BIOCONJUGATE TECHNIQUES

BIOCONJUGATE TECHNIQUES

Greg T. Hermanson
Pierce Biotechnology, Thermo Fisher Scientific, Rockford, IL

ELSEVIER

AMSTERDAM • BOSTON • HEIDELBERG • LONDON
NEW YORK • OXFORD • PARIS • SAN DIEGO
SAN FRANCISCO • SINGAPORE • SYDNEY • TOKYO
Academic Press is an imprint of Elsevier

Acquiring Editor: Janice Audet
Development Editor: Mary Preap
Project Managers: Karen East and Kirsty Halterman
Designer: Alan Studholme

Academic Press is an imprint of Elsevier
32 Jamestown Road, London NW1 7BY, UK
225 Wyman Street, Waltham, MA 02451, USA
525 B Street, Suite 1800, San Diego, CA 92101-4495, USA

Third edition 2013

British Library Cataloguing-in-Publication Data
A catalogue record for this book is available from the British Library

Library of Congress Cataloging-in-Publication Data
A catalog record for this book is available from the Library of Congress

ISBN: 978-0-12-382239-0

Typeset by MPS Limited, Chennai, India
www.adi-mps.com

Back cover review quote reprinted by permission from Macmillan Publishers Ltd: Nature Chemical Biology; Book Review; Francis, M.B.; Updating the bioconjugation catalog: Vol. 4, No. 12, pp. 717, copyright 2008

Printed and bound in China

13 14 15 16   10 9 8 7 6 5 4 3 2 1

On the cover: Molecular model of R-phycoerythrin, a light-harvesting protein from red algae having bright fluorescent properties. The crystal structure was published in Contreras-Martel, C., Martinez-Oyanedel, J., Bunster, M., Legrand, P. Piras, C., Vernede, X., and Fontecilla-Camps, J.C. (2001) Crystallization and 2.2 A resolution structure of R-phycoerythrin from *Gracilaria chilensis*: a case of perfect hemihedral twinning. Acta Crystallogr. Sect.D 57, 52–60. The coordinates are available on the RCSB Protein Data Bank as structure 1eyx. Molecular graphics and analyses were performed with the UCSF Chimera software package. Chimera is developed by the Resource for Biocomputing, Visualization, and Informatics at the University of California, San Francisco (Sanner, M.F., Olson, A.J., Spehner, J.C. (1996) Reduced surface: an efficient way to compute molecular surfaces. Biopolymers 38(3), 305–320).

Dedication

For *Amy and Meghan*, who, since the second edition was published, have now graduated from college and are pursuing careers in biology and medicine. Also for baby *Violet*, who hopefully someday will be able to read and appreciate a future edition of this book.

Contents

9. Cross-Bridges and Cleavable Reagent Systems 387

10. Fluorescent Probes 395

11. (Strept)avidin–Biotin Systems 465

12. Isotopic Labeling Techniques 507

13. Silane Coupling Agents 535

14. Microparticles and Nanoparticles 549

15. Immobilization of Ligands on Chromatography Supports 589

16. Buckyballs, Fullerenes, and Carbon Nanotubes 741

17. Chemoselective Ligation Bioorthogonal Reagents 757

18. PEGylation and Synthetic Polymer Modification 787

Preface to the Third Edition

In the years since the publication of the second edition, the field of bioconjugation has continued to advance at an incredible pace. Since 2008, over 54,000 additional journal publications have appeared in the biological, medical, polymer, material science, and chemistry journals that at least mention the terms "bioconjugate" or "bioconjugation." In addition, many tens of thousands of new links to Internet sites with bioconjugation information also have appeared in this time frame, including sources from academic, corporate, and personal web pages. These journal articles and links describe many new reagents and reactions for forming bioconjugates of all types, including the formation of unique complexes in solution as well as the coupling of molecules to solid-phase surfaces or particles. In addition, exciting new methods are appearing for the application of bioconjugates in highly sensitive assays and detection schemes, for *in vivo* imaging and diagnosis, for therapeutic drug targeting, in the capture and purification of biomolecules, for catalysis and chemical modification, and for vaccine development and immune modulation. These recent advances in bioconjugate techniques have resulted in two new chapters and many new sections and updates throughout the book, as well as the rearrangement and consolidation of chapters to more logically group topics together having common themes.

The third edition also contains two major chapters that were obvious gaps in the previous editions: Chapter 1 is an extensive introduction to the vast field of bioconjugation, while Chapter 15 describes the reagents and techniques used for the immobilization of ligands onto chromatography supports. The new comprehensive introduction to the book begins by describing the basic principles of bioconjugation along with presenting important strategies for designing optimal conjugates for a wide range of applications. Chapter 1 also reviews the major application areas where bioconjugates are being used today and describes the conjugate designs associated with each of these applications.

In addition, the new Chapter 15 contains one of the most extensive overviews of immobilization chemistry ever presented. It begins by reviewing the basic principles of affinity chromatography along with the resins and other solid-phase options used to couple affinity ligands of every type, including proteins, antibodies, enzymes, peptides, nucleic acids, other biomolecules, metal chelates, and organic mimetics. It also presents numerous options for activating supports and the coupling reactions that can be used for covalently attaching ligands onto them. These methods can also be successfully applied to the immobilization of molecules onto any other particle or surface material desired. The resultant affinity supports can be used for capturing target molecules, in the purification of proteins, for studying protein or oligonucleotide interactions, for proteolysis and enzyme catalysis, for removing contaminants from solution, or for developing biosensors and assays for a host of analytes.

Another major change that is immediately noticeable with this edition is the use of full-color illustrations. Literally hundreds of new and updated figures now use color to better illustrate reactions or to show how bioconjugates are being used in applications. While the design of the book may have been radically changed and updated with this edition, it is my hope that the reader will continue to find it useful in the design of new bioconjugates.

Acknowledgments

I thank the thousands of researchers, many of whose names appear in the reference section, who have developed and optimized hundreds of reagents and applications related to the modification, conjugation, and immobilization of biomolecules and affinity ligands. Their work made this book possible. I also want to thank Barb Tanaglia, Sally Etheridge, Crystal Gomez, Heather Flynn, and Brian Weathers for their expert help in obtaining journal references. I also greatly appreciate Craig Smith for reviewing the new material and being supportive of my writing. In addition, I thank Funmilayo Suleman for literally reading the entire second edition and providing helpful feedback on how to make this edition even better. I also thank Peter Bell, Julie Kremer, and Alan Doernberg for providing corporate approval for this entire endeavor. Although Thermo Fisher Scientific did not sponsor the project, the company provided great motivation for me to undertake the effort and complete the third edition.

Finally, special thanks to the one who made it all possible.

Important Information

HEALTH AND SAFETY

This book describes hundreds of reagents, reactions, and applications for use in bioconjugation. Most of the compounds are highly specialized and we have very little information regarding their toxicological properties. At a minimum, bioconjugation reagents should be considered irritants and handled with care. However, the overwhelming majority are known to be reactive and any individual compound or solvent can be corrosive, hazardous, toxic, volatile, flammable, explosive, or otherwise dangerous to personal health and safety. For this reason, the use of any reagent or protocol described in this book should be carried out by taking the appropriate precautions. Before utilizing any of these methods, the user agrees to take complete responsibility and personal liability for any and all risks associated with the reagents and reactions described in this book or within the references cited. Before starting an experiment, the user agrees to reference the appropriate Material Safety Data Sheets (MSDS) relative to every compound or component used in a reaction and to completely understand the properties of the reactions being contemplated. The use of personal protective equipment (PPE), fume hoods, and proper laboratory techniques can ensure safety for both the user and other people in the immediate vicinity. In addition, the disposal of waste materials should be performed according to the appropriate environmental regulations to prevent toxic elements or compounds from entering the water, air, or soil. The inappropriate disposal of excess reagents or reaction byproducts may be harmful to people and the environment.

INTELLECTUAL PROPERTY

Throughout this book I have provided references related to the reagents, reactions, and techniques used in bioconjugation. There are many additional references that can be found by performing the appropriate key word searches on the Internet. However, such knowledge does not necessarily provide the liberty to legally use these reagents and applications for commercial purposes without consideration for existing intellectual property rights. While in some cases pertinent patent references are provided within the book, this is done only to supply additional technical details about the topic being discussed and not to imply anything about freedom to operate.

Today, nearly every important reagent or method reported in the literature has a patent or patent application associated with it, especially if it has potential commercial value. A search of the patent databases, such as the United States Patent and Trademark Office (http://www.uspto.gov/) or the European Patent Office (http://ep.espacenet.com/) for key words or the names of inventors can provide a list of existing issued patents or patent applications related to a bioconjugate technique or compound.

It is the responsibility of the reader to become familiar with the patents and claims that may cover particular compounds, compositions, reactions, or their methods of use in bioconjugate applications. If patents or patent applications exist, it is important that the appropriate permission or a license be obtained from the owner of such intellectual property before exploiting it for commercial use.

CHAPTER

1

Introduction to Bioconjugation

The field of bioconjugation has had a deceptively quiet, but at the same time enormous, impact on science and technology. The use of bioconjugates may not be the focus of many popular press science articles or noticed as important for news stories; nevertheless, the ability to produce discrete bioconjugate complexes having unique properties suitable for a wide variety of applications has become the underlying success story of many research endeavors. Bioconjugation has made possible the discovery of new biomolecules, the elucidation of complex biological processes, and the spawning of entire industries within the medical, diagnostics, life sciences, microelectronics, and material sciences fields. The use of bioconjugates in an almost unending number of applications has also made possible a multibillion dollar worldwide economy that functions to cure diseases and discover the very secrets of life. The process of securing grant money, venture capital investment, or corporate R&D funding often includes the proposed use of novel bioconjugates as key components in reaching technical goals. Indeed, many of the largest pharmaceutical and biotech companies now depend upon bioconjugation to design their future product pipelines and maintain a vital edge over the competition. The "magic bullet" and "targeted drug" concepts often desired in drug development are heavily contingent on the creation of highly specific bioconjugates with therapeutic efficacy toward certain cells, tissues, or disease states. Without the advance of bioconjugate techniques to enable this process, the world would not have anywhere near the invention and innovation that have taken place over the last few decades in the life science and medical fields.

A search for any particular bioconjugate type using PubMed or other major science and technology search engines typically yields at least hundreds or, more likely, thousands of hits. The actual percentage of all publications that use bioconjugates to perform methods critical to scientific discovery may not be determinable precisely, but it is likely to be an overwhelming majority of all life science research being done and published today. In fact, most activities in biological research could not be done easily, if at all, without the use of one or more bioconjugate reagents to assay, detect, track, image, or capture target molecules, or effectively target and treat diseases.

Throughout this book, the techniques of producing and using bioconjugates are presented with a view toward providing options for designing similar conjugates for new or existing applications. This chapter is meant to provide an introduction to aid in understanding the breadth and depth of this field, while acting as a guide to rationally choosing bioconjugate components and reaction strategies for designing effective reagents.

1. WHAT IS BIOCONJUGATION?

In its most fundamental aspect, bioconjugation simply involves the attachment of one molecule to another, usually through a covalent bond, to create a complex consisting of both molecules linked together (Figure 1.1). In most cases, at least one of the molecules is of biological origin or is a fragment or derivative of a biomolecule. In some situations, the conjugate that is formed is entirely synthetic, but its use is directed toward biological or life science applications. When forming such bioconjugates, the process can yield a composite having approximately equal proportions of each component or create a conjugate purposely designed to have more molecules of one component than the other. The final form that the bioconjugate takes is dependent on the

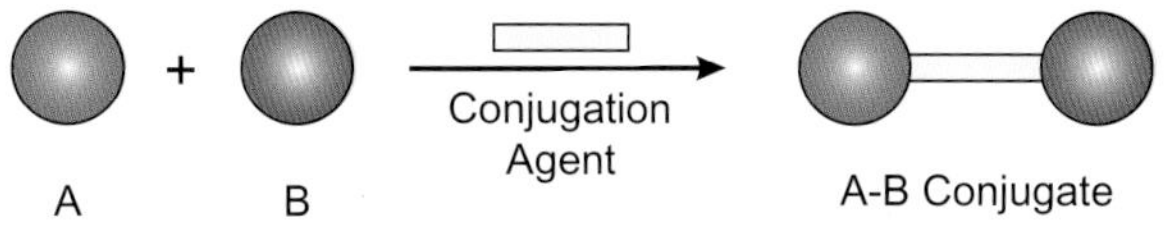

FIGURE 1.1 Forming a basic conjugate often involves the reaction of two molecules and a crosslinking agent that covalently links the components together. In some conjugation schemes an activation agent is used that results in the linking of two molecules without an intervening cross-bridge between them.

Bioconjugate Techniques, Third Edition
DOI: http://dx.doi.org/10.1016/B978-0-12-382239-0.00001-7

desired application and the components and methods used to couple them together. With an understanding of the basic concepts of bioconjugation, this process can be done without difficulty and even become controllable through the appropriate choice of reagents, reactions, and conditions.

The process of making bioconjugates from individual molecules thus creates new complexes having the combined properties of each constituent from which it is made. The result forms novel constructs having characteristics not normally found among naturally occurring substances. For instance, attaching a fluorescent label to an antibody creates a targeting complex that can be used to bind specifically to a desired biomolecule through the antigen binding sites on the antibody, and then, due to the fluorescent properties of the label, the targeted biomolecule can be detected—a feature which would not be possible using the unlabeled antibody alone. Thus, the formation of a useful bioconjugate begins by envisioning the features that are desired in the final complex and then choosing the components necessary to create it. Figure 1.2 illustrates some bioconjugate types that can be made by linking two or more molecules together. These designs are by no means exhaustive; the functionality that can be built into a bioconjugate by connecting two or more molecules

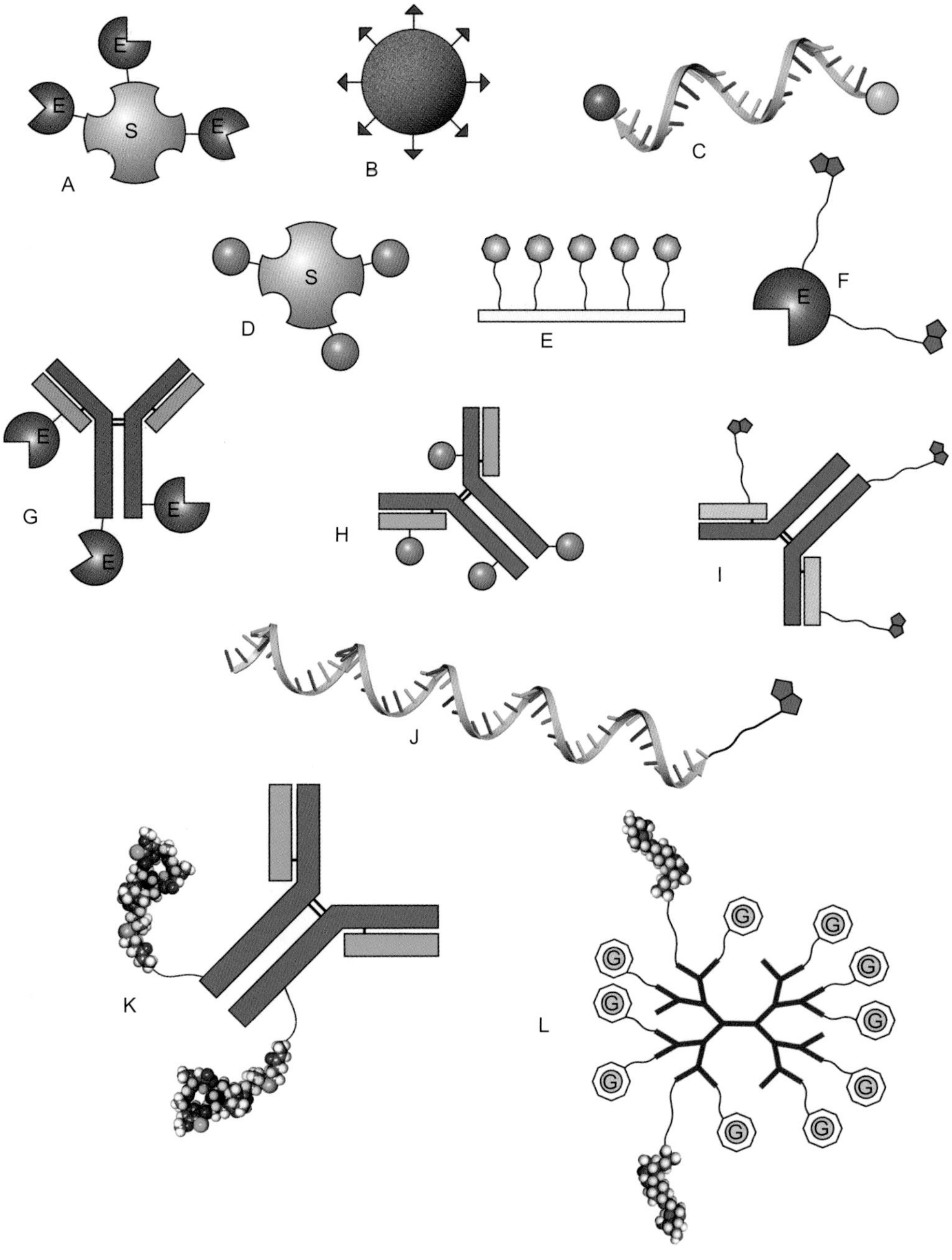

FIGURE 1.2 Some of the common bioconjugate designs often used for life science applications include (A) streptavidin–enzyme conjugate, (B) an immobilized affinity ligand on a particle, (C) an oligo molecular beacon probe containing two fluorescent labels or a fluor and a quencher at each end, (D) fluorescently labeled streptavidin, (E) an affinity ligand attached to a surface, (F) a biotinylated enzyme, (G) an antibody–enzyme conjugate, (H) a fluorescently labeled antibody, (I) a biotinylated antibody, (J) a biotinylated oligo probe, (K) an antibody–drug conjugate, (L) a gadolinium chelate-modified dendrimer containing folate molecules for targeting.

together is limited only by the imagination and the individual components available to assemble it.

The process of creating bioconjugates is typically carried out using reactive crosslinking agents that have been specially designed for this purpose or through the use of an appropriate reactive group on one of the molecules to facilitate the conjugation. The coupling reaction can also be accomplished using secondary activating agents that create an intermediate reactive group on one of the components to be conjugated. Most of the reactive groups used for this process can then be made to couple with specific functional groups on one or more of the molecules to be conjugated, thus linking them together into the final complex. Figure 1.3

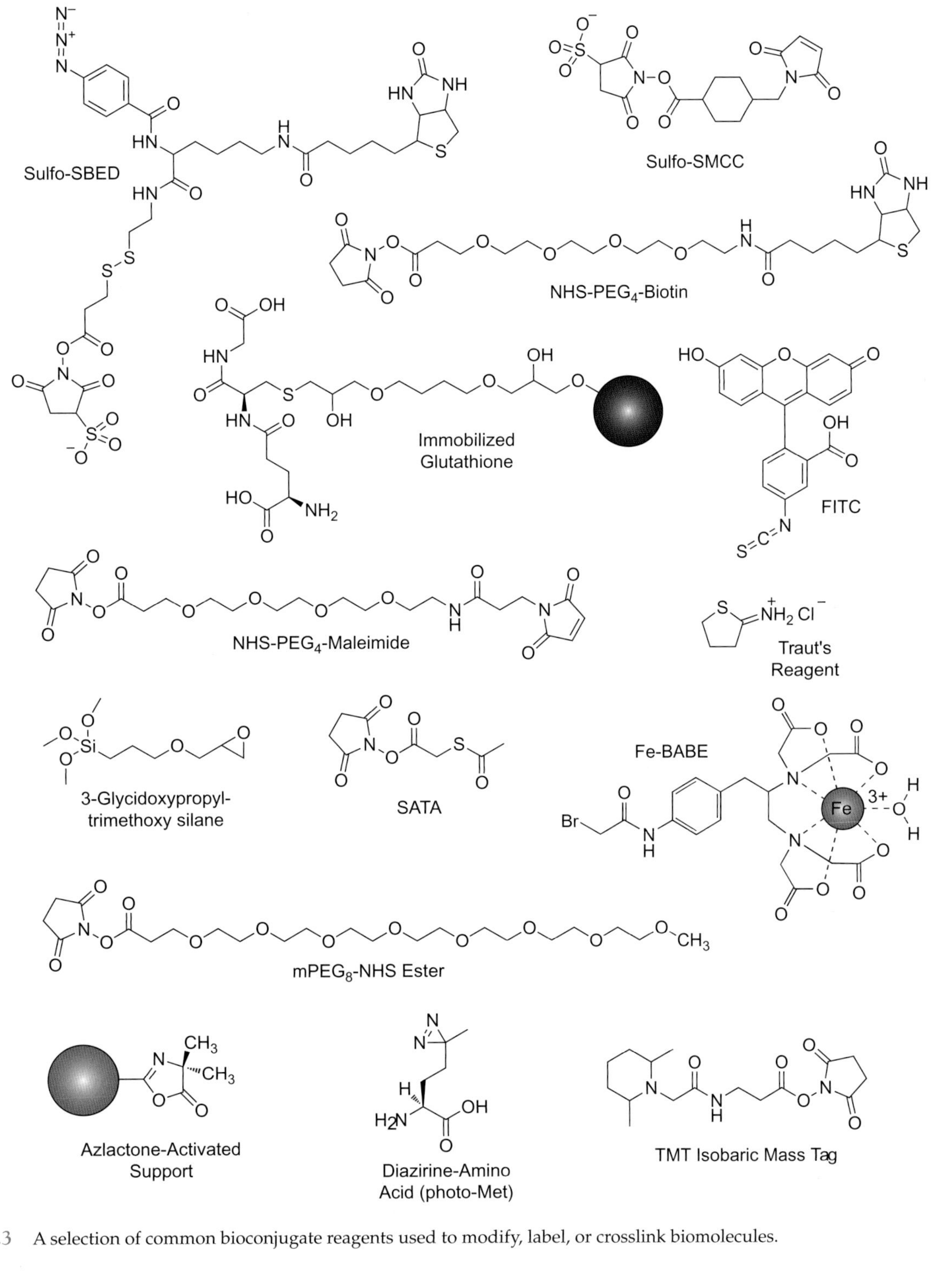

FIGURE 1.3 A selection of common bioconjugate reagents used to modify, label, or crosslink biomolecules.

illustrates some of the major types of reagents used for bioconjugation techniques. These are just a few of the tremendous number of reactive compounds that are available today for modification or conjugation purposes. Many other specific examples of bioconjugation reagents can be found throughout this book.

The reactions commonly utilized to form bioconjugates can be selected from perhaps 1 to 2 dozen principal reactive groups that are commonly used for creating covalent bonds with various functional groups on biomolecules. Of these dozens of choices for reactions, probably less than 10 are used routinely to effect the conjugation of the overwhelming majority of bioconjugates created. The reader will notice that throughout this book time and again the same basic reactions and reagents are used to form diverse conjugates that are useful in a broad range of applications.

The key to forming a successful bioconjugate is to select the proper crosslinking reagents that in turn contain the appropriate reactive groups, which will couple with the chemical functionalities available on the molecules to be linked together. In a sense, it is like choosing the right building blocks needed to construct the final structure of a molecular assembly. Just like a square peg that will not fit into a round hole, improper choice of crosslinkers or reactive groups will not yield the desired bioconjugate or may result in a suboptimal one. Therefore, a basic understanding of the reactions of bioconjugation is essential to understanding how to build a useful conjugate for a particular application. This is not to say one has to become an expert at synthetic organic chemistry to make a successful bioconjugate; rather, one only needs to gain enough knowledge to make the right reagent choices. Matching the correct crosslinking agents or reactive groups with the accessible functional groups on the components to be conjugated is the first step to building a successful and active bioconjugate complex.

Thus, bioconjugates can be constructed by the judicious coupling of two or more individual components to create a multifunctional complex having two or more properties combined into a single macromolecule. The conjugate may contain one or more affinity molecules, which can be used to target, capture, or detect another biomolecule; it may also contain one or more detection molecules or enzymes, toxins, drugs, or other components having some other specific activity or purpose. In addition, some bioconjugates are made using supplementary molecular scaffolds that may not be an active component of the bioconjugate, but are nevertheless important in constructing the final complex. Just like girders in a building, a molecular scaffold can be used to construct the bioconjugate by providing the core structure and attachment points for linking additional molecules into the overall assemblage of parts (Figure 1.4). The scaffold may also allow more copies of a certain molecule to be coupled and therefore become more actively present in the final conjugate than would be possible without its use. Multivalent scaffolds in particular might provide greater numbers of attachment points for coupling detection molecules, thus potentially increasing the detection sensitivity of a bioconjugate in an assay. Therefore, a given bioconjugate might contain components directly linked together to form a complex or it can be made with intermediary molecules which do not have functional activity in the final reagent, but nevertheless contribute to its properties as a whole.

In addition, bioconjugates can be designed to contain small molecule affinity ligands that specifically interact with other proteins, and these proteins in turn can function as scaffolds for increasing conjugate functionality through noncovalent interactions. For instance, biotin can be attached to one molecule and used as an affinity handle to interact with (strept)avidin reagents. The interaction between biotin and (strept)avidin is extremely strong (Chapter 11), and this permits highly specific interactions to occur with the biotinylated conjugate. For instance, forming a bioconjugate by attaching a biotin label to an antibody permits detection of a targeted protein through use of a second bioconjugate, such as a fluorescently labeled (strept)avidin complex. A biotinylated bioconjugate can also be captured and purified from a complex sample solution using an immobilized (strept)avidin particle or resin.

The intermediary use of a biotin–(strept)avidin complex in assays provides another type of bioconjugate scaffold for increasing assay or detection sensitivity. A biotinylated antibody is often used to target an analyte, while a secondary detection conjugate formed using (strept)avidin linked to another detection molecule provides the signaling agent. The docking of the (strept)avidin conjugate onto the biotinylated antibody carries with it more detection molecules than could be coupled to the antibody directly. The use of two conjugates in this assay design therefore provides increased sensitivity beyond that possible using a single antibody-detection conjugate alone.

The differences in function and use between a directly labeled antibody-detection conjugate *versus* a biotinylated antibody plus a secondary labeled (strept)avidin conjugate reflect the fact that the function of a given bioconjugate is usually determined by the individual constituent properties making up the whole. In this sense, a fluorescently labeled antibody can be used and detected directly, while a biotinylated antibody has to be used with another bioconjugate, such as a fluorescently labeled (strept)avidin molecule, to provide the same type of detection capability as the direct bioconjugate (Figure 1.5).

The combined properties of a given bioconjugate—all its various components linked together—thus

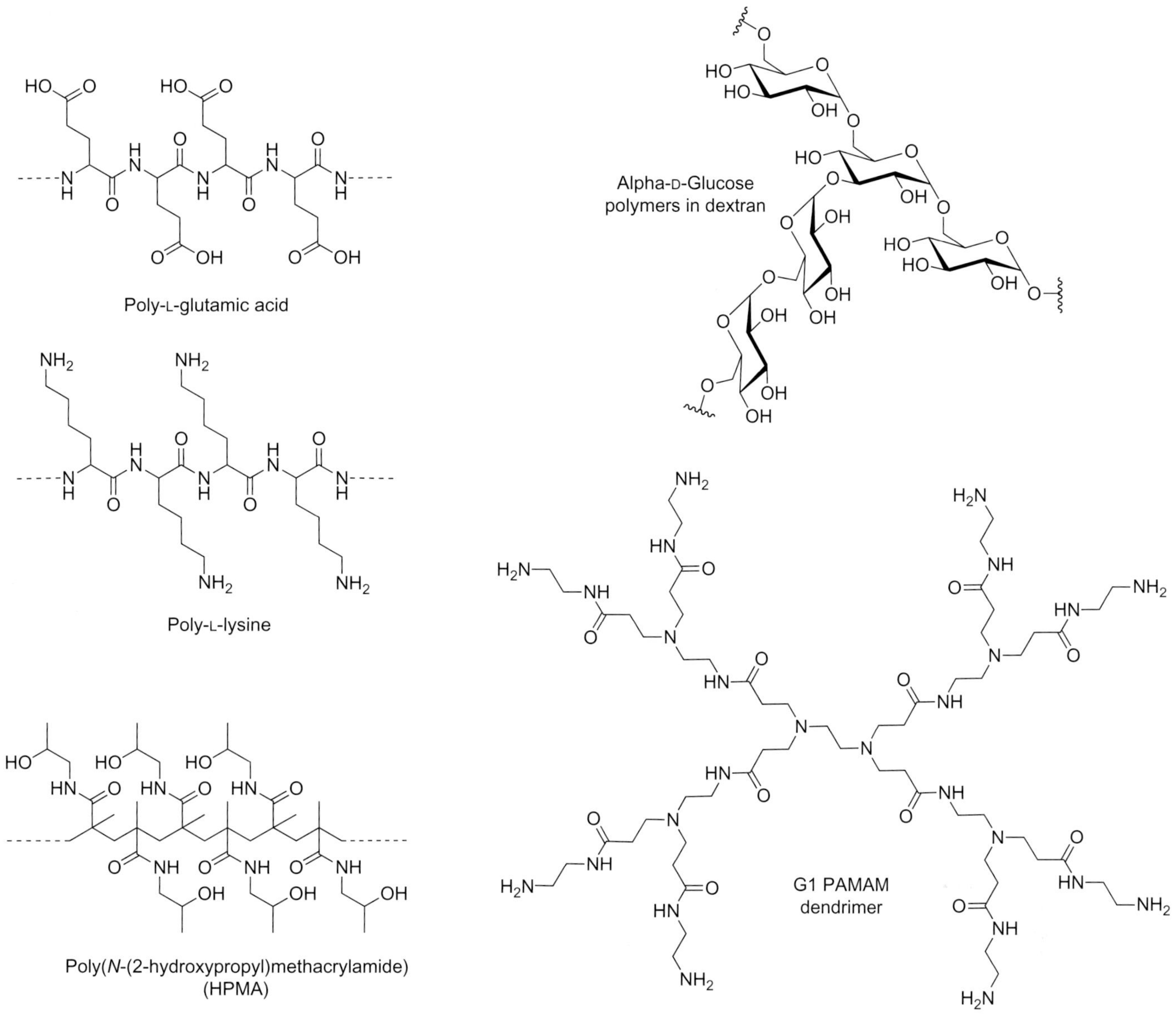

FIGURE 1.4 Polymeric molecular scaffolds often used in bioconjugate design to carry targeting molecules, detection components, or biotherapeutic agents.

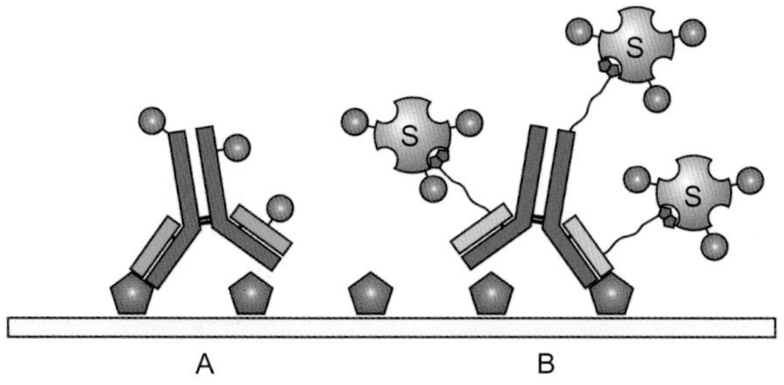

FIGURE 1.5 Comparison of the use of a fluorescently labeled antibody *versus* a biotinylated antibody used with a fluorescently labeled streptavidin conjugate. The multi-layered effect of the streptavidin–biotin interaction provides additional detection molecules able to dock at the site of a target molecule, thus providing potentially higher sensitivity in assays.

govern the properties, applications, and methods for its use. Designing a useful bioconjugate entails envisioning the preferred characteristics and methods of use for the final complex and then choosing the best components and reaction strategies to produce the desired product or products. On the surface, it may seem that the use of a fluorescently labeled antibody produces a simpler and thus a better conjugate than the use of a biotinylated antibody combined with a fluorescently labeled (strept) avidin complex, because a single conjugate is easier to make and use. However, that straightforward conclusion does not tell the whole story. The biotinylated antibody can have multiple biotins on its surface which permit the docking of multiple (strept)avidin molecules on each antibody. If each (strept)avidin molecule also

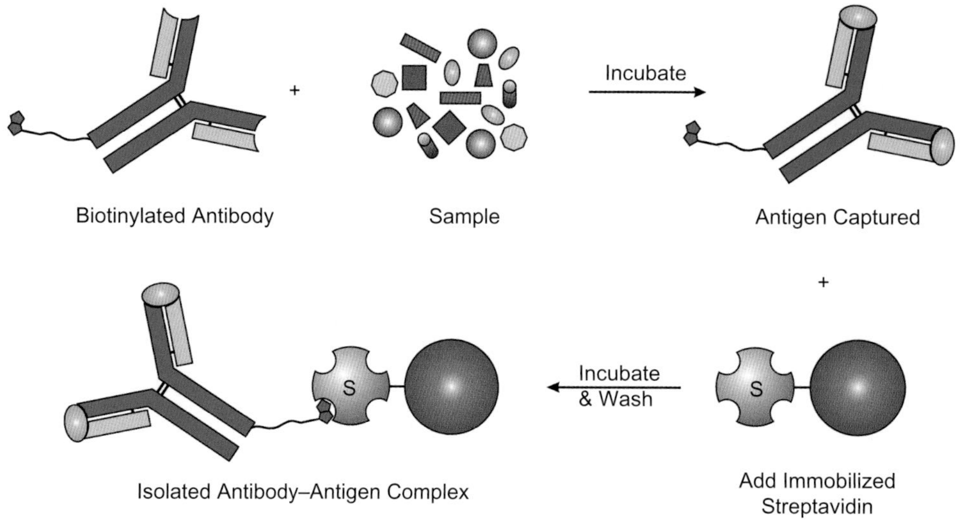

FIGURE 1.6 Immobilized streptavidin used to isolate a biotinylated antibody–target protein complex by immunoprecipitation (IP). A primary labeled antibody is allowed to incubate with a sample containing proteins and the antibody binds to its target. The biotin affinity tag facilitates isolation of this complex by binding to the immobilized streptavidin on an insoluble particle.

contains multiple fluorescent labels, then the resultant complex can possess many more fluorescent molecules than a directly labeled antibody can have attached to it. Therefore, when used in an assay to detect a target molecule, the biotinylated antibody combined with a fluorescently labeled (strept)avidin conjugate can result in much higher fluorescence signals than the fluorescently labeled antibody alone. Multi-layered conjugate complexes often result in greater numbers of molecules being assembled at the site of a targeted analyte, thus recruiting greater conjugate activity at that site as well.

In addition, some bioconjugates like a biotin-modified antibody can have multiple uses. Not only can a biotinylated antibody be used in an assay or detection process but it can also be used to bind to a target protein in a cell lysate and then isolate the target by use of an immobilized (strept)avidin affinity support (Figure 1.6). Thus, the preparation of a single bioconjugate consisting of a biotin-modified antibody can have multiple uses due to the combined properties of its constituent parts. It is clear that envisioning how a bioconjugate ultimately will be used is essential in choosing the optimal bioconjugate to make. The next section will explore this aspect in greater detail.

2. BIOCONJUGATION STRATEGY AND DESIGN

The strategies used to develop bioconjugates for particular applications can be as highly varied as the reactions and components that are used to form such complexes. Bioconjugates can be used in a multitude of applications and in each case the particular bioconjugate that performs best is a result of careful design and is often meticulously optimized for its intended use (see Section 3, this chapter). Each component of the bioconjugate should be considered as being an active part of the entire reagent. Every element has a role to perform that is essential to the functioning of the bioconjugate in its intended application. If any one part performs poorly then the activity or specificity of the entire conjugate is also likely to suffer. For this reason, it is essential to optimize each bioconjugate with its application in mind. This section provides some general guidelines to the design of bioconjugates along with some strategies for choosing the right components and reactions to form the final complex. Reference will also be made to the appropriate chapters within the book for further information regarding functional groups that can be used as well as the conjugation reagents, scaffolds, targeting agents, detection components, polymers, particles, antibodies, and any other molecules that may be incorporated into the final bioconjugation reagent. In addition, it is recommended that the Table of Contents and the Index be consulted to cross-reference the appropriate pages that best relate to the type of bioconjugate being designed and the reactions used to produce it. This section then leads into an extensive review of bioconjugation, especially highlighting how it has been used to create reagents that have become essential to countless fields in science and technology.

2.1. Start with the Application in Mind

The ultimate use of a bioconjugate should be carefully considered in order to properly design its components to be appropriate to the envisioned task they must perform. In addition, the conditions under which the bioconjugates are to be used should be taken into account, because certain components might not perform well in the anticipated medium or environment. For this reason, it is important that when beginning to design any bioconjugate its ultimate application must first be considered before initiating any experimental procedures to actually make the conjugate. If the bioconjugate is to be used in an assay for example, the most appropriate components might consist of a targeting molecule coupled to a detection element, so that the final complex can bind and interact with its target and then be detected. Alternatively, if the final bioconjugate is to be used in the purification of a target molecule, an appropriate conjugate might consist of an insoluble support or particle to which an affinity ligand is attached. Therefore, each bioconjugate application may have radically different conjugate construction requirements.

The design of a new bioconjugate for a unique application might best be explored by first considering the bioconjugates that others have used in similar applications. Section 3 of this chapter should therefore be carefully reviewed as a guide to the bioconjugate types that have been used successfully for published applications. In addition, the entire history of bioconjugation may be consulted through the literature to determine the scope of bioconjugate designs useful in different types of situations. Table 1.1 lists some of the major applications of bioconjugation along with the conjugate styles that have been applied to work with them. This table is not meant to be exhaustive in its presentation of every conjugate that has ever been reported, but only to provide major examples, which can be used as a starting point for the development of new bioconjugates. Since the component options in designing a bioconjugate are vast and the reactions that can be used in producing it are many, the options available in creating a new conjugate can be almost daunting. For this reason, it is extremely useful to gain knowledge of the bioconjugate types and construction details that others have used successfully in similar situations before embarking on a new design. Even with this knowledge, however, it is likely that for a new application, the bioconjugation construction methods will still have to be optimized to work well in the intended application.

For instance, the design of a fluorescently labeled antibody may appear to be quite trivial on the surface, especially with so many years of history in the literature involving the fluorescent modification of antibodies. To prepare this type of conjugate, it seems to be merely a matter of coupling a reactive fluorescent molecule with the desired purified antibody. Using known reaction protocols, such as those described in Chapter 10, one can obtain a labeled antibody that could be used to detect a target molecule. Upon closer inspection, however, the process for creating this conjugate is much more complex. Unless you have experience with similar conjugations using exactly the same fluorescent label that you want to use again, the first variable in the process is to choose an appropriate reactive dye. The choice may be governed by the wavelength of excitation and emission or it may be determined by the general wavelength region that is required for successful use in the desired application. For *in vivo* imaging applications, for instance, the antibody should be labeled with a fluorescent dye that has spectral properties in the far-red to near-infrared (NIR) region of the spectrum, whereas for general cell staining applications the dye could have emission properties anywhere from the UV to the NIR region, which also covers all of the visible wavelengths. Another set of criteria for dye selection in making this type of conjugate is limitations that may exist related to the instrument used for fluorescence detection. For instance, if there is access to a fluorescent microscope with filter sets already available for dyes in the fluorescein (488), Cy3, and Cy5 wavelength ranges, then it would be appropriate to make the antibody conjugate using dyes that correlate with these wavelengths.

In addition, fluorescent dyes may be chosen for antibody conjugation based on their relative hydrophobicity or hydrophilicity. In certain applications, it is far better to choose a dye that is more hydrophilic to prevent antibody aggregation or to reduce the potential for nonspecific binding of the conjugate in an assay. Most dyes have highly hydrophobic, aromatic core structures that can interact nonspecifically with many hydrophobic regions on biomolecules or interact with surfaces used in assay procedures. The use of hydrophilic dyes containing negatively charged sulfonates or neutral PEG chain modifications can dramatically reduce nonspecific binding and enhance assay signal-to-noise ratios. Conversely, the use of dyes with somewhat less water solubility may be appropriate if the antibody conjugate is to be used for *in vivo* imaging or cell-based imaging where efficient penetration of membrane structures is required to reach a desired target. In this case, the use of dyes containing fewer sulfonates (e.g., perhaps no more than two to three negative charges on a large cyanine-type dye) can perform better in the intended application than dyes containing more hydrophilic groups or a greater number of charges. Thus, the performance of the conjugate in its intended application often dictates the best choice of fluorescent dye derivative to use in the conjugate design. Testing of several potential choices to optimize the performance of a dye–antibody conjugate should be carried out in order to obtain the best possible result.

Another important consideration when making a conjugate is the relative ratio of each component in the final complex. In the dye-labeled antibody example, typically it is important to have more than one dye modification on each antibody molecule. This has the effect of increasing the fluorescence intensity of each antibody as

TABLE 1.1 **Bioconjugate Components and Designs Used for Major Applications**

Application	Bioconjugate Components	Bioconjugate Reagents	Bioconjugate Designs
Enzyme Immunoassays	Antibodies; enzymes; scaffolds; biotin; streptavidin; particles; microwells; planar surfaces; antigens; affinity ligands	Heterobifunctional aliphatic crosslinkers; heterobifunctional PEG-based crosslinkers; multifunctional scaffolds; zero length crosslinkers; homobifunctional crosslinkers; thiolation reagents; biotinylation reagents; activated particles, microwells, or surfaces	Antibody–enzyme; biotinylated antibody plus streptavidin–enzyme; biotinylated antibody, biotinylated enzyme, plus streptavidin; antibody–particle or antibody–surface for capture; antibody–polymer–enzyme; streptavidin—polymer–enzyme plus biotinylated antibody; immobilized antigens or affinity ligands
Fluorescent Immunoassays	Fluorescent dyes; antibodies; scaffolds; biotin; streptavidin; phycobiliproteins or tandem dyes; lanthanide chelates; quantum dots; particles; microwells; planar surfaces	Reactive fluorescent dyes; multifunctional scaffolds; reactive phycobiliproteins or reactive tandem dyes; bifunctional lanthanide chelators; activated particles, microwells, or surfaces; heterobifunctional crosslinkers; PEG-based crosslinkers; zero length crosslinkers	Antibody–dye; biotinylated antibody plus streptavidin–dye; antibody–phycobiliprotein; streptavidin–phycobiliprotein; antibody–lanthanide chelate fluor; antibody–quantum dot; antibody–dye-doped-particle; antibody–tandem dye; streptavidin–quantum dot; streptavidin–tandem dye; quencher–fluor–oligonucleotide molecular beacon; quencher–fluor–peptide molecular beacon; inhibitor–dye; antibody–polymer dye; antibody–particle or antibody–surface for capture
Chemiluminescent Immunoassays	Acridinium esters; antibodies; streptavidin; biotin; particles; microwells; planar surfaces	Reactive acridinium ester compounds; biotinylation reagents; activated particles, microwells, or surfaces	Antibody–acridinium ester; biotinylated antibody plus streptavidin–acridinium ester; antibody–particle or antibody–surface for capture
Hybridization Assays, including FISH	Fluorescent dyes; synthetic or genomic oligonucleotide probes; enzymes; biotin; streptavidin; particles; microwells; planar surfaces; digoxigenin (DIG); haptens	Reactive fluorescent dyes; modified or activated oligo probes; activated particles, microwells, or surfaces; biotinylation reagents; DIG-11-dUTP	Oligonucleotide–dye; hapten–labeled oligo; biotinylated oligo; quencher-dye–labeled oligo for molecular beacon; oligo–particle or oligo–surface for capture; reactive oligo probe
ChIP Assays	Oligonucleotide probes; particles; microwells; planar surfaces; zero length crosslinkers; antibodies	Functionalized oligo probes; activated particles, microwells, or surfaces; crosslinking agents	Oligonucleotide–biotin; oligo–particle or oligo–surface for capture; Antibody–enzyme; antibody–dye; streptavidin–enzyme; streptavidin–dye; biotinylated antibody; oligo–particle or oligo–surface for capture
IP and Co-IP Assays	Antibodies; interacting prey protein; particles; microwells; planar surfaces	Activated particles, microwells, or surfaces; crosslinking agents; biotinylation agents	Antibody–particle or antibody–surface for capture; biotinylated antibody; streptavidin–particle or streptavidin–surface for capture
Cellular Imaging and *in vivo* Imaging	Antibodies; enzymes; fluorescent dyes; biotin modifications; streptavidin; nanoparticles; oligonucleotides; peptides; fluorescent proteins; bioluminescent enzymes; split enzymes or fluorescent proteins	Heterobifunctional aliphatic crosslinkers; heterobifunctional PEG-based crosslinkers; multifunctional scaffolds; zero length crosslinkers; homobifunctional crosslinkers; thiolation reagents; biotinylation reagents; activated particles; reactive fluorescent dyes	Antibody–enzyme; biotinylated antibody; antibody–dye; streptavidin–dye; fluorescent active site warhead probes; molecular beacon fluor–peptide–quencher or fluor–oligo–quencher probes
Flow Cytometry	Antibodies; fluorescent dyes; phycobiliproteins; tandem dyes; streptavidin; biotin; quantum dots; nanoparticles; magnetic particles	Heterobifunctional aliphatic crosslinkers; heterobifunctional PEG-based crosslinkers; multifunctional scaffolds; zero length crosslinkers; homobifunctional crosslinkers; thiolation reagents; biotinylation reagents; activated particles; reactive fluorescent dyes	Antibody–dye; biotin–antibody; antibody–phycobiliprotein; streptavidin–dye; streptavidin –phycobiliprotein; antibody–tandem dye; streptavidin–tandem dye; antibody–quantum dot; antibody–magnetic particle; antibody–nanoparticle

Immunohistochemical Staining	Antibodies; enzymes; scaffolds; biotin modifications; streptavidin; fluorescent dyes; nanoparticles; colloidal gold; quantum dots	Heterobifunctional aliphatic crosslinkers; heterobifunctional PEG-based crosslinkers; multifunctional scaffolds; zero length crosslinkers; homobifunctional crosslinkers; thiolation reagents; biotinylation reagents; activated particles; reactive fluorescent dyes	Antibody–enzyme; biotinylated antibody; antibody–dye; streptavidin–dye; streptavidin–enzyme; antibody–gold; antibody–quantum dot
Homogeneous Protein Assays	Antibodies; fluorescent dyes; fluorescent proteins; quenchers; bioluminescent enzymes; split reporter enzymes; split fluorescent proteins; lanthanide chelates; phycobiliproteins; oligonucleotide probes	Reactive fluorescent dyes; reactive lanthanide chelates; reactive phycobiliproteins; reactive oligo probes	Antibody–dye; antibody–lanthanide chelate; fluor–peptide–quencher; fluor–peptide–fluor; antibody–phycobiliprotein; antibody–oligo
Homogeneous DNA Assays	Oligonucleotide probes; fluorescent dyes; quenchers; polymerase and other enzymes; minor groove binding dyes	Modified oligos; reactive fluorescent dyes; reactive quenchers	Dye–oligo–quencher; dye–oligo–dye; dye–primers
Vaccines and Immunogens	Carrier proteins; haptens; antigens; peptides; carbohydrates; protein fragments; synthetic scaffolds	Heterobifunctional aliphatic crosslinkers; heterobifunctional PEG-based crosslinkers; multifunctional scaffolds; zero length crosslinkers; homobifunctional crosslinkers; thiolation reagents	Crosslinked proteins; carrier–peptide; carrier–carbohydrate; carrier–hapten; crosslinked viral particles; carrier-–protein fragments or domains; polymerized antigens
Tumor Targeting	Antibody or antibody fragment; alternative targeting scaffolds; polymers; dendrimers; chemotherapeutic drugs; toxins; enzymes; prodrugs; haptens or ligands	Heterobifunctional aliphatic crosslinkers; heterobifunctional PEG-based crosslinkers; PEGylation agents; multifunctional scaffolds; zero length crosslinkers; homobifunctional crosslinkers; thiolation reagents; spacer arms	Antibody–drug; antibody–enzyme; antibody–polymer–drugs; antibody–polymer–dye–drug; hapten–drug; ligand–drug
Catalytic Transformations	Enzymes; particles; resins; biotin; streptavidin; chemical reactants	Reactive particles; reactive resins; activation agents; crosslinkers	Enzyme–particle; enzyme–resin; biotin–enzyme; streptavidin–resin; reactant–particle
Affinity purification	Beaded porous supports (resins); nonporous particles (microparticles or nanoparticles); membranes; monoliths; surfaces; antibodies; proteins; peptides; small molecule affinity ligands; metal chelates; carbohydrates; organic mimetic ligands	Activation agents; spacer arms; crosslinkers	Primary affinity ligand coupled to a bead, particle, surface, membrane, or monolith; immobilized streptavidin to immobilize biotinylated affinity ligands; protein A or protein G coupled to a support to immobilize antibodies

it is docked upon its intended target. Each fluorophore adds another potential photon emission to each conjugate molecule. However, having more dyes modifying an antibody is not always better than fewer dyes. At a certain substitution level the fluorescent dyes may cause aggregation of the antibody or start to quench the potential fluorescence output due to dye–dye interactions and energy transfer instead of emission. For a given antibody–dye complex there is a certain level of dye substitution that will provide maximal signal in an assay without the potential for precipitation or quenching. That optimal conjugate can only be determined through experimentation by making a number of trial dye–antibody conjugates in small amounts so that each preparation has a slightly different dye substitution level. Testing each of the conjugates for their fluorescence intensity, solution stability, and assay performance will determine which one is best for the intended application.

2.2. Designing the Optimal Bioconjugate

One of the most important aspects in the successful formation of a bioconjugate involves having sufficient knowledge of the structural, chemical, and activity characteristics of the components to be bound together. To design a linking strategy that will form the final complex while still maintaining all the functions of its individual parts is not a trivial matter. A successful process not only involves knowledge of the reactive groups, reagent types, and functional groups present on the biomolecules but also includes an understanding of how modification of a biomolecule can potentially affect its activity. Using current knowledge of protein three-dimensional structure and function, including active site constituents and stability considerations, the reaction strategies for bioconjugation can be designed to maximize the potential activity and function of the final conjugate. With this information, functional groups near binding sites or active centers can often be avoided through careful selection of the proper crosslinking agents or reaction conditions. In some cases, unique functional groups (e.g., bioorthoganol; see Chapter 17) or a functional group with a limited frequency of appearance on a biomolecule (e.g., cysteine thiol groups) can be chosen for conjugations that position the modifications or crosslinks away from the active centers or binding sites, thus helping to preserve activity in the final bioconjugate.

As the technology of bioconjugation advances, certain types of conjugates that have been used for years may be redesigned with newer techniques to provide better functionality. For instance, one of the earliest bioconjugate types consisted of radiolabeled complexes formed by attaching a radioactive atom to a biomolecule. Radioiodination involves the covalent modification of certain organic groups within a molecule with a radioactive iodine-125 atom (or another radiolabel) to form a detectable complex, such as in the modification of tyrosine or histidine residues within proteins or activated aromatic rings within small molecules (Chapter 12). Many antibodies were initially labeled in this way to provide the combination of specific antigen binding with radioactive detection. These early bioconjugates were used in many research and diagnostic assays to detect and quantify target molecules. Beginning in the mid-1970s, however, with the advent of antibody–enzyme conjugates, these new detection complexes started to replace radiolabeled conjugates due to the stability and safety concerns of radioactive compounds and due to the increased performance and sensitivity of enzymatic detection. Radioactive bioconjugates are still used in many important applications, but the main focus for radiolabels now is not for *in vitro* research assays or diagnostic testing but rather for advanced *in vivo* therapeutic targeting and imaging of tumor cells.

The methods and reactions used to make bioconjugates have evolved over the years in more ways than one. Early methods of forming conjugates often involved the use of homobifunctional reactants that almost uncontrollably polymerized the components of the bioconjugate. An example of this type of reaction is the use of glutaraldehyde to form antibody–enzyme conjugates. Glutaraldehyde is an effective crosslinker, but the conjugate that is formed too frequently contains high-molecular-weight components and even partially precipitates due to the oligomerization of antibody and enzyme during the reaction (Figure 1.7). More advanced crosslinkers are now available to remedy this situation and provide more control in the reaction process. For this reason most antibody–enzyme conjugates are no longer made using glutaraldehyde. As the science of bioconjugate techniques progresses, new reagents and reactions continually provide additional options for producing better bioconjugates. Even familiar bioconjugates that have been prepared and applied in research for decades might be improved by redesigning the bioconjugation strategy originally used to make them.

The process of bioconjugation, under the most controlled and optimal conditions, can still result in a number of different structural species making up the final composition. When two components are reacted together and at least one of them has multiple sites for covalent attachment, such as on proteins which contain multiple amines, the resultant bioconjugate will likely have a distribution of molecular weights around an average mass. For instance, the process of modifying a protein using an amine-reactive biotinylation compound can yield an almost Gaussian distribution of the number of biotins coupled per protein, the mean of which is dependent on the molar excess of biotin reagent used over the quantity of protein added at

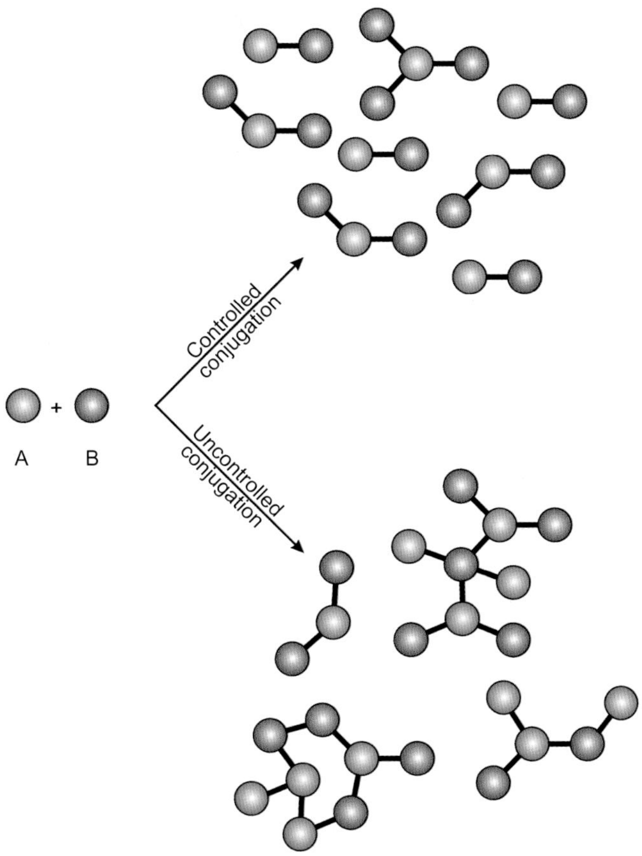

FIGURE 1.7 Oligomerized bioconjugation *versus* controlled bioconjugation. Uncontrolled conjugation can result in high-molecular-weight complexes being formed, which may aggregate or precipitate out of solution over time. Using methods of conjugate formation that limit the size of the resulting complex can enhance stability in solution and increase assay performance in certain applications.

the start of the reaction. For instance, if a 10-fold mole excess of NHS-PEG_4-biotin (Chapter 18) is added to an antibody, the final bioconjugate might have an average of perhaps seven biotins per antibody molecule, but there will also be species present in the population having a range of substitutions from perhaps three biotins to even 12 or more biotins per antibody, which on the high end even exceeds the amount of biotinylation reagent initially added (Adamczyk *et al.*, 1996). There are at least two reasons for this phenomenon: (1) reaction yields are never 100% efficient (especially in this example with competing hydrolysis of the reactive NHS ester group on the biotinylation compound) and therefore some molecules will be more effectively modified than others; and (2) local concentration differences caused when the reaction components are first mixed together may result in some antibody molecules experiencing a higher molar excess of the biotinylation reagent than other antibody molecules experience once the solution is completely homogeneous. Even with such a large distribution of different degrees of biotinylation, if reaction conditions are used that carefully control the amount of modification, it will help to maximize the activity of the conjugate and make production of the final complex more reproducible.

Bioconjugates can also take on many different shapes or structures. The simplest form is a bifunctional conjugate created by reacting two different components together. However, occasionally bioconjugates are made from more than two components, forming trifunctional or even multi-functional complexes that have the combination of three or more distinct properties merged together. One molecule may provide targeting capability to bind to a particular biomolecule or cell constituent, while a second component may be a biopharmaceutical agent useful to treat a certain disease, and a third group might provide detection capability (Figure 1.8). Bioconjugates of this type have been formed using molecular scaffolds that have multiple functional groups, which can be used to link together the components of the final complex. For instance, a bioconjugate of this type could be constructed using a dendrimer polymer (Chapter 8) to link together an antibody, a chemotherapeutic drug, and a fluorescent molecule, which together provide tumor targeting, treatment, and fluorescence detection capability all built into the same complex. In addition, Zaman *et al.* (2006) used a dextran polymer to link the chemotherapeutic drug doxorubicin and the hepatocyte-targeting molecule galactosamine to create a multifunctional bioconjugate for the treatment of cancer.

The use of the proper molecular scaffold can also amplify the activity of one of the components of a bioconjugate. In this regard, a multivalent polymer or dendrimer can be used to couple numerous copies of a detection molecule or enzyme, while in addition attaching a small number of targeting molecules. This can result in a specific targeting complex with extremely sensitive detection capability. An example of this type of conjugate is a gadolinium-chelate dendrimer that can be used as a contrast agent *in vivo* using magnetic resonance imaging (MRI) (Bourne *et al.*, 1996; Bryant *et al.*, 1999). The ability to couple multiple copies of the gadolinium chelate to the dendrimer surface significantly enhances the signal beyond that typical of a directly labeled targeting molecule without the use of a scaffold. Attaching a targeting molecule to this modified dendrimer, such as folic acid which is taken up by tumor cells (Swanson *et al.*, 2008), then provides specificity for imaging cancerous cells. Kojima *et al.* (2000) also used a similar dendrimer scaffold to attach poly(ethylene glycol) (PEG) grafts and noncovalently encapsulate the anti-cancer drug methotrexate for *in vivo* delivery.

Another important aspect of bioconjugate design is the type of linkage used to attach together the components of the conjugate. The linker strategy can

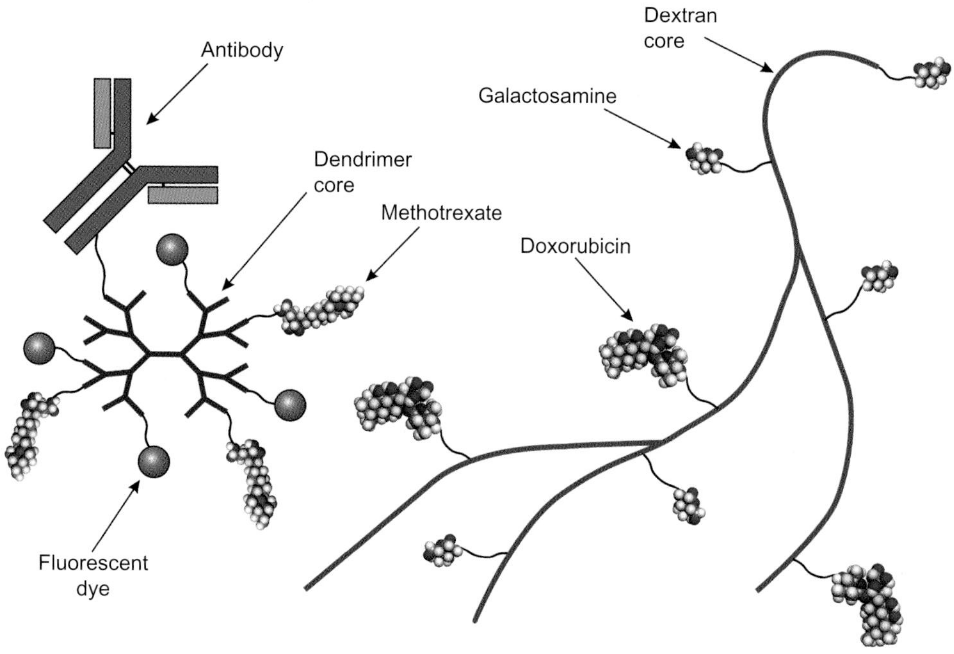

FIGURE 1.8 Examples of multifunctional bioconjugates including the use of molecular scaffolds, such as polymers and dendrimers. Left: A dendritic conjugate containing an antibody for targeting specific epitopes, a chemotherapeutic agent (methotrexate), and a fluorescent dye for imaging. Right: An illustration of a dextran conjugate containing galactosamine residues for targeting hepatocytes and the drug doxorubicin for antitumor activity. The polymeric scaffolds provide multiple sites for conjugation and thus allow more molecules to be attached per conjugate than direct crosslinking could provide.

incorporate multiple reactive groups as well as organic spacer molecules that separate the components and form bridges between the linked molecules in the bioconjugate. The properties of these linker arms can have significant effects on the properties of the final reagent. For instance, some crosslinkers contain hydrophobic spacers consisting of aliphatic or aromatic groups that may work well in forming a bioconjugate, but also might have detrimental effects on conjugate stability or nonspecificity in a given application. Conversely, some linking molecules are hydrophilic, such as PEG-based reagents, and actually increase the solubility of the final bioconjugate in aqueous solution. These property differences between hydrophobic and hydrophilic cross-bridges also apply to modification reagents, such as biotinylation compounds or thiolation reagents, which can affect the properties of intermediates as well as the final bioconjugates. Figure 1.9 illustrates some of the bioconjugation reagents that have similar purposes but differ in spacer arm hydrophilicity. For some applications, however, an extremely hydrophilic conjugate might not be appropriate if the bioconjugate is being designed to pass through cell membrane structures—in which case a more hydrophobic linker may actually perform better. Clearly, the design of every part of a bioconjugate can potentially affect its use and performance in specific applications. This is why there is an emphasis on optimization and testing of more than one bioconjugate design to determine the best possible construction for the desired application.

Even the length and construction of the cross-bridge between two components of a bioconjugate may significantly affect its overall properties. Especially when designing a biotherapeutic agent, the chemical construction of the linkage can have dramatic effects on its *in vivo* efficacy, sometimes in unexpected ways. For instance, Doronina *et al.* (2006, 2008) found that different two-amino acid linkers in the linking arm between an anti-tumor monoclonal antibody and the cytotoxic drug monomethylauristatin F had divergent effects on the toxicity and effectiveness of the bioconjugate. In this case, the construction of the cross-bridge also included a *p*-aminobenzylcarbamate group that was designed to be cleaved by intracellular proteases such as cathepsin B. The presence of this group in the linkage allowed the payload to be released inside the tumor cell, which enhanced its ability to cause cell death. Finally, the authors found that modification of a C-terminal amino acid residue on the drug either improved or impaired potency. Thus, depending on the application, the choice of linker construction should be carefully considered when designing a bioconjugate, including the reactive groups, hydrophilic *versus* hydrophobic cross-bridges, the possibility of incorporating cleavage sites, or the use of complex molecular scaffolds to link the components together. Chapters 9 and 15 discuss various cross-bridge options in greater detail, including the use of many different cleavage groups to release the components of the bioconjugate for specific purposes.

The construction of a bioconjugate using hydrophobic or hydrophilic crosslinkers can also have dramatic effects on reagent stability and nonspecificity, depending on the choice of reagent. Many conjugation schemes involve the modification of one protein with a crosslinker to provide secondary reactive groups and the modification of another protein to create the corresponding functional groups able to link with the

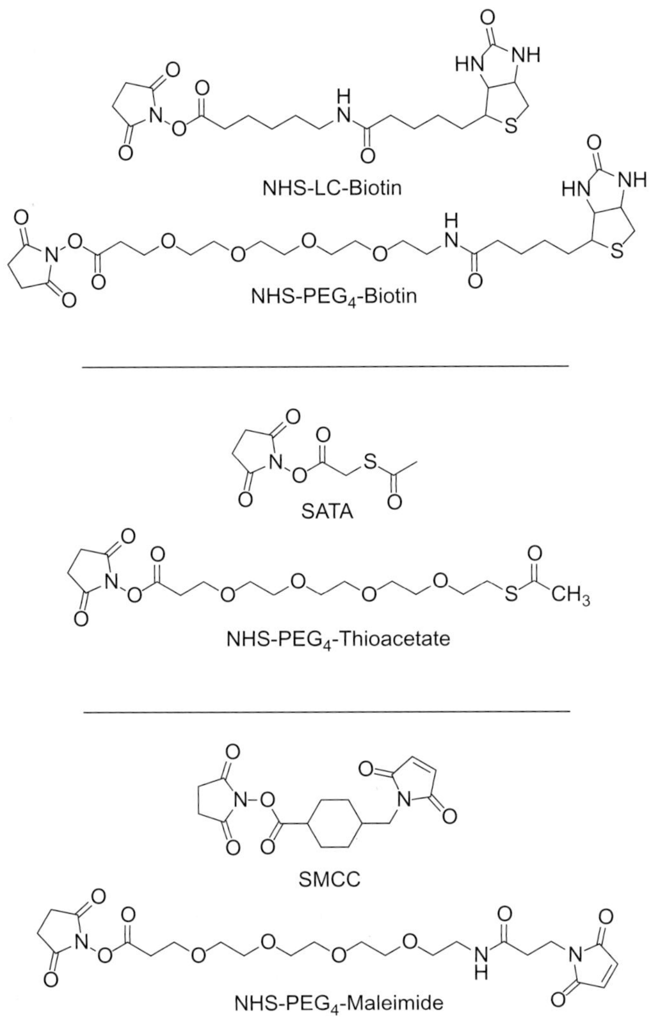

FIGURE 1.9 Hydrophobic *versus* hydrophilic crosslinker designs. Each reagent set provides similar reactivity on both ends, but the pairs of reagents differ by their water solubility and biocompatibility characteristics. Highly hydrophilic reagents such as those made from PEG spacers can offer benefits over aliphatic compounds in their ability to increase conjugate solubility and decrease nonspecific binding with biomolecules.

reactive groups on the first protein. If the initial reactions on the proteins use hydrophobic compounds with aliphatic or aromatic cross-bridges, then both proteins become decorated with these modifications. Such intermediates can contribute considerable hydrophobicity to the individual proteins even before the conjugate is created. Once the two modified proteins are mixed together and the conjugate is formed, the reactive groups or functional groups may be blocked at their ends, but the hydrophobic linker arms still protrude from the proteins. Since these initial modifications typically are done with reagents added in excess to form a multitude of linker arms spread across the protein surfaces, these modifications will still be present in the final bioconjugate. In addition, if every modification site does not result in a protein–protein conjugation event, then in the final complex both proteins will have "dead-end" modifications sticking off their surfaces. Every one of these remaining modification sites potentially can interact noncovalently with hydrophobic pockets on biomolecules (or with each other), thus creating tethers for possible nonspecific binding of the final complex.

These dead-end hydrophobic modification arms can also cause conjugate-to-conjugate interactions, which may result in larger noncovalent complex formation with significant potential for precipitation (Figure 1.10). Antibodies in particular are often succeptible to aggregation in solution even without additional hydrophobic linkers present on their surface. If hydrophobic modification or crosslinking agents are used to form an antibody conjugate, then the result may be to enhance antibody aggregation to the point of significantly lowered stability. Even a single modification reagent used to modify an antibody, such as performing a biotinylation reaction (Chapter 11), can cause immediate or long-term stability issues in aqueous solution. For example, a very popular choice in biotinylation reactions is the use of NHS-LC-biotin (Chapter 11, Section 6.2). This reagent has an extended aliphatic spacer arm between the amine-reactive NHS ester and the biotin group at the other end of the compound. In principle, a long spacer may seem advantageous for increasing the accessibility of a biotinylated antibody to bind with a streptavidin detection conjugate; however, the long aliphatic spacer arms actually cause more problems than advantages. If the biotinylation reagent is reacted in too high a molar excess, the antibody may suffer precipitation almost immediately due to hydrophobic interactions. Even if the biotinylation level is controlled to be at a lower level, the biotinylated antibody will usually slowly aggregate in solution over time through interactions between the biotin tethers on separate biotinylated antibody molecules as well as interactions of the tethers with hydrophobic pockets on the antibodies, thus causing continued activity loss.

By contrast, the use of a hydrophilic biotinylation compound, such as a NHS-PEG$_n$-biotin reagent (Chapter 18, Section 1.3), actually increases the hydrophilicity of a modified antibody after the reaction. Instead of promoting hydrophobic interactions and aggregation, an antibody having PEG-based biotin modifications on its surface becomes more stable in solution and has dramatically lowered nonspecific binding character. Therefore, if water solubility and low nonspecific binding to biomolecules are important criteria in a bioconjugate, it is better to explore the use of hydrophilic crosslinkers and modification reagents in the design strategy.

Another aspect of bioconjugate design to consider is the incorporation of modification agents that can

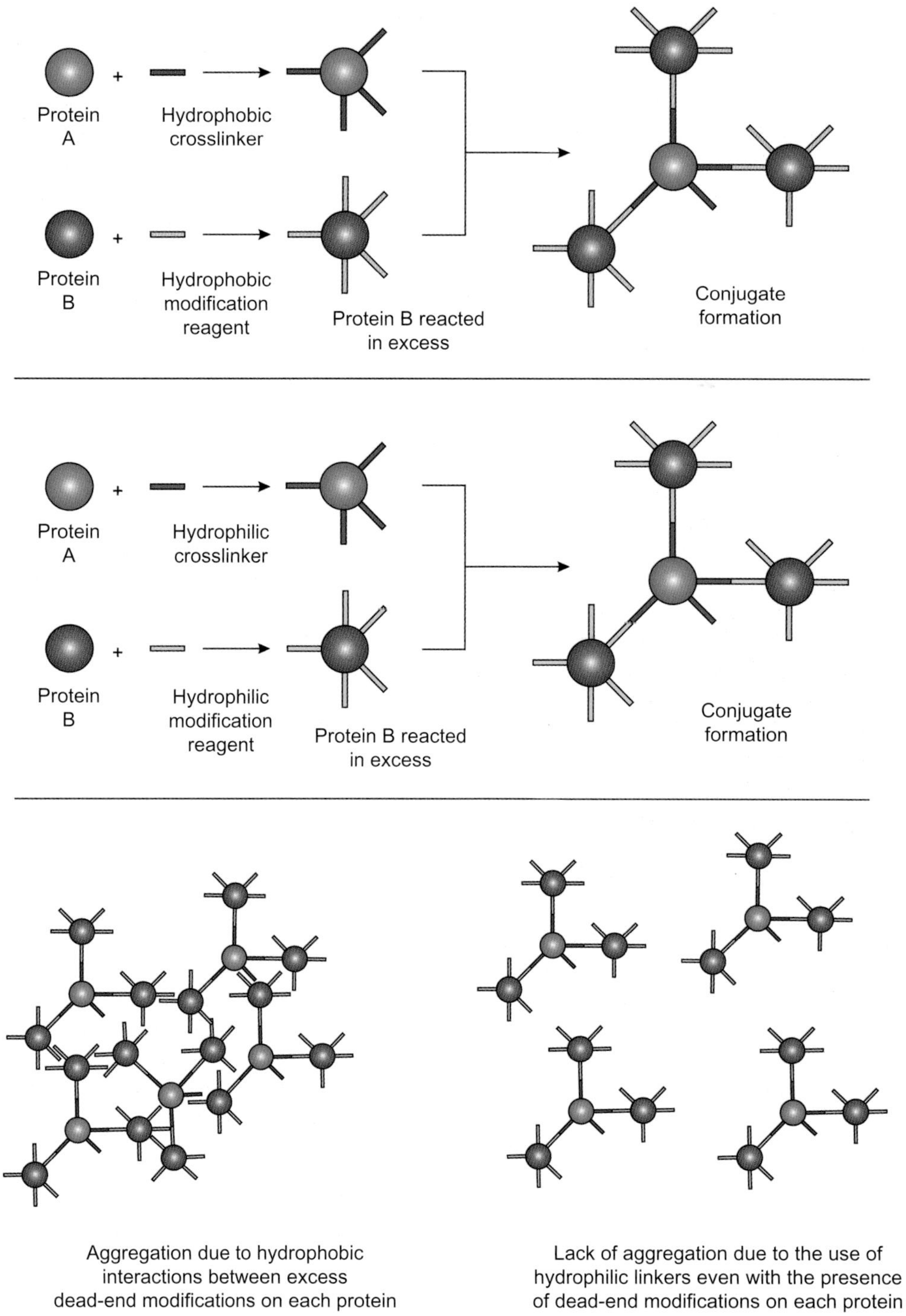

FIGURE 1.10 Conjugates made with hydrophobic reagents *versus* hydrophilic reagents and their propensity to aggregate or remain in solution. Each step in a conjugation process can add hydrophobicity or hydrophilicity to a conjugate. If two proteins are linked together by using crosslinking and modification agents, then their resultant water solubility and biocompatibility are greatly affected by the choice of reagent chosen. If hydrophobic, aliphatic linkers are used, both proteins in the conjugate will display these components on their surfaces, potentially affecting how they behave in solution. If hydrophilic reagents are chosen, such as PEG-based compounds, the molecules are linked by hydrophilic components and both proteins display these reagents on their surfaces. The presence of hydrophobic modifications on protein surfaces may cause aggregation or precipitation of the conjugates over time. It may also cause nonspecific binding characteristics toward molecules in biological samples.

positively affect reagent stability and half-life *in vivo*. One of the most important choices in this area is the use of polyethylene glycol (PEG) modifiers or crosslinkers (Chapter 18). Modification or "PEGylation" of an existing bioconjugate with linear or branched PEG molecules has been shown to enhance water solubility, increase the hydrodynamic volume of a bioconjugate, extend stability and half-life, and reduce immunogenicity and toxicity of drugs *in vivo* (Webster *et al.*, 2007; Jevevar *et al.*, 2010). PEG modification of bioconjugates can also dramatically improve water solubility and stability of reagents *in vitro*. The presence of PEG groups on a bioconjugate surface reduces nonspecific binding to other biomolecules, thus improving signal-to-noise ratios in assays and potentially enhancing detection sensitivity.

One of the most significant developments in PEG-based reagents in recent years is the new discrete molecules of defined polymer length. The early PEG compounds available for protein modification were all based on polydisperse polymer chain lengths in which a distribution of molecular weights was always present in a given reagent preparation. The discrete PEG compounds, by contrast, are based on pure molecular chain lengths, which provide defined reagents of known mass and properties. Discrete PEG-based crosslinkers and modification reagents are now available for creating bioconjugates of every type, and in most cases are better choices in forming conjugates than the corresponding aliphatic linkers. See Chapter 18 for a complete discussion of PEG reagents for modification and crosslinking, both of the polydisperse and discrete variety.

It should be apparent from this brief introduction that the choices involved in preparing bioconjugates are many, and particularly when designing new reagents, it often involves significant optimization to determine the ideal conjugation strategy from an initial concept that will give the best possible performance in an application. Even with experience, sometimes it is not clear from the outset what will be the ultimate bioconjugate design that will yield the best activity or efficacy. For this reason, it is important that a number of bioconjugate reagents and reactions be investigated—at least by initially putting together several options in theory—before deciding on a design strategy and route of synthesis. Especially when preparing bioconjugates for potential therapeutic or *in vivo* use, it is essential to thoroughly investigate the properties and performance of multiple conjugate designs before making conclusions as to the best construct for a given application. Even subtle changes in structure can have significant effects on bioconjugate performance when used for *in vivo* diagnostics or therapeutics.

Conversely, for the preparation of bioconjugates designed for more standard "tried and true" research detection methods or for use in research assays, the choices regarding the components to use and potential reaction strategies that will work are not as daunting as for those that are designed for clinical use. In fact, the rich history of bioconjugate design and applications can help guide the preparation of new conjugates that will work well for almost any intended use. The protocols provided in this book have worked to create bioconjugates of a wide variety of types. They provide viable starting points for the construction of almost any new bioconjugate designs using similar reaction strategies. In the end, however, the most viable bioconjugates are thoroughly optimized to perform best in their intended application and carefully produced to assure reproducible and robust reagents.

3. THE APPLICATIONS OF BIOCONJUGATES

Throughout this book, different bioconjugate types and strategies for making them are presented along with their principal applications for use. These applications generally can be grouped into the six major areas that are described in this section: (1) assay and quantification; (2) detection, tracking, and imaging; (3) purification, capture, and scavenging; (4) catalysis and chemical modification; (5) therapeutics and *in vivo* diagnostics; and (6) vaccines and immune modulation. It will be seen, however, that many conjugates successfully used in one application may also be applied to other areas; thus, many conjugate designs are not exclusive to a single application or method but can be used in diverse applications with equivalent success.

The following sections present an introduction to each of these application areas along with a discussion of the major bioconjugate designs that have been used for them. Each section also points the reader to the relevant chapters where additional information can be obtained on conjugate preparation and use.

3.1. Assay and Quantification

Perhaps the most prevalent use of bioconjugates is in the assay or quantification of target analytes, in which a substance to be measured is present in a complex sample mixture containing many other components. Designing an appropriate bioconjugate and assay strategy to measure such target molecules often involves first creating a two-component conjugate that consists of a specific targeting molecule attached through a covalent bond to a detection molecule. A targeting molecule can be anything that is able to discretely bind and interact with the desired target in the presence of other biological molecules. This can include antibodies, specific interacting proteins, peptides, nucleic acid sequences, substrate analogs, or other affinity ligands that exclusively bind to a portion of the target molecule with high enough affinity to interact and stay bound under the assay conditions.

In the majority of assay applications, the targeting molecule of choice is an antibody or an antibody fragment with specificity for binding a discrete epitope on an analyte. It can also be any other biological or non-biological molecule with biospecific or chemical affinity toward the substance to be measured. Such affinity groups may include saccharides that can bind to certain lectin binding sites, metal chelates that have affinity for certain functional groups (such as His-tagged fusion proteins or phosphate groups on phosphorylated proteins), substrate analogs that are able to bind to the active site on enzymes, and organic ligands that have affinity for receptor binding sites on proteins. Examples of some of these affinity targeting molecules are shown in Figure 1.11.

However, it takes more than just a good affinity molecule to develop an assay. Most molecules with the requisite binding specificity for a target do not in and of themselves possess the needed detection characteristics for quantification. It usually takes a specialized detection molecule coupled to the targeting molecule to provide the resulting complex with sufficient detectability for measurement of the desired target. This is where bioconjugation comes into play.

Targeting molecules such as antibodies have been used for decades to form bioconjugates using detection molecules such as fluorescent labels, enzymes, and chemiluminescent compounds (Figure 1.12). An antibody bioconjugate having this type of detection component can form the basis for making powerful assay systems, which can measure almost any biological molecule in a complex mixture. The readout from these detector molecules typically involves taking a measurement using absorbance, fluorescence, or luminescence. Detection bioconjugates made from a targeting molecule coupled to an enzyme can be used to act on a substrate and convert it into a chromogenic (colored) product or a fluorescent product, or to create an excited intermediate that subsequently emits a photon of light as it decays back to its ground state. There actually are several forms of luminescence signals that can be used in assay designs. Luminescence can be a consequence of the chemical triggering of a chemiluminescent compound or enzymatic turnover of a substrate that generates chemiluminescent or bioluminescent light emission. Ultimately, all of these luminescent detection systems operate through a series of reactions to create an intermediate excited state that then emits light.

Bioconjugates Used in Heterogeneous Immunoassays

To quantify a target molecule using most detection bioconjugates, a separation step must first be performed to capture and isolate the target from all other molecules in the sample solution. Once separated, a detection bioconjugate can be used to measure the amount of target present. However, in order to capture the target, another bioconjugate is required, which consists of a second specific affinity ligand able to interact with the target, bind, and then isolate it from the rest of the sample molecules. This capture step has two important functions: it gets rid of the other non-relevant components of the sample and it enriches the target molecule to enhance the potential signal in an assay.

To make this separation step as simple and rapid as possible, the capture bioconjugate usually is comprised of the capture molecule covalently attached to an insoluble support matrix (Chapters 14 and 15). Immobilization of the appropriate affinity ligand onto an insoluble particle or surface creates a bioconjugate that can be easily separated from soluble molecules simply by washing the affinity support with buffer (Figure 1.13). Upon binding of the target to the immobilized capture molecule, the rest of the unwanted sample components can be separated and washed away, leaving only the bound target molecule. Subsequent application of a detection bioconjugate to this bound target molecule complex followed by another wash step to remove excess detection reagent thus provides the means for quantification. Finally, comparison of the signal obtained from the interaction of the detection bioconjugate with both an unknown amount of target from the sample and that of a similarly treated set of known concentration standards provides the means for an accurate measurement of the amount of target in the original sample.

In the majority of all quantitative measurements for biomolecules, at least two bioconjugates are used in a heterogeneous assay system to measure a particular target molecule. The exact format that the assay takes can be highly variable, depending on the materials available, the measurement instrumentation to be used, and whether a single target is to be measured or multiple targets are to be analyzed simultaneously (multiplexed analysis). Solid phases such as microplates, tubes, planer array surfaces, beaded chromatography supports, microparticles, nanoparticles, and membranes all have been employed to immobilize capture molecules to facilitate a separation step in an assay (Figure 1.14; see Chapters 14 and 15 for immobilization techniques). Such capture conjugates combined with the appropriate detection conjugate can create viable assays for nearly any target, including proteins, nucleic acids, other biological macromolecules, as well as a wide diversity of small organic molecules. The various types of capture and detection bioconjugates used in these assays are the most widely used reagents in life science research, and most scientists do not even think about how the bioconjugates are made when using them to run an assay. This is how common and transparent bioconjugation has become today when doing routine assay procedures.

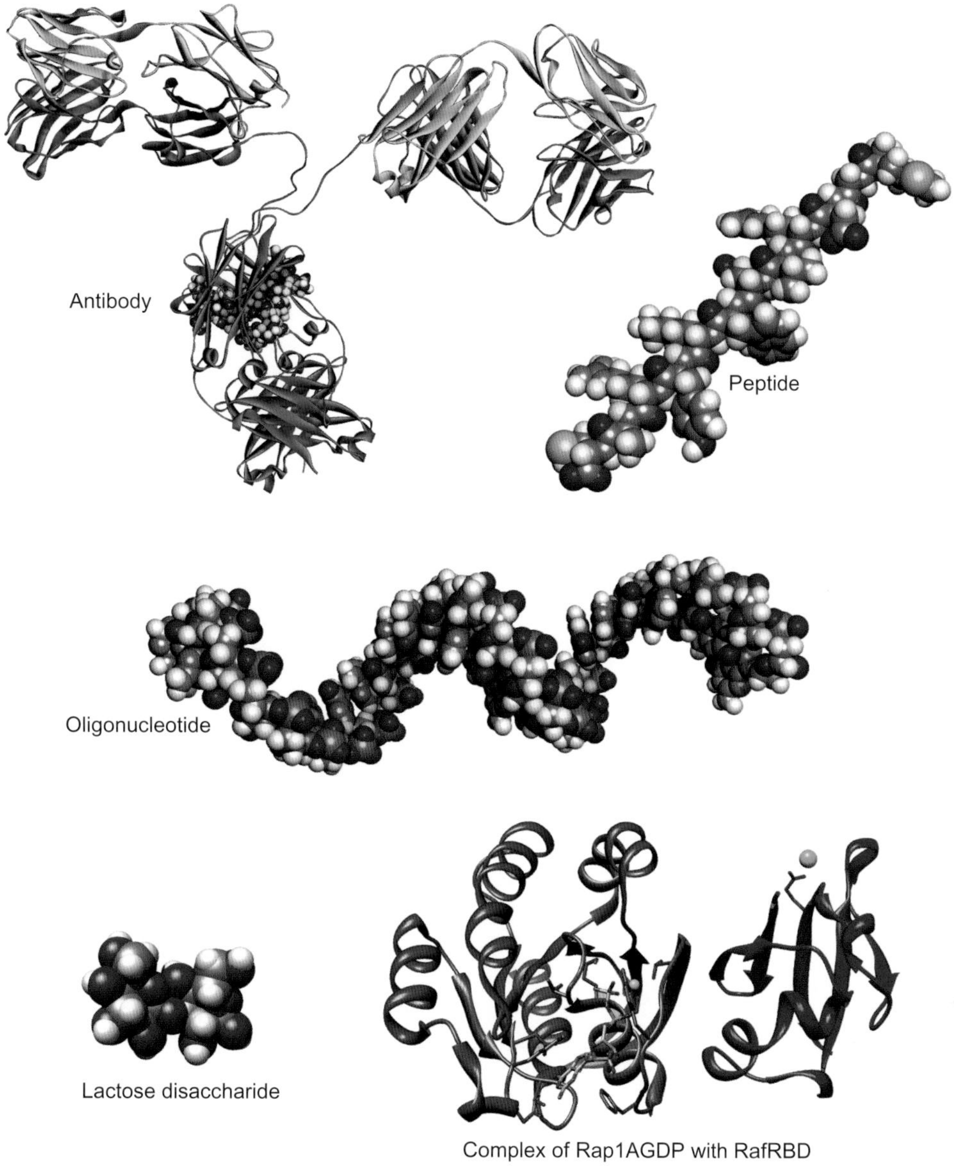

FIGURE 1.11 Affinity molecules commonly used to specifically bind to a target in an assay include the use of antibodies, peptides, oligonucleotides, saccharides or carbohydrates, and proteins that have specific interaction potential with other proteins or molecules. These targeting molecules are often used to form bioconjugates with detection molecules or other components have some specific activity desired in the final complex.

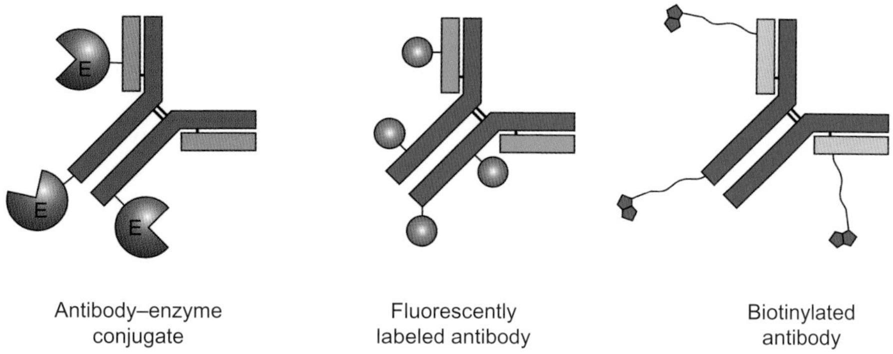

FIGURE 1.12 The most common antibody conjugates used in assays include (l to r) antibody–enzyme conjugates, fluorescently labeled antibodies, and biotinylated antibodies.

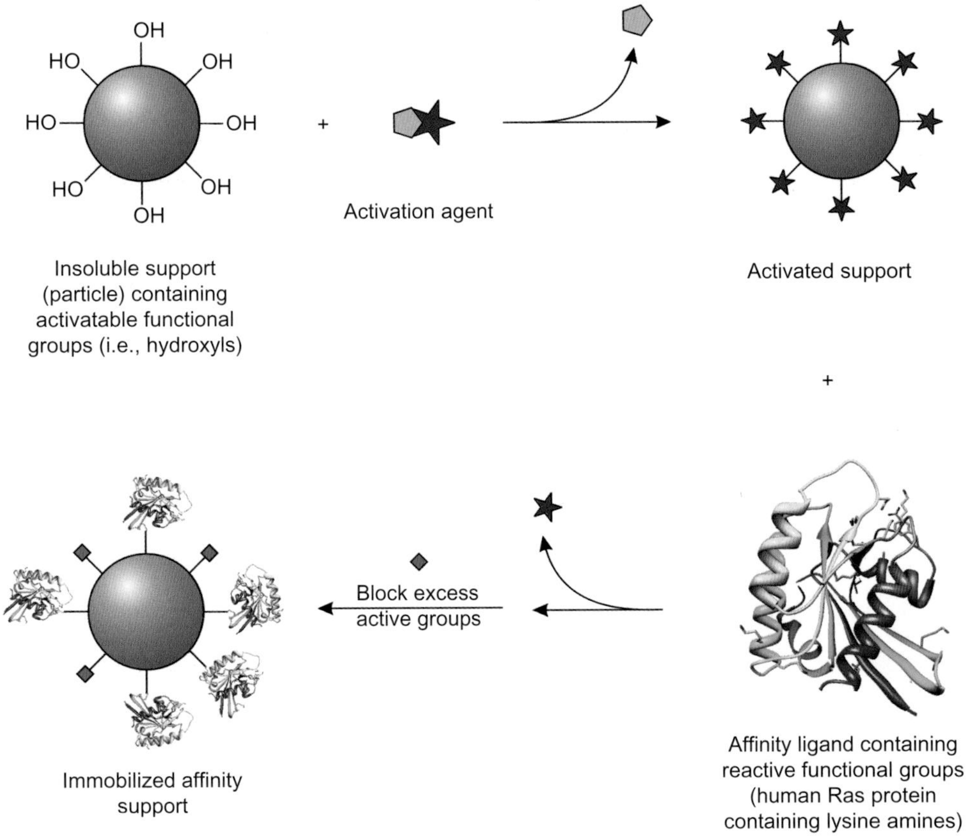

FIGURE 1.13 The immobilization of an affinity ligand onto insoluble supports typically involves activation of the support to create reactive groups and subsequent coupling of the ligand to form covalent bonds, thus linking it to the support.

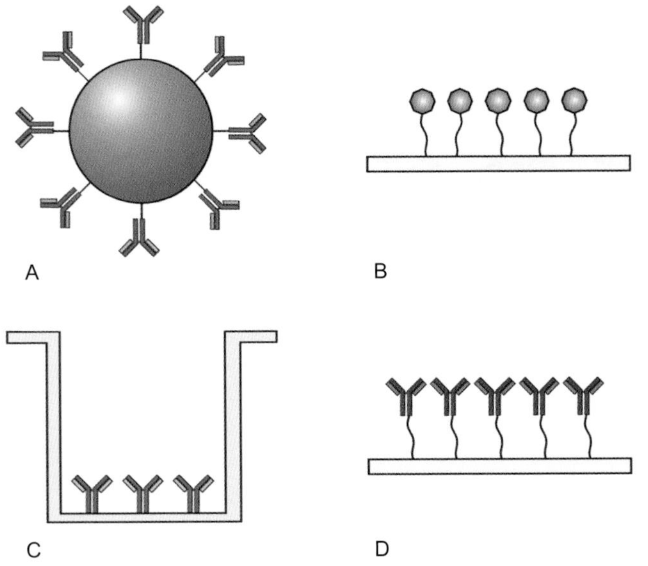

FIGURE 1.14 Some of the most common immobilized capture conjugates used in assays include (A) antibodies coupled to particles, (B) affinity ligands coupled to surfaces, (C) antibodies coated onto microwell plates, and (D) antibodies covalently linked to surfaces.

Bioconjugates for Enzyme-Linked Immunoassays

One of the more common assay designs that makes use of bioconjugates involves microplates with capture antibodies adsorbed or covalently coupled onto the surface for use in an enzyme-linked immunosorbant assay (ELISA) system (Lequin, 2005; Crowther, 2009). The assay is performed by adding samples to the wells of the capture antibody plate and allowing the target analyte (antigen) to bind (Figure 1.15). After a wash step to remove excess sample, the detection process is initiated, which often next includes the addition of an unconjugated primary antibody against the target. The immobilized capture antibody and the primary antibody can be both monoclonals directed against different epitopes on the analyte or polyclonals, or a mixture of both types. After another wash step, the detection antibody–enzyme conjugate is added, which uses a secondary antibody directed against the antibody species type used as the primary antibody. Any analyte-bound primary antibody that bound as a result of the first step will also result in the binding of the secondary antibody–enzyme conjugate. A final wash is then carried out to thoroughly remove any excess unbound detection conjugate before a substrate is added to develop the detectable product of the reaction (i.e., chromogenic, fluorescent, or luminescent signals). In the more rapid designs often used in clinical diagnostic testing, the primary antibody is a part of the detection conjugate to eliminate the extra steps of using a secondary antibody–enzyme (or streptavidin) conjugate in the assay. The only downside to the use of a direct primary antibody–enzyme conjugate is that the sensitivity and minimum detection level of the assay are usually not quite as good as that possible when using an unconjugated primary antibody followed by the secondary antibody–enzyme conjugate. This is because a multi-layering effect of binding the primary antibody plus the secondary antibody–enzyme complex allows for more detection conjugates to bind to each captured analyte than would be possible with just a single primary antibody conjugate alone docking on the analyte.

Another frequent bioconjugate strategy used in immunoassays is the use of a biotinylated primary antibody, which initially is bound to the captured analyte in the assay. To detect this binding event, typically a (strept)avidin conjugate is added that contains the enzyme used to develop the substrate. This multi-layered ELISA detection process is a very common method for assay of antigens using sandwich immunoassays, because the production of a biotinylated primary antibody is simple and the (strept)avidin–enzyme conjugate becomes a universal detection agent for all such assay designs.

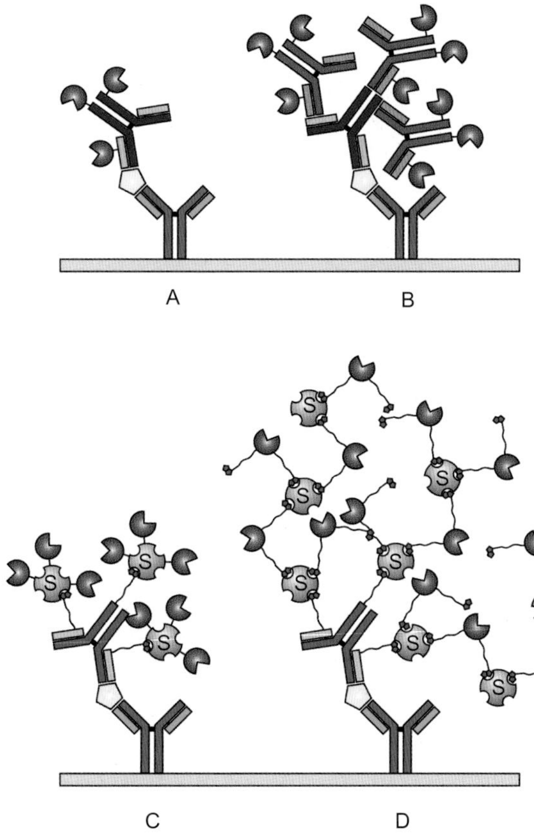

FIGURE 1.15 Typical ELISA methods using capture and detection bioconjugates. (A) a sandwich assay design using a primary antibody-enzyme conjugate, (B) a sandwich assay design using a primary antibody plus a secondary antibody–enzyme conjugate, (C) a sandwich assay using a biotinylated primary antibody and a streptavidin–enzyme conjugate, and (D) a sandwich assay using a biotinylated antibody along with a biotinylated enzyme and a streptavidin bridging molecule. Some methods of using bioconjugates in assay systems can result in better detection capability by recruiting more detection molecules at the site of the measured analyte.

Thus, for ELISA formats of all types the production of a universal detection agent like a secondary antibody–enzyme conjugate or a (stept)avidin–enzyme conjugate can provide the basis for developing virtually any analyte-specific assay. The methods for making such conjugates can be found in Chapters 11, 20, and 22.

Chemiluminescent Acridinium Ester Conjugates

A unique detection bioconjugate that is used in many heterogeneous diagnostic immunoassays consists of a complex between a targeting molecule such as an antibody and an organic chemiluminescent compound, such as an acridinium ester. This type of labeled conjugate can be used in assays that generate a photon of light from each acridinium modification present on the targeting molecule. A trigger solution containing hydrogen peroxide (H_2O_2) is added at the end of a typical immunoassay procedure followed by an addition of NaOH to result in highly alkaline conditions to rapidly oxidize the acridinium ester compound. This oxidation reaction forms an intermediate dioxetanone that subsequently breaks down into the singlet electron excited state (*N*-methylacridone), which then emits a photon of light when it returns to the ground state (Figure 1.16).

The detection limit of an acridinium ester-labeled antibody in an assay is typically in the low fmol range, indicating that targeted analytes can be measured with high sensitivity. One deficiency of using this detection conjugate design, however, is the short lifetime of the chemiluminescence once the label is triggered. For this reason, the trigger solution must be added within an instrument by rapid injection followed by immediate signal measurement to capture the emitted light.

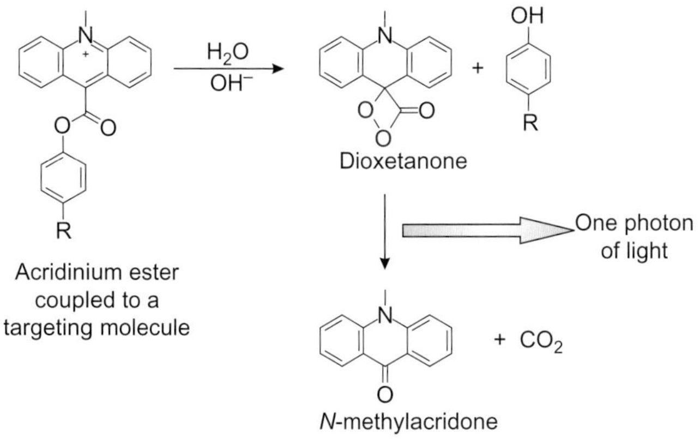

FIGURE 1.16 Acridinium ester bioconjugate and mechanism of light generation. Labeled antibodies made with acridinium esters can be used in a flash chemiluminescent assay system to detect desired antigens.

A recent alternative method of generating the chemiluminescent reaction is a continuous electrochemical production of superoxide using flavin-containing molecules, such as flavin adenine dinucleotide (FAD), flavin mononucleotide, or riboflavin (Aizawa *et al.*, 2006). Another technique uses an initial acridan ester compound, which upon electrochemical-mediated oxidation yields the reactive acridinium ester that can then be further oxidized with peroxide and base to produce light (Wilson *et al.*, 2001). These alternative electrochemical techniques eliminate the need for injection of the trigger solution and extend the overall lifetime of chemiluminescence, but they still require the auxiliary electrochemical system to generate signal.

In addition, unlike chemiluminescent systems that use enzyme–substrate reactions to produce light, the acridinium ester groups attached to the targeting molecule cannot continue to produce light catalytically. Once they are oxidized, the ester is cleaved from the targeting molecule and the acridinium degrades to an *N*-methylacridone derivative that does not get regenerated. Thus, only a single photon of light is emitted per acridinium ester group on each bioconjugate. This contrasts with conjugates made using antibodies linked to HRP or alkaline phosphatase that can use the turnover of a chemiluminescent substrate to produce an ongoing signal. For this reason, the enzyme conjugate systems are inherently more sensitive than the acridinium ester system, and therefore they may be a better choice for routine assays.

Fluorescent Oligonucleotide Probes

Bioconjugates consisting of oligonucleotides having fluorescent labels attached to them for use in hybridization methods can be employed in heterogeneous assays to detect and quantify specific target DNA or RNA sequences. These conjugates can be synthesized during solid-phase oligo synthesis using fluorescently labeled phosphoramidite monomers or the oligo probes may be labeled subsequent to synthesis using a number of reaction strategies, including enzymatic and chemical modification (Chapter 23). Once a fluorescently labeled oligo conjugate is available, it can be used to assay a complementary target sequence after the target has been captured by another oligo attached to a surface or particle. The capture oligo hybridizes to one portion of the targeted genomic DNA or RNA sequence, immobilizing it on the surface while unbound components of the sample are removed, and then a fluorescent detection oligo is added to hybridize to the other end of the targeted sequence. This assay is similar in concept to sandwich antibody assays often used in immunoassay procedures, but here two synthesized oligo conjugates are used to first capture and then detect the target instead of two antibodies (Figure 1.17).

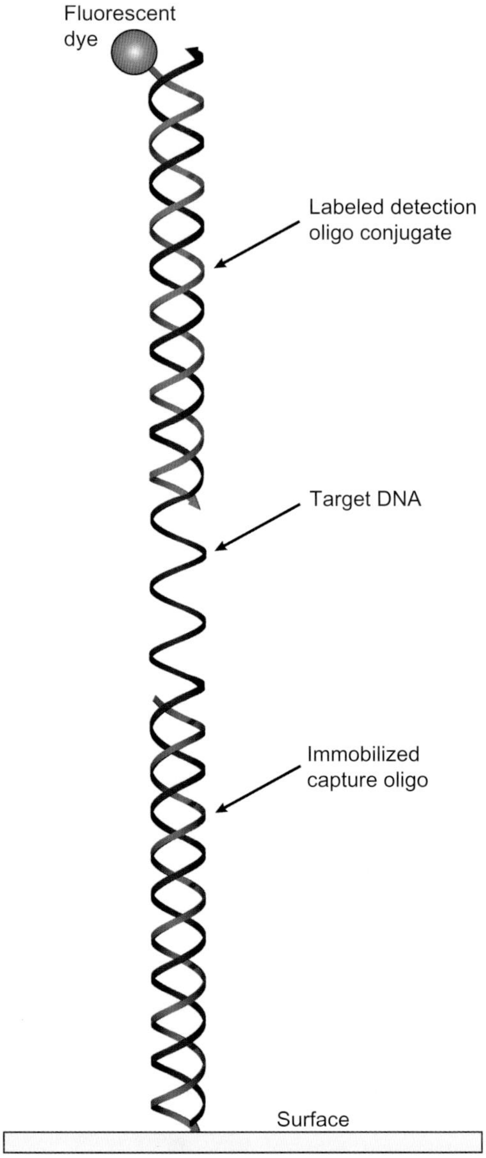

FIGURE 1.17 Fluorescently labeled oligo conjugates and capture oligo used in hybridization assays. The initial conjugate involves the attachment of the capture oligo to the surface, which binds to the targeted genomic DNA. Next a detection oligo conjugate consisting of a fluorescently labeled oligo that can bind to another region of the targeted DNA is used to provide signal for the binding event.

An alternative to the use of fluorescently labeled oligo probes is to prepare conjugates of a targeting oligo that is coupled to an enzyme. Similar to antibody–enzyme bioconjugates used in ELISA systems, the oligo serves as the affinity component that can interact with the complementary DNA- or RNA-targeted sequence and the enzyme functions as the detection component, which can act on a substrate to create a detectable signal—fluorescent, chromogenic, or chemiluminescent.

Hybridization assays can be carried out using many different heterogeneous assay formats. For instance,

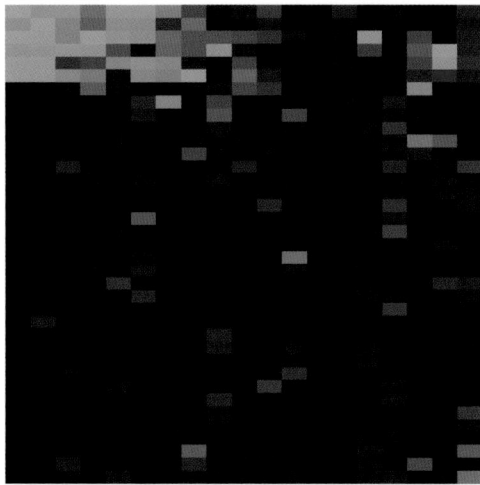

FIGURE 1.18 The fluorescent signals from an oligo array. Spots that have captured specific complementary genomic sequences are detected using a fluorescently labeled detection oligo. *The image was created by Miguel Andrade, as posted on Wikipedia from data available in the public domain at http://www.stembase.ca/ and using the public domain program Cluster from Michael Eisen, which is available from http://rana.lbl.gov/EisenSoftware.htm. The author released the image into the public domain to use for any purpose under the Wikipedia Commons.*

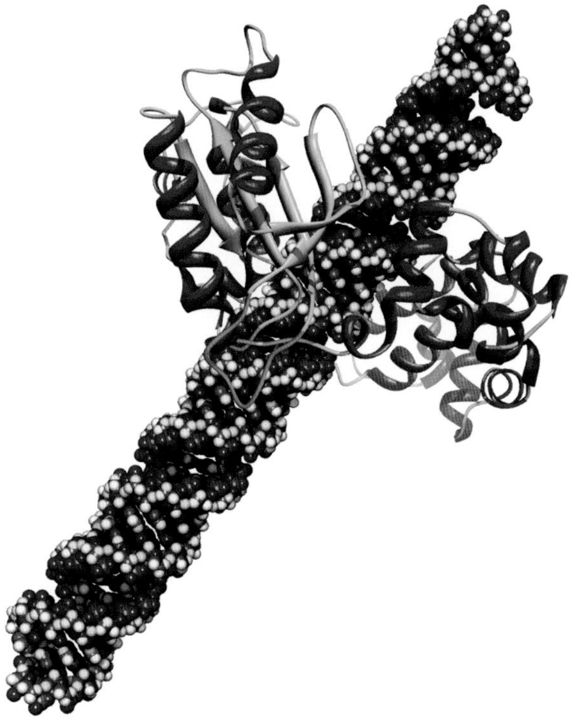

FIGURE 1.19 Selected DNA sequences can be used to capture specific binding proteins in a chromatin immunoprecipitation assay (ChIP), such as the binding of transcription factors and other regulatory proteins. The binding of DNA polymerase beta to a segment of double helix is shown here.

a capture oligo can be immobilized onto the wells of a microplate, onto a microparticle surface, or onto a microchip using reactions suitable for activation of the given surface followed by coupling to an appropriate functional group on the oligo, usually put on the 3′ or 5′ termini (Chapters 14, 15, and 23).

Labeled oligos have become key bioconjugates in the development of multiplexed oligonucleotide assays, which now are used routinely to map critical gene profiles associated with disease or genetic abnormalities. Microchips containing hundreds or thousands of unique oligonucleotides have been produced to simultaneously capture multiple gene targets within microchip boundaries not much larger than a few mm^2. These chips may be probed using fluorescently labeled detection oligos to result in differential fluorescent signals emanating from the oligo array at discrete points corresponding to each captured target. Software algorithms can then be used to quantify the amount of target captured at each array point (Figure 1.18). A number of companies, such as Affymetrix and Roche NimbleGen, have commercialized high-density oligo arrays for use in profiling gene expression and in the diagnosis of disease. The market for Affymetrix products alone in 2011 was $267.5 million (Affymetrix annual report, 2011). Note that this market has been declining as the DNA sequencing market is accelerating upward, which represents a technique that also uses fluorescently labeled nucleotide probes.

Immobilized oligonucleotide probes also are being used for chromatin immunoprecipitation (ChIP) assays. Instead of probing for complementary nucleotide sequences that can bind to the immobilized oligos as in hybridization assays, these bioconjugates are used to capture proteins which specifically interact with them. For instance, transcription factors and polymerases often interact with segments of genomic DNA to turn on genes or replicate the DNA double helix (Figure 1.19). So-called "ChIP-on-chip" arrays can be used to discover the binding regions for specific transcription factors and other regulatory proteins, which interact with certain sequences on genomic DNA and function to control transcription (Buck and Lieb, 2004; Shaked *et al.*, 2008). Oligo capture probes coupled to planar surfaces or particles can be used to isolate these interacting proteins by using the solid phase as a convenient affinity support for removing the non-binding components in a sample prior to analysis (Figure 1.20). Alternatively and more commonly, the regulatory protein-DNA complexes first can be covalently captured within cells using fixation with formaldehyde or other crosslinking agents. Cell lysis is then carried out followed by

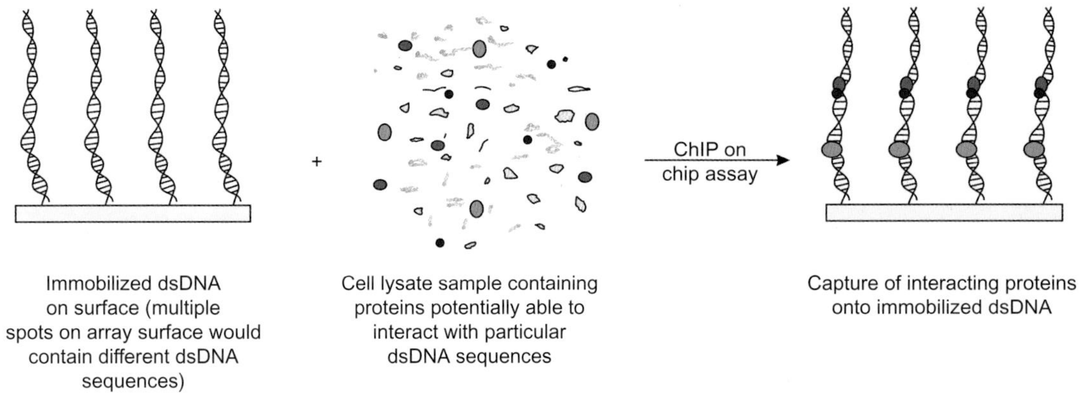

FIGURE 1.20 A ChIP-on-chip array design using immobilized dsDNA to capture interacting proteins is illustrated. A cell lysate sample containing potential binding proteins is incubated with the array and proteins are allowed to bind. These proteins can be eluted and detected by western blotting or mass spec methods.

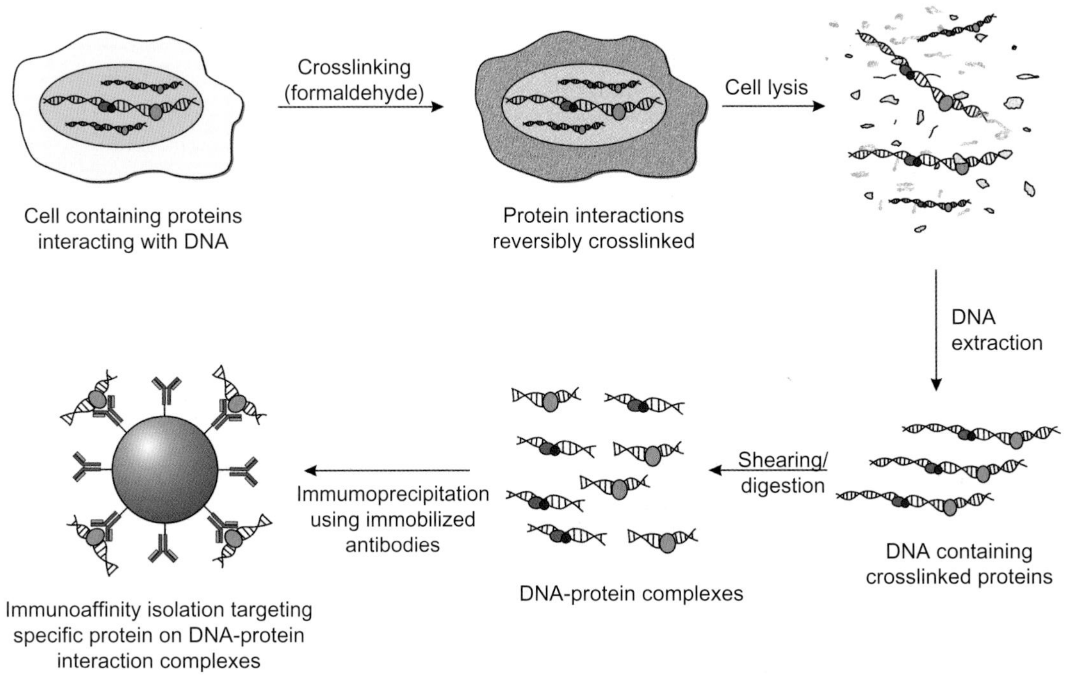

FIGURE 1.21 A ChIP assay design using immobilized antibodies can be used to capture specific interacting proteins onto particles for subsequent analysis by western blotting or mass spectrometry.

disruption of the chromatin using sonication and partial enzymatic digestion, which fragments the DNA into short sequences of 100 to 250bp. These DNA–protein complexes can then be isolated using particles containing immobilized antibodies directed against the regulatory proteins of interest (Figure 1.21). ChIP assays can be used to study gene activation and epigenetic modifications within chromosomes that control protein expression within cells.

Another important application area for oligonucleotide bioconjugates is represented by the technique called "fluorescence *in situ* hybridization" (FISH), described by Pinkel *et al.* (1988). In this method, gene targets can be directly probed within cells or tissue sections using labeled oligo conjugates (Nath and Johnson, 2000; Liehr, 2009). The complementary oligo probes can be prepared by limited digestion of genomic DNA using DNase or made synthetically by selecting the proper sequence regions within the DNA target and synthesizing a complementary strand which will hybridize to it. However, the nucleotide probes used in FISH assays typically are longer than those probes used in hybridization arrays to ensure better specificity and faster binding to the genomic target sequence.

The oligo conjugates used for targeting in FISH can be of several varieties (Figure 1.22). They can be directly labeled with a fluorescent dye (Chapters 10 and 23) or they can first be labeled with another molecule that subsequently can be detected using a fluorescently labeled antibody (Chapter 20) or a fluorescently labeled (strept)avidin conjugate (Chapter 11). One of the more popular oligo labels for antibody targeting is to use the small organic hapten digoxigenin (DIG), which is a plant hormone found only in the leaves of certain plants (foxgloves). Antibodies of high affinity are available against DIG and since it is not found in mammalian cells DIG-labeled oligo probes can be detected within cells with high specificity and low background. An alternative to DIG labeling is to use a biotinylated oligo probe (Chapters 11 and 23), which can be detected using a fluorescently labeled (strept)avidin conjugate. Biotinylated oligonucleotides are common bioconjugates used to probe DNA and RNA sequences in a number of applications.

DIG or biotin modifications can be enzymatically incorporated into DNA probes using dUTP derivatives, typically DIG-11-dUTP or biotin-11-dUTP. These derivatives are commercially available from many suppliers and can be used in PCR labeling procedures or other enzymatic methods to add several DIG or biotin tags into an amplified segment of an oligonucleotide probe (Chapter 23). For instance, terminal transferase can be used to add DIG or biotin dUTP derivatives to the 3′ end of an oligo probe in a template independent manner (Schmitz *et al.*, 1990). This labeling method typically adds multiple copies of modified nucleotides (10–100 copies) to the 3′ end, depending on the concentration of the dUTP derivative used. Alternatively, these tags can be chemically added through conjugation of an amine-modified oligo to the NHS ester derivatives of DIG or biotin (Chapter 11 and 23) (Figure 1.23). The modification of a 5′-amino derivative of an oligonucleotide can add a single biotin handle to the end of an oligo probe, thus avoiding over labeling, which could lead to decreased hybridization efficiency.

Molecular Scaffolds to Increase Assay Sensitivity

In some strategies to further enhance the signal and sensitivity of assays, multivalent scaffolds may be used in bioconjugate construction to increase the number of detection molecules associated with a single targeting molecule. In this approach, the docking

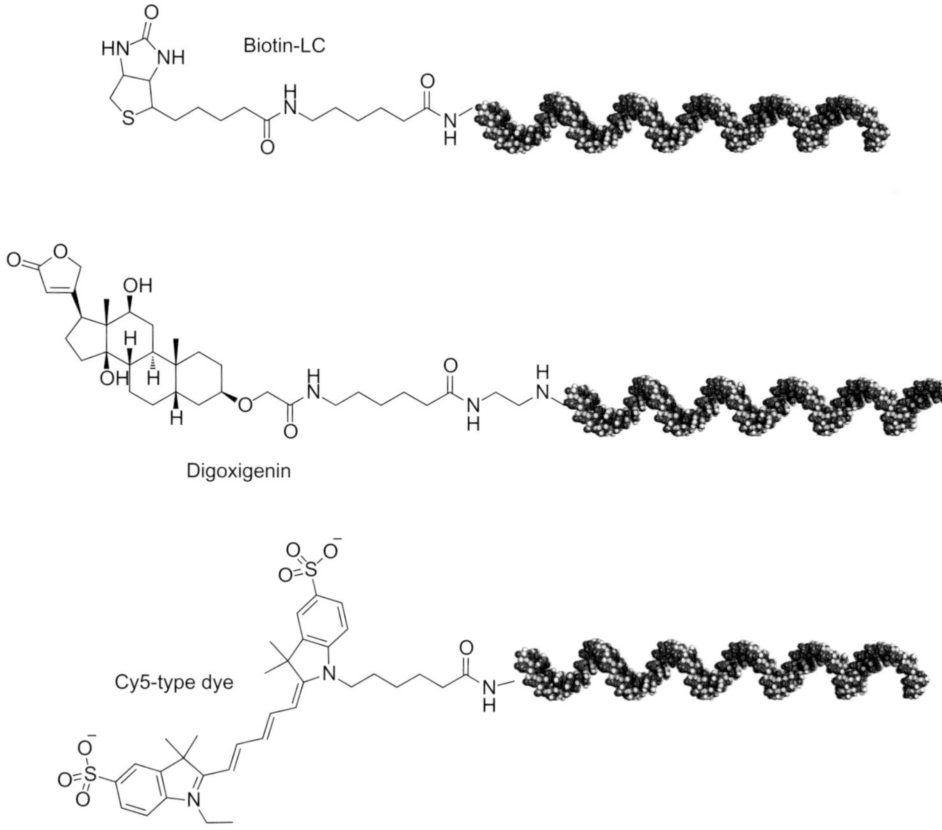

FIGURE 1.22 Fluorescent oligo conjugates typically used in FISH assays include labels such as biotin, digoxigenin, or a fluorescent dye to detect hybridization events. The biotin- and digoxigenin-modified oligos typically are used with streptavidin conjugates or antibody conjugates (DIG) for detection.

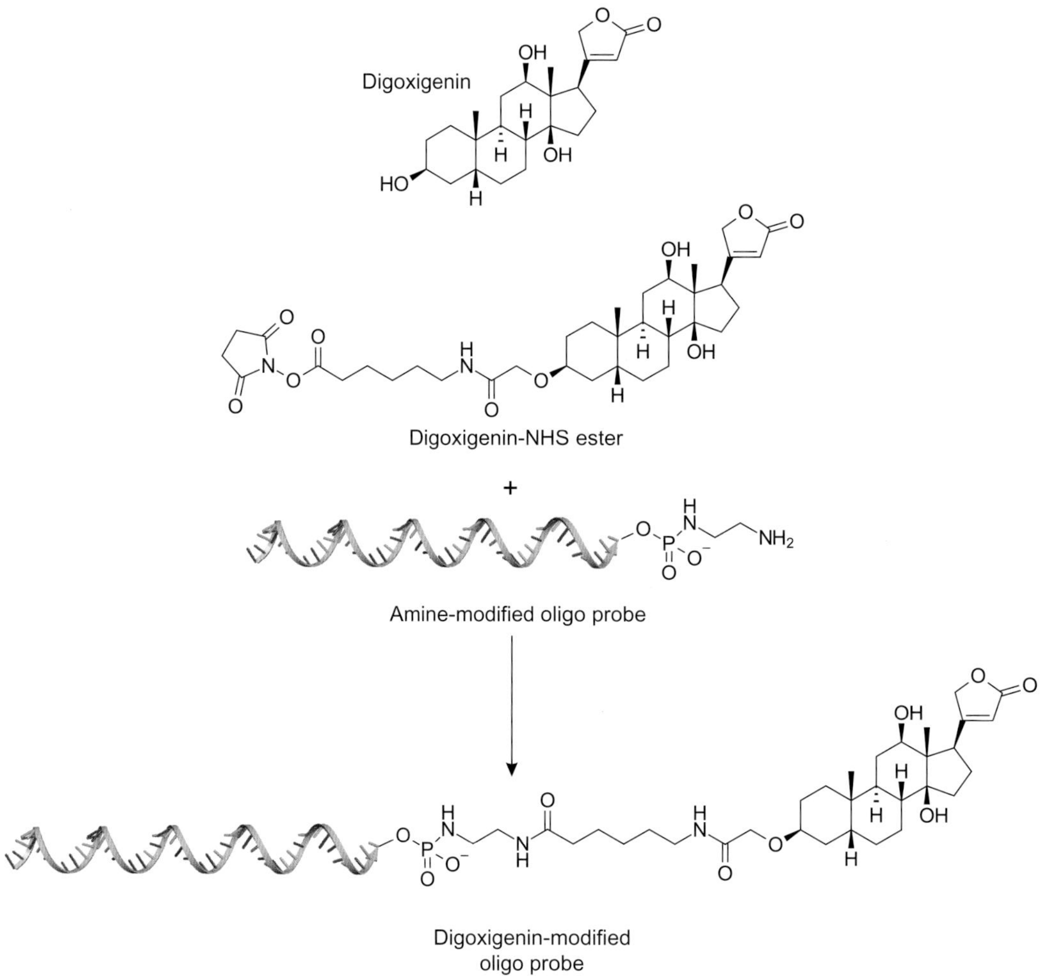

FIGURE 1.23 DIG-NHS ester modification of an amine-modified oligo to create a probe useful in hybridization assays.

of one targeting molecule on its target analyte brings with it a larger number of detection components due to the scaffold design, therefore increasing the resultant sensitivity in a particular assay. Some conjugates that are constructed in this manner to increase signal include scaffold components consisting of dendrimer polymers (Chapter 8), linear or branched polymers (Chapter 18), particles (Chapters 14 and 15), and other interacting molecules including large complexes formed from use of the biotin–(strept)avidin system (Chapter 11) (Figure 1.24). In each system, assay designs can be constructed using advanced bioconjugation techniques that can multiply the detection signal over that normally observed without the use of a scaffold. Such scaffold designs are used not because a simple bioconjugate of a labeled antibody will not work in most assays; rather, the use of auxiliary scaffolds is driven by the need to detect and measure ever decreasing concentrations of target molecules, especially when measuring very low copy numbers of proteins expressed in cells.

Bioconjugates Used in Homogeneous Assays

Specialized bioconjugates can also be developed for use in single-step or homogenous assay formats. Unlike in a heterogeneous assay system, this method involves the mixing of sample and detection reagents together in a single reaction solution and the assay result is obtained without any separation steps being done. With heterogeneous assays, the detection conjugate is bound to its target analyte on an insoluble solid phase support and any excess reagent is removed before a measurement is taken. This removes any nonspecific signal due to unbound conjugate and isolates on the solid phase only the specific signal due to the detection conjugate being bound to the target analyte. Homogeneous or single-step assays have to be designed to allow analyte detection even in the presence of excess detection conjugate. Thus, the bioconjugates used for this type of assay must create a detectable signal if and only if they are bound to their target analyte. The advantage of the homogeneous or single-step assay design is that a test can

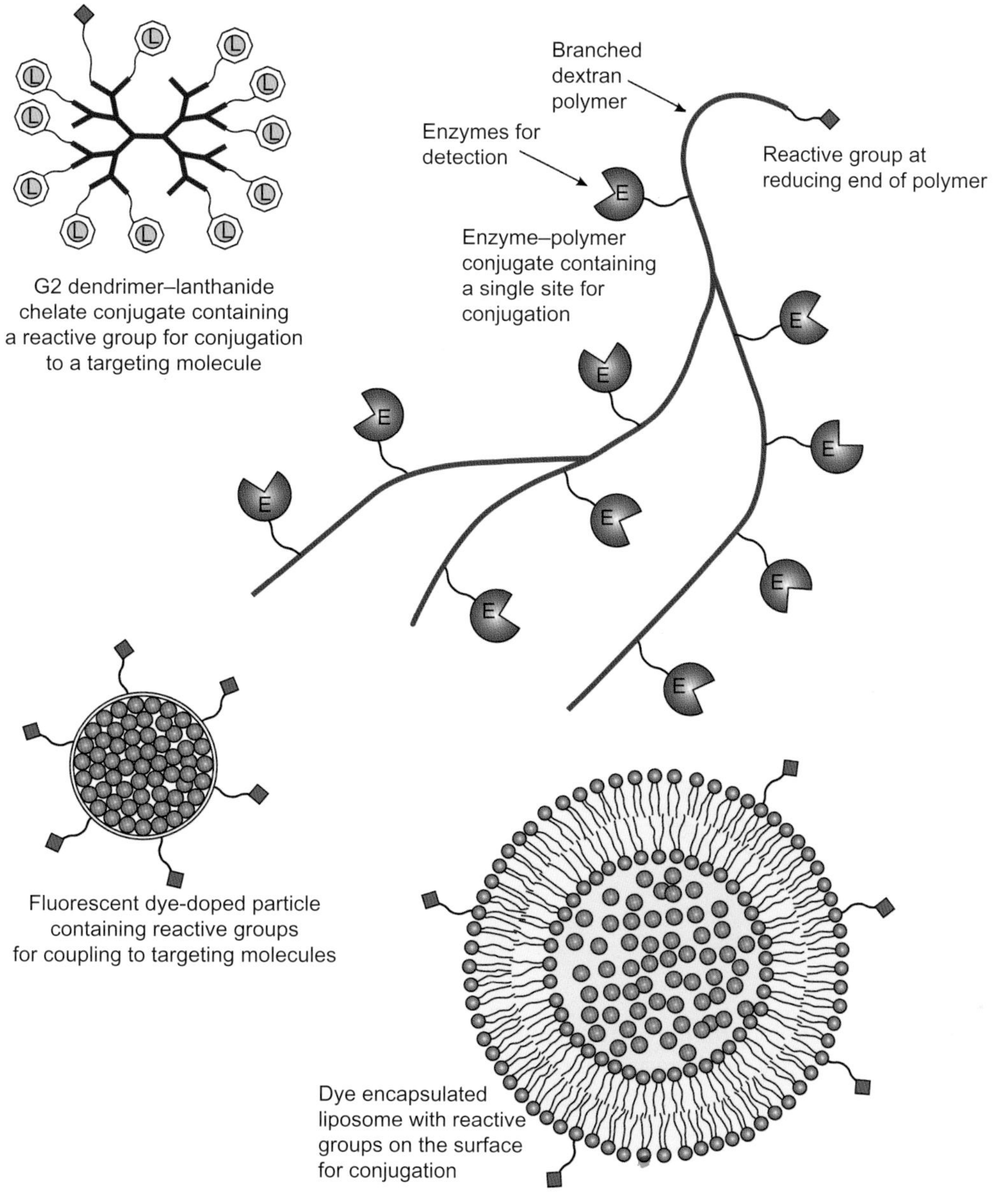

FIGURE 1.24 Scaffolds commonly used to enhance detection in assays through multivalent conjugation sites. The upper left shows a dendrimer containing multiple fluorescent lanthanide chelates for high sensitivity detection plus a reactive group for labeling a targeting molecule, such as an antibody. The upper right shows a dextran polymer linked to numerous enzyme molecules for detection using substrates and a reactive group on the polymer's reducing end for conjugation purposes. The bottom left illustrates a particle containing multiple fluorescent molecules and reactive groups on its surface for bioconjugation. Finally, on the lower right, a liposome-containing dye molecule with reactive groups on its surface for conjugation with targeting molecules is shown. Each of these scaffold designs can be used to generate greater signals in assays or in imaging methods.

be completed almost immediately after the sample and reagents are mixed, and without the need for subsequent washings, reagent additions, or lengthy incubations. As might be imagined, this format is particularly attractive for use in diagnostic testing and drug discovery screening, as both applications are sensitive to the price-per-test expense and the time it takes to complete an assay.

Bioconjugates used in homogeneous assays typically are designed to utilize one of three strategies for assay detection: (1) fluorescence resonance energy transfer (FRET); (2) bioluminescence resonance energy transfer (BRET); or (3) split reporter proteins (Figure 1.25). The first two systems involve non-radiative energy transfer between an emitter of light and an acceptor of light, wherein the acceptor can be either

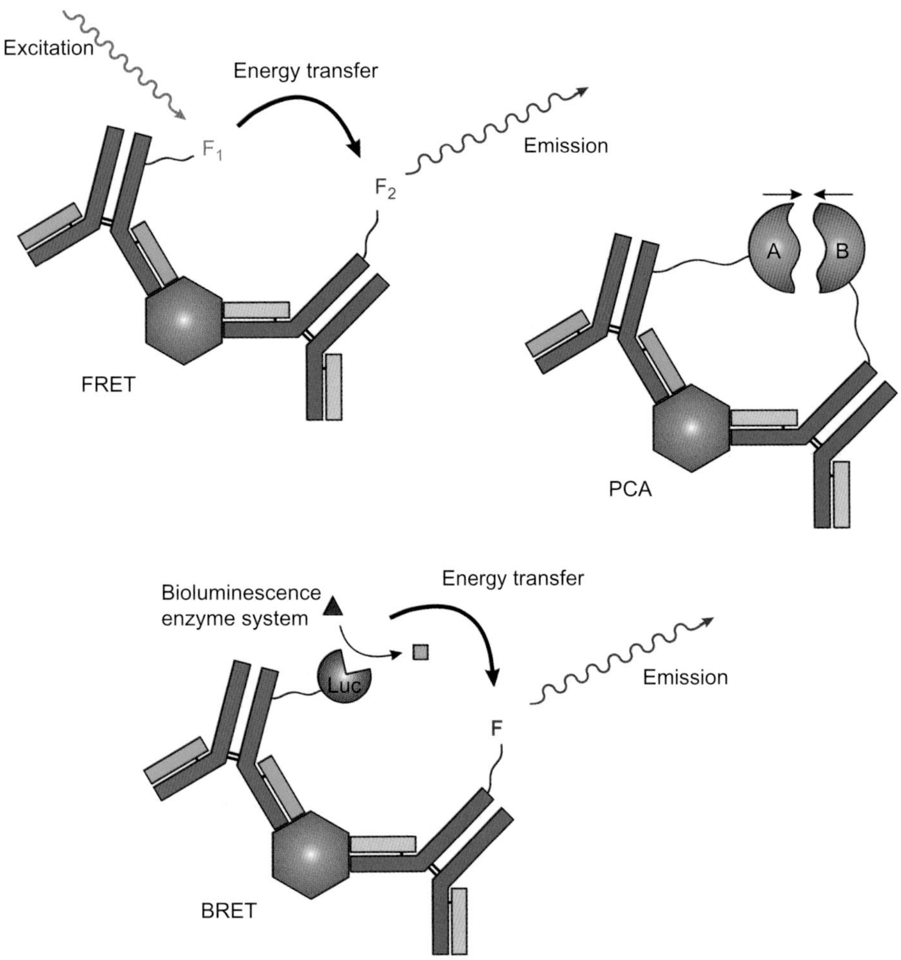

FIGURE 1.25 Basic homogeneous assay formats using bioconjugates. FRET-based systems often use antibodies labeled with two fluorescent molecules that can transfer energy between them if they are held close enough together by docking on a target protein. A protein complementation assay (PCA) uses recombinantly split enzymes or fluorescent proteins that have no activity or fluorescence unless they come together close enough to reconstitute the intact three-dimensional structure of the native protein. Bioluminescence-based systems use energy transfer between a luciferase enzyme and a nearby fluorescent dye to transfer energy from the converted luminescent substrate to the dye for emission at another wavelength. All three designs can facilitate the development of single-step or homogeneous assays without the need for separation steps to detect the specific signals.

an emitter at a different wavelength or a quencher (absorber) of the emitter's energy. Split reporter technology, also called "protein fragment complementation assays" (PCAs), involves the use of truncated portions of proteins that can bind together only when the fragments are positioned and held within an appropriate molecular distance of each other and thus result in restoration of the protein's unique detection characteristics (Michnick, 2003; Michnick *et al.*, 2007). In this system, each fragment of the split reporter produces no detection signal on its own, because it requires the other portion of the molecule to restore native structure and activity.

The strategy fundamental to most homogeneous assay systems involves the use of two different detection components that come together in molecular proximity as they are bound to the desired target. The bioconjugate designs for use in such systems should permit signals to manifest only when the detection reagents are held through tight affinity interactions onto the targeted analyte. This brings the two detectors close enough together to generate an efficient energy transfer or allows for regeneration of split reporter activity, which results in the desired signal. To ensure specificity, any excess detection conjugates that are not bound to an analyte must not contribute to any signal or the assay would have considerable background noise and low sensitivity.

Bioconjugates for FRET Assays

Bioconjugates used in a fluorescence resonance energy transfer (FRET) system are designed to contain fluorescent labels or a fluorescent label combined with a quenching label. A targeting molecule, such as a monoclonal antibody directed against an epitope on the target analyte, is conjugated with one of these molecules while a second targeting molecule, such as another monoclonal antibody against a second epitope on the target analyte, is conjugated with the appropriate complementary detection molecule. For FRET to occur with good efficiency and signal strength, the donor and acceptor detection conjugates must be brought together and held within a molecular distance of less than about 50 Å (5 nm) and preferentially at a distance of less than 20 Å (2 nm). According to the Forster equation (Chapter 10), FRET efficiency decreases with the sixth power of the distance in which the two molecules are separated, so high-proximity assay designs result in the best signals.

Unfortunately, FRET assays using standard organic fluorescent dyes (Chapter 10) usually do not produce strong enough assay signals, because the resultant molecular distances between the fluors can easily be outside the range of efficient energy transfer. A typical IgG targeting molecule, for example, is about 11 nm in diameter (110 Å). Therefore, two fluorescently labeled antibody conjugate pairs of a FRET assay that are used simultaneously to dock on a large protein analyte may result in a complex in which the attached fluorophores are too far away from each other to generate good signals. Energy transfer in such systems usually does not occur with enough efficiency to make homogeneous assays with large analytes possible, unless the two targeting antibodies are designed to interact with target epitopes that are spatially close together.

However, this situation can be overcome for macromolecule assays by using a fluorescent lanthanide chelate label on one of the targeting antibodies (the donor) and a strongly absorbing fluorescence acceptor such as a cyanine dye or a phycobiliprotein on the second antibody. For protein assays, FRET signaling can be done at a greater distance using a lanthanide chelate, typically up to about 8 to 10 nm with good efficiency and signal strength. For instance, in a FRET-based system, a first monoclonal antibody might be conjugated with fluorescent lanthanide chelates having bound europium ions (Chapter 10, Section 9) and a second antibody might be conjugated with either a Cy5-type dye (Chapter 10, Section 8) or a phycobiliprotein, such as allophycocyanin (APC) (Chapter 10, Section 7). When these bioconjugate pairs are combined in an assay and both bind to a target analyte, the fluorescent conjugates are held in close enough proximity to allow efficient energy transfer between their fluorescent labels. Thus, excitation of the lanthanide chelate at 340 nm will cause FRET signaling to the neighboring APC molecules with concomitant emission at 660 nm. Signal emission at 660 nm occurs only in proportion to the amount of analyte present in the sample. Any excess labeled antibody conjugates that do not bind to the target will not undergo FRET signaling, and consequently they will result in no fluorescence emission at 660 nm. Unbound lanthanide conjugates still undergo excitation at 340 nm but only result in sharp signals at the europium chelate's characteristic emission wavelength of about 615 nm but no FRET signaling to the APC conjugates. In this system, a homogeneous assay can be done for a particular target that requires only the mixing of the detection bioconjugates with the sample followed by measurement of the resultant fluorescence at 660 nm, which forms a rapid and powerful system for many assays done in drug discovery (Figure 1.26).

Molecular Beacons and TaqMan Probes

Another application of organic fluorescent labels in FRET assays is their use in molecular beacon-type assays using oligonucleotide probes or peptides. In this case, the organic detection molecules of the FRET system are both attached to the same detection molecule: either an oligo or a peptide. The acceptor in such designs can be another fluorescent dye or a quencher of the first fluorophore's emission. In molecular beacon assays using oligo probes, the acceptor is often a

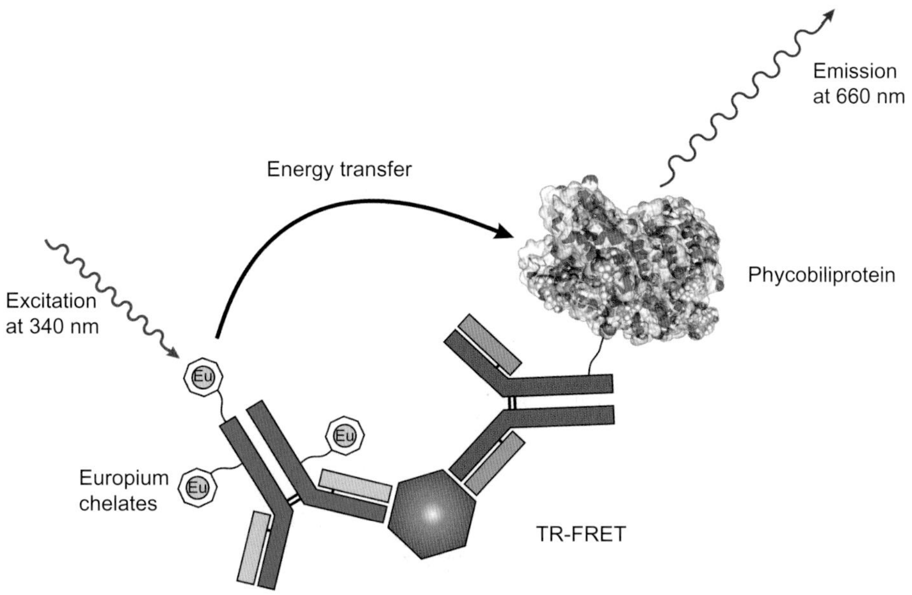

FIGURE 1.26 In a time-resolved FRET assay system (TR-FRET) an Europium chelate modified antibody is used along with a phycobiliprotein (e.g., APC) labeled antibody. Energy transfer between the lanthanide chelate fluorescent label and the fluorescent protein does not occur except when the two labeled antibodies are held close enough together by being bound to a targeted analyte. The efficiency of energy transfer using lanthanide chelates is much better than that observed using standard organic fluorescent dyes.

quencher. Both the fluor and the quencher are labeled on the probes at specific sites along the polymer—typically at the ends. An oligo probe might be labeled with one donor fluor at the 3′-nucleotide end and with the quencher at the 5′-nucleotide end of the sequence. The total length of the oligo probe is long enough to result in inefficient FRET signaling between the two labels if the oligo is stretched out linearly or hybridized to a complementary target sequence. The design of the oligonucleotide probe usually includes a hairpin configuration in which the last few nucleotides at both ends can hybridize to each other, thus bringing the donor and acceptor labels close together as the hairpin forms in solution (Chakravorty *et al.*, 2010; Krasnoperov *et al.*, 2010). In the hairpin configuration, the FRET labels will associate with each other and cause no fluorescence signal, as the fluor is completely quenched by the neighboring quencher molecule. In the presence of specific target sequences, however, the labeled oligo will hybridize to its target, breaking the hairpin structure, and pushing the FRET pairs far enough apart to lose any energy transfer potential. Hybridization to the target consequently causes a fluorescent signal to result, as the fluor-labeled end is no longer in proximity to, and quenched by, the quencher-labeled end (Figure 1.27). The more target that is present in the sample, the greater the fluorescent signal will be in the assay.

Perhaps the most popular molecular beacon-like technology is represented by the TaqMan probes (Roche, Applied Biosystems, and others) that are designed to assay specific nucleotide sequences in real time during a polymerase chain reaction (PCR) amplification process. These assays often are called real-time PCR or quantitative PCR (qPCR) tests. The sensitivity of this technology can measure target gene sequences down to a level of 10 to 100 copies and has a linear assay range of about nine orders of magnitude. There are very few technologies that can tout this degree of sensitivity and linearity in typical assays for any biomolecule. These probes do not use the hairpin structure design discussed previously but are linear oligonucleotide probes labeled at the 5′ nucleotide end by a fluorescent molecule and at the 3′ end by a quencher. The sequence is short enough to maintain nearly a quenched state of the fluor if the probe has not bound to its target genomic sequence; therefore very low background fluorescent signal results from just the probe in solution. However, if the probe binds to its amplified target during PCR, the 5′ nuclease activity of the polymerase enzyme used in the PCR process degrades the bound probe from the 5′ end, releasing the fluorescently labeled nucleotide and generating a characteristic fluorescence emission (Figure 1.28). As the PCR amplification process continues the amount of fluorescent signal increases exponentially, thus resulting in the exquisite sensitivity possible for the assay. Over one million TaqMan type assays have been developed to detect gene expression, mRNA, miRNA, or siRNA in samples from many different species.

The bioconjugate designs used for TaqMan assays have also incorporated minor groove binding molecules to enhance the initial binding to double-stranded target

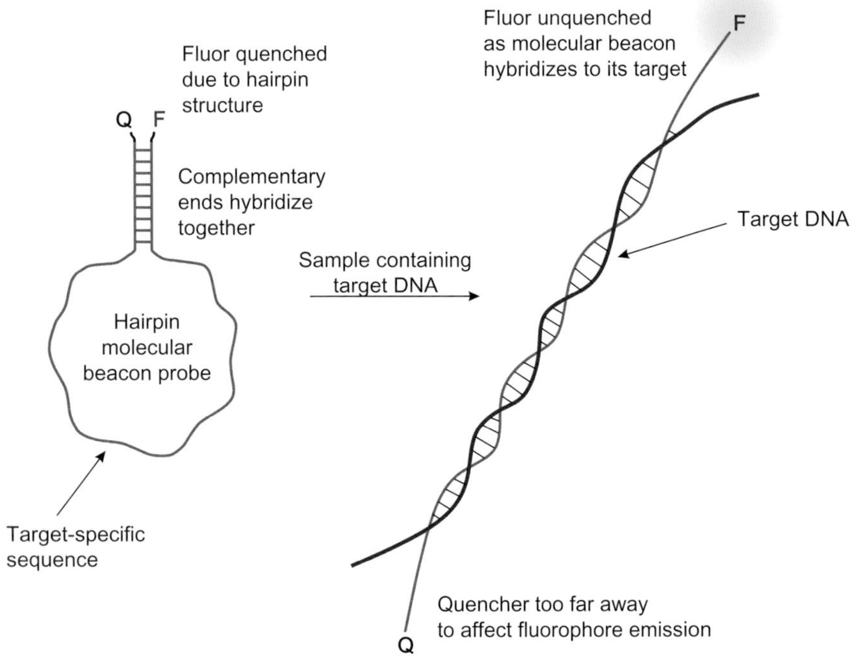

FIGURE 1.27 Oligo hairpin molecular beacon assays using fluor-quencher designs. The complementary region on the labeled probe keeps the fluorescent dye and quencher close enough in solution to eliminate fluorescence in the unbound state. However, once a targeted DNA sequence is hybridized with the probe, the hairpin is opened up and the fluor and quencher are moved apart far enough to allow fluorescence emission to occur. Such molecular beacons allow homogeneous assays to be carried out in solution in a single step.

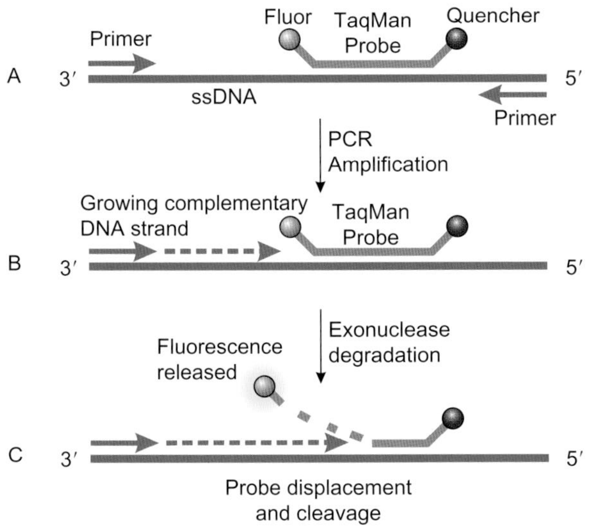

FIGURE 1.28 The TaqMan assay system design involves the use of a fluor–quencher FRET-based oligo probe that has its fluorescence quenched before and during hybridization with a target DNA sequence. Upon PCR amplification, however, the exonuclease activity of the enzyme cleaves off the nucleotides from the hybridized probe starting at the 5′ end labeled with the fluor. As the fluor molecules are released, they are able to emit fluorescence as they move away from the quencher end of the probe. An increase in fluorescence signal therefore results as the targeted DNA sequence is amplified, permitting real-time assays with high sensitivity.

sequences during PCR. Some designs have even used multiple dyes to enable multiplexed qPCR assays. In one such multiplex design, a dual-dye-labeled probe was created using a fluorescein molecule linked to a Cy5.5 dye, which was in turn used to label the 5′ end of the probe (Jothikumar *et al.*, 2009). At the 3′ end, a Black Hole Quencher (BHQ-3) molecule was attached to the terminal nucleotide (Figure 1.29). In solution, the fluorescein dye could be excited by a blue LED at 470nm resulting in FRET signaling to the Cy5.5 dye. However, due to the BHQ-3 quencher attached at the other end, the cyanine dye displays little to no fluorescence emission. Only after the polymerase enzyme has degraded the bound probe during PCR can excitation at 470nm result in emission at the Cy5.5 wavelength of 705nm. Along with the multiple-dye-labeled probe, a second probe labeled only with fluorescein can be used simultaneously during an assay to result in multiplexed signals at both 705nm for the dual-fluor-labeled probe and at 530nm for the fluorescein-labeled probe. Thus, two target sequences can be measured simultaneously in solution using these bioconjugate probes.

Molecular beacon type assays can also be designed using peptide sequences. In this case, the two FRET pairs are labeled onto the N- and C-terminal amino acid residues of a peptide. With peptide molecular beacons, a hairpin design is not possible to bring the ends together as it is with oligo probes. However, if relatively hydrophobic dye structures are used as labels, often they will have a tendency to associate together in solution, thus bringing the ends of the peptide sequence together and quenching any potential fluorescence from the donor fluor. The fluorescent molecule label and the quencher may also be sufficiently close with only a short sequence of amino acids in-between them so as to maintain good quenching characteristics. In this type of probe, the peptide sequence can be chosen to be a substrate for a particular protease of interest. In an assay, the action of the protease will cleave the peptide at a specific site in the center of the sequence and result in the fluor and the quencher being released into solution and thus becoming separated from one another. This releases the FRET quenching and allows the fluorescent molecule to emit light at its characteristic wavelength (Figure 1.30). In an assay, the more of the specific protease that is present and active in solution the greater the fluorescent signal that will be generated from separating the fluor from the quencher. Thus, using a peptide molecular beacon bioconjugate in a single-step homogeneous assay, a protease enzyme can be quantified for activity and estimated for concentration. Stefflova *et al.* (2007) describe the use of different peptide-based molecular beacons for use in cancer imaging *in vivo*.

Bioluminescence Resonance Energy Transfer (BRET)

In a BRET system, similar bioconjugates can be designed for a homogeneous assay as in FRET systems, but in this case using targeting molecules conjugated with a bioluminescent enzyme for the first conjugate and a fluorescent label for the second conjugate (Subramanian *et al.*, 2006). This phenomenon is naturally seen in the sea pansy *Renilla reniformis* which has a luciferase enzyme that emits blue light, but it is associated with a green fluorescent protein (GFP) that accepts the light and re-emits it in the green region of the spectrum. For designing BRET assays, the main criterion is that the fluorescent label must be able to absorb energy emitted by the bioluminescent enzyme–substrate activity and emit it at another wavelength characteristic of the fluor. One of the main advantages of BRET-based assay systems is that they can be used *in vivo* in living cells to study protein interactions. In this design, the bioluminescent enzyme component is fused to one of the two proteins to be studied for potential interactions. The acceptor fluor in this instance consists of a fluorescent protein, such as GFP or one of another color (yellow fluorescent protein [YFP] is particularly popular for this application), which is fused to the second protein of the potential interacting pair. The advantage of this assay design is that both bioconjugates are expressed

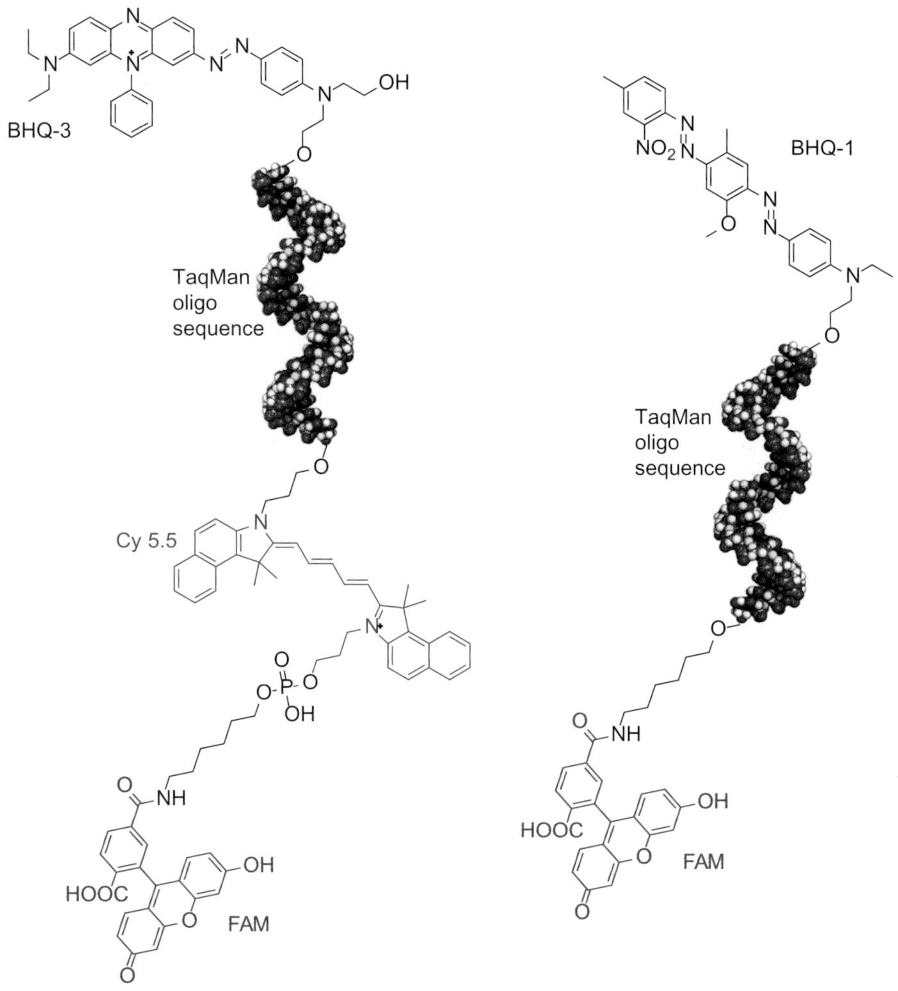

FIGURE 1.29 Multiplex TaqMan probe designs use one fluorescein-quencher probe (right) along with a second probe having a different specificity and containing a fluorescein-Cy5.5 FRET pair of dyes combined with another quencher (left). During an assay both probes can be excited at 488 nm; however the fluorescein-only probe will emit at about 520 nm whereas the fluorescein-Cy5.5 probe will emit in the far red region at the cyanine dye's emission peak due to energy transfer within the FRET system. Thus, two different DNA target sequences can be assayed simultaneously using the same laser excitation source.

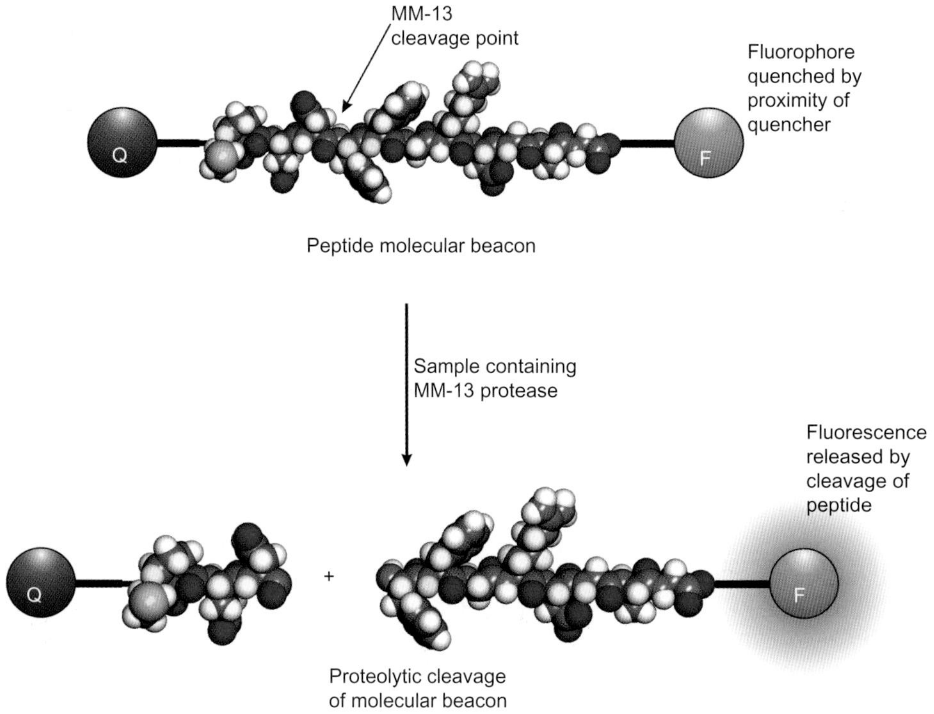

FIGURE 1.30 Peptide molecular beacon assays for protease activity typically are constructed using a specific amino acid sequence that can be cleaved by the intended enzyme target and having one end labeled with a fluorescent dye and the other end labeled with a quencher. Upon cleavage of the probe by the protease enzyme, the fluorescent dye moves far enough away from the quencher to become fluorescent. The degree of fluorescence is indicative of the amount of enzyme present in the sample.

as fusion proteins within the cell, thus poised to be assayed by the simple addition of a bioluminescent substrate. If the bioluminescent enzyme is expressed with its fusion protein partner and is able to act upon the substrate, then light will be generated from this reaction. If this fusion protein has bound with its putative interaction partner, then the fluorescent protein will be held in close enough molecular proximity to be able to accept the energy produced by the bioluminescent enzyme. Therefore, BRET signaling will only occur if the two target proteins are present and held in association from biospecific interactions. This type of assay system has advantages for use with biological systems, because it does not require external illumination to generate the signal, which could cause autofluorescence from biomolecules in the cell or inadvertent excitation of the acceptor fluorescent protein (Figure 1.31).

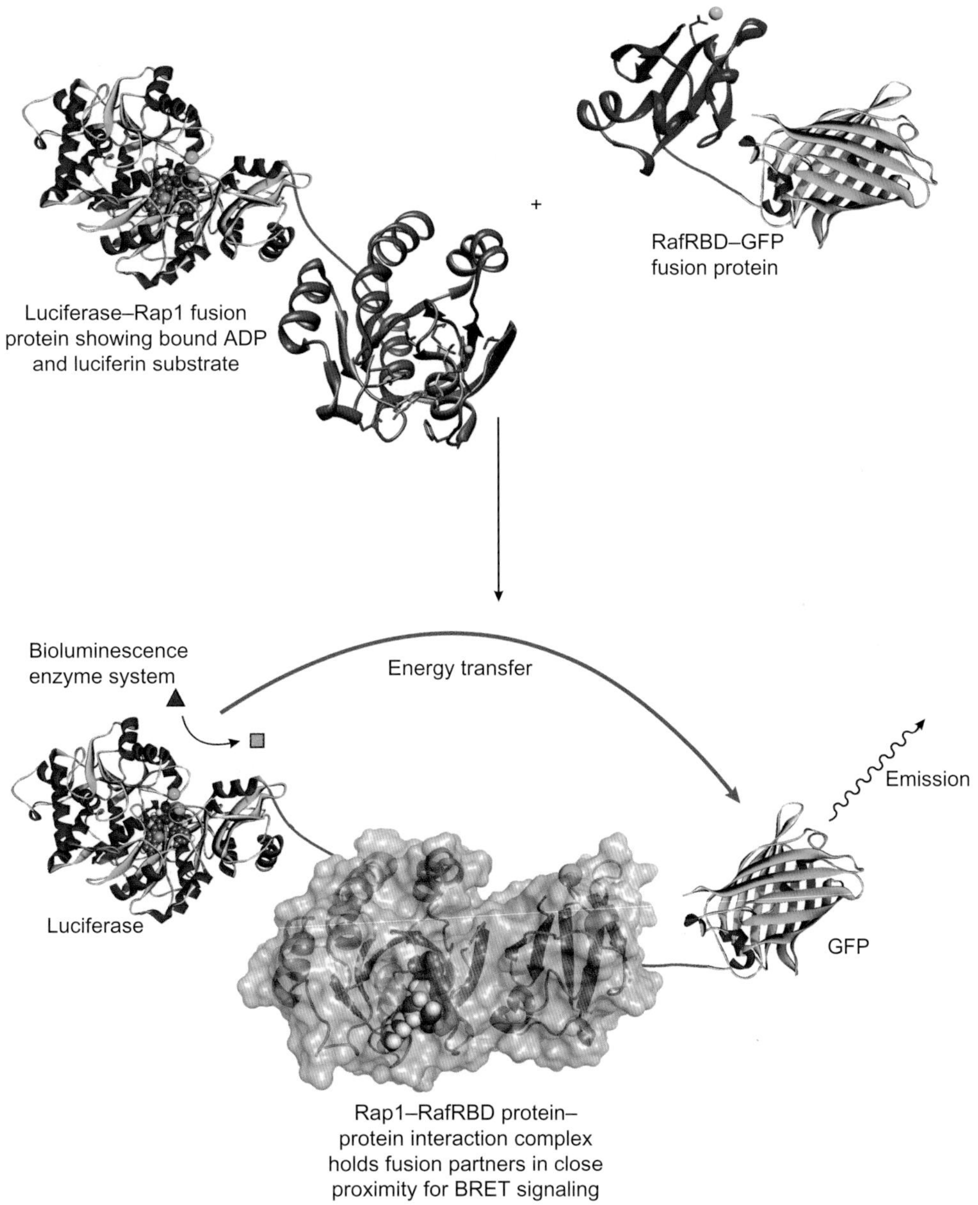

FIGURE 1.31 A bioluminescence resonance energy transfer (BRET) assay system with a bioluminescent enzyme and GFP fusion proteins. In this type of protein–protein assay one of the putative interacting proteins is fusion tagged with a bioluminescent luciferase enzyme (e.g., Rap1) and the other interacting protein (RafRBD) is tagged with a fluorescent protein (GFP). If the two target proteins interact within a cell, then the luciferase and the GFP are held close enough together to allow for the luminescence of the luciferase enzyme/substrate reaction to be transferred to the nearby GFP. If this occurs, then the complex will emit light at the emission peak of GFP and not just at the luminescence wavelength of the luciferase substrate reaction.

Protein Fragment Complementation Assays: Split Reporters

In a split reporter system used for homogeneous assays, the two components can be of several types: (1) an enzyme that acts upon a substrate to produce a chromogenic or fluorescent product; (2) a bioluminescent enzyme that transforms a substrate to produce a photon of light; or (3) a fluorescent protein with characteristic excitation and emission properties, such as the green fluorescent protein (GFP). The fragments of such reporters are designed to be stable but inactive when conjugated to targeting molecules like antibodies. However, once the targeting molecules interact with their targeted analyte in solution or on a solid phase, the coupled reporter fragments are able to interact in a complementation process that restores their native function. For example, in a split firefly luciferase reporter assay, the fragments come together as the two targeting bioconjugates dock upon the analyte and the native bioluminescence enzyme activity is then restored. Enzyme complementation results in a turnover of luciferin substrate in the assay and bioluminescence light output is measured as proportional to the amount of target molecule present in solution. Protein complementation assays (PCAs) are also used to detect interacting proteins within cells, wherein fusion protein complexes similar to the design of the BRET system, each containing a fragment of the split reporter, are detected as a specific interacting complex only if complementation occurs and the unique detection activity of the reporter is restored (Gambhir and Paulmurugan, 2005). Figure 1.32 illustrates the design and function of a split GFP PCA system wherein two labeled antibodies bind to a target protein, resulting in the GFP fragments coming together and restoring fluorescence activity.

Mass Tags for Quantitative Mass Spectrometry

Bioconjugation reagents incorporating stable isotope-labeled structures are becoming common for the quantitative analysis of protein expression using mass spectrometry. Chapter 12, Section 3, describes the major types of mass tags and their protocols for use, including isotopic and isobaric constructs. Mass

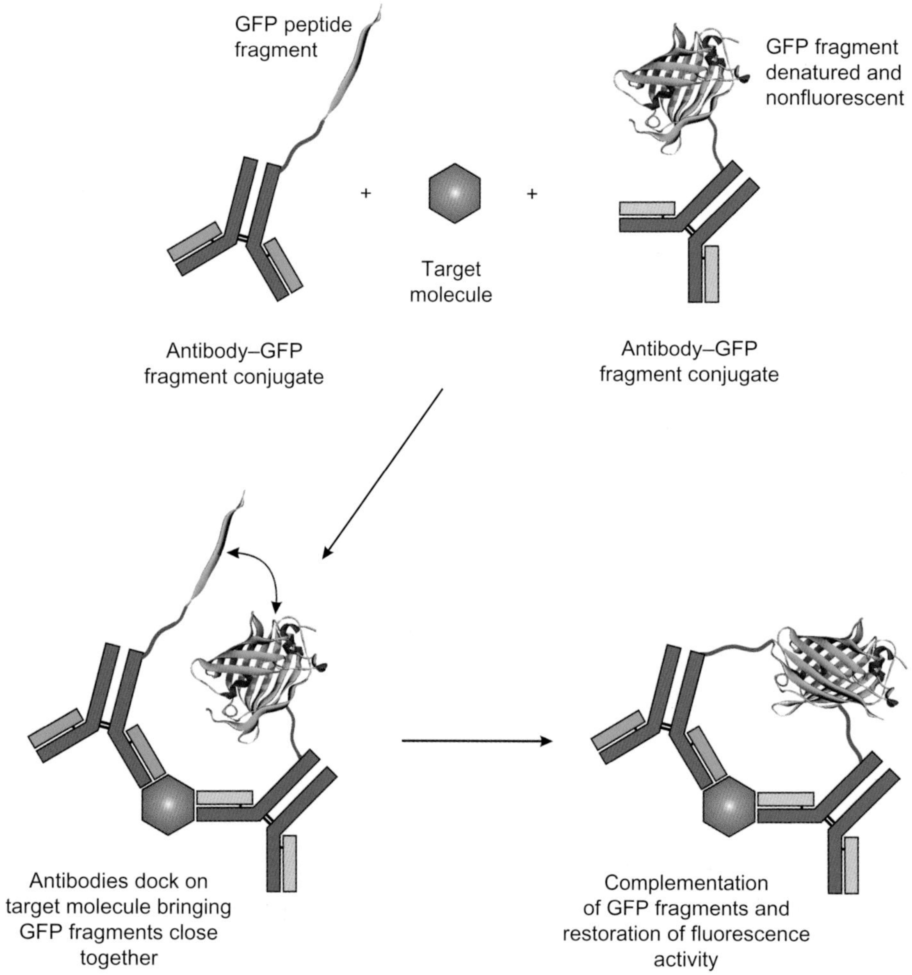

FIGURE 1.32 A protein complementation assay (PCA) can be designed using a recombinantly split GFP. If one antibody is tagged with a fragment of GFP and another antibody is conjugated with the other part of GFP, then bringing the two antibodies together onto a targeted protein will allow the two fragments to recombine and reconstitute its fluorescent properties. This same concept can be used with recombinant fusion proteins to detect protein–protein interactions in cells.

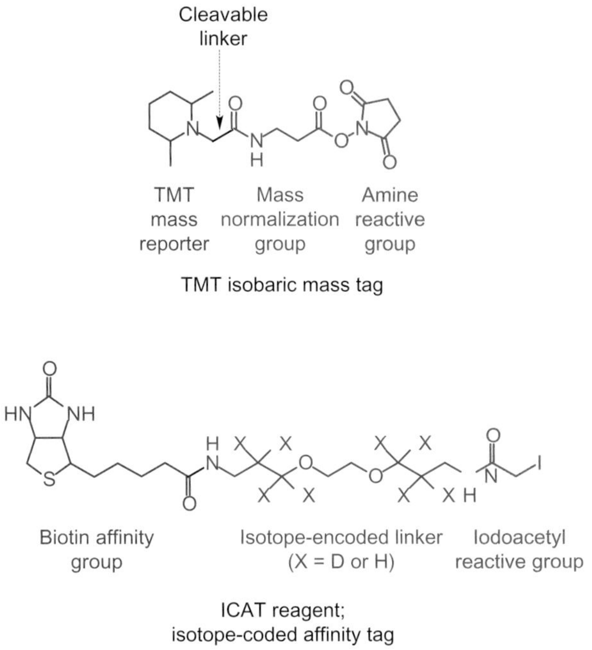

FIGURE 1.33 Examples of isotopic and isobaric mass tag designs. The isobaric mass tag uses a reporter group and a balancing group (mass normalization) to maintain the same total mass within the tag. A set of different isotopically labeled tags can be designed that all have different reporter group masses but the same overall mass. Peptides labeled with these reagents will all elute at the same point during the chromatographic separation, but upon mass spec analysis the different reporter masses can be correlated to different samples, thus allowing multiplexed analysis. An isotope-coded affinity tag, by contrast, has an isotope label region but does not have the properties of the isobaric mass tags. The biotin group allows for enrichment prior to mass spec analysis of labeled peptides.

tags can be used to analyze the expression levels of multiple proteins from a cell sample, from just a few at a time to global proteomic changes across a number of samples simultaneously (Bantscheff *et al.*, 2007). Modification reagents used as mass tags typically contain a mass reporter unit, a spacer, and a reactive group useful for modifying specific sites on peptides or proteins. The mass reporter typically is one of two types, either (1) a simple isotopically labeled compound containing, for instance, deuterium, ^{14}C, or ^{15}N in place of the normal isotopes for hydrogen, carbon, or nitrogen (Zhou *et al.*, 2002); or (2) a more complex isobaric design, wherein there is a mass reporter group that is next to a mass balancing group (Figure 1.33). For simple isotopic mass tags, two samples can be run simultaneously with one sample being modified with the normal isotopic tag and the other sample modified using a tag containing stable heavy isotopes.

In the isobaric form of these reagents, any isotopic changes made to the reporter group are counterbalanced with opposite changes to the balancing unit, thus maintaining the same overall molecular weight of the entire mass tag molecule (Gingras *et al.*, 2007). Isobaric tags allow the labeling of multiple samples using designs with different reporter unit masses, but the same total mass signature. These samples can be mixed in equal ratios prior to mass spec analysis to obtain the relative expression levels for particular proteins within all the samples. Identical peptides labeled with the isobaric tags from different samples will all be contained in the same peak in the first dimension of MS separation; however, in MS^2 where peptide fragmentation occurs to identify the amino acid sequence of each peptide, the reporter groups are also released and used as an identifying mass signature to correlate the peptide sequence to the appropriate sample number. Thus, multiplexed analysis of protein expression is enabled by use of an isobaric mass tag bioconjugation reagent.

Isotopic or isobaric mass tags are used to covalently modify peptides in a protein sample before undergoing mass spec analysis. The mass tag can be designed to contain an amine-reactive, thiol-reactive, or carbonyl-reactive group for covalent modification of peptides at known sites. Each modified protein can then be identified by its combined peptide-plus-tag molecular mass as detected by the mass spectrometer. The tags can also be used to derivatize post-translational modifications on proteins or other biological molecules through specific functional groups. For instance, aminoxy-containing isotopic tags have been used to detect oxidative damage in proteins by reacting with carbonyl groups that are produced on amino acid side chains (Whetstone *et al.*, 2003; Meares *et al.*, 2007).

Mass tags that facilitate relative protein expression analysis between multiple samples can be combined with isotopically labeled peptide standards to permit quantitative analysis of expression levels (Yan and Chen, 2005; Lange *et al.*, 2008). The labeled peptide standard creates a mass spec peak from a known initial concentration added to the sample solution. Comparing the standard peak area to the sample peak area from the same peptide (which is from a known protein) makes possible multiplexed quantitative MS analysis.

Limiting Nonspecificity in Bioconjugate Designs

Another important factor in assay design is to limit the nonspecificity of the bioconjugate as much as possible in the desired application. Many strategies for removing nonspecific binding or background signals include adding components to the assay buffers that can block sites of potential nonspecificity on the immobilized capture support, thus reducing the potential for the detection bioconjugate to bind and generate false signals. Proteins such as BSA, casein, serum, non-fat dried milk, and a host of other proteins or polymers have been used extensively to block nonspecific binding on microplates, membranes, particles, and other assay surfaces.

However, in addition to introducing blocking agents or excipients to solutions, the actual design of the capture or detection bioconjugates may also be altered to limit nonspecificity and improve assay performance. For instance, the use of discrete PEG-based crosslinkers in the formation of the bioconjugates instead of aliphatic-based crosslinkers can dramatically affect assay performance by lowering the potential for nonspecific binding. During modification and conjugation processes, PEG linkers create an extremely hydrophilic surface due to the ethylene oxide spacers on each molecule, which in turn reduces the potential for nonspecific interactions of the conjugate with other molecules. Conjugates designed using hydrophilic instead of hydrophobic crosslinking strategies can improve signal-to-noise ratios in many assay applications. Chapter 18 discusses the reagents and methods used to create such conjugates, which can benefit not only assays but other applications as well.

In forming a typical protein–protein bioconjugate, it is common to have intermediate derivatives prepared that modify and activate both proteins for the subsequent conjugation reaction. For instance, in the preparation of an antibody–enzyme conjugate, often one of the molecules is activated with a heterobifunctional crosslinker, while the other protein is modified to contain an appropriate functional group to couple with the modified first protein. In many traditional bioconjugation procedures this could take the form of reacting the antibody with an aliphatic NHS ester–maleimide crosslinker, such as SMCC. In this reaction, SMCC couples with the amines on the antibody to create amide bonds while the other end of the crosslinker creates thiol reactive sites on the molecule, thus producing terminal maleimide groups sticking out from the surface of the protein (Chapter 20). However, using this approach, the resultant derivatized antibody also contains a number of hydrophobic aliphatic components sticking off its surface (Figure 1.34) and potentially creating sites for nonspecific interactions.

In a secondary reaction, the enzyme molecule typically is modified with an aliphatic thiolation reagent prior to crosslinking with the SMCC-activated antibody. Modification agents such as Traut's reagent or SATA (Chapter 2, Section 4.1) form aliphatic protrusions on the surface of the enzyme to create thiols just as the SMCC modifications made on the antibody create thiol-reactive groups. Figure 1.34 illustrates both the SMCC- modified antibody and an alkaline phosphatase enzyme that is modified using SATA to create terminal thiol groups after deprotection (note: the structure shows the deprotected enzyme with exposed thiol groups). After mixing the SMCC-activated antibody with the thiolated enzyme, the conjugate that is formed may have relative ratios of enzyme-to-antibody of perhaps 2:1 or 3:1, depending on the molar ratio of the components added to the initial reaction

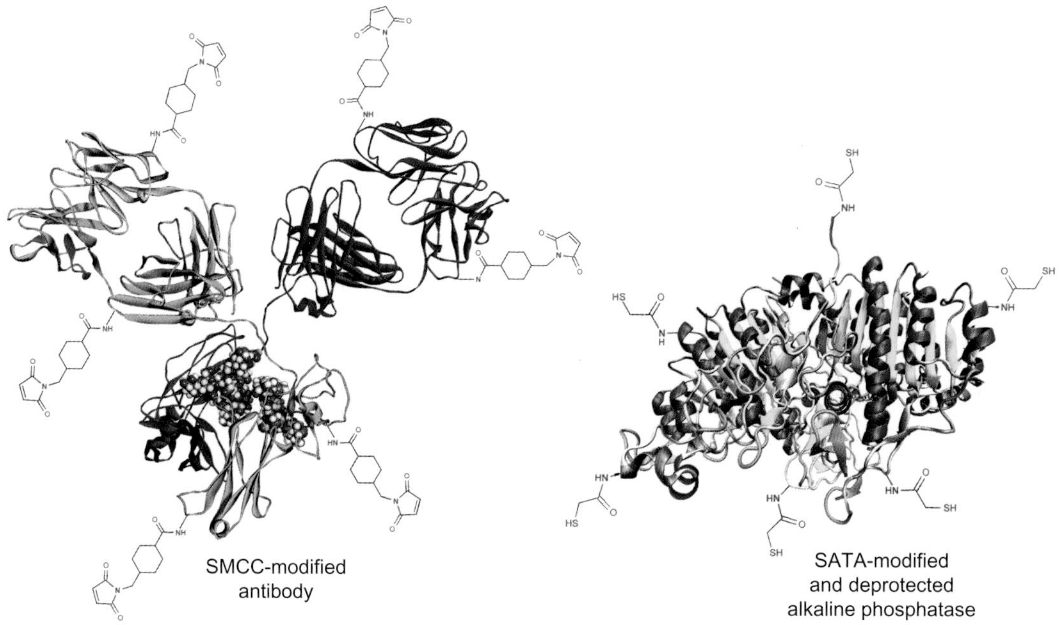

FIGURE 1.34 An antibody modified with aliphatic SMCC linkers and alkaline phosphatase modified with SATA create aliphatic modification sites sticking out of both molecules.

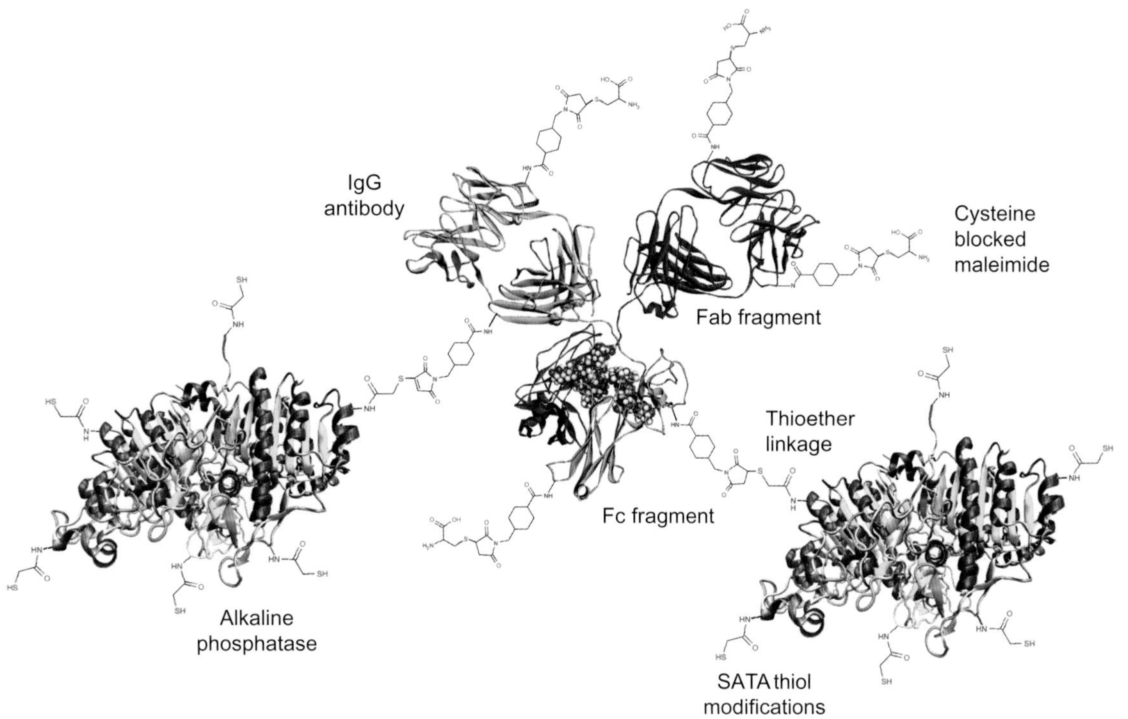

FIGURE 1.35 An antibody–alkaline phosphatase conjugate made using SMCC and SATA. The final conjugate has numerous dead-end linkers sticking out of both components, which can cause nonspecific binding in assays or aggregation of the conjugate in solution.

media. However, after the conjugation reaction is complete many of the modification sites on the antibody and enzyme remain unused. This leaves unreacted SMCC or thiolation modifications on both the antibody and the enzyme protruding off the final conjugate and forming hydrophobic interaction sites (Figure 1.35). Though the reactive groups on the ends of these reagents usually are blocked to prevent subsequent unwanted covalent reactions, the aliphatic groups sticking out from the conjugate remain.

The use of discrete PEG-based crosslinkers and thiolation reagents in place of the standard aliphatic compounds can create a completely different situation. The modification of an antibody with an NHS-$(PEG)_n$ -maleimide compound forms a thiol-reactive intermediate that contains hydrophilic PEG-based spacers. This intermediate actually is more water soluble than the initial antibody, especially when using PEG chains of equal length or longer than PEG_4. Similarly, the modification of the enzyme component with a PEG-based thiolation reagent forms an intermediate containing hydrophilic PEG spacers sticking off the enzyme. Once the activated antibody is mixed with the thiolated enzyme to form the final bioconjugate, the excess modification sites sticking off both components of the complex are all hydrophilic, not hydrophobic. Thus, bioconjugates made from discrete PEG-based reagents end up being more water soluble and less likely to interact nonspecifically with surfaces in assays or bind nonspecifically with other biomolecules present in samples.

3.2. Detection, Tracking, and Imaging

A second major application area for bioconjugate techniques is in the detection, tracking, and imaging of biomolecules. In many ways, this area is closely related to the use of bioconjugates for assay and quantification, but in this case the purpose is to reveal the presence or location of a target molecule, not necessarily to quantify it. Methods in this area typically involve the identification of biomolecules within cells, tissues, or in organisms using bioconjugates composed of a targeting component conjugated to a detection component. It also includes semi-quantitative analysis of specific biomolecules after electrophoretic separation and blotting, such as in the chemiluminescent detection of proteins on western blots using antibody–enzyme conjugates. Bioconjugates used in detection, tracking, and imaging have driven the rapid growth of techniques like immunochemical staining, fluorescence microscopy, and high content analysis (HCA), which have become crucial for advancing an understanding of cellular biology. They also form the most important new technology platforms for performing *in vivo* imaging for diagnostics or therapeutics, which are a significant part of the overall medical imaging market estimated to be close to $6 billion worldwide

(Bergin, 2009). The specific applications of bioconjugates for therapeutics and immune modulation will be discussed in greater detail in Section 2.4 of this chapter.

The most frequently used targeting molecule for detection, tracking, or imaging is an antibody or antibody fragment with specificity toward a desired target molecule in a cell or tissue. Tens of thousands of antibodies now are commercially available with defined specificity and many of them have validated data that prove their utility for a particular application. Conjugation of such antibodies with the appropriate detector molecule can facilitate its use in specific imaging techniques (Chapters 10 and 20). Often, however, the primary antibody is available in quantities too small and at price points too high to economically prepare a conjugate from it. In this case, a secondary antibody conjugate can be used to detect the primary antibody in an imaging application. Another convenient alternative to making a primary antibody conjugate is the use of a biotinylated primary antibody, which can then be detected using a labeled streptavidin conjugate (Chapter 11). The use of secondary antibody conjugates or streptavidin conjugates is the most popular strategy for detecting a primary antibody in detection, tracking, and imaging applications.

In addition to antibodies, targeting molecules can consist of synthetic or small biological affinity binding molecules such as peptides or aptamers as well as active-site probes for enzymes or ligands that are able to interact with certain receptors. Some targeting agents can be molecules that are naturally taken up by cells, such as transferrin, which is taken up more readily by tumor cells than normal cells due to the cells' rapid growth characteristics. Some bioconjugates designed for use in cells have been labeled with short peptide sequences that facilitate rapid uptake by cells, such as trafficking molecules, which shuttle other attached molecules to specific cellular organelles or locations (Morris *et al.*, 2008). Also, certain small molecule drugs with known intracellular targets can be used to deliver a detection reagent to a specific site within a cell (Figure 1.36).

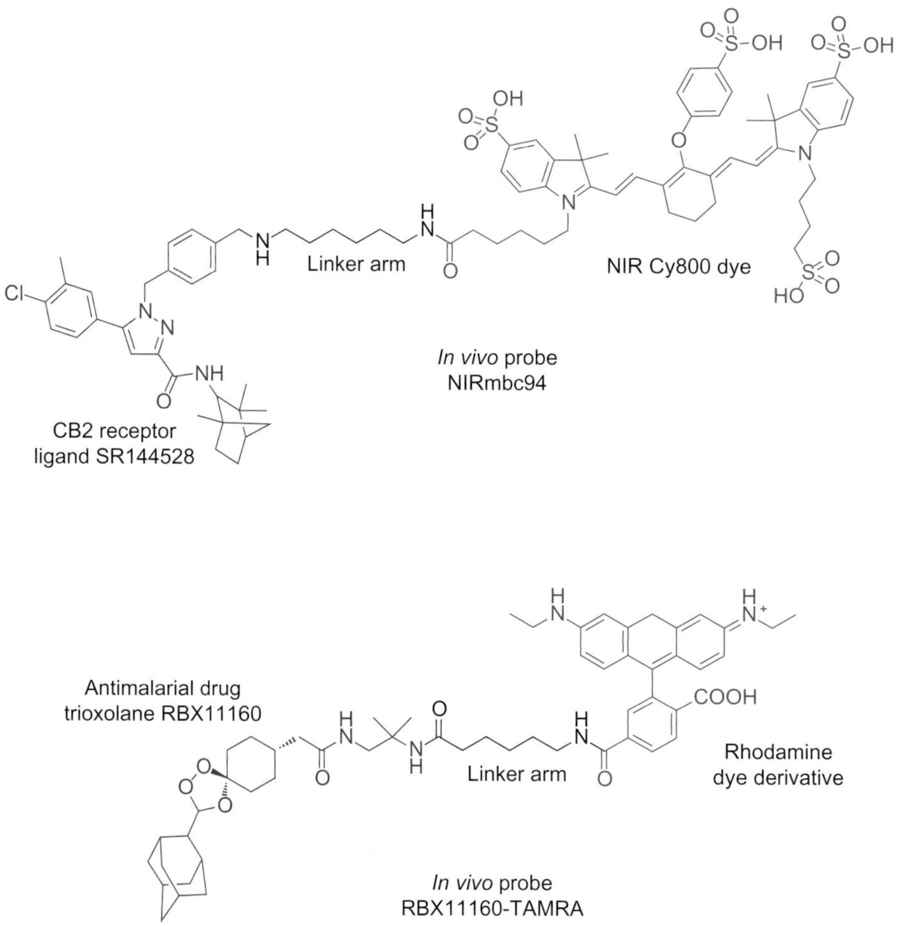

FIGURE 1.36 Two examples of bioconjugates designed for NIR fluorescence *in vivo* imaging. The reagent on the top contains a CB2 receptor affinity ligand that will bind to the receptor sites on cell surfaces and a carbocyanine 800-nm dye with absorbance and emission characteristics well within the NIR region of the spectrum. The probe on the bottom contains an antimalarial drug as the targeting component and a rhodamine dye, which emits within the red region of the spectrum.

Fluorescently labeled pharmacophores represent a novel method for imaging cellular compartments or locals through known drug–target interactions (Uhlemann *et al.*, 2007; Leopoldo *et al.*, 2009). Often, drug candidates have been studied for their mode of action by fluorescently labeling the drug and observing the site of interaction *in vivo*. Using established drug–target interactions is an excellent way of delivering fluorescent signatures directly to specific locations within cells.

The detection module conjugated to detection, tracking, or imaging probes can consist of many different traceable components, such as fluorescent labels, enzymes, radiolabels, high-contrast agents, nanoparticles, intermediary biotin–(strept)avidin reagents, synthetic scaffolds such as dendrimers or fullerenes, carbon nanotubes, or fluorescent fusion proteins (Figure 1.37). Some of these detection modules are identical to those used in bioconjugates for applications in assay and quantification, but some traceable components are unique to the imaging field. For instance, gold-labeled targeting molecules (Chapter 14, Section 6) can be used as high-contrast agents to visualize antigen locations in tissue sections or cells. Detection of such nanoparticle gold conjugates is particularly effective by electron microscopy, as precise locations of each particle can be imaged. Similarly, nanoparticle quantum dots (QDs) (Chapter 10, Section 10) can be used to label a targeting molecule and provide stable and intense fluorescent signals of target molecule locations within cells. QD-labeled antibodies are becoming a popular choice for detection and probing of western blots, because several different conjugates can be used simultaneously to image different protein targets. Each antibody–QD conjugate used in this type of system has its own unique fluorescence emission signature, thus ensuring that each target can be determined independently of the other probes and targets.

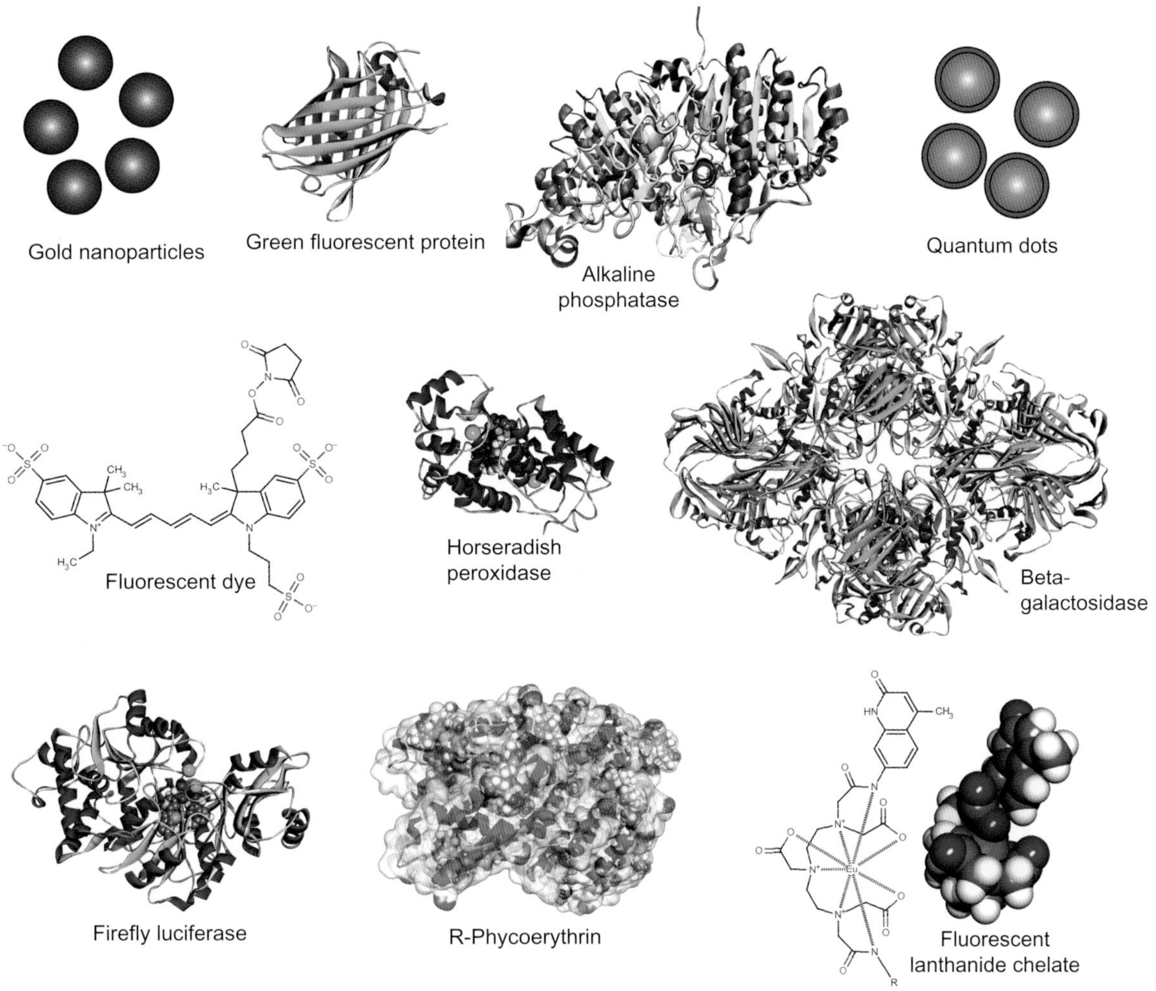

FIGURE 1.37 Common detection agents used in bioconjugates include gold nanoparticles, fluorescent proteins such as GFP, alkaline phosphatase, quantum dot nanocrystals, fluorescent dyes, horseradish peroxidase, beta-galactosidase, luciferase bioluminescence, phycobiliproteins, and fluorescent lanthanide chelates.

Fluorescently Labeled Antibodies and Streptavidin

Perhaps the most common bioconjugate used for cellular imaging is a fluorescently labeled antibody. Fluorescence microscopy has become a standard technique for studying cell biology, and the use of fluor–antibody bioconjugates is an essential tool in this process. Fluorescence detection and the reagents associated with this technique are discussed in Chapter 10. Fluorescent labels consisting of organic molecules, metal chelates, or nanoparticles are able to provide sensitive detection for imaging specific antigen targets within cells (Figure 1.38). A bright fluorescent label conjugated with a highly specific targeting agent, such as an antibody, can yield pertinent information on the biological state of a cell, including expression levels, post-translational modifications on proteins, cellular trafficking (translocation), morphology, and phenotype. The most common antibody used for

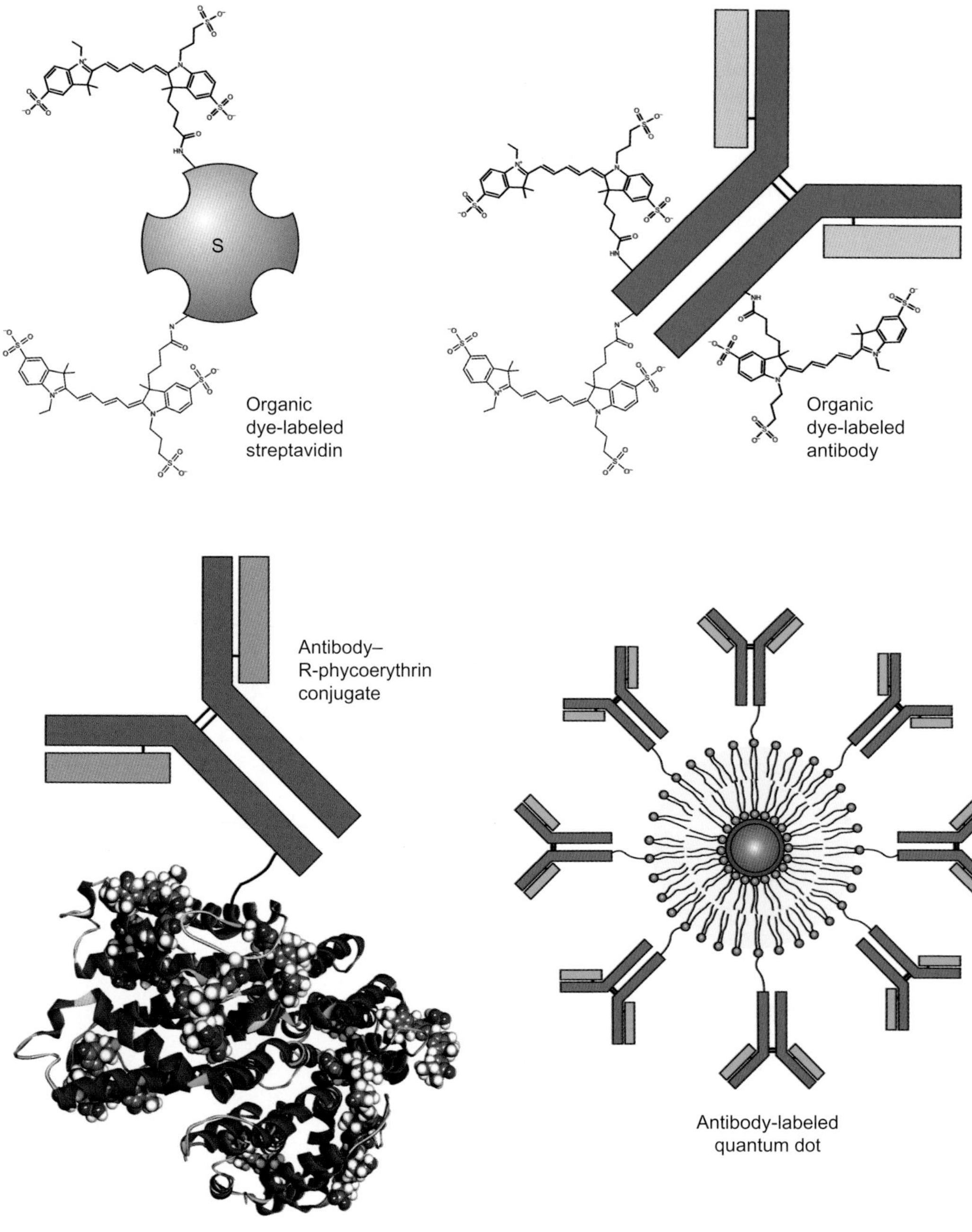

FIGURE 1.38 Fluorescent bioconjugate reagents are among the most common detection constructs used for cellular imaging. Popular conjugates include fluorescently labeled streptavidin, fluorescently labeled antibodies, antibody–phycobiliprotein conjugates, and quantum dot conjugates with specific antibodies attached.

these techniques is a secondary antibody, which can target a primary antibody that has bound to a cellular target. Alternatively, a fluorescently labeled streptavidin bioconjugate can be used to target biotinylated primary antibodies. The use of secondary antibodies or streptavidin conjugates avoids the expense of having to purchase or prepare a fluorescently labeled primary antibody. This layered detection approach using a primary antibody followed by a fluorescently labeled secondary reagent also provides greater fluorescent signal than a primary labeled reagent alone in cellular imaging applications. In practice, most of these fluorescently labeled reagents are used to probe fixed and permeabilized cells and through the permeabilizing treatment are able to accommodate the molecular size of the fluorescent bioconjugate to access virtually any areas within the cells. The only exception to this perhaps involves some antibody–nanoparticle conjugates, which can have diameters greater than 50 nm and thus may have limited access to some small intracellular spaces.

FRET Based Protease Probes

Bioconjugate imaging agents have also been developed to assess the activity of proteases within cells, including the use of molecular beacons (see Section 1.1, this chapter). These are typically fluorescently labeled at one end and labeled with a quencher molecule at the other end, and contain a protease-specific peptide sequence in the middle. These probes generate a fluorescent signal only if the cell has a protease active which has specificity for cleaving the peptide sequence. For instance, the peptide might consist of a caspase-specific sequence containing an aspartic acid cleavage point. This type of bioconjugate probe can be used to determine if a cell is undergoing apoptosis and therefore if caspases are activated (Edgington *et al.*, 2009). Before it encounters a caspase enzyme, the probe essentially is nonfluorescent due to the presence of the quencher molecule attached to the other end of the peptide (FRET quenching). However, as the activated caspases of the apoptotic cascade cleave the labeled peptide probe, the fluor and the quencher are separated and released into the surrounding cytoplasmic solution, and the FRET quenching process no longer occurs. The result is the generation of a fluorescent signal proportional to the amount of caspase enzymes activated within the cell or tissue. This type of molecular beacon assay is a powerful indicator of programmed cell death.

Caspase bioconjugate probes can also be designed with a fluorescent tag, a short peptide recognition sequence for binding to the enzyme, and a reactive warhead end, which covalently couples to the active site upon binding (Ekert, 1999; Amstad, 2000; Bedner, 2000; Smolewski, 2001). This type of bioconjugate has been made using NIR dyes to image tumor cells *in vivo* that are undergoing apoptosis after drug treatment (Griffin *et al.*, 2007; Cursio *et al.*, 2008, 2009; Delgado-Martín *et al.*, 2009; Erman *et al.*, 2009). Other fluorescent dyes can also be incorporated into the probes for use in cell-based imaging or flow cytometry applications. The reactive warhead end of the probe can consist of a fluoromethyl ketone (FMK) group that is highly efficient at covalent coupling to the thiol in the active site of the large subunit of the caspase heterodimer. The warhead end may also be made of a 2,6-difluorophenoxy-methyl ketone group, which has been shown to have even greater binding potential with the active site thiol of caspases (Melnikov *et al.*, 2002; Southerland *et al.*, 2010) (Figure 1.39). Unlike caspase molecular beacon probes, these types of active site conjugation probes do not catalytically generate a fluorescent species due to enzyme cleavage of a peptide, but in this case a fluorescent active site binder stays within the cells while excess probe is removed either by washing (for cell-based imaging) or through removal by the kidneys or liver (for *in vivo* imaging). Animal imaging studies have resulted in excellent tumor cell localization using NIR imaging at wavelengths from 680 to 900 nm, which penetrates tissue to at least several centimeters to excite the bound fluorescent probe.

Interacting Proteins or Domains

Another targeting component that can be used for specific cellular targeting is an interacting protein or a peptide domain, which is able to bind to another protein through native biospecific interactions (Pawson and Nash, 2003). This can also include an interacting protein that only binds to the activated state of the target protein, such as after a phosphorylation event has occurred. Proteins within cellular pathways have specific interacting partners and one of these proteins can be used in this manner to detect its partner. An example of this type of bioconjugate is the use of a fluorescently labeled Src homology 2 (SH2) domain, which is able to recognize and bind to certain phosphotyrosine sites in receptor tyrosine kinases (Hanke and Mann, 2009; Major *et al.*, 2009). In addition, downstream effector molecules or domains have been used to target activated GTPases, such as the Raf1 binding domain that binds to activated Ras or the Pak1 p21 binding domain, which binds to activated Rho (Smith *et al.*, 2008). Using a labeled bioconjugate of an interacting protein can reveal the precise locations of the activated target within the cell and yield information regarding the differences between a diseased cell and a normal one or a treated cell *versus* an untreated one. Yaoi *et al.* (2006) developed a fluorescent bead-based assay using a library of different SH2 domains that can target

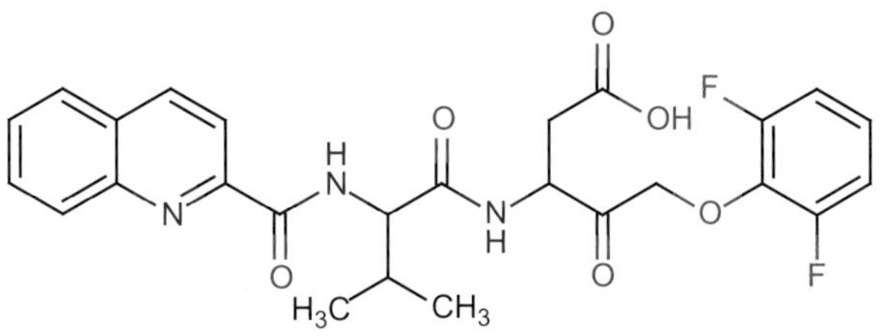

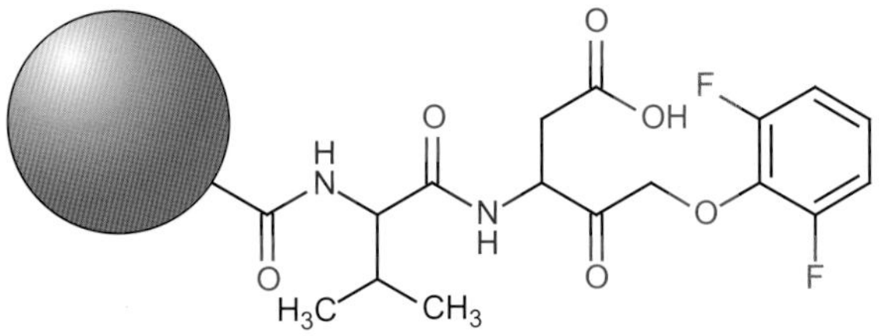

FIGURE 1.39 Two fluorescent warhead probes designed to detect caspase enzymes include a quinoline-labeled valine–aspartic acid (VD) probe, which can bind to the active site on caspases and covalently link to it by reaction of the difluorophenoxy-methyl ketone group. This type of probe can be designed to include NIR fluorescent tags for better signal penetration through tissue.

specific activated tyrosine kinases. This system was able to globally profile tyrosine kinase pathway activation in cell extracts (Figure 1.40).

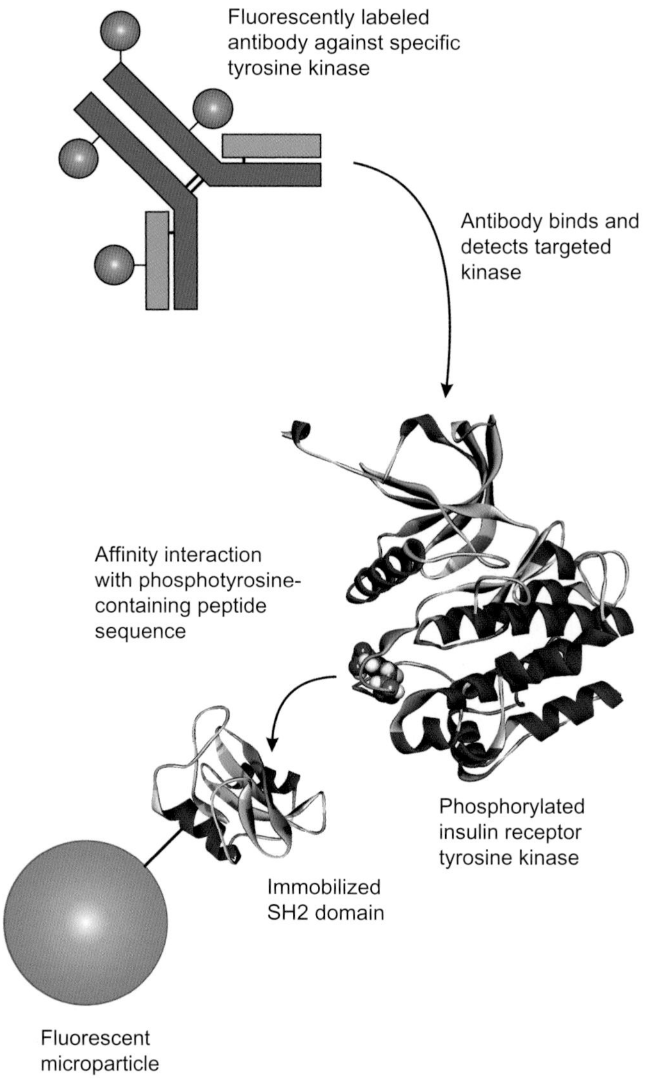

FIGURE 1.40 A fluorescent bead-based affinity assay can be designed using an immobilized protein or protein domain, in this case SH2 domains, to specifically capture phosphorylated tyrosine kinases and detect them using a fluorescently labeled antibody directed against the kinase enzyme.

Probing by Chemoselective Ligation

A recent conjugation strategy that is highly suitable for real-time detection within living cells is the use of *in vivo* labeled proteins or carbohydrates using non-natural amino acid or sugar derivatives (Codelli *et al.*, 2008; Laughlin *et al.*, 2008; de Graaf *et al.*, 2009). These amino acid analogs contain unique reactive groups that can facilitate subsequent bioconjugation with detection or affinity components. Certain amino acids and sugars can be derivatized to contain relatively small functional groups that still allow them to be used with the intracellular translational machinery or carbohydrate biosynthesis enzymes, thus placing these modified residues within newly synthesized proteins or glycans. Cells grown in the presence of a limiting amount of the non-natural amino acid or sugar derivatives will slowly begin incorporating such modifications into proteins, carbohydrates, and glycans. However, an important criteria for success using this strategy is that the reactive groups appended onto the amino acids must be nonreactive with other biological molecules and also be non-toxic to the cells; otherwise, they would react and covalently attach to biomolecules or affect cell viability and potentially cause cell death. Some examples of non-natural amino acids and sugars are shown in Figure 1.41 and include groups such as azides, alkynes, or aryl ketones modified onto the side chains or derivatized off one of the carbon atoms of a monosaccharide. Many other amino acid and sugar derivatives can be used for this purpose; for an excellent review, see Sletten and Bertozzi (2009).

Proteins or glycans containing these non-natural amino acids or sugars can be targeted for detection using reagents containing another reactive group that will only conjugate with the functionality on the

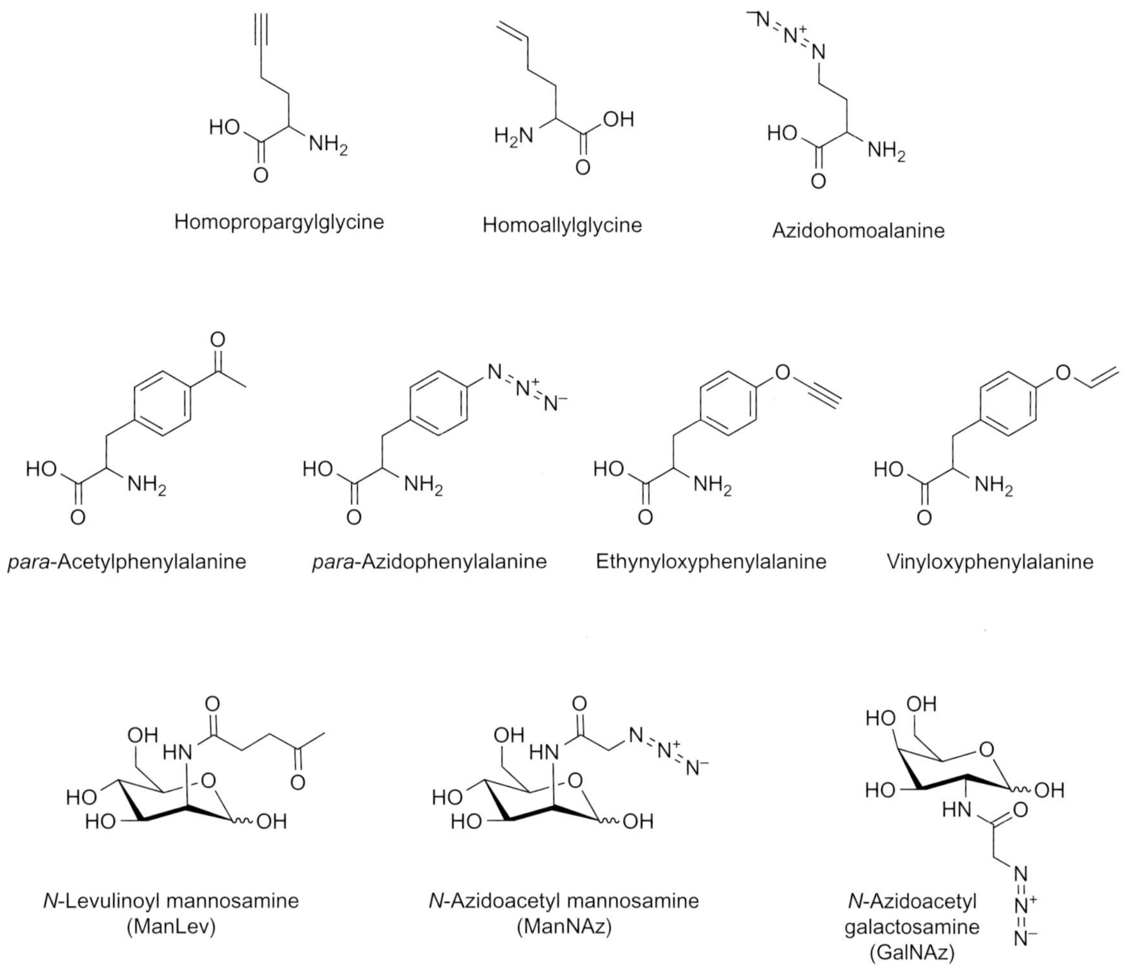

FIGURE 1.41 A number of amino acid and sugar derivatives have been synthesized to contain functional groups able to participate in a chemoselective ligation reaction with a detection reagent. For example, the alkyne-containing amino acids can be used as additives in depleted media to grow cells and incorporate these analogs into proteins within the cells. The alkynes can then be reacted with azide-containing detection reagents to covalent link to these sites by forming a triazole ring. Alternatively, the azide-containing compounds may be incorporated similarly into proteins or carbohydrate components of cells and reacted with an alkyne-containing reagent in a click chemistry reaction or a phosphine derivative in a Staudinger ligation reaction to specifically label the proteins or glycans containing the azido groups.

non-natural component. Such "bioorthogonal" reactions (see Chapter 17) can be used to create bioconjugates within the complex milieu of a cell or a cell lysate without cross-reactivity with other biological components. In this manner, modified glycoproteins have been targeted and imaged using fluorescent dyes containing a complementary bioorthogonal reactive group (Baskin *et al.*, 2007).

There are a number of bioorthogonal reactions that can be used for detection, tracking, and imaging purposes. These include click chemistry, Staudinger ligation, hydrazide–aldehyde reactions, aminoxy–aldehyde reactions, and others (Chapter 17). For detection within live cells, the copper-free click chemistry option is perhaps the best choice, because of its rapid reaction kinetics and the components are relatively nontoxic to cells. The only deficiency of the cyclooctyne chemistry is the potential of the triple bond to react with other nucleophiles within biomolecules, especially free thiols. However, the copper-catalyzed click reaction with an alkyne group can be used to fluorescently label azido amino acids or azido sugars with greater specificity, as long as cell viability is not an issue. Figure 1.42 illustrates the use of a fluorescent rhodamine–alkyne probe to detect azido-labeled glycans on cell surfaces. This particular alkyne compound was originally synthesized by Beatty (2008) and used to image azido amino acids within cells. For imaging within fixed and permeabilized cells, a broad spectrum of other bioorthogonal chemistry approaches may also be used, since cell viability is not an issue.

Proximity Ligation Assays

Detection and imaging within fixed cells can also be done with extraordinarily high sensitivity using unique

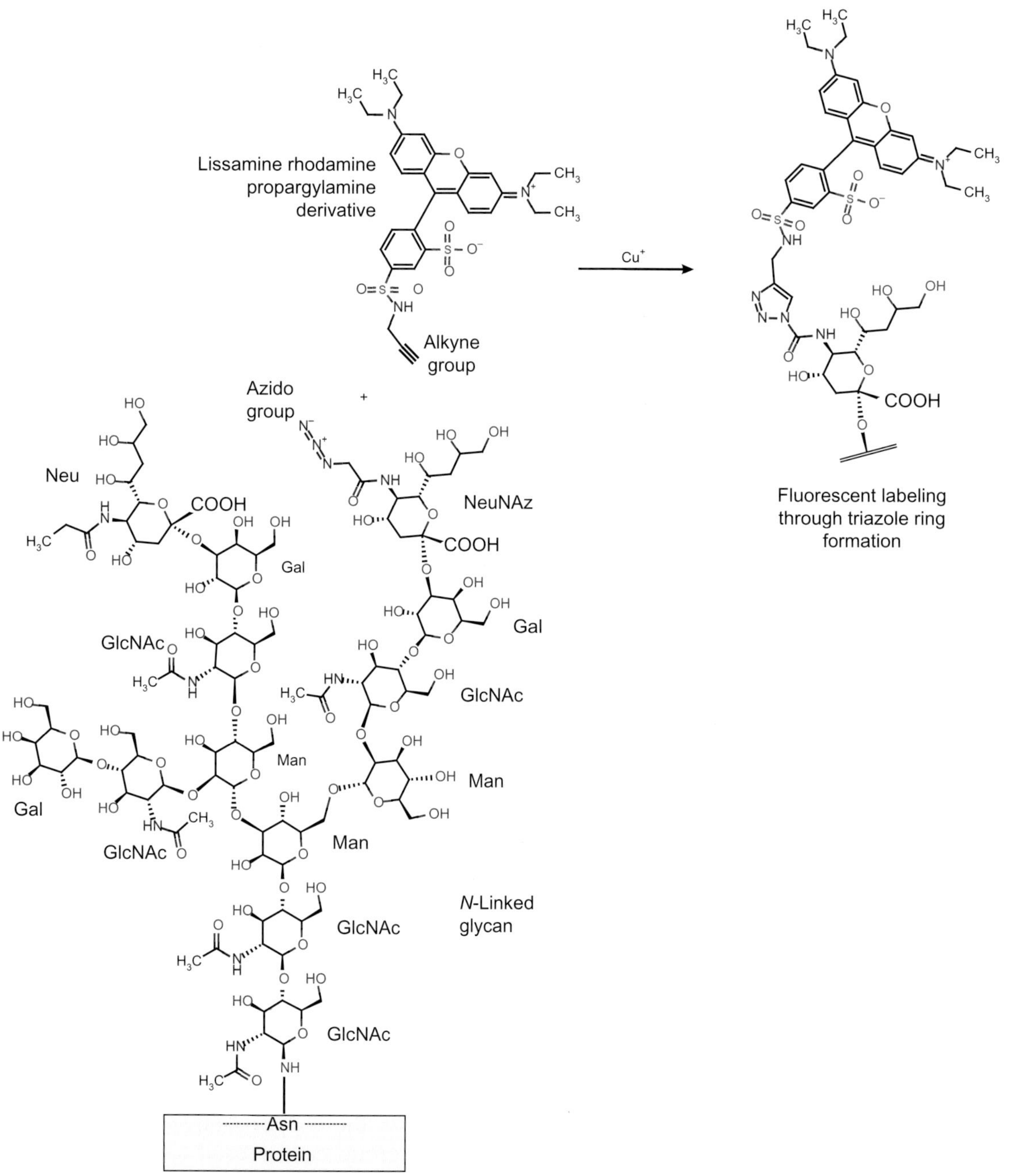

FIGURE 1.42 An illustration of the labeling of an azide-containing glycan on a cell surface glycoprotein or glycolipid with an alkynyl–rhodamine derivative to fluorescently tag the modified carbohydrates.

oligonucleotide bioconjugates in a process called a "proximity ligation assay" (PLA) (Fredricksson *et al.*, 2002; Gullberg *et al.*, 2004; Jarvius *et al.*, 2006; Söderberg *et al.*, 2006). The bioconjugates used in this detection method are capable of visualizing individual cellular events such as discrete protein–protein interactions or the activation of a particular protein through phosphorylation (Jarvius *et al.*, 2007). The reagents used for PLA consist of two main conjugates, each containing a targeting molecule attached to a tether consisting of a terminal synthetic oligonucleotide, which acts as the reporter (Olink Bioscience). The targeting molecules typically are two different secondary antibodies directed against two different primary antibody types (species specific). The primary antibodies that are chosen for an assay are directed against two different targets or epitopes within the cell and can consist of antibodies that are able to bind to particular proteins or interact with any other cellular components desired to be detected. In addition, PLA targeting molecules can

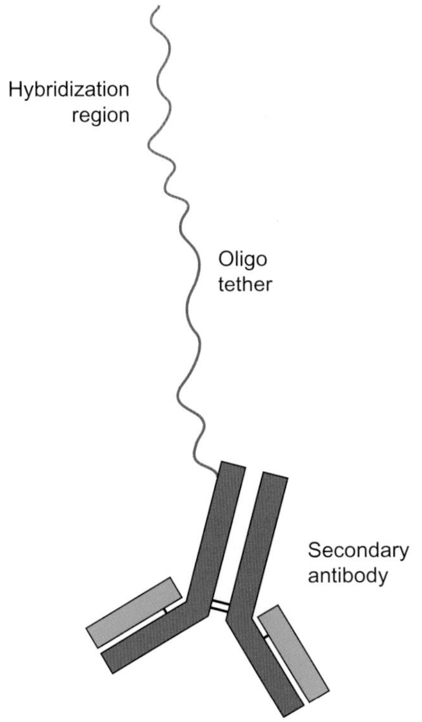

FIGURE 1.43 A proximity ligation assay (PLA) conjugate can be created by covalently linking an oligo tether containing a uniquely designed hybridization region to an antibody specific for a particular target molecule. Secondary antibodies may also be used as part of the bioconjugate to make more universal detection reagents for any appropriate primary antibody.

also be specific oligonucleotide probes that can bind to target DNA or RNA sequences within the cells. Figure 1.43 illustrates the components of a PLA bioconjugate for antibody-based reagents.

The PLA method first involves the binding of the targeting bioconjugates containing the attached oligo tethers with their targets within a cell. If the binding events occur in sufficient proximity, such as in the binding of labeled primary antibodies to a protein within a cell (Figure 1.44), then the oligo tethers on each conjugate are maintained within molecular distance of one another. A second set of synthetic oligos is added to the assay that are able to hybridize with the ends of the oligo tethers and form a circular DNA structure containing some double-stranded features at the points in which they interact with the tethers. The circle that is created is ligated together by adding ligase enzyme and then amplified using polymerase enzyme and deoxynucleotide triphosphates in an isobaric rolling circle amplification process that forms a long repeating complementary oligonucleotide sequence in the immediate vicinity where the initial binding event took place. This amplified piece of DNA is finally detected by adding a third bioconjugate consisting of a fluorescently labeled short oligonucleotide, which is able to hybridize repeatedly along this long amplified strand. The fluorescent signal generated by this detection process is produced through multiple fluor–oligo bioconjugates docking at the site of the targeted molecules, thus resulting in an intensely bright signal of high sensitivity. The resultant fluorescent spots created within a cellular image are about 500 nm in diameter and effectively localize the targeted molecule or interacting proteins with highly discrete signals.

PLA detection methods thus can result in the imaging of individual events within a cell, and the fluorescence signals produced in cell-based imaging amount to numerous discrete spots associated with the amplified DNA at the site of each targeted molecule. Unlike the diffuse signals obtained when using fluorescently labeled antibodies to detect targets within cells, the PLA process produces dots of fluorescence, which can dramatically improve imaging, localizing, and tracking of biomolecules (Figure 1.45). It is likely that this technique permits sensitive assays down to the single molecule detection level for cellular imaging.

Staining with Antibody–Enzyme Conjugates

One of the oldest methods for detecting markers within cells and tissues involves specific immunohistochemical (IHC) staining using antibody–enzyme conjugates. Bioconjugate reagents for IHC were among the first developed for imaging disease from biopsies within thin tissue sections mounted on microscope slides. The pathology market that still extensively uses these bioconjugates for diagnostic testing is crucial for assessing the presence of cancer and other diseases within millions of patients worldwide. The bioconjugates designed for this purpose are constructed virtually the same as those used for ELISA or western blotting applications. The typical targeting components are antibodies, but they must be validated for activity and specificity toward the target of interest. It often takes a trained pathologist to validate the specificity of IHC reagents for targeting and staining the proper antigens within a tissue section. The enzymes usually used for detection are horseradish peroxidase (HRP), alkaline phosphatase, or β-galactosidase, which can all form chromogenic or fluorescent products from turnover of specific substrates. In addition, since the initial substrate is added to the IHC process as a soluble molecule in a buffered solution, the detectable product of the enzyme reaction must be insoluble or it would diffuse away from the site of target interaction. The result of a positive signal in IHC type bioconjugate assays is a dark colorimetric stain or a bright fluorescent signal marking the sites of antigen targets within the tissue section. Figure 1.46 illustrates the use of antibody–enzyme bioconjugates in an IHC staining process.

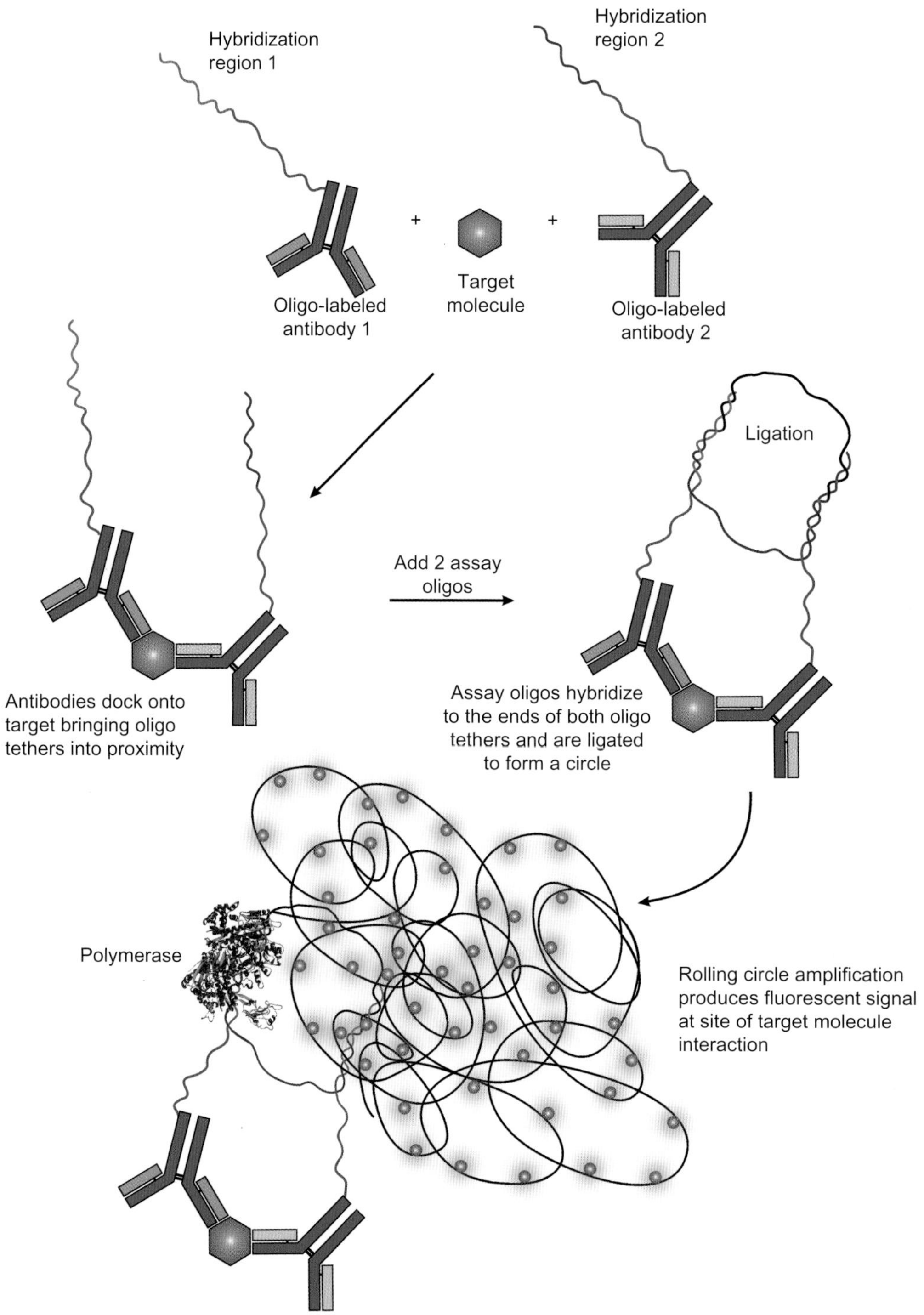

FIGURE 1.44 A proximity ligation assay (PLA) can be done using two oligo-labeled antibodies directed against two separate regions of a target protein (or against two different proteins interacting together). The antibody conjugates are mixed with a sample containing the target and allowed to bind. Assay ligation oligos are then added that are capable of hybridizing to selected regions on the oligo tethers on each antibody. If both antibodies have bound to their target, then the oligo tethers are held in proximity and the ligation oligos can be ligated together to form a circle. At this stage, a polymerase enzyme is added along with the appropriate deoxynucleotide triphosphate molecules to perform an isobaric rolling circle amplification reaction, which replicates the complementary sequence to the circular DNA strand over and over again. Finally, fluorescently labeled detection oligos are added that can hybridize to the amplified DNA and create an intense fluorescent signal at the site of the targeted protein.

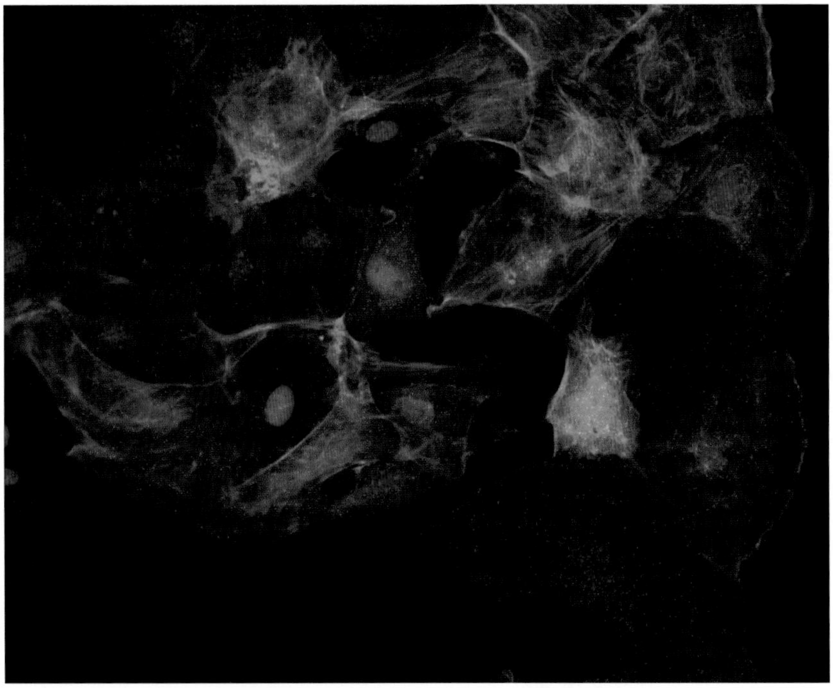

FIGURE 1.45 Fluorescent visualization of protein–protein interactions between SMAD1/2/3 and SMAD4, upon TGF-β stimulation in mouse embryonic fibroblasts. Each red dot represents an interaction event detected by *in situ* PLA® using Duolink®. Blue: counter-staining of nuclei. Green: counter-staining of actin. *Data were collected by Olink Bioscience in collaboration with Katerina Pardali, Uppsala University, Sweden. Image courtesy of Olink Bioscience and used with permission.*

Bioconjugates for Western Blot Detection

In an immunoassay detection procedure similar to IHC staining, western blots can be specifically probed using antibody–enzyme bioconjugates with subsequent highly sensitive chemiluminescent detection. Secondary antibody–enzyme conjugates and streptavidin–enzyme conjugates are the two most common reagents for developing western blots. In this application, a protein sample is separated by molecular weight using SDS polyacrylamide gel electrophoresis (PAGE) and the separated proteins are then transferred from the gel to a membrane for immunochemical probing. The combination of protein separation by mass combined with specific protein targeting and detection using antibodies provides a powerful confirmation for the presence of a particular target.

Typically, a primary antibody is used to initially bind the target protein on the blot. After washing off excess antibody, the secondary antibody–enzyme conjugate is added for the detection process. Alternatively, if the primary antibody is biotinylated, then a streptavidin–enzyme conjugate can be used for detection; however, if the primary antibody is not biotinylated, then a secondary antibody–enzyme conjugate is used. After another thorough wash step, a substrate solution is then added to develop a signal at the site of antibody binding (Figure 1.47). The substrate used for this process is usually one of two types: a precipitating substrate or one that creates a chemiluminescent signal as the result of enzyme turnover. Enzymatic chemiluminescence is by far the most widely used detection scheme in western blotting.

The antibody–enzyme or streptavidin–enzyme bioconjugates used to detect proteins in western blots can be prepared using direct conjugation of the two molecules together (zero length crosslinking) or they can be made with bifunctional crosslinking agents. Zero length conjugation is particularly suitable for conjugations made with HRP, as the carbohydrate on the glycoprotein can be oxidized to contain reactive aldehyde groups, which can then be coupled to the targeting molecule using a reductive amination process (Chapter 4, Section 4, and Chapter 20, Section 1.3). Bioconjugation using crosslinkers can also be done effectively using a number of reactive compounds (Chapters 5 and 6). However, the most effective crosslinkers are heterobifunctional reagents, such as the popular sulfo-SMCC (Chapter 6, Section 1.3), and even better are the hydrophilic PEG-based analogs to this compound (Chapter 18, Section 1.2). For increased sensitivity in western blotting applications, an intermediate scaffold can be used in preparing the detection conjugate, which functions to attach a greater number of enzymes onto each targeting antibody (see previous discussion on scaffold molecules).

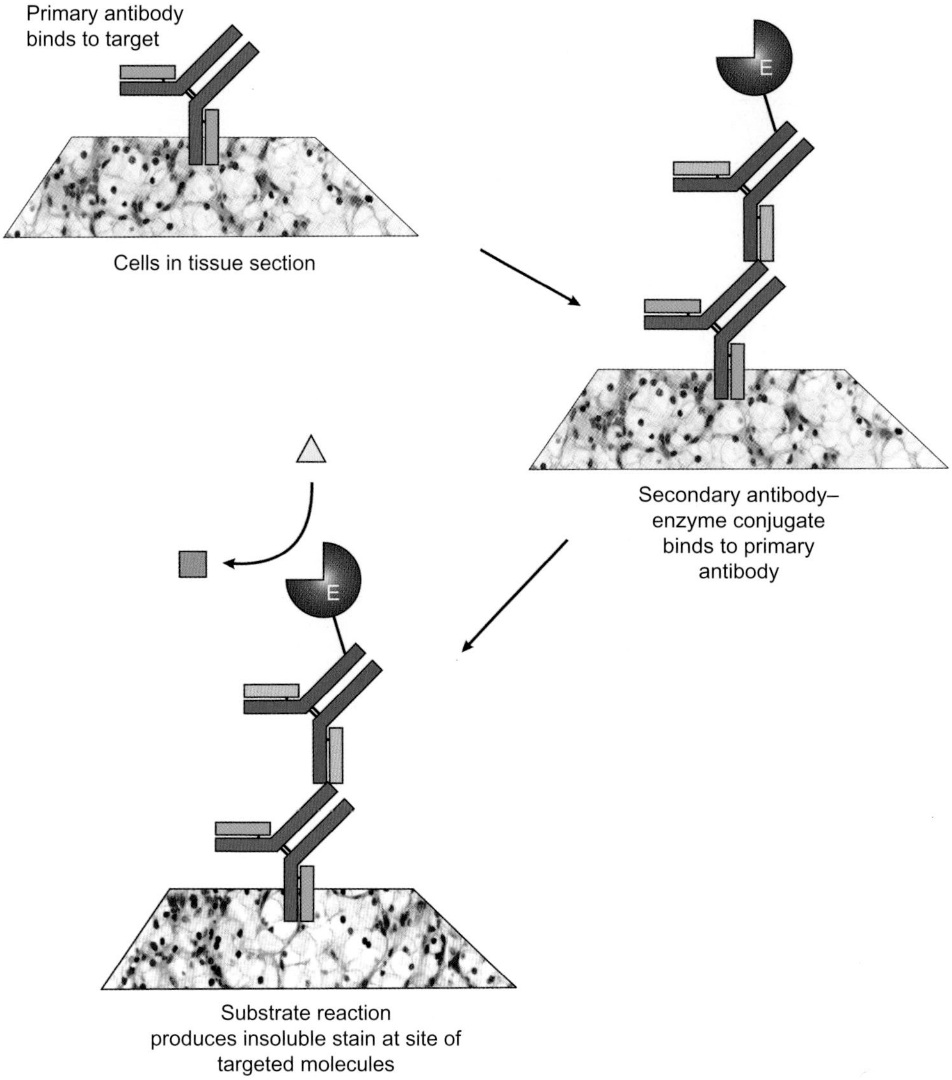
Primary antibody
binds to target
Cells in tissue section
E
Secondary antibody–
enzyme conjugate
binds to primary
antibody
E
Substrate reaction
produces insoluble stain at site of
targeted molecules

FIGURE 1.46 A secondary antibody-enzyme bioconjugate can be used to detect specific antigenic targets within tissue sections. Immunohistochemical (IHC) staining typically involves the binding of a primary antibody to a given target protein within a tissue section. An antibody–enzyme conjugate is then added that contains an antibody able to bind to the antibody type used as the primary antibody, for instance a goat-anti-mouse antibody used to bind to a mouse monoclonal primary antibody. The enzyme part of the conjugate can be used to develop a specific detectable signal by a substrate reaction, such as a chromogenic or fluorescent product at the site of the primary antibody binding event.

Bioconjugates for Super Resolution Microscopy

Some recent exciting developments in imaging are represented by the super resolution microscopy techniques that use specialized bioconjugates containing fluorescent labels. The fluorophores used in these conjugates typically are switchable from an off to an on state and sometimes back again. The technology behind this revolution in resolving ever smaller cellular components actually allow structures to be resolved that would not be possible using simple light microscopy. At one time, it was believed that due to the diffraction limit of light, the fine macromolecular structures within a cell would never be imaged below about half the wavelength of light used. However, a combination of fluorescence technology and software innovation has now made possible the resolution of cellular structures down to the 20- to 50-nm level and sometimes even within the 2- to 20-nm range (Betzig *et al.*, 2006; Chi, 2009; Huang *et al.*, 2009).

The evolution to super resolution imaging occurred through research into single-molecule fluorescence analysis of biomolecules, which are techniques that isolate a single discrete fluorescent conjugate bound to its target from all other bound fluorescent conjugates in the sample (Klar *et al.*, 2000; Kapanidis and Weiss, 2002). It has now been demonstrated that the position of a single fluorescent conjugate can be determined to extremely high accuracy, provided there are enough photons collected

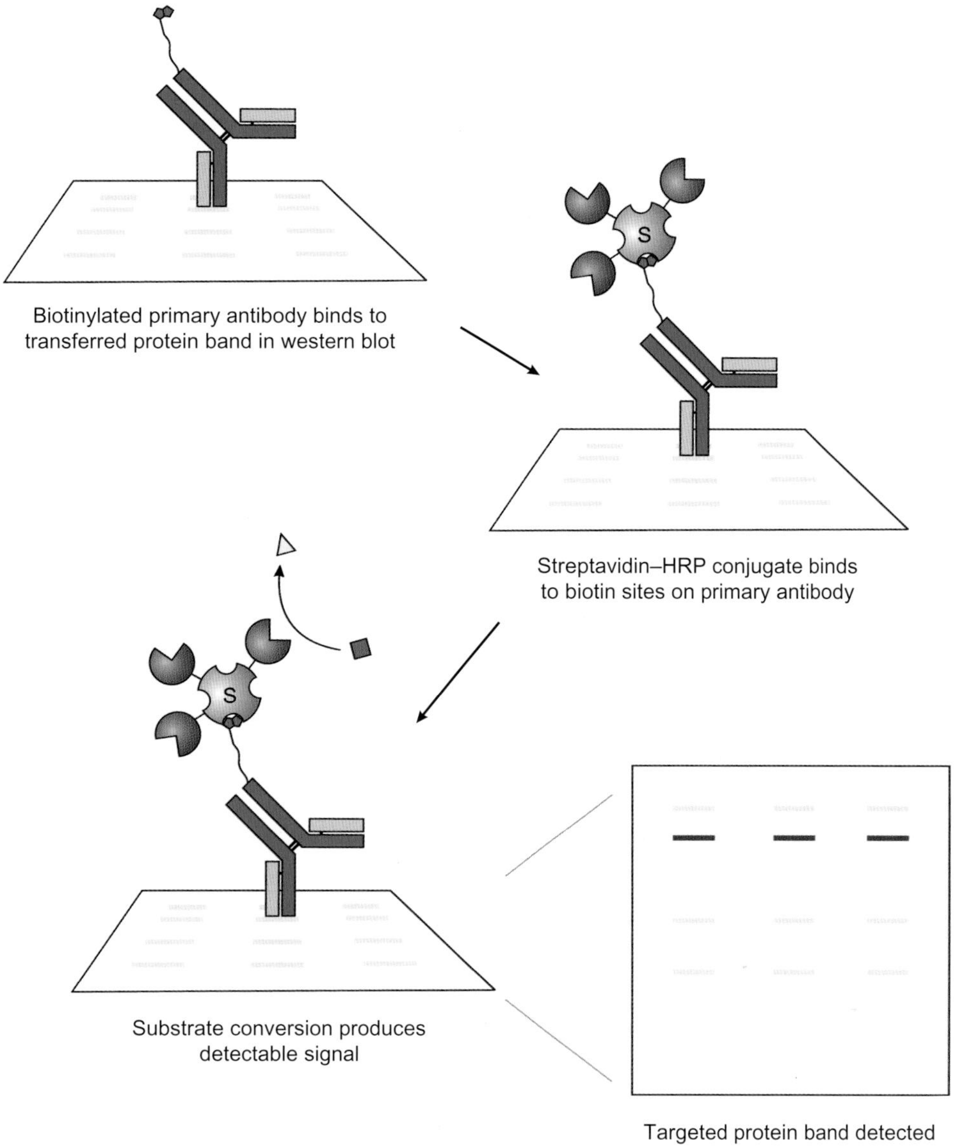
Biotinylated primary antibody binds to transferred protein band in western blot
S
Streptavidin–HRP conjugate binds to biotin sites on primary antibody
S
Substrate conversion produces detectable signal
Targeted protein band detected

FIGURE 1.47 Bioconjugates used for western blotting purposes are among the most common reagents used in everyday laboratory experiments involving biological research. Proteins are separated by SDS polyacrylamide gel electrophoresis and then blotted over onto a membrane, typically nitrocellulose or PVDF. The transferred proteins can be probed using a biotinylated primary antibody specific for the protein of interest. Next, a streptavidin–enzyme conjugate is added that binds to the biotin groups on the primary antibody. Then a substrate reaction is used to develop a detectable signal at the site of the targeted protein, which can be chromogenic, fluorescent, or chemiluminescent in nature. The most common detection method involves the use of the enzyme HRP in the streptavidin conjugate along with a chemiluminescent reaction using a luminol or acridinium substrate system.

from a precise site of emission (van Oijen *et al.*, 1998; Thompson *et al.*, 2002; Yildiz *et al.*, 2003). Using this technique, an extremely high-resolution image of cellular components can then be constructed from repeated localization of individual fluorescent probes.

There are a number of methods used for super resolution microscopy and each one uses unique bioconjugate designs to make optical imaging at the molecular level possible. Some of the more popular techniques include *fluorescence photoactivated localization microscopy* (FPALM or just PALM), *interferometer photoactivated localization microscopy* (iPALM), and *stochastic optical reconstruction microscopy* (STORM). The FPALM, PALM, and STORM techniques are based on nearly identical principles and were developed simultaneously by Rust *et al.* (2006), Hess *et al.* (2006), and Betzig *et al.* (2006).

FPALM, PALM, and iPALM often use bioconjugates containing photoactivatable fluorescent labels that can be switched from a dark state to a bright emission state by exposure to particular wavelengths of light (Heilemann *et al.*, 2005; Betzig *et al.*, 2006; Hess *et al.*, 2006). The fluorescent molecules used for PALM have

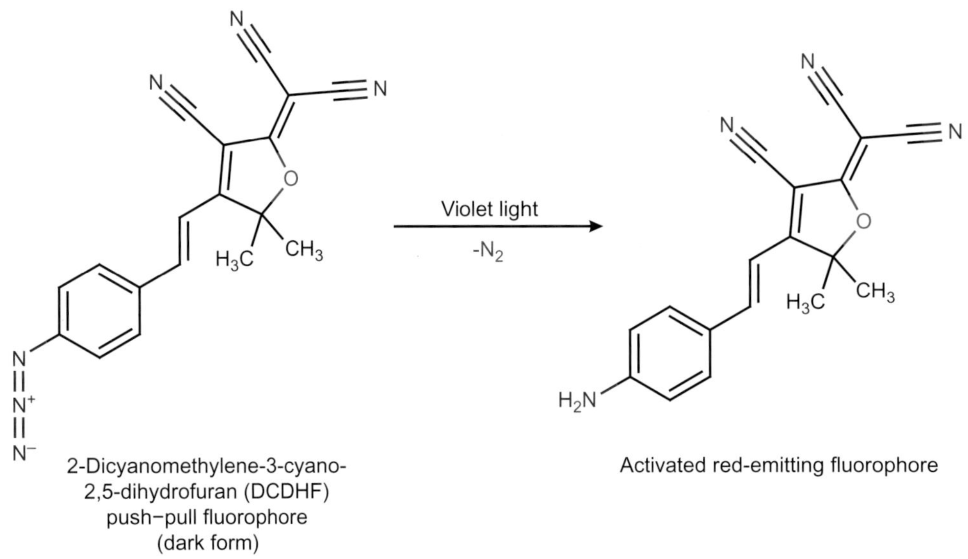

FIGURE 1.48 The photolysis of DCDHF dyes using violet light results in the degradation of the azido phenyl group to form an amine, which creates a red fluorescent dye that can be used to selectively activate bound probe conjugates within a cell for super-high resolution microscopy.

included fluorescent proteins, fluorescent organic dyes (Chapter 10, Sections 1–8), and more recently fluorescent quantum dots (Chapter 10, Section 10) (Pinaud *et al.*, 2010). Initially a probed sample may be completely dark, containing bound labeled conjugates, but none of them capable of fluorescence emission unless activated. In some cases, fluorescent fusion proteins are used that can be switched from a bright to a dark state and back on again. The dark fluorescent labels used in designing bioconjugates for PALM include organic molecules that are quenched by the presence of an azide group attached on an aromatic ring associated with the π-conjugated system of the fluorophore.

One dye family particularly useful for photoactivation and PALM imaging is based on a push–pull design containing an amine donor group on one end and a 2-dicyanomethylene-3-cyano-2,5-dihydrofuran (DCDHF) acceptor group on the other end, which are separated by an electron rich π-conjugated system (Lord *et al.*, 2008, 2009a). The amine group pushes electrons toward the conjugated system, while the cyano groups pull them away at the other end. The electron-withdrawing properties of an azide in place of the amine disrupt the conjugated system and quench the fluorescence while also causing a blue shift in the absorbance spectrum of the dye. The size of the intermediate π system determines the absorption and emission characteristics of the activated fluorescent dye. In super resolution microscopy, the quenched form of DCDHF dyes is relieved when the molecule is irradiated at 407 nm, which results in photolysis of the azide and conversion into the amine (Figure 1.48). This phenylazide group does not ring expand to the dehydroazepine as some other photoreactive phenylazides do, because of the electron withdrawing properties of the cyano groups, which stabilize the intermediate nitrene (Chapter 6, Section 3) (Lord *et al.*, 2008). However, it is possible during nitrene formation that photo-insertion may take place resulting in covalent attachment into C-H or C-C bonds in neighboring biomolecules. This reaction can be used to advantage, as DCDHF dyes could be used to design photoaffinity labeling (PAL) agents that could target a particular active site and become covalently attached and fluorescent upon photoactivation (Lord *et al.*, 2009b). Thus, the fluorescence form of the DCDHF dyes can be revealed with discrete laser pulses at precise locations within a cell during imaging, allowing detection at the single molecule or near single molecule level for highly accurate localization of signals.

The STORM method (Rust *et al.*, 2006), uses fluorescent dye conjugates with targeting molecules (e.g., antibodies), wherein the fluorophores can be switched on and off using exposure to light at particular wavelengths. Unique to this method is the use of dual-labeled conjugates that have two fluorescent probes attached to a single targeting molecule, which are able to initiate a FRET signal between them due to their proximity. It was discovered that certain dyes, for instance the two cyanine dyes Cy3 and Cy5, are able to be turned on and off for dozens of cycles without photobleaching (Figure 1.49). The equivalent dyes that represent more hydrophilic Cy3 and Cy5 derivatives containing additional sulfonate groups may also be used in this type of bioconjugate to decrease nonspecific binding of the dyes to biomolecule structures within cells.

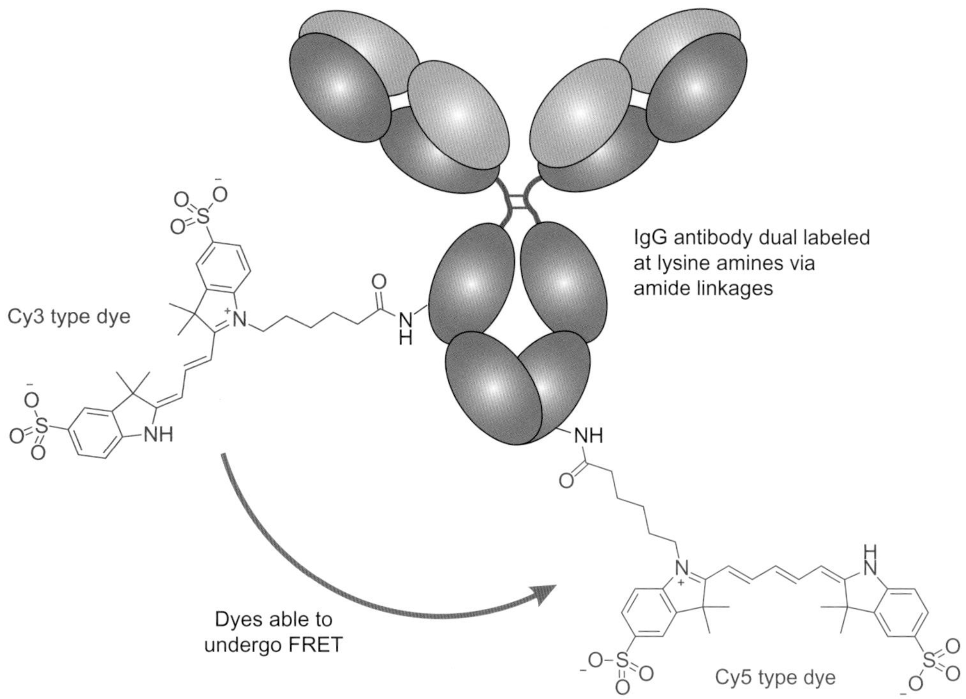

FIGURE 1.49 Dual-labeled antibody conjugates can be constructed to allow FRET signaling between the two carbocyanine dyes attached to a single antibody. This permits on and off switching of the fluorescent emission for use in super resolution microscopy techniques.

The cycling of the emission from dark to bright in such dual-labeled conjugates is essential for isolating one or at most a very limited number of labeled molecules bound to their targets within a cell. Under normal fluorescence imaging conditions using dye-labeled antibodies, the conjugates bound to targets are all excited in mass and what is observed is the average fluorescence signal from many conjugates emitting in combination. In order to determine the position of a single fluorescent probe without interference from other ones, it is necessary to have methods of turning on only a single fluorescent conjugate in the presence of many other conjugates; otherwise, the combined signals would all bleed together and reconstructing their exact positions would be impossible. The dual-labeled conjugates of STORM imaging make this possible by turning on and off the acceptor dye through selective irradiation at certain wavelengths.

In a dual-labeled fluorescent targeting bioconjugate, the acceptor dyes of the FRET pair (e.g., the Cy5-type dye) can be switched to a nonfluorescent state by excitation under non-oxygen-containing conditions in the presence of a thiol. Upon continued excitation at the acceptor dyes' excitation wavelengths, the dyes undergo a reaction to add the thiol compound at a carbon within the polymethine bridge, thus forming a thioether derivative (Figure 1.50). This addition reaction to the double bond reduces a site in the conjugated electron system of each dye to an aliphatic derivative, thus interrupting the conjugated double-bond system and effectively destroying the fluorescence character of the molecules. The Cy5 dyes are therefore no longer able to emit light upon excitation at their normal absorption band and they become dark. However, it was discovered that if the FRET donor dye (e.g., the Cy3-type dye) is excited at its normal excitation peak, it will transfer its energy to the Cy5 dye and restore the acceptor dye's fluorescence characteristics by cleaving off the thiol compound and restoring the original polymethine structure of the dye (Rust *et al.*, 2006). This selective rescue of the acceptor dye fluorescence is the property needed for discrete imaging of single molecule fluorescent conjugates in super resolution microscopy. A discretely focused laser can target and restore fluorescence to dual-labeled conjugates in only a very small area in a cell image, thus ensuring that only one or just a small number of FRET conjugates become fluorescent at a time.

3.3. Purification, Capture, and Scavenging

Another set of important applications where specialized bioconjugates are used involves affinity-mediated enrichment, purification, or scavenging of biomolecules. The immobilization of biospecific affinity ligands onto insoluble resins or other solid support materials makes possible the separation of target molecules from complex solutions, often in a single step (Figure 1.51). Immobilization technology and affinity chromatography originated in the 1960s and '70s using classical

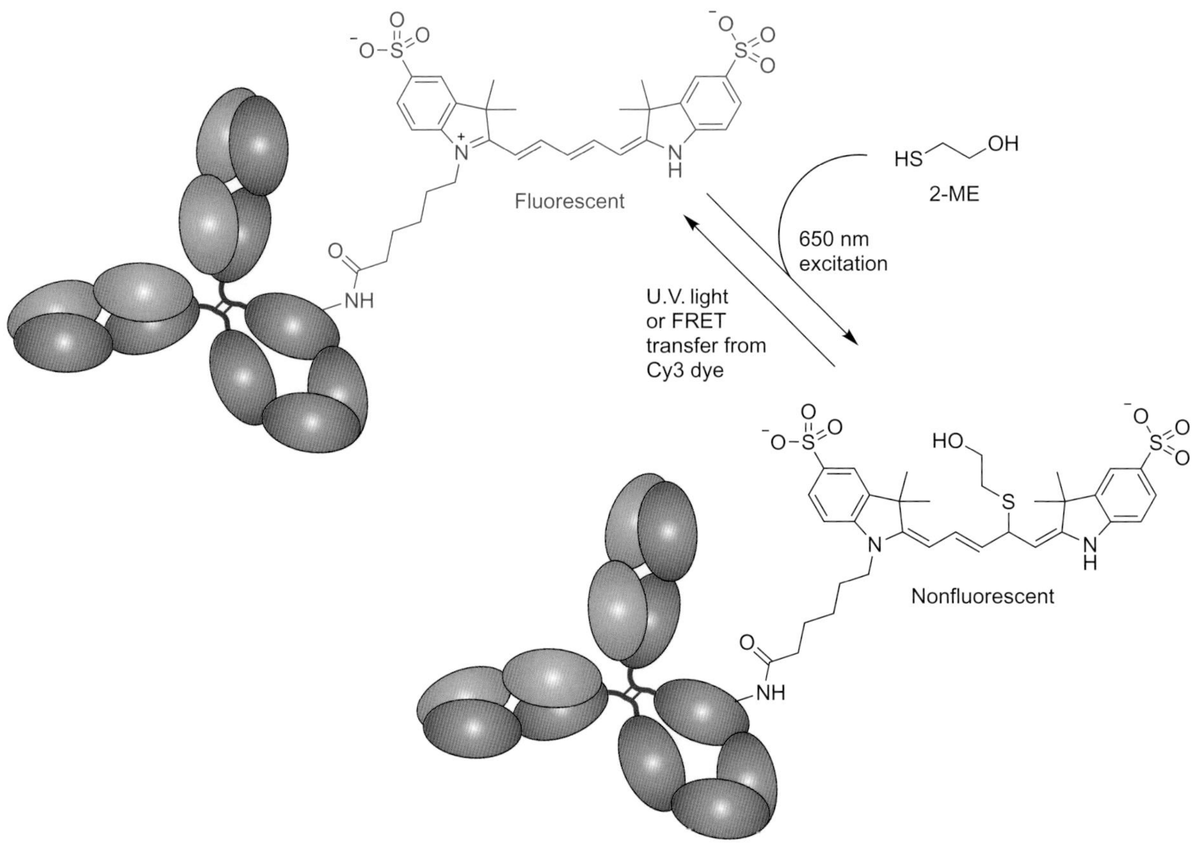

FIGURE 1.50 An antibody labeled with a Cy5-type carbocyanine dye can be switched from an on state to an off state in the presence of a thiol compound, such as 2-mercaptoethanol. Exposure to 650-nm excitation light in the presence of the thiol forms an addition product to the central polymethine bridge, thus interrupting the conjugated electron system and eliminating the fluorescent signal. Subsequent exposure to ultraviolet light causes the thiol to be removed and restores the fluorescence.

porous chromatography supports to purify proteins and other biological molecules by interaction with affinity ligands (Cuatrecasas, 1970; Hermanson *et al.*, 1992; Hage, 1999). The immobilized ligand may be a natural bioaffinity molecule or a synthetic molecule chosen to bind the desired target molecule reversibly so that the affinity support can capture and release the target for isolation. The appropriate affinity ligand typically is covalently attached to the insoluble matrix using immobilization reactions that are similar or identical to those methods used to form soluble bioconjugates in solution (Chapters 14 and 15). Alternatively, affinity ligands can be noncovalently immobilized using an intermediary high-affinity biological interacting pair, an example of which is the streptavidin–biotin interaction. In this technique, a biotinylated molecule or protein (Chapter 15, Section 2.5) is allowed to interact with an immobilized streptavidin support, thus attaching the ligand to the support through the biotin modifications.

Affinity molecules can be immobilized onto a wide range of insoluble support materials, including porous resins and nonporous particles of all shapes and sizes, as well as magnetic particles, planar surfaces, microplates, inside tubes or microfluidic channels, membranes, and porous monolithic structures. The affinity matrix thus obtained can be used to capture out of solution virtually any target molecule in a biological sample, provided the appropriate ligand is immobilized onto it. Today, immobilized affinity ligands are applied in chromatographic separations ranging in scale from microliter quantities of resin used to purify microgram quantities of target molecules to hundreds of liters of resin used in bioprocess purifications to produce kilogram quantities of therapeutic proteins.

Affinity chromatography has also become a vital tool for many proteomics studies investigating the expression, activity, structure, and function of proteins. For instance, an affinity support can be used to enrich low copy target proteins from a cell or tissue lysate to permit detection or characterization by mass spectrometry or other techniques (Zhang *et al.*, 2007; Kota *et al.*, 2009). The appropriate affinity support can be designed that can specifically capture phosphoproteins or phosphopeptides, enrich the glycoprotein fraction from nonglycosylated proteins, isolate peptides containing thiols or disulfides, purify proteins with ubiquitin or other

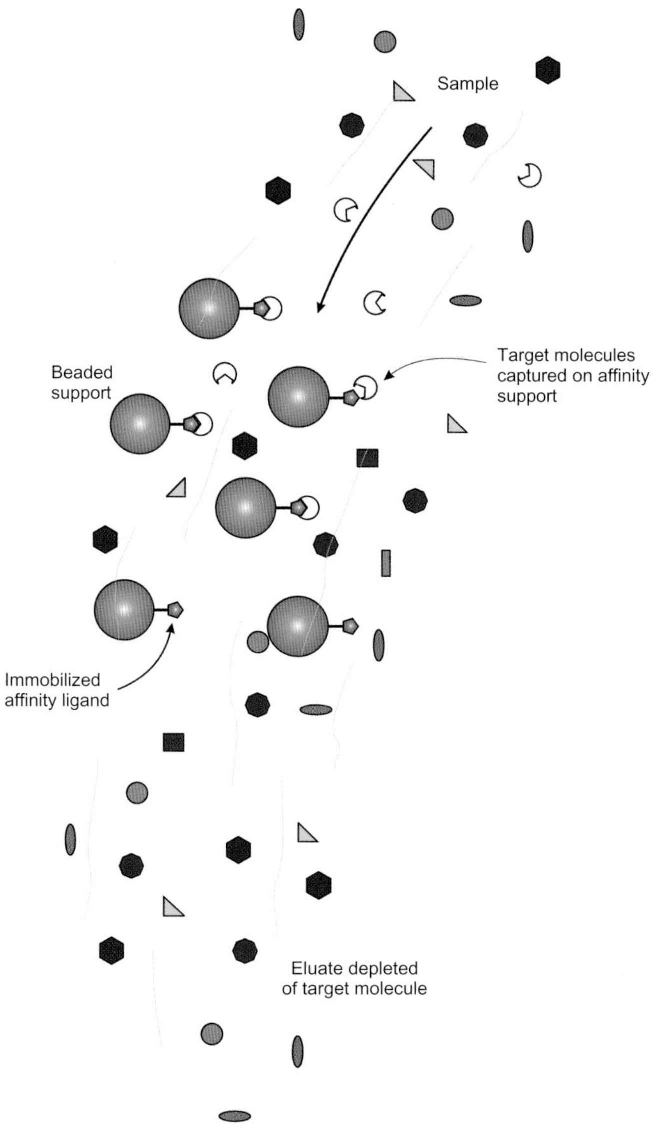

FIGURE 1.51 The principle of affinity chromatography involves the use of an immobilized affinity ligand attached to an insoluble matrix, typically a porous beaded resin that can interact with and capture a desired target molecule from a complex sample. After binding the target, the support is washed free of excess unbound sample and the captured target can then be eluted and isolated in an essentially pure state.

post-translational modifications, or pull out of solution proteins having particular active sites or binding sites able to interact with the immobilized ligand (Azarkana *et al.*, 2007; Patricelli *et al.*, 2007; Wissing *et al.*, 2007; Cravatt *et al.*, 2008).

Purification using Immobilized Affinity Ligands

Biological macromolecules (such as proteins), smaller bioaffinity ligands (such as short peptides, nucleic acids, or carbohydrates), or even synthetic organic molecules (such as substrate analogs or drugs) can be covalently coupled to an insoluble surface or particle to create an affinity matrix for the purpose of specifically binding and interacting with a target molecule in a biological solution. The major use of immobilized affinity ligands is to capture and purify a target biomolecule from an extract, for example a lysate or homogenate. Affinity chromatography has become the primary purification method of choice for isolating biological targets from complex solutions. Auxillary chromatographic methods may also be employed to clean up a biomolecule preparation, but the most powerful specific techniques are separations using immobilized affinity ligands. In fact, the biotechnology and pharmaceutical industries have invested heavily in the purification of biotherapeutic proteins in scales ranging from multi-gram quantities required during clinical trials to tens or hundreds of kilos needed for full production, if a major protein drug candidate is successful and enters the market.

The protein most often targeted for purification using an immobilized affinity ligand is an antibody. At scales appropriate for research separations, μg or mg quantities of IgG are routinely purified from immune serum, mammalian cell culture supernatant, recombinant bacterial fermentation broth, or ascites fluid to isolate a particular antibody having specificity for the desired target. There are several basic types of immobilized affinity supports that can be used for antibody isolation: (1) an immobilized antigen with precise binding specificity for the antibody being isolated; (2) an immobilized immunoglobulin binding protein which is able to bind to the general class of antibody being produced; or (3) an immobilized organic pseudo-affinity ligand having binding characteristics for the antibody type in general (Figure 1.52) (Vljayalakshmi, 1998).

The coupling of an antigen to an insoluble support usually involves conjugation through a functional group on the molecule to a reactive group on the matrix. If the antibody was produced to have specificity for a particular protein (which represents the most common type of antibody specificity), then to create an immobilized affinity support the target protein is simply coupled to the matrix using its available amines, thiols, or occasionally through its carbohydrate. This works especially well if the protein antigen is available in quantities suitable for immobilization and use as a purification tool. Chapter 15 describes the reactions useful in coupling proteins to insoluble support materials and Chapter 20 discusses particular conjugation strategies for use with antibodies, the methods of which are appropriate for both solution-phase conjugation and immobilization to a solid phase.

Purification of specific antibodies from antiserum or from media that may contain other non-relevant IgG molecules usually involves the use of an immobilized antigen column, because this ensures that the antibody of correct specificity is isolated from antibodies having other specificities. However, if the desired antibody is

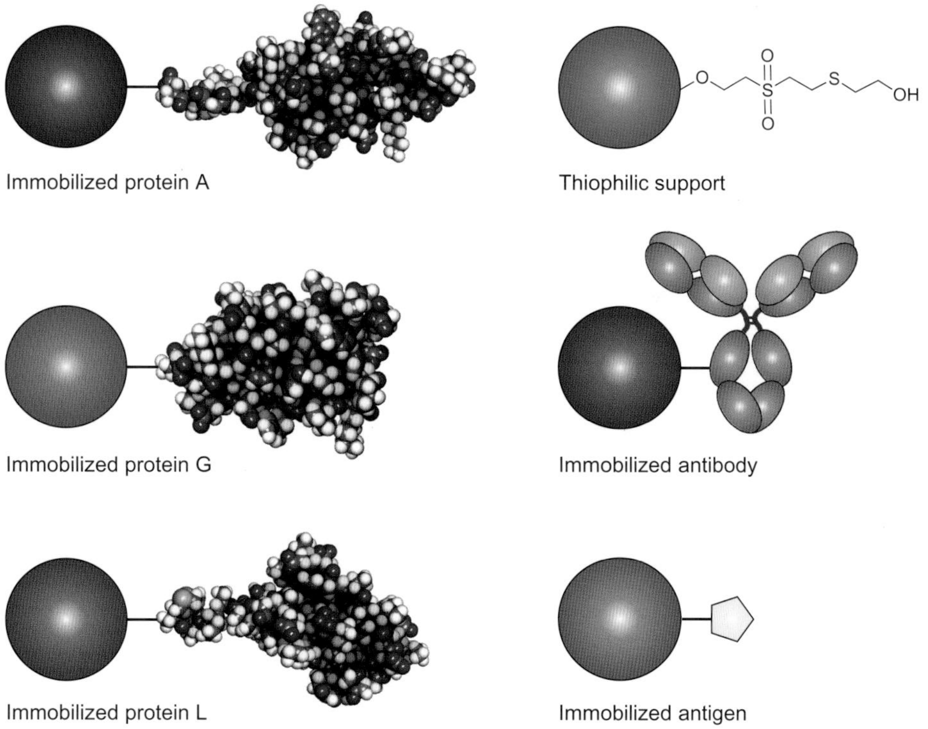

FIGURE 1.52 The most common affinity chromatography supports designed to purify antibody molecules. The immunoglobulin binding proteins protein A and protein G typically are used to purify intact antibodies containing an Fc region, whereas protein L is used with Fab fragments that contain kappa light chains. A thiophilic support binds with high affinity to the Fc region of antibodies in the presence of a lyotropic salt and can be eluted by removing the salt. Immobilized antibodies can be used to purify secondary antibodies in high purity, for instance in the use of an immobilized mouse IgG to purify a goat-anti-mouse antibody. Finally, immobilized antigens are often used to highly purify specific antibodies having discrete antigen binding activity.

present in a solution that contains no other antibodies, such as is typical of ascites fluid when making monoclonal antibodies or of bacterial production of recombinant antibodies, then an immobilized immunoglobulin binding support may be used that can isolate antibodies through its affinity to the Fc or Fab regions of IgG, and not through specific antigen binding site interactions.

For decades, affinity supports prepared using immobilized protein A, protein G, or protein A/G have been used to isolate and purify antibodies from culture supernatants and ascites (Huse *et al.*, 2002; Roque *et al.*, 2007; Bergmann-Leitner *et al.*, 2008). Recombinant protein A in particular has become the most popular immobilized affinity ligand for purification of IgG antibodies in process scale at biopharmaceutical companies. Today, many thousands of liters of immobilized protein A are produced under cGMP conditions to supply the needs of antibody therapeutic production processes worldwide. These quantities will only increase in coming years as many more antibody-based biotherapeutics are making their way through clinical trials.

A third option for the purification of antibodies using an immobilized affinity ligand is the use of a synthetic organic binding molecule. For example, the unique thiophilic support developed in the late 1980s is an affinity

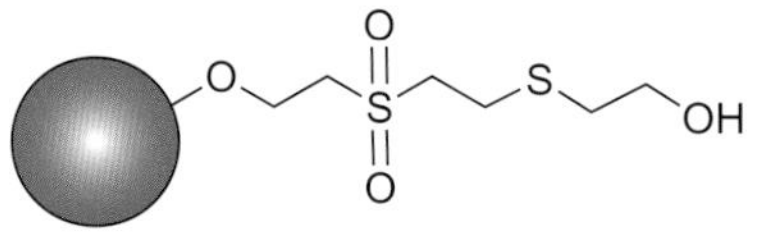

Thiophilic support containing
a thioether next to a sulfone

FIGURE 1.53 A thiophilic affinity support can be prepared by the reaction of 2-mercaptoethanol with a support activated with vinyl sulfone groups. This creates a thioether next to a sulfone group, which is a feature of thiophilic supports having affinity for binding antibody molecules.

matrix having high capacity for binding immunoglobulins (Belew *et al.*, 1987). It was discovered that an affinity ligand containing a thioether group next to a sulfone is able to bind to antibodies in the presence of a lyotropic salt, such as 0.5*M* potassium sulfate (Figure 1.53). The binding presumably occurs in the region of the Fc domains and may involve hydrophobic interactions. A thiophilic support made on an agarose chromatography resin has high capacity for binding IgG at levels of 20 to 50 mg/ml, depending on ligand density. This type of thiophilic support, however, has relatively little binding potential with other proteins that may be present in the antibody solution. Elution of bound antibody is simply accomplished by removing the high salt from the binding buffer, thus eliminating the interaction of the thiophilic groups with the antibody molecules. The only disadvantage of this method is the quantity of salt that must be used to purify large amounts of antibody, therefore creating a disposal issue for biotech companies thinking of using this support in process separations.

Other small organic molecules have been developed to interact with immunoglobulins and mimic the affinity binding of protein A to the Fc region. Some short peptide sequences have been identified using peptide libraries, such as the multimeric ligand TG19318, that are able to interact with IgG in similar fashion to that of protein A (Fassina *et al.*, 1996, 1998; Fassina, 2000). In addition, certain small organic compounds, such as 2-(3-aminophenol)-6-(4-amino-1-naphthol)-4-chloro-*s*-triazine, have been discovered and immobilized on agarose that mimic protein A and have binding characteristics toward immunoglobulins (Lowe *et al.*, 1986; Li *et al.*, 1998; Teng *et al.*, 2000).

Separations using immobilized affinity ligands can be employed to purify many more biological molecules than just antibodies. For instance, immobilized saccharides, carbohydrates, or complex glycans can be used to isolate proteins with binding specificity toward the sugar or sugar sequence of interest. This can permit isolation of lectins and cell–cell interacting proteins, which bind to certain glycan structures (Horlacher and Seeberger, 2006). It can also be used to isolate and characterize cell populations for their ability to interact with specific sugars (El-Boubbou *et al.*, 2010). Intact glycans can be released from glycoproteins or other glycoconjugates using enzymatic or chemical means and subsequently immobilized through their reducing ends to surfaces or particles for analysis of interacting proteins (Kim *et al.*, 2008). Glycan arrays have been produced on a variety of surfaces or particles to allow multiplexed analysis of carbohydrate-interacting proteins from a cell or tissue lysate (Taylor and Drickamer, 2009). The glycan binding properties of the *Arabidopsis* PP2-A1 phloem lectin was studied in this way, using a glycan array for identification of the lectin's carbohydrate specificity (Beneteau *et al.*, 2010). Chapter 15, Section 2.4, describes the reactions of labeling or immobilizing glycans through their reducing ends, which involves coupling to amine-, hydrazide-, or aminoxy-containing compounds to create secondary amine, hydrazone, or oxime linkages, respectively. Figure 1.54 illustrates some of these immobilized saccharide bioconjugates for the capture of interacting proteins.

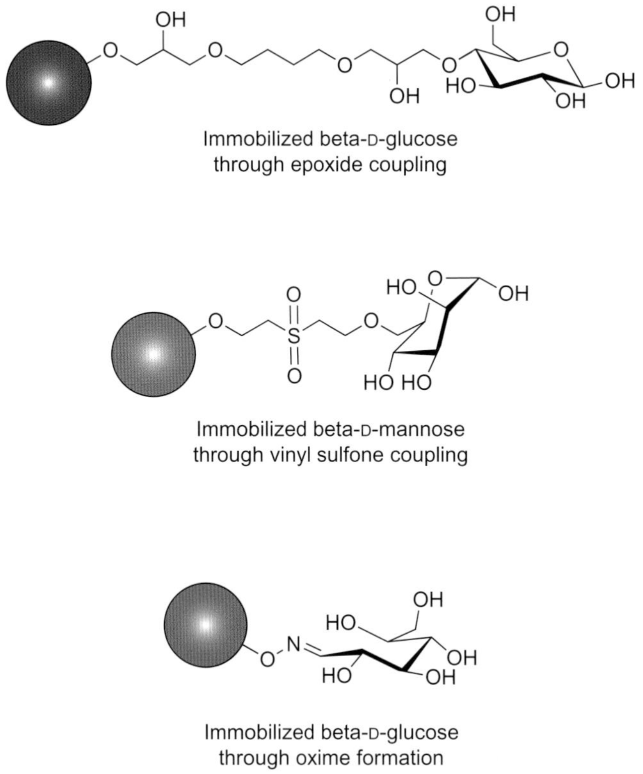

FIGURE 1.54 Immobilized sugars can be created using an epoxy activated or vinyl sulfone activated support to link to hydroxyl groups via ether bonds. In addition, reducing sugars can be coupled to supports using an aminooxy group, which forms an oxime bond with the aldehyde groups.

Conversely, immobilized lectins can be used to isolate certain carbohydrates or glycoproteins based on the specific sugars contained in the target molecule. Many dozens of lectins have been identified having precise saccharide-binding characteristics, any number of which can be immobilized to separate and enrich the portion of the glycome having the requisite carbohydrate structure for interaction (Tateno *et al.*, 2007). In addition, immobilized *m*-aminophenyl boronic acid (Figure 1.55) has been used to bind diol-containing molecules, such as those present in many sugars contained in carbohydrates and glycans (Olajos *et al.*, 2010). The boronic acid affinity ligand coordinates with the adjacent hydroxyl groups to form a covalent cyclic linkage with the carbohydrate, thus capturing sugar-modified biomolecules from solution. Unlike lectin binding, boronate-containing supports are not selective to sugar sequences, but interact with any molecules having glycols or an amine adjacent to a hydroxyl. Immobilized *m*-aminophenyl boronic acid on agarose has been used to isolate glycated hemoglobin in diagnostic testing of diabetics for long-term glucose monitoring (Mallia *et al.*, 1981; Klenk *et al.*, 1982; Frolov and Hoffmann, 2010).

Small molecule substrate analogs or inhibitors have also been used as immobilized affinity ligands to capture or purify certain enzymes from biological solutions. For instance, biomimetic affinity ligands, including immobilized dyes or inhibitors like

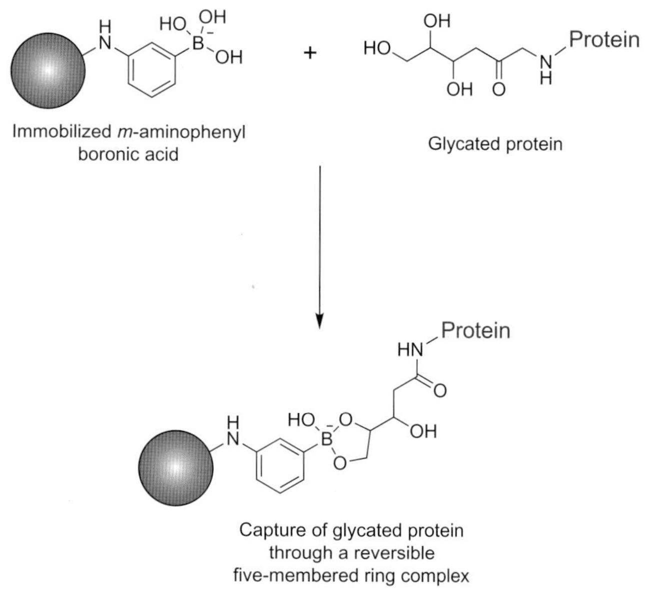

FIGURE 1.55 Immobilized *m*-aminophenyl boronic acid can be used to capture glycated proteins that are the product of nonenzymatic modification caused by glucose, which is present in high amounts in diabetic patients. The boronate group complexes with the glycated protein by forming a ring structure with the diols present in the glucose modification.

benzamidine, an inhibitor of trypsin-like proteases, have been used to isolate these enzymes, including the pharmaceutically important tissue plasminogen activator (t-PA) (Wu and Yu, 2007).

Small molecule biomimetics have been designed to bind selectively to certain proteins or other biomolecules and then immobilized for use as affinity supports in capture and purification of their specific targets (Lowe *et al.*, 2001). In particular, triazine dye molecules or their derivatives have often been used to mimic natural ligands, such as nucleotides, and bind to active sites on enzymes for efficient purification (Clonis *et al.*, 2000). Tsopelas *et al.* (2010) used customized biomimetic affinity chromatography columns to isolate selenium-containing molecules from biological samples, including selenomethionine, selenocystine, selenocystamine, selenourea, dimethyl selenide, and dimethyl diselenide. Gautam (2010) describes peptide biomimetic sequences that are effective in isolating IgM antibodies and also summarizes the history of peptide mimetics for protein purification. Many other proteins or biological targets can be isolated by the appropriately designed biomimetic affinity support, whether the ligand is designed based upon a synthetic organic motif, a biomolecule fragment, or a peptide sequence (Figure 1.56).

Immunoprecipitation Techniques

One of the more popular research applications using immobilized affinity ligands involves antibody-mediated immunoprecipitation. The specificity and bivalent nature of antibodies allows direct targeting through immunoaffinity interactions of proteins and other molecules from biological samples, effectively precipitating them out of the sample solution in an immune complex. If an antibody is coupled to an insoluble support, the resultant bioconjugate can also be used to capture the target molecule and remove the remaining components simply by washing (Figure 1.57). Although it is not strictly immunoprecipitation in the immunological sense, immunoaffinity chromatography on an insoluble support essentially accomplishes the same effect of precipitating the target antigen out of solution.

Polyclonal or monoclonal antibodies can be made for nearly any target protein or biomolecule and used in immunoprecipitation techniques to capture these targets. In practice, there typically are two strategies used to prepare an immunoaffinity support: (1) direct immobilization of the antibody onto the resin or solid phase; or (2) use of an intermediary high-affinity interacting pair, such as the streptavidin–biotin interaction, to capture an antibody after it has interacted with its specific target antigen. The first procedure, direct conjugation, is usually carried out if there is sufficient quantity of the antibody for immobilization, which often is the case with polyclonals. The indirect method of immobilization is sometimes carried out to orient the antigen binding sites for better interaction with the target—in other words, to direct the Fab ends away from the particle or surface—or, alternatively, because the amount of antibody that is available is limited and the price per mg is expensive (Sissona and Castor, 1990).

Direct antibody coupling to an activated support can be done using a number of reactive chemistries that can covalently link to amines, thiols, or even to the carbohydrate portion of an immunoglobulin (Chapters 15 and 20) (Figure 1.58). Immobilization through amines is the most common route and results in a fairly random orientation of the antibodies on the surface of the support. Coupling through antibody thiol groups (Chapter 15, Section 2.2) can also be done to make the attachment point near the hinge region of the antibody. This is particularly useful for immobilization of $F(ab')_2$ fragments to insoluble supports after their disulfides are reduced to prepare Fab′ fragments. This strategy ensures that every conjugation reaction to the matrix with the antibody occurs at the opposite end of the antigen binding site on each monovalent fragment. Finally, polyclonal antibodies and some monoclonals are glycosylated in the Fc region, which can be targeted to facilitate coupling by reductive amination using amine-, hydrazide-, or aminooxy-containing supports, after the carbohydrate has been periodate oxidized to create aldehyde groups (Chapter 15, Section 2.4). This method certainly avoids any coupling to the antibody peptide structure,

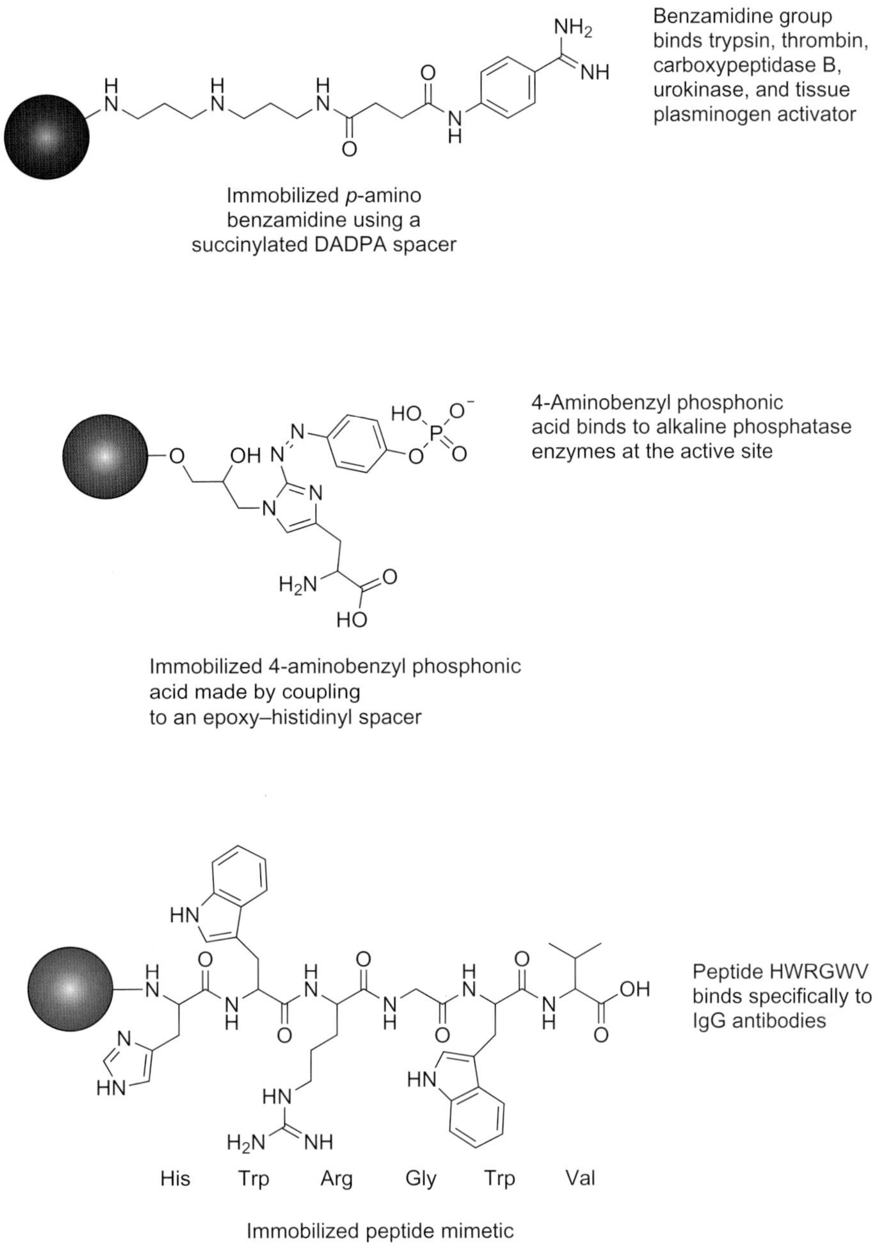
Benzamidine group binds trypsin, thrombin, carboxypeptidase B, urokinase, and tissue plasminogen activator
Immobilized p-amino benzamidine using a succinylated DADPA spacer
4-Aminobenzyl phosphonic acid binds to alkaline phosphatase enzymes at the active site
Immobilized 4-aminobenzyl phosphonic acid made by coupling to an epoxy–histidinyl spacer
Peptide HWRGWV binds specifically to IgG antibodies
His Trp Arg Gly Trp Val
Immobilized peptide mimetic

FIGURE 1.56 Immobilized enzyme inhibitors and biomimetic affinity ligands can be used to target the active sites of specific enzymes or other proteins in biological samples.

thus avoiding the binding sites; however, some carbohydrate-specific antibodies are known to be glycosylated near the antigen binding sites and may even participate in antigen binding through hydrogen bonding (Wright *et al.*, 1991). Thus, immobilization of IgG molecules through their glycans might not be the best strategy in all cases.

All of these methods work well to immobilize antibodies on solid supports through the creation of direct covalent bonds, but coupling through amines may be the simplest strategy, and it almost always results in an antigen binding affinity matrix for immunoprecipitation applications with high activity for capturing target molecules. Chapters 14, 15, and 20 discuss antibody conjugation methods and immobilization reactions that can be used to couple antibodies to solid supports.

Another strategy for performing immunoprecipitation separations is to use immobilized streptavidin as the affinity support and then use biotinylated antibodies in solution to capture targeted antigens (Figure 1.59). This method is convenient in that the streptavidin matrix functions as a universal support for doing virtually any immunoprecipitation procedure using any available biotinylated antibody conjugate. In this case, only one

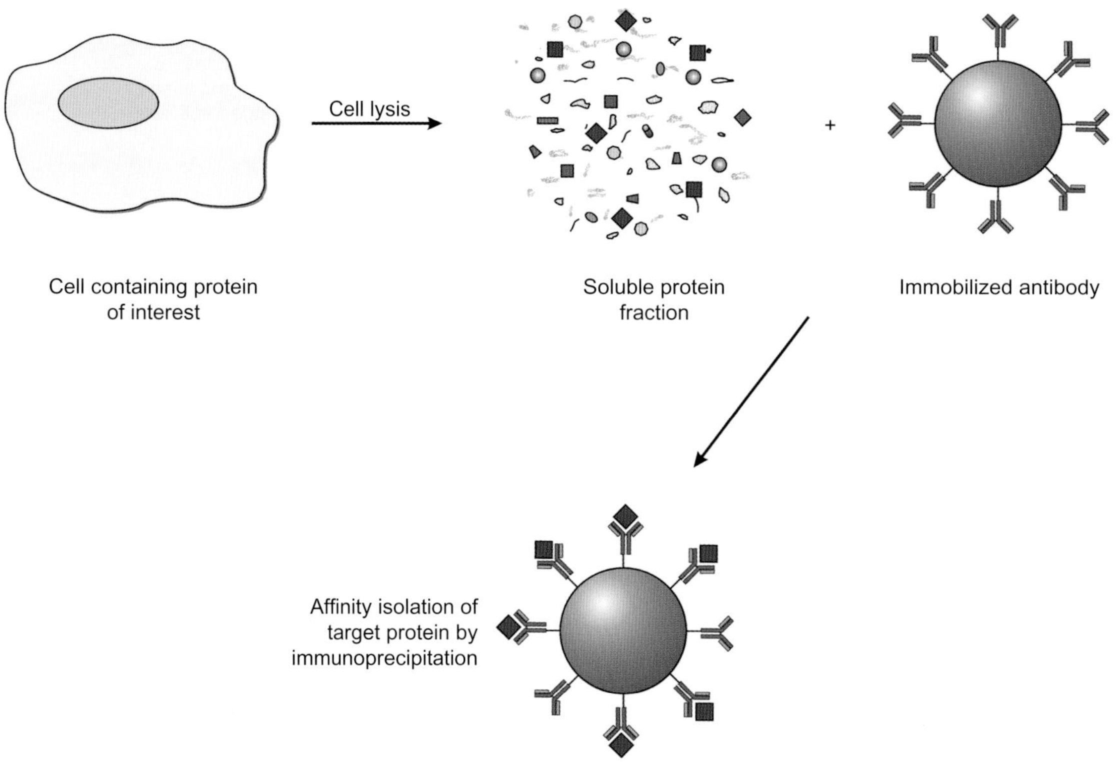

FIGURE 1.57 Immunoprecipitation can be used with an immobilized antibody to capture a specific target protein from a complex sample, such as a cell lysate.

affinity support is needed, immobilized streptavidin, while obtaining or making the desired biotinylated antibodies for use with it is relatively trivial. In addition, the amount of biotinylated antibody required to do an immunoprecipitation assay using this method is minuscule compared to the amount of antibody that would be needed to prepare an affinity support of reasonable capacity by direct immobilization of the antibody.

Another alternative option for the preparation of immobilized antibodies for immunoprecipitation applications is to use immobilized immunoglobulin binding proteins, such as protein A, protein G, or protein A/G, which are bacterial proteins that have affinity for binding IgG primarily through its Fc region. In this approach, an immobilized immunoglobulin binding support is used to capture the primary antibody through its Fc region, thus orienting the IgG molecules with their antigen binding sites free to interact with a targeted antigen (Figure 1.60). Next, this affinity complex is covalently crosslinked using a homobifunctional crosslinking agent to permanently stabilize the interaction. Schneider *et al.* (1982) first used this technique to prepare immobilized antibodies using the crosslinker DNP (Chapter 15, Section 2.6) for immunoprecipitation applications. Since that time, this technique has grown in popularity and evolved to use better crosslinking agents with much better long-term stability. However, the direct immobilization of a primary antibody may still be the best option, because it avoids the potential nonspecificity that can result from having another protein present on the affinity support.

Bioconjugates used for immunoprecipitation techniques can consist of many different solid-phase matrices, such as planar arrays, the wells of microplates, porous chromatography supports like agarose (Chapter 15), or nonporous particles such as latex beads or paramagnetic particles (Chapter 14). Figure 1.61 illustrates some of these bioconjugate types and shows both direct immobilization of an IgG as well as the use of streptavidin or an immunoglobulin-binding protein as intermediaries to immobilize the antibody.

Affinity Capture of Post-Translational Modifications

Affinity supports can be used for enrichment of the proteomic fraction of samples having specific post-translational modifications. The affinity ligand can be designed to target and bind to the modification through a number of different interactions. For instance, one of the principal signaling modifications that occur within cells is phosphorylation. The enzymatic addition of a phosphate group to serine, threonine, and tyrosine -OH groups by specific kinases facilitates activation of particular proteins within dozens of signaling pathways in cells. These

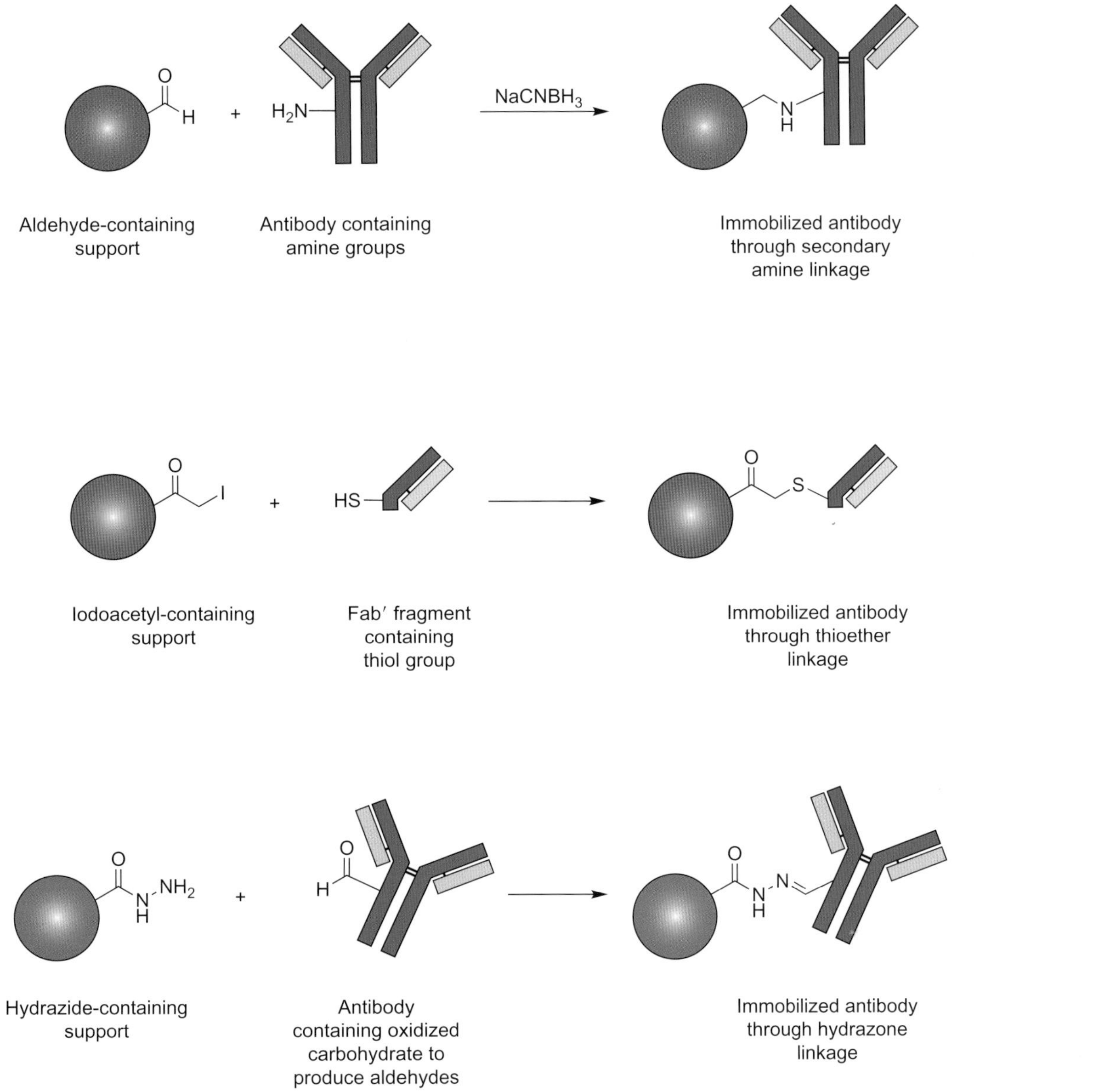

FIGURE 1.58 Three common methods to immobilize antibodies include: (1) the reductive amination coupling of the amines on an antibody to an aldehyde-containing support, such as periodate-oxidized agarose; (2) coupling to thiols on reduced antibodies or antibody fragments with an iodoacetyl-containing support, which binds thiols through a thioether linkage; or (3) using a hydrazide-containing support to couple with antibodies through their carbohydrate after periodate oxidation to create aldehydes.

phosphorylation events can be captured for study by affinity isolation of phosphorylated proteins or peptides using selective affinity matrices able to bind to the amino acid–phosphate modification sites.

There are two main approaches for affinity purification of the phospho-proteome: (1) immunoaffinity chromatography using antibodies with specificity for the phosphopeptide sequence or the phospho–amino acid modification site; or (2) immobilized metal affinity chromatography (IMAC), which selectively coordinates with the phosphate modifications. Immunoaffinity separations have a tendency to work well when using anti-phosphotyrosine antibodies (particularly the monoclonal antibody known as pY20), but the specificity of anti-phosphoserine and anti-phosphothreonine antibodies is not that good. However, some phospho-specific antibodies have specificity for a particular peptide sequence surrounding the phosphorylation site; thus, these antibodies can be used not only to enrich for phosphopeptides in general, but also to capture a unique phosphopeptide representing a phosphorylation event that occurred on a specific protein having that sequence.

Most of the phosphorylation site-specific antibodies available on the market are rather expensive, and for

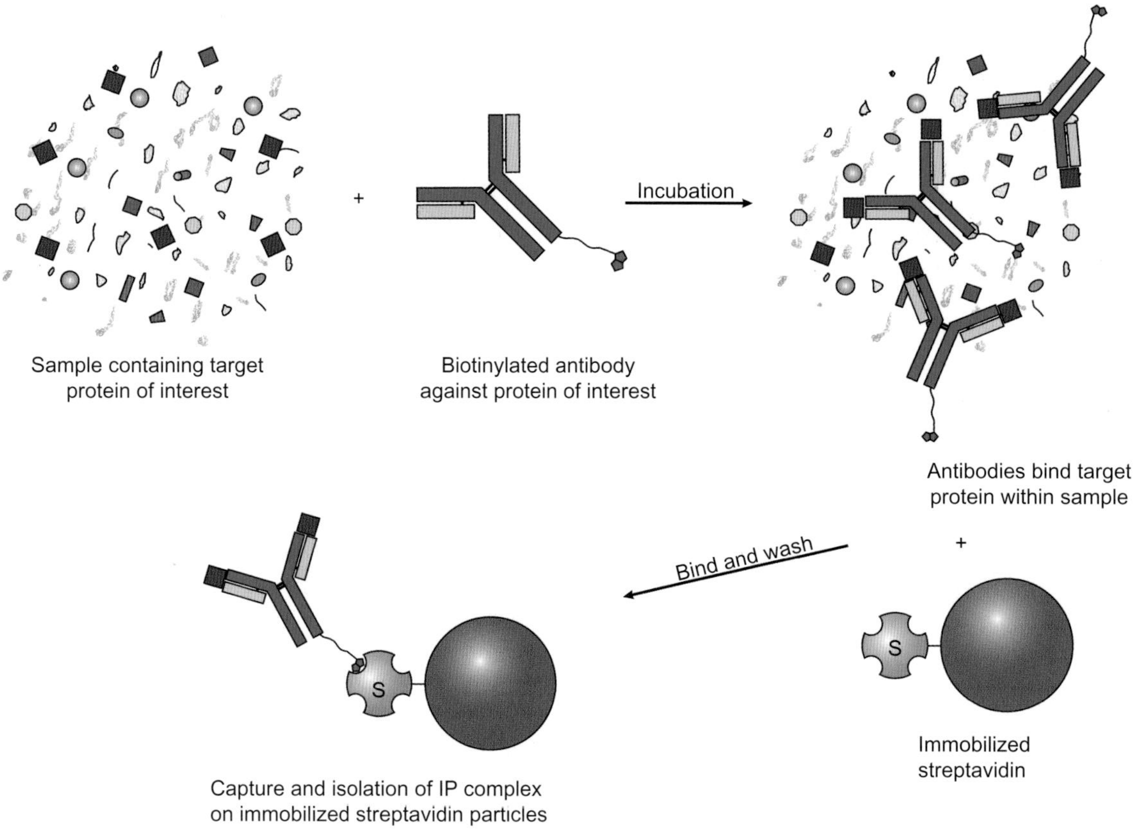

FIGURE 1.59 Immobilized streptavidin can be used in an immunoprecipitation procedure to capture a biotinylated primary antibody that has bound to a target protein in a complex sample. In this procedure, the immobilized streptavidin can be used with any biotinylated antibody and thus becomes a universal reagent for carrying out IP separations.

this reason they are often used in a biotinylated form to interact with a phosphopeptide in solution and then the complex is captured on an immobilized streptavidin column for recovery of the phosphorylated peptide or protein (Figure 1.62).

Serine and threonine phosphorylation can also be targeted for analysis using bioconjugate techniques involving a two-step chemical modification followed by affinity capture (Leitner and Lindner, 2009). Phosphoserine and phosphothreonine groups can be dephosphorylated in the first step of this technique through a beta-elimination reaction that leaves behind a carbon–carbon double bond. This double bond can be substituted in a second reaction with thiol-containing molecules to create a conjugate useful in subsequent detection or isolation procedures. A common strategy in beta-elimination is to label the dephosphorylated peptides with a dithiol derivative, which adds to the double bond to form a thioether linkage. Tags subsequently attached to the free thiol end of the dithiol compound are thus covalently coupled to only phosphorylated peptides, which can then be useful for affinity capture or detection. A common modification technique involves the addition of a biotinylation reagent to the free thiol to modify every phosphopeptide at the point of Ser or Thr phosphorylation (Oda *et al.*, 2001). In this method, each peptide containing a phosphorylation site gets tagged with a biotin group. Enrichment of the biotinylated peptides can then be done on an immobilized streptavidin support to analyze this fraction of the phosphoproteome by mass spec (Figure 1.63). In addition, other tags can be added to the dephosphorylated double bond of serine or threonine to provide fluorescent detection or enhanced ionization properties for mass spec separations. Note, however, that phosphotyrosine does not participate in the beta-elimination reaction and therefore cannot be modified using this strategy. Nevertheless, phosphotyrosine can usually be effectively captured or detected using antibodies, as discussed in the previous paragraphs.

Some limitations of the beta-elimination technique include the combined inefficiency of the dephosphorylation reaction and the subsequent modification of the double bond with a thiol compound as well as the potential for side reactions (Karty and Reilly, 2005; Tinette *et al.*, 2006). None of the reactions provides 100% yields, and under some conditions even non-phosphorylated serine and threonine hydroxyls can

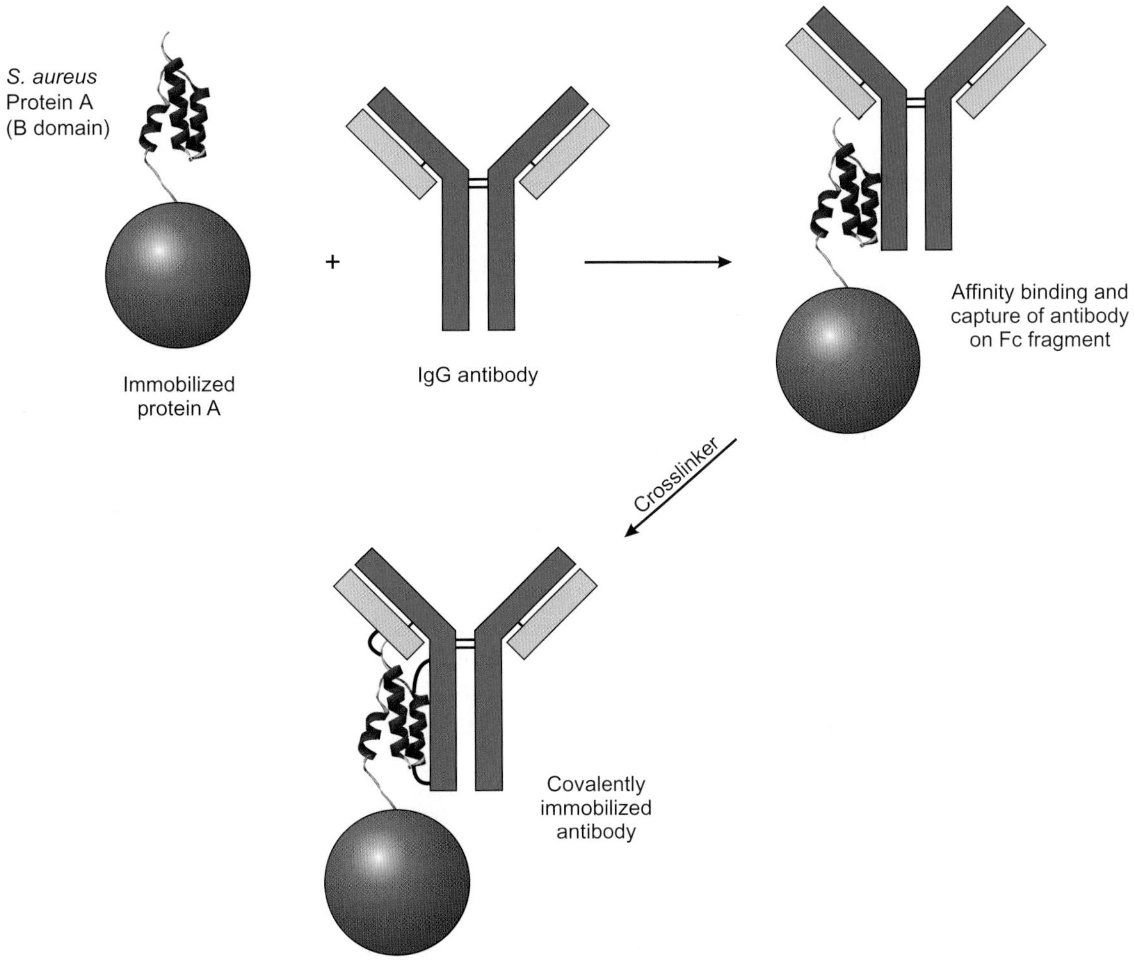

FIGURE 1.60 Immobilized protein A can be used to immobilize an antibody for immunoaffinity chromatography or immunoprecipitation applications by first orienting it with its antigen binding sites pointing outward from the matrix. To stabilize the protein A–antibody interaction a crosslinker is added to create covalent linkages, thus preventing leaching of the antibody during use.

undergo elimination and be modified inadvertently (Li *et al.*, 2003). Therefore, obtaining quantitative information about serine or threonine phosphorylation through beta-elimination is sometimes difficult unless conditions are carefully controlled.

Covalent modification of phospho-amino acid groups in phosphorylated peptides has also been done using diazo compounds (Lansdell and Tepe, 2004; Tepe and Pinnavaia, 2008; Frawley Cass *et al.*, 2009). Diazo carbonyl groups will react with an acidic phosphate group to liberate N_2 and create a phosphoester bond with a phosphopeptide. To avoid side reactions and coupling to carboxylate groups on peptides, a protein digest is first treated with a methanolic HCl solution to block all carboxylic acids and transform them into methyl esters. The subsequent use of an immobilized diazo derivative can function to enrich the phosphorylated proteome through covalent chromatography, specifically coupling to the phospho groups and allowing all other non-phosphorylated peptides to pass through the gel unretarded. This is also called "covalent chromatography," since the target molecule is captured through a covalent linkage. Once phosphopeptides have been isolated, they can be released from the resin for mass spec analysis using acid hydrolysis of the phosphoester linkages (Figure 1.64).

Another powerful way of enriching the phosphoproteome is the use of IMAC with certain chelated metal ions. Porath pioneered the method of metal chelate affinity interactions in the 1970s using immobilized organic metal chelating groups, which can coordinate and hold a metal ion for chromatography (Porath *et al.*, 1975). IMAC supports can be used as pseudo affinity ligands, that is, not truly biospecific but having interaction potential for certain functional groups or amino acid side chains in proteins (for review, see Block *et al.*, 2009). Electron-rich groups that are able to coordinate with the unoccupied sites on the chelated metal ion can bind to the IMAC support and be isolated from the rest of the proteins and

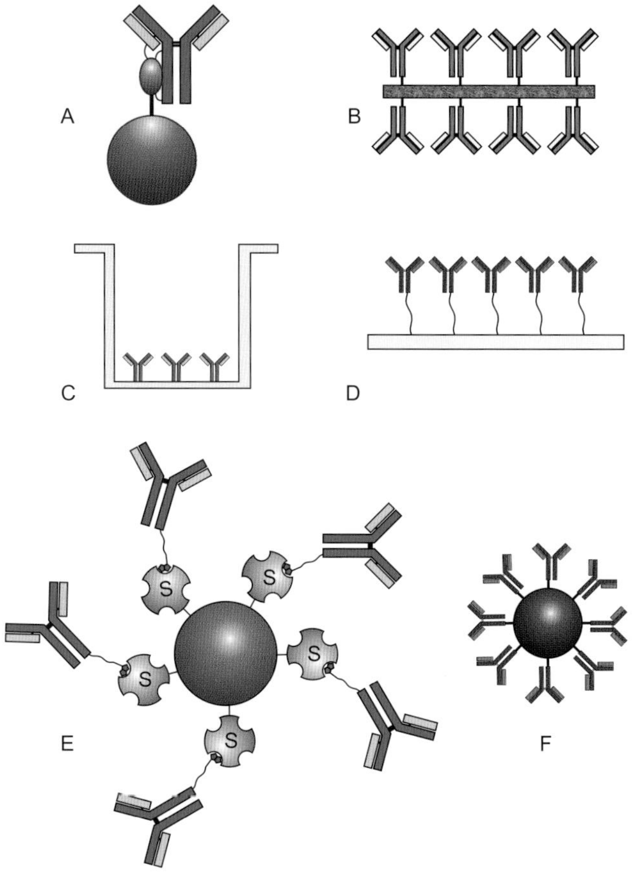

FIGURE 1.61 Solid-phase supports and strategies that are commonly used for the immobilization of antibody molecules include (A) indirectly coupled to a beaded support using an immobilized immunoglobulin binding protein, (B) covalently coupled to a membrane or monolithic support, (C) passive adsorption onto a microwell plate or surface, (D) covalently tethered to a surface via a linker arm, (E) through the use of immobilized streptavidin on particles to couple biotinylated antibodies, and (F) through direct immobilization onto porous and nonporous particles.

other molecules in a sample. Elution of bound proteins is typically done by using a buffer pH shift or through the addition of a counter-ligand, such as imidazole, which can compete for binding to the metal chelate and thus release the bound molecules. Alternatively, the addition of a chelating compound, such as EDTA, can be done to strip off the metal ions from the IMAC resin and elute the bound protein.

IMAC supports containing chelated gallium (Ga III) or iron (Fe III) are particularly useful in the enrichment of the phosphoproteome (Collins *et al.*, 2005; Cantin *et al.*, 2008; Ye *et al.*, 2010). This method is used extensively to isolate phosphopeptides prior to mass spec analysis, a technique which was pioneered by Neville *et al.* in 1997 (Figure 1.65). Since even gallium- or iron-mediated IMAC is not absolutely specific for binding phosphorylation sites on proteins, other amino acid groups can also interact with the support, such as on peptides having an abundance of the acidic residues like aspartic and glutamic acids. For this reason, the bound proteins or peptides are said to be enriched for phosphorylated molecules, but they are not completely purified to homogeneity.

Another major post-translational protein modification that can be targeted for purification using immobilized affinity ligands is glycosylation. Glycomics, or the study of glycoconjugate modification, is perhaps the second most active area in post-translational research today, following phosphorylation. Glycoproteins contain complex carbohydrate structures coupled to certain amino acid sites in either an *N*-glycosidic (asparagine) or an *O*-glycosidic (serine, threonine, or hydroxylysine) linkage (Chapter 2, Section 2). The sugar structure making up the glycan typically is branched and can vary depending on the function of the glycoprotein and the cell of origin. Glycan synthesis is under enzymatic regulatory control in the ER and Golgi apparatus within cells (Varki *et al.*, 2008). Aberrant glycosylation can often occur in disease states, such as cancer, creating tumor markers that are targets for drug discovery and therapeutic development (Brockhausen *et al.*, 1995; Kim *et al.*, 2009). Certain carbohydrate structures containing specific sugar residues can be recognized by carbohydrate-binding proteins (e.g., lectins, selectins, galectins, receptors, many CD cell surface antigens) (Cummings and Smith, 2005; Sakaguchia *et al.*, 2006; Shoseyov *et al.*, 2006; Sharon, 2007). For this reason, lectin bioconjugates have been used as probes of glycan structure or in an immobilized form, to purify glycoproteins (Ueda *et al.*, 2009).

Bioconjugates consisting of immobilized lectins coupled to resins or surfaces can be used to capture carbohydrates and glycans containing the proper sugar sequence, which are able to interact with the lectin binding sites through biospecific affinity contacts (Hirabayashi, 2008). Mixed-bed affinity chromatography supports containing a library of immobilized lectins have also been used to enrich the total glycome from cell or tissue lysates for subsequent analysis by mass spec or other methods (Cummings, 2005). In addition, the immobilization of glycan structures on array surfaces has also been done to assay and study the carbohydrate binding proteins present in samples (Figure 1.66) (Blixt *et al.*, 2004; Smith and Cummings, 2009).

Another important post-translational modification that can be targeted using bioconjugate techniques is nitrosylation. *S*-Nitrosylation occurs when nitric oxide (NO) synthase acts to generate this highly reactive radical gas, which then covalently modifies cysteine thiols in a protein (and the SH group in glutathione) to form a nitrosothiol group (S-NO) (Gaston *et al.*, 2003). This modification is a significant cell signaling event that has downstream effects on a variety of proteins, including

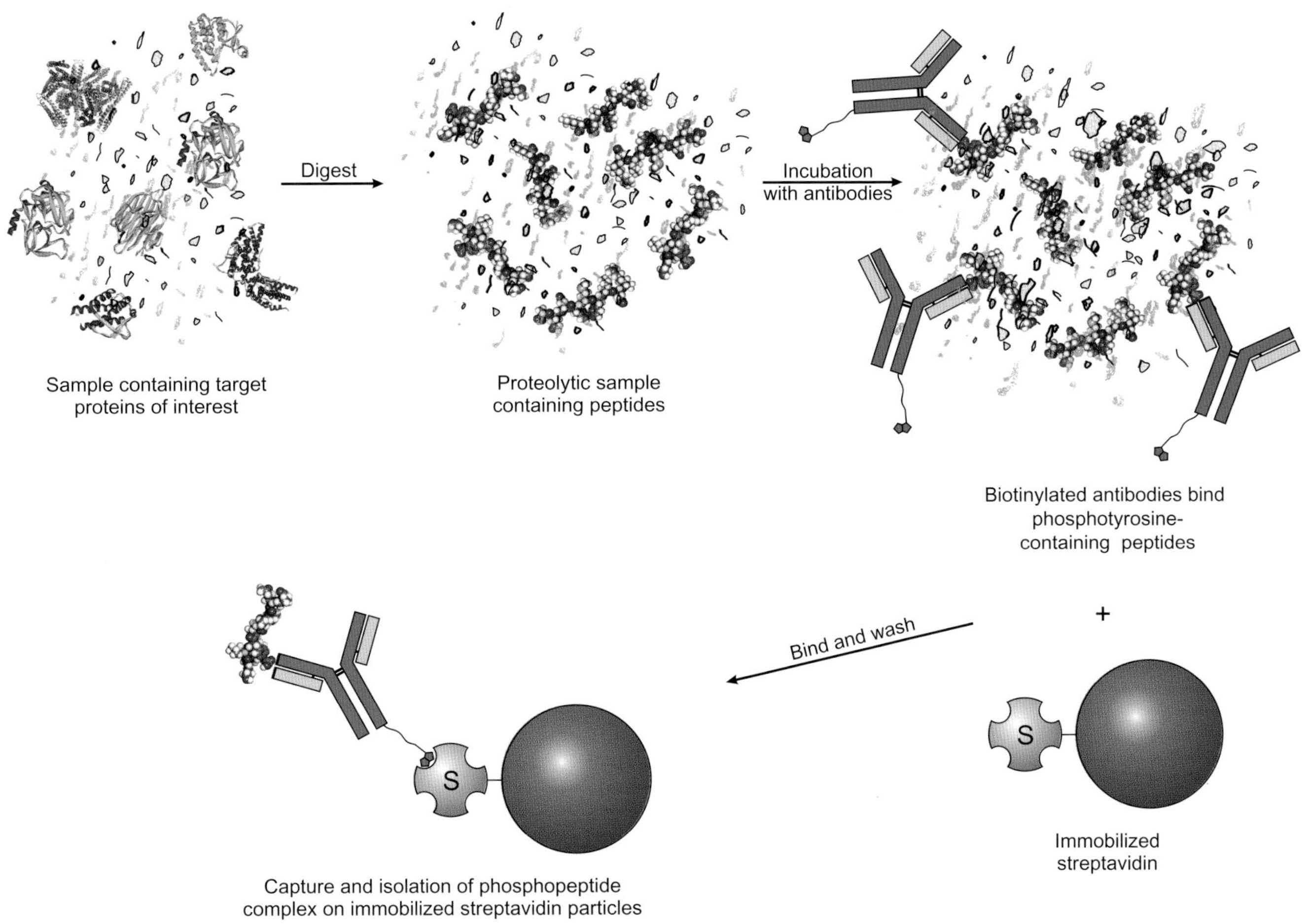

FIGURE 1.62 The enrichment of phosphopeptides in a proteomics workflow can be carried out by digesting a sample containing proteins with a protease (e.g., trypsin) to create peptides, including potentially phosphorylated peptides, incubating the mixture with a biotinylated antibody specific for the phosphopeptide of interest, and then binding the complex to an immobilized streptavidin resin, which captures the biotinylated antibody along with any phosphopeptides it has interacted with.

the control of regulatory proteins in the nucleus, structural and cytoskeletal proteins, signaling proteins, and ion channel proteins in cellular membranes (Stamler *et al.*, 2001).

The *S*-nitrosylated product is a reversible modification that can be cleaved back off *in vitro* using a reducing agent, typically ascorbate. To specifically detect cysteine nitrosylation, the biotin switch assay typically is used, which is a series of reactions that results in the biotinylation of a protein only at the sites of nitrosylation. First all free cysteine thiols are blocked in a sample using an SH-specific modification agent (such as MMTS, iodoacetamide, or *N*-ethylmaleimide; Chapter 2, Section 5.2) in the presence of SDS (to ensure reagent access to buried thiol groups). After removal of excess blocking agent, ascorbate is added to reduce the nitrosothiol group, thus cleaving off the post-translational modification and reforming the thiol. The freed cysteine thiols are then reacted with biotin-HPDP (Chapter 11, Section 6.3) or another SH-reactive biotin compound to result in a biotinylation event occurring at every site of nitrosylation (note: a thiol-reactive, PEG-based biotin compound can also be used to provide hydrophilicity; see Chapter 18, Section 1.3). The biotinylated proteins or peptides that are formed can be detected in western blots using streptavidin detection reagents or purified from a sample using immobilized streptavidin (Figure 1.67).

A similar biotin switch methodology has been developed for use in the measurement of fatty acid *S*-acylation (Drisdel and Green, 2004). This post-translational modification involves proteins conjugated with fatty acid acyl groups that are linked to cysteine thiols, which creates thioester derivatives with hydrophobic lipid chains protruding off. These lipid modifications enable the proteins to strongly associate with cellular membrane structures. This process is also commonly termed *palmitoylation*, although other fatty acids in

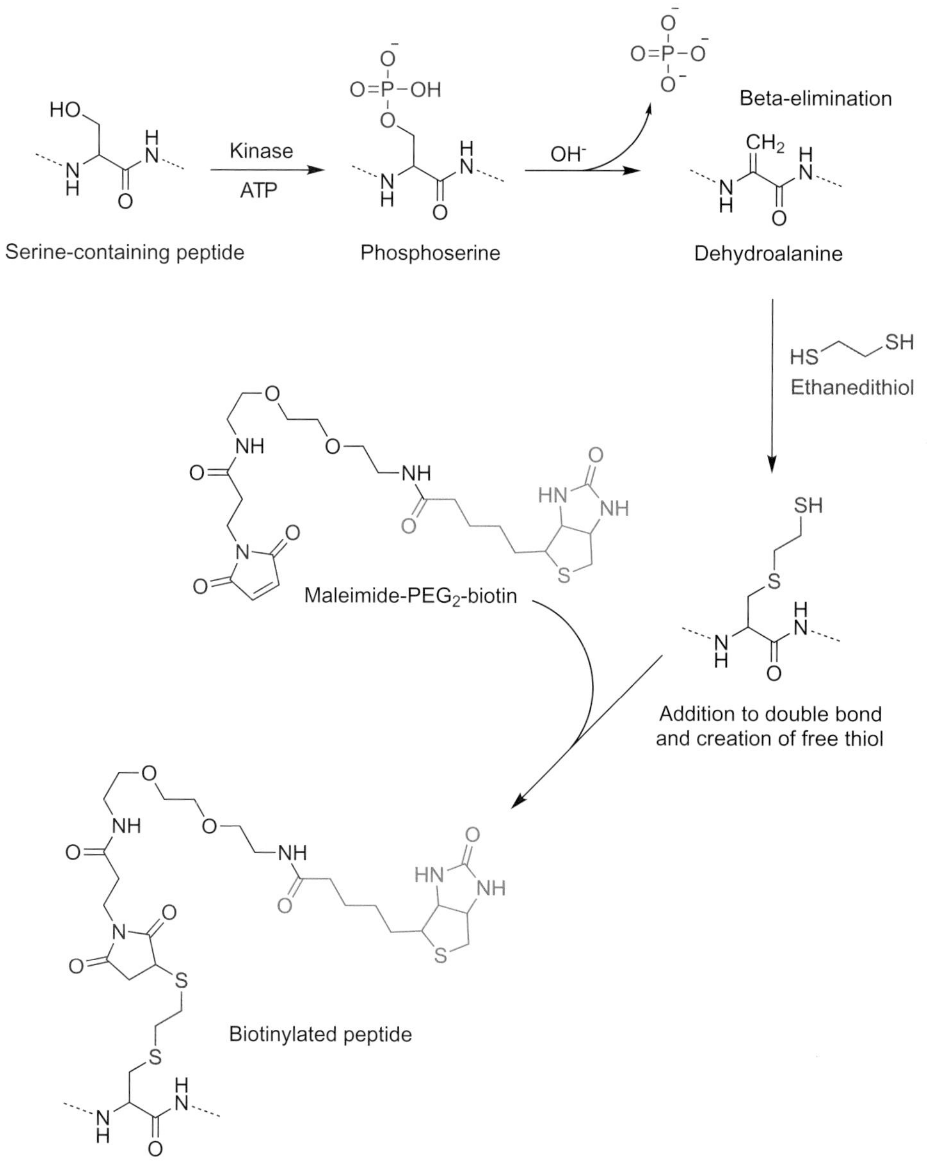

FIGURE 1.63 Phosphopeptides can be modified specifically at the site of phosphorylation using a three-step reaction that first involves a base-catalyzed beta-elimination reaction, which removes the phospho group and leaves a double bond. Modification of this carbon–carbon double bond with ethanedithiol by Michael addition provides a thiol group that can be reacted with a maleimide-PEG$_2$-biotin reagent to biotinylate the created thiol groups. This results in a biotin affinity group being attached at the site of every phosphorylation event on serine, which can be used to analyze the degree of phosphoserine production *in vivo*.

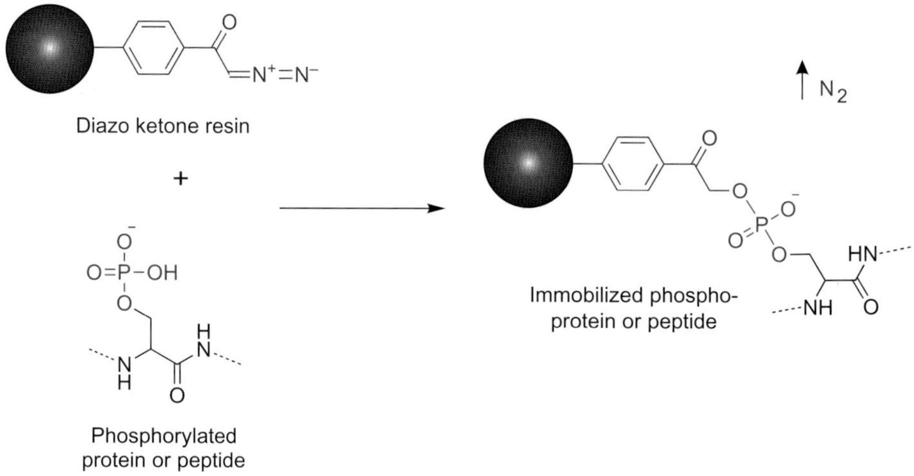

FIGURE 1.64 A diazo ketone containing-support can be used to react with phosphorylated peptides to covalently link to the phosphate groups through a phosphoester bond.

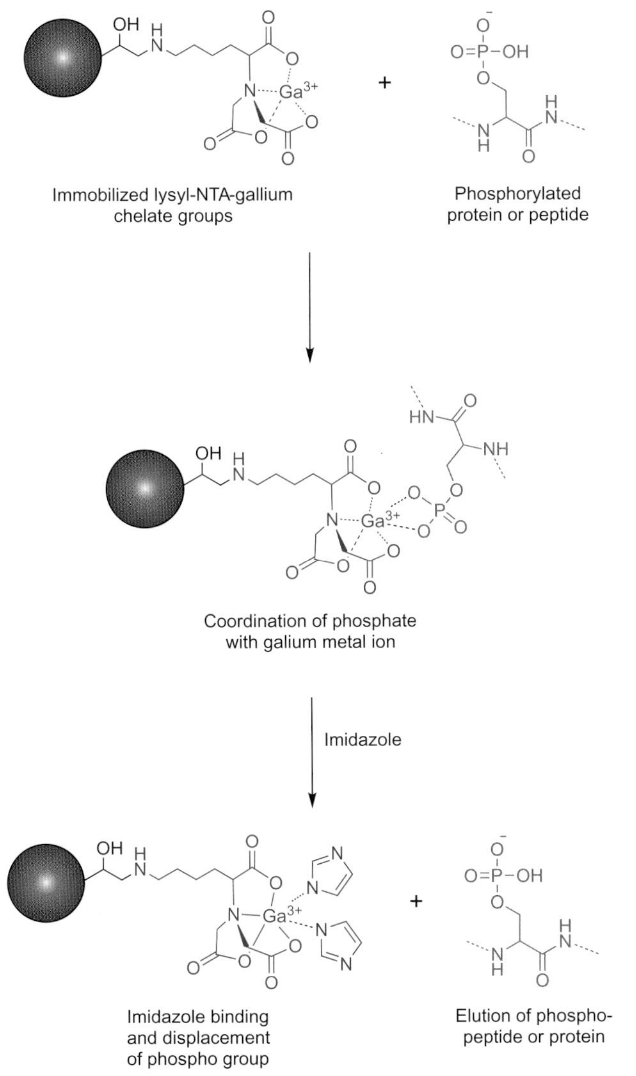

FIGURE 1.65 Immobilized metal chelate affinity chromatography (IMAC) can be used with an NTA-gallium ligand to reversibly bind a phosphorylated protein or peptide. This interaction can facilitate the enrichment of the phospho-proteome from a complex protein sample. Elution of the phosphopeptides can be carried out using a counter-ligand consisting of imidazole, which competes for the coordination sites on the chelated gallium salt.

addition to palmitic acid are involved with post-translational modification onto cysteine thiols. The thioester bond that is formed in *S*-acylation is reversible and can be cleaved using hydroxylamine, thus releasing the fatty acid and reforming the free SH group on the cysteines that were originally modified. If a protein sample containing palmitoylated or *S*-acylated proteins is first treated to block all free cysteines, using for instance *N*-ethylmaleimide and then hydroxylamine is added to cleave the thioester bonds, then all the revealed thiols will derive only from the post-translationally modified sites. Subsequent modification of the released thiols with a thiol-reactive biotinylation agent (Chapter 11, Section 6.3; Chapter 18, Section 1.3) will result in tagging all these sites with a biotin group. The biotin modified proteins can then be isolated on an immobilized streptavidin resin or detected using the appropriate streptavidin conjugates. For instance, fluorescently labeled streptavidin or HRP-labeled streptavidin conjugates can be used to probe western blots for palmitoylated proteins using this assay (Figure 1.68).

Another type of bioconjugate that can be used for affinity capture of phosphorylated target molecules is a biotinylated binding domain, which has the ability to specifically interact with another protein, typically binding to an activated form of the target. An example of such a system is the use of biotinylated SH2 domain fragments that can recognize certain phosphotyrosine residues in tyrosine kinases (Pawson and Nash, 2003; Hanke and Mann, 2009; Major *et al.*, 2009). In cells, a phosphorylation event at a particular site within a tyrosine kinase protein causes activation of the kinase with many downstream effects within signaling pathways. Protein–protein interactions typically occur subsequent to the phosphorylation event to cause certain SH2 domain-containing proteins to approach, interact with the tyrosine kinase, and then get phosphorylated and activated themselves. *In vitro*, a biotinylated SH2 domain molecule can be used to affinity target activated tyrosine kinases and bind to them in a cell lysate. The

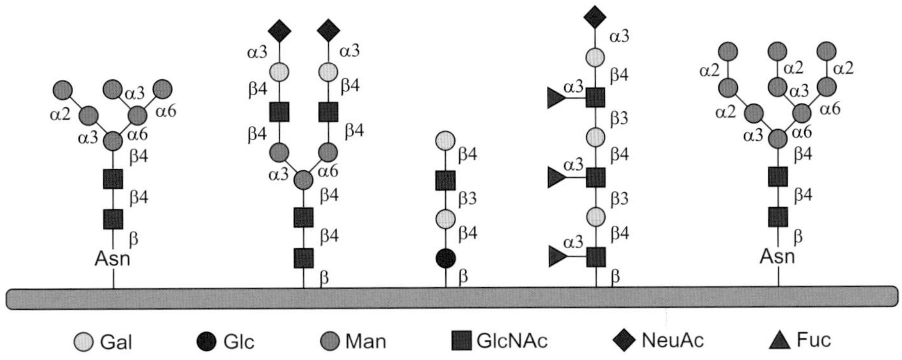

FIGURE 1.66 Immobilized glycan arrays can be created on surfaces that contain a variety of potential glycation forms, which can be used to study carbohydrate binding proteins in complex biological samples.

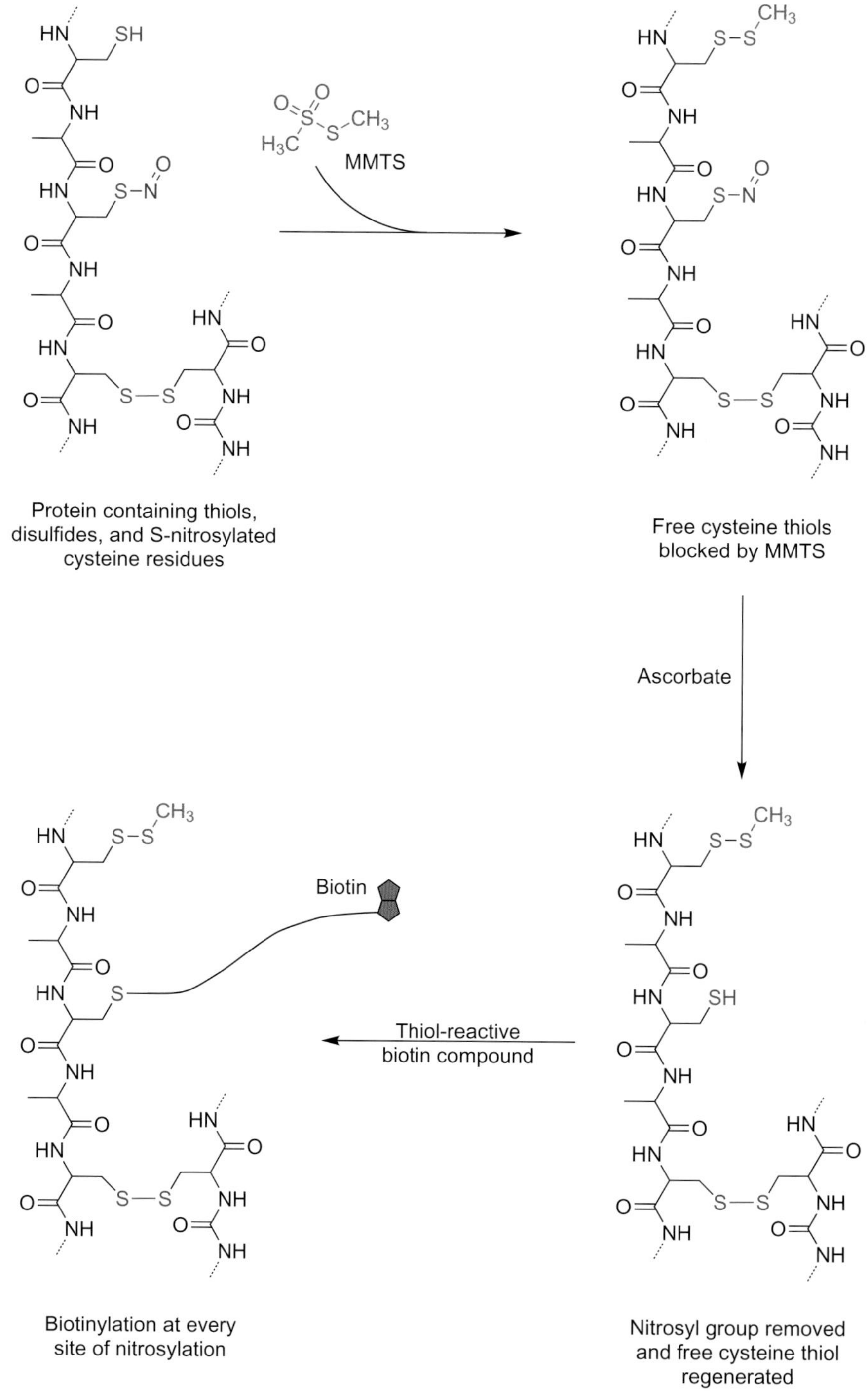

FIGURE 1.67 A post-translational modification event consisting of *S*-nitrosylation can be specifically targeted through modification with a biotin reagent. First, any free thiols in the sample are blocked with MMTS to eliminate interference from endogenous sulfhydryls. Next, the nitrosyl group is removed from the *S*-nitrosylation sites on cysteine using ascorbate treatment. Finally, a thiol-reactive biotinylation agent is added that couples with the freed thiols and provides a biotin affinity group at every site of original nitrosylation. The biotinylated peptides can be captured using an immobilized streptavidin affinity support for further characterization and study.

biotin tag on the SH2 domain can then be used to isolate these interacting complexes through its ability to bind to an immobilized streptavidin support. This "pull down" assay, as it is commonly called, can be used to study the activated tyrosine kinase population in a cell in response to disease or in response to a prospective treatment regime *versus* a normal cell. The enriched population of activated tyrosine kinases can be

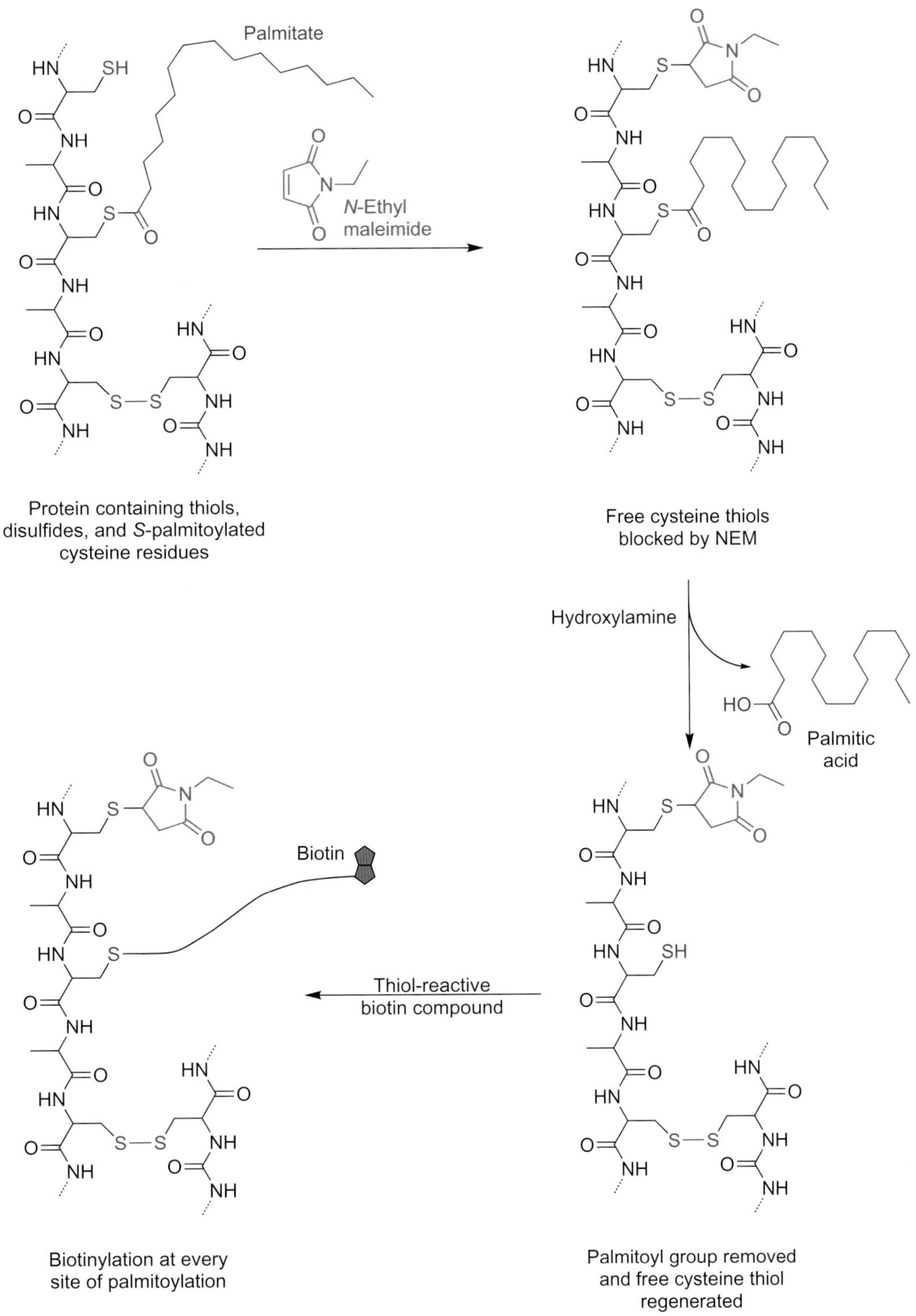

FIGURE 1.68 *S*-Palmitoylated proteins can be specifically biotinylated at each site of modification using a multi-step reaction sequence. First, *N*-ethyl maleimide is used to block all thiol groups in the sample so as not to interfere with subsequent reactions. Next, the palmitic acid groups are removed from cysteine residues using hydroxylamine. Finally, a thiol-reactive biotinylation reagent is used to add a biotin affinity group to each site that was initially palmitoylated. The biotinylated protein or peptide can then be isolated using immobilized streptavidin for further analysis.

analyzed by western blotting or mass spec to determine specific phosphotyrosine activation profiles. Figure 1.40 illustrates a pull-down assay to capture a phosphorylated tyrosine kinase through specific binding to an immobilized SH2 domain molecule, while Figure 1.69 shows the three-dimensional molecular model of an SH2 domain binding to a phosphotyrosine-containing peptide.

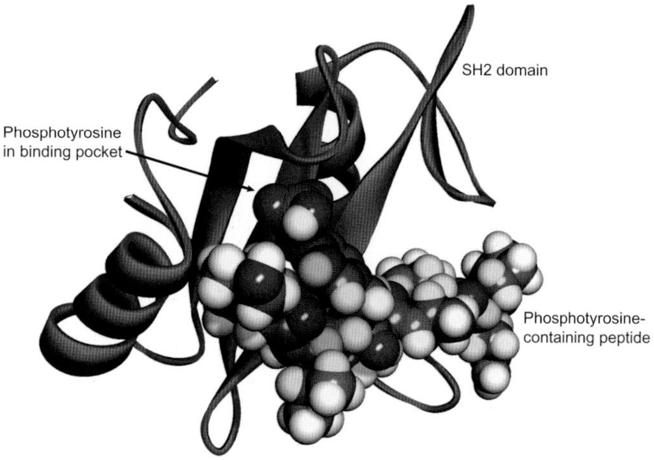

FIGURE 1.69 SH2 domains can be used as affinity molecules to specifically bind to phosphotyrosine-containing peptides. The SH2 domain is shown as the ribbon structure and the phosphorylated peptide as the space filling model.

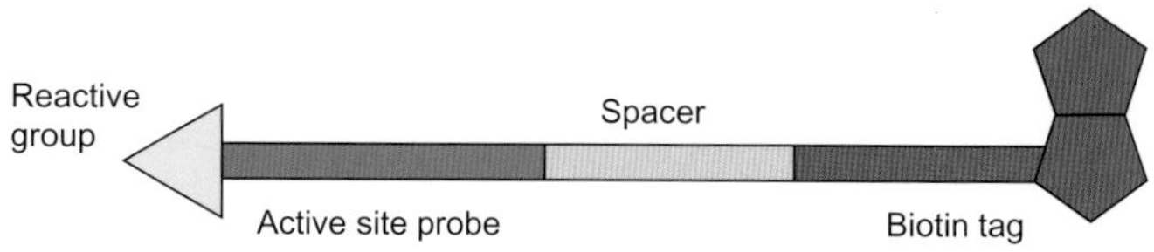

FIGURE 1.70 The general design of an active site warhead probe containing a biotin affinity group. The active site probe portion typically is a peptide binding sequence or an organic inhibitor of a specific enzyme active site. This facilitates the probe to initially interact with the enzyme, after which the reactive group covalently links to a functional group within or near the active center. The biotin handle can then be used to detect the modified enzyme or isolate the complex using streptavidin conjugates.

Affinity Capture using Active Site Binding Probes

Affinity probes containing active site binding components can be used to target and isolate specific enzymes, or enzyme classes, and other macromolecules within a biological sample. In this application, the bioconjugate probe may be an analog of a ligand that binds to a particular receptor or it may be a substrate analog that binds to the catalytic site of an enzyme. In either case, the probe can be designed to contain a detection component for assay or imaging, or an affinity component for purification using a complementary immobilized affinity support. The active site probe portion can consist of a binding molecule that interacts with high affinity but reversibly with the receptor or enzyme, or it can be a reactive probe which covalently couples in the region of the active site once bound. The second type of probe is often said to have a reactive warhead that spontaneously couples to a functional group within molecular reach of the active site.

Active site binding probes can be constructed to contain a terminal biotin group for detection or affinity isolation using streptavidin conjugates or immobilized streptavidin, respectively. Once the warhead end of the probe gets conjugated within or near an active site, the biotin tag can be used to enrich the population of active enzymes from a sample or detect the modified proteins on a western blot (Figure 1.70).

Probes can also be designed on their non-interacting end to contain a bioorthogonal or chemoselective reactive group that does not react with biomolecules, but can be used to subsequently label the probe with a detection molecule or affinity tag once the probe has bound to its target. Activity-based protein profiling (ABPP) has been used to globally assess enzyme activation and catalytic activity in samples (Kam *et al.*, 1993; Abuelyaman *et al.*, 1994; Speers *et al.*, 2003). For instance, Singaravelu *et al.* (2010) used a phenyl sulfonate ester probe containing a terminal alkyne group for labeling a range of enzyme families associated with hepatitis C virus replication. The phenyl sulfonate ester end is able to covalently link to nucleophiles in the active site of many enzymes, while the alkyne group subsequently can be reacted in a click chemistry reaction (e.g., copper-catalyzed Huisgen's 1,3-dipolar cycloaddition; Chapter 17, Section 5) with an azide-containing compound to form a triazole ring linkage. The resultant bioconjugate can be used to capture and enrich the labeled enzymes or detect them in a western blot (Figure 1.71). Activated enzymes isolated from a cell lysate may be analyzed by mass spec to provide a global picture of enzyme status within a cell.

Specific kinase enzyme probes have been developed to permit similar targeting of the activated kinome. ATP analogs can be used that contain an electrophilic anhydride group (acyl phosphate) positioned at the gamma-phosphate of the nucleotide. Such derivatives made from a biotinylation compound containing a terminal carboxylate can be used to form the acyl phosphate group, which creates an ATP binding site warhead that can covalently link to a particular lysine amine immediately adjacent to the nucleotide binding site. This probe is able to label all kinases having their nucleotide binding site accessible and active (Patricelli *et al.*, 2007). Similar nucleotide binding site probes can be constructed using photoreactive groups attached to an ATP derivative or using the reactive group 5′-p-fluorosulfonylbenzoyl adenosine. The fluorosulfonyl group can react with a nearby lysine amine residue similar to the acyl phosphate design. When using a photoreactive group, however, exposure to UV light is the driving force that facilitates the covalent attachment of the probe once it is bound to a nucleotide binding pocket (Karaman *et al.*, 2008). In addition to ATP active site probes, reactive conjugates can be designed to contain ADP, GTP, GDP, or other nucleotides and are useful in probing enzymes that have binding

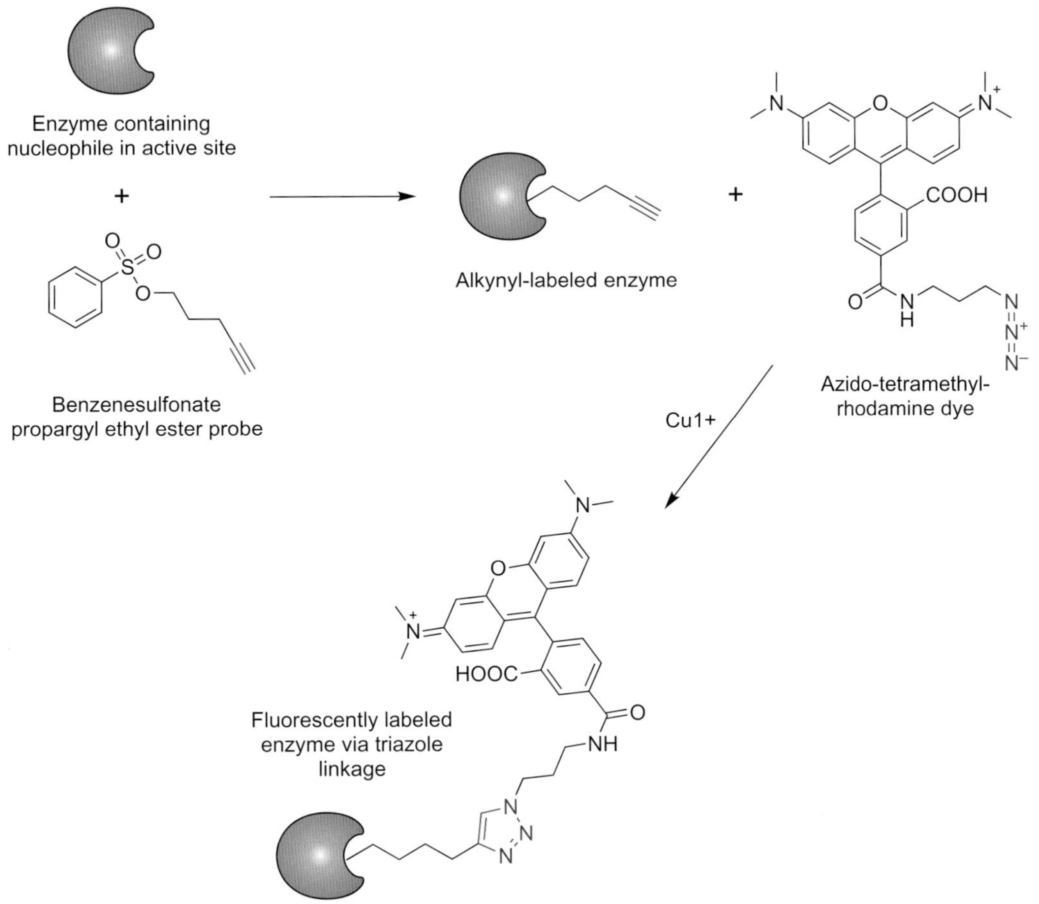

FIGURE 1.71 A benzene sulfonate ester reactive group probe can be used to covalently link to the active site within some enzymes and the alkyne group can be used for labeling with a fluorescent probe by a copper-catalyzed click reaction with an azido-rhodamine dye.

sites for particular nucleotide derivatives (Figure 1.72). Over half of all proteins produced in mammalian cells have a binding pocket for a nucleotide derivative; therefore, ABPP using nucleotide cofactors can be used effectively to analyze much of the proteome.

Other reactive probes can be created to target enzymes of a number of different substrate classes (Cravatt *et al.*, 2008). For instance small bioconjugate probes for serine hydrolases have been constructed using reactive warheads containing phosphonate reactive groups. Serine hydrolases are characterized by a highly conserved serine residue in the active site, which is crucial in forming an intermediate enzyme–substrate ester bond when hydrolyzing ester, thioester, or amide bonds. This serine group can be targeted for labeling by phosphonate derivatives prepared from native enzyme substrates. Examples are fluorophosphonate and aryl phosphonates, which position the phosphonate reactive group adjacent to the active site serine, causing a covalent bond to occur with the serine hydroxyl (Mahrus and Craik, 2005; Pan *et al.*, 2006) (Figure 1.73). Affinity tags or fluorescent labels can also be attached to the substrate–fluorophosphonate warhead for subsequent isolation on affinity supports or detection in western blots (Liu *et al.*, 1999; Kidd *et al.*, 2001; Patricelli *et al.*, 2001; Alexander and Cravatt, 2006). After enrichment of labeled enzymes from a cell sample, the global serine hydrolase activity profile may be analyzed by mass spectrometry.

Activity-based protein profiling using reactive bioconjugates can also be designed to target cysteine proteases, which contain a critical thiol in their active site. Cysteine proteases can be labeled by a bioconjugate warhead containing an epoxide reactive group, a vinyl sulfone, or various ketone derivatives, including diazomethyl ketones, acyloxymethyl ketones, or fluoromethyl ketones (Barrett *et al.*, 1982; Mason *et al.*, 1989; Bromme *et al.*, 1994; Ekert, 1999; Amstad, 2000; Bedner, 2000; Smolewski, 2001; Bogyo *et al.*, 2002; Blum *et al.*, 2005; Kato *et al.*, 2005; Fonovic and Bogyo, 2007). These electrophilic groups react and covalently couple with the active site SH group to form thioether linkages (Figure 1.74). Biotin or fluorescent tags conjugated to the active site probe/warhead combination

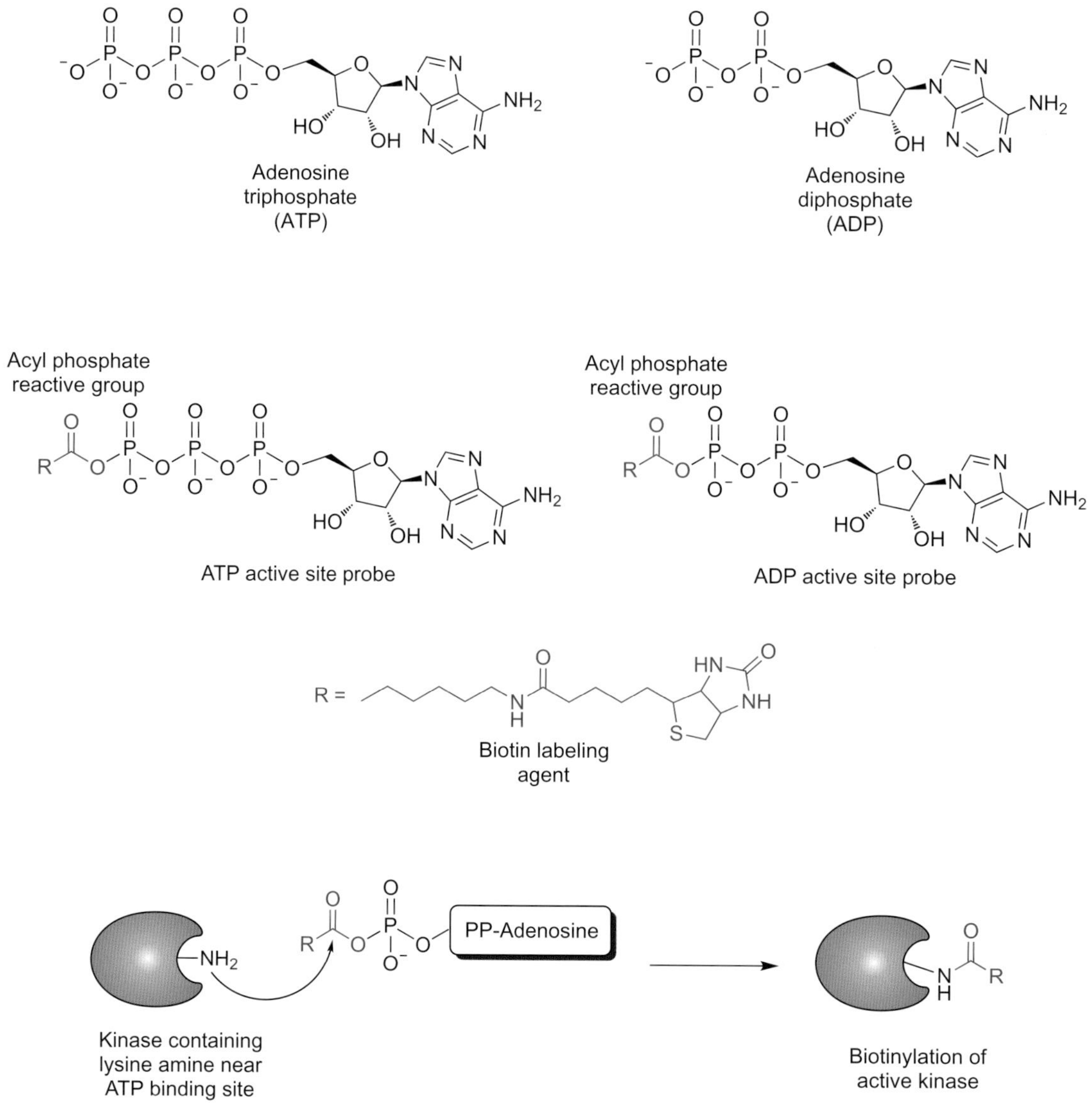

FIGURE 1.72 Active site probes for kinase enzymes consisting of an ATP or ADP analog and containing an amine-reactive acyl phosphate derivative on the terminal phosphate group. As the ATP or ADP probe binds to the nucleotide binding site on these enzymes, it positions the acyl phosphate reactive group near a lysine side-chain amino group. The amine reacts with the acyl phosphate to form a covalent amide bond with the acyl biotin arm while releasing the nucleotide portion, thus biotinylating every enzyme that can bind to the ATP or ADP probes.

can facilitate affinity isolation on immobilized streptavidin or detection using the fluorescent signature. Biotinylated probes can also be detected through the subsequent application of streptavidin–enzyme or streptavidin–dye conjugates.

Unlike serine hydrolases or thiol proteases, metalloprotease enzymes cannot be targeted using an electrophilic reactive group, because there is not an accessible nucleophilic group within their active site. Typically, metalloproteases have a histidine-coordinated metal ion (zinc) in the active site that facilitates a proteolytic hydrolysis reaction through the activation of a water molecule. However, the chelated metal can be targeted using activity-based bioconjugates containing a tight binding peptide substrate with a hydroxamate group to chelate the metal in the active site and a photoreactive group to facilitate covalent coupling (Saghatelian *et al.*, 2004). The incorporation of a benzophenone photoreactive group, for instance, permits photo-crosslinking of the probe to the metalloprotease after exposure to UV light (Figure 1.75). If this type of probe is built with a fluorescent label or a biotin tag the labeled enzyme can be detected within cells or enriched on an immobilized streptavidin support.

Immobilized affinity ligands can also be designed to purify receptor proteins for analysis. Cell surface receptor activation through extracellular ligand binding causes many downstream effects within a cell, including

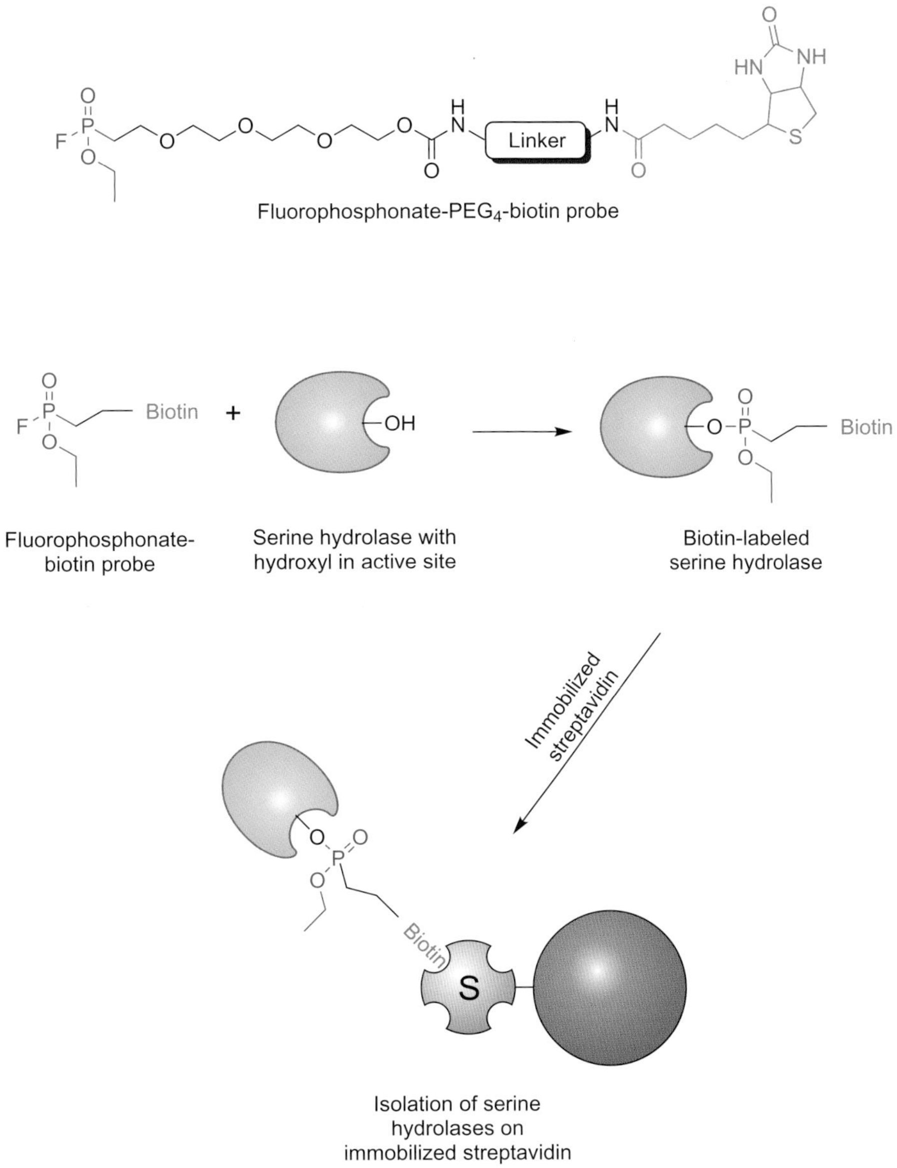

FIGURE 1.73 A serine hydrolase probe can be constructed from a fluorophosphonate-reactive warhead end and a biotin tag. The fluoro–phosphonate group covalently links to the active-site serine hydroxyl group and provides an affinity tag for detection of purification using streptavidin reagents.

enzyme activation, protein translocation and translation, increases in energy production or glucose utilization, stress responses, transmission of action potentials, production of hormones, immune responses, secondary intracellular and extracellular signaling events, and sensing of environmental conditions (e.g., the sense of smell or taste). Many of these biochemical events are mediated through the specific interaction of a small organic ligand with a binding pocket on a receptor protein. Ligands may also be peptides or proteins, which interact specifically with the receptor binding site.

The immobilization of a specific ligand derivative onto an insoluble resin or support material can facilitate the targeting and isolation of corresponding receptors from cell lysates. In addition, the coupling of multiple ligands onto a scaffold can be used to present multivalent interactions that can in turn cause receptor aggregation and activation of signal transduction pathways within cells (Kiessling *et al.*, 2006). The appropriate affinity ligands may be coupled to chromatography supports (Chapter 15), microparticles (Chapter 14), carrier proteins (Chapter 19), dendrimers (Chapter 8), polymers (Chapter 18), or liposomes (Chapter 21) to facilitate receptor isolation or signal activation.

Immobilized affinity ligands and their analogs have been used for years to purify many diverse receptor

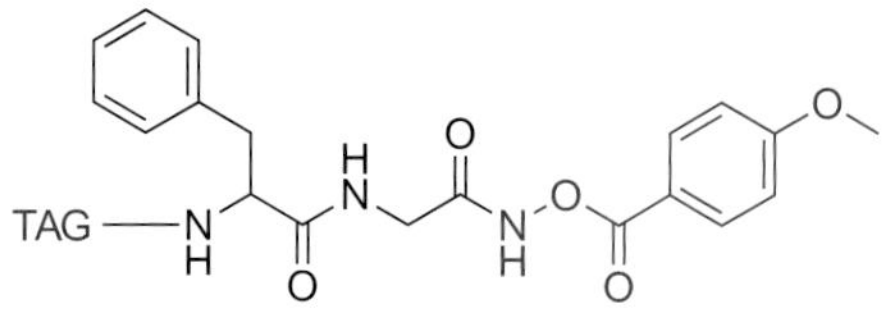

Cathespin inhibitor containing
a thiol-reactive *O*-acyl hydroxamate group

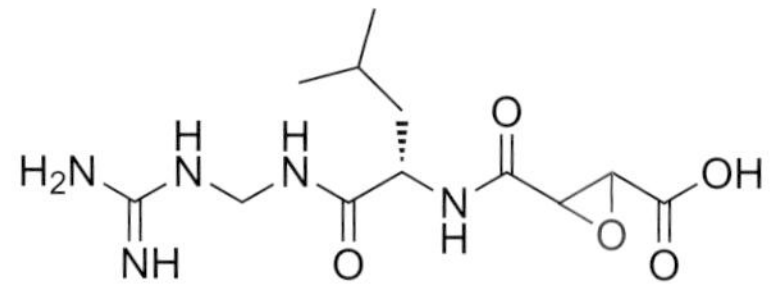

Cathespin inhibitor E-64
containing a thiol-reactive epoxide

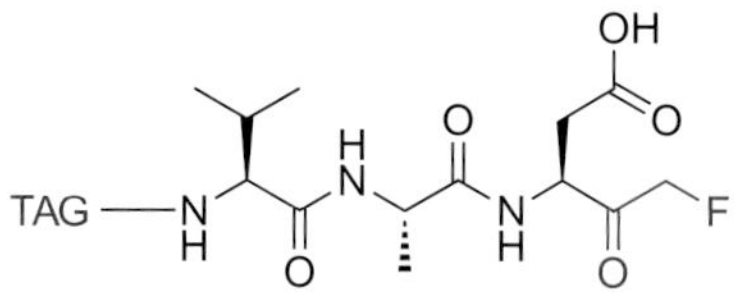

Caspase probe containing
a fluoromethyl ketone reactive group

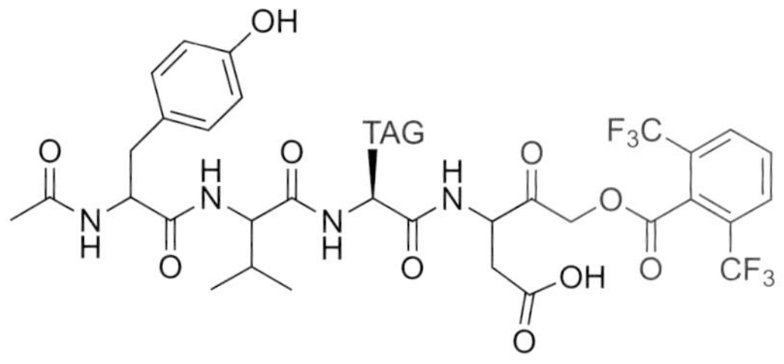

Caspase inhibitor containing an
acyloxymethyl ketone reactive group

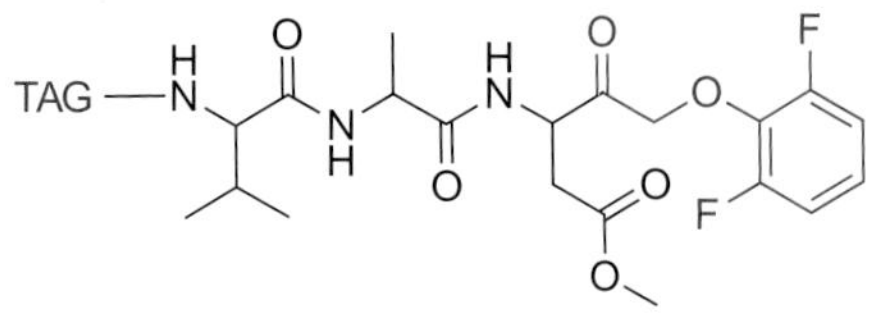

Caspase inhibitor containing an
acyloxymethyl ketone group
(Tag-VAD-OPh probe)

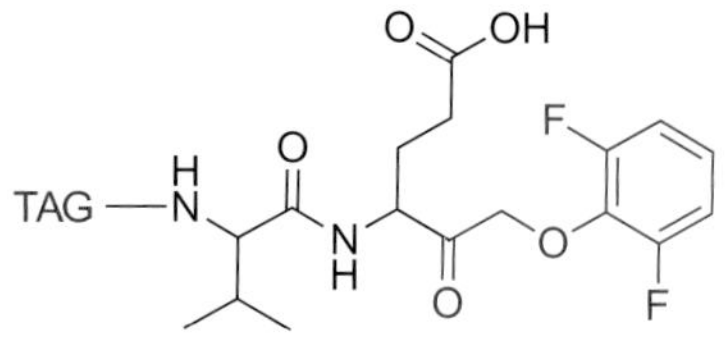

Caspase inhibitor containing an
acyloxymethyl ketone group
(Tag-VE-OPh probe)

FIGURE 1.74 Cysteine protease activity-based probes. Typical tag positions are shown in blue and the reactive groups are shown in red. The amino acid bridge within each probe causes it to bind to the active site of the protease and the warhead end then reacts with the thiol within the binding pocket to form a covalent bond.

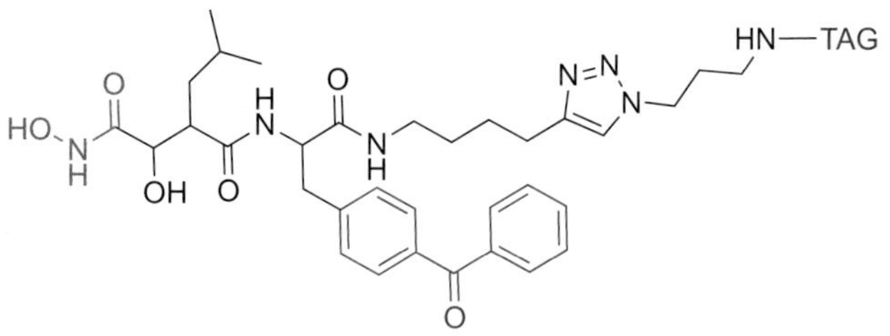

Metalloproteinase inhibitor probe containing
a photoreactive benzophenone group

FIGURE 1.75 Example of a metalloproteinase photoreactive probe. The targeting end of the molecule is shown in green, which contains a hydroxamate group that is able to bind through coordination with the chelated zinc ion in the enzyme active site. After binding, the benzophenone photoreactive group is photolyzed, which results in a covalent linkage with the enzyme near the catalytic site. The tag can be used to detect the enzyme or isolate it for further study.

types for analysis, including benzodiazepine receptors using an immobilized delorazepam derivative (Martini *et al.*, 1981); Fc receptors using an immobilized IgG support (Simister and Rees, 1985); human hepatic ferritin receptors using immobilized ferritin (Adams *et al.*, 1988); and estrogen receptors using immobilized 7β-estradiol, 4-hydroxy tamoxifen, or the synthetic active-site binding compound ICI-182,780 (Nalvarte *et al.*, 2010), to name only a few examples.

In addition, affinity chromatography using immobilized small molecule drugs or drug candidates can be used, followed by mass spec analysis for drug discovery to determine drug interaction targets (Saxena *et al.*, 2009). This technique, often called "capture compound mass spectrometry" (CCMS), can be used to determine pharmacokinetic properties of drugs, specific drug targets and off-target effects, and the toxicological indications of drug candidates (Kroll *et al.*, 2009). Trifunctional designs for CCMS probes have been developed that contain one arm having a drug derivative, a second arm

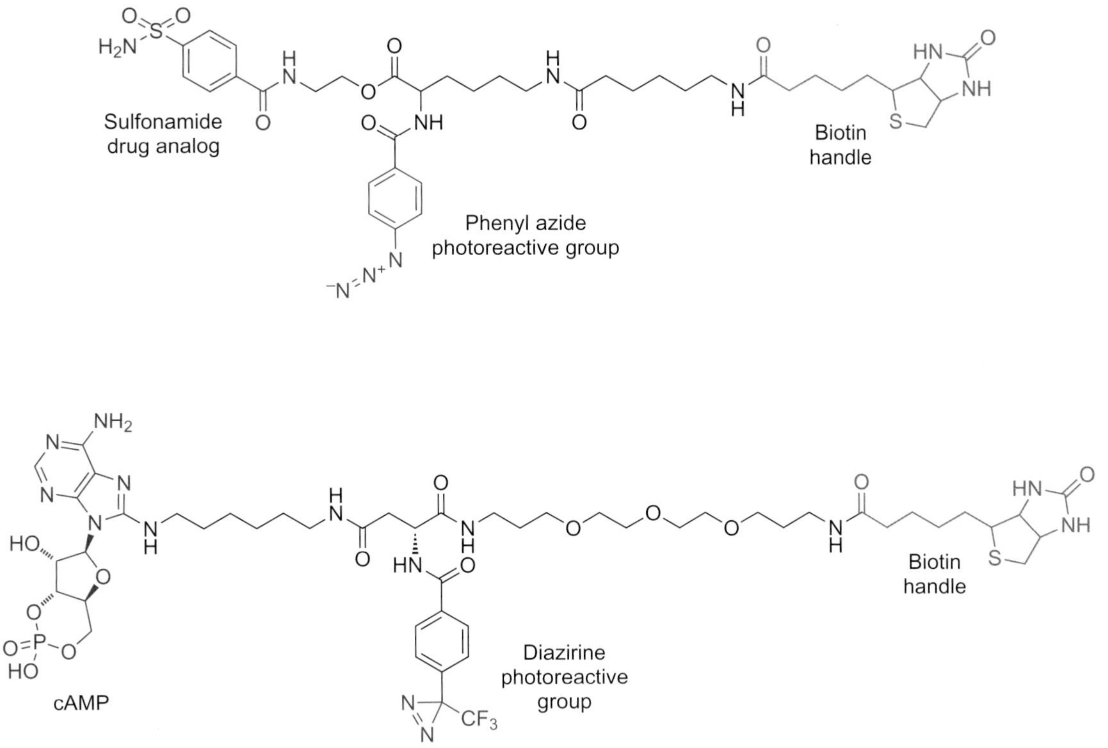

FIGURE 1.76 A CCMS probe containing a trifunctional design. The sulfonamide end binds to carbonic anhydrase and is an analog of a glaucoma drug, the phenyl azide photoreactive group facilitates covalent attachment to the enzyme target upon photolysis, and the biotin handle allows for purification on immobilized streptavidin. The captured molecules can be eluted and analyzed by mass spec.

with a photoreactive warhead, and a third arm with an affinity tag for capture on an affinity support. Köster *et al.* (2007) developed such probes containing a sulfonamide drug analog, a photoreactive phenyl azide group for covalent attachment, and a biotin derivative for affinity capture of the resultant complexes on a streptavidin resin (Figure 1.76). A similar CCMS compound was developed for affinity capture of *S*-adenosyl-L-methionine-dependent methyltransferases (Dalhoff *et al.*, 2010). Its trifunctional probe design contained an *S*-adenosyl-L-homocysteine selectivity group (affinity ligand), a photoreactive phenyl azide, and a biotin tag for isolation on immobilized streptavidin. Other CCMS probes have been designed to contain a GDP selectivity group for capture of GTPases (Luo *et al.*, 2009), a cAMP selectivity group for isolation of cAMP binding proteins (Luo *et al.*, 2009b) (Figure 1.76), and a staurosporine selectivity group for broad binding of kinases (Fischer *et al.*, 2010). The basic trifunctional design of CCMS probes can be used to create numerous affinity probes for study of small molecule binding to proteins, which is especially useful in drug discovery.

Affinity Capture of Recombinant Fusion Proteins

One of the most frequent applications of immobilized affinity ligands is in the capture or isolation of recombinant proteins containing fusion tags. Fusion protein technology has advanced in conjunction with recombinant protein production to allow easy purification of a wide range of expressed proteins by targeting a common peptide tag. In other words, if the recombinantly produced protein also contains another peptide chain attached to its C- or N-terminal end that has known binding characteristics to certain affinity ligands, then the fusion tag can facilitate immobilization, capture, or purification of the expressed protein. The fusion partner thus functions as a universal affinity handle that can be targeted by the proper immobilized ligands regardless of what recombinant protein is being produced.

Fusion protein affinity tags can consist of relatively short peptide segments (i.e., 5 to 8 amino acids) or intermediate length or longer polypeptides consisting of fragments, domains, subunits, or entire proteins having particular active-site binding properties. The shorter tags are often targeted through the use of an immobilized antibody to specifically interact with the fusion tag sequence or through the use of an immobilized metal chelating group (IMAC) (see Table 1.2). Fusion tags such as the FLAG tag, which contains an eight amino acid sequence, and the c-myc tag, which contains 11 amino acids, typically are targeted using monoclonal antibodies specific for their respective sequences. Immobilized immunoaffinity

TABLE 1.2 Common Fusion Protein Affinity Tags and Their Properties

Fusion Tag	Residues	Sequence	Molecular Weight (kDa)	Affinity Support
Poly-His	6–8 (usually 6)	HHHHHH	0.84	Metal chelate (Ni^{2+}-NTA, Co^{2+}-CMA)
HAT (natural His affinity tag)	19 (6 His)	KDHLIHNVHKEFH-AHAHNK	2.31	Metal chelate (Ni^{2+}-NTA, Co^{2+}-CMA)
Glutathione-*S*-transferase (GST)	211	Protein tag	26	Glutathione (GSH)
FLAG tag	8	DYKDDDDK	1.01	Anti-FLAG monoclonal antibody
FLAG repeat	22	DYKDHDGDYKDH-DIDYKDDDDK	2.73	Anti-FLAG monoclonal antibody
c-Myc	11	EQKLISEEDL	1.2	Anti-c-Myc monoclonal antibody
Poly-Arg	5–6	RRRRRR	0.96	Cation exchange resin
S tag	15	KETAAAKFERQHMDS	1.75	*S*-fragment of RNaseA
Calmodulin-binding peptide	26	KRRWKKNFIAVSA-ANRFKKISSSGAL	2.96	Calmodulin
Cellulose-binding domain	27–189	Domains of cellulose binding protein	3–20	Cellulose
Chitin-binding domain	51	TNPGVSAWQVNTAYT-AGQLVTYNGKTYKC-LQPHTSLAGWEPSN- VPALWQLQ	5.59	Chitin
Maltose-binding protein	396	Protein tag	40	Amylose
Streptavidin-binding peptide (SBP)	38	MDEKTTGWRGGHVV-EGLAGELEQLRAR- RLEHHPQGQREP	4.03	Streptavidin
Strep-tag II	8	WSHPQFEK	1.06	Mutant streptavidin (Strep-Tactin)
AviTag	15	GLNDIFEAQKIEWHE (facilitates biotinylation at lysine residue)	1.83	Streptavidin
HA tag (influenza hemagglutinin)	9	YPYDVPDYA (amino acids 98–106)	1.1	Anti-HA monoclonal antibody
DHFR (dihydrofolate reductase)	187	Protein tag (sometimes used as fragment containing 6xHis sequence)	23.8	Anti-DHFR monoclonal antibody

supports thus can be used to isolate any recombinant proteins containing the FLAG or c-myc fusion tags.

Another common fusion tag is a short eight-amino-acid sequence, called Strept-tag II, which binds to the biotin-binding pocket of streptavidin (Schmidt and Skerra, 2007). Typically, a mutant streptavidin protein is used with the Strept-tag II to further enhance the binding affinity of the peptide to the biotin binding region. The Strep-tag II sequence has also been used in dual-tag approaches wherein a recombinant protein is purified through use of two different affinity tags on either of its C- and N-terminal ends (Clontech). This technique, called "tandem affinity purification" (TAP or TAP tag) has important application in the isolation of recombinant proteins (Rigaut *et al.*, 1999; Puig *et al.*, 2001). Purification of an expressed protein using a series of two different affinity chromatography separations directed at the different TAP tags results in increased purity of the desired protein (Figure 1.77). TAP tags have also been used to isolate interacting proteins or protein complexes and even enhance the identification of the interacting partners (Séraphin *et al.*, 2002).

A unique poly-DOPA (dihydroxy phenylalanine) fusion tag has been described that allows the immobilization of a recombinant protein through interactions of the dihydroxy groups with surfaces (Jennissen and Laub, 2007). The tag is expressed as a series of four tyrosine residues at the N-terminal with an intervening serine residue. The tyrosines can be chemically or enzymatically transformed into the DOPA derivatives. This technology mimics the native foot proteins found in mussels, which contain DOPA groups that allow tight noncovalent interactions with mineral, metallic, or organic surfaces in aqueous environments. DOPA

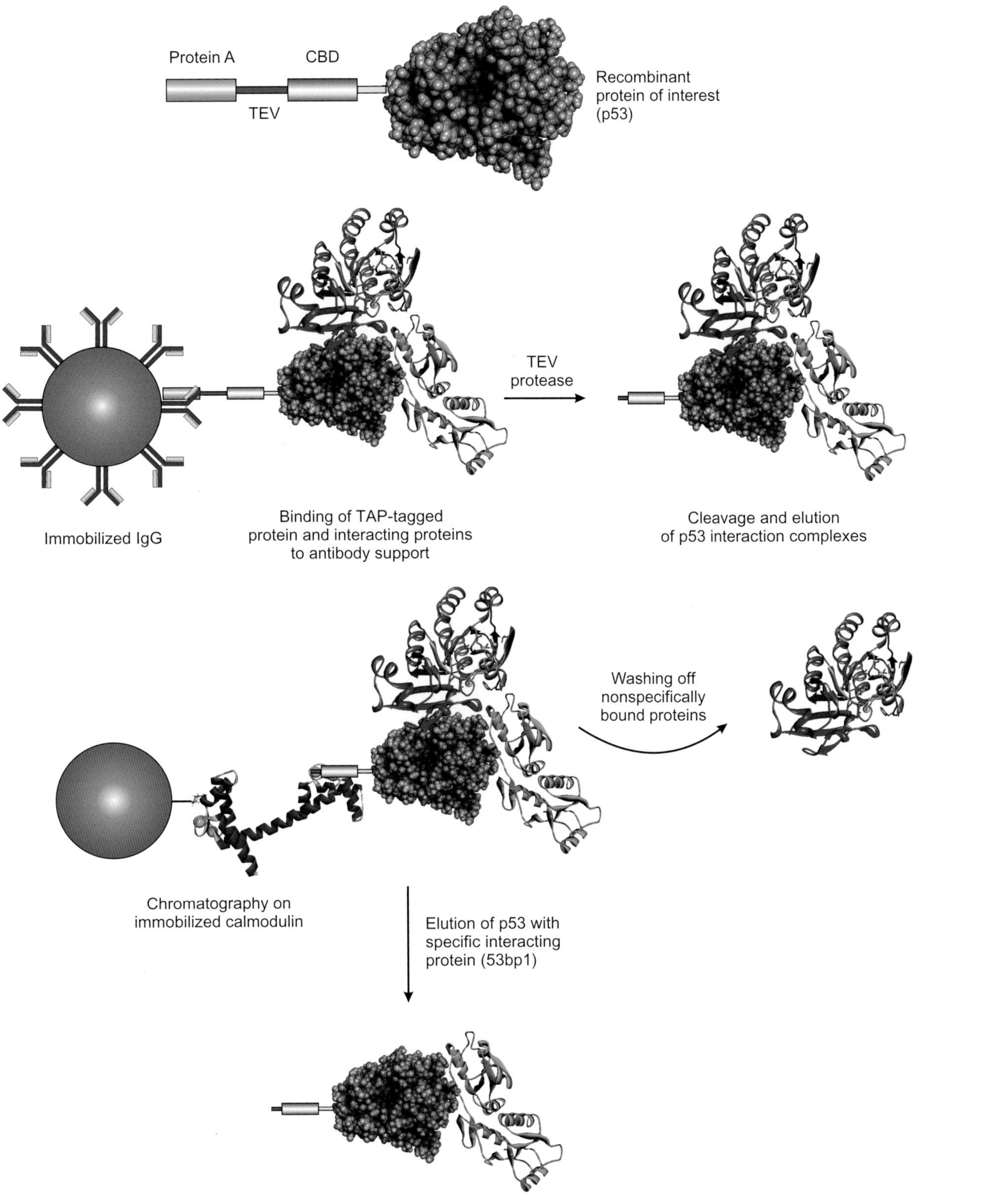

FIGURE 1.77 Tandem affinity purification (TAP) tags make use of a design consisting of two fusion proteins linked by a TEV protease sequence and followed by the protein of interest. This design permits the isolation of true interacting proteins in two chromatography steps by removing any other proteins that may be carried along through nonspecific binding. A common pair of fusion protein tags for this purpose includes the use of protein A to bind immunoglobulins as well as the calmodulin binding domain (CBD) to bind calmodulin. Upon expression of the recombinant protein of interest (in this case, p53) and incubation with a sample to allow binding of putative interacting partners, the fusion tags are used in a series of two affinity steps to aid in the removal of nonspecifically bound proteins that may be interacting with the desired protein interaction complex. Thus, the expressed protein complex is first subjected to affinity purification on immobilized IgG, which binds to the protein A tag and facilitates washing away of the bulk of non-bound proteins. Next TEV protease is added to the affinity resin to cleave the linker between the protein A and CBD fusion partners, thus releasing the recombinant protein complex of interest, which still has the CBD fusion tag attached to it. In a second affinity chromatography step, immobilized calmodulin is used to capture the CBD fusion tag and a second wash step is carried out to remove any contaminating proteins. Elution of the remaining interacting proteins then results in a higher certainty that any bound proteins still remaining are true interacting partners to the protein of interest and not just nonspecific binders.

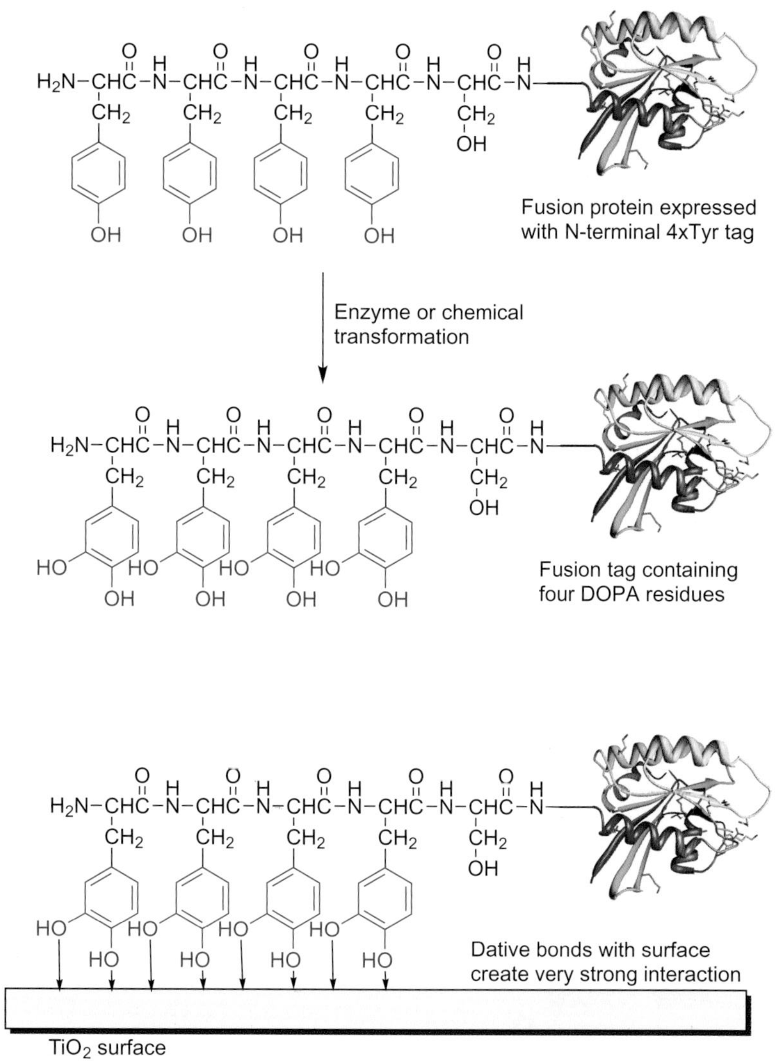

FIGURE 1.78 A fusion protein tag consisting of four tyrosine residues can be transformed into a poly-DOPA tag either enzymatically or chemically. The high density of the hydroxyl groups within this short sequence can provide strong binding potential to metallic surfaces through dative bonds.

groups have been shown to present high-affinity binding with TiO_2 surfaces, which reflect dual coordination of the hydroxy groups providing incredibly high dissociation energies (Lee *et al.*, 2006). A single DOPA residue was found to require in the range of 750 to 800 pN to break its interaction with a TiO_2 surface, thus placing its dissociation energy midway between one of the strongest noncovalent interactions (streptavidin–biotin at ~160 pN) and true covalent bonds, which are on the order of 1,600 pN (note that a single hydrogen bond only requires about 4 pN to dissociate the interaction). Thus, a poly-DOPA-labeled fusion protein can be immobilized onto surfaces with high efficiency and with an effective strength of binding that places it even above the realm of covalent coupling, due to multiple DOPAs interacting simultaneously (Figure 1.78).

Perhaps the most popular fusion tags, however, are the 6xHis or the glutathione *S*-transferase (GST) tags. A 6xHis tag contains a sequence of six histidine residues in a row at one end of the recombinant protein. Occasionally, additional histidine residues are used and sometimes spaced by other non-chelating amino acids (such as in the HAT tag; see Table 1.2). Neighboring histidines are very adept at coordinating transition metals through the unshared pairs of electrons on their side-chain imidazole nitrogens. An immobilized metal chelate containing nickel, cobalt, or gallium with unoccupied coordination sites (or occupied only by water molecules) is commonly used to purify His-tagged proteins or even immobilize them to a surface for analysis.

Recombinant proteins containing a GST fusion partner can be isolated through its native interaction with glutathione (GSH). GSH can be immobilized to an affinity matrix through its thiol group by using an epoxide- reactive group, which forms a thioether upon coupling (Chapter 15, Section 2.3). GSH presented in this manner has good capacity for interactions with GST fusion proteins and can

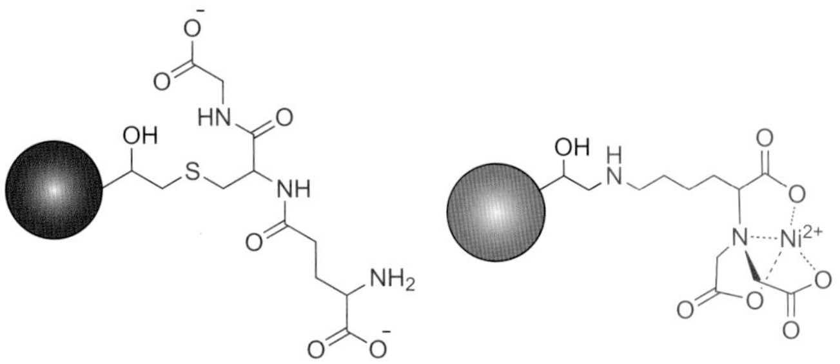

FIGURE 1.79 Two of the most common affinity ligand conjugates for capturing fusion proteins constructed with GST or 6xHis tags are immobilized glutathione and immobilized nickel chelate, respectively.

be used for either isolation or immobilization of the recombinant protein associated with it. Fusion proteins therefore can be used for purification of the desired recombinant protein or for oriented coupling of the protein to a surface or particle. Since the affinity ligand only interacts with the fusion tag, the recombinant protein ends up sticking off the surface with a predictable and reproducible orientation (Figure 1.79). This is an advantage for ensuring that the recombinant protein activity or binding site is available upon immobilization.

Covalent Fusion Tag Technology

A variation on standard fusion tag design is the development of mutant enzyme tags that can bind and covalently bond to a particular substrate molecule. Certain enzymes are designed to cleave substrates through the temporary covalent linkage of an intermediate of the substrate to a residue within their active sites followed by release of the temporary link to free the active site for another substrate. Esterases, such as cutinase for example, are one class of enzyme that undergo such reactions and if they are mutated or used with a special substrate analog, a substrate derivative can be designed to permanently link within the active site through covalent bond formation. Alternatively, some enzymes are classified as suicide enzymes, because they react with a substrate and covalently transfer part of the substrate to a functional group within their active sites, while the remainder of the substrate is cleaved and released. Mutagenesis has been used to mutate amino acid residues around the active site of such enzymes and create a new enzyme with altered activity, which functions to covalently link the substrate to its active site without the subsequent cleavage event taking place. The results of these approaches are enzyme–substrate pairs that are able to covalently couple to the appropriate substrate ligand with high selectivity and efficiency.

Using enzymes and substrates that are able to form a selective covalent bond between one enzyme molecule and one substrate molecule is useful for making reactive fusion proteins, which can then be used to label or immobilize the recombinant protein at a specific site on the fusion partner (Figure 1.80). Examples of this technology include the use of the serine esterase cutinase along with a chlorophosphonate ester substrate (or fluorophosphonate or *p*-nitrophenyl phosphonate) derivative to form a covalent link with the serine in its active site (Hodneland *et al.*, 2002; Nickel *et al.*, 2012); a mutant haloalkane dehalogenase enzyme that covalently links a terminal chloroalkyl substrate derivative to an aspartate carboxylate in its active site to form a stable ester bond (Hata and Nakayam, 2007; Shinohara and Matsubayashi, 2007; Schröder *et al.*, 2009) (HaloTag, Promega); a mutant human O^6-alkylguanine-DNA-alkyltransferase (hAGT) that is able to covalently link a substrate analog containing a terminal benzylguanine group to a cysteine thiol in its active site (Jongsma and Litjens, 2006; Iversen *et al.*, 2008; Engin *et al.*, 2010) (SNAP-tag, NEB); and a second mutant of hAGT that specifically reacts with substrate derivatives containing a benzylcytosine group to the cysteine thiol in the active site (Gautier *et al.*, 2008; Schulz and Köhn, 2008) (CLIP-tag, NEB). The combined use of both the SNAP-tag and CLIP-tag fusion proteins can facilitate multi-protein labeling with different fluorescent substrate derivatives for *in vivo* cellular imaging (Gautier *et al.*, 2008).

Covalent fusion tags allow highly discrete labeling of recombinant proteins at a single site per fusion pair. The mutant enzymes have been engineered to have high-affinity binding constants to their substrate derivatives combined with a high yield of covalent attachment. The result permits selective labeling of expressed fusion proteins in complex solutions containing many other proteins, similar to bioorthogonal chemical labeling techniques. It also allows such fusion proteins to be immobilized onto array surfaces or particles without purification from an expression lysate. Cells expressing a protein containing a covalent fusion tag partner can be grown in microplates and the cells lysed. If the appropriate substrate derivative is immobilized on the plate surface, the expressed protein in the lysate is quickly coupled to the surface through the fusion tag active site, thus immobilizing it for subsequent assay or analysis.

Scavenging of Contaminants or Unwanted Components

Another major application area for immobilized bioconjugate affinity ligands is the removal of contaminants or unwanted components in a biological sample. Often, certain components in a sample extract can interfere with subsequent analysis or inhibit activity of

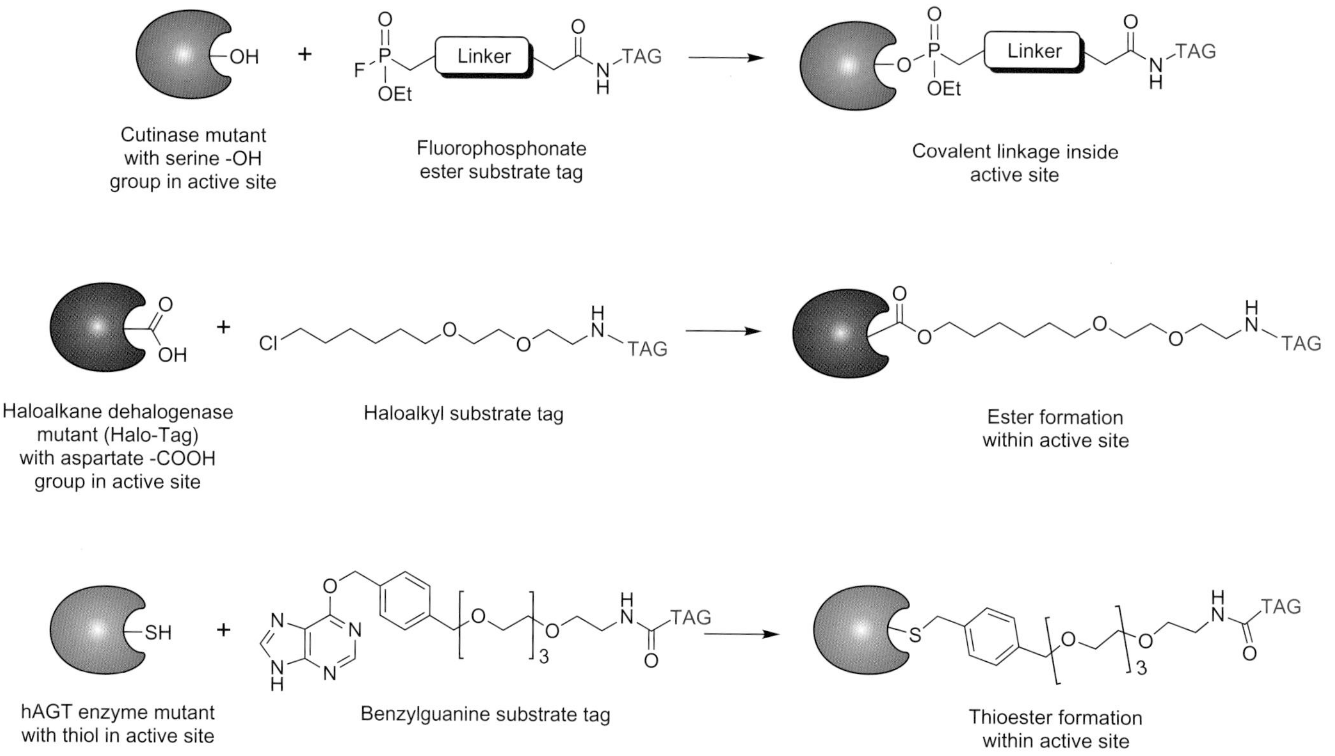

FIGURE 1.80 Three examples of covalent fusion tags consist of mutant versions of the enzymes cutinase, haloalkane dehalogenase, and human O^6-alkylguanine-DNA-alkyltransferase (hAGT). These mutant enzymes react with certain substrates to result in covalent linkages with the key functional groups within their active sites. Substrate derivatives containing detection tags, affinity handles (e.g., biotin), or even immobilized substrates may be used to specifically link to these tags, which can be associated with expressed proteins.

a protein to be studied. Specific removal of such components can be done through affinity chromatography using an appropriate immobilized affinity ligand that interacts selectively with the unwanted molecules. An example of such a scavenging affinity support includes the removal of detergents from protein solutions, especially from cell lysates prepared by detergent-mediated lysis. Detergents are difficult to remove from aqueous solutions, because micelle formation typically restricts their diffusion through dialysis membranes. This also makes it difficult to remove them from proteins by size-exclusion chromatography, because their effective micellular molecular weight is on the same order as proteins or greater.

Affinity supports designed to remove detergents from biological solutions have been developed using hydrophobic interaction chromatography. In this case, the affinity ligand typically is a small hydrophobic molecule or a hydrophobic polymer attached to a resin, which allows the hydrophobic tail of the detergent molecules to bind while proteins pass through (Figure 1.81). In other cases, the entire beaded matrix can be made from a hydrophobic polymer, which can bind to detergents. However, to eliminate the potential for proteins to bind nonspecifically through hydrophobic interactions to these particles, many detergent-removing supports use porous chromatography beads with a rather small molecular exclusion limit (<10 kDa), thus restricting access of the average-sized proteins to interact with the interior of the particles. This design allows proteins to pass around the beads while detergent molecules are able to diffuse within the porous structure and get bound by the hydrophobic ligands.

Variations on this concept have been commercialized by Pierce (Thermo Fisher Scientific) (Extracti-Gel D) as well as Pall Corporation (SDR HyperD resin). Both of these affinity resins contain proprietary hydrophobic molecules immobilized onto the interior spaces of the particles. Another matrix is available made from a polystyrene/divinylbenzene copolymer particle (BioBeads SM-2 from Bio-Rad), but this support can also bind considerable protein molecules nonspecifically due to the hydrophobic surface character of the beads and the fact that the particles are not porous to better exclude proteins from interacting with the hydrophobic surfaces.

Scavenging affinity supports can also be prepared to remove hazardous components from solution, such as in the removal of endotoxins from biological preparations. Lipopolysaccharide (LPS) molecules (also called lipid A) consist of a hydrophilic head group containing

FIGURE 1.81 The principle of detergent removal using hydrophobic interaction chromatography. Porous particles containing hydrophobic ligands coupled to the surface can be used to bind the hydrophobic tails of detergent molecules. If the pore structure is small enough to prevent most proteins from entering the pores and being bound, then the detergents will be removed as the proteins pass around the particles and elute from the column.

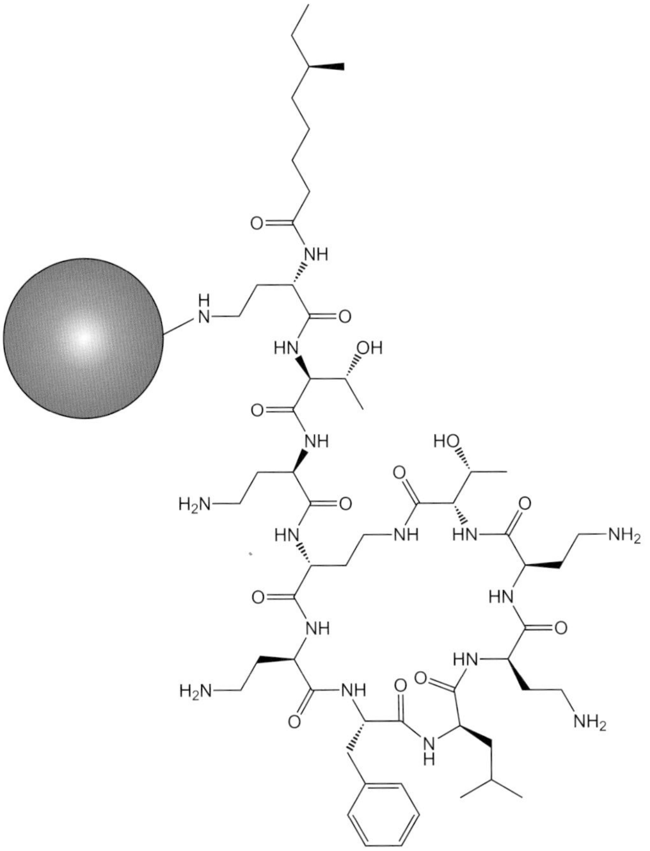

FIGURE 1.82 Immobilized polymyxin B can be used for the removal of endotoxins from solution.

carbohydrate along with two negatively charged phosphate groups as well as seven hydrophobic fatty acid tails attached to the sugar groups. LPS molecules are often components of bacterial membranes with their lipid chains embedded in the bilayer membrane structure. Bacterial contamination of biological materials can cause endotoxins to be present in protein solutions. Endotoxins are pyrogens that cause fever in animals in small doses or even death in larger amounts. Biological solutions containing endotoxins can also cause unwanted cellular effects if they are inadvertently added to cells grown in culture.

Removal of endotoxin can be accomplished through conjugation of polymyxin B to a chromatography resin (Magalhães *et al.*, 2007). Polymyxin contains a large cyclic peptide derivative with a hydrophobic tail that has antibiotic properties and which is able to interact rather specifically with lipid A molecules (Figure 1.82). Protein solutions containing unacceptable amounts of endotoxin can be cleansed from this contaminant by affinity chromatography on an immobilized polymyxin B support. Another specific endotoxin-binding matrix was prepared using an immobilized version of the drug pentamidine, which was found to have similar binding capacity to polymyxin B supports, but is much less expensive (McAllister *et al.*, 2007). In addition, immobilized amino acids have also been found to be effective at removing endotoxins from biological preparations (Wei *et al.*, 2007).

Another type of scavenging affinity support that has found extensive use is an immunoaffinity matrix containing immobilized antibodies that target the most abundant proteins in plasma. Abundant protein removal using immobilized antibodies is often important to uncover and analyze the less abundant protein population in plasma by 2D electrophoresis or mass spectrometry. Blood contains soluble proteins ranging in concentration from albumin at about 80 mg/ml and immunoglobulins at about 20 mg/ml to several other abundant proteins in the mg/ml range to many proteins in the μg/ml range and many thousands more at lower concentrations (down to the low pg/ml levels and below). The range of protein concentrations in blood spans over 10 orders of magnitude. However, the bottom 10% of proteins in plasma in terms of concentration represents over 90% of proteins actually in plasma (in terms of the number of different proteins present). The problem with analyzing the vast majority of proteins in blood is that they are obscured by the ones at high concentration. Removal of the dozen or so most abundant proteins from plasma is beneficial for separating and analyzing proteins at lower concentration without interference.

The preparation of scavenging immunoaffinity supports containing a mixed bed of different immobilized antibodies with specificity toward the most abundant plasma proteins is typically used to remove these high concentration proteins from samples prior to analysis. Each immobilized antibody is prepared individually and then the affinity resins are mixed in proportion to the amount of protein they must remove from a sample. Such affinity supports are available commercially with anywhere from a few antibodies to remove albumin and the major immunoglobulins up to about 12 to 20 different antibodies to eliminate all the most abundant proteins in a single chromatographic step (BAC, Agilent, Pall, Sigma Aldrich, Beckman Coulter). The protein passing through such affinity resins is cleaned of the indicated abundant proteins, but contains all the remaining proteins within the sample (Figure 1.83).

Affinity resins also have been designed to remove viruses from solutions. Immobilization of a Tamiflu analog, for instance, was used to chromatographically isolate influenza A viruses through specific binding to the H1N1 neuraminidase coat proteins (Kimura *et al.*, 2009). Although this was used in a purification mode, the affinity support could also function to remove influenza viral contamination from biological solutions. In addition, Tsao *et al.* (1988) used immobilized cells on polymer membranes as the affinity ligands to capture viruses able to bind to specific cell surface receptors. The support was simple to prepare and highly selective for removing any viruses able to interact with the cells. Immunoaffinity chromatography can also be used to remove contaminating viruses. Detmers *et al.* (2010) developed a specific virus-binding support using camelid antibodies to adenoassociated virus, which is difficult to purify using other techniques.

Another type of scavenging support is one that contains a ligand able to interact and bind to a small contaminant in solution. For instance, immobilized metal chelators can be used to remove metal contamination from aqueous solution, as in remediation processes. This application is different from the use of immobilized metal chelate affinity chromatography (IMAC) for the purification of target molecules (Section 2.3, this chapter). In this case, an uncharged chelating support is used to capture and remove metal ions from solution, whereas in IMAC a pre-charged, metal chelate support is used to affinity capture certain targets having binding ability for the bound metal. To remove metals from various solutions, Duru *et al.* (2001) prepared a composite microparticle containing covalently attached poly(ethylene imine) on a poly(methyl methacrylate) core microsphere. The resultant scavenging affinity support could remove cadmium, copper, and lead ions from aqueous solution with high capacity.

Inorganic solid phases often perform better as a base for producing a metal-scavenging affinity support than polymeric materials, because they are less susceptible to shrinking and swelling or mechanical damage under harsh conditions of use. Silica in particular can be modified with functional silanes to contain a number of metal chelating groups (Chapter 14), which can then be used for metal ion capture. Jal *et al.* (2004) reviewed a number of chelating groups and their immobilization chemistry on silica supports for the selective binding of many metallic ions from solution.

Biopolymer constructs have also been immobilized on solid phases to create a metal scavenging support. Kostal *et al.* (2005) described the use of certain peptide sequences that can be produced by recombinant means and used to effectively remove metal contamination from waste water. For example, a poly-histidine-containing peptide is known to have excellent affinity for chelating a number of metal ions from aqueous solution. A hexahistidine tag on recombinant proteins can be used to purify the protein on an IMAC support (Section 2.3, this chapter), but in this case the poly-histidine ligand itself becomes the chelator to sequester and remove metals. Other peptide sequences known to bind metal ions can also be used for this purpose and immobilized on solid phase supports for heavy metal removal.

3.4. Catalysis and Chemical Modification

One of the most widely used applications of immobilized ligands is to catalyze a reaction in solution. Typically this is done using enzymes, but it may also involve the use of immobilized reactive groups that create certain modifications on target molecules. These immobilized reactants may facilitate specific cleavage of peptide bonds in proteins, hydrolysis of glycosidic bonds in carbohydrates, or the facilitation of chemical reactions to alter the structure of target molecules or synthesize particular substances. Conjugation of such

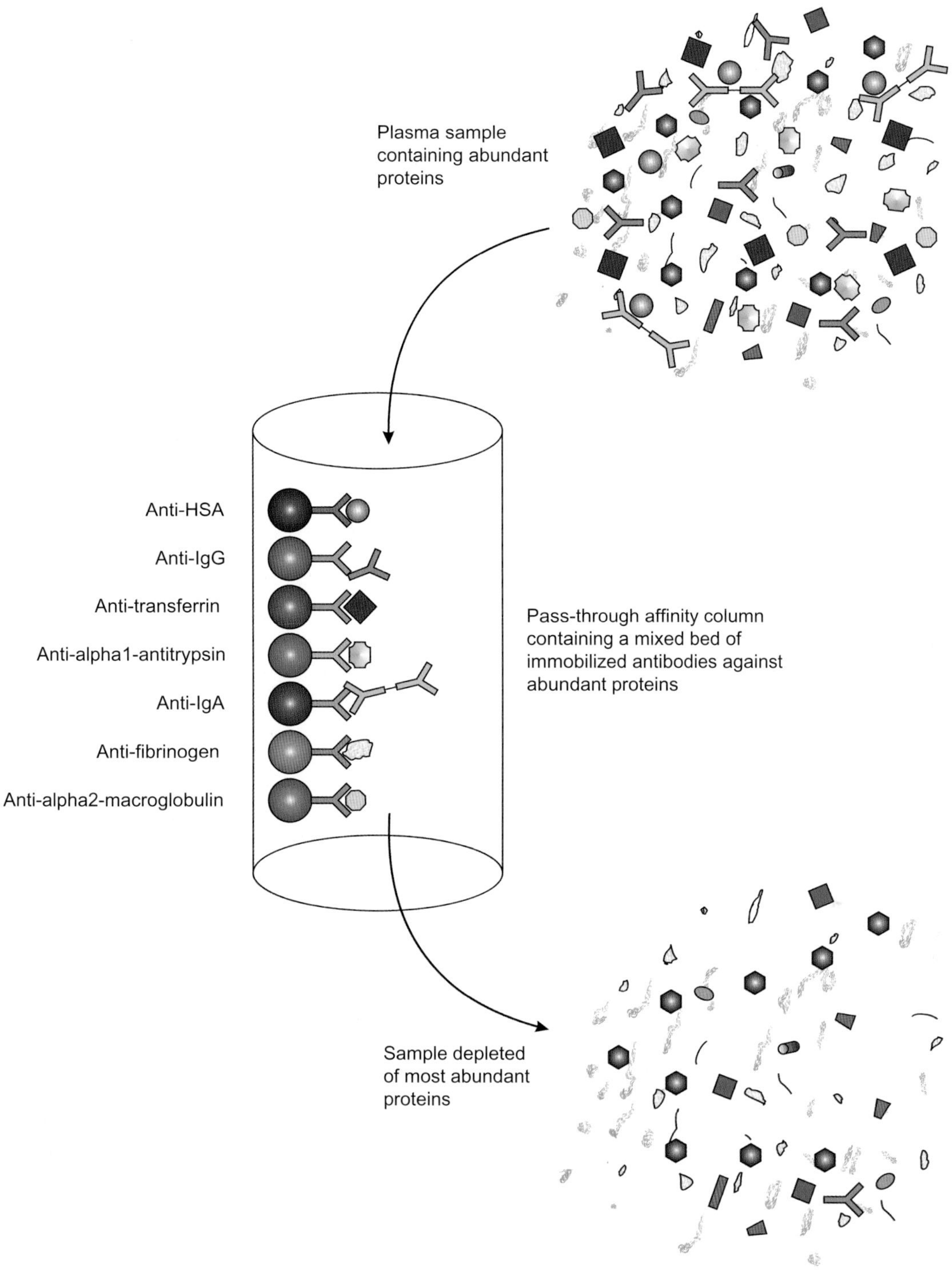

FIGURE 1.83 The principle of abundant protein removal from plasma samples is illustrated. A mixed-bed column is packed with immobilized antibodies against the major proteins in plasma. As a sample is passed through the column, the antibodies remove these predominant proteins from the solution and the remaining plasma proteins pass through essentially depleted of these high-concentration proteins.

reactive agents to solid phases creates a system that permits modification of target molecules without contamination of the solution with the reactive agent. After the reaction, separation of the immobilized catalyst or reactive compound is accomplished simply by filtration and washing of the solid phase to recover the sample solution. Many agents that can catalyze a reaction or react with a target molecule can be covalently attached to resins or surfaces to form an immobilized analog, which has convenient properties for numerous applications.

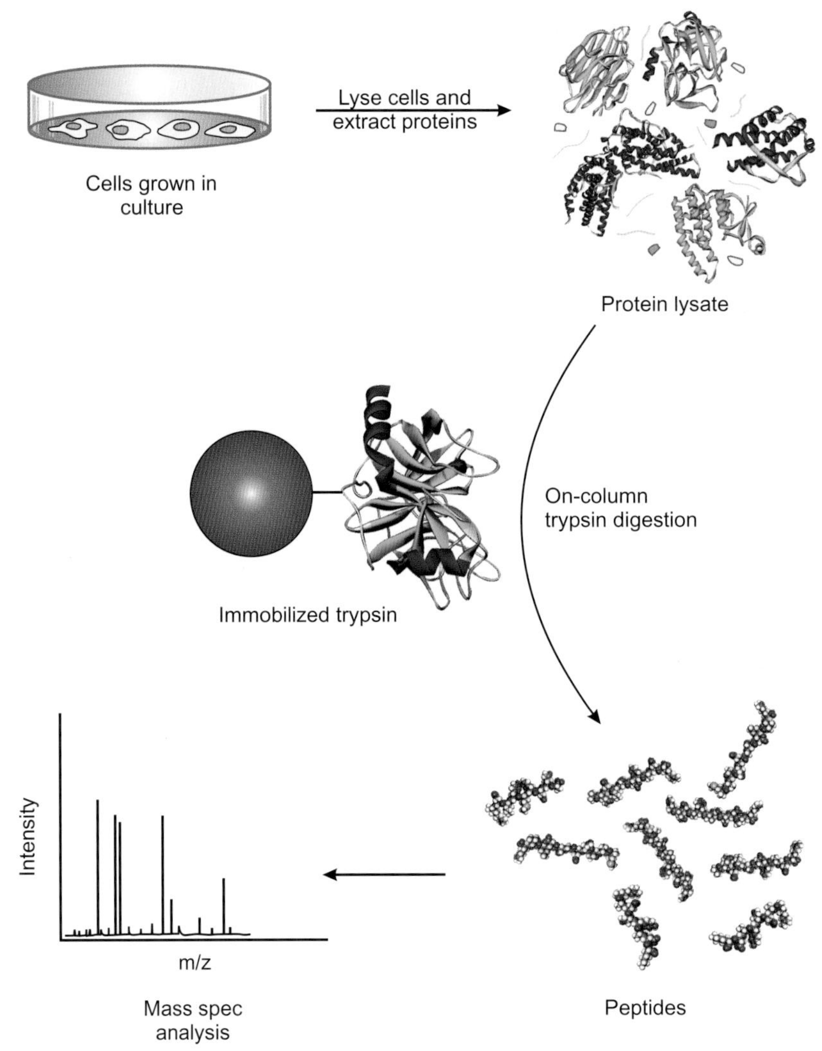

FIGURE 1.84 Immobilized proteases such as trypsin can be used to digest protein samples prior to mass spec analysis of the peptides produced. The use of an immobilized protease prevents contamination of the sample with free protease and its autolysis products.

These reagents are often called "immobilized reactors," and they have found extensive use in the fields of proteomics and biotechnology (Krenkova and Svec, 2009; Ma *et al.*, 2009), in cellulosic ethanol production (Brethauer and Wyman, 2010), in the production of biodiesel fuels (Jegannathan *et al.*, 2008), in the synthesis of organic compounds (Miyazaki and Maeda, 2006; Zhang *et al.*, 2010), and in the remediation and treatment of pollutants (Husain *et al.*, 2009).

Immobilized Proteases for Proteomic Analysis

One of the most common immobilized catalysts used for biological research applications is a protease. Protein hydrolases are enzymes that can cleave peptide bonds at certain sequence-specific locations to form peptides of various lengths from protein samples. If the protease has a high specificity toward cleavage only at precise amino acid sites within a protein, then the peptides formed from the digestion will be of known size and sequence relative to the protein from which they originated. These peptide fragments can be analyzed by mass spec and their masses and sequences compared to established databases, which can then be used to determine the corresponding parent proteins (Figure 1.84). This workflow is the basis for most methods of mass spec proteomics analysis and specific proteases make it possible (Malmström *et al.*, 2007).

Trypsin is a serine protease that cleaves on the C-terminal side of lysine and arginine, and it has become a popular choice for protein digestion prior to mass spectrometry. If trypsin is immobilized through its amine groups to a particle or membrane, a protein sample can be treated for digestion without ultimately contaminating the sample with the enzyme or its autodigestion fragments. Immobilized trypsin on particles has often been exploited in mass spec applications, using it in formats such as spin columns or pipette tips to rapidly digest proteins and subsequently separate the peptide sample from the trypsin reagent (Ota *et al.*, 2007). Additional information on the immobilization of trypsin and other enzymes is provided in Chapter 15.

Immobilized proteases are also used for the cleavage of fusion tags from recombinantly expressed proteins. Many fusion tags are co-expressed with a desired protein to enable solubilization, folding, and/or purification of the target protein (see Table 1.2). However, it may be desirable to remove the fusion tag and thus obtain an expressed native protein structure for study without the tag being present. If a peptide sequence is incorporated between the expressed protein and the fusion tag that can be recognized and cleaved by a highly specific protease, then the tag can be selectively removed. Solution-phase protease treatment as well as the use of immobilized proteases have been employed to cleave fusion tags and recover desired proteins (Kubitzki *et al.*, 2008). The immobilized protease form of these reagents allows recovery of the desired protein to occur without further contaminating the solution with the enzyme, thus providing a more convenient work flow for purification.

Another strategy in the use of enzymes to cleave fusion protein tags is to initially add a soluble enzyme containing an affinity tag to the fusion protein solution. After cleavage of the target fusion protein the enzyme can subsequently be removed using an immobilized affinity column directed at the tag on the enzyme (Pedersen *et al.*, 1999; Jenny *et al.*, 2003; Abdullah *et al.*, 2005; Arnau *et al.*, 2006). This method is more convenient in cases where the amount of protease is limiting and the preparation of an immobilized enzyme support is impractical (Figure 1.85). A simple route to creating this type of workflow is to use a biotinylated protease, which can be removed from the cleaved fusion protein through binding to an immobilized streptavidin support.

Immobilized Reactors in Bioengineering

Immobilized reactors are bioconjugate supports containing an enzyme or some other reactive component that can interact with and convert a target molecule into a desired product. Many of these insoluble reactors have been developed to serve large-scale process operations in fields such as biotechnology, the food industry, organic chemical production, bioremediation, and in the generation of biofuels. Immobilized enzymes in particular can display dramatically improved stability in an immobilized state compared to the corresponding soluble form of the enzyme in solution. Some enzymes such as α-chymotrypsin, chloroperoxidase, glucose oxidase, and lipase have shown stability improvements (including in one case over a thousand-fold gain in half-life of activity loss), increased stability toward organic solvents, and resistance to thermal degradation (Wang *et al.*, 2001; Borole *et al.*, 2004; Kim *et al.*, 2006; Sheldon, 2007).

The coupling chemistry used to attach an enzyme reactor to an insoluble support involves similar reactions to those commonly used to create soluble bioconjugates, except in this case one of the components linked together typically is insoluble and particulate in nature. In most cases, the particle is considerably larger than the enzyme, and thus in the final complex the particle contains numerous copies of the coupled enzyme. For large process operations, the insoluble matrix could consist of a high-fidelity chromatography support, such as agarose or porous beaded polymers typically used in protein purification and having diameters of tens or hundreds of microns, but more often it is a support material that is much less expensive and abundant for scale up.

Some materials that have been used to immobilize reactors for industrial operations include particulate supports made from polymers (e.g., polystyrene), silica (particularly kieselgur or diatomaceous earth), porous and nonporous glass beads, activated carbon, chitosan or chitin, alginate gel, gelatin, polyurethane, agarose, agar, cellulose beads, polyacrylamide, polyvinyl alcohol, alumina, oxides of various transition metals, graphite, many different membrane compositions, and even sand or soil. For a review that contains a comprehensive list of immobilized bioreactor supports see Durán *et al.* (2002).

Both covalent and noncovalent methods have been used to immobilize enzymes on insoluble supports for use as reactors. Noncovalent methods include entrapment in polymer structures as they are formed, encapsulation of enzyme droplets into polymerizing membranes, and passive adsorption onto particle surfaces using one or a combination of hydrophobic interactions, hydrogen bonding, dative bonding, or charge interactions. Covalent bond formation involves the use of a reactive group that can couple with a functional group on the enzyme, such as an amine or a thiol. The covalent coupling strategies useful for the preparation of immobilized reactors are discussed in Chapter 15, which covers general methods of protein and enzyme coupling to insoluble supports.

An example of an immobilized enzymatic reactor is one that is used in the production of ethanol from cellulosic materials. Although most ethanol is produced from corn or sugar cane, which contains large quantities of simple sugars that can feed directly into the fermentation process, the technology developed over recent decades also permits ethanol manufacture from non-edible cellulosic raw materials. Hydrolysis of cellulose

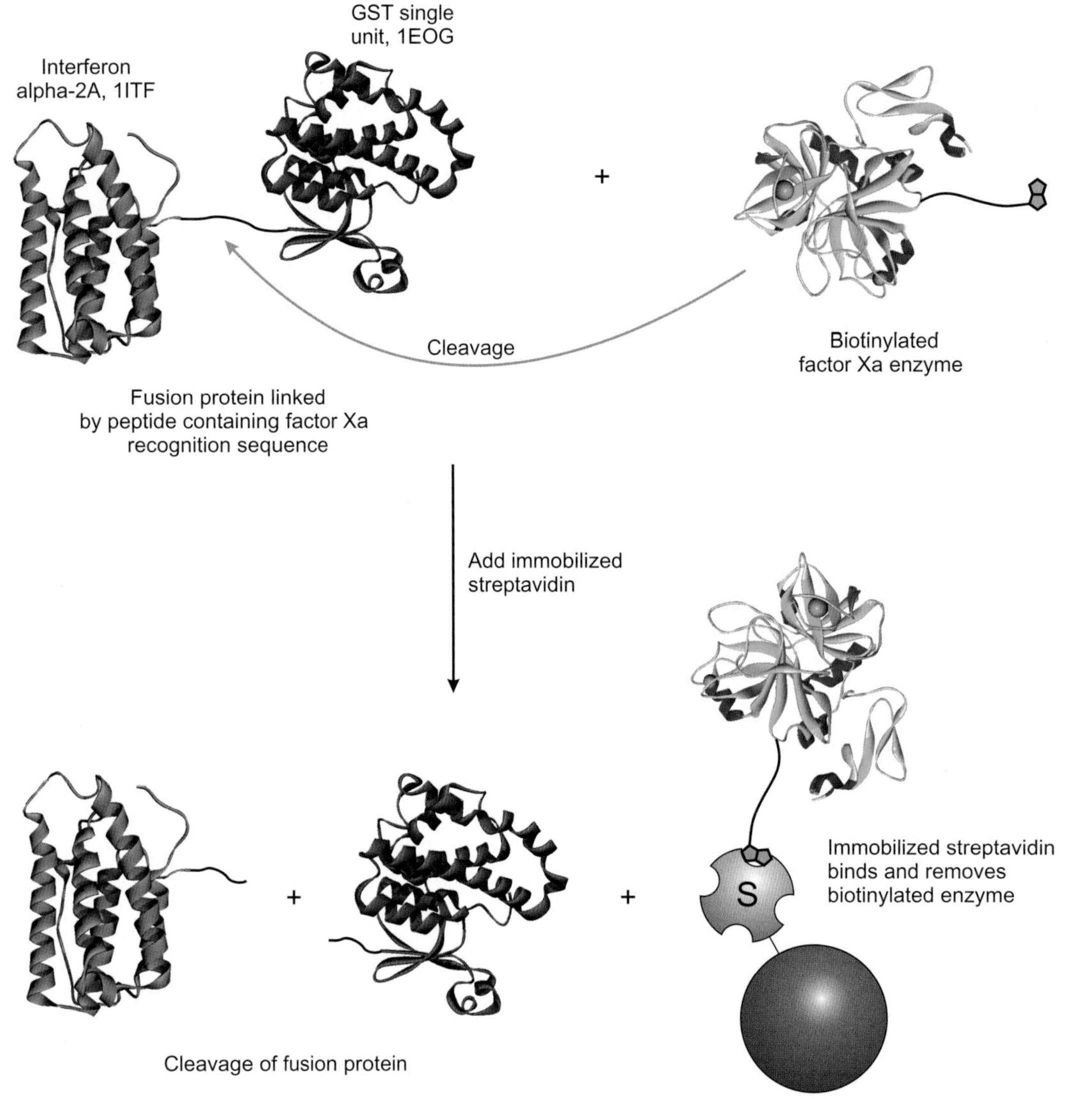

FIGURE 1.85 A biotinylated protease can be used to cleave a peptide sequence between a desired protein and an affinity fusion tag after purification of the complex on an immobilized affinity support. In this example, biotinylated factor Xa is used to cleave the recognition sequence present between interferon alpha-2A and the fusion tag GST after purification on immobilized glutathione. The biotinylated enzyme can then be removed from solution using an immobilized streptavidin support.

or hemicellulose can be carried out using chemical means, but the production can be improved through the use of enzyme reactors. Cellulase and hemicellulase enzymes are often used as alternatives to acid hydrolysis to convert cellulose or hemicellulose to glucose and other sugars in order to enter the fermentation process leading to ethanol (Taherzadeh and Karimi, 2007) (Figure 1.86). Although soluble enzyme can be used successfully in this process, the cost of the enzyme raw material may justify the use of an immobilized reactor for reusability purposes. The main issue in recycling an immobilized cellulase reactor is the difficulty in separating it from the insoluble materials that result from the processing of cellulose and lignin. One solution to this problem involves the use of paramagnetic particles that allow subsequent recovery of the immobilized reactor from the reaction byproducts through the use of a magnetic separation step (Qiu and Li, 2000, 2001; Feng *et al.*, 2006; Liao *et al.*, 2010).

The production of ethanol has also involved the use of immobilized cells as bioreactors, which can be used to convert the sugar raw materials into ethanol in the final step of the process (Anselme and Tedder, 1987; Gulnur *et al.*, 1998; Wai *et al.*, 2008; Menon *et al.*, 2010; Mojović *et al.*, 2010). The use of immobilized yeast cells in bioethanol production allows the saccharification and fermentation processes to go on simultaneously, thus streamlining its manufacture. The use of insoluble

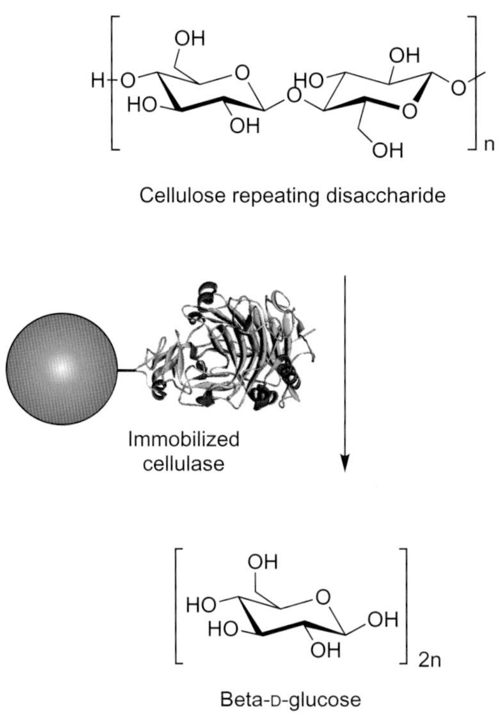

FIGURE 1.86 Cellulase can be used as an immobilized reactor to convert a cellulosic feedstock into beta-D-glucose. This type of reactor is being used in the development of biofuels.

enzymes or cells also eases the removal of these components from the final ethanol product at the completion of the process.

A second type of immobilized bioconjugate reactor involves the use of laccases or tyrosinases (two groups of phenoloxidases) in various process applications, including bioremediation, waste water treatment, wine stabilization, as well as other synthetic or analytical procedures (Durán *et al.*, 2002). These enzymes can be used to catalyze the modification of many phenolic and some non-phenolic aromatic compounds. A phenol oxidation reaction involves the formation of an intermediate aryloxy radical that is subsequently derivatized to a quinone. The quinone can then polymerize through nonenzymatic means (Higuchi, 1989) (Figure 1.87). If one of the sites *para* to the phenolic OH group is occupied, then only a dimer will form from the reaction (Crestini *et al.*, 2010). The products of the polymerization reaction can then be removed from the treated solution by adsorption onto inert polymers (such as Polyclar, which is a commercial polyvinylpolypyrrolidone) or the polymerized products that result can be considered unreactive or nontoxic compared to the initial aromatic substrates (such as in remediation processes). The removal of phenolic compounds from wine can be an important step in reducing a bitter taste, preventing discoloration of white wine, and stabilizing it for long-term shelf life. Immobilized laccases can be used as a reusable reactor in the process of polymerizing the phenolic compounds and removing them to limit the bitterness associated with increased phenol concentration due to the incubation time the must spent in contact with the grape skins (Brenna and Bianchi, 1994; Lante *et al.*, 2000; Minussi *et al.*, 2007).

An immobilized bioreactor conjugate that has found important use in bioprocess operations consists of lipase enzymes coupled to insoluble supports. Lipases are catalysts in the hydrolysis of fatty acid ester bonds, especially those in triglycerides, which can be used to produce oleochemicals in large scale. The preparation of fatty acids from triglycerides can be carried out using traditional chemical processing methods involving high temperatures and pressures. The use of an enzyme catalyst, however, can generate fatty acids at normal temperatures and pressures merely in the presence of excess water for hydrolysis (Malcata *et al.*, 1990; Hermansyah *et al.*, 2007) (Figure 1.88). In addition, under the right conditions, immobilized lipases can be used to effect the reverse reaction and form ester derivatives from fatty acid raw materials. The conversion of fatty acids into their methyl ester or ethyl ester derivatives has been used to produce biodiesel fuel under continuous operations (Shimada *et al.*, 1999; Iso *et al.*, 2001; Noureddini *et al.*, 2005). Transesterification reactions of this type can be performed directly from the triglyceride raw materials in organic solvent with complete retention of enzymatic activity and while avoiding the unwanted production of sulfur oxides and soot common in chemical production processes. Immobilized lipase has also been used to initiate caprolactam polymerization in organic solution (Mei *et al.*, 2003). Novozym 435 is a common commercial lipase (immobilized *Candida antarctica* lipase B) for the preparation of short-chain fats and other esters, including polyesters (Novozymes A/S, Denmark). This immobilized reactor is used extensively in the cosmetics industry and to produce specific fatty acid glycerol esters at relatively low temperatures (60–70°C) compared to standard organic synthesis.

Immobilized enzymes have been used extensively in organic synthesis procedures, especially to effect the creation of complex stereochemical derivatives without resulting in racemic mixtures (Brask, 2009). The use of immobilized enzymes as biocatalytic reactors is growing in popularity due to the realization that many immobilized enzymes can be used in organic environments without denaturing or affecting their catalytic activity. The ability of enzymes to recognize and act on substrates in an exquisitely enantioselective manner creates powerful tools for use in synthesizing drugs for biopharmaceutical applications. Immobilized reactors are currently used in dozens of biocatalytic processes in industry and their use is growing rapidly, especially

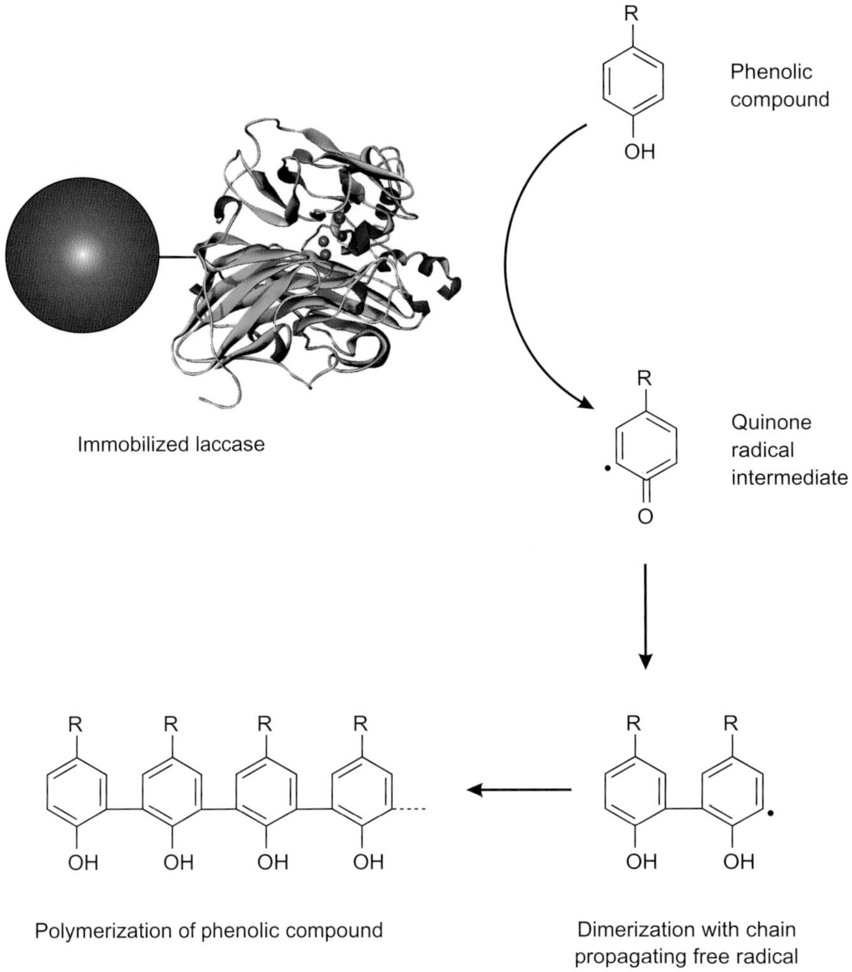

FIGURE 1.87 Immobilized laccase can be used as a reactor to produce a phenolic resin from a phenolic precursor through radical quinone formation, which initiates polymerization.

in situations where classical organic synthesis would prove much more difficult (Wohlgemuth, 2004). Some examples of large-scale biocatalytic processes include the production of high-fructose corn syrup using immobilized glucose isomerase, the transesterification of triglycerides using immobilized lipase, and the production of acrylamide using immobilized nitrile hydratase (Figure 1.89).

In addition to the immobilization of enzyme reactors onto solid phases such as particles, the enzymes themselves have been crosslinked to form insoluble particles without the need to attach them to a secondary support. One common method used for this procedure involves the formation of crosslinked enzyme crystals (CLECs) (St. Clair and Navia, 1992; Govardhan, 1999) or crosslinked enzyme aggregates (CLEAs) (Cao *et al.*, 2000). The method for crosslinking enzyme aggregates is the simplest route, because it does not necessitate the prior formation of micro-crystals. Both CLEC and CLEA formation are typically facilitated by the addition of glutaraldehyde as the crosslinking agent to form the final immobilized reactor. Cao *et al.* (2000) used CLEA formation to prepare insolubilized penicillin G acylase for the synthesis of ampicillin from phenylglycine amide and 6-aminopenicillanic acid (6-APA) (Figure 1.90). The synthesis could occur in a wide range of organic solvents or in aqueous environments with highly selective stereochemical control. The reverse reaction can also be carried out using a CLEA consisting of insolubilized penicillin amidase to form 6-APA in industrial synthesis processes (Brask, 2009).

Immobilized enzyme reactors are also used extensively for analytical determinations. HPLC separations of small molecules can employ post-column detection of substrates using an immobilized enzyme reactor that can act on a substrate and create a fluorescent or chromogenic product, which then can be monitored using standard spectral means. In this application, the enzyme reactor is specific for only one component in the separated sample, and it can provide highly

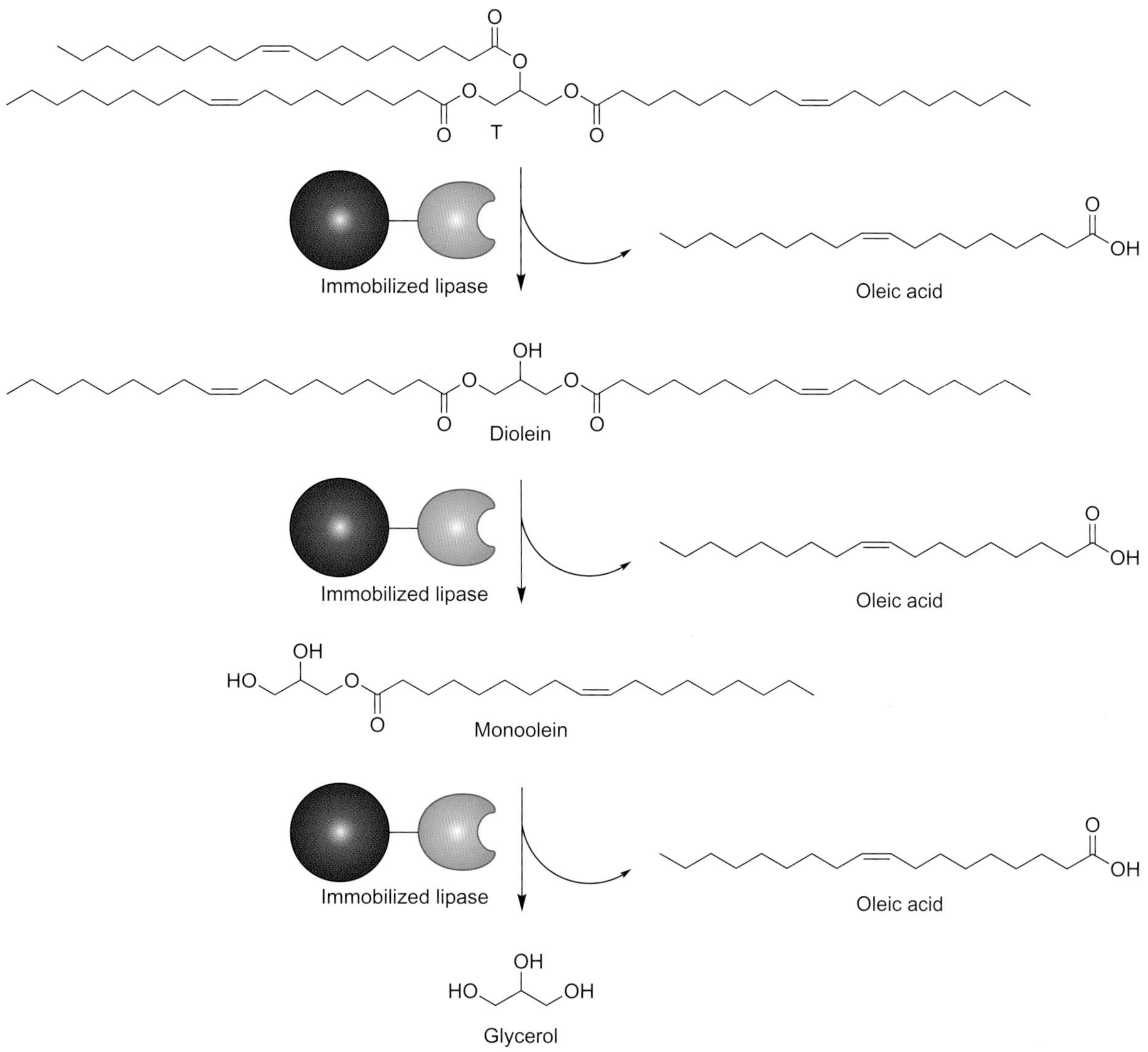

FIGURE 1.88 Immobilized lipase can be used as a reactor in the production of glycerol from a triglyceride feedstock, involving sequences of several reactions to remove the fatty acid acyl chains.

sensitive detection of biological or organic molecules otherwise difficult to analyze (Nagels and Maes, 1995).

In HPLC applications, the enzyme reactor is often immobilized on a silica or controlled pore glass support using an aminopropyl silane derivative (Chapter 13), which is activated with glutaraldehyde for attaching the enzyme through its amines (Girelli *et al.*, 2007). For instance, the use of multiple enzyme reactors in an HPLC system has been successfully developed for the proteomic analysis of phosphopeptides in biological cell extracts (Yamaguchi *et al.*, 2010). An immobilized protease such as trypsin was first used to digest the protein samples passing through the column, and next the peptides produced flowed through an immobilized alkaline phosphatase microreactor to analyze the phosphopeptide fraction. The HPLC enzyme reactor system was fed into a mass spec for final determination of the separated and processed peptides.

HPLC-based microreactors have also been developed for the analysis of a number of small molecules, including sugars or carbohydrates, such as glucose and fructose (Rahman *et al.*, 2008), monosaccharides in cellulose (Olsson *et al.*, 1990), galactose, lactate, ethanol, and L-amino acids (Vojinović *et al.*, 2006), as well as for sulfite analysis in foods (Theisen *et al.*, 2010), and in the determination of drug metabolites (Nicoli *et al.*, 2008). Most of these applications would be difficult to perform if it were not for an immobilized reactor used in line with an HPLC separation column.

Another type of immobilized reactor is a non-enzymatic, single-use support that reacts with a target molecule and transforms it into a desired product. In

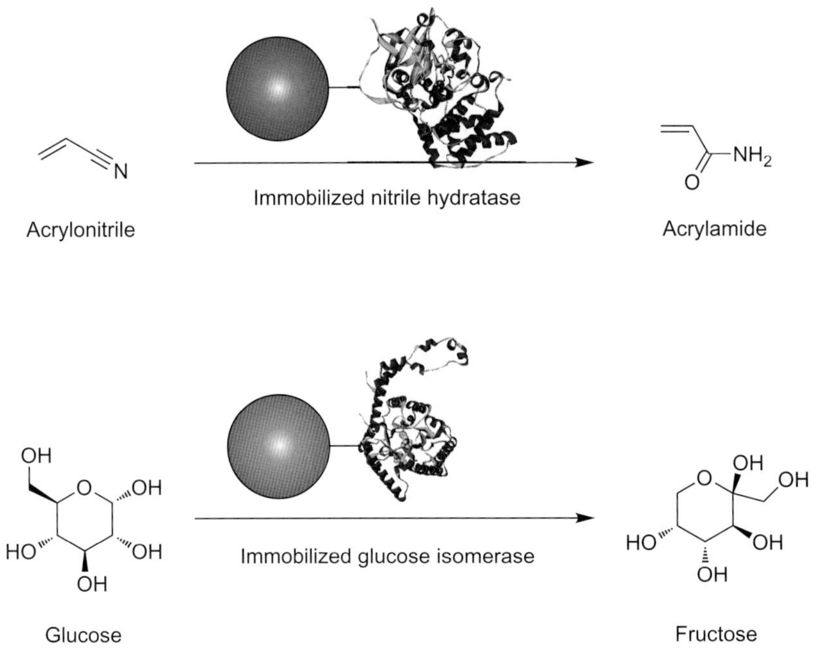

FIGURE 1.89 Two common bioreactor conjugates using enzymes include immobilized nitrile hydratase that can convert acrylonitrile into acrylamide and immobilized glucose isomerase, which can convert glucose into fructose.

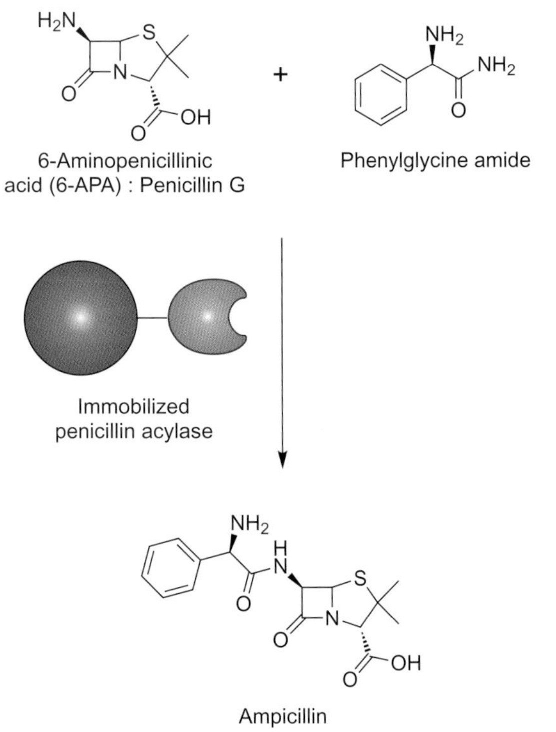

FIGURE 1.90 The antibiotic ampicillin can be synthesized using immobilized penicillin acylase, which acts on the precursors penicillin G and phenylglycine amide to link them together into the final product.

this case, the reactor is not catalytic in nature, but it is capable of spontaneously reacting with a target molecule in solution, and once used it cannot be reused without some sort of regeneration step. An example of this type of support is an immobilized reductant, which can reduce a target molecule to cause disulfide reduction or some other hydrogenation step. Disulfide reducing agents are often used in solution to reduce cysteine disulfides within proteins or peptides (Chapter 2, Section 4.1). Occasionally, it is beneficial to use an immobilized disulfide reducing agent, because it makes it easier to isolate the products of the reaction. For instance, a disulfide-containing short peptide may be reduced with immobilized TCEP (Thermo Fisher): (Chapter 2, Section 4.1) and the thiol-containing peptide is subsequently purified simply by filtering off the solution from the insoluble support. If a soluble disulfide reductant is used, such as DTT, then the purification step would be more difficult, because the molecules may be too close in molecular weight to facilitate simple mass-based separations, such as size exclusion chromatography or dialysis.

Other immobilized reactors have been created to perform certain organic chemical transformations in synthetic procedures, which eliminate the difficulty of subsequent purification steps to purify the product of a reaction. Polymer bound reagents can be designed to react with an organic molecule to produce a certain chemical reaction, while allowing the end product to be separated simply by washing it away from the resin. In

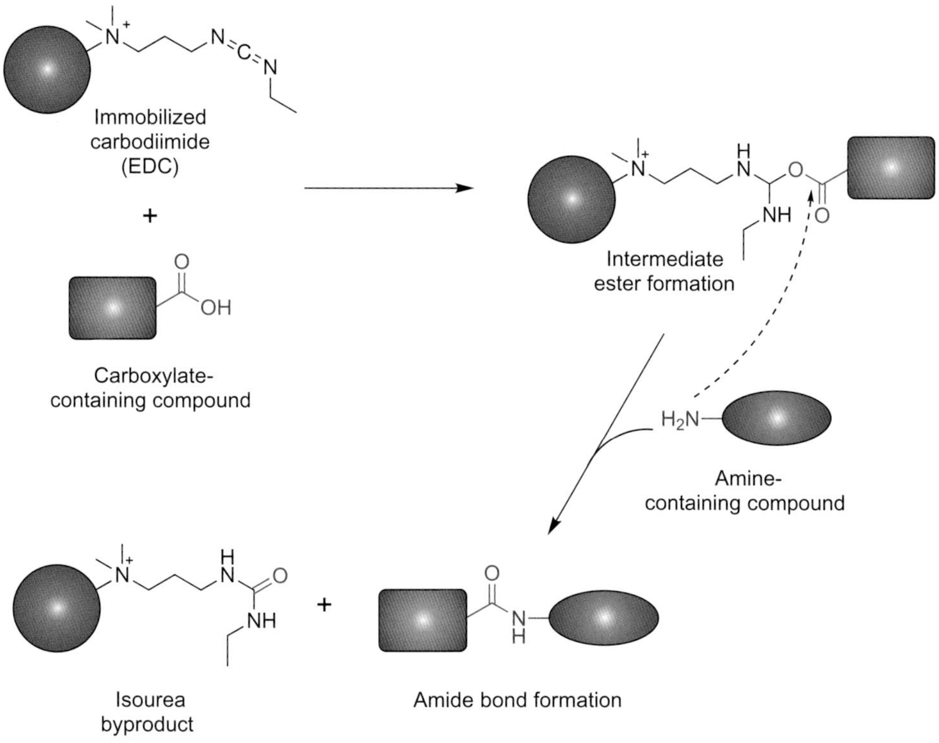

FIGURE 1.91 An immobilized chemical reactant can be used to facilitate a particular synthetic reaction without contaminating the reaction solution with a soluble byproduct. In this case, immobilized EDC is used to form an amide bond between two molecules containing a carboxylate and an amine. Removal of the spent carbodiimide is achieved by simply filtering off the reaction solution from the immobilized reactor.

some circumstances, the immobilized reactor is able to bind to a small molecule to purify it from other contaminants and then release it using a subsequent reaction. This second class of immobilized reactor is also a scavenger, which can be used to isolate a desired molecule or remove a contaminating substance from solution.

Biotage commercializes immobilized reactants for organic synthesis, including solid-phase versions of borohydride and cyanoborohydride for performing reduction of carbonyl compounds, azides, or oximes as well as for reductive amination conjugation of an aldehyde or ketone to hydrazines, hydrazides, and aminoxy groups. The immobilized reductants allow for reactions to be carried out on small molecules without the need for difficult purification schemes to remove the reducing agent. Borohydride or cyanoborohydride coupled to microparticles can simply be removed by centrifugation or filtration, while the product of the reaction is captured in the solution phase.

In addition, amide forming reagents are available in immobilized form to couple a carboxylate-containing molecule to an amine-containing molecule (Biotage). An immobilized carbodiimide creates a convenient reagent for facilitating the conjugation of two small molecules together to form an amide linkage. If an immobilized form of this reagent is not used for small molecule conjugation, a purification strategy would have to be developed to remove the byproducts of the reaction from the final conjugate. With an immobilized carbodiimide, simple removal and washing of the support facilitate removal of the carbodiimide and the isourea byproducts from the desired conjugate (Figure 1.91). The reaction typically is performed in organic solvent, such as dichloromethane and DMF. The reagent may also be used to create reactive esters from carboxylate-containing molecules. In this regard, NHS esters, PFP esters, or HOBt esters (Chapter 15) may be made from a carboxylate-containing compound to activate it to a reactive intermediate that may then be used to conjugate to amine-containing molecules in aqueous solution to form amide bonds.

Immobilized reactors in the form of enzyme catalysts or small reactive molecules can be used to effect many chemical transformations, including ones that would be extremely difficult from a stereochemical perspective if they were done using classical organic synthesis. These immobilized conjugate reagents are currently being used in diverse applications spanning the scale from small lab bench reactions to large industrial production processes. The application of such reagents is only now beginning to become routine across many disciplines and no doubt their use will continue to grow as their advantages become more apparent.

3.5. Therapeutics and *in vivo* Diagnostics

Among the largest and most important application areas for bioconjugate techniques is the field of human therapeutics and diagnostics. Bioconjugation reagents developed for these applications have contributed significantly to the growth of the diagnostic, biotech, and pharmaceutical industries over the last few decades. The advent of targeted therapeutics has almost entirely been made possible by the ability to create designer bioconjugates that can specifically target a disease while limiting its effects on normal tissue. The ideal therapeutic bioconjugate might consist of a targeting component coupled to a drug or an effector molecule, which can deliver an active payload directly to tumor cells, attack invading pathogens, or target other sites within the body for therapy with little to no binding or toxicity toward untargeted cells and organs.

Targeted bioconjugates have also been used for *in vivo* diagnostic procedures to detect and image the sites of tumors and for *in vitro* assays to quantify disease markers or drug concentrations within patients' serum. The specificity and sensitivity of many diagnostic procedures depend on developing the optimal bioconjugate that can interact with specific targets and provide sensitive detection capability.

In addition, the use of designer bioconjugate drugs has had considerable advantages over traditional pharmaceuticals beyond just targeted delivery. The right bioconjugate design can provide increased chemical stability, protection from proteolysis or degradation *in vivo*, increased half-life of the drug within the body, a decrease in off-target effects, a reduction in undesired immunogenicity, increased water solubility, decreased removal by the kidneys or liver, increased specificity to interact with the desired target, and an increase in potential cellular penetration by endocytosis (Monfardini and Veronese, 1998; Veronese and Morpugo, 1999).

Applying the right bioconjugation methods can also result in enhanced properties of more traditional organic pharmaceutical drugs. Small drug compounds have been used for many decades in the treatment of diseases. The preparation of conjugate derivatives of these drugs to create second-generation pharmaceuticals with novel properties is a field that is being explored extensively (Kojima *et al.*, 2000; Kim *et al.*, 2008). For instance, a sparingly soluble organic compound may be rendered much more hydrophilic through selective covalent modification with a water soluble polymer, such as polyethylene glycol (PEG) (Chapter 18). Conjugates of this type can also increase biodistribution and drug availability *in vivo*, while increasing its overall half-life in blood. In addition, dendrimers and other polymers have been favorably investigated for their ability to carry multiple copies of organic drugs to targets *in vivo* better than the administration of the drug without the conjugate carrier (Chapter 8 and Chapter 18, Section 2). Thus, even traditional pharmaceuticals have realized benefits from the design of novel bioconjugate derivatives.

In the following sections some of the major designs and application areas will be discussed for the use of pharmaceutical and diagnostic bioconjugates. These conjugates include compositions made up of antibodies, enzymes, protein toxins, peptides, small molecule organics, radiolabels, metal chelates, inorganic nanoparticles, polymers, and bioaffinity ligands. Bioconjugation for the development of advanced therapeutics and diagnostics is entering into an exciting age of new development, which will no doubt affect the way drugs are designed and disease is detected from this point forward.

Bioconjugates for Cancer Therapy

The strategy in creating therapeutic bioconjugates for cancer is to design a final complex that has high specificity for the intended cells combined with high efficacy in killing the tumor being targeted, but with no off-target effects or toxicity. At first, this goal may seem easy to attain—just conjugate an affinity targeting agent with an anticancer drug component and the result should be an effective biotherapeutic. Unfortunately, even the most logical ideas for constructing anticancer bioconjugates typically run into problems far more complex than such a simple design concept would suggest. Bioconjugates can be rendered ineffective due to unexpected immune system responses including severe allergic reactions, a limited half-life *in vivo* that decreases the availability of the drug, lack of penetration into solid tumors, changes in the antigenic presentation on cancer cells rendering the targeting agent less capable of binding, sequestration of the drug in the liver or other organs, rapid clearance through the kidneys, nonspecific binding to off-target sites, unexpected toxicity or side effects, and a general lack of effectiveness in killing the tumor cells.

Such problems are the reason that the design and development of therapeutic bioconjugates are typically a billion-dollar investment taking seven to ten years to complete. To avoid these negative issues, a bioconjugate destined for use in humans must be developed with precise control over every conceivable aspect of its preparation. For instance, it has been found that relatively minor alterations in the ratio of the molecules making up a final conjugate can result in dramatic changes in specificity and effectiveness. Different bioconjugation methods can also have tremendous effects on the performance of a prospective drug. In some cases, it has been shown that relatively small changes to the cross-bridge formed between a targeting agent and the drug component in the conjugate can have significant effects *in vivo* (Doronina *et al.*, 2006). In other instances, a polymer drug carrier of incorrect type or size can result in rapid kidney ultrafiltration and clearance or undesired polymer accumulation within tissues resulting in toxicity (Seymour *et al.*, 1990; Duncan *et al.*, 2001; Thanou and Duncan, 2003; Jones, 2004).

The challenging aspect of developing anticancer bioconjugates has led to the study of many different conjugate types for their effectiveness in treating the disease. In general, though, the concept of combining a targeting agent with a therapeutic component remains a common theme throughout these bioconjugate designs.

Antibody Targeting for Biotherapeutics

Therapeutic conjugates for cancer often have used antibodies or antibody fragments as the targeting component. The binding specificity of an antibody for an antigen can be used to advantage to target only epitopes on tumor cells while limiting nonspecific interactions with normal cells or tissues. Since the early days of monoclonal antibody technology, the dream of targeted drug therapy has been met with the harsh reality of how difficult it is to effectively utilize even a highly specific antibody *in vivo*. For instance, typical mouse monoclonals can be made to virtually any human tumor epitope, but their effectiveness for human use can be very limited. This is due to the fact that most people quickly develop an antibody response to the monoclonal, which renders the continued use of the antibody futile. To overcome the potential immune response to a mouse monoclonal antibody, particularly the production of what is called human anti-murine antibodies (termed the HAMA response), humanized or chimeric human/mouse monoclonal antibodies and fragments have been developed (Shawler *et al.*, 1985). Such recombinant antibodies can be conjugated to an anti-cancer drug or toxin, bind specifically to a tumor cell, and deliver their chemotherapeutic package directly to the cells at high concentration, thus maximizing the effectiveness of the drug. In addition, it has been shown that chimeric antibodies, or those that contain part human and part mouse sequences, also help to limit the immune response and aid in getting the targeted payload to its intended destination. However, the very best monoclonal antibodies for cancer are those that are completely human in nature, even with respect to their hypervariable reagions. The advent of transgenic mice that have fully human antibody genes has facilitated the development of true human antibodies with high specificity and low immunogenicity *in vivo* (Jakobovits, 1995; Kucherlapati *et al.*, 2012). This development has revolutionalized the use of monoclonal antibodies as targeting agents for the treatment of human disease.

By the end of the first decade of this century, the therapeutic monoclonal antibody market had total worldwide revenues exceeding $35 billion (USD) (from *Therapeutic Monoclonal Antibodies: World Market 2010–2025*, Visiongain, London, 2010). At least 25 antibody-based therapeutic drugs are on the market today, that are approved for use in various regions of the world, with hundreds more in various stages of clinical trials (Carter, 2006; El Bakri *et al.*, 2010). Such biologic drugs as Avastin, Cimzia. Erbitux, Herceptin, Humira, Lucentis, MabThera, Remicade, Synagis, Tysabri, and Xolair employ monoclonal antibodies to successfully treat diseases. Most of the approved antibodies on the market today are not conjugated; they are known as "naked" antibodies and are used to block receptors or neutralize molecules such as epidermal growth factor (EGF), which can prevent activation of its receptor (EGFR) on certain tumor cells (e.g., metastatic colorectal cancer).

The bioconjugate antibody-based therapeutics that have been approved for commercialization include a PEGylated, humanized Fab′ fragment to treat Crohn's disease (Cimzia, UCB Pharma), an anti-CD20 mouse monoclonal that is radiolabeled with I-131 for treatment of non-Hodgkin's lymphoma (GlaxoSmithKline), a mouse IgG1 monoclonal radiolabeled with Y-90 to treat low grade follicular transformed non-Hodgkin's lymphoma (IDEC Pharmaceuticals), a radiolabeled (I-131) chimeric IgG1 monoclonal for the treatment of lung cancer (Shanghai Medipharm Biotech), and a humanized IgG4 against CD33 as an immunotoxin conjugate to treat acute myeloid leukemia (AML) (Celltech/Wyeth).

Cimzia (certolizumab pegol) is a bioconjugate consisting of a humanized monoclonal Fab′ fragment targeting human tumor necrosis factor alpha (TNFα), which is modified with an approximately 40-kDa polyethylene glycol molecule (Connock *et al.*, 2010). The conjugate is formed through the use of a maleimido-PEG derivative that is coupled to a single thiol group on the Fab′ fragment (Figure 1.92). This forms a highly defined complex of approximately 91 kDa, wherein the PEG group functions to increase the hydrodynamic volume of the biologic and maintain the drug in circulation for a longer period *in vivo*. The antibody is able to bind to TNFα and neutralize its activity, which is involved in inflammatory processes. The drug treatment is indicated for the treatment of Crohn's disease and severely active rheumatoid arthritis (Rosa *et al.*, 2010).

Antibody–drug conjugates (ADCs) have also been developed to combine the targeting advantage of a specific antibody with the known anti-tumor chemotherapy of a small-molecule drug (Chari, 2008). These complexes are able to deliver a high dose of the chemotherapeutic agent directly to the site of the cancer cells. The first ADC approved by the U.S. Food and Drug Administration (FDA) was gemtuzumab ozogamicin, which consists of a complex of a monoclonal antibody against CD33 and the drug calicheamicin (Haeuw *et al.*, 2009). The anti-CD33–immunotoxin conjugate was the first approved biologic for antibody-targeted

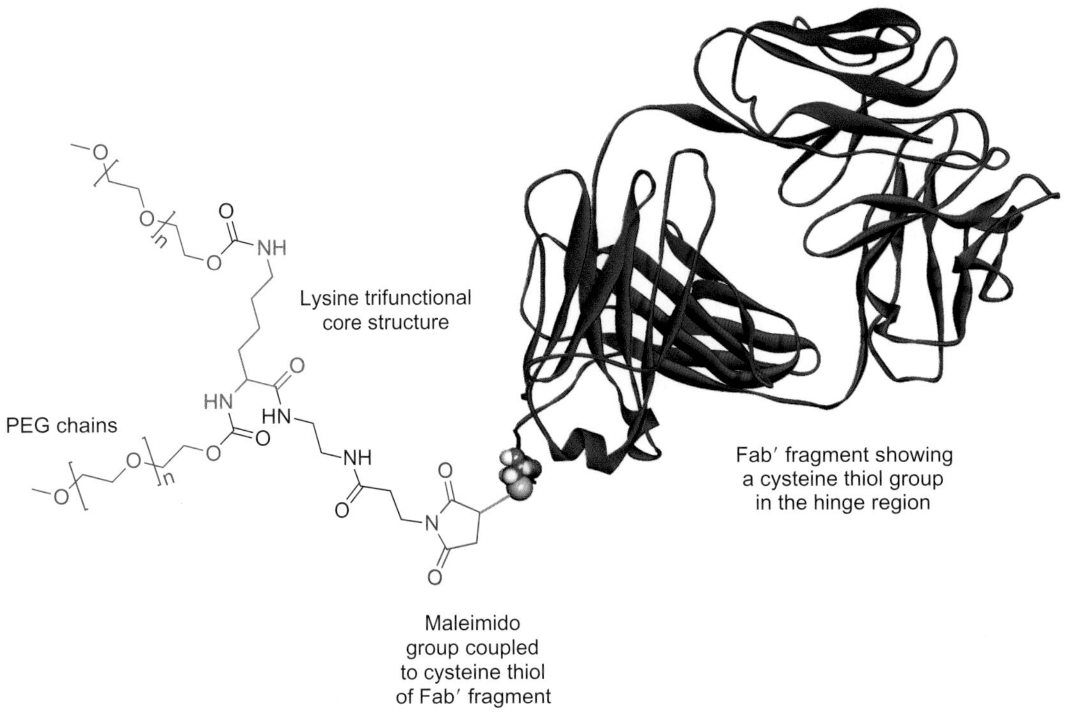

FIGURE 1.92 The structure of the bioconjugate drug Cimzia, which contains a humanized Fab′ fragment directed against TNFα and has been modified with two PEG chains by using a maleimide reactive group to couple with the available thiol in the hinge region.

chemotherapy in cancer, coming to market in the year 2000. The antibody is conjugated with a potent antibiotic that was originally isolated from bacteria in certain clay found in parts of Texas. The anti-tumor agent, calicheamicin, is linked to the humanized monoclonal antibody and commercialized under the name Mylotarg (Figure 1.93). The bioconjugate linkage chemistry proved to be very important for efficacy of this drug. Calicheamicin is a complex di-alkyne-containing molecule that is able to bind to the minor groove of DNA and subsequently cause widespread cleavage of genetic material, which then leads to tumor cell apoptosis. The key to getting the drug payload active at binding DNA was to ensure the release of the calicheamicin after conjugate entry into the cell. Surprisingly, the linkage of calicheamicin to the monoclonal antibody via a disulfide bond actually proved less effective at releasing the antibiotic within the cells than linking through a hydrazone bond using aldehyde and hydrazide reactive groups (Ducry and Stump, 2010). This is an important example of how a thorough exploration of bioconjugation designs can lead to a therapeutic conjugate with the highest possible activity.

Unfortunately, this first ADC to be approved eventually was removed from the market due to its lack of effectiveness. The conjugate proved to be highly heterogeneous, consisting of a broad distribution of calicheamicin substitution levels on the antibody (from 0–8 drugs per antibody) with up to 50% of the antibody present in unconjugated form (Bross *et al.*, 2001). Since the launch and removal of Mylotarg from the market over a dozen other ADCs are in clinical development. Table 1.3 shows the list of current clinical trials involving ADCs in the United States. The future for ADC therapeutics may be much brighter than the initial failure of the first marketed drug would suggest.

The cytotoxic drug component of ADCs can be of many types. Some examples described in the literature include doxorubicin, methotrexate, calicheamicin, two maytansinoids (DM1 and DM4), duocarmycin, and two auristatins (MMAE and MMAF) (McCarron *et al.*, 2005; Beck, 2010). In addition, the methods used to conjugate a drug to an antibody involve many different strategies to build linker arms and covalent bonds that in combination may vitally contribute to the effectiveness of the final complex *in vivo*. Often a cleavable bond and a hydrophilic bridge between the antibody and the drug are present in the final conjugate, which lends flexibility and water solubility to an otherwise rather hydrophobic chemotherapeutic agent. Figure 1.94 illustrates some of these ADC structures along with the exact linker arms and covalent bonds used to couple with the antibody molecules.

The use of antibody bioconjugates has only just begun to realize its true potential for human therapeutics. It is expected that the continued development

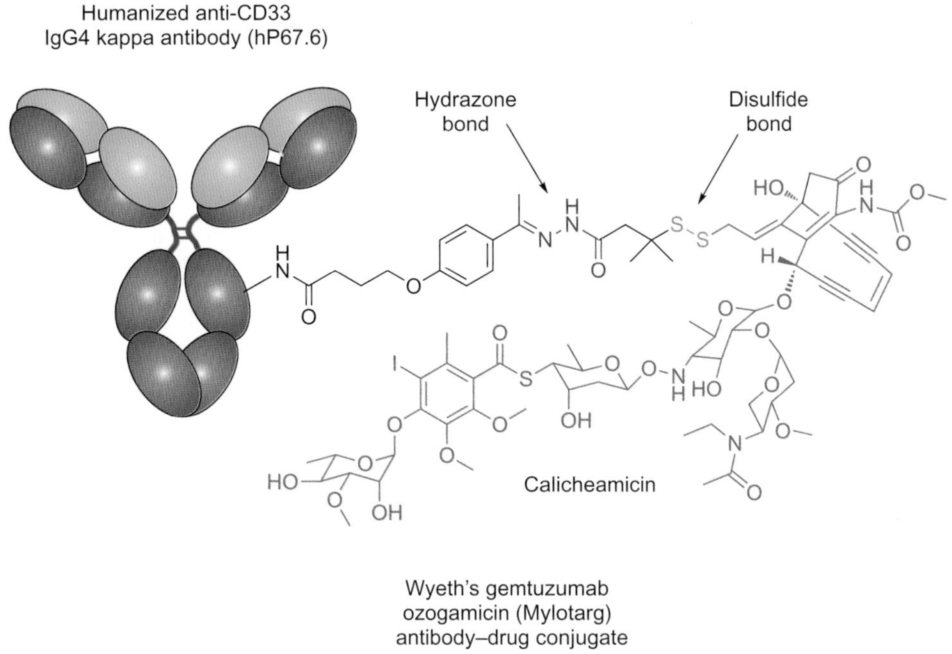

FIGURE 1.93 The antibody–drug conjugate (ADC) gemtuzumab consists of a humanized monoclonal antibody that has the anti-cancer drug calicheamicin attached to it using a linker arm that contains both a hydrazone bond and a disulfide group.

TABLE 1.3 **Antibody–Drug Conjugates for Cancer**

Name; Antibody Type	Company	Drug	Linker Type	Antibody Target	Disease Indication
Gemtuzumab ozogamicin; IgG4	Wyeth & Pfizer	Calicheamicin	Hydrazone and disulfide	CD33	Acute myeloid leukemia
Inotuzumab ozogamicin; IgG4	Wyeth & Pfizer	Calicheamicin	Hydrazone and disulfide	CD22	Non-Hodgkin lymphoma
SAR 3419	Sanofi-Aventis	DM4	Disulfide	CD19	Non-Hodgkin lymphoma
BT-062; IgG4	Biotest	DM4	Disulfide	CD138	Myeloma
IMGN-388	Centocor	DM4	Thioether	Alpha-V integrin	Solid tumors
BIIB015; IgG1	Biogen Idec	DM4	Unknown	Cripto	Breast cancer
Trastuzumab emtansine; IgG1	Genentech	DM1	Thioether	HER2	Breast cancer
Lorvotuzumab mertansine; IgG1	ImmunoGen	DM1	Thioether	CD56	Myeloma
Brentuximab vedotin; IgG1	Seattle Genetics	vcMMAE	Valine–citruline	CD30	Hodgkin lymphoma
Glembatumumab vedotin; IgG2	Celldex	vcMMAE	Valine–citruline	GPNMB	Breast cancer, melanoma
PSMA ADC	Progenics	vcMMAE	Valine–citruline	PSMA	Prostate cancer
ASG-5ME	Agensys	vcMMAE	Valine–citruline	SLC44A4	Pancreatic cancer
SGN-75	Seattle Genetics	mcMMAE	Maleimido–caproic acid	CD70	Non-Hodgkin lymphoma, renal cell carcinoma
MEDI-547; IgG1	MedImmune	mcMMAE	Maleimido–caproic acid	EphA2	Solid tumors
MN	BMS (Medarex)	Auristatin	Unknown	MN	Cancer
MDX-1203	BMS (Medarex)	Duocarmycin	Dipeptide linker	CD70	Non-Hodgkin lymphoma, renal cell carcinoma

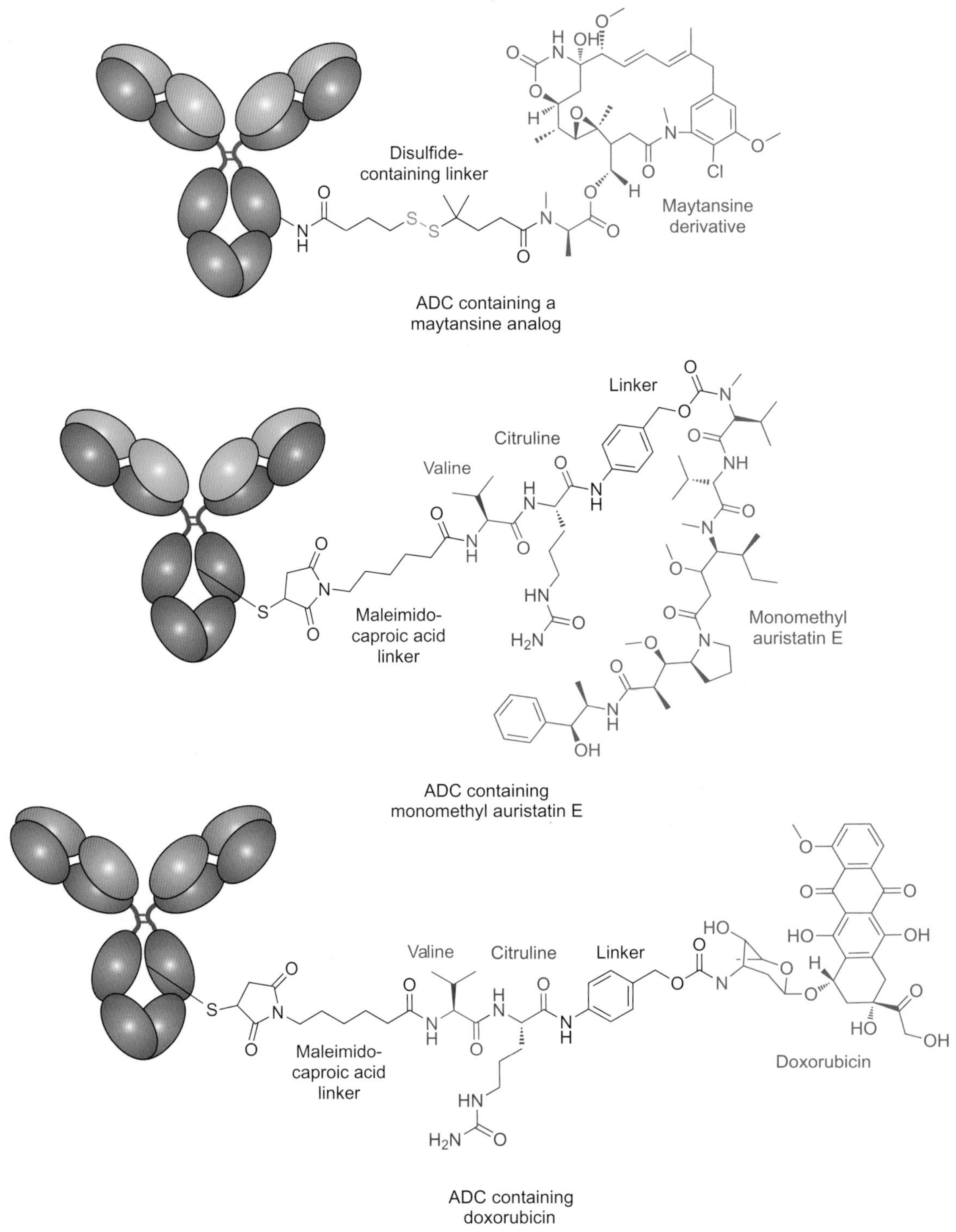

FIGURE 1.94 Examples of ADCs constructed with intact humanized monoclonal antibodies using the drugs maytansine, monomethyl auristatins E, and doxorubicin. Note the construction of the linker arms that may incorporate amino acid derivatives or cleavable groups such as a disulfide bond. Two of the ADC conjugates are linked through thiols on the antibody molecules, which were formed by mild and controlled reduction of disulfides within the IgG structure, while the other one is linked through amine groups on the antibody surface.

of antibodies and antibody conjugates for cancer and other indications will rapidly expand the biologic drug market in the first part of this century. With many hundreds of such bioconjugates currently in research and clinical trials worldwide, it is certain that the number of antibody conjugates on the market will dramatically increase over the next 20 years. We are truly entering the age of targeted therapeutics made possible through rational design using bioconjugate techniques.

Research into antibody bioconjugates for therapeutic purposes has included the use of every conceivable antigen binding construct from intact IgG-type antibodies to many different types of fragments, including $F(ab')_2$, Fab, and single-chain Fv (scFv) (Chapter 20). More elaborate antibody constructs have included the formation of recombinant multivalent hybrid designs having multiple specificities, such as bispecific (diabodies), trispecific (triabodies), tetraspecific (tetrabodies), and even small minibodies, which are constructs containing two of the smallest possible antigen binding fragments (Morrison, 2007; Sharkey *et al.*, 2010). Some of these complexes are formed entirely through recombinant fusion techniques, but others involve the conjugation of fragments having different specificities to form the final multivalent molecule. For instance, trimeric Fab conjugates have been developed using chemical crosslinking of three different Fab fragments onto a central trifunctional core structure to form a trispecific antibody (Hudson and Souriau, 2003).

One unique trispecific cancer therapeutic was made from an IgG fusion of one heavy-/light-chain pair originating from one antibody fused to another heavy-/light-chain pair from another antibody. The result was a fused antibody conjugate with dual antigen binding ability with specificity toward a tumor cell epitope on one Fab arm and a second specificity toward CD3 surface protein on T cells on its other Fab arm, and its Fc region was able to bind to Fc receptors on macrophages, NK cells, or dendritic cells. The purpose of this elaborate trispecific construct was to bring the most important cells of the immune system together at the site of the tumor cell, thus initiating destruction of the cancer cell in the process (Figure 1.95).

Other bispecific constructs have been made by fusing two scFv fragments together that contain two different target antigen specificities. For instance, one scFv end can be designed to bind to a tumor cell epitope, while another scFv end can bind to T-cell surface antigens. One of the most advanced forms of this type of therapeutic agent is called a bispecific T-cell engager (BiTE) (Baeuerle and Reinhardt, 2009). At least two BiTEs are in clinical trials at the time of writing: (1) blinatumomab, which is under development for non-Hodgkin lymphoma as well as B-precursor acute lymphoblastic leukemia (Bargou *et al.*, 2008; Topp *et al.*, 2008); and (2) MT110, which is being developed for lung cancer, squamous cell carcinoma, and gastrointestinal cancer (Maetzel *et al.*, 2009).

In addition, affinity targeting peptides based upon antibody–antigen binding regions within immunoglobulins or even using peptide sequences not originating from antibodies have been designed for therapeutic bioconjugate use. Short peptide sequences that have affinity binding specificity for a target molecule of interest, and which are of low molecular weight, can provide alternatives to traditional antibodies. Small targeting molecules have advantages in that they can provide excellent cell and tissue penetration *in vivo* and are relatively simple to synthesize. Such peptides can be identified through knowledge of native binding site sequences on larger proteins or they can be chosen by screening libraries of potential peptide-binding molecules (Kolonin *et al.*, 2006; Eichler, 2008; Jüse *et al.*, 2010).

Non-antibody-based scaffolds for developing recombinant affinity targeting molecules have also been developed for bioconjugate use (Plückthun, 2009). An alternative scaffold may consist of a protein or protein fragment that can be used as a structural foundation to create a binding region, which can then be recombinantly randomized at particular amino acid sequences to make unique affinity libraries. Such libraries can be screened for their ability to specifically interact with targets, such as biomarkers on tumor cells. Positive hits can be further investigated for targeting suitability and subsequently produced on a large scale by recombinant means, if desired.

Proteins or peptides that have been used as antibody-like scaffolds to develop binding molecules have included the Z-domain of protein A, the Kunitz domain inhibitors (60-amino-acid, single-chain peptide), human pancreatic secretory trypsin inhibitor (55 amino acids), Alzheimer's amyloid β-protein precursor inhibitor, the ecotin bacterial serine protease inhibitor sequence (142 amino acids), the knottin protease inhibitor including the cellulose-binding domain (CBD) peptide sequence (25–35 amino acids), fibronectin III (94 amino acids), and thioredoxin (108 amino acids) (Nygrena and Skerra, 2004). Within these peptide sequences a single loop or multiple neighboring loops are randomly evolved to create libraries of potential affinity binding surfaces that can function in a manner similar to the antigen binding sites on antibodies. Many of these alternative binding scaffolds offer advantages over traditional IgG-type antibodies in that they are mainly single peptide sequences, they have more stable and compact β-sheet or α-helical structures, and they are typically smaller than full-sized immunoglobulins, which has the added advantage of aiding tissue and tumor penetration *in vivo*.

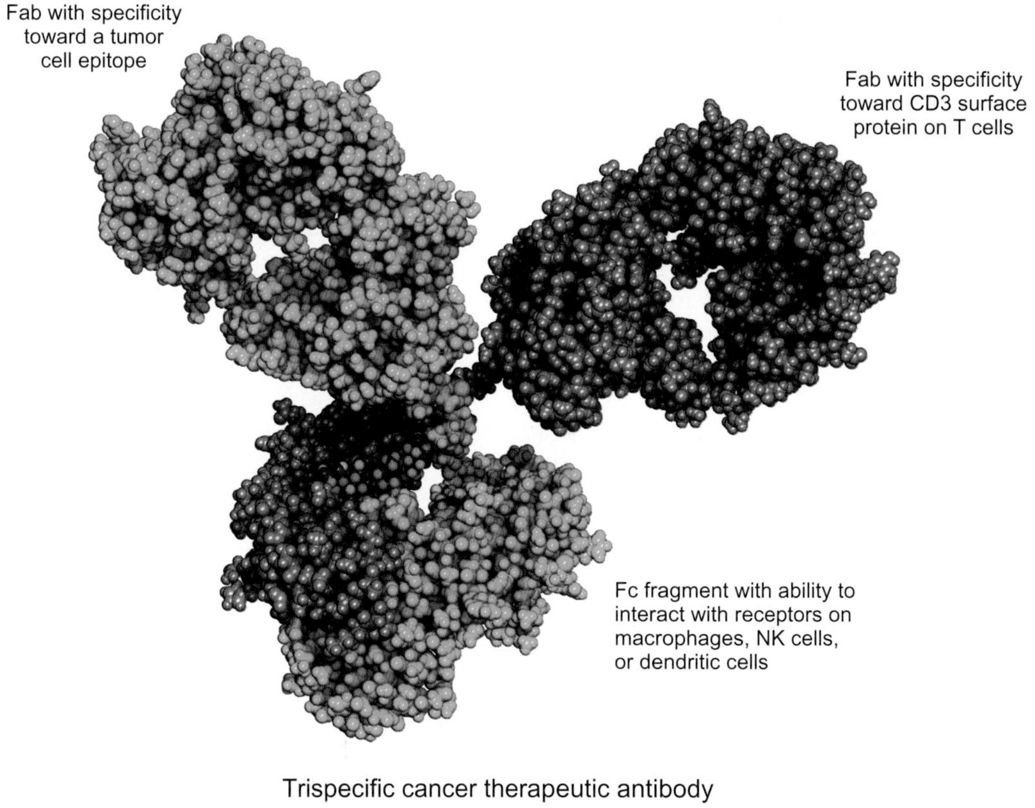

FIGURE 1.95 The design of a trispecific antibody involves the use of a heavy-/light-chain pair from one antibody combined with a heavy-/light-chain pair from another antibody (colored differently in this illustration). The specificities of the two heavy-/light-chain pairs are different to target a tumor cell epitope on one side and markers on T-cell surfaces on the other side. The third interaction potential for this construct occurs through the combined Fc region, which has the ability to interact with receptors on key immune cells. The effect of this conjugate is to bring the important components of the immune system together at the site of cancer cells to cause tumor cell death.

One commercial scaffold construct that has been used to develop diagnostic or therapeutic bioconjugate designs is the Affibody molecule consisting of 2 alpha-helices from the Z-domain of protein A (Tolmachev *et al.*, 2007; Zielinski *et al.*, 2009). This small 6-kDa protein is highly robust, as it resists high temperatures and extremes of pH. Libraries have been produced by randomization of a 13-amino-acid loop within the molecule and screening can be done against over 10 billion Affibody molecules to find specificity toward a desired target.

In addition to domain-based protein scaffolds, short peptide sequences have also been used to develop affinity binding molecules for bioconjugate targeting. One of the first peptide affinity agents to be created was a simple set of linear amino acid sequences that was engineered onto proteins normally present on bacteriophage particles—a technique called "phage display"—which could then be randomized to form a library of potential targeting peptides. First developed by Smith (1985), the method provides a powerful way of creating custom binding peptides with affinity toward desired targets. Screening of this combinatorial peptide library for its potential to bind an antigenic target can yield viable affinity sequences that can be used to conjugate with therapeutic drugs for targeted drug delivery (McCafferty *et al.*, 1990). Libraries of this type can easily be created with selected modification of the minor (pIII gene) or major (pVIII gene) coat proteins located on the ends of phage particles. Using this technique, affinity binding libraries can be made that contain anywhere from 10^9 to 10^{12} potential peptide binders, depending on the length of the peptide sequences (Bratkovic *et al.*, 2005; Lunder *et al.*, 2005). The disadvantage of this type of small peptide targeting molecule is the lack of a constrained structure, as would be present if a loop or multiple loops were used in a larger scaffold. Freedom of motion of small linear peptides in solution often changes the binding characteristics toward targets, the effect of which often depends on the medium in which the peptide is present. For this reason, linear peptide libraries typically are not used to create bioconjugate therapeutics to target epitopes *in vivo*.

Other non-antibody-based bioconjugates have also been designed to carry drugs to tumor cells *in vivo*. In some cases these targeting agents are biological molecules that are preferentially taken up by tumor cells rather than normal cells. For instance, transferrin has been used to effectively target tumors through the cells' inherent property of ingesting high quantities of this protein due to the high growth and division rate of tumor cells (Zheng *et al.*, 2010). In addition, small ligands such as folic acid have frequently been used as targeting agents, because of the dramatic increase in folate receptors on tumor cell surfaces. A folate-containing bioconjugate can be captured by tumor cells much more effectively than by normal cells, thus providing a mechanism for directed delivery of the drug to the desired site without using an antibody (Prabaharan *et al.*, 2009).

Polymeric Scaffolds and Nanoparticles for Biotherapeutic Conjugates

Antibody multimers built on a polymer scaffold have also been made to produce greater binding avidity toward certain antigens and to carry additional chemotherapeutic agents beyond that possible using a single antibody conjugate. For instance, dextran conjugates or liposomal conjugates with antibodies that incorporate multiple copies of an anti-tumor drug or toxin have been developed to treat specific tumor types (Chapters 18 and 20). Many of these polymeric constructs are particulate in nature and have diameters in the low nanometer range, which makes them eminently suitable for *in vivo* delivery. In many instances, the chemotherapeutic properties of traditional drugs administered in a free form can be dramatically improved when incorporated into a polymeric drug delivery system (PDDS) (Kwon, 2005; Uchegbu and Schatzlein, 2006).

A wide variety of polymer carriers have been designed for use as drug transporters, and the type of polymer can be of natural or synthetic origin and self-assembled or synthetically crosslinked (for reviews, see Cho *et al.*, 2008; Duncan, 2003). PDDS constructs have generally fallen into four different structural categories: (1) linear; (2) nanoparticle; (3) branched; and (4) dendritic. Some of them have been used to carry only a chemotherapeutic agent without an associated antibody or other targeting molecule, while others have incorporated affinity binding agents to gain specificity for the particular cell or tissue type being targeted. The chemical structures of some of the major polymers used in the design of drug bioconjugates are shown in Figure 1.96. These polymers have been used alone or in combination with other monomers to form copolymer constructs having a variety of functional groups for conjugation or imparting water solubility or biocompatibility. In some cases, the purpose of the polymer modification is to alter the properties of the attached drug *in vivo* to make it more soluble or extend its half-life in circulation. In other cases, it is to link multiple copies of the drug to one bioconjugate and thus gain additional therapeutic efficacy at the targeted tumor or tissue *in vivo*.

Nanoparticle polymer delivery systems are usually constructed through crosslinking of linear or smaller polymers or alternatively through the self-assembly of amphipathic polymers into micellular structures. Polymer encapsulation into nanoparticles has been used to deliver large numbers of anti-tumor drugs at concentration levels far beyond that normally feasible using drug delivery without a particle transporter. Natural polymer particle constructs can be formed from proteins, inactive viral particles, or even crosslinked carbohydrates. For instance, albumin nanoparticles have been used to carry taxol (Gradishar *et al.*, 2005), a heat-shock protein caged architecture has been used to carry doxorubicin (Flenniken *et al.*, 2006), micellular structures have been used to carry daunorubicin (Rosenthal *et al.*, 2002), and virus nanoparticles were constructed to carry doxorubicin (Manchester and Singh, 2006).

Other polymer constructs for drug delivery have incorporated poly-L-glutamate and poly-L-aspartate, the copolymer HPMA (*N*-(2-hydroxypropyl)-methacrylamide), PLA (poly-L-lactide), dextran and cyclodextrin, chitosan, heparin, PEG (polyethylene glycol), and various advanced dendrimer designs (Khandare and Minko, 2006). In these cases, a targeting agent and the therapeutic drug are conjugated to the polymer through functional groups normally present on the polymer or through short spacer arms extending off from the polymer core, which link the active components to the base structure (Figure 1.97). Regardless of its structure and morphology, the multivalent nature of a polymer backbone creates abundant functional groups for designing selective properties into the final conjugate. Antibodies or other targeting molecules can thus be attached covalently to the PDDS along with therapeutic agents and even detection components to make a multi-functional conjugate for *in vivo* use. Such polymer conjugates can discretely target particular cell types through antibody–antigen interactions or other affinity binding events, such as ligand–receptor interactions. In addition, if a detection molecule is also attached, the resultant complex can be tracked *in vivo* through, for instance, the use of near-IR fluorescent labels. Finally, this type of bioconjugate can carry multiple chemotherapeutic agents to destroy the intended targeted cells.

Polymeric, dendrimer-based constructs have been made that contain multiple functionalities to target, detect, and destroy tumor cells. Singh *et al.* (2008) prepared a G4 polyamidoamine (PAMAM) dendrimer that was able to target tumor cells which have increased numbers of folate receptors by coupling folic acid to the

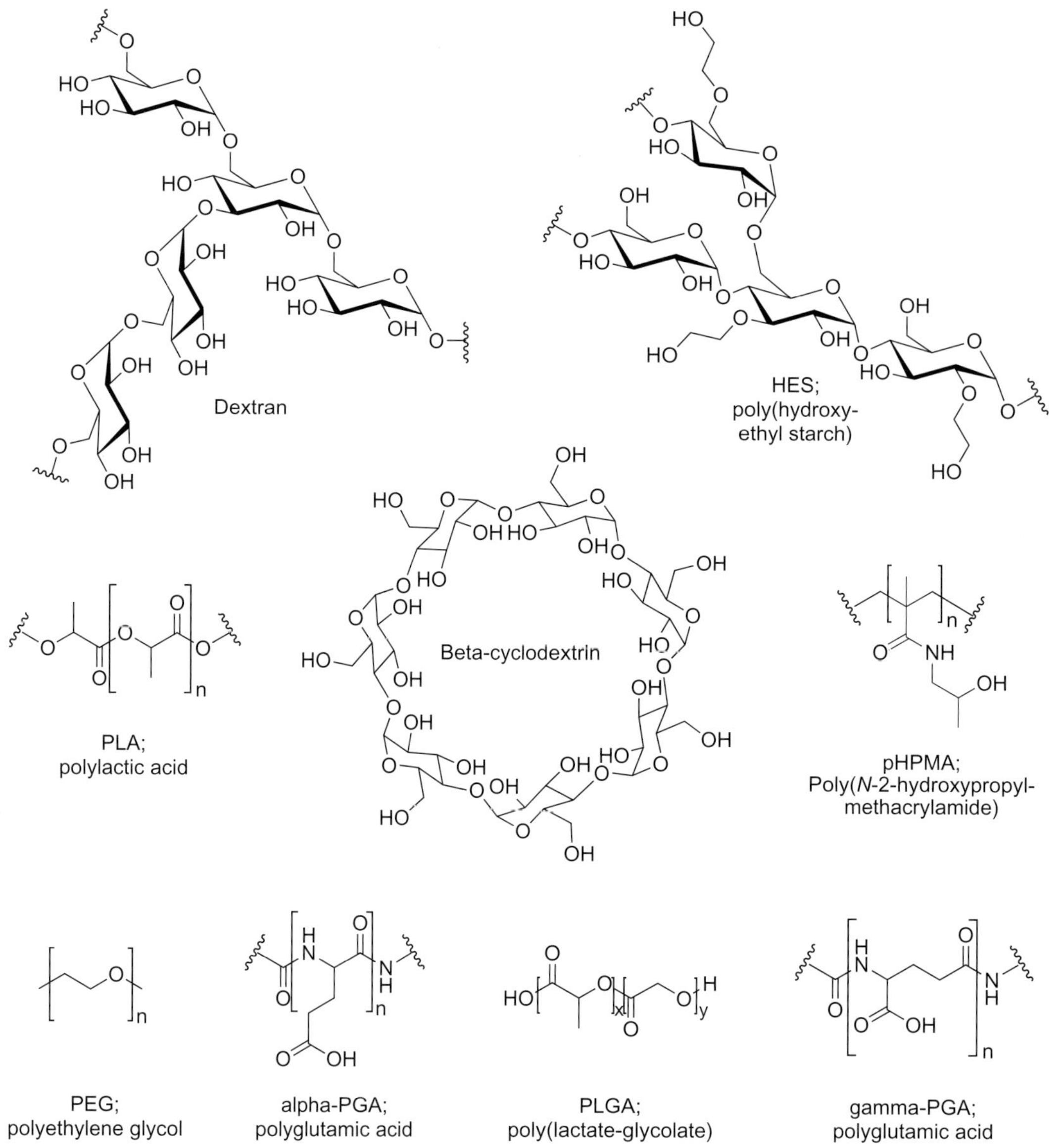

FIGURE 1.96 Some of the most common polymer types used in the design of therapeutic bioconjugates.

pendent amines or through PEG spacers. In addition, the central core of the dendrimer was used as a sequestration carrier to noncovalently encapsulate 5-fluorouracil, a well-known anti-cancer drug. The resultant multifunctional PDDS had the ability to specifically target human epidermoid carcinoma in mice without off-target effects or toxicity.

Liposomes have also been used to create drug delivery vehicles or imaging agents for *in vivo* targeting. A liposome is a synthetic micellular construct containing fatty acids, phospholipids, and other components self-assembled into a bilayer. They are designed to function as a multivalent conjugation scaffold or to sequester active molecules within its bilayer construction for delivery to targets *in vivo* (see Chapter 21). Liposomes can be constructed to contain multiple functional groups or reactive groups on their surface to facilitate conjugation with targeting molecules as well as for the attachment of drugs, enzymes, or imaging agents. Bioconjugates formed using liposome scaffolds can be used to create spherical micelle suspensions of nanometer-sized particles, which can also be filled on the inside of the bilayer with therapeutic agents or dyes for detection (Zhang *et al.*, 2008).

Polymeric nanoparticle-based scaffolds used in biotherapeutic conjugate design have proliferated in recent years and include such constructs as polymeric lipid nanoparticles (PLNs), solid lipid nanoparticles (SLNs), magnetic nanoparticles (MNPs), nanocrystals and quantum dots, amphiphilic polymer vesicles, and phospholipid micelles (Koo *et al.*, 2005; Kingsley *et al.*,

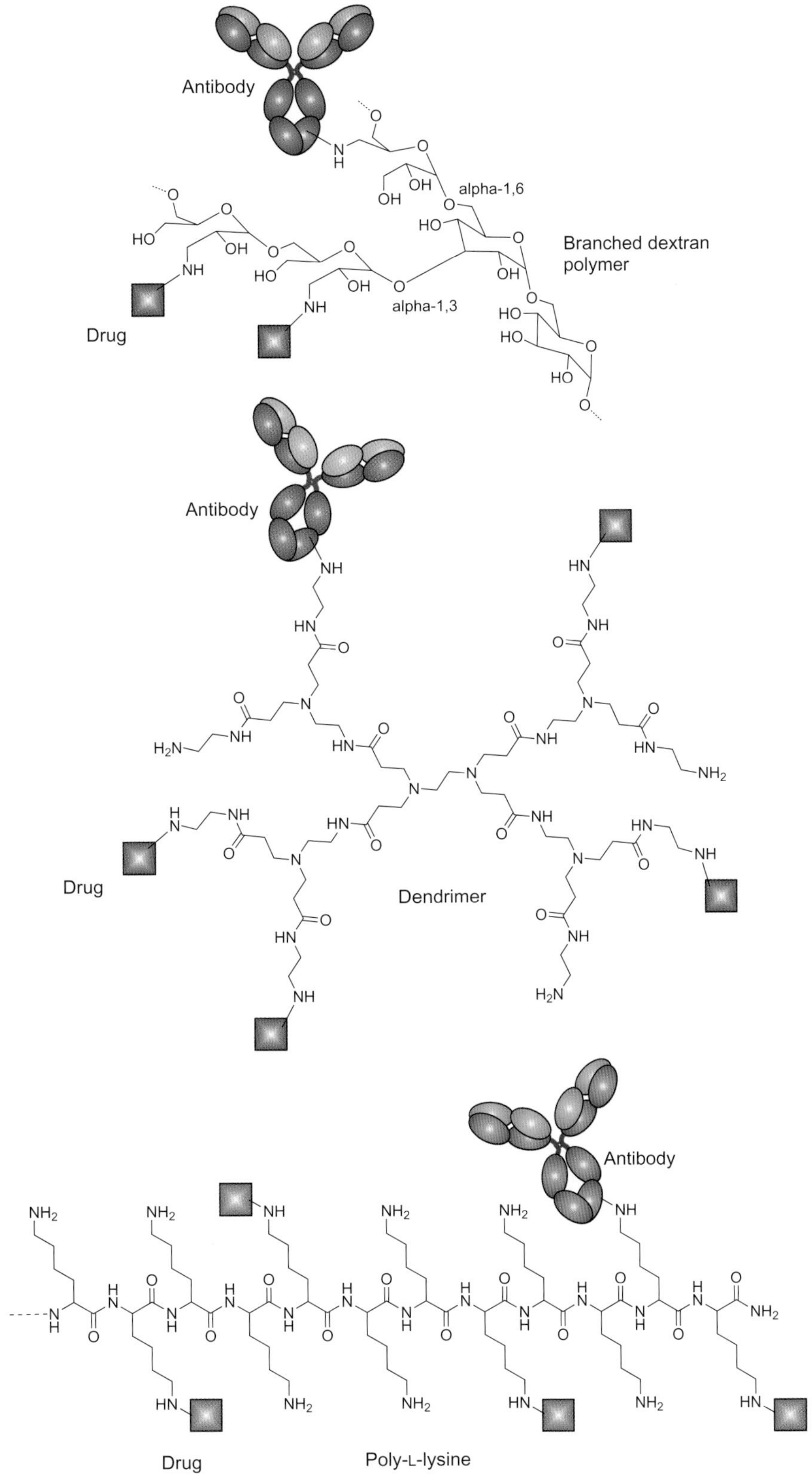

FIGURE 1.97 Polymer-based biotherapeutic conjugates combine the specificity of targeted antibodies with the ability to attach multiple drugs along the multivalent surface of the polymer. The result of this design is to provide a much higher concentration of chemotherapeutic agent at the site of tumor cells than would be possible by linking the drugs directly to the antibodies, as is done with most ADCs.

2006; Caruthers *et al.*, 2007; Ganta *et al.*, 2008; Keck *et al.*, 2008). According to Mishra *et al.* (2010) there have been over 1000 patents or patent applications involving colloidal nanoparticles for use as therapeutic or *in vivo* diagnostic delivery agents.

Antibody-Directed Enzyme Prodrug Therapy

Another antibody-based therapeutic bioconjugate that has been used frequently as an anti-cancer agent is an antibody–enzyme conjugate, wherein the enzyme is chosen for its ability to act on a prodrug substrate and convert it into an active form. Called ADEPT, for "antibody-directed enzyme prodrug therapy," this strategy uses a conjugate that initially has little or no toxic effect on cells (Bagshawe *et al.*, 1988; Neri *et al.*, 1997; Rooseboom *et al.*, 2004). The antibody component is able to interact specifically with the targeted tumor cells through a characteristic biomarker on the cell surface, but neither the enzyme nor the antibody has any immediate therapeutic activity against the tumor. During treatment, the antibody conjugate is administered to the patient, wherein the conjugate finds and binds to the tumor cells, and the excess conjugate is allowed to clear the system before the prodrug is given. The prodrug also has no initial therapeutic activity or cellular toxicity on its own, because of a chemical modification that renders it harmless. However, as the prodrug reaches the tumor cells containing the bound ADEPT conjugate, the enzyme acts on the compound, removing the chemical modification and initiating the drug's anti-tumor activity (Figure 1.98). The prodrug, therefore, has to be designed in anticipation of the enzyme being used in order to obtain successful catalytic removal of its modifying group with concomitant production of the active therapeutic.

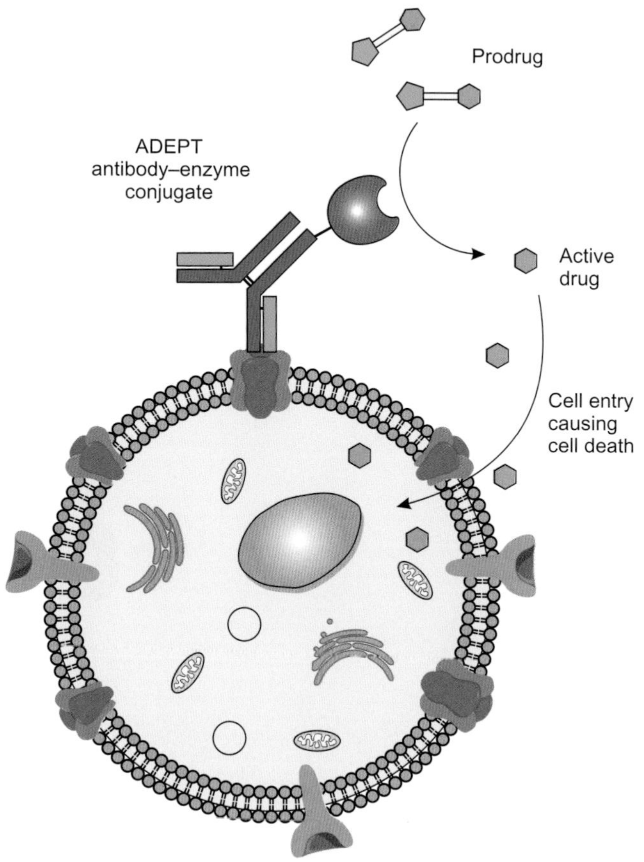

FIGURE 1.98 The principle of an ADEPT antibody–enzyme conjugate is to target the desired cancer cell with the conjugate and then provide an inactive prodrug in circulation, which gets activated only at the tumor cell surface by action of the enzyme. The result of this strategy is to limit the toxicity and off-target effects of the chemotherapeutic agent while maximizing its effect on the tumor cell.

This approach to tumor targeting has several advantages in that it does not use any immediately toxic compounds or conjugates, it involves targeted therapy wherein the antibody–enzyme conjugate only binds to the desired tumor tissue, and the active drug is formed only at the site of each tumor cell thus maximizing its nearby concentration and anti-tumor effectiveness. In addition, since a high concentration of the free drug is attained only around the site where it is needed most, the potential side effects to normal tissue are limited. In addition, the active drug is able to affect tumor cells in the immediate vicinity that may not be specifically labeled with the antibody–enzyme conjugate. This is particularly helpful when epitope changes occur on tumor cells that may eliminate the ability of the antibody to bind to all of the cells within the cancer tissue. Despite epitopic drift, the prodrug molecules will still get converted into the active drug within the tumor tissue and then penetrate throughout the mass to deliver an effective therapeutic dose. Finally, targeted production of a chemotherapeutic agent can avoid the system-wide toxicity of standard chemotherapy, which has numerous sickening side effects on patients.

Some enzymes that have been used in ADEPT bioconjugates include alkaline phosphatase, β-lactamase, carboxypeptidases A and G2, and cytosine deaminase (Senter *et al.*, 1988; Senter, 1990; Senter *et al.*, 1991; Wallace and Senter, 1991; Rodrigues *et al.*, 1995; Vitols *et al.*, 1995; Vrudhula *et al.*, 1995, 1997; Wolfe *et al.*, 1999; Napier *et al.*, 2000; Francis *et al.*, 2002; Harding *et al.*, 2005).

The specific activity of the ADEPT enzyme component ultimately determines the successful design of the associated prodrug used in the therapy. For instance, alkaline phosphatase is particularly useful in removing an organophosphate group from a phenolic prodrug precursor to result in an active drug. One prodrug used for this purpose, etoposide phosphate, is the inactive form of the etoposide anti-cancer agent, which has potent tumor killing properties. The use of

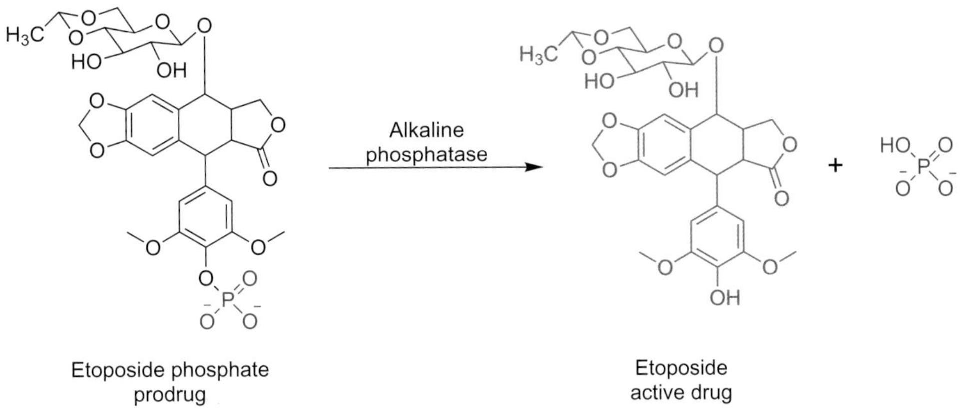

FIGURE 1.99 An example of a prodrug being converted into its active form using the enzyme alkaline phosphatase.

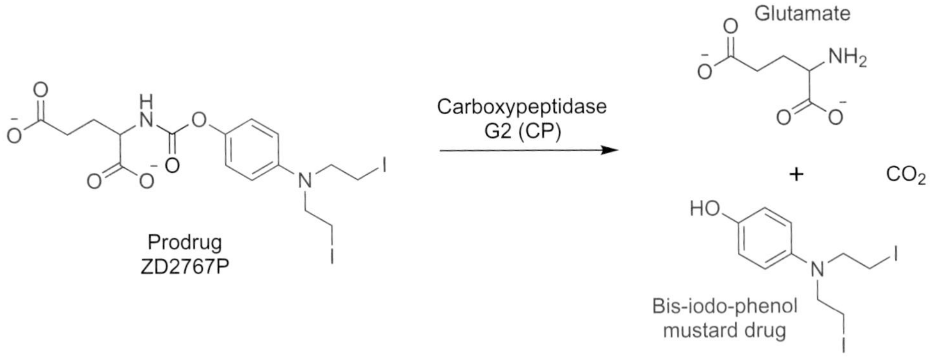

FIGURE 1.100 Carboxypeptidase G2 can be used in an ADEPT conjugate to convert the prodrug ZD2767P into the active chemotherapeutic agent, which is a toxic mustard compound.

the appropriate antibody–alkaline phosphatase ADEPT conjugate can be used to target an antigen on tumor cells, thus bringing the AP enzyme to the cancer tissue in high concentration. After clearance of excess conjugate and administration of etoposide phosphate, the antibody–AP bound to the tumor cells catalytically removes the phosphate groups from the prodrug, resulting in a very high concentration of the chemotherapeutic agent in the surrounding area of the targeted cells (Figure 1.99).

The reaction of an ADEPT biotherapeutic with a prodrug can also be illustrated by the use of an anti-CEA (carcinoembryonic antigen)–carboxypeptidase G2 conjugate and the prodrug ZD2767P as described by Mayer *et al.* (2004). The prodrug ZD2767P is a phenol mustard derivative containing an esterified glutamate residue on the phenolic OH group. Carboxypeptidase G2 can cleave the ester and release the active drug, which is the bifunctional alkylating agent, ZD2767D (4-[*N,N*-*bis*(2-iodoethyl)amino]phenol). The free drug is a nitrogen mustard compound that can rapidly induce cell death in the cells it enters. Figure 1.100 shows the reactions associated with the anti-CEA–G2 conjugate after binding to the targeted tumor cells and administration of the prodrug agent.

ADEPT conjugates using β-lactamase can be used to cleave many types of prodrugs containing β-lactam linkages. β-Lactamases are known to be involved in the metabolic breakdown of certain antibiotics, such as penicillins, carbapenems, monobactams, and cephalosporin, which all contain a lactam ring structure. The enzyme cleaves the lactam ring through hydrolysis, which causes an electron flow toward the associated six-member ring, inducing breakage of any bond at the 3′ substituted position (Kuhn *et al.*, 2004). Due to the enzyme's rapid turnover rate and its ability to act on a wide variety of substrate structures containing a lactam ring, it is now seeing increased use in ADEPT conjugates to convert prodrugs into chemotherapeutic agents. For instance, Wang *et al.* (2001) demonstrated its usefulness in acting upon a cephalosporin-CC-1065 prodrug to ultimately cleave the internal carbamate bond and release the active anti-tumor agent CC-1065 (Reynolds *et al.*, 1986). Figure 1.101 shows the enzymatic conversion of this prodrug to its active form. In a similar strategy, Alderson *et al.* (2006) synthesized the

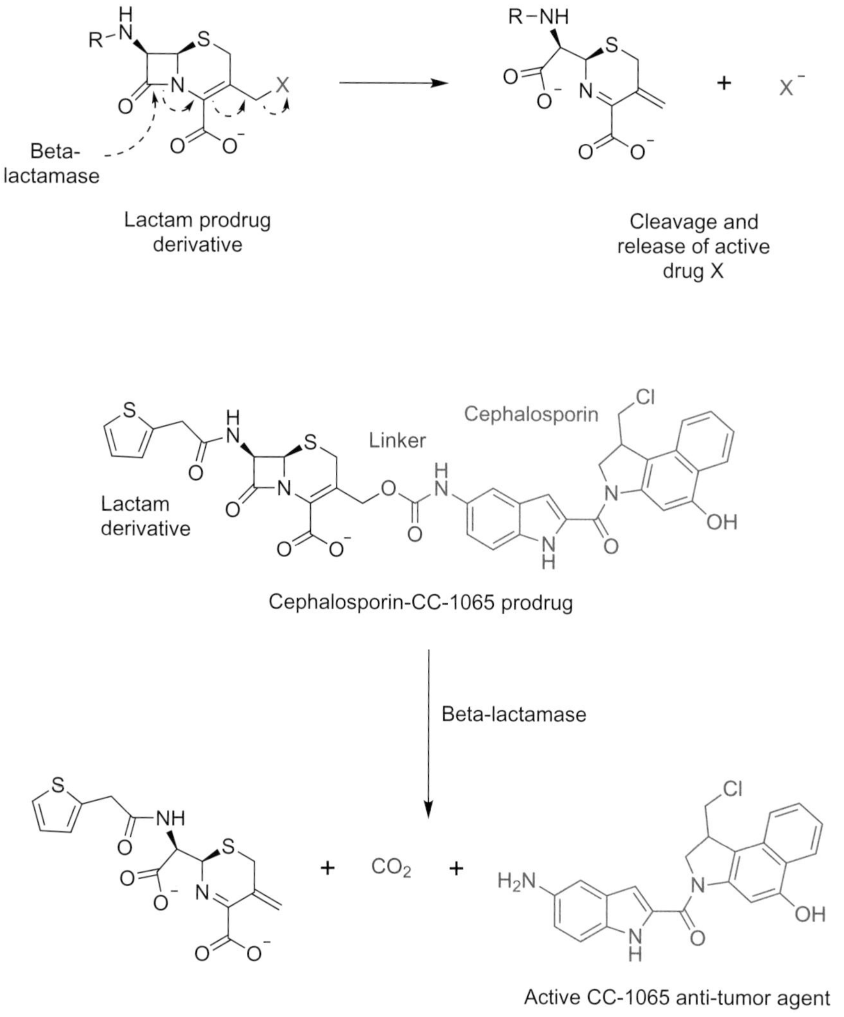

FIGURE 1.101 Beta-lactamase can be used in an ADEPT conjugate to act on lactam-containing prodrugs through a self-immolative process that removes the active drug as a leaving group. The upper reaction scheme illustrates this general concept and the lower reaction shows the specific release of the anti-tumor agent cephalosporin from the lactam precursor derivative.

prodrug GC-Mel, which upon cleavage by a scFv–β-lactamase bioconjugate produced as a fusion protein released the potent anticancer drug melphalan, another nitrogen mustard *bis*-alkylating agent.

Radiolabeled Bioconjugates for Cancer

Radiolabeled antibody conjugates have been prepared by modification with unstable isotopic emitters to create bioconjugates for tumor imaging or to target and destroy cancer cells *in vivo*. The preparation of bioconjugates for radioimmunotherapy involves the attachment of radioisotopes to antibodies or other targeting molecules, such as peptides or small ligands. It can be accomplished using several strategies: (1) direct modification of amino acid side chains in the protein, peptide, or ligand with one or more radioactive atoms; (2) indirect modification of the targeting molecule using a bifunctional chelating compound to coordinate and carry the radioactive isotope; or (3) through the use of a polymeric carrier or nanoparticle attached to the targeting molecule, in which the carrier is modified to contain the radiolabels (Chapter 12) (Figure 1.102). In each case, one or more radionuclides are attached to the targeting molecule to produce a therapeutic agent that can bring to tumor sites a highly concentrated dosage of radiation.

Radiolabeled conjugates for cancer have been used for *in vivo* imaging and for cancer cell destruction, which is called radioimmunotherapy. Some of the radionuclides used to modify peptides or antibodies for tumor imaging or therapy include I-131, Lu-177, In-111, Sm-153, Ho-166, Cu-64, Re-188, Tc-99m, and Y-90 (Weiner and Thakur, 2002). Iodine-131 can be conjugated to antibodies using oxidative techniques or through the use of an iodinatable modification reagent, as described in Chapter 12, Section 2. Modification with other radiolabels, however, typically involves the use of a bifunctional chelating agent, which is able to stably coordinate the isotope and also allow for covalent attachment to protein functional groups. These chelators provide a reactive group that can be used to couple with, for instance, amines on antibodies while the other

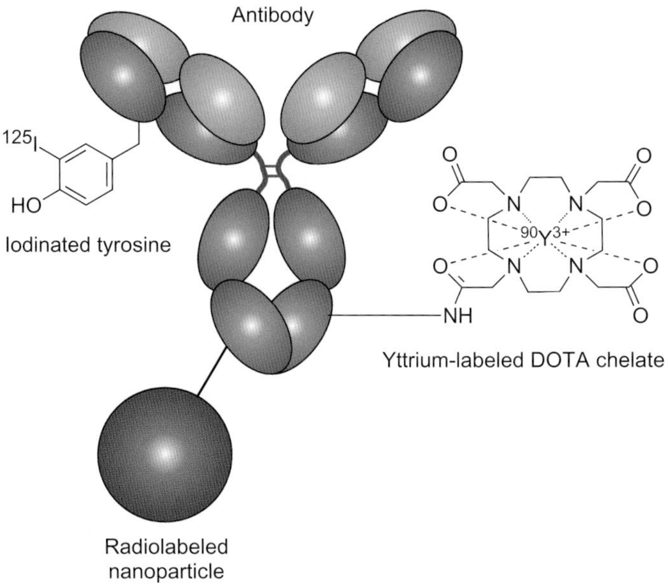

FIGURE 1.102 Several different types of radiolabeled antibodies have been used for therapeutic purposes, including a radioiodinated tyrosine, a radiolabeled nanoparticle conjugate, and a ^{90}yttrium-labeled DOTA chelate conjugate.

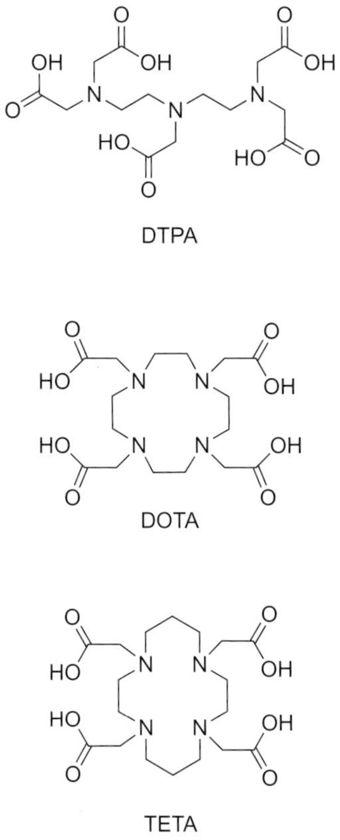

FIGURE 1.103 Three types of chelating groups are shown that are commonly used to make isotopically labeled bioconjugates for radiotherapy or imaging applications.

end of the molecule is designed with a cyclic or wrap-around type of chelation group that can fully coordinate the radionuclide, thus preventing it from leaching from the complex. Some of these bioconjugation chelating agents are illustrated in Figure 1.103 and include the tetraaza-macrocycles DOTA and TETA as well as the multi-armed carboxylate chelator DTPA.

Bioconjugates used in radioimmunotherapy usually employ radionuclides that emit high-energy particles or electromagnetic gamma radiation and are capable of rapid cellular destruction when greatly concentrated at a tumor site. Bioconjugate designs for these agents typically involve the use of one or more chelated radiolabels attached to an antibody targeting molecule, which is directed at tumor cell epitopes. For this type of bioconjugate, one of the most preferred radioisotopes is Y-90, because it emits a high-energy β^- particle and it has a half-life of only 2.7 days. Yttrium-90 labeled antibodies bind to the tumor cells and create extremely high radiation doses right at the site of the cancer mass, thus maximizing its effectiveness for tumor cell destruction and minimizing the effects on other organs and tissues. The first commercial drug of this type was approved for non-Hodgkin lymphoma in 2002 and is marketed under the name Zevalin (Y-90 ibritumomab tiuxetan) (Witzig *et al.*, 2002; Bethge *et al.*, 2010). It consists of an anti-CD20 monoclonal antibody that is modified with a DTPA derivative (tiuxetan), which contains the chelator groups and coordinates the Y-90 radionuclide (Figure 1.104). The chelator is conjugated to the antibody via a thiourea bond formed from the reaction of an isothiocyanate-phenyl-DTPA derivative with amine groups on the antibody molecule. The antibody conjugate is then loaded with fresh Y-90 prior to *in vivo* administration to ensure maximum activity for radioimmunotherapy.

Many other yttrium-90-based radioimmunotherapy conjugates have been investigated in their efficacy in killing tumor cells. DeNardo *et al.* (1997) created a DOTA–peptide–chimeric monoclonal antibody conjugate with a peptide linker that would be cleaved by lysozymes intracellularly. The chimeric antibody was developed by Bristol-Myers Squibb Pharmaceutical Research Institute and consisted of a human Fc region joined to a mouse Fab region, which had antigen binding characteristics toward an integral membrane glycoprotein, which is expressed on breast, colon, ovary, and lung cancer cells. The study found a positive synergistic effect in the use of yttrium-90 conjugate with the anticancer drug taxol, indicating that bioconjugates may also be important adjuncts to standard chemotherapeutic approaches in treating cancer.

In addition, Zang *et al.* (2009) developed an anti-CD25 monoclonal antibody conjugate containing ^{90}Y

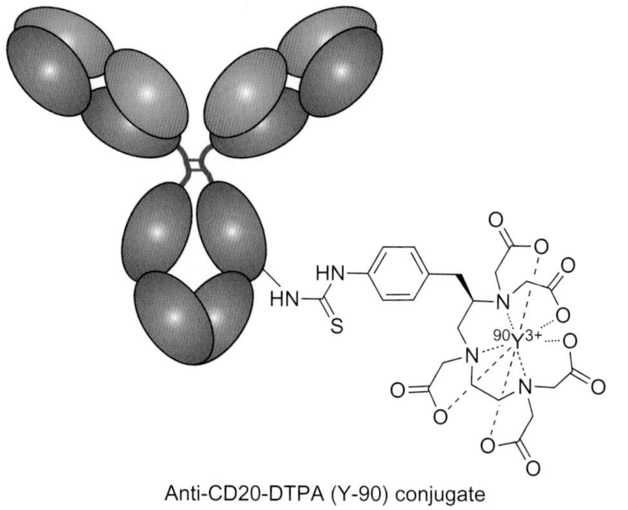

FIGURE 1.104 An antibody labeled with a yttrium-90 chelate can be used to target tumor cells for radioimmunotherapy purposes. This illustration depicts the design of the drug Zevalin, which is approved to treat non-Hodgkin lymphoma.

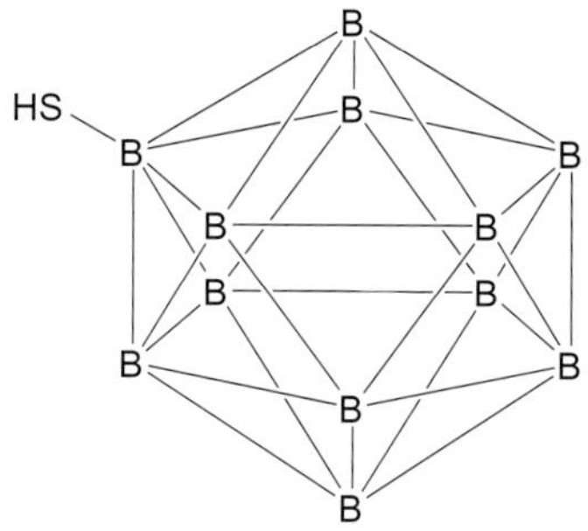

FIGURE 1.105 Borocaptate is a common label for use in boron neutron capture therapy. It consists of 12 boron atoms and a single thiol for convenient conjugation to targeting molecules.

chelated with DTPA derivatives to investigate the treatment of CD25-expressing lymphomas in mice. This radioimmunoconjugate showed efficacy in the treatment of SUDHL-1 lymphoma in a preclinical animal study. Similarly, Shibata *et al.* (2009) used a ^{90}Y–DTPA–anti-CEA (carcinoembryonic antigen) conjugate in a phase I study on patients with advanced CEA-producing solid tumors. The chimeric antibody was labeled with isothiocyanatobenzyl-DTPA and loaded with Y-90 before administration. The major side effect seen in this study that limited the use of the bioconjugate drug was the formation of a human anti-chimeric antibody (HACA) response, which interfered with the targeted delivery of the complex to the tumor.

Boron Neutron Capture Therapy

Boron neutron capture therapy (BNCT) is another application area in which bioconjugates have found important use for the treatment of cancer. The ability of boron to capture neutrons was discovered by Taylor (1935) when he found that bombarding boron-10 with neutrons caused the release of He-4 alpha particles and converted the boron atoms into lithium-7 ions (Slatkin, 1990). Although simple inorganic boron-10-containing salts were found to be useful as neutron capture agents, the nonselective nature of these compounds proved impractical for *in vivo* targeting of tumor cells (Godwin *et al.*, 1955). Second-generation boron compounds were developed and consisted of boron-modified amino acids, but these also suffered from lack of specificity in targeting most cancers, although some compounds do appear to accumulate preferentially in certain brain tumors (Snyder *et al.*, 1958; Radioisotopes in Medicine, World Nuclear Association, 2010).

Perhaps the most convenient form of boron that can be used for bioconjugation was discovered in the early years of boron compound development and consists of a polyhedral borane complex, mercapto-undecahydro-*closo*-dodecaborate (or sodium borocaptate, BSH), which contains 12 boron-10 atoms ($Na_2B_{12}H_{11}SH$). This compact polyhedral compound can be used as a label to provide a high concentration of boron atoms for *in vivo* therapy. Particularly, BSH has a convenient thiol group for conjugation with other molecules, making it ideal for labeling targeting molecules using crosslinkers having thiol-reactive groups (Figure 1.105). Borocaptate has also been modified to contain other functional groups or reactive groups suitable for bioconjugation to targeting molecules, thus establishing it as a powerful tool for BNCT of cancer (Hawthorne and Lee, 2003; Barth *et al.*, 2005).

For BNCT to be effective at killing cancer cells at least 20 μg of B-10 per gram of tumor mass or approximately a billion atoms of B-10 per cell have to be delivered to the tumor site (Barth *et al.*, 2005). This level of concentrated delivery can only be accomplished without side effects through the use of carefully designed bioconjugate drugs containing affinity-targeting agents specific to the tumor type requiring treatment. In some cases, low-molecular-weight targeting molecules have been found to be efficacious, such as a conjugate containing multiple BSH complexes linked to a porphyrin derivative (Evstigneeva *et al.*, 2003; Vicente *et al.*, 2003; Fronczek and Vicente, 2005). Porphyrin molecules are rapidly taken up by tumor cells due to their high metabolic rates, thus providing a mechanism for delivering the boron payload to the cells.

BSH may also be conjugated to monoclonal antibodies or other macromolecule targeting components for the delivery of B-10 to tumor sites (Novick *et al.*, 2002; Wu *et al.*, 2004). The production of these conjugates is rather straightforward, but their use in the treatment of brain cancer may be somewhat restricted due to their limited ability to penetrate the blood–brain barrier and reach the tumor site. However, recent studies in rats have used the monoclonals cetuximab (against wild-type EGFR) or L8A4 (against mutant EGFRvIII), both conjugated with BSH and compared with that of a lower molecular weight conjugate consisting of EGF-BSH (Yang *et al.*, 2009). Best results in that study were obtained using the monoclonal L8A4 as the targeting agent, because it had the greatest specificity for the mutant EGFR expressed on the gliomas.

Other bioconjugates with BSH have been created using various scaffolds to link the boron-10 complexes with antibodies or other targeting molecules. For instance, boronated polylysine was coupled to the carbohydrate groups on polyclonal antibodies to create a conjugate for neutron capture therapy (Novick *et al.*, 2002). The final conjugate was estimated to contain about 6000 atoms of B-10 per antibody molecule, thus making a high-activity complex for targeted delivery directly to tumor cells at levels required to efficiently cause cell death in BNCT procedures.

Photodynamic Therapy

Photodynamic therapy (PDT) involves the use of an organo-metallic photosensitizing agent that combines irradiation by light with available dissolved oxygen within the surrounding media to catalytically produce reactive oxygen species (ROS). The rapid production of ROS at the site of the reaction can cause cell death in proximity to the photosensitizer (Usuda *et al.*, 2006). Small-molecule photosensitizers have been used individually or in conjugated form with other targeting molecules to treat malignant or even certain benign diseases. Many of the photosensitizing agents that have been used for therapy or research involve porphyrin or phthalocyanine derivatives or porphyrin precursors such as 5-aminolevulinic acid (ALA), which is taken up by metabolizing cells (e.g., neoplastic tissue) to create an excess of porphyrin intracellularly. Upon exposure to light in the far-red to NIR region and in the presence of dissolved oxygen, a redox reaction is induced at the site of the coordinated metal ion, which ends up producing highly reactive singlet oxygen ($^1O_2$). The photosensitizer absorbs a photon of light and enters an excited state, which subsequently transitions into an excited triplet state. This intermediate can then undergo redox reactions with nearby molecules and create free radical intermediates, which subsequently can react with oxygen to produce ROS. Since the porphyrin photosensitizer acts as a redox cycle and gets regenerated during the course of the reaction, large amounts of singlet oxygen and other ROS can be produced in the immediate molecular vicinity. Thus, PDT can be very effective at producing a toxic reactive agent right at the site of a tumor or directly within cancer cells. See Chapter 2, Section 1.1, for additional information on the reactions associated with ROS on biomolecules.

At least eight photosensitizers have been approved and commercialized for PDT. Many of these compounds are porphyrin derivatives that are not conjugated with targeting molecules. They are administered and accumulate in membranes and organelles, especially within tumor tissue, because of their rapid growth characteristics. Some commercial photosensitizers consist of ALA derivatives, and these drugs typically are applied topically to skin lesions. Some studies have even found benefits to PDT in the treatment of port wine stain birthmarks (Yuan *et al.*, 2009).

An example of a common drug derivative used for PDT is Photofrin, which consists of a short polymer of porphyrin derivatives linked together via ether bonds (Figure 1.106). Though it has been approved for use with a number of cancers, including non-small cell lung carcinoma, esophageal cancer, and bladder cancer, it and other PDTs suffer from a lack of affinity targeting agent to direct the drug specifically to the desired tumor cells. Side effects can include off-target tissue damage as well as eye and skin sensitivity to light due to the accumulation of porphyrin compounds throughout the body.

In use, the Photofrin compound is reconstituted from a freeze-dried product in 5% dextrose solution or 0.9% NaCl and injected intravenously. The drug is allowed to circulate for up to 40 to 50 hours before irradiating the tumor site with red laser light at a wavelength of 630 nm. The red light penetrates tissue to the tumor containing a high concentration of the Photofrin oligomer. Irradiation begins the ROS production that leads to tumor destruction. However, there are potentially serious side effects to this treatment partly due to the untargeted nature of the treatment. PDT done with advanced bioconjugates containing targeting molecules that limit off-target effects likely will alleviate some of the side effects observed with the use of Photofrin alone.

Another deficiency in the use of many porphyrin derivatives is that they often are best activated with light within the blue or green part of the spectrum. At these wavelengths, light cannot travel far through tissue without being absorbed by biological molecules, especially porphyrin-containing molecules like hemoglobin. This limits the use of some PDT compounds *in vivo* to skin or near-surface lesions. However a porphyrin-like compound called texaphyrin that was first synthesized at the University of Texas (Austin) includes a fifth coordinating nitrogen in its ring structure to chelate a lanthanide ion, such as lutetium or gadolinium (Sessler and Miller, 2000; Young *et al.*, 1996; Sessler and Tomat, 2007). A common derivative of the texaphyrin core structure involves the addition of two short-chain

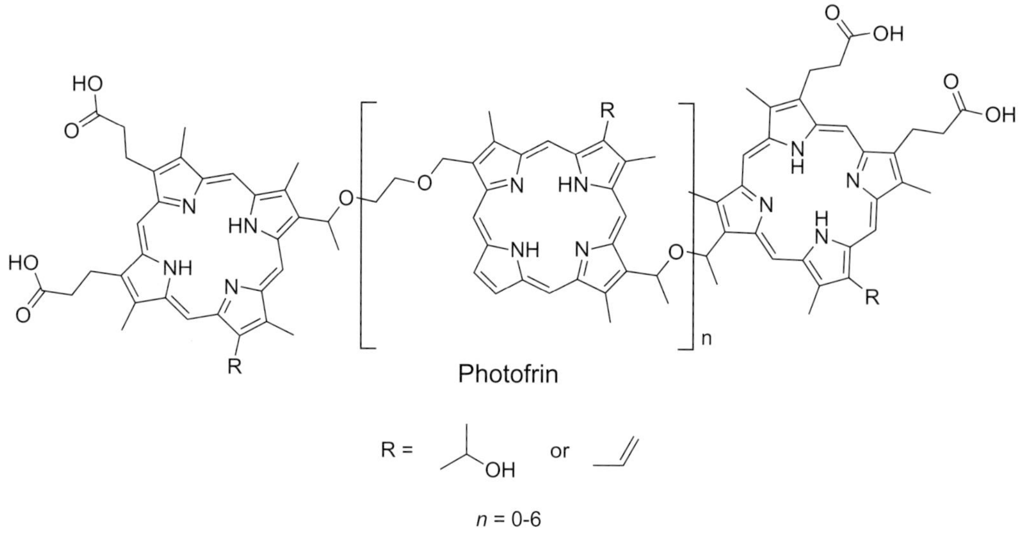

FIGURE 1.106 The structure of Photofrin, which contains multiple porphyrin units and can be used in photodynamic therapy to cause tumor cell death.

PEG arms to improve water solubility for i.v. administration while maintaining the lipophilic nature of the macrocyclic metal chelate (Figure 1.107).

Texaphyrin-based PDT compounds behave spectrally like phthalocyanine derivatives in that they display red-shifted absorbance properties relative to porphyrins. Texaphyrin derivatives can be activated by light in the far-red to NIR region of the spectrum (690–880nm), thus allowing much deeper tissue penetration for therapy. These compounds also have enhanced relaxivity that makes them useful in nuclear magnetic resonance imaging (MRI) for dual-mode detection and therapy (Sessler *et al.*, 1995). Considerable research is being done on Texaphyrin compounds due to their favorable absorbance properties for *in vivo* applications.

To overcome the nonspecific targeting of many standard PDT techniques, photosensitizer compounds can be conjugated to targeting molecules that have high affinity toward the diseased cells being treated. For instance, Staneloudi *et al.* (2007) designed a novel PDT bioconjugate using antibody scFv fragments directed against colorectal tumor cells and coupled with two different porphyrin photosensitizer derivatives, which contained amine-reactive isothiocyanate groups (Figure 1.108). It was found that a reaction ratio of 5:1 (porphyrin-NCS:scFv) was ideal for forming an active scFv–porphyrin conjugate while not affecting the ability of the small antibody fragment to bind antigen—an important factor to consider when working with low-molecular-weight targeting molecules.

A novel photodynamic conjugate was described by Cheng *et al.* (2008) using gold nanoparticles that were PEGylated with thiol-PEG tethers, which functioned as a biocompatible cage to transport a drug to a tumor site

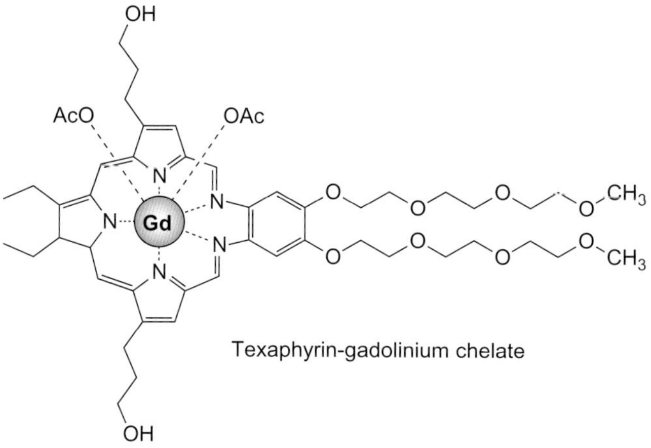

FIGURE 1.107 Texaphyrin is a phthalocyanine-like derivative containing a chelated gadolinium ion and two PEG chains to promote water solubility.

for PDT. The gold nanoparticles are known to have low toxicity and the addition of PEG tethers prevents non-specific binding to proteins *in vivo*. After PEGylation, the gold was further derivatized with a phthalocyanine compound containing a side-chain tertiary amine (Figure 1.109). The unshared pair of electrons on the nitrogen can bind through dative bonds to the gold surface in areas that were not taken up by the PEG-SH tethers, thus forming a water soluble (or colloidal) nanoparticle containing photodynamic groups.

In another example, Liu *et al.* (2009a) developed a small molecule conjugate for targeted PDT consisting of an inhibitor of a biomarker for prostate cancer, the prostate-specific membrane antigen (PSMA), and a porphyrin-like derivative, pyropheophorbide-*a* (Ppa). The

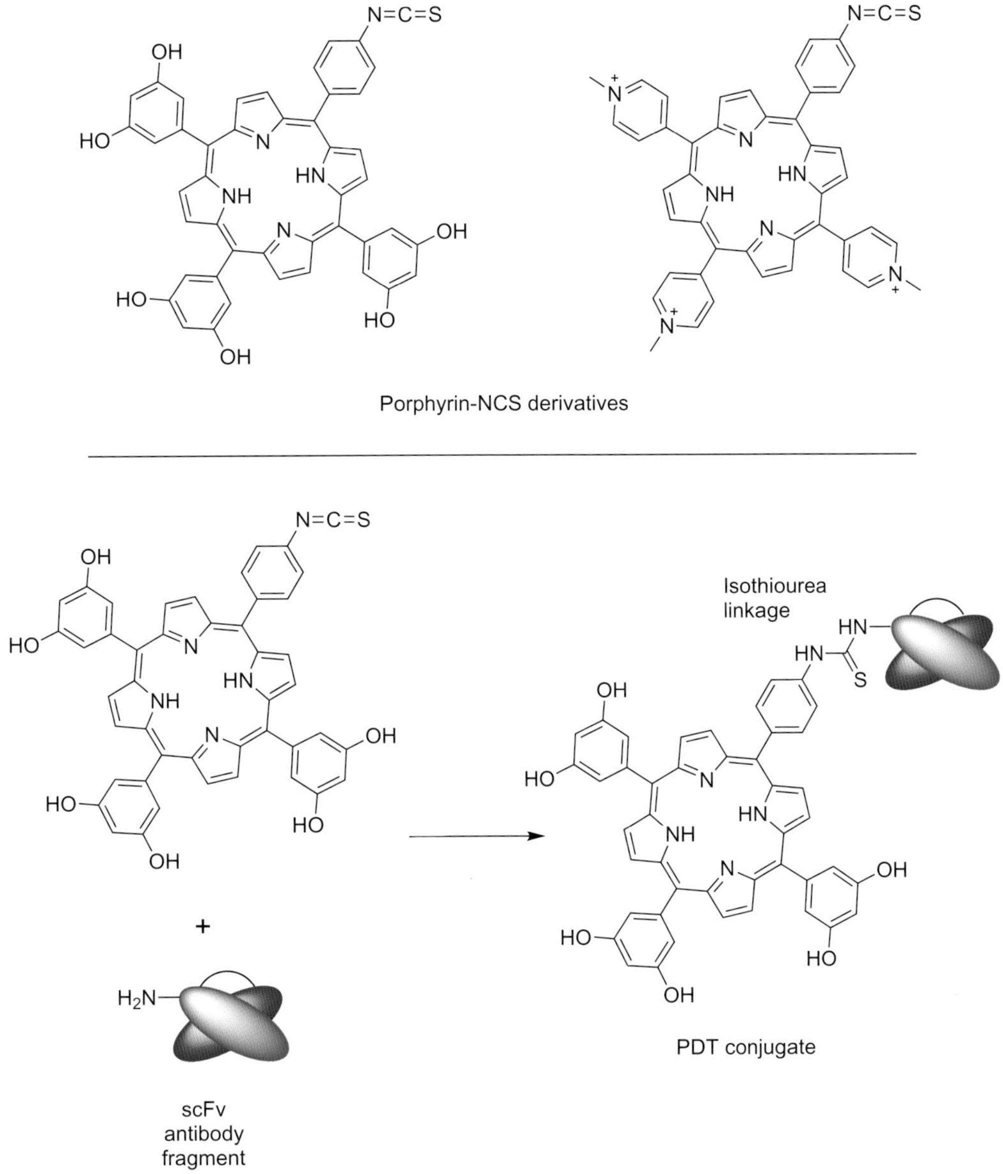

FIGURE 1.108 Reactive porphyrin-isothiocyanate derivatives have been created to provide reactive groups for conjugation of this photosensitizer to antibodies. The conjugate with a scFv monoclonal antibody fragment, made via an isothiourea linkage, was shown to be effective in targeting colorectal tumor cells.

PSMA inhibitor consisted of a peptidomimetic phosphoramidate derivative with a free alpha-amine at one end (termed PSMA inhibitor core 1). The conjugate was made through an amide bond that was formed using an NHS ester on the carboxylate end of Ppa and then reacting it with the amino end on the inhibitor (Figure 1.110). The conjugate was found to specifically target and bind to LNCaP cells in culture through the inhibitor's affinity toward PSMA. Cellular apoptosis was achieved after irradiation as a result of photodynamic destruction of the tumor cells by the generated ROS.

A targeted PDT bioconjugate was also designed by Lu *et al.* (2009) by modifying coagulation factor VII with verteporfin to provide a potential therapeutic for age-related macular degeneration (AMD). Factor VII is a serine protease involved with the initiation of the coagulation cascade. It forms a complex with tissue factor, a protein normally found on the outside endothelial cells of blood vessels and therefore inaccessible to agents in plasma. Upon vascular injury, however, tissue factor becomes available to interact with factor VII in the blood, which starts the enzymatic coagulation cascade process. In AMD involving choroidal neovascularization (CNV), tissue factor is aberrantly present on the inside of the blood vessels. Thus, a bioconjugate consisting of factor VII has affinity to bind only to the

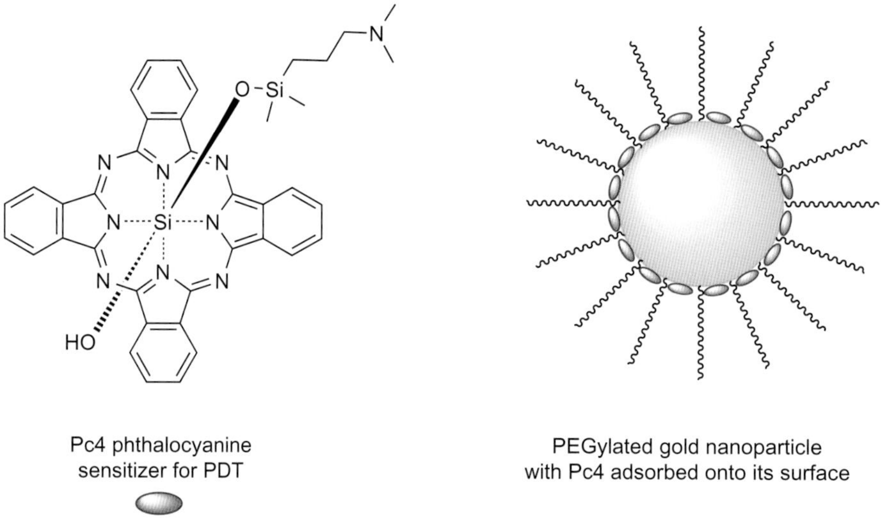

FIGURE 1.109 A nanoparticle sensitizer for photodynamic therapy was created by adsorbing a phthalocyanine compound onto a PEGylated gold particle. The conjugate provided intense singlet oxygen production through the high density of photosensitive groups on each particle surface. The PEG tethers provided water solubility and low nonspecific binding toward biomolecules *in vivo*.

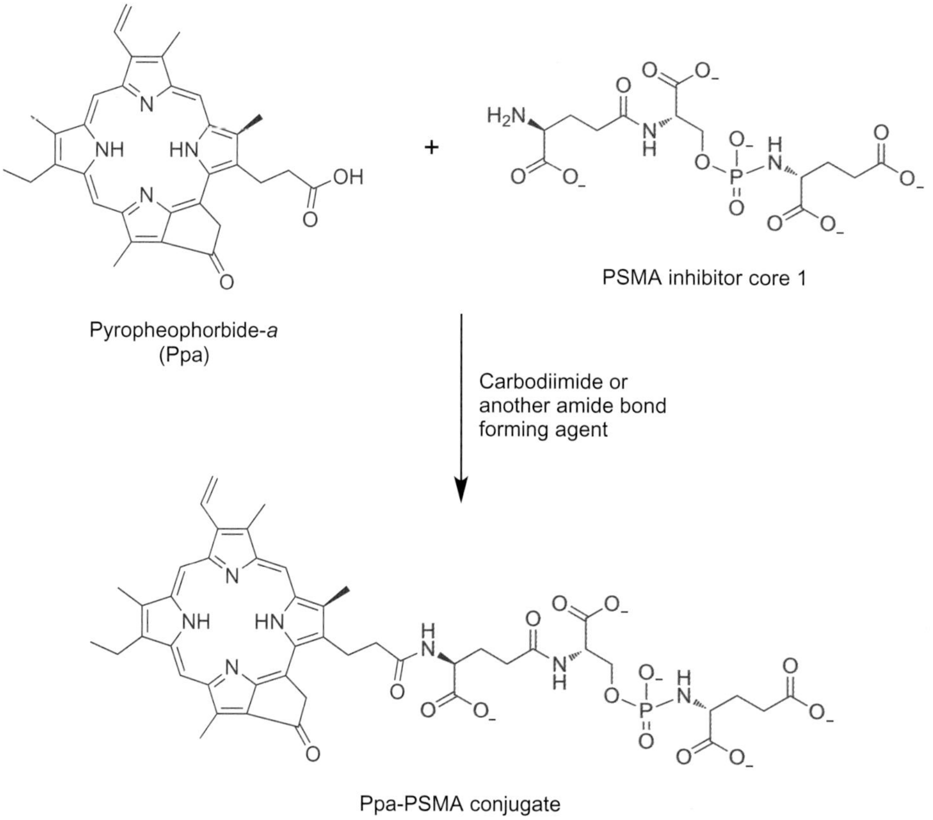

FIGURE 1.110 The conjugation of a porphyrin-like derivative, pyropheophorbide-*a* (Ppa), with the prostate-specific membrane antigen (PSMA) inhibitor core 1 provides targeted photodynamic therapy in the treatment of prostate cancer.

endothelium present in CNV lesions but nowhere else. The porphyrin derivative verteporfin, which contains a carboxylate group, was used in an EDC-mediated carbodiimide reaction (Chapter 4, Section 1) to conjugate with purified recombinant factor VII (Figure 1.111). The resultant PDT bioconjugate was found to be significantly more effective than untargeted reagents in treating AMD using photodynamic methods.

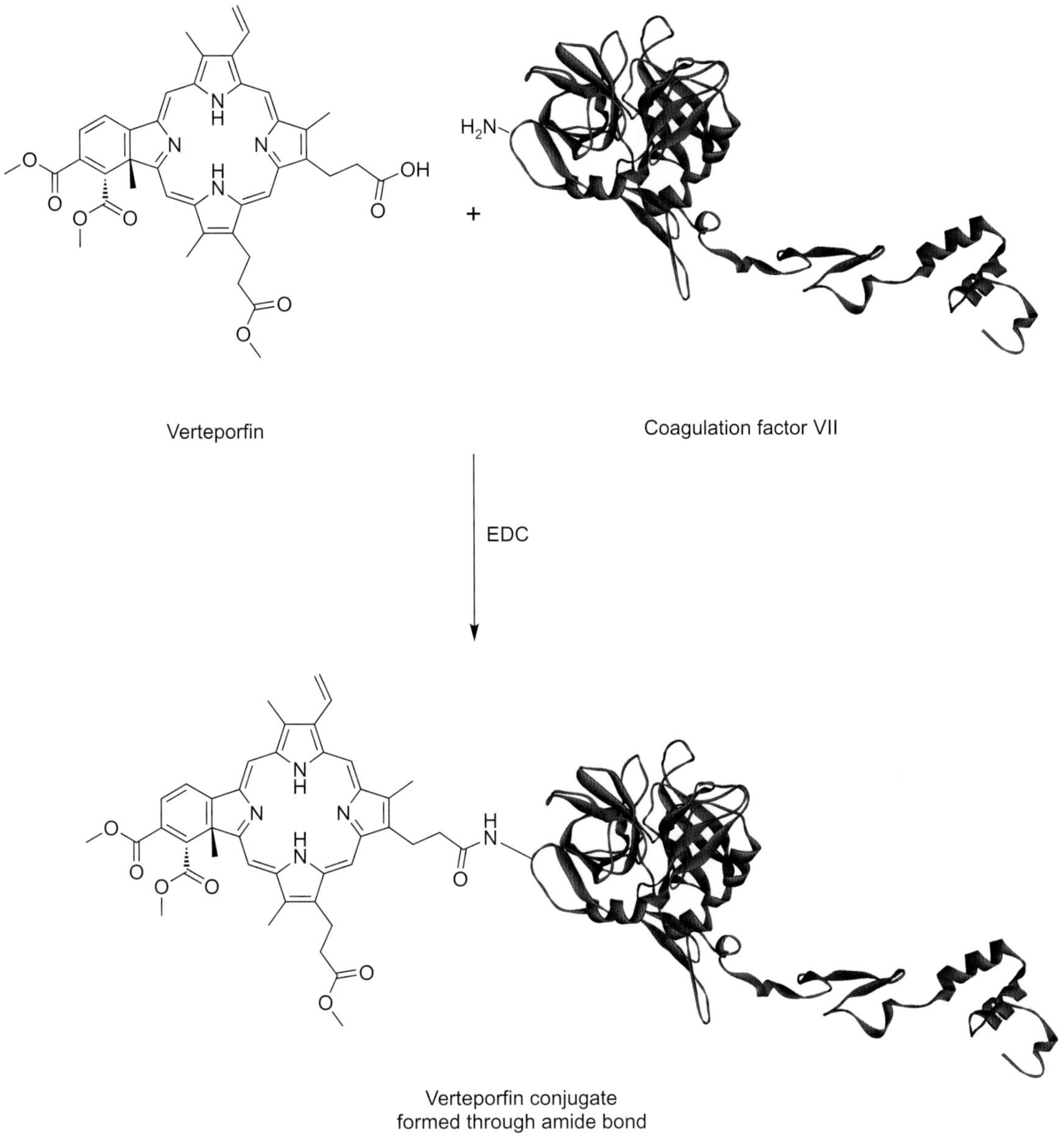

FIGURE 1.111 The conjugation of the photosensitizer verteporfin with protein coagulation factor VII provides a specific PDT agent to treat choroidal neovascularization (CNV) lesions.

In addition, bioconjugates have been designed using the red-shifted porphyrin derivative texaphyrin to provide targeted PDT while also being able to use NIR irradiation for activation *in vivo*. Wei *et al.* (2005) synthesized a conjugate of texaphyrin with the antitumor drug methotrexate (Figure 1.112). This unique conjugate used the tumor uptake capability of the texaphyrin component to better aid in getting methotrexate into tumor cells. This small molecule bioconjugate was designed to overcome the resistance of tumor cells to methotrexate, which can occur very rapidly after treatment. The complex creates the potential for dual-mode therapy using an anti-folate agent combined with PDT.

Diagnostic Bioconjugates for In Vivo Imaging

For decades diagnostic procedures to detect disease have been done *ex vivo* on samples extracted from patients consisting primarily of cells, tissues, or fluids. For example, classical immunohistochemical (IHC) staining methods in pathology have been widely used with antibody conjugates to assess disease states using patient tissue samples obtained through biopsies. These procedures are often unpleasant or painful and can yield differential results depending on how accurately the sampling process was performed. The bioconjugate reagents used for IHC typically are enzyme conjugates wherein an enzyme substrate reaction is used

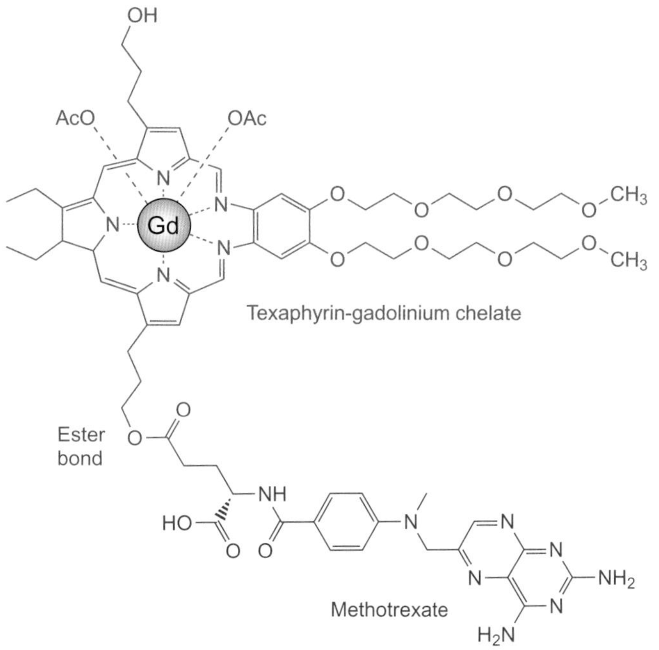

FIGURE 1.112 The conjugation of texaphyrin–gadolinium chelate for PDT and the anti-tumor agent methotrexate offer combined therapy for cancer.

to produce a precipitated chromogenic or fluorescent signal at targeted points within the extracted tissue section. Probing thin tissue sections or cells in this manner has also been done using fluorescently labeled antibodies to directly detect antigens present within the sample (see further discussion on IHC in Section 2.2 of this chapter). However, these procedures are all done *ex vivo* to target and detect diagnostically important pathology within samples extracted from patients in highly invasive procedures.

The current trend is to detect diseased tissues and cells directly within animals or human patients without taking a biopsy or using invasive methods to obtain the critical sample (for review, see Frangioni, 2008). Bioconjugates developed for *in vivo* diagnostic imaging have become a major research area that has grown rapidly over the last two decades. The use of targeted detection with affinity bioconjugates containing highly sensitive labels that can be imaged effectively through skin or tissue already generates billions of dollars annually in health care revenues. The major types of detection bioconjugates currently used for *in vivo* diagnostic procedures consist of chelated radionuclides, high-contrast agents, nanoparticles, and organic fluorescent labels. Many of these imaging bioconjugates also incorporate polymeric constructs (e.g., dendrimers) as scaffolds that can be used to couple multiple copies of detection molecules for enhanced sensitivity. Today, the majority of clinical *in vivo* imaging involves the use of contrast agents and radiopharmaceuticals along with instrumentation for magnetic resonance imaging (MRI) or positron emission tomography (PET); however, fluorescence detection is also predicted to increase dramatically over the next decade to become the third leading detection method for *in vivo* diagnostics (*Biologic Imaging Reagents: Technologies and Global Markets*, BCC Research, Wellesley, MA, 2009, pp. 124, 144). The following sections discuss the bioconjugates used in these three primary detection modes for *in vivo* imaging applications.

Bioconjugates for Radio Imaging

The overwhelming majority of radiopharmaceuticals prepared for *in vivo* imaging involve the use of Tc-99m-labeled compounds (Liu, 2004). This radiolabel is used in millions of diagnostic procedures every year. Tc-99m emits gamma radiation at approximately the same wavelength as X-rays used in diagnostic imaging; therefore, conventional medical instrumentation can often be used for detection. Many of the commercial Tc-99m complexes are not conjugates with antibodies or peptide targeting agents, but are chelator organo-metal complexes that are able to target certain sites within the body due to their small size, charge, or hydrophobicity. The design of Tc-99m chelators is highly diverse due to the complex redox chemistry of technetium (Liu *et al.*, 1997). A number of commercialized radiopharmaceuticals currently on the market are shown in Figure 1.113. One such compound is represented by the drug Cardiolyte (Bristol-Myers Squibb), which consists of a technetium complex of six molecules of 2-methoxy isobutyl isonitrile complexed with one Tc ion to form $Tc99m[MIBI]_6^+$ coordination complex (^{99m}Tc-Sestamibi, Figure 1.113). This diagnostic agent is indicated for detecting coronary artery disease and is used frequently in imaging procedures to noninvasively detect atherosclerosis.

Other popular small-isotope conjugates used for *in vivo* imaging consist of F-18-labeled reagents, which are used in positron emission tomography (PET). The F-18 radionuclide emits a positron, which is the antiparticle counterpart to an electron. As the positron is emitted within tissue, it soon encounters an electron at which point they both get annihilated, producing two high-energy photons (gamma emission). The emitted photons are detected in the PET scanning instrumentation, which frequently also incorporates a simultaneous X-ray CT scan or MRI that can produce 3D images of the body. Upon decay of F-18 the stable isotope O-18 is formed, which has no further radio-emission characteristics.

The most popular F-18-containing reagent is 2-deoxy-2-[^{18}F]fluoro-D-glucose (abbreviated ^{18}F-FDG) (Figure 1.114). This compound is preferentially taken up by cells undergoing rapid metabolism, such as is

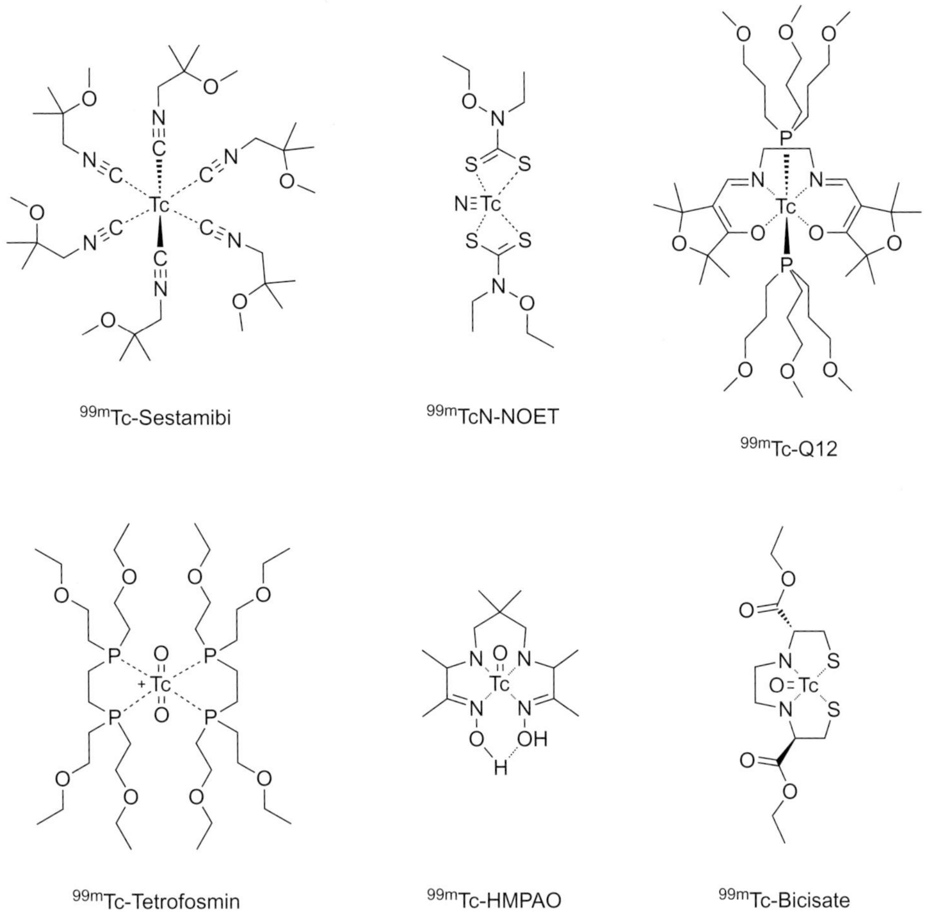

FIGURE 1.113 The major commercial Tc-99m complexes used for radioimaging.

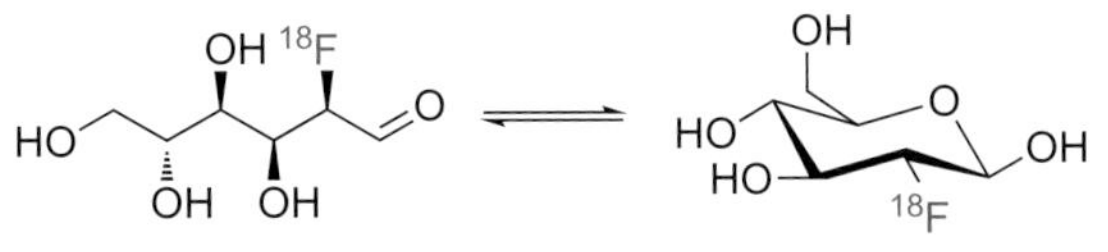

2-deoxy-2-[^{18}F]fluoro-D-glucose

FIGURE 1.114 ^{18}F-labeled deoxy glucose is one of the major radiolabels used in PET imaging.

common in tumor cells. In ^{18}F-FDG, the 2-hydroxyl group normally present in glucose is missing; thus, once it enters a cell, it cannot enter the glycolysis pathway. Within the cell, however, it does get phosphorylated at its carbon-6 position, which prevents it from leaving the cell. Therefore, once it enters cells undergoing extensive glycolysis, it is trapped there and can be imaged until its positron emission capability decays. Upon decay, the 6-phospho-^{18}F-FDG becomes 6-phospho-^{18}O-FDG, which effectively transforms the molecule into glucose-6-phosphate containing the stable heavy isotope of oxygen, wherein it then gets metabolized normally within the cell without toxic effects.

Other conjugates of F-18 are being investigated for PET imaging purposes. Vandenberghe *et al.* (2010) reported on a phase two clinical trial involving an F-18 modified flutemetamol probe being developed by GE Healthcare. This compound is a thioflavin derivative of a previous C-11 isotopic probe of Aβ amyloidosis plaques (called ^{11}C-Pittsburgh compound B [PiB]), which is an important indicator of Alzheimer's disease (Rabinovici *et al.*, 2007; Nelissen *et al.*, 2009). The C-11 probe required more expensive processes for preparation of the unstable isotope, but the newer ^{18}F-flutemetamol is more widely available and can be used with any PET imaging instruments.

Another F-18 affinity probe has been developed to image tumors containing upregulated $\alpha_v\beta_3$ integrin receptors. This conjugate consists of an F-18 modified galactose group attached to the common integrin ligand recognition sequence Arg-Gly-Asp (RGD) (Haubner *et al.*, 2004). The ^{18}F-galacto-RGD affinity probe binds to rapidly dividing cells, such as those involved in angiogenesis and metastasis (Beer *et al.*, 2007) (Figure 1.115). Similar probes having enhanced *in vivo* characteristics were prepared using a dimeric RGD binding site

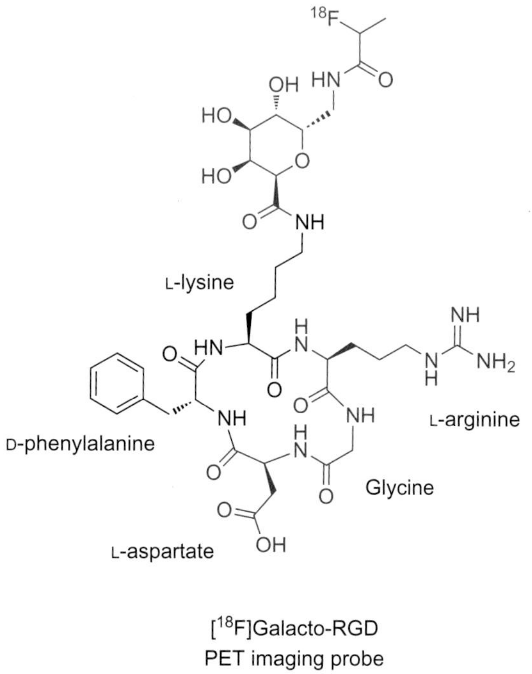

FIGURE 1.115 A targeted PET imaging probe has been constructed by conjugating ^{18}F-galactose (red) to a cyclic peptide containing the amino acid sequence RGD (blue), which binds specifically to integrin receptors.

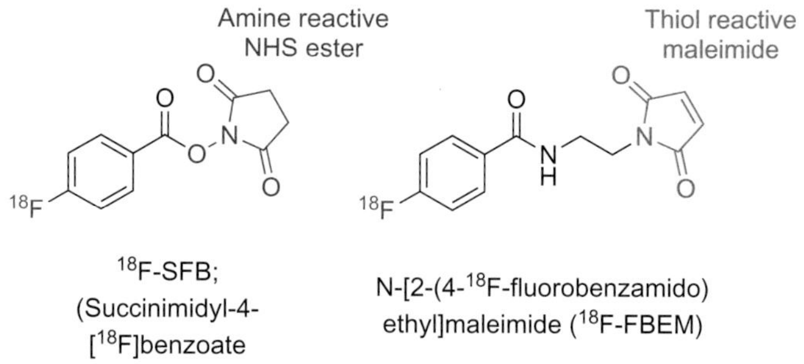

FIGURE 1.116 Labeling reagents have been made that allow the addition of ^{18}F groups to targeting molecules, such as antibodies, for *in vivo* PET imaging applications. The compounds react with amines using an NHS ester group (l) or with thiols using a maleimide group (r).

and a PEGylated arm leading to the ^{18}F modification (Cai *et al.*, 2005).

Other probe conjugates containing the positron emission isotope F-18 may be constructed for PET imaging by coupling a compound containing the radionuclide to an affinity molecule that can bind *in vivo* to the desired diagnostic target. Cai *et al.* (2006) prepared two modification reagents to add an F-18 label to amine-containing molecules (using an NHS ester) or to thiol-containing molecules (using a maleimide group) (Figure 1.116). De Bruin *et al.* (2005) also reported a maleimide-containing F-18 compound for protein and peptide labeling. These reagents provide flexibility in the design of custom PET imaging probes for diagnostic use. In addition, an F-18 labeled antibody mimic (Affibody) was developed against the epidermal growth factor receptor 2 (HER2) to target cells over-expressing this protein, as is common in some breast and lung tumors. Affibody can be used in PET imaging and for imaging guided surgical techniques (Kramer-Marek *et al.*, 2008). The bioconjugate probe was prepared by reacting the modification reagent N-[2-(4-[^{18}F]fluorobenzamido)ethyl]maleimide with a thiol-modified anti-HER2 Affibody.

Nanoparticles labeled with multiple F-18 isotopes have also been used in bioconjugates with a targeting agent for *in vivo* PET imaging (Devaraj *et al.*, 2009). The particle was constructed from a superparamagnetic core that was coated with dextran and labeled with F-18 using short PEG spacer arms with terminal azido groups for a click chemistry-based reaction scheme. The probe could be used for PET imaging with 200 times greater sensitivity than MRI imaging with iron oxide nanoparticles alone.

There have also been reports on the development of F-18 labeling reagents for the modification of oligonucleotide probes for *in vivo* imaging. Kuhnast *et al.* (2003, 2004) reported the synthesis of a bromoacetamide reactive group on two types of ^{18}F-tagged small organic reagents for labeling DNA by alkylation. In addition, Mercier *et al.* (2011) used a click chemistry-based reaction with an azide and an alkyne (see Chapter 17) to couple an F-18 reagent with an siRNA probe for therapeutic *in vivo* inhibition of protein synthesis with simultaneous PET imaging capability (Figure 1.117).

It has been shown that bioconjugates incorporating any one of the three most popular radionuclides, Y-90, F-18, or I-131, can be imaged using rather inexpensive optical imaging devices (Liu *et al.*, 2010), because their photon emission is intense enough to be detected using CCD cameras. The emission wavelengths of these isotopes actually encompass a broad region through the visible and NIR, which is the standard range used for most imaging instruments; therefore, most commercial imagers are suitable for rapid imaging, especially when working with animals for research purposes.

Bioconjugates for High-Contrast Imaging

High-contrast *in vivo* imaging using magnetic resonance imaging (MRI) techniques has become a common way to noninvasively detect pathophysiological abnormalities in individuals with disease or injury. MRI can provide high-definition 3D images of internal structures through the use of a combination of an intense magnetic field and radiofrequency radiation. Images are formed through detection of the energy released from hydrogen atom relaxation in tissue after application of the magnetic field followed by a pulse of radio waves

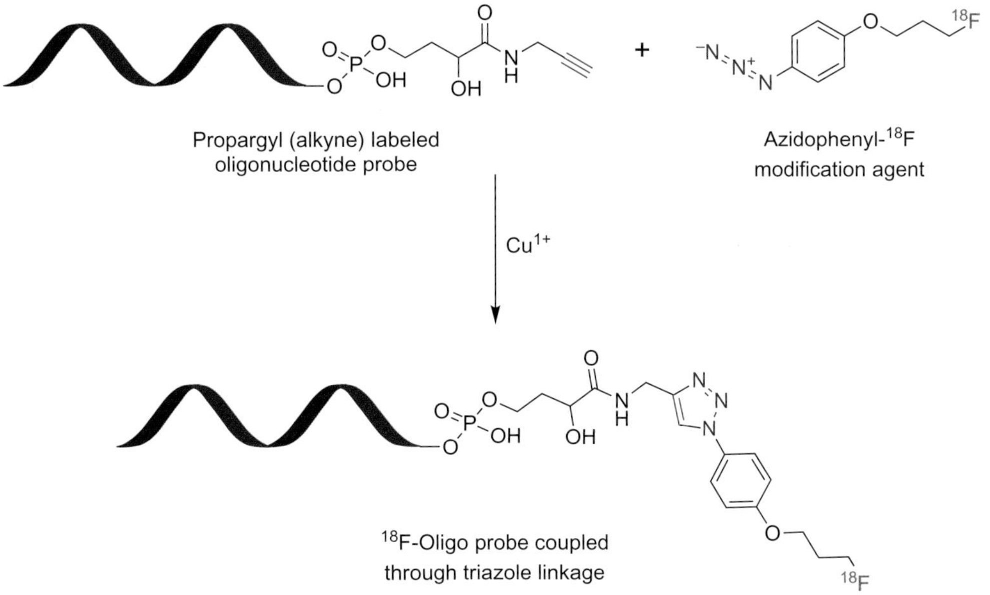

FIGURE 1.117 An azido ^{18}F labeling reagent can be used to modify alkyne-containing molecules using the copper-catalyzed click reaction, which forms a triazole linkage.

tuned to the proper frequency (Sands and Levitin, 2004; Strijkers *et al.*, 2007).

Lanthanide contrast agents have been used in MRI procedures to increase the contrast between differences in tissue types and thus improve the images obtained through scans. In particular, chelates of the trivalent, paramagnetic gadolinium or dysprosium ions have been developed for use with or without conjugated targeting agents to aid in the production of high-resolution images (Rocklage and Watson, 1993). The chemistry of the lanthanide metals, including gadolinium, and their complexes with caged bifunctional chelating agents such as DOTA have recently been reviewed (Sherry *et al.*, 2009).

At the time of writing, there are nine gadolinium-based contrast agents (GBCAs) approved for diagnostic use in the United States and Europe. All of these agents are low- molecular-weight constructs containing a single gadolinium ion coordinated in the core of an organic chelate. Products such as Magnevist and Omniscan consist of a linear chelating molecule built from a DTPA derivative (Chapter 12, Section 1.1), while the product ProHance is a DOTA derivative having a cyclic caged structure to coordinate the metal ion (Chapter 12, Section 1.2).

Bioconjugates of GBCAs with targeting molecules have also been developed to form enhanced delivery agents able to target disease *in vivo* with high specificity toward a particular tissue or cell or to have greater half-life in circulation. One simple conjugate consists of a GBCA–albumin complex that, once injected IV, functions to maintain greater circulating drug half-life to better image fine vasculature (Niemi *et al.*, 1991). This conjugate provides a significant increase in half-life over the use of small organic chelating compounds in an unconjugated state. However, other bioconjugate designs have incorporated modified targeting molecules such as monoclonal antibodies to perform enhanced imaging of tumors or other structures *in vivo* (Göhr-Rosenthal *et al.*, 1993; Ramakrishnan *et al.*, 2008). Figure 1.118 shows the bioconjugation reactions involved in creating the Gd-DTPA-anti-VEGFR2 antibody as described by Jun *et al.* (2010). Chapter 12, Section 1, describes the bioconjugation of suitable reactive chelating compounds with targeting molecules of all types to form specific high-contrast imaging agents that can bind to targeted cells and tissues *in vivo*. Once charged with a lanthanide metal ion such as gadolinium, the resultant conjugate can be used as a MRI contrast agent, which can enhance the detection of diseased cells and other tissue structures.

To increase the sensitivity of MRI even further using contrast agents, polymeric scaffolds have been used to permit greater numbers of chelated gadolinium ions associated with each targeting agent. For instance, Swanson *et al.* (2008) used a G5 PAMAM dendrimer (see Chapter 8) to create a conjugate containing multiple Gd(III)-DOTA chelate groups (>50) in addition to folic acid groups (~4.5) for targeting folate receptors on tumor cells. The resultant bioconjugate was retained in tumor cells and functioned in MRI to give greater relaxivity (more than four times that of the commercial agent Omniscan) and higher contrast sensitivity compared to other Gd chelates that lack the dendritic core.

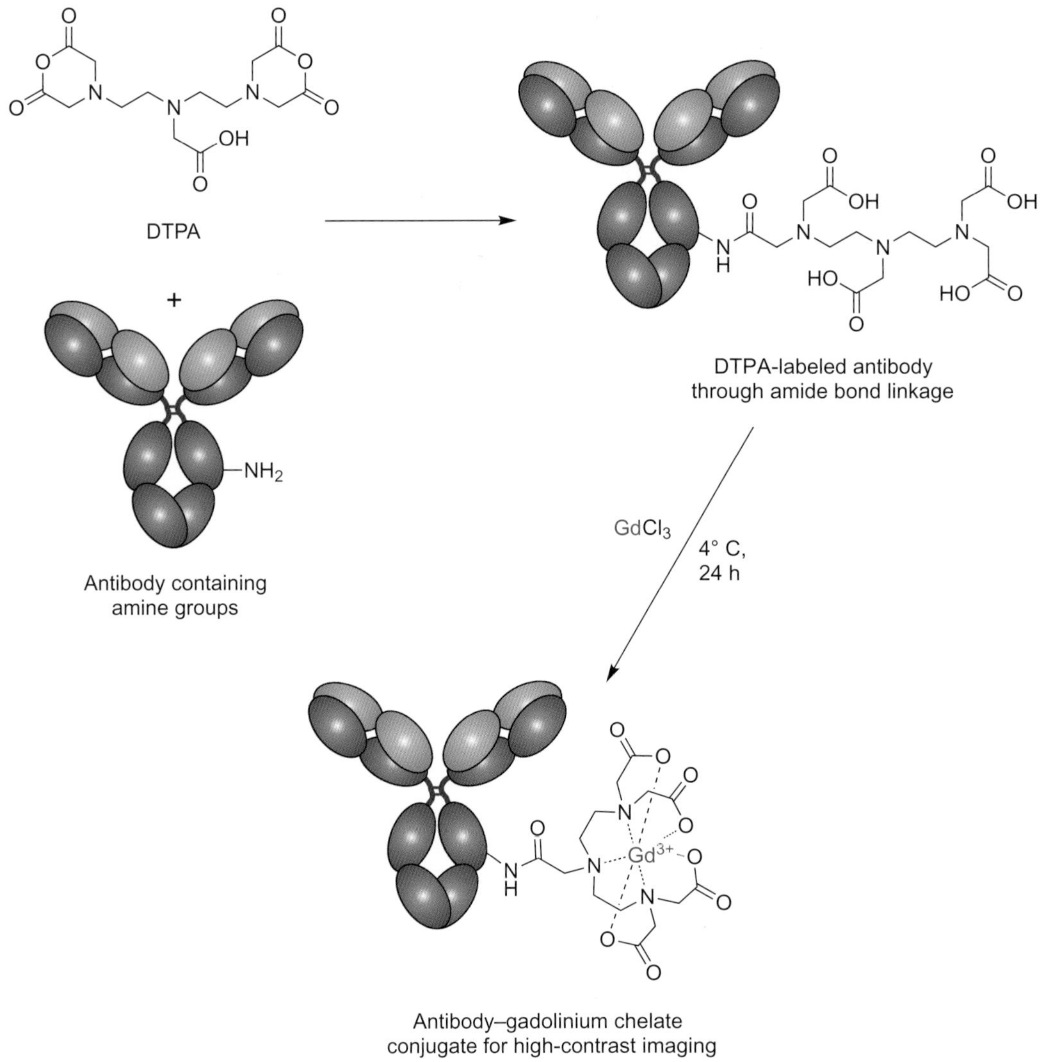

FIGURE 1.118 The chelating compound DTPA can be reacted with the amine groups on antibodies to form an amide bond derivative that is able to bind gadolinium ions useful for high-contrast imaging applications. The conjugate provides targeted delivery of the imaging agent directly to tumor cells *in vivo*.

A similar dendritic contrast agent conjugate was created using a G8 dendrimer with maximal modification of its terminal amines with a DTPA derivative for loading with gadolinium (Kobayashi *et al.*, 2003). The greater size of this G8 dendritic complex compared to the G5 dendrimer used above permitted MRI of the entire lymphatic system in mice to detect various disease states or infections. Thus, the use of a dendrimer-based scaffold can provide increased sensitivity for imaging due to the multivalent nature of the polymer plus it can be modulated for size to permit targeting of the desired organ or cell.

NIR Fluorescent Conjugates for In Vivo Imaging

Diagnostic procedures involving *in vivo* imaging often depend on the creation of a bioconjugate containing a targeting molecule coupled to a detection component, which can be observed from a vantage point outside the body. For instance, the exact location of tumors may be detected using bioconjugates designed to localize highly sensitive imaging agents right at the site of the cancer cells. The majority of *in vivo* imaging is done using PET or MRI with radionuclides or high-contrast agents, respectively (see previous sections). However, one of the latest trends in diagnostic imaging is to make use of fluorescence detection, which can have the advantages of avoiding the use of radioactive substances and extraordinarily expensive instrumentation. In fact, fluorescence-based *in vivo* imaging procedures can make use of relatively portable detection devices that can be used in clinics or even during surgical procedures to image tumors in real time as operations are taking place (De Grand and Frangioni, 2003; Gioux *et al.*, 2005; Troyan *et al.*, 2009).

Far-red or NIR fluorescent conjugates can be used to visualize cells directly through skin or tissue (Hilderbrand and Weissleder, 2010). NIR fluorescent dyes attached to antibodies or other targeting molecules can be excited through biological matter, because biomolecules typically have very low absorption potential for light in this unique spectral window between 680nm and 900nm (the "biological window"). Any attempt to use fluorescent probes that have lower excitation and emission wavelengths, such as through the UV or visible regions of the spectrum, will encounter severe interference from biomolecules with high absorption characteristics. Naturally absorbing biomolecules include porphyrin-containing proteins, such as the highly abundant protein hemoglobin and the cytochromes, as well as aromatic residues in proteins and small molecules like melanin, lipids, and riboflavin (Figure 1.119). These ubiquitous absorbing components within biosystems are the primary reason why visible light fluorescence imaging cannot work through tissues, consequently restricting light penetration to only a few millimeters at best. In addition, wavelengths of light above about 1000nm also are not very favorable for *in vivo* imaging, because water and many biomolecules begin to absorb strongly at the higher end of the NIR spectrum (Cheong *et al.*, 1990; Chen *et al.*, 2004; Khullar *et al.*, 2009).

The signals emanating from NIR fluorescent probes conjugated to the appropriate targeting molecules can be combined with imaging technology sensitive to higher wavelengths than visible light and detected through at least several centimeters of tissue *in vivo* (Rao *et al.*, 2007). At these wavelengths, the only remaining limitation to tissue penetration is light scattering, which is also minimized at higher wavelengths but still decreases signal with the distance traveled through tissues. Thus, targeted NIR probes can be used to detect the sites of tumors for diagnostic purposes or, using the right bioconjugate designs, they can also be used to detect the activity of certain enzymes indicative of tumorogenesis or apoptotic events.

NIR bioconjugate probes are simpler to use compared to radionuclide probes, because the difficulties encountered in production, safety, and stability of radioisotopes are eliminated through use of simple organic fluorophore labels. Where radiolabeled probes can be dangerous to handle and may be toxic *in vivo*, NIR fluorescent probes typically have low toxicity and require no extraordinary precautions in handling or disposal. Although the high-contrast probes for MRI and the positron emitting probes for PET imaging still are the most widely used detection platforms for *in vivo* applications, the NIR fluorescence field is growing rapidly.

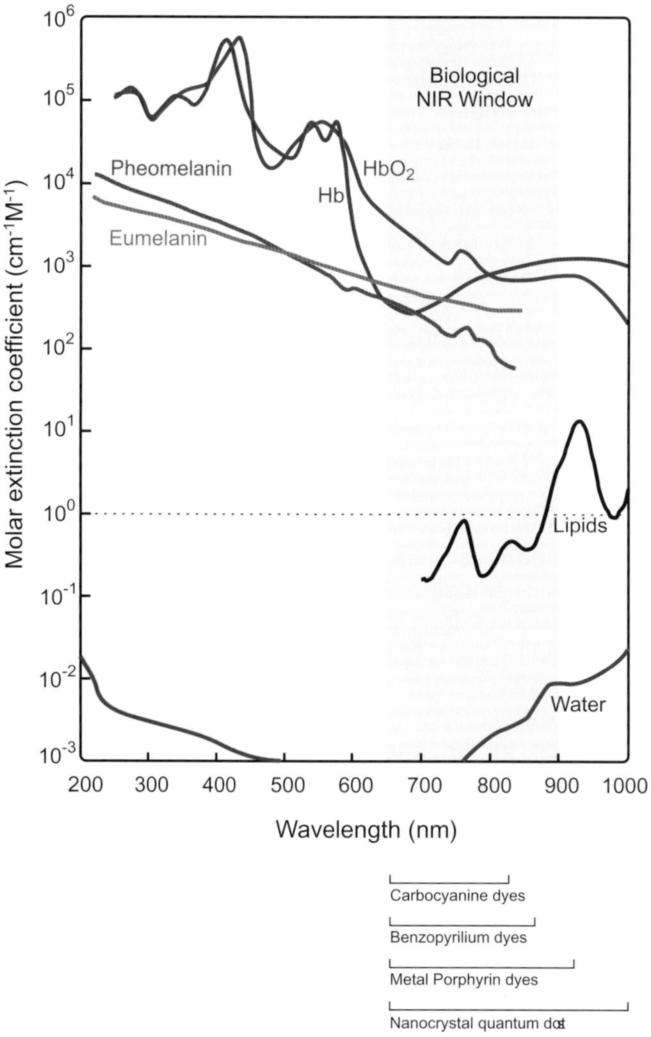

FIGURE 1.119 The biological NIR window in which the absorbance interference due to biomolecules is at a minimum and therefore fluorescent probes can be observed directly through tissues and fluids. *The spectral properties of hemoglobin and water were adapted with permission from Kobayashi, H., Ogawa, M., Alford, R., Choyke, P.L., and Urano, Y. (2010) New strategies for fluorescent probe design in medical diagnostic imaging. Chem. Rev. 110, 2620–2640, figure 1; Copyright 2010, American Chemical Society. The spectral properties of lipids, eumelanin, and pheomelanin were adapted from the Wikipedia page Near-infrared window in biological tissue (http://en.wikipedia.org/wiki/Near-infrared_window_in_biological_tissue; author: Zhun310) and licensed for any use by the Creative Commons Attribution-Share Alike 3.0 Unported license.*

Several strategies can be used to target and fluorescently image biomarkers *in vivo* (Rao *et al.*, 2007). In a common targeting design, antibodies or proteins having specific affinity toward a desired protein or biomolecule can be labeled with a NIR fluorescent dye to create the active bioconjugate. In a similar design, antibody fragments, small peptides, or organic ligands having affinity for an active site or receptor binding site on a protein can be used as targeting agents and similarly labeled with a NIR dye. In both of these designs the imaging bioconjugate is inherently fluorescent and the *in vivo* target becomes progressively labeled with the probe as the complex circulates throughout the cardiovascular system.

In an alternative design approach, probes can be made that are not immediately fluorescent, but generate a NIR signal upon action by a targeted enzyme. For example, probes can be designed to detect specific protease activities *in vivo*, and they function to generate fluorescent signals only upon conversion by the targeted enzyme. In this approach, a short peptide substrate containing an amino acid sequence that is able to be recognized and cleaved by a targeted enzyme *in vivo* is dual labeled with a fluorescent molecule and a quencher to create a FRET (fluorescence resonance energy transfer) signaling agent. The initial probe is dark due to the action of the quencher molecule being held in proximity to the fluorescent molecule—any energy absorbed by the fluor is transferred to the quencher without radiative emission taking place. However, when the peptide is cleaved by the targeted protease, the quenching effect on the probe is relieved as the quencher and fluor are able to diffuse apart, and a fluorescence signal is then generated in the immediate region of the enzymatic activity. Thus, this type of probe is called a peptide molecular beacon, which produces a fluorescent signal in response to the amount of specific protease detected. Regardless of the imaging probe design, the conjugate should be constructed to find its target *in vivo* and create a NIR signal directly at the point of the cells or tissues being monitored. Figure 1.120 illustrates some of the main types of NIR fluorescent probes.

Perhaps the most widely used NIR fluorescence probes are based on carbocyanine dyes that contain either a pentamethine (5-carbon) or heptamethine (7-carbon) bridge. Carbocyanine dyes also typically have aromatic ring structures at both ends of the bridge, which can be of various compositional types (see Chapter 10, Section 8). The most effective dyes for *in vivo* imaging correspond to derivatives of the standard Cy5.5 or Cy7 core structures, which have excitation and emission characteristics within the range of 680 to 900 nm. In addition, the alternative benzopyrillium dyes within the NIR range have also been found successful for imaging applications (Czerney *et al.*, 2005) (Figure 1.121). Derivatives of these dyes, such as those containing water-solubilizing sulfonates and amine-reactive NHS esters or thiol-reactive maleimide groups, provide conjugation ability with biomolecules to create targeted *in vivo* NIR probes of many types (Thermo Fisher, Dyomics).

As an alternative NIR platform, quantum dots (QDs) have also been used for *in vivo* imaging purposes in animals. The intense fluorescence and high photostability of QDs are attractive for use in extended imaging applications. Like organic fluorescent probes, they can be designed to have spectral properties in the NIR region, which produces intense fluorescence in the ideal wavelength range for tissue penetration. They can also be conjugated to targeting molecules to provide specific interaction with biomarkers *in vivo*. However, unlike organic fluors, their alloy construction, which uses heavy metals with known toxic properties, has restricted their utility to non-human diagnostic imaging applications (Chapter 10, Section 10).

Bioconjugate targeting complexes containing affinity molecules linked to NIR dyes have been used to image the cardiovascular system, tissue, organs, and cells directly within patients. The types of conjugates that have been created are highly varied depending on the biomarker being targeted. For instance, the short peptide sequence RGD (Arg-Gly-Asp) has been conjugated to a Cy5.5 dye to specifically bind cell surface integrin $\alpha_v\beta_3$ receptors, which are upregulated in many tumors, including glioblastoma (Chen *et al.*, 2004) (Figure 1.122). Goldshaid *et al.* (2010) created a multi-purpose *in vivo* imaging conjugate using a cyclic RGD peptide for affinity targeting with a bacteriochlorophyll derivative that displayed photodynamic and fluorescent properties. The probe could be used both to localize breast tumor necrotic domains containing the $\alpha_v\beta_3$ integrin receptors by NIR imaging and to treat them using PDT. Other groups have used multimeric RGD constructs to improve the binding to receptors and increase imaging signals (Cheng *et al.*, 2005; Ye *et al.*, 2006). Previously, this same small peptide targeting agent was used along with a [^{18}F]galactose label to successfully detect tumors *in vivo* using PET imaging. The ability to use the same targeting component with NIR fluorescence imaging provides a powerful detection scheme, which can be used in situations where PET imaging instrumentation is not available or its use is impractical.

NIR imaging conjugates can also consist of fluorescently labeled antibodies that can target and bind to specific antigens *in vivo*. Bioconjugate probes containing intact IgG-type antibodies are often too large to effectively penetrate cells or the blood brain barrier, but they can be quite useful at targeting surface antigens on cells or tissue structures. The use of antibody fragments, such as recombinant scFv fragments, labeled with NIR dyes can provide bioconjugates small enough in size to access tumor cells throughout the body (Ramjiawan *et al.*, 2000). The imaging of xenograft tumors in mice or in tumor cells was carried out using a recombinant Affibody–NIR fluorescent conjugate targeted at epidermal growth factor receptors (EGFRs), which often are dramatically upregulated on the surfaces of cancer cells (Gong *et al.*, 2010; Miao *et al.*, 2010). The NIR dyes used in these studies consisted of a Cy5.5 molecule, which was a carbocyanine derivative having

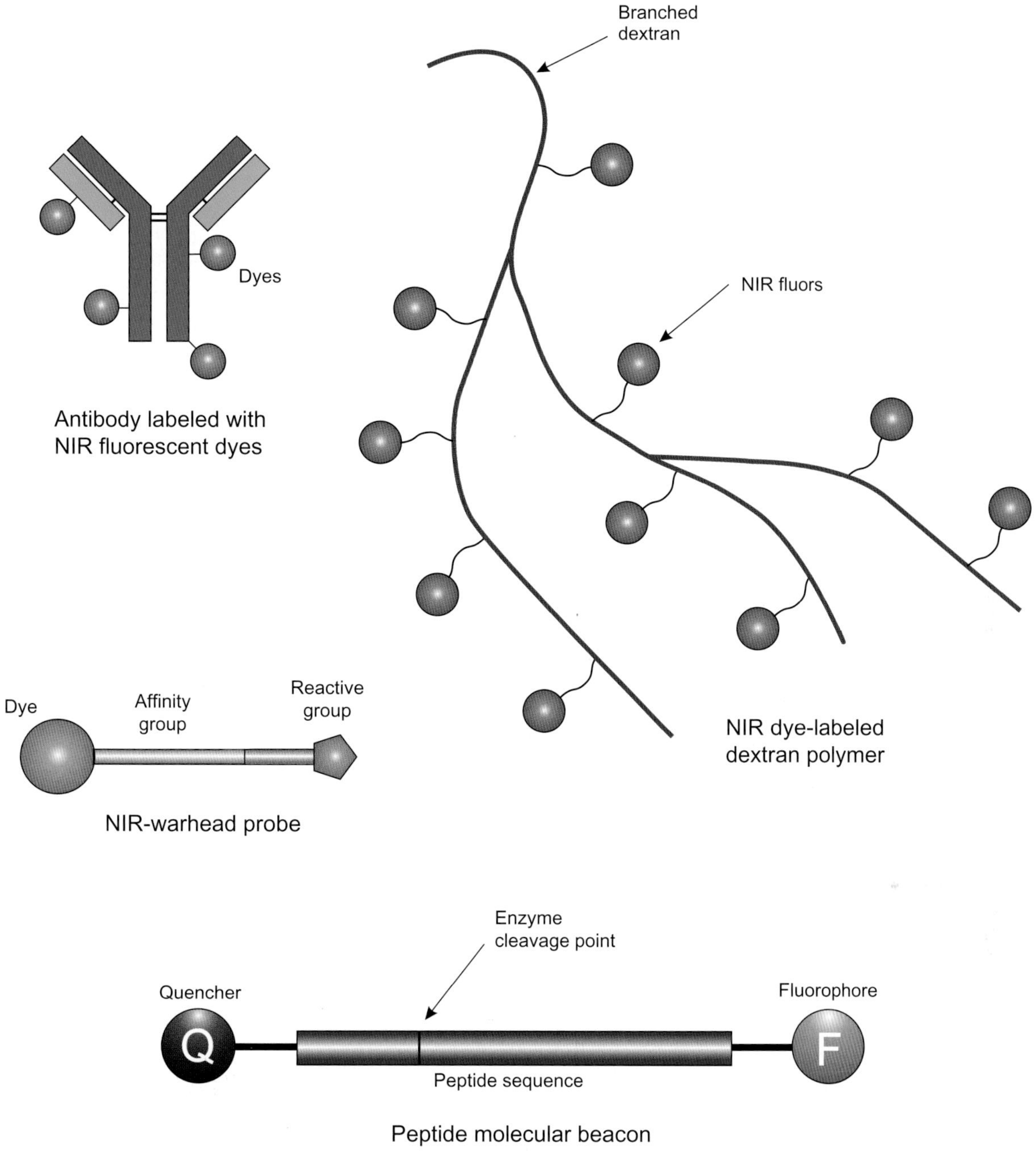

FIGURE 1.120 Probe designs for NIR *in vivo* imaging applications can use a number of conjugate forms. Fluorescent dye-labeled antibodies are perhaps the most common way of targeting an epitope *in vivo*. In addition, polymeric scaffold conjugates have been used to increase the number of dyes delivered to a particular site. NIR warhead probes are used to target an active site on an enzyme while the associated reactive group covalently attaches to a functional group near or within the binding region and the dye provides the detectable fluorescent signal. A molecular beacon probe typically includes two dyes attached to either end of a short peptide that can FRET signal between each other when the peptide is intact. The use of a quencher in this design results in a dark probe upon administration. Once the specific peptide sequence is cleaved by a targeted protease, however, the quencher moves away from the dye and fluorescence is restored.

spectral properties that produced emission near 800 nm (IRDye800CW, LI-COR), well into the region where biomolecule absorption will not interfere with *in vivo* imaging. Monoclonal antibodies have also been conjugated with QDs to successfully target cancer using animal models; however, the large size of nanoparticle conjugates typically limits them to cell surface epitopes, which are more easily accessed by the vascular system (Gao *et al.*, 2004).

Imaging bioconjugates can also be constructed by using antibodies that previously have been validated as therapeutics for targeting specific tumor antigens

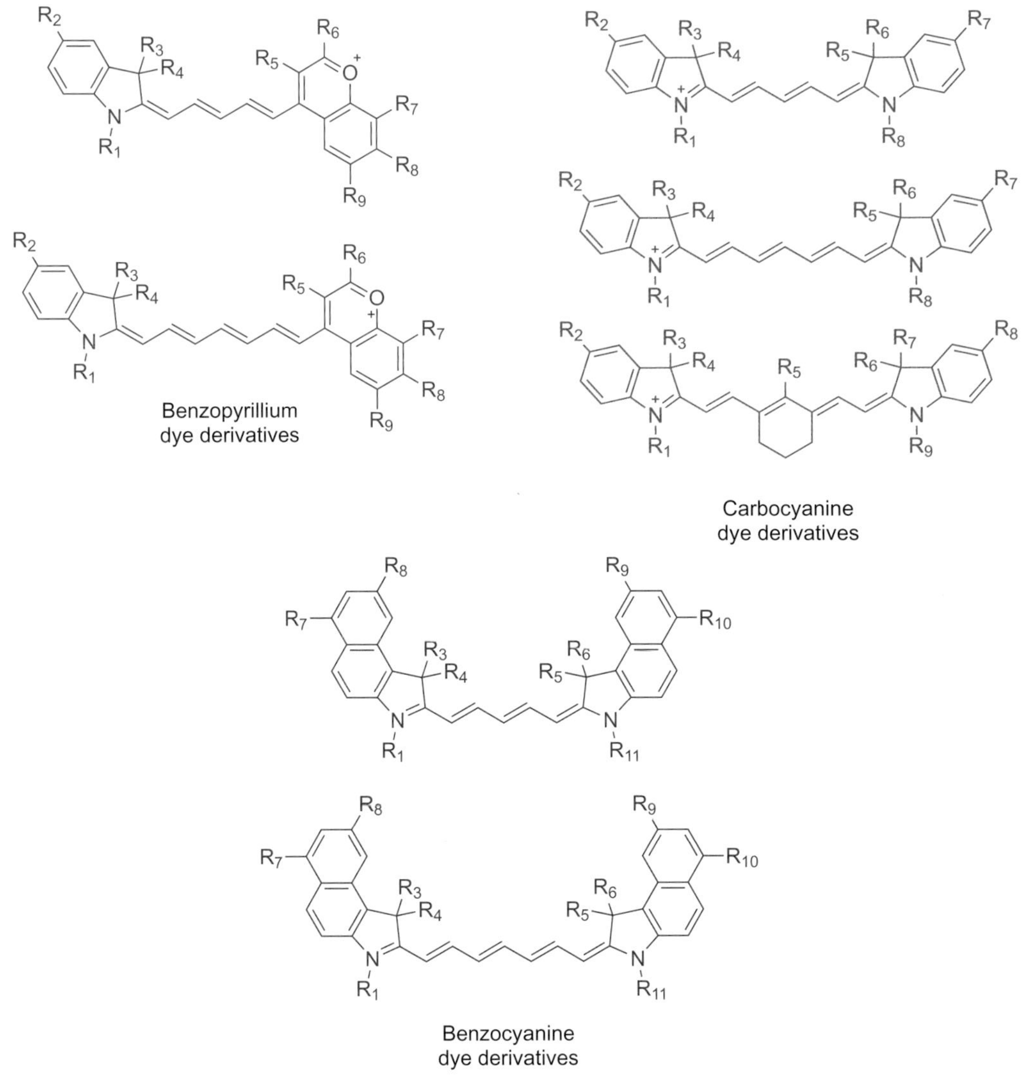

FIGURE 1.121 Some dye types that are commonly used for NIR *in vivo* imaging purposes.

in vivo. For instance, Portnoy *et al.* (2011) reported on the use of a liposome conjugate containing the NIR fluorescent probe indocyanine green with the anti-EGFR antibody Cetuximab. Selective binding of the bioconjugate to colon carcinoma cells that over-express EGFR was used to successfully image the tumor in cell based assays. It was also found that the dye had greater fluorescent signal when associated with the liposome than when using the corresponding dye directly conjugated with the antibody.

Other NIR fluorescent constructs can be designed with quenched dyes and certain peptide sequences to image cells expressing particular proteases. The enzymes are able to cleave the peptide linker, release the quenching on the fluorescent label, and produce a NIR signal within the associated cells (Weissleder *et al.*, 1999; Funovics *et al.*, 2003; McIntyre *et al.*, 2004). Enzyme-activatable, NIR fluorogenic probes are capable of highly sensitive imaging due to the catalytic nature of signal generation upon encounter with the targeted enzyme (Tung, 2004). Unlike the use of affinity targeting probes that bind to an epitope and carry with it fluorescent dyes but produce no additional fluorescent products upon binding, enzyme-sensitive molecular beacon probes continually generate fluorescence in proximity to the protease so long as additional substrate is available (Rao *et al.*, 2007). For instance, using this type of conjugate, cells can be detected that produce an abundance of cathepsin enzymes, which are associated with many tumor types (Nomura and Katunuma, 2005). In this case, a cathepsin-specific amino acid sequence is used as the linker in the molecular beacon design, which can get cleaved in the presence of the enzyme.

Alternatively, cells that produce caspases might be imaged using a similar conjugate design but another

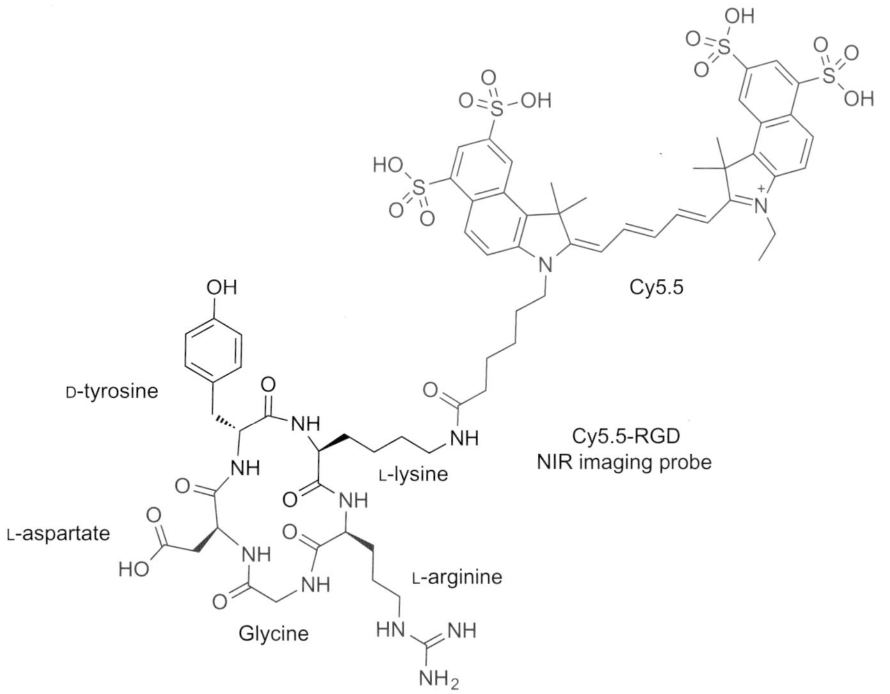

FIGURE 1.122 A conjugate of the integrin binding peptide RGD with the NIR imaging dye Cy5.5 provides a targeted probe for *in vivo*.

amino acid sequence in the core structure, which can then be used as an indicator of apoptosis in cancer cells after a successful treatment regimen (see previous section on FRET peptide probes in this chapter). The design of this type of imaging probe usually consists of a molecular beacon with a fluorescent molecule at one end of the peptide and a quencher molecule on the other side of the cleavage point in the amino acid sequence (Edgington *et al.*, 2009; Rai *et al.*, 2010) (Figure 1.123). In solution, the probe generates no fluorescence signature, because the quencher molecule is held close enough to the fluor to effectively absorb any of its emission energy. As activated caspases of the apoptotic cascade within a cell cleave the labeled peptide probe, the tether linking the fluor and the quencher is hydrolytically broken and the FRET quenching process no longer occurs. This proteolytic cleavage event generates a fluorescent signal proportional to the amount of caspase enzyme activated in the cell or tissue. The fluorescent signal is thus an indicator of apoptotic cell death and can be used as a sign of a potentially effective tumor treatment by a drug candidate.

A major challenge of small molecular beacon conjugates for use *in vivo*, such as protease probes, is their rapid clearance from circulation, which limits the time available for them to interact with the targeted proteases. Their small size, however, can also be an advantage for being able to penetrate cell membranes and interact with intracellular targets. For extracellular protease detection, such as targeting of matrix metalloproteinases (MMPs), larger poly-amino acid scaffolds or PEG-modified conjugates have been used to increase the probe size and therefore better retain it within the vascular system for longer periods (Bremer *et al.*, 2001). In addition, Zhu *et al.* (2011) designed a bioconjugate probe for MMPs that allowed real-time video imaging *in vivo* for up to 24 hours post injection. In the construction of this probe, it was found that modification of the molecular beacon with discrete $mPEG_{12}$ chains resulted in an optimal signal for imaging (Figure 1.124). See Chapter 18 for further discussion on the use and advantages of short, discrete PEG compounds in bioconjugate design.

Carbocyanine dyes have also been used in a quenched FRET probe system without the use of a nonfluorescent absorbing partner as the energy acceptor. Weissleder *et al.* (1999) used a poly-L-lysine polymer that was modified with approximately 92 mPEG molecules along with 11 Cy5.5 NIR fluorescent labels. The mPEG modifications provided increased solubility and *in vivo* stability, thus extending the half-life of the probe in circulation. After modification, the lysine polymer backbone still contained enough unmodified lysines to act as an efficient substrate for proteases. In addition, the high number of Cy5.5 dyes attached to the polymer caused dye–dye interactions resulting in quenching, which caused a 15-fold lowering of fluorescence signal compared to the free dye. Once the poly-lysine

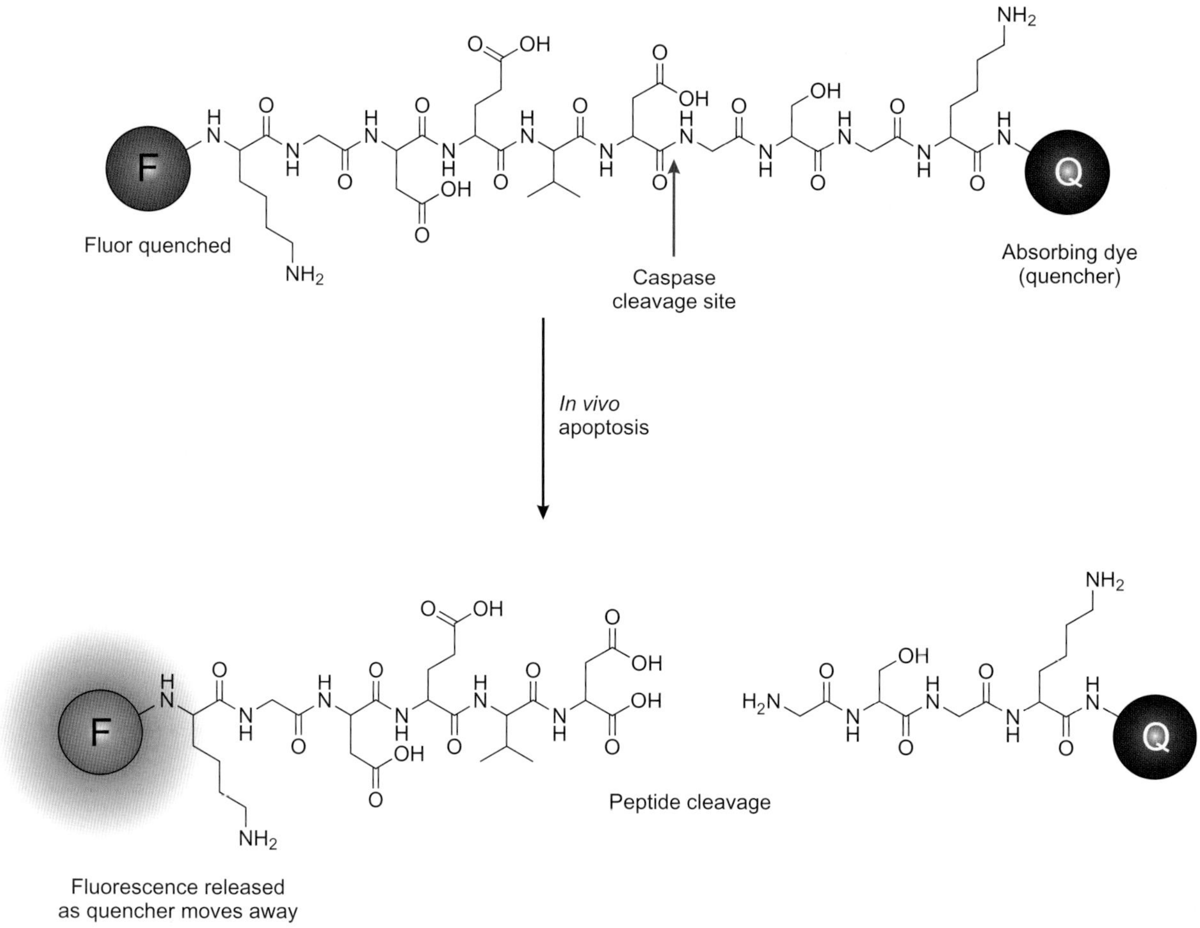

FIGURE 1.123 A molecular beacon probe for caspase enzymes can be constructed from a caspase-specific peptide sequence with a NIR fluorescent dye on one end and a quencher on the other end. When the probe is cleaved in the presence of caspase enzymes the dye will be released and its fluorescent signal restored.

chain was proteolytically cleaved by an enzyme *in vivo*, the probe liberated dye molecules on short peptide segments, which relieved the fluorescence quenching and resulted in an increase in signal at the site of the cells producing the proteases (Figure 1.125) A similar design was used to detect caspase-1 *in vivo* by conjugating the cleavable peptide substrate Gly-Trp-Glu-His-Asp-Gly-Lys containing the NIR fluorescent probe Cy5.5 to a partially PEGylated poly-L-lysine polymer (Messerli *et al.*, 2004).

Zhang *et al.* (2009) designed an activatable NIR caspase-3 probe by using two derivatives of a heptamethine carbocyanine dye (Cy7), which could efficiently FRET quench between them and thus only yield a fluorescent signal if enzyme was present. The dyes were linked on the ends of a core DEVD amino acid sequence (Asp-Glu-Val-Asp), which is known to be specific for caspase-3. Upon cleavage of the short peptide, an *in vivo* fluorescence image could be detected using an excitation laser at 785 nm with emission at 805 nm.

In another bioconjugate design for targeting proteases *in vivo*, a specific peptide sequence can be labeled with a fluorescent molecule at one end and also contain a reactive group at the other end. Unlike enzyme activatable probes that catalytically produce a NIR fluorescent signal upon cleavage, these NIR probes are able to bind to the active site of a targeted enzyme and link covalently to a nearby functional group, usually a nucleophile, thus permanently tagging the enzyme with a single fluorescent label (Griffin *et al.*, 2007; Cursio *et al.*, 2008, 2009; Delgado-Martín *et al.*, 2009; Erman *et al.*, 2009). The peptide sequence typically used in a covalent probe of this type is an inhibitor of the enzyme active site and not susceptible to cleavage by the enzyme. In addition, this "warhead" probe design must contain a reactive group that does not just indiscriminately couple to other proteins or functional groups present on biomolecules, but only reacts if it is held in proximity to a nucleophile at the targeted enzyme active site.

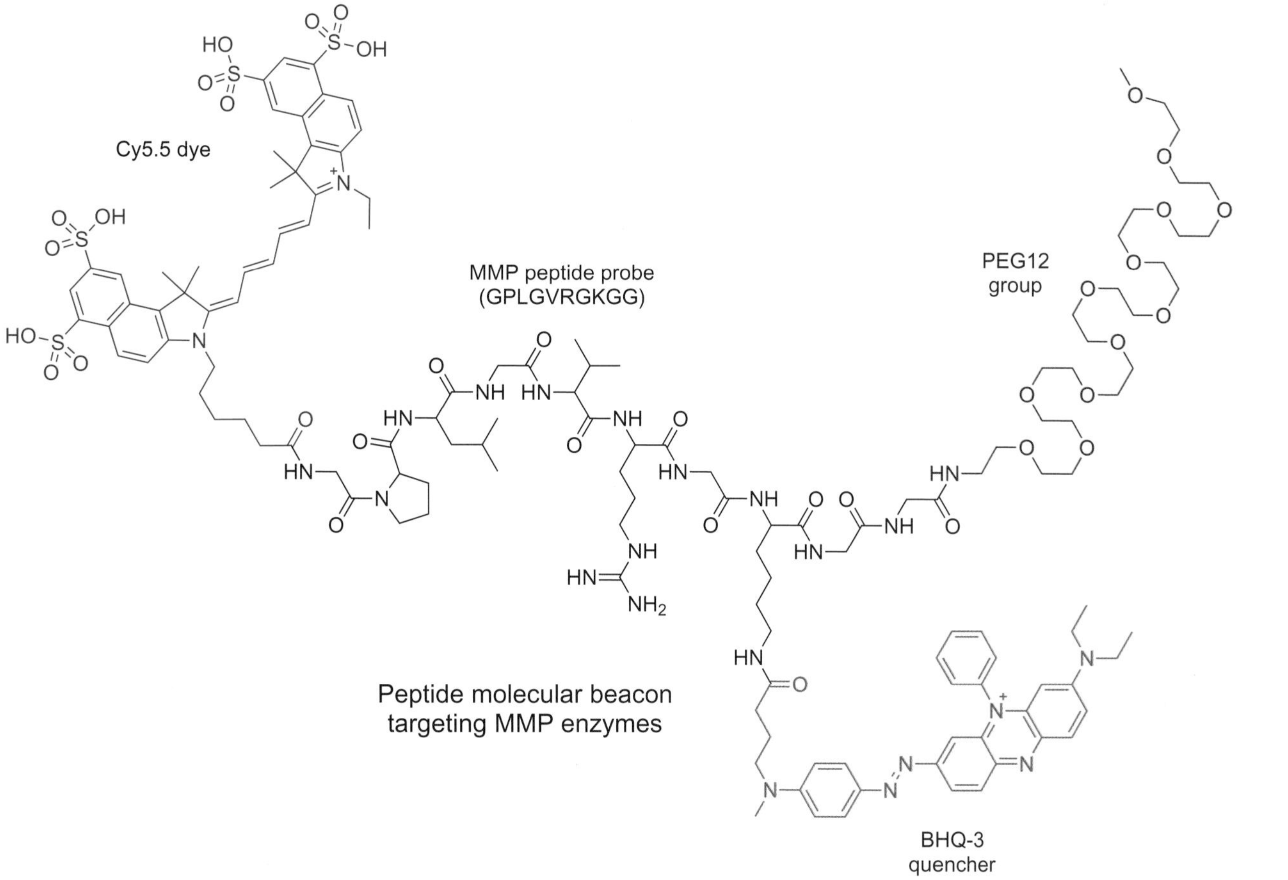

FIGURE 1.124 A conjugate for targeting MMP activity within cells or *in vivo* has been constructed using a peptide sequence combined with the fluorescent dye Cy5.5 and the quencher molecule BHQ-3. Also added to this conjugate was a PEG group to provide water solubility and enhanced life time in circulation *in vivo*.

Warhead NIR probes have been designed to contain reactive groups such as epoxide, fluoromethyl ketone (FMK), chloromethyl ketone (CMK), aldehyde, and difluorophenoxy methyl ketone (OPh) (Blum *et al.*, 2009). Examples of these probes are shown in Figure 1.126. Since the NIR fluorescent label is not quenched in this probe design, excess probe that does not covalently link to the enzyme must be allowed to wash free of the targeted area. For *in vivo* imaging this is done merely by extending the time of interaction for several hours to permit the circulatory system to clear excess probe. For cell-based imaging applications, the cells are washed with buffer to remove excess reagent before imaging the bound probe, which has remained within the cells through covalently linking with the enzyme target.

3.6. Vaccines and Immune Modulation

A conjugate vaccine is a therapeutic agent designed to stimulate the immune system to recognize and attack a potential pathogen invader or a diseased cell. Conjugate vaccines can also be used to generate protective IgG antibodies against allergy-related antigens to prevent the binding of IgE and modulate an allergic reaction. Such immunogen conjugates are an advanced form of the original vaccines that consisted of inactivated or attenuated viruses or bacteria cells, which have been administered to healthy individuals for decades to develop immunity toward many dangerous diseases. Since the very beginnings of this technology with the cholera vaccine in the late 1800s, vaccine therapy has progressed from the use of inactivated whole pathogens to more elaborate designer hapten-carrier immunogen conjugates that build protective immunity using key peptide sequences or other antigenic epitopes from viruses or bacteria. Immunogen conjugates typically consist of a carrier molecule that has a number of haptens linked onto its surface. Methods for the preparation of hapten-carrier immunogen conjugates are described in Chapter 19.

Immunogen conjugates can take on many different forms depending on the type of carrier molecule chosen and the target epitope selected to link to the carrier. The final complex is the immunogen used for immunization

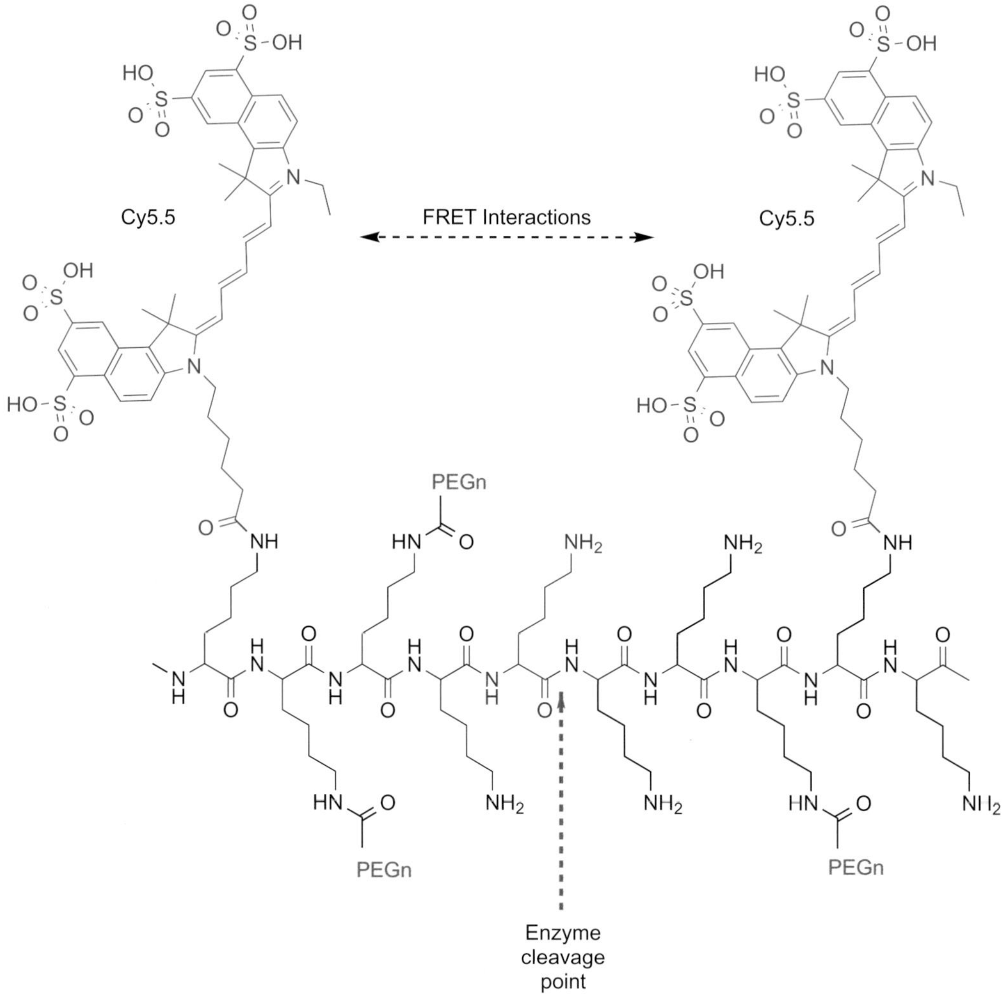

FIGURE 1.125 A FRET quenching molecular beacon style probe can be created by using two fluorescent dyes that are attached very close to each other along the peptide backbone. In this case, the emission from one fluor will be absorbed by the other fluor, thus effectively quenching its fluorescence signature. Once the peptide is cleaved by an enzyme, however, the fluorescent dyes are able to move away from each other to restore fluorescence.

and it usually consists of a large macromolecular core (the "carrier") and a number of small molecule haptens attached to it through various covalent conjugation strategies. Suitable carriers for the attachment of haptens can consist of foreign proteins, synthetic polymers, or lipid micelles that have functional groups or reactive groups on their surface to facilitate a conjugation reaction.

Protein carriers that are commonly used include bovine serum albumin (BSA), keyhole limpet hemocyanin (KLH), thyroglobulin (THY), ovalbumin (OVA), tetanus toxoid, diphtheria toxoid, and tuberculin purified protein derivative (PPD). The two carriers KLH and BSA are widely used for routine immunogen preparation; however, PPD has been found to elicit high titer specificities to coupled antigens with a higher percentage of antigen-specific antibodies being produced from polyclonal or monoclonal production (De Silva *et al.*, 1999).

Synthetic carrier constructs have also been used in the design of conjugate vaccines or immunogens. For instance, a popular peptide-like carrier consists of a polylysine core typically containing three or eight branched amino acids to which peptide haptens may be attached at each pendent side-chain amine group. This design is called the "multiple antigenic peptide" (MAP) system and the final conjugate typically has up to eight copies of peptide antigens attached to each core structure.

The hapten component of a conjugate vaccine is usually much smaller than the carrier and can consist of protein fragments, short peptide sequences, carbohydrates or glycans, or other primary immunogenic portions of a targeted pathogen. Similarly, to prepare immunogen conjugates toward allergens, key epitopes are coupled to the carrier, which are known to elicit strong allergenic responses *in vivo*. In all cases, the carrier molecule is designed to aid in developing an

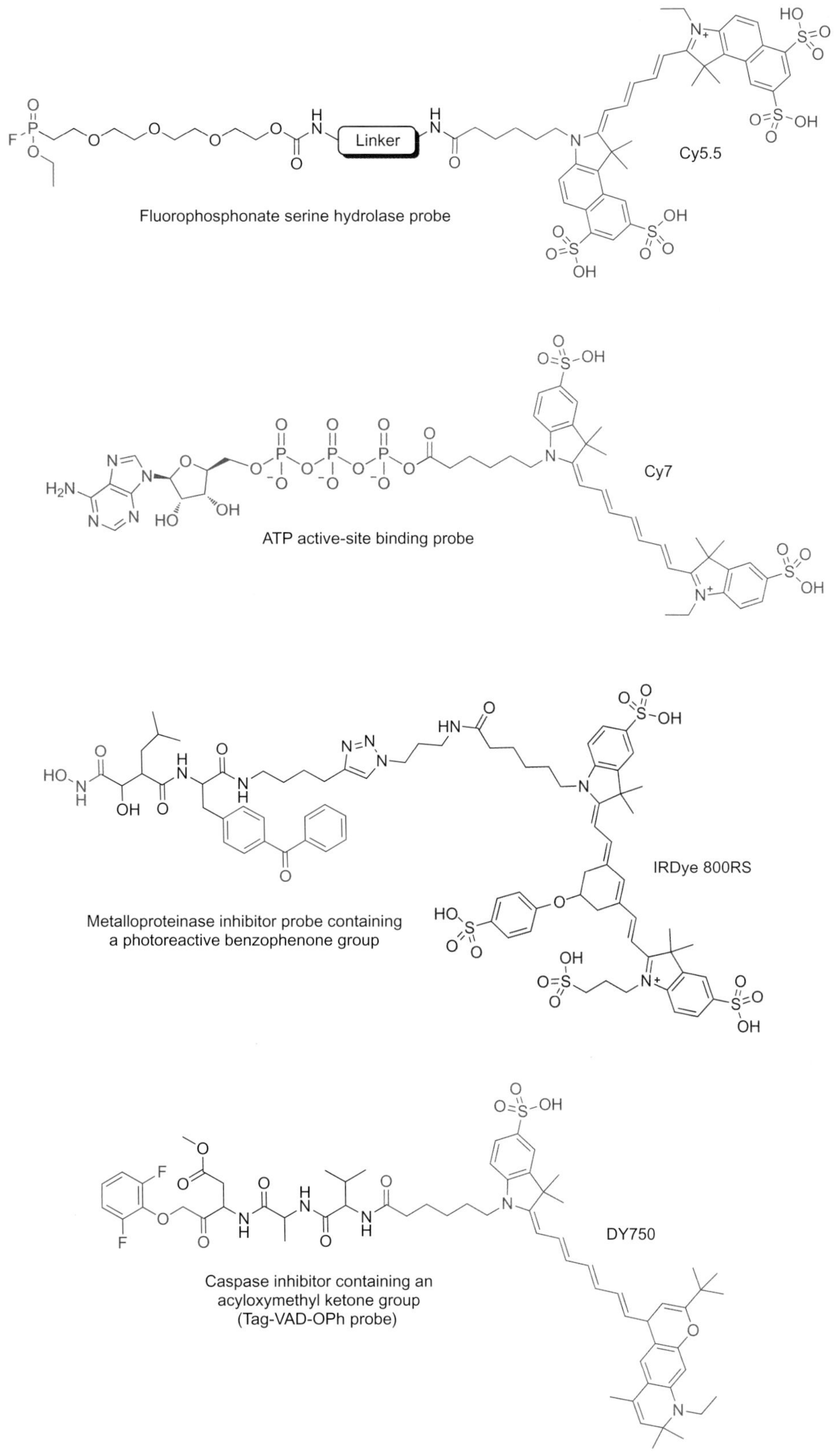

FIGURE 1.126 NIR fluorescent active site probes can be constructed from an affinity warhead type probe containing an affinity group for an active site on a particular enzyme and a reactive group for covalently linking it to the enzyme once bound. The fluorescent dyes used in these conjugates have absorbance and emission properties between about 680 nm and 820 nm, within the NIR window for *in vivo* imaging application.

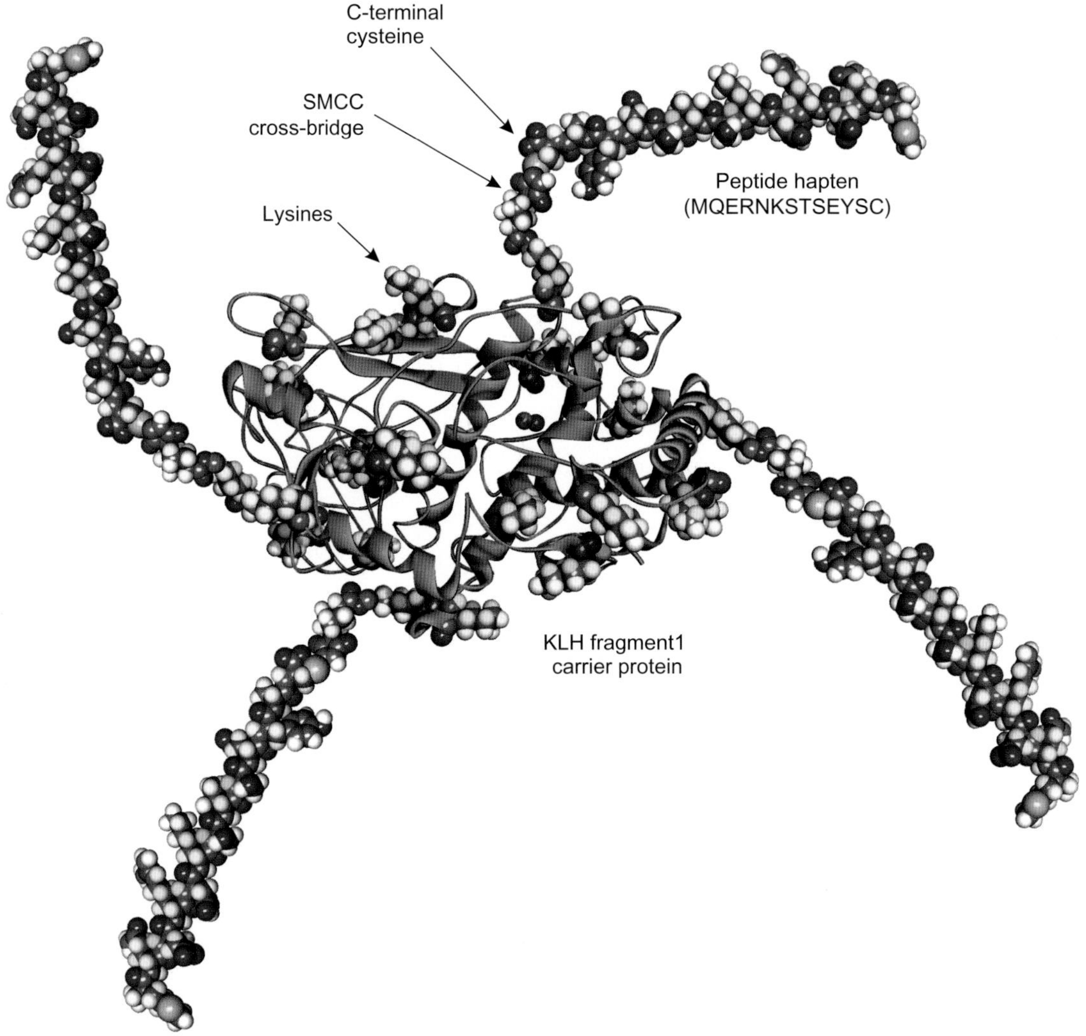

FIGURE 1.127 Vaccine or immunogen conjugates typically are prepared by conjugating a peptide hapten to a carrier protein such as keyhole limpet hemocyanin (KLH), which provides the immunogenicity needed to elicit an immune response to the peptide. Peptides can be synthesized to contain a terminal cysteine thiol group for convenient conjugation using SMCC-activated KLH.

immunogenic response to the bound hapten, whereas separately or alone the hapten would have little or no immunogen response. The basic design of an immunogen conjugate is illustrated in Figure 1.127, which shows peptide haptens coupled to the KLH functional unit 1 carrier protein through SMCC crosslinks.

The research into conjugate vaccines is growing, as the design allows safe and effective therapeutics to be developed without the risks associated with attenuated pathogens or intact allergens. For example, an effective vaccine against *Streptococcus pneumonia* bacteria was developed using the most common strains of the pathogen observed in the United States but instead of using all the different strains in the form of intact bacteria and attenuating them into a mixed cocktail, the key bacteria capsule polysaccharides were isolated and coupled to a nontoxic recombinant mutant of diphtheria toxin, CRM_{197}. The unique carbohydrates were conjugated to the amines on the carrier protein through reductive amination using their reducing ends to form the final immunogen complexes (Chapter 19, Section 8). These polysaccharide immunogen conjugates were then mixed to create the final vaccine, which in its various forms contains the carbohydrate molecules from 7, 10, 13, or 23 different bacteria strains. This type of conjugate vaccine was developed and marketed by a number of companies, including Wyeth Lederle Vaccines, Pfizer, and GlaxoSmithKline under the trademarks PCV7, PPV23, Prevnar, Synflorix, or Prevnar 13. The efficacy and safety of these conjugate vaccines were reviewed by DeStefano *et al.* (2008).

The first instance of a conjugate vaccine developed using the capsular polysaccharides of a bacterium was against *Haemophilus influenza* type b, which proved to be very successful, especially in young children (Peltola, 2000). This approach to vaccine development was extended to other bacterial pathogens, including those approved to prevent meningococcal disease (Maiden and Spratt, 1999; Miller *et al.*, 2001; Mueller *et al.*, 2006; Trotter and Ramsay, 2007). There are many polysaccharide-encapsulated pathogens that might be successfully prevented using immunogen conjugates consisting of protein–carbohydrate immunogens. The theory of how these conjugate vaccines work effectively is that the polysaccharide chain is able to interact with multiple B-cell receptors on the immune cell surface, thus crosslinking the receptors and stimulating the production of immunoglobulins and memory B cells (Pollard *et al.*, 2009). A direct comparison of the protein conjugate vaccine *versus* the use of free polysaccharide vaccine demonstrated the superior results of the conjugate vaccine in chronic obstructive pulmonary disease (Dransfield *et al.*, 2009).

The latest development in the use of conjugate vaccines for the prevention of infections is the use of combination vaccines to treat and prevent two or more different bacterial infectious agents simultaneously. For instance, a combination investigational vaccine was created through pooling of two polysaccharide–tetanus toxoid conjugates made from the polysaccharides of *Haemophilus influenza* type b and *Neisseria meningitides* serogroups C and Y (Nolan *et al.*, 2011). The results of a clinical study showed that the combination vaccine worked equally well compared to the use of separate conjugate vaccines. Thus, there may be significant potential to create a pan-bacteria vaccine using bioconjugate techniques to produce effective immunogen conjugates.

Cancer Vaccines

Another area where conjugate vaccines are beginning to find use is in the development of vaccines to prevent or treat certain cancers. Cancer immunotherapy provides a powerful alternative to the commonly used treatments including surgery, radiotherapy, and chemotherapy. If the immune system can be activated against certain tumor specific biomarkers that are not normally present on cells or tissues, it may be possible to provoke an innate immune response consisting of antibodies and lymphoid cells acting in concert to destroy the tumor (Chaudhuri *et al.*, 2009). It is known that the immune system has a potent ability to rapidly destroy tissue when activated. For instance, acute transplant rejection due to a mismatch of human leukocyte antigen (HLA) on a xenograft organ can occur within hours and result in total tissue destruction (Frohn *et al.*, 2001). The use of conjugate vaccines to target cancer is an active area of research to get the immune system to similarly react to the presence of tumor tissue.

There are at least three ways in which research into cancer vaccines is being applied: as a prophylactic to prevent the development of future tumor types (known as cancer preventive vaccines); as a vaccine to prevent infections by certain cancer-causing viruses (also a cancer preventative); or as a treatment to stimulate the immune system to respond to existing tumors (known as cancer treatment vaccines) (Giarelli, 2007). Effective vaccine therapy against cancer is greatly dependent upon the cellular immune response to unique tumor antigens, primarily on tumor cell surfaces. Changes in cancerous cells with respect to proteins displayed on outer membrane surfaces or changes in post-translational modifications on cell surfaces, particularly glycans, can result in the appearance of tumor-specific biomarkers, which can be used in designing effective immunotherapy (Pashov *et al.*, 2010).

Tumor-specific biomarkers have been used in active immunotherapy to stimulate a patient's own natural immune system to respond to the presence of tumor cells. The goal of such treatment is to induce the formation of an initial immune response combined with immunological memory to continue the attack upon cancer cells beyond the time frame of the treatment regimen. The ideal treatment with cancer vaccines will continue to protect against tumor recurrence without the need for additional treatment (Gulley *et al.*, 2007; Kipp and McNeel, 2007; Marrari *et al.*, 2007; Rescigno *et al.*, 2007).

Using this approach, conjugate vaccines have been developed to prevent or treat cancers using tumor-specific epitopes in the conjugate vaccine design. In this case, a biomarker characteristic of a tumor type is used to make an immunogen complex to stimulate the immune system to recognize the cancer cells as danger and attack them. The first cancer vaccine of this design was approved by the FDA in April of 2010 and is manufactured by Dendreon under the name Provenge. The key bioconjugate used in this treatment consists of prostatic acid phosphatase (PAP) covalently linked to the protein granulocyte–macrophage colony-stimulating factor (GM-CSF). The PAP component is a biomarker present on the majority of prostate cancer cells while GM-CSF is able to stimulate the immune system to better display antigens on antigen-presenting cells (APCs). In use, the PAP–GM-CSF conjugate functions as the immunogen in an *ex vivo* treatment of APCs isolated from the blood of the prostate cancer patient (Kipp and McNeel, 2007; Marrari *et al.*, 2007; Higano *et al.*, 2009; Kantoff *et al.*, 2010). It is believed that the mode of action of this conjugate vaccine is that PAP–GM-CSF gets taken up by the APCs during the *ex vivo* procedure and gets processed into small peptides, some of which end up being displayed on the cell surfaces in the binding pocket of the HLA complexes. Once the APCs are re-administered into the patient, these peptide-HLA complexes are able to interact with receptors on B cells and T cells, thus activating the key components of the immune system to respond with a humoral (antibody) and cellular (T lymphocyte) attack on the prostate tumor cells.

There are a number of other ongoing clinical studies that use *ex vivo* treatment of APCs with a conjugate immunogen in active immunotherapy procedures (Baron-Bodo *et al.*, 2005; Goldman and DeFrancesco, 2009). These include the use of the HER2-based immunogen Lapuleucel-T, a fusion conjugate of the HER2/neu antigen linked to GM-CSF, which is designed to treat breast cancer (Park *et al.*, 2007; Peethambaram, *et al.*, 2009), as well as a vaccinia virus–prostate-specific antigen (PSA) immunogen conjugate to treat metastatic prostate cancer (Kantoff *et al.*, 2010).

In addition to unique protein biomarkers on the surfaces of cancer cells, tumors often display aberrant glycosylation patterns on normally glycosylated proteins or glycolipids. It has been observed that some cancer patients develop antibodies to tumor-associated carbohydrate antigens (TACAs), which have actually been associated with longer survival rates among these patients (Vollmers and Brandlein, 2009). It has been speculated that these tumor-specific antibodies keep the cancer somewhat in check even though complete tumor destruction may not be possible simply through a humoral response. Aberrant glycosylation on tumor cells may provide viable epitopes for targeting in cancer immunotherapy just as unique carbohydrates associated with pathogens have been used successfully in the development of conjugate vaccines against various bacteria. Bioconjugation will no doubt contribute heavily to the continued development of these new cancer vaccines.

The National Cancer Institute lists a number of ongoing clinical trials investigating the use of cancer vaccines (*Cancer Vaccines*, NCI Factsheet, U.S. Department of Health and Human Services, August 4, 2010, p. 5). For instance, one of the drug studies involving metastatic melanoma in brain cancer is using a complex between Gp96, a non-polymorphic inducible heat-shock protein (HSP), and antigenic peptides that originate from patient tumor cells. Heat shock proteins are among the most abundant intracellular proteins and they are adept at binding to antigen-presenting cells (Amato, 2008). Purification of Gp96 from tumor biopsies carries along with it the critical peptide epitopes characteristic of the patient's cancer cells. Gp96 is a peptide shuttle protein that can elicit powerful T-cell responses against the tumor peptide epitopes it has bound to it, resulting in protective immunity. Tumor-derived heat-shock protein–peptide complexes (HSPPCs) are examples of noncovalent immunogen complexes having potent anti-tumor characteristics, which can be used against many tumor types (Tosti *et al.*, 2009; Eton *et al.*, 2010). The clinical trial using HSPPC-96 utilizes a conjugate vaccine called Vitespen (formerly Oncophage) that could end up being the first of many autologous vaccines, which are derived from tumor biopsies of individual cancer patients.

Another route to preventing cancer through vaccines is to target certain viruses that have been linked to the development of particular types of cancer. Three vaccines have already been introduced that prevent viral infections that are known to lead to cancer in certain instances. Human papillomavirus (HPV) conjugate vaccine has been introduced by two companies, Merck and GlaxoSmithKline, under the drug names Gardasil and Cervarix, respectively. Both vaccines consist of a multi-protein complex that contains virus coat proteins made recombinantly and self-assembled into virus-like particles (VLPs). These complexes are non-covalently held together by protein–protein interactions, which naturally take place when virus particles are assembled *in vivo*. However, VLPs are non-pathogenic since they do not contain any genetic information, just the outer envelope or capsid structural proteins. VLP complexes are strongly immunogenic and elicit excellent T-cell and B-cell immune responses in immunized patients (Chackerian and Schiller, 2012).

A third virus preventative vaccine was introduced in 1981 against hepatitis B virus (HBV), which aids in the prevention of hepatocellular carcinoma, which is a type of liver cancer (Chang *et al.*, 1997). The initial vaccine consisted of inactivated virus, but it was replaced in 1986 with a recombinant form consisting of the hepatitis B surface antigen (HBsAg), which is one of the envelope proteins within the virus particle (Merck, Recombivax). The purified HBsAg is crosslinked with formaldehyde, making a large complex that is then used with adjuvant to create the final vaccine.

Other vaccines that have a cancer preventative aspect include hepatitis C virus (for the prevention of hepatocellular carcinoma), Epstein–Barr virus (prevents Burkitt lymphoma, non-Hodgkin lymphoma, Hodgkin lymphoma, and nasopharyngeal carcinoma), human T-cell lymphotropic virus 1 (prevents T-cell leukemia), *Helicobacter pylori* (prevents stomach cancer), schistosomes (a parasite that causes bladder cancer), and liver flukes (a parasite that causes cholangiocarinoma, a type of liver cancer) (obtained from http://www.cancer.gov/cancertopics/factsheet).

Immunogen Conjugates in the Production of Antibodies

Immunogen conjugates have also been used extensively for the production and harvesting of specific antibodies in animals. Generation of polyclonal and monoclonal antibodies has been carried out using carrier conjugates with short peptide sequences, protein fragments or domains, glycans, carbohydrates, small organic compounds, and other molecules of biological or non-biological origin. Even small neurotransmitters and biogenic amine compounds can be conjugated to suitable carriers to make highly specific antibodies against them (Huisman *et al.*, 2009).

The preparation of hapten-carrier conjugates can be carried out through a number of routes (Chapter 19) and the result is an immunogenic complex that can produce a specific antibody response to the coupled hapten within an animal. After immunization, boosting, and harvesting of the antisera, the specific antibody can then be isolated using affinity chromatography techniques and applied toward many different applications in biological targeting and detection (see previous sections, in this chapter; Harlow and Lane, 1988, 1999). The use of immunogen conjugates in the generation of antibodies has been instrumental in the growth of the antibody business for life science research applications to at least $3 billion in annual revenues shared among dozens of antibody producers, while the market for diagnostic antibodies is approximately $10 billion per annum, and the market for therapeutic antibodies is estimated to be at least $53 billion (*Life Science Tools and Reagents: Global Markets*, BCC Research, Wellesley, MA, 2011, p. 74).

Conjugate vaccines for immunization therapy and immunogen conjugates for antibody production both have many advantages. For *in vivo* therapeutic use, conjugate vaccines can be significantly less toxic than the corresponding use of attenuated pathogens. The conjugation of key immunogenic epitopes from a bacteria or virus to a carrier protein entirely avoids the use of whole biological pathogens, which can have unexpected negative effects in patients. In addition, the use of immunogen conjugates to produce antibodies in animals can be carried out using highly specific peptide sequences, polysaccharide sequences, or other small molecule haptens that can generate the desired antibody binding specificity with minimal cross-reactivity toward other epitopes. Antibodies can also be generated to discrete fragments of biological molecules or pathogens that otherwise in their native form would be highly toxic *in vivo*, thus entirely avoiding any animal toxicity side effects.

4. SUMMARY

It is amazing that the field of bioconjugation has affected nearly every technology area, including having significant impact on life science research, diagnostics, therapeutics, organic chemistry, material research, and many other specialized disciplines within the sciences. The ability to produce unique functional conjugates can make possible the creation of a virtually limitless selection of reagents having properties of enormous diversity. The use of such reagents can facilitate quantification, detection, purification, synthesis, imaging, therapy, diagnosis, or almost any other purpose the imagination can conceive.

The chapters of this book are designed to cover the chemistry, reagents, and applications of bioconjugation with a view toward making the techniques understandable and practical. They were assembled and outlined based upon reagent types, common reactions, or by an underlying focus on a particular application or topic. Information is liberally cross-referenced to other areas within the book where the same reactions or similar reagents can be used to make additional bioconjugates for different fields of use. The best way to find all of the available information on a specific reagent, reaction, bioconjugate, or application is to refer to the Table of Contents and the Index. Often, information presented on a topic in one chapter can be supplemented by additional discussion in another chapter; the end result is hopefully to paint a more complete picture of the growing world of bioconjugate techniques.

CHAPTER

2

Functional Targets for Bioconjugation

Modification and conjugation techniques are dependent on two interrelated chemistries: the reactive functionalities present on the various crosslinking or derivatizing reagents and the functional groups present on the target macromolecules to be modified. Without both types of functional groups being available and chemically compatible, the process of derivatization would be impossible. Reactive functionalities on crosslinking reagents, tags, and probes provide the means to specifically label certain target groups on ligands, peptides, proteins, carbohydrates, lipids, synthetic polymers, nucleic acids, and oligonucleotides. Knowledge of the basic mechanisms by which the reactive groups couple to target functionalities provides the means to intelligently design a modification or conjugation strategy. Choosing the correct reagent systems that can react with the chemical groups available on target molecules forms the basis for successful chemical modification.

The process of designing a derivatization scheme that works well in a given application is not as difficult as it may seem at first glance. A basic understanding of about a dozen reactive functionalities that are commonly present on modification and crosslinking reagents combined with knowledge of about half that many functional target groups can provide the minimum skills necessary to plan a successful experiment.

Fortunately, the principal reactive functionalities commonly encountered on bioconjugate reagents are now present on scores of commercially obtainable compounds. The resource that this arsenal of reagents provides can assist in solving almost any conceivable modification or conjugation problem. The following sections describe the predominant targets for these reagent systems. The functionalities discussed are found on virtually every conceivable biological molecule, including amino acids, peptides, proteins, sugars, carbohydrates, polysaccharides, nucleic acids, oligonucleotides, lipids, and complex organic compounds. A careful understanding of target molecule structure and reactivity provides the foundation for the successful use of all of the modification and conjugation techniques discussed in this book.

1. MODIFICATION OF AMINO ACIDS, PEPTIDES, AND PROTEINS

Protein molecules are perhaps the most common targets for modification or conjugation techniques. As the mediators of specific activities and functions within living organisms, proteins can be used *in vitro* and *in vivo* to perform certain tasks. Having enough of a protein that can bind a particular target molecule can result in a way to detect or assay the target providing the protein can be followed or measured. If such a protein does not possess an easily detectable component, it can often be modified to contain a chemical or biological tracer to allow detectability. This type of protein complex can be designed to retain its ability to bind its natural target, while the tracer portion can provide the means to find and measure the location and amount of target molecules.

Detection, assay, tracking, or targeting of biological molecules using the appropriately modified proteins are the main areas of application for modification and conjugation systems. The ability to produce a labeled protein having specificity for another molecule provides the key component for much of biological research, clinical diagnostics, and human therapeutics.

In this section, the structure, function, and reactivity of amino acids, peptides, and proteins will be discussed with the goal of providing a foundation for successful derivatization. The interplay of amino acid functionality and the three-dimensional folding of polypeptide chains will be seen as forming the basis for protein activity. Understanding how the attachment of foreign molecules can affect this tenuous relationship, and thus alter protein function, ultimately will create a rational approach to protein chemistry and modification.

Bioconjugate Techniques, *Third Edition*.
DOI: http://dx.doi.org/10.1016/B978-0-12-382239-0.00002-9

1.1. Protein Structure and Reactivity

Amino Acids

Peptides and proteins are composed of amino acids polymerized together through the formation of peptide (amide) bonds. The peptide-bonded polymer that forms the backbone of polypeptide structure is called the α-chain. The peptide bonds of the α-chain are rigid planar units formed by the reaction of the α-amino group of one amino acid with the α-carboxyl group of another (Figure 2.1). The peptide bond possesses no rotational freedom due to the partial double-bond character of the carbonyl–amino amide bond. The bonds around the α-carbon atom, however, are true single bonds with considerable freedom of movement.

The sequence and properties of the amino acid constituents determine protein structure, reactivity, and function. Each amino acid is composed of an amino group and a carboxyl group bound to a central carbon, termed the α-carbon. Also bound to the α-carbon are a hydrogen atom and a side chain unique to each amino acid (Figure 2.2). There are 20 common amino acids found throughout nature, each containing an identifying side chain of particular chemical structure, charge, hydrogen bonding capability, hydrophilicity (or hydrophobicity), and reactivity. The side chains do not participate in polypeptide formation and are thus free to interact and react with their environment.

Amino acids may be grouped by type depending on the characteristics of their side chains. There are seven amino acids that contain aliphatic side chains, which are relatively nonpolar and hydrophobic: glycine, alanine, valine, leucine, isoleucine, methionine, and proline (Figure 2.3). Glycine is the simplest amino acid, its side chain consisting of only a hydrogen atom. Alanine is next in line, possessing just a single methyl group for its side chain. Valine, leucine, and isoleucine are slightly more complex with three or four carbon branched-chain constituents. Methionine is unique in that it is the only reactive aliphatic amino acid, containing a thioether group at the terminus of its hydrocarbon chain. Proline is actually the only *imino* acid. Its side chain forms a pyrrolidine ring structure with its α-amino group. Thus, it is the only amino acid containing a secondary α-amine. Due to its unique structure, proline often causes severe turns in a polypeptide chain. Proteins rich in proline, such as collagen, have tightly formed structures of high density. Collagen also contains a rare derivative of proline, 4-hydroxyproline, found in only a few other proteins. Proline, however, cannot be accommodated in normal α-helical structures, except at the ends where it may create the turning point for the chain. Poly-proline α-helical structures have been formed, but the structural characteristics of these artificial polypeptides are quite different from native protein helices.

Phenylalanine and tryptophan contain aromatic side chains that, like the aliphatic amino acids, are also relatively nonpolar and hydrophobic (Figure 2.4). Phenylalanine is unreactive toward common derivatizing reagents, whereas the indolyl ring of tryptophan is quite reactive, if accessible. The presence of tryptophan

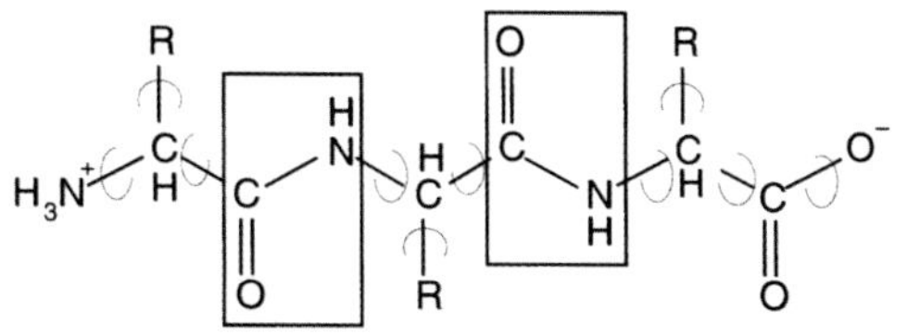

FIGURE 2.1 Rigid peptide bonds link amino acid residues together to form proteins. Other bonds within the polypeptide structure may exhibit considerable freedom of rotation.

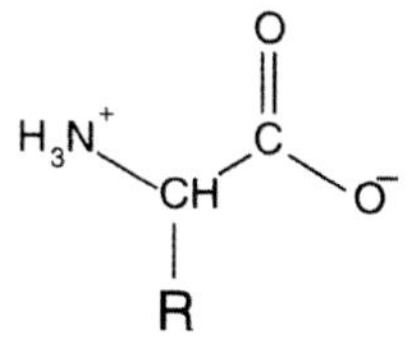

FIGURE 2.2 Individual amino acids consist of a primary (α) amine, a carboxylic acid group, and a unique side-chain structure (R). At physiological pH, the amine is protonated and bears a positive charge, while the carboxylate is ionized and possesses a negative charge.

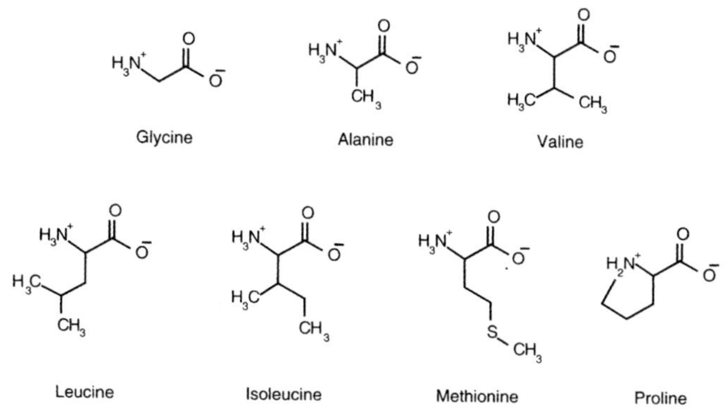

FIGURE 2.3 Common aliphatic amino acids.

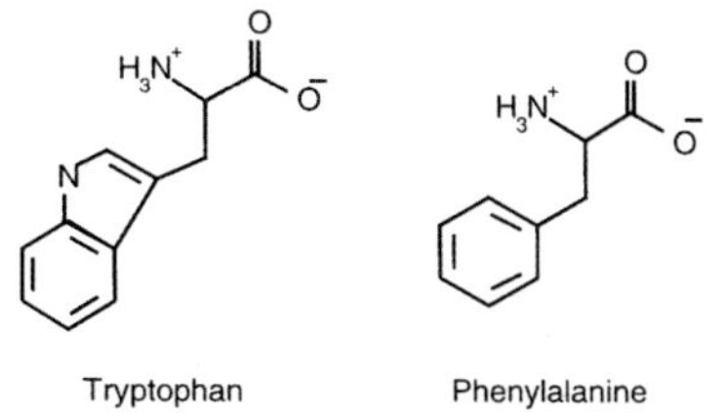

FIGURE 2.4 The two aromatic amino acids that have nonpolar and nonionizable side-chain groups.

in a protein contributes more to its total absorption at 275 to 280 nm on a mole-per-mole basis than any other amino acid. The phenylalanine content, however, adds very little to the overall absorbance in this range.

All of the aliphatic and aromatic hydrophobic residues are often located at the interior of protein molecules or in areas that interact with other nonpolar structures such as lipids. They usually form the hydrophobic core of proteins and are not readily accessible to water or other hydrophilic molecules.

There is another group of amino acids that contains relatively polar constituents and are thus hydrophilic in character. Asparagine, glutamine, threonine, and serine (Figure 2.5) are usually found in hydrophilic regions of a protein molecule, especially at or near the surface where they can be hydrated with the surrounding aqueous environment. Asparagine, threonine, and serine are often found post-translationally modified with carbohydrate in *N*-glycosidic (asp) and *O*-glycosidic linkages (thr and ser). Though these side chains are enzymatically derivatized in nature, the hydroxyl and amide portions have relatively the same nucleophilicity as that of water and are therefore difficult to modify with common reagent systems under aqueous conditions.

The most significant amino acids for modification and conjugation purposes are the ones containing ionizable side chains: aspartic acid, glutamic acid, lysine, arginine, cysteine, histidine, and tyrosine (Figure 2.6). In their unprotonated state, each of these side chains can be potent nucleophiles to engage in addition reactions (see the discussion on nucleophilicity below).

Both aspartic and glutamic acids contain carboxylate groups that have similar ionization properties to the C-terminal α-carboxylate. The theoretical pK_a of the β-carboxyl of aspartic acid (3.7–4.0) and the γ-carboxyl of glutamic acid (4.2–4.5) are somewhat higher than the α-carboxyl groups at the C-terminal of a polypeptide chain (2.1–2.4). At pH values above their pK_a, these groups are generally ionized to negatively charged carboxylates. Thus, at physiological pH, they contribute to the overall negative charge contribution of an intact protein (see following section).

Carboxylate groups in proteins may be derivatized through the use of amide bond-forming agents or through active ester or reactive carbonyl intermediates (Figure 2.7). The carboxylate actually becomes the acylating agent to the modifying group. Amine-containing nucleophiles can couple to an activated carboxylate to give amide derivatives. Hydrazide compounds react similarly to amines. While a thiol group is reactive toward an activated carboxylate and results in a thioester linkage, it forms relatively unstable derivatives, which can exchange with other nucleophiles such as amines or hydrolyze in aqueous solutions.

Lysine, arginine, and histidine have ionizable amine-containing side chains that, along with the N-terminal α-amine, contribute to a protein's overall net positive charge. Lysine contains a straight four-carbon chain terminating in a primary amine group. The ε-amine of lysine differs in pK_1 from the primary α-amines by having a slightly higher ionization point (pK_a of 9.3–9.5 for lysine *versus* pK_a of 7.6–8.0 for α-amines). At pH values lower than the pK_a of these groups, the amines are generally protonated and possess a positive charge. At pH values greater than the pK_a, the

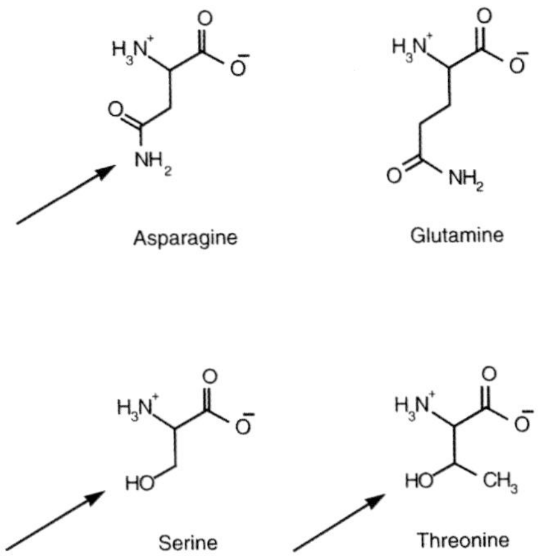

FIGURE 2.5 The four amino acids with polar, uncharged side chains. The arrows show the attachment points for carbohydrate that may be present in post-translational modifications on glycoproteins.

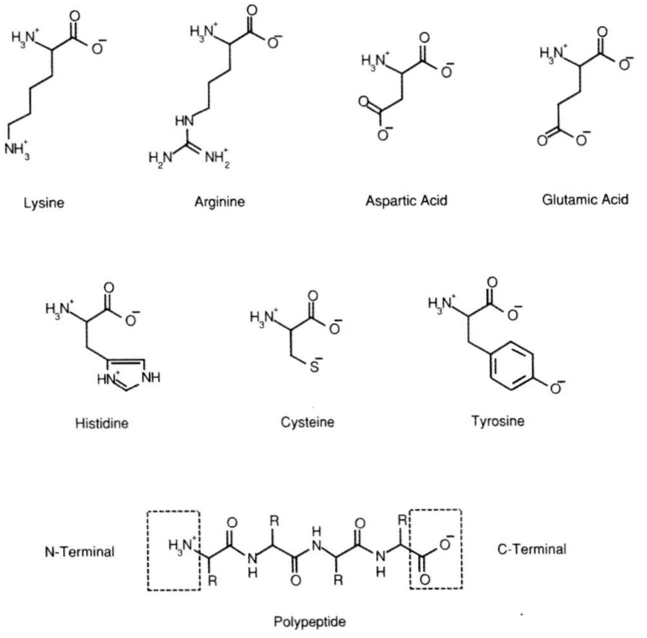

FIGURE 2.6 The amino acids with ionizable side chain groups possess some of the most important functional groups for bioconjugate applications. The C- and N-terminals of each polypeptide chain also are included in this group.

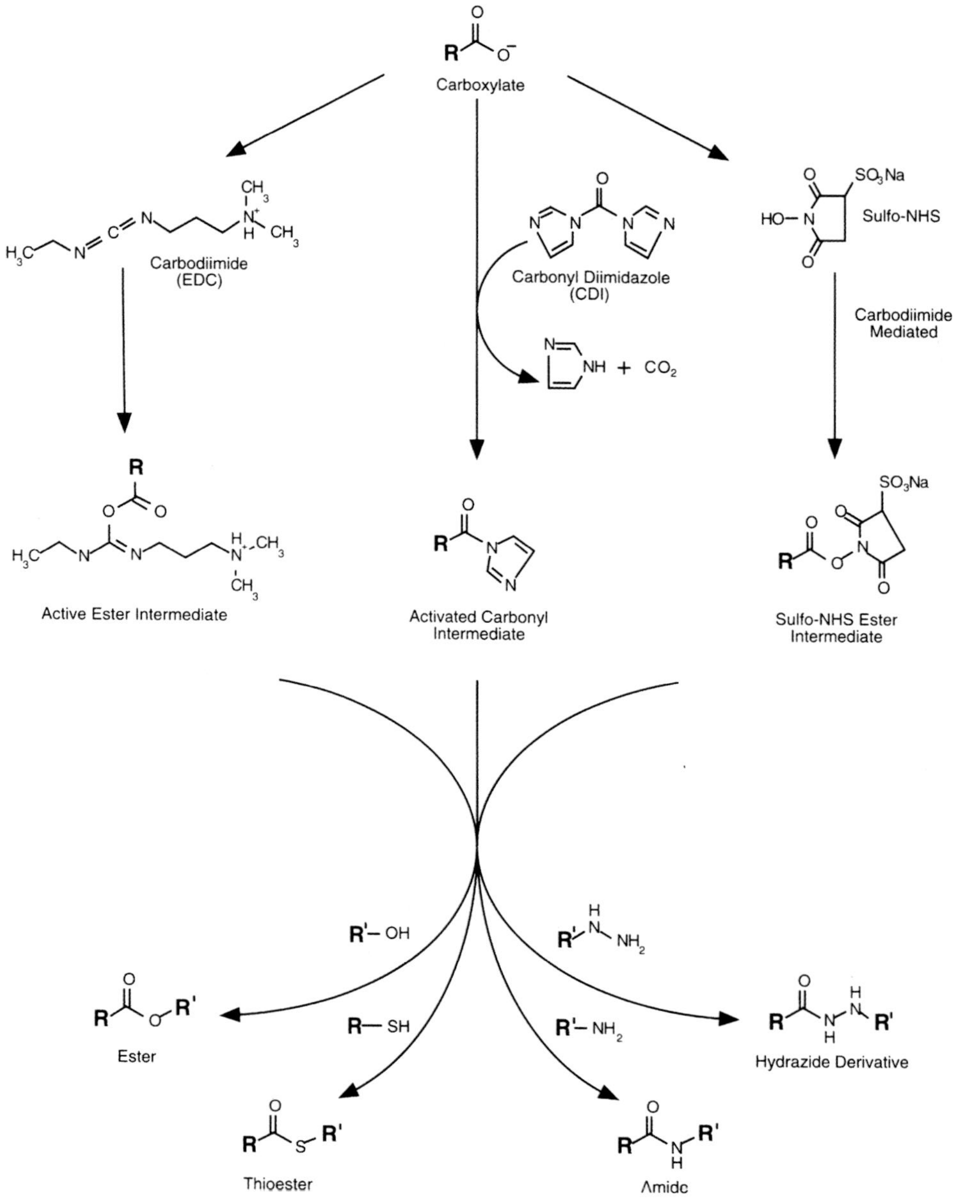

FIGURE 2.7 Derivatives of carboxylic acids can be prepared through the use of active intermediates that react with target functional groups to give acylated products.

amines are unprotonated and contribute no net charge. Arginine contains a strongly basic chemical constituent on its side chain called a guanidino group. The ionization point of this residue is so high ($pK_a > 12.0$) that it is virtually always protonated and carriers a positive charge. The histidine side chain is an imidazole ring that is potentially protonated at slightly acidic pH values ($pK_a = 6.7–7.1$). Thus, at physiological pH, these residues contribute to the overall net positive charge of an intact protein molecule.

The amine-containing side chains in lysine, arginine, and histidine typically are exposed on the surface of proteins and can be derivatized with ease. The most important reactions that can occur with these residues are alkylation and acylation (Figure 2.8). In alkylation, an active alkyl group is transferred to the amine nucleophile with loss of one hydrogen. In acylation, an active carbonyl group undergoes addition to the amine. Alkylating reagents are highly varied and the reaction with an amine nucleophile is difficult to generalize. Acylating reagents, however, usually proceed through a carbonyl addition mechanism as shown in Figure 2.9. The imidazole ring of histidine is also an important reactive species in electrophilic reactions, such as in iodination using radioactive ^{125}I or ^{131}I (Chapter 12, Section 2).

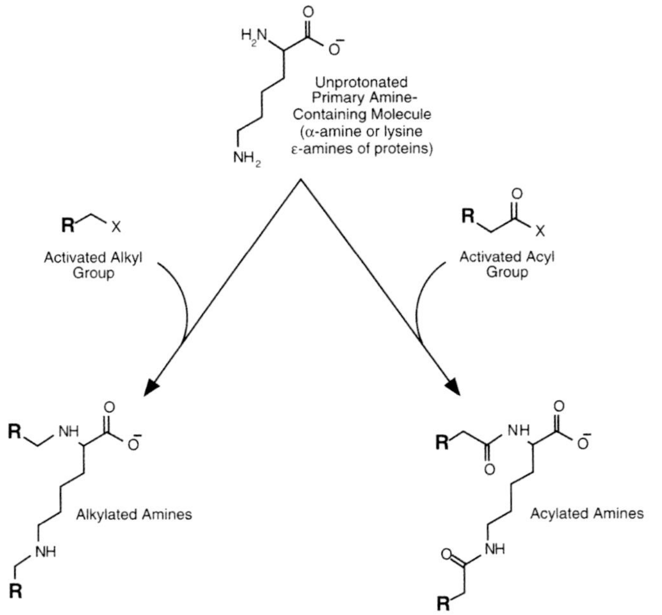

FIGURE 2.8 Derivatives of amines can be prepared from acylating or alkylating agents to give amide, secondary amine, or tertiary amine bonds.

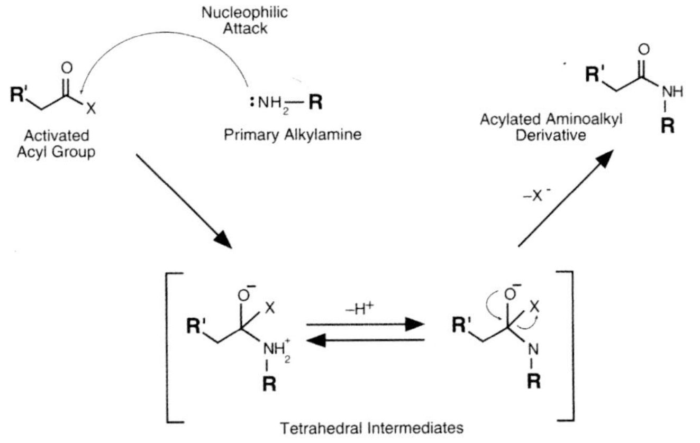

FIGURE 2.9 The mechanism of acylation proceeds through the attack of a nucleophile, generating a tetrahedral intermediate, which then goes on to form the product.

Cysteine is the only amino acid containing a sulfhydryl group. At physiological pH, this residue is normally protonated and possesses no charge. Ionization only occurs at high pH (pK_a = 8.8–9.1) and results in a negatively charged thiolate residue. The most important reaction of cysteine groups in proteins is the formation of disulfide crosslinks with another cysteine molecule. Cysteine disulfides (called cystine residues) are often key points in stabilizing protein structure and conformation. They frequently occur between polypeptide subunits, creating a covalent linkage to hold two chains together. Cysteine and cystine groups are relatively hydrophobic and usually can be found within the core of a protein. For this reason, it is often difficult to fully reduce the disulfides of large proteins without a deforming agent present to open up the inner structure and make them accessible (see Section 4.1).

Cysteine sulfhydryls and cystine disulfides may undergo a variety of reactions, including alkylation to form stable thioether derivatives, acylation to form relatively unstable thioesters, and a number of oxidation and reduction processes (Figure 2.10). Derivatization of the side chain sulfhydryl of cysteine is one of the most important reactions of modification and conjugation techniques for proteins.

Tyrosine contains a phenolic side chain with a pK_a of about 9.7 to 10.1. Due to its aromatic character, tyrosine is second only to tryptophan in contributing to a protein's overall absorptivity at 275 to 280 nm. Although the amino acid is only sparingly soluble in water, the ionizable nature of the phenolic group makes it often appear in hydrophilic regions of a protein—usually at or near the surface. Thus, tyrosine derivatization proceeds without much need for deforming agents to further open protein structure.

Tyrosine may be targeted specifically for modification through its phenolate anion by acylation, through electrophilic reactions such as the addition of iodine or diazonium ions, and by Mannich condensation reactions. The electrophilic substitution reactions on tyrosine's ring all occur at the *ortho* position to the –OH group (Figure 2.11). Most of these reactions proceed effectively only when tyrosine's ring is ionized to the phenolate anion form.

In summary, protein molecules may contain up to nine amino acids that are readily derivatizable at their side chains: aspartic acid, glutamic acid, lysine, arginine, cysteine, histidine, tyrosine, methionine, and tryptophan. These nine residues contain eight principal functionalities with sufficient reactivity for modification reactions: primary amines, carboxylates, sulfhydryls (or disulfides), thioethers, imidazoles, guanidinyl groups, and phenolic and indolyl rings. All of these side-chain functionalities in addition to the N-terminal α-amino and the C-terminal α-carboxylate form the full complement of polypeptide reactivity within proteins (Figure 2.12).

Nucleophilic Reactions and the pI of Amino Acid Side Chains

Ionizable groups within proteins can exist in one of two forms: protonated or unprotonated. Carboxylate groups below their pK_a values exist in the protonated state and are therefore in the conjugate acid form and carry no charge. However, at pH values above the pK_a of the carboxylic group, the acid is ionized and therefore unprotonated to a negative charge. This same relationship is true of the –OH group on the phenol ring of tyrosine. At pH values below its pK_a, tyrosine's side

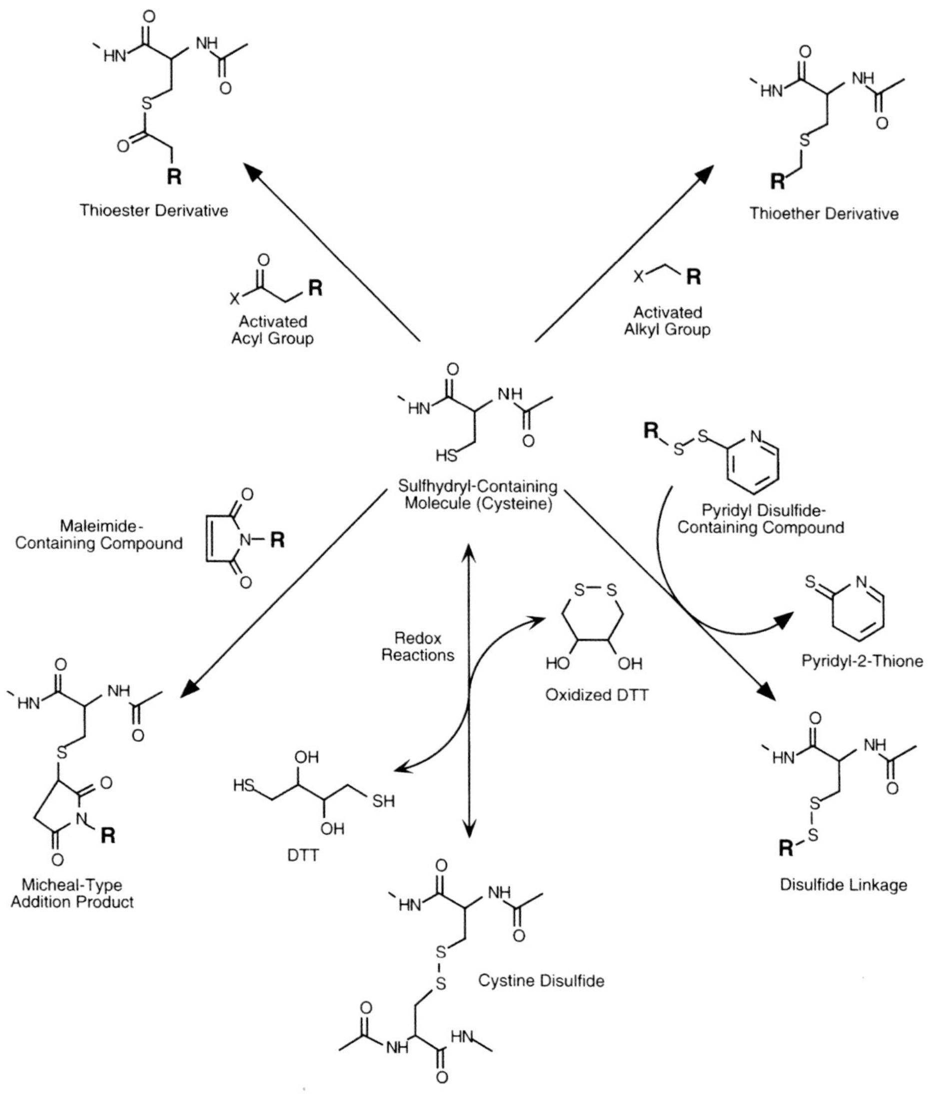

FIGURE 2.10 Sulfhydryl groups may undergo a number of additional reactions, including acylation and alkylation. Thiols also may participate in redox reactions, which generate reversible disulfide linkages.

chain is uncharged. Above the pK_a, however, the hydrogen ionizes off, leaving a negatively charged phenolate. Conversely, amine nucleophiles below their pK_a values are in a protonated state and possess a positive charge. At pH values above the pK_a of the amino group, it is then ionized and unprotonated to neutrality.

Each type of ionizable group in proteins will have a unique pK_a based upon the theoretical value for the amino acid and modulated from that value by its own surrounding microenvironment. Minute environmental changes will cause amine-containing residues at different structural locations to have different ionization potentials, even if the groups are otherwise chemically identical.

Thus, the actual pK_a of each ionizable group within protein molecules may range considerably lower or higher than the theoretical values as the microenvironment of individual groups changes. Identical side chains in differing parts of a protein molecule may have widely varying pK_a values depending on the immediate chemical milieu. Such factors as the presence of other amino acid side chains in the vicinity, salts, buffers, temperature, ionic strength, and other effects of the solvent medium all play crucial roles in creating microenvironmental changes that affect the ionization potential of these groups (Tanford and Hauenstein, 1956; Schewale and Brew, 1982).

The Henderson–Hasselbalch equation (1) explains the relationship of pH and pK_a to the relative ratios of protonated (acid) and unprotonated (base) forms of an ionizable group. Note that the ionized form of such a group does not have to possess a negative charge, as in the case of unprotonated primary amines. Indeed,

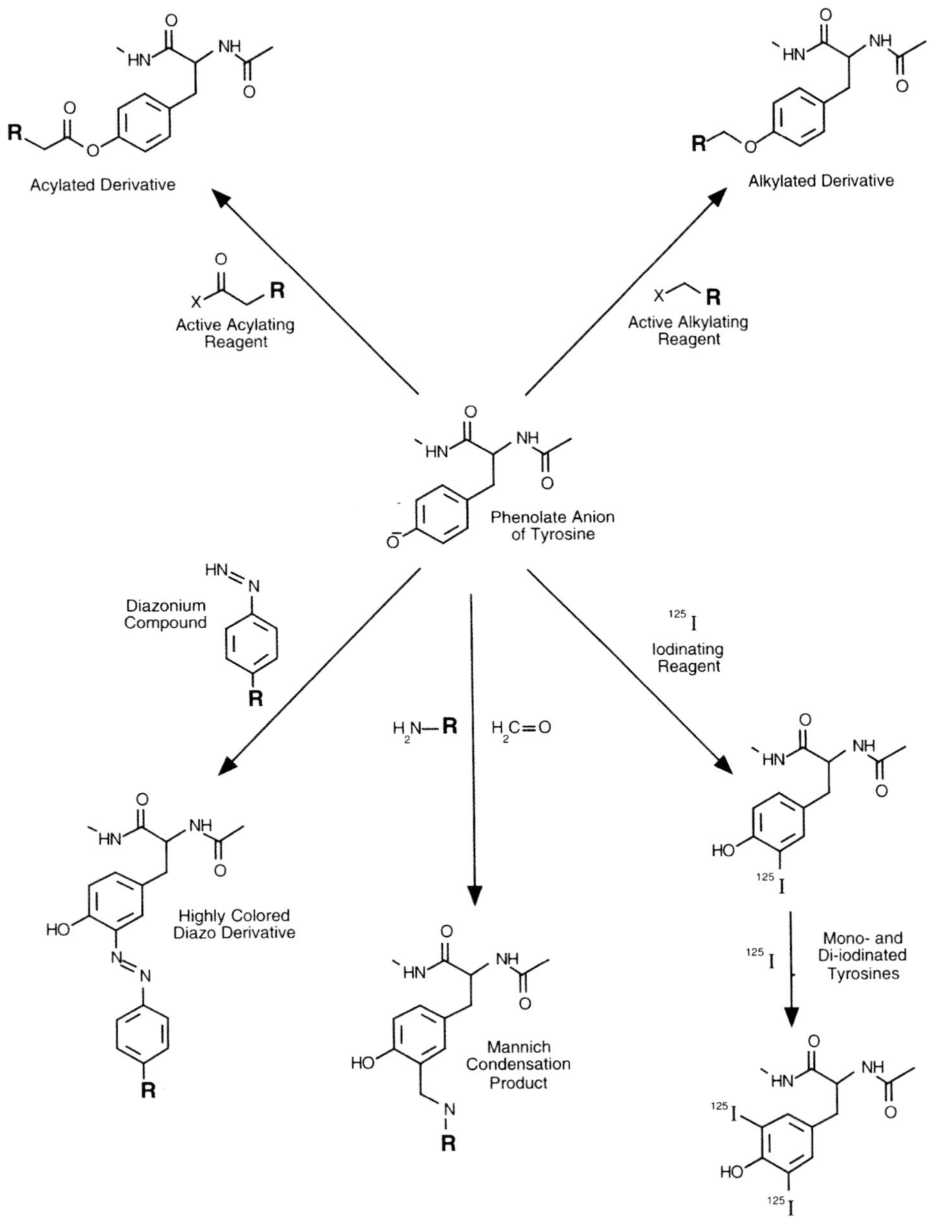

FIGURE 2.11 Tyrosine residues are subject to nucleophilic and electrophilic reactions. The unprotonated phenolate ion may be alkylated or acylated using a variety of bioconjugate reagents. Its aromatic ring also may undergo electrophilic addition using diazonium chemistry or Mannich condensation, or be halogenated with radioactive isotopes such as ^{125}I.

in that instance it is the protonated amine that bears a charge of positive one. According to the mathematical implications of this equation, an ionizable group at its pK_a value is exactly 50% ionized. This means that aspartic acid side chains placed in a medium with a pH equal to its pK_a should have half of its carboxylates ionized to a negative charge and half of them unionized with no charge.

$$pH = pK_a + \log\{[base]/[acid]\} \quad (1)$$

Further implications of this equation are that at one pH unit below or above the pK_a, an ionizable group will be 91% unionized (protonated) or 91% ionized (unprotonated), respectively. Two pH units below or above translate to a 99% unionized or 99% ionized state.

The absolute ratio of protonated : unprotonated forms will change from this theoretical approach based upon the microenvironment each group experiences. The reactivity of amino acid side chains is directly related to them being in an unprotonated or ionized state. Many reactions of modification and conjugation

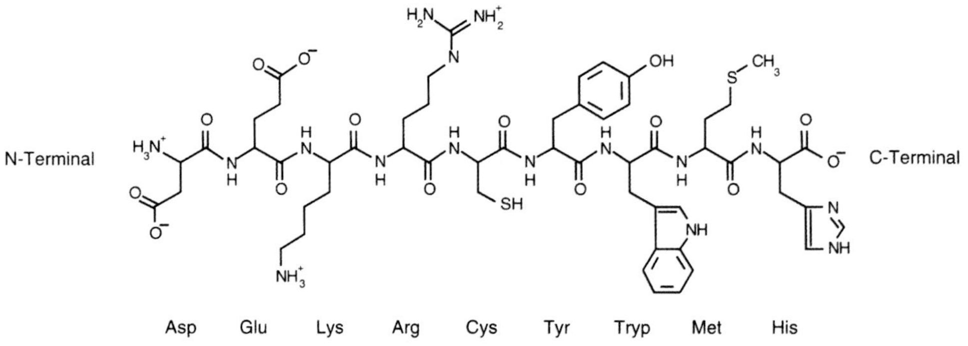

FIGURE 2.12 The more important polypeptide functional groups are represented by these nine amino acids. Bioconjugate chemistry may occur through the C- and N-terminals of each polypeptide chain, the carboxylate groups of aspartic and glutamic acids, the ε-amine of lysine, the guanidino group of arginine, the sulfhydryl group of cysteine, the phenolate ring of tyrosine, the indol ring of tryptophan, the thioether of methionine, and the imidazole ring of histidine.

occur efficiently only when the nucleophilic species is in an ionized form. As the unprotonated form increases in concentration, the relative nucleophilicity of the ionizable group increases. Many of the reactive groups commonly used for protein modification will couple in greater yield as the pH of the reaction is raised closer to the pK_a of the ionizable target. However, continuing to increase the pH beyond the pK_a may not be necessary for increased yield, and may even be detrimental, because many reactive groups will begin to lose activity through hydrolysis at high pHs.

A nucleophile is any atom containing an unshared pair of electrons or an excess of electrons able to participate in covalent bond formation. Nucleophilic attack at an atomic center of electron deficiency or positive charge is the basis for many of the coupling reactions that occur in chemical modification. Thus, an uncharged amine group is a more powerful nucleophile than the protonated form bearing a positive charge. Likewise, a negatively charged carboxylate has greater nucleophilicity than its uncharged, protonated conjugate acid form. In addition, an unprotonated thiolate, bearing a negative charge (RS^-), is a much more powerful nucleophile than its protonated, uncharged sulfhydryl form.

According to the theory of nucleophilicity (Edwards and Pearson, 1962; Bunnett, 1963; Pearson *et al.*, 1968), the relative order of nucleophilicity relative to the major groups in biological molecules can be summarized as follows:

$$R{-}S^- > R{-}SH$$

$$R{-}NH_2 > R{-}NH_3^+$$

$$R{-}COO^- > R{-}COOH$$

$$R{-}O^- > R{-}OH$$

$$R{-}OH = H{-}OH$$

and finally,

$$R{-}S^- > R{-}NH_2 > R{-}COO^- = R{-}O^-$$

Using these relationships, it is obvious that the strongest nucleophile in protein molecules is the sulfhydryl group of cysteine, particularly in the ionized, thiolate form. Next in line are the amine groups in their uncharged, unprotonated forms, including the α-amines at the N-terminals, the ε-amines of lysine side chains, the secondary amines of histidine imidazolyl groups and tryptophan indole rings, and the guanidino amines of arginine residues. Finally, the least potent nucleophiles are the oxygen-containing ionizable groups including the α-carboxylate at the C-terminal, the β-carboxyl of aspartic acid, the γ-carboxyl of glutamic acid, and the phenolate of tyrosine residues.

According to the theoretical pK_a values for the ionizable side chains of amino acids, nucleophilic substitution reactions involving primary amines or sulfhydryl groups on proteins should not be efficient below a pH of about 8.5 (Table 2.1). In practice, however, reactions can be carried out with these groups in high yield at pH values not much higher than neutrality. This discrepancy relates to the changes in pK_a due to microenvironmental effects experienced by the residues within the three-dimensional structure of the protein molecule. In reality, the ε-amine groups on lysine side chains within proteins, having theoretical pK_as of over 10, nonetheless exist in sufficient quantity in an unprotonated form even at a pH of 7.2 that modification easily occurs.

One important point should be noted, however. The changes that occur in the pK_a of ionizable groups in protein molecules due to microenvironmental effects sometimes make it difficult to select certain residues for modification simply by careful modulation of reaction pH. For instance, at least in theory, overlap of the pK_a range for sulfhydryls and amine-containing residues

TABLE 2.1 **pK_a of Ionizable Amino Acids**

Group Location	Functionality	pK_a Range
α-Amine; N-terminus	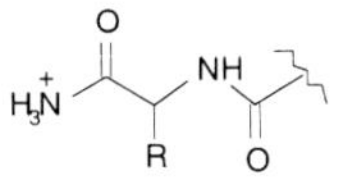	7.6–8
Lysine's ε-amine	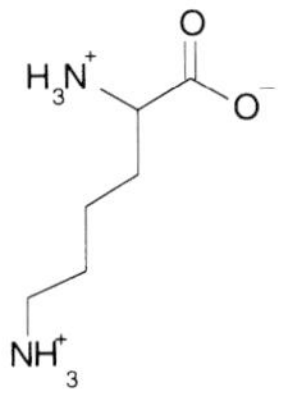	9.3–9.5
Histidine's imidazolyl nitrogen	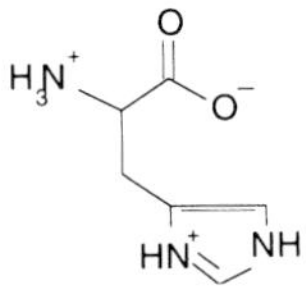	6.7–7.1
Arginine's guanidinyl group	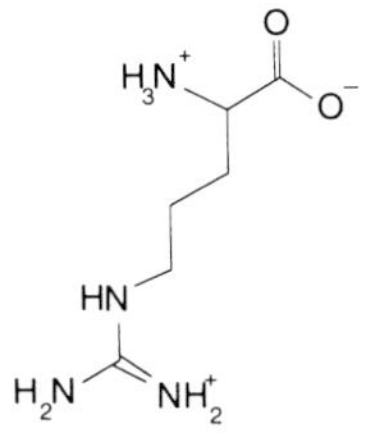	>12
Tyrosine's phenolic hydroxyl	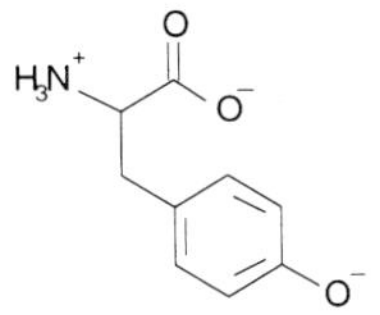	9.7–10.1
α-Carboxyl; C-terminus	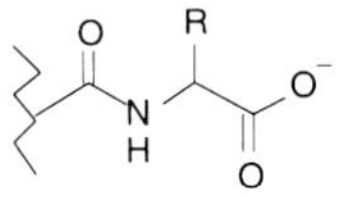	2.1–2.4
Aspartic acid's γ-carboxyl	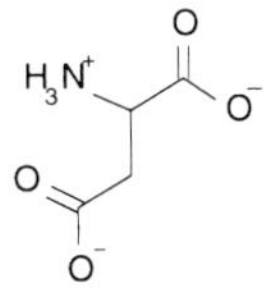	3.7–4
Gutamic acid's γ-carboxyl	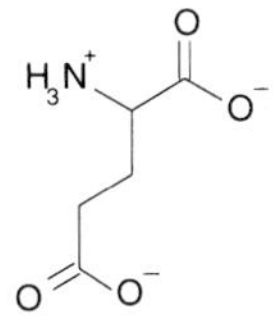	4.2–4.5
Cysteine's sulfhydryl	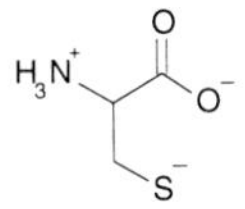	8.8–9.1

would eliminate any chance of directing a reaction toward –SH groups solely by adjusting the pH of the reaction medium. However, because of the microenvironmental changes that occur in complex biomolecules, pH sometimes can be used along with the right reactive group to target thiols without amine modification. Thus, in practice, to effectively site-direct a modification reaction, the proper choice of reactive group and reaction conditions can result in highly discrete conjugation to certain sites within proteins.

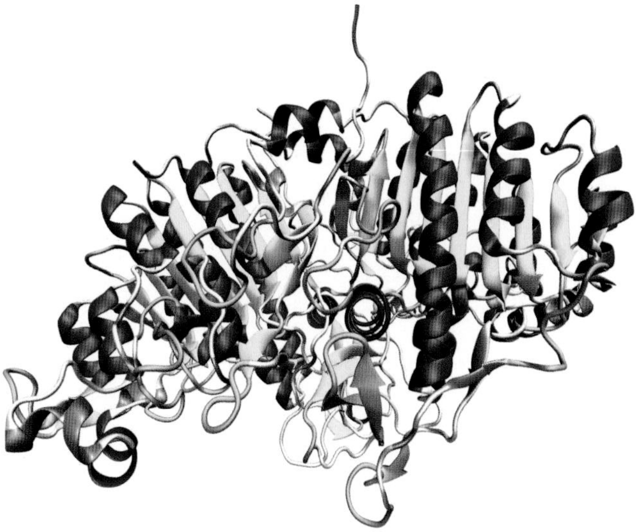

FIGURE 2.13 The α-chain structure of alkaline phosphatase illustrates the complex nature of polypeptide structure within proteins (Kim and Wyckoff, 1991).

Secondary, Tertiary, and Quaternary Structure

Amino acids are linked through peptide bonds to form long polypeptide chains. The *primary* structure of protein molecules is simply the linear sequence of each residue along the α-chain. Each amino acid in the chain interacts with surrounding groups through various weak, noncovalent interactions and through its unique side-chain functionalities. Noncovalent forces such as hydrogen bonding and ionic and hydrophobic interactions combine to create each protein's unique organization.

The sequence and types of amino acids and the way that they are folded provide protein molecules with specific structure, activity, and function. Ionic charge, hydrogen bonding capability, and hydrophobicity are the major determinants for the resultant three-dimensional structure of protein molecules. The α-chain is twisted, folded, and formed into globular structures, α-helicies, and β-sheets based upon the side-chain amino acid sequence and weak intramolecular interactions such as hydrogen bonding between different parts of the peptide backbone (Figure 2.13). Major secondary structures of proteins such as α-helicies and β-sheets are held together solely by massive hydrogen bonding created through the carbonyl oxygens of peptide bonds interacting with the hydrogen atoms of other peptide bonds (Figure 2.14).

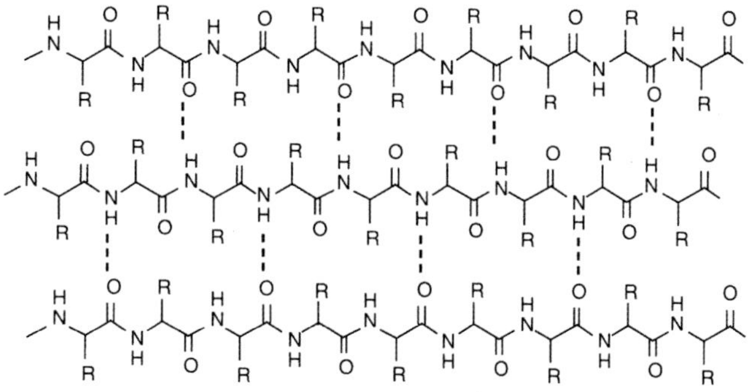

FIGURE 2.14 Secondary structures within proteins may be stabilized through hydrogen bonding between adjacent α-chains, forming β-sheet conformations.

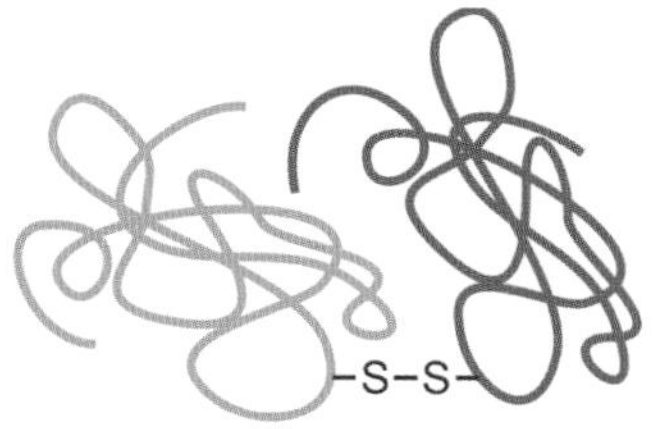

FIGURE 2.15 Polypeptide chains may be bound together through disulfide linkages occurring between cysteine residues within each subunit.

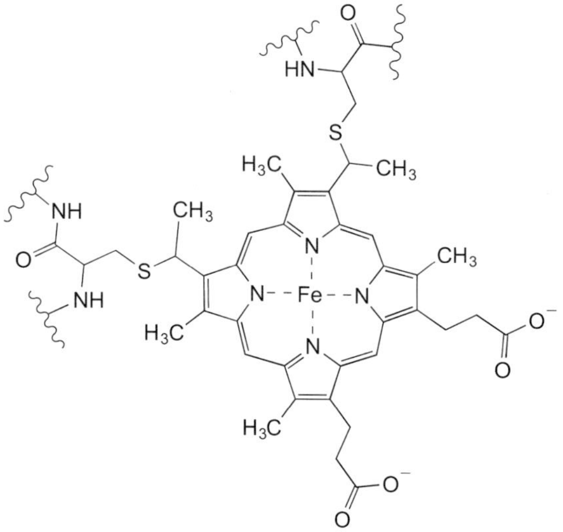

FIGURE 2.16 The heme ring of cytochrome *c* is a non-amino-acid, prosthetic group bound to the protein through two cysteine residues.

In addition, negatively charged residues may become bonded to positively charged groups through ionic interactions. Nonpolar side chains may attract other nonpolar residues and form regions of hydrophobicity to the exclusion of water and other ionic groups. Occasionally, disulfide bonds are also found holding different regions of the polypeptide chain together. All of these forces combine to create the *secondary* structure of proteins, which is the way the polypeptide chain folds in local areas to form larger, sometimes periodic structures.

On a larger scale, the unique folding and structure of one complete polypeptide chain is termed the *tertiary* structure of protein molecules. The difference between local secondary structure and complete polypeptide tertiary structure is arbitrary and sometimes of little practical difference.

Larger proteins often contain more than one polypeptide chain. These multi-subunit proteins have a more complex shape, but are still formed from the same forces that twist and fold the local polypeptide. The unique 3-dimensional interaction between different polypeptides in multi-subunit proteins is called the *quaternary* structure. Subunits may be held together by noncovalent contacts, such as hydrophobic or ionic interactions, or by covalent disulfide bonds formed from the cysteine residue of one polypeptide chain being crosslinked to a cysteine sulfhydryl of another chain (Figure 2.15).

Thus, aside from the covalently polymerized α-chain itself, the majority of protein structure is determined by weaker, non-covalent interactions that potentially can be disturbed by environmental changes. It is for this reason that protein structure can be easily disrupted or denatured by fluctuations in pH or temperature or by substances that can alter the structure of water, such as detergents or chaotropes.

Not surprisingly, chemical modification to the amino acid constituents of a polypeptide chain may also cause significant disruption in the overall three-dimensional structure of a protein. If amino acid residues critical to folding near functionally important regions are modified with chemical groups that change the charge, hydrophilicity, or hydrogen bonding character of the polypeptide chain, protein structure may be altered and activity may be compromised. This concept will be discussed further in subsequent sections.

Prosthetic Groups, Cofactors, and Post-Translational Modifications

Proteins may contain structures other than polypeptide chains that are important for biological function. Prosthetic groups and cofactors are small organic compounds that are sometimes tightly bound to a protein and aid in forming the active center. A prosthetic group is usually carried within the three-dimensional protein structure in a firm-fitting pocket or even attached through covalent bonds, such as the heme ring associated with cytochrome *c* molecules which is bonded through thioether linkages with adjacent cysteine residues (Figure 2.16). Cofactors, by contrast, may be bound only transiently to proteins during periods of activity. Enzymes often require cofactors to act as donors or acceptors of chemical groups that are added to or cleaved from a substrate molecule. Some common cofactors are ATP, ascorbic acid, coenzyme A, NAD, NADP, FAD, FMN, and biotin. Sometimes, the enzyme cofactor is also an energy source for the catalytic reaction, as in the case of ATP dependent reactions.

Frequently, metal ions are associated with the prosthetic group or cofactor. Heme rings usually contain a chelated iron atom. Occasionally, however, these metals are merely bound within folded polypeptide regions with no additional organic constituents required. Many metal ions are known to participate in enzymatic activity. One or more of the ions of Na, K, Ca, Zn, Cu, Mg, and Mn, as well as Co and Mo, are often required by enzymes to maintain activity.

Prosthetic groups and cofactors, whether organic or metallic, may be removed from a protein to create an inactive *apo* protein or enzyme. Loss of these groups may occur through environmental changes, such as removing metal ions from solution or adding denaturants to unfold protein structure. In many cases, simply re-introducing the needed group into the surrounding medium can restore full activity.

In addition to small organic molecules or metal ions, proteins may have other components tightly associated with them. Nucleoproteins, for instance, contain noncovalently bound DNA or RNA, as in some of the structural proteins of viruses. Lipoproteins contain associated lipids or fatty acids and may also carry cholesterol, as in the high-density and low-density lipoproteins in serum.

During modification or conjugation reactions, prosthetic groups and other associated molecules may be lost or damaged. Metal ions temporarily may be removed by the inclusion of a chelating agent added to maintain sulfhydryl stability during coupling through the –SH groups of a protein. To restore activity after conjugation, it is necessary to remove the chelator and add the required metal salts. Other changes to the prosthetic carriers may not be so easily corrected. For instance, heme-containing molecules are sensitive to the presence of agents that can form a coordination complex with or modify the oxidation state of the chelated metal ion. Some reagent systems may permanently inactivate the heme-containing protein.

Thus, loss of activity can occur not only through changes to the amino acid constituents of a protein, but through prosthetic group or cofactor loss or damage as well. Most of these potential difficulties can be overcome through careful selection of the reaction conditions and through knowledge of the cofactor dependencies that are critical to the activity of the protein being modified.

Post-translational modifications to protein structure are covalent changes that occur as the result of controlled enzymatic reactions or due to chemical reactions not under enzymatic regulation. One of the most common cellular modifications performed on proteins after ribosomal synthesis is glycosylation. Proteins newly synthesized on ribosomes may be transported to the Golgi apparatus, where specific glycosyl transferases catalyze the coupling of carbohydrate residues to the polypeptide chains. Glycoproteins and mucoproteins are formed by the coupling of polysaccharides through *O*-glycosidic linkages to serine, threonine, or hydroxylysine and through *N*-glycosidic linkages with the amide side-chain group of asparagine.

The structure of most glycoprotein carbohydrate is branched with the sugars mannose, *N*-acetylglucosamine (GlcNAc), sialic acid, glactose, and L-fucose being prevalent. Asparagine-linked polysaccharides are well characterized and are known to be constructed of a core unit consisting of three mannose residues and two *N*-acetylglucosamine residues. The GlcNAc residues are bound to the Asp side-chain amide nitrogen through a β1 linkage (Kornfield and Kornfield, 1985). The three mannose groups then usually form the first branch point in the oligosaccharide chain (Section 2.1.2).

The content by weight of carbohydrate in glycoproteins may vary from only a few percent to over 50% in some proteins in mucous secretions. Although the function of the polysaccharide in most glycoproteins is unknown, in some cases it may provide hydrophilicity, recognition, and points of non-covalent interaction with other proteins through lectin-like affinity binding.

The presence of carbohydrate on protein or peptide molecules can provide important points of attachment for modification or conjugation reactions. Coupling exclusively through polysaccharide chains can often direct the reaction away from active centers or critical points in the polypeptide chain, thus preserving activity. Polysaccharides can be specifically targeted on glycoproteins through mild sodium periodate oxidation. Periodate cleaves adjacent hydroxyl groups in sugar residues to create highly reactive aldehyde functionalities (Chapter 3, Section 4.4). The level of periodate addition can be adjusted to selectively cleave only certain sugars in the polysaccharide chain. For instance, a concentration of 1-m*M* sodium periodate at temperatures less than 4°C specifically oxidizes sialic acid residues to contain aldehydes, leaving all other monosaccharides untouched. Increasing the concentration to 10-m*M* and carrying out the reaction at room temperature, however, will cause oxidation of other sugars in the carbohydrate chain, including galactose and mannose. The generated aldehydes can then be used in coupling reactions with amine-or hydrazide-containing molecules to form covalent linkages. Amines can react with formyl groups under reductive amination conditions using a suitable reducing agent such as sodium cyanoborohydride. The result of this reaction is a stable secondary amine linkage (Chapter 3, Section 5.3).

Alternatively, hydrazides spontaneously react with aldehydes to form hydrazone linkages, although the addition of a reducing agent increases the efficiency of the reaction (Chapter 3, Section 5.1).

Another form of post-translational modification that may add carbohydrate to a polypeptide is nonenzymatic glycation. This reaction occurs between the reducing ends of sugar molecules and the amino groups of proteins and peptides. See Section 2.1 in this chapter for further details and the reaction sequence behind this modification.

Protecting the Native Conformation and Activity of Proteins

The goal of most protein modification or conjugation procedures is to create a stable product with good retention of the native state and activity. Ideally, any derivatization should result in a protein that performs exactly as it would in its unmodified form, but with the added functionality imparted by whatever is conjugated to it. Thus, an antibody molecule tagged with a fluorophore should retain its ability to bind to antigen and also have the added functionality of fluorescence.

One of the best ways to ensure retention of activity in protein molecules is to avoid carrying out chemistry at the active center. The active center is that portion of the protein where ligand, antigen, or substrate binding occurs. In simpler terms, the active center (or active site) is that part that has specific interaction with another substance (Means and Feeney, 1971). For the preparation of enzyme derivatives, it is important to protect the site of catalysis where conversion of substrate to product happens. For instance, when working with antibody molecules, it is crucial to stay away from the two antigen binding sites.

The best chemical procedures avoid the active site by selecting functional groups away from that area or by protecting the site through the incorporation of additives. In some cases, the inclusion of substrates, cofactors, ligands, inhibitors, or antigens in the modification reaction will protect the active site. Addition of the appropriate substance can bind the active site and mask it from modification by crosslinking agents. In enzyme derivatization procedures, this is often just a matter of adding a reversible inhibitor or substrate analog. For instance, when working with alkaline phosphatase merely carrying out the reaction in phosphate buffer protects the active center from chemical modification, since phosphate ions bind in the catalytic site. With trypsin, the incorporation of benzamidine similarly masks and protects the active site.

However, protecting the antigen binding sites on an antibody molecule by using this method is often more difficult. Inclusion of antigen to mask the binding sites is effective in blocking these areas, but it may also cause irreversible crosslinking of the antigen to the antibody. This is especially true when the antigen is a peptide or a protein having the same chemical functionalities as the antibody. Any modification reactions that are directed at the antibody may modify the antigen as well. Therefore, only use this method if the antigen is lacking in the chemical targets that are going to be used on the antibody. For instance, if the polysaccharide chains on the antibody are targeted for modification, then using a protein antigen that does not contain carbohydrate to block the antigen binding sites may work well.

An equally effective method of protecting the activity of a protein is by using site-directed reactions that result in modifications away from the active center. In some cases, specific functionalities are known to be present only at restricted sites within the three-dimensional structure of a protein. If these functionalities are not present close to the active site, then using them exclusively for modification reactions should ensure good retention of activity. For instance, sulfhydryl groups or carbohydrate chains are often present in limited quantity and in specific regions on a protein. Selecting reagent systems that target these groups ensures derivatization only at restricted sites within the protein molecule, thus potentially avoiding the active center.

Surprisingly, the goal of some protein crosslinking schemes is to somewhat alter the native presentation of the conjugate. This is especially true in hapten–carrier conjugation as used for immunogen or vaccine preparation. In this case, the main objective is to modify the environment of the hapten to create an immunological response *in vivo*. A hapten is usually a small molecule that is not able to generate an immune response on its own, but can react with the products of such a response once generated. Most often these products are antibodies having binding specificity for the hapten.

The complexities involved in achieving a successful conjugation strategy are best illustrated in the problems and concerns dealing with hapten–carrier conjugation. In order to produce the initial immune response to a small molecule, the hapten is typically coupled to a larger protein that can generate a response on its own. In simple terms, the larger carrier protein confers immunogenicity to the smaller hapten. The native presentation of the hapten is altered toward the immune system, thus creating the immune response.

The site of attachment of the hapten to the carrier and the nature of the crosslinker are both important to the specificity of the resultant antibodies generated against it. For proper recognition, the hapten must be coupled to the carrier with the appropriate orientation. For an antibody subsequently to recognize the free hapten without the attached carrier, the hapten–carrier conjugate must present the hapten in an exposed and accessible form. Optimal orientation is often achieved by directing the crosslinking reaction to specific sites on the hapten molecule. With peptide haptens, this is typically carried out by attaching a terminal cysteine residue during synthesis. This provides a free thiol group on one end of the peptide for conjugation to the carrier. Crosslinking through this group provides hapten attachment only at one end, therefore ensuring consistent orientation.

In hapten–carrier conjugation, the goal is not to maintain the native state or stability of the carrier, but to present the hapten in the best possible way to the immune system. In reaching this goal, the choice of conjugation chemistry may control the resultant titer, affinity, and specificity of the antibodies generated against the hapten. It may be important in some cases to choose a crosslinking agent containing a spacer arm long enough to present the antigen in an unrestricted fashion. It may also be important to control the density of the hapten on the surface of the carrier. Too little hapten substitution may result in little or no response. A hapten density that is too high may actually cause immunological suppression and decrease the response. In addition, the crosslinker itself may generate an undesired immune response. Fortunately, for the majority of hapten–carrier conjugation problems, a few main crosslinking techniques provide a workable compromise to solving all these concerns and ultimately generating an effective immune response (Chapter 19).

Oxidation of Amino Acids in Proteins and Peptides

The modification of amino acids in proteins and peptides by oxidative processes plays a major role in the development of disease and in aging (Kim *et al.*, 1985; Tabor and Richardson, 1987; Halliwell and Gutteridge, 1989, 1990; Stadtman, 1992). Tissue damage through free-radical oxidation is known to cause various cancers, neurological degenerative conditions, pulmonary problems, inflammation, cardiovascular disease, and a host of other problems. Oxidation of protein structures can alter activity, inhibit normal protein interactions, modify amino acid side chains, cleave peptide bonds, and even cause crosslinks to form between proteins.

Due to their abundance in cells relative to other biological molecules, proteins are one of the primary targets of oxidation *in vivo*. However, sometimes oxidation reactions involving proteins and peptides are thought of solely as the creation of disulfides from thiols on cysteine residues. This is certainly an important form of oxidation that can affect protein structure and function or even cause problems relevant to bioconjugation reactions. The presence of an accessible free thiol on a protein in an aqueous solution can be highly unstable to rapid oxidation unless precautions are taken to prevent disulfide formation. Dissolved oxygen and other potentially catalytic components, such as certain metal salts, can quickly result in disulfides being formed within a protein or between different protein molecules.

From a broader perspective, protein oxidation can result in covalent modification at many sites other than just at cysteine thiols. The earliest reports on protein oxidation date from the first decade of the 20th century, but it took many more years to characterize these reactions and their products (Dakin, 1906).

The significance of protein oxidation became paramount with the advent of recombinant protein biologics used as human therapeutics. Careful characterization of protein stability is essential to maintaining the efficacy of protein pharmaceuticals. If even a single side-chain amino acid residue becomes oxidized, then a protein therapeutic may not have the same activity *in vivo* as the unmodified protein.

Oxidation of proteins can result from exposure to oxidative species from many sources: reactive oxygen intermediates caused by metabolic reactions within cells (mitochondrial electron transport function and certain enzymes, such as oxidases, peroxidases, and P-450 enzymes), from the byproducts of oxidative stress reactions in cells (Sayre *et al.*, 2001), or through the presence of strongly oxidizing compounds within a solution—all of these can contribute to selective damage or modification to protein structures. Some examples of chemical agents that can oxidatively modify proteins include hydrogen peroxide (H_2O_2) and other peroxy compounds, such as perborate and peroxycarbonate; hydroperoxyl radical ($HO_2\cdot$); superoxide anion ($O_2^-\cdot$); singlet oxygen ($^1O_2$); hydroxyl radical ($\cdot OH$), periodate (IO_4^-); metal salts in the presence of oxygen species, such as those of iron (Fe^{3+} and Fe^{2+}) and copper (Cu^{2+}); ozone (O_3); peroxynitrite ($ONOO^-$); hypobromous acid (HOBr); hypochlorous acid (HOCl); performic acid (HC(O)OOH); trichloromethylperoxyl radical ($CCl_3OO\cdot$); under the right conditions, metal-chelating compounds, such as porphyrins, texaphyrins, and FeBABE; and gamma radiation and ultraviolet light. For additional information, see Winterbourn and Kettle, 2000; Baynes and Thorpe, 2000; Greenacre and Ischiropoulos, 2001; Halliwell and Gutteridge, 1989, 1990; and Stadtman, 1992.

Singlet oxygen ($^1O_2$) differs from the predominant oxygen molecule in that O_2 is in the ground state or triplet state and its outer two unshared electrons have parallel spins (sometimes designated $^3O_2$), which is nearly unreactive toward other molecules, while singlet oxygen has increased energy and has its outer electrons transformed into an opposite spin orientation, which is highly reactive. Superoxide ($O_2^-\cdot$) is different from singlet oxygen in that it is a reduced form of oxygen having an extra unpaired electron, called a radical. The presence of the radical makes superoxide extremely

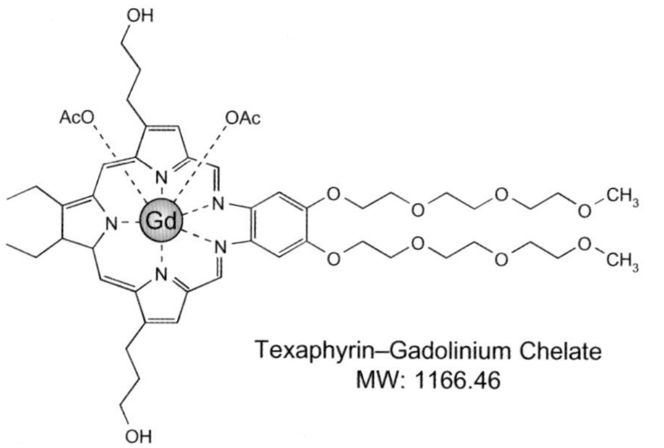

FIGURE 2.17 The texaphyrin–gadolinium chelate structure used as a photosensitizer and MRI contrast agent in the detection and treatment of cancer.

reactive and highly damaging to proteins and other biological molecules.

Singlet oxygen and superoxide, in addition to their modifying effects on proteins, are also important reactive oxygen species in biological applications, as they are intermediates used in some detection methods and in photodynamic therapy (PDT) for the treatment of cancer. One of the more common compounds used in PDT is Photofrin, which is a mixture of oligomers consisting of ether and ester linkages that combine up to eight porphyrin groups (Misawa *et al.*, 2005). The generation of reactive oxygen species takes place by irradiation with 630-nm wavelength laser light, which also penetrates the skin effectively during therapy. Photoactivation of the Photofrin molecule causes radical initiation to form porphyrin-excited states. Transfer of electrons from the porphyrin groups to molecular oxygen then generates the highly reactive singlet oxygen species. Subsequent radical reactions can also form superoxide and hydroxyl radicals, all of which severely damage tissue in the region of the tumor and ultimately cause cancer cell death.

Another compound used to generate reactive oxygen species for PDT is texaphyrin, which contains a metal-chelating ring structure resembling a porphyrin group (Figure 2.17). Typically, a gadolinium atom is chelated in the texaphyrin center and this complex becomes both a photosensitizing agent and an MRI contrast agent to better visualize tumor locations for irradiation therapy (Donnelly *et al.*, 2004).

The reactive oxygen species involved with protein oxidation can be generally categorized according to their relative reactivity as follows:

$$HO\cdot,\ HO_2\cdot > O_2^{-}\cdot > ROOH,\ H_2O_2 > {}^1O_2,\ ClO^-,\ BrO^- > O_2$$

Thus, radicals are the most reactive and destructive of protein structure, followed by peroxy derivatives, singlet oxygen, and other oxygen compounds. The oxidative reactivity of some of these oxygen species is so high that just contact of the pure compound with paper or cotton fabrics can cause combustion (e.g., superoxide).

In vitro studies of protein oxidation indicate that virtually all proteins and peptides are susceptible to damage by the radicals $\cdot OH$ and $O_2^{-}\cdot$. Analysis of protein modification products after oxidation indicates the presence of altered molecular weight (either fragmentation or oligomerization), altered net charge, tryptophan destruction, and the formation of tyrosine dimers (Davies, 1987). Even in the presence of very low concentrations of oxidants (nM), SDS polyacrylamide gel electrophoresis of proteins can indicate multiple bands of higher and lower molecular weight due to oxidative damage.

Transition metals in solution can catalyze the formation of reactive oxygen species that are particularly damaging to proteins and other biomolecules. In a series of reactions, reduced transition metals, such as Fe^{2+} and Cu^{1+}, can be oxidized by oxygen to produce superoxide and ultimately undergo a Fenton reaction to create hydroxyl radicals (Kim *et al.*, 1985). Transition metal-chelating groups can accelerate this reaction, as demonstrated in the process of hydroxyl radical footprinting of protein interactions using EDTA chelates of iron (see discussion on the reagent FeBABE in Chapter 24, Section 4).

Production of superoxide: $Fe^{2+} + O_2 \rightarrow Fe^{3+} + O_2^{-}\cdot$

Production of hydrogen peroxide:
$$2O_2^{-}\cdot + 2H^+ \rightarrow H_2O_2 + O_2$$

Production of hydroxyl radical:
$$Fe^{2+} + H_2O_2 \rightarrow Fe^{3+} + OH\cdot + OH^-$$

The potential sites of oxidation within a protein molecule include the peptide backbone and the side-chain amino acid groups. Hydrogen atom abstraction at the alpha carbon of an amino acid chain can occur upon reaction with an oxidative species to form a radical intermediate. Subsequent reaction can result in peptide bond cleavage and fragmentation of the protein structure, often forming carboxylic acids or carbonyls (aldehydes or ketones). This is the basic mechanism of fragmentation caused by the bifunctional chelating reagent FeBABE. When used in the presence of H_2O_2 and ascorbic acid, polypeptides will fragment in the neighborhood of interacting proteins.

Amino acid side chains can undergo oxidation through hydrogen abstraction, elimination, or by addition reactions. In the presence of oxygen, aliphatic amino acids usually experience oxidation to a peroxy intermediate, which causes either hydrogen atom abstraction or an elimination reaction resulting in the formation of carbonyls, hydroxyls, or other peroxides (Requena *et al.*, 2001) (Figure 2.18). Aromatic amino acids typically undergo addition reactions following exposure to strong oxidants. An example of this type of reaction is the nitrosation of tyrosine groups in the presence of a peroxynitrite ($ONOO^-$) to create *o*-nitrotyrosine (see Figure 2.19).

After exposure to an oxidant, the potential types of oxidation products in proteins and peptides can be extensive (Stadtman and Levine, 2000). Cysteine and methionine undergo a variety of sulfur oxidation reactions to yield cysteine disulfides, methionine sulfoxide, methionine sulfone, and sulfonate products (e.g., cysteic acid) (Ghesquiere *et al.*, 2011) (Figure 2.20). Oxidation with performic acid can be used purposely to convert methionine and cysteine in peptides and proteins to more stable products prior to acid hydrolysis and amino acid analysis. Cysteine and methionine are perhaps the most sensitive amino acids to oxidation, and for this reason they are an early indicator of oxidative damage to proteins.

Tyrosine is also easily modified through addition reactions due to the ring activating nature of its phenolic group. Using oxidants, tyrosine's ring can be chlorinated or iodinated, undergo nitrosation or hydroxylation, and even form tyrosine–tyrosine cross-links. The last product can be formed purposely by use of a peroxidase in the presence of hydrogen peroxide, and this type of reaction has been studied extensively in the manufacture of phenolic polymer resins (Dordick, 1991). In addition, Fancy *et al.* (1996) as well as Fancy and Kodadek (1997, 1998) have applied the oxidation of tyrosine to form dityrosine to the study of protein–protein interactions using nickel-chelated 6xHis-tagged fusion proteins in oxidative environments.

Nitrogen-containing side chains in amino acids can be altered by oxidation forming chloramines or even become deaminated. The result is often the formation of carbonyls (e.g., aldehydes) and hydroxyls. Lee *et al.* (2006)

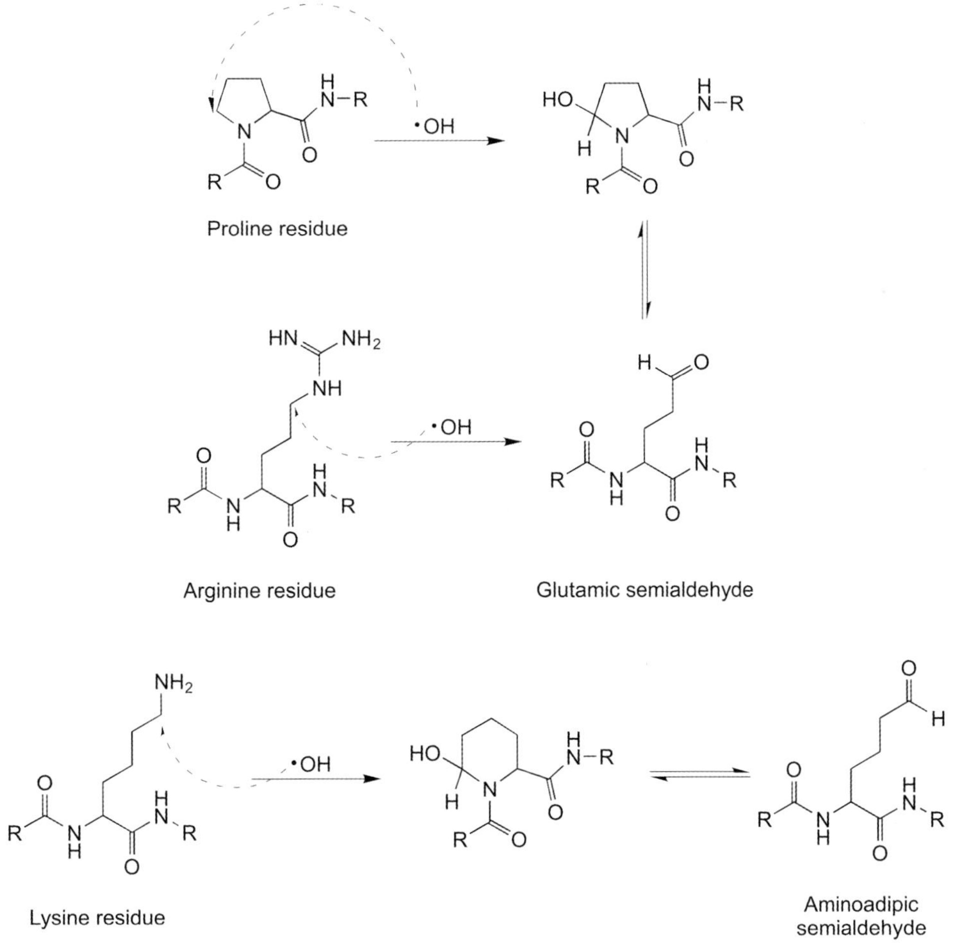

FIGURE 2.18 Reaction of proline, arginine, and lysine residues with hydroxyl radical results in oxidation of side-chain structures that form carbonyls. Both arginine and proline oxidation will result in the same product being formed.

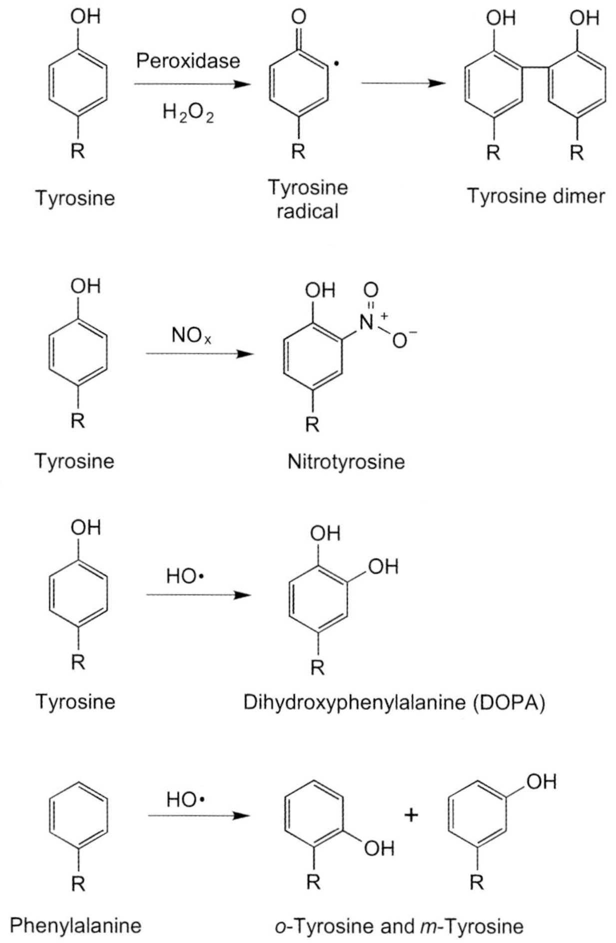

FIGURE 2.19 Tyrosine and phenylalanine residues can undergo oxidation to modify their phenyl side chain groups. Tyrosine can form covalent dimers that link two side chains together *via* a radical reaction. Both tyrosine and phenylalanine can be modified by oxidation to add oxygen-containing groups directly to their aromatic ring.

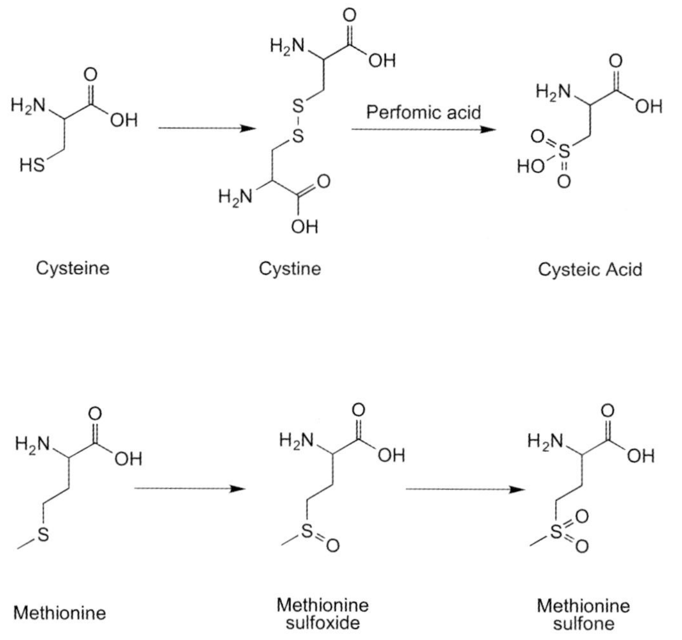

FIGURE 2.20 Cysteine and methionine are highly susceptible to oxidation reactions. Cysteine thiols can forms disulfide linkages with other cysteine groups or be oxidized to cysteic acid. Methionine is oxidized very easily to the sulfoxide or sulfone products.

found that Fe-EDTA-mediated oxidation of human serum albumin resulted in extensive aldehyde and ketone formation from modification of lysine, arginine, histidine, proline, threonine, and aspartic and glutamic acids. Some groups will oxidize and convert to another amino acid altogether. For instance, histidine can be converted to asparagine and proline to hydroxyproline, and tyrosine can be changed to dihydroxyphenylalanine (DOPA) through oxidation reactions.

It is obvious that the oxidation of protein molecules can have detrimental effects on protein structure and function. However, there are some unique methods in bioconjugation wherein controlled and purposeful oxidation is carried out to study protein–protein interactions (Chapter 24, Section 4).

Unfortunately, there are no universal methods to detect all types of protein oxidation, because the products formed can be so diverse in nature. However, some forms of protein oxidation can be assayed using chemical modification (Davies *et al.*, 1999; Shacter, 2000). In particular, the formation of carbonyl groups on proteins can be targeted using the reagent 2,4-dinitrophenylhydrazine (DNPH). This compound reacts with aldehydes to form 2,4-dinitrophenylhydrazone derivatives, which create chromogenic modifications that can be detected at high sensitivity in microplate assays or western blot analysis (Buss *et al.*, 1997; Winterbourn *et al.*, 1999).

In addition, a method involving mass spec analysis to determine carbonyl formation as a result of protein oxidation was developed using a novel mass tag. The carbonyl-specific Element-Coded Affinity Mass Tag (O-ECAT) can covalently couple to aldehyde or ketone oxidation sites using an aminoxy group to form an oxime (Lee *et al.*, 2006; see also Chapter 12, Section 3). The ECAT mass tag consists of a bifunctional metal-chelating group that coordinates a lanthanide metal ion of specific mass. Proteins that have been oxidized to contain carbonyls can be labeled with this reagent and the exact sites of modification determined by analyzing the mass spec signature of the labeled peptides after proteolysis.

Solvent Accessibility of Functional Targets in Proteins

Proteins are highly complex, folded polypeptide chains consisting of at least 20 different amino acids that are strung together in unique sequences, which relate to structure and function. Particular amino acids in proteins may be further modified post-translationally to contain a wide variety of covalent modifications normally found in native proteins. The way in which a peptide chain is wrapped and folded governs each amino acid's relative exposure to the outside environment, but post-translational modifications can also obscure the protein surface from easy access to the solvent environment.

Amino acid side chains are the primary effectors of the three-dimensional structure of a protein, because their properties vary depending on the presence of charged groups, uncharged polar components, aliphatic chains, aromatic rings, and groups able to form hydrogen bonds with other amino acid residues. The relative hydrophilicity or hydrophobicity of an amino acid side chain is a major factor in determining whether the group will be found on the surface of a globular protein or buried within its globular structure.

However, just considering the individual properties of each amino acid type is not enough to determine its accessibility to the surrounding aqueous environment. There have been many attempts at developing analytical models with predictive value for determining buried or surface accessible amino acids in a folded polypeptide chain (Vasicek *et al.*, 2012). These studies have concluded fractional assignments for each residue that relate to its accessible surface area (ASA) or its solvent exposed area (SEA).

In most cases, there are general trends that emerge from theoretical studies in which hydrophilic amino acids are more likely to be found on the surface of a protein and hydrophobic amino acids are more likely to be inside its three-dimensional structure, but we already knew this intuitively so this conclusion is not surprising. However, a real-life study of the positions of amino acids in proteins whose structures are known is more revealing. The data for Figure 2.21 were calculated from 55 proteins in the Brookhaven database by Bordo and Argos (1991), and the graph was derived from the analysis as presented by the Jena Image Library of Biological Macromolecules (http://www.imb-jena.de/IMAGE_AA.html). Although most of these structures were determined using crystallographic means and thus the proteins are "frozen" in a single structural state, the results are revealing as to how often particular amino acids are accessible to the surrounding solvent.

Three levels of SEA are presented in the graph for each amino acid, which corresponds to areas in Å^2 accessible to the solvent environment: greater than 30 Å^2 for highly accessible amino acids, between 10 Å^2 and 30 Å^2 for medium accessibility, and less than 10 Å^2 for those residues that are relatively not accessible to the solvent. Only the SEA for each amino acid of >30 Å^2 is shown in the plotted data. The graph shows that the polar amino acids such as serine, threonine, asparagine, glutamine, and tyrosine often have large areas accessible to the solvent, as do the charged amino acids aspartic acid, glutamic acid, lysine, histidine, and arginine. Surprisingly, proline also falls in the highly accessible

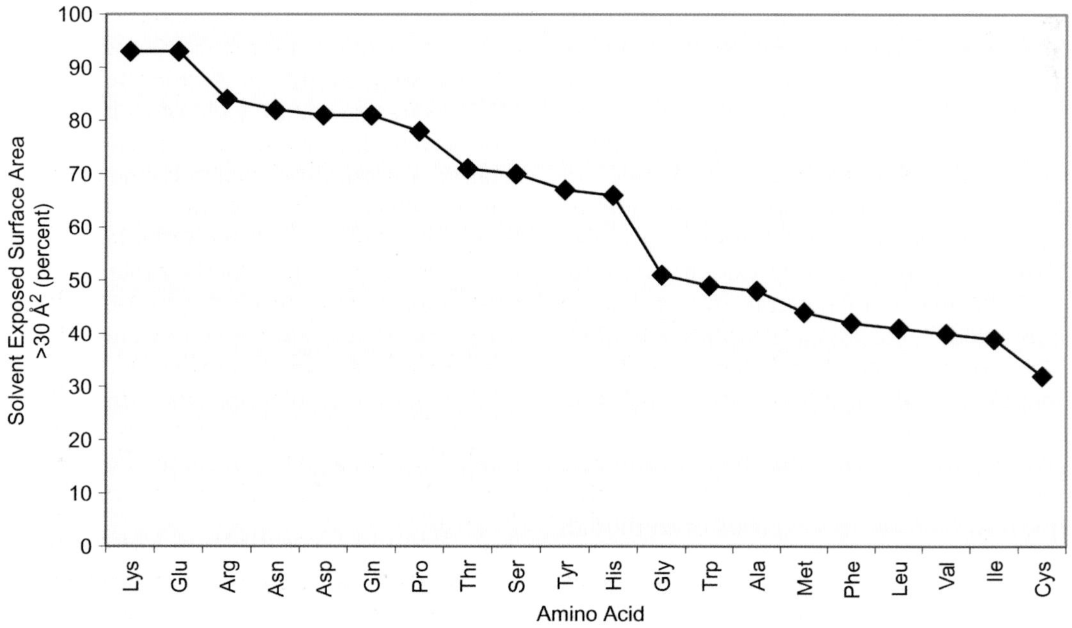

FIGURE 2.21 Comparison of the solvent exposed surface area of amino acids in proteins. Data are plotted as a percentage of each amino acid in a protein having greater than a 30-Å^2 exposure to the aqueous environment. Charged and polar amino acids are seen to have the most solvent exposure, while uncharged, aromatic, or aliphatic amino acids have the least exposure.

group, which is not as expected, because it does not carry a charge, nor is it a highly polar amino acid. However, proline does have a unique characteristic that may explain its appearance on the surface of proteins: it cannot freely rotate about its imino group as other amino acids in a peptide chain can do at amide bonds. This effect results in a kink in the polymer (called a beta-turn), and these sharp turns in a peptide backbone probably occur most often near the surface. Thus, proline is found to be frequently accessible to the solvent environment despite its hydrophobic nature.

The nonpolar amino acids glycine, alanine, valine, leucine, isoleucine, methionine, phenylalanine, and cysteine have lower exposure to the solvent environment than charged or polar residues. However, the frequency at which these groups are found to have an SEA of greater than 30 Å^2 is much higher than one would expect based solely upon consideration of their hydrophobicity. In fact, nearly 30 to 50% of the time nonpolar amino acids in a protein can be found at the surface.

At the two extremes, lysine is observed as the amino acid most accessible on the surface of proteins while cysteine is the least exposed amino acid. The inaccessibility of cysteine probably stems from the fact that disulfides are typically buried within the polypeptide structure of proteins, whether they are intrachain or in nature, and proteins rarely contain many reduced cysteine thiols.

It is clear from these data that proteins have complex hydrophilic and hydrophobic regions on their surfaces that determine their potential interactions, binding sites, and active centers. For bioconjugation purposes, targeting of an amino acid even with a high SEA for modification or crosslinking may not result in every residue being modified that is theoretically present in a protein based only on knowledge of its amino acid composition. Even when coupling to very polar or charged groups, such as lysine, there are varying degrees of accessibility to a given reagent, because of the complex folding of the polypeptide chains at the protein surface.

Figure 2.22 shows the globular structure of an immunoglobulin (IgG) Fc region to illustrate this point. In this space-filling model, the lysine residues are highlighted in yellow to easily show their locations within the two polypeptides of the heavy chains. Notice that some of the ε-amino groups at the ends of the side chains are protruding far out into the solvent and are therefore highly accessible for modification. Some of these groups, however, are less exposed even though they are still near the surface, and a few lysines are seen to be between the heavy chain regions where it would be difficult to modify them due to crowding.

Figure 2.64 in this chapter provides data to validate this effect. The reaction of the thiolating reagent SATA with IgG resulted in only a percentage of the available lysines being modified. As the molar ratio of SATA to IgG was increased, the yield of lysine modification actually became lower. This result can be explained by the relative accessibility of each lysine in the immunoglobulin structure. Some residues are easily accessible and they get modified with high yield even with low molar ratios of SATA-to-IgG. As the molar ratio is increased, it gets more difficult to modify those lysines that are less accessible to the solvent environment or are partially obscured by another polypeptide chain. Thus, the solvent accessibility of particular amino acids is a major factor in whether they can be effectively targeted and modified with a given bioconjugate reagent.

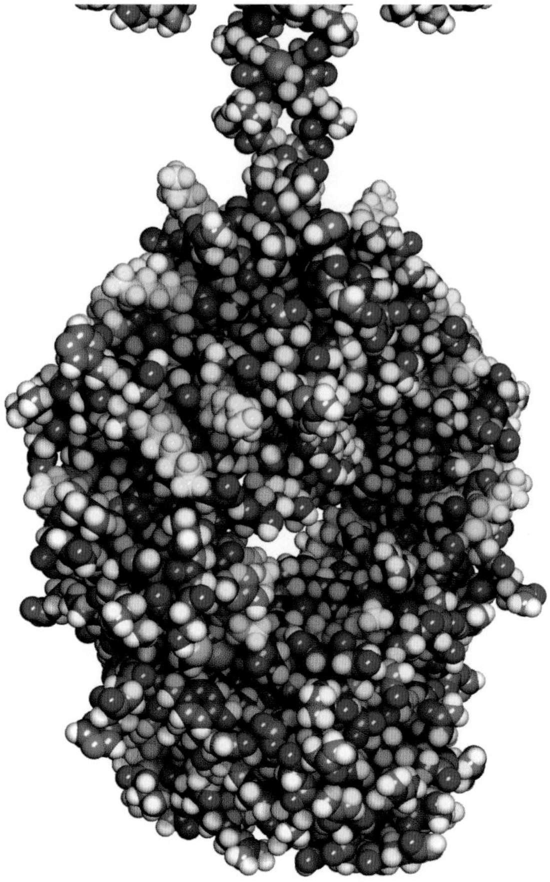

FIGURE 2.22 The solvent accessibility of lysine residues in the Fc region of an antibody is illustrated by highlighting the lysine groups in yellow. Some lysine ε-amine groups are extremely accessible to conjugation, while others are only partially exposed, thus making it difficult to modify all of them in bioconjugation reactions.

1.2. Protein Crosslinking Methods

The crosslinking of two proteins using a simple homobifunctional reagent (Section 2.2) can potentially result in a broad range of conjugates being produced (Avramease, 1969). The reagent may initially react with either one of the proteins, forming an active intermediate. This activated protein may then form crosslinks with the other protein or with another molecule of the same protein. The activated protein may also react intramolecularly with other functionalities on part of its own polypeptide chain. Other crosslinking molecules may continue to react with these conjugated species to form various mixed products, including severely polymerized proteins that may fall out of solution (Figure 2.23).

The problems of indeterminate conjugation products are amplified in single-step reaction procedures using homobifunctional reagents (Chapter 5). Single-step procedures involve the addition of all reagents at the same time to the reaction mixture. This technique provides the least control over the crosslinking process and invariably leads to a multitude of products, only a small percentage of which represent the desired or optimal conjugate. Excessive conjugation may cause the formation of insoluble complexes that consist of very high-molecular-weight polymers. For example, one-step glutaraldehyde conjugation of antibodies and enzymes (Chapter 20, Section 1.2) often results in significant oligomers and precipitated conjugates. To overcome this shortcoming, multi-step reaction procedures have been developed using both homobifunctional and heterobifunctional reagents (Chapters 5 and 6). Controlled, multi-step conjugation protocols alleviate the polymerization problem and form relatively low-molecular-weight, soluble antibody–enzyme complexes (Chapter 20, Section 1.1).

In two-step protocols, one of the proteins to be conjugated is reacted or "activated" with a crosslinking agent and excess reagent and byproducts are removed. In the second stage, the activated protein is mixed with the other protein or molecule to be conjugated, and the final conjugation process occurs (Figure 2.24).

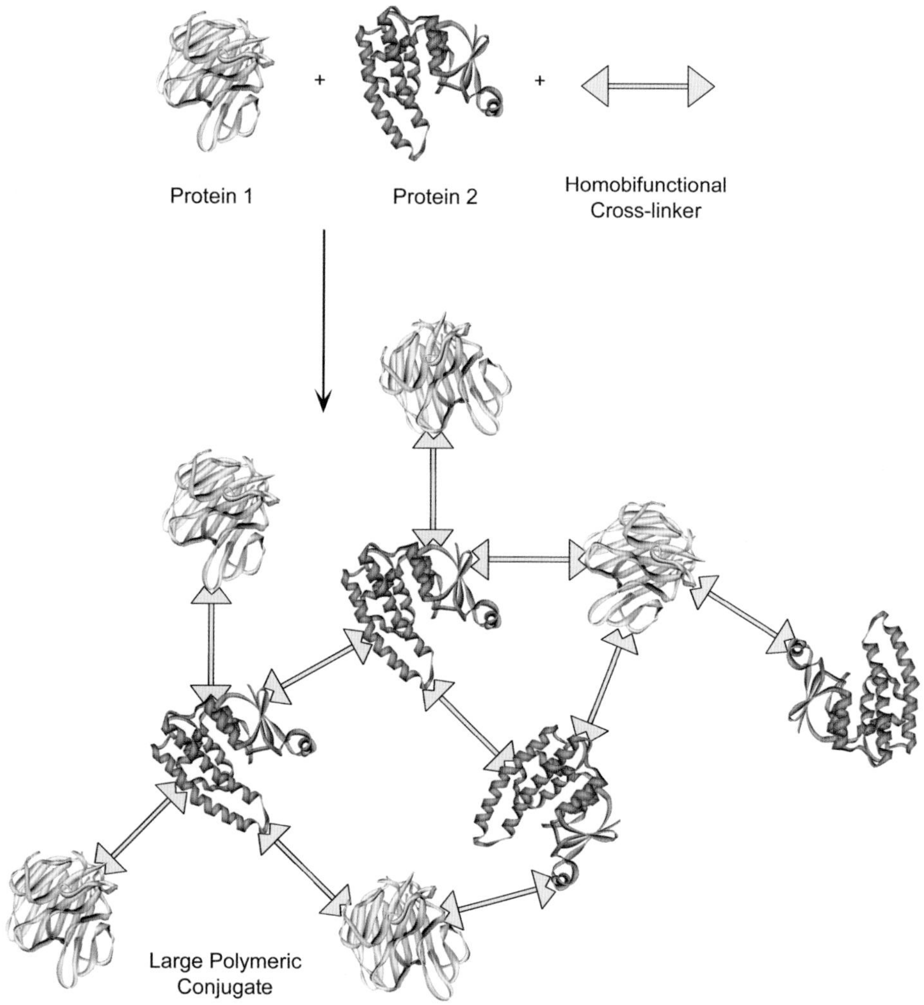

FIGURE 2.23 Protein crosslinking reactions carried out using homobifunctional reagents can result in large polymeric complexes of multiple sizes and indefinite structure.

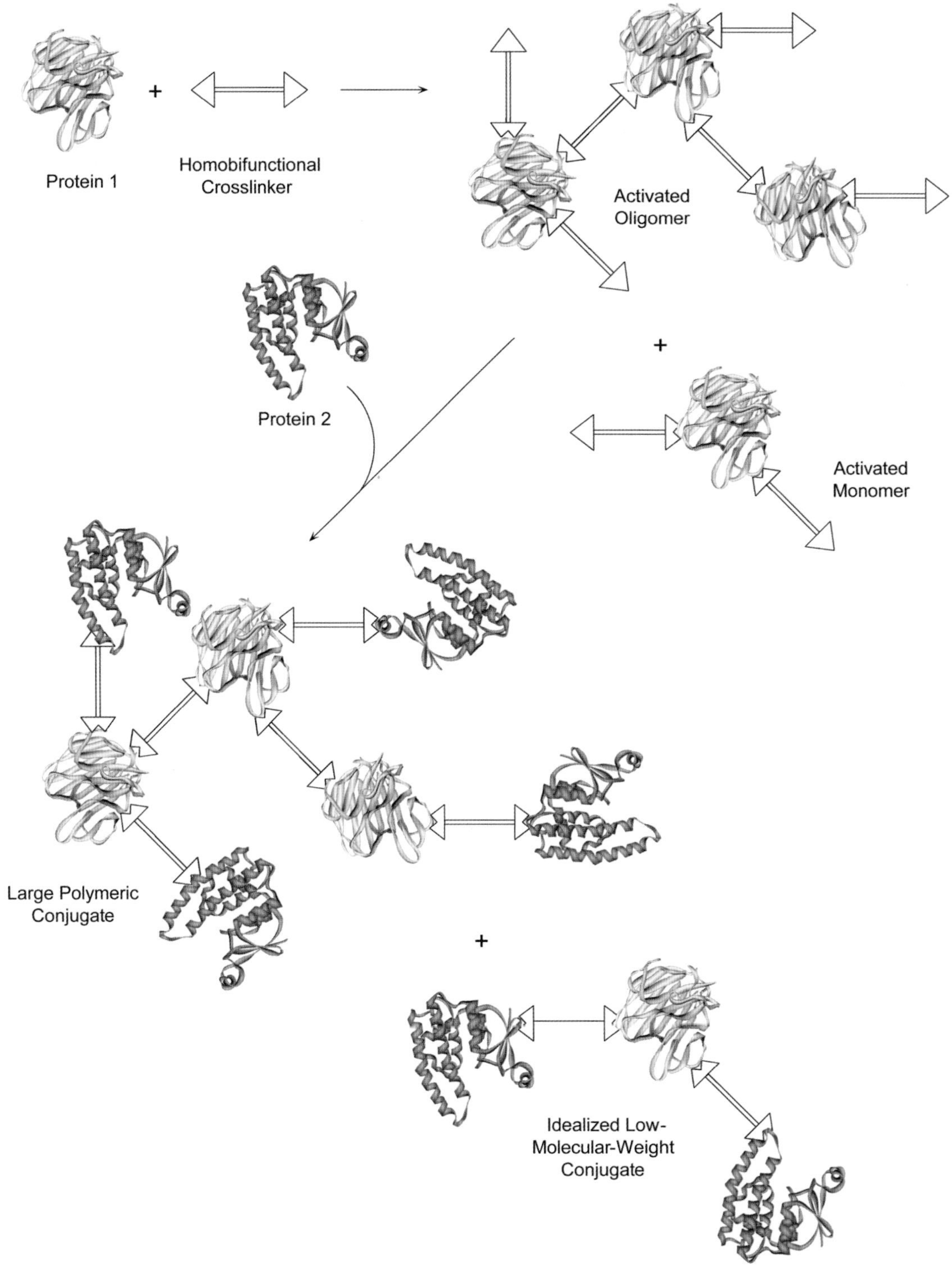

FIGURE 2.24 A two-step protocol using a homobifunctional crosslinking agent offers more control than single-step methods, but still may result in oligomer formation.

The use of homobifunctional reagents in two-step protocols still creates many of the problems associated with single-step procedures, because the first protein can crosslink and polymerize with itself long before the second protein is added. Homobifunctional reagents by definition have the same reactive group on both ends of the crosslinking molecule. Since the protein to be activated has target functionalities on every molecule that can couple with the reactive groups on the crosslinker, both ends of the reagent can potentially react.

This inherent potential to uncontrollably polymerize is unfortunately characteristic of all homobifunctional reagents, even in multi-step protocols.

The greatest degree of control in crosslinking procedures is afforded using heterobifunctional reagents (Section 2.3). Since a heterobifunctional crosslinker has different reactive groups on either end of the molecule, each side can be directed specifically toward different functional groups on proteins. Using a multi-step conjugation protocol with a heterobifunctional reagent can allow one macromolecule to be activated, excess crosslinker removed, and then a second macromolecule added to induce the final linkage. Directed conjugation will occur as long as the first protein that is activated does not have groups able to couple with the second end of the crosslinker, whereas the second molecule does possess the correct functionalities.

Occasionally, the second protein does not naturally have the target groups necessary to couple with the second end of the crosslinker. In such cases, a specific functionality can usually be created to make the conjugation successful (Chapter 2, Section 4). In such three-step systems, the first protein is activated with the heterobifunctional reagent and purified away from excess crosslinker. The second protein is then modified to contain the specific target groups required for the second stage of the conjugation. Finally, in step three, the two modified proteins are mixed to cause the coupling reaction to take place (Figure 2.25).

Two- and three-step protocols using heterobifunctional crosslinkers are often designed around amine-reactive and sulfhydryl-reactive chemical reactions. Many of these reagents utilize NHS esters on one end for coupling to amine groups on the first protein and

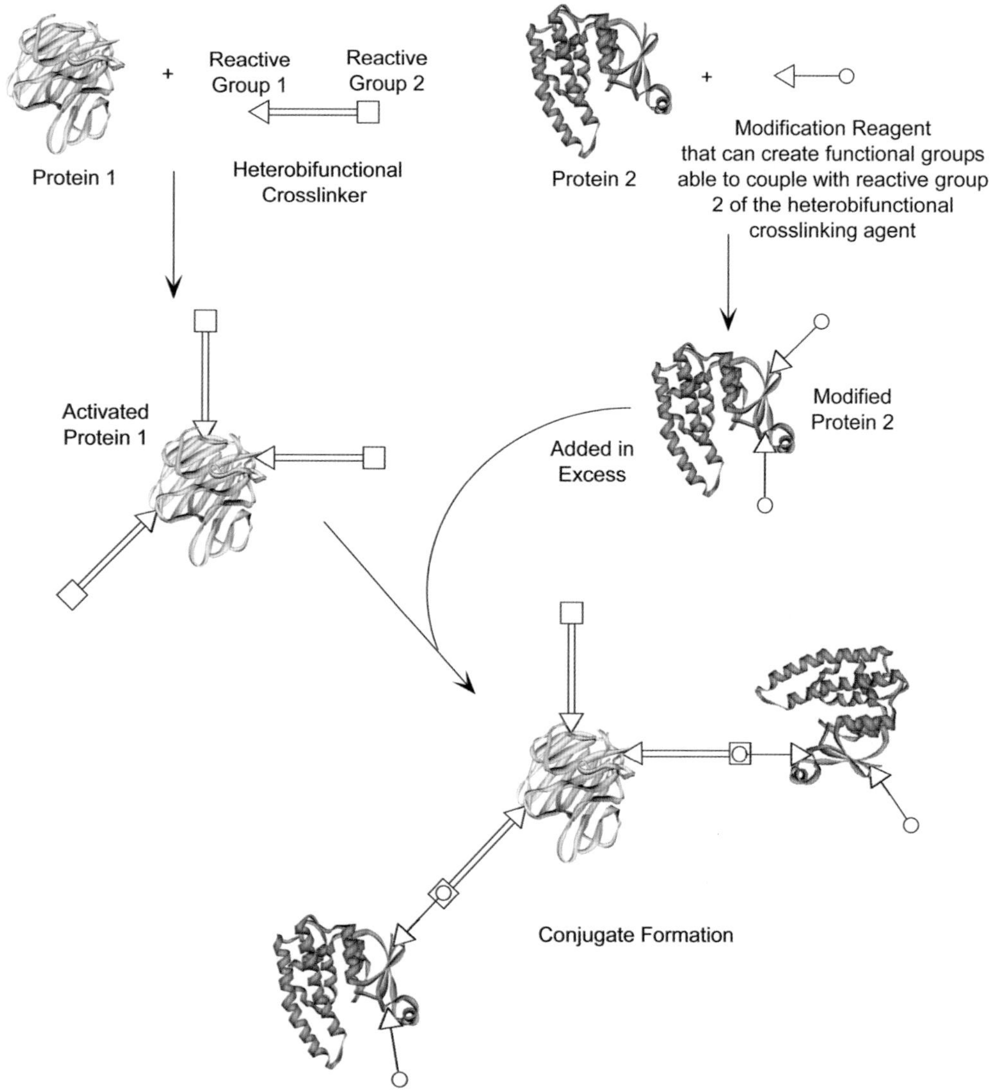

FIGURE 2.25 Heterobifunctional crosslinking agents used in multi-step protocols result in the best control over the products formed.

maleimide groups on the other end that can react with sulfhydryls on the second protein. The NHS ester end is reacted with the first protein to be conjugated, forming an activated intermediate containing reactive maleimide groups. Fortunately, the maleimide end of such crosslinkers is relatively stable to degradation, thus the activated protein can be isolated without loss of sulfhydryl coupling ability. Additionally, if the second protein does not contain indigenous sulfhydryls, these can be created by an abundance of methods (Chapter 2, Section 4.1). After mixing the maleimide-activated protein with the sulfhydryl-containing protein, conjugation can only occur in one direction.

Control of the products of conjugation will generally increase as the protocols progress from single-step to multi-step reactions. Likewise, control of the chemistry of conjugation increases as the reagent systems evolve from simple homobifunctional to site-directed heterobifunctional. It may appear to be a paradox, but often as the method of conjugation gets more complex the result is less potential for side reactions and therefore fewer products being formed. Therefore, multi-step processes using advanced heterobifunctional reagents are the best combination to ensure that the protein conjugate formed is indeed the one desired.

2. MODIFICATION OF SUGARS, POLYSACCHARIDES, AND GLYCOCONJUGATES

The basic units of food energy for cells and living organisms consist of polysaccharides or simple sugars, principally glucose and its derivatives. Biological molecules themselves often contain carbohydrate or are made exclusively of such components. Complex carbohydrate "trees" frequently project off the surface of cells, providing specific points of attachment or sites of recognition. Lipids and proteins that contain these components may possess them to give identity or partial hydrophilicity to their parent structures.

Many of the macromolecules that are the subject of modification or conjugation reactions contain significant proportions of carbohydrate. Reactions can be designed to directly target these polysaccharide portions, either selectively modifying them with small, detectable compounds or using them as conjugation bridges to couple with other macromolecules. The reactivity of carbohydrate molecules in such derivatizations is an important factor in the success of many bioconjugate techniques.

This section describes the basic chemical attributes of carbohydrate molecules. Principal sites of reactivity on carbohydrates are discussed with the aim of developing a rational approach to using them in modification and conjugation procedures.

2.1. Carbohydrate Structure and Functionality

Carbohydrates are characterized by the presence of polyhydroxylic aldehyde or polyhydroxylic ketone structures or polymers made of such units. Sugars and polysaccharides have definite three-dimensional structures that are important for many biological functions. They are hydrophilic and thus easily accessible to aqueous reaction mediums. The chemistry of bioconjugation using carbohydrate molecules begins with an understanding of the building blocks of polysaccharide molecules.

Basic Sugar Structure

The simplest carbohydrate, called a monosaccharide, is composed of a structure that cannot be hydrolyzed to simpler polyhydroxylic compounds. A disaccharide is a carbohydrate that contains two of these basic units, and a polysaccharide contains many polyhydroxylic monomers.

A monosaccharide that contains an aldehyde group is called an aldose, and one that contains a ketone group is a ketose. Monosaccharides are further classified by the number of carbon atoms they contain. Thus, a five-carbon sugar is known as a pentose and a six-carbon sugar, a hexose. All monosaccharides containing accessible aldehyde or ketone functionalities are reducing sugars—that is, they are able to reduce Fehling's or Tollen's reagent.

The aldehyde or ketone group of monosaccharides can undergo an intramolecular reaction with one of its own hydroxyl groups to form a cyclic, hemiacetal, or hemiketal structure, respectively (Figure 2.26). In aqueous solutions, this cyclic structure actually predominates. The open-chain aldehyde or ketone form of monosaccharides is in equilibrium with the cyclic form, but the open structure exists less than 0.5% of

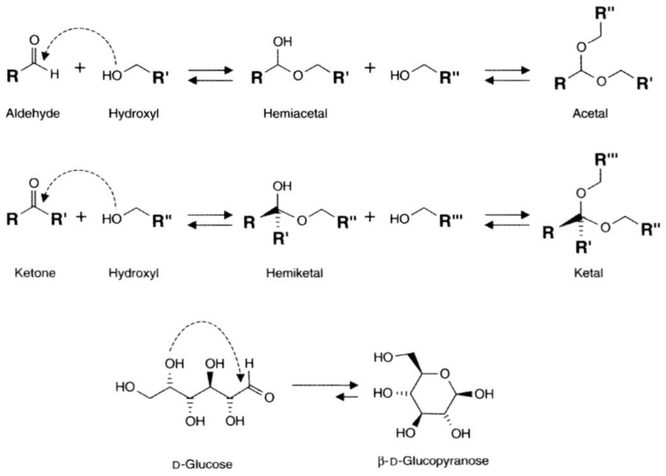

FIGURE 2.26 Carbonyl groups and hydroxyls may react to form acetal or ketal products. Sugars naturally undergo these reactions to form ring structures in aqueous solution.

the time in aqueous environments. It is the open form that reduces Fehling's or Tollen's reagent. However, due to this predominance of the cyclic structure of monosaccharides, they do not have the capability of reacting with bisulfite or Schiff's reagent, as do normal unblocked aldehydes and ketones. Thus, the carbonyl functionalities of sugars have reduced reactivity, because of hemiacetal and hemiketal formation.

Figure 2.27 shows the structures of some of the most common monosaccharide molecules: D-glyceraldehyde, D-erythrose, D-ribose, D-arabinose, D-xylose, D-glucose, D-glucosamine, *N*-acetyl-D-glucosamine, D-mannose, D-galactose, D-galactosamine, and *N*-acetyl-D-galactosamine of the aldose family and dihydroxyacetone, D-ribulose, D-fructose, and D-*N*-acetylneuraminic acid of the ketose family. Formation of the cyclic structure of each of these sugars can result in one of two stereoisomers, designated α and β, depending on the orientation of the aldehyde group or ketone group during hemiacetal formation. For aldoses, the α form is drawn in the standard Haworth projection with the number 1 carbon hydroxyl pointing down. For ketoses, the α form consists of the number 2 carbon hydroxyl pointing down. All the common monosaccharide structures shown in Figure 2.27 are in the β-stereoisomer form.

Since in aqueous solutions the cyclic form of monosaccharides is in equilibrium with their corresponding open forms, the α and β structures continually interconvert. At equilibrium, one form usually predominates. For instance, glucose dissolved in water consists of about a 2:1 ratio of β-D-glucose to α-D-glucose. Although their chemical constituents are identical, the biochemical properties between the α and β forms can be quite different. Monosaccharides linked together to form disaccharides and polysaccharides cannot continue to interconvert and are therefore frozen in the α or β forms. Changing one monosaccharide in a complex carbohydrate to its opposite stereoisomer form can produce radical structural changes in the polysaccharide chain and significantly alter its biochemical properties.

Sugar Functional Groups

Monosaccharide functional groups consist of either a ketone or an aldehyde, several hydroxyls, and the possibility of amine, carboxylate, sulfate, or phosphate groups as additional constituents. Amine-containing sugars may possess a free primary amine, but often are modified to the *N*-acetyl derivative, such as the *N*-acetylglucosamine residue of chitin. Sulfate-containing monosaccharides frequently are found in certain mucopolysaccharides, including chondroitin

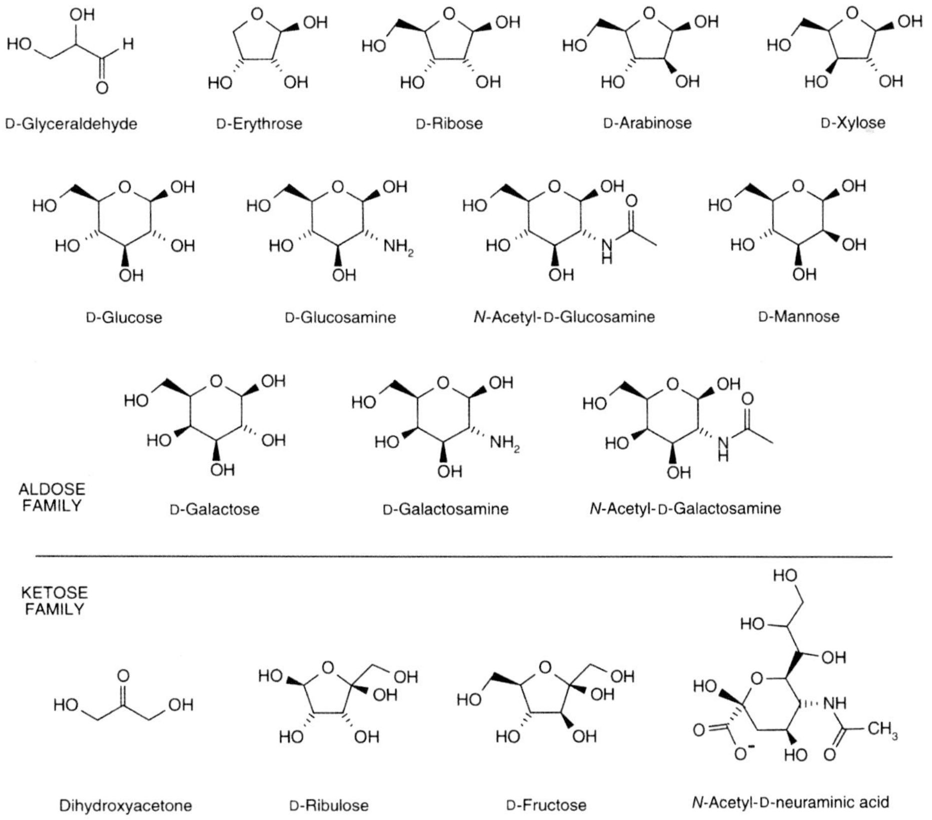

FIGURE 2.27 Common monosaccharides of the aldose and ketose families found in biological molecules.

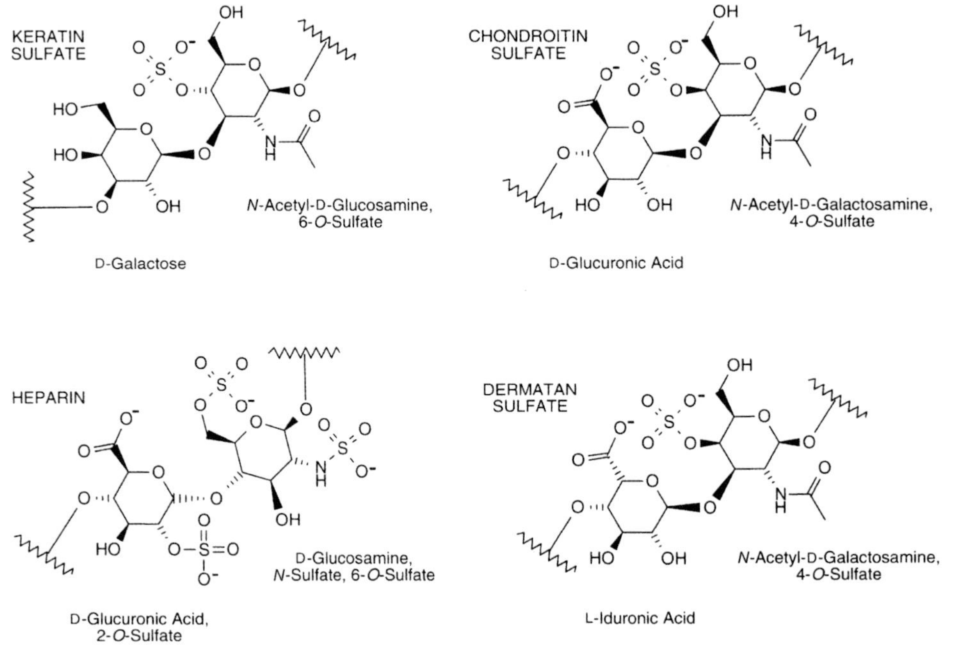

FIGURE 2.28 Common sulfonated polysaccharides of biological origin.

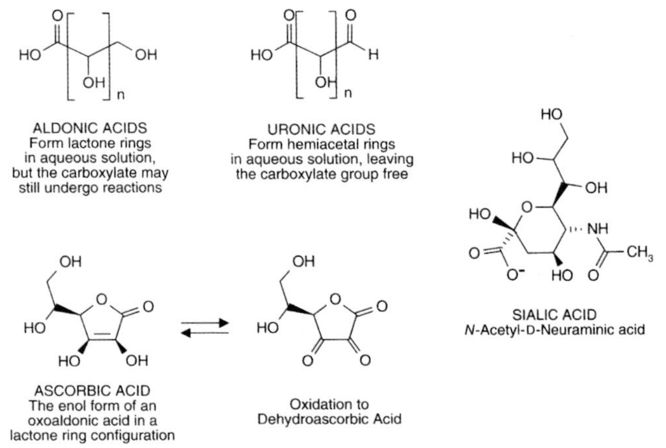

FIGURE 2.29 Monosaccharides containing carboxylate groups. Sialic acid is often found at the terminal residues of polysaccharides within glycoproteins.

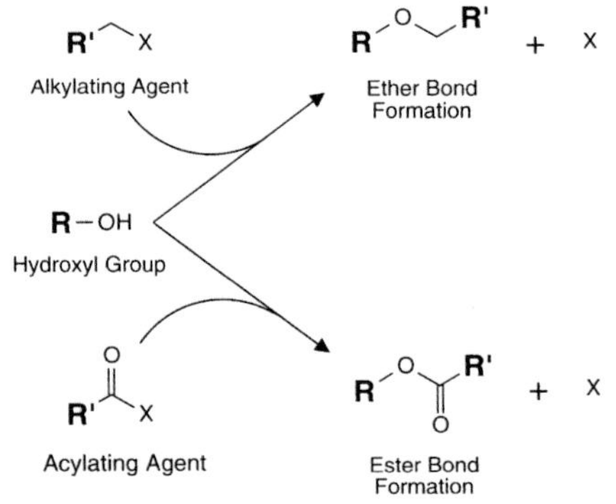

FIGURE 2.30 Hydroxyl groups within sugar residues may undergo alkylation or acylation reactions, forming ether or ester linkages.

sulfate, dermatan sulfate, heparin sulfate, and keratin sulfate (Figure 2.28). Carboxylate-containing sugars include sialic acid as well as many aldonic, uronic, oxoaldonic, and ascorbic acid derivatives (Figure 2.29). Phosphate-containing monosaccharides are almost exclusively created in metabolic processes involving energy utilization, such as in the production of glucose-1-phosphate formed during glycogen breakdown and glucose-6-phosphate produced during glycolysis. Perhaps the most common phosphate sugar derivative, however, is the 5′-phosphate of D-ribose or D-2-deoxyribose found as a repeating component of RNA and DNA, respectively.

Modification and conjugation reactions can be designed to target many of these functionalities. Sugar hydroxyl groups, for example, may be derivatized by acylating or alkylating reagents, similar to the principal reactions of primary amines (Section 1). However, acylation of a hydroxyl group usually creates an unstable ester derivative that is subject to hydrolysis in aqueous solution. An exception to this is acylation by a carbonylating reagent such as carbonyldiimidazole (CDI) (Chapter 3, Section 4.2) or *N,N′*-disuccinimidyl carbonate (DSC) (Chapter 3, Section 4.3), which can produce stable carbamate linkages after subsequent conjugation with an amine-containing molecule. By contrast, alkylating reagents, such as alkyl halogen compounds (Chapter 3, Section 4.6), typically form more stable ether bonds after reaction with hydroxyls. Figure 2.30

shows the reactions associated with alkylation and acylation of hydroxyl residues.

Carbohydrates containing hydroxyl groups on adjacent carbon atoms may be treated with sodium periodate (Section 4.4, this chapter) to cleave the associated diol carbon–carbon bond and oxidize the hydroxyls to reactive formyl groups (Bobbitt, 1956). Modulating the concentration of sodium periodate can direct this oxidation to exclusively modify sialic acid groups (using 1-m*M* concentration at temperatures <4°C) or to convert all available diols to aldehydes (using 10-m*M* or greater concentrations at room temperature). Specific monosaccharide residues may be targeted with selective sugar oxidases to generate similar aldehyde functions only at discrete points within a complex polysaccharide structure (Section 4.4, this chapter) (Avigad, 1962; Gahmberg, 1978). The creation of formyl groups in this manner may be carried out on purified polysaccharide molecules, as in the case of soluble dextrans (Chapter 18, Section 2.2), or may be selectively performed on carbohydrate constituents of glycoproteins and other glycoconjugates. Once formed, aldehyde groups may be covalently coupled with amine-containing molecules by reductive amination using sodium cyanoborohydride (Chapter 4, Section 4) (Dottavio-Martin and Ravel, 1978; Cabacungan *et al.*, 1982).

The native reducing ends of carbohydrates may also be conjugated to amine-containing molecules by reductive amination. The reaction, however, is typically less efficient than using periodate-created aldehydes, since the open structure is in low concentration in aqueous solutions compared to the cyclic hemiacetal form. The reaction is usually allowed to continue for a week or more to reach good yields of coupling. Proteins may be modified to contain carbohydrate using this procedure (Gray, 1974, 1978; Baues and Gray, 1977; Schwartz and Gray, 1977). See Section 4.6 of this chapter for a more complete discussion of methods for the introduction of saccharide or glycan groups into proteins or other molecules.

The reducing ends of oligosaccharides can be modified with β-(*p*-aminophenyl)ethylamine to yield terminal arylamine derivatives (Jeffrey *et al.*, 1975; Zopf *et al.*, 1975). The aromatic amines can then be diazotized for coupling to active hydrogen-containing molecules, such as the tyrosine phenolic residues in proteins (Zopf *et al.*, 1978a,b). Alternatively, the arylamines may be transformed into isothiocyanate derivatives for coupling to amine-containing molecules, such as proteins (Smith *et al.*, 1978). The aromatic amine may also be used to conjugate the modified oligosaccharide directly with amine-reactive crosslinking agents or probes.

Another potential reaction of created or native aldehyde groups on carbohydrates is with hydrazide functionalities to form hydrazone linkages. Hydrazide-containing probes or crosslinking reagents may be conjugated with periodate-oxidized polysaccharides or with the reducing ends of sugars. The hydrazone bonds may be reduced with sodium cyanoborohydride to more stable linkages (Chapter 3, Section 5.1). The reduction step is recommended for long-term stability of crosslinked molecules. An example of this modification strategy is the use of biotin-hydrazide (Chapter 11, Section 6.4) or hydrazide-PEG$_4$-biotin (Chapter 18, Section 1.3) to label specifically glycoproteins at their carbohydrate locations.

Reducing sugars can be detected by reaction with phenylhydrazine to yield a hydrazone product, except the result of the reaction is not what one might imagine given the structure of aldoses and ketoses. Glucose, for example, can react with phenylhydrazine to yield the anticipated 1-phenylhydrazone derivative. In an excess of phenylhydrazine, however, the reaction continues to yield a 1,2-phenylhydrazone product, called an osazone, with concomitant production of aniline and ammonia (Figure 2.31). Exactly how the number 2 hydroxyl group gets oxidized to react with another molecule of phenylhydrazine is not entirely clear, but probably proceeds through an enol intermediate. This reaction is typical of all α-hydroxy aldehydes and α-hydroxy ketones, not just those occurring in carbohydrate molecules. Thus, glucose, mannose, and fructose all yield the same osazone product upon reaction with phenylhydrazine, since the stereochemical differences about carbons 1 and 2 are eliminated. Reversal of the phenylhydrazone linkage with an excess of benzaldehyde yields an osone, a 1-aldehyde-2-keto-derivative of the sugar. Many simple hydrazide-containing reagents probably are capable of forming similar 1,2-hydrazone

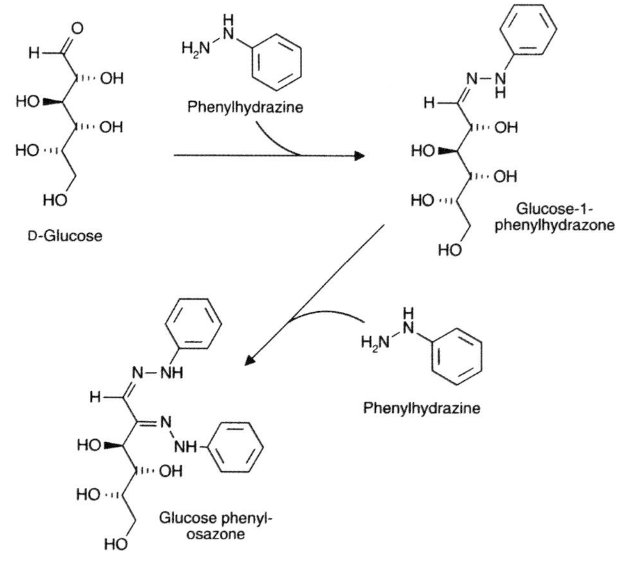

FIGURE 2.31 Phenylhydrazine can react with aldehyde or ketone groups within carbohydrates to give detectable products.

derivatives with reducing sugars, provided their size does not cause steric difficulties.

Polysaccharides, glycoproteins, and other glycoconjugates therefore may be specifically labeled on their carbohydrate portions by creating aldehyde functionalities and subsequently derivatizing them with another molecule containing an amine or a hydrazide group. This route of derivatization is probably the most common way of modifying carbohydrates.

The hydroxyl residues of polysaccharides may also be activated by certain compounds that form intermediate reactive derivatives containing good leaving groups for nucleophilic substitution. Reaction of these activated hydroxyls with nucleophiles such as amines results in stable covalent bonds between the carbohydrate and the amine-containing molecule. Activating agents that can be employed for this purpose include carbonyl diimidazole (Chapter 3, Section 4.2, and Chapter 3, Section 3), certain chloroformate derivatives (Chapter 3, Section 4.3), tresyl- and tosyl-chloride, cyanogen bromide, divinylsulfone, cyanuric chloride (Chapter 18, Section 2.1), disuccinimidyl carbonate (Chapter 5, Section 1.7), and various *bis*-epoxide compounds (Chapter 3, Section 1.8). Refer also to Chapters 14 and 15 for the use of many of these activators in the coupling of affinity ligands to particles or chromatography supports. Such activation procedures are typically carried out in nonaqueous solutions (i.e., dry dioxane, acetone, DMF, or DMSO) to prevent hydrolysis of the active species. While many pure polysaccharides can tolerate these organic environments, many biological glycoconjugates cannot. Thus, these methods are suitable for activating pure polysaccharides such as dextran, cellulose, agarose, and other carbohydrates, but are not appropriate for modifying sugar residues on glycoproteins. Many of these hydroxyl-activating reagents can also be used to activate polysaccharide chromatography supports and other hydroxyl-containing synthetic polymers such as polyethylene glycol or hydroxylic particles (Chapters 14, 15, and 18). For a complete treatment of polysaccharide chromatographic support activation through hydroxyl groups, see Chapter 15 and Hermanson *et al.* (1992). For a description of the activation of soluble polysaccharides and synthetic polymers, see Chapter 18.

While the hydroxyl groups on carbohydrate molecules are nucleophilic in aqueous solution, they are approximately equal to water in relative nucleophilicity. Since the majority of reactive functionalities on bioconjugation reagents are dependent upon nucleophilic reactions to initiate covalent bond formation, specific hydroxyl group modification is usually not possible in aqueous solution—especially with other biomolecules displaying stronger nucleophilic groups as well (e.g., amines and thiols). In many instances, hydrolysis of the active groups on crosslinking reagents occurs faster than hydroxyl group modification, due to the relative high abundance of water molecules compared to the amount of carbohydrate hydroxyls present. In some cases, even if modification does occur, the resultant bond may be unstable. For instance, NHS esters (Chapter 3, Section 1.4) can react with hydroxyls to form ester linkages, which are themselves unstable to hydrolysis.

Anhydrides, such as acetic anhydride (Sections 4.2 and 5.1, this chapter), may react with carbohydrate hydroxyls even in aqueous environments to form acyl derivatives. The reaction, however, is reversible by incubation with hydroxylamine at pH 10 to 11.

Epoxide-containing reagents, such as the homobifunctional 1,4-(butanediol) diglycidyl ether (Chapter 5, Section 7.1), can react with polysaccharide hydroxyl groups to form stable ether bonds. *bis*-Epoxy compounds have been used to couple sugars and polysaccharides to insoluble matrices for affinity chromatography (Sundberg and Porath, 1974; see also Chapter 15, Section 2.3). The reaction of epoxides, however, is not specific for hydroxyl groups and they will cross-react with amines and sulfhydryls if these functionalities are present.

Hydroxyl groups on carbohydrates may be modified with chloroacetic acid to produce a carboxylate functionality for further conjugation purposes (Plotz and Rifai, 1982). In addition, endogenous carboxylate groups, such as those in sialic acid residues and aldonic-or uronic-acid-containing polysaccharides, may be targeted for modification using typical carboxylate modification reactions (Chapter 3, Section 3). However, when these polysaccharides are part of macromolecules containing other carboxylic acid groups such as glycoproteins, the targeting will not be specific for the carbohydrate alone. Pure polysaccharides containing carboxylate groups may be coupled to amine-containing molecules by use of the carbodiimide reaction (Chapter 4, Section 1). The carboxylate is activated to an *O*-acylisourea intermediate, which is in turn attacked by the amine compound. The result is the formation of a stable amide linkage with loss of one molecule of isourea.

Carbohydrate molecules containing amine groups, such as D-glucosamine, may easily be conjugated to other macromolecules using a number of amine-reactive chemical reactions and crosslinkers (Chapters 3 to 6). Some polysaccharides containing acetylated amine residues, such as chitin which contains *N*-acetylglucosamine, may be deacetylated under alkaline conditions (Jeanloz, 1963) to free the amines (forming chitosan in this case).

Amine functionalities may also be created on polysaccharides (Section 4.3, this chapter). The reducing ends of carbohydrate molecules (or generated aldehydes) may be reacted with small diamine compounds to yield short alkylamine spacers that can be used for subsequent conjugation reactions. Hydrazide groups may be similarly created using *bis*-hydrazide compounds (Sections 4.5 and 4.6, this chapter).

Phosphate-containing carbohydrates that are stable, such as the 5′-phosphate of the ribose derivatives of oligonucleotides, may be targeted for modification using a carbodiimide-facilitated reaction (Section 4.3, this chapter). The water soluble carbodiimide EDC (1-ethyl-3-(3-dimethylaminopropyl)carbodiimide) can react with the phosphate groups to form highly reactive phosphoester intermediates. These intermediates can react with amine-or hydrazide-containing molecules to form stable phosphoramidate bonds.

Polysaccharide and Glycoconjugate Structure

Aldose monosaccharide units are frequently bound together through the number 1 carbon hydroxyl group of one sugar to another sugar's number 4 or 6 hydroxyl group, forming a complete acetal linkage. Two monosaccharides coupled in this fashion are termed a disaccharide. Numerous monosaccharides bound together to form a chain are called a polysaccharide. The most abundant polysaccharides in nature, starch and cellulose, consist of glucose bound together in α-1,4, β-1,4, and, to a lesser extent, α-1,6 acetal linkages (Figure 2.32). While the hemiacetal, cyclic structure of individual sugars shows some reversibility under equilibrium conditions, the acetal linkage between two monosaccharides is quite stable, only hydrolyzing under severe pH extremes.

Similarly, ketose sugars participate in polysaccharide formation by reaction of their anomeric carbon with a hydroxyl of another monosaccharide to create a ketal linkage. The acetal and ketal bonds within polysaccharides are termed *O*-glycosidic linkages.

Hemiacetal hydroxyl groups of carbohydrate molecules may also be coupled to amine-containing molecules to form *N*-glycosidic linkages, such as those in nucleic acids and oligonucleotides.

Polysaccharides may or may not have reducing power, depending on the way they are linked together and whether the terminal, potentially reducing end is available. The structure of simple disaccharides can illustrate this point. Of the most common disaccharides, sucrose and lactose, sucrose is a non-reducing sugar since β-D-fructose is linked through its reducing C-2 hydroxyl, and lactose remains a reducing sugar, since the terminal glucose is linked to β-D-galactose through its C-4 hydroxyl, leaving its reducing end free (Figure 2.33).

Polysaccharide synthesis is under enzymatic control, but does not occur from a template as in protein synthesis. For this reason, each molecule of a particular polysaccharide will have its own unique molecular weight. The molecular weight of a carbohydrate polymer is usually expressed as an average. Starch

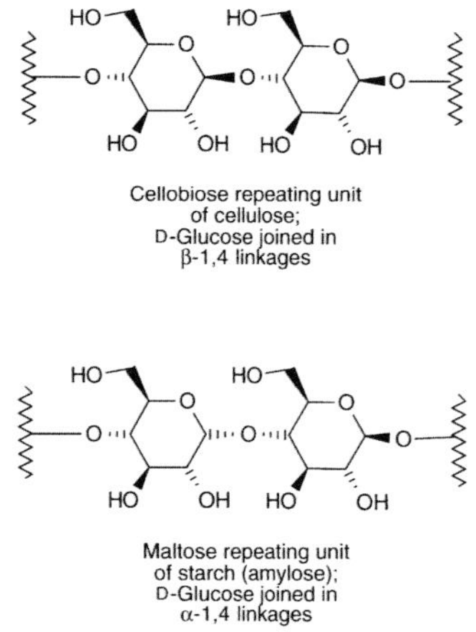

FIGURE 2.32 The repeating units of cellulose and starch, two of the most common polysaccharides in nature.

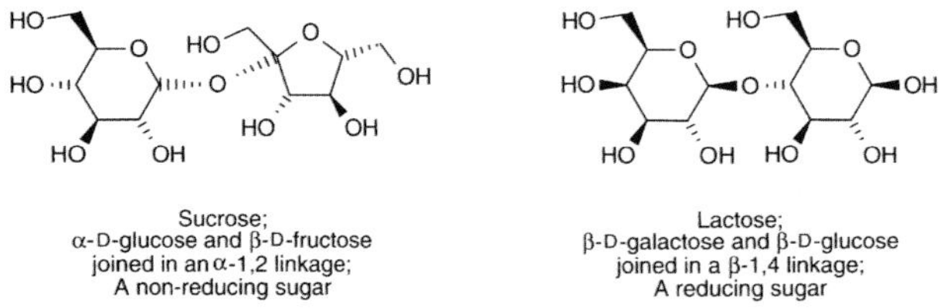

FIGURE 2.33 Comparison of a reducing and a nonreducing disaccharide.

or cellulose chains, for example, may vary by several hundred thousand in their molecular weights between individual molecules. For an excellent review of carbohydrate chemistry, see Binkley (1988).

Due to their polyhydroxylic structures, all carbohydrates are polar and will possess associated water molecules in aqueous solution, but they may not be fully water soluble. Large polysaccharides such as cellulose form intricate matrices created from extensive hydrogen bonding. Neighboring monosaccharide units hydrogen bond within the same chain, while neighboring polymers form interchain hydrogen bonds between hydroxyls. The three-dimensional structure of a carbohydrate to a large extent is determined by these hydrogen bonds—sometimes resulting in sheeted or helical structures, as in the triple helix of agarose polysaccharide chains. Water will be intimately associated in this internal arrangement, but the overall multi-polymer structure is often too large to allow for complete water solubility. For a review, see Preis (1980).

Polysaccharide solubility in aqueous solutions is usually dependent on polymer size and its allied three-dimensional structure. Even water-insoluble carbohydrates may be solubilized by controlled hydrolysis of *O*-glycosidic linkages to create smaller polysaccharide molecules. Thus, cellulose may be solubilized by heating in an alkaline solution until the polymers are broken up sufficiently to reduce their average molecular weight. Many such soluble forms of common polysaccharides are available commercially.

Carbohydrate also is an important constituent of many biological molecules. Polysaccharides may be found covalently conjugated to proteins and lipids, forming glycoproteins, proteoglycans, glycolipids, and lipopolysaccharides. Such glycoconjugates (glycans) are produced in the cell through controlled, enzymatic processes. With proteins, the modification occurs after translational synthesis of the polypeptide chain at the ribosome.

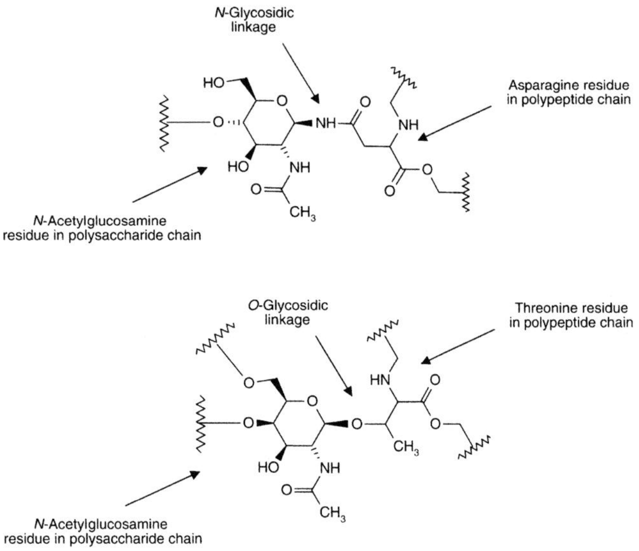

FIGURE 2.34 Common attachment points for polysaccharide chains on glycoproteins.

Proteins newly synthesized on ribosomes may be transported to the Golgi apparatus where specific glycosyl transferases catalyze the coupling of monosaccharides to the polypeptide chains. Glycoproteins and mucoproteins are formed by the coupling of polysaccharides through *O*-glycosidic linkages to serine, threonine, or hydroxylysine in addition to *N*-glycosidic linkages with the amide side-chain group of asparagine (Figure 2.34). For reviews of glycoconjugate structure and function, see Hynes (1987), Lennarz (1980), Jentoft (1990), Steer and Ashwell (1986), and the entire issue of *Science*, Vol. 291, March 23, 2001 (special edition on carbohydrates and glycobiology). The GlycomeDB also may be consulted to provide exact structures of known carbohydrates and glycans (Ranzinger *et al.*, 2010).

The structure of most glycoprotein carbohydrate consists of a complex, branched heteropolysaccharide with the sugars mannose, *N*-acetyl glucosamine, sialic acid, galactose, and *L*-fucose being prevalent. Asparagine-linked polysaccharides are well characterized and are known to be constructed of a core unit consisting of three mannose residues and two *N*-acetyl-glucosamine (GlcNAc) residues. The GlcNAc residues are bound to the Asp side-chain amide nitrogen through a β1 linkage (Kornfield and Kornfield, 1985). The three mannose groups then usually form the first branch point in the oligosaccharide chain. Figures 2.35 and 2.36 illustrate

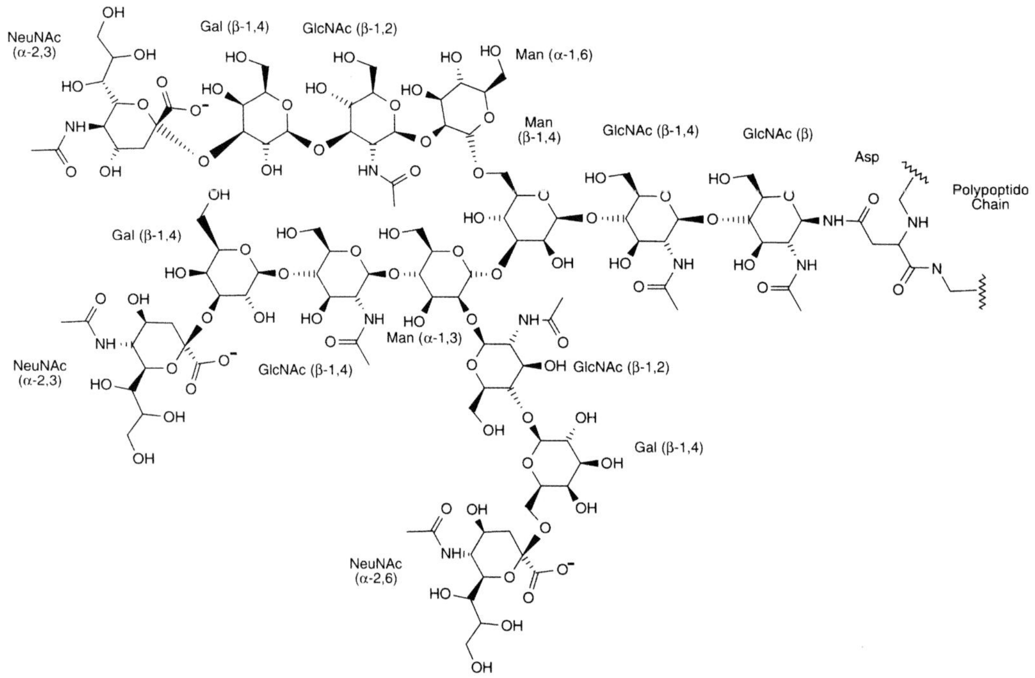

FIGURE 2.35 The complex structure of an asparagine-linked polysaccharide. Note the branched nature of the polymer with terminal sialic acid residues on each chain.

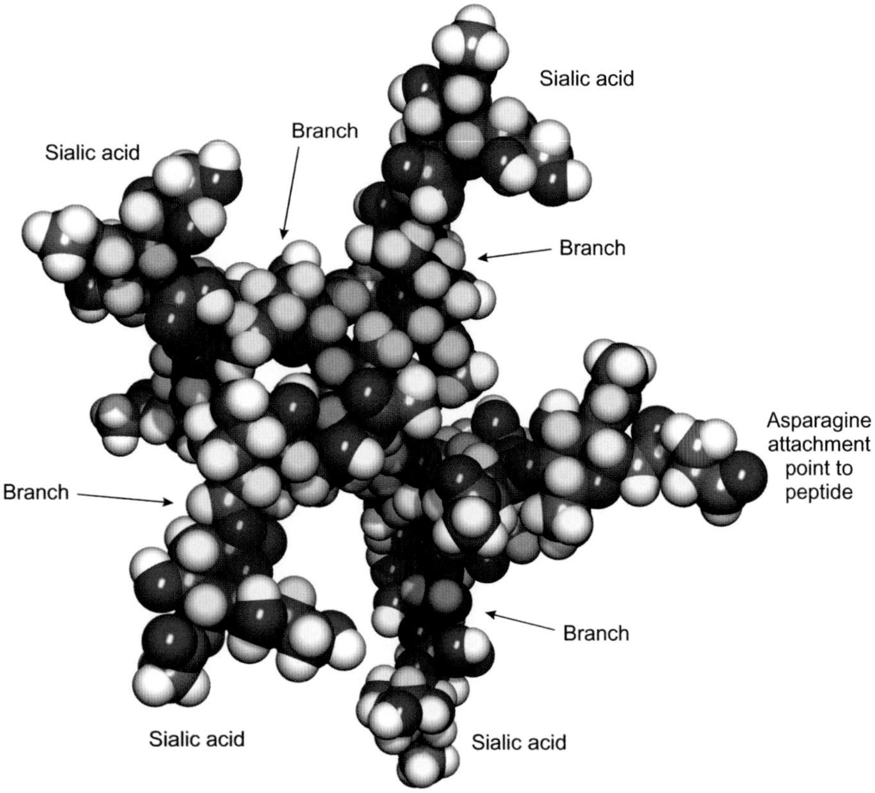

FIGURE 2.36 A space-filling model of an *N*-linked glycan showing four branch points.

the chemical makeup of a typical *N*-linked glycan and a space-filling model of a glycan's molecular structure. Much of the detailed structural knowledge of glycoconjugates is developed using controlled chemical or enzymatic degradation of the polysaccharides followed by analysis by gas chromatography and mass spectrometry (Vliegenthart *et al.*, 1983; Sweeley and Nunez, 1985; McCleary and Matheson, 1986; Biermann and McGinnis, 1989).

The content by weight of carbohydrate in glycoproteins may vary from only a few percent to as much as 70% in some proteins in mucous secretions. Although the exact function of the polysaccharide in most glycoproteins is unknown, in some cases it may provide hydrophilicity, recognition, and points of non-covalent interaction with other proteins through lectin affinity binding. Glycosylation also contributes to the correct folding of proteins after translation, probably by ensuring that certain amino acid regions end up at the surface of the protein structure. In addition, extensive polysaccharide modification is helpful in preventing proteolytic digestion of the underlying polypeptide chain.

Another form of post-translational modification that may add carbohydrate to a polypeptide is non-enzymatic glycation. This reaction occurs between the reducing ends of sugar molecules and the amino groups of proteins and peptides. The aldehyde group of a reducing sugar first forms a reversible Schiff base linkage with the α-amino or ε-amino groups of the protein. This bond can then undergo an Amadori rearrangement to form a stable ketoamine derivative (Figure 2.37). The result is a blocked amine containing a sugar derivative with available hydroxyl residues. This reaction commonly occurs with proteins continually exposed to reducing sugars, such as glucose in blood. The measurement of glycated hemoglobin is a clinically important parameter in the management of diabetes mellitus. Increases in the blood sugar level in diabetes are the cause of concomitant increases in the level of non-enzymatic glycation of blood proteins. Measuring the relative amount of glycated hemoglobin provides the physician with information concerning a diabetic patient's blood glucose control.

2.2. Carbohydrate and Glycan Conjugation Methods

The presence of carbohydrate on biomolecules provides important points of attachment for modification and conjugation reactions. Coupling only through polysaccharide chains often can direct the reaction away from active centers or critical points in protein

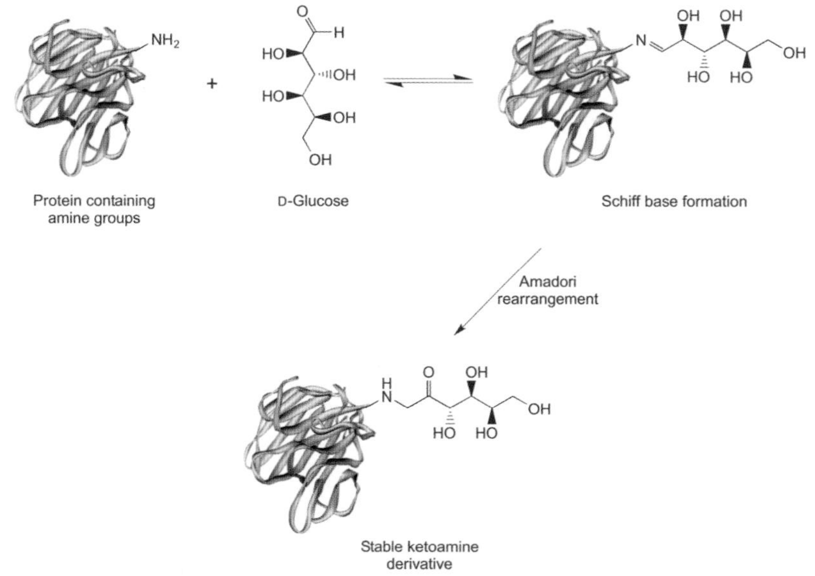

FIGURE 2.37 A reducing sugar may modify protein amine groups through Schiff base formation followed by an Amadori rearrangement to give a stable ketoamine product. Glucose is a common *in vivo* modifier of blood proteins through this process.

molecules, thus preserving activity. Crosslinking strategies involving polysaccharides or glycoconjugates usually involve a two- or three-step reaction sequence. If no reactive functionalities other than hydroxyl groups are present on the carbohydrate, then the first step is to create sufficiently reactive groups to couple with the functional groups of a second molecule.

Perhaps the easiest way specifically to target polysaccharides on glycoproteins is through mild sodium periodate oxidation. Periodate cleaves the carbon–carbon bond connecting adjacent hydroxyl groups in sugar residues to create highly reactive aldehyde functionalities (Section 4.4, this chapter). The level of oxidant addition can be adjusted to cleave selectively only certain sugars in the polysaccharide structure. A concentration of 1-m*M* sodium periodate at 0 to 4°C oxidizes sialic acid residues to aldehydes, leaving all other monosaccharides untouched. Increasing the concentration to 10-m*M* at room temperature, however, will cause oxidation of other sugars in the carbohydrate, including galactose and mannose residues in glycans. The generated aldehydes can then be used in coupling reactions with amine- or hydrazide-containing molecules to form covalent linkages. Amines react with formyl groups under reductive amination conditions when a suitable reducing agent is used, such as sodium cyanoborohydride. The result of this reaction is a stable secondary amine, linkage (Chapter 3, Section 5.3). Hydrazides spontaneously react with aldehydes to form hydrazone linkages, although the addition of a reducing agent greatly increases the efficiency of the reaction and the stability of the bond (Chapter 3, Section 5.1).

Oxidized glycoconjugates are usually stable enough to be stored in a freeze-dried state without loss of activity prior to a subsequent conjugation reaction, provided the protein itself is stable to lyophilization. Storage in solution, however, may cause slow polymerization if the molecule also contains amine groups, as in glycoproteins. Sometimes the protein can be treated to block its amines prior to periodate oxidation, as in the procedure often used with the enzyme horseradish peroxidase (HRP) (Chapter 22, Section 1.1), thus eliminating the potential for self-conjugation. Even in the absence of amines, periodate-oxidized HRP may polymerize due to the Mannich reaction (Chapter 3, Section 5.5).

If the second molecule to be coupled to the oxidized glycoconjugate already has the requisite amines or hydrazide groups, then directly mixing the two components together in the presence of a reductant is all that is needed to form the conjugate. This is an example of a two-step procedure. However, if the second molecule possesses none of the appropriate functionalities for coupling, then modifying it to contain amine or hydrazide groups must be carried out prior to the conjugation reaction (see Section 4.3 and 4.5), which results in a three-step protocol. The use of other functionalities (either indigenous or created) on polysaccharide molecules to effect a crosslinking reaction can be carried out in similar two- or three-step strategies.

Occasionally, it is important to conjugate a polysaccharide-containing molecule to another molecule while retaining, as much as possible, the carbohydrate's original chemical and three-dimensional structure. For instance, in the preparation of immunogen conjugates by coupling a polysaccharide molecule to a carrier, care should be taken to preserve the structure of the carbohydrate to ensure antibody recognition of the native molecule. In this case, periodate oxidative techniques may not be the best choice to effect crosslinking due to the potential for extensive ring opening throughout the chain. Under controlled conditions, however, where periodate is carefully used in limiting quantities, this method has proved successful in creating oligosaccharide–carrier conjugates (Anderson *et al.*, 1989).

Retention of native carbohydrate structure is also important in applications that utilize the conjugated polysaccharide in binding studies with receptors or lectins. In these cases, the carbohydrate should be modified at limited sites, preferentially only at its reducing end. Section 4.6 of this chapter discusses glycan conjugation techniques in greater detail.

3. MODIFICATION OF NUCLEIC ACIDS AND OLIGONUCLEOTIDES

The nucleic acid polymers DNA and RNA form the most basic units of information storage within cells. The conversion of DNA's unique information code into RNA and proteins is the fundamental step in controlling all cellular processes. Targeting segments of this encoded data with labeled probes that are able to bind to specific genetic regions allows detection, localization, or quantification of discrete oligonucleotide sequences. This targeting capability is made possible by the predictable nature of nucleic acid interactions. Despite the complexity of the genetic code, the base-pairing process that causes one oligonucleotide to bind to its complementary sequence is rather simple to predict and decipher. Nucleic acids are the only type of complex biological molecule wherein their binding properties can be fully anticipated and incorporated into synthetic oligonucleotide probes. Thus, a short DNA segment can be synthetically designed and used to target and hybridize to a complementary DNA strand within much larger chromosomal material or extracted genomic DNA. If the small oligonucleotide is labeled with a detectable component that does not interfere in the base-pairing process, then the targeted DNA can be identified or assayed.

Bioconjugate techniques involving nucleic acids are becoming one of the most important application areas of crosslinking and modification chemistry. With the secrets of the genetic code now revealed by such mammoth efforts as the Human Genome Project, knowledge of the DNA sequence that governs specific protein expression is leading to diagnostic tests able to assess the presence of critical genetic markers associated with certain disease states. To test for particular target sequences, complementary oligonucleotide probes are used that possess conjugated enzymes, fluorophores, haptens, radiolabels, or other such groups which can be used to detect a hybridization signal. Such oligonucleotide conjugates can be used to discover target sequences in blots, electrophoresis gels, tissues, cells immobilized to surfaces, or in solution.

The power and advantages of assessing cellular processes at their most fundamental level is propelling the science of oligonucleotide probe detection into one of the most prominent positions in bioconjugate chemistry. Oligonucleotide arrays containing hundreds or thousands of tests are now carried out routinely to monitor different aspects of genetic information—all with the use of specific oligonucleotide probes.

In this section, the chemistry and structure of nucleic acids and oligonucleotides are discussed with a view to creating functional conjugates with detectable molecules. The corresponding strategies and protocols associated with DNA or RNA modification and conjugation can be found in Chapter 23.

3.1. Polynucleotide Structure and Functionality

Nucleic acid polymers are characterized by the types of base residues present and the structure of their sugar backbone. The bases are nitrogenous ring compounds consisting of either purine or pyrimidine derivatives. A purine is a fused-ring compound containing one six-membered ring attached to a five-membered ring, whereas a pyrimidine consists of a single six-membered ring structure (Figure 2.38).

Nucleic acids can contain of any one of three kinds of pyrimidine ring systems (uracil, cytosine, or thymine) or two types of purine derivatives (adenine or guanine). Adenine, guanine, thymine, and cytosine are the four main base constituents found in DNA. In RNA molecules, three of these four bases are present, but with thymine replaced by uracil to make up the fourth. Some additional minor derivatives are found in messenger RNA (mRNA), transfer RNA (tRNA), and ribosomal RNA (rRNA), particularly the N^4,N^4-dimethyladenine and N^7-methylguanine varieties.

Nucleic acid sugar residues are attached to the associated base units in an *N*-glycosidic bond, involving the

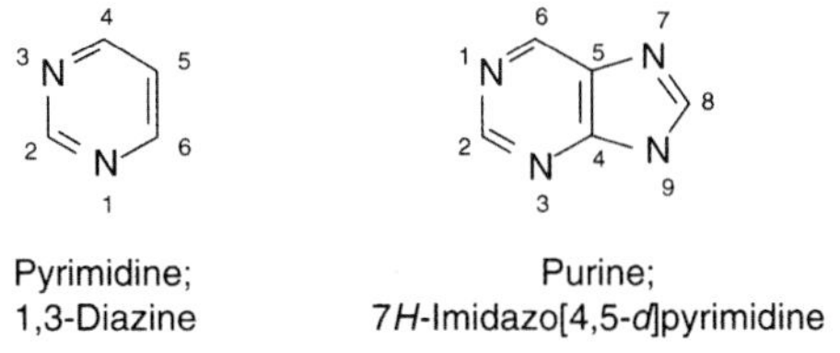

FIGURE 2.38 The pyrimidine and purine ring structures common to nucleic acids.

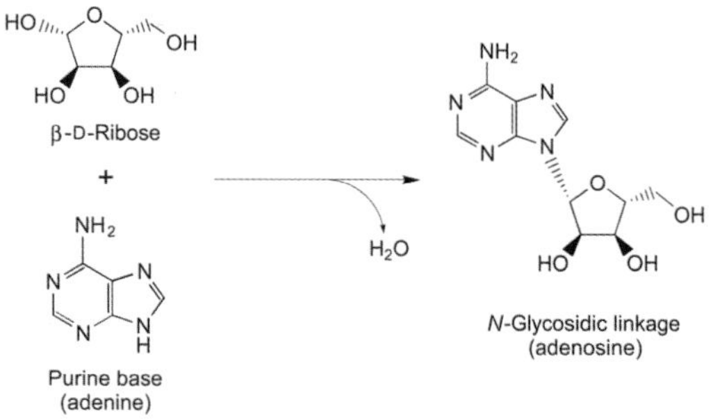

FIGURE 2.39 The formation of an *N*-glycosidic bond links the base unit of nucleic acids to the associated ribose derivative.

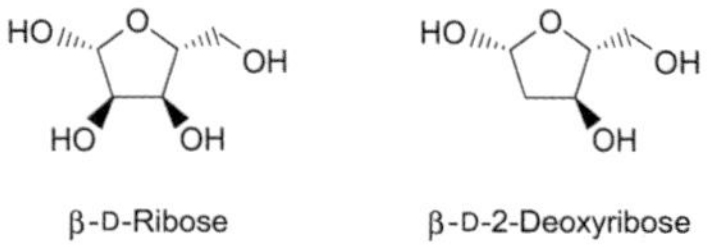

FIGURE 2.40 The two forms of sugar residues commonly found in nucleic acids. β-D-Ribose is the sugar constituent of RNA, while β-D-2-deoxyribose is a component of DNA.

TABLE 2.2 Nucleic Acid Nomenclatures

Base Name	Nucleoside Name[a] (Base + Sugar)	Nucleotide Name[b] (Base + Sugar+Phosphate)
Adenine	Adenosine	Adenosine monophosphate (AMP)
Guanine	Guanosine	Guanosine monophosphate (GMP)
Cytosine	Cytidine	Cytidine monophosphate (CMP)
Thymine	Thymidine	Thymidine monophosphate (TMP)
Uracil	Uridine	Uridine monophosphate (UMP)

[a]*For deoxyribose nucleosides, add "deoxy" before the nucleoside name; for example, adenosine becomes deoxyadenosine.*
[b]*For the presence of two phosphate groups, the names are changed to diphosphate; for three phosphate groups, the terminology is triphosphate.*

number 1 nitrogen of pyrimidine bases or the number 9 nitrogen of purines directly linked to the number 1 carbon of the monosaccharide derivative (Figure 2.39). The sugar group consists of either a β-D-ribose unit (found in RNA) or a β-D-2-deoxyribose unit (in DNA) (Figure 2.40). In mRNA and rRNA, a minor sugar derivative, a 2′-*O*-methylribosyl group, also is found.

The nomenclature of nucleic acid chemistry further characterizes the structure of the associated groups. A *nucleoside* contains only a base group and an attached sugar. A *nucleotide* consists of a base and a sugar plus a phosphate group. At this point, the naming system gets somewhat confusing due to the fact that the nucleoside name is a derivative of the base name. Table 2.2 shows this relationship and their associated abbreviations (which are simpler to remember).

In each nucleotide monomer of DNA or RNA molecules, a phosphate group is attached to the C-5 hydroxyl of each sugar residue in an ester (anhydride) linkage. These phosphate groups in turn are linked in diester bonds to neighboring sugar groups of adjacent nucleotides through their 3′-ribosyl hydroxyl to create the oligonucleotide polymer backbone (Figure 2.41). Thus, the phosphate–sugar repeating unit produces the linear sequence within the DNA or RNA structure, while the four types of base units protrude out from this backbone, creating the unique code making up the genetic information.

Nucleotide Functional Groups

Chemical attachment of a detectable component to an oligonucleotide forms the basis for constructing a sensitive hybridization reagent. Unfortunately, the methods developed to crosslink or label other biological molecules such as proteins do not always apply to nucleic acids. The major reactive sites on proteins involve primary amines, sulfhydryls, carboxylates, or phenolates—groups that are relatively easy to derivatize. RNA and DNA contain none of these functionalities. They also are relatively unreactive directly with many of the common bioconjugate reagents discussed in Part II.

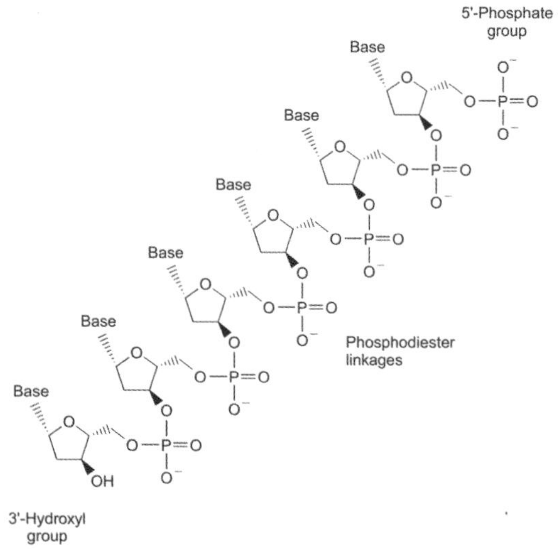

FIGURE 2.41 Polynucleotides are formed through phosphodiester bonds linking the associated sugar groups together. In DNA, the 3′-hydroxyl of one deoxyribose unit is bound to the 5′-hydroxyl of the next, creating direction in the polymer backbone.

However, there are particular sites that can be modified on the bases, sugars, or phosphate groups of nucleic acids to produce derivatives able to couple with a second molecule. The chemistry is almost entirely unique to DNA and RNA work, but once mastered the process of conjugation can be carried out with the same ease as with protein molecules.

The following sections discuss the major constituents of oligonucleotides with special emphasis on the chemical sites useful for bioconjugation.

Cytosine, Thymine, and Uracil Residues

The pyrimidine base units cytosine, thymine, and uracil contain six-membered nitrogenous ring structures with various points of unsaturation. Thymine and uracil are similar, containing the same double bond between carbons 5 and 6 and the same two ketone groups on C-2 and C-4 of the ring, but differ only in the presence of a methyl group on the number 5 carbon of thymine. Cytosine, by contrast, contains an additional site of unsaturation between carbons 3 and 4 as well as an amine group on C-4 instead of a ketone (Figure 2.42).

Figure 2.43 indicates major sites of reactivity within the ring structures for nucleophilic displacement reactions. Cytosine, thymine, and uracil all react toward nucleophilic attack at the same two sites, the C-4 and C-6 positions. The presence of powerful nucleophiles, even at neutral pH, can lead to significant base modification or cleavage with pyrimidine residues (Debye, 1947). For instance, hydrazine spontaneously adds to the 5,6-double bond, initiating further ring reactions, which eventually leads to oligonucleotide degradation. A similarly strong nucleophile, hydroxylamine is almost entirely specific for modifying pyrimidines. It too can add to the 5,6-double bond, creating a 6-hydroxylamino derivative. In general, the pyrimidines can undergo reactions at the 5,6-double bond leading to a stable modification at the C-5 position (Figure 2.44).

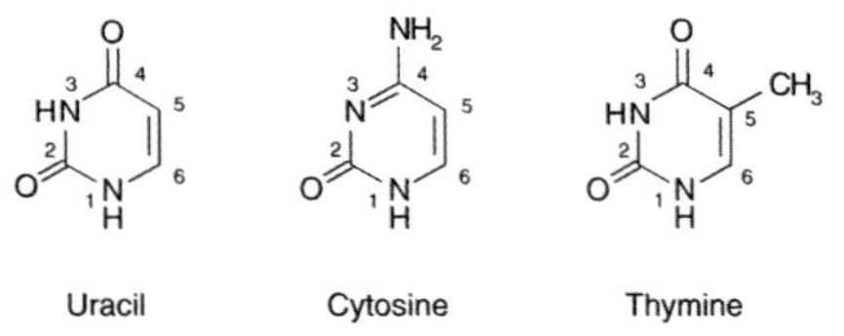

FIGURE 2.42 The three pyrimidine bases common to nucleic acid construction. Cytosine and thymine are found in DNA, while in RNA uracil residues replace thymine. The associated sugar groups are bound in *N*-glycosidic linkages to the N-1 nitrogen.

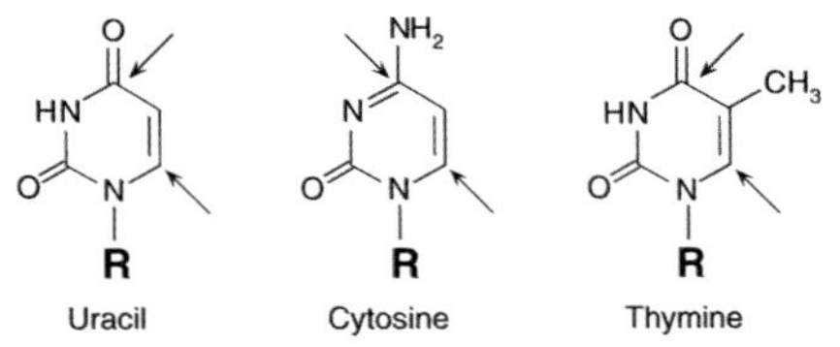

FIGURE 2.43 Pyrimidine bases ate subject to nucleophilic displacement reactions primarily at the C-4 and C-6 positions.

Addition of a nucleophile to the C-6 position of cytosine often results in fascile displacement reactions occurring at the N-4 location. With hydroxylamine attack, nucleophilic displacement causes the formation of an N^4-hydroxy derivative. A particularly important reaction for bioconjugate chemistry, however, is that of nucleophilic bisulfite addition to the C-6 position. Sulfonation of cytosine can lead to two distinct reaction products. At acid pH wherein the N-3 nitrogen is protonated, bisulfite reaction results in the 6-sulfonate product followed by spontaneous hydrolysis. Raising the pH to alkaline conditions causes effective formation of uracil. If bisulfite addition is carried out in the presence of a nucleophile, such as a primary amine or hydrazide compound, then transamination at the N-4 position can take place instead of hydrolysis (Figure 2.45). This is an important mechanism for adding spacer arm functionalities and other small molecules to cytosine-containing oligonucleotides (see Chapter 23, Section 2).

Electrophilic reagents can also modify the pyrimidine rings of nucleic acids. Alkylation and acylation reactions can take place at several sites on all three bases. Figure 2.46 illustrates the principal locations where electrophilic attack can occur. In particular, the heteroatoms (oxygen and nitrogen) are the best positions of high electron density, thus functioning as nucleophiles in reaction processes. Of the pyrimidine residues, however, it is the N-3 position of cytosine derivatives that is the most susceptible to alkylation. Reactions can occur with ethylenimine compounds (Section 4.3, this chapter), alkyl halogens (Chapter 3, Section 2.1), epoxides (Chapter 3, Section 1.8), and many other strong alkylating agents (for a review, see Brown, 1974).

Acylation reactions can be carried out at the nucleophilic sites on pyrimidines using activated forms of carboxylic acids. Acylation of functional groups in nucleotides is typically used for protection during synthesis (Reese, 1973). However, for bioconjugate

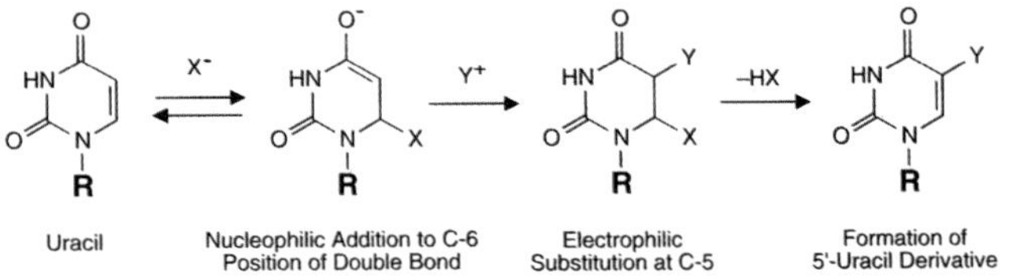

FIGURE 2.44 Nucleophilic addition at C-6 of the pyrimidine double bond can cause electrophilic substitution to occur at the C-5 position.

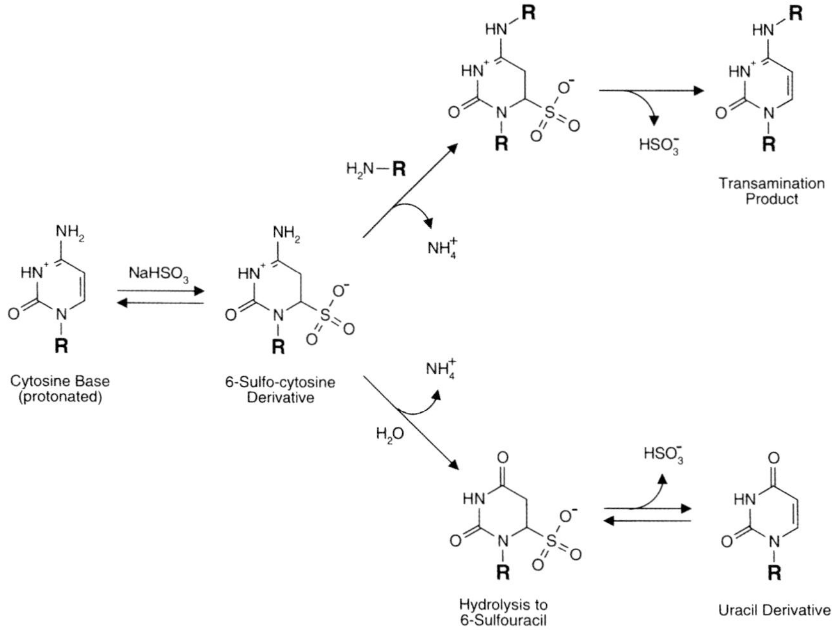

FIGURE 2.45 Reaction of bisulfite with cytosine bases is an important route of derivatization. It can lead to uracil formation or, in the presence of an amine- (or hydrazide)-containing compound, transamination can occur, resulting in covalent modification.

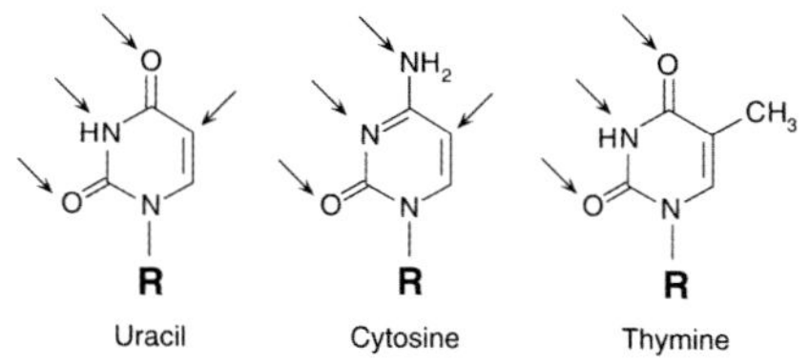

FIGURE 2.46 Potential sites of electrophilic attack on pyrimidine bases.

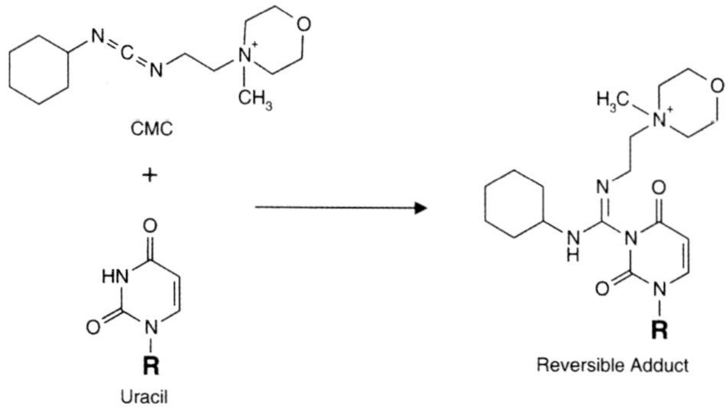

FIGURE 2.47 The carbodiimide CMC can react with the N-3 nitrogen to yield a reversible product.

applications, the reactivity of native groups on pyrimidines is not as great as that obtained using an amine-terminal spacer derivative, such as those described in Chapter 23, Section 2.1. Yields and reaction rates are typically low for direct acylation or alkylation of pyrimidine bases, especially in aqueous environments.

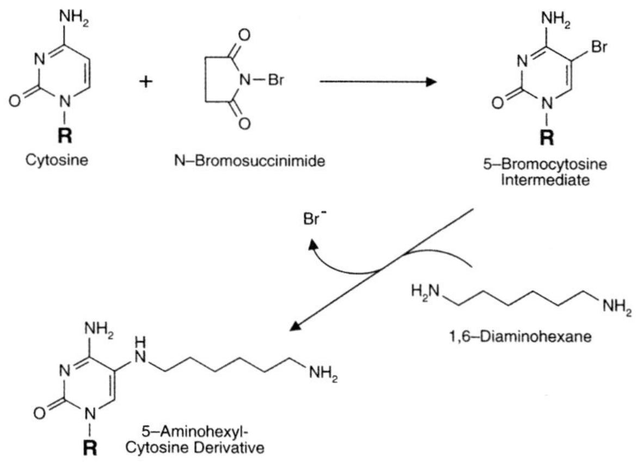

FIGURE 2.48 Cytosine bases are susceptible to bromination at the C-5 double bond position, resulting in active intermediates capable of reacting with amine nucleophiles.

The N-3 position of uracil can also be modified with carbodiimide reagents. In particular, the water soluble carbodiimide CMC (1-cyclohexyl-3-(2-morpholinoethyl) carbodiimide, as the metho *p*-toluene sulfonate salt) can react with the N-3 nitrogen at pH 8 to give an unstable, charged adduct. The derivative is reversible at pH 10.5, regenerating the original nucleic acid base (Figure 2.47). Cytosine is unreactive in this process.

Halogenation of pyrimidine bases may be carried out with bromine or iodine. Bromination occurs at the C-5 of cytosine, yielding a reactive derivative, which can be used to couple diamine spacer molecules by nucleophilic substitution (Figure 2.48) (Traincard *et al.*, 1983; Sakamoto *et al.*, 1987; Keller *et al.*, 1988). Other pyrimidine derivatives are also reactive to bromine compounds at the C-5 position. Either an aqueous solution of bromine or the compound *N*-bromosuccinimide can be used for this reaction. The brominated derivatives can then be used to couple amine-containing compounds to the pyrimidine ring structure (Chapter 23, Section 2.1).

Other reactions characterized for pyrimidine residues include mercuration at C-5 of cytosine or uracil (Hopman *et al.*, 1986), cycloaddition to the 5,6-double bond of thymine and uracil (Cimino *et al.*, 1985), and thiolation at the C-4 amino group of cytosine (Malcom and Nicolas, 1984).

Adenine and Guanine Residues

The purine bases of nucleic acids are constructed of a two-ring system made from a pyrimidine-type, six-membered ring fused with a five-membered imidazole ring. Adenine and guanine are present in both RNA and DNA. They differ in their six-membered ring structures by an additional point of unsaturation between C-6 and N-1 (in adenine) and by the presence of amine or ketone groups attached to C-2 or C-6 (Figure 2.49). Attachment to ribose or deoxyribose in nucleosides is made through an *N*-glycosidic linkage at N-9 of the imidazole ring on either purine.

As in the case of pyrimidine bases discussed previously, adenine and guanine are subject to nucleophilic displacement reactions at particular sites on their ring structures (Figure 2.50). Both compounds are reactive with nucleophiles at C-2, C-6, and C-8, with C-8 being the most common target for modification. However, the purines are much less reactive to nucleophiles than the pyrimidines. Hydrazine, hydroxylamine, and bisulfite—all important reactive species with cytosine, thymine, and uracil—are almost unreactive with guanine and adenine.

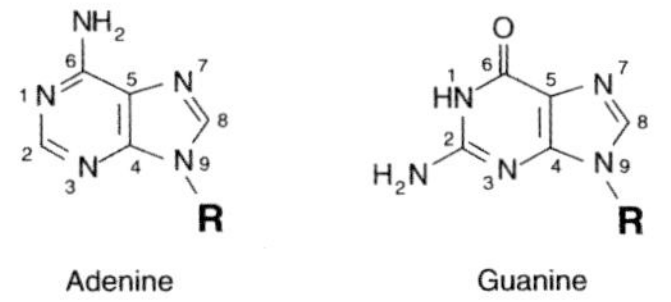

FIGURE 2.49 The structures of the common purine bases of RNA and DNA. The associated sugar groups are bound in *N*-glycosidic linkages to the N-9 position.

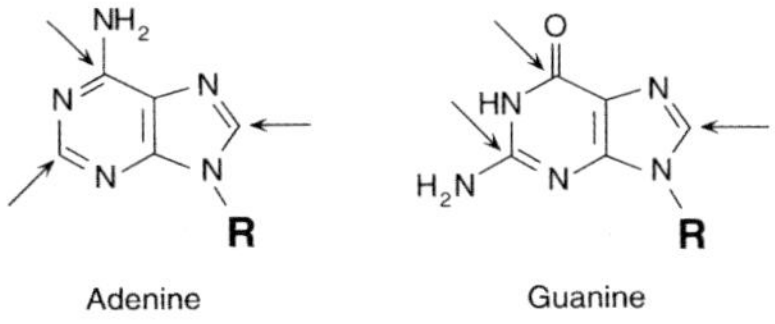

FIGURE 2.50 Primary nucleophilic displacement sites on purine bases.

With purines, reaction with electrophilic species is the most important route to derivatization. Figure 2.51 identifies the major sites of electrophilic attack on adenine and guanine. On both bases it is the heteroatoms that make up the majority of sites. Alkylation reactions thus can occur at N-1, N-3, and N-7 in adenine or N-3 and N-7 in guanine. However, the greatest location of electron density (nucleophilicity) occurs at N-7 on the imidazole ring of guanine, followed by N-1 of adenine. According to Brown (1974), the order of reactivity of nucleosides toward alkylation by esters of strong acids is guanine > adenosine > cytidine >> uridine (nearly unreactive).

As with pyrimidines, the water soluble carbodiimide CMC may react with guanine derivatives to give a reversible adduct at N-1 (Figure 2.52). Raising the pH to highly alkaline conditions regenerates the purine group. Adenine residues, however, display no reactivity in this process.

One of the most important reactions of purines is the bromination of guanine or adenine at the C-8 position. It is this site that is the most common point of modification for bioconjugate techniques using purine bases (Figure 2.53). Either an aqueous solution of bromine or the compound *N*-bromosuccinimide can be used for this reaction. The brominated derivatives can then be used to couple amine-containing compounds to the pyrimidine ring structure by nucleophilic substitution (Chapter 23, Section 2.1).

Adenine may also undergo an additional reaction at its C-6 amine group using a Fischer–Dimroth

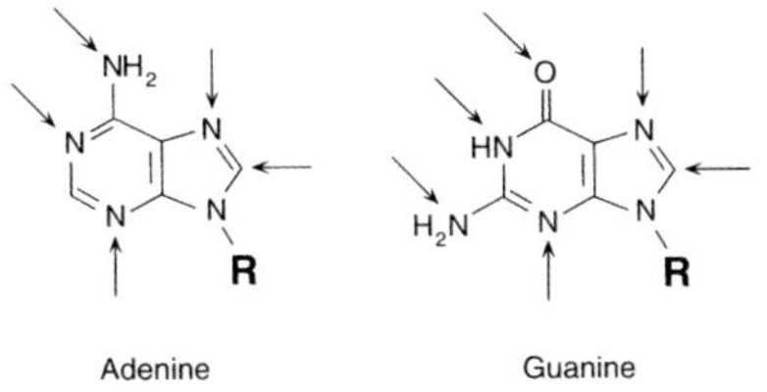

FIGURE 2.51 Electrophilic attack can occur at a number of sites on both purine bases.

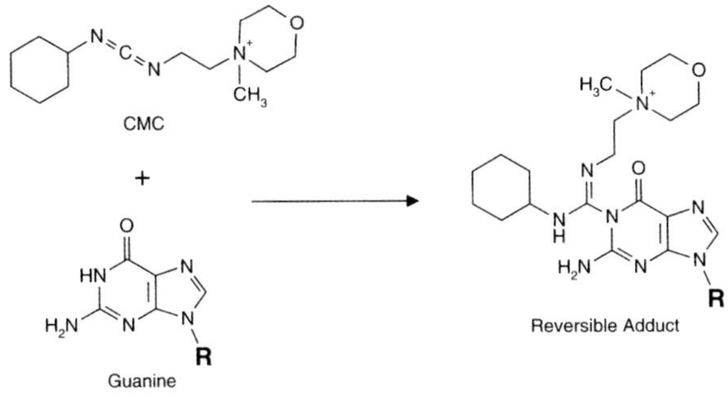

FIGURE 2.52 The carbodiimide CMC can react with guanine at the N-1 position to form a reversible complex.

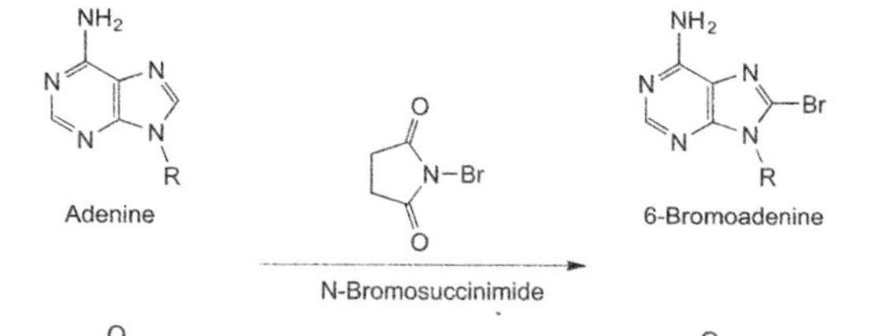

FIGURE 2.53 The purine bases are subject to bromination reactions at the C-8 position, forming an important reactive intermediate for derivatization purposes.

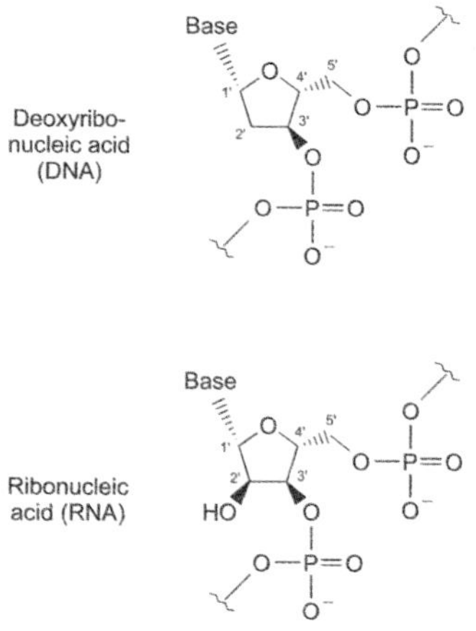

FIGURE 2.55 The similar structures of DNA and RNA basic units.

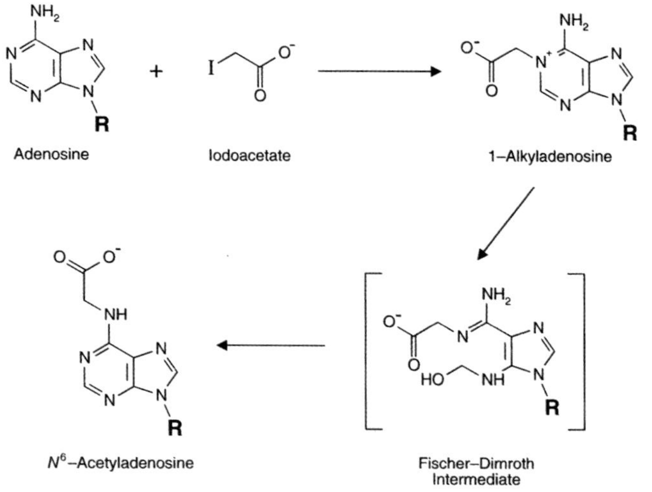

FIGURE 2.54 Alkylation reactions can occur at the N-1 position of adenosine, resulting in a Fischer–Dimroth rearrangement to yield an N^6 derivative.

rearrangement mechanism. Alkylation at N-1 can result in a rearrangement to give the C-6 alkylated product. The reaction at N-1 usually requires extended time to obtain good yields. For instance, alkylation with iodoacetic acid takes 5 to 10 days at pH 6.5. Under alkaline conditions and elevated temperatures, the six-membered ring is then broken and reformed, resulting in the 6-aminoalkylated product containing a terminal carboxylate group (Figure 2.54). The resultant acid can be used in further derivatization reactions to facilitate conjugate formation (Lowe, 1979).

An additional reaction reported for adenine involves the coupling of glutaraldehyde to the 6-amino group (Matthews and Kricka, 1988). However, reaction at this group with electrophilic reagents such as those discussed in Section 2 proceeds more slowly than that possible using a primary aliphatic amine. In general, bioconjugate chemistry carried out with nucleic acid bases involves the formation of an intermediate derivative containing a spacer arm terminating in an amine, sulfhydryl, or carboxylate to obtain acceptable reactivity and yields.

Sugar Groups

The sugar portion of oligonucleotides is a five-carbon pentose occurring in one of two forms. In RNA, it is β-D-ribose in a ring structure. In DNA, the monosaccharide is β-D-2-deoxyribose, wherein the number 2′ carbon of the ring lacks a hydroxyl group. An individual nucleotide will have its 1′ hydroxyl group of the ribose unit tied up in an *N*-glycosidic bond with the associated base and its C-5 hydroxyl group bound to phosphate in an ester linkage. If the nucleotide is of the deoxy form, then the only remaining hydroxyl is on the 3′ carbon of the sugar unit. Ribonucleic acids, by contrast, contain a diol group formed from the two hydroxyls on the 2′- and 3′-carbons of ribose (Figure 2.55). Polymers of nucleic acids are created through diester phosphate bonds, mainly connected between the 5′ hydroxyl of one sugar group and the 3′ hydroxyl of the next adjacent sugar. Thus, DNA contains no hydroxyl groups except the single one at the 3′ terminal of each strand. RNA has one hydroxyl at each nucleotide sugar unit and a diol group at the 3′ end.

Conjugation or modification reactions may be done through the 3′ hydroxyl group of deoxyribonucleic acids or the 2′,3′-diol of ribonucleic acids. Hydroxyls may be targeted for coupling using strong alkylating agents under alkaline conditions. Epoxide compounds (Chapter 5, Section 7) are particularly effective at modifying hydroxyl groups. The most common method of conjugation through nucleotide sugar units, however, is periodate oxidation of the adjacent hydroxyls of ribonucleic acids. Treatment with periodate breaks the carbon–carbon bond between the two hydroxyl residues and creates two aldehyde groups (Seela and Waldeck, 1975). A procedure for oxidizing carbohydrates with sodium periodate can be found in Section 4.4, this chapter. This method can be used to create RNA conjugates through directed coupling only at the 3′ end or to immobilize ribonucleic acids such as ATP to insoluble supports for affinity chromatography (Lowe, 1979).

Phosphate Groups

The phosphate groups of nucleotides are joined to the 5′ hydroxyl group of the sugar component in an ester or anhydride linkage. Several forms of nucleoside phosphate compounds are possible, containing up to three esterified phosphate groups polymerized off the ribose or deoxyribose unit. The presence of these groups contributes an overall negative charge to the nucleotide—minus two for the terminal phosphate group and minus one for each internal phosphate under alkaline conditions. Multiple esterified phosphates contain considerable potential energy from their easily hydrolyzed anhydride bonds. This energy is the basis for many biochemical transformations in biological systems. It is the triphosphate form of nucleosides that is utilized in DNA and RNA synthesis *in vivo*. However, nucleoside triphosphates and diphosphates such as ATP and ADP have numerous contributions to cellular metabolism beyond just oligonucleotide construction. Controlled hydrolysis of their multiple phosphate ester bonds releases energy for many biological operations. Other derivatives of nucleoside phosphate compounds provide cofactors for enzymes (such as coenzyme A) or are involved in signal transduction processes (such as cyclic AMP [cAMP]). Figure 2.56 shows some of these common nucleoside phosphate derivatives.

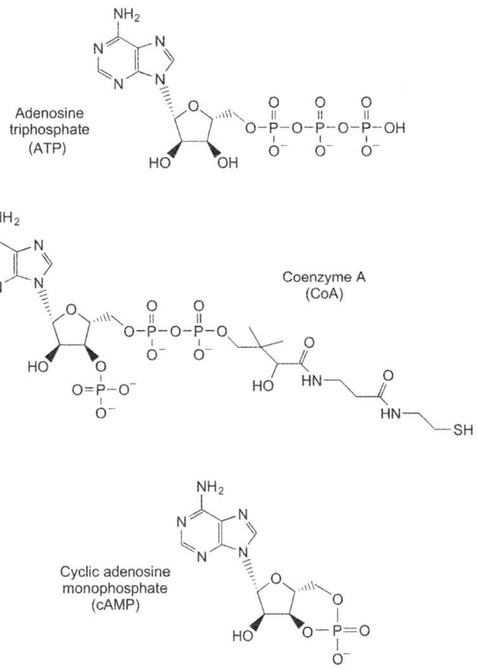

FIGURE 2.56 Nucleotide derivatives have additional functions *in vivo* beyond their role in oligonucleotide construction.

The phosphate groups of nucleotides may be targeted for modification reactions using condensation agents such as carbodiimides. In aqueous environments, EDC (Chapter 4, Section 1.1) may be used to couple amine-containing compounds to the terminal phosphate group of an oligonucleotide, forming a phosphoramidate linkage. In DNA or RNA chains, the internal phosphate groups do not react under the pH conditions of the modification. In this way, the 5′-phosphate group may be specifically targeted for modification or conjugation, thus avoiding potential interference with hydrogen bonding interactions with complementary polynucleotide strands. Chapter 23, Sections 2.1 and 2.2 describe the use of this reaction in bioconjugate applications.

Another phosphate modification procedure that is effective at adding detectable components to oligonucleotide probes is to use a phosphoramidite derivative. The common method of automated oligonucleotide synthesis is to use phosphoramidite chemistry to add nucleotides to the growing sequence. A functionalized phosphoramidite nucleotide derivative can be added at particular points in the synthetic process to create labeled probes of known structure. Non-nucleotide phosphoramidites may also be used to produce modified probes containing fluorescent molecules, biotin, chelating groups, or spacer groups with amines for further derivatization. Most of these techniques require an automated DNA synthesizer. The methods of DNA modification during synthesis have been reviewed and are beyond the scope of this book (Beaucage and Iyer, 1993).

RNA and DNA Structure

The nucleotides forming RNA or DNA molecules are linked together in phosphodiester bonds with sugar–phosphate repeating units. The esters are directionally linked between the 3′ hydroxyl of one ribosyl group and the 5′ hydroxyl of the next. The fundamental step in cellular DNA synthesis involves the reaction of a deoxynucleoside triphosphate group with the 3′ end of an existing chain. The nucleotide sequence of a new strand is enzymatically controlled by use of a complementary chain as a template. Each new nucleotide addition is facilitated by the energy released through hydrolysis of two phosphates from the triphosphate group of the incoming nucleoside. The resulting succession of nucleotides encodes the message for protein synthesis, with each three-base code signaling a particular amino acid in a polypeptide sequence.

Nucleotide bases projecting from the sugar–phosphate backbone of a polynucleotide are able to interact with other strands through hydrogen bonding. Hydrogen bonding can occur between cytosine and guanine base

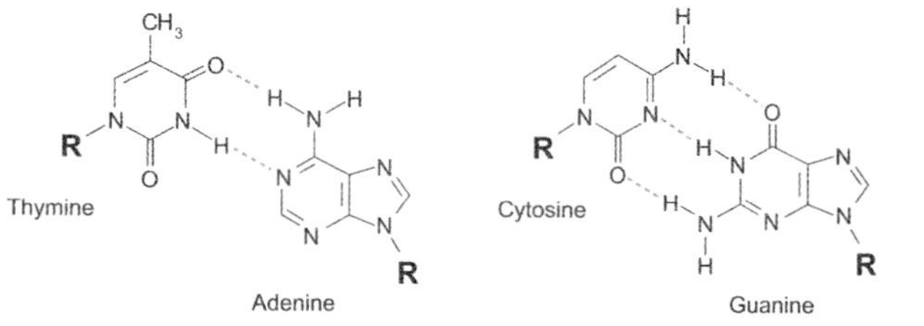

FIGURE 2.57 Base-pairing can occur between complementary bases in opposing oligonucleotide strands. These predictable interactions form the basis for using synthetic oligonucleotide probes to target particular DNA sequences.

units in different strands of DNA through interaction of the C-2 ketone oxygen, the N-3 nitrogen, and C-4 amine groups of cytosine with the C-2 amine, N-1 nitrogen, and the C-6 ketone oxygen of guanine. In a similar fashion, thymine (or uracil) residues can hydrogen bond with adenine groups through the N-3 nitrogen and C-4 ketone oxygen of thymine interacting with the N-1 nitrogen and C-6 amine of adenine (Figure 2.57).

This specific base-pairing capability of oligonucleotides defines the structure of complementary DNA molecules. In the classic Watson–Crick model, two complementary DNA strands interact in an antiparallel fashion to form a right-handed double helix. Thus, one chain runs in the 3′-to-5′ direction while the complementary chain runs in the 5′-to-3′ direction through the helical structure. This standard double helix, now called the B form, occurs often in aqueous solution and is the most stable structure under physiological conditions (Figure 2.58). However, there are several other forms that double-stranded DNA can take in solution. Another right-handed helical construction, the A form, can occur under nonaqueous conditions and is more compact than the B form. A completely different DNA structure, the Z form, is a left-handed helix that can occur in some segments containing an abundance of alternating pyrimidines and purines. Short segments of Z structure have been found in some cells. Finally, some rare DNA sequences can form triple-helical regions through normal and non-Watson–Crick base pairing.

Unlike the double-stranded nature of DNA, RNA molecules usually occur as single strands. This does not mean they are unable to base-pair as DNA can. Complementary regions within an RNA molecule often undergo base-pairing and form complex tertiary structures, even approaching the three-dimensional nature of proteins. Some RNA molecules, such as transfer RNA (tRNA), possess several helical areas and loops as the strand interacts with itself in complementary sections. Other hybrid molecules such as the enzyme RNase P contain protein and RNA portions. The RNA part is highly complex, with many circles, loops, and helical regions creating a convoluted structure.

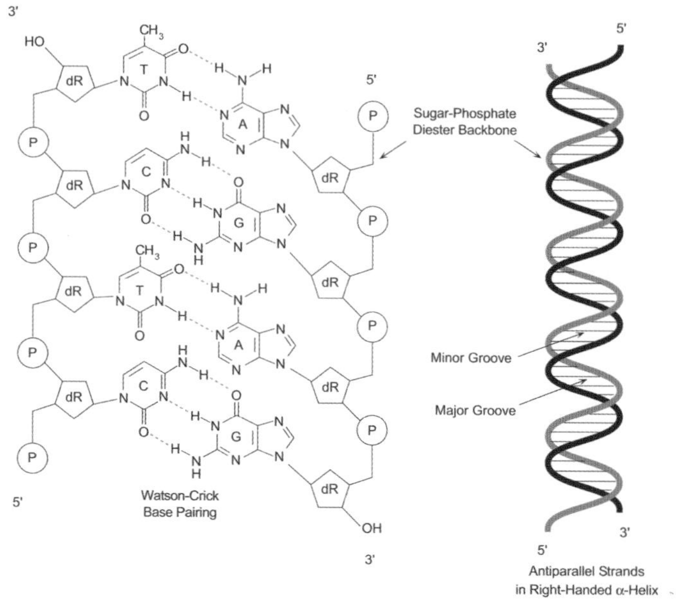

FIGURE 2.58 The classic Watson–Crick DNA double helix is formed through base-pairing interactions between two antiparallel strands. In physiological conditions, the two strands take on an α-helical shape with about 10bp per turn of the helix. The phosphate–sugar backbone of the helix faces outward, while hydrogen bonding between opposing bases occurs in the middle of the wrapped strands. This configuration creates minor and major grooves between the phosphate–sugar backbones, potentially exposing the internal bases to interactions with other molecules.

The predictable nature of DNA and RNA base pairing make their interactions the most defined of any biological system. The specific affinity of one strand for its complementary sequence makes it possible to target genetic markers with extreme accuracy. Synthetic segments of RNA or DNA can be used to detect or quantify their complementary targets, even in highly dilute environments containing many other oligonucleotide molecules. If the oligonucleotide probe is labeled with a highly detectable component, then specific base-pairing interactions can be assayed. This ability has created an extensive utilization of labeled probes in molecular biology. Detection of target DNA or RNA can be carried out in cells, tissue sections, blots, and electrophoresis gels; after amplification by polymerase chain reaction (PCR techniques); or in solution. The ability to detect single-copy genes through the use of labeled oligonucleotide probes will make this field one of the leading application areas for bioconjugate techniques.

3.2. Polynucleotide Crosslinking Methods

The unique properties of oligonucleotides create crosslinking options that are far different from any other biological molecule. Nucleic acids are the only major class of macromolecule that can be specifically duplicated *in vitro* by enzymatic means. The addition of modified nucleoside triphosphates to an existing DNA strand by the action of polymerases or transferases allows addition of spacer arms or detection components at random or discrete sites along the chain. Alternatively, chemical methods that modify nucleotides at selected functional groups can be used to produce spacer arm derivatives or activated intermediates for subsequent coupling to other molecules.

Thus, both chemical and enzymatic derivatization techniques can be used to form oligonucleotide probes of high activity in hybridization assays. The main consideration for successful polynucleotide crosslinking, as in other bioconjugate applications, is to avoid probe inactivation during the modification or conjugation process. Since the purpose in constructing a DNA or RNA probe is to hybridize to a complementary oligonucleotide through hydrogen bond interactions, any derivatization procedure that significantly interferes with Watson–Crick base pairing should be avoided. This means that a large amount of base derivatization along a polynucleotide chain has potential for causing obstructions in the hybridization process, sometimes dramatically reducing or eliminating base-pairing efficiency. In general, base modifications within an oligonucleotide probe should be limited to no more than about 30 to 40 sites per 1000 bases to maintain hybridization ability.

By contrast, derivatization at the ends of an oligo or at the sugar–phosphate backbone usually produces little interference in base pairing. Conjugates may be created by enzymatic polymerization of functionalized nucleoside triphosphates off the 3′ end or by chemical modification of the 5′ phosphate group with minimal to no interference in hybridization potential. The application of these strategies to creating labeled oligonucleotide probes is discussed in Chapter 23.

4. CREATING SPECIFIC FUNCTIONALITIES

It is often desirable to alter the native structure of a macromolecule to provide functional targets for modification or conjugation. The use of most reagent systems requires the presence of particular chemical groups to effect coupling. For instance, heterobifunctional crosslinkers contain two different reactive species that are directed against different functionalities. One target molecule has to contain chemical groups able to react with one end of the crosslinker, while the other target molecule must contain groups able to react with the other end. Occasionally, the required chemical groups are not present on one of the target molecules and must be created. This can usually be done by reacting an existing chemical group with a modification reagent that contains or produces the desired functionality upon coupling. Thus, an amine can be "changed" into a sulfhydryl or a carboxylate can be altered to yield an amine simply by using the appropriate reagent.

This same type of modification strategy can also be used to create highly reactive groups from functionalities of rather low reactivity. For instance, carbohydrate chains on glycoproteins can be modified with sodium periodate to transform their rather unreactive hydroxyl groups into highly reactive aldehydes. Similarly, cystine or disulfide residues in proteins can be selectively reduced to form active sulfhydryls, or 5′ phosphate groups of DNA can be transformed to yield modifiable amines.

Alternatively, spacer arms can be introduced into a macromolecule to extend a reactive group away from its surface. The extra length of a spacer can provide less steric hindrance to conjugation and often yields more active complexes.

The use of modification reagents to create specific functionalities is an important technique to master. In one sense, the process is like using building blocks to construct on a target molecule any desired functional groups necessary for reactivity. The success of many conjugation schemes depends on the presence of the correct chemical groups. Care should be taken in choosing a modification strategy, however, since some chemical changes will radically affect the native structure and activity of a macromolecule. A protein may lose its capacity to bind a specific ligand. An enzyme may lose the ability to act upon its substrate. A DNA probe may no longer be able to hybridize to its complementary target. In many cases, the potential for inactivation relates to changing conformational structures, blocking active sites, or modifying critical functional groups. Trial and error and careful literature searches are often necessary to optimize any modification tactic.

4.1. Introduction of Sulfhydryl Residues (Thiolation)

The sulfhydryl group is a popular target in many modification strategies. Crosslinking agents that have more than one reactive group often employ a sulfhydryl-reactive functionality at one end to direct the conjugation reaction to a particular part of a target macromolecule. The frequency of sulfhydryl occurrence in proteins or other molecules is usually low (or nonexistent) compared to other groups such as amines or carboxylates. The use of sulfhydryl-reactive chemistries thus can restrict modification to only a limited number of sites within a target molecule. Limiting modification greatly increases the chances of retaining activity after conjugation, especially in sensitive proteins such as some enzymes. Unfortunately, sulfhydryl groups often need to be generated (from reduction of indigenous disulfides) or created (from use of the appropriate thiolation reagent systems). The following sections describe the most popular techniques for creating these functionalities. Some of these reagent systems are specifically designed to form –SH groups, while others are crosslinkers that can also serve the dual purpose of sulfhydryl-generating agents.

Sulfhydryl groups are susceptible to oxidation and formation of disulfide crosslinks. To prevent disulfide bond formation, remove oxygen from all buffers by degassing under vacuum and bubbling an inert gas (e.g., nitrogen) through the solution. In addition, EDTA (0.01–0.1-*M*) may be added to buffers to chelate metal ions, preventing metal-catalyzed oxidation of sulfhydryls. Some proteins of serum origin (particularly BSA) contain so much contaminating metal ions (presumably iron from hemolyzed blood) that 0.1-*M* EDTA is required to prevent this type of oxidation.

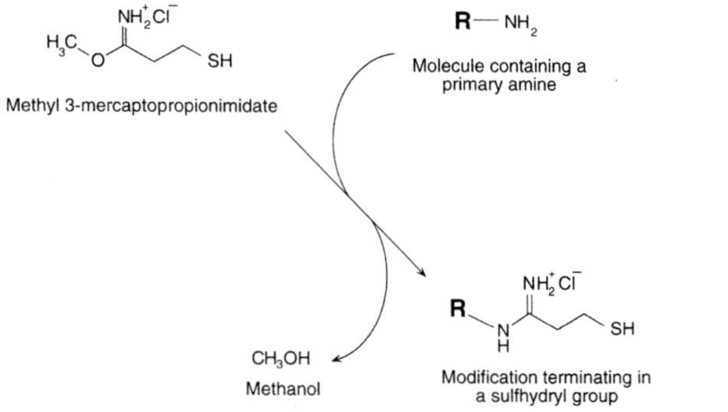

FIGURE 2.59 Thiolation of an amine-containing compound with methyl 3-mercaptopropionimidate. The modification preserves the positive charge on the primary amine.

Modification of Amines with 2-Iminothiolane (Traut's Reagent)

Perham and Thomas (1971) originally prepared an imidoester compound containing a thiol group, methyl 3-mercaptopropionimidate hydrochloride. The imidoester group can react with amines to form a stable, charged linkage (Chapter 3, section 1.11), while leaving a sulfhydryl group available for further coupling (Figure 2.59). Traut *et al.* (1973) subsequently synthesized an analogous reagent containing one additional carbon, methyl 4-mercaptobutyrimidate. Later, this compound was found to cyclize as a result of the sulfhydryl group reacting with the intrachain imidoester, forming 2-iminothiolane (Jue *et al.*, 1978). The cyclic imidothioester can still react with primary amines in a ring-opening reaction that regenerates the free sulfhydryl (Figure 2.60).

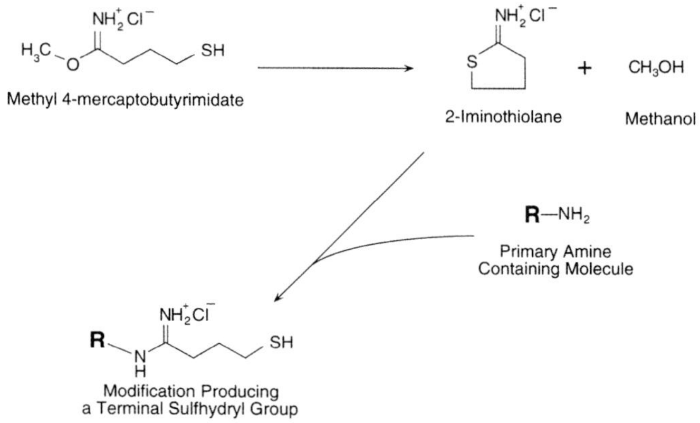

FIGURE 2.60 Methyl 4-mercaptobutyrimidate forms 2-iminothiolane, which can react with a primary amine to create a sulfhydryl group. The modification preserves the positive charge of the original amine.

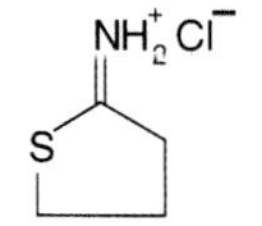

Traut's Reagent;
2-Iminothiolane
MW 137.6
8.1 Å

Traut's reagent is fully water soluble and reacts with primary amines in the range of pH 7 to 10. The cyclic imidothioester is stable to hydrolysis at acid pH values, but its half-life in solution decreases as the pH increases beyond neutrality. However, even at pH 8.0 in 25-m*M* triethanolamine the rate of sulfhydryl formation without added primary amine was found to be negligible. Upon addition of dipeptide amine, the reagent reacted quickly as evidenced by the production of Ellman's reagent color. The rate of reaction can also be followed by 2-iminothiolane's absorbance at 248 nm (λ_{max}; $\varepsilon = 8840\,M^{-1}cm^{-1}$). As the cyclic imidate reacts with amines, its absorbance at this wavelength decreases. With addition of the dipeptide glycylglycine, the starting absorbance of a solution of Traut's reagent decreased over 80% within 20 minutes (Jue *et al.*, 1978). Thus, protein modification with 2-iminothiolane is very efficient and proceeds rapidly at slightly basic pH.

At high pH (10.0), Traut's reagent is also reactive with aliphatic and aromatic hydroxyl groups, although the rate of reaction with these groups is only about 0.01 that of primary amines. In the absence of amines, however, carbohydrates such as agarose or cellulose membranes can be modified to contain sulfhydryl residues (Alagon and King, 1980). Polysaccharides modified in this manner are effective in covalently crosslinking antibodies for use in immunoassay procedures.

Proteins modified with 2-iminothiolane are subject to disulfide formation upon sulfhydryl oxidation. This can cause unwanted conjugation, potentially precipitating the protein. The addition of a metal-chelating agent such as EDTA (0.01–0.1-*M*) will prevent metal-catalyzed oxidation and maintain sulfhydryl stability. In the presence of some serum proteins (e.g., BSA) a 0.1-*M* concentration of EDTA may be necessary to prevent metal-catalyzed oxidation, presumably due to the high contamination of iron from hemolyzed blood.

Traut's reagent has been used successfully in the investigation of ribosomal proteins (Sunn *et al.*, 1974; Jue *et al.*, 1978; Kenny *et al.*, 1979; Lambert *et al.*, 1983; Blattler *et al.*, 1985), RNA polymerase (Hillel and Wu, 1977), progesterone receptor subunits (Birnbaumer *et al.*, 1979), and in the synthesis of enzyme-labeled DNA hybridization probes (Ghosh *et al.*, 1990). It is an excellent thiolation reagent for use in the preparation of immunotoxins (Section 3.3). It also has been used to modify and introduce sulfhydryls into oligosaccharides from asparagine-linked glycans (Tarentino *et al.*, 1993).

Side reactions other than oxidation to disulfides can also occur using Traut's reagent. Once an amine on a protein is modified with 2-iminothiolane, the terminal thiol can recyclize by attacking the amidine carbon (Figure 2.61). This can then rearrange into an iminothiolane derivative, which effectively ties up the thiol (Singh *et al.*, 1996; Mokotoff *et al.*, 2001). Proteins and other molecules thiolated using Traut's reagent can lose substantial amounts of available thiol to recyclization in just hours. For this reason, the thiolated product of a Traut's reaction should be used immediately in a conjugation reaction to avoid significant loss of activity.

Protocol

1. Prepare the protein or macromolecule to be thiolated in a non-amine-containing buffer at pH 8.0. For the modification of ribosomal proteins (often cited in the literature) use 50-m*M* triethanolamine hydrochloride, 1-m*M* $MgCl_2$, 50-m*M* KCl, pH 8.0. The magnesium and potassium salts are for stabilization of some ribosomal proteins. If other proteins are to be thiolated, the same buffer may be used without added salts for stabilization. Alternatively, 50-m*M* sodium phosphate, 0.15-*M* NaCl, pH 8.0, or 0.1-*M* sodium borate, pH 8.0, may be used. For the modification of polysaccharides, use 20-m*M* sodium borax, pH 10, to produce reactivity towards carbohydrate hydroxyl residues. Dissolve the protein to be modified at a concentration of 10 mg/ml in the reaction buffer of choice. Lower concentrations may also be used with a proportional scaling back of added 2-iminothiolane.
2. Dissolve the Traut's reagent (Thermo Fisher) in water at a concentration of 2 mg/ml (makes a 14.5-m*M* stock solution). The solution should be used immediately. For the modification of IgG at a concentration of 10 mg/ml using a 10-fold molar excess of Traut's reagent, add 45.8 µl of the stock solution to each milliliter of the protein solution.
3. React for 1 h at room temperature (a 4°C reaction temperature may be used successfully as well).
4. Purify the thiolated protein from unreacted Traut's reagent by gel filtration using your buffer of choice (e.g., 20-m*M* sodium phosphate, 0.15-*M* NaCl, 1-m*M* EDTA, pH 7.2). The addition of EDTA to this buffer helps to prevent oxidation of the sulfhydryl groups and the resultant disulfide formation. After purification, use the thiolated protein immediately in a conjugation reaction to avoid the recyclization of the free thiol with concomitant decrease in thiol availability.
5. The degree of –SH modification may be determined using the Ellman's assay (Section 4.1, this chapter).

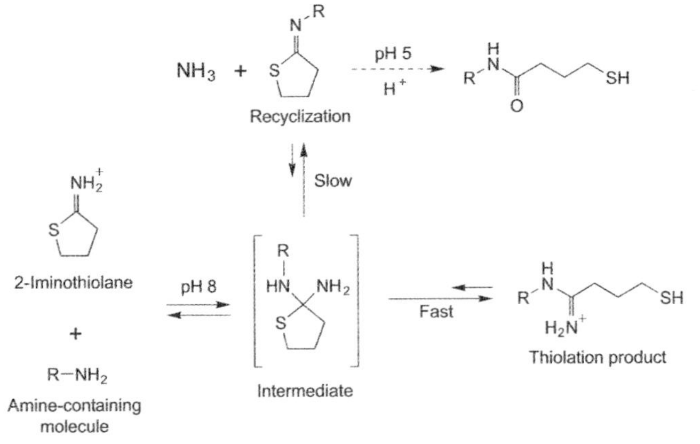

FIGURE 2.61 Traut's reagent can undergo side reactions after the modification of an amine-containing molecule. The terminal thiol group can recyclize to create another iminothiolane derivative that effectively ties up the thiol.

When 2-iminothiolane is used to modify proteins in tandem with 4,4′-dipyridyl disulfide, a protected sulfhydryl can be introduced in a single step (King *et al.*, 1978). The simultaneous reaction between a protein, 2-iminothiolane, and 4,4′-dipyridyl disulfide yields a modification containing pyridyl disulfide groups. The pyridyl disulfide subsequently may be reduced with DTT to yield a free sulfhydryl. Pyridyl disulfides are also highly reactive toward sulfhydryls through disulfide interchange (Chapter 3, Section 2.6). The protocol is a modification of the method of King *et al.* (1978). 2-Iminothiolane is used to modify proteins in tandem with 4,4′-dipyridyl disulfide, in which a protected sulfhydryl can be introduced in a single step (King *et al.*, 1978). The simultaneous reaction between a protein, 2-iminothiolane, and 4,4′-dipyridyl disulfide yields a modification containing pyridyl disulfide groups. The pyridyl disulfide subsequently may be reduced with DTT to yield a free sulfhydryl. Pyridyl disulfides are also highly reactive toward sulfhydryls through disulfide interchange (Chapter 6,

Section 1.1). The following protocol is a modification of the method of King *et al.* (1978).

Protocol

1. Dissolve 1–10 mg of a protein to be modified in 1.0 ml of 0.025-*M* sodium borate, pH 9.0.
2. Dissolve 2-iminothiolane in 0.025-*M* sodium borate to a concentration of 0.02-*M*.
3. Dissolve 4,4′-dipyridyl disulfide at a concentration of 2 mg/ml in acetonitrile.
4. Add 0.2 ml of (3) and 1.0 ml of (2) to the protein solution.
5. React for 2 h at room temperature.
6. Purify the modified protein by gel filtration or dialysis.

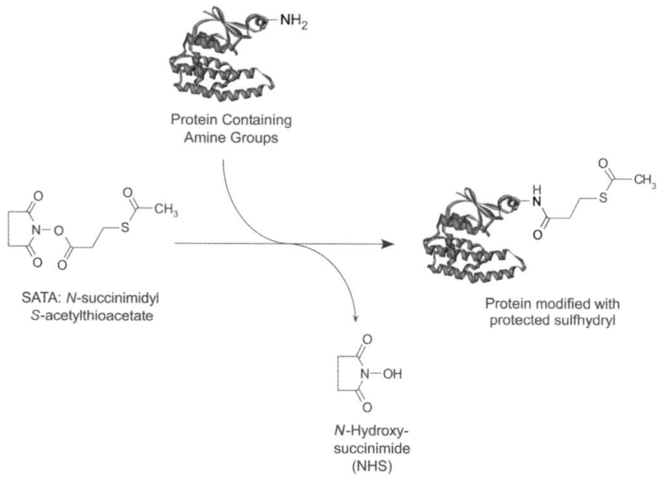

FIGURE 2.62 SATA can react with available amine groups in proteins and other molecules *via* its NHS ester end to form protected sulfhydryl derivatives. The illustrated protein is glutathione-*S*-transferase (E.C.2.5.1.18) (Ji *et al.*, 1995).

Occasionally, a protein modified in this manner will begin to precipitate as the reaction proceeds. Stopping the reaction earlier or adding a smaller quantity of modifying reagents may limit this effect.

Modification of Amines with SATA

A versatile reagent for introducing sulfhydryl groups into proteins is SATA, *N*-succinimidyl *S*-acetylthioacetate (Duncan *et al.*, 1983). The active NHS ester end of SATA reacts with amino groups in proteins and other molecules to form a stable amide linkage (Figure 2.62) (Chapter 3, Section 1.4). The modified protein then contains a protected sulfhydryl that can be stored without degradation and subsequently deprotected as needed with an excess of hydroxylamine (Figure 2.63). Since the protecting group can be removed without adding disulfide reducing agents like DTT, disulfides indigenous to the native protein will not be affected. This is an important consideration if disulfides are vital to activity, such as in the case of antibodies and some protein toxins.

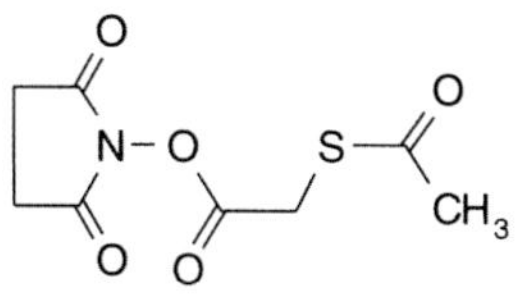

SATA;
N-succinimidyl
S-acetylthioacetate
MW 231.2

SATA is often used to form antibody–enzyme conjugates utilizing maleimide-containing heterobifunctional crosslinking agents. Most polyclonal antibody molecules may be modified to contain up to about six SATA

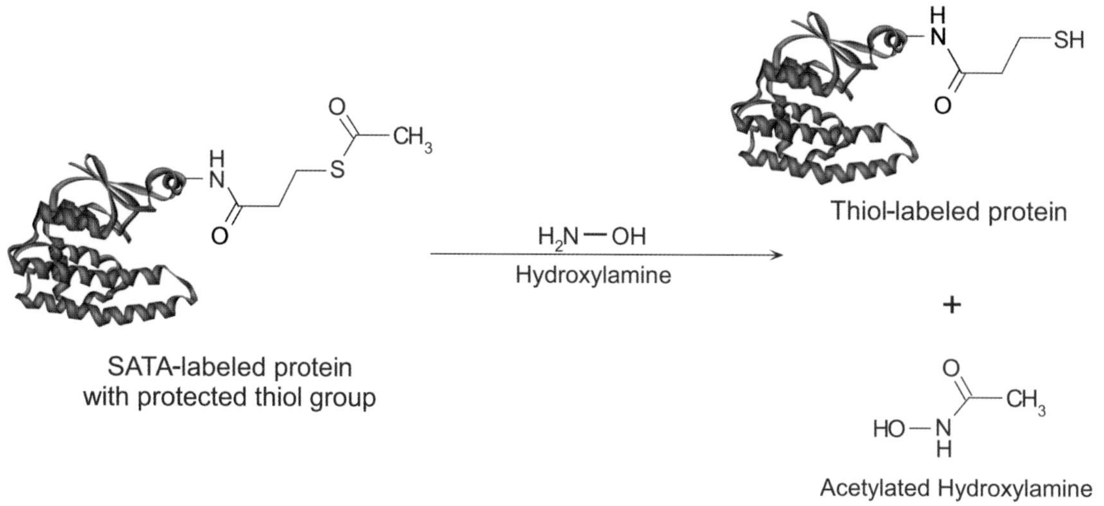

FIGURE 2.63 Deprotection with hydroxylamine of the acetylated thiol of SATA-modified proteins yields a free sulfhydryl group.

molecules per immunoglobulin with minimal effect on antigen binding activity. Some sensitive monoclonal antibodies, however, may be susceptible to modification and should be tested on a case-by-case basis. The modified antibody may then be deprotected and reacted with a maleimide-activated enzyme to form a conjugate useful in immunoassays (Chapter 20, Section 1.1). Conjugates formed using SATA are usually of low molecular weight with very few high-molecular-weight oligomers. They also maintain a bivalent antibody structure, ensuring a conjugate containing two antigen binding sites. This is an advantage over reduction schemes that break the antibody molecule into two heavy–light chain pairs to create sulfhydryls, since disulfide cleavage yields antibody fragments with only one antigen binding site.

SATA has been used to form conjugates with avidin or steptavidin with excellent retention of activity (Chapter 11, Section 3.1). It has also been used in the formation of a therapeutically useful toxin conjugate with recombinant CD4 (Ghetie *et al.*, 1990), to study syntaxin proteins (Amessou *et al.*, 2007), to prepare biospecific antibodies (Lenderfer *et al.*, 2001), and to make a unique polylysine conjugate as a vehicle for drug delivery (Sakharov *et al.*, 2001).

SATA is freely soluble in many organic solvents. In use, it is typically dissolved as a stock solution in DMSO, DMF, or methylene chloride, and then an aliquot of this solution is added to an aqueous reaction mixture containing the protein to be modified.

The thiolation method described below is generally applicable for the modification of proteins with SATA, particularly for subsequent conjugation with a maleimide-activated secondary protein. The degree of modification described usually yields 3 to 4 moles of –SH groups per mole protein when thiolating immunoglobulins. Other macromolecules containing primary amines may be modified using a similar procedure. The degree of modification observed with other molecules may vary depending on the number of available primary amines and their relative reactivity. The molar ratio of SATA to immunoglobulin added to a reaction for the modification of rabbit polyclonal IgG *versus* the degree of sulfhydryl incorporation is illustrated in Figure 2.64 (Sykaluk, 1994).

The following protocol represents a generalized method for protein thiolation using SATA. For comparison purposes, contrast the variation of this SATA modification method as outlined in Chapter 20, Section 1.1, for use in the preparation of antibody–enzyme conjugates.

Protocol

1. Dissolve the protein to be thiolated at a concentration of 1–5 mg/ml in 50-m*M* sodium phosphate, pH 7.5, containing 1–10-m*M* EDTA. Other non-amine-containing buffers such as borate, HEPES, and bicarbonate may also be used as the reaction medium. The effective pH for the NHS ester modification reaction is in the range of 7.0 to 9.0, but environments closer to neutrality will limit the hydrolysis of the ester.
2. Dissolve the SATA reagent (Thermo Fisher) in DMSO at a concentration of 65 m*M* (15 mg/ml). Note: DMSO should be handled in a fume hood.
3. Add 10 μl of the SATA solution to each milliliter of protein solution.
4. Mix and react for 30 min at room temperature.
5. Separate modified protein from unreacted SATA and reaction byproducts by dialysis against 50-m*M* sodium phosphate, pH 7.5, containing 1-m*M* EDTA or by gel filtration on a Sephadex G-25 column (Pharmacia) using the same buffer.
6. Deprotect the acetylated –SH groups as needed by adding 100 μl of 0.5-*M* hydroxylamine hydrochloride in 50-m*M* sodium phosphate, 25-m*M* EDTA, pH 7.5, to each milliliter of the SATA modified protein solution.
7. Mix and react for 2 h at room temperature.
8. Purify the sulfhydryl-modified protein by dialysis against 50-m*M* sodium phosphate, 1-m*M* EDTA, pH 7.5, or by gel filtration on a Sephadex G-25 column using the same buffer.

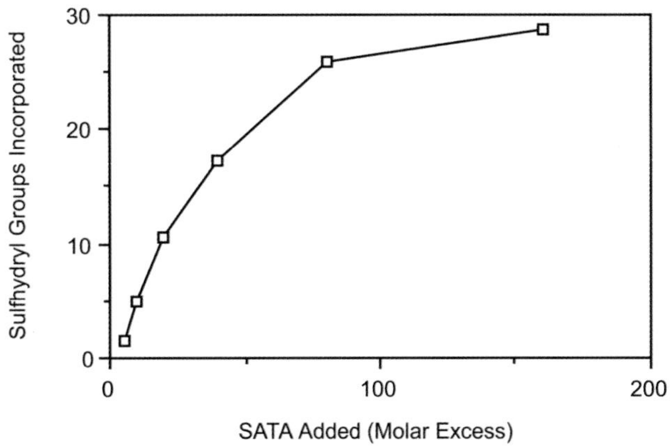

FIGURE 2.64 SATA modification of rabbit polyclonal IgG with the resultant sulfhydryl incorporation level.

The deacetylated protein should be used immediately to prevent loss of sulfhydryl content through disulfide formation. The degree of –SH modification may be determined by performing an Ellman's assay (Section 4.1).

Modification of Amines with SATP

SATP, succinimidyl acetylthiopropionate, is an analog of SATA (Section 4.1) containing one additional carbon atom in length (Fuji *et al.*, 1985). The compound retains all the advantages of a protected sulfhydryl, including stability of the modified protein and selective release of the protecting group with hydroxylamine to free the sulfhydryl as needed (Figure 2.65). SATP is soluble in DMF

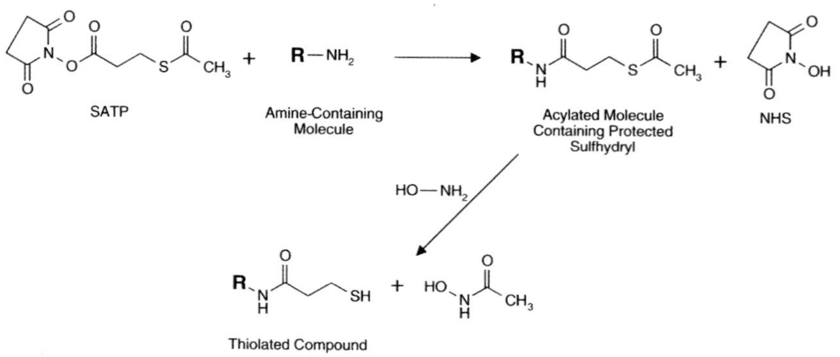

FIGURE 2.65 SATP reacts with amine-containing proteins or other molecules *via* its NHS ester end to create protected sulfhydryl derivatives in a manner similar to that of SATA. Deprotection can be done with hydroxylamine to free the thiol.

and methylene chloride. It is usually first solubilized in organic solvent and an aliquot added to an aqueous solution containing the macromolecule to be modified. It is particularly useful in adding an N-terminal –SH group at the completion of peptide synthesis.

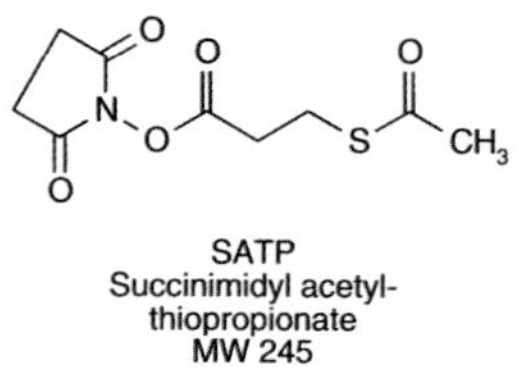

Protocol

1. Dissolve the protein or peptide to be thiolated at a concentration of 10 mg/ml in 50-m*M* sodium phosphate, pH 7.5, containing 1-m*M* EDTA. Other non-amine-containing buffers such as borate, HEPES, and bicarbonate may also be used as the reaction medium. The effective pH for the NHS ester modification reaction is in the range of 7.0 to 9.0. Conditions closer to neutral pH will limit the degree of NHS ester hydrolysis during the reaction.
2. Dissolve the SATP reagent (Molecular Probes) in DMF at a concentration of 65 m*M* (16 mg/ml). Note: DMF should be handled in a fume hood.
3. Add 10 μl of the SATP solution to each milliliter of protein or peptide solution.
4. Mix and react for 30–60 min at room temperature (or 2–4 h at 4°C).
5. Separate modified protein from unreacted SATP and reaction byproducts by dialysis against 50-m*M* sodium phosphate, pH 7.5, containing 1-m*M* EDTA or by gel filtration using a desalting column with the same buffer. If a peptide of low molecular weight is being modified, careful gel filtration using a matrix having a low exclusion limit will separate the peptide from the reaction byproducts, but the separation should be carried out on an automated system to accurately capture the peaks. In this case, use either Sephadex G-25 or Sephadex G-10 for the chromatography.
6. Deprotect the acetylated –SH groups as needed by adding 100 μl of 0.5-*M* hydroxylamine hydrochloride in 50-m*M* sodium phosphate, 25-m*M* EDTA, pH 7.5, to each milliliter of the SATP-modified protein solution.
7. Mix and react for 2 h at room temperature.
8. Purify the sulfhydryl-modified protein by dialysis against 50-m*M* sodium phosphate, 1-m*M* EDTA, pH 7.5, or by gel filtration on a Sephadex G-25 column using the same buffer. Again, if a peptide of low molecular weight is being modified, use careful gel filtration for purification.

The deacetylated protein should be used immediately to prevent loss of sulfhydryl content through disulfide formation. The degree of –SH modification may be determined by performing an Ellman's assay (Section 4.1, this chapter).

Modification of Amines with SPDP

SPDP, *N*-succinimidyl 3-(2-pyridyldithio)propionate, is one of the most popular heterobifunctional crosslinking agents (Chapter 6, Section 1.1). The NHS ester end of SPDP reacts with amine groups to form an amide linkage, while the 2-pyridyldithiol group at the other end can react with sulfhydryl residues to form a disulfide linkage (Carlsson *et al.*, 1978). The crosslinker is used extensively to form immunotoxin conjugates for *in vivo* administration (Chapter 20, Section 3.2.1). The reagent is also useful in creating sulfhydryls in proteins and other molecules. Once modified with SPDP, a protein can be treated with DTT (or other disulfide reducing agents; see Section 4.1, this chapter) to release the pyridine-2-thione leaving group and form the free sulfhydryl (Figure 2.66). The terminal –SH group can then be used to conjugate with any crosslinking agents containing sulfhydryl-reactive groups, such as maleimide or iodoacetyl functionalities (for covalent conjugation) or 2-pyridyldithiol groups (for reversible conjugation).

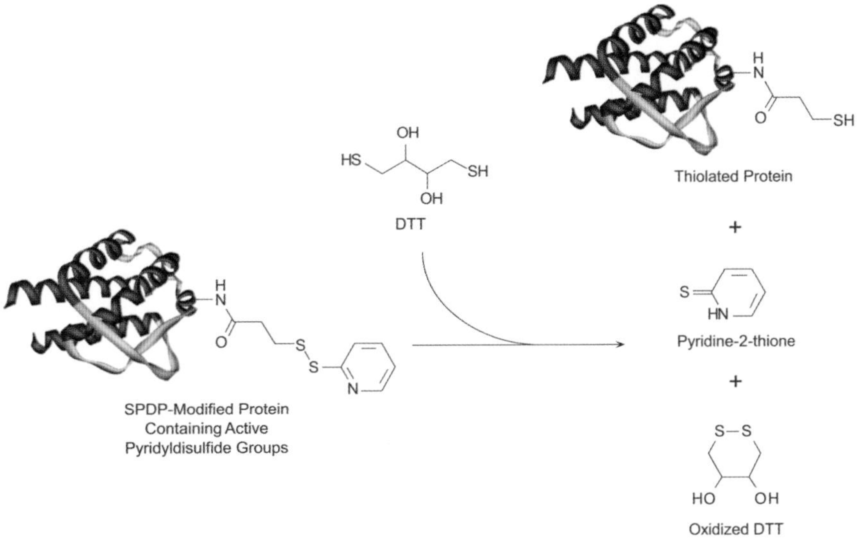

FIGURE 2.66 SPDP-modified proteins can be reduced with DTT to yield free sulfhydryl groups for conjugation.

There are three forms of SPDP analog currently available commercially (Thermo Fisher): the standard SPDP, a long-chain version designated LC-SPDP, and a water soluble, sulfo-NHS form also containing an extended chain, called Sulfo-LC-SPDP (Chapter 6, Section 1.1). The main disadvantage in using SPDP to create sulfhydryls is the necessity of using a reducing agent to remove the pyridine-2-thione group. Reducing agents will also affect indigenous disulfides within a protein molecule, cleaving and reducing them. This method therefore works well for proteins containing no sulfhydryls or no disulfides that are critical to function, but it may cause loss of activity or subunit breakdown in proteins containing essential disulfides.

The following procedure is similar to the method of Cumber *et al.* (1985), but with some modifications.

Protocol

1. Dissolve the protein or macromolecule to be thiolated at a concentration of 10 mg/ml in 50-m*M* sodium phosphate, 0.15-*M* NaCl, pH 7.2. Other non-amine-containing buffers such as borate, HEPES, and bicarbonate may also be used in this reaction. The effective pH for the NHS ester modification reaction is in the range of 7.0 to 9.0, but environments closer to neutrality will limit ester hydrolysis.
2. Dissolve SPDP (Thermo Fisher) at a concentration of 6.2 mg/ml in DMSO (makes a 20-m*M* stock solution). Alternatively, LC-SPDP may be used and dissolved at a concentration of 8.5 mg/ml in DMSO (also makes a 20-m*M* solution). If the water soluble Sulfo-LC-SPDP is used for the reaction, a stock solution in water may be prepared just prior to adding an aliquot to the protein solution. In this case, prepare a 10-m*M* solution of Sulfo-LC-SPDP by dissolving 5.2 mg/ml in water. Since an aqueous solution of the crosslinker will degrade by hydrolysis of the sulfo-NHS ester, it should be used quickly to prevent significant loss of activity. If a sufficiently large amount of protein will be modified to allow accurate weighing of Sulfo-LC-SPDP, the solid may be added directly to the reaction mixture without preparing a stock solution in water.
3. Add 25 μl of the stock solution of either SPDP or LC-SPDP in DMSO to each milliliter of the protein to be modified. If Sulfo-LC-SPDP is used, add 50 μl of the stock solution in water to each milliliter of protein solution.
4. Mix and react for at least 30 min at room temperature. Longer reaction times, even overnight, will not adversely affect the modification.
5. Purify the modified protein from reaction byproducts by dialysis or gel filtration using 50-m*M* sodium phosphate, 0.15-*M* NaCl, pH 7.2.
6. To release the pyridine-2-thione leaving group and form the free sulfhydryl, add DTT at a concentration of 0.5 mg DTT per mg of modified protein. A stock solution of DTT may be prepared to make it easier to add it to a small amount of protein solution. In this case, dissolve 20 mg of DTT per milliliter of 0.1-*M* sodium acetate, 0.1-*M* NaCl, pH 4.5. Add 25 μl of this solution per mg of modified protein. Release of pyridine-2-thione can be followed by its characteristic absorbance at 343 nm ($\varepsilon = 8.08 \times 10^3 M^{-1} cm^{-1}$).
7. Mix and react at room temperature for 30 min.

8. Purify the thiolated protein from excess DTT by dialysis or gel filtration using 50-m*M* sodium phosphate, 0.15-*M* NaCl, 1-m*M* EDTA, pH 7.2. The modified protein should be used immediately in a conjugation reaction to prevent sulfhydryl oxidation and formation of disulfide crosslinks.

Modification of Amines with SMPT

SMPT, succinimidyloxycarbonyl-α-methyl-α-(2-pyridyldithio)toluene, contains an NHS ester end and a pyridyldisulfide end similar to SPDP, but its hindered disulfide makes conjugates formed with this reagent more stable (Thorpe *et al.*, 1987) (Chapter 6, Section 1.2). The reagent is especially useful in forming immunotoxin conjugates for *in vivo* administration (Chapter 20, Section 3.2.1). A water soluble analog of this crosslinker containing an extended spacer arm is also commercially available as Sulfo-LC-SMPT (Thermo Fisher).

SMPT or Sulfo-LC-SMPT may be used as thiolation reagents by first reacting its NHS ester end with an amine-containing molecule and then releasing the pyridine-2-thione leaving group with DTT to free the sulfhydryl (Figure 2.67). The disadvantage of this approach is the necessity of using a reducing agent to create the –SH group modification. This method of thiolation should only be used if there are no disulfides in the target molecule that are critical to function. If a reductant cannot be used, choose a thiolation method that does not need DTT treatment, such as the use of Traut's reagent or SATA (Section 4.1, this chapter).

Since SMPT is not soluble in aqueous solutions it must first be dissolved in organic solvent and an aliquot of this stock solution transferred to the reaction solution. The reagent is soluble in DMF and DMSO, but is much more stable in solutions of acetonitrile. A stock solution of SMPT in acetonitrile may be kept frozen without loss of activity. The NHS ester of SMPT is also extraordinarily stable to hydrolysis in water. Even when an SMPT/acetonitrile aliquot is added to an aqueous solution and stored at room temperature, SMPT will only lose about 5% of its activity after 16h. By contrast, other NHS esters usually have half-lives measured in minutes or hours in aqueous environments, depending on the pH.

Sulfo-LC-SMPT is not as stable as SMPT. The sulfo-NHS ester is more susceptible to hydrolysis in aqueous solutions and the pyridyl disulfide group is more easily reduced to the free sulfhydryl. Stock solutions of Sulfo-LC-SMPT may be prepared in water, but should be used immediately to prevent loss of amine coupling ability.

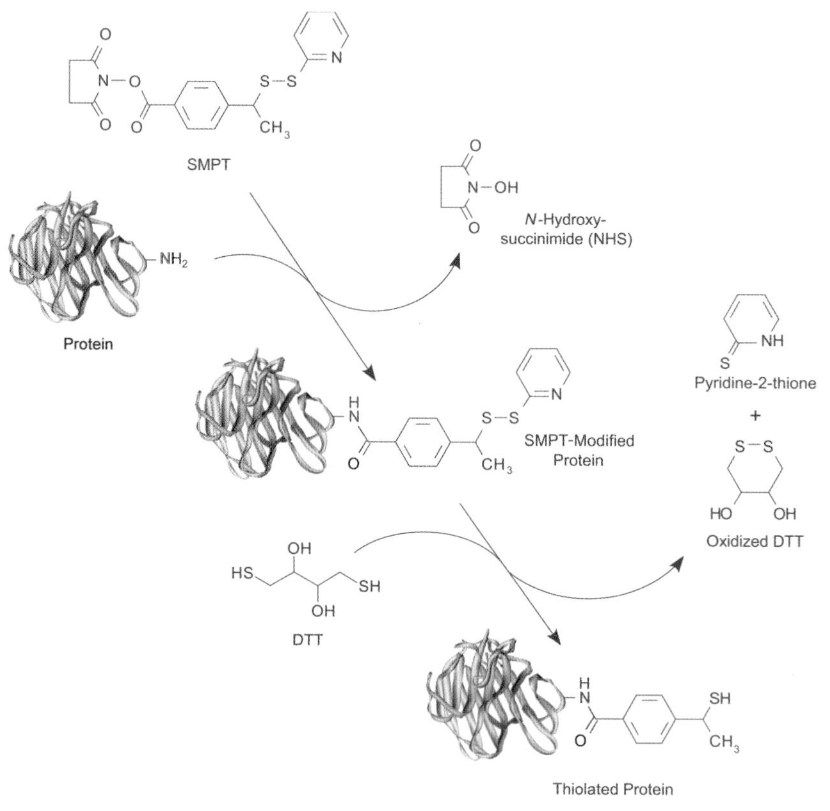

FIGURE 2.67 SMPT can be used to modify the amine groups of proteins to form disulfide intermediates. The disulfides can be reduced with DTT to create free thiols for subsequent conjugation purposes.

Protocol

1. Dissolve the protein or macromolecule to be thiolated at a concentration of 10 mg/ml in 50-m*M* sodium phosphate, 0.15-*M* NaCl, pH 7.2. Other non-amine-containing buffers such as borate, HEPES, and bicarbonate may also be used as the reaction medium. The effective pH for the NHS ester modification reaction is in the range of 7.0 to 9.0, but conditions close to neutrality will limit ester hydrolysis.
2. Dissolve SMPT (Thermo Fisher) at a concentration of 7.7 mg/ml in acetonitrile (makes a 20-m*M* stock solution). Alternatively, the water soluble Sulfo-LC-SMPT may be used and dissolved at a concentration of 6.0 mg/ml in water (makes a 10-m*M* solution). This should be carried out just prior to adding an aliquot to the thiolation reaction. Since an aqueous solution of the crosslinker will degrade by hydrolysis of the sulfo-NHS ester, it should be used quickly to prevent significant loss of activity. If a sufficiently large amount of protein will be modified to allow accurate weighing of Sulfo-LC-SMPT, the solid may be added directly to the reaction mixture without preparing a stock solution in water, but this is not recommended with most reactions.
3. Add 25 µl of the stock solution of SMPT in acetonitrile to each milliliter of the protein to be modified. If Sulfo-LC-SMPT is used, add 50 µl of the stock solution in water to each milliliter of protein solution.
4. Mix and react for at least 30 min at room temperature. Longer reaction times, even overnight, will not adversely affect the modification.
5. Purify the modified protein from reaction byproducts bydialysis or gel filtration using 50-m*M* sodium phosphate, 0.15-*M* NaCl, pH 7.2.
6. To release the pyridine-2-thione leaving group and form the free sulfhydryl, add DTT at a concentration of 0.5 mg DTT per mg of modified protein. A stock solution of DTT may be prepared to make it easier to add it to a small amount of protein solution. In this case, dissolve 20 mg of DTT per milliliter of 0.1-*M* sodium acetate, 0.1-*M* NaCl, pH 4.5. Add 25 µl of this solution per mg of modified protein. Release of pyridine-2-thione can be followed by its characteristic absorbance at 343 nm ($\varepsilon = 8.08 \times 10^3 M^{-1} cm^{-1}$).
7. Mix and react at room temperature for 30 min.
8. Purify the thiolated protein from excess DTT by dialysis or gel filtration using 50-m*M* sodium phosphate, 0.15-*M* NaCl, 1-m*M* EDTA, pH 7.2. The modified protein should be used immediately in a conjugation reaction to prevent sulfhydryl oxidation and formation of disulfide crosslinks.

Modification of Amines with N-Acetyl Homocysteine Thiolactone

N-Acetyl homocysteine thiolactone (also called citiolone or 2-acetamido-4-mercaptobutyric acid) is a cyclic derivative of homocysteine containing a blocked α-amino group. The compound can react with primary amines in a ring opening reaction to create free sulfhydryl modifications (Figure 2.68). It was originally used as a reagent for insolubilizing antibodies. Later, it was immobilized on an amine-containing matrix to form a disulfide reducing support for cleaving cystine residues in peptides and proteins (Eldjarn and Jellum, 1963; Jellum, 1964) (see *Use of Disulfide Reductants*, this section). The thiolation reaction of amine-containing macromolecules proceeds much like the reaction for 2-iminothiolane. Nucleophilic attack occurs at the carbonyl, cleaving the thiolactone and producing an amide linkage with the target molecule, while at the same time creating the free sulfhydryl (Benesch and Benesch, 1956, 1958). *N*-Acetyl homocysteine is soluble in aqueous buffers.

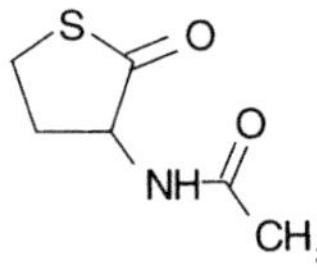

N-Acetyl Homocysteine Thiolactione
MW 159

Thiolation of peptides and other small molecules containing amines proceeds easily with *N*-acetyl

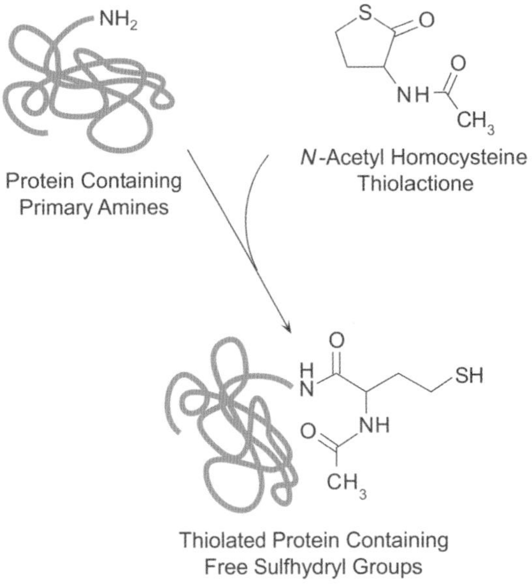

FIGURE 2.68 *N*-Acetyl homocysteine thiolactone spontaneously reacts with amine groups on proteins to create sulfhydryl groups.

homocysteine thiolactone. However, protein modification often results in much lower yields unless the reaction is carried out for extended periods at pH 10 to 11.

It has been found that silver ions catalyze the thiolation process with proteins, allowing the reaction to be completed rapidly at physiological pH (Benesch and Benesch, 1958). The addition of an equal molar concentration of $AgNO_3$ forms an insoluble complex with the thiolactone, and this in turn reacts with protein amines.

Protocol

1. Dissolve the amine-containing molecule to be thiolated at a concentration of 10 mg/ml in cold (4°C) 1-*M* sodium bicarbonate (reaction buffer). For proteins, dissolve them in deionized water at a pH of 7.0 to 7.5, at room temperature. Note: The presence of some buffer salts, such as phosphate or carbonate, is incompatible with silver nitrate.
2. Add *N*-acetyl homocysteine thiolactone (Aldrich) to the bicarbonate reaction mixture to obtain a concentration representing a 10- to 20-fold excess over the amount of amines present. For protein thiolation, add the same molar excess of thiolactone reagent to the water reaction medium, and then slowly add an equivalent molar quantity of silver nitrate ($AgNO_3$). Maintain the pH at 7.0 to 7.5 with periodic addition of NaOH.
3. For the bicarbonate reaction, gently mix for 20 h at 4°C. For the silver-catalyzed reaction, continue the reaction for 1 h or until the silver complex has fully dissolved.
4. To remove the silver mercaptide formed from the facilitated protein thiolation reaction, add an excess of thiourea to convert all the silver into a soluble $Ag(thiourea)_2^+$ complex and free the sulfhydryl modifications.
5. Remove unreacted *N*-acetyl homocysteine thiolactone and reaction byproducts by gel filtration or dialysis against 10-m*M* sodium phosphate, 0.15-*M* NaCl, 10-m*M* EDTA, pH 7.2. Other buffers suitable for individual protein stability may be used as desired. For the silver nitrate-containing reaction, removal of the silver–thiourea complex may be carried out by adsorption onto Dowex 50, and the protein subsequently eluted from the resin by 1-*M* thiourea. Removal of the thiourea may then be carried out by gel filtration or dialysis.

Including EDTA in the final preparation inhibits metal-catalyzed oxidation of the sulfhydryl groups to disulfides. The modified peptide or protein should be used immediately to assure full sulfhydryl reactivity.

Modification of Amines with SAMSA

S-Acetylmercaptosuccinic anhydride, or SAMSA, is an amine-reactive reagent containing a protected sulfhydryl much like SATA, described previously. The anhydride portion opens in response to the attack of an amine nucleophile, yielding an amide linkage (Klotz *et al.*, 1962; Weston *et al.*, 1980). The ring opening reaction, however, does produce a free carboxylate group that lends a negative charge to the modified molecule where once there may have been a positive charge (Figure 2.69). This charge reversal may affect the conformation and activity of some sensitive proteins. After the initial modification step, releasing the acetylated sulfhydryl-protecting group with hydroxylamine forms the thiolated derivative.

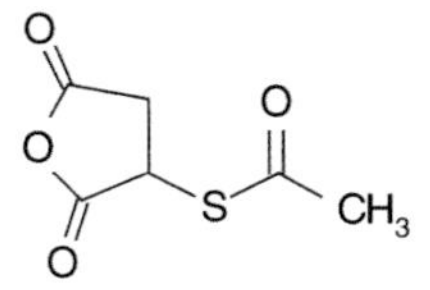

SAMSA;
S-Acetylmercaptosuccinic
Anhydride
MW 174

Protocol

1. Dissolve the protein or other amine-containing macromolecule in 0.1-*M* sodium phosphate, 0.15-*M* NaCl, pH 7.5, at a concentration of 5 mg/ml.
2. Dissolve SAMSA in DMF at a concentration of 25 mg/ml.
3. Add 20 µl of the stock SAMSA solution to each milliliter of the protein solution, with mixing.
4. React at room temperature for 30 min.
5. Remove excess reagent and reaction byproducts by dialysis or gel filtration using 0.1-*M* sodium phosphate, 0.15-*M* NaCl, 10-m*M* EDTA, pH 7.5. For chromatographic separation, use a desalting gel filtration support such as the Zeba desalting spin columns (Thermo Fisher) or the equivalent. The SAMSA-modified protein may be stored at −20°C until needed.
6. To deprotect the acetylated sulfhydryl group of SAMSA-modified proteins, add 100 µl of 0.5-*M* hydroxylamine hydrochloride in 50-m*M* sodium phosphate, 25-m*M* EDTA, pH 7.5, to each milliliter of protein solution.
7. Mix and react for 2 h at room temperature.
8. Purify the sulfhydryl-modified protein by dialysis against 50-m*M* sodium phosphate, 1-m*M* EDTA, pH 7.5, or by gel filtration on a Sephadex G-25 column using the same buffer.

The deacetylated protein should be used immediately to prevent loss of sulfhydryl content through disulfide formation. The degree of –SH modification may be

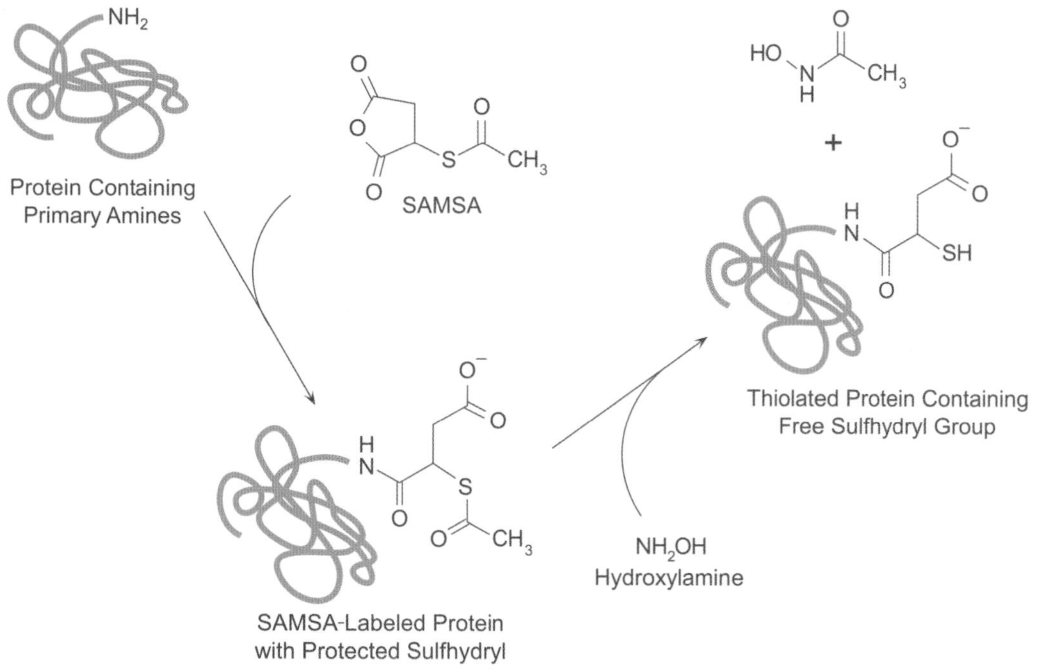

FIGURE 2.69 SAMSA is an anhydride compound containing a protected thiol. Reaction with protein amine groups yields amide bond linkages. Deprotection of the acetylated thiol produces free sulfhydryl groups for conjugation.

determined by performing an Ellman's assay (see *Ellman's Assay for the Determination of Sulfhydryls*, this chapter).

Modification of Aldehydes or Ketones with AMBH

AMBH (2-acetamido-4-mercaptobutyric acid hydrazide) is a unique hydrazide derivative that can thiolate aldehydes and ketones to form reactive sulfhydryl groups (Taylor and Wu, 1980). It is particularly useful in converting oxidized carbohydrates to contain a thiol. In this respect, glycoproteins or other carbohydrate and diol-containing molecules may be treated with sodium periodate under relatively mild conditions to form aldehyde residues (Section 4.4, this chapter). The aldehydes readily react with the hydrazide groups of AMBH to form hydrazone linkages, leaving a free terminal sulfhydryl residue to use in further conjugation reactions (Figure 2.70).

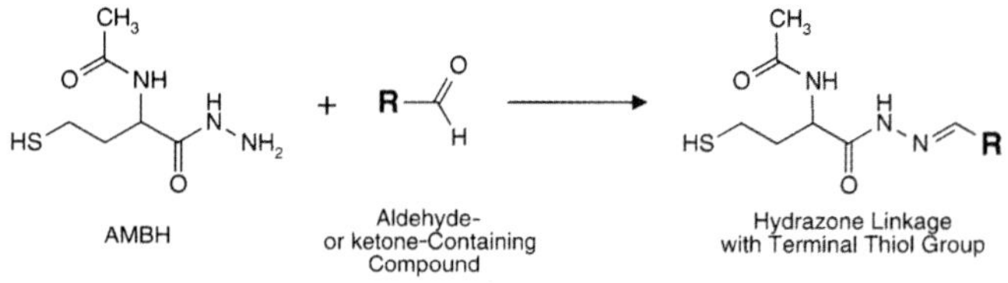

FIGURE 2.70 AMBH is a hydrazide-containing compound that reacts with carbonyl groups to form hydrazone bonds. The free thiol can be used for subsequent conjugation reactions.

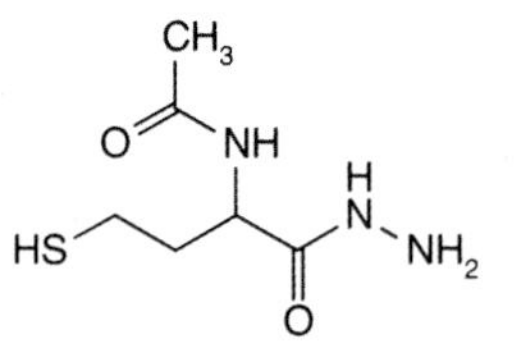

AMBH
2-Acetamido-4-mercaptobutyric acid hydrazide
MW 191

Protocol

1. Dissolve an aldehyde-containing macromolecule to be modified (e.g., a periodate oxidized glycoprotein) in 0.01-*M* sodium phosphate, 0.15-*M* NaCl, pH 7.4, containing 1-m*M* EDTA. A suitable concentration range for a protein is 1–10 mg/ml.
2. Add a 10-fold molar excess of AMBH (pre-dissolved in ethanol) (Molecular Probes) over the expected amounts of aldehydes to be modified.
3. React for 2 h at room temperature.
4. Purify the modified protein by gel filtration.

Modification of Carboxylates or Phosphates with Cystamine

Cystamine is decarboxylated cystine [or 2,2′-dithiobis(ethylamine)], a small disulfide-containing molecule with primary amines at both ends. This versatile reagent can be used in several conjugation techniques. Cystamine may be used to introduce sulfhydryl residues in proteins, nucleic acids and other molecules, or as the active species in disulfide exchange crosslinking reactions, or in reversible conjugation procedures. The reagent can be used to create sulfhydryl groups in proteins or other molecules by first conjugating one of its terminal amino groups with the carboxylates on a target molecule using a carbodiimide reaction (Chapter 3, Section 1.12 and Chapter 4, Section 1). Subsequent reduction of the disulfide group liberates the free sulfhydryl (see *Ellman's Assay for the Determination of Sulfhydryls*, this section) (Figure 2.71). This same modification procedure can also be used to introduce sulfhydryl residues at the 5′ phosphate group of DNA (Chu *et al.*, 1986; Ghosh *et al.*, 1990). The carbodiimide activates the phosphate and the amines of cystamine may then react with this active species to form a phosphoramidate bond (Chapter 23, Section 2.2) (Figure 2.72). Specific labeling of DNA probes only at the 5′ end is possible using this technique.

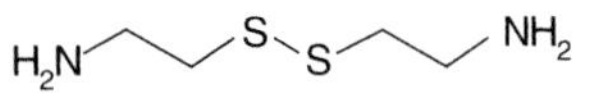

Cystamine;
2,2'-dithiobis(ethylamine)
MW 152

The carbodiimide of choice used to couple cystamine to carboxylate- or phosphate-containing molecules is most often the water soluble carbodiimide, EDC (1-ethyl-3-(3-dimethylaminopropyl)carbodiimide hydrochloride; Chapter 4, Section 1.1). This reagent rapidly reacts with carboxylates or phosphates to form an active ester intermediate, which is highly reactive toward primary amines. The reaction is efficient from pH 4.7 to 7.5, and a variety of buffers may be used, providing they do not contain competing groups.

Cystamine is also used as an activating reagent for disulfide exchange reactions. In this procedure, the reagent is used to modify one of two proteins to be conjugated. The cystamine-modified protein is then mixed with the other protein that contains, or is thiolated to contain, a sulfhydryl group. By disulfide exchange, the sulfhydryl-containing molecule cleaves the disulfide of the cystamine-modified protein, releasing 2-mercaptoethylamine and forming a disulfide crosslink (Figure 2.73).

Using this approach, EGF has been successfully conjugated by disulfide exchange to the A chain of diphtheria toxin (Shimisu *et al.*, 1980). A cystaminyl derivative of insulin could also be conjugated to the A chain of diphtheria toxin by this method (Miskimins and Shimizu, 1979). Other references to disulfide exchange using cystamine include Oeltmann and Forbes (1981) and Bacha *et al.* (1983), who prepared antibody–toxin and peptide–toxin conjugates, respectively.

Finally, cystamine may be used to conjugate two macromolecules through its terminal amine groups. In this case, the internal disulfide bridge remains intact, forming a reversible conjugate of the two molecules through reduction of the disulfide bond. Using this approach, the first molecule is modified with cystamine by use of the EDC reaction. A second molecule is then reacted with the free amines of cystamine on the

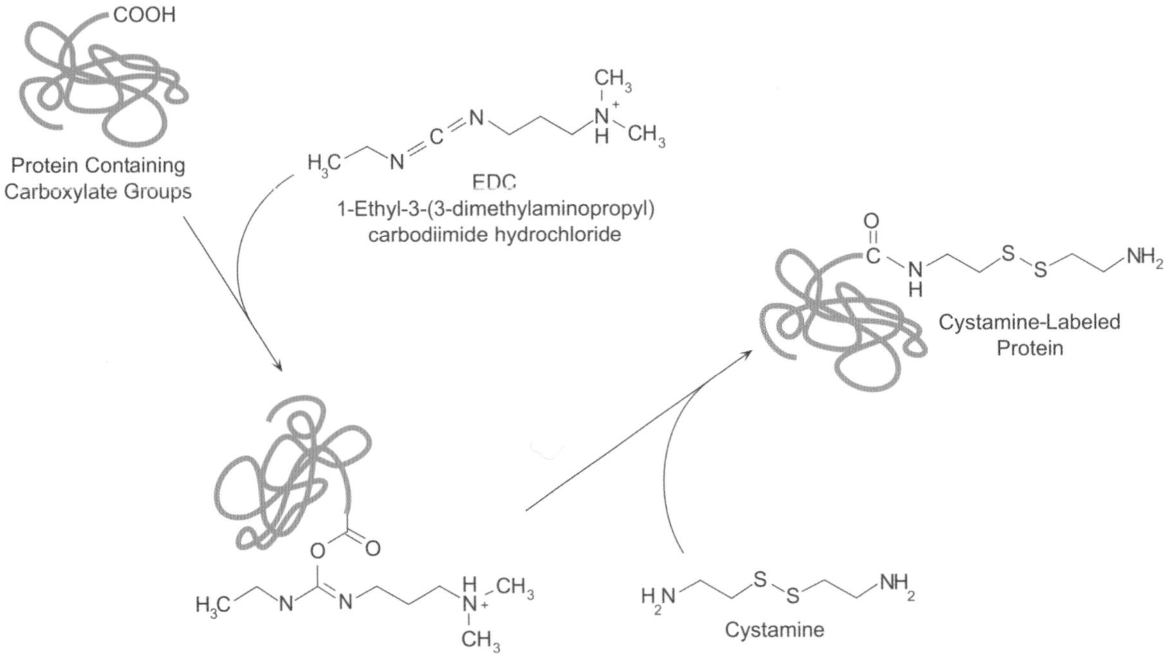

FIGURE 2.71 Cystamine may be used to label protein carboxylate groups using the water soluble carbodiimide EDC.

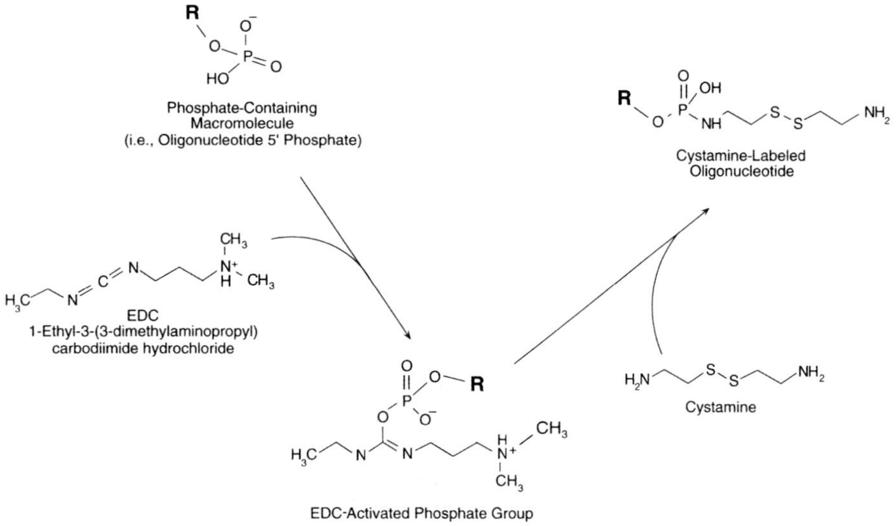

FIGURE 2.72 Cystamine may be used to label phosphate groups, such as at the 5′ end of oligonucleotides, *via* a carbodiimide reaction using EDC. The resultant phosphoramidate linkage is a common way to modify oligonucleotides at the 5′ end.

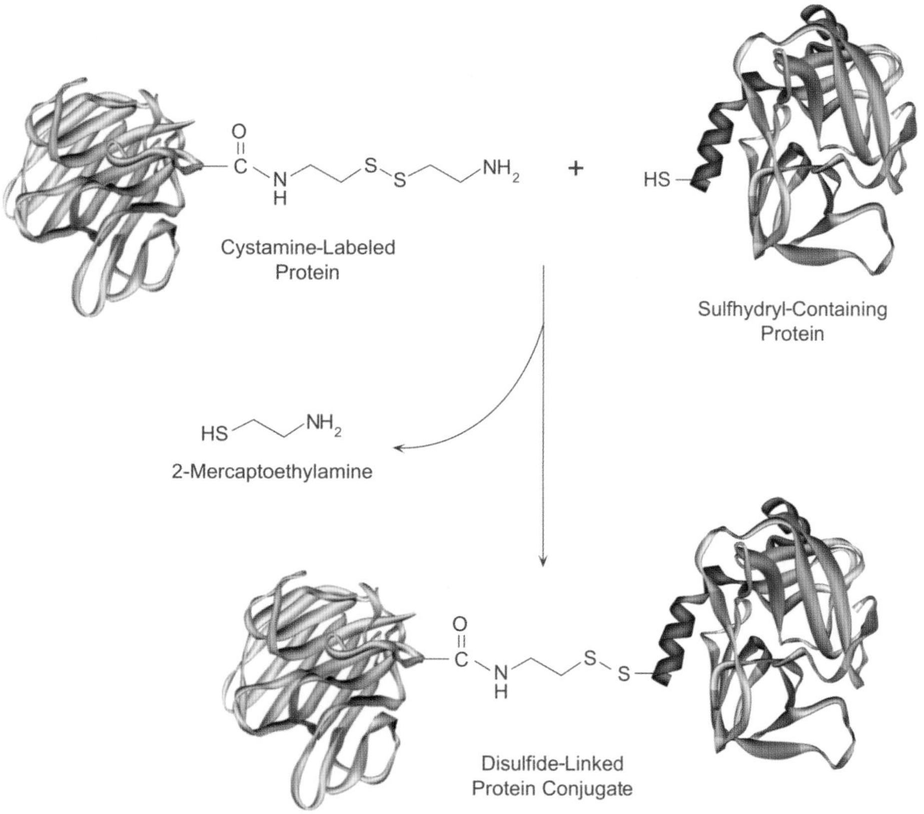

FIGURE 2.73 The disulfide group of a cystamine-modified protein may undergo disulfide interchange reactions with another sulfhydryl-containing protein to yield a disulfide-linked conjugate.

first molecule by use of an amine-reactive chemistry. Typically, this reaction scheme is used if the first molecule initially contains no reactive amines and the second molecule is often an amine-reactive fluorescent tag or other probe. For instance, DNA probes may be cystamine-modified through their 5′ phosphate group using this method and amine-reactive biotin labels subsequently attached. The biotin label is then reversible by virtue of the cystamine cross-bridge through simple disulfide reduction (Chapter 23, Section 2.2).

Modification of Proteins with Cystamine

The following protocol is useful for the modification of proteins with cystamine with subsequent reduction to create the free sulfhydryl.

Protocol

1. Dissolve the protein to be modified at a concentration of 10 mg/ml in a buffer having a pH between 4.7 and 7.5. Avoid buffers or other components containing competing groups to the carbodiimide reaction (i.e., carboxylates or amines). For the lower pH conditions, 0.1-*M* MES, pH 4.7 works best. For a physiological pH environment, 0.1-*M* sodium phosphate, 0.15-*M* NaCl, pH 7.2, will also give good incorporation of cystamine. For other concentrations of protein in solution, proportionally adjust the amount of reagents added.
2. Dissolve cystamine (Aldrich) in the reaction buffer at a concentration of 2.25 mg/ml (10-m*M*). Add an aliquot of this solution to the protein solution to be modified. Use about a 10- to 20-fold molar excess of cystamine over the amount of protein present. For a protein of MW 100,000 at a concentration of 10 mg/ml, add 10 µl of the stock cystamine solution to each milliliter of protein solution to obtain a 10-fold molar excess.
3. Add EDC (Thermo Fisher) to the solution prepared in (2) to obtain at least a 5-fold molar excess over the amount of cystamine present. React for 2 h at room temperature.
4. Separate excess cystamine and EDC (and reaction byproducts) from the modified protein by dialysis or gel filtration using 10-m*M* sodium phosphate, 0.15-*M* NaCl, pH 7.2. A desalting column may be used for the gel filtration procedure (e.g., Zeba spin columns from Thermo Fisher).
5. To reduce the disulfide groups, add DTT at a concentration of 0.5 mg DTT per mg of modified protein. A stock solution of DTT may be prepared to make it easier to add it to a small amount of protein solution. In this case, dissolve 20 mg of DTT per milliliter of 0.1-*M* sodium acetate, 0.1-*M* NaCl, pH 4.5. Add 25 µl of this solution per mg of modified protein.
6. Mix and react at room temperature for 30 min.
7. Purify the thiolated protein from excess DTT by dialysis or gel filtration using 50-m*M* sodium phosphate, 0.15-*M* NaCl, 1-m*M* EDTA, pH 7.2. The modified protein should be used immediately in a conjugation reaction to prevent sulfhydryl oxidation and formation of disulfide crosslinks.

Modification of Nucleic Acids and Oligonuleotides with Cystamine

DNA or RNA also may be modified with cystamine at the 5′ phosphate group using a carbodiimide reaction. See Chapter 23, Section 2.2, for a complete discussion of the labeling protocol.

Use of Disulfide Reductants

One of the most convenient ways of generating sulfhydryl groups is by reduction of indigenous disulfides. Many proteins contain cystine disulfides that are not critical to structure or activity. In some cases, mild reducing conditions can free one or more –SH groups for conjugation or modification purposes. The creation of free sulfhydryls in this manner allows for site-directed modification at a limited number of locations within the protein molecule.

This method of creating sulfhydryls for conjugation purposes should be avoided, however, if the indigenous disulfides are important for maintaining native structure and activity. Disulfides are often the point of attachment for subunits within a protein molecule. The cystine bonds may be crucial for maintaining quaternary integrity. Reduction may cause a protein to break up into two or more subunits with little or no remaining activity. Disulfides may also be critical for retention of ligand binding activity. Deformation of an active site may occur if important disulfides are reduced. In these cases, the best mode of thiolation is through the use of a reagent system that does not require a disulfide reducing agent, such as 2-iminothiolane or SATA (see previous discussion, this section).

Occasionally, even a protein containing critical disulfides can be partially reduced to yield a useful thiolated derivative. IgG molecules contain disulfide groups that hold together the two heavy chains as well as disulfides holding the light chain–heavy chain pairs together. Selective reduction of some or all of the hinge region disulfides between the heavy chains can result in a divalent or even a monovalent antibody molecule that still maintains its antigen binding capability. Reductants such as DTT, 2-mercaptoethylamine, 2-mercaptoethanol, or tris(2-carboxyethyl)phosphine (TCEP) in a non-denaturing environment can be used at low concentrations to perform this type of partial cleavage. The thiol-containing antibody that is generated can then be successfully conjugated with enzymes or other molecules through the sulfhydryl residue(s) in the exposed hinge region (Chapter 20, Section 1.1).

Disulfide reductants are also used to investigate protein structural properties. In this case, retention of activity is not the critical issue, but complete reduction of all disulfides is paramount. The standard method of carrying out protein subunit molecular weight determinations by SDS polyacrylamide gel electrophoresis often depends on complete disulfide reduction. When total reduction is necessary, the reductants must also contain a deforming agent to unfold protein tertiary structure. This is typically done by including high concentrations of denaturants such as 8-*M* urea or guanidine or detergents such as SDS. Under severely deforming conditions, proteins unfold, exposing internal disulfides to the reducing agent. Without these added reagents to deform native protein structure, many buried disulfides would remain unaffected by the reductants.

The following reducing agents represent the most popular options for cleaving disulfide bonds. Their properties and use vary widely. The decision as to which reagent is best is often governed by the molecule being reduced and the potential application. Careful review of these properties may sway the success or failure of a conjugation protocol.

CLELAND'S REAGENT: DTT AND DTE

Dithiothreitol (DTT) and dithioerythritol (DTE) are the *trans* and *cis* isomers of the compound 2,3-dihydroxy-1,4-dithiolbutane. The reducing potential of these versatile reagents was first described by Cleland in 1964. Due to their low redox potential (−0.33 V) they are able to reduce virtually all accessible biological disulfides and maintain free thiols in solution despite the presence of oxygen. The compounds are fully water soluble with very little of the offensive odor of the 2-mercaptoethanol they were meant to replace. Since Cleland's original report, literally thousands of references have cited the use of mainly DTT for the reduction of cystine and other forms of disulfides.

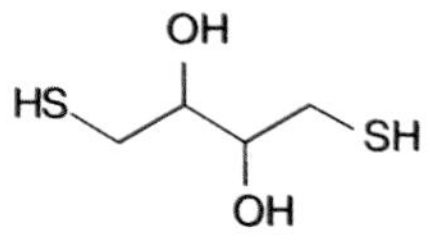

DTT;
Dithiothreitol
MW 154.25

The unique characteristics of DTT and DTE are mainly reflected in their ability to form intramolecular ring structures upon oxidation. Disulfide reductants such as 2-mercaptoethanol, 2-mercaptoethylamine, glutathione, thioglycolate, and 2,3-dimercaptopropanol cleave disulfide bonds in a two-step reaction that involves the formation of a mixed disulfide (Figure 2.74). In the second stage of the reducing process, the mixed disulfide is cleaved by another molecule of reductant, freeing the sulfhydryl and forming a dimer of the reducing agent through the formation of an intermolecular disulfide bond. For simple reductants containing only one thiol, the equilibrium for disulfide exchange is nearly equivalent for the reductant and target protein. Thus, monothiol compounds are usually required in extreme excess to drive the reaction to completion.

The presence of two sulfhydryl groups in DTT and DTE, however, allows the formation of a favored cyclic disulfide during the course of target protein reduction (Figure 2.75). This drives the equilibrium toward the reduction of target disulfides. Therefore, complete reduction is possible with much lower concentrations of DTT or DTE than when using monothiol systems.

As with all reductants, DTT and DTE will reduce disulfides only if they are accessible. The three-dimensional structure of a protein molecule often contains disulfides buried deep in the inner structure of

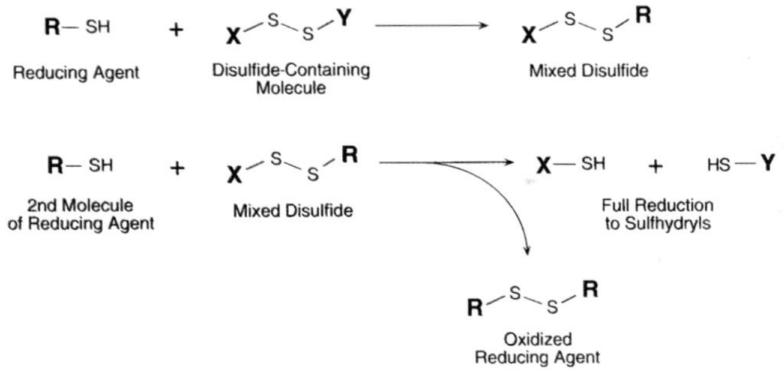

FIGURE 2.74 Thiol-containing disulfide reductants reduce disulfide groups through a multi-step process producing a mixed disulfide intermediate.

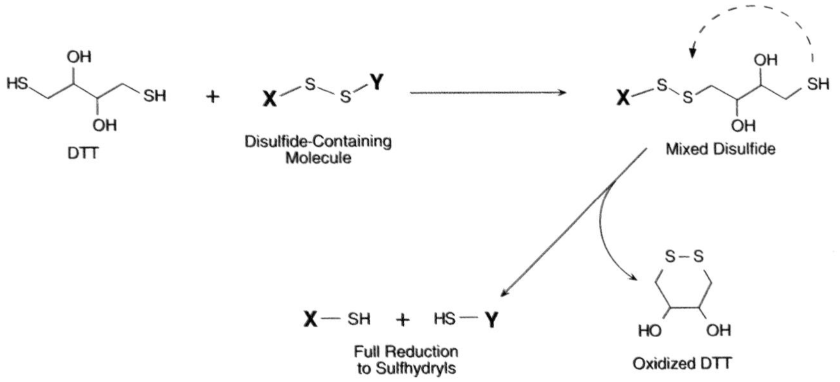

FIGURE 2.75 DTT is highly efficient at reducing disulfides, since a single molecule can reduce the intermediate mixed disulfide by forming a ring structure.

the polypeptide chains. A protein retaining its native conformation is frequently protected from complete reduction. In the absence of denaturants such as urea, guanidine, or SDS, DTT is not capable of reducing all available disulfides within some proteins (Bewley *et al.*, 1968; Bewley and Li, 1969). For instance, at moderate concentrations of DTT and no denaturants, limited cleavage of disulfides in antibody molecules can result in reducing mainly the bonds between the heavy chains of the immunoglobulin. This produces two half-antibody molecules, each containing one antigen binding site and free sulfhydryls in the hinge region. This limited reduction process can be used to site-direct sulfhydryl-reactive conjugation reagents away from the antigen binding sites, thus preserving activity (de Rosario *et al.*, 1990). However, using an appropriate concentration of deforming agents, DTT efficiently reduces all protein disulfides in the antibody and allows subunit separation for analysis (Konigsberg, 1972).

In a comparative study of disulfide reducing agents, it was determined that use of the relatively strong reductants DTT and TCEP required only 3.25 and 2.75 mole equivalents per mole equivalent of antibody molecule to achieve the reduction of two interchain disulfide bonds between the heavy chains of a monoclonal IgG (Sun *et al.*, 2005). This limited reduction strategy retains intact bispecific antibody molecules while providing discrete sites for conjugation to thiols.

DTT may also be used to cleave disulfide containing modification and crosslinking reagents. For thiolation procedures, DTT may be used to remove a dithiopyridyl group or cleave other disulfides to produce a free sulfhydryl. In this case, the presence of a denaturant is usually not required to access and reduce the disulfide of the modification reagent. Similarly, disulfides of crosslinking agents may be reduced after two macromolecules have been conjugated to release them as desired. This technique is often used to analyze receptor–ligand interactions or to discover how two proteins associate *in vivo*.

COMPLETE REDUCTION OF DISULFIDES IN PROTEIN MOLECULES USING DTT

Protocol

1. Dissolve a disulfide-containing protein or peptide at a concentration of 1–10 mg/ml in 6-*M* guanidine hydrochloride, 0.01-*M* sodium phosphate, 0.15-*M* NaCl, pH 7.4. Alternative denaturant conditions may be used (e.g., 8-*M* urea or 2.3% (w/w) SDS) along with any other buffer salts and pH values desired. A pH between 7.0 and 8.1 usually works best.
2. Add DTT to a final concentration of 10–100-mm.
3. Incubate for 2 h at room temperature. For some buried disulfides to become exposed and fully reduced, it may be necessary to heat the solution (in a capped test tube) at 50°C for 30 min. Some procedures use a 2-min incubation in a boiling water bath to completely denature the protein.
4. For removal of excess DTT, a protein of molecular weight greater than 5000 may be isolated by gel filtration using a desalting column. To maintain the stability of the exposed sulfhydryl groups, include 1–10-m*M* EDTA in the chromatography buffer. The presence of oxidized DTT can be monitored during elution by measuring the absorbance at 280 nm. The protein should elute in the first peak and the DTT reaction products in the second peak.

USE OF DTT TO CLEAVE DISULFIDE-CONTAINING CROSSLINKING AGENTS

The following method may be used to reduce the disulfide bonds of some crosslinking agents, thus cleaving conjugated proteins. This procedure will reduce the pyridyl disulfide group of SPDP to create a thiolated species (see previous discussion in this section and Chapter 6, Section 1.1). It may also be used to partially reduce the indigenous disulfides in some protein molecules. In this regard, low concentrations of DTT under non-denaturing conditions have been used to selectively reduce the disulfides between the heavy chains of immunoglobulin G (Edelman *et al.*, 1968; Sun *et al.*, 2005). Without an added denaturant to open up the polypeptide chain, internally buried disulfides will typically remain unreduced.

Protocol

1. Dissolve a crosslinked protein or peptide that has been conjugated with the use of a disulfide-containing crosslinker at a concentration of 1–10 mg/ml in 0.01-*M* sodium phosphate, 0.15-*M* NaCl, pH 7.4. Alternative buffer conditions and pH values may be used; however, a pH between 7.0 and 8.1 usually works best.
2. Add DTT to a final concentration of 1–10-mm.
3. Incubate for 2 h at room temperature.
4. For removal of excess DTT, a protein of molecular weight greater than 5000 may be isolated by gel filtration using a desalting resin. To maintain the stability of the exposed sulfhydryl groups, include 10-m*M* EDTA in the chromatography buffer. The presence of oxidized DTT can be monitored during elution by measuring the absorbance at 280 nm. The protein should elute in the first peak and the DTT reaction products in the second peak.

2-MERCAPTOETHANOL

2-Mercaptoethanol is one of the most common agents used for disulfide reduction. Sometimes referred to as β-mercaptoethanol, it is a clear, colorless liquid with an extremely strong odor. All operations with this chemical should be performed in a well-ventilated fume hood. The reduction of protein disulfides with 2-mercaptoethanol proceeds rapidly *via* a two-step process involving an intermediate mixed disulfide (Figure 2.76). Due to its strong reducing properties, the reagent is used most often when complete disulfide reduction is required. It can also be used to cleave disulfide-containing crosslinking agents. Usually a concentration of 0.1-*M* 2-mercaptoethanol will cleave a disulfide-containing crosslinker and liberate conjugated proteins (Chapter 9, Section 1).

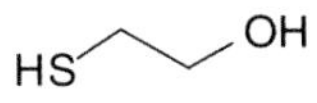

2-ME;
2-Mercaptoethanol
MW 78.13

2-Mercaptoethanol is used as a reducing additive in a number of biochemical reagents. It is used as a reductant for a Gram-negative bacteria lysis buffer (Schwinghamer, 1980; Scopes, 1982), as the second dimensional equilibration buffer for 2-D electrophoresis (Dunbar, 1987), as the sample reducing buffer for SDS polyacrylamide gel electrophoresis (Laemmli, 1970), and as a participant in the *o*-phthalaldehyde (OPA) reaction for the detection of primary amines (Jones and Gilligan, 1983).

PROTOCOL FOR PREPARATION AND USE OF A GRAM-NEGATIVE BACTERIA LYSIS BUFFER

1. Prepare a solution consisting of 2.5 ml glycerol, 100 μl of 10% Triton X-100 (Thermo Fisher Surfact-Amps X-100), and 10 μl 2-mercaptoethanol.
2. Add 10 g of wet packed cells to the lysis buffer and stir vigorously for 30 min.
3. Add 30 ml of an extraction buffer consisting of 20-m*M* potassium phosphate, pH 7.0, 1-m*M* EDTA, 0.2 mg/ml lysozyme, and 10 μg/ml DNase I.
4. Add 5 mg PMSF dissolved in 0.5 ml acetone and 0.1 mg pepstatin A.
5. Centrifuge for 20 min at 15,000 *g*. Recover the extracted, solubilized material in the supernatant.

PROTOCOL FOR PREPARATION AND USE OF THE SECOND-DIMENSION EQUILIBRATION BUFFER FOR 2-D GELS The following procedure relates to electrophoretic protocols where the first dimension is developed by isoelectric focusing (in tube gels) and the second dimension is a size-exclusion separation by SDS polyacrylamide electrophoresis in a slab gel.

1. Add 4.0 g SDS and 20 ml of 10% glycerol to 150 ml of 0.125-*M* Tris, pH 6.8, and adjust the final volume to 200 ml. Once dissolved, add a few crystals of bromophenol blue, mix, and pass the solution through a 0.2-μm filter. For storage, freeze in 10- to 15-ml aliquots.
2. Immediately before use, add 2-mercaptoethanol to a final concentration of 0.5–0.8%.
3. Incubate the first dimensional electrophoresis tube gel in this reducing buffer for 15 min. Drain off excess buffer and electrophorese in the second dimension.

SDS SAMPLE BUFFER FOR RUNNING ELECTROPHORESIS SIZE SEPARATIONS UNDER REDUCING CONDITIONS

1. Dissolve 2.0 g of SDS, 0.75 g Tris base, and 10 ml of glycerol in 90 ml of water. Adjust the pH to 6.8 and bring the final volume to 100 ml.
2. To a small aliquot of the above buffer, add 2-mercaptoethanol to obtain a final concentration of 2–5%. Only 200 μl of this buffer typically is required to treat and reduce about 10–500 μg of protein. Solubilize the protein sample in this buffer.
3. Incubate in a sealed tube at 95°C for 5–10 min or in a boiling water bath for 1–2 min. Electrophorese immediately.

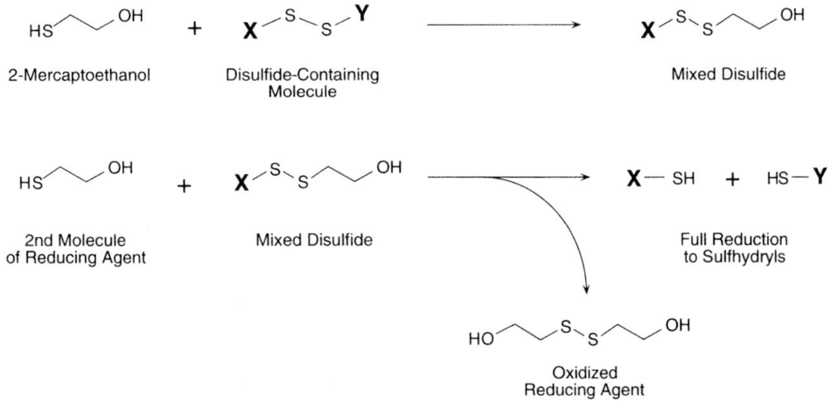

FIGURE 2.76 The reduction of disulfides by 2-mercaptoethanol proceeds through a mixed disulfide intermediate.

O-PHTHALALDEHYDE SOLUTION FOR THE FLUORESCENT DETECTION OF PRIMARY AMINES (SEE SECTION 4.3, OPA)

1. Add 3 ml of the detergent Brij-35 (as a 30% solution) and 2 ml of 2-mercaptoethanol to 950 ml of Fluoraldehyde Reagent Diluent (all reagents from Thermo Fisher).
2. Dissolve 0.5–0.8 g of *o*-phthalaldehyde (OPA) crystals in about 10 ml of methanol.
3. Mix the OPA solution with the solution from (1) and store under nitrogen in sealed glass bottles at 4°C. The addition of an aliquot of this solution to a sample containing primary amines will yield an intense blue fluorescence.

2-MERCAPTOETHYLAMINE

2-Mercaptoethylamine (also called aminoethanethiol) is a disulfide reducing agent that has found widespread application in the partial reduction of immunoglobulin molecules. The reagent is supplied as a solid in the hydrochloride form (Thermo Fisher) and possesses very little of the sulfhydryl odor of 2-mercaptoethanol. When used under non-denaturing conditions, 2-mercaptoethylamine can cleave the disulfide bonds between the heavy chains of IgG. This directed reduction is important for generating sulfhydryls while preserving antigen binding activity.

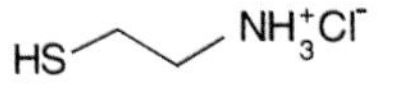

2-MEA;
2-Mercaptoethylamine Hydrochloride
MW 113.62

The complex structure of an antibody molecule creates two antigen binding sites from the interaction of the hypervariable regions on both the heavy and light chains. For this reason, heavy–light chain pairing must remain intact during any modification procedure to ensure that antigen binding activity is retained. In addition, it is important that any chemistry take place away from the antigen binding sites so they are not sterically blocked by modification reagents or by subsequent conjugation steps. 2-Mercaptoethylamine can be used to cleave disulfides primarily in the hinge region of IgG—away from the antigen binding sites—thus preserving the disulfides that hold the heavy and light chains together (Yoshitake *et al.*, 1979). It can also be used to reduce $F(ab')_2$ fragments, because they still retain the hinge region disulfides of intact IgG (Figure 2.77).

Once reduced with 2-mercaptoethylamine, immunoglobulins will often be cleaved in half, forming two heavy chain–light chain molecules of MW 75–80,000 and each containing one antigen binding site. These half molecules of IgG will possess reactive sulfhydryls in the hinge region that can be used in conjugation protocols with sulfhydryl-reactive crosslinking reagents. For instance, a reduced antibody may be used to make a conjugate with a maleimide-activated enzyme, forming a reagent useful in immunoassays (Chapter 20, Section 1.1). Similarly, $F(ab')_2$ fragments may be reduced to yield two molecules, each containing an antigen binding site. Making conjugates with this low molecular weight fragment can dramatically

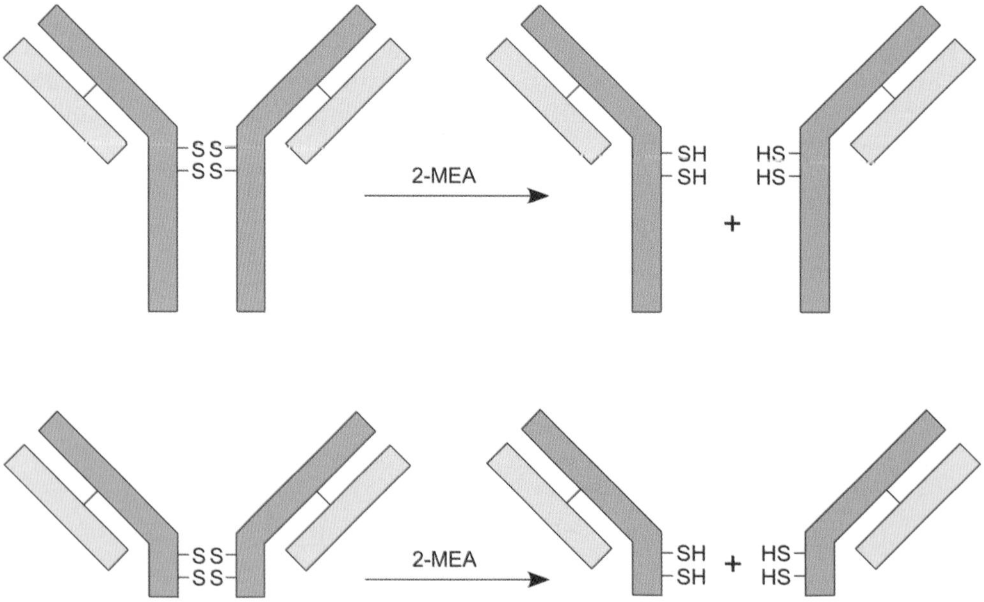

FIGURE 2.77 Disulfide reducing agents such as 2-mercaptoethylamine can be used to cleave the disulfide bonds in the hinge region of antibody molecules. Either intact IgG molecules or $F(ab')_2$ fragments may be reduced in this manner to yield monofunctional antigen binding fragments.

reduce background in assay systems or provide access to antigens restricted to higher molecular weight conjugates made with intact antibody (such as in immunohistochemical staining techniques).

The use of a 500-fold molar excess of 2-mercaptoethylamine over the concentration of antibody presence was found to result in a partially reduced antibody in which two disulfides were reduced to yield four thiols (Sun *et al.*, 2005). This strategy can be used to retain a biospecific antibody construct for subsequent discrete conjugation at the hinge region between the heavy chains.

Protocol

1. Dissolve the antibody to be reduced at a concentration of 10 mg/ml in 20-m*M* sodium phosphate, 0.15-*M* NaCl, pH 7.4, containing 1–10-m*M* EDTA.
2. To each milliliter of the antibody solution, add 6 mg of 2-mercaptoethylamine hydrochloride (final concentration is 50-m*M*). Mix to dissolve. Alternatively, to limit the degree of disulfide reduction, add a 500-fold molar excess of 2-mercaptoethylamine over the concentration of antibody present.
3. Incubate the solution in a sealed tube for 90 min at 37°C.
4. Purify the reduced IgG from excess 2-mercaptoethylamine and reaction byproducts by dialysis or gel filtration using a desalting resin. All buffers should contain 1–10-m*M* EDTA to preserve the free sulfhydryls from metal-catalyzed oxidation. The sulfhydryl-containing antibody may now be used in conjugation protocols that use –SH-reactive heterobifunctional crosslinkers (Chapter 6, Section 1).

TCEP

The reduction of disulfide bonds with trivalent phosphines has been known for some time (Levison *et al.*, 1969; Ruegg *et al.*, 1977; Kirley, 1989). Unfortunately, trialkylphosphines are generally water insoluble, undergo autoxidation, and are extremely odious.

To overcome these issues, the water soluble tris(2-carboxyethyl)phosphine (TCEP) was synthesized and successfully used to cleave organic disulfides to sulfhydryls in water (Burns *et al.*, 1991). The advantage of using this phosphine derivative in disulfide reduction as opposed to previous ones is its excellent stability in aqueous solution, its lack of reactivity with other common functionalities in biomolecules, and its freedom from odor.

The reaction of TCEP with biological disulfides proceeds with initial cleavage of the S–S bond followed by oxidation of the phosphine (Figure 2.78). The stability of

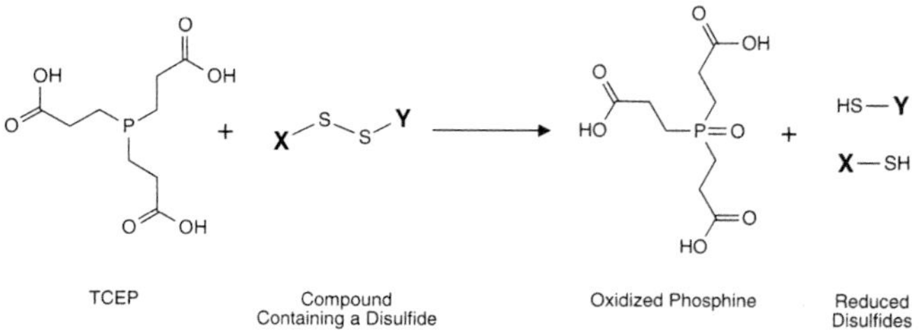

FIGURE 2.78 TCEP reduction of disulfides proceeds without the use of thiol compounds.

the phosphine oxide bond that is formed in this process is great enough to prevent reversal of the reaction. Since this reaction is performed without any added –SH compounds, subsequent conjugation with the generated sulfhydryl groups can be carried out without removal of excess TCEP or reaction byproducts (provided the conjugation step does not involve disulfide exchange reactions, such as with the active disulfide-containing reagent SPDP; Chapter 6, Section 1.1).

Although TCEP is capable of rapidly and quantitatively reducing simple organic disulfides in solution, it requires the presence of a deforming agent to fully reduce all disulfides in proteins. Without opening up the internal disulfides in many protein molecules, TCEP will not be able to reduce them. For complete reduction of IgG, it was found that 20-m*M* TCEP and 5 min of boiling were needed (Hines, 1992). Partial reduction, however, is possible for some more accessible disulfides in protein using aqueous buffers at room temperature. For instance, the use of a 2.75-fold molar excess of TCEP over the concentration of a monoclonal IgG resulted in the reduction of only two disulfide bonds in the hinge region of the antibody, leaving all other disulfides intact (Sun *et al.*, 2005).

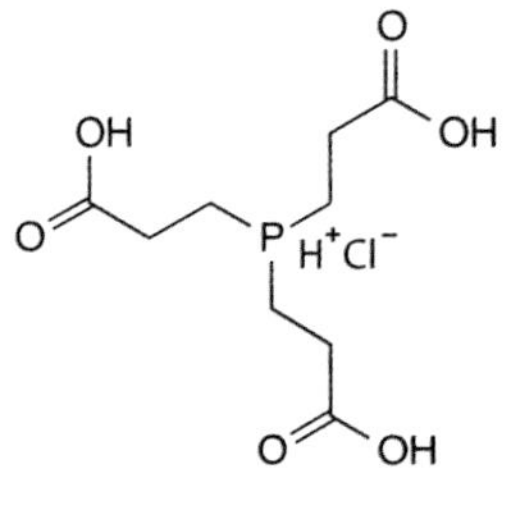

TCEP
Tris(2-carboxyethyl)phosphine
(hydrochloride)
MW 250.19

PROTOCOL FOR THE COMPLETE REDUCTION OF DISULFIDE BONDS WITHIN PROTEIN MOLECULES

1. Dissolve the protein to be reduced at a concentration of 1–10 mg/ml in 20-m*M* sodium phosphate, 0.15-*M* NaCl, pH 7.4. Other buffers and pH values may also be used. A strong denaturant may be added (6-*M*

guanidine or 8-*M* urea) to this solution to promote protein unfolding and make buried disulfides more accessible.

2. Add TCEP to a final concentration of 20-m*M*.
3. Place in a sealed tube and incubate in a boiling water bath for 5 min. If a denaturant was included in the buffer from (1), then high temperature may not be necessary. Alternatively, incubate the sample at 50°C for 30 min.
4. To remove excess TCEP and reaction byproducts, dialyze the solution or purify by gel filtration using a buffer containing 1–10-m*M* EDTA.

PROTOCOL FOR PARTIAL REDUCTION OF PROTEIN DISULFIDES OR FOR CLEAVING DISULFIDE-CONTAINING MODIFICATION REAGENTS

1. Dissolve the protein to be reduced at a concentration of 1–10 mg/ml in 20-m*M* sodium phosphate, 0.15-*M* NaCl, pH 7.4. Other buffers and pH values may also be used. Do not add a denaturant to unfold protein structure.
2. Add TCEP to a final concentration of 20-m*M*. For partial reduction of antibody disulfides in the hinge region while maintaining a biospecific IgG molecule, add TCEP in a 2.75-fold molar excess over that of the antibody concentration.
3. Incubate for 2 h at room temperature or 37°C.
4. To remove excess TCEP and reaction byproducts, dialyze the solution or purify the protein by gel filtration using a buffer containing 1–10-m*M* EDTA.

IMMOBILIZED DISULFIDE REDUCTANTS

Many extracellular proteins like immunoglobulins, protein hormones, serum albumin, pepsin, trypsin, ribonuclease, and others contain one or more indigenous disulfide bonds. For functional and structural studies of proteins, it is often necessary to cleave these disulfide bridges. Disulfide bonds in proteins are commonly reduced with small, soluble mercaptans, such as DTT, TCEP, 2-mercaptoethanol, thioglycolic acid, cysteine, etc. High concentrations of mercaptans (molar excess of 20- to 1000-fold) are usually required to drive the reduction to completion.

Cleland (1964) showed that dithiothreitol (DTT) and dithioerythritol (DTE) are superior reagents in reducing disulfide bonds in proteins (see previous discussion, this section). DTT and DTE have low oxidation–reduction potential and are capable of reducing protein disulfides at concentrations far below that required with 2-mercaptoethanol. However, even these reagents have to be used in approximately 20-fold molar excess in order to get close to 100% reduction of a protein.

An immobilized disulfide reductant usually consists of an insoluble beaded support material such as agarose that has been modified with a small ligand containing a terminal sulfhydryl group. The presence of densely coupled sulfhydryl groups on the matrix creates enormous disulfide reducing potential. Simply mixing a solution of a disulfide-containing peptide or protein with the immobilized reductant efficiently breaks any disulfide linkages and creates free sulfhydryls. This is done without extraneous sulfhydryl contamination by the reductant, as in the case of soluble reductants.

The use of immobilized disulfide reductants thus has the following advantages over solution-phase agents:

1. Immobilized disulfide reductants can be used to reduce all types of biological disulfides without liberating product or byproducts contaminants.
2. Soluble components that interfere with the assay of free thiol groups are not present if immobilized disulfide reductants are used.
3. Small molecules containing disulfide bonds (such as cystine-containing peptides) may be reduced and isolated simply by removing the immobilized reductant. Separation of reduced molecules from reductant is much more difficult if a soluble reducing agent is used with low molecular weight disulfides.
4. Immobilized disulfide reductants can easily be regenerated and reused many times.

Immobilized dihydrolipoamide (thioctic acid) (Gorecki and Patchornick, 1973, 1975) and immobilized *N*-acetyl-homocysteine thiolactone (Eldjarn and Jellum, 1963; Jellum, 1964) are the two most commonly used immobilized disulfide reductants. In addition, immobilized TCEP provides a reducing matrix that is free of thiols (Thermo Fisher). Such immobilized reductants can successfully be used to reduce many types of biological disulfides, including small molecules such as oxidized glutathione and bovine insulin. They are particularly convenient to reduce peptide disulfides prior to conjugation, which may be necessary even with peptides labeled at their end with a cysteine group, as the sulfhydryl may become oxidized over time and form an inter-peptide disulfide bridge.

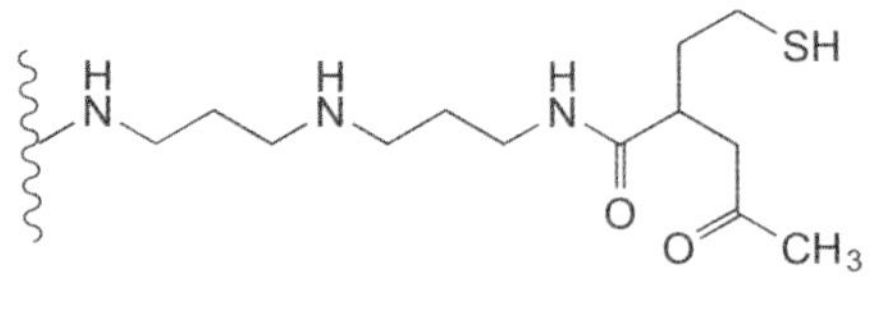

Immobilized *N*-acetyl homocysteine attached to a diaminodipropylamine spacer

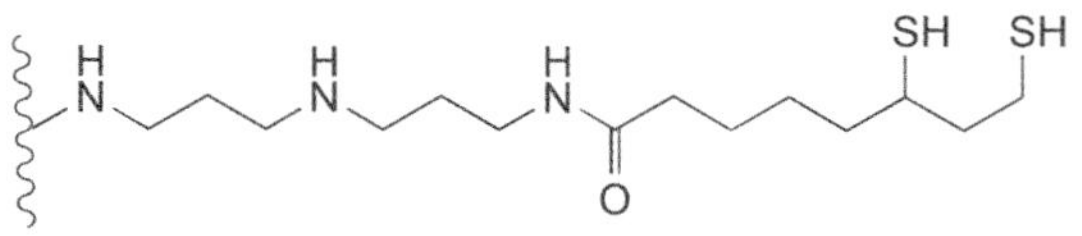

Immobilized dihydrolipoamide attached to a diaminodipropylamine spacer

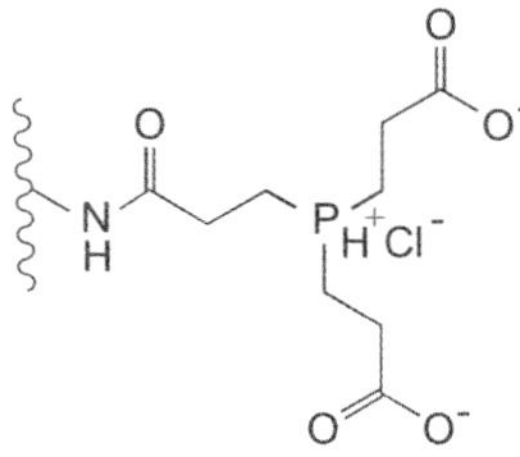

Immobilized TCEP attached via an amide bond

Immobilized disulfide reductants may be synthesized as described in Hermanson *et al.* (1992) or by using the immobilization methods described in Chapter 15, or they may obtained commercially (Thermo Fisher).

REDUCTION OF PEPTIDES USING IMMOBILIZED REDUCTANTS

Note: For optimal reduction of peptides, the following steps should be performed at room temperature.

1. Pack an immobilized reductant gel (2 ml settled gel) in a disposable polypropylene column and wash with 5 ml of 0.1-*M* sodium phosphate buffer, pH 8.0, containing 1-m*M* EDTA (equilibration buffer).
2. Prepare the sulfhydryl column by washing with a disulfide reducing agent. Apply 10 ml of freshly made 10-m*M* DTT solution (15.4 mg of DTT dissolved in 10 ml of equilibration buffer). This treatment converts the immobilized ligands into a fully reduced form (free –SH groups).
3. Wash the column with 20 ml of equilibration buffer-1 to remove free DTT.
4. Apply to the column 1.0 ml of peptide solution (dissolved in equilibration buffer) to be reduced. Normally, small peptides (molecular weight less than or equal to that of insulin) require no deforming agent (denaturant) such as guanidine to be completely reduced.
5. After the sample has completely entered into the gel bed, wash the column with 9 ml of equilibration buffer, while collecting 1.0-ml fractions.
6. Monitor the elution of reduced peptide from the column by measuring the absorbance at 280 nm (if peptide absorbs at this wavelength) as well as by performing an Ellman's assay (Section 4.1, this chapter) for thiol groups using a small aliquot (10–20 μl) of each collected fraction
7. Regenerate the sulfhydryl-containing support by following steps 2 and 3 above. Such columns can be regenerated and reused at least ten times without any significant decrease in the reductive capacity.
8. Store the column in 0.02% sodium azide at 4°C.

REDUCTION OF PROTEINS USING IMMOBILIZED REDUCTANTS

Note: For optimal reduction of proteins, the following steps must be performed at room temperature.

1. Pack an immobilized reductant gel (2 ml) in a disposable polypropylene column and wash with 5 ml of 0.1-*M* sodium phosphate buffer, pH 8.0, containing 1-m*M* EDTA (equilibration buffer-1).
2. Prepare the sulfhydryl column by washing with a disulfide reducing agent. Apply 10 ml of freshly made 10-m*M* DTT solution (15.4 mg of DTT dissolved in 10 ml of equilibration buffer-1).
3. Wash the column with 10 ml of equilibration buffer-1 and 10 ml of 0.1-*M* sodium phosphate buffer, pH 8.0, containing 1-m*M* EDTA and 6-*M* guanidine hydrochloride (equilibration buffer-2) to remove free DTT.
4. Apply to the column 1.0 ml of protein solution (dissolved in equilibration buffer-2) to be reduced. The inclusion of a denaturant in the solution deforms the protein structure so that inner disulfides are available to the immobilized reductant. Without the presence of guanidine or another deforming agent (e.g., urea, SDS), only partial reduction of the protein is possible.
5. After the sample has completely entered the gel bed, incubate the column at room temperature for 1 h.
6. Wash the column with 9 ml of equilibration buffer-2 while 2-ml fractions are collected.
7. Monitor elution of reduced protein from the column by measuring the absorbance at 280 nm as well as by performing an Ellman's assay for thiol groups (Section 4.1, this chapter) using a small aliquot (50–100 μl) of each collected fraction.
8. Regenerate the sulfhydryl-containing column by following steps 2 and 3 above. Such columns can be regenerated and reused at least 10 times without any significant decrease in the reductive capacity.
9. Store the column in 0.02% sodium azide at 4°C.

SODIUM BOROHYDRIDE

Perhaps the simplest route to the reduction of disulfide groups in peptides is the use of sodium borohydride ($NaBH_4$). This common reducing agent often used in organic synthesis is able to specifically reduce disulfides to free thiols without affecting any of the other major functional groups in proteins. Gailit (1993) developed a protocol for borohydride reduction, which avoids any purification steps to remove the reducing agent after the reaction. Thus, peptides reduced by this protocol can be used immediately in bioconjugate applications without additional steps.

Protocol

1. Dissolve the peptide to be reduced in a buffer at pH 8 to 10. Sodium phosphate or sodium bicarbonate at 0.1-*M* work well. The optimal pH range for borohydride activity is alkaline; therefore, avoid using buffers at neutral pH.
2. Add sodium borohydride (Aldrich) to the peptide solution to obtain a final concentration of 0.1-*M*. Generation of hydrogen bubbles will occur as the borohydride is dissolved.
3. Incubate at room temperature for 30–60 min.
4. Adjust the pH of the reaction to pH 4.0 using dilute HCl. Incubate for 10 min to ensure the complete destruction of excess borohydride. Hydrogen bubbles again will be evolved from the solution.
5. Readjust the pH to the optimal value for the bioconjugate application to be carried out using the generated thiols. Use the reduced peptide immediately to prevent reoxidation of the thiols to disulfides.

ELLMAN'S ASSAY FOR THE DETERMINATION OF SULFHYDRYLS

Ellman's reagent, 5,5′-dithio*bis*(2-nitrobenzoic acid) (DTNB), reacts with thiols under slightly alkaline conditions to release the highly-chromogenic compound, 5-thio-2-nitrobenzoic acid (TNB) (Ellman, 1959; Riddles, 1979) (Figure 2.79). The reagent contains a disulfide bond between two TNB groups, and reacts with free sulfhydryls to create a mixed disulfide product. The target of the reaction is the unprotonated, conjugate base form of the thiol, R–S⁻. At pH 8.0, the release of one TNB group per available thiol provides a yellow-colored product with an extinction coefficient at 412 nm of 13,600 $M^{-1} cm^{-1}$. The increase in absorbance at this wavelength is directly proportional to the concentration of sulfhydryls in solution. Correlation to a standard curve of known sulfhydryl concentrations allows accurate measurement of thiol content in unknown samples.

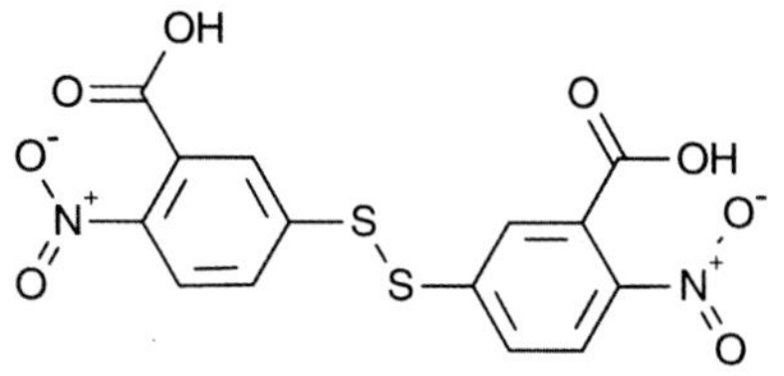

Ellman's Reagent
5,5'-Dithio-*bis*-(2-nitrobenzoic acid)
MW 396.4

Ellman's reagent has been used not only for the determination of sulfhydryls in proteins and other molecules, but also as a pre-column derivatization reagent for the separation of thiol compounds by HPLC (Kuwata *et al.*, 1982), in the study of thiol-dependent enzymes (Tsukamoto and Wakil, 1988; Alvear *et al.*, 1989; Masamune *et al.*, 1989), and to create sulfhydryl-reactive chromatography supports for the coupling of affinity ligands (Jayabaskaran *et al.*, 1987) (Chapter 15, Section 2.2). Another important use of the compound is in the assessment of conjugation procedures using sulfhydryl-reactive crosslinking agents (Chapter 6, Section 1).

Depending on the conditions, an Ellman's assay can detect as little as 10-nm cysteine concentration. The linearity can extend into the m*M* range, making the test extremely flexible for different sample situations.

Protocol

1. Dissolve Ellman's reagent (Thermo Fisher) in 0.1-*M* sodium phosphate, pH 8.0, at a concentration of 4 mg/ml.
2. Prepare a set of standards by dissolving cysteine in 0.1-*M* sodium phosphate, pH 8.0, at an initial concentration of 2-m*M* (3.5 mg/ml) and serially diluting this solution (1:1) with reaction buffer down to at least 0.125-m*M*. This will produce five solutions

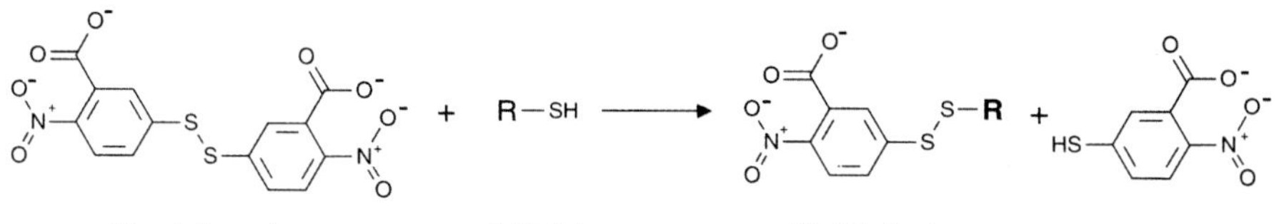

FIGURE 2.79 The reaction of Ellman's reagent with a sulfhydryl group releases the chromogenic TNB anion, which can be quantified by its absorbance at 412 nm.

of cysteine for generating a standard curve. If a more dilute concentration range is required, continue to serially dilute until a set of standards in the desired range is obtained.

3. Label a set of tubes according to the standards and samples to be used. Add 250 µl of each standard and sample to the appropriate tubes. If the samples are in a buffer that may significantly change the pH of the reaction buffer, the samples should be buffer-exchanged or dialyzed into 0.1-*M* sodium phosphate, pH 8.0, before running the assay.
4. Add 50 µl of Ellman's reagent to each standard and sample tube. Mix well.
5. Incubate at room temperature for 15 min.
6. Measure the absorbance of each solution at 412 nm.
7. Plot the absorbance *versus* cysteine concentration for each of the standards. Determine the sulfhydryl concentration of the samples by comparison to the standard curve.

4.2. Introduction of Carboxylate Groups

Modification of various functional groups in macromolecules with the following types of reagents will introduce carboxylate functions for further derivatization purposes. Amines, sulfhydryls, histidine, and methionine side chains are readily modified to contain short molecules terminating in a carboxylic acid. The short chain can serve as a spacer to enhance steric accommodations and the terminal carboxylate group can facilitate subsequent couplings with amines or hydrazides. The introduction of carboxylates also affects the overall charge characteristics or pI of the molecule being derivatized. The modification of amine residues by acylation with anhydrides not only eliminates the positive charge contribution of the protonated amine, but also adds the negative charge contribution of the acid. The result may be a change of minus two in net charge per group modified. While the reactions involved in such derivatizations are conducted under relatively mild conditions, severe alterations in net charge may cause some macromolecules, like proteins, to denature or lose activity. In addition, if the group being modified happens to be critical for active center operation then functionality may be compromised regardless of conditions. While the following reactions are facile and efficient, it should be kept in mind that in certain instances modification may lead to inactivity.

Modification of Amines with Anhydrides

Acid anhydrides, as their name implies, are formed from the dehydration reaction of two carboxylic acid

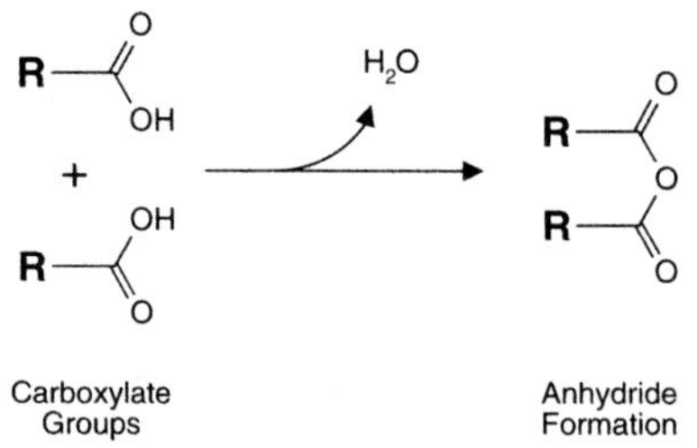

FIGURE 2.80 Anhydrides are created from two carboxylate groups by the removal of one molecule of water.

groups (Figure 2.80). Anhydrides are highly reactive toward nucleophiles and are able to acylate a number of the important functional groups of proteins and other macromolecules. Upon nucleophilic attack, the anhydride yields one carboxylic acid for every acylated product. If the anhydride was formed from monocarboxylic acids, such as acetic anhydride, then the acylation occurs with release of one carboxylate group. However for dicarboxylic acid anhydrides, such as succinic anhydride, upon reaction with a nucleophile the ring structure of the anhydride opens, forming the acylated product modified to contain a newly formed carboxylate group. Thus, anhydride reagents may be used to both block functional groups and to convert an existing functionality into a carboxylic acid.

Protein functional groups able to react with anhydrides include the α-amines at the N-terminals, the ε-amine of lysine side chains, cysteine sulfhydryl groups, the phenolate ion of tyrosine residues, and the imidazolyl ring of histidines. However, acylation of cysteine, tyrosine, and histidine side chains forms unstable complexes that are easily reversible to regenerate the original group. Only amine functionalities of proteins are stable to acylation with anhydride reagents (Fraenkel-Conrat, 1959; Smyth, 1967).

Another potential site of reactivity for anhydrides in protein molecules is modification of any attached carbohydrate chains. In addition to amino group modification in the polypeptide chain, glycoproteins may be modified at their polysaccharide hydroxyl groups to form ester derivatives. Esterification of carbohydrates by acetic anhydride, especially cellulose, is a major industrial application for this compound. In aqueous solutions, however, esterification may be a minor product, since the oxygen of water is about as strong a nucleophile as the hydroxyls of sugar residues.

The major side reaction to the desired acylation product is hydrolysis of the anhydride. In aqueous solutions anhydrides may break down by the addition of one molecule of water to yield two carboxylate groups. The presence of an excess of the anhydride in the reaction medium is usually enough to minimize the effects of competing hydrolysis.

Since both hydrolysis and acylation yield the release of carboxylic acid functionalities, the medium becomes acidic during the course of the reaction. This requires either the presence of a strongly buffered environment to maintain the pH or periodic monitoring and adjustment of the pH with base as the reaction progresses.

SUCCINIC ANHYDRIDE

Succinic acid is a four-carbon molecule with carboxylic acid groups on both ends. The anhydride has a five-atom cyclic structure that is highly reactive toward nucleophiles, especially amines. Attack of a nucleophile at one of the carbonyl groups opens the anhydride ring, forming a covalent bond with that carbonyl and releasing the other to create a free carboxylic acid (Klotz, 1967). Succinylation of positively charged amino groups of proteins and other molecules thus creates amide bond derivatives and converts the cationic site into a negatively charged carboxylate (Figure 2.81). Succinylated proteins often experience dramatic changes in their three-dimensional structure.

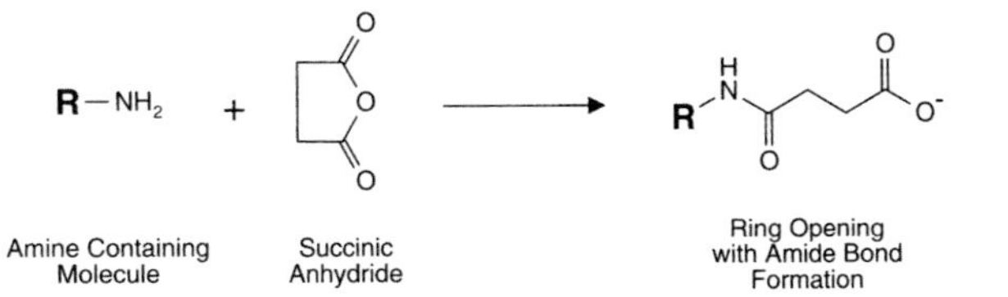

FIGURE 2.81 Succinic anhydride reacts with primary amine groups in a ring-opening process, creating an amide bond and forming a terminal carboxylate.

Subunits may dissociate (Klotz and Keresztes-Nagy, 1962), enzymatic activity may be compromised (Riordan and Valle, 1963, 1964), and the molecular radius and viscosity may be increased (Habeeb *et al.*, 1958). Other effects on protein conformation and function have also been studied (Meighen *et al.*, 1971; Shiao *et al.*, 1972; Shetty and Rao, 1978).

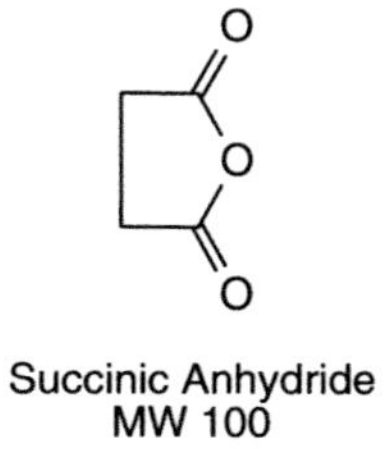

Succinic anhydride may also react with protein phenolate side chains of tyrosine residues and the –OH group of aliphatic hydroxy amino acids (Figure 2.82). The phenolate ester derivatives are unstable above pH 5.0, whereas the serine and threonine esters are more stable but may be cleaved by treatment with hydroxylamine at basic pH (Gounaris and Perlman, 1967).

A succinylated casein derivative that has nearly all its amines blocked can be used as a substrate in protease assays (Hatakeyama, 1992). As the casein is degraded by a protease, free amines are created from α-chain cleavage and release of α-amino groups. The creation of amines can be monitored by an amine detection reagent such as trinitrobenzene sulfonic acid (TNBS; Section 4.3, this chapter). The procedure forms the basis for a highly sensitive assay for protease activity.

Succinylated derivatives of nucleic acids may be prepared by reaction of the anhydride with available –OH groups. The reaction forms relatively stable ester

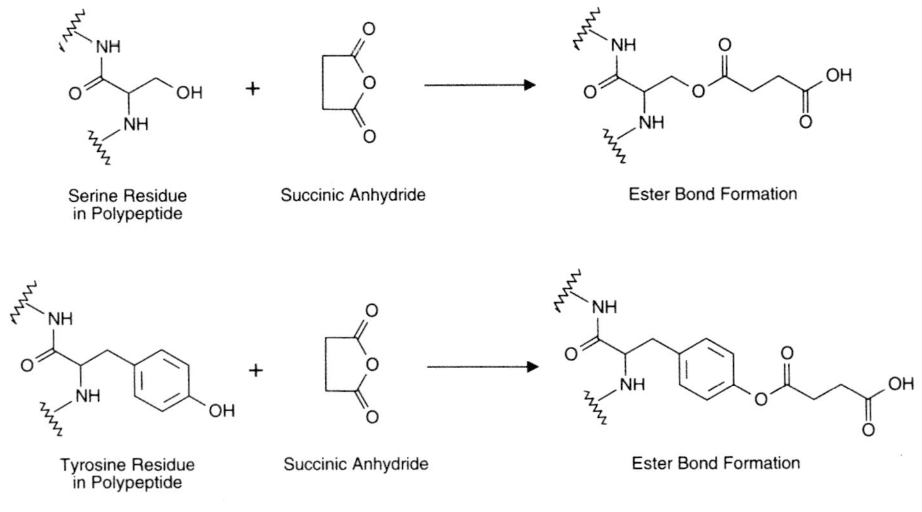

FIGURE 2.82 The hydroxyl group of serine residues and the phenolate ring of tyrosine groups may be modified with succinic anhydride to produce relatively unstable ester bonds. In aqueous conditions these reactions are minor due to competing hydrolysis by water.

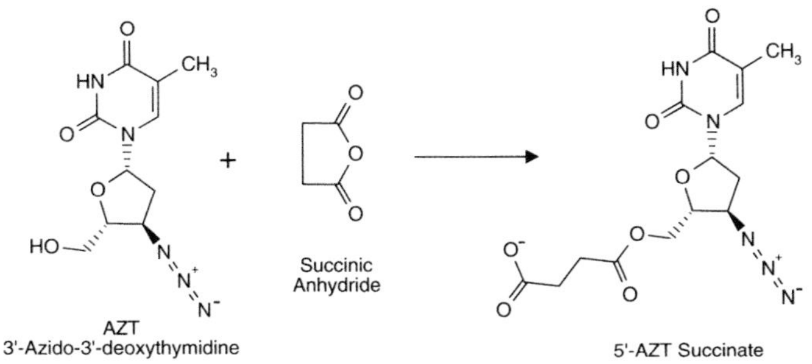

FIGURE 2.83 Succinic anhydride has been used in nonaqueous conditions to modify the 5′-hydroxyl group of nucleic acid derivatives such as AZT.

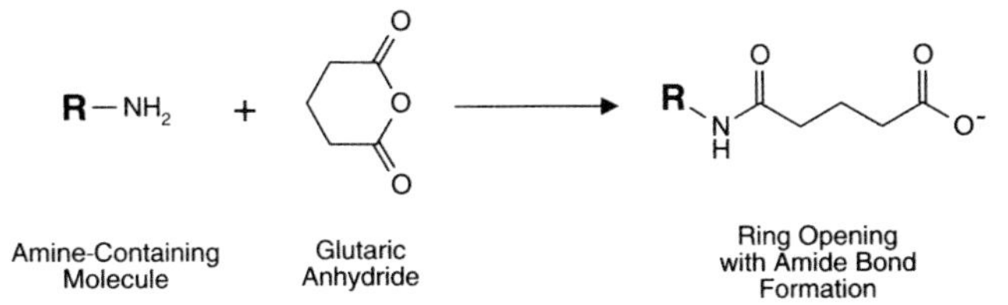

FIGURE 2.84 Glutaric anhydride reacts with amines in a ring-opening process to create an amide bond linkage and a terminal carboxylate group.

derivatives that create carboxylates on the nucleotide for further conjugation or modification (Figure 2.83). This method has been used in nucleic acid synthesis (Matteucci and Caruthers, 1980) and to derivatize nucleotide analogs such as AZT (Tadayoni, 1993).

Succinic anhydride is also a convenient extender for creating spacer arms on chromatography supports. Supports derivatized with amine-terminal spacers may be succinylated to totally block the amine functionalities and form terminal carboxylic acid linkers for coupling amine-containing affinity ligands (Cuatrecasas, 1970). See Chapter 15 for a complete discussion of immobilization techniques.

Molecules modified with succinic anhydride to create terminal carboxylate functionalities may be further conjugated to amine-containing molecules by use of amide bond-forming reagents such as carbodiimides (Chapter 4, Section 1).

Protocol

1. Dissolve (or suspend in the case of insoluble polymers or support materials) the amine-containing molecule to be succinylated in a buffer having a pH between 6.0 and 9.0. Higher pH buffers will cause the reaction to occur faster and result in more amines in an unprotonated state. Suitable buffer salts include sodium acetate, sodium phosphate, and sodium carbonate in a 0.1–1.0-*M* concentration. Avoid buffers containing primary amine groups such as Tris. Alternatively, the substance may be dissolved in water and the pH maintained in the proper range by periodic addition of NaOH. This is conveniently done by means of a pH stat. Even in buffered reactions, the pH should be monitored to prevent severe acidification of the reaction solution, which could damage the molecule being modified.
2. Add a quantity of succinic anhydride to the reaction medium to provide at least a 5–10 molar excess of reagent over the amount of amines to be modified. Even greater molar excess may be required for total blocking of all the amines of some proteins. When adding solid succinic anhydride, multiple additions may be carried out to maintain solubility of the reagent in the reaction solution. The anhydride may also be dissolved in dry dioxane before addition to aid in dissolution.
3. React for at least 1–2h at room temperature. To ensure complete blocking of all amine groups, the reaction may be continued overnight.
4. Remove excess reactants from the succinylated molecule by dialysis, gel filtration, or some other suitable method. The efficiency of amine modification may be assessed by use of the TNBS test for amines (Section 4.3, this chapter). A negative test for amines indicates complete succinylation.

GLUTARIC ANHYDRIDE

Glutaric acid is a linear, five-carbon molecule with carboxylic acid groups on both ends. It contains one more carbon in length than the similar compound succinic acid. The anhydride of glutaric acid forms a cyclic structure containing six atoms. Attack of a nucleophile, such as an amino group, on one of the carbonyl groups of glutaric anhydride opens the ring, forming an amide linkage and liberating the other carboxylic acid (Figure 2.84). Reaction with the phenolate of tyrosine or the sulfhydryl group of cysteine forms unstable linkages (an ester and a thioester, respectively) that can easily hydrolyze. As with succinic anhydride, however, aliphatic hydroxyl groups such as those of serine and threonine may be modified with glutaric anhydride to create more stable ester bonds (see above).

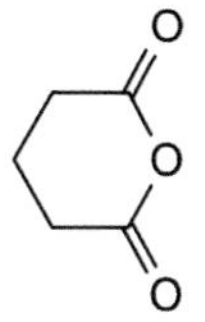

Glutaric Anhydride
MW 114

PROTOCOL The procedure for the modification of amine-containing compounds with glutaric anhydride is identical to that described for succinic anhydride, above.

MALEIC ANHYDRIDE

Maleic acid is a linear, four-carbon molecule with carboxylate groups on either end similar to succinic acid, but with a double bond between the central carbon atoms. The anhydride of maleic acid is a cyclic molecule containing five atoms in its ring. Although the reactivity of maleic anhydride is similar to other such reagents like succinic anhydride, the products of maleylation are much more unstable toward hydrolysis, and the site of unsaturation lends itself to additional side-reactions. Acylation products of amino groups with maleic anhydride are stable at neutral pH and above, but they readily hydrolyze at acid pH values (around pH 3.5) (Butler *et al.*, 1967). Maleylation of sulfhydryls and the phenolate of tyrosine are even more sensitive to hydrolysis.

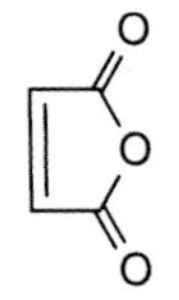

Maleic Anhydride
MW 98

As with other cyclic anhydrides, the acylation of an amine residue will proceed with elimination of the potential positive charge of the amine and addition of the negative charge created by the anhydride ring opening (Figure 2.85). Thus, a molecule can undergo a change of minus two in net charge per site of maleylation. Proteins extensively modified with maleic anhydride may spontaneously dissociate into subunits or experience a general opening of their three-dimensional structures (Sia and Horecker, 1968; Uyeda, 1969).

The double bond of maleic anhydride may undergo free-radical polymerization with the proper initiator. Polymers of maleic anhydride (or copolymers made with another monomer) are commercially available (Polysciences). They consist of a linear hydrocarbon backbone (formed from the polymerization of the vinyl groups) with cyclic anhydrides repeating along the chain. Such polymers are highly reactive toward amine-containing molecules.

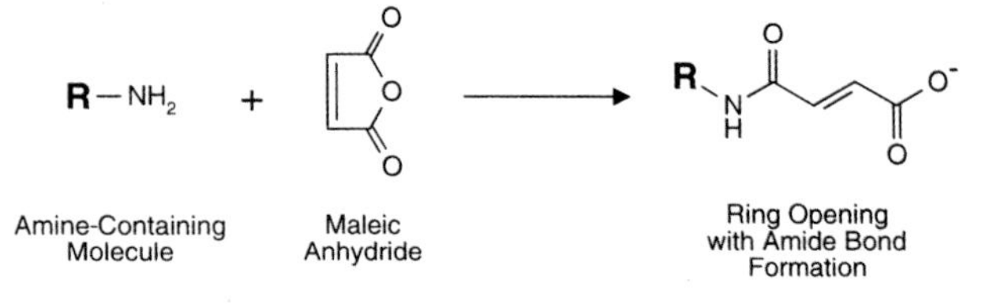

FIGURE 2.85 Maleic anhydride reacts with amine groups in a ring-opening process to create carboxylate derivatives.

Maleic acid imides (maleimides) are derivatives of the reaction of maleic anhydride and ammonia or primary amine compounds. The double bond of a maleimide may undergo an alkylation reaction with a sulfhydryl group to form a stable thioether bond (Chapter 3, Section 2.2). Maleic anhydride may presumably undergo the same reaction with cysteine residues and other sulfhydryl compounds.

Proteins derivatized with maleic anhydride exhibit an increase in their absorptivity at wavelengths below 280 nm, due to the addition of the unsaturated carbon–carbon bond. The extent of maleylation may be estimated by measuring the absorbance increase before and after modification (Freedman *et al.*, 1968).

PROTOCOL Modification of amines with maleic anhydride is done essentially the same as that described for succinic anhydride, except the pH of the reaction should be kept alkaline (pH 8–9) at all times to prevent unwanted de-acylation. Deblocking of maleylated amines can be accomplished according to the following procedure of Butler *et al.* (1967).

1. Adjust the pH of the maleylated protein or other molecule to pH 3.5 with formic acid and aqueous NH_3.
2. Incubate the solution at 37°C for 30 h.
3. Stop the deblocking reaction by the addition of NaOH to raise the pH back to neutrality.

CITRACONIC ANHYDRIDE

Citraconic anhydride (or 2-methylmaleic anhydride) is a derivative of maleic anhydride that is even more reversible after acylation than maleylated compounds. At alkaline pH values (pH 7–8) the reagent effectively

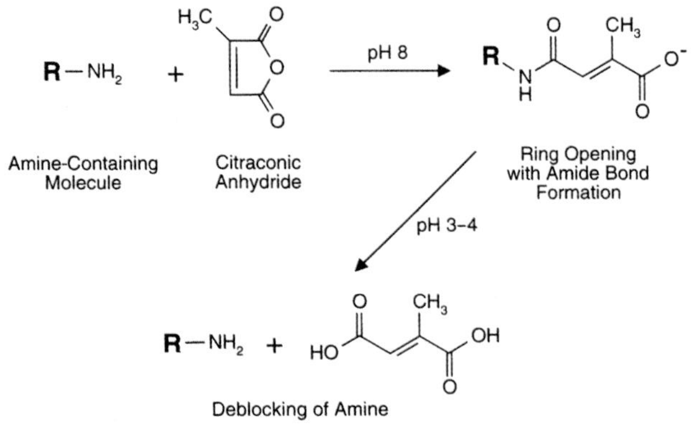

FIGURE 2.86 Citraconic anhydride can be used to block amine groups reversibly. The amide bond derivative is unstable to acidic conditions.

reacts with amine groups to form amide linkages and a terminal carboxylate. However, at acid pH (3–4), these bonds rapidly hydrolyze to release citraconic acid and free the amine (Figure 2.86) (Dixon and Perham, 1968; Habeeb and Atassi, 1970; Klapper and Klotz, 1972; Shetty and Kinsella, 1980). Thus, citraconic anhydride has been used to temporarily block amine groups while other parts of a molecule are undergoing derivatization. Once the modification is complete, the amines can then be unblocked to create the original structure.

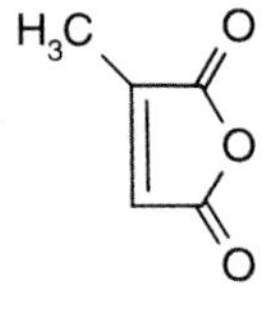

Citraconic Anhydride MW 113

Acid-labile, heterobifunctional crosslinking reagents have been synthesized using 2-methylmaleic anhydride at one end (Blattler *et al.*, 1985). Amines can be reacted with the anhydride end under alkaline conditions to form amide linkages. The other end, containing another functionality, in this case a maleimide group, is then made to react with a sulfhydryl-containing molecule. After the conjugation is complete, the citraconylamide end can be specifically released by lowering the pH.

Citraconic anhydride has also been used to reverse the effects of formalin fixation in tissue sections. Namimatsu *et al.* (2005) found that heating deparafinized tissue sections in a dilute solution of citraconic anhydride broke the formaldehyde crosslinks and restored antigen recognition of proteins within the samples. In addition, the compound has been used to stabilize certain enzymes and improve their catalytic performance (Rodrigues *et al.*, 2011).

Citraconic anhydride is a toxic liquid that should be handled with extreme care in a fume hood. Avoid contact with skin or eyes and inhalation of vapors.

Protocol

1. Dissolve the amine-containing molecule to be modified in a buffer having a pH of between 8 and 9. Maintenance of this pH range is necessary due to the high tendency of citraconylamides to hydrolyze at lower pH. Suitable buffer salts include sodium phosphate and sodium carbonate in a 0.1- to 1.0-*M* concentration. Avoid buffers containing primary amine groups such as Tris. Also avoid thiol reducing agents containing –SH groups, as these may be acylated by the anhydride. Alternatively, the substance may be dissolved in water and the pH maintained in the proper range by periodic addition of NaOH. This is conveniently done by means of a pH stat.
2. Add a quantity of citraconic anhydride to the reaction medium to provide at least a 5–10 molar excess of reagent over the amount of amines to be modified. Even greater molar excesses may be required for total blocking of all the amines of some proteins.
3. React for at least 1–2 h at room temperature. To ensure complete blocking of all amine groups, the reaction may be continued overnight.
4. Remove excess reactants from the citraconylated molecule by dialysis or gel filtration. The efficiency of amine modification may be assessed by use of the TNBS test for amines (Section 4.3, this chapter). A negative test for amines indicates complete modification.

To remove the citraconic modifications and free the amine groups, the protein may be treated in one of two ways:

1. Adjust the pH of the citraconylated molecule to 3.5–4.0 by addition of acid. Incubate at room temperature overnight or for at least 3 h at 30°C.
   Or
2. Treat the citraconylated molecule with 1-*M* hydroxylamine at pH 10 for 3 h at room temperature.

Modification of Sulfhydryls with Iodoacetate

Iodoacetate (and bromoacetate) can react with a number of functional groups within proteins: the sulfhydryl group of cysteine, both imidazolyl side-chain nitrogens of histidine, the thioether of methionine, and the primary ε-amine group of lysine residues and N-terminal α-amines (Gurd, 1967). The relative rate of reaction with each of these residues is generally dependent on the degree of ionization and thus the pH at which the modification is carried out. The exception to this is methioninyl thioethers, which react rapidly at nearly all pH values above about 1.7 (Vithayathil and Richards, 1960). The reaction products of these groups with iodoacetate are illustrated in Figure 2.87. The only reaction resulting in one definitive product is that of the alkylation of cysteine sulfhydryls, giving the carboxymethylcysteinyl derivative (Cole *et al.*, 1958). Histidine groups may be modified at either nitrogen atom of their imidazolyl side chain. Both mono-substituted derivatives and di-substituted products of the imidazole ring are possible (Crestfield *et al.*, 1963). With primary amine groups such as in the side chain of lysine residues, the products of the reaction are the secondary amine (monocarboxymethyllysine) or the tertiary amine derivative (dicarboxymethyllysine). Methionine thioether groups give the most

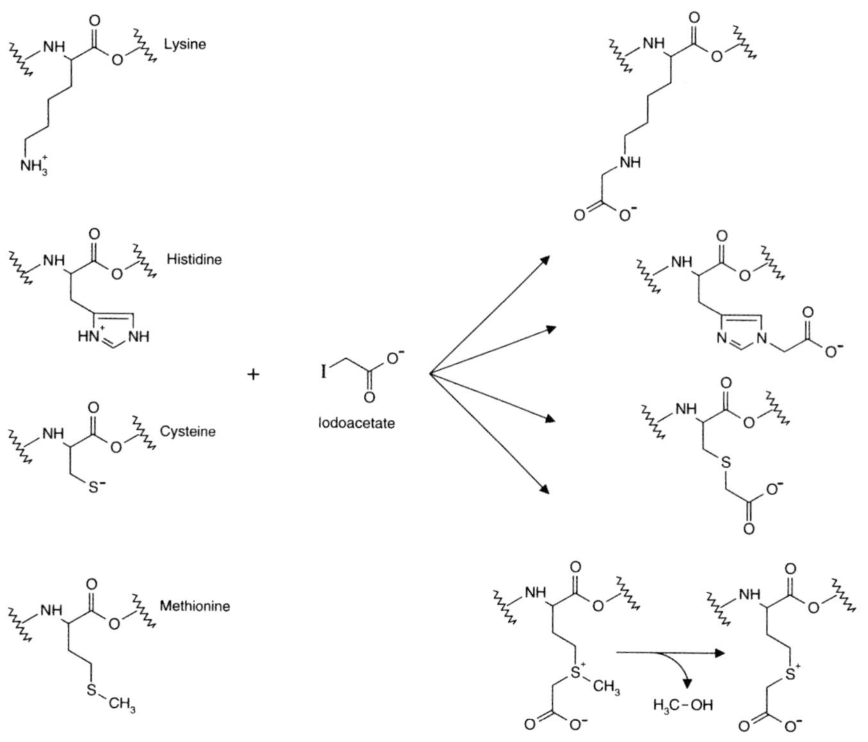

FIGURE 2.87 Iodoacetate can modify a number of amino acid side chains in proteins, forming alkylated derivatives containing a terminal carboxylate.

complicated products, some of which rearrange or decompose unpredictably. The only stable derivative of methionine is where the terminal methyl group is lost to form carboxymethylhomocysteine, the same product as the reaction of iodoacetate with homocysteine (Gundlach *et al.*, 1959).

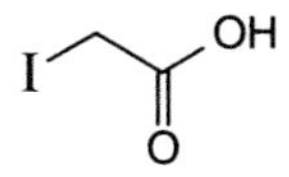

Iodoacetate
MW 185.9

The relative reactivity of α-haloacetates toward protein functionalities is sulfhydryl > imidazolyl > thioether > amine. Among halo derivatives the relative reactivity is I > Br > Cl > F, with fluorine being almost unreactive. The α-haloacetamides have the same trend of relative reactivities, but will obviously not create a carboxylate functional group. The acetamide derivatives typically are used only as blocking agents.

Thus, iodoacetate has the highest reactivity toward sulfhydryl cysteine residues and may be directed specifically for –SH modification. If iodoacetate is present in limiting quantities (relative to the number of sulfhydryl groups present) and at slightly alkaline pH, cysteine modification will be the exclusive reaction. The specificity of this modification has been used in the design of heterobifunctional crosslinking reagents, where one end of the crosslinker contains an iodoacetamide derivative and the other end contains a different functionality directed at another chemical target (see SIAB, Chapter 6, Section 1.5).

Protocol

1. Dissolve the sulfhydryl-containing protein or macromolecule to be modified at a concentration of 1–10 mg/ml in 50-m*M* Tris, 0.15-*M* NaCl, 5-m*M* EDTA, pH 8.5. EDTA is present to prevent metal-catalyzed oxidation of sulfhydryl groups. The presence of Tris, an amine-containing buffer, should not affect the efficiency of sulfhydryl modification. Not only do amines generally react more slowly than sulfhydryls, the amine in Tris buffer is of particularly low reactivity. If Tris does pose a problem, however, use 0.1-*M* sodium phosphate, 0.15-*M* NaCl, 5-m*M* EDTA, pH 8.0.
2. Add iodoacetate to a concentration of 50-m*M* in the reaction solution. Alternatively, add a quantity of iodoacetate representing a 10-fold molar excess relative to the number of –SH groups present. An estimation of the sulfhydryl content in the protein to be modified can be accomplished by performing an Ellman's assay (Section 4.1, this chapter). Readjust the pH if necessary. To aid in adding a small quantity of iodoacetic acid to the reaction,

a concentrated stock solution may be made in the reaction buffer, the pH re-adjusted, and an aliquot added to the protein solution to give the desired concentration.

3. Mix and react for 2h at room temperature. To avoid the possibility of methionine modification, limit the reaction to 30min.
4. Purify the modified protein from excess iodoacetate by dialysis or gel filtration.
5. An Ellman's assay comparing the unmodified protein to the iodoacetylated protein may be carried out to assess the degree of modification.

Modification of Sulfhydryls with BMPA

BMPA is *N*-β-maleimidopropionic acid (or 3-maleimidopropionic acid), which contains a thiol-reactive maleimide group at one end and a carboxylate group on the other end (Rich, 1975; Moroder, 1983, 1987). The compound is the acid precursor to the short, heterobifunctional crosslinker 3-maleimidopropionic acid *N*-hydroxysuccinimide ester (BMPS; Chapter 6, Section 1.15).

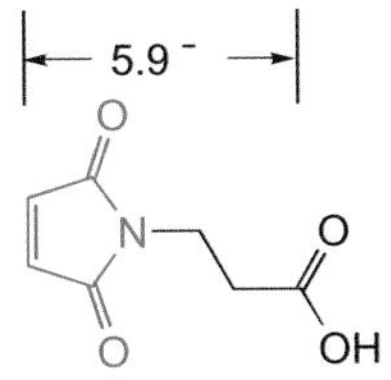

BMPA
N-β-Maleimidopropionic acid
(3-Maleimidopropionic acid)
MW: 169.13

Like BMPS, BMPA is spontaneously reactive toward sulfhydryls through its maleimide, but unlike BMPS it must be activated using a carbodiimide, such as EDC, to couple to amines or hydrazides through its carboxylic acid end. In its use as a blocking or modification agent for sulfhydryls, BMPA may be reacted with a thiol-containing molecule to form a stable thioether bond. The blocking of thiols takes place in buffered aqueous conditions from slightly acidic to moderately basic pH by addition to the double bond of the maleimide group. The final product, which then contains the short propionic acid spacer, terminates in a negatively charged carboxylate. Thus, thiols can be modified and transformed into carboxylate-containing molecules using this reagent.

BMPA has also been used as a crosslinking agent in a number of applications, including the preparation of peptide–protein conjugates for immunogens (Iglesias *et al.*, 2001), to prepare immunomodulating adducts (Gemeiner *et al.*, 1992; Cruz *et al.*, 2001), in the preparation of novel trifunctional compounds containing the metal-chelating group lysine nitrilotriacetic acid (NTA) for conjugation with His-tagged proteins (Meredith *et al.*, 2004), for the immobilization of proteins onto surfaces (Jung and Wilson, 1996), to create thiol-reactive quantum dots for labeling biomolecules (Evident Technologies, Inc., website protocols, 2005), to form albumin–insulin conjugates for slow-release drugs (Shechter *et al.*, 2005), and to synthesize thiol-reactive luminescent chelates for time-resolved fluorescence applications (Chen and Selvin, 1999).

Figure 2.88 shows the reactions of BMPA for the modification of thiols and a subsequent reaction using an EDC-mediated amide bond formation for coupling

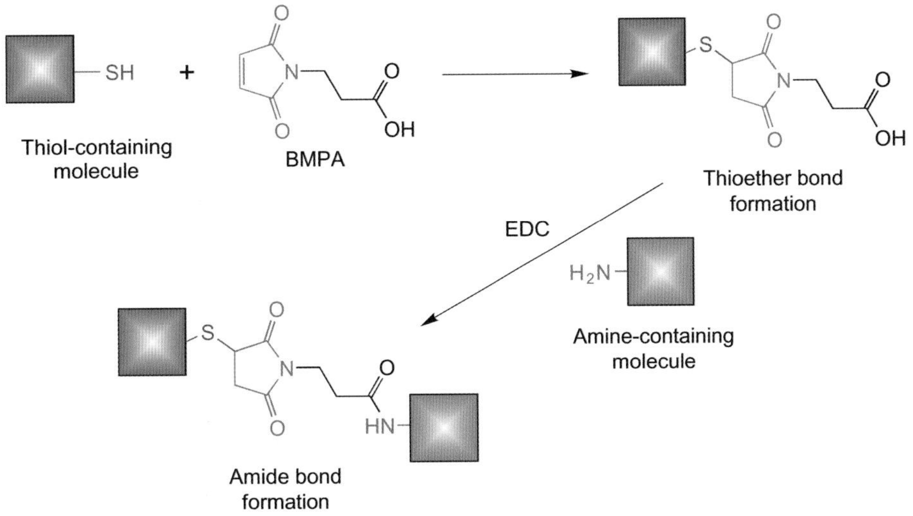

FIGURE 2.88 The maleimide group of BMPA reacts with a thiol-containing molecule to result in a modification having a terminal carboxylate group. Amine-containing molecules can then be conjugated to the carboxylate using a carbodiimide reaction with EDC.

to its carboxylate end. The maleimide–thiol reaction proceeds at physiologic pH to form a stable thioether linkage with sulfhydryl-containing molecules. The combination of EDC and sulfo-NHS (Chapter 4, Section 1.2) may also be used to form an intermediate sulfo-NHS ester, which can enhance the yield of amide bond formation. Cysteine groups in proteins and peptides may be permanently blocked using this reagent, yielding a modification that terminates in the negatively charged carboxylate.

The use of BMPA to block a thiol and create a terminal carboxylate is illustrated in the following protocol. The protocol relates to the modification of proteins, but similar reaction conditions can be used to modify other thiol-containing molecules or surfaces. See Chapter 6, Section 1.15, for additional information related to BMPA and its amine-reactive counterpart, BMPS.

Protocol

1. Dissolve a thiol-containing protein in phosphate buffered saline (PBS), pH 6.5 to 7.5, at a concentration of 1–10 mg/ml. Disulfides may be reduced to yield free thiols using DTT, TCEP, or other reducing agents, but reductants containing sulfhydryls should be completely removed by dialysis or desalting prior to reaction with BMPA.
2. Dissolve BMPA in DMSO or DMF to prepare a stock solution at a higher concentration such that adding an aliquot of this solution to the protein solution will result in the desired molar excess of the maleimide over the concentration of thiols present.
3. Add a quantity of BMPA to the protein solution to obtain at least a 5-fold molar excess of maleimide reagent over the amount of thiol present in the protein. The final concentration of organic solvent in the protein solution should not exceed 10% to prevent protein precipitation. Mix thoroughly to dissolve.
4. React for 2 h at room temperature.
5. Purify the modified protein from reactants and reaction byproducts by dialysis or gel filtration.

Modification of Hydroxyls with Chloroacetic Acid

Chloroacetic acid can be used to transform a rather unreactive hydroxyl into a carboxylate group that can be used in a variety of conjugation reactions. The reaction proceeds under basic conditions, yielding a stable ether bond terminating in a carboxymethyl group (Figure 2.89) (Plotz and Rifai, 1982; Brunswick *et al.*, 1988). Side reactions will occur with other nucleophiles, such as amines, if they are present in the molecule to be modified. The reagent is used most often to modify pure polysaccharides or hydroxyl-containing polymers that contain no other functionalities.

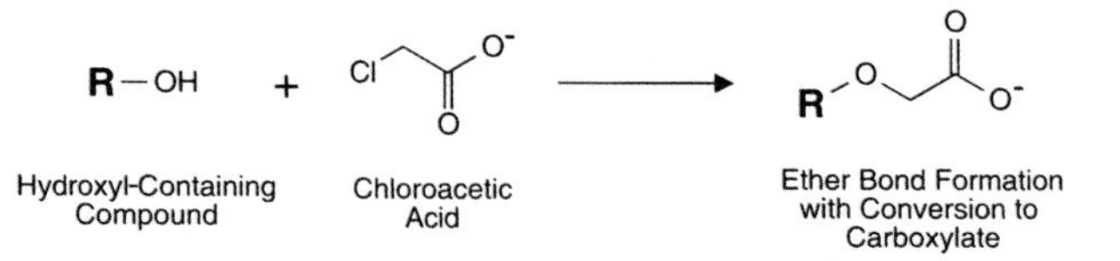

FIGURE 2.89 Chloroacetic acid can be used to create a carboxylate group from a hydroxyl.

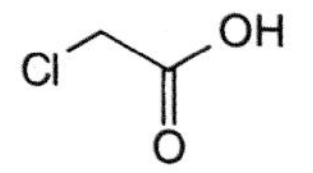

The following protocol illustrates the modification of a dextran polymer with chloroacetic acid.

Protocol

1. In a fume hood, prepare a solution consisting of 1-*M* chloroacetic acid in 3-*M* NaOH.
2. Immediately add dextran polymer to a final concentration of 40 mg/ml. Mix well to dissolve.
3. React for 70 min at room temperature with stirring.
4. Stop the reaction by adding 4 mg/ml of solid NaH_2PO_4 and adjusting the pH to neutral with 6-*N* HCl.
5. Remove excess reactants by dialysis.

4.3. Introduction of Primary Amine Groups

Primary amine groups on proteins consisting of N-terminal α-amines and lysine side-chain ε-amines are typically present in abundant quantities for modification or conjugation reactions. Occasionally, however, a protein or peptide will not contain sufficient amounts of available amines to allow for an efficient degree of coupling to another molecule or protein. For instance, horseradish peroxidase (HRP), a popular enzyme employed in the preparation of antibody conjugates, only possesses two free amines that can participate in conjugation protocols. Creating additional amines on HRP allows for higher amounts of modification and thus produces more active conjugates.

Other non-protein molecules, such as nucleic acids and oligonucleotides, may not normally possess primary amines of sufficient nucleophilicity to react with common modification reagents. The ability to add amine functionalities to these molecules is sometimes the only route to successful conjugation. Creating amines at specific sites within these molecules allows for site-directed modification at known positions, thus better ensuring active conjugates once formed.

The following reagents and techniques can be used to directly transform carboxylates or sulfhydryls into reactive amine functional groups. In addition, sugars, polysaccharides, or glycan-containing macromolecules may be modified to contain amines after mild periodate activation to form aldehyde groups or through modification at the reducing end of a carbohydrate.

Modification of Carboxylates with Diamines

Carboxylic acids may be covalently modified with short compounds containing primary amines at either end to form amide linkages. The result of such alterations is to block the carboxylates and form terminal amino groups. Reacting the diamine in high molar excess ensures that only one end of the compound couples to each carboxylate and does not crosslink the molecules being modified. Amide bond formation may be accomplished by several methods including carbodiimide-mediated coupling (Chapter 4, Section 1), active ester intermediates such as *N*-hydroxysuccinimide esters (Chapter 3, Section 1.4), and the use of carbonylating compounds such as *N,N'*-carbonyldiimidazole (Chapter 4, Section 3). A combination of the water soluble carbodiimide EDC and sulfo-NHS is also an efficient way of creating amide linkages (Chapter 4, Section 1.2).

Diamines that can be used for aminoalkylation include ethylene diamine, 1,3-diaminopropane, 3,3'-iminobispropylamine (also known as diaminodipropylamine), 1,6-diaminohexane, and the Jeffamine-type compounds containing a hydrophilic chain consisting of polyethylene- or polypropylene-oxide (formerly from Texaco Chemical Co., now Huntsman Corporation). Ethylene diamine is perhaps the most popular choice for protein carboxylate modification. Its short chain length ensures minimal steric effects and virtually no hydrophobic interactions. Diaminodipropylamine provides a longer spacer arm and has been used extensively as a bridging molecule for coupling carboxylate-containing ligands to insoluble supports (Hermanson *et al.*, 1992). The long hydrocarbon chain of 1,6-diaminohexane, however, may induce hydrophobic effects and probably should be avoided. The longest diamine of the group is the Jeffamine compound. Its chain is extremely hydrophilic and should function as an excellent modifier of carboxylates when a longer spacer is desired.

In addition, there are other diamine spacers containing discrete PEG chains available with one end blocked using either a t-BOC group or a CBZ-amido group (Quanta BioDesign). These diamine spacers are extremely hydrophilic due to the presence of a PEG_3 or PEG_{11} cross-bridge unit. Since one end of these compounds is masked with a reversible blocking group commonly used in peptide synthesis, the free amine end can be conjugated to a carboxylic acid without the possibility of crosslinking. The blocked end can then be removed with TFA (for t-BOC groups) or by reduction using hydrogen in the presence of a Pd/C catalyst (for CBZ amido groups). This type of deblocking reaction should only be used with molecules that can tolerate these conditions, such as organic molecules and short peptides. Complex proteins, however, may be denatured or lose activity under those conditions.

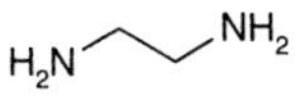

Ethylenediamine
MW 60

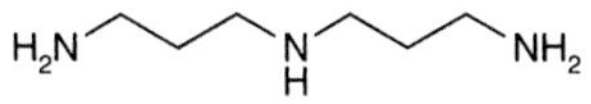

3,3'-Iminobispropylamine
(diaminodipropylamine)
MW 131

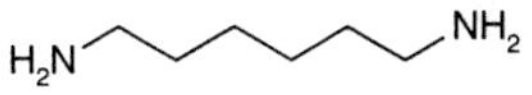

1,6-Diaminohexane
MW 116

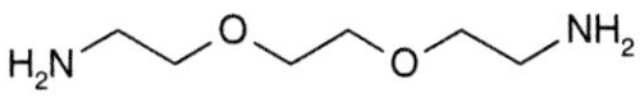

Jeffamine EDR-148
MW 148

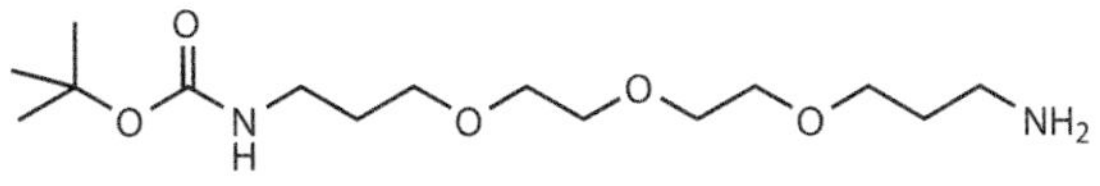

Mono-N-t-boc-amido-PEG_3-amine
MW 320.42

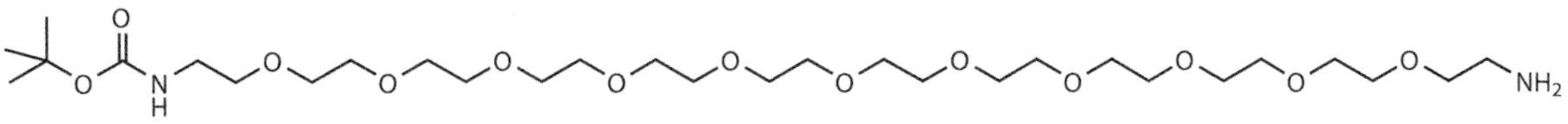

Mono-N-t-boc-amido-PEG_{11} amine
MW 644.79

Diamine modification of proteins can have dramatic effects on the net charge of the molecule, usually significantly raising the pI from the native state. The amide linkage eliminates the negative potential of the carboxylate and the terminal amine adds the potential for a positive charge. Thus, diamine modification may have the net effect of changing the overall charge by plus two for every carboxylate residue coupled. Proteins heavily modified with diamines may exhibit vital changes in activity due to the alteration of microenvironmental charge at each site of modification. In some cases, native conformation may be changed and activity completely lost.

Raising the pI of macromolecules can also significantly alter the immune response toward them upon *in vivo* administration. Cationized proteins (those modified with diamines to increase their net charge or pI) are known to generate an increased immune response compared to their native forms (Muckerheide *et al.*, 1987a,b; Apple *et al.*, 1988; Domen *et al.*, 1988). The use of cationized BSA as a carrier protein for hapten conjugation can result in a dramatically higher antibody response toward a coupled hapten (Chapter 19).

The following protocol using the carbodiimide EDC is an efficient way of modifying protein carboxylates with diamines either to increase the amount of amines present for further conjugation or to create a cationized protein having an increased net charge (Figure 2.90). Note that glycoproteins containing sialic acid may be modified at this sugar's COOH group in addition to coupling at C-terminal, aspartic acid, and glutamic acid functions on the polypeptide chain. Other carboxylate-containing macromolecules may be modified using this procedure as well.

Protocol

1. Dissolve the protein to be modified at a concentration of 1–10 mg/ml in 0.1-*M* MES, pH 4.7 (coupling buffer). Other buffers may be used as long as they do not contain groups that can participate in the carbodiimide reaction. Avoid carboxylate- or amine-containing buffers such as citrate, acetate, glycine, or Tris. Higher pH conditions may be used up to about pH 7.5 (in sodium phosphate buffer) without severely affecting the yield of modification. The protein in solid form also may be added directly to the diamine solution prepared in (2).
2. Dissolve the diamine chosen for modification at a concentration of 1-*M* made up in the coupling buffer. If a free-base form of diamine is used, then the solution will become highly alkaline upon dissolution. This operation will also generate heat—the solution process being highly exothermic. The easiest way to dissolve such a diamine is to initially add the correct amount to a beaker containing a quantity of crushed ice equal to the final solution volume desired. The ice should be made from deionized water or the equivalent to maintain purity. All operations should be carried out in a fume hood. Next, add an equivalent weight of concentrated HCl and mix. As the mixing becomes complete, the ice will almost totally melt and provide nearly the correct final solution volume. Finally, add an amount of MES buffer salt to bring its concentration to 0.1-*M*, and adjust the solution pH to 4.7. In some cases, the dihydrochloride form of the diamine is commercially available and can be used to avoid such unpleasant pH adjustments. For instance, ethylene diamine dihydrochloride is available from Aldrich. It can be added to the 0.1-*M* MES buffer without a significant change in pH.
3. Add the protein solution to an equal volume of diamine solution and mix. Alternatively, the solid protein can be dissolved directly in the diamine

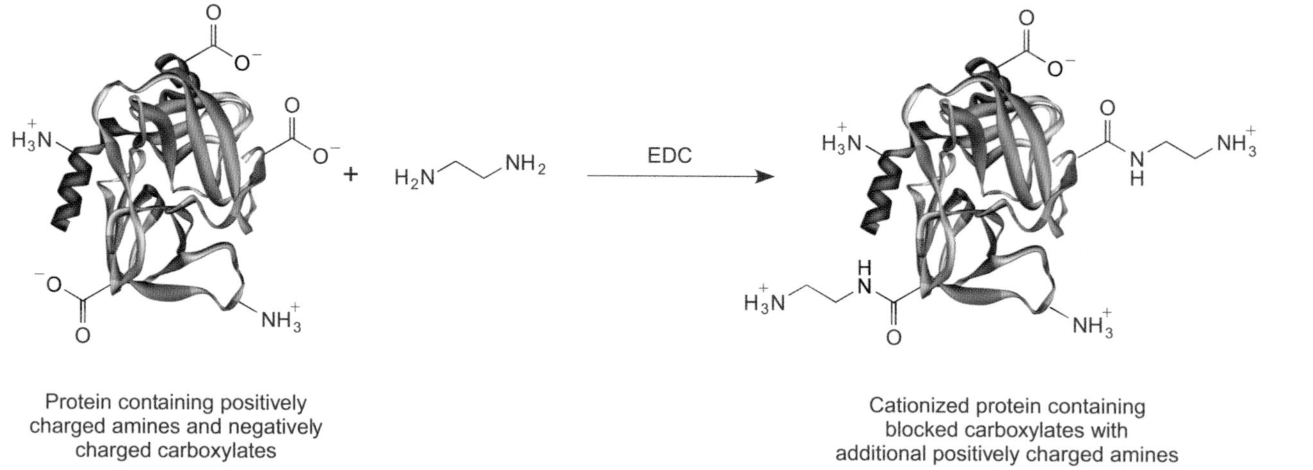

FIGURE 2.90 Cationization of protein molecules can be carried out using ethylene diamine to modify carboxylate groups using a carbodiimide reaction process.

solution (after pH adjustment) at the indicated concentration.

4. Add EDC (1-ethyl-3-(3-dimethylaminopropyl) carbodiimide hydrochloride; Thermo Fisher) to a final concentration of 2 mg/ml in the reaction solution. To aid in the addition of a small amount of EDC, a higher concentration stock solution may be prepared in water and an aliquot added to the reaction to give the proper concentration. Since EDC is labile in aqueous solutions, the stock solution must be made quickly and used immediately.
5. React for 1–2 h at room temperature.
6. Purify the modified protein by extensive dialysis against 0.02-*M* sodium phosphate, 0.15-*M* NaCl, pH 7.4 (PBS), or another suitable buffer.

The changes that occur in the pI of a protein modified with diamines may be assessed by isoelectric focusing or by general electrophoresis based upon relative migration due to charge. A cationized protein will possess a higher pI value or migrate further toward the anode than its native form. Using the above protocol typically alters the net charge of bovine serum albumin from a native pI of 4.9 to the highly basic range of pI 9.5 to over pI 11.0.

Modification of carboxylate groups with diamines may also be carried out in organic solvent for those molecules insoluble in aqueous buffers. Some peptides are quite soluble in solvents such as DMF and DMSO, but relatively insoluble in water. Such molecules may be reacted in these solvents with the carbodiimide DCC (dicyclohexyl carbodiimide) using the same basic reactant ratios as given previously for EDC in aqueous solutions (Chapter 4, Section 1.4).

Modification of Sulfhydryls with N-(β-Iodoethyl) Trifluoroacetamide [Aminoethyl-8]

The conversion of sulfhydryl groups on cysteine residues or other molecules to amine-containing groups may be accomplished by aminoethylation with *N*-(β-iodoethyl) trifluoroacetamide (Schwartz *et al.*, 1980). The haloalkyl group specifically reacts with sulfhydryls to form the aminoalkyl derivative in one step. Under the conditions of the reaction, the trifluoroacetate amine-protecting group spontaneously hydrolyzes to expose the free primary amine without the need for a secondary deblocking step (Figure 2.91). This reagent is commercially available from Thermo Fisher Chemical under the name Aminoethyl-8™.

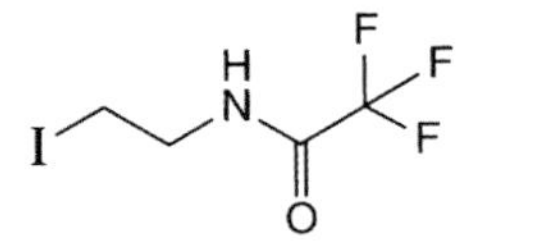

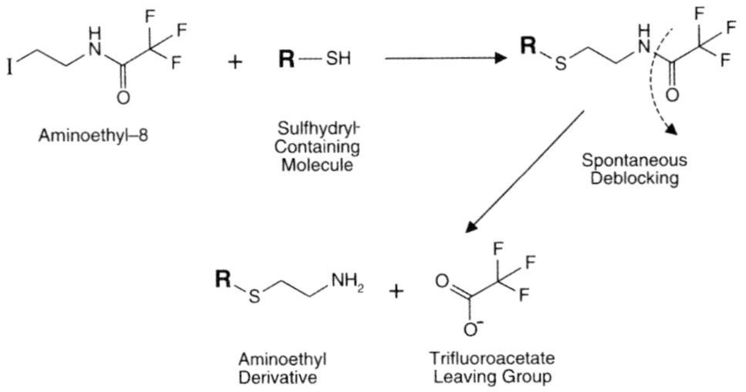

FIGURE 2.91 Aminoethyl-8 can be used to transform a sulfhydryl group into an amine. The intermediate spontaneously undergoes deblocking to release the primary amine group.

Aminoethyl-8 has an advantage over ethylenimine modification (next section), due to the potential polymerization of ethylenimine in aqueous solutions. Such polymers are highly cationic and may nonspecifically block protein molecules by coating them. The specificity of Aminoethyl-8 for sulfhydryls makes it an optimum choice for modification.

For small molecules containing sulfhydryls or for low molecular weight peptides containing cysteine residues, modification may proceed without deforming agents. However, for intact proteins containing both disulfides and free sulfhydryls, a denaturant and a disulfide reducing agent may be required to open buried or structurally inaccessible groups if complete modification is desired.

Protocol

1. Dissolve the protein to be aminoalkylated at a concentration of 1–10 mg/ml in 6-*M* guanidine hydrochloride, 0.2-*M* *N*-ethylmorpholine acetate, pH 8.1. All water used in preparing buffers should be deoxygenated by boiling followed by cooling and bubbling with nitrogen. Small molecules that do not require denaturants to expose internal disulfides or sulfhydryls may be modified without using guanidine treatment.
2. Add dithiothreitol (DTT) to obtain a 20-fold molar excess over the amount of disulfides present.
3. React for 4 h at room temperature, maintaining a blanket of nitrogen over the solution.
4. Adjust the pH to 8.6 with NaOH, and heat the solution to 50°C.
5. Add a quantity of Aminoethyl-8 in methanol to equal a 25-fold molar excess over the amount of sulfhydryl present (including the amount of DTT added). The solution in methanol should be made concentrated enough so that only a small amount of methanol has to be added to the reaction solution (i.e., no more than 10% of the final volume). A second addition of modifying agent may be made after 1 h to drive

the reaction more completely toward total –SH aminoalkylation.

6. React for 3h at 50°C.
7. Purify the modified protein or other macromolecule by gel filtration or dialysis. Occasionally, complete modification with Aminoethyl-8 will cause precipitation of the protein.

Modification of Sulfhydryls with Ethylenimine

The cyclic compound ethylenimine reacts with protein sulfhydryl groups causing ring opening and forming the aminoalkyl derivative, *S*-(2-aminoethyl)cysteine (Raftery and Cole, 1963, 1966) (Figure 2.92). Under physiological conditions ethylenimine is virtually specific for sulfhydryls with no cross-reactivity toward other protein functionalities. At acid pH, a small degree of reactivity occurs with methionine residues, forming *S*-(2-aminoethyl) methionine sulfonium ion (Schroeder *et al.*, 1967). Since aminoethylated cysteine groups resemble the side-chain structure of lysine residues, except for the replacement of one methylene group with a thioether, these modifications make them susceptible to tryptic hydrolysis, although at an abbreviated rate (Plapp *et al.*, 1967; Wang and Carpenter, 1968).

Ethylenimine
MW 43

Ethylenimine may be used to introduce additional sites of tryptic cleavage for protein structural studies. In this case, complete sulfhydryl modification is usually desired. Proteins are treated with ethylenimine under denaturing conditions (6- to 8-*M* guanidine hydrochloride) in the presence of a disulfide reductant to reduce any disulfide bonds before modification. Ethylenimine may be added directly to the reducing solution in excess (similar to the procedure for Aminoethyl-8 described previously) to totally modify the –SH groups formed.

The disadvantage of using ethylenimine for protein modification stems from the fact that in the presence of water slow formation of polyethylenimine occurs. The polymer is highly positively charged at physiological pH and can interact strongly with protein molecules, masking sites of potential sulfhydryl modification. Also, the polymer may have terminal aziridine residues (Chapter 3, Section 2.3), making it reactive and potentially forming a covalent attachment with the protein (Dermer and Ham, 1969).

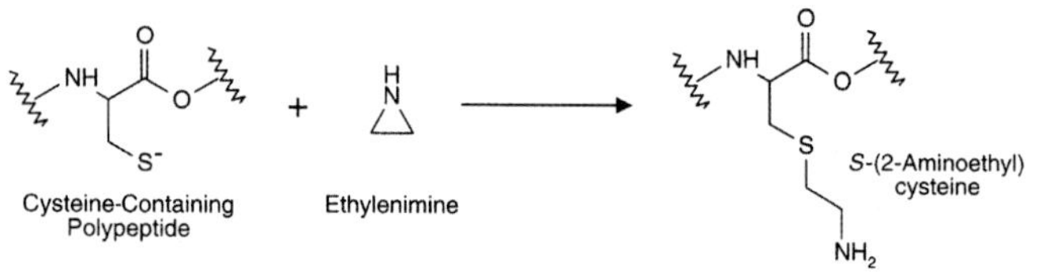

FIGURE 2.92 The small compound ethylenimine can react with sulfhydryls to form aminoethyl derivatives.

Modification of Sulfhydryls with 2-Bromoethylamine

2-Bromoethylamine may undergo two reaction pathways in its modification of sulfhydryl groups in proteins (Figure 2.93). In the first scheme, the thiolate anion of cysteine attacks the number 2 carbon of 2-bromoethylamine to release the halogen and form a thioether bond (Lindley, 1956). This straightforward reaction mechanism is similar to the modification of sulfhydryls with iodoacetate (Section 4.2, this chapter). In a two-step, secondary process, 2-bromoethylamine is converted under alkaline conditions to the cyclic ethylenimine derivative by the intramolecular attack of its primary amine on the number 2 carbon, causing release of the halogen and ring formation (Cole, 1967). Ethylenimine then goes on to react with the sulfhydryl to form the aminoalkylated derivative (as described in the previous section). The two-step reaction is slower than direct aminoalkylation by either 2-bromoethylamine or ethylenimine.

2-Bromoethylamine
MW 123.92

The selectivity of protein alkylation using 2-bromoethylamine was investigated for use in adding a stable isotopically labeled ^{13}C at cysteine residues (Marincean *et al.*, 2012). It was documented that the reaction at cysteine thiols was energetically favored over modification of lysine amines by 15.4 kJ mol^{-1} at alkaline pH values.

Protocol

1. Dissolve the protein or peptide to be aminoalkylated at cysteine sulfhydryls in 0.5-*M* sodium carbonate. If cystine disulfides are present, add a 10- to 25-fold molar excess of DTT to fully reduce them to free sulfhydryls.

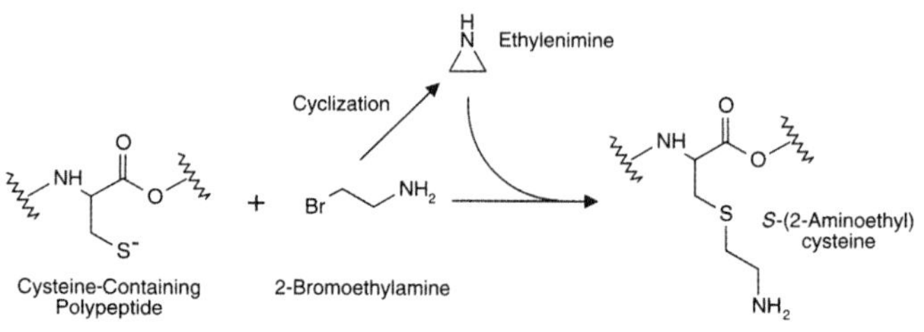

FIGURE 2.93 2-Bromoethylamine can be used to transform a thiol into an amine. The reaction may proceed through the intermediate formation of ethylenimine, yielding an aminoethyl derivative.

2. Add a quantity of 2-bromoethylamine to obtain a 10-fold molar excess over the number of sulfhydryls present in the sample, including any added DTT.
3. React overnight at room temperature.
4. Purify the modified protein by gel filtration or dialysis.

Modification of Sulfhydryls with 2-Aminoethyl-2′-Aminoethanethiolsulfonate (AEAETS)

Thiolsulfonate-containing compounds can react with thiols with release of the sulfonate end of the molecule to yield disulfide derivatives. The modification reagent 2-aminoethyl-2′-aminoethanethiolsulfonate, or AEAETS, reacts with a sulfhydryl with release of taurine (2-aminoethanesulfonate) to form a 2-aminoethyl-dithiol derivative (Figure 2.94). AEAETS can be used to block cysteine residues in proteins and form derivatives containing positively charged amines.

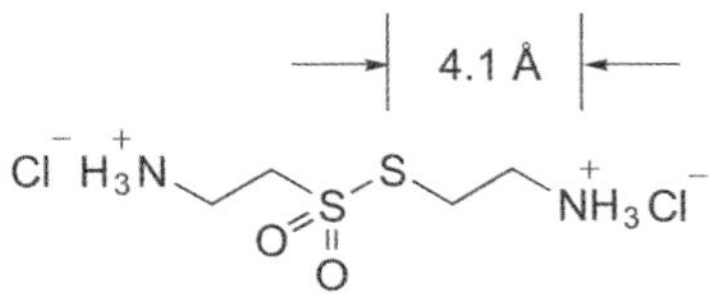

AEAETS
2-Aminoethyl-2'-aminoethanethiolsulfonate dihydrochloride
Mol. Wt.: 257.20

The basic reactions of thiolsulfonates have been known for decades (Field *et al.*, 1961, 1964), but more recently they have been applied to the study of protein interactions by site-directed modification of native cysteines or through modification of cysteines introduced at particular points in proteins by mutagenesis. Such studies have yielded insights into the structure and binding site characteristics of proteins (Kirley *et al.*, 1989). Pascual *et al.* (1998) used AEAETS to probe the acetylcholine receptor from the extracellular side of the membrane in order to investigate the molecular accessibility and electrostatic potential within the open and closed channel.

The AEAETS reaction with thiols is similar to that of sodium tetrathionate (Section 5.2, this chapter) and the methanethiosulfonate (MTS) or disulfide exchange compounds described in Chapter 3, Section 2.6. All of these reagents form disulfides upon reaction with sulfhydryls, and the modifications can subsequently be reversed using disulfide reducing agents, like DTT or TCEP. AEAETS is moisture sensitive and hydrolyzes slowly in aqueous solution, cleaving to release mecaptoethylamine and taurine. Also, avoid contact with oxidizing agents, such as peroxide, as these will oxidatively cleave and inactivate the reagent.

AEAETS is freely soluble in aqueous buffers and in DMF or DMSO to about 100 mg/ml. A stock solution may be made in organic solvent and a small aliquot transferred to a buffered reaction medium to initiate the reaction. The reaction of AEAETS with thiols takes place rapidly (within minutes) provided the group is accessible. Since cysteine is the least accessible amino acid in proteins (Section 2.1, this chapter), globular proteins may contain sulfhydryls that are buried or not fully accessible to the surrounding aqueous environment, and these may react slowly or not at all. The modification of cysteine thiols may be carried out in 10-m*M* Hepes, pH 7.5, and containing 150-m*M* NaCl using 10–100 μ*M* AEAETS in the final reaction medium.

Modification of Carbohydrates with Diamines

Carbohydrates or oligosaccharides may be modified to contain primary amino groups by selective reaction with a diamine compound. Several reaction pathways may be used to accomplish this modification. In some cases, a particular carbohydrate may contain sugar residues that possess potential amine coupling groups without prior derivatization to form such functionalities. For example, if carboxylate-containing sugars are present, such as sialic or uronic acid (Figure 2.95), then direct modification with a diamine is possible using the carbodiimide coupling protocol described previously in this section.

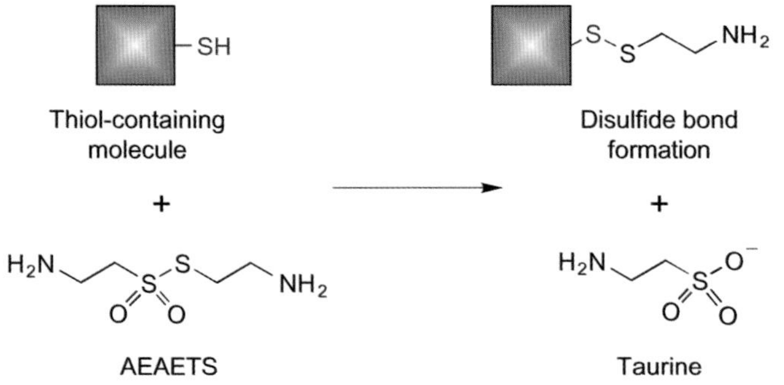

FIGURE 2.94 AEAETS reacts with thiol groups to form a disulfide modification that terminates in a primary amine.

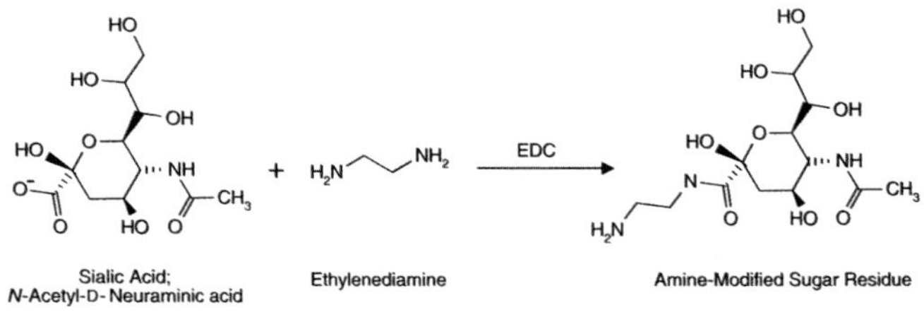

FIGURE 2.95 Carboxylate-containing sugars may be modified with diamines using a carbodiimide-mediated reaction to create available amine groups for subsequent conjugation.

If carboxylates are lacking in the carbohydrate molecule, then indigenous hydroxyls may be utilized to create aldehydes for coupling diamines by one of two routes. The simplest method of creating amine-reactive groups in sugar molecules is by oxidation using sodium periodate (Section 4.4, this chapter). Periodic acid cleaves adjacent hydroxyls to form highly reactive aldehyde groups (Rothfus and Smith, 1963). At a concentration of 1-m*M* in the cold, sodium periodate specifically cleaves only at the adjacent hydroxyls between the number 7, 8, and 9 carbon atoms of sialic acid residues (Van Lenten and Ashwell, 1971; Wilchek and Bayer, 1987). The product is the formation of one aldehyde group on the number 7 carbon and liberation of two molecules of formaldehyde. The sialic acid aldehyde then can be coupled with diamines by Schiff base formation and reductive amination (Chapter 3, Section 5.3, and Chapter 4, Section 4).

Oxidation of polysaccharides using a 10-m*M* or greater concentration of sodium periodate will result in the cleavage of adjacent diol-containing carbon–carbon bonds on other sugars besides just sialic acid residues. Glycoproteins and polysaccharides may be modified using this procedure to form multiple formyl functionalities for coupling diamines or other amine-containing molecules.

In some instances, reducing sugars are present that can be reductively aminated without prior periodate treatment. A reducing end of a monosaccharide, a disaccharide, or a polysaccharide chain may be coupled to a diamine by reductive amination to yield an aminoalkyl derivative bound by a secondary amine linkage (Figure 2.96). Also see Section 4.6, this chapter, for an extensive discussion on carbohydrate modification techniques.

An alternative to the use of chemical means to create formyl groups is the specific modification afforded by sugar oxidases (Section 4.4, this chapter). For instance, galactose oxidase may be reacted with a carbohydrate-containing terminal D-galactose or *N*-acetyl-D-galactosamine residues to transform the C-6 hydroxyl group into an aldehyde (Avigad *et al.*, 1962). Subsequent reaction with a diamine yields the desired amine modification.

The appropriate protocols for diamine modification of various carbohydrate or glycoprotein derivatives may be found in the indicated sections.

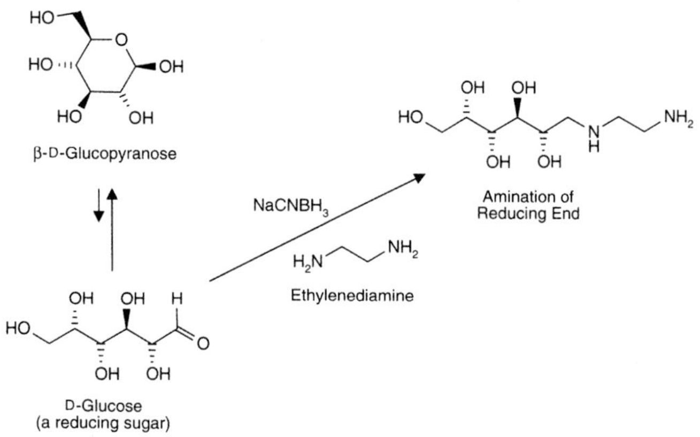

FIGURE 2.96 Reducing sugars may be aminated with diamines in the presence of sodium cyanoborohydride to produce amine modifications.

Modification of Alkylphosphates with Diamines

Alkylphosphate groups can be made to react with diamines to form aminoalkylphosphoramidate modifications. The primary amine thus formed may then be used to conjugate with other molecules containing amine-reactive groups. In this sense, DNA or RNA may be modified with a diamine at the 5′ phosphate group mediated by a carbodiimide reaction. N-substituted carbodiimides can react with a phosphate group to form highly reactive phosphodiester derivatives that are extremely short-lived in aqueous solution (Chapter 4, Section 1) (Figure 2.97). This active species can then react with a nucleophile such as a primary amine to form a phosphoramidate bond (Chu *et al.*, 1986). The process is analogous to the activation of a carboxylate by a carbodiimide with subsequent coupling to an amine-containing molecule to form an amide linkage (Williams and Ibrahim, 1981).

In most procedures, the water soluble carbodiimide EDC (1-ethyl-3-(3-dimethylaminopropyl)carbodiimide hydrochloride) is the most effective mediator of this reaction. Both EDC and its reaction byproducts are fully soluble in aqueous buffers and can be easily separated from the modified aminoalkylphosphate (Chapter 3, Section 1.1).

In some methods, the reaction is carried out in a two-step process by first forming an intermediate, reactive phosphorylimidazolide by EDC conjugation in an imidazole buffer. Next, the diamine, in this case cystamine, is reacted with the activated oligonucleotide, causing the imidazole to be replaced by the amine and creating a phosphoramidate linkage (Chu *et al.*, 1986). An easier protocol was described by Ghosh *et al.* (1990) in which the oligo, cystamine, and EDC were all reacted together in an imidazole buffer. A modification of this method developed by Zanocco (1993) is described in Chapter 23, Section 2.1.

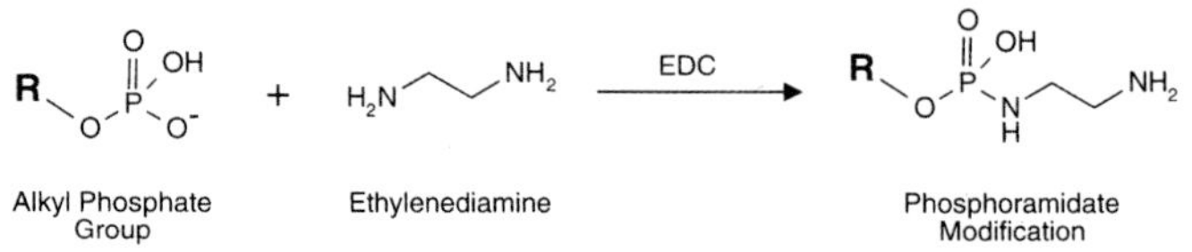

FIGURE 2.97 Phosphate groups may be modified to possess amines by a carbodiimide reaction in the presence of a diamine.

Modification of Aldehydes with Ammonia or Diamines

Aldehyde groups can be converted into terminal amines by a reductive amination process with ammonia or a diamine compound. The reaction proceeds by initial formation of a Schiff base interaction—a dehydration step yielding an imine derivative (or an aldimine). Reduction of the Schiff base with sodium cyanoborohydride or sodium borohydride produces the primary amine (in the case of ammonia) or a secondary amine derivative terminating in a primary amine (for a diamine compound) (Figure 2.98).

This simple strategy can be used to add amine residues to polysaccharide molecules after formation of aldehydes by periodate or enzymatic oxidation (Section 4.4, this chapter). Thus, glycoconjugates or carbohydrate polymers such as dextran may be derivatized to contain amines for further conjugation reactions.

The reaction occurs rapidly at alkaline pH (7–10), with higher pH values resulting in better yields due to faster Schiff base formation. To ensure complete conversion of available aldehydes to amines, add the ammonia or diamine compound to the reaction in at least a 10-fold molar excess over the expected number of formyl groups present. Diamines that are commonly used for this process include ethylene diamine, diaminodipropylamine (3,3′-iminobispropylamine), 1,6-diaminohexane, the Jeffamine derivative EDR-148 containing a hydrophilic, 10-atom chain (Texaco Chemical Co.), and other PEG-based diamines having one end masked with a reversible blocking agent (see previous discussion of discrete PEG diamines in this section).

Introduction of Arylamines on Phenolic Compounds

Compounds having phenol ring structures, such as tyrosine residues in proteins, can often be derivatized to contain aromatic amine groups through a two-stage reaction process. First, the phenolic ring is nitrated with tetranitromethane in aqueous solution to add a nitro group *ortho* (or *para*, if available) to the hydroxyl. This type of modification can be used to detect tyrosine residues by the strong absorptivity of the unprotonated (at pH 9), 3-nitrophenolate ring at 428 nm (extinction coefficient = 4200 $M^{-1}cm^{-1}$) (Sokolovsky *et al.*, 1967). The method has been used to quantify the tyrosine content in porcine trypsinogens and trypsins and to modify a variety of other proteins (Vincent *et al.*, 1970; Lundblad, 1991).

Nitration of the tyrosine rings in the four binding pockets of avidin or streptavidin can be done to increase the steric hinderance within the biotin binding sites (Morag *et al.*, 1996). This process yields chromogenic proteins that have reduced binding affinity for biotin, thus allowing elution of biotinylated molecules under mild conditions.

The nitrophenol group of nitrated molecules may also be reduced to an aminophenyl derivative in alkaline conditions with the use of sodium dithionite (sodium hydrosulfite, $Na_2S_2O_4$) (Pojer, 1979). Dithionite is often used to convert aromatic nitro groups into aryl amines in organic synthesis procedures (Nakayama *et al.*, 2011). The aryl amine groups can then be used to conjugate with amine-reactive crosslinking reagents to label peptides or proteins at their tyrosine side chains. In addition, this strategy can be a route to creating modifiable amine groups on aromatic molecules other than just tyrosine. For instance, the Bolton–Hunter reagent (Chapter 12, Section 2.5) can be used to modify amine groups on proteins, leaving a phenolic end that is typically used as a site for radioiodination. However, such a derivative could also be used to create an arylamine for further transformation into a highly reactive diazonium group for coupling to tyrosines or phenolic functionalities in other molecules (Figure 2.99) (Chapter 3, Section 6.1).

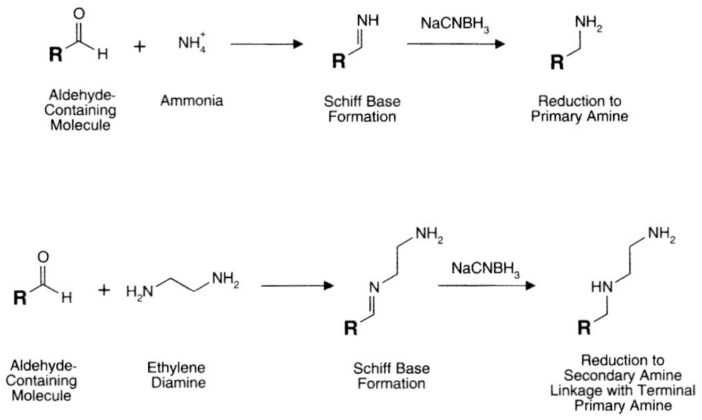

FIGURE 2.98 Aldehydes may be transformed into primary amines by reaction with ammonia or a diamine in the presence of a reducing agent.

Protocol

1. Dissolve the protein containing tyrosine residues (or another phenolic macromolecule) in 0.02-*M* sodium phosphate, 0.15-*M* NaCl, pH 7.4, at a concentration of 2–4 mg/ml.
2. With stirring, add to each milliliter of the protein solution, 20 μl of 0.15-*M* tetranitromethane in 95% ethanol (Sigma). Make the addition in small aliquots if more than several milliliters of solution are to be derivatized. **Note**: all operations with tetranitromethane should be carried out in a fume hood with extreme care, as this compound is sensitive to heat, friction, and shock or impact.
3. React for 1 h at room temperature.
4. Quench the reaction by immediate gel filtration using a column of Sephadex G-25 (Pharmacia).

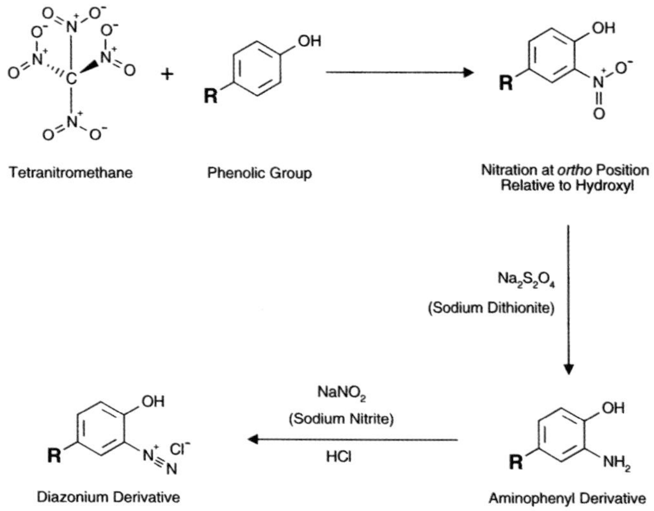

FIGURE 2.99 Phenolic compounds, such as the side chain of tyrosine residues, may be modified to contain an amine group by nitration followed by reduction to the aminophenyl derivative.

Equilibrate the column and perform the chromatography using 0.2-*M* sodium borate, pH 9.0, so that the protein will be at the proper pH for the reduction step. After the separation, a determination of the modification level may be made by measuring its absorbance at 428 nm.

5. Add sufficient sodium dithionite to bring the final concentration in the reaction medium to 0.1-*M*.
6. React for 1 h at room temperature.
7. Purify the aminophenyl derivative by gel filtration or dialysis.

The formation of a diazonium group from the arylamine derivative can be carried out by treatment with sodium nitrite in HCl (see protocol in Chapter 10, Section 6).

Amine Detection Reagents

There are several methods available for the detection or measurement of amine groups in proteins and other molecules. Accurate determination of target amine groups in molecules before or after modification may be important for assessing reaction yield or suitability for subsequent crosslinking procedures. The following methods use commercially available reagents and are easily employed to detect primary amines with simple spectrophotometric measurement.

TNBS

Molecules containing primary amines or hydrazide groups can react with 2,4,6-trinitrobenzenesulfonate (TNBS) to form a highly chromogenic derivative (Figure 2.100). This reaction may be used to assay the amine content of compounds by measuring the absorbance of the orange-colored product at 335 nm.

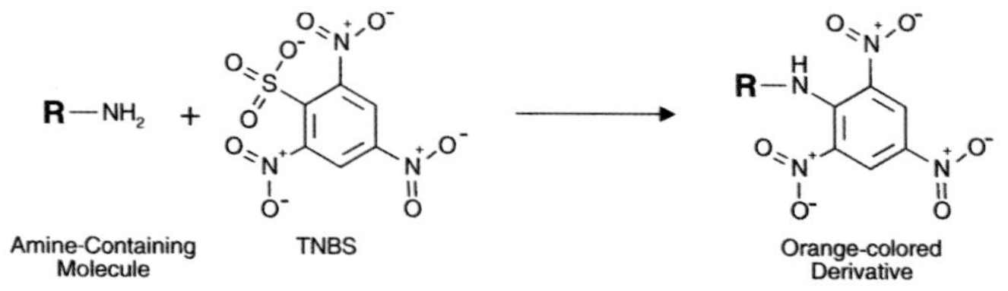

FIGURE 2.100 TNBS may be used to detect or quantify amine groups through the production of a chromogenic derivative.

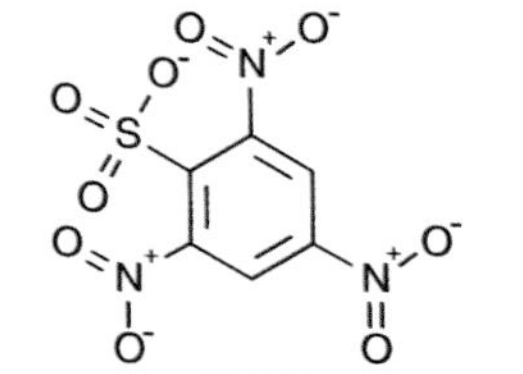

TNBS;
Trinitrobenzene sulfonic acid
MW 293

TNBS has been used to measure the free amino groups in proteins (Habeeb, 1966; Kakade and Liener, 1969); as a qualitative check for the presence of amines, sulfhydryls, or hydrazides (Inman and Dintzis, 1969); and to specifically determine the number of ε-amino groups of L-lysine in carrier proteins (Sashidhar *et al.*, 1994). It is also a convenient reagent to quantitatively or qualitatively assess the formation or blocking of amine groups during bioconjugation reactions (Tripathi *et al.*, 2011).

The following protocol may be used for the measurement of amines in soluble molecules, such as proteins or other macromolecules.

Protocol

1. Dissolve or dialyze the molecule to be assayed into 0.1-*M* sodium bicarbonate, pH 8.5, at a concentration of 20–200 μg/ml (for large molecules like proteins) or 2–20 μg/ml (for small molecules like amino acids).
2. Dissolve TNBS in 0.1-*M* sodium bicarbonate, pH 8.5, at a concentration of 0.01% (w/v). Prepare fresh. **Note**: TNBS may be prepared as a stock solution in ethanol at a concentration of 1.5%. This solution is stable for long-term storage and may be diluted as needed in the bicarbonate buffer to the required concentration.
3. Add 0.5 ml of TNBS solution to 1 ml of each sample solution. Mix well.
4. Incubate at 37°C for 2 h.
5. Add 0.5 ml of 10% SDS and 0.25 ml of 1-*N* HCl to each sample.
6. Measure the absorbance of the solutions at 335 nm. Determination of the number of amines present in a particular sample may be achieved by

comparison to a standard curve generated by use of an amine-containing compound (i.e., an amino acid) dissolved at a series of known concentrations in the bicarbonate sample buffer and assayed under identical conditions.

OPA

OPA (*o*-phthaldialdehyde) is an amine detection reagent that reacts in the presence of 2-mercaptoethanol to generate a fluorescent product (for preparation, see Section 4.1, 2-Mercaptoethanol, this chapter) (Figure 2.101). The resultant fluorophore has an excitation wavelength of 360 nm and an emission point at 455 nm. OPA can be used as a sensitive detection reagent for the HPLC separation of amino acids, peptides, and proteins (Fried *et al.*, 1985). It is also possible to measure the amine content in proteins and other molecules using a test-tube or microplate format assay with OPA. Detection limits are typically in the μg/ml range for proteins.

Protocol

1. Prepare a series of standards, preferably consisting of serial dilutions of the substance to be measured, dissolved in water or non-amine-containing buffer. The concentration range of the standards can be anywhere between about 500 ng/ml and 1 mg/ml.
2. Prepare the samples dissolved in water or non-amine-containing buffer at an expected concentration level that falls within the standard curve range. The assay can tolerate the presence of most buffer salts, denaturants, and detergents. However, the standard curve should be run in the same buffer environment as the samples to obtain consistent response.
3. To a set of labeled tubes, add 2 ml of OPA reagent (Thermo Fisher) and 200 μl of the appropriate standard or sample. Mix well. If using a microplate format, scale back these quantities 10-fold to fit in the microwells.
4. Measure the fluorescence of each sample and standard using an excitation wavelength of 360 nm and an emission wavelength of 450 nm (or using a filter close to the 436- to 455-nm range).
5. Determine the concentration of the samples by comparison to the standard curve. Since the assay measures the presence of amine groups, the results may be correlated to the relative amount of amines available.

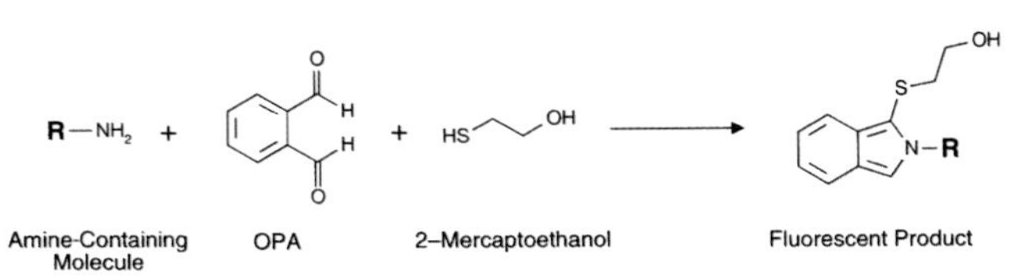

FIGURE 2.101 OPA reacts with amines to form a fluorescent product.

4.4. Introduction of Aldehyde Residues

The formation of an aldehyde group on a macromolecule can produce an extremely useful derivative for subsequent modification or conjugation reactions. In their native state, proteins, peptides, nucleic acids, and oligonucleotides contain no naturally occurring aldehyde residues. There are no aldehydes on amino acid side chains, none introduced by post-translational modifications, and no formyl groups on any of the bases or sugars of DNA and RNA. To create reactive aldehydes at specific locations within these molecules opens the possibility of directing modification reactions toward discrete sites within the macromolecule.

There are two basic ways of introducing aldehyde residues in biological macromolecules: (1) oxidation of carbohydrates or molecules containing diols; and (2) modification of available amino groups with reagents that contain or produce aldehydes. In both cases, aldehydes can be created that will allow easy conjugation to amine-containing molecules by Schiff base formation and reductive amination (Chapter 3, Section 5.3, and Chapter 4, Section 4). The following sections describe these methods.

Periodate Oxidation of Glycols and Carbohydrates

Carbohydrates and other biological molecules that contain polysaccharides, such as glycoproteins, can be specifically modified at their sugar residues to produce reactive aldehyde functionalities. With proteins, this method often allows modification to occur only at specific locals, usually away from critical active centers or binding sites.

Periodate oxidation is perhaps the simplest route to transforming the relatively unreactive hydroxyls of sugar residues into amine-reactive aldehydes. Periodate cleaves carbon–carbon bonds that possess adjacent hydroxyls, oxidizing the –OH groups to form highly reactive aldehydes (Bobbit, 1956; Rothfus and Smith, 1963). Terminal *cis*-glycols result in the loss of one carbon atom as formaldehyde and the creation of an aldehyde group on the former number 2 carbon atom. Varying the concentration of sodium periodate during the oxidation reaction gives some specificity with regard to what sugar residues are modified. Sodium periodate at a concentration of 1-m*M* at near 0°C specifically cleaves only at the adjacent hydroxyls between carbon atoms 7, 8, and 9 of sialic acid residues (Van Lenten and Ashwell, 1971; Wilchek and Bayer, 1987). The product is the formation of one aldehyde group on the number 7 carbon and liberation of two molecules of formaldehyde (Figure 2.102).

Since sialic acid is a frequent terminal sugar constituent of the polysaccharide trees on glycoproteins,

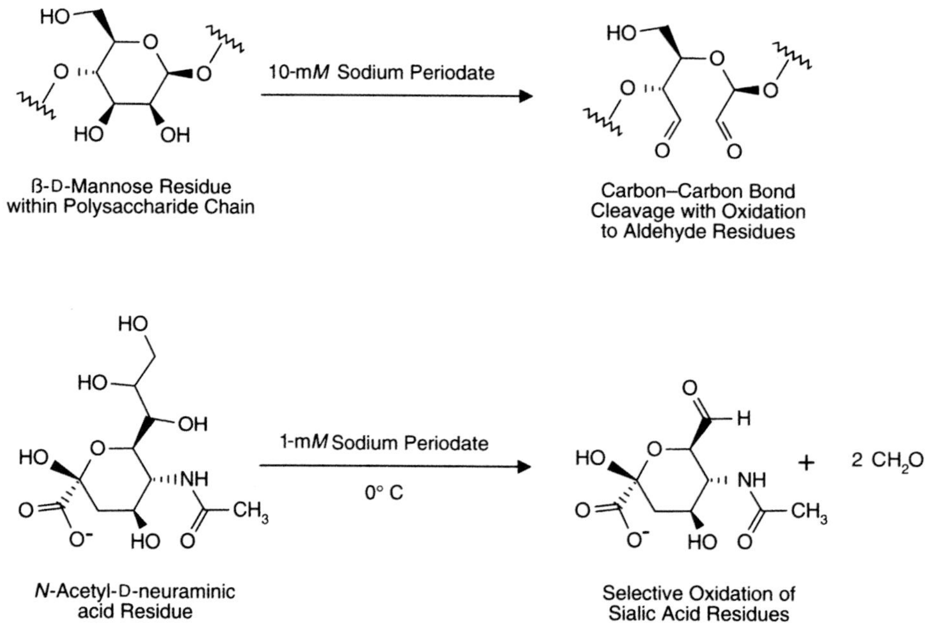

FIGURE 2.102 The reaction of sodium periodate with sugar residues can produce aldehydes for conjugation reactions.

this method selectively forms reactive aldehydes on the most accessible parts for subsequent modifications. The carbohydrate polymer of a protein provides a long spacer arm that can be used to conjugate another large macromolecule, such as a second protein, with few steric problems.

Oxidation of polysaccharides using 10-m*M* or greater concentrations of sodium periodate at room temperature results in the cleavage of adjacent hydroxyl-containing carbon–carbon bonds on other sugars besides just sialic acid residues (Lotan *et al.*, 1975). High concentrations of periodate result in sugar ring opening and the creation of many aldehydes on each polysaccharide tree. The oxidation of dextran polymers has been investigated and found to involve the formation of several aldehyde-containing products depending on the diol location being oxidized (Maia *et al.*, 2011).

Using these methods, carbohydrate-containing proteins may be altered to contain aldehydes for conjugation with other proteins or for detection using hydrazide-containing fluorescent probes (Chapter 10). The aldehydes thus formed can then be coupled to other amine-containing molecules by Schiff base formation and reductive amination (Chapter 3, Section 5.3 and Chapter 4, Section 4). For instance, the enzyme horseradish peroxidase (HRP) can be activated with periodate for conjugation with antibodies (Nakane and Kawaoi, 1974). Alternatively, such reactive aldehyde groups may be conjugated to hydrazide-containing molecules to form hydrazone bonds (Chapter 5, Section 8, Chapter 10, Section 1, and Chapter 22, Section 2.2). Cell surface polysaccharides may be probed with hydrazide-containing reagents for sialic acid groups or total glycoconjugates. Glycoproteins or glycopeptides in solution may also be tagged in this manner. Gangliosides and other glycolipids may also be modified with hydrazide reagents (Spiegel *et al.*, 1982).

Protocol

1. The glycoprotein or diol-containing molecule is dissolved in deionized water or a buffer at physiological pH. Sodium phosphate buffer (0.01–0.1-*M*), pH 7.0, is an appropriate choice. When oxidizing cell surface glycoconjugates, use a buffer suitable for cellular stability requirements. Avoid amine-containing buffers such as Tris and glycine, because they may interact with the aldehyde groups as they are formed. For glycoproteins in solution, a concentration range of 1–10 mg/ml will produce acceptable results in this procedure. For sialic acid modification, place the sample in ice to cool to near 0°C.
2. Dissolve sodium periodate (MW 213.91) in water at a concentration of 10 mg/ml (0.046-*M*). Protect from light. To obtain approximately a 1-m*M* concentration of sodium periodate in the reaction solution (suitable for oxidizing only sialic acid residues), add 21.8 µl of this stock solution to each milliliter of the glycoprotein solution to be oxidized. Maintain the solution on ice. For general oxidation of carbohydrates other than just sialic acid, add 218 µl of the stock solution to obtain an approximate final concentration of 10-m*M* periodate in the reaction. Use room temperature conditions for general

carbohydrate oxidation. Wrap the vial containing the reaction solution with aluminum foil to protect from light. The use of an amber vial also is suitable for this purpose.
3. React for 15–30 min at room temperature.
4. Quench the reaction by the addition of 0.1 ml of glycerol per milliliter of reaction solution. Alternatively, the reaction may be stopped by immediate gel filtration on a desalting column. If a dextran-based resin is used for the chromatography, the support itself will react with sodium periodate to quench excess reagent. Alternatively, *N*-acetylmethionine may be added to quench the reaction, because the thioether of the methionine side chain will react with periodate to form sulfoxide or sulfone products (Geoghegan and Stroh, 1992). In addition, sodium sulfite (Na_2SO_3) was used by Stolowitz *et al.* (2001) to quench the periodate oxidation of HRP in solution. To quench the reaction with cellular samples, wash the cells with buffer to remove remaining traces of periodate.

Oxidase Modification of Sugar Residues

Another method of forming aldehyde groups on carbohydrates and glycoproteins involves the use of specific sugar oxidases. These enzymes only affect the monosaccharide they are specific toward, leaving other sugar residues within polysaccharides unaffected. Probably the most often used oxidase for this purpose is galactose oxidase, which can form C-6 aldehydes on terminal D-galactose or *N*-acetyl-D-galactose residues (Avigad *et al.*, 1962) (Figure 2.103). When galactose residues are penultimate to sialic acid residues, another enzyme, neuraminidase, must be used to remove the sialic acid sugars and expose galactose as the terminal residue (Wilchek and Bayer, 1987). The specificity of using glycosidases to create aldehyde residues on carbohydrates may be the method's greatest advantage. However, the use of a simple chemical reagent such as sodium periodate may still be the easiest way to create aldehydes on carbohydrates (Section 4.4, this chapter).

The following protocol was used by Wilchek and Bayer (1987) to label cell-surface galactose residues.

Protocol

1. Prepare a 5% cell suspension in an appropriate buffer. Avoid amine-containing buffers, as these will interact with aldehydes.
2. Add 0.05 units of *Vibrio cholerae* neuraminidase and 5 units of galactose oxidase per milliliter of cell suspension.
3. Incubate for 60 min at 37°C.
4. Wash cells with PBS to remove excess enzymes.

Modification of Amines with NHS-Aldehydes (SFB and SFPA)

Succinimidyl *p*-formylbenzoate (SFB) (SFB) and succinimidyl *p*-formylphenoxyacetate (SFPA) are amine-reactive reagents that contain terminal aldehyde residues. Their NHS ester ends react with primary amines in proteins and other molecules at pH 7 to 9 to yield amide linkages (see Chapter 3, Section 1.4, and Chapter 17, Section 2) (Figure 2.104). The resulting formyl derivatives may be utilized to couple to other amine or hydrazine-containing molecules (Kraehenbuhl *et al.*, 1974; Galardy *et al.*, 1978). In particular, SFB can be used to produce aldehyde groups on alkaline phosphatase for conjugation with 5′-hydrazide-modified DNA for use in hybridization assays (Chapter 23, Section 2.4) (Ghosh *et al.*, 1989). It has also been used to couple antibodies to sensor surfaces to produce AFM tips (Wildling *et al.*, 2011). The aryl aldehyde group or SFB does not as readily form Schiff base interactions with amines as aliphatic aldehydes. However, it can be used with aminooxy-,

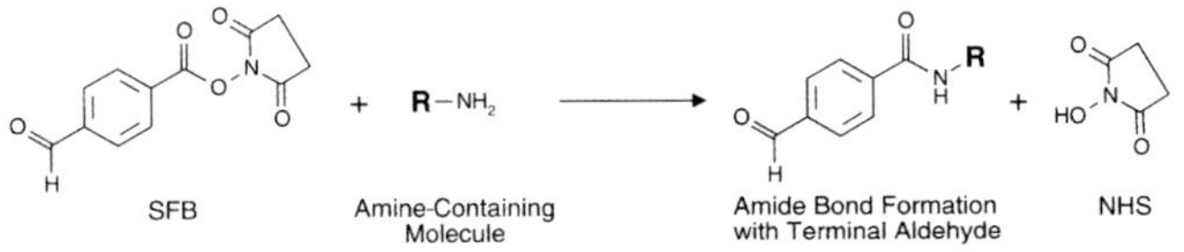

FIGURE 2.104 SFB reacts with primary amines to form amide bond derivatives containing aldehyde groups.

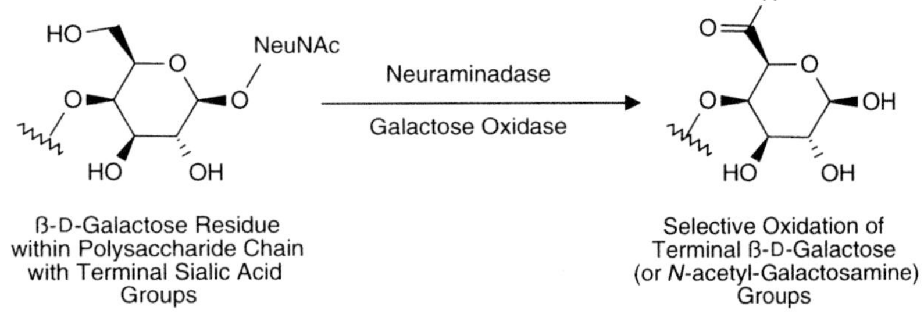

FIGURE 2.103 Galactose oxidase may be used to transform specifically the C-6 hydroxyl group of galactose into an aldehyde.

hydrazide-, and hydrazine-containing reagents in a chemoselective reaction to form crosslinks in the presence of complex biological samples (see Chapters 15 and 17). SFB and SFPA are insoluble in water, but may be predissolved in DMF or acetonitrile before adding a small quantity to an aqueous reaction mixture. Both reagents contain aromatic phenyl rings and have absorptivity at wavelengths less than 300nm. Their structures may contribute a significant degree of hydrophobicity to macromolecules being modified, especially if high-density couplings are achieved. For this reason, modified proteins and other soluble molecules may have a tendency to precipitate if modification is carried out too heavily. The optimal amount of modification may have to be adjusted to maintain solubility in each application.

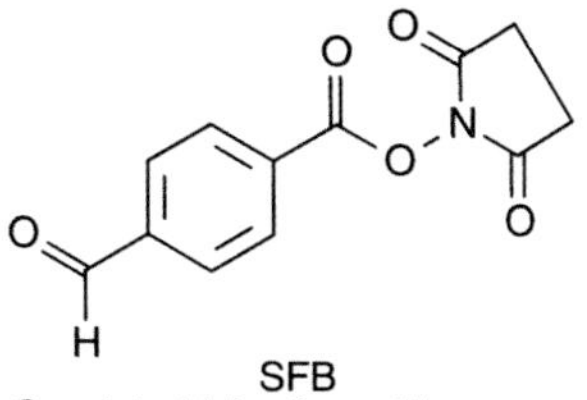

SFB
Succinimidyl-*p*-formyl benzoate
MW 247

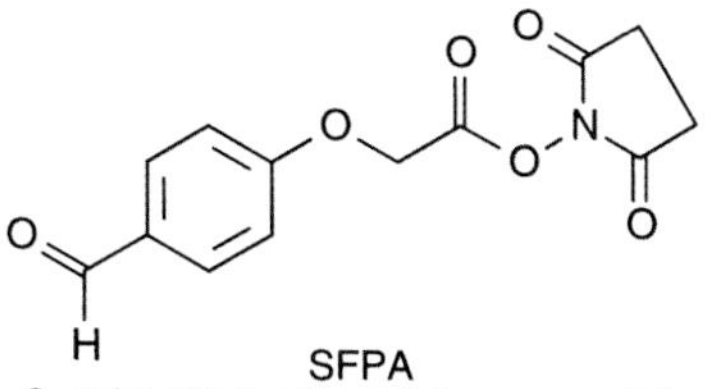

SFPA
Succinimidyl-*p*-formylphenoxyacetate
MW 277

Protocol

1. Dissolve a macromolecule containing amine groups at a concentration of 1–10mg/ml in a buffer having a pH of 7.0–9.0 (e.g., 0.1-*M* sodium phosphate, pH 7.5). Higher pH conditions will increase the hydrolysis rate of the NHS ester. Avoid amine-containing or nucleophilic buffers such as Tris, glycine, or imidazole (see Chapter 3, Section 1.4).
2. Dissolve SFB (Thermo Fisher, Solulink) or SFPA (Molecular Probes) in DMF. The concentration should be such that a small aliquot can be added to the reaction medium to obtain at least a 10-fold molar excess of modifying reagent over the amount of amines to be modified. Add no more than 100μl of the modifier/DMF solution to each milliliter of the macromolecule solution prepared in (1).
3. React for 2h at room temperature.
4. Purify the modified macromolecule from excess reagent and reaction byproducts by dialysis or gel filtration.

Modification of Amines with glutaraldehyde

Amino groups on proteins may be reacted with the *bis*-aldehyde compound glutaraldehyde to form activated derivatives able to crosslink with other proteins. The reaction mechanism for this modification proceeds by one of several possible routes. In the first option, one of the aldehyde ends can form a Schiff base linkage with ε-amines or α-amines on proteins to leave the other aldehyde terminal free to conjugate with another molecule. Alternatively, a cyclic glutaraldehyde derivative or polymer may undergo vinyl addition reactions to create stable secondary amine bonds, leaving the aldehydes exposed for potential secondary reductive amination reactions. Finally, a cyclized form of glutaraldehyde may also react with the ε-amines of two neighboring lysine side chains to form a quaternary pyridinium crosslink (Figure 2.105). See Chapter 15, Section 2.1, for a more extensive explaination of the many reactions glutaraldehyde can undergo in aqueous solutions.

Schiff base interactions between aldehydes and amines typically are not stable enough to form irreversible linkages. These bonds may be reduced with sodium cyanoborohydride or a number of other suitable reductants (Chapter 3, Section 5) to form permanent secondary amine bonds. However, proteins crosslinked by glutaraldehyde without reduction nevertheless show stabilities unexplainable by simple Schiff base formation. The stability of such unreduced glutaraldehyde conjugates has been postulated to be due to the vinyl addition mechanism, which does not depend on the creation of Schiff bases.

Glutaraldehyde modification readily proceeds at alkaline pH. The higher the pH, the more efficient is Schiff base formation. Using a reductant like sodium cyanoborohydride that does not affect the aldehyde groups, while efficiently transforming Schiff bases into a secondary amines, provides the best possible yields. In many cases, the degree of glutaraldehyde-induced crosslinks is so severe that conjugate precipitation occurs. This is especially well documented in antibody–enzyme conjugation schemes employing this reagent (Chapter 20, Section 1.2).

Glutaraldehyde can also be used to create aldehydes on amine-containing polymers. The use of this reagent in derivatizing chromatography supports and other soluble polymers is well known (Hermanson *et al.*, 1992). See Chapter 15, Section 2.1, for updated protocols on the use of glutaraldehyde for immobilizing affinity ligands on chromatography supports.

The following protocol may be used as the first stage of a two-step glutaraldehyde conjugation reaction. In this initial reaction, glutaraldehyde modification converts available protein amines into reactive aldehyde groups (and possibly sites for Michael-type addition

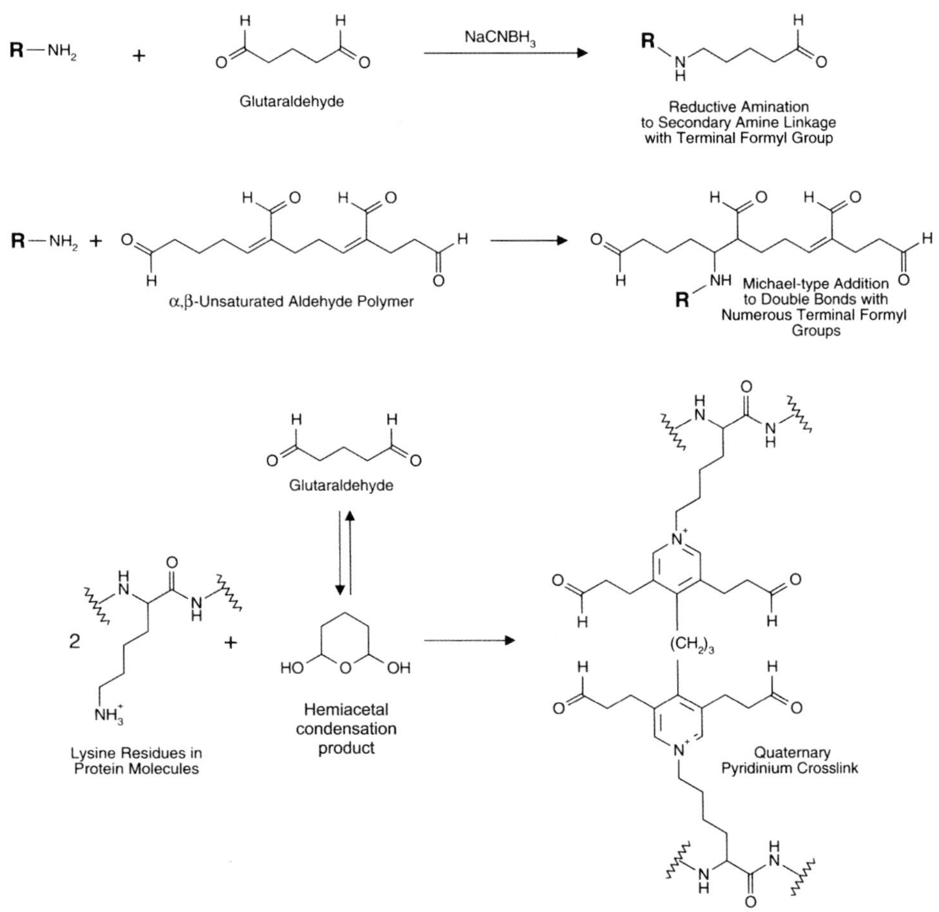

FIGURE 2.105 Glutaraldehyde can undergo complex reactions with amine groups, resulting in aldehyde-containing derivatives that can be used in conjugation reactions.

reactions). The subsequent addition of a second protein or another amine-containing molecule causes the activated protein to crosslink with the amines and forms a conjugate. Glutaraldehyde may also be used in single-step conjugation procedures where the aldehyde-modified protein is not isolated before addition of a second protein. In single-step conjugations both proteins to be crosslinked are together in solution and glutaraldehyde is added to effect crosslinking (Chapter 19, Section 7).

Protocol

1. Dissolve the protein or other amine-containing macromolecule to be modified at a concentration of 1–10 mg/ml in a buffer having a pH from 7 to 10. The higher the pH, the more efficiently Schiff base formation will occur. Phosphate, borate, and carbonate buffers at 0.01–0.1-*M* are acceptable. Avoid amine-containing buffers like Tris and glycine, since they will react with glutaraldehyde.
2. Add a quantity of glutaraldehyde equal to a 10-fold molar excess over the amount of amines to be modified. A typical concentration of glutaraldehyde in the reaction mixture is 1.25%. In some cases, trial experiments will have to be performed to check for solubility of the resultant modified protein. Scale back the quantity of glutaraldehyde added if precipitation occurs.
3. React for at least 2 h at 4°C.
4. Quickly isolate the modified protein by gel filtration using a desalting resin.

In some cases, the modified protein may be stored for long periods before conjugation with another amine-containing molecule by immediate freezing and lyophilization. If stability is a problem, however, the modified protein should be conjugated immediately.

Periodate Oxidation of N-Terminal Serine or Threonine Residues

Sodium periodate can be used to form aldehydes on unmodified N-terminal serine or threonine residues in proteins and peptides (Geoghegan and Stroh, 1992). Periodate cleaves carbon–carbon bonds that both possess primary or secondary hydroxyls or amines (i.e., diols or 2-amino alcohol groups). If a primary hydroxyl is present, such as in the case of N-terminal serine residues, then the reaction liberates formaldehyde and forms an aldehyde group (an α-*N*-glyoxylyl) at the end of the peptide (Figure 2.106). This reaction can be used to direct bioconjugation to a site-specific point on biomolecules, provided that there are no other periodate-oxidizable groups within the protein structure. A synthetic peptide designed to have an N-terminal serine or threonine residue can provide a site of coupling at the end of the chain. This strategy is a viable alternative to the incorporation of a cysteine group for bioconjugation at the end of a peptide.

If other oxidizable groups are present in a protein, such as carbohydrates or sensitive amino acid side chains (Section 2.1, this chapter), then this method should be avoided, because modification can occur at sites other than just the N-terminal. This method has been used successfully to conjugate tags with small peptides (Geoghegan *et al.*, 1993), to attach fluorescent probes to enzyme substrates (Gaertner *et al.*, 1994), to effect the conjugation of lactamase to a Fab′ antibody fragment (Mikolajczyk *et al.*, 1994), for the coupling of antibodies to liposomes (Koning *et al.*, 1999), and to couple a PEG polymer to the amino terminus of proteins (Gaertner and Offord, 1996). The reaction has also been used to selectively label N-terminal serine-containing peptides in the development of a fluorescence caspase assay (Rahman *et al.*, 2012).

When using sodium periodate to oxidize an N-terminal serine or threonine residue on a large molecule like a protein, excess oxidant can be removed simply by dialysis or size exclusion chromatography. However, when using periodate to oxidize a low molecular weight peptide, it can become problematic to remove excess reactant by size separation methods alone. For this reason, the addition of a reducing agent may be used to scavenge any remaining periodate, so long as the reductant chosen does not also reduce the aldehydes that have been formed. Geoghegan and Stroh (1992) used *N*-acetylmethionine for this purpose, because the thioether of the methionine side chain can readily react with periodate to form sulfoxide or sulfone products, but it will not affect the aldehyde groups formed at the end of the peptide chains or interfere

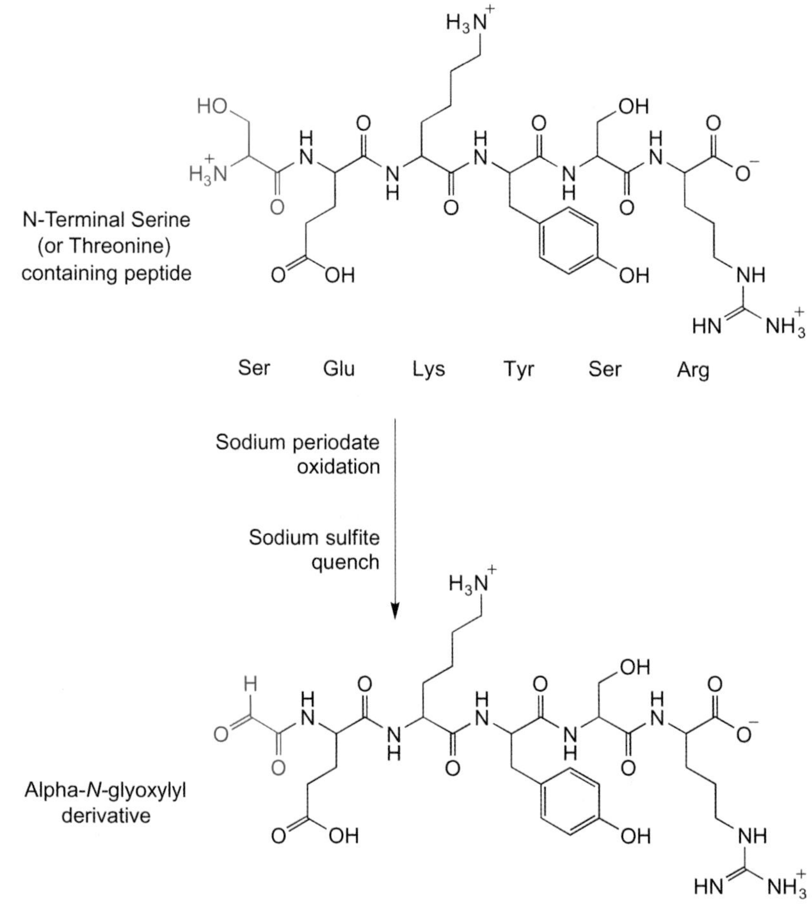

FIGURE 2.106 An N-terminal serine or threonine residue can be oxidized with sodium periodate to produce an aldehyde group. The reaction can be quenched with sodium sulfite to eliminate excess periodate.

with subsequent coupling reactions. A less expensive reagent, sodium sulfite (Na_2SO_3), was used by Stolowitz *et al.* (2001) to quench the periodate oxidation of HRP in solution, and ultimately this may prove to be the best choice for stopping the reaction.

Conjugation of molecules to periodate-oxidized N-terminal serine or threonine residues in peptides is best carried out using hydrazine or hydrazide reagents to avoid potential cross-reactions with lysine amino groups in the peptide structure. The aldehyde group preferentially reacts with the hydrazino group even in the presence of other amines to form a hydrazone bond (Figure 2.107). After the reaction, the hydrazone may be reduced with sodium cyanoborohydride to stabilize the linkage.

The following protocol is based on the methods of Geoghegan and Stroh (1992) and Stolowitz *et al.* (2001).

Protocol

1. Dissolve the peptide containing an N-terminal serine or threonine group at a concentration of at least 2 mg/ml in 0.04-*M* sodium phosphate, pH 7.0. Higher concentrations of peptides or proteins may be used without modification to the rest of the protocol, because the amount of periodate used in the reaction is in sufficient molar excess, even when low molecular weight peptides are being oxidized. Peptides that are initially insoluble at physiological pH may first be dissolved at higher concentration in 0.01% TFA before adding a small aliquot to the buffer. Adjust the pH back to neutral, if needed.
2. With mixing, add sodium periodate to a final concentration of 2.5-m*M*. Periodate should be pre-dissolved as a stock solution at a higher concentration in buffer or water, and then a small aliquot added to the peptide solution to start the reaction. Pre-dissolving the periodate will facilitate immediate dissolution in the final reaction medium without the creation of regions of high concentration, as would occur if solid sodium periodate was added directly to the peptide solution. Protect all solutions containing periodate from exposure to light.
3. React for 3 min at room temperature. Longer reactions increase the likelihood of oxidative damage to other amino acids within the peptide structure.
4. Quench the oxidation by the addition to the peptide solution of at least a 4-fold molar excess of *N*-acetylmethionine or sodium sulfite over the concentration of periodate in the reaction mixture. Pre-dissolve the quencher in buffer at a higher concentration prior to adding an aliquot of it to the reaction solution. React for 10 min.
5. The oxidized peptide may be reacted with an amine- or hydrazide-containing molecule by reductive amination to conjugate with the newly formed aldehyde at the N-terminal (for protocols see the following section, this chapter; Chapter 3, Sections 5.1–5.3; and Chapter 17, Section 2). For conjugation with amine-containing molecules, the peptide must not have any other competing amines (e.g., lysine residues) present or else ring formation or peptide-to-peptide coupling may occur. If lysines are present within the peptide, then the use of a hydrazide (or hydrazine) conjugation process will eliminate interference from lysine amines.

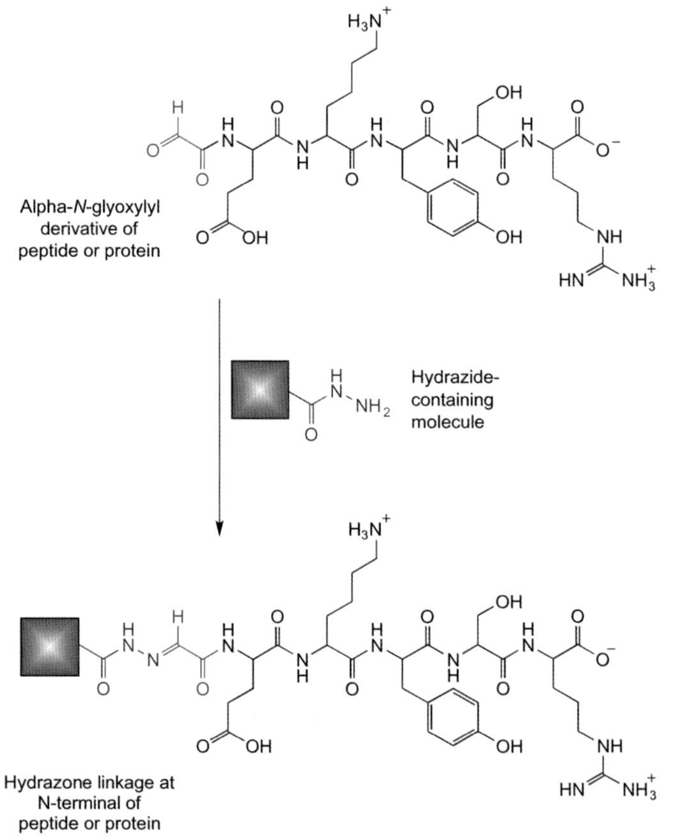

FIGURE 2.107 The N-terminal aldehyde group on a peptide formed from periodate oxidation of serine or threonine residues can be conjugated with a hydrazide-containing molecule to produce a hydrazone bond.

4.5. Introduction of Hydrazine or Hydrazide Functionalities

Hydrazide-containing reagents can be used for probing or conjugation of carbonyl-containing compounds, including macromolecules possessing aldehydes and ketones. Fluorescent or enzymatic probes containing hydrazide functionalities can be used to assay or label carbohydrates, glycoproteins, the polysaccharide portion of cell surfaces, gangliosides, and glycoconjugates on blots (Lotan, 1975; Hurwitz *et al.*, 1980; Spiegal *et al.*, 1982; Gershoni *et al.*, 1985;

Wilchek and Bayer, 1987). Multivalent forms of hydrazide reagents created by modifying enzymes, ferritin, and polymers such as dextran and polypeptides with *bis*-hydrazides can be used to target formyl groups with high avidity and sensitivity (Roffman *et al.*, 1980; Kaplan *et al.*, 1983).

The creation of hydrazide probes is often based on the derivatization of a detectable molecule with a *bis*-hydrazide compound. Although hydrazine itself (in the form of hydrazine hydrate) can be used in a methanolic solution to modify activated carboxylate molecules forming hydrazides, the availability of the bifunctional hydrazides provides a built-in spacer to accommodate greater steric accessibility.

The following protocols make use of the compounds adipic acid dihydrazide and carbohydrazide to derivatize molecules containing aldehydes, carboxylates, and alkylphosphates. The protocols are applicable for the modification of proteins, including enzymes, soluble polymers such as dextrans and poly-amino acids, and insoluble polymers used as micro-carriers or chromatographic supports. See Chapter 15, Section 2.8, for a discussion on the use of hydrazide-containing spacers in the preparation of affinity supports.

The addition of hydrazide groups into macromolecules containing aldehydes, carboxylates, or alkylphosphates has the effect of increasing the pI or net charge. In the case of carboxylates or alkylphosphates, blocking these groups with hydrazide compounds eliminates the negative charge contribution of the original functionality and adds a potential positive charge contribution due to the terminal hydrazide. The consequence of raising the pI of a macromolecule can have dramatic effects on the molecule's conformation and activity or on its relative nonspecificity in assay systems due to the presence of additional positive charge. For instance, the modification of avidin with adipic acid dihydrazide by coupling through the protein's carboxylate groups significantly increases the net charge of an already highly cationic molecule, and therefore increases its overall cross-reactivity in avidin–biotin assays (Chapter 11, Section 5).

Modification of Aldehydes with bis-Hydrazide Compounds

Aldehyde-containing macromolecules will react spontaneously with hydrazide compounds to form hydrazone linkages. The hydrazone bond is a form of Schiff base that is more stable than the Schiff base formed from the interaction of an aldehyde and an amine. The hydrazone, however, may be reduced and further stabilized by the same reductants utilized for reductive amination purposes (Chapter 3, Section 5.4). The addition of sodium cyanoborohydride to a hydrazide–aldehyde reaction drives the equilibrium toward formation of a stable covalent complex. Mallia (1992) found that adipic acid dihydrazide derivatization of periodate oxidized dextran (containing multiple formyl functionalities) proceeds with much greater yield when sodium cyanoborohydride is present.

The reaction of an excess of adipic acid dihydrazide with aldehyde groups present on glycoproteins or other molecules will result in modified proteins containing alkylhydrazide groups (Figure 2.108). Another *bis*-hydrazide compound, carbohydrazide, may also be employed with similar results, except the spacer afforded through its use is considerably shorter. Target aldehydes may be created on macromolecules according to the protocols described in Section 4.4, this chapter. Thus, glycoproteins and other molecules containing polysaccharides may be periodate-oxidized to contain formyl groups and then modified with a *bis*-hydrazide compound to create the hydrazide-activated reagent. Modification of proteins through glycan residues obviates the blocking of negatively charged carboxylates and only adds a limited number of hydrazides at discrete points on the molecule. The enzyme HRP is conveniently modified with hydrazide functionalities using this approach (Chapter 22, Section 2.4).

Protocol

1. Dissolve a macromolecule (such as a protein) containing aldehyde functionalities at a concentration of about 1–10 mg/ml in 100-m*M* sodium citrate, 150-m*M* NaCl, pH 6.0. Alternatively, a buffered solution at a pH of about 7–8.5 will also work well in this protocol. Phosphate, carbonate, borate, or similar buffers adjusted to this pH range work well. Avoid amine-containing buffers (e.g., glycine or Tris) or other components containing strong nucleophiles, since these may react with the aldehydes. To modify a molecule to contain aldehyde groups, see Section 4.4.
2. Add a quantity of adipic acid dihydrazide or carbohydrazide (Aldrich) to the protein solution to obtain at least a 10-fold molar excess over the amount of aldehyde functionality present. High molar ratios are necessary to avoid protein conjugation during the reaction process. If the concentration of aldehydes is unknown, the addition of 32 mg adipic acid dihydrazide per milliliter of the protein solution to be modified should work well.
3. React for 2 h at room temperature. While hydrazone formation does not require the addition of a reductant to create a linkage, including sodium cyanoborohydride in the reaction considerably increases the yield and stability of bonds formed. If the presence of a reducing agent will not cause harm

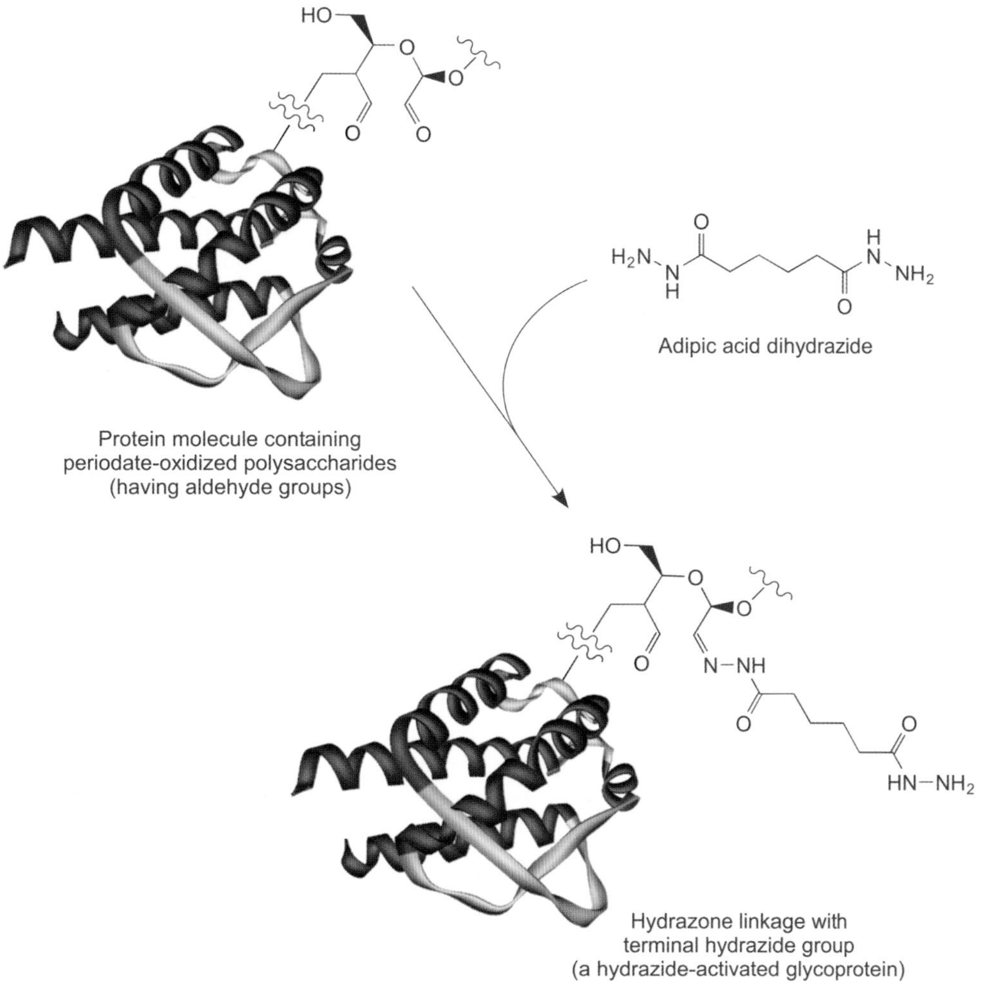

FIGURE 2.108 Glycoproteins that have been treated with sodium periodate to produce aldehyde groups can be further modified with adipic acid dihydrazide to result in a hydrazide derivative.

to the macromolecule being modified, the addition of 10 μl of 5-*M* sodium cyanoborohydride (Sigma) per milliliter of reaction solution may be carried out. Caution: Cyanoborohydride is extremely toxic. All operations should be performed with care in a fume hood. Also, avoid any contact with the reagent, as the 5-*M* solution is prepared in 1-*N* NaOH.

4. Purify the modified protein by dialysis or gel filtration using a desalting resin.

Hydrazide-activated proteins are stable for long-term storage at 4°C in the presence of a preservative (0.05% sodium azide) or in a frozen or lyophilized state.

Modification of Carboxylates with bis-Hydrazide Compounds

Carboxylic acids may be covalently modified with adipic acid dihydrazide or carbohydrazide to yield stable diacyl hydrazide bonds, which also contain extending terminal hydrazide groups. Hydrazide functionalities do not spontaneously react with carboxylate groups the way they do with aldehyde groups (Section 4.5, this chapter). In this case, the carboxylic acid must first be activated with another compound that makes it reactive toward nucleophiles. In organic solutions, this may be accomplished by using a water-insoluble carbodiimide (Chapter 4, Section 1.4) or by creating an intermediate active ester, such as an NHS ester (Chapter 3, Section 1.4).

In aqueous solutions, the easiest method for forming this type of bond is to use the water soluble carbodiimide EDC (Chapter 4, Section 1.1). For proteins and other water soluble macromolecules, EDC reacts with their available carboxylate groups to form an intermediate, highly reactive *O*-acylisourea. This active ester species may further react with nucleophiles such as a hydrazide to yield a stable imide product (Figure 2.109).

Most proteins contain an abundance of carboxylic acid groups from C-terminal functionalities and aspartic and glutamic acid side chains. These groups are

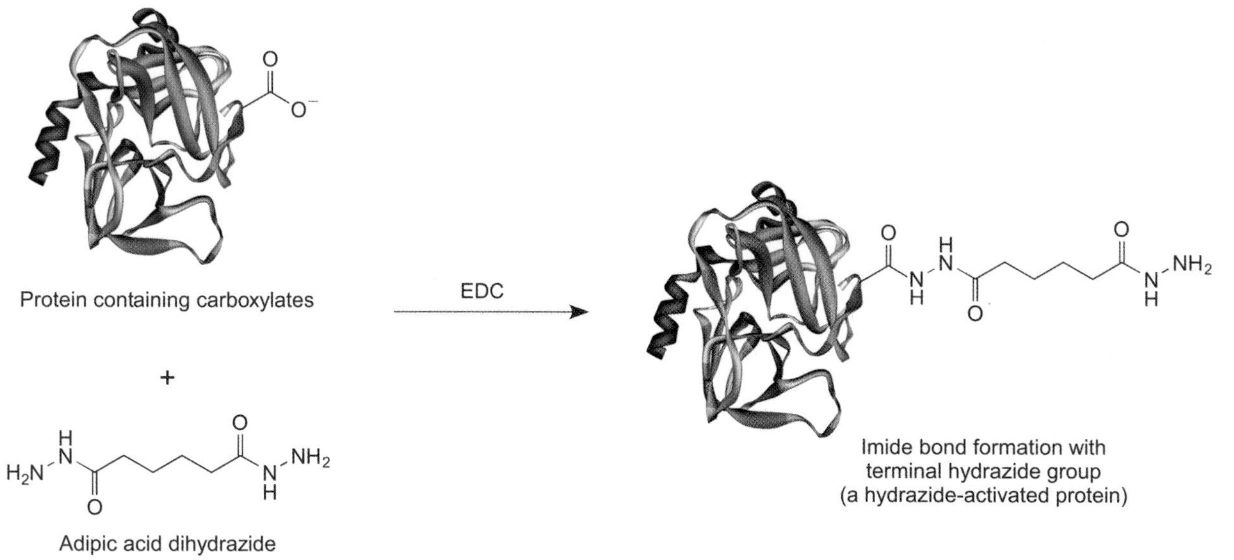

FIGURE 2.109 Carboxylate groups on proteins may be modified with adipic acid dihydrazide in the presence of a carbodiimide to produce hydrazide derivatives.

readily modified with *bis*-hydrazide compounds to yield useful hydrazide-activated derivatives. Both carbohydrazide and adipic acid dihydrazide have been employed in forming these modifications using the carbodiimide reaction (Wilchek and Bayer, 1987).

Protocol

1. Dissolve 32 mg of adipic acid dihydrazide per milliliter of 0.1-*M* sodium phosphate, 0.15-*M* NaCl, pH 7.2.
2. Dissolve 5 mg of the protein or other macromolecule to be modified per milliliter of the above solution.
3. Add 16 mg EDC and react at room temperature for 2–4 h.
4. Purify the modified protein by dialysis or gel filtration using a desalting resin.

Modification of Amines with SANH, SHNH, or SHTH

The introduction of hydrazine or hydrazide functional groups may be performed using the bifunctional crosslinkers that are a part of the hydrazine–aldehyde chemoselective ligation reagents, which are described in Chapter 17, Section 2 (Wong, 1991; Hartmann *et al.*, 2002; Zhong *et al.*, 2003; Kozlov *et al.*, 2004). These are amine-reactive compounds that all have NHS esters on one end and which form amide bonds when coupled to primary amines, such as on the side chain of lysine. SANH (succinimidyl 4-hydrazinonicotinate acetone hydrazone) and SHNH (succinimidyl hydraziniumnicotinate hydrochloride) are both hydrazine derivatives of nicotinic acid, while SHTH (succinimidyl 4-hydrazidoterephthalate hydrochloride) is a hydrazide derivative of terephthalate. Hydrazines and hydrazides react with carbonyl compounds, such as aldehydes and ketones, to form hydrazone bonds. The hydrazone linkage formed between a hydrazine and an aldehyde is much more stable than that formed between a hydrazide and an aldehyde. Thus, hydrazine–aldehyde bonds typically do not require further reduction to stabilize the linkage, as is often the case with hydrazide–aldehyde linkages. For both hydrazide and hydrazine groups, however, the bond that is formed with ketones is typically weaker than the hydrazone formed with aldehydes.

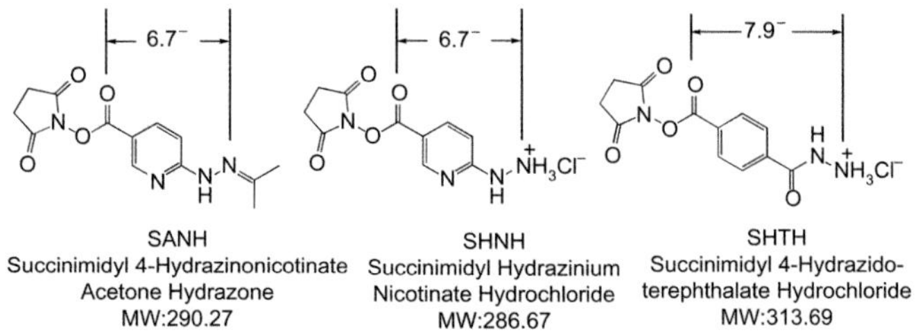

The reaction of SANH, SNHN, or SHTH with an amine-containing molecule proceeds at physiological pH or at slightly basic pH conditions to form an amide bond (Figure 2.110). Reported reaction conditions used the following buffers to modify proteins, amine-modified DNA, or amine-modified surfaces: (a) 0.1-*M* sodium borate buffer, pH 8.1–8.4; or (b) 0.1-*M* sodium phosphate, 0.15-*M* NaCl, pH 7.2–7.4. Whereas NHS esters are reactive throughout this range, the use of pH buffers closer to neutrality will provide optimal yields while reducing the amount of competing hydrolysis as compared to higher pH conditions. Avoid the use of amine or thiol components in the reaction medium, as

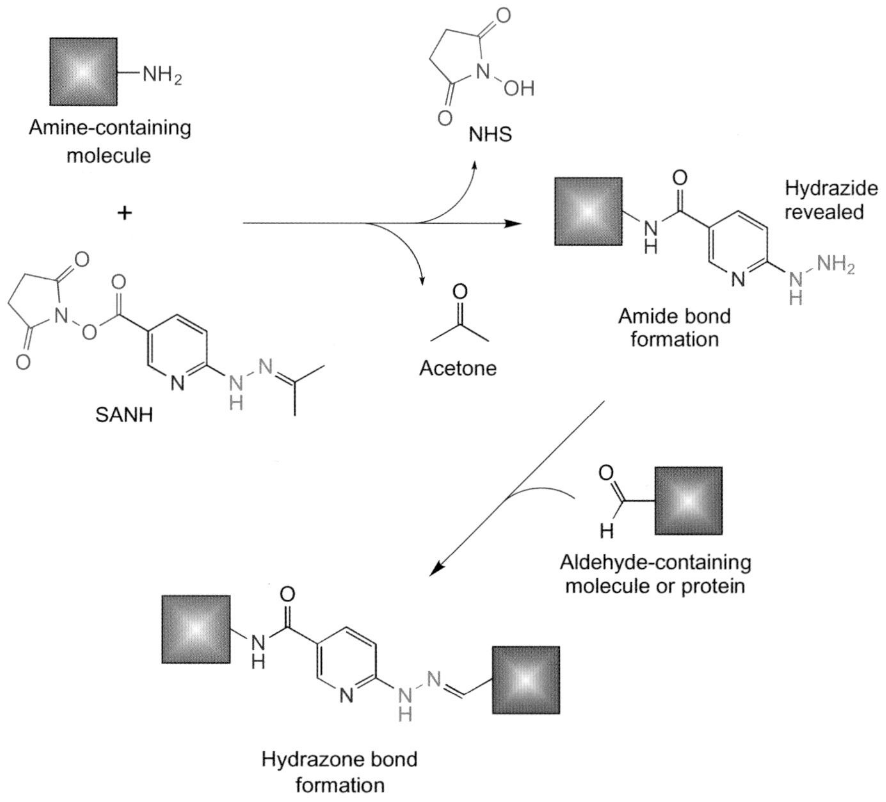

FIGURE 2.110 The reaction of SANH with an amine-containing molecule results in an amide bond derivative that terminates in a protected hydrazine group. Reaction with an aldehyde-containing molecule results in release of the acetone protecting group and formation of a stable hydrazone bond.

these will react with the NHS ester (e.g., avoid Tris buffer, glycine, DTT, or imidazole).

The following protocol describes a general method for modifying amines with SANH, SNHN, or SHTH and assaying for the incorporation of hydrazino groups in the final derivative.

Protocol

A. MODIFICATION OF AMINE GROUPS ON PROTEINS WITH SANH, SHNH, OR SHTH

1. Dissolve SANH, SHNH, or SHTH in DMF to prepare a stock solution at a concentration of 2.0–4.0 mg in 100–200 µl. Use highly pure and dry solvent (H_2O content<0.1% or treat with molecular sieves) to prevent hydrolysis of the NHS ester.
2. Dissolve the protein to be modified in 0.1-*M* sodium phosphate, 0.15-*M* NaCl, pH 7.2–7.4, at a concentration of at least 1–10 mg/ml.
3. With mixing, add an aliquot of the crosslinker solution to the protein solution such that the desired molar excess of reagent over the protein present in the reaction is attained. For most applications, molar ratios in the range of 5:1 to 20:1 will work best to generate a number of hydrazine groups on the protein. Maintain the final percentage of DMF in the reaction mixture at less than 10% to avoid precipitation of protein.
4. React at room temperature for at least 30–60 minutes or 2–3 h at 4°C.
5. Purify the modified protein away from excess reagent and reaction byproducts by gel filtration using a desalting column or dialysis.
6. Calculate the protein concentration in the final preparation using its absorbance at 280 nm or a colorimetric method, such as the Coomassie assay. (Note: The presence of hydrazine or hydrazide groups on the protein will interfere with the BCA assay for total protein concentration.)

B. MEASUREMENT OF HYDRAZINE MODIFICATION LEVEL The number of hydrazine groups per protein molecule can be determined by reacting a small portion of the hydrazine-modified protein with *p*-nitrobenzaldehyde, which forms a chromogenic product upon formation of the hydrazone derivative (Figure 2.111).

1. Prepare a 0.5-m*M* *p*-nitrobenzaldehyde buffer by initially dissolving the compound at a higher concentration in organic solvent (e.g., methanol, DMF, DMSO) and then adding the appropriate

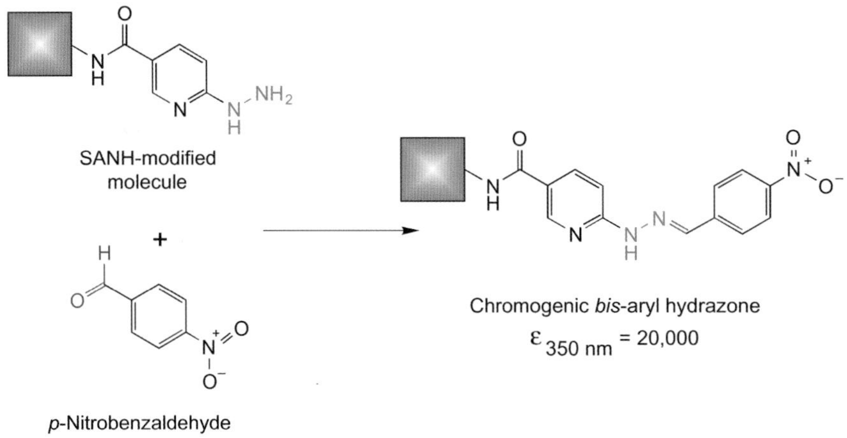

FIGURE 2.111 An SANH-modified molecule can be detected and measured by reaction with *p*-nitrobenzaldehyde, which forms a chromogenic derivative with a characteristic absorbance at 350 nm.

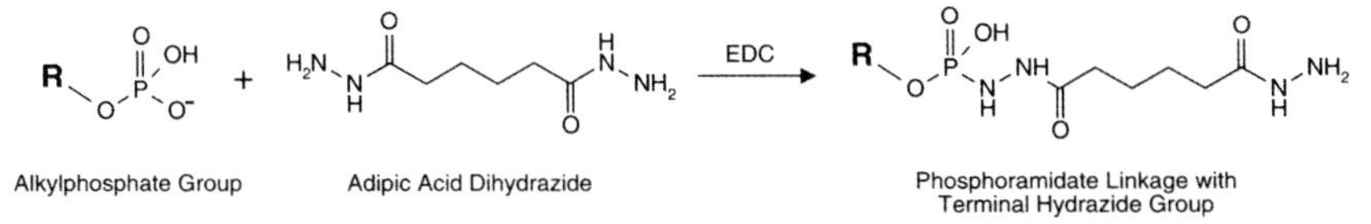

FIGURE 2.112 Phosphate groups can be modified with adipic acid dihydrazide in the presence of a carbodiimide to produce hydrazide derivatives. This is a common modification route for the 5′ phosphate group of oligonucleotides.

aliquot to 100-m*M* acetate, pH 5.0, to result in the final concentration.

2. Add an aliquot of the hydrazine-modified protein solution to the *p*-nitrobenzaldehyde solution and incubate at 37°C for 1 h or at room temperature for 2 h. To ensure accuracy, determine the linear response range of the test by adding a series of different concentrations of the hydrazine-modified protein solution to the *p*-nitrobenzaldehyde buffer. This is achieved by preparing a set of serial dilutions of the protein solution and adding an equal volume of each dilution to an aliquot of the *p*-nitrobenzaldehyde solution in separate tubes. Determine the absorbance of each solution at 380 nm *versus* a blank prepared by the addition of an equal aliquot of buffer alone.
3. Calculate the substitution level of hydrazine groups into the protein by determining the hydrazine concentration using the molar extinction coefficient of the resultant *p*-nitrobenzyl derivative (22,000 $M^{-1} cm^{-1}$) and dividing this value by the protein concentration (moles/liter). Any sample falling within the linear response range of the test can be used for this calculation.

Modification of Alkylphosphates with bis-Hydrazide Compounds

Alkylphosphate groups such as those present at the 5′ end of RNA and DNA molecules may be specifically modified with *bis*-hydrazide compounds. Mediated by the addition of the water soluble carbodiimide EDC and imidazole, adipic acid dihydrazide or carbohydrazide will react with the phosphate group in a two-step process to form phosphoramidate bonds with short linker arms containing terminal hydrazides (Figure 2.112) (Ghosh *et al.*, 1989). In the first stage, EDC activates the phosphate group, forming a short-lived, but highly reactive phosphodiester species, which in turn reacts with a molecule of imidazole to form a longer-lived, active phosphorimidazolide. The second stage involves addition and attack of the hydrazide nucleophile, releasing imidazole and forming the phosphoramidate bond. In a modification of the two-stage reaction, Zanocco (1993) developed a single-pot reaction in which the alkylphosphate molecule is reacted in the presence of EDC, imidazole, and the bis-hydrazide compound. The modification reaction proceeds rapidly at room temperature.

Protocol

1. Weigh out 1.25 mg of the carbodiimide EDC (1-ethyl-3-(3-dimethylaminopropyl)carbodiimide hydrochloride; Thermo Fisher) into a microfuge tube.
2. Add to the tube 7.5 μl of RNA or DNA containing a 5′ phosphate group. The concentration of the oligonucleotide should be 7.5–15 nmol/μl or total of about 57–115.5 μg. Also, immediately add 5 μl of 0.25-*M* adipic acid dihydrazide or carbohydrazide dissolved in 0.1-*M* imidazole, pH 6.0. Because EDC is labile in aqueous solutions, the addition of the oligo and *bis*-hydrazide/imidazole solutions should occur quickly.

3. Mix by vortexing, then place the tube in a microcentrifuge and spin for 5 min at maximal rpm.
4. Add an additional 20 μl of 0.1-*M* imidazole, pH 6.0. Mix and react for at least 2 h at room temperature. The additional buffer prevents pH drift during the carbodiimide reaction.
5. Purify the hydrazide-labeled oligo by gel filtration on a desalting resin using 10-m*M* sodium phosphate, 0.15-*M* NaCl, 10-m*M* EDTA, pH 7.2. The hydrazide-containing probe may now be used to conjugate with a molecule containing an aldehyde-reactive group.

4.6. Introduction of Saccharide or Glycan Groups

The modification of proteins with sugar groups occurs *in vivo* through both enzymatic and non-enzymatic processes. Approximately 1% of proteins encoded in the genomes of mammals are enzymes that are involved with carbohydrate production or modification. Many of these enzymes digest carbohydrate in foods to provide energy for cellular metabolism, but others are involved with the controlled modification of proteins or other biomolecules to create complex polysaccharide structures. This process results in carbohydrates, called glycans, covalently attached to proteins at discrete locations on only certain amino acid residues within a polypeptide sequence (Section 2, this chapter). The presence of carbohydrate modifications on proteins has a pronounced effect on biological activity *in vivo*.

Non-enzymatic modification of proteins with saccharides also occurs *in vivo* through uncontrolled glycation of lysine amines with the reducing end of sugars, especially glucose. This reaction results in the formation of an initial Schiff base with a subsequent rearrangement to form a stable ketoamine derivative. The non-enzymatic glycation reaction has been studied extensively as a result of it being a major factor in the development of the complications associated with diabetes (for a review, see Singh *et al.*, 2001).

In vitro modification of protein can be performed synthetically to add specific sugars or complex carbohydrates to proteins for further bioconjugation or for subsequent study of the glycan derivative *in vivo*. Investigations of the effect of these synthetic carbohydrate–protein conjugates (neoglycoproteins) on the immune response date back many decades with the diazonium-mediated coupling of aminophenol glucosides to study type-II and type-III pneumonia polysaccharides (Goebel *et al.*, 1932). More recently, conjugation of carbohydrates to protein carriers has been carried out to illicit a specific immune response to glyco-antigens of infectious diseases or tumor cells (Toyokuni and Singhal, 1995; Koganty *et al.*, 1996; Ragupathi *et al.*, 1997; Pozsgay, 1998; Mawas *et al.*, 2002; Karsten *et al.*, 2004). Synthetic peptide–glycan conjugates have also been prepared by conjugation of carbohydrates to peptide sequences that can be presented by MHC (major histocompatibility complex) molecules to enhance the immune response against the carbohydrate component (Kihlberg and Magnusson, 1996). For an excellent review of glycoconjugation, see Davis (1999).

Sugar residues can also be used to modify a protein, molecule, or surface for subsequent use in a bioconjugation procedure or to increase the hydrophilicity of the modified molecule. For bioconjugation purposes, a sugar group can be added to facilitate the covalent conjugation of another molecule. Since many saccharides contain diols that can be oxidized by periodate to create aldehydes, certain sugars can be used after glycol oxidation to couple with amine-containing molecules by reductive amination. For instance, the amine group on the monosaccharide glucosamine can be coupled to an amine-reactive surface or to a protein through its carboxylate groups using EDC (Chapter 4, Section 1.1). The glucosamine-modified surface or molecule subsequently can be oxidized by sodium periodate to create aldehydes (Sections 2 and 4.4, this chapter). This reactive intermediate can be conjugated to another amine-containing molecule to create the final conjugate through reductive amination (Chapter 3, Section 5).

Periodate oxidation of carbohydrates should be avoided, however, if generating an immune response against the saccharide component is desired or antibody recognition needs to be retained against the carbohydrate. Woodward *et al.* (1985) demonstrated that periodate oxidation of glycans effectively destroys antibody binding if the specificity of the antibody is truly toward the carbohydrate. The assay of antibody binding to a carbohydrate with and without periodate oxidation is often used to demonstrate antibody specificity. A specific antibody will fail to bind to periodate oxidized carbohydrate, but will bind to the non-oxidized glycan.

The modification of molecules with saccharides also has the effect of increasing the hydrophilicity of the resultant complex due to the presence of multiple hydroxyl groups. Native glycan modification of proteins functions in much the same manner, because the carbohydrate "tree" becomes fully hydrated through the ability of the hydroxyls to hydrogen bond with the surrounding water molecules. In addition, post-translational modification of proteins helps polypeptides fold properly by assuring that certain regions are maintained near the solvent surface. Chemical modification with saccharides thus can be done to increase the hydrophilicity and solubility of proteins or other molecules in aqueous solution.

Some reports even indicate that the conjugation of proteins or peptides with carbohydrates can dramatically increase their activity compared to that of the native state (Susaki *et al.*, 1998). Carbohydrates can also provide a protective effect on modified peptides toward proteolytic digestion (Rudd *et al.*, 1994) or mask recognition of a peptide by the immune system (Harding *et al.*, 1993). The creation of neoglycoproteins thus can affect the activity of peptides and proteins, which are not normally glycosylated *in vivo*. Complex glycans can now be obtained through automated synthesis with amino groups on their reducing ends for ready conjugation with surfaces or proteins (Krock *et al.*, 2012).

The following sections describe several examples of saccharide modification for the purpose of bioconjugation, the study of glycan function, to prepare immunogens, or to increase the water solubility of a modified molecule.

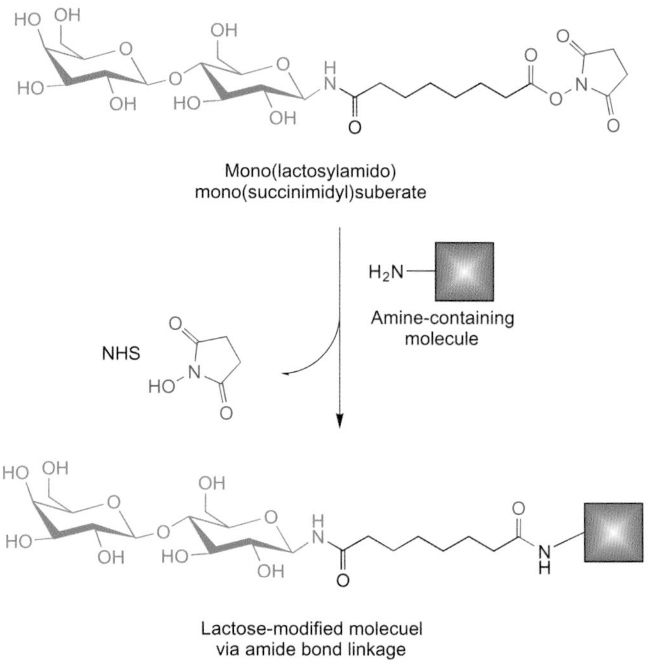

FIGURE 2.113 The NHS ester-suberate derivative of lactose can be used to add lactose groups to amine-containing molecules. The reaction results in the formation of amide bonds containing terminal lactose groups.

Modification of Amines with Mono(lactosylamido) Mono(succinimidyl)suberate

The amine-reactive compound mono(lactosylamido) mono(succinimidyl)suberate contains a lactose group at the end of a suberate bridge, which terminates at the other end in an NHS ester. Vetter *et al.* (1995) prepared this reagent by reaction of the corresponding glycosylamine derivative with disuccinimidyl suberate (DSS). The starting lactosylamine derivative was prepared *via* a 5-day reaction of the reducing end of lactose with aqueous ammonia solution in sodium carbonate. Other similar 1-amino, 1-deoxy sugar derivatives may be synthesized in a like manner for subsequent bioconjugation. The final product after reaction with DSS, the mono-NHS ester derivative of lactose containing an 8-carbon suberate spacer, can be used to modify proteins or other amine-containing molecules with a lactose group. The NHS ester spontaneously reacts with an amine at neutral or slightly basic pH values to form an amide bond (Figure 2.113). The presence of the lactose disaccharide provides a hydrophilic modifying group that increases the water solubility of the resultant conjugate. Other glycosylamine derivatives may be prepared using a similar strategy to modify biomolecules or surfaces with specific saccharide compounds.

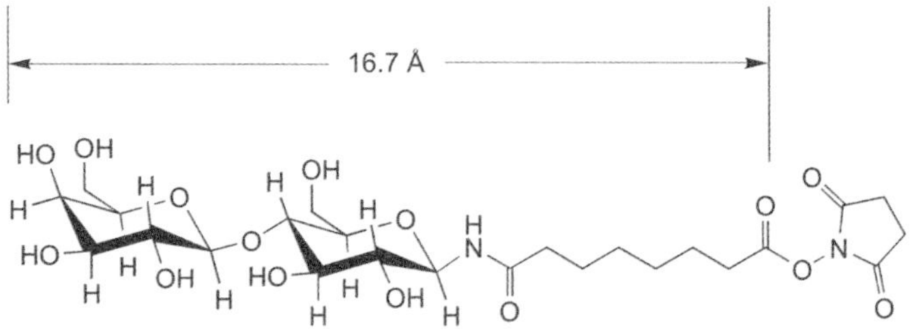

Mono(lactosylamido)
mono(succinimidyl)suberate
MW 594.56

The following protocol describes the modification of a protein with mono(lactosylamido) mono(succinimidyl)suberate. The reagent is available from Thermo Fisher. The use of this reagent to couple to amine-containing surfaces, such as polystyrene beads, has also been performed using similar reaction conditions (Vetter *et al.*, 1995). It was also used to develop an electrochemical binding assay for cholera toxin by affinity interaction with the lactose group conjugated to daunomycin (Kuramitz *et al.*, 2011).

Protocol

1. Dissolve mono(lactosylamido) mono(succinimidyl) suberate in dry DMF to prepare a concentrated solution from which an aliquot may be taken and added to a final aqueous reaction medium. The compound is extremely soluble in DMF, and solutions of 100 mg/ml may be prepared. The use of dry solvent is essential to prevent hydrolysis of the NHS ester. However, make only enough of this stock solution so that a small amount added to the protein reaction will provide the appropriate molar excess desired for the modification reaction.
2. Prepare the protein to be modified in a non-amine-containing buffer at a slightly basic pH (i.e., avoid Tris or imidazole). The use of 0.1-*M* sodium phosphate, 0.15-*M* NaCl, pH 7.2 works well for NHS ester reactions. The concentration of the protein in the reaction buffer may vary from μg/ml to mg/ml, but highly dilute solutions will result in less efficient

modification yields. A protein concentration from 1 to 10 mg/ml works well in this reaction.

3. With mixing, add a quantity of the mono(lactosylamido) mono(succinimidyl)suberate in dry DMF to the protein solution to result in a 10- to 20-fold molar excess of reagent over the amount of protein present. Depending on the desired application for the lactosyl-modified protein, several different molar ratios of reactant-to-protein may have to be tried to optimize the resulting modification level.
4. React for 30–60 min at room temperature with gentle mixing.
5. Purify the modified protein away from reactants and reaction byproducts using dialysis or size exclusion chromatography.

Modification of Amine or Hydrazide Molecules by Carbohydrates and Glycans

The modification of proteins, surfaces, or other molecules with reducing sugars through a reductive amination process is still perhaps the most common method for glycosyl addition. A saccharide or glycan molecule containing a reducing end typically has a masked carbonyl group (i.e., aldehyde or ketone) that can be reacted with an amine or hydrazide in the presence of a reducing agent (such as sodium cyanoborohydride or borane compounds) to form a secondary amine or a reduced hydrazone bond (Figure 2.114). However, since most reducing ends of sugars or glycans exist mainly in an acetal (or ketal) ring form, while only the open form with the exposed aldehyde can participate in a reductive amination reaction, the rate of modification by this process can be slow and the yield low.

To help overcome the predominance of the cyclic acetal form, which inhibits the desired amination reaction, functional groups other than amines (having greater reactivity) have been used, along with elevated temperatures and high concentrations of reducing agent, to form more efficiently the initial Schiff base and thus drive the coupling reaction to completion. In particular, hydrazide or hydrazine groups have been used successfully to modify reducing glycans and saccharides with or without a reducing agent present (Rothenberg *et al.*, 1993; Toomre and Varki, 1994;

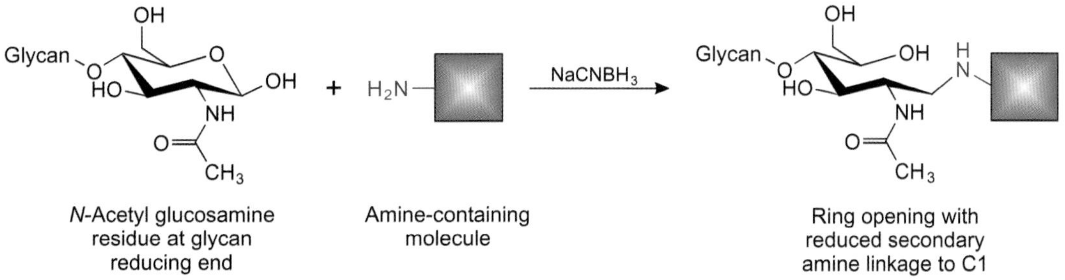

FIGURE 2.114 The reaction of a reducing sugar with an amine-containing compound in the presence of sodium cyanoborohydride results in ring opening with the formation of a secondary amine derivative.

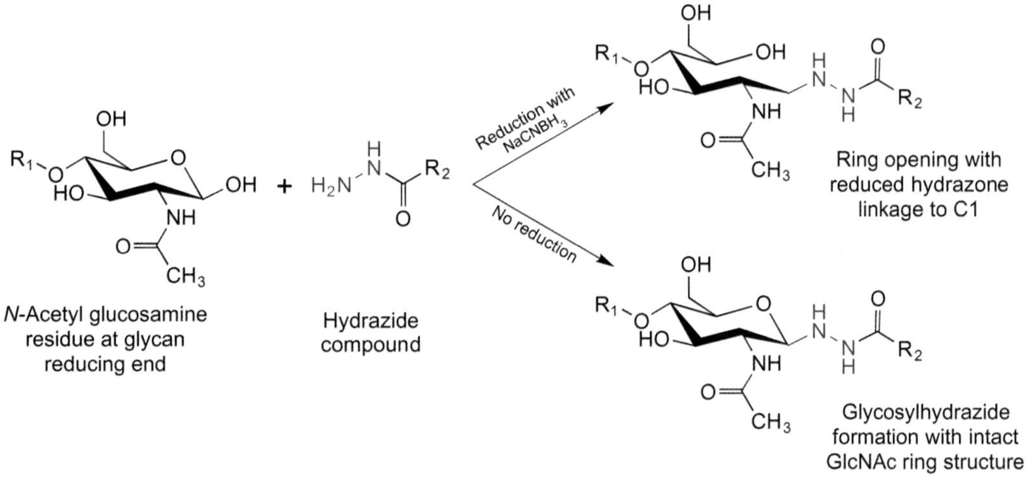

FIGURE 2.115 The reaction of a reducing sugar with a hydrazide-containing molecule can proceed by one of two routes depending on whether a reducing agent is present. If the reaction is performed in the presence of sodium cyanoborohydride, then ring opening will occur followed by the formation of a reduced hydrazone linkage. If no reducing agent is present, then the reaction gives a glycosylhydrazide derivative with retention of the ring structure of the sugar group.

Leteux *et al.*, 1998; Srikrishna *et al.*, 2001) (Figure 2.115). Whereas reactions performed with primary amines at room temperature may take days to reach acceptable coupling yields, the modification of hydrazino groups with the reducing end of a sugar can be done in hours at 60 to 80°C. Also, the modification of glycans with hydrazines can be carried out without the addition of cyanoborohydride, because the resultant hydrazone linkage is more stable compared to the Schiff base formed between an aldehyde and an amine. Although when coupling hydrazino compounds to the reducing end of a glycan, the addition of a reductant is often performed to drive the reaction to completion and further stabilize the hydrazone, reduction of the terminal saccharide can cause changes in the binding potential of some lectins that recognize the core sugar structure (Leteux *et al.*, 1998).

The reversible nature of the hydrazone bond formation between the reducing end of a glycan and an immobilized hydrazide group was exploited to enrich glycans prior to analysis (Yang and Zhang, 2012). In this case, hydrazide beads were used to temporarily bind release glycans from glycoproteins and wash off any non-glycan molecules from the sample. Subsequent elution by hydrolysis of the hydrazone resulted in a pure glycan fraction, which can be analyzed by mass spectrometry.

The following protocol may be used to conjugate the available reducing end of a saccharide or glycan with an amine, hydrazide, or hydrazine group. These functional groups may be present on a variety of molecules, such as biotin groups for labeling or molecules having fluorescent properties, which allows detection of the glycan derivative. For specific modification details, see the methods of Rothenberg *et al.* (1993), Toomre and Varki (1994), Leteux *et al.* (1998), and Srikrishna *et al.* (2001). The following method is based on the optimized conditions as determined by Bigge *et al.* (1995). The use of organic solvents should be performed in a fume hood.

Protocol

1. Dissolve a carbohydrate, saccharide, or glycan sample having a free reducing end in 0.1-*M* sodium acetate, pH 5.0. (Note: Glycans may be released from glycoconjugates by hydrazinolysis using pure hydrazine or by endoglycosidase treatment with PNGase F.) Alternative coupling conditions that can be used for the modification reaction include 30% glacial acetic acid in DMSO (v/v) or acetic acid/pyridine (1:2, v/v). The use of DMSO or pyridine often facilitates solubilization of a greater range of carbohydrates or glycans than aqueous buffers. The presence of acetic acid has been found to accelerate the reductive amination reaction when the organic solvent conditions are used (Bigge *et al.*, 1995). The concentration of the carbohydrate should be 5–100 μ*M* for glycans. For modification of other more abundant carbohydrates, higher concentrations can be used if required. For glycan modification of biomolecules that are not compatible with organic solvents, such as proteins, the glycan initially may be solubilized in DMSO and then an aliquot added to the aqueous reaction buffer.
2. Add to the glycan solution the molecule to be labeled containing an available amine, hydrazine, or hydrazide group. For small molecule derivatization, the final concentration of the nucleophile in the glycan solution should be about 0.3-*M* to result in maximal efficiency of labeling. For protein modification, an aqueous reaction buffer should be used, and the protein should be as concentrated as possible.
3. Add to the reaction mixture a quantity of reducing agent (e.g., sodium cyanoborohydride or borane dimethylamine, BDA) to give a final concentration of 1.0-*M*.
4. When using nonaqueous reaction conditions, incubation should be carried out for 1–2 h at 60–80°C. For reactions in an aqueous environment with temperature-sensitive molecules, the reaction may be performed at room temperature or 37°C. In this case, the reaction time should be extended to at least 24 h. Longer reaction times are not unusual when modifying carbohydrates by reductive amination at ambient temperature. For instance, the coupling of heparin through its reducing end to a solid phase containing a hydrazide group takes up to 72 h at room temperature to obtain maximal yield.
5. Purify the modified glycan from reactants and reaction byproducts by dialysis, gel filtration, or ion-exchange chromatography, depending on the size and type of the molecule being modified by the carbohydrate. For instance, Rothenberg *et al.* (1993), fractionated glycans modified with a biotin-diaminopyridine derivative from excess biotin compound by size exclusion chromatography on a 1.5×48 cm Toyopearl HW40S column equilibrated with 50% acetonitrile/10-m*M* sodium acetate.

Labeling Glycans with Fluorescent 2-Aminopyridine, 2-Amino Benzamide, or Anthranilic Acid

The ability to label glycans released from glycoproteins and other glycoconjugates is important for tracking complex carbohydrates through purification or analysis, such as HPLC, electrophoresis, and mass spec. Glycan molecules typically do not have spectrally detectable groups and therefore benefit by being labeled with a detectable component for easy assay and detection. Small, amine-containing fluorescent compounds have been found to be particularly useful in this regard. The compounds 2-aminopyridine, benzamide, and anthranilic acid (2-aminobenzoic acid) have been used to label the reducing end of glycans and other carbohydrates.

Glycosylated proteins and other glycoconjugates may be treated to release the modifying carbohydrate component, making available the reducing end for conjugation. *N*-linked glycans may be released from glycoconjugates by PNGase F or PNGase A or released by anhydrous hydrazine and regenerated to reducing oligosaccharides, thus yielding free reducing ends for bioconjugation. In addition, the enzymes Endo H and Endo F release *N*-linked glycans by cleaving the GlcNAc(β1-4) GlcNAc core, thereby resulting in one less GlcNAc moiety at the reducing end. *O*-linked glycans may be released using *O*-glycanase treatment, or by mild hydrazinolysis followed by regeneration to the reducing oligosaccharides. *O*-linked glycans may also be released by non-reductive beta-elimination. However, one should avoid strong reductive glycan-releasing methods using sodium borohydride, as these may reduce the C1 aldehyde on the core sugar group and make it unreactive for further derivatization.

Bigge *et al.* (1995) described a method to label released glycans by reductive amination with the small fluorescent compounds 2-amino benzamide and anthranilic acid (2-aminobenzoic acid). The result is a secondary amine bond with the C1 carbon atom of the reducing end of the glycan (Figure 2.116). The derivatives provide stable fluorescent modifications on glycans, which can be used for detection during electrophoresis, separation, and purification techniques (such as HPLC). The 2-amino benzamide label was found to work best for chromatographic separations, enzymatic modifications, and mass spec analysis, while the anthranilic acid derivative was more suitable for electrophoretic separations, because its negative charge produced sharper bands.

Rothenberg *et al.* (1993) described the synthesis of a very useful fluorescent glycan labeling agent that contains a biotin group. Diaminopyridine was reacted with sulfo-NHS-biotin to form 2-amino-(6-amidobiotinyl)pyridine

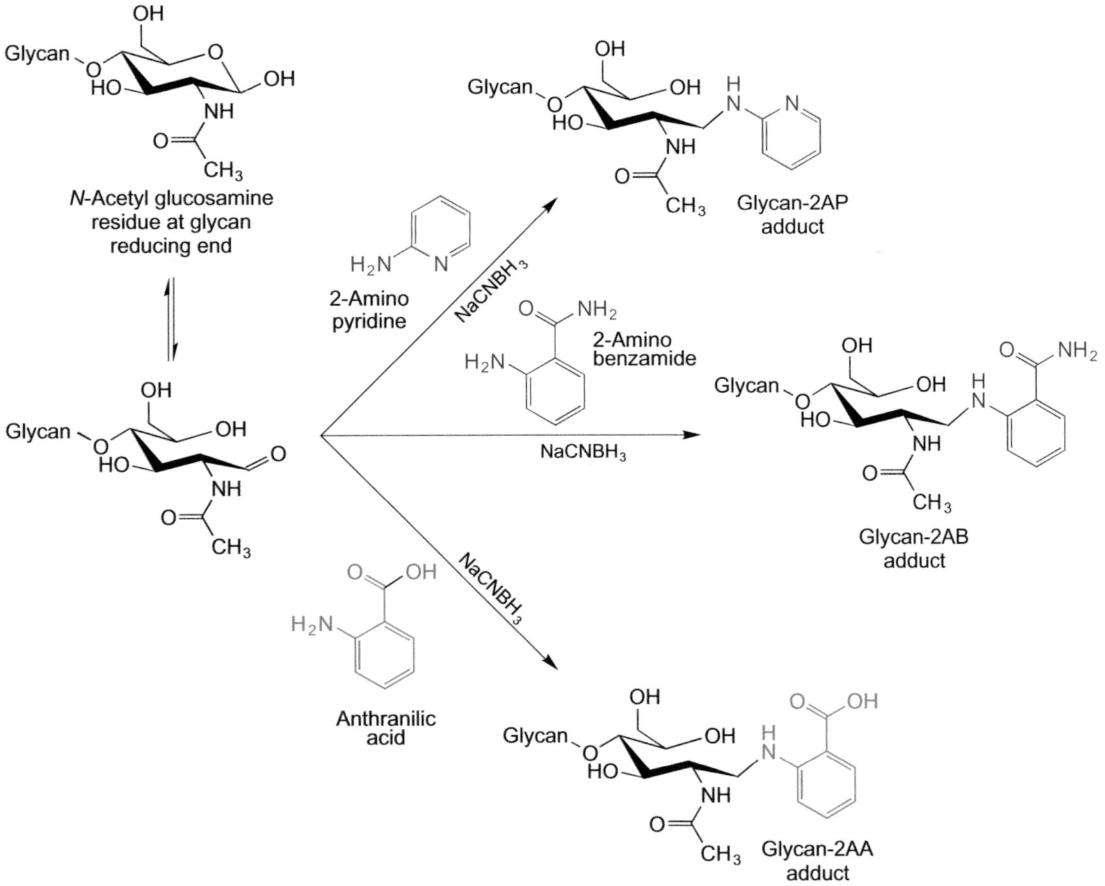

FIGURE 2.116 Released glycans can be labeled with small fluorescent compounds containing amines for subsequent detection upon chromatographic separation. In the presence of sodium cyanoborohydride these compounds react at the reducing end of glycans to form secondary amine derivatives with characteristic spectral properties.

(BAP). This reagent contains the fluorescent diaminopyridine group for detection and a biotin handle for purification or immobilization using its strong interaction with streptavidin. BAP can be used to label the reducing ends of glycans and other carbohydrates by reductive amination to form a secondary amine linkage with the aminopyridine group.

The fluorescent 2-aminopyridine group may also be used without a biotin handle for modification of glycans for detection, similar to the use of anthranilic acid and 2-amino benzamide. In this case, the tag just functions as a fluorescent label, which can be used to track glycans during purification or analysis. Modification of glycans with the bifunctional 2,6-diaminopyridine group using reductive amination produces a fluorescently labeled carbohydrate with an available amine for further coupling to surfaces or other biomolecules. Care should be taken, however, when using amine-reactive reagents to modify such amino glycans, because many electrophilic compounds cross-react with hydroxyls on the sugar molecules.

A similar fluorescent biotin label to the BAP reagent was created by Leteux *et al.* (1998) for labeling glycans without reduction of the reducing end of the monosaccharide. Biotinyl-L-3-(2-naphthyl)-alanine hydrazide (BNAH) contains a strongly UV absorbing naphthalene group, which has fluorescent characteristics, and a hydrazide group for conjugation to the reducing end of glycans. This compound was found to label effectively glycans at the reducing end, while retaining the unreduced nature of the carbohydrate, thus preserving critical protein interactions, which may recognize this region of the sugar.

The reductive amination protocol for coupling these types of small tags to glycans can be found in the previous section.

Synthesis of Glycosylamines for Conjugating Glycans

Reducing sugars and carbohydrates or glycans containing an unmodified reducing end may be derivatized in a simple reaction to provide an amine group on C1 for further conjugation. The anomeric hydroxyl group at the reducing end of such sugars can be converted into an amino group by reaction in an aqueous, saturated solution of ammonium carbonate. Kochetkov amination, as it is called (Likhosherstov *et al.*, 1986), can be used to modify a wide variety of reducing carbohydrates and glycans, including neutral and charged monosaccharides, disaccharides, and oligosaccharides (Kallin *et al.*, 1989; Urge *et al.*, 1991, 1992; Manger *et al.*, 1992; Cohen-Anisfeld and Langburg, 1993). A more recent analysis of the reaction mechanisms of glycosylamine formation identified a number of routes to product formation (Ghadban *et al.*, 2012). Once an amine is created on the anomeric carbon, subsequent labeling or coupling reactions can be performed using standard amine-reactive conjugation agents, for instance fluorescently labeling with a dye molecule (Song *et al.*, 2011).

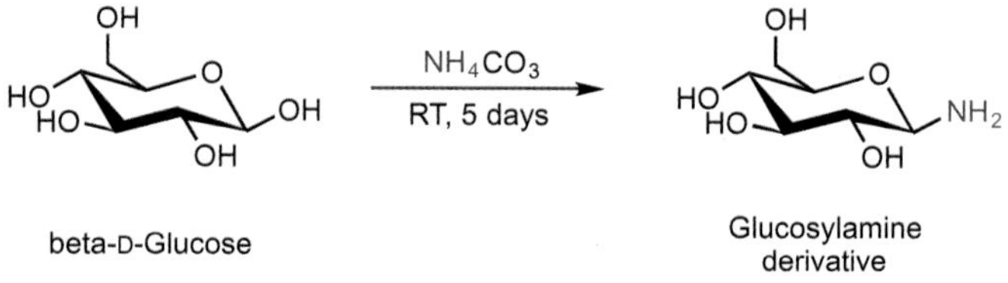

FIGURE 2.117 Glucosylamine derivatives can be prepared at the reducing end of glycans or other reducing carbohydrates by reaction with ammonium carbonate. The resultant amine derivative can be used to conjugate the carbohydrate with other proteins or molecules without disturbing the cyclic character of the reducing end.

Modification using this protocol yields saccharides that are highly stereochemically pure, with typical reactions giving >95% glycosylamines of the β-anomer structure (Figure 2.117). The resulting primary amine group can be further coupled with amine-reactive bioconjugation reagents or used for carbohydrate immobilization onto surfaces or insoluble supports. The major side reaction of amination is the dimeric diglycosylamine derivative, which has two sugar groups attached to a single secondary amine in the middle. This byproduct can be present in up to 10% of the resulting mass of product formed by the reaction.

The following protocol for synthesis of glycosylamines from reducing sugars is based on the method of Likhosherstov *et al.* (1986).

Protocol

1. Prepare a solution of the saccharide to be aminated at a concentration of up to 1% (w/v) in an aqueous solution of saturated ammonium carbonate. Even higher concentrations of saccharides may be used if the carbohydrate being modified is abundant and inexpensive.
2. React with stirring at room temperature for up to 5 days. Over the course of the reaction, add to the solution solid ammonium carbonate at a rate of about 40 mg per mg of saccharide to assure continued saturation.
3. After the reaction, freeze the solution and lyophilize to remove excess ammonium carbonate. Complete removal of volatile salt can be accomplished by re-dissolving the solid in warm methanol. After the completion of CO_2 evolution, dry the saccharide by evaporation under vacuum. Removal of ammonium carbonate is essential, as the ammonium ion will interfere with any subsequent conjugations attempted with the glycosylamine derivative.

5. BLOCKING OR PROTECTING GROUPS

It is often necessary to block specific groups on macromolecules to prevent them from participating in modification or conjugation reactions. In most blocking procedures, a chemical group is covalently coupled to an undesired functional group on the macromolecule to mask or eliminate its reactivity. In this sense, the modification is performed with a compound that is relatively inert in whatever application the macromolecule is intended for use. The blocking agent is usually a small organic compound containing a functional group able to couple with the group to be masked. The blocking molecule may contain another functional group of its own, converting the blocked group into a chemical function of another type, but this conversion is all right, providing the newly created function does not interfere in subsequent reactions or applications.

In some cases, a blocking procedure is performed to direct a conjugation reaction to discrete sites in a macromolecule. In other instances, blocking a group on one of two macromolecules can prevent self-polymerization and promote the desired intermolecular conjugation. For instance, HRP can be blocked with an amine-specific coupling reagent prior to periodate oxidation to prevent the reactivity of its two amino groups during subsequent conjugation with an antibody molecule (Chapter 20, Section 1.3).

In other uses of blocking reagents, proteins dissociated into subunits by the use of denaturants and disulfide reductants may be prevented from re-association or oxidation of their sulfhydryls by blocking the –SH groups with the appropriate reagent. Alternatively, sulfhydryls may be blocked on a protein prior to activation with a heterobifunctional crosslinking agent that contains both amine-reactive and sulfhydryl-reactive ends. The amine-reactive end will couple to the amines of the protein without reaction of the sulfhydryl-reactive end. This can prevent oligomer formation during the activation process and thus ensure that the sulfhydryl-reactive function is available for conjugation with the desired molecule.

Controlled functional studies of a protein's active center may also be carried out by blocking specific groups and observing its effect on activity. Often, this blocking procedure is performed through the use of a reversible blocking agent to subsequently regenerate activity, thus demonstrating that the affect was directed at functionalities present in the active site (Perham and Jones, 1967).

Blocking may also be carried out to quench further modification or conjugation through a targeted functional group. After a conjugation reaction, excess functional groups may be masked from nonspecifically reacting with other molecules. For instance, periodate-oxidized glycoproteins may still contain aldehyde groups after conjugation with another protein by reductive amination. Blocking the aldehydes with a small amine-containing molecule prevents unwanted reactions from occurring when the conjugate is used in an assay or targeting operation. This is also true of excess sulfhydryl groups that may undergo disulfide interchange with other sulfhydryl molecules subsequent to a conjugation reaction. Blocking these groups with the appropriate reagent prevents this type of side-reaction from occurring.

Blocking of amine groups on proteins has also been used to create a sensitive reagent for measuring protease activity (Hatakeyama, 1992). With nearly all the primary amines of casein blocked, an amine detection reagent such as trinitrobenzene sulfonic acid (TNBS) will react only minimally with the protein and form its typical orange derivative. As proteases cleave the protein, however, primary α-amines are created from cleavage of the α-chain peptide bonds, and TNBS then can react with them. The more protease activity present, the more color is formed.

The choice and application of a specific blocking reagent can produce a modified macromolecule with unique and useful properties. Many of the common blocking reagents are discussed in this section. Beyond the scope of this book, however, is a discussion of the numerous blocking agents used in peptide or nucleic acid synthesis to temporarily block specific reactive groups during growth of the polymer chain.

5.1. Blocking or Protecting Amine Groups

The amine functionalities most commonly found in macromolecules are primary amines such as those at the N-terminal of polypeptide chains (α-amines) and the side-chain ε-amino groups of lysine residues. Several acylation reagents can effectively block these primary amines, some of which are reversible under the right conditions. It should be noted that the cyclic anhydrides mentioned in this section react with amino groups to form amide bonds, opening the anhydride ring and effectively transforming the amine function into a carboxylate. There are additional compounds described in Section 4.2 (this chapter) that also create carboxylates from amines, but in this section the two cyclic anhydrides discussed, maleic anhydride and citraconic anhydride, are both reversible and designed more for temporary masking than permanent blocking. For more stable blocking of amines, the compounds sulfo-NHS acetate and acetic anhydride are the best choices.

Sulfo-NHS-Acetate

Sulfo-NHS acetate is the *N*-hydroxysulfosuccinimide ester of acetic acid. The NHS ester end provides high reactivity with the amino groups of proteins at a pH

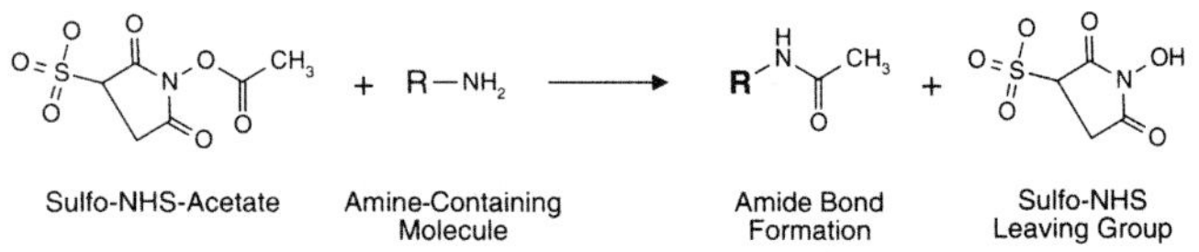

FIGURE 2.118 Sulfo-NHS acetate may be used to block amine groups, forming permanent amide bond derivatives.

range of 7 to 9, acylating the amines and forming nonreversible acetamide modifications (Figure 2.118). The sulfonate derivative of the NHS ester provides good water solubility to the reagent. Thus, the compound can be added directly to an aqueous solution of the protein to be blocked, or a stock solution may be prepared and a small aliquot added to the reaction medium. Stock solutions should be dissolved rapidly and used immediately. In aqueous solutions, the main competing reaction is hydrolysis of the active ester to release nonreactive sulfo-NHS and acetic acid. The use of a 10- to 50-fold molar excess of sulfo-NHS acetate over the molar amount of groups to be blocked should provide good yields of acylated amines. Reaction buffers should contain no extraneous amines that could cross-react with the sulfo-NHS acetate. Avoid Tris, glycine, and imidazole-containing buffers. Phosphate, borate, or bicarbonate buffers work well at a concentration of 0.05–0.1-*M*. React for at least 1 h at room temperature.

It should be noted that complete blocking of all amines on proteins with sulfo-NHS acetate may cause precipitation or loss of native structure and function. The acetate modifications are uncharged and relatively hydrophobic, which may decrease solubility of proteins or other molecules.

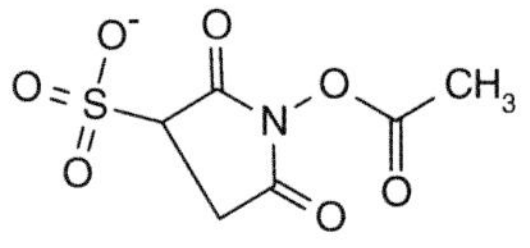

Sulfo-NHS Acetate
MW 259.17

Protocol

1. Dissolve the protein or other amine-containing macromolecule at a concentration of 1–10 mg/ml in 0.1-*M* sodium phosphate, 0.15-*M* NaCl, pH 7.5.
2. Add a 25-molar excess of sulfo-NHS acetate over the amount of amines present in the sample. If the precise amount of amines is not known, adding an equal mass of reagent to the mass of protein will provide a large excess of reactivity to completely block all amines.
3. React at room temperature for at least 1 h.
4. Purify the modified protein by dialysis or gel filtration.

FIGURE 2.119 Acetic anhydride reacts with amines to form amide bond derivatives.

Acetic Anhydride

Acetic anhydride is the only monocarboxylic acid anhydride that is important in modification reactions. The acetylation of the amino groups of proteins can be made relatively specific if the reaction is carried out in saturated sodium acetate, since the *O*-acetyltyrosine derivative is unstable to an excess of acetate ions (Fraenkel-Conrat, 1959). The tyrosine derivative rapidly hydrolyzes in alkaline reaction conditions, even in the absence of added acetate buffer (Uraki *et al.*, 1957; Smyth, 1967). Treatment with hydroxylamine also cleaves any *O*-acetyltyrosine modifications, forming acetylhydroxamate, which can be followed by its purple complex with Fe^{+3} at 540 nm (Balls and Wood, 1956).

At physiological pH values, acetylation of amine groups proceeds rapidly, requiring less than an hour to go to completion (Figure 2.119).

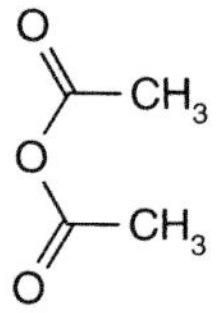

Acetic Anhydride
MW 102

Protocol

1. Dissolve the macromolecule to be modified at a concentration of 1–10 mg/ml in a buffered solution having a pH between 6.5 and 7.5. Avoid amine-containing buffers such as glycine and Tris. Sodium phosphate buffer at a concentration of 0.1-*M* works well. The addition of an equal volume of a saturated solution of sodium acetate may be carried out to prevent tyrosine derivatization.
2. Cool the solution on ice. With stirring, add an amount of acetic anhydride equal to the mass of macromolecule to be modified. Alternatively, add a 10-fold molar excess of acetic anhydride over the amount of amines present. The addition of the anhydride slowly or in several aliquots over the course of 1 h will ensure good yield of acetylation.
3. React with stirring for at least 1 h while cooling in an ice bath.
4. Purify the acetylated macromolecule by gel filtration or dialysis.

Citraconic Anhydride

Citraconic anhydride (or 2-methylmaleic anhydride) is a derivative of maleic anhydride that is reversible after acylation of amine groups. At alkaline pH values (pH 7–8) the reagent reacts with amines to form amide linkages with an extending terminal carboxylate. However, at acid pH (3–4), these bonds rapidly hydrolyze to release citraconic acid and free the amine (Dixon and Perham, 1968; Habeeb and Atassi, 1970; Klapper and Klotz, 1972; Shetty and Kinsella, 1980). Thus, citraconic anhydride is useful in temporarily blocking amine groups while other parts of a molecule are undergoing derivatization. Once the modification is complete, the amines can then be unblocked to create the original structure. See Section 4.2 for additional information and a protocol for modification of proteins with citraconic anhydride.

Maleic Anhydride

Maleic acid is a linear four-carbon molecule with carboxylate groups on both ends and a double bond between the central carbon atoms. The anhydride of maleic acid is a cyclic molecule containing five atoms. Although the reactivity of maleic anhydride is similar to other cyclic anhydrides, the products of maleylation are much more unstable toward hydrolysis, and the site of unsaturation lends itself to additional side-reactions. Acylation products of amino groups with maleic anhydride are stable at neutral pH and above, but they readily hydrolyze at acid pH values around pH 3.5 (Butler *et al.*, 1967). Maleylation of sulfhydryls and the phenolate of tyrosine are even more sensitive to hydrolysis. Thus, maleic anhydride is an excellent reversible blocker of amino groups to temporarily mask them from reactivity while another reaction is being performed. For additional information and a protocol for the modification of proteins with this reagent, see Section 4.2.

5.2. Blocking or Protecting Sulfhydryl Groups

The sulfhydryl group is among the most highly reactive of nucleophiles found in biological macromolecules. Cysteine sulfhydryls in proteins undergo covalent reactions rapidly with most of the reactive groups utilized in modification and conjugation reagents. To prevent modification from occurring at these sites, it is often necessary to use a blocking agent that ties up the sulfhydryl and renders it inert toward further reactions.

There are two types of sulfhydryl blocking agents: permanent and reversible. The permanent ones form thioether linkages that do not readily break down. The reversible ones form disulfide bonds that are susceptible to cleavage by the addition of the appropriate reducing agent. Reversible sulfhydryl blockers can be used to temporarily mask an –SH group from modification while a reaction is performed at another site. This is especially useful when the sulfhydryl forms a critical part of the active center of a protein. After the final modification is complete, the blocking agent can be removed to regenerate activity.

N-Ethyl Maleimide

N-Ethyl maleimide (NEM) (NEM) is an alkylating reagent that reacts with sulfhydryls to form stable thioether bonds (Smyth *et al.*, 1960). Maleimide reactions are specific for sulfhydryl groups in the pH range of 6.5 to 7.5 (Smyth *et al.*, 1964; Gorin *et al.*, 1966; Heitz *et al.*, 1968; Partis *et al.*, 1983) (see Chapter 3, Section 2.2). At higher pH values some cross-reactivity with amino groups takes place (Brewer and Riehm, 1967). One of the carbons adjacent to the double bond undergoes nucleophilic attack by the thiolate anion to generate the addition product (Figure 2.120). When sufficient quantities of –SH groups are being blocked, the reaction may be followed spectrophotometrically by the decrease in absorbance at 300 nm as the double bond reacts and disappears. The result is a stable, inert derivative that terminates in the ethyl group. NEM is useful for permanently blocking sulfhydryl residues in proteins and other macromolecules. It has been used for blocking sulfhydryl-containing reagents that interfere in a glucose oxidase assay system (Haugaard *et al.*, 1981).

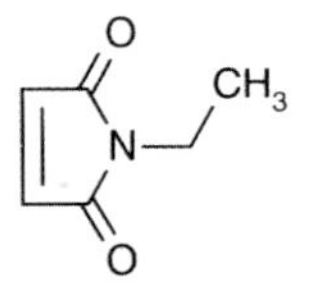

N-Ethylmaleimide
MW 125.12

Protocol

1. Dissolve the macromolecule containing sulfhydryl groups to be blocked in a buffer having a pH of 6.5 to 7.5. Sodium phosphate (0.01–0.1-*M*) at pH 7.2 works well. Avoid amine-containing buffers, since an excess of amines may cause some reactivity with the maleimide groups. Also, avoid the presence of sulfhydryl-containing disulfide reductants such as

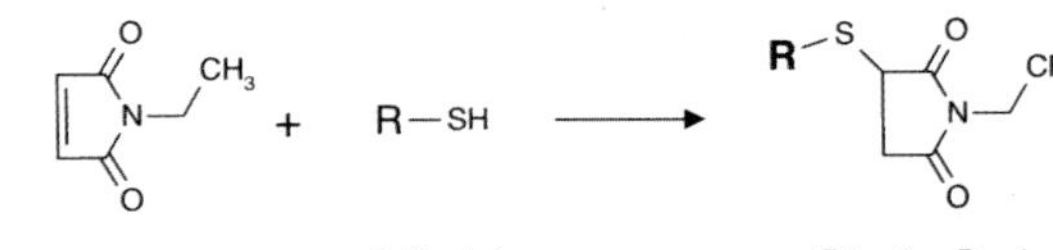

FIGURE 2.120 The reaction of *N*-ethylmaleimide with sulfhydryl groups yields a thioether derivative, permanently blocking the thiol.

DTT or 2-mercaptoethanol, which will rapidly react with NEM.

2. Add at least a 10-fold molar excess of NEM over the amount of sulfhydryls present in the reaction. Alternatively, add an equal mass of NEM to the amount of macromolecule present. To facilitate the addition of a small quantity of reagent, a more concentrated stock solution may be prepared in buffer and an aliquot added to the reaction medium. Make the stock solution up fresh, and use it immediately to prevent loss of activity due to maleimide group breakdown.
3. React for 2h at room temperature.
4. Purify the modified protein by gel filtration or dialysis.

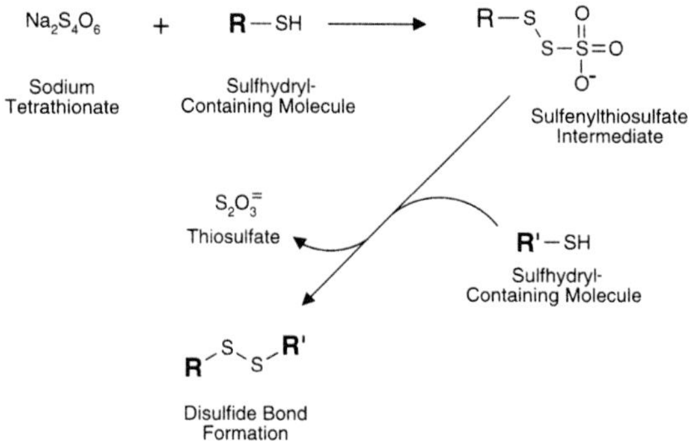

FIGURE 2.121 Sodium tetrathionate reacts with thiols to form reactive sulfenylthiosulfate intermediates. Another sulfhydryl-containing molecule may couple to this active group to create a disulfide linkage.

Iodoacetate Derivatives

Iodoacetate (and bromoacetate) can react with several nucleophilic functional groups within proteins. Their relative reactivity toward protein functionalities is sulfhydryl > imidazolyl > thioether > amine. Among α-haloacetate derivatives the relative reactivity is I > Br > Cl > F, with fluorine being almost unreactive. The α-haloacetamides have the same trend of relative reactivities, but will obviously not create a charged carboxylate functional group. The acetamide derivatives typically are used only as blocking reagents. The bond formed from the reaction of iodoacetamide and a sulfhydryl group is a stable thioether linkage that is not reversible under normal conditions.

Thus, iodoacetamide has the highest reactivity toward cysteine sulfhydryl residues and may be directed specifically for –SH blocking. If iodoacetamide is present in limiting quantities (relative to the number of sulfhydryl groups present) and at slightly alkaline pH, cysteine modification will be the exclusive reaction. For additional information on α-haloacetate reactivities and a protocol for blocking, see Section 4.2.

Sodium Tetrathionate

Sodium tetrathionate ($Na_2S_4O_6$) is a redox compound that under the right conditions can facilitate the formation of disulfide bonds from free sulfhydryls. The tetrathionate anion reacts with a sulfhydryl to create a somewhat stable active intermediate, a sulfenylthiosulfate (Figure 2.121). Upon attack of the nucleophilic thiolate anion on this activated species, the thiosulfate ($S_2O_3^=$) leaving group is removed and a disulfide linkage forms (Pihl and Lange, 1962). The reduction of tetrathionate to thiosulfate *in vivo* was a subject of early study (Chen *et al.*, 1934; Theis and Freeland, 1940).

$Na_2S_4O_6$

Sodium Tetrathionate
MW 270.22

Depending on the proximity of cysteine sulfhydryl groups in proteins, intrachain and interchain disulfide formation is possible upon reaction with tetrathionate. When neighboring sulfhydryl groups are not close enough to create disulfide linkages, the sulfenylthiosulfate modification is sufficiently stable to temporarily block exposed –SH groups. For sulfhydryls present in the active centers of enzymes, tetrathionate may lead to reversible inactivation (Parker and Allison, 1969). Thus, the reagent may be used to protect certain sulfhydryl residues during modification reactions performed elsewhere on a protein. Using this approach, the enzyme ficin may be temporarily protected with tetrathionate during modification, conjugation, or immobilization reactions performed through its amine groups (Liener and Friedenson, 1970). Subsequent treatment with thiol-containing disulfide reducing agents frees the sulfenylthiosulfate and regenerates the sulfhydryl with enzymatic activity. The following protocol is an adaptation of Englund *et al.* (1968), used in the purification of ficin.

Protocol

1. The macromolecule containing sulfhydryl residues to be blocked or protected is dissolved in a buffer suitable for its individual stability requirements. The blocking process may be performed on a purified protein or during the early stages of a purification process to protect sulfhydryl active centers from oxidation. PBS buffers containing 1-m*M* EDTA work well.
2. Add sodium tetrathionate to obtain a final concentration of 10-m*M*.
3. React for 1h at room temperature.
4. Excess tetrathionate may be removed by dialysis or gel filtration.
5. To remove the sulfenylthiosulfate blocking group, add a 300-fold excess of DTT over the amount of

blocked sulfhydryls present. Alternatively, add DTT to obtain a final concentration of 0.01 to 0.1-*M*. Cysteine also may be utilized to regenerate some enzymes to full activity.

6. Incubate for 2h at room temperature.
7. For removal of excess DTT, a protein of molecular weight greater than 5000 may be isolated by gel filtration using a desalting resin. To maintain the stability of the exposed sulfhydryl groups, include 10-m*M* EDTA in the chromatography buffer. The presence of oxidized DTT can be monitored during elution by measuring the absorbance at 280nm. The protein should elute in the first peak and the DTT reaction products in the second peak.

Methyl Methanethiosulfonate

Methyl methanethiosulfonate (MMTS) is a small reversible blocking agent for sulfhydryl groups (Thermo Fisher, Toronto Research). It reacts with free thiols to form a dithiomethane modification with release of sulfinic acid (Figure 2.122). The sulfinic acid component decomposes into volatile products, which do not affect the disulfide formed from the MMTS reaction. Alkylthiosulfonates will react rapidly with thiols under mild conditions at physiological pH. The MMTS compound is a liquid at 10.6-*M* concentration and is conveniently added to a reaction medium by pipette. The only caveat in this regard is that the reagent is more viscous than water and will fall to the bottom of most aqueous buffers and form a two-phase solution. Vigorous mixing is necessary to completely cause dissolution in the reaction buffer or medium. Complete thiol modifications of available cysteine residues in proteins can be achieved even using relatively dilute (μ*M* to m*M*) concentrations of the reagent. Typically, MMTS need only be added in several-fold molar excess over the quantity of thiols present to result in stoichiometric sulfhydryl group blocking. Reactions can be performed in organic solvent, aqueous buffers, or a mixture of organic/aqueous solutions, whatever is suitable for the sulfhydryl compound being modified.

MMTS-mediated modification of thiols is reversible by use of disulfide reductants. Reducing agents such as DTT, 2-mercaptoethanol, or TCEP will cleave the dithiomethane modification groups to restore the original sulfhydryl. The reagent has been used to identify the cysteine residues important for organic cation transport in oocytes (Sturm *et al.*, 2007), to study the peptide loading complex within MHC class I (Santos *et al.*, 2007), for investigations of the Zn^{21}-dependent redox switch in an intracellular interface channel (Wang *et al.*, 2007), and to study how disulfide isomerization functions to switch tissue factor from coagulation to cell signaling (Ahamed *et al.*, 2006). MMTS has also been used to block cysteine thiol groups prior to the determination of *S*-nitrosylation within proteins (Sangwung *et al.*, 2012).

MMTS is a popular thiol blocking agent that functions similar to sodium tetrathionate in forming reversible disulfide derivatives (previous section). This reactive group has also been used as the basis of creating sulfhydryl-reactive crosslinking agents, such as the trifunctional compounds MTS-ATF-Biotin and MTS-ATF-LC-Biotin (Chapter 24, Section 3.2). In addition, it has been used to form thiol modification reagents to

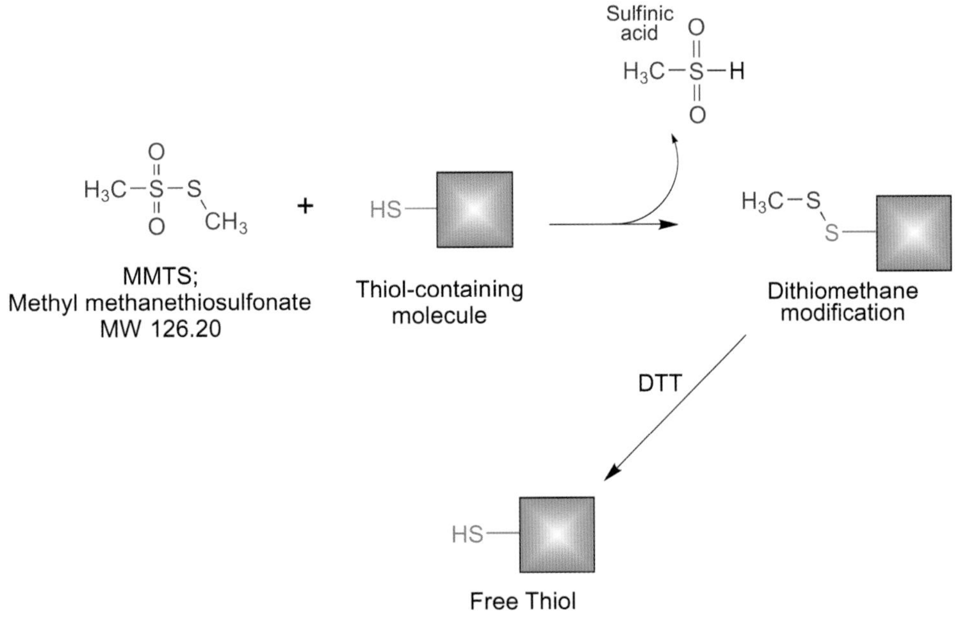

FIGURE 2.122 2,2′-Dipyridyl disulfide reacts with thiols to form an active pyridyl disulfide intermediate.

study site-directed mutagenesis, including how small modifications might affect protein folding or protein interactions (Toronto Research).

Ellman's Reagent

Ellman's reagent or DTNB, is a compound useful for the quantitative determination of sulfhydryls in solution (Ellman, 1958, 1959). The disulfide of Ellman's reagent readily undergoes disulfide exchange with a free sulfhydryl to form a mixed disulfide and release of one molecule of the chromogenic substance 5-sulfido-2-nitrobenzoate, also called 5-thio-2-nitrobenzoic acid (TNB). The intense yellow color produced by the TNB anion can be measured by its absorbance at 412 nm ($\varepsilon = 1.36 \times 10^4 M^{-1} cm^{-1}$ at pH 8.0). Since each sulfhydryl present generates one molecule of TNB per molecule of Ellman's reagent, direct quantitation is easily performed. This reagent has been used to measure the sulfhydryl content in peptides, proteins, and tissue samples (Anderson and Wetlaufer, 1975; Riddles *et al.*, 1979). See Section 4.1 in this chapter for the use of Ellman's reagent in the determination of sulfhydryl groups.

The same reaction between Ellman's reagent and the sulfhydryls of macromolecules can be used to temporarily block available –SH groups by the formation of a mixed disulfide bond. Treatment of a sulfhydryl-containing protein with an excess of Ellman's reagent blocks the accessible sulfhydryls with the TNB group, allowing chemistries to be performed on other functionalities. Studies have shown that the rate of Ellman's reaction with the thiol groups in proteins is dependent on their accessibility (Damjanovich and Kleppe, 1966; Colman, 1969). The addition of a disulfide reducing agent then cleaves the TNB group and regenerates the free sulfhydryl. Enzymes containing sulfhydryls in their active sites may be reversibly blocked using this technique to preserve activity after modification or conjugation. Deblocking then restores catalytic activity in most instances.

Protocol

1. Dissolve the protein to be blocked at a concentration of 1–10 mg/ml in 0.1-*M* sodium phosphate, pH 8.0.
2. Dissolve the Ellman's reagent at a concentration of 4 mg/ml in 0.1-*M* sodium phosphate, pH 8.0.
3. Mix the protein solution with an equal volume of the Ellman's reagent solution and react for 15 min at room temperature.
4. Purify the modified protein from excess Ellman's reagent and reaction byproducts by dialysis or gel filtration. A measurement of sulfhydryl content may be carried out by reading the absorbance of the modification reaction at 412 nm ($\varepsilon = 1.36 \times 10^4 M^{-1} cm^{-1}$) *versus* a series of sulfhydryl standards treated in the same manner (e.g., cysteine).

To deblock the TNB modified sulfhydryl residues, treat the protein with an excess of DTT according to the protocol described in Section 4.1, DTT (this chapter).

Dipyridyl Disulfide Reagents

The similar reagents, 4,4′-dipyridyl disulfide (Grassetti and Murray, 1967) and 2,2′-dipyridyl disulfide (Brocklehurst, 1974), react in an analogous manner to Ellman's reagent, both forming pyridyl disulfide bonds with free sulfhydryls and releasing a molecule of either pyridine-4-thione or pyridine-2-thione, respectively (Figure 2.122). Both leaving groups are measurable spectrophotometrically at 324 nm (pyridine-4-thione) or 343 nm (pyridine-2-thione) to quantify the amount of sulfhydryl modification. The reagent 2,2′-dipyridyl disulfide is useful for creating sulfhydryl-reactive cross-linking agents, such as SPDP (Chapter 6, Section 1.1) and for activating particles or chromatography supports for the coupling of thiol-containing ligands (Chapters 14 and 15). Both reagents may be used to temporarily block sulfhydryl groups in macromolecules or to activate –SH groups for coupling to another sulfhydryl-containing molecule. The pyridine disulfide-modifying group can react with a sulfhydryl to form a disulfide linkage. The pyridine disulfide may also be cleaved with an excess of disulfide reducing agents, such as DTT, making it a reversible blocking agent.

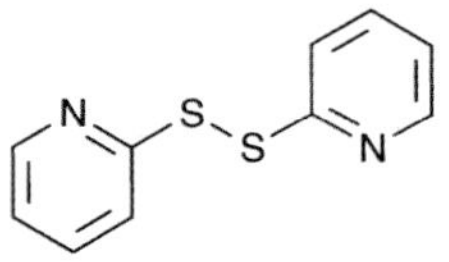

2,2′-Dipyridyl disulfide

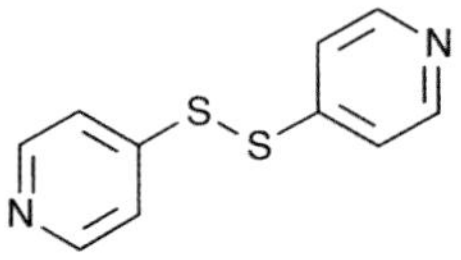

4,4′-Dipyridyl disulfide

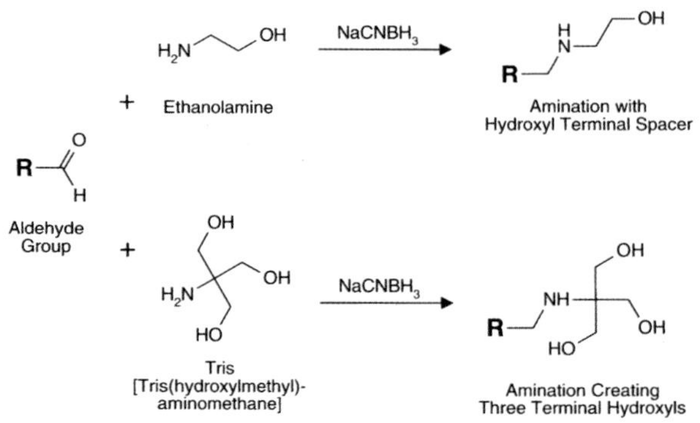

FIGURE 2.123 Aldehyde groups may be blocked with Tris or ethanolamine using a reductive amination process.

Unfortunately, 2,2′-dipyridyl disulfide is relatively insoluble in aqueous buffers. The use of this compound to modify molecules usually involves prior dissolution in an organic solvent such as acetone and then performing the blocking reaction in an aqueous/organic mixture. Many proteins will not tolerate high concentrations of organic solvents without precipitation.

The 4,4′-dipyridyl disulfide can be used in aqueous solutions, but it has been found that modification of proteins with this reagent yields rapid disulfide bond formation. Only when 2-iminothiolane is used in tandem with 4,4′-dipyridyl disulfide can 4-dithiopyridyl groups be introduced into proteins (King *et al.*, 1978) (Section 4.1, this chapter). This is due to disulfide interchange reactions predominating without the addition of 2-iminothiolane.

For one-step methods, the use of Ellman's reagent (previous section) to yield a similar reversible sulfhydryl blocking group is probably a better choice with protein molecules.

5.3. Blocking or Protecting Aldehyde or Ketone Groups

Aldehyde groups are useful in facilitating modification or conjugation reactions, easily forming secondary amine linkages with amine-containing molecules in reductive amination procedures or hydrazone linkages with hydrazide-containing molecules. Macromolecules modified to contain aldehyde groups for use in these reactions (see Section 4.4) should be treated after conjugation to remove any excess formyl functionalities. The blocking step prevents subsequent nonspecific interactions when a conjugate is used in assay or targeting applications.

Reductive Amination with Tris or Ethanolamine

One of the simplest methods for blocking aldehyde functionalities involves reductive amination with a small amine-containing molecule. The best such blockers do not have extra functionalities that may create additional sites of reactivity after blocking. Tris and ethanolamine are ideal in this regard. They both contain primary amines that readily react with aldehydes in the presence of a reductant, and they both possess relatively inert hydroxyl groups that maintain hydrophilicity after coupling. Reductive amination (Chapter 2, Section 5.3, and Chapter 4, Section 4) facilitated by the use of sodium cyanoborohydride can quickly block residual aldehyde groups and transform them into unreactive hydroxyls of low nonspecific binding potential (Figure 2.123).

Protocol

1. Dissolve the macromolecule containing aldehydes to be blocked (e.g., a glycoprotein that has been oxidized with sodium periodate to create formyl groups) at a concentration of 1–10 mg/ml in 0.1-*M* Tris buffer, pH 8.0. Alternatively, dissolve the macromolecule in 0.1-*M* sodium phosphate containing 0.1-*M* ethanolamine, pH 8.0. The use of other buffers having a pH between 7 and 10 will work as well, but the Tris or ethanolamine concentrations should be maintained in high excess to efficiently block all the aldehyde residues.
2. Add 10 μl of 5-*M* sodium cyanoborohydride in 1-*N* NaOH (Aldrich) per milliliter of the macromolecule solution volume prepared in (1). *Caution: Highly toxic compound. Use a fume hood and be careful to avoid skin contact with this reagent.*
3. React for 15 min at room temperature.
4. Purify the derivatized macromolecule by dialysis or gel filtration using a buffer suitable for the nature of the substance being modified.

Oxime Formation with Hydroxylamine

Another convenient option for blocking aldehyde groups is the use of a small aminooxy compound to react spontaneously with aldehydes (or ketones) to form oxime bonds (aldoxime or ketoxime bonds with aldehydes or ketones, respectively). Hydroxylamine is the simplest aminooxy molecule, and the addition of this compound to an aldehyde-containing molecule will block the carbonyl groups and form oximes (Figure 2.124). An oxime bond is a stable linkage under normal conditions, but it may be cleaved under certain circumstances (e.g., in the presence of sodium dithionite) (Pojer, 1979) or by heating in the presence of an inorganic acid (which is typically too harsh for use with many biomolecules). The formation of the oxime reaction product may be accelerated in aqueous solution by the addition of aniline to the reaction, which acts as a catalyst by forming an intermediate imine that goes on to react with the aminooxy compound (hydroxylamine) to make the oxime (Cordes and Jencks, 1962; Dirksen *et al.*, 2006a,b; Dirksen and Dawson, 2008; Byeon *et al.*, 2010).

Hydroxylamine has also been used for the cleavage of certain peptide bonds in proteins, particularly asparagine–glycine bonds (Smith, 2002). For efficient cleavage, the typical reaction conditions require alkaline

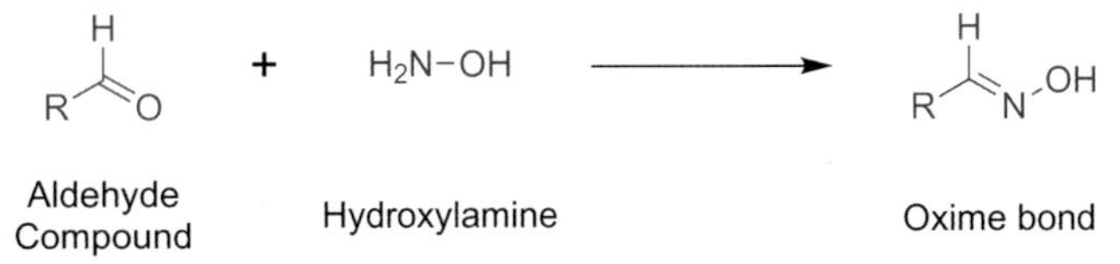

FIGURE 2.124 Reaction of hydroxylamine with an aldehyde to form an oxime.

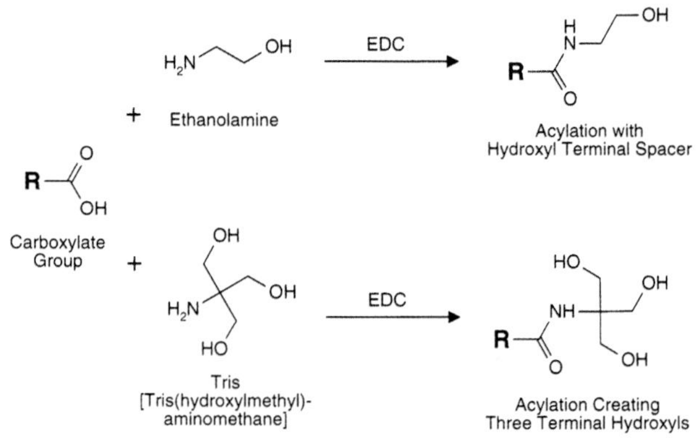

FIGURE 2.125 Carboxylate groups may be blocked with Tris or ethanolamine using a carbodiimide-mediated process.

pH (9.0) along with heating to at least 40°C for 4h. The use of hydroxylamine as a blocking agent at acid pH and under ambient temperature conditions should not adversely affect protein stability. Hydroxylamine should not be used, however, with heme-containing proteins, as the molecule will irreversibly coordinate with the central metal ion and cause inactivation. It should also be completely removed from serine proteases, as it has been shown to accelerate enzyme activity by acting as a nucleophilic agent within the active site (Nakai *et al.*, 2012).

Protocol

1. Dissolve an aldehyde-containing protein to be blocked at a concentration of 1–10 mg/ml in 1 ml of 0.1-*M* sodium acetate, 0.15-*M* NaCl, pH 5.5 (reaction buffer).
2. In a fume hood, prepare a stock solution consisting of 1-*M* hydroxylamine in reaction buffer by adding 0.33 g (0.27 ml) of hydroxylamine to about 7 ml of reaction buffer. Adjust the pH of the solution back down to 5.5 with acid and add sufficient additional reaction buffer to bring the final volume up to 10 ml.

   Note: If using hydroxylamine hydrochloride, add 694 mg to about 8 ml of reaction buffer and adjust the pH back to 5.5 with base if necessary, then add buffer to make a final volume of 10 ml.
3. In a fume hood, add 18 μl of aniline catalyst to the protein solution with stirring. This results in approximately a 0.1-*M* aniline solution in the modification reaction.
4. Add 27 μl of the 1-*M* hydroxylamine stock solution to the protein solution while stirring (results in approximately a 0.1-*M* solution).
5. React for 2–4 h at room temperature with constant mixing by end-over-end rotation in a sealed container.
6. Remove excess reactants by dialysis or size exclusion chromatography using a buffer of choice.

5.4. Blocking or Protecting Carboxylate Groups

The presence of unwanted carboxylate groups in macromolecules may easily be blocked by the use of a small amine-containing molecule coupled *via* the carbodiimide procedure (Chapter 3, Section 1.1).

Tris or Ethanolamine Plus EDC

Tris or ethanolamine are excellent choices for blocking procedures involving carboxylic acid groups, since they contain hydrophilic hydroxyls that mask the carboxylate and create an inert modification with low nonspecific binding potential. Using the water soluble carbodiimide EDC to facilitate this reaction, the carboxylate is activated by forming an intermediate *O*-acylisourea. The amine-containing compound then reacts with this active species to create a stable amide linkage (Figure 2.125).

Protocol

1. Dissolve the macromolecule containing carboxylate groups to be blocked at a concentration of 1–10 mg/ml in 0.1-*M* MES, pH 4.7, containing 0.1-*M* Tris or ethanolamine. Other conditions may be used to perform this reaction. See Chapter 4, Section 1, for further details.
2. Add 10 mg of EDC per ml of the solution prepared in (1).
3. React for 2–4 h at room temperature.
4. Purify by gel filtration or dialysis.

CHAPTER

3

The Reactions of Bioconjugation

Every chemical modification or conjugation process involves the reaction of one functional group with another, resulting in the formation of a covalent bond. The creation of bioconjugate reagents with spontaneously reactive or selectively reactive functional groups forms the basis for simple and reproducible crosslinking or tagging of target molecules. Of the hundreds of reagent systems described in the literature or offered commercially, most utilize common organic chemical principles that can be reduced down to a few dozen or so primary reactions. An understanding of these basic reactions can provide insight into the properties and use of bioconjugate reagents even before they are applied to problems in the laboratory.

This section is designed to provide a general overview of activation and coupling chemistry. Some of the reagents discussed in this chapter are not themselves crosslinking or modification compounds, but may be used to form active intermediates with another functional group. These active intermediates can subsequently be coupled to a second molecule that possesses the correct chemical constituents, which allows bond formation to occur.

Ultimately, this section is meant to function as a ready-reference database for learning or review of bioconjugate chemistry. In this regard, a reaction can quickly be found, a short discussion of its properties and use understood, and a visual representation of the chemistry of bond formation illustrated. What this section is not meant to be is an exhaustive discussion on the theory or mechanism behind each reaction, nor a review of every application in which each chemical reaction has been used. For particular applications where the reactions are employed, cross-references are given to other sections in this book or to outside literature sources.

Table 3.1 is a compendium of the major reactions used in bioconjugation along with the appropriate reaction partners, functional groups, activation agents or catalysts, the bonds formed as a result of the reactions, and in some cases the reactive group intermediate or new functional group which is formed. Each reaction is indicated by reading across a row of the table. For instance, in row one, an amine can react with an isothiocyanate group to give an isothiourea bond. In another example further down the table, an aminooxy group can react with aldehydes or ketones to give an oxime bond. The index may be used to look up the reactions or terms used in this table and find the associated chapters within the book where the topic is discussed in more detail.

1. AMINE REACTIONS

Reactive groups able to couple with amine-containing molecules are by far the most common functional groups present on crosslinking or modification reagents. An amine-coupling process can be used to conjugate with nearly all protein or peptide molecules as well as a host of other macromolecules. The primary coupling reactions for modification of amines proceed by one of two routes: acylation or alkylation (Chapter 2, Section 1.1). Most of these reactions are rapid and occur in high yield to give stable amide or secondary amine bonds.

1.1. Isothiocyanates

Isothiocyanates can be formed by the reaction of an aromatic amine with thiophosgene (Rifai and Wong, 1986). The group reacts with nucleophiles such as amines, sulfhydryls, and the phenolate ion of tyrosine side chains (Podhradsky *et al.*, 1979). The only stable product of these reactions, however, is with primary amine groups. Therefore, isothiocyanate compounds are almost entirely selective for modifying ε-amino groups in lysine side chains and N-terminal α-amines in proteins or primary amines in other molecules (Jobbagy and Kiraly, 1966). The reaction involves attack of the nucleophile on the central, electrophilic carbon of the isothiocyanate group (Reaction 3.1). The resulting electron shift and proton loss create a thiourea linkage

Bioconjugate Techniques, Third Edition.
DOI: http://dx.doi.org/10.1016/B978-0-12-382239-0.00003-0

TABLE 3.1 **The Major Reactions of Bioconjugation**

Functional Group	Reactive Group	Secondary Reactive Group	Activation Agent or Catalyst	Bond Formed	Reactive or Functional Group Formed
Amine	Isothiocyanate			Isothiourea	
Amine	Isocyanate			Isourea	
Amine	Acyl azide			Amide	
Amine	NHS ester			Amide	
Amine	Sulfonyl chloride			Sulfonamide	
Amine					
Amine	Tosyl ester			Secondary amine	
Amine	Tresyl ester			Sulfonamide	
Amine	Aldehyde			Schiff base	
Amine	Aldehyde		Sodium cyanoborohydride	Secondary amine	
Amine	Aldehyde	Active hydrogen compound	Aniline (Mannich reaction catalyst for targeting tyrosine)	Secondary amino-methyl-	
Amine	Amine		Formaldehyde	Secondary amino-methyl-amino-	Quaternary ammonium salt
Amine	Epoxide			Secondary amine	
Amine	Carbonate			Carbamate	
Amine	Aryl halide			Aryl amine	
Amine	Haloacetyl or alkyl halide			Secondary amine	
Amine	Imido ester			Amidine	
Amine	Carboxylate		Carbodiimide	Amide	Carbodiimide active ester
Amine	Alkyl phosphate (e.g., 5′-phosphate of oligonucleotide)		Carbodiimide plus imidazole	Phosphoramidate	
Amine	Anhydride			Amide	
Amine	Fluorophenyl ester			Amide	
Amine	HOBt ester			Amide	
Amine	Hydroxymethyl phosphine			Secondary amine	
Amine	*O*-methylisourea			Guanidine	
Amine	DSC				NHS carbamate
Amine	NHS carbamate			Isourea	
Amine	Glutaraldehyde			Schiff base or secondary amine	Aldehyde, hemiacetal, or double bond
Amine	Activated double bond			Secondary amine	
Amine	Cyclic hemiacetal			Secondary amine	
Amine	NHS carbonate			Carbamate	
Amine	Imidazole carbamate			Carbamate	
Amine	Acyl imidazole			Amide	
Amine	Methylpyridinium ether			Secondary amine	
Amine	Azlactone			Amide	
Amine	Cyanate ester			Isourea	
Amine	Cyclic imidocarbonate			Substituted imidocarbonate	
Amine	Chlorotriazine			Secondary amine	
Amine	Dehydroazepine			Secondary amine	
Amine	6-sulfo-cytosine derivative			4-Amino derivative of cytosine	
Aryl amine			Sodium nitrite, HCl		Diazonium group
Thiol	Haloacetyl or alkyl halide			Thioether	
Thiol	Maleimide			Thioether	
Thiol	Aziridine			Thioether	
Thiol	Aryl halide			Aryl thioether	
Thiol	Pyridyl disulfide			Disulfide	
Thiol	2,2′-dipyridyl disulfide or 4,4′-dipyridyl disulfide				Pyridyl disulfide
Thiol	TNB-thiol			Disulfide	
Thiol	Ellman's reagent				TNB-thiol
Thiol	Peroxide (other oxidants)			Disulfide, sulfonate	

(Continued)

TABLE 3.1 (Continued)

Functional Group	Reactive Group	Secondary Reactive Group	Activation Agent or Catalyst	Bond Formed	Reactive or Functional Group Formed
Thiol	Vinylsulfone			Thioether	
Thiol	Metal surface			Dative bond	
Thiol	Phenylthioester (on C-terminal peptides for native chemical ligation)			Thioester (followed by S→N shift to amide on N-terminal of peptide)	
Thiol	Cisplatin			Thioether (thio-platinum bond)	
Thiol	Activated double bond			Thioether	
Carboxylate	Diazoalkanes or diazoacetyl compounds			Ester	
Carboxylate			CDI		Acyl imidazole
Carboxylate			Carbodiimide		Carbodiimide active ester
Carboxylate			Carbodiimide plus NHS or sulfo-NHS		NHS ester or sulfo-NHS ester
Carboxylate			DSC		NHS ester
Carboxylate			TSTU		NHS ester
Hydroxyl	Epoxide			Ether	
Hydroxyl			CDI		Imidazole carbonate
Hydroxyl			DSC		NHS carbonate
Hydroxyl	Haloacetyl or alkyl halide			Ether	
Hydroxyl	Isocyanate			Carbamate	
Aldehyde	Amine			Schiff base	
Aldehyde	Amine		Sodium cyanoborohydride	Secondary amine	
Aldehyde	Amine	Active hydrogen compound (e.g., phenol compound or tyrosine)	Aniline (Mannich reaction catalyst to target tyrosines)	Secondary amine	
Aldehyde	Hydrazide (hydrazine)		Aniline	Hydrazone	
Aldehyde	Aminooxy compound		Aniline	Oxime	
Active hydrogen compound	Diazonium			Diazo bond	
Active hydrogen compound	Amine		Aldehyde ± aniline	Secondary amine (Mannich reaction)	
Active hydrogen compound	I_2		Oxidizing agent	Iodinated compound (R-I)	
Active hydrogen compound	Halogenated phenyl azide		UV light	Nonspecific insertion at active hydrogen site	
Active hydrogen compound	Phenyl azide	Amine	UV light	Secondary amine or low yield nonspecific insertion at active hydrogen site	
Active hydrogen compound	Benzophenone		UV light	Nonspecific insertion at active hydrogen site	
Active hydrogen compound	Anthraquinone		UV light	Nonspecific insertion at active hydrogen site	
Active hydrogen compound	Diazo derivatives		UV light	Nonspecific insertion at active hydrogen site	
Active hydrogen compound	Diazirine derivatives		UV light	Nonspecific insertion at active hydrogen site	
Thymine base	Psoralen derivative		UV light	Cycloaddition at 5,6-double bond of thymine	

(*Continued*)

TABLE 3.1 **(Continued)**

Functional Group	Reactive Group	Secondary Reactive Group	Activation Agent or Catalyst	Bond Formed	Reactive or Functional Group Formed
Diene	Alkene			Cycloalkene (4 + 2 cycloaddition)	
Hydrazine (or hydrazide)	Aldehyde			Hydrazone	
	Aminooxy	Aldehyde or ketone		Oxime	
	Azide	Alkyne	Cu^{1+}	Triazole (3 + 2 cycloaddition)	
	Azide	Alkene	Cu^{1+}	Triazoline (3 + 2 cycloaddition)	
	Azide	Triphenyl phosphine with electrophilic trap (Staudinger ligation)		Aryl amide	
	Azide	Aryl ester triphenyl phosphine derivative (traceless Staudinger ligation)		Amide bond	
	Phenyl boronic acid	Salicylhydroxamate		Boronate-containing 6- or 5-member ring	
Alkyl phosphate (e.g., 5′-phosphate of oligonucleotide)			Carbodiimide plus imidazole		Phosphoryl imidazolide
Cytosine base			Sodium bisulfite		6-Sulfo-cytosine
Guanine base			*N*-Bromosuccinimide		8-Bromo-guanine
Guanine base	Diazo benzoyl derivative			8-(*p*-diazobenzoyl)-guanine derivative	
Diol			Periodate		Cleavage to form two aldehydes
Disulfide			DTT, 2-mercaptoethanol, TCEP,		Cleavage to form two sulfhydryls (thiols)
Diazo bond			Sodium dithionite		Cleavage to form two aryl amines
Ester bond			Hydroxylamine, OH^-		Cleavage to form a carboxylate and a hydroxyl
Sulfone bond			OH^-		Cleavage to form a sulfonate and a hydroxyl
Acyl hydrazone bond			Hydrazide		Cleavage to form a hydrazone and a hydrazide
o-Nitrophenyl group			UV light		Cleavage to form a methyl ketone and an amine

between the isothiocyanate-containing compound and the amine with no leaving group involved.

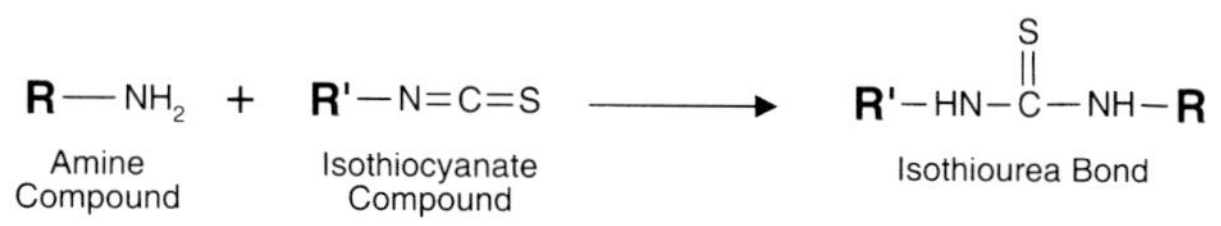

(REACTION 3.1)

Isothiocyanate compounds react best at alkaline pH values where the target amine groups are mainly unprotonated. Many reactions are carried out in 0.1-*M* sodium carbonate buffer at pH 9.0. Reaction times vary from 4 to 24h at 4°C. Rana and Meares (1990) found that by reacting isothiocyanate-containing chelates at pH 7 they could selectively modify a monoclonal antibody only at its N-terminal α-amines while leaving lysine amines unmodified. This is an excellent method for selectively modifying only a single site on a protein or peptide molecule. Since the isothiocyanate group is relatively unstable in aqueous conditions, reagents containing this function should be stored desiccated at refrigerator or freezer temperatures.

1.2. Isocyanates

Isocyanates are similar to the isothiocyanates discussed above, except an oxygen atom replaces the sulfur. An isocyanate can be formed from the reaction of an aromatic amine with phosgene (Rifai and Wong, 1986). The group can also be created from acyl azides by treatment at 80°C in the presence of an alcohol (Section 2.6.1.5). Under these conditions, the acyl azide rearranges to form an isocyanate. Isocyanates can react with amine-containing molecules to form stable isourea linkages (Reaction 3.2). The reactivity of isocyanates is greater than that of isothiocyanates, but for the same reason their stability can be a problem. Many commercial suppliers of bioconjugate reagents have found isocyanate compounds too unstable to offer them for sale, since moisture rapidly decomposes them, releasing CO_2 and leaving an aromatic amine.

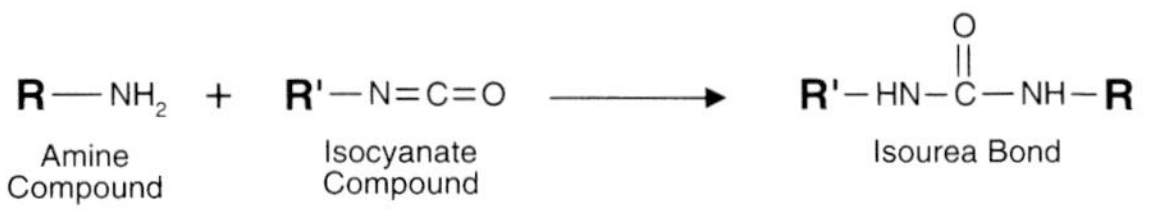

(REACTION 3.2)

Isocyanate-containing reagents can also be used to crosslink or label hydroxyl-containing molecules. Recently, a heterobifunctional compound containing an isocyanate group on one end and a maleimide group on the other end was reported (Annunziato *et al.*, 1993). PMPI, or *p*-maleimidophenyl isocyanate, can be used to conjugate hydroxyl-containing compounds such as polysaccharides with sulfhydryl-containing molecules (available from Thermo Fisher).

1.3. Acyl Azides

Acyl azides are activated carboxylate groups that can react with primary amines to form amide bonds. The azide function is a good leaving group similar to the *N*-hydroxysuccinimide group of NHS ester compounds. An acyl azide can be formed by treatment of a hydrazide with sodium nitrite at 0°C (Lowe and Dean, 1974). A coupling reaction with an amine group occurs by attack of the nucleophile at the electron-deficient carbonyl group (Reaction 3.3). Optimum conditions for the reaction are a pH range of 8.5 to 10 in buffers which contain no competing amines or other nucleophiles.

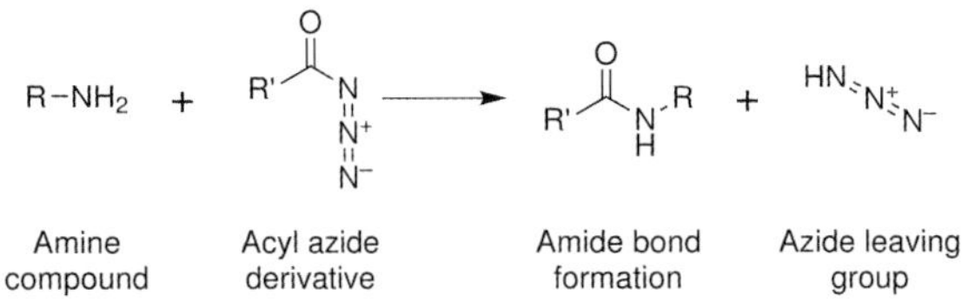

(REACTION 3.3)

The major competing reaction in acyl azide coupling is hydrolysis. The higher the pH of the reaction medium, the faster becomes the reactivity, which is true with regard to both amine reactivity and hydrolysis. Cross-linkers or modification reagents containing this compound must be kept dry to preserve activity. Reactions are complete in 2 to 4 h at room temperature.

1.4. NHS Esters

An *N*-hydroxysuccinimide (NHS) ester is perhaps the most common activation chemistry for creating reactive acylating agents. NHS esters were first introduced as reactive ends of homobifunctional crosslinkers (Bragg and Hou, 1975; Lomant and Fairbanks, 1976). Today, the great majority of amine-reactive crosslinking or modification reagents commercially available utilize NHS esters. An NHS ester may be formed by the reaction of a carboxylate with NHS in the presence of a carbodiimide. To prepare stable NHS ester derivatives, the activation reaction must be performed in nonaqueous conditions using water-insoluble carbodiimides or condensing agents, such as DCC, or, better, an uronium-based activation agent (Chapter 4, Section 1.4, and Chapter 15, Section 2.1).

NHS or sulfo-NHS ester-containing reagents react with nucleophiles with release of the NHS or sulfo-NHS leaving group to form an acylated product (Reaction 3.4). The reaction of such esters with a sulfhydryl or hydroxyl group may not yield stable conjugates, forming thioesters or ester linkages, respectively. Both of these bonds can potentially hydrolyze in aqueous environments or exchange with neighboring amines to form amide bonds. Histidine side-chain nitrogens of the imidazolyl ring may also be acylated with an NHS ester reagent, but they hydrolyze very rapidly in aqueous environments (Cuatrecasas and Parikh, 1972). Thus, the presence of imidazole in reaction buffers only serves to increase the hydrolysis rate of the active ester. Reaction with primary and secondary amines, however, creates stable amide and imide linkages, respectively, that do not readily break down. Thus, in protein molecules, NHS ester crosslinking reagents couple principally with the α-amines at the N-terminals and the ε-amines of lysine side chains.

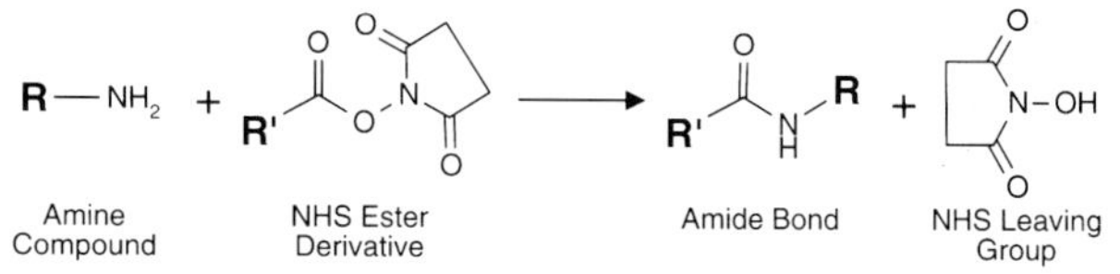

(REACTION 3.4)

Recent studies with NHS ester crosslinking agents using mass spec analysis have determined that the specificity of their reaction with proteins is dependent on the side-chain amino acid functionalities that may be present (Kalkhof and Sinz, 2008). In addition to the expected reactivity with N-terminal α-amines and lysine side-chain ε-amines, NHS esters can react and couple with tyrosine, serine, and threonine –OH groups. This is especially true if there is a neighboring histidine residue present in the polypeptide sequence (Mädler *et al.*, 2009). The histidine will rapidly react with the NHS ester, but since the intermediate acyl imidazole is highly unstable and also reactive toward nucleophiles, it ends up reacting with the nearby –OH group to form an ester linkage. This reaction does not take place as readily in aqueous solution without a histidine group present, because of competition with water, which is in much higher concentration and can promote hydrolysis of the NHS ester rather than ester formation with serine, threonine, or tyrosine.

NHS esters may also be formed *in situ* to react immediately with target molecules in aqueous reaction media. Using the water soluble carbodiimide EDC (Chapter 4, Section 1.1) a carboxylate-containing molecule can be transformed into an active ester by reaction in the presence of NHS or sulfo-NHS (*N*-hydroxysulfosuccinimide) (Chapter 4, Section 1.2). Sulfo-NHS esters are hydrophilic active groups that couple rapidly with amines on target molecules with the same specificity and reactivity as NHS esters (Staros, 1982). Unlike NHS esters that are relatively water insoluble and must be first dissolved in organic solvent before being added to aqueous solutions, sulfo-NHS esters are relatively water soluble and longer lived and hydrolyze more slowly in water. In the presence of amine nucleophiles that can attack at the electron-deficient carbonyl of the active ester, the sulfo-NHS group rapidly leaves, creating a stable amide linkage with the amine compound. Sulfhydryl and hydroxyl groups will also react with such active esters, but the products of such reactions, thioesters and esters, are unstable in aqueous environments or in the presence of amine nucleophiles.

NHS esters have a half-life on the order of hours under physiological pH conditions. However, hydrolysis and amine reactivity both increase with increasing pH. At 0°C at pH 7.0, the half-life is typically 4 to 5 h (Lomant and Fairbanks, 1976). At pH 8.0 at 25°C it falls to 1 h (Staros, 1988), and at pH 8.6 and 4°C the half-life is only 10 min (Cuatrecasas and Parikh, 1972). The rate of hydrolysis may be monitored by measuring the increase in absorptivity at 260 nm as the NHS leaving group is cleaved. The molar extinction coefficient of the NHS group in solution is $8.2 \times 10^3 M^{-1}cm^{-1}$ in Tris buffer at pH 9.0 (Carlsson *et al.*, 1978), but somewhat decreases to $7.5 \times 10^3 M^{-1}cm^{-1}$ in potassium phosphate buffer at pH 6.5 (Partis *et al.*, 1983). Unfortunately, the relatively low sensitivity of this absorptivity measurement does not allow for determining the rate of reaction in an actual crosslinking procedure.

To maximize the modification of amines and minimize the effects of hydrolysis, maintain a high concentration of protein or other target molecule in the reaction medium. By adjusting the molar ratio of crosslinker to target molecule(s), the level of modification and conjugation may be controlled to create an optimal product. Water insoluble crosslinkers containing NHS esters may be reacted in organic solvents, eliminating the hydrolysis problem, provided the target molecule is soluble and stable in such environments. For nonaqueous reactions, an organic base (proton acceptor) is typically added, such as triethylamine or 4-(dimethylamine)pyridine (DMAP).

1.5. Sulfonyl Chlorides

Sulfonyl chlorides are reactive sulfonic acid derivatives similar in properties and reactivity to acid chlorides of carboxylates. The sulfonic acid group, however, is a highly hindered molecule, containing a tetrahedral configuration of substituents. The attack of a nucleophile on a sulfonyl chloride involves temporary formation of a pentavalent intermediate which is highly crowded and unstable. Unlike the capability of using other condensing agents such as carbodiimides when preparing amide linkages between carboxylate groups and amines, sulfonic acids are too hindered to allow such bulky active intermediates to be formed. Thus, the main activation chemistry employed with sulfonates is to create the sulfonyl chloride derivative. Reaction of a sulfonyl chloride compound with a primary amine-containing molecule proceeds with loss of the chlorine atom and formation of a sulfonamide linkage (Reaction 3.5).

(REACTION 3.5)

Sulfonic acids (sulfonates) are frequent constituents of fluorescent probes to aid in water solubility of the core dyes (Chapter 10, Section 1). The sulfonyl chloride derivative allows simple conjugation of these molecules with proteins and other amine-containing compounds. The derivative is prepared by reaction of the sulfonate with thionyl chloride or phosphorus pentachloride in nonaqueous conditions. The reaction of a sulfonyl chloride with an amine proceeds under alkaline pH conditions (typically carried out at pH 9–10). It may also be performed in organic solvent for the modification of water-insoluble compounds. Hydrolysis is the major competing reaction in aqueous environments, although the overall rate of sulfonyl chloride reactivity and hydrolysis is less than that of the corresponding acid chlorides of carboxylates. However, sulfonyl chloride-containing reagents should be stored under nitrogen or in a desiccator to prevent breakdown by moisture.

1.6. Tosylate Esters

Reactive groups consisting of tosylate esters can be formed from the reaction of 4-toluenesulfonyl chloride (also called tosyl chloride or TsCl) with a hydroxyl group to yield the sulfonyl ester derivative (Whitaker *et al.*, 2006). The sulfonyl ester is an electrophilic reactive group that can couple with nucleophiles to produce a covalent bond. This can result in a secondary amine linkage with primary amines, a thioether linkage with sulfhydryl groups, or an ether bond with hydroxyls. Tosylate esters are commonly used in organic synthesis and in the immobilization of affinity ligands onto insoluble support materials or surfaces (see Chapter 15, Section 2.1). An interesting aspect of this reaction is the replacement of the original hydroxyl that is activated with TsCl with the reacting nucleophile, which is being coupled.

Nucleophilic attack on the reactive sulfonate ester actually occurs at the carbon atom adjacent to the ester (Cremlyn, 1996; James and Cremlyn, 2002). This causes cleavage of the sulfonyl ester and release of the leaving group, which is the tosyl sulfonate (Figure 3.1). The reaction occurs under basic conditions using aqueous buffers in the pH range of 8.5 to 10. The reactivity of nucleophiles typically is $-SH \rightarrow -NH_2 > -OH$, and the optimal pH for the reaction increases for these functional groups in the same order. The reactions may also be carried out in nonaqueous conditions using an organic base to catalyze the reaction process. Suitable bases include pyridine, triethylamine (TEA), diisopropylethylamine (DIEA), or dimethylaminopyridine (DMAP).

1.7. Aldehydes and Glyoxals

Carbonyl groups such as aldehydes, ketones, and glyoxals can react with amines to form Schiff base intermediates which are in equilibrium with their free forms. The interaction is pH dependent, being more efficient at low pH and especially efficient at high pH conditions. Certain compounds, particularly some reducing sugars, may undergo an Amadori rearrangement after Schiff base formation to a stable ketoamine structure

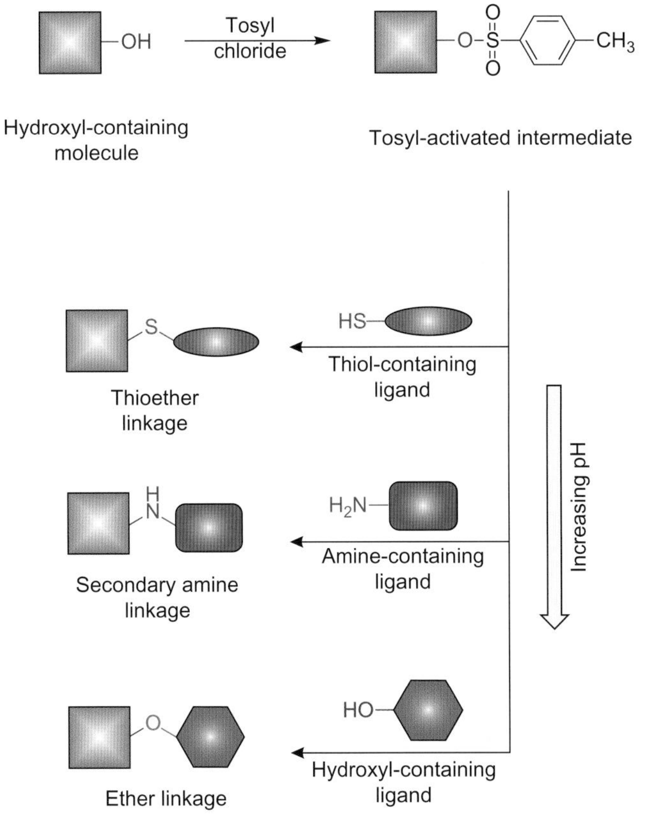

FIGURE 3.1 Tosyl chloride can be used to activate a hydroxyl group in nonaqueous conditions to create a reactive tosyl ester. The sulfonyl ester can react with thiols, amines, and hydroxyls to form thioether, secondary amine, and ether linkages, respectively. In general, an increase in pH from neutral to highly alkaline is needed to efficiently couple with thiols, amines, and hydroxyls.

(Chapter 2, Section 2.1). This occurs *in vivo* as glucose modifies amine-containing components in the blood to form glycated derivatives. Such modification is thought to be related to aging and is a signal for protein and cellular regeneration.

The rather labile Schiff base interaction can be chemically stabilized by reduction. The addition of sodium borohydride or sodium cyanoborohydride to a reaction medium containing an aldehyde compound and an amine-containing molecule will result in reduction of the Schiff base intermediate and covalent bond formation, creating a secondary amine linkage between the two molecules (Reaction 3.6).

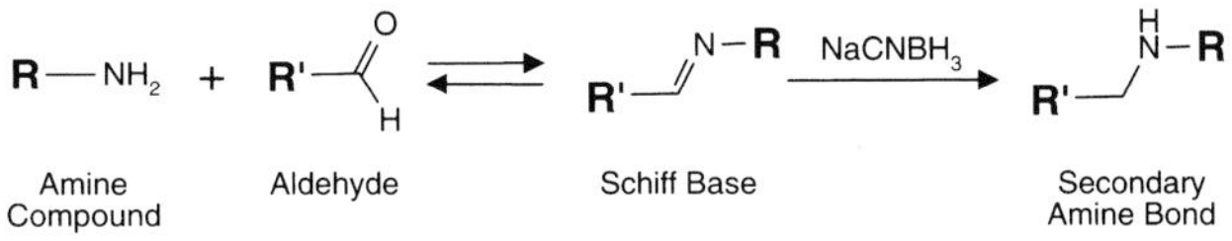

(REACTION 3.6)

Although both borohydride and cyanoborohydride have been used for reductive amination purposes, borohydride will reduce the reactive aldehyde groups to hydroxyls at the same time it converts any Schiff bases present to secondary amines. Cyanoborohydride, by contrast, is a milder reducing agent. It has been shown to be at least five times milder than borohydride in reductive amination processes with antibodies (Peng *et al.*, 1987). While cyanoborohydride does not reduce aldehydes, it is very effective at Schiff base reduction. Thus, higher yields of conjugate formation can be realized using cyanoborohydride instead of borohydride. Other reducing agents have also been explored for reductive amination processes, including various amine boranes and ascorbic acid (Cabacungan *et al.*, 1982; Hornsey *et al.*, 1986). See Chapter 4, Section 4, for a protocol for reductive amination coupling.

1.8. Epoxides and Oxiranes

An epoxide or oxirane group will react with nucleophiles in a ring-opening process. The reaction can take place with primary amines, sulfhydryls, or hydroxyl groups to create secondary amine, thioether, or ether bonds, respectively. During the coupling process, ring opening forms a β-hydroxy group on the epoxy compound (Reaction 3.7). The reaction of the epoxide functionalities with hydroxyls requires high pH conditions, usually in the range of pH 11 to 12. Amine nucleophiles react at more moderate alkaline pH values, typically needing buffer environments of at least pH 9.0. Sulfhydryl groups are the most highly reactive nucleophiles with epoxides, requiring a buffered system closer to the physiological pH range of 7.5 to 8.5 for efficient coupling.

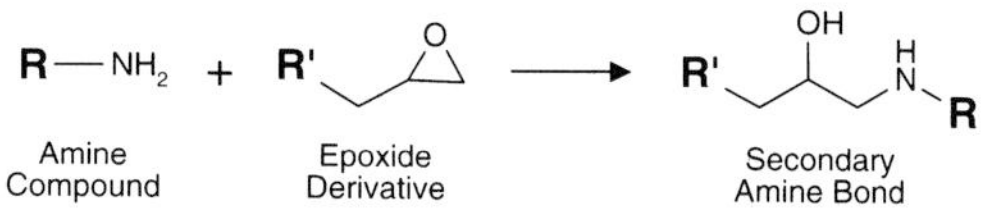

(REACTION 3.7)

The principal side reaction to epoxide coupling is hydrolysis. Particularly at acid pH values, the epoxide ring can hydrolyze to form adjacent hydroxyls. This diol can be oxidized with periodate to create a terminal aldehyde residue with loss of one molecule of formaldehyde (Chapter 2, Section 4.4). The aldehyde then can be used in reductive amination reactions. The reaction of an epoxide group with an ammonium ion generates a terminal primary amine group that can also be used for further derivatization.

1.9. Carbonates

Carbonates are diester derivatives of carbonic acid formed from its condensation with hydroxyl compounds. These groups may be created from the reaction of a bifunctional carbonic acid compound like phosgene or carbonyl diimidazole (Chapter 4, Section 3) with two alcohols. Carbonates can rapidly react with nucleophiles to form carbamate linkages, which are extremely stable bonds (Reaction 3.8). A commonly used bifunctional carbonate compound, disuccinimidyl carbonate, can be used to activate hydroxyl-containing molecules to form amine-reactive succinimidyl carbonate intermediates (Section 4.3, this chapter). This carbonate activation procedure can be used with great success in coupling polyethylene glycol (PEG) to proteins and other amine-containing molecules (Chapter 18, Section 2.1).

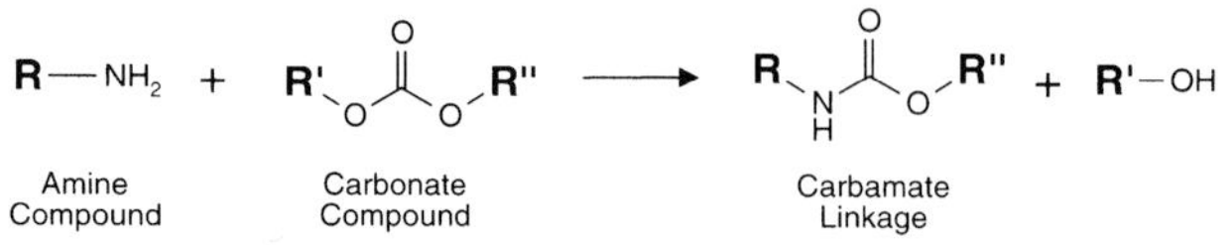

(REACTION 3.8)

Nucleophiles, such as the primary amino groups of proteins, can react with the succinimidyl carbonate functional groups to give stable carbamate (aliphatic urethane) bonds. The linkage is identical to that obtained through CDI activation of hydroxyl groups with subsequent coupling of amines (Chapter 4, Section 3, and Chapter 15, Section 2.1). However, the reactivity of the succinimidyl carbonate is much greater than that of the imidazole carbamate formed as the active species in CDI activation. A succinimidyl carbonate group may hydrolyze in aqueous solution to release NHS and CO_2, essentially regenerating the underivatized hydroxyl. Carbonates formed from esterification of two alcohol groups similarly hydrolyze to release CO_2 plus the original hydroxyl compounds.

The coupling reaction of a carbonate functional group with an amine is best carried out in slightly alkaline pH (7–9) and in the absence of any competing amine or sulfhydryl components.

1.10. Arylating Agents

Aryl halide compounds such as fluorobenzene derivatives can be used to form covalent bonds with amine-containing molecules like proteins. The reactivity of aryl halides, however, is not totally specific for amines. Other nucleophiles such as thiol, imidazolyl, and phenolate groups of amino acid side chains can also react (Zahn and Meinhoffer, 1958). Conjugates formed with sulfhydryl groups are reversible by cleaving with an excess of thiol (Shaltiel, 1967).

Fluorobenzene-type compounds have been used as functional groups in homobifunctional crosslinking agents (Chapter 5, Section 4). Their reaction with amines involves nucleophilic displacement of the fluorine atom with the amine derivative, creating a substituted aryl amine bond (Reaction 3.9). Detection reagents incorporating reactive aryl chemistry include 2,4-dinitrofluorobenzene and trinitrobenzenesulfonate (Eisen *et al.*, 1953). These compounds form colored complexes with target amine groups. The relative rate of reactivity for aryl compounds is F > Cl~Br > sulfonate.

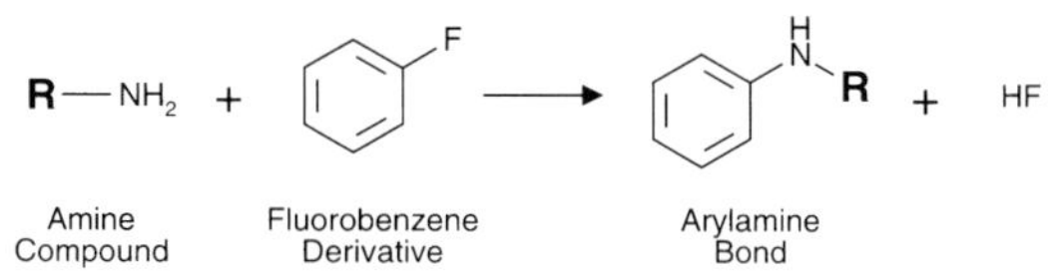

(REACTION 3.9)

1.11. Imidoesters

The imidoester (or imidate) functional group is one of the most specific acylating agents available for modifying primary amines. Unlike most other coupling chemistries, imidoesters possess minimal cross-reactivity toward other nucleophilic groups in proteins. The α-amines and ε-amines of proteins may be targeted and crosslinked by reacting with homobifunctional imidoesters at a pH of 7 to 10 (optimal pH 8–9). The product of this reaction, an imidoamide (or amidine) (Reaction 3.10), is protonated and thus carries a positive charge at physiological pH (Liu *et al.*, 1977; Kiehm and Ji, 1977; Ji, 1979; Wilbur, 1992).

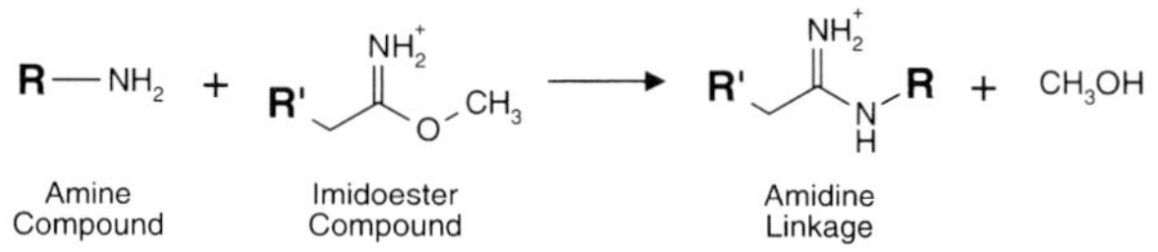

(REACTION 3.10)

The amidine bond formed is quite stable at acid pH; however, it is susceptible to hydrolysis and cleavage at high pH. A typical reaction condition for using imidate crosslinkers is a buffer system consisting of 0.2-*M* triethanolamine in 0.1-*M* sodium borate, pH 8.2. After conjugating two proteins with a bifunctional imidoester crosslinker, excess imidoester functional groups may be blocked with ethanolamine.

1.12. Carbodiimides

Carbodiimides are zero-length crosslinking agents used to mediate the formation of an amide or phosphoramidate linkage between a carboxylate group and an amine or a phosphate and an amine, respectively (Hoare and Koshland, 1966; Chu *et al.*, 1986; Ghosh *et al.*, 1990). They are called zero-length reagents because in forming these bonds no additional chemical structure is introduced between the conjugating molecules.

N-substituted carbodiimides can react with carboxylic acids to form highly reactive, *O*-acylisourea derivatives that are extremely short-lived (Reaction 3.11). This active species can then react with a nucleophile such as a primary amine to form an amide bond (Reaction 3.12) (Williams and Ibrahim, 1981). Other nucleophiles are also reactive. Sulfhydryl groups may attack the active species and form thioester linkages, although these are not as stable as the bond formed with an amine.

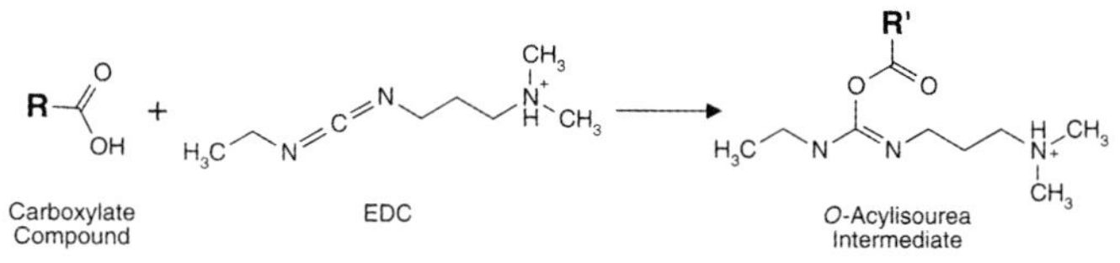

(REACTION 3.11)

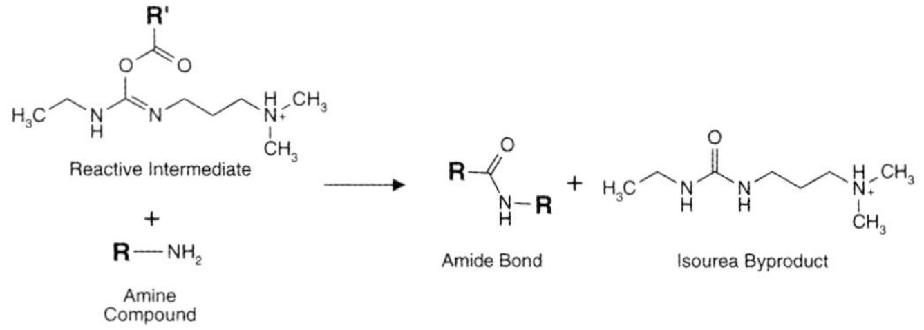

(REACTION 3.12)

Hydrazide-containing compounds can also be coupled to carboxylate groups using a carbodiimide-mediated reaction. Using bifunctional hydrazide reagents, carboxylates can be modified to possess terminal hydrazide groups able to conjugate with other carbonyl compounds (Chapter 5, Section 8).

In addition, oxygen atoms may act as the attacking nucleophile, such as those in water molecules. In aqueous solution, hydrolysis by water is the major competing reaction, both inactivating EDC itself and cleaving off the activated ester intermediate, forming an isourea, and regenerating the carboxylate group (Gilles *et al.*, 1990).

Nakajima and Ikada (1995) investigated the reactions of EDC amide bond formation in aqueous solution using a hydrogel of poly(acrylic acid) to contribute the carboxylate groups and ethylenediamine or benzylamine as the amine functional groups. Their results indicate that carboxylate activation occurs most effectively at pH 3.5 to 4.5, while amide bond formation occurs with highest yield at pH 4 to 6. However, the maximal rate of hydrolysis of EDC occurs at acidic pH values with increasing stability of the carbodiimide in solution at or above pH 6.5. When working with proteins and peptides, experience indicates that EDC-mediated amide bond formation effectively occurs between pH 4.5 and 7.5. Buffer systems using MES or phosphate may be used to stabilize the pH during the course of the reaction. For additional information on specific carbodiimides used in bioconjugate chemistry, see Chapter 4, Section 1.

Molecules containing phosphate groups, such as the 5′ phosphate of oligonucleotides, may also be conjugated to amine-containing molecules by using a carbodiimide-mediated reaction (Chapter 23, Section 2.1). The carbodiimide activates the phosphate to an intermediate phosphate ester similar to its reaction with carboxylates (Chapter 4, Section 1). In the presence of an amine, the ester reacts to form a stable phosphoramidate bond (Reaction 3.13).

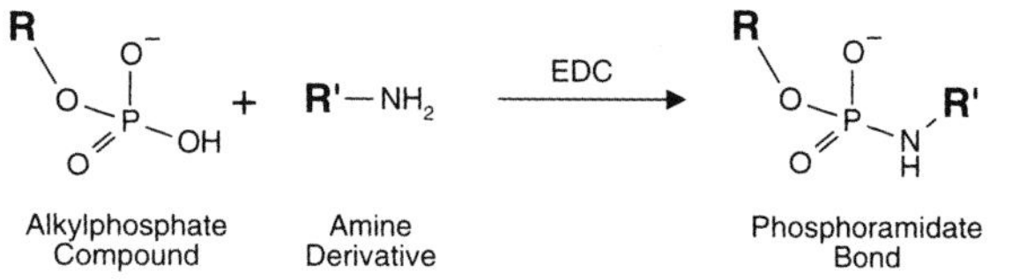

(REACTION 3.13)

1.13. Anhydrides

Acid anhydrides, as their name implies, are formed from the dehydration reaction of two carboxylic acid groups. Anhydrides are highly reactive toward nucleophiles and are able to acylate a number of the important functional groups of proteins and other macromolecules. Upon nucleophilic attack, the anhydride yields one carboxylic acid for every acylated product. If the anhydride was formed from monocarboxylic acids, such as acetic anhydride, then the acylation occurs with release of one carboxylate group. However, for dicarboxylic acid anhydrides, such as succinic anhydride, upon reaction with a nucleophile the ring structure of the anhydride opens, forming the acylated product modified to contain a newly formed carboxylate group (Reaction 3.14). Thus, anhydride reagents may be used to both block functional groups and to convert an existing functional group into a carboxylic acid.

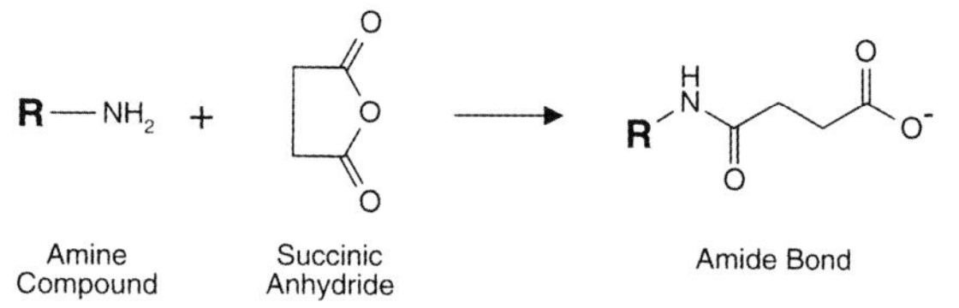

(REACTION 3.14)

Protein functional groups able to react with anhydrides include the α-amines at the N-terminals, the ε-amine of lysine side chains, cysteine sulfhydryl groups, the phenolate ion of tyrosine residues, and the imidazolyl ring of histidines. However, acylation of cysteine, tyrosine, and histidine side chains forms unstable complexes that are easily reversible to regenerate the original group. Only amine functional groups of proteins are stable to acylation with anhydride reagents, forming amide bonds (Fraenkel-Conrat, 1959; Smyth, 1967).

Another potential site of reactivity for anhydrides in protein molecules is modification of any attached carbohydrate chains. In addition to amino group modification in the polypeptide chain, glycoproteins may be modified at their polysaccharide hydroxyl groups to form esterified derivatives. Esterification of carbohydrates by acetic anhydride, especially cellulose, is a major industrial application for this compound. In aqueous solutions, however, esterification will be a minor product, since the oxygen of water is about as strong a nucleophile as the hydroxyls of sugar residues.

The major side reaction to the desired acylation product is hydrolysis of the anhydride. In aqueous solutions, anhydrides may break down by the addition of one molecule of water to yield two unreactive carboxylate groups. The presence of an excess of the anhydride in the reaction medium usually is used to minimize the effects of competing hydrolysis.

Since both hydrolysis and acylation result in the release of carboxylic acid functionalities, the medium becomes acidic during the course of the reaction. This requires either the presence of a strongly buffered environment to maintain the pH or periodic monitoring and adjustment of the pH with base as the reaction progresses.

1.14. Fluorophenyl Esters

Another type of carboxylic acid derivative that reacts with amines consists of the ester of a fluorophenol compound, which creates a group capable of forming amide bonds with proteins and other molecules. Several types of fluorophenyl esters have been used as reactive groups: a pentafluorophenyl (PFP) ester, a tetrafluorophenyl (TFP) ester, and a sulfo-tetrafluorophenyl (STP) ester. All of these derivatives have similar reactivity with amines, but the uncharged ones are hydrophobic and the sulfonated one is negatively charged in aqueous solution, thus providing water solubility to active ester compounds (Gee *et al.*, 1999) (Figure 3.2).

Fluorophenyl esters react with amine-containing molecules at slightly alkaline pH values to give the same amide bond linkages as NHS esters (Reaction 3.15). However, in most cases, the fluorophenyl ester compound will display better stability toward hydrolysis in aqueous solution. It has been reported that a TFP ester has over twice the half-life in basic pH buffers (pH ~8) than a corresponding NHS ester on the same compound (Molecular Probes).

Fluorophenyl ester compounds can be coupled to amines at a pH range of 7 to 9, with 0.1-*M* sodium bicarbonate, pH 8, a suggested reaction medium.

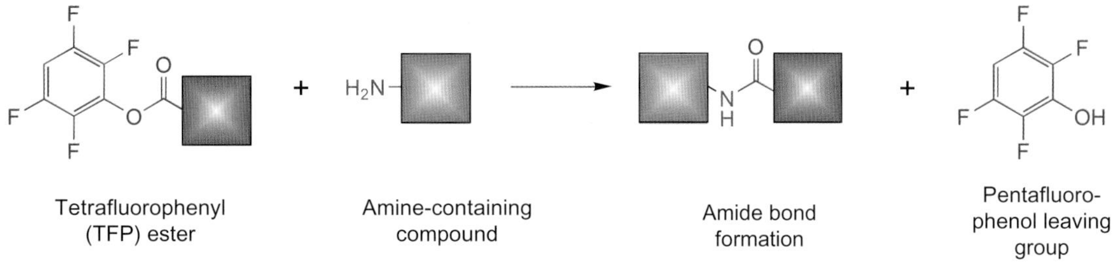

(REACTION 3.15)

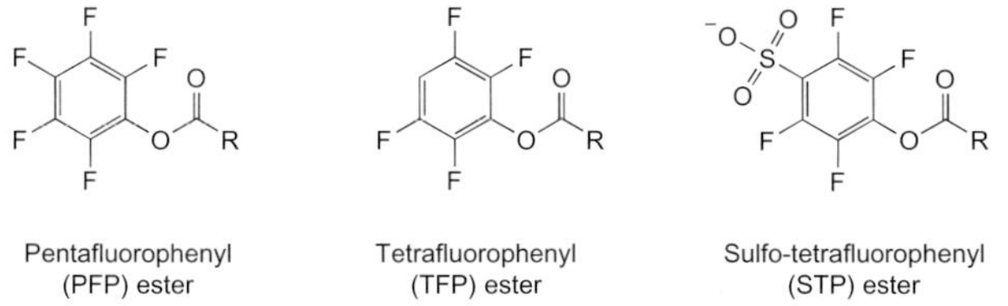

FIGURE 3.2 Three types of fluorophenyl esters have been used for coupling to amine-containing molecules. The pentafluorophenyl and tetrafluorophenyl esters are relatively hydrophobic and typically have better stability toward hydrolysis in aqueous solution than NHS esters. The sulfo-tetrafluorophenyl ester is water soluble due to the negatively charged sulfonate group, and it provides better solubility to associated crosslinkers or bioconjugation reagents similar to that of a sulfo-NHS ester group.

1.15. Hydroxymethyl Phosphine Derivatives

Phosphine compounds are often thought of only in terms of having reductant properties, especially in biological applications. However, there are classes of phosphine derivatives with hydroxymethyl group substitutions that can also act as bioconjugation agents for coupling or crosslinking purposes. Tris(hydroxymethyl) phosphine (THP) and β-[tris(hydroxymethyl)phosphino] propionic acid (THPP; Thermo Fisher) are small trifunctional compounds that spontaneously react with nucleophiles, such as amines, to form covalent linkages (Henderson *et al.*, 1994; Katti, 1996; Katti *et al.*, 1999). Nucleophiles react with the hydroxymethyl arms of THP and THPP by attack on the electron-deficient carbon atom next to the oxygen with loss of water to form secondary or tertiary amine bonds (Reaction 3.16).

Both THP and THPP are stable in aqueous solution, as the only potential product of hydrolysis is the reformation of the hydroxymethyl groups. It is unusual for an amine-reactive functional group to have long-term stability in water or buffer, which makes these reagents uniquely suitable for creating reactive surfaces or reactive molecules for subsequent conjugation with proteins or other amine-containing compounds. Hydroxylic chromatographic supports have also been activated with hydroxymethyl phosphine derivatives for immobilization of enzymes (Petach *et al.*, 1994).

Hydroxymethyl phosphines are susceptible to oxidation to form the phosphine oxide derivative. Therefore,

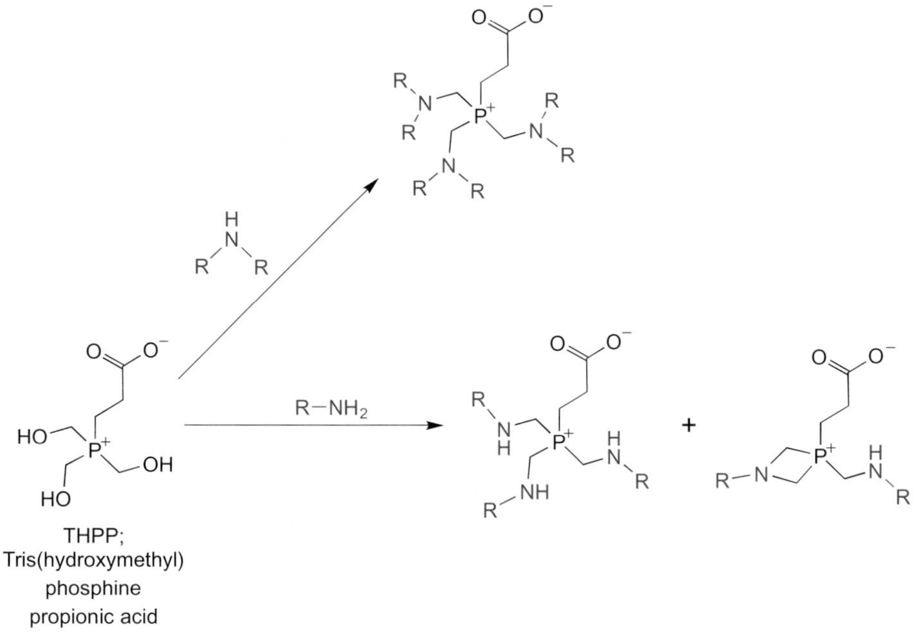

(REACTION 3.16)

avoid excess oxygen, oxidizing agents, or azide compounds, which react with phosphines in the Staudinger reaction (Chapter 17, Section 6). In addition, metallic surfaces can be modified via the phosphine group to result in hydroxymethyl group substitutions.

1.16. Guanidination of Amines

The addition of a guanidino group to amine-containing molecules can be performed using the compound *O*-methylisourea (as the hemisulfate salt). Guanidination has been used to increase the ionization potential of lysine-containing peptides for greater sensitivity in mass spec analysis (Brancia *et al.*, 2000). The process has also been used to add stable isotope labels (^{15}N) to tryptic peptides (Cristea *et al.*, 2004). Upon trypsin digestion, protein samples are cleaved at arginine and lysine residues to yield peptide fragments containing these amino acids at their C-terminal. The guanidine group of arginine is known to aid in the ionization of peptides for MS analysis. The reaction of *O*-methylisourea hemisulfate with lysine ε-amino groups produces homoarginine (Reaction 3.17), which ionizes far better than lysine and aids in the detection of these peptides (Warwood *et al.*, 2006). The addition of a guanidine group to a lysine residue adds 42.02 Daltons to the resultant modified peptide, which needs to be taken into account for mass spec purposes.

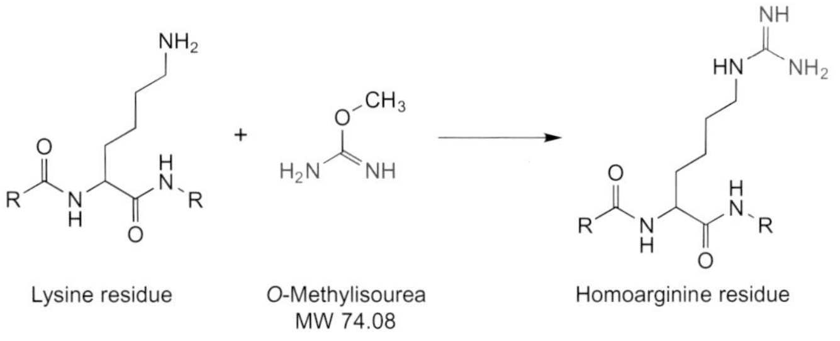

(REACTION 3.17)

Protocols for guanidination reactions typically use basic conditions to deprotonate all the lysine ε-amino groups, which is necessary to achieve efficient yields. The amount of N-terminal α-amine labeling is minimal, but may occur to some extent, especially for peptides containing an N-terminal glycine residue (Beardsley and Reilly, 2002). The initial protocols for guanidination usually use reaction times of several hours, but optimization has resulted in decreasing this to 5 to 10 min at elevated temperature. The following protocol is based on the method of Beardsley and Reilly (2002).

Protocol

1. Dissolve 50 mg of *O*-methylisourea hemisulfate in 51 μl of water.
2. Prepare 5 μl of a peptide solution to undergo guanidination at a concentration of 1 pmol/μl and add 5.5 μl of 7-*N* ammonium hydroxide.
3. Add 1.5 μl of the *O*-methylisourea hemisulfate solution to the peptide solution with mixing.
4. React for 5–10 min at 65°C in an oven.
5. Stop the reaction with the addition of 15 μl of 10% TFA (v/v).

2. THIOL REACTIONS

Reactive groups able to couple with sulfhydryl-containing molecules are perhaps the second most common functional groups present on crosslinking or modification reagents. Especially in the design of heterobifunctional crosslinkers, sulfhydryl-reactive groups are frequently present on one of the two ends. The other end of such crosslinkers is often an amine-reactive group that is coupled to a target molecule before the sulfhydryl-reactive end, due to the labile nature of amine acylation chemistries. The primary coupling reactions for modification of sulfhydryls proceed by one of two routes: alkylation or disulfide interchange. Many of the reactive groups that undergo these reactions are stable enough in aqueous environments to allow a two-step conjugation strategy to be used (Chapter 6, Section 1). Once initiated, most of these reactions are rapid and occur in high yield to give stable thioether or disulfide bonds.

2.1. Haloacetyl and Alkyl Halide Derivatives

Three forms of activated halogen derivatives can be used to create sulfhydryl-reactive compounds: haloacetyl, benzyl halides that react through a resonance activation process with the neighboring benzene ring, and alkyl halides that possess the halogen β to a nitrogen or sulfur atom, as in N- and S-mustards. In each of these compounds, the halogen group is easily displaced by an attacking nucleophilic substance to form an alkylated derivative with loss of HX (where X is the halogen and the hydrogen comes from the nucleophile). Haloacetyl compounds and benzyl halides typically are iodine or bromine derivatives, whereas the halo-mustards mainly employ chlorine and bromine forms (see Chapter 5, Section 10, for examples of homobifunctional reagents that employ reactive halogen groups). Iodoacetyl groups have also been used successfully to couple affinity ligands to chromatography supports (Chapter 15, Section 2.2).

Although the primary utility of active halogen compounds is to modify sulfhydryl groups in proteins or other molecules, the reaction is not totally specific. Iodoacetyl (and bromoacetyl) derivatives can react with a number of functional groups within proteins: the sulfhydryl group of cysteine, both imidazolyl side chain nitrogens of histidine, the thioether of methionine, and the primary ε-amine group of lysine residues and N-terminal α-amines (Gurd, 1967). The relative rate of reaction with each of these residues is generally dependent on the degree of ionization and thus the pH at which the modification is performed. The exception to this rule is methioninyl thioethers which react rapidly at nearly all pH values above 1.7 (Vithayathil and Richards, 1960). The only reaction resulting in one definitive product is that of the alkylation of cysteine sulfhydryls, giving the carboxymethylcysteinyl derivative (Cole *et al.*, 1958) (Reaction 3.18). Histidine groups may be modified at either nitrogen atom of their imidazolyl side chain, thus producing the possibility of either mono-substituted or di-substituted products (Crestfield *et al.*, 1963). With primary amine groups such as in the side chain of lysine residues, the products of the reaction are the secondary amine, monocarboxymethyllysine, or the tertiary amine derivative, dicarboxymethyllysine. Methionine thioether groups give the most complicated products, some of which rearrange or decompose unpredictably. The only stable carboxy derivative of methionine is where the terminal methyl group is lost to form carboxymethylhomocysteine, the same product as the reaction of iodoacetate with homocysteine. For a complete illustration of these reactions, see Chapter 2, Section 4.2.

(REACTION 3.18)

The relative reactivity of α-haloacetates toward protein functional groups is sulfhydryl > imidazolyl > thioether > amine. Among halo derivatives the relative reactivity is I > Br > Cl > F, with fluorine being almost unreactive. The α-haloacetamides have the same trend of relative reactivities, but will create a terminal amide group not a terminal carboxylate.

Thus, iodoacetate has the highest reactivity toward sulfhydryl cysteine residues and may be directed specifically for –SH modification. If iodoacetate is present in limiting quantities (relative to the number of sulfhydryl groups present) and at slightly alkaline pH, cysteine modification will be the exclusive reaction. The specificity of this modification has been used in the design of heterobifunctional crosslinking reagents, where one end of the crosslinker contains an iodoacetamide derivative and the other end contains a different functionality directed at another chemical target (see SIAB, Chapter 6, Section 1.5).

2.2. Maleimides

Maleic acid imides (maleimides) are derivatives of the reaction of maleic anhydride and ammonia or an amine derivative. This functional group is a popular constituent of many heterobifunctional crosslinking agents (Chapter 6). The double bond of maleimides may undergo an alkylation reaction with sulfhydryl groups to form stable thioether bonds. Maleimide reactions are specific for thiols in the pH range of 6.5 to 7.5 (Smyth *et al.*, 1964; Gorin *et al.*, 1966; Heitz *et al.*, 1968; Partis *et al.*, 1983). At pH 7.0, the reaction of the maleimide with sulfhydryls proceeds at a rate 1000 times greater than its reaction with amines. At higher pH values, some cross-reactivity with amino groups takes place (Brewer and Riehm, 1967). One of the carbons adjacent to the maleimide double bond undergoes nucleophilic attack by the thiolate anion to generate the addition product (Reaction 3.19). When sufficient quantities of –SH groups are being alkylated, the reaction may be followed spectrophotometrically by the decrease in absorbance at 300 nm as the double bond reacts and disappears.

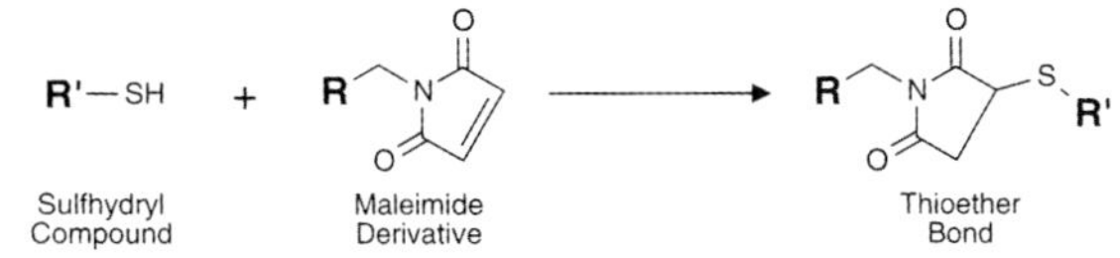

(REACTION 3.19)

The maleimide group may also undergo hydrolysis to an open maleamic acid form that is unreactive toward sulfhydryls (Chapter 19, Section 5). Hydrolysis may also occur after sulfhydryl coupling to the maleimide. This ring opening reaction typically happens faster the higher the pH becomes. Hydrolysis is also dependent on the type of chemical group next to the maleimide function. For instance, the cyclohexane ring of SMCC (Chapter 6, Section 1.3) provides increased stability to maleimide hydrolysis, probably due to its steric effects and its lack of aromatic character. However, the adjacent phenyl ring of MBS allows much greater rates of hydrolysis to occur at the maleimide ring (Chapter 6, Section 1.4).

2.3. Aziridines

An aziridine reactive group is a small ring system composed of one nitrogen and two carbon atoms. The highly hindered nature of this heterocyclic ring gives it strong reactivity toward nucleophiles. Sulfhydryls will react with aziridine-containing reagents in a ring-opening process, forming thioether bonds (Reaction 3.20). The simplest aziridine compound, ethylenimine, can be used to transform available sulfhydryl groups into amines (Chapter 2, Section 4.3).

(REACTION 3.20)

The reaction of an aziridine with a thiol is highly specific at slightly alkaline pH values. In aqueous solution, the major side reaction is hydrolysis.

Substituted aziridines have been used to form homobifunctional and trifunctional crosslinking agents, although their use has been limited (Ross, 1953; Alexander, 1954). The functional group has found use, however, in the design of the fluorescent probe dansyl aziridine (5-dimethylaminonaphthalene-2-sulfonyl aziridine) (Johnson *et al.*, 1978; Grossman *et al.*, 1981).

2.4. Acryloyl Derivatives

Reactive double bonds are capable of undergoing additional reactions with sulfhydryl groups. A popular example of this type of functional group is the maleimide group (Section 2.2, this chapter). However, derivatives of acrylic acid or methacrylic acid (more stable) are also able to participate in this reaction, although the rate of sulfhydryl addition is somewhat slower than that of maleimides. The reaction of an acryloyl compound with a sulfhydryl group occurs with the creation of a stable thioether bond (Reaction 3.21).

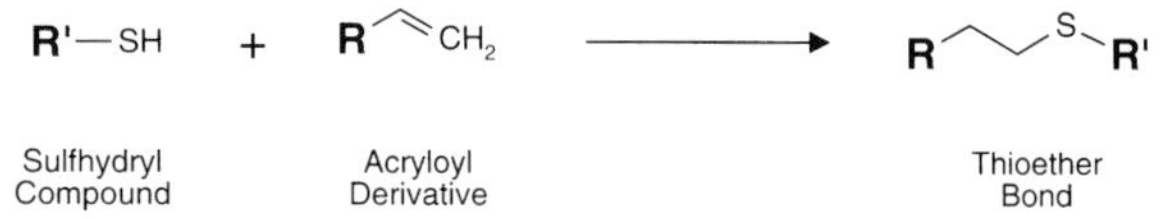

(REACTION 3.21)

Although acryloyl crosslinking agents have not been common, the reactive group has found use in the design of the sulfhydryl-reactive fluorescent probe, 6-acryloyl-2-dimethylaminonaphthalene (acrylodan; Molecular Probes) (Epps *et al.*, 1992; Yem *et al.*, 1992).

2.5. Arylating Agents

Arylating agents are reactive aromatic compounds containing a constituent on the ring that can undergo nucleophilic substitution. The most common arylating agents are derivatives of benzene which possess either halogen or sulfonate groups on the ring. The presence of electron-withdrawing constituents, such as nitro groups, increases the reactivity of the replaceable group. Although aryl halides are commonly used to modify amine-containing molecules to form aryl amine derivatives, they also react quite readily with sulfhydryl groups.

Fluorobenzene-type compounds have been used as functional groups in homobifunctional crosslinking agents (Chapter 5, Section 4). Their reaction with nucleophiles involves bimolecular nucleophilic substitution, causing the replacement of the fluorine atom with the sulfhydryl derivative and creating a substituted aryl bond (Reaction 3.22). Conjugates formed with sulfhydryl groups are reversible by cleaving with an excess of thiol (such as DTT) (Shaltiel, 1967). Detection reagents incorporating reactive aryl chemistry include 2,4-dinitrofluorobenzene and trinitrobenzenesulfonate (Eisen *et al.*, 1953). The relative rate of reactivity for aryl compounds is F > Cl~Br > sulfonate.

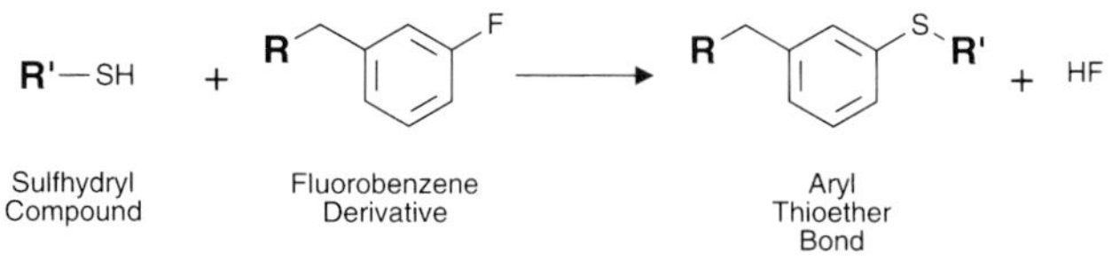

(REACTION 3.22)

2.6. Thiol–Disulfide Exchange Reagents

Compounds containing a disulfide group are able to participate in disulfide exchange reactions with another thiol. The disulfide exchange (also called interchange) process involves attack of the thiol at the disulfide, breaking the -S–S- bond, with subsequent formation of a new mixed disulfide comprising a portion of the original disulfide compound (Reaction 3.23). The reduction of disulfide groups to sulfhydryls in proteins using thiol-containing reductants proceeds through the intermediate formation of a mixed disulfide (Chapter 2, Section 4.1). If the thiol is present in excess, the mixed disulfide can go on to form a symmetrical disulfide consisting entirely of the thiol reducing agent—thus completely reducing the original disulfide to free sulfhydryls. If the thiol reductant is not present in large enough excess, the mixed disulfide product is the end result.

(REACTION 3.23)

Crosslinking or modification reactions using disulfide exchange processes form disulfide linkages with sulfhydryl-containing molecules. These bonds are reversible using disulfide reducing agents. Thus, conjugates may be created and later released for analysis by incubation with DTT or other disulfide reductants (e.g., TCEP). The disulfide bond within these crosslinks also permits important reactions to occur *in vivo*, such as the release of the toxin component of immunotoxin conjugates, allowing the cytotoxic portion to penetrate target cells and cause cell death (Chapter 20, Section 3).

Disulfide exchange reactions occur over a broad range of conditions—from acid to basic pH—and in a wide variety of buffer constituents. Most crosslinking reactions involving disulfide exchange are carried out under physiological conditions or those most appropriate to maintain stability of the protein or other molecule being modified.

Pyridyl Disulfides

A pyridyl dithiol is perhaps the most popular type of thiol-disulfide exchange functional group used in the construction of crosslinkers or modification reagents. Pyridyl disulfides can be created from available primary amines on molecules through the reaction of 2-iminothiolane in tandem with 4,4′-dipyridyl disulfide (King *et al.*, 1978). For instance, the simultaneous reaction among a protein, 2-iminothiolane, and 4,4′-dipyridyl disulfide yields a modification containing reactive pyridyl disulfide groups in a single step (Chapter 2, Section 4.1).

A pyridyl disulfide will readily undergo an interchange reaction with a free sulfhydryl to yield a single mixed disulfide product. This is due to the fact that the pyridyl disulfide contains a leaving group that is easily transformed into a non-reactive compound not capable of participating in further mixed disulfide formation. Thus, the thiol–disulfide exchange reaction can be controlled to occur with only one-half of the original disulfide compound. For instance, a reagent system containing a pyridyl disulfide group, such as SPDP (Chapter 6, Section 1.1), is able to react with sulfhydryl groups by releasing the electron-stabilized compound pyridine-2-thione (Reaction 3.24). Since the leaving group does not possess a free thiol, it cannot undergo disulfide exchange with another molecule of the attacking sulfhydryl compound. Thus, only one end of the reagent has potential for becoming attached to the sulfhydryl-containing molecule.

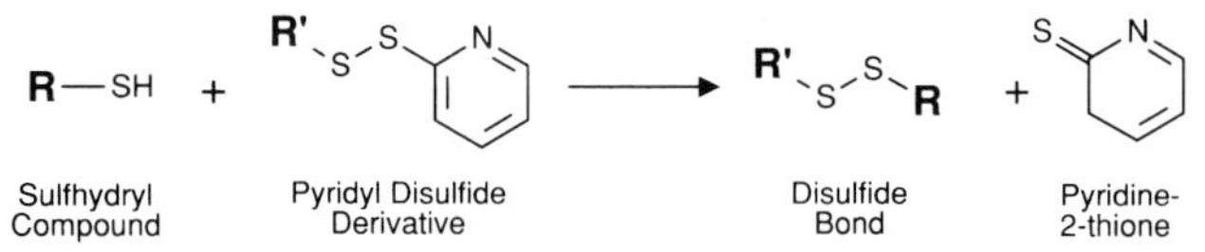

(REACTION 3.24)

Pyridyl-dithiol containing crosslinking and modification reagents are highly efficient in forming disulfide bonds with sulfhydryl-containing molecules. Newer reagents containing a hydrophilic PEG crossbridge in the structure provide increased water solubility and decreased nonspecific binding character (Chapter 18, Section 1.2), which is an advantage in many assay and detection applications for the resultant conjugates. In addition, during the course of a pyridyl disulfide reaction, the pyridine-2-thione leaving group has unique spectral properties that allow the measurement of sulfhydryl coupling by monitoring the increase in absorbance at 343 nm ($\varepsilon = 8.08 \times 10^3 M^{-1} cm^{-1}$). Once a disulfide linkage is formed, it may be cleaved using standard disulfide reducing agents (Chapter 2, Section 4.1). The activation of chromatography supports to contain pyridyl disulfide reactive groups and their subsequent coupling with thiol-containing affinity ligands is described in Chapter 15, Section 2.2.

TNB-Thiol

Sulfhydryl groups activated with the leaving group 5-thio-2-nitrobenzoic acid can be used to couple free thiols by disulfide interchange similar to pyridyl disulfides, as discussed previously. A TNB–thiol-activated species may be created by reaction of a sulfhydryl group with Ellman's reagent, 5,5′-dithio-*bis*(2-nitrobenzoic acid), or DTNB, a compound useful for the quantitative determination of sulfhydryls in solution (Ellman, 1958, 1959) (Chapter 2, Section 4.1). The disulfide of Ellman's reagent readily undergoes disulfide exchange with a free sulfhydryl to form a mixed disulfide with concomitant release of one molecule of the chromogenic substance 5-sulfido-2-nitrobenzoate, also called 5-thio-2-nitrobenzoic acid (TNB). The TNB–thiol group can again undergo interchange with a sulfhydryl-containing target molecule to yield a disulfide crosslink. Upon coupling with a sulfhydryl compound, the TNB group is released (Reaction 3.25). The intense yellow color produced by the TNB anion can be measured by its absorbance at 412 nm ($\varepsilon = 1.36 \times 10^4 M^{-1} cm^{-1}$ at pH 8.0). Since each sulfhydryl which is coupled generates one molecule of TNB per molecule of Ellman's reagent, the possibility for quantifying the reaction exists.

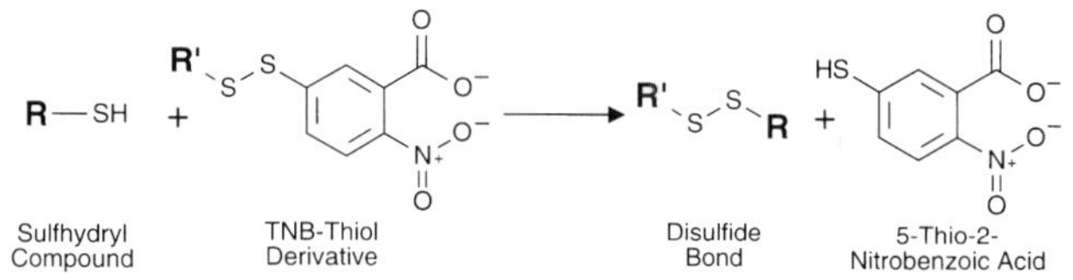

(REACTION 3.25)

Disulfide exchange with a TNB–thiol group occurs efficiently at physiological to slightly alkaline pH conditions. Avoid the presence of disulfide reducing agents, as these will cleave the TNB group and prevent specific coupling.

Disulfide Reductants

Disulfide reduction by the use of disulfide interchange can be performed using thiol-containing compounds such as TCEP, DTT, 2-mercaptoethanol, or 2-mercaptoethylamine (Chapter 2, Section 4.1). The formation of free sulfhydryls from a disulfide group occurs in two stages. First, one molecule of the reducing agent undergoes disulfide exchange, cleaving the disulfide and forming a new, mixed disulfide. In the next stage, a second molecule of the thiol cleaves the mixed disulfide, releasing a free sulfhydryl and forming a molecule of oxidized reducing agent (Reaction 3.26).

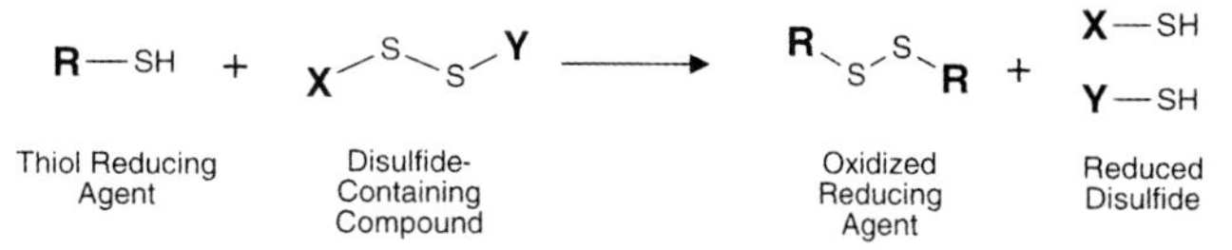

(REACTION 3.26)

Disulfide reduction occurs over a broad pH range and in a variety of buffer environments. The reaction can be performed in denaturants, chaotropic agents, detergents, and in high salt conditions.

2.7. Vinyl Sulfone Derivatives

A vinyl sulfone group can be used to conjugate with nucleophiles, especially thiol groups, in aqueous solution and under mild conditions (Masri and Friedman, 1988). In addition to thiols, they can react with amines and hydroxyls under higher pH conditions. As opposed to a maleimide group, the vinyl sulfone group is not as strong an electrophile, but efficiently couples with thiols at slightly alkaline pH values to give stable β-thiosulfonyl linkages (Reaction 3.27). The product of the reaction of a thiol with a vinyl sulfone gives a single stereoisomer structure, unlike conjugation with maleimides, which produces two potential stereoisomers. In addition, crosslinkers and modification reagents containing a vinyl sulfone can be used to activate surfaces or molecules to contain thiol-reactive groups. These vinyl sulfone groups are stable in aqueous solution for extended periods, as they are not subject to hydrolysis at neutral pH. Thus, they retain excellent coupling potential for thiol-containing proteins or other molecules even when used in aqueous buffer conditions.

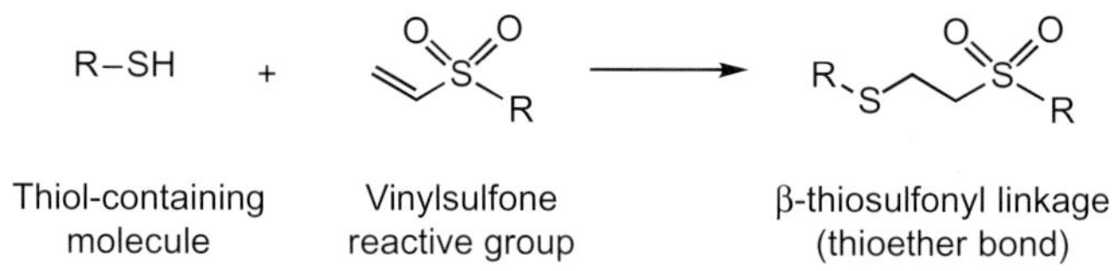

(REACTION 3.27)

Vinyl sulfone-activated chromatography supports have long been used for coupling affinity ligands that contain thiols or other nucleophiles (Porath, 1974) (see Chapter 15, Section 2.2). This reactive group has also been used to activate PEG polymers for modification of thiol-containing molecules (Morpurgo *et al.*, 1996). There are now homobifunctional and heterobifunctional crosslinking agents commercially available that use the vinyl sulfone reactive group (Thermo Fisher and Molecular Biosciences).

2.8. Metal–Thiol Dative Bonds

Thiol-containing molecules can interact with metal ions and metal surfaces to form dative bonds. Dative bonds are also known as coordinate covalent bonds. They differ from normal covalent linkages, because they are formed by two electrons coming from a single atom, instead of two atoms each sharing one electron. In a coordinate bond formed with a thiol, the unshared pair of electrons on the sulfur atom is able to form a dative bond with a metal atom. In this sense, even disulfides are able to datively link to a metal surface without prior reduction to thiols (Reaction 3.28).

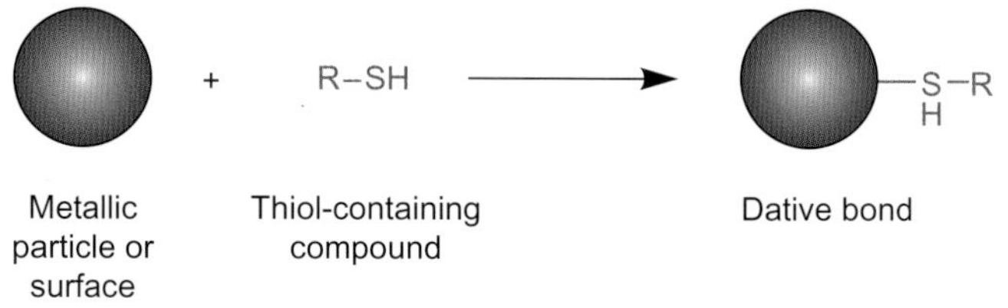

(REACTION 3.28)

Other atoms containing a lone pair of electrons are also effective at forming coordinate bonds. Oxygen- and nitrogen-containing organic molecules are often used to chelate metal ions, such as in various lanthanide chelates (Chapter 10, Section 9), bifunctional metal chelating compounds (Chapter 12, Section 1), and FeBABE (Chapter 24, Section 4.1). In addition, amino acid side chains and prosthetic groups in proteins frequently form bioinorganic motifs by coordinating a metal ion as part of an active center (Degtyarenko, 2000). Thiol organic compounds, however, are used routinely to coat metallic surfaces or particles to form biocompatible layers or create functional groups for further conjugation of biomolecules.

Thiol ligand modification in particular has been used extensively to create water soluble quantum dots (Sapsford *et al.*, 2006) and gold nanoparticles having reduced nonspecific binding character to the metallic surface and to form surface groups for coupling proteins and other affinity molecules (Chapter 10, Section 10, and Chapter 14, Section 6). For instance, thiol-containing aliphatic/PEG linkers have been used to form self-assembled monolayers (SAMs) on planar gold surfaces and particles (Prime and Whitesides, 1991).

A monodentate thiol–metal bond is not as strong as a true covalent linkage. These bonds are subject to displacement by other molecules containing thiols or other atoms with a lone pair of electrons. The thiol may also oxidize off the surface if exposed to oxygen in air or oxygen dissolved in aqueous solutions. Instead of monodentate compounds, the use of multidentate molecules can dramatically increase the stability of a coordination bond. A bidentate DOPA ligand, for instance, was found to have far greater bonding adhesion to a

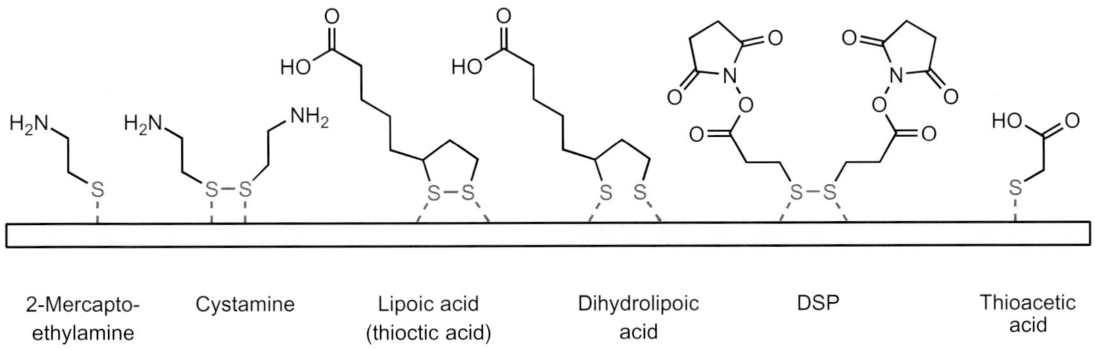

FIGURE 3.3 A number of small thiol-containing molecules have proven useful for modification of gold or metallic surfaces. The dithiol derivatives provide better dative bond stability and cannot be displaced easily by competing thiols or oxidation. Most thiol-containing compounds used for surface modification also contain terminal functional groups or reactive groups for coupling affinity ligands.

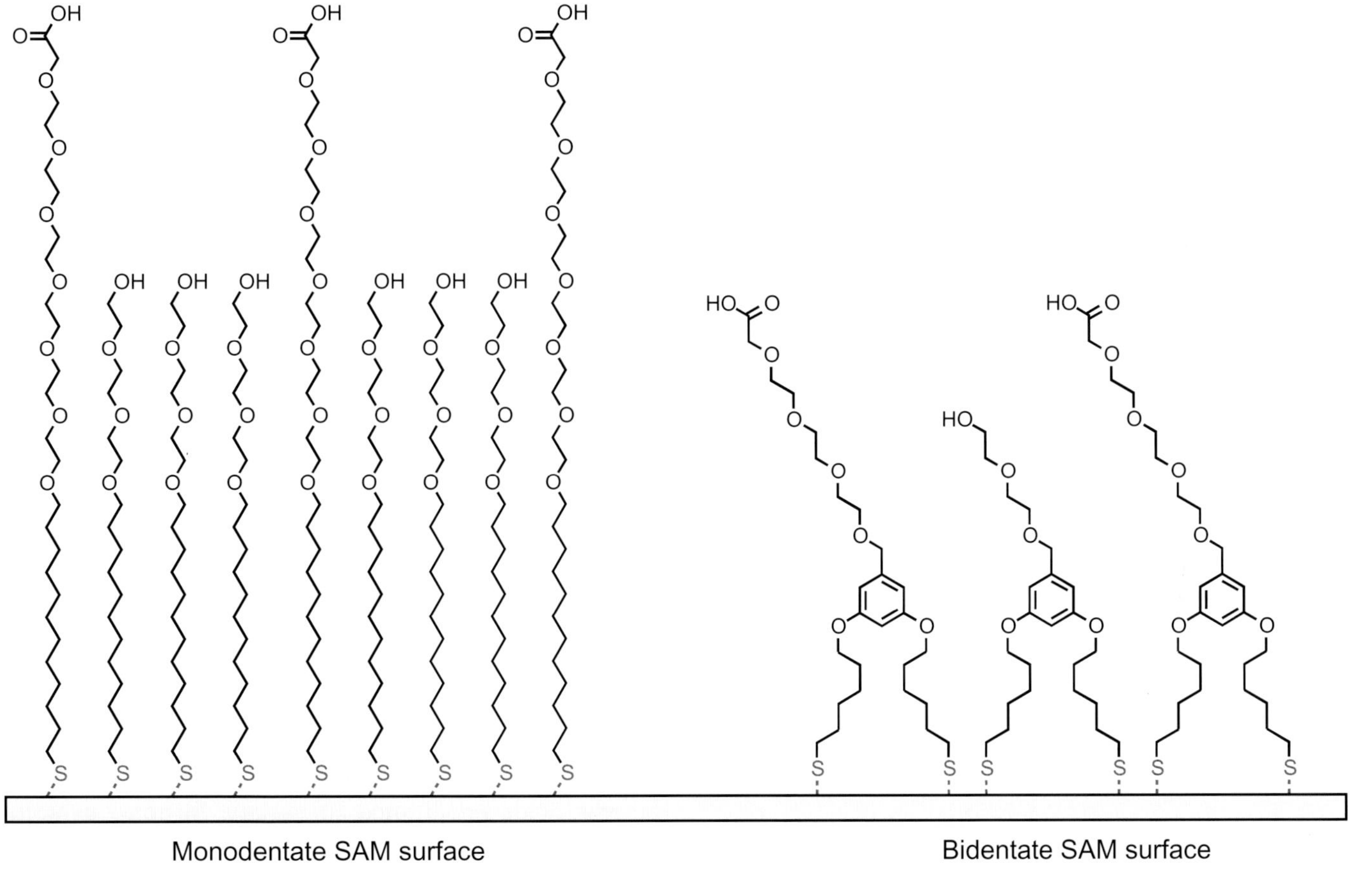

FIGURE 3.4 SAM surface modification has been performed using monothiol and dithiol compounds containing PEG linkers. Useful coatings typically contain mainly PEG–hydroxyl or PEG–monomethyl ether linkers that provide a biocompatible lawn, which prevents nonspecific binding of proteins to the metallic surface. About 10% of the surface modifications are performed using a longer carboxylate-containing thiol-PEG linker that provides sites for attachment of affinity ligands.

surface than monodentate ligands and nearly as much as a covalent bond (Lee *et al.*, 2006).

To increase the strength of thiol dative linkages, the application of multivalent thiols has been used to increase the overall strength of a single molecule tether bound to a metal. Bidentate thiol ligands that have been designed for metal surface modification include lipoic acid (thioctic acid) derivatives (Cheng and Brajter-Toth, 1992; Willey *et al.*, 2004; Hahn *et al.*, 2007), dithiobis(succinimidyl)propionate (DSP) modifications (Grubor *et al.*, 2004) (Chapter 5, Section 1.1), and a dithiol linker built from a central phenyl ring, which contains a PEG spacer for hydrophilicity on the SAM surface (Spangler *et al.*, 2004) (Figures 3.3 and 3.4).

2.9. Native Chemical Ligation

Native chemical ligation involves reactions that are very similar to intein splicing and peptide ligation found in certain protein expression systems. A peptide having a C-terminal thioester reacts with an N-terminal cysteine residue in another peptide to undergo a trans-thioesterification reaction, which results in the formation of an intermediate thioester with the cysteine thiol. However, due to the proximity of the neighboring α-amine group on cysteine, a subsequent nucleophilic attack of the electron-rich nitrogen on the ester carbonyl results in an S→N shift; this then forms a native amide (peptide) bond (Reaction 3.29).

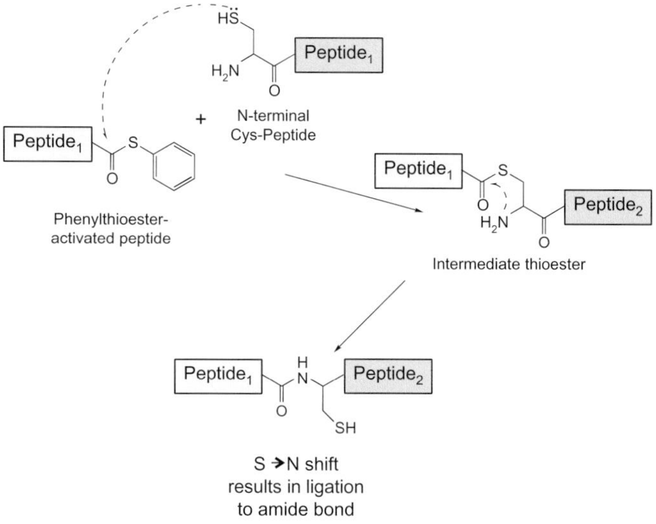

(REACTION 3.29)

Native chemical ligation is effective at conjugating two peptides together in aqueous solution to form a longer peptide. The reaction proceeds at physiological pH under mild conditions without any additional additives, except for the presence of the thioester-containing peptide and the cysteine-containing peptide. In addition, other thioester compounds may be used in this reaction to label specifically N-terminal cysteine peptides through amide bond formation. For instance, a biotin thioester derivative may be used to add a biotin group to a peptide or a fluorescent label may be added by the same process. Thus, native chemical ligation is an excellent way of discretely labeling peptides only at their N-terminal. See Chapter 17, Section 7, for additional information on the use of this reaction.

2.10. Cisplatin Modification of Methionine and Cysteine

Platinum complexes have long been used as tumor therapeutic agents for their ability to bind to DNA at guanine and adenine residues and interfere with

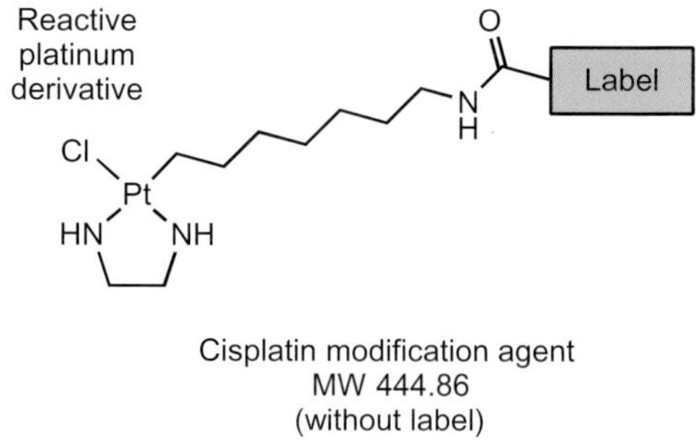

FIGURE 3.5 The general design of a cisplatin modification agent consists of the reactive cisplatin group and a short linker that typically terminates in a detectable label.

transcription (Repta and Long, 1980; Eastman, 1987; Reedijk *et al.*, 1987). The reaction of Platinum II compounds with nucleic acids has been studied in great detail (Anin *et al.*, 1992), and the reactive group has been applied to bioconjugation labeling reagents (van Belkum *et al.*, 1994; Heetebrij *et al.*, 2003).

Cisplatin reagents also react with certain amino acid residues in proteins. Rapid coupling occurs with methionine and cysteine residues over a broad pH range (2–9), while slower reaction kinetics results in modification of histidine imidazole ring with an optimal pH at 8.0. The relative rate of reaction of cisplatin derivatives for sulfur nucleophiles is over 100 times more than their reaction with nitrogen nucleophiles (Hay and Porter, 1999). The basic structure of cisplatin labeling reagents is shown in Figure 3.5, and Figure 3.6 illustrates the modification reactions of a cisplatin derivative with amino acid or nucleic acid targets. The reactive group is stable in aqueous conditions and covalent bonds formed with it are stable to typical conditions used in biological assays and detection procedures. Reagent systems are available from Kreatech.

3. CARBOXYLATE REACTIONS

Chemical groups that specifically react with carboxylic acids are limited in variety. In aqueous solutions, the carboxylate functionality displays rather low nucleophilicity. For this reason, it is unreactive with the great majority of bioconjugate reagents which couple through a nucleophilic addition process.

Several important chemistries, however, have been developed that allow conjugation through a carboxylate group. The following sections briefly describe these reactions.

3.1. Diazoalkanes and Diazoacetyl Compounds

Diazomethane and other diazoalkyl derivatives have long been used to label carboxylate groups for analysis (Herriott, 1947; Riehm and Scheraga, 1965). A major

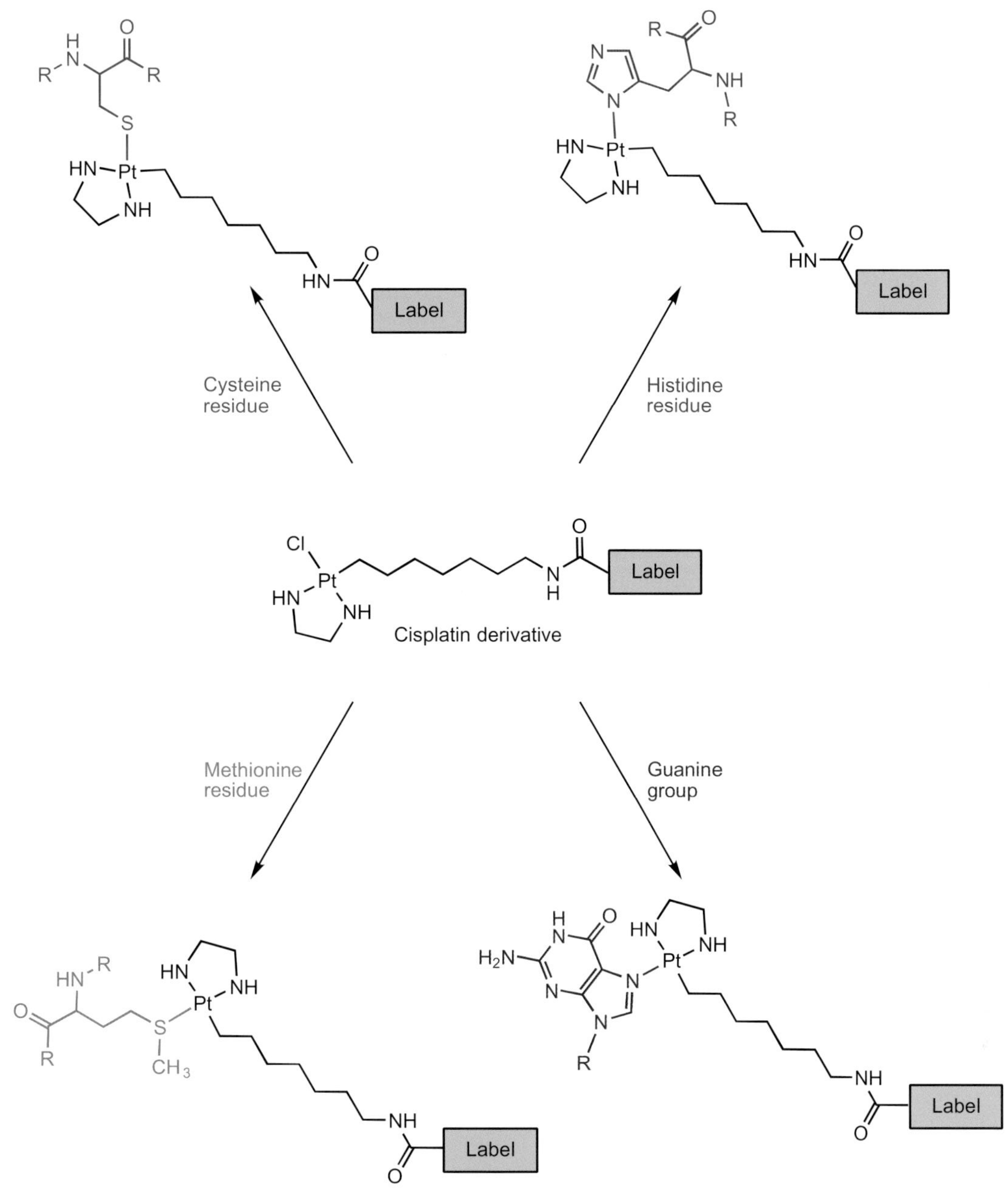

FIGURE 3.6 The cisplatin reactive group can covalently couple to methionine-, cysteine-, and histidine-containing peptides or proteins. It also reacts with guanine groups to form a covalent modification on the N_7 nitrogen.

application of such reagents has been in the HPLC analysis of low-molecular-weight compounds such as fatty acids (DeMar *et al.*, 1992). Several coumarin derivatives containing stable, carboxylate-reactive diazoalkane functionalities are also available for fluorescent labeling of target molecules (Molecular Probes) (Ito and Sawanobori, 1982; Ito and Maruyama, 1983).

Diazoalkanes and diazoacetyl compounds (amides and esters) are spontaneously reactive with carboxylate groups without addition of other reactants or catalysts. The reaction mechanism involves attack of a negatively charged carboxylate oxygen atom on a protonated diazoalkyl group, liberating nitrogen gas and forming a covalent linkage (Reaction 3.30).

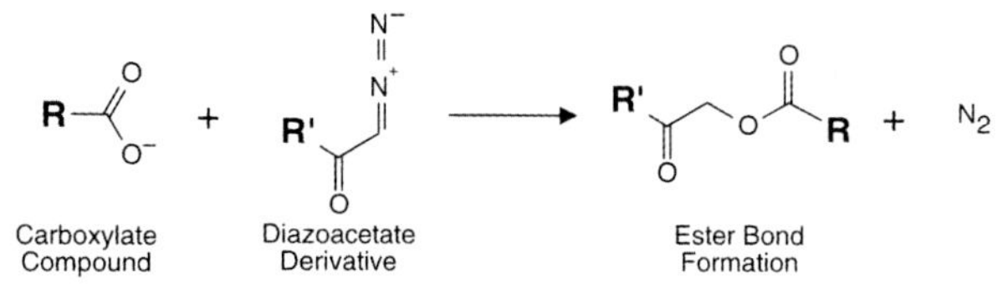

(REACTION 3.30)

The reaction with carboxylates occurs over a range of pH values, but is optimal at pH 5.0. Unfortunately, the diazoalkyl compounds will cross-react with sulfhydryl groups at this pH. At higher pH conditions, the reaction is even less specific due to reaction with other nucleophiles. In aqueous solution, the most-likely side reaction is hydrolysis.

3.2. N,N′-Carbonyl Diimidazole

N,N′-Carbonyl diimidazole (CDI) is an active carbonylating agent that contains two acylimidazole leaving groups (Chapter 4, Section 3, and Chapter 15, Section 2.1). CDI reacts with carboxylic acids under nonaqueous conditions to form *N*-acylimidazoles of high reactivity (Reaction 3.31). The active intermediate forms in excellent yield due to the driving force created by the liberation of carbon dioxide and imidazole (Anderson, 1958). An active carboxylate can then react with amines to form amide bonds or with hydroxyl groups to form ester linkages (Reaction 3.32). The reaction has been used successfully in peptide synthesis (Paul and Anderson, 1960, 1962). In addition, activation of a styrene/4-vinylbenzoic acid copolymer with CDI was used to immobilize the enzyme lysozyme through its available amino groups to the carboxyl groups on the matrix (Bartling *et al.*, 1973).

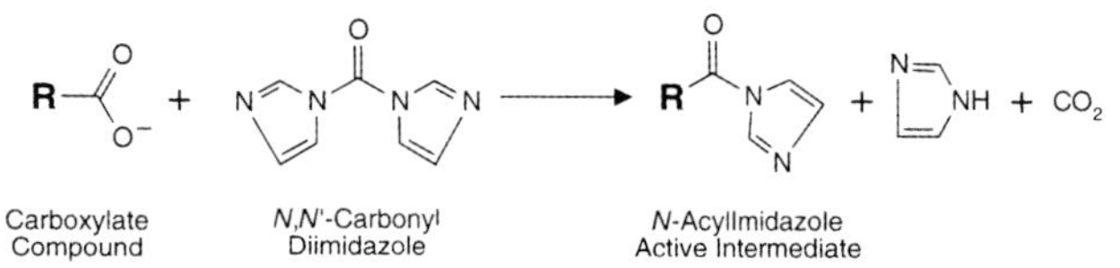

(REACTION 3.31)

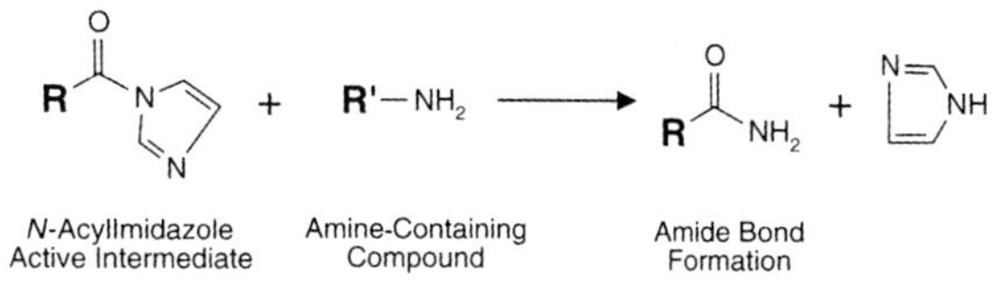

(REACTION 3.32)

CDI functions as a zero-length crosslinker if the activated species is a carboxylic acid, because the attack of another nucleophile liberates the imidazole leaving group. The conjugation reaction can be performed in organic solvent or aqueous conditions, depending on the solubility of the nucleophile. For aqueous coupling of *N*-acylimidazoles to amine-containing compounds, optimal conditions include an alkaline pH environment from about pH 7 to 9 and in buffers containing no amines (avoid Tris or imidazole).

3.3. Carbodiimides

Carbodiimides function as zero-length crosslinking agents capable of activating a carboxylate group for coupling with an amine-containing compound. There are several major types of carbodiimide reagents commonly available that can be used for organic or aqueous reactions, depending on their individual solubility characteristics (Chapter 4, Section 1). The water soluble reagents are used mainly for biological conjugations involving proteins and other macromolecules. The water-insoluble carbodiimides can be used in peptide synthesis or for the synthesis of other organic compounds.

Carbodiimides are used to mediate the formation of amide or phosphoramidate linkages between a carboxylate and an amine or a phosphate and an amine, respectively (Hoare and Koshland, 1966; Chu *et al.*, 1986; Ghosh *et al.*, 1990). Regardless of the type of carbodiimide, the reaction proceeds by the formation of an intermediate *O*-acylisourea that is highly reactive and short-lived in aqueous environments. The attack of an amine nucleophile on the carbonyl group of this ester results in the loss of an isourea derivative and formation of an amide bond (see Reactions 3.11 and 3.12). The major competing reaction in water is hydrolysis.

4. HYDROXYL REACTIONS

Hydroxyl-reactive chemical compounds include not only those modification agents able to directly form a stable linkage with an –OH group, but also a broad range of reagents that are designed to temporarily activate the group for coupling with a secondary functional group. Many of the chemical methods for modifying hydroxyls originally were developed for use with chromatography supports in the coupling of affinity ligands. Some of these same chemical reactions have found application in bioconjugate techniques for crosslinking a hydroxyl-containing molecule with another substance, usually containing a nucleophile. For instance, carbohydrate-containing molecules such as polysaccharides or glycoproteins can be coupled through their sugar residues using hydroxyl-specific reactions. In addition, polymers and other organic compounds containing hydroxyls (such as PEG) may be conjugated with another molecule using these chemistries.

4.1. Epoxides and Oxiranes

An epoxide or oxirane group can react with nucleophiles in a ring-opening process. The reaction can take place with primary amines, sulfhydryls, or hydroxyl groups to create secondary amine, thioether, or ether bonds, respectively. See Section 1.7 (this chapter) for further information on this reaction.

4.2. N,N′-Carbonyl Diimidazole

N,N′-Carbonyl diimidazole (CDI) is an active carbonylating agent that contains two acylimidazole leaving groups (Chapter 4, Section 3, and Chapter 15, Section 2.1). The compound can react with a carboxylate to form an active acylimidazole group capable of coupling with amine-containing molecules (Section 3.2). However, CDI can also react with hydroxyl groups to create a reactive intermediate. If CDI is used to activate a hydroxyl functional group, the reaction proceeds quite differently from its reaction with carboxylates. The active intermediate formed by the reaction of CDI with an –OH group is an imidazolyl carbamate (Reaction 3.33). Attack by an amine releases the imidazole, but not the carbonyl. Thus, hydroxyl-containing molecules may be coupled to amine-containing molecules with the result of a one-carbon spacer, forming stable urethane (*N*-alkyl carbamate) linkages (Reaction 3.34). This coupling procedure has been applied to the activation of hydroxyl-containing chromatography supports for the immobilization of amine-containing affinity ligands (Bethell *et al.*, 1979, 1979, Hearn *et al.*, 1983) and also to the activation of polyethylene glycol for the modification of amine-containing macromolecules (Beauchamp *et al.*, 1983).

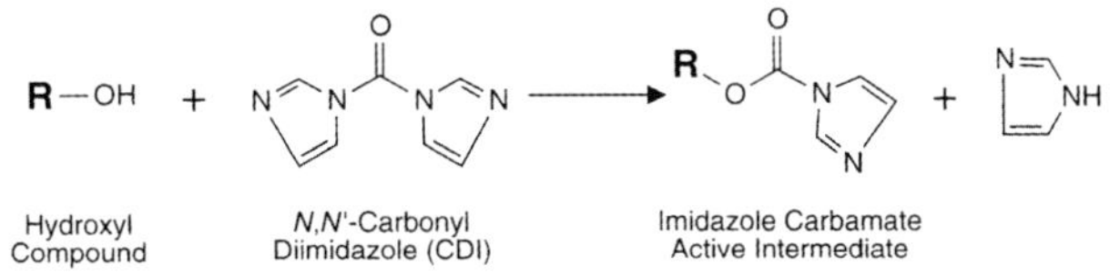

(REACTION 3.33)

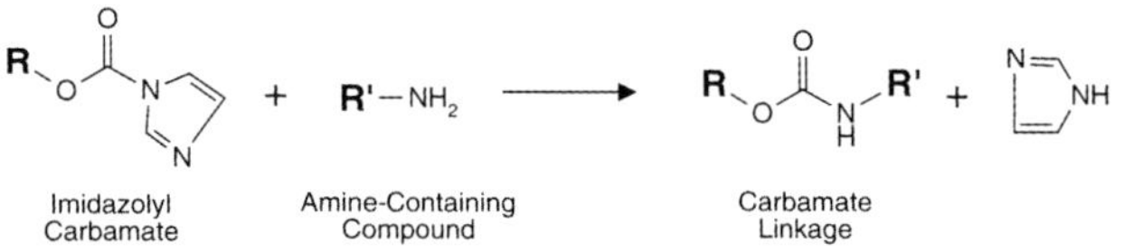

(REACTION 3.34)

4.3. N,N′-Disuccinimidyl Carbonate or N-Hydroxysuccinimidyl Chloroformate

N,N′-Disuccinimidyl carbonate (DSC) consists of a carbonyl group containing, in essence, two NHS esters. The compound is highly reactive toward nucleophiles. In aqueous solutions, DSC will hydrolyze to form two molecules of *N*-hydroxysuccinimide (NHS) with release of one molecule of CO_2. In nonaqueous environments, the reagent can be used to activate a hydroxyl group to a succinimidyl carbonate derivative (Reaction 3.35). DSC-activated hydroxylic compounds can be used to conjugate with amine-containing molecules to form stable crosslinked products (Reaction 3.36). The linkage created from this reaction is a urethane derivative or a carbamate bond, displaying excellent stability.

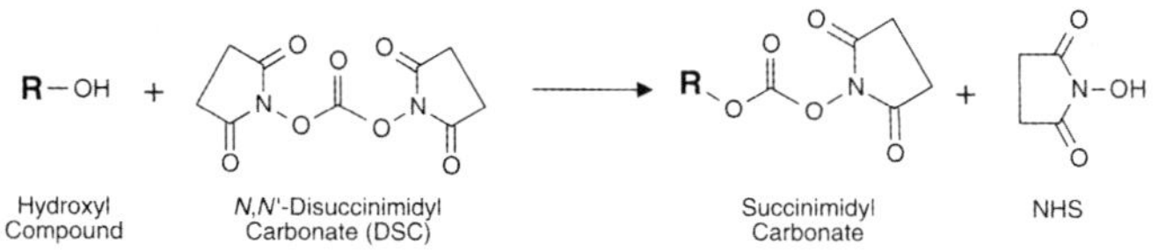

(REACTION 3.35)

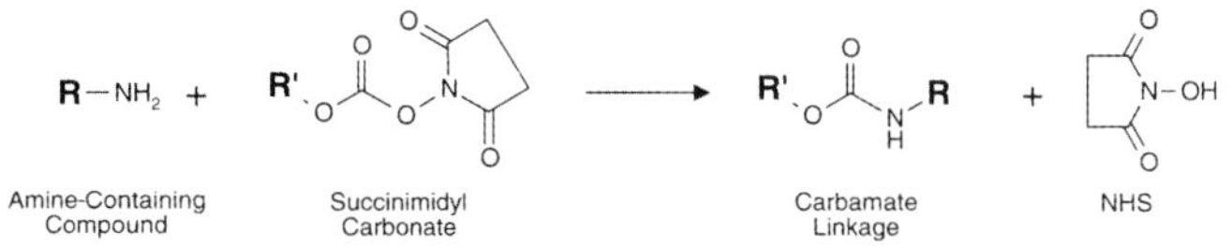

(REACTION 3.36)

A related reagent, *N*-hydroxysuccinimidyl chloroformate is also a bifunctional carbonyl derivative containing an NHS ester and a acid chloride. In aqueous solutions, the compound is unstable to hydrolysis, rapidly breaking down to NHS, CO_2, and HCl. In nonaqueous environments, however, NHS-chloroformate may be used to activate a hydroxyl group similar to DSC. Reaction of the chloroformate with a hydroxylic residue forms the same succinimidyl carbonate derivative as the reaction of DSC with –OH groups (Reaction 3.37). Subsequent conjugation with an amine-containing compound yields a carbamate linkage. The bond is identical to that formed from the reaction of CDI-activated hydroxyls with amine-containing compounds (see previous section).

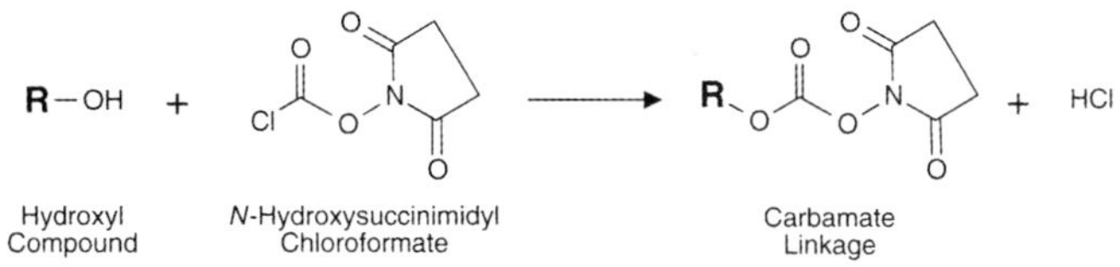

(REACTION 3.37)

4.4. Oxidation with Periodate

Sodium periodate can be used to oxidize hydroxyl groups on adjacent carbon atoms, forming reactive aldehyde residues suitable for coupling with amine- or hydrazide-containing molecules. The reaction occurs with two adjacent secondary hydroxyls to cleave the carbon–carbon bond between them and create two terminal aldehyde groups (Reaction 3.38). When one of the adjacent hydroxyls is a primary hydroxyl, reaction with periodate releases one molecule of formaldehyde and leaves a terminal aldehyde residue on the original diol compound (Reaction 3.39). These reactions can be used to generate crosslinking sites in carbohydrates or glycoproteins for subsequent conjugation of amine-containing molecules by reductive amination (Chapter 2, Section 4.4, and Chapter 4, Section 4). Sodium periodate also reacts with 2-aminoethanol derivatives—compounds containing a primary amine and a secondary hydroxyl group on adjacent carbon atoms. Oxidation cleaves the carbon–carbon bond, forming a terminal aldehyde group on the side that had the original hydroxyl residue (Reaction 3.40). This reaction can be used to create reactive aldehydes on N-terminal serine residues of peptides (Geoghegan and Stroh, 1992).

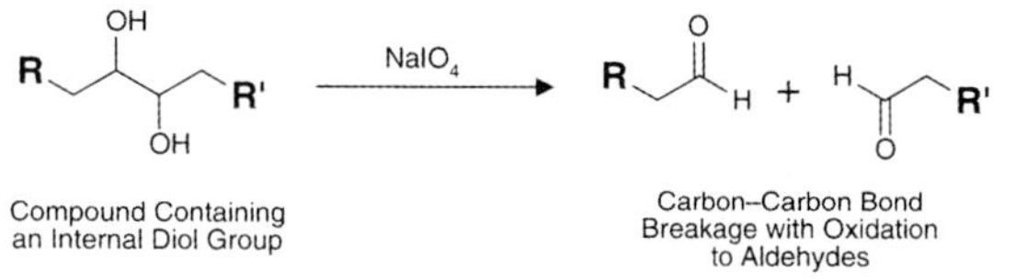

(REACTION 3.38)

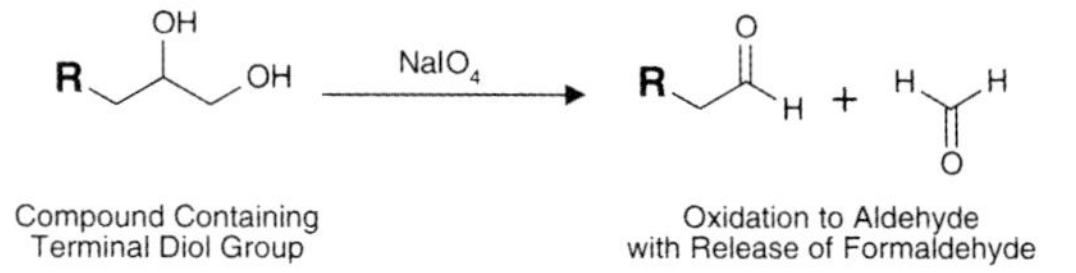

(REACTION 3.39)

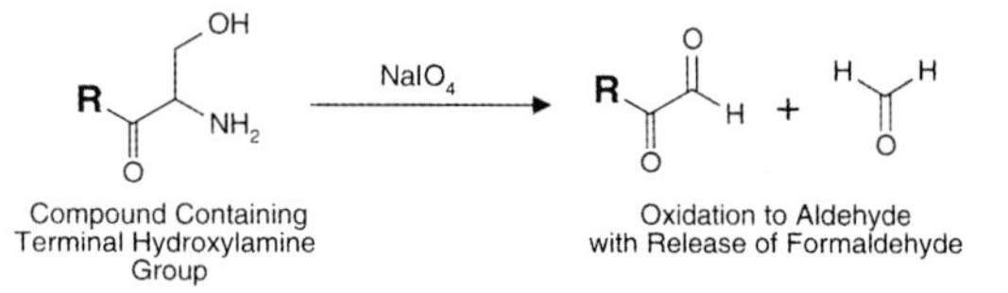

(REACTION 3.40)

4.5. Enzymatic Oxidation

Certain enzymes may be used to oxidize hydroxyl-containing carbohydrates to create aldehyde groups (Chapter 2, Section 4.4). For example, the reaction of galactose oxidase on terminal galactose or *N*-acetyl-D-galactose residues proceeds to form C-6 aldehyde groups on polysaccharide chains (Reaction 3.41). These groups can then be used for conjugation reactions with amine- or hydrazide-containing molecules.

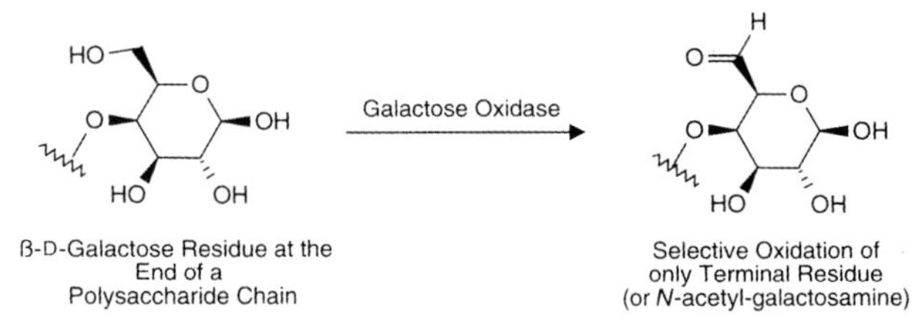

(REACTION 3.41)

4.6. Alkyl Halogens

Reactive alkyl halogen compounds can be used to specifically modify hydroxyl groups in carbohydrates, polymers, and other molecules. Chloro- or bromo-derivatives of short alkyl chains containing an electron-withdrawing second functional group on their other end (typically a carboxylate group) can be used to form spacer arms useful for conjugation with another substance. Brunswick *et al.* (1988) used chloroacetic acid to modify the hydroxyl groups of dextran, forming the carboxymethyl derivative (Reaction 3.42). The carboxylates may then be coupled with amine-containing molecules using a carbodiimide reaction scheme. In a somewhat similar approach, Noguchi *et al.* (1992) prepared a carboxylate spacer arm by reacting 6-bromohexanoic acid with a dextran polymer (Chapter 18, Section 2.2).

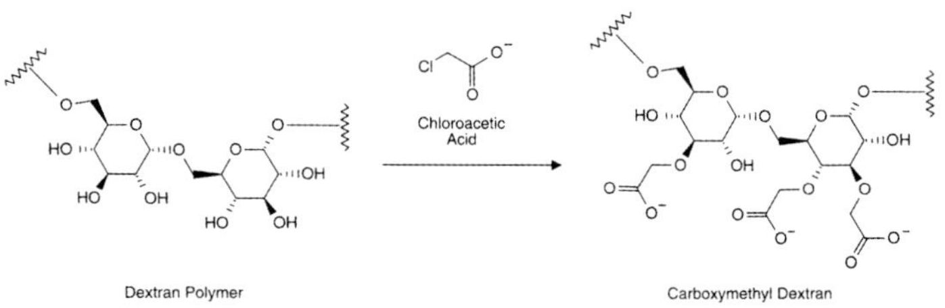

(REACTION 3.42)

Modification of hydroxyl groups with such compounds can be performed in 3- to 10-*M* NaOH by reacting from 25°C to 40°C for 1.5 to 4 h.

4.7. Isocyanates

Isocyanates can be formed from the reaction of an aromatic amine with phosgene (Rifai and Wong, 1986). They can also be created from acyl azides by treatment at 80°C in the presence of an alcohol (Chapter 10, Section 5). In the transformation, the acyl azide group rearranges to form an isocyanate that can react with hydroxyl-containing molecules to form a urethane (carbamate) linkage (Reaction 3.43). The reactivity of isocyanates is excellent, but for the same reason their stability can be a problem. In storage, moisture decomposes them, releasing CO_2 and leaving an aromatic amine in its place. In aqueous environments, the aromatic amine can react with another molecule of isocyanate to form a urea derivative (Annunziato *et al.*, 1993).

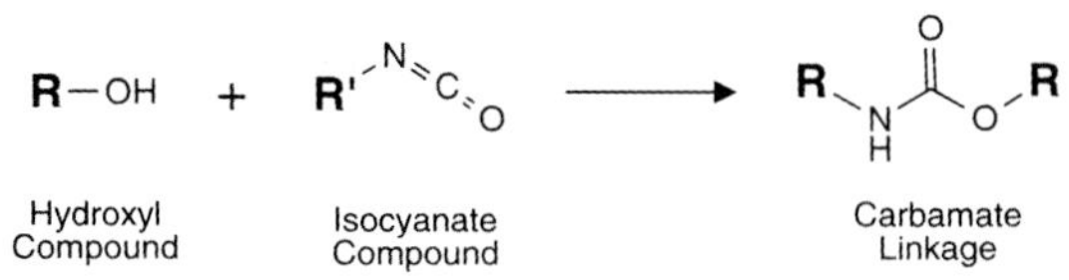

(REACTION 3.43)

Isocyanate-containing reagents can be used to crosslink or label hydroxyl-containing molecules, including polysaccharides. Carbohydrate modification can be performed without the need for prior oxidation of sugar residues with periodate to form reactive aldehydes, as is common in many protocols (Chapter 2, Section 4.4). The reaction occurs best at alkaline pH values (e.g., pH 8.5). Many coupling protocols avoid the hydrolysis problem by performing the reaction in organic solvent (i.e., DMSO).

Annunziato *et al.* (1993) have reported on the synthesis and use of a novel heterobifunctional crosslinking reagent containing a hydroxyl-reactive isocyanate group on one end and a sulfhydryl-reactive maleimide group on the other end. The compound can be useful in labeling hydroxylic molecules for subsequent conjugation with thiol-containing molecules.

5. ALDEHYDE AND KETONE REACTIONS

Aldehyde and ketone groups are important reactive sites in molecules for many bioconjugate strategies. Although some pharmacological agents contain ketones, these groups are not usually present in proteins and other biological macromolecules. Even when a molecule does not contain these functionalities, however, they may be created through a number of processes (Chapter 4, Section 4.4). The following sections discuss the major reactions that can be performed with aldehydes and ketones to modify or crosslink molecules containing them.

5.1. Hydrazine and Hydrazide Derivatives

Derivatives of hydrazine, especially the hydrazide compounds formed from carboxylate groups, can react specifically with aldehyde or ketone functional groups in target molecules. Reaction with either group creates a hydrazone linkage (Reaction 3.44)—a type of Schiff base. This bond is relatively stable if it is formed with a ketone, but somewhat labile if the reaction is with an aldehyde group. However, the reaction rate of hydrazine derivatives with aldehydes is typically faster than the rate with ketones. Hydrazone formation with aldehydes, however, results in much more stable bonds than the easily reversible Schiff base interaction of an amine with an aldehyde. To further stabilize the bond between a hydrazide and an aldehyde, the hydrazone may be reacted with sodium cyanoborohydride to reduce the double bond and form a secure covalent linkage.

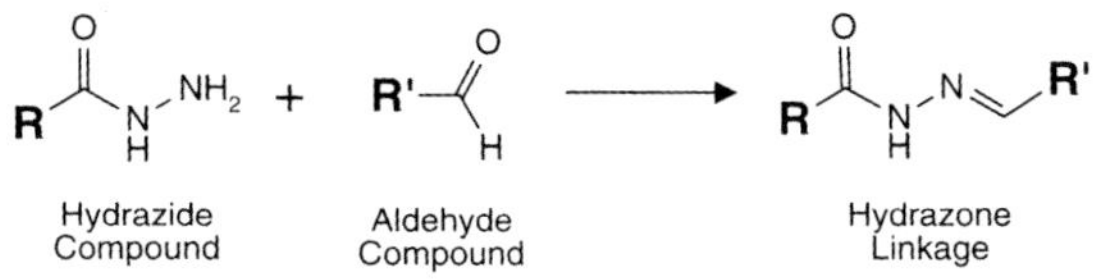

(REACTION 3.44)

5.2. Schiff Base Formation

Aldehydes and ketones can react with primary and secondary amines to form Schiff bases, a dehydration reaction yielding an imine (Reaction 3.45). However, Schiff base formation is a relatively labile, reversible interaction that is readily cleaved in aqueous solution by hydrolysis. The formation of Schiff bases is enhanced at alkaline pH values, but they are still not stable enough to use for crosslinking applications unless they are reduced by reductive amination (see below).

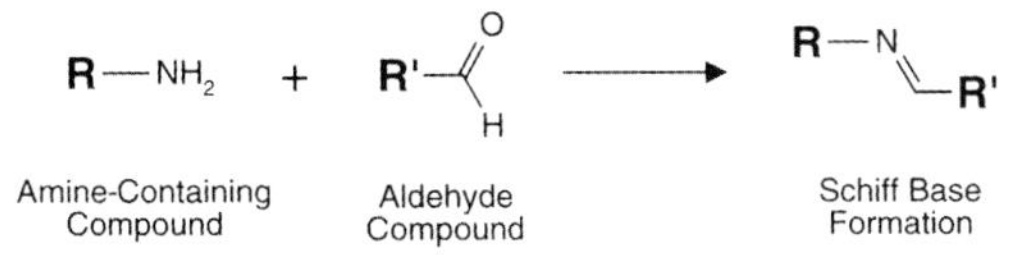

(REACTION 3.45)

The reaction of dicarbonyl compounds, such as glyoxal or phenylglyoxal, with a guanidinyl group, such as that of an arginine residue, proceeds to yield a more stable linkage due to the formation of a cyclic derivative (Reaction 3.46).

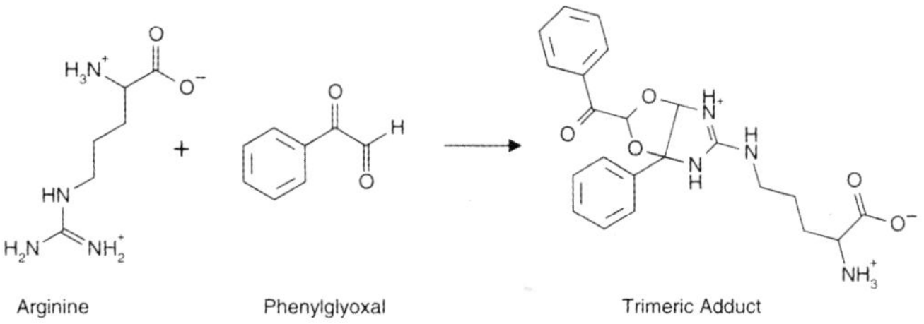

(REACTION 3.46)

5.3. Reductive Amination

Reductive amination (or alkylation) may be used to conjugate an aldehyde- or ketone-containing molecule with an amine-containing molecule. Schiff base formation between aldehydes and amines occurs readily in aqueous solutions, especially at elevated pH. This type of linkage, however, is not stable unless reduced to secondary or tertiary amine bonds. A number of reducing agents can be used to specifically convert the Schiff base interaction into an alkylamine linkage (Reaction 3.47). Once reduced, the bonds are highly stable and will not readily hydrolyze in aqueous environments. The use of reductive amination to conjugate an aldehyde-containing molecule to an amine-containing molecule results in a zero-length crosslinking procedure where no additional spacer atoms are introduced between the molecules (Section 2.1.4). Reaction of ammonia or a diamine compound with an aldehyde by reductive amination is a method of creating a primary amine functional group (Chapter 2, Section 4.3).

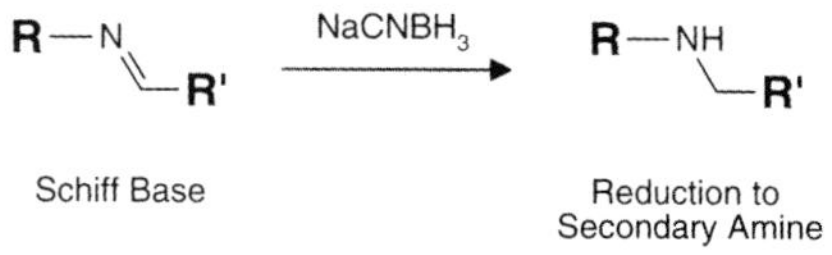

(REACTION 3.47)

5.4. Aminooxy Derivatives

Aminooxy groups (also called aminoxy or alkoxyamine) contain a terminal primary amine group next to oxygen (–ONH_2), wherein the oxygen atom can be attached to a linker arm of a bioconjugation reagent or attached to a solid phase such as a particle or surface. The chemoselective ligation reaction that occurs between an aldehyde group and an aminooxy group yields an oxime linkage (aldoxime) that has been used in many bioconjugation reactions, as well as in the coupling of ligands to insoluble supports including surfaces (see Chapter 15, Section 2.4) (Thumshirn *et al.*, 2003; Poethko *et al.*, 2004; Liu *et al.*, 2007; Colombo and Bianchi, 2010). This reaction is also quite efficient with ketones to form an oxime called a ketoxime. The simplest form of this reaction has been known for over a century and occurs with aldehydes or ketones as they react with the small compound hydroxylamine (Reaction 3.48).

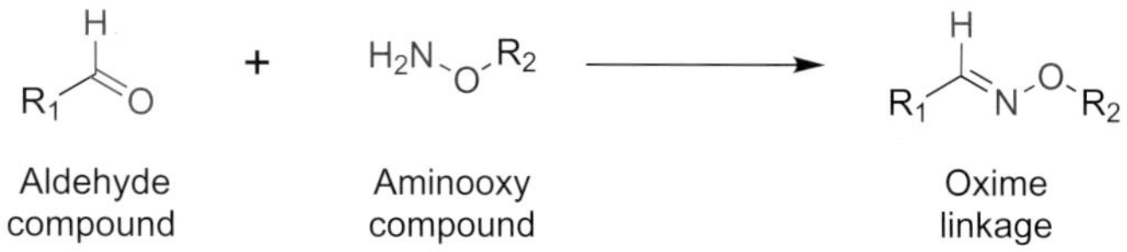

(REACTION 3.48)

Oxime linkages are very stable bonds that, unlike in the case of a hydrazone bond between a hydrazide group and an aldehyde, do not require further stabilization through reduction to eliminate leakage. The formation of the oxime can be further accelerated through the use of aniline as a catalyst. The aniline aryl amine group will first react with an aldehyde to form an intermediate Schiff base with the aldehyde, which then undergoes efficient attack by the aminooxy group to result in loss of aniline and formation of the oxime linkage. Additional information on the use of this reaction can be found in Chapter 15.

5.5. Mannich Condensation

Aldehydes may participate in a condensation reaction with an amine compound and a substance containing a sufficiently active hydrogen, yielding an alkylated derivative that effectively crosslinks the two molecules through the carbonyl group of the aldehyde. Strictly speaking, the Mannich reaction consists of the condensation of formaldehyde (or sometimes another aldehyde) with ammonia, in the form of its salt, and another compound containing an active hydrogen. Instead of using ammonia, however, this reaction can be carried out with primary or secondary amines, or even with amides. An example is illustrated in the condensation of phenol, formaldehyde, and a primary amine salt (Reaction 3.49).

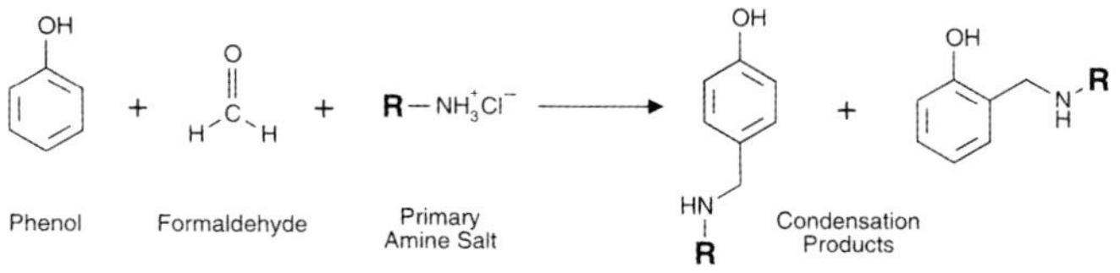

(REACTION 3.49)

The Mannich reaction provides an often superior alternative to diazonium conjugation (Section 6.1, this chapter), because of the disadvantages inherent in the instability of both the diazonium group and the resultant diazo linkage. By contrast, conjugations performed through the use of a Mannich condensation process result in stable covalent bonds.

The crosslinking scheme using this method can make use of the native ε- and N-terminal amines on proteins as the source of primary amine for the condensation reaction. Added to the conjugation reaction is formaldehyde and the desired molecule to be coupled containing an appropriately active hydrogen. The Mannich reaction should not be used for molecules containing both an amine and a reactive hydrogen, since polymerization may occur. It is especially useful for preparing hapten–carrier conjugates when the hapten contains no other available functionalities suitable for crosslinking, but does contain an active hydrogen (Chapter 19, Section 6.2).

A modification of the Mannich reaction was described by Joshi *et al.* (2004) which uses aniline in the reaction solution as a catalyst to accelerate the formation of the condensation product. This alternative reaction scheme is more fully described in Chapter 15, Section 2.7. The aniline catalyzed Mannich reaction may be used to specifically target tyrosine-containing peptides for covalent conjugation at the *ortho* position relative to its phenolic ring –OH group (see Figures 15.103 and 15.104 in Chapter 15).

6. ACTIVE HYDROGEN REACTIONS

Many compounds contain reactive (or replaceable) hydrogens that are able to participate in conjugation procedures using certain chemical reactions. These hydrogens typically are associated with aromatic systems wherein an electron donating group activates positions on the ring toward substitution reactions. At such carbons, the hydrogen is easily displaced by an attacking electrophilic group able to form a new covalent linkage. Several common modification reactions are used in bioconjugate chemistry to label or crosslink molecules at active hydrogen sites. The following three sections discuss these chemical reactions.

6.1. Diazonium Derivatives

Diazonium groups react with active hydrogen sites on aromatic rings to give covalent diazo bonds. Generation of a diazonium functional group is usually made from an aromatic amine by reaction with sodium nitrite under acidic conditions at 0°C (Chapter 2, Section 4.3, and Chapter 19, Section 6.1). The highly reactive and unstable diazonium is reacted immediately with an active hydrogen-containing compound at pH 8 to 10. In general, at pH 8.0 the diazonium group will react principally with histidinyl residues, attacking the electron rich nitrogens of the imidazole ring. At higher pH, the phenolic side chain of tyrosine groups can be modified (Reaction 3.50). The reaction proceeds by electrophilic attack of the diazonium group toward the electron-rich points on the target molecules. Phenolic compounds are modified at positions *ortho* and *para* to the aromatic hydroxyl group. For tyrosine side chains, only the *ortho* modification is available.

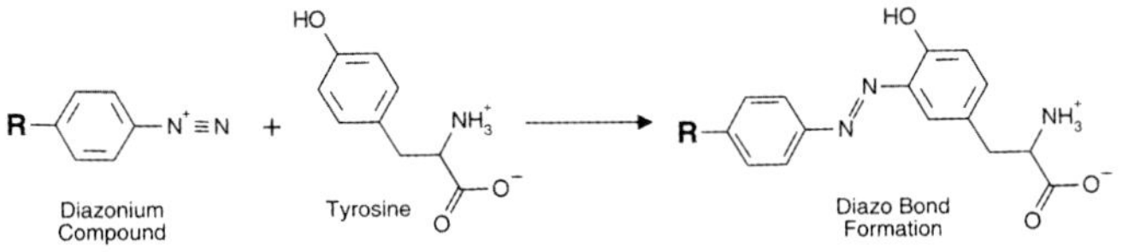

(REACTION 3.50)

Crosslinking using diazonium compounds usually creates deeply colored products characteristic of the diazo bonds. Occasionally, the conjugated molecules may turn dark brown or even black. The diazo linkages are reversible by addition of 0.1-*M* sodium dithionite in 0.2-*M* sodium borate, pH 9.0. Upon cleavage, the color of the complex is lost.

6.2. Mannich Condensation

The Mannich reaction consists of the condensation of an active hydrogen-containing compound with an amine-containing compound in the presence of formaldehyde. See Section 5.5 (this chapter) for additional details.

6.3. Iodination Reactions

Radioiodination involves the substitution of radioactive iodine atoms for reactive hydrogen sites in target molecules. The process usually involves the action of a strong oxidizing agent to transform iodide ions into a highly reactive electrophilic iodine compound (typically I_2 or a mixed halogen species such as ICl). Formation of this electrophilic species leads to the potential for rapid iodination of aromatic compounds containing strong activating groups, such as aryl compounds. In particular, aromatic constituents that have electron-donating groups can sufficiently activate the carbons on the ring to undergo electrophilic substitution reactions. Therefore, phenols, aniline derivatives, or alkyl anilines that contain OH, NH_2, or NHR constituents, respectively, are very susceptible to being iodinated. In proteins, this translates into tyrosine side-chain phenolic groups and histidine side-chain imidazole groups (Reaction 3.51). See Chapter 12, Section 2, for further details on iodination reactions.

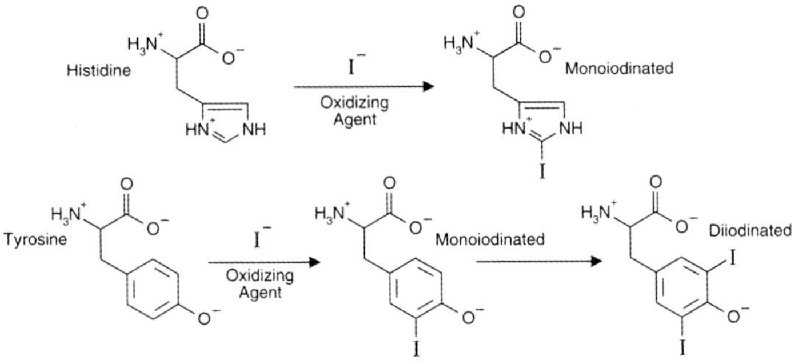

(REACTION 3.51)

7. PHOTOCHEMICAL REACTIONS

Photoreactive groups can be induced to couple with target molecules by exposure to UV light. Until they are photolyzed, photosensitive functional groups are relatively nonreactive in typical thermochemical processes. For this reason, reagents designed with a photoreactive group can be used in highly controlled reactions. The labeling reaction can be induced by a UV flash at predetermined points in an experimental protocol. For instance, covalent bond formation can be initiated after binding of photo-labeled ligands to receptors or after some other biochemical process has taken place. In this regard, photoreactive chemistry has become an important device for numerous bioconjugate applications. The following sections describe the major photosensitive groups that can be used in the design of modification or crosslinking reagents. Chapter 5, Section 5; Chapter 6, Sections 3–7; Chapter 7; and Chapter 11, Section 6.5, describe the reagents that utilize these functional groups.

7.1. Aryl Azides and Halogenated Aryl Azides

The most popular type of photosensitive functional group is the aryl azide derivative. Upon photolysis, phenyl azide groups form short-lived nitrenes that react rapidly with the surrounding chemical environment (Gilchrist and Rees, 1969). Nitrenes can insert nonspecifically into chemical bonds of target molecules, including undergoing addition reactions with double bonds and insertion reactions into active hydrogen bonds at C–H and N–H sites. Abundant evidence, however, indicates that the photolyzed intermediates of aryl azides principally undergo ring expansion to create nucleophile-reactive dehydroazepines. Instead of inserting non-selectively at active carbon–hydrogen bonds, dehydroazepines have a tendency to react preferentially with nucleophiles, especially amines (Reaction 3.52).

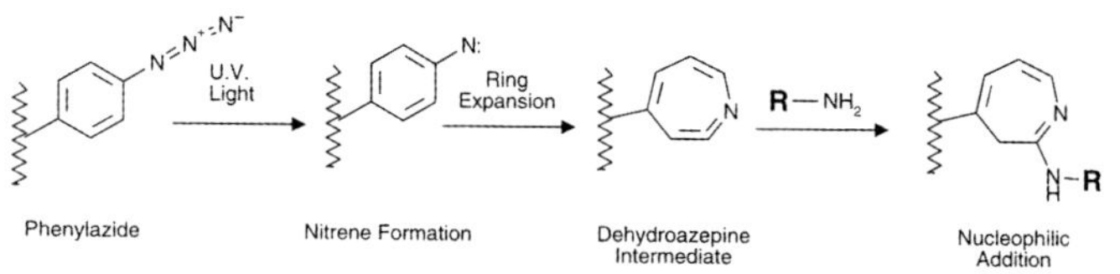

(REACTION 3.52)

However, some investigators have shown that aryl azides that possess a perfluorinated ring structure or are substituted completely with halogen atoms are quite efficient at forming the desired nitrene intermediate (Keana and Cai, 1990; Cai *et al.*, 1993; Schnapp and Platz, 1993; Schnapp *et al.*, 1993; Soundararajan *et al.*, 1993; Yan *et al.*, 1994). The ring substitution prevents ring expansion after nitrene formation, thus allowing the reactive intermediate to survive long enough to react with target molecules. Halogenated phenyl azides undergo the insertion reactions that were typically attributed to unsubstituted aryl azides in the past (Reaction 3.53).

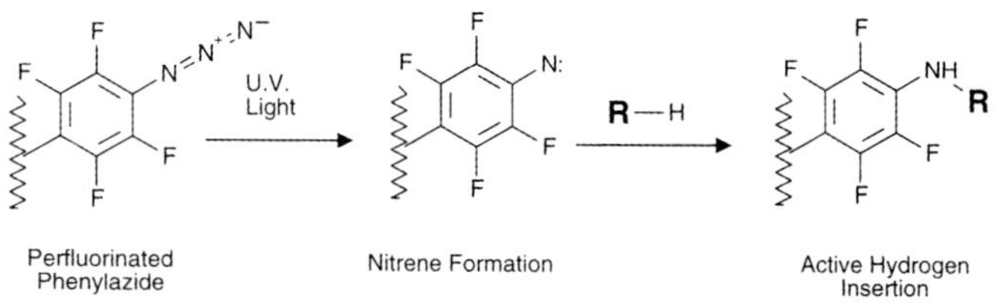

(REACTION 3.53)

7.2. Benzophenones

A photoreactive group consisting of a benzophenone residue photolyzes upon exposure to UV light to give a highly reactive triplet-state ketone intermediate (Walling and Gibian, 1965). Similar to the reactive nitrene of photolyzed phenyl azides, the energized electron of an activated benzophenone can insert in hydrogen–carbon bonds and other active groups to give covalent linkages with target molecules (Reaction 3.54). Unlike phenyl azides, however, the decomposition or decay of the photoactivated species does not yield an inactive compound. Instead, benzophenones that have become deactivated without forming a covalent bond can once again be photolyzed to an active state. As a result of this multiple-activation characteristic, a benzophenone reagent has more than one chance to form a covalent bond with its intended target. Thus, it typically gives much higher yields of photo-crosslinking than comparable phenyl azide crosslinkers.

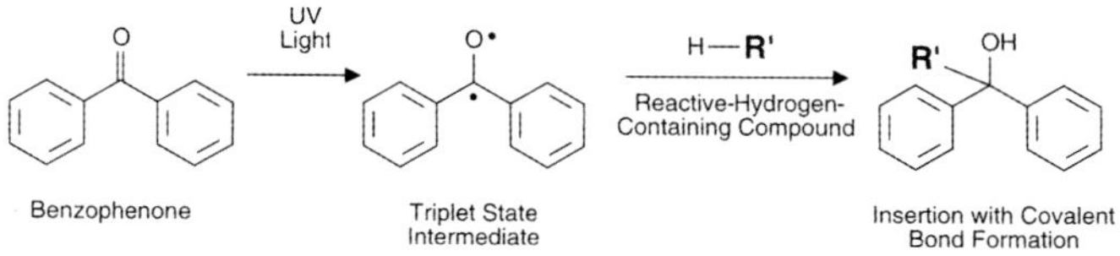

(REACTION 3.54)

The use of a benzophenone photoactivatable group in the design of bioconjugate reagents is rare. Two sulfhydryl-reactive ones incorporating a maleimide group and an iodoacetyl group opposite the benzophenone are described in Chapter 6, Sections 4.3, and 4.4. A newer benzophenone modification reagent containing a water soluble PEG spacer is described in Chapter 18, Section 1.3.

7.3. Anthraquinones

Anthraquinone groups are highly photoreactive by exposure to long UV light in the range of 340 to 360 nm. Unlike photoreactive groups that form intermediate nitrene or carbine precursors after photoactivation, anthraquinones react by a radical generation process, which is much more efficient in coupling to C–H substrates. Photoactivation results in a highly reactive, excited species that involves the formation of a triplet n,π*-state, which becomes a powerful electron acceptor. If an organic substrate is present containing a reactive C–H bond, then the excited anthraquinone is able to cause rapid proton abstraction, resulting in the formation of a reduced, phenoxy radical intermediate with a second radical formed on the hydrogen-donating substrate. This radical pair can then react to covalently link the anthraquinone group to the substrate, forming an ether linkage and effectively immobilizing the reagent to the substrate (Brennan and Beutel, 1969; Koch *et al.*, 2000) (Figure 3.7).

Modification or crosslinking reagents containing an anthraquinone group can be made from the 2-carboxylic acid derivative (Kumar *et al.*, 2004) (Exiqon). A spacer arm can be added to the carboxylate to terminate in a reactive group for bioconjugation or an affinity group (e.g., biotin) for interaction with other molecules.

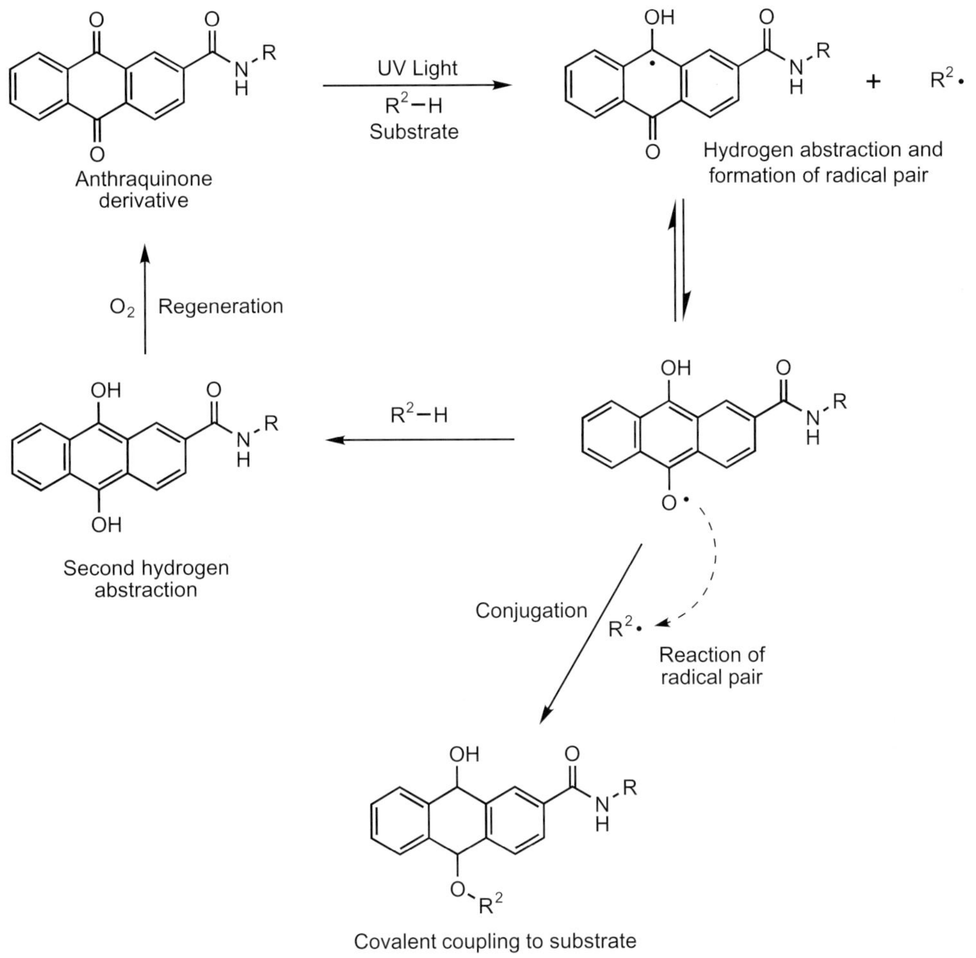

FIGURE 3.7 Anthraquinone derivatives can photoreactively couple to substrates by means of a free radical generation process. The reactive intermediate can also be regenerated back to the initial anthraquinone by proton abstraction and oxidation, resulting in the possibility of again being photolyzed and successfully coupled to the substrate.

Anthraquinone photoreactive linking reagents can be used to modify any surfaces or particles containing C–H groups, including polymer-based microplates, slides, and particles. Inorganic surfaces can be modified with an organosilane compound (Chapter 13) and then further reacted using an anthraquinone compound; however, the benefit of this strategy may be negated by choice of the proper reactive group on the silane. Reactions with surfaces are usually performed with anthraquinone concentrations in aqueous buffers ranging from 100 ng/ml to about 2 μg/ml. Protect all photoreactive reagents and solutions from light until ready to photoactivate them.

7.4. Certain Diazo Compounds

Certain diazo compounds can be photolyzed with UV light to generate highly reactive carbenes (Reaction 3.55). Similar to nitrenes, carbenes can insert into active C–H or N–H bonds or add to double bonds, forming covalent linkages with target molecules (Gilchrist and Rees, 1969). Few diazo photoreactive reagents have been synthesized, probably due to their tendency to react with water molecules after photoactivation, thus severely decreasing coupling yields with intended molecules. One heterobifunctional crosslinker, PNP–DTP, containing an amine-reactive end and a photosensitive diazotrifluoropropionate group is available (Chapter 6, Section 3.12).

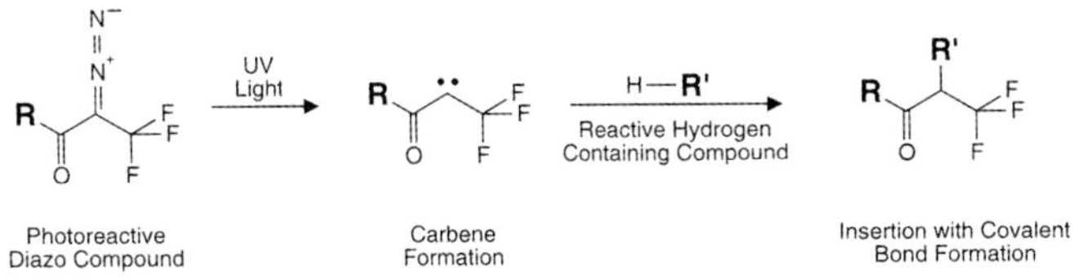

(REACTION 3.55)

Diazopyruvates are another class of photoreactive diazo compounds that have a unique coupling mechanism (Chapter 6, Section 3.11). The diazo functional

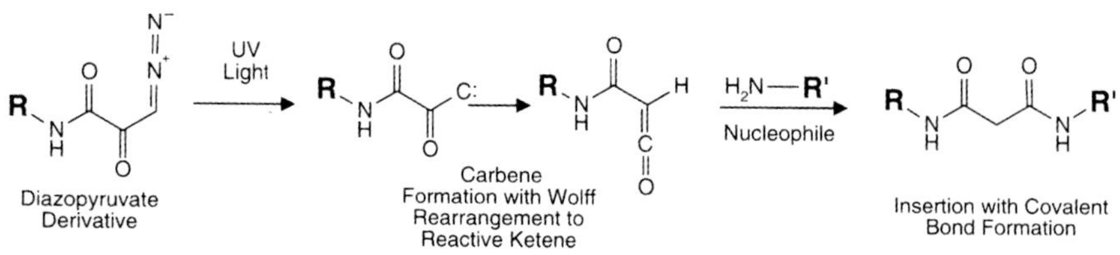

(REACTION 3.56)

group can by photolyzed by exposure to irradiation at 300 nm, forming a highly reactive carbene, which can undergo a Wolff rearrangement to produce a ketene amide intermediate. In the presence a nucleophilic species on a target molecule, the ketene can undergo an acylation reaction to form a stable malonic acid derivative. The photolyzed product thus can couple to hydrazide- or amine-containing targets to form covalent linkages (Reaction 3.56).

7.5. Diazirine Derivatives

Diazirine compounds are similar in their photoreactivity to diazo groups, forming highly reactive carbene intermediates upon exposure to UV light of about 360 nm (Reaction 3.57). Diazirines consist of a three-member ring system containing two nitrogen atoms connected through a double bond. First developed by Smith and Knowles (1973), the photosensitive diazirine is perhaps second in popularity to phenyl azides in the design of photoreactive crosslinking agents.

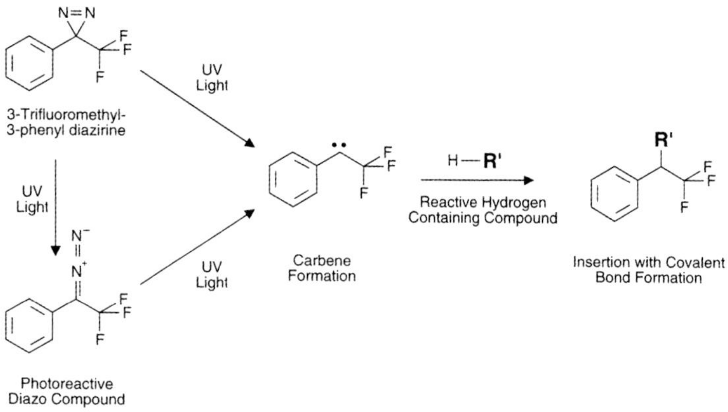

(REACTION 3.57)

Some diazirines, particularly the 3-trifluoromethyl-3-aryldiazirines, can rearrange upon photolysis to a linear diazo derivative, similar in structure to the photosensitive end of the crosslinker PNP–DTP (Chapter 6, Section 3.12). These isomerized products themselves can be photolyzed to the reactive carbene.

Carbene generation from photolysis of diazirine compounds leads to efficient insertion into C–H or N–H bonds and also causes addition reactions with points of unsaturation within target molecules. Diazirine-containing photoaffinity probes have been used to study numerous ligand–receptor interactions (Bergmann *et al.*, 1994). Heterobifunctional crosslinkers containing a diazirine photosensitive group has also been used to attach macromolecules to surfaces such as polystyrene and glass (Collioud *et al.*, 1993). In addition, diazirine-containing photoreactive amino acid analogs, photo-leucine and photo-methionine, have been developed to study protein interactions within cells (Suchanek *et al.*, 2005).

7.6. Psoralen Compounds

Psoralen, or derivatives of 9-methoxy-7*H*-furo[3,2-*g*]chromen-7-one tricyclic ring structures, are used as photoreactive groups in crosslinkers, biotinylation compounds, and nucleic acid probes. Psoralens have been used for many years as photochemotherapy agents for treatment of psoriasis and vitiligo (Smith and Barker, 2006). Psoralens react when exposed to UV light in the range of 320 to 400 nm to form an excited triplet state intermediate that can insert in certain double-bond structures, especially at the 5,6-double bond of thymine bases.

Psoralens can react by two different routes upon photoactivation (Parsons, 1980; Pathak, 1984). The first route is through the well-known photoreaction mechanism that principally involves intercalation within double-stranded DNA or RNA with the formation of adducts with adjacent thymine bases. The furan-side and pyrone-side rings in psoralen can both form cycloaddition products with the 5,6-double-bond of thymine to create a crosslink between two DNA strands (Reaction 3.58) or, to a lesser extent, within double-strand regions of RNA.

Psoralen can also undergo reactions with oxygen to produce reactive oxygen species, including the formation of singlet oxygen ($^1O_2$), superoxide anion ($O_2^{\cdot -}$), and hydroxyl radicals ($^{\cdot}OH$). The production of reactive oxygen species by psoralen derivatives can damage biological molecules and structures. See Chapter 2, Section 1.1, for additional information on the oxidation of amino acids.

Psoralen–biotin compounds have been used to label double-stranded DNA for detection using streptavidin reagents (Henriksen *et al.*, 1991; Wygrecka *et al.*, 2007). The compound psoralen–PEG_3–biotin is commercially available for this purpose (Thermo Fisher). Crosslinking

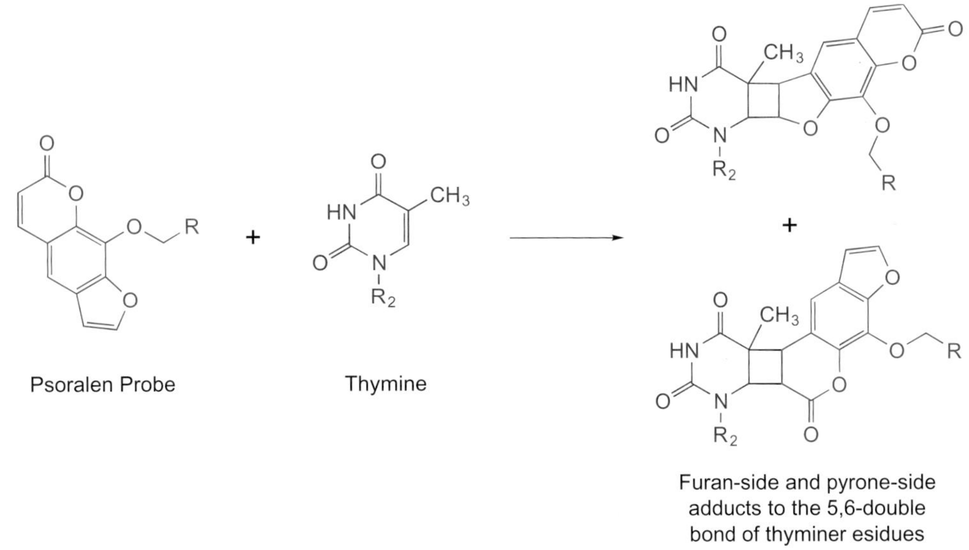

(REACTION 3.58)

agents can also be built using a photoreactive psoralen ring system at one end. The reagent succinimidyl-[4-(psoralen-8-yloxy)]butyrate (SPB) contains an NHS ester to covalently link a psoralen group to proteins or other amine-containing molecules (Thermo Fisher, Molecular Biosciences). Oser *et al.* (1988) created a lanthanide chelate with a psoralen group to label DNA with a time-resolved fluorescent probe. Psoralen derivatives can also be coupled to polymeric surfaces by a photoreaction process. Elsner and Mouritsen (1994) used psoralen linkers for modification of the surface of microplates to create sites for covalent binding of affinity molecules.

8. CYCLOADDITION REACTIONS

The following sections briefly describe three cycloaddition reactions that can be used to form bioconjugates. These reactions represent highly specific reactant pairs that have a chemoselective nature, meaning they mainly react with each other and not other functional groups, such as those found on biomolecules. For a complete discussion of chemoselective ligation reactions, see Chapter 17.

8.1. Diels–Alder Reaction

The Diels–Alder reaction consists of the covalent coupling of a diene with an alkene to form a six-membered ring complex. This process has been used extensively in organic synthesis, but only recently has it been applied to bioconjugation reactions (Hill *et al.*, 2001). This reaction proceeds at room temperature or slightly elevated temperature conditions (30°C) to give the 2 + 4 cycloaddition product, a hexane ring containing a single double bond. The reaction can be performed using a maleimide group as the alkene derivative and a hexadienyl group as the diene (Reaction 3.59). The reaction process can give 90 to 95% yields in 1 to 18h.

See Chapter 17, Section 1, for additional information concerning the use of the Diels–Alder reaction in bioconjugation applications.

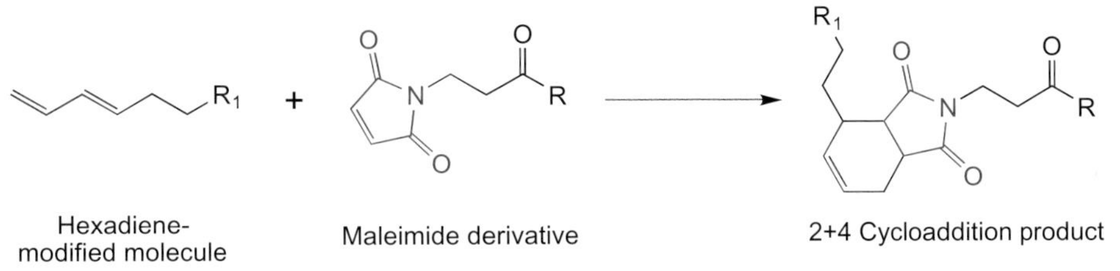

(REACTION 3.59)

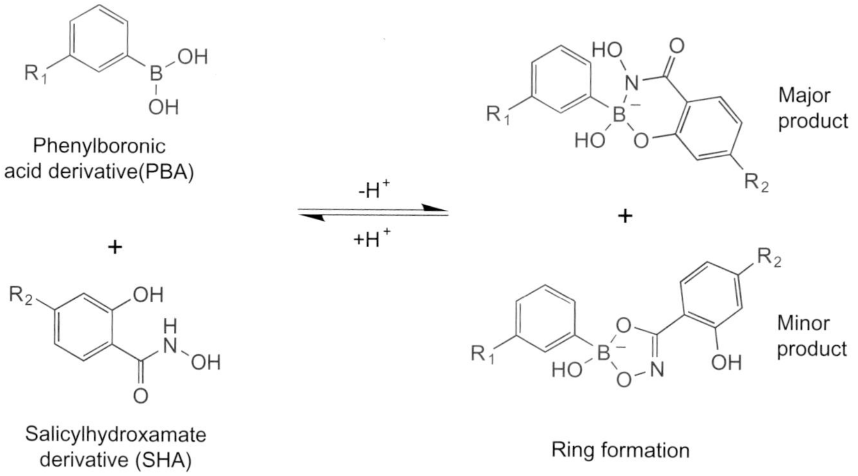

(REACTION 3.60)

8.2. Complex Formation with Boronic Acid Derivatives

Boronic acid derivatives are able to form ring structures with other molecules having neighboring functional groups consisting of 1,2- or 1,3-diols, 1,2- or 1,3-hydroxy acids, 1,2- or 1,3-hydroxylamines, 1-2- or 1,3-hydroxyamides, 1,2- or 1,3-hydroxyoximes, as well as various sugars containing these species (Weith at al., 1970; Rosenberg and Gilham, 1971; Rosenberg *et al.*, 1972; Pace and Pace, 1980; Singhal *et al.*, 1980). The products of these reactions are five- or six-membered heterocyclic rings, which in some cases are reversible with a change in pH or by the addition of a counter-ligand having competing functional groups.

Typically, the boronic acid group is part of an aminophenyl boronic acid derivative, and this group has been used for bioconjugation and affinity chromatography purposes (Burnett *et al.*, 1980; O'Shannessy and Quarles, 1987). A common partner for a phenyl boronic acid group in bioconjugation is the salicylhydroxamic acid (SHA) group (Chapter 17, Section 4) (Reaction 3.60).

8.3. Click Chemistry: Cu^{1}-Promoted Azide–Alkyne [3 + 2] Cycloaddition

Click chemistry refers to the reaction between an azido functional group and an alkyne to form a [3 + 2] cycloaddition product, a five-membered triazole ring. This reaction has been used for many years in organic synthesis to form heterocyclic rings. Normally, the click reaction requires high temperatures, and this was the main reason that it was not used as a bioconjugation tool. However, it was discovered that in aqueous solutions and in the presence of Cu(I), the reaction kinetics are dramatically accelerated to provide high yields even at room temperature and ambient pressures (TornØe *et al.*, 2002; Rostovtsev *et al.*, 2002; Sharpless *et al.*, 2005).

The advantage of the click reaction for bioconjugation is that the reactant pair is not reactive with any other functional group encountered in biological systems. This property of bioorthogonality provides extreme selectivity for bringing together azide and alkyne derivatives to form triazoles even in complex biological samples.

Reaction 3.61 shows the reaction of an alkyne with an azide to form a triazole ring in the presence of a Cu(I) catalyst. See Chapter 17, Section 5, for additional details on the use of this conjugation reaction.

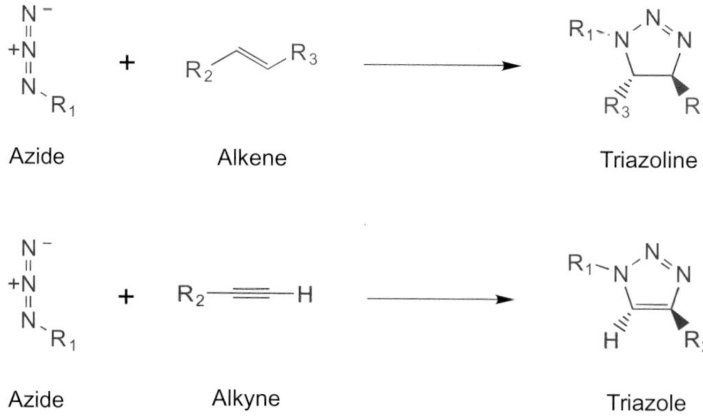

(REACTION 3.61)

CHAPTER

4

Zero-Length Crosslinkers

The smallest available reagent systems for bioconjugation are the so-called zero-length crosslinkers. These compounds mediate the conjugation of two molecules by forming a bond containing no additional atoms. Thus, one atom of a molecule is covalently attached to an atom of a second molecule with no intervening linker or spacer. In many conjugation schemes, the final complex is bound together by virtue of chemical components that add foreign structures to the substances being crosslinked. In some applications, the presence of these intervening linkers may be detrimental to the intended use. For instance, in the preparation of hapten–carrier conjugates the complex is formed with the intention of generating an immune response to the attached hapten. Occasionally, a portion of the antibodies produced by this response will have specificity for the crosslinking agent used in the conjugation procedure. Zero-length crosslinking agents eliminate the potential for this type of cross-reactivity by mediating a direct linkage between two substances.

The reagents described in this section can initiate the formation of three types of bonds: an amide linkage made by the condensation of a primary amine with a carboxylic acid, a phosphoramidate linkage made by the reaction of a organic phosphate group with a primary amine, and a secondary or tertiary amine linkage made by the reductive amination of a primary or secondary amine with an aldehyde group. Therefore, using these reagent systems, substances containing amines can be conjugated with other molecules containing phosphates or carboxylates. Alternatively, substances containing amines can be crosslinked to molecules containing formyl groups. All of the reactions are quite efficient, and, depending on the reagent chosen and the desired application, they may be performed in aqueous or nonaqueous environments.

1. CARBODIIMIDES

Carbodiimides are used to mediate the formation of amide linkages between carboxylates and amines or phosphoramidate linkages between phosphates and amines (Hoare and Koshland, 1966; Chu *et al.*, 1986; Ghosh *et al.*, 1990). They are probably the most popular type of zero-length crosslinker in use, being efficient in forming conjugates between two protein molecules, between a peptide and a protein, between an oligonucleotide and a protein, between a biomolecule and a surface or particle, or any combination of these with small molecules. There are two basic types of carbodiimides: water soluble and water insoluble. The water soluble ones are the most common choice for biochemical conjugations, because most macromolecules of biological origin are soluble in aqueous buffer solutions. Not only is the carbodiimide itself able to dissolve in the reaction medium, but the byproduct of the reaction, an isourea, is also water soluble, facilitating easy purification. Water-insoluble carbodiimides, by contrast, are used frequently in peptide synthesis and other conjugations involving molecules soluble only in organic solvents. Both the organic-soluble carbodiimides and their isourea byproducts are insoluble in water.

1.1. EDC

EDC (or EDAC; 1-ethyl-3-(3-dimethylaminopropyl) carbodiimide hydrochloride) is the most popular carbodiimide used for conjugating biological substances containing carboxylates and amines. In fact, it may also be the most frequently used crosslinking agent of all. Its application in particle and surface conjugation procedures along with NHS (*N*-hydroxysulfosuccinimide) or sulfo-NHS is nearly universal (Chapter 14) and this fact makes it the most common bioconjugation reagent in use today. EDC is water soluble, which allows for its direct addition to a reaction without prior organic solvent dissolution. Both the reagent itself and the isourea formed as the byproduct of the crosslinking reaction are water soluble and may be removed easily by dialysis or gel filtration (Sheehan *et al.*, 1961, 1965). The reagent is, however, labile in the presence of water, especially in acidic solutions. In alkaline solutions,

Bioconjugate Techniques, Third Edition.
DOI: http://dx.doi.org/10.1016/B978-0-12-382239-0.00004-2

the reagent is more stable, but it reacts much more slowly. The bulk chemical should be stored desiccated at −20°C. Warm the bottle to room temperature before opening to prevent condensation occurring that will cause decomposition of the reagent over time. A concentrated solution of EDC in water may be prepared to facilitate the addition of a small molar amount to a reaction, but the stock solution should be dissolved rapidly and used immediately to prevent extensive loss of activity.

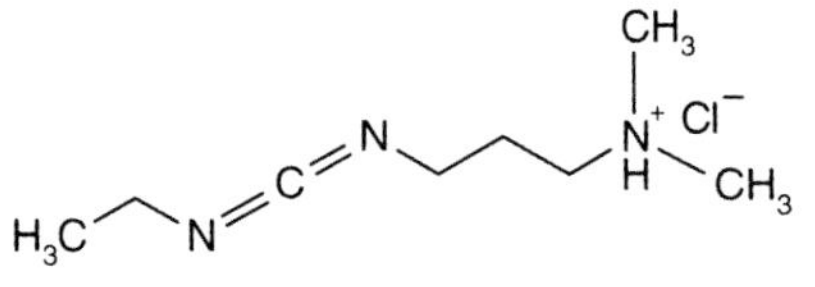

EDC
1-Ethyl-3-(3-dimethylaminopropyl)
Carbodiimide Hydrochloride
MW 191.7

A variety of chemical conjugates may be formed using EDC Chu *et al.*, 1976, 1977, 1982; Yamada *et al.*, 1981; Chase *et al.*, 1983, provided one of the molecules contains an amine and the other a carboxylate group. N-substituted carbodiimides can react with carboxylic acids to form highly reactive *O*-acylisourea intermediates (Figure 4.1). This active species can then react with a nucleophile such as a primary amine to form an amide bond (Williams and Ibrahim, 1981). Other nucleophiles are also reactive. Sulfhydryl groups may attack the active species and form thiol ester linkages, although these are not as stable as the bond formed with an amine. In addition, oxygen atoms may act as the attacking nucleophile, such as those in water molecules. In aqueous solutions, hydrolysis by water is the major competing reaction, cleaving off the activated ester intermediate, forming an isourea, and regenerating the carboxylate group (Gilles *et al.*, 1990).

Nakajima and Ikada (1995) investigated the reactions of EDC amide bond formation in aqueous solution using hydrogels of acrylic acid- or maleic acid-containing polymers or other carboxylate molecules to contribute the activatable groups and ethylenediamine or benzylamine as the amine functional groups to be conjugated. Their results indicate that carboxylate activation occurs most effectively with EDC at pH 3.5 to 4.5, while amide bond formation occurs with highest yield in the range of pH4 to 6. However, EDC hydrolysis occurs maximally at acidic pH values with increasing stability of the carbodiimide in solution at or above pH 6.5. When working with proteins and peptides, experience indicates that EDC-mediated amide bond formation effectively occurs between pH 4.5 and 7.5. Beyond this pH range, however, the coupling reaction occurs more slowly with lower yields.

An EDC-mediated reaction to form an amide bond in aqueous solution involves a number of potential side reactions that can occur in addition to the desired conjugation product. The preferred reaction route is facilitated first by protonation of one of the nitrogens on the imide group of EDC, which results in the formation of an intermediate carbocation on the central carbon

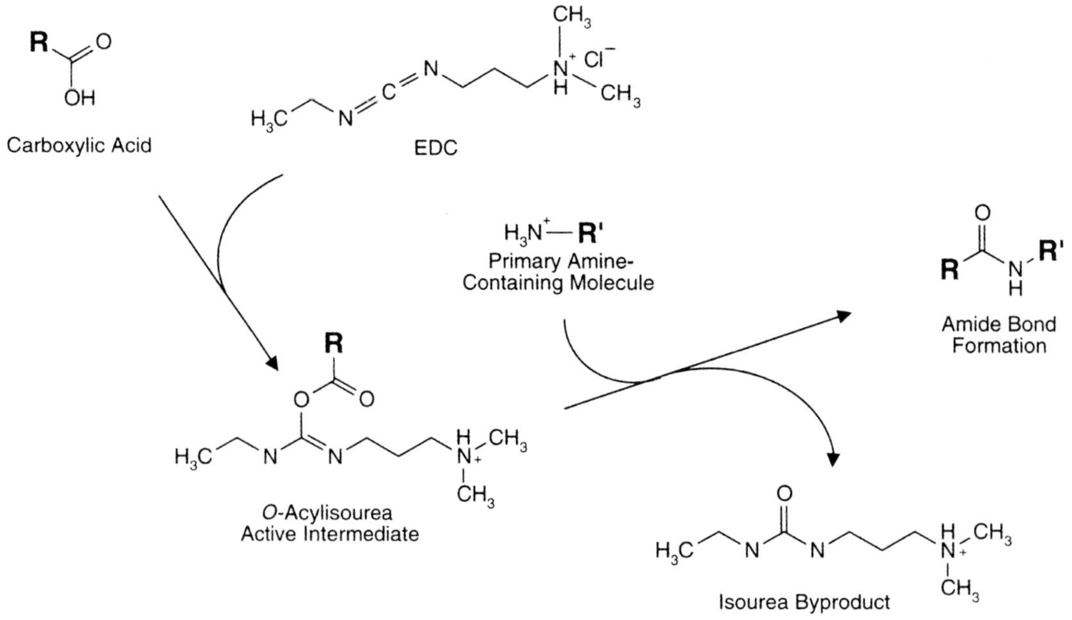

FIGURE 4.1 EDC reacts with carboxylic acids to create an active-ester intermediate. In the presence of an amine nucleophile, an amide bond is formed with release of an isourea byproduct.

atom. At this point, the modified carbodiimide can itself hydrolyze to form an inactive isourea that no longer can participate in the reaction process. It can also react with an available ionized carboxylate group to create the desired *O*-acylisourea reactive ester intermediate (Figure 4.2). This ester again can accept another proton to form a second carbocation on the central carbon atom, and it is this form of the reactive ester that can go on to react with an amine to create an amide bond. However, this intermediate also can undergo a hydrolysis event to yield the same undesired isourea derivative, which again inactivates the compound. Therefore, there are two stages in the reaction sequence in which hydrolysis can occur and inhibit the desired product formation (Nakajima and Ikada, 1995). If hydrolysis does not occur, there are at least three subsequent reactions that can happen, including the desired amide bond formation. If a neighboring carboxylate group is in close proximity to the *O*-acylisourea ester, it may react with it and form an anhydride intermediate. This especially can occur in polymers containing repeating carboxylate groups, such as in polymethacrylate, where the primary intermediate reactive group formed from EDC may be anhydrides (Wang *et al.*, 2011). Fortunately, an anhydride is also reactive with amine groups, so the desired amide bond formation can still occur with at least one of the two carboxylates making up the anhydride. Indeed, anhydride formation may result in higher yields of amide bond formation in certain instances (Nakajima and Ikada, 1995). In addition, if EDC is in large excess over the amount of carboxylates present, then the intermediate ester may exist for a longer period and potentially it can rearrange by reacting with the neighboring secondary amines in the carbodiimide and thus form an *N*-acylisourea derivative, which is inactive and permanently attaches the EDC derivative to the carboxylate compound. Finally, the EDC-reactive ester can also react with the desired amine-containing molecule and form an amide bond. Given this degree of potential side reactions, it is amazing that EDC-mediated amide bond formation can be done with reproducibility, especially when scaling up reactions in production processes. The propensity for EDC to undergo side reactions may be a reason that high variability has been reported using the carbodiimide for particular conjugation reactions (Young *et al.*, 2004).

To complicate matters even further, the presence of both carboxylates and amines on one of the molecules to

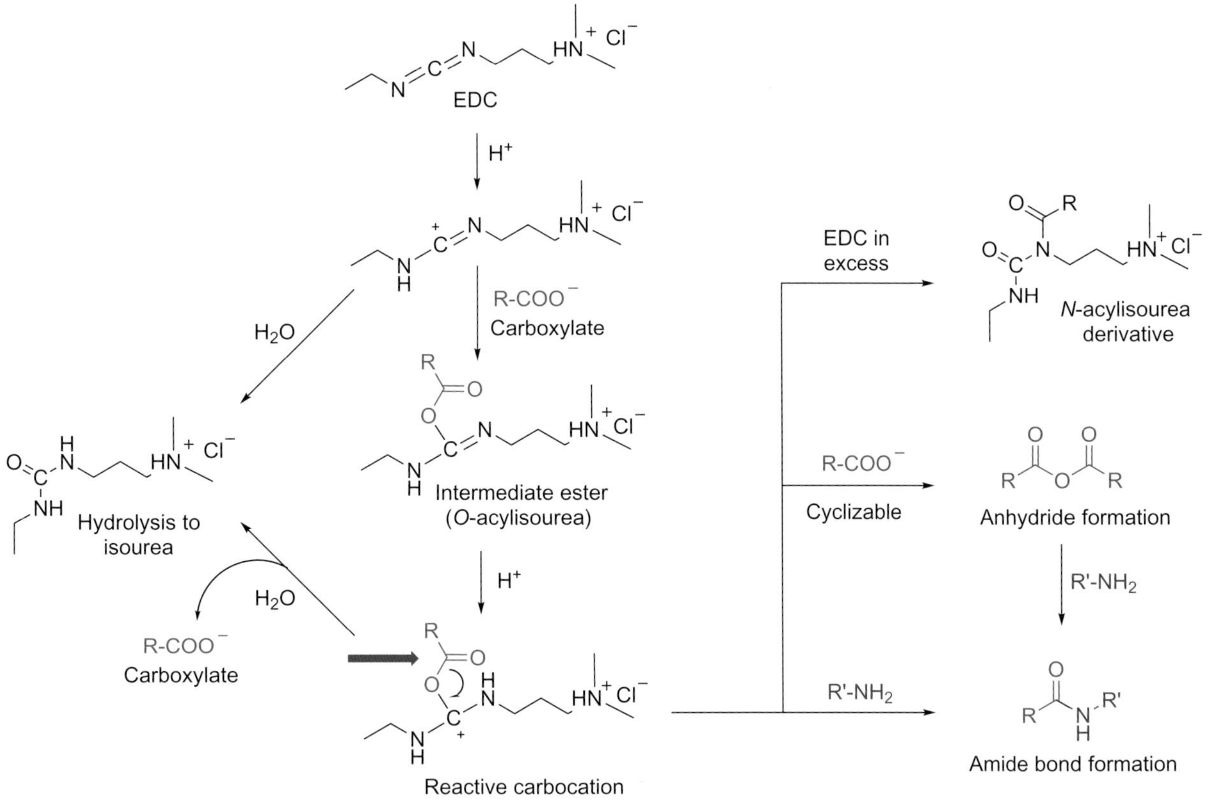

FIGURE 4.2 EDC can undergo a number of potential side reactions that compete with amide bond formation. In aqueous solution, hydrolysis may occur at two points in the reaction path to decrease the yield of the desired conjugate in addition to the possible formation of a permanent EDC complex to create an *N*-acylisourea derivative. This complex reaction pathway often makes EDC conjugations difficult to reproduce in large-scale conjugations.

be conjugated with EDC may result in self-polymerization, because the substance can then react with another molecule of its own kind instead of the desired target. For instance, when conjugating peptides to carrier proteins using EDC, the peptide usually contains both a carboxylate and an amine. The result typically is peptide polymerization in addition to coupling to the carrier (see Chapter 19, Section 3). For this type of immunogen conjugation, polymerization is not usually detrimental to its use, because polymerized peptide is also immunogenic. However, for other crosslinking applications where it may be more desirable to avoid oligomer formation, the use of a carbodiimide may not be the best choice of reagent, especially if one of the molecules being conjugated contains both a carboxylate and an amine.

EDC has been used extensively in the conjugation of amine-containing molecules to carboxylate particles, including small microparticles and nanoparticles (Chapter 14). Most references to the use of EDC describe the optimal reaction medium to be at a pH from 4.7 to 6.0. However, the carbodiimide reaction occurs effectively up to at least pH 7.5 without significant loss of yield. Conjugations performed under mildly alkaline pH conditions (e.g., pH 8.5) can also be carried out to limit the polymerization of proteins, while still facilitating the coupling of a carboxylate-containing molecule at a lower substitution level per protein. See Chapter 19, Section 3, for additional information on the properties of EDC conjugation using small peptides coupled to carrier proteins.

Some procedures recommend the use of water as the solvent in an EDC reaction, while the pH is maintained constant by the addition of HCl. Buffered solutions are more convenient, because the pH does not have to be monitored during the course of the reaction. For acidic pH conjugations, MES [2-(*N*-morpholino)ethane sulfonic acid] buffer at 0.1-*M* works well. When carrying out neutral pH reactions, a phosphate buffer at 0.1-*M* is appropriate. Any buffers may be used that do not interfere with the reaction, but avoid amine- or carboxylate-containing buffer salts or other components in the medium that may react with the carbodiimide.

There are some additional side reactions that may occur when using EDC with proteins. In addition to reacting with carboxylates, EDC itself can form a stable complex with exposed sulfhydryl groups (Carraway and Triplett, 1970). Tyrosine residues can react with EDC, most likely through the phenolate ionized form of its side chain (Carraway and Koshland, 1968). The imidazolyl group of histidine may react with sulfo-NHS esters, resulting in an active carbonyl imidazole group which subsequently hydrolyzes (Cuatrecasas and Parikh, 1972). Finally, EDC may promote unwanted polymerization due to the usual abundance of both amines and carboxylates on protein molecules.

The following protocol is a generalized description of how to conjugate a small amine- or carboxylate-containing molecule to a protein. The protocol may be modified by changing the pH, buffer salts, and ratios of reactants to obtain the desired product. Specific protocols utilizing EDC in selected conjugation applications may be found in Chapters 8, 14, 15, 19, 21, and 23. In some cases, the parameters of this generalized protocol may have to be modified to retain solubility or activity of the resulting conjugate. For instance, coupling hydrophobic molecules to the surface of proteins often causes partial or complete precipitation. This problem may be somewhat alleviated by decreasing either the amount of EDC or the amount of the hydrophobic molecule added to the reaction, thus resulting in a lower density of substitution. Protocols on the use of EDC to couple proteins or other molecules to carboxylated nanoparticles and microparticles may be found in Chapter 14 and Chapter 10, Section 10.

Protocol

1. Dissolve the protein to be modified at a concentration of 10 mg/ml in one of the following reaction media: (a) water; (b) 0.1-*M* MES, pH 4.7 to 6.0; or (c) 0.1-*M* sodium phosphate, pH 7.3. NaCl may be added (i.e., 0.15-*M*) if desired. If lower or higher concentrations of the protein are used, adjust the amounts of the other reactants added as necessary to maintain the correct molar ratios. For the preparation of a peptide–protein immunogen conjugate, a 200 μl solution of the carrier protein at a concentration of 10 mg/ml in 0.1-*M* MES, pH 4.7, usually works well.
2. Dissolve the molecule to be coupled in the same buffer used in step 1. For small molecules, add them to the reaction in at least a 10-fold molar excess to the amount of protein present. If possible, the molecule may be added directly to the protein solution in the appropriate excess. Alternatively, dissolve the molecule in the buffer at a higher concentration, and then add an aliquot of this stock solution to the protein solution. In the example of preparing a peptide–protein conjugate, dissolve the peptide in 0.1-*M* MES, pH 4.7, at a concentration of up to 2 mg/500 μl.
3. Add the solution prepared in step 2 to the protein solution to obtain at least a 10-fold molar excess of small molecule-to-protein. In the case of the peptide–protein immunogen conjugate, add the 500 μl of peptide solution to the 200 μl of protein solution.
4. Add EDC (Thermo Fisher) to the above solution to obtain at least a 10-fold molar excess of EDC to the protein. Alternatively, a 0.5- to 0.1-*M* EDC concentration in the reaction mixture usually works well. To make it easier to add the correct quantity of EDC, a higher concentration stock solution may

be prepared if it is dissolved and used immediately. To prepare the peptide–protein conjugate, add the solution from step 3 to 10 mg of EDC in a test tube. Mix to dissolve. If this ratio of EDC to peptide or protein results in precipitation, then scale back the amount of carbodiimide addition until a soluble conjugate is obtained. For some proteins, as little as 0.1 times this amount of EDC may have to be used to maintain solubility.

5. React for 2 h at room temperature.
6. Purify the conjugate by gel filtration or dialysis using the buffer of choice (for many conjugates 0.01-*M* sodium phosphate, 0.15-*M* NaCl, pH 7.4, is appropriate). If some turbidity has formed during the conjugation procedure, it may be removed by centrifugation or filtration. When using EDC to prepare immunogen conjugates, the presence of some precipitated material is usually not of concern, because precipitated immunogens are often more immunogenic than soluble proteins.

1.2. EDC Plus Sulfo-NHS

The water soluble carbodiimide EDC may be used to form active ester functionalities with carboxylate groups using the water soluble compound *N*-hydroxysulfosuccinimide (sulfo-NHS) (Thermo Fisher). Sulfo-NHS esters are hydrophilic reactive groups that couple rapidly with amines on target molecules (Staros, 1982; Denney and Blobel, 1984; Kotite *et al.*, 1984; Jennings and Nicknish, 1985; Ludwig and Jay, 1985; Beth *et al.*, 1986; Donovan and Jennings, 1986; Anjaneyulu and Staros, 1987). Unlike many non-sulfonated NHS esters that can be relatively water insoluble and must first be dissolved in organic solvent before being added to aqueous solutions, sulfo-NHS esters typically are water soluble and longer lived and don't hydrolyze quite as quickly in water. However, in the presence of amine nucleophiles that can attack at the carbonyl group of the ester, the sulfo-NHS group rapidly leaves, creating a stable amide linkage with the amine. Sulfhydryl and hydroxyl groups will also react with such active esters, but the products of such reactions, thioesters and esters, are relatively unstable compared to an amide bond.

The advantage of adding sulfo-NHS to EDC reactions is to increase the solubility and stability of the active intermediate, which ultimately reacts with the attacking amine. EDC reacts with a carboxylate group to form an active ester (*O*-acylisourea) leaving group. Unfortunately, this reactive complex is slow to react with amines and can hydrolyze in aqueous solutions, having a rate constant measured in seconds (Chapter 14, Section 4) (Hoare and Koshland, 1967). If the target amine does not find the active carboxylate before it hydrolyzes, the desired coupling cannot occur. This is especially a problem when the target molecule is in low concentration compared to water, as in the case of protein molecules. In addition, Nakajima and Ikada (1995) found that if a carboxylate-containing compound can form an anhydride from the *O*-acylisourea intermediate reactive ester, then the yield of amide bond formation is increased. In a similar approach, forming a sulfo-NHS ester intermediate from the reaction of the hydroxyl group on sulfo-NHS with the EDC active-ester complex dramatically increases the resultant amide bond formation. Since the concentration of added sulfo-NHS usually is much greater than the concentration of target molecule, the reaction preferentially proceeds through the more efficient sulfo-NHS ester intermediate. However, the final product of this two-step reaction is identical to that obtained using EDC alone: The activated carboxylate reacts with an amine to give a stable amide linkage (Figure 4.3).

EDC/sulfo-NHS coupled reactions are highly efficient and usually increase the yield of conjugation significantly over that obtainable solely with EDC. Staros *et al.* (1986) showed that the addition of just 5-m*M* sulfo-NHS to the EDC coupling of glycine to keyhole limpet hemocyanin increased the yield of derivatization about 20-fold as compared to using EDC alone. This technique can also be used to create activated proteins containing sulfo-NHS esters (Grabarek and Gergely, 1990). A protein can be incubated in the presence of EDC/sulfo-NHS, the active ester form isolated and then mixed with a second protein or other amine-containing molecule for conjugation. This two-step process allows the active species to form on only one protein, thus gaining greater control over the conjugation reaction (Figure 4.4).

In addition to the potential side reactions of EDC as mentioned previously (Section 1.1, this chapter), the additional efficiency obtained by the use of a sulfo-NHS intermediate in the process may cause other problems. In some cases, the conjugation actually may be too efficient to result in a soluble or active complex. Particularly when coupling some peptides to carrier proteins, the use of EDC/sulfo-NHS often causes severe precipitation of the conjugate. Scaling back the amount of EDC/sulfo-NHS added to the reaction may be performed to solve this problem. However, eliminating the addition of sulfo-NHS altogether may have to be done in some instances to preserve the solubility of the final product.

An EDC/sulfo-NHS reaction using carboxylated particles to couple amine-containing ligands has the added advantage of creating an intermediate negatively charged particle surface, which helps maintain strong repulsion between particles and prevents aggregation during the reaction. Many types of carboxylated particles have been used in this procedure to couple biomolecules in high yield (Bartczak and Kanaras, 2011).

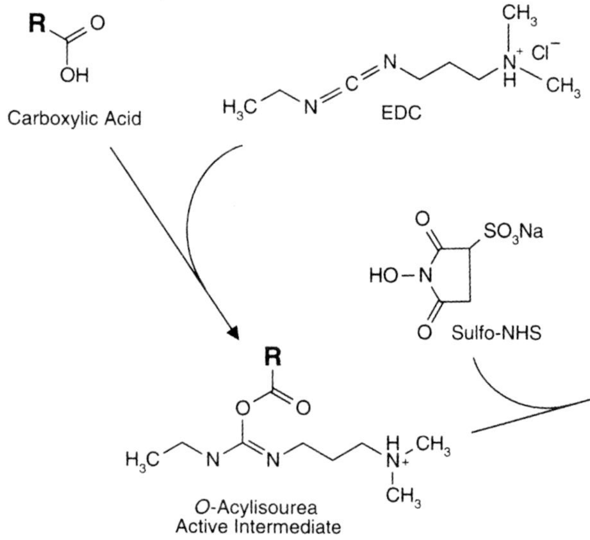

FIGURE 4.3 The efficiency of an EDC-mediated reaction may be increased through the formation of a sulfo-NHS ester intermediate. The sulfo-NHS ester is more effective at reacting with amine-containing molecules. Thus, higher yields of amide bond formation may be realized using this two-step process as opposed to using a single-step EDC reaction.

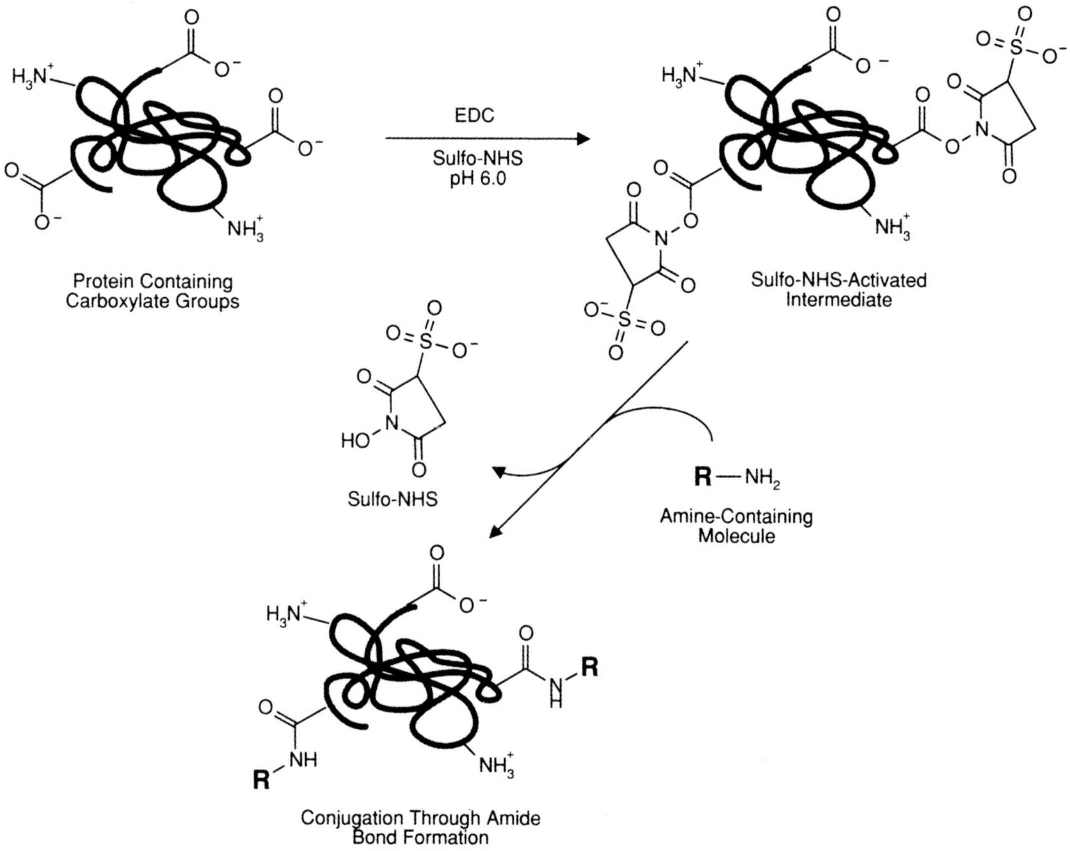

FIGURE 4.4 EDC may be used in tandem with sulfo-NHS to create an amine-reactive protein derivative containing active ester groups. The activated protein can couple with amine-containing compounds to form amide bond linkages.

The following protocol is a generalized description of how to incorporate sulfo-NHS ester intermediates in EDC conjugation procedures. For specific applications of this technology, the amount of each reagent and unconjugated species may have to be adjusted to obtain an optimal conjugate. See also Chapter 14 and Chapter 10, Section 10, for protocols using EDC/sulfo-NHS in the coupling of proteins to particles and quantum dots, respectively.

Protocol

1. Dissolve the protein to be modified at a concentration of 1 to 10 mg/ml in 0.1-*M* sodium phosphate, pH 7.4. NaCl may be added to this buffer if desired. For the modification of keyhole limpet hemocyanin (KLH; Thermo Fisher) as described by Staros *et al.* (1986), include 0.9-*M* NaCl to maintain the solubility of this high-molecular-weight protein. If lower or higher concentrations of the protein are used, adjust the amounts of the other reactants as necessary to maintain the correct molar ratios.
2. Dissolve the molecule to be coupled in the same buffer used in step 1. For small molecules, add them to the reaction in at least a 10-fold molar excess over the amount of protein present. If possible, the molecule may be added directly to the protein solution in the appropriate excess. Alternatively, dissolve the molecule in the buffer at a higher concentration, and then add an aliquot of this stock solution to the protein solution.
3. Add the solution prepared in step 2 to the protein solution to obtain at least a 10-fold molar excess of small molecule to protein.
4. Add EDC (Thermo Fisher) to the above solution to obtain at least a 10-fold molar excess of EDC over the amount of protein present. Alternatively, a 0.05- to 0.1-*M* EDC concentration in the reaction usually works well. Also, add sulfo-NHS (Thermo Fisher) to the reaction to bring its final concentration to 5-m*M*. To make it easier to add the correct quantity of EDC or sulfo-NHS, higher concentration stock solutions may be prepared if they are dissolved and used immediately. Mix to dissolve. If this ratio of EDC/sulfo-NHS to peptide or protein results in precipitation, scale back the amount of addition until a soluble conjugate is obtained.
5. React for 2 h at room temperature.
6. Purify the conjugate by gel filtration or dialysis using the buffer of choice (for many conjugates 0.01-*M* sodium phosphate, 0.15-*M* NaCl, pH 7.4, is appropriate). If some turbidity has formed during the conjugation procedure, it may be removed by centrifugation or filtration.

A modification of a two-step protocol (Grabarek and Gergely, 1990) for the activation of proteins with EDC/sulfo-NHS and subsequent conjugation with amine-containing molecules is given below. The variation in the pH of activation from that described above provides greater stability for the active ester intermediate. At pH 6.0, the amines on the protein will be protonated and therefore be less reactive toward the sulfo-NHS esters that form. In addition, the hydrolysis rate of the esters is dramatically slower at slightly acid pH. Thus, the active species may be isolated in a reasonable time frame without significant loss in conjugation potential. To quench the unreacted EDC, 2-mercaptoethanol is added to form a stable complex with the remaining carbodiimide, according to Carraway and Triplett (1970). In the following protocol, sulfo-NHS is used instead of NHS so that active ester is more water soluble and ester hydrolysis is slowed (Anjaneyulu and Staros, 1987; Thelen and Deuticke, 1988).

Protocol

1. Dissolve the protein to be activated in 0.05-*M* MES, 0.5-*M* NaCl, pH 6.0 (reaction buffer), at a concentration of 1 mg/ml.
2. Add to the solution in step 1 a quantity of EDC and sulfo-NHS (Thermo Fisher) to obtain a concentration of 2-m*M* EDC and 5-m*M* sulfo-NHS. To aid in aliquoting the correct amount of these reagents, they may be quickly dissolved in the reaction buffer at a higher concentration, and then a volume immediately pipetted into the protein solution to obtain the proper molar quantities.
3. Mix and react for 15 min at room temperature.
4. Add 2-mercaptoethanol to the reaction solution to obtain a final concentration of 20-m*M*. Mix and incubate for 10 min at room temperature. Note: If the protein being activated is sensitive to this level of 2-mercaptoethanol, instead of quenching the reaction chemically the activation may be terminated by desalting (step 5).
5. If the reaction was quenched by the addition of 2-mercaptoethanol, the activated protein may be added directly to a second protein or other amine-containing molecule for conjugation. Alternatively, or if no 2-mercaptoethanol was added, the activated protein may be purified from reaction by-products by gel filtration using a desalting resin. The desalting operation should be performed rapidly to minimize hydrolysis and recover as much active ester functionality as possible. The use of centrifugal spin columns of some sort may afford the greatest speed in purification (Thermo Fisher). After purification, add the activated protein to the second molecule for conjugation. The second protein or other

amine-containing molecule should be dissolved in 0.1-*M* sodium phosphate, pH 7.5. This will bring the pH of the coupling medium above pH 7.0 to initiate the active ester reaction.

6. React for at least 2 h at room temperature.
7. Remove excess reactants by gel filtration or dialysis.

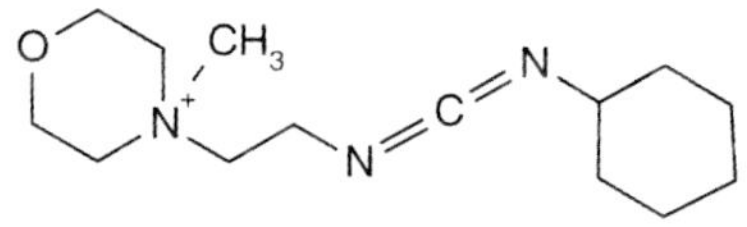

CMC
1-Cyclohexyl-3-(2-morpholinoethyl)carbodiimide
MW 423.58
(as the metho-p-toluene sulfonate salt)

1.3. CMC

CMC, or 1-cyclohexyl-3-(2-morpholinoethyl) carbodiimide (usually synthesized as the metho *p*-toluene sulfonate salt) (Aldrich), is a water soluble reagent used to form amide bonds between one molecule containing a carboxylate and a second molecule containing an amine. The presence of the positively charged morpholino group creates its water solubility. Along with EDC (Section 1.1, this chapter), CMC is the only other water soluble carbodiimide commonly available for biological conjugations. It was first utilized in peptide synthesis (Sheehan and Hlavka, 1956) and found to be superior to other coupling agents used at the time (Ondetti and Thomas, 1965). It has also been used for the quantitative modification and estimation of total carboxyl groups in protein molecules (Hoare and Koshland, 1967) and for investigating the secondary structure of nucleic acids (Metz and Brown, 1969). Another early application area of CMC relates not to solution-phase crosslinking of two molecules but to coupling of ligands to insoluble support materials for use in affinity chromatography (Lowe and Dean, 1971; Marcus and Balbinder, 1972; Schmer, 1972). It has also been used to conjugate antibodies to quantum dots (East *et al.*, 2011).

CMC reacts with carboxylate groups by addition of the carboxyl across one of its diimide bonds, resulting in the characteristic active ester, *O*-acylisourea intermediate common to all carbodiimide mechanisms. Nucleophilic attack on this intermediate yields the acylated product—usually an amide bond, resulting from the reaction with a primary amine (Figure 4.5). However, carbodiimide chemistry does create several potential side-reactions. Sulfhydryl groups may react with CMC to form a stable covalent complex unreactive toward further conjugation. The reagent may also react with phenols, alcohols, and other nucleophiles to quench the crosslinking reaction. In aqueous solutions, hydrolysis of the carbodiimide and the active ester are by far the most frequent side-reactions. Reaction of the ester with water molecules regenerates the carboxylate and releases a soluble isourea by-product.

CMC should be able to participate in the two step reaction using a sulfo-NHS ester intermediate similar to EDC, however there are no reports in the literature to this effect. Protocols for the use of this reagent in biological crosslinking applications should be essentially the same as those given previously for EDC, except substituting a molar equivalent quantity of CMC. See Section 1.1 and 1.2 in this chapter for additional information concerning carbodiimide reactions.

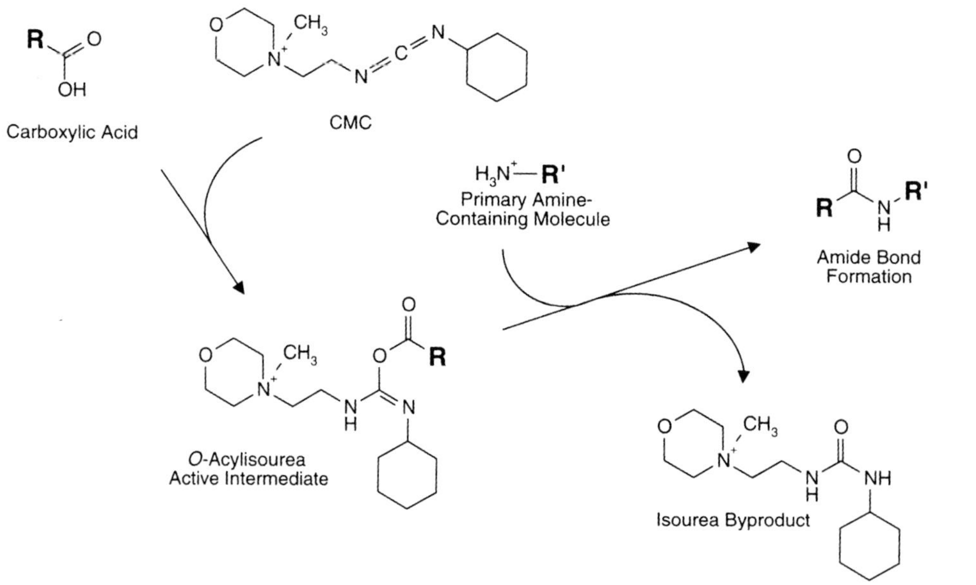

FIGURE 4.5 The water soluble carbodiimide CMC reacts with carboxylates to form an active-ester intermediate. In the presence of amine-containing molecules, amide bond formation can take place with release of an isourea byproduct.

1.4. DCC

DCC (dicyclohexyl carbodiimide) is one of the most frequently used coupling agents, especially in organic synthesis applications. It has been used for peptide synthesis since 1955 (Sheehan and Hess, 1955) and continues to be a popular choice for creating peptide bonds (Barany and Merrifield, 1980). DCC is water-insoluble, but it has been used in 80% DMF for the immobilization of small molecules onto carboxylate-containing chromatography supports for use in affinity separations (Larsson and Mosbach, 1971; Gutteridge and Robb, 1973; Lowe *et al.*, 1973). In addition to forming amide linkages, DCC has been used to prepare active esters of carboxylate-containing compounds using NHS or sulfo-NHS (Staros, 1982). Unlike the EDC/sulfo-NHS reaction described in Section 1.2 (this chapter), active ester synthesis with DCC typically is carried out in organic solvent and therefore does not have the hydrolysis problems of water soluble EDC-formed esters. Thus, DCC is most often used to synthesize active ester containing crosslinking and modifying reagents and not to perform biomolecular conjugations.

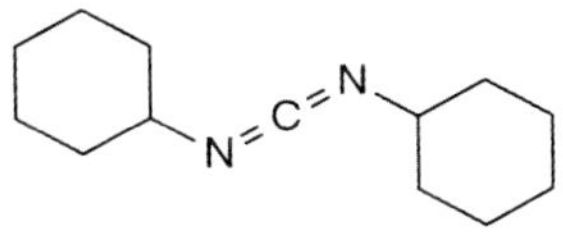

DCC
N,N'-Dicyclohexyl carbodiimide
MW 206.32

DCC is a waxy solid that is often difficult to remove from a bottle. Its vapors are extremely hazardous to inhalation and to the eyes. It should always be handled in a fume hood. The isourea byproduct of a DCC-initiated reaction, dicyclohexyl urea (DCU) (Figure 4.6), is also water insoluble and must be removed by organic solvent washing. For synthesis of peptides or affinity supports on insoluble matrices this is not a problem, because washing of the support material can be performed without disturbing the conjugate coupled to the support. For solution-phase chemistry, however, reaction products must be removed by solvent washings, precipitations, or recrystallizations.

A potential undesirable effect of DCC coupling reactions is the spontaneous rearrangement of the *O*-acylisourea to an inactive *N*-acylurea (Stewart and Young, 1984) (Figure 4.7). The rate of this rearrangement is dramatically increased in aprotic organic solvents, such as DMF. Another potential problem with using DCC to create active ester, such as in the synthesis of NHS esters, is a side reaction that can occur with NHS to create a bifunctional crosslinking agent in solution (Chapter 15, Section 2.1) (Wilchek and Miron, 1987). If a reaction involving the carbodiimide DCC with added NHS is used to create the NHS ester groups on a carboxylate support under nonaqueous conditions, as is also typical for synthesizing most NHS ester-containing bioconjugation reagents, the formation of a *bis*-NHS derivative of β-alanine can be created at the same time that NHS esters are being formed with the carboxylate spacers on the matrix. This side reaction results from the initial interaction of DCC with the hydroxyl on NHS to

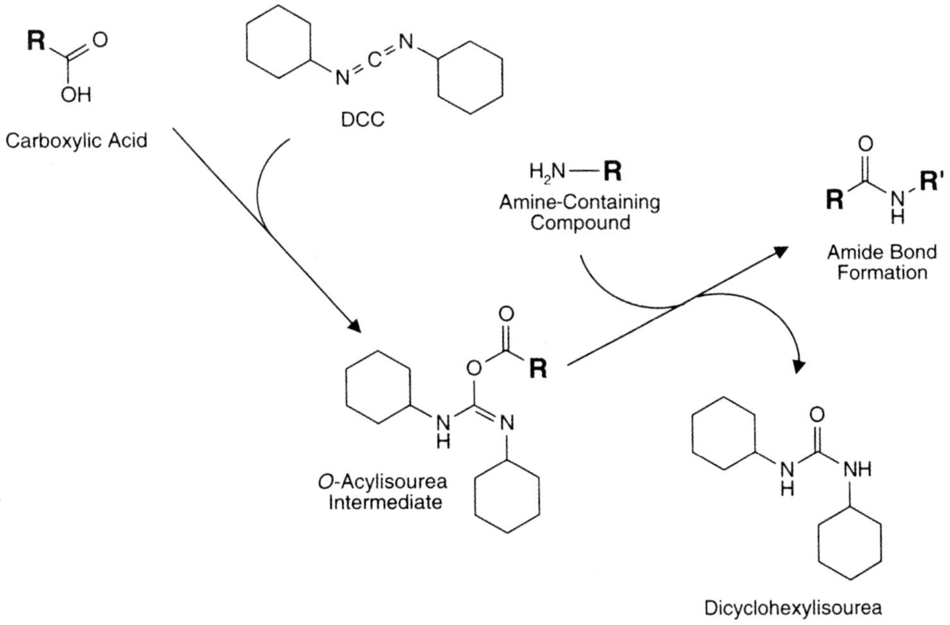

FIGURE 4.6 The organic-soluble carbodiimide DCC is often used to create amide bonds, especially between water-insoluble compounds.

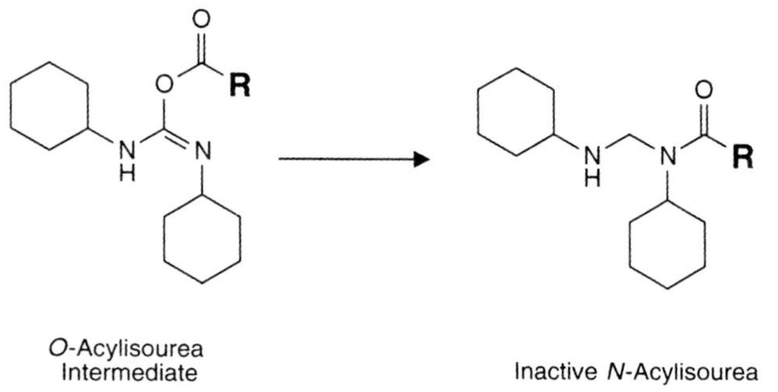

FIGURE 4.7 The active-ester intermediate formed from the reaction of DCC with a carboxylate group may undergo rearrangement to an inactive *N*-acylisourea product.

form a carbodiimide-activated NHS group. This ability of carbodiimides to react with hydroxyl groups correlates to the reported activation of hydroxyls on chitosan using the carbodiimide EDC (Chiou and Wu, 2004).

The activation efficiency of DCC is extraordinarily high, especially in anhydrous solutions that do not have competing hydrolysis problems. *O*-Acylisourea-activated carboxylates may undergo two side reactions that form other active groups. If DCC is added to an excess of a carboxylate-containing molecule without the presence of an amine-containing target, then the activated carboxylate may react with another carboxylic acid to form a symmetrical anhydride (Figure 4.8). The formation of an anhydride intermediate may be a frequent mechanism an route to the creation of an amide bond with an amine, especially under anhydrous conditions (Rebek and Feitler, 1974; Nakajima and Ikada, 1995). In addition, a DCC-activated carboxylate may react with an amino acid to form an azlactone (Figure 4.9) (Coleman *et al.*, 1990). Both the anhydride and the azlactone will react with amines to form covalent amide linkages. However, the ring-opening reaction of an azlactone will form a different product than the zero-length crosslinking result of coupling directly to an amine-containing molecule (Figure 4.10) (Chapter 15, Section 2.1). Finally, if DCC is used in the presence of NHS under anhydrous conditions, a bifunctional crosslinking product may result from a series of reactions that activate and then open up the NHS ring and form amine-reactive groups at both ends of a beta-alanine cross-bridge. See Chapter 15, Figure 15.42, for an illustration of this reaction.

1.5. DIC

DIC, or diisopropyl carbodiimide, is another water-insoluble amide bond-forming agent that has advantages over DCC (Section 1.4, this chapter). It is a liquid at room temperature and is therefore much easier to dispense than DCC. The byproducts of

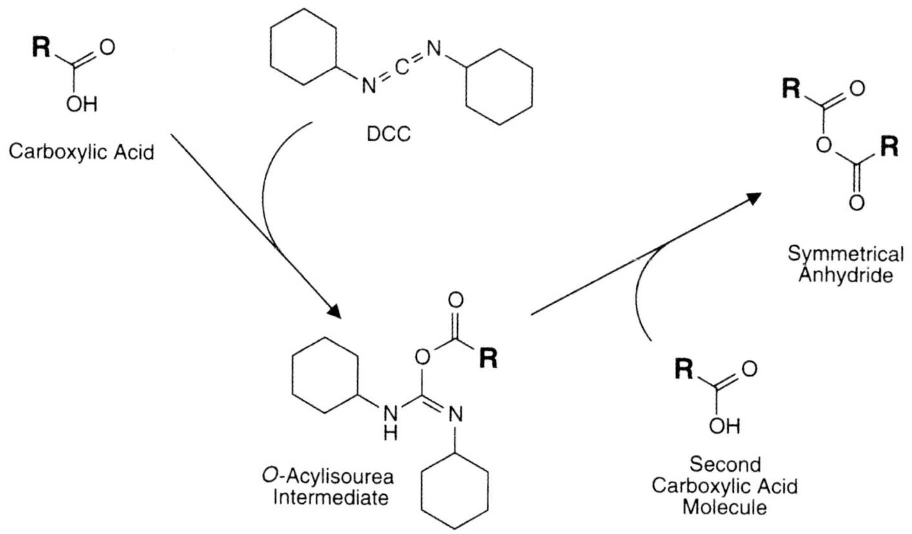

FIGURE 4.8 The reaction of DCC with a carboxylate compound in excess may create anhydride products in the absence of nucleophiles.

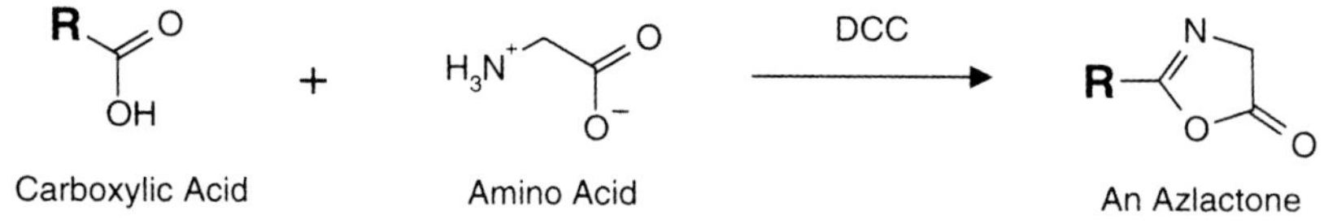

FIGURE 4.9 A DCC-mediated reaction with a carboxylate group in the presence of a small amino acid may form azlactone rings.

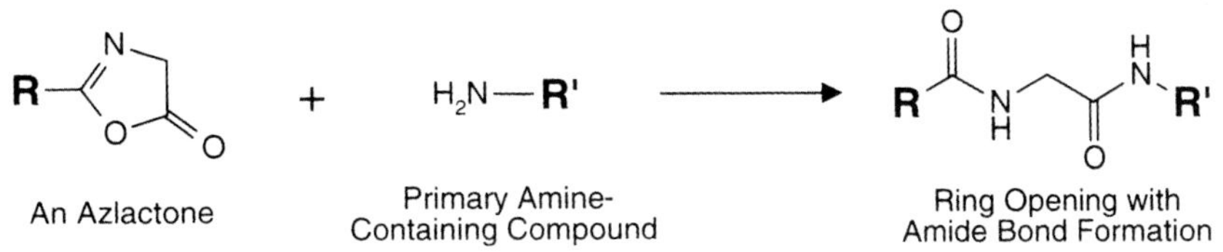

FIGURE 4.10 An azlactone reacts with an amine group through a ring-opening process, creating an amide bond linkage with the attacking nucleophile.

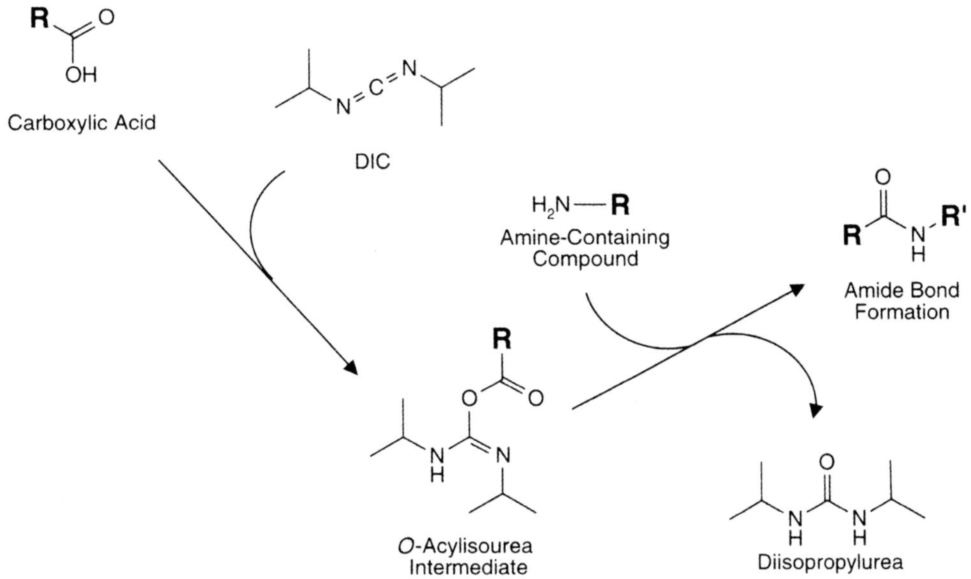

FIGURE 4.11 The symmetrical carbodiimide DIC reacts with carboxylates to form active-ester intermediates able to couple with amine-containing compounds to form amide bond linkages.

its activation reaction with a carboxylate, diisopropylurea and diisopropyl-*N*-acylurea, are more soluble in organic solvents than the DCU byproduct of a DCC reaction. DIC reacts similarly to DCC, forming an active *O*-acylisourea intermediate with a carboxylic acid group (Figure 4.11). This active species may then react with a nucleophile such as an amine to form an amide bond. Presumably, all the possible side reactions that DCC may undergo are also possible with DIC, although it is not well documented.

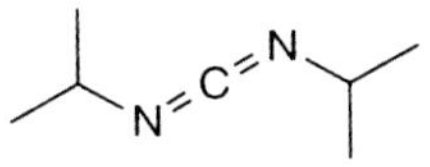

DIC
Diisopropyl carbodiimide
MW 126.2

2. WOODWARD'S REAGENT K

Woodward's reagent K is *N*-ethyl-3-phenylisoxazolium-3′-sulfonate, a zero-length crosslinking agent able to cause the condensation of carboxylates and amines to form amide bonds (Woodward *et al.*, 1961; Woodward and Olofson, 1961). The reaction mechanism involved in activating a carboxylate includes the conversion of the reagent under alkaline conditions to a reactive ketoketenimine. This intermediate then reacts with a carboxylate to create an enol ester. The enol ester is highly susceptible to nucleophilic attack. The reaction with an amine proceeds to amide bond formation with loss of the inactive diketo derivative (Figure 4.12). In aqueous solution, the major side reaction is hydrolysis, which occurs rapidly (Dunn and Affinsen, 1974). Although Woodward's reagent K has been used successfully for conjugation applications with proteins and other molecules to form amide linkages (Boyer, 1986; Pikuleva and Turko, 1989), its mechanism of reaction was called into question by Johnson and Dekker (1996), who found that the compound reacted with cysteine and histidine groups in *E. coli* L-threonine dehydrogenase, not the available aspartate or glutamate groups. Woodward's reagent K is available from Fluka.

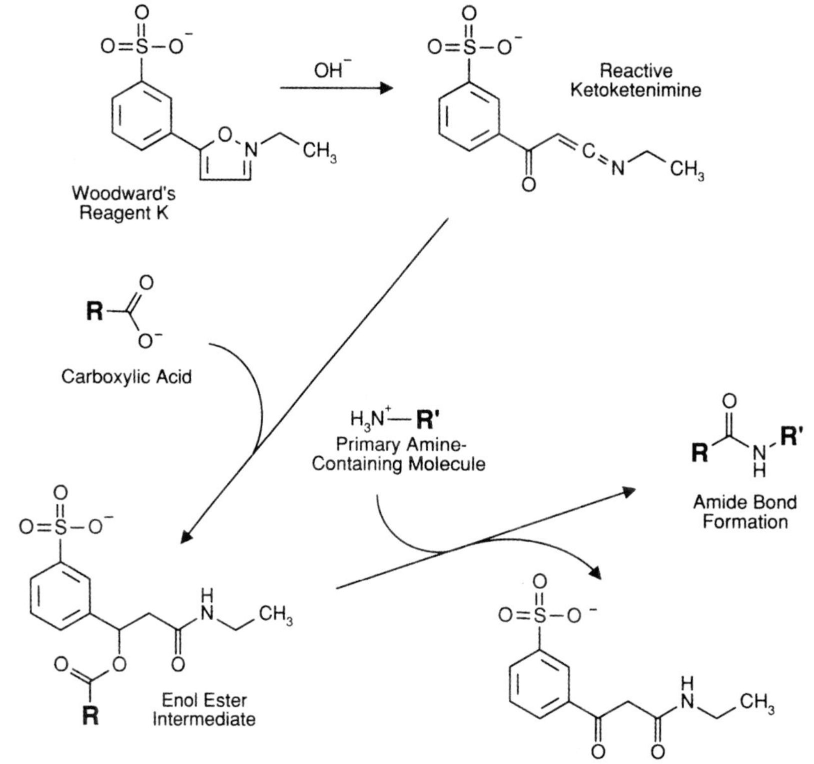

FIGURE 4.12 Woodward's reagent K undergoes a rearrangement in alkaline solution to form a reactive ketoketenimine. This active species can react with a carboxylate group to create another active group, an enol ester derivative. In the presence of amine nucleophiles, amide bond formation takes place.

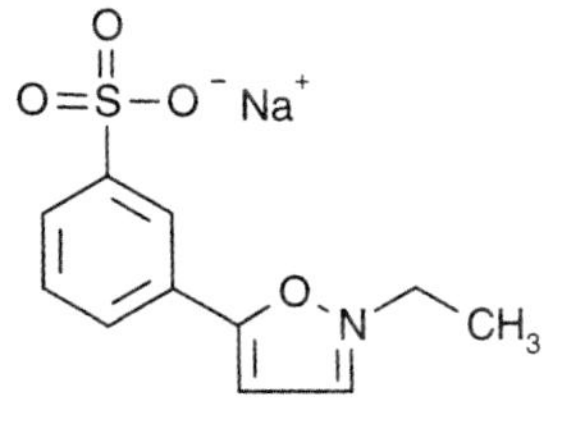

Woodward's Reagent K
N-Ethyl-5-phenylisoxazolium-3'-sulfonate, sodium salt
MW 176

3. N, N'-CARBONYL DIIMIDAZOLE

CDI, or *N,N'*-carbonyl diimidazole, is a highly active carbonylating agent that contains two acylimidazole leaving groups (Aldrich). The result is that CDI can activate carboxylic acids or hydroxyl groups for conjugation with other nucleophiles, creating either zero-length amide bonds or one-carbon-length *N*-alkyl carbamate linkages between the crosslinked molecules. Carboxylic acid groups react with CDI to form *N*-acylimidazoles of high reactivity. The active intermediate forms in excellent yield due to the driving force created by the liberation of carbon dioxide and imidazole (Anderson, 1958). The active carboxylate can then react with amines to form amide bonds or with hydroxyl groups to form ester linkages (Figure 4.13). Both reaction mechanisms have been used successfully in peptide synthesis (Paul and Anderson, 1960, 1962). In addition, activation of a styrene/4-vinylbenzoic acid copolymer with CDI was used to immobilize the enzyme lysozyme through its available amino groups to the carboxyl groups on the matrix (Bartling *et al.*, 1973). Other carboxylate-containing polymers have also been activated with CDI to form bioconjugates useful even in therapeutic applications (Peng *et al.*, 2011).

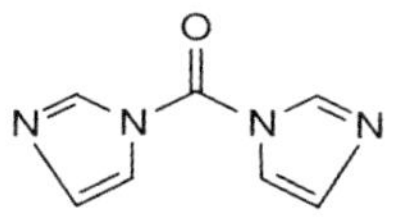

CDI
N,N'-Carbonyldiimidazole
MW 162

CDI functions as a zero-length crosslinker if the activated species is a carboxylic acid, because the attack of another nucleophile liberates the imidazole leaving group. However, if CDI is used to activate a hydroxyl functional group, the reaction proceeds quite differently. The active intermediate formed by the reaction of CDI with an OH group is an imidazolyl carbamate

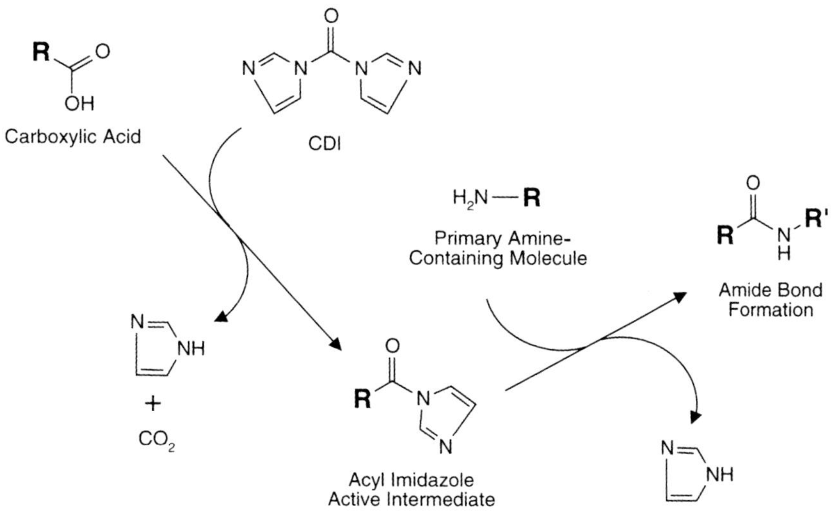

FIGURE 4.13 CDI reacts with carboxylate groups to form an active acylimidazole intermediate. In the presence of an amine nucleophile, amide bond formation can take place with release of imidazole.

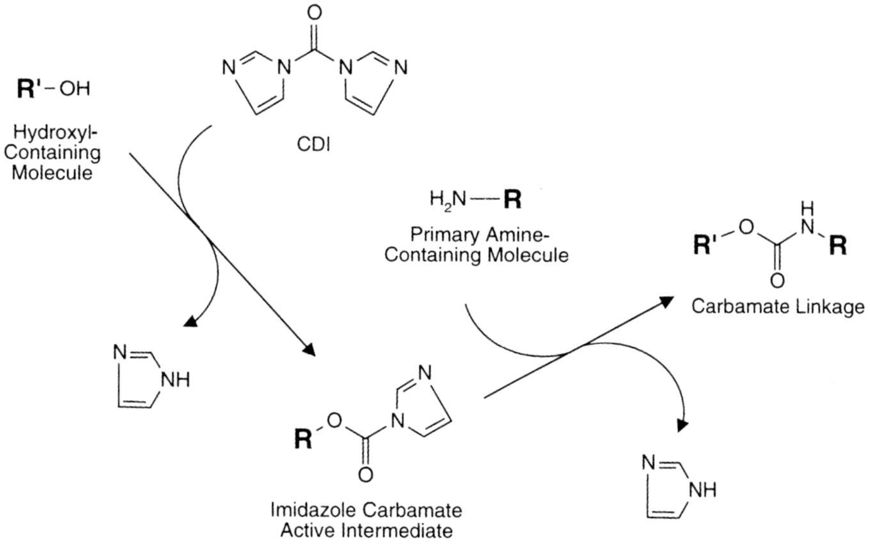

FIGURE 4.14 CDI reacts with hydroxyl groups to form an active imidazole carbamate intermediate. In the presence of amine-containing compounds, a carbamate linkage is created with loss of imidazole.

(Figure 4.14). Attack by an amine releases the imidazole, but not the carbonyl. Thus, a hydroxyl-containing molecule may be coupled to an amine-containing molecule with the result of a one-carbon spacer and formation of a stable urethane (*N*-alkyl carbamate) linkage. This coupling procedure has been applied to the activation of hydroxyl-containing chromatography supports for the immobilization of amine-containing affinity ligands (Chapter 15, Section 2.1; Chapter 14, Section 4.2) (Bethell *et al.*, 1979; Hearn *et al.*, 1979, 1983; Hermanson *et al.*, 1992) and also to the activation of polyethylene glycol for the modification of amine-containing macromolecules (Chapter 18, Section 2.1) (Beauchamp *et al.*, 1983).

CDI-activated hydroxyls may also undergo a side reaction to form active carbonates. This occurs when an imidazolyl carbamate reacts with another hydroxyl group before the second hydroxyl has had a chance to get activated with CDI. Particularly with adjacent hydroxyls on the same molecule, this can be a problem if a defined reactive species is desired. Any carbonates formed, however, are still reactive toward amines to create carbamate linkages.

Formation of the activated species, whether with a carboxylate or a hydroxyl, must take place in nonaqueous environments due to the rapid breakdown of CDI by hydrolysis. Even in solvents containing small amounts of water, CDI quickly hydrolyzes to

CO_2 and imidazole. It is best to use solvents with less than 0.1% water to prevent extensive CDI breakdown. Characteristic bubble formation is an indication of reagent hydrolysis, although CO_2 also is released upon reaction with a carboxylic acid. Activation of carboxylates or hydroxyls may be performed in dry organic solvents such as acetone, dioxane, DMSO, THF, DMAC, or DMF. If an excess of CDI is used during the activation step, it should be removed before adding the active intermediate to an amine-containing molecule for conjugation. Alternatively, equal molar quantities of CDI and the molecule to be activated may be mixed to form the active species. After about an hour of activation, add an equivalent molar quantity of the amine-containing target molecule to be conjugated.

Aqueous reaction conditions that result in the best conjugation yields using CDI usually reflect the relative pK_a of the nucleophilic amine being coupled. Proteins are best coupled to CDI-activated supports or molecules in an environment at least one pH unit above their pI values. Frequently, the greatest coupling yields occur in alkaline buffers within the range of pH 8.0 to 10.0. In aqueous solutions, CDI-activated carboxylates or hydroxyls will hydrolyze and slowly lose activity. *N*-Acylimidazoles hydrolyze by loss of imidazole and regenerate the original carboxylate. The imidazole carbamate active species hydrolyzes by loss of CO_2 and imidazole, regenerating, in this case, the original hydroxyl group. CDI-activated carboxylic acids hydrolyze faster in aqueous solutions than CDI-activated hydroxyls; however, both reactive intermediates undergo increasing hydrolysis with increasing pH.

Conjugation or immobilization reactions using CDI may also be performed in organic solutions (Chapter 15, Section 2.1). This is a distinct advantage if the reactants are not very soluble in aqueous environments. In addition, organic coupling will not result in concomitant loss of activity due to hydrolysis as water-based reactions, thus nonaqueous reactions will usually provide greater yields.

A protocol for the use of CDI in the activation of poly(ethylene glycol) is discussed in Chapter 18, Section 2.1, while CDI activation procedures for particles are described in Chapters 14 and 15.

4. SCHIFF BASE FORMATION AND REDUCTIVE AMINATION

Aldehydes and ketones can react with primary and secondary amines to form Schiff bases. A Schiff base is a relatively labile bond that is readily reversed by hydrolysis in aqueous solution. The formation of Schiff bases is enhanced at alkaline pH values, but they are still not completely stable unless reduced to secondary

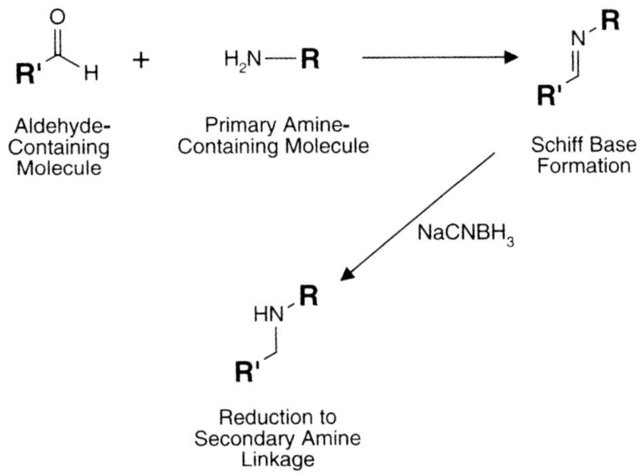

FIGURE 4.15 Carbonyl groups such as in aldehydes and ketones can react with amine nucleophiles to form reversible Schiff base intermediates. In the presence of a suitable reductant, such as sodium cyanoborohydride, the Schiff base is stabilized to a secondary amine bond.

or tertiary amine linkages (Figure 4.15). A number of reducing agents can be used to convert specifically the Schiff base into an alkylamine linkage. Once reduced, the bonds are highly stable. The use of reductive amination to conjugate an aldehyde-containing molecule to an amine-containing molecule results in a zero-length crosslink where no additional spacer atoms are introduced between the molecules.

Reductive amination (or alkylation) may be used to conjugate an aldehyde- or ketone-containing molecule with an amine-containing molecule. The reduction reaction is best facilitated by the use of a reducing agent such as sodium cyanoborohydride, because the specificity of this reagent is toward the Schiff base structure and will not affect the original aldehyde groups. By contrast, sodium borohydride is also used in this reaction, but its strong reducing power rapidly converts any aldehydes not yet reacted into nonreactive hydroxyls, effectively eliminating them from further participation in the conjugation process. Borohydride may also affect the activity of some sensitive proteins, whereas cyanoborohydride is gentler, successfully preserving the activity of even some labile monoclonal antibodies. Cyanoborohydride has been shown to be at least five times milder than borohydride in reductive amination processes with antibodies (Peng *et al.*, 1987). Other reducing agents that have been explored for reductive amination include various amine boranes and ascorbic acid (Cabacungan *et al.*, 1982; Hornsey *et al.*, 1986).

Immobilization by reductive amination of amine-containing biological molecules onto aldehyde-containing solid supports has been used for quite some time (Sanderson and Wilson, 1971). The reaction proceeds with excellent efficiency (Domen *et al.*, 1990). The optimum pH for the reaction is alkaline, although good yield can be realized from pH 7 to 10. At the high end of this range (pH 9–10), the formation of the Schiff bases is more efficient, and the yield of conjugation or immobilization reactions can be dramatically increased (Hornsey *et al.*, 1986) (see Chapter 15, Section 2.1).

The introduction of aldehyde functional groups into proteins and other molecules can be accomplished by a number of methods (Chapter 2, Section 4.4). Glycoproteins may be oxidized at their carbohydrate residues using sodium periodate or by using a specific sugar oxidase. Amine groups may be modified to produce a formyl group by reacting with NHS–aldehyde compounds (Chapter 17, Section 2) or *p*-nitrophenyl diazopyruvate. The following generalized protocol assumes that the requisite groups are present on the two molecules to be conjugated.

Protocol

1. Dissolve the amine-containing protein to be conjugated at a concentration of 1 to 10 mg/ml in a buffer having a pH between 7 and 10. Higher pH reactions will result in greater yield of conjugate formation. Suitable buffers include 0.1-*M* sodium phosphate, 0.15-*M* NaCl, pH 7.2; 0.1-*M* sodium borate, pH 9.5; or 0.05-*M* sodium carbonate, 0.1-*M* sodium citrate, pH 9.5. Avoid amine-containing buffers like Tris.
2. Add a quantity of the aldehyde-containing molecule to the solution in step 1 to obtain the desired molar ratio for conjugation. For instance, if the amine-containing protein is an antibody and the aldehyde-containing protein is an enzyme such as horseradish peroxidase (HRP), a typical molar ratio for the reaction might be 2 to 4 moles of HRP per mole of antibody.
3. Add 10 μl of 5-*M* sodium cyanoborohydride in 1-*N* NaOH (Aldrich) per ml of the conjugation solution volume. *Caution: Highly toxic compound. Use a fume hood and be careful to avoid skin contact with this reagent.*
4. React for 2 h at room temperature.
5. To block unreacted aldehyde sites, add 20 μl of 3-*M* ethanolamine (pH adjusted to desired value with HCl) per ml of the conjugation solution volume. React for 15 min at room temperature.
6. Purify the conjugate by dialysis or gel filtration using a buffer suitable for the nature of the proteins being crosslinked.

CHAPTER

5

Homobifunctional Crosslinkers

The first crosslinking reagents used for modification and conjugation of macromolecules consisted of bireactive compounds containing the same functionality at both ends (Hartman and Wold, 1966). Most of these homobifunctional reagents were symmetrical in design with a carbon chain spacer connecting the two identical reactive ends (Figure 5.1). Like molecular rope, these reagents could tie one protein to another by covalently reacting with the same common groups on both molecules. Thus, the lysine ε-amines or N-terminal amines of one protein could be crosslinked to the same functionalities on a second protein simply by mixing the two together in the presence of the homobifunctional reagent.

The ability to so easily link two proteins or other molecules having different binding specificities or catalytic activities opened the potential for creating a new universe of unique and powerful reagent systems for use in assay and targeting applications. The variety and reactivity of homobifunctional reagents multiplied dramatically throughout the 1970s and 1980s. Today, there are dozens of commercially available crosslinkers possessing almost every length and reactivity desired.

The main disadvantage, however, of using simple homobifunctional reagents is the potential for creating a broad range of poorly defined conjugates (Avramease, 1969). When crosslinking two proteins, for example, the

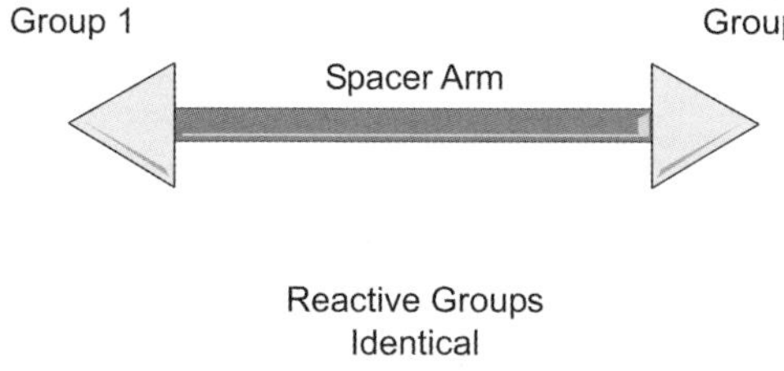

FIGURE 5.1 The general design of a homobifunctional crosslinking agent. The two reactive groups are identical and typically are located at the ends of an organic spacer arm. The length of the spacer may be designed to accommodate the optimal distance between two molecules to be conjugated.

reagent may react initially with either one of the proteins, forming an active intermediate. This activated protein may form crosslinks with the second protein or react with another molecule of the same type. It also may react intramolecularly with other functional groups on part of its own polypeptide chain. In addition, other crosslinking molecules may continue to react with these intermediates to form various mixed oligomers, including severely polymerized products that may even precipitate (see Chapter 2, Section 1.2).

The problem of poorly defined conjugation products is exacerbated in single-step reaction procedures using homobifunctional reagents. Single-step procedures involve the addition of all reagents at the same time to the reaction mixture. This technique provides the least control over the crosslinking process and invariably leads to a multitude of products, only a small percentage of which represent the desired conjugate. Excessive conjugation may cause the formation of insoluble complexes that consist of very high-molecular-weight polymers. For example, one-step glutaraldehyde conjugation of antibodies and enzymes (Chapter 20, Section 1.2) often results in significant oligomers and precipitated conjugates. To overcome this shortcoming, two-step reaction procedures have been developed using homobifunctional reagents. Controlled, two-step conjugation protocols somewhat alleviate the polymerization problem with homobifunctional reagents, but can never totally avoid it.

In two-step protocols, one of the proteins to be conjugated is reacted with the homobifunctional reagent and excess crosslinker and byproducts are removed. In the second stage, the activated protein is mixed with the other protein or molecule to be conjugated, and the final conjugation process occurs (Figure 5.2).

One potential problem of such two-step procedures is hydrolysis of the activated intermediate before addition of the second molecule to be conjugated. For instance, *N*-hydroxysuccinimide (NHS) ester homobifunctional reagents hydrolyze in aqueous buffers and may degrade before the second stage of the crosslinking is initiated. In addition, the use of homobifunctional reagents in

Bioconjugate Techniques, Third Edition.
DOI: http://dx.doi.org/10.1016/B978-0-12-382239-0.00005-4

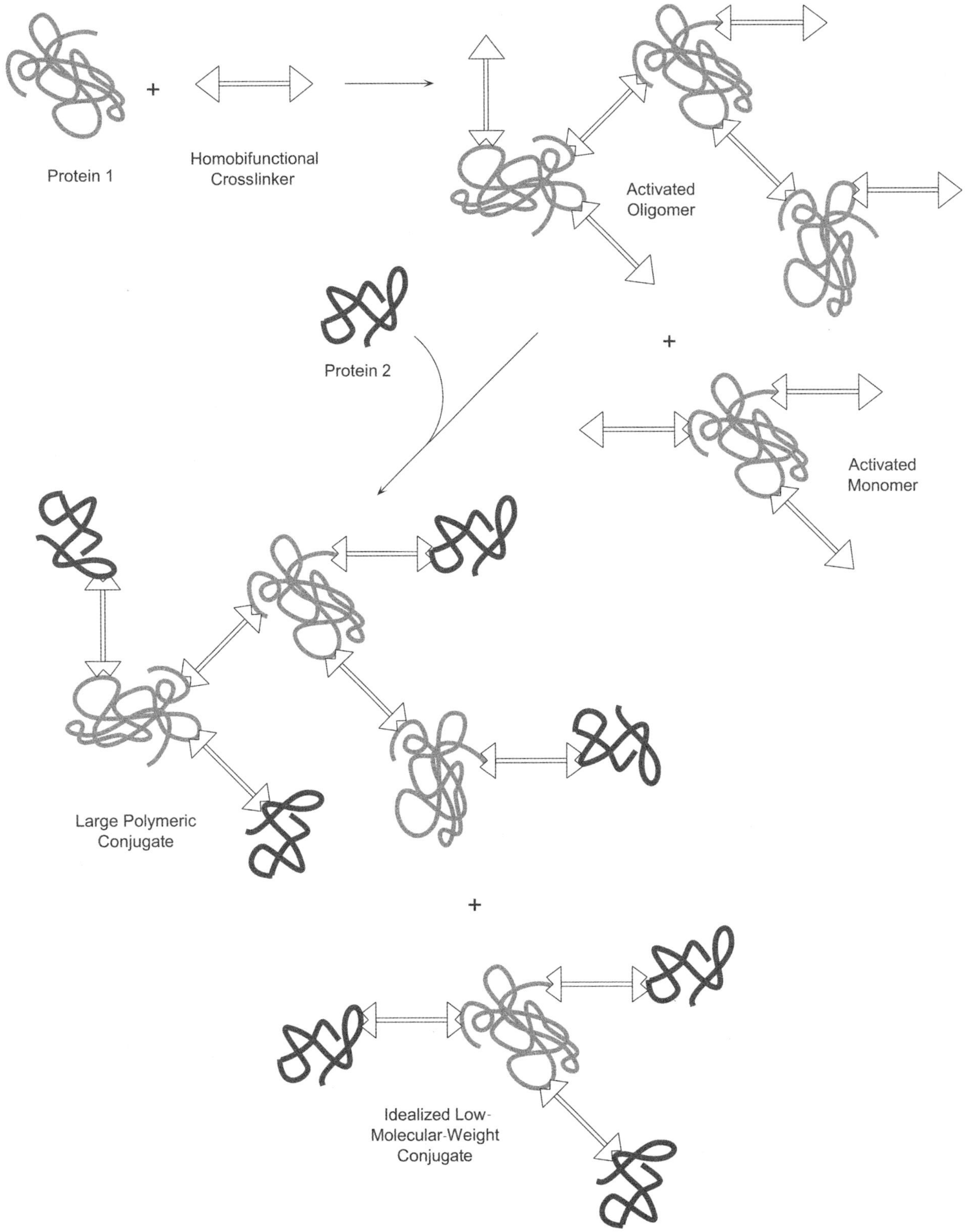

FIGURE 5.2 Homobifunctional crosslinkers may be used in a two-step process to conjugate two proteins or other molecules. In the first step, one of the two proteins is reacted with the crosslinker in excess to create an active intermediate. After removal of remaining crosslinker, a second protein is added to effect the final conjugate. Two-step reaction schemes somewhat limit the degree of polymerization obtained when using homobifunctional reagents, but can't entirely prevent it.

two-step protocols still produces many of the problems associated with single-step procedures, because the first protein can crosslink and polymerize with itself long before the second protein is added. Since the first protein to be activated has target functional groups on every molecule that can couple with both reactive groups of the crosslinker, both ends of the reagent potentially can react. This inherent capacity to polymerize uncontrollably is unfortunately a characteristic of all homobifunctional reagents, even in multi-step protocols.

Although their shortcomings in this regard are clearly recognized, homobifunctional reagents continue to be popular choices for all kinds of conjugation applications, including in the preparation of immobilized antibodies used for biosensor surfaces (Hansberry and Clark, 2012) and in the generation of complex bispecific antibodies (Ellerman and Scheer, 2011). The fact is, in many crosslinking functions, they work well enough to form effective conjugates. Even glutaraldehyde-mediated antibody–enzyme conjugates are still commonly utilized in everything from research to diagnostics.

The particular crosslinkers discussed in this section are the types most often referred to in the literature or are commercially available. Many other forms of homobifunctional reagents containing almost every conceivable chain length and reactivity are mentioned in the scientific literature.

1. HOMOBIFUNCTIONAL NHS ESTERS

Carboxylate groups activated with *N*-hydroxysuccinimide (NHS) esters are highly reactive toward amine nucleophiles. In the mid-1970s, NHS esters were introduced as reactive ends of homobifunctional crosslinkers (Bragg and Hou, 1975; Lomant and Fairbanks, 1976). Their excellent reactivity at physiological pH quickly established NHS esters as viable alternatives to the imidoesters predominating at the time (Section 2, this chapter).

Unfortunately, many NHS ester-containing crosslinkers are insoluble in aqueous buffers. Most protocols involve dissolving the compound at a relatively high concentration in an organic solvent and transferring the required quantity into the reaction medium. Prior dissolution helps to maintain at least some solubility in the buffered crosslinking environment. Most of the time, however, the addition of an organic-solvent-solubilized crosslinker into a buffered solution will result in a microprecipitate that slowly goes into solution during the course of reaction.

In the early 1980s, Staros prepared a derivative of NHS that aids in the water solubility of NHS ester crosslinkers (Staros, 1982). *N*-Hydroxysulfosuccinimide (sulfo-NHS) esters possess a negatively charged sulfonate group on carbon number 2 or 3 of the succinimido ring (Figure 5.3).

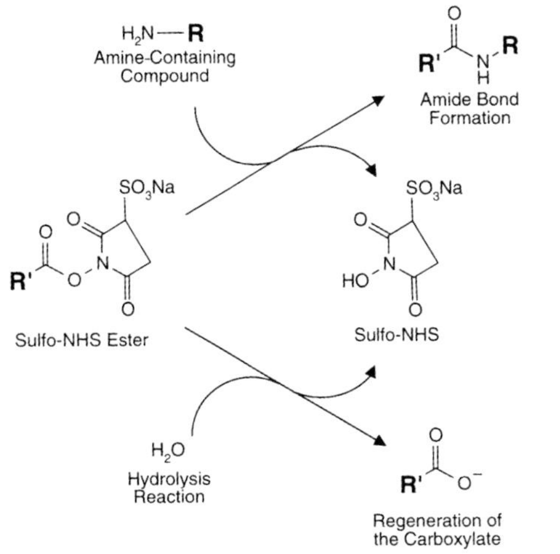

FIGURE 5.3 In aqueous solution, a sulfo-NHS ester can either couple to an amine group to form an amide bond or react with water to hydrolyze back to a carboxylate. Both processes release the sulfo-NHS leaving group.

Crosslinking reagents containing sulfo-NHS esters have half-lives of hydrolysis of the order of hours, sometimes even better than their NHS ester analogs (Anjaneyula and Staros, 1987). The sulfo-NHS ester often lends enough charge and polarity to a crosslinker to provide water solubility and thus eliminate the need for organic solvent dissolution. In addition, sulfo-NHS crosslinkers may be used for surface-only modification of membranes and cells, since they are more hydrophilic and will not penetrate the lipid environment of the membrane (Robertson *et al.*, 1986). By contrast, many of the more hydrophobic NHS ester crosslinkers can be used to traverse the cell membrane and modify intracellular components.

NHS- or sulfo-NHS ester-containing homobifunctional crosslinkers react with nucleophiles to release the NHS or sulfo-NHS leaving group and form an acylated product. The reaction of such esters with a sulfhydryl or hydroxyl group is possible but does not yield stable conjugates, as it forms thioesters and ester linkages which may hydrolyze in aqueous environments. Histidine side-chain nitrogens of the imidazolyl ring also may be acylated with an NHS ester reagent, but they too hydrolyze rapidly (Cuatrecasas and Parikh, 1972). Reaction with primary and secondary amines, however, creates stable amide and imide linkages, respectively, that don't readily break down. In protein molecules, NHS ester crosslinking reagents primarily react with the α-amines at the N-terminals and the abundant ε-amines of lysine side chains.

NHS ester crosslinking reactions in aqueous solutions offer the potential for hydrolysis as well as the desired amide bond formation. Crosslinkers containing NHS esters have fairly good stability in aqueous solutions, despite their susceptibility to attack and breakdown by water. Studies done on the NHS ester-containing homobifunctional reagent, dithiobis(succi-nimidylpropionate) (DSP), indicate that the activated carboxylates have half-lives on the order of hours at physiological pH. However, hydrolysis and amine reactivity both increase with increasing pH. At 0°C at pH 7.0, the half-life of the crosslinking reagent DSP is 4 to 5h (Lomant and Fairbanks, 1976). At pH 8.0 at 25°C it falls to about 1h (Staros, 1988), and at pH 8.6 and 4°C the half-life is only 10min (Cuatrecasas and Parikh, 1972).

The rate of hydrolysis may be monitored by measuring the increase in absorption at 260nm as the NHS leaving group is cleaved. The molar extinction coefficient of the NHS group in solution is $8.2 \times 10^3 M^{-1} cm^{-1}$ in Tris buffer at pH 9.0 (Carlsson *et al.*, 1978), but somewhat decreases to $7.5 \times 10^3 M^{-1} cm^{-1}$ in potassium phosphate buffer at pH 6.5 (Partis *et al.*, 1983). Unfortunately, the sensitivity of assay using absorption usually does not allow for measuring the rate of reaction in an actual crosslinking procedure.

To maximize the modification of amines and minimize the effects of hydrolysis, maintain a high concentration of protein or other target molecule. By adjusting the molar ratio of crosslinker to target molecule(s), the level of modification and conjugation may be controlled to create an optimal product.

The reaction buffer chosen for the conjugation reaction should be free of extraneous amines. Avoid Tris or glycine buffers. Also, avoid imidazole buffers, since the nitrogens of the imidazole ring may react with the active ester and then quickly hydrolyze. The effect is that imidazole only acts to catalyze the hydrolysis process. The pH of the reaction should be in the range of 7 to 9 to promote the unprotonated state of primary amines, which is the nucleophilic species that most effectively attacks the activated carbonyl group. Dissolve NHS ester crosslinkers that are insoluble in water in an organic solvent such as DMF or DMSO prior to addition to the reaction medium. Sulfo-NHS crosslinkers may be added directly to the reaction mixture or pre-dissolved in buffer at a higher concentration before adding an aliquot to the reaction. Aqueous stock solutions should be used immediately to prevent extensive hydrolysis of the active esters.

NHS ester crosslinking reagents also may be used in organic solvent-based reactions without the competing hydrolysis problem provided the target molecules are soluble and stable in such environments. In this case, both molecules to be conjugated must be soluble in the solvent. DMF, DMSO, acetone, and dioxane are examples of solvents that can be utilized as the reaction medium. Refer to any published solubility data on the crosslinking reagent of choice to see which solvents are most appropriate.

1.1. DSP and DTSSP

Lomant's reagent [(dithiobis(succinimidylpropionate), or DSP]) is a homobifunctional NHS ester crosslinking agent containing an eight-atom spacer 12Å in length (Lomant and Fairbanks, 1976) (Thermo Fisher). It is symmetrically constructed around a central disulfide group that is cleavable after conjugation with typical disulfide reducing agents (Chapter 2, Section 4.1).

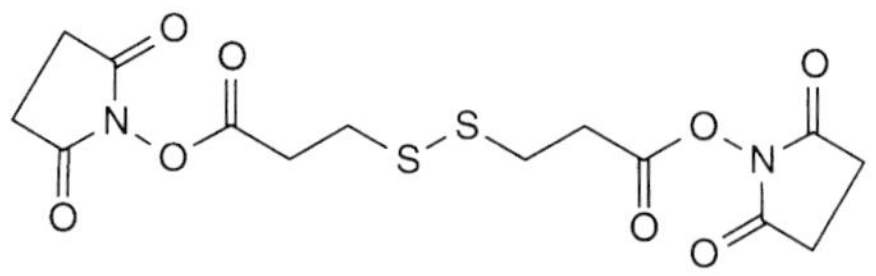

DSP
Dithiobis(succinimidylpropionate)
MW 404.42
12 Å

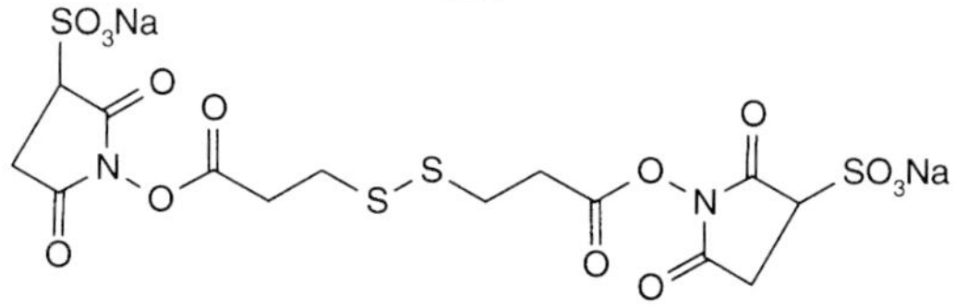

DTSSP
3,3'-Dithiobis(sulfosuccinimidylpropionate)
Water Soluble
MW 608.51
12 Å

DSP is water insoluble and must be pre-dissolved in an organic solvent before addition to a conjugation reaction. Concentrated stock solutions may be prepared in DMF or DMSO and an aliquot added to a buffered reaction medium. The NHS ester reaction occurs most efficiently at pH 7 to 9, with hydrolysis of the active species accelerating the greater the pH. The crosslinking buffers should be free of amine-containing components other than the target molecules to be conjugated. Avoid Tris or glycine buffers. A reaction buffer consisting of 0.1-*M* sodium phosphate, 0.15-*M* NaCl, pH 7.2 to 7.5 works well for most applications involving the crosslinking of two purified proteins. The relatively high concentration of sodium phosphate is used to prevent pH drift downward during the course of the reaction. For *in vitro* crosslinking of cellular components such as membrane proteins, a more dilute PBS buffer containing isotonic saline is more appropriate. Since DSP is a hydrophobic reagent, it is able to penetrate the cell membrane and conjugate membrane components. For this reason, it has become quite popular for use in investigating the interactions of membrane proteins.

DSP reacts with ε-amine groups on the side chains of lysine residues or the α-amine at the N-terminal of proteins to form amide linkages. Amine-containing macromolecules may be reversibly crosslinked with this reagent and later cleaved with dithiothreitol (DTT) or 2-mercaptoethanol (Figure 5.4). For reductive cleavage of conjugated molecules, add 10- to 50-m*M* DTT, and incubate

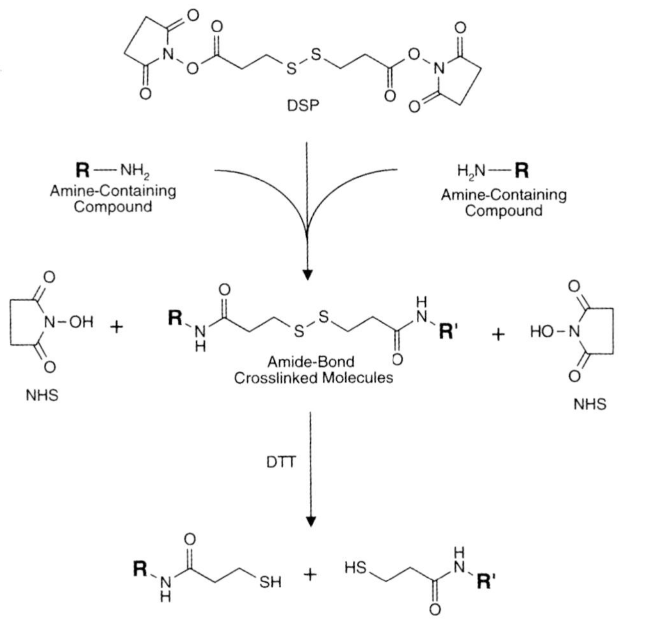

FIGURE 5.4 The reaction of DSP with amine-containing molecules yields amide bond crosslinks. The conjugates may be cleaved by reduction of the disulfide bond in the cross-bridge with DTT.

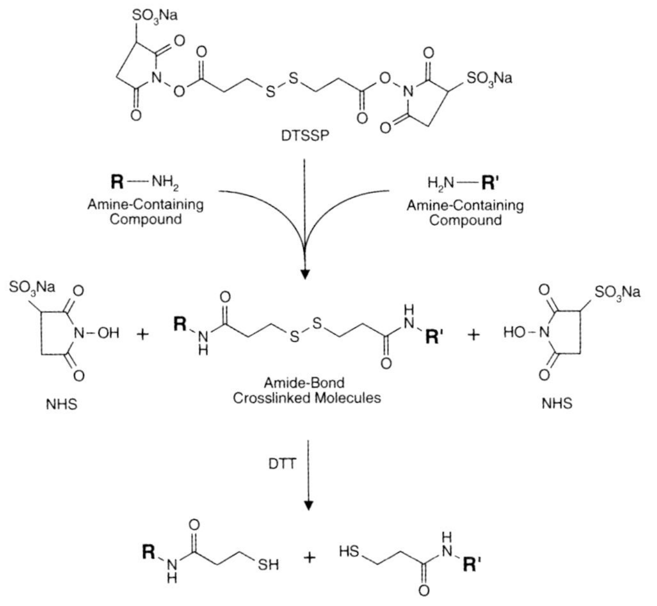

FIGURE 5.5 DTSSP can form crosslinks between two amine-containing molecules through amide linkages. The conjugates may be cleaved by disulfide reduction using DTT.

at 37°C for 30 min. Alternatively, the conjugate may be reduced prior to electrophoresis using SDS sample buffer with 5% 2-mercaptoethanol at elevated temperatures.

Lomant's reagent is one of the most popular of all crosslinking agents, especially for the investigation of protein interactions. Hordern *et al.* (1979) used it to investigate the spatial relationships in the capsid polypeptides of the *Mengo* virion. It has been used to study the interactions of proteins involved with active transport (dePont *et al.*, 1980; Joshi and Burrows, 1990), to identify crosslinks involving cytochrome P-450 (Baskin and Yang, 1982), to characterize cell-surface receptors for colony-stimulating factor (Park *et al.*, 1986), to determine various membrane antigens by crosslinking to specific monoclonal antibodies (Hamada and Tsuro, 1987), to study prothrombin self-association (Tarver *et al.*, 1982), to investigate chemotaxis in *E. coli* (Chelsky and Dahlquist, 1980), for the molecular identification of receptors for vasoactive intestinal peptide in rat intestinal epithelium (Laburthe *et al.*, 1984), to study the crosslinking of the affinity purified CCAAT transcription factor α-CP1 (Kim and Sheffrey, 1990), and to fix tissue for immunostaining procedures (Xiang *et al.*, 2004). The dithiol bridge in DSP has also been used to activate metal surfaces or metal particles through dative bonding (Grubor *et al.*, 2004). The presence of the disulfide group creates two dative bonds in each linker, which significantly increases the stability of the surface bonds. DSP's two NHS ester groups then can be used to couple amine-containing molecules to the metallic surface. See Chapter 3, Section 2.8, for additional information.

The sulfo-NHS version of DSP, dithiobis(sulfosuccinimidylpropionate) or DTSSP, is a water soluble analog of Lomant's reagent that can be added directly to aqueous reactions without prior organic solvent dissolution (Staros, 1982). DTSSP still contains the disulfide center portion that is cleavable with the proper reducing agents, and the sulfo-NHS ends have virtually the same reactivity as DSP (Figure 5.5). Due to its hydrophilicity, however, DTSSP will not penetrate cellular membranes as does its more hydrophobic analog, DSP. It is therefore an excellent choice for the crosslinking of cell-surface components without affecting intracellular substances.

DTSSP reportedly has been used to crosslink the extracytoplasmic domain of the anion exchange channel in human erythrocytes (Staros and Kakkad, 1983), for the characterization of a ribosomal complex in *B. subtilis* (Caufield *et al.*, 1984), to investigate ascites hepatoma cytokeratin filaments (Knoller *et al.*, 1991), to study the B lymphocyte Fc receptor for IgE (Waugh *et al.*, 1989), and to crosslink platelet glycoproteins (Jung and Moroi, 1983). Both DSP and DTSSP have been used extensively to study protein interactions and to investigate the oligomeric state of many proteins within cells (Sampathkumar *et al.*, 2012).

1.2. DSS and BS^3

Disuccinimidyl suberate (DSS) is an amine-reactive, homobifunctional, NHS ester, crosslinking reagent that produces an eight-atom bridge (11.4 Å) between conjugated molecules (Figure 5.6) (Thermo Fisher). Its hydrocarbon

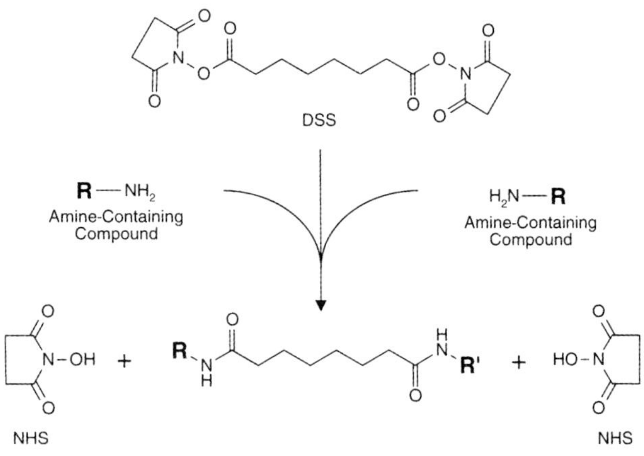

FIGURE 5.6 DSS reacts with two amine-containing molecules to form amide bond crosslinks. The crossbridge is noncleavable.

chain is non-cleavable, so crosslinks formed are irreversible. Many of the reported applications of DSS involve investigations of receptor–ligand binding on cell surfaces using radiolabeled molecules. The crosslinker is hydrophobic and must be solubilized in organic solvent prior to addition to a conjugation reaction. Pre-dissolving in dry dioxane, DMF, or DMSO may be done at higher concentration and then an aliquot added to the aqueous reaction medium as needed. The final concentration of the organic solvent in the buffered reaction should not exceed 10% to avoid precipitation of biomolecules. Stock solutions should be prepared fresh. DSS is membrane permeable and is therefore useful for intracellular and intramembrane conjugations. The optimum conditions for the crosslinking reaction are a pH range of 7 to 9 using buffers and other salts which contain no amines. Avoid the use of Tris or glycine. A phosphate buffer (PBS) at physiological pH works well. See Section 1 for additional information on NHS ester reactions.

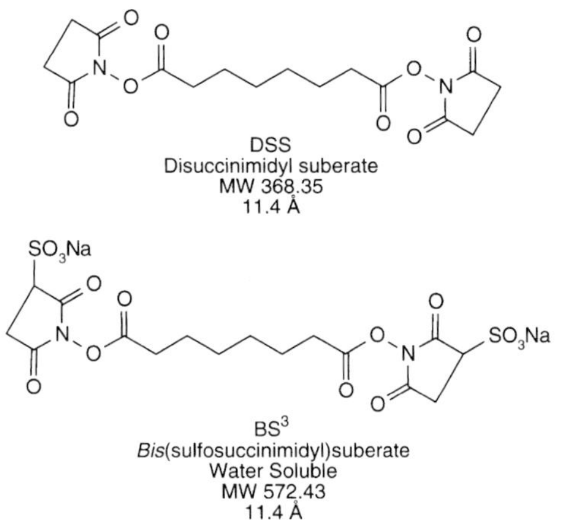

Reported applications of DSS include crosslinking the A and B subunits of ricin (Montesano *et al.*, 1982), studying human somatotropin and the components of the lactogenic binding sites of rat liver (Caamano *et al.*, 1983), crosslinking CSF-1 to its cell-surface receptor (Morgan and Stanley, 1984), studying angiotensin II interactions with its receptor (Petruzzelli *et al.*, 1985), crosslinking of vasoactive intestinal peptide to its receptor on human lymphoblasts (Wood and O'Dorisio, 1985), investigating insulin-dependent protein kinases (Petruzzelli *et al.*, 1985), identifying a cellular receptor for TNF (Kull *et al.*, 1985), affinity crosslinking of atrial natriuretic factor in aorta membranes (Vandelen *et al.*, 1985), studying the receptor for human interferon (Rashidbaigi *et al.*, 1986), crosslinking of endorphin to membranes rich in opioid receptors (Helmeste *et al.*, 1986), immunoprecipitation studies of the crosslinked complex of parathyroid hormone with its receptor (Wright *et al.*, 1987), binding of human interferon Y to its receptor (Novick *et al.*, 1987), identifying the erythropoietin receptor on Friend virus-infected erythroid cells (Sawyer *et al.*, 1987), and binding of the p75 peptide to an interleukin 2 receptor (Tsudo *et al.*, 1987).

Bis(sulfosuccinimidyl) suberate (BS^3) is an analog of DSS that contains sulfo-NHS esters on both carboxylates. The affect of the negative charges provided by the sulfonate groups lends water solubility to the compound. Prior organic solvent dissolution (before addition to a reaction) is not necessary. The hydrophilicity of BS^3 also makes it membrane impermeable. Therefore, cell labeling with BS^3 results in hydrophilic-region modification and crosslinking, targeting surface functionalities, whereas DSS is capable of targeting hydrophobic regions within the membrane structure itself. As with DSS, BS^3 is non-cleavable, and thus all crosslinks formed are irreversible. The reactivity of the sulfo-NHS esters is identical to NHS esters, being highly reactive toward amines in the pH range of 7 to 9.

Reported applications of BS^3 include crosslinking of the β-endorphin–calmodulin interaction (Staros, 1982), crosslinking of the extracellular domain of intact human erythrocytes' anion exchange channel (Staros and Kakkad, 1983), crosslinking of hepatoma cytokeratin filaments (Ward *et al.*, 1985), investigating the β-lymphocyte Fc receptor for IgE (Lee and Conrad, 1985; Staros *et al.*, 1987), crosslinking of the tripeptide Arg–Gly–Asp to an adhesion receptor on platelets (Souza *et al.*, 1988), crosslinking of the large and small subunits of cytochrome b_{559} (Knoller *et al.*, 1991), and for general receptor–ligand crosslinking (Waugh *et al.*, 1989). See also Dihazi and Sinz (2003), Koller *et al.* (2004), and Law *et al.* (2002) for additional applications of BS^3 in proteomic applications. Also see Ishmael *et al.* (2005) and Longshaw *et al.* (2004) for additional applications involving DSS in studying protein interactions.

Both DSS and BS^3 continue to be used extensively to investigate protein interactions and the structure of individual proteins within cellular systems (Fioramonte *et al.*, 2012).

1.3. DST and Sulfo-DST

Disuccinimidyl tartarate (DST) is a homobifunctional NHS ester crosslinking reagent that contains a central diol that is susceptible to cleavage with sodium periodate (Thermo Fisher). DST forms amide linkages with α-amines and ε-amines of proteins or other amine-containing molecules (Figure 5.7). The reagent is fairly insoluble in aqueous buffers, but it may be pre-dissolved in THF, DMF, or DMSO prior to addition of an aliquot to a reaction. Optimal conditions for reactivity include a pH range of 7 to 9 with no extraneous amines present that may cross-react with the NHS esters. Avoid Tris, glycine, or imidazole buffers. Subsequent to conjugating proteins with DST, the crosslinks may be broken for analysis by treatment with 0.015-*M* sodium periodate.

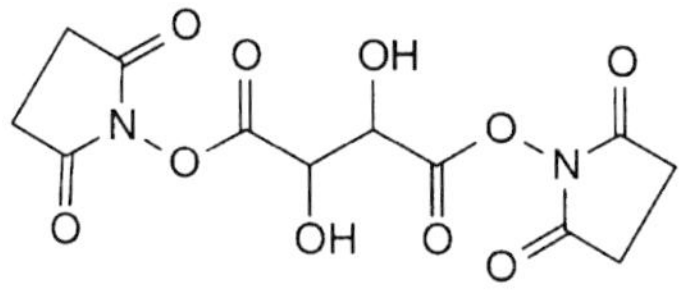

DST
Disuccinimidyl tartarate
MW 344.24
6.4 Å

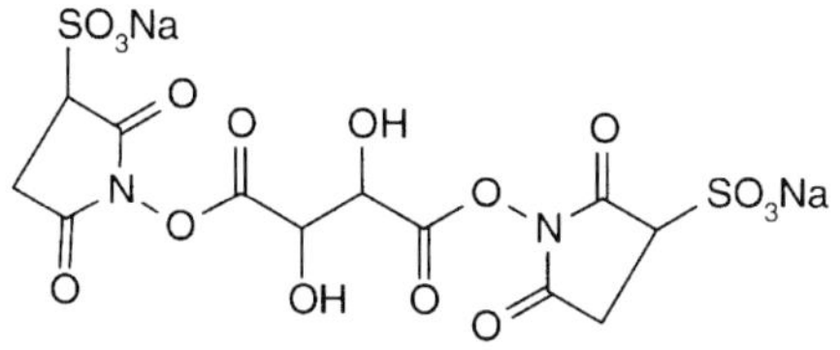

Sulfo-DST
Disulfosuccinimidyl tartarate
Water Soluble
MW 548.34
6.4 Å

Reported applications of DST include the crosslinking of ubiquinone cytochrome *c* reductase (Smith *et al.*, 1978), characterization of the cell-surface receptor for colony-stimulating factor (Park *et al.*, 1986), investigation of the Ca^{+2}, Mg^{+2} activated ATP of *E. coli* (Bragg and Hou, 1980), and characterization of human properdin polymers (Farries and Atkinson, 1989).

Disulfosuccinimidyl tartarate (sulfo-DST) is an analog of DST that contains sulfo-NHS esters. The negatively charged sulfonate groups contribute enough hydrophilicity to provide water solubility for the reagent without the need for organic solvent dissolution before adding it to a crosslinking reaction. The conditions for conjugation are otherwise identical to DST. DST and sulfo-DST also have been used to study protein–lipid complexes (Predescu *et al.*, 2001), to investigate the binding protein of corticotropin-releasing factor (Jahn *et al.*, 2002), and to investigate the acidic C-terminal domain of Rna1p (Haberland *et al.*, 1997).

DST and sulfo-DST also have been used to stabilize fragile cells by crosslinking membrane proteins (Deng, 2011).

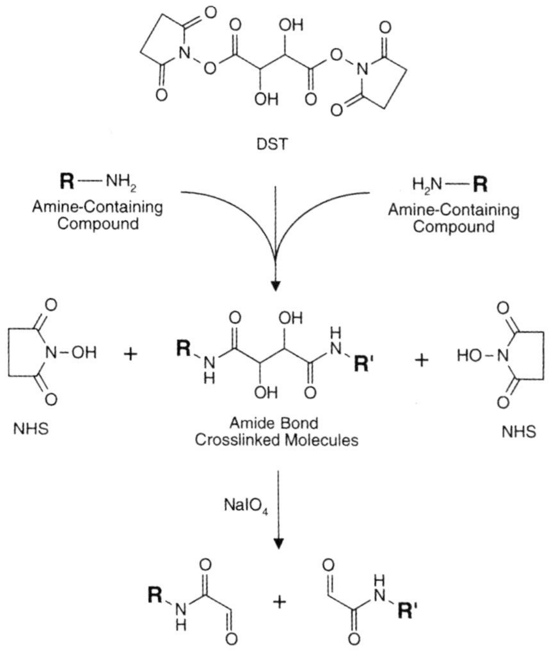

FIGURE 5.7 DST may be used to crosslink amine-containing molecules, forming amide bond linkages. The central diol of the crossbridge is cleavable by treatment with sodium periodate.

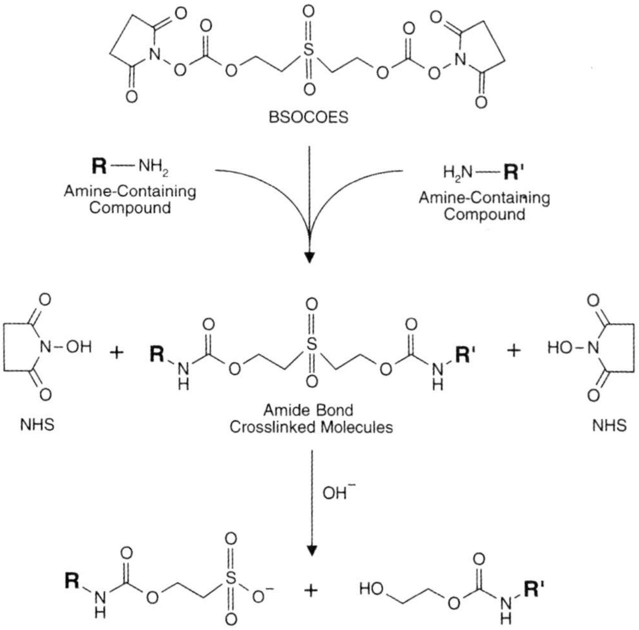

FIGURE 5.8 BSOCOES reacts with amine-containing molecules to create amide bond crosslinks. The internal sulfone group is cleavable under alkaline conditions.

1.4. BSOCOES and Sulfo-BSOCOES

BSOCOES [*bis*[2-(succinimidyloxycarbonyloxy)ethyl] sulfone] is a water-insoluble, homobifunctional NHS ester crosslinking reagent that contains a central sulfone group, which is cleavable under alkaline conditions (Figure 5.8)

(Thermo Fisher). The two NHS ester ends are reactive with amine groups in proteins and other molecules to form stable amide linkages. Once proteins are crosslinked using this reagent, they may be dissociated for analysis by raising the pH to 11.6 and incubating for 2h at 37°C. The sulfone group is base labile under these conditions, and the conjugate cleaves at the center of the bridge.

BSOCOES is a hydrophobic crosslinker and therefore must be dissolved in organic solvent prior to its addition to an aqueous reaction medium. Preparing a stock solution in DMF or DMSO and then adding an aliquot to the crosslinking reaction is recommended. Do not exceed a concentration of more than 10% organic solvent in the buffered reaction to avoid protein precipitation.

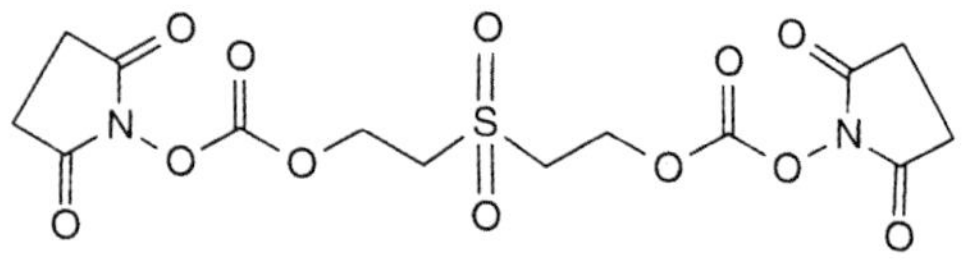

BSOCOES
Bis[2-(succinimidooxycarbonyloxy)ethyl]sulfone
MW 436.36
13 Å

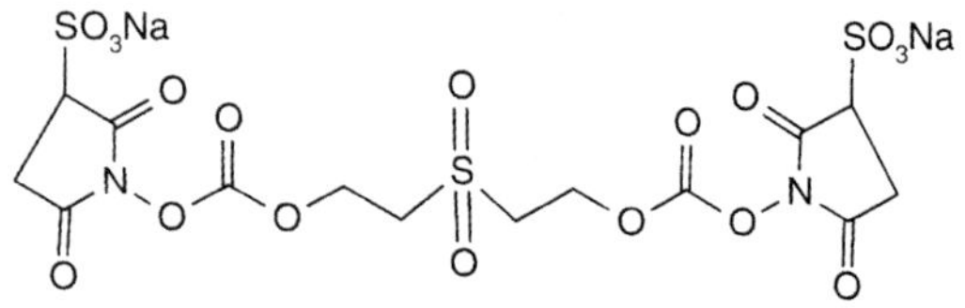

Sulfo-BSOCOES
Bis[2-(sulfosuccinimidooxycarbonyloxy)ethyl]sulfone
Water Soluble
MW 640.46
13 Å

Reported applications of BSOCOES include studying the polypeptide antigens on lymphocyte cell surfaces (Zarling *et al.*, 1980), crosslinking labeled β-endorphin to its opioid receptors (Howard *et al.*, 1985), and isolation and characterization of calcitonin receptors in rat kidney (Bouizar *et al.*, 1986).

A water soluble version of this reagent is also available. Sulfo-BSOCOES [*bis*[2-(sulfosuccinimidooxycarbonyloxy) ethyl]sulfone] is built on the same chemical structure as BSOCOES, but contains the negatively charged sulfonate groups on both of its succinimido rings. The presence of the sulfonates provides enough charge and hydrophilicity to lend water solubility to the entire reagent. Thus, it may be added directly to aqueous reactions at concentrations of up to 10-m*M*. Prior dissolution in organic solvent, however, may provide solubility at greater concentrations in aqueous solutions.

Additional applications of BSOCOES and sulfo-BSOCOES include the study of factors influencing lysosomal membrane permeabilization (Zhao *et al.*, 2012), the study of nanotopography-induced changes in mesenchymal stem cells (Yim *et al.*, 2010), investigations of the cellular and subcellular distribution of the type II vasopressin receptor (Fenton *et al.*, 2007) and TNF-alpha (Grinberg *et al.*, 2005), and studying mechanisms in the control of plasmid replication (Das *et al.*, 2005).

1.5. EGS and Sulfo-EGS

Ethylene glycolbis(succinimidylsuccinate) (EGS) is a homobifunctional crosslinking agent that contains NHS ester groups on both ends (Thermo Fisher). Its central bridge is constructed from an ethylene glycol group esterified on either side with succinic acid, the terminal carboxylates of which are activated by forming *N*-hydroxysuccinimide esters. The two NHS esters are amine reactive, forming stable amide bonds between crosslinked molecules within a pH range of about 7 to 9. Avoid amine-containing buffers such as Tris or glycine, since they will cross-react with the NHS esters. Imidazole also should be avoided, because it has the effect of catalyzing the hydrolysis of the NHS ester groups. The internal structure of EGS provides two cleavable ester sites that may be broken at pH 8.5 by incubation with 1-*M* hydroxylamine for 3 to 6h at 37°C (Abdell *et al.*, 1979) (Figure 5.9). Thus, conjugates produced from the EGS crosslinking of the specific interaction of proteins or other molecules subsequently may be cleaved with hydroxylamine for analysis.

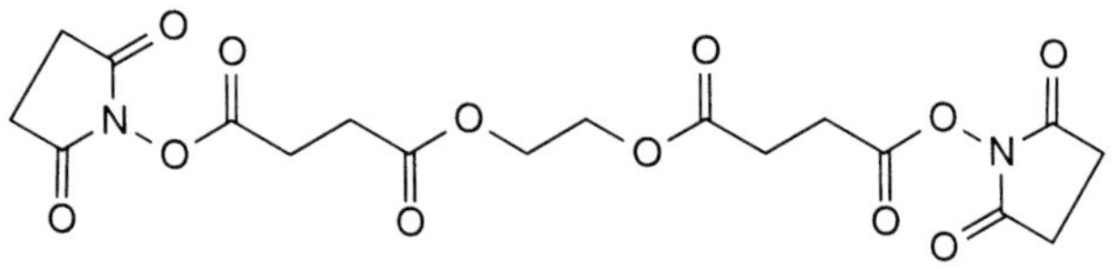

EGS
Ethylene glycolbis(succinimidylsuccinate)
MW 456.37
16.1 Å

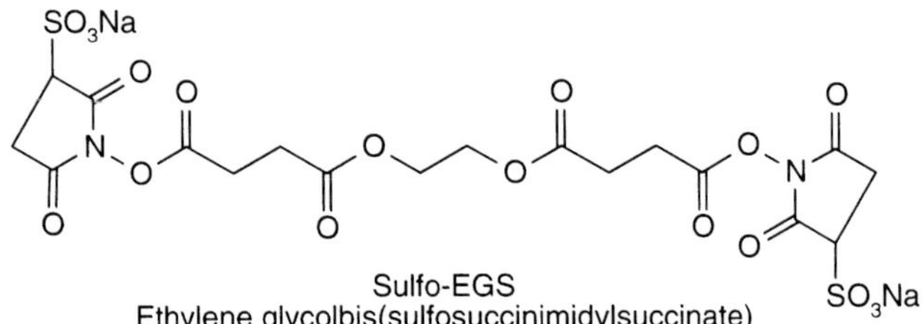

Sulfo-EGS
Ethylene glycolbis(sulfosuccinimidylsuccinate)
MW 660.47
16.1 Å

EGS is not very water insoluble and must be dissolved in an organic solvent prior to its addition to an aqueous reaction. Prepare a concentrated solution of EGS in DMF or DMSO and add an aliquot of the stock solution to the reaction. Do not exceed a concentration of more than about 10% organic solvent in the aqueous reaction buffer or precipitation of buffer salts or protein may occur.

Reported applications of EGS include crosslinking studies of cytochrome P-450 (Baskin and Yang, 1980), studies of the conjugation of tumor necrosis factor with lymphotoxin (Browning and Ribolini, 1989),

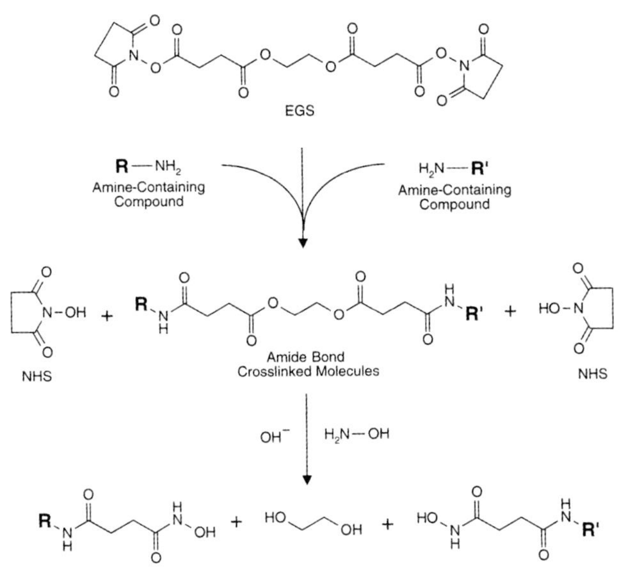

FIGURE 5.9 EGS reacts with amine-containing molecules to form amide-linked conjugates. The ester groups within its crossbridge are cleavable under alkaline conditions using hydroxylamine.

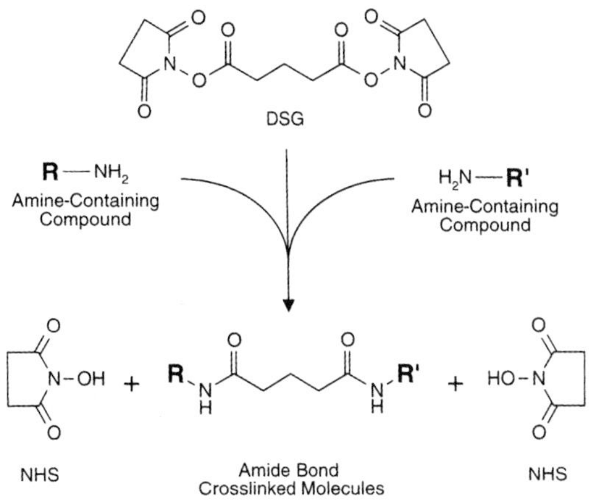

FIGURE 5.10 DSG is a noncleavable crosslinker that can react with two amine-containing molecules to form amide bonds.

the conversion of a gonadotropin-releasing hormone antagonist to an agonist (Conn *et al.*, 1982a), preparation of a conjugate of gonadotropin-releasing hormone with an agonist (Conn *et al.*, 1982b), covalent crosslinking of vasoactive peptide to its lymphoblast receptors (Wood and O'Dorisio, 1985), and study of bombesin receptors in Swiss 3T3 cells (Millar and Rozengur, 1990).

A water soluble analog of EGS is also available commercially (Thermo Fisher). Sulfo-EGS, or ethylene glycolbis (sulfosuccinimidylsuccinate), contains negatively charged sulfonate groups on its NHS rings. The resultant charge and hydrophilicity of this modification provide water solubility to the entire compound so that prior dissolution in organic solvent is not necessary.

EGS and sulfo-EGS also have been used to study the surface loop motion in FepA (Scott *et al.*, 2002), to characterize the high-affinity copper transporter in *Saccharomyces cerevisiae* (Pena *et al.*, 2000) to study of protein interactions and large protein complexes (Petrotchenko *et al.*, 2005), and to study the activation of Syk through dimerization (Hughes *et al.*, 2010).

1.6. DSG

Disuccinimidyl glutarate (DSG) is a water-insoluble, homobifunctional crosslinker containing amine-reactive NHS esters at both ends (Thermo Fisher). The active esters react with amino groups in protein molecules in the pH range of 7 to 9 to form amide linkages. DSG is a non-cleavable reagent, forming stable five-carbon bridges between amine-containing molecules (Figure 5.10).

DSG should be dissolved in an organic solvent prior to addition to an aqueous reaction medium. Suitable solvents include DMF and DMSO. To initiate a reaction, add an aliquot of the organic solution to the buffered medium containing the molecules to be crosslinked. Reaction buffers should not contain any competing amine compounds such as Tris or glycine, as these will cross-react with the active esters. Avoid imidazole-containing buffers, also, since they catalyze the hydrolysis of NHS esters.

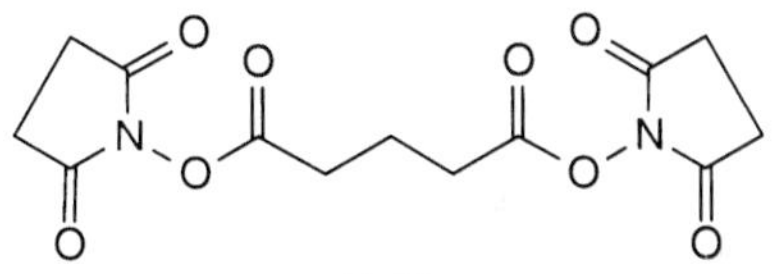

DSG
Disuccinimidyl glutarate
MW 326.26
7.7 Å

The reported applications of DSG include receptor–ligand studies by covalent crosslinking of their complexes (Waugh, 1989); the capture of protein interactions on protein array surfaces by DSG crosslinking (MacBeath, 2007); studying TorT, a member of a periplasmic binding protein family (Baraquet *et al.*, 2006); in the investigation of left-helical conformation of L-DNA for analyzing biomarkers (Hauser *et al.*, 2006); and studying the covalently bound terminal proteins at the 5′ ends of telomeres of *Streptomyces* linear plasmids (Tsai *et al.*, 2011).

1.7. DSC

N,N′-Disuccinimidyl carbonate (DSC) is the smallest homobifunctional NHS ester crosslinking reagent available (Aldrich). It is, in essence, merely a carbonyl group containing two NHS esters. The compound is highly reactive toward nucleophiles. In aqueous solutions, DSC will rapidly hydrolyze to form two molecules of *N*-hydroxysuccinimide (NHS) with release of CO_2. In nonaqueous environments, it can react with two amine groups to form a substituted urea derivative with loss of two molecules of NHS. The reagent also can be used in anhydrous organic solvents to activate a hydroxyl group to an amine-reactive succinimidyl carbonate derivative (Chapter 15, Section 2.1). This procedure is commonly used to activate poly(ethylene glycol) for conjugation with proteins and other molecules (Chapter 18, Section 2.1). In this sense, DSC-activated hydroxylic compounds can be used to conjugate with an amine-containing molecule to form a stable derivative (Figure 5.11). The linkage created from this reaction is a urethane derivative or a carbamate bond, displaying excellent stability.

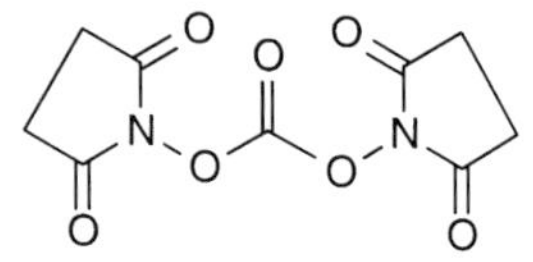

N,N'-Disuccinimidyl Carbonate (DSC)
MW 256.17

Activation of hydroxyl groups with DSC can be done in acetone or dioxane by reacting for 4 to 6h at room temperature. Subsequent conjugation with amine-containing molecules is done in organic or aqueous solutions. For buffered reactions, the optimal conditions include a pH range of 7 to 9 using common buffer salts (avoid amine-containing components, such as Tris). React for at least 4h at room temperature or up to overnight at 4°C.

DSC also is used to activate hydroxylic particles for coupling to amine-containing ligands (Miron and Wilchek, 1993). It also has been used to activate carboxylate-containing molecules for coupling to amine-containing dendrimers to make targeted detection agents (Biswas *et al.*, 2012). For methods involving particle conjugation using this homobifunctional compound, see Chapter 14, Section 4.2, and Chapter 15, Section 2.1.

2. HOMOBIFUNCTIONAL IMIDOESTERS

Crosslinking compounds containing imidoesters at both ends are among the oldest homobifunctional reagents used for protein conjugation (Hartman and Wold, 1966). The imidoester (or imidate) functionality is one of the most specific acylating groups available for the modification of primary amines, with minimal cross-reactivity toward other nucleophilic groups in proteins. The α-amines and ε-amines of proteins may be targeted and crosslinked by reacting with homobifunctional imidoesters at a pH of 7 to 10 (optimal pH 8–9). The product of this reaction, an imidoamide (or amidine) (Figure 5.12), is protonated and thus carries a positive charge at physiological pH (Kiehm and Ji, 1977; Liu, *et al.*, 1977; Ji, 1979; Wilbur, 1992). The result is no alteration of the charge characteristics of the crosslinked proteins, since the amines being modified were themselves protonated and originally contributed to the overall positive charge of the molecule.

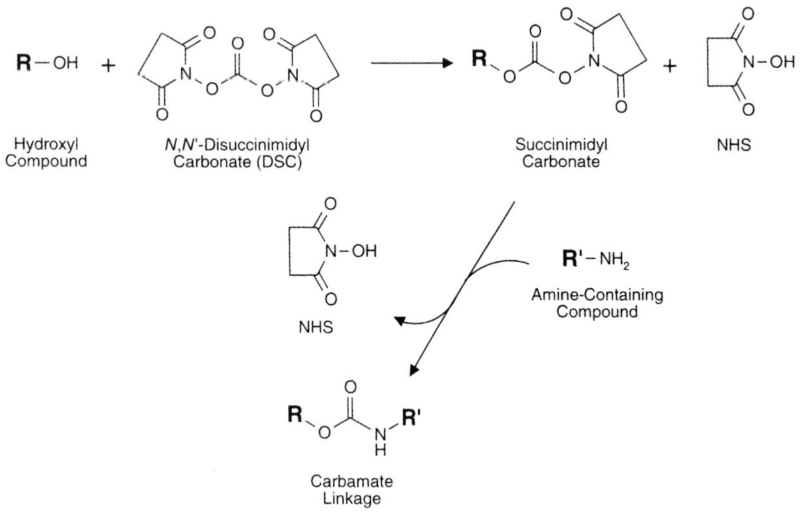

FIGURE 5.11 DSC can react with hydroxyl groups to create a succinimidyl carbonate intermediate that is highly reactive toward nucleophiles. In the presence of an amine-containing molecule, the active species can form stable carbamate linkages.

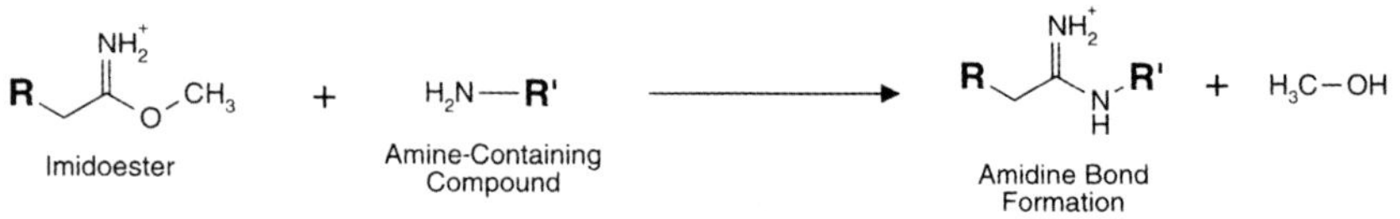

FIGURE 5.12 Imidoesters react with amine groups to form amidine bonds, which are positively charged at physiological pH.

Imidoesters can therefore preserve the microenvironment within the vicinity of the crosslink bridge, possibly retaining native structure and activity better than reagents that modulate the net charge of a protein.

The amidine bond is quite stable at acid pH; however, it is susceptible to hydrolysis and cleavage at alkaline pH. Derivatized proteins may be assayed by amino acid analysis after acid hydrolysis without loss of imidate modifications.

Imidoester crosslinkers are highly water soluble, but undergo continuous degradation due to hydrolysis. The half-life of the imidate functionality is typically less than 30 min, especially in the alkaline conditions of the reaction medium (Hunter and Ludwig, 1962; Browne and Kent, 1975). Concentrated stock solutions may be prepared before addition of a small amount to a conjugation reaction, but they should be dissolved rapidly and used immediately.

The following list of homobifunctional imidoesters represent compounds that are commonly used for protein crosslinking and are currently available from commercial sources.

2.1. DMA

Dimethyl adipimidate (DMA) is a short-chain, homobifunctional crosslinking agent containing imidoesters at both ends (Thermo Fisher). After reaction with amine groups on target molecules, the compound creates a non-cleavable, six-atom bridge with terminal amidine bonds (Figure 5.13). DMA is water soluble and may be added directly to a crosslinking reaction or predissolved at higher concentration before addition of an aliquot to the reaction medium. Stock solutions should be used immediately to prevent breakdown by hydrolysis. Reaction buffers having a pH of 8 to 9 are optimal. Avoid buffers containing primary amines (glycine and Tris), since these will cross-react with the imidoester groups. Borate or bicarbonate buffers adjusted to the optimum pH range work well. The addition of (or the exclusive use of) 0.1- to 0.2-*M* triethanolamine is often carried out to help catalyze the coupling reaction.

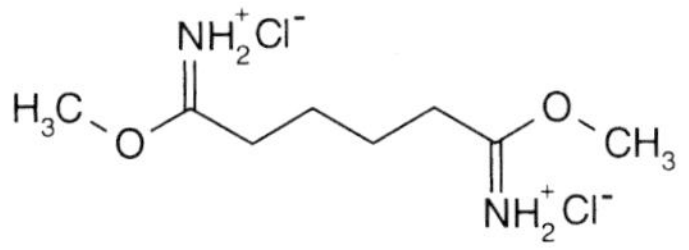

DMA
Dimethyl adipimidate dihydrochloride
MW 245.15
8.6 Å

Reported applications of DMA include the crosslinking of bovine pancreatic ribonuclease A (Hartman and Wold, 1967), treatment of erythrocyte membranes to reduce the effects of sickle cell anemia (Waterman *et al.*, 1975), conjugation and analysis of the outer membrane proteins of *Neisseria gonorrhoeae* (Newhall *et al.*, 1980), protein structural studies of bovine α-crystalline (Siezen *et al.*, 1980), crosslinking of hemoglobin *S* (Pennathur-Das *et al.*, 1982), and forming S-carbomethoxyvaleramidine during hydrolysis of DMA (Mentzer *et al.*, 1982). The compound has also been used to study uranyl–antibody interactions by atomic force microscopy (Odorico *et al.*, 2007), to produce antibody–drug conjugates for the treatment of non-Hodgkin lymphoma (Polson *et al.*, 2007), to investigate the subcellular distribution of the type II vasopressin receptor in kidney (Fenton *et al.*, 2007), and to study protein–chromatin interactions (Tian *et al.*, 2011).

2.2. DMP

Dimethyl pimelimidate (DMP) is a homobifunctional crosslinking agent that has imidoester groups on either end (Thermo Fisher). The imidoesters are amine reactive to give stable amidine linkages with target molecules. The seven-atom bridge created by DMP crosslinks is non-cleavable and positively charged at physiological pH due to the protonated amidine bonds (Figure 5.14). The reagent is water soluble and may be reacted with

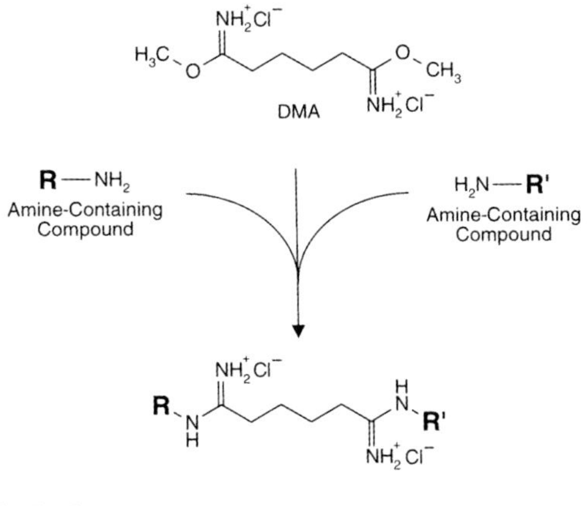

FIGURE 5.13 DMA can crosslink two amine-containing molecules to form charged amidine linkages.

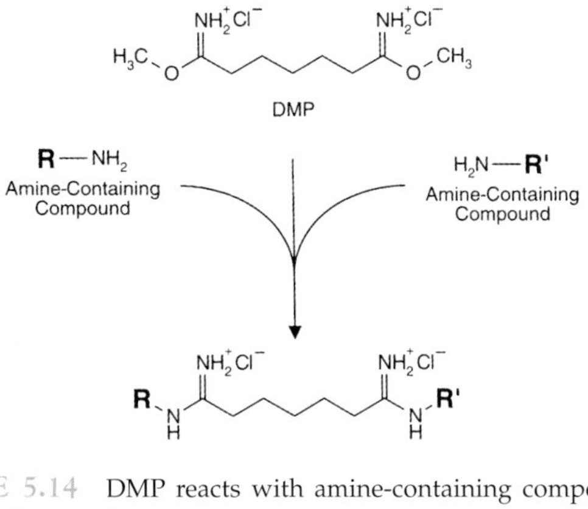

FIGURE 5.14 DMP reacts with amine-containing compounds to form amidine bonds.

proteins or other amine-containing macromolecules at a pH of 8 to 9 in aqueous media. Use buffers that contain no amine groups that may cross-react with the imidoesters. Avoid glycine or Tris buffers.

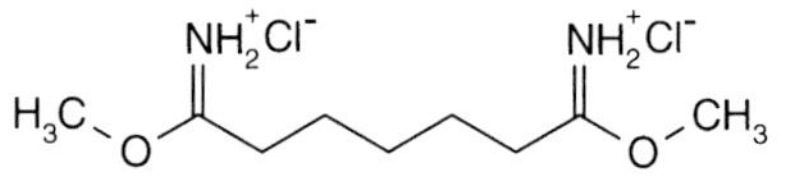

DMP
Dimethyl pimelimidate dihydrochloride
MW 259.18
9.2 Å

In protein crosslinking studies, DMP has been used to examine the subunit structure of muscle pyruvate kinase (Davies and Kaplan, 1972), for the crosslinking of lactose synthetase (Brew *et al.*, 1975), and to conjugate a fluorescent derivative of α-lactalbumin to galactosyltransferase (O'Keefe *et al.*, 1980). The reagent has also found use in the immobilization of antibody molecules to insoluble supports containing bound protein A (Schneider *et al.*, 1982) or protein G (Tozawa *et al.*, 2012). The antibody molecules are first allowed to interact with the coupled protein A, orienting them with their antigen-binding sites facing away from the matrix. DMP is then added to covalently anchor the antibodies to the protein A, forming a permanent immunoaffinity matrix.

DMP also has been used to study the interaction between the endoplasmic reticulum and microtubules (Ogawa-Goto *et al.*, 2007), the conformational changes in the outer arm dynein of *Chlamydomonas* in response to calcium (Sakato *et al.*, 2007), and secretion of the adipocyte-specific secretory protein adiponectin (Wang *et al.*, 2007).

2.3. DMS

Dimethyl suberimidate, DMS, is a homobifunctional crosslinking agent containing amine-reactive imidoester groups on both ends. The compound is reactive toward the ε-amine groups of lysine residues and N-terminal α-amines in the pH range of 7 to 10 (pH 8–9 is optimal). The resulting amidine linkages are positively charged at physiological pH, thus maintaining the positive charge contribution of the original amine. DMS creates eight-atom bridges between conjugated molecules that are not cleavable (Figure 5.15).

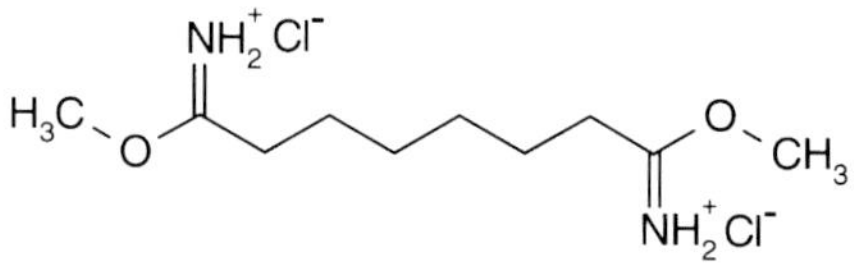

DMS
Dimethyl suberimidate dihydrochloride
MW 273.2
11 Å

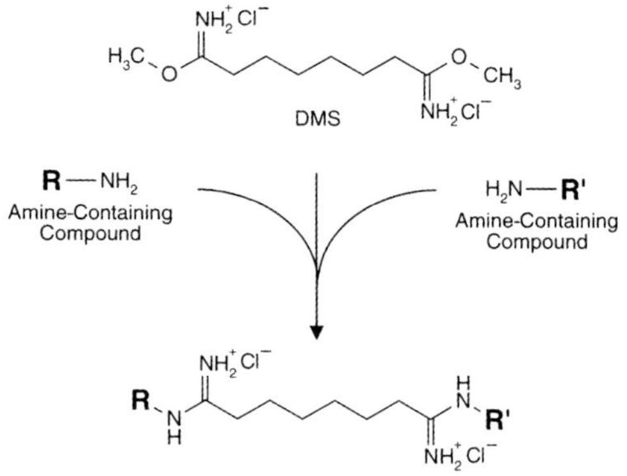

FIGURE 5.15 The reaction of DMS with amine-containing molecules yields amidine linkages.

DMS reportedly has been used as a tissue fixative for light and electron microscopy (Hassell and Hand, 1974), to study the subunit structure of oligomeric proteins (Davies and Stark, 1970), to investigate ATPase activity (Adolfson and Moudrianokis, 1976), to study crosslinking ribonuclease A (Wang *et al.*, 1976), in binding studies of nerve growth factor to its receptor (Pulliam *et al.*, 1975), to study red cell shape (Mentzer and Lubin, 1979), to study crosslinking of glycogen phosphorylase *b* (Hajdu *et al.*, 1979), to study crosslinking of APO low-density lipoproteins (Ikai and Yanagita, 1980), to study the mechanism of binding of multivalent immune complexes to Fc receptors (Dower *et al.*, 1981), to investigate the quaternary structure of the pyruvate dehydrogenase multienzyme complex of *Bacillus stearothermophilus* (Packman and Perham, 1982), in crosslinking studies of the protein topography of rat liver microsomes (Baskin and Yang, 1982), in affinity crosslinking studies of the protein topography of rat liver microsomes (Pfeuffer *et al.*, 1985), and to study the quantitative chemical crosslinking of CAD protein (Lee *et al.*, 1985).

DMS also has been used to study the interaction between the *Helicobacter pylori* accessory proteins HypA and UreE (Benoit *et al.*, 2007) as well as to identify the interaction between the Ca^{2+}-binding protein S100A11 and the Ca^{2+}- and phospholipid-binding protein annexin A6 (Chang *et al.*, 2007) and the recognition of the 70S ribosome by the RNA degradosome (Tsai *et al.*, 2012).

2.4. DTBP

Dimethyl 3,3′-dithiobispropionimidate (DTBP) is a homobifunctional, reversible crosslinking agent containing imidoester groups on both ends (Thermo Fisher). The compound, commonly called Wang and Richards' reagent, is water soluble, and reacts with amines in the pH range of 7 to 10 (optimum pH 8–9) to produce amidine linkages (Wang, 1974; Wang and Richards, 1974). Conjugated molecules subsequently may be cleaved by

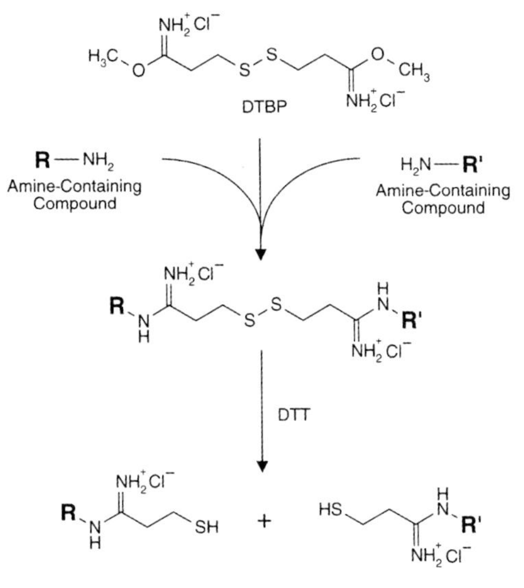

FIGURE 5.16 DTBP reacts with amine-containing molecules to form charged amidine bonds. The internal disulfide group can be cleaved with DTT to release the conjugate.

reduction of the internal disulfide group of the eight-atom bridge (Figure 5.16). Crosslinked molecules may be analyzed by one- or two-dimensional electrophoresis, making use of the easy reversibility of the disulfide bond.

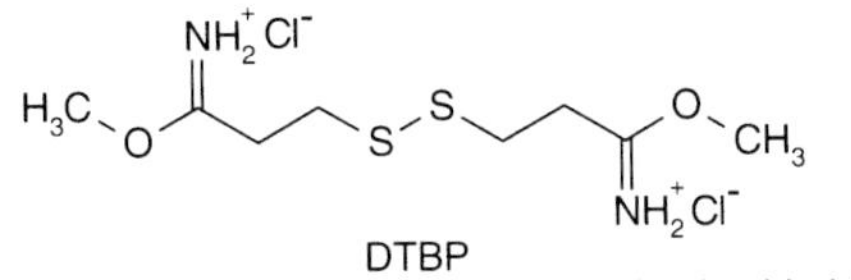

DTBP
Dimethyl-3,3'-dithiobispropionimidate dihydrochloride
MW 309.28
11.9 Å

Reported applications include studying protein–protein interactions with paramyxoviruses (Markwell and Fox, 1980), inhibition of adenylate cyclase activity (Young, 1979), red cell shape (Mentzer and Lubin, 1979), crosslinking phytochrome to its putative receptor (Yu and Schweinberger, 1979), the subunit structure of Na, K-ATPase (DePont, 1979), Newcastle disease virus proteins (Nagai *et al.*, 1978), thylakoid membrane proteins (Novak-Hofer and Siegenthaler, 1978), glutamate dehydrogenase–amino transferase complexes (Fahien et al., 1978), vesicular stomatitis virus proteins (Mudd and Swanson, 1978), hemagglutinin of influenza virus (Wiley *et al.*, 1977), pig heart lactate dehydrogenase crystals (Bayne and Ottesen, 1977), beef heart mitochondrial coupling factor 1 (Baird and Hammes, 1977), chloroplast coupling factor 1 (Baird and Hammes, 1976), rabbit muscle skeletal sarcoplasmic reticulum protein (Louis *et al.*, 1977), human hemoglobin and erythrocyte membrane proteins (Wang and Richards, 1974, 1975; Miyakawa *et al.*, 1978), subunit interface of the E. coli ribosome (Cover *et al.*, 1981), rat liver 60S ribosomal subunits (Uchiumi *et al.*, 1980), proteins in avian sarcoma and leukemia viruses (Pepinsky *et al.*, 1980), outer membrane proteins of *Neisseria gonorrhoeae* (Dade and Johnston, 1980; Newhall *et al.*, 1980), monooxygenase enzymes (Baskin and Yang, 1980), cytochrome P-450 and reduced NAD phosphate-cytochrome P-450 reductase (Baskin and Yang, 1980b), crosslinking initiation factor IF2 to proteins in 70S ribosomes of E. coli (Heinmark *et al.*, 1976), sheep red blood cell membranes (Brandon, 1980), conjugation of F-actin to skeletal muscle myosin subfragment 1 (Labbe *et al.*, 1982), decreased staining of proteins after electrophoresis (Leffak, 1983), and the molecular association of IA antigens after T- and B-cell interaction (Shivdasani and Thomas, 1988).

DTBP also has been used to investigate the dimerization and actin-bundling properties of villin (George *et al.*, 2007), to study the interaction of the Mre11 complex with RPA (Olson *et al.*, 2007), to study gamma–secretase complex assembly (Spasic *et al.*, 2007), and to study the multi-protein assembly of Kv4.2, KChIP3, and DPP10 (Jerng *et al.*, 2005).

3. HOMOBIFUNCTIONAL SULFHYDRYL REACTIVE CROSSLINKERS

Crosslinking agents that contain homobifunctional sulfhydryl-reactive groups at either end fall into two general categories: those that form permanent bonds with available sulfhydryls and those that create reversible linkages. Reactive groups yielding permanent links with sulfhydryls usually form thioether bonds that are quite stable. Those that result in disulfide bonds can be cleaved with the use of a disulfide reducing agent like DTT (Chapter 2, Section 4.1). Mercurial-based coupling groups also can be reversed with reducing agents.

Many varieties of homobifunctional, sulfhydryl-reactive crosslinkers have been synthesized and described in the literature. Some have been based on *bis*-mercurial salts (Edelhoch *et al.*, 1953; Edsall *et al.*, 1954; Kay and Edsall, 1956; Singer *et al.*, 1960; Mandy *et al.*, 1961). Such mercurial reactive groups also have been used in reversible covalent chromatography applications to purify thiol-containing proteins (Cuatrecasas, 1970, 1972; Ruiz-Carrillo and Allfrey, 1973). Other homobifunctional sulfhydryl-reactive reagents have been based on forming a mixed disulfide active group with TNB (5-thio-2-nitrobenzoic acid). Reaction of the TNB active group with a sulfhydryl-containing macromolecule results in a reversible disulfide linkage (Gaffney *et al.*, 1983; Willingham and Gaffney, 1983). Active groups consisting of *bis*-thiosulfonates also have been used to create SH-reactive crosslinkers (Bloxham and Sharma, 1979; Bloxham and Cooper, 1982). The thiosulfonate groups react with available sulfhydryls to form disulfide linkages, in this case with loss of the sulfonate groups. All of these disulfide crosslinks are cleavable with disulfide reducing agents.

A number of *bis*-alkyl halide-reactive groups have been used to create homobifunctional sulfhydryl-reactive crosslinkers (Husain and Lowe, 1968; Ozawa, 1967; Wilchek and Givol, 1977). These react with sulfhydryls to create stable, nonreversible thioether bonds. Similar thioether bond formation has been realized using various *bis*-maleimide derivatives (Moore and Ward, 1956; Kovacic and Hein, 1959; Tawney *et al.*, 1961; Fasold *et al.*, 1963; Simon and Konigsberg, 1966; Zahn and Lumper, 1968; Freedberg and Hardman, 1971; Chang and Flaks, 1972; Wells *et al.*, 1980; Heilmann and Holzner, 1981; Sato and Nakao, 1981; Moroney *et al.*, 1982; Partis, 1983; Chantler and Bower, 1988; Srinivasachar and Neville, 1989). Sulfhydryls add to the double bond of the maleimide to create a thioether linkage.

The differences within these families of reagents generally relate to the length of the spacer or bridging portion of the molecule. Occasionally, the cross-bridge portion itself is designed to be cleavable by one of a number of methods (Chapter 9). The great majority of homobifunctional, sulfhydryl-reactive crosslinkers mentioned in the literature are not readily available from commercial sources and would have to be synthesized to make use of them. The ones listed in this section are obtainable from Thermo Fisher.

3.1. DPDPB

DPDPB, or 1,4-di-[3′-(2′-pyridyldithio)propionamido] butane, is a homobifunctional crosslinking agent that contains sulfhydryl-reactive dithiopyridyl groups on both ends. These coupling groups are identical to the sulfhydryl-reactive end of the popular heterobifunctional crosslinking agent, SPDP (Chapter 6, Section 1.1). Available thiols on proteins and other molecules can react with the pyridyl disulfide groups to form disulfide linkages with release of pyridine-2-thione (Figure 5.17). Conjugation of two macromolecules with DPDPB results in a 14-atom spacer of approximately 16 Å in length. Release of two molecules of pyridine-2-thione during the crosslinking reaction may be followed by their characteristic absorbance at 343 nm ($\varepsilon = 8.08 \pm 0.3 \times 10^3 M^{-1} cm^{-1}$) (Stuchbury *et al.*, 1975).

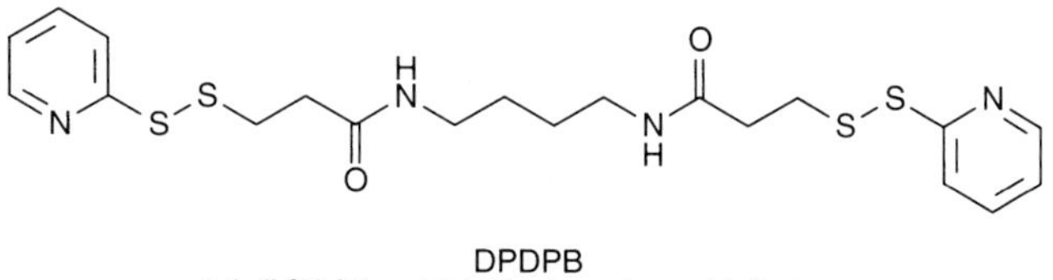

DPDPB
1,4-di-[3′-(2′-pyridyldithio)propionamido]butane
MW 482.7
19.9 Å

DPDPB is insoluble in aqueous solutions and should be initially dissolved in an organic solvent prior to addition of a small aliquot to a buffered reaction medium. Preparation of a stock solution in DMSO at a concentration of 25 m*M* DPDPB works well. The addition of an aliquot of this stock solution to the conjugation reaction should not result in more than about 10% organic solvent by volume in the buffered mixture or protein precipitation may occur.

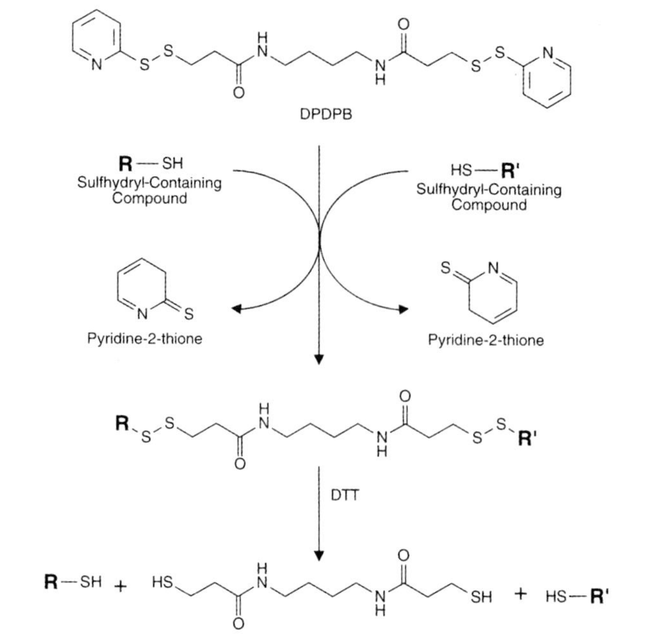

FIGURE 5.17 DPDPB is a sulfhydryl-reactive crosslinker that forms disulfide bonds with thiol-containing molecules. The conjugates may be disrupted using a disulfide reducing agent such as DTT.

DPDPB has two absorbance maxima, one peak at 237 nm ($\varepsilon = 1.2 \times 10^4 M^{-1} cm^{-1}$) and another at 287 nm ($\varepsilon = 8.8 \times 10^3 M^{-1} cm^{-1}$) (Traut *et al.*, 1989; Zecherle, 1990;). Reduction of the pyridyldithio groups causes a shift in the absorbance characteristics of the molecule, such that the peak at 237 nm is shifted to 272 nm and the peak at 287 nm is shifted to 343 nm. This absorbance shift correlates to the release of the pyridine-2-thione groups.

DPDPB has been used to produce liposomal multilamellar vaccines (Moon *et al.*, 2012) and to study the endocytosis of cadherin from intracellular junctions (Troyanovsky *et al.*, 2006), the subunit arrangement in the flagellar rotor assembly (Lowder *et al.*, 2005), and the disease-associated mutations in myelin proteolipid protein in the ER (Swanton *et al.*, 2005). DPDPB can be used to conjugate reduced antibody molecules to β-D-galactosidase using essentially the same protocol as that described by O'Sullivan *et al.* (1979).

3.2. BMH

Homobifunctional crosslinking compounds containing maleimide groups on both ends have been used

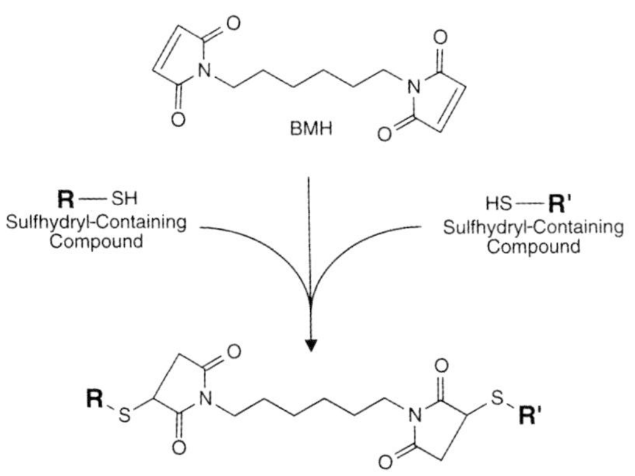

FIGURE 5.18 BMH contains two maleimide groups specific for crosslinking sulfhydryl-containing molecules. The thioether bonds that are formed are stable.

for quite some time (Moore and Ward, 1956; Kovacic and Hein, 1959; Tawney *et al.*, 1961; Fasold *et al.*, 1963; Simon and Konigsberg, 1966; Zahn and Lumper, 1968; Freedberg and Hardman, 1971; Chang and Flaks, 1972; Wells *et al.*, 1980; Heilmann and Holzner, 1981; Sato and Nakao, 1981; Moroney *et al.*, 1982; Partis, 1983; Chantler and Bower, 1988; Srinivasachar and Neville, 1989).

Bismaleimidohexane (BMH) is a homobifunctional reagent containing a non-cleavable, 6-atom spacer between terminal maleimides (Thermo Fisher). The maleimide groups can react with sulfhydryls to form stable thioether linkages (Figure 5.18). Crosslinks formed with this reagent create a 16.1-Å cross-bridge between conjugated macromolecules. The reaction takes place optimally from pH 6.5 to 7.5. Within this pH range, the reaction is very specific for sulfhydryls. At higher pH values, cross-reactivity with amino groups may occur (see Chapter 3, Section 2.2).

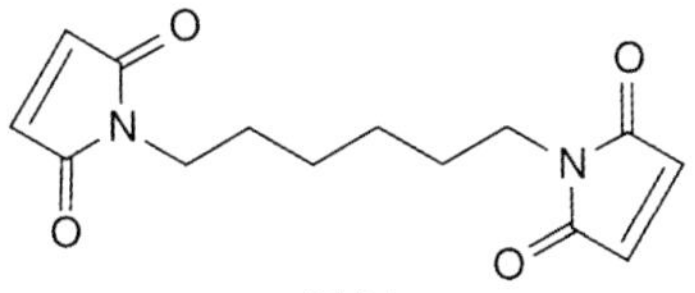

BMH
Bismaleimidohexane
MW 276.29
16.1 Å

BMH has been used to study the initiation of Bak oligomerization (Dai *et al.*, 2011) and to study the binding of SUMO conjugating enzyme to the PR-Set7 histone methyltransferase (Spektor *et al.*, 2011).

4. DIFLUOROBENZENE DERIVATIVES

Difluorobenzene derivatives are small homobifunctional crosslinkers that react with amine groups.

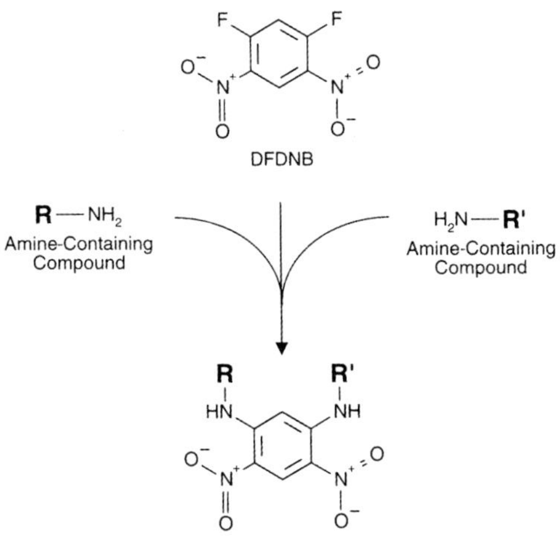

FIGURE 5.19 DFDNB is a small crosslinker able to form covalent bonds between amine-containing molecules. The aromatic fluorine atoms are readily displaced by nucleophiles.

Conjugation using these compounds results in bridges only about 3Å in length, potentially providing information concerning very close interactions between macromolecules.

4.1. DFDNB

DFDNB is the acronym for an aryl halide-containing compound having the structural names 1,5-difluoro-2,4-dinitrobenzene or 1,3-difluoro-4,6-dinitrobenzene (Thermo Fisher). The reagent contains two reactive fluorine atoms that can couple to amine-containing molecules, yielding stable arylamine bonds (Figure 5.19). However, the reactivity of aryl halides is not totally specific for amines. Thiol, imidazolyl, and phenolate groups of amino acid side chains also can react (Zahn and Meinhoffer, 1958). Conjugates formed with sulfhydryl groups, however, are reversible by cleaving with an excess of thiol (such as DTT) (Shaltiel, 1967). The compound is especially useful in crosslinking cellular membrane proteins, since it is able to penetrate the hydrophobic regions of the lipid bilayer.

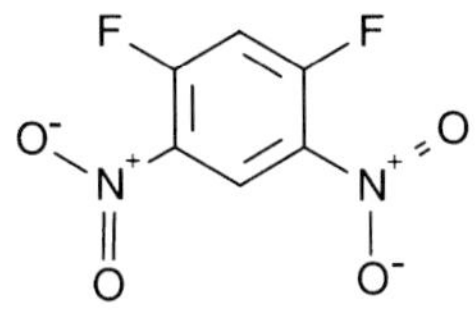

DFDNB
1,5-Difluoro-2,4-dinitrobenzene
MW 204.1
3 Å

Difluorobenzene reagents have been used for crosslinking phospholipids in human erythrocyte membranes (Berg *et al.*, 1965; Marfey and Tsai, 1975),

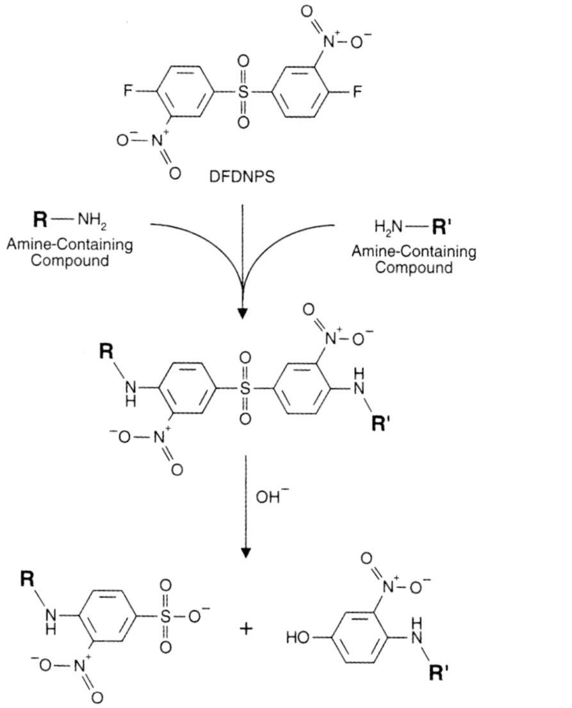

FIGURE 5.20 DFDNPS reacts with amine-containing molecules to form arylamine crosslinks. The central sulfone group in the cross-bridge can be cleaved under alkaline conditions.

conjugation of small peptides to the carrier protein albumin (Tager, 1976), studying the interaction of proteins in the myelin membrane (Golds and Braun, 1978), crosslinking cytochrome oxidase subunits (Kornblatt and Lake, 1980), and studying the conformational effects of calcium on troponin C (Kareva *et al.*, 1986). DFDNB also has been used to investigate the conformational changes in *Chlamydomonas* outer arm dynein in response to calcium ions (Sakato *et al.*, 2007), to study the dimerization properties of villin (George *et al.*, 2007), to characterize the intraflagellar transport complex B core (Lucker *et al.*, 2005), and to study peptide binding to human vascular endothelial growth factor (VEGF) (Marquez *et al.*, 2012).

4.2. DFDNPS

DFDNPS (4,4′-difluoro-3,3′-dinitrophenylsulfone) is a di-aryl halide reagent containing a central sulfone group (Figure 5.20). The aromatic fluorines are reactive with amines, sulfhydryls, phenolates, and imidazolyl groups of proteins (see previous section). The reaction with amines forms stable arylamine linkages. The reaction with sulfhydryl groups, however, is reversible by treatment with an excess of thiol. The central sulfone group provides cleavability through hydrolysis with base at pH 11 to 12 at 37°C (Wold, 1961, 1972; Zahling *et al.*, 1980).

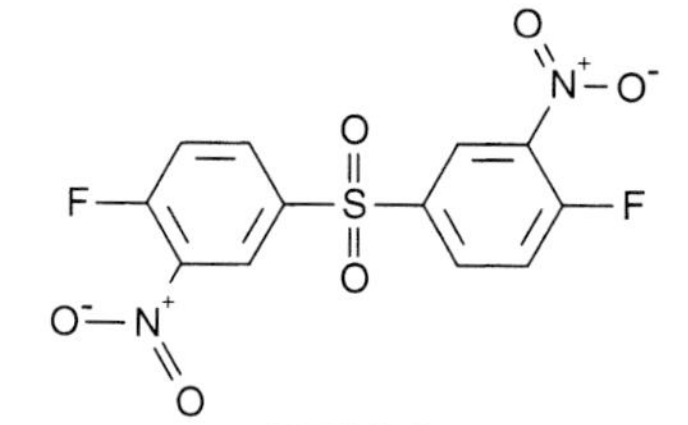

DFDNPS
4,4'-Difluoro-3,3'-dinitrodiphenylsulfone

5. HOMOBIFUNCTIONAL PHOTOREACTIVE CROSSLINKERS

Although there are a number of photosensitive coupling chemistries that have been used in modification and conjugation reactions (Chapter 3, Section 7), it has been primarily aryl azides that have found application in homobifunctional crosslinkers. The photolysis reaction requires exposure of the phenyl azide to a bright light source at a wavelength of 265 to 275nm (Ji, 1979). If the aromatic ring contains a nitro group *meta* to the azide functionality, then photolysis can occur at higher wavelengths (300–460nm). The photolysis process initially forms a highly reactive aryl nitrene, but these quickly undergo ring expansion to create a dehydroazepine. This active species principally reacts with nucleophiles, rather than inserting in C–H or N–H bonds or adding to double bonds. Thus, instead of non-selective coupling into nearly any part of a molecular structure, aryl azides ultimately react with primary amines more than any other functionality (Schnapp *et al.*, 1993).

Reported structures for homobifunctional aryl azides include a biphenyl derivative and a naphthalene derivative (Mikkelsen and Wallach, 1976), a biphenyl derivative containing a central, cleavable disulfide group (Guire, 1976), and a compound containing a central 1,3-diamino-2-propanol bridge between phenyl azide rings that are nitrated (Guire, 1976). The only commercially available homobifunctional photoreactive crosslinker is BASED.

5.1. BASED

Bis-[β-(4-azidosalicylamido)ethyl]disulfide (BASED) is a homobifunctional photoreactive crosslinking agent containing phenyl azide groups at both ends (Thermo Fisher). Its central bridge contains a cleavable disulfide bond that may be broken after conjugation with the appropriate reducing agent (Chapter 2, Section 4.1). The aryl azides are salicylate derivatives that contain hydroxylic functions that activate the ring toward electrophilic reactions. Thus, the phenolic rings are modifiable with ^{125}I using traditional oxidative radioiodination reagents. Prior to the photoreactive conjugation step, the crosslinker may be iodinated with Iodogen or Iodobeads (Chapter 12, Sections 2.2 and 2.3).

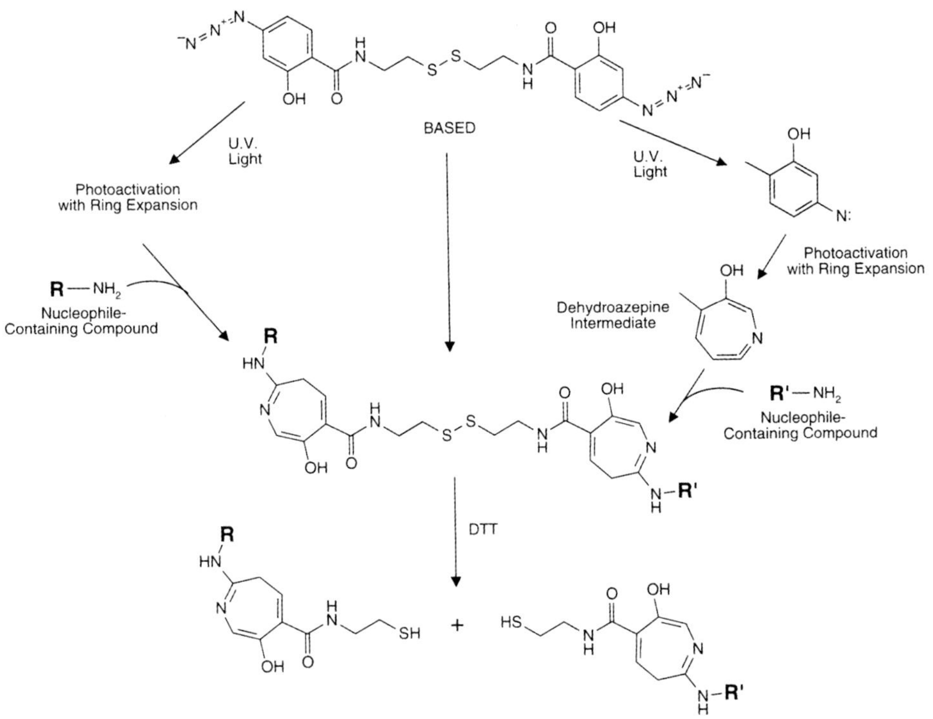

FIGURE 5.21 BASED can react with molecules after photoactivation to form crosslinks with nucleophilic groups, primarily amines. Exposure of its phenyl azide groups to UV light causes nitrene formation and ring expansion to the dehydroazepine intermediate. This group is highly reactive with amines. The cross-bridge of BASED is cleavable using a disulfide reducing agent.

After two proteins are crosslinked, cleavage of the conjugate with DTT releases the link but maintains a radiolabel on each of the molecules (Figure 5.21).

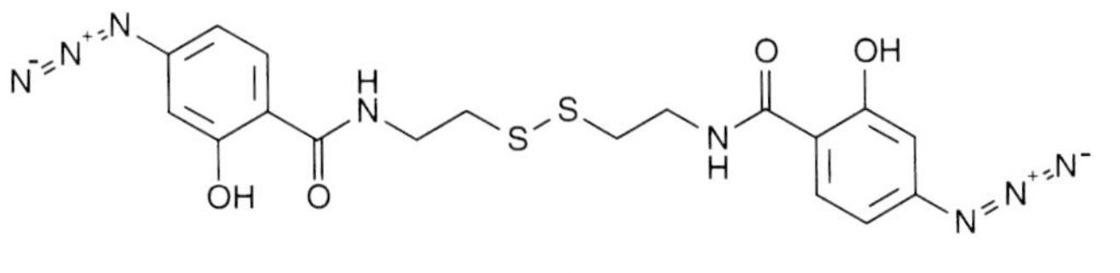

BASED
Bis-[β-(4-azidosalicylamido)ethyl]disulfide
MW 474.54

6. HOMOBIFUNCTIONAL ALDEHYDES

Numerous *bis*-aldehyde reagents have been used for the conjugation of biomolecules. Nearly every small organic compound containing two aldehyde groups has been at least tried in crosslinking reactions. The repertoire of available homobifunctional aldehydes ranges from the single-carbon formaldehyde (Section 6.1) through the two-carbon atom glyoxal (Brooks and Klamerth, 1968), the three-carbon malondialdehyde (Cater, 1963), the four-carbon succinaldehyde (Cater, 1963), the popular five-carbon glutaraldehyde (Section 6.2), the six-carbon adipaldehyde as well as its α-hydroxy derivative (Fein and Filachione, 1957; Seligsberger and Sadlier, 1957; Cater, 1963; Richard and Knowles, 1968; Hopwood, 1969), and several pyridoxal-polyphosphate derivatives that are internally cleavable with acid or base (Shimomura and Fukui, 1978; Benesch and Kwong, 1988).

By far, the two most popular *bis*-aldehyde reagents are formaldehyde and glutaraldehyde.

6.1. Formaldehyde

Formaldehyde is the smallest crosslinking reagent available that does not create a zero-length bridge between two molecules (Chapter 4). Although technically not a homobifunctional reagent, it undergoes crosslinking reactions as though it possessed two functional groups. In concentrated aqueous solutions, it can form the low-molecular-weight polymers typically observed in formalin preparations. In dilute solutions, it exists mainly in its monomeric state. Older solutions of formaldehyde may contain precipitated polymer that often can be resolubilized by heating. Commercial preparations of formaldehyde may be obtained as a 37% solution stabilized against polymerization by the addition of methanol.

Conjugation reactions using formaldehyde may proceed by one of two routes: the Mannich reaction or *via* an immonium cation intermediate. The Mannich reaction consists of the condensation of formaldehyde (or sometimes another aldehyde) with ammonia (in the form of its salt) and another compound containing an active hydrogen. For a review of this reaction mechanism, see Burns and Jacobsen (2011) and Adams *et al.* (1942). Instead of ammonia, however, this reaction can be performed with primary or secondary amines, or even with amides. An example of this is illustrated in

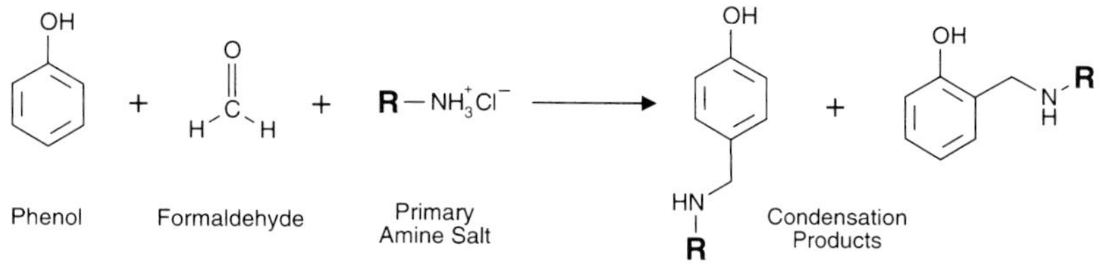

FIGURE 5.22 The Mannich reaction occurs between an active hydrogen-containing compound (phenol) and an amine-containing molecule in the presence of an aldehyde (formaldehyde). The condensation reaction forms stable crosslinks.

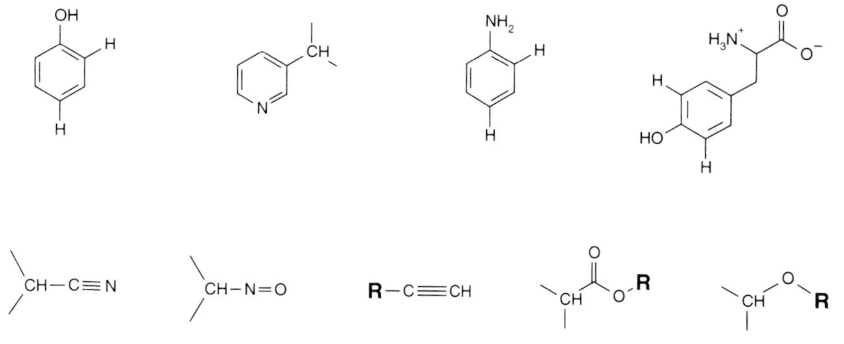

FIGURE 5.23 Examples of active hydrogen-containing compounds that can participate in the Mannich reaction. The points of reactivity are shown by the hydrogen atoms.

Figure 5.22 by the condensation of phenol, formaldehyde, and a primary amine salt (the active hydrogens are shown underlined). Figure 5.23 shows some active hydrogen-containing functional groups that can participate in the Mannich reaction.

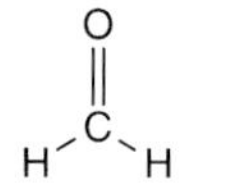

Formaldehyde
MW 30

The Mannich reaction can be used for the immobilization of certain drugs, steroidal compounds, dyes, or other organic molecules that do not possess the typical nucleophilic groups able to participate in traditional coupling reactions (Chapter 15, Section 2.7) (Hermanson *et al.*, 1992). It also can be used to conjugate hapten molecules to carrier proteins when the hapten contains no convenient nucleophile for conjugation (Chapter 19, Section 6.2) (Liao *et al.*, 2011). In this case, the carrier protein contains the primary amines and the hapten contains at least one sufficiently active hydrogen to participate in the condensation reaction.

To obtain acceptable yields, the Mannich reaction must be done at elevated temperatures. Incubation at 37 to 57°C for at least 2 to 24h usually is required to complete the reaction. Addition of formaldehyde is done by adding an aliquot of a 37% solution to the reaction to obtain about a 10- to 100-fold molar excess over the amount of active hydrogen-containing compound to be conjugated. Thermo Fisher has designed a kit for the conjugation of haptens to carrier proteins using the Mannich reaction mechanism.

A secondary reaction pathway also is possible in formaldehyde-facilitated conjugations. Formaldehyde may react with a primary amine to form a quaternary ammonium salt. This intermediate spontaneously reacts to create a highly active immonium cation with loss of one molecule of water (Blass *et al.*, 1965; Ji, 1983). The immonium cation is reactive toward nucleophiles in proteins and other molecules, including amines, sulfhydryls, phenolic groups, and imidazole nitrogens. The reaction yields methylene bridges between two nucleophiles, binding macromolecules with a one-carbon linker (Figure 5.24).

It is obvious that the Mannich reaction pathway and the immonium ion mechanism may occur simultaneously, especially at conditions of room temperature or greater. Formaldehyde-facilitated crosslinking reactions between molecules that both contain nucleophiles probably occur primarily by the immonium ion pathway, since the Mannich reaction proceeds at a slower rate. In addition, the Mannich reaction will cause nondescript polymerization between molecules that possess both active hydrogens and amine groups. It is best to utilize the Mannich reaction only when one of the molecules contains no nucleophilic groups but at least one active hydrogen, and the other molecule contains a primary or secondary amine.

Formaldehyde also can be used to study protein interactions in cells or tissue sections by crosslinking and capturing protein complexes. Chapter 24, Section 1.3, describes this method and contains a protocol for use.

6.2. Glutaraldehyde

Glutaraldehyde is the most popular *bis*-aldehyde homobifunctional crosslinker in use today. However, a glance at glutaraldehyde's structure is not indicative of the complexity of its possible reaction mechanisms. Reactions with proteins and other amine-containing

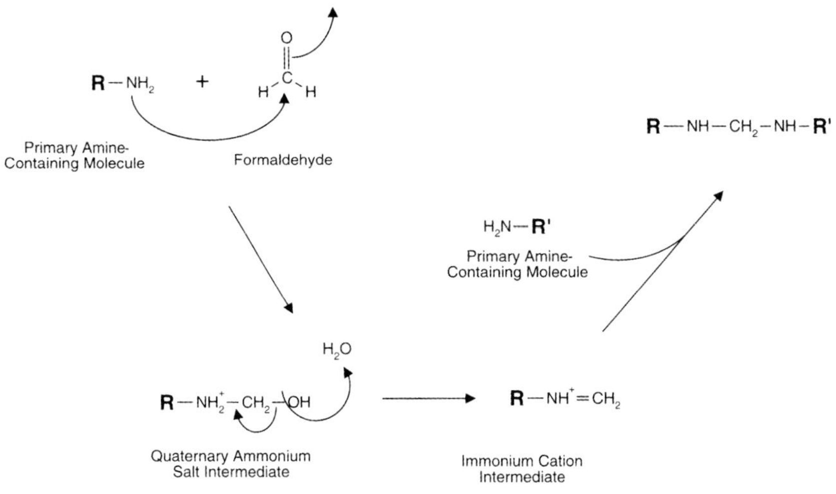

FIGURE 5.24 Two amine-containing molecules can be crosslinked by formaldehyde through formation of a quaternary ammonium salt with subsequent dehydration to an immonium cation intermediate. This active species then can react with a second amine compound to form stable secondary amine bonds.

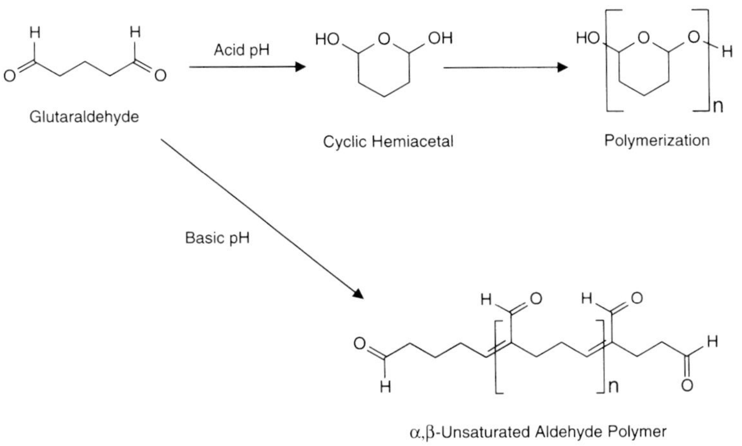

FIGURE 5.25 Glutaraldehyde in aqueous solution may polymerize at either acid or basic pH.

molecules would be expected to proceed through the formation of Schiff bases. Subsequent reduction with sodium cyanoborohydride or another suitable reductant would yield stable secondary amine linkages (Chapter 3, Sections 4.4 and 5.3). This reaction sequence certainly is possible, but other crosslinking reactions also occur.

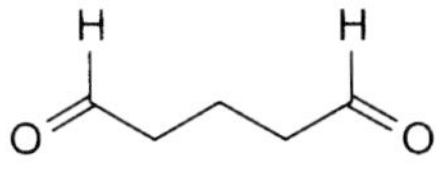

Glutaraldehyde
MW 100.11

Glutaraldehyde in aqueous solutions can form polymers containing points of unsaturation due to aldol formation ((Figure 5.25) (Chapter 15, Section 2.1) (Hardy *et al.*, 1969, 1976; Monsan et al., 1975). Such α,β-unsaturated glutaraldehyde polymers are highly reactive toward nucleophiles, especially primary amines. Reaction with a protein results in alkylation of available amines, forming stable secondary amine linkages. These glutaraldehyde-modified proteins still may react with other amine-containing molecules either through the Schiff base pathway or through addition at other points of unsaturation (Figure 5.26). The proposed reaction mechanism of conjugation using these polymer conjugates may explain the stability of proteins crosslinked by glutaraldehyde that has not been reduced. Schiff base formation alone would not yield stable crosslinked products without reduction. In addition, a number of other potential reactions of glutaraldehyde in aqueous solution also can contribute to its stable crosslinking ability. These include reactions involving hemiacetal rings sometimes combined with aldol formation products, which can couple to amine groups without the formation of Schiff base linkages (see Chapter 15, Section 2.1, for an in-depth discussion of these reactions).

Crosslinking using glutaraldehyde polymers can be difficult to reproduce and scale up. Since the exact glutaraldehyde state in solution, including its potential polymer size and structure, is difficult to determine, the exact nature of the conjugates formed by this method may be indeterminable as well. The age of a

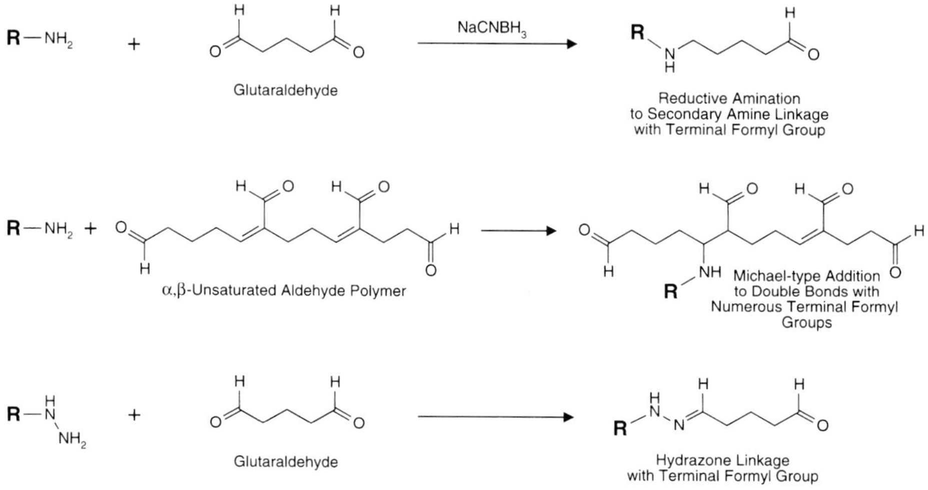

FIGURE 5.26 Glutaraldehyde may react by several routes to form covalent crosslinks with amine-containing molecules.

glutaraldehyde solution is another variable, because the older the solution the more polymer could be formed. Fresh glutaraldehyde often will not yield the same results as aged solutions. Some methods to control the glutaraldehyde activation and coupling process have been successfully carried out for the immobilization of affinity ligands (Chapter 15, Section 2.1); therefore, it also may be possible to use such methods to better control conjugate formation in solution.

A third method of using glutaraldehyde in conjugation reactions is through its ability to react rapidly with hydrazide groups. A molecule containing hydrazide functionalities or modified to contain them (Chapter 2, Section 4.5) can be conjugated with another molecule containing either amines or hydrazides. Glutaraldehyde will react with the hydrazide groups to form hydrazone linkages. When two macromolecules are crosslinked in solutions that contain multiple sites of conjugation, the multivalent hydrazone bonds will be strong enough to create a stable conjugate. If a small molecule is involved, however, reduction of the hydrazone with sodium cyanoborohydride is recommended to produce a leak-resistant bond.

Glutaraldehyde has been used extensively as a homobifunctional crosslinking reagent, especially for antibody–enzyme conjugations (Avrameas, 1969; Avrameas and Ternynck, 1971) and to produce vaccine immunogens (De Filette *et al.*, 2011; Chapter 19, Section 7). To help overcome its tendency to form large-molecular-weight polymers upon crosslinking two proteins, a two-step protocol often is employed. In this regard, one protein first is reacted with glutaraldehyde and purified away from excess reagent. The second protein then is added to effect the conjugate formation. See the introduction to this chapter and Chapter 2, Section 1.2 as well as Chapter 20, Section 1.2 for additional information on the use of glutaraldehyde and two-step crosslinking procedures.

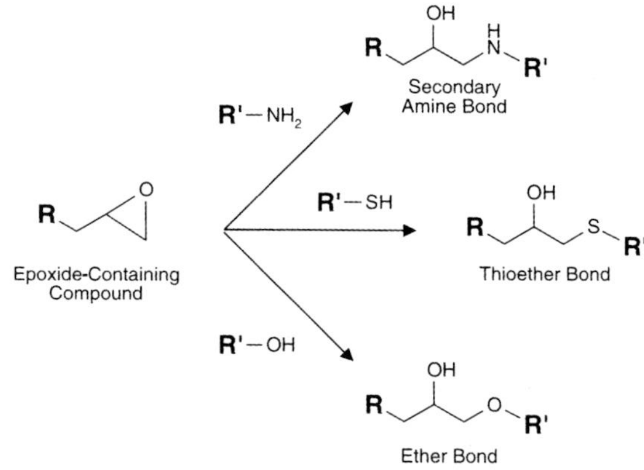

FIGURE 5.27 Epoxide groups are reactive toward sulfhydryls, amines, and hydroxyls.

7. *BIS*-EPOXIDES

Homobifunctional compounds containing epoxide groups on both ends can be used to crosslink molecules containing nucleophiles, including amines, sulfhydryls, and hydroxyls. The reaction proceeds with epoxide ring opening to create secondary amine, thioether, or ether bonds with these functional groups (Figure 5.27). During the ring-opening process, a β-hydroxy group is created. Hydrolysis of the epoxy function without coupling to a nucleophile yields adjacent hydroxyls that can be oxidized with sodium periodate to create reactive aldehydes (Figure 5.28). Epoxide groups, however, are quite stable in aqueous environments around neutral pH or above. They are reactive at alkaline pH values toward other nucleophilic molecules, while they can be hydrolyzed to a diol at acid pH.

Certain *bis*-epoxide reagents have been used to activate hydroxylic matrices for coupling ligands containing amine, sulfhydryl, or hydroxyl groups for affinity chromatography purposes (Chapter 15, Section 2.3) (Hermanson *et al.*, 1992). Conjugation reactions involving proteins have

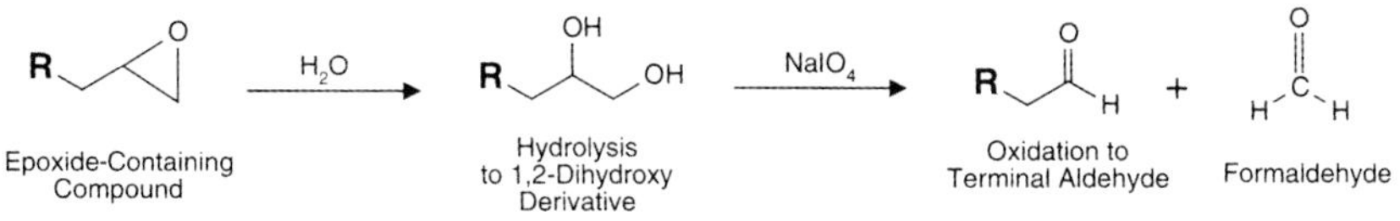

FIGURE 5.28 Hydrolysis of epoxy groups forms 1,2-dihydroxy derivatives that can be oxidized with periodate to create reactive aldehydes.

also been performed using epoxide crosslinkers, but not to the extent of their use in immobilization.

Bis-Epoxy compounds that have been used for crosslinking purposes vary mainly in their chain length, ranging from the 4-carbon bridge of 1,2:3,4-diepoxybutane (Kohn *et al.*, 1966; Skold, 1983), the 6-carbon spacer of 1,2:5,6-diepoxyhexane (Fearnley and Speakman, 1950), the 7-atom bridge of *bis*(2,3-epoxypropyl)ether (Kohn *et al.*, 1966), to the 12-atom spacer of the popular 1,4-(butanediol) diglycidyl ether (discussed below) (Sundberg and Porath, 1974; Porath, 1976). Longer-chain polymeric *bis*-epoxide compounds also have been utilized in collagen crosslinking experiments (Murayama, 1988).

7.1. 1,4-Butanediol Diglycidyl Ether

The most commonly used homobifunctional epoxide compound is 1,4-butanediol diglycidyl ether. The reagent can react with hydroxyls, amines, or sulfhydryl groups to produce ether, secondary amine, or thioether bonds, respectively. The reaction of the epoxide functionalities with hydroxyls requires high pH conditions, usually in the range of pH 11 to 12. Amine nucleophiles react at more moderate alkaline pH values, typically needing buffer environments of at least pH 9. Sulfhydryl groups are the most highly reactive nucleophiles with epoxides, requiring a buffered system in the range of only pH 7.5 to 8.5 for efficient coupling.

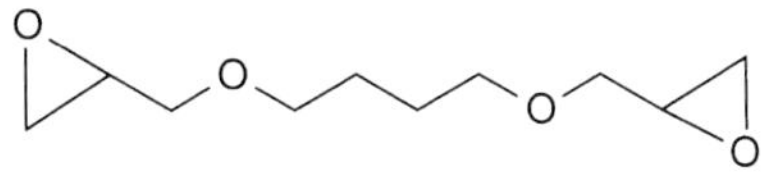

1,4-Butanediol Diglycidyl Ether
MW 202.25

1,4-Butanediol diglycidyl ether is a viscous liquid having a density of 1.45 at 20°C. It is a hygroscopic, corrosive compound with a displeasing odor and should be handled with care in a fume hood. Aqueous solutions of the bis-epoxide usually possess a characteristic oily film on their surfaces, indicating the limited solubility of the reagent.

An example of the use of 1,4-butanediol diglycidyl ether for the activation of soluble dextran polymers is given in Chapter 18, Section 2.2. One end of the *bis*-epoxide reacts with the hydroxylic sugar residues of dextran to form ether linkages, which terminate in epoxy functionalities. The epoxides of the activated derivative then can be used to couple additional molecules containing nucleophilic groups to the dextran backbone.

bis-Epoxy compounds also have been used to couple affinity ligands to particles and chromatography supports. Amine- or hydroxyl-containing supports can be reacted with one end of 1,4-butanediol diglycidyl ether (in excess) to result in particles containing an available terminal epoxide group for further immobilization reactions ((Chapter 15, Section 2.3) (Goyal *et al.*, 2011).

8. HOMOBIFUNCTIONAL HYDRAZIDES

Homobifunctional crosslinking agents containing hydrazide groups at both ends can be used to conjugate molecules containing carbonyl or carboxyl groups. In one scheme, a *bis*-hydrazide compound can be reacted with the carboxylate groups of a protein in the presence of the water soluble carbodiimide 1-ethyl-3-[3-dimethylaminopropyl]carbodiimide hydrochloride (EDC), to yield acyl hydrazide linkages containing terminal alkyl hydrazides. The hydrazide-activated protein then can be used to conjugate with a glycoprotein that had been previously oxidized with sodium periodate to generate reactive aldehyde residues. The resulting hydrazone bonds can be further stabilized by reducing with sodium cyanoborohydride to give the secondary amine linkage.

These techniques have been used to target, detect, or assay glycoproteins in solution or on cell surfaces by using hydrazide-activated enzymes, avidin, or streptavidin (Chapter 11, Section 5) (Bayer and Wilchek, 1990; Bayer *et al.*, 1987, 1990) and to form conjugates with glycoproteins.

bis-Hydrazide-containing molecules can also be used to activate soluble polymeric substances containing aldehyde groups. For instance, dextran may be periodate-oxidized to create numerous formyl functionalities on each molecule. Subsequent reaction with a homobifunctional hydrazide in large excess results in a hydrazide-activated polymer having multivalent binding capability toward aldehydes or ketones (Chapter 18, Section 2.2). Insoluble support matrices suitable for affinity chromatography have been activated in a similar fashion to create the hydrazide derivative (O'Shannessy and Wilchek, 1990).

8.1. Adipic Acid Dihydrazide

The dihydrazide derivative of adipic acid (Aldrich) is perhaps the most popular homobifunctional hydrazide compound in use. The reagent provides a 10-atom bridge between crosslinked molecules after conjugation. Adipic dihydrazide (ADH) is a solid that is soluble in aqueous solutions, but may need to be moderately heated to create concentrated solutions. Aldehyde-containing substances may be modified with this reagent to form hydrazone bonds with alkyl hydrazide spacers suitable for reaction with other formyl-containing molecules (Figure 5.29). In this sense, affinity chromatography matrices have been activated with ADH to produce a hydrazide derivative for coupling to aldehyde-containing ligands (Chapter 15, Section 2.4) (O'Shannessy and Wilchek, 1990), enzymes have been modified at available carboxylate groups using an EDC-facilitated reaction to create hydrazide-activated derivatives appropriate for targeting oxidized glycoproteins (Bayer *et al.*, 1987), and the biotin-binding proteins avidin and streptavidin have been activated with *bis*-hydrazides to assay glycoconjugates using biotinylated enzymes (Bayer and Wilchek, 1990; Bayer *et al.*, 1990). The crosslinker has also been utilized to study the carbohydrate portion of yeast acid phosphatase (Kozulic *et al.*, 1984) and to prepare vaccine conjugates of the capsular polysaccharide of *Salmonella typhi* coupled to the carrier *Pseudomonas aeruginosa* recombinant exoprotein A (Kossaczka *et al.*, 2012).

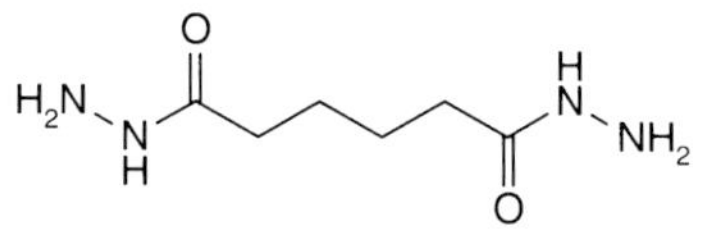

Adipic Acid Dihydrazide
MW 174.2

Protocols for the use of ADH in the modification of aldehyde or carboxylate functionalities can be found in Chapter 2, Section 4.5, and Chapter 11, Section 5.

8.2. Carbohydrazide

Carbohydrazide (carbonic dihydrazide or 1,3-diaminourea) is a small homobifunctional reagent containing reactive hydrazide groups on both ends. Its lack of an internal aliphatic bridge, as found in adipic dihydrazide, gives the compound excellent solubility characteristics in aqueous solutions. Carbohydrazide is freely soluble in water, but practically insoluble in ethanol and other organic solvents. The two hydrazide functional groups of the molecule can react with aldehyde or ketone groups to form hydrazone linkages. When reacted in excess with a molecule containing carbonyl groups, carbohydrazide modification results in short derivatives terminating in available hydrazides (Figure 5.30). The compound has been used to modify microplate wells that have been graft polymerized with glycidyl methacrylate to form surfaces that would couple antibodies through their carbohydrate portions (Allmer *et al.*, 1990; Brillhart and Ngo, 1991) and also used to activate chromatography supports for the immobilization of aldehyde-containing ligands (Chapter 15, Section 2.4). In addition, carbohydrazide has been used as part of the linker arm to create a fucose-specific conjugation scheme for attaching glycan carbohydrates to a vascular-targeting monoclonal antibody (Zuberbuhler *et al.*, 2012). Although its use for protein modification has not been realized, carbohydrazide may be a superior alternative to adipic dihydrazide due to its hydrophilicity. Its only disadvantage may be in its shorter bridge (5-atom spacer as opposed to ADH's 10-atom bridge).

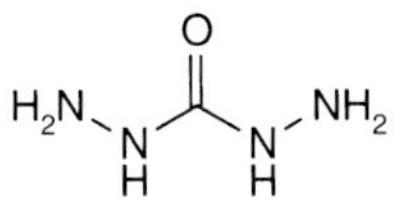

Carbohydrazide
MW 90.09

A protocol for the use of carbohydrazide in the modification of aldehyde or carboxylate functional groups can be found in Chapter 2, Section 4.5, and its potential use for the activation of particles or chromatography supports is described in Chapter 15, Section 2.4.

9. *BIS*-DIAZONIUM DERIVATIVES

Diazonium groups react with active hydrogens on aromatic rings to give covalent diazo bonds. Generation of a diazonium-reactive group is usually done from an aromatic amine by reaction with sodium nitrite under acidic conditions at 0°C (see Chapter 2, Section 4.3, and

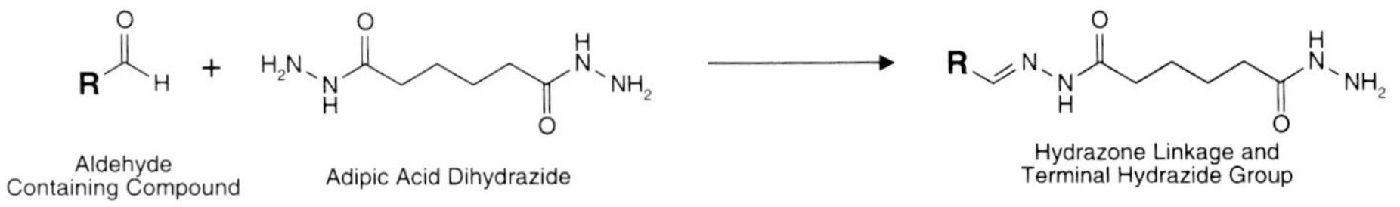

FIGURE 5.29 Adipic acid dihydrazide spontaneously reacts with aldehydes to form hydrazone linkages.

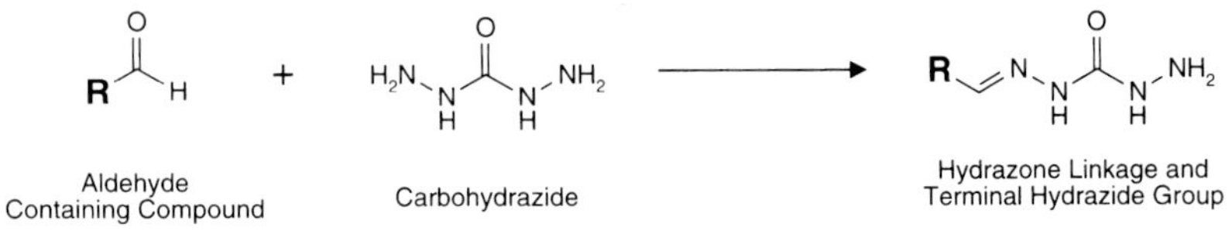

FIGURE 5.30 Carbohydrazide can be used to transform an aldehyde residue into a hydrazide group.

Chapter 3, Section 6.1). The highly reactive and unstable diazonium is reacted immediately with an active hydrogen-containing compound at pH 8 to 10. In general, at pH 8.0 the diazonium group will react principally with histidine residues, attacking the electron-rich nitrogens of the imidazole ring. At higher pH, the phenolic side chain of tyrosine groups can be modified. The reaction proceeds by electrophilic attack of the diazonium group toward the electron-rich points on the target molecules. Phenolic compounds are modified at positions *ortho* and *para* to the aromatic hydroxyl group. For tyrosine side chains, only the *ortho* modification is available and reactive.

bis-Diazonium compounds are useful in crosslinking molecules containing no other convenient functional groups such as amines, carboxylates, or sulfhydryls. Conjugations done using these compounds usually create deeply colored products characteristic of the diazo bonds. Occasionally, the conjugated molecules may turn dark brown or even black. The diazo linkages are reversible by addition of 0.1-*M* sodium dithionite in 0.2-*M* sodium borate, pH 9.0. Upon cleavage, the color of the complex is lost.

9.1. *o*-Tolidine, Diazotized

o-Tolidine, or 3,3′-dimethylbenzidine, is a *bis*-aromatic-amine-containing compound that can be readily diazotized to a homobifunctional diazonium crosslinker by reaction with sodium nitrite (Figure 5.31). The reagent is typically used in a one-step conjugation reaction wherein two active hydrogen-containing molecules are crosslinked through the addition of *o*-tolidine immediately after it has been diazotized under acidic conditions by reacting with sodium nitrite (Figure 5.32). pH adjustment to alkaline conditions after diazonium formation rapidly causes crosslinking to occur. The diazotized form of *o*-tolidine must be used quickly due to its instability in aqueous solutions. The reagent has been used to couple active hydrogen-containing haptens to carrier proteins to form immunogens suitable for the production of antibodies (Chapter 19, Section 6.1).

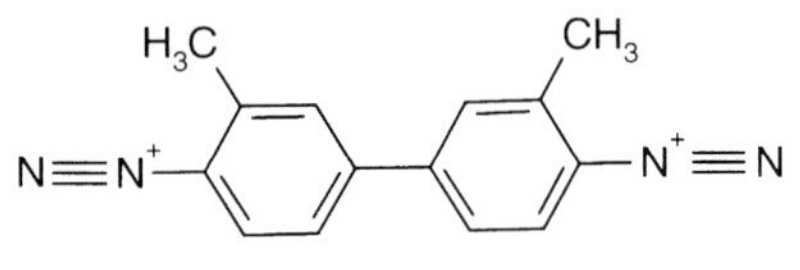

bis-Diazotized *o*-Tolidine

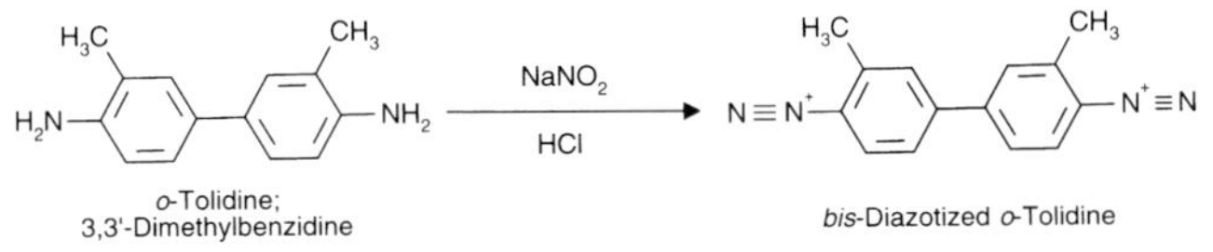

FIGURE 5.31 Reaction of *o*-tolidine with sodium nitrite in the presence of HCl yields a highly reactive diazo-derivative.

o-Tolidine is a benzidine derivative that should be considered a potential carcinogen. Handling should be done with proper safety precautions and with the use of a fume hood to avoid breathing in any dust particles. The reagent is sparingly soluble in water, but is more soluble under the dilute acidic conditions necessary for activation to a diazonium derivative.

9.2. *bis*-Diazotized Benzidine

Benzidine, or *p*-diaminodiphenyl, may be diazotized with sodium nitrite to form a homobifunctional diazonium crosslinking agent useful in conjugating active hydrogen-containing molecules (Figure 5.33). The coupling reaction proceeds *via* electrophilic attack on atoms containing extractable hydrogens. Particularly reactive are the phenolic side chains of tyrosine residues and the imidazole rings of histidine groups.

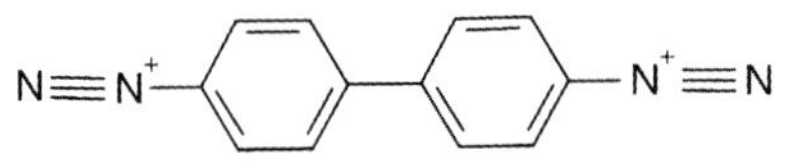

bis-Diazotized Benzidine

Benzidine is a known carcinogen and should be handled with extreme caution (Fourth Annual Report on Carcinogens; NTP 85-002, 1985, p. 37). The solid and its vapors may be rapidly absorbed through skin. Protective clothing and the use of a fume hood are recommended. The compound is only sparingly soluble in water as the free base. The dihydrochloride form, however, is soluble in water and ethanol.

bis-Diazotized benzidine has been used to create active hydrogen-reactive spacer arms on chromatographic

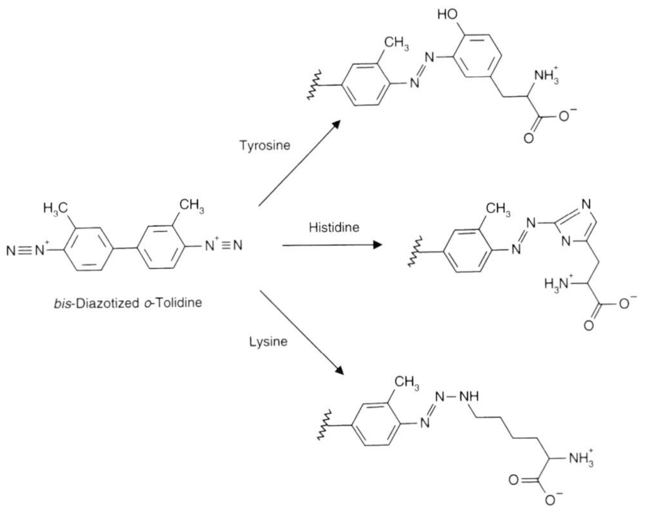

FIGURE 5.32 *bis*-Diazotized tolidine can form crosslinks with proteins through available tyrosine, histidine, or lysine residues.

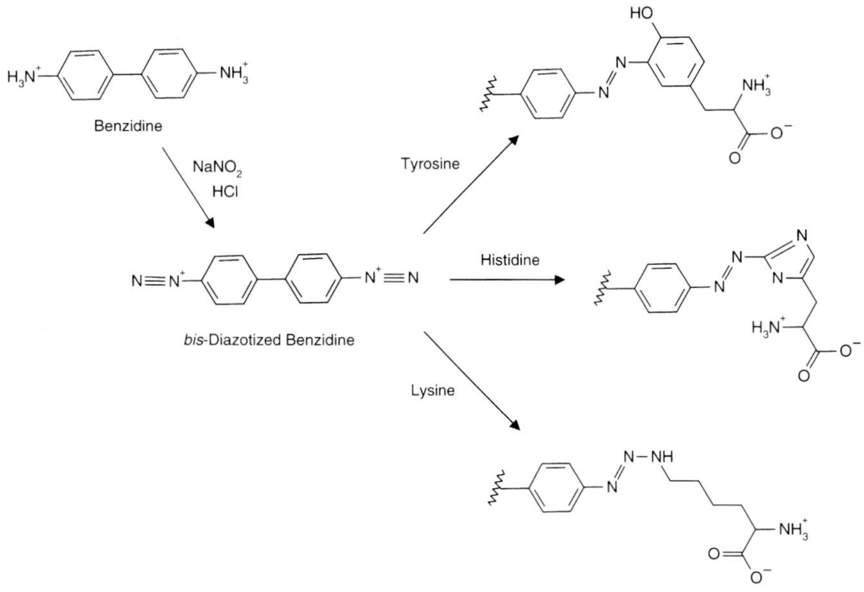

FIGURE 5.33 Benzidine can be diazotized with sodium nitrite and HCl for reaction with proteins through their tyrosine, histidine, or lysine side-chain groups.

matrices (Silman *et al.*, 1966; Lowe and Dean, 1971). The compound may be used similarly to *o*-tolidine for the conjugation of active hydrogen-containing molecules (see Section 9.1, and Chapter 3, Section 6.1).

10. *BIS*-ALKYLHALIDES

Homobifunctional reagents containing reactive halogen groups on both ends are capable of crosslinking sulfhydryl-, amino-, or histidine-containing molecules by nucleophilic substitution. Three forms of activated halogen functionalities can be used to create these reagents: haloacetyl derivatives (see Chapter 3, Section 2.1), benzyl halides that react through a resonance activation process with the neighboring benzene ring, and alkyl halides that possess the halogen β to a nitrogen or sulfur atom, as in *N*- and *S*-mustards. Haloacetyl compounds typically are iodo- or bromo-derivatives, the simplest of which are 1,3-dibromoacetone (Husain and Lowe, 1968) and various iodoacetyl derivatives of short diamine alkyl spacers (Ozawa, 1967) (Figure 5.34). Benzyl halides also are usually iodo or bromo derivatives, whereas the halo-mustards mainly employ chloro- and bromo-forms (Figure 5.35).

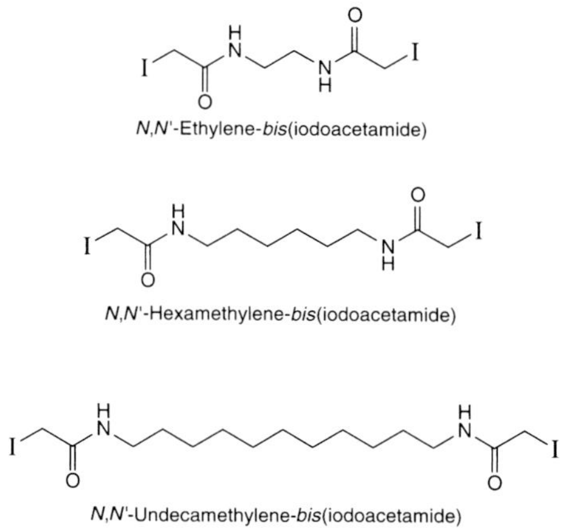

FIGURE 5.34 Several varieties of iodoacetylated diamine compounds have been investigated for crosslinking proteins through sulfhydryl groups.

Reactive halogen crosslinkers are mainly specific for sulfhydryl groups at physiological pH, however at more alkaline pH values they can readily cross-react with amines and the imidazole nitrogens of histidine residues. Some reactivity with hydroxyl-containing compounds also may be realized, particularly with trichloro-*s*-triazine (TsT) derivatives under alkaline conditions (see Chapter 15, Section 2.3).

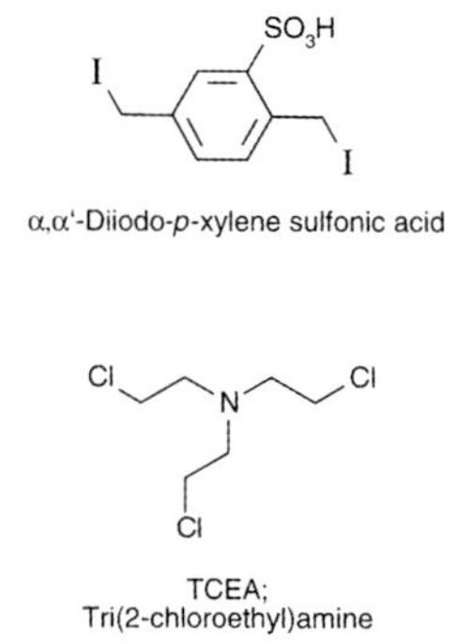

FIGURE 5.35 Benzylhalides and halo-mustards can be used as crosslinking agents reactive toward sulfhydryl groups.

Most of the *bis*-alkyl halides referenced in the literature are unavailable commercially and therefore must be synthesized. Some key references to the preparation and use of these compounds for the crosslinking of sulfhydryl-containing proteins and other molecules include Goodlad (1957), Segal and Hurwitz (1976), Ewig and Kohn (1977), Wilchek and Givol (1977), Prestayko *et al.* (1981), Luduena *et al.* (1982), Hiratsuka (1988), and Aliosman *et al.* (1989).

CHAPTER

6

Heterobifunctional Crosslinkers

Heterobifunctional conjugation reagents contain two different reactive groups that can couple to two different functional targets on proteins and other macromolecules (Figure 6.1). For example, one part of a crosslinker may contain an amine-reactive group, while another portion may consist of a sulfhydryl-reactive group. The result is the ability to direct the crosslinking reaction to selected parts of target molecules, thus garnering better control over the conjugation process.

Heterobifunctional reagents can be used to crosslink proteins and other molecules in a two- or three-step process that limits the degree of polymerization often obtained using homobifunctional crosslinkers (Chapter 2, Section 1.2, and Chapter 5, Section 2.2). In a typical conjugation scheme, one protein is modified with a heterobifunctional compound using the crosslinker's most reactive or most labile end. The modified protein is then purified from excess reagent by gel filtration or rapid dialysis. Most heterobifunctionals contain at least one reactive group that displays extended stability in aqueous environments, therefore allowing purification of an activated intermediate before adding the second molecule to be conjugated. For instance, an

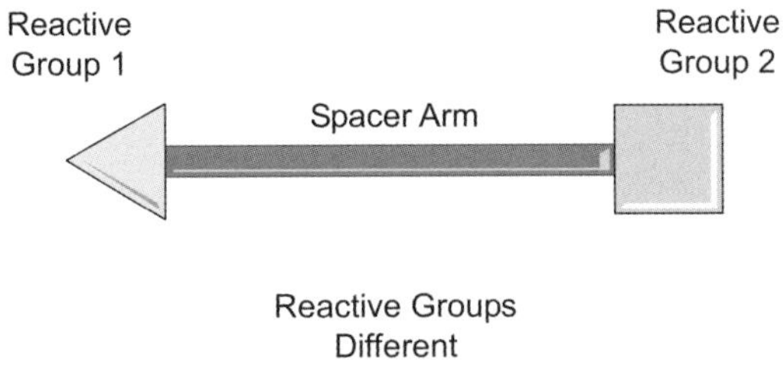

FIGURE 6.1 The general design of a heterobifunctional crosslinking agent includes two different reactive groups at either end and an organic cross-bridge of various length and composition. The cross-bridge may be constructed of chemically cleavable components for selective disruption of conjugates.

N-hydroxysuccinimide (NHS ester–maleimide heterobifunctional (for example, see Section 1.3, this chapter) can be used to react with the amine groups of one protein through its NHS ester end (the most labile functionality), while preserving the activity of its maleimide functionality. Since the maleimide group has greater stability in aqueous solution than the NHS ester group, a maleimide-activated intermediate may be created. After a quick purification step, the maleimide end of the crosslinker can then be used to conjugate to a sulfhydryl-containing molecule.

Such multi-step protocols offer greater control over the resultant size of the conjugate and the molar ratio of components within the crosslinked product. The configuration or structure of the conjugate can be regulated by the degree of initial modification of the first protein and by adjusting the amount of second protein added to the final conjugation reaction. Thus, low- or high-molecule-weight conjugates may be obtained to better fashion the product toward its intended use.

Heterobifunctional crosslinking reagents may also be used to site-direct a conjugation reaction toward particular parts of target molecules. Amines may be coupled on one molecule while sulfhydryls or carbohydrates are targeted on another molecule. Directed coupling is often important in preserving critical epitopes or active sites within macromolecules. For instance, antibodies may be coupled to other proteins while directing the crosslinking reaction away from the antigen binding sites, thus maximizing antibody activity in the conjugate.

Heterobifunctional reagents containing one photoreactive end may be used to insert nonselectively into target molecules by UV irradiation. Ligands having specific affinity toward a receptor may be labeled with a photoreactive crosslinker, allowed to interact with its target, and then photolyzed to permanently label the receptor at its binding site. The photoreactive group is stable until exposed to high-intensity light at UV

Bioconjugate Techniques, *Third Edition*.
DOI: http://dx.doi.org/10.1016/B978-0-12-382239-0.00006-6

wavelengths. Photoaffinity labeling techniques are an important investigative tool for determining binding site characteristics.

The third component of all heterobifunctional reagents is the cross-bridge or spacer that ties the two reactive ends together. Crosslinkers may be selected based not only on their reactivities, but also on the length and type of cross-bridge they possess. Some heterobifunctional families differ solely in the length of their spacer. The nature of the cross-bridge may also govern the overall hydrophilicity of the reagent. For instance, polyethylene glycol (PEG)-based cross-bridges create hydrophilic reagents that provide water solubility to the entire heterobifunctional compound (see Chapter 18, Section 1). A number of heterobifunctionals contain cleavable groups within their cross-bridges, lending greater flexibility to the experimental design. A few crosslinkers contain peculiar cross-bridge constituents that actually affect the reactivity of their functional groups. For instance, it is known that a maleimide group that has an aromatic ring immediately next to it is less stable to ring opening and loss of activity than a maleimide that has an aliphatic ring adjacent to it. In addition, conjugates destined for use *in vivo* may have different properties depending on the type of spacer on the associated crosslinker. Some spacers may be immunogenic and cause specific antibody production to occur against them. In other instances, the half-life of a conjugate *in vivo* may be altered by the choice of cross-bridge, especially when using cleavable reagents.

The following heterobifunctional reagents are organized according to their reactivities. The majority are commercially available and well-documented in the literature as to their properties. Additional heterobifunctional compounds are described in Chapter 17 (Chemoselective Ligation; Bioorthogonal Reagents) and Chapter 18, Section 1 (Discrete PEG Reagents).

1. AMINE-REACTIVE AND SULFHYDRYL-REACTIVE CROSSLINKERS

Perhaps the most popular heterobifunctional reagents are those which contain amine-reactive and sulfhydryl-reactive ends. The amine-reactive group is usually an active ester, most often an NHS ester, while the sulfhydryl-reactive portion may be one of several different functional groups. The amine-reactive end of these crosslinkers is typically an acylating agent possessing a good leaving group that can undergo nucleophilic substitution to form an amide bond with primary amines. The sulfhydryl-reactive portion, by contrast, is usually an alkylating agent that is capable of creating either thioether or disulfide linkages with sulfhydryl-containing molecules. Depending on the chemistry chosen, linkages with a sulfhydryl-containing molecule may be either permanent covalent bonds or reversible disulfide bonds that can be cleaved by use of a suitable disulfide reductant.

The active ester chemistry of the amine-reactive end of these crosslinkers is characteristically the most labile functional group, being susceptible to rapid hydrolysis under the aqueous conditions of a conjugation reaction. The sulfhydryl-reactive group, however, is usually much more stable to breakdown in aqueous environments. Therefore, these reagents typically are used in multi-step conjugation protocols wherein one protein or molecule is first modified through its amines to yield a sulfhydryl-reactive intermediate. After removal of excess crosslinker by gel filtration, a second protein or molecule containing a sulfhydryl group is added to effect the final conjugation. The stability of the sulfhydryl-reactive end of these crosslinkers allows greater control over the crosslinking process than is possible with single-step procedures.

1.1. SPDP, LC-SPDP, and Sulfo-LC-SPDP

SPDP, *N*-succinimidyl 3-(2-pyridyldithio)propionate, is one of the most popular heterobifunctional crosslinking agents available. The activated NHS ester end of SPDP reacts with amine groups in proteins and other molecules to form an amide linkage. The 2-pyridyldithiol group at the other end reacts with sulfhydryl residues to form a disulfide linkage with thiol-containing molecules (Carlsson *et al.*, 1978) (Figure 6.2). The crosslinker is used extensively to form enzyme conjugates for use in immunoassays or in labeled DNA probe techniques. It is also frequently used for the preparation of immunotoxin conjugates for *in vivo* administration (Chapter 20, Section 3). In addition, the reagent is effective in creating sulfhydryls on proteins and other molecules (Chapter 2, Section 4.1). Once modified with SPDP, a protein can be treated with dithiothreitol (DTT) (or another disulfide-reducing agent) to release the pyridine-2-thione leaving group and form a free sulfhydryl. The terminal SH group can then be used to conjugate with any crosslinking agents containing sulfhydryl-reactive groups, such as maleimide or iodoacetyl (both for covalent conjugation) or 2-pyridyldithiol groups (for reversible conjugation).

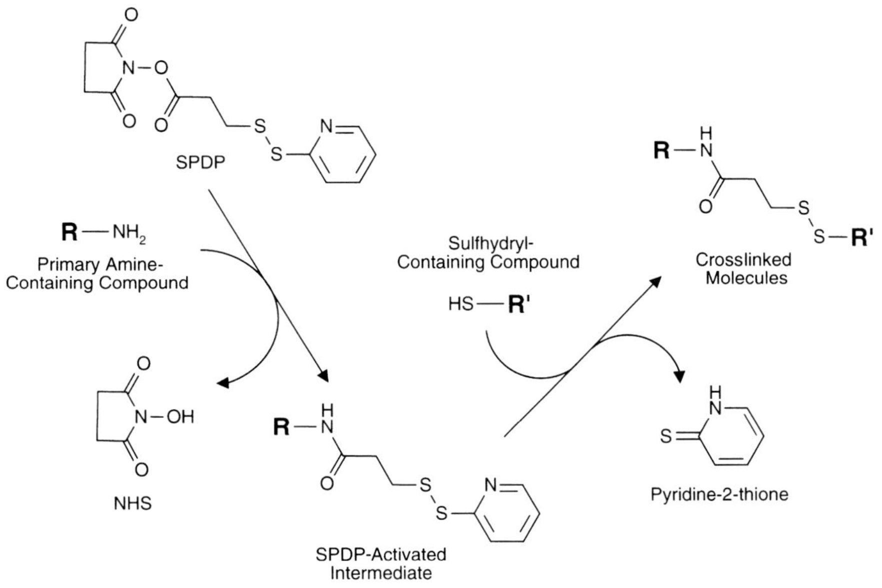

FIGURE 6.2 SPDP can react with amine-containing molecules through its NHS ester end to form amide bonds. The pyridyl disulfide group can then be coupled to a sulfhydryl-containing molecule to create a cleavable disulfide bond.

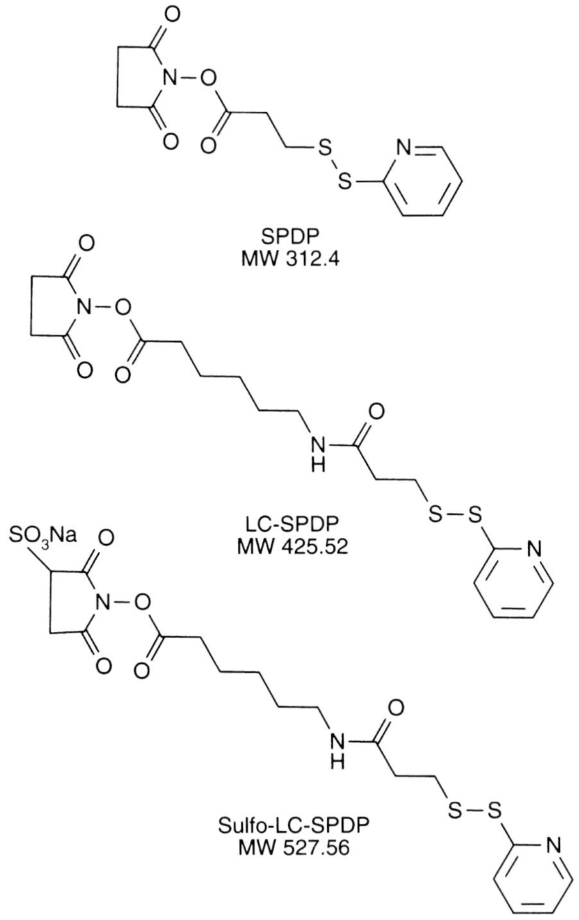

There are three forms of SPDP analogs currently available commercially (Thermo Fisher): the standard SPDP, a long-chain version designated LC-SPDP, and a water soluble, sulfo-NHS form also containing an extended chain, called sulfo-LC-SPDP. Both the standard SPDP and the LC-SPDP are insoluble in aqueous solutions and must first be solubilized in DMSO prior to addition to the reaction solution. The sulfo-LC-SPDP may be solubilized directly in water or buffer. The long-chain versions extend the length of the crosslinker for those applications that require greater accessibility to react with sterically hindered functional groups. Since many sulfhydryl residues are found below the surface of a protein structure in more hydrophobic domains, the longer spacer arm of the LC versions may be more effective in conjugations with these groups. The deficiency in having long aliphatic spacers, however, is that the reagent becomes much more hydrophobic. Care should be taken not to over modify proteins in order to avoid the potential for precipitation.

SPDP or its analogs have been used in many conjugation applications, including the preparation of peptide-based immunoconjugates for prostate cancer therapy (Rege *et al.*, 2007), the preparation of antibody conjugates for a time-resolved fluorescence assay system (Liang *et al.*, 2007), in the study of bone morphogenetic protein type 2 receptor in pulmonary hypertension (Reynolds *et al.*, 2007), and to form conjugates of cationic cell-permeable peptides with siRNA for delivery into targeted cells (Lee *et al.*, 2012).

The following procedure is a suggested multi-step protocol involving the activation of one protein by modification of its amines through the NHS ester end of SPDP, purification of this active intermediate, and subsequent addition of a sulfhydryl-containing molecule for conjugation via the remaining pyridyl disulfide group.

Protocol

1. Dissolve a protein or macromolecule containing primary amines at a concentration of 10 mg/ml in 50-m*M* sodium phosphate, 0.15-*M* NaCl, pH 7.2. Other non-amine-containing buffers such as borate, HEPES, and bicarbonate may also be used in this reaction. Avoid sulfhydryl-containing components in the reaction mixture as these will react with the pyridyl disulfide end of SPDP. The effective pH for the NHS ester modification reaction is in the range of 7 to 9, but hydrolysis will increase at the higher end of this range.
2. Dissolve SPDP at a concentration of 6.2 mg/ml in DMSO (makes a 20-m*M* stock solution). Alternatively, LC-SPDP may be used and dissolved at a concentration of 8.5 mg/ml in DMSO (also makes a 20-m*M* solution). If the water soluble sulfo-LC-SPDP is used, a stock solution in water may be prepared just prior to adding an aliquot to the thiolation reaction. In this case, prepare a 10-m*M* solution of sulfo-LC-SPDP by dissolving 5.2 mg/ml in water. Since an aqueous solution of the crosslinker will degrade by hydrolysis of the sulfo-NHS ester, it should be used quickly to prevent significant loss of activity. If a sufficiently large amount of protein will be modified, the solid may be added directly to the reaction mixture without preparing a stock solution in water to allow accurate weighing of sulfo-LC-SPDP.
3. Add 25 μl of the stock solution of either SPDP or LC-SPDP in DMSO to each ml of the protein to be modified. If sulfo-LC-SPDP is used, add 50 μl of the stock solution in water to each ml of protein solution.
4. Mix and react for at least 30 min at room temperature. Longer reaction times, even overnight, will not adversely affect the modification.
5. Purify the modified protein from reaction byproducts by dialysis or gel filtration using 50-m*M* sodium phosphate, 0.15-*M* NaCl, 10-m*M* EDTA, pH 7.2. Alternatively, centrifugal spin columns containing a desalting resin may be used for rapid purification (Thermo Fisher).
6. Add a sulfhydryl-containing protein or other molecule to the purified SPDP-modified protein to effect the conjugation reaction. Molecules lacking available sulfhydryl groups may be modified to contain them by a number of methods (Chapter 2, Section 4.1). The amount of this second protein added to the reaction should be governed by the desired molar ratio of the two proteins in the final conjugate. The conjugation reaction should be done in the presence of at least 10-m*M* EDTA to prevent metal-catalyzed sulfhydryl oxidation.

1.2. SMPT and Sulfo-LC-SMPT

SMPT, succinimidyloxycarbonyl-α-methyl-α-(2-pyridyldithio) toluene, is a heterobifunctional crosslinking agent that contains an amine-reactive NHS ester on one end and a sulfhydryl-reactive pyridyl disulfide group on the other end. SMPT is therefore an analog of SPDP that differs only in its cross-bridge, which contains an aromatic ring and a hindered disulfide group (Thorpe *et al.*, 1987; Ghetie *et al.*, 1990). The spacer arm of SMPT is slightly longer than SPDP (11.2 Å *versus* 6.8 Å), but the presence of the benzene ring and an α-methyl group adjacent to the disulfide sterically hinders the structure sufficiently to provide increased half-life of conjugates *in vivo*.

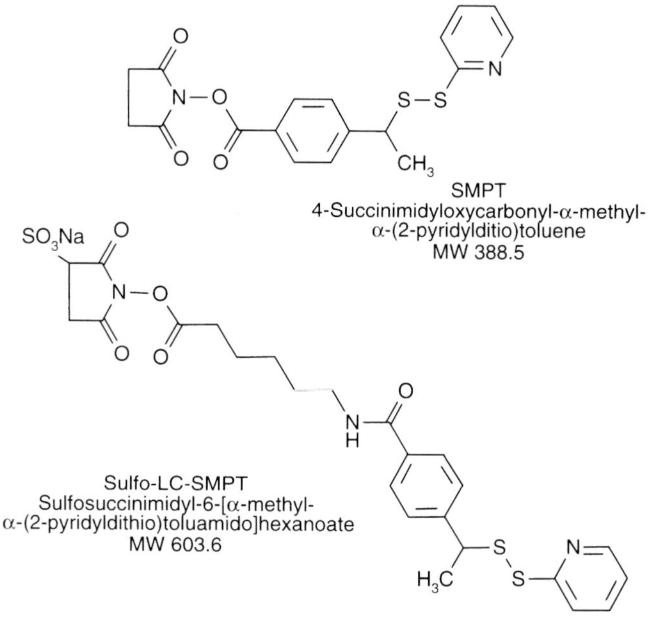

SMPT
4-Succinimidyloxycarbonyl-α-methyl-α-(2-pyridylditio)toluene
MW 388.5

Sulfo-LC-SMPT
Sulfosuccinimidyl-6-[α-methyl-α-(2-pyridyldithio)toluamido]hexanoate
MW 603.6

Conjugation reactions performed using SMPT often proceed by a multi-step protocol involving modification of one protein through its amine groups to create a pyridyl disulfide-activated intermediate. Since SMPT is not soluble in water, the reagent is first solubilized in DMF or DMSO and an aliquot of this stock solution added to the reaction. The NHS ester end of the reagent reacts with ε- and N-terminal amine groups to create stable amide linkages. After removal of excess crosslinker by gel filtration or dialysis, a second protein containing a

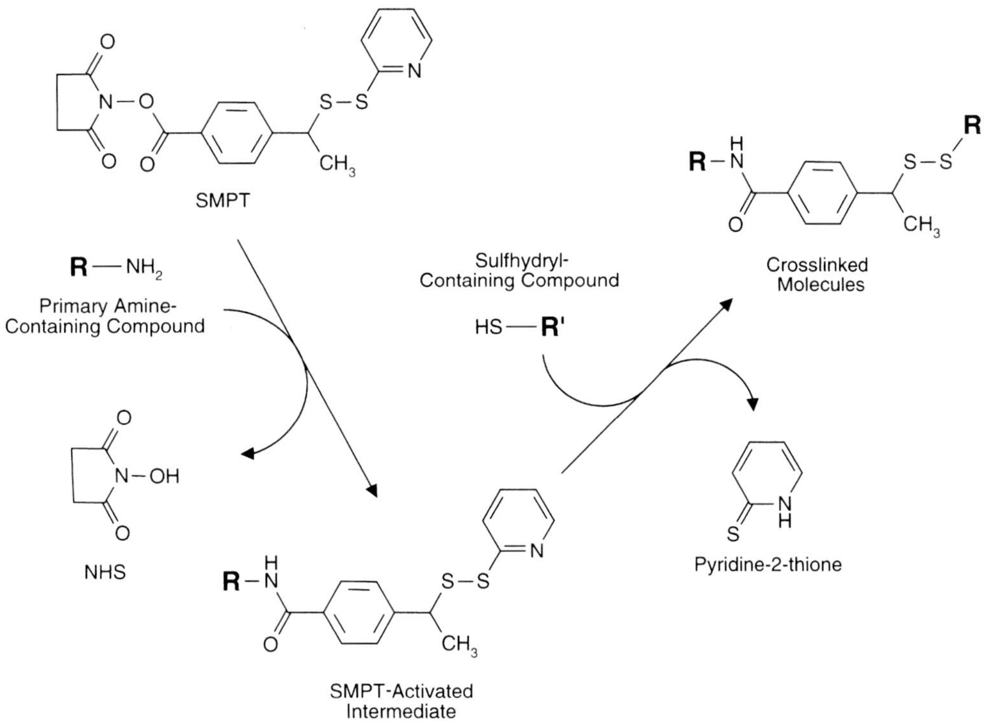

FIGURE 6.3 SMPT can form crosslinks between an amine-containing molecule and a sulfhydryl-containing compound through amide and disulfide linkages, respectively. The hindered nature of the disulfide group provides better stability toward reduction and cleavage.

sulfhydryl group is added to effect the final conjugation (Figure 6.3). The resultant protein–protein crosslink contains a disulfide bond that is susceptible to cleavage by reduction, although more slowly due to the hindered nature of the cross-bridge.

SMPT is often used for the preparation of immunotoxin conjugates that contain a monoclonal antibody directed against some cell-surface antigen (usually a tumor-associated antigen) crosslinked to a protein toxin molecule. It has been shown that a cleavable linkage between the antibody and toxin molecules helps to ensure a potent immunotoxin (Lambert *et al.*, 1985). Increased cytotoxicity is typically observed for immunotoxin conjugates containing cross-bridge disulfides as opposed to non-cleavable linkages. Cleavability presumably facilitates the release of the toxin from the antibody after the conjugate has bound to the cell surface. However, the disulfide bonds formed from some crosslinkers, such as SPDP, are readily reduced and cleaved *in vivo*—often before they reach their target. The hindered disulfide of SMPT has distinct advantages in this regard. Thorpe *et al.* (1987) showed that SMPT conjugates had approximately twice the half-life *in vivo* as SPDP conjugates.

A water soluble analog of SMPT, called sulfo-LC-SMPT, or sulfosuccinimidyl-6-[α-methyl-α-(2-pyridyldithio) toluamido]hexanoate, is available from Thermo Fisher. The sulfo–NHS ester end of the reagent provides the water solubility due to the negative charge of the sulfonate group. While sulfo-LC-SMPT has the same chemical reactivity as SMPT, its cross-bridge contains an additional 6-aminocaproic acid spacer providing a 20-Å crosslink as opposed to the 11.2-Å length of SMPT. The reactivity and use of sulfo-LC-SMPT are essentially the same as that of SMPT, except that the reagent may be added directly to aqueous reaction media or pre-dissolved in water. A stock solution made in water should be used immediately to prevent extensive NHS ester hydrolysis.

SMPT or sulfo-LC-SMPT has been used to develop conjugates for *in vivo* delivery of siRNA to hepatocytes (Rozema *et al.*, 2007), in preparing an anti-CD25-immunotoxin conjugate (Mielke *et al.*, 2007), and in preparing conjugates for selective depletion of donor lymphocytes in stem cell transplantation (Solomon *et al.*, 2005). The crosslinkers have also been used to form "stealth" liposomes carrying chemotherapeutics through the conjugation of anti-tumor antibodies (Manjappa *et al.*, 2011).

1.3. SMCC and Sulfo-SMCC

SMCC, succinimidyl-4-(*N*-maleimidomethyl)cyclohexane-1-carboxylate, is a heterobifunctional reagent with significant utility in crosslinking proteins, particularly in the preparation of antibody–enzyme and hapten–carrier conjugates (Hashida and Ishikawa, 1985;

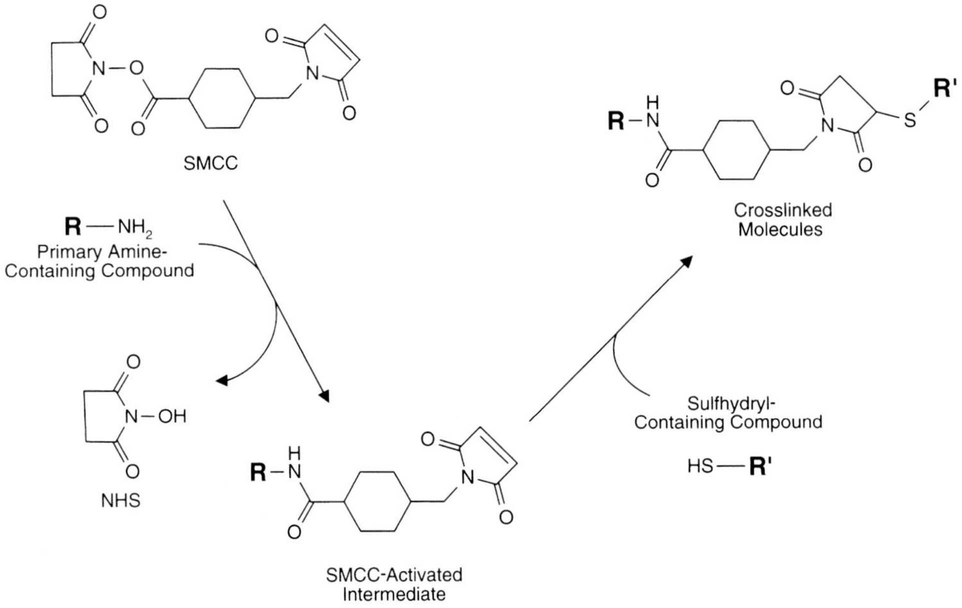

FIGURE 6.4 SMCC reacts with amine-containing molecules to form stable amide bonds. Its maleimide end may then be conjugated to a sulfhydryl-containing compound to create a thioether linkage.

Dewey *et al.*, 1987). In fact, it may be the most popular crosslinker ever designed for protein conjugation purposes. The NHS ester end of the reagent can react with primary amine groups on proteins to form stable amide bonds. The maleimide end of SMCC is specific for coupling to sulfhydryls when the reaction pH is in the range of 6.5 to 7.5 (Smyth *et al.*, 1964) (Figure 6.4).

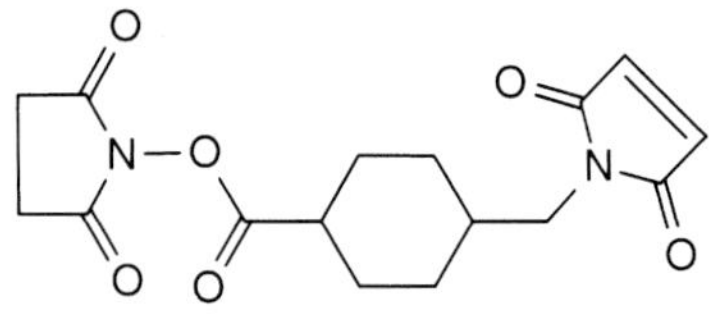

SMCC;
Succinimidyl 4-(*N*-maleimidomethyl)-
cyclohexane-1-carboxylate
MW 334.33
11.6 Å

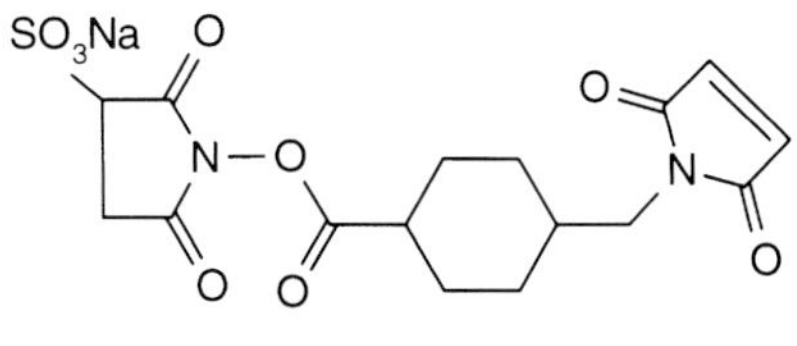

Sulfo-SMCC;
Sulfosuccinimidyl 4-(*N*-maleimidomethyl)-
cyclohexane-1-carboxylate
Water Soluble
MW 436.37
11.6 Å

At pH 7, the reaction of the maleimide group with sulfhydryls proceeds at a rate 1000-times greater than its reaction with amines. At more alkaline pH values, however, its reaction with amines becomes more evident. The maleimide end may also undergo hydrolysis to an open maleamic acid form that is unreactive toward sulfhydryls. Hydrolysis may occur after sulfhydryl coupling to the maleimide, as well. This ring-opening reaction typically happens faster at higher pH values. However, the maleimide group of SMCC displays unusual stability up to pH 7.5. The increased stability of SMCC's maleimide group may be due to it not being attached directly to an aromatic ring structure. By contrast, some maleimide-containing reagents, such as *N,N'-o*-phenylenedimaleimide and *N,N'*-oxydimethylenedimaleimide are far less stable under these conditions. Reportedly, only 4% of the maleimide groups of SMCC will decompose at neutral pH within 2h at 30°C (Ishikawa *et al.*, 1983). For this reason, proteins may be modified with SMCC to form relatively long-lived, maleimide-activated intermediates. The SMCC derivative may then be freeze-dried to provide a stock preparation of sulfhydryl-reactive protein.

SMCC is frequently used to prepare hapten–carrier or antibody–enzyme conjugates. In both applications, one of the molecules is activated (usually the carrier or the enzyme) with the crosslinker, purified to remove excess reagents, and then mixed with the sulfhydryl-containing second molecule to make the final conjugate. Published applications using SMCC are numerous, but include conjugation of glucose oxidase to rabbit antibodies (Yoshitake *et al.*, 1979), crosslinking

Fab′ fragments to horseradish peroxidase (Imagawa *et al.*, 1982; Yoshitake *et al.*, 1982a,b; Ishikawa *et al.*, 1983; Uto *et al.*, 1991), coupling anti-digoxin $F(ab')_2$ fragments to β-galactosidase (Freytag *et al.*, 1984), preparing conjugates of alkaline phosphatase and human IgG $F(ab')_2$ fragments (Mahan *et al.*, 1987), and preparing immunogens (Peeters *et al.*, 1989). SMCC has also been used to prepare antibody–drug conjugates targeted to CD79 for treatment of non-Hodgkin lymphoma (Polson *et al.*, 2007), to make antibody-conjugated, radiolabeled carbon nanotubes for tumor targeting (McDevitt *et al.*, 2007), and to prepare peptide–ovalbumin conjugates to study nucleus-to-cytoplasm shuttling of human acireductone dioxygenase (Gotoh *et al.*, 2007). In addition, Hoppmann *et al.* (2011) used sulfo-SMCC to prepare Affibody–albumin bioconjugates for HER2-positive cancer targeting *in vivo*.

Since SMCC is a water-insoluble crosslinker, it must first be dissolved in organic solvent (DMSO or DMF) before adding it to a protein to be modified. In some cases, addition of even a small amount of organic solvent to a protein solution may be detrimental to activity. To be safe, no more than 10 to 20% solvent should be present in the aqueous reaction medium.

Sulfo-SMCC, sulfosuccinimidyl-4-(*N*-maleimidomethyl) cyclohexane-1-carboxylate, is a water soluble analog of SMCC that possesses a negatively-charged sulfonate group on it *N*-hydroxysuccinimide ring. The charge gives just enough polarity to the molecule to provide water solubility at a level of at least 10 mg/ml at room temperature. This allows direct addition of the reagent to reaction mixtures without prior dissolution in organic solvent. The crosslinker is known to be soluble at a concentration of at least 10-m*M* in the following buffers: (a) 50-m*M* sodium acetate, pH 5.0; (b) 50-m*M* sodium borate, pH 7.6; and (c) 0.1-*M* sodium phosphate, pH 6 to 7.5. Aqueous stock solutions may be prepared using sulfo-SMCC, but these should be dissolved rapidly and used immediately to prevent extensive loss of sulfo–NHS coupling ability due to hydrolysis. Concentrated aqueous stock solutions (up to about 50 mg/ml) may be made by heating for a few minutes under hot running water. Quickly cool to room temperature before using. However, to avoid the potential of activity loss by hydrolysis, even sulfo-SMCC may be dissolved in DMSO prior to adding a small aliquot to an aqueous reaction.

The following is a generalized protocol for the activation of a protein with sulfo-SMCC with subsequent conjugation to a sulfhydryl-containing second molecule or protein. Specific examples of the use of this crosslinker to make antibody–enzyme or hapten–carrier conjugates may be found in Chapter 20, Section 1.1, and Chapter 19, Section 5, respectively.

Protocol

1. Dissolve 10 mg of a protein or other macromolecule to be activated with sulfo-SMCC in 1 ml of 0.1-*M* sodium phosphate, 0.15-*M* NaCl, pH 7.2.
2. Weigh out 2 mg of sulfo-SMCC and add it to the above solution. Mix gently to dissolve. To aid in measuring the exact quantity of crosslinker, a concentrated stock solution may be made in water (or DMSO) and an aliquot equal to 2 mg transferred to the reaction solution. If a stock solution is made, it should be dissolved rapidly and used immediately to prevent extensive hydrolysis of the active ester. As a general guideline of addition for a particular protein activation, the use of a 40- to 80-fold molar excess of crosslinker over the amount of protein present usually results in good activation levels.
3. React for 1 h at room temperature with periodic mixing.
4. Immediately purify the maleimide-activated protein by applying the reaction mixture to a desalting column packed with a desalting resin. The use of a centrifugal spin column may provide faster separations (Thermo Fisher). Do not use dialysis to purify the solution, since the maleimide activity will be lost over the time course required to complete the operation. To obtain good separation between the protein peak (eluting first) and the peak representing excess reagent and reaction byproducts (eluting second), the applied sample size should be no more than 8% of the column bed volume. If complete separation of the activated protein from excess crosslinker is not obtained, then the maleimide content contributed from contaminating crosslinker may prevent subsequent conjugate formation. Perform the chromatography using 0.1-*M* sodium phosphate, 0.15-*M* NaCl, pH 7.2. Collect 1 ml fractions and pool the peak containing the protein. At this point, the maleimide-activated protein may be used immediately in a conjugation reaction with a sulfhydryl-containing protein or other molecule or freeze-dried to preserve the maleimide activity.
5. To effect the conjugation reaction, mix the maleimide-activated protein at the desired molar ratio with a sulfhydryl-containing molecule dissolved in 0.1-*M* sodium phosphate, 0.15-*M* NaCl, pH 7.2. The purified protein from step 4 may be concentrated if necessary using centrifugal concentrators, but this should be done quickly to avoid extensive loss of activity. The molar ratio of addition depends on the desired conjugate to be obtained. For instance, if coupling a sulfhydryl-containing small molecule to a protein, the molecule might be added in excess to the amount of maleimide activity present on the protein if a conjugate linking all reactive sites is desired. In such a case, at least a 10- to 100-fold molar excess

may be appropriate (see Chapter 19, Section 5). However, if preparing protein–protein conjugates, as in the case of antibody–enzyme conjugates, the ratio of maleimide-activated protein to the sulfhydryl-containing protein is a matter of choice. Often, when coupling enzymes to antibodies, the enzyme is in molar excess over the antibody (see Chapter 20, Section 1.1). Typical molar ratios of enzyme-to-antibody can range from 2:1 to 7:1.

6. React for 2 to 24h at room temperature or 4–24h at 4°C.
7. The conjugate may be isolated by gel filtration if the molecular weight of the complex is sufficiently different from that of the unconjugated molecules.

1.4. MBS and Sulfo-MBS

MBS, *m*-maleimidobenzoyl-*N*-hydroxysuccinimide ester, is a heterobifunctional crosslinking agent containing an NHS ester on one end and a maleimide group on the other end. The NHS ester can react with primary amines in proteins and other molecules to form stable amide bonds, while the maleimide end nearly exclusively reacts with sulfhydryl groups to create stable thioether linkages (Figure 6.5). These characteristics allow highly controlled conjugation reactions to be performed with MBS using two or three step processes. In this sense, the NHS ester end of the reagent is typically reacted with the first protein to be crosslinked, forming a maleimide-activated intermediate. The maleimide group is more stable to breakdown by hydrolysis than the NHS ester, so the activated intermediate can be quickly purified from excess crosslinker and reaction byproducts before adding it to the sulfhydryl-containing second molecule. However, due to the aromatic ring adjacent to its maleimide functional group, MBS displays less stability toward maleimide ring opening than SMCC (Section 1.3, this chapter). Unlike SMCC, MBS is therefore not recommended for preparing freeze-dried, maleimide-activated proteins, since during the processing necessary to purify and stabilize the derivative much activity can be lost by hydrolysis.

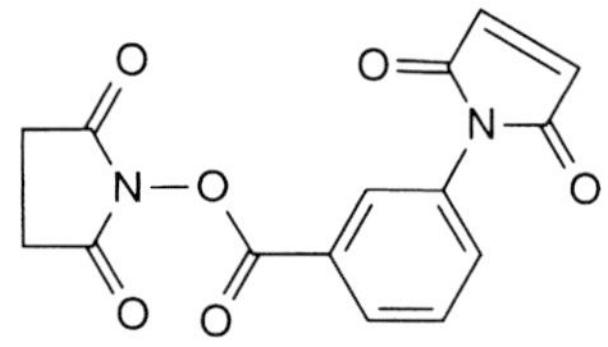

MBS;
m-Maleimidobenzoyl-N-hydroxy-succinimide ester
MW 314.2
9.9 Å

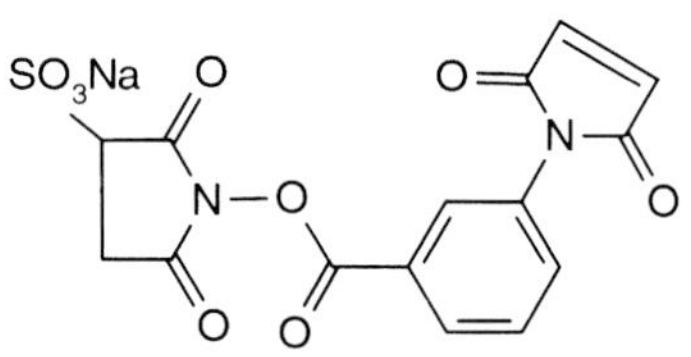

Sulfo-MBS;
m-Maleimidobenzoyl-N-hydroxy-sulfosuccinimide ester
MW 416.24
9.9 Å

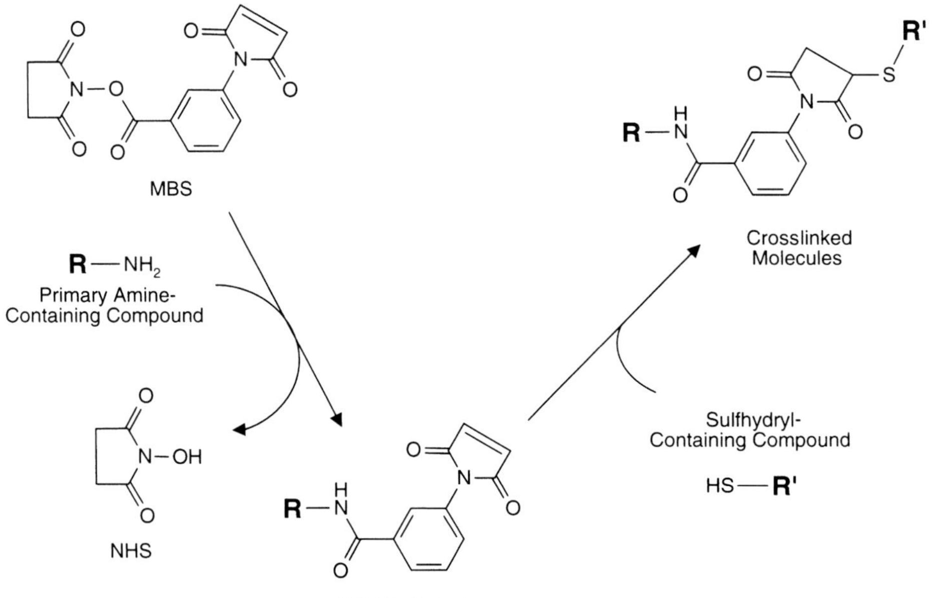

FIGURE 6.5 The two-step conjugation procedure for the MBS crosslinking of an amine-containing molecule with a sulfhydryl-containing molecule.

MBS contains a benzoic acid derivative as its cross-bridge, thus lending considerable hydrophobicity to the entire molecule. Since the reagent is water insoluble, it must first be dissolved in organic solvent before adding it to an aqueous reaction medium. Making a concentrated stock solution of MBS in DMF or DMSO allows transfer of a small amount to a conjugation reaction (total concentration of the organic solvent should not exceed 10% in the reaction buffer). When these solvents are used, a micro-emulsion typically is formed in the aqueous solution, which provides crosslinker efficiently to the conjugating species. The reagent is also readily permeable to membrane structures due to its hydrophobic nature.

Sulfo-MBS, *m*-maleimidobenzoyl-*N*-hydroxysulfosuccinimide ester, is a water soluble analog of MBS that contains a negatively charged sulfonate group on its NHS ring (Martin and Papahadjopoulos, 1982; Aithal *et al.*, 1988). The negative charge lends enough hydrophilicity to the crosslinker to allow direct addition of the reagent to aqueous reaction media without prior dissolution in organic solvents. Sulfo-MBS has the identical reactivity of MBS.

MBS was one of the first and most popular of the family of NHS ester–maleimide heterobifunctionals (Kitagawa and Aikawa, 1976). It has been used extensively to produce antibody–enzyme and other enzyme conjugates (Kitagawa *et al.*, 1978; Freytag *et al.*, 1984; O'Sullivan *et al.*, 1979), in the preparation of hapten–carrier immunogens (Liu *et al.*, 1979; Lerner *et al.*, 1981; Kitagawa *et al.*, 1982; Niman *et al.*, 1985; Chamberlain *et al.*, 1989; Edwards *et al.*, 1989; Miller *et al.*, 1989; Swanson *et al.*, 1991), and for making immunotoxin conjugates (Youle and Nevelle, 1980; Dell'Arciprete *et al.*, 1988; Myers *et al.*, 1989). Additional applications include investigations of carnitine palmitoyltransferase 1 (Faye *et al.*, 2007), preparation of a targeting conjugate containing Cyt1Aa toxin directed at myeloma cells (Cohen *et al.*, 2007), and development of an improved method for coupling synthetic peptide haptens to carrier proteins (Lateef *et al.*, 2007). Ji *et al.*, (2012) also used MBS to prepare an arginine–glycine–aspartic acid (RGD) peptide conjugated to albumin nanoparticles for targeted delivery to pancreatic cancer cells.

The generalized protocol for performing a multi-step conjugation reaction with MBS or sulfo-MBS is similar to that described for SMCC (Section 1.3, this chapter). Specific examples may be found in the cited references.

1.5. SIAB and Sulfo-SIAB

SIAB, *N*-succinimidyl(4-iodoacetyl)aminobenzoate, is a heterobifunctional crosslinker containing amine-reactive and sulfhydryl-reactive ends (Weltman, 1983). The NHS ester of SIAB can couple to primary amine-containing molecules, forming stable amide linkages (Chapter 3, Section 1.4). The other end contains an iodoacetyl group that is specific for coupling to sulfhydryl residues, creating stable thioether bonds (Chapter 3, Section 2.1). The aminobenzoate cross-bridge is a hydrophobic spacer that helps the reagent become fully permeable to membrane structures.

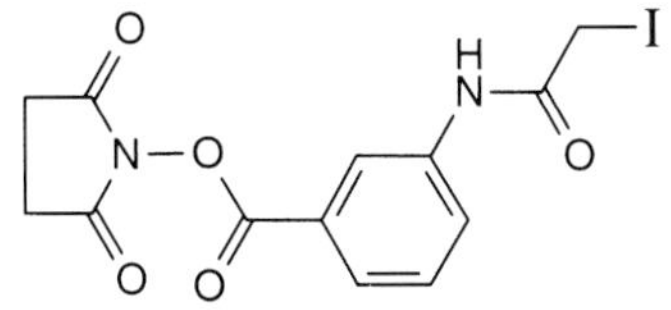

SIAB;
N-Succinimidyl(4-iodoacetyl)-
aminobenzoate
MW 402.15
10.6 Å

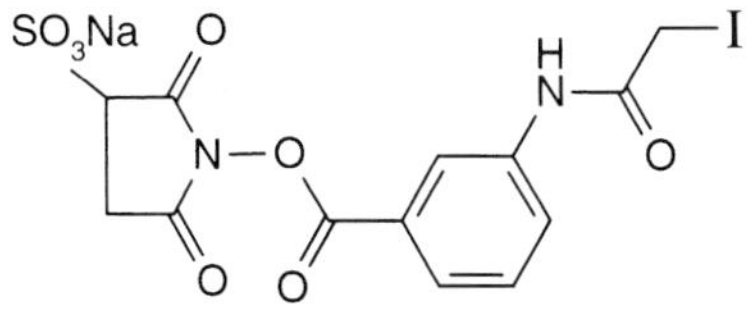

Sulfo-SIAB;
Sulfo-succinimidyl(4-iodoacetyl)-
aminobenzoate
MW 504.2
10.6 Å

Since SIAB is water insoluble, it must be dissolved first in organic solvent prior to addition to an aqueous reaction medium. The most commonly used solvents for this purpose include DMSO and DMF. Typically, a concentrated stock solution is prepared in one of these solvents and an aliquot added to the protein conjugation solution. Long-term storage of the reagent in these solvents is not recommended, however, due to slow uptake of water and breakdown of the NHS ester end.

Conjugations done with SIAB usually proceed by a multi-step process. Because the crosslinker's NHS ester end is its most labile functionality, an amine-containing protein or molecule is reacted first to create an iodoacetyl-activated intermediate (Figure 6.6). This iodoacetyl derivative is stable enough in aqueous solution to allow purification of the derivatized protein from excess reagent and other reaction byproducts without significant loss of activity. The only consideration is to protect the iodoacetyl derivative from light, which may generate iodine and reduce the activity of the intermediate. Finally, the modified protein is mixed with a sulfhydryl-containing molecule to effect the conjugation through a thioether bond. The result of such two-step procedures is to direct the coupling toward only sulfhydryls on the second molecule while avoiding the polymerization problems that can occur with single-step protocols. Conjugations done with SIAB should avoid buffer components containing amines (i.e., Tris, glycine, or imidazole) or sulfhydryls

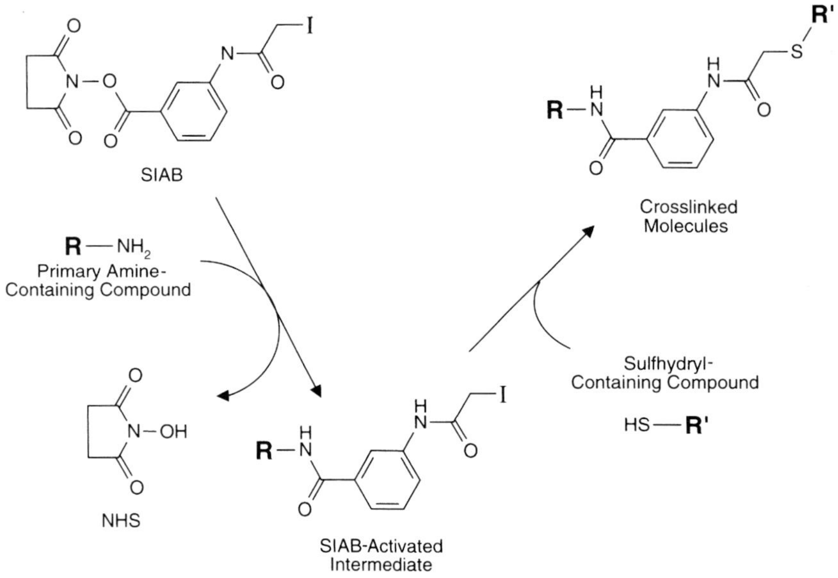

FIGURE 6.6 SIAB may be used to modify an amine-containing molecule for subsequent conjugation to a sulfhydryl-containing molecule.

(e.g., DTT, 2-mercaptoethanol, cysteine), since these will compete with the desired crosslinking reactions.

Sulfo-SIAB, sulfosuccinimidyl(4-iodoacetyl)aminobenzoate, is a water soluble analog of SIAB that contains a negatively charged sulfonate on its NHS ring. The negative charge lends enough hydrophilicity to the entire molecule to provide good solubility in aqueous solutions (up to about 10-m*M*). Sulfo-SIAB may be added directly to reaction mediums without prior dissolution in organic solvent, or solutions that are more concentrated may be made in water before transfer of an aliquot to the reaction to facilitate easy addition of small quantities. Aqueous stock solutions should be dissolved rapidly and used immediately to avoid excessive hydrolysis of the NHS ester.

SIAB and sulfo-SIAB have been used to make a high-capacity RNA affinity column for the purification of human IRP1 and IRP2 (Allerson *et al.*, 2003), to couple antibodies or Fab fragments to amine-modified microparticles (Härmä *et al.*, 2000), in the attachment of oligonucleotides to surfaces for detection arrays (Adessi *et al.*, 2000), and in the preparation of polymer bioconjugates for targeted drug delivery (Viadya *et al.*, 2011).

The following protocol illustrates the use of SIAB in preparing antibody–enzyme conjugates using β-galactosidase.

Protocol

1. Dissolve a specific antibody to be conjugated at a concentration of 10 mg/ml in 50-m*M* sodium borate, 5-m*M* EDTA, pH 8.3 (reaction buffer).
2. Dissolve SIAB (Thermo Fisher) in DMSO at a concentration of 1.4 mg/ml. Alternatively, dissolve sulfo-SIAB in deionized water at a concentration of 1.7 mg/ml. Prepare fresh and use immediately. Protect from light.
3. Add 100 μl of the SIAB stock solution to each ml of the antibody solution. Mix gently to dissolve.
4. React for 1 h at room temperature in the dark.
5. Purify the modified antibody by gel filtration on a desalting resin. Spin columns may be used to speed the separation process (Thermo Fisher). Perform the chromatography using the reaction buffer. To obtain good separation, apply sample at a ratio of no more than 8% of the total column gel volume. Monitor the eluting peak by using a small aliquot of each fraction and reacting it with a protein detection reagent such as Coomassie Protein Assay Reagent (Thermo Fisher) in a microplate. This avoids exposure of the entire modified protein fractions to UV light from a spectrophotometer, which could inactivate the iodoacetyl group. Collect the first peak eluting from the column, which contains the protein.
6. Add β-galactosidase to the activated antibody solution at a ratio of 4 mg of enzyme per mg of antibody.
7. React for 1 h at room temperature in the dark.
8. To block any remaining iodoacetyl sites, add cysteine to a final concentration of 5-m*M* and react for an additional 15 min at room temperature.
9. Purify the conjugate by gel filtration using a buffer of choice (e.g., PBS, pH 7.4).

1.6. SMPB and Sulfo-SMPB

SMPB, succinimidyl-4-(*p*-maleimidophenyl)butyrate, is a heterobifunctional analog of MBS (Section 1.4, this

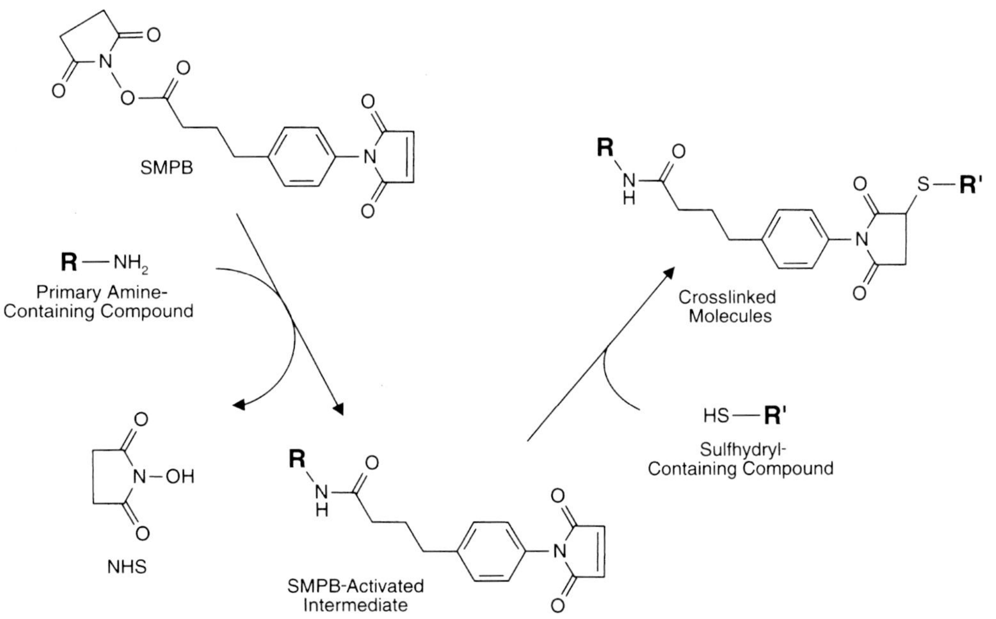

FIGURE 6.7 SMPB may be used in a two-step procedure to conjugate an amine-containing molecule to a sulfhydryl-containing compound, forming amide and thioether bonds, respectively.

chapter) containing an extended cross-bridge (Thermo Fisher). The reagent has an amine-reactive NHS ester on one end and a sulfhydryl-reactive maleimide group on the other end (Figure 6.7). Conjugates formed using SMPB thus are linked by stable amide and thioether bonds. A comparison with SPDP-produced conjugates concluded that SMPB formed more stable complexes that survive *in vivo* for longer periods (Martin and Papahadjopoulos, 1982).

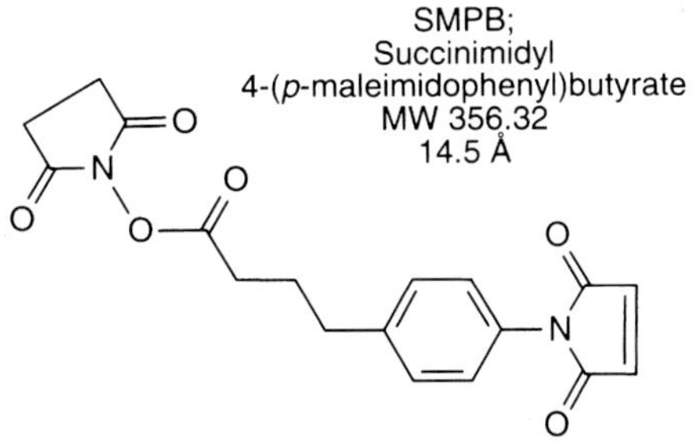

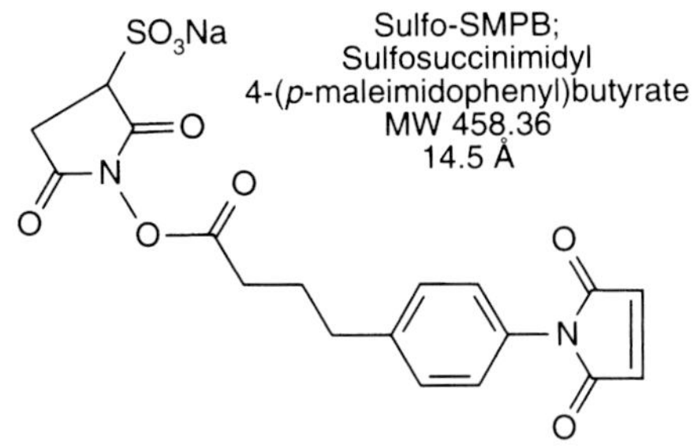

Conjugation reactions performed with SMPB are typically multi-step procedures, wherein a protein is modified through its amine groups, purified to remove excess reagent, and then mixed with a sulfhydryl-containing molecule to effect the final conjugation. The maleimide group of SMPB is highly specific for coupling to sulfhydryl-containing proteins and other molecules, thus directing the conjugation to discrete points on the second molecule. This maleimide is, however, more labile to ring opening in aqueous solution than the maleimide group of SMCC due to its proximity to an aromatic ring. Therefore, the first protein modified with SMPB (to obtain a maleimide-activated intermediate) should be purified quickly to prevent extensive activity loss from hydrolysis and maleimide ring opening.

SMPB contains a hydrophobic cross-bridge and relatively nonpolar ends, which allows the reagent to permeate membrane structures. Due to its water-insolubility, it must be dissolved in an organic solvent prior to adding an aliquot to a reaction mixture. The solvents DMF and DMSO work well for this purpose. A concentrated stock solution prepared in these solvents allows for easy addition of a small amount to a conjugation reaction. Long-term storage in these solvents is not recommended due to slow water pickup and possible hydrolysis of the NHS ester end.

A water soluble analog to SMPB, called sulfo-SMPB [sulfosuccinimidyl-4-(*p*-maleimidophenyl)butyrate] contains a negatively charged sulfonate group which lends considerable hydrophilicity to the molecule (Thermo Fisher). Sulfo-SMPB may be added directly to aqueous reaction mixtures without prior dissolution in organic solvent. Concentrated stock solutions made in water should be dissolved quickly and used immediately to prevent hydrolysis of the NHS ester.

SMPB or sulfo-SMPB have been used to conjugate preformed vesicles and Fab′ fragments in a liposome carrier study (Martin and Papahadjopoulos, 1982), to attach insulin molecules to reconstituted Sendai virus envelopes (Gitman *et al.*, 1985), to target loaded virus envelopes by covalently attaching insulin molecules to receptor-depleted cells (Gitman *et al.*, 1985b), to form alkaline phosphatase–Fab′ fragment conjugates for an ELISA (Teale and Kearney, 1986), to prepare peptide–protein immunogen conjugates (Iwai *et al.*, 1988), to study the transport of the variant surface glycoprotein of *Trypanosome brucia* (Bangs *et al.*, 1986), and to prepare immunotoxin conjugates (Myers *et al.*, 1989). SMPB or sulfo-SMPB have also been used to modify glass slides for coupling thiol-modified oligonucleotides (Zhang *et al.*, 2006), to investigate the protein decorin (Zhu *et al.*, 2005), to create conjugates with β-galactosidase that can cross the blood–brain barrier (Zhang and Pardridge (2005), and to develop a microRNA delivery system using virus-like particles (Pan *et al.*, 2012).

1.7. GMBS and Sulfo-GMBS

GMBS, *N*-(γ-maleimidobutyryloxy)succinimide ester, is a heterobifunctional crosslinking agent that contains an NHS ester on one end and a maleimide group on the other (Fujiwara *et al.*, 1988) (Thermo Fisher). Its internal cross-bridge contains a linear four-carbon spacer, resulting in 10.2-Å crosslinks between conjugated molecules (Figure 6.8). GMBS is water insoluble and therefore must be dissolved in organic solvent. Typically, a concentrated stock solution is prepared in DMF or DMSO just before use, and then an aliquot of the solution is transferred to the aqueous reaction medium. The result is the formation of a micro-emulsion that effectively supplies crosslinker to the aqueous phase.

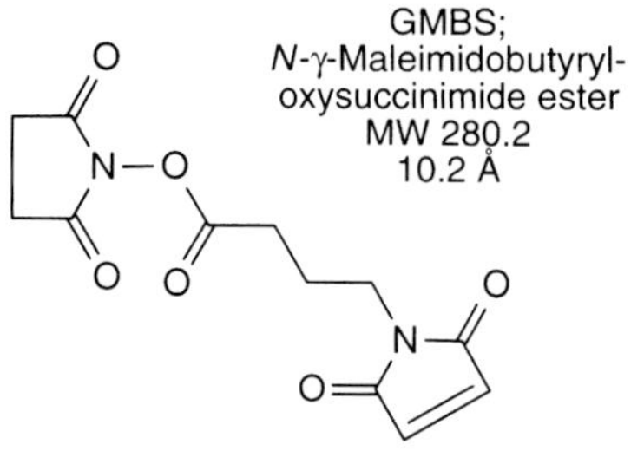

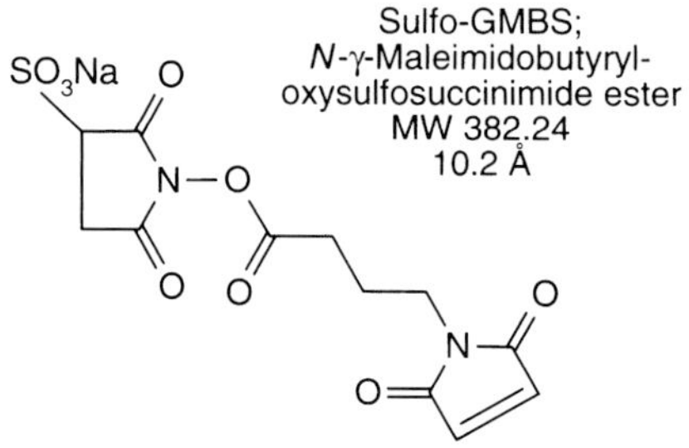

GMBS can be used in multi-step conjugation protocols wherein an amine-containing molecule or protein is first modified via the NHS ester end (its most labile reactive group) to create a stable amide bond. The derivative at this point contains reactive maleimide groups able to couple with the available sulfhydryl groups on a second protein or molecule. This active intermediate is then purified to remove excess reagent

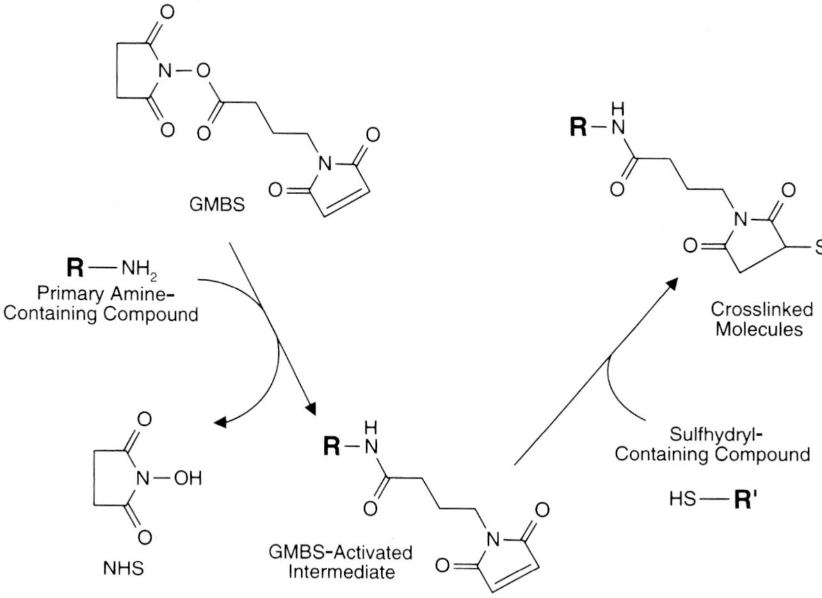

FIGURE 6.8 The reaction of GMBS with an amine-containing molecule yields a maleimide-activated intermediate that can then be used to crosslink with a sulfhydryl-containing compound.

and reaction byproducts, and immediately added to the sulfhydryl-containing molecule to effect the final conjugation.

The maleimide group of GMBS is adjacent to an aliphatic spacer, so its stability toward ring opening should be better than crosslinkers like MBS which contain adjacent aromatic groups. Hydrolysis of the maleimide group results in loss of sulfhydryl coupling capability. However, GMBS is not as stable as the hindered maleimide group of SMCC, since the cyclohexane ring of that reagent inhibits hydrolysis and ring opening.

Sulfo-GMBS, *N*-(γ-maleimidobutyryloxy)sulfosuccinimide ester, is a water soluble analog of GMBS containing a negatively charged sulfonate group on its NHS ring (Thermo Fisher). The charge provides enough hydrophilicity to allow at least 10-m*M* concentrations of the crosslinker to be made in aqueous reaction mediums. Its reactivity is identical to that of GMBS.

GMBS or sulfo-GMBS have been used for studying carnitine palmitoyltransferase 1 in its formation of a complex within the outer mitochondrial membrane (Faye *et al.*, 2007), for investigating protein organization of the postsynaptic density (Liu *et al.*, 2006), for studying the structure and dynamics of rhodopsin (Jacobsen *et al.*, 2006), and to produce conjugates of cell-penetrating peptides for delivery of antisense oligomers (Moulton, 2012).

The protocol for using GMBS or sulfo-GMBS in protein–protein crosslinking applications is similar to that of SMCC or sulfo-SMCC (see Section 1.3).

1.8. SBAP

SPAB is succinimidyl-3-(bromoacetamide)propionate, a short amine- and thiol-reactive reagent built from a central beta-alanine core. Inman *et al.* (1991) used this compound to modify BOC-protected lysine at its epsilon amine group in order to incorporate bromoacetyl reactive groups into synthesized peptides. SPAB may be used similar to SIAB in procedures for crosslinking proteins or other molecules containing amines or thiols.

The NHS ester end of the molecule can be used to modify proteins by reaction at pH 7 to 9 in phosphate, HEPES, or borate buffers. Use a buffer concentration of 50- to 100-m*M* to prevent pH drift to acidic values as the reaction takes place. SPAB can be dissolved in DMSO as a concentrated stock solution before adding an aliquot to the aqueous reaction medium. After purification of the modified protein by gel filtration or dialysis, the bromoacetyl group on the first protein may be coupled to a thiol-containing protein by reaction at pH 8.5 in borate buffer. The bromoacetyl group on the crosslinker and the intermediate activated protein should be protected from light to prevent degradation. In addition, reducing agents should be avoided (especially those containing thiols) to prevent inactivation and coupling to the reactive group prior to coupling with the desired thiol-containing protein.

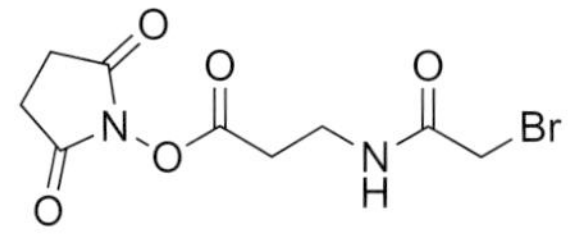

SBAP;
Succinimidyl-3-(bromoacetomido)propionate
MW: 307.10

1.9. SIA

The shortest amine-reactive and thiol-reactive crosslinker available is SIA, which is simply the NHS ester of iodoacetate, also called succinimidyl iodoacetate or NHS-iodoacetate. After conjugation, the crosslinked molecules or proteins are held together by only a 1.5-Å spacer represented by the two-carbon length of the acetate core. When conjugating proteins using this reagent, it is unlikely that there will be any cross-bridge-related artifacts introduced into the conjugate, such as extreme hydrophobicity or immunogenicity. The reactivity and use of SIA are similar to those of SIAB, which has identical reactive groups but is built on a longer cross-bridge.

The NHS ester end of SIA is typically reacted with an amine-containing protein or molecule first, because it is the most labile end of the reagent. During handling and use, SIA should be protected from light to prevent degradation of the iodoacetyl group. In addition, avoid the presence of reducing agents (particularly thiol-containing compounds), which will react with the iodoacetyl end and inactivate it. The compound can be dissolved before use in dry DMSO as a concentrated stock solution and then an aliquot of this solution is added to an aqueous reaction mixture to initiate the amine modification reaction using the NHS ester end. After the first protein is modified, the excess crosslinker is removed by size exclusion chromatography or dialysis before the addition of a thiol-containing protein to link the two molecules together.

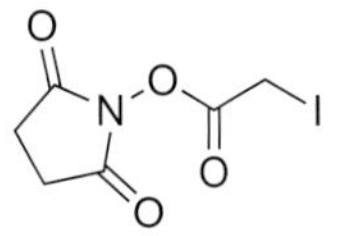

SIA;
Succinimidyl iodoacetate
(NHS-iodoacetate)
MW: 283.02

Thorpe *et al.* (1984) used SIA to conjugate intact ricin to an antibody and by virtue of the short linkage arm was able to effectively block the galactose binding site on the B chain. In addition, Rector *et al.* (1978) used this reagent to activate IgG for conjugation with thiolated proteins in a highly controlled reaction, which produced conjugates of defined composition. More recently, Choi *et al.* (2011) used SIA to prepare a RGD labeled metalloproteinase 2 probe as an antitumor agent, and McCarthy *et al.* (2012) used it to prepare a thrombus targeting nanoparticle conjugated with recombinant tissue plasminogen activator (tPA).

2. CARBONYL REACTIVE AND SULFHYDRYL REACTIVE CROSSLINKERS

A relatively new set of heterobifunctional crosslinking agents now are available which contain a carbonyl-reactive group on one end and a sulfhydryl-reactive functionality on the other end. The main utility of these reagents is in conjugating carbohydrate-containing molecules, such as glycoproteins, to sulfhydryl-containing molecules. Both polysaccharide residues and sulfhydryl groups are usually present on proteins in limiting quantities and at discrete sites. In certain cases, conjugation through these groups can direct the coupling reaction away from critical active centers or binding sites, thus preserving activity of the proteins after crosslinking. A prime example of the advantages of this type of directed coupling can be seen when conjugating antibody molecules to other proteins, such as enzymes. The carbohydrate residues of immunoglobulin molecules often occur on the Fc portion, away from the antigen binding sites. Coupling procedures which direct the crosslinking reaction to parts on the antibody removed from the antigen binding sites have the best chance of retaining activity after conjugate formation. However, some antibodies do contain glycosylation sites in the Fab region of the molecule, thus making conjugation strategies through carbohydrates less certain as to their affect on antigen binding activity (Endo *et al.*, 1995; Mattu *et al.*, 1998).

The carbonyl-reactive group on these crosslinkers is a hydrazide that can form hydrazone bonds with aldehyde residues. To utilize this functional group with carbohydrate-containing molecules, the sugars must first be mildly oxidized to contain aldehyde groups by treatment with sodium periodate. Oxidation with this compound will cleave adjacent carbon–carbon bonds which possess hydroxyl groups, as are abundant in polysaccharide molecules (Chapter 2, Sections 2 and 4.4).

Two types of sulfhydryl-reactive functions are available on these reagents: pyridyl disulfide groups and maleimide groups. The pyridyl disulfide group will react with a sulfhydryl residue to create a disulfide bond. This linkage is reversible by treatment with a disulfide reducing agent. Reaction of a maleimide group with a sulfhydryl, however, forms a permanent thioether bond of good stability. Thus, either reversible or permanent conjugates may be designed using these heterobifunctionals.

2.1. MPBH

MPBH, or 4-(4-*N*-maleimidophenyl)butyric acid hydrazide, is a heterobifunctional crosslinking agent that contains a carbonyl-reactive hydrazide group on one end and a sulfhydryl-reactive maleimide on the other end (Thermo Fisher). The cross-bridge between the two functional ends provides a 17.9-Å spacer. The hydrazide group is produced as the hydrochloride salt. The reagent as a whole has good water solubility. It can be dissolved in 0.1-*M* sodium acetate, pH 5.5, up to a concentration of 327 mg/ml. It is also freely soluble in DMSO and may be stored as a concentrated stock solution in this solvent without degradation.

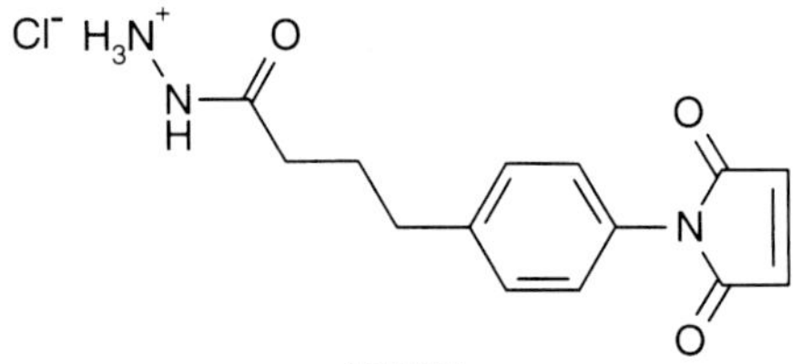

MPBH
4-(4-*N*_Maleimidophenyl)butyric acid hydrazide hydrochloride
MW 309.75
17.9 Å

The maleimide group of MPBH is adjacent to an aromatic ring and thus may exhibit instability to hydrolysis in aqueous solutions, especially at alkaline pH. Hydrolysis opens the maleimide ring and destroys its coupling ability with sulfhydryls. However, both reactive ends of the crosslinker are stable enough to survive a multistep coupling protocol without extensive loss of activity. Thus, a sulfhydryl-containing protein or molecule may be modified via the maleimide end of MPBH, the derivative purified by gel filtration to remove excess reactants, and then mixed with a glycoprotein (that has previously been oxidized to provide aldehyde residues) to effect the final conjugation (Figure 6.9). The opposite approach is also possible: modification of the glycoprotein first, purification, and subsequent mixing with a sulfhydryl-containing molecule. With this second option, however, the purification step should be carried out quickly to prevent extensive hydrolysis of the maleimide group.

MPBH has been used to conjugate CD4 without loss of biological activity (Chamow *et. al.*, 1992), to prepare a polyethyleneimine–lipid conjugate as a pH-sensitive carrier for gene delivery (Sawant *et al.*, 2012), and in the development of a multiplexed bead-based assay for profiling glycosylation patterns (Li *et al.*, 2011).

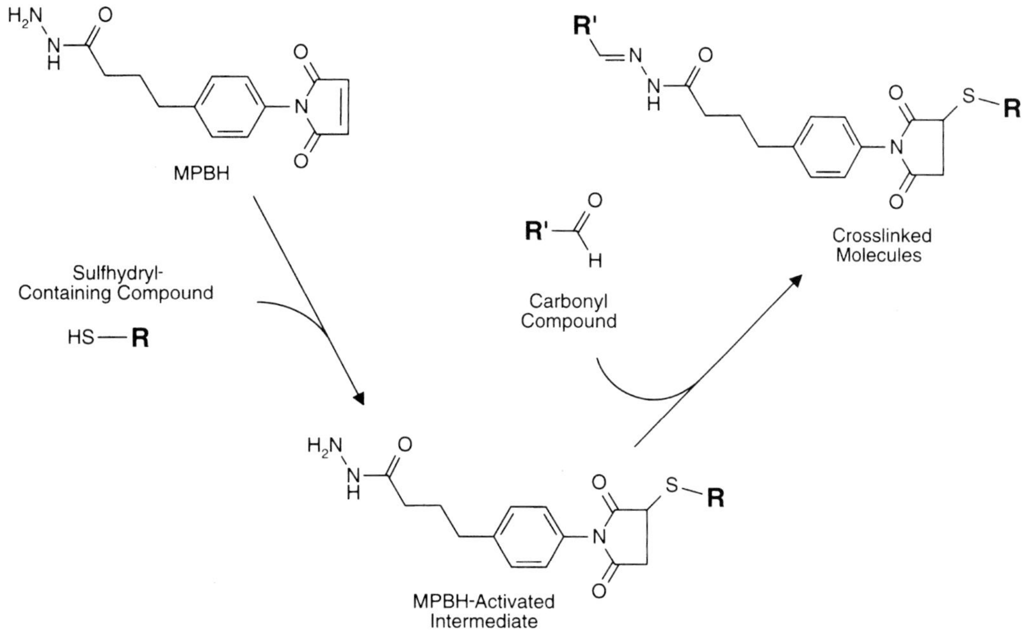

FIGURE 6.9 MPBH reacts with sulfhydryl-containing molecules through its maleimide end to produce thioether linkages. Its hydrazide group can then be used to conjugate with carbonyl-containing molecules (such as periodate oxidized carbohydrates that contain aldehydes) to give hydrazone bonds.

2.2. M_2C_2H

M_2C_2H, 4-(*N*-maleimidomethyl)cyclohexane-1-carboxyl-hydrazide, is a heterobifunctional crosslinking agent that contains a carbonyl-reactive hydrazide group on one end and a sulfhydryl-reactive maleimide group on the other end (Thermo Fisher). The reagent is similar to MPBH (described previously), but the maleimide group on M_2C_2H is expected to be more stable in aqueous solutions, since it is adjacent to an aliphatic cyclohexane ring instead of an aromatic phenyl group. In this sense, the cross-bridge of M_2C_2H is nearly identical to that of SMCC, which contains one of the most stable maleimide groups known. The hydrophobic, hindered environment of the cyclohexane ring should provide similar stability advantages to this reagent. Reaction of the maleimide group with a sulfhydryl residue results in the formation of a stable thioether bond.

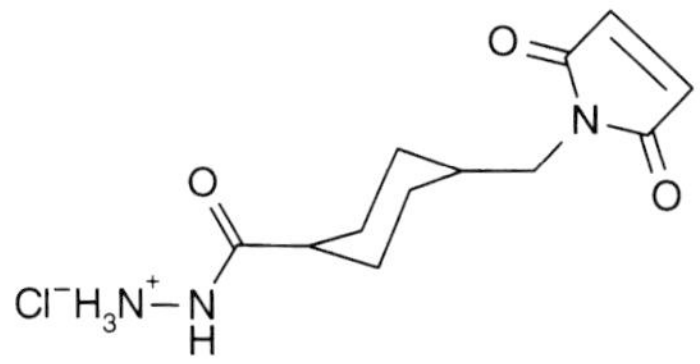

M_2C_2H
4-(*N*-Maleimidomethyl)cyclohexane-1-carboxyl-hydrazide hydrochloride
MW 287.75
15.1 Å

On the other end of the crosslinker, the hydrazide functional group can react with periodate-oxidized carbohydrate molecules or with the reducing end of carbohydrates to form hydrazone linkages (Chapter 2, Sections 2 and 4.5). Thus, glycoproteins can be targeted specifically at their polysaccharide chains, avoiding crosslinking at active sites, which can lead to activity losses (Figure 6.10). The crosslinker has been used to produce polymer bioconjugates for drug delivery (Vaidya *et al.*, 2011).

M_2C_2H is slightly soluble in aqueous solutions, reportedly having a maximal solubility of 3.2 mg/ml in 0.1-*M* sodium acetate at pH 5.5. It is also soluble in organic solvents, which allows for the preparation of concentrated stock solutions to be made prior to addition of a small aliquot to an aqueous reaction mixture. The crosslinker is particularly stable in acetonitrile.

2.3. PDPH

PDPH, 3-(2-pyridyldithio)propionyl hydrazide, is a heterobifunctional reagent that possesses a carbonyl-reactive hydrazide group on one end and a sulfhydryl-reactive pyridyl disulfide group on the other end (Thermo Fisher). Thus, sulfhydryl-containing proteins or other thiol molecules may be conjugated to carbohydrate-containing molecules (after treatment of the polysaccharide portion with sodium periodate to create aldehyde residues) (Figure 6.11). Using this crosslinker,

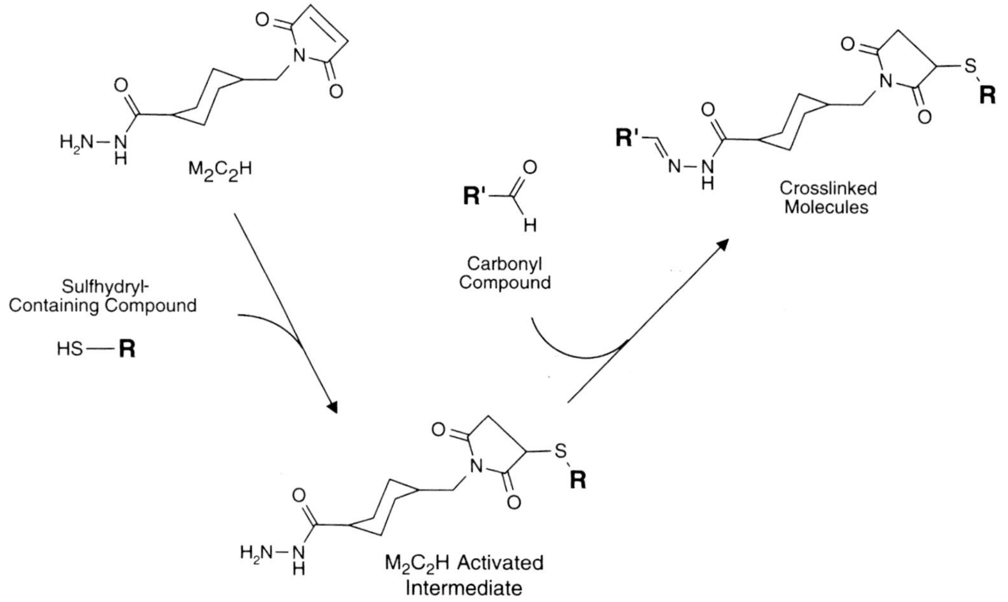

FIGURE 6.10 M_2C_2H can be used to crosslink a sulfhydryl-containing molecule with an aldehyde-containing compound. Glycoproteins may be conjugated using this reagent after treatment with sodium periodate to form reactive aldehyde groups.

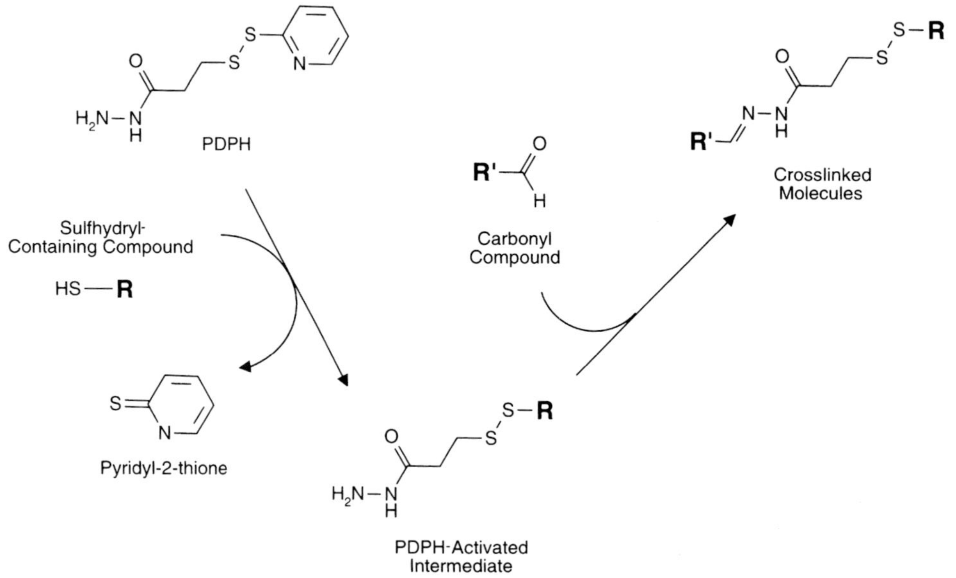

FIGURE 6.11 PDPH reacts with thiol-containing compounds through its pyridyl disulfide end to form reversible disulfide linkages. Its hydrazide end may then be subsequently conjugated with an aldehyde-containing molecule to form hydrazone bonds. Glycoproteins may be crosslinked using this approach after periodate activation to generate aldehyde groups.

glycoproteins can be coupled specifically through their carbohydrate chains, in many cases better avoiding active centers or binding sites than when coupling through abundant polypeptide groups like amines. Since the pyridyl disulfide group reacts with sulfhydryls to create disulfide bonds, the crosslinked proteins can be cleaved by reduction with DTT (Chapter 2, Section 4.1).

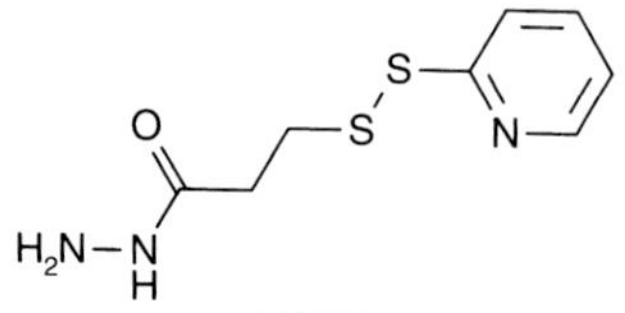

PDPH
3-(2-Pyridyldithio) propionyl hydrazide
MW 229.31
9.2 Å

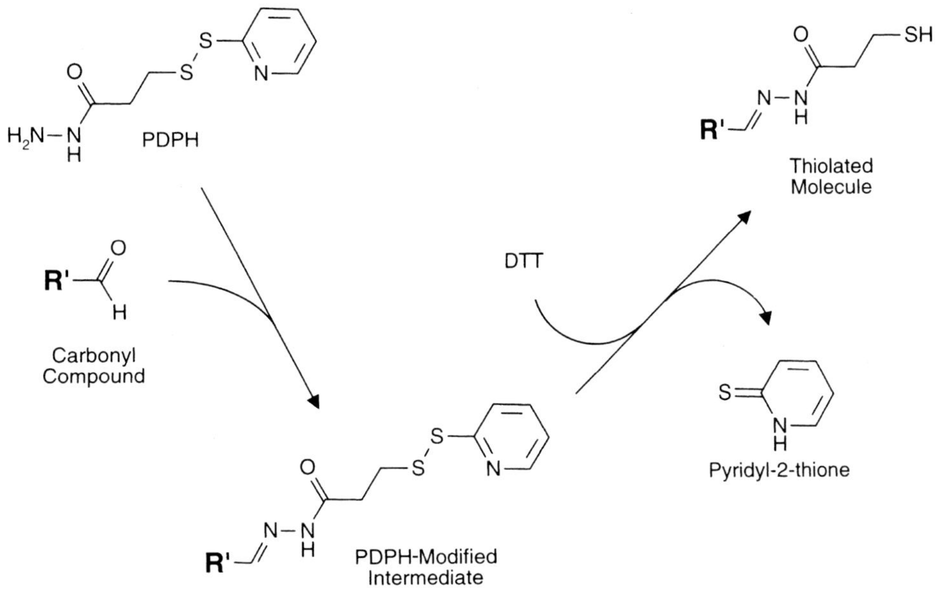

FIGURE 6.12 PDPH may be used to add a sulfhydryl group to an aldehyde-containing molecule. After reacting its hydrazide end with the aldehyde to form a hydrazone bond, the pyridyl disulfide may be reduced with DTT to create a free thiol.

PDPH may also be used as a thiolation reagent to add sulfhydryl functional groups to carbohydrate molecules. The reagent can be used in this sense similar to the protocol described for AMBH (Chapter 2, Section 4.1). After modification of an oxidized polysaccharide with the hydrazide end of PDPH, the pyridyl group is removed by treatment with DTT, leaving the exposed sulfhydryl (Figure 6.12).

PDPH is soluble in 0.1-*M* sodium acetate, pH 5.5, at a maximal concentration of 14.2 mg/ml. The reagent is particularly stable in acetonitrile for preparation of concentrated stock solutions.

PDPH has been used in the preparation of immunotoxin conjugates (Zara *et al.*, 1991). It has also been used to create a unique conjugate of nerve growth factor (NGF with an antibody directed against the transferrin receptor OX-26, which could traverse the blood–brain barrier (Friden, 1993). Labeling of antibody molecules with PDPH at oxidized polysaccharide sites followed by reduction to free the sulfhydryl has been used to form a technetium-99 m complex for radiopharmaceutical use (Ranadive *et al.*, 1993) (Chapter 12, Section 1.5). PDPH has also been used to study the uptake of rsCD4 across the blood–brain barrier (Walus *et al.*, 1996), to prepare an NGF conjugate (Bäckman *et al.*, 1996), to study novel engineered cell surface receptors (Lee *et al.*, 1999), and to prepare siRNA conjugates with poly (D,L-lactic-co-glycolic acid) (PLGA) via a cleavable disulfide bond (Lee *et al.*, 2011).

3. AMINE-REACTIVE AND PHOTOREACTIVE CROSSLINKERS

An important class of heterobifunctional reagents is the photoreactive crosslinkers that have one end that can be photolyzed to initiate coupling. Photoreactive crosslinkers may be designed to utilize any one of a number of photosensitive groups, including aryl azides, fluorinated aryl azides, benzophenones, anthraquinones, certain diazo compounds, and diazirine derivatives (Chapter 3, Section 7.5). The best photoreactive groups are stable in aqueous solution in the dark, and may be activated at the desired time by a pulse of light at the appropriate wavelength. The other end of these heterobifunctionals usually contains a spontaneously reactive functionality that will couple rapidly with certain groups present on target molecules. This secondary functionality is sometimes called *thermoreactive* to differentiate it from the photoreactive end and to emphasize its ready reactivity or sometimes its labile nature in aqueous environments. The thermoreactive end is typically amine reactive, sulfhydryl reactive, carbonyl reactive, carboxylate reactive, or arginine reactive. Still another class of photoreactive heterobifunctionals may use a biotin handle at one end to crosslink specifically, but noncovalently, with avidin or streptavidin molecules (Chapter 11, Section 6.5).

Photoreactive groups can be categorized by the reactive species that is generated upon photolysis. The most popular type of photosensitive group, an aryl azide derivative, forms a short-lived nitrene that reacts extremely rapidly with the surrounding chemical environment (Gilchrist and Rees, 1969). Subsequent evidence, however, indicates that the photolyzed intermediates of aryl azides can undergo ring expansion to create nucleophile-reactive dehydroazepines. Instead of inserting nonselectively at active carbon–hydrogen bonds, dehydroazepines have a tendency to react preferentially with nucleophiles, especially amines (Figure 6.13). However, some investigators have shown that aryl azides that possess a perfluorinated ring structure or are substituted completely with halogen atoms are quite efficient at forming the desired nitrene intermediate (Keana and Cai, 1990; Cai *et al.*, 1993; Schnapp and Platz, 1993; Schnapp *et al.*, 1993; Soundararajan *et al.*, 1993; Yan *et al.*, 1994). A few crosslinking reagents now use halogen-substituted phenyl azides to provide greater efficiency of photoreactive insertion into target molecules (Chapter 24, Sections 2.2 and 3.2).

One advantage of aryl azide photoreactive crosslinkers is that they have a relatively low energy of activation, which is optimal in the long UV region. In addition, many aryl azides possess nitro groups on their associated aromatic ring structures. These electron-withdrawing groups tend to increase the optimal wavelength for photolysis upwards close to the 350-nm range. The benefit of this approach is that relatively low light exposure at the higher energy UV wavelengths avoids potential bond breakage that may occur with some sensitive compounds upon photolysis. In addition, some biological molecules can undergo crosslinking reactions upon irradiation with UV light of less than 300 nm (e.g., DNA).

Other phenyl azide-containing reagents possess hydroxyl groups on their aromatic rings. These electron-donating groups activate the ring system to allow electrophilic substitution reactions to occur on the crosslinker prior to its use. A major application of this

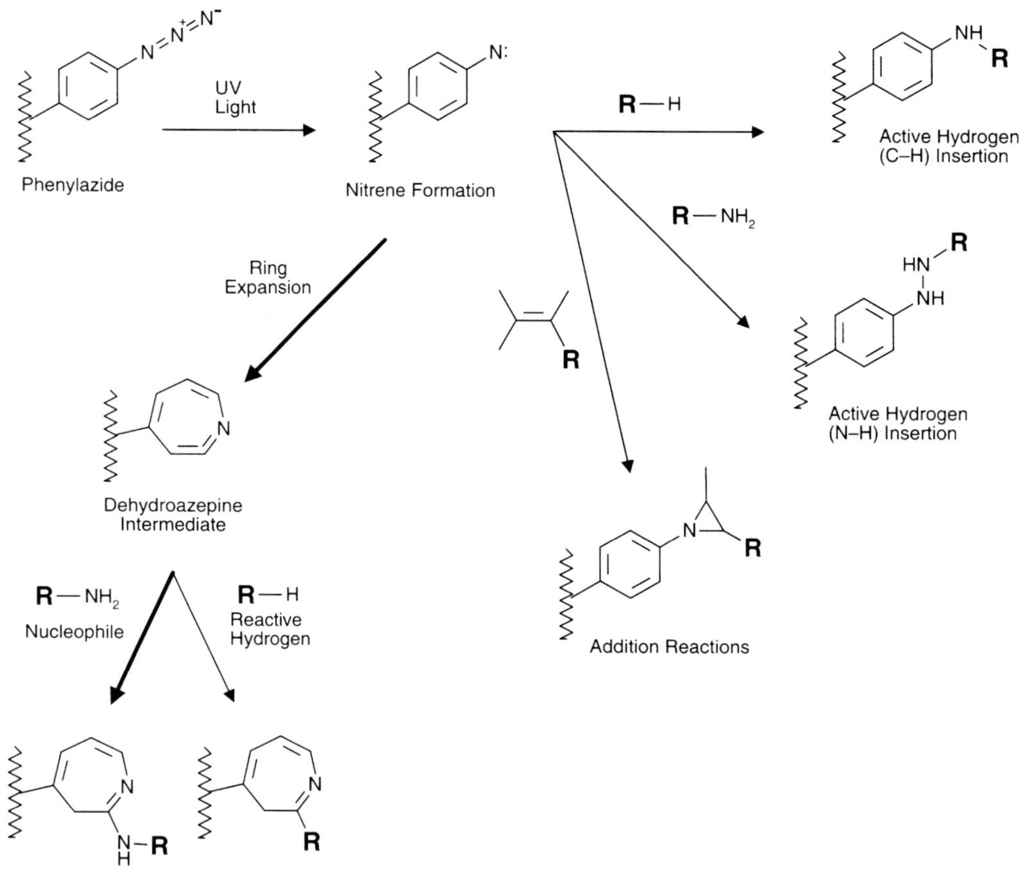

FIGURE 6.13 Photoactivation of a phenyl azide group with UV light results in the formation of a short-lived nitrene. Nitrenes may undergo a number of reactions, including insertion into active carbon–hydrogen or nitrogen–hydrogen bonds and addition to points of unsaturation in carbon chains. The most likely route of reaction, however, is to ring-expand to a dehydroazepine intermediate. This group is highly reactive toward nucleophiles, especially amines.

ability is to radioiodinate the photoreactive end, thus permitting crosslinking and detection of proteins within samples.

Suitable light sources for photolyzing include sunlamps manufactured by a number of companies, such as Philips Ultrapnil MLU 300 W, General Electric sunlamp RSM 275 W, or National Self-Ballasted BHRF 240–250 V 250 W W-P lamp. Irradiation for 15 min with such lamps while the sample is cooled in an ice bath will result in good photolysis of photoreactive crosslinkers and modification reagents. In addition, many long-wavelength UV light sources with irradiation capability at about 366 nm work well.

Although photoreactive aryl azides are relatively inert to thermochemical reactions prior to photolysis, they are not stable in the presence of sulfhydryl compounds which can reduce the azide functionality to an amine with concomitant release of N_2. Avoid, therefore, the use of reductants such as DTT or 2-mercaptoethanol before the photoreaction step, as these can react with the aryl azide within minutes (Staros *et al.*, 1978). Also, avoid amine-containing buffer components such as Tris or glycine, because of the potential for nucleophilic reactivity with the photolyzed dehydroazepine intermediate formed from photolysis of unsubstituted phenyl azides.

Of the following amine-reactive and photoreactive crosslinkers, the overwhelming majority use an aryl azide group as the photosensitive functional group. Only a few use alternative photoreactive chemistries, particularly perfluorinated aryl azide, benzophenone, or diazo compounds. For general background information on photoreactive crosslinkers, see Das and Fox (1979), Kiehm and Ji (1977), Vanim and Ji (1981), and Brunner (1993).

3.1. NHS–ASA, Sulfo-NHS–ASA, and Sulfo-NHS–LC–ASA

NHS–ASA (*N*-hydroxysuccinimidyl-4-azidosalicylic acid) is a heterobifunctional reagent containing an NHS ester on one end and a photoreactive aryl azide group on the other (Thermo Fisher). The amine-reactive NHS ester can be reacted with proteins or other primary amine-containing molecules to yield a photosensitive derivative suitable for probing biological interaction sites. Upon photolysis with a long UV light source, the aryl azide end is activated to covalently complex with closely associated target molecules (Figure 6.14). The small cross-bridge of NHS–ASA is built from a salicylate derivative that contains a hydroxyl group on the aromatic ring. The ring-activating nature of this group provides an iodination site on the crosslinker to allow tracking of modified molecules (Ji and Ji, 1982) (Chapter 12, Section 2.5).

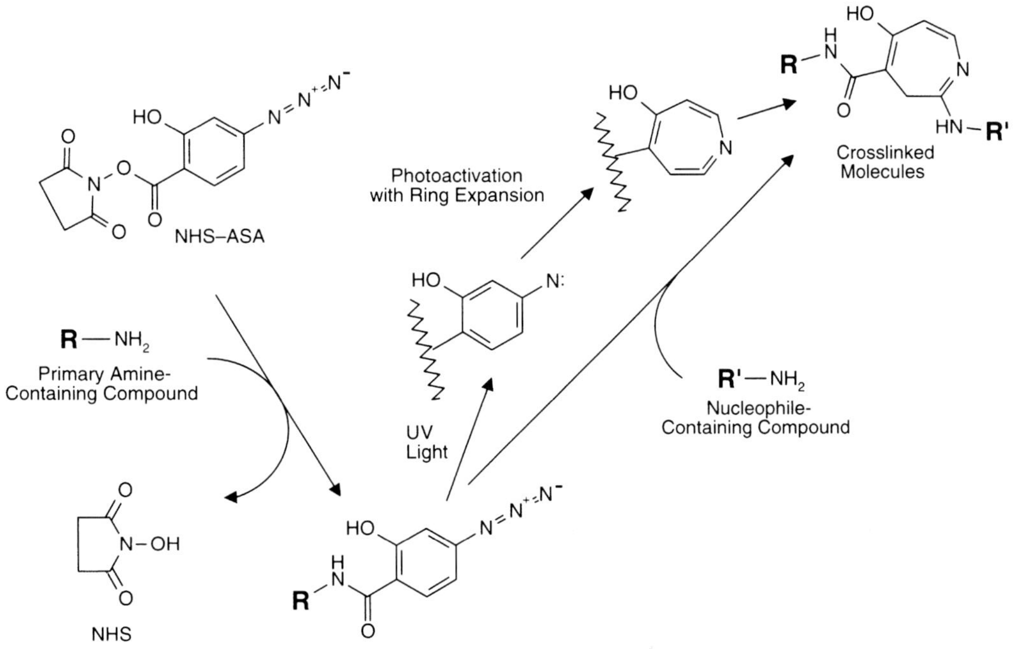

FIGURE 6.14 NHS–ASA reacts with amine-containing compounds to form stable amide linkages. Photoactivation with UV light results in ring expansion to a dehydroazepine intermediate, which can react with amines to form covalent bonds.

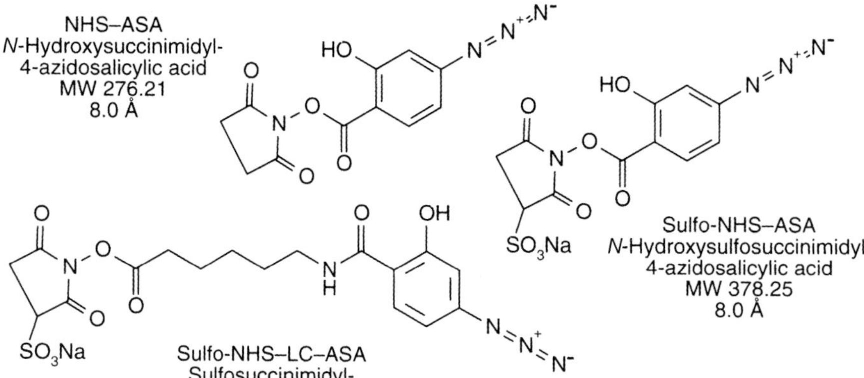

NHS–ASA
N-Hydroxysuccinimidyl-4-azidosalicylic acid
MW 276.21
8.0 Å

Sulfo-NHS–ASA
N-Hydroxysulfosuccinimidyl-4-azidosalicylic acid
MW 378.25
8.0 Å

Sulfo-NHS–LC–ASA
Sulfosuccinimidyl-(4-azidosalicylamido)hexanoate
MW 491.41
18.0 Å

Reported applications of NHS–ASA include photoaffinity labeling of ^{125}I–ASA–Con A to erythrocyte ghosts (Ji and Ji, 1982), derivatization of human choriogonadotropin with ^{125}I–NHS–ASA with photo-initiated crosslinking of the α-β dimer (Ji *et al.*, 1985), radiolabeling of D-glucose and conjugation of the sugar to the human erythrocyte monosaccharide transporter protein (Shanahan *et al.*, 1985), photoaffinity labeling of a bacterial sialidase (van der Horst *et al.*, 1990), identification of the peptide binding site of DnaK (Zhang and Walker, 1996), and characterization of the vitamin B-12 receptor (Nguyen, 1999).

Two analogs of NHS–ASA that provide alternative physical characteristics are available. Sulfo-NHS–ASA is a water soluble version of the crosslinker that contains a negatively charged sulfonate group on its NHS ring. Sulfo-NHS–LC–ASA also has the water-solubility advantage provided by a sulfonate, but it possesses a longer cross-bridge made from a 6-aminocaproic acid chain in its internal structure. The longer spacer increases the potential distance between conjugated molecules, thus allowing more flexibility in the experimental design. Both analogs are still iodinatable to provide radiolabeling capability.

3.2. SASD

SASD (sulfosuccinimidyl-2-(*p*-azidosalicylamido) ethyl-1,3′-dithiopropionate) is a heterobifunctional crosslinker containing a photoreactive group and an amine-reactive NHS ester (Thermo Fisher). The NHS ring possesses a negatively-charged sulfonate group which lends water solubility to the reagent. The cross-bridge of SASD contains a central disulfide group that provides cleavability after conjugation. Reaction with a disulfide reducing agent such as DTT breaks the disulfide bond and releases the crosslinked molecules. The photosensitive end of SASD is built from a salicylic acid derivative which contains a ring-activating hydroxyl group. Due to the presence of this group, the crosslinker can be radiolabeled with ^{125}I prior to a conjugation reaction. Iodination occurs *ortho* or *para* to the hydroxyl group on the phenyl ring, next to the aryl azide function (Figure 6.15) (Chapter 12, Section 2.5).

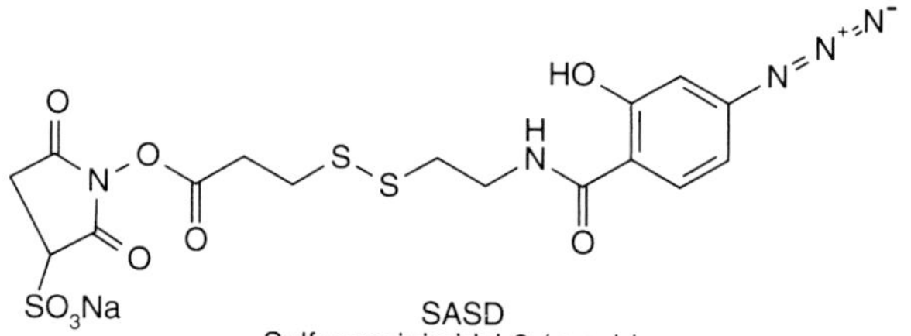

SASD
Sulfosuccinimidyl-2-(*p*-azido-salicylamido)ethyl-1,3'-dithiopropionate
MW 541.51
18.9 Å

The combination of radiolabeling and cleavability provides the ability to detect the fate of the protein that retains the radiolabel after disulfide reduction. Thus, for investigations involving biomolecular interactions, a purified protein can be labeled with SASD through its amine groups via the NHS ester end of the crosslinker, allowed to interact *in vivo* with unknown target proteins, and photolyzed to effect a crosslink with these unknown substances. Subsequently the complex can be localized in the cell or effectively isolated by following the radiolabel. Alternatively, the conjugate can be cleaved by reduction, which results in the label being transferred to the unknown interacting protein, and its fate determined or the identity of the unknown protein revealed through the radiolabel (Figure 6.16). Such label transfer reagents are described in more detail in Chapter 24, Section 3.

Reported applications of SASD involve modification of lipopolysaccharide (LPS) molecules and studying their interaction with albumin and an antibody directed against LPS (Wollenweber and Morrison, 1985), identification of the murine interleukin-3 receptor and an *N*-formyl peptide receptor (Sorenson *et al.*, 1986), crosslinking of factor V and Va to iodinated peptides (Chattopadhyay *et al.*, 1992), and a comparison of radiolabeling techniques for the crosslinker (Shephard *et al.*,

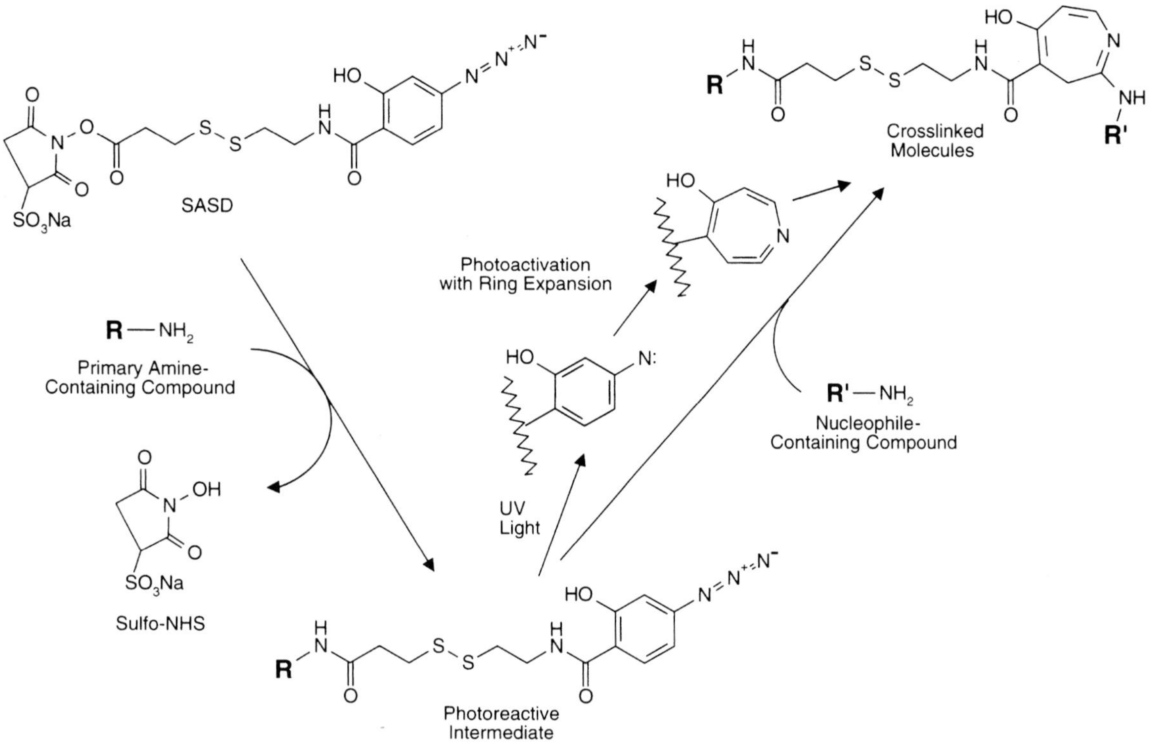

FIGURE 6.15 SASD is a photoreactive crosslinker that can be used to modify amine-containing compounds through its NHS ester end and subsequently photoactivated to initiate coupling with nucleophiles (after ring expansion to an intermediate dehydroazepine derivative). The crosslinks may be selectively cleaved at the internal disulfide group using DTT.

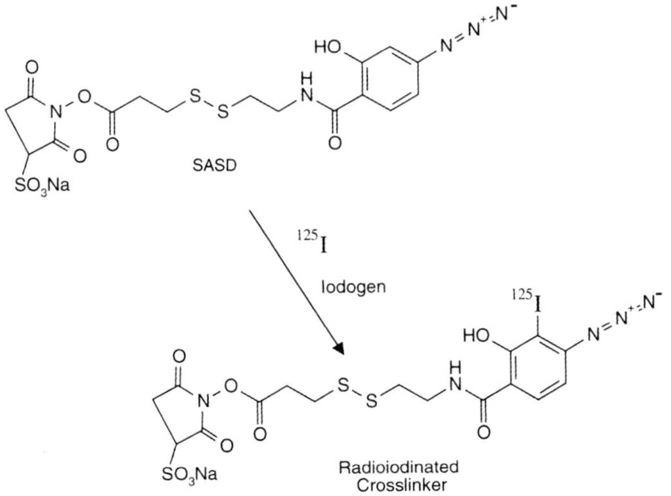

FIGURE 6.16 The hydroxyl group on the phenyl azide ring of SASD may be iodinated with ^{125}I to allow radiolabeling studies to be done on photolyzed conjugates.

1988). Other studies have involved the investigation of protein interactions using the label transfer nature of radioiodinated SASD (Gupta *et al.*, 2005; Lindersson *et al.*, 2005; LeFebvre *et al.*, 2006) and the development of photoaffinity probes for studying carbohydrate biology (Yu *et al.*, 2012).

The best radiolabeling technique for SASD is to use the Iodogen method (Shephard *et al.*, 1988) described in Chapter 12, Section 2.3. The following suggested protocol for using SASD was based on the method described in the Thermo Fisher Catalog.

Protocol

The following operations should be performed using standard safety procedures for working with radioactive compounds. All steps involving SASD prior to initiation of the photoreaction should be carried out protected from light to avoid loss of phenyl azide activity. The radiolabeling procedure should be performed quickly to prevent excessive loss of NHS ester activity due to hydrolysis.

1. Radiolabel 55 nmol of SASD using Iodogen (Thermo Fisher) and 40 μCi $Na^{125}I$ for 30 s. Do not use Chloramine-T, since termination of the iodination reaction with this reagent involves addition of a reducing agent which may cleave the disulfide bonds of the crosslinker.
2. Terminate the iodination by removing the SASD solution from the Iodogen reagent using a transfer pipette. Be careful not to carry any solid Iodogen reagent with the transfer. Since free radioactive iodine may still be present in the solution, it may be necessary to add an iodine scavenger to prevent

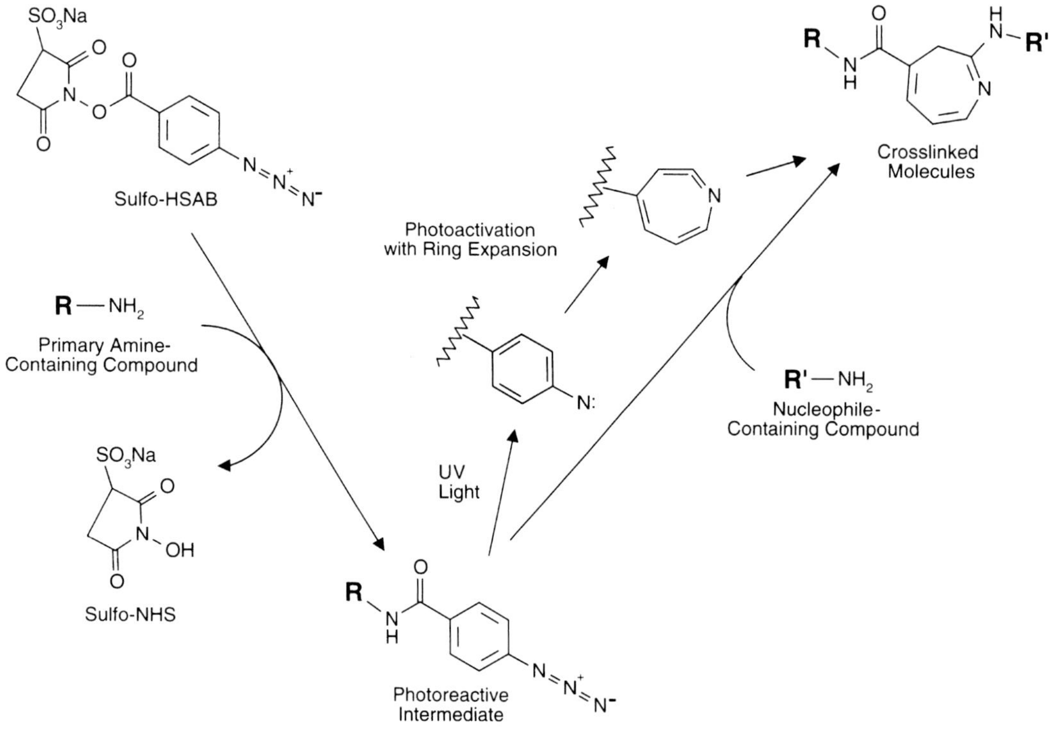

FIGURE 6.17 Sulfo-HSAB is a short photoreactive crosslinker that can be used to modify amine-containing molecules through its NHS ester end to form amide linkages. After photoactivation, the phenyl azide group can react with amines to create a covalent bond.

the possibility of radiolabels being incorporated into the proteins being crosslinked. Suitable scavengers include tyrosine or *p*-hydroxyphenylacetic acid. Adding these compounds in molar excess to the amount of iodine present will prevent any secondary modifications from occurring. Immediately add the radiolabeled SASD solution to the equivalent of 16 nmol of a protein to be modified. The protein should be dissolved previously in a minimum quantity of 0.1-*M* sodium borate, pH 8.4 (conjugation buffer). The more concentrated the protein, the more efficient will be the modification reaction.

3. React for 30 min to create the SASD derivative, coupled through the NHS ester reactive group of the crosslinker onto available amine groups of the protein (forming amide bonds).
4. Purify the modified protein by desalting using a desalting resin and performing the chromatography using a buffer of choice. Pool fractions containing protein. The protein should be radiolabeled at this point and also contain photoreactive phenyl azide groups from the SASD modification.
5. Add the SASD-modified protein to a second protein or other molecule to be conjugated. After mixing, expose the solution to long-wave UV light for 10 to 15 min at room temperature to effect the conjugation. The solution may be kept on ice to prevent over heating of sensitive proteins.

3.3. HSAB and Sulfo-HSAB

HSAB (*N*-hydroxysuccinimidyl-4-azidobenzoate) is a heterobifunctional reagent containing an amine-reactive NHS ester on one end and a photoreactive phenyl azide group on the other end (Thermo Fisher). The small cross-bridge, built from a benzoic acid group, provides crosslinking ability at short intermolecular distances. Reaction of one protein *via* the NHS ester end of the crosslinker provides a stable derivative that can be incubated with a target molecule and then photolyzed to effect the final conjugation (Figure 6.17).

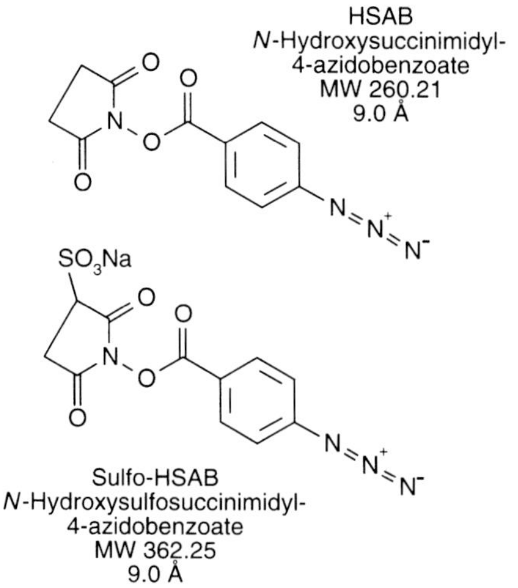

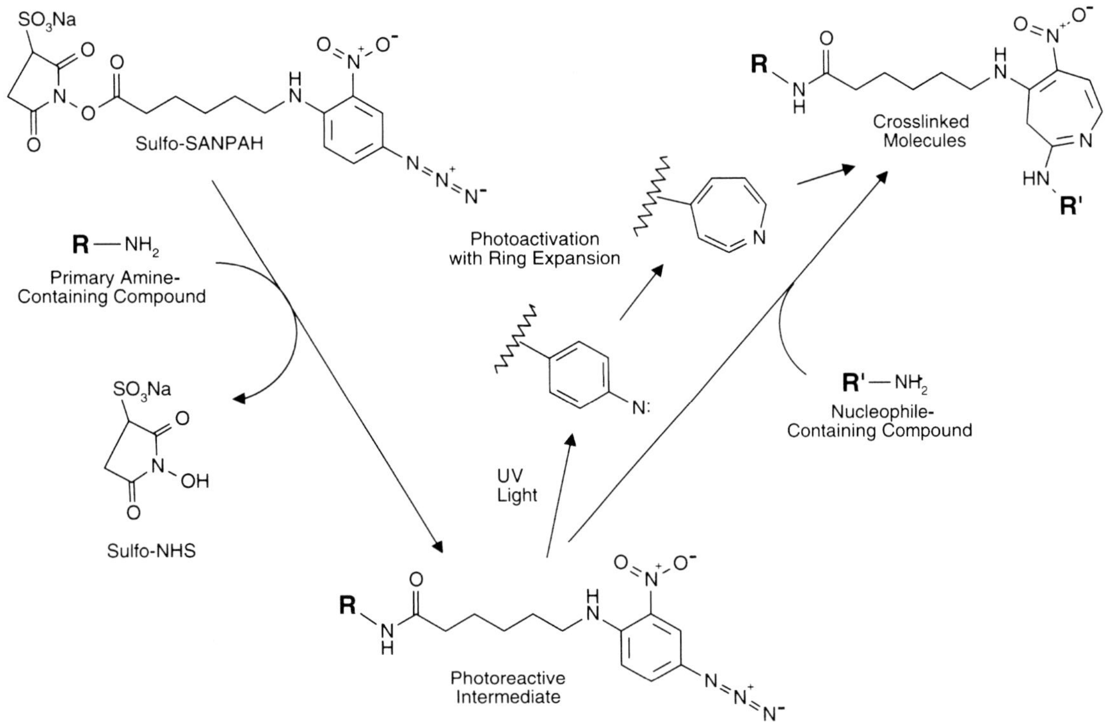

FIGURE 6.18 The reaction sequence of crosslinking with sulfo-SANPAH first involves derivatizing an amine-containing molecule using its NHS ester end to create an amide bond. Exposure to UV light then causes ring expansion to the dehydroazepine derivative, which can couple with amines to form the final conjugate.

Reactions performed with HSAB should involve dissolution of the crosslinker in organic solvent prior to addition to an aqueous reaction medium. DMSO or DMF are suitable solvents to prepare concentrated stock solutions. Protect all solutions from light to avoid loss of photoreactive phenyl azide groups prior to the desired point of photolysis.

Reported applications of HSAB include photoaffinity labeling of peptide hormone binding sites (Galardy *et al.*, 1974), photoaffinity labeling of the insulin receptor with derivatized insulin analog (Yeung *et al.*, 1980), identifying nerve growth factor receptor proteins in sympathetic ganglia membranes (Massague *et al.*, 1981), labeling of the hormone receptor of both α and β subunits of human choriogonadotropin (Ji and Ji, 1981), isolation of *in situ* crosslinked ligand–receptor complexes (Ballmer-Hofer *et al.*, 1982), crosslinking vasoactive intestinal polypeptide to its receptors on intact human lymphocytes (Wood and O'Dorisio, 1985), conjugation of monoclonal antibodies to nanoparticle polymers (Le *et al.*, 2012), and conjugation of proteins to poly(butyl cyanoacrylate) nanoparticles (Reukov *et al.*, 2011).

Sulfo-HSAB, *N*-hydroxysulfosuccinimidyl-4-azidobenzoate, is a water soluble analog of HSAB possessing a negatively charged sulfonate group on its NHS ring. This crosslinker may be added directly to aqueous reaction media without prior dissolution in organic solvent. To aid in the addition of small quantities of the reagent, a concentrated solution of sulfo-HSAB may be made in water and then an aliquot added to the reaction. Aqueous stock solutions should be dissolved quickly and used immediately to prevent extensive hydrolysis of the NHS ester.

HSAB and sulfo-HSAB have also been used to investigate serum amyloid A (Cai *et al.*, 2007), the functional role of C-terminal sequence elements in the transporter associated with antigen processing (Ehses *et al.*, 2005), and the kinetics of intermolecular interactions during cytochrome *c* protein folding (Nishida *et al.*, 2004).

3.4. SANPAH and Sulfo-SANPAH

SANPAH (*N*-succinimidyl-6-(4′-azido-2′-nitrophenylamino)hexanoate) is a heterobifunctional crosslinking agent containing an NHS ester and a photoreactive phenyl azide group (Thermo Fisher). The NHS ester end can react with amine groups in proteins and other molecules, forming stable amide linkages. The photoreactive end is sensitive to long UV light, being selectively activated to a highly reactive nitrene or dehydroazepine intermediate. Either of these photolyzed species can couple to molecules within van der Walls contact, rapidly forming covalent bonds (Figure 6.18). The cross-bridge of SANPAH is a non-cleavable 6-aminohexanoic acid derivative, which provides a long spacer between conjugated molecules.

The phenyl azide group also contains a nitro group on the ring that has the effect of increasing the wavelength of optimal photolysis. Exposure to light at a wavelength in the range of 320 to 350 nm promotes the photoreaction process. SANPAH is a water-insoluble crosslinker that will permeate membrane structures efficiently. The reagent should be dissolved in DMSO or DMF prior to addition of an aliquot to an aqueous reaction medium.

SANPAH
N-Succinimidyl-6-(4'-azido-2'-nitrophenylamino)hexanoate
MW 390.95
18.2 Å

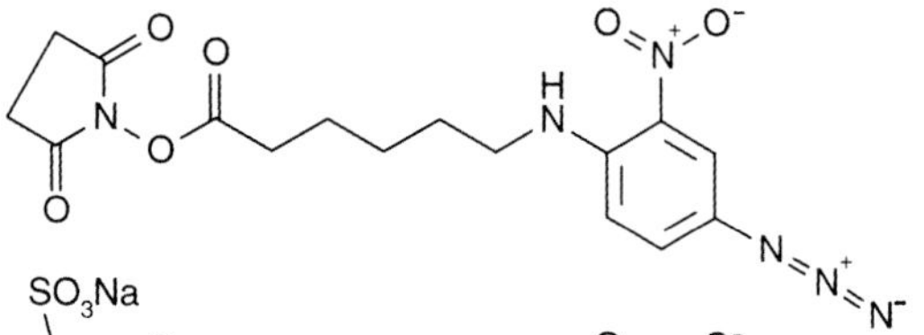

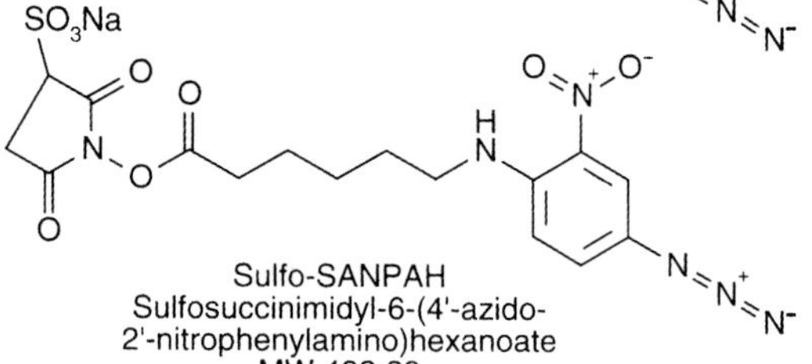

Sulfo-SANPAH
Sulfosuccinimidyl-6-(4'-azido-2'-nitrophenylamino)hexanoate
MW 492.39
18.2 Å

Reported applications of SANPAH include the crosslinking of ligand–receptor complexes *in situ* (Ballmer-Hofer *et al.*, 1982), preparing photoactivatable glycopeptide derivatives for site-specific labeling of lectins (Baenziger and Fiete, 1982), photoaffinity labeling of the *N*-formyl peptide receptor binding site of intact human polymorphonuclear leukocytes (Schmitt *et al.*, 1983), and the crosslinking of vasoactive intestinal peptide to receptors on intact human lymphoblasts (Wood and O'Dorisio, 1985).

A water soluble version of this crosslinker also exists. Sulfo-SANPAH (sulfosuccinimidyl-6-(4′-azido-2′-nitrophenylamino)hexanoate) contains the negatively charged sulfonate group on its NHS ring, lending greater hydrophilicity to the compound. SANPAH and sulfo-SANPAH have also been used for investigations into endothelial cell spreading and adhesion (Wallace *et al.*, 2007), to study the behavior of pre-osteoblastic cells (Khatiwala *et al.*, 2006), to develop a real-time microscopic method for studying biomolecular interactions (Sasuga *et al.*, 2006), and to prepare ligand-modified agarose gels to study fibronectin- and collagen-mimetic ligands (Connelly *et al.*, 2011).

3.5. ANB-NOS

N-5-Azido-2-nitrobenzoyloxysuccinimide (ANB-NOS) is a photoreactive, heterobifunctional crosslinker containing an amine-reactive NHS ester group (Thermo Fisher). Its cross-bridge consists of a benzoic acid derivative, allowing molecules to be conjugated at relatively short 7.7-Å distances apart. The phenyl ring of ANB-NOS contains a nitro group that has the effect of shifting the optimal wavelength of activation to longer UV regions. The photoreaction is initiated by exposure to light in the range of 320 to 350 nm. Without the presence of the nitro group, activation would occur at much lower wavelengths, around 265 to 275 nm—wavelengths that potentially can damage biological molecules when exposed under high-photon irradiation. ANB-NOS typically is used to label an amine-containing protein or molecule by its NHS ester end. The resultant derivative is allowed to interact with other molecules that potentially can bind specifically to it and is photolyzed to effect the final conjugation, capturing any interacting partners (Figure 6.19).

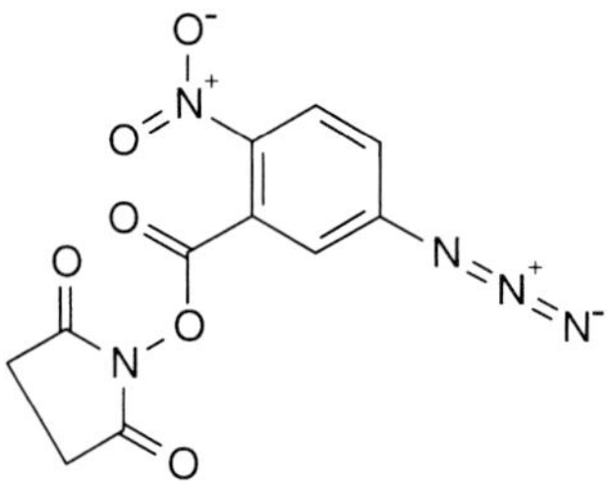

ANB–NOS
N-5-Azido-2-nitrobenzoyloxy-succinimide
MW 305.21
7.7 Å

Reported applications using this reagent include crosslinking of the aggregation state of cobra venom phospholipase A_2 (Lewis *et al.*, 1977) and conjugation of the signal sequence of nascent preprolactin to a polypeptide of the signal recognition particle (Krieg *et al.*, 1986). In addition, ANB-NOS has been used to capture interacting proteins in a two-step process (Nadeau and Carlson, 2007), to study the self-association of the yeast TATA-binding domain (Adams *et al.*, 2004), and to investigate the peptide-binding cleft of major histocompatibility complex (MHC) class I molecules (Park *et al.*, 2003).

3.6. SAND

Sulfosuccinimidyl-2-(*m*-azido-*o*-nitrobenzamido)-ethyl-1,3′-dithiopropionate (SAND) is a heterobifunctional reagent containing an amine-reactive sulfo-NHS ester at one end and a photoreactive phenyl azide group on the other end (Thermo Fisher). The presence of the sulfonate group on the NHS ring lends water solubility to the reagent due to its negative charge in aqueous solutions. In addition, the phenyl azide group contains a nitro constituent which shifts the

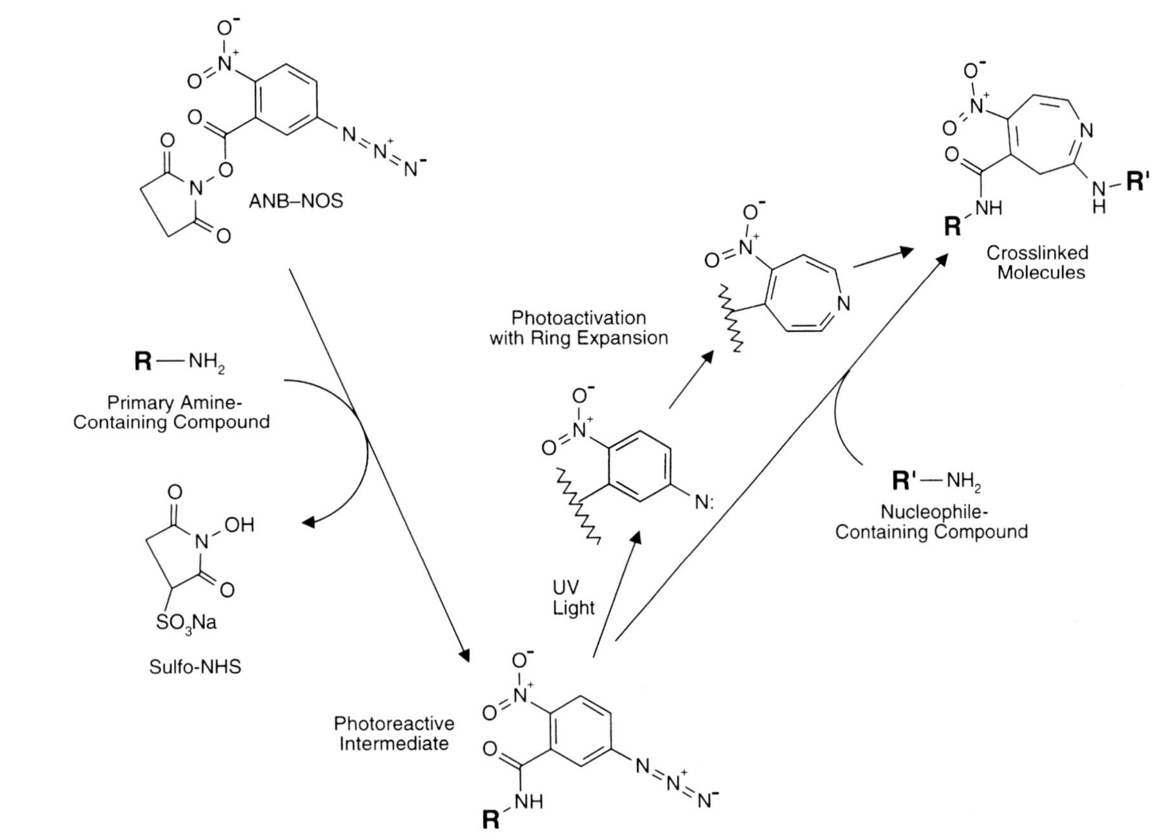

FIGURE 6.19 The NHS ester of ANB-NOS reacts with amines to form amide bonds. Subsequent photoactivation of the complex with UV light causes phenyl azide ring expansion and reaction with neighboring amines.

optimal range of photoactivation toward higher wavelengths—into the 320- to 350-nm region, thus decreasing the potential of photolytic damage to other sensitive groups that may be present during crosslinking. The extended cross-bridge of SAND (18.5 Å) provides a long spacer arm to accommodate even relatively distant sites between interacting molecules. The presence of a disulfide bond within the cross-bridge infers that the reagent is also cleavable by the use of a disulfide reductant, allowing the potential for disruption of the crosslinks after purification of the conjugate.

In use, SAND is first reacted with an amine-containing protein or other molecule—being careful to protect the photoreactive group from inadvertent degradation by exposure to excessive room light or sun. The modified intermediate is then allowed to interact with a target molecule. Finally, the photolyzing process is performed to effect a nonselective crosslink between the modified molecule and any target molecules within van der Waals distance to the crosslinker (Figure 6.20). Its use may be similar to that reported for sulfo-SANPAH, and its cleavability similar to that reported for SADP. For a more detailed discussion on the use of photoreactive crosslinkers to capture protein–protein interactions, see Chapter 24, Sections 2 and 3.

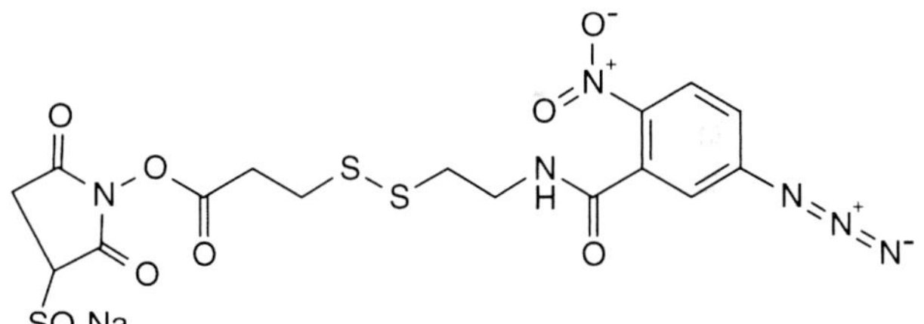

SAND
Sulfosuccinimidyl-2-(*m*-azido-*o*-nitrobenzamido)-ethyl-1,3'-dithiopropionate
MW 570.52
18.5 Å

3.7. SADP and Sulfo-SADP

SADP, N-succinimidyl-(4-azidophenyl)1,3′-dithiopropionate, is a photoreactive heterobifunctional crosslinker that is cleavable by treatment with a disulfide reducing agent (Thermo Fisher). The crosslinker contains an amine-reactive NHS ester and a photoactivatable phenyl azide group, providing specific, directed coupling at one end and nonselective insertion capability at the other end.

SADP is first used to modify a protein via its amine groups through the reactive NHS ester end of the crosslinker. After allowing for interaction of the modified

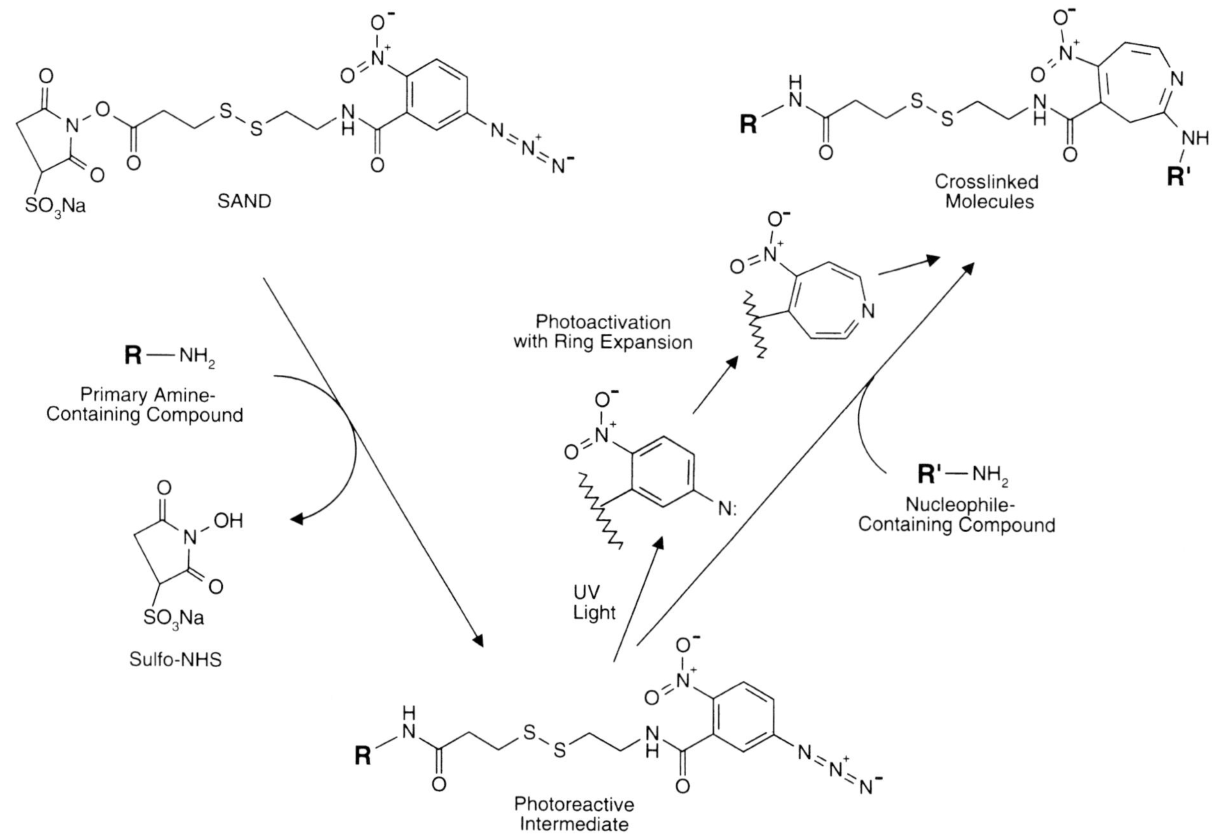

FIGURE 6.20 SAND can be used to modify amine-containing molecules and then photo-initiate crosslinking to another amine-containing molecule via a ring-expansion process. The conjugates may be disrupted by reduction of the cross-bridge disulfide with DTT.

protein with target molecules, the photoreactive group is used to couple with any molecules within van der Waals distance. The photolysis reaction requires UV exposure in the range of 265 to 275nm to form the final linkage. The presence of the disulfide group in SADP's cross-bridge allows disruption of crosslinks with 50-m*M* DTT after the conjugation reaction is complete (Figure 6.21).

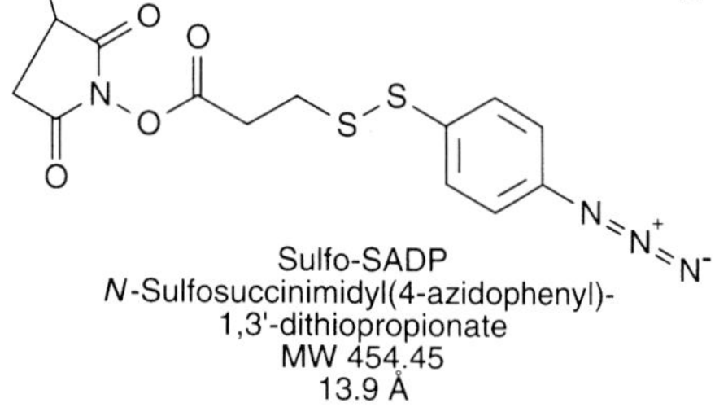

SADP is hydrophobic and should be dissolved in organic solvent prior to addition of a small aliquot to an aqueous reaction. Concentrated stock solutions can be prepared in dry DMSO or DMF. Final concentration of the organic solvent in a crosslinking reaction should not exceed about 10% to prevent protein precipitation or denaturation.

Reported applications of SADP include the crosslinking of Con A to receptors on human erythrocyte membranes (Vanin and Ji, 1981), site-specific labeling of lectins using modified glycopeptides (Baenziger and Fiete, 1982), conjugation of a mouse cell-surface polypeptide with a Sendai virion envelope on newly infected cells (Zarling *et al.*, 1982), and crosslinking of platelet glycoprotein Ib (Jung and Moroi, 1983).

Sulfo-SADP is a water soluble analog of SADP which contains a negatively charged sulfonate group on its NHS ring. The reagent may be added directly to aqueous reaction mixtures without prior dissolution in an organic solvent. Concentrated stock solutions prepared in water should be used immediately to prevent extensive hydrolysis of the sulfo–NHS ester group.

SADP or sulfo-SADP have also been used to study the phenylalanine–methionine–arginine–phenylalanine–

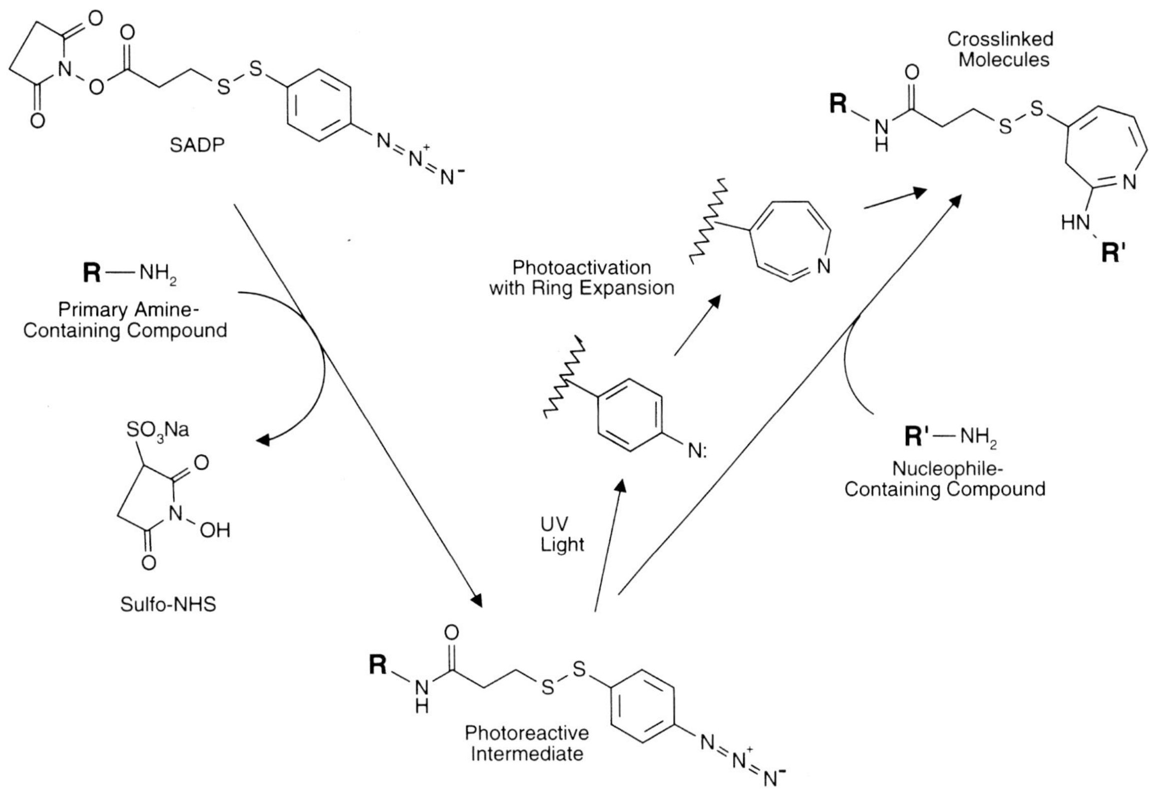

FIGURE 6.21 SADP reacts with amines via its NHS ester end to produce amide bonds. The modified molecule may then be photoactivated to create a nucleophile-reactive dehydroazepine intermediate able to covalently couple with amine-containing compounds.

amide-activated sodium channel (Coscoy *et al.*, 1998), various apolipoprotein E isoforms (Mann *et al.*, 1995), the high-affinity phenylalkylamine Ca^{2+} antagonist binding protein from guinea pig (Moebius *et al.*, 1994), the interaction of non-histone proteins with nucleosome core particles (Reeves and Nissen, 1993), and the interactions among cytochromes P-450 in the endoplasmic reticulum (Alston *et al.*, 1991). See Chapter 24 for methods of using photoreactive heterobifunctional crosslinkers to study protein interactions.

3.8. Sulfo-SAPB

Sulfo-SAPB, sulfosuccinimidyl 4-(*p*-azidophenyl) butyrate, is a photoreactive heterobifunctional reagent containing an amine-reactive sulfo-NHS ester at one end (Thermo Fisher). The crosslinker is similar in design to sulfo-HSAB (Section 3.3, this chapter), but it contains a three-carbon-longer cross-bridge. The sulfo–NHS ester provides water solubility to the reagent due to the negative charge of the sulfonate group. The phenyl-azide end can be photolyzed by exposure to UV light in the wavelength range of 265 to 275 nm (Figure 6.22). Sulfo-SAPB has been used in the bioconjugation of polymer scaffolds to create biotherapeutic agents for targeted drug delivery (Vaidya *et al.*, 2011). The commercial availability of the reagent provides additional options for spacer length to study the interactions between two proteins or other molecules.

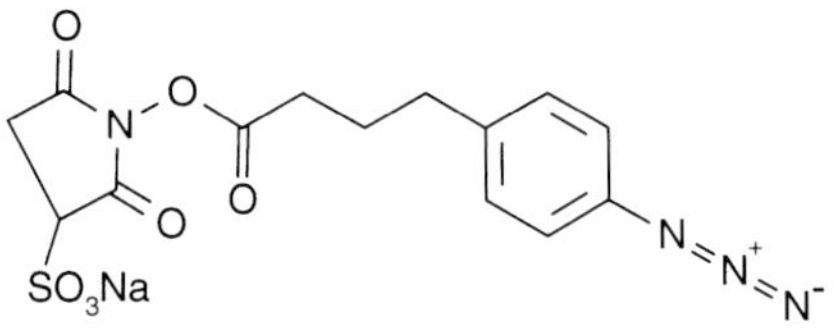

Sulfo-SAPB
Sulfosuccinimidyl-
4-(*p*-azidophenyl)butyrate
MW 404.32
12.8 Å

3.9. SAED

Sulfosuccinimidyl 2-(7-azido-4-methylcoumarin-3-acetamide)ethyl-1,3′-dithiopropionate (SAED) is a photoreactive heterobifunctional crosslinking agent that also contains a fluorescent group (Thermo Fisher). The sulfo-NHS ester end of the reagent reacts with primary amines in proteins and other molecules to form stable amide linkages. The photoreactive end is an AMCA derivative (Chapter 10, Section 3) containing a light-sensitive azide group on the aromatic ring. Photolyzing with light in the long UV range will result

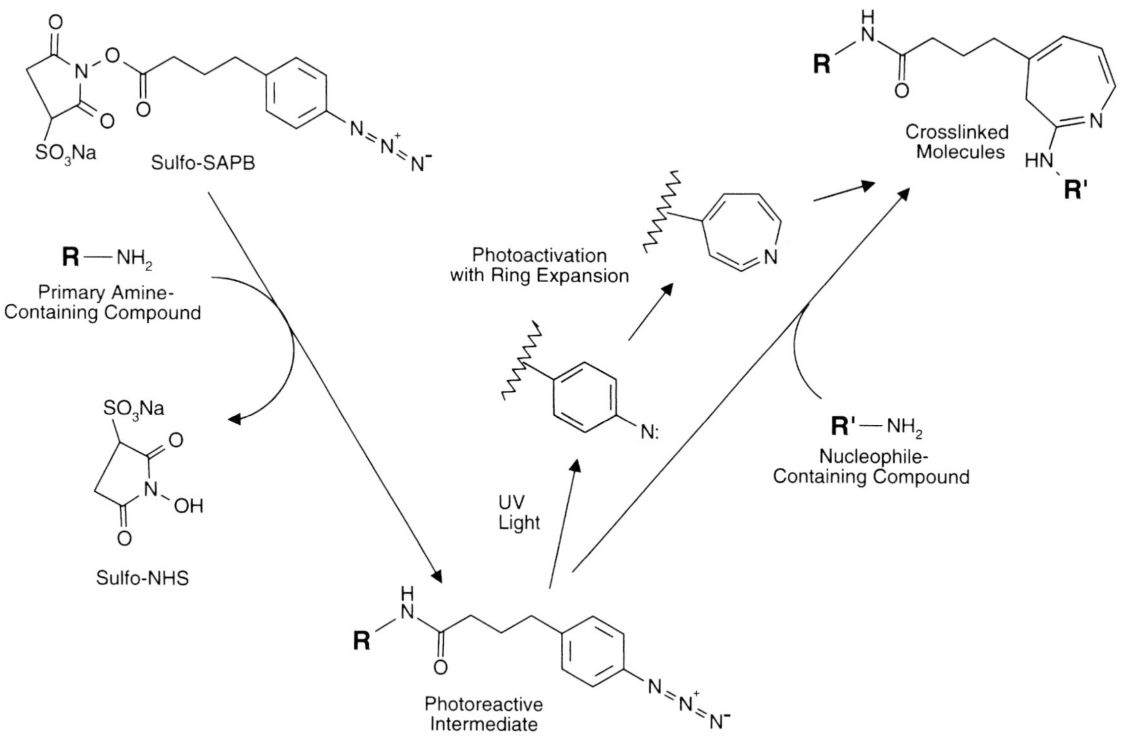

FIGURE 6.22 The reaction of sulfo-SAPB with an amine group is performed first to form an amide bond derivative through its NHS ester end. Subsequent exposure to UV light causes the phenyl azide group to ring expand to a highly reactive dehydroazepine, which can couple to nucleophiles, such as amines.

in nonselective bond formation with nucleophiles and active carbon–hydrogen bonds within van der Waals distance (Figure 6.23).

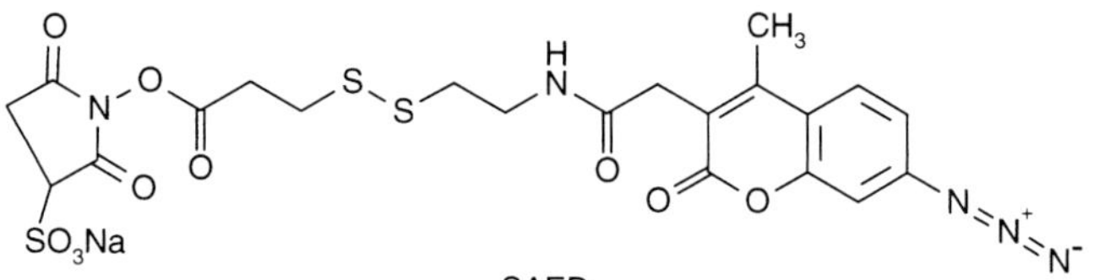

SAED
Sulfosuccinimidyl-2-(7-azido-4-methyl coumarin-3-acetamide)ethyl-1,3'-dithiopropionate
MW 621.6
23.6 Å

SAED is a relatively large crosslinker containing a long (22.5-Å) cross-bridge. The central portion of its cross-bridge contains a disulfide bond, making the reagent susceptible to cleavage with disulfide reducing agents. The aromatic character of the coumarin derivative creates a maximal UV absorptivity at 327 nm with an extinction coefficient of 18,200 $M^{-1}cm^{-1}$ for a 1-mg/ml solution in acetonitrile:water (15:2 v/v). The extinction coefficient at 298 nm for the same concentration of SAED in the identical solvent is 13,625 $M^{-1}cm^{-1}$.

SAED contains a sulfo-NHS ester with a negatively charged sulfonate group on its ring. The presence of this negative charge does lend some expected water solubility to the reagent (3 mg/ml at room temperature), but because of the reagent's large size, it does not provide the same water solubility benefits as with other smaller crosslinkers. It is also sparingly soluble in acetonitrile (2.5 mg/ml), but only if a small amount of water is present (15:2 acetonitrile:water, v/v). However, SAED is very soluble in DMSO and DMF (about 50 mg/ml). Stock solutions may be prepared in dry DMSO or DMF while maintaining fairly good stability of the reagent's functional groups. The addition of a small quantity of these stock solutions to an aqueous reaction medium facilitates the amine-modification process via the sulfo-NHS ester end of the crosslinker. The final concentration of organic solvent in the aqueous reaction should not exceed 10% to avoid protein denaturation and precipitation. Protect all solutions of the crosslinker from light to prevent premature activation of the photoreactive group.

The coumarin derivative of SAED is not fluorescent until the photolysis reaction is initiated. A protein modified with SAED will fluoresce after activation with UV light whether or not the photoreactive end actually couples to the intended target, since breakdown of the azide group on the ring is all that is required to initiate fluorescence. Thus, the level of SAED incorporation into a macromolecule may be assessed by the resultant coumarin fluorescence after separation of the derivative from excess reagent. Native AMCA has an excitation

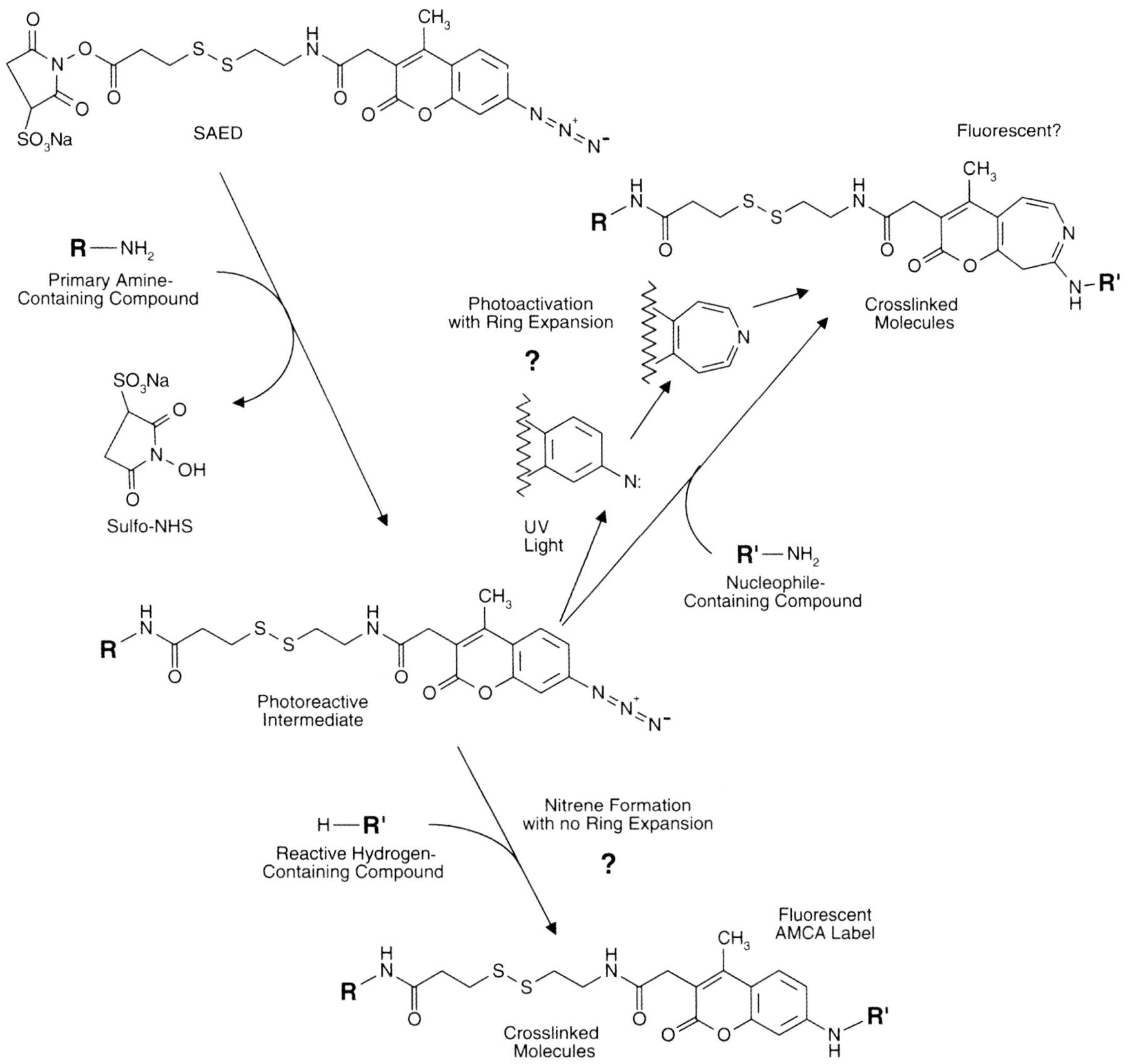

FIGURE 6.23 SAED can be used to modify amine-containing molecules through its NHS ester end. Subsequent exposure to UV light causes bond formation with nearby nucleophilic groups, such as amines. The photosensitive phenyl azide group is created on the aromatic ring of an AMCA fluorophore. Before photoactivation, the azide group makes the crosslinker nonfluorescent. After photoactivation, however, the azide group is either lost by N_2 generation or couples to a target molecule. Either way, the AMCA portion becomes fluorescent to allow tracking of the conjugate. The photoreaction may occur through ring expansion to an intermediate dehydroazepine or might happen through nitrene formation.

optimum at 345 to 350nm and an emission wavelength range of 440 to 460nm. The quantum yield of SAED may change somewhat upon its attachment to macromolecules due to fluorescent quenching; however, the coumarin tag will still remain fluorescently active even after crosslinking.

Since the crosslinker is cleavable, SAED provides a means of fluorescent transfer of the coumarin tag to a second molecule which interacts with the initially modified protein (Figure 6.24). For example, soybean trypsin inhibitor (STI) was labeled with SAED and then allowed to interact with trypsin. After photoreactive crosslinking of the two interacting molecules, the complex was reduced with DTT, breaking the conjugate and transferring the fluorescent tag to trypsin near the STI binding site (Thevenin *et al.*, 1991). This type of fluorescent label transfer reagent is important for studying unknown interacting proteins, because the unknown protein can be detected and isolated by the tag after cleavage of the complex.

In another study, SAED was used to investigate the role of the foot protein moiety of the triad and its relationship to Ca^{2+} release from sarcoplasmic reticulum (Kang *et al.*, 1991). Modification of poly-L-lysine (a Ca^{2+} release inducer) and neomycin with the crosslinker was carried out, followed by subsequent incubation with the

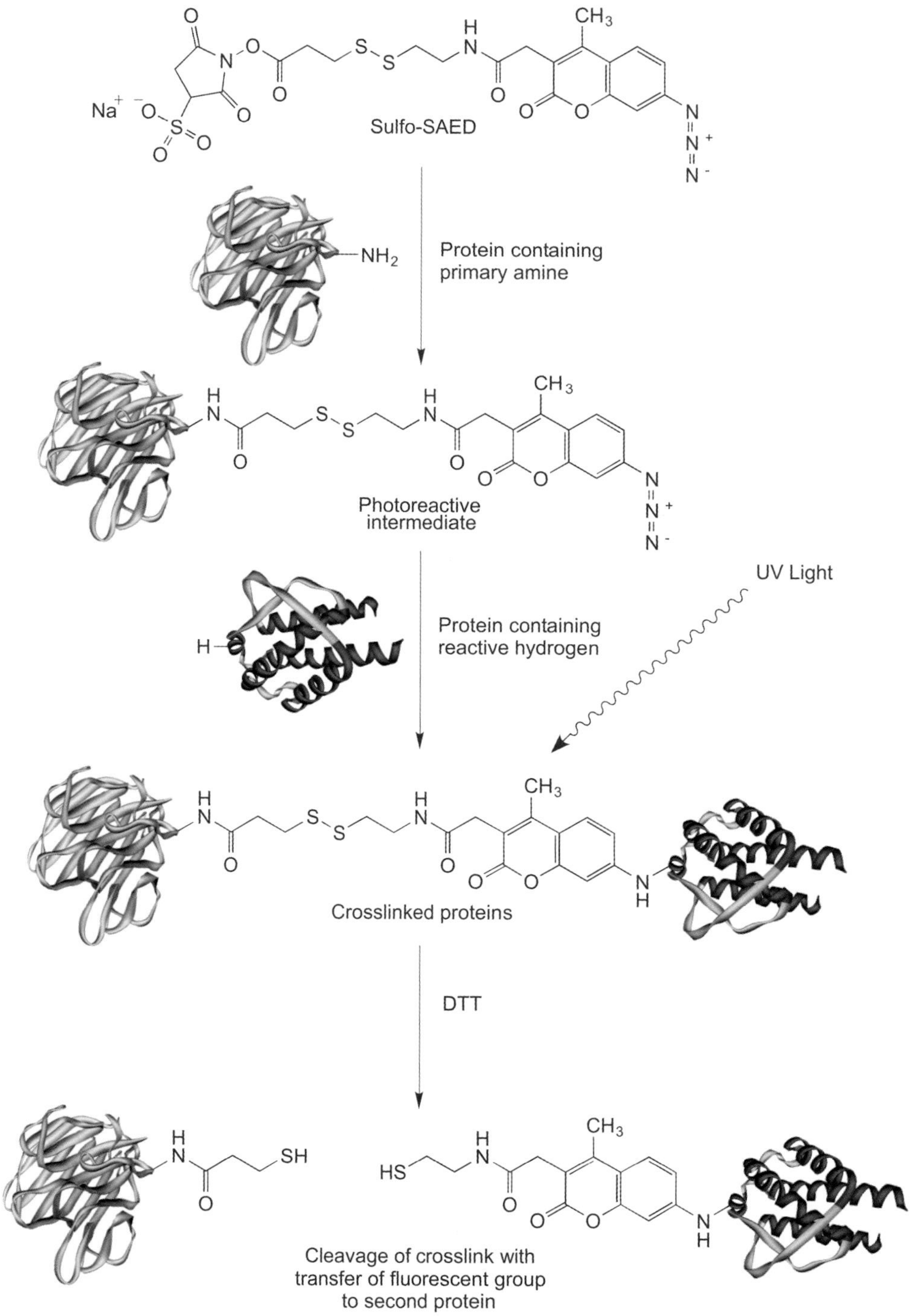

FIGURE 6.24 SAED may be used to transfer the fluorescent AMCA label from the first molecule modified with the crosslinker to the second molecule crosslinked with it by reduction of its internal disulfide bond. Thus, unknown target molecules may be fluorescently tagged to follow them *in vivo*.

foot protein and photoreactive conjugation. Cleavage of the crosslinks with a disulfide reductant allowed transfer of the fluorescent tag to the foot protein in areas near the binding sites. Fluorescent monitoring of conformational changes within the protein upon varying the Ca^{2+} concentration was then possible.

Since the photoreactive crosslinking step with SAED occurs rapidly upon exposure to even bright light within the visible spectrum, UV lamps are not required. However, special care should be taken to protect the reagent from exposure to light before the photolysis reaction is initiated. The solid should be stored in amber

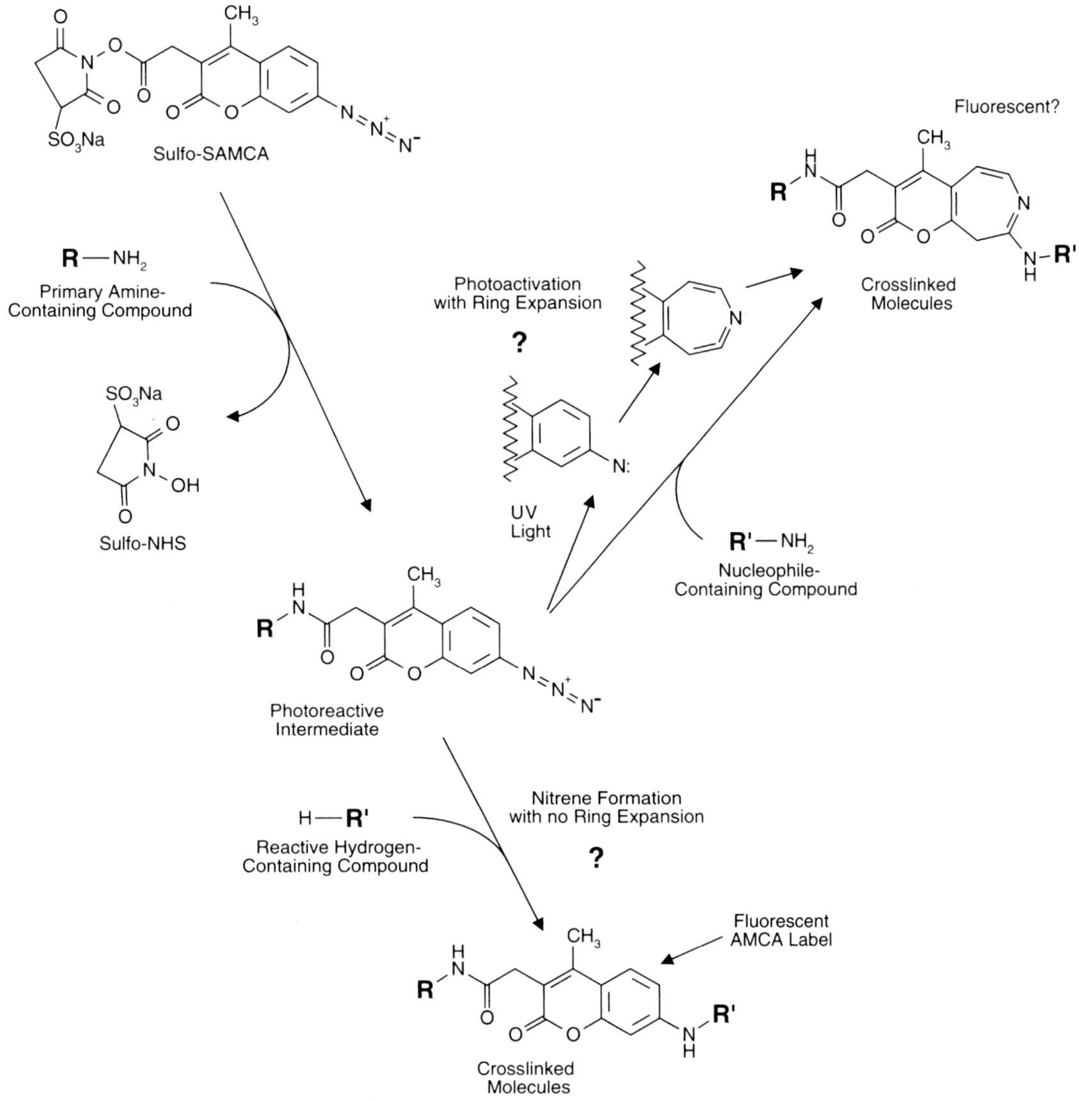

FIGURE 6.25 Sulfo-SAMCA can be used to modify amine-containing molecules through its NHS ester end. Subsequent exposure to UV light causes bond formation with nearby nucleophilic groups, such as amines. The photosensitive phenyl azide group is created on the aromatic ring of an AMCA fluorophore. Before photoactivation, the azide group makes the crosslinker nonfluorescent. After photoactivation, however, the azide group either is lost by N_2 generation or couples to a target molecule. Either way, the AMCA portion becomes fluorescent to allow tracking of the conjugate. The photoreaction may occur through ring expansion to an intermediate dehydroazepine or might happen through nitrene formation.

bottles and any stock solutions prepared in organic solvent should be wrapped to exclude light. In addition, the initial derivatization of an amine-containing molecule should be done in the dark in wrapped containers.

Additional applications of SAED include study of the ryanodine receptor (Yano *et al.*, 2005; Mochizuki *et al.*, 2007) and investigating the protein organization of the postsynaptic density (Liu *et al.*, 2006).

3.10. Sulfo-SAMCA

Sulfo-SAMCA, sulfosuccinimidyl 7-azido-4-methylcoumain-3-acetate, is a heterobifunctional reagent similar in design to SAED (Section 3.9, this chapter) (Thermo Fisher). One end of the crosslinker contains an amine-reactive sulfo-NHS ester, while the other end is an AMCA derivative (Chapter 10, Section 3) that contains a photosensitive phenyl azide group. Unlike SAED, however, sulfo-SAMCA contains a short noncleavable cross-bridge (12.8 Å) where the active ester functionality is constructed directly off the carboxylate group of AMCA without any other intervening spacer groups. Conjugated molecules will retain the fluorescent label, thus providing detectability to the complexes formed (Figure 6.25). However, since crosslinks formed with this reagent are not cleavable, sulfo-SAMCA cannot function as a fluorescent label transfer agent in the fashion of SAED.

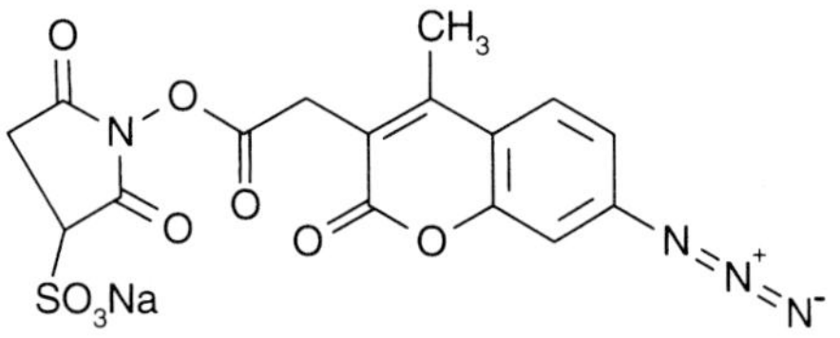

Sulfo-SAMCA
Sulfosuccinimidyl-7-azido-4-methylcoumarin-3-acetate
MW 458.34
12.8 Å

3.11. *p*-Nitrophenyl Diazopyruvate

Diazopyruvates represent a unique class of photoreactive reagents that are not often used in heterobifunctional crosslinker design. The *p*-nitrophenyl ester derivative of diazopyruvate provides amine-reactive, acylating potential, while the photosensitive group can be activated with UV light to generate reactive aldehydes. More specifically, the diazo functionality can by photolyzed by exposure to irradiation at 300 nm, forming a highly reactive carbene which can undergo a Wolff rearrangement that produces a ketene amide intermediate. In the presence of a nucleophilic species on a target molecule, the ketene can undergo an acylation reaction to form a stable malonic acid derivative. The photolyzed product thus can couple to hydrazide- or amine-containing targets to form covalent linkages (Figure 6.26).

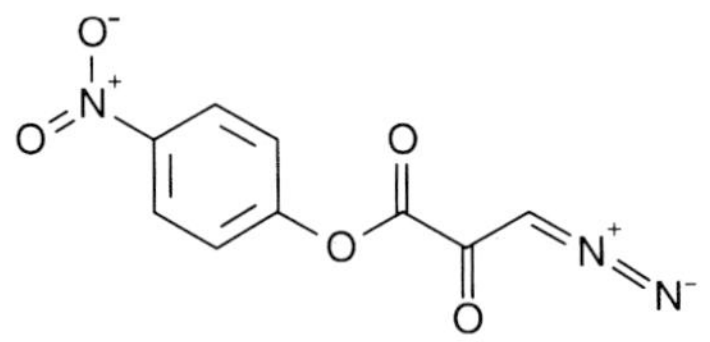

pNPDP
p-Nitrophenyl diazopyruvate
MW 235

p-Nitrophenyl diazopyruvate (Invitrogen) is relatively insoluble in water or aqueous buffers, but may be pre-dissolved in DMF before adding an aliquot of the stock solution to an aqueous reaction mixture. All solutions of the reagent should be carefully protected from light to prevent premature photolysis. *p*-Nitrophenyl diazopyruvate has an absorbance maximum at 390 nm with a molar extinction coefficient of about 19,000 $M^{-1}cm^{-1}$ in methanol.

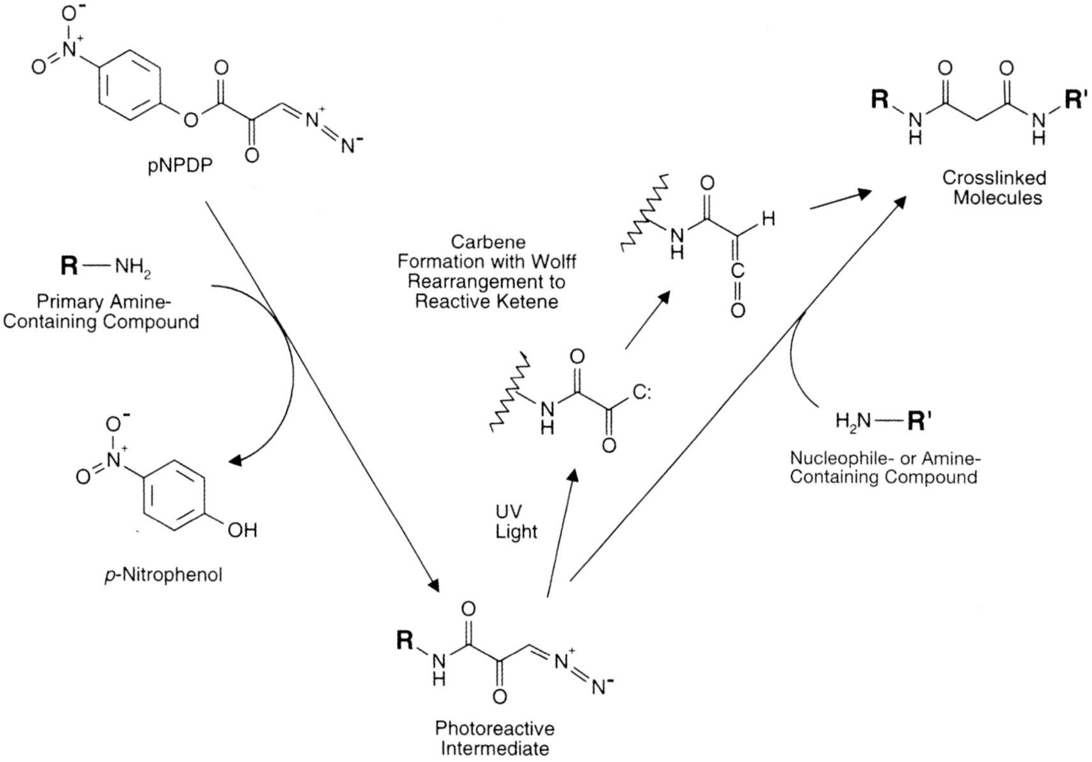

FIGURE 6.26 pNPDP reacts with amine-containing compounds by its *p*-nitrophenyl ester group to form amide bonds. After photoactivation of the diazo derivative with UV light, a Wolff rearrangement to a highly reactive ketene intermediate occurs. This group can couple to nucleophiles such as amines.

p-Nitrophenyl diazopyruvate has been used in the photoreactive crosslinking of calmodulin with adenylate cyclase from bovine brain (Harrison *et al.*, 1989), to crosslink aldolase (Goodfellow *et al.*, 1989), as a bonding agent for tissue containing type-I collagen (Givens *et al.*, 2003) or to bond to corneal tissue (Timberlake *et al.*, 2005), and in the photoreactive coupling of DNA to paramagnetic beads (Penchovsky *et al.*, 2000).

3.12. PNP–DTP

PNP–DTP, *p*-nitrophenyl-2-diazo-3,3,3-trifluoropropionate, is a photoreactive heterobifunctional crosslinker that contains an amine-reactive group on one end and a photosensitive diazo group on the other (Chowdhry *et al.*, 1976) (Thermo Fisher). *p*-Nitrophenyl esters react similarly to NHS esters (Chapter 5, Section 1, and Chapter 3, Section 1.4) but in this case with *p*-nitrophenol as the leaving group upon reaction with a nucleophile (Figure 6.27). Amine-containing target molecules such as proteins can be modified with this reagent to form amide bond derivatives possessing photoactivatable functionalities. The reagent is small enough to probe deep within the active centers of receptor molecules and other sites of biomolecular interactions.

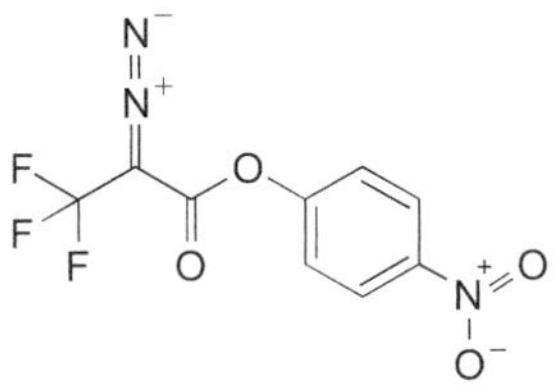

PNP-DTP;
p-Nitrophenyl-2-diazo-
3,3,3-trifluoropropionate
Mol. Wt.: 275.14

PNP–DTP has been used to photoaffinity label the thyroid hormone nuclear receptors in intact cells by preparing a derivative of 3,5,3′-triiodo-L-thyronine with the crosslinker (Pascual *et al.*, 1982; Casanova *et al.*, 1984). Effective photoreactive conjugation was found to occur after irradiation with UV light at 254 nm or 310 nm.

4. SULFHYDRYL-REACTIVE AND PHOTOREACTIVE CROSSLINKERS

The benefits of nonselective photoreactive crosslinking can be merged with the directed coupling ability of sulfhydryl-reactive functionalities to create heterobifunctional reagents possessing greater utility than the standard amine- and photoreactive agents discussed previously. Having a sulfhydryl-reactive group on one end of the crosslinker allows the initial conjugation to take place at more discrete sites on proteins and other molecules before irradiation to effect the final photosensitive reaction.

The following reagents contain a variety of sulfhydryl-reactive groups, including iodoacetyl derivatives, maleimide compounds, and pyridyl disulfide chemistries. The iodoacetyl and maleimide functions form permanent thioether bonds with target molecules containing free sulfhydryls. The pyridyl disulfide derivative reacts with SH groups to form reversible disulfide linkages, which can be cleaved with disulfide reducing agents, like DTT.

The photoreactive end of the following crosslinkers also varies from the traditional aryl azide group to the newer benzophenone and fluorinated aryl azide derivatives. The fluorinated phenyl azide functional groups will photolyze to true nitrenes without the ring expansion side reaction characteristic of aryl azides. The result is that fluorinated aryl azides more effectively insert into active carbon–hydrogen bonds, rather than potentially undergoing nucleophilic reactions like phenyl azides. In addition, benzophenone groups generally have higher degrees of bond formation with the intended target molecule compared to the yields obtained using traditional phenyl azides, due to their ability to be repeatedly photolyzed without breakdown of the precursor to an inactive form.

The number of commercially available crosslinkers for sulfhydryl- and photoreactive conjugations provide enough variety to design successful experiments in photolabeling, such as studying active centers and macromolecular interactions.

4.1. ASIB

ASIB, 1-(*p*-azidosalicylamido)-4-(iodoacetamido)butane, is a heterobifunctional crosslinker containing a sulfhydryl-reactive iodoacetyl group on one end and a photosensitive phenyl azide group on the other end (Thermo Fisher). The phenyl azide ring is substituted with a ring-activating hydroxyl group which provides the ability to radioiodinate the compound before the conjugation reaction is performed. Since both the iodoacetyl and phenyl azide functionalities are relatively stable in aqueous solutions, the steps involved in iodination and crosslinking do not detrimentally affect the subsequent reactivity of the reagent. All operations should be protected from light, however, to prevent premature photolysis before the desired crosslinking reaction is initiated. The cross-bridge of the reagent provides an 18.8-Å spacer between crosslinked molecules.

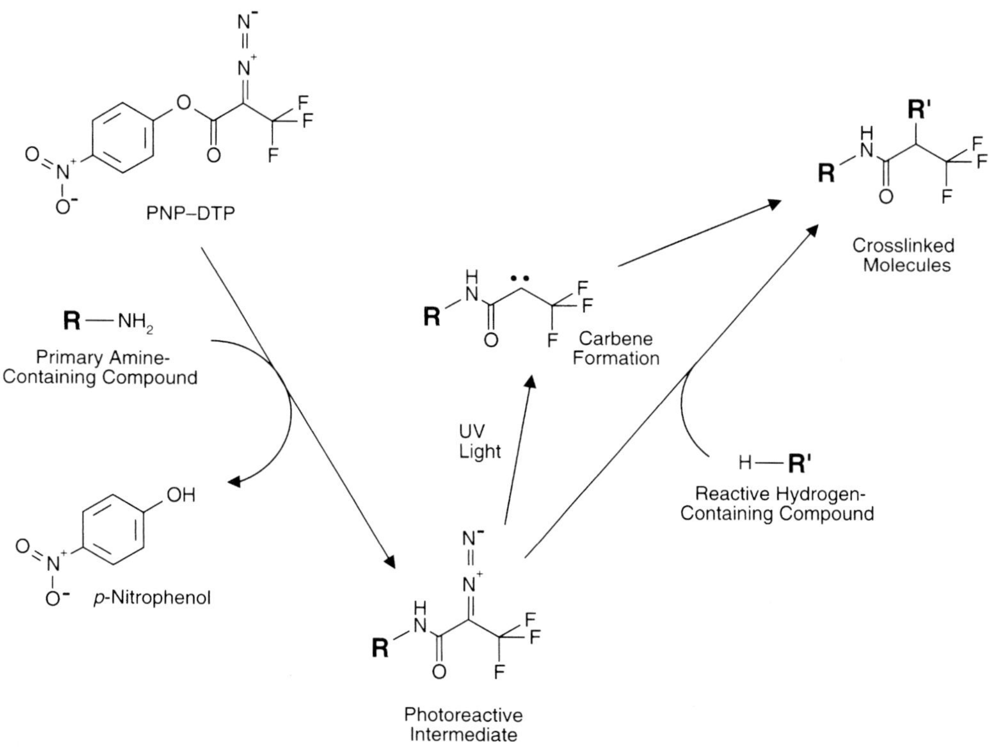

FIGURE 6.27 PNP–DTP can modify amine-containing molecules through its *p*-nitrophenyl ester group to form amide bonds. Exposure of its photosensitive diazo group with UV light generates a highly reactive carbene that can insert into active C—H or N—H bonds.

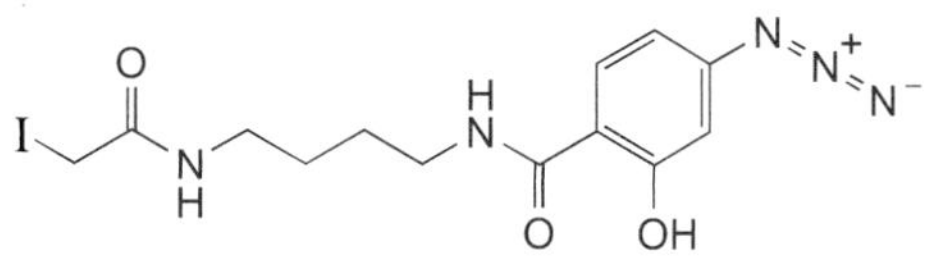

The reaction of ASIB with sulfhydryl-containing molecules can be performed at mildly alkaline pH with excellent specificity. Higher pH conditions may cause cross-reactivity with amines. Photolyzing with UV light may result in immediate reaction of the nitrene intermediate with a target molecule within van der Waals distance, or may result in ring expansion to the nucleophile-reactive dehydroazepine. The ring-expanded product is reactive primarily with amine groups (Figure 6.28).

4.2. APDP

APDP, N-[4-(*p*-azidosalicylamido)butyl]-3′-(2′-pyridyldithio) propionamide, is a radioiodinatable, heterobifunctional crosslinking agent that contains a sulfhydryl-reactive pyridyl disulfide group on one end and a photosensitive phenyl azide on the other end (Thermo Fisher). Radioiodinatable crosslinkers eliminate the need to radiolabel one of the reacting proteins, thus avoiding potential activity losses due to modification of important residues (Chapter 12, Section 2.5). They also allow radiolabeling of unknown target molecules which interact with the initially modified protein. APDP reacts with sulfhydryl-containing proteins and other molecules to form a reversible disulfide bond. If the crosslinker is radiolabeled prior to conjugation, cleavage of the disulfide group with DTT after crosslinking effectively transfers the iodinated portion to the secondary, photocoupled protein. This radiolabel transfer process allows tracking of a specific receptor or other interacting species after conjugation with its complementary ligand (Figure 6.29). APDP thus falls into the general category of label transfer reagents that can be used to study protein interactions (Chapter 24, Section 3).

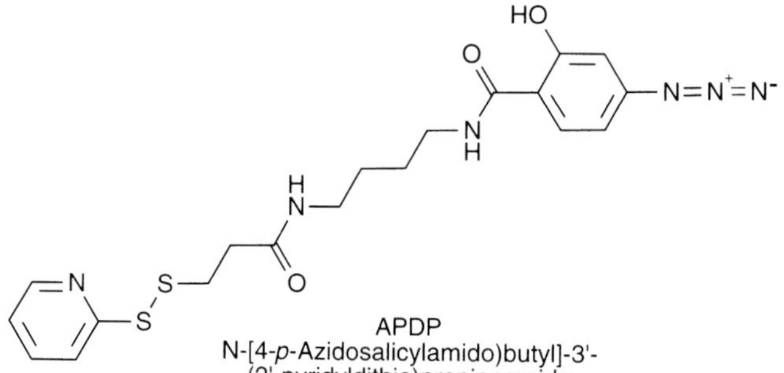

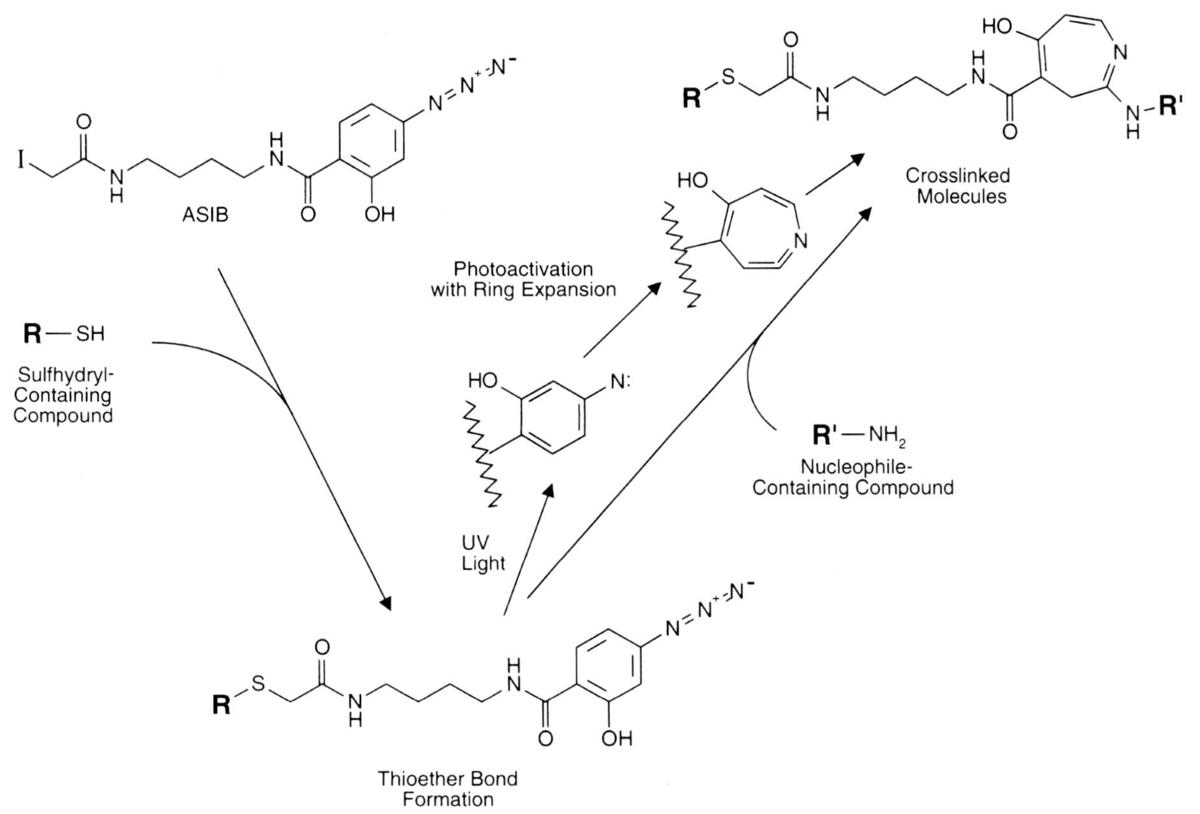

FIGURE 6.28 ASIB can react with sulfhydryl-containing molecules through its iodoacetate group to form thioether linkages. Subsequent exposure to UV light causes a ring-expansion process to occur, creating a highly reactive dehydroazepine intermediate that can couple to amine-containing molecules.

The reactions of APDP are similar to that of the reported compound *N*-(4-azidophenyl)thiophthalimide, a non-radioiodinatable crosslinker (Moreland *et al.*, 1988). Both the phenyl azide group and the pyridyl disulfide portion are stable in aqueous environments prior to the crosslinking reaction. The initial modification with a sulfhydryl-containing protein should be performed protected from light to preserve the activity of the photosensitive group. Avoid, also, in the reaction medium disulfide reducing agents that can react with the pyridyl disulfide group as well as inactivate the phenyl azide portion.

The cross-bridge of APDP provides a long, 21.02-Å spacer that is able to reach distant points between two interacting molecules. Cleavage of the crosslink with a disulfide reducing agent regenerates the original sulfhydryl-modified protein without leaving any other chemical groups behind. The remainder of the crosslinker stays attached to the second, interacting protein.

APDP is soluble in DMSO and DMF, but almost insoluble in acetone or water. Stock solutions may be prepared in DMSO or DMF and a small aliquot added to an aqueous reaction mixture. Do not exceed 10% organic solvent in the buffered reaction. Both functionalities of APDP will react in a variety of salt conditions and pH values. For reaction with a sulfhydryl-containing protein, a buffer at physiological pH containing a chelating agent to protect the free sulfhydryl groups from metal-catalyzed oxidation is recommended (i.e., 0.01- to 0.1-*M* sodium phosphate, 0.15-*M* NaCl, pH 7.2, containing 10-m*M* EDTA).

Iodination of the crosslinker may be performed according to the procedures discussed in Chapter 12, Section 2.5, or performed similar to that described for SASD (Section 3.2, this chapter).

4.3. Benzophenone-4-Iodoacetamide

A photoreactive group consisting of a benzophenone residue photolyzes upon exposure to UV light to give a highly reactive triplet-state ketone intermediate (Walling and Gibian, 1965). Similar to the reactive nitrene of photolyzed phenyl azides, the energized electron of an activated benzophenone can insert in active hydrogen–carbon bonds and other reactive groups to give covalent linkages with target molecules. Unlike phenyl azides, however, the decomposition or decay of the photoactivated species does not yield an inactive compound. Instead, benzophenones that have become deactivated without forming a covalent bond can once

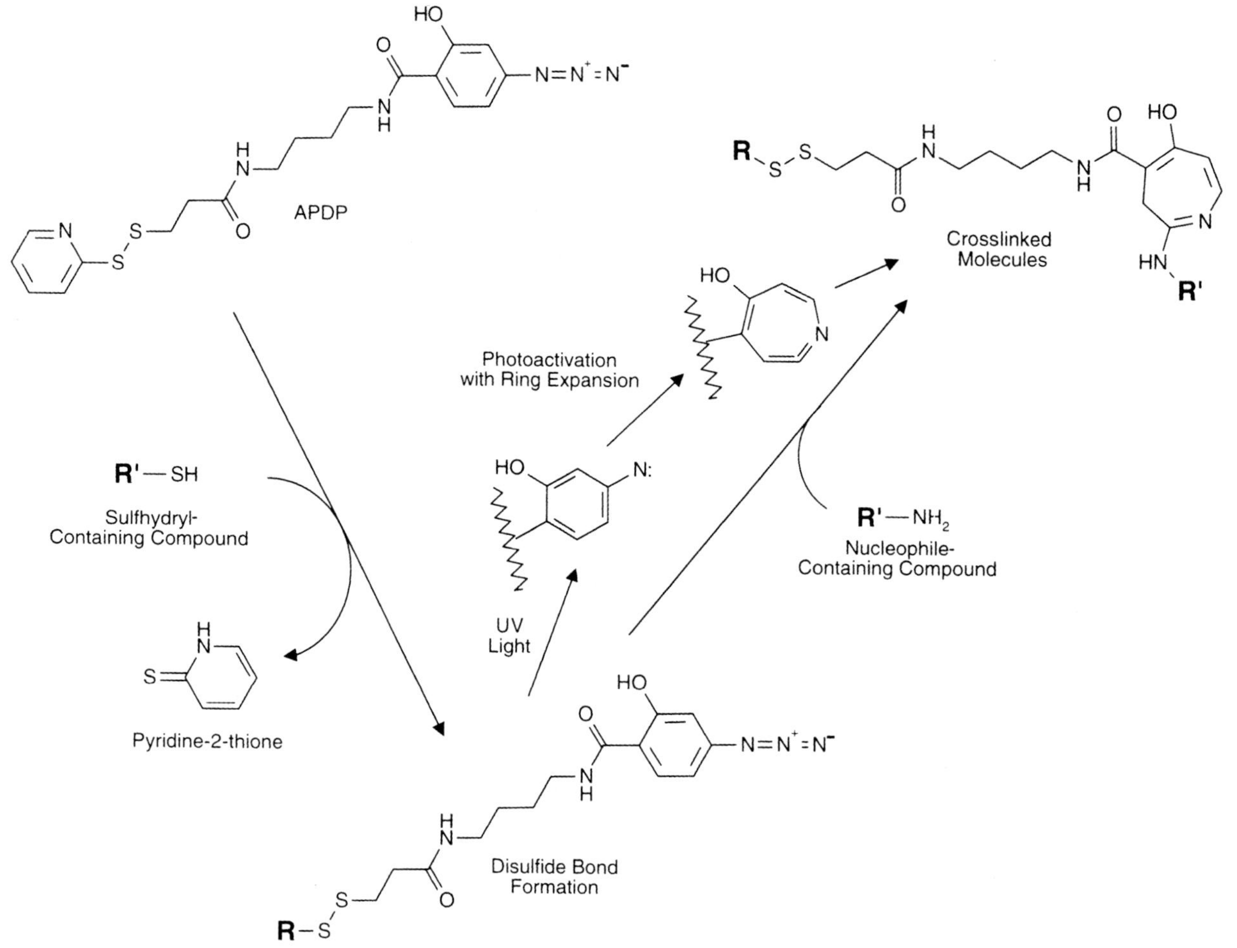

FIGURE 6.29 APDP can modify sulfhydryl-containing compounds through its pyridyl disulfide group to form disulfide bonds. Its phenyl azide end can then be photolyzed with UV light to couple with nucleophiles via a ring expansion process. The disulfide group of the crosslink can be selectively cleaved using DTT.

again be photolyzed to an active state. The results of this multiple-activation characteristic are more than one chance to form a crosslink with the intended target and much higher yields of photo-crosslinking.

The heterobifunctional crosslinker benzophenone-4-iodoacetamide is a photoreactive reagent containing a sulfhydryl-reactive iodoacetyl derivative at one end and a benzophenone group on the other end (Invitrogen) (Hall and Yalpani, 1980; Tao *et al.*, 1984; Lu and Wong, 1989). The iodoacetyl group has similar reactivity to the same group on the heterobifunctional reagent SIAB (Section 1.5, this chapter). Under alkaline conditions (pH 8–9), the iodoacetyl reaction is highly specific for sulfhydryl residues in proteins and other molecules, forming stable thioether linkages. The initial modification reaction of sulfhydryl-containing compounds should be performed protected from light to avoid premature photolysis of the benzophenone functionality. Also, avoid thiol-containing reducing agents in the sample, as these will react with the iodoacetyl group. After purification of the benzophenone-modified protein from excess reagent (by dialysis or gel filtration), it is mixed with a sample containing a second target molecule (e.g., a cell lysate) to allow an interaction to take place and then photolyzed with UV light to effect the final crosslink (Figure 6.30). Since repeated photolysis of the benzophenone species is possible, the yield of such conjugation reactions can be significantly higher than using other photoreactive groups. One report indicated that crosslinking with chymotrypsin approached 100% efficiency (Campbell and Gioannini, 1979).

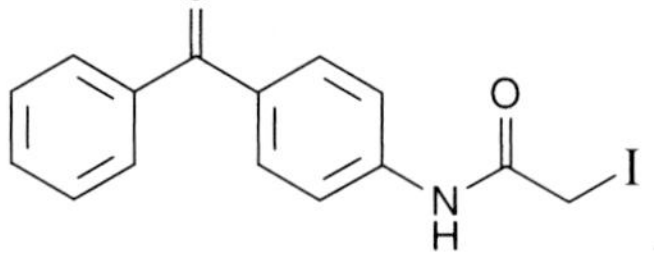

Benzophenone-4-iodoacetamide
MW 365

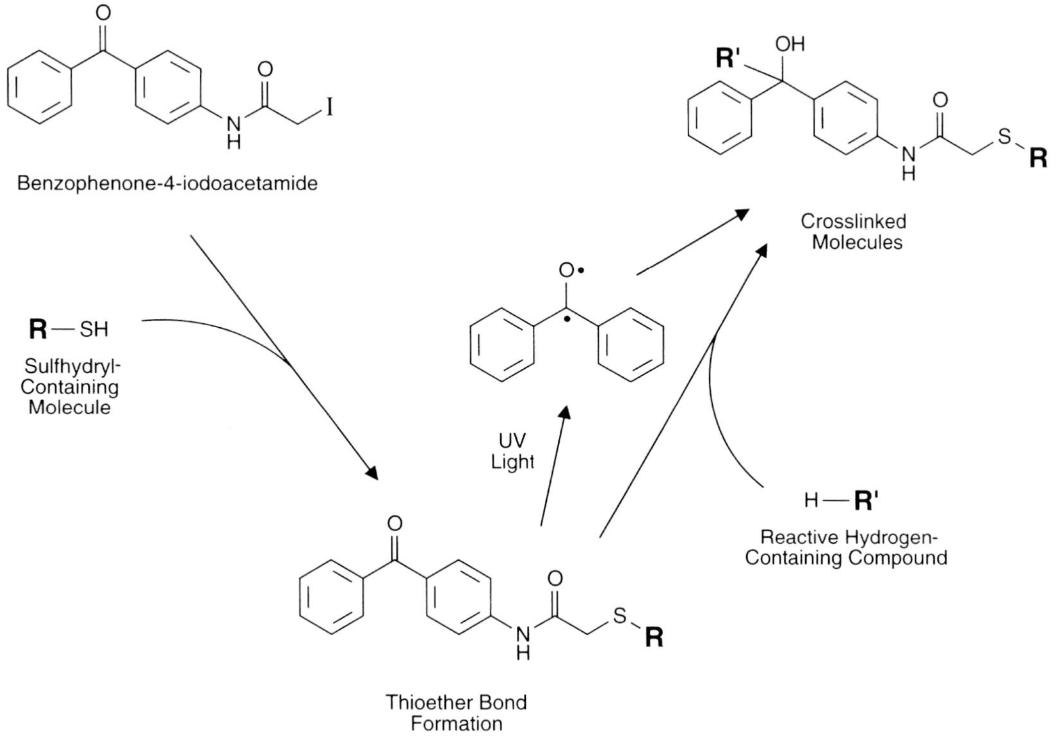

FIGURE 6.30 Benzophenone-4-iodoacetamide reacts with sulfhydryl-containing compounds to give thioether linkages. Subsequent photoactivation of the benzophenone residue gives a highly reactive triplet-state ketone intermediate. The energized electron can insert in active C—H or N—H bonds to give covalent crosslinks.

Benzophenone-4-iodoacetamide is water insoluble and should be pre-dissolved in DMF or another organic solvent prior to adding an aliquot to an aqueous reaction mixture. Stock solutions may be prepared and stored successfully if protected from light.

Benzophenone-4-iodoacetamide has been used to study the 100-kDa U5 snRNP protein (hPrp28p) and its interactions (Ismaili *et al.*, 2001).

4.4. Benzophenone-4-maleimide

Benzophenone-4-maleimide is a heterobifunctional photoreactive crosslinker that has sulfhydryl reactivity similar to benzophenone-4-iodoacetamide discussed in the previous section (Invitrogen). In this case, the sulfhydryl-reactive portion is provided by the presence of a maleimide group that couples to thiols by addition to the double bond (Chapter 3, Section 2.2). The maleimide group is specific for sulfhydryls under physiological conditions, and the reaction results in a thioether linkage that is quite stable. Sulfhydryl-containing proteins or other molecules modified with this reagent may be used in photoaffinity labeling studies to investigate the specific interactions between two molecules. After mixing a modified protein with a sample, the solution may be photolyzed to create a covalent crosslink between any interacting species. UV photolysis of the benzophenone group results in a highly reactive triplet-state intermediate which can rapidly insert or add to organic components within van der Waals distance (Figure 6.31). Decay of the active-state intermediate returns the photosensitive group to its original chemical form, thus allowing repeated photoactivations without losing the potential for coupling to its intended target.

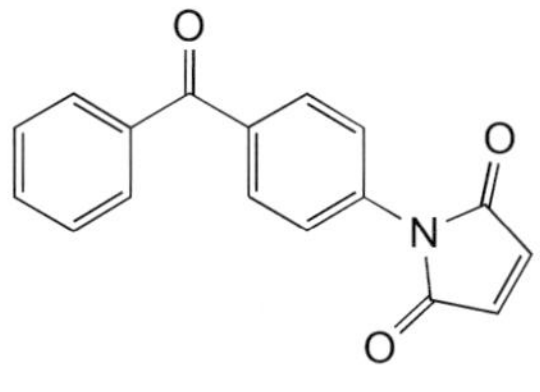

Benzophenone-4-maleimide
MW 277.27

Benzophenone-4-maleimide is water insoluble and should be pre-dissolved in DMF or another organic solvent prior to adding an aliquot to an aqueous reaction mixture. Stock solutions may be prepared and stored successfully if protected from light. The hydrophobicity

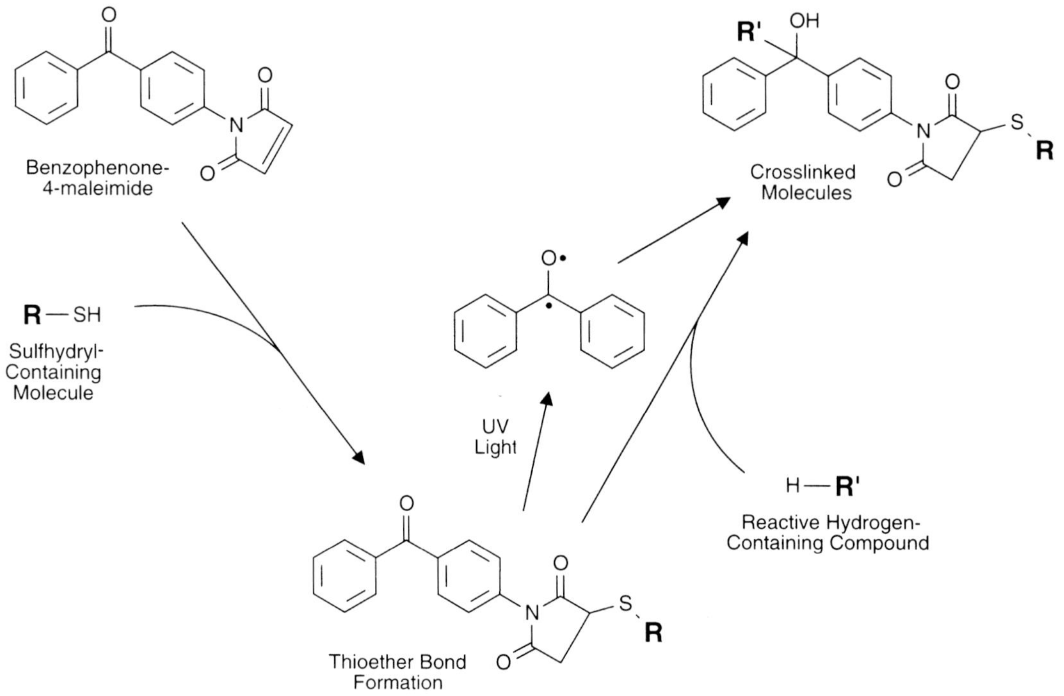

FIGURE 6.31 Benzophenone-4-maleimide can couple to thiol-containing molecules to form stable thioether bonds. Exposure of the benzophenone group to UV light causes transition to a triplet-state ketone of high reactivity for insertion into C—H or N—H bonds.

and bulkiness of the benzophenone group may cause insolubility problems in the initial protein that is modified if the derivatization is carried out at too high a level. Fortunately, the use of a sulfhydryl-reactive reagent can limit the degree of derivatization, since thiol groups usually are present in lower quantities and in more discrete locations than amines.

5. CARBONYL-REACTIVE AND PHOTOREACTIVE CROSSLINKERS

Crosslinking reagents containing a photoreactive group on one end and a carbonyl-reactive group on the other end are rare. The use of an amine group on one end of a photosensitive heterobifunctional reagent has been described (Drafler and Marinetti, 1977; Das and Fox, 1979; Gorman and Folk, 1980), but the presence of a hydrazide is required for spontaneous reactivity toward carbonyls. The following compound is the only commercially available reagent containing a phenyl azide photoreactive group and a hydrazide functional group.

5.1. ABH

ABH, *p*-azidobenzoyl hydrazide, is a small, heterobifunctional crosslinker containing a photoreactive phenyl azide group on one end and a hydrazide functionality on the other end (Thermo Fisher). The hydrazide can react with carbohydrate-containing molecules after oxidation with sodium periodate (Chapter 2, Section 4.4) to create aldehyde residues. The reaction forms a hydrazone linkage. Thus, glycoproteins may be specifically labeled on their polysaccharide chains for subsequent investigation of their interaction with receptor molecules (Figure 6.32). In this sense, lectin–carbohydrate interactions may be studied through direct modification of the sugar groups at or adjacent to the binding site. Other amine- or sulfhydryl-reactive probes may not be suitable for such studies due to the lack of amine or sulfhydryl groups near enough to a polysaccharide structure.

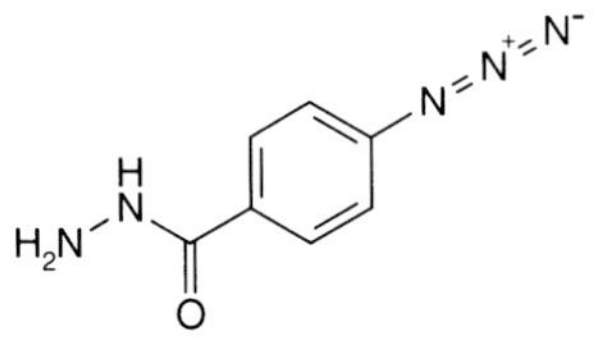

ABH
p-Azidobenzoyl hydrazide
MW 177.17
11.9 Å

The cross-bridge of ABH consists of a benzoic acid derivative and thus provides a short spacer between conjugated molecules. After ABH modification of a glycoprotein and incubation with a potential target

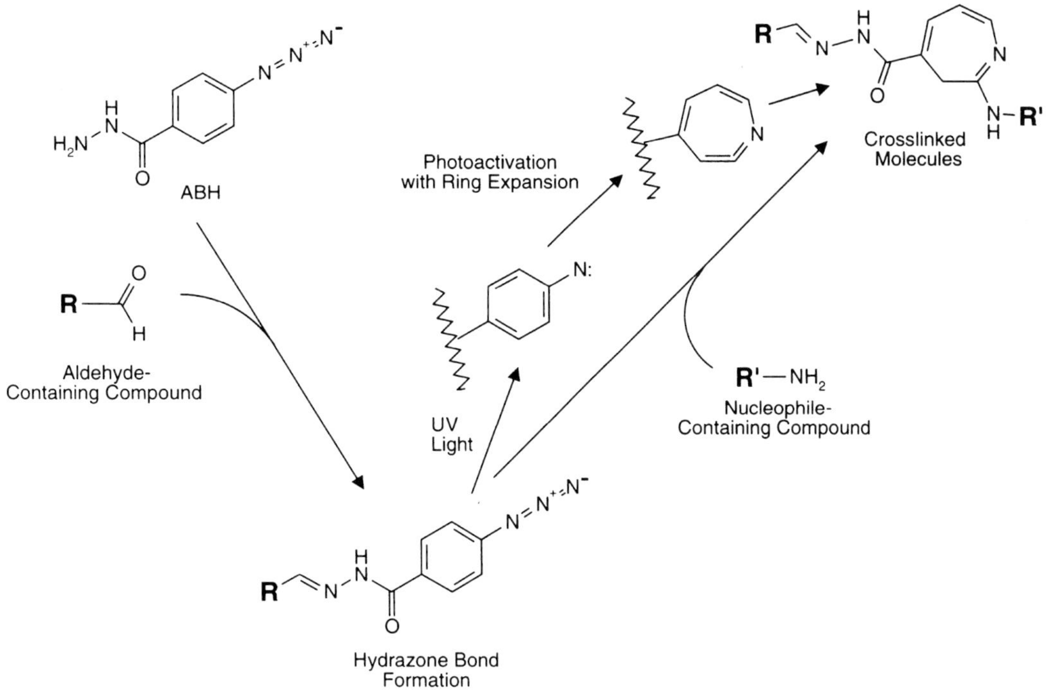

FIGURE 6.32 ABH reacts with aldehyde-containing compounds through its hydrazide end to form hydrazone linkages. Glycoconjugates may be labeled by this reaction after oxidation with sodium periodate to form aldehyde groups. Subsequent photoactivation with UV light causes transformation of the phenyl azide to a nitrene. The nitrene undergoes rapid ring expansion to a dehydroazepine that can couple to nucleophiles, such as amines.

molecule, the solution may be photolyzed with UV light to initiate the final crosslink. Prior to photolysis, the reagent and all modified species should be protected from light to prevent degradation of the phenyl azide group.

ABH is relatively insoluble when directly added to water or buffer, and therefore it should be pre-dissolved in DMSO prior to addition of an aliquot to an aqueous reaction medium. Stock solutions at a concentration of 50-m*M* ABH in DMSO work well. Since both reactive groups on ABH are stable in aqueous environments as long as the solution is protected from light, a secondary stock solution may be made from the initial organic preparation by adding an aliquot to the hydrazide reaction buffer (0.1-*M* sodium acetate, pH 5.5) (O'Shannessy and Quarles, 1985; O'Shannessy *et al.*, 1984). Make a 1:10 dilution of the ABH/DMSO solution in the reaction buffer. This solution may be stored in the dark at 4°C without decomposition.

6. CARBOXYLATE-REACTIVE AND PHOTOREACTIVE CROSSLINKERS

A carboxylate-reactive crosslinking compound typically contains a primary amine functional group that can be coupled to a carboxylic acid group in a protein or other molecule through the use of a suitable activating agent, such as a carbodiimide. The carbodiimide forms an active ester intermediate that then reacts with the amine to create an amide bond (Chapter 4, Section 1). Reported use of diazoalkyl derivatives that spontaneously react with carboxylates have been tried with fluorescent probes, but not yet applied to heterobifunctional crosslinking agents (DeMar *et al.*, 1992; Schneede and Ueland, 1992) (Chapter 3, Section 3.1). The following heterobifunctional reagent is the only carboxylate-reactive photosensitive crosslinker currently available commercially.

6.1. ASBA

ASBA, 4-(*p*-azidosalicylamido)butylamine, is a carboxylate-reactive crosslinking agent containing a primary amine on one end and a photosensitive phenyl azide group on the other (Thermo Fisher). The crosslinker is not spontaneously reactive with carboxylates, but must be used with another activating agent that facilitates bond formation. For instance, it can be used in conjunction with a carbodiimide or other such reagent system that can initiate covalent bond formation with a carboxylic acid. A water soluble carbodiimide like EDC (Chapter 4, Section 1.1) is able to activate the carboxylates on a target molecule, forming active ester

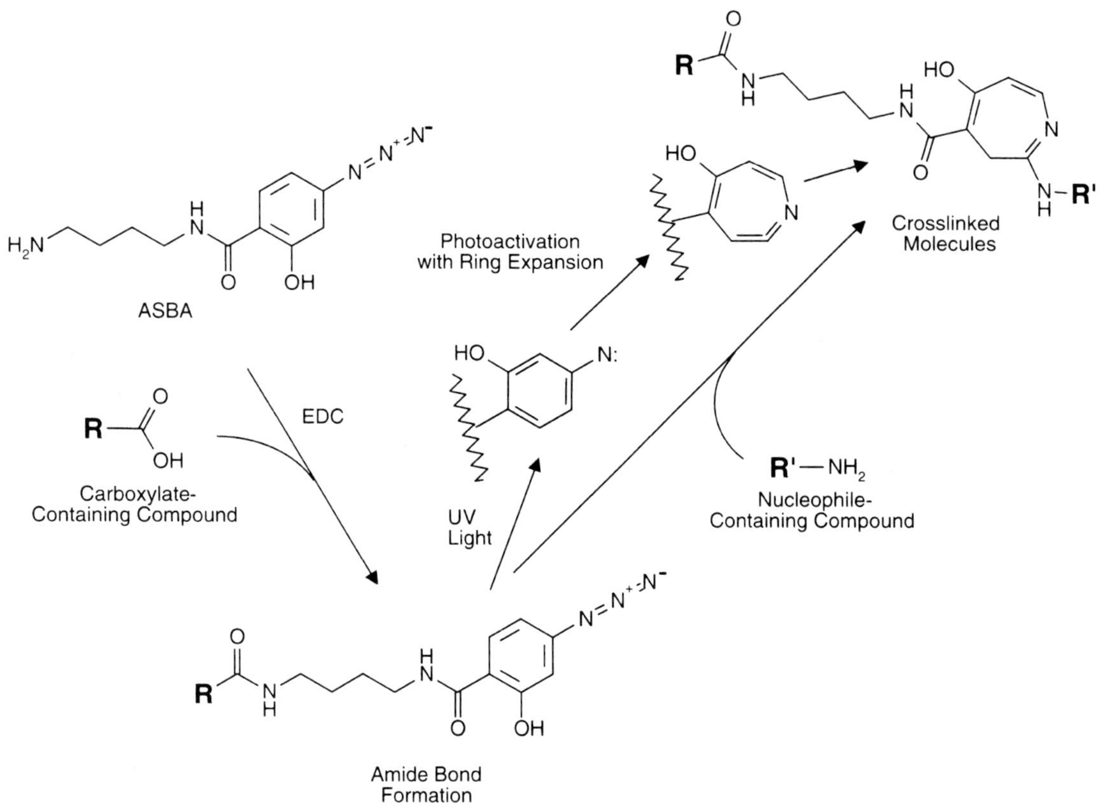

FIGURE 6.33 ASBA contains a primary amine group that may be conjugated to carboxylate compounds using the carbodiimide EDC. Subsequent exposure to UV light initiates the photoreaction leading to covalent crosslinks.

intermediates (Figure 6.33). In the presence of ASBA, derivatization will occur, resulting in amide bond formation and thus leading to modification of the carboxylate-containing molecule with a photoreactive group.

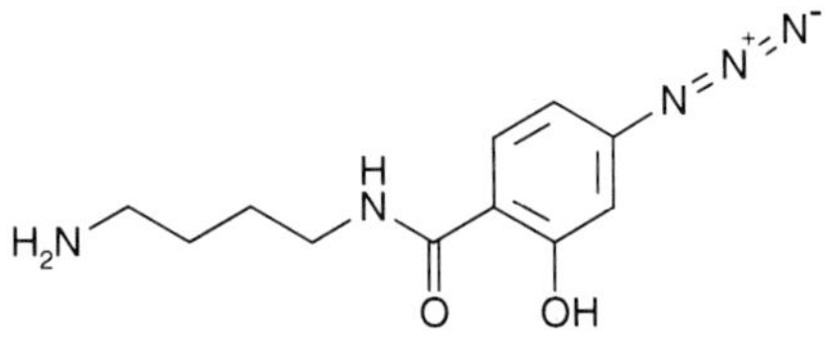

ASBA
4-(*p*-Azidosalicylamido) butylamine
MW 249.27
16.3 Å

The cross-bridge of ASBA provides a reasonably long spacer (16.3 Å). The phenyl azide portion is constructed from a salicylic acid derivative and thus possesses a ring-activating hydroxyl group. The presence of this group allows radioiodination of the ring prior to crosslinking (Chapter 12, Section 2.5). Before the photolyzing step is initiated, the reagent should be handled in the dark or protected from light to avoid decomposition of the phenyl azide group.

ASBA has been used to identify parasite adhesive proteins (Gowda *et al.*, 2007), for active site-directed labeling of glucosidase I (Romaniouk *et al.*, 2004), and to study interactions with the proteasome (Qureshi *et al.*, 2003).

7. ARGININE-REACTIVE AND PHOTOREACTIVE CROSSLINKERS

The guanidinyl group on arginine's side chain can be specifically targeted by the use of 1,2-dicarbonyl reagents, such as the diketone group of glyoxal (Chapter 3, Section 5.2). Under alkaline conditions, this type of group can condense with the guanidinyl residue to form a Schiff base-like complex. The presence of other chemical compounds in the reaction can cause further structural rearrangements, such as stabilization by boronate (Pathy and Smith, 1975). Derivatives such as phenylglyoxal and *p*-nitrophenylglyoxal can be used to block or quantitatively determine the amount of arginine in a protein (Yamasaki *et al.*, 1981). Studies have shown that if the reaction is performed with a 2:1 ratio of glyoxal compound to arginine residues, then the modification that

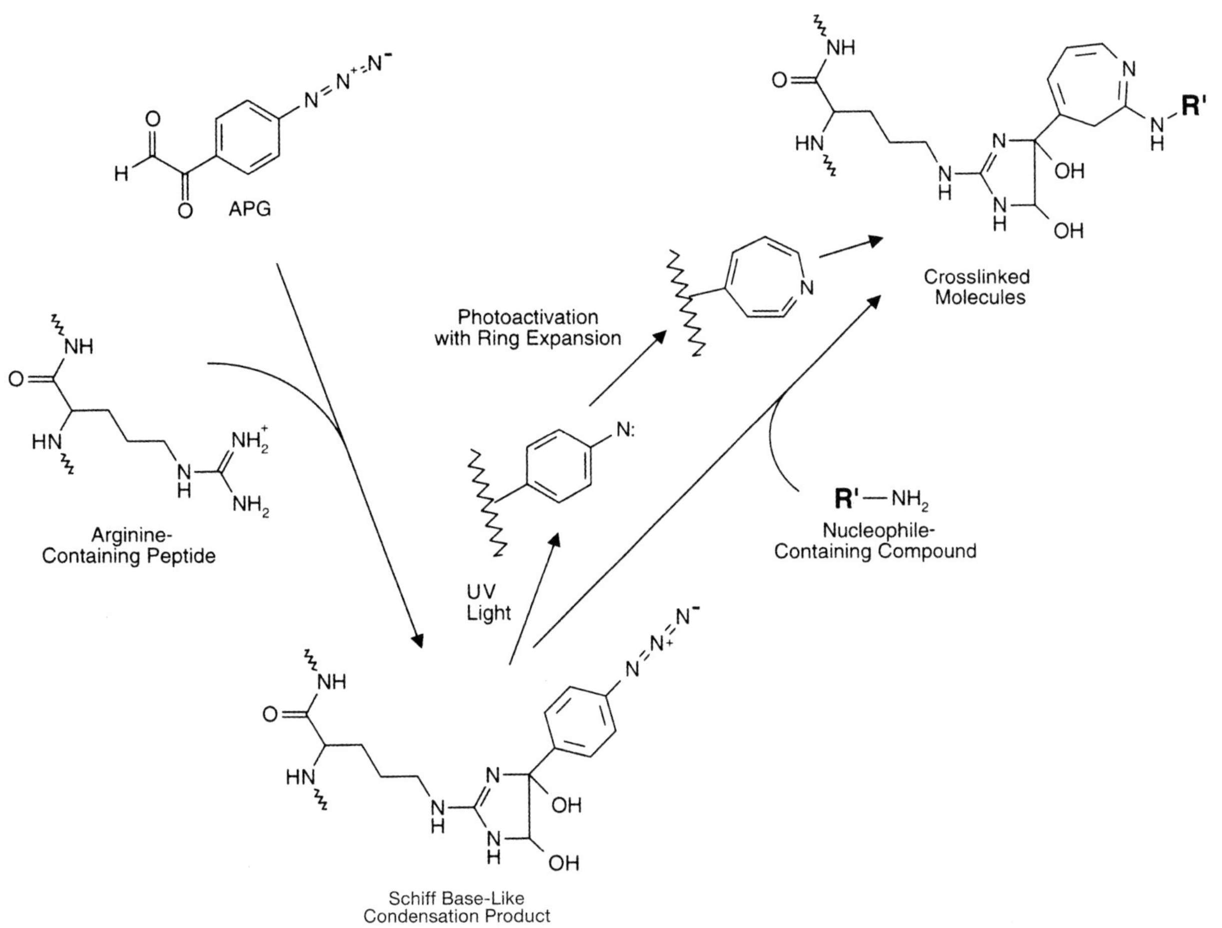

FIGURE 6.34 APG can be used to label specific arginine residues in proteins, producing stable, cyclic Schiff base-like bonds with the side-chain guanidino groups. Photoactivation with UV light then causes ring expansion of the phenyl azide group, initiating covalent bond formation with amines.

results is reversible (Takahashi, 1968). However, if the modification is performed at a 1:1 stoichiometry, then it is irreversible (Konishi and Fujioka, 1987).

The ability to direct conjugation or modification specifically through arginine residues using this chemistry has been exploited in the availability of the only photoreactive glyoxal derivative, APG.

7.1. APG

APG, *p*-azidophenyl glyoxal, is a heterobifunctional crosslinker containing an arginine-specific diketone group on one end and a photosensitive phenyl azide group on the other end (Thermo Fisher). The reagent is a derivative of phenylglyoxal, a compound long used as an arginine guanidinyl modifier. Reaction of APG with proteins at pH 7 to 8 results in selective modification of arginine, leaving photoreactive groups available for subsequent crosslinking with interacting molecules (Figure 6.34). Exposure to UV light effects the final crosslink. The cross-bridge of an APG crosslink is only 9.3 Å in length, allowing proximity interactions to be studied or the irreversible labeling of arginine areas in proteins.

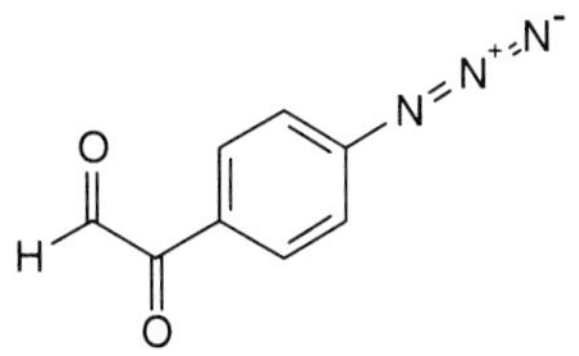

APG
p-Azidophenyl glyoxal
MW 193.16
(as the monohydrate)
9.3 Å

APG has been used to investigate the binding step in collagen phagocytosis (Chong *et al.*, 2007); to inhibit bovine heart lactic dehydrogenase, egg white lysozyme, horse liver alcohol dehydrogenase, and yeast alcohol dehydrogenase (Ngo *et al.*, 1981); to crosslink RNA–protein interactions in *E. coli* ribosomes (Politz *et al.*, 1981), and to identify regions of brome mosaic virus coat protein chemically crosslinked *in situ* to viral RNA (Sgro *et al.*, 1986).

CHAPTER

7

Trifunctional Crosslinkers

Trifunctional crosslinkers represent a relatively small but important category of bioconjugation reagents, possessing three different reactive or complexing groups per molecule. The design of this type of reagent is more elaborate than multifunctional crosslinkers built off polymers such as polyaldehyde dextran (Chapter 18, Section 2.2) or small organic molecules like trichloro-s-triazine (Chapter 18, Section 2.1) which merely contain more than two groups of the same functionality per molecule. The trifunctional approach incorporates elements of the heterobifunctional concept wherein two ends of the linker contain reactive groups able to couple with two different functional groups on target molecules. A trifunctional reagent, however, has a third arm terminating in yet another group able to specifically link to a third chemical or biological target.

A convenient molecule from which to build trifunctionals is the amino acid, L-lysine. Its three functional groups, α-carboxy, α-amino, and ε-amino, can be derivatized independently to contain three arms. Each arm can be designed to terminate in a complexing group able to participate in a particular type of conjugation reaction or affinity interaction.

The initial attempts at producing trifunctional reagents used biocytin as the core compound. Biocytin is the lysine derivative of biotin having its valeric acid side chain amide-bonded to the ε-amino group of the amino acid (Chapter 11, Section 6.2). Thus, crosslinkers built on this compound have one of their trifunctional arms ending in a biotin label which is able to specifically complex with avidin or streptavidin probes. The creation of two additional reactive arms from the α-carboxy and α-amino groups of biocytin results in the completed trifunctional. The following two crosslinkers are examples of this approach.

1. 4-AZIDO-2-NITROPHENYLBIOCYTIN-4-NITROPHENYL ESTER

Wedekind *et al.* (1989) designed a trifunctional reagent for studying the hormone binding site of the insulin receptor. The crosslinker, 4-azido-2-nitrophenylbiocytin-4-nitrophenyl ester (ABNP) contains a nitrophenyl ester group that can react with amine functions in proteins and peptides, similar to the reaction of *N*-hydroxysuccinimide (NHS) esters with amines (Chapter 3, Section 1.4). This group can be used to modify a ligand (such as insulin) prior to its binding to a specific receptor molecule. The second chemically reactive functional group on ABNP is a photosensitive phenyl azide group capable of being activated by exposure to UV light. After the labeled ligand is allowed to interact with its receptor, forming an interaction complex, the mixture is photolyzed to effect a covalent attachment point (Figure 7.1). The third arm of the trifunctional reagent is the biotin handle (from biocytin). This component allows the complex to be purified by affinity chromatography on immobilized avidin or immobilized streptavidin (Chapter 15, Section 2.5). Alternatively, the biotin group can be used to visualize the binding of the ligand to its receptor using labeled avidin or streptavidin reagents (Chapter 11, Sections 1–5).

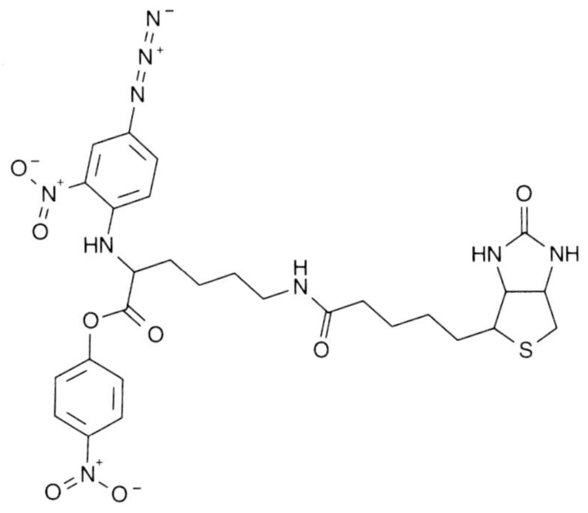

4-Azido-2-nitrophenylbiocytin-4-nitrophenyl ester

ABNP is soluble in dimethylformamide (DMF) but insoluble directly in aqueous solution. Insulin

Bioconjugate Techniques, Third Edition.
DOI: http://dx.doi.org/10.1016/B978-0-12-382239-0.00007-8

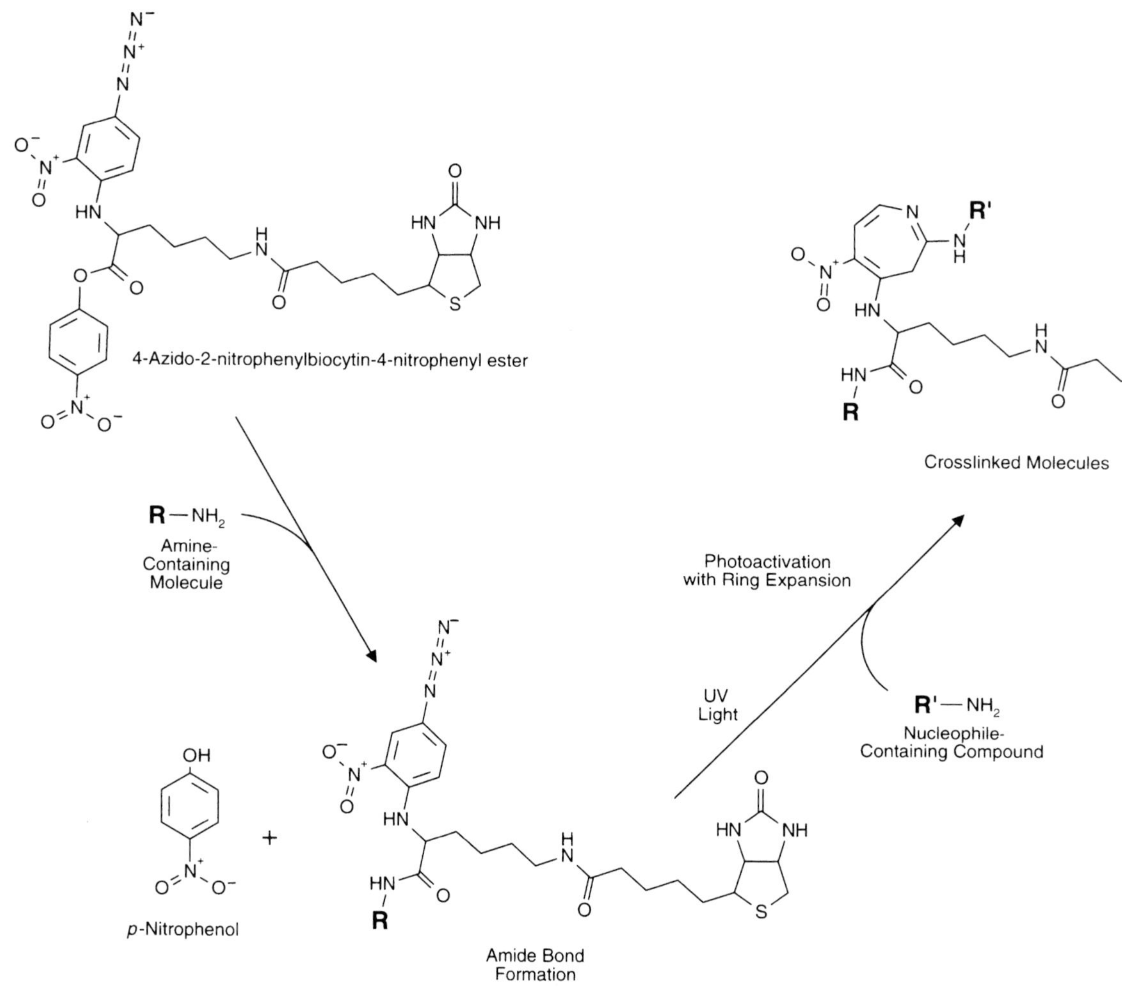

FIGURE 7.1 The Wedekind trifunctional crosslinker can react with amine groups via its *p*-nitrophenyl ester to form amide bond linkages. The phenyl azide group then can be photoactivated with UV light to generate covalent bond formation with a second molecule. The biotin side chain provides binding capability with avidin or streptavidin probes.

labeling was done in DMF:water at a ratio of 9:1. For molecules not soluble in organic solvent, such as proteins, the trifunctional may first be dissolved in DMF and a small aliquot added to an aqueous reaction medium. The nitrophenyl ester reactive group can be coupled to amine groups at alkaline pH (7–9) and in buffers containing no extraneous amines (avoid Tris). Unfortunately, ABNP is not commercially available at the time of this writing.

2. SULFO-SBED

Another trifunctional crosslinking agent is sulfo-SBED or sulfosuccinimidyl-2-[6-(biotinamido)-2- (*p*-azido-benzamido) hexanoamido]ethyl-1,3′-dithiopropionate, developed by Ed Fujimoto at Pierce Chemical (now Thermo Fisher). Similar to ABNP discussed previously, sulfo-SBED is built on a biocytin backbone. Thus, one arm of the trifunctional compound consists of a biotin handle that can be used for purification or detection purposes using avidin or streptavidin probes. The chemically reactive groups of sulfo-SBED include a sulfo-NHS ester and a phenyl azide group. The sulfo-NHS ester provides amine-coupling capability, forming amide bond linkages with target molecules (Chapter 3, Section 1.4). The phenyl azide is photosensitive and may be activated by exposure to UV light at wavelengths >300 nm. Most phenyl azides react by ring expansion to dehydroazepines with subsequent reactivity toward nucleophiles, especially amines (Chapter 3, Section 7.1) (Figure 7.2).

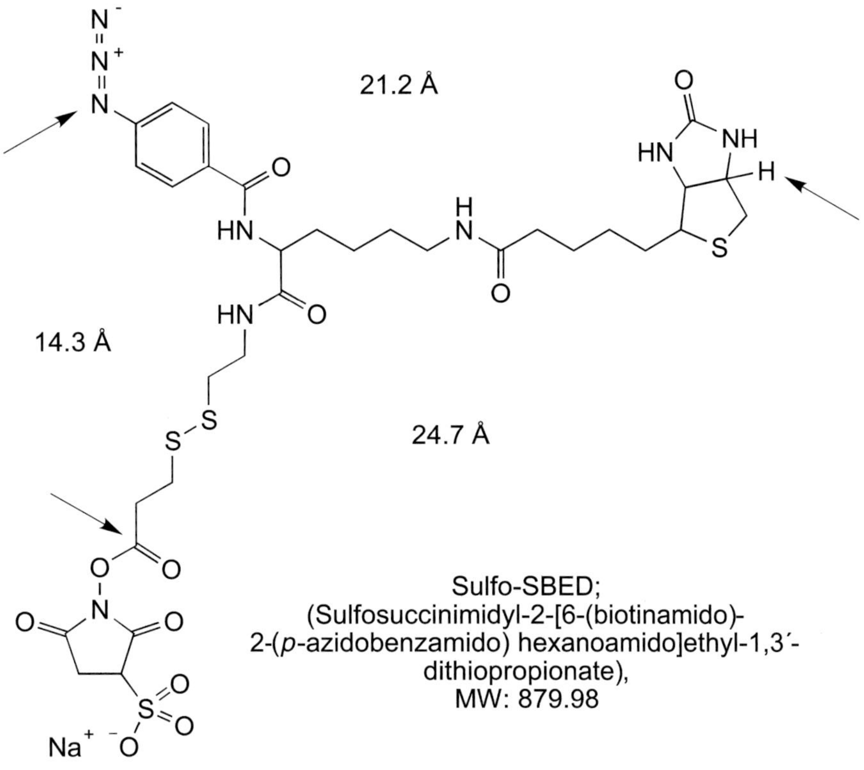

Sulfo-SBED;
(Sulfosuccinimidyl-2-[6-(biotinamido)-
2-(*p*-azidobenzamido) hexanoamido]ethyl-1,3′-
dithiopropionate),
MW: 879.98

The sulfo-NHS ester of sulfo-SBED is negatively charged and provides a degree of water solubility (about 5-m*M* maximum concentration) for the entire molecule. Limited water solubility is all that can be expected due to the large size of the trifunctional, most of it consisting of relatively hydrophobic structures. However, the reagent is much more soluble in organic solvents such as DMF (170 m*M*) and dimethyl sulfoxide (DMSO) (125 m*M*). Concentrated stock solutions may be prepared in these solvents prior to addition of a small aliquot to an aqueous reaction mixture.

Since the active ester end of the molecule is subject to hydrolysis (half-life of about 20 min in phosphate buffer at room temperature conditions), it should be coupled to an amine-containing protein or other molecule before the photolysis reaction is carried out. During the initial coupling procedure, the solutions should be protected from light to avoid decomposition of the phenyl azide group. The degree of derivatization should be limited to no more than a 5- to 20-fold molar excess of sulfo-SBED over the quantity of protein present to prevent possible precipitation of the modified molecules. For a particular protein, studies may have to be done to determine the optimal level of modification.

An additional feature of sulfo-SBED is the presence of a cleavable disulfide group in the cross-bridge of the NHS ester arm of the molecule. After a conjugation reaction has taken place, the complexes first may be purified using immobilized avidin or immobilized streptavidin and then the conjugates released by treatment with a disulfide reducing agent. This allows analysis of the complexed molecules, for example, after the binding of a ligand to its receptor. Alternatively, the disulfide group may be cleaved after interaction and capture of unknown proteins, thus transferring the biotin label to the prey protein. The unknown interacting protein may then be detected or purified using the biotin tag. Such label transfer procedures are important options for studying protein–protein interactions (Chapter 24).

Since sulfo-SBED has three functional arms, the length of each portion should be considered when doing conjugation studies involving interacting proteins. The biotin handle has an effective length of 19.1 Å, including the side-chain length for the lysine component. The sulfo-NHS ester arm is approximately 13.7 Å long, measuring from the same point in the lysine group. The phenyl azide arm is the shortest, only 9.1 Å long. The structure for sulfo-SBED shown in this section includes molecular distance measurements somewhat different from these numbers in that the total distances between the three arms are given, which reflects the intramolecular distances between the terminal reactive groups or interacting group on biotin (indicated by the arrows).

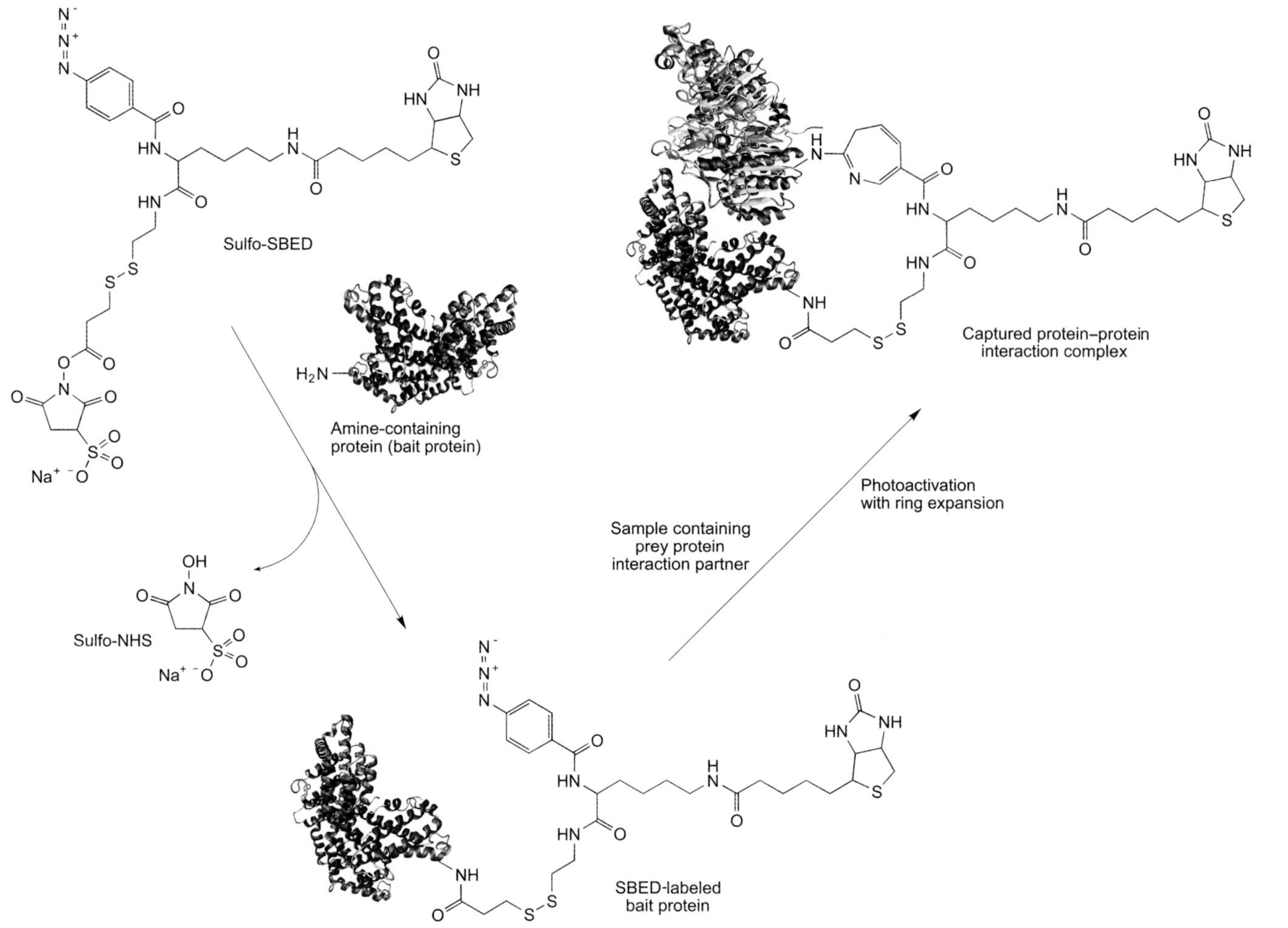

FIGURE 7.2 The trifunctional reagent sulfo-SBED reacts with amine-containing bait proteins via its NHS ester side chain. Subsequent interaction with a protein sample and exposure to UV light can cause crosslink formation with a second interacting protein. The biotin portion provides purification or labeling capability using avidin or streptavidin reagents. The disulfide bond on the NHS ester arm provides cleavability using disulfide reductants, which effectively transfers the biotin label to an unknown interacting protein.

Sulfo-SBED has been used to confirm the interaction between human frataxin and the scaffold protein ISU with mass spec detection (Watson *et al.*, 2012), to investigate the binding of receptor tyrosine kinase TrkB to the neural cell adhesion molecule (NCAM) (Cassens *et al.*, 2010), and in a host of other applications designed to identify protein interactions (see Chapter 24, Section 3.1, for additional information on the use of sulfo-SBED in capturing interacting partners).

The following suggested protocol was developed by Barb Olson at Thermo Fisher for the labeling of soybean trypsin inhibitor with its subsequent complexation with trypsin. Modifications to this procedure may have to be done to study other proteins.

Protocol

1. Dissolve 5 mg of soybean trypsin inhibitor (STI) in 0.5 ml 0.1-*M* sodium phosphate, 0.15-*M* NaCl, pH 7.2.
2. In a fume hood, dissolve 1.12 mg of sulfo-SBED in 25 μl of DMSO. Prepare fresh and protect from light.
3. Add 11 μl of the sulfo-SBED solution to the STI solution. Mix well.
4. React for 30 min at room temperature or for 2 h at 4°C.
5. If some precipitation occurs, clarify the solution by centrifugation using a microfuge. Remove excess reactant by gel filtration using a desalting resin.
6. Mix the purified sulfo-SBED-modified STI with 5 mg of trypsin dissolved in 0.1-*M* sodium phosphate, 0.15-*M* NaCl, pH 7.2.
7. Incubate at room temperature for 3.5 min to allow the specific binding of the two molecules to occur.
8. Photolyze the solution with long UV light (about 365 nm) at a distance of about 5 cm for 15 min. This process may be done with the solution on ice to prevent heating of the sample.

Isolation of complexed molecules may be done by affinity chromatography using a column of immobilized avidin or immobilized streptavidin. Cleavage of the disulfide bond of the crosslinker may be done by treatment with 50-m*M* dithiothreitol (DTT). For additional information on the use of sulfo-SBED in the study of protein interactions, see Chapter 24, Section 3.1.

3. MTS–ATF–BIOTIN AND MTS–ATF–LC-BIOTIN

MTS–ATF–Biotin and MTS–ATF–LC-Biotin are trifunctional crosslinkers similar in design to sulfo-SBED discussed previously, but in addition to the biotin handle, they contain a thiol-reactive group and an enhanced photoreactive, perfluorinated phenyl azide group. The two reagents differ only in the length of the cross-bridge in the photoreactive arm, with the LC version containing an extended aminocaproyl spacer. Relative to the spacing possible between the reactive groups on these compounds, the LC version therefore provides nearly twice the maximal molecular distance over its shorter analog (21.8 Å *versus* 11.1 Å). Thus, interacting proteins may be captured through use of either a long or short crosslink, depending on the optimal distances between the proteins—or at least to the nearest thiol on the bait protein.

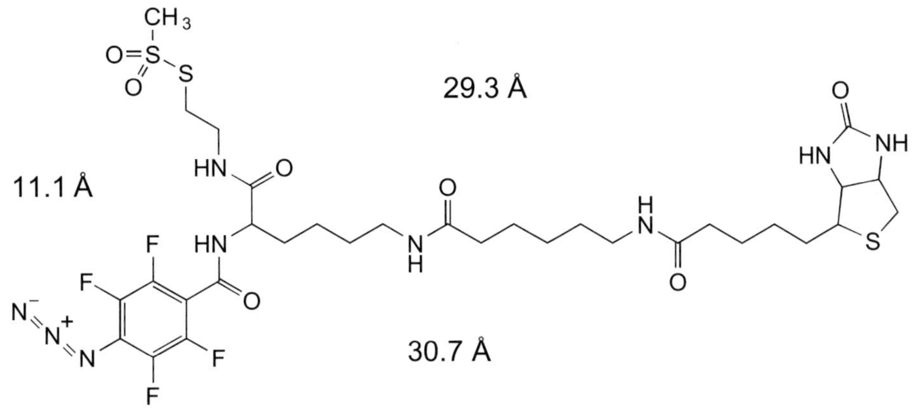

MTS–ATF–Biotin
(Methanethiosulfonate-azidotetrafluoro–Biotin)
M.W. 839.95

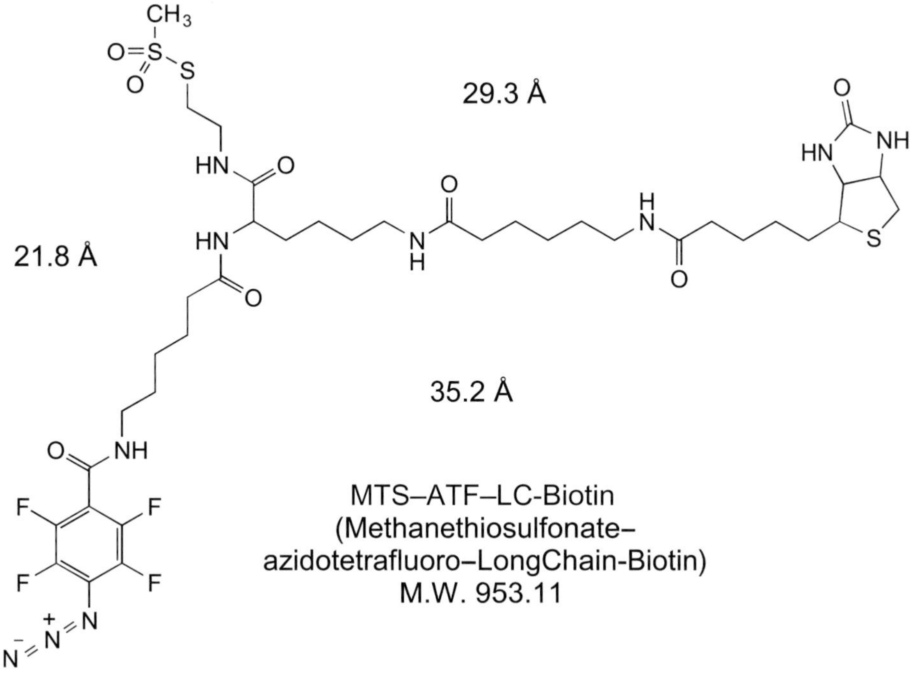

MTS–ATF–LC-Biotin
(Methanethiosulfonate-azidotetrafluoro–LongChain-Biotin)
M.W. 953.11

Both MTS–ATF–Biotin and MTS–ATF–LC-Biotin contain a methanethiolsulfonate group (MTS) on one arm, which is able to couple with thiols. This reaction proceeds with loss of the methyl sulfonate leaving group (sulfinic acid) and forms a disulfide linkage (Figure 7.3). Unlike a pyridyl disulfide group, however, which also reacts with thiols to form disulfide linkages, the MTS group is unstable to hydrolysis in aqueous solution,

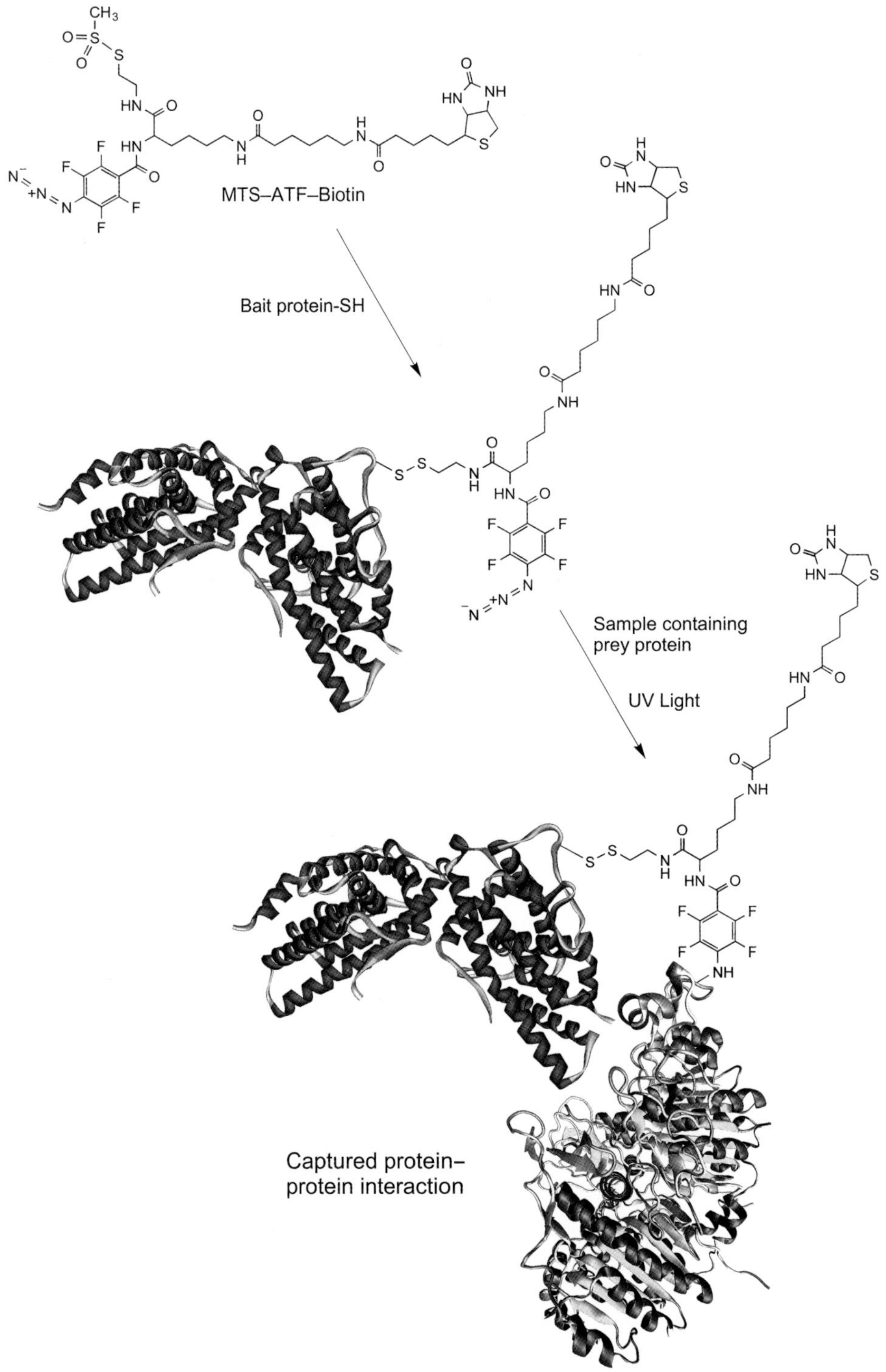

FIGURE 7.3 Mts–Atf–Biotin can be used to label bait proteins at available thiol groups using the methanethiolsulfonate group, which forms a disulfide linkage after reaction. The modified protein is then allowed to interact with a protein sample and photoactivated with UV light to cause a covalent crosslink with any interacting proteins. Cleavage of the disulfide bond effectively transfers the biotin label to the unknown interacting protein.

especially if other strong nucleophiles are present. Therefore, most MTS compounds dissolved or brought into PBS buffer at physiological pH will hydrolyze with a half-life on the order of 10 to 15 min. However, they also have very rapid reactivity with thiols (Stauffer and Karlin, 1994; Holmgren *et al.*, 1996; Liu *et al.*, 1996). The reaction of an MTS group with a thiol on a bait protein can take place with high yields in just a matter of minutes. Both of these trifunctional label transfer compounds are hydrophobic, so their MTS reactivity in the aqueous phase may be somewhat slower than corresponding hydrophilic MTS reagents.

For use in studying protein interactions, these compounds are first reacted with a thiol on a purified bait protein to form a disulfide bond. Since the reagents are water insoluble, they must be dissolved in an organic solvent such as DMF or DMSO and then an aliquot added to the bait protein in an aqueous buffer to initiate the reaction. Once modified, the biotinylated bait protein is then incubated with a sample containing potentially interactive prey proteins. After an incubation period, initiating the photoreaction by exposure to UV irradiation captures the interacting proteins. Any interaction complexes thus formed can be isolated or detected using the biotin handle. In addition, the disulfide bond formed with the bait protein during the crosslinking reaction can be reduced to cleave the conjugates and transfer the biotin label to the unknown interacting prey proteins. This is a powerful way of labeling unknown interacting proteins for subsequent analysis.

MTS–ATF–Biotin and MTS–ATF–LC-Biotin have been used to investigate protein interactions involving the SufBCD Fe-S scaffold complex binding with SufA for Fe-S cluster transfer (Chahal *et al.*, 2009), in the characterization of carriers and receptors of the Lewis^X glycan (Katagihallimath, 2008), and for study of the piccolo NuA4 catalyzed acetylation of nucleosomal histones (Huang and Tan, 2012). A protocol for the use of these compounds in the study of protein interactions can be found in Chapter 24, Section 3.2.

4. HYDROXYMETHYL PHOSPHINE DERIVATIVES

Although phosphine compounds are often used as disulfide reducing agents in bioconjugate chemistry (Chapter 2, Section 4.1), there are classes of organo-phosphine reagents containing three hydroxymethyl groups that can act as trifunctional bioconjugation agents for coupling or crosslinking purposes. Tris(hydroxymethyl) phosphine (THP) and β-[tris(hydroxymethyl)phosphino] propionic acid (THPP) (Thermo Fisher) are small trifunctional compounds that spontaneously react with nucleophiles, such as amines, to form covalent linkages (Henderson *et al.*, 1994; Katti, 1996; Katti *et al.*, 1999). Nucleophiles react with the hydroxymethyl arms by attack on the electron-deficient carbon atom with loss of water to form secondary or tertiary amine bonds (Figure 7.4).

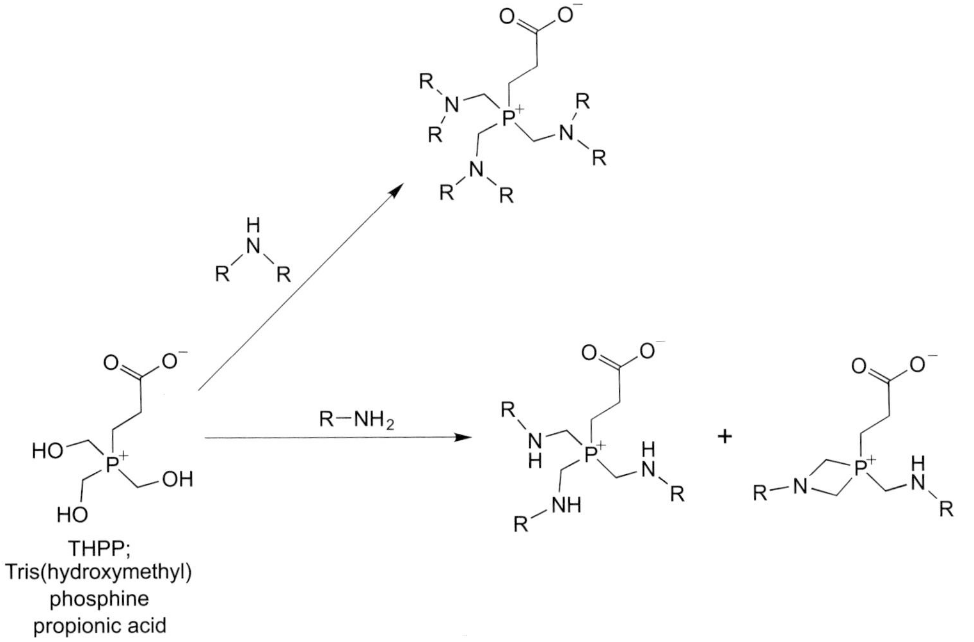

FIGURE 7.4 THPP reacts with amine-containing molecules to form secondary or tertiary amine bonds.

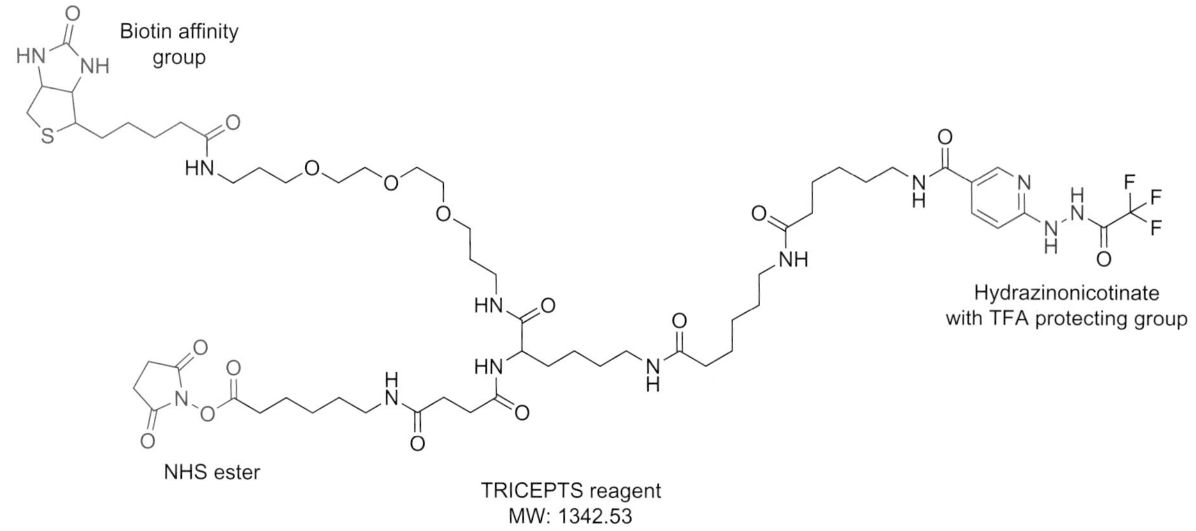

FIGURE 7.5 The TRICEPTS reagent contains an amine-reactive NHS ester to couple with a ligand that can interact with a cell surface receptor. The hydrazinonicotinate end of the compound is able to form hydrazone bonds with glycans on these receptors if the carbohydrate has been previously oxidized with sodium periodate to form aldehyde residues. The trifluoroacetyl protecting group on the hydrazine spontaneously hydrolyzes off during this reaction. Thus, receptors that are able to bind to the labeled ligand become permanently labeled with the TRICEPTS reagent and now contain a biotin affinity handle for detection or purification purposes. Receptors so labeled can be detected using streptavidin conjugates or enriched on immobilized streptavidin columns and analyzed by mass spectrometry.

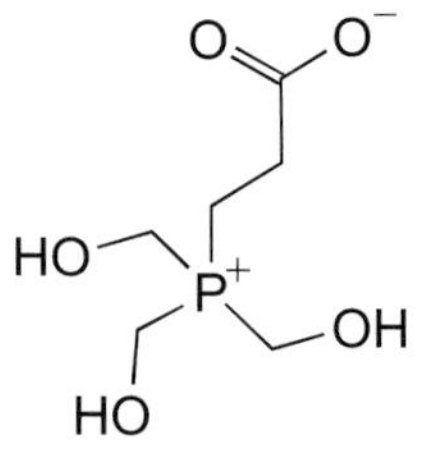

THPP;
Tris(hydroxymethyl)
phosphine propionic acid

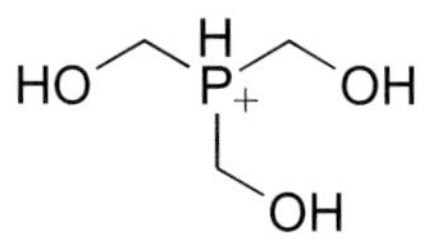

THP;
Tris(hydroxymethyl)
phosphine

THP has also been used as a reducing agent in organic synthesis for some time and it is able to form coordination complexes with metals to create potent hydrogenation catalysts (James and Lorenzini, 2010). Both THP and THPP are stable in aqueous solution, as the only potential product of hydrolysis is the reformation of the hydroxymethyl groups. It is unusual for an amine-reactive functional group to have long-term stability in water or buffer, which makes these reagents uniquely suitable for creating reactive surfaces or reactive molecules for subsequent conjugation with proteins or other amine-containing compounds. Hydroxylic chromatographic supports have also been activated with hydroxymethyl phosphine derivatives for immobilization of enzymes (Petach *et al.*, 1994).

Hydroxymethyl phosphines are susceptible to oxidation to form the phosphine oxide. Therefore, avoid excess oxygen, oxidizing agents, or azide compounds, which react with phosphines in the Staudinger reaction (Chapter 17, Section 6). In addition, metallic surfaces can be modified via the phosphine group to result in hydroxymethyl group substitutions.

5. TRICEPTS REAGENT

Frei *et al.* (2012) reported on a new trifunctional crosslinker design, which they termed TRICEPTS, a descriptive name that depicts its three arms capable of binding three different components (Figure 7.5). Similar in design to the trifunctional reagents described previously in Sections 1 to 3, this compound was specifically constructed to capture interacting molecules associated with receptor–ligand binding events. One arm of this reagent contains a terminal NHS ester group that can be used to label a protein or ligand that interacts with a receptor through the formation of an amide linkage. A second arm of the compound contains a trifluoroacetyl-protected hydrazine group that is capable of binding to glycans on cell surface receptors after they have been treated by mild oxidation with sodium periodate to produce aldehydes (Chapter 2, Section 4.4). The hydrazine group can spontaneously form a hydrazone linkage with an aldehyde on these glycoproteins, thus locking the ligand to its receptor after the interaction has occurred. The third arm of the TRICEPTS compound contains a long biotin affinity group that can then be used to detect or isolate the interacting ligand–receptor complexes using streptavidin conjugates or immobilized streptavidin.

The TRICEPTS reagent was used to successfully detect interactions between ligands and receptors using labeled insulin, transferrin, apelin, and epidermal growth factor (EGF), as well as therapeutic antibodies able to interact with receptors on tumor cells. Fluorescently labeled streptavidin could be used to detect the crosslinked complex on cell surfaces. In addition, the complexes captured on immobilized streptavidin could be analyzed by mass spec to determine the identities of the receptors being bound.

CHAPTER

8

Dendrimers and Dendrons

1. DENDRIMER CONSTRUCTION

The science of nanotechnology has employed many different constructs having low nanometer dimensions, including inorganic scaffolds, biological macromolecules, and various forms of polymers and particles. One of the most defined nanoparticle constructs is a unique polymeric assemblage called a dendrimer. First described in 1983 by Tomalia and Dewald in the application to U.S. patent 4,507,466 and again in 1985 by Tomalia *et al.* and Newkome *et al.*, dendrimers are monodisperse, globular macromolecules grown by successive synthetic steps from a central core molecule. Each step, called a generation, adds a distinct layer to the previous one so that a dendrimer grows out from the core like the branches of a tree. In fact, the name dendrimer comes from the Greek word for tree, or "dendron." The result is a polymeric molecule having a fractal dimensional quality with properties and shape that are determined by the types of monomers used to grow the branches.

Each step in dendrimer synthesis occurs independently of the other steps; therefore, a dendrimer can take on the characteristics defined by the chemical properties of the monomers used to construct it. Dendrimers thus can have almost limitless properties depending on the methods and materials used for their synthesis. Characteristics can include hydrophilic or hydrophobic regions, charged or uncharged parts, the presence of functional groups or reactive groups, metal-chelating properties, core/shell dissimilarity, electrical conductivity, hemispherical divergence, biospecific affinity, and photoactivity, or the dendrimers can be selectively cleavable at particular points within their structure.

Dendrimers have been used for many diverse applications within the biological, chemical, polymer, and nanotechnology fields. Some major applications include their use as multivalent bioconjugation scaffolds, for enhancement of signals in assays, to solubilize hydrophobic molecules in aqueous environments by internal entrapment, to functionalize surfaces and particles for conjugation, as transfection agents for cells, to create targeted therapeutic constructs for the treatment of disease, as carriers of affinity ligands, and as additives for other polymer mixtures. For excellent reviews of dendrimer technology, see Fréchet and Tomalia (2002), as well as Boas *et al.* (2006) and Menjoge *et al.* (2010).

The synthesis and structure of a dendrimer can be illustrated by the well-known poly(amidoamine) type (called PAMAM), which describes the monomers making up the complete polymer. The synthesis starts from a core diamine (or ammonia) molecule. The diamine can be of various lengths and spacer arm properties and even contain cleavable components. Typically, the core is a short diamine, such as ethylenediamine. The first reaction that is done to form a PAMAM dendrimer is reacting the core with methyl acrylate to form the Michael addition product, a tetra-methyl ester branched molecule. Next ethylenediamine is again added in large excess to the tetra-methyl ester intermediate, which undergoes amidation to form the tetra-amidoethylamine generation-0 (G-0) product containing four pendent amines. Another round of methyl acrylate addition followed by ethylenediamine yields the eight-amine generation 1 (G-1) PAMAM dendrimer (Figure 8.1).

Similar successive additions of methyl acrylate and ethylenediamine result in progressively higher-generation dendrimers, which branch out to larger diameters and contain greater numbers of amines on their surface. For dendrimers of the classic PAMAM type, each additional generation results in doubling the number of pendent amine groups decorating its surface, because alkylation of each terminal amine can be done twice with two molecules of methyl acrylate. Since each step in dendrimer synthesis adds greater branching of monomer units as they grow out from the core, the design has become known as "starburst" dendrimers. Figure 8.2 illustrates graphically how the growth of dendrimers built from a bifunctional core results in ever more branching as the generational size increases. The corresponding G-3 and G-4 PAMAM dendrimer chemical structures are shown in Figures 8.3 and 8.4.

Bioconjugate Techniques, Third Edition.
DOI: http://dx.doi.org/10.1016/B978-0-12-382239-0.00008-X

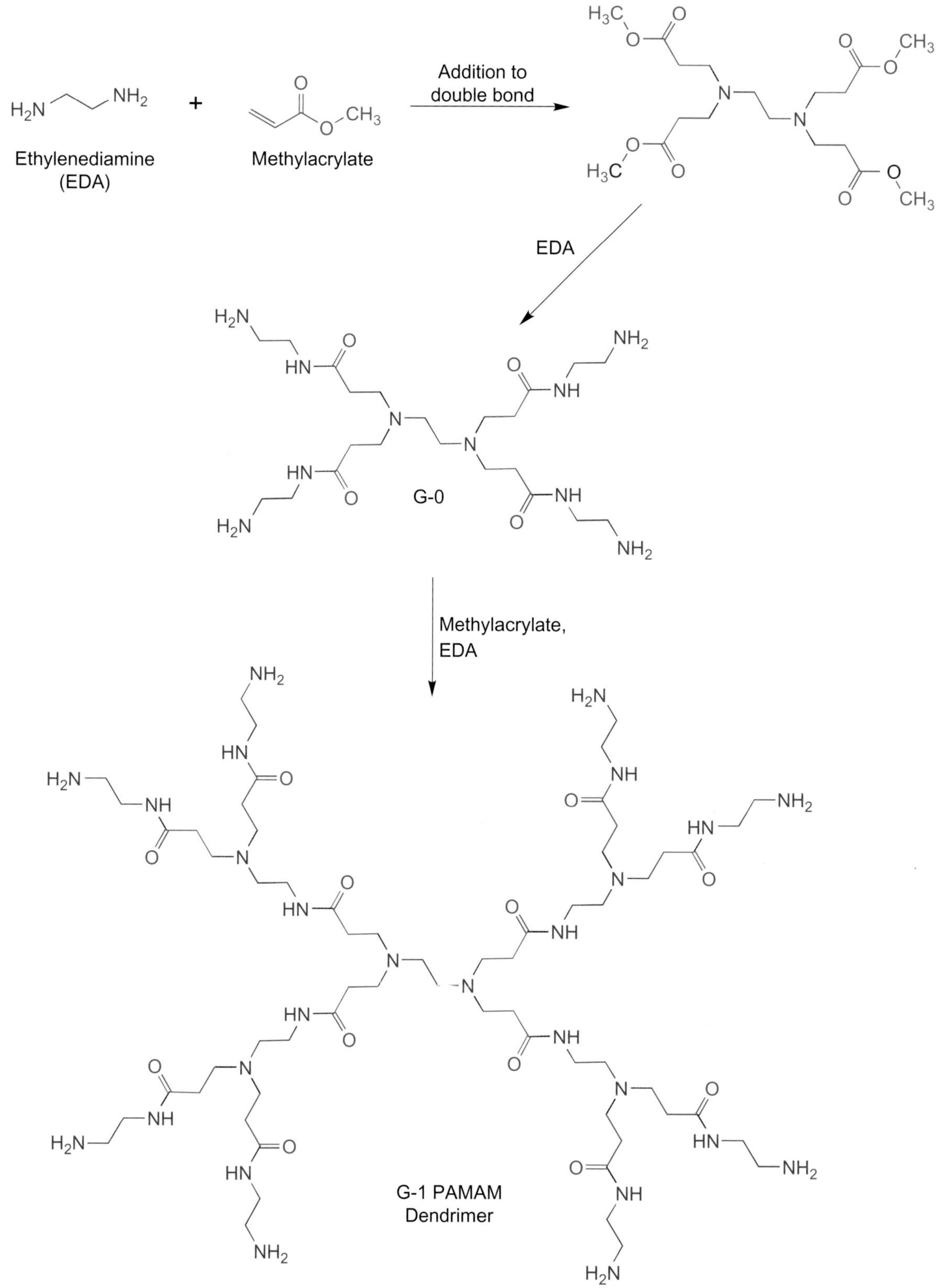

FIGURE 8.1 The synthesis of a PAMAM-type dendrimer proceeds from a diamine core (e.g., ethylene diamine, or EDA) by initial addition of the amines to the double bonds of methacrylate. Subsequent reaction of the methyl ester groups with EDA produces a G-0 dendrimer with four pendent amine groups. Another round of methacrylate and EDA additions results in a G-1 PAMAM dendrimer containing eight primary amines.

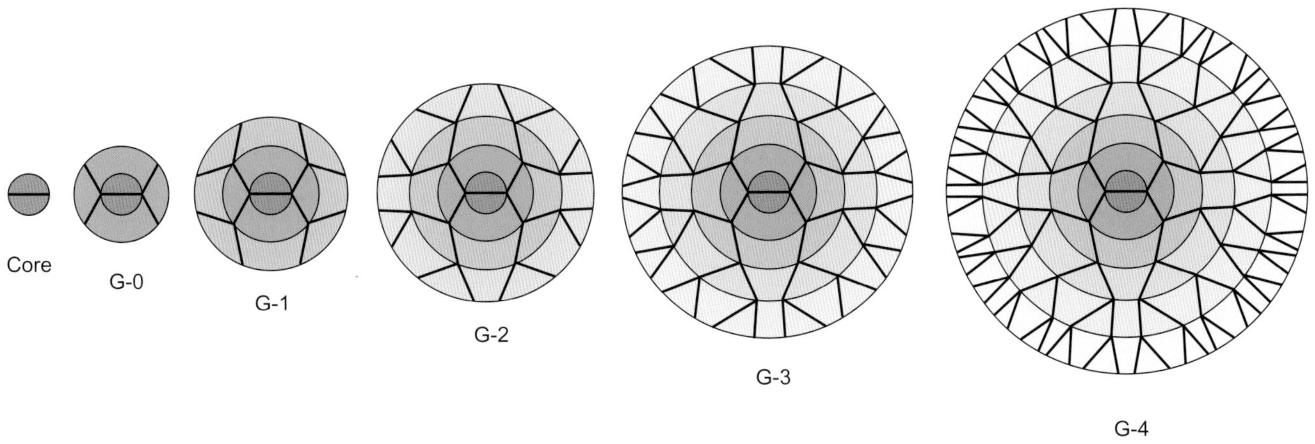

FIGURE 8.2 A graphical illustration of the growth of dendrimer structures from an initial bifunctional core to a G-4 dendrimer containing 64 terminal groups on its outer surface.

Although a two-dimensional depiction of this dendrimer structure may look like the molecule has nearly perfect circular symmetry, its true three-dimensional structure actually appears more asymmetrical for lower-generation dendrimers and like a complex globular protein (Figure 8.5) for generations above G-4.

There are two main methods of synthesizing dendrimers: (1) the divergent method, which involves building the core outward in successive steps or generations as described above for PAMAM dendrimers; and (2) the convergent method (Hawker and Fréchet, 1990; Hodge, 1993; Grayson and Fréchet, 2001), which consists of building a single-branched tree as it would grow out from the core in the divergent method, but in this case, after synthesis of individual trees, they are linked to the core structure as single units (Figure 8.6). Thus, convergent dendrimers are constructed from the outside in. One advantage of the convergent synthesis strategy is that different dendritic starting materials (dendrons) can be built and combined to form a segment-block dendrimer or a layer-block dendrimer, which consists of polymers of different types within the same dendrimer structure. The only major disadvantage of using the convergent approach to making dendrimers is that the result is limited to rather small dendritic molecules, because as the size of the building blocks increases, steric crowding prevents efficient reaction of all the dendrons with the core.

A third method of constructing dendrimers is through self-assembly of engineered building blocks. When the blocks are put together under the correct conditions in solution, they spontaneously assemble into dendritic structures, thus dramatically simplifying an often tedious, multi-step production process (Walter and Malkroch, 2012). For instance, dendrimers have been assembled through use of chelating components, which assemble into dendritic structures upon addition of the appropriate metal ions (Denti *et al.*, 1992; Balzani *et al.*, 1996; Kawa and Fréchet, 1998). Oligonucleotide dendrimers also have been formed by using intelligently designed sequences that hybridize to other oligos in such a way that dendritic molecules are spontaneously created in solution (Genisphere technology). In addition, dendrimers have been made from dendron trees containing interior groups that can self-assemble through hydrogen bonding with groups on neighboring trees, thus forming the complete dendrimer as the adjacent groups interact at the core (Hudson *et al.*, 1997; Percec *et al.*, 1998).

Finally, dendrimers have been synthesized using solid-phase peptide synthesis resins, wherein the core is linked to the resin and the half-dendrimer (dendron) is built out from it in sequential steps (March *et al.*, 1996; Swali *et al.*, 1997; Wells *et al.*, 1998). The advantage of this method is the ease with which reactants are added and the growing dendron is purified from reaction by-products. This method also can be used to add terminal peptides onto the dendron for affinity-targeting purposes. Monaghan *et al.* (2001) used this approach to make peptide–dendrimer conjugates for the investigation of integrin binding.

One of the most important advances in synthesizing dendrimers is the use of the cycloaddition reaction between azides and alkynes, which has become known as "click chemistry" (Chapter 17, Section 5). The copper I-catalyzed conjugation reaction forms cyclic 1,2,3-triazoles in very high yield. Using the appropriate monomers, the click chemistry-mediated assemblage of dendrimers has provided dramatic improvement to both the yield and the ease of synthesis and thus

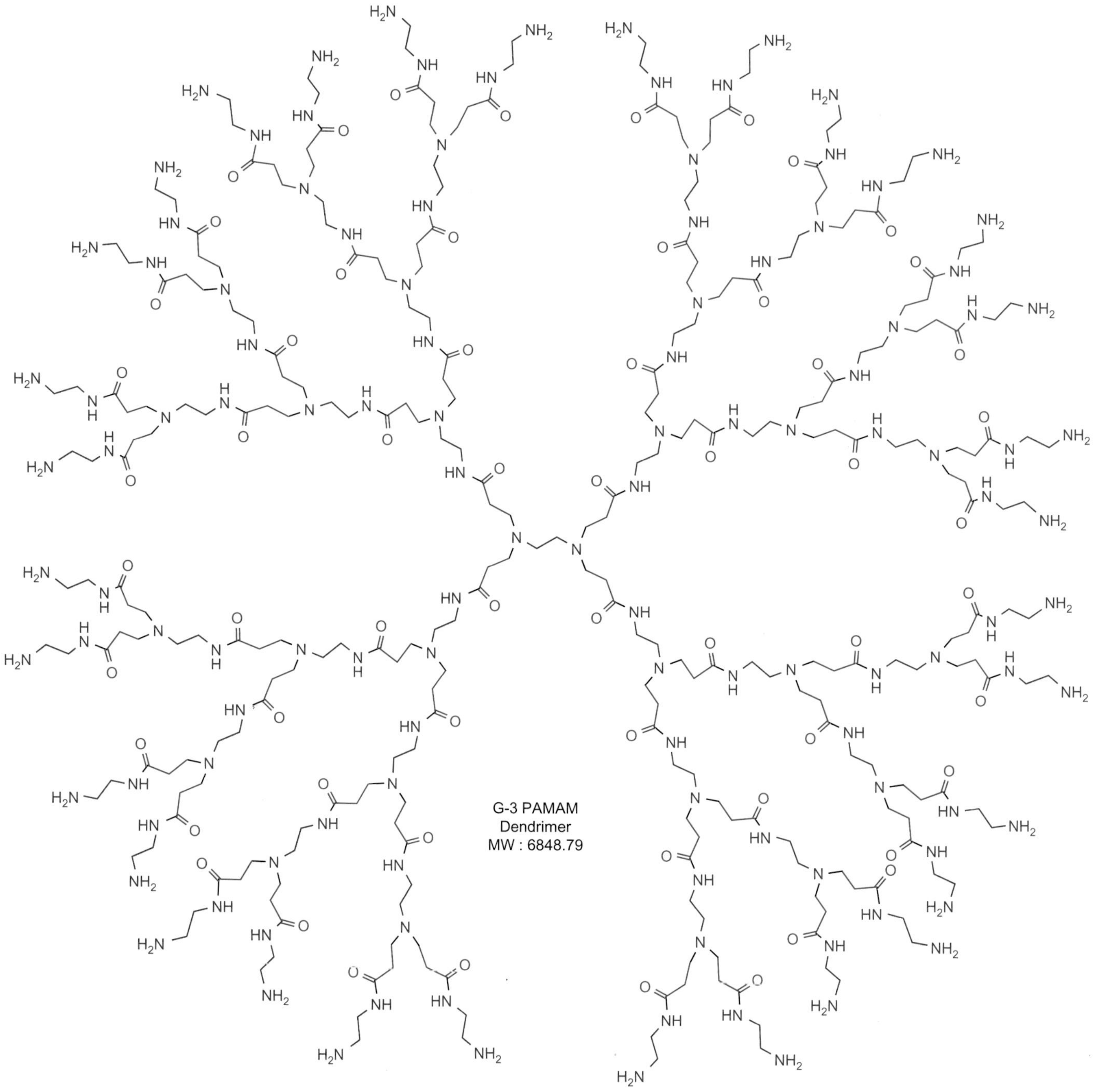

FIGURE 8.3 The chemical structure of a G-3 PAMAM dendrimer.

has dramatically decreased the cost of making bulk amounts of dendritic molecules (Rouhi, 2004; Wu *et al.*, 2004; Joralemon *et al.*, 2005; Lee and Kim, 2005; Wu *et al.*, 2005; Huang et al., 2012) (Figure 8.7). Even unsymmetrical dendrimers can be prepared using a convergent synthesis approach between propargyl-functionalized PAMAM dendrons and azido-functionalized PAMAM dendrons (Lee *et al.*, 2007).

Regardless of how they are made, the higher the dendrimer generation the greater the density of its branching becomes. Dendrimers of small size have an internally open configuration that freely permits the flow of small molecules within their inner structure. As dendrimers increase in diameter from G-0 through G-7, their appearance and size become more and more similar to those of globular proteins of corresponding mass.

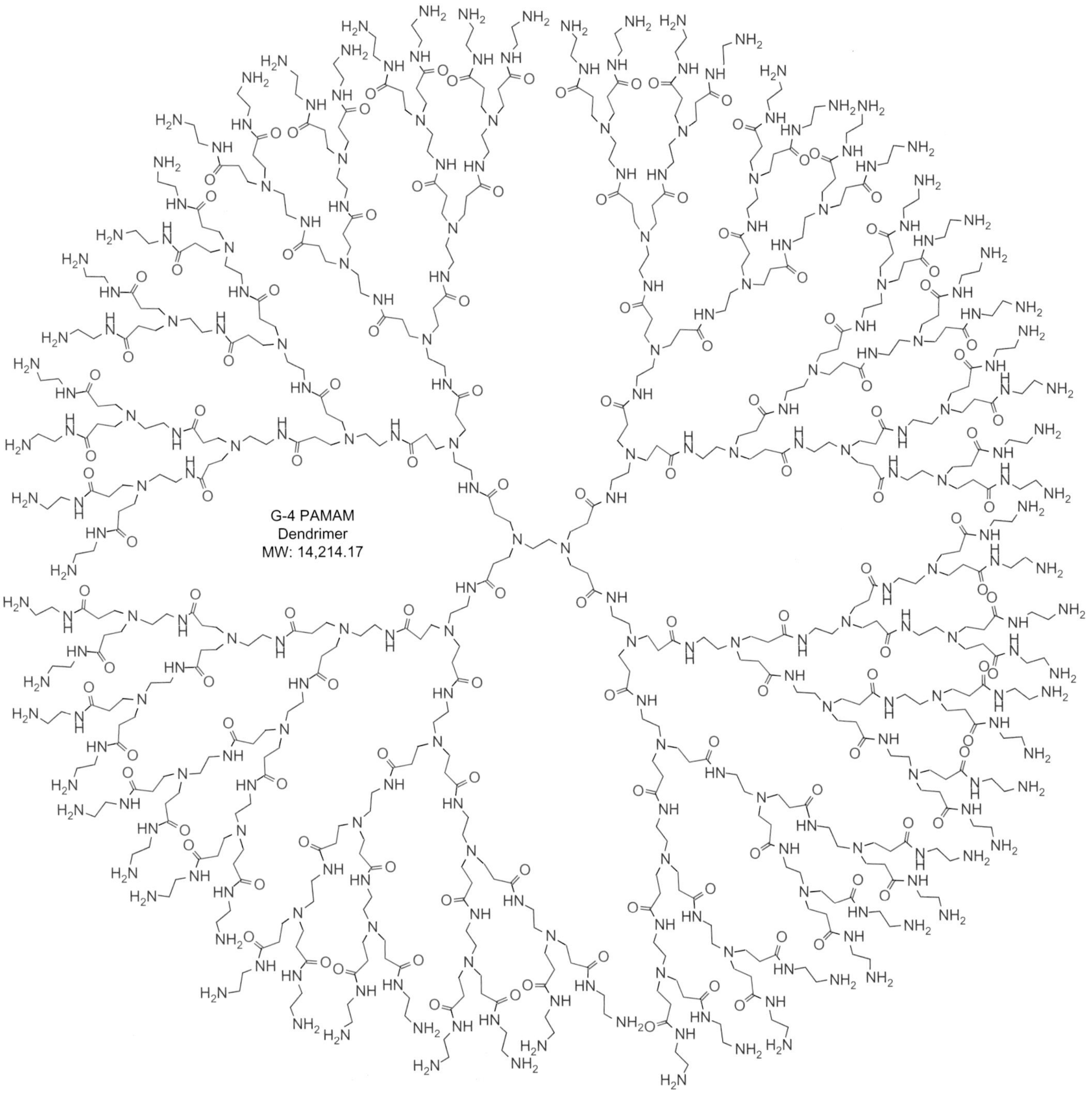

FIGURE 8.4 The chemical structure of a G-4 PAMAM dendrimer.

For instance, a G-0 PAMAM dendrimer has a molecular weight of 517 and a diameter of about 1.5 nm, whereas a G-7 dendrimer has a mass of 116,493 and a diameter of 8.1 nm, which are very similar in mass and diameter to peptides and proteins of comparable sizes (Eichman *et al.*, 2001). Unlike proteins, however, the surfaces of dendrimers become increasingly dense as their size increases. This is due to the doubling of branches and pendent groups on the outer surface for each generation increase in size. As the dendrimer generation and size increase, the molecules become more symmetrical and spherical in shape due to the dense outer branch packing. At a certain point, surface crowding becomes so great that no further access to the internal structure is possible and the dendrimer becomes a rigid ball. As dendrimer size increases, it also becomes increasingly

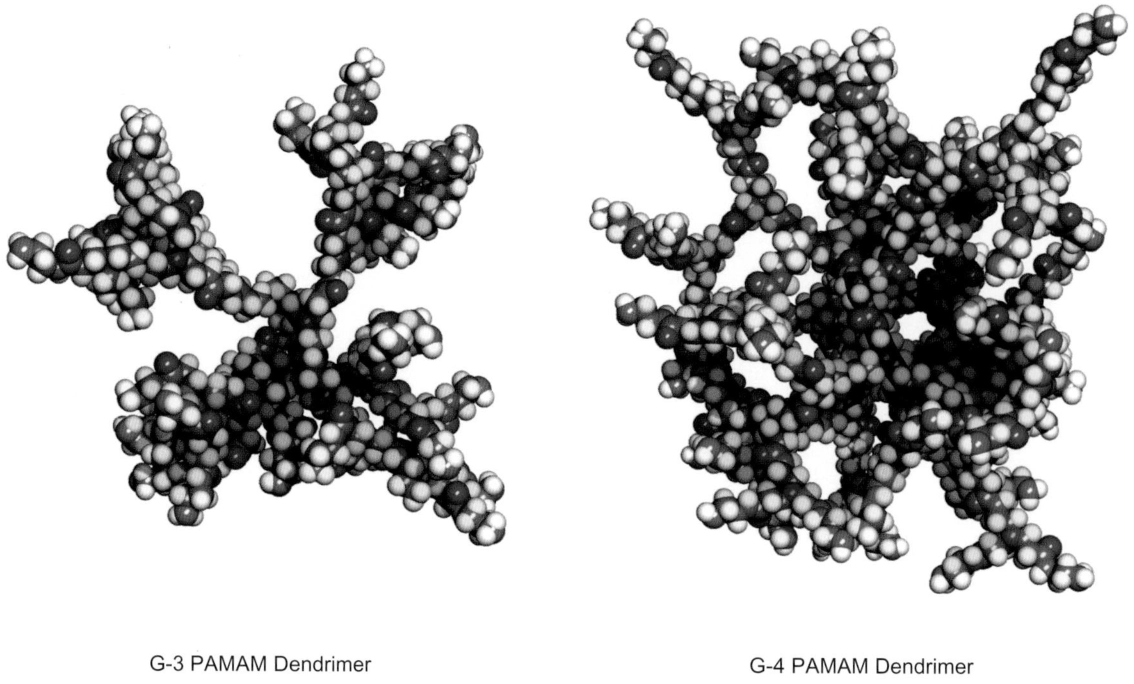

FIGURE 8.5 The G-3 and G-4 PAMAM structures shown in Figures 8.3 and 8.4 are illustrated here as space-filling molecular models. The structures were energy minimized and then allowed to go through molecular dynamics to give the three-dimensional configurations seen here. The G-4 level begins to show similarity to globular proteins as the number of pendent arms becomes great enough to create an almost fully filled core.

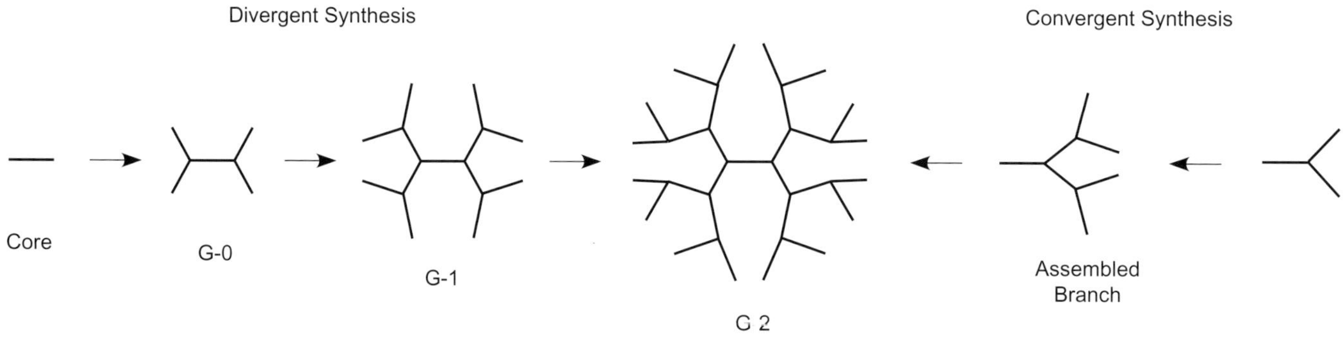

FIGURE 8.6 The divergent and convergent synthesis methods of dendrimer formation.

difficult to add another layer of monomers to the surface due to steric hindrance.

In general, dendrimers of size G-0 through G-3 have open, asymmetric, and flexible structures with effectively no protected internal areas, due to a large freedom of motion in their branches, and they can readily accommodate additional covalent attachments to their surfaces. See Figures 8.3 to 8.5 for illustrations of the 2D and 3D structure of a G-3 and G-4 dendrimer, in which G-3 dendrimers can be seen to have a great deal of internal space. However, G-4 through G-6 dendrimers display a more globular structure and contain internal void areas, which can hold guest molecules, such as delivery agents or drugs. Above this size, G-7 and greater, dendrimers are more like solid spheres or particles with inaccessible interiors and highly dense surfaces, and they appear upon imaging much like polymer nanoparticles of similar size (Chaper 14).

Unlike linear or ordinary branched polymers, dendrimers display intrinsically low viscosity, even at high mass. As standard polymeric molecules increase in mass and size, their viscosity normally increases

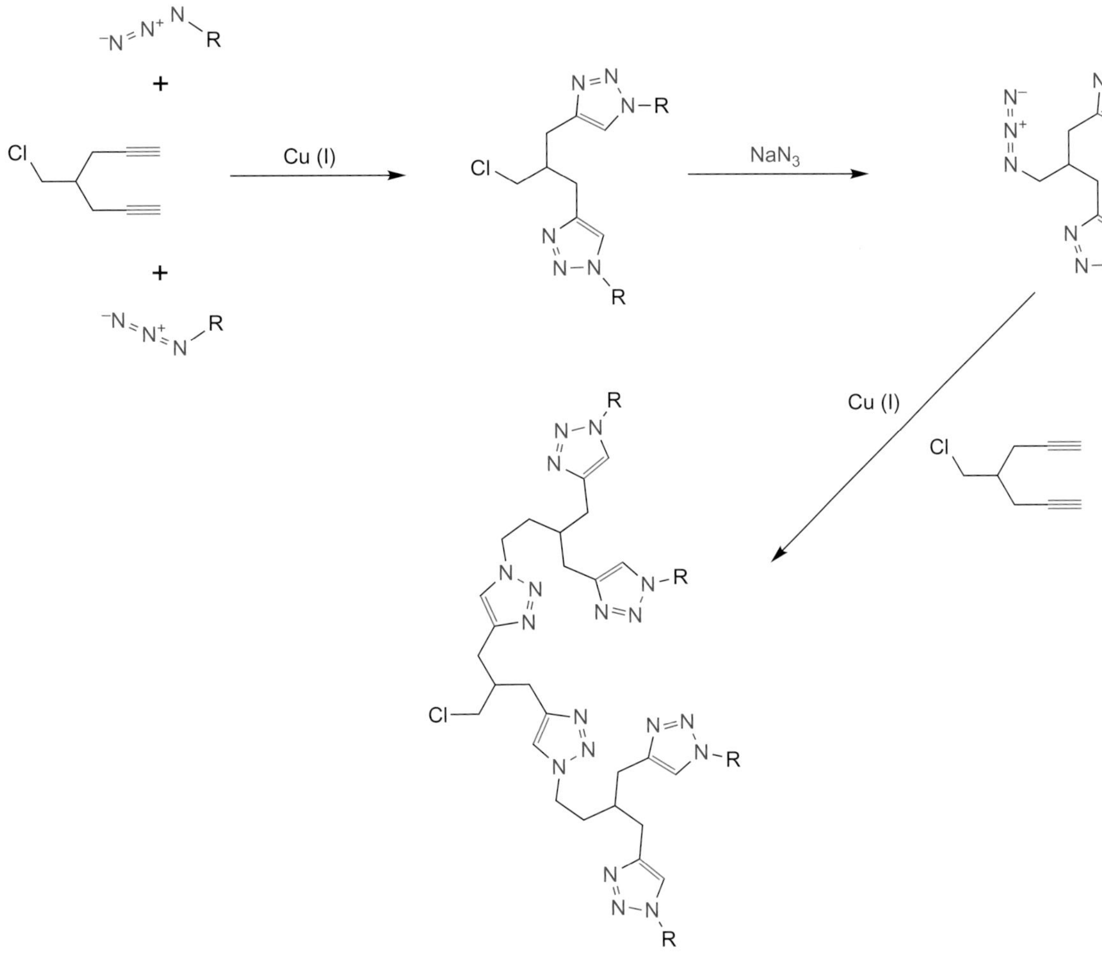

FIGURE 8.7 The synthesis of dendrimer molecules using click chemistry proceeds with high yield. Each step results in the cycloaddition reaction between azide-containing molecules and alkyne molecules to form triazole linkages.

continually. With dendrimers, viscosity increases only up to about the fourth generation, after which it actually begins to decline (Mourey, 1992; Fréchet, 1994). In addition, with control over the type of pendent groups that adorn the surface, dendrimers can maintain high solubility regardless of size.

Dendrimer molecules of the mid-size range have been found to be excellent carriers of guest molecules for solubilization and drug delivery. Wang *et al.* (2012) studied the host–guest chemistry of dendrimer–cyclodextrin conjugates in detail using NMR to determine what types of molecules are best contained within the structures for drug delivery purposes. In addition, certain groups added to the surface of such dendrimers can aid in the entrapment of molecules within the dendrimer cores. Such a construct, called a dendrimer box, can be designed to release the guest molecules upon certain conditions being met, such as pH-facilitated hydrolysis or a photoreaction of the surface capping groups. In this case, the capping groups are reversible and thus the dendrimer can be induced to release its cargo by cleavage (Jansen *et al.*, 1994; Jansen and Meijer, 1995). Miklis *et al.* (1997) used molecular dynamics to investigate the encapsulation of rose bengal molecules in a dendrimer box, which was formed by the coupling of tBOC-L-Phe cap molecules to the 64 terminal primary amines of a G-5 poly(propyleneimine) dendrimer. It was discovered that without the capping groups, the rose bengal molecules were in equilibrium between the solvent, surface, and interior of the dendrimer. However, forming a dendrimer box with the tBOC-L-Phe caps stably was found to keep the guest molecules within the dendrimer interior without leakage.

In some cases, a dendrimer box can be designed to provide a slow release of a drug by capping the branched structure with hydrophilic PEG groups. The internal structure is designed to effectively dissolve the organic drug and sequester it, while the PEG capping groups provide extreme dendrimer solubility and inhibit the movement of the drug out of the interior dendritic space. Liu *et al.* (2000) used this design to entrap indomethacin within the hydrophobic branches of a dendrimer built from phenyl-group-containing monomers and provided an $mPEG_{16}$ cap to allow a time-release effect for the drug *in vivo*.

In addition to the PAMAM variety, many different chemical constituents have been used as monomer units to build a dendritic molecule. Some have used trifunctional aromatic units, rigid monomers to keep the branches from bending, polyethers, polyhydroxyls, heterocyclic compounds, etc. In some dendrimer types, the interior structure is hydrophobic and can be used potentially to carry small organic molecules, while the surface groups are made to be hydrophilic to promote overall solubility. The potential variety of dendrimer construction is limited only by the imagination of organic or inorganic building blocks that can be conceived and linked together.

2. CONJUGATION TO DENDRIMERS

Many of the applications of dendrimers involve the covalent coupling of other molecules to the dendrimer surface or to points within the branched structure. These attached molecules can function as detection agents, affinity ligands, targeting components, radioligands, imaging agents, or pharmaceutically active compounds. The methods used for dendrimer conjugation are similar to the procedures used with other macromolecules and particles. The essential elements in designing a dendrimer conjugate are the functional groups that are present on the dendrimer and the functional groups on the molecule to be coupled.

Dendrimers are commercially available containing a variety of dendron structures and pendent groups on their surface (Dendritic Nanotechnologies; Dendritech). Some of the functionalities available allow for bioconjugation reactions to be done, but others are designed to create certain solubility properties and cannot be used directly for coupling other ligands. Figure 8.8 shows some of the available surface groups for traditional PAMAM-based dendrimers, including amine, carboxylate, hydroxyl, methyl ester, mPEG, and a hydrophobic C_6 chain. Selections of different diamine cores also are available with these surface functionalities, including the chain lengths C_2, C_4, C_6, and C_{10}, and a disulfide cleavable cystamine core. Most dendrimers of the PAMAM type are commercially available in sizes up to G-6, but can be custom ordered in higher generations, which are more difficult to synthesize and more expensive.

The Priostar dendrimers originally available from Dendritic Nanotechnology (now Starpharma) are created using a new manufacturing process and provide greater flexibility in dendrimer design and much better stability in the final product. These dendrimers are stable at room temperature, unlike PAMAM dendrimers, and can withstand extremes in pH or temperature without hydrolyzing or decomposing. In addition, the dendritic structures can be formed to have internal functional groups along their branches for conjugation, including secondary amines and hydroxyls. Therefore, fluorescent molecules or other organic molecules can be attached at internal locations, leaving the surface groups available for additional conjugation sites.

Priostar dendrimers are available with amine, hydroxyl, carboxylate, or epoxy functionalities on their

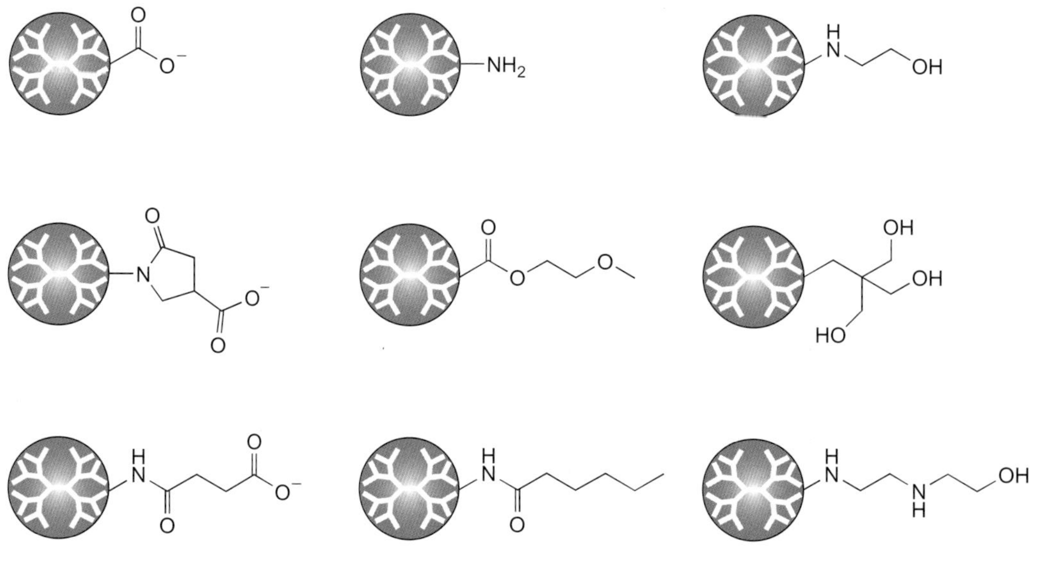

FIGURE 8.8 The pendent groups available on the surface of dendrimer molecules are highly varied. Some groups provide functional or reactive groups for bioconjugation, while other groups create unique solubility characteristics for the dendrimer.

surface for bioconjugation reactions. Also, due to the monomers used for synthesis, each generation of this dendrimer contains more surface functionalities than the corresponding PAMAM dendrimer. For example, PAMAM dendrimers with ethylenediamine cores have 4, 8, 16, and 32 pendent groups present for generations G-0, G-1, G-2, and G-3, respectively. By contrast, Priostar dendrimers have 4, 12, 30, and 100 pendent functionalities on their surface for G-0, G-1, G-2, and G-3, respectively. The result is much greater surface functionality at much lower generation number and size; therefore, surface functionality can be maximized without growing the dendrimer so large that it no longer can accommodate guest molecules within its core structure.

Other than the epoxy groups available on one Priostar dendrimer type and a methyl ester available on a PAMAM dendrimer, the commercial suppliers generally do not offer a selection of spontaneously reactive dendrimers for bioconjugation purposes. For this reason, most of the applications published for coupling biomolecules to dendrimers have used various modification or activation steps to create the appropriate reactive groups for conjugation (e.g., Leon *et al.*, 1996).

Due to the multivalent nature of dendrimers, the first consideration for conjugating molecules to them is to decide how many modifications should occur on the surface. For some molecules, maximizing the ligand:dendrimer modification ratio may be desirable. An example is in creating sugar–dendrimer derivatives to interact with carbohydrate binding proteins on cell surfaces. Since many sugar–lectin associations are of low affinity, creating a dendrimer conjugate having numerous sugar molecules attached to its surface is advantageous to form multiple interaction points. This approach results in the sugar–dendrimer complex binding to the cell surface with higher avidity than a single sugar derivative would be able to achieve.

However, for other bioconjugation applications, the optimal number of molecules attached to a dendrimer may have to be determined by experimentation. Too many modifications may result in decreased activity of the final conjugate as compared to a similar conjugate made without the use of a dendrimer. For instance, numerous fluorescent molecules can be attached to a G-3 amine-containing dendrimer to provide an enhanced fluorescent conjugate, which is brighter than a single fluorescent molecule for labeling proteins. However, if too many fluorescent molecules are attached, fluorescence quenching may take place and obviate any benefit the use of a multivalent dendrimer may provide toward signal enhancement. Therefore, for any given dendrimer conjugate preparation, some thought must be given to optimizing the number of modifications (or the ligand:dendrimer ratio) to obtain the best possible conjugate activity in the intended application.

In addition, if a dendrimer is to be labeled with one molecule and then ultimately attached to another molecule to form the complete complex, then the second conjugation step also must be planned from the beginning. Are some surface groups going to be used for coupling the first molecule and then the remaining groups used for coupling to the second one or will a disulfide dendritic core be used for coupling to the second molecule after cleavage of the modified dendrimer? Such decisions will affect the conjugation strategies used with dendrimers and often govern the usefulness of the resultant conjugate.

Dendrimer conjugates are being used more frequently in the development of multivalent constructs for delivering drugs to discrete sites *in vivo*. Dendrimers have been used to target cancer cells using biotin–dendrimer conjugates (Yang *et al.*, 2009), modified with folate and PEG polymers to carry DNA to tumor cells (Arima *et al.*, 2012), and loaded with DTPA–gadolinium chelates and folate in a PEG-dendrimer for high-contrast imaging of tumors (Chen *et al.*, 2010). In addition, a doxorubicin–dendrimer conjugate was designed to carry increased dosages of the chemotherapeutic agent to cancer cell targets (Chandra *et al.*, 2011).

The following methods of linking molecules to dendrimers present options for deciding the best reactions to exploit in creating a conjugate. In all cases, the ratio of reactants and the nature of the final conjugate should be carefully considered. In the end, running a series of trial conjugations to optimize the final conjugate will result in a method that is appropriate to the intended application and consistent in performance from batch to batch.

2.1. Coupling to Amine-Containing Dendrimers

PAMAM amine-containing dendrimers have surfaces that are adorned with many primary amine groups, the number of which is dependent on the generational size of the dendritic structure. For instance, a G-3 PAMAM dendrimer has 32 amine groups on the outer ends of its branches, a G-4 has 64 amines, and a G-5 has 128 amines. As long as the dendrimer size is within the mid-range where crowding of surface components is not severe, then the reactivity of these pendent amines is much greater than the amines on a globular protein. This is due to the fact that with protein molecules solvent accessibility of amines is dependent on the relative exposure of the side-chain lysine amines to the environment. For many globular proteins, a percentage of amines are highly accessible and react quickly, but for another population of amines, they are somewhat buried below the surface polypeptide structure and do not react as readily (Chapter 2, Section 1.1).

With dendrimers, the pendent amine groups are all approximately equally accessible and very reactive for bioconjugation or modification reactions (Fréchet, 1994). The only potential limitation for coupling to a dendrimer surface would be steric crowding, which would restrict the number of large molecules from coupling to every functional group on the dendrimer. For the coupling of small molecules such as sugars or biotin derivatives, the total number of possible modifications will approach the total number of amine groups on the surface, providing the dendrimer is not so large that there is already severe surface crowding of the branches. For globular protein coupling, however, the number of potential coupling points may be much less than the total number of functional groups available, because each protein molecule will overlap and block some functional groups as it is coupled to the dendrimer surface.

Modification of Amine-Containing Dendrimers with Sulfo-NHS-LC-SPDP

The amines of a PAMAM dendrimer may be reacted with any heterobifunctional crosslinker that contains an amine-reactive group on one side and another end having different reactivity. The result will form a reactive dendrimer that is useful for coupling proteins and other ligands, that have functional groups able to form a covalent linkage with the second reactive group on the crosslinker. Sulfo-NHS–LC-SPDP contains a sulfo-NHS ester that will couple with the pendent amine groups on the dendrimer and a pyridyl disulfide at its other end to couple with thiol-containing ligands (Chapter 6, Section 1.1). Reaction of this reagent with an amine-containing dendrimer yields a reactive intermediate containing a long aliphatic spacer arm and terminating in reactive pyridyl disulfide groups (Figure 8.9). A disulfide linkage can be formed by reaction of the pyridyl disulfide end with a thiol-containing ligand, which is reversible by reduction with DTT or TCEP.

Another strategy using this crosslinker is purposely to reduce the pyridyl disulfide end after dendrimer modification to create a free thiol on the dendrimer. Singh (1998) used this process to thiolate dendrimers for subsequent coupling with protein molecules containing a group reactive to thiols, such as sulfo-SIAB (or sulfo-SMCC)-activated antibodies. After the initial activation, SPDP-PAMAM dendrimers may be stored indefinitely either lyophilized or frozen, because the thiol-reactive group is stable to hydrolysis or degradation. The modified dendrimers also may be treated with DTT to release pyridine-2-thione and create free sulfhydryls on the dendrimer surface. A thiolated dendrimer should be kept in the presence of at least 10-m*M* EDTA and used immediately for bioconjugation to prevent oxidation of the thiols, which could form oligomers over time if disulfides are formed between dendrimer molecules.

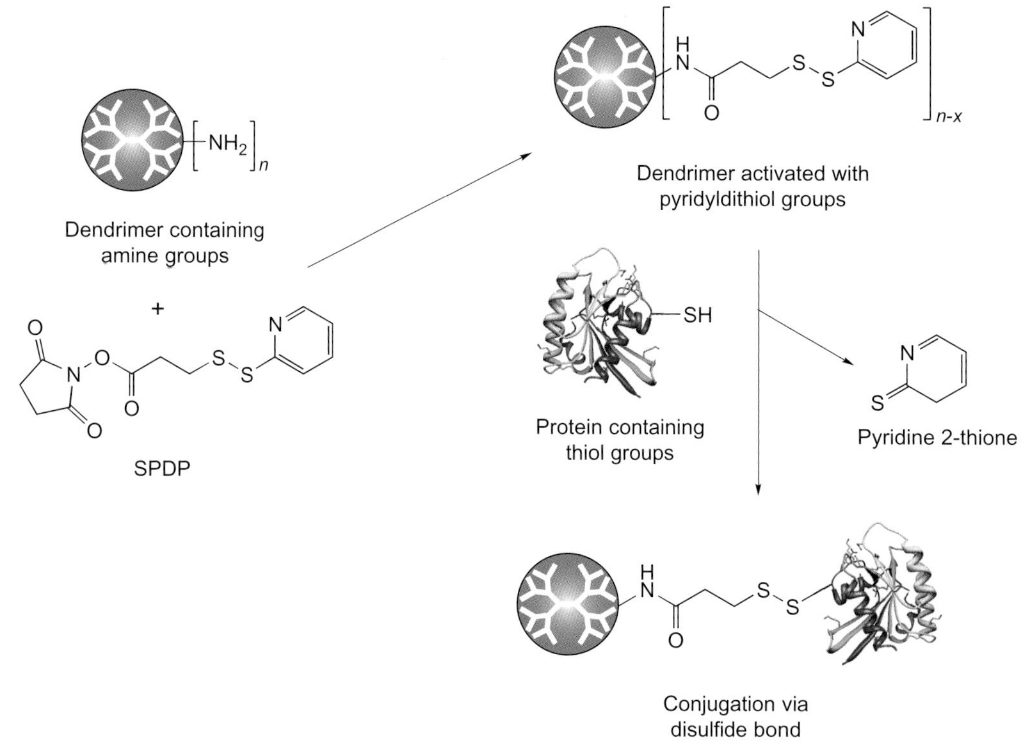

FIGURE 8.9 Amine-containing dendrimers can be activated with SPDP to create thiol-reactive derivatives. Alternatively, the pyridyl dithiol group may be reduced to create free thiols on the dendrimer surface for subsequent conjugation.

Use of sulfo-NHS–LC-SPDP or other heterobifunctional crosslinkers to modify PAMAM dendrimers may be done along with the use of a secondary conjugation reaction to couple a detectable label or another protein to the dendrimer surface. Patri *et al.* (2004) used the SPDP activation method along with amine-reactive fluorescent labels (FITC or 6-carboxytetramethylrhodamine succinimidyl ester) to create an antibody conjugate that also was detectable by fluorescent imaging. Thomas *et al.* (2004) used a similar procedure and the same crosslinker to thiolate dendrimers for conjugation with sulfo-SMCC-activated antibodies. In this case, the dendrimers were labeled with FITC at a level of five fluorescent molecules per G-5 PAMAM molecule. In addition, Bosnjakovic *et al.* (2012) prepared a dendrimer-based immunosensor for detection of tumor necrosis factor-alpha-cytokine by activation with SPDP.

Reaction of a G-3 PAMAM dendrimer with only a 4- to 10-fold molar excess of sulfo-NHS–LC-SPDP was found to yield a partially modified surface containing only a few pyridyl disulfide groups and leaving the rest of the amines available for further conjugation (Singh, 1998). To create seven to ten thiols per dendrimer, the reaction can be carried out using a 20-fold excess of sulfo-NHS–LC-SPDP. The following protocol should be optimized to incorporate the level of thiolation best suited for the application of the final conjugate.

Protocol

1. Dissolve the amine-containing PAMAM dendrimer in methanol or a buffered aqueous medium at a pH of 7 to 9 (e.g., 50-m*M* sodium phosphate, pH 7.5) and at a concentration of at least 10 mg/ml. Note that Singh (1998) used a concentration of 110 mg/ml in methanol, but other dendrimer concentrations should work equally well. For nonaqueous reactions, the addition of a proton acceptor may aid in driving the reaction to maximal yields (e.g., triethylamine or dimethylaminopyridine).
2. Dissolve sulfo-NHS–LC-SPDP at a concentration of 20-m*M* (5.2 mg/ml) in DMSO or water. If water is used, the solution must be used immediately to prevent hydrolysis of the sulfo-NHS ester.
3. With mixing, add an aliquot of the crosslinker to the dendrimer solution to provide the desired molar excess of reagent. For many applications, less than 10 pyridyl disulfide groups are needed per dendrimer molecule; therefore, molar ratios in the range of 5–20×excess of crosslinker over the amount of dendrimer present typically are used.
4. React with mixing for at least 30 min at room temperature. Longer reactions may be used without problems.
5. Purify the derivatized dendrimer using gel filtration (size exclusion chromatography) on a desalting column or through use of ultrafiltration spin-tubes (for G-4 and above). For smaller dendrimers, the derivatives may be purified by repeated precipitation from a methanolic solution by addition of ethyl acetate, dioxane, or benzene. The SPDP-dendrimer may be dried by lyophilization (if in water or buffer) or by solvent evaporation in vacuo (if the precipitation method was used).

   The SPDP-modified dendrimer of step 5 may be further derivatized with an amine-reactive fluorescent molecule for use in fluorescence detection applications. The fluorophore modification may be done prior to coupling an antibody or other molecules to the dendrimer for targeting purposes. The following steps illustrate the procedure used to obtain the fluorescently labeled dendrimer and then to use the SPDP-modified dendrimer to form thiols on the surface through reduction or to link to another molecule containing a thiol. Step 6 is optional for adding fluorescence detection capabilities, and the protocol of either step 7 or 8 may be used to conjugate the SPDP-reactive group to another protein.
6. Dissolve the purified SPDP-modified dendrimer of step 5 in 50-m*M* sodium phosphate, 0.15-*M* NaCl, pH 7.5, or in DMSO at a concentration of at least 10 mg/ml. Add a 10–20×molar excess of an amine-reactive fluorescent molecule (i.e., NHS-rhodamine or a hydrophilic NHS-Cy5 derivative—see section on fluorescent probes). React with mixing for 1 h at room temperature. Purify the fluorescently labeled SPDP-modified dendrimer using gel filtration or ultrafiltration. Follow the method of either step 7 or 8 to conjugate the dendrimer to another protein or molecule.
7. To create a thiolated dendrimer, the pyridyl disulfide groups may be reduced by addition of 50-m*M* DTT or TCEP in 50-m*M* sodium phosphate, 0.15-*M* NaCl, 10-m*M* EDTA, pH 7.5, which will release pyridine-2-thione groups and leave sulfhydryls on the dendrimer surface. Remove excess reducing agent by gel filtration on a column of Sephadex G-25 or the equivalent. The thiolated dendrimer should be used immediately to conjugate to a protein or another molecule containing a thiol-reactive group, such as a maleimide or iodoacetyl group.
8. Alternatively, a thiol-containing protein may be directly conjugated to the SPDP-modified dendrimer to create a disulfide linkage. Add a sulfhydryl-containing protein or other molecule to the purified SPDP-modified dendrimer in 50-m*M* sodium phosphate, 0.15-*M* NaCl, 10-m*M* EDTA, pH 7.5. The amount of the thiol-containing protein to be added to the dendrimer should be determined experimentally to be optimal for the intended application of the

conjugate. Typically, an excess of thiol-containing protein over the number of SPDP groups per dendrimer is added to ensure that the number of proteins coupled will efficiently utilize the number of modifications initially made using sulfo-NHS–LC-SPDP. Thus, controlling the molar ratio used in the initial crosslinker reaction will control the molar ratio of protein-to-dendrimer in the final conjugate.

NHS–PEG$_n$–Maleimide Coupling to Amine-Containing Dendrimers

An alternative method to the use of sulfo-NHS–LC-SPDP for coupling thiol-containing proteins or antibodies to PAMAM dendrimers is to use a heterobifunctional crosslinker containing an amine-reactive NHS ester and a thiol-reactive maleimide group (Chapter 6, Section 1). Unlike the pyridyl disulfide reaction with a sulfhydryl as in the SPDP protocol described previously, a maleimide group forms a stable thioether linkage with a sulfhydryl-containing ligand, which is not cleavable by reduction.

A common choice of crosslinker for this type of reaction is sulfo-SMCC, which has been used extensively for antibody conjugation (Chapter 20, Section 1.1) and can also be used to conjugate molecules to dendrimers. Perhaps a better option for dendrimer conjugation, however, is to use a similar crosslinker design, but one that contains a hydrophilic PEG spacer arm to promote dendrimer hydrophilicity after modification. Derivatization of an amine dendrimer with an NHS–PEG$_n$–maleimide can create an intermediate that is coated with water soluble PEG spacers. This modification helps to mask any potential for nonspecific interactions that the PAMAM surface may have, while providing terminal thiol-reactive maleimides for coupling ligands (Figure 8.10). Rong and Reinhard (2012) used this strategy in developing dendrimer reagents to study the lateral dynamics of ErbB1-enriched membrane domains in live cells.

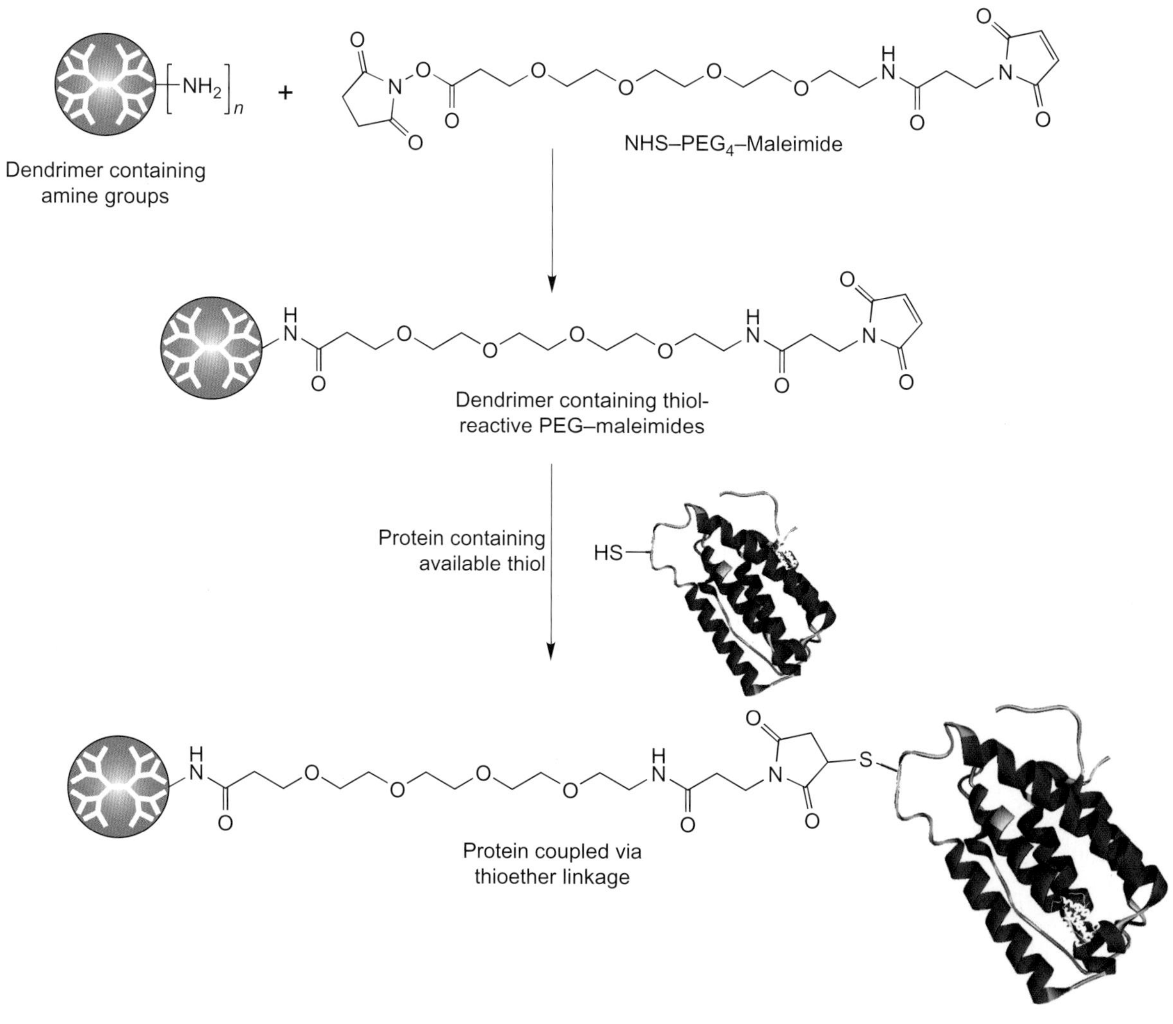

FIGURE 8.10 An NHS–PEG–maleimide compound can be used to functionalize dendrimers to provide a hydrophilic spacer terminating in thiol-reactive groups. Thiol-containing proteins then can be conjugated to this reactive intermediate to form covalent thioether bonds.

If an NHS–PEG$_n$–maleimide compound is used for this type of activation and coupling, the intermediate maleimide-activated dendrimer should be quickly purified of excess crosslinker and reaction byproducts and immediately used to couple ligand. This is due to the fact that the maleimide may hydrolyze in aqueous solution at a higher rate than when using a SMCC-type crosslinker, because of the extreme hydrophilicity of the PEG spacer arm compared to the cyclohexane spacer of SMCC.

NHS–PEG$_n$–maleimide crosslinkers are available in a number of spacer lengths depending on the size of the polymer chain in the PEG arm. Long-chain crosslinkers of this type use PEG polymers of molecular weight approximately 2000 to 5000, containing from about $n = 45$ to over $n = 100$ repeating polyethylene oxide units. Shukla *et al.* (2003) used PEG modifications to increase the half-life of folate receptor-targeted dendrimers for boron neutron capture therapy. However, a major deficiency of such full-length PEG polymers is that they are extremely polydisperse and consist of a broad range of polymer lengths, which makes reproducing the size of the conjugate difficult to achieve. Modifying a dendrimer with this type of polymeric crosslinker results in high variability of the length of the PEG chains displayed on its surface, which may be detrimental for coupling some ligands. In addition, the use of full-length PEG polymers perhaps should be avoided if the resultant hydrodynamic volume of the dendrimer derivative becomes unacceptably large for some applications.

Alternatively, shorter, discrete crosslinkers containing PEG spacers of known chain length are now available (Thermo Fisher, Quanta BioDesign; see Chapter 18, Section 1.2). These reagents are designed to contain an exact number of PEG units, typically from between 2 repeating units to 24 repeating units. The PEG spacer chain increases the overall water solubility of modified molecules and decreases nonspecific binding potential of the final conjugate. The following protocol may be used with any of these discrete PEG crosslinkers with the appropriate adjustments in the quantity of reagent added to take into account differences in molecular weight due to the PEG length. The longer of these discrete NHS–PEG–maleimide crosslinkers will provide the greatest degree of hydrophilicity after modification of an amine-containing dendrimer surface. Intermediate-length NHS–PEG–maleimides probably provide a sufficient combination of hydrophilicity while maintaining a smaller conjugate size. One caution should be noted when using PEG-based reagents with amine-containing dendrimers. The polyether crossbridge of PEG compounds has the ability to hydrogen bond to the dendrimer surface, especially when using dendrimers with a high density of amines. Reactions performed at higher levels of PEG-containing reagents may be done to overcome this tendency. Alternatively, the amine surface may be partially blocked or derivatized with another molecule prior to using the PEG compound to avoid hydrogen bond interactions.

Protocol

1. Dissolve 10 mg of an amine-containing dendrimer into 1 ml of 50-m*M* sodium phosphate, 0.15-*M* NaCl, pH 7.5 (coupling buffer), with mixing.
2. Dissolve NHS–PEG$_6$–maleimide (MW 601.6) into DMSO at a concentration of 20 m*M*. Short, PEG-type crosslinkers often exist as a thick oily mass, and preparing the solution typically involves dissolving an entire vial of the compound into DMSO to determine accurately the required concentration. Use only dry DMSO to avoid hydrolysis of the NHS ester.
3. Add 50 μl of the NHS–PEG$_6$–maleimide solution to the 1-ml dendrimer solution and mix thoroughly to dissolve. This represents approximately a 14-fold molar excess of crosslinker over the quantity of dendrimer present, if a G-3 PAMAM dendrimer is used with an ethylenediamine core. The optimum molar ratio of crosslinker-to-dendrimer should be determined experimentally for best performance of the resultant conjugate in its intended application. If enough material is available, carrying out a series of experiments at different mole ratios of crosslinker-to-dendrimer will help to optimize the resultant conjugate.
4. React for 1 h at room temperature with mixing.
5. Purify the derivatized dendrimer from excess crosslinker and reaction byproducts using gel filtration (size exclusion chromatography) on a 10-ml desalting column or through use of ultrafiltration spin-tubes. If a dendrimer of at least G-3 size is being modified, the separation should be done on a support with an exclusion limit of no more than 2500 to 5000 Daltons to avoid losing the derivatized molecule through the membrane or not obtaining sufficient separation during the chromatography.
6. Add 1 to 10 mg of a protein or antibody containing an available thiol group to the purified, modified dendrimer from step 5. Alternatively, add the protein to be coupled to the dendrimer suspension in an amount equal to 1–10×molar excess over the estimated number of maleimide groups on the modified dendrimer. The amount of maleimide functionality may be determined using the protocol in Chapter 19, Section 5. Creating thiol groups from disulfides in proteins may be done according to the procedures in Chapter 2, Section 4.1. Alternatively, the use of a thiolation reagent may be done to add

thiols to the protein surface for coupling to the maleimide groups. The optimal amount of protein to be added to the dendrimer should be determined experimentally.

7. React the protein with activated dendrimer for 2h at room temperature with mixing. At the completion of the reaction, cysteine may be added at 50 m*M* to block excess maleimide reactive sites that are not coupled with protein.
8. Purify the conjugate and remove excess protein by gel filtration using a column with an exclusion limit that is able to accommodate both the protein being conjugated and the dendrimer–protein conjugate. If the conjugate elutes in the void volume while the protein is retained in the pores, this also will result in sufficient separation to purify the conjugate. Store the dendrimer conjugate frozen (especially if it's a PAMAM-type) or lyophilized. The addition of a stabilizing excipient to the freeze-dried conjugate may be done to protect the coupled protein during lyophilization (e.g., sucrose).

Coupling Glycoproteins to Amine-Dendrimers by Reductive Amination

The amines on the surface of PAMAM-type and other amine-containing dendrimers may be used to couple to aldehyde groups in other molecules, including those formed after periodate oxidation of sugar groups in glycosylated proteins. Mild treatment of glycoproteins with sodium meta periodate results in cleavage of diol carbon–carbon bonds with concomitant oxidation of the hydroxyls to aldehyde groups. These aldehydes can be used to conjugate the proteins to amine-containing dendrimers through Schiff base formation and reduction to secondary amine linkages using sodium cyanoborohydride (Chapter 2, Section 4.4, and Chapter 3, Section 5) (Figure 8.11).

The following protocol involves the conjugation to an amine-containing dendrimer of a periodate-oxidized glycoprotein, such as an antibody that has been treated to produce aldehydes according to the protocols in Chapter 2, Section 4.4. This type of conjugation reaction to dendrimers may be used even after the dendrimer

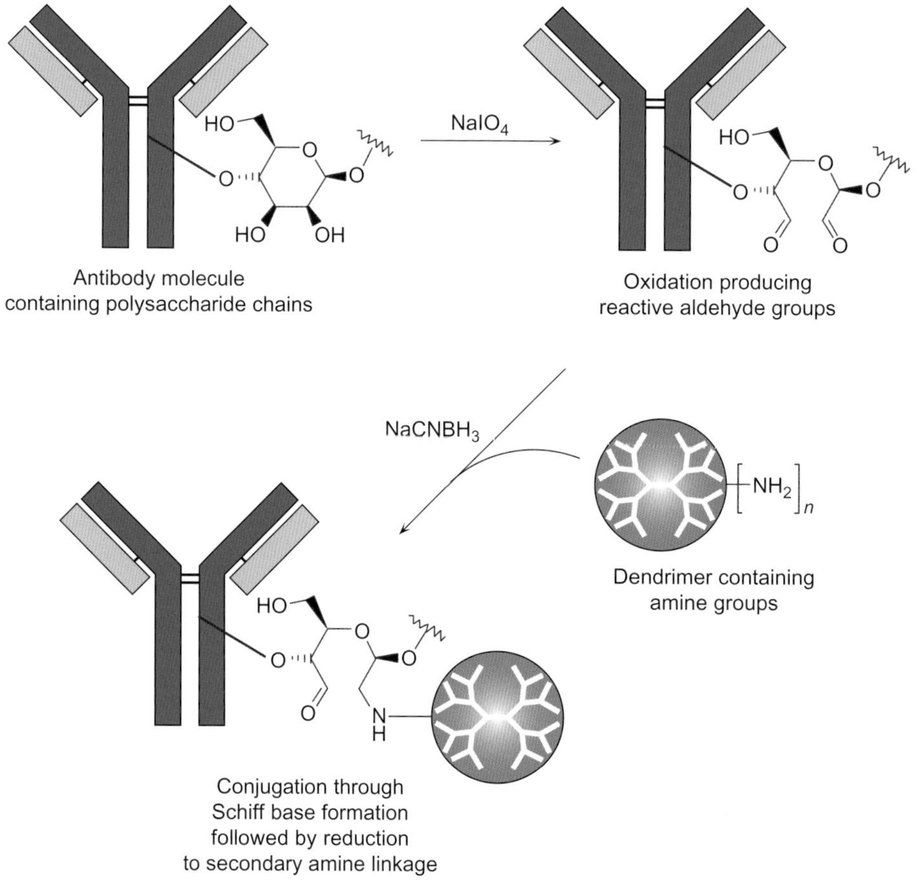

FIGURE 8.11 Oxidation of glycoproteins with periodate, such as glycosylated antibodies, results in the formation of aldehyde groups that can be used for conjugation to dendrimers containing amine groups. Reductive amination with sodium cyanoborohydride results in coupling *via* secondary (or tertiary) amine bonds.

has been initially modified with a limited number of heterobifunctional crosslinking molecules. Thus, an amine-containing dendrimer first may be reacted with crosslinkers such as sulfo-NHS–LC-SPDP or NHS–PEG_6–maleimide as described previously, and the remaining amines on the dendrimer surface used to couple with an oxidized glycoprotein. This would create a complex that contained a covalently linked glycoprotein with thiol-reactive groups available for further conjugation with another molecule containing a sulfhydryl group. Singh (1998) used this technique to produce a dendrimer conjugate containing alkaline phosphatase for detection and a Fab′ fragment of an antibody directed against creatine kinase MB isoenzyme. The phosphatase enzyme was coupled through its carbohydrates by reductive amination and the Fab′ fragment was coupled through a thiol group.

Protocol

1. Dissolve an amine-containing dendrimer in 50-m*M* sodium phosphate, 0.15-*M* NaCl, pH 7.5, at a concentration of at least 10 mg/ml. Note: The use of a buffer at pH 9–10 (e.g., 0.1-*M* sodium carbonate) for the initial Schiff base formation (step 2) will result in higher efficiency of conjugation. However, if a higher pH is used during this first stage, then the pH must be adjusted back down to more neutral conditions before the addition of reductant (step 4), as the reducing agent is not effective in the higher pH environment.
2. Add a quantity of a periodate-oxidized glycoprotein to the dendrimer solution (made according to Chapter 2, Section 4.4) to provide a molar ratio of protein-to-dendrimer of 1:1 to 10:1. Mix to dissolve. The optimal level of protein addition should be adjusted to provide maximal performance and activity in the intended application. If another protein also is to be attached to the dendrimer to make the final complex, then limiting the density of the periodate-oxidized protein may be necessary to leave room on the surface for additional protein coupling. In addition, some proteins with multiple glycosylation sites may cause dendrimer oligomerization if the conjugation reaction is done with too low of a protein-to-dendrimer ratio.
3. React for 2 h at room temperature with mixing.
4. In a fume hood, add 10 μl of 5-*M* sodium cyanoborohydride (Sigma) per ml of reaction solution. Caution: Cyanoborohydride is extremely toxic. All operations should be done with care in a fume hood. Also, avoid any contact with the reagent, as the 5-*M* stock solution is dissolved in 1-*N* NaOH. If a higher pH buffer was used for the Schiff base formation, then adjust the solution to pH 7.5 before adding the cyanoborohydride.
5. React for 30 min at room temperature in a fume hood.
6. Block unreacted aldehydes by the addition of 50 μl of 1-*M* ethanolamine, pH 7.5, per ml of reaction solution. Block for 30 min at room temperature with mixing.
7. Purify the conjugate from unreacted protein or unreacted dendrimer using gel filtration chromatography with a matrix having an exclusion limit appropriate to accommodate the size of the molecules being separated (e.g., a HiPrep 16/60 column packed with Sephacryl S-200 HR, GE Healthcare).

Blocking Amines on PAMAM Dendrimers

The number of amine groups on dendrimers of medium to large size (G-3 and above) often is many more than what is required to prepare a conjugate. Since amines can become protonated and carry a positive charge even under physiological conditions, they may interact nonspecifically with biomolecules. Ionic interactions can be a significant source of high backgrounds in assays or create off-target effects if using a dendrimer conjugate in the development of *in vivo* applications (cell-based or whole animal). In particular, nonspecific binding to the positively charged amine groups on dendrimers, if left unblocked, can be cytotoxic to normal cells *in vivo* especially if the conjugate is designed to carry chemotherapeutic components that can be used to kill cancer or diseased cells (Patri *et al.*, 2002; Kukowska-Latallo *et al.*, 2005). To overcome the positive charge character of amine-containing dendrimer excess amine groups can be covalently blocked to eliminate the possibility of ionic interactions. This blocking process can occur before activation of the dendrimer with a crosslinker or after modification, providing that the blocking agent used to couple with the remaining amines does not affect the activation chemistry.

Singh (1998) used succinylation with succinic anhydride to block excess amines and convert them into negatively charged carboxylates, which often have lower interaction potential with biomolecules than positive charges. Thomas *et al.* (2004) used acetic anhydride to acetylate and block the amines of a G-4 PAMAM dendrimer before or after activation with sulfo-NHS–LC-SPDP. Similarly, Patri *et al.* (2004) also used acetic anhydride to block amines on a G-5 PAMAM dendrimer prior to activation with sulfo-NHS–LC-SPDP, but the blocking step was done to eliminate about 80 of the 128 amine groups before effecting the conjugation with an antibody. In all these instances, the amine groups on the dendrimer are converted into amides, which carry no charge at physiological pH.

Reactions with succinic anhydride or acetic anhydride to block dendrimer amines can be done in aqueous or methanolic solution. If organic solvent is used for the reaction, then it is typical to include triethylamine as a proton acceptor, which helps drive the reaction. Such reactions, however, cannot be done to dendrimer amines once a protein containing amines also has been conjugated, as the protein too will get modified.

Another method of blocking excess amines on dendrimers is to use small, hydrophilic blockers that contain hydroxyl or ether groups. Islam *et al.* (2005) used the short epoxy compound glycidol to modify amine groups to reduce nonspecific interactions in a dendrimer conjugate. The primary amine groups can be alkylated a maximum of two times with this reagent in methanol, resulting in four hydroxyl groups for each amino functionality (Shi *et al.*, 2007). The reaction with glycidol, however, was shown to be not as efficient in creating a homogeneous product as acylation with either succinic anhydride or acetic anhydride (Shi *et al.*, 2005). This likely is due to the formation of branched polymers off the dendrimer using glycidol (see Chapter 15, Section 2.1) as opposed to a single acylation product using either anhydride compound. Glycidol modification of an amine-containing dendrimer is done according to the following protocol.

Protocol

1. Dissolve the amine-containing dendrimer to be modified in methanol at a concentration of 10 mg/ml (all operations are done in a fume hood).
2. Add glycidol to the dendrimer solution in a 4-fold molar excess over the number of amines to be blocked (Shi *et al.*, 2007). The number of amines may be calculated from the generation number of the dendrimer and the degree of blocking desired. Obtaining the optimal molar ratio for a particular conjugate application may have to be done by experimentation using a series of different glycidol-to-dendrimer molar ratios.
3. Allow the reaction to continue for 24 h in the fume hood with mixing.
4. Dialyze the modified dendrimer against PBS buffer and then water to remove excess reactants.

Another option to limit the nonspecific binding character of amine-containing dendrimer is the use of PEG compounds. Amine-reactive PEG derivatives can be used to covalently link to excess amines on dendrimers and create a hydrophilic tether, which can dramatically limit nonspecific interactions with the exposed surface. For this purpose, PEG reagents containing an activated carboxylate (NHS ester) on one end and a methyl ether group (mPEG) on the other end work well (Chapter 18, Section 1.4). NHS–mPEG derivatives that are spontaneously reactive to the dendrimer surface amines will form amide bonds and eliminate the positive charge character of the pendent amino groups. Such PEG compounds may be a superior option to either succinylation or acetylation of amines to eliminate charge, because the PEG component also adds considerable hydrophilicity and low interaction potential to the resultant dendrimer derivative (Figure 8.12). PEG also is often added to dendrimers to increase their half-life in circulation for *in vivo* targeting. For instance, Zhu *et al.* (2012) created a multivalent dendrimer probe containing the integrin-binding RGD peptide and the antitumor drug doxorubicin and also modified with PEG to modulate its *in vivo* behavior.

Acylation reactions to block amine groups on PAMAM dendrimers with anhydride compounds are done in a similar manner to glycidol modification. The following protocol is based on the method of Majoros *et al.* (2005).

Protocol

1. Dissolve an amine-containing dendrimer in methanol at a concentration of 15 mg/ml with mixing. All operations should be done in a well-ventilated fume hood. A G-5 dendrimer was used in this reaction by Majoros *et al.* (2005) that contained 110 available amine groups.
2. Add to the dendrimer solution a quantity of acetic anhydride that represents a molar ratio of anhydride-to-amines of 0.72 : 1.0 (680 µl of acetic anhydride was added for the G-5 dendrimer). Using a molar quantity of anhydride that is less than the amount of amines present on the dendrimer ensures that only a portion of the amines will become blocked, so that further modification remains possible.
3. Add a quantity of triethylamine to the solution so that a 25% molar excess over the amount of anhydride will result (1.25 ml for the G-5 dendrimer).
4. React for 2 h at room temperature with mixing (in a fume hood).
5. Extensively dialyze the reaction mixture against water or buffer to remove excess reactants.

Preparation of Sugar–Dendrimer Derivatives

The multifunctional nature of an amine-containing dendrimer can be used to advantage to mimic multidentate interactions of molecules with cell surfaces or virus particles. For instance, the binding affinity of carbohydrate binding proteins (lectins) for individual sugar molecules typically is weak, of the order of $10^6 M^{-1}$. In the native method used to increase the binding strength of these interactions, lectins on cell surfaces usually engage in multi-point attachments with carbohydrates or glycans. The conjugation of sugars to the pendent amine groups on dendrimers provides a scaffold for similar multi-site interactions with lectins, which effectively increases the avidity of the resultant

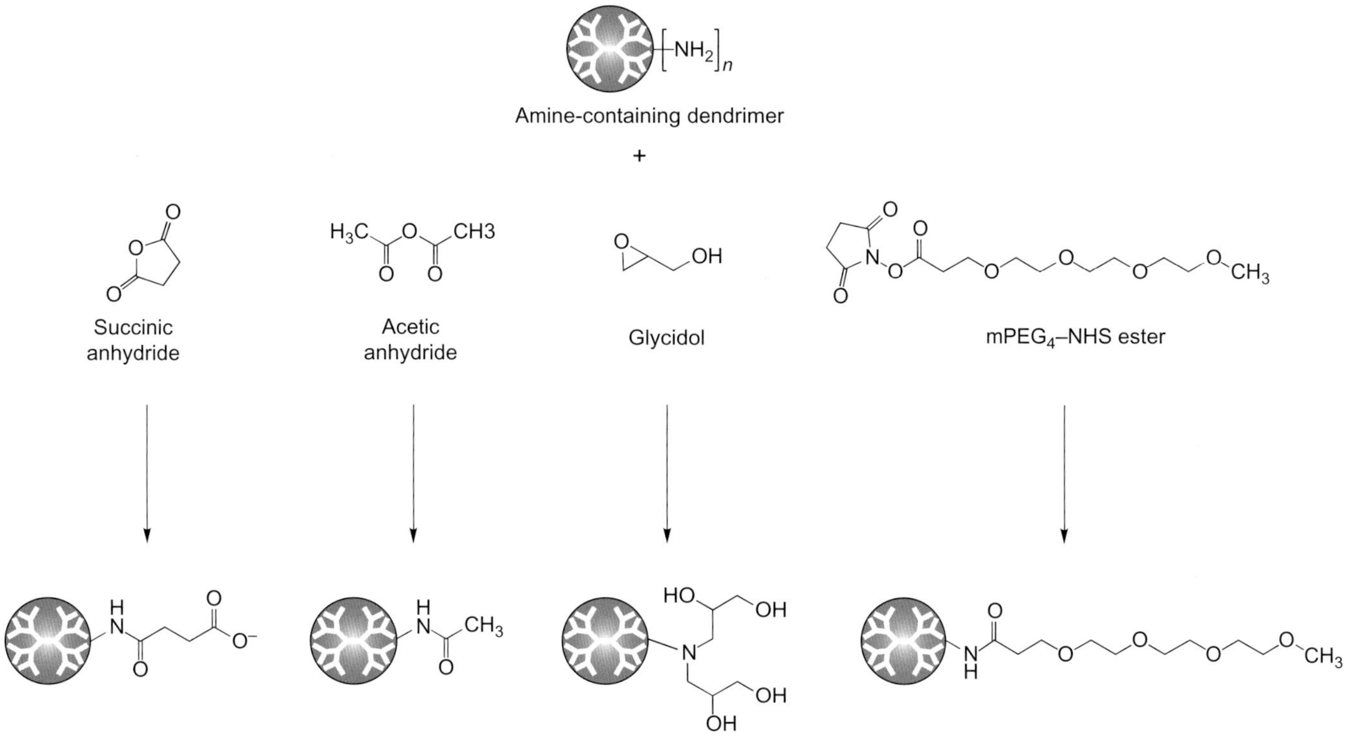

FIGURE 8.12 Amine-containing dendrimers can be modified using a number of common reactive modification agents. Excess amine groups can be blocked using acetic anhydride, glycidol, or an NHS–mPEG compound. Amines also can be converted into carboxylates using succinic anhydride.

binding complex (Aoi *et al.*, 1995; Bertozzi and Kiessling, 2001). This design is structurally similar to the multiple tree-branched structures of glycans on glycoproteins. The terminal sugar groups on such glycoconjugates are capable of interacting with more than one binding site or more than one receptor on cell surfaces. This effectively turns a single low-affinity interaction into a high-affinity binding event, which forms the basis for many life processes, including cellular recognition, adhesion, transport, and cell signaling (Clark and Wilson, 1988; Sharon and Lis, 1989). For a review of glycobiology, see the entire issue of *Science*: Carbohydrates and Glycobiology, Vol. 291, March 23, 2001, pp. 2263–2502.

A series of different mannose-dendrimers was synthesized to investigate their interaction with concanavalin A (Con A) (Woller and Cloninger, 2002; Woller *et al.*, 2003). It was discovered that the sugars on the dendrimer surface were able to bind to the Con A binding sites just like free methyl mannose in solution. However, as the size of the dendrimer increased and a number of multivalent mannose residues became available for binding to multiple Con A interaction sites, the affinity of the interaction dramatically increased. For a G-3 mannose–dendrimer derivative, the sugar complex was about 45 times more active than methyl mannose in solution. For larger-sized mannose–dendrimer complexes, the increase in activity of binding was up to 660-fold greater than free methyl mannose. However, the interaction potential for the mannose–dendrimer derivatives also was shown to be dependent on the degree of mannose loading. For large dendrimers, steric crowding of sugar molecules on the dendrimer surface decreased its binding activity toward Con A if the mannose loading was greater than about 50% of the amines modified. Thus, both dendrimer size and the level of modification must be carefully considered when designing sugar–dendrimer conjugates.

An amine-containing dendrimer's multivalent surface functionalities can accommodate a combination of carbohydrate modifications along with other groups placed on the surface to provide other activities. For instance, Lo Conte *et al.* (2011) prepared glycodendrimers by activating the surface amines to provide free-radical thiol-ene sites for coupling thiolated sugars by addition to carbon–carbon double bonds. The covalent attachment of mono- or disaccharides was easily done by irradiation at 365 nm for 1 h. On the same dendrimer, they also introduced PEG groups and glutathione molecules to create a G-4 dendrimer with the ability to avoid the reticuloendothelial system *in vivo*.

Such sugar–dendrimer complexes ("sugar balls") have been used to inhibit the interactions of viruses with cell surfaces. Many viruses bind to particular carbohydrate residues on cell surfaces, which in turn facilitate their entry into cells and the resultant infection process. A virus particle presents a multi-dentate surface consisting of many carbohydrate-binding proteins able to interact with multiple cell surface carbohydrates. The surface of a dendrimer that is modified with sugar molecules is able to mimic the glycan-rich field on the surface of a cell. In this way, sugar–dendrimer complexes are able to interfere with virus binding to cellular targets. For instance, Borges *et al.* (2005) used dendrimers modified with a carbohydrate usually found on immune cell surface glycosphingolipids to function as an inhibitor of HIV-1 infection. The polyvalent nature of the sugar–dendrimer derivative was able to bind effectively to the virus particles with high avidity, thus preventing the virus from binding to the carbohydrates on the immune cell surface (Borges and Schengrund, 2005). The sugar–dendrimer complex in the vaginal ointment Vivagel, developed by Starpharma, is undergoing clinical trials as a preventative for HIV infection (Halford, 2005). The dendrimer formulation also is said to be active against other STDs, including chlamydia, herpes simplex virus, hepatitis B, and human papilloma virus.

Sugar–dendrimer complexes also may be used as affinity ligands for purification of carbohydrate- or sugar-binding proteins. Szwergold *et al.* (2001) used a polyvalent fructose–dendrimer derivative immobilized on a chromatography matrix to purify human fructosamine-3-kinase. The ligand was coupled to the support through excess amine groups on the dendrimers, which were not coupled with sugar molecules, and gel functioned with high affinity toward binding the enzyme out of complex samples.

The synthesis of sugar–dendrimer derivatives may be done using a number of reaction strategies. Aoi *et al.* (1995) used the lactone derivatives of lactose and maltose to react with the terminal amine groups on G-2 to G-4 PAMAM dendrimers. The reaction of a lactone with an amine involves nucleophilic attack on the lactone carbonyl with ring opening to form an amide bond (Figure 8.13). The conjugation was done in organic solvent (DMSO), but the reaction also may be done in aqueous buffers under alkaline conditions.

The preparation of sugar–dendrimer conjugates may encompass a variety of modification levels and sizes, depending on the generation of dendrimer used and the density of sugar molecules modifying its surface. Interactions with lectins vary according to the optimal sugar-to-sugar distance within the final complex. For instance, it is known that the three binding sites on the mammalian hepatic lectin for interaction with galactose and *N*-acetylgalactosamine are arranged on the protein in a triangle with a separation of 1.5 nm, 2.2 nm, and 2.5 nm (Lee *et al.*, 1984). Aoi *et al.* (1995) found that the molecular distances between sugar molecules on a G-3 PAMAM dendrimer prepared according to their protocol was between 1.3 and 2.9 nm. Thus, the sugar ball dendrimer had an optimal inter-sugar spacing to interact with the binding sites on the targeted lectin.

The following protocol represents the method of Aoi *et al.* (1995) for the coupling of lactone sugars to amine-dendrimers in organic solvent.

Protocol

1. In a fume hood, dissolve 270 mg of a G-4 dendrimer (PAMAM or other amine containing) in 2 ml of dry DMSO. Maintain a nitrogen blanket over the reaction to prevent oxidation. The use of a three-necked flask and a heating mantle is recommended to control mixing and the proper temperature.
2. Add with mixing a 300-fold molar excess of a sugar lactone derivative, such as *O*-β-D-galactopyranosyl-(1→4)-D-glucono-1,5-lactone (2.6 g) dissolved in 3 ml of DMSO.
3. React at 40°C for 9 h or overnight with stirring.
4. After cooling to room temperature, the sugar–dendrimer derivative may be precipitated with a large volume of methanol. The precipitate may be purified from reaction products by dialysis against water or buffer using a membrane with a molecular weight cutoff of 3500 Daltons. The final product may be stored frozen or lyophilized to a white powder.

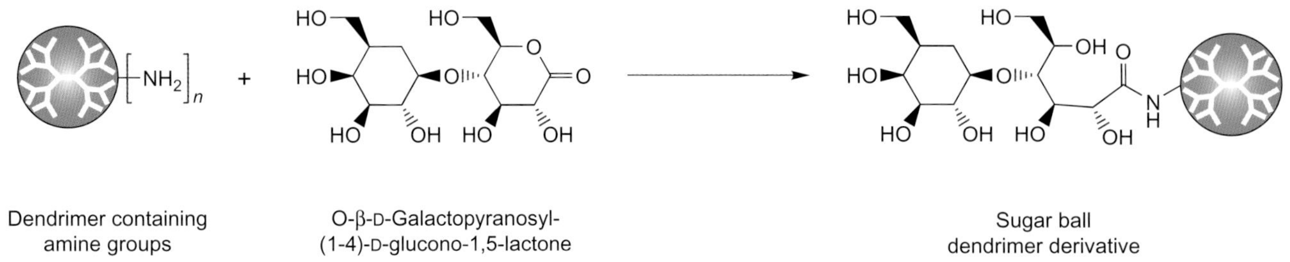

FIGURE 8.13 Lactone sugar derivatives can be used to react with amine-containing dendrimers, which results in coupling *via* amide bonds. The "sugar ball" dendrimers then can be used to specifically bind to carbohydrate binding proteins.

The modification of amine-containing dendrimer with lactose also can be done using mono(lactosylamido) mono(succinimidyl)suberate (Thermo Fisher). The amine-reactive compound contains a lactose group at the end of a suberate bridge and terminates at the other end in an NHS ester (Vetter *et al.*, 1995). The NHS ester spontaneously reacts with the amine groups on a dendrimer at neutral or slightly basic pH values to form an amide bond (Figure 8.14). The presence of the lactose disaccharide provides a hydrophilic modifying group that maintains the water solubility of the resultant conjugate.

Protocol

1. Dissolve mono(lactosylamido) mono(succinimidyl) suberate in dry DMF to prepare a concentrated stock solution. The compound is extremely soluble in DMF, and solutions of 100 mg/ml may be prepared.
2. Prepare the amine-containing dendrimer to be glycosylated in a buffer at a slightly basic pH (avoid amine-containing buffers, such as Tris or imidazole). The use of 0.1-*M* sodium phosphate, 0.15-*M* NaCl, pH 7.2, works well for NHS ester reactions. The concentration of the dendrimer in the reaction buffer should be at least 10 mg/ml. Other dendrimer concentrations also will work, but highly dilute solutions will result in less efficient modification yields.
3. With mixing, add a quantity of the mono(lactosylamido) mono(succinimidyl) suberate in dry DMF to the dendrimer solution to result in at least a 10- to 20-fold molar excess of reagent over the quantity of amines to be modified in the dendrimer. Depending on the desired application for the lactosyl-modified dendrimer, several different molar ratios may have to be tried to optimize the resulting modification level.
4. React for 30 to 60 min at room temperature with gentle mixing.
5. Purify the modified dendrimer away from reactants and reaction by-products using dialysis or size exclusion chromatography.

Another method for coupling carbohydrates or sugar derivatives to amine-containing dendrimer is to take advantage of the reducing end of sugars to reductively aminate the amines on the dendrimer surface. This reaction will form stable secondary amine linkages between carbon-1 of the sugars and the dendrimer (Figure 8.15). Saccharides with reducing ends have carbonyl groups that can be coupled to an amine-containing dendrimer in the presence of a reducing agent. However, since most reducing ends of sugars or glycans exist mainly in an acetal (or ketal) ring form, and only the open form with the exposed aldehyde can participate in a reductive

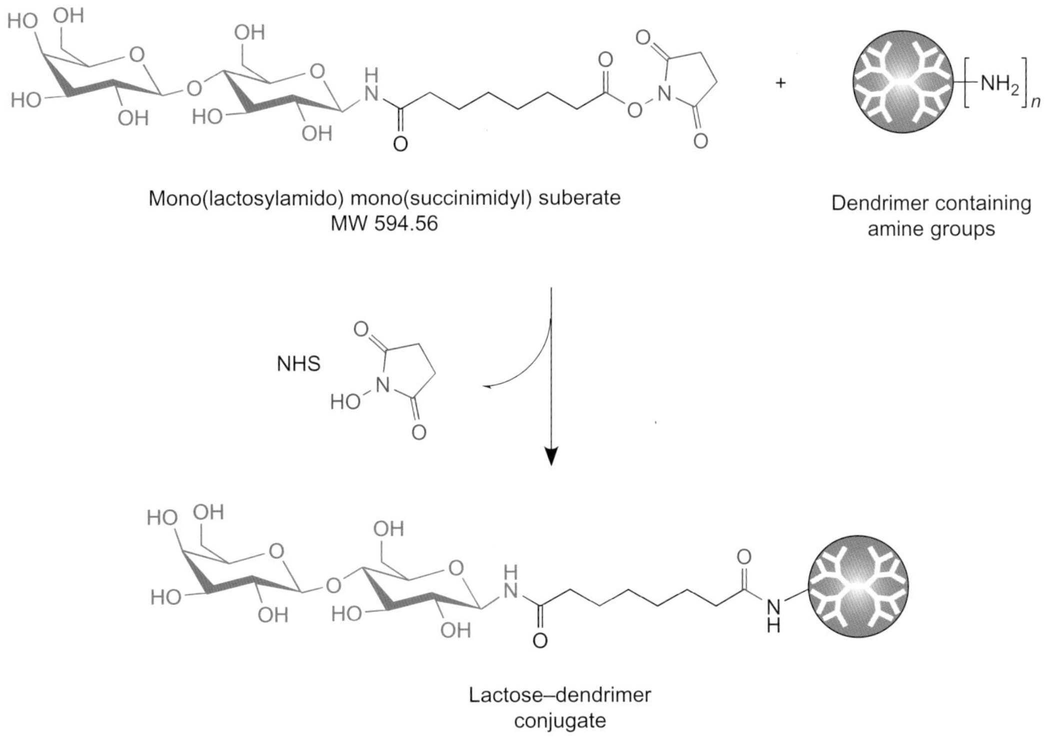

FIGURE 8.14 The reaction of an amine-containing dendrimer with mono(lactosylamido) mono(succinimidyl) suberate results in the attachment of a lactose unit *via* an amide bond.

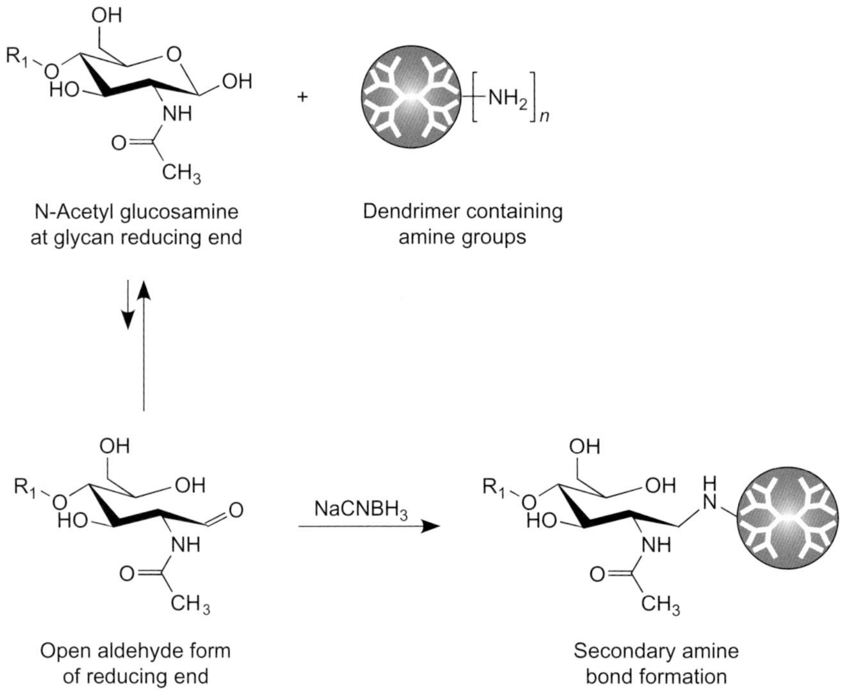

FIGURE 8.15 The reducing end of a glycan or a carbohydrate can be used to conjugate to an amine-containing dendrimer by reductive amination, which results in the formation of a secondary amine linkage.

amination reaction, the rate of modification by this process can be slow. The open form of a reducing sugar is available at any given time in only a small percentage of the total saccharide present, usually far less than 1%. For this reason, reactions carried out with primary amines at room temperature may take days to reach acceptable coupling yields. Typically, such reactions are done at elevated temperatures and over the course of at least several days, or sometimes weeks, to reach completion.

The following method for carbohydrate conjugation to dendrimers may be used to couple a variety of reducing sugars to amine-containing dendrimer, including saccharides, longer-chain carbohydrates, and even complex glycans after release from a protein (see Chapter 2, Section 4.6).

Protocol

1. Dissolve a carbohydrate, saccharide, or glycan sample having a free reducing end in 0.1-*M* sodium acetate, pH 5.0. Alternative coupling conditions that can be used for the modification reaction include 30% glacial acetic acid in DMSO (v/v) or acetic acid/pyridine (1:2, v/v), if the dendrimer is soluble in these solutions. The use of DMSO or pyridine often facilitates solubilization of a greater range of carbohydrates or glycans than aqueous buffers. The presence of acetic acid has been found to accelerate the reductive amination reaction when the organic solvent conditions are used (Bigge *et al.*, 1995). A well-ventilated fume hood should be used for all organic reactions. The concentration of the carbohydrate should be 5 to 100 μ*M* for glycans, but for the modification of dendrimers with other more abundant carbohydrates higher concentrations should be used. For modification of amine-containing dendrimer that are not soluble in the recommended organic solvents, the glycan or carbohydrate initially may be solubilized in DMSO and then an aliquot added to the aqueous reaction buffer.
2. Add to the glycan solution the amine-containing dendrimer to be labeled. Dendrimer concentrations of at least 10 mg/ml will help to increase the reaction kinetics, although the reaction proceeds slowly, mainly due to the limited quantity of open-form reducing sugar available for coupling at any given time.
3. In a fume hood, add to the reaction mixture a quantity of reducing agent (e.g., sodium cyanoborohydride or borane dimethylamine [BDA]) to give a final concentration of 1.0 *M*. The high concentration of reductant will aid in accelerating the reaction to completion. Cyanoborohydride is extremely toxic and should not be handled outside of a fume hood. Use proper protective clothing when handling such compounds.
4. When using nonaqueous reaction conditions, incubate for 1 to 2 h at 60 to 80°C. For reactions in an aqueous environment, the reaction may be carried out at room temperature or 37°C. In this case, the reaction time should be extended to

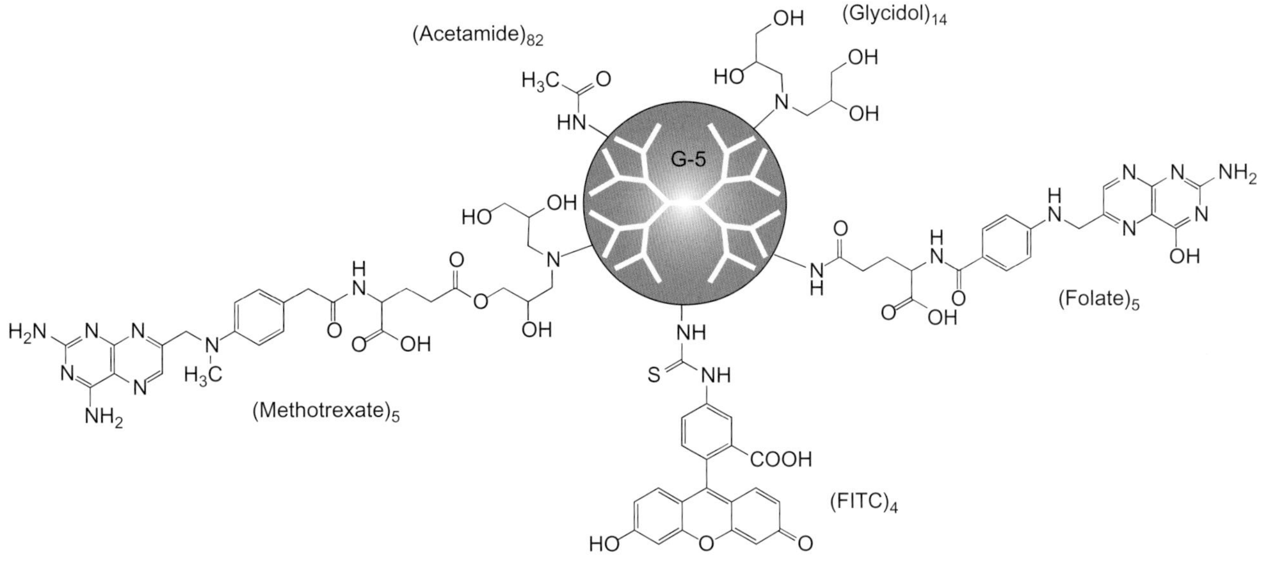

FIGURE 8.16 The creation of a tumor-targeting dendrimer conjugate can take advantage of the multivalent character of the dendrimer polymer. This figure illustrates the attachment of five different groups to an amine-containing dendrimer to produce a chemotherapeutic construct.

at least 24h. Longer reaction times are not unusual when modifying carbohydrates by reductive amination at ambient temperature.

5. Purify the modified dendrimer from reactants and reaction by-products by dialysis, gel filtration, or ion-exchange chromatography.

Synthesis of other reactive groups on sugars for conjugation with amine-containing dendrimer also has been done with success. For example, André *et al.* (1999) used the *p*-isothiocyanato derivative of *p*-aminophenyl-β-D-lactoside to couple with the amine groups on a G-3 PAMAM dendrimer, forming isothiourea linkages between the sugar and dendrimer. Davis *et al.* (2005) used glycosyl derivatives containing the thiol-reactive group methanethiosulfonate to couple with dendrons modified with thioacetic acid to contain terminal thiol groups. This reactive functionality was again used to conjugate to thiol-containing proteins to create glyco-dendrimer-protein reagents. Woller and Cloninger (2001) used another isothiocyanate derivative of mannose to couple with amine-containing dendrimer of several different generations. Finally, Kensinger *et al.* (2004a) synthesized a thiopropionic acid derivative of galactose and coupled it to amine-containing dendrimer using the amide bond-forming agent HATU (Aldrich) in acetonitrile. The galactose functionalized dendrimers were used to study their binding to HIV-1 gp120 protein (Kensinger *et al.*, 2004b).

Conjugation of Carboxylate Organic Molecules to Amine-Containing Dendrimers

Amine-containing dendrimer molecules of various generations have been used as carriers for small organic molecules, such as chemotherapeutic agents or small-ligand, cell-targeting compounds, such as folic acid (Quintana *et al.*, 2002 Shukla *et al.*, 2003; Islam *et al.*, 2005; Arima *et al.*, 2012; Choi *et al.*, 2012). The presence of folic acid on the dendrimer surface allows the drug complex to bind with over-expressed folate receptors on certain tumor cells (Ross *et al.*, 1994). Adding a chemotherapeutic drug to the folate–dendrimer conjugate creates a toxic payload deliverable directly to the cancer cells *in vivo*.

Majoros *et al.* (2005) synthesized such dendrimer conjugates using the carbodiimide EDC to activate the carboxylate group of folic acid or methotrexate for reaction with the amines on the PAMAM dendrimer surface. A G-5 amine-containing dendrimer first was partially blocked by acetylation according to the previously described procedure. The result was that about 82 of the available amines were capped, leaving about 28 amines still free for coupling with other molecules. In some derivatives, a fluorescent probe (FITC) also was conjugated to the partially blocked dendrimers to provide a detectable tag for following binding to cells. Then both folic acid and methotrexate were conjugated to the labeled dendrimer in separate reactions to yield the final conjugate. Careful tuning of the molar ratios used during each reaction resulted in a modification level per dendrimer of four FITC labels, four folic acids, and five methotrexate groups. In some cases, glycidol also was used to block about 14 of the amine groups and form hydrophilic hydroxylated regions on the dendrimer surface to promote water solubility. Figure 8.16 illustrates the final conjugate composition with a G-5 dendrimer. Although the conjugate structure indicates a single glycidol monomer at each coupling site on the dendrimer, it is likely that some polymerization of glycidol also occurs during the derivatization step (see Chapter 15, Section 2.1). These conjugates were extensively studied by HPLC to determine properties and modification levels (Islam *et al.*, 2005).

Hong *et al.* (2007) investigated the binding affinity of such folate-dendrimer derivatives toward the folate binding protein (FBP), typically over-expressed on cancer cells. Using SPR analysis, they were able to measure the binding constants of a series of folate–dendrimer conjugates containing different levels of folate modification (from 2 to 14 folates/dendrimer). The results indicate that the combined avidity of multiple folates on a single dendrimer enhanced the binding to FBP by up to 5 orders of magnitude (2500 to 170,000-fold enhancement). This is the first definitive study on the benefits of multivalent dendrimer drugs as compared to the same drug interactions individually in solution. Using this approach to drug targeting, Kukowska-Latallo *et al.* (2005) found that a dendrimer conjugate improved the therapeutic response toward tumor cells in animal models of human epithelial cancer.

Similar dendrimer conjugates for therapeutic application also have been designed for boron neutron capture therapy in the treatment of cancer (Barth *et al.*, 1994; Newkome *et al.*, 1994; Qualmann *et al.*, 1996). In one design, a G-3 PAMAM dendrimer was modified with folate and 12 to 15 polyhedral decaborate clusters (Alam *et al.*, 1989) along with an average of 1 to 1.5 PEG_{2000} units to make the therapeutic conjugate (Shukla *et al.*, 2003). The folate targeting component was not attached to the surface amines, but to the end of an additional amino–PEG_{800} unit. The PEG groups also facilitated increased *in vivo* half-life and prevented immediate uptake of the complex into the liver.

Most of these conjugation reactions involve amide bond formation between a small, organic molecule containing a carboxylate and the amines on the dendrimer surface. There are two potential reaction strategies for creating such conjugates: aqueous or nonaqueous reactions. For the coupling of folic acid to PAMAM dendrimers, Majoros *et al.* (2005) used an organic solvent environment according to the following protocol.

Protocol

1. Dissolve 33 mg of folic acid in a mixture of 24 ml DMF and 8 ml DMSO at room temperature.
2. Add a 14-fold molar excess of EDC (200 mg) with mixing.
3. React for 1 h at room temperature. The product of this reaction forms an amine-reactive ester on folate for coupling to the dendrimer.
4. Dissolve an amine-containing dendrimer in water at a concentration of at least 4.5 mg/ml. Majoros *et al.* (2005) used 403 mg of a partially acetylated G-5 dendrimer that contained 82 blocked amines and 28 available amines for coupling, dissolved in 90 ml of water.
5. Slowly add with mixing, the activated folic acid from step 3 to the amine-containing dendrimer solution to give a final molar ratio of folate-to-dendrimer of 5.5 : 1.
6. React for 1 h at room temperature with mixing.
7. Purify the modified dendrimer from excess reactants and reaction byproducts by dialysis against water or buffer.

The same type of modification with carboxylate molecules can be done in aqueous solution using EDC. If the ligand to be coupled only has a single carboxylate with no amines or other nucleophiles present, then the dendrimer and ligand may be dissolved at a similar molar ratio in aqueous buffer and EDC added to facilitate the coupling reaction.

Amine-containing dendrimers also have been used to quantify phosphorylated peptides for phosphoproteome analysis by mass spec (Tao *et al.*, 2005; for a review, see Nita-Lazar, 2011). In this novel application, proteins undergoing analysis are reduced and alkylated with iodoacetamide to block sulfhydryls, then trypsin digested. The peptides are then desalted and lyophilized. The dried peptides are methylated in methanolic HCl to eliminate interference by carboxylate groups and then the solvent is removed under vacuum. The mixture of blocked peptides next is reacted with a G-5 PAMAM dendrimer using 50-m*M* EDC and 100-m*M* imidazole in 0.2-*M* MES buffer, pH 6.0 (see the method described in Chapter 23, Section 2.1). The reaction activates the phosphoryl groups on the phosphorylated peptides, which in turn react with the amines on the dendrimers to create phosphoramidate linkages (Figure 8.17). The conjugated phosphopeptides can be separated from non-phosphorylated peptides by simple filtration and washing. Finally, the isolated phosphopeptides can be cleaved from the dendrimer using 10% TFA treatment and analyzed by mass spec.

Epoxy Activation of Amine-Containing Dendrimers

Epoxy groups can be formed on the surface of amine-containing dendrimers by reaction with either *bis*-epoxide compounds or epibromohydrin (or epichlorohydrin). Modification with these compounds forms terminal epoxide groups that can be used for subsequent conjugation with amine-, thiol-, or hydroxyl-containing ligands. Singh (1998) described a simple procedure using epibromohydrin to form an epoxy-activated PAMAM dendrimer for conjugation with thiol-modified alkaline phosphatase enzymes (Figure 8.18).

The following protocol is based on the method of Singh (1998). Other amine-containing dendrimers besides the PAMAM type, such as the PrioStar dendrimers from Starpharma may be used as well.

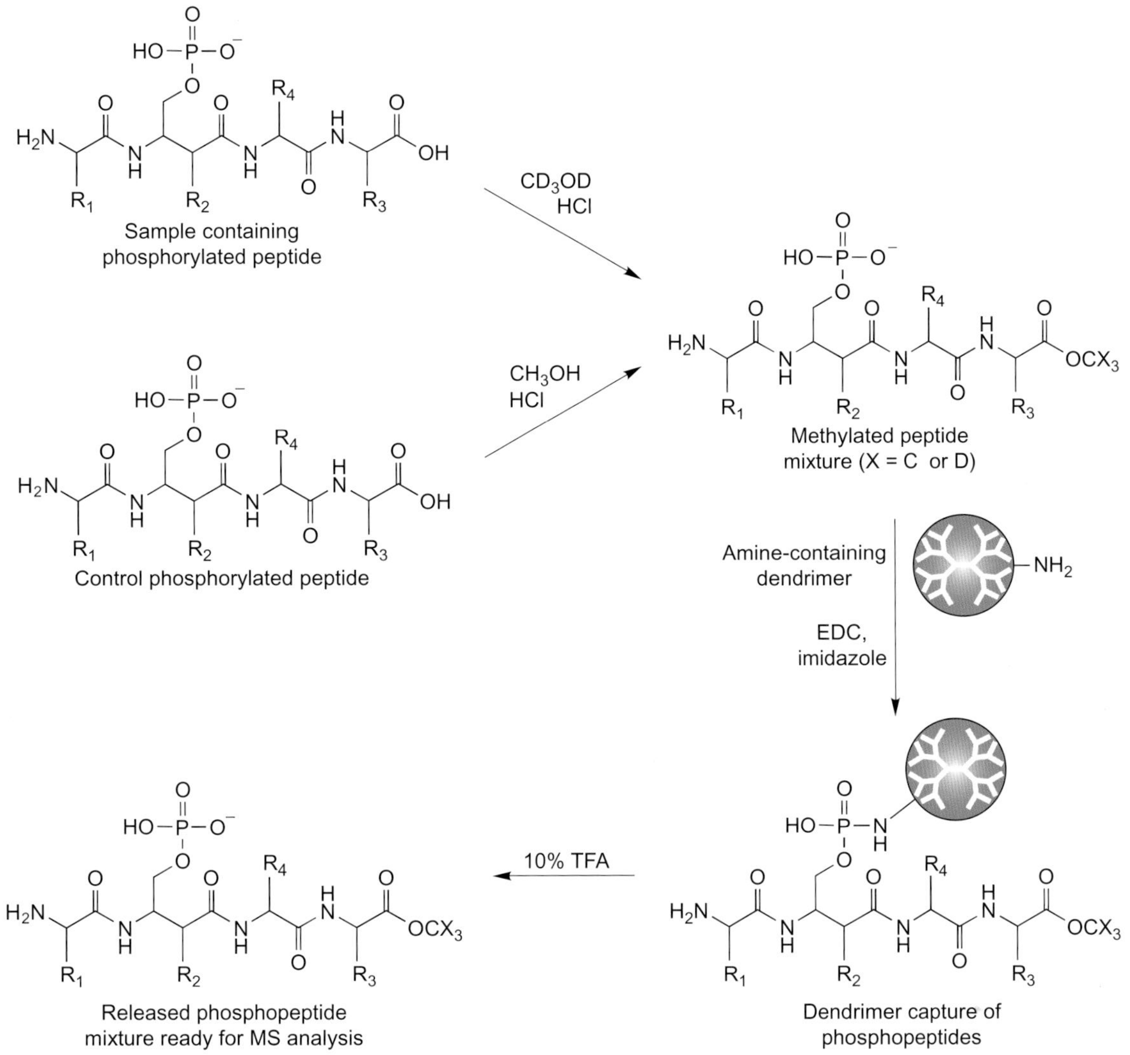

FIGURE 8.17 Phosphopeptides can be separated for mass spec analysis using an amine-containing dendrimer. A control sample and a test sample first are methylated to block their carboxylate groups and then mixed together in equal amounts. An amine-containing dendrimer then is used to capture the phosphorylated peptides by conjugation with the phospho groups using EDC and imidazole to form phosphoramidate bonds. After separation of the phosphopeptide–dendrimer conjugates using size exclusion chromatography, the phosphopeptides are cleaved off using TFA and analyzed by mass spec.

Protocol

1. Dissolve a dendrimer in 50% methanol/water at a concentration of at least 10 mg/ml. Singh (1998) used about 165 mg of a G-5 PAMAM dendrimer dissolved in 1 ml of the alcohol/water mixture. Use a fume hood for all operations.
2. Add 200 mg of sodium carbonate per ml of the dendrimer solution prepared in step 1 and a quantity of epibromohydrin equal to a 285-fold molar excess over the amount of dendrimer present. The large excess of reagent ensures that the amines get fully blocked during the reaction and will prevent any amines from crosslinking by reaction with the epoxy ends of the modified dendrimer.
3. React for 4 h with mixing in a fume hood.
4. Dilute the reaction mixture 10-fold with 50% methanol/water and purify the epoxy-activated dendrimer by repeated ultrafiltration using a membrane with a molecular weight cutoff able to retain the size of dendrimer being activated. Repeat the dilution and concentration steps until the filtrate is neutral in pH.
5. The epoxy-activated dendrimer may be conjugated to thiol-containing proteins by reaction in 50-m*M* sodium phosphate, pH 7.2. The reaction can be done at 4°C or at room temperature for 8 to 16 h to form thioether linkages.
6. Purify the conjugate by gel filtration to separate protein–dendrimer complexes from excess protein or dendrimer.

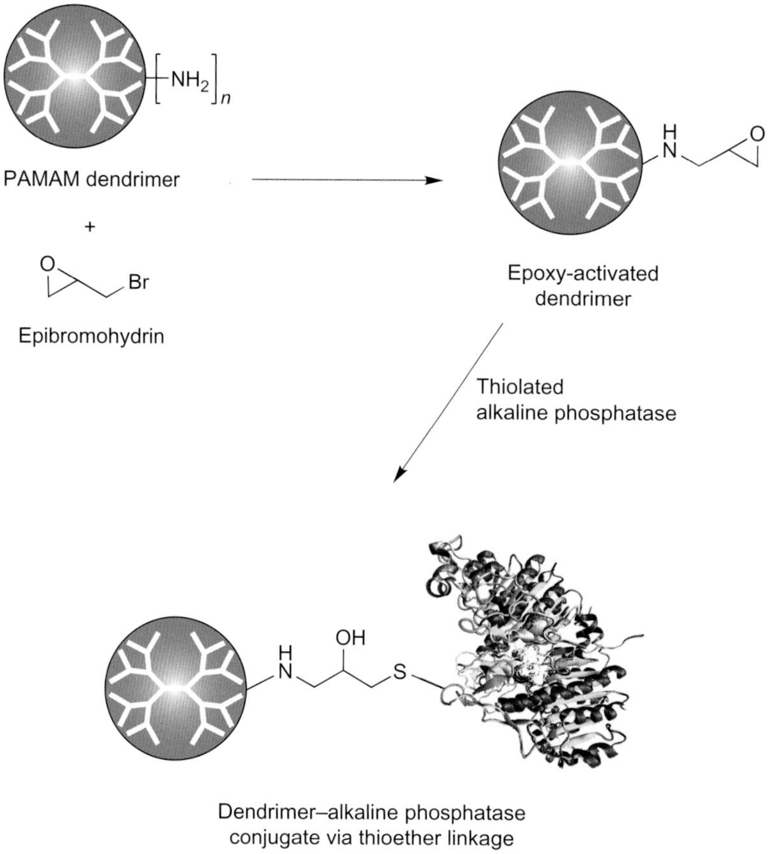

FIGURE 8.18 Amine-containing dendrimers can be activated with epibromohydrin to result in the formation of reactive epoxy groups on the dendrimer surface. This reactive intermediate then can be used to conjugate with thiol-containing proteins, such as thiolated alkaline phosphatase. The reaction results in the formation of a thioether bond.

Biotinylation of Amine-Containing Dendrimers

Amine-containing dendrimers can be modified to contain one or more biotin groups for interaction with avidin or streptavidin reagents. The polyvalent nature of a dendrimer permits the formation of a biotin multimeric structure for potential enhancement of detection in avidin–biotin assays. Since a biotin–dendrimer scaffold allows many biotin-binding proteins to dock simultaneously and form larger complexes, these conjugates can recruit more detection molecules to bind at the site of an analyte. This can have direct effect on the sensitivity of immunoassays, such as fluorescence detection, enzyme linked immunoadsorbent assays (ELISAs), and western blotting procedures.

This type of biotinylated dendrimer-based signal amplification technique has been done to increase the detection of genomic DNA in suspension arrays (Borucki *et al.*, 2005), wherein a DNA dendrimer was biotinylated multiple times and also modified to contain a targeting oligo sequence. The DNA dendrimer, which in that case was a construct consisting of partially hybridized oligonucleotide sequences branching out from a central core, was modified to contain as many as 850 to 900 biotin molecules on its surface along with at least one targeting oligonucleotide sequence. After genomic DNA was allowed to bind to a capture oligo on a fluorescence microparticle, the biotinylated dendrimer was added and it bound to the genomic DNA via hybridization with the targeting oligo. Finally, streptavidin–phycoerythrin detection conjugate was added and it then was bound to the biotin groups on the dendrimer (Figure 8.19). The resultant fluorescent signal was amplified beyond that possible using a simple biotinylated oligo directly.

A similar type of biotin-dendritic multimer also was used to boost sensitivity in DNA microarray detection by 100-fold over that obtainable using traditional avidin–biotin reagent systems (Stears, 2000; Striebel *et al.*, 2004). With this system, a polyvalent biotin dendrimer is able to bind many labeled avidin or streptavidin molecules that may carry enzymes or fluorescent probes for assay detection. In addition, if the biotinylated dendrimer and the streptavidin detection agent are added at the same time, then at the site of a captured analyte, the biotin–dendrimer conjugates can form huge multidendrimer complexes wherein avidin or streptavidin

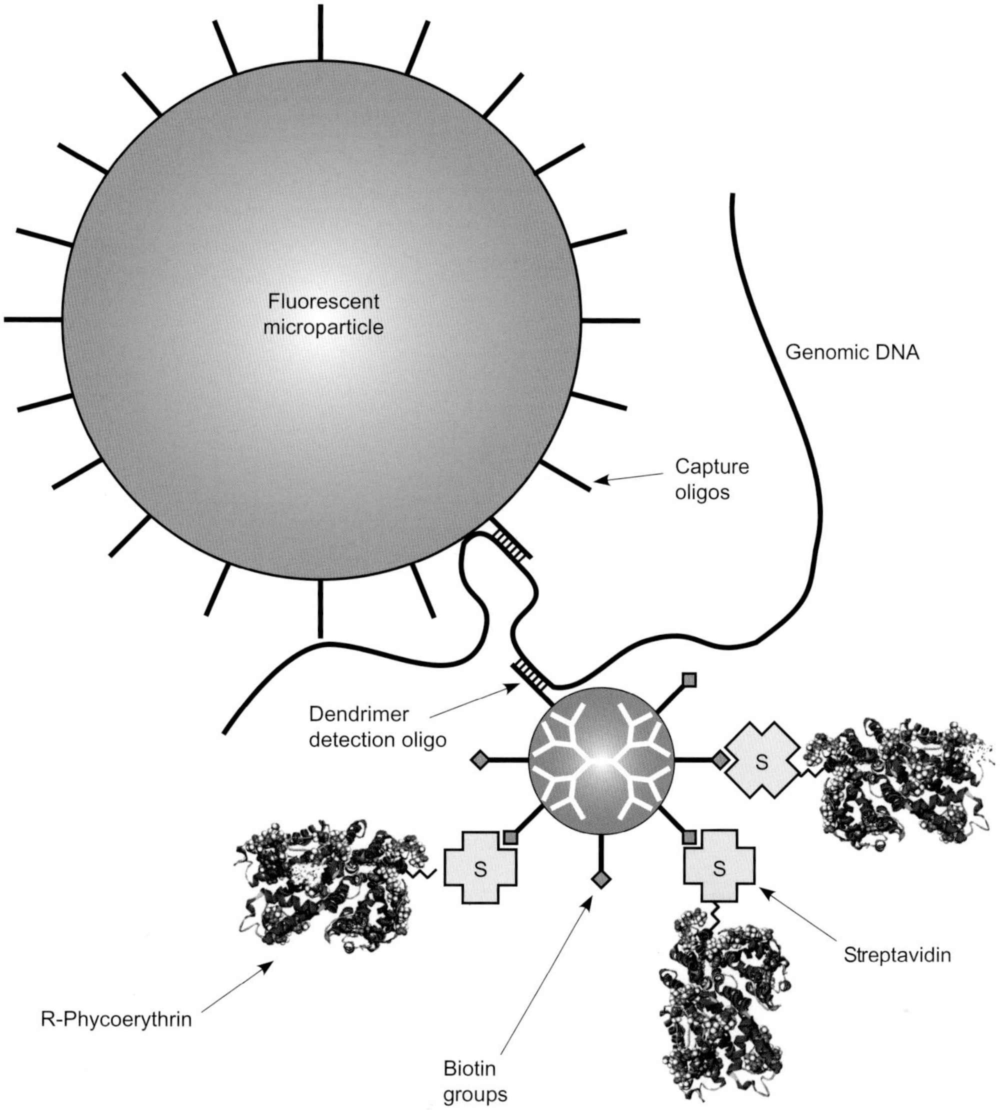

FIGURE 8.19 Biotinylated dendrimers can be used to enhance signals in assay procedures using the (strept)avidin–biotin interaction. In this example, a fluorescent microsphere containing capture oligos is able to interact with target DNA while a biotinylated dendrimer also containing detection oligo sequences binds to the genomic DNA target. The interaction of the dendrimer is detected using streptavidin–phycoerythrin conjugates, which are intensely fluorescent.

detection reagents bridge between more than one dendrimer. Thus, the use of multivalent biotin–dendrimers can become universal enhancers of DNA hybridization assays or immunoassay procedures.

Another example of an immunoassay enhancement using biotinylated dendrimers involves a novel detection technique called carbonylmetallo-immunoassay (CMIA). This technology involves the use of an NHS-4-pentynoate-(dicobalt hexacarbonyl) transition metal chelate labeling reagent or (n^5-cyclopentadienyl) iron dicarbonyl (n^1-*N*-maleimidato) group, which can be detected using Fourier transform infrared spectroscopy (Salmain *et al.*, 1991, 1992; Philomin *et al.*, 1994; Varenne *et al.*, 1992, 1995; Vessières *et al.*, 1999). These compounds can be coupled to amine groups on PAMAM dendrimers to yield amide linkages. Salmain *et al.* (2002) developed a dendrimer-based signal enhancement method to increase the sensitivity of the IR detection chelate. A G-4 PAMAM dendrimer containing 64 primary amine groups was labeled with NHS-biotin and the iron chelate to yield a complex containing 45 chelate groups and 4 biotins (Figure 8.20). The ratios in the final dendrimer conjugate were of course dependent on the molar ratios in the initial reaction mixture and could be controlled through careful planning of the conjugation process.

Biotinylated dendrimers also have been used to develop targeting conjugates for therapeutic use in

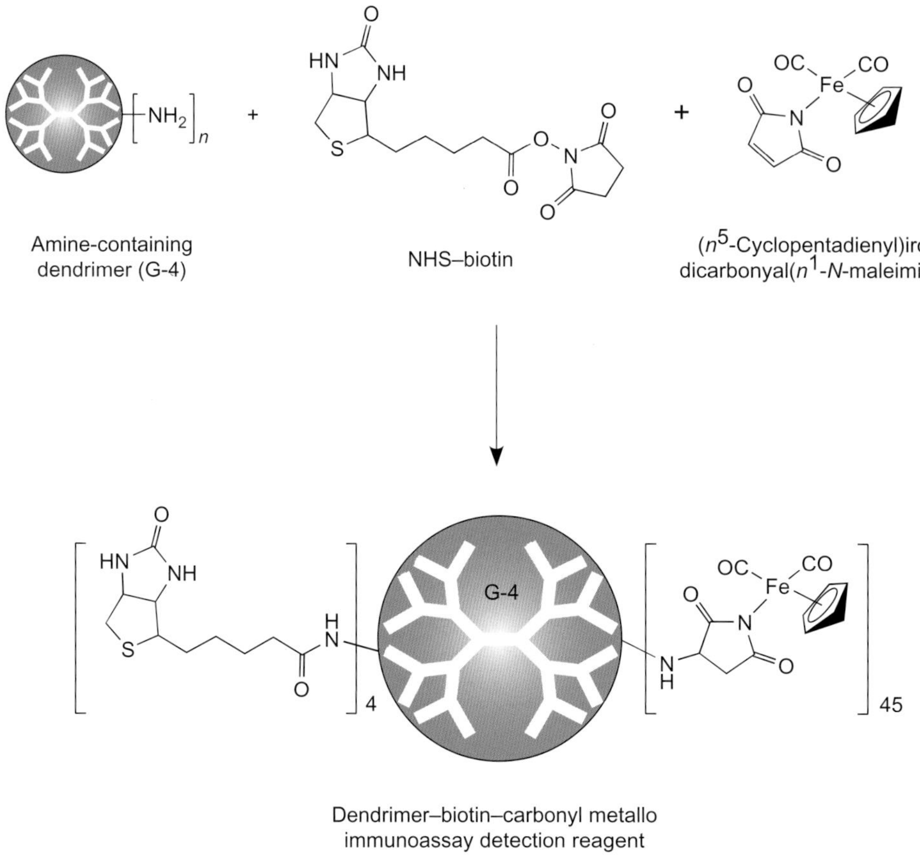

FIGURE 8.20 The multivalent surface of dendrimers can be used to couple biotin groups and labels for detection in immunoassays. One such conjugate was made by coupling NHS–biotin and a maleimido–iron chelate to an amine-containing dendrimer for use in an unique carbonyl metallo assay method.

targeting cancer cells. Wilbur *et al.* (1998) studied several different PAMAM generations in the coupling of a water soluble biotin analog with unique biotinidase insensitivity. The biotin–dendrimer conjugates were compared with biotin trimers or biotin tetramers in their ability to recruit radionuclide labeled streptavidin complexes after binding to a streptavidin–antibody conjugate already bound to antigen on a tumor cell surface. It was found that the biotinylated dendrimer increased the number of radionuclide-streptavidin binding events by a factor of 4.

The efficiency of biotinylation also was determined relative to the size of the amine-containing dendrimer (Wilbur *et al.*, 1998). The reactions were performed in DMF with triethylamine added as a proton acceptor, so competing hydrolysis of the NHS ester on the biotin compound was not an issue. In reacting each generation of dendrimer with a large excess of biotinylation agent, it was found that G-0, G-1, G-2, and G-3 PAMAM dendrimers could all be completely biotinylated by modification of all of the pendent amine groups. Only when reaching the size of a G-4 dendrimer was the biotinylation yield less than the number of total amine groups present (51 of 64 possible amines modified). This indicates that, unlike with protein labeling, the reactivity and accessibility of available amines on dendrimers remain high even when working with mid-sized molecules.

Mamede *et al.* (2003) developed a biotinylated G-4 PAMAM dendrimer containing multiple DTPA chelating groups for use in radiotherapy of intraperitoneally disseminated tumors. The G-4 dendrimer was reacted with sulfo-NHS–LC-biotin at a molar ratio that resulted in approximately 2.5 biotin molecules per dendrimer. This conjugate then was conjugated with 2-(*p*-isothiocyanatobenzyl)-6-methyldiethylenetriaminepentaacetic acid to give about 52 DTPA chelating groups per dendrimer. Finally, the complex was loaded with radioactive ^{111}In and also incubated with avidin to result in two to three avidin molecules per dendrimer binding to the biotin groups on the surface. The positive charge of the avidin protein molecules facilitated entry into the tumor cells, while the radioactive cargo killed the cells. The use of dendrimers in cancer therapeutics, including biotinylated ones, has been recently reviewed (Guo and Shi, 2012).

The biotinylation of amine-containing dendrimer may be accomplished using either an organic reaction environment or an aqueous medium. For modification of PAMAM dendrimers with a biotinidase-resistant biotin compound, Wilbur *et al.* (1998) performed the reaction in DMF with triethylamine as catalyst (proton acceptor). The following protocol illustrates this type of procedure using the biotinylation reagent NHS–PEG_4–biotin, which closely compares to the biotinidase insensitive compound used in the published procedure (also see Chapter 18, Section 1.3).

Protocol

1. In a fume hood, dissolve an amine-containing dendrimer in DMF at a concentration of at least 10 mg/ml with mixing. For the modification of a G-3 dendrimer, Wilbur *et al.* (1998) used approximately 240 mg of dendrimer dissolved in 10 ml of DMF. Other concentrations will work well in this procedure.
2. Add 15.7 µl of triethylamine (0.12 m*M*) to the dendrimer solution with mixing.
3. Dissolve a quantity of NHS–PEG_4–biotin (MW 588.67) in the dendrimer solution with mixing to bring the final reagent-to-dendrimer ratio in the reaction medium to at least 1.25 times greater than the molar amount of amines to be modified. To saturate completely a G-3 dendrimer initially containing 32 primary amine groups, add at least 8 mg of biotinylation reagent for each mg of dendrimer being modified. Using this molar ratio of biotin-to-dendrimer in the reaction will result in a high modification yield of all the amine groups on the dendrimer, especially for G-1 through G-3 size dendrimers. If a lower modification level is desired, then scale back the amount of biotinylation reagent added accordingly. Note: The most important consideration is optimizing the molar ratio of biotinylation reagent to dendrimer to obtain the desired modification level in the final conjugate. This entirely depends on the intended application of the biotinylated dendrimer, and a series of reactions containing different biotinylation ratios may have to be done to determine the best ratio to use.
4. React for a minimum of 1 h at room temperature with mixing. Longer reaction times may be used, particularly if a maximal modification level of biotin is desired.
5. Purify the biotinylated dendrimer by diluting it with an equal volume of water and then using dialysis, ultrafiltration, or size exclusion chromatography.

Another method that can be used to biotinylate an amine-containing dendrimer is to perform a similar reaction in aqueous buffer conditions. The following protocol is based on the methods of Tomalia *et al.* (1998) and Mamede *et al.* (2003).

Protocol

1. Dissolve 10 mg (0.7 µmol) of a G-4 PAMAM dendrimer in 1 ml of 0.1-*M* sodium phosphate, pH 9.0.
2. Add a 3-fold molar excess of biotinylation reagent over the molar quantity of dendrimer present. For the use of sulfo-NHS–LC-biotin (MW 556), this represents the addition of 2.1 µmol or 1.16 mg. This reaction ratio will result in a modification level of about 2.5 biotin groups per dendrimer. Other molar ratios also may be used, depending on the desired level of modification and the intended use for the conjugate.
3. React for 30 to 60 min at room temperature with mixing.
4. Purify the biotin-dendrimer using size exclusion chromatography on a desalting matrix or by use of ultrafiltration (e.g., centrifugal concentrators).

Fluorescent Labeling of Amine Dendrimers

The multivalent nature of dendrimers can be used to advantage for signal enhancement in numerous fluorescence detection schemes. A broad selection of hydrophilic organic fluorescent probes is available as amine-reactive derivatives (Chapter 10), many of which may be used to form conjugates with amine-containing dendrimers. Coupling multiple fluorescent labels to each dendrimer molecule creates a complex with increased luminescence or brightness for use in fluorescence assays (Tomalia *et al.*, 1998). Even fluorescent environmental probes have been found to have enhanced activity if conjugated to dendrimer core structures (Albertazzi *et al.*, 2011). In addition, combining a fluorescently labeled dendrimer with the ability to link it to targeting molecules, such as antibodies or streptavidin, provides a reagent system having superior detection capability compared to the use of fluorescently labeled proteins. Manduchi *et al.* (2002) demonstrated that a fluorescent dendrimer could be a more sensitive detection reagent than the use of directly labeled primary antibodies or indirect immunoassay methods using labeled secondary antibodies for microarray assay methods.

For instance, a dendrimer easily can be coupled with a large number of fluorescent dyes and still provide additional coupling sites for biotinylation. The only limitation to the number of fluorescent modifications is if fluorescence quenching starts to take place, in which case no further modifications will result in increased signal. A series of such conjugates using different levels of fluorophore modification should be done to determine the optimal level of dye-to-dendrimer before quenching occurs.

A fluorescent, biotinylated dendrimer of this type then can be used in a (strept)avidin–biotin assay to

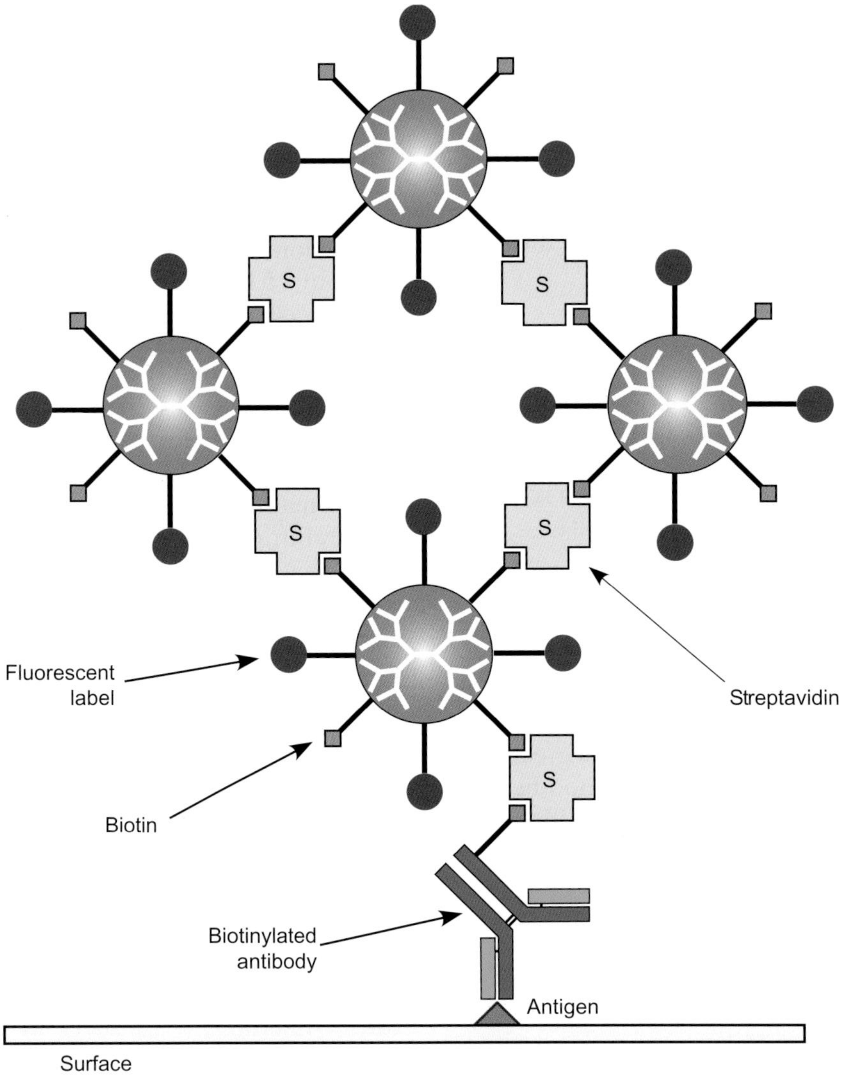

FIGURE 8.21 Dendrimers that are fluorescently labeled as well as biotinylated create enhanced detection reagents for use in (strept)avidin–biotin-based assays. Large complexes containing multiple fluorescent dendrimers can bind to antigens and form a highly sensitive detection system that exceeds the detection capability of fluorescently labeled antibodies.

detect analytes. Having more than one biotin group on such a conjugate allows it to interact with multiple streptavidin molecules, thus forming a megameric complex with a biotinylated primary antibody bound at the site of an analyte. In this way, many more fluorescent molecules are recruited to the point of detection than is possible using directly labeled proteins (Figure 8.21). In fact, for detection of DNA targets, dendrimer-based fluorescence enhancement methods have been identified as an important route to increasing signal and avoiding polymerase chain reaction (PCR) or other amplification methods (Nilsen *et al.*, 1997; Vogelbacker *et al.*, 1997; Kricka, 1999).

Fluorescent labels on dendrimers also are convenient to track the interaction of functionalized dendrimers for cell targeting. After the conjugation of targeting groups such as antibodies or folic acid on a dendrimer along with radiolabels or toxic components, there are still amine groups available to add fluorescent labels for detection purposes (Minard-Basquin *et al.*, 2003; Patri *et al.*, 2004; Thomas *et al.*, 2005). Such complexes may be assessed as to their specificity by monitoring the fluorescent signal as the dendrimers bind to target cells.

Dendrimers bearing certain fluorescent molecules attached on their periphery have been shown to be environmentally sensitive probes for the presence of certain metal ions or to changes in pH (Balzani *et al.*, 2000; Paola *et al.*, 2005). In addition, PAMAM dendrimers modified with the relatively hydrophobic dye Oregon Green 488 were shown to be a more effective transfection agent for anti-sense oligonucleotides than the dendrimer alone—plus the complex could be tracked within the cell due to the fluorescence of the dye (Yoo and Juliano, 2000).

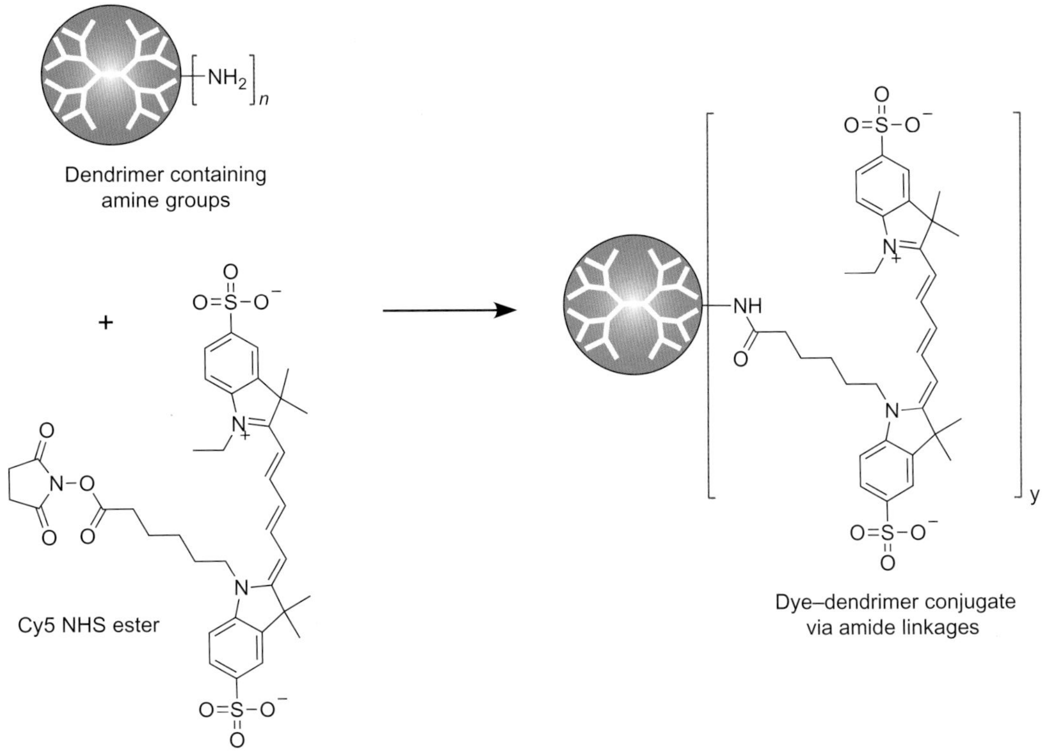

FIGURE 8.22 Fluorescent dyes such as an amine-reactive Cy5 derivative can be coupled to amine-containing dendrimer at relatively high substitution levels to create intensely fluorescent detection agents. If the dendrimer also is derivatized to contain an affinity group or a targeting group then specific fluorescent detection at high sensitivity can be realized.

Specific procedures for the conjugation of fluorescent labels to amine-containing molecules can be found in Chapter 10. The following protocol describes one such reaction for an amine-containing dendrimer with the fluorescent cyanine reactive probe, DyLight 650 NHS ester. This dye has spectral properties that are nearly identical to the original Cy5-type dye, but the DyLight one has negatively charged sulfonate groups, which makes the resultant compound extremely water soluble. Coupling reactions with other fluorescent dyes may be performed similarly, but in each case optimization of the molar ratio of dye-to-dendrimer should be performed to produce the best conjugate for its intended application. Note that higher levels of dye incorporation do not necessarily correlate to brighter fluorescent conjugates, because fluorescence energy transfer and quenching may occur between dye molecules as the density of dyes increases on the dendrimer surface. However, it appears that fluorescent dye substitution on dendrimers can be done at a higher density than on proteins before such quenching effects occurs. Figure 8.22 illustrates the reaction between an amine-containing dendrimer and a disulfonated Cy5 NHS ester derivative. Although the exact structure of the DyLight 650 NHS ester is not published, it is similar to the known Cy5 structure but contains additional sulfonate groups to increase water solubility.

Protocol

1. In a fume hood, dissolve the DyLight 650 NHS ester at a concentration of 10 mg/ml in dry DMF.
2. Dissolve the amine-containing dendrimer to be modified in DMF or buffer (50-m*M* sodium borate, pH 8.5) at a concentration of at least 10 mg/ml. Avoid the use of amine-containing buffers for an aqueous reaction, such as Tris or imidazole, as these will react with the NHS ester on the dye and prevent conjugation to the dendrimer. For reactions done in organic solvent, add triethylamine to a final concentration of 1.25-times greater than the amount of reactive dye to be added to the solution.
3. Add a quantity of the DyLight 650 dye to the dendrimer solution to provide at least a 1.25-fold molar excess of dye over the amount of dendrimer present (for nonaqueous reactions) or a 6- to 15-fold molar excess for aqueous reactions. Mix well to dissolve. The optimal amount of dye added should be determined experimentally by preparing a series of conjugates using different molar ratios of dye-to-dendrimer.
4. React for 1 h at room temperature with mixing.
5. Remove unreacted dye and reaction byproducts from the modified dendrimer by gel filtration or dialysis, using a molecular weight cutoff suitable for the size of the dendrimer.

3. DENDRIMER–CHELATE DERIVATIVES FOR IMAGING APPLICATIONS

Imaging agents consisting of metal chelating compounds have been used extensively for the modification of targeting agents for the *in vivo* detection of specific cells, organs, or vascular systems. The proper chelating group can coordinate securely a radioactive element or a contrast-enhancing agent, which can be imaged using radio-imaging techniques or magnetic resonance imaging (MRI), respectively. Many such polydentate chelating agents are described in Chapter 12, Section 1. These compounds are designed to have bifunctional characteristics in that one end is reactive for coupling to another molecule and the other end contains the chelating group.

The polyvalent nature of dendrimers has been investigated as vehicles for carrying multiple chelator groups to enhance signals in various imaging applications (Barthand Soloway, 1994; Yoo *et al.*, 1999; Kobayashi *et al.*, 2000, 2001; Sato *et al.*, 2001; Klemm *et al.*, 2012). In addition, in certain chelate–dendrimer constructs, excess amines on the dendrimer surface can aid in the cellular uptake process through charge-mediated endocytosis.

Roberts *et al.* (1990) conjugated porphyrins to PAMAM dendrimers as carriers to link them to antibody molecules for *in vivo* targeting. The porphyrin derivative was a *N*-(4-nitrobenzyl)-5-(4-carboxylphenyl)-10,15,20-tris (4-sulfophenyl)porphine containing one carboxylate group, which could be coupled to the amine groups on the dendrimer using the EDC/NHS method to form an amide bond. This porphyrin compound was effective at chelating copper ions, especially ^{67}Cu, which has nuclear decay properties ideal for potential use as a chemotherapeutic agent.

One of the more common chelating groups used for imaging applications is diethylenetriamine pentaacetic acid (DTPA). This compound is available in amine-reactive form as the anhydride or as a isothiocyanatobenzyl derivative, such as 2-(*p*-isothiocyanatobenzyl)-6-methyldiethylenetriaminepentaacetic acid chelate (1B4M; MW 555) (Kobayashi *et al.*, 1999). There are various isothiocyanatobenzyl-DTPA derivatives available that avoid conjugation of one of the chelator's carboxylate groups, which would be necessary using the anhydride form (Mirzadeh *et al.*, 1990; Brechbiel and Gansow, 1991; Behr *et al.*, 1999; Michel *et al.*, 2002; Brouwers *et al.*, 2003; Macrocyclics, Inc.). Any of these derivatives can be used to modify an amine-containing dendrimer to contain multiple chelating groups on its surface (Figure 8.23). Mamede *et al.* (2003) used the SCN-Bzl-DTPA compound 1B4M to modify a biotinylated G-4 PAMAM dendrimer to contain approximately 52 chelating group substitutions. The conjugate then was loaded with ^{111}In and complexed with avidin for investigation of its use for radiotherapy applications.

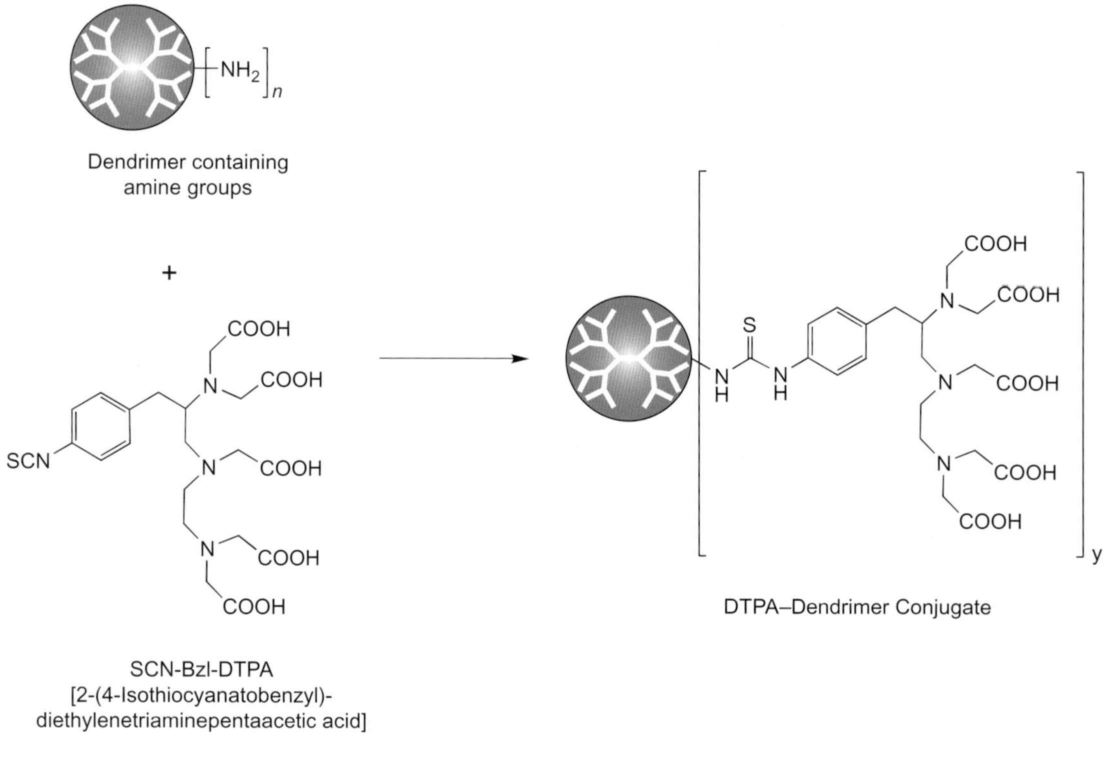

FIGURE 8.23 Chelating groups such as the isothiocyanate derivative of DTPA can be used to create multivalent chelating complexes with amine-containing dendrimer. Such complexes are able to coordinate multiple metal ions for detection, imaging, or radioimmunotherapy purposes.

Kobayashi *et al.* (2003) similarly prepared a DTPA-modified PAMAM dendrimer using the same isothiocyanate derivative for use in MRI imaging. A G-8 amine-dendrimer was reacted with a 1024-fold excess of chelating compound, resulting in a heavily loaded complex. In this case, the resultant conjugate was charged with ^{153}Gd to create a high-intensity contrast agent for *in vivo* imaging use in animals.

The following protocol for the modification of an amine-containing dendrimer with an SCN-Bzl-DTPA chelator is based on the literature references previously cited. Dendrimers of other generations will work well in this procedure provided that the molar ratios of reactants are adjusted for the size of dendrimer being used and the substitution level desired.

Protocol

1. Dissolve 10 mg of a G-4 amine-containing dendrimer in 1 ml of 0.1-*M* sodium carbonate buffer, pH 9.0.
2. Add a quantity of the SCN-Bzl-DTPA bifunctional chelating agent to obtain the desired molar excess of label over the amount of dendrimer present. The optimal ratio may be determined experimentally by preparing a series of dendrimer–chelate conjugates using different molar ratios and choosing the one that works the best in the intended application.
3. React for at least 1 h at room temperature with mixing. Some reactions have been done for up to 24 h at elevated temperatures (e.g., 40°C).
4. Purify the DTPA–dendrimer using dialysis, size exclusion chromatography, or spin-tube concentrators having a molecular weight cutoff of 5000 Daltons.

Other chelate–dendrimer constructs may be prepared using alternative bifunctional chelating agents according to Chapter 12, Section 1.

4. DENDRIMER DERIVATIVES AS SURFACE MODIFICATION AGENTS

The extremely branched nature of dendrimers provides multiple functionalities for creating high-capacity surfaces for coupling affinity molecules. The presence of numerous pendent groups on dendrimers can be used to advantage both in facilitating attachment to the surface and in providing a biocompatible coating for protein coupling. For instance, dendrimer modification of gold surfaces, both planer and nanoparticle, have the potential to create a matrix for high-density ligand coupling and low nonspecificity in assays. However, the dendrimer type and the characteristics of the pendent functionalities come into play in forming the optimal surface environment. PAMAM or amine-containing dendrimers of other types could result in relatively high nonspecific binding, particularly if unmasked terminal amines provide sites for charge interactions with other proteins (Yoon *et al.*, 2002; Hong *et al.*, 2003a, 2004).

Amine-containing dendrimers also can be used to pattern biological molecules on gold surfaces (Hong *et al.*, 2003b). In this application, PAMAM dendrimers were printed onto a SAM surface prepared using 11-mercaptoundecanoic acid to create sites for linking biomolecules. The carboxylate groups on the termini of the decanoic acid groups were activated with carbodiimide (EDC) and petafluorophenol to create reactive PFP esters (Chapter 3, Section 1.14). A PAMAM dendrimer solution in methanol (0.7 n*M*) was printed onto this activated SAM surface by contact printing. Reaction between the PFP esters on the SAM molecules and the amines on the PAMAM molecules covalently coupled the dendrimers to the surface via amide bond linkages. The dendrimers then were biotinylated with sulfo-NHS–biotin, washed, and allowed to interact with fluorescently labeled avidin, which was bound to the surface through interaction with the biotin groups. The fluorescent label was used to detect the bound protein and assess the efficiency of dendrimer surface functionalization.

Alternatively, PAMAM dendrimers can be linked to glass surfaces containing aldehyde functionalities through Schiff base formation and reductive amination using sodium cyanoborohydride. Hong *et al.* (2003b) patterned dendrimers on aldehyde slides and then blocked excess aldehyde groups using a 2-h incubation with 1-*M* 2-(2-aminoethoxy)ethanol. The result was covalently linked dendrimers on the slides containing an abundance of dendritic amines for further conjugation.

Hong *et al.* (2004) also found that modification of PAMAM dendrimers with a short PEG linker arm could act to reduce nonspecificity caused by the amines on the dendrimer modified surface. An azido-PEG$_3$-amine spacer was activated with nitrophenyl carbamate to yield an activated intermediate that could be used to modify the amines on the dendrimer (Figure 8.24). Reaction at high molar ratio resulted in about 61 PEG–azido spacers on the dendrimer. Reduction of the azido group to an amine using triphenylphosphine in THF provided the dendrimer–PEG–amine derivative for surface modification. The added presence of the PEG spacer arm reduced nonspecific binding to the dendrimer surface despite the fact that both the PAMAM and PEG-modified dendrimer both had terminal amino groups present. Another bioconjugation route to using this azido–PEG–dendrimer is to make an alkyne-containing protein or ligand and couple it to the modified dendrimer using click chemistry. The reaction of an azide with an alkyne in the presence of Cu(I) results in a cycloaddition reaction, which forms a triazole ring. See Chapter 17, Section 5, for more details on the click reaction.

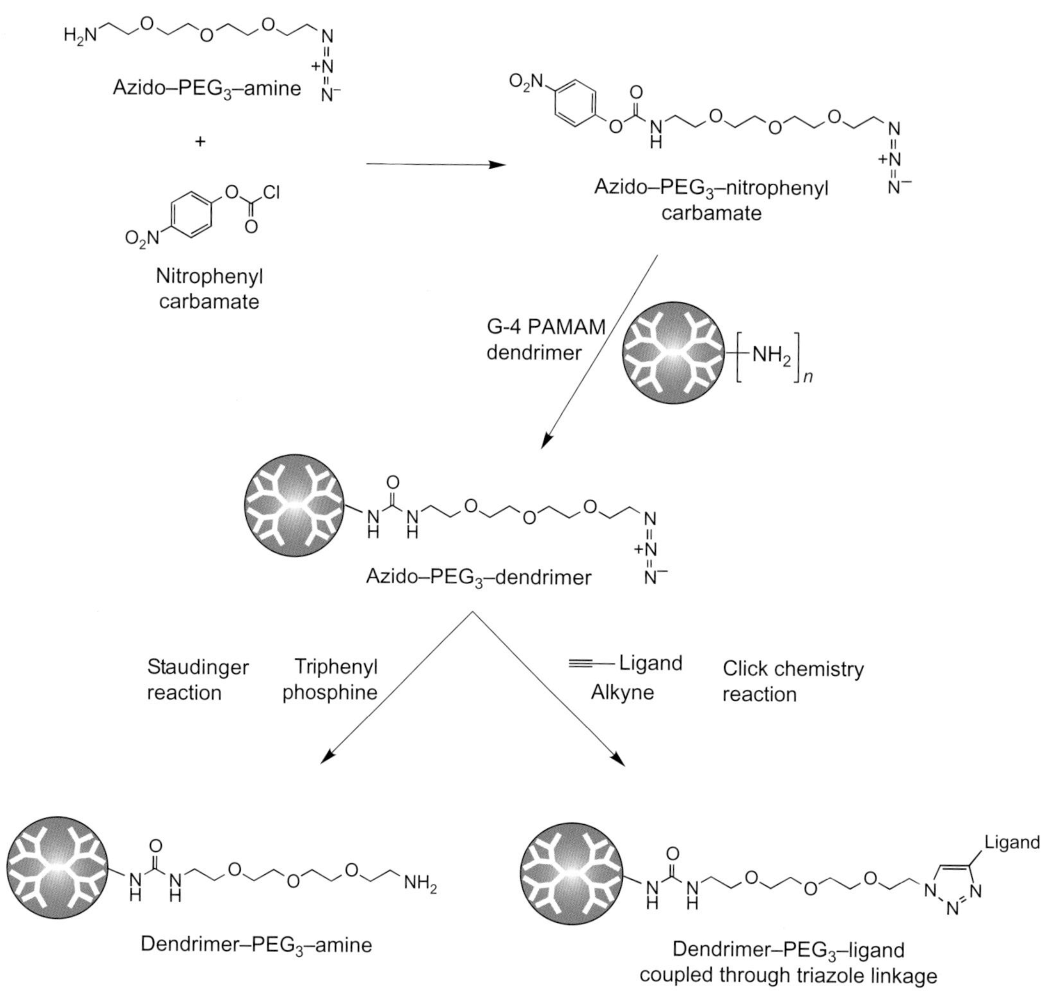

FIGURE 8.24 Azido–PEG–amine linkers can be coupled to amine-containing dendrimers by formation of an intermediate reactive nitrophenyl carbamate group. This compound then can be reacted with the amines on a dendrimer to form an azido–dendrimer derivative that can be used in chemoselective ligation strategies. The azide groups can be reacted with alkyne derivatives in a click chemistry reaction to couple ligands through a triazole linkage. Alternatively, the azido groups can undergo a Staudinger reaction with a phosphine compound to give the amino–PEG–dendrimer derivative. In addition, a Staudinger ligation reaction can be done using a phosphine derivative that would result in a covalent amide bond linkage (see Chapter 17, Section 4).

In another interesting route to creating dendrimer-functionalized surfaces, Pathak *et al.* (2004) activated the silanol groups on glass slides using CDI (Chapter 4, Section 3). The resultant reactive imidazole carbamate groups could be coupled to amine-containing dendrimers to form covalent carbamate linkages directly to the surface. Studying the effect of surface modification with generations 1 to 5 of a series of poly(propyleneimine) dendrimers resulted in the conclusion that dendrimers of G-3 size and above created surfaces of very high binding activity for coupling affinity ligands, with a G-4 dendrimer giving the greatest binding potential. This method may provide an excellent modification strategy for immobilizing affinity ligands onto planer and spherical silica surfaces. Once the amine-containing dendrimer surface was created, coupling proteins to it was done using the heterobifunctional crosslinker, sulfo-GMBS, which formed a thiol-reactive surface.

Benters *et al.* (2002) used three different coupling strategies to coat glass slides with amine-containing dendrimers for subsequent immobilization of oligonucleotides for preparing DNA microarrays. Cleaned slides were initially modified with 3-aminopropyltriethoxysilane (APTS) or glycidyloxypropyltrimethoxysilane (GOPTS) to create functional groups for further modification. The APTS-modified slides subsequently were reacted either with glutaric anhydride to form terminal carboxylate groups or with 1,4-phenylenediisothiocyanate (PDITC) to create reactive isothiocyanate groups on the surface. The carboxylate-containing slides were activated using the carbodiimide DCC along with NHS in DMF to form reactive NHS ester groups for coupling the amine-containing dendrimer.

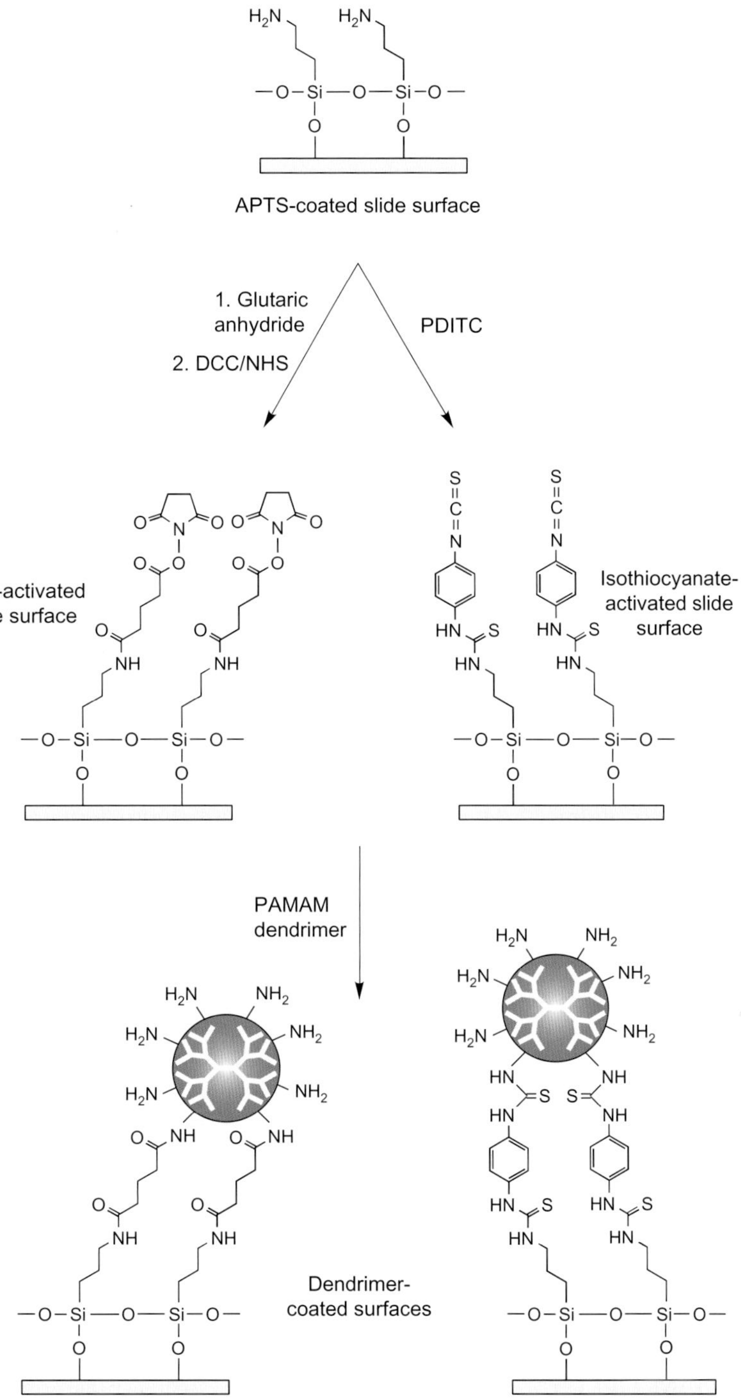

FIGURE 8.25 The multivalent nature of dendrimers can be used to add increased functionality to surfaces. Aminopropyl silane surfaces can be activated with either PDITC or through use of a cyclic anhydride plus DCC/NHS to give amine-reactive surfaces. These reactive surfaces can be used to couple amine-containing dendrimer to provide a high density of amine groups on the surface for further bioconjugation.

G-4 PAMAM dendrimers were attached to each of the three activated slides using a 100-mg/ml solution in methanol. The result formed covalently attached dendrimer layers on the slide surfaces, which then were again activated for the coupling of oligonucleotides. The dendrimer activation process was similar to that of the APTS surface to form reactive NHS esters. The dendrimers first were modified with glutaric anhydride and then activated using DCC and NHS in DMF. This process formed amine reactive NHS esters on the dendrimers, which then could be used to immobilize directly 5′-amine modified oligos (Figure 8.25).

Dendrimer-coated slides prepared using the methods of Benters *et al.* (2002) were found to provide significantly greater signal in fluorescent DNA hybridization assays than conventional slides prepared by coupling DNA directly to silane-or poly-L-lysine-modified surfaces. Clearly, dendrimers provide a 3D surface with greater functionality and biocompatibility for producing high-activity arrays.

Le Berre *et al.* (2003) also investigated the use of dendrimers to provide enhanced coupling capability and increased signal for DNA microarrays. In this application, the dendrimers consisted of a core of hexachlorocyclo-triphosphazene ($N_3P_3Cl_6$), which terminated in aryl aldehyde groups on the surface (Launay *et al.*, 1994, 1997; Slomkowski *et al.*, 1999). A G-4 aldehyde dendrimer containing 96 pendent aldehydes was coupled to a slide surface after modification with APTS to contain amines. Multiple Schiff base interactions with the amine groups on the derivatized slide effectively immobilized the dendrimers. Next, the spotting of amine-modified DNA to the aldehyde-dendrimer surface with overnight incubation followed by reduction with sodium cyanoborohydride resulted in covalently linking both the dendrimers to the surface and the oligos to the aldehyde groups through secondary amine bonds. A comparison of the dendrimer-coupled DNA slides to 11 commercially available activated glass slides indicated that the dendrimer slide provided among the highest fluorescence intensities in hybridization assays and 10- to 100-fold higher detection sensitivity than conventional slides.

Dendrimers also can provide increased capacity to immobilize capture antibodies for use in ELISA assays. Bosnjakovic *et al.* (2012) coupled PAMAM dendrimers to microplate wells using a PEG-based spacer arm and then used the amino groups on the dendrimer to couple the antibodies. The dendrimer-based immunoassay platform provided enhanced capture and detection for tumor necrosis factor-alpha cytokine as compared to normal ELISA capture plates prepared by simple adsorption of the antibody onto the surface.

5. DENDRIMER FLUORESCENT QUANTUM DOTS

Dendrimers can be used to effectively coat and passivate fluorescent quantum dots to make biocompatible surfaces for coupling proteins or other biomolecules. In addition, the ability of dendrimers to contain guest molecules within their 3D structure also has led to the creation of dendrimer–metal nanoclusters having fluorescent properties. In both applications, dendrimers are used to envelop metal or semiconductor nanoparticles that possess fluorescent properties useful for biological detection.

Huang and Tomalia (2005) used PAMAM dendrimers to coat gold nanoparticles or CdSe/CdS core/shell fluorescent quantum dots by preparing a disulfide-core dendrimer (using cystamine). The dendrimer was then succinylated to create terminal carboxylate groups, its core reduced with DTT, and the thiol dendron used to modify quantum dots by dative bonding to the particle surface. The result was an organized polymeric coating on the gold particles or quantum dots that terminated in multiple carboxylate groups for conjugation of biomolecules (Figure 8.26). The negative charges on the dendron terminals provided charge repulsion to maintain the colloidal stability of the small nanoparticles in solution, while the polyvalent nature of the dendrons made available an abundance of coupling sites for conjugation.

Testing of G-1, G-2, and G-3 dendrimers in this application provided insight into the density of surface modification needed to passivate completely the particles and prevent aggregation. The G-1 dendron was insufficient in this regard, but both the G-2 and G-3 dendrons were big enough to create a surface barrier, which resulted in excellent colloidal stability of the particles in solution.

Zheng and Dickson (2002) created a new type of fluorescent dendrimer construct by sequestering small nanoclusters of silver within hydroxyl-terminated G-2 or G-4 PAMAM dendrimers (16 hydroxyls on G-2 and 64 hydroxyls on G-4). The internal structure of a dendrimer is known to interact with charged silver ions in solution (Varnavski *et al.*, 2001). Without adding a reducing agent, such as sodium borohydride, which is typically used to form silver nanoparticles, it was discovered that the silver ion dendrimer complex could be photoactivated to cause silver reduction to elemental silver within the internal structure of the dendrimers. The resulting very small dendrimer/silver nanoclusters displayed strong fluorescence with absorption bands at 345 nm and 430 nm and a broad emission curve extending from slightly less than 500 nm to nearly 700 nm. By contrast, if borohydride reduction was used, larger silver nanoparticles were formed (3–7 nm) within the dendrimers which displayed no fluorescence characteristics but only strong plasmon absorption at 398 nm. Thus, small fluorescent nanoclusters containing only up to about 8 atoms of silver were formed by photoactivation and stably sequestered within the G-2 or G-4 interior. These properties were in agreement with studies on silver clusters of 2 to 8 atoms, which have been shown to have size-dependent fluorescence characteristics (Tani and Murofushi, 1994).

The broad emission band displayed by these silver/dendrimer constructs actually was found to consist of five overlapping fluorescent peaks caused by individual silver/dendrimer complexes. Each of these complexes

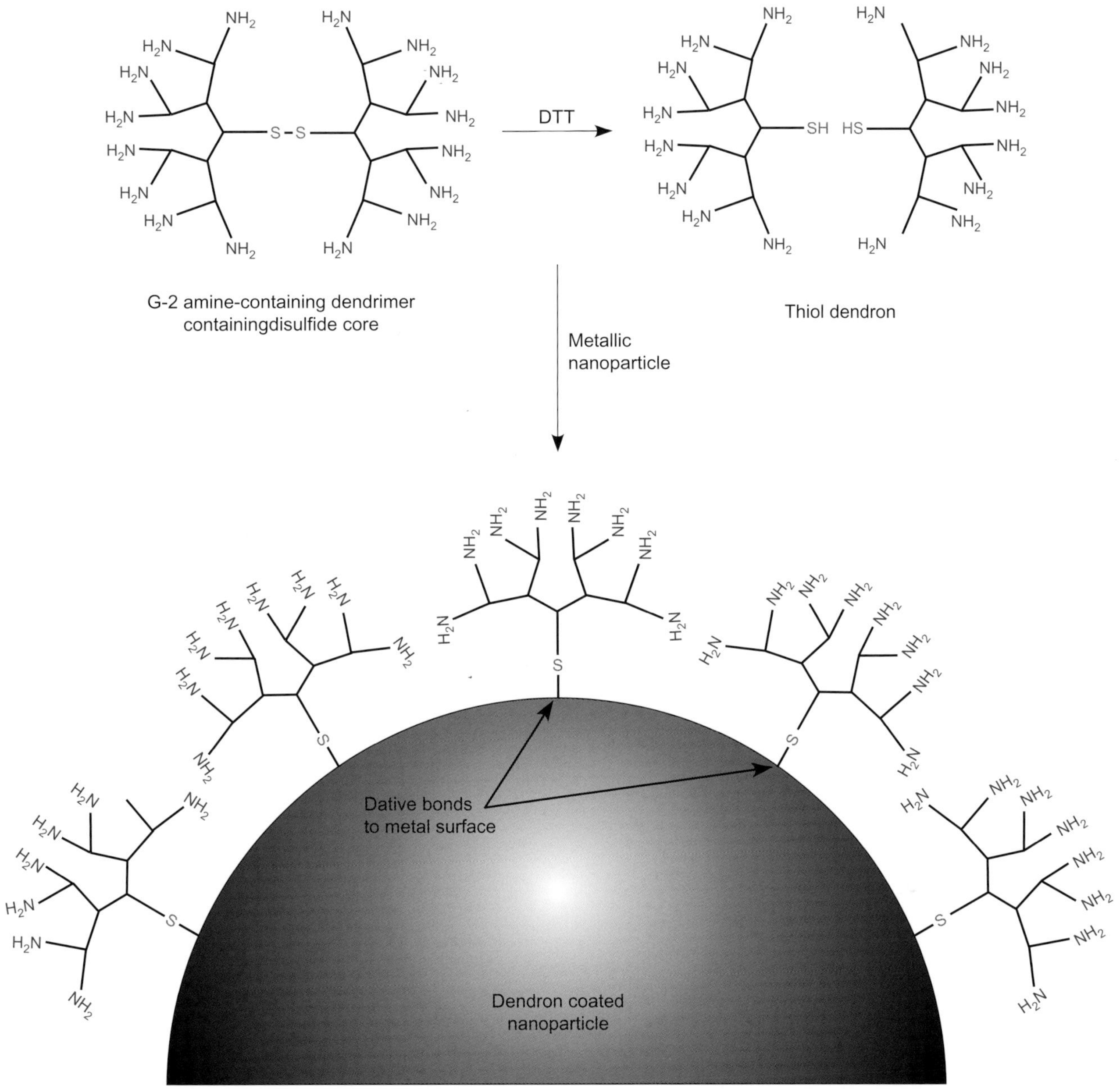

FIGURE 8.26 Dendrimers made with a disulfide-containing core can be reduced to produce dendrons having free thiol groups for surface modification. Dative binding of these thiol dendrons to gold or metallic surfaces can provide a high density of amine groups for coupling proteins or other molecules.

evidently contained a uniquely sized silver nanocluster, which resulted in an individual emission peak. Therefore, all the silver/dendrimer complexes together in solution presented a combined average of these five discrete emission peaks, and thus displayed the broad emission band covering nearly 200nm in width across the spectrum.

Lesniak *et al.* (2005) also describe the preparation and use of similar dendrimer/silver nanoclusters using G-5 PAMAM dendrimers terminated with amino, hydroxyl, or carboxyl functionalities. In this method, 25 Ag (I) ions were entrapped within each dendrimer and photoactivated by irradiation with UV light. This reduced the silver ions to Ag (0) metal nanoclusters totalling 25 atoms per dendrimer. The resultant silver/dendrimer nanocomposites were fluorescent with excitation in the range of 300 to 400nm and emission in the range of 400 to 500nm. Investigations into the use of the complexes as fluorescent detection reagents for cell-based imaging proved positive, with the properties closely matching the expected characteristics of the parent dendrimer for cell uptake and cytotoxicity.

The polyvalent nature of dendrimer molecules also has been used to modify semiconductor quantum dot surfaces to provide increased bioconjugation capacity or to use a PAMAM dendrimer's surface positive charge to enhance the entry of the nanoparticle into cells (Higuchi *et al.*, 2011). In addition, dendrimer-functionalized quantum dots have been used to develop fluorescent imaging agents for tumor targeting *in vivo* (Akin *et al.*, 2012). The conjugation methods that can be used for quantum dots are discussed in greater detail in Chapter 10, Section 10.

CHAPTER

9

Cross-Bridges and Cleavable Reagent Systems

Crosslinking and modification reagents may possess functionality other than their reactivity toward certain chemical groups. In particular, the cross-bridge of the molecule can be designed to contain constituents that allow cleavage of the reagent after use. Occasionally, the coupling reaction itself provides linkages susceptible to subsequent cleavage. Why would you want to break a conjugate apart after having formed it? The ability can be very important in studies involving the biospecific interactions between two molecules, especially if only one of the molecules is known or characterized. Cleavability allows the conjugation reaction to be verified through identification of the crosslinked molecules after conjugation and purification of the complex. Precise points of modification can be determined by mass spec analysis after cleavage. In some cases, purification of unknown target molecules is facilitated by the ability to cleave the crosslinking bonds after isolation of the complex. For instance, a protein modified near its binding site with a photoreactive heterobifunctional crosslinker can be incubated with its receptor or specific binding molecule, a covalent linkage then can be formed by exposure to UV light, and the complex analyzed by subsequent cleavage of the crosslink.

In another example, ligands can be biotinylated with a cleavable biotinylation reagent and then incubated with receptor molecules. The resulting complex can be isolated by affinity chromatography on immobilized (strept)avidin. Final purification of the ligand–receptor can be accomplished by cleaving the biotin modification sites while the complex is still bound to the support. The receptor complex thus can be eluted from the column without the usual harsh conditions required to break the avidin–biotin interaction.

The ability to cleave a crosslink can also provide a means of transferring a label from one protein to another. For instance, a photoreactive heterobifunctional crosslinker that is iodinatable and cleavable can be used to tag an unknown receptor molecule after conjugation. For example, the crosslinker SASD (Chapter 6, Section 3.2) can be iodinated before it is employed in a crosslinking reaction. It is then used to modify a protein through its amine-reactive NHS ester end and purified from excess crosslinker. After incubation of the modified protein with specific binding molecules (e.g., other proteins) and photoreactive crosslinking, the conjugate can be broken by reduction of the disulfide group within the cross-bridge of the reagent. Since the radiolabeled part of the crosslinker is now attached to the unknown interacting molecule, the tracer is effectively transferred from the initially modified molecule. Thus, unknown interacting molecules can be tracked after their binding to an SASD-labeled substance.

Another crosslinker, SAED (Chapter 6, Section 3.9), can be used in a similar fashion, but instead of transferring a radioactive label, it contains a fluorescent portion that is transferred to a binding molecule after cleavage. Similarly, sulfo-SBED routinely is used to study protein interaction. Cleavage of a disulfide bridge after capture of interacting proteins results in transfer of a biotin label to the unknown prey protein (Chapter 24, Section 3.1). The biotin modification can then be used to detect or isolate the unknown interactor for subsequent identification.

The ability to break a crosslink can be an important feature of a modification or conjugation reagent. This chemistry typically is built into the cross-bridge or reactive ends of a reagent using disulfides, glycol groups, diazo bonds, esters, sulfone groups, acylhydrazones, nitrophenyl groups, or acetal linkages. The following sections describe these chemical characteristics and their respective cleavage conditions.

1. CLEAVAGE OF DISULFIDES BY REDUCTION

The formation of a disulfide linkage between crosslinked molecules is an important option for many conjugation chemistries. Examples of reagents that have

Bioconjugate Techniques, Third Edition.
DOI: http://dx.doi.org/10.1016/B978-0-12-382239-0.00009-1

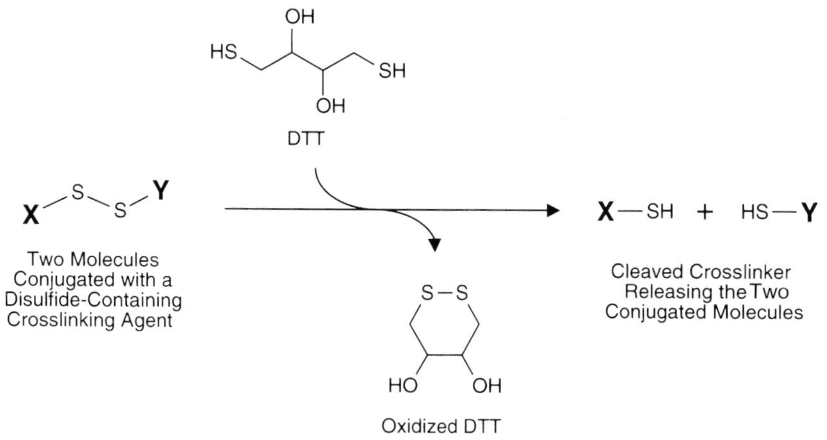

FIGURE 9.1 Cleavage of disulfide-containing crosslinking compounds can be achieved using a reducing agent such as DTT. Reduction causes the conjugates to break apart into their original components with each component containing a portion of the crosslinker that terminates in a thiol group.

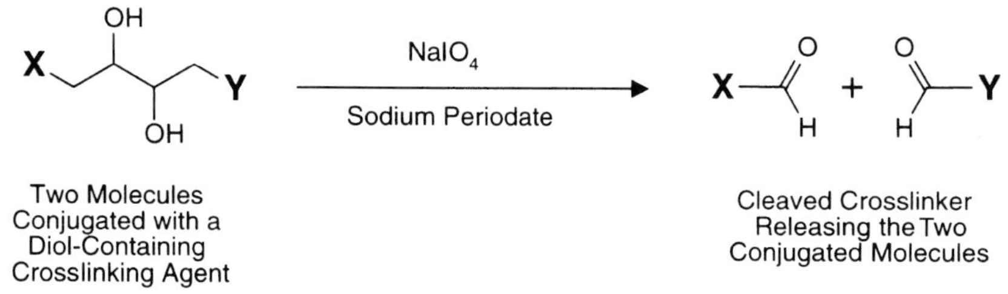

FIGURE 9.2 Crosslinkers containing a diol group in their cross-bridge design may be cleaved by oxidation with sodium periodate.

this capability include the pyridyl disulfide-containing heterobifunctionals like SPDP (Chapter 6, Section 1.1) and SMPT (Chapter 6, Section 1.2). Other non-sulfhydryl-reactive crosslinkers still may possess a disulfide group within their cross-bridge construction. The presence of such disulfide groups, whether designed in the crosslinker or created as a product of their reactions, allows for specific cleavage of the complex or modified molecule after conjugation. Disulfide bonds can be broken by a number of methods (Chapter 2, Section 4.1), utilizing either direct hydrogenolysis by a strong reductant such as sodium borohydride or through a disulfide interchange process with a compound containing one or more free sulfhydryls (Figure 9.1).

Cleavage of disulfide bonds is easily achieved by incubation with a reducing agent at a level of 10- to 100-m*M* concentration. If the disulfides in the crosslinks are the only ones present in the complexed molecules, then reduction will yield unconjugated molecules—one or both of which will contain a portion of the crosslinker, and on both molecules a free sulfhydryl will be created. Caution should be used with this method of cleavage, however, if other disulfides are present in the conjugated molecules. Some protein disulfides, for instance, may also be affected by the reduction step. Complete cleavage of all disulfides in crosslinked proteins by inclusion of unfolding agents (e.g., guanidine) may yield additional protein fragments of lower molecular weight due to subunit disassociation.

2. PERIODATE-CLEAVABLE GLYCOLS

Crosslinking agents can be designed to contain adjacent carbon atoms possessing hydroxyl groups. Cross-bridges containing such diols or glycol residues can be constructed from the inclusion of an internal tartaric acid spacer or similar compound in their synthesis (e.g., DST, Chapter 5, Section 1.3). These groups can easily be cleaved by oxidation with sodium periodate (Chapter 2, Section 4.4). Treatment with 15-m*M* periodate at physiological pH will break the carbon–carbon bond between the glycol portion, oxidizing each hydroxyl to an aldehyde, and cleaving the associated crosslinked molecules (Figure 9.2). Under these conditions, glycosylated portions of glycoproteins or other carbohydrate-containing molecules will also be affected, forming additional aldehyde groups. In some cases, the production of aldehyde residues may cause secondary reactions to occur, especially Schiff base formation with available amine groups. To avoid unexpected crosslinks that may form through such intermolecular Schiff base formation, after oxidation and removal of excess periodate, Tris or ethanolamine may be included to tie up the aldehydes and stabilize the cleaved products.

Sodium periodate may also affect tryptophan residues in some proteins. The oxidation of tryptophan can result in activity losses if the amino acid is an essential component of the active site. For instance, avidin and streptavidin may be severely inactivated by treatment

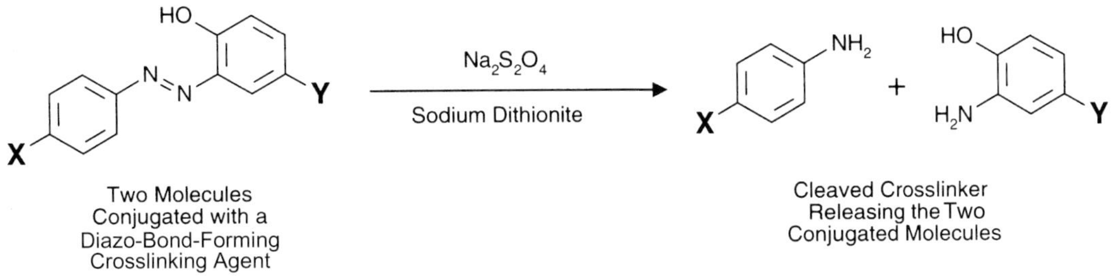

FIGURE 9.3 Crosslinking agents that form diazo bonds may be cleaved using sodium dithionite.

with periodate, since tryptophan is important in forming the biotin-binding pocket. In addition, many other amino acid residues are susceptible to oxidation by periodate (Chapter 2, Section 1.1). Limiting the time of oxidation is important to restricting oxidation to diol groups while not affecting other protein structures.

The use of periodate as a cleavage agent does have advantages, however. Unlike the use of cleavable crosslinkers that contain disulfide bonds which require a reductant to break the conjugate, cleavage of diol-containing crosslinks with periodate typically preserves the indigenous disulfide bonds and tertiary structure of proteins and other molecules. As a result, with most proteins bioactivity usually remains unaffected after mild periodate treatment.

3. DITHIONITE-CLEAVABLE BONDS

Crosslinking compounds containing diazo bonds within their structures can be specifically cleaved with dithionite (Jaffe *et al.*, 1980). In addition, crosslinks formed by the reaction of a diazonium compound (Chapter 3, Section 6.1) with a tyrosine residue can be broken using this reagent. Sodium dithionite (also called sodium hydrosulfite) reduces the diazo linkage, breaking the bond between the nitrogens and leaving a primary amine on both fragments of the crosslinker (Figure 9.3). The reaction is usually carried out in alkaline conditions; a 25-min incubation with 0.1-*M* dithionite in 0.2-*M* sodium borate, pH 9, works well. As the diazo bonds are broken, any color associated with the reagent will disappear.

Dithionite is also capable of cleaving oxime linkages that are formed as the result of the reaction of an aldehyde and aminoxy group (Pojer, 1979). Thus, crosslinks formed between two proteins or other molecules using this reaction strategy can be specifically broken using sodium dithionite.

In a novel application, Fonović *et al.* (2007) designed dithionite-cleavable, activity-based probes for the covalent modification of thiol or serine proteases. In many cell extracts DTT is added to maintain a reducing environment; therefore, cleavable probes containing disulfides would be incompatible with the extract buffer system. One probe design contains an active site binding region with an epoxide warhead group, which can link to the thiol within the binding pocket of a cysteine protease through a thioether bond. The other end of the probe contains a biotin tag to allow purification of labeled enzymes on a streptavidin affinity chromatography support. The crossbridge of this probe contains a diazo group, which can be cleaved using dithionite (Figure 9.4). The design facilitates mild elution of isolated enzymes from the streptavidin column without the typical severe denaturing conditions required to break the biotin–streptavidin interaction.

4. HYDROXYLAMINE-CLEAVABLE ESTERS

Hydroxylamine is a powerful nucleophile which, under alkaline conditions, is effective in breaking ester bonds. Crosslinking agents containing esterified spacer components can be cleaved after undergoing a conjugation reaction by incubation with 0.1-*M* hydroxylamine, pH 8.5, for 3 to 6h at 37°C (Abdella *et al.*, 1979). The reaction results in the formation of an amide derivative on one fragment of the cleaved crosslinker and a hydroxyl group on the other fragment (Figure 9.5). Thioester bonds are also susceptible to cleavage under these conditions. Thioesters may be broken with the production of an amide and a sulfhydryl group on either side of the crosslinker fragments.

An example of a hydroxylamine–cleavable reagent is EGS (Chapter 5, Section 1.5) which contains two ester bonds made by the esterification of ethylene glycol with succinic acid. Cleavage with hydroxylamine yields two fragments terminating with an amide bond and concomitant release of ethylene glycol. Mass spec evidence of the cleavage products from an EGS crosslink indicates that the cleaved parts can undergo cyclization to form a succinimide group on both fragments (Petrotchenko *et al.*, 2005).

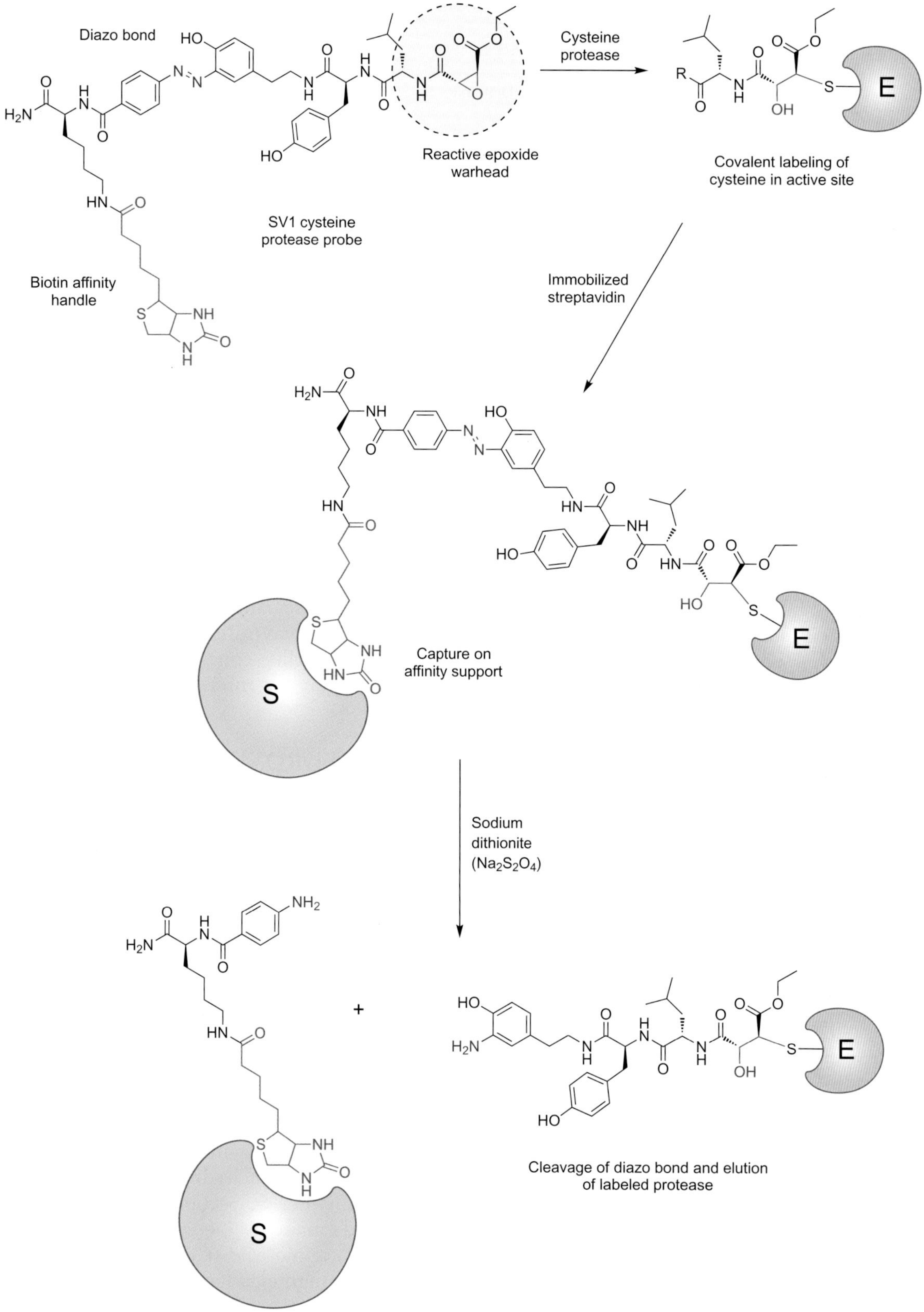

FIGURE 9.4 Use of an active site warhead probe containing a dithionite-cleavable diazo bond.

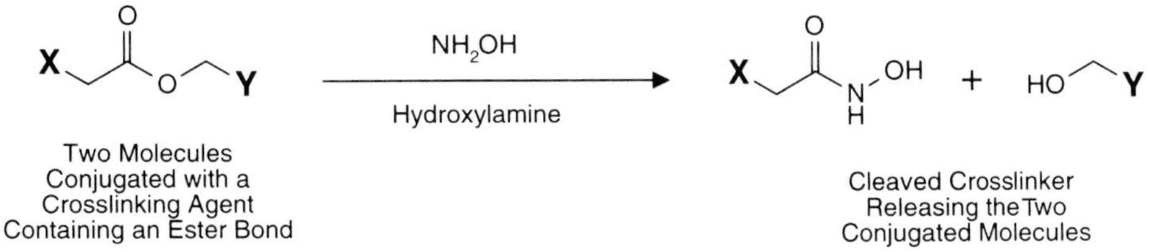

FIGURE 9.5 Crosslinkers containing an ester group in their cross-bridge are susceptible to cleavage under alkaline conditions using hydroxylamine.

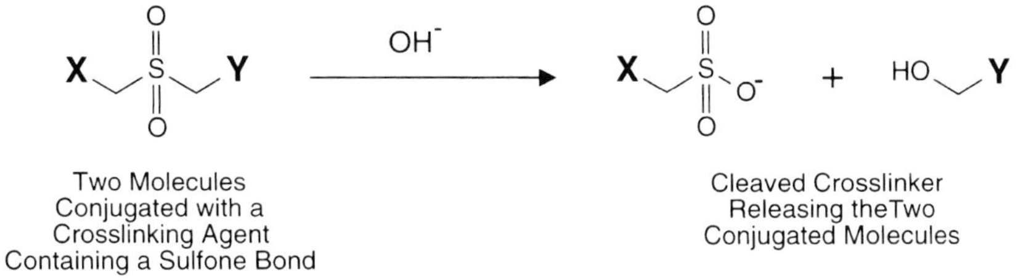

FIGURE 9.6 Crosslinkers that have an internal sulfone group in their cross-bridge may be cleaved using base.

5. BASE LABILE SULFONES

The presence of a sulfone group in a crosslinking reagent can allow for cleavage of a conjugate through hydrolysis of the linkage under basic conditions. In 0.1-*M* sodium phosphate, adjusted to pH 11.6 by addition of Tris base, containing 6-*M* urea, 0.1% SDS, and 2-m*M* DTT, sulfone groups were successfully cleaved after incubation at 37°C for 2h (Zarling *et al.*, 1980). In that study, peptide antigens on the surface of lymphocyte receptors were crosslinked with the homobifunctional, amine-reactive reagent BSOCOES (Chapter 5, Section 1.4), purified, and cleaved for analysis. The presence of urea, SDS, and DTT was not absolutely necessary for breaking the sulfone bond; rather, they served to disrupt completely protein/peptide structure for complete dissociation of the complex.

In addition to BSOCOES, the amine-reactive, bisfluorobenzene reagent DFDNPS (Chapter 5, Section 4.2) also contains an internal sulfone group that is easily cleaved under basic conditions (Wold, 1961, 1972). Hydrolysis of the sulfone yields two crosslinker fragments, one terminating in a sulfonic acid group and the other containing a hydroxyl group (Figure 9.6).

6. ACYL HYDRAZONE-CLEAVABLE LINKERS

A common method of targeting active sites within proteins is to use a biotin-containing probe that can covalently link to a functional group within or near the active center of an enzyme or receptor protein. Once active proteins are labeled in a complex sample such as a cell extract, the biotin handle is used as an affinity tag that can interact with immobilized streptavidin to facilitate purification (Chapter 1, Section 3.3, and Chapter 15, Section 2.5). This is an efficient capture method to isolate the modified proteins, but subsequent elution from the streptavidin support often results in severe denaturation of bound proteins. To overcome this problem, various cleavable biotin tags have been designed which contain, for instance, disulfides or esters that can be cleaved using reducing agents or hydroxylamine under alkaline conditions, respectively. Recently, Park *et al.* (2009) introduced an acylhydrazone-containing biotin probe that can be cleaved under very mild conditions. A single acylhydrazone linkage in the cross-bridge between the biotin tag and the active site probe allows cleavage and elution from a streptavidin affinity support in the presence of an excess of a hydrazide-containing reagent (Figure 9.7).

Hydrazone bonds have been used extensively in the design of bioconjugates for use as drug delivery agents (Kale and Torchilin, 2007), for chemoselective ligation reactions (Chapter 17, Section 2), and to immobilize affinity ligands on particles or chromatography supports (Chapter 14, Section 4.2, and Chapter 15, Section 2.4). A hydrazone bond is reasonably stable at neutral or slightly alkaline pH values, but at slightly acidic pH (4–5) it is labile to hydrazide exchange reactions. Under these conditions, the addition of a hydrazide derivative will cleave the hydrazone bond in the cross-bridge of an enzyme probe, resulting in hydrazide exchange and covalent modification of the labeled enzyme with the cleavage reagent. Thus, the active site probe cannot

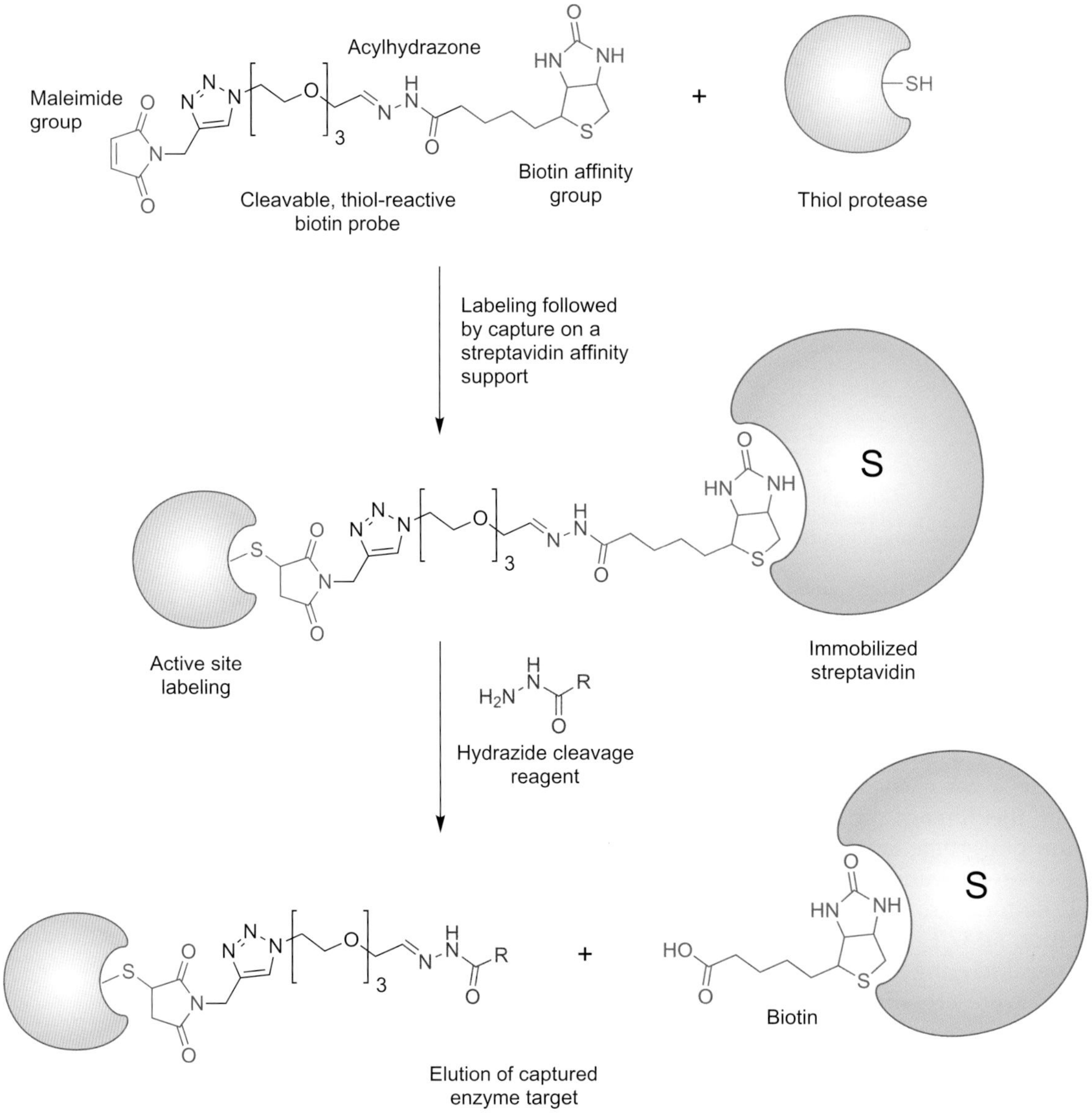

FIGURE 9.7 Cleavage of biotin acylhydrazone active site probe using a labeled hydrazide reagent.

only facilitate labeling of active enzymes and isolation from complex proteomic mixtures, but further labeling with detection components, if desired, during the cleavage reaction. Capture of a biotin probe–enzyme complex on an immobilized streptavidin support can be followed by release and labeling of the enzyme with, for instance, a fluorescent tag containing a hydrazide group. The fluorescently labeled enzyme can then be used in detection experiments to image or track the enzyme.

7. PHOTOCLEAVABLE LINKERS

The ability to cleave cross-bridges through the use of light is an important option that allows for controlled release of protecting groups or for rapid, timed release of part of a conjugate component from a covalent complex. Photolabile groups can be incorporated into the design of standard bioconjugate reagents and used in normal fashion to create crosslinked, modified, or immobilized molecules. Photocleavable groups have been used in the study of protein–protein interactions and in drug discovery and development, especially as the trigger for the release of "caged" ligands or prodrugs that have therapeutic activity only upon photocleavage (Gorman and Prestwich, 2000).

Reagents containing a photocleavage site can be of many types, including crosslinkers or modification agents, which can have reactive groups at either end or contain an affinity group on one end. The cross-bridge

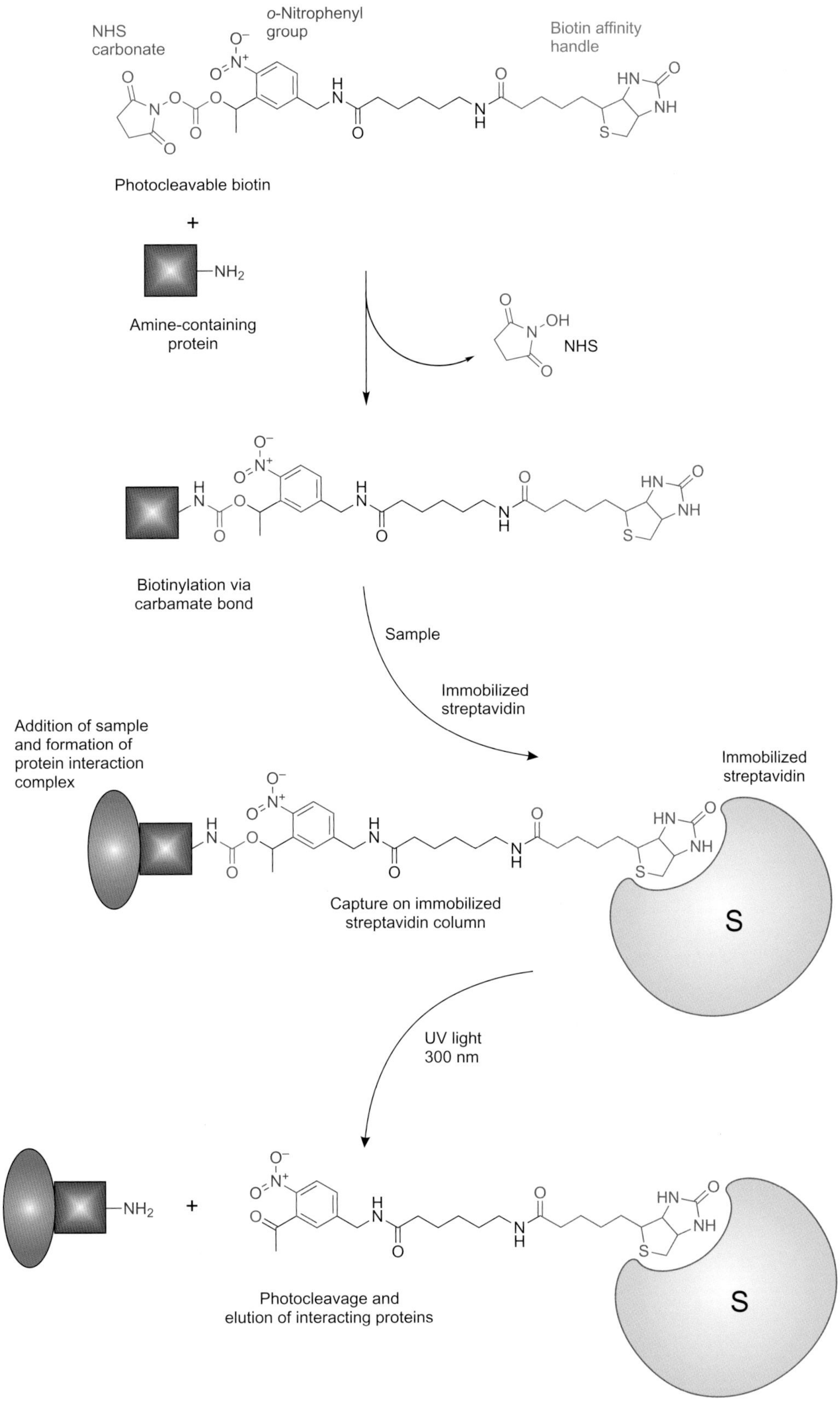

FIGURE 9.8 Photocleavable biotin reactions with an amine-containing protein or molecule and subsequent photocleavage of the cross-bridge to release the biotinylated molecule.

of such compounds remains stable so long as it is not exposed to light strong enough and at the right wavelength to cause cleavage (normal room light poses no problems for brief exposures). Olenjnik *et al.* (1995) designed biotin derivatives containing an internal *ortho*-nitrophenyl group within the spacer arm, which also contains a thermoreactive aryl carbonate ester (before coupling) or an aryl carbamate bond (after coupling). Upon exposure of this cross-bridge with UV light at 300 nm, this group undergoes a photoreaction leading to bond breakage at the aryl carbamate linkage with concomitant cleavage of the biotin group from the rest of the molecule (Rothschild *et al.*, 1999) (Figure 9.8). The photocleavage reaction is rapid and complete within about 5 min of exposure to UV light.

The advantage of incorporating photocleavable sites into a biotinylation agent (or an active site probe containing biotin) is that the resultant biotin conjugate can be captured on an immobilized streptavidin support and subsequently eluted by exposure to light. This process can facilitate the isolation of a group of interacting proteins that can be studied in their native form, because elution can be carried out without the use of strongly denaturing conditions typically associated with the use of the biotin–streptavidin interaction.

CHAPTER

10

Fluorescent Probes

Labels, tags, and probes are relatively small modifying agents that can be used to modify proteins, nucleic acids, and other molecules. These compounds often contain groups that provide, either directly or indirectly, sensitive detectability by virtue of some intrinsic chemical or atomic property such as fluorescence, chemiluminescence, absorptivity, radioactivity, or affinity toward another biomolecule. Most probes can be designed to contain a reactive group capable of coupling to the functional groups of biomolecules, thus making conjugation possible. After modification of a protein, for instance, the probe becomes covalently attached, thus permanently tagging it with a unique detectable property, which can be used to assay or track the conjugate as it interacts with other biomolecules.

Fluorescent labels used in this manner can provide tremendous sensitivity due to their property of repeated light emission upon excitation. Proteins, nucleic acids, and other molecules can be labeled with fluorescent probes to provide highly sensitive reagents for numerous *in vitro* assay procedures. For instance, fluorescently tagged antibodies can be used to probe cells and tissues for the presence of particular antigens and then be detected through the use of fluorescence microscopy techniques. Since each probe has its own fluorescence emission character, more than one labeled molecule—each tagged with a different fluorophore—can be used at the same time to detect two or more target molecules.

A fluorescent molecule has the ability to absorb photons of energy at one wavelength and subsequently emit energy at another wavelength. The absorption process is also called excitation, since the quantum energy levels of some of the compound's electrons increase with photon uptake. The absorption band usually is not isolated at a discrete photon energy level but is spread out over a range of wavelengths with at least one peak of maximal absorbance within this excitation region. The extinction coefficient (ε, expressed as $M^{-1}cm^{-1}$) at the absorbance peak maximum is a unique characteristic of each fluorophore under a given environmental condition (such as the solvent it is dissolved in).

The excess energy of an excited fluorophore can be lost as heat or through collisions with adjacent molecules or released as photons of light as the electrons return to the lower, ground-state energy level (Figure 10.1). This process of light emission occurs in less than 10^{-4}s after excitation and is known as fluorescence. Fluorescence takes place from the lowest excited singlet state as the electron collapses back down to the ground state and releases a photon. According to Stokes' law, the emission wavelength is always longer and thus of lower energy than the wavelength of excitation (Kawamura, Jr., 1977). The ratio of total photon emission over the entire range of fluorescence to the total photon absorption is called the quantum yield (QY). Quantum yield values range from 0 to 1. The larger the QY value the more efficient the photon emission or luminescence. For a particular fluorophore under fixed environmental conditions, both its extinction coefficient and its quantum yield are fundamental characteristics of its photochemical behavior.

Not only should a fluorescent compound suitable for analytical studies have a high QY, but also its fluorescence emission spectrum should be separated sufficiently from its excitation spectrum to ensure good signal isolation. A fluorophore's Stokes shift is a measure of the separation of its maximal absorbance wavelength from its emission wavelength maxima (Figure 10.2). The greater the Stokes shift, the better the signal isolation and therefore less interference from Rayleigh-scattered excitation light.

The majority of reported fluorophores of practical use in labeling biomolecules contain an aromatic ring system as the generator of luminescence. In general, as the conjugated electron system gets larger, the emission wavelength is shifted to the red or longer wavelengths. Also, the extinction coefficient and QY of larger conjugated systems typically are greater than those of small aromatic compounds. Fluorescent compounds emitting in the far red or near infrared (NIR) region typically have extinction coefficients greater than $100{,}000\,M^{-1}cm^{-1}$ and sometimes greater than $200{,}000\,M^{-1}cm^{-1}$, whereas dyes that emit in the blue region of the spectrum usually have extinction coefficients of less than $25{,}000\,M^{-1}cm^{-1}$.

Bioconjugate Techniques, Third Edition.
DOI: http://dx.doi.org/10.1016/B978-0-12-382239-0.00010-8

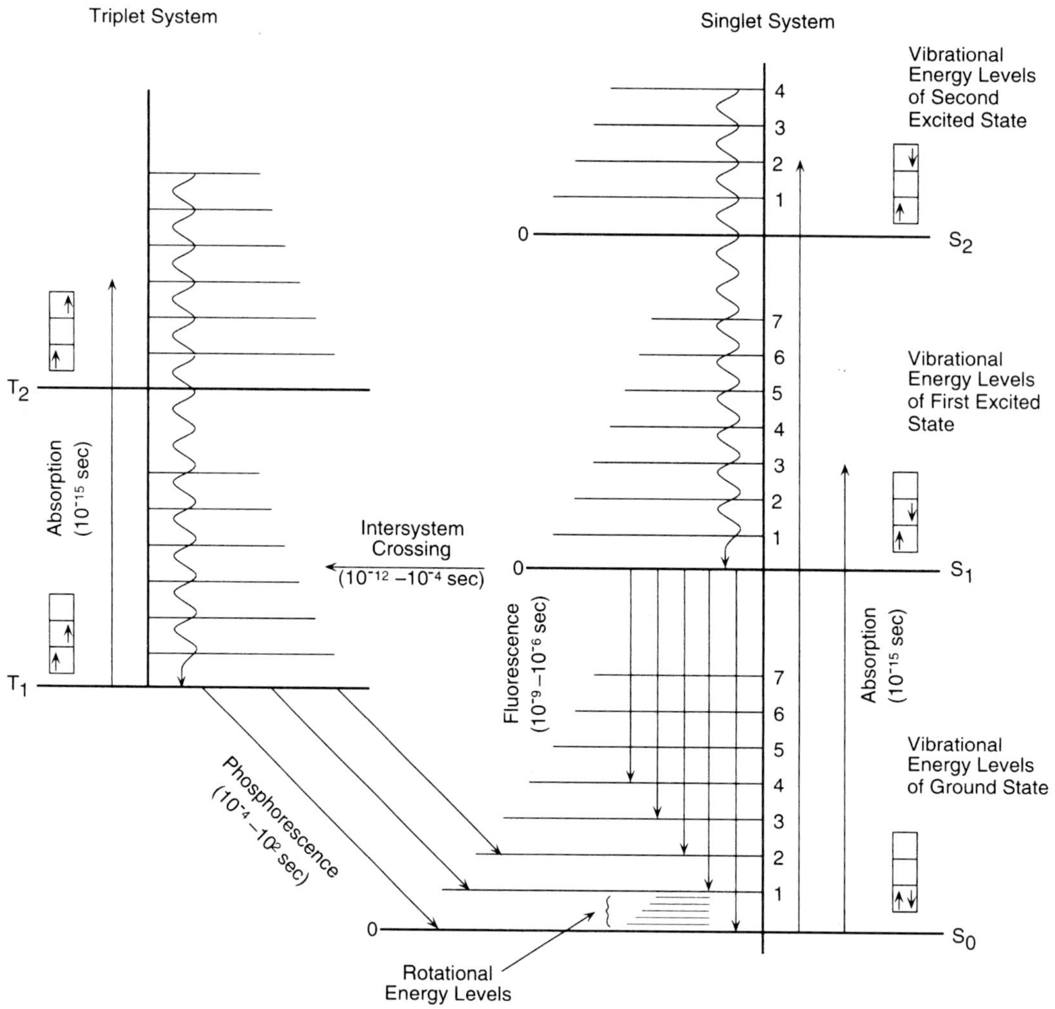

FIGURE 10.1 Energy diagram showing the transition states involved in the absorption and decay of electromagnetic energy. Energy may be released through heat or internal collisions, transferred to other molecules, or emitted as photons of light. Fluorescence occurs from within the singlet system as light energy is released, returning the electrons to the ground state. Phosphorescence occurs from the triplet system and involves a longer emission process at lower energies than that of fluorescence.

Aromatic ring constituents can have a pronounced affect on fluorescence. The presence of ring activators or electron donating groups (e.g., *ortho* and *para* directors) generally can increase QY, while the presence of electron-withdrawing groups generally reduces QY. An example of this effect can be found in the photoreactive hetero-bifunctional crosslinker SAED (Chapter 6, Section 3.9). Before photoactivation, SAED possesses a photoreactive phenyl azide group on its AMCA-derived end. This electron-withdrawing group quenches the fluorescence of the compound so that the AMCA dye does not behave as it characteristically does in the underivatized state. Upon photolysis, however, the phenyl azide either gets coupled to a target molecule or breaks down to an amine. In either case, the presence of an amine (or its derivative) eliminates the fluorescence-quenching effects of the azide and restores the coumarin group's fluorescent character.

Another potential ring constituent having a dramatic effect on fluorescence is the presence of heavy atoms. Aromatic rings possessing heavy atoms typically diminish QY by enhancing the probability of the excited singlet state going on to triplet transition. Energy decay from a triplet excited state causes phosphorescence instead of fluorescence. The phosphorescent band is located at longer wavelengths (and thus lower energies) relative to the fluorescent spectrum. The energy transition to the triplet state is therefore in direct contention with fluorescence, and has the effect of decreasing overall luminescence. Halogen substitution on aromatic-ring-containing fluorophores also can have a quenching effect, which generally follows the size of the atom: I > Br > Cl > F. Depending on the dye type and other constituents on the aromatic system, the presence of heavier halogen atoms on a fluorescent aromatic ring typically may cause a decrease

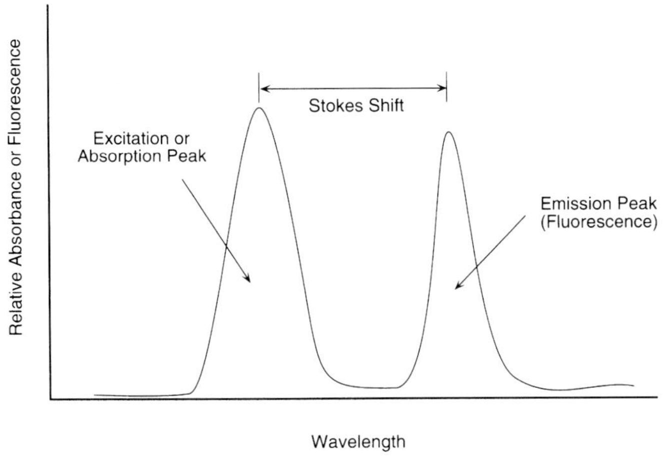

FIGURE 10.2 Typical spectral scan of a fluorescent compound showing its absorbance peak or wavelengths of most efficient excitation and its emission peak or wavelengths where light emission occurs. The Stokes shift is the distance in nanometers between the absorbance peak and the emission peak. The larger the Stokes shift, the less interference that will occur from excitation light when measuring fluorescence emission.

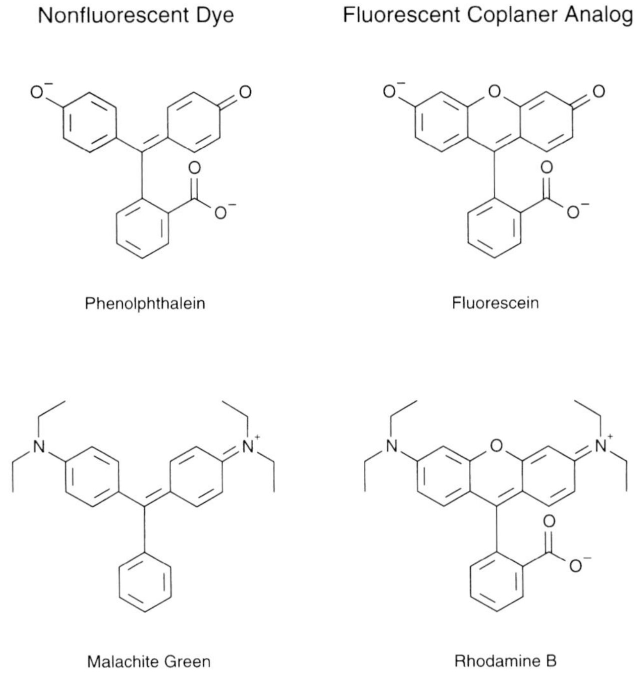

FIGURE 10.3 Fluorescent character in organic compounds is often determined by the presence of a planar aromatic ring system. The fluorescent compounds differ from the nonfluorescent ones only in the closure of their central ring system, which produces a conjugated system having the required constraints to create a planar triple-ring configuration.

in fluorescence or eliminate it altogether. However, the carbocyanine dye Alexa 680 has a bromine atom on its outer pyridinium ring and it still retains intense fluorescence character. However the presence of an iodine atom nearly guarantees complete elimination of fluorescence (Turro *et al.*, 2009). Thus, radioiodination of a fluorescent ring structure likely is not possible without dramatically decreasing (or eliminating) its QY.

The aromatic ring systems of most of the following fluorophores consist of polycyclic structures that may also contain a conjugated bridge between ring systems. To maintain fluorescence in such compounds, it is important that the entire system be coplanar or the rings be in the same dimensional plane. In fact, the differences between some non-fluorescent chromogenic dyes and their corresponding, structurally similar fluorophores are minor, but the planer nature of the fluorophores gives them their luminescent properties. For instance, Figure 10.3 illustrates the similarity of the dyes phenolphthalein and malachite green to the nearly identical fluorescent molecules fluorescein and rhodamine, respectively. The only difference among these dyes and fluorophores is the presence of the oxygen bridge between the upper phenyl rings which creates the rigid aromatic system that constricts the molecule to a planer shape, thus conferring luminescent qualities.

Fluorescent compounds are sensitive to changes in their chemical environment. Alterations in media pH, buffer components, solvent polarity, or dissolved oxygen can affect and quench the QY of a fluorescent probe (Bright, 1988). The presence of absorbing components in solution which absorb light at or near the excitation wavelength of the fluorophore will also have the effect of decreasing luminescence. In addition, noncovalent interactions of the probe with other components in solution can inhibit rotational freedom and quench fluorescence. In this regard, the binding of an anti-fluorescein antibody to the fluorescent dye completely abolishes its fluorescence. Also, environmental ozone levels have been shown to affect the fluorescence of some dye molecules, particularly cyanine-type dyes used in microarray gene measurements (Branham *et al.*, 2007) (Section 8).

Over-labeling a macromolecule with a fluorescent probe also causes decreases in QY due to dye–dye interactions. Quenching caused by energy transfer interactions between fluorescent molecules often occurs as the level of probe substitution reaches about 8 to 10 fluorophores per protein. Not only can the emission intensity decrease severely at high substitution levels, but the degree of nonspecific binding caused by the number of aromatic groups attached to the biomolecule can increase severely. Fluorophore self-quenching at high substitution or concentration levels can be due to energy transfer from excited-state molecules to ground-state dimers (Chen and Knutson, 1988). Hydrophobic dyes are particularly prone to such aggregation and quenching effects in aqueous buffers.

One method of reducing dye–dye interactions is to add negative charge character to the fluorescent molecule. This is typically performed by the addition of sulfonate groups on aromatic rings or on alkyl tethers built off the central dye structure. The high negative charge character of the sulfonate groups creates like-charge repulsion between dye molecules, thus dramatically reducing their tendency to interact. The presence of three to four sulfonate groups on a probe usually creates a very hydrophilic molecule with lower nonspecific binding character than in biological solutions and greater QY than the corresponding dye without sulfonates.

Another approach to creating hydrophilic dyes is to add discrete PEG tethers to selected regions of the dye structure. The excellent water-solubilizing properties of PEG modifications can mask the core hydrophobicity of polycyclic aromatic fluorescent molecules, thus limiting dye–dye interactions and dramatically decreasing nonspecific binding toward biomolecules in a sample. The combination of PEG and sulfonate modifications on certain dye structures can create an optimum dye with maximal QY and excellent hydrophilicity. Many modern carbocyanine and xanthene type dyes incorporate water-solubilizing groups to avoid insolubility of dye conjugates and reduce background signals in fluorescent staining procedures. See Chapter 18 for a discussion of discrete PEG compounds that can be used for this purpose.

The following sections describe some of the most popular fluorescent probes for use in labeling biomolecules. These fluorophores are available from a number of manufacturers in several different forms, with a variety of reactive groups able to couple to specific functional groups on target molecules or with different substituents added to increase QY or hydrophilicity. By far, the most often used fluorescent core structures are water-soluble derivatives of cyanine, fluorescein, rhodamine, and coumarin—probably in that order. However, new fluorescent molecules are being developed and reported continually. For instance, there is growing interest in the use of lanthanide chelates for time-resolved fluorescence (Section 9) or quantum dot nanocrystals (Section 10) for highly stable fluorescence in multiplex detection schemes. Careful study of each dye's individual properties can lead to the successful labeling of a biomolecule for any anticipated application.

Suggested references on fluorescent spectroscopy and the use of fluorescent probes include Pansare *et al.* (2012), Lakowicz (1991, 1999, 2006), Bright (1988), Dewey (1991), McGown and Warner (1990), Ploem and Tanke (1987), Darzynkiewicz and Crissman (1990), Haugland (1991), and Waggoner (1990).

1. FLUORESCEIN DERIVATIVES

Fluorescein and its derivatives represent one of the most popular of all fluorescent labeling agents. Its fluorescent character is created by the presence of a multi-ring aromatic xanthene core structure, which creates the rigid planar nature of its upper, fused three-ring system (Figure 10.4). In its most elementary form, the molecule has a molecular weight of about 332 Da, but most modification reagents based on fluorescein are derivatives of this basic structure prepared through substitutions off the No. 5 or 6 carbons of its lower ring. The derivatives provide reactivity toward particular functional groups in biomolecules, allowing rapid labeling of proteins and nucleic acids.

Fluorescein has an effective excitation wavelength range of about 488 to 495 nm, closely matching the photon emission for an argon laser (488 nm). Some fluorescein derivatives are sold using the 488 reference in their name to reflect this excitation characteristic (e.g., DyLight 488 from Thermo Fisher and Alexa 488 from Invitrogen). The dye's emission spectrum occurs between 518 and 525 nm, depending on the derivative chosen. Under ideal conditions, its QY can be as high as 0.75; however, its fluorescent intensity fades when it is dissolved in aqueous buffers, exposed to light, or stored for extended periods (Kawamura, Jr., 1977). In environments below pH 7.0, fluorescein's QY is significantly quenched. In addition, when derivatives of fluorescein are conjugated to proteins, the degree of fluorescent quenching can be as high as 50%, especially for relatively hydrophobic forms of the dye. Fluorescein derivatives having multiple sulfonates on them helps to alleviate the quenching problem when labeling at higher density but does not completely eliminate it. Even so, the fluorophore usually maintains excellent detectability in assay systems.

The following sections describe the most important fluorescein derivatives having reactive groups commonly used to label biomolecules.

1.1. Amine-Reactive Fluorescein Derivatives

Two general forms of amine-reactive fluorescein probes are available. Both of them react under alkaline

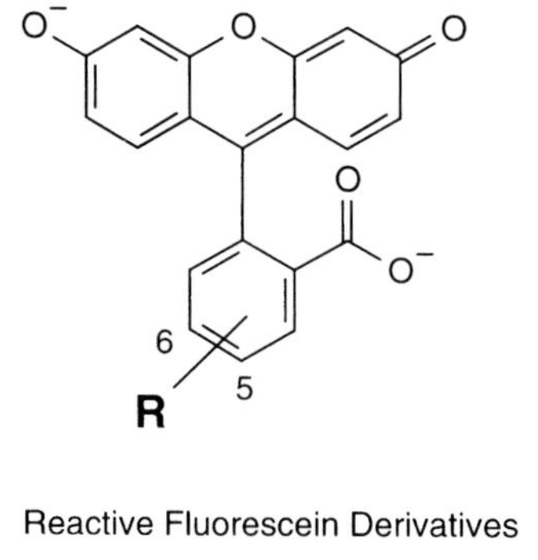

FIGURE 10.4 Fluorescein derivatives often are produced through modification at the C-5 or C-6 positions on the lower aromatic ring.

conditions with primary amines in proteins and other molecules to form stable, highly fluorescent derivatives.

1.2. Fluorescein Isothiocyanate (FITC)

FITC is one of the most popular fluorescent probes ever created. An isothiocyanate derivative of fluorescein is synthesized by modification of its lower ring at the 5- or 6-carbon positions. The two resulting isomers are nearly identical in their reactivity and spectral properties, including excitation and emission wavelengths and intensities. Their chemical differences, however, may affect the separation of modified proteins from excess reagent or the analysis of tagged molecules by electrophoresis. For this reason, most manufacturers purify the carbon-5 derivative as the FITC reagent of choice.

Isothiocyanates react with nucleophiles such as amines, sulfhydryls, and the phenolate ion of tyrosine side chains (Podhradsky *et al.*, 1979). The only stable product, however, is with primary amine groups, and so FITC is almost entirely selective for modifying ε- and N-terminal amines in proteins (Jobbagy and Kiraly, 1966). The reaction involves attack of the nucleophile on the central, electrophilic carbon of the isothiocyanate group (Figure 10.5). The resulting electron shift creates a thiourea linkage between FITC and the protein with no leaving group.

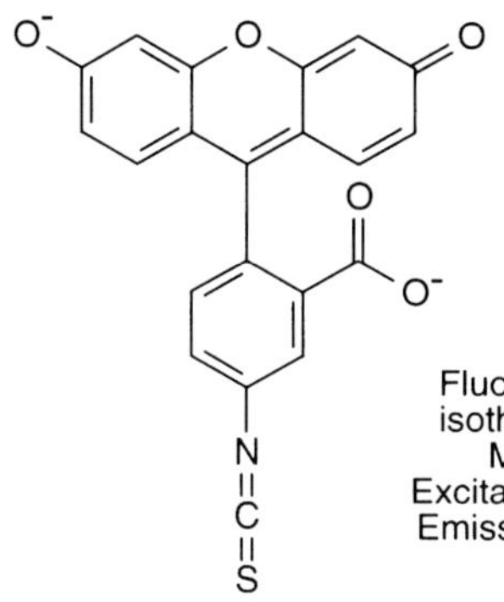

FITC;
Fluorescein-5-isothiocyanate
MW 389
Excitation 494 nm
Emission 520 nm

FITC can be dissolved in DMF as a concentrated stock solution prior to its addition to an aqueous reaction mixture. This may make aliquoting small quantities of the compound easier. The reagent is water soluble above pH 6. The isothiocyanate group is reasonably stable in aqueous solution for short periods, but will degrade. FITC also can break down and lose activity upon storage. It is best, therefore, to use fresh reagent for modification purposes. Storage should be under desiccated conditions, protected from light, and at −20°C.

The fluorescent properties of FITC include an absorbance maximum at about 495 nm and an emission wavelength of 520 nm. Fluorescent quenching of the molecule is possible. Under concentrated conditions, fluorescein-to-fluorescein interactions result in energy transfer and self-quenching, which reduce the luminescence yield. This phenomenon can occur with fluorescein-tagged molecules as well. If derivatization of a protein is carried out at too high a level, the resultant quantum yield of the conjugate will be depressed. Typically, modifications of proteins involve adding no more than eight to ten fluorescein molecules per protein molecule, with a four to five substitution level considered optimal.

FITC has been used in numerous applications involving fluorescence detection. Antibodies or their fragments can be labeled to detect antigens in cells, tissue sections, blots, or on surfaces (Clausen, 1988). Tagging molecules with FITC is also useful in detecting proteins after electrophoretic separations (Strottmann *et al.*, 1983), for microsequencing analysis of proteins and peptides (Muramoto *et al.*, 1984), in analysis of molecules using capillary zone electrophoresis (Cheng and Dovichi, 1988), and in tracking and detecting molecules involved in various bio-interactions (Friedman and Ball, 1989; Burtnick, 1984).

The level of fluorescein modification in a macromolecule can be determined by measuring its absorbance at or near its characteristic excitation maximum (~498 nm). The number of fluorochrome molecules per molecule of protein is known as the F/P ratio. This value should be measured for all derivatives prepared with fluorescent tags. The ratio is important in predicting the behavior

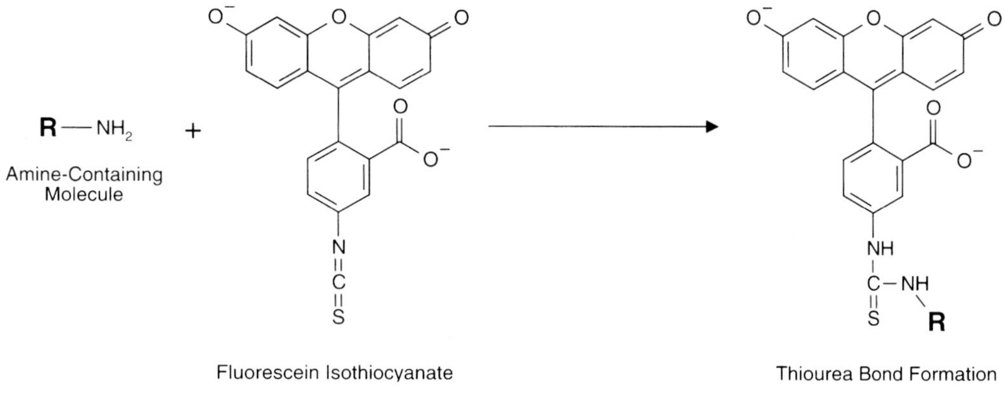

FIGURE 10.5 FITC reacts with amine-containing compounds to produce an isothiourea linkage.

of antibodies labeled with FITC (Hebert *et al.*, 1967; Beutner, 1971). Using the known extinction coefficient of FITC in solution at pH 13 ($\varepsilon_{498\,nm}$ = 8.1–8.5 × 10^4) (McKinney *et al.*, 1964; Jobbagy and Jobbagy, 1972) a determination of derivatization level can be made after excess FITC is removed. At pH 7.8, the absorbance of protein-coupled FITC decreases by 8% (Dalen and Haaijman, 1974).

Despite the fact that FITC is one of the oldest reactive dyes used for labeling biomolecules, it still is being used extensively for detection purposes (Lui *et al.*, 2011; Segal *et al.*, 2011; Pushpanathan *et al.*, 2012).

A general protocol for the modification of proteins, particularly immunoglobulins, with FITC is given below. Slight modifications to the amount of reagent added to the reaction may be made to optimize the F/P ratio.

Protocol

1. Prepare a protein solution in 0.1-*M* sodium carbonate, pH 9.0, at a concentration of at least 2 mg/ml.
2. In a darkened lab, dissolve FITC (Thermo Fisher) in dry DMSO at a concentration of 1 mg/ml. Do not use old FITC, as breakdown of the isothiocyanate group over time may decrease coupling efficiency. Protect from light by wrapping in aluminum foil or using amber vials.
3. In a darkened lab, slowly add 50 to 100 µl of FITC solution to each milliliter of protein solution (at 2-mg/ml concentration). Gently mix the protein solution as the FITC is added.
4. React for at least 8 h at 4°C in the dark.
5. The reaction may be quenched by the addition of ammonium chloride to a final concentration of 50-m*M*. Some protocols also include at this point the addition of 0.1% xylene cylanol and 5% glycerol as a photon absorber and fluorescence stabilizer, respectively (Petrou *et al.*, 2002). React for a further 2 h to stop the reaction by blocking remaining isothiocyanate groups.
6. Purify the derivative by gel filtration using a PBS buffer or another suitable buffer for the particular protein being modified. The use of a desalting resin with low exclusion limits works well. To obtain complete separation, the column size should be 15 to 20 times the size of the applied sample. Fluorescent molecules often nonspecifically stick to the gel filtration support, so reuse of the column is not recommended.

1.3. NHS–Fluorescein and NHS–LC-Fluorescein

NHS–fluorescein is another amine-reactive fluorescent probe that contains a carboxy-succinimidyl ester group off the No. 5 or 6 carbons on fluorescein's lower ring structure (Khanna and Ullman, 1980; Vigers *et al.*, 1988; Brinkley, 1992). The 5- and 6-isomers are virtually identical in their reactivity and fluorescent characteristics. Similar to FITC (above), NHS–fluorescein can be used to label proteins and other macromolecules that contain primary amine groups. This reagent is more stable than FITC, especially in storage. The NHS ester reaction proceeds rapidly at slightly alkaline pH values, resulting in a stable, amide-linked derivative (Chapter 3, Section 1.4; Figure 10.6).

NHS-Fluorescein;
5-Carboxyfluorescein,
succinimidyl ester
MW 457
Excitation 491 nm
Emission 518 nm

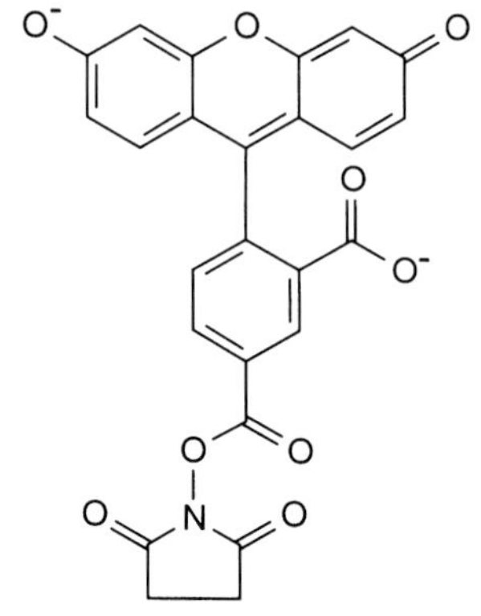

NHS-LC-Fluorescein;
6-(Fluorescein-5-carboxamido)
hexanoic acid, succinimidyl ester
MW 587
Excitation 494 nm
Emission 519 nm

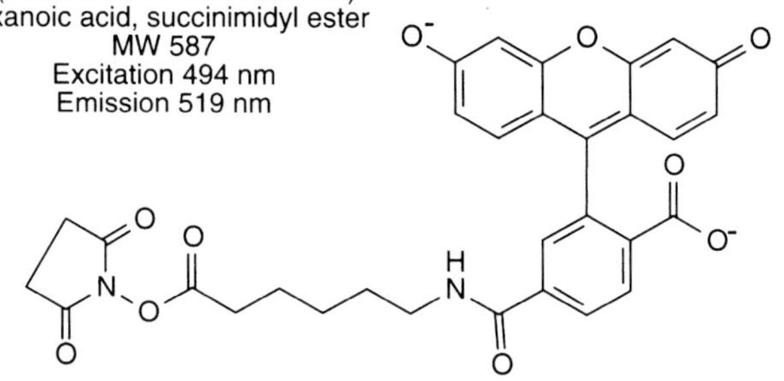

The fluorescent properties of NHS–fluorescein are similar to FITC. The wavelength of maximal absorbance or excitation for the reagent is 491 nm and its emission maximum is 518 nm, exhibiting a visual color of green (Sheehan and Hrapchak, 1980). Its molar extinction coefficient at 491 nm in a pH 8.0 buffer environment is 66,000 $M^{-1}\,cm^{-1}$. Other components in solution as well as the pH can change this value.

NHS–fluorescein is insoluble directly in aqueous solution and should be dissolved in organic solvent prior to addition of a small aliquot to a buffered reaction medium. Concentrated stock solutions may be prepared in DMSO or DMF. Such solutions are relatively stable if protected from light. Reaction conditions should be maintained at the optimal reactivity for NHS esters—pH 7 to 9.

NHS–LC-fluorescein (Invitrogen) is an analog of NHS–fluorescein that contains a 6-aminocaproic acid spacer group, extending the NHS ester group away from the fluorescein portion. The longer length of the coupling arm may decrease steric hindrance around the fluorescent head of the molecule, thus reducing any

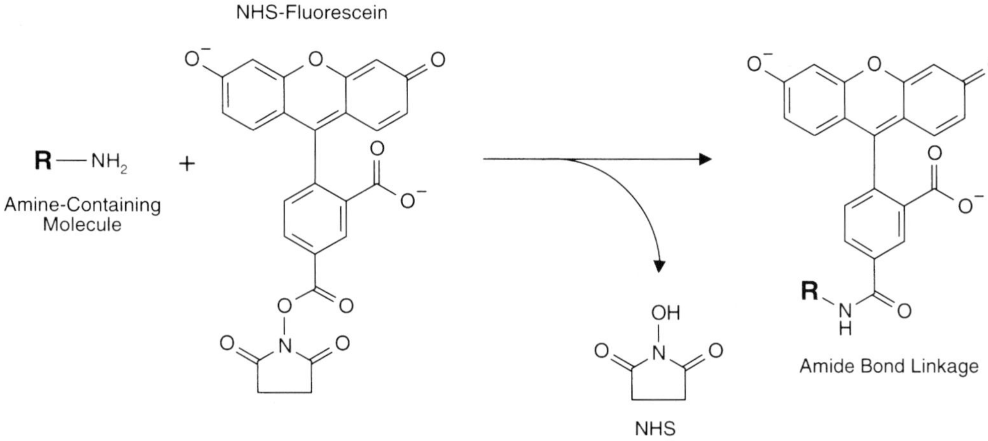

FIGURE 10.6 NHS–fluorescein reacts with amine-containing compounds via its NHS ester group to form amide bonds.

fluorescence quenching due to its attachment to a macromolecule. All other properties of the long-chain version are virtually identical to that described above for NHS–fluorescein.

Sulfonated versions of NHS–fluorescein dyes also are available that contain negative charges to make the molecule more water soluble and avoid nonspecific interactions with biomolecules (Thermo Fisher and Invitrogen). These 488-type dyes are a better choice than the original fluorescein dyes, because of their advantages when working with proteins and other biological samples, including greater brightness, lower nonspecific binding, and less tendency to quench. The NHS ester reactions of the sulfonated dyes are identical to those of the non-sulfonated types, but the sulfonated ones can be labeled on proteins at higher densities without fluorescence quenching or protein precipitation.

NHS–fluorescein has been used to label proteins for fluorescence detection (Domashevskiy *et al.*, 2012), to modify dendrimers (Biswas *et al.*, 2012), for labeling peptides to monitor their interaction with surfaces (Jeong *et al.*, 2011), and to develop assays based on fluorescence polarization (Novitsky and Sloyer, 2012).

The following procedure is a suggested method for using NHS-fluorescein to label immunoglobulins.

Protocol

1. Dissolve the IgG to be labeled in 50-m*M* sodium bicarbonate buffer, pH 8.5, at a concentration of 10 mg/ml.
2. Dissolve 0.5 mg of NHS-fluorescein (Thermo Fisher) in 0.5 ml DMSO. Protect from light.
3. In a darkened lab, slowly add 50 to 100 μl of the NHS–fluorescein solution to the antibody solution, while mixing. Protect from light by wrapping the reaction vessel in aluminum foil.
4. React for 2 h on ice.
5. Remove unreacted NHS–fluorescein and reaction by-products by gel filtration or dialysis. Continue to protect all labeled protein solutions from light.

A spectrophotometric assessment of the F/P ratio should be carried out after purification of the tagged antibody. The measurement of absorbance at 495 nm (for fluorescein) divided by the absorbance at 280 nm should be between 0.3 and 1.0 to obtain a good fluorescent derivative of acceptable activity and low background. This usually translates into a ratio of about four to seven fluorescein molecules per protein molecule.

1.4. Sulfhydryl-Reactive Fluorescein Derivatives

Fluorescein derivatives containing a sulfhydryl-reactive group off the lower ring structure are available to direct the modification reaction to more limited sites on target molecules. Coupling through sulfhydryls instead of amines can help to avoid active centers in proteins, thus preserving activity in the fluorescent probe complex. Sulfhydryl reaction sites can be naturally available through free cysteine side chains, generated by reduction of disulfides, or created by the use of thiolation reagents (Chapter 2, Section 4.1).

The first two compounds discussed in this section are truly sulfhydryl reactive, using the common iodoacetyl and maleimide functionalities, respectively. The third derivative, however, is not reactive directly with sulfhydryl groups but contains a protected sulfhydryl which, after deprotection, can be used to react with other sulfhydryl-reactive crosslinkers.

1.5. 5-(and 6)-Iodoacetamido-Fluorescein

The iodoacetamido derivatives of fluorescein possess a sulfhydryl-reactive iodoacetyl group (Chapter 2,

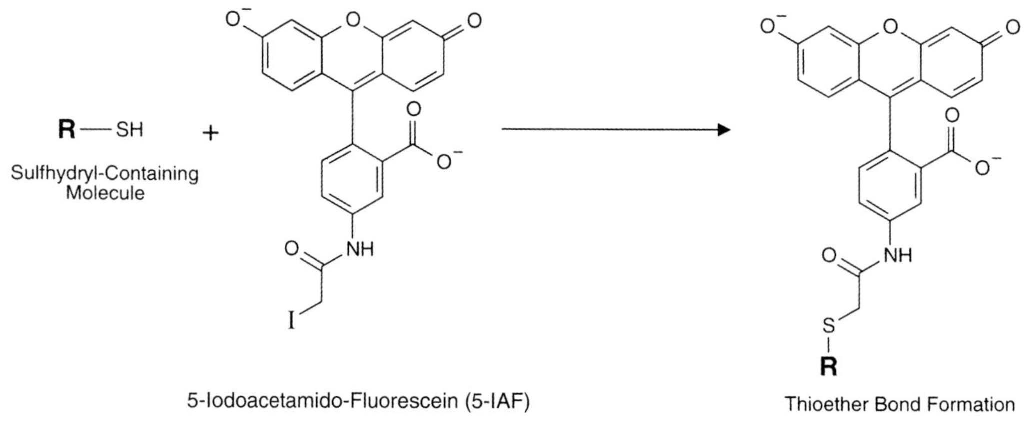

FIGURE 10.7 5-IAF can be used to modify sulfhydryl-containing molecules, creating stable thioether linkages.

Section 4.2, and Chapter 3, Section 2.1) at either the 5- or 6-carbon position on their lower ring. The isomers are commercially available in purified form, since some reactivity and specificity differences between the 5- and 6-derivatives toward various sulfhydryl sites in proteins may be observed. Both iodoacetamido derivatives are among the most intense fluorophores available for labeling biomolecules due to high QY.

The iodoacetyl group of both isomers reacts with sulfhydryls under slightly alkaline conditions to yield stable thioether linkages (Figure 10.7). They do not react with unreduced disulfides in cystine residues or with oxidized glutathione (Gorman *et al.*, 1987). The thioether bonds will be hydrolyzed under conditions necessary for complete protein hydrolysis prior to amino acid analysis.

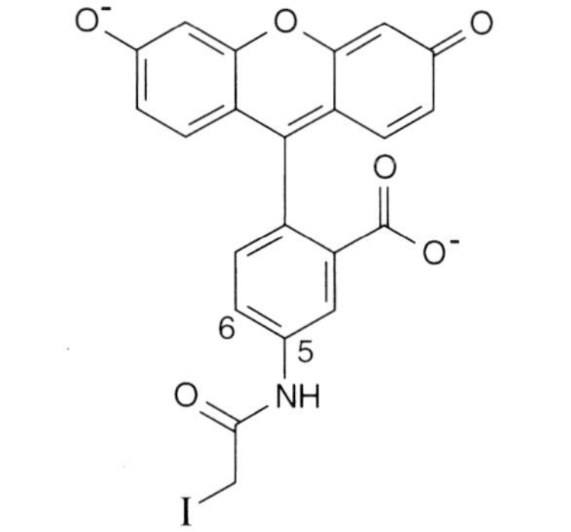

5- (and 6)-Iodoacetamido-Fluorescein
MW 515.26
Excitation: 490–495 nm
Emission: 515–520 nm
ε at 492 nm = 80,000–85,000 $M^{-1}cm^{-1}$

Care must be taken to protect these reagents from light, not only to maintain the fluorescent yield of the fluorescein probe, but also to protect the iodoacetyl group from light-catalyzed breakdown. Iodoacetamidofluorescein is soluble in DMF and also in aqueous solutions maintained above pH 6.0. Concentrated stock solutions may be prepared in DMF prior to the addition of a small aliquot to a reaction mixture. Protect solutions from light by wrapping in aluminum foil and working in subdued light.

The spectral properties of these derivatives are similar to native fluorescein. The excitation maximum occurs at about 490 to 495 nm and its emission peak at 515 to 520 nm, producing light in the green region of the spectrum. The extinction coefficient of 5-iodoacetamido-fluorescein at its wavelength of maximum absorbance, 491 nm, is 82,000 $M^{-1}cm^{-1}$ (pH 9), whereas the extinction coefficient of 6-iodoacetamido-fluorescein is 77,000 $M^{-1}cm^{-1}$ at 493 nm (pH 9).

The 5-iodoacetamido derivative of fluorescein (5-IAF) has been used to label numerous proteins and other biomolecules, including actin (Plank and Ware, 1987), myosin (Aguirre *et al.*, 1986), troponin (Greene, 1986), hemoglobin (Hirsch *et al.*, 1986), and sulfhydryl-containing proteins separated by SDS electrophoresis (Gorman, 1984).

The 6-iodoacetamido derivative (6-IAF) has been used to label myosin (Ando, 1984), actin (Konno and Morales, 1985), microtubule-associated proteins (Scherson *et al.*, 1984), and histones (Cocco *et al.*, 1986).

Iodoacetamido-fluorescein continues to be used to tag thiol-containing molecules for fluorescence detection. Abate-Pella *et al.* (2012) used the dye to investigate Ras-mediated signal transduction, Maki (2011) used it to label human p53 to study its import and export from the nucleus, and the dye was also used to analyze alpha-1-acid glycoprotein isoforms (Garrido-Medina *et al.*, 2012).

The following protocol for labeling proteins with 5-IAF is adapted from Gorman (1987). It is a bit unusual

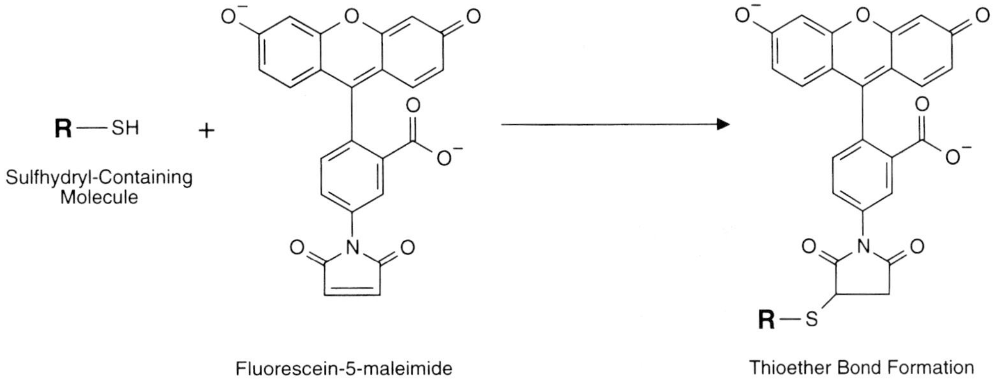

FIGURE 10.8 Fluorescein-5-maleimide can be used to modify sulfhydryl groups in proteins and other molecules, forming thioether bonds.

in that it involves reduction of disulfides with DTT and immediate reaction with 5-IAF in excess without removal of excess reductant. The procedure can be changed to include a gel filtration step after disulfide reduction to remove excess DTT, but in any case, it should be optimized for each protein to be modified. An alternative to the use of DTT to produce sulfhydryls is thiolation with a compound that can generate free thiols upon reaction with a protein (Chapter 2, Section 4.1).

Protocol

1. Dissolve a protein containing disulfide residues at a concentration of 5 to 10mg/ml in 0.1-*M* ammonium carbonate containing 1% SDS and 20-m*M* DTT. Note: The presence of detergent may be eliminated if certain disulfides can be reduced in the protein without completely denaturing it, such as in the reduction of antibodies in the hinge region. If only partial reduction is done, then the amount of DTT should be reduced to about a 3-fold molar excess over the concentration of antibody present (Sun *et al.*, 2005). For this type of labeling the reaction buffer should be 50-m*M* sodium borate, pH 8.5.
2. For complete reduction of all disulfides in the presence of a denaturant, react for 16h at 0°C and 2h at room temperature. For partial reduction of disulfides, the reaction time may be reduced to 2h at 37°C, particularly for antibody thiol reduction, if only partial reduction of thiols in the hinge region is done.
3. Add a 5-fold molar excess of 5-IAF (Thermo Fisher) over the amount of DTT present. The fluorescent probe may be solubilized in DMF prior to addition of a small aliquot to the reaction mixture. Do not exceed 10% DMF in the final aqueous solution.
4. React for 2h at room temperature in the dark.
5. To recover a protein that has been completely reduced and labeled, precipitate the protein by the addition of 9 volumes of methanol at −20°C. Collect the protein pellet by centrifugation at 8000*g* for 15min (4°C). If partial reduction was done followed by labeling with 5-IAF, then purify the labeled protein by gel filtration or dialysis using a molecular weight cutoff of about 5000 Daltons.

An alternative protocol for labeling sulfhydryl-containing proteins that does not require DTT reduction can be found in a method adapted from Ando (1984). When preparing any fluorescently labeled protein, optimization of the dye-to-protein ratio is important to obtain the best performance in the intended application.

Protocol

1. Prepare a 20-m*M* 6-IAF (Thermo Fisher) solution by dissolving 10.3mg per milliliter of DMF. Prepare freshly and protect from light.
2. Dissolve the protein to be modified at a concentration of 5 to 10mg/ml in 20-m*M* TES, pH 7.0. TES is 2-[tris(hydroxymethyl)methylamino]-1-ethanesulfonic acid.
3. Slowly add 25 to 50μl of the 6-IAF solution to each milliliter of the protein solution while mixing.
4. React for 2h at 4°C in the dark.
5. Remove excess reactant and reaction byproducts by gel filtration using a desalting resin or dialysis.

1.6. Fluorescein-5-Maleimide

Fluorescein-5-maleimide is a fluorescent probe containing a sulfhydryl-reactive maleimide group on it lower ring structure. The modification of sulfhydryl-containing molecules under physiological pH conditions will result in stable thioether bonds (Chapter 3, Section 2.2) (Figure 10.8). The derivative thus possesses fluorescent properties closely characteristic of fluorescein molecules: excitation wavelength = 490 nm; emission wavelength = 515 nm, in the green spectral region. Conjugates prepared by fluorescein-5-maleimide are among the most intensely fluorescent probes available. The reactivity of the maleimide group is similar to that of the iodoacetyl derivative discussed previously.

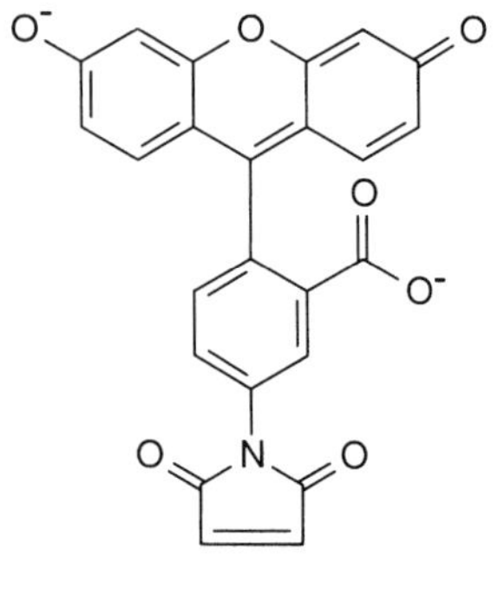

Fluorescein-5-maleimide
MW 427
Excitation: 490–495 nm
Emission: 515–520 nm
ε at 490 nm = 83,000 $M^{-1}cm^{-1}$

Thus, this reagent can be used to fluorescently label proteins and other biomolecules containing free sulfhydryl residues. If there are no SH groups available, their creation can be accomplished by reduction of indigenous disulfides or through the use of various thiolation reagents (Chapter 2, Section 4.1).

Fluorescein-5-maleimide is slightly soluble in aqueous solutions above pH 6 (~1-m*M* concentration). It may be dissolved in DMF at higher concentrations and a small addition of this solution made to an aqueous reaction mixture to initiate labeling. Do not exceed 10% DMF in the reaction buffer to avoid protein precipitation. At pH 8, the reagent has an extinction coefficient at 490 nm of about 78,000 $M^{-1}cm^{-1}$.

Fluorescein-5-maleimide has been used in numerous applications, including labeling the transmembrane glycoprotein H-2K^k on both the N- and C-terminal regions to investigate the structure of the molecule (Cardoza *et al.*, 1984), for the determination of two different conformations of the protein actin (Konno and Morales, 1985), in the study of a bacterial sensory receptors (Falke *et al.*, 1988), in the structural mapping of chloroplast coupling factor (Snyder and Hammes, 1984, 1985), for localization of the stilbenedisulfonate receptor on human erythrocytes (Rao *et al.*, 1979), to investigate the calcium-dependent ATPase protein structure of sarcoplasmic reticulum (Bigelow and Inesi, 1991), and to study the movement of tRNA during peptide bond formation on ribosomes (Odom *et al.*, 1990).

Fluorescein-5-maleimide has also been used to study the assembly dynamics of mycobacterium tuberculosis (Chen *et al.*, 2007), to study monomers and dimers of NhaA Na^+/H^+ antiporter of *E. coli* (Rimon *et al.*, 2007), to investigate the regulation of the protein disulfide proteome by mitochondria (Yang *et al.*, 2007), and to study the folded state of fibronectin type III domains (Lemmon *et al.*, 2011).

Protocol

1. Dissolve a sulfhydryl-containing protein or other macromolecule at a concentration of 1 to 10 mg/ml in 20-m*M* sodium phosphate, 0.15-*M* NaCl, pH 7.2. Other buffers within the range of pH 6.5 to 7.5 may be used as long as they do not contain extraneous sulfhydryls.
2. Dissolve fluorescein-5-maleimide (Thermo Fisher, Invitrogen) in DMF at a concentration of 10-m*M* (4.25 mg/ml). Protect from light.
3. In subdued lighting conditions, add 25 to 50 μl of the fluorescein solution to each milliliter of protein solution while mixing. Alternatively, determine the exact molar quantity of protein present and add a 25-fold molar excess of fluorescein-5-maleimide solution.
4. React for 2 to 4 h at room temperature in the dark. The reaction also may be performed at 0 to 4°C, but allow at least 8 h for completion.
5. Immediately purify the derivative using gel filtration on a desalting resin. Protect the solutions from light during the chromatography.

1.7. SAMSA–Fluorescein

SAMSA–fluorescein, 5-((2-(and 3)-*S*-(acetylmercapto) succinoyl)amino)fluorescein, is a fluorescent probe containing a protected sulfhydryl group. In its protected state, the compound is unreactive. The acetyl protecting group can be removed by treatment with dilute NaOH at pH 10.0 (Figure 10.9). The resulting free sulfhydryl derivative can be used to label thiol-reactive crosslinkers or to couple with sulfhydryl residues on proteins and other molecules. After activating proteins with crosslinkers containing terminal maleimide, pyridyl disulfide, or iodoacetyl groups, SAMSA–fluorescein can be used to assess the level of modification. For instance, a maleimide-activated protein that has been derivatized with succinimidyl-4-(*N*-maleimidomethyl) cyclohexane-1-carboxylate (SMCC) could be reacted with this reagent to yield a fluorescein derivative that can be assayed spectrofluorometrically for its level of fluorescence. Using the molar extinction coefficient for SAMSA–fluorescein (ε at 495 nm = 80,000 $M^{-1}cm^{-1}$), the molar level of incorporation of the label can be calculated. This determination directly correlates to the original level of maleimide groups present on the protein.

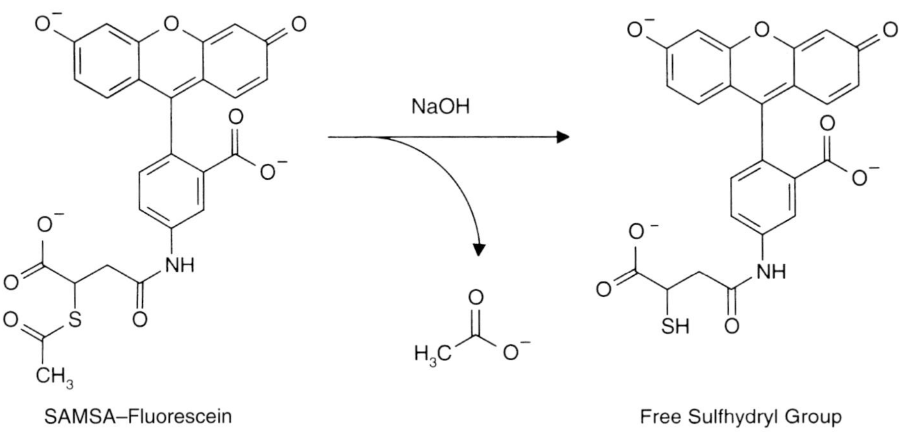

FIGURE 10.9 SAMSA–fluorescein contains a protect thiol that can be deblocked by treatment with hydroxylamine. The reagent then can be used to modify molecules containing sulfhydryl-reactive groups.

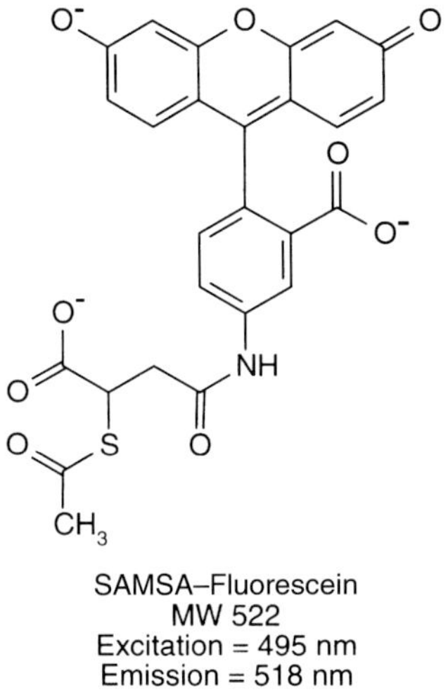

The dye has been used to investigate the uptake mechanisms of a mannan nanogel in macrophages (Ferreira *et al.*, 2012), to optimize the signal-to-noise ratios in making signal-enhanced Raman spectroscopy (SERS) tags (Fabris, 2012), and to tag silver nanoparticles for the development of a therapeutic to treat multi-drug resistant cancer (Liu *et al.*, 2012).

SAMSA–fluorescein is an orange solid compound. Dissolved in buffer at pH 9.0, its maximal wavelength of absorption or excitation is at 495 nm, while its emission wavelength maximum is 520 nm. The reagent and all solutions and derivatives made from it are light sensitive and should be stored in the dark. SAMSA–fluorescein is soluble in aqueous solutions above pH 6.0, but it can be dissolved in DMF to prepare a concentrated stock solution prior to adding a small amount to a buffered reaction mixture.

Protocol

1. To deprotect the acetylated sulfhydryl, dissolve the desired amount of SAMSA–fluorescein (Invitrogen) in 100-m*M* NaOH, pH 10.0, at a concentration of 10 mg/ml.
2. React for 15 min at room temperature.
3. Lower the pH to 7 to 8 by the addition of solid sodium phosphate.
4. Add the required amount of deprotected SH-fluorescein to a protein or other macromolecule that had been modified to contain a sulfhydryl-reactive group. Use a 5- to 10-fold molar excess of SH-fluorescein to the expected amount of sulfhydryl reactivity present.
5. React for 2 h at room temperature, protected from light.
6. Remove excess fluorescent probe by gel filtration using a desalting resin.
7. Measure the absorbance of the derivative at 495 nm. Determine the level of fluorophore incorporation by using its molar extinction coefficient.

1.8. Aldehyde-/Ketone- and Cytosine-Reactive Fluorescein Derivatives

Hydrazide groups directly react with aldehyde and ketone groups to form relatively stable hydrazone linkages (Chapter 3, Section 5.1). Two fluorescein derivatives are commonly available that contain hydrazide groups off their No. 5 carbons on the lower ring structure. Both may be used to fluorescently label aldehyde- or ketone-containing molecules. Although most biomolecules do not contain aldehyde or ketone groups in their native state, carbohydrates, glycoproteins, RNA, and other molecules containing sugar residues can be oxidized with sodium periodate to produce reactive formyl groups. The use of modification reagents that generate aldehydes upon coupling to a molecule also can be used to produce a hydrazide-reactive site (Chapter 2, Section 4.4).

DNA and RNA may be modified with hydrazide-reactive probes by reacting their cytosine residues with bisulfite to form reactive sulfone intermediates. These derivatives undergo transamination to couple hydrazide- or amine-containing probes (Draper and Gold, 1980) (Chapter 23, Section 2.3).

1.9. Fluorescein-5-thiosemicarbazide

Fluorescein-5-thiosemicarbazide (Invitrogen) is a hydrazide derivative of fluorescein that can spontaneously react with aldehyde- or ketone-containing molecules to form a covalent, hydrazone linkage (Figure 10.10). It also can be used to label cytosine residues in DNA or RNA by use of the bisulfite activation procedure (Chapter 23, Section 2.1). The resulting fluorescent derivative exhibits an excitation maximum at a wavelength of 492 nm and a maximal emission wavelength of 519 nm when dissolved in buffer at pH 8.6. In the same buffered environment, the compound has an extinction coefficient of approximately 78,000 $M^{-1}cm^{-1}$ at 492 nm.

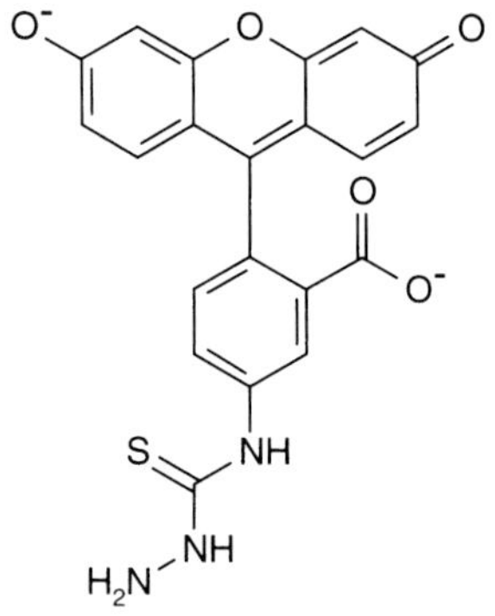

Fluorescein-5-thiosemicarbazide
MW 421
Excitation: 492 nm
Emission: 516 nm
ε at 492 nm = 85,000 $M^{-1}cm^{-1}$

Fluorescein-5-thiosemicarbazide is soluble in DMF or in buffered aqueous solutions at pH values above 7.0. The reagent may be dissolved in DMF as a concentrated stock solution before adding a small aliquot to an aqueous reaction medium. The compound itself and all solutions made with it should be protected from light to avoid decomposition of its fluorescent properties.

This hydrazide derivative of fluorescein has been used in a number of applications, including site-directed labeling of antibodies through their carbohydrate chains (Duijndam *et al.*, 1988); labeling thrombin and anti-thrombin (Atha *et al.*, 1964), Na^+/K-ATPase glycoprotein (Lee and Fortes, 1985), and periodate-oxidized RNA (Odom *et al.*, 1980, 1984; Ferguson and Yang, 1986; Friedrich *et al.*, 1988); determining carbonyl groups in proteins; and detecting oxidized glycoproteins in gels (Ahn *et al.*, 1987). It also has been used to investigate the molecular basis of RNA recognition (Pagano *et al.*, 2007), in a non-invasive visualization method for assessment of carbonylated protein (Fujita *et al.*, 2007), for the detection of protein carbonyls in aging liver tissue (Chaudhuri *et al.*, 2006), and to label polysaccharides to image as they are transported into cells (Zhang *et al.*, 2011).

The following protocols are generalized for the labeling of cell surface glycoproteins or glycoproteins in solution. Some optimization may be necessary to achieve the best level of fluorescent modification for each particular application.

Protocol for Labeling Cell Surfaces

1. Add 10^6 to 10^8 cells per ml in a PBS solution (10-m*M* sodium phosphate, 0.15-*M* NaCl, pH 7.4) containing 1-m*M* sodium periodate and incubate on ice for 30 min in the dark. This level of periodate addition will target the oxidation only to sialic acid residues (Chapter 2, Section 4.4). If additional sites of glycosylation also are to be labeled, increase the periodate concentration to 10-m*M* and perform the reaction at room temperature in the dark.

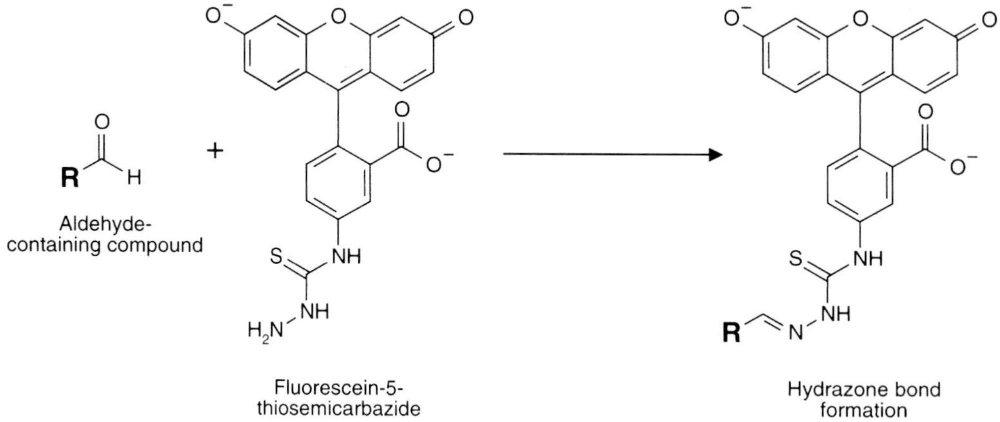

FIGURE 10.10 Fluorescein-5-thiosemicarbazide reacts with aldehyde groups to produce hydrazone linkages.

2. Centrifuge and wash cells several times with PBS. Some protocols include a quench step wherein excess sodium periodate is eliminated by addition of glycerol. This can be accomplished by adding 3 volumes of 0.1-*M* glycerol in PBS prior to centrifugation.
3. Resuspend cells in PBS containing 0.5 mg/ml fluorescein-5-thiosemicarbazide.
4. Incubate 30 min in the dark at room temperature.
5. To reduce the hydrazone bonds to more stable linkages, cool the cell suspension to 0°C and add an equal volume of 30-m*M* sodium cyanoborohydride in PBS. Incubate for 40 min. Note: If the presence of a reducing agent is detrimental to protein activity, eliminate the reduction step. In most cases, the hydrazone linkage is stable enough for fluorescent labeling experiments.
6. Centrifuge and wash cells extensively with PBS.

Protocol for Labeling Glycoproteins in Solution

1. Dissolve the glycoprotein(s) to be labeled in ice-cold 1-m*M* sodium periodate, 10-m*M* sodium phosphate, 0.15-*M* NaCl, pH 7.4, for the exclusive oxidation of sialic acid residues. For general carbohydrate oxidation, increase the periodate concentration to 10-m*M* in PBS at room temperature.
2. React for 30 min on ice (for sialic acids) or at room temperature (for other polysaccharide residues).
3. Remove excess reactant by gel filtration using a desalting resin with PBS, pH 7.4. Some protocols use a quenching agent to remove excess periodate prior to gel filtration. This can be done by adding glycerol to a final concentration of 0.1-*M*.
4. To the purified, oxidized glycoprotein(s), add fluorescein-5-thiosemicarbazide to a final concentration of 0.5 mg/ml.
5. React for 30 min at room temperature in the dark.
6. To reduce the hydrazone bonds to more stable linkages, cool the solution to 0°C and add an equal volume of 30-m*M* sodium cyanoborohydride in PBS. Incubate for 40 min. Note: If the presence of a reducing agent is detrimental to protein activity, eliminate the reducing step. In most cases, the hydrazone linkage is stable enough for fluorescent labeling experiments.
7. Purify the fluorescently labeled glycoprotein(s) by gel filtration using a desalting resin.

1.10. 5-(((2-(Carbohydrazino)methyl)thio) acetyl)-aminofluorescein

Another hydrazine derivative of fluorescein, 5-(((2-(carbohydrazino)methyl)thio)acetyl)-aminofluorescein (invitrogen), contains a longer spacer arm off the No. 5 carbon atom of its lower ring than fluorescein-5-thiosemicarbazide, described previously. The reagent can be used to react spontaneously with aldehyde- or ketone-containing molecules and form a hydrazone linkage (Figure 10.11). It also can be used to label cytosine residues in DNA or RNA by use of the bisulfite activation procedure (Chapter 23, Section 2.1). The resulting fluorescent derivative exhibits a maximal excitation at 490 nm and a maximal luminescence emission peak at 516 nm when dissolved in buffer at pH 8.0. In the same buffered environment, the compound has an extinction coefficient of approximately 75,000 $M^{-1}cm^{-1}$ at 490 nm.

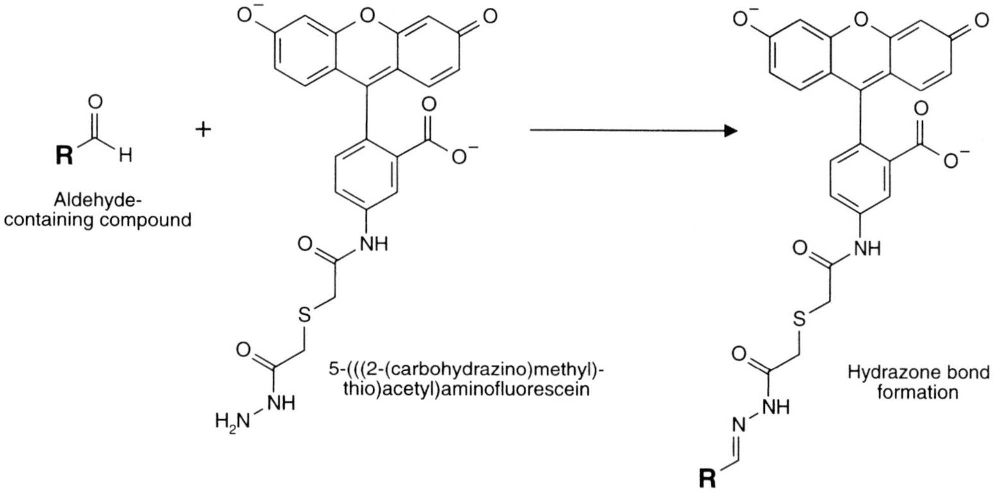

FIGURE 10.11 This carbohydrazide-containing fluorescein derivative can be used to modify aldehyde-containing molecules. Glycoconjugates may be labeled with this reagent after treatment with sodium periodate to produce aldehydes on the carbohydrate portion.

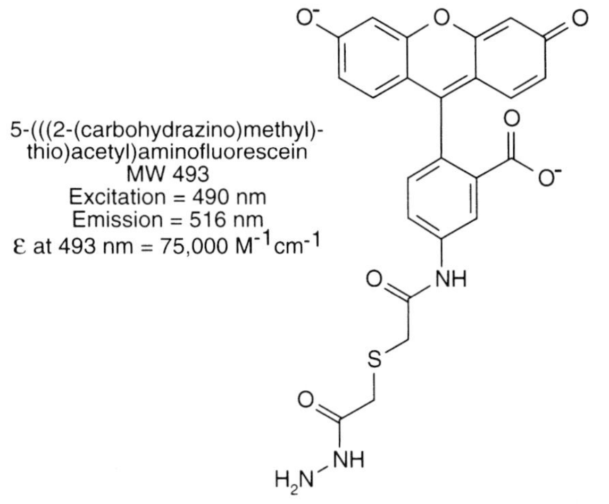

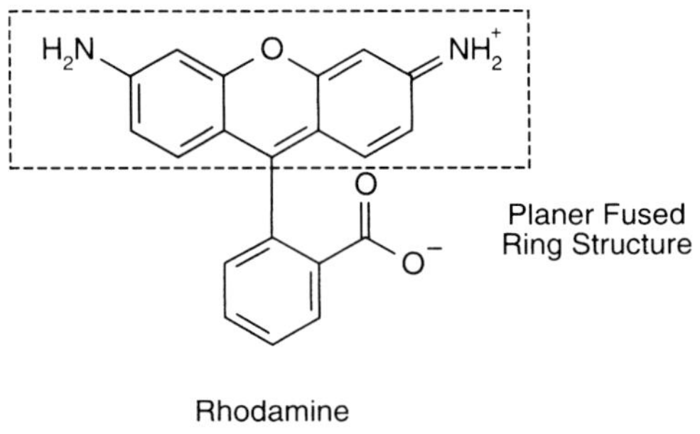

FIGURE 10.12 The basic structure of rhodamine derivatives with the xanthene triple ring system at its core.

The fluorescent probe, 5-(((2-(carbohydrazino)methyl) thio)acetyl)-aminofluorescein, is soluble in DMF or in buffered aqueous solutions at pH values above 7.0. The reagent may be dissolved in DMF as a concentrated stock solution before adding a small aliquot to an aqueous reaction medium. The compound itself and all solutions made with it should be protected from light to avoid decomposition of its fluorescent properties.

The methods for using this reagent in labeling glycoproteins on cell surfaces or in solution are similar to those described for fluorescein-5-thiosemicarbazide, above.

2. RHODAMINE DERIVATIVES

Rhodamine and its derivatives are popular fluorescent probes for labeling all types of biomolecules. Their fluorescent character is created by the presence of a planar, multi-ring aromatic xanthene core structure, similar to the core of fluorescein, but in this case with nitrogen atoms replacing the oxygens on the outer rings (Figure 10.12). Fluorescent modification reagents based on rhodamine are derivatives of this basic structure. Activated rhodamine probes have reactive groups prepared through substitutions off the numbers 5 or 6 carbon atoms of its lower ring. These derivatives provide reactivity toward particular functional groups in biomolecules, allowing rapid labeling of proteins and nucleic acids. Other alterations to the basic rhodamine structure modulate its fluorescent character, creating more intense or stable fluorophores, or changing its wavelength of excitation and emission toward the red region. Many such derivatives are now commercially available.

The tetramethylrhodamine derivative, for instance, has two methyl groups attached to each nitrogen on its outer rings. Activated forms of tetramethylrhodamine are among the most common derivatives of rhodamine used for fluorescent labeling. Another useful derivative is rhodamine B, which contains two ethyl groups on each nitrogen as well as a carboxylate group at the No. 3 position on its lower ring. Rhodamine 6G adds two methyl groups on the outer rings as well as an ethyl ester group off rhodamine B's carboxylate. Rhodamine 110 contains no substituents on the upper nitrogens and only the carboxylate on the lower ring. Sulforhodamine B possesses rhodamine B's two ethyl groups on each nitrogen of the upper rings, but has two sulfonates at the number 3 and 5 positions of its lower ring. This derivative is often called Lissamine™ rhodamine B—Lissamine being a trademark of Imperial Chemical Industries. Another popular derivative of rhodamine, sulforhodamine 101, goes by the name of Texas Red™ (a trademark of Invitrogen). This derivative has intense luminescent properties that take it the farthest into the red region of the spectrum. The basic structures of these rhodamine derivatives are shown in Figure 10.13. The corresponding commercially available rhodamine fluorophores usually contain additional reactive groups, on the No. 5 or 6 carbons of the lower ring, to permit coupling to target molecules (Invitrogen, Thermo Fisher).

Rhodamine derivatives have effective excitation wavelengths within the visible light spectrum from the low- to high-500-nm range, depending on the particular derivative. Their associated emission wavelengths occur from the mid- to high-500-nm range—with Texas Red derivatives typically emitting at over 600 nm—within the orange-to-red visible spectrum. The QY of rhodamine derivatives is generally less than that of fluorescein—only about 25%. However, its fluorescent intensity fades more slowly than fluorescein when it is dissolved in buffers, exposed to light, or stored for extended periods. In addition, its orange-to-red luminescence is in stark contrast to the green of fluorescein. Thus, these two types of probes form an ideal pair for use in double staining techniques, especially in fluorescent microscopy.

The following sections describe the most important rhodamine derivatives commonly used to label biomolecules.

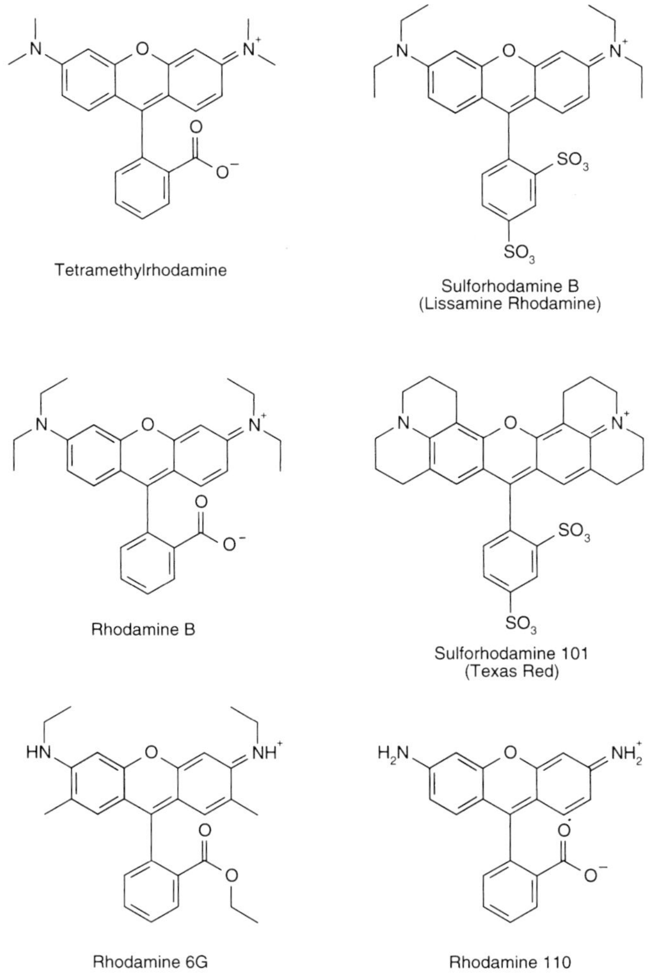

FIGURE 10.13 The primary rhodamine derivatives useful for fluorescent labeling.

2.1. Amine-Reactive Rhodamine Derivatives

Four forms of amine-reactive rhodamine probes are commonly available. Two of them are based on the tetramethyl derivatives of the fundamental rhodamine structure, one is based on the sulforhodamine B or Lissamine derivative, and the last is the sulforhodamine 101 or Texas Red-type of derivative. All of them react under alkaline conditions with primary amines in proteins and other molecules to form stable, highly fluorescent complexes.

2.2. Tetramethylrhodamine-5-(and 6)-isothiocyanate (TRITC)

TRITC is one of the most popular fluorescent probes available. The isothiocyanate derivative of tetramethylrhodamine is synthesized by modification of its lower ring at the 5- or 6-carbon positions. The two resulting isomers are almost identical in their reactivity but slightly different in their spectral properties, including excitation and emission wavelengths and intensities. The chemical differences in the isomers, however, may affect the separation of modified proteins from excess probe or the analysis of tagged molecules by electrophoresis. For this reason, most manufacturers offer the mixed isomers as well as the purified 5- or 6-isothiocyanate derivatives individually.

Isothiocyanates react with nucleophiles such as amines, sulfhydryls, and the phenolate ion of tyrosine side chains (Podhradsky *et al.*, 1979). The only stable product, however, is with primary amine groups, and so TRITC is almost entirely selective for modifying ε- and N-terminal amines in proteins. The reaction involves attack of the nucleophile on the central, electrophilic carbon of the isothiocyanate group (Figure 10.14). The resulting electron shift creates a thiourea linkage between TRITC and the protein with no leaving group.

Tetramethylrhodamine-
5-isothiocyanate (TRITC)
MW 444
Excitation = 544 nm
Emission = 570 nm
ε at 544 nm = 100,000 $M^{-1}cm^{-1}$

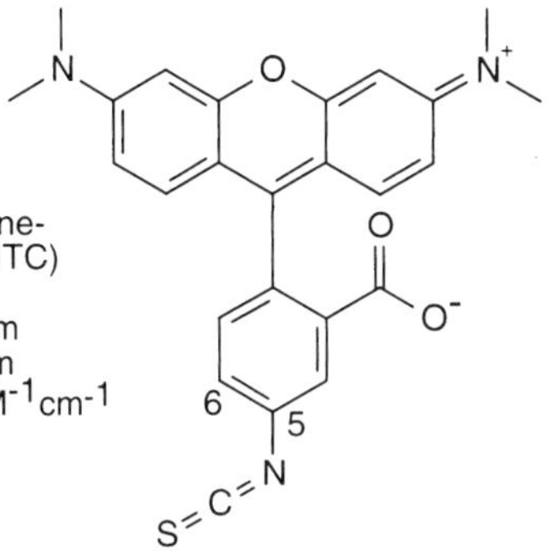

TRITC is relatively insoluble in water, but it can be dissolved in DMF or DMSO as a concentrated stock solution prior to its addition to an aqueous reaction mixture. The isothiocyanate group is reasonably stable in aqueous solution for short periods, but will degrade by hydrolysis. TRITC also is more stable to photobleaching than FITC (Section 1), and its absorption and emission spectra are less sensitive to environmental conditions, such as pH. It is best, however, to use only fresh reagent for modification purposes. Storage should be under desiccated conditions, protected from light, and at −20°C.

The fluorescent properties of TRITC (mixed isomers) include an absorbance maximum at about 544 nm and an emission wavelength of 570 nm. Fluorescent quenching of the molecule is possible. Under concentrated conditions, rhodamine-to-rhodamine interactions result in self-quenching which reduces its luminescence yield. This phenomenon can occur with TRITC-tagged molecules, as well. If derivatization of a protein is done at too high a level,

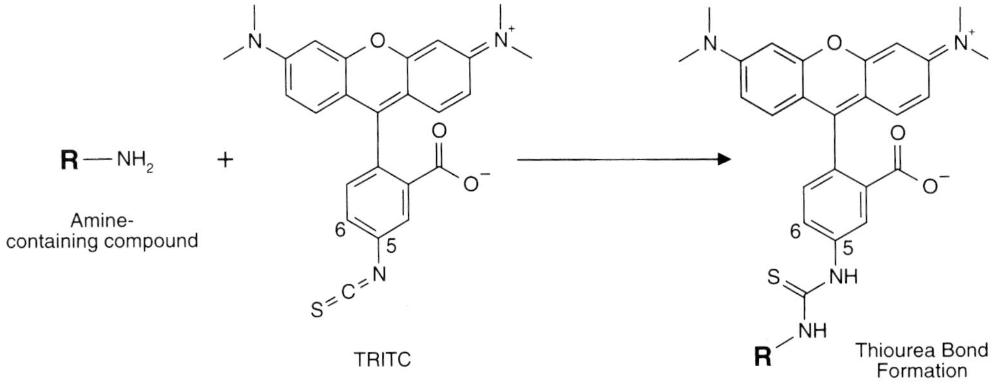

FIGURE 10.14 TRITC reacts with amine-containing molecules to create an isothiourea linkage.

the resultant QY of the conjugate will be depressed from expected values. Typically, modifications of proteins involve adding no more than eight to ten rhodamine molecules per molecule of protein, with a four to five substitution level considered optimal.

TRITC has been used in numerous applications involving fluorescence detection, including double-staining techniques with fluorescein-labeled probes (Mossberg and Ericsson, 1990), the synthesis of fluorescently labeled DNA probes (Smith *et al.*, 1985), as a label in homogeneous immunoassay systems (Nithipatikom and McGown, 1987), investigating specific interactions of proteins with cell surfaces (Hochman *et al.*, 1988), and as an important fluorescent tag of antibodies in immunohistochemical staining techniques (Davidson and Hilchenbach, 1990). The fluorescent dye also has been used to investigate dynein-dependent transport of virus proteins (Ramanathan *et al.*, 2007), the trafficking of the prion protein (Campana *et al.*, 2007), the distribution of FAT1 isoforms in migrating cells (Braun *et al.*, 2007), and the development of a dual immunofluorescence system for confocal microscopy (Entwistle and Noble, 2011). Many thousands of additional references cite the use of TRITC in labeling molecules in fluorescence detection applications.

The level of TRITC modification in a macromolecule can be determined by measuring its absorbance at or near its characteristic absorption maximum (~575 nm). The number of fluorochrome molecules per molecule of protein is known as the F/P ratio. This value should be measured for all derivatives prepared with fluorescent tags. The ratio is especially important in predicting the behavior of antibodies labeled with TRITC. For a TRITC-labeled protein, the ratio of its absorbance at 575 nm to 280 nm should be between 0.3 and 0.7.

A general protocol for the modification of proteins, particularly immunoglobulins, with TRITC is given below. Modifications to the amount of reagent added to the reaction may be done to optimize the F/P ratio.

Protocol

1. Prepare a protein solution in 0.1-*M* sodium carbonate, pH 9.0, at a concentration of at least 2 mg/ml.
2. In a darkened lab, dissolve TRITC (Thermo Fisher) in dry DMSO at a concentration of 1 mg/ml. Do not use old TRITC, as breakdown of the isothiocyanate group over time may decrease coupling efficiency. Protect from light by wrapping in aluminum foil or using amber vials.
3. In a darkened lab, slowly add 50 µl of TRITC solution to each milliliter of protein solution. Gently mix the protein solution as the TRITC is added.
4. React for at least 8 h at 4°C in the dark or 2 to 4 h at room temperature.
5. The reaction may be quenched by the addition of ammonium chloride to a final concentration of 50-m*M*. Some protocols also include at this point the addition of 0.1% xylene cylanol and 5% glycerol as a photon absorber and protein stabilizer, respectively. React for a further 2 h to stop the reaction by blocking remaining isothiocyanate groups.
6. Purify the derivative by gel filtration using a PBS buffer or another suitable buffer for the particular protein being modified. The use of a desalting resin (e.g., Excellulose, Thermo Fisher) or similar matrices with low exclusion limits works well. To obtain complete separation, the column size should be 15 to 20 times the size of the applied sample. Fluorescent molecules often nonspecifically stick to gel filtration supports, so reuse of the column is not recommended.

2.3. NHS–Rhodamine

NHS–rhodamine is an amine-reactive fluorescent probe that contains a carboxy-succinimidyl ester group off the No. 5 or 6 carbons on rhodamine's lower ring structure (Kellogg *et al.*, 1988). The 5- and 6-isomers are virtually identical in their reactivity and fluorescent

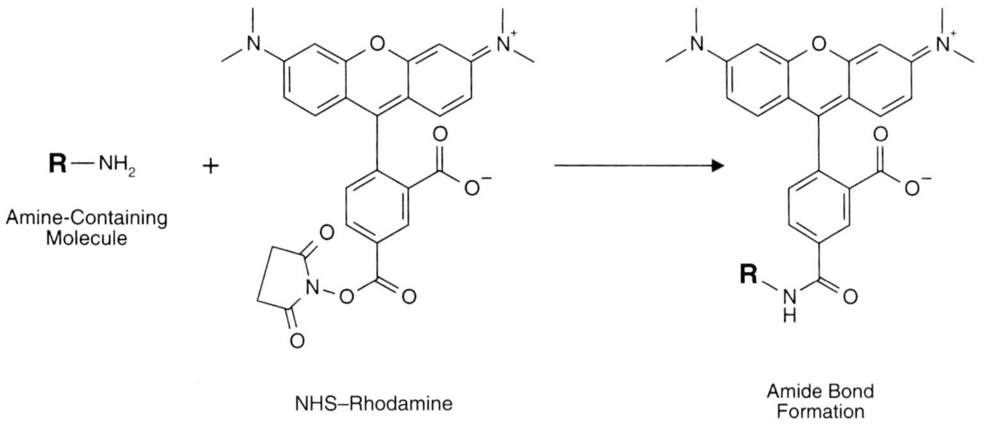

FIGURE 10.15 NHS–rhodamine can be used to label amine-containing molecules such as proteins by reaction with its NHS ester group.

characteristics. Similar to TRITC (described previously), NHS–rhodamine can be used to label proteins and other macromolecules that contain primary amine groups. The isomeric forms of the fluorescent probe are available in mixed and purified forms (Invitrogen, Thermo Fisher). The pure forms are particularly important for labeling nucleic acid probes that will be separated by electrophoresis (Chehab and Kan, 1989). The NHS ester labeling reaction proceeds rapidly at slightly alkaline pH values, resulting in a stable, amide-linked derivative (Chapter 3, Section 1.4; Figure 10.15).

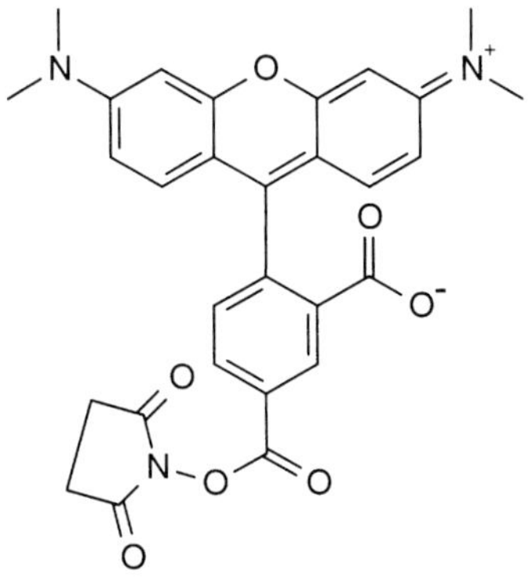

NHS–Rhodamine;
5-Carboxytetramethyl-rhodamine, succinimidyl ester
MW 528
Excitation = 546 nm
Emission = 579 nm
ε at 546 nm = 100,000 $M^{-1}cm^{-1}$

The fluorescent properties of NHS–rhodamine are similar to TRITC. The wavelength of maximal absorbance or excitation for the reagent is 544 nm and its emission maximum is 576 nm, exhibiting a visual color of orange–red. Its molar extinction coefficient at 546 nm in a methanol environment is 63,000 $M^{-1}cm^{-1}$. Other components in solution as well as the pH (in aqueous buffers) can change this value.

NHS–rhodamine is insoluble directly in aqueous solution and should be dissolved in organic solvent prior to addition of a small aliquot to a buffered reaction medium. Concentrated stock solutions may be prepared in DMSO or DMF. Such solutions are relatively stable for short periods if protected from light, but should be prepared fresh. Reaction conditions should be maintained at the optimal reactivity for NHS esters, which is pH 7 to 9.

NHS–rhodamine has been used in numerous applications, including the detection of specific DNA sequences (Chehab and Kan, 1989), studying the behavior of microtubules and actin filaments in living *Drosophila* embryos (Kellogg *et al.*, 1988), investigation into the light-initiated breakup of microtubules (Vigers *et al.*, 1988), studying ω-conotoxin-sensitive channels in neurons (Jones *et al.*, 1989), investigating growth cones during axon elongation (Tanaka and Kirschner, 1991), and studying the pathways of mitotic spindle assembly *in vitro* (Sawin and Mitchison, 1991). NHS–rhodamine has also been used to perform fluorescent studies of membrane protein complexes (Wittig *et al.*, 2007), to investigate the cross-bridge between microtubules (Li *et al.*, 2007), to study a multi-enzyme network in digestion (Delcroix *et al.*, 2006), and to label self-assembling CpG DNA nanoparticles for antigen delivery (Rattanakiat *et al.*, 2012).

The following generalized protocol relates to the labeling of IgG with NHS–rhodamine. Optimization of the level of rhodamine incorporation may have to be done with other proteins or other macromolecules.

Protocol

1. Dissolve an immunoglobulin to be labeled in ice-cold, 50-m*M* sodium bicarbonate, pH 8.5, at a concentration of 10 mg/ml.
2. Dissolve NHS–rhodamine at a concentration of 1 mg/ml in DMSO. Protect from light.

3. In a darkened lab, slowly add 50 to 100 μl of the NHS–rhodamine solution to each milliliter of the antibody solution with mixing. Wrap the vessel with aluminum foil to protect from light.
4. Place the sample on ice and react for 2 h.
5. Remove unreacted NHS–rhodamine and reaction byproducts by gel filtration or dialysis.

2.4. Lissamine™ Rhodamine B Sulfonyl Chloride

The Lissamine form of rhodamine B consists of diethyl modifications on the two nitrogens of the upper rings of the basic rhodamine molecule as well as two sulfonate groups added at the 3- and 5-carbon positions of the lower ring. Lissamine rhodamine B sulfonyl chloride is an amine-reactive reagent made by converting the No. 5 sulfonate group to a reactive sulfonyl halide. Reaction with proteins and other amine-containing molecules results in the formation of sulfonamide bonds (Figure 10.16). Lissamine is a trademark of Imperial Chemical Industries.

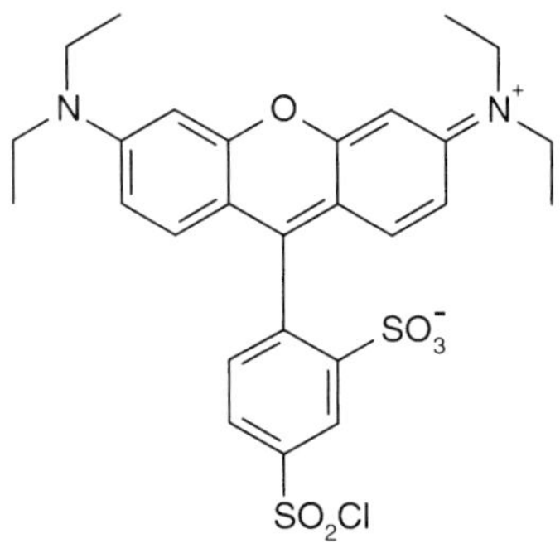

Lissamine Rhodamine B
Sulfonyl Chloride
MW 577
Excitation = 556 nm
Emission = 576 nm
ε at 556 nm = 100,000 $M^{-1}cm^{-1}$

The spectral characteristics of protein conjugates made with Lissamine rhodamine B derivatives are of longer wavelength than those of tetramethylrhodamine—more toward the red region of the spectrum. In addition, modified proteins have better chemical stability and are somewhat easier to purify than those made from TRITC (discussed previously). Lissamine derivatives also make more photostable probes than the fluorescein derivatives (Section 1).

Lissamine rhodamine B sulfonyl chloride is relatively insoluble in water, but may be dissolved in DMF prior to the addition of a small aliquot to an aqueous reaction. Do not dissolve in DMSO, as sulfonyl chlorides will readily react with this solvent (Boyle, 1966). The compound has a maximal absorptivity at 556 nm with an extremely high extinction coefficient of up to 93,000 $M^{-1}cm^{-1}$ (in methanol) in highly purified form. Its emission maximum occurs at 576 nm, emitting red luminescence.

A sulfonyl chloride rapidly reacts with amines on proteins and other molecules to form stable sulfonamide bonds. It also may react with tyrosine OH groups, aliphatic alcohols, thiols, and imidazole groups (such as histidine side chains). Conjugates of sulfonyl chlorides with sulfhydryls and imidazole rings are unstable, while esters formed with alcohols are subject to nucleophilic displacement (Nillson and Mosbach, 1984; Scouten and Van der Tweel, 1984). The only stable derivative with proteins is therefore the sulfonamide, formed by reaction with ε-lysine and N-terminal amines. Optimal conditions for coupling are non-amine-containing buffers in the pH range of 9 to 10. Phosphate, bicarbonate, or borate buffers are recommended for the modification reaction. Avoid the presence of other nucleophiles that can cross-react with the sulfonyl chloride (e.g., amine-containing components or sulfhydryl-containing reducing agents). In aqueous solutions, hydrolysis is a competing reaction, but occurs more slowly with sulfonyl halides than with the corresponding acid chlorides of carboxylate groups. Unreacted, hydrolyzed probe is

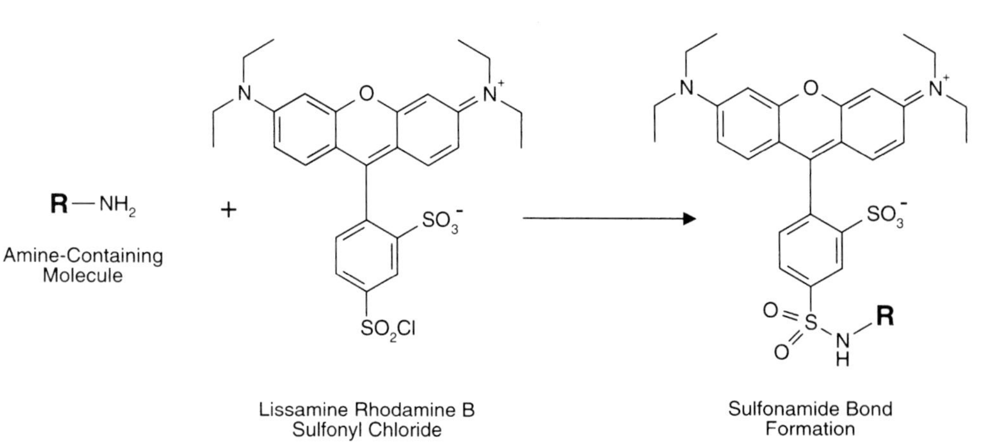

FIGURE 10.16 Lissamine rhodamine B sulfonyl chloride reacts with amine-containing molecules to produce stable sulfonamide bonds.

the water soluble sulforhodamine B fluorophore that is easily removed by gel filtration or dialysis.

Lissamine rhodamine B sulfonyl chloride has been used in numerous applications, including multiple-labeling techniques in microscopy (Wessendorf, 1990), for confocal microscopy techniques (Tsien and Wagoner, 1990), in the study of fibronectin receptors (Duband *et al.*, 1988), for investigations into microtubule and intermediate filament association (Geiger and Singer, 1980), for the labeling of glycoconjugates (Wilchek *et al.*, 1980), for studying regulation of the Na^+,K^+–ATPase system (Sipe *et al.*, 1991), and for investigations into the redox potential within mitochondria (Chazotte and Hackenbrock, 1991). The dye has also been used to study the shape of giant unilamellar vesicles as model plasma membranes (Gudheti *et al.*, 2007), to investigate the activity of xyloglucan xylogucosyl transferase (Hrmova *et al.*, 2007), to study the switching of neurofilaments between mobile and stationary states (Trivedi *et al.*, 2007), and in the synthesis of bioactive fluorescent analogues of the marine alkaloid discorhabdin C (Lam *et al.*, 2012). Hundreds of additional biological studies have used this fluorescent probe.

The following protocol is a general guide for labeling biological macromolecules with Lissamine rhodamine B sulfonyl chloride. Optimization of the fluorophore incorporation level (F/P ratio) may have to be done for specific labeling experiments.

Protocol

1. Dissolve the amine-containing macromolecule to be labeled (i.e., a protein) in 0.1-*M* sodium carbonate/bicarbonate buffer, pH 9.0, at a concentration of 1 to 5 mg/ml.
2. Dissolve Lissamine rhodamine B sulfonyl chloride (Invitrogen) in DMF at a concentration of 1 to 2 mg/ml. Protect from light and use immediately.
3. In a darkened lab and with gentle mixing, slowly add 50 to 100 µl of the fluorophore solution to the protein solution.
4. React for 1 h at room temperature in the dark.
5. Remove excess fluorophore by gel filtration using a desalting resin or by dialysis.

2.5. Texas Red Sulfonyl Chloride

Texas Red sulfonyl chloride is the active halogen derivative of sulforhodamine 101. This important derivative of the basic rhodamine molecule possesses dual aliphatic rings off the upper-ring nitrogens and sulfonate groups on the No. 3- and 5-carbon atoms of its lower ring component. The sulfonyl chloride group can react with primary amines in proteins and other molecules to form stable sulfonamide bonds (Figure 10.17). The group, however, can hydrolyze in the presence of moisture. For this reason, only fresh Texas Red sulfonyl chloride should be used for modification experiments.

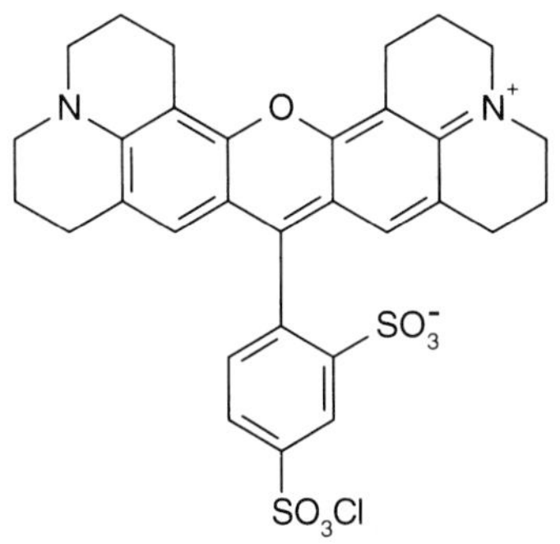

Texas Red Sulfonyl Chloride
MW 577
Excitation = 556 nm
Emission = 576 nm
ε at 556 nm = 93,000 $M^{-1}cm^{-1}$

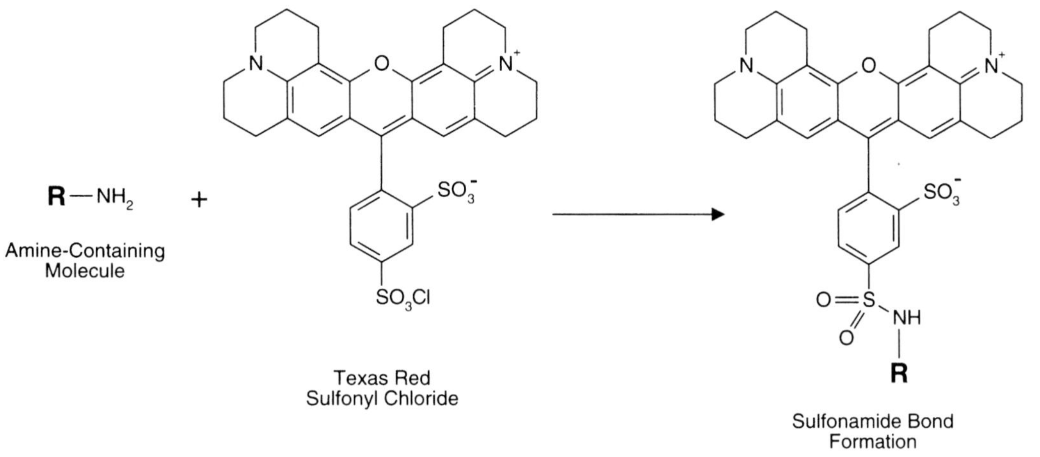

FIGURE 10.17 Texas Red sulfonyl chloride can be used to label amine-containing molecules through sulfonamide bond formation.

The intense Texas Red fluorophore has a QY that is inherently higher than the tetramethylrhodamine or Lissamine rhodamine B derivatives. Texas Red's luminescence is shifted maximally into the red region of the spectrum, and its emission peak only minimally overlaps with that of fluorescein. This makes Texas Red derivatives one of the best choices of labels for use in double-staining techniques.

Texas Red sulfonyl chloride has a maximal excitation at 589 nm and a maximum emission at 615 nm when dissolved in methanol. The extinction coefficient of the compound dissolved in acetonitrile is 85,000 $M^{-1}cm^{-1}$ at 596 nm. The only disadvantage of this fluorophore is its poor excitation by the standard argon laser at 488 nm. However, since both Texas Red and fluorescein are weakly excited by an argon laser at 514 nm, it makes them fairly good pairs for use in laser confocal microscopy or flow cytometry (Mossberg and Ericsson, 1990). The fluorophore is particularly appropriate for excitation by the 568-nm line produced by an argon–krypton mixed laser used on some confocal devices. Compared to other rhodamine derivatives, Texas Red fluorophores display low background in staining techniques and are among the most photostable probes available.

This rhodamine dye derivative has been used to produce purine-scaffold fluorescent probes (Taldone *et al.*, 2011), to study glomerular sieving by use of a dye-labeled albumin (Saleh *et al.*, 2012), and to investigate synaptic activity using patch-clamp chips (Martina *et al.*, 2011).

Texas Red sulfonyl chloride is soluble in DMF or acetonitrile and may be dissolved as a concentrated stock solution in either solvent prior to the addition of a small aliquot to an aqueous reaction medium. Avoid the use of DMSO, as sulfonyl chlorides react with this solvent (Boyle, 1966). The solid and all solutions made from it must be protected from light to avoid photodecomposition. Prepare the stock solution immediately before use.

A sulfonyl chloride group rapidly reacts with amines in the pH range of 9 to 10 to form stable sulfonamide bonds. Under these conditions, it also may react with tyrosine OH groups, aliphatic alcohols, thiols, and histidine side chains. Conjugates of sulfonyl chlorides with sulfhydryls and imidazole rings are unstable, while esters formed with alcohols are subject to nucleophilic displacement (Nillson and Mosbach, 1984; Scouten and Van der Tweel, 1984). The only stable derivative with proteins therefore is the sulfonamide, formed by reaction with ε-lysine and N-terminal amines. For coupling, the reaction media should use only non-amine-containing buffers, such as phosphate, borate, or bicarbonate (avoid Tris, imidazole, or glycine).

A suggested protocol on the use of this fluorescent probe is described below. Optimization may be necessary to achieve the best level of fluorescent modification (F/P ratio) for a particular application.

Protocol

1. Dissolve the protein or macromolecule to be labeled in 0.1-*M* sodium carbonate, pH 9.0, at a concentration of 1 to 5 mg/ml.
2. Dissolve Texas Red sulfonyl chloride (Thermo Fisher, Invitrogen) in acetonitrile at a concentration of 20 mg/ml. Prepare fresh and protect from light. Use a fume hood for all operations using organic solvents.
3. In subdued lighting conditions, add 50 μl of the Texas Red sulfonyl chloride solution to each milliliter of the protein solution. Mix well.
4. React for 1 h at room temperature.
5. Remove excess fluorophore and reaction byproducts by gel filtration using a desalting resin or by dialysis.

Determine the level of fluorophore incorporation (the F/P ratio) by measuring the absorbance of the labeled protein at 520 nm and 280 nm. Labeled proteins having a 520 nm/280 nm ratio of absorbency of 0.3 to 0.8 should perform well in most applications (Titus *et al.*, 1982).

At the time of writing, over 15,000 references cited the use of Texas Red for labeling biological molecules or in bioconjugate detection applications.

2.6. Sulfhydryl-Reactive Rhodamine Derivatives

Rhodamine derivatives containing a sulfhydryl-reactive group off the lower ring structure are available to direct the modification reaction to more limited sites on target molecules. Coupling through sulfhydryls instead of amines can help to avoid active centers in proteins, thus preserving activity in the fluorescent probe complex. Sulfhydryl reaction sites can be naturally available through free cysteine side chains, generated by reduction of disulfides, or created by the use of thiolation reagents (Chapter 2, Section 4.1).

2.7. Tetramethylrhodamine-5-(and 6)-iodoacetamide

The iodoacetamido derivatives of tetramethylrhodamine possess a sulfhydryl-reactive iodoacetyl group (Chapter 2, Section 4.2, and Chapter 3, Section 2.1) at either the 5- or 6-carbon position on their lower ring. The isomers are commercially available only in mixed form (Invitrogen), but some reactivity and specificity differences between the purified 5- and 6-derivatives

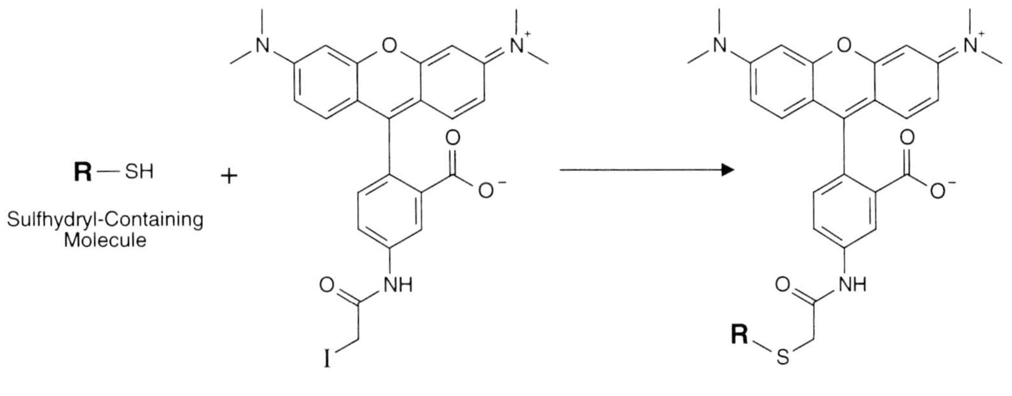

FIGURE 10.18 This iodoacetamide derivative of tetramethylrhodamine can be used to label sulfhydryl groups to form a thioether bond formation.

toward various sulfhydryl sites in proteins may be observed (Ajtai *et al.*, 1992).

The iodoacetyl group of both isomers reacts with sulfhydryls under slightly alkaline conditions to yield stable thioether linkages (Figure 10.18). They do not react with unreduced disulfides in cystine residues or with oxidized glutathione (Gorman *et al.*, 1987). The thioether bonds are hydrolyzed under conditions necessary for complete protein hydrolysis prior to amino acid analysis.

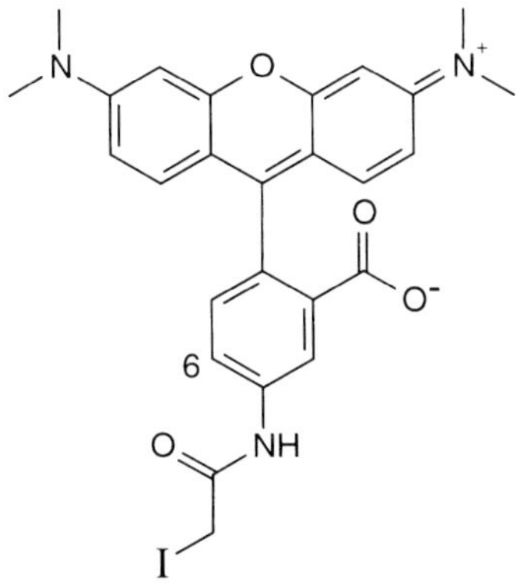

Tetramethylrhodamine-5-iodoacetamide
MW 569
Excitation = 540 nm
Emission = 567 nm
ε at 540 nm = 76,000 $M^{-1}cm^{-1}$

Care must be taken to protect these reagents from light, not only to maintain the fluorescent yield of the rhodamine derivative, but also to protect the iodoacetyl group from light-catalyzed breakdown. Tetramethylrhodamine-5-(and-6)-iodoacetamide is soluble in DMF and DMSO. Concentrated stock solutions may be prepared in these solvents prior to addition of a small aliquot to an aqueous reaction mixture. Protect solutions from light by wrapping in aluminum foil and working in subdued light. Quenching reactions with cysteine, glutathione, or mercaptosuccinic acid will sometimes facilitate removal of unconjugated fluorophore by dialysis or gel filtration.

The spectral properties of these derivatives are similar to native rhodamine. The excitation maximum occurs at about 543 nm and its emission peak at 567 nm, producing light in the orange–red region of the spectrum. The extinction coefficient of tetramethylrhodamine-5-(and-6)-iodoacetamide in methanol at its wavelength of maximum absorptivity, 542 nm, is 81,000 $M^{-1} cm^{-1}$.

The fluorescent probe has been used extensively to label numerous proteins and other biomolecules, including actin (Glacy, 1983; Wang, 1985; Meige and Wang, 1986), myosin light chains (Mittal *et al.*, 1987), α-actin (Simon and Taylor, 1988; Stickel and Wang, 1988), blood coagulation factor Va (Isaacs *et al.*, 1986), and histones (Murphy *et al.*, 1982). The dye has also been used to study conformational changes in proteins (Heuck *et al.*, 2007), the binding region of protein C on factor Va (Yegneswaran *et al.*, 2007), how flavonoids affect actin functions (Boehl *et al.*, 2007), and the stretching of rigor cross-bridges in skeletal muscle fibers (Ushakov *et al.*, 2011). Hundreds of additional publications cite the use of this dye for various biological detection applications.

The following protocol for labeling proteins with tetramethylrhodamine-5-(and-6)-iodoacetamide represents a general guideline. The procedure should be optimized for each macromolecule being labeled to obtain the best F/P ratio to produce intense fluorescence and high activity in the final complex.

Protocol

1. Prepare a 20-m*M* tetramethylrhodamine-5-(and-6)-iodoacetamide solution by dissolving 11.3 mg per milliliter of DMF. Prepare fresh and protect from light.
2. Dissolve the protein to be modified at a concentration of 5 to 10 mg/ml in 50-m*M* sodium phosphate, pH 7.5.
3. Slowly add 25 to 50 μl of the tetramethylrhodamine-5-(and-6)-iodoacetamide solution to each milliliter of the protein solution while mixing.
4. React for 2 h at 4°C in the dark.
5. Remove excess reactant and reaction byproducts by gel filtration using a desalting resin or by dialysis.

2.8. Aldehyde-/Ketone- and Cytosine-Reactive Rhodamine Derivatives

Hydrazide groups can be coupled directly to aldehydes and ketones to form relatively stable hydrazone linkages (Chapter 3, Section 5.1). Two rhodamine derivatives are commonly available that contain a sulfonyl hydrazine group off their No. 5 carbon on the lower ring structure (Invitrogen). They are based on the Lissamine and Texas Red structures and may be used to label aldehyde- or ketone-containing molecules with an intensely fluorescent probe. Although most biomolecules don't contain aldehyde or ketone groups in their native state, carbohydrates, glycoproteins, RNA, and other molecules containing sugar residues (or diols) can be oxidized with sodium periodate to produce reactive formyl groups. The use of modification reagents that generate aldehydes upon coupling to a molecule also can be used to produce a hydrazide-reactive site (Chapter 2, Section 4.4).

DNA and RNA may be modified with hydrazide-reactive probes by reacting their cytosine residues with bisulfite to form reactive sulfone intermediates. These derivatives undergo transamination to couple with hydrazide- or amine-containing probes (Draper and Gold, 1980) (Chapter 23, Section 2.1).

2.9. Lissamine Rhodamine B Sulfonyl Hydrazine

Lissamine rhodamine B sulfonyl hydrazine is a hydrazide derivative of sulforhodamine B that can spontaneously react with aldehyde- or ketone-containing molecules to form a covalent, hydrazone linkage (Figure 10.19). It also can be used to label cytosine residues in DNA or RNA by use of the bisulfite activation procedure (Chapter 23, Section 2.1). The resulting fluorescent derivative exhibits an excitation maximum at a wavelength of 556 nm and a maximal emission wavelength of 580 nm when dissolved in methanol. In the same solvent, the compound has an extinction coefficient of approximately 75,000 $M^{-1}cm^{-1}$.

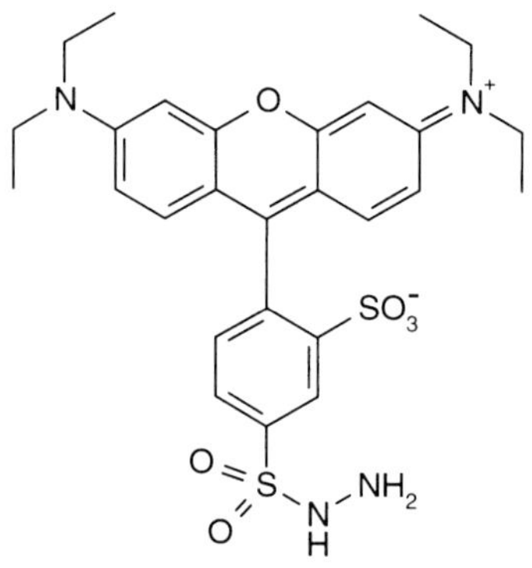

Lissamine Rhodamine B
Sulfonyl Hydrazine
MW 573
Excitation = 560 nm
Emission = 585 nm
ε at 560 nm = 95,000 $M^{-1}cm^{-1}$

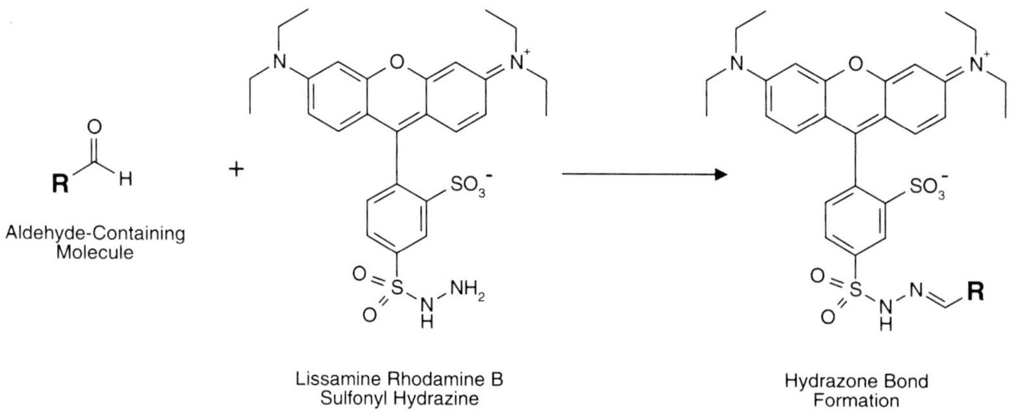

FIGURE 10.19 Lissamine rhodamine B sulfonyl hydrazine reacts with aldehyde groups to form hydrazone linkages.

Lissamine rhodamine B sulfonyl hydrazine is soluble in DMF. The reagent may be dissolved in this solvent as a concentrated stock solution before adding a small aliquot to an aqueous reaction medium. The compound itself and all solutions made with it should be protected from light to avoid decomposition of its fluorescent properties.

This dye has been used to create a dual-wavelength staining method for glycoproteins in polyacrylamide gels (Chiang *et al.*, 2011), to fluorescently label a PEG-phosphatidyl ethanolamine-based polymer for incorporation into liposomes (Biswas *et al.*, 2011), and to track liposomal fusion with cells (Dutta *et al.*, 2011).

Generalized protocols for the use of hydrazine probes reactive toward aldehyde residues can be found in Section 1 of this chapter. These procedures are directed at the labeling of cell surface glycoproteins or glycoproteins in solution. Substitution of Lissamine rhodamine B sulfonyl hydrazine for the fluorescein-5-thiosemicarbazide reagent described in that section can be done without difficulty. Some optimization may be necessary to achieve the best level of fluorescent modification for each particular application.

2.10. Texas Red Hydrazide

Texas Red hydrazide is a derivative of Texas Red sulfonyl chloride made by reaction with hydrazine (Invitrogen). The result is a sulfonyl hydrazine group on the No. 5 carbon position of the lower ring structure of sulforhodamine 101. The intense Texas Red fluorophore has a QY that is inherently higher than either the tetramethylrhodamine or Lissamine rhodamine B derivatives of the basic rhodamine molecule. Texas Red's luminescence is shifted maximally into the red region of the spectrum, and its emission peak only minimally overlaps with that of fluorescein. This makes derivatives of this fluorescent probe among the best choices of labels for use in double staining techniques.

The hydrazide derivative can be used to modify aldehyde- or ketone-containing molecules, including cytosine residues using the bisulfite activation procedure described in Chapter 23, Section 2.1. The sulfonyl hydrazine group of Texas Red hydrazide reacts with aldehydes or ketones in target molecules to form hydrazone bonds (Figure 10.20). Carbohydrates and glycoconjugates can be specifically labeled at the polysaccharide portion if the required aldehydes are first formed by periodate oxidation or another such method (Chapter 2, Section 4.4).

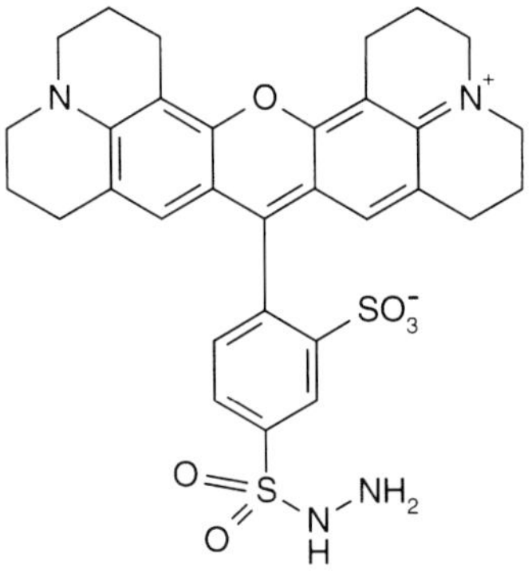

Texas Red Hydrazide
MW 621
Excitation = 580 nm
Emission = 604 nm
ε at 580 nm = 80,000 $M^{-1}cm^{-1}$

Texas Red hydrazide has a maximal excitation wavelength of 580 nm and a maximum emission at 604 nm when dissolved in methanol. Its extinction coefficient in the same solvent is 80,000 $M^{-1}cm^{-1}$ at 580 nm. The only disadvantage of this fluorophore is its poor excitation by an argon laser at 488 nm. However, since both

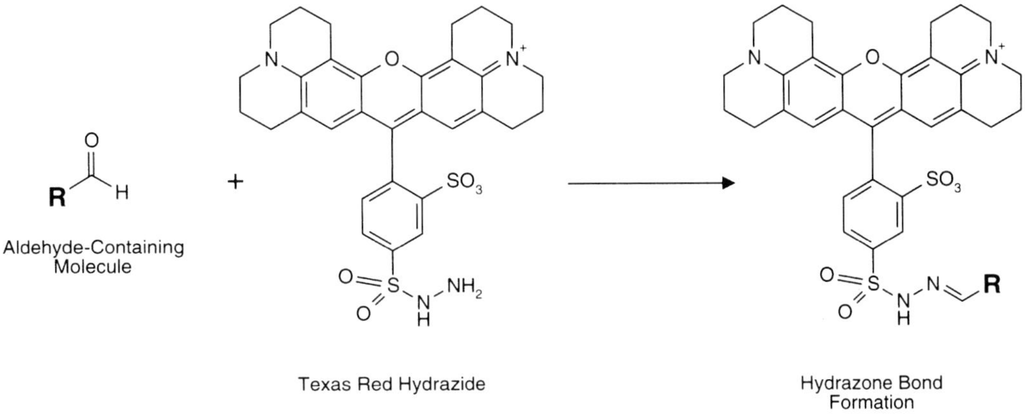

FIGURE 10.20 Texas Red hydrazide reacts with aldehydes to create hydrazone bonds.

Texas Red and fluorescein are weakly excited by an argon laser at 514 nm, it makes them fairly good pairs for use in laser confocal microscopy or flow cytometry (Mossberg and Ericsson, 1990). The fluorophore is particularly appropriate for excitation by the 568-nm line produced by an argon–krypton mixed laser used on some confocal devices. White-light illumination also can be used to excite the dye, while its emission is detected using an appropriate filter. Compared to other rhodamine derivatives, Texas Red fluorophores display low background in staining techniques and are among the most photostable probes available.

Texas Red hydrazide is soluble in DMF and may be dissolved as a concentrated stock solution in this solvent prior to the addition of a small aliquot to an aqueous reaction medium. The solid and all solutions made from it must be protected from light to avoid photodecomposition. Prepare the stock solution fresh immediately before use. A suggested protocol on the use of this fluorescent probe may be obtained by following the method outlined for fluorescein-5-thiosemicarbazide in Section 1 of this chapter. Optimization may be necessary to achieve the best level of fluorescent modification (F/P ratio) for a particular application.

This dye has been used to study the cellular internalization of various proteins and molecules (Ratelade *et al.*, 2011), in a fucose-specific labeling procedure for a vascular-targeting antibody (Zuberbuhler *et al.*, 2012), and in the fluorescent labeling of carbon nanotubes (Yoshimura *et al.*, 2011).

3. COUMARIN DERIVATIVES

Coumarin (2H-1-benzopyran-2-one) is a naturally occurring substance found in tonka beans, lavender oil, and sweet clover (Merck Index **11**:2563) (Figure 10.21). Many of its derivatives are highly fluorescent compounds that are widely studied (Schimitschek *et al.*, 1974; Ernsting *et al.*, 1982; Eschrich and Morgan, 1985; Jones *et al.*, 1984, 1985; Baranowska- Kortylewicz and Kassis, 1993a,b). The 7-amino-4-methylcoumarin derivatives have excellent fluorescent properties useful for labeling biological molecules with a detectable tag

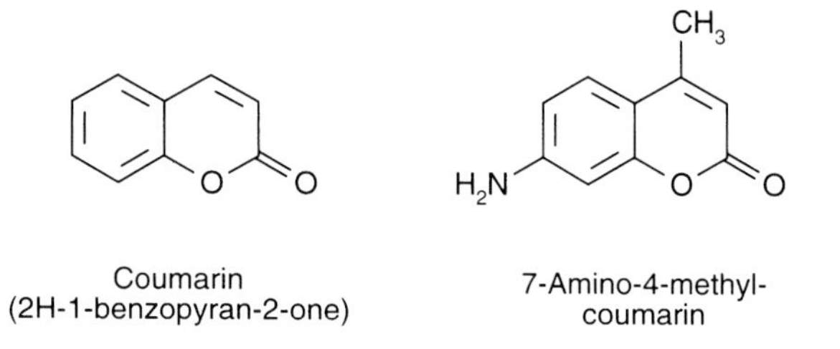

FIGURE 10.21 The basic structural characteristics of coumarin fluorophores.

(Uchino *et al.*, 1979; Bos, 1981). Particularly, the 3-acetic acid derivative of this molecule, known as AMCA, provides a carboxylate group from which to create easily reactive groups suitable for coupling to proteins and other molecules.

Aminomethylcoumarin derivatives possess intense fluorescent properties within the blue region of the visible spectrum. Their emission range is sufficiently removed from other common fluorophores that they are excellent choices for double-labeling techniques. In fact, coumarin fluorescent probes are very good donors for excited-state energy transfer to fluoresceins.

The following sections describe the most popular derivatives of aminomethylcoumarin used to label proteins and other biological macromolecules.

3.1. Amine-Reactive Coumarin Derivatives

Three main forms of amine-reactive AMCA probes are commonly available. One of them is simply the free acid form of AMCA, which can be used to couple to amine-containing molecules using the carbodiimide reaction (Chapter 4, Section 1). The other two are active-ester derivatives of AMCA—the water-insoluble NHS ester and the water soluble sulfo-NHS ester forms—both of which spontaneously react with amines to create stable amide linkages. All of them react under mild conditions with primary amines in proteins and other molecules to form highly fluorescent derivatives.

3.2. AMCA

AMCA, or 7-amino-4-methylcoumarin-3-acetic acid, is a fluorescent probe that exhibits a spectacular blue fluorescence (Khalfan *et al.*, 1986) (Thermo Fisher). AMCA absorbs light at a wavelength of 345 nm and luminesces in the range of 440 to 460 nm. Its emission wavelength is in a region that does not overlap with the emission spectra of other major fluorescent probes. This makes double-staining techniques particularly effective with this fluorophore. AMCA also has pronounced stability toward photobleaching, retaining its full fluorescence more than three times longer than fluorescein-based probes. AMCA derivatives, and labeled molecules fluoresce with a bright-blue color upon excitation. This color is easily visualized using fluorescent microscopes or imagers. The blue emission color avoids problems of autofluorescence associated with high background. In addition, the large Stokes shift of the molecule minimizes interference from Rayleigh light scatter effects during excitation. The fluorescent intensity of AMCA is not affected by changes in pH in the range of 3 to 10. This is in marked contrast to other fluorescent probes, such as fluorescein, which display considerable variability in their emission spectra with pH.

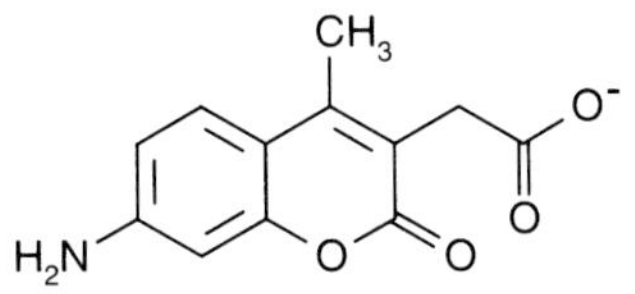

AMCA
7-Amino-4-methyl-coumarin-3-acetic acid
MW 233
Excitation = 345–350 nm
Emission = 440–460 nm

AMCA may be coupled to amine-containing molecules through the use of the carbodiimide reaction using EDC (Chapter 4, Section 1.1). EDC will activate the carboxylate on AMCA to a highly reactive *O*-acylisourea intermediate. Attack by a nucleophilic primary amine group on the carbonyl of this ester results in the formation of an amide bond (Figure 10.22). Derivatization of AMCA off its carboxylate group causes no major effects on its fluorescent properties. Thus, proteins and other macromolecules may be labeled with this intensely blue probe and easily detected by fluorescence microscopy and other techniques.

3.3. AMCA–NHS and AMCA–Sulfo-NHS

AMCA–NHS, succinimidyl-7-amino-4-methylcoumarin-3-acetic acid, is an amine-reactive derivative of AMCA containing an NHS ester on its carboxylate group (Thermo Fisher). The result is reactivity directed toward amine-containing molecules, forming amide linkages with the AMCA fluorophore (Figure 10.23). Proteins labeled with AMCA show little to no effect on the isoelectric point of the molecule.

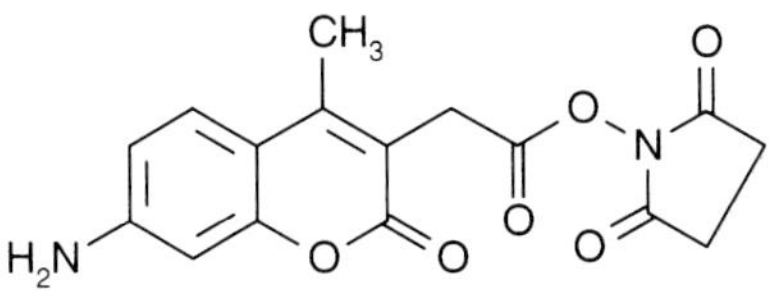

AMCA–NHS
Succinimidyl-7-amino-4-methylcoumarin-3-acetic acid
MW 330

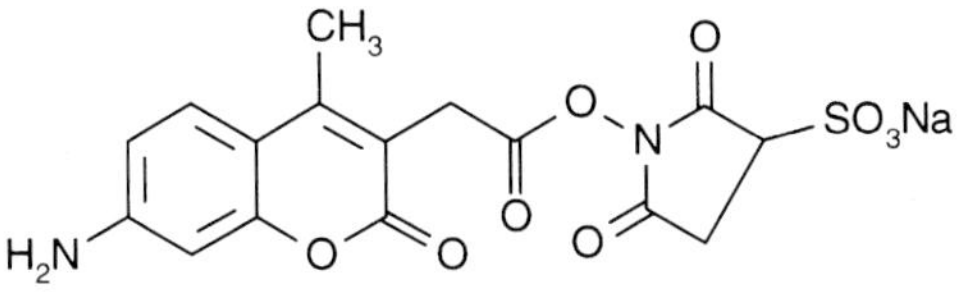

AMCA–Sulfo-NHS
Sulfosuccinimidyl-7-amino-4-methylcoumarin-3-acetic acid
MW 431

Reaction of AMCA–NHS with proteins proceeds efficiently in the pH range of 7 to 9. Avoid buffers containing amines which can compete in the coupling reaction, such as Tris or glycine, and avoid imidazole buffers since they promote hydrolysis of the NHS ester.

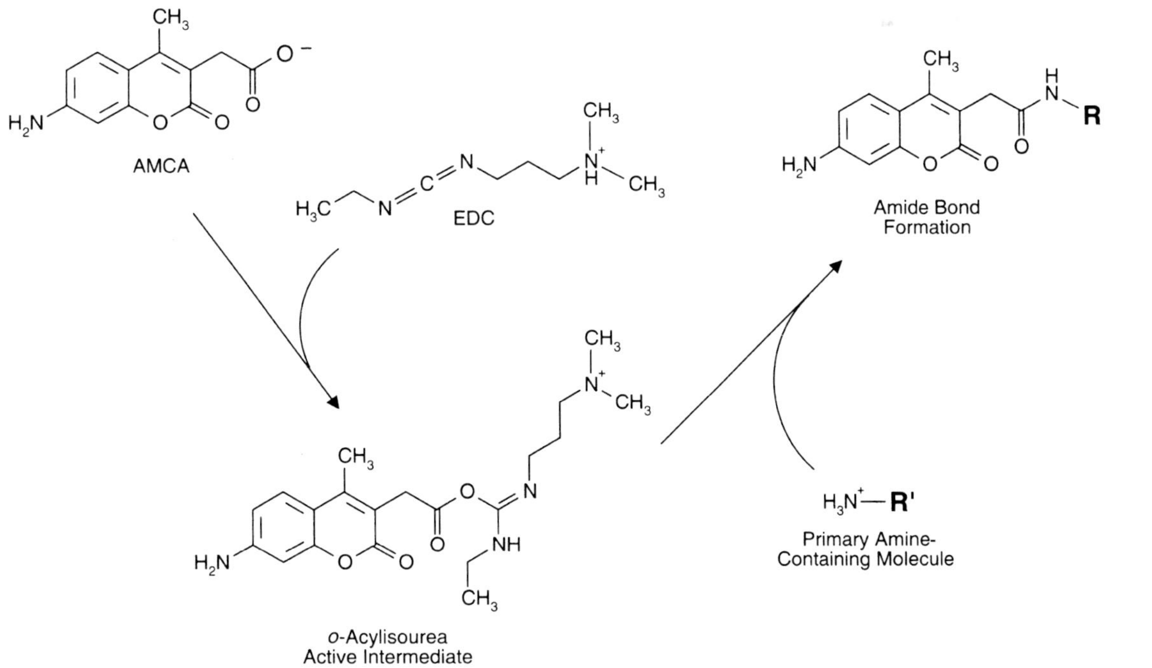

FIGURE 10.22 AMCA may be linked to amine-containing molecules through its carboxylate group using a carbodiimide reaction with EDC.

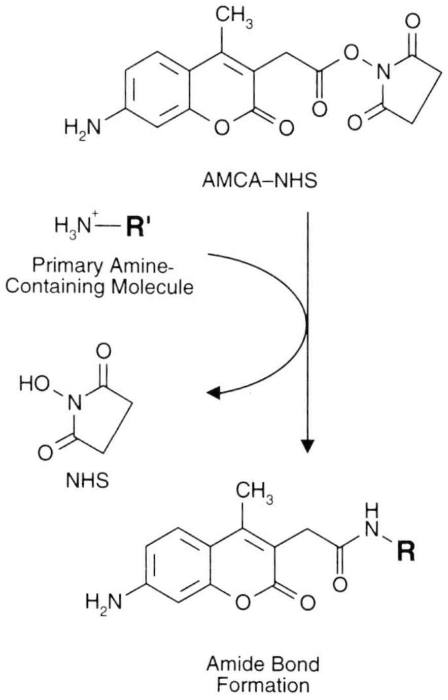

FIGURE 10.23 AMCA–NHS reacts with amines to form amide bonds.

AMCA–NHS is relatively insoluble in aqueous buffers. The compound must first be dissolved in organic solvent prior to adding a small aliquot to the reaction mixture. A concentrated stock solution may be prepared in DMSO and stored up to 2 weeks refrigerated or frozen without loss of activity. The solid and all solutions of AMCA–NHS should be protected from light to avoid photobleaching of the fluorophore.

AMCA–sulfo-NHS is an analog of AMCA–NHS which contains a sulfonate group on its NHS ring (Thermo Fisher). The negative charge of this group provides enough polarity to promote water solubility for the entire reagent in its activated form. The reactivity and properties of AMCA–sulfo-NHS are identical to those of AMCA–NHS.

These reactive dyes have been used for the modification of peptides (Mentinova *et al.*, 2012), to study the trafficking and release of *Leishmania* metacyclic HASPB on macrophage invasion (MacLean *et al.*, 2012), to investigate the structure and activity relationships of quinolone-type molecules against *Trypanosoma brucei* (Hiltensperger *et al.*, 2012), and as a labeled secondary antibody for general immunofluorescence detection assays (De Oliveira *et al.*, 2011).

Preparing an optimal fluorescent conjugate is largely dependent upon the degree of modification with the label. The following protocol is generalized for the labeling of a protein with AMCA–NHS. For particular labeling experiments, it is often necessary to vary the amount of fluorophore added to the reaction mixture to obtain the best combination of protein activity and fluorescent intensity in the conjugate. Too much label and nonspecific binding or fluorescent quenching may result; too little label and the complex will not possess enough fluorescent intensity to be sufficiently detectable.

Protocol

1. Dissolve the protein to be modified in 50-m*M* sodium borate, pH 8.5, at a concentration of 10 mg/ml. Other buffers may be used for an NHS ester reaction, including 0.1-*M* sodium phosphate, pH 7.5 (Chapter 3, Section 1.4).
2. Dissolve AMCA–NHS (Thermo Fisher) in DMSO at a concentration of 2.6 mg/ml. Protect from light.
3. In subdued lighting conditions, slowly add 50 to 100 μl of the AMCA–NHS stock solution to each milliliter of the protein solution, with gentle mixing.
4. React for 1 h at room temperature in the dark.
5. Remove excess reagent and reaction byproducts by gel filtration using a desalting resin. The sample volume should be no more than about 5 to 8% of the column volume.

The F/P ratio of the purified, labeled protein may be determined by measuring the absorbance at 345 nm and 280 nm. Ratios between 0.3 and 0.8 usually produce labeled molecules having acceptable levels of fluorescent intensity and good retention of protein activity. AMCA-labeled proteins may be lyophilized without significant loss of fluorescence. The addition of bovine serum albumin (15 mg/ml) or another such stabilizer is often necessary to retain solubility of the freeze-dried, labeled protein after reconstitution.

3.4. Sulfhydryl-Reactive Coumarin Derivatives

Two derivatives of aminomethylcoumarin are available for labeling sulfhydryl-containing molecules. The ability to label SH groups in proteins provides a means of directing the modification reaction to a limited number of sites, possibly avoiding active centers or binding regions better than when using amine-reactive probes. The first sulfhydryl-reactive probe discussed in this section makes use of a pyridyl disulfide group on the AMCA derivative and forms a reversible disulfide bond with thiol-containing molecules. The second probe is an iodoacetyl compound made from a diethyl and aminophenyl derivative of the basic aminomethylcoumarin structure, which forms a stable thioether linkage with thiol groups.

3.5. AMCA–HPDP

AMCA–HPDP is *N*-[6-(7-amino-4-methylcoumarin-3-acetamido)hexyl]-3′-(2′-pyridyldithio)propionamide.

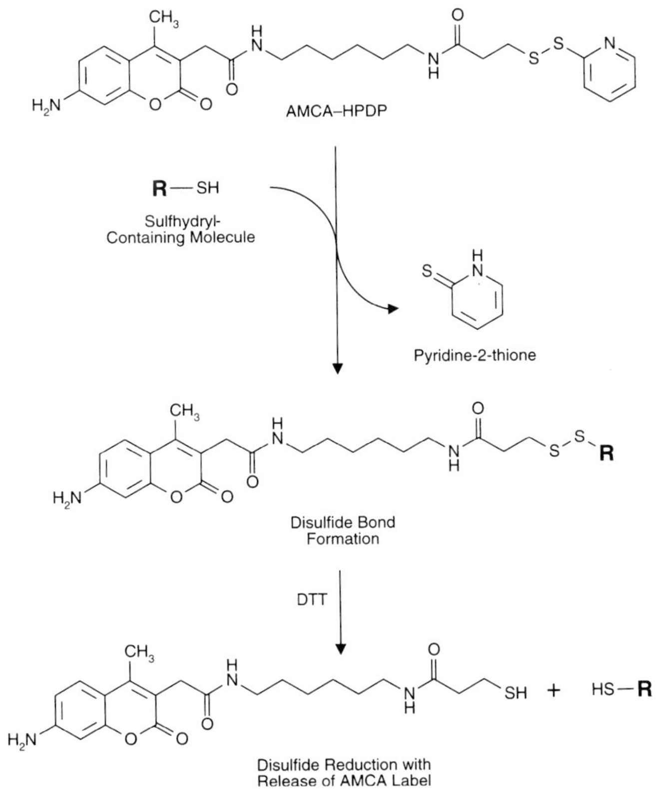

FIGURE 10.24 AMCA–HPDP reacts with sulfhydryl groups through its pyridyl disulfide end to form reversible disulfide bonds.

It is formed from AMCA plus a 1,6-diaminohexyl spacer off the carboxylate that has been additionally modified at its terminal end with SPDP (Chapter 6, Section 1.1). The result is a long spacer arm terminating in a pyridyl disulfide group reactive toward free sulfhydryl residues. The reaction of this group with a thiol creates a disulfide bond between the AMCA fluorophore and the molecule being modified. Thus, the fluorescent tag can be specifically cleaved by reduction with DTT or other disulfide reducing agents (Figure 10.24).

The required sulfhydryl residues can be naturally occurring on a protein, created by reduction of cystine crosslinks or by thiolation (Chapter 2, Section 4.1). For the labeling of antibody molecules, mild reduction with 2-mercaptoethylamine, DTT, or TCEP results in free sulfhydryl groups in the hinge region. Labeling in this area is advantageous to direct the modification away from antigen binding regions. Sulfhydryl residues also may be created on oligonucleotides without difficulty (Chapter 23, Section 2.2).

Although the reaction of a sulfhydryl-containing molecule with AMCA–HPDP results in the release of the chromogenic leaving group, pyridine-2-thione, using it to quantify the extent of modification may be difficult, because it absorbs at 343nm, which is in the same region as AMCA itself (345nm). The emission range of the AMCA probe is about 440 to 460nm, in the blue region of the spectrum. It has been used as a fluorescent label in the study of *S*-nitrosylation (Lopez-Sanchez *et al.*, 2012; Raju *et al.*, 2012).

The following protocol is a suggested method for labeling a protein with AMCA–HPDP. It is assumed that the presence of a sulfhydryl on the protein has been documented or created. The reaction conditions can be carried out in a variety of buffers between pH 6 and 9. Avoid the presence of extraneous sulfhydryl-containing compounds (such as disulfide reductants) that will compete in the reaction. The inclusion of EDTA in the modification buffer prevents metal-catalyzed sulfhydryl oxidation. Optimization for a particular labeling experiment should be performed to obtain the best level of fluorophore incorporation.

Protocol

1. Dissolve the sulfhydryl-containing protein to be labeled in 0.1-*M* sodium phosphate, 0.15-*M* NaCl, 10-m*M* EDTA, pH 7.2, at a concentration of 10mg/ml.
2. Dissolve AMCA–HPDP in DMSO at a concentration of 0.5mg/ml. Protect from light.
3. In subdued lighting conditions, add 50 to 100μl of the AMCA–HPDP stock solution to each milliliter of sulfhydryl-containing protein solution. Mix.
4. React for 1h at room temperature in the dark with occasional mixing.
5. Remove excess fluorophore and reaction by-products by gel filtration using a desalting resin.

To determine the F/P ratio of the labeled protein, measure the absorbance of the purified preparation at 345nm and 280nm. Ratios of 345nm/280nm within the range of 0.3 to 0.8 usually result in fluorescent

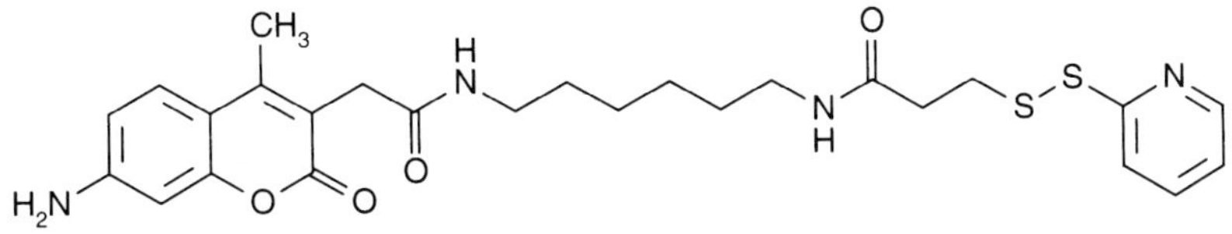

AMCA–HPDP
N-[6-(7-amino-4-methylcoumarin-3-acetamido)hexyl]-3'-(2'-pyridyldithio)propionamide
Excitation = 345 nm
Emission = 440-460 nm

conjugates with a good balance of high-intensity luminescence, low nonspecific binding, and excellent retention of biological activity within the protein component.

3.6. DCIA

DCIA is 7-diethylamino-3-[(4′-(iodoacetyl)amino) phenyl]-4-methylcoumarin, a derivative of the basic aminomethylcoumarin structure that contains a sulfhydryl-reactive iodoacetyl group and a diethyl substitution on its amine. This particular coumarin derivative is among the most fluorescent UV-excitable iodoacetamide probes available (Invitrogen) (Sippel, 1981).

The iodoacetyl group of DCIA reacts with sulfhydryls under slightly alkaline conditions to yield stable thioether linkages (Figure 10.25). They do not react with unreduced disulfides in cystine residues or with oxidized glutathione (Gorman *et al.*, 1987).

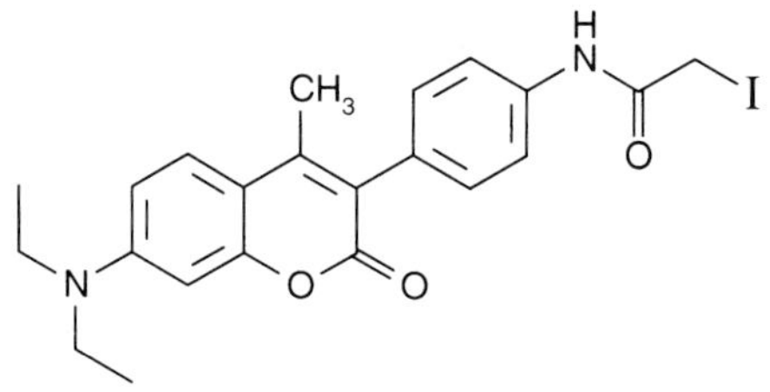

DCIA
7-Diethylamino-3-((4'-(iodoacetyl)-amino)phenyl)-4-methylcoumarin
MW490
Excitation = 382 nm
Emission = 472 nm
ε at 382 nm = 33,000 $M^{-1}cm^{-1}$

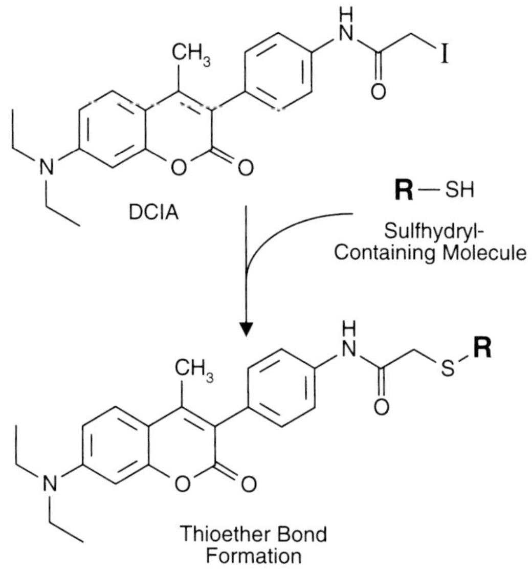

FIGURE 10.25 DCIA through its iodoacetamide group can modify sulfhydryls in proteins and other molecules to form thioether linkages.

Care must be taken to protect iodoacetyl reagents from light, not only to maintain the fluorescent yield of the coumarin component, but also to protect the iodoacetyl group from light-catalyzed breakdown. DCIA is soluble in DMF and DMSO. Concentrated stock solutions may be prepared in either solvent prior to addition of a small aliquot to a reaction mixture. Protect solutions from light by wrapping vessels in aluminum foil and working in subdued light.

The spectral properties of this fluorophore are similar to those of other coumarin derivatives. The excitation maximum occurs at about 382 nm and its emission peak at 472 nm, producing light in the blue region of the spectrum. The extinction coefficient of DCIA at its wavelength of maximum absorbance, 382 nm, is 33,000 $M^{-1}cm^{-1}$ (in methanol).

DCIA has been used to label numerous proteins and other biomolecules, including phospholipids (Silvius *et al.*, 1987); to study the interaction of mRNA with the 30S ribosomal subunit (Czworkowski *et al.*, 1991); in the investigation of cellular thiol components by flow cytometry (Durand and Olive, 1983); in the detection of carboxylate compounds using peroxyoxalate chemiluminescence (Grayeski and DeVasto, 1987); for general sulfhydryl labeling (Sippel, 1981); and in the development of advanced FRET methods to study protein–lipid interactions (Fernandes *et al.*, 2012).

A general protocol for the use of DCIA for fluorescently labeling proteins that contain sulfhydryl residues may be obtained by following the method discussed for AMCA–HPDP (previous section). After purification of the labeled protein, the F/P ratio of fluorophore incorporation may be determined by measuring its 382 nm/280 nm absorbance ratio.

3.7. Aldehyde- and Ketone-Reactive Coumarin Derivatives

Hydrazide groups can be coupled directly to aldehyde and ketone groups to form relatively stable hydrazone linkages (Chapter 3, Section 5.1). One AMCA derivative is commonly available that contains a hydrazine group modification on its carboxylate. Although most biomolecules don't contain aldehyde or ketone groups in their native state, carbohydrates, glycoproteins, RNA, and other molecules that contain sugar residues can be oxidized with sodium periodate to produce reactive formyl groups. The use of modification reagents that generate aldehydes upon coupling to a molecule also can be used to produce a hydrazide-reactive site (Chapter 2, Section 4.4).

DNA and RNA may be modified with hydrazide-reactive probes by reacting their cytosine residues with bisulfite to form reactive sulfone intermediates. These derivatives can undergo transamination reactions with hydrazide- or amine-containing probes to yield covalent bonds (Draper and Gold, 1980) (Chapter 23, Section 2.1).

3.8. AMCA–Hydrazide

AMCA–hydrazide is 7-amino-4-methylcoumarin-3-acetyl hydrazide, a hydrazine derivative off the carboxyl group of the basic AMCA molecule (Thermo Fisher). AMCA-based fluorophores are highly stable toward photobleaching. Molecules labeled with this probe retain their fluorescent intensity over three times longer than a fluorescein label when exposed to light. In addition, AMCA derivatives exhibit a large Stokes shift of over 100 nm, thus they are minimally affected by Rayleigh scattering effects during excitation. The blue light emitted by these labels is in a region of the spectrum well removed from the emission characteristics of other major fluorescent probes. This means that double-staining techniques easily can be used with an AMCA label. AMCA also exhibits little luminescence dependency on pH over the range of pH 3 to 10.

The hydrazide derivative of AMCA can be used to modify aldehyde- or ketone-containing molecules, including cytosine residues using the bisulfite activation procedure described in Chapter 23, Section 2.1. AMCA–hydrazide reacts with these target groups to form hydrazone bonds (Figure 10.26). Carbohydrates and glycoconjugates can be labeled specifically at their polysaccharide portion if the required aldehydes are first formed by periodate oxidation or another such method (Chapter 2, Section 4.4). The dye has been used to study the UV photodissociation of chromophore-labeled oligosaccharides (Ko and Brodbelt, 2011) and to label reactive peptides in site-specific tagging methods (Eldridge and Weiss, 2011).

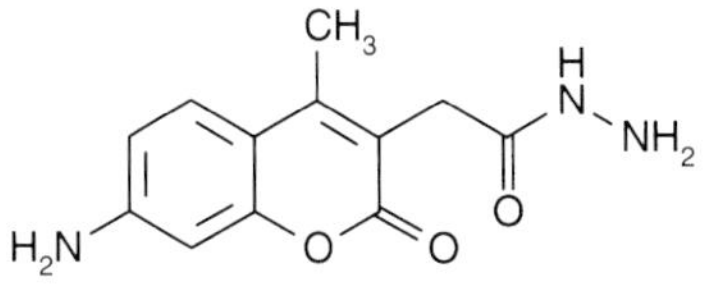

AMCA–Hydrazide
7-Amino-4-methylcoumarin-3-acetyl hydrazide
MW 247. 1
Excitation = 345 nm
Emission = 440–460 nm

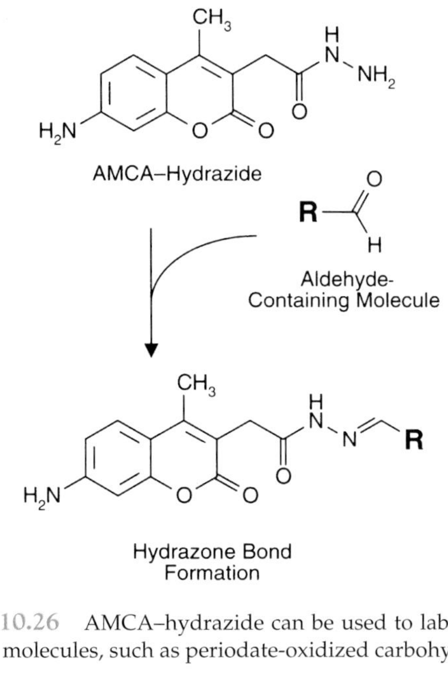

FIGURE 10.26 AMCA–hydrazide can be used to label aldehyde-containing molecules, such as periodate-oxidized carbohydrates.

AMCA–hydrazide has a maximal excitation wavelength of 345 nm and a maximum emission wavelength in the range of 440–460 nm. A solution of AMCA in PBS at a concentration of 16.7 ng/ml (71.61 nmoles/ml) gives an absorbance at 345 nm of about 1.28. This translates into a molar extinction coefficient at this wavelength of about 13,900 $M^{-1}cm^{-1}$. Different solvents and conditions may alter this value somewhat.

AMCA–hydrazide is soluble in DMSO or DMF and may be dissolved as a concentrated stock solution in either of these solvents prior to the addition of a small aliquot to an aqueous reaction medium. The solid and all solutions made from the fluorophore must be protected from light to avoid photo-decomposition. Freshly prepare the stock solution immediately before use. A suggested protocol on the use of this fluorescent probe may be obtained from the following method on the labeling of periodate-oxidized IgG. Optimization may be necessary to achieve the best level of fluorescent modification (F/P ratio) for a particular application.

Protocol for Oxidation of IgG Carbohydrate Residues with Sodium Periodate

1. Dissolve the antibody to be labeled in 0.1-*M* sodium phosphate, 0.15-*M* NaCl, pH 7.5, at a concentration of at least 10 mg/ml. The immunoglobulin must be glycosylated to work in this procedure.
2. Dissolve sodium periodate in water to a final concentration of 100-m*M*. Protect from light. Add 0.1 ml of this stock periodate solution to each milliliter of the antibody solution.
3. React for 15 min at room temperature, protected from light.
4. Quench the reaction by addition of 0.1 ml of glycerol per milliliter of reaction volume, mix, and then react for an additional 15 min. Remove excess reagents by gel filtration using a desalting column, and perform the chromatography using the phosphate buffer.
5. Adjust the concentration of IgG in the purified preparation to 1 mg/ml by the addition of 0.1-*M* sodium phosphate, 0.15-*M* NaCl, pH 7.5.

Protocol for Modification of Oxidized IgG with AMCA–Hydrazide

1. Dissolve AMCA–hydrazide in DMF at a concentration of 0.4 mg/ml. Protect from light.
2. Add 50 to 100 µl of the AMCA–hydrazide stock solution to each milliliter of the oxidized antibody solution. Note: At a level of 50 µl probe addition, polyclonal human IgG will be modified at a level that gives an F/P ratio of about 0.113. Since the labeling occurs only at the oxidized carbohydrate sites, the fluorophore incorporation typically is less than that observed when using amine-reactive probes.
3. React for 30 min at room temperature in the dark.
4. Remove excess fluorophore by dialysis or gel filtration using a desalting resin. Protect the labeled immunoglobulin from light.

Determine the F/P ratio by measuring the absorbance at 345 nm and 280 nm.

4. BORON DIPYRROMETHENE DERIVATIVES

Fluorescent dye compounds constructed from a core structure consisting of boron-dipyrromethene tricyclic ring system (e.g., 4-bora-3a, 4a-diaza-*s*-indacene) are a class of dyes all characterized by a bound boron atom in their center (Loudet and Burgess, 2007; Arroyo *et al.*, 2009; Schmitt *et al.*, 2009; Tram *et al.*, 2009; Boens *et al.*, 2012). These derivatives often go by the trade name BODIPY (Life Technologies), and a broad variety of compounds have been reported that have a wide range of fluorescence properties throughout the visible region of the spectrum. They are noted for their sharp excitation and emission peaks, but they also have a relatively small Stokes shift and can be sensitive to dye–dye quenching effects and environmental conditions, which cause changes in brightness and QY.

BODIPY fluorophores are a class of probes based on the fused, multi-ring structure, 4,4-difluoro-4-bora-3a,4a-diaza-*s*-indacene (Figure 10.27) (Invitrogen) (U.S. patent 4,774,339). This fundamental molecule can be modified, particularly at its 1, 3, 5, 7, and 8 carbon positions, to produce new fluorophores with different characteristics. The modifications cause spectral shifts in its excitation and emission wavelengths and can provide sites for chemical coupling to label biomolecules.

These dyes have been used for a wide range of applications, including in the development of FRET probes (Shao *et al.*, 2011), in the formation of light-harvesting metal-organic frameworks (Lee *et al.*, 2011), in the construction of dye-sensitized solar cells (Kolemen *et al.*, 2011), as labels of nucleotide derivatives to study DNA methylation (Falck *et al.*, 2011), in the study of the dissociation constant of the streptavidin–biotin interaction (Raphael *et al.*, 2011), and conjugated to targeting molecules for photodynamic therapy (Kamkaew *et al.*, 2012).

The BODIPY derivatives typically have high extinction coefficients and excellent QY, often greater than 0.8. Their spectral characteristics are relatively insensitive to changes in pH. Luminescent changes with shifts in pH usually are due to reconfiguration of a fluorophore's π-electron cloud if an atom on the ring system becomes protonated or unprotonated. Since the BODIPY structure lacks an ionizable group, alterations in pH have no effect on its spectral attributes.

The emission spectra of BODIPY derivatives normally display narrow bandwidths, providing intensely fluorescent labels for biomolecules. Unfortunately, they also have very small Stokes shifts, typically on the order of only 10 to 20 nm. Excitation at the optimal wavelength may cause some interference in measurements at the emission wavelength due to light scattering or cross-over from the wide bandwidth of the excitation source. The dyes usually require excitation at sub-optimal wavelengths to prevent this problem.

The following sections discuss the major BODIPY derivatives that are reactive toward particular functional groups in proteins and other molecules.

4.1. Amine-Reactive Boron Dipyrromethene Dyes

A number of BODIPY derivatives that contain reactive groups able to couple with amine-containing molecules are commonly available. The derivatives either contain a carboxylate group, which can be reacted with an amine in the presence of a carbodiimide to create an amide bond, or an NHS ester derivative of the carboxylate, which can react directly with amines to form amide linkages. The three discussed in this section are representative of this amine-reactive BODIPY family. The two NHS ester derivatives react under alkaline conditions with primary amines in molecular targets to form stable, highly fluorescent derivatives.

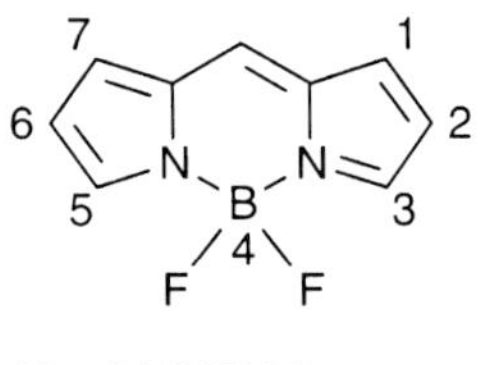

The BODIPY Structure
4,4-Difluoro-4-bora-
3a,4a-diaza-*s*-indacene

FIGURE 10.27 The basic structure of boron dipyrromethene (BODIPY) fluorophores.

The carboxylate derivative can be coupled to an amine using the EDC/sulfo-NHS reaction discussed in Chapter 4, Section 1.2.

The only disadvantage of using BODIPY fluorophores to label amines in macromolecules is the tendency for fluorescence quenching to occur if too many sites on one molecule are modified. Especially with proteins, using an amine-reactive probe usually results in multiple sites being modified on each molecule. All fluorophores experience some quenching effect if the degree of substitution is high, because dye–dye interactions are possible that can transfer energy from an excited-state fluorophore to a ground-state fluorophore before luminescence occurs. BODIPY probes, however, are especially notorious for dye–dye quenching effects. For this reason, the manufacturer (Invitrogen) recommends that the amine-reactive BODIPY probes only be used to modify substances that have the potential for just one substitution per molecule. In this sense, BODIPY fluorophores are particularly well-suited for tagging DNA probes at the 5′ end or lipid molecules on their head groups. Oligonucleotides modified to contain an amine on their 5′ phosphate group (Chapter 23, Section 2.1) are good candidates for labeling with this fluorophore. Other BODIPY probes that contain reactivity toward non-amine functionalities such as sulfhydryls or polysaccharides may be more effective at labeling macromolecules like proteins, since these groups occur at more limited sites within the molecules and the modification level can be better controlled.

4.2. BODIPY FL C_3-SE

BODIPY FL C_3-SE is 4,4-difluoro-5,7-dimethyl-4-bora-3a,4a-diaza-*s*-indacene-3-propionic acid, succinimidyl ester (Invitrogen). The derivatization to the base BODIPY molecule results in fluorescent properties which mimic fluorescein in its emission wavelength. The molecule thus emits light in the green region of the spectrum. The NHS ester on its propionic side chain provides amine reactivity, resulting in amide bond linkages with modified molecules (Figure 10.28).

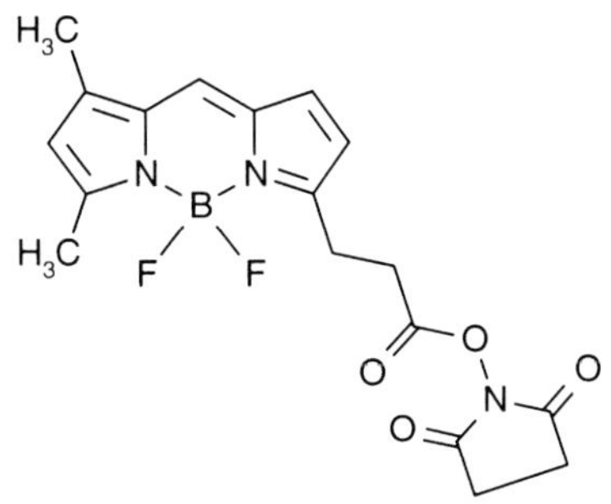

BODIPY FL C_3-SE
4,4-Difluoro-5,7-dimethyl-4-bora-3a,4a-diaza-*s*-indacene-3-propionic acid, succinimidyl ester
MW 389
Excitation = 502 nm
Emission = 510 nm
ε at 502 nm = 77,000 $M^{-1}cm^{-1}$

This fluorophore has an excitation maximum at 502 nm and an emission maximum at 510 nm. The small Stokes shift of only 8 nm creates some difficulty in discrete excitation without contaminating the emission measurement with scattered or overlapping light. The extinction coefficient of the molecule in methanol is about 77,000 $M^{-1}\,cm^{-1}$ at 502 nm.

BODIPY FL C_3-SE is insoluble in aqueous solution, but may be dissolved in DMF or DMSO as a concentrated stock solution prior to addition of a small aliquot to a reaction. For aqueous reactions, a pH range of 7 to 9 is optimal. Avoid amine-containing buffers. The reaction also may be done in organic solvent.

Since BODIPY fluorophores are easily quenched if substitutions on a molecule exceed a 1:1 stoichiometry, modification of proteins with this fluorophore probably will not yield satisfactory results. However, for labeling molecules which contain only one amine group, BODIPY FL C_3-SE will give intensely fluorescent derivatives.

4.3. BODIPY 530/550 C_3

BODIPY 530/550 C_3 is 4,4-difluoro-5,7-diphenyl-4-bora-3a,4a-diaza-*s*-indacene-3-propionic acid (Invitrogen).

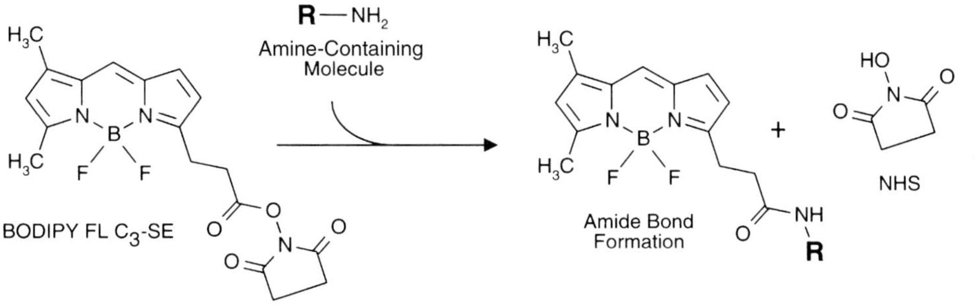

FIGURE 10.28 The terminal NHS ester on the linker arm of this BODIPY derivative can be used to modify amine-containing molecules, forming amide bond linkages.

This derivative of the basic BODIPY structure contains two phenyl rings off the No. 5 and 7 carbon atoms and a propionic acid group on the No. 3 carbon atom. The carboxylate group may be used to attach the fluorophore to amine-containing molecules via a carbodiimide reaction to create an amide bond. The substituents on this BODIPY fluorophore result in alterations to its spectral properties, pushing its excitation and emission maximums up to higher wavelengths.

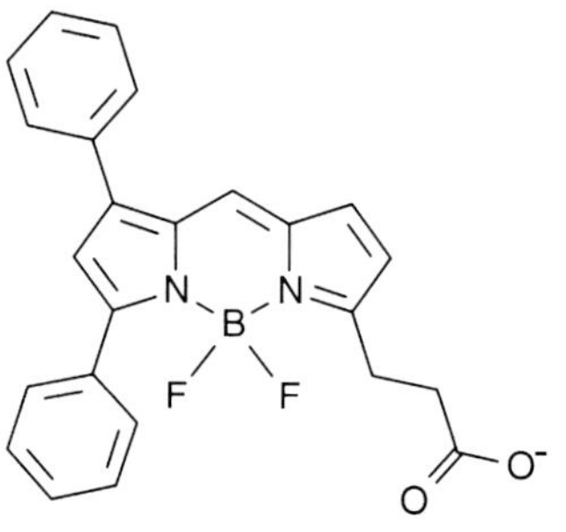

BODIPY 530/550 C_3
4,4-Difluoro-5,7-diphenyl-4-bora-3a,4a-diaza-*s*-indacene-3-propionic acid
MW 416
Excitation = 535 nm
Emission = 552 nm
ε at 535 nm = 62,000 $M^{-1}cm^{-1}$

The excitation maximum for the molecule occurs at 535 nm and its emission at 552 nm. Its Stokes shift is slightly greater than some of the other BODIPY fluorophores, producing a 17-nm separation between excitation and emission peaks. BODIPY 530/550 C_3 has an extinction coefficient in methanol of about 62,000 $M^{-1}cm^{-1}$ at 535 nm.

BODIPY 530/550 C_3 is insoluble in aqueous solution, but it may be dissolved in DMF or DMSO as a concentrated stock solution prior to addition of a small aliquot to a reaction. Coupling to amine-containing molecules may be done using the EDC/sulfo-NHS reaction as discussed in Chapter 4, Section 1.2 (Figure 10.29). However, modification of proteins with this fluorophore probably will not yield satisfactory results, since BODIPY fluorophores are easily quenched if substitutions on a molecule exceed a 1:1 stoichiometry. For labeling molecules which contain only one amine group, such as DNA probes modified at the 5′ end to contain an amine (Chapter 23, Section 2.1), BODIPY 530/550 C_3 will give intensely fluorescent derivatives.

4.4. BODIPY 530/550 C_3-SE

BODIPY 530/550 C_3-SE is 4,4-difluoro-5,7-diphenyl-4-bora-3a,4a-diaza-*s*-indacene-3-propionic

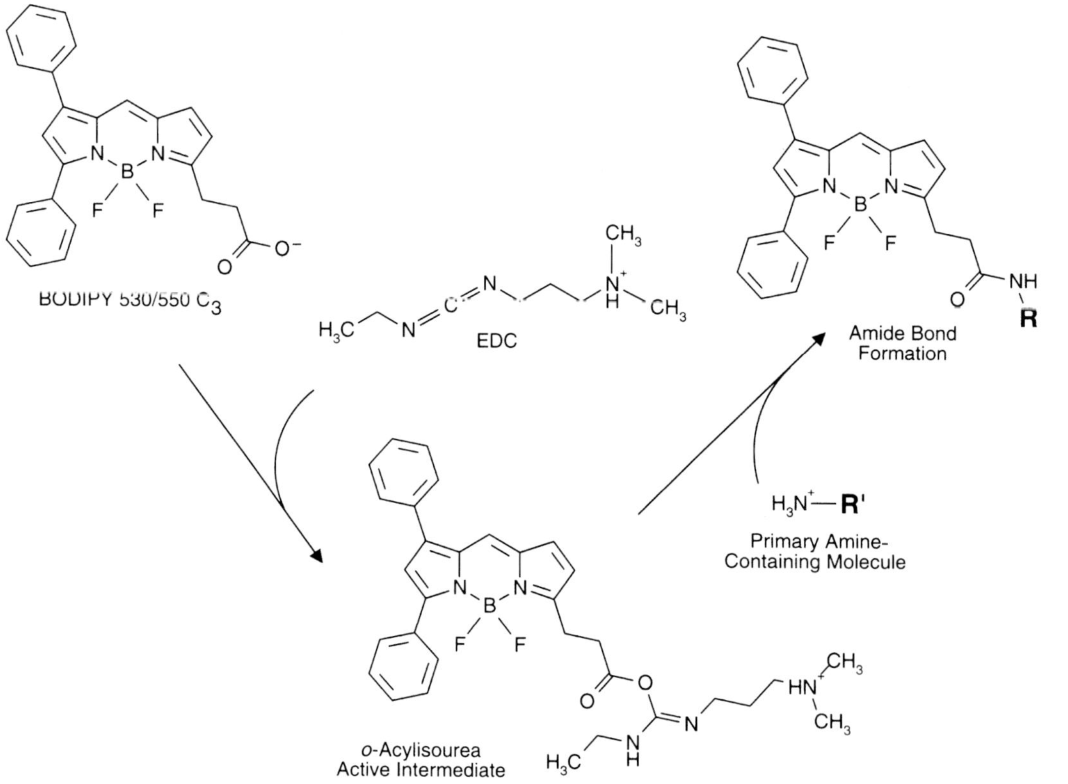

FIGURE 10.29 This BODIPY fluorophore contains a carboxylate group that can be attached to amine-containing molecules using a carbodiimide reaction.

acid, succinimidyl ester (Invitrogen). The compound is an analog of BODIPY 530/550 C3 that contains an active NHS ester on its propionic acid side chain (Chapter 3, Section 1.4). The ester reacts with primary amines to form stable amide bonds.

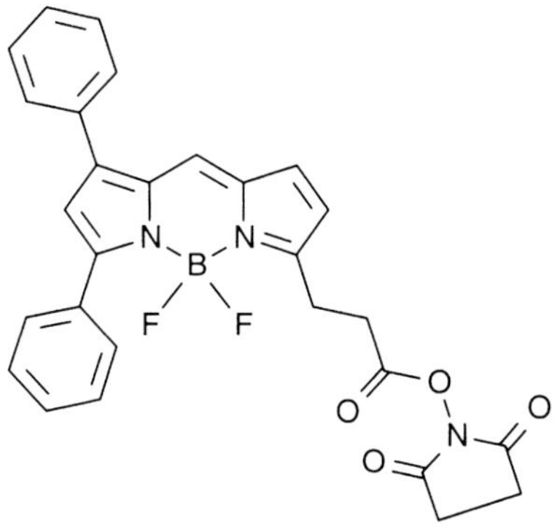

BODIPY 530/550 C_3-SE
4,4-Difluoro-5,7-diphenyl-4-bora-3a,4a-diaza-s-indacene-3-propionic acid, succinimidyl ester
MW 513
Excitation = 533 nm
Emission = 550 nm
ε at 533 nm = 70,000 $M^{-1}cm^{-1}$

The excitation maximum for BODIPY 530/550 C_3-SE occurs at 533 nm and its emission at 550 nm. Its Stokes shift is relatively small and may not be enough to avoid completely problems of excitation-light interference in emission measurements. The molecule has an extinction coefficient in methanol of about 70,000 $M^{-1}cm^{-1}$ at 533 nm.

BODIPY 530/550 C_3-SE is insoluble in aqueous solution, but may be dissolved in DMF or DMSO as a concentrated stock solution prior to addition of a small aliquot to a reaction. Coupling to amine-containing molecules proceeds by nucleophilic attack at the carbonyl group, release of the NHS leaving group, and formation of an amide linkage (Figure 10.30). The reaction may be done in buffered environments having a pH range of 7 to 9. However, modification of proteins with this fluorophore may not yield satisfactory results, since BODIPY fluorophores are easily quenched if substitutions on a molecule result in a high molar ratio of incorporation. For labeling molecules which contain only one amine group, such as DNA probes modified at the 5′ end to contain an amine (Chapter 23, Section 2.1), BODIPY 530/550 C_3-SE will give intensely fluorescent derivatives.

4.5. Aldehyde-/Ketone-Reactive Boron Dipyrromethene Dyes

Hydrazide groups react with aldehyde and ketone groups to form hydrazone linkages (Chapter 3, Section 5.1). Three BODIPY derivatives are available that contain a hydrazine group modification of carboxylate side chains. Biomolecules such as proteins that do not normally possess aldehyde residues can be modified to contain them by a number of chemical means (Chapter 2, Section 4.4).

In addition, DNA and RNA may be modified with hydrazide-containing fluorophores by a transamination reaction of their cytosine residues using bisulfite as a catalyst (Chapter 23, Section 2.1) (Draper and Gold, 1980).

4.6. BODIPY 530/550 C_3 Hydrazide

BODIPY 530/550 C_3 hydrazide is 4,4-difluoro-5,7-diphenyl-4-bora-3a,4a-diaza-*s*-indacene-3-propionyl hydrazide, a derivative of the basic BODIPY structure that contains two phenyl rings off the No. 5 and 7 carbon atoms and a propionic acid hydrazide group on the No. 3 carbon atom (Invitrogen). The hydrazide group reacts with aldehyde- or ketone-containing molecules to form hydrazone linkages (Figure 10.31). The compound may be used to label glycoproteins or other carbohydrate-containing molecules

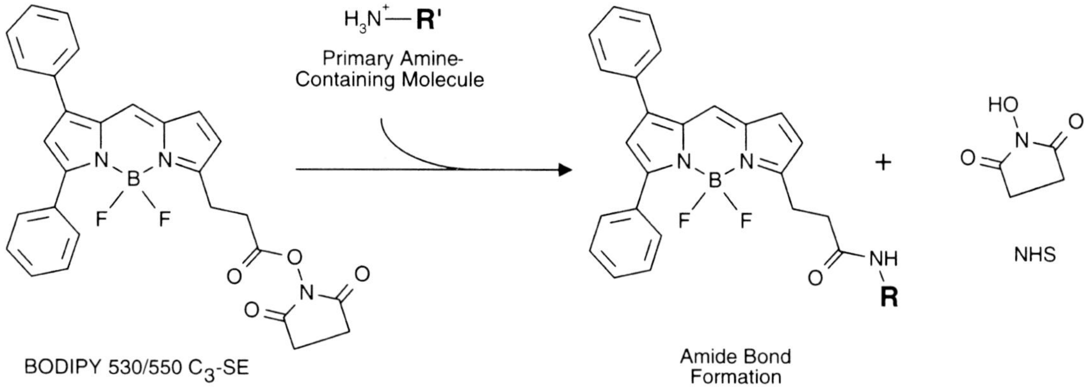

FIGURE 10.30 The NHS ester group of this BODIPY compound provides amine reactivity, which can be used to label proteins through amide bond linkages.

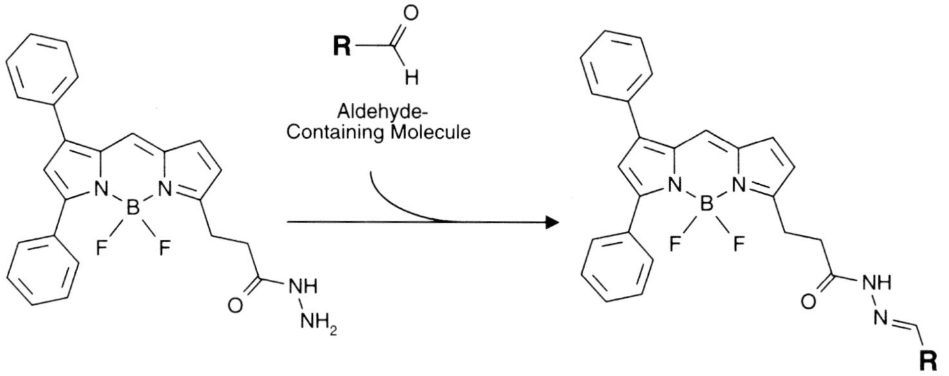

BODIPY 530/550 C_3 Hydrazide

Hydrazone Bond Formation

FIGURE 10.31 The side-chain hydrazide group of this BODIPY derivative can be used to label aldehyde-containing molecules. Glycoconjugates may be labeled after oxidation of carbohydrates with sodium periodate to produce the required aldehydes.

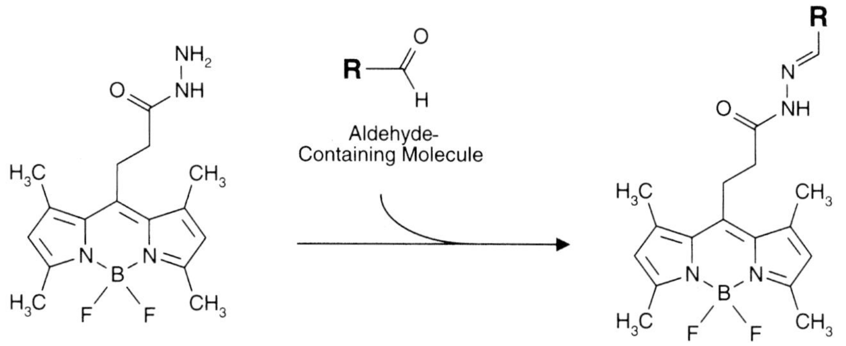

BODIPY 493/503 C_3 Hydrazide

Hydrazone Bond Formation

FIGURE 10.32 Reaction of this BODIPY fluorophore with aldehyde groups creates hydrazone linkages.

after oxidation of their polysaccharide portions with sodium periodate to yield aldehydes.

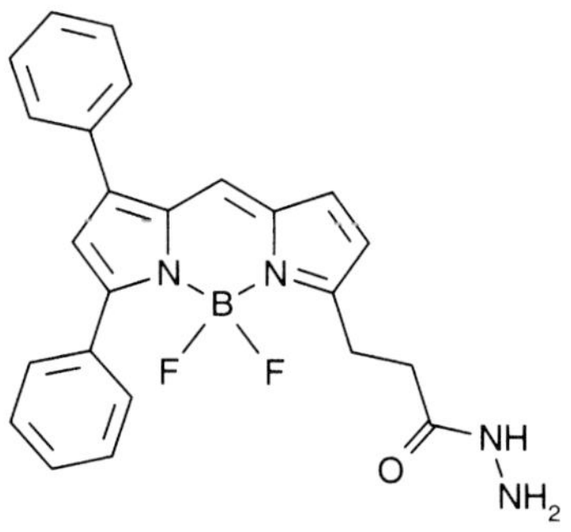

BODIPY 530/550 C_3 Hydrazide
4,4-difluoro-5,7-diphenyl-4-bora-3a,4a-diaza-*s*-indacene-3-propionyl hydrazide
MW 430
Excitation = 534 nm
Emission = 551 nm
ε at 534 nm = 79,000 $M^{-1}cm^{-1}$

The excitation maximum for BODIPY 530/550 C_3 hydrazide occurs at 534 nm and its emission at 551 nm. The molecule has an extinction coefficient in methanol of about 79,000 M^{-1} cm^{-1} at 534 nm.

BODIPY 530/550 C_3 hydrazide is insoluble in aqueous solution, but may be dissolved in DMF or methanol as a concentrated stock solution prior to addition of a small aliquot to a reaction. Coupling to aldehyde-containing molecules occurs rapidly with the formation a hydrazone linkage. The reaction may be done in buffered environments having a pH range of 5 to 10. However, modification of glycoproteins with this fluorophore may not yield satisfactory results, since BODIPY fluorophores are easily quenched if substitutions on a molecule result in a high molar ratio of incorporation.

4.7. BODIPY 493/503 C_3 Hydrazide

BODIPY 493/503 C_3 hydrazide is 4,4-difluoro-1,3,5,7-tetramethyl-4-bora-3a,4a-diaza-s-indacene-8-propionyl hydrazide (Invitrogen). Unlike BODIPY 530/550 C_3 hydrazide, this BODIPY derivative contains substituents that shift to lower wavelengths the spectral characteristics of its fluorescence. The molecule is highly reactive toward aldehyde-containing compounds, including glycoproteins which have been oxidized with sodium periodate to create the requisite groups (Figure 10.32).

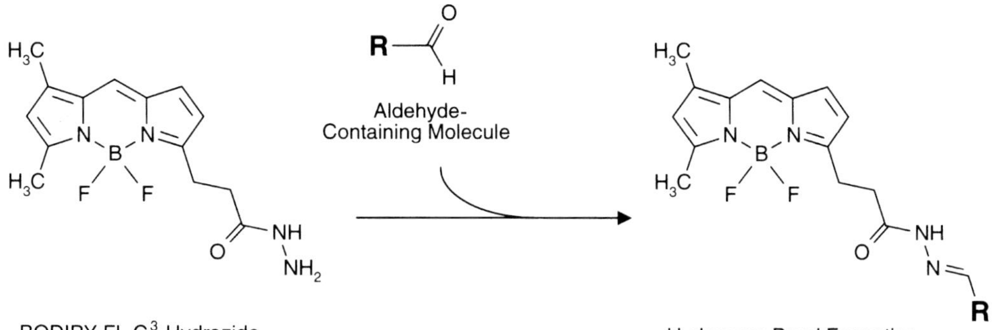

BODIPY FL C^3 Hydrazide

Hydrazone Bond Formation

FIGURE 10.33 Modification of aldehyde-containing molecules can be done through this BODIPY derivative's hydrazide group.

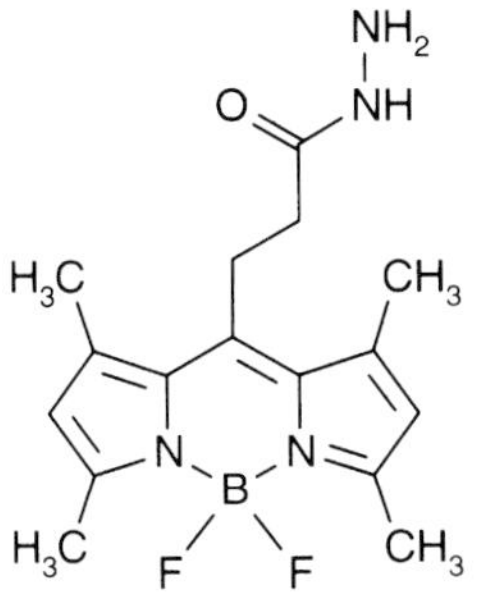

BODIPY 493/503 C_3 Hydrazide
4,4-difluoro-1,3,5,7-tetramethyl-4-bora-3a,4a-diaza-*s*-indacene-8-propionyl hydrazide
MW 334
Excitation = 498 nm
Emission = 506 nm
ε at 498 nm = 92,000 $M^{-1}cm^{-1}$

The excitation maximum for BODIPY 493/503 C_3 hydrazide occurs at 498nm and its emission at 506nm. Since this is an extremely small Stokes shift, it may be difficult to avoid completely problems of excitation light scattering interference in critical emission measurements unless sub-optimal excitation wavelengths are used. The molecule has an extinction coefficient in methanol of about 92,000 $M^{-1} cm^{-1}$ at 493 nm.

BODIPY 493/503 C_3 hydrazide is insoluble in aqueous solution, but may be dissolved in DMF or DMSO as a concentrated stock solution prior to addition of a small aliquot to a reaction mixture. Coupling to aldehyde-containing molecules occurs rapidly with the formation of a hydrazone linkage. The reaction may be done in buffered environments having a pH range of 5 to 10. However, modification of glycoproteins with this fluorophore may not yield satisfactory results, since BODIPY fluorophores are easily quenched if substitutions on a molecule result in a high molar ratio of incorporation. Limiting the modification level by the reaction to no more than a 2- to 4-fold molar excess of probe to the amount of glycoconjugate present may overcome this quenching problem.

4.8. BODIPY FL C_3 Hydrazide

BODIPY FL C_3 hydrazide is 4,4-difluoro-5,7-dimethyl-4-bora-3a,4a-diaza-*s*-indacene-3-propionyl hydrazide (Invitrogen). Unlike the two BODIPY hydrazide derivatives discussed above, this derivative contains substituents that produce luminescent characteristics similar to that of fluorescein, particularly with regard to fluorescing in the green region of spectrum. The molecule is highly reactive toward aldehyde-containing compounds, including glycoproteins that have been oxidized with sodium periodate to create the requisite aldehyde groups (Figure 10.33).

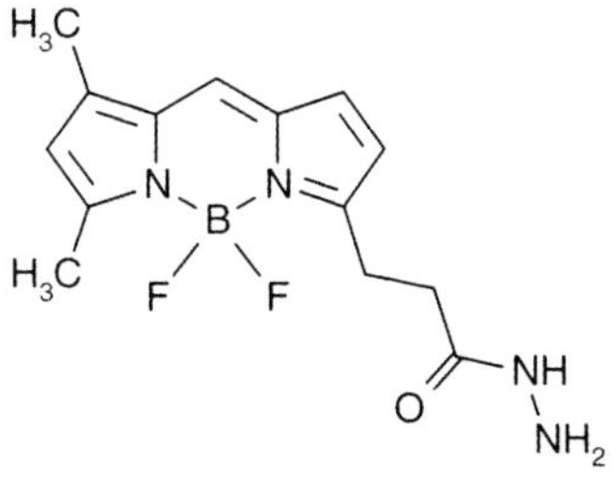

BODIPY FL C_3 Hydrazide
4,4-difluoro-5,7-dimethyl-4-bora-3a,4a-diaza-*s*-indacene-3-propionyl hydrazide
MW 306
Excitation = 503 nm
Emission = 510 nm
ε at 503 nm = 71,000 $M^{-1}cm^{-1}$

The excitation maximum for BODIPY FL C_3 hydrazide occurs at 503nm and its emission at 510nm. The extremely small Stokes shift makes it difficult to avoid problems of excitation-light interference in critical emission measurements unless sub-optimal excitation wavelengths are used. The molecule has an extinction coefficient in methanol of about 71,000 $M^{-1} cm^{-1}$ at 503 nm.

BODIPY FL C_3 hydrazide is insoluble in aqueous solution, but may be dissolved in DMF or methanol as a concentrated stock solution prior to addition of a small aliquot to a reaction mixture. Coupling to aldehyde-containing molecules occurs rapidly with the formation of a hydrazone linkage. The reaction may be done in buffered environments having a pH range of 5–10. However, modification of glycoproteins with this fluorophore may result in fluorescent quenching effects if substitutions on a molecule are at a high molar ratio of incorporation. Limiting the modification level by reacting no more than a 2- to 4-fold molar excess of probe to the amount of glycoconjugate present may help to overcome the quenching problem.

4.9. Sulfhydryl-Reactive Boron Dipyrromethene Dyes

Three BODIPY derivatives are available for labeling sulfhydryl-containing molecules. The ability to label SH groups in proteins with sulfhydryl-reactive probes provides a means of directing the modification reaction to a more limited number of sites than occurs when using amine-reactive chemistries. Directed coupling potentially can avoid active centers or binding regions. The first two sulfhydryl-reactive probes discussed in this section make use of iodoacetyl derivatives off the basic BODIPY molecule. The third probe is a bromomethyl derivative that also has good reactivity toward sulfhydryls.

4.10. BODIPY FL IA

BODIPY FL IA is *N*-(4,4-difluoro-5,7-dimethyl-4-bora-3a,4a-diaza-*s*-indacene-3-propionyl)-*N′*-iodoacetylethylenediamine, an intensely fluorescent derivative of the basic BODIPY structure which is useful in modifying sulfhydryl groups (Invitrogen). The iodoacetyl group reacts with SH groups in proteins and other molecules to form a stable thioether linkage (Figure 10.34). The reactive group is at the end of a reasonably long spacer arm, providing enough length to avoid steric problems in modifying sulfhydryls not easily accessible at the surface of macromolecules.

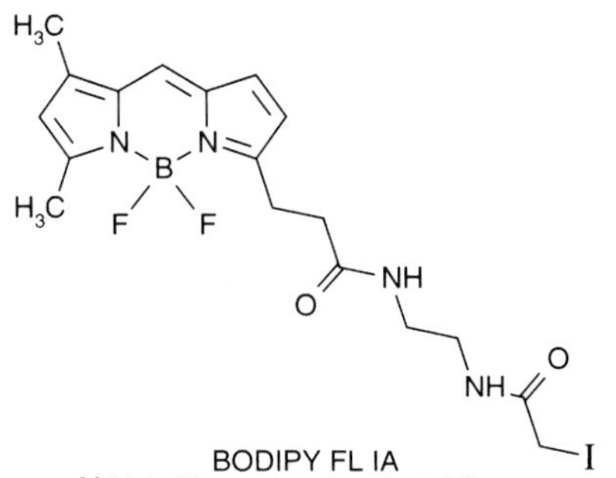

BODIPY FL IA
N-(4,4-difluoro-5,7-dimethyl-4-bora-3a,4a-diaza-*s*-indacene-3-propionyl)-*N'*-iodoacetylethylenediamine
MW 502
Excitation = 504 nm
Emission = 510 nm
ε at 504 nm = 79,000 $M^{-1}cm^{-1}$

The spectral characteristics of BODIPY FL IA somewhat mimic the green luminescence of fluorescein, thus the origin of the FL designation in its name. The excitation maximum for the probe occurs at 504nm and its emission at 510nm. The extremely small Stokes shift makes it difficult to avoid problems of excitation-light interference in critical emission measurements unless sub-optimal excitation wavelengths below its excitation maximum are used. The molecule has an extinction coefficient in methanol (containing 1% sodium acetate and 1% 2-mercaptoethanol) of about 79,000 $M^{-1}cm^{-1}$ at 504nm.

BODIPY FL IA is insoluble in aqueous solution, but may be dissolved in DMF or DMSO as a concentrated stock solution prior to addition of a small aliquot to a reaction mixture. Coupling to sulfhydryl-containing molecules occurs rapidly with the formation of a thioether linkage. The reaction may be done in 50-m*M* sodium borate, 5-m*M* EDTA, pH 8.3. The main consideration is to protect the iodoacetyl derivative from light which may generate iodine and reduce the reactivity of the probe. In addition, to avoid the fluorescence quenching effects that are often a problem with BODIPY probes, react no more than a 2- to 4-fold molar excess of probe to the amount of sulfhydryl groups present. Oligonucleotides containing a sulfhydryl modification at their 5′ ends (Chapter 23,

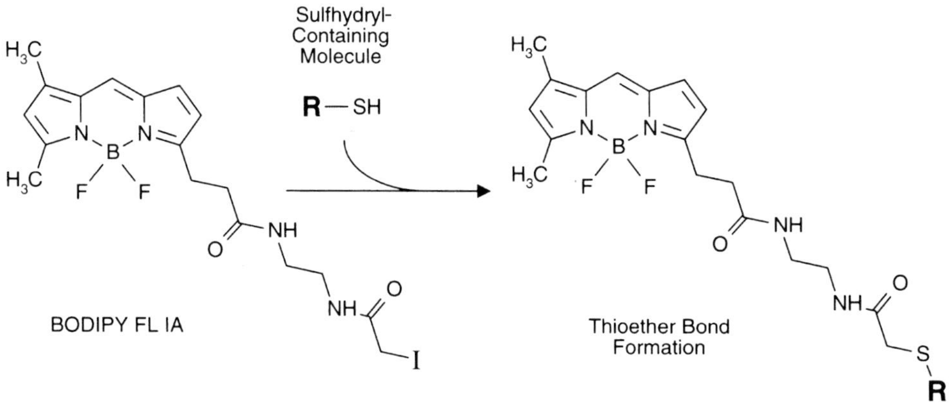

FIGURE 10.34 The long linker arm of this BODIPY derivative contains a sulfhydryl-reactive iodoacetamide group that can couple with a thiol group in proteins and other molecules to form a thioether bond.

Section 2.2) may be coupled with BODIPY FL IA, yielding highly fluorescent probes.

4.11. BODIPY 530/550 IA

BODIPY 530/550 IA is *N*-(4,4-difluoro-5,7-diphenyl-4-bora-3a,4a-diaza-*s*-indacene-3-propionyl)-*N'*-iodoacetylethenediamine, a derivative similar to that of BODIPY FL IA, but containing two phenyl groups rather than two methyl substituents in the No. 5 and 7 positions. This change in structure results in modulation of the spectral characteristics such that its excitation and emission wavelengths and its Stokes shift are all increased. The spacer arm off the No. 3 carbon atom of the basic BODIPY core contains a terminal iodoacetyl group, which reacts with sulfhydryl groups in proteins and other macromolecules to create stable thioether linkages (Figure 10.35).

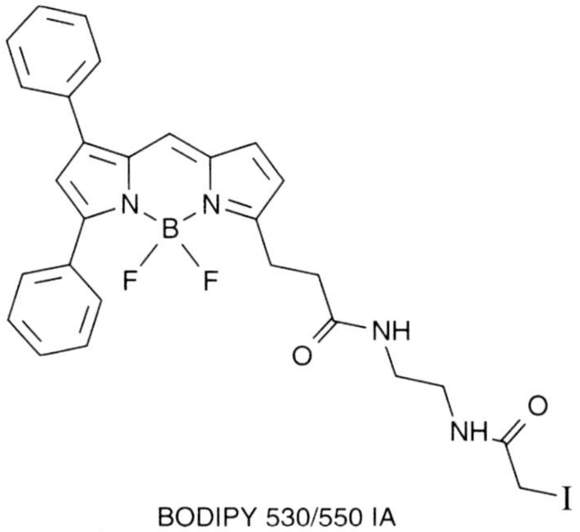

BODIPY 530/550 IA
N-(4,4-difluoro-5,7-diphenyl-4-bora-3a,4a-diaza-*s*-indacene-3-propionyl)-*N'*-iodoacetylethylenediamine
MW 626
Excitation = 534 nm
Emission = 552 nm
ε at 534 nm = 69,000 $M^{-1}cm^{-1}$

The excitation maximum for BODIPY 530/550 IA occurs at 534nm and its emission at 552nm. The Stokes shift is greater than the "FL" BODIPY derivatives, having an 18nm differential between excitation and emission wavelengths. The molecule has an extinction coefficient in methanol (containing 1% sodium acetate and 1% 2-mercaptoethanol) of about 69,000 $M^{-1}cm^{-1}$ at 534nm.

BODIPY 530/550 IA is insoluble in aqueous solution, but may be dissolved in DMF or DMSO as a concentrated stock solution prior to addition of a small aliquot to a reaction mixture. Coupling to sulfhydryl-containing molecules occurs rapidly with the formation a thioether linkage. The reaction may be done in 50-m*M* sodium borate, 5-m*M* EDTA, pH 8.3. The main consideration is to protect the iodoacetyl derivative from light which may generate iodine and reduce the reactivity of the probe. To limit the degree of fluorescent quenching in the resultant conjugate, the probe should be reacted at no more than a 2- to 4-fold molar excess over the amount of target molecule present.

4.12. Br-BODIPY 493/503

Br-BODIPY 493/503 is 8-bromomethyl-4,4-difluoro-1,3,5,7-tetramethyl-4-bora-3a,4a-diaza-*s*-indacene, a small BODIPY derivative containing a short, sulfhydryl-reactive bromomethyl group. The modifications to the core molecule result in modulation of its spectral characteristics such that, after conjugation, its excitation and emission wavelengths are reduced somewhat from other BODIPY probes. The reagent can be coupled to SH-containing molecules to produce a thioether linkage (Figure 10.36).

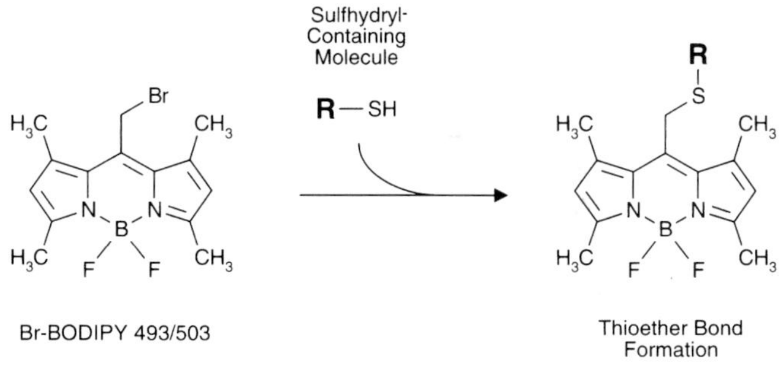

FIGURE 10.36 Br-BODIPY can be used to modify sulfhydryl-containing molecules to form thioether linkages.

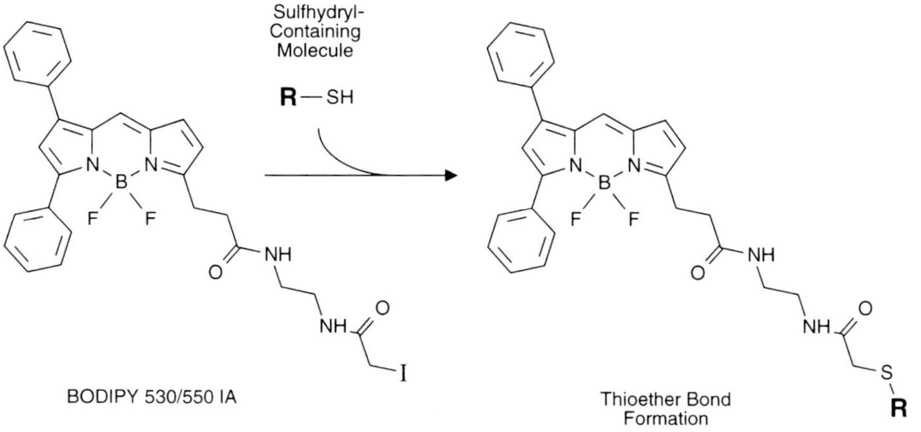

FIGURE 10.35 The iodoacetamide group of this BODIPY fluorophore can react with sulfhydryl-containing molecules to form thioether linkages.

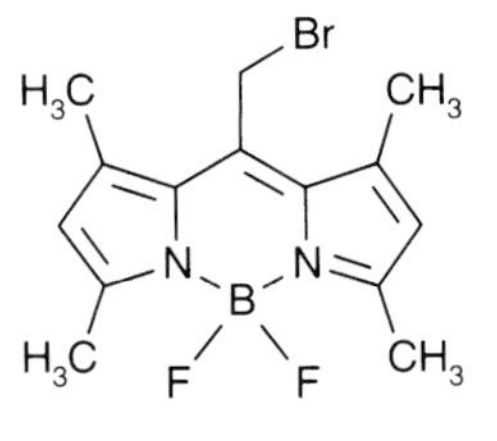

Br-BODIPY 493/503
8-Bromomethyl-4,4-difluoro-1,3,5,7-tetramethyl-4-bora-3a,4a-diaza-*s*-indacene
MW 341
Excitation = 515 nm
Emission = 525 nm
ε at 515 nm = 55,000 $M^{-1}cm^{-1}$

The excitation maximum for Br–BODIPY 493/503 is 515nm and its emission occurs at 525nm when dissolved in methanol. Upon coupling to a sulfhydryl compound, however, the excitation wavelength of the adduct decreases to 493nm and its emission drops to 503nm. The very small 10-nm Stokes shift may be a problem, particularly in avoiding interference due to of excitation-light scattering in critical emission measurements. Sub-optimal excitation wavelengths below the excitation maximum may be used to reduce extraneous light contamination. The molecule has an extinction coefficient in methanol of about 55,000 M^{-1} cm^{-1} at 515nm.

Br-BODIPY 493/503 is insoluble in aqueous reaction mixtures, but may be dissolved in DMF or DMSO as a concentrated stock solution prior to addition of a small amount to a buffered solution. Coupling to sulfhydryl-containing molecules is rapid, leading to the formation a thioether linkage. The reaction may be done in 50-m*M* sodium borate, 5-m*M* EDTA, pH 8.3. An important consideration is to protect the iodoacetyl derivative from light which may generate iodine and reduce the reactivity of the probe.

5. PYRENE DERIVATIVES

Fluorescent dyes having a core structure based on a pyrene ring system (or four fused benzene rings) are characterized as having strong emission characteristics in the blue region of the spectrum. Pyrene derivatives can be rather hydrophobic and unsuitable for use in bioconjugation unless they contain sulfonate groups on the rings, which provide dramatically increased water solubility. One family of pyrene dyes, termed Cascade Blue (Life Technologies), is available in a broad range of different forms. Other pyrene-type dyes are available from Thermo Fisher and Dyomics. All of these compounds also have various reactivities toward functional groups on biomolecules, which allows labeling of virtually any targeting molecule for use in assays. Pyrene-type dyes have been used for a wide range of applications, including in the study of polymer dynamics (Beija *et al.*, 2011), to investigate biopolymer structure (Cantor, 2012), in the labeling and detection of nanoparticle constructs (Kainz *et al.*, 2011), in conjugates of oligonucleotide probes (Juskowiak, 2011), and in the labeling of antibodies for use in cellular imaging (Guo *et al.*, 2011).

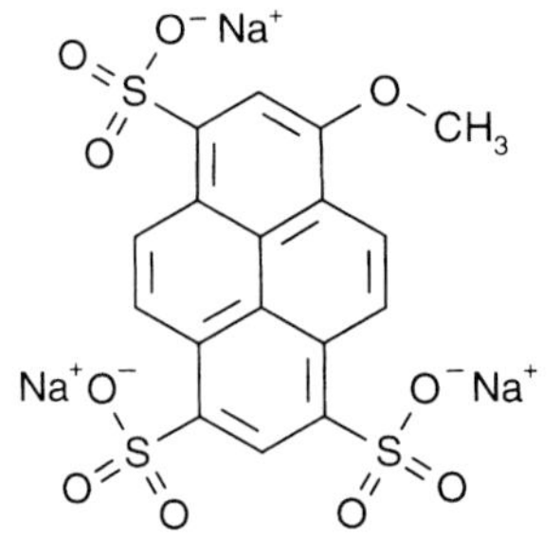

8-Methoxypyrene-1,3,6-trisulfonic acid, trisodium salt

FIGURE 10.37 The basic structure of pyrene-based fluorophores.

Cascade Blue derivatives are fluorescent probes having strong luminescence in the blue region of the spectrum (Invitrogen). The basic Cascade Blue molecule is derived from a trisulfonated pyrene backbone (Figure 10.37) (Whitaker *et al.*, 1991). Other sulfonated pyrene derivatives are commercially available from other sources as well (e.g., Thermo Fisher). This dye type is a reactive analog of 8-methoxypyrene-1,3,6-trisulfonic acid, which is a blue fluorescent neural tracer. The fluorophore emits light in a region removed from the luminescent signal of fluorescein or Lucifer Yellow, making it a good choice for multi-labeling applications. The dye can be used along with Lucifer Yellow CH and sulforhodamine 101 for three-color mapping of neuronal components and processes. Cascade Blue derivatives have relatively high absorptivity, good QY (typically about 0.54), excellent water solubility due to the presence of the negatively charged sulfonate groups, and good photostability. Labeling proteins and other macromolecules with Cascade Blue derivatives can be done with little fluorescent quenching due to dye–dye interactions.

The following sections discuss the most important Cascade Blue derivatives that are available for covalent modification purposes.

5.1. Amine-Reactive Cascade Blue Acetyl Azide

One Cascade Blue derivative is available for creating linkages with amine-containing molecules. The acetyl azide functionality of this reagent reacts with primary amines at ambient temperatures or below to create amide bond derivatives (Lanier and Recktenwald, 1991; Oparka *et al.*, 1991). At elevated temperatures (80°C in DMF), the acetyl azide group rearranges to form an

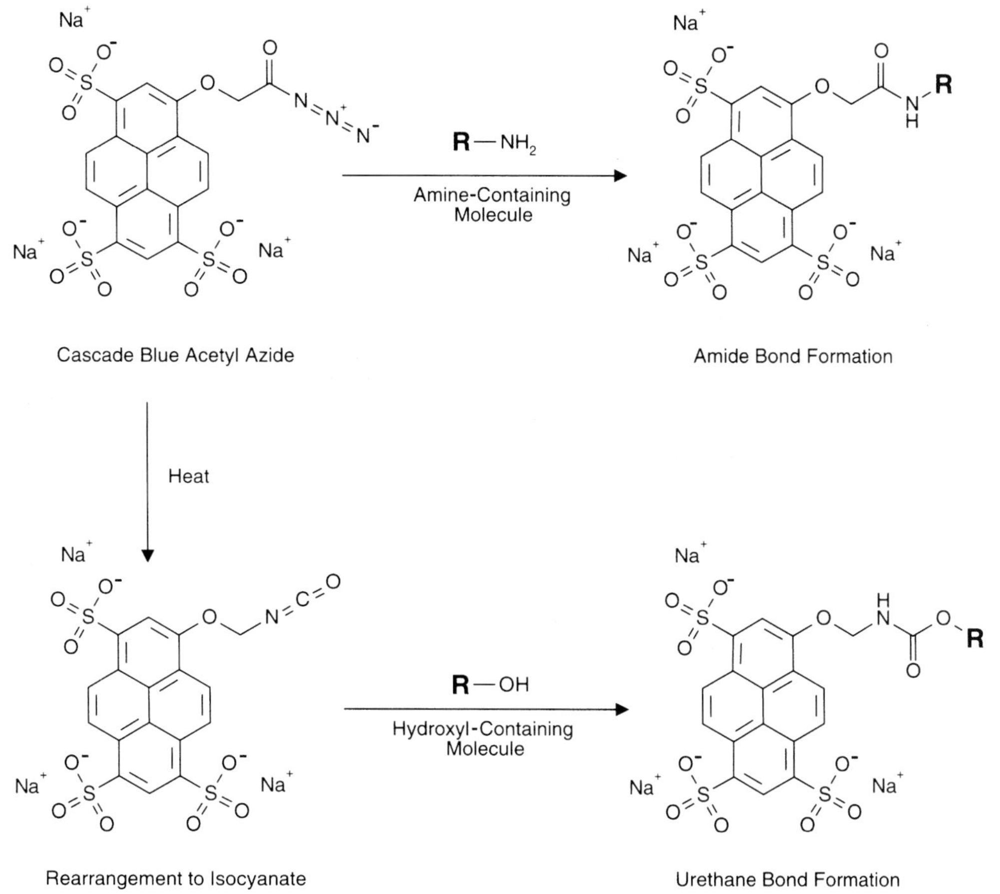

FIGURE 10.38 The acetyl azide group of this Cascade Blue derivative has dual functions. It can react with amine groups to form amide bonds, or it can be converted to an isocyanate at elevated temperature to couple with hydroxyl functional groups, which creates a carbamate linkage.

isocyanate that can react with hydroxyl-containing molecules to form a urethane linkage (Figure 10.38). The Cascade Blue urethane derivatives of macromolecules are extremely fluorescent and can be detected down to femtogram quantities (Takadate *et al.*, 1985). This dye has been used in a fluorescence lifetime assay to analyze the complex between cytochrome *b* and an anti-p22^{phox} antibody (Taylor *et al.*, 2012).

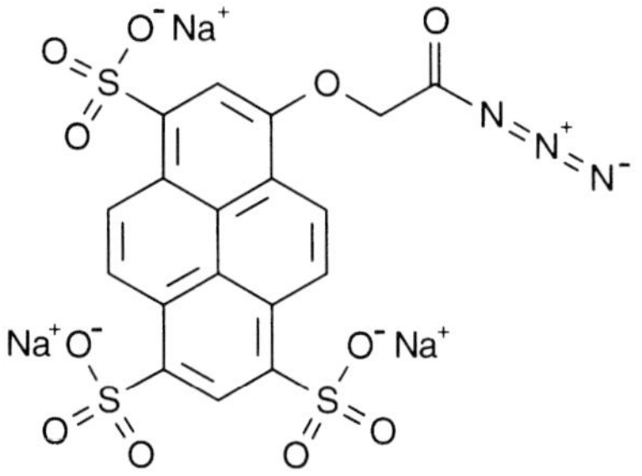

Cascade Blue Acetyl Azide
MW 607
Excitation = 375, 400 nm
Emission = 410 nm
ε at 375 nm = 27,000 $M^{-1}cm^{-1}$

This fluorophore has excitation maxima at 375 nm and 400 nm and an emission maximum at 410 nm. The small Stokes shift may create some difficulty in discrete excitation without contaminating the emission measurement with scattered or overlapping light. The extinction coefficient of the molecule in water is about 27,000 $M^{-1}cm^{-1}$. Cascade Blue and Lucifer Yellow derivatives can be simultaneously excited by light of less than 400 nm, resulting in two-color detection at 410 nm and 530 nm.

Cascade Blue acetyl azide is soluble in aqueous solution, but the reactive azide group will hydrolyze and should be used immediately in a conjugation reaction. A concentrated stock solution may be prepared in water and dissolved quickly, and an aliquot quickly added to a buffered reaction medium. For aqueous reactions, a pH range of 7 to 9 is optimal. Avoid amine-containing buffers.

5.2. Carboxylate-Reactive Cascade Blue Cadaverine and Cascade Blue Ethylenediamine

Cascade Blue cadaverine and Cascade Blue ethylenediamine both contain a carboxamide-linked diamine

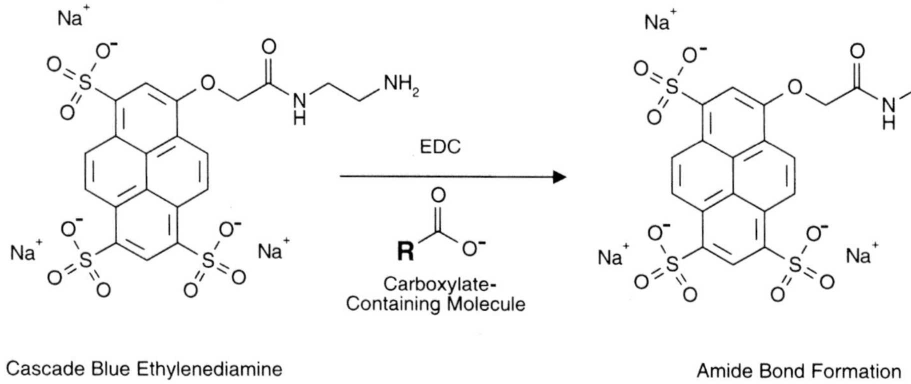

FIGURE 10.39 The terminal primary amine group on the linker arm of this Cascade Blue derivative can be coupled to carboxylate-containing molecules using a carbodiimide reaction.

spacer off the 8-methoxy group of the pyrene trisulfonic acid backbone. The cadaverine version contains a 5-carbon spacer, while the ethylenediamine compound has only a 2-carbon arm. Both can be coupled to carboxylic acid-containing molecules using a carbodiimide reaction (Chapter 4, Section 1). Since Cascade Blue derivatives are water soluble, the carbodiimide EDC can be used to couple these fluorophores to proteins and other carboxylate-containing molecules in aqueous solutions at a pH range of 4.5 to 7.5. The reaction forms amide bond linkages (Figure 10.39).

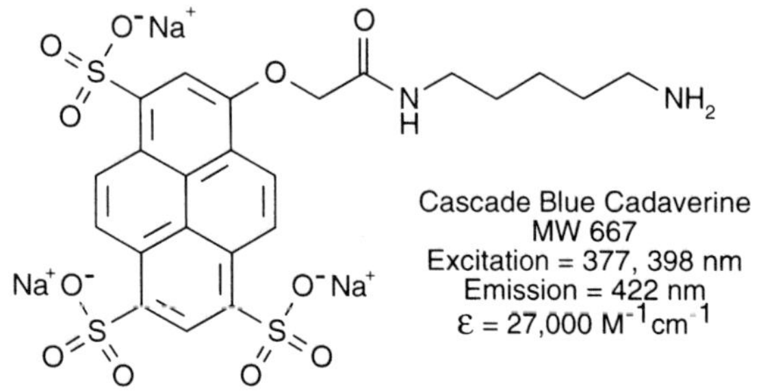

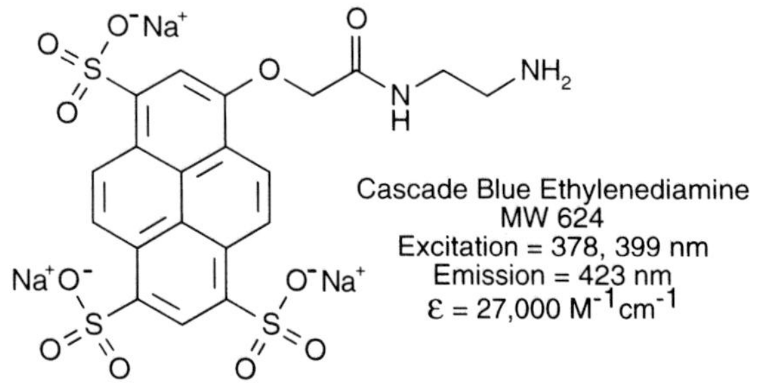

These fluorophores have excitation maxima at 377 to 378 nm and at 398 to 399 nm and emission maxima at 422 to 423 nm. The extinction coefficient of the molecules in water is about 27,000 $M^{-1}cm^{-1}$. The Cascade Blue derivatives can be used along with Lucifer Yellow derivatives and simultaneously excited by light of less than 400 nm, resulting in two-color detection at 422 nm and 530 nm.

Cascade Blue diamine derivatives are soluble in aqueous solution. A concentrated stock solution may be prepared in water and dissolved quickly, and an aliquot immediately added to a buffered reaction medium. For aqueous reactions, 0.1-*M* MES, pH 4.7 to 6.5, may be used to stabilize the pH during the coupling process. Avoid amine- or carboxylate-containing buffers such as Tris or glycine, since these can compete with the coupling reaction.

5.3. Aldehyde-/Ketone-Reactive Cascade Blue Hydrazide

Cascade Blue hydrazide is a carboxy-hydrazine derivative of the 8-methoxy group on the pyrene trisulfonic acid fluorophore. Hydrazide groups react with aldehyde and ketone groups to form relatively stable hydrazone linkages (Chapter 3, Section 5.1) (Figure 10.40). Although most biomolecules don't contain aldehyde or ketone groups in their native state, carbohydrates, glycoproteins, RNA, and other molecules that contain sugar residues can be oxidized with sodium periodate to produce reactive formyl groups. Modification reagents that generate aldehydes upon coupling to a molecule also can be used to produce hydrazide-reactive sites (Chapter 2, Section 4.4).

In addition, DNA and RNA may be modified with hydrazide-reactive probes by reaction of their cytosine residues with bisulfite to form reactive sulfone intermediates (Chapter 23, Section 2.1). These derivatives can undergo transamination reactions with hydrazide- or

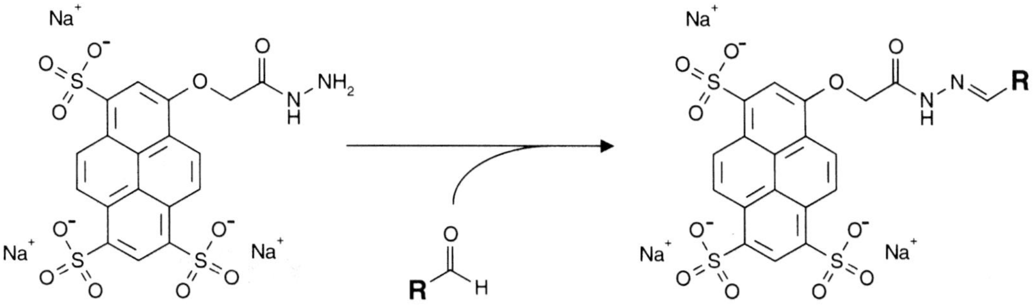

Cascade Blue Hydrazide — Aldehyde-Containing Molecule — Hydrazone Bond Formation

FIGURE 10.40 Cascade Blue hydrazide can be used to modify aldehyde-containing molecules to form hydrazone bonds, including modification of the reducing end of carbohydrates or in periodate-oxidized sugars.

amine-containing probes to yield covalent bonds (Draper and Gold, 1980). This dye has been used to investigate the integrin-mediated adhesion and proliferation of human MSCs (Krishna *et al.*, 2011), to analyze the vascular sequestration of fluorescent probes in plant cells (Oparka and Hawes, 2011), and to study the photostimulation of the dendritic arbor (Yang *et al.*, 2011).

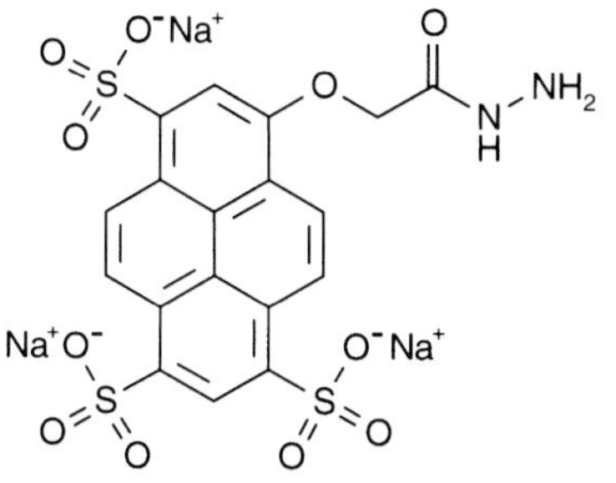

Cascade Blue Hydrazide
MW 645
Excitation = 400 nm
Emission = 420 nm
ε at 400 nm = 31,000 $M^{-1}cm^{-1}$

This fluorophore has an excitation maximum at 400 nm and an emission maximum at 420 nm. The extinction coefficient of the molecule in aqueous solution at pH 7 is about 31,000 $M^{-1}cm^{-1}$. Cascade Blue hydrazide and Lucifer Yellow derivatives can be excited simultaneously by light of less than 400 nm, resulting in two-color detection at 420 nm and 530 nm.

Cascade Blue hydrazide is soluble in aqueous solution, and it should be stable if protected from light. A concentrated stock solution of the reagent may be prepared in water and an aliquot added to a buffered reaction medium to facilitate the transfer of small quantities. For aqueous reactions, a pH range of 5 to 9 will result in efficient hydrazone formation.

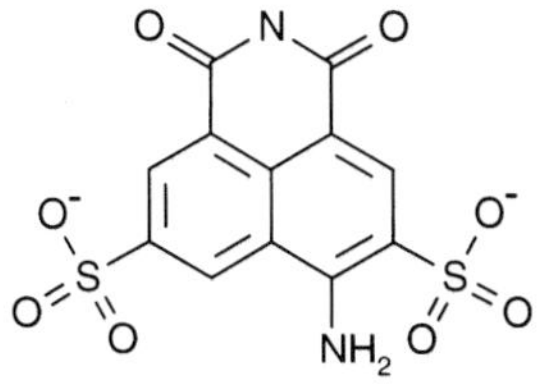

3,6-Disulfonate-4-amino-naphthalimide

FIGURE 10.41 The basic structure of naphthalimide-based (Lucifer Yellow) fluorophores.

6. NAPHTHALIMIDE DERIVATIVES

Fluorescent dyes based on a core structure of naphthalimide have been used as conjugates, probes, and tracers of biological systems for decades. The sulfonate groups typically seen modifying the naphthalene outer rings provide water solubility to an otherwise hydrophobic aromatic bicyclic ring system. Many reactive naphthalimide compounds have been commercialized under the name Lucifer Yellow since its introduction in 1978 and they have been used in a wide range of applications (Hanani, 2011). As their popular name implies, the spectral characteristics of these dyes fall within the yellow region of the spectrum.

Lucifer Yellow derivatives are used extensively for cytochemical staining applications, especially in neurophysiology (Stewart, 1981a,b). The fluorophores are 3,6-disulfonate 4-aminonaphthalimide derivatives that can be further modified at their imide nitrogen to contain reactive groups suitable for conjugation with biomolecules (Figure 10.41). Cell staining with membrane-impermeant Lucifer Yellow dyes is usually carried out by osmotic shock, microinjection, or pinocytosis (Swanson *et al.*, 1987). Derivatives containing amines or hydrazide groups are fixable with

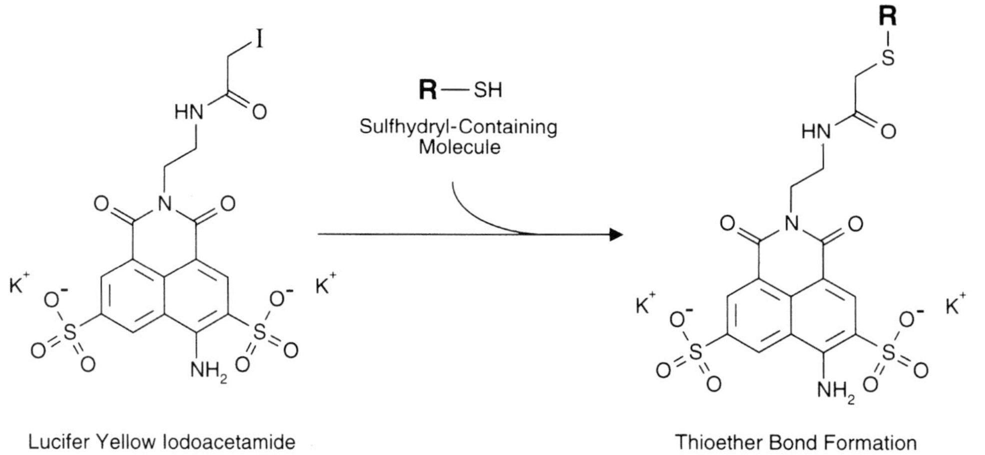

FIGURE 10.42 Lucifer Yellow iodoacetamide can be used to label sulfhydryl-containing molecules, forming thioether bonds.

formaldehyde or glutaraldehyde, coupling to nearby proteins or other amine-containing molecules intracellularly during the reaction. After periodate oxidation, glycoconjugates on cell surfaces or in solution may be labeled with the hydrazide derivative (Stewart, 1978). Sulfhydryl-containing molecules may be tagged with the iodoacetamide derivative.

Lucifer Yellow probes are water soluble to at least 1.5%. The absorbance maximum of the derivatives occurs about 426 to 428 nm with an emission peak at about 530 to 535 nm, in the yellow region of the spectrum. The quantum yield of Lucifer dyes is about 0.25. The good intensity of luminosity from these dyes makes the detection of small quantities of labeled molecules possible intracellularly. The fluorescent conjugates are readily visible in living cells at concentrations that are nontoxic to cell viability. The low molecular weight and water solubility of these dyes allow passage of labeled compounds from one cell to another, potentially revealing molecular relationships between cells.

6.1. Sulfhydryl-Reactive Lucifer Yellow Iodoacetamide

One Lucifer Yellow derivative is available for labeling sulfhydryl-containing molecules. Lucifer Yellow iodoacetamide is a 4-ethyliodoacetamide derivative of the basic disulfonate aminonaphthalimide fluorophore structure (Invitrogen). The iodoacetyl groups react with SH groups in proteins and other molecules to form stable thioether linkages (Figure 10.42). The dye has been used as a FRET acceptor to study dityrosine in the juvenile hormone-binding protein (JHBP) (Bystranowska *et al.*, 2012), for structural analysis of a yeast prion strain (Marcelino-Cruz, 2011), for end-labeling of oligonucleotides (Zearfoss and Ryder, 2012), and to study oligomerization of αA- and αB-crystallins (Wu *et al.*, 2012).

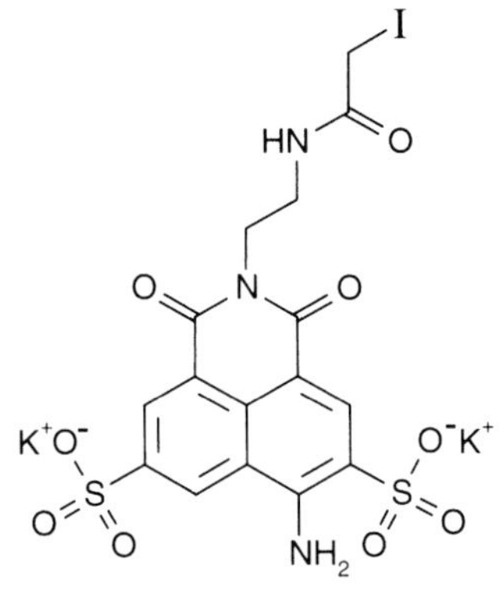

Lucifer Yellow Iodoacetamide, dipotassium salt
MW 649
Excitation = 426 nm
Emission = 530 nm
ε at 426 nm = 13,000 $M^{-1}cm^{-1}$

The spectral characteristics of Lucifer Yellow iodoacetamide produce luminescence at somewhat higher wavelengths than the green luminescence of fluorescein, thus the yellow designation in its name. The excitation maximum for the probe occurs at 426 nm and its emission at 530 nm. The rather large Stokes shift makes sensitive measurements of emission intensity possible without interference by scattered excitation light. The 2-mercaptoethanol derivative of the fluorophore has an extinction coefficient at pH 7 of about 13,000 $M^{-1}cm^{-1}$at 426 nm.

Lucifer Yellow iodoacetamide is soluble in aqueous solution due to its negatively charged sulfonate groups. A concentrated stock solution may be prepared in water prior to the addition of a small aliquot to a reaction mixture. Coupling to sulfhydryl-containing molecules occurs rapidly with the formation a thioether linkage. The reaction may be performed in 50-m*M* sodium borate, 5-m*M* EDTA, pH 8.3. The main consideration is

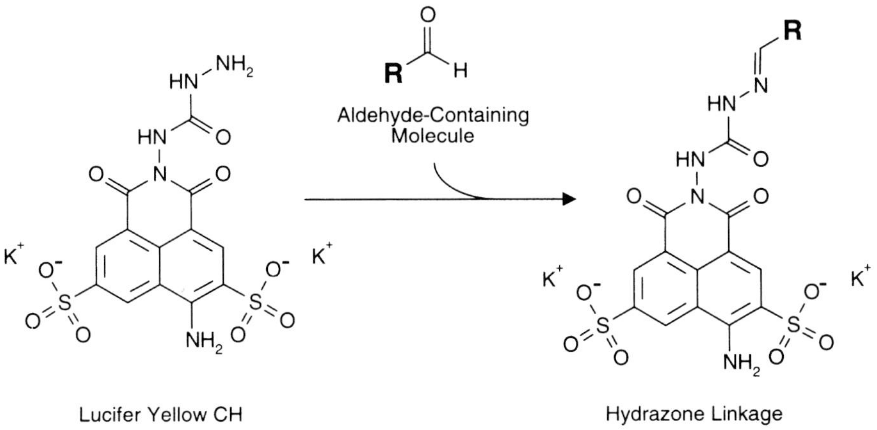

FIGURE 10.43 The hydrazide group of this Lucifer Yellow derivative can react with aldehyde-containing molecules to form hydrazone bonds.

to protect the iodoacetyl derivative from light, which will generate iodine and reduce the reactivity of the probe. The reaction may be limited to sulfhydryls (avoiding any amine derivatization) by maintaining a low molar excess of probe to the amount of sulfhydryl groups present. In addition, oligonucleotides containing a sulfhydryl modification at their 5′ ends (Chapter 23, Section 2.2) may be coupled with Lucifer Yellow iodoacetamide, yielding highly fluorescent, yellow probes.

6.2. Aldehyde-/Ketone-Reactive Lucifer Yellow CH

Lucifer Yellow CH is a carbohydrazide derivative of the basic disulfonate aminonaphthalimide fluorophore structure (Invitrogen). Hydrazide groups react with aldehyde and ketone groups to form relatively stable hydrazone linkages (Chapter 3, Section 5.1) (Figure 10.43). Although most biomolecules do not contain aldehyde or ketone groups in their native state, carbohydrates, glycoproteins, RNA, and other molecules that contain sugar residues can be oxidized with sodium periodate to produce reactive formyl groups. The use of modification reagents that generate aldehydes upon coupling to a molecule also can be used to produce hydrazide-reactive sites (Chapter 2, Section 4.4). In addition, DNA and RNA may be modified with hydrazide-reactive probes by reacting their cytosine residues with bisulfite to form reactive sulfone intermediates (Chapter 23, Section 2.1). These derivatives can undergo transamination reactions with hydrazide- or amine-containing probes to yield covalent bonds (Draper and Gold, 1980).

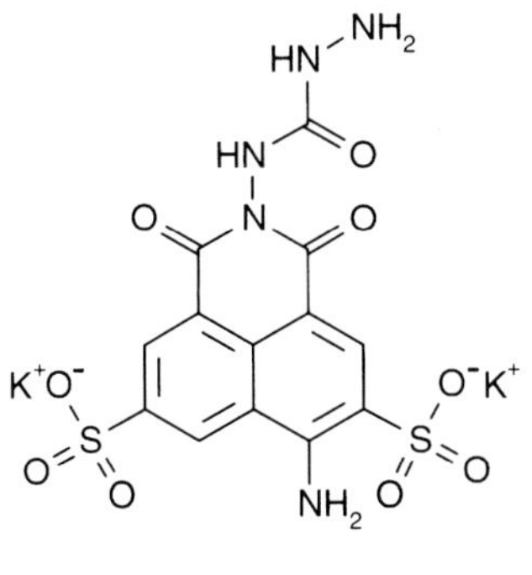

Lucifer Yellow CH
MW 522 (potassium salt)
MW 457 (lithium salt)
MW 479 (ammonium salt)
Excitation = 428 nm
Emission = 533-535 nm
ε at 428 nm = 12,000 $M^{-1}cm^{-1}$

Lucifer Yellow CH is commonly used as a neuronal tracer by staining cells and then fixing them with formaldehyde or glutaraldehyde. It also can be used to label glycoproteins or glycolipids on cell surfaces after periodate oxidation (Spiegel *et al.*, 1983; Lee and Fortes, 1985). The labeling of oxidized ribonucleotides and gangliosides can be performed similarly (Spiegel *et al.*, 1985; Sun *et al.*, 1988).

This fluorophore has an excitation maximum at 428 nm and an emission maximum at 534 nm. The extinction coefficient of the molecule in aqueous solution is about 12,000 $M^{-1}cm^{-1}$. Cascade Blue hydrazide and Lucifer Yellow CH derivatives can be excited simultaneously by light of less than 400 nm, resulting in the possibility for two-color detection at 420 nm and 534 nm.

Lucifer Yellow CH is soluble in aqueous solution, and it should be stable for a while if protected from light. The reagent is available as three different salts of the sulfonate groups. The ammonium salt of the fluorophore is soluble to a level of 9% in water, while the lithium and potassium salts have solubilities of 5% and 1%, respectively. A concentrated stock solution of the fluorophore may be prepared in water and an aliquot added to a buffered reaction medium to facilitate the transfer of small quantities. For aqueous reactions, a pH range of 5 to 9 will result in efficient hydrazone formation with aldehyde or ketone residues.

TABLE 10.1 **Properties of the Phycobiliproteins**

Property	B-Phycoerythrin	R-Phycoerythrin	C-Phycocyanin	Allophycocyanin
Source	*Porphyridium cruentum*	*Gastroclonium coulteri*	*Anabaena variabilis*	*Anabaena variabilis*
Subunit structure	$(\alpha\beta)_6\varepsilon$	$(\alpha\beta)_6\gamma$	$(\alpha\beta)_2$	$(\alpha\beta)_3$
Molecular weight	240,000	240,000	72,000	110,000
Pigment content (bilin groups)	34	34	4	6
Absorbance maximum	546 nm	566 nm	614 nm	650 nm
Molar extinction coefficient	2.4×10^6	2.0×10^6	5.8×10^5	7.0×10^5
Emission maximum	575 nm	574 nm	643 nm	660 nm

7. PHYCOBILIPROTEIN DERIVATIVES

Phycobiliproteins are intensely fluorescent proteins that function as components in the photosynthetic apparatus of eukaryotic blue–green algae and cyanobacteria (Glazer, 1981; Samarakoon *et al.*, 2012). The proteins are found as aggregates in phycobilisome particles near the chlorophyll regions (Kronick, 1986). In the native state, phycobiliproteins do not fluoresce; rather, excitation energy is designed to be efficiently transferred to chlorophyll molecules for utilization in synthetic processes within the cell. Once purified, however, excitation energy is released from phycobiliproteins as strong luminosity. The fluorescent quantum efficiencies of these proteins can be as high as 0.98, far better than most synthetic probes (Grabowski and Gantt, 1978). In addition, each biliprotein contains multiple chromophoric bilin prosthetic groups, conferring extremely high absorbance coefficients to each protein molecule. B-Phycoerythrin, for example, typically contains 34 chromophoric groups giving an effective, combined extinction coefficient at 545 nm of $2.4 \times 10^6 M^{-1} cm^{-1}$ (Glazer and Hixson, 1977). The strong absorption bands are in the visible region of the spectrum, extending from the green to the far red wavelengths. These absorption spectra extend over a broad range of potential excitation wavelengths, allowing for versatility in the excitation source employed and creating large Stokes shifts, thus minimizing interference from Rayleigh-scattered light (Loken, 1987).

Due to the presence of multiple fluorescent groups in each phycobiliprotein, conjugates of these molecules form extraordinarily luminescent probes. Labeling of macromolecules with phycobiliprotein derivatives can provide absorption coefficients 30-fold higher than labeling with small, synthetic fluorophores. Their ability to be monitored by fluorescing in the red region of the spectrum decreases potential interferences from indigenous biological fluorescence. The protected bilin (tetra-pyrrole) prosthetic groups are not easily affected by their external environment. They are not readily quenched by conjugation to another molecule or affected by other components in solution. The prosthetic group orientation within the protein molecules enables fluorescence to take place independent of pH or ionic strength. The excellent solubility of phycobiliproteins in aqueous solution allows easy chemical manipulation for modification or conjugation reactions, and their hydrophilic nature provides low nonspecific binding character in fluorescent detection applications.

There are three main classes of phycobiliproteins, differing in their protein structure, bilin content, and fluorescent properties. These are phycoerythrin, phycocyanin, and allophycocyanin. There are two main forms of phycoerythrin proteins commonly in use: B-phycoerythrin isolated from *Porphyridium cruentum* and R-phycoerythrin from *Gastroclonium coulteri*. There also are three main forms of pigments found in these proteins: phycoerythrobilin, phycourobilin, and phycocyanobilin (Glazer, 1985; Richa *et al.*, 2011). The relative content of these pigments in the phycobiliproteins determines their spectral properties. All of them, however, have extremely high absorption coefficients ranging from a magnitude of 10^5 to 10^6 and excellent QYs ranging from 0.51 up to 0.98.

The spectral properties of four major phycobiliproteins used as fluorescent labels can be found in Tables 10.1 and 10.2. The bilin content of these proteins ranges from a low of 4 prosthetic groups in C-phycocyanin to the 34 groups of B- and R-phycoerythrin. Phycoerythrin derivatives can, therefore, be used to create the most intensely fluorescent probes possible using these proteins. The fluorescent yield of the most luminescent phycobiliprotein molecule is equivalent to about 30 fluoresceins or 100 rhodamine molecules. Streptavidin–phycoerythrin conjugates, for example, have been used to detect as few as 100 biotinylated antibodies bound to receptor proteins per cell (Zola *et al.*, 1990).

Conjugation of phycobiliproteins to targeting components such as antibodies, avidin, biotin, or other molecules preserves the binding or activity of the attached

TABLE 10.2 **Spectral Properties of Phycobiliproteins**

Phycobiliprotein	Molecular Weight	Absorption Max (nm)	EC ($M^{-1}cm^{-1}$)	Emission Max (nm)	Fluorescence QY
B-phycoerythrin	240,000	546, 565	2,410,000	575	0.98
R-phycoerythrin	240,000	496, 546, 565	1,960,000	578	0.82
Allophycocyanin	104,000	650	700,000	660	0.68

EC, extinction coefficient; QY, quantum yield.

constituent and does not alter the spectral characteristics of the bilin prosthetic groups. Common heterobifunctional crosslinking agents can be used to create phycobiliprotein conjugates, including SPDP (Chapter 6, Section 1.1), SMCC (Chapter 6, Section 1.3), and SMPB (Chapter 5, Section 1.6) (Oi *et al.*, 1982). These crosslinkers react with amine groups on the phycobiliproteins, producing activated intermediates able to couple with sulfhydryl-containing molecules. Thiolation reagents such as 2-iminothiolane, *N*-succinimidyl-*S*-acetylthioacetate (SATA), and SAMSA (Chapter 2, Section 4.1) can be used to create thiols on the secondary molecule to effect the final coupling reaction.

Glazer and Stryer (1983) reported on the preparation and use of tandem phycobiliprotein conjugates wherein B-phycoerythrin is crosslinked to allophycocyanin to create an energy donor–acceptor pair. The B-phycoerythrin component can be excited at 545 nm, emitting energy at 575 nm that can be accepted by allophycocyanin (APC), which in turn emits light at 660 nm. The result is a large shift in the spectral characteristics from that of the individual proteins, increasing the effective Stokes shift to over 100 nm. Conjugation of such tandem pairs to other proteins can create superior fluorescent reagents.

Phycobiliproteins can also be modified with reactive organic fluorescent probes to produce tandem dyes having modulated emission properties. In this sense, dyes having excitation wavelengths that overlap with the emission wavelength of a phycobiliprotein can be used to extend the emission of the tandem complex farther into the red region. For instance, a Cy5-type dye (see Section 8) can be used to modify R-phycoerythrin (RPE) to produce a conjugate that can be excited at the normal wavelengths used for RPE, but emit light at the Cy5 range of 660 to 665 nm instead of the typical 575-nm emission for the phycobiliprotein itself. Fluorescence resonance energy transfer (FRET) of excitation energy from the bilin units in RPE to the Cy5 dyes modifying the protein results in the red-shifted emission characteristics of the tandem dye construct. In an optimized tandem dye conjugate, the fluorescence emission of the phycobiliprotein is almost entirely eliminated by the FRET signaling to the organic dye (Tian and Pappas, 2011; Tian, 2012).

Preparation of a series of phycobiliprotein tandem dyes allows multiplexed analysis of different targets in a sample. In addition, since RPE can be excited by the argon-ion laser at 488 nm, a fluorescein-labeled probe can be used concurrently with RPE alone and RPE-tandem conjugates to create a multiplexed system of different fluorescent probes that can be used simultaneously. Table 10.3 shows the different combinations of dyes that can be used in this type of assay with RPE and APC.

The following protocol is a generalized method for creating sulfhydryl-reactive phycobiliprotein reagents for coupling to SH containing molecules. The procedure uses the heterobifunctional crosslinker SPDP (Chapter 6, Section 1.1). Other amine- and sulfhydryl-reactive crosslinking agents may be used in a similar manner, such as sulfo-SMCC or the hydrophilic NHS-PEG_n-maleimide crosslinkers (Chapter 18). The discrete PEG-based crosslinkers will create conjugates of phycobiliproteins that have low nonspecific binding character due to the biocompatible nature of the PEG groups.

7.1. Protocol

1. Dialyze the phycobiliprotein into 50-m*M* sodium borate, 0.3-*M* NaCl, pH 8.5. Note: Commercial preparations of these proteins come as an ammonium sulfate suspension. After dialysis, adjust the protein solution to a concentration of 1 mg/ml. Higher protein concentrations may be used, but the amount of crosslinking reagent added to each milliliter of the reaction should be proportionally scaled up as well. Protect the protein solution from undue exposure to light.
2. Dissolve SPDP at a concentration of 6.2 mg/ml in DMSO (makes a 20-m*M* stock solution). Alternatively, LC-SPDP may be used and dissolved at a concentration of 8.5 mg/ml in DMSO (also makes a 20-m*M* solution). If the water soluble sulfo-LC-SPDP is used, a stock solution in water may be prepared just prior to adding an aliquot to the reaction. In this case, prepare a 10-m*M* solution of sulfo-LC-SPDP by dissolving 5.2 mg/ml in water. Since an aqueous solution of the crosslinker will degrade by hydrolysis of the sulfo-NHS ester, it should be used quickly to prevent significant loss of activity. If a sufficiently

TABLE 10.3 Tandem Dye Fluorescence Properties

	Excitation (nm)		
	488	568	647/650
Emission (nm)	Argon-Ion Laser (488)	Krypton-Argon Laser (568)	Krypton-Argon (647) or Red Diode Laser (650)
519	DyLight 488 or Alexa 488		
575	R-Phycoerythrin (RPE)	R-Phycoerythrin (RPE)	
590		Lissamine Rhodamine	
615–630	RPE-Texas Red, RPE-Alexa 610, RPE-DyLight 594, or RPE-DyLight 610	RPE-Texas Red, RPE-DyLight 594 or RPE-DyLight 610	
660–665	RPE-Alexa 647 or RPE-DyLight 649	RPE-DyLight 649	DY649, Allophycocyanin (APC)
702–709	RPE-DyLight 682	RPE-DyLight 680	APC Alexa 680 or APC-DY-682
719–735	RPE-Alexa 700 or RPE-DY701	RPE-Alexa 700 or RPE-DY701	APC-Alexa 700 or APC-DY701
770–780	RPE-Alexa 750 or RPE-DY752	RPE-Alexa 750 or RPE-DY752	APC-Alexa 750 or APC-DY752

large amount of phycobiliprotein will be modified, the solid may be added directly to the reaction mixture without preparing a stock solution in water to allow accurate weighing of Sulfo-LC-SPDP.

3. Add 25 μl of the stock solution of either SPDP or LC-SPDP in DMSO to each milliliter of the protein solution. If sulfo-LC-SPDP is used, add 50 μl of the stock solution in water to each milliliter of protein solution.
4. Mix and react for at least 30 min at room temperature. Longer reaction times, even overnight, will not adversely affect the modification.
5. Purify the modified protein from reaction byproducts by dialysis using a membrane with a low-molecular-weight cutoff or gel filtration using desalting resin and a buffer consisting of 50-m*M* sodium phosphate, 0.15-*M* NaCl, 10-m*M* EDTA, pH 7.2.

The SPDP-activated phycobiliprotein may be reacted with a sulfhydryl-containing protein to create a fluorescent conjugate linked through disulfide bonds.

8. CYANINE DYE DERIVATIVES

One of the most popular fluorescent dye types for labeling biomolecules is built from two cationic, aromatic, nitrogenous ring structures linked by an unsaturated polymethine bridge. One of the rings must have a quanternized nitrogen atom possessing a positive charge. The ring structure types can be highly varied, from five or six membered heterocycles to multiple fused ring systems. The polymethine bridge to a great extent determines the fluorescence character of the dye as well as contributing to the naming convention that distinguishes each dye within a family of similar structures. Thus, the general structure of all cyanine dyes can be represented by

$$X\text{–}(CH{=}CH)_n\text{–}CH{=}Y$$

where X and Y are the nitrogenous rings at both ends of the polymethine bridge and n can vary from 0 to 3. The name of a cyanine dye often has a number following it, which reflects the number of carbon atoms in the polymethine chain. Thus, if $n = 0$, the dye is called a monomethine cyanine dye; if $n = 1$, it is called a tricyanine or Cy3; if $n = 2$, it is a pentacyanine or Cy5; and if $n = 3$, it is a heptacyanine or Cy7.

The longer the polymethine bridge in a cyanine dye, the higher the absorbance and emission wavelengths become. In general, for each incremental increase of n, the absorbance and emission characteristics of the dye increase by about 100 nm. Thus, Cy3 dyes typically display excitation and fluorescence in the mid-500 nm range, Cy5 has spectral properties in the mid-600 nm range, and Cy7 in the mid-700 nm range.

Though the bridge structure significantly affects the spectral character of a cyanine dye, the heterocyclic rings also can contribute heavily to its properties. For instance, the structure of the ring systems can highly affect the absorptivity and brightness of a particular cyanine dye. Some heterocycle structures, such as the commonly used indol rings of many commercial cyanine compounds, have extremely high extinction coefficients and thus provide intensely bright fluorescent labels. Other constituents on the heterocyclic rings can provide a blue shift or a red shift in the spectral characteristics of a dye. This is the basis for creating intermediate Cy dyes having fluorescent properties between the standard ones. For instance, Cy5.5 falls between the emission wavelengths of Cy5 and Cy7, completely due to an alternative fused ring structure (e.g., benzo-indolium groups; see Figure 10.44). The relative fluorescence effects of different ring systems and numerous possible substitutions on these rings often can be predicted from decades of investigation and data (Michaela, 2002). Thus, many commercial suppliers of these dyes have fine-tuned their properties to make them more suitable for certain applications.

The nonreactive base structures of cyanine dyes (or carbocyanines) have been used for many years as components in photographic emulsions to increase the range and sensitivity of film and also in CD-R and DVD-R optical disks to record digital information. A major innovation came when Ernst *et al.* (1989) and Waggoner *et al.* (1993) recognized that cyanine dyes would make excellent labels for fluorescence detection and, for this reason, they synthesized reactive dye derivatives that could then be covalently attached to proteins and other molecules.

The nitrogenous ring structures at each end of a cyanine dye can be the same (symmetrical structure) or different (unsymmetrical), depending on the heterocycle type and constituents on the rings. Reactive cyanine dyes nearly always contain unsymmetrical structures due to the presence of a reactive arm on one end to facilitate conjugation to biomolecules or the presence of one or more negatively charged groups designed to increase water solubility.

The earliest cyanine dye labels were relatively hydrophobic due to the lack of hydrophilic groups or having only a minimal number of charged groups on the molecules. These dyes, while intensely fluorescent, often cause aggregation or precipitation of labeled proteins, especially if more than just a few fluorescent molecules are attached to a single biomolecule. Dye–dye interactions due to ring stacking or hydrophobic interactions also cause fluorescence quenching, because energy transfer can take place between dye molecules, negating the emission of light.

To make cyanine dyes more biocompatible for protein labeling, sulfonate groups typically are added to the ring systems or at the ends of short spacer arms protruding from the base dye structure (Mujumdar *et al.*, 1993; Southwick *et al.*, 2005). The addition of sulfonate groups provides a negative charge character that helps solubilize the cyanine dye in aqueous solution and prevents dye–dye interactions through like-charge repulsion. In general, the more sulfonates that a cyanine dye possesses, the greater will be its hydrophilicity and the lower its tendency to bind nonspecifically to hydrophobic structures on proteins or other molecules. Cyanine dyes of this type having from two to four sulfonate groups are available commercially, with three or four sulfonates per dye giving the best results for bioconjugation purposes in aqueous solution.

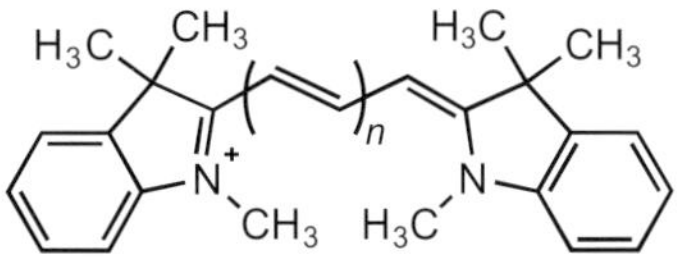

Indolium-based cyanine dye

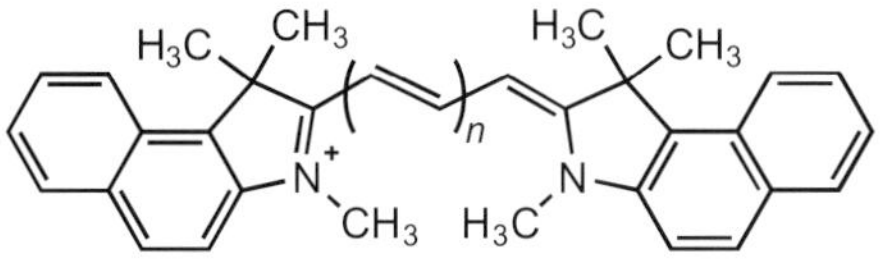

Benzo-indolium-based cyanine dye

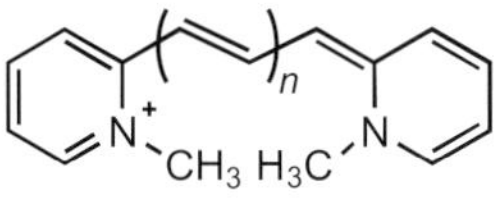

Pyridinium-based cyanine dye

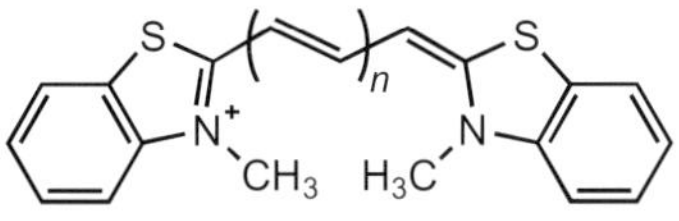

Thiazolium-based cyanine dye

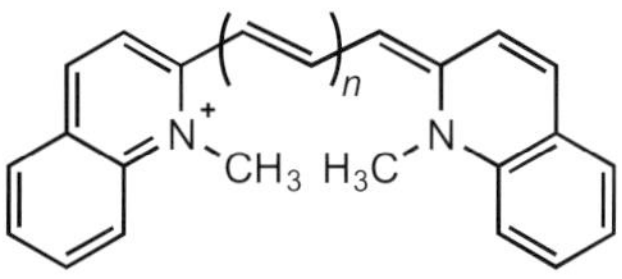

Quinolinium-based cyanine dye

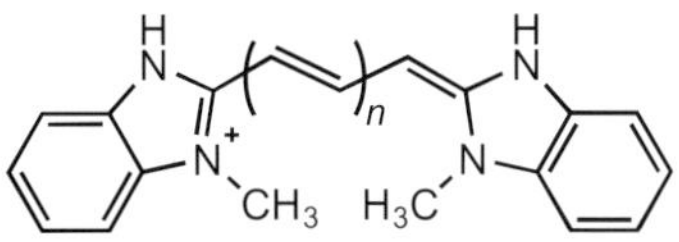

Imidazolium-based cyanine dye

FIGURE 10.44 The common nitrogenous ring structures of cyanine-type dyes. Indolium-based dye derivatives are the most frequently used fluorescent labels due to the high extinction coefficient of the indol groups. Antibodies and proteins labeled with these dyes can provide bright fluorescent signals for highly sensitive detection assays.

Cyanine dyes also are used as labels for oligonucleotide probes. Unlike the hydrophilic cyanine dyes valuable for protein labeling, the use of dye–phosphoramidite compounds to synthesize DNA or RNA probes typically requires the use of more hydrophobic dye structures to make them compatible with the solvents and reactions of oligonucleotide synthesis. Thus, indol cyanines containing few or no sulfonates are used in these applications to label oligos for applications such as array detection, hybridization assays, and RT-PCR.

In addition, small unsymmetrical cyanine compounds have been designed that bind to the minor groove of DNA to measure DNA concentration or stain DNA by fluorescence. These derivatives often contain a positive charge character with short crescent shapes to wrap around the helical structure of double-stranded oligonucleotides. As minor groove binding dyes interact with DNA, they typically undergo a slight twist in their molecular structure, which dramatically increases fluorescence QY and brightness. For example, the minor groove binding dye BOXTO displays a 300-fold increase in fluorescence when binding to double-stranded DNA (Karlsson *et al.*, 2003).

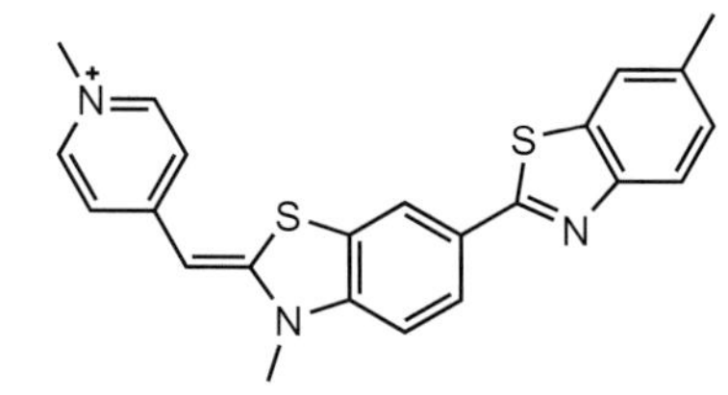

BEBO
MW 402.55

Minor groove binding cyanine dyes

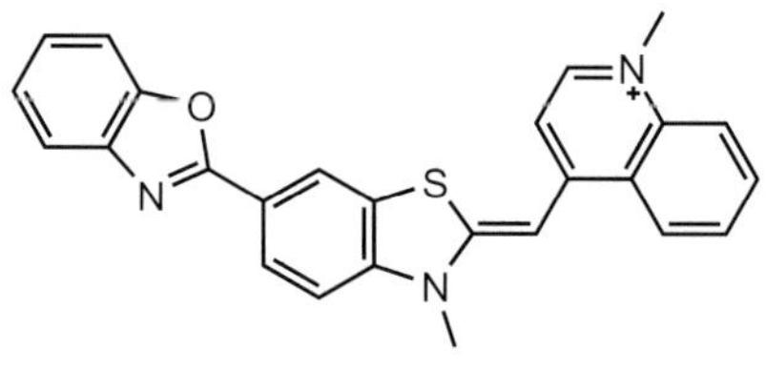

BOXTO
MW 422.52

Most companies selling cyanine dyes do not reveal their exact structures. This is likely due to each company keeping proprietary the small synthetic tweaks that create unique fluorescence properties for their dyes. However, some structures are available through published documents, such as patents and early publications (Leung *et al.*, 2005). Figure 10.45 illustrates some of these structures, which may not reflect precisely what any one company actually offers today but does give an idea of the types of modifications that can be made to add water solubility and reactivity.

Cyanine dyes are commercially available in a number of different structural configurations containing appropriate reactive groups to couple with many of the major functional groups (Thermo Fisher, GE Healthcare, Invitrogen, Sigma, Dyomics, Atto-Tec, and Denovo Biolabels). Most of these dyes contain sulfonate groups to increase their biocompatibility, but careful comparisons should be made, as most companies do not provide structures. The following sections describe some of the reactive cyanine dye reagents and the protocols used to label proteins and other molecules.

8.1. Amine-Reactive Cyanine Dyes

Amine-reactive cyanine dyes typically contain an NHS ester group on the end of a short hydrocarbon spacer for attachment to biomolecules. The NHS ester can react with amine groups on proteins to form amide bond linkages (Figure 10.46). This reaction is efficient at physiological pH or under slightly more alkaline conditions. While poly-sulfonated dye molecules are very water soluble, it is best first to prepare a stock solution in organic solvent to prevent NHS ester hydrolysis prior to adding a small aliquot to a reaction mixture.

The following protocol for protein labeling with an NHS-ester cyanine dye is based on the Thermo Fisher instructions for use of DyLight dyes. When preparing a fluorescently labeled protein the degree of modification should be optimized to provide maximal fluorescence signal while not affecting the activity or binding potential of the protein for other molecules. In fact, dye loadings of greater than about 8 fluorescent labels per protein will result in fluorescence quenching due to energy transfer between dye molecules. Higher loadings will not result in greater fluorescence signals, only greater potential to interfere with protein activity. Therefore, limit the molar excess of dye over protein to a level that will provide less than about eight substitutions per protein molecule.

Protocol

1. Dissolve a protein or antibody to be labeled with an NHS-ester cyanine dye in 0.1-*M* sodium phosphate buffer, pH 7.2 to 7.5, at a concentration of 1 to 10 mg/ml. The higher the concentration of protein, the greater will be the yield of the labeling reaction. Other buffers may also be used, including borate buffer or carbonate buffer, but higher pH reactions will result in greater hydrolysis of the NHS ester and may decrease labeling efficiency. Avoid amine-containing buffers such as Tris or imidazole. Also, avoid the addition of thiol reducing agents or

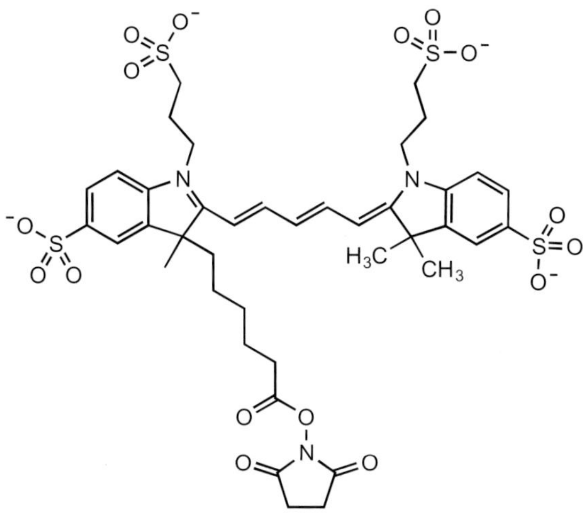

Tetrasulfonyl-Cy5-NHS ester

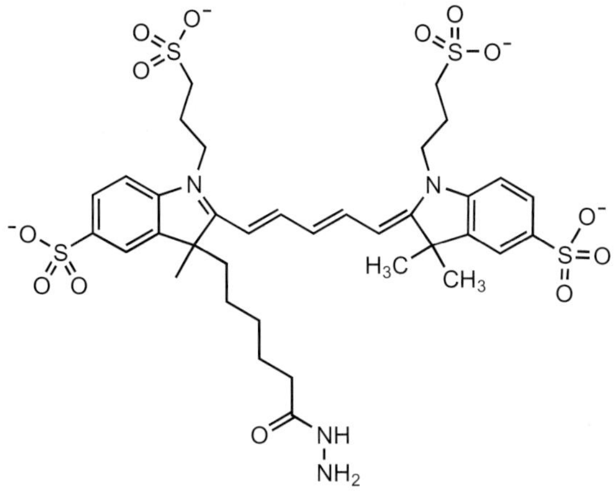

Tetrasulfonyl-Cy5-Hydrazide

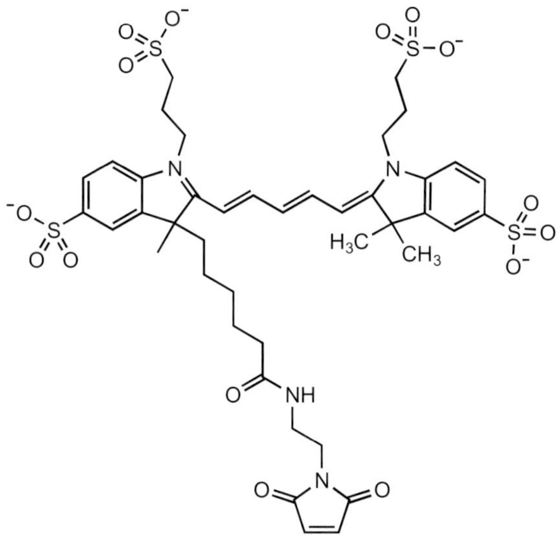

Tetrasulfonyl-Cy5-Maleimide

FIGURE 10.45 Typical Cy5 derivatives contain negatively charged sulfonate groups for water solubility and a side chain terminating in a reactive group for covalent coupling. The most common reactive groups are NHS esters for labeling amines, maleimide groups for coupling to thiols, and hydrazide groups for labeling aldehydes.

glycerol, as these also will react with the NHS ester. If the protein concentration is below 1 mg/ml, the molar excess of dye added to the reaction will have to be greatly increased to maintain yield of the reaction.

2. In a fume hood, dissolve the NHS-ester cyanine dye in DMF at a concentration of 10-m*M*. Protect all dye solutions from light.
3. With mixing, add a quantity of the dye solution to the protein solution to provide the desired molar excess of dye over protein. For instance, for an antibody dissolved in buffer at 10 mg/ml, the addition of 66 μl of dye solution will give a 10-fold molar excess of dye over protein. Note: For dyes that are rather hydrophobic in character, the mole excess of dye over the amount of protein present may have to be scaled back to prevent precipitation of the conjugate or dye–dye quenching from occurring (e.g., 5-fold mole excess). For highly hydrophilic dye

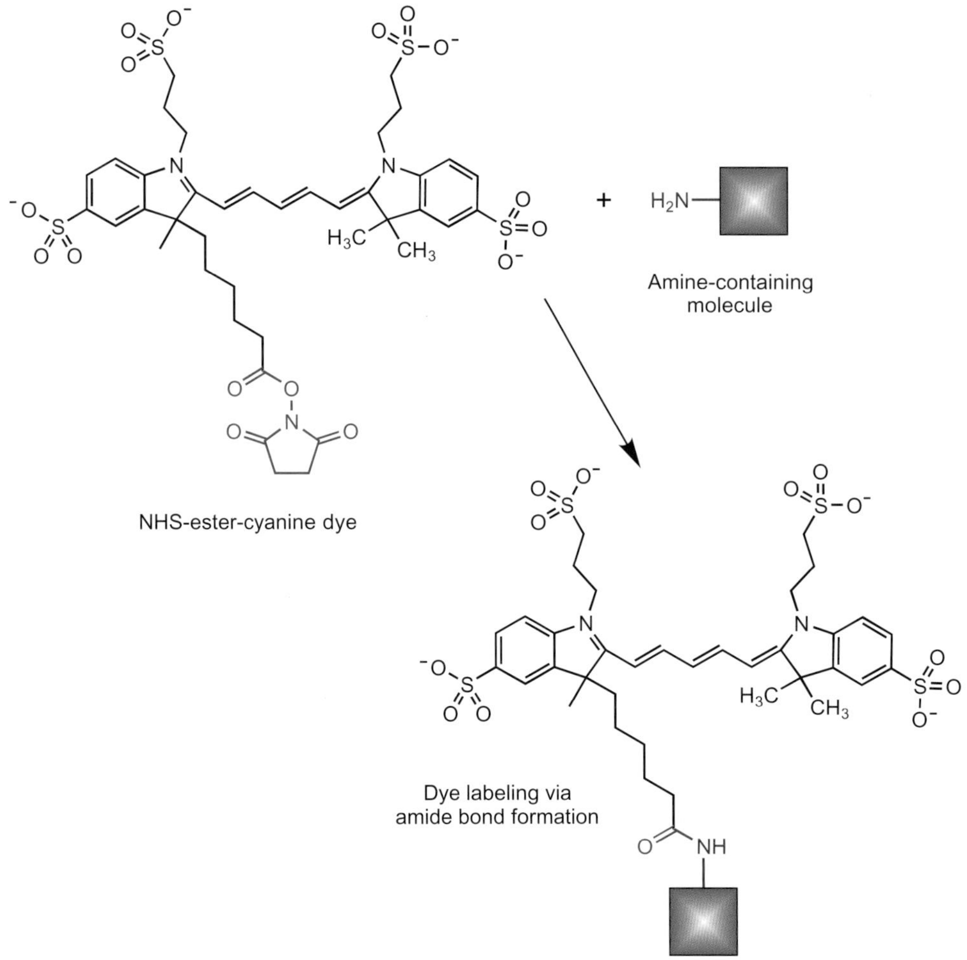

FIGURE 10.46 An NHS-ester-containing cyanine dye can be used to label amine-containing proteins or other molecules to form amide bonds.

derivatives, such as those containing at least three to four sulfonate groups or PEG modifications, then a 10-fold mole excess is appropriate.

4. React for 1 h at room temperature with gentle mixing.
5. Purify the labeled protein from excess dye and reaction by-products using dialysis or gel filtration on a desalting resin.

8.2. Thiol-Reactive Cyanine Dyes

Cyanine dyes that have a thiol-reactive group are typically maleimide derivatives. Maleimide groups react with sulfhydryls under neutral pH conditions to form a thioether linkage (Figure 10.47). Since thiols are present in proteins in limited amounts, dye labeling through these groups often will result in modifications that occur only at discrete locations on the molecule. Labeling thiol groups also may be done using peptides that have been synthesized with terminal cysteine residues or using oligonucleotides containing a 5′ thiol group.

The pH of a maleimide conjugation reaction should be in the range of 6.5 to 7.5 to ensure specificity toward thiol groups. Higher pH conditions will begin to result in cross-reactions with amines through a Michael addition process. Coupling to proteins can be done through disulfide reduction or through use of a thiolation reagent (Chapter 2, Section 4.1). Reduction of disulfides may be performed using immobilized TCEP (Thermo Fisher), which effectively forms free thiols on proteins and peptides while not contaminating a reaction with the reducing agent.

The following protocol is based on the methods recommended by Thermo Fisher for use of the cyanine dye DyLight 649. Antibody reduction is based on the methods of Sun *et al.* (2005) and results in partially reduced bispecific immunoglobulin containing available thiols in the hinge region for labeling.

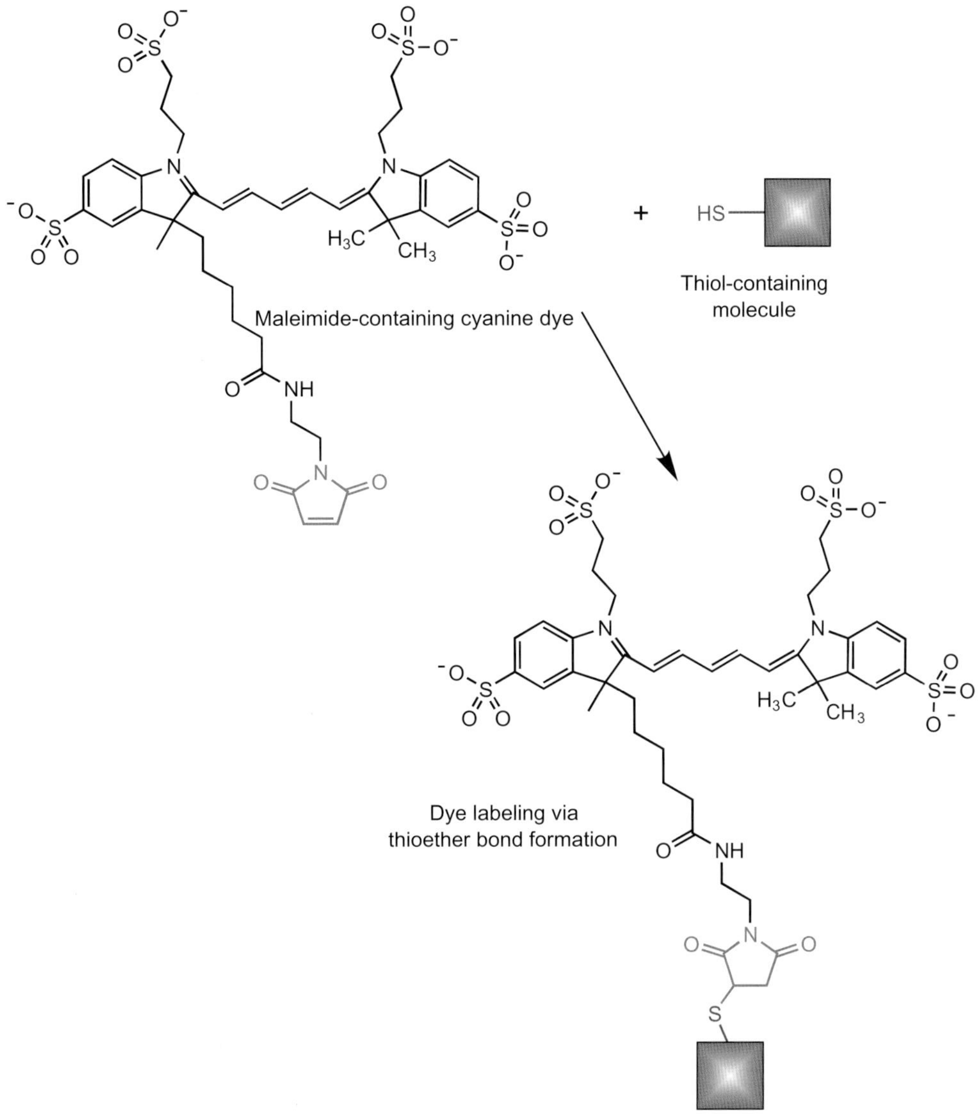

FIGURE 10.47 A maleimide-containing cyanine dye can be used to label thiol-containing molecules to form thioether bonds.

Protocol for Partial Reduction of IgG Antibody

1. Dissolve an IgG antibody in 50-m*M* sodium phosphate, 150-m*M* NaCl, 10-m*M* EDTA, pH 7.2 (reaction buffer), at a concentration of 1 to 10 mg/ml.
2. Dissolve TCEP in reaction buffer at a concentration of 10-m*M* and readjust the pH if necessary.
3. Add a quantity of the TCEP solution to the antibody solution with mixing to achieve a 2.75 molar excess of the reducing agent over the amount of antibody present.
4. Incubate for 2 h at 37°C.

The reduced antibody can be used immediately to label with a maleimide-based cyanine dye using the following protocol.

Protocol for Labeling Reduced Antibody with Thiol-Reactive Cyanine Dye

5. In a fume hood, dissolve the maleimide-containing cyanine dye in DMF at a concentration of 10-m*M*. Protect all dye solutions from light.
6. With mixing, add a quantity of the dye solution to the antibody solution to provide the desired molar excess of dye. For instance, for an antibody dissolved in buffer at 10 mg/ml, the addition of 66 μl of dye solution will give a 10-fold molar excess of dye over protein.
7. React for 2 h at room temperature with gentle mixing.
8. Purify the labeled protein from excess dye and reaction byproducts using dialysis or gel filtration on a desalting resin. Protect the labeled protein from light to avoid photobleaching the dye and losing fluorescence intensity.

8.3. Carbonyl-Reactive Cyanine Dyes

Cyanine-type dyes containing hydrazide functional groups may be used to label molecules containing carbonyl groups, especially aldehydes (Figure 10.48). Aldehydes may be created on carbohydrates, sugars, and glycans by periodate oxidation, which cleaves the carbon–carbon bonds between diols to form reactive formyl groups. Thus, glycoproteins and other glycoconjugates can be labeled specifically with a cyanine fluorescent dye only through carbohydrate components, which often avoids binding sites or active centers. Hydrazide-containing cyanine dyes also may be used to detect glycoproteins in cells, tissues, gels, or on western blots after periodate oxidation. In this regard, bands containing glycoproteins can be detected in a sample separately from the rest of the non-glycosylated protein pool (Thermo Fisher).

Glycan molecules that have been released from a protein by enzymatic means also can be labeled at their reducing ends by a hydrazide-containing cyanine dye. This reaction results in a single fluorescent label on the C-1 carbon of the innermost sugar of the glycan tree. The labeled glycan can be tracked through separation steps or detected for its specific interactions with carbohydrate binding proteins (see Chapter 2, Section 4.6).

Hydrazide-based cyanine dyes are reactive with common formaldehyde fixatives for cell and tissue studies. This enables these dyes to function as general stains for protein-rich areas within cells, and they get crosslinked into place by the formaldehyde reaction process.

The following protocol relates to the labeling of glycoproteins with hydrazide-containing cyanine dyes. Similar methods can be used to detect glycoproteins in other applications, such as within electrophoresis gels or on blots. For methods related to the labeling of glycans at their reducing ends with hydrazide reagents, see Chapter 11, Section 6.7.

Protocol

1. Dissolve a glycoprotein to be labeled in 50-m*M* sodium phosphate, pH 7 (reaction buffer), at a concentration

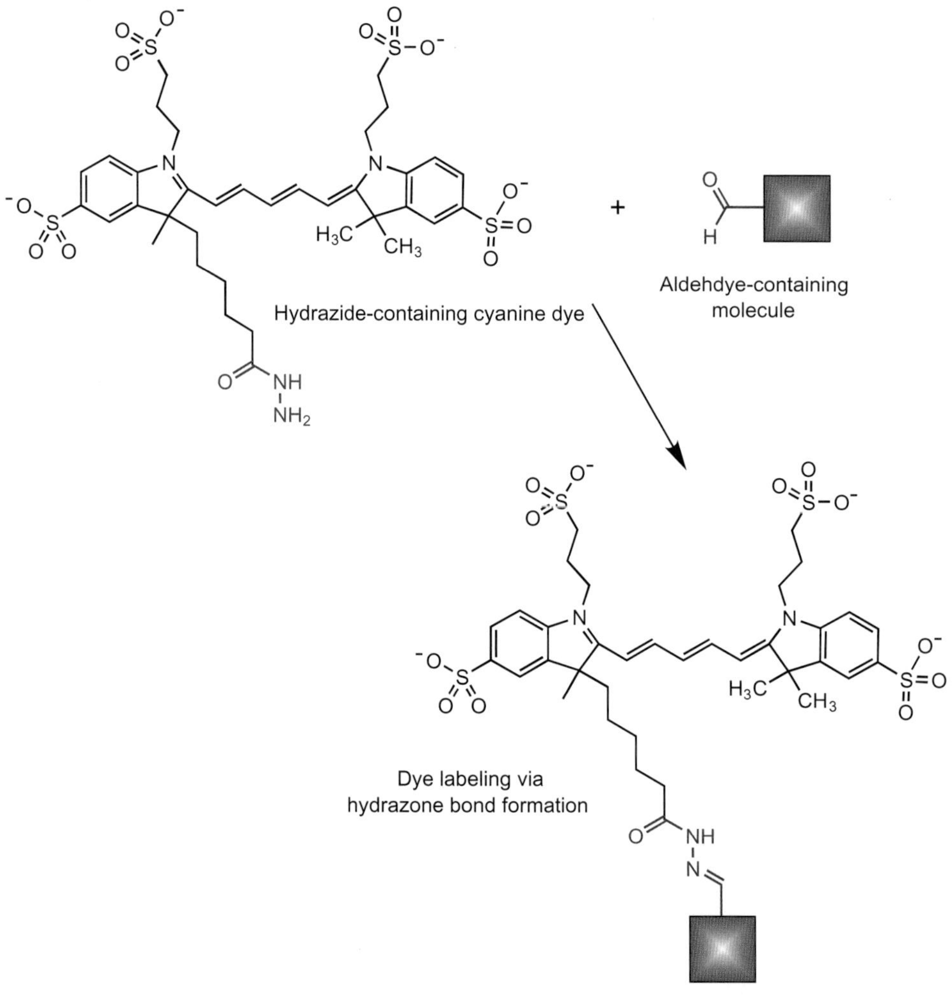

FIGURE 10.48 A cyanine dye containing a hydrazide group can be used to label glycans at their reducing end or other reducing sugars, forming a hydrazone linkage. Glycoproteins also can be labeled after periodate oxidation to form aldehyde groups.

of 1 to 10 mg/ml. For sialic acid modification, place the sample in ice to cool to near 0°C.

2. Dissolve sodium metaperiodate in reaction buffer at a concentration of 10 mg/ml (0.046-*M*). Protect from light. To obtain approximately a 1-m*M* concentration of sodium periodate in the reaction solution (suitable for oxidizing only sialic acid residues), add 21.8 µl of this stock solution to each milliliter of the glycoprotein solution to be oxidized. Maintain the solution on ice. For general oxidation of carbohydrates other than just sialic acid, add 218 µl of the stock solution to obtain an approximate final concentration of 10-m*M* periodate in the reaction. Use room-temperature conditions for general carbohydrate oxidation. Wrap the vial containing the reaction solution with aluminum foil to protect from light. The use of an amber vial is also suitable for this purpose.
3. React for 15 to 30 min at room temperature.
4. Quench the reaction by immediate gel filtration on a desalting column. If a dextran-based resin is used for the chromatography, the support itself will react with sodium periodate to quench excess reagent. Alternatively, *N*-acetylmethionine may be added to quench the reaction, because the thioether of the methionine side chain will react with periodate to form sulfoxide or sulfone products (Geoghegan and Stroh, 1992). In addition, sodium sulfite (Na_2SO_3) was used by Stolowitz *et al.* (2001) to quench the periodate oxidation of HRP in solution. To quench the reaction with cellular samples, wash the cells with buffer to remove remaining traces of periodate.
5. In a fume hood, dissolve the hydrazide-cyanine dye in DMF at a concentration of 10-m*M*. Protect all dye solutions from light.
6. With mixing, add a quantity of the dye solution to the oxidized antibody solution to provide the desired molar excess of dye. For instance, for an antibody dissolved in buffer at 10 mg/ml, the addition of 66 µl of dye solution will give a 10-fold molar excess of dye over protein.
7. React for 2 hours at room temperature with gentle mixing.
8. Purify the labeled protein from excess dye using dialysis or gel filtration on a desalting resin. Protect the labeled protein from light to avoid photobleaching the dye and losing fluorescence intensity.

9. LANTHANIDE CHELATES FOR TIME-RESOLVED FLUORESCENCE

Lanthanide metals are sometimes called rare earths and are located in period 6 of the periodic table of elements with atomic numbers 57 to 71, beginning with lanthanum (although there is some discrepancy over exactly where they start and end). Due to their unique electronic properties, lanthanide ions display luminescent properties with very sharp emission peaks. Unfortunately, the metal ions alone have extremely low extinction coefficients of no more than about 1–10 $M^{-1}cm^{-1}$, which makes them poorly fluorescent. The dim nature of lanthanide luminescence can be overcome by associating the metal ion with a chromophore group that can act as a light antenna. Lanthanide metals can form nine coordination bonds with ligands, creating three-faced centered trigonal prism structures. Coordinated in this way in an organic chelate structure that also contains a strongly absorbing chromophore nearby, the dimness of lanthanide metal luminescence can be transformed into an extraordinarily bright complex. The appropriate antenna group is one that can absorb energy and transfer it to the lanthanide metal, which then emits light at specific wavelengths, depending on which lanthanide element is in the coordination complex. For a review of chelating compounds for lanthanide luminescence, including their use in the development of time-resolved immunoassays, see Hagan and Zuchner (2011), Mathis and Bazin (2011), and Arnaud and Georges (2003). Some examples of antenna-chelate structures are shown in Figure 10.49. Figure 10.50 illustrates the principle of lanthanide luminescence.

Fluorescent labels built from lanthanide chelating groups of this type are extremely useful in biological applications due to their large Stokes shift (nearly 290 nm), no overlap between excitation and emission peaks, and very long fluorescent lifetimes. When bound by a chelating group that completely envelops and interacts with most of the lanthanide's nine possible coordination sites, the metal also is protected from any solvent-quenching effects that may reduce luminescence (Pietraszkiewicz *et al.*, 1993). If the antenna-chelating group also contains a reactive group, then the lanthanide–chelate complex can be covalently linked as a label on targeting molecules or affinity ligands. The result is one of the most intensely fluorescent probes available for biological detection applications.

The fluorescence lifetimes of lanthanide chelates are among the longest of all fluorescent compounds. Whereas most organic fluors or inorganic nanocrystals have lifetimes measured in the nanosecond or picosecond range, lanthanide chelates typically emit light for microseconds to milliseconds. The advantage of this long-lived emission is that time-gated measurements can be taken, which detect lanthanide luminescence long after emission from other organic fluors has decayed to zero. In complex biological samples, there are often many organic components that have fluorescent properties, especially in the lower visible or UV regions of the spectrum. When using lanthanide chelates as fluorescent probes and labels, any indigenous

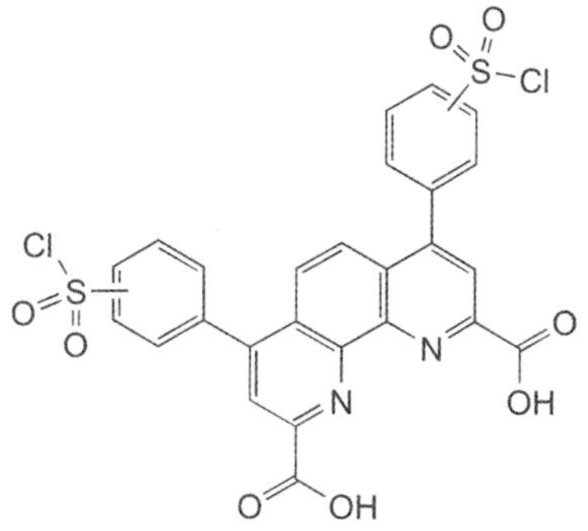

BCPDA;
4,7-*bis*-(chlorosulfophenyl)-
1,10-phenanthroline-2,9-dicarboxylic acid

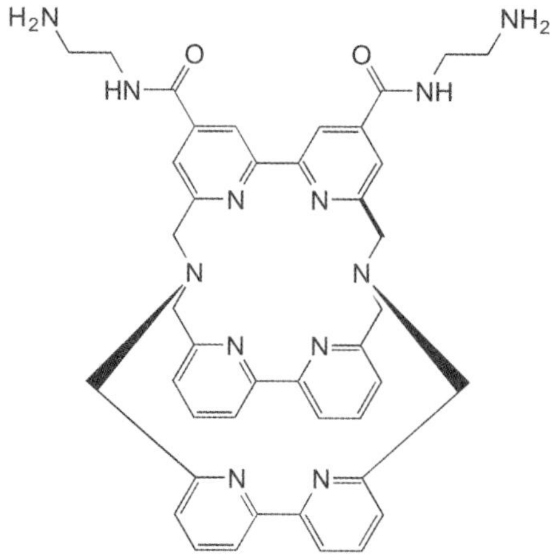

TBP;
trisbipyridine cryptate

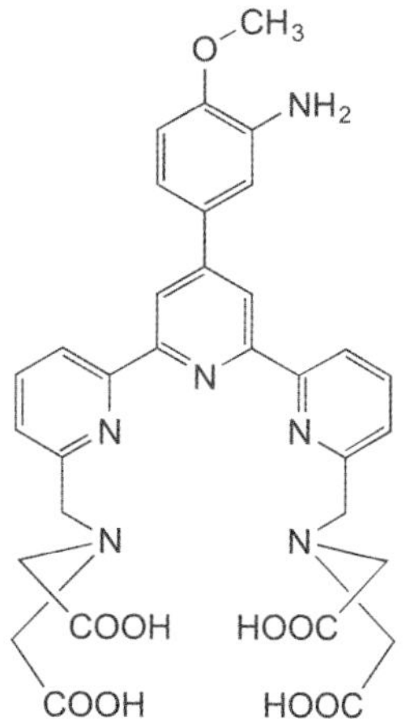

TMT;
4'-(3-isothiocyanato-4-methoxyphenyl)-
6,6"-*bis*[*N,N*-*bis*(carboxymethyl)
aminomethyl)-2,2';6',2"-terpyridine

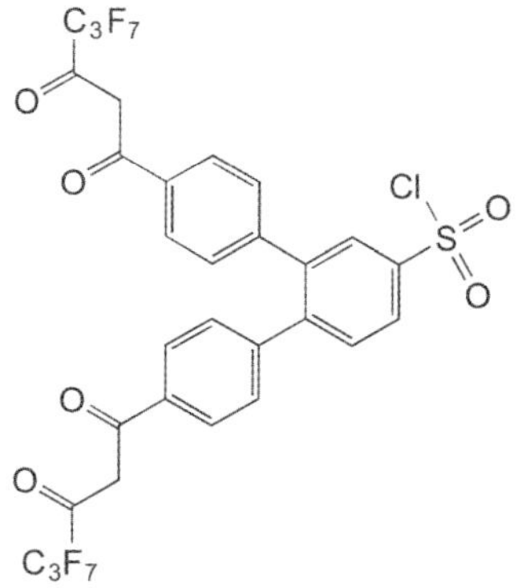

BHHCT;
4,4'-*bis*(1",1",1",2",2",3",3"-
heptafluoro-4",6"-hexanedion-6"-yl)
chlorosulfo-*o*-terphenyl

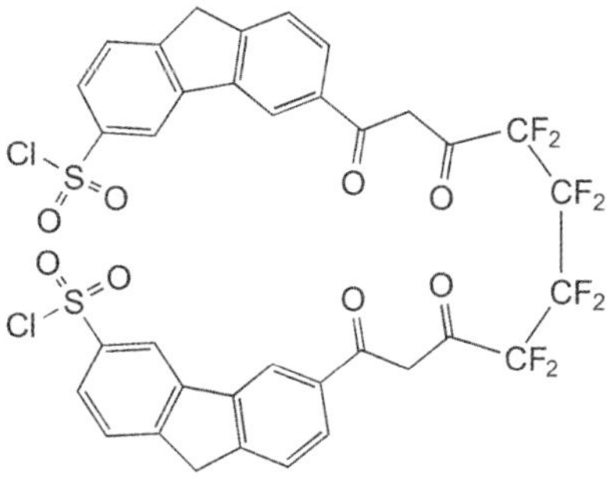

BCOT;
1,10-*bis*(8'-chlorosulfo-
dibenzothiophene-2'-yl)-4,4,5,5,6,6,7,7, -
octafluorodecane-1,3,8,10-tetraone

FIGURE 10.49 Examples of antenna–chelator compounds that have been used for lanthanide luminescence.

sample fluorescence can be eliminated before the chelate emission is measured. This "time-resolved" luminescence technique virtually abolishes background fluorescence and makes detection and assays using lanthanide chelates among the most sensitive fluorescence measurements possible. For reviews of time-resolved fluorescence using lanthanide chelates in biological assays, see Patsenker *et al.* (2011), Hagan and Zuchner (2011), Mathis and Bazin (2011), Soini and Kojola (1983), Soini *et al.* (1990), Diamandis and Christopoulos (1990), Hemmila (1985), Diamandis (1993), and Hemmila (1998).

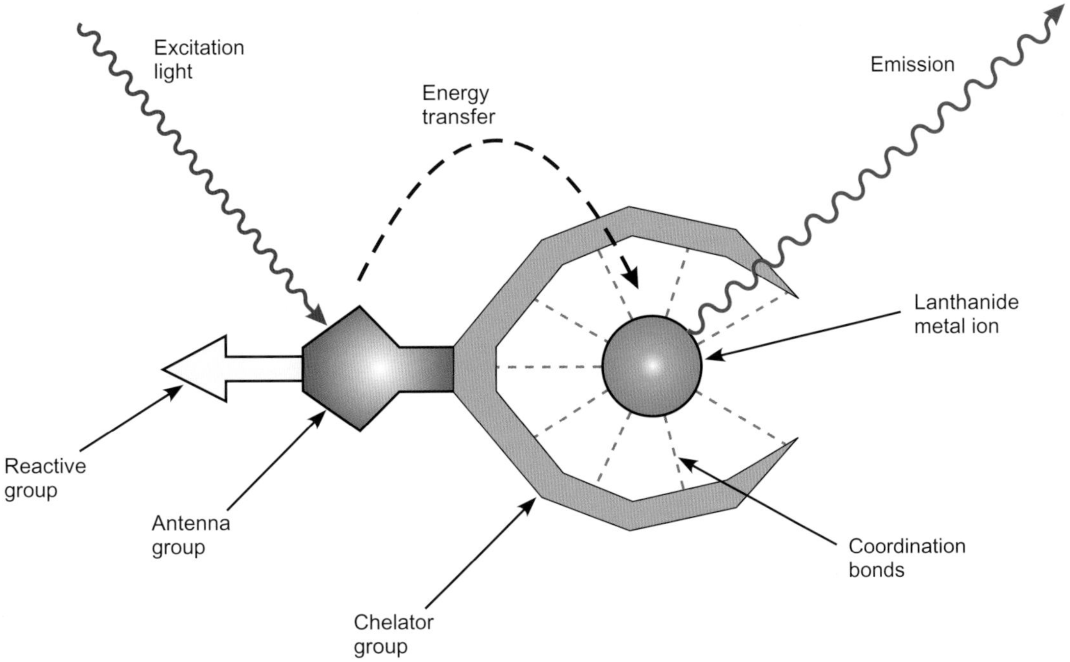

FIGURE 10.50 The principle of enhanced lanthanide luminescence using antenna–chelator compounds. The organic dye antenna group absorbs energy and transfers it to the lanthanide metal ion coordinated within the chelate structure. The best chelating groups fully coordinate the metal and protect it from the quenching effects of water. The lanthanide atom then emits light at a longer wavelength, which is characteristic of the lanthanide metal. Lanthanide chelates also contain a reactive group that can be used to attach covalently the label to targeting molecules.

Different lanthanide metals also produce different emission spectrums and different intensities of luminescence at their emission maximums. Therefore, the relative sensitivity of time-resolved fluorescence also is dependent on the particular lanthanide element complexed in the chelate. The most popular metals along with the order of brightness for lanthanide chelate fluorescence are europium(III) > terbium(III) > samarium (III) > dysprosium(III). For instance, Huhtinen *et al.* (2005) found that lanthanide chelate nanoparticles used in the detection of human prostate antigen produced relative signals for detection using europium, terbium, samarium, and dysprosium of approximately 1.0:0.67:0.16:0.01, respectively. The emission wavelength for each of these lanthanides is also different. Terbium has a main peak of luminescence in the range of 545nm, dysprosium at about 575nm, europium at about 615nm, and samarium at approximately 645nm (although there are other emission bands for all of these lanthanide metals as well).

In assays, europium chelates usually produce the greatest sensitivity in an assay, followed closely by terbium chelates. Another advantage of europium is that it has the longest fluorescent lifetime of all the lanthanides, making it the best choice for time-resolved applications. For these reasons, most applications of lanthanide fluorescence use europium or terbium chelates to modify biological molecules, as they provide the brightest conjugates achievable when using this technology.

Another advantage of lanthanide chelates is their lack of self-quenching effects. Since the excitation and emission peaks do not overlap for a given lanthanide metal, the chelates do not quench each other if modified at high density on other molecules or immobilized close together on surfaces or particles. Therefore, creating complexes having multiple lanthanide chelating groups is a strategy that can dramatically increase the total fluorescence of a detection conjugate. This property differs from organic fluorescent reagents, as dyes usually start to quench if as few as six to eight fluorescent labels modify a single protein. By contrast, it is possible to form polymer or particle labels containing dozens, hundreds, or even thousands of lanthanide chelating groups, which are capable of increasing the fluorescent signal in assays equal to the sum of the total number of groups present. Using this approach, Huhtinen *et al.* (2005) created nanoparticle labels containing hundreds of fluorescent lanthanides. Similarly, Scorilas (2000) created polyvinylamine polymer chains with multiple europium chelates and also added biotin labels to form huge complexes with streptavidin in solution. Such reagents dramatically increase the lanthanide luminescence signal in immunoassays or other detection applications beyond that possible with standard organic fluors.

Lanthanide chelates also can be used in fluorescence resonance energy transfer (FRET) applications with other fluorescent probes and labels (Figure 10.51).

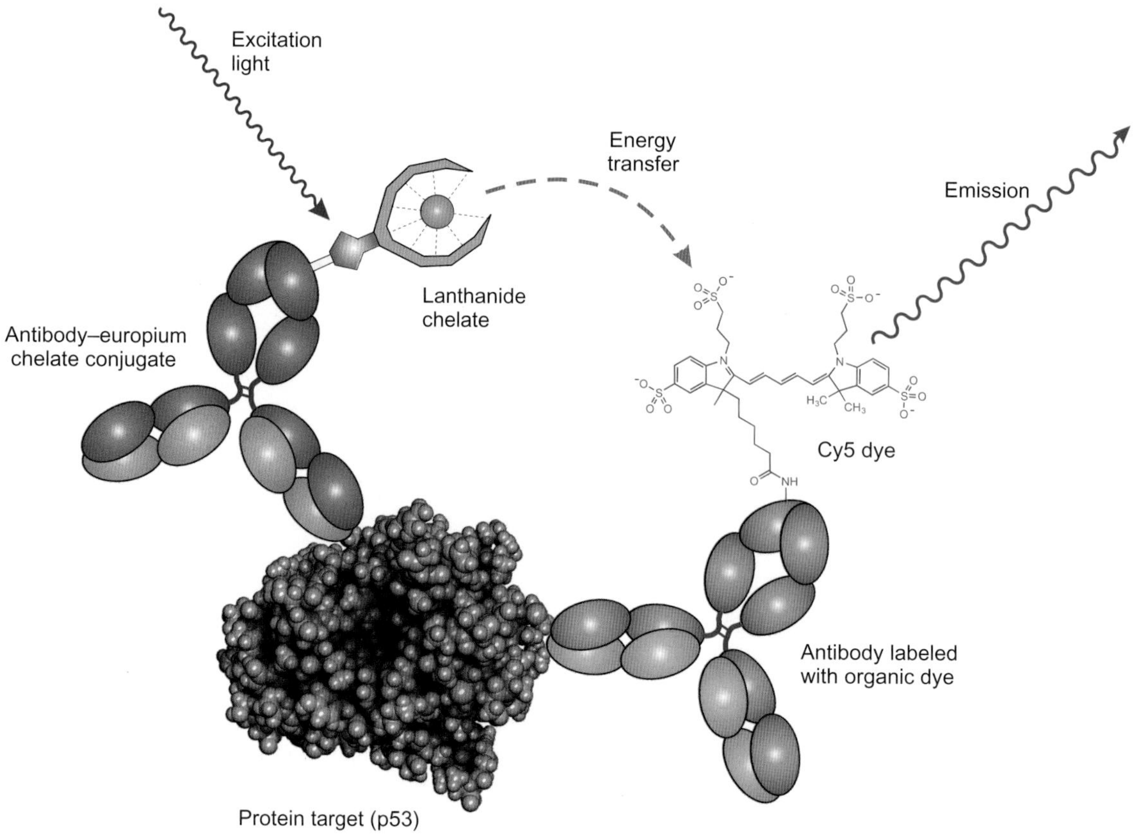

FIGURE 10.51 Time-resolved FRET assay systems involve energy transfer between the lanthanide chelate and an organic dye (or a fluorescent phycobiliprotein) that are brought together as two labeled molecules bind to an analyte. In this illustration, an antibody labeled with a lanthanide chelate is used along with a Cy5-labeled antibody to detect a protein target in solution. Excitation of the lanthanide label results in energy transfer and excitation of the cyanine dye only if they are held within close enough proximity to allow efficient FRET to occur. Under these conditions, excitation of the lanthanide chelate results in cyanine dye emission, which will not occur if the labeled antibodies have not bound to a target.

In this application, the time-resolved (TR) nature of lanthanide luminescent measurements can be combined with the ability to tune the emission characteristics through energy transfer to an organic fluor (Comley, 2006). TR-FRET, as it is called, is a powerful method to develop rapid assays with low background fluorescence and high sensitivity, which can equal the detection capability of enzyme assays (Selvin, 2000).

FRET signaling is limited by the distance requirements for energy transfer between fluorescent molecules. The donor molecule must be constrained within the immediate molecular vicinity of the acceptor fluorescent molecule. FRET systems can be described in terms of the Förster radius, which is the distance at which energy transfer efficiency is 50% (Förster, 1948; Lakowicz, 1999). For energy transfer between organic fluors, the Förster radius is usually no more than about 15 to 20 Å. This means that for FRET signaling to occur with high retention of fluorescence yield, the donor and acceptor molecules must be in extreme proximity, held much closer than the radius of the average globular protein.

Lanthanide chelates have distinct advantages in FRET systems, because their Förster radius can be on the order of 80 to 100 Å. This means that two biological molecules coming together in solution, one labeled with a donor lanthanide chelate and the other labeled with an acceptor organic fluor, usually are positioned close enough to undergo efficient energy transfer. Thus, lanthanide TR-FRET assays can be developed that are completely homogeneous in nature, taking advantage of both the long lifetime of fluorescence and the efficient energy transfer characteristics of a long Förster radius. For instance, an antibody labeled with a europium chelate can be used along with another antibody labeled with an organic fluorescent label able to absorb energy from the lanthanide. If both antibodies recognize different epitopes on a target antigen, then a solution-phase assay can be designed wherein the FRET signal is observed only if both antibodies are bound to the antigen. The excess labeled antibodies in solution still will be free to diffuse throughout the sample and thus not be held in enough proximity to undergo energy transfer. The result is low background interference, even though excess fluor is not washed away.

To make a luminescence resonance energy transfer (LRET) system work, the proper choice of acceptor fluor must be made for the lanthanide chelate. The acceptor must have an excitation band corresponding to or overlapping an emission band of the lanthanide label, so that excitation energy may be transferred. Although energy transfer is never 100% efficient as would be evident by a complete disappearance of the lanthanide emission band, the efficiency can be high. To maximize the yield of fluorescence transfer, the acceptor should also be an efficient absorber with a high extinction coefficient and should have bright fluorescent properties (high QY) to ensure good signal in the final assay application. In this regard, one of the best acceptor molecules for lanthanide luminescence is the phycobiliprotein allophycocyanin (APC). APC often is paired with europium chelates to create LRET assays. The protein's multiple fluorescent bilin groups enhance its ability to receive energy transfer from the luminescence of the donor europium. In a europium–APC system, excitation in the UV range with subsequent energy transfer results in emission in the red region of the spectrum, giving an effective Stokes shift of over 300 nm. Combine the FRET signal with a time-gated measurement and the result is a highly sensitive assay with virtually no interference from other biological or organic molecule fluorescence.

Organic fluorescent dyes with the appropriate spectral properties can also be paired with lanthanide chelates in FRET systems. For instance, many rhodamine dyes and the cyanine dye Cy5 have ideal excitation wavelengths for receiving energy from a nearby europium chelate. The LeadSeeker assay system from GE Healthcare incorporates various Cy5-labeled antibodies for developing specific analyte assays. In addition, when using a terbium chelate is used as the donor, a Cy3 fluorescent dye can be used in assays as the acceptor.

Antenna–chelator molecules for lanthanide luminescence typically have a reactive group to allow conjugation with targeting molecules to be useful for TR-FRET in biological applications. Some of the reactive groups that have been designed for polyaminocarboxylate chelates include the amine-reactive isothiocyanate derivatives (Li and Selvin, 1997), thiol-reactive maleimide groups, and pyridyl disulfide groups (Chen and Selvin, 1999). Alternatively, fluorescent lanthanide chelates can be embedded into nanoparticles at high density and form intensely bright particle-labeling agents for biomolecule detection (Ju *et al.*, 2011).

Two of the more common chelate structures are built from an EDTA or DTPA backbone (Chapter 12, Section 1). Depending on the method of antenna group addition, the DTPA chelate structure can have a maximum of eight coordination groups consisting of carboxylates and amines to hold the lanthanide atom. An EDTA chelating group can have up to six coordination groups per chelate molecule. The remaining coordination sites are taken up by water molecules in aqueous solution, which can potentially quench luminescence, because they are effective acceptors of excitation energy from the lanthanide. Both EDTA and DTPA chelating groups with antenna molecules do perform well in luminescent assays. However, it is better to have a chelating group that can coordinate with all or a majority of the nine coordination bonds on a lanthanide metal ion. This avoids the potential for quenching effects from other chelators in solution or from water itself. In this regard, reagents built from a DTPA chelating group are potentially better than those based on EDTA. For a review of stable lanthanide chelate constructs useful for luminescent assays, see Mathis and Basin (2011).

In some cases, the initial lanthanide chelating group may not have an antenna group built into its structure. One of the early commercial applications of europium luminescence in a TR-FRET-based assay system used an isothiocyanatophenyl–EDTA or an isothiocyanatophenyl–DTPA derivative to modify proteins (Delfia system from Wallac). This system uses an isothiocyanate chelate group to carry the lanthanide and attach it to antibodies and other molecules, but it does not have an antenna group. In an assay, this system requires the use of a secondary fluorescence enhancer to create the lanthanide fluorescence signal. This secondary reagent is typically β-naphthoyltrifluoroacetone, which when added in excess at the end of an assay extracts the lanthanide metal from the nonfluorescent chelate and forms another soluble chelate that is highly fluorescent.

More advanced designs incorporate efficient antenna groups directly into the chelating structure of the label. For instance, the terpyridine-*bis*(methylenamine) tetraacetic acid (TMT) chelator of europium developed by Toner *et al.* (1993) has multiple pyridine rings and two iminodiacetic acid groups on each side to create a chelator having nine coordination sites. The pyridine groups provide the antenna structure and also supply nitrogen atoms with unshared pairs of electrons to coordinate with the lanthanide metal ion. The result is a protected lanthanide that cannot be quenched easily by water or other matrix components. Complexed with europium, this label can be excited at 340 nm and provide sharp emission peaks at wavelengths of 589, 599, 618 (maximum emission), 623, 651, 689, 697, and 702 nm. The largest peaks in the >600 nm range can optimally excite Cy5 dyes, such as the hydrophilic DyLight 649 (Section 8), and create an effective TR-FRET system for homogeneous assay design.

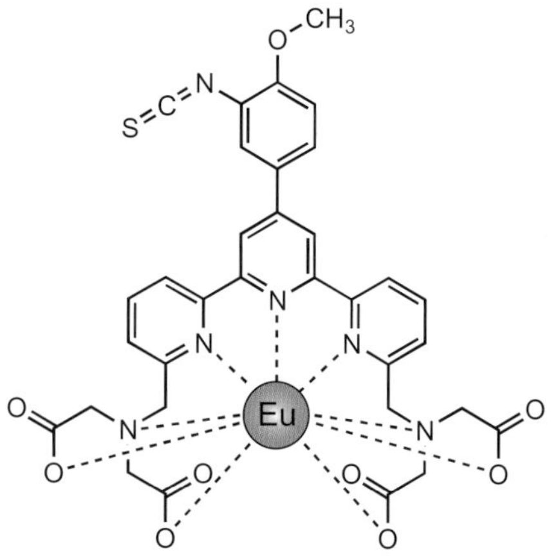

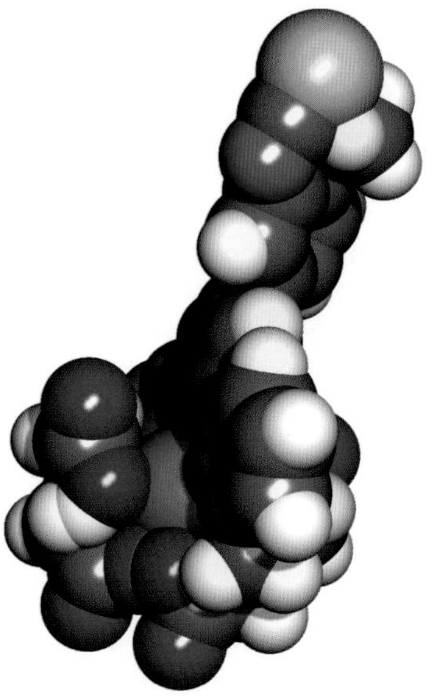

The TMT chelator containing an amine-reactive isothiocyanate group can be used to label antibodies, proteins, or other molecules using the following protocol (Figure 10.52).

Protocol

1. In a fume hood, dissolve the TMT chelator in a 1:1 mixture of DMF:DMSO at a concentration of 10mg/ml.
2. Dissolve a protein to be labeled in 0.1-*M* sodium bicarbonate, pH 9.0, at a concentration of 1 to 10mg/ml.
3. With mixing, add a quantity of the TMT chelator to the protein solution to provide a 5- to 10-fold molar excess of reagent. Optimization may have to be done to determine the best ratio of chelator to protein for the intended application.
4. React at 4°C overnight with gentle mixing.
5. Purify the labeled protein by gel filtration or dialysis.

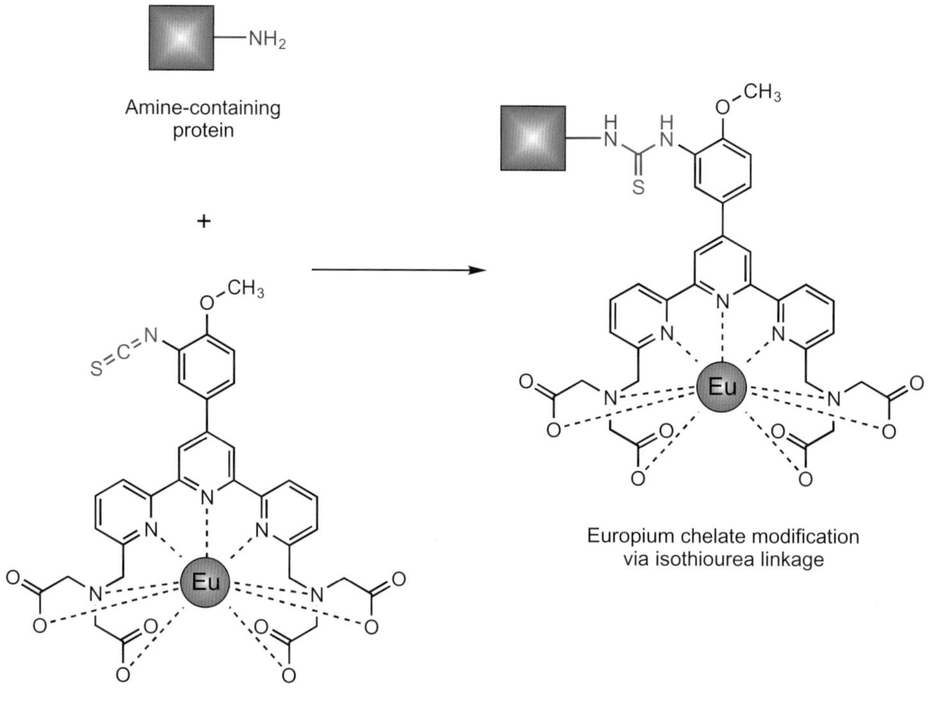

FIGURE 10.52 The isothiocyanate group of a TMT europium chelate can be used to label amine-containing proteins and other molecules to result in an isothiourea bond.

As in the structure of the TMT chelator group, pyridine derivatives have long been known to be enhancers of lanthanide luminescence (Thomas *et al.*, 1978). One such compound, dipicolinic acid (DPA), contains two carboxylates on both sides of a pyridine core, providing three coordination sites per molecule to complex with lanthanides. In solution, three DPA molecules can coordinate with one lanthanide ion to fully surround the metal and protect it from quenching by water or other molecules. Each DPA pyridine group can absorb energy in the UV region and transfer it to the central lanthanide, which results in strong emission at characteristic wavelengths.

Lamture and Wensel (1995) synthesized a unique modification of the basic DPA structure to make the chelator bifunctional and thus able to react with functional groups on other molecules. This reactive DPA reagent, 4-(iodoacetamido)-2,6-dimethylpyridine dicarboxylate, contains an iodoacetyl group on the C4 of the pyridine ring, thus making it particularly reactive with thiol groups and amines at higher pH. This derivative was used to create a large polymeric chelating structure from poly-L-lysine that contained 50 to 100 DPA units along its length. Loading this complex with Tb(III) ions resulted in highly intense luminescence. The remaining amines on poly-L-lysine finally were succinylated to create carboxylates for coupling to proteins via carbodiimide conjugation. Figure 10.53 shows DPA and iodoacetyl-DPA derivatives; Figure 10.54 shows the DPA-poly-L-lysine derivative succinylated.

Another antenna group that is particularly effective for lanthanide luminescence is the carbostyril derivatives (7-amino-4-methyl-2(1H)-quinolinone), which can be attached to many chelating compounds *via* amide bond formation with the 7-amino group to a carboxylate (Figure 10.55) (Hemmila and Laitala, 2011). A carbostyril–DTPA derivative has intense luminescence for use in TR-FRET-based assays for drug discovery. The 7-aminoquinolinone structure is a common choice for designing fluorescent lanthanide chelates (Soini and Lövgren, 1987; Sammes and Yanhioglu, 1996; Selvin, 2002; Selvin, 2003). The synthesis of various carbostyril derivatives is described in Ge and Selvin (2004). The 7-aminoquinolinone group first is put onto DTPA by reaction with one of the chelator's anhydride rings. Linking the chelator to proteins or other molecules is done through the anhydride on the opposite end by forming an amide-linked spacer arm. In some cases, sulfonate groups or other hydrophilic components on the carbostyril structure can provide increased hydrophilicity for modifying biomolecules.

When used with europium or terbium ions, a carbostyril-based lanthanide chelate can be excited at 340 nm and provide sharp characteristic emission bands for transfer of energy to the appropriate acceptor fluor. Similar to the TMT chelator described previously, luminescence from terbium FRET signals well with Cy3 dyes, and luminescence from europium can be used with APC or Cy5 dyes. Other fluorescent dyes that have similar excitation and emission ranges to these can also be used as acceptors in TR-FRET assays. For instance, terbium chelates can be used with fluorescein and tetramethyl rhodamine dyes, if the appropriate filters are used for measuring emission.

The potential for creating an expressed protein time-resolved luminescence system was realized in the

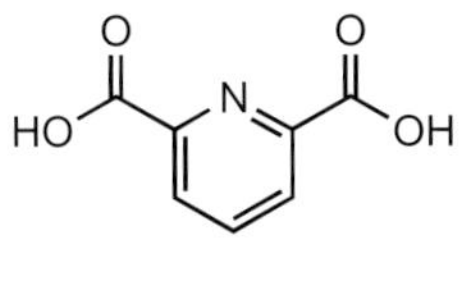

Dipicolinic acid (DPA)
MW 167.12

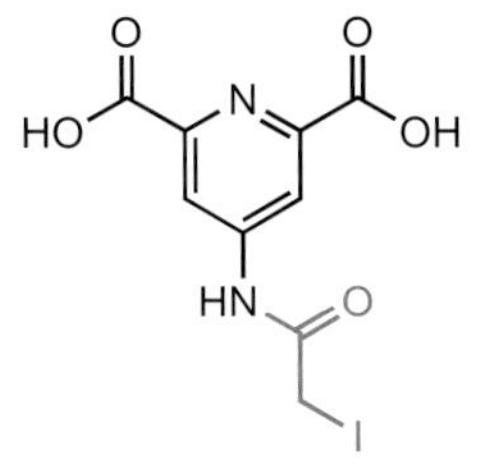

DPA-iodoacetate;
4-(iodoacetamido)-2,6-dimethylpyridine dicarboxylate
MW 350.07

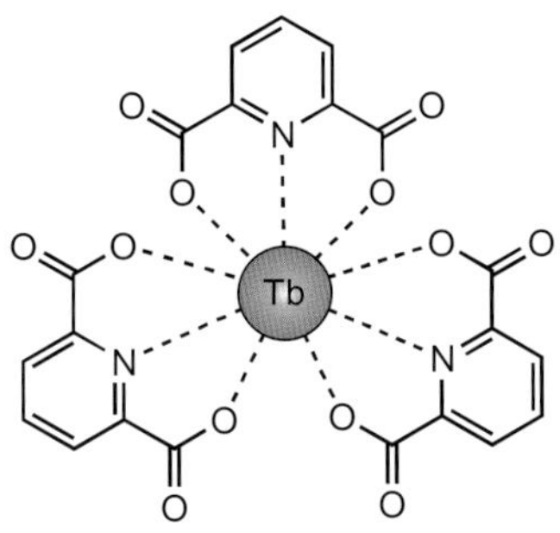

DPA-Terbium chelate

FIGURE 10.53 DPA derivatives have been used as potent enhancers of lanthanide luminescence. Three DPA groups can coordinate with a terbium ion. The iodoacetate derivative of DPA has been used to label covalently molecules for lanthanide luminescence.

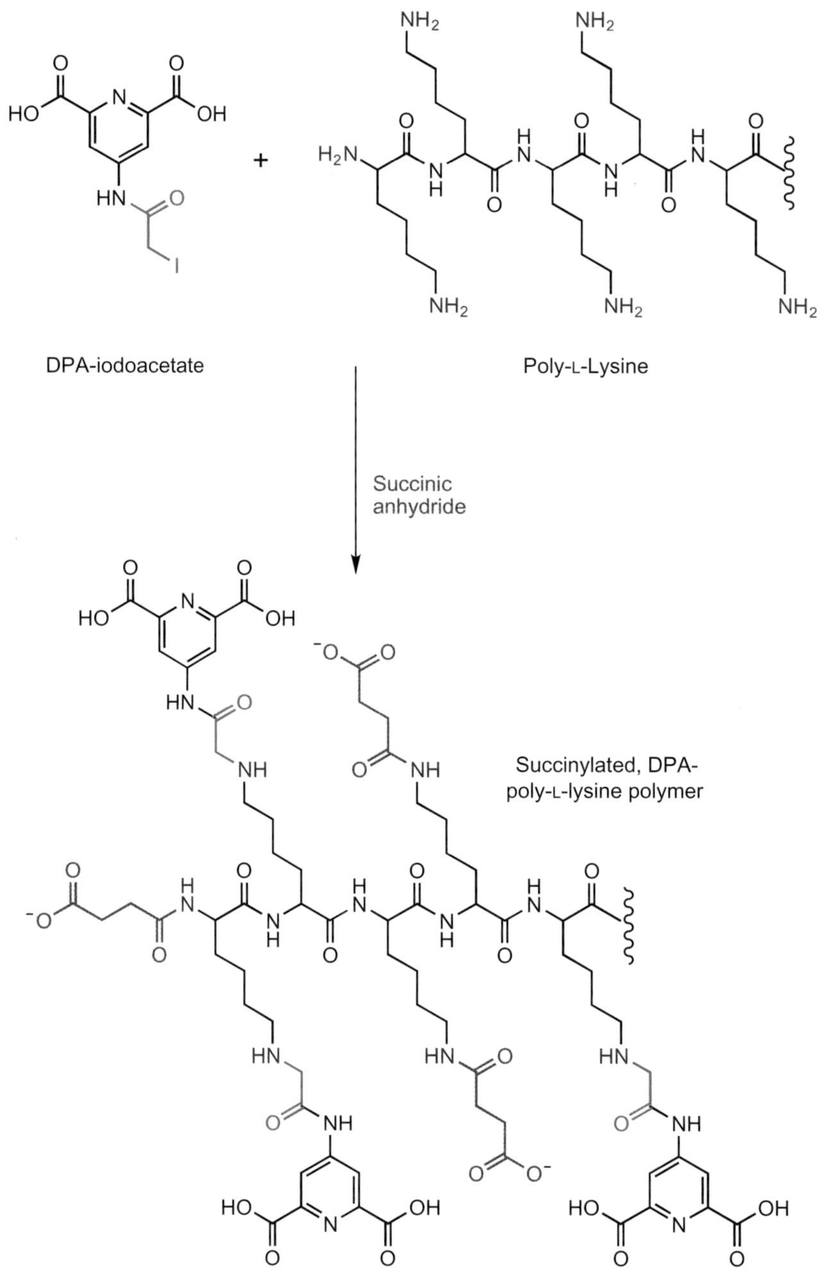

FIGURE 10.54 The iodoacetamide derivative of DPA has been used to create a chelating polymer of lanthanide metals using poly-L-lysine as the backbone to attach multiple chelators along its length.

development of a novel lanthanide chelating complex derived from a metal-binding peptide sequence. This sequence, called a lanthanide binding tag (LBT), can be used as a fusion tag in recombinant protein expression (Franz *et al.*, 2003; Nitz *et al.*, 2003; Wöhnert *et al.*, 2003; Lim and Franklin, 2004; Goda *et al.*, 2007). To create this luminescent tag, a peptide sequence of about 15 to 20 amino acids representing the basic structural domain of a calcium-binding loop from calmodulin was mutated to tightly coordinate a terbium ion. Luminescence occurs through excitation in the UV of a nearby tryptophan residue with energy transfer to the chelated lanthanide. Recombinant mutagenesis of the original calcium-binding domain resulted in optimizing the dissociation constant for terbium to within the nanomolar range, thus the metal ion is tightly associated with the peptide tag (Martin *et al.*, 2005). Recombinant proteins containing this small fusion tag can be detected using time-resolved fluorescence techniques. Such *in vivo* expression of a luminescent lanthanide tag can be used to study protein interactions using LRET or track individual proteins within cells (Sculimbrene and Imperiali, 2006).

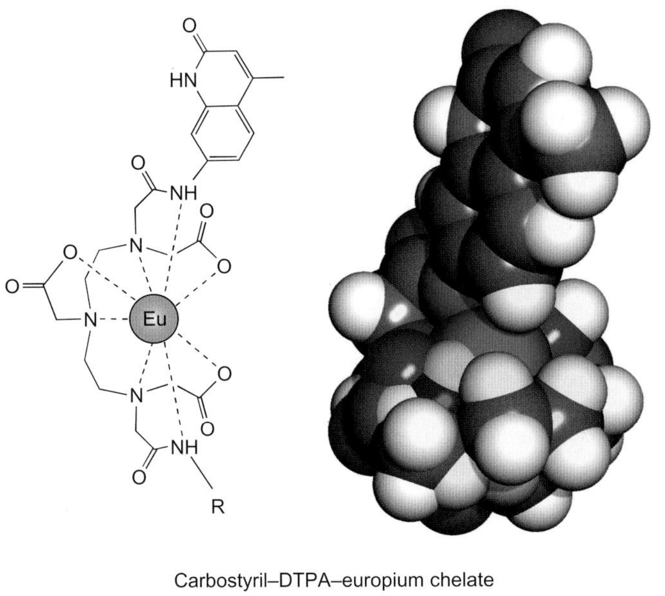

FIGURE 10.55 A carbostyril–DTPA–europium chelator is a strong enhancer of lanthanide luminescence. The chemical structure and three-dimensional structure of the chelator are shown. The metal ion is well protected at the center of the coordinating groups of the chelator.

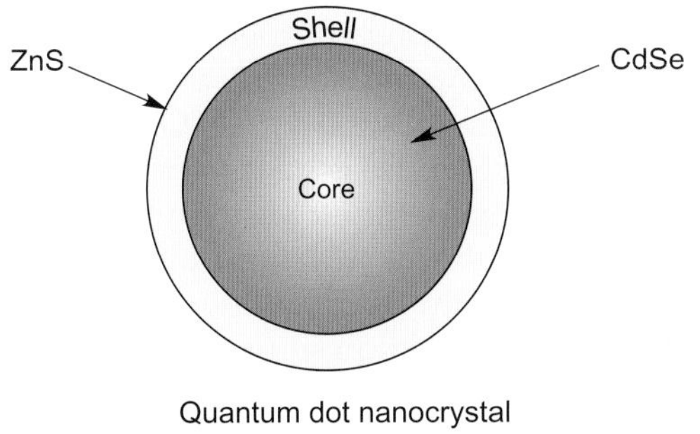

FIGURE 10.56 The structure of a typical QD nanocrystal includes a semiconductor alloy core surrounded by a shell consisting of a different alloy structure. Early QD compositions involved the use of a CdSe core capped with a ZnS shell, but many different alloy compositions have been used and are possible.

10. QUANTUM DOT NANOCRYSTALS

10.1. Properties of Quantum Dots

Quantum dots (QDs) are nanoparticles typically made of a semiconductor metal alloy arranged in a spherical crystalline core and capped with a shell consisting of a second metal alloy composition (Figure 10.56). The size of this raw core/shell construct is usually less than 10 nm in diameter or about the same size as many globular protein molecules. Upon exposure to light at the appropriate wavelength range, QDs are able to absorb a photon of energy, which results in the excitation of an electron within the core. The excited electron is confined within the nanocrystal, because its core diameter is less than the exciton Bohr radius, which thus leads to quantum confinement. The shell structure aids in this confinement and prevents the electron from tunneling out of the core and escaping into the outer medium or undergoing non-radiative deactivation. The size and shape of a QD govern the discrete energy levels that the excited state electron can attain within it, thus dots can be tuned to have desired electronic properties by careful adjustment of core diameter and composition. Upon return of the electron to its ground state, a radiative QD emits a photon of light, the wavelength of which is dependent on the alloy material type and its diameter (Alivisatos, 1996). QD size *versus* fluorescence is shown in Figure 10.57.

As a result of their unique optical and electronic properties, particularly their ability to fluoresce at discrete wavelengths directly proportional to their sizes and material compositions, QDs have found use in many fields, including electronics, biology, medicine, and even cosmetics. The first attempts to modify their surface characteristics to make them water soluble and biocompatible eventually led to their use as fluorescent labels for biomolecules in many applications (Rogach *et al.*, 1996; Bruchez *et al.*, 1998; Chan and Nie, 1998).

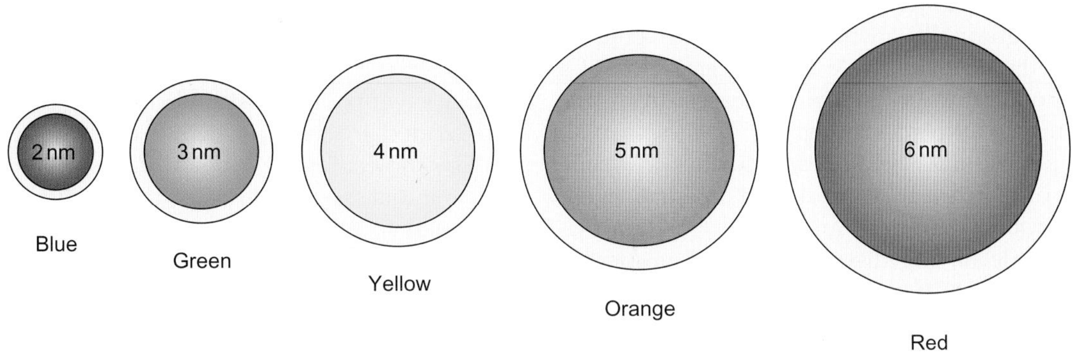

FIGURE 10.57 The size of a QD directly affects its emission wavelength. Careful control of nanocrystal diameter during the manufacturing process can result in discrete QD populations having emission properties ranging from the blue to the red region of the visible spectrum.

QDs have a number of advantages over other fluorescent molecules such as organic dyes, including: (1) resistance to photobleaching, which allows them to be imaged over long periods without loss of fluorescence; (2) narrow, nearly symmetrical emission peaks with no red-shift tail typical of organic fluors, thus creating a bright fluorescence signal at characteristic wavelengths (3) a broad absorbance band, which increases almost exponentially toward shorter wavelengths with extremely high extinction coefficients (10^5–$10^7\,M^{-1}\,cm^{-1}$); (4) the ability to excite at a single wavelength an entire family of QDs having different emission characteristics, thus providing multiplexed assay capability; (5) the capacity to design QDs with emission characteristics ranging from the low visible wavelengths to well within the IR region; and (6) the potential for relatively high quantum yields of fluorescence (0.65–0.95 for CdSe).

The emission properties of QDs can be adjusted based upon core diameter and nanoparticle composition. Nanoparticle diameters are typically carefully controlled during manufacture to be between 2 and 10 nm. In addition, the band-gap energy or energy of fluorescence emission is inversely proportional to the diameter of the QD particle. Thus, the smaller the particle, the more blue-shifted is its emission; the larger the QD, the more red-shifted is its emission bands. QDs also have an intrinsic color to their solutions that corresponds to the size of the particles and their fluorescence emission characteristics. However, to create a single particle population with a tight fluorescence emission pattern, the diameter of the particles must be controlled to well within a nanometer. The emission peak width is directly proportional to the size distribution of a particle population. This makes manufacturing reproducible QDs a constant challenge for most suppliers that rely on size to control fluorescence properties.

However, as opposed to the difficulty of tuning emission properties by particle diameter, QD alloy composition instead may be adjusted independent of size to control the wavelength of emission for a given particle population. In a QD having a concentration gradient composition, the concentration of an alloy of a first semiconductor gradually increases from the core to the surface of the particle, while the concentration of a second semiconductor gradually decreases from the core to the surface (Nie and Bailey, 2007). A third semiconductor type also may be added to further fine tune the emission properties. By careful adjustment of these semiconductor concentration gradients, QD populations can be made having discrete emission properties without changing the particle size. Therefore, tuning QD spectral characteristics can be done using a single particle size and by making selective changes to the alloy composition. This avoids the difficulty in manufacturing particles of uniform size, because all particle populations can have the same size but only vary in their relative semiconductor gradient concentrations to attain particles having discrete fluorescence character.

The material types making up the core of a quantum dot also affect the range of emission wavelengths that can be attained. For common material types, the ranges of emission wavelengths that can be achieved by adjustment of particle diameter or composition are: CdSe = ~470–660 nm, CdTe = ~520–750; InP = ~620–720 nm; PbS > 900 nm; and PbSe > 1000 nm.

Quantum dots have been made using a number of techniques. A common method to make bulk quantities of particles involves carrying out colloidal suspension synthesis in organic solvent with nucleation of semiconductor metals under high-temperature conditions (Murray *et al.*, 1993; Hines and Guyot-Sionnest, 1996; Dabbousi *et al.*, 1997). In one such process, a solvent such as octadecene is stirred at constant rate and heated to >300°C at which point solutions containing the semiconductor metals are injected. The metals at first decompose under high heat and then recombine to form alloys consisting of nanoparticle seeds, which grow to create the QDs. The reaction time determines the size of the nanoparticles and thus their spectral properties. Detergent molecules are often added to coat the resulting nanoparticles and prevent their aggregation during nucleation. Originally, the solvent and detergent molecule used for making QDs was TOPO (tri-*n*-octylphosphine oxide), which ends up coating the particles with the phosphine component interacting with the semiconductor surface and the alkyl chains pointing out into the organic solution (Figure 10.58). Other additives, such as stearic or oleic acid, function similarly. The raw particles thus prepared are hydrophobic and not dispersible in aqueous solution.

To use QDs in biological applications, the particles must be rendered biocompatible by coating with a hydrophilic layer that masks the surface, thus preventing aggregation and nonspecific binding. This is not a trivial problem, as the successful commercialization of QDs for biomolecule labeling took at least 5 years from the time the first two papers appeared in *Science* describing water soluble particles for bioconjugation (Bruchez *et al.*, 1998; Chan and Nie, 1998). The fact is, these early particles were not very soluble in aqueous environments and tended to clump together or bind nonspecifically with biomolecules.

The initial modifications carried out to covalently link molecules to QDs need to displace the TOPO or detergent coating on the raw nanocrystal surface with a new organic derivative imparting water solubility. The first attempts at making biocompatible QDs all involved the use of simple monothioacids, such as mercaptoacetic acid (thioglycolic acid), which can link to the shell through thiol dative bonding and provide a terminal

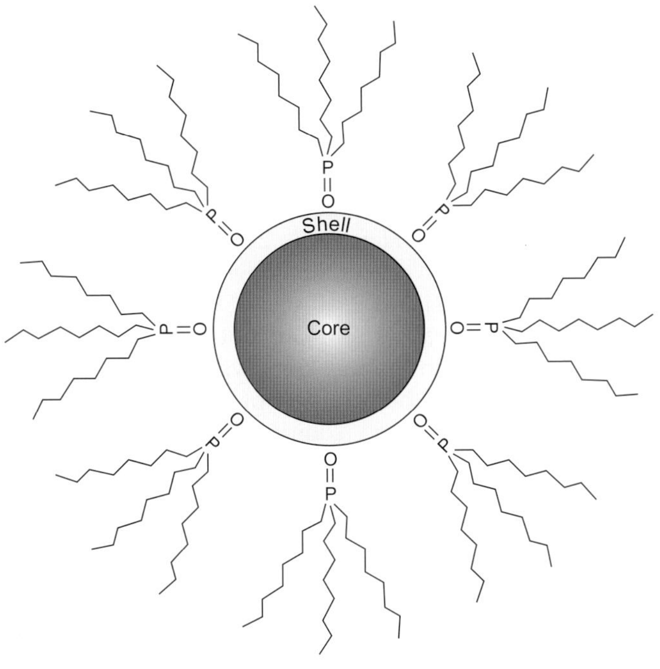

FIGURE 10.58 QDs made using the TOPO process typically have a layer of these molecules associated with their outer surface. The TOPO groups must be displaced and replaced by water soluble groups to provide biocompatibility for bioconjugation purposes.

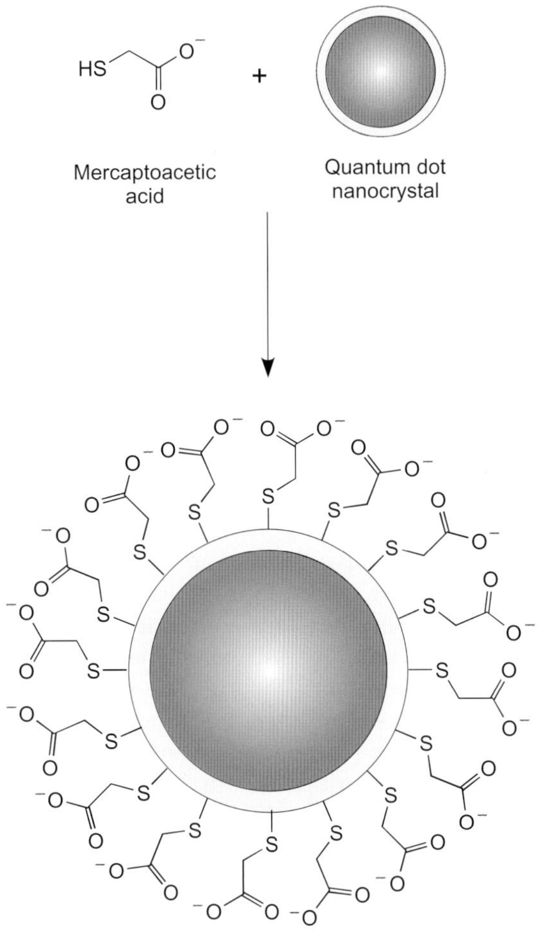

FIGURE 10.59 One of the first methods of preparing water-soluble QDs was to use mercaptoacetic acid modification of the nanocrystal surface. This resulted in a negative charge on the surface of each dot that provided like-charge repulsion of particles suspended in aqueous solution. The carboxylate group also could be used for conjugation with amine-containing molecules.

carboxylate for further conjugation (Figure 10.59). However, monothiol linkers can easily oxidize back off QDs and leave behind surface gaps, which become hydrophobic sites for particle clumping and nonspecific binding to biomolecules. A better approach is to use a dithiol compound, which forms two dative bonds per linker on the QD surface. This makes the linkage to the QD resistant to oxidation and prevents nonspecific surface gaps from forming. One such dithiol compound that is particularly useful is dihydrolipoic acid (DHLA), which contains a carboxylate group on the other end. QDs modified with DHLA can subsequently be modified with PEG groups to provide increased hydrophilicity of the surface or directly linked to proteins *via* electrostatic interactions or through an EDC-mediated reaction (Mattoussi *et al.*, 2000; Uyeda *et al.*, 2003; Medintz *et al.*, 2004; Anikeeva *et al.*, 2006; Clapp *et al.*, 2006). The functionalization of nanoparticles for bioimaging applications has been reviewed (Erathodiyil and Ying, 2011).

The negative charge character of DHLA-modified QDs has been used to link noncovalently positively charged proteins, such as avidin (Goldman *et al.*, 2002) or a recombinant protein containing a positively charged fusion peptide. In this regard, the highly positive leucine zipper peptide has been used as a fusion tag (Goldman *et al.*, 2002) as well as a penta-histidine peptide tag (Medintz *et al.*, 2003). Combinations of positively charged fusion proteins and avidin have also been used to control the resultant biotin-binding density on a DHLA-QD surface for use in live cell imaging (Jaiswal *et al.*, 2004).

One major advantage of DHLA modification is that the diameter of the QD remains as small as possible, while still creating a water soluble particle. QDs of <10-nm diameter have been created using this process and were successfully used to image intracellular proteins (Jaiswal *et al.*, 2003). The methods involved in conjugation of DHLA-QDs with peptides, proteins, and DNA have been reviewed (Prasuhn *et al.*, 2011).

Another type of simple surface modification involves noncovalent coating of the QDs with detergents or lipids. The hydrophobic tails of these molecules bind to the QD particle, while the hydrophilic portions interact with the aqueous phase and render the dots dispersible. Still other surface modification schemes use polymeric coatings containing multiple binding points to the QD, thus eliminating the possibility for leaching. Coatings containing PEG spacers can also be used to create a highly hydrophilic layer on top of the semiconductor surface. All of these modification strategies provide QDs that are water soluble (or dispersible and stable in suspension) and that contain functional groups for covalent attachment of proteins or other affinity molecules.

Masking the surface of semiconductor QD particles can also be done by adding another inorganic layer to the outer shell alloy structure. This layer can take the form of a silica coating formed by the reaction of a silane derivative with the shell (Darbandi and Nann, 2005). For instance, a pure silica surface can be created by controlled polymerization of the raw nanocrystals with tetraethyl orthosilicate (TEOS), which forms a siliceous sphere with silanol groups on the outer surface. Another organosilane derivative that is appropriate for use with metallic particles is mercaptopropyl-tris-hydroxy-silane. The thiol groups on the silane compounds datively bind to the surface while the hydroxy-silane groups polymerize to form a new silica coating. The resultant silanol-containing surface can then be functionalized using other organosilane compounds containing functional groups or reactive groups for further conjugation with biomolecules (see Chapter 13 for organosilane reagents and protocols). The only disadvantage of this approach is the increasingly greater particle diameter that results from building successive layers on the initial QD core, which may inhibit their use for probing within cells or tissues (Figure 10.60).

Water soluble QDs are now available from a number of manufacturers (Invitrogen, Evident Technologies, and Crystalplex). Each supplier uses their own proprietary methods of surface pacification to create biocompatible particles. Even coated quantum dot clusters are available that contain hundreds of particles bound together in a polymer matrix (Crystalplex). These form intensely bright labels for biomolecules, because the nanocrystals do not quench when clustered together at high density.

Most QD surfaces for biological applications contain negatively charged carboxylates for conjugation with amine-containing molecules *via* a carbodiimide reaction with EDC and (sulfo)NHS (Chapter 4, Section 1). The negative charges on the QD surface prevent particle aggregation through like-charge repulsion. An alternative method of creating water-dispersible dots is to form a hydrophilic coating that carries along with it a layer of hydration consisting of hydrogen-bonded water molecules. This often is done using hydroxylic polymers or PEG modifications. This too prevents aggregation due to the high energy needed to remove the bound water layer.

QDs have been used successfully in many biological applications that exploit their best properties of brightness, photostability, and multiplex capability. There are many publications that use QDs for in-cell or whole-organism-based imaging, including tracking of targets within cells (Dahan *et al.*, 2003), gene localization within chromosomes (Xiao and Barker, 2004), embryo developmental monitoring (Dubertret *et al.*, 2002), tumor imaging *in vivo* (Gao *et al.*, 2004), and multiplexed imaging and assays (Medintz *et al.*, 2003; Wu *et al.*, 2003), including FRET signaling (Han *et al.*, 2001). For a review on the use of QDs for cancer imaging and treatment, see Cassette *et al.* (2012) and Vashist *et al.* (2006).

However, QD labels still are not without potential problems. Although the raw dots typically are less than 10 nm in diameter, the addition of thick surface layers for biocompatibility and conjugation can increase the hydrodynamic radius considerably. Most particles with polymer coatings are in the 20- to 50-nm range in diameter, which often limits their use for cell-based detection due to their inability to easily penetrate cells and diffuse freely to intracellular targets. Most cell imaging applications with this type of QD involve cell surface staining or transport within cells by endocytosis, which limits particle access to other areas within the cell. In addition, nonspecific binding still plagues some QD probes when used with complex biological samples.

Another potential deficiency of QDs is the toxic nature of their metallic composition. Most particles contain at least one known toxic metal (e.g., cadmium) or contain alloys with unknown toxilogical properties. Cadmium-based QDs exposed to UV light for long periods release cadmium ions, which are highly toxic to cells (Derfus *et al.*, 2004). The initial proposal that QDs would be ideal as fluorescent probes for *in vivo* diagnostic imaging may not be fully realized due to the potential for heavy metal toxicity in humans. Even the use of QDs for *in vitro* research purposes must be carried out with care, as the solutions should be regarded as hazardous waste and disposed of according to standards for handling heavy-metal-contaminated solutions.

Another potential difficulty with using quantum dots relates to their special spectral characteristics. Excitation of QDs optimally occurs in the low region of the spectrum, typically below 400 nm, while emission usually is measured in regions that can be hundreds of nanometers away from the excitation wavelength. Unfortunately, many instruments for imaging or fluorimetry still do not contain lasers and filter sets that exactly match QD excitation and emission patterns. Most instruments in use today were initially designed for organic fluors with matched filter sets for such common dye derivatives as fluorescein, rhodamine, and the cyanine dyes. Using QDs with these instruments may mean exciting at a non-optimal, higher wavelength than is recommended to obtain full brightness.

However, the most important issue with quantum dot fluorescence is their tendency to blink or to be

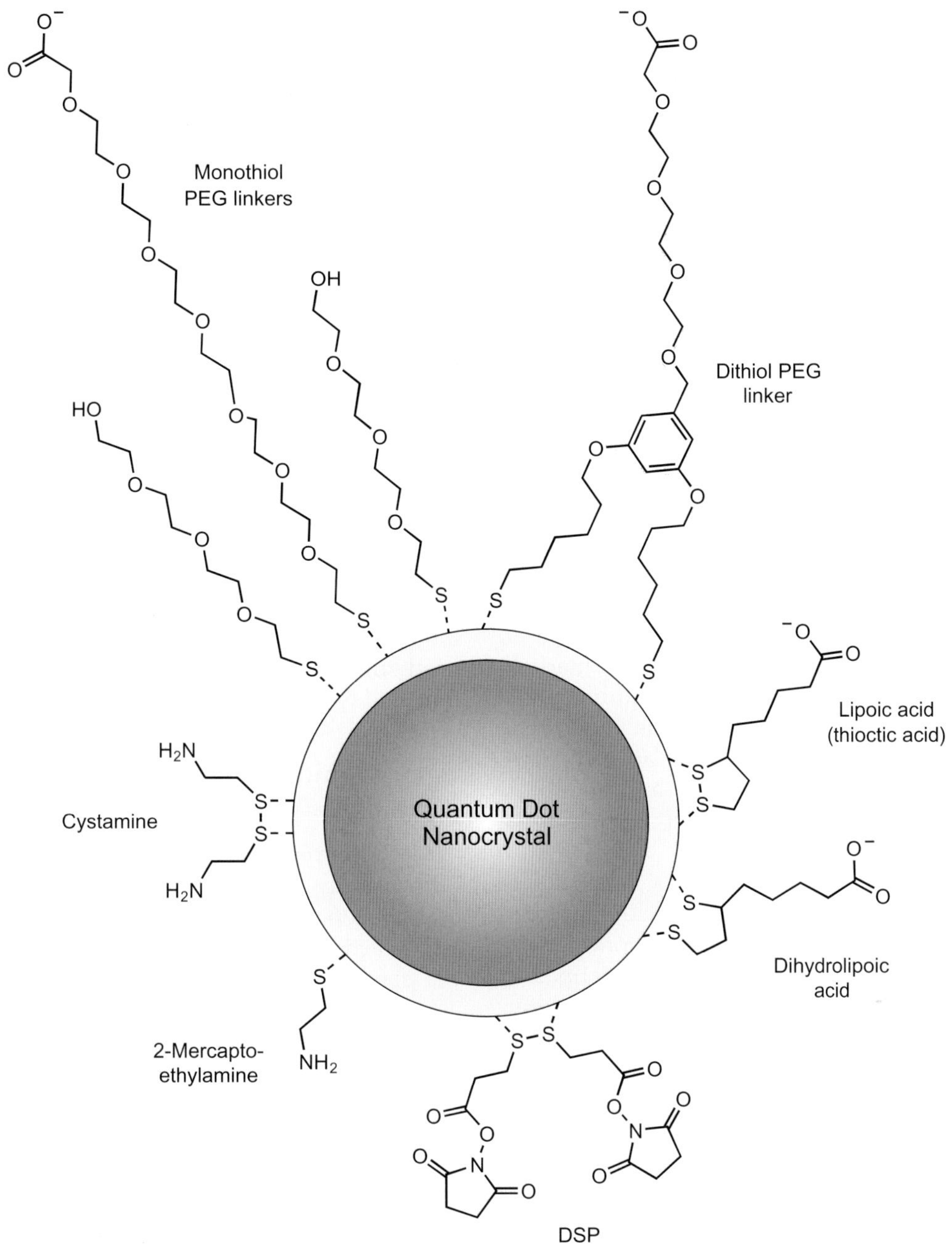

FIGURE 10.60 Many different thiol-containing linkers can be used to prepare water soluble QDs. The monothiol compounds suffer from the deficiency of being easily oxidized or displaced off the surface, thus creating holes for potential nonspecific binding. The dithiol linkers are superior in this regard, as they form highly stable dative bonds with the semiconductor metal surface that do not get displaced. The PEG-based linkers are especially effective at creating a biocompatible surface for conjugation with biomolecules.

completely dark and not fluoresce at all. Individual QDs can undergo an on/off cycle that results in a dark period of no light emission after a photon has been emitted (Nirmal *et al.*, 1996; Efros and Rosen, 1997). This blinking can be observed with a frequency on the order of milliseconds and can be problematic when imaging at the single-dot resolution or if using QDs for flow cytometry purposes is important. When imaging a larger population of QDs, the problem of blinking will be overcome by the fact that at any given moment many of the particles will be emitting light and thus contributing to the overall signal. Blinking will lower the apparent quantum yield for the combined QD population, but it will not eliminate signal entirely. In addition, certain solution additives used at relatively high concentration (i.e., 100-m*M* DTT or 2-mercaptoethanol) may serve to limit the blinking phenomena (Hohng and Ha, 2004).

A more severe issue with QDs, however, is the problem of dark or nonradiant dots in aqueous solution. Yao *et al.* (2005) documented that a significant fraction of commercially available QD particles in a population can be entirely dark. The nonradiant properties are probably due to defects in their nanocrystalline structure that occurred during manufacture. The percentage of dark dots varies for each sample, but they can represent 44 to 47% of all dots in a population. Thus, the use of QD conjugates for biomolecule imaging may mean that nearly half of all particles in an assay do not contribute to the resultant fluorescence signal.

10.2. Conjugation to Quantum Dots

Many antibodies, proteins, and other targeting or affinity ligands have been conjugated to QDs for biological applications. Antibodies to tumor markers have been used to image cancer cells *in vivo* (Tada *et al.*, 2007), fluoroimmunoassays have been developed using antibody-conjugated QDs (Goldman *et al.*, 2005; Tian *et al.*, 2012), peptide–QD conjugates have been made to target proteins *in vivo* (Ness *et al.*, 2003), antibody–QD conjugates containing tumor toxic agents have been designed to image and kill tumor cells (Gao *et al.*, 2004; Tavares *et al.*, 2011; Pöselt *et al.*, 2012), and sugar-QD conjugates have been made to detect carbohydrate binding proteins (Babu *et al.*, 2007).

The conjugation of proteins and other molecules to QDs involves standard coupling reactions with the added caveat related to the potential difficulties of working with particles. Many of the coupling strategies described in Chapter 14 for dealing with nanoparticles and microparticles are valid for use with QDs, but it is best when using commercially available particles to pay close attention to the manufacturer's suggested protocols.

When designing a protein–QD conjugate, it is also important to consider the optimal number of proteins to be coupled per particle. In some applications, a low ratio of protein-to-QD may result in the highest signal for fluorescence detection. This is often the case for antibody–QD conjugates where two to three antibodies coupled per particle are sufficient to target antigens. However, other applications may require a high density of protein on the QD surface. For instance, as a result of investigations into the interaction of CD8 with HLA A2 complexes, it was found to be important to create multiple contacts between the HLA molecules and the CD8 molecules on cell surface membranes. Therefore, up to 12 HLA A2 complexes per QD were coupled (the maximal possible to fit on the particle surface) to ensure the greatest interaction potential with cells (Anikeeva *et al.*, 2006). Additionally, when preparing (strept)avidin–QD conjugates, it may be desirable to maximize the biotin-binding ability of the complex to interact with as many biotinylated molecules as possible. Commercial streptavidin–QD conjugates typically have at least 5 to 10 proteins coupled per particle. Since each QD conjugate will have its own unique use, the conjugation process should be optimized to perform best in its intended application.

As is the case with most nanoparticles, buffer and salt compatibility should be taken into account when working with QDs. Small charged particles maintain colloidal stability by like-charge repulsion, which prevents aggregation by van der Waals or hydrophobic attraction. Any salt or buffer constituents that neutralize or eliminate the surface charge on the QDs will cause clumping or precipitation. Some buffer additives will also cause loss of fluorescence intensity and should be avoided in concentrations above a certain level. Consult the supplier's guidelines for buffer suggestions to maintain particle stability.

The following methods are based on those cited in the literature or in company instruction manuals for coupling molecules to fluorescent nanoparticles. The coupling of unique proteins or other molecules to QD surfaces may need further optimization of reactant ratios as well as time and temperature to obtain the best conjugates.

10.3. Conjugation of Proteins to Quantum Dots using EDC

Carbodiimide coupling to carboxylate-containing QDs usually involves the use of EDC in a single-step or two-step process to form an amide bond. If a one-step reaction is done, the QD is activated with EDC in the presence of an amine-containing molecule, such as a protein. Many protocols use this method, but it can result in protein polymerization in addition to coupling, because proteins contain both carboxylates and amines. A two-step protocol results in better control of the reaction (Figure 10.61). In the first step, EDC is used in the presence of sulfo-NHS to activate the carboxylates on the particles to intermediate sulfo-NHS esters. After a quick separation step to remove excess reactants, the activated QDs are added to the protein solution to be coupled. This then results in amide bond formation without polymerization of the protein in solution. See Chapter 4, Section 1, and Chapter 14, Section 4.2, for additional information on this process.

The following protocol describes the coupling of amine-containing proteins to carboxylated QDs using a single-step EDC reaction, as recommended by several manufacturers.

Protocol

1. Prepare a carboxylated QD solution in 10-m*M* sodium borate, pH 7.4 (reaction buffer), at a concentration of 1 μ*M*. The supplier of QDs usually will provide the reagent concentration as a molar

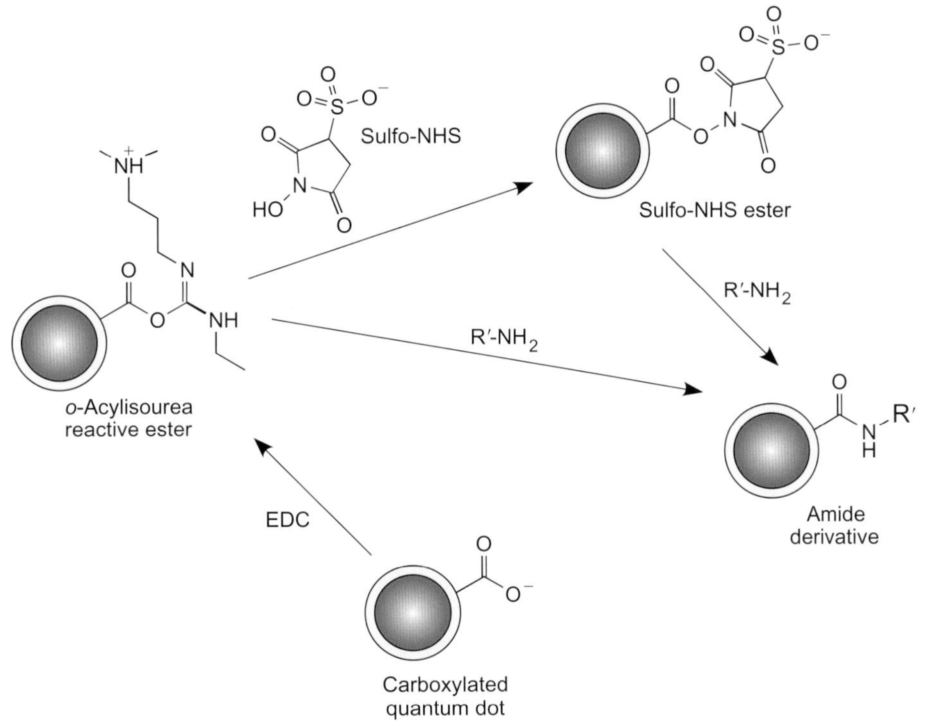

FIGURE 10.61 QDs containing carboxylate groups can be coupled to amine-containing proteins or other molecules using the EDC/sulfo-NHS reaction to form amide bond linkages. The intermediate sulfo-NHS ester is negatively charged and will help maintain particle stability due to like charge repulsion between particles.

quantity, which treats each particle as though it was a single molecule. A typical QD solution as obtained from a manufacturer may be about 8 μ*M* starting concentration.
2. Dissolve the protein to be conjugated to the QD in reaction buffer at a concentration of 1 to 10 mg/ml.
3. Add a quantity of the protein solution to the QD solution with mixing to obtain the desired molar excess of protein over the concentration of nanoparticles. Using a 1- to 20-fold molar excess typically works well, but optimization should be carried out to determine the best ratio for a particular application.
4. Prepare a solution of EDC in water at a concentration of 10 mg/ml. Immediately add 57 μl of this solution to the protein/QD solution. Mix well.
5. React for 2 h at room temperature with gentle mixing.
6. Filter the solution through a 0.2-μm filter (low protein binding type) to remove any precipitated protein or particles.
7. Separate excess protein from the protein–QD conjugate by use of ultrafiltration spin columns using an exclusion limit appropriate for allowing passage of the unconjugated protein, but retention of the protein–QD conjugate. For proteins of molecular weight below 100 kD, a membrane of this cutoff will work well.

QD nanoparticles containing carboxylate groups also may be reacted in a two-step EDC/sulfo-NHS reaction to couple proteins and other molecules containing both amines and carboxylates. This type of reaction is designed to remove excess EDC activating agent before addition of protein, so protein polymerization cannot occur.

Protocol

1. Prepare a QD solution in 25-m*M* PIPES buffer, pH 7.0 (reaction buffer), at a concentration of 100 μg/ml.
2. Prepare a protein solution in reaction buffer at a concentration of 1 mg/ml.
3. Prepare a solution of 20-m*M* EDC, 50-m*M* sulfo-NHS in water immediately before use.
4. To each milliliter of QD solution, add 50 μl of the EDC/sulfo-NHS stock solution. Maintain the pH at 7.0 by the addition of base, if necessary. Small volume reactions may be controlled using a pH stat.
5. React for 10 min at room temperature with gentle mixing.
6. Add 1.4 μl of 2-mercaptoethanol to each milliliter of the reaction mixture to quench excess EDC. Sonicate the QD solution several times to maintain particle dispersion.
7. Add 5 μl of the protein solution to each 100 μg quantity of activated QDs.

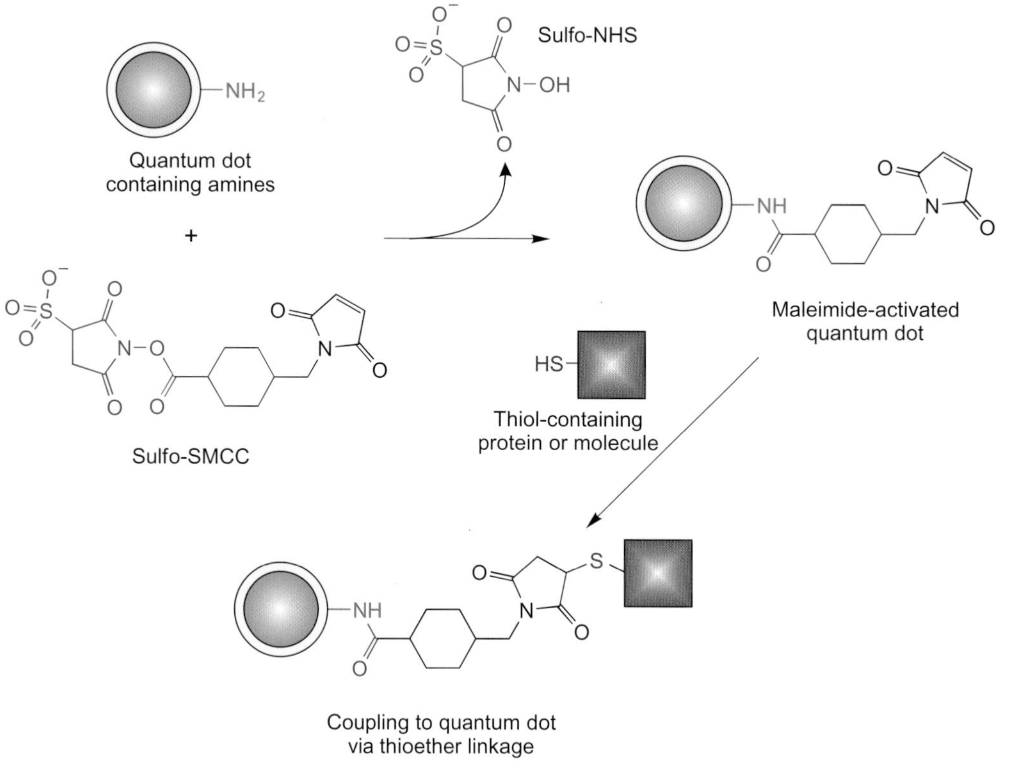

FIGURE 10.62 Sulfo-SMCC can be used to conjugate amine-containing QDs with thiol-containing proteins or other molecules using a two-step coupling procedure.

8. React for 60 min at room temperature with mixing.
9. Remove excess reactants and block remaining sulfo-NHS ester sites by dialysis against 50-m*M* Tris, pH 7.4. Use a membrane cutoff appropriate to allow passage of the protein being coupled, but retention of the protein–QD conjugate.

10.4. Conjugation to Quantum Dots Using Sulfo-SMCC

Sulfo-SMCC is a heterobifunctional crosslinking agent containing an amine-reactive NHS ester on one end and a thiol-reactive maleimide group on the other end (Chapter 6, Section 1.3). Amine-containing QDs may be activated with sulfo-SMCC to contain sulfhydryl-reactive maleimides for conjugation with thiol-containing proteins or other molecules (Figure 10.62).

An alternative to the use of an aliphatic crosslinker like sulfo-SMCC is to use a hydrophilic PEG-based one, which displays greater biocompatibility and lower nonspecific binding character. Particularly, the NHS–PEG$_n$–maleimide crosslinkers described in Chapter 18 may be a superior choice in designing QD conjugates of low background and high sensitivity for biomolecule-detection purposes.

The following protocol illustrates the process of activating an amine-containing QD with sulfo-SMCC and coupling a thiol-containing molecule in a second step.

Protocol

1. Prepare a 200-μl solution of amine-containing QDs at a concentration of 2.5 μ*M* in 50-m*M* sodium phosphate, pH 7.4 (reaction buffer). This represents 0.5 n*M* of QD in 200 μl buffer.
2. Add 1 mg of sulfo-SMCC to the QD solution with mixing to dissolve the crosslinker.
3. React for 60 min at room temperature with mixing.
4. Remove excess crosslinker and reaction by products by gel filtration on a desalting resin. The QD fraction is identified by its characteristic fluorescence. This operation should be done quickly to limit the degree of maleimide hydrolysis. The resulting particles contain reactive maleimide groups for coupling to a thiol-containing protein or other molecule.
5. Prepare a protein or antibody containing an available thiol group in reaction buffer at a concentration of 1 to 10 mg/ml. Add a quantity of this protein solution to the purified nanoparticle suspension to attain the desired molar excess of protein over the concentration of activated QDs. Typically, a 1- to 20-fold molar excess of protein over the QD concentration works well. The optimal amount of protein to be added should be determined experimentally by considering the best performance of the fluorescent conjugate in its intended application. Tada *et al.* (2007) created a monoclonal anti-HER2 antibody–QD conjugate at a

level of three antibodies per nanoparticle to target tumors in mice. Creating thiol groups on proteins or peptides may be done from disulfides by reduction. Alternatively, a thiolation reagent may be used to add thiols to the protein surface for coupling (see the protocols in Chapter 2, Section 4.1).

6. React with mixing for 2 h at room temperature. At the completion of the reaction, cysteine may be added at 50-m*M* to block excess maleimide reactive sites.
7. Purify the QD conjugate using gel filtration or ultrafiltration using a micro-spin device. The use of a molecular weight cutoff for gel filtration that will accommodate both the conjugate and the not-coupled protein is appropriate (e.g., Superdex-200 resin). For ultrafiltration, use a membrane cutoff that will retain the conjugate but permit the not-coupled protein to pass through.

CHAPTER

11

(Strept)avidin–Biotin Systems

One of the most popular methods of noncovalent conjugation is to make use of the natural strong binding of (strept)avidin for the small molecule biotin. The strength of the (strept)avidin–biotin interaction has made it a useful tool in specific targeting applications and assay design. Since each (strept)avidin molecule contains a maximum of four biotin-binding sites, the interaction can be used to enhance the signal strength in immunoassay systems.

Modification reagents that can add a functional biotin group to proteins, nucleic acids, and other molecules now come in many shapes and reactivities (Section 6, this chapter, and Chapter 18, Section 1.3). Depending on the functionality present on the biotinylation compound, specific reactive groups on antibodies or other proteins may be modified to create a (strept)avidin binding site. Amines, carboxylates, sulfhydryls, and carbohydrate groups can be specifically targeted for biotinylation through the appropriate choice of biotin derivative. In addition, photoreactive biotinylation reagents (Section 6.5, this chapter) are used to add nonselectively a biotin group to molecules containing no convenient functional groups for modification. In this manner, oligonucleotide probes often are modified for detection with (strept)avidin conjugates (Chapter 23, Section 2.3).

The following sections discuss the concept and use of the (strept)avidin–biotin interaction in bioconjugate techniques. Preparation of biotinylated molecules and (strept)avidin conjugates also are reviewed with suggested protocols. For a discussion of the major biotinylation reagents, see Section 6, this chapter, and Chapter 18, Section 1.3.

1. THE (STREPT)AVIDIN–BIOTIN INTERACTION

Avidin is a glycoprotein found in egg-whites that contains four identical subunits of 16,400 Daltons each, giving an intact molecular weight of approximately 66,000 (Green, 1975). Each subunit contains one binding site for biotin, or vitamin H, and one oligosaccharide modification (Asn-linked). The tetrameric protein is highly basic, having a pI of about 10. The biotin interaction with avidin is among the strongest noncovalent affinities known, exhibiting a dissociation constant of about 1.3×10^{-15}M. Tryptophan and lysine residues in each subunit are known to be involved in forming the binding pocket (Gitlin *et al.*, 1987, 1988).

The tetrameric native structure of avidin is resistant to denaturation under extreme chaotropic conditions. Even in 8-*M* urea or 3-*M* guanidine hydrochloride the protein maintains structural integrity and activity (Green, 1963). When biotin is bound to avidin, the interaction promotes even greater stability to the complex. An avidin–biotin complex (ABC) is resistant to breakdown in the presence of up to 8-*M* guanidine at pH 5.2. A minimum of 6- to 8-*M* guanidine at pH 1.5 is required for inducing complete dissociation of the avidin–biotin interaction complex (Cuatrecasas and Wilchek, 1968; Bodanszky and Bodanszky, 1970). Since the subunits in avidin are not held together by disulfide bonds, conditions that cause denaturation also result in subunit disassociation.

The strength of the noncovalent avidin–biotin interaction along with its resistance to breakdown makes it extraordinarily useful in bioconjugate chemistry. Biotinylated molecules and avidin conjugates can "find" each other under the most extreme conditions to bind and complex together. The biospecificity of the interaction is similar to antibody–antigen or receptor–ligand recognition, but on a much higher level with respect to affinity constants. Variations in buffer salt, pH, the presence of denaturants or detergents, and extremes of temperature will not prevent the interaction from occurring (Ross *et al.*, 1986).

The only disadvantage to the use of avidin is its tendency to bind nonspecifically with components other than biotin due to its high pI and carbohydrate content. The strong positive charge on the protein causes ionic interactions with more negatively charged molecules, especially cell surfaces. In addition, carbohydrate

Bioconjugate Techniques, Third Edition.
DOI: http://dx.doi.org/10.1016/B978-0-12-382239-0.00011-X

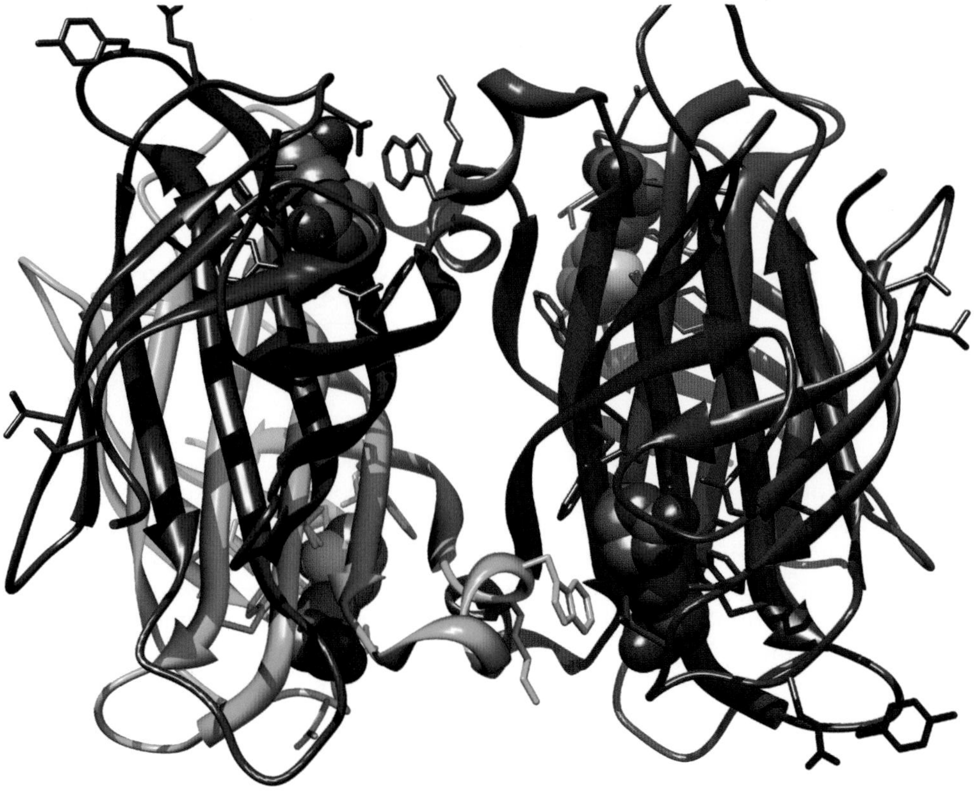

binding proteins on cells can interact with the polysaccharide portions on the avidin molecule to bind them in regions devoid of targeted biotinylated molecules. These nonspecific interactions can lead to elevated background signals in some assays, preventing the full potential of the avidin–biotin amplification process to be realized.

Streptavidin is a biotin-binding protein similar to avidin, but it is of bacterial origin and originates from *Streptomyces avidinii*. Due to streptavidin's structural differences, however, it can overcome some of the nonspecific binding deficiencies of avidin (Chaiet and Wolf, 1964). Similar to avidin, streptavidin contains four subunits, each with a single biotin-binding site. After some postsecretory modifications, the intact tetrameric protein has a molecular mass of about 60,000 Daltons, slightly less than that of avidin (Bayer *et al.*, 1986, 1989).

The primary structure of streptavidin is considerably different than that of avidin, despite the fact that they both bind biotin with similar avidity. The molecular model illustrates the four identical subunits of streptavidin as colored ribbons with four biotins shown as space-filling molecules bound deep within the binding pockets of each polypeptide chain (Le Trong *et al.*, 2011) (PDB structure 3RY1). The variation in the amino acid sequence from avidin results in a much lower isoelectric point for streptavidin (pI 5–6) compared to the highly basic pI of 10 for the egg-white protein. Moderation in the overall charge of streptavidin substantially reduces the amount of nonspecific binding due to ionic interaction with other molecules. Of additional significance is the fact that streptavidin is not a glycoprotein, thus there is no potential for binding to carbohydrate receptors. These factors lead to better signal-to-noise ratios in assays using streptavidin–biotin interactions than those employing avidin–biotin.

Both avidin and streptavidin can be conjugated to other proteins or labeled with various detection reagents without loss of biotin-binding activity. The biotin-binding proteins can also be immobilized onto surfaces, chromatography supports, microparticles, and nanoparticles for use in coupling biotinylated molecules (Xie *et al.*, 2011; Williams *et al.*, 2012; Yu *et al.*, 2012). Streptavidin is slightly less soluble in water than avidin, but both are extremely robust proteins that can tolerate a wide range of buffer conditions, pH values, and chemical modification processes. Bioconjugate techniques can utilize the ε- or N-terminal amines on these proteins for direct conjugation or employ modification reagents to transform their existing functional groups into other reactive groups (Chapter 2, Section 4).

In this chapter and throughout this book, the use of the term "(strept)avidin" is meant to infer that either avidin or streptavidin can be used in the associated protocols, conjugates, and applications. However, due to its enhanced properties, streptavidin has effectively replaced native avidin in most capture or detection methods designed for use with biomolecules. There are some chemically modified avidin preparations, such as NeutrAvidin (Thermo Fisher), that have eliminated its negative properties through deglycosylation and reducing its pI through covalent modification of charged residues. These modified avidin preparations perform much better than native avidin and for certain applications even perform better than streptavidin.

2. USE OF (STREPT)AVIDIN–BIOTIN INTERACTIONS IN ASSAY SYSTEMS

The specificity of biotin binding to (strept)avidin provides the basis for developing assay systems to detect or quantify analytes. Biotinylated molecules can be targeted in complex mixtures by using the appropriate (strept)avidin conjugates. If the biotinylated component has affinity for binding a particular antigen, then the antigen can be located through the use of a (strept)avidin conjugate containing a detectable molecule. A series of (strept)avidin–biotin interactions can be built upon each other—utilizing the multivalent nature of each tetrameric (strept)avidin molecule—to further enhance the detection capability for the target. Immunoassays built on the layering of (strept)avidin–biotin interactions can result in amplified sensitivities and lower limits of detection than assays using antibodies directly targeting an analyte (He *et al.*, 2012).

A common application for (strept)avidin–biotin chemistry is in immunoassays. The specificity of antibody molecules provides the targeting capability to recognize and bind particular antigen molecules. If there are biotin labels on the antibody, it creates multiple sites for the binding of (strept)avidin. If (strept)avidin is in turn labeled with an enzyme, fluorophore, etc., then a very sensitive antigen-detection system is created. The potential for more than one labeled (strept)avidin to become attached to each antibody through its multiple biotinylation sites is the key to dramatic increases in assay sensitivity over that obtained through the use of antibodies directly labeled with a detectable tag.

There are several basic immunoassay designs that make use of the enhanced sensitivity afforded by the (strept)avidin–biotin interaction. Most of these assays use conjugates of (strept)avidin with enzymes, such as horseradish peroxidase (HRP) or alkaline phosphatase, although other labels (such as fluorophores) can be used as well. In the simplest assay design, called the labeled avidin–biotin (LAB) system (Figure 11.1), a biotinylated antibody is allowed to incubate and bind with its target antigen. Next, a (strept)avidin–enzyme conjugate is introduced and allowed to interact with the available biotin sites on the bound antibody. Just as in other enzyme-linked immunosorbent assay (ELISA) tests, substrate development then provides the chemical detectability necessary to quantify the antigen (Guesdon *et al.*, 1979).

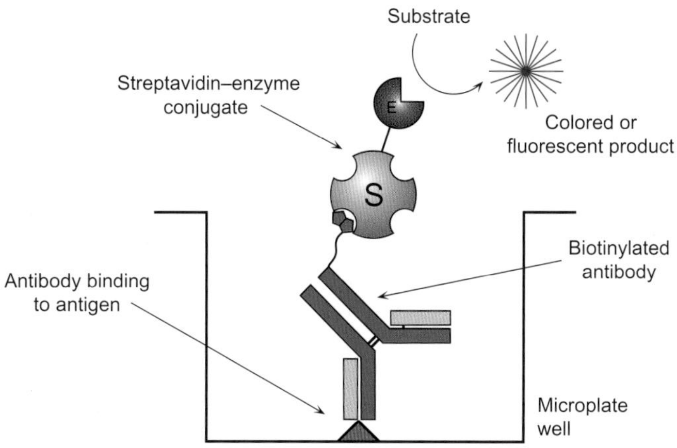

FIGURE 11.1 The basic design of the labeled avidin–biotin (LAB) assay system.

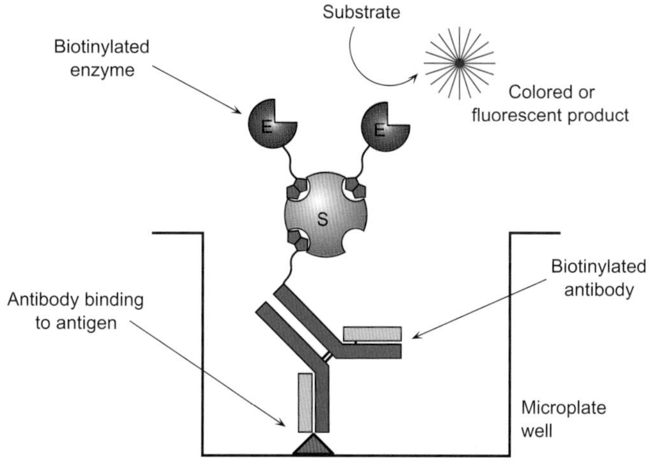

FIGURE 11.2 The basic design of the bridged avidin–biotin (BRAB) assay system.

In a slightly more complex design, the bridged avidin–biotin (BRAB) system uses (strept)avidin's multiple biotin-binding sites to create an assay of potentially higher sensitivity than that of the LAB assay. Again the biotinylated antibody is allowed to bind to its target, but in the next step an unmodified (strept)avidin is introduced to bind with the biotin binding sites on the antibody. Finally, a biotinylated enzyme is added to provide a detection vehicle (Figure 11.2). Since the bound (strept)avidin still has additional biotin-binding sites available, the potential exists for more than one biotinylated enzyme to interact with each bound (strept)avidin. In some cases, sensitivity can be increased over that of the LAB technique by using this bridging ability of (strept)avidin (Seydack, 2008; Jedrychowski *et al.*, 2009).

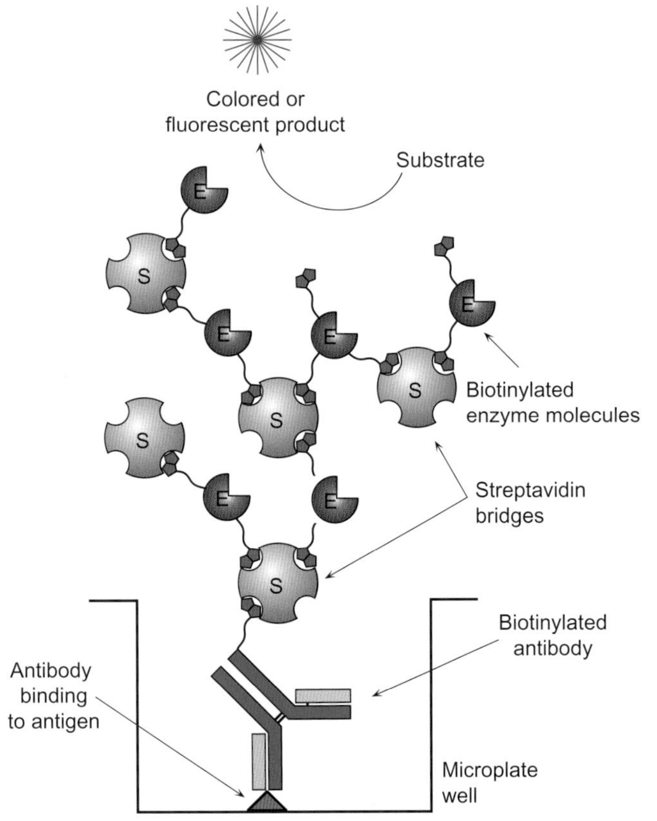

FIGURE 11.3 The assay design of the avidin–biotin complex (ABC) system.

A modification on this theme can be used to produce one of the most sensitive enzyme-linked assay systems known. The ABC system (for avidin–biotin complex) increases the detectability of antigen beyond that possible with either the LAB or BRAB designs by forming a polymer of biotinylated enzyme and (strept)avidin before its addition to an antigen-bound, biotinylated antibody (Bayer *et al.*, 1988). When (strept)avidin and a biotinylated enzyme are mixed together in solution in the proper proportion, the multiple binding sites on (strept)avidin create a linking matrix to form a high-molecular-weight complex. If the biotinylated enzyme is not in large enough excess to block all the binding sites on (strept)avidin, then additional sites will still be available on this complex to bind a biotinylated antibody which is bound to its complementary antigen. The large complex provides multiple enzyme molecules to enhance the sensitivity of detecting antigen (Figure 11.3). Thus, the ABC procedure is currently among the highest-sensitivity methods available for immunoassay work. It is a common assay technique used for microplate-based ELISA assays and for immunohistochemistry (IHC) procedures (Hadisaputri *et al.*, 2011; Inouye *et al.*, 2012; Li *et al.*, 2012).

Similar techniques can be used to devise (strept)avidin–biotin assay systems for detection of nucleic acid hybridization. DNA probes labeled with biotin can be detected after they bind their complementary DNA target through the use of (strept)avidin-labeled complexes (Bugawan *et al.*, 1990; Lloyd *et al.*, 1990). Direct detection of hybridized probes can be accomplished, similar to the LAB method, by incubating with a (strept)avidin–enzyme conjugate followed by substrate development. BRAB-like and ABC-like assays can also be utilized to further enhance a DNA probe signal (Chapter 23, Section 2.3).

Non-enzyme assay systems can be designed with the (strept)avidin–biotin interaction, as well. Fluorescently labeled (strept)avidin molecules can be used to detect a biotinylated molecule after it has bound its target. In fact, a single preparation of a fluorescent (strept)avidin derivative can be used as a universal detection reagent for any biotinylated targeting molecule. The main application of this technique is in cytochemical staining wherein the fluorescence signal is used to localize antigen or receptor molecules in cells and tissue sections. In addition, detection of analytes on arrays is commonly carried out using fluorescently labeled (strept)avidin conjugates to bind to biotinylated primary antibodies interacting with specific targets on the array surface.

Other tags or probes can be coupled to (strept)avidin and used in a similar fashion. For instance, radiolabeled (strept)avidin can be employed as a universal detection reagent in radioimmunoassay designs (Wojchowski and Sytkowski, 1986) (Chapter 12, Section 2). (Strept)avidin labeled with ^{125}I can be used to localize biotinylated monoclonal antibodies directed against tumor cells *in vivo* for imaging purposes (Paganelli *et al.*, 1988). Chemical tags such as in hydrazide–(strept)avidin derivatives can be made to site-direct (strept)avidin's interaction toward oxidized carbohydrate residues for specific detection of glycoconjugates (Section 5, this chapter). Colloidal gold-labeled (strept)avidin can be used as highly sensitive detection reagents for microscopy techniques (Cubie and Norval, 1989; Krager *et al.*, 2012) (Chapter 14, Section 6.6). Finally, cytotoxic substances coupled to (strept)avidin can be used to direct cell-killing activity toward a tumor-cell-bound, biotinylated monoclonal antibody (or other targeting molecule) for cancer therapy (Hashimoto *et al.*, 1984; Haugland and Bhalgat, 2008) (Chapter 20, Section 3).

Universal detection reagents can also be constructed through biotinylation techniques. Modification of immunoglobulin-binding proteins with biotin tags, for instance, creates a reagent useful for the general assay of antibody molecules. In this sense, biotinylated protein A or biotinylated protein G can be used to detect the binding of any primary IgG to its antigen target (provided there is no other antibody molecule presence to cause nonspecific binding of the protein A component). Subsequent addition of a labeled (strept)avidin molecule binds to the biotinylated protein A, completing the formation of a detection complex (Jagannath and Sehgal, 1989).

To develop assay systems using the (strept)avidin–biotin interaction, it is first necessary to produce the associated (strept)avidin conjugates and/or biotinylated components. When the LAB technique is employed, the (strept)avidin conjugate is made using crosslinking agents, not biotinylation reagents, in order to maintain the binding capacity of the (strept)avidin tetramer toward other biotinylated molecules. In the BRAB assay system, (strept)avidin is left unconjugated and acts merely as the multivalent bridging molecule, while both the targeting molecule and the detection molecule are biotinylated. The components for the ABC assay are identical to the BRAB system.

The following sections discuss the main techniques used to make (strept)avidin conjugates and various biotinylated components. Section 6 of this chapter and Chapter 18, Section 1.3, should be consulted for a complete overview of biotinylation reagents.

3. PREPARATION OF (STREPT)AVIDIN CONJUGATES

Conjugates of (strept)avidin with other protein molecules must be prepared to design systems using the LAB assay technique. Suitable protein molecules attached to (strept)avidin either possess indigenous detectability, such as in the case of ferritin or phycobiliproteins, or possess catalytic activity (enzymatic) that can be utilized to produce a detectable substrate product. The majority of conjugation procedures for making (strept)avidin–protein conjugates use the amines, sulfhydryls, or carbohydrates on each protein as functional groups for crosslinking.

Perhaps the most common conjugates of (strept) avidin involve attaching enzyme molecules for use in ELISA systems. As in the case of antibody–enzyme conjugation schemes (Chapter 20, Section 1), by far the most commonly used enzymes for this purpose are horseradish peroxidase and alkaline phosphatase. Other enzymes such as β-galactosidase and glucose oxidase are used less often, especially with regard to assay tests for clinically important analytes (Chapter 22).

Other proteins commonly crosslinked to (strept)avidin are chromogenic or fluorescent molecules, such as ferritin or phycobiliproteins (Chapter 10, Section 7). These conjugates can be used in microscopy techniques to stain and localize certain antigens or receptors in cells or tissue sections.

The following sections discuss three main methods for preparing these types of (strept)avidin–protein conjugates: (1) the use of an *N*-hydroxysuccinimide (NHS) ester–maleimide heterobifunctional crosslinker, (2) the use of the carbohydrate on glycoproteins for reductive amination coupling, and (3) employing the old technique of homobifunctional crosslinking with glutaraldehyde.

3.1. NHS Ester–Maleimide-Mediated Conjugation Protocols

Heterobifunctional crosslinking agents can be used to control the degree of protein conjugation, thus limiting polymerization and controlling the molar ratio of each component in the final complex (Chapter 6). Particularly useful heterobifunctionals include the amine- and sulfhydryl-reactive NHS ester–maleimide crosslinkers discussed in Chapter 6, Section 1. Chief among these is succinimidyl-4-(*N*-maleimidomethyl)cyclohexane-1-carboxylate (SMCC) or sulfo-SMCC (Chapter 6, Section 1.3), which contains a reasonably long spacer and a relatively stable maleimide group due to the adjacent cyclohexane ring in its cross-bridge.

Conjugations carried out with SMCC usually involve up to three steps. In the first stage, one of the proteins is modified at its amine groups *via* the NHS ester end of the crosslinker to form amide linkages, which upon modification then create derivatives that terminate in reactive maleimide groups. If the other protein to be conjugated does not contain sulfhydryl residues necessary to react with the maleimide-activated protein, it must be modified to contain them (Chapter 2, Section 4.1). Finally, the two reactive components are mixed together in the proper ratio to effect the conjugation reaction (Shuvaev *et al.*, 2011; Lillich *et al.*, 2012).

For the preparation of (strept)avidin–enzyme conjugates, either protein may first be modified with SMCC and the other one modified to contain –SH groups. Since (strept)avidin does not possess any free sulfhydryls—and the disulfides present in (strept)avidin are inaccessible to easy reduction—it must be modified with either a crosslinker or with a thiolation agent before conjugation. If the enzyme employed contains free sulfhydryls in its native state, such as β-galactosidase, then it is convenient to activate (strept)avidin with SMCC and simply add the sulfhydryl-containing protein to it for conjugation. If the enzyme does not contain free sulfhydryls (as is the case with alkaline phosphatase or horseradish peroxidase), then the choice of which component gets maleimide-activated and which gets thiolated is up to the individual.

The following protocol describes the activation of (strept)avidin with sulfo-SMCC and its subsequent

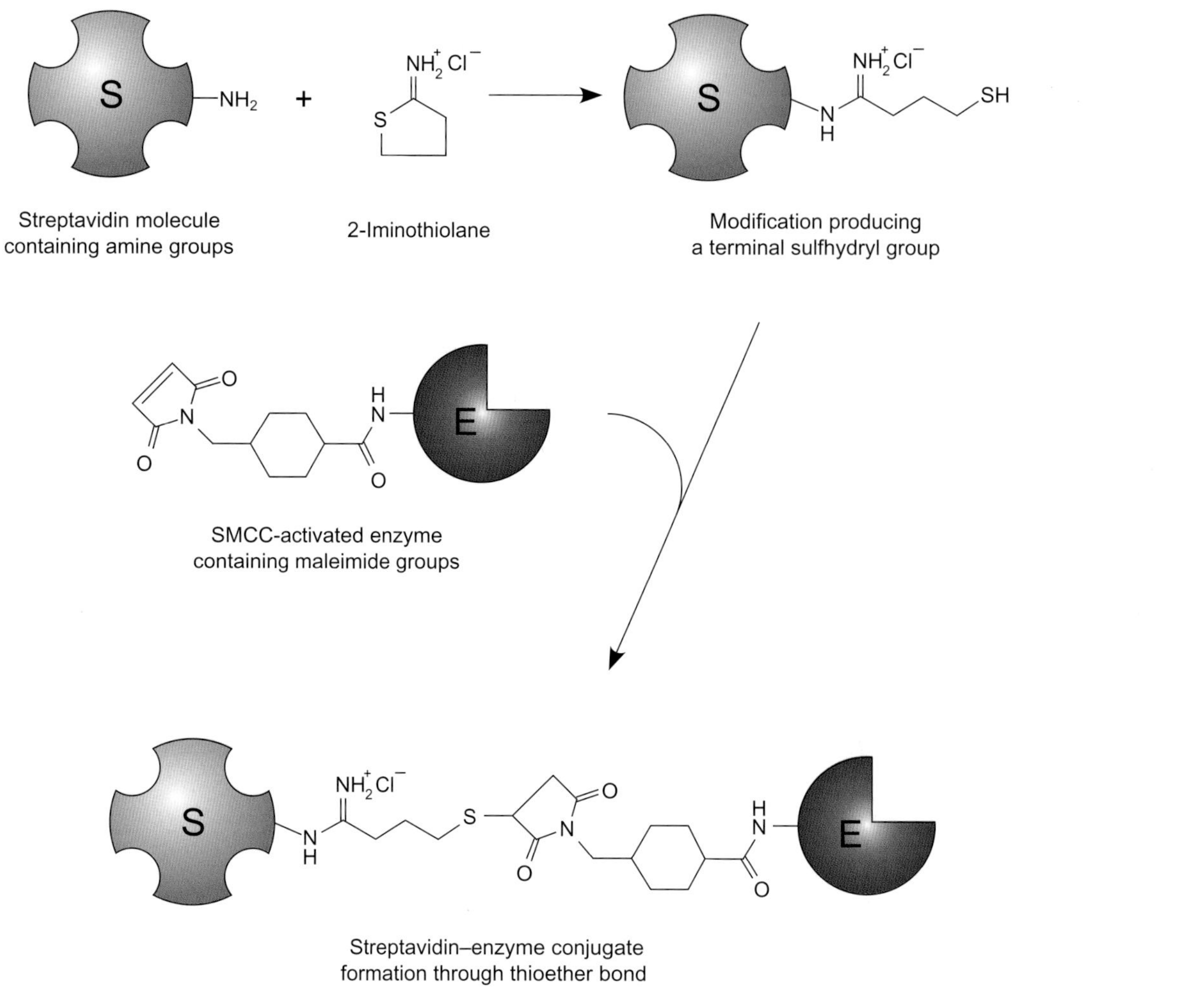

FIGURE 11.4 Streptavidin can be modified with 2-iminothiolane (Traut's reagent) to produce sulfhydryl groups. Subsequent reaction with a maleimide-activated enzyme produces a thioether-linked conjugate.

conjugation with an enzyme modified to contain sulfhydryls using *N*-succinimidyl-*S*-acetylthioacetate (SATA) (Chapter 2, Section 4.1). A method for the opposite approach, wherein the enzyme is activated with SMCC and the (strept)avidin component is thiolated, is presented immediately after this protocol. This strategy may be the most common approach to forming these conjugates (Figure 11.4). In addition, since there are enzymes commercially available that are activated with SMCC (Thermo Fisher), their use may be the easiest solution to forming conjugates.

Protocol for the Conjugation of SMCC-Activated (Strept)avidin with Thiolated Enzyme

ACTIVATION OF (STREPT)AVIDIN WITH SMCC

1. Dissolve (strept)avidin (Thermo Fisher) in 0.1-*M* sodium phosphate, 0.15-*M* NaCl, pH 7.2, at a concentration of 10 mg/ml.
2. Add 1.0 mg of sulfo-SMCC (Thermo Fisher) to each ml of (strept)avidin solution. Mix to dissolve.
3. React for 30 to 60 min at room temperature. Since maleimide groups are labile in aqueous solution, extended reaction times should be avoided.
4. Immediately purify the maleimide-activated (strept)avidin away from excess crosslinker and reaction byproducts by gel filtration on a desalting resin. A spin column will facilitate the most rapid purification. Use 0.1-*M* sodium phosphate, 0.15-*M* NaCl, pH 7.2, as the chromatography buffer. Pool the fractions containing protein (the first peak eluting from the column). After elution, adjust the protein concentration to 10 mg/ml for the conjugation reaction (centrifugal concentrators work well for this step). At this point, the maleimide-activated (strept)avidin may be frozen and lyophilized to preserve its maleimide activity. The modified protein is stable for at least 1 year in a

freeze-dried state. If kept in solution, the maleimide-activated (strept)avidin is labile and should be used immediately to conjugate with a thiolated enzyme following the procedure described below.

MODIFICATION OF ENZYME WITH SATA

If β-galactosidase is used to conjugate with an SMCC-activated (strept)avidin, then there is no need to thiolate the enzyme, since it contains sulfhydryls in its native state (Fujiwara, 1988; Sivakoff and Janes, 1988). For conjugations using horseradish peroxidase, alkaline phosphatase, or glucose oxidase, however, thiolation is necessary to add the necessary sulfhydryls.

1. Dissolve the enzyme to be modified in 0.1-*M* sodium phosphate, 0.15-*M* NaCl, pH 7.2, at a concentration of 10 mg/ml.
2. Prepare a stock solution of SATA (Thermo Fisher) by dissolving it in DMSO at a concentration of 13 mg/ml. Use a fume hood to handle the organic solvent.
3. Add 25 μl of the SATA stock solution to each ml of 10 mg/ml enzyme solution. For different concentrations of enzyme in the reaction medium, proportionally adjust the amount of SATA addition; however, do not exceed 10% DMSO in the aqueous reaction medium.
4. React for 30 min at room temperature.
5. To purify the SATA-modified enzyme perform a gel filtration separation using a desalting resin or dialyze against 0.1-*M* sodium phosphate, 0.15-*M* NaCl, pH 7.2, containing 10-m*M* EDTA. Purification is not absolutely required, since the following deprotection step is carried out using hydroxylamine at a significant molar excess over the initial amount of SATA added. Whether a purification step is carried out or not, at this point, the derivative is stable and may be stored under conditions which favor long-term enzyme activity.
6. Deprotect the acetylated sulfhydryl groups on the SATA-modified enzyme according to the following protocol:
   a. Prepare a 0.5-*M* hydroxylamine solution in 0.1-*M* sodium phosphate, pH 7.2, containing 10-m*M* EDTA.
   b. Add 100 μl of the hydroxylamine stock solution to each ml of the SATA-modified enzyme. Final concentration of hydroxylamine in the enzyme solution is 50-m*M*.
   c. React for 2 h at room temperature.
   d. Purify the thiolated enzyme by gel filtration on a desalting resin using 0.1-*M* sodium phosphate, 0.1-*M* NaCl, pH 7.2, containing 10-m*M* EDTA as the chromatography buffer. To obtain efficient separation between the thiolated protein and excess hydroxylamine and reaction byproducts, the sample size applied to the column should be at a ratio of no more than 5% sample volume to the total column volume. Collect 0.5-ml fractions. Pool the fractions containing protein by measuring the absorbance of each fraction at 280 nm.

PRODUCTION OF CONJUGATE

1. Immediately mix the thiolated enzyme with an amount of maleimide-activated (strept)avidin to obtain the desired molar ratio of enzyme-to-(strept) avidin in the conjugate. Use of a 4 : 1 (enzyme : avidin) molar ratio in the conjugation reaction usually results in high-activity conjugates suitable for use in many enzyme-linked immunoassay procedures employing the LAB approach.
2. React for 30 to 60 min at 37°C or 2 h at room temperature. The conjugation reaction may also be carried out at 4°C overnight.

A variation of the above method can be used, wherein the enzyme is first activated with SMCC and conjugated to a thiolated (strept)avidin molecule. This approach is probably the most common way of preparing (strept)avidin–enzyme conjugates, and since the activated enzymes are readily available (Thermo Fisher) it may also be the easiest.

Protocol for the Conjugation of SMCC-Activated Enzymes with Thiolated (Strept)avidin

ACTIVATION OF ENZYME WITH SULFO-SMCC

The following protocol describes the activation of horseradish peroxidase (HRP) with sulfo-SMCC. Other enzymes may be activated in a similar manner. The activated enzyme possesses maleimide groups that are relatively unstable in aqueous solution. Therefore, the thiolation reaction should be coordinated with the activation process so that the final conjugation can be carried out immediately. Note: If activated enzymes are obtained (Thermo Fisher), this step may be eliminated.

1. Dissolve HRP in 0.1-*M* sodium phosphate, 0.15-*M* NaCl, pH 7.2, at a concentration of 10 mg/ml.
2. Add 3.3 mg of sulfo-SMCC (Thermo Fisher) to each ml of the HRP solution. Mix to dissolve and react for 30 min at room temperature. Alternatively, two equal additions of crosslinker may be carried out—the second one after 15 min of incubation—to obtain even more efficient modification.
3. Immediately purify the maleimide-activated HRP away from excess crosslinker and reaction byproducts by gel filtration on a desalting column. Use 0.1-*M* sodium phosphate, 0.15-*M* NaCl, pH 7.2, as the chromatography buffer. HRP can be

observed visually as it flows through the column due to the color of its heme ring. Pool the fractions containing the HRP peak. After elution, adjust the HRP concentration to 10 mg/ml for the conjugation reaction. At this point, the maleimide-activated enzyme may be frozen and lyophilized to preserve its maleimide activity. The modified enzyme is stable for at least one year in a freeze-dried state. If kept in solution, the maleimide-activated HRP should be used immediately to conjugate with thiolated (strept) avidin following the protocols outlined below.

THIOLATION OF (STREPT)AVIDIN

1. Dissolve (strept)avidin in 0.1-*M* sodium phosphate, 0.15-*M* NaCl, pH 7.2, at a concentration of 10 mg/ml.
2. Prepare a stock solution of SATA by dissolving it in DMSO at a concentration of 13 mg/ml. Use a fume hood to handle the organic solvent.
3. Add 25 µl of the SATA stock solution to each ml of 10 mg/ml (strept)avidin solution. For different concentrations of protein in the reaction medium, proportionally adjust the amount of SATA addition; however, do not exceed 10% DMSO in the aqueous reaction medium.
4. React for 30 min at room temperature.
5. To purify the SATA-modified (strept)avidin use gel filtration on a desalting column or dialyze against 0.1-*M* sodium phosphate, 0.15-*M* NaCl, pH 7.2, containing 10-m*M* EDTA. At this point, the derivative is stable and may be stored under conditions which favor long-term (strept)avidin activity.
6. Deprotect the acetylated sulfhydryl groups on the SATA-modified protein according to the following protocol:

   a. Prepare a 0.5-*M* hydroxylamine solution in 0.1-*M* sodium phosphate, pH 7.2, containing 10-m*M* EDTA.
   b. Add 100 µl of the hydroxylamine stock solution to each ml of the SATA-modified (strept)avidin. Final concentration of hydroxylamine in the solution is 50-m*M*.
   c. React for 2 h at room temperature.
   d. Purify the thiolated protein by gel filtration on Sephadex G-25 using 0.1-*M* sodium phosphate, 0.1-*M* NaCl, pH 7.2, containing 10-m*M* EDTA as the chromatography buffer.

CONJUGATION OF SMCC-ACTIVATED ENZYME WITH THIOLATED (STREPT)AVIDIN

7. Immediately mix the SMCC-activated enzyme with an amount of thiolated (strept)avidin to obtain the desired molar ratio of enzyme-to-(strept)avidin in the conjugate. Use of a 4:1 (enzyme:avidin) molar ratio in the conjugation reaction usually results in high-activity conjugates suitable for use in many enzyme-linked immunoassay procedures employing the LAB approach.
8. React for 30 to 60 min at 37°C or 2 h at room temperature. The conjugation reaction also may be done at 4°C overnight.

3.2. Conjugation Using Periodate Oxidation/Reductive Amination

Glycoproteins may be conjugated with another amine-containing protein through the process of periodate oxidation and reductive amination. Periodate oxidation of polysaccharide components on the glycoprotein results in the formation of reactive aldehyde residues by cleavage of carbon–carbon bonds and oxidation of the associated adjacent hydroxyls (Chapter 2, Section 4.4). Conjugation with another protein may be achieved by reacting the aldehydes with amines to form intermediate Schiff bases with subsequent reduction using sodium cyanoborohydride to create stable secondary amine bonds.

This method of conjugation is particularly well-suited for coupling HRP or ferritin with (strept)avidin (Dhar *et al.*, 2012). Both HRP and (strept)avidin are glycoproteins that can be oxidized with sodium periodate to generate aldehydes. Thus, HRP–(strept)avidin and ferritin–(strept)avidin may be prepared by reductive amination. Ferritin is a large, complex protein of molecular weight 750,000. Its structure is comprised of a protein shell of diameter approximately 12 nm that surrounds a micelle core consisting of ferric hydroxide of about 6-nm diameter. This core contains more than 2000 iron atoms, making the protein extremely electron dense and thus perfect for electron microscopy applications. The properties of HRP are described in Chapter 22, Section 1.

The following protocol is adapted from Bayer *et al.* (1976).

Protocol for the Conjugation of (Strept)avidin with Ferritin Using Reductive Amination

1. Dissolve (strept)avidin in 0.1-*M* sodium acetate, 0.15-*M* NaCl, pH 4.5, at a concentration of 3 mg/ml.
2. Dissolve ferritin in 0.1-*M* sodium acetate, 0.15-*M* NaCl, pH 4.5, at a concentration of 100 mg/ml.
3. Add 1 ml of ferritin solution to every 5 ml of (strept) avidin solution. Chill on ice.
4. Dissolve sodium periodate in water at a concentration of 100-m*M*. Prepare fresh and protect from light.
5. Add 110 µl of sodium periodate solution to each ml of (strept)avidin/ferritin solution.
6. React for 3 h on ice with periodic mixing. Protect from light.

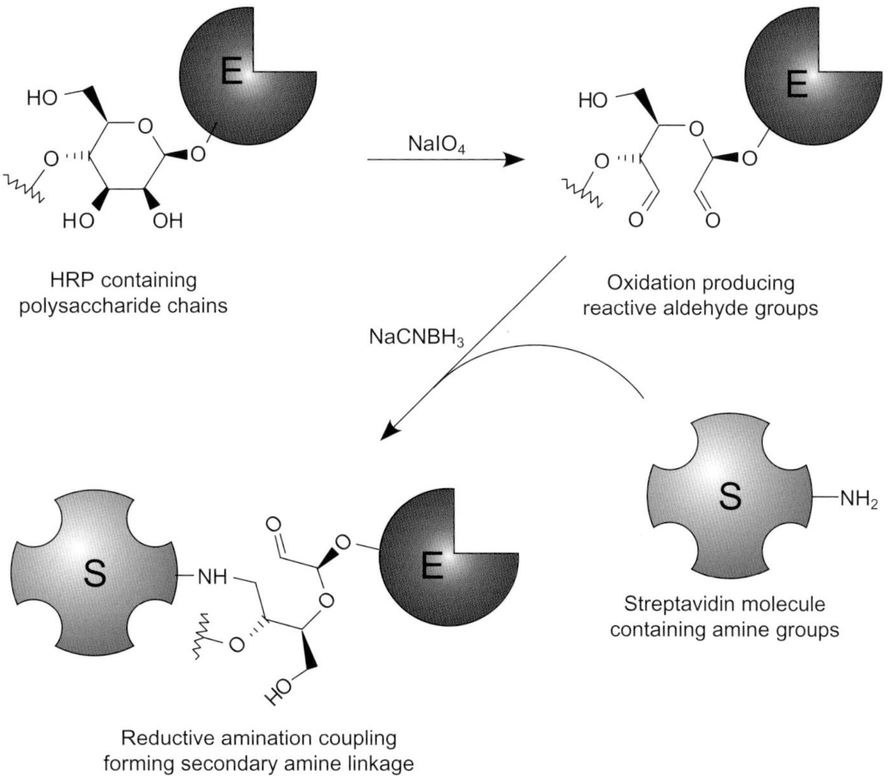

FIGURE 11.5 Oxidation of the polysaccharide components of HRP produces reactive aldehyde groups. Conjugation to streptavidin may then be achieved by reductive amination.

7. Remove excess periodate by gel filtration on a column of Sephadex G-25 or by overnight dialysis against 50-m*M* sodium borate, 0.15-*M* NaCl, pH 8.5.
8. Dissolve 10 mg of sodium borohydride in 1 ml of 10-m*M* NaOH. Prepare fresh. Add 83 µl of this reducing solution to each ml of (strept)avidin/ferritin solution.
9. React for 1 h on ice.
10. Remove excess reductant by gel filtration using a column of Sephadex G-25 or by extensive dialysis against 20-m*M* sodium phosphate, 0.15-*M* NaCl, pH 7.4.

Conjugation of HRP by reductive amination can be achieved by oxidizing the carbohydrate on the enzyme and subsequently coupling to the amines on (strept)avidin (Figure 11.5).

Protocol for the Preparation of (Strept)avidin–HRP by Reductive Amination

OXIDATION OF HRP WITH SODIUM PERIODATE

1. Dissolve HRP in water or 0.01-*M* sodium phosphate, 0.15-*M* NaCl, pH 7.2, at a concentration of 10 to 20 mg/ml.
2. Dissolve sodium periodate in water at a concentration of 0.088-*M*. Protect from light.
3. Immediately add 100 µl of the sodium periodate solution to each ml of the HRP solution. This results in an 8-m*M* periodate concentration in the reaction mixture. Mix to dissolve. Protect from light.
4. React in the dark for 15 min at room temperature. A color change will be apparent as the reaction proceeds—changing from the brownish/gold color of concentrated HRP to green. Longer reaction times will result in a decrease in HRP enzymatic activity.
5. Immediately purify the oxidized enzyme by gel filtration using a column of Sephadex G-25. The chromatography buffer is 0.01-*M* sodium phosphate, 0.15-*M* NaCl, pH 7.2. Collect 0.5-ml fractions and monitor for protein at 280 nm. HRP may also be detected by its absorbance at 403 nm. In oxidizing large quantities of HRP, the fraction collection process may be done visually by pooling the colored HRP peak as it comes off the column.
6. Pool the fractions containing protein. Adjust the enzyme concentration to 10 mg/ml for the conjugation step. The periodate-activated HRP may be stored frozen or freeze-dried for extended periods without loss of activity. However, do not store the preparation in solution at room temperature or 4°C, since precipitation will occur over time due to self-polymerization.

CONJUGATION OF PERIODATE-OXIDIZED HRP WITH (STREPT)AVIDIN

1. Dissolve (strept)avidin at a concentration of 10 mg/ml in 0.2-*M* sodium bicarbonate, pH 9.6, at room temperature. The high pH buffer will result in very efficient Schiff base formation and conjugation

with the highest possible incorporation of enzyme molecules per (strept)avidin molecule. To produce lower molecular weight conjugates (using less efficient Schiff base formation conditions), dissolve the proteins at a concentration of 10 mg/ml in 0.1-*M* sodium phosphate, 0.15-*M* NaCl, pH 7.2.

2. The periodate-oxidized HRP (prepared above) is finally purified using 0.01-*M* sodium phosphate, 0.15-*M* NaCl, pH 7.2. For conjugation using the lower-pH-buffered environment, this HRP preparation can be used directly at 10 mg/ml concentration. For conjugation using the higher pH carbonate buffer, dialyze the HRP solution against 0.2-*M* sodium carbonate, pH 9.6 for 2 h at room temperature prior to use.
3. Mix the (strept)avidin solution with the enzyme solution at a ratio of 1:6.6 (v/v). Since (strept)avidin has a molecular weight of about 66,000 and HRP's molecular weight is 40,000, this ratio of volumes will result in a molar ratio of HRP:(strept)avidin equal to 4:1. For conjugates consisting of greater enzyme-to-(strept)avidin ratios, proportionally increase the amount of enzyme solution as required. Typically, molar ratios of 2:1 to 10:1 (enzyme:avidin) give acceptable conjugates useful in a variety of ELISA techniques.
4. React for 2 h at room temperature to form the initial Schiff base interactions.
5. In a fume hood, add 10 µl of 5-*M* sodium cyanoborohydride (Sigma) per ml of reaction solution. Caution: Cyanoborohydride is extremely toxic. All operations should be carried out with care in a fume hood. Also, avoid any contact with the reagent, as the 5-*M* solution is prepared in 1-*N* NaOH.
6. React for 30 min at room temperature (in a fume hood).
7. Block unreacted aldehyde sites by addition of 50 µl of 1-*M* ethanolamine, pH 9.6, per ml of conjugation solution. Approximately 1-*M* ethanolamine solution may be prepared by addition of 300 µl ethanolamine to 5 ml of deionized water. Adjust the pH of the ethanolamine solution by addition of concentrated HCl, keeping the solution cool on ice.
8. React for 30 min at room temperature.
9. Purify the conjugate from excess reactants by dialysis or gel filtration using Sephadex G-25. Use 0.01-*M* sodium phosphate, 0.15-*M* NaCl, pH 7.0, as the buffer for either operation. Use a fume hood, since cyanoborohydride will be present in some of the fractions.

3.3. Glutaraldehyde Conjugation Protocol

Glutaraldehyde is one of the oldest homobifunctional reagents used for protein conjugation. It reacts with amine groups to create crosslinks by one of several routes (Chapter 5, Section 6.2, and Chapter 15, Section 2.1). In aqueous solution, the molecule undergoes a number of transformations through hemiacetal and aldol products to create a variety of reactive intermediates. Although glutaraldehyde is a bifunctional aldehyde molecule, it doesn't react at both ends as one might expect through aldehyde–amine Schiff base formation and reductive amination. The complex activation and conjugation reactions described in Chapter 15 for the use of glutaraldehyde to immobilize affinity ligands onto chromatography supports can be referenced as a basis for the conjugation reactions that can occur for bioconjugation in solution. The reagent is highly efficient at protein conjugation, but it also has a tendency to form high-molecular-weight polymers due to its bifunctional or multifunctional nature. Single-step protocols using glutaraldehyde are particularly notorious at resulting in some degree of insoluble protein oligomers (Porstmann *et al.*, 1985). Two-step methods somewhat alleviate this problem, but the potential for conjugate precipitation is still present.

Preparation of (strept)avidin conjugates with other proteins can be accomplished using either a one- or two-step glutaraldehyde procedure. Both methods may result in some degree of oligomer formation; however, the two-step protocol may keep insoluble material to a minimum. Although the following procedures are described using particular proteins, they may be used as a general guide for coupling enzymes, ferritin, phycobiliproteins, or other detectable proteins to (strept)avidin. Some optimization may be necessary to obtain the best yield of active conjugate.

Protocol for the One-Step Glutaraldehyde Conjugation of Ferritin to (Strept)avidin

This protocol is adapted from Bayer and Wilchek (1980).

1. Prepare a solution containing 5 mg/ml (strept)avidin and 25 mg/ml ferritin in 0.02-*M* sodium phosphate, 0.15-*M* NaCl, pH 7.4, at room temperature. Note: For the coupling of other proteins to (strept)avidin, their concentration may be reduced from the 25 mg/ml stated for ferritin.
2. In a fume hood, add 10 µl of 25% glutaraldehyde per ml of (strept)avidin/ferritin solution. Mix well.
3. React for 1 h at room temperature.
4. To reduce the resultant Schiff bases and any excess aldehydes, add sodium borohydride to a final concentration of 10 mg/ml.

Note: Some protocols do not call for a reduction step. The addition of borohydride at this level may result in disulfide bond cleavage and loss of protein activity in some cases. As an alternative to reduction, add 50 µl of 0.2-*M* lysine in 0.5-*M* sodium carbonate, pH 9.5, to each ml of the conjugation reaction to block excess reactive sites. Block for 2 h at room temperature. Other amine-containing small molecules may be substituted for lysine—such as glycine, Tris buffer, or ethanolamine.

5. Reduce for 1 h at 4°C.
6. To remove any insoluble polymers that may have formed, centrifuge the conjugate or filter it through a 0.45-µm filter. Purify the conjugate by gel filtration or dialysis using PBS, pH 7.4.

A two-step glutaraldehyde protocol may result in lower molecular weight conjugates, thus limiting the degree of insoluble material formed during the cross-linking process. The following protocol is adapted from Avrameas (1969).

Protocol for the Two-Step Glutaraldehyde Conjugation of Enzymes to (Strept)avidin

1. Dissolve the enzyme at a concentration of 10 mg/ml in 0.1-*M* sodium phosphate, 0.15-*M* NaCl, pH 6.8.
2. Add glutaraldehyde to a final concentration of 1.25%.
3. React overnight at room temperature.
4. Purify the activated enzyme from excess glutaraldehyde by gel filtration (using Sephadex G-25) or by dialysis against PBS, pH 6.8.
5. Dissolve (strept)avidin at a concentration of 10 mg/ml in 0.5-*M* sodium carbonate, pH 9.5. Mix the activated enzyme with the (strept)avidin solution at the desired molar ratio to effect the conjugation. Mixing the equivalent of 1 to 2 moles of enzyme per mole of (strept)avidin usually results in acceptable conjugates.
6. React overnight at 4°C.
7. To reduce the resultant Schiff bases and any excess aldehydes, add sodium borohydride to a final concentration of 10 mg/ml.

Note: Some protocols avoid a reduction step, as it can lead to disulfide bond cleavage and detrimental effects on protein activity. As an alternative to reduction, add 50 µl of 0.2-*M* lysine in 0.5-*M* sodium carbonate, pH 9.5, to each ml of the conjugation reaction to block excess reactive sites. Block for 2 h at room temperature. Other amine-containing small molecules may be substituted for lysine—such as glycine, Tris buffer, or ethanolamine.

8. Reduce for 1 h at 4°C.
9. To remove any insoluble polymers that may have formed, centrifuge the conjugate or filter it through a 0.45-µm filter. Purify the conjugate by gel filtration or dialysis using PBS, pH 7.4.

4. PREPARATION OF FLUORESCENTLY LABELED (STREPT)AVIDIN

Fluorophore modification of (strept)avidin creates a reagent system that can be used to detect and localize biotinylated targeting molecules. The application of such reagents in immunohistochemical staining techniques is significant (Bonnard *et al.*, 1984). A biotinylated antibody directed against a particular tissue antigen can be allowed to bind its target *in situ*, and then a fluorescently tagged (strept)avidin may be added to bind and visualize the antibody-bound antigenic sites by luminescence. Individual cellular structures can be labeled in similar assay strategies and detected by fluorescent microscopy or cell sorting techniques (Sternberger, 1986; Abou-Samra *et al.*, 1990). Biotinylated targeting molecules like antibodies can have some nonspecific binding potential due to the presence of a hydrophobic biotin tag. However, if hydrophilic biotinylation compounds are used in the modification reaction (such as those containing a PEG spacer; see Chapter 18, Section 1.3), then the degree of nonspecific binding in the conjugate can be kept to a minimum. The multivalent nature of (strept)avidin's biotin-binding sites combined with the potential of more than one biotin tag per antibody creates a system of much greater potential sensitivity than when using fluorescently modified antibodies alone. The complex formed from the (strept)avidin–biotin interaction amplifies the fluorescent signal beyond that capable in directly labeled antibody techniques.

Double labeling systems can also be developed using the (strept)avidin–biotin interaction. If two primary antibodies directed against separate antigenic determinants are labeled, one with biotin and the other with another detection component (such as a fluorophore, enzyme, gold particles, etc.), then both may be used to simultaneously localize different antigens in tissue sections. The biotinylated antibody may subsequently be detected by the addition of a fluorescently labeled (strept)avidin reagent. An example of a double label (strept)avidin–biotin detection system is that of Feller *et al.* (1983). A pair of tonsil antigens was visualized using two monoclonal antibodies, one fluorescein-labeled, and the other biotinylated. The biotinylated antibody was detected by using a phycoerythrin-labeled (strept)avidin conjugate (Section 4.4, this chapter, and Chapter 10, Section 7). Even triple-labeling systems may be developed using this strategy (van Dongen *et al.*, 1985).

The following sections present suggested protocols for labeling (strept)avidin with selected fluorophores. Other fluorescent probes may be constructed using the reagents and methods discussed in Chapter 10.

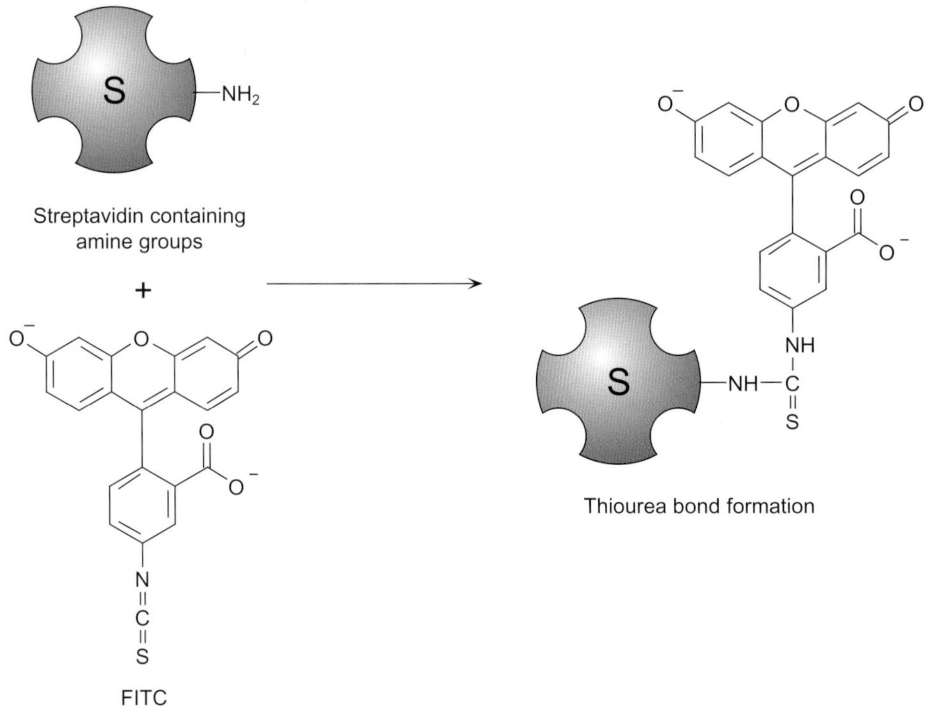

FIGURE 11.6 The reaction of FITC with streptavidin produces a fluorescent probe *via* isothiourea bonds.

4.1. Modification with FITC

Fluorescein isothiocyanate (FITC) has been one of the most common fluorescent labels used to modify proteins and other biomolecules (Chapter 10, Section 1). The isothiocyanate group reacts with amines in protein molecules to form a stable thiourea linkage (Figure 11.6). (Strept)avidin may be tagged with this reagent to yield highly fluorescent derivatives useful both in single-staining and double-staining techniques (Bayer and Wilchek, 1980; Bakkus *et al.*, 1989; Szabo *et al.*, 1989). Optimal modification levels for fluorescein are in the range of 3 to 8 fluorophores per (strept)avidin molecule. Lower incorporation levels will result in low luminescence and poor sensitivity. Higher levels may cause fluorescein–fluorescein quenching effects, resulting in decreased fluorescence. Too high a modification level may also result in nonspecific binding of the derivatized proteins to non-targeted components in assay systems or even result in precipitation of the modified protein due to hydrophobic aggregation of the dyes.

Although FITC and other reactive fluorescein derivatives are still widely used to label (strept)avidin and other proteins, better fluorescence yield and stability will be obtained if one of the newer hydrophilic fluorescein dyes is used, such as DyLight 488 (Thermo Fisher). See Chapter 10, Section 1, for additional details on labeling proteins with fluorescein.

Protocol

1. Dissolve (strept)avidin in 0.1-*M* sodium carbonate, pH 9.5, at a concentration of 2 to 4 mg/ml.
2. Dissolve FITC in DMF at a concentration of 2 mg/ml. Protect from light.
3. Add 50 to 100 μl of the FITC solution to each ml of the (strept)avidin solution.

Note: The optimal level of fluorescein modification should be determined experimentally to obtain a bright conjugate without the tendency to precipitate or fluorescently quench due to over-labeling. For relatively hydrophobic dyes such as FITC, this may mean reacting the (strept)avidin with no more than a 5-fold mole excess of dye to obtain about two to three fluorescent labels per molecule. Excessive labeling also causes nonspecific binding of dye-labeled conjugates in assays and detection applications.

4. React overnight at 4°C in the dark.
5. Remove excess fluorescein by gel filtration using a column of Sephadex G-25.

4.2. Modification with Lissamine Rhodamine B Sulfonyl Chloride

Rhodamine derivatives are popular probes to use in tandem with fluorescein labels. The Lissamine derivatives of rhodamine (Chapter 10, Section 2) are intensely

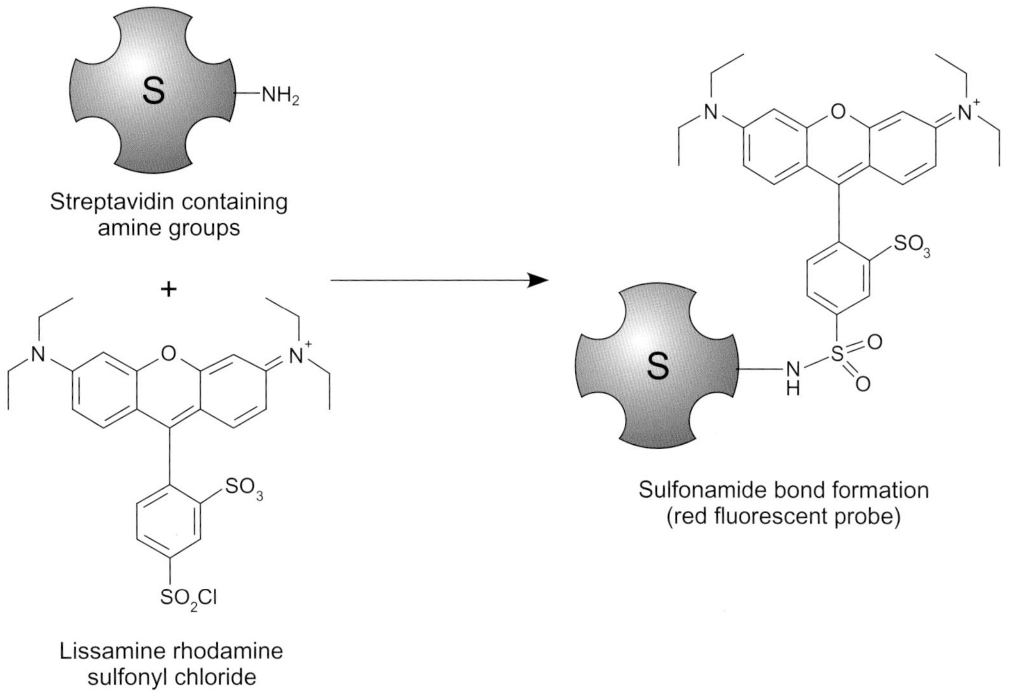

FIGURE 11.7 Streptavidin can be labeled with Lissamine rhodamine sulfonyl chloride to form a fluorescent probe.

fluorescent, strongly emitting in the red region of the spectrum. The red luminescence of Lissamine rhodamine contrasts sharply with the green emission of fluorescein derivatives. Lissamine rhodamine B sulfonyl chloride can be used to modify proteins at their ε- and N-terminal amine functional groups. The resultant derivatives are linked through stable sulfonamide bonds, resulting in rhodamine's fluorescent character being incorporated into the modified molecules. (Strept)avidin derivatives of this fluorophore are particularly popular for use in fluorescent assay systems (Figure 11.7).

Protocol

1. Dissolve (strept)avidin in 0.1-*M* sodium carbonate/bicarbonate buffer, pH 9.0, at a concentration of 1 to 5 mg/ml.
2. Dissolve Lissamine rhodamine B sulfonyl chloride in DMF at a concentration of 1 to 2 mg/ml. Protect from light and use immediately. Do not use DMSO as the solvent, as sulfonyl chlorides react with it.
3. In a darkened lab and with gentle mixing, slowly add 50 to 100 μl of the fluorophore solution to each ml of the (strept)avidin solution.

Note: Determine the optimal level of dye labeling experimentally to avoid over-labeling, which can cause fluorescent quenching, nonspecific binding in assays, or even precipitation of the conjugate. For hydrophobic dyes, a reaction including no more than about a 5-fold mole excess of dye over the amount of (strept)avidin in solution may result in the best performing conjugate.

4. React for 1 h at room temperature in the dark.
5. Remove excess fluorophore by gel filtration using a column of Sephadex G-25 or by dialysis.

Modification of (strept)avidin with Texas Red sulfonyl chloride may be done similarly, except the fluorophore is first dissolved in acetonitrile prior to addition to the aqueous reaction mixture.

4.3. Modification with AMCA–NHS

AMCA derivatives possess fluorescent properties within the blue region of the visible spectrum (Chapter 10, Section 3). Their emission range is well removed from other common fluorophores, making them excellent choices for use in double-labeling techniques, for example with fluorescein-labeled molecules. Coumarin-based fluorescent probes are very good donors for excited-state energy transfer to fluorescein dye derivatives. AMCA–NHS reacts with amine-containing molecules to result in stable amide-bond derivatives (Figure 11.8). (Strept)avidin may be labeled with this reagent to give probes useful for immunohistochemical staining of biotinylated targeting molecules. AMCA-labeled proteins are fairly stable to photoquenching and exhibit a large Stokes shift, allowing sensitive measurements to be made without interference from scattered excitation light.

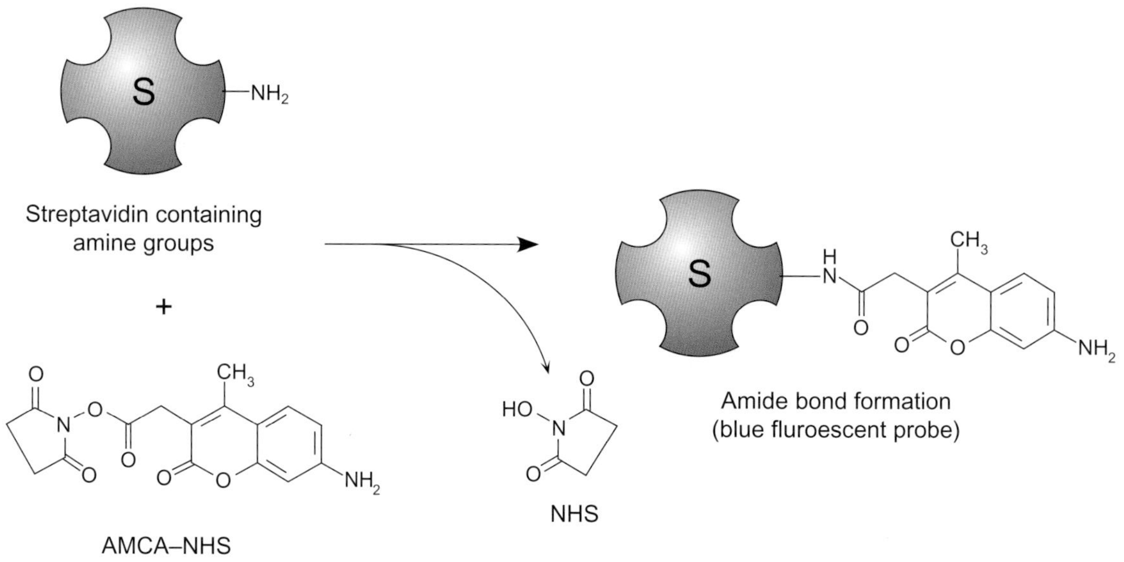

FIGURE 11.8 AMCA–NHS reacts with the amine groups of streptavidin to produce amide bonds.

Protocol

1. Dissolve (strept)avidin) in 50-m*M* sodium borate, pH 8.5, at a concentration of 10 mg/ml. Other buffers may be used for an NHS ester reaction, including 0.1-*M* sodium phosphate, pH 7.5 (Chapter 3, Section 1.4).
2. Dissolve AMCA–NHS (Thermo Fisher) in DMSO at a concentration of 2.6 mg/ml. Protect from light.
3. In subdued lighting conditions, slowly add 50–100 µl of the AMCA–NHS stock solution to each ml of the (strept)avidin solution, with gentle mixing.

Note: The optimal level of fluorescent labeling should be determined experimentally to obtain the best conjugate in the intended application. Avoid over-labeling as this may cause conjugate precipitation or non-specific binding in assays. Relatively hydrophobic dyes may perform best if the substitution level of the conjugate is not greater than about three to five dyes per (strept)avidin molecule.

4. React for 1 h at room temperature in the dark.
5. Remove excess reagent and reaction byproducts by gel filtration using a column of Sephadex G-25 or by dialysis.

4.4. Conjugation with Phycobiliproteins

Phycobiliproteins are incredibly fluorescent due to their multiple chromophoric bilin prosthetic groups, conferring extremely high absorbance coefficients to each protein molecule (Chapter 10, Section 7). Conjugates of these biliproteins with targeting molecules form extraordinarily luminescent probes. Labeling with phycobiliprotein derivatives can provide absorption coefficients 30-fold higher than labeling with small, synthetic fluorophores. Their ability to be monitored by fluorescing in the red region of the spectrum decreases potential interferences from indigenous biological fluorescence. Phycoerythrin-labeled (strept)avidin probes can be used in double-staining procedures with a fluorescein-labeled antibody, detecting two antigens in the same tissue section simultaneously by excitation at 488 nm (Feller *et al.*, 1983; Agata *et al.*, 2012).

The bilin content of these fluorescent proteins ranges from a low of 4 prosthetic groups in C-phycocyanin to the 34 groups of B- and R-phycoerythrin. Phycoerythrin derivatives, therefore, can be used to create the most intensely fluorescent probes possible using these proteins. (Strept)avidin–phycoerythrin conjugates, for example, have been used to detect as few as 100 biotinylated antibodies bound to receptor proteins per cell (Zola *et al.*, 1990).

Conjugates of (strept)avidin with these fluorescent probes may be prepared by activation of the phycobiliprotein with SPDP to create a sulfhydryl-reactive derivative, followed by modification of (strept)avidin with 2-iminothiolane or SATA (Chapter 2, Section 4.1) to create the free sulfhydryl groups necessary for conjugation. The protocol for SATA modification of (strept)avidin can be found in Section 3.1, this chapter. The procedure for SPDP activation of phycobiliproteins can be found in Chapter 10, Section 7. Reacting the SPDP-activated phycobiliprotein with thiol-labeled (strept)avidin at a molar ratio of 2:1 will result in highly fluorescent biotin-binding probes.

Other fluorescent probes may also be used to label (strept)avidin molecules for detection of biotinylated

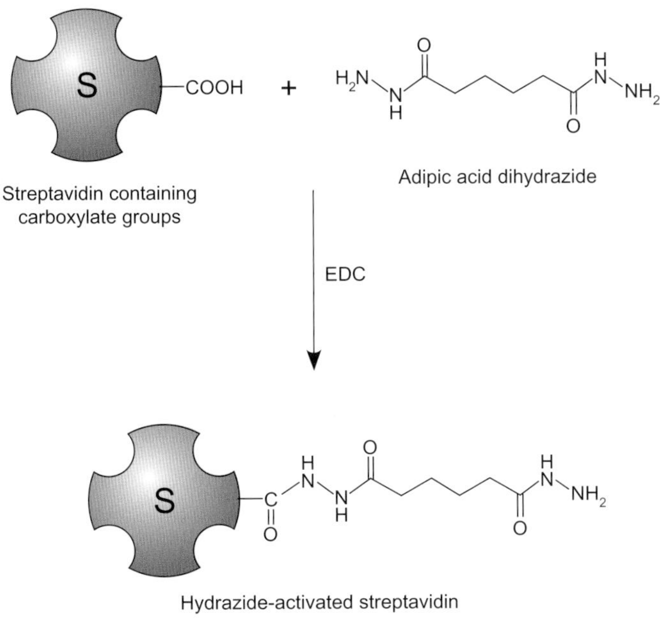

FIGURE 11.9 Reaction of adipic acid dihydrazide with streptavidin produces a hydrazide derivative that is highly reactive toward periodate-oxidized polysaccharides.

targeting molecules. Chapter 10 reviews many additional fluorescent labels, such as quantum dots, lanthanide chelates, and cyanine dye derivatives, all of which may be used in similar protocols to create detection conjugates for (strept)avidin–biotin-based assays.

5. PREPARATION OF HYDRAZIDE ACTIVATED (STREPT)AVIDIN

Hydrazide groups can react with aldehydes or ketones to form hydrazone linkages (Chapter 3, Section 5.1). Proteins may be labeled with hydrazide residues by reaction of their indigenous carboxylate groups with *bis*-hydrazine compounds such as adipic acid dihydrazide or carbohydrazide (Chapter 5, Section 8). A carbodiimide-mediated reaction between the protein and the *bis*-hydrazine reagent forms diimide-bond derivatives terminating in hydrazide groups (Figure 11.9). (Strept)avidin labeled with adipic acid dihydrazide can form the basis of a carbohydrate detection system using the (strept)avidin–biotin interaction (Bayer *et al.*, 1987, 1990; Bayer and Wilchek, 1990). Glycoconjugates in tissue sections, cells, or blots may be treated with sodium periodate or galactose oxidase to create aldehyde groups on the associated sugar components. Introduction of hydrazide-activated (strept)avidin causes hydrazone bonds to form between the hydrazides and aldehydes, thus specifically targeting glycoproteins and other carbohydrate-containing molecules. Subsequent detection with a biotinylated enzyme allows precise localization of glycoconjugates. Detection in a single step using this strategy is possible using preformed complexes of hydrazide-activated (strept)avidin and a biotinylated enzyme (Figure 11.10).

The activation of (strept)avidin with adipic dihydrazide may be done using the method of Bayer *et al.* (1987). A summary of this protocol is given below.

Protocol

1. Dissolve 160 mg of adipic acid dihydrazide (Aldrich) in 5 ml of 0.1-*M* sodium phosphate, pH 6.0. Some heating of the tube under a hot-water tap may be required to help solubilize the compound. Cool to room temperature.
2. Dissolve 50 mg of (strept)avidin in the adipic acid dihydrazide solution.
3. Add 160 mg of the water soluble carbodiimide EDC (Thermo Fisher) (Chapter 4, Section 1.1) to the solution, and mix to dissolve.
4. React for 4 h at room temperature.
5. Dialyze against PBS, pH 7.2, to remove excess reagent and reaction byproducts.

Hydrazide-activated (strept)avidin may be stored as a freeze-dried preparation without loss of activity.

6. BIOTINYLATION TECHNIQUES

The highly specific interaction of (strept)avidin with the small vitamin biotin can be a useful tool in designing assay, detection, and targeting systems for biological analytes (see Sections 1 and 2). The extraordinary affinity of (strept)avidin's interaction with biotin allows biotin-containing molecules in complex mixtures to be discretely bound with (strept)avidin conjugates. If the (strept)avidin–biotin complex contains detection components, then the targeted analytes can be located or quantified. This assay concept is made possible through the ability of biotin to be covalently attached to other targeting molecules, such as antibodies. In this sense, biotin derivatives may be prepared which contain reactive portions able to couple with particular functional groups in proteins and other molecules. Biotin modification of secondary molecules, called *biotinylation*, results in covalent derivatives containing one or more bicyclic biotin rings extending from the parent structure. These biotinylation sites are still capable of binding avidin or streptavidin with the specificity and nearly the same avidity of free biotin in solution. Since the biotin components are relatively small, macromolecules can be modified with these reagents without significantly affecting their physical or chemical properties (Della-Penna *et al.*, 1986). Proteins, carbohydrates, lipid molecules, and nucleic acids can be modified to

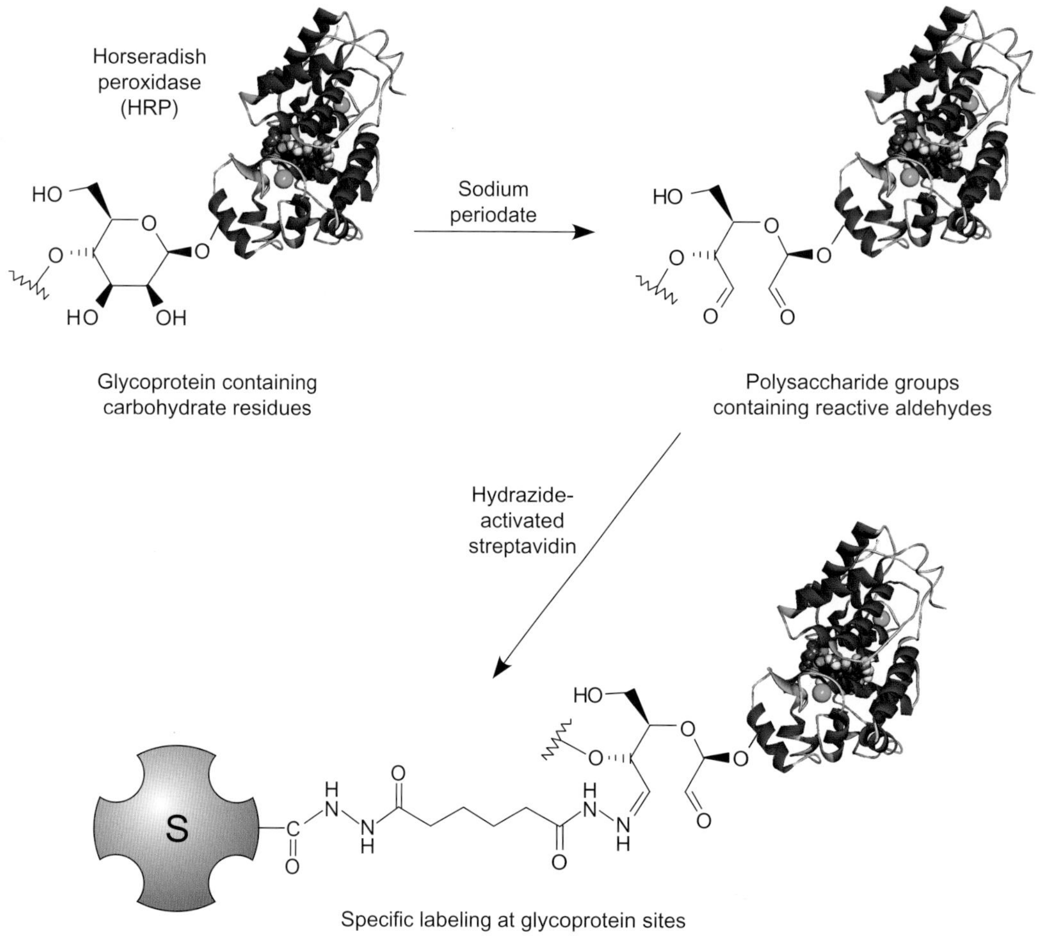

FIGURE 11.10 Glycoproteins can be oxidized with sodium periodate to generate aldehyde residues. These may be specifically labeled using a hydrazide–streptavidin derivative through hydrazone bond formation. Subsequent detection may be done using biotinylated enzymes.

contain one or more biotins able to strongly interact with (strept)avidin. The technique of biotinylation is made easier through the commercial availability of a range of different biotin derivatives having a number of important reactivity and property characteristics useful in (strept)avidin–biotin chemistry.

The basic design of a biotin labeling compound is illustrated in Figure 11.11. Common to all such modification reagents is the presence of the bicyclic biotin ring at one end of the structure and a reactive group at the other end that can be used to couple with other molecules. Biotinylation reagents also possess various cross-bridges or spacer groups built off the valeric acid side chain of the molecule. Since the binding sites for biotin on avidin and streptavidin are pockets buried about 9Å beneath the surface of the proteins, spacers can affect the accessibility of biotinylated compounds for efficiently binding avidin or streptavidin conjugates (Green *et al.*, 1971). In some applications, the use of a long spacer arm in the biotinylation reagent will result in the greatest potential assay sensitivity. The rate of binding of an avidin or streptavidin probe to a biotinylated molecule is also affected by the length of spacer in the biotinylation reagent. When longer spacers are utilized to make biotinylated macromolecules, it can potentially result in a 5-fold greater rate of streptavidin interaction (Bonnard *et al.*, 1984).

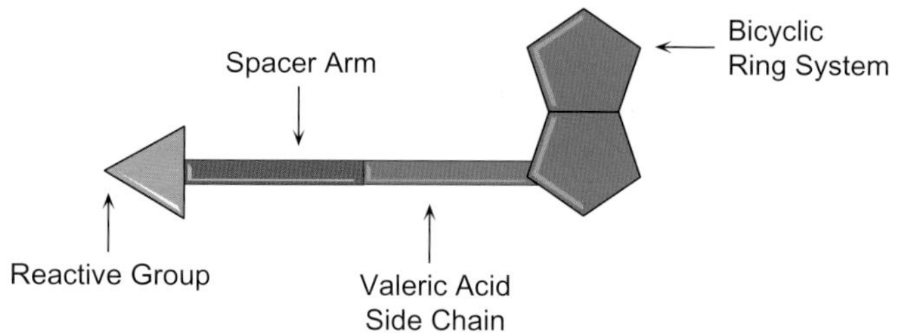

FIGURE 11.11 The basic design of a biotinylation reagent includes the bicyclic rings and valeric acid side chain of D-biotin at one end and a reactive group to couple with target groups at the other end. Spacer groups may be included in the design to extend the biotin group away from modified molecules, thus ensuring better interaction capability with avidin or streptavidin probes.

Biotinylation reagents containing spacers can be subdivided into several categories: (1) aliphatic spacers, (2) PEG-based spacers, and (3) cleavable or non-cleavable. The aliphatic spacer arms typically consist of hydrocarbon chains, which may be substituted with amide bonds or disulfide linkages. These spacers increase the length of the biotin group from a modified protein, but they may also confer considerable hydrophobicity to the biotinylation agent and molecules labeled with them. In fact, many biotinylated proteins may have a tendency to aggregate in solution or even precipitate due to the overall increased hydrophobicity of the biotin modifications. By contrast, biotinylation compounds built using hydrophilic PEG-based spacers have excellent water-solubility properties and have little or no tendency to aggregate or precipitate in solution. When using biotinylation agents having aliphatic spacers, such as the popular NHS–LC-biotin, it is therefore advantageous to keep the level of biotin modification at a minimum to avoid nonspecific binding, aggregation, or precipitation caused by the reagent. Antibodies biotinylated using such biotin compounds often do well if there are only one to three biotin groups per antibody molecule; however, higher modification levels can quickly cause stability problems as well as nonspecific binding in assays. This problem is circumvented by use of the hydrophilic biotin compounds containing a PEG spacer arm. Even over-biotinylation of an antibody with a PEG-based reagent usually won't result in precipitation, especially if the length of the PEG spacer is at least PEG_4 or greater. These hydrophilic biotins are discussed in detail in Chapter 18, Section 1.3.

Another variable to consider in choosing biotinylation reagents is the use of a biotin analog such as iminobiotin that has a moderated affinity constant in its binding of avidin or streptavidin (Section 6.2, this chapter). Analogs may be useful if release of the (strept)avidin–biotin bond is important for isolating a targeted analyte. Using native biotin, the interaction with avidin is so strong that up to 6- to 8-*M* guanidine at pH 1.5 is required to break the bond, possibly causing extensive denaturation of any other complexed molecules. By contrast, iminobiotinylated molecules can be released simply by adjusting the pH down to 4.

The following sections discuss some of the more common biotinylation reagents available for modification of proteins and other biomolecules. Each biotin derivative contains a reactive portion (or can be made to contain a reactive group) that is specific for coupling to a particular functional group on another molecule. Careful choice of the correct biotinylation reagent can result in directed modification away from active centers or binding sites, and thus preserve the activity of the modified molecule.

6.1. Determination of the Level of Biotinylation

It is often important to determine the extent of biotin modification after a biotinylation reaction is complete. Measuring biotin incorporation into macromolecules can aid in optimizing a particular (strept)avidin–biotin assay system. It can also be used to ensure reproducibility in the biotinylation process. The most common method of measuring the degree of biotinylation makes use of the HABA dye assay (Green, 1965). HABA is 4′-hydroxyazobenzene-2-carboxylic acid. In the absence of biotin, the dye is capable of specifically forming noncovalent complexes with (strept)avidin at its biotin-binding sites. Upon binding to (strept)avidin in aqueous solution, HABA exhibits a characteristic absorption band at 500 nm ($\varepsilon = 35{,}500\,M^{-1}cm^{-1}$, expressed as per mole of HABA bound). The addition of biotin to this complex results in displacement of HABA from the binding site, since the affinity constant of the (strept)avidin–biotin interaction ($1.3 \times 10^{15}\,M^{-1}$) is much greater than that for (strept)avidin–HABA ($6 \times 10^{6}\,M^{-1}$). As HABA is displaced, the absorbance of the complex decreases proportionally. Thus, the amount of biotin present in the solution can be determined by plotting the (strept)avidin–HABA absorbance at 500 nm *versus* the absorbance modulation with increasing concentrations of added biotin. Comparing an unknown biotin-containing sample to this standard response curve, can result in the determination of the biotin concentration in the sample.

Since a biotinylated molecule potentially is able to interact with (strept)avidin at its biotin-binding sites just as strongly as biotin in solution, the degree of biotinylation may be determined using the HABA method as well. Comparison of the response of a biotinylated protein, for example, with a standard curve of various biotin concentrations allows calculation of the molar ratio of biotin incorporation.

Two variations of the HABA–dye assay for biotinylated proteins are possible. In one approach, the biotinylated protein is digested using the enzyme pronase prior to doing the assay. The digestion process breaks the protein into small fragments, some of which possess biotin modifications. The digestion is done to eliminate any sterically hindered biotinylation sites from not being able to interact with (strept)avidin. The second approach merely uses the intact biotinylated protein in the assay, assuming that the HABA assay results will then provide a truer picture of the level of *accessible* biotin sites on the molecule. Pronase addition obviously is not necessary for assessing biotinylated molecules which are not proteins.

The following protocol describes both of these HABA-based tests for determining the level of biotinylation.

Protocol

1. Dissolve (strept)avidin in 0.05-*M* sodium phosphate, 0.15-*M* NaCl, pH 6.0, at a concentration of 0.5 mg/ml. A total of 3 ml of the (strept)avidin solution is required to create a standard curve using known concentrations of biotin and an additional 3 ml is needed for each sample determination.
2. Dissolve the HABA dye (Sigma) in 10-m*M* NaOH at a concentration of 2.42 mg/ml (10-m*M*). Prepare about 100 µl of the HABA solution for each 3-ml portion of (strept)avidin solution required.
3. Dissolve the biotinylated protein to be measured in 0.05-*M* sodium phosphate, 0.15-*M* NaCl, pH 6.0, at a concentration of 10 to 20 mg/ml. The amount required is about 100 µl of sample per determination.
4. Dissolve D-biotin in 0.05-*M* sodium phosphate, 0.15-*M* NaCl, pH 6.0, at a concentration of 0.5-m*M*.
5. For the proteolytic digestion procedure, dissolve pronase in water at a concentration of 1% (w/v).
6. If pronase digestion of the biotinylated protein is to be carried out, heat 100 µl of the sample at 56°C for 10 min, then add 10 µl of the pronase solution. Allow the sample to digest enzymatically at room temperature overnight. If no pronase digestion is desired, simply use the biotinylated protein solution prepared in step three without further treatment.
7. To construct a standard curve of various biotin concentrations, first zero a spectrophotometer at an absorbance setting of 500 nm with sample and reference cuvettes filled with 0.05-*M* sodium phosphate, 0.15-*M* NaCl, pH 6.0. Remove the buffer solution from the sample cuvette and add 3 ml of the (strept)avidin solution plus 75 µl of the HABA dye solution. Mix well and measure the absorbance of the solution at 500 nm. Next add 2-µl aliquots of the biotin solution to this (strept)avidin–HABA solution, mix well after each addition, and measure and record the resultant absorbance change at 500 nm. With each addition of biotin, the absorbance of the (strept) avidin–HABA complex at 500 nm decreases. The absorbance readings are plotted against the amount of biotin added to construct the standard curve.
8. To measure the response of the biotinylated protein sample, add 3 ml of the (strept)avidin solution plus 75 µl of the HABA dye to a cuvette. Mix well and measure the absorbance of the solution at 500 nm. Next, add a small amount of sample to this solution and mix. Record the absorbance at 500 nm. If the change in absorbance due to sample addition was not sufficient to obtain a significant difference from the initial (strept)avidin–HABA solution, add another portion of sample and measure again. Determine the amount of biotin present in the protein sample by using the standard curve. The number of moles of biotin divided by the moles of protein present gives the number of biotin modifications on each protein molecule.

6.2. Amine-Reactive Biotinylation Agents

Amine-reactive biotinylation reagents contain reactive groups off biotin's valeric acid side chain that are able to form covalent bonds with primary amines in proteins and other molecules. Two basic types are commonly available: NHS esters and carboxylates. NHS esters spontaneously react with amines to form amide linkages (Chapter 3, Section 1.4), whereas carboxylate-containing biotin compounds can be coupled to amines *via* a carbodiimide-mediated reaction using EDC (Chapter 4, Section 1.1).

D-Biotin and Biocytin

D-Biotin (hexahydro-2-oxo-1H-thieno[3,4-d]imidazole-4-pentanoic acid) is a naturally occurring growth factor present in small amounts within every cell. It is a key component in numerous processes involving carboxylation reactions, wherein it functions as a cofactor and transporter of CO_2 (coenzyme R). Biotin is mainly found covalently attached to lysine ε-amine groups of proteins *via* its valeric acid side chain. The compound was originally discovered through symptoms of deficiency caused by eating too many raw egg-whites. Biotin (or vitamin H) was found to be complexed and inactivated by the egg-white protein avidin (Boas, 1927; du Vigneaud, 1940). Treatment with additional vitamin H alleviated the symptoms.

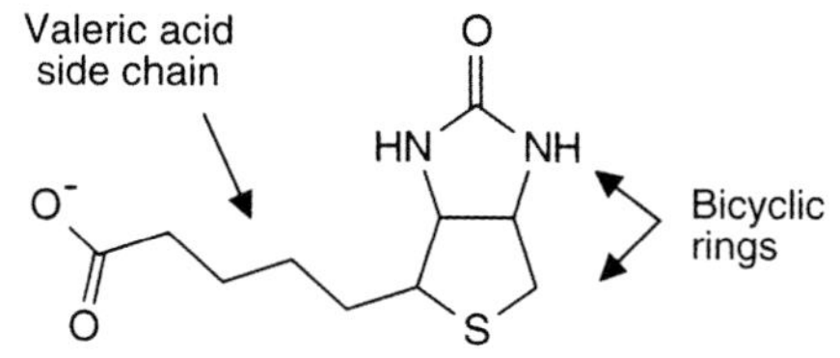

D-Biotin
MW 244.31

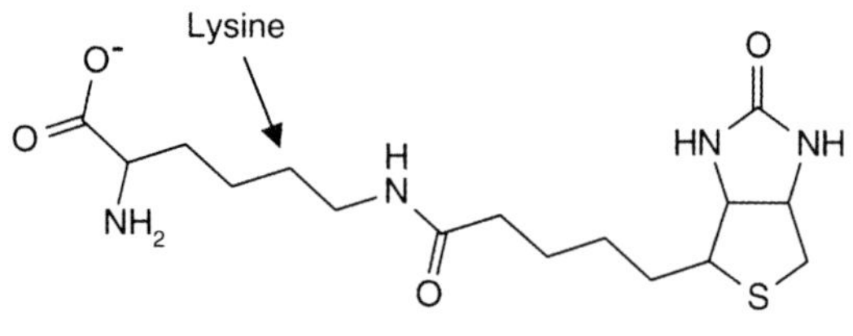

Biocytin
MW 372.48

The interaction of biotin with the proteins avidin and streptavidin is among the strongest noncovalent affinities known (Ka = $10^{15}M^{-1}$). The binding occurs between the bicyclic ring of biotin and a pocket within each of the four subunits of the proteins. The valeric acid portion is not directly involved with the interaction with avidin (Green, 1975; Wilchek and Bayer, 1988), but elimination of its carboxylate group or changing the amide linkage in biotinylation compounds can affect its affinity for (strept)avidin. Amide derivatives of biotin's valeric acid group, however, can be made without interfering with its high-affinity interactions. This characteristic allows modification of the valeric acid side chain without affecting the binding potential toward avidin or streptavidin.

D-Biotin is thus the basic building block for constructing biotinylation reagents. The molecule may be attached directly to a protein *via* its valeric acid side chain or derivatized at this carboxylate with other organic components to create spacer arms and various reactive groups. Reaction of biotin with primary amine groups on proteins can be done using the water soluble carbodiimide EDC (Chapter 4, Section 1.1). EDC activates the carboxylate to create a highly reactive, intermediate ester. This ester can then couple to amines to form stable amide bond derivatives (Figure 11.12). Biotinylated molecules thus formed retain the ability to bind avidin or streptavidin with high affinity.

The only potential deficiency in using D-biotin to directly modify a protein is the relatively short spacer arm afforded by the indigenous valeric acid group. Some applications may require longer spacers to maintain good binding potential toward avidin or streptavidin.

Biocytin is ε-*N*-biotinyl-L-lysine, a derivative of D-biotin containing a lysine group coupled at its ε-amino side chain to the valeric acid carboxylate. It is a naturally occurring complex of biotin that is typically found in serum and urine, and probably represents breakdown products of recycling biotinylated proteins. The enzyme biotinidase specifically cleaves the lysine residue and releases the biotin component from biocytin (Ebrahim and Dakshinamurti, 1986, 1987).

Biocytin has been used extensively as a labeling reagent for intracellular components within neurons (Horikawa and Armstrong, 1988; King *et al.*, 1989; Izzo, 1991; Granata and Kitai, 1992). It is particularly good for anterograde tracing studies in the central nervous system, since it can be easily injected into neurons using micropipettes. Subsequent visualization of biocytin locations may be achieved using a (strept)avidin–enzyme conjugate (Section 3, this chapter).

Biocytin should not be used in a carbodiimide reaction to modify proteins or other molecules, since it contains both a carboxylate and an amine group. A carbodiimide-mediated reaction, as suggested for D-biotin previously, would cause self-conjugation and polymerization of this reagent.

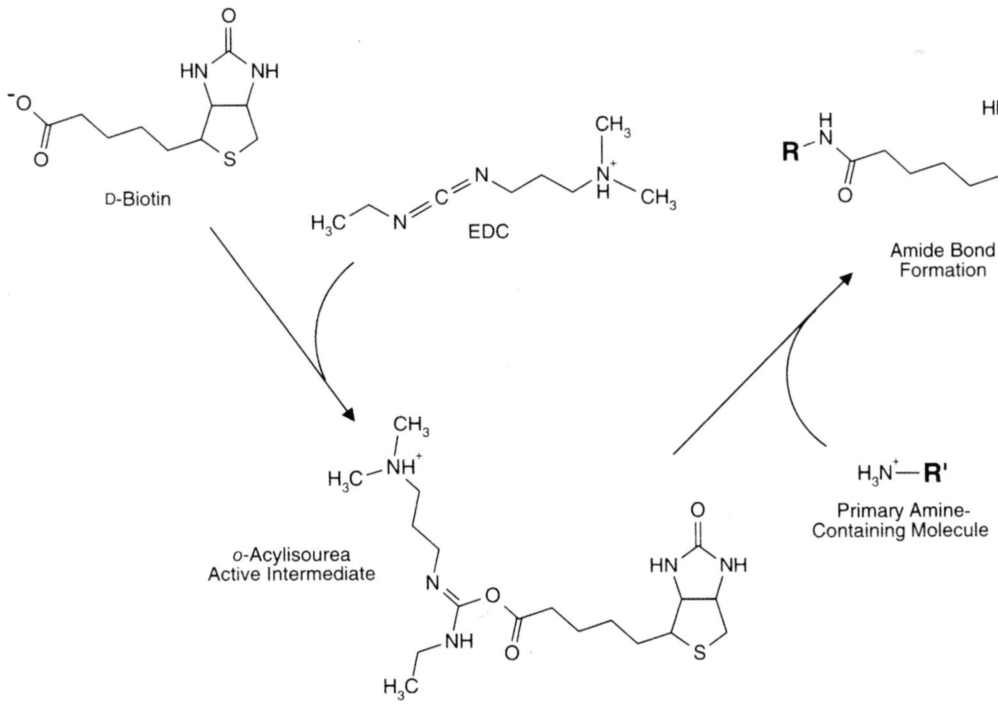

FIGURE 11.12 D-Biotin can be directly coupled to amine-containing molecules using the water soluble carbodiimide EDC to form an amide bond linkage.

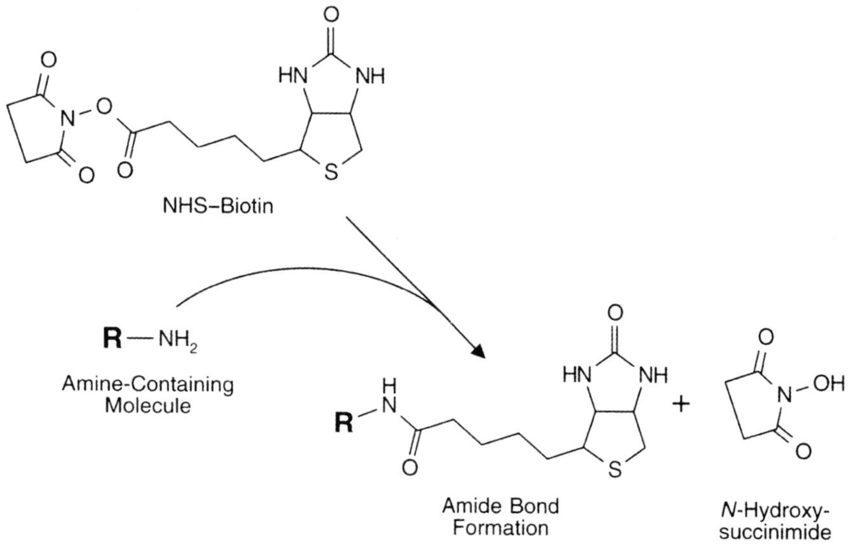

FIGURE 11.13 The active ester group of NHS–biotin reacts with amine-containing compounds to form amide bond linkages.

Biocytin, however, can form the basis for constructing trifunctional crosslinking reagents (Chapter 7). The lysine component of the molecule contains a free carboxylate and an α-amine group that can be used to build spacers and reactive groups for crosslinking purposes. The biotin component is the third arm of the trifunctional system, retaining its ability to bind (strept)avidin probes after conjugation has occurred at its other two ends. Such a trifunctional derivative has been used to study the hormone binding site of the insulin receptor (Wedekind *et al.*, 1989). This compound, 4-azido-2-nitrophenyl-biocytin-4-nitrophenyl ester, contains an amine-reactive group and a photoreactive phenyl azide functionality (Chapter 7, Section 1). The nitrophenyl ester reacts with amines on proteins and other molecules to from stable amide linkages. Once a molecule is modified in this manner, it contains both a photosensitive group and a biotin handle for conjugation and detection, respectively. Interaction of the modified protein with another protein followed by subsequent photolysis with UV light will result in covalent crosslinking. Localization and detection of the crosslinked molecules can then be done using an avidin or streptavidin conjugate. Another trifunctional compound, sulfo-SBED (or sulfosuccinimidyl-2-[6-(biotinamido)-2-(pazidobenzamido) hexanoamido] ethyl-1,39-dithiopropionate), is also based on a biocytin core (Chapter 7, Section 2). Additional information on the properties and use of these trifunctional biotin compounds based upon biocytin is described in Chapter 7 and in Chapter 24, Section 3, related to the study of protein interactions.

NHS–Biotin and Sulfo-NHS–Biotin

The valeric acid carboxylate of D-biotin may be activated to an NHS ester for direct modification of amine groups in proteins and other molecules. NHS esters react by nucleophilic attack of an amine on the carbonyl group, releasing the NHS group and forming a stable amide linkage (Chapter 3, Section 1.4) (Figure 11.13). NHS–biotin is the simplest biotinylation reagent available. Modification reactions are carried out under mildly alkaline conditions, and they usually result in a high efficiency of biotin incorporation.

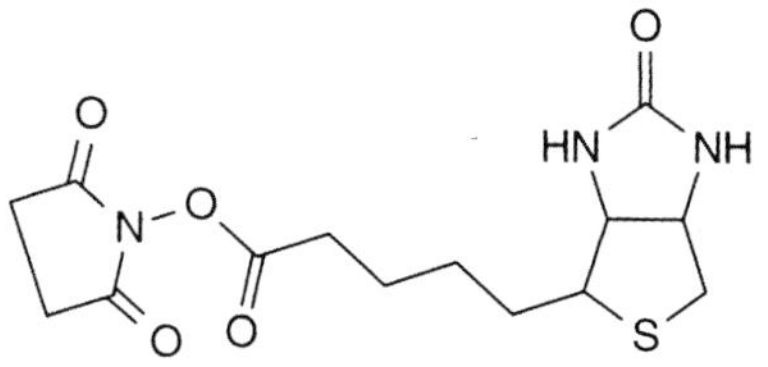

NHS–Biotin
MW 341.38
13.5 Å

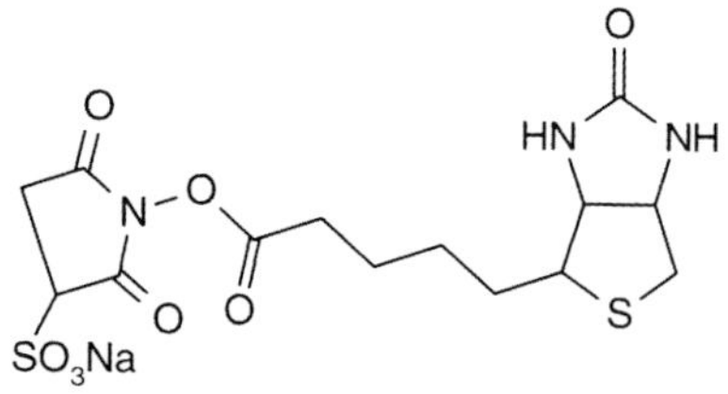

Sulfo-NHS–Biotin
MW 443.42
13.5 Å

NHS–biotin is insoluble in aqueous environments. It must be dissolved first in organic solvent as a concentrated stock solution and an aliquot added to an aqueous reaction medium to facilitate dissolution. Organic solvents such as dimethylformamide (DMF) or dimethyl sulfoxide (DMSO) are suitable for this purpose. Addition of an NHS–biotin solution to a reaction should not exceed a level of about 10% organic solvent in the buffer to avoid protein precipitation problems or precipitation of the buffer salt itself. Once added to the reaction medium, NHS–biotin may appear as a cloudy or hazy suspension, indicating incomplete solubility. However, such micro-dispersions are still effective at modification, often driving the bulk of the reagent into solution as the NHS ester reacts. Biotinylation of peptides or other molecules that are water insoluble may be carried out completely in organic solvent. For example, insulin can be biotinylated with NHS–biotin in an organic medium (Hofmann *et al.*, 1977).

A water soluble analog of NHS–biotin containing a negatively charged sulfonate group on its NHS ring structure is also available. Sulfo-NHS–biotin may be added directly to aqueous reactions without the need for organic solvent dissolution. A concentrated stock solution may be prepared in water to facilitate the addition of a small quantity to a reaction, but hydrolysis of the NHS ester will occur at a rapid rate, so the solution must be used immediately. This reagent is widely used to biotinylate antibodies and proteins for subsequent detection using streptavidin-based conjugates or for capture on immobilized streptavidin resins, especially in pull-down assays (Lei *et al.*, 2012; Wu *et al.*, 2012) (Chapter 15).

The only disadvantage to the use of NHS–biotin or sulfo-NHS–biotin is the lack of a long spacer group off the valeric acid side chain. Since the binding site for biotin on avidin and streptavidin is somewhat below the surface of the proteins, some biotinylated molecules may not interact as efficiently with (strept)avidin as when longer cross-bridges are used (Green *et al.*, 1971; Bonnard *et al.*, 1984).

NHS esters of D-biotin have been used in many applications, including the biotinylation of rat IgE to study receptors on murine lymphocytes (Lee and Conrad, 1984), in the development of an immunochemical assay for a post-synaptic protein and its receptor (LaRochelle and Froehner, 1986), in the study of plasma membrane domains by biotinylation of cell surface proteins in *Dictyostelium disoideum* amoebas (Ingalls *et al.*, 1986), and for the detection of blotted proteins on nitrocellulose membranes after transfer from polyacrylamide electrophoresis gels (LaRochelle and Froehner, 1986b).

The following protocol is a generalized method for the biotinylation of a protein using sulfo-NHS–biotin.

Protocol

1. Dissolve the protein to be biotinylated in 0.1-*M* sodium phosphate, 0.15-*M* NaCl, pH 7.2 to 7.5, at a concentration of 1 to 10 mg/ml.
2. Immediately before use, dissolve sulfo-NHS–biotin (Thermo Fisher) in water at a concentration of 20 mg/ml. Alternatively, the compound may be dissolved in organic solvent to prevent hydrolysis prior to a reaction (i.e., dry DMF or DMSO). Adjust the concentration and quantity of this stock solution to be prepared according to the amount of reagent needed to biotinylate the desired amount of protein. If prepared in water, the sulfo-NHS–biotin stock solution must be used immediately, since the NHS ester is subject to hydrolysis in aqueous environments.
3. With mixing, add a quantity of the sulfo-NHS–biotin solution to the protein solution to obtain a 5- to 10-fold molar excess of biotinylation reagent over the quantity of protein present. For instance, for an immunoglobulin (MW 150,000) at a concentration of 10 mg/ml, 10 μl of a sulfo-NHS–biotin solution (containing 8×10^{-4} mmoles) should be added per ml of antibody solution to obtain a 6-fold molar excess. For more dilute protein solutions (e.g., 1–2 mg/ml), an increased amount of biotinylation reagent may be required (e.g., >10-fold molar excess) to obtain similar incorporation yields as when using more concentrated protein solutions.

Note: To obtain biotinylation densities in the range of one to five biotins per protein, use no more than a 5-fold mole excess of biotinylation reagent over the concentration of protein present in the reaction. Some optimization may have to be done to obtain the best biotinylation level for use in the anticipated application.

4. React for 30 to 60 min at room temperature.
5. Purify the biotinylated protein from excess reagent and reaction byproducts by gel filtration using a desalting resin or by dialysis against PBS.

Determination of the degree of biotinylation can be done using the HABA assay (Section 6.1, this chapter).

NHS–LC-Biotin and Sulfo-NHS–LC-Biotin

NHS–LC-biotin is a derivative of D-biotin containing a spacer arm off the valeric acid side chain, terminating in an NHS ester. The compound is also known as succinimidyl-6-(biotinamido)hexanoate or NHS-X-biotin. The 6-aminocaproic acid spacer provides greater length between a covalently modified molecule and the bicyclic biotin rings. The total distance from an attached molecule to the biotin component is about 22.4 Å, significantly greater than the 13.5-Å length of NHS–biotin without a spacer arm. This increased distance can result in better binding potential for avidin or streptavidin probes, because the binding sites on these proteins are buried relatively deep inside the surface plane.

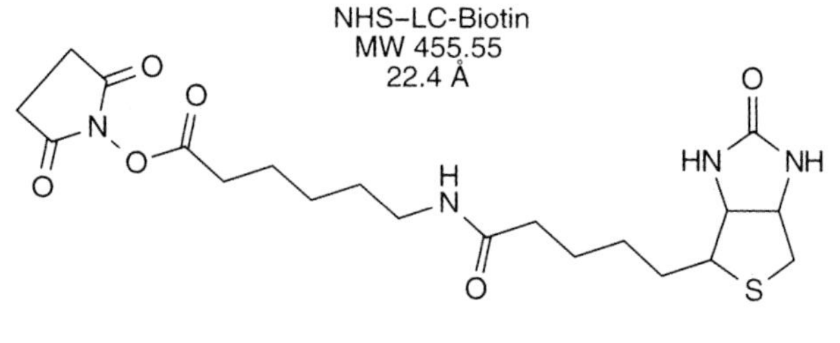

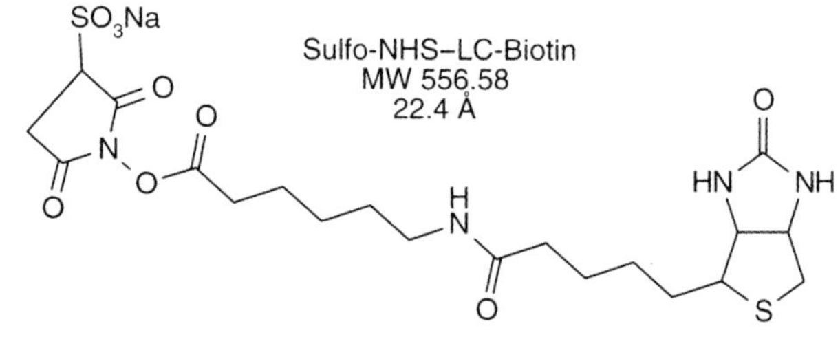

The NHS ester end of NHS–LC-biotin reacts with amine groups in proteins and other molecules to form stable amide-bond derivatives (Figure 11.14). Optimal reaction conditions are at a pH of 7 to 9, but the higher the pH the greater will be the hydrolysis rate of the ester. Avoid amine-containing buffers that will compete in the acylation reaction. NHS–LC-biotin is insoluble in aqueous reaction conditions and must be solubilized in organic solvent prior to the addition of a small quantity to a buffered reaction. Preparation of concentrated stock solutions may be carried out in DMF or DMSO. Nonaqueous reactions may also be carried out with this reagent for the modification of molecules insoluble in water. The molar ratio of NHS–LC-biotin to a protein in a reaction can be from about 2:1 to about 50:1, with higher levels resulting in better incorporation yields (Gretch *et al.*, 1987).

In a study comparing NHS–LC-biotin with two other derivatives of biotin, NHS–SS-biotin (Section 6.2, this chapter) and biotin hydrazide (Section 6.4, this chapter), it was found that modification through amines on monoclonal antibodies resulted in 2.5 times more activity in binding a streptavidin–agarose affinity column than when modification of carbohydrate residues using hydrazide conjugation chemistry was done (Gretch *et al.*, 1987). This was probably due to the greater abundance of amino groups over polysaccharide residues on these antibodies or the limited accessibility of the glycan between the heavy chains in the Fc region.

NHS–LC-biotin can be used to add a biotin tag to monoclonal antibodies directed at certain tumor antigens. The biotinylated monoclonals are allowed to bind to the tumor cell surfaces *in vivo*, and subsequent administration of an avidin or streptavidin conjugate can form the basis for inducing cytotoxic effects or creating traceable complexes for use in imaging techniques (Hnatowich *et al.*, 1987).

The reagent has also been used in a unique tRNA-mediated method of labeling proteins with biotin for nonradioactive detection of cell-free translation products (Kurzchalia *et al.*, 1988), in creating one- and two-step noncompetitive avidin–biotin immunoassays (Vilja, 1991), for immobilizing streptavidin onto solid surfaces using biotinylated carriers with subsequent use in a protein avidin–biotin capture system (Suter and Butler, 1986), and for the detection of DNA on nitrocellulose blots (Leary *et al.*, 1983).

Sulfo-NHS–LC-biotin, a water soluble analog of NHS–LC-biotin, is also available (Thermo Fisher), which contains a negatively charged sulfonate group on its NHS ring structure. The presence of the negative

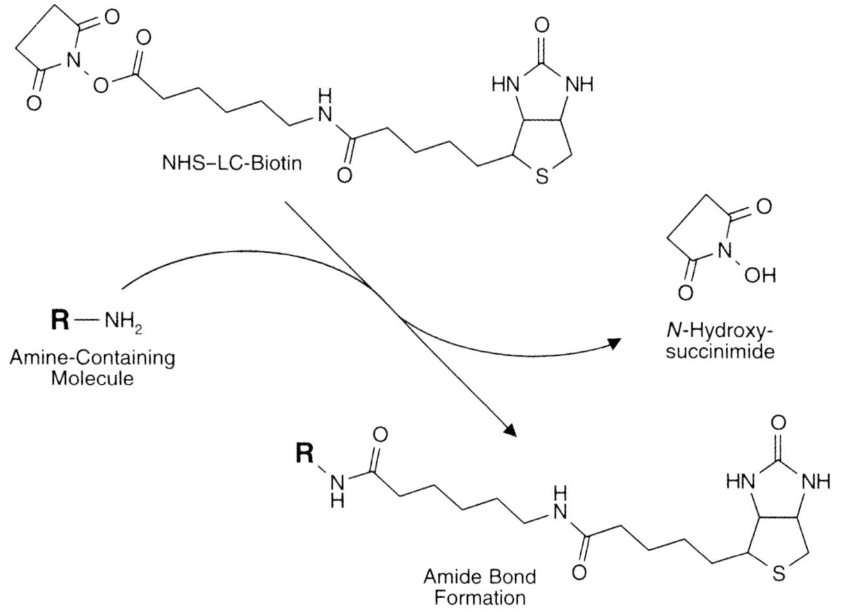

FIGURE 11.14 NHS–LC-biotin provides an extended spacer arm to allow greater distance between the biotin rings and a modified molecule. Reaction with amines forms amide linkages.

charge creates enough polarity within the molecule to allow direct solubility in aqueous reaction mediums. All other properties of the sulfonated version of the reagent are the same as those of NHS–LC-biotin. Sulfo-NHS–LC-biotin has been used to develop a dual-labeling method for virus imaging (Liu *et al.*, 2012), as a biotinylation agent having reduced membrane permeability for cell labeling (Strassberger *et al.*, 2011), and to develop a pull-down assay to investigate pore opening in ion channels (Tolino *et al.*, 2011).

Although NHS–LC-biotin and sulfo-NHS–LC-biotin are very popular reagents for biotinylation, they both result in hydrophobic aliphatic biotin modifications on proteins and antibodies. Unfortunately, these groups have a tendency to aggregate in aqueous solution and may cause protein precipitation or loss of activity over time, even when using low amounts of biotin substitution. Many commercial applications of biotinylated antibodies, for instance, use a low substitution level of biotin to avoid the aggregation issue as much as possible. As an alternative, the use of more hydrophilic PEG-based biotin compounds of approximately the same spacer length are a better choice for maintaining water solubility of modified proteins (Chapter 18, Section 1.3).

The following protocol is a suggested method for the biotinylation of proteins with either NHS–LC-biotin or sulfo-NHS–LC-biotin.

Protocol

1. Dissolve the protein to be biotinylated in 0.1-*M* sodium phosphate, 0.15-*M* NaCl, pH 7.2 to 7.5, at a concentration of 10 mg/ml.
2. Dissolve NHS–LC-biotin (Thermo Fisher) in dry DMF at a concentration of 40 mg/ml. This stock solution is stable for reasonable periods, although long-term storage is not recommended. For use of the water soluble sulfo-NHS–LC-biotin, a stock solution may be prepared in either organic solvent or water. If a solution in water is made to facilitate the addition of a small quantity of reagent to a reaction, then the solution should be prepared quickly and used immediately to prevent hydrolysis of the NHS ester. Sulfo-NHS–LC-biotin may be dissolved in water at a concentration of 20 mg/ml.
3. Add 50 μl of the NHS–LC-biotin solution in DMF to each ml of the protein solution in two aliquots apportioned 10 min apart. Alternatively, add a quantity of the sulfo-NHS–biotin solution prepared in water to the protein solution to obtain a 5- to 10-fold molar excess of biotinylation reagent over the quantity of protein present. For instance, for an immunoglobulin (MW 150,000) at a concentration of 10 mg/ml, 10 μl of the sulfo-NHS–biotin solution (8×10^{-4} mmoles) should be added per ml of antibody solution to obtain a 6-fold molar excess.

Note: To obtain a lower level of biotinylation (better to maintain solubility and avoid aggregation of biotinylated proteins), react the biotin compound at a level of no more than a 5-fold mole excess over the amount of protein present in the reaction medium.

4. React for a total of 30 to 60 min at room temperature or several hours at 4°C.
5. Remove unreacted biotinylation reagent and reaction byproducts by gel filtration using a desalting resin or dialysis against PBS.
6. Assay for the level of biotin incorporation using the HABA dye procedure (Section 6.1, this chapter).

NHS–Iminobiotin

NHS–iminobiotin is *N*-hydroxysuccinimido-2-iminobiotin, the guanidino analog of NHS–biotin that has a lower affinity constant for binding avidin or streptavidin. Iminobiotin replaces the 2-oxo-imidazole upper ring structure of D-biotin with a 2-imino-imidazole structure, causing moderated interaction with the avidin or streptavidin binding sites. This biotin analog can be used in situations requiring mild dissociation of the (strept)avidin-biotin complex. Normally, breaking the (strept)avidin–biotin interaction requires 6- to 8-*M* guanidine hydrochloride at a pH of 1.5, an environment too severe for most proteins to maintain native structure or recover activity. Iminobiotin, by contrast, can be bound to avidin or streptavidin at a pH wherein the guanidino group is unprotonated and thus uncharged. Binding occurs at pH values above 9.5 (typically done with good affinity at pH 11), and elution can be accomplished simply by changing the pH to 4.0—an environment that protonates the 2-imino group and creates a positive charge, thus effectively dissociating the interaction (Figure 11.15).

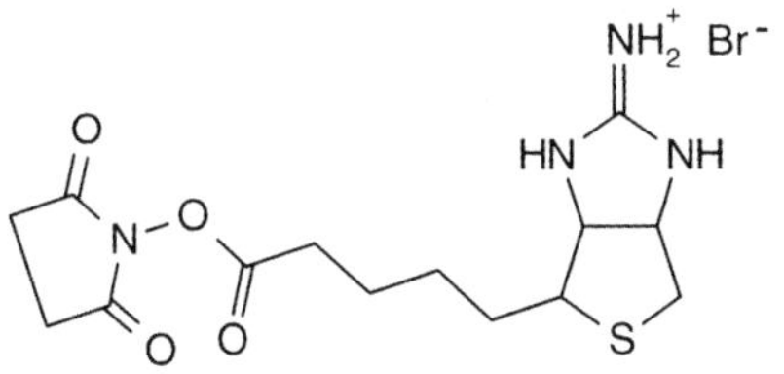

NHS–Iminobiotin
N-Hydroxysuccinimido-iminobiotin hydrobromide
MW 421.32
13.5 Å

NHS–iminobiotin can be used to label amine-containing molecules with an iminobiotin tag, providing reversible binding potential with avidin or streptavidin. The NHS ester reacts with proteins and other amine-containing molecules to create stable

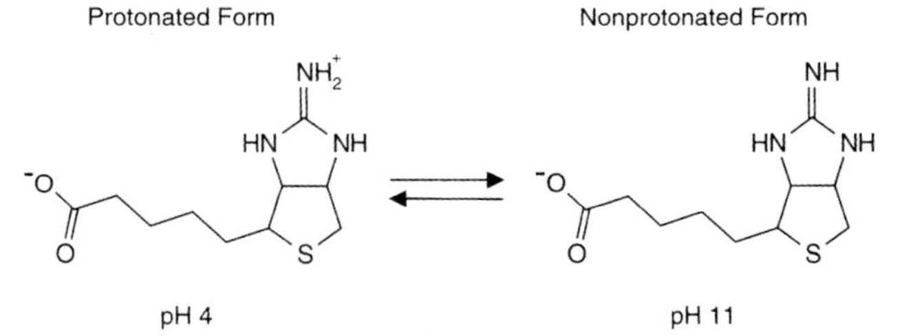

FIGURE 11.15 At pH 4, the protonated form of iminobiotin does not interact with the binding sites on avidin or streptavidin. At pH 11, the imino group is unprotonated and regains binding capability toward these proteins.

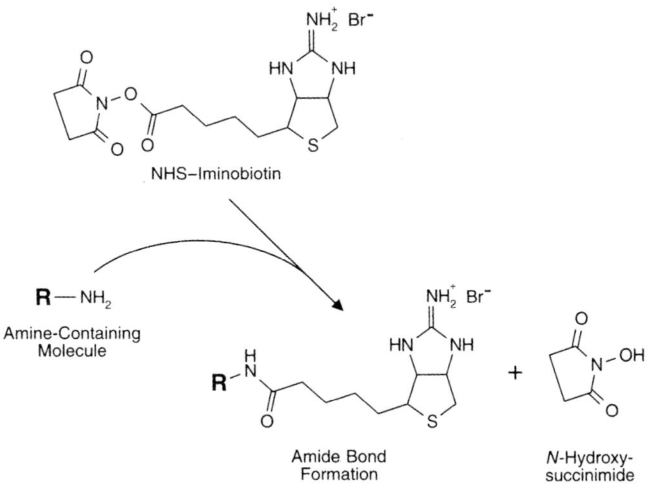

FIGURE 11.16 NHS–iminobiotin can be used to label amine-containing molecules, creating amide linkages.

amide bond derivatives (Figure 11.16). An iminobiotinylated molecule can then be used to target and purify other components in biological samples. For instance, a targeting molecule, such as an antibody, can be iminobiotinylated and allowed to bind its target in complex mixtures (such as tissue sections, cell extracts, or homogenates). The antibody–antigen complex subsequently can be purified using an affinity column of immobilized avidin with binding at pH 10 to 11 and simple elution at pH 4 (Orr, 1981; Zeheb *et al.*, 1983). The relatively mild elution condition allows recovery of the bound antigen without exposure to severely denaturing conditions.

The iminobiotin–avidin interaction can also be utilized in the opposite approach. Immobilized iminobiotin affinity columns can be used to purify avidin- or streptavidin-containing complexes under mild elution conditions (Hofmann *et al.*, 1980).

NHS–iminobiotin is insoluble in aqueous solution. It can be dissolved in organic solvent (DMF) prior to addition of a small aliquot to a buffered reaction medium. Do not exceed 10% DMF in the reaction to avoid protein precipitation problems. Optimal conditions for protein derivatization include non-amine-containing buffers at a pH of 7 to 9. The following protocol is a suggested method for labeling antibodies with NHS–iminobiotin. Some optimization may have to be carried out for particular derivatization needs.

Protocol

1. Dissolve the antibody to be modified in 50-m*M* sodium borate, pH 8.0, at a concentration of 5 mg/ml.
2. Dissolve NHS–iminobiotin in DMF at a concentration of 1 mg/ml. Prepare fresh.
3. Add 100 μl of the NHS–iminobiotin solution to each ml of the antibody solution. Mix well to dissolve.

Note: In the beginning, some turbidity may be present in the reaction due to incomplete dissolution of the NHS–iminobiotin. The solution may look cloudy or have a micro-particulate suspension present. This is normal for many water-insoluble reagents when added to an aqueous solution in an organic solvent. As the reaction takes place, the NHS–iminobiotin will be driven into solution, both by coupling to the protein and by hydrolysis of the NHS ester.

Note: The level of biotinylation of the protein to be modified should be controlled to prevent over-labeling, which can result in precipitation of the conjugate and nonspecific binding in the intended application. Reacting at a mole excess of no more than about 5-fold over the amount of protein present may result in the best performing biotinylated proteins, especially when using relatively hydrophobic biotin compounds. Some experimentation may be necessary to determine the optimal conjugate for a particular application.

4. React for 30 to 60 min at room temperature or for 3 h at 4°C.
5. Remove unreacted NHS-iminobiotin and reaction byproducts by dialysis or gel filtration using a desalting resin.

Sulfo-NHS–SS-Biotin

Sulfo-NHS–SS-biotin (also known as NHS–SS-biotin) is sulfosuccinimidyl-2-(biotinamido)ethyl-1,3-dithiopropionate, a long-chain cleavable biotinylation reagent that can be used to modify amine-containing proteins and other molecules (Thermo Fisher). The cross-bridge of the compound provides a 24.3-Å spacer arm that creates plenty of distance between the modified molecule and the biotin end. Using a long-chain biotinylation reagent can increase the efficiency of biotinylated molecules to bind avidin or streptavidin conjugates, thus enhancing the potential sensitivity of assay systems.

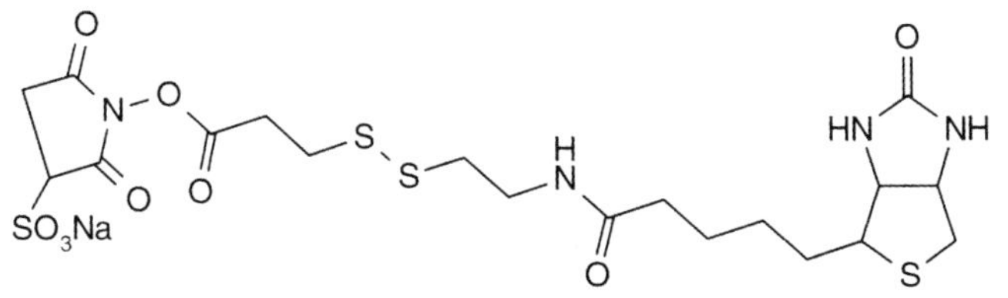

Sulfo-NHS–SS-Biotin
Sulfosuccinimidyl-2-(biotinamido)-ethyl-1,3-dithiopropionate
MW 606.7
24.3 Å

After molecules modified with sulfo-NHS–SS-biotin are allowed to interact with avidin or streptavidin probes, the complexes can be cleaved at the disulfide bridge by treatment with 50-m*M* DTT. Reduction releases the biotinylated molecule from the avidin or streptavidin capture reagent without breaking the (strept)avidin interaction. The use of disulfide biotinylation reagents thus provides much gentler conditions to break the complex than would be required if the avidin–biotin interaction itself were disrupted (which dissociates only at 6- to 8-*M* guanidine, pH 1.5).

The use of a cleavable biotinylation reagent also provides a means to purify targeted molecules using affinity chromatography on a column of immobilized avidin or streptavidin. For instance, an antibody modified with sulfo-NHS–SS-biotin can be allowed to bind its target in complex mixtures (such as tissue sections, cell extracts, or homogenates). The antibody–antigen complex subsequently can be isolated using an affinity column of immobilized avidin or streptavidin in a "pull-down assay" (Knezevic *et al.*, 2011; Hayashi, 2012). Elution from the column with DTT breaks the disulfide bonds, releasing the antibody and its bound antigen. The isolation of herpes virus proteins (Gretch *et al.*, 1987) and the recovery of DNA binding proteins (Shimkus *et al.*, 1985) were both carried out using this approach. Other methods of immunoprecipitation using non-cleavable biotinylation agents result in the inability to recover the captured proteins except under severely denaturing conditions.

Due to the presence of the negatively charged sulfonate group, sulfo-NHS–SS-biotin is a water soluble biotinylation reagent that may be added directly to aqueous reactions without prior dissolution in organic solvent. For the addition of small quantities of reagent, the compound may be dissolved in water and an aliquot transferred to the reaction medium. If an aqueous stock solution of sulfo-NHS–SS-biotin is prepared, it must be dissolved rapidly and used immediately to prevent hydrolysis of the active ester. The NHS ester reaction forms stable amide linkages with amine-containing proteins and other molecules (Figure 11.17). Optimal conditions for the NHS ester reaction include a

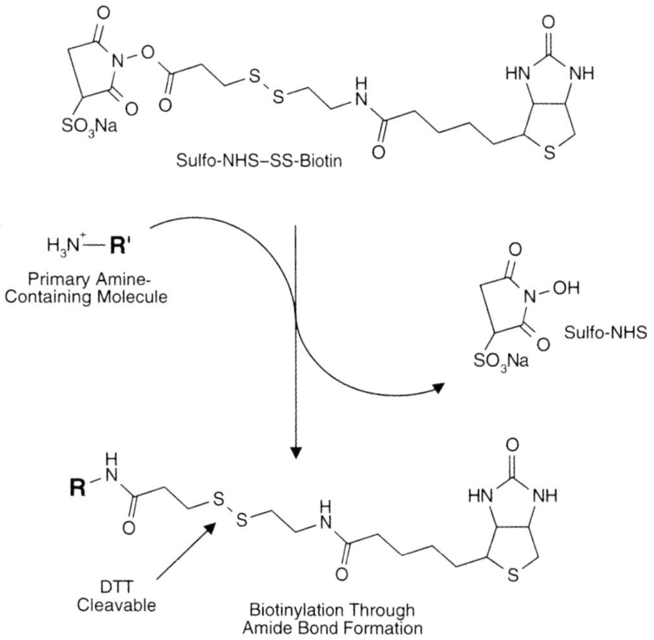

FIGURE 11.17 Sulfo-NHS–SS-biotin reacts with amine groups to form amide bonds. The biotin group can be later cleaved off the modified molecule by reduction of its internal disulfide linkage.

pH of 7 to 9, avoidance of any amine-containing buffers or other components that may compete in the reaction (including imidazole buffers which catalyze hydrolysis of these esters), and avoidance of reducing agents that could cleave the disulfide bridge.

The following protocol is a suggested method for biotinylating antibody molecules with sulfo-NHS–SS-biotin. Some optimization may have to be done with each application to ensure good biotin incorporation with retention of antigen binding activity. Other proteins and amine-containing molecules may be biotinylated using similar conditions.

Protocol

1. Dissolve the antibody to be biotinylated in 50-m*M* sodium bicarbonate, pH 8.5, at a concentration of 10 mg/ml. Other buffers and pH conditions between pH 7 and 9 can be used as long as no amine-containing buffers like Tris are present. Avoid also the presence of disulfide reducing agents that can cleave the disulfide group of the biotinylation reagent.
2. Add 0.3 mg of sulfo-NHS–SS-biotin (Thermo Fisher) to each ml of the antibody solution. To measure out small amounts of the biotinylation reagent, it may first be dissolved in water (or DMF) at a concentration of at least 1 mg/ml. Immediately transfer the appropriate amount to the antibody solution.

Note: This level of sulfo-NHS–SS-biotin addition represents about an 8-fold molar excess over the amount of antibody present. This should result in a molar incorporation of approximately two to four biotins per immunoglobulin molecule. Some optimization of the biotinylation level may have to be carried out to obtain the best performing conjugate for a particular application.

3. React for 30 to 60 min at room temperature or for 2 to 4 h at 4°C.
4. Remove unreacted biotinylation reagent and reaction byproducts by dialysis or gel filtration using a desalting resin.

Sulfo-NHS–SS-biotin can also be used to label cell surface proteins for subsequent detection or isolation using (strept)avidin reagents. The negative charge character of the compound prior to its reaction with an amine on a protein prevents it from penetrating the cell membrane bilayer. Thus, proteins on the outer surface of the cell can be specifically tagged with a biotin group. The disulfide cross-bridge of sulfo-NHS–SS-biotin allows recovery of labeled proteins after capture on an immobilized (strept)avidin support. Reduction of the disulfide using DTT or TCEP releases the proteins without the severe denaturing conditions usually required to break the (strept)avidin–biotin interaction. This allows isolation of cell surface proteins under non-denaturing conditions for subsequent analysis (Schuberth *et al.*, 1996; DeBlaquiere and Burgess, 1999; Ellerbroek *et al.*, 2001; Jang and Hanash, 2003).

The following protocol is based on the method of Thermo Fisher, as found in the instructions for the cell surface biotinylation kit.

Protocol

1. Grow cells in four T75-cm^2 flasks until they are 90 to 95% confluent.
2. Remove the media and wash the cells twice with 8 ml of cold 0.1-*M* sodium phosphate, 0.15-*M* NaCl, pH 7.2 (PBS). Note: This buffer contains a high buffer salt content to stabilize the pH during the biotinylation reaction. Do not allow the cells to remain in contact with it for more than 5 s to prevent detachment from the flask surface.
3. Dissolve 12 mg of sulfo-NHS–SS-biotin in 48 ml of cold PBS and immediately add 10 ml of the solution to each flask containing the washed cells.
4. React with gentle rocking for 30 min at 4°C.
5. Quench the reaction by the addition of 1 ml of 1-*M* Tris, pH 7.2.
6. Scrape the cells from each flask and transfer them into a 50-ml tube. Wash each flask using a single 10-ml portion of 0.025-*M* Tris, 0.15-*M* NaCl, pH 7.2, and add the solution to the scraped cells.
7. The isolated cells may be lysed using standard mechanical or detergent methods and the biotinylated cell surface proteins analyzed or isolated using (strept)avidin reagents.

6.3. Sulfhydryl-Reactive Biotinylation Agents

Sulfhydryl-reactive biotinylation reagents allow modification at cysteine –SH groups or at sites of specific thiolation within proteins and other molecules. Targeting sulfhydryls for modification, as opposed to amines, usually results in more limited derivatization, often away from active centers or binding sites. Directed coupling of biotin in this manner can aid in preserving activity. For instance, antibodies may be treated by reduction of their disulfide groups (mainly in the hinge region), which forms free sulfhydryls that are removed from the antigen-binding sites (Chapter 20, Section 1.1). Biotinylation at these sites produces a derivative that can bind efficiently to both antigen and (strept)avidin probes without steric hindrance.

Sulfhydryl groups also can be added to 5′-phosphate end of DNA probes (Chapter 23, Section 2.2). Biotinylation at these sites avoids disruption of base pairing with complementary DNA targets, since the point of modification is restricted to a single end position on the oligonucleotide.

The following sections discuss three sulfhydryl-reactive biotinylation reagents that utilize maleimide-, pyridyl disulfide-, and iodoacetyl-reactive groups, respectively. The maleimide and iodoacetyl options produce nonreversible, covalent thioether linkages with target –SH groups. The pyridyl disulfide chemistry results in disulfide bonds that are reversible through cleavage with a reducing agent.

Biotin–BMCC

Biotin–BMCC is 1-biotinamido-4-[4′-(maleimidomethyl) cyclohexane-carboxamido]butane, a biotinylation reagent containing a maleimide group at the end of an extended spacer arm (Thermo Fisher). The maleimide end reacts with sulfhydryl groups in proteins and other molecules to form stable thioether linkages (Figure 11.18). The reaction is highly specific for –SH groups in the range of pH 6.5 to 7.5. The long spacer arm (32.6 Å) provides more than enough distance between modified molecules and the bicyclic biotin end to allow efficient binding of avidin or streptavidin probes.

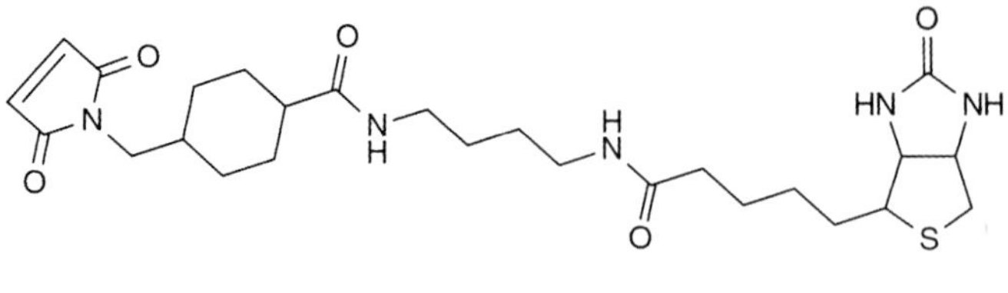

Biotin–BMCC
1-Biotinamido-4-[4'-(maleimidomethyl)-cyclohexane-carboxamido]butane
MW 533.69
32.6 Å

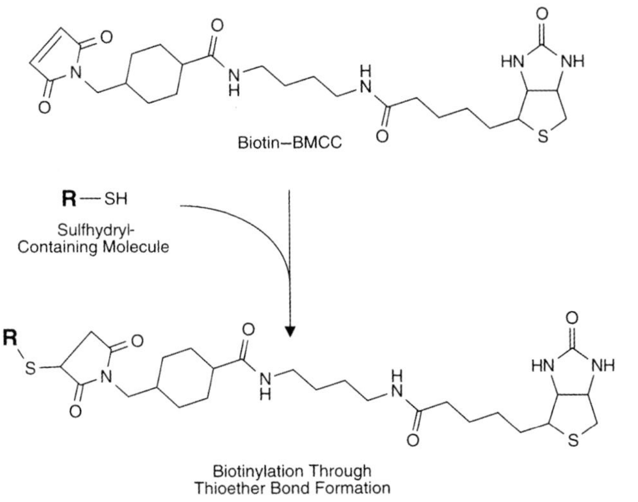

FIGURE 11.18 Biotin–BMCC provides sulfhydryl reactivity through its terminal maleimide group. The reaction creates a stable thioether linkage.

The reagent is similar to another maleimide-containing biotinylation reagent, 3-(*N*-maleimidopropionyl) biocytin, a compound used to detect sulfhydryl-containing molecules on nitrocellulose blots after SDS-electrophoresis separation (Bayer *et al.*, 1987). Biotin–BMCC should be useful in similar detection procedures.

Biotin–BMCC is insoluble in water and must be dissolved in an organic solvent prior to addition to an aqueous reaction mixture. Preparing a concentrated stock solution in DMF or DMSO allows transfer of a small aliquot to a buffer reaction. The upper limit of biotin–BMCC solubility in DMSO is approximately 33-m*M* or 17 mg/ml. In DMF, it is only soluble to a level of about 7-m*M* (4 mg/ml). Upon addition of an organic solution of the reagent to an aqueous environment (do not exceed 10% organic solvent in the aqueous medium to prevent protein precipitation), biotin–BMCC may form a micro-emulsion. This is normal, and during the course of the reaction the remainder of the compound will be driven into solution as it couples or hydrolyzes.

The required sulfhydryl groups for biotin–BMCC modification may be indigenous in molecules, formed through reduction of disulfides or created by the use of thiolation reagents (Chapter 2, Section 4.1). At physiological pH, the rate of the maleimide reaction toward sulfhydryls is almost 1000-fold faster than its reaction toward amines. However, at higher pH values the maleimide will couple to amines quite readily (Wu *et al.*, 1976; Ishi and Lehrer, 1986). Maleimides can also undergo a ring-opening hydrolysis reaction which increases in rate with pH, effectively inactivating the reactive group for thiol coupling.

Biotin–BMCC has been used to investigate palmitoylation (Fairbank, 2012), to study the heat-shock response in yeast (Wang *et al.*, 2012), and to study *S*-nitrosylation of proteins (Cheng *et al.*, 2012).

The following protocol is a suggested method for modifying sulfhydryl-containing proteins with biotin–BMCC. Some optimization of biotinylation levels may have to be carried out for particular applications.

Protocol

1. Dissolve the protein to be biotinylated (containing one or more free sulfhydryls) in 0.1-*M* sodium phosphate, 0.15-*M* NaCl, 10-m*M* EDTA, pH 6.5 to 7.5, at a concentration of 2.5 mg/ml.
2. Dissolve biotin–BMCC (Thermo Fisher) in DMSO at a concentration of 5 mg/ml.
3. Add 100 µl of the biotin–BMCC solution to each ml of the protein solution. Mix well.
4. React for at least 2 h at room temperature.
5. Remove excess biotinylation reagent and reaction byproducts by dialysis or gel filtration using a desalting resin.

Biotin–HPDP

Biotin–HPDP is *N*-[6-(biotinamido)hexyl]-3′-(2′-pyridyldithio)propionamide (Thermo Fisher). The reagent contains a 1,6-diaminohexane spacer group which is attached to biotin's valeric acid side chain. The terminal amino group of the spacer is further modified *via* an amide linkage with the acid precursor of SPDP (Chapter 6, Section 1.1) to create a terminal, sulfhydryl-reactive group. The pyridyl disulfide end of biotin–HPDP can react with free thiol groups in proteins and other molecules to form a disulfide bond with loss of pyridine-2-thione (Figure 11.19). This leaving group may be monitored by its characteristic absorbance at 343 nm to assess the level of biotinylation. However, since its extinction coefficient is rather low (about $8 \times 103\,M^{-1}cm^{-1}$), small-scale biotinylation reactions may not be quantifiable using this technique.

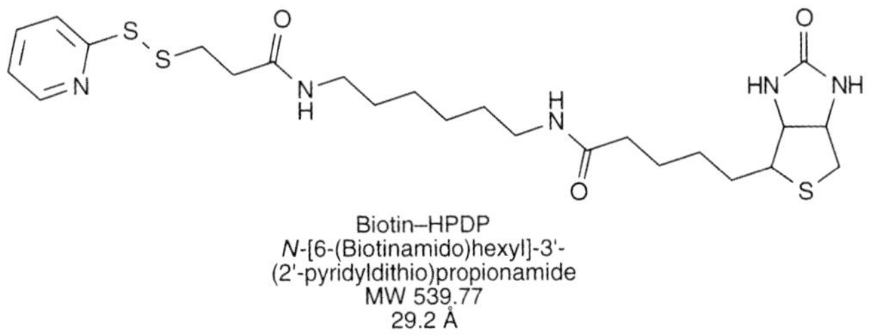

Modifications carried out with biotin–HPDP produce biotinylated compounds with long spacer arms (29.2 Å), which typically provide good binding efficiency with avidin or streptavidin probes. After coupling to sulfhydryl-containing molecules, the biotin–HPDP component can be cleaved back off by treatment with

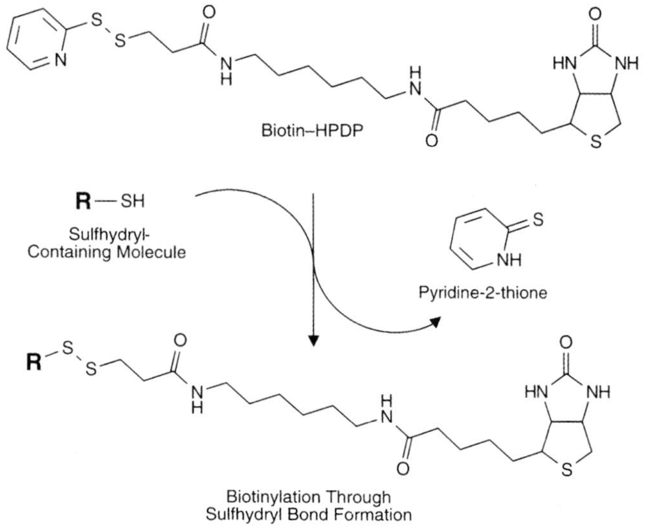

FIGURE 11.19 Biotin–HPDP reacts with sulfhydryl-containing molecules through its pyridyl disulfide group, forming reversible disulfide bonds. The biotin group may be released from modified molecules by reduction with DTT.

disulfide reducing agents, such as DTT. Breaking this bond releases the biotin modifications and regenerates the original sulfhydryl-containing molecule. This cleavability also provides a means of recovering target complexes after purification of the biotinylated molecules by affinity chromatography on immobilized avidin or streptavidin. Thus, biotin–HPDP-modified antibodies directed against some specific cellular antigen can be used to aid in the isolation of targeted components using affinity chromatography (immunoprecipitation) followed by elution with a disulfide reductant.

Using a similar approach, C1q has been modified with biotin–HPDP and allowed to interact with its specific receptor. Subsequent purification of the C1q receptor was accomplished through binding to immobilized streptavidin (Chapter 15) and subsequent cleavage of the disulfide bridge of the biotinylation reagent using DTT (Ghebrehiwet *et al.*, 1988). Similar pull-down assays continue to be done with biotin–HPDP to study protein interactions in cells (Seth and Stamler, 2011) or to analyze post-translational modifications (Lee *et al.*, 2011).

Biotin–HPDP is water insoluble and therefore must be dissolved in an organic solvent prior to addition to an aqueous reaction medium. Suitable solvents include DMSO and DMF. Concentrated stock solutions may be prepared in DMSO and a small aliquot transferred to a buffered reaction solution. Do not add more than 10% organic solvent to the aqueous reaction to prevent precipitation or denaturation of biological molecules. After addition, a micro-emulsion may result. This is normal for many water-insoluble reagents. The solution usually will become clearer during the course of the reaction. Optimal conditions for the disulfide interchange reaction include a pH range of 6 to 9 in buffer systems that do not contain any extraneous sulfhydryl compounds or reducing agents such as DTT, 2-mercaptoethanol, or TCEP. If reducing agents are used to create sulfhydryls in the protein to be biotinylated, these must be completely removed by dialysis or gel filtration before reacting with biotin-HPDP.

A suggested protocol for the use of biotin–HPDP in the modification of sulfhydryl-containing proteins follows. Similar procedures may be used when biotinylating other molecules containing thiols.

Protocol

1. Dissolve the sulfhydryl-containing protein to be biotinylated in 0.1-*M* sodium phosphate, 0.15-*M* NaCl, 10-m*M* EDTA, pH 7.2, at a concentration of at least 2 mg/ml.
2. Dissolve biotin–HPDP (Thermo Fisher) in DMSO at a concentration of 4-m*M* (2.1 mg/ml).
3. Add 100 µl of the biotin–HPDP stock solution to each ml of the protein solution. Mix well.
4. React for 90 min at room temperature.
5. Purify the biotinylated protein by gel filtration using a desalting resin or by dialysis. The PBS/EDTA buffer described in step 1 is suitable for either operation.

Iodoacetyl–LC-Biotin

Iodoacetyl–LC-biotin is *N*-iodoacetyl-*N*-biotinylhexylenediamine, a sulfhydryl-reactive biotinylation agent (Thermo Fisher). The reagent contains a 1,6-diaminohexane spacer group that is attached to biotin's valeric acid side chain. The terminal amino group of the spacer is further modified *via* an amide linkage with an iodoacetyl group to provide the sulfhydryl reactivity. Coupling to sulfhydryl-containing proteins or other molecules creates nonreversible thioether bonds (Figure 11.20). Modifications carried out with iodoacetyl–LC-biotin produce biotinylated compounds with sufficiently long spacer arms (27.1 Å), which help to ensure excellent binding potential with avidin or streptavidin probes.

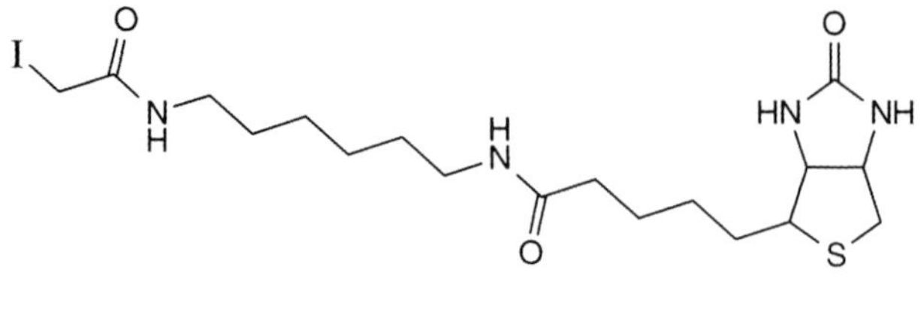

Iodoacetyl–LC-Biotin
N-Iodoacetyl-N-biotinylhexylenediamine
MW 510.42
27.1 Å

Iodoacetyl–LC-biotin is water insoluble and therefore must be dissolved in an organic solvent prior to

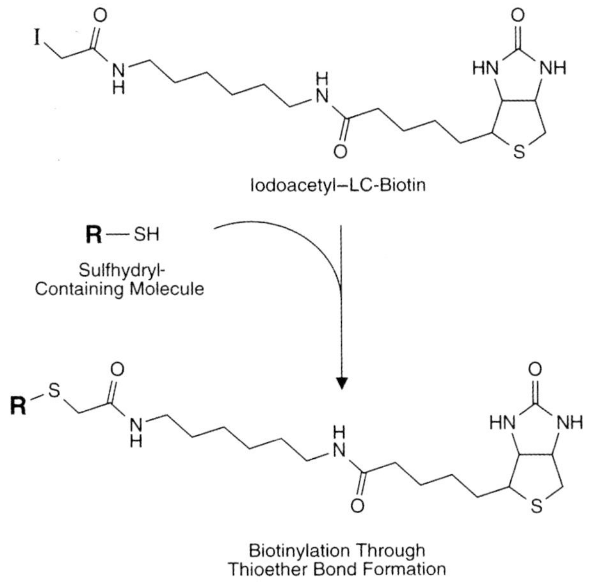

FIGURE 11.20 This biotinylation reagent reacts with sulfhydryl groups through its iodoacetamide end to form thioether bonds.

addition to an aqueous reaction medium. Suitable solvents include DMSO and DMF. Concentrated stock solutions may be prepared in DMSO and a small aliquot transferred to a buffered reaction solution. Do not add more than 10% organic solvent to the aqueous reaction to prevent precipitation or denaturation of biological molecules. After addition, a micro-emulsion may result. This is normal for many water-insoluble reagents. The solution usually will become clear during the course of the reaction. Optimal conditions for coupling using iodoacetyl-containing reagents include a pH range of 7.5 to 8.5 in buffer systems that do not contain any extraneous sulfhydryl compounds. In addition, protect all solutions containing iodoacetyl–LC-biotin from light, since photolysis may cause liberation of iodine, degrading the activity of the compound and possibly causing modification of tyrosine or histidine residues by iodination.

Iodoacetyl–LC-biotin has been used to localize the SH1 thiol of myosin by use of an avidin–biotin complex visualized by electron microscopy (Sutoh *et al.*, 1984), to determine the spatial relationship between SH1 and the actin binding site on the myosin subfragment-1 surface (Yamamoto *et al.*, 1984), to study markers for pancreatic cancer (Zhu *et al.*, 2012), and to assay protein–DNA interactions (Ritzefeld and Sewald, 2012).

The following protocol is a suggested method for biotinylating sulfhydryl-containing proteins using iodoacetyl-LC-biotin. The required sulfhydryl groups may be provided through reductive cleavage of disulfide bonds or by the use of thiolation reagents (Chapter 2, Section 4.1). Other molecules may be modified with iodoacetyl–LC-biotin using similar techniques.

Protocol

1. Dissolve the sulfhydryl-containing protein to be biotinylated in 50-m*M* Tris, 0.15-*M* NaCl, 10-m*M* EDTA, pH 8.3, at a concentration of 4 mg/ml.
2. Dissolve iodoacetyl–LC-biotin (Thermo Fisher) in DMF at a concentration of 4-m*M* (2 mg/ml). Protect from light.
3. Add 50 μl of the iodoacetyl–LC-biotin solution to each ml of the protein solution. Mix well. This level of addition represents a 3.28-fold molar excess of biotinylation reagent over the quantity of protein present if the protein has a molecular weight of 67,000 and possesses one sulfhydryl. Adjustments to the amount of reagent addition may have to be made to be appropriate for other proteins of different molecular weight. Consideration of the number of sulfhydryls present per protein molecule should also be carried out. React the biotinylation reagent at no more than a 3- to 5-fold molar excess over the amount of sulfhydryls present to ensure specificity of the iodoacetyl group for only –SH groups. Higher ratios of reagent-to-protein may cause reaction with amine groups present on the protein.
4. React for 90 min in the dark at room temperature.
5. Remove excess reactants and reaction byproducts by dialysis or gel filtration using a desalting resin.

6.4. Carbonyl- or Carboxyl-Reactive Biotinylation Agents

Hydrazide- or amine-containing biotinylation compounds can be used to modify carbonyl or carboxyl groups on other molecules. Hydrazides spontaneously react with aldehydes or ketones to give hydrazone linkages. The reaction may be further accelerated by the addition of an aniline catalyst (see Chapter 15, Section 2.4, for additional information on this reaction mechanism). The resulting hydrazone bonds may be further stabilized by reduction with sodium cyanoborohydride. The amine-containing biotinylation reagents (or the hydrazide ones) may be coupled to carboxylate groups using a carbodiimide reaction (Chapter 4, Section 1.1). In addition, amine- or hydrazide-containing biotinylation reagents may be coupled to cytosine residues in DNA or RNA by transamination catalyzed by bisulfite (Chapter 23, Section 2.3).

Biotin–Hydrazide and Biotin–LC-Hydrazide

Biotin–hydrazide is *cis*-tetrahydro-2-oxothieno[3,4-d]-imidazoline-4-valeric acid hydrazide, the hydrazine derivative of D-biotin off its valeric acid carboxylate (Thermo Fisher). The hydrazide functionality reacts with aldehyde and ketone groups to give hydrazone linkages. Although formyl groups are not common in biological molecules, they may be created by oxidation of diols with sodium periodate (Chapter 2, Section 4.4). Thus, glycoconjugates may be targeted specifically at their sugar residues. Biotinylation of these oxidized carbohydrates with biotin–hydrazide produces modifications that may be away from active centers or binding sites on proteins (Figure 11.21). Particularly, immunoglobulins may be biotinylated with this reagent at their polysaccharide groups which typically are present in the Fc region of the IgG molecule. Directed modification in this manner avoids the antigen-binding sites at the ends of the heavy and light chains, thus preserving antibody activity and allowing avidin or streptavidin probes to dock without blocking or interfering with antigen binding (although care should be taken in this respect, as some antibodies contain carbohydrate near their antigen-binding sites).

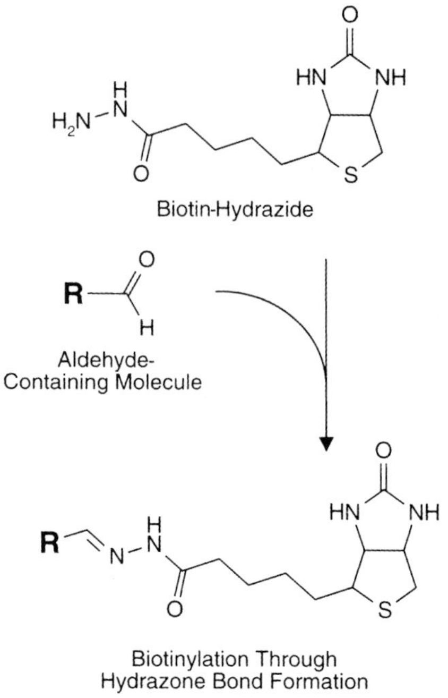

FIGURE 11.21 Biotin–hydrazide can be used to label aldehyde-containing molecules, creating hydrazone bonds.

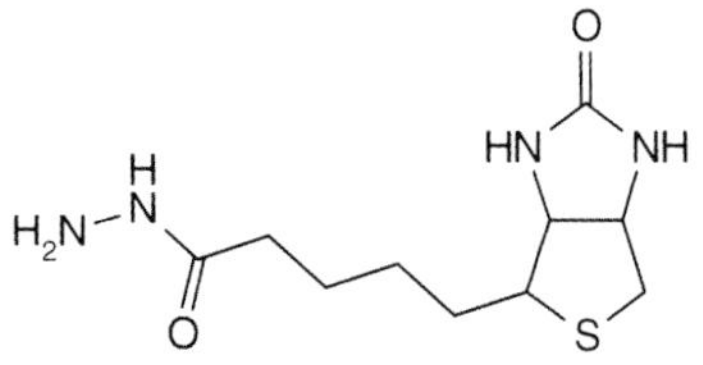

Biotin-Hydrazide
MW 258.34
15.7 Å

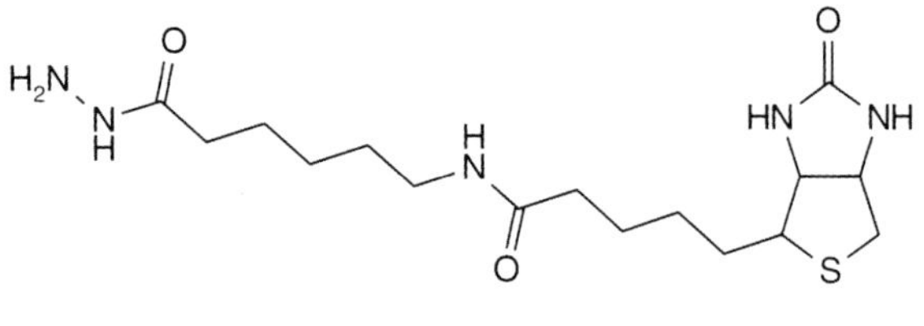

Biotin–LC-Hydrazide
MW 371.5
24.7 Å

Biotin–hydrazide may also be used to couple with carboxylate-containing molecules. Hydrazides can be coupled with carboxylic acid groups by using the carbodiimide reaction (Chapter 4, Section 1.1). The carbodiimide activates a carboxylate to an *o*-acylisourea intermediate. Biotin–hydrazide can react with this intermediate *via* nucleophilic addition to form a stable covalent bond.

Biotin–hydrazide has been used to biotinylate antibodies at their oxidized carbohydrate residues (O'Shanessy *et al.*, 1984, 1985, 1987; Hoffman and O'Shannessy, 1988), to modify the low-density lipoprotein (LDL) receptor (Wade *et al.*, 1985), to biotinylate nerve growth factor (NGF) (Rosenberg *et al.*, 1986), and to modify cytosine groups in oligonucleotides to produce probes suitable for hybridization assays (Reisfeld *et al.*, 1987) (Chapter 23, Section 2.3).

An analog of this biotinylation reagent with a longer spacer arm also exists. Biotin–LC-hydrazide contains a 6-aminocaproic acid extension off its valeric acid group (Thermo Fisher). The increased length of this spacer (24.7 Å) provides more efficient interaction potential with avidin or streptavidin probes, possibly increasing the sensitivity of assay systems. The reactions of biotin–LC-hydrazide are identical to those of biotin–hydrazide.

Both biotin–hydrazide and biotin–LC-hydrazide can be used to identify sites of carbonylation in biomolecules, such as those caused by oxidative stress. These sites typically involve some formation of aldehyde and ketone groups, which can be specifically tagged using hydrazide-containing biotinylation agents. The hydrazide group forms a hydrazone bond at the sites of carbonylation, thus allowing detection using streptavidin conjugates or isolation by using immobilized streptavidin in pull-down assays (Frohnert *et al.*, 2011; Curtis *et al.*, 2012).

The following protocol describes the use of biotin–hydrazide to label glycosylated proteins at their carbohydrate residues. Control of the periodate oxidation level can result in specific labeling of sialic acid groups or general sugar residues (Chapter 2, Section 4.4).

Protocol

1. Dissolve a periodate-oxidized glycoprotein (i.e., antibodies—see Chapter 20, Section 1.3) in 0.1-*M* sodium phosphate, 0.15-*M* NaCl, pH 7.4, at a concentration of 2 mg/ml. Note: The buffer, 0.1-*M* sodium acetate, pH 5.5, is typical of literature references for reaction of a hydrazide compound

with an aldehyde-containing molecule to form a hydrazone linkage. Alternative buffer conditions using higher pH values also work well. Physiological pH conditions with the use of a reducing agent such as sodium cyanoborohydride (step four) produce the most efficient labeling yields when using hydrazide-containing reagents.

2. Add biotin–hydrazide or biotin–LC-hydrazide to a final concentration of 5-m*M*.
3. React for 2h at room temperature.
4. To reduce the hydrazone bonds to more stable linkages, cool the solution to 4°C and add an equal volume of 30-m*M* sodium cyanoborohydride in PBS. Incubate for 40min. Note: If the presence of a reducing agent is detrimental to protein activity, eliminate this step. In most cases, the hydrazone linkage is stable enough for avidin–biotin detection experiments.
5. Remove excess reactants by dialysis or gel filtration using a desalting column.

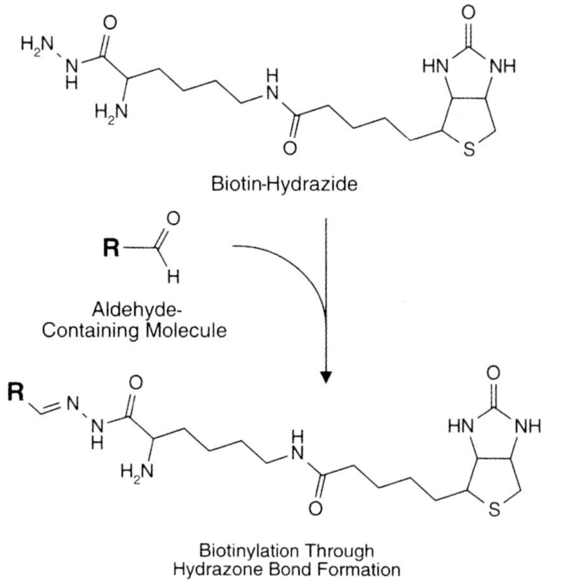

FIGURE 11.22 Biocytin hydrazide reacts with aldehyde-containing molecules to form hydrazone bonds.

Biocytin Hydrazide

Another biotinylation reagent that can spontaneously couple with aldehyde- or ketone-containing molecules is biocytin hydrazide (Thermo Fisher). Produced by forming the hydrazine derivative of biocytin—a lysine–biotin complex often found naturally in serum (Section 6.2, this chapter)—the compound has better solubility in aqueous solutions than either biotin–hydrazide or biotin–LC-hydrazide discussed previously. The solubility enhancement of biocytin hydrazide is due to the presence of lysine's α-amino group, which is protonated and positively charged at physiological pH. The reagent can be used to label carbohydrate-containing molecules, such as glycoproteins, after they have been oxidized to contain reactive aldehydes (Chapter 2, Section 4.4). The hydrazide group forms a hydrazone linkage with the aldehydes, thus directing the biotinylation reaction toward the polysaccharide regions of glycoconjugates (Figure 11.22).

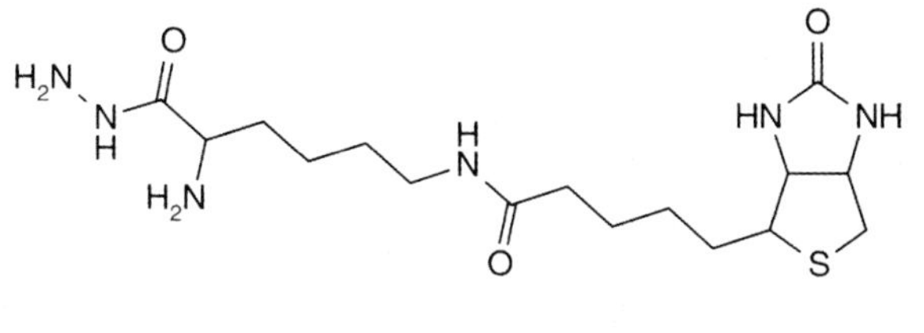

Biocytin Hydrazide
MW 386.51

Biocytin hydrazide was used to label specifically sialic acid residues, galactose residues, and for general sugar modification (Bayer *et al.*, 1988). The galactose residues were oxidized using galactose oxidase after treatment with neuraminidase (Chapter 2, Section 4.4). The use of this approach for labeling glycoproteins *in situ* was found to be optimal, due to the other potential side-reactions that may occur when using sodium periodate.

The reactivity and use of biocytin hydrazide is similar to that described for biotin–hydrazide in Section 6.4, this chapter. The following protocol for labeling glycoproteins at oxidized carbohydrate (galactose) sites is from Bayer and Wilcheck (1992).

Protocol

1. Dissolve the glycoprotein to be labeled in 0.1-*M* sodium phosphate, 0.15-*M* NaCl, pH 7.4, containing 1-m*M* $CaCl_2$ and 1-m*M* $MgCl_2$ (labeling buffer), at a concentration of 1mg/ml.
2. Dissolve biocytin hydrazide (Thermo Fisher) in 0.1–*M* sodium phosphate, 0.15-*M* NaCl, pH 7.4 (PBS), at a concentration of 20mg/ml.
3. To each ml of glycoprotein solution, add 30μl of neuraminidase (1 unit/ml as supplied by Behringwerke AF), then 30μl of galactose oxidase (previously dissolved at 100 units/ml in the labeling buffer of step one), and finally 100μl of the biocytin hydrazide solution.
4. React for 2h at 37°C.
5. Remove unreacted reagents by dialysis or gel filtration.

5-(Biotinamido)pentylamine

The D-biotin derivative, 5-(biotinamido)pentylamine, contains a 5-carbon cadaverine spacer group attached to the valeric acid side chain (Thermo Fisher).

The compound can be used in a carbodiimide reaction process to label carboxylate groups in proteins and other molecules, forming amide bond linkages (Chapter 4, Section 1). However, the main use of this biotinylation reagent is in the determination of factor XIIIa or transglutaminase enzymes in plasma, cell, or tissue extracts.

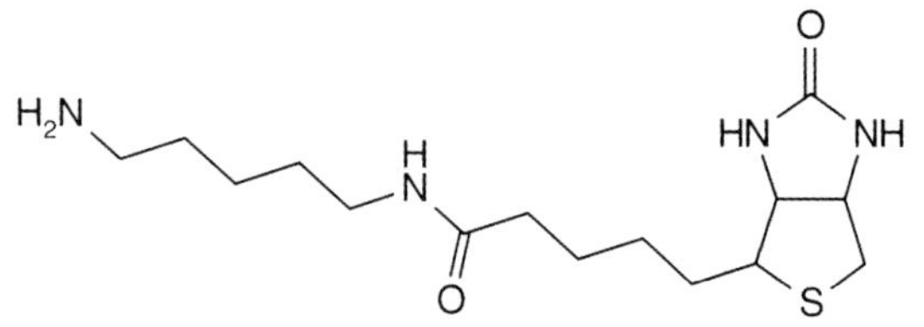

5-(Biotinamido)pentylamine
MW 328.48

Factor XIII, also known as plasma transglutaminase, is an enzyme of the blood coagulation cascade. It is activated by thrombin and calcium to factor XIIIa, at which point it catalyzes covalent crosslinks between the ε-amine group of lysine side chains and the γ-glutamyl side chain of glutamine residues. Abnormal levels of factor XIII in plasma are clinically important, being associated with cancer, liver or renal dysfunction, or various bleeding disorders. The assay of transglutaminase activity therefore is important for investigating the activity and function of this enzyme as it relates to post-translational protein modification as well as various disease states.

5-(Biotinamido)pentylamine is able to participate in the acyltransferase reaction, becoming covalently attached to protein substrates at their glutamine residues (Figure 11.23). Lee *et al.* (1988) used this biotinylation reagent to quantify factor XIII in plasma (D'Eletto *et al.*, 2012; Kuper *et al.*, 2012). Transglutaminase activity resulted in the modification of an *N,N'*-dimethylcasein substrate which was subsequently detected by an avidin–biotin assay procedure. The assay may be carried out in microplates using wells coated with the substrate protein and quantifying the enzyme activity with streptavidin–alkaline phosphatase (Slaughter *et al.*, 1992). Jeon *et al.* (1989) subsequently applied the assay to the measurement of transglutaminase activity in cells. Components biotinylated in cellular systems also can be isolated by use of affinity chromatography on immobilized avidin (Lee *et al.*, 1992).

6.5. Photoreactive Biotinylation Agents

Biotin derivatives containing a photoreactive group provide nonselective biotinylation potential at certain reactive hydrogen sites or nucleophilic groups. They can be used to incorporate an avidin-binding biotin group into molecules that do not possess amines, sulfhydryls, or other easily modifiable functional groups. Many of these photoreactive derivatives

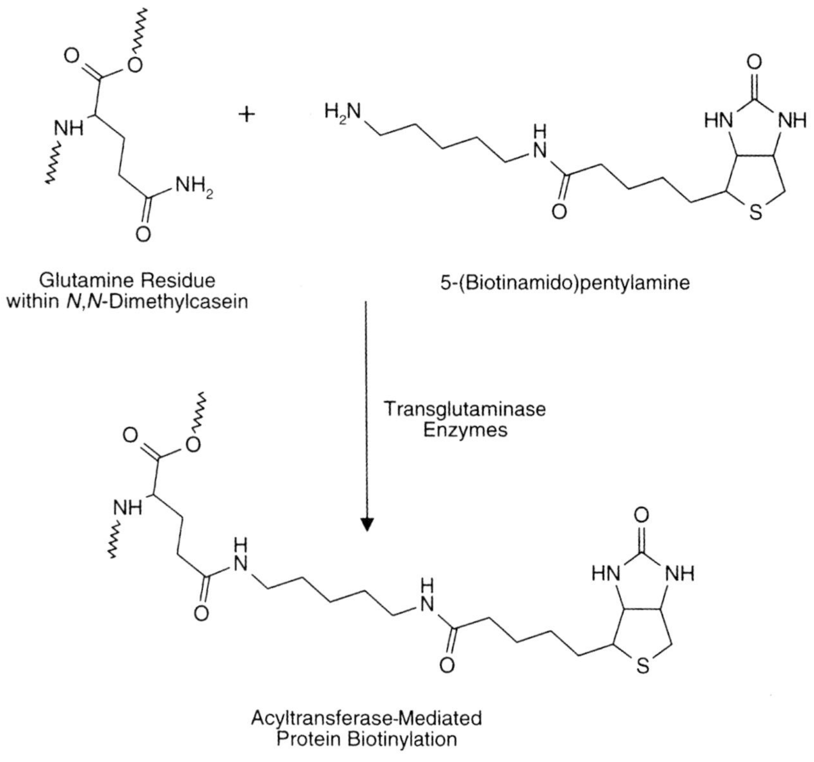

FIGURE 11.23 5-(Biotinamido)pentylamine can be used to label glutamine residues in proteins by enzymatic action of transglutaminase.

utilize the phenyl azide-type of photosensitive group, which can be activated by exposure to UV light to an intermediate nitrene or the nucleophile-reactive dehydroazepine (Chapter 3, Section 7.1, and Chapter 6, Section 3). However, additional photoreactive groups that are also useful include a psoralen ring system and a benzophenone group. A psoralen-based biotinylation agent is presented in this section, while a benzophenone-containing biotin compound with a water soluble PEG spacer is discussed in Chapter 18, Section 1.3.

Photobiotin

Perhaps the most common photoreactive biotin derivative is *N*-(4-azido-2-nitrophenyl)-aminopropyl-*N′*-(*N*-d-biotinyl-3-aminopropyl)-*N′*-methyl-1,3-propanediamine, simply called photoactivatable biotin or photobiotin (Forster *et al.*, 1985) (Thermo Fisher). The compound contains a 9-atom diamine spacer group on the biotin valeric acid side chain at one end, while the other end of the spacer terminates in an aryl azide reactive group. The presence of a nitro group on the phenyl azide ring allows for photoactivation at higher UV wavelengths approaching the visible region of the spectrum, thus avoiding potential breakdown of biological molecules through UV exposure. Photolyzing with light at a wavelength of 350 nm causes rapid activation with nitrene formation. The nitrene can couple to replaceable hydrogen sites in target molecules, add to double bonds within van der Waals distance, or undergo ring expansion to the dehydroazepine. If ring expansion occurs, the principal target group for coupling is a nucleophile, such as a primary amine (Figure 11.24).

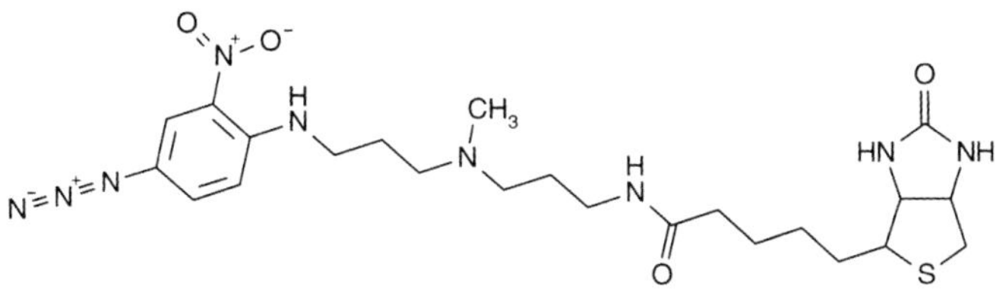

Photobiotin
N-(4-azido-2-nitrophenyl)-aminopropyl-
N′-(*N*-d-biotinyl-3-aminopropyl)-*N′*-methyl-1,3-propanediamine
MW 533.65
30.0 Å

Photobiotin has been used to biotinylate numerous macromolecules, including proteins and nucleic acids. The biotinylation of alkaline phosphatase was carried out with complete retention of activity (Forster *et al.*, 1985). Tubulin was labeled with photobiotin and detected on dot blots down to a level of 10 pg of sample using an avidin–enzyme conjugate (Lacey and Grant, 1987). DNA and RNA were labeled for use in hybridization assays (Forster *et al.*, 1985; Keller, 1989). For instance, photobiotin-modified probes have been used to detect flavivirum RNA in infected cells (Khan and Wright, 1987), to detect single-copy genes and low-abundance mRNA (McInnes *et al.*, 1987), for the diagnosis of barley yellow dwarf virus (Habili *et al.*, 1987), to assay luteinizing hormone β mRNA in individual gonadotropes (Childs *et al.*, 1987), to perform DNA mapping using a cross-hybridization technique (Chetrit *et al.*, 1989), to create patterns of fluorescent streptavidin molecules on dextran-coated surfaces (Ahn *et al.*, 2011), and to form multilayer films on titanium dioxide substrates (Weng *et al.*, 2011).

Photobiotin can be dissolved in water or buffer at a concentration of 1 mg/ml and stored in the dark at −20°C until needed. As long as no exposure to light is permitted, the compound is stable for at least 1 year under these conditions.

The protocol for modifying DNA probes with photobiotin can be found in Chapter 23, Section 2.3. It is based on the method of Forster *et al.* (1985). The following method is a suggested protocol for the modification of proteins using a photoreactive biotin derivative. Some optimization may be necessary to obtain the best incorporation levels.

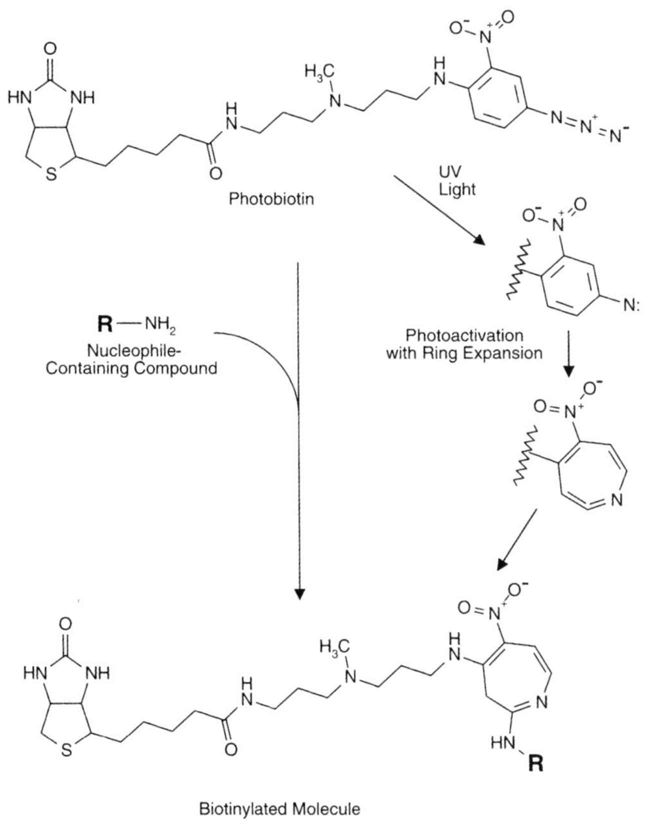

FIGURE 11.24 Photobiotin can be made to couple spontaneously with nucleophiles by exposure to UV light. The phenyl azide ring undergoes ring expansion to a highly reactive dehydroazepine intermediate, which can react with amines.

Protocol

1. Dissolve the protein to be biotinylated at a concentration of at least 1 mg/ml in water or dilute buffer at neutral pH.
2. In subdued light, dissolve photobiotin (Thermo Fisher) in water at a concentration of 1 mg/ml.
3. Add a quantity of photobiotin solution to the protein solution to give at least a 5-fold molar excess of biotinylation reagent.
4. Place in an ice bath and irradiate from above (about 10 cm away) for 15 min using a sunlamp (such as Philips Ultrapnil MLU 300 W, General Electric sunlamp RSM 275 W, or National Self-Ballasted BHRF 240–250 V 250 W W-P lamp).
5. Remove excess photobiotin by dialysis or gel filtration using a desalting column.

Psoralen–PEG$_3$–Biotin

Psoralen–PEG$_3$–biotin is a photoreactive biotinylation reagent containing a psoralen group at one end and a triethylene glycol (PEG-based) spacer in the middle (Thermo Fisher). This compound is water soluble due to the presence of the hydrophilic PEG arm. It is able to photo-insert into double-stranded DNA and to a lesser extent into double-stranded regions of RNA. The reaction occurs upon exposure to UV light in the range of 320 to 400 nm, which forms an excited triplet state intermediate that can insert in certain double-bond structures, especially at the 5,6-double bond of thymine bases.

The psoralen ring system can intercalate within double-stranded DNA or RNA and induce the formation of adducts with adjacent thymine bases (Stutz *et al.*, 2011) (Figure 11.25). The furan side and pyrone side of the tricyclic rings in psoralen can both form cycloaddition products with the 5,6-double bond of thymine residues, which results in crosslinks between the DNA strands with a PEG–biotin label sticking out.

Psoralen–PEG$_3$–biotin has been used to label double-stranded DNA for detection using (strept)avidin reagents (Henriksen *et al.*, 1991; Wygrecka *et al.*, 2007). The psoralen photoreactive group provides better insertion yields than typical phenyl azide-based systems, such as the standard photobiotin probe discussed previously in this section.

Protocol

1. Dissolve the DNA sample to be modified at a concentration of 20 to 100 μg/ml in 10-m*M* Tris, 1-m*M* EDTA, pH 7.4. Note: The sample may be heated to denature and solubilize genomic DNA and then cooled to form dsDNA for modification.
2. Dissolve the psoralen–PEG$_3$–biotin reagent in DMF at a concentration of 20-m*M* (use a fume hood). Protect from light.
3. Add a quantity of the psoralen–PEG$_3$–biotin solution to the DNA solution to result in a final concentration of 200 μ*M*. Mix well.
4. Expose the solution to long-wavelength UV light at about 365 nm (Philips TL 20 W/09 UV light works well) for 10 to 30 min. The solution may be cooled on ice to prevent heating during the irradiation process.
5. Precipitate the sample to remove unreacted biotinylation reagent by adding 0.1-*M* potassium acetate and ethanol (1 : 2 ratio). Centrifuge and wash the biotinylated DNA pellet with ethanol, then dry it under nitrogen. The purified sample may be dissolved in water or buffer.

6.6. Active Hydrogen-Reactive *p*-Aminobenzoyl Biocytin, Diazotized

p-Aminobenzoyl biocytin contains a 4-aminobenzoic acid amide derivative off the α-amino group of biocytin (Section 6.2, this chapter) lysine residue (Thermo Fisher). The aromatic amine can be treated with sodium nitrite in dilute HCl to form a highly reactive diazonium group (Figure 11.26), which is able to couple with active hydrogen-containing compounds. A diazonium reacts rapidly with histidine or tyrosine residues within proteins, forming covalent diazo

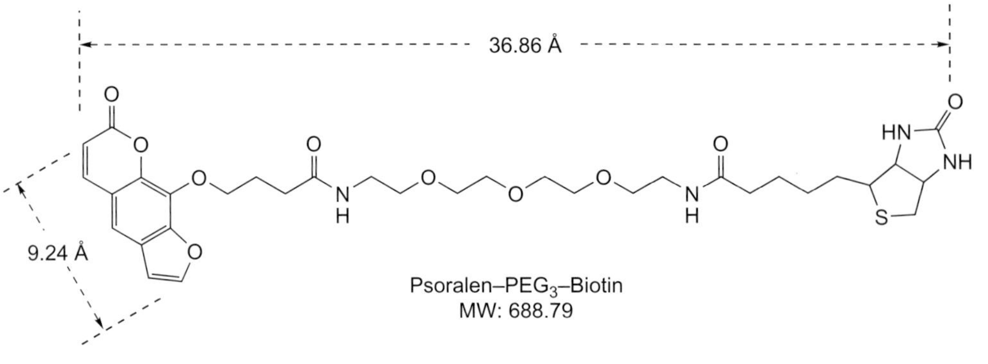

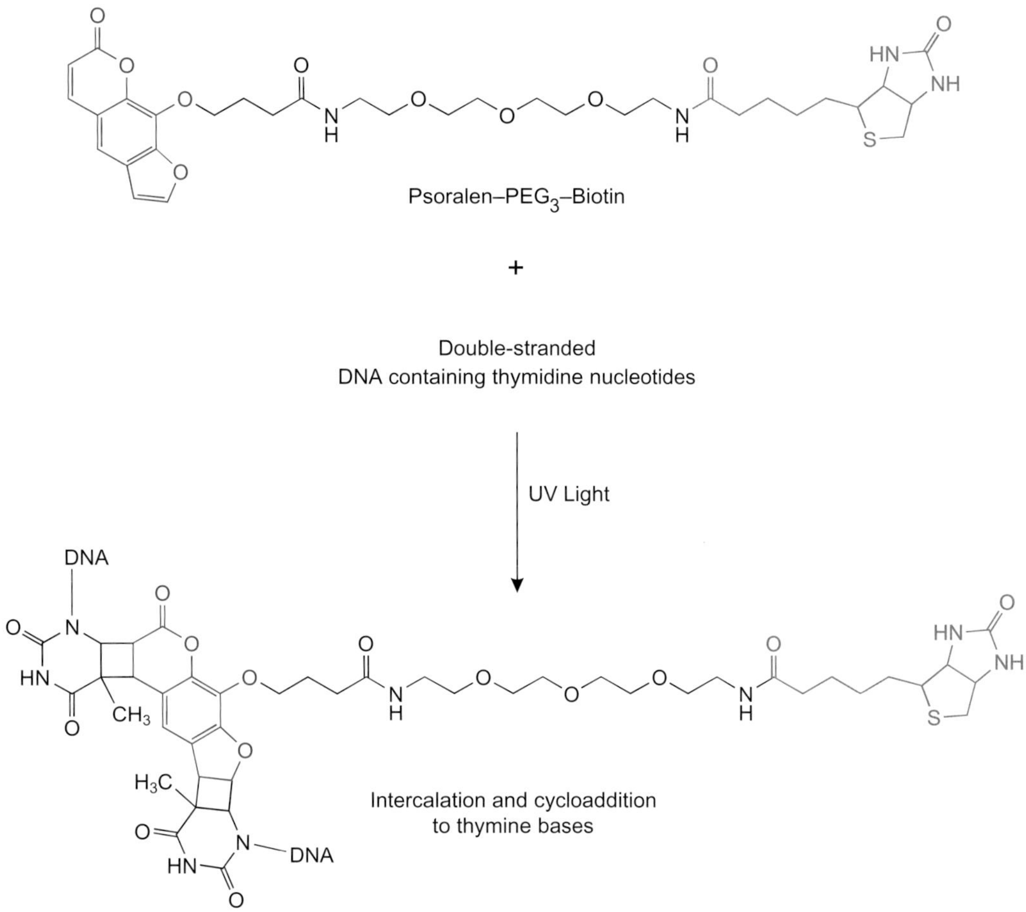

FIGURE 11.25 The photoreactive compound psoralen–PEO$_3$–biotin can intercalate into double-stranded DNA or RNA segments and covalently link to thymine bases *via* a photoreaction process.

bonds (Wilchek *et al.*, 1986) (Figure 11.27). It can also react with guanidine residues within DNA at position eight of the base (Rothenberg and Wilchek, 1988) (Figure 11.28). Biotinylation *via* diazo linkages is reversible by treatment with a 10-fold molar excess of $Na_2S_2O_4$ (sodium dithionite) in 50-m*M* Tris, pH 8.5 (Gorecki *et al.*, 1971) (Chapter 3, Section 6.1, and Chapter 5, Section 9).

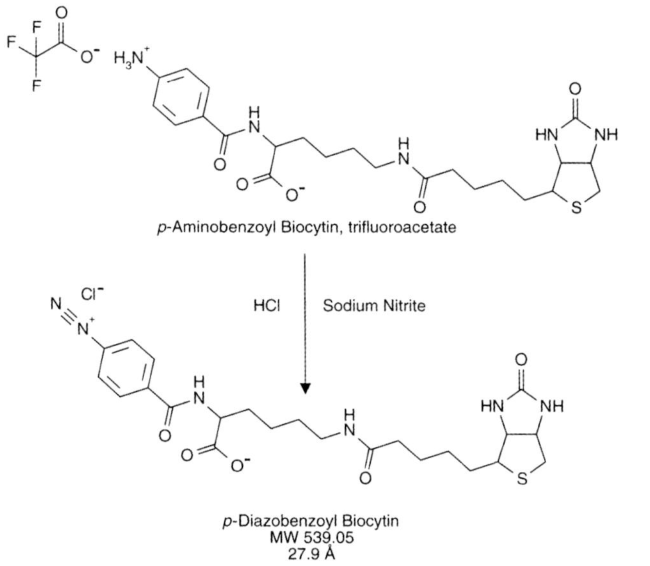

FIGURE 11.26 The aminophenyl group of this biotin derivative can be transformed into a diazonium reactive group by treatment with sodium nitrite in dilute HCl.

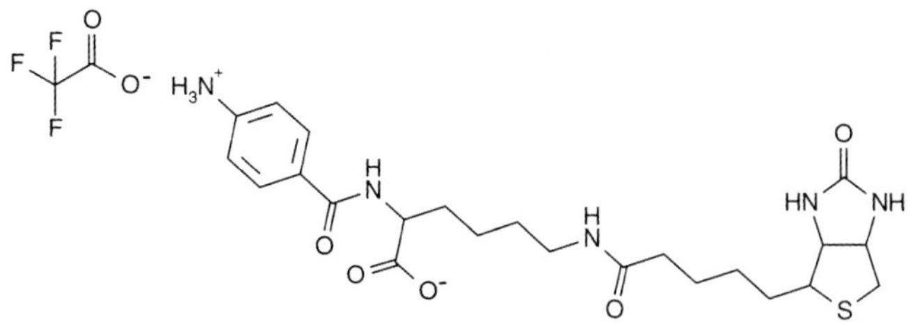

p-Aminobenzoyl Biocytin,
trifluoroacetate salt
MW 605.63

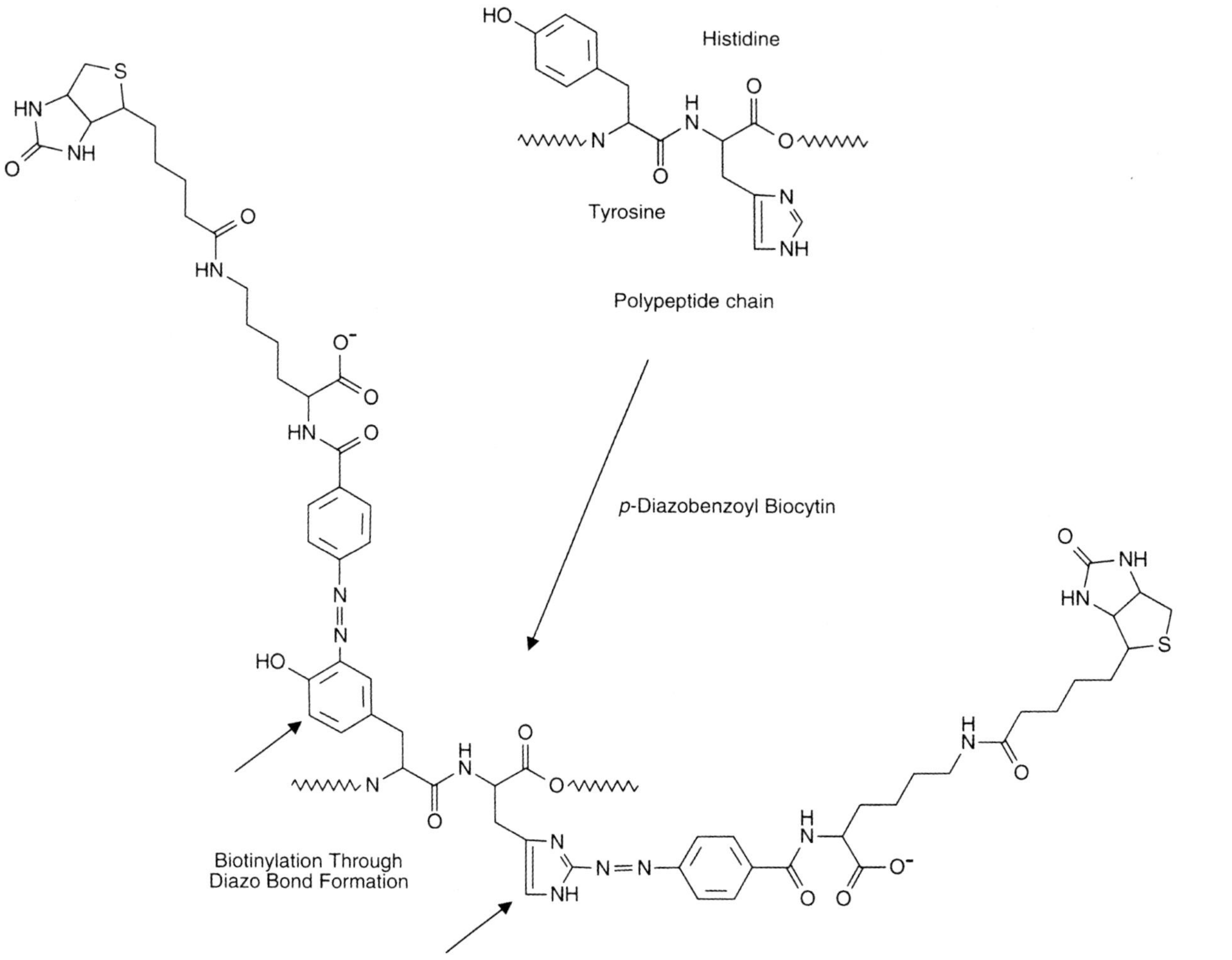

FIGURE 11.27 The diazonium group of *p*-diazobenzoylbiocytin can react with tyrosine or histidine residues in proteins to form diazo bonds.

The following procedure describes the process for creating a diazonium derivative of *p*-aminobenzoyl biocytin along with the subsequent coupling of the activated species to a protein or a nucleic acid probe.

Protocol for Formation of the Diazonium Derivative

1. Dissolve 2 mg of *p*-aminobenzoyl biocytin (Thermo Fisher) in 40 µl of 1-*N* HCl (concentration of 50 mg/ml). Cool the solution on ice.
2. Dissolve 7.7 mg of sodium nitrite in 1 ml of ice-cold water. Prepare fresh.
3. Mix 40 µl of the *p*-aminobenzoyl biocytin solution with 40 µl of the sodium nitrite solution.
4. React for 5 min on ice to create the diazonium derivative.
5. Stop the reaction by the addition of 35 µl of 1-*N* NaOH. Use immediately for biotinylation.

Protocol for Biotinylation of Proteins on Blots Using the Diazonium Derivative of p-Aminobenzoyl Biocytin

1. Dilute the diazonium derivative of *p*-aminobenzoyl biocytin with 0.2-*M* sodium borate, pH 8.4, to a concentration of 10 µg/ml.
2. Transfer proteins onto a nitrocellulose membrane using any appropriate procedure, including dot blotting the protein solution onto the surface.
3. Incubate the membrane with the biotin derivative at a ratio of 1 ml/cc^3 of membrane.
4. React for 1 h at room temperature.
5. Wash the membrane thoroughly with 0.1-*M* Tris, 0.15-*M* NaCl, pH 7.5.
6. Block nonspecific sites on the membrane with an appropriate blocking component (such as BSA) and detect the biotinylated proteins using an avidin or streptavidin conjugate.

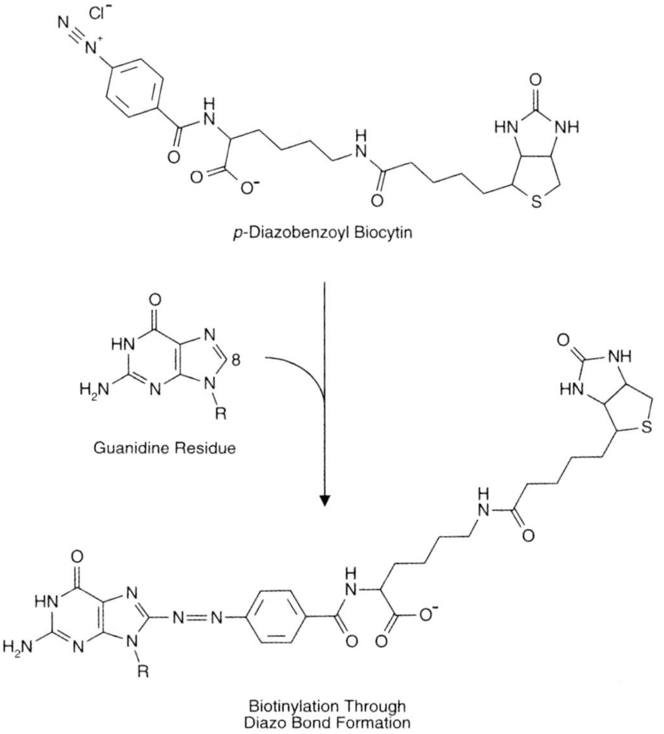

FIGURE 11.28 The diazonium group of p-diazobenzoylbiocytin can couple to the C-8 position of guanidine bases in nucleic acids, forming diazo bonds.

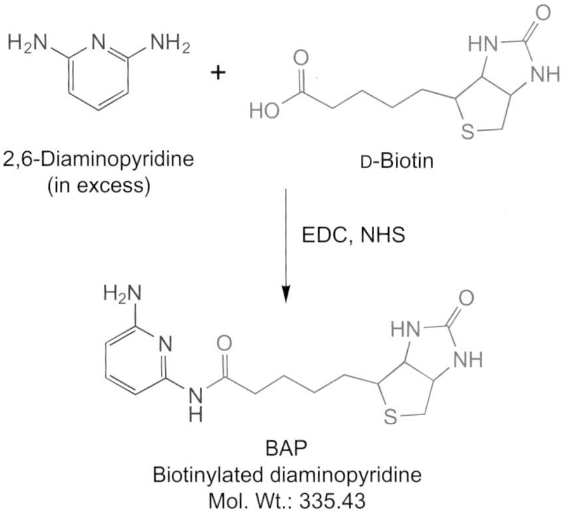

FIGURE 11.29 The synthesis of BAP can be done by the reaction of an excess of diaminopyridine with biotin in the presence of EDC and NHS.

6.7. Glycan Biotinylation Reagents

Biotinylated oligosaccharides are convenient probes of carbohydrate interactions, because the biotin label can be captured or detected using an avidin or streptavidin derivative. For instance, immobilized streptavidin can be used to purify glycoconjugates that have been labeled with a biotin group, potentially isolating glycoproteins or carbohydrate binding proteins (see Chapter 15). Enzyme-labeled or fluorescently labeled avidin or streptavidin can be used to probe for biotin-labeled carbohydrates in cells or tissue samples. In addition, a biotinylated glycan can be displayed on avidin or streptavidin to make an immunogen for developing specific antibodies to the carbohydrate.

Complex glycans on glycoproteins or other carbohydrate-bearing molecules can be modified with a biotinylation reagent using a number of reaction strategies. Oxidation with sodium meta periodate can be used to create aldehyde residues from diols on sugars, and this technique has been used to specifically modify sialic acids on glycans by reductive amination with a biocytin-hydrazide compound (see Section 6.4, this chapter) (Bayer *et al.*, 1988). Other procedures make use of released glycans from glycoproteins or other glycoconjugates, which contain reducing ends upon cleavage. The reducing ends can then be reacted with amine- or hydrazide-containing biotinylation compounds to couple with the open aldehyde group at the reducing end, thus forming a hydrazone linkage. The hydrazone bond may be reduced to stabilize the bond (recommended when reacting with amine-containing biotin compounds) or left as the unreduced hydrazone, which typically is carried out when coupling with hydrazide-biotin compounds. Alternatively, the reducing end of a carbohydrate can be reacted with an amine to form a glycosylamine derivative without opening the hemiacetal ring, thus better preserving the native structure of a glycan, which is important in some studies involving protein interactions.

A recent addition to the methods of glycan biotinylation makes use of the Staudinger ligation reaction with a phosphine–biotin derivative (see also Chapter 17, Section 6). Carbohydrates containing azide derivatives have been modified with this biotin compound to probe for glycoconjugates *in vivo*. This reaction is particularly useful for doing cell-based assays, because the ligation reaction is completely orthogonal to any biological reactions or interactions.

The following sections describe fluorescent biotinylation reagents that can be used to study carbohydrate function and interactions.

Biotinylated Aminopyridine

BAP (biotinylated aminopyridine or 2-amino-(6-amidobiotinyl)pyridine) is a derivative of D-biotin made by reacting the NHS ester of this vitamin with 2,6-diaminopyridine (DAP) in large molar excess, typically carried out using a carbodiimide EDC/NHS reaction (Figure 11.29). The resultant compound has fluorescent

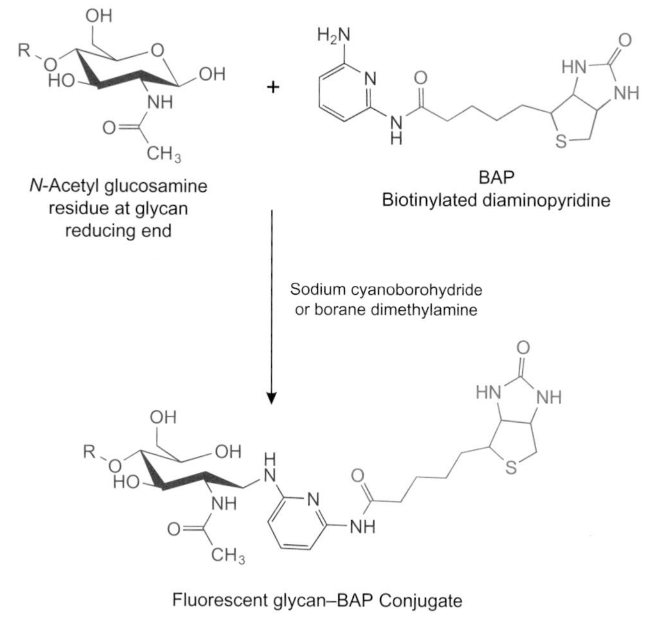

FIGURE 11.30 BAP can be used to label the reducing end of released glycans by reductive amination in the presence of a reducing agent.

properties due to the present of the aminopyridine ring, and its remaining free amine group may be used to modify reducing saccharides and glycans by reductive amination (Figure 11.30). BAP can be used to label oligosaccharides under mild conditions and without the carbohydrate structural degradation that results using periodate oxidation of carbohydrates. After modification, the glycans or carbohydrates can be analyzed by chromatography, electrophoresis, or mass spectrometry (Harvey, 2011; Nakano *et al.*, 2011).

Rothenberg *et al.* (1993) demonstrated the utility of BAP for highly sensitive fluorescence detection and separation of oligosaccharides by reverse-phase HPLC, with limits of detection down to about the 50-femtomole level (low picomole levels if using a cuvette reader with a 1-cm path length). Toomre and Varki (1994) subsequently published an improvement on the synthesis and use of the BAP reagent. In addition, the biotin group of BAP-labeled glycans can be used to create neoglycoproteins by interaction with tetrameric avidin or streptavidin molecules. The resultant glyco-complexes have been shown to be potent immunogens for evoking an IgG immune response in mice toward the glycan components (Srikrishna, 2001).

BAP-modified glycans also can be used to probe for receptors or binding proteins, which can then be detected by use of streptavidin conjugates or isolated by affinity chromatography on immobilized streptavidin or immobilized monomeric avidin (Thermo Fisher). The use of immobilized monomeric avidin is convenient, because the biotinylated glycans can be released by elution with acid pH or by using a solution containing biotin. BAP–carbohydrate adducts have been shown to be high affinity binders of both streptavidin and avidin, despite the relatively short spacer afforded by the diaminopyridine–biotin linker (Toomre and Varki, 1994).

The biotin group of BAP makes the compound somewhat hydrophobic, but attached to glycans the conjugates should display good water solubility due to the abundance of hydroxyl groups in addition to potentially having other charged groups on the sugars. BAP solubility in water was reported to be about 1 mg/ml, but with the addition of less than 1% DMSO, this solubility can be increased more than 10-fold (Toomre and Varki, 1994).

Fluorescence of the diaminopyridine group allows detection of conjugates down to the picomole range, with excitation and emission maxima at 345 nm and 400 nm, respectively. For detection of BAP and its conjugates, the optimal buffer environment is less than pH 5, because its fluorescent properties are pH dependent. A preferred buffer is sodium acetate at pH 4.

After BAP conjugation to saccharides or glycans, separation of the conjugates and unreacted BAP can be fluorescently followed using size exclusion chromatography (SEC) on a TSK-G3000PW column, which successfully resolves most of the lower molecular weight conjugate species, including single-sugar adducts through three-sugar carbohydrates. BAP–glycan conjugates containing more than three sugars elute early in the separation and do not resolve into discrete peaks, as smaller adducts do. Unreacted BAP elutes last in the SEC separation.

Alternatively, separations can be carried out by anion-exchange chromatography using a column packed with the HPLC support TSK-DEAE-2SW, which effectively resolves negatively charged carbohydrates, such as those containing sialic acid residues. BAP–glycan conjugates will elute according to their degree of negative charge character. Sulfate-containing sugars in general will interact more strongly with the matrix and have longer retention times than those containing only carboxylates.

BAP may be prepared by the reaction of NHS–biotin with DAP. The biotinylation compound is commercially available (Thermo Fisher) or it may be formed *in situ* by reaction of D-biotin with EDC and NHS. An optimized protocol for preparation of the reactive intermediate ester and the final BAP compound can be determined from Rothenberg *et al.* (1993) with modifications by Toomre and Varki (1994). Basically, a solution of 0.3-*M* DAP is prepared in 40 ml of 50-m*M* MES, pH 6.5, and 10 ml of 0.1-*M* D-biotin in DMSO is added. The reaction is initiated by the addition of EDC and NHS to a final concentration of 150-m*M* and 50-m*M*, respectively. The reaction is allowed to continue overnight at room temperature with mixing before purification of BAP on a C_{18} sample prep cartridge. The reaction mixture is applied to the cartridge and reaction byproducts and

DAP removed by washing with water and 10% acetonitrile. BAP is finally eluted in high purity by washing with 50% acetonitrile.

The conjugation of BAP to oligosaccharides can be achieved by the following protocol based on the method of Toomre and Varki (1994).

Protocol

1. Dissolve an oligosaccharide or glycan having a reducing end to be modified in 2:1 pyridine/glacial acetic acid (vol/vol) with a total reaction volume of 10 to 100 μl. If the carbohydrate initially is insoluble in the reaction solution, a prior dissolution in a minimal amount of DMSO or water can be carried out and then an aliquot transferred to the reaction medium.
2. Add to the solution a 50-fold molar excess of BAP over the estimated amount of carbohydrate present in the reaction mixture.
3. Heat at 80°C in a sealed Reactivial (Thermo Fisher) for 1 h.
4. Add to the reaction mixture an equal volume of the reducing agent borane-dimethylamine (BDA) complex, which was previously prepared in the reaction buffer at a concentration of 125 mg/ml.
5. React for another hour at 80°C.
6. Purify the BAP–glycan conjugate from unreacted glycans by use of a C_{18} sample prep cartridge, as described above for the synthesis of BAP. Separation of excess BAP from the conjugates may be carried out by SEC or anion-exchange chromatography, depending on the size of carbohydrates being modified and their intrinsic charge. Follow the separations by visualization of BAP fluorescence with a hand-held UV lamp or through the use of a fluorescence detector.

Biotinyl-L-3-(2-Naphthyl)-Alanine Hydrazide (BNAH)

BNAH is a biotin–hydrazide derivative containing a UV-absorbing and fluorescent naphthalene group (biotinyl-L-3-(2-naphthyl)-alanine hydrazide) (Leteux *et al.*, 1998). Unlike BAP described previously, BNAH has a hydrazide group for coupling to the reducing end of carbohydrates, instead of an amine. While both groups can be successfully conjugated to an aldehyde of a released glycan, the biotin–amine compounds require a reducing agent to stabilize the resultant hydrazone bond. BNAH may be coupled to reducing sugars without reduction, since the linkage formed between a hydrazide and an aldehyde is much more stable than that with an amine. The modified glycans or carbohydrates can be detected using the fluorescent properties of the naphthalene group or captured by immobilized streptavidin for further analysis (Harvey, 2011).

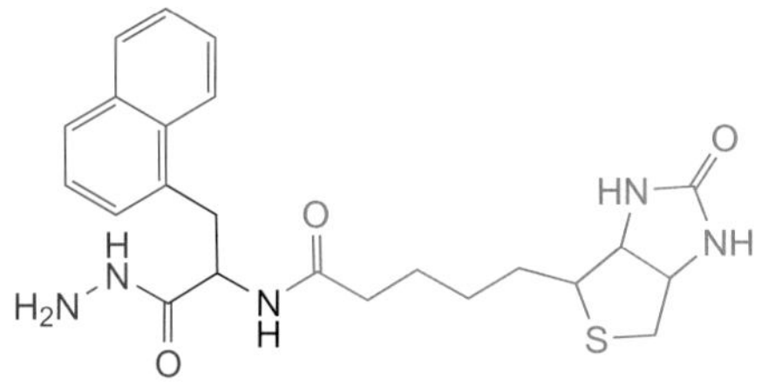

BNAH
Biotinyl-L-3-(2-naphthyl)-alanine hydrazide
Mol. Wt.: 455.57

In some cases, the ability to modify glycans at the reducing end without reduction preserves the carbohydrate's native structure sufficiently to allow interactions with proteins that would otherwise not interact if the bond were reduced. Therefore, depending on the ultimate use of the biotinylated carbohydrate, using a hydrazide-mediated conjugation process can have advantages over the use of amine–biotin compounds.

In the case of BNAH, however, it was determined that the resultant linkage with the reducing end of an oligosaccharide or glycan was not a hydrazone bond, but a glycosylhydrazide derivative, which preserves the pyranose ring structure of the sugar (Figure 11.31). This finding is the main reason a BNAH-modified glycan effectively displays a near-native conformation at the reducing end. In addition, the biotin label in this configuration is reversible and can be released by incubation under acidic conditions, thus allowing recovery of the carbohydrate.

Another advantage of the BNAH derivative is that the conjugation reaction with reducing sugars can be carried out in aqueous conditions and in an environment that permits carbohydrate and biotinylation reagent solubility.

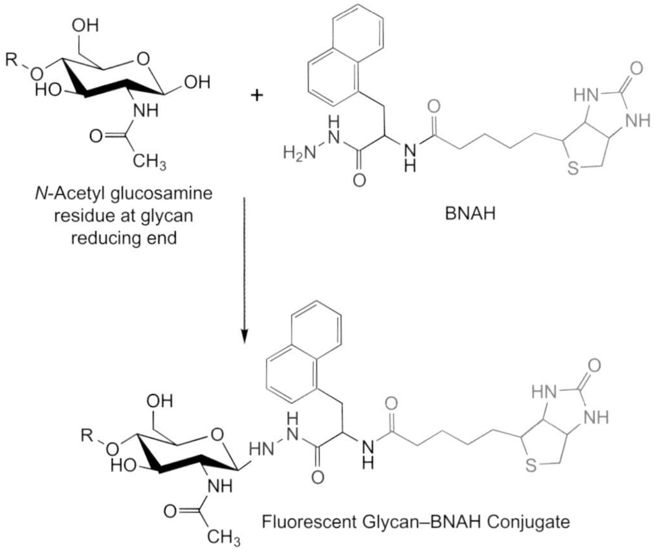

FIGURE 11.31 BNAH contains a hydrazide group that can be used to label the reducing end of released glycans through the formation of a hydrazone bond.

Modified carbohydrates may be stored for at least one year at −20°C in a solution of water/methanol (9:1, v/v) without degradation.

The following protocol is based on the method of Leteux *et al.* (1998).

Protocol

1. Dissolve 100 nmol of the carbohydrate to be modified and 500 nmol of BNAH in 25 μl of methanol (20-m*M* BNAH solution).
2. Evaporate the solution to dryness and re-dissolve in 25 μl an acidic solution of methanol/water/acetic acid (74:8:8, v/v) or in a neutral solution consisting of methanol/water (9:1, v/v), depending on the relative solubility of the carbohydrate.
3. React for at least 5 h at 60°C if using the acidic reaction solution or for a total of 16 h at 60°C if using the neutral solution.
4. Purification of the conjugates may be achieved by reverse-phase HPLC separation. Dry the reaction solution under a nitrogen stream and reconstitute in a minimum volume of acetonitrile/water (1:1, v/v). Apply the sample to a 5-μm C_{18}-silica HPLC column (250 × 4.6 mm, Nucleosil). Elute with a gradient of water to acetonitrile at a flow rate of 1 ml/min over a time course of 30 min. Free BNAH and BNAH–glycan derivatives can be monitored by absorbance at 275 nm. The conjugate peak will also be positive for carbohydrate by reaction with orcinol, which can be detected by spray after spotting a small eluted sample on a TLC plate.

Biotin–PEG$_3$–Phosphine

Saxon and Bertozzi (2000) reported on the synthesis and use of a novel biotinylation compound containing a phosphine group for coupling to azide containing molecules. The reagent has a biotin handle at one end, a triethylene-glycol (PEG) diamine spacer imparting increased water solubility in the middle, and a 3-(diphenylphosphino)-4-(methoxycarbonyl)benzamide group on the other end. Biotin–PEG$_3$–phosphine reacts with azide derivatives of amino acids, sugars, and cross-linkers to form an intermediate aza-ylide, which spontaneously rearranges in aqueous solution to create a stable amide bond (Figure 11.32).

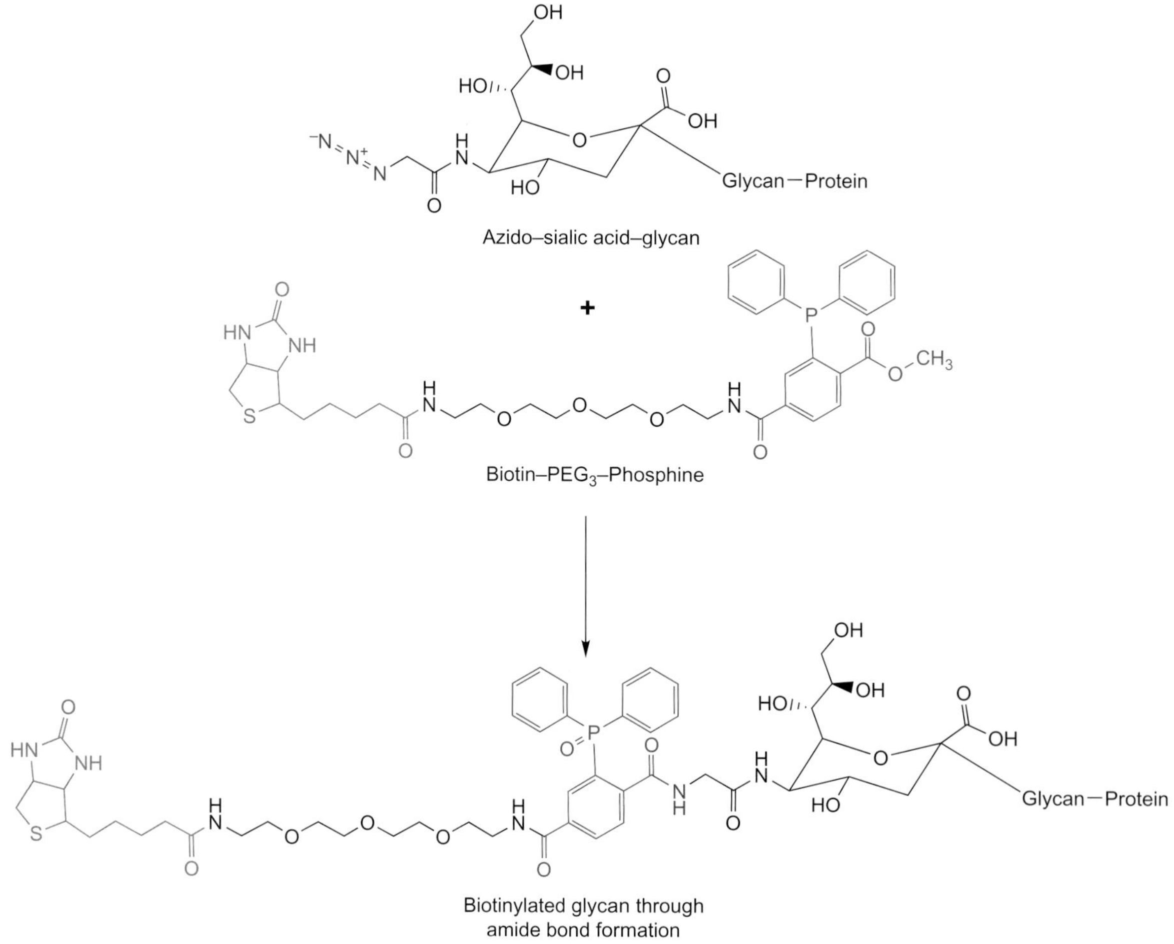

FIGURE 11.32 Azido–sialic acid-containing glycans can be labeled *in vivo* with biotin–PEG–phosphine using the Staudinger ligation reaction, which forms an amide bond.

This reaction is a modified Staudinger ligation that can be used to target azide-containing glycans or proteins *in vivo*.

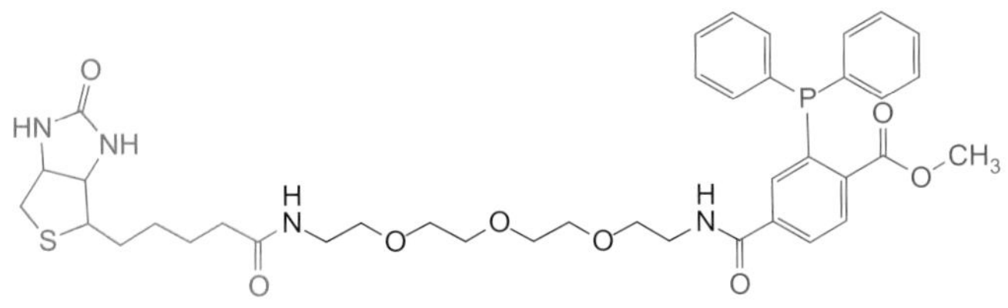

Biotin–PEG_3–Phosphine
Mol. Wt.: 764.87

Cells grown in the presence of azide analogs of certain amino acids or sugars will incorporate these derivatives into proteins or carbohydrates through enzymatic synthesis using the native cell machinery. Azides thus displayed on biomolecules are unreactive with other substances typically found within the cell, but the azide derivatives may be targeted for conjugation using the modified Staudinger reaction (see Chapter 17, Section 6, for additional information on this reaction and its use in biological labeling).

Biotin–PEG_3–phosphine can be used to label glycans or proteins that have been modified to contain azide groups. The beautiful specificity of this reaction and its lack of toxic side reactions or additives make it suitable for use in living organisms, including cells and animals. The reaction is completely orthogonal to functional groups and reactions found in living systems, so the biotinylation process proceeds with no cross-reactions with other biomolecules. In addition, no cell toxicity has been observed due to the phosphine or the phosphine oxide byproduct of the coupling reaction. The phosphine also does not appear to be capable of reducing disulfides within proteins.

Prescher *et al.* (2004) have shown that mice fed with the peracetylated azido–mannose sugar derivative Ac_4ManNAz efficiently convert it through deacetylation by cytosolic esterases into an azido–sialic acid (SiaNAz), which then gets incorporated enzymatically into cell surface glycans. Biotinylation of these aberrant carbohydrates provides a method of detecting or isolating glycoproteins or other glycoconjugates (Vainauskas *et al.*, 2012). For instance, fluorescently labeled streptavidin can be used to image the cell surface structures containing the biotinylated azido–sugar derivatives. Alternatively, immobilized streptavidin or immobilized monomeric avidin can be used to purify biotinylated azido–glycoconjugates after cell lysis and provide a method for studying glycoprotein function and interactions.

The following protocol is based on the methods of Saxon and Bertozzi (2000) and Prescher *et al.* (2004).

Protocol

1. Treat and grow cells in the presence of 20-μ*M* Ac_4ManNAz for at least 3 days. The azide–mannose derivative may be solubilized in 70% DMSO as a more concentrated stock solution and then an aliquot added to the media containing the cells. Alternatively, mice may be treated with the azide–sugar derivative in aqueous DMSO at a level of 100 to 300 mg/kg, using an injection of 200 μl administered intraperitoneally daily for 7 days.
2. When working with cells, first wash them several times with PBS, pH 7.4, to remove any remaining Ac_4ManNAz from the media. Alternatively, if working with animals, isolate the tissue type desired and prepare the cells to be labeled in PBS, pH 7.4.
3. To label cell surface azide–glycans, the cells are reacted for 1 h using a final concentration of 1-m*M* biotin–PEG_3–phosphine reagent dissolved in PBS, pH 7.4.
4. Wash the cells several times with PBS, pH 7.4, to remove excess biotinylation compound.
5. The labeled glycans may be analyzed by cell sorting after staining with fluorescently modified streptavidin. Alternatively, the cells may be lysed and the labeled glycans isolated using immobilized streptavidin.

Additional biotinylation reagents are discussed in Chapter 18, Section 1.3, which describes newer hydrophilic biotin compounds containing PEG spacers. These reagents offer significant advantages over the more traditional aliphatic compounds, because the pronounced water solubility of the PEG cross-bridge can prevent aggregation of biotinylated molecules. With some longer chain biotin compounds that contain hydrophobic hydrocarbon spacers, proteins can precipitate or lose activity over time due to the insolubility of the biotin modifications. Biotinylated antibodies are particularly susceptible to aggregation and loss of antigen binding ability if they are modified using hydrophobic biotin compounds. The PEG-based biotin reagents show better solubility and longer stability than their corresponding aliphatic biotinylation compounds of equivalent size.

CHAPTER

12

Isotopic Labeling Techniques

1. BIFUNCTIONAL CHELATING AGENTS AND RADIOIMMUNOCONJUGATES

Monoclonal antibodies provide extremely high antigen specificity that can be useful as cancer-targeting and therapeutic reagents *in vivo* (Waldmann, 1991; Dougan and Dranoff, 2012; Modjtahedi *et al.*, 2012). Radiolabeled monoclonals are currently undergoing developmental clinical trials for their use in the diagnosis or treatment of cancer. The antigen binding specificity of the radioimmunoconjugate provides the targeting capability to localize in tumor sites, while the associated radiolabel provides cytotoxic properties or detectability for imaging applications (Schlom, 1986) (see also Chapter 1, Section 3.5).

Iodine-131 was among the first radioactive isotopes used for radioimmunoconjugate preparation (Order, 1982; Regoeczi, 1984). Since the earliest studies on the efficacy of radiotherapy, additional isotopes have been employed, such as iodine-125, bismuth-212, yttrium-90, yttrium-88, technetium-99m, copper-67, rhenium-188, rhenium-186, galium-66, galium-67, indium-111, indium-114m, indium-115, and boron-10.

There are several methods commonly used to label monoclonal antibodies with radionuclides. In a direct labeling process, a radioactive atom is attached to functional groups on the antibody without the use of an intervening chemical spacer. For instance, radioiodination can be done through modification of tyrosine side chains using established techniques and reagents (see Section 2). Another direct method uses indigenous sulfhydryl groups or those formed through disulfide reduction to couple covalently certain metal nuclides (Holmberg and Meurling, 1993; Ranadive *et al.*, 1993). Thiolation reagents also can be used in this regard to create the requisite SH groups by modification of other protein functional sites (Joiris *et al.*, 1991) (Chapter 2, Section 4.1).

Indirect methods of protein labeling with radiolabels utilize organic compounds able to chelate metal ions in a coordination complex. Bifunctional chelating agents (BCAs), as they are called, contain a reactive group for coupling to proteins or other molecules and a strong metal chelating group for complexing certain radioactive metals (Lattuada *et al.*, 2011; Maruk *et al.*, 2011). Their extensive use with monoclonal antibodies that are able to target specific cellular antigens has resulted in important radiopharmaceutical applications for the diagnosis and treatment of cancer (Wessels and Rogus, 1984; Meares, 1986; Otsuka and Welch, 1987; Hnatowich, 1990; Liu and Wu, 1991; Subramanian and Meares, 1991). The BCAs may be loaded with the radioactive metal before or after their conjugation with a monoclonal antibody (Frytak *et al.*, 1991). If they are loaded with radionuclides prior to modifying the antibody, the BCA–metal pair is called a preformed complex (Kasina *et al.*, 1991).

The following sections describe the major methods and BCAs used to create radioimmunoconjugates.

1.1. DTPA

DTPA is diethylenetriaminepentaacetic anhydride, a bifunctional chelating agent containing two amine-reactive anhydride groups. The compound reacts with N-terminal and ε-amine groups of proteins to form amide linkages. The anhydride rings open to create multivalent, metal-chelating arms able to bind metals tightly in a coordination complex (Figure 12.1) (Hnatowich *et al.*, 1982). Metal-chelate bonds are created through dative interactions with the unshared pair of electrons on each oxygen and nitrogen atom on DTPA, thus creating the potential for eight coordination sites with metal ions (from three nitrogens and five oxygens). Targeting molecules modified with DTPA can be used to carry radioactive or nonradioactive metal ions for *in vivo* imaging purposes, including PET/CT scanning and high-contrast imaging (Krausz *et al.*, 2011; Nayak *et al.*, 2011).

Optimal reaction conditions for antibody modification with DTPA are neutral to slightly alkaline pH environments containing no extraneous amines. A pH of 7 to 8 may be used with buffering provided by phosphate or bicarbonate buffers at 0.1-*M*. Since two anhydride groups are present on each DTPA molecule there is potential for creating crosslinks between two

Bioconjugate Techniques, Third Edition.
DOI: http://dx.doi.org/10.1016/B978-0-12-382239-0.00012-1

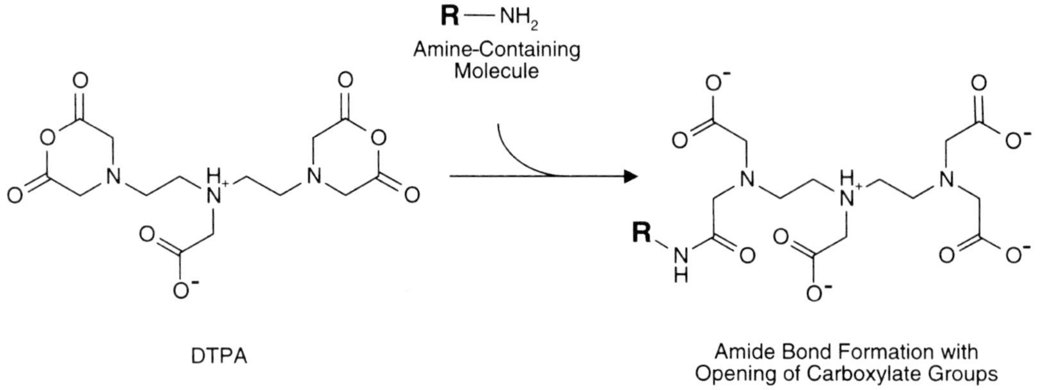

FIGURE 12.1 DTPA reacts with amine-containing molecules via ring opening of its anhydride groups to create amide bond linkages. The potential also exists for both anhydride groups to react and cause crosslinking of modified molecules, which is undesirable.

amine-containing molecules. Conjugation of antibody through DTPA crosslinks may be a major reason some immunoglobulins lose antigen-binding activity after modification (Lanteigne and Hnatowich, 1984). Optimization of the amount of protein present and the quantity of DTPA added to the reaction may have to be done to avoid this type of crosslinking and polymerization.

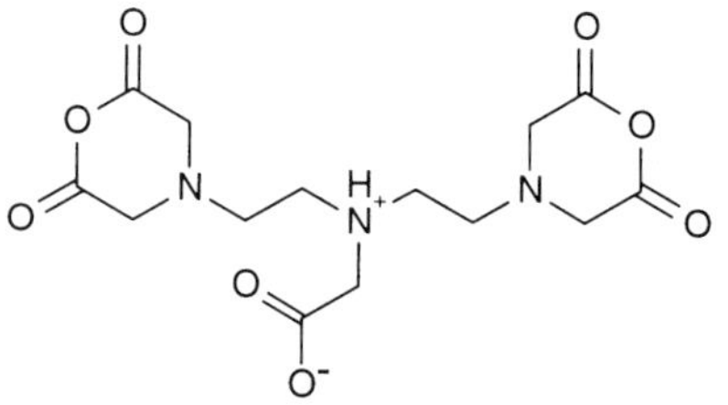

DTPA
Diethylenetriamine-
pentaacetic anhydride
MW 357. 33

DTPA also can be used to modify amine-containing polymers, such as poly-L-lysine, to create a chelating polymer possessing multiple metal binding sites (Trubetskoy *et al.*, 1993). Subsequent polymer modification of antibodies provides much higher radioactivity per molecule than if DTPA is directly coupled to the protein. Such chelating polymers can introduce into proteins as many as 100 DTPA residues, each one able to hold one radiolabel, for each 55,000-Dalton poly-L-lysine chain modified (Torchilin *et al.*, 1993). Directed coupling to the antibody through only the N-terminal of the poly-L-lysine chain limits the modification to a single point along the polymer, thus avoiding crosslinking or multi-site attachment that can affect antibody activity (Slinkin *et al.*, 1991). This approach can dramatically increase the radioactivity level at tumor sites, thus increasing cytotoxicity or enhancing imaging capability. Similar DTPA–polymer constructs can be formed using PAMAM dendrimers (Chapter 8), which can provide an amine-containing scaffold that can hold dozens of chelating groups, depending on its size.

While DTPA has been used extensively as a BCA to prepare radiopharmaceutical reagents, newer metal chelators such as those discussed below may show greater promise for *in vivo* applications.

1.2. DOTA, NOTA, and TETA

Cyclic chelating compounds have been developed that have both nitrogens and carboxylates and are able to tightly coordinate metal ions useful for *in vivo* imaging in diagnostics and drug discovery (Brechbiel, 2008; Stasiuk *et al.*, 2012). Such metal chelates attached to antibodies and other targeting agents have widespread utility in the major imaging modalities, including MRI (chelates of Gd), PET (chelates of ^{64}Cu and ^{68}Ga), and SPECT (using ^{99m}Tc). Three of the most commonly used cyclic, bifunctional chelating agents are DOTA, NOTA, and TETA.

DOTA is 1,4,7,10-tetraazacyclododecane-*N*,*N′*,*N″*, N‴-tetraacetic acid, a chelating ring structure containing four acetic acid carboxylate groups off the four nitrogens of its 12-atom cyclic structure. C- or N-functionalized derivatives of this basic structure produce a BCA capable of modifying proteins and binding radioactive metal ions in strong coordination complexes of up to eight dative bonds (Cox *et al.*, 1990; Renn and Meares, 1992). Perhaps the simplest method of DOTA functionalization is through modification of one of its carboxylates to contain a short spacer terminating in a reactive group capable of being coupled to proteins (Li and Meares, 1993). However, modification of carbons on its cyclic backbone also is possible, which preserves all of the carboxylates for coordinating with a metal ion (Moi and Meares, 1988; Brechbiel *et al.*, 1993). Complexes of metal ions with DOTA have been studied in detail (Sherry *et al.*, 1989; Aime *et al.*, 1992).

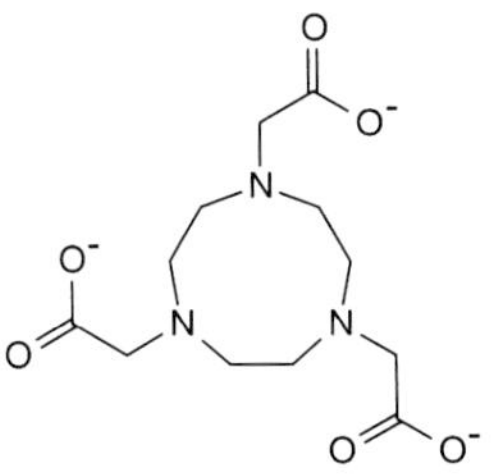

NOTA
1,4,7-Triazacyclononane-*N*, *N'*, *N''*-triacetic acid

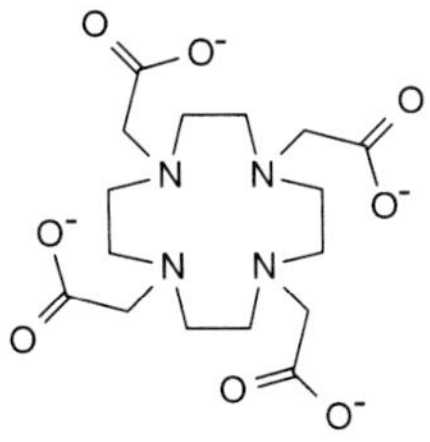

DOTA
1,4,7,10-Tetraazacyclododecane-*N*, *N'*, *N''*, *N'''*-tetraacetic acid

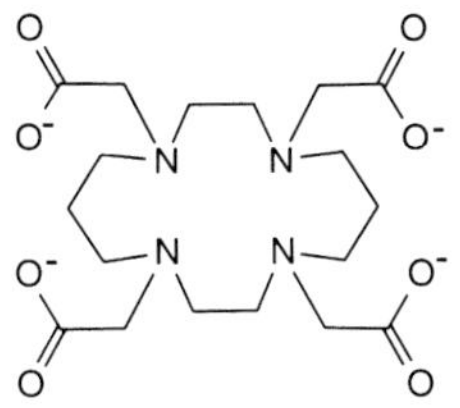

TETA
1,4,8,11-Tetraazacyclotetradecane-*N*, *N'*, *N''*, *N'''*-tetraacetic acid

A BCA similar to DOTA is NOTA, which is 1,4,7-triazacyclononane-*N*,*N'*,*N''*-triacetic acid. NOTA contains a smaller ring structure in comparison to DOTA and only has three chelating carboxylate groups and three nitrogens, creating a maximal coordination potential for six dative bonds with metal ions. Some reports indicate that NOTA is superior to DOTA (Zhang *et al.*, 2011). Synthesis of functional derivatives can be done through C- or N-modifications, creating reactive groups able to couple to proteins and other molecules (Cox *et al.*, 1990). Cox and co-workers (1990) have prepared an (S)-lysine derivative of a benzamide-protected, C-substituted NOTA. Since coupling antibodies with NOTA through an amide linkage leaves only two free carboxylic acids, potentially making the reagent pentadentate for chelating purposes, metal complexes formed with this reagent may have lower stability than those containing greater numbers of chelating groups.

A third compound in the same category as DOTA and NOTA is TETA, which is 1,4,8,11-tetraazacyclotetradecane-*N*,*N'*,*N''*,*N'''*-tetraacetic acid. TETA contains a larger ring structure than the other two BCAs and has four chelating carboxylate groups and four nitrogens. Similar to the other two chelators, C- and N-functionalized derivatives can be prepared with TETA (Brechbiel *et al.*, 1993). In addition, a novel *p*-bromoacetamidobenzyl–TETA derivative could be used to label antibodies through sulfhydryl groups and securely bind radioactive copper for probing biological systems *in vivo* (Moi *et al.*, 1985).

1.3. DTTA

DTTA is N^1-(*p*-isothiocyanatobenzyl)-diethylenetriamine-N^1,N^2,N^3,N^3-tetraacetic acid. This BCA contains four carboxylate groups and three nitrogens that can hold metals tightly in a coordination complex of seven dative bonds. The compound is especially good at chelating lanthanide-series elements, such as europium, samarium, terbium, and dysprosium (Appel *et al.*, 2011). Unlike the previous bifunctional chelating agents which are used to prepare radiopharmaceutical reagents, this one is used primarily for complexing metals to form fluorescent probes for time-resolved fluoroimmunoassays (Hemmila, 1988) (see Chapter 10, Section 9). The most commonly used lanthanides for this purpose are europium (Eu^{3+}), terbium (Tb^{3+}), and samarium (Sm^{3+}). Proteins modified with DTTA and complexed with lanthanide metal ions form the basis for unique fluorescent probes possessing long-lived signals upon excitation.

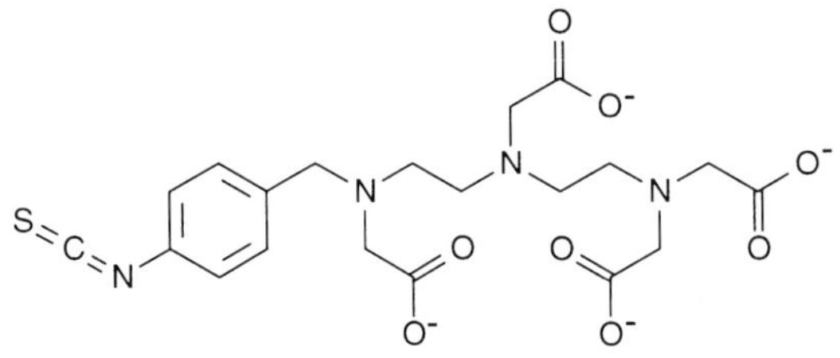

DTTA
N^1-(*p*-isothiocyanatobenzyl)-diethylenetriamine-N^1, N^2, N^3, N^3-tetraacetic acid

The isothiocyanate group of DTTA reacts with primary amines in proteins and other molecules to form stable thiourea bonds (Figure 12.2) (Mukkala *et al.*, 1989). The reagent is water soluble and can be reacted under relatively mild conditions (in 0.1-*M* sodium carbonate, pH 9.0). The isothiocyanate group is reasonably stable in aqueous solution for short periods, but will degrade. Best results will be obtained if fresh DTTA is used. The reaction involves attack of the nucleophile on the central, electrophilic carbon of the isothiocyanate group. The resulting electron shift creates a thiourea linkage between the chelating compound and the protein with no leaving group. Modification of antibodies can be done without loss of significant antigen binding activity, even when up to 10 to 15 DTTA chelates are substituted per immunoglobulin molecule (Stahlberg *et al.*, 1993). Diagnostic assays using DTTA–Eu^{3+} chelates are commercially available employing the DELFIA® (Wallac Oy, Turku, Finland) time-resolved fluoroimmunoassay system.

1.4. DFA

DFA or deferoxamine is *N'*-[5-[[4-[[5-(acetylhydroxyamino)pentyl]amino]-1,4-dioxobutyl]hydroxyamino]pentyl]-*N*-(5-aminopentyl)-*N*-hydroxybutanediamide, a naturally occurring product isolated from *Streptomyces pilosus*. Its native activity is forming iron complexes, but it is also very proficient at forming coordination

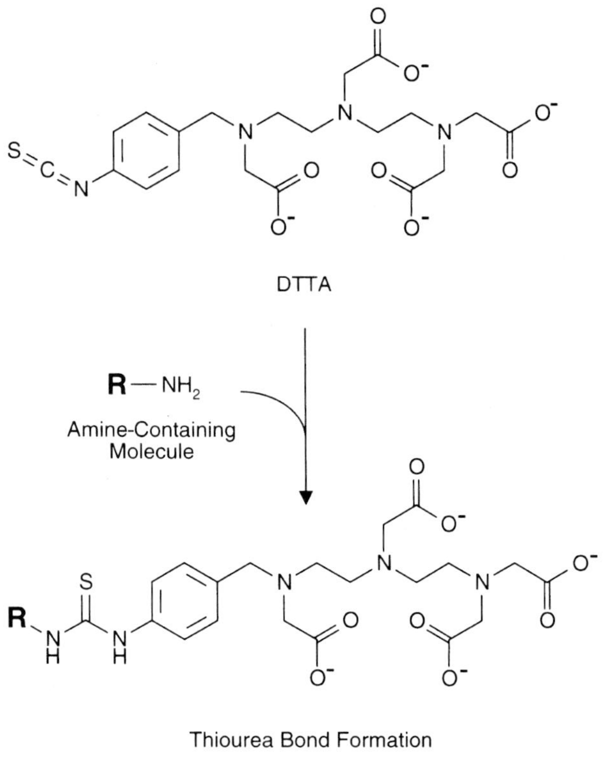

FIGURE 12.2 The isothiocyanate group of DTTA can react with amine-containing molecules to form isothiourea bonds.

chelates with other metals, particularly lanthanides such as galium-68 (Motta-Hennessy *et al.*, 1985) and ^{99m}Tc (Maruk *et al.*, 2011). Radiopharmaceutical agents for imaging can be produced by modification of targeting molecules with DFA complexes containing radioactive metals. The amine group of DFA can be utilized for direct labeling of antibodies and other proteins through coupling with available carboxylates using carbodiimide-mediated conjugation with EDC (Chapter 4, Section 1.1). The use of amine-reactive, homobifunctional crosslinkers (Section 2.2) also can be employed to modify proteins with DFA at their amino groups.

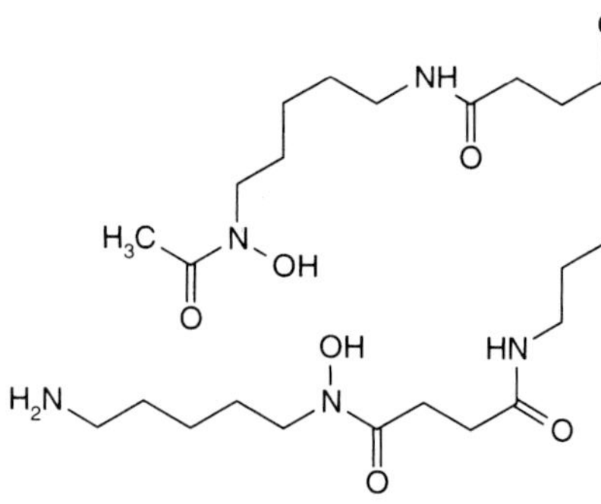

DFA
Deferoxamine
MW 560.71

DFA–polymer conjugates can be made containing multiple chelating groups along the length of the polymer. The polymer backbone utilized for this synthesis can be either activated dextran (Torchilin *et al.*, 1989) or succinylated poly-L-lysine (Slinkin *et al.*, 1990; Torchilin *et al.*, 1993). These DFA–polymer constructs can be attached to antibody molecules through additional functional groups on the polymer. Chelating polymers can provide much higher signals than direct attachment of DFA to antibodies, since each coupled polymer derivative can possess dozens of chelated radioactive metals. In addition, high substitution levels of DFA directly coupled to antibodies can significantly affect activity by denaturation or blocking of the antigen binding sites.

1.5. Use of Thiolation Reagents for Direct Labeling to Sulfhydryl Groups

Proteins containing sulfhydryl residues can be labeled with a radioactive element by direct coordination to the SH group through a dative bond (Chapter 3, Section 2.8), avoiding entirely the use of a bifunctional chelating agent. Particularly, reduced sulfhydryls in antibody molecules can be coupled with ^{99m}Tc to yield thiol–metal derivatives (Thakur and DeFulvio, 1991; Rhodes, 1991). However, cleavage of disulfide linkages within the antibody can lead to activity losses and fragmentation (Pimm *et al.*, 1991). The required sulfhydryl groups can be introduced into antibodies without disulfide reduction through the use of a thiolation reagent that modifies amine residues within the antibody (Joiris, 1991). Thiolating agents such as 2-iminothiolane or SATA provide efficient ways of introducing multiple sulfhydryl groups for this type of radiopharmaceutical preparation (Chapter 2, Section 4.1).

Site-directed thiolation at carbohydrate residues within the Fc region of antibody molecules may prove to be the best choice for SH group introduction while maintain antigen-binding activity. Ranadive and coworkers (1993) used the heterobifunctional crosslinking agent PDPH (3-(2-pyridyldithio)propionyl hydrazide) (Chapter 6, Section 2.3) to react specifically with oxidized polysaccharide components of monoclonals. The polysaccharide chains are treated first with sodium periodate (Chapter 2, Section 4.4) to generate reactive aldehyde residues. PDPH then is coupled to these aldehydes via its hydrazide end to create stable hydrazone linkages. The other end of the crosslinker, containing a pyridyl disulfide group, is reduced with DTT under mild conditions (25-m*M* DTT, pH 4.5, 30 min) to produce the free sulfhydryl groups. Since the thiolation occurs only at carbohydrate locations within the antibody, the modification has a better chance of being away from the antigen binding sites, thus preserving

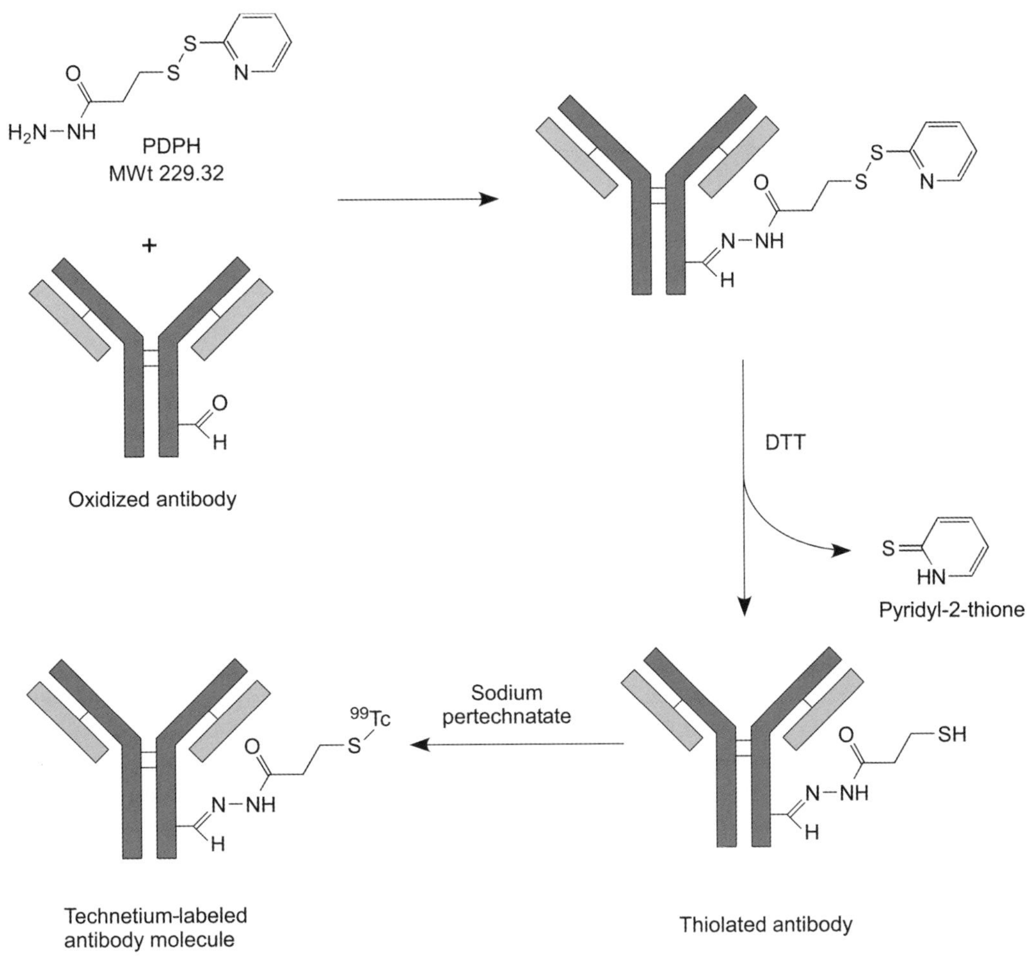

FIGURE 12.3 Antibody molecules oxidized with sodium periodate to create aldehyde groups on their polysaccharide chains can be modified with PDPH to produce thiols after reduction of the pyridyl disulfide. Direct labeling of the sulfhydryls with ^{99}Tc produces a radioactive complex.

immunoglobulin activity. Subsequent treatment with sodium pertechnatate yields the ^{99m}Tc derivative on the sulfhydryl groups (Figure 12.3).

1.6. FeBABE

FeBABE (pronounced "iron-babe") is Fe(III) (S)-1-(*p*-bromoacetamido-benzyl)ethylene diamine tetraacetic acid, a bifunctional chelating agent designed to be a hydroxyl radical probe to footprint interacting domains between proteins (Rana and Meares, 1990, 1991). The bromoacetyl group reacts with thiols to form a thioether bond. The EDTA chelating group coordinates Fe^{3+} to create a redox active complex that is able to form reactive oxygen species in aqueous solution. Iron–EDTA chelates are effective at generating hydroxyl radicals (·OH) that can react with peptide bonds to oxidize and cut them non-selectively. The reaction is initiated with the addition of peroxide and ascorbate, which catalyzes the peptide cleavage reaction. Thus, one protein labeled with FeBABE can be allowed to interact with a protein-binding partner and the cleavage reaction initiated, which results in cuts in the peptide structure immediately surrounding the region where FeBABE is attached. The pattern of peptide cutting can be used to determine the area on both interacting proteins that constitutes the binding region, because all other peptide regions will be affected by the oxidation reaction.

FeBABE has been used to study the interactions of the RNA polymerase RbpA (Hu *et al.*, 2012), to map the protein–protein interactions of ATP-dependent chromatin remodelers (Hota *et al.*, 2012), to investigate the binding of the Swi2/Snf2 remodeller Mot1 in complex with its substrate TBP (Wollmann *et al.*, 2011), and to study the interaction of transcriptional factor SoxS and a subunit of RNA polymerase (Zafar *et al.*, 2011).

Chapter 24, Section 4.1, provides additional information about the FeBABE chelator and its use in studying protein interactions.

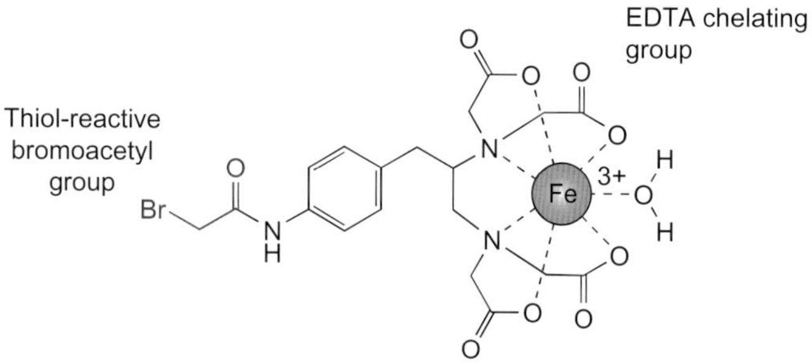

FeBABE
Fe(III)(S)-1-(*p*-bromoacetamido-benzyl)
ethylene diamine tetraacetic acid
MW589.15

2. IODINATION REAGENTS

Modification of proteins and other molecules with a radioactive element provides a means of detection that can be extremely sensitive for assay, localization, and imaging applications. Among the most common radiolabels for biological studies are ^{14}C, ^{32}P, ^{35}S, ^{3}H, and the isotopes of iodine, ^{124}I, ^{125}I, and ^{131}I. The unstable isotopes of carbon, phosphorus, sulfur, and hydrogen are all β emitters, releasing subatomic particle radiation consisting of either positrons or electrons. To measure labeled molecules containing β emitters often necessitates tedious sample manipulation including tissue homogenization and mixing with scintillation cocktails for subsequent counting.

The radioactive isotopes of iodine, by contrast, are both γ emitters, providing a much easier route to measurement than β-particle-emitting radioisotopes. High-energy electromagnetic radiation can be detected directly without the need for intermediate scintillation cocktails. Iodine-131 was the first unstable iodine isotope to be used for labeling protein molecules (Li, 1945; Pressman and Keighley, 1948). The ^{131}I isotope decays by both β^- (electron) and γ emission. The specific activity of this element can be as high as 6550Ci/mmol, providing extraordinary sensitivity for detecting labeled molecules.

Iodine-125 decays by electron capture followed by γ emission. However, the maximum energy of ^{125}I electromagnetic energy emission can be as little as 1/10 to 1/3 that of ^{131}I (Wilbur, 1992; Powsner, 1994). The greater energy intensity of ^{131}I emission actually can be a disadvantage, since γ rays emanating from it are more penetrating, requiring increased precautions and greater protective equipment. In addition, the relatively short half-life of ^{131}I (8.1 days) as compared to ^{125}I (60 days) necessitates that labeled compounds be prepared more often, since activity losses will be severe upon storage. Because ^{125}I is not a particle emitter, its use *in vivo* for imaging applications limits radiation damage to the immediate surrounding proteins, cells, and tissues.

$$I_3^- \rightleftharpoons I_2 + I^-$$

$$I_2 + H_2O \rightleftharpoons H_2OI^+ + I^-$$

FIGURE 12.4 Iodide anion in aqueous solution undergoes an equilibrium reaction process to form the reactive H_2OI^+ species.

Iodine-124 also is being used routinely for developing imaging probes for radiopharmacological evaluation (Koehler *et al.*, 2010).

These factors make ^{125}I the iodine label of choice for radiolabeling biological molecules. Its commercial availability from a number of suppliers at relatively low cost further adds to its popularity. Even though it has lower specific activity than ^{131}I, iodine-125 still provides much greater sensitivity than ^{14}C, ^{32}P, ^{35}S, or ^{3}H in labeling biomolecules. In fact, the use of a radioactive iodine label can create probes that have 150-fold more sensitivity than tritiated molecules and as much as 35,000-times the detectability of ^{14}C-labeled molecules (Bolton and Hunter, 1986).

Radioiodination is the process of chemically modifying a molecule to contain one or more atoms of radioactive iodine. Early studies on protein modification determined that iodine in aqueous solution formed a reactive ion, H_2OI^+ (Figure 12.4), that is capable of modifying tyrosine side chains, the imidazole groups of histidine, and either modifying sulfhydryl groups or catalyzing their oxidation to disulfides (Figure 12.5). Most methods now utilize a chemical agent to create the reactive iodine species, thus driving the reaction at much greater rates.

There are two main methods of radioiodination that are commonly employed to modify proteins and other molecules: (1) direct labeling of the desired protein or other target molecule in the presence of an oxidizing agent; or (2) indirect labeling of the desired molecule by first labeling an intermediate compound which is then used to perform the final modification. Direct labeling methods are by far the most common, and the chemistries used in this process have been reviewed (Regoeczi, 1984). For a review of radioiodination methods and chemistry, see Eisenhut and Mier (2011).

The prevailing procedures for direct coupling of ^{125}I to a protein or other molecule are through the use of oxidizing agents. The *in situ* preparation of an electrophilic radioiodine species is fundamental to the ability to modify certain reactive sites within the desired molecules. The most common oxidizing compounds are *N*-haloamine derivatives, such as *N*-chlorotoluenesulfonamide (chloramine-T) or 1,3,4,6-tetrachloro-3α,6α-diphenylglycouril (Iodogen). In most instances, such compounds do not harm the proteins being labeled, although reaction times should be carefully controlled to prevent over-labeling or oxidative damage. A secondary method of producing an

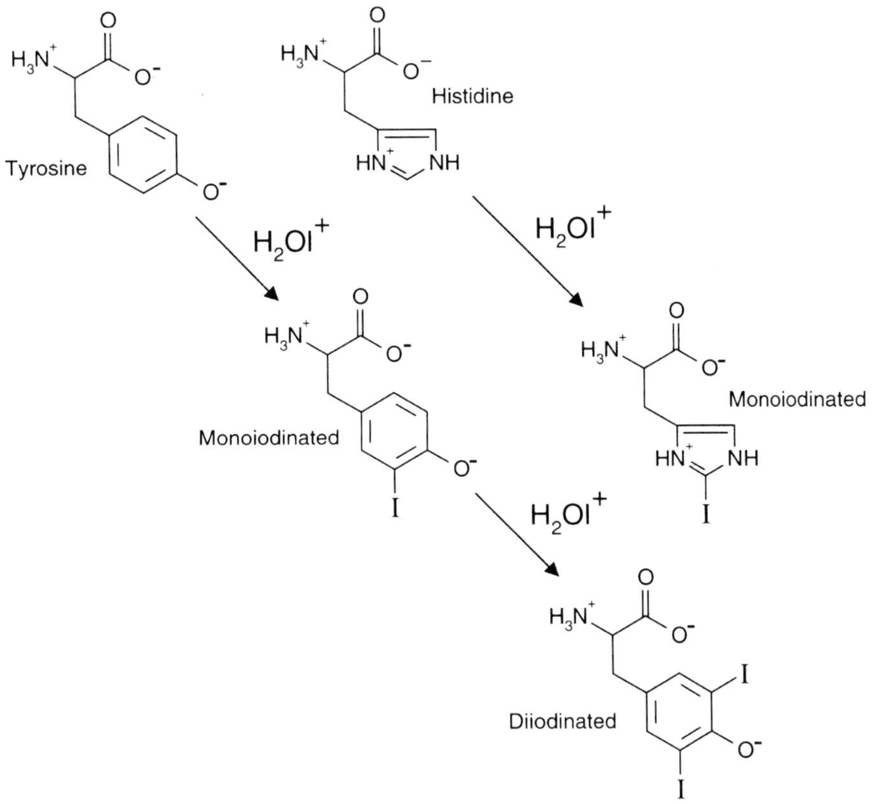

FIGURE 12.5 The iodination of tyrosine or histidine residues in proteins by H_2OI^+.

oxidative effect is to use an enzyme-driven system. The glucose oxidase/lactoperoxidase reaction creates reactive iodine through the production of hydrogen peroxide from glucose with the subsequent action of peroxidase to form I_2 from I^-.

Formation of the electrophilic halogen species leads to the potential for rapid reaction with compounds containing strongly activating groups, such as in activated aryl compounds. Particularly, substances containing aromatic ring structures that have substituents on the ring which are electron donating can sufficiently activate the carbons on the ring to undergo electrophilic substitution reactions. Therefore, phenols, aniline derivatives, or alkyl anilines that contain OH, NH_2, or NHR constituents, respectively, are very susceptible to being iodinated. In proteins, this translates into tyrosine side-chain phenolic groups and histidine side-chain imidazole groups. Crosslinking compounds or modification reagents containing ring-activated groups also are capable of being iodinated.

The addition of a radioactive iodine atom to a protein molecule typically has little effect on the resultant protein activity, unless the active center is modified in the process. The size of an iodine atom is relatively small and does not result in many steric problems with large molecules. The sites of potential protein modification are tyrosine and histidine side chains. Tyrosine may be modified with a total of two iodine atoms per phenolate group, whereas histidine can incorporate one iodine atom. Sulfhydryl modification at cysteine residues is typically unstable.

The result of iodination at tyrosine groups can alter the spectral characteristics of the protein in solution (Hughes, 1950). The typical protein absorbency at 280nm can shift to a maximum at about 305 to 315nm due to the addition of iodine atoms to the phenolate ring of tyrosine. The degree of absorbance shift is dependent on how many iodine atoms are incorporated into the protein and whether they result in mainly mono-iodotyrosine or di-iodotyrosine formation. In addition, as the level of iodination increases, the solubility of a protein in aqueous solution can dramatically decrease until complete insolubility results in proteins with high numbers of tyrosines.

Thus, controlling the degree of iodination is an important consideration both in choosing the oxidant used and in controlling the time of reaction. Typically, most radiohalogenations are done in a time period of 30s to as long as 30min. Optimization may have to be performed to determine the correct time of use for a particular modification reaction. Termination of the iodination reaction may be done through addition of a reducing agent, such as sodium metabisulfite. Bisulfite reduces the electrophilic iodine species to unreactive iodide, effectively stopping the modification process.

The following sections discuss the major radioiodination reagents available for direct labeling as well as the main crosslinkers or modification reagents used for indirect labeling techniques.

2.1. Chloramine-T

Chloramine-T, or *N*-chlorotoluenesulfonamide (Sigma), has been one of the most popular oxidizing reagents used for radioiodination techniques since its introduction by Greenwood *et al.* in 1963. It has strong oxidizing properties that readily lead to the formation of the required electrophilic halogen species that result in iodine incorporation into target molecules. The reactions of chloramine-T are well documented, being suitable for both macromolecular protein iodination and small-molecule modification (Wilbur, 1992). It also can be used to modify molecules with other radioactive halogen elements, such as isotopes of bromine and astatine (Hadi *et al.*, 1979; Mazaitis *et al.*, 1981).

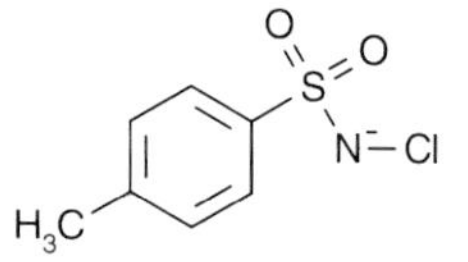

Chloramine-T
N-chlorotoluenesulfonamide

The reaction of chloramine-T with iodide ion in solution results in oxidation with subsequent formation of a reactive, mixed halogen species, ICl (Figure 12.6). Either ^{125}I or ^{131}I can be used in this reaction. The ICl then rapidly reacts with any sites within target molecules that can undergo electrophilic substitution reactions. Within proteins, any tyrosine and histidine side-chain groups can be modified with iodine within 30 s to 30 m. Since chloramine-T is a water soluble reagent and the reaction is done completely in the solution phase, higher incorporation of radioactive iodine can be obtained than when using insoluble or immobilized oxidants (see subsequent sections). However, a greater yield of specific radioactivity does not always translate into a better radiolabeled probe. Chloramine-T, being a strong oxidant with rapid reaction rates, can easily over-label a target molecule or cause oxidative damage to sensitive proteins (Lee and Griffiths, 1984). The reaction may be quenched with a reductant, usually done by addition of sodium metabisulfite. Although chloramine-T is still widely used, alternative iodination reagents that are insoluble or immobilized on insoluble supports may provide milder reaction conditions and be more controllable. Hussien *et al.* (2011) compared chloramine-T with Iodogen (Section 2.3) in the radioiodination of etodolac, an imaging agent for inflammation. Both iodination agents performed similarly; however, the concentration of Iodogen required to result in similar iodination yields was less than required for chloramine-T. Both agents had an optimum pH for the reaction at about pH 7 and

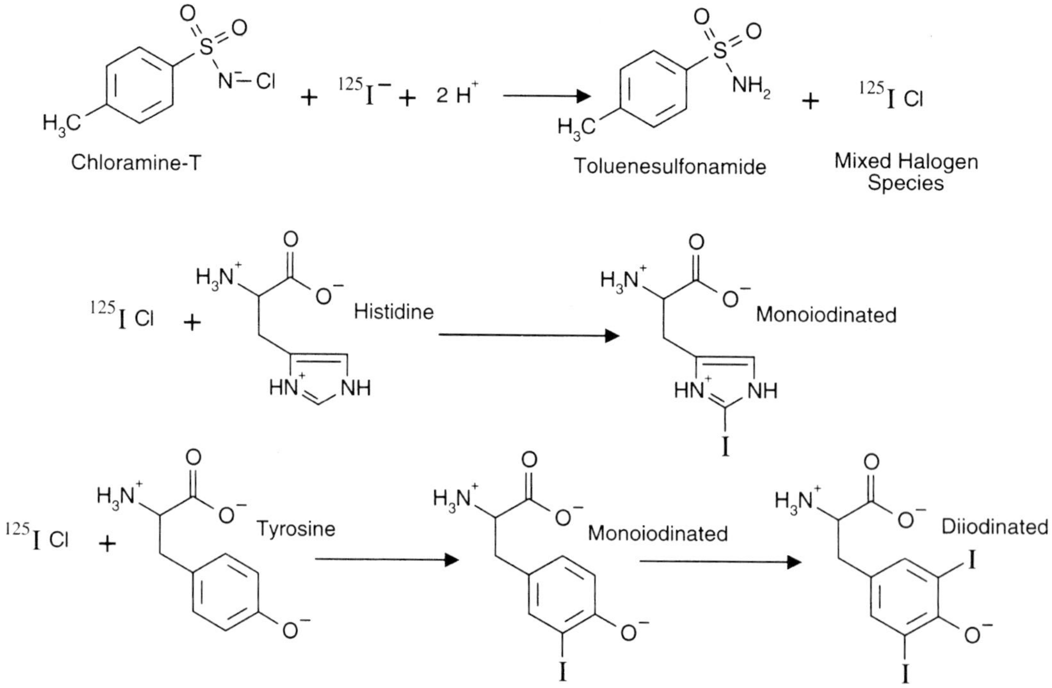

FIGURE 12.6 The strong oxidant chloramine-T can react with iodide anions in aqueous solution to form a highly reactive mixed halogen species. ^{125}ICl then can modify tyrosine and histidine groups in proteins to form radiolabeled products.

resulted in about 83 to 88% yield of iodine incorporation into etodolac.

The following protocol is representative of those found in the literature for iodination of protein molecules using chloramine-T.

Protocol

Caution: Handle all radioactive substances according to the radiation safety regulations instituted at each facility approved to handle such materials. Use adequate precautions to protect personal safety and the environment. Dispose of radioactive waste only by following approved guidelines.

1. Dissolve chloramine-T in 50-m*M* sodium phosphate, pH 7.0, at a concentration of 4 mg/ml. Prepare fresh. Approximately 25 µl of this solution (100 µg) is required to iodinate 5 µg of a protein.
2. Dissolve sodium metabisulfite in 50-m*M* sodium phosphate, pH 7.0, at a concentration of 12.6-m*M* (240 µg/100 µl). Prepare fresh. Approximately 100 µl of this solution is required for a 5-µg protein iodination.
3. Obtain fresh $Na^{125}I$ and adjust its concentration to approximately 0.5 µmCi/µl. Two microliters of this solution are required to iodinate 5 µg of protein.
4. Add to a suitable reaction vial 25 µl of a solution consisting of 5 µg of a protein dissolved in 50-m*M* sodium phosphate, pH 7.0. Mix using a small magnetic stirring chip.
5. Add 2 µl of the $Na^{125}I$ solution (about 1 mCi) to the protein in the vial. Seal the vial using a screw-cap septum that can be penetrated with a syringe.
6. Using a syringe, add 25 µl of the chloramine-T solution and continue to mix for at least 30 s. Longer reaction times can be used, but the solution-phase iodination usually proceeds very rapidly.
7. Add 100 µl of the sodium metabisulfite solution to the iodination reaction to stop it. Stir for 10 s.
8. Purify the iodinated protein from excess reactants by gel filtration using a desalting resin. The column may be pre-treated by passing a solution of bovine serum albumin (BSA) through it to eliminate nonspecific binding sites that could cause significant protein loss in small-sample applications.

2.2. IODO-BEADS

IODO-BEADS (Thermo Fisher) are an immobilized preparation of a chloramine-T analog, consisting of nonporous, polystyrene beads of diameter 1/8″ that have been derivatized to contain *N*-chlorobenzenesulfonamide groups (as the sodium salt) (refer to U.S. patents 4,448,764 and 4,436,718). During the manufacturing process, the hydrophobic nature of the polystyrene is changed to a rather hydrophilic surface due to the chlorosulfonamide modifications. The surface character results in excellent protein recoveries (typically greater than 90%). The oxidizing capability of IODO-BEADS is limited to surface reactions on the outer shell of the nonporous polystyrene ball. The effect is to reduce the rate of iodine incorporation into macromolecules from the extremely rapid 30-s reaction of soluble chloramine-T to a more relaxed pace of about 2 to 15 min to obtain a similar yield. This also creates a milder oxidizing environment, thus minimizing the potential for protein degradation or activity loss. A slower iodination process allows more control over the level of iodine derivatization. Often, tyrosine iodination can be limited to mono-iodo forms, avoiding the detrimental effects on solubility or activity that excessive modification can cause.

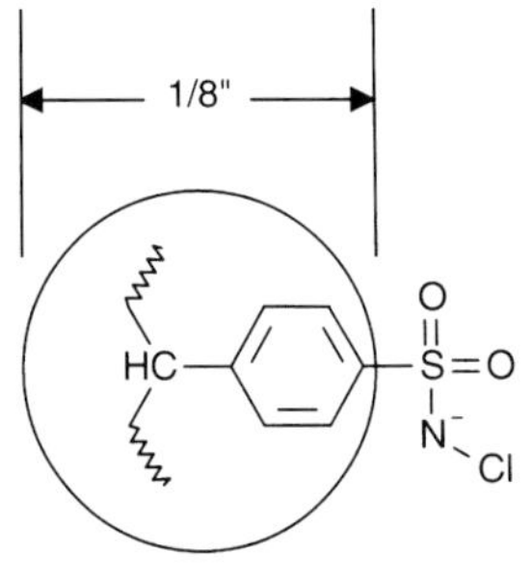

IODO-BEADS Containing *N*-Chlorobenzenesulfonamide Groups on a Polystyrene Backbone

Markwell (1982) reported that the reaction mechanism for creating the electrophilic iodine species may be somewhat different for IODO-BEADS than other oxidizing agents. It was demonstrated that the active component remained at or near the surface of the beads during the course of the iodination process. Markwell speculated that an intermediate reactive species, *N*-iodobenzenesulfonamide, is formed from substitution of the chlorine atoms on the bead (Figure 12.7). It is possibly that this intermediate is involved in the direct iodination of target molecules that approach the bead surface.

IODO-BEADS may be used in a variety of buffer salts and in the presence of detergents or denaturants without affecting iodination. The reagent, however, is susceptible to inactivation by reducing agents such as disulfide reductants, and it can be inactivated by moisture upon storage. Also, avoid organic solvents that can dissolve or affect the surface characteristics of polystyrene, such as dimethylformamide (DMF) or dimethyl sulfoxide (DMSO).

To determine the optimal reaction time for a particular radioiodination, 5-µl aliquots of the reaction medium can be removed every 30 seconds, diluted 1:20,000 with 20-m*M* Tris, 1-m*M* ethylenediaminetetraacetic acid (EDTA), pH 7.4, containing 0.5 mg/ml BSA

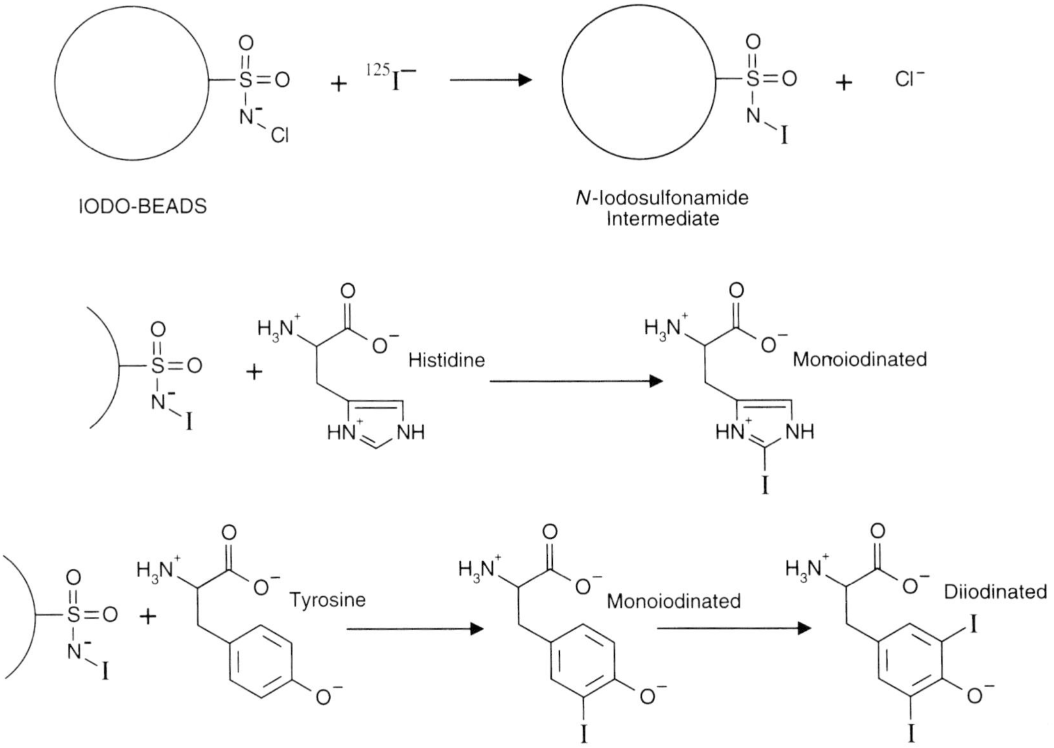

FIGURE 12.7 IODO-BEADS contains immobilized chloramine-T groups that can react with radioactive iodide in aqueous solution to form a highly reactive intermediate. The active species may be an iodosulfonamide derivative, which then can iodinate tyrosine or histidine residues in proteins.

as a carrier protein. Finally, precipitate a small amount of the diluted aliquot with trichloroacetic acid (TCA, 60%), centrifuge to recover the pellet, wash the pellet once with TCA, and measure the amount of radioactivity in the pellet and supernatant using a gamma counter. The reaction period representing optimal radiolabel incorporation should be used for subsequent radioiodination reactions.

Directing the iodination reaction toward histidine residues in proteins, as opposed to principally tyrosine modification, is possible simply by increasing the pH of the IODO-BEADS reaction from the manufacturer's recommended pH 7.0 to pH 8.2 (Tsomides *et al.*, 1991). No reducing agent is required to stop the iodination reaction as is the case with chloramine-T and other methods. Simple removal of the bead(s) from the reaction is enough to eliminate the iodination process. The mild nature of the IODO-BEADS iodination reaction can result in better recovery of active protein than using soluble oxidants (Lee and Griffiths, 1984).

IODO-BEADS are being used extensively for the radioiodination of peptides, proteins, and other molecules for *in vivo* imaging and therapeutic purposes (Bertrand *et al.*, 2010; Koehler *et al.*, 2010; Ha *et al.*, 2012).

Each bead is able to iodinate up to 500 μg of tyrosine-containing protein or peptide. This translates into an oxidative capacity of about 0.55 μmoles per bead. The rate of reaction can be controlled by changing the number of beads that are used and altering the sodium iodide concentration added to the reaction. Reaction volumes of 100 to 1000 μl are possible per bead. The following protocol is suggested for iodinating proteins. Optimization should be done to determine the best incorporation level to obtain good radiolabel incorporation with retention of protein activity.

Protocol

Caution: Handle all radioactive substances according to the radiation safety regulations instituted at each facility approved to handle such materials. Use adequate precautions to protect personal safety and the environment from contamination. Dispose of radioactive waste by following approved guidelines.

1. Wash one or more IODO-BEADS (Thermo Fisher) with the iodination buffer of choice. Buffers containing 0.1-*M* sodium phosphate or 0.1-*M* Tris at slightly acidic to slightly alkaline pH work well. A buffer consisting of 0.1-*M* sodium phosphate, pH 6.5, will give the highest possible reaction rates and yields.
2. Add the desired number of beads to a solution of carrier-free $Na^{125}I$ in iodination buffer at a

concentration level of about 1 mCi per 100 μg of protein to be modified. The total reaction volume should be 100 to 1,000 μl per bead.

3. Add from 5 μg to 500 μg of a tyrosine-containing peptide or protein dissolved in iodination buffer to the reaction mixture.
4. React for 2 to 15 min at room temperature. Reactions carried out at 4°C are possible, but will result in slightly lower incorporation of iodine.
5. Stop the reaction by removing the solution from the beads. This can be done by simply pipetting the solution away from the beads or by physically removing the beads. The beads may be washed once with iodination buffer to ensure complete recovery of protein. Exact timing of the reaction is important to obtain reproducible results.
6. Remove excess ^{125}I from the iodinated protein by gel filtration using a desalting resin.

2.3. Iodogen

Iodogen (Thermo Fisher), first described by Fraker and Speck in 1978, is 1,3,4,6-tetrachloro-3α,6α-diphenylglycouril, an *N*-haloamine derivative with oxidizing properties similar to those of IODO-BEADS and chloramine-T. The compound is insoluble in aqueous solution, thus making it a type of solid-phase radioiodination reagent. However, unlike IODO-BEADS wherein the oxidizing group is immobilized on another support material, Iodogen must be plated out on the surface of a reaction vessel prior to iodination. Due to the reagent's stability, the plated reaction vessels can be prepared well in advance and stored in a desiccator until needed (Marwell and Fox, 1978). Alternatively, pre-coated tubes now are available (Thermo Fisher).

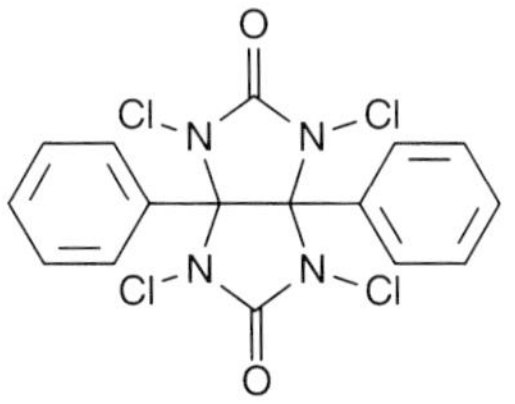

Iodogen
1,3,4,6-Tetrachl ro-
3α,6α-diphenylglycouril
MW 432.09

The reaction of Iodogen with iodide ion in solution results in oxidation with subsequent formation of a reactive, mixed halogen species, ICl (Figure 12.8); ^{124}I, ^{125}I, or ^{131}I can be used in this reaction. The ICl then rapidly reacts with any sites within target molecules that can undergo electrophilic substitution reactions. Within proteins, any tyrosine and histidine side-chain groups can be modified with iodine within 30 s to 30 m. In addition, crosslinking or modification reagents possessing phenyl rings with activating groups present (e.g., electron-donating constituents, such as OH or NH_2) can be iodinated using Iodogen. The incidence of side reactions appears to be negligible.

Since Iodogen is insoluble in aqueous solution and is plated on the surface of the vessel during the iodination reaction, it is possible to stop the reaction simply by removing the aqueous phase. The plating technique is important for the successful use of this reagent. Failure to plate properly the Iodogen reagent on the surface of the reaction vessel may cause the oxidizing agent to become suspended in the reaction medium. However, even with well-plated vessels, there is some potential that a portion of the iodinating reagent can break off in small pieces and contaminate the aqueous phase. For this reason, it is not advisable to stop the iodination reaction only by removing the supernatant as in the case of Iodobeads (Section 2). To be certain that the iodination has stopped, an aliquot of sodium metabisulfite can be added to ensure complete cessation of the oxidative process. Alternatively, immediate separation of the iodinated protein from the reactants by gel filtration can be used to stop the reaction (any suspended particles of Iodogen will be filtered out on the top of the gel).

Specific radioactivity of 1×10^5 cpm of ^{125}I per microgram of protein easily can be obtained using Iodogen. Iodination efficiencies are typically 60% or better and may be controlled by regulating the amount of I^- concentration added to the reaction.

When iodinating intact cells, Iodogen can be used to radiolabel the outer cell surface proteins or be directed more toward the inner membrane areas simply by modulating the reaction conditions. Membrane proteins in hydrophobic regions can be labeled to a greater extent by including a small excess of carrier iodide, using high salt conditions, or employing detergents to disrupt the membrane integrity. Cell surface hydrophilic proteins may be preferentially labeled by not including components that increase cell permeability, by the use of carrier-free iodide, and using short reaction times (Markwell and Fox, 1978).

Hussien *et al.* (2011) compared iodogen to chloramine-T for relative effectiveness and yield of radioiodination of etodolac, a molecule used for inflammation imaging *in vivo*. The results indicated that the yield of iodination was slightly higher for iodogen (88% *versus* 83%) and the reaction required less iodogen than chloramine-T to result in similar yields. Iodogen is being used frequently to prepared radiolabeled peptides, proteins, and other probes for imaging or therapeutic use (Kuntner *et al.*, 2011; Tran *et al.*, 2011; El-Bary *et al.*, 2012).

The following protocol describes the use of Iodogen for the radioiodination of proteins and peptides.

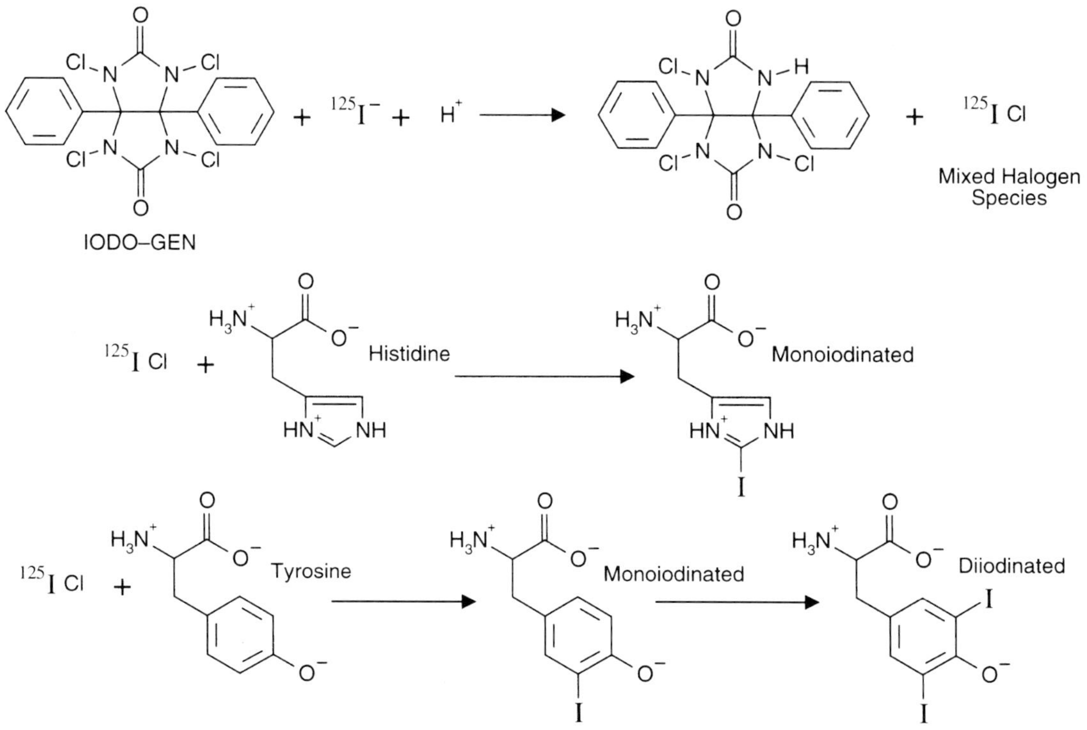

FIGURE 12.8 Iodogen is a water-insoluble oxidizing agent that can react with $^{125}I^-$ to form a highly reactive mixed halogen species, ^{125}ICl. This intermediate can add radioactive iodine atoms to tyrosine or histidine side chain rings.

Protocol

Caution: Handle all radioactive substances according to the radiation safety regulations instituted at each facility approved to handle such materials. Use adequate precautions to protect personal safety and the environment from contamination. Dispose of radioactive waste by following approved guidelines.

1. In a fume hood, dissolve 10 to 100 μg of Iodogen (Thermo Fisher) in 100 to 500 μl of chloroform, methylene chloride, or DMSO. The use of 10 μg of Iodogen per 100 μg of protein or 10^7 cells to be iodinated will result in good incorporation yields.
2. Add the Iodogen solution to a clean, dry, glass reaction vessel in an amount needed for the quantity of protein to be labeled. Slowly evaporate the solvent in the vessel using a stream of dry nitrogen or other inert gas. Do not use a strong gas jet, since rapid evaporation or turbulence in the solvent solution will cause uneven Iodogen distribution with possible clumping. Do not merely leave the vessel to dry in a hood, since contaminants or moisture may get into the reagent film. If done properly, the plating process should leave a film of Iodogen on the inner surface of the vessel that is difficult to see—looking like a slight clouding of the glass. After solvent evaporation, seal the container with nitrogen and store in a desiccator until needed.
3. Dissolve the protein to be iodinated in a buffer compatible with its known biological stability. Conditions ranging from pH 4.4 to pH 9 and temperatures from 0°C to 37°C can be used with good results. The amount of protein to be labeled should be contained in a volume of 100 μl or less. The sample buffer should not contain reducing agents, antioxidants, 2-mercaptoethanol, DTT, cysteine, glycerol, high detergent concentrations, or anything that may interfere with the iodination reaction or dislodge the plated Iodogen.
4. Rinse the plated reaction vessel once with sample buffer to remove any loose particles of Iodogen that may not be strongly adhered to the surface of the glass.
5. Add carrier-free $Na^{125}I$ to the reaction vessel in a ratio of about 500 μCi per 100 μg protein.
6. React for 10 to 15 min at room temperature. Optimization of the reaction time and the amount of ^{125}I added to the reaction may have to be done to obtain the best radioactivity incorporation and retention of protein activity.
7. Remove the sample from the reaction vessel. This process should terminate the iodination reaction, unless small Iodogen particles break off from the sides of the vessel. To ensure safe handling, carrier NaI may be added to the reaction mixture to a final concentration of 1-m*M*.
8. Remove excess reactants by gel filtration using a desalting resin.

2.4. Lactoperoxidase-Catalyzed Iodination

An enzyme-catalyzed process also may be used to form reactive iodine species capable of iodinating proteins and other molecules (Marcholonis, 1969; Morrison and Bayse, 1970). The enzymatic approach utilizes lactoperoxidase in the presence of H_2O_2 to oxidize $^{125}I^-$ to I_2. The iodine thus formed may react with tyrosine or histidine sites within proteins, forming radiolabeled complexes (Fu *et al.*, 2011; Hadjikakou *et al.*, 2011; Amiri *et al.*, 2012). Unlike the use of chemical oxidants for iodination, the enzymatic reaction is very pH dependent—the optimum being between pH 6 and 7. If H_2O_2 is directly added to the reaction medium, it must be highly pure with no stabilizing agents such as metals, since they inhibit the oxidation process.

An alternative to direct addition of H_2O_2 is to form it *in situ* through the use of a second enzymatic reaction. Enzymobeads (originally from Bio-Rad, but no longer commercially available) used immobilized lactoperoxidase along with immobilized glucose oxidase to create the necessary oxidative environment. The glucose oxidase reaction transformed added glucose in the iodination medium to the required H_2O_2. As it was formed, the lactoperoxidase (coupled in tandem to the same beads) would catalyze the formation of I_2 (Figure 12.9). The immobilized enzymes create an iodination environment that is more oxidatively gentle than direct addition of a soluble chemical oxidant like chloramine-T.

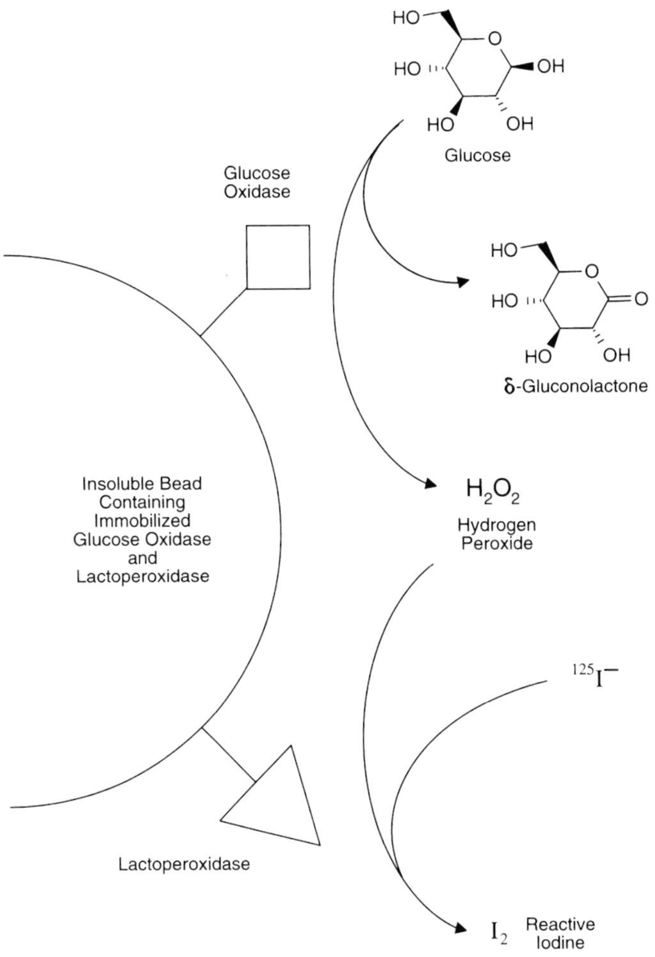

FIGURE 12.9 The immobilized glucose oxidase/lactoperoxidase system radioiodinates proteins through the intermediate formation of hydrogen peroxide from the oxidation of glucose. H_2O_2 then reacts with iodide anions to form reactive iodine (I_2). This efficiently drives the formation of the highly reactive H_2OI^+ species that is capable of iodinating tyrosine or histidine residues (see Figure 12.2).

2.5. Iodinatable Modification and Crosslinking Agents

Radioiodination can be performed using an indirect approach that utilizes a radiolabeled crosslinking or modification reagent which is then used to label the target molecule. One advantage of indirect labeling over direct modification of tyrosine or histidine residues in proteins is to be able to control the iodination to occur with functional groups other than just using indigenous amino acids. In addition, the ability to add a radiolabeled modification agent to a molecule can facilitate the radioactive tagging of substances that normally do not have radioiodinatable sites. Another major advantage of using iodinatable modification agents is to eliminate the potential for oxidative damage to sensitive biological molecules, as may occur when an oxidant is used in direct iodination procedures.

For instance, an amine-reactive modification reagent can be radiolabeled and subsequently used to couple with ε- and N-terminal amines on a protein molecule. The protein is not exposed to oxidative conditions, and the level of radiolabeling can be discretely controlled by the molar ratio of modification reagent addition. The use of iodinatable crosslinking reagents can similarly provide radioactive tags incorporated into conjugates at the time of formation. In addition, iodinatable bioconjugation reagents that react with groups such as sulfhydryls, aldehydes, or other functionalities of limited occurrence in proteins or other macromolecules, can be used to direct the point of radiolabeling to areas away from active centers or binding sites, thus better preserving biological activity. Finally, some photoreactive crosslinking agents can be iodinated, used to label a targeting molecule, photolyzed at the point of binding to its target, and the cross-bridge of the resulting complex chemically cleaved, resulting in the transfer the radiolabel to the targeted component (Chapter 6, Section 3). This process can be used to follow the targeted molecule *in vivo* or in cellular systems.

Direct iodination of proteins and other molecules does not provide the range of experimental options available through indirect labeling. The main disadvantage of the indirect labeling process is the additional steps needed to prepare the radiolabeled crosslinker or modification reagent before iodination of the desired molecule. The following sections discuss some of the major indirect iodination methods, including the reagents available for doing such procedures.

Bolton–Hunter Reagent

Bolton and Hunter (1973) developed the reagent *N*-succinimidyl-3-(4-hydroxyphenyl) propionate (SHPP) for the indirect radioiodination of proteins and other macromolecules (Thermo Fisher). The NHS ester end of the molecule reacts with amine groups in target molecules to form stable amide bond derivatives (Chapter 3, Section 1.4). The other end of the reagent contains a phenolic group that is ideally suited for modification with ^{125}I. Iodination of the phenol group occurs *ortho* to the hydroxyl, thus accommodating either one or two iodine substitutions per molecule (Figure 12.10).

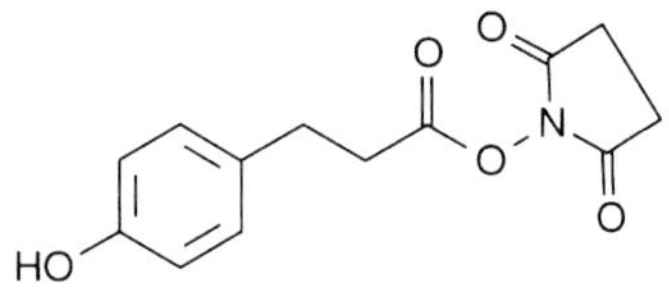

Bolton–Hunter Reagent
N-Succinimidyl-3-
(4-hydroxyphenyl)propionate
MW 263.26

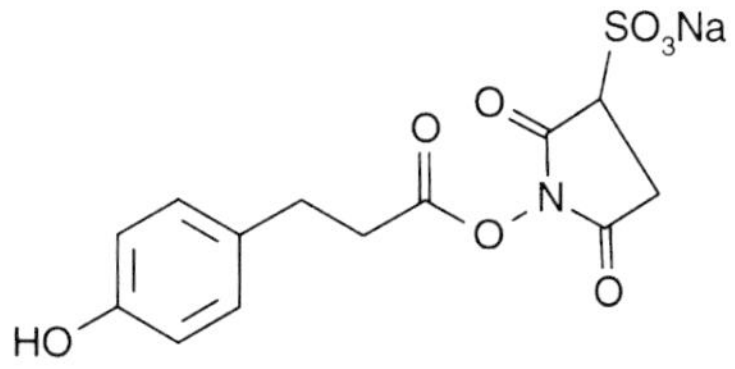

Water-Soluble Bolton–Hunter Reagent
N-Sulfosuccinimidyl-3-
(4-hydroxyphenyl)propionate
MW 365.30

The use of the Bolton–Hunter reagent to incorporate radioactive labels into proteins results in at least as good incorporation of ^{125}I as direct labeling procedures using an oxidant. In many cases, the degree of radioiodine labeling can be much greater than that possible by direct labeling of tyrosine residues, because the total number of amines in a protein (from N-terminal and lysine side chains) is typically significantly more than the number of tyrosine residues present. The major advantage of the indirect approach, however, is that non-tyrosine-containing proteins also may be iodinated. In addition, substances sensitive to oxidant exposure can be labeled without loss of activity or structural degradation. Ultimately, any molecule containing an available amino group can be radioiodinated with SHPP, even if it does not contain a strongly activated aromatic ring system to allow direct iodine substitution. For reviews of protein modification using radiolabeled Bolton–Hunter reagent, see Langone (1980, 1981) and Wilbur (1992). For its use in labeling cellular components, see Katz *et al.* (1982) and Davies and Palek (1981). The reagent has found widespread use in radioiodination of many different compounds and proteins (Waentig *et al.*, 2011; Zhou *et al.*, 2011).

SHPP is relatively insoluble in aqueous environments and must be dissolved in an organic solvent prior to addition to a reaction medium. Suggested solvents include dioxane and DMSO that are low in water content to avoid hydrolysis of the NHS ester.

A water soluble version of the original Bolton–Hunter reagent is also available, called sulfo-SHPP or sulfosuccinimidyl-3-(4-hydroxyphenyl)propionate (Thermo Fisher). This compound contains a negatively charged sulfonate group on the NHS ring structure, which provides enough hydrophilicity to allow direct addition to aqueous reaction mediums.

The following procedure describes the iodination process for the Bolton–Hunter reagent and its subsequent use for the radiolabeling of protein molecules. Modification of other macromolecules can be carried out using the same general method. For particular labeling applications, optimization of the level of iodine incorporation may have to be done to obtain the best specific radioactivity with retention of biological activity.

Protocol

Caution: Handle all radioactive substances according to the radiation safety regulations instituted at each facility approved to handle such materials. Use adequate precautions to protect personal safety and the environment from contamination. Dispose of radioactive waste by following approved guidelines.

1. Dissolve SHPP (Thermo Fisher) in dry dioxane or DMSO at a concentration of 0.5 mg/ml. Prepare fresh. If sulfo-SHPP is used, dissolution in organic solvent is unnecessary, although it may better facilitate the addition of a small quantity to the aqueous reaction and also prevent hydrolysis prior to initiating the iodination reaction.
2. Dissolve chloramine-T (Sigma) in 50-m*M* sodium phosphate, pH 7.5 (reaction buffer), at a concentration of 100 μg per 25 μl of buffer. Prepare fresh.
3. Prepare 0.5 mCi $Na^{125}I$ dissolved in 2 μl of reaction buffer. Prepare fresh.
4. In a suitable reaction vessel (such as a glass ReactiVial, Thermo Fisher), add 2 μl of the SHPP solution and 2 μl of the $Na^{125}I$ solution to 25 μl of reaction buffer. Once the Bolton–Hunter reagent is in an aqueous environment, the NHS ester end

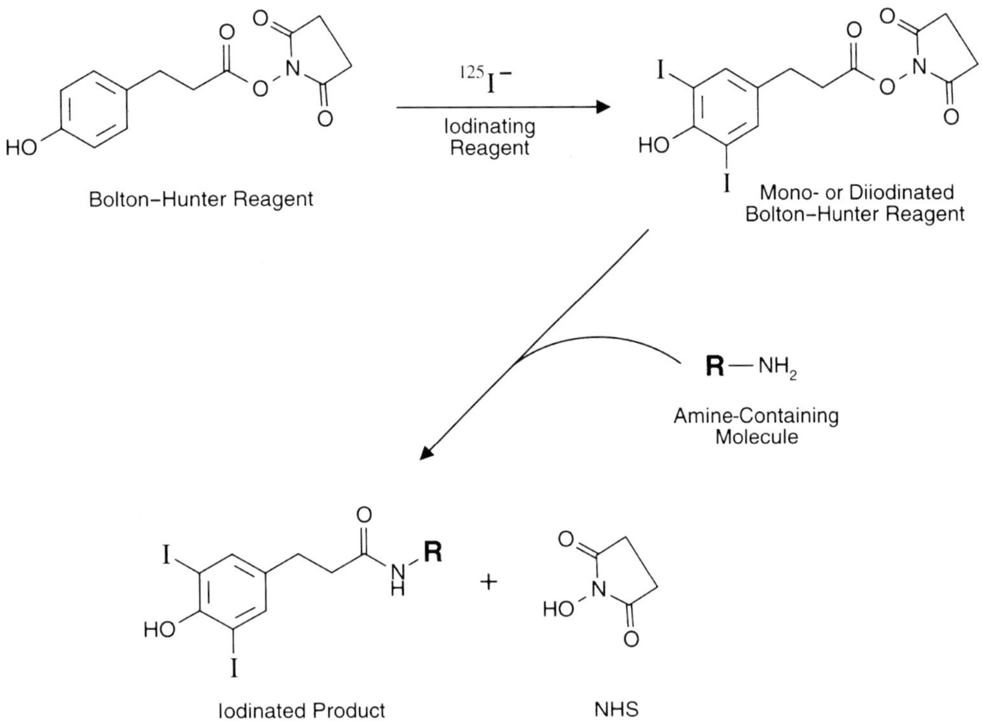

FIGURE 12.10 The Bolton–Hunter reagent may be radioiodinated at its phenolic ring structure prior to reaction with an amine-containing molecule to form an amide bond modification.

of the compound will hydrolyze; therefore, all aqueous handling of the reagent from this point on should be done quickly to preserve enough amine-coupling activity to label the protein after the iodination reaction.

5. Immediately add 25 µl of the chloramine-T solution to the SHPP solution. Mix well.
6. React for 15 s with mixing.
7. Add to the iodination reaction 5 µl of DMF and 100 µl of benzene. Mix to extract the iodinated Bolton–Hunter reagent into the organic phase.
8. Remove the aqueous phase and transfer the organic phase into a clean glass vial or tube.
9. Remove the organic solvent by evaporation using a steady, but gentle, stream of nitrogen.
10. Dissolve the equivalent of 250 ng of a protein to be labeled in 2 to 2.5 µl of ice-cold 50-m*M* sodium borate, pH 8.5.
11. Add the protein solution to the dried, iodinated Bolton–Hunter reagent.
12. React for 2 h on ice. An overnight reaction may be performed at 4°C.
13. Remove excess labeled Bolton–Hunter reagent by gel filtration or dialysis.

The Bolton–Hunter reagent also may be used to modify a molecule prior to the iodination reaction. In this case, an amine-containing protein or other molecule is coupled via the NHS ester end of the reagent to form an amide-bond derivative. This derivative is then iodinated using any of the iodination reagents discussed in this section. This approach can be useful in preparing stable Bolton–Hunter derivatives that can be stored for extended periods until requiring iodination, eliminating the relatively short half-life of ^{125}I-labeled probes.

Iodinatable Bifunctional Crosslinking Agents

Bifunctional crosslinking agents containing an activated aromatic ring system may be radioiodinated using similar procedures as that described previously for the Bolton–Hunter reagent. Certain conjugation compounds have been designed with the potential for radiolabeling in mind. For instance, there are a number of photoreactive phenyl azide crosslinkers that possess an activating hydroxyl group on their phenyl rings. The phenolic group provides sites of facile iodination *ortho* and *para* to the hydroxyl.

The heterobifunctional crosslinker ASBA (4-(*p*-azidosalicylamido)butylamine) (Chapter 6, Section 6.1) is an example of this type of iodinatable photoreactive reagent. The phenyl azide group may be radiolabeled using any standard iodination process (described previously) before coupling of its primary amine end to a carbonyl group on a macromolecule. After allowing the modified molecule to bind to a target, the complex may be photolyzed and the covalent conjugate detected by its radioactivity.

The homobifunctional photoreactive BASED (Chapter 5, Section 5.1) has two photoreactive phenyl azide groups, each of which contains an activating hydroxyl. Radioiodination of this crosslinker can yield one or two iodine atoms on each ring, creating an intensely radioactive compound. Crosslinks formed between two interacting molecules are reversible by disulfide reduction, thus allowing traceability of both components of the conjugate.

An extremely versatile iodinatable heterobifunctional is APDP (*N*-[4-(*p*-azidosalicylamido) butyl]-3′-(2′-pyridyldithio)propionamide) (Chapter 6, Section 4.2). One end of the crosslinker can couple with sulfhydryl-containing molecules, while the other end is a nonselective photoreactive phenyl azide. Again, the phenyl ring contains an activating hydroxyl group, providing radioiodination capability. Modification of a sulfhydryl-containing molecule may be done after iodination of the reagent. After the labeled molecule is allowed to interact with a target molecule, the photoreactive process can be initiated to form a covalent conjugate. Subsequently, the crosslinks may be cleaved using a disulfide reducing agent, thus transferring the radiolabel to the second molecule.

SASD (sulfosuccinimidyl-2-(*p*-azidosalicylamido) ethyl-1,3′-dithiopropionate) (Chapter 6, Section 3.2) behaves in a similar manner, except that it contains an amine-reactive end that can be coupled to proteins and other molecules. Its photoreactive end can be iodinated using any of the radioiodination reagents discussed previously. Just as in the case of APDP, SASD crosslinks can be cleaved by a disulfide reductant to transfer the radioactive component to a second molecule.

Finally, the small amine-reactive and photoreactive crosslinker NHS–ASA (Chapter 6, Section 3.1) can be iodinated to provide a non-cleavable radioactive conjugate.

Figure 12.11 shows the iodination products resulting from labeling these reagents with ^{125}I. Any of the iodination reagents described previously can be used to radiolabel these compounds prior to their incorporation into target molecules. However, the insoluble iodination reagents are probably the best choice, since the separation of radiolabeled compound from excess oxidant simply involves removing the solution.

There are many other compounds that have been investigated for their use in indirect radiolabeling of

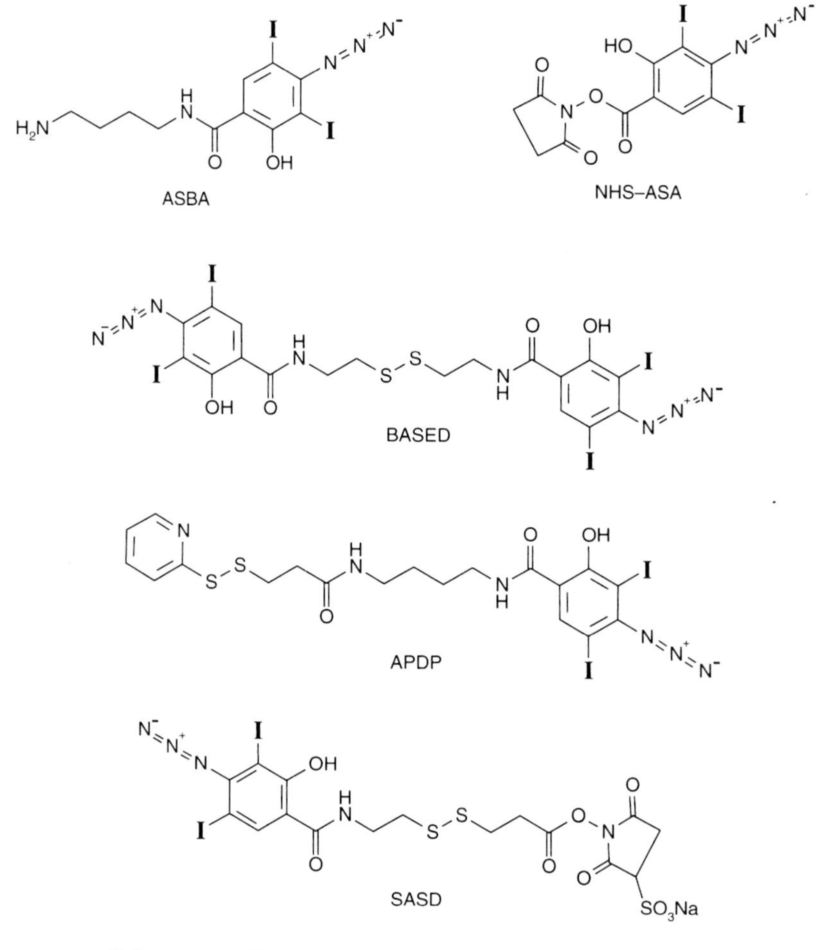

FIGURE 12.11 Some common crosslinking agents that are capable of being radioiodinated. The sites of iodination are shown in bold.

proteins. For an excellent overview of these chemical reactions, see Wilbur (1992).

3. MASS TAGS AND ISOTOPE TAGS

Mass spectrometry has become one of the most important tools for analyzing proteins in complex biological samples. The ability to separate proteins and peptides at high resolution has made possible the simultaneous identification of hundreds of proteins within samples (for reviews, see Bantscheff *et al.*, 2012; Sabidó *et al.*, 2012; Siuzdak, 2006; Hamdan and Righetti, 2005; and Gingras *et al.*, 2005). Proteins can be analyzed for their presence or compared between samples for their relative expression levels. One cell population treated with a drug candidate, for instance, can be compared by mass spectrometry to another sample as control to assess the effect of the drug on expression levels of certain proteins.

There are several ways that proteins can be analyzed using mass spectrometry. Whole proteins can be separated using the electrospray ionization (ESI) technique or by using matrix-assisted laser desorption/ionization (MALDI). Both of these methods inject intact proteins into the mass spectrometer, ionize them, and separate the resultant charged components by their individual mass/charge ratios. These methods work well for small and medium-sized proteins in samples of low complexity, but analysis of larger proteins or highly complex samples is difficult. Alternatively, proteins first can be proteolytically digested using an enzyme such as trypsin and then the peptide products analyzed by mass spectrometry. This proteolysis method is more universally applicable, because analysis of peptide fragments allows mass spectrometry separation to be done for all proteins regardless of their original intact mass prior to digestion. The peptides typically are subjected next to a higher energy secondary mass spectrometry separation that fragments them into their component amino acids, which then can be identified by their masses. Samples then are analyzed by correlation of the peptide amino acid sequences to online databases of mass spectrometry information, which can identify the parent protein that each peptide came from.

However, the interpretation of mass spectrometry data on whole samples can be daunting, especially when analyzing proteolytically digested samples, which results in many times more species to analyze per protein than intact proteins. In order to reduce the complexity of sample analysis, a number of techniques have been developed to fractionate the proteome prior to mass spectrometry separation. For instance, two-dimensional electrophoresis can separate proteins both by charge and by molecular weight and allow picking of only certain spots for subsequent mass spectrometry analysis. However, 2D electrophoresis is severely limited in its sensitivity for picking up low- or medium-copy proteins (Gygi *et al.*, 2000). Alternatively, affinity separations on resins or surfaces can be carried out to capture only those proteins having certain epitopes or chemical characteristics, such as post-translational modifications. In addition, nanoliter HPLC separations in one or two dimensions can be done to fractionate the peptides in complex samples by charge, size, or hydrophobicity before being injected into the mass spec.

Another major technique to simplify the analysis of protein samples is to use mass tags. Mass tags are modification reagents that contain a reactive group for coupling to biomolecules and another component of known mass, which behaves predictably upon MS separation. The mass tag may also contain a functional group for capture and separation on an affinity support, which permits further fractionation of the proteome. MS analysis of mass tagged peptides can be done by focusing only on those peptides that contain an additional mass component representing the tag's known mass contribution. Thus, all other peaks on the MS spectrum can be ignored, which greatly reduces the complexity of the sample. Mass tag reagents have been developed with reactive groups to modify specifically only certain low frequency amino acids within proteins. For instance, a thiol-reactive iodoacetyl group on a mass tag can be used to modify only those peptides having cysteine residues, thus removing from the analysis window all other peptides not containing cysteine.

The design of mass tags can also be combined with stable isotope labels to create more than one mass unit for each tag type (Schneider and Hall, 2005; Helsens *et al.*, 2011). For example, certain hydrogen atoms on one mass tag can be replaced with deuterium atoms on another derivative. Everything else on the tag is identical except for the isotope substitutions. Thus, the two mass tag analogs will differ in molecular weight by exactly the mass difference represented by the isotopic substitutions. Such tags can be used to modify a test sample with the stable isotope tag *versus* a control sample modified with the normal tag. If the two samples are then combined and analyzed by mass spectrometry, their signal peaks generated from the tagged peptides will differ in mass units by the isotopic mass differences in the two tags. Identification of the peptides from both samples is done by looking for peptide peak pairs differing by the characteristic mass amount, therefore greatly reducing the complexity of sample analysis and allowing simultaneous investigation of two samples. In this way, a test sample's protein expression levels can be compared to a control sample by measuring the different areas of the paired peptide peaks. The ability to analyze protein expression in two samples is vitally important to drug discovery and life science research applications studying the proteome.

Mass tags also can be broad spectrum in their modification properties to derivatize all peptides as they are formed upon proteolysis. For instance, one of the simplest mass tagging systems is to use the oxygen isotope ^{18}O in the water ($H_2{}^{18}O$) used during the enzymatic digestion of a protein sample (Miyagi and Rao, 2007). Upon hydrolysis by trypsin, the resultant C-terminal carboxylates that are formed each incorporate two ^{18}O atoms. Thus, peptides formed from ^{18}O digestion will be four mass units heavier than peptides formed by proteolysis using normal water. Mass spectrometry analysis of this difference can identify the peptide pairs resulting from a control sample and a test sample run simultaneously.

Other broad-spectrum mass tag modification agents are designed to modify all amine groups and yield tags on every peptide at their N-terminal amines. For instance, small molecule tags using deuterium-labeled tags and regular hydrogen-labeled ones, such as in the use of isotopically labeled propionic anhydride (Zappacosta and Annan, 2004), provide differentiation in the mass spectrometry signals of peptides from test samples and controls. To eliminate interference, side chain lysine amines are blocked by guanidination with *O*-methylisourea hemisulfate and cysteine thiols are blocked with iodoacetamide (Leitner and Lindner, 2004). Some mass tag reagents of this type are able to differentiate peptides from six to ten samples analyzed at the same time (see section on isobaric tags, this chapter).

The following sections describe some of the major mass tag types and discuss the general protocols for their use.

3.1. ICAT Reagents

Isotope-coded affinity tags (ICATs) are bifunctional mass tagging agents containing a reactive group on one end of the molecule and an affinity capture group on the other end (Aebersold, 2003; Gygi *et al.*, 1999; Roepstorff, 2012) (Figure 12.12). In addition, a portion of the tag can contain stable isotope substitutions, usually designed to be in the cross-bridge between the reactive group and the biotin handle. The original ICAT reagent contained eight deuterium atom substitutions on the outer ends of an ethylene oxide spacer. The isotope tagged version thus differs from its normal atom analog by exactly 8 mass units.

Most ICAT-style compounds contain a thiol-reactive iodoacetyl group on one end and a biotin handle on the other end of a spacer arm (Figure 12.13). Reagents of this type are highly specific for reacting with cysteine thiols in proteins to result in stable thioether modifications containing a terminal biotin group (Figure 12.14).

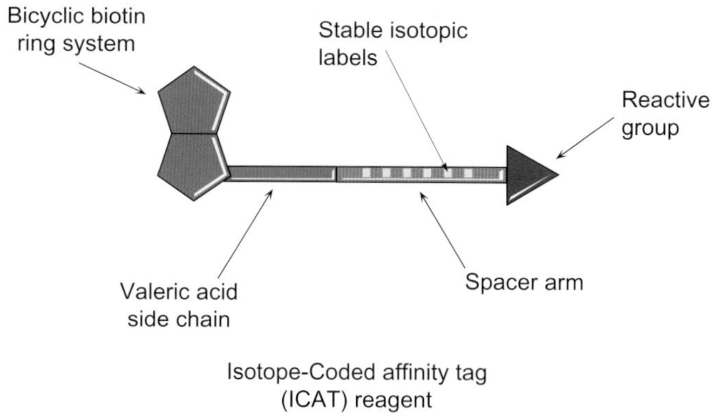

FIGURE 12.12 The general design of an ICAT reagent consists of a biotinylation compound with a spacer arm containing stable isotope substitutions. The reactive group is used to label proteins or peptides at particular functional groups and the biotin affinity tag is used to isolate labeled molecules using immobilized (strept)avidin.

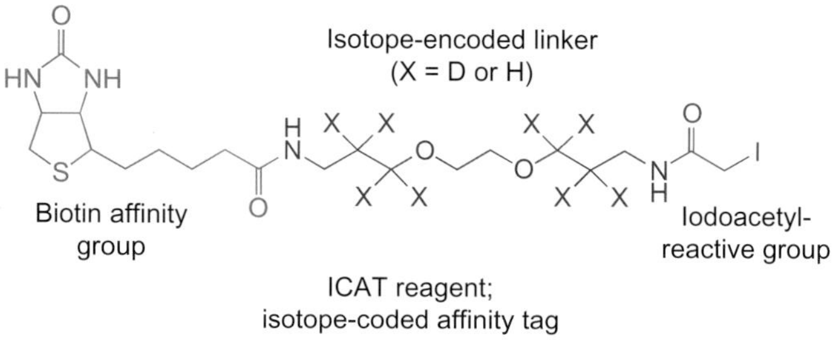

FIGURE 12.13 The original design of the ICAT compound. The iodoacetyl group provides reactivity with thiol groups. The isotopically labeled spacer arm typically is substituted with eight deuterium atoms.

After enzymatic digestion, modified peptides then can be isolated using immobilized (strept)avidin, which specifically binds only to those peptides containing the biotin tag and allows the other peptides to be discarded. Thus, the sample complexity can be reduced to analyze only peptides that contain a cysteine residue, which in the human proteome represents about 26.6% of the total tryptic peptides in a sample. This translates into the ability to cover 96.1% of all the proteins in the human proteome by targeting only cysteine-containing peptides (Zhang *et al.*, 2004; Yan and Chen, 2005).

ICAT reagents can be used to compare two different samples by mass spectometry analysis. For instance, one cell population can be treated with a drug candidate, while another one remains untreated and acts as a control. Alternatively, one cell population can represent a disease state and the control population is the normal cell line. After cell lysis, the proteins in each sample are denatured and reduced to make available all of the cysteine thiols for modification. One sample is then reacted with the heavy atom ICAT reagent, while the other sample is reacted with the normal

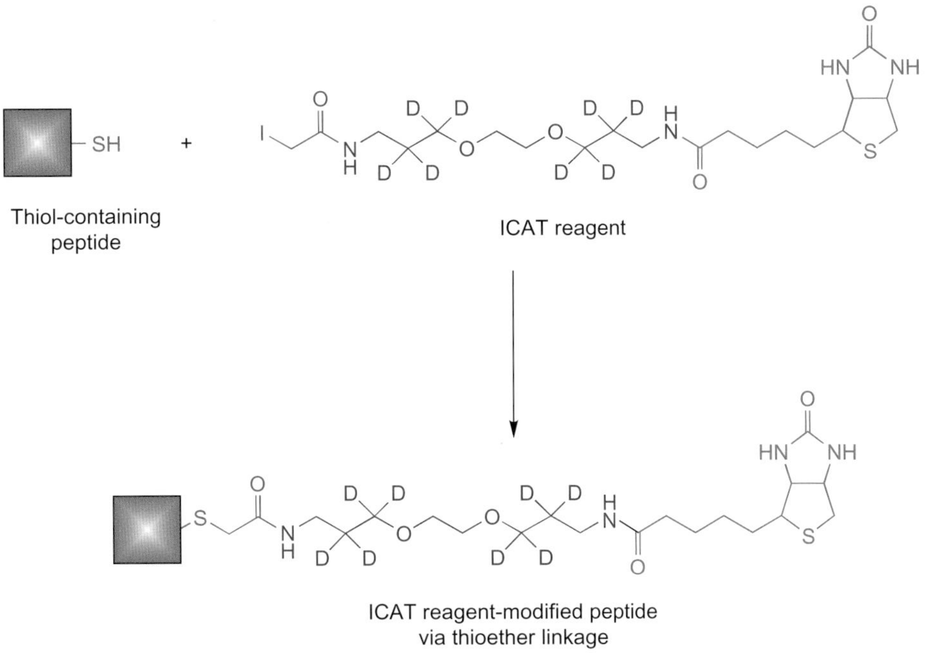

FIGURE 12.14 The ICAT reagent reacts with cysteine-containing peptides to form a thioether bond.

isotope compound. The two samples are then combined and enzymatically digested with trypsin to generate peptide fragments, some of which will contain ICAT-labeled cysteine groups. This combined peptide sample is affinity separated on an immobilized (strept) avidin column (or monomeric avidin column), which binds biotin-labeled peptides from both sample populations equally. After removal of the non-biotinylated peptides by washing the column followed by elution of the ICAT-labeled peptides, the sample is subjected to capillary reverse-phase chromatography leading into ESI or MALDI mass spectrometry analysis. The final HPLC separation again reduces the complexity of the sample set by further fractionating the peptides based on comparative hydrophobicity. In the MS spectrum, the relative peptide concentrations are determined by comparing all peaks separated by exactly the mass unit differential between the heavy atom mass tag and the normal atom mass tag. Each peptide sequence is then identified by fragmentation of the peptides into amino acid ions in a second-dimension MS separation (MS/MS). Comparison of the amino acid sequence of each peptide peak to known sequence databases can identify the protein from which it came. Thus, the resultant peptide peak ratios are directly proportional to the relative amounts of the corresponding proteins present in the cell population.

The original ICAT design was found to have a number of deficiencies that often prevented the reagent from providing acceptable MS results. First, the deuterium isotope-labeled compound has a tendency to behave differently than the normal hydrogen isotope-labeled compound during reverse-phase separation (Regnier *et al.*, 2002). If the labeled peptides that are identical except for the presence or absence of a D_8-ICAT modification do not elute at precisely the same point in an HPLC separation, then the MS analysis will not provide the corresponding peak pairs necessary for quantification. To solve this problem, a second-generation ICAT compound was designed containing ^{13}C isotopic substitutions instead of deuterium atoms. This type of reagent facilitates precise chromatographic separation of the labeled peptides and thus gives far superior performance upon MS analysis.

A second problem in the original ICAT design relates to the presence of the biotin tag. The biotinylated peptides often give undesirable fragmentation patterns during MS/MS analysis, which interferes with the smooth identification of peaks. Removing the biotin tag before mass spectrometry analysis therefore would be beneficial to interpreting the MS results. Another issue with using a biotin tag is the elution step from the immobilized (strept)avidin column. Only under severely denaturing conditions is the interaction between biotin and (strept)avidin disrupted. However, even when using such conditions, the bound peptides do not always get released reproducibly from the column. The result is inefficient recovery of labeled peptides, which directly translates into a lack of precision in the MS data. Even using an immobilized monomeric avidin column does not completely solve this problem, because this affinity support sometimes has higher-affinity binding sites or binds non-biotinylated peptides nonspecifically. To solve these issues, new cleavable ICAT designs were created that contain a bond within the cross-bridge that can be chemically broken (Li *et al.*, 2003). After binding to the (strept)avidin column, elution can be accomplished by cleaving the biotin arm, not by breaking the (strept)avidin–biotin interaction. The cleavage site can consist of a disulfide group within the cross-bridge (Turecek, 2002) that can be reduced for elution from the (strept)avidin column or it can consist of an acid cleavable linker arm (e.g., a carbamate bond) within the ICAT structure (Fauq *et al.*, 2006) (Figure 12.15). Both methods dramatically improve the recovery of labeled peptides from the affinity column and thus provide increased precision in the samples leading into the LC–MS analysis. The ^{13}C-labeled, acid-cleavable ICAT reagent has been used to identify successfully low-level protein expression in highly complex samples (Hansen *et al.*, 2003).

In another ICAT design, termed a "catch-and-release" tag, the compound contains a constrained, sterically hindered disulfide linkage with bulky alkyl groups on both sides. The hindered nature of the disulfide makes it stable to standard protein reduction procedures, but it can be specifically reduced upon the addition of tris(2-carboxyethyl)phosphine (TCEP) (Gartner *et al.*, 2007) (Figure 12.16). This allows proteins to be labeled with the catch-and-release ICAT compound that have undergone reduction using dithiothreitol (DTT) to cleave protein disulfides but not affect the disulfide group in the reagent cross-bridge. Only after capture of labeled peptides on a (strept)avidin column is the cross-bridge cleaved by the addition of TCEP and the labeled peptide recovered.

A variation on the ICAT mass tag concept was made by immobilizing the label on a solid phase (Zhou *et al.*, 2002). Using this design, cysteine-containing peptides are modified directly on a beaded insoluble support. After washing away non-cysteine peptides, the linked

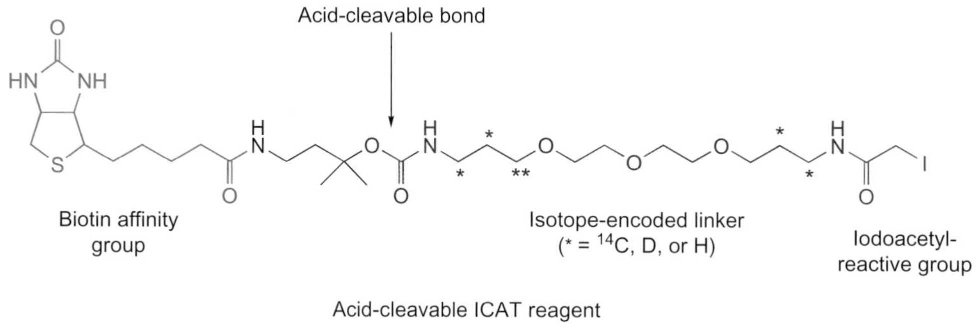

FIGURE 12.15 A more advanced ICAT design uses an acid-cleavable spacer arm to facilitate elution of labeled peptides from a (strept)avidin affinity column. The use of ^{14}C isotopes instead of deuterium labels permits precise reverse-phase separations prior to mass spectrometry that show no elution peak time differences between isotope-labeled and normal atom-labeled peptides.

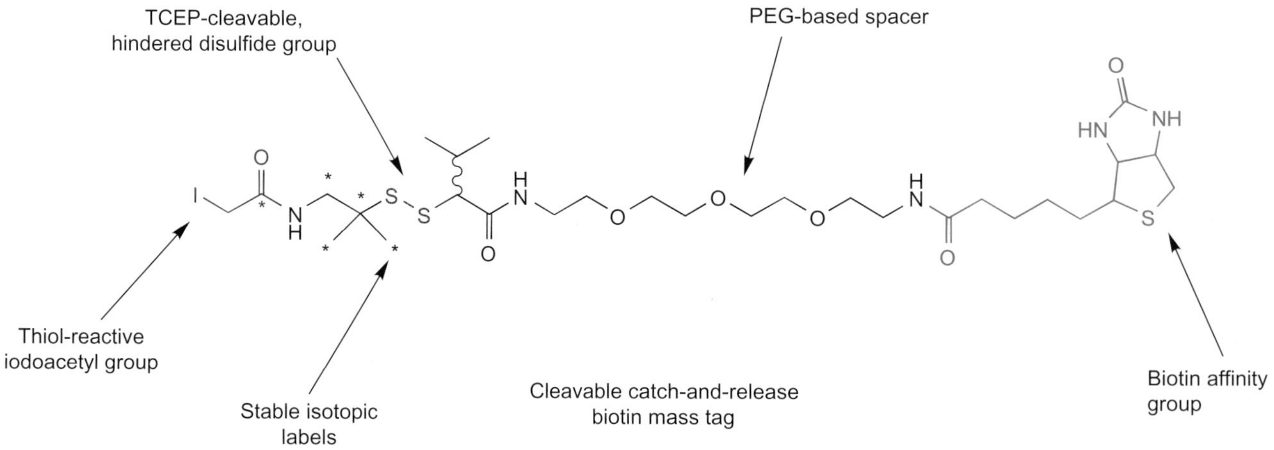

FIGURE 12.16 A catch-and-release ICAT design incorporates a gem-methyl group and an isopropyl group on either side of a disulfide bond within its spacer arm. The hindered disulfide permits the use of standard reducing gel electrophoresis conditions using DTT without reduction. After purification on a (strept)avidin affinity column, however, the disulfide group can be cleaved with TCEP, which provides recovery of the labeled peptides prior to mass spectrometry separation.

peptides can be cleaved from the matrix by use of a photo-cleavable group and eluting off the peptides with an isotope tag modification. This approach results in cleaner and more efficient isolation of tagged peptides and simplifies the ICAT labeling process (Figure 12.17).

Another novel mass tag design involves a spectrally visible ICAT variant developed to include a fluorescent group for detection purposes (Lu *et al.*, 2004). Like the original ICAT reagent, the VICAT compound includes a thiol-reactive iodoacetyl group, a cleavable cross-bridge, an isotopically labeled portion, and a biotin handle. It also has another arm, however, that contains the chromogenic label, which is detectable by absorption at 493 nm and emission at 503 nm. This group allows detection of peptides in samples separated by chromatographic or electrophoretic means. Quantification of the fluorescent tag in isolated peptides can provide absolute information regarding the level of proteins present in a cell.

ICAT-type compounds are designed to enrich for peptides containing one particular amino acid residue, usually cysteine. The affinity capture step removes other non-cysteine peptides and thus reduces the complexity of the MS data set. ICAT reagents can also be designed with a different reactive group that is able to covalently couple to other amino acid groups (or even sites of post-translation modification) to change the selectivity of the peptide population being analyzed. However, it is best to target amino acids or functional groups present in limited amounts within proteins, otherwise the tag may capture more peptides than could be conveniently measured by mass spectrometry. Han *et al.* (2007) developed a hydrazide–ICAT compound to identify proteins modified by 2-alkenals derived from

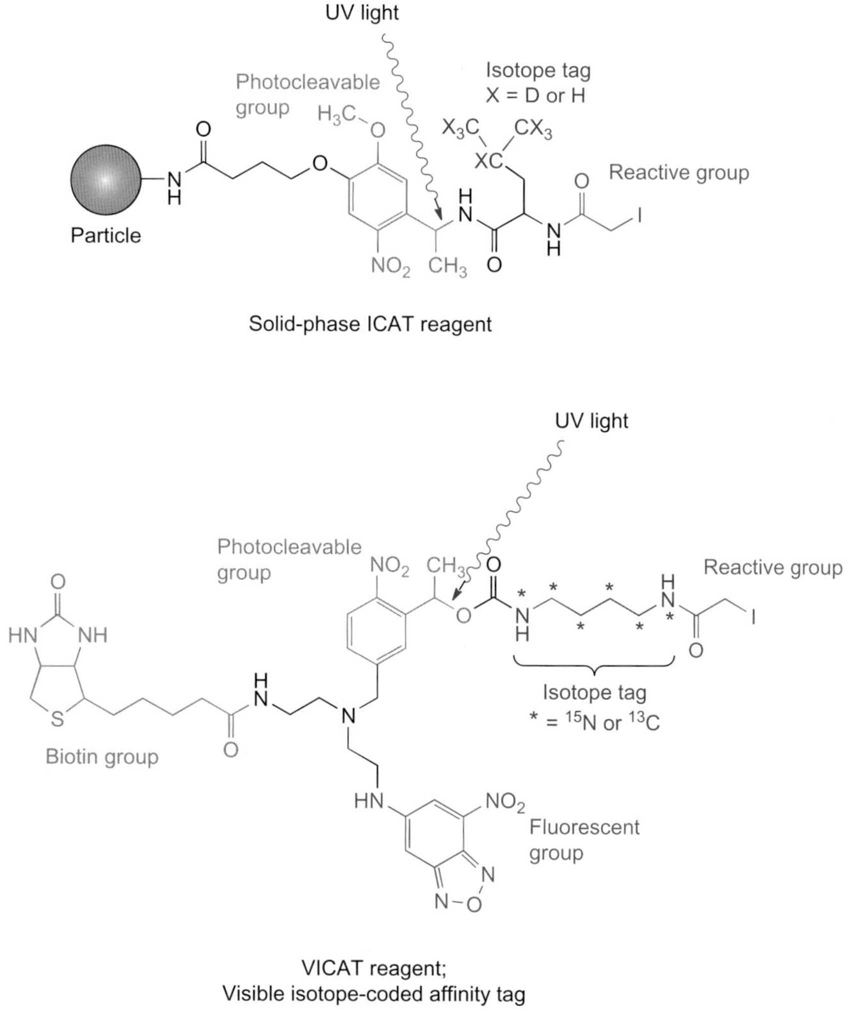

FIGURE 12.17 The solid-phase ICAT reagent provides a thiol-reactive iodoacetyl group to capture cysteine peptides, a spacer containing stable isotopic labels, and a photocleavable group that can release the captured peptides for mass spectrometry analysis. The VICAT mass tag is a solution-phase labeling agent that also has a photocleavable site to release isolated peptides from a (strept)avidin affinity resin. This compound adds a fluorescent group to better detect labeled peptides as they are being isolated from a sample.

lipid peroxidation (LPO). This type of mass tag should be useful for the study of other oxidative changes on proteins, such as those resulting in aldehyde or ketone modifications (see Chapter 2, Section 1). The ICAT reagents are being used for analyzing the global differential expression of proteins in particular cell lines (Helsens *et al.*, 2011), to map protein interaction surfaces (Crane, 2011), and to discover biomarkers related to specific disease states (Abonnenc and Mayr, 2012).

The following protocol describes the use of an acid-cleavable ICAT reagent, currently available from Applied Biosystems. This is not meant to be a detailed method describing every aspect concerning the use of mass spectrometry, but only to describe the modification reaction of the ICAT compound with proteins.

Protocol

1. Grow cells to 70 to 80% confluence and harvest by scraping the cells into 5 ml PBS, 5-m*M* EDTA, pH 7.4. Aliquot cell counts of approximately 2.5–5 × 10^6 for processing. Lyse cells using a detergent lysis buffer (e.g., Poppers, Thermo Fisher) or by mechanical means. Centrifuge and discard the cellular debris. Measure the total protein concentration using the BCA assay (Thermo Fisher) and adjust the protein concentration to 1.5 mg/ml using 50-m*M* Tris, 0.1% SDS, pH 8.0. Separately process a test sample and a control sample made up of different cell populations.
2. Reduce disulfides in the two protein samples by the addition of 2 µl of 50-m*M* TCEP (Thermo Fisher) to each 100-µl aliquot of protein solution. Cover and boil the samples for 10 min in a water bath to completely denature and reduce the proteins. Avoid the use of thiol-containing reductants, such as DTT, as these will react with the iodoacetyl group on the ICAT compound.
3. Dissolve one vial of heavy isotope, cleavable ICAT reagent (Applied Biosystems) in 20 µl acetonitrile (use a fume hood for handling solvents). Dissolve a second vial containing the normal isotope ICAT compound in 20 µl acetonitrile. Vortex mix each vial to dissolve.
4. Add 100 µg of the control protein solution to one vial of dissolved normal isotope ICAT reagent. Mix to dissolve. Add 100 µg of the test protein solution to one vial of dissolved heavy isotope ICAT reagent. Mix to dissolve.
5. React both solutions for 2 h at 37°C.
6. Combine the test sample with the control sample in a single vial. Mix well.
7. Prepare a solution of TCPK–trypsin in 50-m*M* ammonium bicarbonate, pH 8.0, at a concentration of 100 ng/µl. Add 10 µl of the trypsin solution to every 10 µg of combined, labeled protein solution from step 6. Incubate at 37°C for 12 to 16 h or overnight with mixing.
8. Centrifuge the digested peptide mixture to remove any insoluble material. The biotinylated peptides then are purified on 20 µl immobilized monomeric avidin column. The column is first primed with elution buffer (0.4% TFA in 30% acetonitrile) and then washed with binding buffer (100-m*M* ammonium bicarbonate, pH 8.0). The sample is applied and washed through with binding buffer until all not-bound peptides are completely removed. The acetonitrile/TFA elution is subsequently used to cleave off the biotin group from the eluted peptides.

Note: Additional fractionation may be done to reduce further the complexity of the sample, such as the use of ion-exchange chromatography. The eluted, labeled peptides finally are analyzed by LC–MS/MS.

3.2. ECAT Reagents

Element-coded affinity tags represent a new type of isotope tag for mass spectrometry analysis (Corneillie *et al.*, 2003, 2004; Whetstone *et al.*, 2003; Meares *et al.*, 2007). This system uses a bifunctional metal chelate group that securely coordinates a lanthanide metal ion. A reactive group also is present for the modification of certain amino acids in proteins, which typically consists of a bromoacetyl group. This group reacts similarly to an iodoacetyl group and forms thioether linkages with cysteine thiols. The ECAT design includes a DOTA chelating group (Section 1.2) containing four nitrogen atoms and four carboxylates to complex with any of the lanthanide series metal ions *via* eight coordination bonds (Figure 12.18). Simply by using different lanthanide elements within the complex the result will be a unique set of isotope tags having different mass signatures by MS. At least in theory, up to 15 different

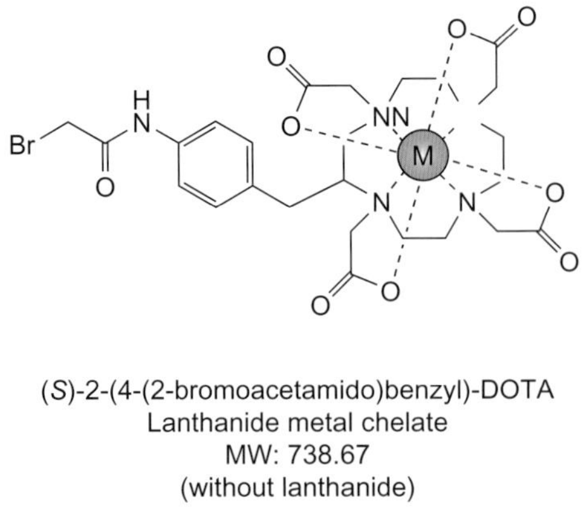

FIGURE 12.18 The ECAT mass tag consists of a DOTA metal chelate group that can coordinate a lanthanide metal ion and a bromoacetyl group for coupling to cysteine-containing proteins.

ECAT mass tag compounds could be created by using all of the different lanthanide metals representing elements 57 to 71. In addition, the lanthanides are naturally mono-isotopic in that they occur mainly in nature with only a single isotope. Only cerium contains a high percentage of another isotope in nature (88% ^{140}Ce and 11% ^{142}Ce); all the other lanthanides are >97% a single isotope. This is important for mass spectrometry separations, as the resultant peaks will not contain extraneous mass signatures due to multiple isotopes.

The ECAT reagent bromoacetamidobenzyl–DOTA (BAD) can be used like an ICAT tag to label only those peptides containing cysteine residues (Figure 12.19), thus reducing the total number of peptides having to be analyzed in a mass spectrometry separation. Unlike the ICAT reagent design, the ECAT compound does not contain a biotin handle for affinity separation. Instead, a monoclonal antibody has been developed with specificity for the DOTA chelate containing a bound lanthanide metal. The antibody will recognize any lanthanide element bound in the chelate and thus function as an affinity ligand for separating ECAT-labeled peptides. The affinity of the monoclonal for ECAT chelate structure allows highly stringent washes to be done prior to elution of the labeled peptides. The affinity column typically is washed with low-pH, high-salt, high-pH, and 30% acetonitrile before elution is done. This removes all traces of nonspecifically bound peptides, so they can't interfere with the mass spec analysis. The ECAT-labeled peptides are then eluted with 20% acetonitrile containing 0.1% TFA.

The ECAT system has an advantage over ICAT reagents in that it is available with more mass signatures than is achievable using ^{13}C- or ^{2}H-labeled tags, which are difficult to synthesize. In addition, it is not hampered by the binding idiosyncrasies of a biotin group interacting with (strept)avidin or the fragmentation problems a biotin tag gives on MS analysis. The ECAT chelate does not generate fragmentation products during the mass spec analysis. Also, the mass defect characteristic of lanthanide metals results in a mass signature upon MS separation that occurs in a relatively unoccupied region of the m/z spectrum (Schneider and Hall, 2005). Thus, identification of ECAT-labeled peptides potentially is simpler than using ICAT reagents.

Another variant of ECAT reagent technology has been developed to analyze the products of protein oxidation (Lee *et al.*, 2006). Called O-ECAT, for "oxidation-dependent, carbonyl-specific element coded affinity tag," the compound contains the same DOTA lanthanide chelating group, but instead of a thiol-reactive bromoacetyl group it has an aldehyde- or ketone-reactive aminoxy group (Figure 12.20). The aminoxy functional group can covalently link to aldehydes or ketones to give an oxime bond that is stable under aqueous conditions (Figure 12.21).

The O-ECAT reagent is a superior alternative to the use of 2,4-dinitrophenylhydrazine (DNPH) (Chapter 2, Section 1.1) in the study of protein oxidation. DNPH modification produces detectable complexes, but it does not provide information as to what amino acids are involved. O-ECAT modifies carbonyl end products of protein oxidation and in addition can provide exact

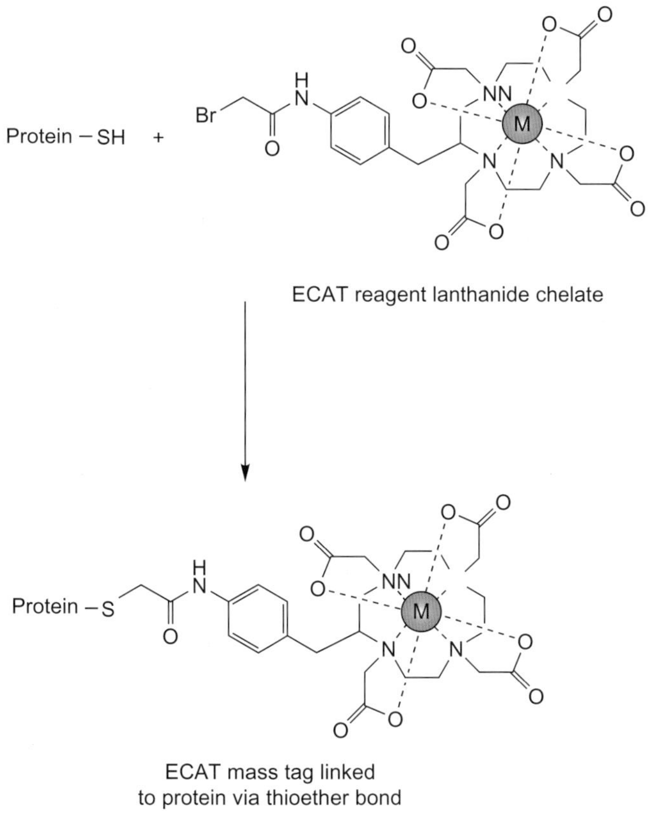

FIGURE 12.19 Reaction of the ECAT reagent with a cysteine-containing protein results in the formation of a stable thioether bond.

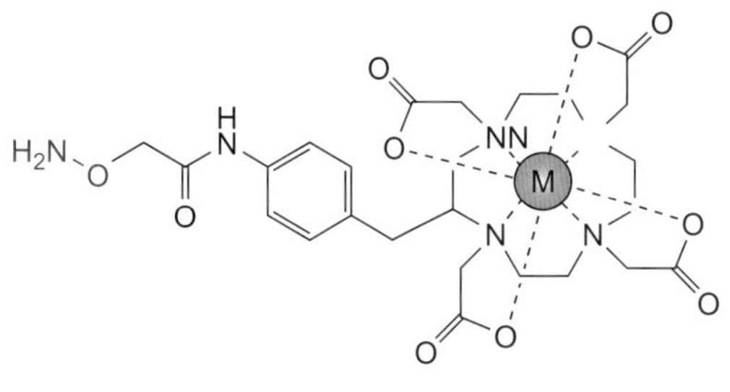

FIGURE 12.20 The O-ECAT reagent structure contains a DOTA chelating group and a terminal aminooxy group for coupling to aldehyde and ketone sites of oxidation within biological molecules.

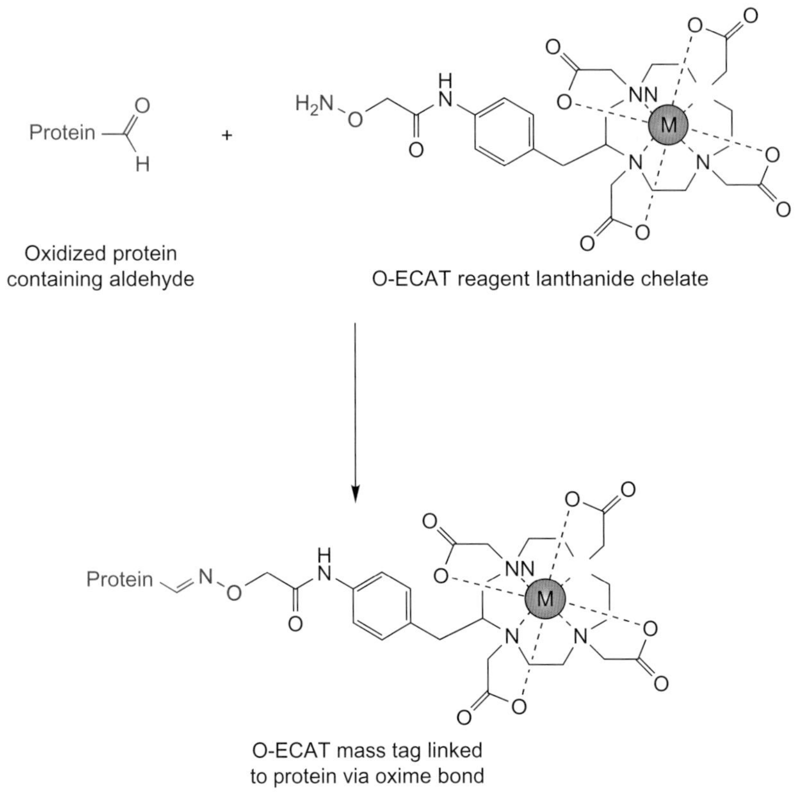

FIGURE 12.21 The O-ECAT mass tag can covalently link to any oxidized proteins containing aldehydes, forming an oxime bond.

information as to the amino acids that were oxidized. Mass spectrometry analysis of modified proteins performed after proteolysis gives the exact amino acid sequences including the sites of O-ECAT reagent modification. The same antibody that is specific for the metal chelate portion of the standard ECAT reagent also can be used to capture and detect the O-ECAT-labeled proteins or peptides. Thus, O-ECAT-modified proteins can be detected in western blots or the sites of oxidation quantified using ELISA-based assays. The technology is being used to study carbonylation of biomolecules as a result of oxidative stress and other oxidation processes in cells and tissues (Møller *et al.*, 2011; Rao and Møller, 2011).

3.3. Isobaric Mass Tags

The use of mass tagging reagents to analyze proteomic data has greatly improved the ability to compare samples for protein expression differences. However, a major limitation of the ICAT procedure (Section 3.1) is that it can only compare two samples simultaneously, usually a test and a control. Even with the ECAT design (Section 2) using multiple lanthanide metals to make a series of different mass tag signatures, it is difficult to extend the analysis to multiple samples, because of the sheer number of peaks that result in the mass spectrometry separation.

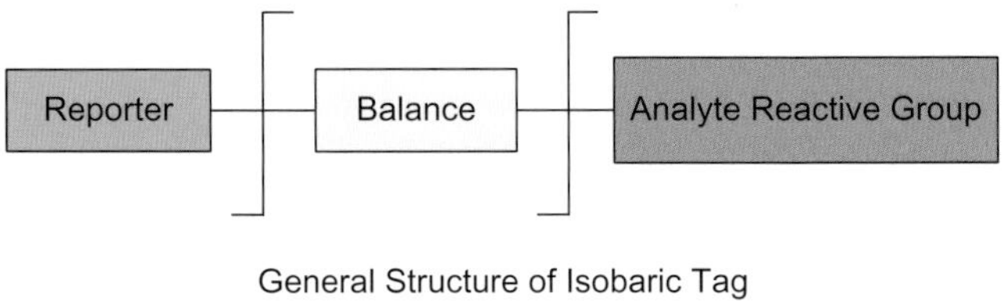

FIGURE 12.22 The general structure of an isobaric mass tag reagent. The reactive group facilitates coupling to discrete sites on peptides, such as amines. The reporter group creates a unique mass signal in MS^2 analysis and its total mass is exactly balanced by the balance group by changing the stable isotopic labels to provide the opposite mass differential as that of the reporter group. The result is that all isobaric tags have the same initial molecular mass, but upon fragmentation in MS^2 the reporter group is released and it provides the unique mass signal to identify the sample being analyzed.

A new type of mass tag extends the benefits of stable isotopic labeling of peptides for mass spectrometry to the analysis of multiple samples and multiple proteins separated simultaneously within the same mass spectrometry run. An isobaric tag consists of a reactive group for coupling to peptides followed by an isotopically labeled mass normalization group, a cleavable linker, and another isotopically labeled group, called a mass reporter (Figure 12.22). For every isotopic substitution in the mass reporter region, the mass normalization group has an inverse mass substitution. Using this balancing process, the total molecular mass of the entire tag always stays the same—thus the name "isobaric tag." A series of isobaric labels can be created by careful selection of the isotopic substitutions in the reporter group, which are exactly balanced in the normalization group.

If there are enough potential isotopic substitution sites in the isobaric tag design, a set of four to ten tags can be created in which each has a different reporter mass, but all of them have the same total molecular weight. Proteome Sciences and Thermo Fisher, Applied Biosystems (now ABSciex), and PerkinElmer/Agilix (now discontinued) each have developed isobaric tag sets based on the reporter group/normalization group blueprint for multiplexed sample analysis. Figure 12.23 shows examples of isobaric tag design that were obtained from company advertising, scientific publications, or the associated patents. Most of the tags also contain a tertiary amine group or a guanidino group to aid in the ionization of the labeled peptide and provide better mass spectrometry signals.

In use, a protein sample is first proteolytically digested and labeled with an individual isobaric tag. The most common isobaric tag design contains an amine-reactive NHS ester group to label each peptide at its N-terminus. Therefore, within a given sample all of the different peptides after labeling will have a unique isobaric tag modifying them with a characteristic reporter group mass. Unlike the use of an ICAT tag that targets only cysteine-containing peptides, modification of the N-terminus of every peptide ensures 100% coverage of the proteome. If multiple samples are being analyzed, then each peptide sample is separately labeled with a different isobaric tag in the series. After labeling, all of the samples then are combined and separated chromatographically to reduce the total sample complexity.

Since each isobaric tag in a set is structurally identical except for its isotopic substitution pattern, all of them perform identically with regard to peptide modification and chromatographic separation. Multiple peptide samples then can be labeled separately with different tags in a series. The samples can then be combined, subjected to fractionation typically done by multidimensional protein identification technology (MudPIT) (Lin *et al.*, 2003), and injected into a mass spectrometer for analysis. As each peak comes off the MudPIT separation system and goes into the mass spectrometer, it contains the same sequences of labeled peptides from each sample that happen to elute under the instant conditions of salt strength and solvent addition being used at that moment. The only difference in the peptide mixtures within a given chromatographic peak, then, is the type of isobaric label attached to them, which is indicative of the sample it came from. Thus, the peptides representing a particular sample will all be labeled with a tag having a characteristic reporter group mass. Another set of peptides with the same amino acid sequence but coming from another sample will be labeled with a different isobaric tag having another reporter group mass signature.

It is only upon tandem mass spectrometry analysis that the isobaric labels can be distinguished and the peptides identified. For instance, if four samples were each labeled with a different isobaric tag, in the first dimension of a MS separation a given peptide from each sample will appear in the same peak, because the peptides will all have the same sequences and the isobaric tags labeling them will all have the same mass signatures. Therefore, their mass:charge ratios will all be the same and they will be indistinguishable at this point. However, upon MS/MS separation wherein additional energy is used to promote fragmentation of the labeled peptides within the peak, usually by collision-induced dissociation (CID) or electron capture dissociation (ECD), the peptides will break down into their respective charged amino acids and the isobaric tags will be cleaved to release their reporter groups. From the peaks that result from this second-stage MS separation both the peptide sequence and the sample from which it came can be identified from the amino acids and the mass of the respective reporter groups.

There are many publications describing the development and use of isobaric tags. For instance, Thompson *et al.* (2003) described the development of tandem mass tags based on an isotopically labeled peptide design. A later iteration of this concept uses commercially available isotopes of alanine to form the reporter group, followed by a piperazine ring (as the charge-carrying group), a second alanine used as the mass normalization group, and a proline residue that functions as the electrospray cleavable linker (Thompson *et al.*, 2007). Ross *et al.* (2004) used a set of four isobaric tags designed around an *N*-methyl piperazine group to study the global protein expression of a wild-type yeast strain *versus* strains defective in certain pathways. These tags are part of the iTRAQ (isotope tags for relative and absolute quantitation) system from Applied Biosystems (now ABSciex). Figure 12.23 shows some of the major structural characteristics of commercial isobaric tags. Figure 12.24 illustrates the design of a multiplex isobaric set that uses a piperidine ring as the reporter group, showing all of the various isotopic modifications done to balance the reporter and normalization group to create six unique tags. The use of tandem mass tags (TMTs; Proteome Sciences and Thermo Fisher) in relative protein quantitation has been reviewed (Dayon and Sanchez, 2012).

Isobaric labels thus permit quantitative information regarding protein expression levels in multiple samples analyzed simultaneously by MS. The multiplexed capability of these reagents allows the measurement of

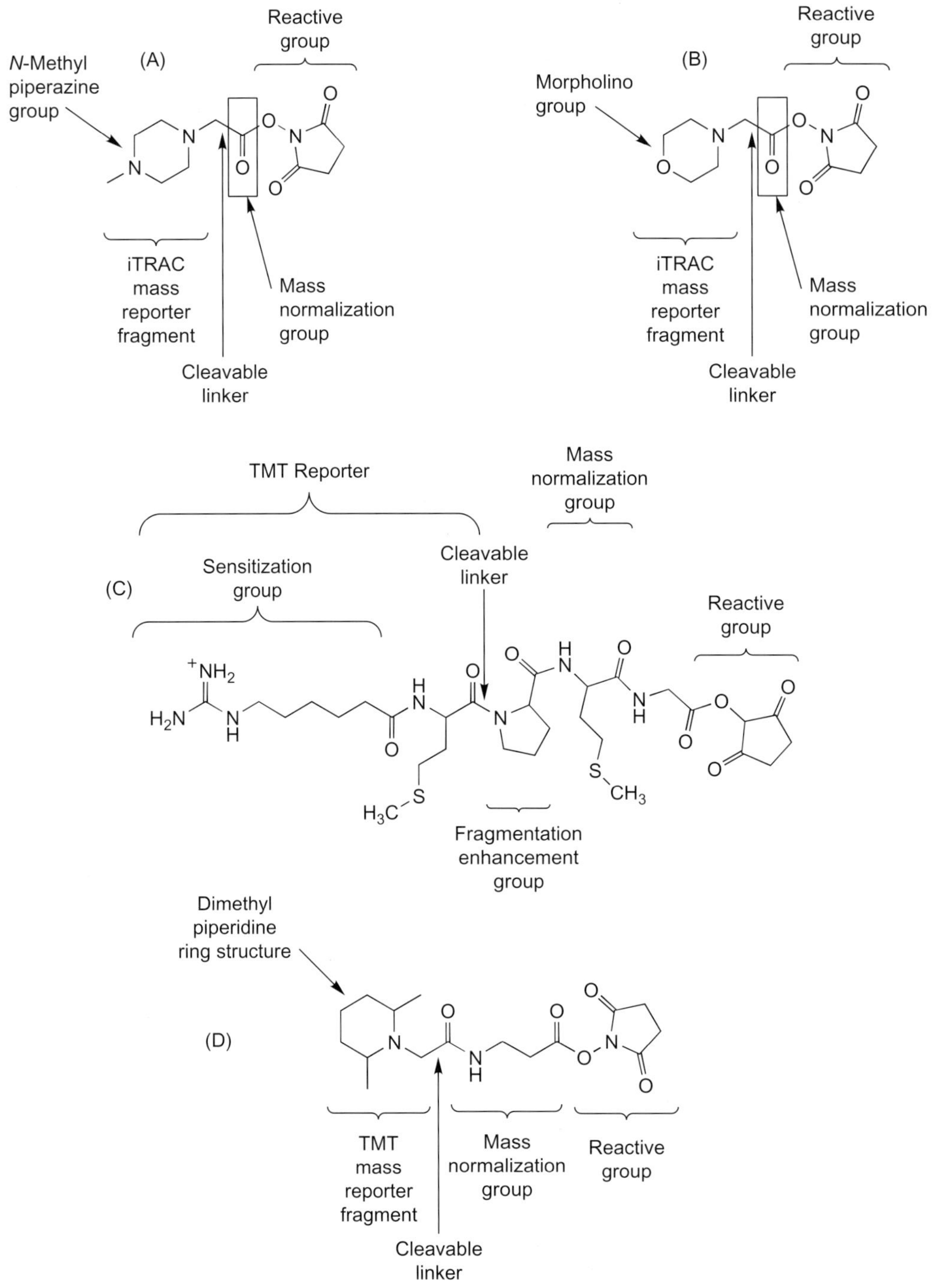

FIGURE 12.23 A number of isobaric tags have been developed having different structural motifs. (A and B) Isobaric tags used as part of the iTRAC reagents (Applied Biosystems). (C) An early design of isobaric tags using a core of amino acid derivatives (Proteome Sciences). (D) The TMT tag design as commercialized by Proteome Sciences. All of these isobaric tags contain a sensitization group to enhance mass spectrometry detection, which is represented either by a tertiary amine that can be protonated to carry a positive charge or a guanidino group.

peptides and proteins in diseased samples, treated samples, and normal samples all in the same experiment. In addition, since all peptides from a given protein get labeled at their N-termini, the MS analysis generates more than one peptide signal, which can be used to confirm protein identity with greater confidence than using a cysteine label, like ICAT.

The protocol for using isobaric tags differs from that described previously for the ICAT or ECAT type of reagents. In the following method, the proteins are

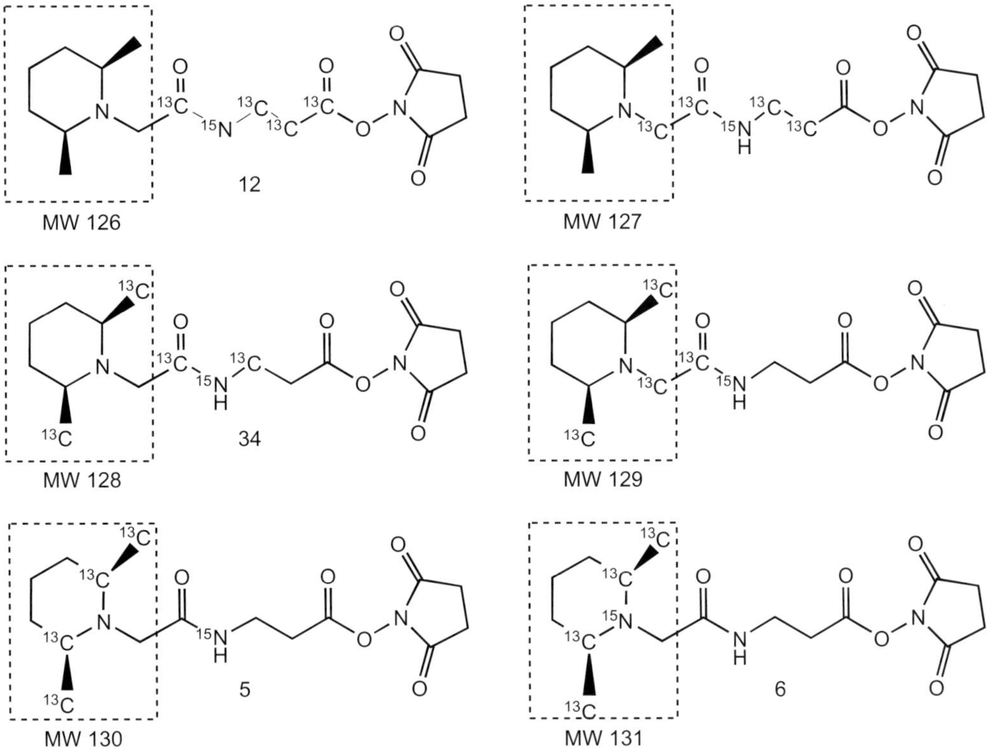

FIGURE 12.24 Isobaric tags allow multiplexed analysis of different samples by changing the mass of the reporter group and balancing that change by the opposite change in the balance group. This figure illustrates the 6-plex tag system from Proteome Sciences, which contains six different reporter groups (boxed areas). All of the reagents contain amine-reactive NHS esters for modifying lysine side chains in proteins and peptides.

denatured and the disulfides reduced and then alkylated to block them permanently. This eliminates disulfide re-association and also prevents the isobaric tags from forming thioester modification with cysteine thiols. Next, the proteins are digested with trypsin and then modified with an isobaric tag. Each sample is labeled with a different isobaric compound so that the samples can be differentiated upon MS/MS analysis.

The following protocol illustrates the modification reaction and the handling of the protein samples, but it is not meant to be instructive of mass spectrometry techniques.

Protocol

1. Grow cells to 70 to 80% confluence and harvest by scraping the cells into 5 ml of 0.1-*M* sodium borate, pH 7.5. Avoid the use of amine-containing buffers, as these will react with the NHS esters on the isobaric tags. Aliquot cell counts of approximately 2.5–5 × 10^6 for processing. Lyse cells using a detergent lysis buffer (e.g., Poppers, Thermo Fisher) or by mechanical means. Centrifuge and discard the cellular debris. Measure the total protein concentration using the BCA assay (Thermo Fisher) and adjust the protein concentration to 1 mg/ml using 0.1-*M* sodium borate, 0.1% SDS, pH 7.5. A total of 100 µg of each protein sample can be used in this protocol. Separately process test samples and a control sample made up of different cell populations. Simultaneous measurement can be done on a total number of samples equal to the number of different isobaric tags available.
2. Reduce disulfides in the protein sample by the addition of 2 µl of 50-m*M* TCEP (Thermo Fisher) to each 100-µl aliquot of protein solution (final concentration 1-m*M*). Cover and boil the samples for 10 min in a water bath to completely denature and reduce the proteins. Alternatively, reduction may be done at 60°C for 1 h. Avoid the use of thiol-containing reductants, such as DTT, as these will react with the thiol blocking agent used in the next step.
3. Add 6 µl of iodoacetamide to each sample solution and react with mixing for 30 min at room temperature.
4. Prepare a solution of TCPK–trypsin in 0.1-*M* sodium borate, pH 7.5, at a concentration of 400 ng/µl. Add 10 µl of the trypsin solution to each sample and incubate at 37°C for 12 to 16 h or overnight with mixing.

5. Dissolve the isobaric tagging reagents in acetonitrile or ethanol at a concentration of 50-m*M* (or according to the manufacturer's recommendations). Use a fume hood to handle organic solvents.

6. Add a quantity of the appropriate isobaric tag solution to each sample to provide a final concentration of 10 to 20-m*M*. This quantity of reagent will ensure a large molar excess of reagent over the concentration of peptides present in order to modify completely all peptides at their N-terminus. Note that the ε-amino groups of lysine residues also will be modified by this procedure. React for 1 h at room temperature.

7. To eliminate acylation products at tyrosine residues, add 1 μl of 15-*M* hydroxylamine solution in water to each protein sample and incubate for 30 min at 37°C (Zappacosta *et al.*, 2006).

8. Combine the contents of each labeled sample into a single tube and mix by vortexing, then centrifuge.

9. Before LC–MS/MS analysis, the sample must be cleaned up to remove excess salts, SDS, reducing agent, and cell lysis buffer components. This typically is done by using a strong cation exchange matrix. Protocols may be found for this procedure in the instruction booklets related to the use of isobaric tagging reagents (e.g., Thermo Fisher, ABSciex).

CHAPTER

13

Silane Coupling Agents

A silane compound is a monomeric silicon-based molecule containing four constituents. Since silicon is in the same family as carbon on the periodic table, it too is able to form covalent bonds with four other atoms. However, it is less electronegative than carbon and undergoes reactions that are unique compared to typical organic compounds. Silanes that contain at least one bonded carbon atom are called organosilanes. Organosilanes can also have hydrogen, oxygen, or halogen atoms directly attached to the silicon atom core. Some of these derivatives are highly reactive and can be used to form covalent linkages with other molecules or surfaces (Mittal, 2009; Borup and Weissenbach, 2010). The process of adding a silane coating (R_3Si–) to a particle or surface is called silanization or sometimes silation. (Note: The process of adding a silyl group to a functional group on a molecule is called silylation, which is a derivatization method used extensively to separate molecules in gas chromatography or as a temporary protecting group in organic synthesis.) The more useful organosilanes for silanization purposes are those that contain a functional organic component in addition to one or more silane-reactive groups. The organic portion also may contain a reactive group, which allows conjugation of the organosilane molecule to other organic compounds. Conversely, the silane-reactive groups typically are unreactive toward organic molecules, but can covalently couple to certain inorganic substrates. The advantage of this type of functional silane derivative is to promote the bonding of an organic molecule to an inorganic particle, surface, or substrate.

The general structure of a functional silane coupling agent is shown in Figure 13.1. The organic arm typically has a structure that terminates in a functional group or reactive component, which facilitates the covalent linkage to another organic molecule. The other part consists of the silane-reactive groups attached directly to the silicon atom and can be of several types. They may comprise simply a hydrogen atom (called a silicon hydride); a halogen–silicon derivative, such as a chlorine atom (called a chlorosilane); an –OH group (called a silanol); or groups containing methyl ether or ethyl ether organic constituents (called a methoxy- or ethoxysilane, respectively).

The general reactions of silane coupling agents toward inorganic substrates are illustrated in Figure 13.2. The reaction mechanism and formation of a reactive intermediate can be different from that usually encountered with organic reactive groups. Unlike most other reactive groups discussed in this book, the most common silane coupling groups (alkoxy) often must first undergo hydrolysis to form a reactive intermediate, which then couples to the substrate. The hydrolysis of an alkoxysilane produces a silanol, which actually is the highly reactive form necessary for coupling to inorganic surface hydroxyls.

The initial reaction that occurs in solution or near an inorganic substrate is condensation of the silane coupling agents together to form a polymer matrix linked together by –Si–O–Si– bonds. Concurrently, the growing silane network interacts with the inorganic substrate through the formation of a hydrogen bonding network with the –OH groups on its surface. Another condensation reaction then occurs, usually requiring heat or vacuum to remove water, which results in the formation of covalently linked organosilane polymer

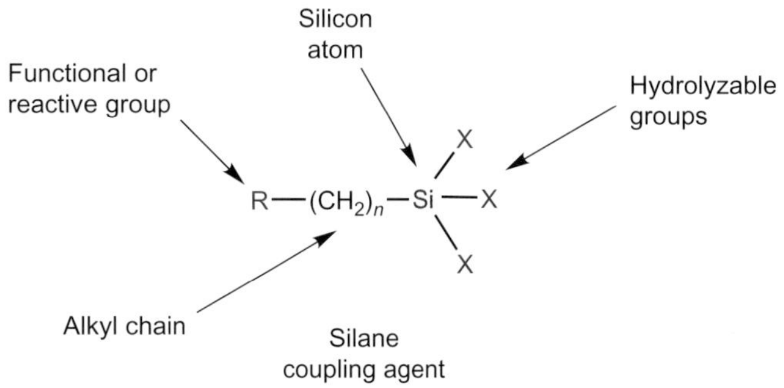

FIGURE 13.1 The general structure of a silane coupling agent includes a functional group or reactive group at the end of an organic spacer. This alkyl chain is attached to the central silicon atom, which also has up to three hydrolysable groups attached to it.

Bioconjugate Techniques, Third Edition.
DOI: http://dx.doi.org/10.1016/B978-0-12-382239-0.00013-3

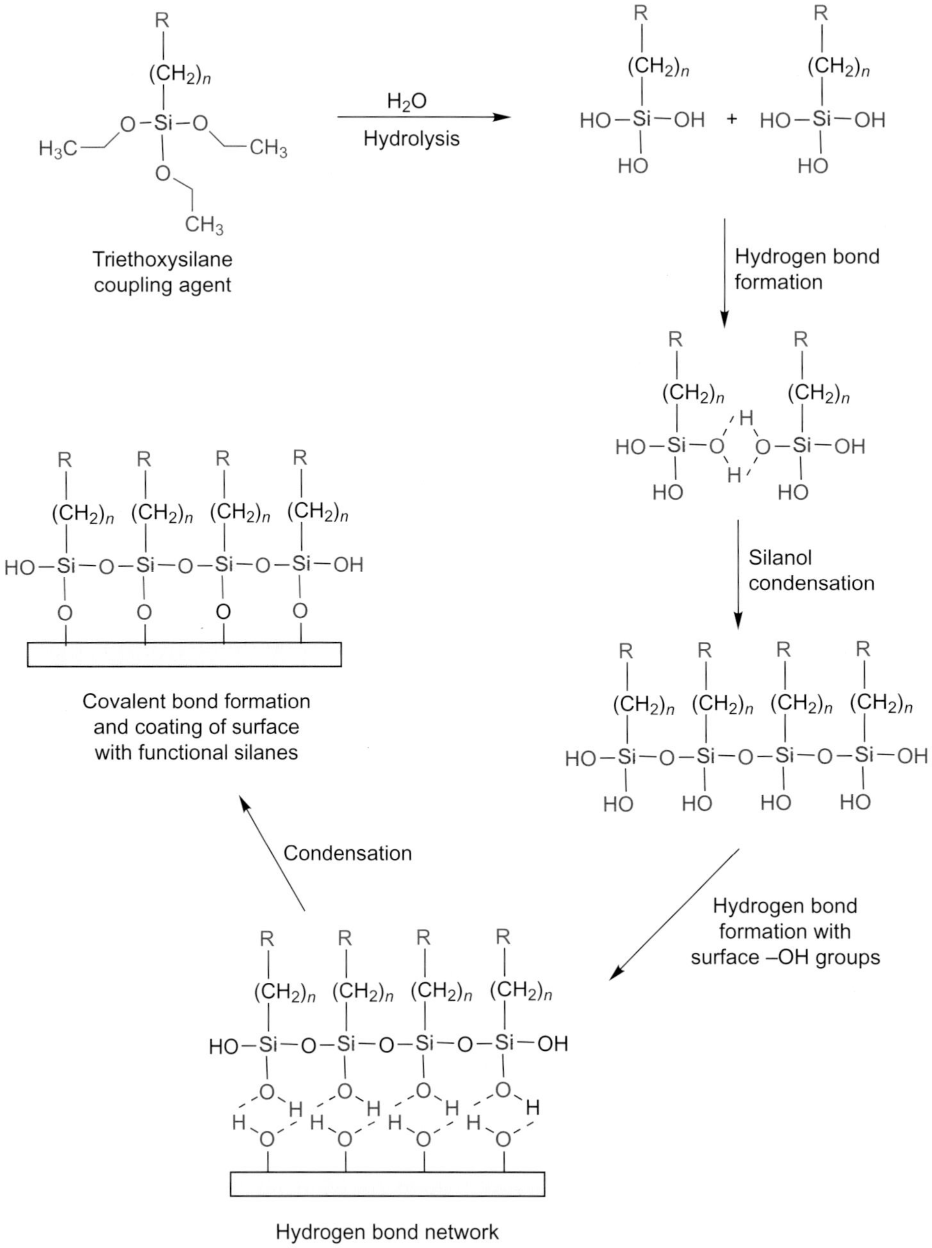

FIGURE 13.2 The reactions involved with the coupling of an organosilane compound to an inorganic surface containing available –OH groups. The first step in the reaction involves the hydrolysis of the alkoxysilane groups to form highly reactive silanols. The silanols undergo hydrogen bonding with other silanols in solution and on the substrate, resulting in a network of associated organosilane derivatives. A condensation reaction then takes place to form a polymerized coating of the organosilane on the surface of the substrate.

to the surface, forming stable siloxane linkages (or oxane bonds). The organosilane coating applied in this manner does not result in a monolayer, but a thicker polymer layer with the reactive organic components extending out from the surface. The thickness of the organo–siloxane layer is dependent on the concentration of silane coupling agent used in the reaction and the amount of water present in the solution. Gelest reports that deposition of a silane coupling agent using a 0.25% aqueous solution will result in a layer approximately three to five silanes thick.

Alkoxysilanes are used for modification of surfaces due to the instability of silanol derivatives, which will spontaneously hydrogen bond and conjugate together in solution. The formation of an alkoxy derivative of the corresponding silanol acts as a stabilizer to prevent silanol polymerization. However, alkoxysilanes typically are unreactive with substrate hydroxyl groups under ambient temperature conditions. Ethoxysilanes are virtually unreactive to substrate –OH groups without prior hydrolysis. Methoxysilanes are the most reactive, but only react very slowly at room temperature. However, under the right conditions, both methoxy- and chlorosilane groups are sufficiently reactive to couple directly with inorganic substrate functionalities without prior hydrolysis. For instance, the addition of a catalyst to the reaction to increase the hydrogen bonding capability of substrate hydroxyls has been done to increase the reaction rate for methoxysilanes (Kanan *et al.*, 2002). Also, chlorosilanes and methoxysilanes can be reacted in an organic solvent without the presence of water (i.e., in THF, toluene, or hydrocarbon solvents), if performed under refluxing conditions to drive the reaction to completion. In this case, a siloxane polymer network does not form in solution, because no hydrolysis occurs to form silanols on the silane coupling agents. Therefore, instead of a thick polymer layer forming on the substrate as in aqueous reactions, a monolayer results where each organosilane is coupled directly to the substrate *via* a siloxane bond. The reactive organic components stick out from this monolayer for eventual conjugation with biomolecules or other molecules.

Therefore, functional silanes in a sense are bifunctional compounds that can be used to attach one substance to another. They have been used for many years as adhesive agents to promote the bonding of an organic layer to an inorganic layer. Some examples are in the coupling of inorganic metals, particles, fiberglass, or fillers to organic polymers, plastics, rubber, or resins. The uses for functional silanes thus are extensive and span many industries, such as automotive, building materials, coatings, and electronics, to name but a few. For reviews of silane coupling agents and their uses, see Borup and Weissenbach (2010), Mittal (2009), Plueddemann (1991), and VanDerVoort *et al.* (1996).

Some common inorganic substrates for use with silane coupling agents in approximate order of efficiency and stability for modification include silica, quartz, glass, and the oxides of aluminum, copper, tin, titanium, iron, chromium, zirconium, nickel, and zinc. All of these substrates can have functional inorganic –OH groups on their surface that react with the silanols on the silane coupling agents to form siloxane bonds. Sometimes the surfaces require prior treatment to form the –OH groups and remove contaminants, which will interfere with the silanization process. Glass, for instance, often needs to be treated with acid (5% HCl) for several hours to remove non-binding metal ions, especially sodium, potassium, and calcium, which are ubiquitous in the environment. In addition, treatment with a mixture of 25% sulfuric acid and 15% hydrogen peroxide (piranha solution) for about 30 min is carried out to create a high density of hydroxyl functionalities suitable for silane modification. Glass slides also can be cleaned and washed prior to modification with a silane with DMSO, ethanol, and water, and then etched using 10% NaOH (w/w) in water for 1 h.

The distinctive bifunctional nature of silane coupling agents has led to their application in bioconjugate chemistry. There are many silane coupling agents commercially available that contain functional groups or reactive groups that can be used to covalently link biomolecules to inorganic substrates (Dow Corning, Gelest). Inorganic substrates treated with a suitable silane coupling agent subsequently can be used to couple antibodies, proteins, oligonucleotides, or other biomolecules containing the appropriate chemical groups for reaction. By judicious choice of the proper organo-functional component, a silane coupling agent can be used to design virtually any bioconjugation complex. The following references provide examples of how silane coupling agents have been used to modify surfaces and provide covalent linkage sites for coupling biomolecules or organic substituents: Debs *et al.* (2009), Iijima *et al.* (2009), Lin *et al.* (2011), Zhao *et al.* (2011), and Smirnov *et al.* (2012).

1. SILANE REACTION STRATEGIES

The reaction techniques that can be used with functional silane coupling agents are varied. Reactions can be performed in aqueous solution, entirely in organic solvent, in organic solutions containing a small amount of water, and even in the vapor phase. They also can be done at room temperature or under elevated temperature conditions. The choice of reaction strategy often is governed by the type of substrate initially being modified by the silane compound and the inorganic reactive groups on the silane. For instance, if the inorganic substrate can be treated by mixing or suspending it in a solution of functional silane, such as with particle coatings, then this may be simplest option for silanization. Another choice may be to dip the substrate into the silane solution, as is sometimes done with glass slides and other small surfaces that can be handled in trays or baskets. Alternatively, complex surfaces, such as those often encountered with devices, may be treated in a chamber wherein the silane compound is volatilized by heating or by being placed under vacuum. This option usually results in the uniform modification of all surfaces with a thin layer of functionalized silane by vapor phase deposition.

The following sections describe the major organosilane reaction strategies used to modify surfaces and provide suggested protocols, which may be used with the functional silane compounds described later in this chapter. Many organosilane compounds are initially only sparingly soluble in aqueous solution. Therefore, solutions containing silanes in either 100% aqueous or in a water/organic mixture containing a large amount of the aqueous component often will appear as insoluble two-phase systems. As the alkoxy groups hydrolyze, the silanols will dissolve into the aqueous solution and the solution will become clear. However, if the solution is allowed to sit long enough, the silanols will react, form siloxane linkages, and polymerize in solution. At this point, the solution again will become cloudy due to these large polymers.

1.1. Aqueous/Organic Solvent Deposition

This method involves the deposition of functional silane onto an inorganic substrate using a dilute solution of water in organic solvent. The small amount of water at acid pH facilitates the hydrolysis of the alkoxy groups on the silane to form the reactive silanols necessary for coupling. The method of silane solution contact with the substrate may be accomplished by suspension stirring, dipping, spinning, or spraying the silane solution.

Protocol

All operations should be carried out in a well-ventilated fume hood. Use care not to inhale vapors or get reactive silanes on your skin or in your eyes.

1. Prepare a solution containing 3 to 5% water in ethanol (v/v) and adjust the pH to 4.5 to 5.5 with acetic acid.
2. Dissolve a silane coupling agent in the acidic water/ethanol solution with stirring to a final concentration of 2 to 5% (v/v). Allow hydrolysis to occur for 5 min at room temperature to form reactive silanols.
3. Contact the inorganic substrate with the silane solution for as brief as 2 min to as long as several hours, depending on the degree of organosilane polymer deposition desired on the surface. Benters *et al.* (2002) performed the reaction on glass slides in a stirring bath for 2 h. Optimization of the reaction time should be performed to determine the best performance of the modified substrate in its intended application.
4. Wash the substrate several times with ethanol or the water/ethanol mixture to remove excess silane compound.
5. Cure the organosilane-modified substrate by incubation at 110°C for 30 min or at room temperature in a low-humidity environment. Curing under vacuum also will aid in the removal of water and the formation of siloxane bonds.

1.2. Aqueous Deposition

Functional organosilanes can be applied to substrates directly from aqueous solutions, provided the silane compound is soluble in water.

Protocol

1. Dissolve an alkoxysilane in water at a concentration of 0.5 to 2.0% (v/v). If the compound is not very soluble in water, a nonionic detergent can be added to the solution at 0.1% to promote solubility.
2. Adjust the solution with acetic acid to pH 5.5.
3. Contact the substrate with the silane solution from 2 min to as long as 1 h, depending on the degree of organosilane polymer deposition desired on the surface. Optimization of the reaction time should be done to determine the best performance of the modified substrate in its intended application.
4. Remove excess solution and cure at 110 to 120°C for 30 min to remove traces of water and form the siloxane bonds.
5. Wash the substrate with water or buffer.

1.3. Organic Solvent Deposition

This method of silanization, which uses organic solvent without the addition of water, is suitable for highly reactive silane derivatives, such as chlorosilanes, aminosilanes, and methoxysilanes. This procedure will not work for ethoxysilanes, as these compounds are not reactive enough without prior hydrolysis to create the silanol. This method is convenient for use in silica particle modification and for the functionalization of metallic nanoparticles having the requisite –OH groups present (see Chapter 14, Section 5).

Protocol

1. In a fume hood, dissolve the organosilane coupling agent containing chloro-, amino-, or methoxysilane-reactive groups in toluene, THF, or a hydrocarbon solvent at a concentration of 5% silane.
2. To ensure a thin monolayer of silane modification on the substrate, first remove all traces of water by drying it at 150°C for 4 h. This step is especially important for nanoparticle modification, as the high surface area of the particle population can hold a lot of water by hydrogen bonding. Note that if the substrate contains water, a thicker polymer layer of silane will build up on it due to hydrolysis of the reactive groups to silanols close to the surface of the substrate. After drying, wash the particles with solvent several times by centrifugation before resuspending them in the silane solution. For other substrates, submerge them in the silane solution with mixing.
3. React with gentle stirring for 12 to 24 h by heating to reflux in a fume hood.
4. Cool the reaction and wash the substrate with solvent to remove excess silane reagent and reaction byproducts. The modified substrate may be dried or washed into aqueous buffer for further conjugation with biomolecules.

1.4. Vapor-Phase Deposition

Many silane coupling agents can be applied to substrates by volatilization in an enclosed chamber under heat or vacuum. In this approach, the substrate is placed within the chamber in a fashion to allow for vapor-phase molecules to access all areas that are to be derivatized. This method is commonly used for silanizing glass slides or substrates that are difficult to suspend in a silane solution. Slides are often placed in racks within the chamber and all surfaces get modified with silane in a uniform manner. Vapor deposition also uses a very small amount of organosilane compound compared to what may be necessary to fully immerse a substrate device in solution.

Protocol

1. In an enclosed chamber made of glass or acrylic, such as a vacuum chamber, place the substrate to be modified in such a manner as to expose the surfaces to the vapor phase.
2. Place the organosilane coupling agent in a small reservoir under or next to the substrate in the chamber. If volatilization by heating is to be done, an explosion-proof heating device should be used to maintain the silane solution temperature at 50°C or above. Alternatively, apply a vacuum to the chamber until the silane compound begins to volatilize. The vacuum tubing may be clamped off to maintain vacuum within the chamber once a sufficient level of evacuation has been reached. Vapor-phase deposition usually requires that the silane have at least a 5-mm vapor pressure to achieve adequate concentration in the atmosphere within the chamber.
3. React for 4 to 24 h within the chamber to result in a uniform coating of the substrate surface with organosilane. Often surfaces are coated overnight to complete the reaction.

2. FUNCTIONAL SILANE COMPOUNDS

Functional silane compounds containing an organo-functional or organo-reactive arm can be used to conjugate biomolecules to inorganic substrates. The appropriate selection of the functional or reactive group for a particular application can allow the attachment of proteins, oligonucleotides, whole cells, organelles, or even tissue sections to substrates. The organosilanes used for these applications include functional or reactive groups such as hydroxyl, amino, aldehyde, epoxy, carboxylate, thiol, and even alkyl groups to bind molecules through hydrophobic interactions.

The following sections discuss the organosilane compounds commonly used for conjugation to inorganic surfaces. These reagents and many other silane derivatives are available from a number of commercial sources, which include Dow Corning, Gelest, and Aldrich, among others.

2.1. 3-Aminopropyltriethoxysilane (APTS) and 3-Aminopropyltrimethoxysilane

These two silane coupling agents are among the most popular choices for creating a functional group on an inorganic surface or particle. Both reagents contain a short organic 3-aminopropyl group, which terminates in a primary amine. The only difference in these compounds is the silane-reactive portion that contains either a triethoxy group or a trimethoxy group. Thus, the reagents display differences in reactivity toward substrate –OH groups related to the relative reactivity of the alkoxysilane functionalities. The trimethoxy compound is more reactive and can be deposited on a substrate using 100% organic solvent without the presence of water to promote hydrolysis of the alkoxy groups prior to coupling. In this case, the organic solvent deposition protocol described in the previous section can be used to covalently modify substrates with a layer of aminosilane compounds. The advantage of this process is that a thinner, more controlled deposition of the silane can be made to create a monolayer of aminopropyl groups on the surface.

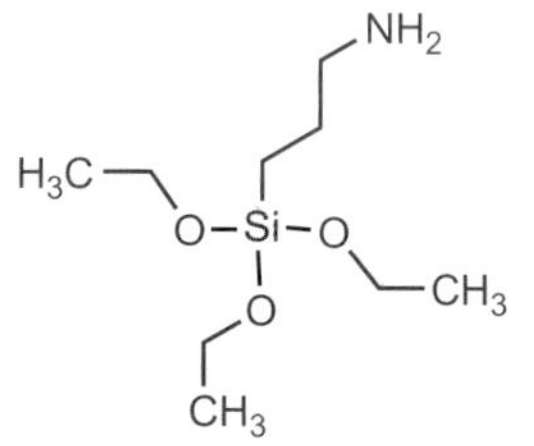

APTS;
3-Aminopropyltriethoxysilane
MW 221.37

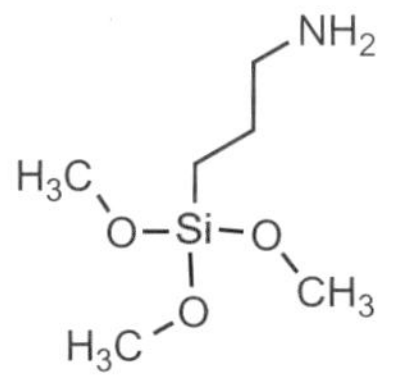

3-Aminopropyltrimethoxysilane
MW 179.29

When using APTS (the triethoxy version), the reaction must occur in at least a partially aqueous environment. The ethoxy groups are not reactive enough to couple spontaneously with the –OH groups on an inorganic surface without prior hydrolysis to form silanols. This is typically done in 5% water in ethanol that has been acidified with acetic acid to pH 4.5 to 5.5. The protocol described in the previous section for aqueous solvent deposition can be used with success to modify surfaces or particles with this reagent (Figure 13.3). This reaction process results in a layer containing about three to eight organosilanes in thickness, which effectively masks the underlying inorganic substrate with aminopropyl functionalities.

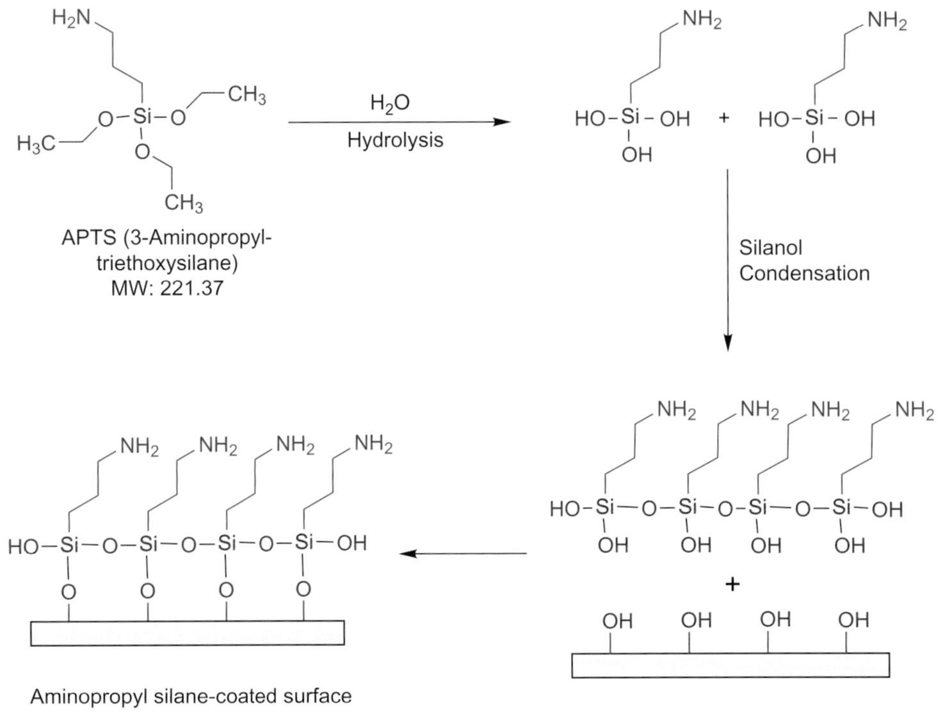

FIGURE 13.3 The deposition of APTS on inorganic substrates results in the formation of a covalent coating containing primary amine groups.

Once deposited on a substrate, the alkoxy groups form a covalent polymer coating with the primary amine groups sticking off the surface and available for subsequent conjugation. Carboxyl- or aldehyde-containing ligands may be directly coupled to the aminopropyl groups using a carbodiimide reaction (Chapter 4, Section 1) or reductive amination (Chapter 4, Section 4), respectively. Alternatively, surfaces initially derivatized with an aminopropylsilane compound can be modified further with spacer arms or crosslinkers to create reactive groups for coupling affinity ligands or biomolecules. For instance, the amine groups may be derivatized with an NHS–PEG$_n$–azide compound for use in click chemistry or Staudinger ligation reactions for linking proteins or other biomolecules (Chapter 17, Sections 5 and 6, and Chapter 18, Section 1.2). This modification forms extremely hydrophilic PEG spacers on the surface that terminate in alkyl azide groups for conjugation with alkyne or phosphine derivatives (Figure 13.4).

Other crosslinking agents that contain an amine-reactive group on one end also may be used to modify and activate the APTS-modified substrate. Surfaces may be designed to contain, for instance, reactive hydrazine or aminooxy groups for conjugation with carbonyl-containing molecules, such as aldehydes formed through periodate oxidation of carbohydrates or natively present at the reducing end of sugars and glycans (Chapter 17, Sections 2 and 3, and Chapter 2, Section 4.6).

Benters *et al.* (2002) used two approaches to modify APTS surfaces. In one instance, the amine groups were acylated using glutaric anhydride to create carboxylate functionalities, which were then activated with NHS/DCC to form the NHS ester. This derivative could be used to couple amine-containing proteins and other molecules *via* amide bond formation. In a second activation strategy, the aminopropyl groups on the surface were activated with 1,4-phenylenediisothiocyanate (PDITC) to create terminal isothiocyanate groups for coupling amines. Both methods resulted in the successful coupling of amine–dendrimers to silica surfaces for use in arrays (see Chapter 8, Section 4).

APTS has been used extensively to modify inorganic surfaces to produce organic functional groups for further coupling with ligands or additional modification reactions. The application of this reagent has involved modification of quantum dots (Chapter 10) (Li *et al.*, 2012), coating ZnO nanoparticles (Zhao *et al.*, 2012), synthesizing SiO_2-coated Fe_3O_4 nanoparticles with amine groups (Khosroshahi *et al.*, 2012), and developing mesoporous silica nanoparticles suitable for anticancer drug delivery (Yanes and Tamanoi, 2012).

Amine surfaces prepared using an aminosilane compound can be modified to contain carboxylate groups using the following protocol involving the reaction with an anhydride, such as succinic anhydride or glutaric anhydride (Figure 13.5). After modification, the

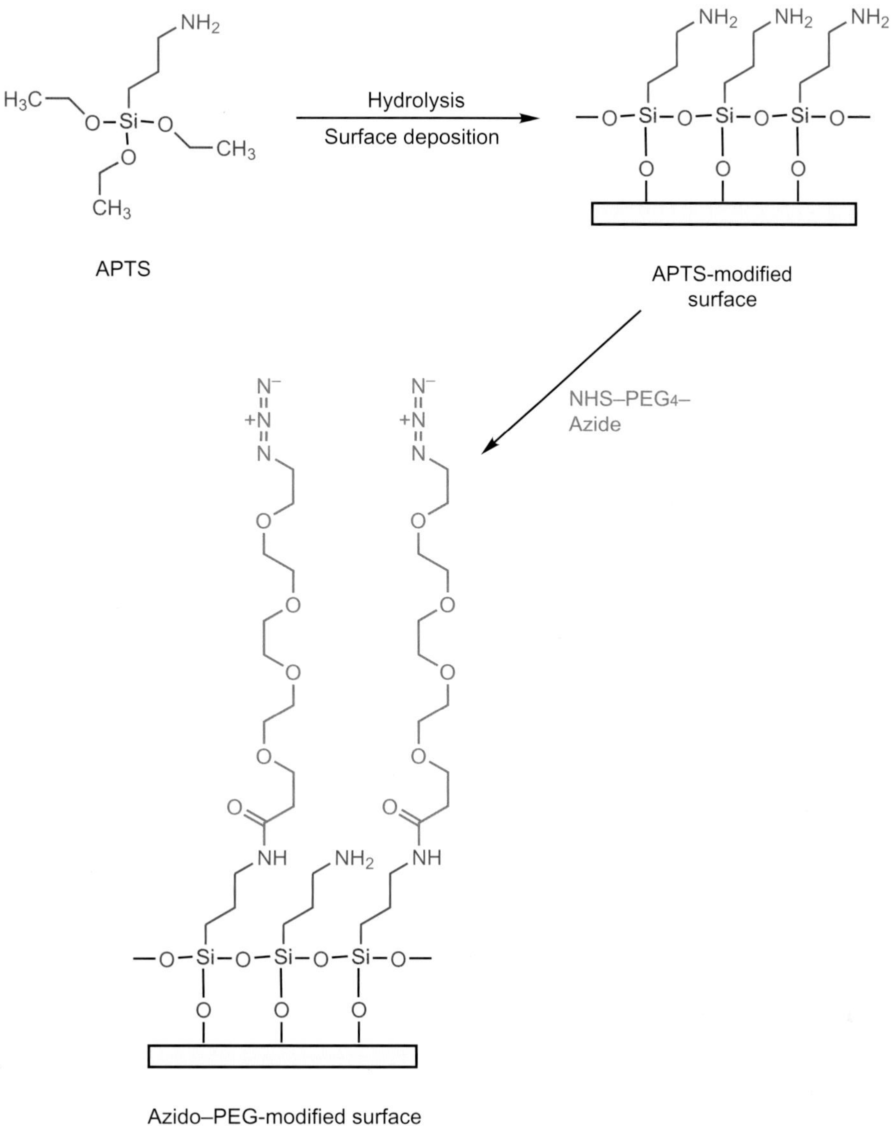

FIGURE 13.4 APTS-modified surfaces may be further derivatized with amine-reactive crosslinkers to create additional surface characteristics and reactivity. Modification with NHS–PEG_4–azide forms a hydrophilic PEG spacer terminating in an azido group that can be used in a click chemistry or Staudinger ligation reaction to couple other molecules.

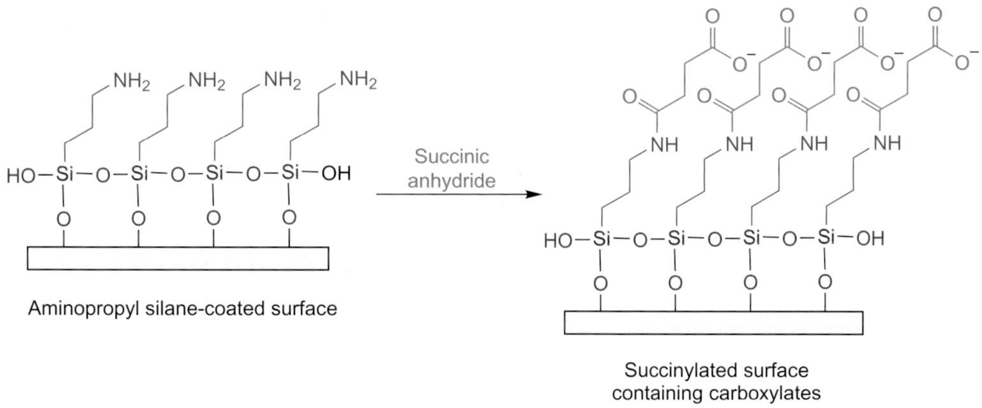

FIGURE 13.5 Modification of an APTS surface with glutaric anhydride creates terminal carboxylates for coupling of amine-containing ligands.

carboxylates then can be used to couple amine-containing molecules using a carbodiimide reaction with EDC plus sulfo-NHS (Chapter 4, Section 1.2).

Protocol

1. Dissolve glutaric anhydride (succinic anhydride also may be used) (Chapter 2, Section 4.2) in DMF at near saturation (use a fume hood). Add triethylamine to a concentration of at least 1 mg/ml to function as a proton acceptor (base).
2. Apply the anhydride solution to the aminosilane-modified surface (typically done by emersion) and mix by stirring.
3. React 2–4 h at room temperature.
4. Wash the carboxylated surface with DMF and then with water. Dried surfaces are stable indefinitely.

The carboxyl groups on the surface may be activated to an NHS ester for coupling with amine-containing biomolecules according to the following protocol, based on the method of Benters *et al.* (2002).

Protocol

1. In a fume hood, dissolve *N*-hydroxysuccinimide (NHS) in DMF (highly pure and dried over molecular sieves) at a concentration of 115 mg/ml (1 *M*) along with 206 mg/ml (1 *M*) *N,N*′-dicyclohexylcarbodiimide (DCC; Chapter 4, Section 1.4).
2. Add the NHS/DCC solution to the carboxylated substrate to immerse fully the surface. Mix by stirring.
3. React for 1 to 2 h at room temperature.
4. Wash the surfaces with DMF. The NHS-activated surface may be used to couple amine-containing molecules in a buffer at physiological pH (7.2–7.4) using 50-m*M* sodium phosphate.

Aminosilane surfaces also may be activated by use of a bifunctional crosslinker to contain reactive groups for subsequent coupling to biomolecules. In one such reaction, *N,N*′-disuccinimidyl carbonate (DSC) was used to react with the amines on a slide surface and create terminal NHS–carbonate groups, which then could be coupled to amine-containing molecules (Niemeyer, 2004) (Figure 13.6). The following procedure is based on this method.

Protocol

1. Dissolve 1.5 g of DSC (Chapter 5, Section 1.7) and 5 ml of diisopropylethylamine (DIEA) in 145 ml of dry acetone (in a fume hood).
2. Add the DSC solution to the aminosilane-slides (or other aminosilane-modified surface) to immerse fully the substrate in the solution and mix by stirring.
3. React for 2 h at room temperature.
4. Wash the modified surfaces with dry acetone to remove excess reactants and reaction by-products. Dry under nitrogen. The NHS–carbonate groups on the surface are stable to storage under desiccated conditions (package under nitrogen with a desiccant).
5. Amine-containing molecules, such as proteins or amine-modified oligonucleotides, are coupled to the NHS–carbonate-activated surface by dissolving them in 50-m*M* sodium phosphate, pH 7.4, at a concentration of at least 1 to 10 mg/ml (for proteins). Spotting of biomolecules onto the activated surface may be done to create an array. Since this process usually is done using only microliter quantities or less of solution per spot, the activated surface should be kept in a humidity chamber to prevent evaporation during the coupling reaction.
6. React for 2 to 4 h at room temperature.
7. Wash the slides with coupling buffer to remove excess reactants.

2.2. Carboxyethylsilanetriol

Silane coupling agents containing carboxylate groups may be used to functionalize a surface with carboxylic acids for subsequent conjugation with amine-containing molecules. Carboxyethylsilanetriol contains an acetate organo group on a silanetriol inorganic

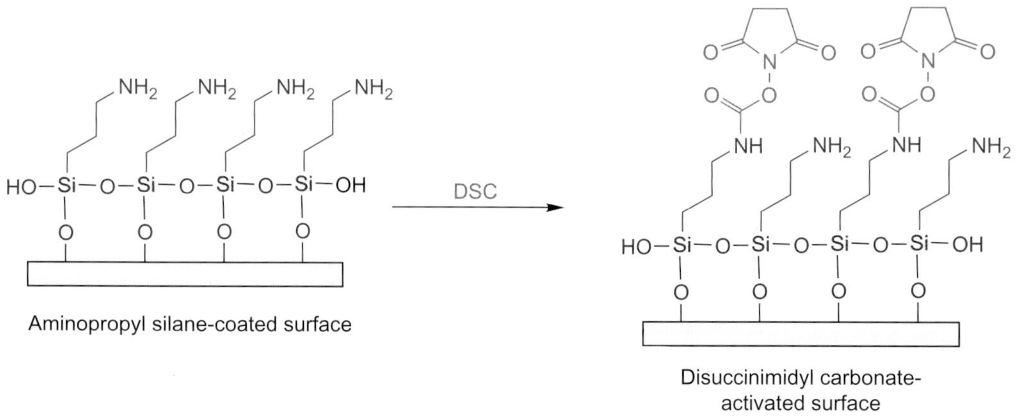

FIGURE 13.6 APTS-modified surfaces can be activated with DSC to form amine-reactive succinimidyl carbonates for coupling proteins or other amine-containing molecules.

reactive end (Gelest). The silanetriol component is reactive immediately with inorganic –OH substrates without prior hydrolysis of alkoxy groups, as in the case with most other silanization reagents (Figure 13.7). This compound is supplied in a 25% aqueous solution as the monosodium salt, and it can be used just by diluting into water or buffer.

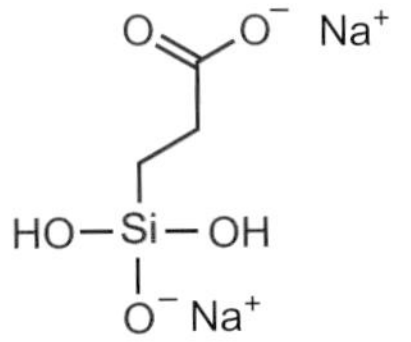

Carboxyethyl-
silanetriol, disodium salt
MW 196.14

Carboxyethylsilanetriol has been used to add carboxylate groups to fluorescent silica nanoparticles to couple antibodies for multiplexed bacteria monitoring (Wang *et al.*, 2007; Cai *et al.*, 2012; Subbiahdoss *et al.*, 2012) (see Chapter 14, Section 5). This reagent can be used in similar fashion to add carboxylate functionality to many inorganic or metallic nano-materials, which also will create negative charge repulsion to maintain particle dispersion in aqueous solutions. Covalent coupling to the carboxylated surface then can be done by activation of the carboxylic acid groups with a carbodiimide to facilitate direct reaction with amine-containing molecules or to form intermediate NHS esters (Chapter 4, Section 1).

The following protocol for using carboxyethylsilanetriol is based on the method of Wang *et al.* (2007). Other metallic particles or surfaces may be modified with this silane compound in like manner.

Protocol

1. Suspend 10 mg of silica particles in 1 ml of 10-m*M* sodium phosphate, pH 7.4, with gentle mixing.
2. Add 20 μl of the carboxyethylsilanetriol (as the 25% aqueous solution) to the particle suspension with mixing.
3. React at room temperature for 3 to 4 h with mixing.
4. Wash the modified particles several times with buffer using centrifugation for separation. A final wash with 10-m*M* MES, pH 5.5, is done to prepare the sample for coupling amine-containing molecules using a carbodiimide reaction. For suggested protocols, see Chapter 14, Section 4.2, entitled *Coupling to Carboxylate Particles*.

2.3. N-(Trimethoxysilylpropyl)Ethylenediamine Triacetic Acid (TMS-EDTA)

Another useful silanization reagent containing carboxylates actually is an effective chelator of metal ions. *N*-(Trimethoxysilylpropyl)ethylenediamine triacetic acid (TMS–EDTA) contains a trimethoxy group for coupling to –OH groups on inorganic substrates and an EDTA group for coordinating metals (minus one carboxylate, which instead functions as the linking arm attached to the silicon atom). This compound can be used to coat substrates and provide a metal-chelating functional group for use in affinity separations.

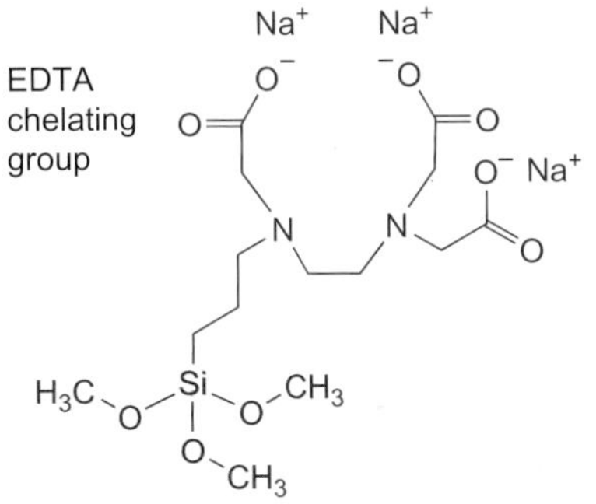

TMS–EDTA;
N-(Trimethoxysilylpropyl)ethylene-
diamine triacetic acid, trisodium salt
MW 462.41

Immobilized metal affinity chromatography (IMAC) has been used for decades as an affinity method for targeting certain functional groups in proteins or other

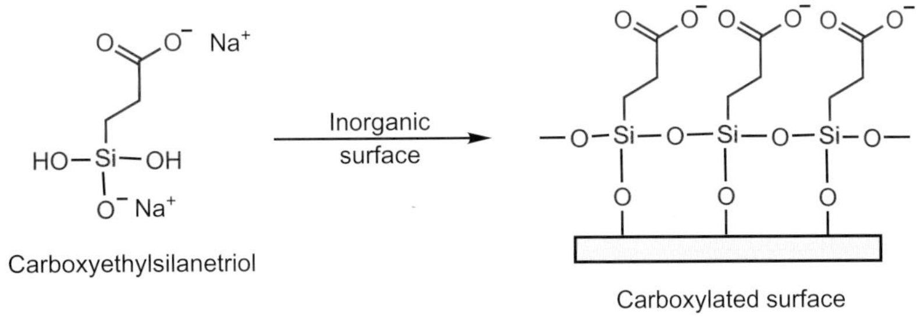

FIGURE 13.7 Carboxylethylsilanetriol can be used to modify an inorganic substrate to containing carboxylate groups for coupling amine-containing ligands.

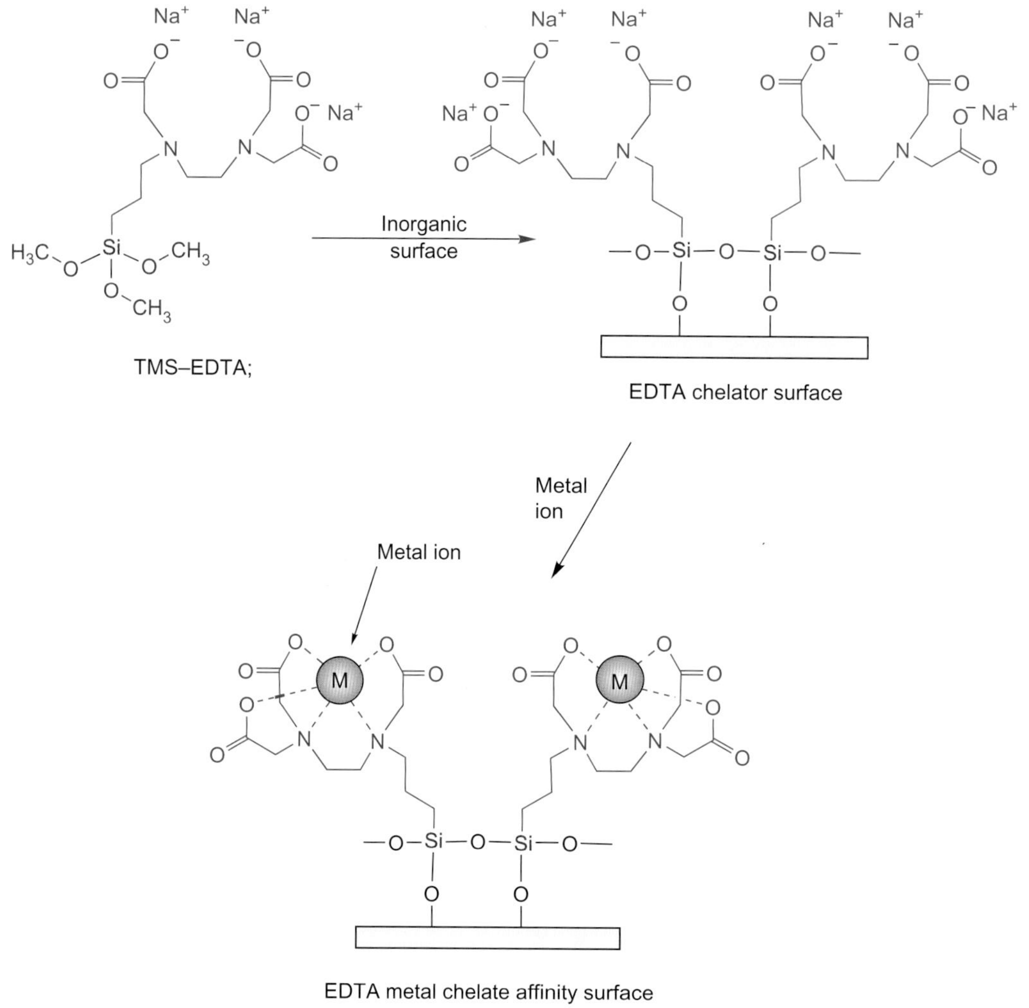

FIGURE 13.8 TMS–EDTA can be used to modify an inorganic substrate to contain EDTA chelating groups for complexation with metal ions.

biomolecules (for reviews, see Hage, 1999; Lopatin and Varlamov, 1995; Winzerling *et al.*, 1992; Porath, 1992; and Hermanson *et al.*, 1992). Various metal ions can be chelated by EDTA affinity groups to provide directed targeting of biological groups such as phosphate modifications or histidine-rich areas in proteins. Thus, phosphorylation sites in proteins can be bound using a silane–EDTA-modified surface containing, for instance, gallium (Posewitz and Tempst, 1999). Alternatively, if the chelating groups are charged with nickel or cobalt, the binding of His-tagged proteins can be done.

The preparation of particles or surfaces that are able to capture specifically a fraction of the proteome using metal affinity separations makes possible analysis of distinct protein populations by mass spectrometry (Zhou *et al.*, 2000). The reactions and use of TMS–EDTA in modifying inorganic surfaces and coordinating metals for affinity chromatography is shown in Figure 13.8.

The coating of surfaces with TMS–EDTA can be done using the general protocol described previously in this chapter entitled Aqueous/Organic Solvent Deposition. The acidified water/organic solvent environment hydrolyzes the methoxy groups to silanols, which then polymerize and covalently link to inorganic surface –OH groups. After a surface has been modified with a chelating ligand, it is washed thoroughly with metal-free water or buffer. Finally, an aqueous solution of the appropriate metal salt is contacted with the surface and the EDTA groups will bind the metal for use in affinity separations.

It should be noted that the chelator group in TMS–EDTA is a pentadentate ligand, which will coordinate five bonds with an associated metal ion. Common IMAC ligands use tridentate or tetradentate ligands that have a greater number of available coordination sites for interaction with target molecules. Iminodiacetic acid-based chelators interact with metal ions through three bonds, typically leaving a maximum of three coordination sites left for affinity binding. Nitrilotriacetic acid–lysine (NTA) chelating groups interact with metal ions through four coordination bonds, leaving two sites available for affinity interactions. However, the TMS–EDTA ligand holds metal ions through five coordination bonds, thus typically having only one remaining for interacting with other molecules. This property may result in lower affinities with biological molecules and change the capture behavior for specific metal interaction sites on proteins. Careful testing should be done for each application envisioned for this metal-chelating silane coupling agent. TMS–EDTA-modified particles have been used without a chelated metal ion to capture heavy metal ions for cleaning metal contaminated water (Addy *et al.*, 2012).

2.4. 3-Glycidoxypropyltrimethoxysilane (GOPTS) and 3-Glycidoxypropyltriethoxysilane

Two very useful silane modification agents are glycidoxy compounds containing reactive epoxy groups. Surfaces covalently coated with these silane coupling agents can be used to conjugate thiol-, amine-, or hydroxyl-containing ligands, depending on the pH of the reaction (Chapter 3, Section 4.1). Thus, 3-glycidoxypropyltrimethoxysilane (GOPTS) or 3-glycidoxypropyltriethoxysilane can be used to link inorganic silica or other metallic surfaces containing –OH groups with biological molecules containing any three of these major functional groups. GOPTS has been used most often in bioconjugation applications, but the triethoxysilane compound also may be used in similar protocols (Figure 13.9).

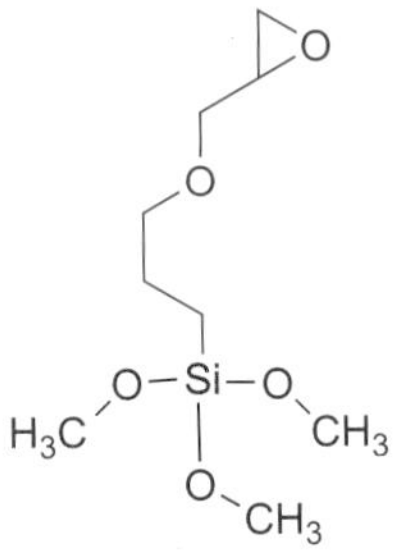

GOPTS;
3-Glycidoxypropyl-
trimethoxysilane
MW 236.34

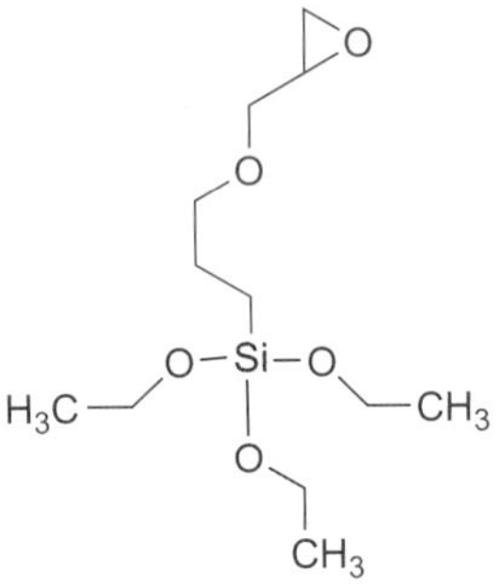

3-Glycidoxypropyl-
triethoxysilane
MW 278.42

The reaction of the epoxide with a thiol group yields a thioether linkage, whereas reaction with a hydroxyl gives an ether and reaction with an amine results in a secondary amine bond. The relative reactivity of an epoxy group is thiol > amine > hydroxyl, and this is reflected by the optimal pH range for each reaction. In this case, the lower the reactivity of the functional group the higher the pH required to drive the reaction efficiently. Chapter 15, Section 2.3, contains a thorough discussion on coupling affinity ligands to epoxy activated supports.

GOPTS has been used to create a high-density PEG surface on glass slides for use in arrays (Piehler *et al.*, 2000). It also has been used in the development of a high-throughput analyzer using biochip technology on aluminum oxide sheets (FitzGerald *et al.*, 2005), in the activation of glass surfaces for detection of antibodies specific for hepatitis B and C viruses (Duan *et al.*, 2005), and in the covalent attachment of biomolecules onto silicon nitride planar waveguide surfaces (Psarouli *et al.*, 2011).

The following protocol is based on these methods. All operations should be done in a fume hood, including wearing proper protective clothing.

Protocol

1. Prepare glass slides by washing with acid (5% HCl) for several hours to remove non-binding metal ions, especially sodium, potassium, and calcium. Treatment with a mixture of 25% sulfuric acid and 15% hydrogen peroxide (piranha solution) for about 30 min is done to create a high density of hydroxyl functionalities suitable for silane modification. Glass slides also can be cleaned and washed prior to modification with a silane with DMSO, ethanol, and water, and then etched using 10% NaOH (w/w) in water for 1 h.
2. Prepare a GOPTS solution in *o*-xylene or 95% ethanol at a concentration of 2% (v/v). If the organic solvent is used, add 2 mg/ml *N*-ethyldiisopropylamine (DIPEA) as base.
3. Immerse the glass slides in the GOPTS solution and mix by stirring.
4. React at 37°C (for the ethanol solution) or 55°C (for the *o*-xylene solution) for at least 5 to 6 h with mixing.
5. Wash slides thoroughly with solvent and then dry in an oven at 135°C for 1 h (explosion-proof oven). The slides are now ready to couple ligands through their epoxy groups.

For protocol suggestions on conjugation to epoxy groups, see Chapter 3, Sections 1.8 and 4.1. Also, see Chapter 14, Section 4.2, for a method to attach affinity ligands to nanoparticle or microparticle surfaces that have been activated with epoxide groups.

2.5. Isocyanatopropyltriethoxysilane (ICPTES)

Isocyanate groups are extremely reactive toward nucleophiles and will hydrolyze rapidly in aqueous solution (Chapter 3, Sections 1.2 and 4.7). They especially are useful for covalent coupling to hydroxyl groups under nonaqueous conditions, which is appropriate for conjugation to many carbohydrate ligands.

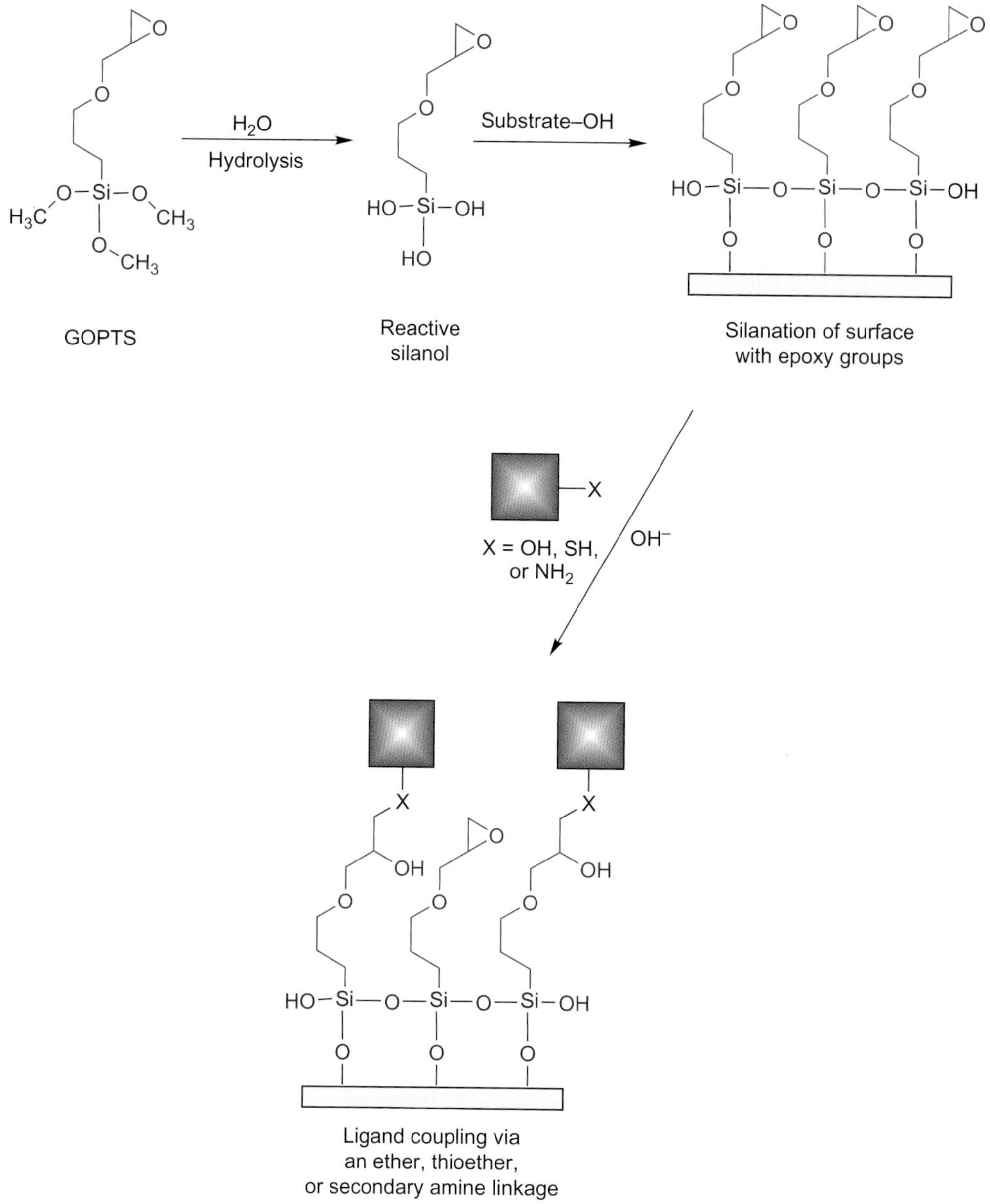

FIGURE 13.9 Epoxy-containing silane coupling agents form reactive surfaces that can be used to couple amine-, thiol-, or hydroxyl-containing ligands.

Isocyanatopropyltriethoxysilane (ICPTES) contains an isocyanate group at the end of a short propyl spacer, which is connected to the triethoxysilane group useful for attachment to inorganic substrates. Silanization can be accomplished in dry organic solvent to form reactive surfaces while preserving the activity of the isocyanates.

An isocyanate reacts with amines to form isourea linkages and with hydroxyls to form carbamate (urethane) bonds. Both reactions can take place in organic solvent to conjugate molecules to inorganic substrates (Figure 13.10). The solvent used for this reaction must be of high purity and should be dried using molecular sieves prior to adding the silane compound.

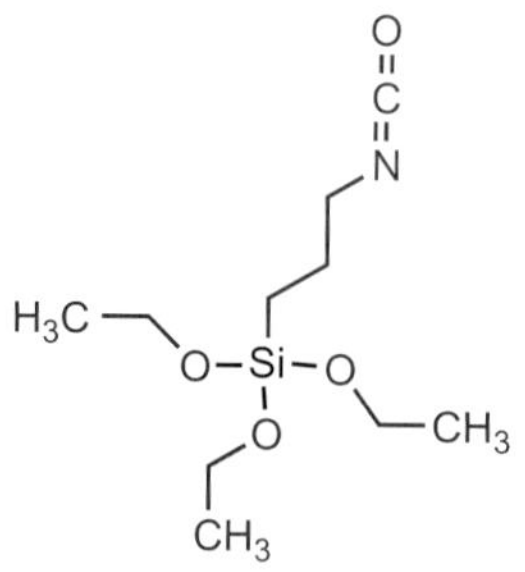

ICPTES;
Isocyanatopropyl-
triethoxysilane
MW 247.36

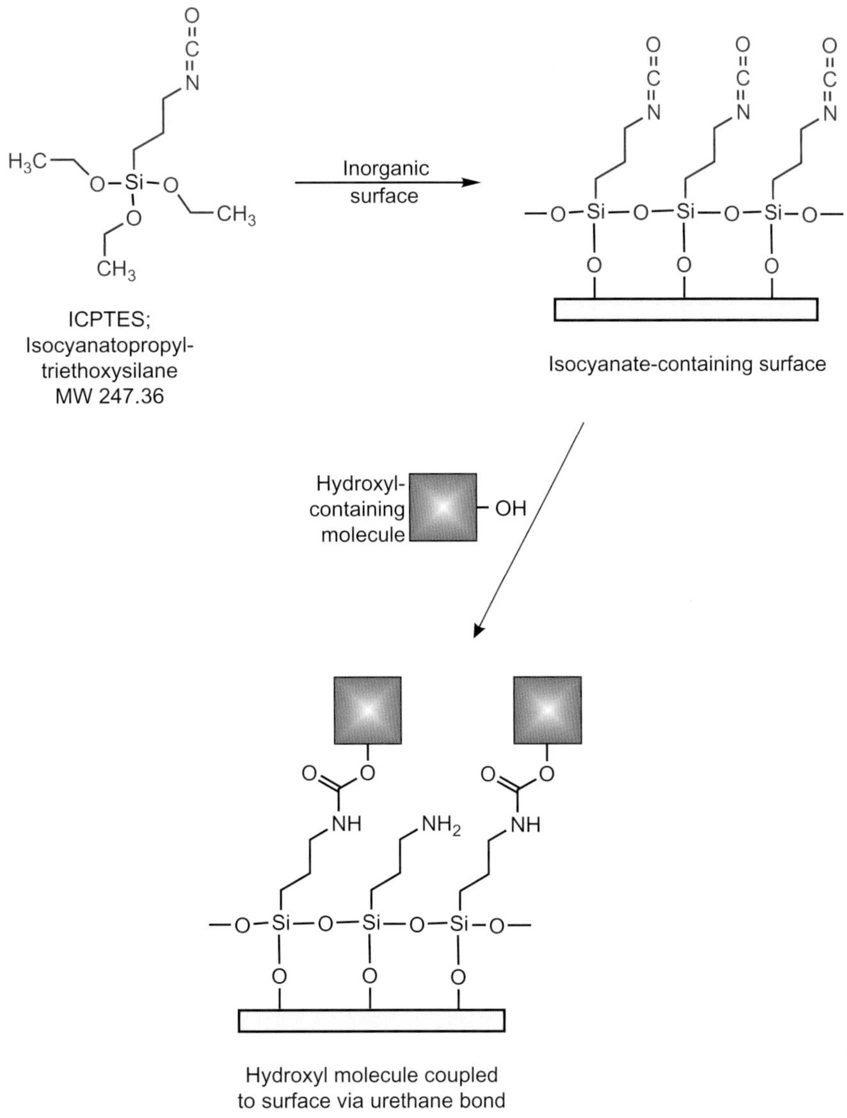

FIGURE 13.10 The isocyanate-containing silane coupling agent can be used to couple hydroxyl-containing molecules to inorganic surfaces. The reactions should be carried out in dry organic solvent to prevent hydrolysis of the reactive group.

Silva *et al.* (2005) used ICPTES to create novel chitosan–siloxane hybrid polymers by coupling the isocyanate groups to the functional groups of the carbohydrate and forming a silica polymer using the triethoxysilane backbone. Boev *et al.* (2005) used this functional silane coupling agent to prepare CdS fluorescent nanoparticles in a urea–silicate matrix. ICPTES and APTS have been used in combination to create organically modified silica xerogels through carboxylic acid solvolysis that formed hybrid materials with luminescent properties (Fu *et al.*, 2006). In addition, Wu *et al.* (2011) used ICPTES to modify carbon nanotubes, Fu *et al.* (2011) created a glucose biosensor by immobilizing an enzyme using it, and Yu *et al.* (2011) developed a magnetic NIR luminescent nanocomposite using the silane compound.

The following protocol is based on the method of FitzGerald *et al.* (2005), who used the technology to develop aluminum oxide-based biochips for high-throughput analysis.

Protocol

1. In a fume hood, dissolve ICPTES in anhydrous toluene at a concentration of 5% (v/v) and also containing 1% DIPEA base.
2. Clean and prepare an inorganic substrate to remove unreactive metal salts and to create –OH sites for coupling the silane compound. For glass, this can be done using acid or base washes. Formation of –OH groups on glass typically is done using piranha solution, which is 25% sulfuric acid and

15% hydrogen peroxide. For cleaning aluminum oxide sheets, use an alkaline detergent solution and apply sonication for 1 h, then wash with water.

3. Treat the inorganic substrate with the ICPTES solution by immersion and react at 50°C overnight with mixing.
4. Wash the derivatized substrate with toluene at least twice and then rinse a final time with acetone.
5. Dry the modified substrate at 120°C for 30 min in an explosion-proof oven.

Many other functional silane coupling agents are available from commercial suppliers, including hydroxyl, aldehyde, acrylate and methacrylate, and anhydride compounds. Substrate modification procedures similar to those discussed above can be used with these reagents to link a biomolecule to an inorganic surface or particle.

CHAPTER

14

Microparticles and Nanoparticles

The use of particles in biological assays and other applications dates back many decades (Singer and Plotz, 1956). Assays and detection systems made by coupling affinity ligands to particles of nanometer or micrometer diameter have been used in diagnostic tests and numerous other research procedures. Latex microspheres and gold nanoparticles were perhaps the earliest examples of solid-phase spheres used for these purposes. With the recent explosion in nanotechnology, the relevant particle size has shrunk by 2 to 3 orders of magnitude and the applications for particles have dramatically expanded. Nanoparticles and microparticles are used in agglutination tests and assays, particle capture ELISA methods, lateral flow tests, solid-phase assays, scintillation proximity assays, polymerase chain reaction (PCR) tests, superparamagnetic-based assays and magnetic separation systems and biosensors; as enhancers of Raman spectral signals; in light-scattering assays; in drug delivery applications; and as fluorescent labels or stains for detecting biological molecules (Bulte and Modo, 2008; Carregal-Romero *et al.*, 2012; Otsuka *et al.*, 2012; Wang *et al.*, 2012; Yurt *et al.*, 2012).

1. PARTICLE TYPES

The types of particles used in biological applications are extremely varied. Most often, they are nonporous in nature and spherical in shape. However, the material science revolution in nanotechnology has provided particle types and compositions of almost limitless shape and size, including spherical, amorphous, or aggregate particles, as well as elaborate geometric shapes like rods, tubes, cubes, triangles, and cones. In addition, new symmetrical organic constructs have emerged in the nanometer range that include fullerenes (e.g., Buckyballs), carbon nanotubes, and dendrimers, which are highly defined synthetic structures used as bioconjugation scaffolds in various applications. These specialized nanoparticle-like constructs are discussed in more detail in Chapter 8; Chapter 10, Section 10; and Chapter 16.

The chemical composition of particles can be just as varied as their shape. Commercial particles can consist of polymers or copolymers, inorganic constructs, metals, semiconductors, superparamagnetic composites, biodegradable constructs, synthetic dendrimers, and dendrons. Often, both the composition of a particle and its shape govern its suitability for a particular purpose. For instance, composite particles containing superparamagnetic iron oxide typically are used for small-scale affinity separations, especially for cell separations followed by flow cytometry analysis or fluorescence-activated cell sorting (FACS). Core-shell semiconductor particles, by contrast, are the basis for quantum dot nanoparticle labels, which provide intensely fluorescent detection reagents.

The original polymeric latex particles are still widely used for separation and detection. Polymers provide a matrix that can be swollen for embedding other molecules in their core, such as organic dyes or fluorescent molecules. Even nanoparticle quantum dots can be incorporated into larger latex particles to form highly fluorescent composite microparticles.

Such fluorescent or dyed latex particles are useful in multiplexed detection systems using suspension arrays (e.g., Luminex technology) (Armstrong *et al.*, 2000). In this application, dyes are incorporated within the beads having a range of different spectral properties to create a series of different colored particles. In addition, changing the amount of dye molecules within the beads or blending two or more dyes in a single particle population can form a gradient of different color compositions. Particle subpopulations of a particular color or emission wavelength and intensity can then be used to identify the type of target being measured in a suspension assay. Mixing together in a single solution such particle subpopulations having different colors permits multiple targets to be assayed simultaneously, wherein each particle type is identified by its color and correlated to the analyte being targeted. Flow-cytometry-based instruments are then used to detect and measure the particle color and assay result.

Bioconjugate Techniques, Third Edition.
DOI: http://dx.doi.org/10.1016/B978-0-12-382239-0.00014-5

Dyed particles also are commonly used in diagnostic lateral flow tests (like the common home pregnancy test), as the colors can be seen with the eye without the need for special detectors. In this type of assay, antibodies or antigens are coupled to the dyed particles and a sample solution applied to the test strip carries them along within a membrane. The particles then are captured at points in the membrane that represent either a control or a positive sample result. Large numbers of color particles docking at these points within the membrane create the visual lines associated with these disposable tests.

Polymeric particles can be constructed from a number of different monomers or copolymer combinations. Some of the more common ones include polystyrene (traditional "latex" particles), poly(styrene/divinylbenzene) copolymers, poly(styrene/acrylate) copolymers, polymethylmethacrylate (PMMA), poly(hydroxyethyl methacrylate) (pHEMA), poly(vinyltoluene), poly(styrene/butadiene) copolymers, and poly(styrene/vinyltoluene) copolymers. In addition, by mixing into the polymerization reaction combinations of functional monomers, one can create reactive or functional groups on the particle surface for subsequent coupling to affinity ligands. One example of this is a poly(styrene/acrylate) copolymer particle, which creates carboxylate groups within the polymer structure, the number of which is dependent on the ratio of monomers used in the polymerization process.

Many of the common polymer particle types present a surface with hydrophobic character, such as the poly-aromatic styrene-containing ones. These in particular are better used with modifications on the particle surface to add charge or hydrophilicity, which masks the underlying polymer core. Some constructs of this type provide a grafted hydrophilic surface to limit the degree of nonspecific binding and form a more biocompatible particle that will not denature or bind protein through unwanted hydrophobic interactions. Some polymer particles are made from hydrophilic monomers and display better biocompatibility. Poly-HEMA is very hydrophilic due to an abundance of hydroxyl groups, and its properties are extremely biocompatible without further surface derivatization.

Inorganic particles are used extensively in various bioapplications, too. Gold nanoparticles long have been used as detection labels for immunohistochemical (IHC) staining and lateral flow diagnostic testing. These dark, dense particles provide single particle detection capability for sensitive staining techniques, especially in microscopy applications. In addition, the ease with which clumps of gold particles can be seen visually makes their use in diagnostic strip tests a very popular alternative to dyed latex particles (see Section 6 for a thorough review of gold conjugation methods).

Invariably, the use of particles in bioapplications involves the attachment of affinity capture ligands to their surface, by either passive adsorption or covalent coupling. The coupling of an affinity ligand to such particles creates the ability to bind selectively biological targets in complex sample mixtures. The affinity particle complexes can thus be used to separate and isolate proteins or other biomolecules or to specifically detect the presence of these targets in cells, tissue sections, lysates, or other complex biological samples.

The reactions used for coupling affinity ligands to nanoparticles or microparticles are basically the same as those used for bioconjugation of molecules or for immobilization of ligands onto surfaces or chromatography supports (see Chapter 15). However, with particles, size can be a major factor in how a reaction is performed and in its resultant reaction kinetics. Since particle types can vary from the low-nanometer diameter to the micron size, there are dramatic differences in how such particles behave in solution and how the density of reactive groups or functional groups affects reactions.

2. PARTICLE CHARACTERISTICS AND STABILITY

Particle size, surface composition, and density directly affect how a particle behaves in suspension. This in turn affects coupling protocols, especially in the handling and washing techniques used for particles during the conjugation process. Larger particles of micron size will generally settle over time just in normal gravity. As particle size decreases, however, a point is reached where a true colloidal suspension may occur, wherein the particles will not separate, no matter how long they sit in suspension. This typically happens when particle size gets to about 100 nm, and Brownian motion causes water molecules to collide with particles with high enough force-to-mass ratios to prevent them from settling under gravity. Many dense particles of less than 100 nm, such as silica, can still be separated from solution using a bench-top centrifuge; however, as particles approach the size of biological macromolecules, or around 10 nm, an ultracentrifuge would be required for separation. For a comparison of particle sizes and how they contrast to biomolecules, see Figures 14.1 and 14.2.

The forces acting on particles in suspension can mainly be explained by the Derjaguin, Landau, Verwey, and Overbeek (DLVO) theory, which describes the attractive van der Waals forces and repulsive electrostatic forces affecting their stability (Derjaguin and Landau, 1941; Verwey and Overbeek, 1952). A corollary to this theory indicates that for hydrophobic

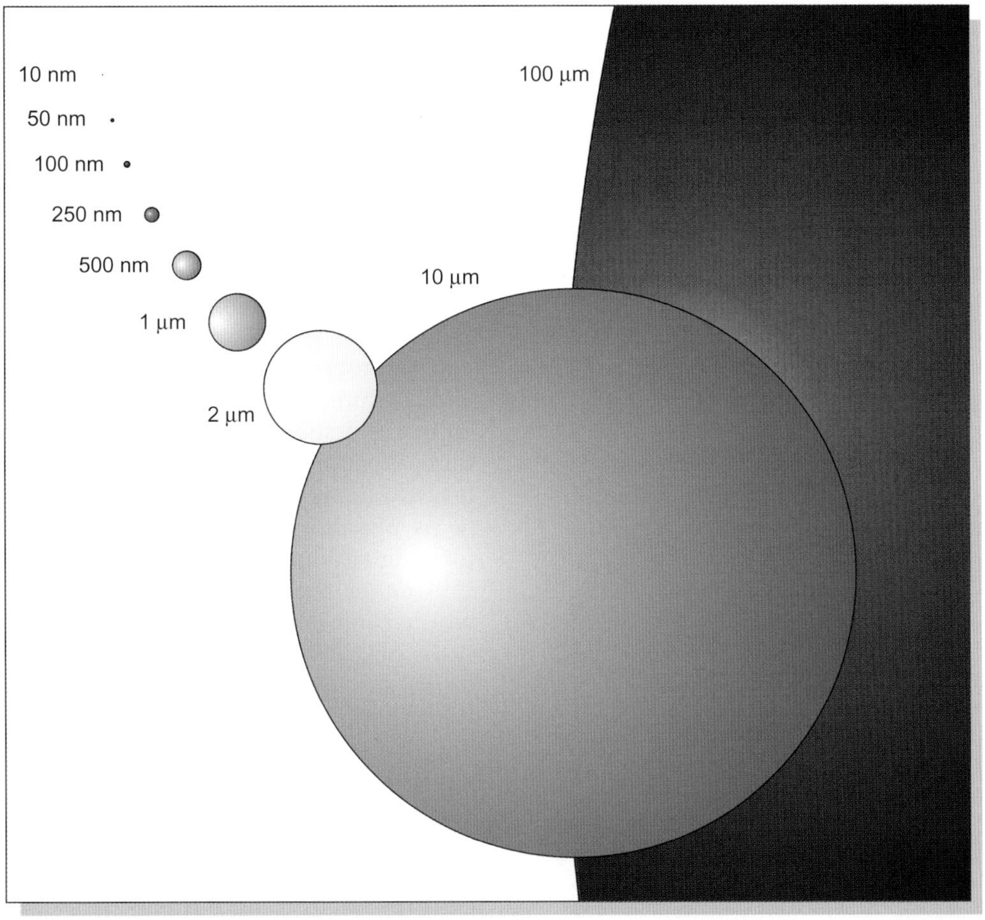

FIGURE 14.1 Particles commonly used in biological applications can range in size over three orders of magnitude, from as small as biological macromolecules (~10 nm) to approximately the diameter of cells and beyond (>10 μm). The diameter of a particle population can dramatically affect its behavior in solution.

surfaces, adding carboxylates can result in a stabilization of particles in solution and a prevention of aggregation through charge repulsion. Of course, this is true provided the pH of the solution is maintained above about pH 5, so that the carboxylates are not protonated, and their negative charge character is maintained. Often, without the presence of surface charge to create significant repulsive effects, many particles of commercial or research interest would aggregate and fall out of suspension rather quickly due to hydrophobic surface interactions and van der Waals forces (electrostatic attractions). The smaller the particle, the more significant the van der Waals attractive forces may become. See Figure 14.3.

By contrast, relatively hydrophilic particles like those made of pHEMA may maintain colloidal stability even at small size due to the "repulsive" effects of a water of hydration layer, which forms around each particle in aqueous solution and is energetically difficult to remove or penetrate. A closely bound water layer is apparently the result of hydrogen bonding and dipole interactions with polar surface groups, such as the hydroxyls on pHEMA particles. Many such hydrophilic particles can display stability at high salt concentrations, even if DLVO theory may predict aggregation (Molina-Bolívar, *et al.*, 1998).

The degree and type of surface charge can dramatically affect particle behavior in suspension. Latex particles that have a very low density of charged groups, for instance, may still present hydrophobic polymer surface areas large enough to cause instability. The type of charged groups on the particle surface can directly affect the magnitude of charge repulsion. In particular, the activation of carboxylate particles using an EDC/sulfo-NHS two-step reaction (Chapter 4, Section 1.2) will temporarily replace the negatively charged carboxylates with negatively charged sulfonates. The sulfonate groups on the sulfo-NHS ester intermediates create a stronger negative charge on the particle surface than the original carboxylates. In some cases, the increase in negative charge repulsion can result in an inability to pellet the particles by centrifugation after the activation step, even if the particles could be separated by centrifugation before activation. This was true for 40-nm carboxylated silica nanoparticles in our hands (unpublished observations).

The charge repulsion effects between particles can be severely affected by the buffer and salt composition of the solution they are suspended in. Charges can be eliminated or neutralized by ionizable groups being protonated or unprotonated or by the concentration of ions in solution. For instance, lowering the pH of an aqueous solution below the pK_a of the surface carboxylates will result in them being protonated. With most particles, especially ones having hydrophobic surfaces, this will cause particle aggregation due to loss of surface negative charge. Similarly, a high salt concentration can effectively mask the charge character of a carboxylated particle by having too many positively charged ions associated with the surface negative charges. Most particle types that are stable in suspension due to like-charge repulsion can be made to aggregate if the pH is changed or the buffer or salt concentration is too high. Part of the challenge of successfully working with small particles is to maintain optimal solution characteristics to keep the particles dispersed throughout the conjugation process. This includes all activation, coupling, and washing steps that are used to conjugate an affinity ligand and subsequently use it in its intended application. Therefore, for each particle type being used to conjugate a ligand, one should first become knowledgeable of its physical and solution characteristics before performing any coupling reactions.

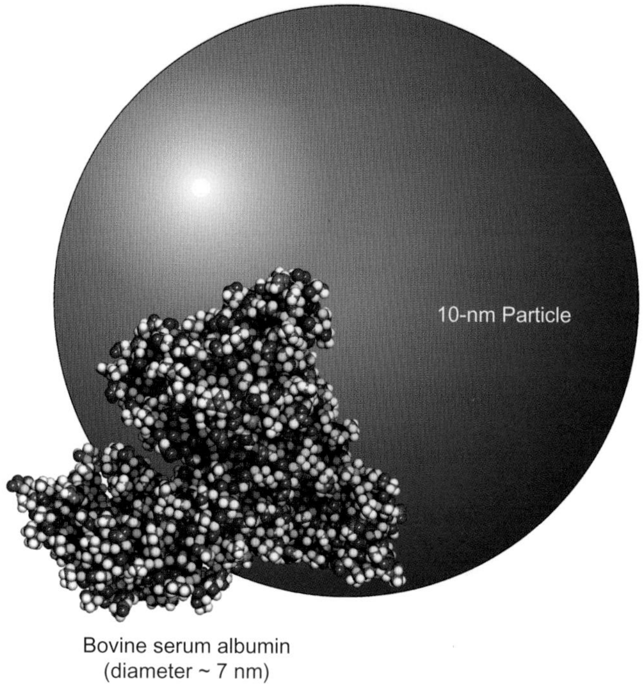

FIGURE 14.2 Comparison of the relative size of a 10-nm particle and the diameter of a typical protein, bovine serum albumin (BSA), at about 7 nm across its largest axis.

Another aspect of particle handling that should be worked out before actually coupling an affinity ligand is the best procedure to wash the particles and remove unreacted ligands or byproducts of a coupling reaction. With small quantities of particles, most often this can be done by centrifugation or, if working with very small nanoparticles, by gel filtration or size exclusion chromatography. Membrane filtration is also another option, especially for larger micron-sized particles. If the solution volume and quantity of particles are large enough, then tangential flow filtration is a viable alternative to other methods of separation.

Although many particle types can, in theory, be separated from solution by centrifugation, some particles may become irreversibly aggregated upon pelletting and should not be centrifuged under any circumstances. The relative hydrophobicity and degree of charge on a particle's surface directly affect their tendency to aggregate, and this property to a large extent governs whether the particle population can be resuspended successfully after centrifugation. In particular,

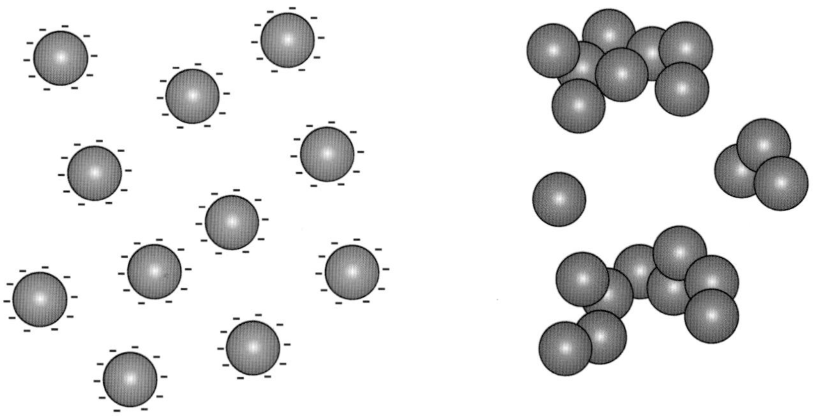

FIGURE 14.3 Small particles are often stabilized in aqueous solution by like-charge repulsion, which prevents particle aggregation and precipitation.

particles that contain higher densities of like surface charge, which functions to repulse and keep individual particles away from one another, typically can be resuspended successfully after centrifugation. To aid in resuspension, centrifuged particles should be subjected to sonication, by using either a bath sonicator or a sonic probe to disrupt the pellet. Vortex mixing should usually be avoided, as it is not very effective at re-dispersing small particles pelletted by centrifugation. Before using centrifugation or any other separation technique, it is best to contact the manufacturer to see whether they offer any guidelines for handling the particles.

3. PARTICLE CONCENTRATION

If individual particles in suspension are considered the equivalent of discrete molecules, then the molar concentration of a given particle suspension can be calculated based on the known particle diameter, density, and the mass of particles present. This allows particles to be treated similarly to other biomolecules with respect to determining concentration for conjugation purposes. However, there are important differences that should be recognized when working with particles as opposed to working with soluble macromolecules, like proteins. Since common commercial particles can vary in size from the molecular range (approximating the size of an antibody or ~10-nm diameter) to a scale 1000 times larger (or approaching the size of a cell at 10 µm), a change in diameter affects the concentration of particles as well as the effective concentration of surface functional groups present in suspension. Also potentially affected are the dispersion characteristics of particles as their size is changed (Suttiponparnit *et al.*, 2011).

In general, as particle size decreases, the molar concentration of particles in a constant volume of solution increases (for a given mass of particles). For instance, a 1-mg quantity of 1-µm latex microspheres represents far fewer particles than a 1-mg amount of 50-nm nanoparticles. Thus, the effective molar concentration of nanoparticles in solution will be much greater than the concentration of the same mass of microparticles (if both are suspended at the same mass quantity and in same volume of solution). In addition, as the diameter decreases for a given mass of particles, the ratio of a particle's surface area to mass increases. This means that the total surface area available for conjugation on the nanoparticles is much greater than the total surface area present on the microparticles. If both particles contain the same functional groups on their surfaces for coupling affinity ligands (i.e., carboxylates), then for the same mass of particles the effective concentration of these functional groups in solution is much greater for the nanoparticles than the concentration of the same groups in a given solution for the microparticles (assuming both have about the same surface density or "parking area" of the carboxylate functional groups).

Thus, conjugation reactions performed with nanoparticles should take into account a potentially greater reactivity than the same reactions performed using microparticles, due to the higher effective concentration of functional groups present in solution for the nanoparticles. As particle size decreases and particle concentrations increase, the available surface area increases, and the effective concentration of reactive groups increases along with it. All of these factors must be considered when optimizing conjugation reactions to particles. This means that an optimal protocol for coupling proteins to 1-µm carboxylated microspheres will most likely have to be re-optimized for use with carboxylated nanoparticles.

The following sections discuss many of the major particle types and provide bioconjugation options for the coupling of ligands to the surface of functionalized particles. Some additional nanoparticle constructs, including gold particles, dendrimers, carbon nanotubes, Buckyballs and fullerenes, and quantum dots are discussed more fully elsewhere (see Section 6 in this chapter, as well as Chapter 8; Chapter 10, Section 10; and Chapter 16).

4. POLYMERIC MICROSPHERES AND NANOSPHERES

Perhaps the most common particle type used for bioapplications is the polymeric microsphere or nanosphere, which consists basically of a spherical, nonporous, "hard" particle made up of long, entwined linear or crosslinked polymers. Creation of these particles typically involves an emulsion polymerization process that uses vinyl monomers, sometimes in the presence of divinyl crosslinking monomers (Kumacheva and Garstecki, 2011). Larger microparticles are usually built from successive polymerization steps through growth of much smaller nanoparticle seeds. The investigation of particle surfaces by high-resolution electron microscopy can often reveal the presence of numerous entangled, amorphous nanoparticles making up the morphology of much larger nanospheres or microspheres. Instead of being smooth spheres, the surfaces of most polymeric particles are actually torturous and made up of many craters, pits, and crevasses, which may appear featureless under lower-resolution imaging.

Some of the most common forms of polymeric particles consist of polystyrene or copolymers of styrene, like styrene/divinylbenzene, styrene/butadiene, styrene/acrylate, or styrene/vinyltoluene. Other common

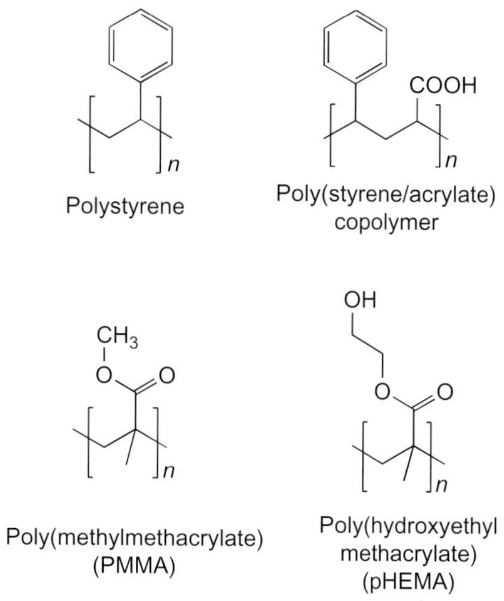

FIGURE 14.4 Some of the most common particles consist of these polymers.

polymer supports include polymethylmethacrylate (PMMA), polyvinyltoluene, poly(hydroxyethyl methacrylate) (pHEMA), and the copolymer poly(ethylene glycol dimethacrylate/2-hydroxyethylmetacrylate) [poly(EGDMA/HEMA)] (Ayhan *et al.*, 2002) (Figure 14.4). The types of monomers used to form the polymers ultimately govern the final particle characteristics. If the monomers are hydrophobic, such as those that form most of the styrene-based particles, then the resultant particle surfaces will be hydrophobic, as well. The inclusion of hydrophilic monomers, including charged groups and polar constituents, will create more hydrophilic character on the surface. In addition, since many particles are polymerized in the presence of detergents, particles typically have detergent molecules non-covalently adsorbed on their surfaces, unless they have been purposely removed by extensive cleaning after polymerization.

Polymeric particles traditionally have been called "latex" beads or spheres, probably from the classic definition of an "emulsion of rubber or plastic globules in water." However, due to the polymeric diversity of these particles, even those that consist of a styrene base, it is best to identify the particle type by its exact chemical composition.

Polymeric particles can also be further diversified by their surface compositions. Even traditional polystyrene particles contain additional groups on their surfaces due to the polymerization reaction used to create them. For instance, nearly all such particles contain some negatively charged sulfonate groups, which are the result of the catalyst, persulfate, used to initiate the polymerization process. In addition, many particles contain non-covalently adsorbed detergent molecules that can add amphipathic character to the particle surface, depending on the type of detergent used during the polymerization reaction. Detergents can be removed by extensive washing after the manufacturing process, but this may be a source of variability from manufacturer to manufacturer.

Another way of altering surface characteristics is through the purposeful addition of secondary polymers onto a polymeric core by graft copolymerization, adsorptive coating, or covalent attachment of another polymer type (Charleux *et al.*, 2011; Pichot *et al.*, 2011). Adding hydrophilic coatings onto hydrophobic particle cores is often carried out to promote biocompatibility and limit nonspecific interactions with proteins or other biomolecules. Examples of hydrophilic coat polymers that have been used for this purpose include poly(ethylene glycol) (PEG), poly(vinyl alcohol) (PVA), poly(acrylic acid), poly(methacrylic acid), poly(acrylamide), and poly(vinyl pyrrolidone). PEGylation of nanoparticles has been used to create biocompatible surfaces for imaging and therapy applications *in vivo* (Jokerst *et al.*, 2011). Hydrophilic modification of hydrophobic particle cores can aid in blocking the nonspecific binding potential that the core may have toward proteins or other biological molecules. It can also alleviate the tendency of particles to aggregate in aqueous solution by creating a hydrated shell around each particle, which helps keep them separated and in suspension. For example, Harper *et al.* (1995) described the copolymerization of a poly(ethylene oxide) methacrylate monomer (PEG-type) with styrene to create polystyrene particles with hydrophilic surfaces for the attachment of cells.

Covalent attachment of hydrophilic spacer arms to particles is another way of masking the hydrophobic character of the underlying surface while making a particle more biocompatible. Spacers of this type are typically short organic molecules containing a number of polar groups along the chain which are able to interact with the surrounding aqueous environment and limit non-specific protein adsorption on the particles. A common spacer arm construction consists of a PEG chain that may also include a short hydrophobic straight chain linker, which attaches to the particle surface. To provide an attachment site for affinity ligands to the spacer, a functional group can be included on the outer end of the PEG polymer, such as a carboxylate group. These functional groups provide sites for covalent coupling of affinity ligands, including proteins, while the unmodified PEG chains create a hydrophilic "lawn" to limit nonspecific interactions. The best configuration of this spacer type is to have about 10% of the spacers terminate in a carboxylate group while the rest of them terminate in a PEG hydroxyl or a PEG methyl ether group (mPEG). This design has been used with success

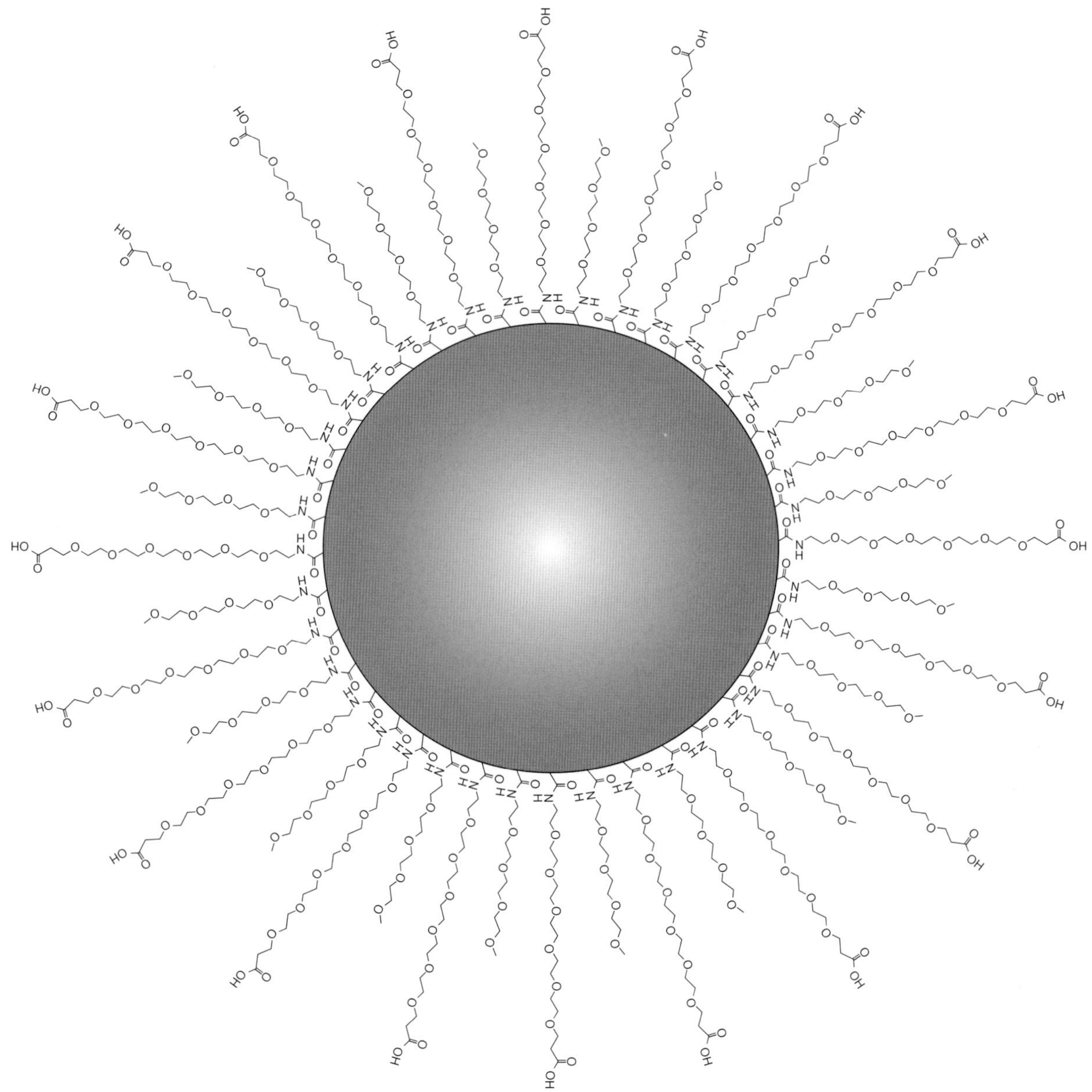

FIGURE 14.5 A method of making particles biocompatible includes the use of PEG-based spacers. A lawn of mPEG molecules is interspersed with some longer PEG chains that terminate in carboxylate groups for coupling amine-containing molecules. The result is an extremely hydrophilic surface with low non-specific binding character.

with other surface types, such as gold nanoparticles or planar surfaces (Prime and Whitesides, 1991), and it can be adapted to polymer particles with little difficulty (Figure 14.5).

4.1. Passive Adsorption

One of the simplest methods of attaching biomolecules to hydrophobic polymeric particles is the use of passive adsorption. Some of the earliest examples related to the use of particles in immunoassays include the use of non-covalently adsorbed antibody or antigen onto latex microspheres. Protein adsorption onto hydrophobic particles takes place through strong interactions of nonpolar or aromatic amino acid residues with the surface polymer chains on the particles with concomitant exclusion of water molecules. Since proteins usually contain hydrophobic core structures with predominately hydrophilic surfaces, their interaction with hydrophobic particles must involve significant conformational changes to create large-scale hydrophobic contacts. These conformational changes often result in complete denaturation of the first protein layer to be adsorbed onto the particles. Subsequent protein molecules which bind and add to this initial layer are adsorbed through protein–protein binding or aggregation events, which ultimately result in the formation of protein clusters in which the outer layers of protein have more of a native conformation and activity (Butler, 2000a, 2000b).

Due to this tendency of proteins to form multilayer coatings on hydrophobic surfaces, the amount of excess protein beyond the particle saturation point (theoretical monolayer density for the protein type) added during adsorption processes should be limited. If extremely large excesses of protein are added to particles, highly unstable constructs may result which may continually leach off loosely adsorbed protein. Most protocols recommend a protein concentration of about 3- to 10-times excess of protein over the calculated monolayer concentration for the particles used. An estimate of the maximal amount of a given protein that can bind as a monolayer to a particle surface can be determined by the fact that bovine serum albumin (BSA) (MW 67 kDa) has an adsorption capacity of about 3 mg/m^2 of particle surface. A larger protein like an antibody (IgG; MW 150 kDa) has a maximal adsorption density of about 2.5 mg/m^2 of particle surface. To translate this into a real-world example, 1 g of 1-µm polystyrene microspheres can adsorb about 18 mg bovine serum albumin (BSA) or about 15 mg IgG (Cantarero *et al.*, 1980). Knowledge of a particle's surface area per gram will permit reasonable estimates of maximal adsorption capacity for a given protein relative to the size of an antibody or albumin molecule.

Passive adsorption has been studied extensively as it relates to the immobilization of antibody molecules onto microplates (polystyrene) or microspheres (hydrophobic beads of various compositions). Although the denaturing effect of passive adsorption of proteins onto surfaces was known since 1956 when Bull studied the adsorption of albumin onto glass, it was not widely recognized as being detrimental until a number of subsequent studies were published. See in particular Butler *et al.* (1992), which investigates the physical and functional behavior of capture antibodies adsorbed on polystyrene; Butler J.E. (2000a), which provides a summary of the effects of the adsorption of various proteins on hydrophobic surfaces; Cantarero *et al.* (1980), which discusses the adsorptive characteristics of proteins for polystyrene surfaces; and Butler *et al.* (1997), which compares the effect of passive adsorption on the antigen specificity of immobilized antibody molecules.

The detrimental effects on biomolecules as a result of passive adsorption onto hydrophobic surfaces can be dramatic. Butler *et al.* (1993) determined that over 90% of monoclonal antibodies and about 75% of polyclonal antibodies against fluorescein were denatured upon adsorption onto polystyrene surfaces. In addition, Bagchi and Birnbuam (1980) observed that IgG was optimally adsorbed to poly(vinyltoluene) particles at a pH value near the antibody's pI, indicating that the interactions were entirely hydrophobic in nature and did not depend on charge interactions. Once the antibody was adsorbed, however, it was tightly bound to the surface, provided that the pH was maintained at the initial adsorption point. If any pH cycling was done, such as may occur with some wash steps, the adsorbed antibody had a tendency to leach off the surface, demonstrating the non-covalent nature of the interaction.

Protein adsorption to hydrophobic polymer particles should thus be done at or near the isoelectric point of the particular protein to ensure that the highest density of protein is attached to the particles. Proteins at their isoelectric point exist in the most collapsed conformation possible, because charge repulsion effects do not play as significant a role on overall globular structure. Upon adsorption under isoelectric conditions, each protein then can bind to the polymer surface at maximal density. Therefore, most suggested buffer cocktails used for passive adsorption recommend pH conditions somewhere in the pI range of the protein. Since many proteins have a range of different biological modifications (post-translational) resulting in a range of pI values, some optimization may need to be done to determine the best pH conditions to use for an adsorption process.

Another factor to consider is the amount of protein available for adsorption. If expensive antibody in very small amounts is used to passively adsorb to polymeric particles, there is often not enough antibody available to saturate the particle surface. In this case, another carrier or blocking protein may be added to the mixture to take up otherwise unoccupied sites on the particle surface. This will prevent particle aggregation that may occur if low concentrations of protein are used in the adsorption process. Carrier proteins frequently used for this purpose include albumin or polyclonal IgG pools, such as bovine gamma globulin. Other standard protein blocking agents may also be used for blending with the desired antibody, such as casein, non-fat dried milk proteins, fish serum, etc.

The polymeric particle composition also plays a significant role in the amount of protein adsorbed and its stability and activity after immobilization. In general, copolymer blends in which a hydrophobic monomer is polymerized with a charged monomer (such as acrylic acid or methacrylate) create a surface that has both hydrophobic character along with areas of charge. Proteins adsorbed onto these copolymer particles interact with the surface through hydrophobic interactions and positive–negative charge attraction. The result is less protein denaturation upon adsorption and better retention of activity after immobilization. A comparison of protein adsorption onto pure polystyrene and several other copolymer particle types by Bale *et al.* (1989) resulted in the following order of binding affinity: polystyrene > polystyrene/PMMA copolymer > PMMA > polystyrene/polyacrylic acid copolymer.

Thus, pure hydrophobic polymer compositions adsorb proteins very strongly but have the greatest tendency to denature protein at the initial protein–surface interface. Copolymers containing some polarity or negative charge character bind protein less avidly, but result in better retention of activity. The inclusion of some polar hydrophilic groups within an overall hydrophobic surface structure has been used with success for polystyrene microplates, as most of the so-called high binding plates contain a population of polar constituents on the well surfaces, allowing both hydrophobic and charge or dipole interactions to occur.

The following protocol for passive adsorption is based on methods reported for use with hydrophobic polymeric particles, such as polystyrene latex beads or copolymers of the same. Other polymer particle types may also be used in this process, provided they have the necessary hydrophobic character to promote adsorption. For particular proteins, conditions may need to be optimized to take into consideration maximal protein stability and activity after adsorption. Some proteins may undergo extensive denaturation after immobilization onto hydrophobic surfaces; therefore, covalent methods of coupling onto more hydrophilic particle surfaces may be a better choice for maintaining native protein structure and long-term stability.

Protocol

1. Dilute the particles to a mass concentration of 10 mg/ml (1%) in coating buffer at a pH near the pI of the protein being adsorbed. If particle aggregation becomes a problem, the initial particle concentration may be reduced to 5 mg/ml. Some typical coating buffer suggestions include: (a) 10-m*M* sodium phosphate, 0.15-*M* NaCl, pH 7.4 (PBS); (b) 50-m*M* sodium borate, pH 8.5; (c) 50-m*M* sodium acetate, pH 3.6 to 5.6; (d) 25-m*M* MES, pH 6.1; or (e) 50-m*M* sodium bicarbonate, pH 8.5 to 9.5. NaCl may be added to any of these buffers at a final concentration of 0.15-*M* to promote protein stability, if required. However, avoid the use of detergents or chaotropic agents, as these will compete with protein adsorption or reduce the strength of hydrophobic interactions, thus limiting the yield.
2. Dissolve the protein to be adsorbed in coating buffer at a concentration such that when the solution is mixed with the particle suspension, the appropriate protein concentration is reached to obtain an excess of about 3 to 10 times over the maximal monolayer density for that protein on the particles (see previous discussion). Note: A good strategy is to perform a protein titration study, wherein a range of protein concentrations are tried with a given particle quantity. For instance, using a 300-nm latex particle, a suitable set of trial experiments would employ protein concentrations in the range of about 10 to 200 μg protein per mg particles. Plotting the amount of protein charged to the coating reaction *versus* the amount bound will provide data to identify the optimal protein concentration to be used.
3. With rapid mixing, add protein solution to the particle suspension. For small volumes, mixing can be performed by pulling the solution up and down in a pipette tip or by vortexing. For larger volumes, a stir bar or paddle may be used. Some protocols recommend adding the particles to the protein solution to obtain the best initial mixing rate, thus ensuring even particle coating throughout the particle suspension. The key is rapid and efficient mixing to create a homogeneous mixture of protein and particles, so adsorption can occur uniformly on each particle.
4. Mix the suspension for 1 h at room temperature.
5. Remove excess protein by centrifugation or tangential flow filtration. Many particles may be pelletted successfully using a tabletop centrifuge, especially if the particle size is above about 150 nm in diameter. If particle clumping is a problem upon pelletting or the particle size is too small to be effectively pelletted, then filtration is the better alternative. Centrifuge and wash the particles at least twice with coating buffer to ensure complete removal of non-bound protein. Complete re-suspension of the particles may be accomplished by the use of sonication, especially through the use of a sonic probe dipped into the solution.
6. Resuspend the particles to a final concentration of 1% in coating buffer containing a preservative. Avoid changing the composition of the storage buffer from that used to coat the protein, as any pH or compositional changes often result in elution of some of the adsorbed protein.
7. Determine the amount of adsorbed protein on the particles by using a suitable protein assay technique, such as the bicinchoninic acid (BCA) protein assay (Thermo Fisher).

4.2. Covalent Coupling to Polymeric Particles

Many particle types contain functional groups that are built into the polymer backbone and displayed on their surface. The quantity of these groups can vary widely depending on the type and ratios of monomers used in the polymerization process or the degree of secondary surface modifications that have been

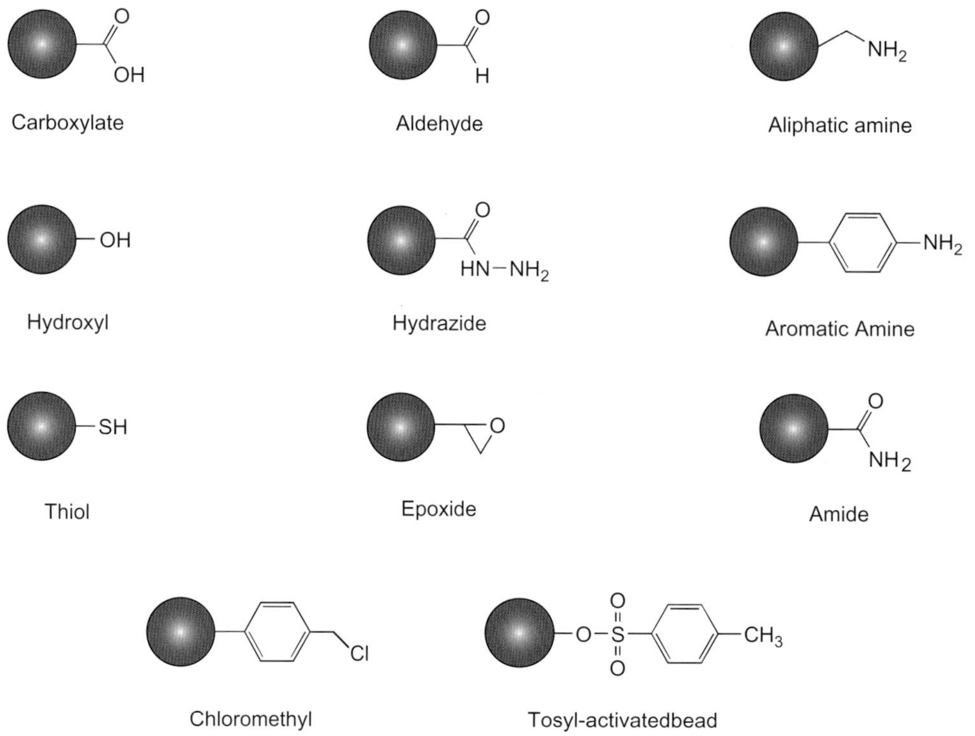

FIGURE 14.6 Common functional groups or reactive groups on particles that provide the ability to couple proteins or other ligands.

performed. Some common particle functionalities are shown in Figure 14.6. Many of these functionalized particles can be used to couple covalently biomolecules through the appropriate reaction conditions (Illum and Jones, 1985; Arshady, 1993). For each type of particle, manufacturers may offer several different densities of functional groups for different applications. The modification of particle surface characteristics and the functionalization of particles for covalent attachment of affinity ligands or targeting molecules requires the use of activation and coupling reactions to produce the most optimal particle derivatives for the intended application (Mout *et al.*, 2012). The following sections describe the reactive groups and immobilization methods that can be used with microparticles and nanoparticles. The reader also should refer to Chapter 15, which describes additional immobilization techniques and similar procedures for the coupling of affinity ligands onto porous chromatography supports.

Coupling to Carboxylate Particles

One of the more common particle types is the carboxylate particle, which is typically created by copolymerization or grafting with acrylic acid or methacrylic acid monomers. However, the characteristics of carboxylate particles can vary between manufacturers and particle sources due to the differences in surface density of carboxylate groups present. A measurement of chemical density on particle surfaces is called the "parking area," referring to the average area in $Å^2$ occupied by each functional group. A brief survey of carboxylate particles from different manufacturers indicates that the average carboxylate parking area can vary widely from about $10\,Å^2$ ($0.1\,nm^2$) to over $125\,Å^2$ ($1.25\,nm^2$). One carboxylate group every $125\,Å^2$ is equivalent to about one carboxylate for every square having dimensions of $11.1\,Å \times 11.1\,Å$ ($1.11\,nm \times 1.11\,nm$), or the space that might be occupied by one small molecule on the particle surface. Note: The amino acid lysine occupies a space of about $8.7 \times 4.8\,Å$, whereas a macromolecule such as albumin (67 kDa) has an effective molecular diameter of about 70 Å (7 nm), and an IgG molecule (150 kDa) has a diameter of about 110 Å (11 nm). A lone carboxylate functional group sticking off the surface of a particle has a diameter of about 2.2 Å. At a density of one carboxylate every $125\,Å^2$, there still is considerable uncharged polymer surface exposed, which could be a source for nonspecific binding, especially if the underlying polymer structure has hydrophobic character. A protein coupled to a carboxylate on such a particle could unfold through interactions with the exposed hydrophobic polymer surface and become denatured. Conversely, at the high end of carboxylate density, there would be about one carboxylate present for every $10\,Å^2$ ($0.1\,nm^2$) of particle surface. For the covalent coupling of proteins, this high-density surface

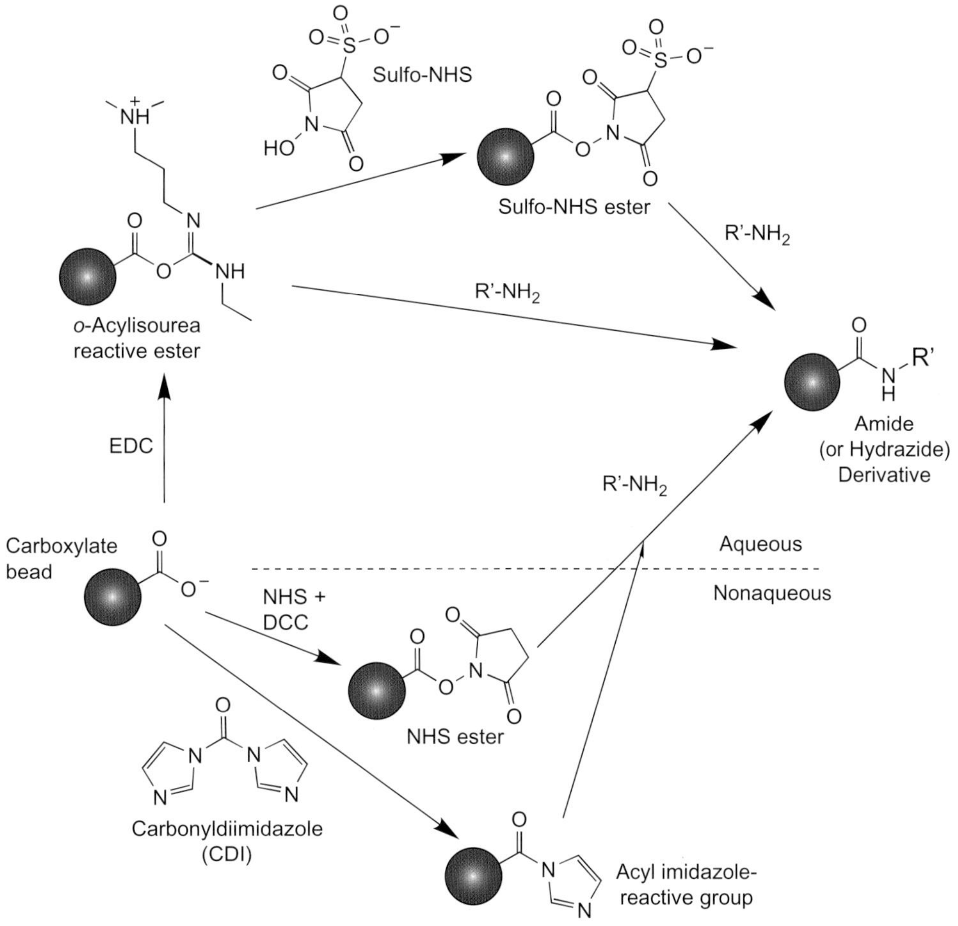

FIGURE 14.7 Carboxylate particles can be coupled to amine-containing molecules using a number of reaction strategies. The most frequently used method involves an aqueous two-step coupling process using EDC and NHS or sulfo-NHS to form an amide bond with a protein or other molecules. It proceeds through an intermediate (sulfo)NHS ester, which has better reactivity for coupling amines than the initial EDC reactive ester. Organic solvent activation processes using NHS esters or acyl imidazole reactive groups can also be used with solvent-stable particles to result in the same product with an amine-containing molecule.

would provide more reactive sites for conjugation, while more effectively masking the underlying polymer surface with negative charges. This may help to maintain a stable particle suspension through negative charge repulsion as well as prevent nonspecific binding or protein denaturation on the surface.

Carboxylate particles can be activated by a number of strategies that yield reactive intermediates capable of coupling with nucleophiles in proteins and other biomolecules. Figure 14.7 illustrates some of the reactions that can be used for coupling amine-containing molecules to carboxylate particles, all of which form stable amide linkages with the particle. The water soluble reactions can be carried out in entirely aqueous conditions, both for activation and coupling. For some reactions, however, the activation process must be carried out under nonaqueous conditions to prevent hydrolysis of the activator or active intermediate. Activation in organic solvent is an option for particles that remain stable in nonaqueous environments. Some polymer particles, especially non-crosslinked ones, may unacceptably swell or partially dissolve in organic solvent and become damaged by such treatment. Polymer particles prepared using a crosslinking monomer, however, can typically tolerate organic solvents without damage.

By far the most common reaction strategy for coupling proteins and other amine-containing molecules to carboxylate particles is through an aqueous, carbodiimide-mediated process using EDC (1-ethyl-3-(3-dimethylaminopropyl)carbodiimide). This reaction can be performed by use of a single-step coupling method (having all components in the reaction at once) or by use of a two-step reaction that better controls the immobilization course (see Chapter 4, Section 1.2, for additional information). Of all the crosslinking methods used in bioresearch applications today, this relatively simple coupling procedure is the most frequent conjugation reaction carried out with proteins and other amine-containing molecules for immobilization onto carboxylate particles. During this reaction, the carboxylate groups on the particles first are activated with the water soluble carbodiimide to create intermediate EDC esters. These esters are reactive directly with amines on proteins or other molecules to create amide bond linkages, which is the desired outcome. However, there are other side reactions that can occur. In aqueous solution, the intermediate ester might hydrolyze back to a carboxylate without coupling ligand. Also, if the density of carboxylates is high on the polymeric particles (e.g., small parking area), then the EDC esters can also undergo reaction with a neighboring carboxylate group to form anhydrides (Wang *et al.*, 2011), but these groups also are very reactive toward amines (Chapter 4, Section 1.1). EDC activation of carboxylate particles also can be used with the addition of NHS or sulfo-NHS in a two-step reaction, which results in the formation of secondary intermediate reactive groups, which are NHS esters or sulfo-NHS esters, respectively. The formation of these secondary esters results in a more stable intermediate in aqueous solution than the one initially formed with EDC therefore, the subsequent coupling reaction with proteins typically proceeds with higher yield than with the use of EDC alone (Figure 14.7). In addition, by forming the secondary (sulfo)NHS esters, excess EDC can be removed from the particles before adding protein, thus preventing carbodiimide-mediated protein polymerization due to the presence of both amines and carboxylates on most proteins (Staros, 1982; Borque *et al.*, 1994; Bonfield *et al.*, 2005).

For particle conjugation, it is important to maintain a repulsive force between particles to stabilize the colloidal suspension, even during the activation and coupling reactions. For this reason, the use of sulfo-NHS instead of NHS in this reaction results in an intermediate ester that is strongly negatively charged. The sulfonates on the sulfo-NHS esters actually are more effective at keeping particles from aggregating than the original carboxylates; therefore, the reactive intermediate particles easily can be purified from excess reactants and then mixed with protein for coupling. If the non-sulfonated NHS is used in this reaction, the uncharged intermediate NHS ester may cause unacceptable particle aggregation, depending on the particle type, although uncharged NHS has been used widely with success.

The following protocols for coupling amine-containing molecules to carboxylated particles represent viable starting points for optimizing a method that works best for the particular molecule or protein being immobilized. The single-step method using EDC alone is appropriate for use in coupling molecules having one or more amines present without any carboxylates. It works especially well for small organic molecules, such as haptens, 5′-amino-modified oligonucleotides, and amine-containing steroid derivatives (Hager, 1974; Quash *et al.*, 1978; Nathan and Cohn, 1981; Fuller *et al.*, 2006). If the molecule being coupled has both amines and carboxylates, such as proteins, then it is best to use the two-step method, which eliminates excess EDC before the addition of ligand; otherwise, unacceptable ligand polymerization may take place.

Single-Step EDC Coupling Protocol

1. Wash particles (e.g., 100 mg of 1-μm carboxylated latex beads) into coupling buffer (i.e., 50-m*M* MES, pH 6.0, or 50-m*M* sodium phosphate, pH 7.2; buffers with pH values from pH 4.5 to 7.5 may be used with success; however, as the pH increases the reaction rate will decrease). Suspend the particles in 5 ml coupling buffer. The addition of a dilute detergent solution may be carried out to increase particle stability (e.g., final concentration of 0.01% SDS). Avoid the addition of any components containing carboxylates or amines (such as acetate, glycine, Tris, imidazole, etc.). Also, avoid the presence of thiols (e.g., dithiothreitol (DTT), 2-mercaptoethanol), as these will react with EDC and effectively inactivate it.
2. Dissolve the amine-containing ligand to be coupled in 5 ml coupling buffer at a concentration sufficient to provide a 1- to 10-fold molar excess of ligand over the maximal calculated carboxylate group concentration for the amount and type of beads used. For particle manufacturers reporting a carboxylate concentration in meq/g, this is equivalent to μmoles/mg.
3. Combine the ligand solution with the particle suspension and mix thoroughly.
4. Add 100 mg EDC and mix to dissolve. To facilitate faster dissolution, EDC may be dissolved immediately before use as a concentrated stock solution in reaction buffer and then an aliquot of this solution added to the particle suspension to obtain the correct final concentration.
5. React at room temperature 2 to 4 h with mixing.
6. Wash the beads and resuspend them in coupling buffer containing 100-m*M* of an amine-containing, hydrophilic quenching molecule to block excess reactive sites (i.e., ethanolamine or Tris).
7. Wash beads and store in an appropriate buffer containing a preservative.

Two-Step EDC/Sulfo-NHS Coupling Protocol

An alternative to this procedure was used by Kulin *et al.* (2002) for coupling antibodies to carboxylated microspheres, which provides different buffer conditions and activation with EDC without the use of sulfo-NHS or NHS.

Activation

1. Wash particles (e.g., 100 mg of 1-μm carboxylated latex beads) into coupling buffer (50-m*M* MES, pH 6.0). Suspend in 5 ml coupling buffer. The addition of a dilute detergent solution may be done to increase bead stability and prevent clumping (e.g., 0.01% SDS). Avoid the addition of any components containing carboxylates or amines (such as acetate, glycine, Tris, imidazole, etc.). Also, avoid the presence of thiols (e.g., DTT, 2-mercaptoethanol), as these will react with EDC and effectively inactivate it.
2. Add 100 mg of EDC and 100 mg of sulfo-NHS. Mix to dissolve. To facilitate faster dissolution, EDC and sulfo-NHS may be dissolved immediately before use as a concentrated stock solution in reaction buffer and then an aliquot of this solution added to the particle suspension to obtain the correct final concentration.
3. React for 15 min at room temperature.
4. Quickly wash beads 2 times with coupling buffer using centrifugation and resuspend using a sonic probe in 5 ml of the same buffer.

Coupling

5. Dissolve protein to be coupled in 5 ml coupling buffer at a concentration sufficient to provide 1- to 10-fold molar excess of ligand over the maximal calculated monolayer concentration for the amount and type of beads used. For particle manufacturers reporting a carboxylate concentration in meq/g, this is equivalent to μmoles/mg. The optimal protein concentration should be optimized. Note: Too low a protein concentration may result in particle crosslinking. For coupling of expensive antibodies that may not be available in enough quantity to reach the optimal molar ratio on the particles, another protein (i.e., bovine gamma globulin or BSA) may be added to take up remaining reactive sites.
6. Combine the protein solution with particles and mix thoroughly.
7. React at room temperature 2 to 4 h with mixing.
8. Wash beads with coupling buffer and resuspend in the same buffer containing 100-m*M* of an amine-containing, hydrophilic quenching molecule to block excess reactive sites (i.e., ethanolamine or Tris).
9. Wash beads and resuspend in an appropriate buffer for storage.

Coupling to Amine Particles

Primary amine-containing polymeric particles are available from a number of manufacturers and have either aliphatic or aryl amine groups on their surface. Occasionally, a particle type may have secondary or tertiary amines present, but these should be avoided for covalent coupling, as primary amines typically give better reaction yields than secondary amines and tertiary amines are unreactive.

Primary amine particles may be used for covalent immobilization using a number of reaction routes (Figure 14.8). A carbodiimide-mediated coupling process may be used, as described above for carboxylate particles, but this time it is done by activation of a carboxylate group on the ligand and subsequent amide bond formation with the amines on the particles. A single-step EDC reaction strategy will work well for small ligands containing one or more carboxylates with no amines. However, for carbodiimide-mediated protein coupling to amine particles, the protein should first be activated with EDC and sulfo-NHS according to the method of Grabarek and Gergely (1990) and then the activated intermediate added to the amine particles for conjugation (see Chapter 4, Section 1.2, for the activation protocol). This type of two-step reaction will prevent protein polymerization during the coupling process. Add a quantity of activated protein to the amine particles to result in a 1- to 10-times molar excess over the total molar quantity of amines present on all the particles used in the reaction.

Coupling to Amine Particles Using Crosslinking Agents

Another possible route to coupling ligands to amine particles is to use a bifunctional crosslinking agent to react with the amines and provide another reactive group at the other end to couple with the ligand. In this approach, virtually any reactive group desired can be formed on the particles (Trindade and da Silva, 2011). Two strategies can be used with crosslinking agents: (1) use of a homobifunctional reagent that contains the same reactive groups on either end, or (2) use of a heterobifunctional compound that contains different reactive groups on each end.

A homobifunctional amine-reactive compound can be used initially to modify the amine groups on particles, while leaving the remaining amine-reactive groups available to couple with ligands. This type of reaction must be carried out with the crosslinker in great excess to prevent polymerization of the amine particles themselves. There must be enough crosslinker present during the activation stage to avert the free end from attaching to another particle before that particle's amines are modified with other crosslinkers.

Once the amine particles are modified with a homobifunctional crosslinker, the excess reagent is removed so that when the ligand is added, it doesn't become polymerized, particularly if it has more than one amine (e.g., proteins). Homobifunctional crosslinkers containing NHS esters, imidoesters, or aldehyde groups on each end can be used for this type of coupling process. The following protocol illustrates this method using glutaraldehyde.

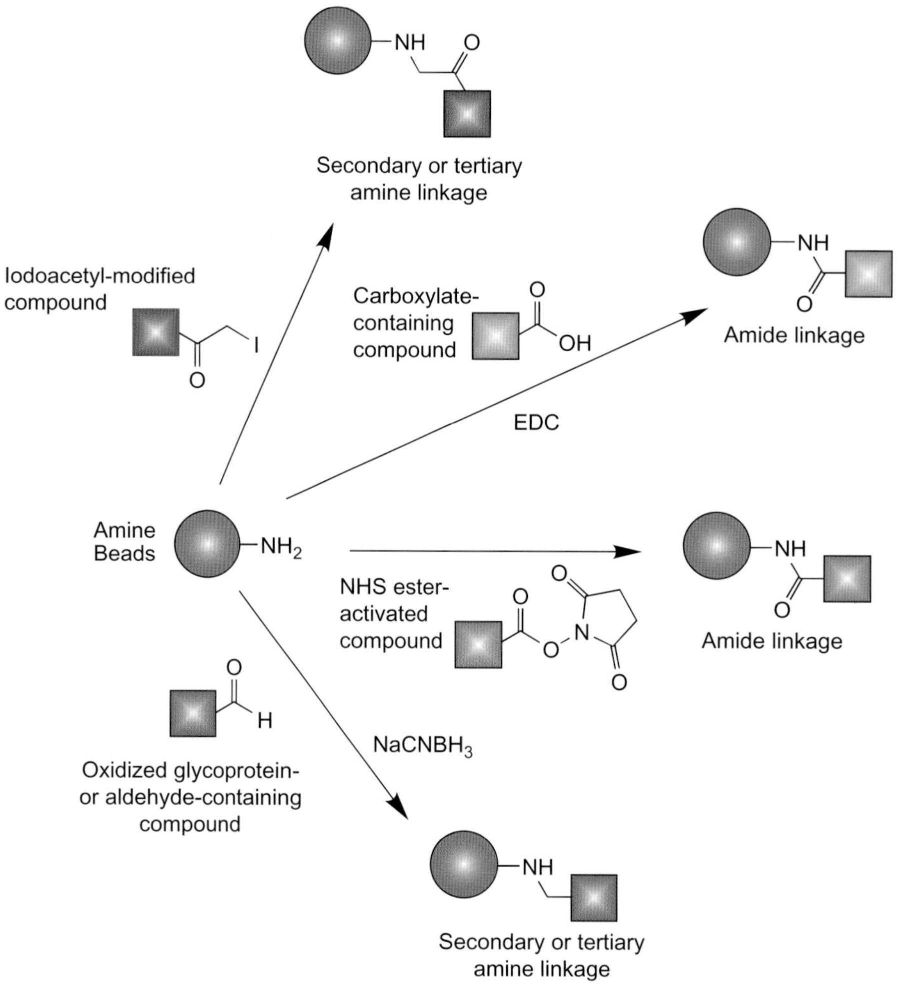

FIGURE 14.8 Amine-containing particles can be conjugated using alkylation or acylation reactions to result in secondary or tertiary amine linkages or amide bonds.

Glutaraldehyde

In one of the simplest methods, a homobifunctional amine-reactive crosslinker may be used to activate the surface of an amine particle for coupling with an amine-containing ligand. This has been done with success using glutaraldehyde or polyglutaraldehyde (Rembaum *et al.*, 1976; Rembaum *et al.*, 1978; Margel *et al.*, 1979; Kaplan *et al.*, 1983). More recent studies have found that glutaraldehyde can undergo intra- and intermolecular transformations in aqueous solution which involve the production of a number of reactive intermediates (see Chapter 15, Section 2.1, for a detailed discussion). In particular, two main forms of glutaraldehyde can be successfully used to activate particles for coupling amine-containing affinity ligands. In particular, the formation of a cyclic hemiacetal ring or a hemiacetal/aldol bicyclic ring structure create reactive derivatives that are useful for particle activation (Figure 14.9). Despite its potential shortcomings, glutaraldehyde is still being used to immobilize many different ligands onto particles of every size (Xu *et al.*, 2011; Alahakoon *et al.*, 2012).

The following protocol describes a two-step coupling method using glutaraldehyde on amine particles that makes use of either of these glutaraldehyde forms; it is based on the procedures of Betancor *et al.* (2006). The reaction mechanism can be controlled by modulating the pH of the reaction as well as the concentration of glutaraldehyde in solution. The monomeric hemiacetal activation method involves the use of highly dilute glutaraldehyde in a relatively brief activation process. The dimeric hemiacetal/aldol activation method uses a more concentrated glutaraldehyde solution for an extended activation time.

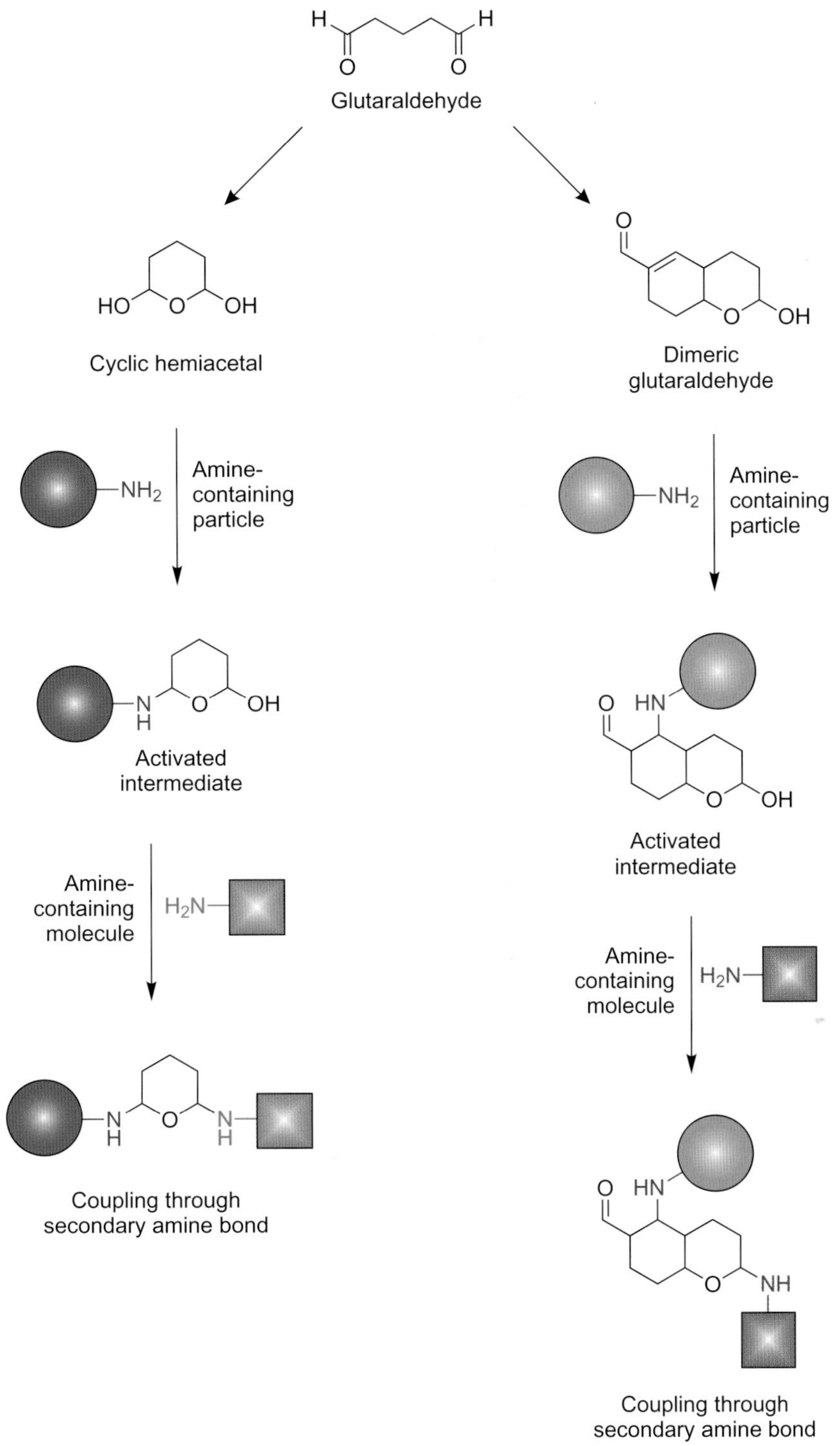

FIGURE 14.9 Glutaraldehyde-mediated immobilization onto amine particles can be performed with some control using two main reaction routes. The use of a cyclic hemiacetal glutaraldehyde species in aqueous solution can be used to activate the amino groups by coupling with one of the anomeric carbons on the ring and forming a reactive intermediate. After washing off excess reagent, the activated particles can be used to immobilize an amine-containing protein or ligand through secondary amine linkages. A similar reaction scheme can be carried out using a dimeric glutaraldehyde species that first reacts with the amine groups on particles by addition to the double bond. The final amine-containing ligand then is immobilized through a secondary amine bond to the anomeric carbon of the hemiacetal ring.

Protocol

1. Wash 100 mg of amine particles 3 times with 0.1-m*M* sodium phosphate, pH 7.0 (activation buffer).
2. After the final wash, suspend the particles in 10 ml of coupling buffer containing 0.5% glutaraldehyde (for monomeric activation) or 15% glutaraldehyde (for dimeric activation) and mix well to create a homogeneous suspension.
3. React with mixing for 1 h (for monomeric) or 15 h (for dimeric) at room temperature.
4. Wash the particles with coupling buffer at least several times using centrifugation to remove excess glutaraldehyde. Resuspend in 5 ml of coupling buffer (25-m*M* sodium phosphate, pH 7.0). Note: The activated particles can be stored at 4°C in water until used.
5. Add the protein to be coupled to the particle suspension in an amount equal to 1- to 10-times molar excess over the calculated particle monolayer for the protein type to be coupled. Mix thoroughly to dissolve. Low concentrations of protein may result in particle aggregation, because a single protein molecule can react with more than one particle.
6. React with mixing at room temperature for 2 to 4 h.
7. Add to the particle suspension a final concentration of 0.2-*M* glycine (or another amine-containing quench molecule, such as ethanolamine or Tris). The blocking agent will couple to any remaining glutaraldehyde reactive sites. Excess aldehyde groups on the support also may be block by the addition of 50-m*M* hydroxylamine, which will react with the aldehydes to create stable oxime linkages.
8. Remove excess protein and reactants by washing with coupling buffer at least three times using centrifugation. Storage of the modified particles may be done at 4°C in a suitable buffer containing a preservative.

SPDP Coupling to Amine Particles

Another crosslinker-based method that has been used for coupling proteins to amine particles involves the use of *N*-succinimidyl 3-(2-pyridyldithio)propionate (SPDP). This reagent contains an amine-reactive NHS ester and a thiol-reactive pyridyl disulfide group (Chapter 6, Section 1.1). Amine-containing particles can be activated by reaction with SPDP to form thiol-reactive derivatives (Xie *et al.*, 2011). Thiol-containing proteins, such as partially disulfide-reduced antibodies, can be coupled to the activated particles in a two-step reaction. An alternative method outlined by Illum and Jones (1985) that was based on Barbet *et al.* (1981) couples SPDP-modified amine particles to antibodies that were also modified with SPDP and then the pyridyl disulfide group reduced to form thiols. Mixing the thiolated antibody with the SPDP-activated particles will result in covalent coupling *via* disulfide linkages (Figure 14.10).

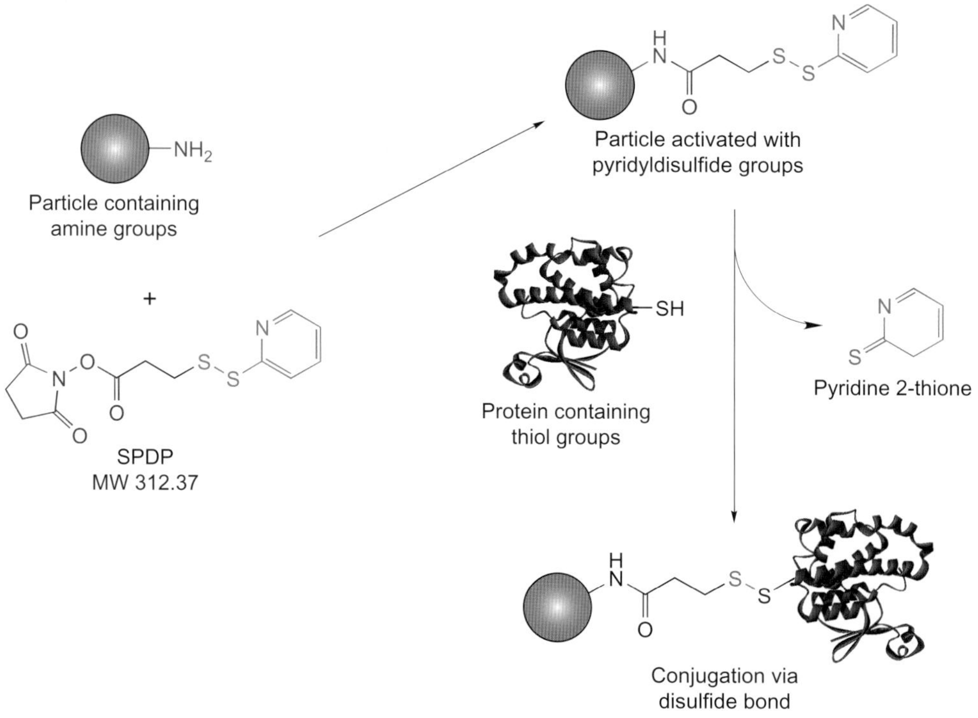

FIGURE 14.10 The crosslinker SPDP can be reacted with amine particles to create thiol-reactive pyridyl disulfide groups on the surface. Thiol-containing proteins or other thiol molecules can be reacted with these activated particles to result in disulfide linkages, which are reversible by reduction.

Unlike the use of homobifunctional crosslinkers, heterobifunctional compounds usually do not have to be used in large excess with amine particles to prevent aggregation. This is due to the fact that only one of the ends of the crosslinker can react with the amines on the particles.

Protocol

1. Wash 10 mg of amine particles into 10-m*M* sodium phosphate, pH 7.2, using centrifugation. Resuspend the particles in 1 ml of the same buffer.
2. Dissolve SPDP in dimethylformamide (DMF) at a concentration of 6.2 mg/ml (makes a 20-m*M* stock solution). Add 50 μl of the SPDP solution to the 1-ml particle suspension and mix to dissolve. Note: The small quantity of DMF in a polymeric particle suspension should not affect particle stability, even if the polymer type is susceptible to swelling in pure DMF. Other particle types, such as metallic or silica based, usually are not affected by organic solvent addition, unless their surfaces are non-covalently coated with a dissolvable polymer.
3. React for 30 min at room temperature with mixing.
4. Wash particles with coupling buffer at least three times using centrifugation and resuspend in the same buffer using a sonic probe to fully disperse the particles.
5. Add 1 to 10 mg of a protein or antibody containing an available thiol group to the particle suspension. Alternatively, add the protein to be coupled to the particle suspension in an amount equal to 1- to 10-times molar excess over the calculated monolayer for the protein type to be coupled. The optimal amount of protein to be added should be determined experimentally. Creating thiol groups from disulfides in proteins may be achieved by using a reducing agent or through the use of a thiolation reagent (Chapter 2, Section 4.1).
6. React with mixing for 2 h at room temperature. At the completion of the reaction, cysteine may be added at 50-m*M* to block excess pyridyl disulfide reactive sites.
7. Remove excess protein and reactants by washing with coupling buffer at least three times. Storage of particles may be done in a suitable buffer at 4°C containing a preservative.

NHS–PEG$_n$–Maleimide Coupling to Amine Particles

An alternative method for coupling thiol-containing proteins or antibodies to amine particles is to use a heterobifunctional crosslinker containing an amine-reactive NHS ester at one end and a thiol-reactive maleimide group on the other end (Chapter 6, Section 1). Unlike the pyridyl dithiol reaction with a sulfhydryl as in the SPDP protocol described previously, a maleimide group forms a stable thioether linkage with a sulfhydryl-containing ligand, which is not cleavable by reduction.

A common choice of crosslinker for this type of reaction is sulfo-SMCC (sulfosuccinimidyl-4-(*N*-maleimidomethyl)cyclohexane-1-carboxylate), which has been used extensively for antibody conjugation (Chapter 6, Section 1.3). However, perhaps a better option for particle conjugation is to use a similar crosslinker design, but one which contains a hydrophilic PEG spacer arm to promote particle hydrophilicity after modification. The modification of an amine particle with an NHS–PEG$_n$–maleimide reagent can create a surface that is essentially coated with PEG spacers (Chapter 18, Section 1.2). This helps to mask any hydrophobic character that the particle surface may have, while providing terminal thiol-reactive maleimides for coupling ligands (Figure 14.11).

If an NHS–PEG$_n$–maleimide compound is used for this type of activation and coupling, the intermediate maleimide-activated particle should be quickly washed free of excess crosslinker and used to couple ligand immediately. This is due to the fact that the maleimide hydrolyzes in aqueous solution at a higher rate than that of a maleimide on an SMCC-type crosslinker, because of the extreme hydrophilicity of the PEG spacer arm compared to the cyclohexane spacer of SMCC.

NHS–PEG$_n$–maleimide crosslinkers are available in a number of spacer lengths depending on the size of the polymer chain in the PEG component (Chapter 18, Section 1.2). Long-chain crosslinkers of this type use PEG polymers of molecular weight approximately 2000 to 5000 Daltons, containing from about 45 to over 100 repeating polyethylene oxide units (Chapter 18, Section 2.1). These full-length polymers are polydisperse and actually consist of a broad range of polymer lengths, which makes reproducibility of the crosslinker size nearly impossible to achieve. Modifying a particle with this type of polymer diversity causes variability in the length of the crosslinkers displayed on the surface, which may be detrimental for coupling some ligands.

However, shorter, discrete crosslinkers containing PEG spacers of known chain length are now available (Thermo Fisher, Quanta BioDesign). These reagents are designed to contain an exact number of PEG units, typically from between 2 repeating units and 24 repeating units. The following protocol may be used with any of these compounds with the appropriate adjustments in the quantity of crosslinker added to the reaction to take into account differences in molecular weight due to the PEG length. The longer of these discrete NHS–PEG$_n$–maleimide crosslinkers will provide the greatest degree of hydrophilicity after modification of an amine particle surface (see Chapter 18 for additional details on discrete PEG compounds.

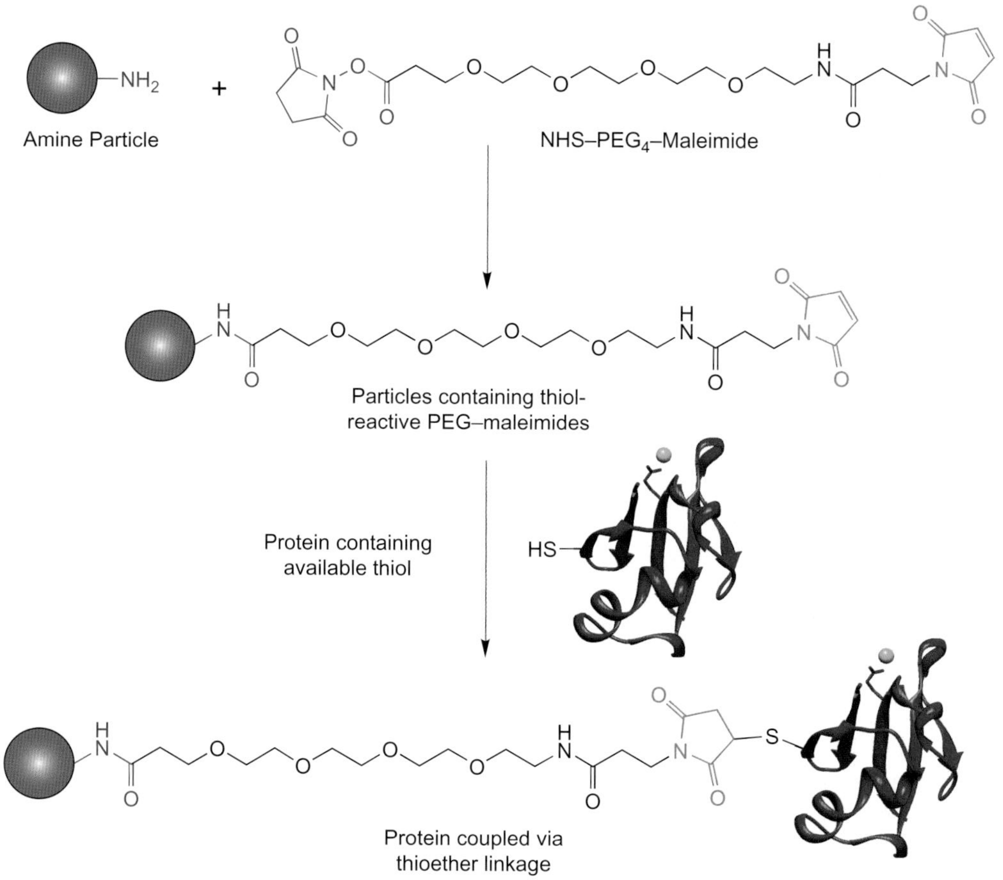

FIGURE 14.11 The modification of amine-containing particles with NHS–PEG_4–maleimide produces hydrophilic PEG spacers containing terminal thiol-reactive groups. Coupling of thiol-containing proteins then results in the formation of thioether linkages.

Protocol

1. Wash 10mg of amine particles into 10-m*M* sodium phosphate, pH 7.2 (coupling buffer) using centrifugation. Resuspend the particles in 1ml of the same buffer.
2. Dissolve NHS–PEG_6–maleimide (MW 601.6) into DMSO at a concentration of 20-m*M*. PEG type crosslinkers often exist as a thick oily mass, and preparing the solution may involve dissolving an entire vial of the compound into DMSO to accurately determine the required concentration.
3. Add 50µl of the NHS–PEG_6–maleimide solution to the 1ml particle suspension and mix thoroughly to dissolve.
4. React for 1h at room temperature with mixing.
5. Quickly wash the particles with coupling buffer at least twice using centrifugation.
6. Add 1 to 10mg of a protein or antibody containing an available thiol group to the particle suspension. Alternatively, add the protein to be coupled to the particle suspension in an amount equal to 1- to 10-times molar excess over the calculated monolayer for the protein type to be coupled. The optimal amount of protein to be added should be determined experimentally. Creating thiol groups on proteins or peptides may be done from disulfides by reduction. Alternatively, a thiolation reagent may be used to add thiols to the protein surface for coupling (see the protocols in Chapter 2, Section 4.1).
7. React with mixing for 2h at room temperature. At the completion of the reaction, cysteine may be added at 50-m*M* to block excess maleimide reactive sites.
8. Remove excess protein and reactants by washing with coupling buffer at least three times using centrifugation. Storage of the particles may be done in a suitable buffer at 4°C containing a preservative.

Coupling to Hydroxyl Particles

Polymeric particles containing hydroxyl groups are often created from copolymers or composites of poly(hydroxyethyl methacrylate) (pHEMA), frequently with other more rigid polymer cores, such as polystyrene (Figure 14.12). pHEMA particles have surfaces that contain an abundance of primary hydroxyls, which tend to produce favorable hydrophilic surface characteristics (Ahmad *et al.*, 2003; Tauer *et al.*, 2005). The

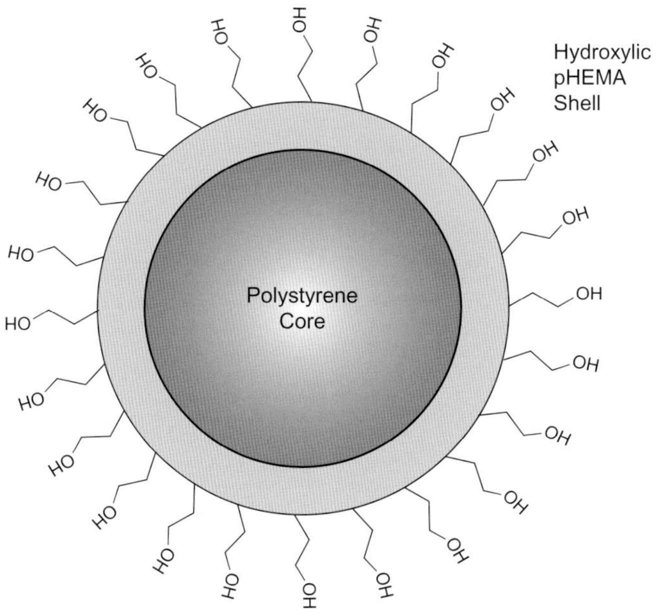

FIGURE 14.12 A core/shell polymeric particle made of a hydrophobic polystyrene core that is capped with a hydrophilic pHEMA shell, which contains numerous hydroxyl groups for hydrophilicity and biocompatibility.

hydroxyls can hydrogen bond with a layer of water molecules in aqueous solution, which forms an interface between individual particles and stabilizes them against aggregation, even in the presence of relatively high salt concentrations. The typical lack of interaction potential of biomolecules with pHEMA particles usually lowers nonspecific binding potential compared to particles of more hydrophobic polymer construction. This enhanced hydrophilicity of pHEMA particles translates into a high degree of biocompatibility, which is important for decreasing background in particle-based assays and in preventing denaturation of immobilized proteins on the particle surface. Horák *et al.* (2011) used a copolymerization process with HEMA and glycidyl methacrylate in the presence of Fe_3O_4 to produce magnetic nanoparticles suitable for affinity targeting. Cui *et al.* (2012) used a similar copolymer preparation to produce pHEMA-containing magnetic microspheres for the immobilization of lipase enzymes.

Although hydroxyls are not spontaneously reactive toward functional groups on biomolecules, they can be activated for covalent coupling by a number of known reaction mechanisms. Most of the reactions that can result in covalent attachment of ligands to hydroxylic or pHEMA particles originated in the development of immobilization technology for affinity chromatography using larger hydroxyl-containing porous beads (for reviews, see Chapter 15 and Hermanson *et al.*, 1992).

Most activation strategies for hydroxylic particles are carried out under nonaqueous conditions, because the activating agent and the intermediate reactive group are typically susceptible to hydrolysis. A convenient method of activation is to form a reactive carbonyl group on the hydroxyl particle using compounds such as carbonyl diimidazole (CDI) (Bethell *et al.*, 1979) or disuccinimidyl carbonate (DSC) (Miron and Wilchek, 1993). These activating agents create imidazole carbamates (using CDI) or NHS–carbonates (using DSC) on the particle surface, which are then spontaneously reactive toward amines (see Figures 14.13 and 14.14). After washing away excess activating agent in organic solvent, the particles are centrifuged to remove most solvent and then resuspended in aqueous buffer containing the amine ligand to be coupled (e.g., a protein). These activation and coupling reactions can also be used to prepare reactive dextran-coated silica nanoparticles for a variety of uses (Schulz *et al.*, 2011).

The imidazole carbamate group is more stable to hydrolysis in aqueous buffer than the NHS–carbonate group, which is similar in reactivity to an NHS ester. However, this means that the imidazole carbamate is also slower to react and couple with amines. NHS–carbonate reactions usually go to completion within 1 to 2h at room temperature, whereas imidazole carbamates typically require higher pH conditions and overnight incubations to get maximal yield of ligand coupling.

The following protocols involve the activation of hydroxylic pHEMA particles in organic solvent using either CDI or DSC and the subsequent coupling of amine-containing molecules using aqueous buffer conditions. Various solvents may be used for the activation step as long as the activation agents are soluble in them and the particles do not get damaged due to solvent exposure. Only water-miscible solvents should be used to facilitate exchange into and out of aqueous conditions. However, the solvents should be anhydrous to prevent hydrolysis of the activating agent or the subsequent reactive group formed on the particles after activation.

Protocol for Activation Using CDI

This method is derived from that of Bethell *et al.* (1979) and Colvin *et al.* (1988). See Figure 14.15.

1. Solvent exchange pHEMA particles into anhydrous THF using centrifugation and resuspension with sequential exchange into greater percentages of THF in water until the particles are suspended in 100% THF. Wash several times with anhydrous THF to eliminate any remaining traces of water, letting the particles agitate in solvent between each centrifugation step. After the final THF wash centrifuge the particles and remove excess solvent.

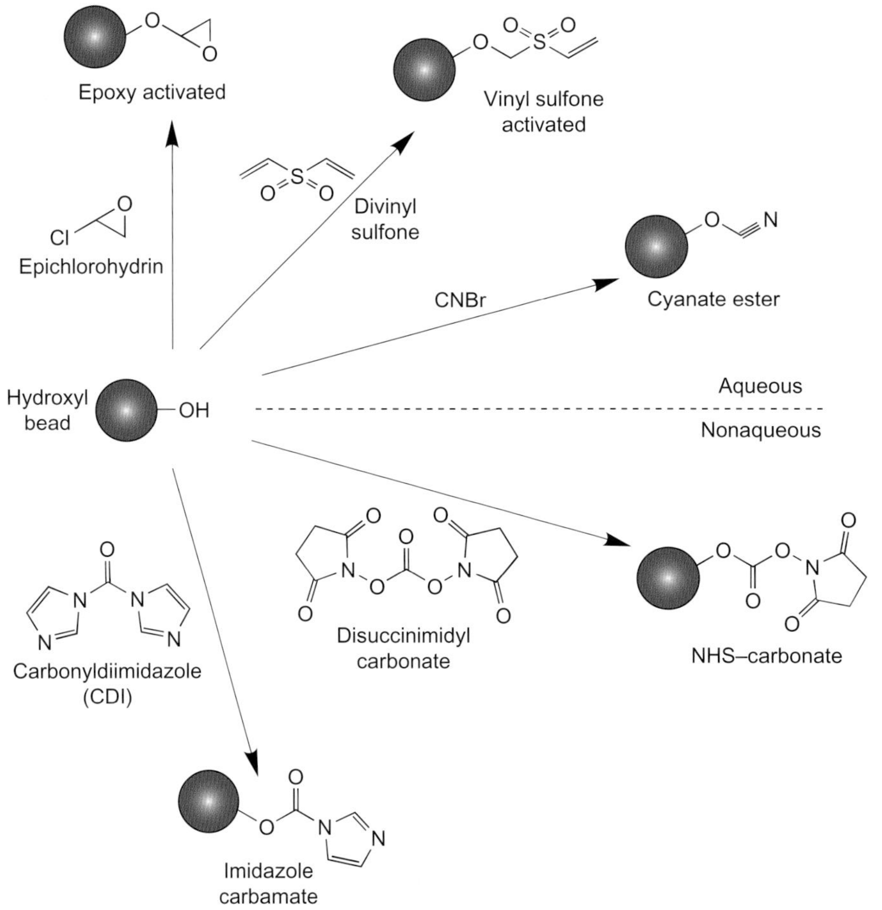

FIGURE 14.13 Hydroxyl-containing particles can be activated for coupling ligands using a number of strategies, which involve either aqueous or nonaqueous reactions. Epoxy and vinyl sulfone activation procedures provide reactive groups able to couple with amine-, thiol-, or hydroxyl-containing ligands. Cyanogen bromide activation and the CDI and DSC methods provide reactive groups for coupling amines.

2. Resuspend the particles as a 5% suspension in anhydrous THF containing CDI at a concentration of 50 mg/ml (0.3-*M*).
3. React with mixing for 2 h at room temperature.
4. Wash the activated particles three times with anhydrous THF to remove excess CDI and reaction byproducts. After the final wash, remove the solvent and perform a quick wash with ice-cold water to remove most traces of solvent in the particle pellet. Finally, resuspend the particles at 10 mg/ml in cold 0.1-*M* sodium phosphate, pH 8.2, or 0.1-*M* sodium carbonate, pH 9.5 (coupling buffer). The higher pH coupling buffer will result in greater reactivity of the imidazole carbamate and greater coupling yields for proteins.
5. Add 1 to 10 mg of a protein or antibody containing an available thiol group to the particle suspension in coupling buffer and mix to dissolve. Alternatively, add the protein to the particle suspension in an amount equal to 1- to 10-times molar excess over the calculated monolayer for the protein type to be coupled. The optimal amount of protein to be added should be determined experimentally.
6. React with mixing for at least 18 h at 4°C. Longer reaction times (24–48 h) may be necessary when using the pH 8.2 coupling buffer. Room temperature reactions will increase the reaction rate.
7. Add ethanolamine to the particle suspension at a final concentration of 0.1-*M* to quench any remaining active groups and react with mixing for several hours.
8. Centrifuge and wash the particles at least three times with buffer to remove unreacted protein and ethanolamine. Finally, suspend the particles in a suitable storage buffer containing a preservative.

Protocol for Activation Using DSC

This method is derived from that of Miron and Wilchek (1993) for the activation of hydroxyl groups on PEG molecules and Wilchek and Miron (1985) for the activation of agarose chromatography beads (see Chapter 15). Many types of hydroxyl-containing

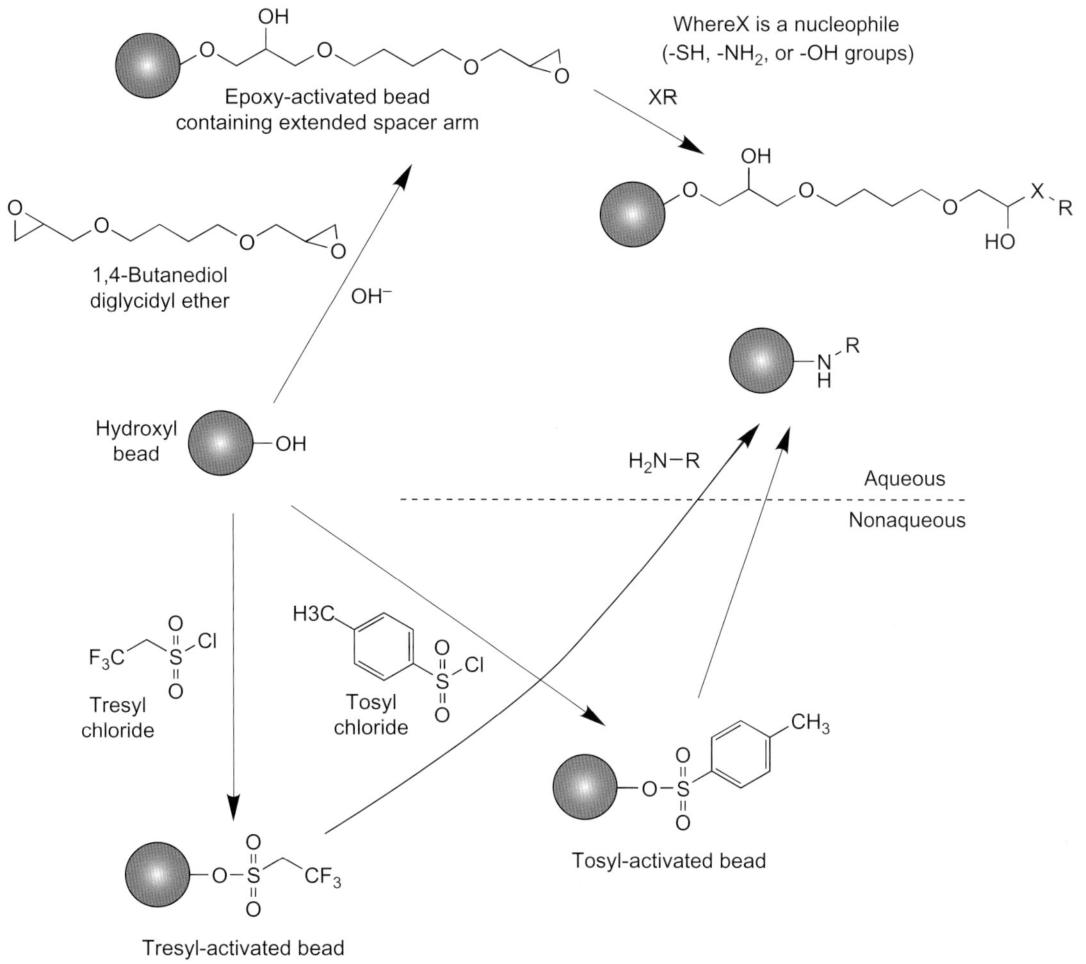

FIGURE 14.14 Additional hydroxyl particle activation methods include *bis*-epoxide modification, tosyl activation, and tresyl activation methods. The tosyl chloride and tresyl chloride activation procedures must be carried out in dry organic solvent, but the coupling of an amine-containing ligand can be performed either in organic solvent or aqueous buffer.

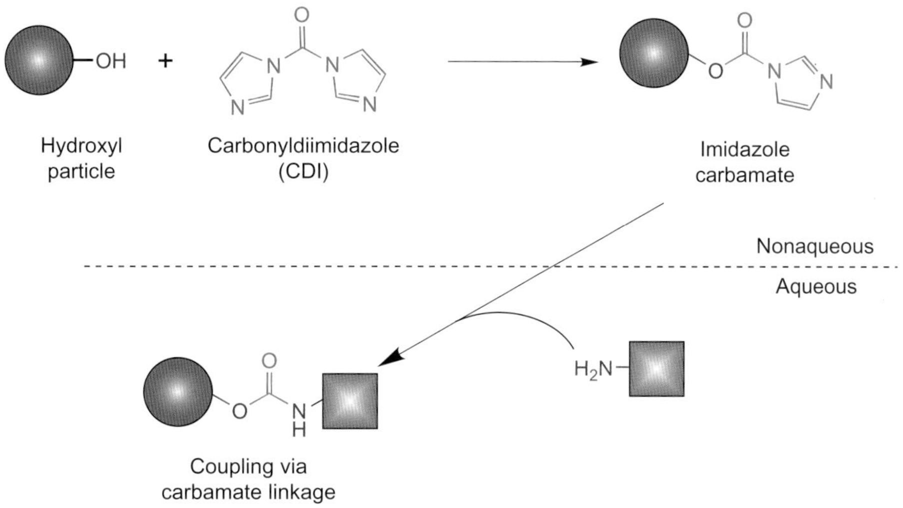

FIGURE 14.15 CDI can be used to activate hydroxyl particles in organic solvent and then the intermediate reactive imidazole carbamate brought into aqueous solution for coupling amine-containing ligands.

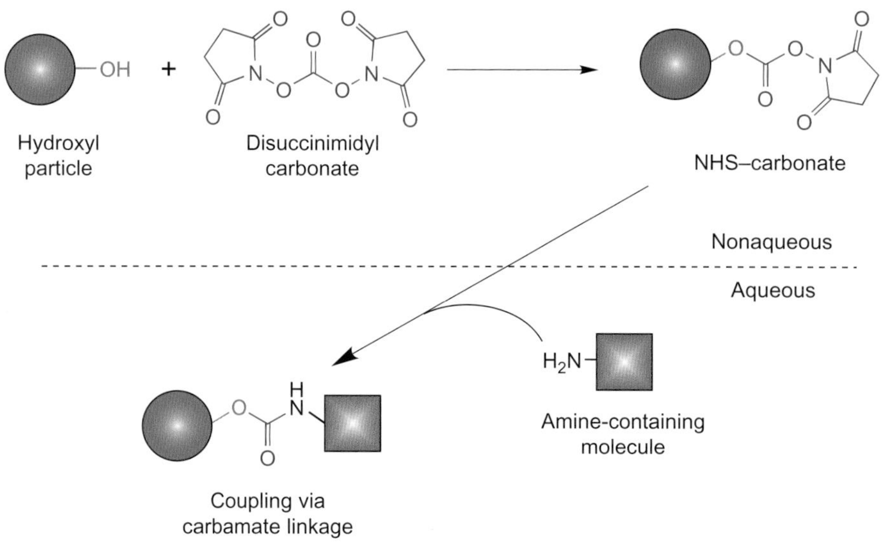

FIGURE 14.16 DSC can be used to activate hydroxyl particles to a reactive NHS–carbonate derivative. The subsequent coupling of amine-containing ligands can be carried out in either organic solvent or aqueous conditions.

particles may be used in this procedure, provided that the solvent used for the activation step does not deleteriously affect particle integrity. See Figure 14.16.

1. Solvent exchange pHEMA particles into anhydrous acetone, dioxane, acetonitrile, THF, or DMF (having very low amine content) using centrifugation and resuspension with sequential exchange into greater percentages of solvent in water until the particles are suspended in 100% organic solvent. Wash several times with anhydrous solvent to eliminate any remaining traces of water, letting the particles agitate in solvent between each centrifugation step. After the final solvent wash, centrifuge the particles and remove excess solvent.
2. Resuspend the particles as a 5% suspension in anhydrous solvent containing DSC at a concentration of 50 mg/ml (0.2-*M*).
3. React with mixing for 2 h at room temperature.
4. Wash the activated particles three times with anhydrous solvent to remove excess DSC and reaction byproducts. After the final wash, remove the solvent and perform a quick wash with ice-cold water to remove most traces of solvent in the particle pellet.
5. Immediately resuspend the particles at a concentration of 10 mg/ml in 0.1-*M* sodium phosphate, pH 7.2 (coupling buffer), containing 1 to 10 mg of a protein or antibody and mix to dissolve. Alternatively, add the protein to the particle suspension in an amount equal to 1- to 10-times molar excess over the calculated monolayer for the protein type to be coupled. The optimal level of protein to be added should be determined experimentally.
6. React with mixing for at least 2 h at room temperature or twice as long at 4°C.
7. Add ethanolamine to the particle suspension at a final concentration of 0.1-*M* to quench any remaining active groups and react with mixing for 1 h. Other amine-containing quenchers may be used, too, such as Tris buffer. Note: DSC-activated sites on the particles that completely hydrolyze will revert back to the original hydroxyls.
8. Centrifuge and wash the particles at least three times with buffer to remove unreacted protein and quenching agent. Finally, suspend the particles in a suitable storage buffer containing a preservative.

Protocol for Activation Using Cyanogen Bromide

Cyanogen bromide can be used to activate hydroxyl groups on particles to create reactive cyanate esters, which then can be coupled to amine-containing ligands to form an isourea bond (Figure 14.17). CNBr activation also can produce cyclic imidocarbonate groups, which are less reactive than the cyanate ester, but can form imidocarbonate bonds. The exact reactive species formed by the reaction is dependent on the structure of the hydroxylic support being activated (Kohn and Wilchek, 1982).

Unlike the previous methods for activation of hydroxyls on particles, cyanogen bromide is used under aqueous conditions, thus eliminating the need for organic solvents. This method originated in the activation and coupling of ligands to agarose supports for affinity chromatography (Axen *et al.*, 1967), and it was subsequently used to couple antibodies to hydroxyl-containing latex particles (Yen *et al.*, 1979) and more

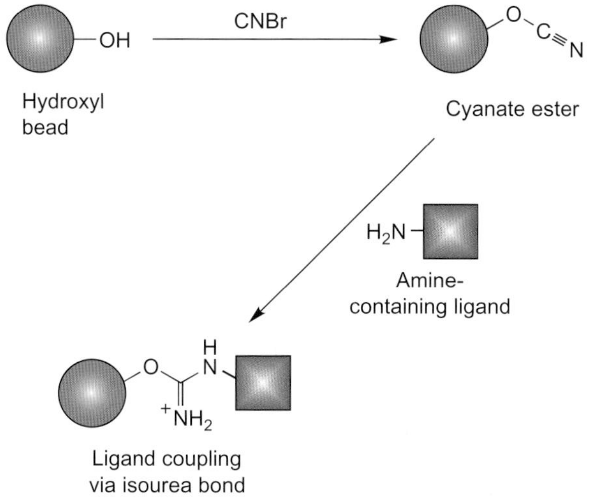

FIGURE 14.17 Cyanogen bromide can be used to activate a hydroxyl particle to a reactive cyanate ester, which can then be used to couple amine-containing ligands.

recently to immobilize protein A onto pHEMA particles (Denizli *et al.*, 1995).

Cyanogen bromide is an extremely toxic chemical and should be used only in a well-ventilated fume hood while using the appropriate personal protective gear. The following protocol is based on the method of March *et al.* (1974), as recommended by Bangs Laboratories.

All operations should be performed in a fume hood.

1. Wash 100 mg of hydroxyl-containing particles two times using centrifugation with 0.1-*M* sodium carbonate, pH 8.5 (coupling buffer). After the second wash, resuspend the particles at 10 mg/ml in 2-*M* sodium carbonate (activation buffer; no pH adjustment necessary).
2. Dissolve an amount of protein in 10 ml of coupling buffer equal to 1- to 10-times excess over the calculated monolayer for the type of particles being used.
3. In a fume hood, dissolve 1 g of cyanogen bromide in 0.5 ml of acetonitrile (highly toxic!).
4. Add a drop of the cyanogen bromide solution at a time to the particle suspension with constant mixing at room temperature. The entire solution should be added to the particles over the course of about 10 s.
5. Activate the particles with mixing for exactly 2 min at room temperature.
6. Quickly wash the particles with ice-cold deionized water and then with a volume of cold coupling buffer. Resuspend the particles in the protein solution prepared in step 2.
7. React for 24 h at 4°C with mixing.
8. Wash the particles with coupling buffer and block excess reactive groups by resuspending in 50-m*M* ethanolamine, pH 9.0. React for 1 h at room temperature with mixing.
9. Thoroughly wash the particles with storage buffer (e.g., PBS, pH 7.5, or other suitable buffer) and resuspend them at 10 mg/ml in storage buffer containing a preservative.

Coupling to Hydrazide Particles

Hydrazide particles can be made from carboxylate particles by modification with a *bis*-hydrazide compound using the carbodiimide reaction with EDC. Suitable bifunctional hydrazides include the small carbohydrazide compound or the longer adipic dihydrazide (Chapter 5, Section 8). A *bis*-hydrazide compound is reacted with a carboxylate particle population in large excess to prevent particle polymerization during the reaction. The resultant hydrazide particles may be used to couple to carbonyl-containing ligands, such as carbohydrates or glycans at their reducing ends or after the formation of aldehydes on carbohydrates using oxidation with sodium periodate (Chapter 2, Section 4.4).

Perhaps a better design for a *bis*-hydrazide compound to modify carboxylate particles would include a short PEG spacer arm between the two hydrazide groups. This type of linker would result in a hydrophilic surface due to the presence of the PEG spacers, while providing the terminal hydrazide functionality necessary for coupling to carbonyl compounds. Unfortunately, this type of compound is not currently available, so the aliphatic *bis*-hydrazides are the only choice.

Another route to the formation of a hydrazide on a surface is to use an aldehyde-containing particle (such as HEMA/acrolein copolymers) and subsequently modify the aldehydes to form hydrazone linkages with *bis*-hydrazide compounds, which then can be stabilized by reduction with sodium cyanoborohydride (Chapter 3, Section 5). The resulting derivative contains terminal hydrazides for immobilization of carbonyl ligands (see Figure 14.18).

Hydrazide-containing particles provide functional groups for the coupling of aldehyde or ketone containing ligands through a dehydration reaction to form hydrazone linkages (a type of Schiff base). However, a single hydrazone bond between a ligand and the particle surface may not provide enough stability to prevent leaching of ligand due to hydrolysis. There are two routes to overcome this instability: (1) reduce the hydrazone linkage using sodium cyanoborohydride, or (2) create multiple hydrazone linkages between the ligand and the particle surface. Multi-site attachment provides sufficient ligand stability, because not all the hydrazones will hydrolyze simultaneously to release ligand, and when one hydrazone bond breaks it will have enough time to reform before the other

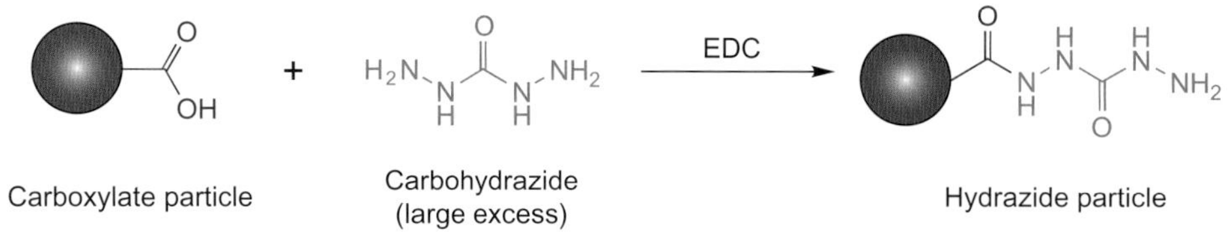

FIGURE 14.18 Carboxylate particles or aldehyde particles can be modified with carbohydrazide in excess to create a hydrazide particle that can be used to couple with aldehyde-containing molecules.

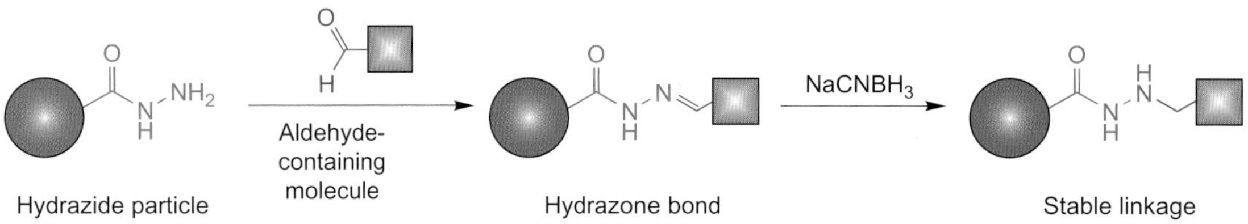

FIGURE 14.19 Aldehyde-containing molecules, such as periodate oxidized carbohydrates or glycoproteins, can be coupled to hydrazide particles to form a hydrazone bond. This bond can be further stabilized by reduction with sodium cyanoborohydride.

hydrazones hydrolyze. Thus, glycosylated proteins coupled after oxidation to hydrazide particles will most likely be stable due to the presence of more than one aldehyde group, but small ligands containing only a single carbonyl group should probably be treated with cyanoborohydride to stabilize the hydrazone bond.

The reactions involved with coupling carbonyl-containing ligands to hydrazide particles originated with the activation and coupling chemistry associated with the preparation of affinity chromatography supports (O'Shannessy and Wilchek, 1990). The application of this strategy to hydrazide-containing microparticles or nanoparticles is straightforward and will work well so long as accommodation is given to particle stability during the process. Using this method, a broad range of carbonyl-containing ligands can be coupled, such as reducing sugars or carbohydrates containing a reducing end, glycans after release from proteins, glycoproteins and other glycoconjugates, and small organic compounds containing an aldehyde or ketone group (Figure 14.19). Horak *et al.* (1999) used hydrazide–pHEMA particles to couple oxidized HRP in good yield and retention of enzymatic activity. Hydrazide particles are also a standard technique for the immobilization of antibodies through their carbohydrate portions after periodate oxidation to create aldehyde residues (Li and Ng, 2012).

The following protocol describes the oxidation of carbohydrate (glycans) on antibody molecules to form aldehydes and the subsequent coupling to hydrazide particles.

Protocol

1. Dissolve the antibody to be coupled in 10-m*M* sodium phosphate, 0.15-*M* NaCl, pH 7.5, at a concentration of 10 mg/ml. The antibody must be glycosylated to work in this procedure.
2. Prepare a solution of 0.1-*M* sodium periodate in water. Protect from light.
3. With mixing, add 0.1 ml of the periodate solution to each milliliter of the antibody solution.
4. React for 30 min at room temperature, protected from light.
5. The reaction may be quenched by the addition of 0.1 ml of glycerol per milliliter of reaction or by the addition of sodium bisulfite to a final concentration of 10-m*M*.
6. Remove excess reactants from the reaction mixture using size exclusion chromatography on a column of Sephadex G25 (or equivalent).
7. Wash 100 mg of hydrazide particles two times with 10 ml of 10-m*M* sodium phosphate, 0.15-*M* NaCl, pH 7.5 (coupling buffer), using centrifugation.

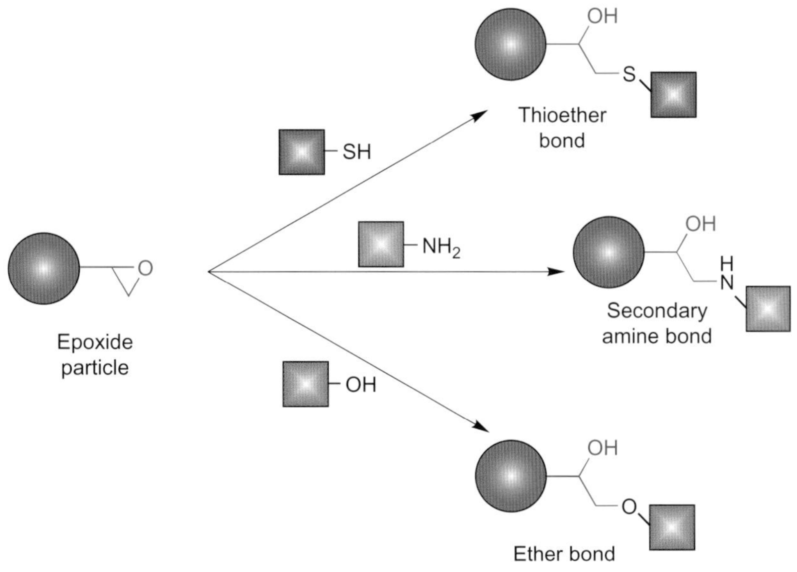

FIGURE 14.20 Particles containing reactive epoxy groups can be coupled with amine-, thiol-, or hydroxyl-containing molecules.

8. After the final wash, resuspend the particles at a concentration of 10 mg/ml in coupling buffer and add an appropriate amount of the solution from step 6, which contains the purified, oxidized antibody. The amount of oxidized antibody to add to the particles should be about 1 to 10 times over the amount of the calculated monolayer for the particle type used. Note: For 100 mg of 1-μm hydrazide particles, a monolayer equivalent of antibody will be about 1.5 mg, so the total amount added should be in the range of 1.5 to 15 mg for a 1- to 10-times excess.
9. React with mixing for at least 6 h at room temperature or overnight at 4°C.
10. Wash the beads at least several times with coupling buffer and resuspend in the same buffer containing 0.05 to 0.1% of a blocking molecule (such as BSA, gelatin, non-fat dried milk, PEG, PVP, etc.). The best blocking agent may be found by experimentation and testing of the antibody-coupled particles in the intended application.
11. Wash the particles several times with coupling buffer and store in an appropriate buffer containing a preservative at 4°C.

Coupling to Epoxy Particles

Polymeric particles containing epoxide groups can be used to couple thiol-, amino-, or hydroxyl-containing ligands *via* a ring opening reaction that is facilitated under alkaline conditions. This reactive group can be used to couple proteins, nucleic acids, sugars and carbohydrates, and other organic molecules containing these functionalities. Epoxide groups can be introduced into polymeric particles through free-radical copolymerization with oxirane-containing vinyl monomers, such as allyl glycidyl ether or glycidyl methacrylate, or they may be introduced by surface modification using a *bis*-epoxide compound, such as 1,4-butanediol diglycidyl ether (Sundberg and Porath, 1974). Horák *et al.* (2011) used a polymeric magnetic particle made from HEMA and glycidyl methacrylate to immobilize streptavidin through the epoxide groups through a secondary amine linkage. In addition, the polymerization of a glycidyl methacrylate shell onto an existing polymer particle can be performed by seeded emulsion polymerization, which creates reactive epoxides on the surface of otherwise unreactive core particles (Omer-Mizrahi and Margel, 2009). Epoxy activation also has been used extensively to immobilize affinity ligands onto porous beaded chromatography supports (see Chapter 15, Section 2.3), and it can be used with equal success to couple ligands to microparticles and nanoparticles.

Epoxide-containing particles can be used to couple thiol-containing ligands at slightly basic pH (pH 7.5–8.5), amine-containing ligands at higher pH values (pH 9–11), and hydroxyl-containing ligands at very high alkaline conditions (pH > 11). The following protocol can be used to couple thiol-, amine-, or hydroxyl-containing ligands with the proper pH adjustment of the carbonate coupling buffer (Figure 14.20).

Protocol

1. Wash 100 mg of epoxy particles with coupling buffer (i.e., 0.1-*M* sodium carbonate, pH 10, for coupling amine-containing ligands). Use higher pH conditions if coupling hydroxylic molecules and lower pH for coupling thiol-containing ligands. Suspend the particles at a 5% solution in coupling buffer.

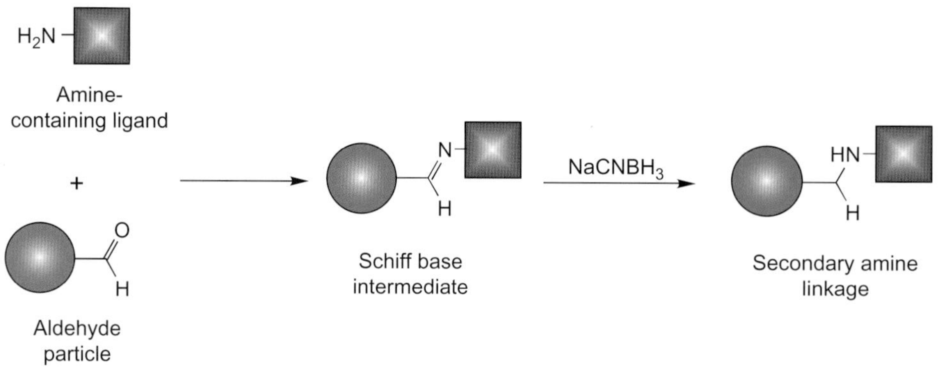

FIGURE 14.21 Aldehyde particles can be reacted with amine-containing proteins or other molecules to form intermediate Schiff bases, which can be stabilized by reduction with sodium cyanoborohydride to form secondary amine linkages.

2. With mixing, add to the particle suspension a quantity of ligand dissolved in coupling buffer in an amount that represents a 1- to 10-times excess over the molar quantity of epoxide groups present on the particles.
3. React at room temperature (for sensitive ligands) or at 45 to 60°C (for more stable ligands) for at least 20h with mixing.
4. Block excess epoxy groups by the addition of cysteine to a final concentration of 50mM. Other small molecules can be used for blocking, provided they will efficiently react with the excess epoxides and not result in a modification that could interfere with the subsequent use of the particles. Continue the reaction with mixing for at least 2h.
5. Wash the particles thoroughly with coupling buffer and then into a more moderate pH storage buffer containing a preservative.

Coupling to Aldehyde Particles

Polymeric particles containing aldehydes are produced by two general routes: copolymers containing an aldehyde monomer (e.g., acrolein) or through periodate oxidation of diols incorporated onto the surface of particles. HEMA/acrolein derivatives are examples of copolymer aldehyde particles (Kumakura and Kaetsu, 1984; Chang *et al.*, 1986; Colvin *et al.*, 1988), which are hydrophilic due to the large number of hydroxyl groups and provide aldehydes for covalent attachment of amine-containing ligands. Epoxy particles also can be used to create surface aldehydes by opening up the epoxide ring through acid hydrolysis and subsequent oxidation of the resultant diols by sodium periodate (Schiel *et al.*, 2006).

Aldehyde particles are spontaneously reactive with hydrazine, hydrazide, or aminooxy derivatives, forming hydrazone or oxime linkages (see Chapter 15, Section 2.4) upon Schiff base formation (Abdelrahman, 2011). Reactions with amine-containing molecules, such as proteins, can be carried out through a reductive amination process using sodium cyanoborohydride (Figure 14.21).

Protocol

1. Wash 10mg of aldehyde particles 3 times with 10-m*M* sodium phosphate, pH 7.4 (coupling buffer). Buffers of higher pH value e.g., carbonate buffer at pH 10) will result in more efficient Schiff base formation with amine-containing molecules than neutral pH conditions.
2. After the final wash, suspend the particles at 5 to 10mg/ml in coupling buffer and add a protein to be coupled to the particle suspension in an amount equal to 1- to 10-times molar excess over the calculated monolayer for the protein type to be coupled. (Note: It takes about 18mg of BSA or 15mg of IgG to saturate 1g of 1-μm particles, and more protein if the particles are smaller.) Mix thoroughly to dissolve. Low concentrations of protein may result in particle aggregation, because a single protein molecule can react and bridge more than one particle.
3. Incubate with mixing for 2 to 4h at room temperature.
4. Add to the particle suspension a quantity of sodium cyanoborohydride in water to bring the final concentration up to 10-m*M*. If a high pH buffer was used for the initial incubation between the particles and the protein, perform a quick wash with 10-m*M* sodium phosphate, pH 7.4, to bring the pH down to a point in which the reducing agent is active. Mix for 30min at room temperature. The reducing agent will convert all the resultant Schiff bases into stable secondary amine linkages.
5. Add to the particle suspension a quenching molecule (such as glycine, ethanolamine, or Tris) to give a final concentration of 0.2-*M*. The blocking agent will couple to any remaining aldehyde reactive sites.

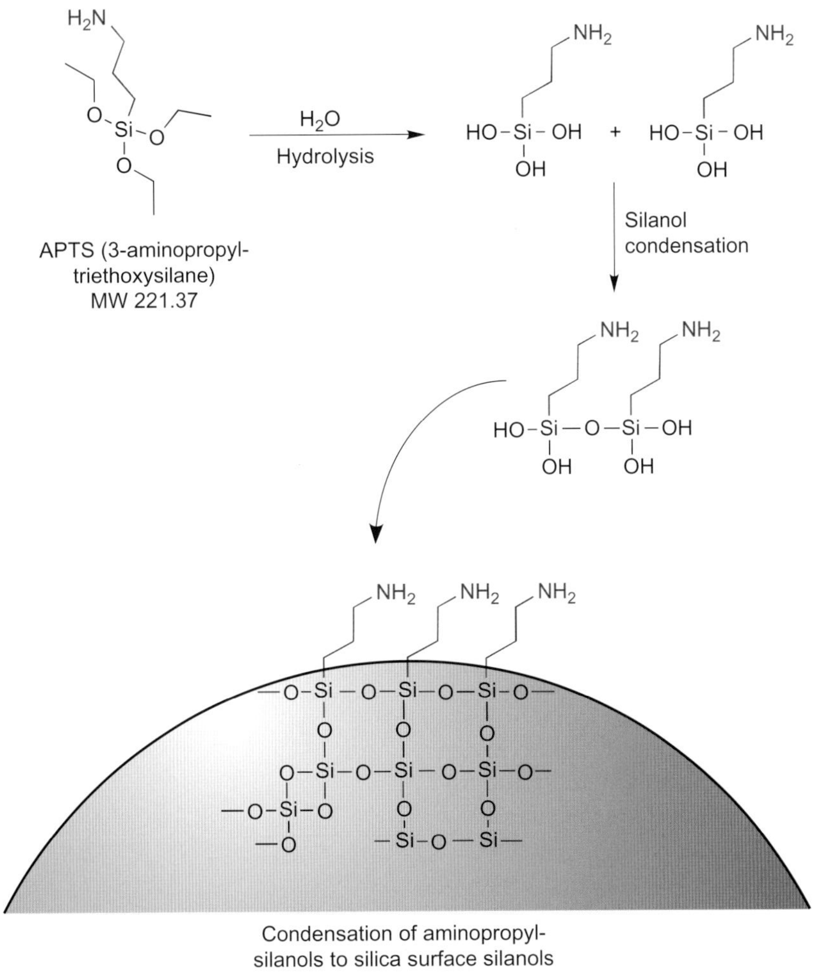

FIGURE 14.22 Silica particles can be functionalized to contain amine groups by reaction with APTS, which coats the particles with a silane layer containing numerous aminopropyl groups on the surface for further conjugation.

6. Remove excess protein and reactants by washing with coupling buffer at least three times using centrifugation. The particles may be stored in a suitable buffer containing a preservative at 4°C.

5. SILICA PARTICLES

The use of silica particles in bioapplications began with the publication by Stöber *et al.* in 1968 on the preparation of monodisperse nanoparticles and microparticles from a silica alkoxide monomer (e.g., tetraethyl orthosilicate, or TEOS). Subsequently, in the 1970s, silane modification techniques provided silica surface treatments that eliminated the nonspecific binding potential of raw silica for biomolecules (Regnier and Noel, 1976). Derivatization of silica with hydrophilic, hydroxylic silane compounds can thoroughly passivate the surface and make possible the use of both porous and nonporous silica particles in all areas of bioapplications (Schiel *et al.*, 2006). Silica nanoparticles or microparticles are routinely synthesized to contain functional groups for the covalent attachment of affinity ligands (del Campo *et al.*, 2005; Pattnaik, 2011).

The modification of silica particles to provide sites for coupling affinity ligands can be done similarly through covalent derivatization of the surface with a functional silane containing a side-chain reactive group or functional group (see Chapter 13). For instance, reaction of a silica particle with 3-aminopropyltriethoxysilane (APTS) under the appropriate conditions coats the surface with primary amino groups for conjugation with electrophilic groups (Pattnaik, 2011; Sen and Bruce, 2012). Reagents containing alkoxy silane groups can condense, particularly after hydrolysis to create reactive silanols, with the OH groups on standard silica particles to form stable siloxane linkages (or oxane bonds) (Figure 14.22). This reaction is typically catalyzed by heat or through the use of at least a partially aqueous environment, which forms the requisite silanols from the alkoxy groups on the silane compound. Unlike other alkoxy silanes, trimethoxy silane compounds can react directly with particle silanols without the need for high-temperature conditions or hydrolysis to form siloxane linkages. The many options available in silane functional groups for surface modification can provide a broad range of silica particle properties for subsequent coupling of biomolecules or other affinity ligands (VanDerVoort *et al.*, 1996; Bruce and Sen, 2005; Gartmann *et al.*, 2010; Thanh and Green, 2010; Rother *et al.*, 2011).

Silica particles have some advantages over polymer particles. They have a higher density than polymeric particles (1.96 g/cm^3 *versus* 1.05 g/cm^3) and thus they can be washed using centrifugation, even when working with nanometer-sized particles. Typically, silica particles as small as 30 to 40 nm still can be separated from suspension using a benchtop microfuge, making handling and processing of silica particles potentially much simpler than polymeric particles of the same size.

Another advantage of using inorganic silica particles over polymer particles is that they do not shrink or swell when exposed to aqueous or nonaqueous environments. In the case of silica, nonaqueous solvents may be used for activation reactions without concern that softening or dissolving of polymeric structures will damage the particles, as silica does not dissolve in organic solvent environments.

A potential disadvantage of silica-based particles, however, is the tendency for the siloxane linkages between silicon atoms within the particle to dissolve hydrolytically, especially under alkaline conditions above pH 8.0. Surface treatment with organosilane derivatives can somewhat stabilize particles to hydrolysis, but long-term exposure to highly alkaline environments should still be avoided. It also is recommended that at least 0.05-*M* NaCl be maintained in the solution to increase stability of the particles in aqueous environments. The best stability for silica is obtained at neutral or acidic pH conditions containing a low concentration of salt.

5.1. Fluorescent Silica Particles

Silica particles have been exploited in virtually every assay or detection strategy that polymer particles have been used in for bioapplications purposes. Fluorescent dye-doped silica nanoparticles have been developed by a number of groups that have similar fluorescence characteristics to quantum dot nanocrystals (Chapter 10, Section 10). Fluorescent silica nanoparticles can be synthesized less expensively than quantum dots due to the fact that the silica particles incorporate standard organic dyes (Ow *et al.*, 2005; Wang *et al.*, 2006; Saleh *et al.*, 2011) and are not dependent on making reproducible populations of semiconductor particles with precise diameters to tune emission wavelengths.

The preparation of fluorescent silica particles can be carried out using a number of strategies. Santra *et al.* (2001) described a water-in-oil emulsion using detergent-mediated reverse micelle formation and controlled hydrolysis of TEOS to create mono-disperse silica nanoparticles (see also Arriagada and Osseo-Asare, 1995). Adding the water soluble fluorescent dye, Tris(2,2′-bipyridyl) dichlororuthenium (II) hexahydrate (Ru(II)bpy$_3^{2+}$) to this emulsion resulted in dye molecules being entrapped within the silica particle structure as it formed (Figure 14.23) (also see Chapter 24, Section 4.2, for additional properties of Ru(II)bpy$_3^{2+}$). Subsequent functionalization of the surface with silane derivatives can be done to facilitate ligand immobilization.

These Ru(II)bpy$_3^{2+}$ fluorescent silica nanoparticles have been used to detect single bacterial cells using

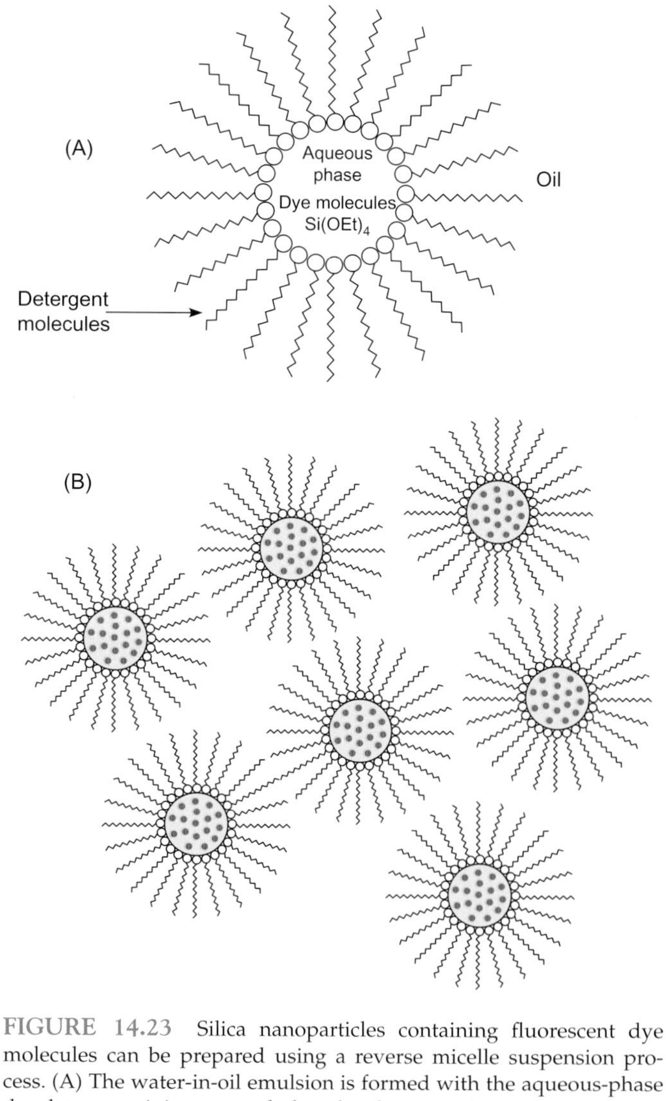

FIGURE 14.23 Silica nanoparticles containing fluorescent dye molecules can be prepared using a reverse micelle suspension process. (A) The water-in-oil emulsion is formed with the aqueous-phase droplets containing tetraethyl orthosilicate and dye molecules in detergent. (B) The final particles contain entrapped dye within the silica particle matrix, creating highly fluorescent particles.

antibodies conjugated to the surface after functionalization with trimethoxysilyl–propyldiethylenetriamine followed by succinylation to create terminal carboxylates. Specific antibody molecules against *E. coli* O157 were then coupled to this modified fluorescent particle using the carbodiimide method with EDC and NHS (Zhao *et al.*, 2004).

Another type of fluorescent silica particle was formed from silica bubbles created on the surface of gold nanoparticles (Liz-Marzan *et al.*, 1996; Makarova *et al.*, 1999). Fluorescein isothiocyanate (FITC) was adsorbed onto the gold nanoparticle surface and then reacted with 3-aminopropyltrimethoxy silane. The isothiocyanate groups on the adsorbed dye molecules coupled to the amine groups on the silane as it polymerized on the gold surface, effectively forming a silica shell around the gold particle. The gold core was then dissolved by reaction with cyanide ions to leave behind the silica nanobubbles filled with water, which also left the fluorescent molecules attached on the inner surface (Figure 14.24).

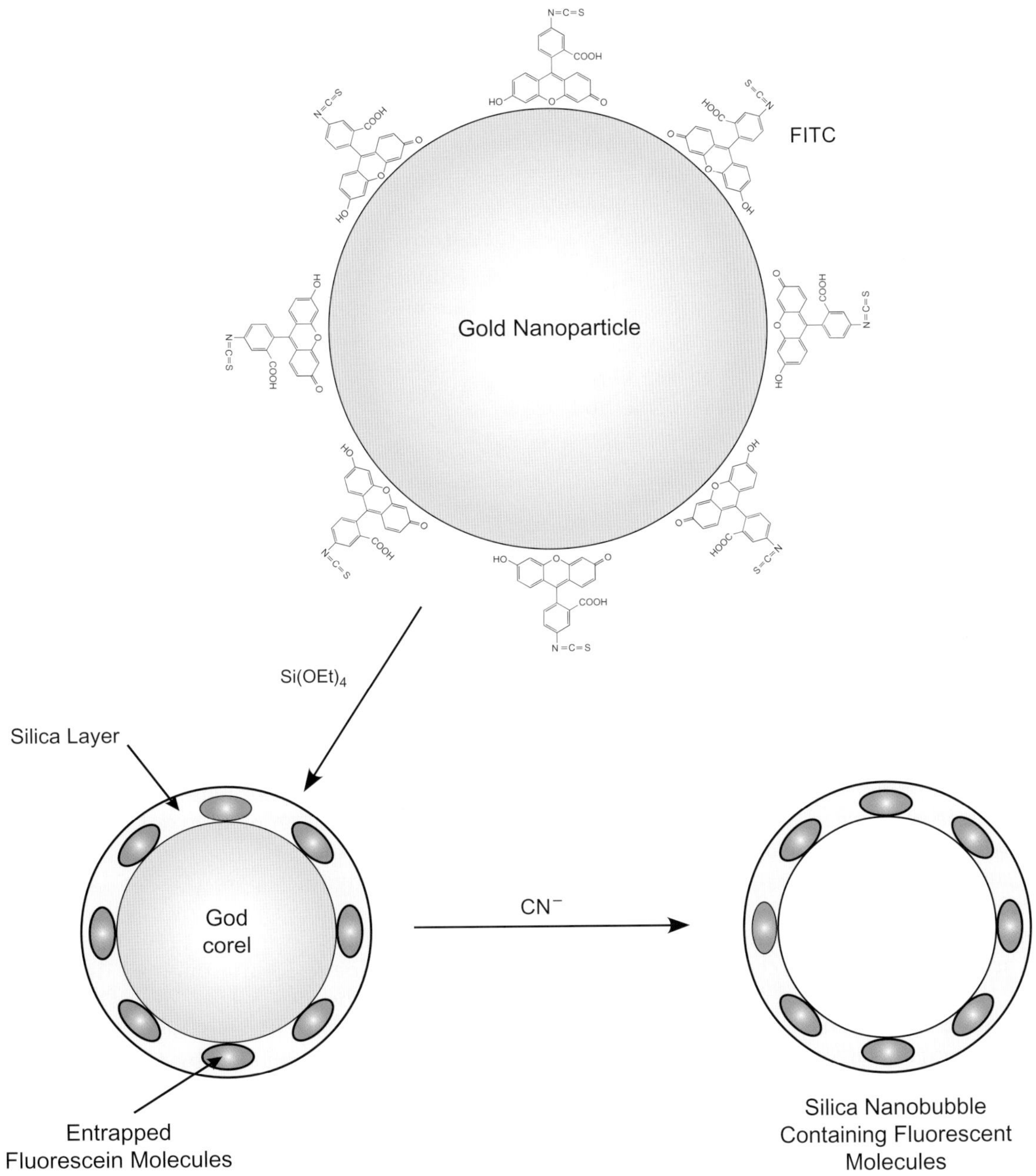

FIGURE 14.24 Fluorescent silica nanobubbles have been created using gold nanoparticle seeds that are initially coated by adsorption with a fluorescent dye. The particles then are capped by a layer of silica by polymerizing tetraethyl orthosilicate, which entraps the dye molecules within it. Finally, the gold core is dissolved by reaction with cyanide, leaving behind hollow fluorescent silica nanobubbles.

Fluorescent silica nanoparticles, called FloDots, were created by Yao *et al.* (2006) by two synthetic routes. Hydrophilic particles were produced using a reverse micro-emulsion process, wherein detergent micelles formed in a water-in-oil system form discrete nanodroplets in which the silica particles are formed. The addition of water soluble fluorescent dyes resulted in the entrapment of dye molecules in the silica nanoparticle. In an alternative method, dye molecules were entrapped in silica using the Stöber process, which typically results in hydrophobic particles. Either process resulted in luminescent particles that can then be surface modified with functional silanes to contain appropriate functionalities for coupling to affinity ligands or biomolecules. Ru(II)bpy_3^{2+} or standard organic fluorescent dyes can be incorporated into such silica particles with high efficiency. In one example, a 70-nm particle containing Ru(II)bpy_3^{2+} was found to be equal in fluorescence intensity to 39 quantum dots having an emission at 605 nm (Yao *et al.*, 2006). Another silica nanoparticle prepared using a rhodamine dye contained about 1290 molecules of the dye entrapped within each particle, which produced intensely fluorescent labels.

In a different method of producing dye-doped silica particles, van Blaaderen and Vrij (1992) and Verhaegh and van Blaaderen (1994) developed a method to covalently link amine-reactive dyes to 3-aminopropyltriethoxysilane (APTS) and then polymerize the resultant conjugate with tetraethyl orthosilicate (TEOS) in mixtures of ammonia, water, and ethanol to form fluorescent "organosilica" spheres. The dye molecules could be dispersed throughout the entire particle, contained in the core, or contained within discrete spherical regions within the particles. Either hydrophilic or hydrophobic particles could be created, depending on the surface treatment used subsequent to particle formation. Fluorescent particles consisting of either fluorescein or rhodamine dyes made by this method were shown to be susceptible to photobleaching similar to the organic dyes in solution.

Finally, fluorescent silica microspheres have been created by use of organosilane compounds forming particles in the presence of CdTe nanocrystals (quantum dots; see Chapter 10). This process loads numerous fluorescent nanoparticles into each silica microparticle (Liu and He, 2012). Using different colors of quantum dots for different preparations resulted in the formation of a range of different microparticles that had a broad spectrum of fluorescent characteristics. A single fluorescent quantum dot/silica particle was far brighter than individual fluorescent nanocrystals or any other organic dye used for detection applications.

The following protocol is based on the creation of fluorescent silica core/shell particles using the method of van Blaaderen and Vrij (1992).

Protocol for Formation of Dye Silica Core Particles

1. React APTS (11.5 mg) with FITC (10.6 mg) in anhydrous ethanol (1 ml) with mixing for 12 h (protect from light). The use of anhydrous conditions will prevent the hydrolysis of the APTS alkoxy groups, which would cause premature condensation.
2. Prepare a solution consisting of 75 ml of ethanol containing 8.5 ml of ammonia.
3. To the stirring ethanol/ammonia solution, add 3.3 ml of TEOS along with the completed reaction solution from step 1, which is now the conjugate of fluorescein and APTS linked through the amino group on the alkyl silane.
4. The reaction is allowed to continue for 24 h with slow stirring.
5. Wash the resultant core particles twice with the ethanol/ammonia solution using centrifugation.

Protocol for Formation of the Silica Shell

1. Dilute the particles to 1.2 l using a 10 : 1 mixture of ethanol:ammonia. Add 28 ml of TEOS to the particle suspension and mix to dissolve.
2. React for 24 h with slow mixing.
3. Wash the particles with the ethanol/ammonia solution several times using centrifugation. Finally, wash the particles into the desired solution (ethanolic or aqueous) without ammonia present to prevent silica hydrolysis upon storage.

Ow *et al.* (2004) developed an improved method of incorporating fluorescent molecules into silica particles using a modified Stöber synthesis, which resulted in both enhanced fluorescence and photostability of the encapsulated dyes. In this two-stage procedure, reactive organic dyes first are conjugated to a silane derivative and condensed to form a polysiloxane-dye-rich nanoparticle core structure. No TEOS is added at this point. These dyed core nanoparticles are then used as seed for the addition of silica sol-gel monomers to condense around the core and create a silica network (shell) around the dye-rich core (Figure 14.25). Nyffenegger *et al.* (1993) used a similar process to form fluorescein particles by first coupling FITC to 3-aminopropyltrimethoxy silane to form a thiourea derivative and then condensing the dye silane derivative with tetramethoxysilane to form the dyed silica particles.

The method developed by Ow and coworkers permits control over the size of the particles and allows the incorporation of virtually any organic dye into silica, provided it can first be conjugated to a silane derivative to form the core. Fluorescent particles made by this procedure may be made as monodisperse populations having diameters from less than 10 nm to over 1-μm spheres, with a high degree of precision. The core

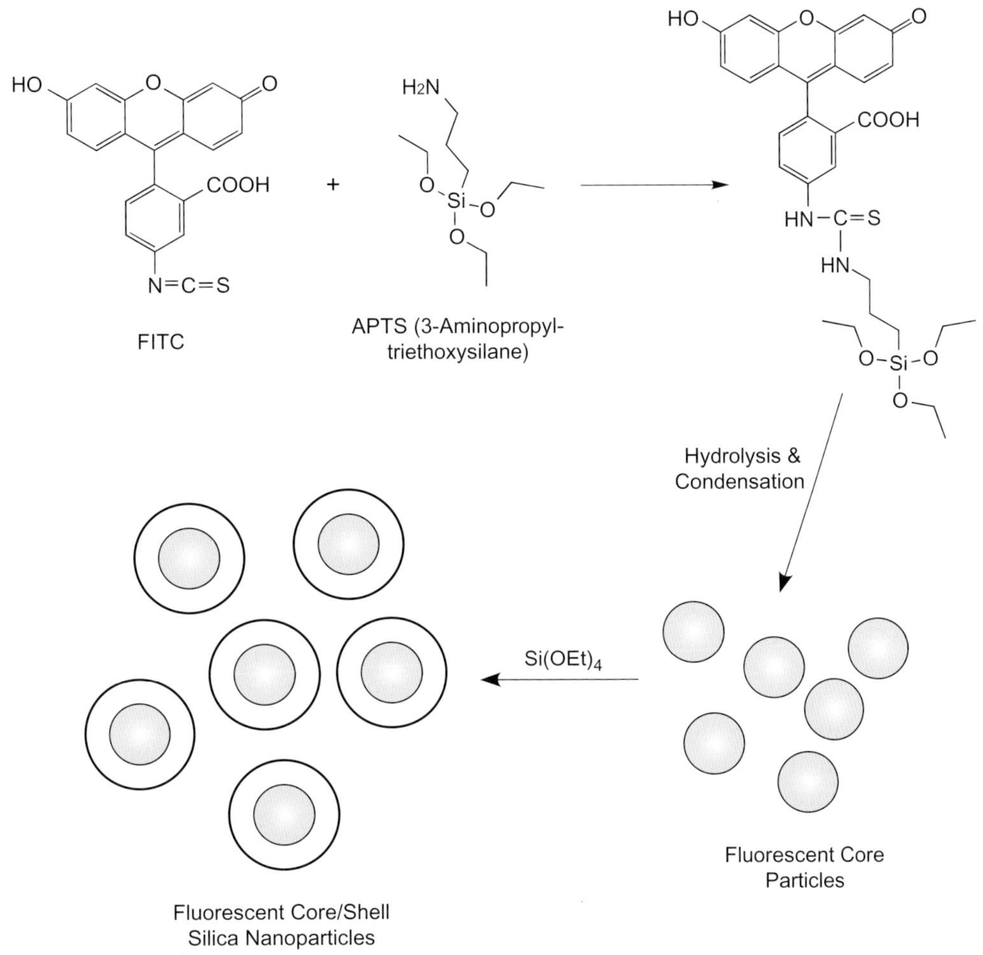

FIGURE 14.25 The preparation of highly controlled fluorescent silica nanoparticles can be done by first polymerizing APTS that has been covalently modified with an amine-reactive dye to form fluorescent core particles. The core then is capped by a shell of silica by polymerization of TEOS. The shell layer can be derivatized further with silane coupling agents to provide functional groups for conjugation.

size can also be varied by changing the concentration of the dye–silane derivative during the condensation process. The procedure for making these particles is similar to that described above using the method of van Blaaderen and Vrij (1992), but without the addition of TEOS for creation of the dye-silane core. Exact procedures are given in Wiesner *et al.* (2006).

In the preparation of 15-nm core-shell fluorescent silica particles, Ow *et al.* (2004) reported that the naked core (2.2 nm) alone produced a fluorescence intensity of less than the free dye in solution, presumably due to dye quenching. However, upon addition of the outer silica shell around the core, the brightness of the particles increased to 30 times that of the free dye (using tetramethylrhodamine-5-(and 6)-isothiocyanate (TRITC)). They speculate that the shell may protect the core from solvent effects, as evidenced by a lack of spectral shift upon changing the solvent in which the particles are suspended.

The enhanced photophysical properties of these fluorescent core-shell silica nanoparticles make them potentially as useful as semiconductor quantum dots for bioconjugation purposes. In this case, standard, commercially available, organic dyes can be incorporated into the silica nanoparticles to provide a range of emission properties as diverse as those available using the dyes alone. In addition, once encapsulated in the particles, the dyes display much better photostability and increased fluorescence (brightness) compared to the free dyes in solution. A major advantage of silica-based particles is that they are known to have greater biocompatibility than quantum dots in that they are nontoxic and hydrophilic and can be conjugated to proteins and other targeting molecules with relative ease. The ability to add any desired fluorescence characteristic to such particles simply by choosing the appropriate organic dye or metal chelate luminescent molecule makes dye-doped silica nanoparticles especially useful for bioapplications.

5.2. Silane Functionalization of Silica Particles

Surface functionalization of silica particles or fluorescent silica particles is typically done using functional alkyl silanes (Douroumis *et al.*, 2012; Lin *et al.*, 2012). The process may be used to add a reactive group to the surface of the particles for spontaneous coupling to biomolecules or it may be used to add the appropriate nucleophilic group to the surface, such as an amine or a carboxylate. Silane modification chemistry is discussed in more detail in Chapter 13.

The following protocol for modification of silica nanoparticles is based on the method of Zhao *et al.* (2004), which describes the addition of amine functionalities using trimethoxysilyl–propyldiethylenetriamine. Other functional silane modifications may be performed similarly.

Protocol

1. Add 32 mg of silica nanoparticles (fluorescent or plain) to 20 ml of 1-m*M* acetic acid containing 1% trimethoxysilyl–propyldiethylenetriamine with stirring. Other concentrations of silane derivatives used for particle modification typically range from 1% to 5% (w/w). Optimization of this concentration may have to be carried out for a particular silica particle size and type.
2. React with mixing for 30 min at room temperature.
3. Wash the amine-derivatized particles at least three times with water using centrifugation to remove excess reactants.
4. Store the particles in a suitable buffer at neutral or slightly acidic pH containing a preservative.
   The amine-modified silica particles may be used to couple with carboxylate-containing ligands using a carbodiimide reaction. Similar coupling protocols may be used as that previously described for amine-containing polymer particles (Section 4, this chapter). The amine particles may also be further derivatized by reaction with succinic anhydride to create carboxylated particles for coupling to proteins or other amine-containing ligands.
5. To prepared the succinylated carboxylate derivative of the amine particles, wash the particles with water and then into DMF.
6. Suspend the amine particles in DMF containing 10% succinic anhydride.
7. React for 6 h under nitrogen gas with mixing.
8. Wash the carboxylated particles at least three times with DMF by centrifugation. Resuspend in water and wash three times with water to remove DMF. Store the particles in water or a suitable buffer at neutral or slightly acidic pH.

Carboxylated silica particles may be coupled with amine-containing ligands, such as proteins, using a carbodiimide reaction with EDC. A protocol similar to that previously described for coupling to carboxylate polymer particles may be used. The following protocol is based on the method of Zhao *et al.* (2004), which was used for immobilizing monoclonal antibodies to *E. coli* O157.

Protocol

1. Suspend the carboxylated silica particles from step 8, above, in 10 ml of 0.1-*M* MES, pH 6.8.
2. With mixing, add 500 mg of EDC and 500 mg of NHS (or sulfo-NHS).
3. React for 25 min at room temperature with stirring.
4. Quickly wash the activated particles with water using centrifugation to remove excess reactants. Resuspend the washed particles in 10 ml of 0.1-*M* sodium phosphate, pH 7.3.
5. Add a quantity of protein or antibody to the activated particles representing a 1- to 10-times excess of protein over the calculated monolayer for the type of particles used. For coupling a limiting amount of antibody to the activated fluorescent particles, something that may be desirable when using expensive monoclonals, the reaction should be carried out in very dilute particle suspension (i.e., 0.1 mg/ml) to prevent aggregation of particles due to the possibility of one antibody reacting with more than one particle. Zhao *et al.* (2004), used an antibody concentration of just 5 μg/ml to create antibacterial fluorescent particles.
6. React for 2 to 4 h at room temperature.
7. Add a blocking agent, such as a non-relevant protein (e.g., BSA) to a final concentration of 1% to mask any nonspecific binding sites and to couple with any remaining reactive groups on the silica particle surface. This is important especially if a limiting amount of antibody was initially reacted with the particles in step 5. React for 30 min to 1 h at room temperature.
8. Wash the particles several times with PBS, pH 7.2, to remove excess protein and reaction by products. Store the particles in neutral or slightly acidic buffer containing a preservative.

6. GOLD PARTICLES

As early as the first decade of the twentieth century colloidal gold sols containing particles of less than 10 nm were produced by chemical means (Zsigmondy, 1905). However, the application of these inorganic suspensions to protein labeling did not occur until 1971 when Faulk and Taylor invented the immunogold staining procedure. Since that time, the labeling of targeting molecules, especially proteins, with gold nanoparticles has revolutionized the visualization of cellular or tissue components by electron microscopy (Horisberger *et al.*, 1975; Horisberger, 1979). The silver enhancement technique further broadened the application of gold labeling to include light microscopy (Holgate *et al.*, 1983). The electron-dense and visually opaque nature of gold labels also provided excellent detection qualities for such techniques such as blotting, flow cytometry, cytochemical staining, and hybridization assays (Jackson *et al.*, 1990; Gee *et al.*, 1991). Double- or triple-labeling systems have been constructed using immunogold methods in tandem with immunoenzymatic techniques to detect more than one antigen at the same time (Gillitzer *et al.*, 1990). Today, biosensing applications of gold nanoparticles have grown dramatically (Zeng *et al.*, 2011), and the methods of production and bioconjugation using gold are important aspects of particle-based methods to understand.

This section discusses the properties of gold particles as well as the common methods of labeling proteins and other biomolecules with them. The cited references should be consulted to obtain protocols for using these protein–gold complexes in assay and detection systems.

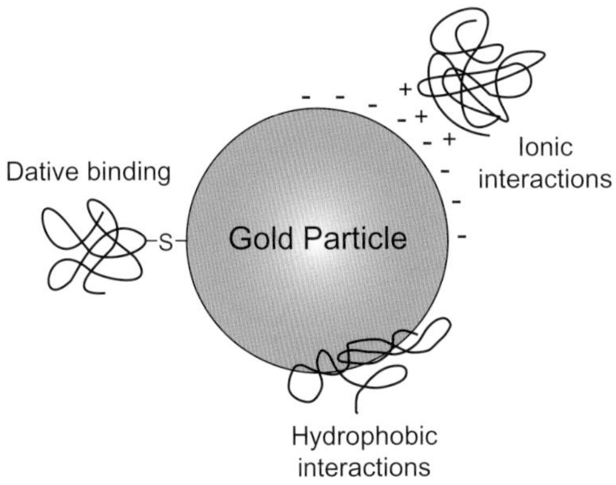

FIGURE 14.26 Protein binding to gold particles can occur through several types of interactions.

6.1. Properties and Use of Gold Conjugates

Colloidal gold suspensions consist of small granules (spheroids) of this transition metal in a stable, uniform dispersion. Viewed under the light or electron microscope, they appear as solid spheres of dense material. In electron microscopy the gold particles are visible as dense, dark markers usually black in appearance. In light microscopy, they can appear as light dots on a darker background due to the high reflectance of the particles or as an orange–red coating where they are localized in large conglomerates on cells or tissues. Colloidal gold particles act as efficient nuclei for deposition of silver, thus markedly enhancing their detection under light microscopy (Danscher and Rytter-Norgaard, 1983). The same silver–gold combination also provides increased sensitivity in blotting applications (Moeremans *et al.*, 1984).

Most preparations of colloidal gold consist of particles varying in diameter from about 5 nm to around 150 nm. The methods of forming small-particle gold suspensions of known diameter are discussed in Section 2.

The labeling of macromolecules with naked gold particles proceeds through a number of rather poorly understood processes. Preparing stable protein–gold complexes depends on several interactions: (1) the electronic attraction between the negatively charged gold particles and the abundant positively charged sites on the protein molecule; (2) an adsorption phenomenon involving hydrophobic pockets on the protein binding to the metal surface; and (3) the potential for binding of the gold with accessible sulfhydryl, oxygen, or nitrogen atoms *via* their unshared pairs of electrons to form dative bonds (see Chapter 3, Section 2.8) (Figure 14.26).

Deryagin and Landau (1941) and Verwey and Overbeek (1948) working independently developed a theory of the behavior of colloidal systems that aids in understanding macromolecular labeling with gold particles. Called the DLVO theory from the initials of the four authors, it views the particles in a sol as consisting of two components producing opposite effects in aqueous suspension. The overlap of the electrical double layer of each particle causes a negative charge on the surface, leading to particle–particle repulsion and stabilizing the sol from aggregation. The other phenomenon is electromagnetic in nature and leads to the potential for van der Waals attraction between the metal surface and other molecules.

In the colloidal suspension, a balance exists between the negative charge repulsion and the attractive forces which could cause coagulation. As particles approach each other, an energy barrier must be traversed to overcome the repulsive character of the negative surface and enter the region of van der Waals attraction. This barrier can be breached by the addition of electrolytes to the solution that can mask the negative charge on each particle. At a certain concentration of electrolytes, the colloid will begin to collapse as the gold particles adsorb onto one another, forming large aggregates and ultimately falling out of suspension.

Electrolyte-mediated coagulation forms the basis for creating all gold conjugates with other molecules. If macromolecules such as proteins are present in the colloidal suspension as the electrolyte concentration is raised to surpass the negative repulsion effects, then adsorption will occur with the protein molecules instead of with other gold particles. Thus, in place of aggregation and collapse of the suspension, labeling occurs.

The most common electrolyte additions in protein–gold labeling are NaCl or buffer salts. If no macromolecules are present, the addition of NaCl would itself cause gold particle coagulation. The aggregation is accompanied by a color change from orange–red to red–violet or blue (Roth and Binder, 1978), and it may be quantified spectrophotometrically by the change in absorbance at 580 nm (Horisberger *et al.*, 1976).

In practice, the addition of a protein to a gold sol will result in spontaneous adsorption on the surface of the gold particles due to electrostatic, hydrophobic, and van der Waals interactions. To prepare labeled proteins, initially the gold suspension is rapidly mixed while a quantity of protein is added. As the gold is bound to the protein molecules, a decrease in the absorbance at 580 nm occurs as the gold particles become stabilized and less coagulated. To check for the completeness of the adsorption process and to determine if the gold particles are totally blocked, a portion of the sol can be removed and an aliquot of NaCl added. If coagulation occurs upon addition of salt (increase in $A_{580\,nm}$), more protein should be added to completely stabilize the sol. Finally, many protocols further stabilize the colloidal suspension after protein binding by the addition of polyethylene glycol (PEG) or an immunochemical blocking agent, such as BSA or a solution of dried milk. These blocking agents completely mask any remaining sites of potential gold–gold or gold–protein interactions, thus preventing aggregation or nonspecific binding during assays.

To produce acceptable gold probes, it is often a common practice to add the minimum quantity of protein needed to prevent NaCl-induced aggregation plus about 10 to 20% excess (Horisberger and Rosset, 1977; De Mey *et al.*, 1981). Other investigators have reported that the addition of large excesses of protein to the amount of gold present yields conjugates of higher specific activity (Tokuyasu, 1983; Tinglu *et al.*, 1984). However, there is some evidence that overloading may cause leaching of loosely bound protein (Horisberger and Clerc, 1985).

As in any conjugation procedure, optimization of the ratios of reactants must be carried out to obtain the best probes. In labeling proteins with gold particles, several parameters should be considered: (1) the pI of the protein, (2) the pH of the adsorption process, and (3) the quantity of protein charged to the labeling reaction. It is generally believed that most proteins can be made to adsorb maximally at or near their isoelectric point (Norde, 1986). This is the pH of net electrical neutrality for a protein, wherein any electrically induced repulsive or attractive forces are balanced. For many proteins, especially antiserum-derived immunoglobulins, the average pI is a broad band encompassing a range of pH values. Thus, a polyclonal antibody preparation may possess an average pI much different than a particular purified monoclonal.

Geoghegan (1988) determined that as the pH of the adsorption reaction increased beyond the pI range, the percentage of IgG bound to gold particles decreased. However, for high-pI immunoglobulins, coupling at basic pH values increased the coupling yield. Geoghegan also noted that the more immunoglobulin that was charged to the adsorption process, the more ended up being coupled, although the percentage bound would decrease.

Thus, while definite standards for the ratio of protein to gold are not universally agreed upon, the efficiency of the process can be improved by following these general guidelines: (1) perform the adsorption reaction at a pH within the range of the pI of the protein being modified or at slightly higher pH; (2) charge an amount of protein to the gold particles that is slightly more (by about 10%) than necessary to maintain colloidal stability upon addition of NaCl; (3) avoid high overloads of protein, since this may promote subsequent leaching of bound material; (4) evaluate the degree of adsorption and the relative coagulation of the gold particles by measuring the absorbance of the solution at 580 nm; and (5) optimize each protein–gold conjugate as to colloidal stability and retention of activity.

An approximation of the correct amount of protein to be added to a gold sol to maintain stability of the colloid can be carried out using the following protocol (Slot and Geuze, 1984).

Protocol

1. Add 0.25 ml of the gold suspension to separate tubes containing 25 µl of different concentrations of the protein to be adsorbed. The amount of protein required to stabilize 1 ml of most gold sols is in the microgram range. The protein concentrations should be from about 10 µg/100 µl to about 150 µg/100 µl. Mix well.
2. After about 1 min, add 0.25 ml of 10% NaCl to the gold/protein suspension. Mix well.
3. Monitor the stability of the gold sol by its color or by the absorbance of the mixture at 580 nm. As long as the colloid continues to turn blue, and thus forms gold aggregates with addition of electrolyte, the amount of protein added is not sufficient to stabilize the suspension. This condition translates into a decrease in the absorbance at 580 nm. When the concentration of protein added is enough to stabilize the colloidal suspension, the solution no longer changes color (the absorbance at 580 nm no longer decreases).

4. The amount of protein added at the stabilization point plus 10% should be used to produce the final protein–gold conjugate.

The use of gold probes in detection systems has a number of advantages. The ability to label macromolecules with a range of gold particle sizes makes it possible to visualize the probe under a variety of microscopic conditions. Gold avoids all the disadvantages of radioactive labels, while being much more stable to quenching or fading than fluorescent probes or enzymatically developed substrate chromophores. A gold-labeled tissue, cell, or blot will maintain its record of staining on a permanent basis. Under sufficient magnification, an assessment of the degree of antigen labeling can be made simply by counting the number of gold particles present per unit area of cell or tissue mass. This cannot be done with other labeling systems, since chemical stains develop an amorphous quality that does not allow differentiation of individual molecules. Finally, gold probes are essentially non-toxic and relatively inexpensive to use.

A variety of biological molecules can be labeled with gold particles. Proteins are perhaps the most common gold probes; toxins, antibodies, immunoglobulin-binding proteins such as protein A, enzymes, lectins, avidin and streptavidin, lipoproteins, and glycoproteins have all been labeled with colloidal gold to form highly sensitive reagents. In addition, polymers, hormones, carbohydrates, and lipids have been gold-labeled for various applications. Small hapten molecules co-adsorbed with adjuvant peptides to gold particles make extraordinary immunogen complexes, producing polyclonal antibody responses having very high titers (Pow and Crook, 1993).

Very small gold particles can even be derivatized to contain specific chemical reactive groups for covalent coupling to macromolecules. For instance, an NHS ester-containing gold particle of 1.4 nm is manufactured by Nanoprobes (Stony Brook, NY). Presumably, such derivatives are formed by adsorption of chemically reactive polymers or by dative binding with a sulfhydryl-containing modification reagent.

The following sections discuss the preparation of colloidal gold suspensions of various particle sizes and their use in labeling proteins for detection purposes. Gold-labeled molecules and proteins are available from a number of manufacturers (Janssen, E-Y Labs, and Nanoprobes).

6.2. Preparation of Mono-Disperse Gold Suspensions for Protein Labeling

Mono-disperse colloidal gold suspensions useful for labeling macromolecules can be produced by a variety of chemical methods. Three main procedures have become common for making particles that fall into predictable particle-size ranges. All of them use reductive processes on chloroauric acid ($HAuCl_4$) to create the spheroidal gold particles. In general, the greater the power and concentration of the reducing agent, the smaller the resultant particles will be.

To create large-particle colloidal gold dispersions, chloroauric acid is normally treated with sodium citrate. The result is a particle range of about 15 to 150 nm, depending on the concentration of citrate utilized (Horisberger and Rosset, 1977; Horisberger, 1979; Pow and Morris, 1991). Medium-sized gold particles of between 6-nm and 15-nm diameter (average 12 nm) are formed by treatment with sodium ascorbate as the reductant (although some procedures use trisodium citrate at higher concentrations than the sodium citrate used for making large particles) (Horisberger and Tacchini-Vonlanthen, 1983; Albrecht, 1989). The smallest gold particles (<5 nm diameter) are created by reduction with either yellow or white phosphorus (Zsigmondy, 1905; Faulk and Taylor, 1971; Horisberger and Rosset, 1977; Pawley and Albrecht, 1988). Particles as small as 2 nm may be created by reduction with sodium borohydride (Bonnard *et al.*, 1984). Other more esoteric methods of reducing gold salts to form nanoparticles also have been developed, although their routine use is not yet established (Ryu *et al.*, 2011; Sharma *et al.*, 2012).

The following protocols for creating colloidal gold sols are adaptations from the above-cited articles. To obtain reproducible preparations, extreme care should be taken in making each batch to maintain the same reagent concentrations, temperatures, and times for the reactions. In each preparation, a color change is noted as the chloroauric acid is reduced from its initial state to the final gold sol. The initial color is typically a brown, purple–red, or dark blue, depending on the reductant used and other conditions. The final color of the mono-disperse colloidal gold preparation is typically red.

Preparation of 2-nm Gold Particle Sols

1. Prepare 1 ml of a 4% $HAuCl_4$ solution in deionized water.
2. Add 375 μl of the chloroauric acid solution plus 500 μl of 0.2-*M* K_2CO_3 to 100 ml deionized water, cooled on ice to 4°C. Mix well.
3. Dissolve sodium borohydride ($NaBH_4$) in 5 ml of water at a concentration of 0.5 mg/ml. Prepare fresh.
4. Add 5 1-ml aliquots of the sodium borohydride solution to the chloroauric acid/carbonate suspension with rapid stirring. A color change from bluish-purple to reddish-orange will be noted as the additions take place.
5. Stir for 5 min on ice after the completion of sodium borohydride addition.

Preparation of 5-nm Gold Particle Sols

1. Prepare 7 ml of a 1% $HAuCl_4$ solution in deionized water.
2. Add 6.25 ml of the chloroauric acid solution plus 5.8 ml of 0.1-*M* K_2CO_3 to 500 ml deionized water. Mix well.
3. In a fume hood, prepare a saturated solution of white phosphorus in diethyl ether, then dilute 1 part of the saturated phosphorus solution with 4 parts of diethyl ether.
4. Add 4.16 ml of the diluted phosphorus solution to the chloroauric acid/carbonate solution with mixing.
5. React at room temperature for 15 min.
6. Bring the mixture to a boil and reflux until the color of the suspension turns from brownish to red. This should take no more than about 5 min.
7. Cool the sol to room temperature.
8. The pH of the suspension will be around 6. Adjustments to more alkaline conditions for adsorbing macromolecules of higher pI may be done by addition of 0.1-*M* K_2CO_3 with stirring. Monitor pH of the sol using a gel-filled electrode (Orion Research, No. 9115, Cambridge, MA) (Geoghegan *et al.*, 1980). After pH adjustment, the gold should be used immediately for complexing with a protein or other macromolecule.

Preparation of 12-nm Gold Particle Sols

1. Prepare 5 ml of a 1% $HAuCl_4$ solution in deionized water.
2. Add 4 ml of the chloroauric acid solution plus 4 ml of 0.1-*M* K_2CO_3 to 100 ml deionized water. Mix well and cool the solution on ice.
3. With rapid mixing of the chloroauric acid/carbonate solution, quickly add 1 ml of a 7% sodium ascorbate solution prepared in water. Maintain the solution cooling in an ice bath. Higher temperatures will create larger particle sizes. The color of the solution at this point will turn to a purple–red.
4. Adjust the volume of the reaction to 400 ml with deionized water.
5. Bring the mixture to a boil and reflux until the color of the suspension turns from purple–red to red.
6. Cool the sol to room temperature.
7. The pH of the suspension will be around 6. Adjustments to more alkaline conditions for adsorbing macromolecules of higher pI may be done by addition of 0.1-*M* K_2CO_3 with stirring. Monitor pH of the sol using a gel-filled electrode (Geoghegan *et al.*, 1980). After pH adjustment, the gold should be used immediately for complexing with a protein or other macromolecule.

Preparation of 30-nm Gold Particle Sols

1. Prepare 1 ml of a 4% $HAuCl_4$ solution in deionized water.
2. Add 0.5 ml of the chloroauric acid solution to 200 ml of deionized water and bring to a boil while mixing.
3. Add to the boiling, rapidly mixing solution of chloroauric acid 3 ml of a 1% sodium citrate solution.
4. Reflux for 30 min. The color of the suspension will change from dark blue to red as the mono-disperse colloidal gold particles are formed.
5. Cool to room temperature.

Any of the particle sols prepared above may be used to adsorb macromolecules to create gold probes. To concentrate the suspensions, the solutions may be filtered through a small-pore filter. Centrifugation also may be done. Each protein–gold complexation should be optimized for the proper amount of protein to add to maintain stability of the colloid. This can be done according to the method described in Section 1.

6.3. Preparation of Protein A–Gold Complexes

Protein A–gold probes (as well as other immunoglobulin-binding proteins adsorbed to gold) have been used to visualize antibody binding to antigenic sites in tissue sections, cells, and blots (Hearn, 1987; Jemmerson and Agre, 1987; Lethias *et al.*, 1987; Bendayan and Garzon, 1988; Yokota, 1988; Bendayan, 1989; Herbener, 1989; Roth *et al.*, 1989; Stump *et al.*, 1989; Hoefsmit *et al.*, 2011; Mannweiler *et al.*, 2011). Gold labeling of immunoglobulin-binding proteins provides "universal" probes for detection of any antibody–antigen interaction (Figure 14.27). Thus, only one gold-labeled reagent need be prepared to visualize many different immunochemical targeting procedures. This avoids having to make antibody–gold probes for each specific immunoglobulin used. The downside of this approach, however, is the potential nonspecificity of protein A in binding other antibodies that may be present within the sample.

Protocol

1. Determine the minimum amount of protein A required to stabilize the colloidal gold sol being used. The colloidal suspension should be adjusted, if needed, with 0.1-*M* K_2CO_3 to pH 6 to 7. Measure the pH of the sol using a gel-filled electrode. Determining the stabilization amount of protein A can be achieved according to the method described in Section 1.
2. Mix a stabilizing amount of protein A plus an additional 10% with the appropriate volume of colloidal gold. For example, Herbener (1989) mixed

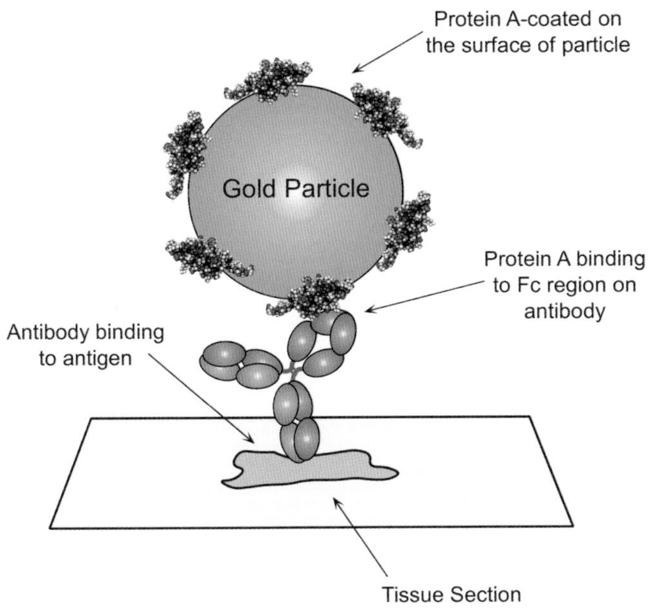

FIGURE 14.27 Antigens may be detected in cells or tissue sections through the use of protein A-coated gold particles. The binding of a specific primary antibody to its target antigen can be localized by the immunoglobulin binding capability of protein A, which occurs in the Fc region of the antibody.

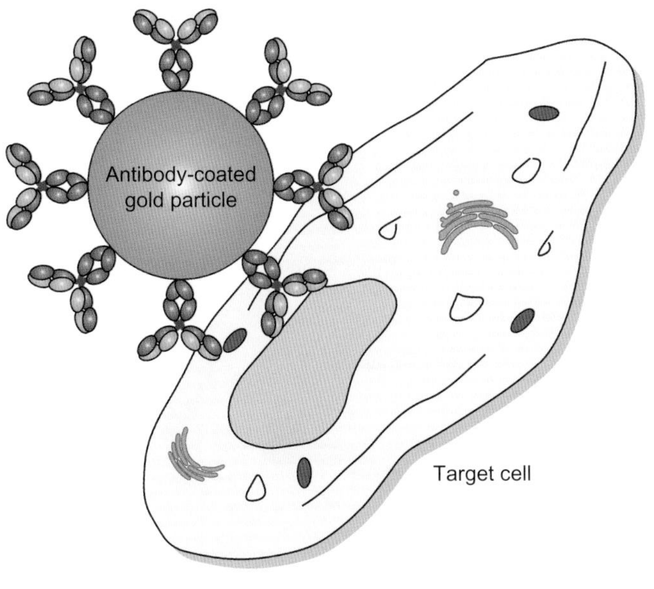

FIGURE 14.28 Antibodies coated on colloidal gold particles can be used to detect specific antigens in cells.

10 ml of a 14-nm gold particle sol at pH 6.9 with 0.3 mg of protein A dissolved in 0.2 ml water. Mix well.

3. After 1 min, add 250 μl of 1% polyethylene glycol (PEG; molecular weight 20,000) per 10 ml of gold sol used. The PEG helps to stabilize further the sol against aggregation.
4. Stir for an additional 5 min.
5. To remove excess protein A, centrifuge the preparation at a minimum of 50,000 *g* for 30 min to several hours (4°C), depending on the size of the particles and the amount of solution. Discard the supernatant, and resuspend the protein A–gold pellet in 0.01-*M* sodium phosphate, pH 7.4, containing 1% PEG.

6.4. Preparation of Antibody–Gold Complexes

Immunocytochemical staining with antibody–gold probes is a powerful way to detect, localize, and quantify antigen molecules in tissue sections and cells (Figure 14.28). Antibody–gold particles also have been used as chromogenic sensors in immunochromatographic strip tests (or lateral flow assays) (Preechakasedkit *et al.*, 2011). Metabolic processes can be followed, epitope mapping of the structural characteristics of macromolecules can be performed, and detection of pathogens or other foreign substances within cells can be accomplished using gold-labeled antibodies (Ellis *et al.*, 1988; Albrecht *et al.*, 1989; Cramer *et al.*, 1989; De Waele *et al.*, 1989; Nielsen *et al.*, 1989; Martinez-Ramon *et al.*, 1990; van den Brink *et al.*, 1990; Thiruppathiraja et al., 2011).

The optimal coupling pH for an antibody should be determined by measurement of the relative pI range of the immunoglobulin. Many antibodies, however, adsorb best at a pH of 8 to 9. The optimal level of protein addition to the gold sol to prevent aggregation should be determined according to the method of Section 6.1. In addition, bovine serum albumin (BSA) is often added instead of PEG (see the protein A coupling procedure, described previously) to further stabilize the antibody–gold suspension.

Protocol

1. Determine the minimum amount of antibody required to stabilize the colloidal gold sol being used. The colloidal suspension should be adjusted, if needed, with 0.1-*M* K_2CO_3 or NaOH to pH 8 to 9. Measure the pH of the sol using a gel-filled electrode. Determining the stabilization amount of antibody can be done according to the method described in Section 1.
2. Mix a stabilizing amount of antibody plus an additional 10% with the appropriate volume of colloidal gold. For example, Geoghegan (1988) found that an addition of 10 to 14 μg of antibody per milliliter of gold colloid resulted in stable preparations. Mix well after addition of antibody to the gold suspension.

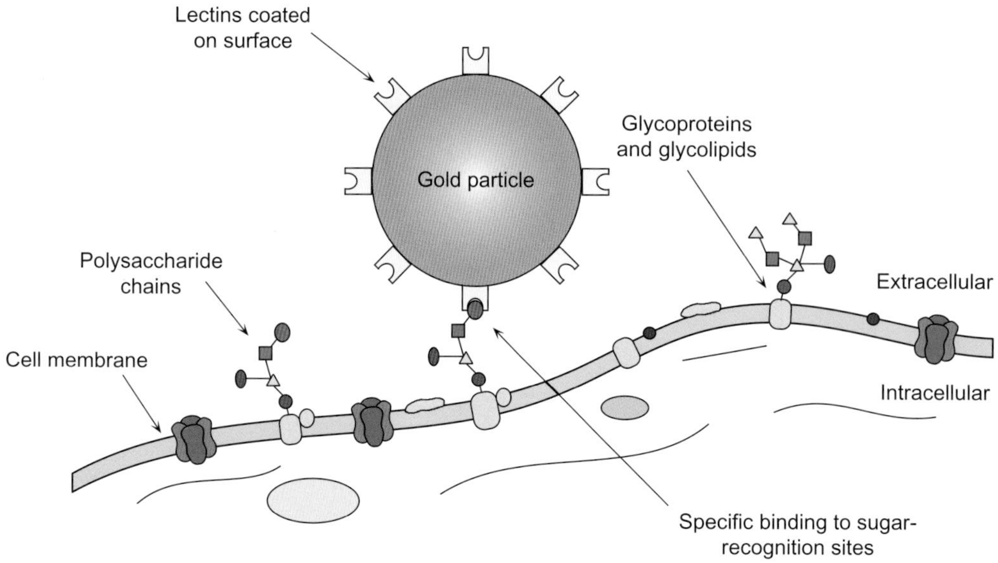

FIGURE 14.29 Lectins coated on gold particles can be used to detect specific carbohydrate sequences in cell surface glycoconjugates.

3. After 1 min, add a quantity of 10% BSA to bring the concentration to 0.25% in the antibody–gold suspension. The BSA helps to further stabilize the sol against aggregation and also blocks nonspecific binding sites. Alternatively, PEG may be added according to step 3 of Section 3.
4. Stir for an additional 5 min.
5. To remove excess IgG, centrifuge the preparation at a minimum of 50,000 *g* for 30 min to several hours (4°C), depending on the size of the particles and the amount of solution. Discard the supernatant, and resuspend the antibody–gold pellet in 0.01-*M* sodium phosphate, pH 7.4, containing 0.25% BSA (or 1% PEG, as desired).

6.5. Preparation of Lectin–Gold Complexes

Lectins, or proteins with specific binding sites for carbohydrates, can be used as targeting molecules to localize particular glycoconjugates such as glycoproteins or glycolipids on cell surfaces (Figure 14.29). Labeled with gold particles, lectins are important probes for detection of cell-surface components, intracellular receptors, and in immunological or biochemical assay procedures (Bog-Hansen *et al.*, 1978; Nicolson, 1978; Kimura *et al.*, 1979; Roth, 1983, 2011; Benhamou *et al.*, 1988; Nakajima *et al.*, 1988; Sánchez-Pomales *et al.*, 2012).

The following generalized protocol is an adaptation for the labeling of 15-nm gold particles with *Aplysia* gonad lectin, as described by Benhamou *et al.* (1988). Each lectin–gold preparation will have its own unique pH optimum and ratio of lectin-to-gold for the absorption process.

Protocol

1. Determine the minimum amount of lectin required to stabilize the colloidal gold sol being used. The colloidal suspension should be adjusted, if needed, with 0.1-*M* K_2CO_3 or NaOH to a pH equal to or slightly above the pI of the lectin being used. For *Aplysia* gonad lectin, the optimal pH for adsorption was determined to be 9.5. Nakajima *et al.* (1988) include pI conditions for a number of different lectins. Measure the pH of the sol using a gel-filled electrode. Determining the stabilization amount of lectin can be done according to the method described in Section 1.
2. Mix a stabilizing amount of lectin plus an additional 10% with the appropriate volume of colloidal gold. For example, Benhamou *et al.* (1988) found that an addition of 5 µg of lectin per milliliter of gold colloid resulted in stable preparations. However, in their final lectin–gold conjugate preparation a 5-fold increase in this ratio (25 µg lectin/ml gold) was used to fully stabilize the sol. Mix well after addition of lectin to the gold suspension.
3. After 1 min, add 250 µl of 1% polyethylene glycol (PEG; molecular weight 20,000) per 10 ml of gold sol used. The PEG helps to further stabilize the sol against aggregation.
4. Stir for an additional 5 min.
5. To remove excess lectin (particularly important if the 5-fold excess ratio is used), centrifuge the preparation at a minimum of 50,000 *g* for 30 min to several hours (4°C), depending on the size of the particles and the amount of solution. Discard the supernatant, and resuspend the lectin–gold pellet in 0.01-*M* sodium phosphate, pH 7.4, containing 1% PEG.

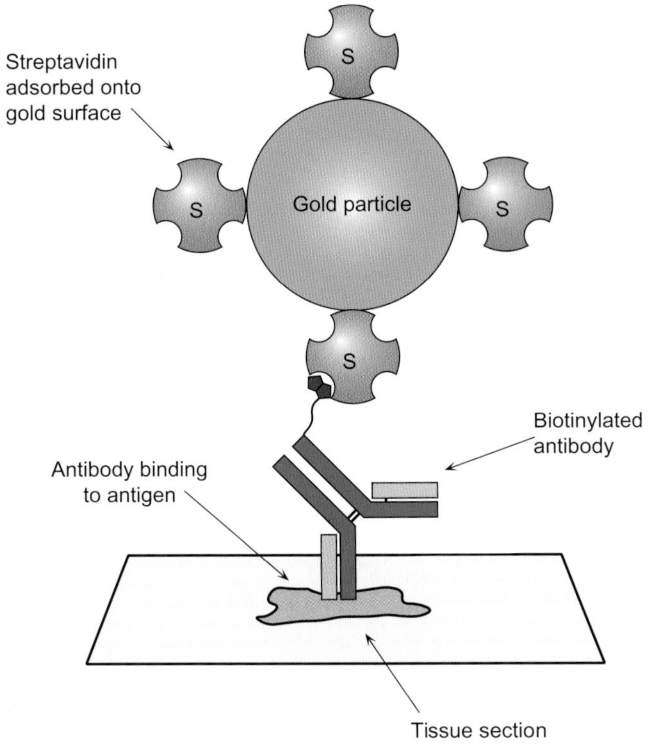

FIGURE 14.30 Streptavidin-coated gold particles can be used to detect biotinylated antibodies that are bound to specific antigenic determinants.

6.6. Preparation of (Strept)avidin–Gold Complexes

Avidin– or streptavidin–gold conjugates can be used to detect, localize, or quantify the binding of biotinylated molecules in cells, tissue sections, or blots (Bonnard *et al.*, 1984; Morris and Saelinger, 1984; Bronckers *et al.*, 1987; Gillitzer *et al.*, 1990; Ishida-Yamamoto *et al.*, 2012; Krager *et al.*, 2012) (Figure 14.30). These reagents are similar to the use of protein A–gold complexes in detecting immunoglobulins (Section 3) in that they are "universal" for detecting any biotin-labeled molecules. Thus, targeting molecules need not be directly modified with gold, only biotinylated so that they are able to interact with avidin– or streptavidin–gold conjugates. See Chapter 11 for additional information on (strept)avidin–biotin techniques, including conjugation and biotinylation protocols.

The following protocol is based on the method of Morris and Saelinger (1984) for the labeling of succinylated avidin with gold particles of 5.2 nm diameter. Succinylated avidin was used to reduce the pI of the protein, thus eliminating nonspecific binding due to the strong positive charge of the native tetramer. Alternatively, streptavidin or NeutrAvidin (Thermo Fisher) can be used in a similar immobilization protocol to coat gold particles without the need for using a succinylated derivative.

Protocol

1. Prepare a 200-ml gold sol by using white phosphorus reduction as described in Section 2.
2. Prepare 5 ml of a 1-mg/ml succinylated avidin solution by dissolving the protein in 50-m*M* sodium phosphate, pH 7.5.
3. With stirring, add the succinylated avidin solution to the colloidal gold suspension at room temperature.
4. React for 30 min with constant mixing.
5. Remove excess protein by centrifugation at a minimum of 50,000 *g* for several hours.
6. Suspend the succinylated avidin–gold pellet in 50-m*M* Tris, 0.15-*M* NaCl, pH 7.5, containing 0.5 mg/ml PEG (molecular weight 20,000).
7. Centrifuge again under the same conditions to ensure complete removal of non-adsorbed protein.
8. Resuspend the pellet in Tris buffer containing PEG and store at 4°C.

A similar protocol has been used by Bonnard *et al.* (1984) in the preparation of streptavidin–gold probes.

Protocol

1. Prepare 20 ml of a gold sol by the white phosphorus method described in Section 6.2.
2. Dissolve streptavidin in 0.1-*M* sodium phosphate buffer, pH 7.4, at a concentration of 1 mg/ml.
3. With stirring, add the streptavidin solution to the gold suspension. Immediately add 200 µl of 1-*M* sodium bicarbonate.
4. React for 10 min at room temperature.
5. To stabilize further the colloid, add 200 µl of 2% PEG 6000.
6. Centrifuge the streptavidin–gold suspension at a minimum of 50,000 *g* for several hours to remove excess protein solution.
7. Resuspend the pellet in 0.1-*M* sodium phosphate, 0.02% PEG, pH 7.4, and store at 4°C.

CHAPTER

15

Immobilization of Ligands on Chromatography Supports

Affinity chromatography has become one of the most powerful techniques available for the isolation of biological molecules. Since the beginnings of this technology in the late 1960s and early 1970s the application of affinity separations using immobilized affinity ligands has grown to affect nearly every aspect of life science research and has even been used in many other scientific disciplines. The basic concept of affinity chromatography involves the use of a biospecific or chemically specific interaction between an immobilized ligand and a desired target molecule to selectively bind and interact with the target even in complex solutions containing many other components. This specific affinity interaction is able to capture the target while removing contaminants or other molecules in a solution and in a single step enrich or purify the targeted molecule away from all other molecules that cannot bind the ligand (Figure 15.1).

Immobilized affinity ligands have been used for many purposes, including the isolation or purification of proteins and other biological molecules, in the capture and study of interacting proteins, for removal of contaminants from biological solutions, for enzymatic or chemical catalysis, and for analytical separations involving the assay of a targeted molecule (Calleri *et al.*, 2011; KumaraSwamy *et al.*, 2011; Sheshagirirao *et al.*, 2011; Zeng *et al.*, 2011; Cheung *et al.*, 2012; Vuignier *et al.*, 2012). Chapter 1, Sections 3.3 and 3.4, describe many of these application techniques in the general introduction to bioconjugation. In this chapter, the techniques of immobilization of affinity ligands will be described, including the matrices available for coupling, the activation chemistry used to form reactive groups on these supports, and the coupling chemistry commonly used to couple a wide variety of ligand types. Many of the methods are based upon those described in Hermanson *et al.* (1992), but they are significantly updated here to reflect the latest developments in chromatography support materials and coupling techniques.

The immobilization of affinity ligands onto insoluble support materials allows the creation of a specific affinity matrix that has binding specificity toward a desired target molecule. The insoluble support typically consists of a biocompatible material that can be modified to covalently link to the affinity ligand. The word "matrix" often is used to describe any material to which a biospecific ligand may be attached covalently. In most cases, the matrix is insoluble in a biological solution containing the target molecule. In some cases, however, the matrix may consist of a soluble polymer that can be modified to contain an affinity ligand.

The most common matrix type used for affinity chromatography consists of porous, beaded materials of generally spherical design that can accommodate biological macromolecules within their porous structure. This type of support material is typically used for chromatographic purposes in the purification of proteins and other biological molecules in volumes ranging from the bench-top research scale to the bioprocess scale. These materials often are referred to as affinity "supports," "resins," "beads," or "gels." Affinity chromatography supports of this type that are used in purification processes often have particle diameters ranging from about 50 to 150 μm and easily settle out of suspension when left standing without mixing. Matrices of this diameter can be used in packed columns or batch operations in scales ranging from microliters or milliliters for bench chromatography to literally hundreds or thousands of liters in size, which are exploited for many process chromatography applications in biotech and pharmaceutical companies.

In other applications, affinity supports can be prepared using nonporous beads consisting of particles at the low-end micron- or even nanometer-diameter range. Affinity supports prepared with such small

Bioconjugate Techniques, *Third Edition*.
DOI: http://dx.doi.org/10.1016/B978-0-12-382239-0.00015-7

particles also can be used in packed columns for chromatography purposes, such as in HPLC or UPLC separations, but most often they are used to create insoluble affinity supports for use in batch separations or in analytical techniques. The immobilization of antibodies or other affinity ligands onto microspheres, for example, has been extensively used to develop heterogeneous immunoassays to measure the concentration of proteins or other target molecules in biological samples. Superparamagnetic microspheres can also be coupled with affinity ligands to form a separation system that can capture a target molecule and then be rapidly separated from solution through the application of a magnetic field. Such affinity supports are widely used in clinical diagnostic autoanalyzers to measure important analytes in patient samples. See Chapter 14 for a discussion on the coupling methods typically used with these very small particles to produce affinity supports.

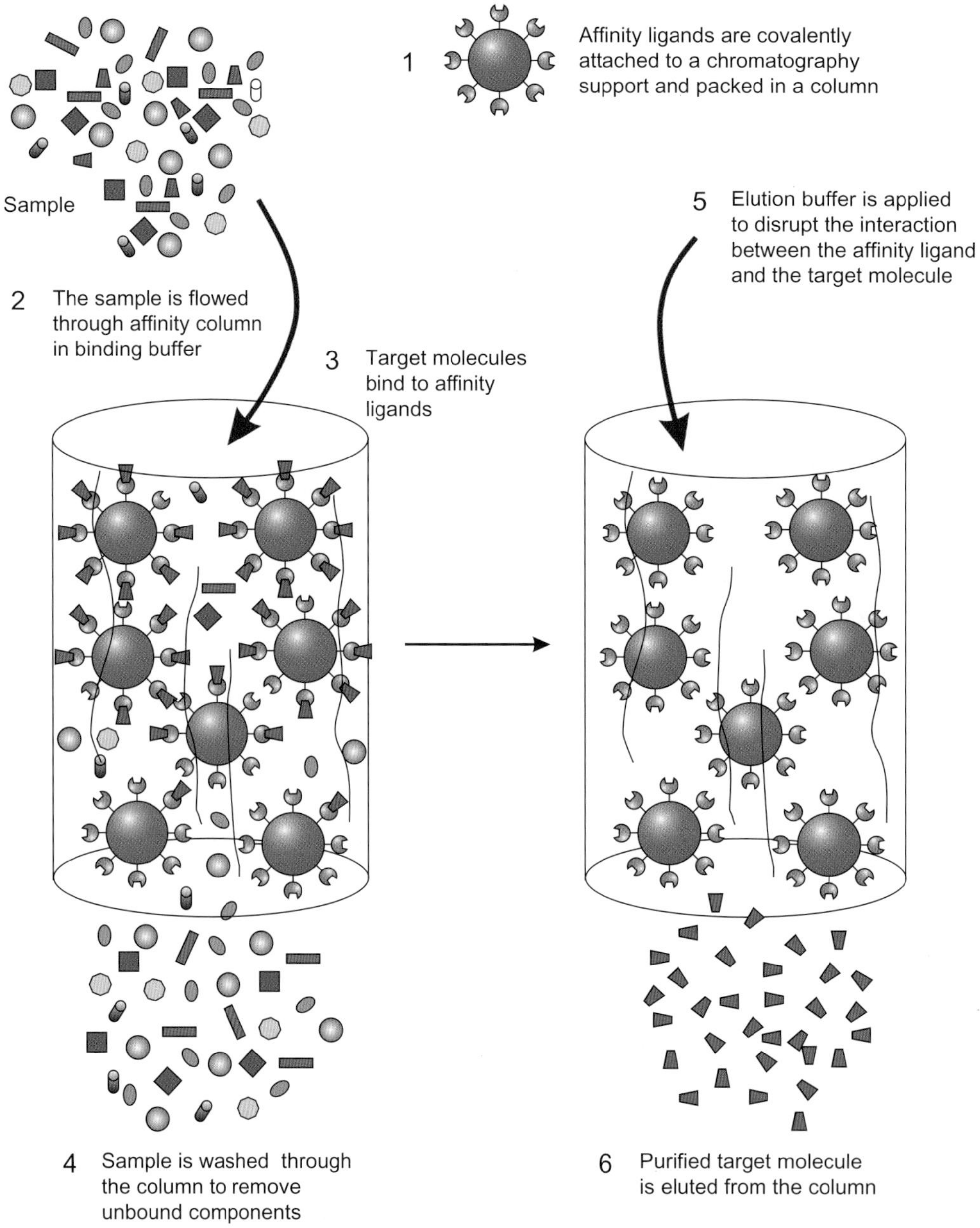

FIGURE 15.1 The principal of affinity chromatography in a column format is depicted in this illustration. An immobilized ligand is attached to an insoluble support material which is typically beaded in nature and the affinity resin is packed into a column. The column is equilibrated with binding buffer and a sample containing a target molecule which has affinity for the ligand is passed through the column. The target molecule binds to the immobilized ligands and is retained in the column while any not-bound material is washed off. An elution buffer is then added to the column, which disrupts the interaction between the ligand and the target molecule and allows the purified target to be isolated.

1. SUPPORT MATERIALS USED FOR AFFINITY CHROMATOGRAPHY

Affinity supports used for chromatography purposes can be highly diverse and have a wide range of properties. They can consist of natural polymers, synthetic organic supports, inorganic ceramic particles, naturally occurring constructs such as diatomaceous earth, membranes or surfaces, or composite constructions consisting of two or more of these materials. The most widely used matrix materials are greater than 10 μm in size and consist of natural polymers such as agarose or synthetic polymeric beads that are formed from hydrophilic monomers. Affinity supports of less than 10 μm in diameter commonly consist of copolymer derivatives or silica-based supports. Even extremely small nanoparticles made of metallic spheres or semiconductor alloys also are important substrates for the immobilization of affinity ligands; however, the coupling of ligands to nonporous particles of less than about 5 μm in size is described in Chapter 14, while this chapter exclusively focuses on the coupling reactions commonly used to make affinity chromatography supports on larger porous particles, membranes, and monolithic constructs.

For use with biological molecules, affinity supports must have a base matrix that is both hydrophilic and biocompatible so that the matrix itself does not bind nonspecifically to molecules in a sample. Therefore, to prepare a highly specific affinity chromatography support, the matrix must have two characteristics that are important to the performance of the final affinity complex: (1) it must be inherently low in nonspecific binding; and (2) it must contain reactive groups or functional groups that can be activated to couple the desired affinity ligand. The following sections summarize the major matrix types typically used for affinity chromatography and fulfill these criteria for biocompatibility and functionality.

1.1. Natural Product Chromatography Supports

Agarose Supports

One of the most common support materials is agarose, which is the naturally occurring major polysaccharide component found in agar, extracted from certain red seaweeds (class Rhodophyta, red algae). It consists of alternating residues of D-galactose and 3,6-anhydro-L-galactopyranose (Figure 15.2), with the occasional presence of repeating D-galactose side chains in α-1,4 linkages (initially linked to the C6 hydroxyl of D-galactose of the main chain). Some of the L-galactose residues are not in the anhydro form, thus the polymer may contain a quantity of alternating D-galactose and L-galactose residues. The C6 –OH group of the D-galactose units can be methylated in some preparations, as can the C2 –OH group of the L-galactose units on the main chains. The main polymer strands can be

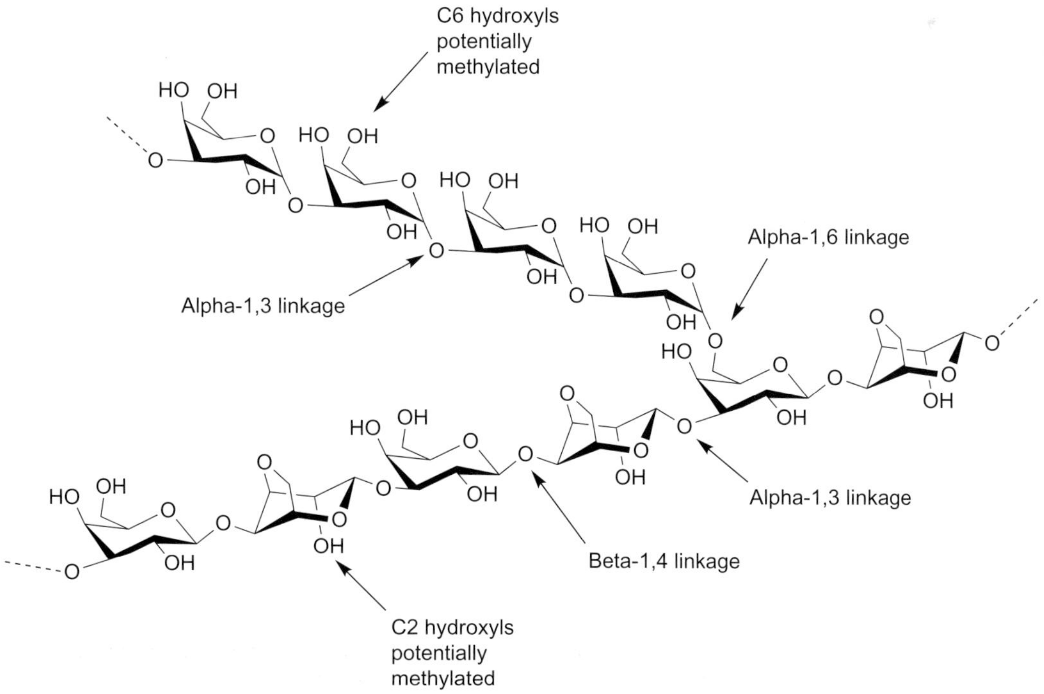

FIGURE 15.2 The repeating disaccharide structure of an agarose polymer is shown, which contains abundant hydroxyl groups that confer excellent hydrophilicity to chromatography supports made of this natural carbohydrate.

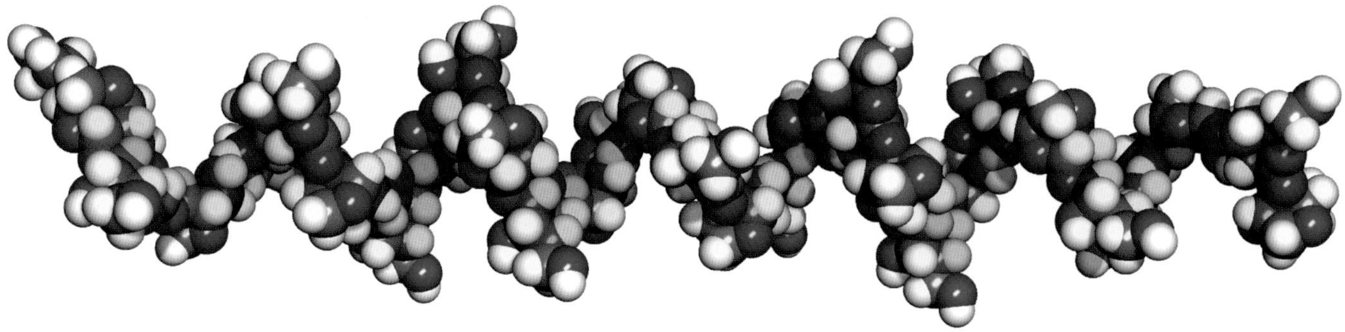

FIGURE 15.3 A space-filling, twisted helical structure of a single agarose polymer strand is illustrated in this figure. In agarose chromatography supports multiple double helices are often associated in super-helical structures which include triple helices. These polymer structures are naturally stabilized through numerous hydrogen-bonding sites along the polysaccharide chains. In crosslinked agarose, the structure is further stabilized by covalent attachments between the chains, which prevent disruption due to melting or denaturation.

further associated together in a left-handed, double-helical construction held together by hydrogen bonds, and these double helices can be further associated with two other double helices in a large wound complex. Two such complexes subsequently can be wrapped around each other into larger supermolecular helices, which are all stabilized by a huge network of hydrogen bonding between chains. Figure 15.3 illustrates the natural helical arrangement of a single agarose polymer strand. The double and higher order helical polymers arrange themselves together through a complex structure that is extraordinarily gelatinous in nature and when formed into beads. The agarose component takes up only about 2 to 6% of the particles by weight, while the remainder is water which hydrates and swells the aerogel structure. The agarose polymer strands weave throughout each particle to form a highly porous gelatinous bead that has a translucent appearance under a microscope and a spongy consistency. Since the large hydrodynamic volume of the support forms the bulk of each bead, the pore structure of agarose is huge compared to other chromatography supports. For this reason, agarose can have an exclusion limit for macromolecules well into the millions of Daltons.

In addition, the carbohydrate nature of an agarose support provides an uncharged, very hydrophilic environment for optimal use with biomolecules, which naturally has very low nonspecific binding character toward proteins. Commercial beaded agarose supports are typically crosslinked with small bifunctional reagents such as 2,3-dibromopropanol, epichlorohydrin, or divinyl sulfone to increase the stability of the matrix and prevent melting under elevated temperature or changing structure under denaturing conditions. Non-crosslinked agarose supports also can be used for chromatography, but the chemical, thermal, and mechanical stability of the crosslinked agarose supports is far superior for use in affinity separations.

The aerogel structure of agarose chromatography supports creates a large inner particle environment that accommodates even the very largest macromolecules; however, due to its gel-like construction and compressible nature agarose easily can be damaged. The most critical factors that may cause bead damage include mechanical disruption such as grinding particles against surfaces, using too high a flow rate during chromatography which may result in an excessive pressure drop which crushes the beads, the use of solvents that alter the internal structure of the beads by disrupting hydrogen bonding, or drying the particles without the use of an excipient. Agarose actually can be freeze dried and stabilized against particle damage through the addition of a sugar molecule such as lactose, which takes up the hydrogen bonding interactions upon loss of water. Some reactive agarose supports are stabilized for commercial purposes by freeze drying in the presence of 15% lactose and will rehydrate rapidly without difficulty or damage to the particles. Other nonreducing sugars may also be used for this purpose, as the multiple hydroxyl groups on the sugars effectively stabilize the agarose polymers and prevent large-scale collapse which could lead to irreversible shrinkage and disruption of the spherical nature of the beads.

Crosslinked agarose can be used with a large number of water-miscible solvents such as dimethyl formamide (DMF), dimethyl sulfoxide (DMSO), tetrahydrofuran (THF), dimethyl acetamide (DMAC), acetone, dioxane, methanol, ethanol, or pyridine, provided the exchange from water into the solvents is done slowly and sequentially with increasing concentrations of solvent. The recommended procedure for exchanging agarose into a solvent is to first wash with water, then with 30% of a solvent/water (v/v) solution, followed by a 70% solvent/water (v/v) solution, and finally washing the beads with 100% solvent. For complete removal of water from within the porous structure of agarose the final wash with solvent should be extensive (at least 10 bed volumes of solvent). Many activation methods of agarose are performed in a nonaqueous environment using one of these solvents and it is essential that all water be removed to prevent hydrolysis of the activating agent or hydrolysis of the resulting reactive groups.

The handling of agarose chromatography supports during washing, activation, ligand coupling, and mixing procedures should be done carefully to avoid particle damage. Although crosslinked agarose is very robust and can be used with many different reactions, there are a number of caveats that should be considered to avoid gel damage. When filtering agarose to wash with water or solvents, care should be taken to avoid drying out the support during the filtration process. The use of a fritted glass filter funnel using vacuum filtration perhaps is the easiest process for filtration and washing, because the particles will not clog the filter and neither will they get lost around a piece of filter paper as they could if a Büchner funnel was used. As the aqueous or solvent solution is filtered through the beads a packed bed of agarose will form on the fritted funnel. As the solution filters through the support and reaches the top of the bed, the vacuum should be removed and the filtration stopped to avoid drying out the top of the agarose cake.

In addition, beaded agarose supports should never be used with a stirring bar, because during the mixing process the bar can grind the beads against the vessel and cause particle damage. For mixing during activation or coupling procedures, gentle end-over-end mixing or the use of an overhead paddle stirrer should be performed depending on the volume of gel being derivatized. Neither of these methods will cause mechanical damage to agarose particles.

The techniques involved with the immobilization of affinity ligands on agarose are very similar to bioconjugation reactions used for the attachment of one molecule to another. Typically, chromatography supports are handled in bulk using batch methods for washing, mixing, activation, or coupling procedures. However, once an affinity support is prepared on agarose by coupling a ligand, any subsequent affinity separations can be carried out using batch methods or by column chromatography while the support is packed in a column.

Agarose chromatography supports can be used with a wide range of activation and coupling reactions. The support mostly contains secondary hydroxyls that originate from the repeating disaccharide composition of the polymer strands along with hydroxyls that may have formed from the crosslinking process if it was done with epoxide-containing reagents. Epoxide crosslinking may also form some primary hydroxyls from unreacted epoxides, which subsequently underwent ring opening to form diols. In addition, the principal crosslinker used to form crosslinked agarose matrices, namely 2,3-dibromopropanol, will add a primary hydroxyl to each site of its reaction within the agarose polymers.

The hydroxyls on agarose may be activated in nonaqueous solution to form reactive electrophilic derivatives that are able to react with nucleophiles on affinity ligands to be coupled. Agarose may also be activated in aqueous solution using sodium periodate, which cleaves diols at the associated carbon–carbon bonds to form primary aldehydes. The primary linear chain of agarose only contains diols if some of the 3,6-anhydro-L-galactose are not in the anhydro form and instead exist as L-galactose residues. Since all of the commercial crosslinked agarose chromatography resins can be periodate oxidized to a considerable activation level, there must be large amounts of L-galactose present in all agarose preparations. Section 2 in this chapter describes many of these activation and coupling reactions along with examples for the immobilization of important affinity ligands.

Agarose chromatography supports are among the most popular beaded matrices for affinity separations. General Electric (formerly Pharmacia and Amersham) manufactures a broad range of basic crosslinked agarose beads suitable for separations ranging from bench scale to process chromatography. GE's agarose supports, sold under the Sepharose name, are the most widely used for biopharmaceutical production, including the manufacture of recombinant antibodies destined for therapeutic use. Other manufacturers, such as Sterogene and Agarose Bead Technologies (ABT), also supply the base gel in one or more forms for use in producing affinity chromatography supports. Many suppliers of affinity media use these base supports to activate and couple affinity ligands. Important commercial suppliers of activated agarose supports as well as the associated affinity ligand derivatives include GE, Thermo Fisher, Bio-Rad, Qiagen, and Sigma, among many other companies.

Cellulose Supports

Cellulose is a natural polysaccharide polymer containing hundreds or thousands of repeating units of D-glucose linked through β-1,4 glycosidic bonds (Figures 15.4 and 15.5) (Svec, 2002). It is derived from plants, being a primary component of the cell walls, and is commonly used in many consumer products, from biofuels to fiber in foods, paper, and insulation. Cellulose chromatography supports can consist of two main varieties, either an amorphous, fibrous, or crystalline particulate form with limited porosity or a spherical beaded form having a range of different porosities. Amorphous cellulose has long been used as a rather crude chromatography support for separations of various organic molecules and occasionally of biological macromolecules. Its use for affinity chromatography applications has been limited by its poor flow characteristics and rather inadequate performance in protein-based separations due to its lack of porosity and capacity.

Cellulose chromatography media are also available in a spherical beaded form with porosities suitable for affinity separations dealing with proteins and other macromolecules. Natural cellulose supports such as the fibrous variety typically have tightly packed polysaccharide chains, which limit the maximal porosity that is achievable. The use of regenerated cellulose provides a more hydrated and less tightly packed form of cellulose and thus permits larger pore sizes to be formed when creating beads. Porous cellulose beads are typically made by preparing a solution of a cellulose derivative and dispersing it into another solvent system that is immiscible. Rapid stirring of this mixture forms small droplets that allow the cellulose to re-solidify into spherical beads the size of the droplets. Chisso Corporation (JNC) manufactures a wide range of regenerated cellulose beads under the Cellufine name having two different porosities for use in biomolecule separations. Some of these supports include a number of immobilized affinity ligands for the separation of proteins, and they are available in quantities suitable for research chromatography to process scale separations. One matrix in particular, the GCL-2000 gel filtration support, is an underivatized particle having a size exclusion limit of 2 million Daltons, which is eminently appropriate for protein separations. This base cellulose matrix may be used in many of the activation and coupling reactions that are described in this chapter for the immobilization of affinity ligands on hydroxylic supports.

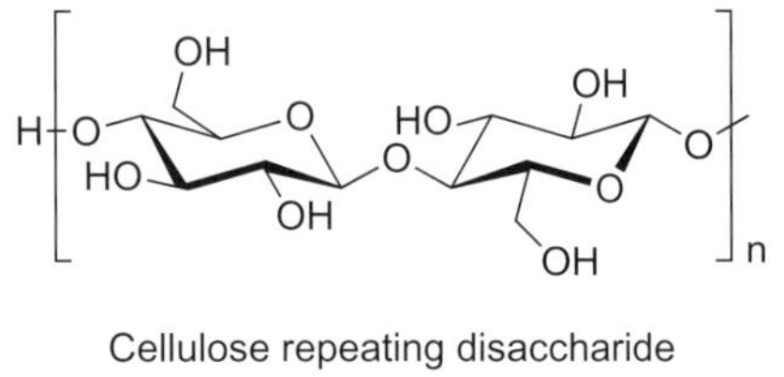

FIGURE 15.4 The repeating saccharide structure of cellulose, which contains numerous glucose residues linked through at least two types of glycosidic bonds. Neighboring chains are held together through extensive hydrogen bonding networks to stabilize its macromolecular structure.

Another beaded cellulosic media available for affinity chromatography that includes particle sizes ranging from about 30 μm up to 0.5 mm in diameter is produced by Iontosorb in the Czech Republic and distributed under the trade name Perloza. These supports consist of regenerated cellulose with no crosslinks, but nonetheless they have excellent mechanical stability even with large pore structures, which can accommodate proteins and other macromolecules. Reactive derivatives of the Perloza support are available containing cyanuric chloride or tosyl groups for the immobilization of affinity ligands.

A different type of cellulose matrix that has been used for affinity chromatography separations is available in the form of membranes. Cellulose membranes typically have pore sizes in the low-micron range

FIGURE 15.5 The three-dimensional structure of a single cellulose polymer is shown, demonstrating the abundant hydroxyl groups that provide hydrophilicity and functional groups for activation and coupling of affinity ligands.

(e.g., 0.45 μm) that easily accommodate proteins and are often used for filtration or clarification of biological solutions. Cellulosic membranes may be activated in a similar manner to cellulose beads and then covalently modified to contain affinity ligands for use in chromatographic separations. Hybrid cellulose membranes (containing other polymers) or pure cellulose membranes have both been used for the immobilization of ligands for affinity separations. For instance, a regenerated cellulose microporous membrane was activated and coupled with protein A to form an affinity support for the purification of antibodies (Boi *et al.*, 2008). The affinity membrane was found to have excellent capacity and could capture IgG from polyclonal human antibody solutions in good yield even at high flow rates. Membranes offer a monolithic alternative to bead-based resins for use in higher flow conditions and they require less maintenance to maintain a column for purification purposes. Although beaded resins still dominate the chromatography industry, membrane-based separations are definitely gaining more interest for their potential advantages in rapid affinity separations.

Inorganic Supports

Another support material often used for affinity separations consists of porous inorganic particles of extremely rigid construction and consisting of a variety of possible compositions. Some of the most common matrices in this category include porous glass, silica, alumina, ceramic supports, zeolites, and even the use of naturally occurring siliceous shells of unicellular organisms, called diatomaceous earth (or kieselgur; originating from aquatic plants called diatoms).

A commonly used support for affinity chromatography applications is controlled pore glass (CPG). CPG is typically made from borosilicate glass under high temperature resulting in the segregation of the silicate from the borate-rich portions. Subsequent rapid cooling and acidic etching of the particles causes the removal of the borate phase, which in turn results in the formation of craters and pore structures throughout the remaining silicate matrix. CPG supports having pore diameters in the range of ≥1000 Å (≥100 nm) are large enough to allow affinity chromatography for the separation of proteins and other macromolecules. CPG supports are extremely rigid and will hold up to high linear flow rates in packed column beds. However, the glass particles also are breakable, so that care should be taken in handling, mixing, and packing columns, as fines can form easily from particle-on-particle grinding. For this reason, avoid the use of magnetic stir bars or harsh mixing conditions. Paddle stirring under moderate mixing rates works well to preserve the integrity of the CPG particles, as does the use of a round-bottom flask in a rotary evaporator under slow rotational speeds.

Another stability issue with glass particles is the potential to dissolve under alkaline conditions. CPG supports are stable at acid or neutral pH, but will dissolve at pH values equal to or higher than pH 8.0. Buffered conditions containing 50- to 150-m*M* NaCl should be used for all aqueous conditions with CPG beads, because some degradation can occur even at neutral pH in unbuffered water. Conversely, CPG is capable of tolerating a wide range of nonaqueous solvent conditions without the shrinking and swelling effects of most polymeric supports. It also withstands high-temperature conditions for use with any derivatization reactions requiring elevated temperatures.

Naked glass particles have considerable potential for nonspecific binding of biomolecules, especially proteins and oligonucleotides, due to the preponderance of silicic acid groups (–Si–OH) on the surface. For this reason, the base CPG support first needs to be modified to create a layer that is not heavily charged with silicic acid functionalities, which will make it more appropriate for chromatography applications. Typically, this is achieved through silanization of the surface with a functional silane compound that contains uncharged groups and preferentially hydrophilic characteristics. Chapter 13 discusses the modification of particles and surfaces with functional silanes for all kinds of bioconjugation and immobilization applications. CPG beads may be modified with reactive or hydroxylic functionalities to block the silicic acid character and allow subsequent coupling of affinity ligands. Once blocked and functionalized, the degree of nonspecific binding toward proteins and oligonucleotides should drop to negligible.

A recommended first step in modifying CPG is to silanize the surface using an epoxy silane, which can be used directly to couple ligands, or alternatively the epoxide groups can be hydrolyzed to form hydrophilic diols (Figure 15.6). The following protocol is based on the methods of Weetal (1969) and Regnier *et al.* (1976). Similar protocols may be used to coat the particles with other functional silanes containing alternative reactive groups or functional groups for immobilization purposes. An important factor in the use of functional silanes to derivatize porous silica or glass particles is that the method must result in only a thin layer of silane coating or the pores will be clogged with silane polymer. In particular, if the reaction is performed in aqueous solution, which is often used for the silane modification of surfaces or nonporous particles, then the resultant high degree of silanol polymerization will quickly reduce porosity. Silylation reactions performed under nonaqueous conditions will limit the degree of modification to only a uniformly thin layer throughout the inner and outer surfaces of the CPG particles, thus preserving the pore structure of the support while efficiently coating the raw glass core.

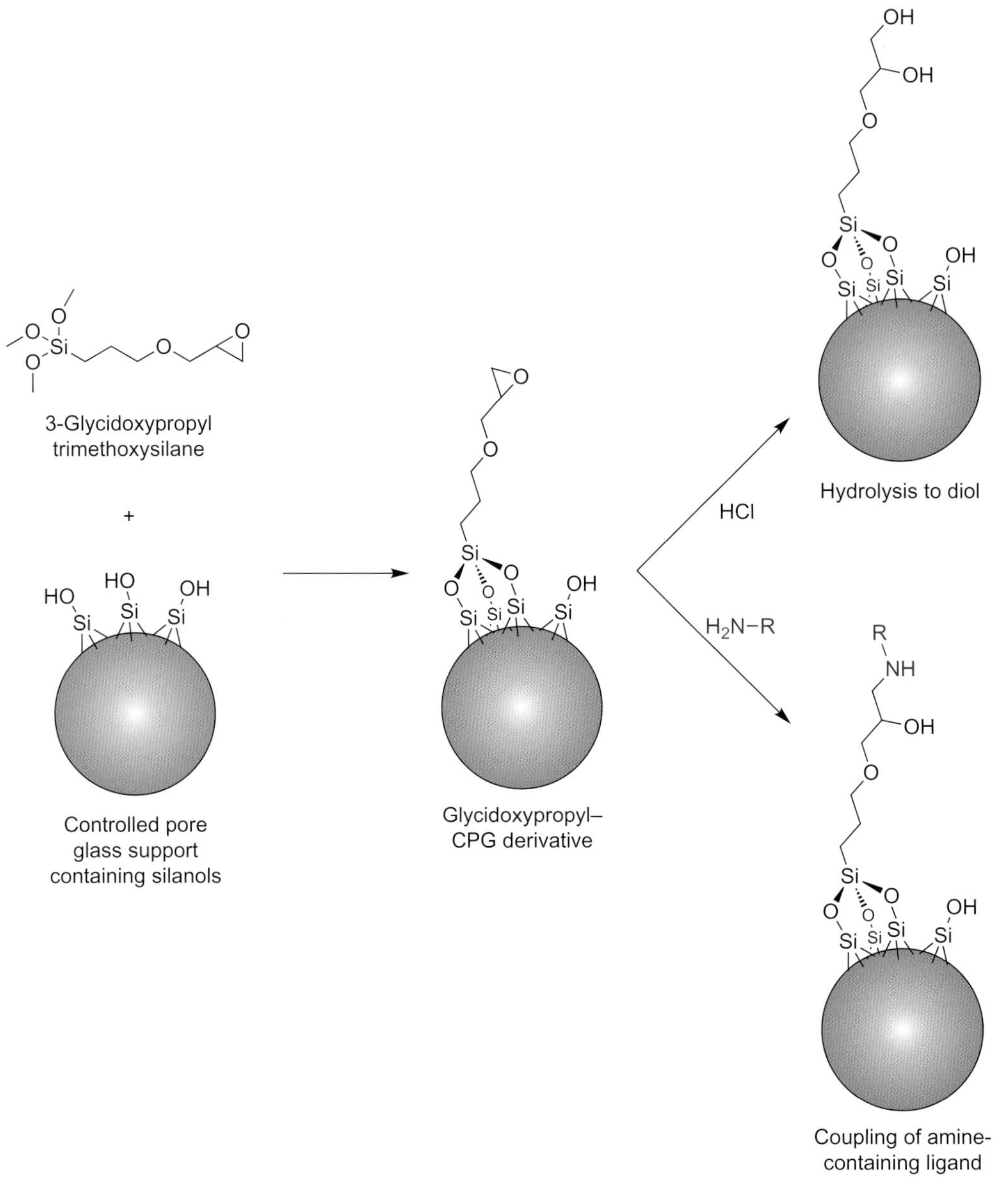

FIGURE 15.6 The reaction of 3-glycidoxypropyl trimethoxysilane with CPG and hydrolysis of the epoxide groups to yield diols or coupling of an amine-containing ligand to the epoxy groups to yield a secondary amine linkage.

Other silica or inorganic particles containing –OH groups may be modified using the following protocol. For the modification of particles less than about 30 µm, the filtration and washing process should be changed from vacuum filtration to centrifugation or tangential flow filtration to prevent clogging of filters or loss of particles through filter pads that are too coarse to prevent the particles from going through the pores.

Protocol

1. Add a quantity of CPG particles to a round-bottom, rotary evaporator flask and add a quantity of 50-m*M* HCl to cover the matrix and provide enough volume to suspend them during rotation. Place the flask in a rotary evaporator and slowly initiate rotation while placing the flask under vacuum. The HCl protonates the silicic acid groups and the vacuum removes entrapped air within the pores of the matrix. Mix for 1 h at room temperature, while intermittently breaking and reestablishing the vacuum.
2. Remove the particles from the flask and transfer them to a fritted glass filter funnel or a Buchner funnel containing a glass fiber filter pad. Drain the HCl solution and then wash the particles with 5 volumes of deionized water followed by washing into acetone using sequentially increasing concentrations of acetone in water. Filtration may be carried out by application of a vacuum to a filter

flask to increase the flow through the filter pad. Wash with at least 10 volumes of pure acetone to remove the last traces of water. Allow the particles to dry on the filter while applying a vacuum to aid in the removal of acetone by filtration and evaporation. Gently break up the filter cake using a plastic spatula to better disperse the particles and facilitate drying.

3. Transfer the particles to the round-bottom rotary evaporator flask and cover them in toluene. Again rotate the flask for an hour while pulling a vacuum to remove entrapped air. Next, heat the slurry in the rotary evaporator unit to 40 to 60°C while continuing the application of vacuum to begin distillation of the toluene. Continue the distillation process until water stops azeotropically distilling along with the toluene. If the last traces of water are not completely removed, the silylation process will result in hydrolysis of the trimethoxy groups and rapid polymerization of the functional silane, which will clog the pores of the CPG support. Cool the flask and remove excess toluene from the support by filtration.
4. Prepare a solution containing 10% (v/v) 3-glycidoxypropyl trimethoxysilane (Acros Organics) and a total volume equal to the volume of particles to be treated. Transfer the support back into the rotary evaporator flask and add the silane solution to the particles to resuspend them. React overnight at room temperature with rotation at slow speed.
5. Transfer the particles to the filter funnel and remove excess silane solution. Wash extensively with dry acetone to remove remaining traces of silane and toluene. Finally, filter off the acetone and dry the particles under vacuum. The coated, epoxy functional particles may be used directly for coupling an affinity ligand or may be treated as in step 6 to hydrolyze the epoxide group to a hydrophilic diol. Protocols for the coupling of amine-, thiol-, or hydroxyl-containing ligands to epoxide-reactive groups may be found elsewhere in this chapter.
6. Transfer the modified CPG support into a clean rotary evaporator flask and add a quantity of 50-m*M* HCl to cover the particles. Place under vacuum to remove entrapped air with slow rotation. Heat the flask to 50 to 60°C for 1 to 2 h to hydrolyze the epoxide groups into diols. The hydroxyl groups may be used in a number of different activation reactions to couple affinity ligands. For instance, the adjacent hydroxyl groups may be oxidized with sodium periodate to yield a reactive aldehyde group for coupling to amine-containing molecules.

Glass supports treated in this manner to yield a diol coating on the surface will have low nonspecific binding to proteins or other biomolecules. The silylation process coats the inner and outer surfaces of the particles and prevents biomolecule interactions with the silicic acid functionalities. However, it is known that silica- and glass-based chromatography supports give higher potential nonspecific binding character toward protein-containing samples than agarose supports, presumably due to some exposure of the silica backbone (Ghose *et al.*, 2007). Alternative silane treatments may also be carried out with CPG supports to create other functionalities for further derivatization. In particular, an aminopropyl silane may be used to add amine functionalities to a glass surface, which may be used for direct coupling of carboxylate-containing ligands or to build hydrophilic spacer arms prior to coupling an affinity molecule. In addition, other support materials such as silica and alumina may be treated in a similar manner to create functional matrices suitable for affinity chromatography. See Chapter 13 for further information regarding functional silane compounds and their reactions with substrates.

Commercial affinity supports prepared on glass particles include a high-capacity protein A or protein G matrix, which has a high density of immunoglobulin-binding proteins immobilized on its surface (Prosep-A or Prosep-G from Millipore or Trisopor-Protein A from BioCat). These supports can tolerate high flow rates during chromatographic purification of antibodies, because the base particle is extremely robust and incompressible. Binding capacities of over 45 mg/ml for antibodies are not unusual on this type of affinity support material.

Composite Gel–Ceramic Supports

Novel chromatographic support materials have also been created from the use of porous, inorganic particles that have been coated with a gelatinous natural polymer. The core ceramic particle provides an incompressible substrate that can take the rigors of large columns and rapid flow rates, while the gel coating creates a hydrophilic surface that is low in nonspecific binding character. This type of support was originally developed by BioSepra and commercialized under the trade name HyperD. BioSepra subsequently underwent a series of acquisitions and sales until the company was ultimately sold to Pall Corporation in 2004. U.S. patents 5,234,991 (1987) and 5,268,097 (1993) describe the preparation and use of these "gel-in-a-shell" supports, which consist of a base inorganic mineral oxide bead containing a hydrogel-filled pore structure (Figure 15.7). The type of core particle used for the HyperD supports is a porous silica bead called Spherosil, which was originally manufactured by Rhone-Poulenc but is now produced by Pall Corporation.

The hydrogel that is used to fill and coat the internal surfaces should ideally be a long linear polymer with a molecular weight of at least 10^4 Daltons. It also

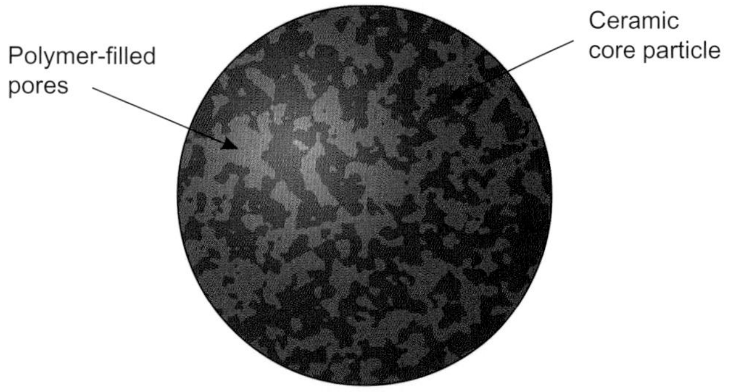

FIGURE 15.7 An illustration of a gel–ceramic composite support material. The ceramic core forms the structural spheroid shape of the support, and it has also very large pore structures to accommodate macromolecules. These pores are filled with polymeric gels, such as agarose, to create a hydrophilic porous resin, which combines the advantages of the polymer with the robust and rigid nature of the ceramic core to give a superior chromatographic support useful for process scale affinity chromatography.

should be hydrophilic to produce a biocompatible surface and be partially cationic in nature in order to stick to the core inorganic support material, which has a negative charge character. Most often the polysaccharide derivative is aminated to contain tertiary or quaternary amine derivatives along the length of the polymer. Examples of polymers that can be used in this fashion include DEAE–dextran and QAE–dextran derivatives, which contain multiple positively charged amines. It is believed that the positively charged polymer interacts with the negatively charged mineral oxide bead surface and forms multi-point, strong noncovalent interactions that are essentially irreversible. For instance, in the modification of CPG or silica supports, the negatively charged silicic acid groups likely interact with the cationic groups of the polymer, creating a large number of charge interactions along each polymer molecule. The polymer-coated surface also can be treated with a crosslinker such as a *bis*-epoxide or epichlorohydrin to further stabilize the hydrogel and provide a composite gel–ceramic particle, which is stable to strong acid or base without disrupting the polymer coating.

The use of aminated dextran, starch, or agarose as the gel layer for coating the particles can result in a poly-hydroxylic support material that can be activated to contain reactive groups capable of coupling affinity ligands. Methods to produce this type of composite gel–ceramic particle can be used successfully on such mineral oxide supports as glass, silica, alumina, titanium, and magnesium. The final "gel-in-a-shell" construction combines the benefits of the rigid ceramic core with the hydrophilicity and immobilization capacity of the biocompatible polymer (Xia *et al.*, 2011). The inorganic phase typically consists of particles with huge pore structures that allow for fast mobile-phase access that eliminates the diffusion limitations often seen with standard porous beaded supports. The large pores allow rapid sample distribution throughout the interior structures of the beads, thus overcoming the drop-off in capacity usually observed as flow rates through a column increase.

Activation and coupling methods for the immobilization of affinity ligands on these composite supports can involve any of the methods described in this chapter for the activation of hydroxylic functional groups. This includes the use of aqueous periodate oxidation to form aldehyde groups, which can then be used to couple amine-containing ligands such as proteins through the use of a reductive amination reaction process. This activation and coupling reaction process is described elsewhere in this chapter in the section entitled *Amine-Reactive Immobilization Methods*.

1.2. Synthetic Polymeric Chromatography Supports

Perhaps the most diverse category of affinity chromatography supports is represented by resins that consist of various types of synthetic polymers. Synthetic supports are produced from the controlled polymerization of various functional monomers and crosslinking monomers in a stirred solution that produces beaded particles having a porous structure. In some cases, polymeric supports may consist of a combination of natural polymers and synthetic polymers. For instance, an agarose–polyacrylamide composite support may be made that combines the hydrophilic functionality of a polysaccharide with the tight pore structure of a polyacrylamide particle, thus creating a matrix that can be easily derivatized to contain affinity ligands.

Synthetic polymeric supports can be designed to have a variety of different physical characteristics that are appropriate for chromatographic separations (Gokmen and Du Prez, 2012). As compared to natural supports, synthetic matrices may have superior physical and chemical stability, while still being able to withstand the high flow rates typically seen in process chromatographic separations. Polymeric supports also can withstand enzymatic degradation or microbial contamination better than some naturally occurring polymers, which may be ideal substrates for bacterial growth. Porous polymeric chromatography supports designed by using reactive or functional monomers can easily be coupled directly with affinity ligands or activated to contain a reactive group for immobilization. Many commercially available synthetic polymeric supports contain primary or secondary hydroxyls and are extremely hydrophilic and low in nonspecific binding. In addition, hydroxylic-containing supports can be activated by a number of different immobilization chemistries and subsequently coupled with a wide variety of affinity molecules.

Polyacrylamide Supports

Polyacrylamide chromatography supports are typically made by the copolymerization of two monomers: acrylamide and *N,N'*-methylene bisacrylamide, first described by Hjerten and Mosbach (1962) (Figure 15.8). The support can have a variety of pore sizes depending on the degree of crosslinking monomers added to the polymerization reaction. The greater the amount of bisacrylamide present in polyacrylamide, the smaller the pore structure in the resultant beads. Bio-Rad offers a broad range of polyacrylamide resins that differ in the pore size and exclusion limit of the particles. These supports are classic gel filtration (size exclusion chromatography, or SEC) matrices that can separate proteins based on their molecular weight. However, they may also be used as supports to couple affinity ligands for purification of selected target molecules. For working with proteins, a size exclusion limit for the polyacrylamide beads should be chosen that will allow the protein molecules to enter the pores and interact with immobilized affinity molecules.

Polyacrylamide resins are typically supplied dry and must be hydrated in aqueous solution with gentle agitation prior to use. Weigh out an appropriate amount of dry beads and slowly add them with mixing to a quantity of aqueous buffer equal to about twice the anticipated volume of hydrated gel.

Polyacrylamide supports are characteristically soft in nature and gelatinous in appearance. They also display low nonspecific binding character toward biomolecules, rather good pH and buffer stability, and because they are totally synthetic, the supports are not able to support microbial growth. In addition, since polyacrylamide does not contain carbohydrate components, the support is ideal for use in the affinity purification of carbohydrate-binding proteins. For this reason, affinity supports prepared on polyacrylamide containing immobilized sugars, disaccharides, or small polysaccharides often have been used for the specific purification of lectins.

A major deficiency of polyacrylamide-type resins is their inability to support acceptable flow rates to permit larger-scale separations to take place in a reasonable time frame. Due to the compressibility of polyacrylamide, maximal linear flow rates are often limited to no more than 10 to 15 cm/h, which is far less than 10% of the flow rate that can be expected from agarose supports or other polymeric matrices. In addition, polyacrylamide beads have a tendency to shrink and swell with different solvents, buffers, and salt concentrations. Severe reduction in particle size can occur in organic solvents and

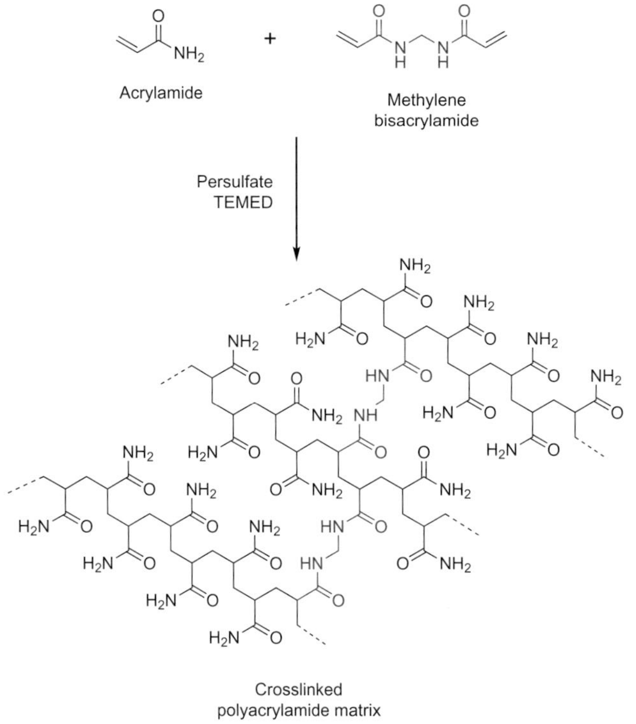

FIGURE 15.8 Polyacrylamide supports are made from the copolymerization of acrylamide and the crosslinking monomer *N,N'*-methylene bisacrylamide. The ratio of crosslinking monomer to acrylamide monomer controls the degree of porosity in the final support material.

this characteristic limits the range of chemical reactions that can be done to immobilize affinity ligands on them. Shrinkage of the particles also causes a collapse of the internal pore structures, which in this case may restrict coupling of ligands only to the outer surfaces.

Chromatography particles made of polyacrylamide do not naturally contain reactive groups or functional groups suitable for the immediate attachment of affinity ligands. Derivatives of polyacrylamide are typically made through partial hydrolysis of the amide bond linkages within the gel or through a process of transamidination carried out at elevated temperatures using amine-containing spacer arms or ligands. Using such strategies a number of methods have been developed to produce the requisite functional or reactive groups needed for immobilization of a wide range of affinity ligands.

Modification of polyacrylamide through transamidination involves the addition of a high concentration of an amine-containing ligand and heating while gently mixing. For instance, a diamine-containing spacer molecule may be reacted in large excess with the polyacrylamide support while mixing at 90°C and the amide bonds within the matrix will be broken and reformed with a linkage to the amine group on the

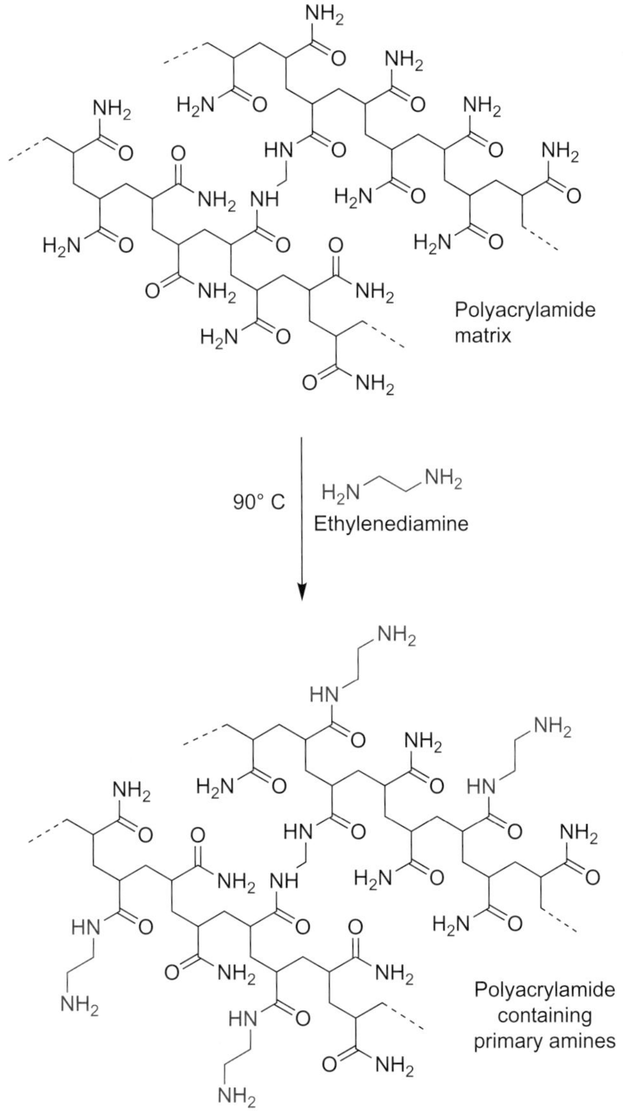

FIGURE 15.9 Polyacrylamide supports may be modified with ethylenediamine using a transamidination reaction at high temperature to produce amine groups on the support for further reactions.

spacer (Figure 15.9). A high density of such modifications may be obtained using a variety of amine- or diamine-containing molecules. In addition, hydrazide groups may be formed on the polyacrylamide support by reacting the gel at 50°C with an excess of hydrazine. (Note: Hydrazine is extremely toxic and explosive—use a fume hood and use proper personal protective equipment to handle safely.)

The following protocol describes the modification of polyacrylamide beads with ethylenediamine (EDA) using thermal transamidination, as disclosed by Inman and Dintzis (1969). A similar procedure may be used to create hydrazide groups using hydrazine substitution. All operations using EDA or hydrazine should be done in a fume hood using appropriate protective equipment.

PROTOCOL

1. Set up a three-necked, round-bottom flask in a fume hood with a heating mantle and paddle stirring rod and overhead stirring motor. Add 500 ml of EDA to the flask and stir the solution while heating to 90°C. Monitor the temperature using a thermometer inserted into one of the three necks in the flask.
2. Stir the EDA solution rapidly while slowly adding 25 g of dry polyacrylamide beads. The beads should be added in small amounts to avoid clumping of the particles. The beads will swell in the hot EDA solution as the reaction occurs.
3. Stir the solution for 5 h at 90°C.
4. Remove the heating mantle and allow the gel slurry to cool to room temperature with continued stirring. After cooling, add an equal volume of crushed ice made from deionized water to chill the slurry as it comes into contact with aqueous solution, because the EDA solution is highly basic and its dissolution into water is exothermic.
5. Transfer the gel slurry with the crushed ice to a fritted glass filter funnel or Büchner funnel containing a glass fiber filter pad and wash the matrix with 0.2-*M* NaCl in 1-m*M* HCl. Add additional crushed ice to prevent severe heating of the gel during the initial wash steps. Wash the gel until the excess EDA has been removed and the gel has decreased in pH to neutrality or acidic conditions. After the unreacted EDA has been washed out of the support, wash thoroughly with 0.2-*M* NaCl for at least 5 bed volumes. The TNBS test for amines may be used to check for the presence of EDA in the washings and to check a small volume of derivatized gel for successful coupling (Chapter 2, Section 4.3). An orange color is indicative of the presence of amines.

Trisacryl Supports

Another polymeric support made for chromatographic operations is Trisacryl, a matrix that was originally produced by IBF Biotechnics but is now available from Pall Corporation as a result of several acquisitions. The polymeric support gets its name from the fact that a key monomer used in the copolymerization process is made from the amide derivative of acrylic acid with Tris buffer (tris(hydroxymethyl)aminomethane), the actual chemical name for which is *N*-acryloyl-2-amino-2-hydroxymethyl-1,3-propanediol. This acrylamide derivative is reacted along with the crosslinking dimer *N*,*N*′-diallyltartradiamide or another crosslinking dimer, *N*,*N*′-methylene bisacrylamide, to give the final beaded, porous polymer known as Trisacryl (Figure 15.10). The Tris component of the polymeric mixture provides strong hydrophilicity due to the three

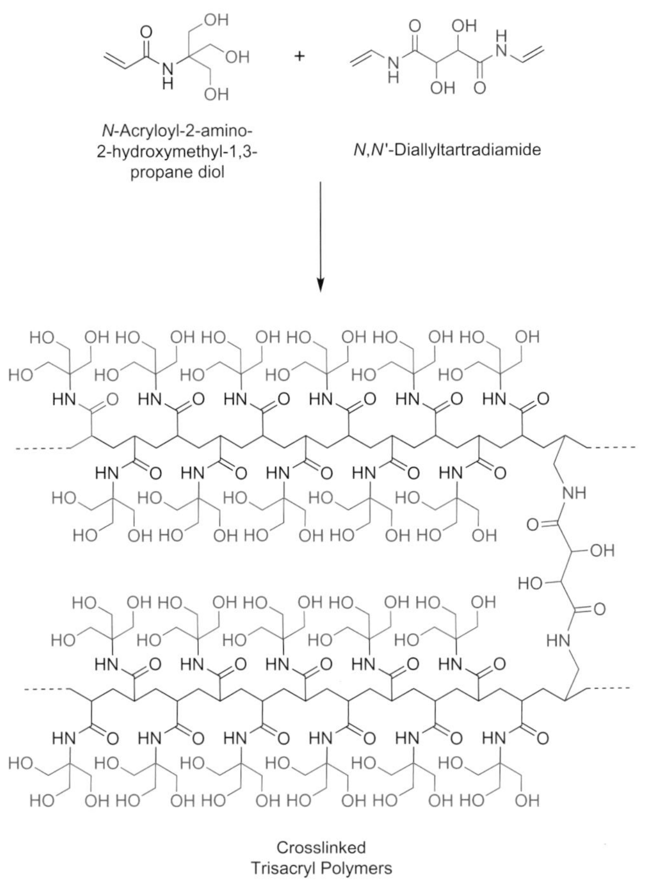

FIGURE 15.10 Trisacryl supports are made by the copolymerization of a vinyl monomer containing a tris-hydroxymethyl group and the crosslinking monomer *N,N'*-diallyltartradiamide. The support contains a high density of hydrophilic hydroxyl groups, which can be used for activation and immobilization reactions to couple affinity ligands.

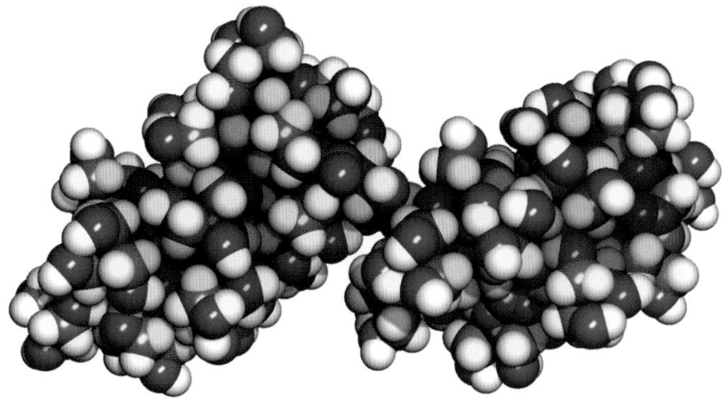

FIGURE 15.11 The three-dimensional, space-filling molecular model of a portion of the Trisacryl polymeric structure showing a single crosslink.

hydroxyl groups present on each monomer unit, while the crosslinking monomer is added in an optimized ratio to control the final porosity and mechanical strength of the particles (Batista-Viera *et al.*, 2011). The main Trisacryl polymers are highly dense structures three-dimensionally with an outer presentation of hydroxyl groups completely surrounding each linear chain and periodic crosslinking monomers bridging the chains to create a stable matrix (Figure 15.11).

The primary hydroxyls within Trisacryl create abundant functional groups for activation and coupling of affinity ligands. They also help to prevent nonspecific binding to biological molecules, thus making affinity supports made on Trisacryl have the potential to be highly specific without secondary "matrix effects." Limiting the degree of crosslinking within the matrix can produce a support with large pores having an exclusion limit exceeding 10 million Daltons, which makes it imminently fitting for protein separations. This type of beaded support, called Trisacryl GF-2000, is available in two particle size ranges that can be chosen to reflect the needs of the affinity chromatography application. The smaller particle size range, 40 to 80 μm, can be used for analytical to medium-scale separations wherein tighter peak shapes are desired from packed columns used along with bench-scale liquid chromatography systems, for instance FPLC-type instruments. Alternatively, a larger particle size range is available, 80 to 160 μm, that can be used for process scale separations where faster flow rates are necessary for greater efficiencies in the isolation of biomolecules. Both supports have identical porosities, so the ultimate choice of particle size should be made based upon the intended application needs.

Unlike plain polyacrylamide resins described previously, Trisacryl supports have much greater rigidity and are not as gelatinous in nature. For this reason, columns packed with Trisacryl have better stability to changes in mechanical pressure and flow rates used during chromatography; therefore affinity resins made using this support can be used without difficulty to produce large quantities of proteins or other biomolecules in process separation procedures. The chemical stability of Trisacryl also is excellent, as it can be used in pH environments from pH 1 to 11 without breakdown due to hydrolysis. At highly basic pH values, however, the support material may hydrolyze by cleavage of the polymer amide bonds; therefore, only brief exposure to NaOH (0.2*N*) for sanitization and regeneration should be done to prevent matrix damage.

The abundant hydroxyl groups within Trisacryl can be used for activation and immobilization of affinity ligands. The activation reactions typically form intermediate electrophilic reactive groups that can be used to couple with affinity ligands containing nucleophilic groups, such as amines, thiols, and other hydroxyls. Examples of activation reagents that can be used with Trisacryl include cyanogen bromide (CNBr), carbonyl

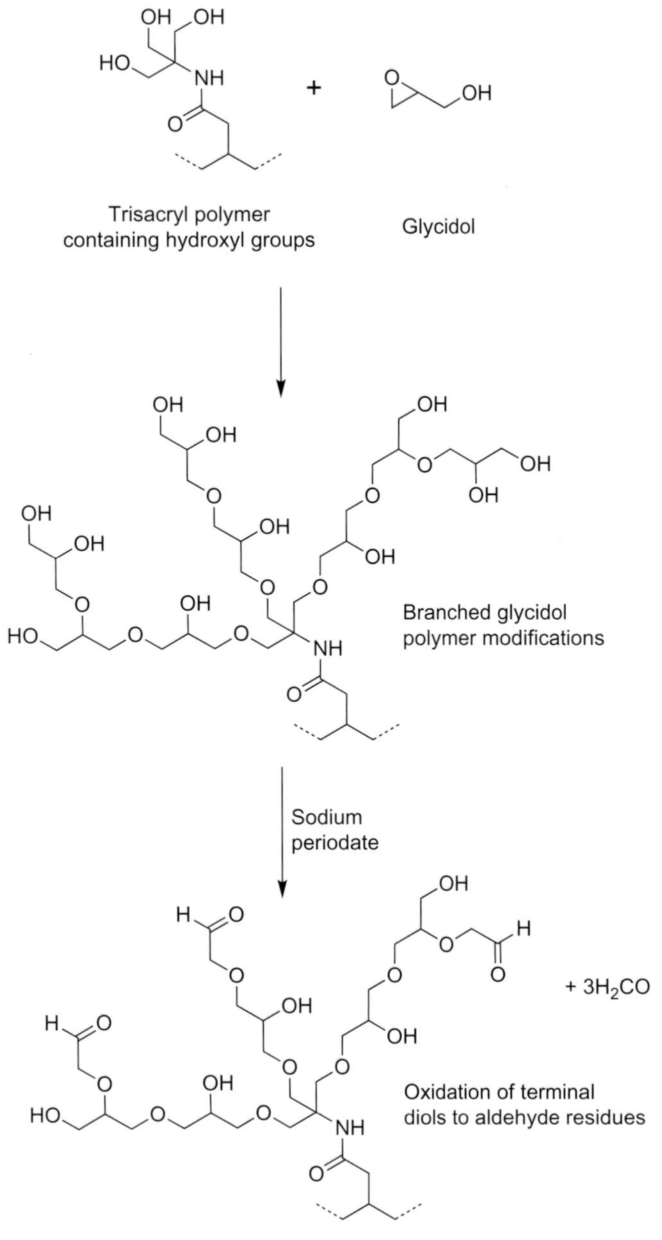

FIGURE 15.12 Glycidol modification of Trisacryl supports followed by periodate oxidation to form aldehydes useful for coupling amine-containing affinity ligands. The glycidol polymers formed within the porous structure of the support can be controlled as to their density and size by modulating the mole excess of glycidol added to the reaction.

diimidazole (CDI), tresyl chloride, tosyl chloride, divinyl sulfone, and *bis*-epoxides such as 1,4-butanediol diglycidyl ether (see corresponding sections later in this chapter). The only caveat to keep in mind with regard to the activation of Trisacryl is that sodium periodate treatment to form aldehydes, which is a common route to form reactive groups on polysaccharide-containing supports, cannot be used successfully. This is due to the fact that the hydroxyls on Trisacryl come from the Tris groups along the polymer strands and are not on adjacent carbon atoms, so there are no periodate-oxidizable groups present within the matrix. One alternative strategy to create aldehyde groups that has been found to be effective is to first react the matrix with glycidol (a monofunctional epoxide–alcohol; Figure 15.12), and then periodate-oxidize the resultant gel, which now contains diols, to form reactive aldehydes (see corresponding section, this chapter).

Sephacryl Supports

Sephacryl supports were originally developed by Pharmacia and are now sold by GE Healthcare Life Sciences as a result of a series of company acquisitions and mergers. This support is actually a composite of a naturally occurring polysaccharide derivative and a synthetic polymer. The porous, beaded support is made from the copolymerization of allyl dextran and *N,N′*-methylene bisacrylamide wherein the dextran carbohydrate is crosslinked by the bisacrylamide, which forms a strong hydrophilic matrix consisting of a multitude of glucose residues (Figure 15.13). The relative porosity of the matrix can be controlled by the amount of the components, especially the crosslinking monomer, added to the polymerization reaction. Sephacryl is available in a range of different molecular weight exclusion limits, and the largest of these, Sephacryl S-300 or S-400 with exclusion limits of 1500 and 8000 kDa, respectively, are appropriate for use as affinity supports as they can accommodate interacting proteins within their pore structure without difficulty.

Some Sephacryl supports have particle sizes that are rather small for use in gravity columns or for process-scale separations, because the matrix was originally designed to be a gel filtration or size exclusion type of support, which provides high resolution on FPLC- or HPLC-based applications. However, the Sephacryl HR series of supports has a particle size distribution that is larger than the Sephacryl SF series and this allows it to be used for moderate-scale affinity separations (25- to 75-μm diameter particles).

Sephacryl supports are supplied pre-swollen as an aqueous suspension containing a preservative. The gel is relatively stable to mechanical stresses during handling, so it can be mixed and washed using standard filter funnels to exchange the matrix into the appropriate solvent or buffer for activation and coupling. Stirring should be performed using a paddle stirrer or by rotation, however, and not using a stir bar to avoid grinding and breakdown of the support. Also avoid drying the support to prevent collapse of the particle, which may be irreversible without the presence of an excipient such as lactose in high concentration.

Immobilization reactions performed on dextran-based gels like Sephacryl can take advantage of the

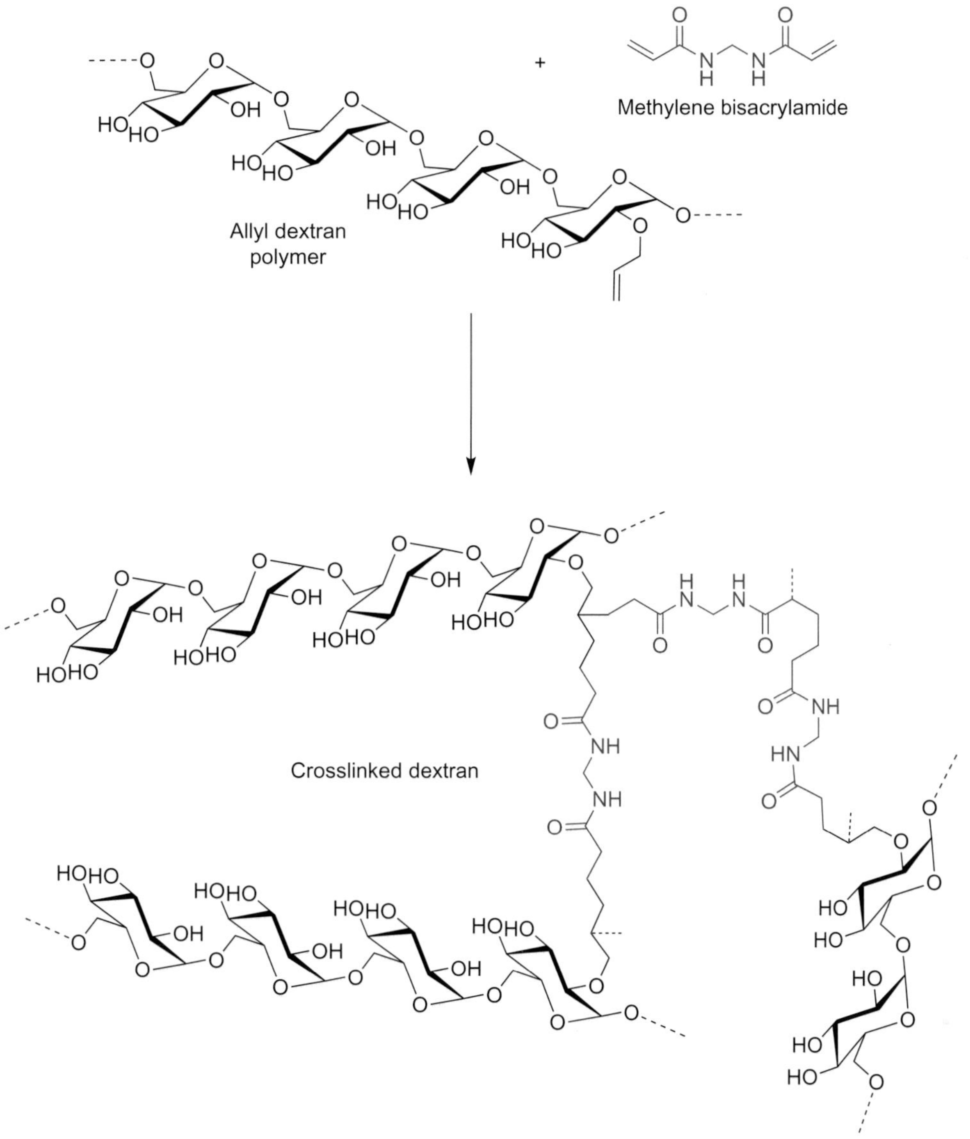

FIGURE 15.13 The chemical nature of Sephacryl supports, which are formed from the copolymerization of *N,N'*-methylene bisacrylamide and allyl dextran to create a crosslinked dextran matrix having good physical stability.

abundant quantity of hydroxyl groups on the glucose residues within the matrix. Since the bisacrylamide crosslinks provide increased physical stability, the support will tolerate washing into environments including 100% organic solvents, which are often used for activation reactions prior to coupling an affinity ligand. In this regard, activation reactions using CDI, tresyl chloride, and tosyl chloride can be carried out to create intermediate reactive groups that then can be used in organic solvent or transferred back into aqueous buffers for coupling to amine-containing ligands. In addition, aqueous-phase activation methods also can be used with Sephacryl, including periodate oxidation of the diols on glucose to form aldehydes, activation with divinyl sulfone, or activation with *bis*-epoxide-based compounds such as 1,4-butanediol diglycidyl ether. See the corresponding sections in this chapter for the experimental protocols related to activation and coupling using these methods.

Ultrogel AcA Supports

Originally, Ultrogel supports were produced by IBF but as a result of acquisitions they are now sold by Pall Corporation. The supports consist of agarose that is copolymerized with acrylamide monomers to produce a particle that has the biocompatible advantage of agarose and the size exclusion and controlled porosity capability of polyacrylamide-based supports (Figure 15.14). The result is a range of size exclusion matrices that are superior to standard polyacrylamide

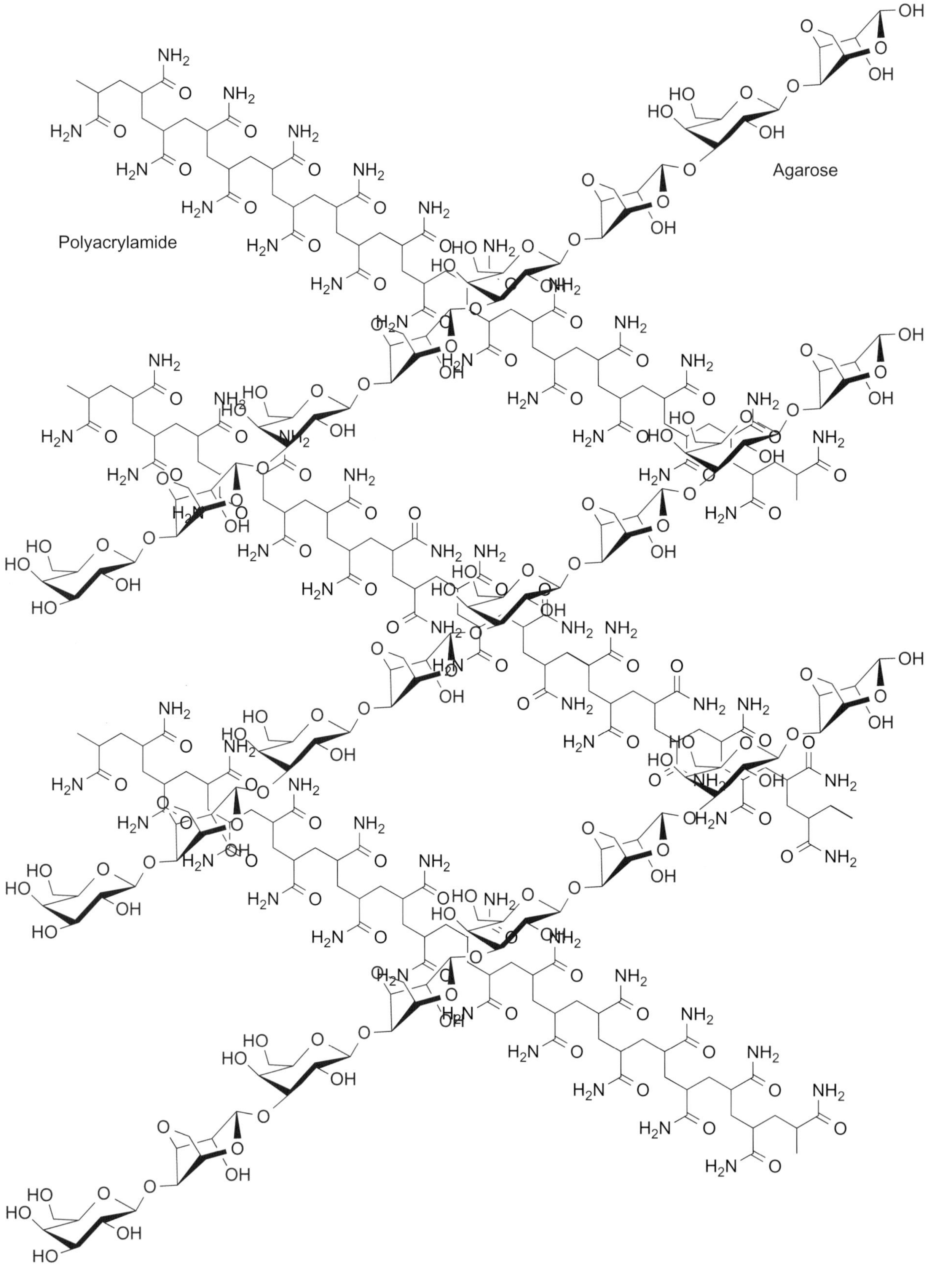

FIGURE 15.14 Ultrogel supports contain a combination of agarose polymers and polyacrylamide to create a complex network within the particles.

supports, but have smaller pore structures than supports made of agarose alone. In the preparation of the beads, the percentage of polyacrylamide in the support governs the final porosity, with a higher percentage of agarose providing a higher exclusion limit of the final support. A particular Ultrogel support containing 2% agarose and 2% polyacrylamide provides an exclusion limit of about 3000 kDa, which is sufficient for most affinity chromatography applications involving proteins. Unfortunately, Ultrogel supports do not utilize crosslinked agarose in the preparation and thus they may have some solvent sensitivity in addition to the potential to be damaged by high concentrations of denaturants or extremes of pH.

Ultrogel is also available as a paramagnetic particle in which iron oxide (Fe_3O_4) has been incorporated into the core of the polymer. The resultant bead can be used in magnetic separations wherein the particles can be removed from solution by the application of a magnetic field. The porosity of this particle is not as great as that of the chromatographic support in that it only has an exclusion limit of 200 kDa, which will likely reduce the overall capacity of the matrix for binding proteins. However, it will likely have superior binding capacity compared to the use of nonporous magnetic particles, which are often used in small-scale affinity separations.

Ultrogel supports should not be used with activation methods that require nonaqueous conditions, as solvents will damage the support. However, the matrix may be activated in aqueous conditions using CNBr, divinyl sulfone, or epoxide activation methods with success. It is anticipated that activation with *bis*-epoxides will create internal crosslinks and help to stabilize the support as compared to the base support. See the appropriate sections in this chapter that describe these activation protocols.

UltraLink Azlactone Supports

The UltraLink resins that are available from Thermo Fisher Scientific (Pierce) were originally developed at 3M in the mid-1990s and commercialized under the Emphaze name. This support material consists of the copolymerization of vinyldimethyl azlactone (oxazolone) with the crosslinking monomer, *N,N*′-methylene bisacrylamide (Figure 15.15). The azlactone monomer creates a particle containing reactive groups that is immediately useful in the coupling of amine-containing ligands, such as proteins. Therefore, the base support is unusual in that it comes already activated for the immobilization of affinity ligands, which is a feature designed into the matrix from the start. The amount of azlactone-reactive groups can be controlled simply by altering the ratio of the reactive monomer to the crosslinking monomer. Beads have been synthesized that contain from 100 μmoles/ml gel to 300 μmoles/ml of azlactone functionality. The typical azlactone reactivity in UltraLink resin is sufficient for the immobilization of over 30 mg/ml of protein A or over 20 mg/ml of human IgG, which is in excess of the levels usually required for protein-based affinity ligands.

The UltraLink support in its active form comes as a dry powder that, upon addition to an aqueous buffer, swells to become fully hydrated beads with a particle size of about 50 to 80 μm and a spherical appearance under a microscope. The beads typically hydrate in aqueous solution to yield a support having a total swollen volume of about 10 ml/g of particles. The fully swollen particles have pore structures of at least 500 Å with a total surface area of about 250 m^2/g. The porous internal structure of the beads is large enough to accommodate interacting proteins, including antibodies, thus making the support very appropriate for affinity chromatography applications dealing with proteins. Once the particle is hydrated, the azlactone-reactive groups are susceptible to hydrolysis and so they should be added directly to a ligand solution for coupling to avoid losing reactivity. Although the azlactone groups will hydrolyze with ring opening to terminal carboxylates (Figure 15.16), the rate of hydrolysis is relatively slow compared to other electrophilic groups commonly used for immobilization. The rate of coupling to an amine-containing ligand takes place with maximal yield usually within 1 to 2 h at room temperature.

Azlactone groups are reactive toward nucleophiles (e.g., $-NH_2$, –SH, –OH), but with the greatest coupling potential being with amines at basic pH values. The reaction proceeds with nucleophilic attack on the azlactone carbonyl group causing ring opening and rearrangement to an amide bond linkage with the amine-containing ligand and forming a short spacer arm to the matrix (Figure 15.17). Small amine-containing ligands may be coupled in aqueous or organic solution, provided that the organic solvent chosen allows the beads to fully swell upon addition to the ligand solution, so that the internal pore structures may be accessed during the immobilization reaction (see Table 15.1).

The UltraLink polymeric support has excellent mechanical properties for rapid chromatographic separations, being able to tolerate high linear flow rates while still maintaining high binding capacities. The support will tolerate flow rates several times higher than highly crosslinked agarose supports, thus making it useable in large columns under rapid flow without collapse of the particles. The particle structure permits high-pressure chromatography, with typical maximal pressure drops across the column exceeding 1000 psi before bead damage is observed. In practice, such extreme pressure conditions are never required, because very high flow rates can be realized under

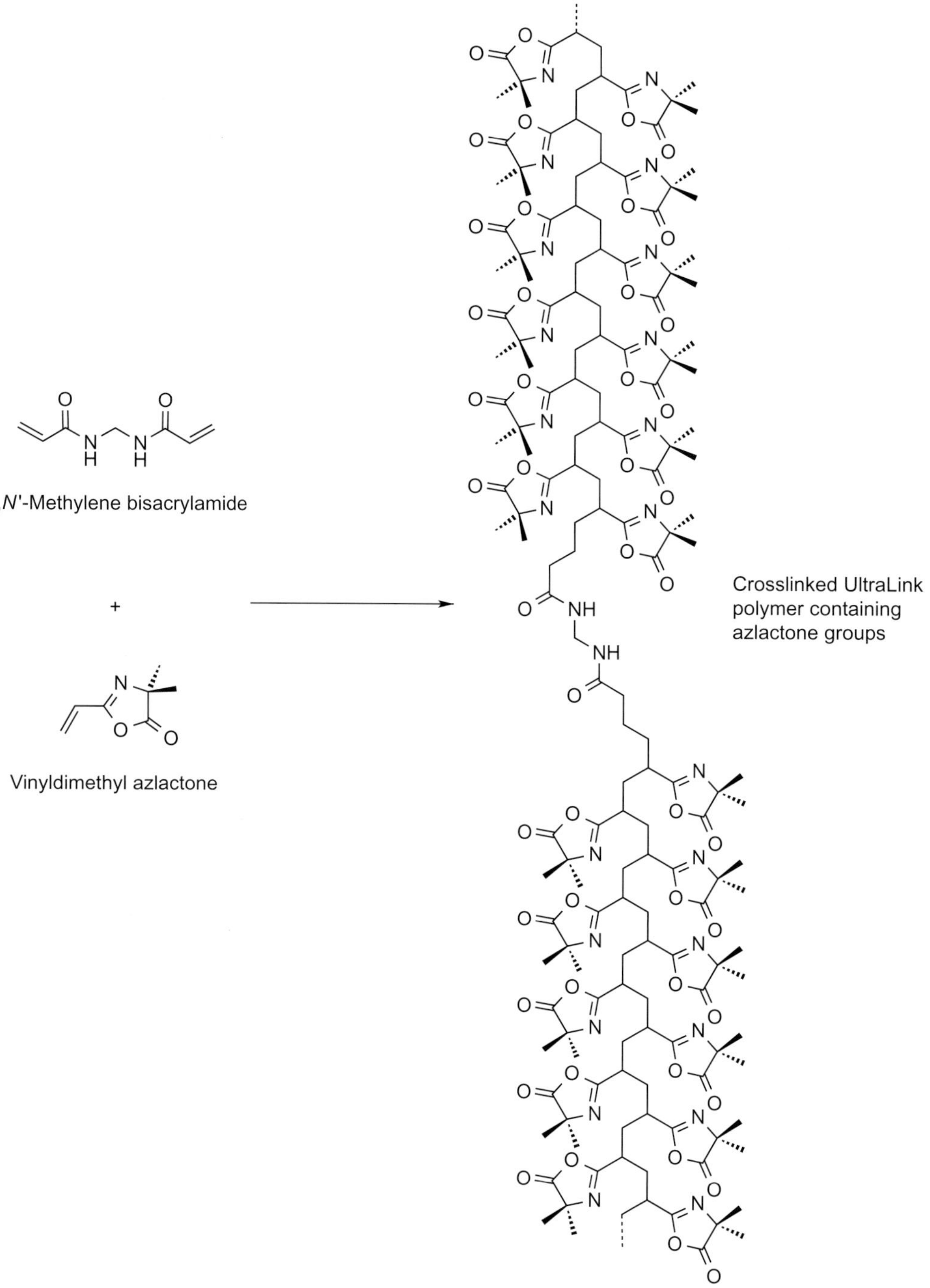

FIGURE 15.15 Ultralink resins are prepared by the copolymerization of *N,N'*-methylene bisacrylamide and the reactive monomer vinyldimethyl azlactone. The ratio of these two monomers used in the polymerization process controls the crosslinking and the degree of azlactone functionality in the final support.

pressures of far less than 100 psi. In addition, studies performed with derivatized UltraLink supports have demonstrated the excellent chemical stability of the base copolymer construction as well as the stability of the covalent bond formed with amine-containing ligands. The polymeric matrix is stable to degradation in extremes of pH (1–14) and heat, as well as being resistant to harm in highly denaturing conditions or high salt or buffer conditions.

For use of UltraLink azlactone-activated supports in the coupling of affinity ligands, the dry matrix should be accurately weighed to result in a known volume of hydrated gel in the coupling buffer. Approximately 0.1 g of beads will be hydrated to a volume of 1 ml of gel,

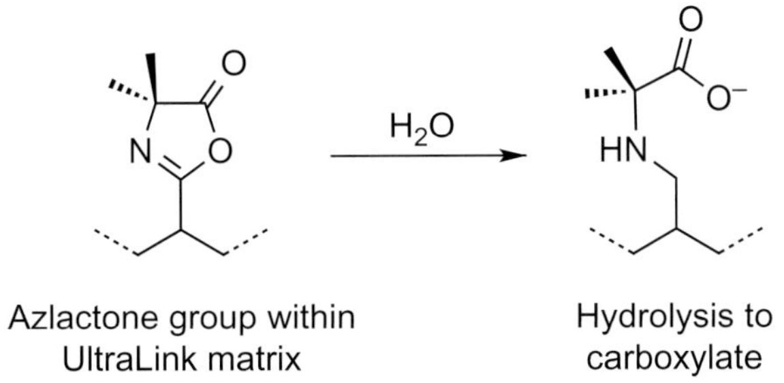

FIGURE 15.16 Azlactone groups can hydrolyze in aqueous solution, which results in ring opening and the formation of a carboxylate group at the end of a short spacer.

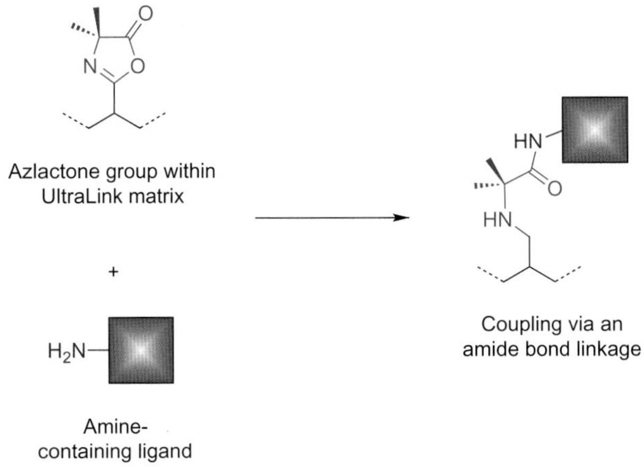

FIGURE 15.17 An amine-containing ligand can react with an azlactone group by attack at the electrophilic carbon atom of the carbonyl to result in ring opening and formation of an amide bond linkage with the ligand.

TABLE 15.1 **UltraLink Swelling Properties in Various Solvents**

	Solvent	Solubility Parameter (cal/ml)	Volume (ml/gm)	Relative Swelling[2] (%)
Dry particles	–	–	4.5	0
Aqueous Solvents	Water	23.4	8.10	100
	1-*N* HCl	–	8.08	99.4
	1-*N* NaOH	–	7.98	96.6
	1-*M* NaCl	–	7.90	94.4
Nonaqueous Solvents	Dimethyl sulfoxide (DMSO)	12.0	8.50	111.1
	Methanol	14.5	8.33	106.4
	Formamide	17.2	8.25	104.2
	Dimethyl formamide (DMF)	12.1	8.25	104.1
	Ethanol	12.7	8.08	99.4
	Acetonitrile	11.9	8.00	97.2
	Isopropanol (IPA)	11.5	7.98	96.7
	Acetone	9.9	7.82	92.2
	Tetrahydrofuran (THF)	9.1	7.40	80.6
	Methyl ethyl ketone (MEK)	9.3	7.07	71.4
	2-Ethylhexanol	9.5	6.70	61.1
	Ethyl acetate	9.1	6.50	55.6
	Dichloromethane (DCM)	9.7	6.26	48.8
	Methyl isobutyl ketone	8.4	5.5	27.8
	Heptane	7.4	5.15	18.1
	Triethylamine (TEA)	7.4	5.15	18.1
	Decane	6.6	4.95	12.5
	Toluene	8.9	4.95	12.5
	Tetrachloroethylene (TCE)	9.3	4.85	9.7

[2]*Relative percent swelling equals the amount of swelling in a solvent from dry beads relative to the amount of swelling in water = ([ml in solvent] − [ml of dry particles]) × 100/([ml in water] − [ml of dry particles]) = ([ml in solvent] − 4.5 ml) × 100/3.6 ml.*

but the exact swelling factor is production-lot dependent, so the user should consult the lot-specific information before use. The amount of coupling reaction solution to use with a given amount of dried particles should be equal to at least twice the amount of the hydrated volume of beads that is anticipated after swelling. This is due to the fact that an equal amount of solution to the hydrated bead volume will be completely taken up within the pore volume of particles during the hydration process and a proportion of at least 50% slurry of gel in the reaction solution is needed to allow proper mixing during the reaction.

The dried UltraLink particles are supplied with a small amount of Triton X-100 coating within the particles to facilitate rapid uptake of water into the pore structures without entrapment of air. For this reason, there will be an inherent absorbance at 280 nm to the reaction medium due to the release of detergent from the support. Unfortunately, this interference will prevent the determination of the amount of protein coupled during the reaction simply by measuring the difference in absorbance at 280 nm of the solution before and after coupling, as can be done with many other coupling methods. Protein coupling can, however, be determined using the BCA protein assay reagent (Thermo Fisher) to measure the amount of immobilized protein directly on the surface of the beads after immobilization.

The handling of UltraLink supports can be performed using spatulas for manipulation, paddle stirrers or rotators for mixing, and standard filter funnels with membranes or fritted glass filters for washing. The matrix is very robust and is not easily damaged in use. The methods used for the immobilization of proteins or other affinity ligands onto azlactone-activated supports are described in subsequent sections within this chapter.

Toyopearl HW Supports

Toyopearl supports are comprised of a synthetic polymeric matrix that is commercially available from Tosoh Bioscience. The matrix is made from the copolymerization of glycidyl methacrylate, polyethylene glycol (PEG), and pentaerythritol dimethacrylate to form a complex structure that is hydrophilic and rich in hydroxyl groups for activation and coupling of affinity ligands (Figure 15.18). The polymerization of the two acrylate-containing monomers is combined with the covalent linkage of the PEG–OH groups to the glycidyl epoxides, thus forming pentaerythritol and PEG crosslinks within the porous polymer network making up the bead internal topography. The PEG chains are very hydrophilic and certainly contribute heavily to the biocompatible nature of the support, while the pentaerythritol dimethacrylate crosslinks add rigidity by locking the polymer strands together.

Toyopearl-type supports have moderate mechanical stability and are able to tolerate pressures of at least 45 psi (3 bar) without bead collapse. This makes possible reasonable linear flow rates and large column configurations for process-scale separations or automated separations using an FPLC or HPLC system for smaller scale operations. A type of Toyopearl support containing a higher degree of crosslinking, and thus greater rigidity for higher pressure resistance, is available under the designation "PW," but this matrix has lower capacity than the HW resins for protein purification applications. The larger-particle-size versions of the support will be able to achieve higher linear flow rates at lower pressure drops across a column bed and thus be capable of reach higher flow rates than the smaller particle grades. The matrix is also quite robust to handling, mixing, and chemical reactions carried out during activation and coupling procedures. Toyopearl supports are able to withstand extremes in pH (2–12), high salt or buffer concentrations, the presence of denaturants, and a solvent exchange into nonaqueous conditions without suffering damage to the particles.

The wide variety of Toyopearl supports available commercially come pre-swollen as thick slurries in aqueous solution. A wide selection of different particles sizes and porosities is available for different applications, depending on the resolution of separation desired and the size of the column being used. Typically, smaller particles will be more appropriate for analytical separations, while larger particle diameters are more suitable for bench-scale to process-scale separations. The different exclusion limits available reflect the fact that the matrix was originally developed for gel filtration or size exclusion chromatography and is still used for these applications. Small particle sizes of the support can be obtained that consist of 20 to 40-μm spherical beads and are designated "S" grade (for "superfine"), while medium particles are in the range of 40 to 90 μm ("M" grade), larger particles that have diameters of about 90 to 120 μm (designated "C" grade for "coarse"), and very large particles of diameters ranging from 100 to 300 μm (designated "EC" grade for "extra coarse"). The particle size distribution generally gets broader as the particle grade and diameter increase, thus giving the smaller-particle-sized supports the potential to provide greater peak resolution in gel filtration chromatographic separations. In general, the larger particles provide less back pressure for a given flow rate and are more amenable to large-scale chromatography, while the smaller particles provide greater resolution and are more appropriate for analytical separations where peak resolution is important.

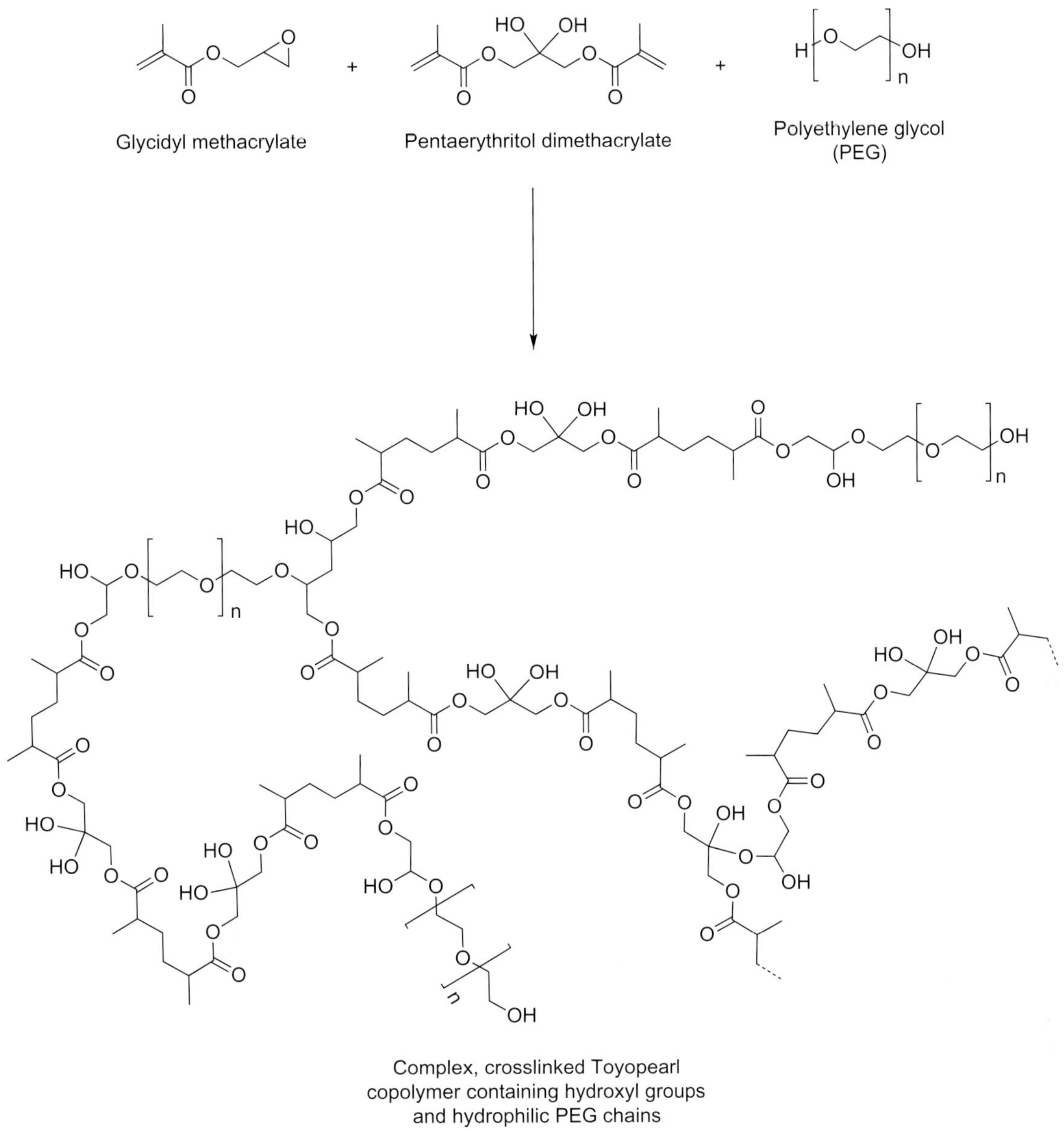

FIGURE 15.18 Toyopearl supports are constructed from the copolymerization of glycidyl methacrylate, the crosslinking monomer pentaerythritol dimethacrylate, and polyethylene glycol (PEG). The epoxy monomer is also able to provide crosslinks within the support with neighboring hydroxyl groups and also incorporate PEG chains within the structure by reacting with the epoxides, which provides significant hydrophilicity. The support also contains an abundance of primary and secondary hydroxyl groups that can be used to couple affinity ligands.

For affinity chromatography use, the Toyopearl support having an appropriate exclusion limit (correlating to its maximal pore size) may be chosen to best accommodate the molecules being separated. For protein separations, typically the supports having the larger exclusion limits will provide the best opportunity for protein–ligand interactions in affinity separations, especially when two large proteins have to interact within the pores. For routine affinity purifications, the Toyopearl HW-65 or HW-75 resins will provide the best performance, as the exclusion limit of these gels are up to 5000 kDa, with a fractionation range for size exclusion chromatography listed as approximately 50 to 5000 kDa. Pores of this size will allow even the largest macromolecules to interact within the matrix, including use in applications involving the capture of interacting proteins using co-immunoprecipitation procedures.

Toyopearl supports contain mainly secondary hydroxyl groups within its structure due to the presence of the pentaerythritol crosslinks as well as resulting from the creation of secondary hydroxyls upon epoxide group coupling with another hydroxyl within the web

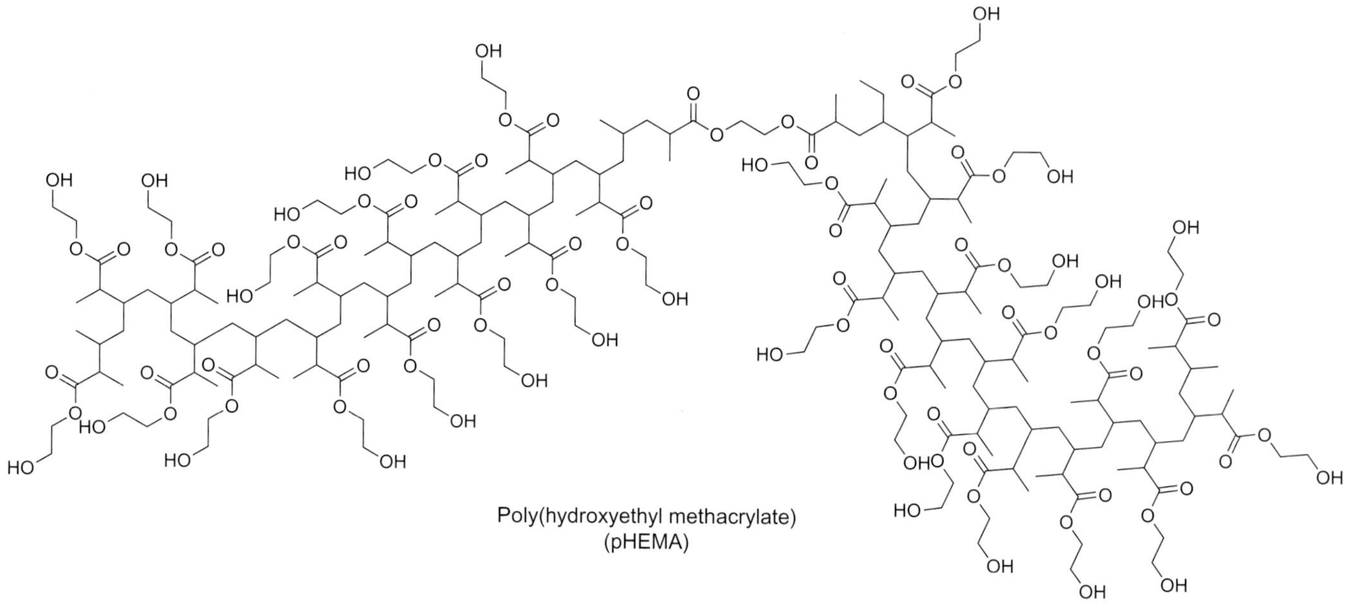

FIGURE 15.19 Poly(hydroxyethyl methacrylate) (pHEMA) supports can be made from the copolymerization of 2-hydroxyethyl methacrylate (HEMA) and the crosslinking monomer ethylene dimethacrylate. The result is a support that is very hydrophilic and contains numerous primary hydroxyl groups, which can be activated for the coupling of affinity ligands.

of polymers. There is also some potential for primary hydroxyls being present if epoxide hydrolysis occurred without coupling to another hydroxyl group or some of the PEG–OH groups remained uncoupled at one end within the gel. Using any of these hydroxyls, the support can be activated through any of the activation reactions designed to create an electrophilic reactive group. This includes the use of such common activation reagents as CNBr, CDI, tresyl chloride, tosyl chloride, *bis*-epoxide, and divinylsulfone. However, the support cannot be directly oxidized with sodium periodate to give aldehydes, as in the case of agarose-containing matrices; nevertheless, an intermediate modification with glycidol can be carried out to create polymer grafts on every hydroxyl within the gel that terminate in diols, which afterward can be oxidized to provide abundant aldehydes for immobilization through reductive amination (similar to the procedure previously mentioned for Trisacryl supports). The initial glycidol derivatization process results in the small epoxide coupling to internal hydroxyls on the support and forming a terminal diol as the epoxide ring opens. Additional glycidol molecules then can react with these newly formed hydroxyls, thereby creating polymer grafts. Subsequent oxidation with periodate then creates an aldehyde at the end of each glycidol chain with simultaneous loss of one molecule of formaldehyde. Coupling of amine-containing ligands on this derivative can be achieved through Schiff base formation followed by reductive amination to form a secondary amine linkage (see Section 2.1).

Separon HEMA Supports

Copolymer particles made of 2-hydroxyethyl methacrylate have been used to create highly hydrophilic chromatography supports that have abundant primary hydroxyls, which can be used for activation and coupling of affinity ligands. This synthetic support made from the copolymerization of HEMA and ethylene glycol dimethacrylate was developed decades ago in Czechoslovakia and commercialized by Tessek, Ltd., under the name Separon HEMA (Figure 15.19). The dimethacrylate crosslinking monomer adds rigidity to the HEMA polymer, and the ratio of crosslinker to HEMA in the final matrix controls the relative porosity of the support as well. A molecular model of a portion of the crosslinked HEMA polymer is shown in Figure 15.20, which illustrates the abundant hydroxyl groups present within the matrix. The particles are spherical and are available in sizes of 10-μm and 60-μm average particle diameters. HEMA supports also come in a range of porosities from 40 to 2000 kDa that provide abundant choices of porosity for size exclusion chromatography applications, and the degree of porosity can be chosen based upon the size of the molecules being separated. Particles having the largest pore size should be selected for general affinity chromatography separations using macromolecules, because the interior of the particles is able to better accommodate interactions between large biomolecules.

The high density of primary hydroxyl groups on the HEMA support allows convenient derivatization with

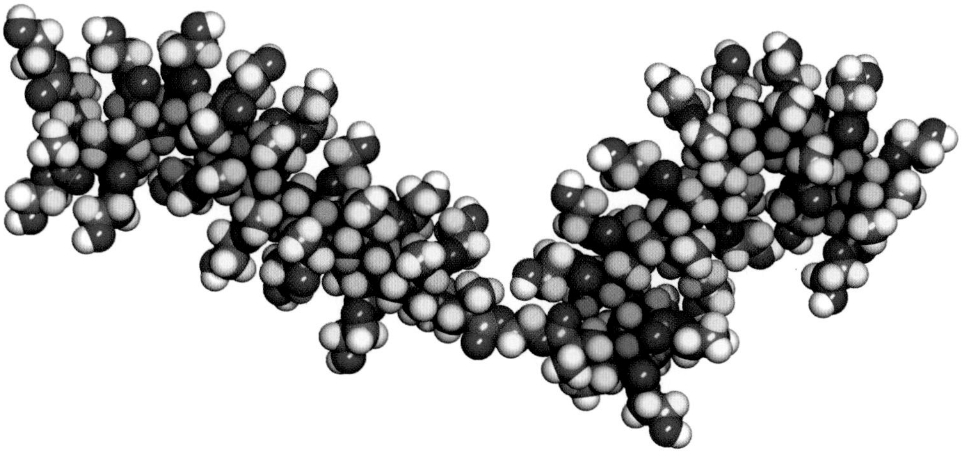

FIGURE 15.20 The three-dimensional, space-filling molecular model of a portion of a pHEMA support showing the numerous hydroxyethyl groups and a crosslink made from ethylene dimethacrylate.

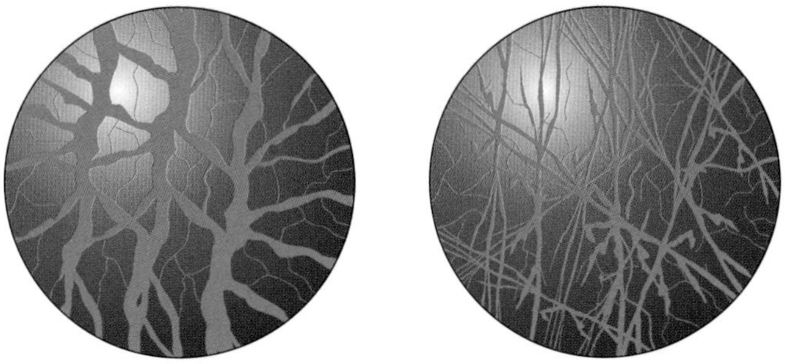

FIGURE 15.21 An illustration of the difference between a perfusive chromatography support containing large through-pores and a typical diffusive chromatography support which contains mainly diffusion-limited access to the inner pore structure of each bead.

typical activation reagents to form reactive, electrophilic intermediates capable of coupling affinity ligands. All of the common activation methods for hydroxyls described in this chapter may be used with HEMA, except for the use of periodate activation, which requires diols to form aldehydes, and this support does not contain any periodate-oxidizable hydroxyls.

Although HEMA-based supports contain esters from each hydroxyethyl group linked to the methacrylate monomers, the ester bonds are extremely stable due to the configuration containing a tertiary alpha-carbonyl ester. It does not easily hydrolyze nor does it react with nucleophiles such as amines. The base support is stable to extremes of pH (2–12) and will tolerate NaOH sanitization procedures without breakdown of the polymeric structure. Two activated versions of the HEMA support are available from Tessek for the immobilization of affinity ligands: epoxy and vinyl sulfone. However, other activation methods may be designed from the base support for custom coupling purposes. In particular, glycidol modification may be carried out to create grafted polymers within the matrix that contain terminal diols on each chain. As described previously in the section on Toyopearl supports, these diols can be periodate oxidized to form terminal aldehydes for use in reductive amination immobilization procedures (see the corresponding section under activation methods in this chapter).

UNOsphere Supports

Another innovative polymeric chromatography support is produced by Bio-Rad under the name UNOsphere. The base matrix construction is unique in that the beads are said to contain "through-pores" that are a minimum of 0.5 μm in diameter with virtually no pores of less than 0.1 μm. This is reminiscent of Perceptive Biosystems' HPLC supports that were developed during the 1980s, which also are claimed to have larger through-pores, which make faster flow rates possible for analytical separations (Jungbauer, 1996; Gallant, 2004). The through-pores allow the mobile phase to enter into the internal spaces of the bead structure by convective flow rather than just by diffusion. In theory, this permits faster on and off rates for interactions with immobilized ligands than is normally possible using supports that only have convective flow around the beads (Figure 15.21). The majority of chromatography resins restrict internal bead access by diffusional processes originating from outside the particles. In perfusion supports, the larger through-pores inside the particles allow convective fluid flow to reach sites deep within each bead, thus increasing the effective volume of the support that can be used for affinity interactions during rapid chromatographic operations. The theory of intraparticle flow through perfusive supports and monoliths has been studied in great detail (Hamaker and Ladisch, 1996).

The UNOsphere polymeric resin is made from the copolymerization of functional and nonfunctional monomers in addition to a portion consisting of cross-linking monomer to form the final matrix. The resultant beaded support has a particle size range of about 45 to 90 μm with a mean diameter of 60 μm. The spherical particles form a chromatography support that is analogous in its internal pore structure to some of the monolithic-type supports that are formed by polymerization of monomers directly within a column to produce a single continuous cylindrical matrix. It is also similar to the use of affinity membranes that contain a torturous pore structure with large and small pores where both convective flow and diffusion take place simultaneously throughout the matrix.

The large cavernous structures of through-pores within UNOsphere resins are constructed by careful selection of polymerization conditions, as described in U.S. patent 6,423,666 by Liao and Hjerten (2002). Just like most other polymeric supports, the particles are formed by suspension polymerization using a monofunctional monomer that may contain functional groups or reactive groups. For instance, the proper choice of monomer can yield a beaded chromatography resin containing ion-exchange functionalities, hydrophobic groups, or reactive derivatives able to spontaneously couple affinity ligands. One key to forming the large pore structure is controlled use of a crosslinking monomer during the polymerization reaction. Particularly, it has been found to be important to keep the concentration of crosslinker to a mole fraction of about 0.3 to 0.4 relative to the total monomer mixture. The construction of the through-pores is accomplished by the addition of some hydrophobic character to the monomers as well as the use of additives that modify the polarity of the aqueous medium during polymerization. Additives such as sulfate or other lyotropic agents can be used in addition to the potential use of hydrophilic polymers such as PEG, dextran, or methyl cellulose.

The base support can be used in process purification applications that involve large column sizes or high linear flow rates (up to 600 cm/h) without particle collapse or encountering severe back pressures. The polymeric support also can be sanitized using 0.1-*N* NaOH or regenerated after use with denaturants such as 6-*M* guanidine. A study of IgG purification on a UNOsphere support containing immobilized protein A was published by Perez-Almodovar and Carta (2009).

A reactive derivative of the UNOsphere support is available that contains epoxide groups, which can be used to immobilize affinity ligands containing amines, thiols, or hydroxyl groups (see the section on activation and coupling reactions, this chapter). It is supplied dry to ensure stability of the reactive group upon storage. The support contains approximately 50 to 132 μmoles/g of reactive epoxy groups, and upon addition to an aqueous solution hydration will cause the matrix to swell to approximately 5.5 to 8.0 ml/g of dry resin. The dry, activated support can be added directly to a ligand solution for coupling. The optimal pH range for coupling is 9 to 13 in a buffered solution containing at least 10-m*M* buffer salts to prevent pH changes during the reaction. Thiols can be coupled to the support at the lower pH, amines can be coupled in the middle of this range, and hydroxyl-containing ligands require pH 13 for efficient coupling. UNOsphere supports are not available in a hydroxylic form that could be used for custom activation and coupling of affinity ligands.

Eupergit Supports

Certain polymeric supports have found use for the immobilization of enzymes, which are often used as immobilized reactors in the manufacture of foods, antibiotics, and other organic compounds. Many of these supports are used in ton quantities annually for large-scale production processes. One such support that has found more use as an immobilized reactor than in affinity chromatography is Eupergit. These porous beads are made of methacrylate derivatives and were originally manufactured by Rohm Pharma Polymers, which was a unit within the Specialty Acrylics division of Degussa, AG in Darmstadt, Germany, but since 2007 this technology became a part of Evonik Industries. Eupergit is made by a copolymerization of glycidyl methacrylate, allyl glycidyl ether, methacrylamide, and the crosslinking agent *N,N'*-methylene *bis*-methacrylamide. The Eupergit C support that contains reactive epoxide groups has become very common for use in the preparation of immobilized reactors (Boller *et al.*, 2002; Berrio *et al.*, 2007) (see Chapter 1, Section 3.4, for a discussion of immobilized reactor applications). Unfortunately, after Rohm Pharma's merger with Evonik Industries, the Eupergit support was discontinued; however, some distributors still sell small quantities of it for research purposes (Sigma Aldrich).

Sepabeads, ReliZyme, and ReliSorb Supports

Although Eupergit is no longer manufactured commercially (see previous section), other polymeric supports continue to be used to prepare immobilized reactors for production purposes. One such material is called Sepabeads and is produced by Resindion, which is a subsidiary of Mitsubishi Chemical. This support material consists of a porous polymer made entirely of polymethacrylate which is formed into spherical white opaque beads. Sepabeads is available in a number of different formats including two different pre-activated supports containing epoxide groups ready to couple ligands, two amino functionalized supports containing different length spacer arms, a diol derivative that can be periodate oxidized to create aldehydes for coupling amine-containing ligands, and two derivatives especially designed for hydrophobic interaction

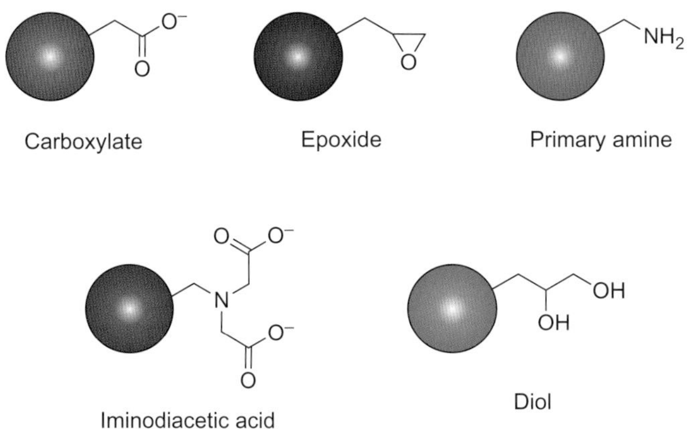

FIGURE 15.22 The derivatives of Resindion chromatography supports that are available commercially.

chromatography: a C_4 butyl and a C_{18} (Figure 15.22). The particle size of Sepabeads is available in two different size ranges: an "S" grade with a particle size distribution of 100 to 300 µm and an "M" grade having a particle size distribution of 200 to 500 µm.

The polymethacrylate Sepabeads support has an average pore diameter of approximately 10 to 20 nm, which is rather small for typical affinity chromatography resins. The average IgG antibody is about 11 nm in diameter, so large proteins may not be accessible to the smaller pore openings. However, a more recent iteration of the Sepabeads-type polymer was introduced under the name ReliZyme, and this material contains average pore sizes in the range of 40 to 60 nm, which are more amenable to protein chromatography applications. All of the derivatives available for the Sepabeads products also are available in the larger particle ReliZyme beads.

Resindion's Sepabeads and ReliZyme particles, however, were developed with a view toward their use as immobilized reactors rather than for applications in affinity chromatography. Most enzyme reactors are prepared through adsorption or covalent immobilization of enzymes onto solid-phase supports, and their subsequent use is to act as catalysts in the transformation of substrates into desired products. According to the company, an enzyme can be immobilized onto the Sepabeads or ReliZyme supports using one of four different approaches: (1) passive adsorption, (2) ionic interaction, (3) crosslinking, and (4) covalent coupling with a reactive group on the particles. Enzyme reactors have been prepared successfully using all of these strategies, although the most robust materials are formed using covalent coupling, which provides low leaching potential and a high degree of reusability.

A third resin material developed by Resindion is also potentially suitable for affinity chromatography applications. The ReliSorb supports are polymethacrylate spherical particles available in three different particle size ranges: 50–150 µm, 75–200 µm, and 200–500 µm, with corresponding mean diameters of 90 µm, 120 µm, and 300 µm. The larger average particle size of the ReliSorb resins permits high-flow chromatography over a wide range of linear flow rates. The spherical particles are extremely rigid with low swelling characteristics when hydrated in aqueous solvent. In addition, the pore structure of this support is in the range of 40 to 50 nm, thus making it useable for affinity chromatography separations using proteins and other biological macromolecules. The linear velocity of the largest particle size support can exceed 10 m/h with a pressure drop of only 0.5 bar/m of column bed height.

The ReliSorb particles also can be used in an expanded bed chromatography application, because the high specific gravity of the particles (>1.1 g/ml) permits controlled bed expansion when used in a reverse-flow mode. A packed bed of resin will expand from 100% to 200% over the linear flow rate of 200 cm/h to 300 cm/h, when used in an expanded bed separation system. This important characteristic allows crude samples to be separated on a bed of ReliSorb matrix without clogging the column. These properties are similar to the Streamline support materials from GE Healthcare.

The ReliSorb supports are available in a number of different derivatives including epoxide-activated, an iminodiacetic acid form for immobilized metal chelate chromatography, a diol derivative that can be periodate oxidized to form aldehyde groups, two amine functionalized resins, three different supports for hydrophobic interaction chromatography including a phenyl, butyl, and a C18 support material, as well as a number of derivatives for ion-exchange chromatography. The epoxide- and amine-containing supports may be used for custom immobilization of affinity ligands containing amines, thiols, hydroxyls, and carboxylates. The diol-containing support may also be activated using a number of different reaction strategies described later in this chapter (Section 2.1), which permits further customization of the support for particular applications.

Poly(Vinyl Alcohol) Supports

Crosslinked poly(vinyl alcohol) supports have been used as resins in column-based separations, particularly involving high-performance liquid chromatography (HPLC) (Battinelli *et al.*, 1996) (Figure 15.23). Poly(vinyl alcohol) (PVA) copolymers are typically made from the polymerization of poly(vinyl acetate) followed by the removal of the acetate ester by hydrolysis to create the PVA derivative. Itagaki *et al.* (1989) described the production of porous PVA particles for use in chromatography applications. The beads are formed by dissolving the linear PVA polymer in an aqueous

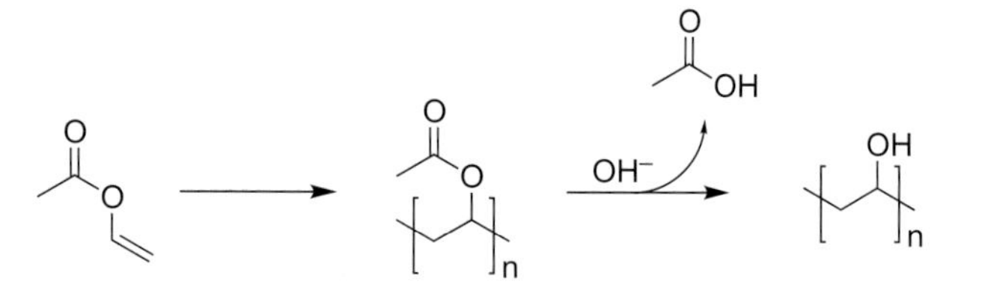

FIGURE 15.23 The reactions involved in the synthesis of a poly(vinyl alcohol) (PVA) support, which contains a huge density of hydroxyl groups throughout its structure.

solution containing a salt and dispersing the solution into a non-miscible organic solvent with stirring. The gelatinous polymer beads that form within the aqueous phase are stabilized through crosslinking with a bifunctional crosslinker, such as glutaraldehyde. One popular form of PVA copolymer is the Carbowax supports that are typically made from a combination of polyethylene glycol (PEG), polydimethylsiloxane, and PVA (note that Carbowax is a trade mark of Dow Chemical, which refers to their line of products incorporating PEG compounds). These resins are often used for separating compounds in analytical gas chromatography or HPLC through differential interactions *via* hydrogen bond formation with the ether oxygens and hydroxyls making up the beaded matrix. Supports made of PVA are extremely hydrophilic due to the abundant –OH groups. Derivatives of these supports have been prepared through esterification or by activation of the hydroxyl followed by covalent linkage of affinity ligands. For instance, a reverse-phase HPLC support was prepared by esterification of dodecanoic acid onto the hydroxyl groups to form a C18 column able to separate compounds based on hydrophobic interactions (Battinelli *et al.*, 1996).

Composite chromatography supports containing an inner core of a hydrophobic matrix and an outer shell of PVA have been created to make a biocompatible hydrogel for protein separations (Leonard *et al.*, 1995). A core particle made from porous polystyrene/divinyl benzene copolymer as modified through adsorption of PVA followed by stabilization by crosslinking. The resultant resin retained its ability to be used for size exclusion chromatography while the PVA outer layer provided hydrophilicity and decreased nonspecific binding toward proteins.

PVA polymers have also been used as carriers in the formation of drug conjugates. Kakinoki *et al.* (2008) created a paclitaxel–PVA conjugate to a linear 80-kDa PVA water soluble synthetic polymer to deliver therapeutic doses of the drug to tumor cells *in vivo*. The aqueous solubility of paclitaxel was dramatically enhanced through conjugation to the extremely hydrophilic PVA.

Membrane Supports

The use of porous membranes for the separation of proteins and other biological materials has existed for decades and remains an important component for purification, concentration, diafiltration, and removal of particulates including viruses and bacteria contaminants. Membranes are used routinely in processes ranging from small research-scale operations to huge bioprocess-scale production procedures in the pharmaceutical, biotechnology, and food industries. Synthetic and inorganic membrane supports are available with a broad range of porosity characteristics and structural compositions. Membranes may be constructed with large pores that permit macromolecules to freely pass through without being retarded or they may be made with smaller pore characteristics for selected filtration of molecules above a certain molecular size (e.g., ultrafiltration membranes). The internal pore structure of polymeric membranes is formed by the torturous entanglement of molecular strands, which leaves gaps and winding pathways through its three dimensional structure.

At a microscopic scale, the internal structure of a membrane is similar to that of a macroscopic sponge. The inner pores can absorb an aqueous solution within its open spaces and allow flow of a mobile phase directly through them at much faster linear flow rates than is possible using beaded chromatography supports. A packed volume of porous particles typically allows the mobile phase to freely move through it only by traveling around the beads, whereas the interior porous structure of the particles is accessed by diffusion. The process of diffusion is slower than perfusive flow and delays the interaction of molecules traveling in the mobile phase with the internal structure of the porous beads by the time needed for molecules within the mobile phase to diffuse into and out of the particles. By contrast, most of a membrane's internal pore structure is directly accessible by a flowing mobile phase with minimal need for diffusion to reach inner structures. The dominance of convective flow through membranes with minimal dependence on diffusional mass transfer is a key advantage of membranes over packed-bed chromatography methods. In principle, this allows for much faster affinity interactions between the mobile phase and a ligand attached to the membrane's surface than is possible using columns of porous particles.

However, membranes often have lower functional density than beaded supports, which can lead to a lower capacity for an affinity molecule to bind with a desired target. To overcome this potential deficiency, composite constructions of particles embedded within membranes have been made to combine the higher capacity of beaded chromatography supports with the faster flow rates possible with membrane supports.

FIGURE 15.24 An SEM picture of the internal structure of a porous PET membrane-particle composite construction. The nanoparticles are covalently attached to the membrane inner pore structure and provide increased capacity for affinity separations or enzyme reactor applications. Reprinted with permission from Ulbricht, M. (2006) Advanced functional polymer membranes. *Polymer* 47(2006): 2217–2262; copyright © 2006 Elsevier.

Figure 15.24 shows the inner pore structure of a track-etched PET membrane (polyethylene terephthalate) containing large 1-μm pores and having nanoparticles covalently attached to the surfaces within it (Ulbricht, 2006). The particles provide significantly increased capacity for coupling enzymes used as reactors for chemical transformations, which occur in the mobile phase as it passes through the membrane.

Membranes are available having a broad selection of porosities ranging from those designed for ultrafiltration of biomolecules to ones more appropriate for filtration of particulates or microbial contamination. Ultrafiltration membranes can be selected to have pore sizes that can allow the passage of molecules below a particular molecular weight range while generally excluding molecules above that point. The molecular weight cutoff value of membranes actually is an approximate figure, not an exact exclusion limit, but it provides useful information about what biomolecules potentially can access the interior porous structure of the support and pass through. To use membranes for affinity separations, the selection of molecular weight cutoff must be one that is high enough to accommodate the intended molecules being targeted so they can bind an immobilized affinity ligand attached within the membrane's structure. Thus, most membrane-based affinity supports utilize membranes having large pore sizes in the micron range that allow proteins to pass through, for instance >0.2 μm. This cutoff range is essential to provide internal space for protein–protein interactions to occur within the internal membrane structures while also allowing for rapid mobile phase flow.

Commercial polymeric membranes are available in a wide range of chemical compositions, including hydrophobic or hydrophilic polymers, having functional groups on the matrix available for chemical modification, containing anionic or cationic groups for ion-exchange separations, or having affinity ligands for capture of specific target molecules. Similar to beaded chromatography supports, polymer membranes can be made from naturally occurring polysaccharides or synthetic polymers. Some of the more popular polymeric compositions consist of cellulose (regenerated), nitrocellulose, cellulose acetate, polyamide (e.g., nylon), polyimide, polyester, polyvinylidene difluoride (PVDF), polyacrylonitrile, polysulfone, polyethylene, polypropylene, poly(vinyl alcohol) (PVA, crosslinked), pHEMA (crosslinked), and a number of copolymers or composites of these polymers (Ulbricht, 2006) (Figure 15.25).

Nitrocellulose and PVDF have long been popular for use with proteins, especially in western blotting procedures in which electrophoretically separated proteins are non-covalently transferred and adsorbed onto the membranes and probed immunochemically using antibody conjugates. These membrane supports also can be used in affinity separations using non-covalently adsorbed proteins as capture ligands to selectively purify target molecules. However, to prepare more stable covalent attachment of affinity ligands it is necessary to have a functional group available on the membrane that can be activated and used to couple a protein or other affinity molecule. In this regard, membranes containing amines, carboxylates, or hydroxyl groups provide the greatest flexibility in the selection of activation and coupling chemistries available to immobilize affinity ligands. Hydroxyl-containing membranes are particularly desirable for use in affinity separations. Unmodified cellulose (including regenerated cellulose) membranes in addition to PVA and pHEMA membranes or composites provide excellent hydrophilicity

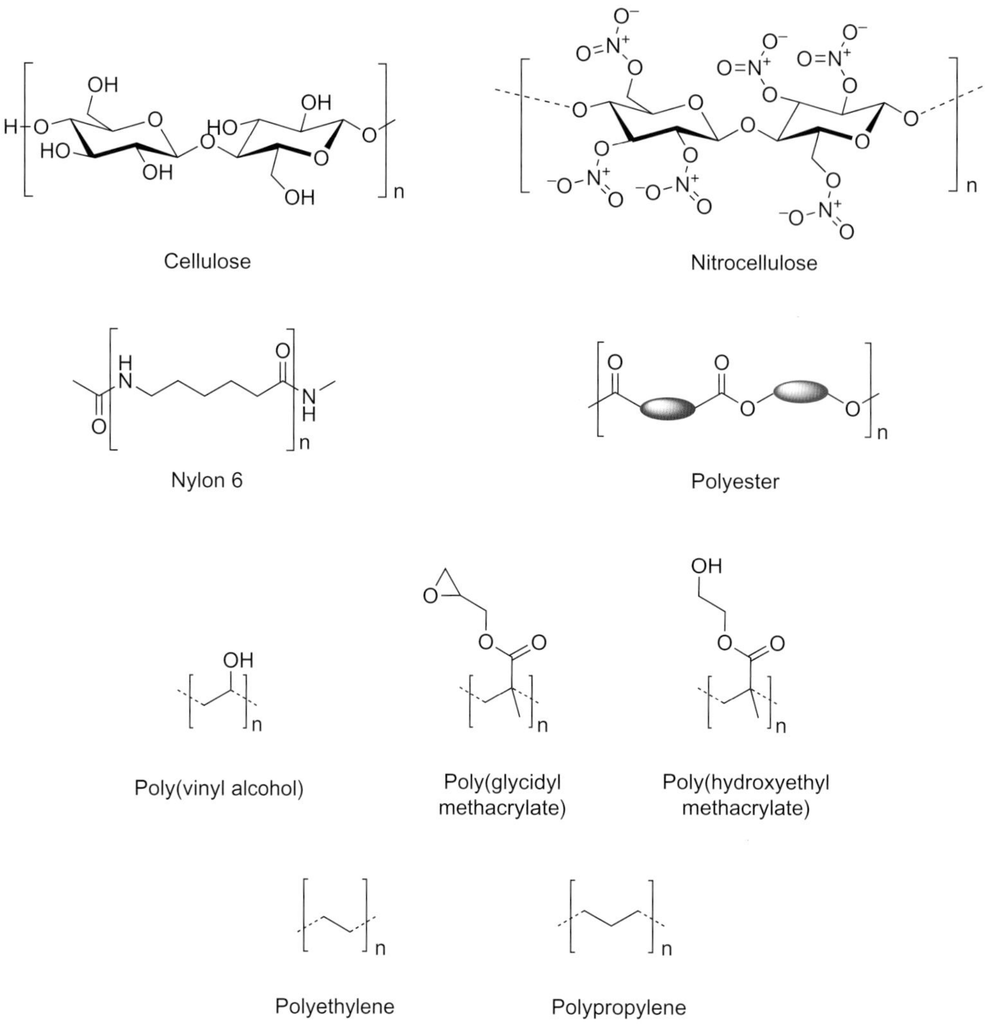

FIGURE 15.25 The polymer structures of common membrane compositions.

and low nonspecific binding with abundant hydroxyls for coupling ligands. Composites containing naturally occurring hydroxylic polymers such as chitosan or dextran also provide abundant functionalities for activation and coupling ligands. Another benefit of hydroxylic membranes is that they do not generate residual charge character from the remaining functional groups after coupling an affinity ligand, as may be the case with amine-containing or carboxylate-containing membranes, which could create considerable positive or negative charges, respectively.

An important monomer type that has been used to prepare affinity membranes is glycidyl methacrylate, which contains an epoxide group that can be used to immobilize affinity ligands through available thiols, amines, or hydroxyl groups. Copolymers of glycidyl methacrylate can be prepared with other hydrophilic monomers to provide the membrane composition of choice. The terminal epoxy groups on poly(glycidyl methacrylate) membranes also can be hydrolyzed by acid treatment to create terminal diols, which then can be periodate-modified to form aldehydes. The aldehydes can subsequently be used for coupling amine-containing ligands by reductive amination (see the corresponding section in this chapter on reductive amination and also Chapter 3, Section 5, and Chapter 4, Section 4). For instance, Akgöl *et al.* (2002) prepared a membrane consisting of poly(hydroxyethyl methacrylate-co-glycidyl methacrylate) that contained hydroxyl groups from the pHEMA component and epoxy groups from the polymerized glycidyl monomers (Figure 15.26). This membrane was used to couple the enzyme cholesterol oxidase for use as an immobilized reactor. In addition, Teke and Baysal (2007) used a base membrane consisting of nylon-6 (polyamide) and then grafted glycidyl methacrylate onto it using benzophenone as an initiator to create a reactive epoxide-containing membrane, which was used for coupling urease.

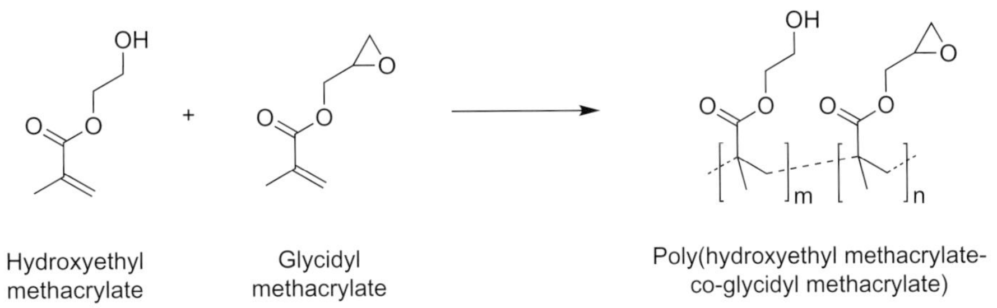

FIGURE 15.26 The structure of a poly(hydroxyethyl methacrylate-co-glycidyl methacrylate) membrane.

Polymer grafting onto existing substrates has been an important route to the creation of reactive groups or functional groups on the surface of a non-functionalized membrane. Grafting onto a block copolymer substrate can introduce novel characteristics into a membrane that normally could not be created through the original membrane synthesis (Koguma *et al.*, 2000). The process of grafting can create new polymer arms or tentacles coming off of the original membrane surface (Figure 15.27). For instance, the grafting of new polymer strands onto the internal structure of porous membranes has been used to increase the coupling capacity of membranes for immobilizing affinity ligands (Muller, 1990; van Reis and Zydney, 2007). Covalent grafting can be done *via* solution-phase polymerization of vinyl monomers in the presence of a membrane substrate and an initiator. Many routes to initiating this type of grafting reaction have been described and some of them can be used to graft synthetic polymers onto natural polymer surfaces (Jenkins and Hudson, 2001). Once the polymerization reaction is initiated, a new graft copolymer will be formed within the solution and can be attached to the existing membrane through any remaining vinyl groups left over from its initial synthesis. Most often, however, linking groups have to be created on the original block polymer surface to facilitate covalent attachment of the new polymer. Covalent graft copolymer modification of membranes and surfaces may use irradiation (i.e., using electron beam or Co^{60}) to form radicals on the existing substrate, which can provide initiation sites for graft polymer growth (Gürsel *et al.*, 2008). New polymer grafts may also be created by linking a polymer tentacle to a membrane using traditional conjugation chemistry. Gacal *et al.* (2006) used the Diels–Alder chemoselective ligation reaction (Chapter 17) to covalently attach PEG or poly(methyl methacrylate) polymers to a polystyrene substrate, thus creating novel surfaces with completely different properties from that of the original polymer. Noncovalent grafting also can be performed through the adsorptive interaction of a functional polymer with the existing membrane substrate, which instead of forming polymer arms, forms a layer of the new polymer onto the initial membrane, in some cases filling the inner pores (Yang *et al.*, 2011). This type of composite membrane is similar in construction to the "gel-in-a-shell" structure of agarose-filled ceramic particles (described previously in this chapter).

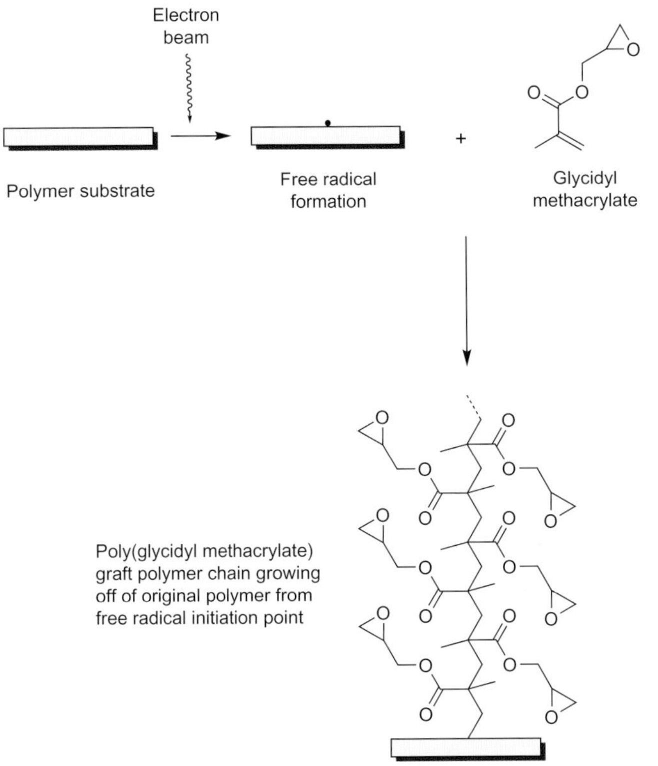

FIGURE 15.27 General illustration of grafting onto an existing polymer substrate. Many different polymer types can be grafted onto substrates to create new properties or add functional groups for coupling affinity ligands.

Membrane-affinity supports have also been formed by the combination of porous beaded chromatography particles, which have been entrapped or grafted within the tangled polymer strands of a membrane (Borneman, 2006). This composite construction of porous beads within a membrane scaffold creates a chromatography support having the high potential convective flow rates characteristic of polymeric membranes plus the high capacity and increased surface area present within the internal pores of particles. Particle-loaded nonwoven fibrous membranes were developed by 3M (Markell *et al.*, 1994) and commercialized under the trade name Empore. These membranes contain either derivatized polystyrene particles or silica particles within an inert 10% polytetrafluoroethylene (PTFE) matrix and are used primarily for solid-phase extraction (SPE). Another fibrillated membrane/particle construct was made by incorporating 3M Emphaze beads (UltraLink, see previous section, this chapter) into a PTFE membrane, which provides azlactone-reactive groups for coupling affinity ligands (Haddad *et al.*, 2010). A similar membrane-entrapped particulate technology was developed by Millipore and used to produce chromatography supports in pipette tips (Zip Tips) (Kopaciewicz *et al.*, 2000). In addition, particle-embedded membranes have been used for the immobilization of enzymes to use as reactors, forming supports of high activity due to the capacity enhancement made possible by the entrapped particles (Ulbricht, 2006). An SEM picture of a nanoparticle-embedded membrane was shown previously in Figure 15.24 of this chapter. All of these constructs provide the ease of use and stability that are typical of membranes with the higher capacity of packed bed chromatography supports. Unlike packed beds, however, the particle-loaded membranes maintain better matrix stability by holding the particles in place and not being susceptible to bed disruption, which can lead to non-uniform flow characteristics.

Particle-loaded membranes have been used in affinity separations by immobilizing ligands on the embedded chromatography particles. Virtually any affinity separations that can be performed using beaded supports can be accomplished using particle/membrane composites. For instance, immunoaffinity applications using immobilized antibodies have been used successfully in SPE techniques to target and capture small antigens out of complex solutions, such as drugs in serum samples (Dombrowski *et al.*, 1998; Delaunay *et al.*, 2000).

Inorganic membranes also can be used for affinity separations. Membranes made of porous inorganic materials have been prepared from the same base compositions as beaded chromatography supports, including porous glass, silica, alumina, and ceramic materials. Many of these membranes are more rigid than those made from polymeric materials and can be molded during construction to fit the shape of a desired flow path or device. An exception to the usual rigidity of most inorganic membranes is glass fiber materials, which often are used in filtration applications and come in flexible disks of all sizes. Inorganic membranes may be surface modified to contain functional groups or reactive sites for coupling affinity ligands. The methods used to introduce sites for covalent coupling on inorganic substrates often make use of functional silanes to form a polymeric network containing side chains, which terminate in the functional group or reactive group of choice. These reagents and their uses are described in Chapter 13.

Monolithic Supports

Chromatographic monolithic supports are similar in construction to membranes, but rather than polymerizing monomers into a thin membranous sheet, the polymerization process forms a single continuous matrix that is contained within a three-dimensional shape, which acts as a mold. The final monolith thus can be constructed within a column shaped like a cylinder for standard chromatography purposes or formed within another type of mold that would result in a custom matrix shape. Early pseudo monoliths were constructed from stacked membrane sheets or from rolls of membranes, which formed larger beds of a discontinuous matrix. True monolithic supports, however, are constructed of a homogeneous porous matrix that takes on a desired shape dependent upon the mold in which it is contained. One of the first references to the name *monolith* with regard to a chromatography support was by Noel *et al.* (1993), who they prepared a continuous matrix of cellulose that took on the consistency of a sponge-like material.

However, explorations into the use of a continuous-phase matrix for chromatography separations occurred as early as the late 1960s and early 1970s, wherein uniform separation media were prepared from such materials as pHEMA, poly(ethyleneglycol methacrylate), foamed polyurethane, and other polymers for potential use in separations (Kubin *et al.*, 1967; Ross and Jefferson, 1970; Hileman *et al.*, 1973). Since that time, a wide variety of molded polymers have been used to form monolithic porous polymers useful for all types of separations, including immobilized affinity ligand applications (Svec and Fréchet, 1999; Svec *et al.*, 2003; Neff and Jungbauer, 2011; Shin *et al.*, 2011; Arrua *et al.*, 2012).

An example of a commercial monolith support is made by BIA Separations (Austria). The "Convective Interaction Media" produced by BIA is made in the form of disks and cylinders (Champagne *et al.*, 2007). The composition of the polymer monolith consists of a homogeneous unit made of polymethacrylate derivatives (Jungbauer *et al.*, 2008). The polymerization process produces large through-pores of 1.5-μm diameter with interconnecting pores throughout the support, which permits interactions between even the largest proteins or biological macromolecules within the media (Zmak *et al.*, 2003; Etzel and Bund, 2011; Neff and Jungbauer, 2011). The BIA monolith supports are available in popular affinity derivatives, including protein A,

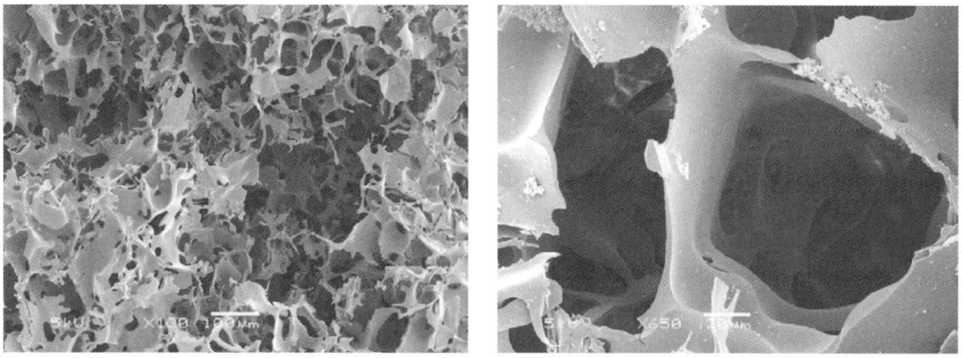

FIGURE 15.28 SEM images of the supermacroporous structure of a Protista monolithic cryogel support. Reprinted with permission from Arvidsson, P., Plieva, F.M., Savina, I.N., Loxinsky, V.I., Fexby, S., Bulow, L., Galaev, I.Y., and Mattiasson, B. (2002) Chromatography of microbial cells using continuous supermacroporous affinity and ion-exchange columns. *J. Chromatogr.* 977: 27–38; copyright © 2002 Elsevier.

protein G, and metal chelate. In addition, an epoxy-activated version is offered for direct immobilization of amine-containing affinity ligands as well as a carboxy and an amino form for coupling of custom affinity molecules.

Protista International AB also commercialized a monolithic chromatography matrix that contains unusually large pore structures. This "supermacroporous" support was produced by the polymerization of monomers in aqueous solution within a column at temperatures that cause frozen sections to develop within the forming polymer (Arvidsson *et al.*, 2002, 2003). The freezing of the aqueous regions causes very large pores to form consisting of cavernous spaces of 10- to 200-μm diameter throughout the "cryogel" monolith, thus creating a series of connected through-pores for rapid convective flow within the support (Arvidsson *et al.*, 2002, 2003) (Figure 15.28). A variety of monomers can be used to create the monolith, including derivatives containing ion-exchange groups, functional groups for modification, or reactive groups for immobilization of affinity ligands. A hydrophilic hydroxylic monolith was created using 2-hydroxyethyl methacrylate to form a cryogel containing abundant hydroxyl groups for subsequent activation and coupling (Savina *et al.*, 2007). Other cryogel compositions have also been formed using other acrylamide or methacrylate monomers (Persson *et al.*, 2004; Savina *et al.*, 2006; Plieva *et al.*, 2008).

The porous structure of the cryogel monolith even allows cell separations to be performed directly within the monolithic support using immobilized affinity ligands specific for certain cell types. For instance, Dainiak *et al.* (2005) used an IMAC support to separate different microbial cells by their ability to bind to the metal chelate ligands. Similarly, Kumar *et al.* (2003) used immobilized protein A to capture lymphocytes within a monolithic cryogel support, and Arvidsson *et al.* (2002) used immunoaffinity and anion-exchange monolith supports to capture microbial cells (Figure 15.29).

Polymeric monolithic supports are chemically similar to the corresponding membranes or spherical beads made from the same polymers—only the shape and size of the support are different. Therefore, the techniques used to activate and immobilize affinity ligands on a monolith's polymer surface are nearly identical to those used for other support materials of equivalent chemical composition. For an illustration of the polymeric structures of typical monolithic supports, see the previous overview of beaded chromatography supports in this chapter. The major difference between a monolith and a beaded support in the methods of ligand coupling is how the matrix is handled and washed during the activation and coupling reactions. Particulate supports can be manipulated as suspensions and washed after each reaction step using filtration devices, but a monolithic support is usually handled directly within the column or mold in which it was formed. For instance, if a column containing a porous monolith consisting of a pHEMA matrix is to be coupled with a protein affinity ligand, then each step in the immobilization process would be carried out by pumping the reagents through the column in a chromatography mode involving a series of sequential additions, incubations, and washing steps. This chromatographic reaction scheme must be carried out carefully to ensure that all parts of the monolith are treated equally and end up with the same density of reactive groups and immobilized ligands. This is not a trivial point, because reactants are added in the mobile phase and are pumped through from one end of the column and out the other end. As the reactions occur, the concentration of the reactants change as the solution passes through the matrix. This is true even in radial flow chromatography operations where the mobile phase travels from the center of a cylindrical shaped monolith through the side walls. In general, as reactions take place within the monolithic support, the

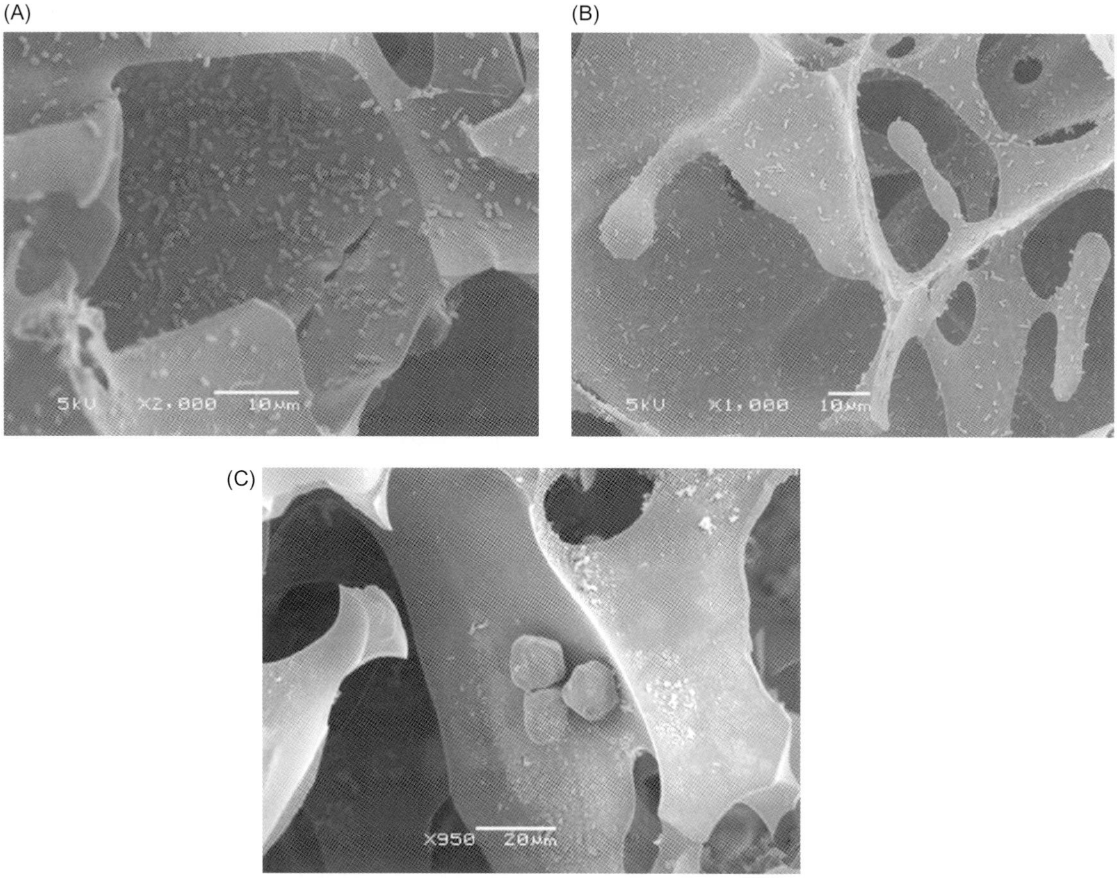

FIGURE 15.29 SEM images of supermacroporous monolith supports capturing *E. coli* cells using an anion-exchange group on the matrix (A and B) and lymphocytes from blood samples using immobilized protein A (C). Reprinted with permission from Arvidsson, P., Plieva, F.M., Savina, I.N., Loxinsky, V.I., Fexby, S., Bulow, L., Galaev, I.Y., and Mattiasson, B. (2002) Chromatography of microbial cells using continuous supermacroporous affinity and ion-exchange columns. *J. Chromatogr*. 977: 27–38, copyright © 2002 Elsevier; and Kumar, A., Plieva, F.M., Galaev, I.Y., and Mattiasson, B. (2003) Affinity fractionation of lymphocytes using a monolithic cryogel. *J. Immunol. Meth*. 283: 185–194, copyright © 2003 Elsevier.

reactant concentration in the mobile phase will decrease as flow travels through the support and any byproducts of the reactions that are released into the solution will increase as the reagent front passes through.

Thus, the rates and yields of the activation and coupling reactions will decrease as the mobile phase flows through the monolith polymer. This could potentially result in dramatic differences in the amount of activation and ligand coupling that occurs from the top of the monolithic column bed to the bottom, if consideration of this effect was not taken into account. To overcome this phenomenon, enough reactant solution must be passed through the matrix at each step in the activation and coupling process to fully saturate the monolithic support from top to bottom. One method of accomplishing this is to prepare a large excess of reagent solution and continually recirculate the solution through the column until the reaction period is complete. Alternatively, an excess of reactant solution may be passed through the monolith and then the support allowed to incubate in the solution of the reactant for an extended time to fully drive the reaction to completion. This process also can be used when coupling affinity ligands to an activated monolith, in which the support is gently mixed in the ligand coupling solution after an initial amount of ligand has been passed through the column to saturate all the internal pore structures. This is the general strategy recommended by BIA Separations for immobilizing ligands on an epoxy-activated monolith made from poly(glycidyl methacrylate).

2. ACTIVATION AND COUPLING OF AFFINITY LIGANDS TO CHROMATOGRAPHY SUPPORTS

The matrix and activation choices available for the preparation of immobilized affinity ligands are two of the most important considerations for creating an optimal affinity support. The selection of an affinity ligand is obviously equally important, but for this discussion it is assumed that the target molecule to be captured and the affinity ligand to be immobilized have already been identified by the user. Matching an affinity ligand with the best possible matrix material as well as choosing the optimal coupling strategy for immobilizing the ligand are both critically important factors for ensuring high specificity and low nonspecific binding in the final chromatography support.

One of the first criteria in choosing an appropriate immobilization strategy is that the method used to activate and couple a ligand to a given support must be compatible with both the ligand and the matrix. In one regard, this means that the chemistry of activation must be chosen to be compatible in its reactivity with the type of functional groups present on the support material. For example, if the support is a polymer that contains abundant hydroxyl groups, which is a common characteristic for many matrices (see previous section), then an activation method should be chosen that will yield an activated intermediate based upon the presence of an initial hydroxyl. In addition, the reactive intermediate created from activating the support must be appropriate for coupling to the available functional groups on the ligand to be immobilized. For instance, an electrophilic reactive group that spontaneously couples to an amine may be an optimal choice for matrix activation when immobilizing an amine-containing affinity ligand, such as a protein.

Another important consideration is the reaction medium that is used to activate the chromatography support. Some activation methods require nonaqueous conditions in pure organic solvents, which potentially may harm a support material by causing collapse or shrinkage of particles with consequential changes to internal pore structures. Determining if a desired support can be exchanged from an aqueous environment into 100% solvent and back again without severe shrinkage or damage often is a prerequisite to deciding on an optimal activation strategy. For instance, a cross-linked agarose support can be washed into anhydrous acetone (or other water-miscible solvents) for activation and taken back into aqueous coupling buffer for immobilization of a protein without irreversible damage occurring to the particles. However, a beaded polyacrylamide matrix that was initially swollen in water will collapse by being exchanged into acetone, which will so severely constrict the pores within each particle that activation and coupling cannot occur efficiently throughout the internal regions of the support.

When choosing an activation method, consideration also should be made to residual functional groups left over after activation and coupling of an affinity ligand. Most activation chemistries alter the initial characteristics of a matrix to form the intermediate reactive groups. If this activated intermediate successfully couples with a ligand, then the desired covalent bond will form, which often replaces part of the reactive group with the coupled ligand. However, if ligand coupling does not occur at every reactive site on the support, then there will likely be reactive groups or hydrolysis products remaining after the reaction. The presence of such residual groups, if they are active, will still have coupling potential toward molecules in a sample when the affinity support is put to use. If any reactive groups remain after immobilizing an affinity ligand, the potential exists to covalently couple molecules that are not desired to be bound to the support and thus create sites for nonspecific interactions. Additionally, if hydrolysis of reactive groups occurs at the same time as ligand coupling, then another functional group may be formed in the process which represents the hydrolysis product. In some cases, this may generate charge characteristics, such as negatively charged carboxylates being left over after a reaction. Residual charged groups may cause nonspecific interactions with sample molecules independent of the desired affinity interactions. For these reasons, there is typically a blocking step associated with the ligand-coupling process that eliminates any residual reactive groups or blocks potential sites of nonspecific binding by capping them with a nonrelevant molecule. In some instances, the blocking of residual charged groups may not be possible after ligand coupling, because the ligand itself may have similar groups present and therefore would be modified in the process. In the end, the most important factor is to consider all of the reactions involved with support activation and ligand coupling to decide the best combination of reactions to produce an optimal affinity support.

Other important things to consider when choosing an activation method include the hazardous nature of the reagents required (personal safety and waste disposal issues), the reaction rate and yield when coupling a ligand, whether the reaction is reproducible and controllable under the chosen conditions, and if the process can be scaled up and performed in large batch sizes with similar efficiency to bench-scale experiments. Some activation and coupling methods work well in small scale, but are difficult to handle and control in process-scale operations. For instance, cyanogen bromide activation, despite its hazardous nature, is relatively simple to perform in a fume hood using milliliter quantities of a support; however, attempts to scale up this process to activate many liters of a support can result in significant problems related to environmental hazards and the difficulties in rapidly activating, washing, and using the activated resin. Therefore, each prospective activation strategy should be carefully considered as to its relative ease of use and scalability in production before deciding upon a method of choice for coupling a desired affinity ligand.

In addition, the quantity of the support to use in the activation and coupling reactions is dependent on how much target molecule ultimately needs to be captured when using the affinity support in a chromatographic separation. The amount of support to use is also governed by how much affinity ligand is available for coupling. For abundant affinity ligands, an excess of affinity support can be prepared without significant cost considerations; however, for expensive ligands such as some primary antibodies or recombinant proteins, the ligand cost and availability may be the governing factors in deciding how much of an affinity support to prepare. Most of the protocols in this section involve the activation of from 1 ml to 100 ml of a support material. The amount of material actually used to produce a desired immobilized ligand may be varied from the given protocols simply by proportionally adjusting the reagents used in the reactions.

If a protein is to be immobilized onto a matrix, ideally it should be dissolved in the coupling buffer at concentrations of at least in the low mg/ml levels. Many proteins to be used as affinity ligands perform well if the immobilization reaction is performed at concentrations of about 3 to 6 mg/ml. In some cases, when maximal binding capacity is desired for the affinity support to capture a target molecule, a protein ligand may be coupled at concentrations much higher than this range, even >10 mg/ml. Most coupling reactions usually provide high yields of protein immobilization when using concentrations of up to about 20 mg/ml during the immobilization reaction. Immunoglobulin-binding proteins such as protein A are typically coupled to chromatography supports at the high end of this concentration range (>6 mg/ml) to maximize the binding capacity of the resultant affinity support for purifying antibodies.

However, regardless of the type of affinity support being prepared, it is recommended that experimental optimization be done for the coupling ligand concentration and the reaction conditions to provide the best possible affinity support performance for binding a desired target molecule. If some proteins are immobilized at too high a density on the support, then non-specific binding may occur toward other molecules in a sample, or alternatively it may result in difficulty in eluting a bound target molecule due to extremely high affinity or avidity effects. Conversely, if too low a quantity of protein ligand is immobilized, the capacity and efficiency of binding to the desired target molecule may be unacceptably low. Of course, if only a small amount of a ligand is available for coupling, then using reaction concentrations down to the μg/ml levels may be carried out with success, but along with the realization that the resultant binding capacity of the support will be lower as well.

If a small amine-containing ligand is to be immobilized on the support, its concentration in the coupling buffer should be at least three times greater than the level of active groups on the matrix to obtain decent yields. Many activated supports may contain about 20 to 40 μmoles/ml gel of reactive groups; therefore a small amine-containing ligand might be reacted at a level of 60 to 120 μmoles/ml gel. Higher concentrations will result in higher yields of coupling, but optimization of the resultant ligand density should be carried out to obtain the best performance of the affinity support in its intended application.

A slightly different situation results if a diamine spacer molecule is to be coupled to an activated support to result in a free terminal amine. To avoid the reaction of the support with both ends of the linker the molecule should be dissolved in the coupling buffer at very high concentration. For instance, a 1.0- to 1.5-*M* concentration of ethylene diamine will maximize coupling yield while preventing both ends of the compound from coupling to the support.

It is interesting to note that when coupling small amine-containing ligands or larger proteins to microparticles or nanoparticles (see Chapter 14) the most frequently used immobilization technique involves activation of carboxylates on the particle surface using the carbodiimide EDC (with or without the addition of NHS or sulfo-NHS), which is then immediately used to react with the ligand. However, for larger particles used in chromatographic separations, the carbodiimide reaction is typically applied only when immobilizing small amine-containing molecules or diamine spacers, and then only in cases in which the molecules do not also contain carboxylates and thus have no potential for polymerization during the reaction. Most coupling methods involving chromatographic supports make use of a much wider variety of activation and coupling reactions to facilitate the immobilization of affinity ligands than those typically used with microparticles or nanoparticles. This statement does not mean that the methods described in this chapter cannot be applied to small particles—indeed they can with the appropriate changes to accommodate the differences in handling and washing, which typically involve centrifugation or magnetic separations as opposed to simple bulk filtration as can be done with chromatography supports. The reactions discussed in this chapter can be applied successfully to any immobilization requirement, whether the solid phase is a planar surface, a microparticle, a nanoparticle, a membrane, or a larger porous bead.

This section describes the activation methods commonly used to create reactive groups on chromatography support materials. All of these reactions have been used successfully for many years and have abundant publications that describe the coupling of specific affinity ligands of all types. This section also contains suggested ligand coupling protocols following each activation procedure. All of these reactions should be performed with caution. Care should be taken to use the appropriate personal protective equipment to guard against exposure to potentially toxic compounds or flammable solvents. Most activation methods should be carried out in a fume hood wearing gloves and a lab coat as a minimum. Occasionally, a full-face shield or an organic mask also should be used to prevent chemical exposure to hazardous compounds.

2.1. Amine-Reactive Immobilization Methods

Periodate Oxidation and Reductive Amination

One of the most popular methods of immobilizing affinity ligands on chromatography supports involves the formation of aldehyde groups on the matrix followed by a reductive amination coupling reaction to covalently link amine-containing molecules. Sodium *meta*-periodate is an oxidizing agent that can be used to create aldehydes on chromatography supports that contain diols. Hydroxyl groups present on adjacent carbon atoms are a common feature in many resins made from polysaccharides, such as beaded agarose, cellulose supports, or cellulosic membranes. Diols may also be formed on synthetic polymeric support materials by careful choice of the monomers used during manufacture or from the selected chemical modification of a support after production. For instance, glycidyl methacrylate monomers can be used to prepare chromatography matrices that will yield epoxide groups in the final beaded support or membrane. Hydrolysis of the epoxides under acidic conditions will generate terminal diols on the carbon atoms that originally made up the epoxide rings. In addition, the crosslinking of some support materials with epoxide-containing agents, such as the use of epichlorohydrin or 1,4-butanediol diglycidyl ether to stabilize and strengthen beaded agarose supports, may create additional periodate-oxidizable diols from leftover, hydrolyzed epoxide groups (Figure 15.30).

Diols can even be formed on support materials containing non-periodate-oxidizable hydroxyl groups through modification with the small epoxide-containing compound glycidol (Hoyer and Shainoff, 1980; Shainoff, 1980; Guisán, 1988; Guisán *et al.*, 1997). The epoxy group of this compound reacts with hydroxyl groups under alkaline conditions to form an ether bond with the associated formation of a terminal diol (Figure 15.31). Reaction of glycidol in excess with hydroxylic supports results in the polymerization of the compound within the matrix through continued reaction with the hydroxyls produced as the epoxide couples. This may form branched polyglycerol chains in which each branch terminates in a diol, the formation of which dramatically increases the hydrophilicity of the matrix while forming the requisite periodate oxidizable components to create aldehydes for coupling affinity ligands. The amount of glycidol used in this process determines the ultimate size of the polymers formed and the potential affect on the internal porosity of the support, which can be significantly reduced even in highly porous gels (Eriksson, 1987). Glycidol modification has been used to immobilize and stabilize enzymes through multipoint attachment to the matrix, and the level of modification and subsequent activation to aldehydes can be controlled to produce between about 7 and 70 µmoles of reactive groups/ml gel (Suh *et al.*, 2003; Tardioli *et al.*, 2011; Nwagu *et al.*, 2012).

Periodate oxidation of diols proceeds by cleavage of the associated carbon–carbon bond with concomitant oxidation of the hydroxyl groups to aldehydes. If the diol consists of secondary hydroxyls present within the sugar rings of a polysaccharide polymer, it will result in two aldehydes being formed per oxidation event along with scission of the ring; however, if the diols are terminal—in other words, containing one primary hydroxyl group adjacent to a secondary hydroxyl, such as those created from epoxide hydrolysis or glycidol modification—an oxidation reaction would yield one molecule of formaldehyde (released) and one aldehyde remaining on the support at every diol site. In addition, if a series of three hydroxyls occur on neighboring carbons, such as in glucose-containing polymers like dextran, then treatment with periodate would result in the formation of one molecule of formaldehyde (from oxidation and loss of the central carbon atom) and two aldehydes generated from oxidation and cleavage of the glucose ring (Figure 15.32). The oxidation of polysaccharide polymers can be highly complex and has been studied extensively (Rinaudo, 2010). Periodate-oxidized supports prepared in this manner can be washed free of excess periodate (and formaldehyde) and stored for long periods without loss of aldehyde functionality.

Aldehydes (or ketones) can react with amine groups in aqueous solution to create reversible Schiff base linkages, the formation of which is enhanced at alkaline pH values. Schiff base formation is a dehydration reaction that can reversibly undergo hydrolysis in aqueous solution to reform the carbonyl group and the amine. However, Schiff bases can be stabilized and made essentially permanent by reduction to a secondary amine bond (Figure 15.33). Reducing agents that are suitable for this process include sodium cyanoborohydride and certain borane compounds (Borch *et al.*, 1971;

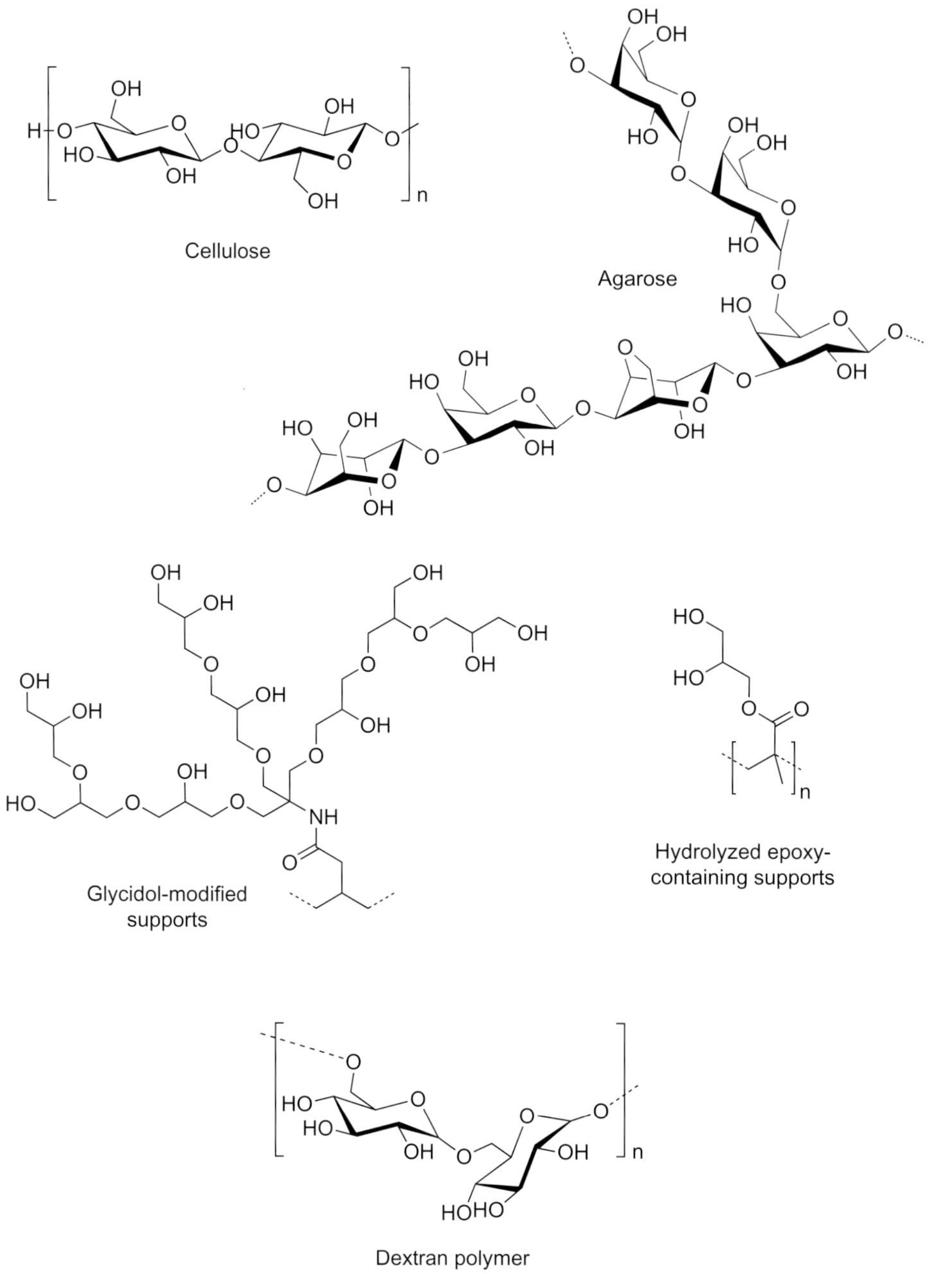

FIGURE 15.30 Polymeric supports and modified supports that contain hydroxyl groups suitable for activation and coupling of affinity ligands.

Jentoft and Dearborn, 1979; Baxter and Reitz, 2002). Stults *et al.* (1989) showed that up to 50 mg of BSA could be immobilized per gram of a crosslinked agarose-beaded support using pyridine borane as the reducing agent. Sodium borohydride may also be used to reduce a Schiff base, but it is about 5-fold stronger as a reducing agent than cyanoborohydride, and thus it could also reduce the aldehyde reactive groups used to create the initial Schiff base and it could reduce and cleave disulfides within protein molecules (Peng *et al.*, 1987). Cyanoborohydride will not normally appreciably reduce carbonyls, except under acidic conditions at pH 3 to 4 (Lane, 1972). Therefore, under slightly more alkaline conditions (pH 6–8), cyanoborohydride can be used to help drive the immobilization reaction to completion—reducing the Schiff base interactions as they form—whereas stronger reductants may also reduce one of the starting reactants, thus potentially reducing yields.

It has been shown that the coupling efficiency and the resultant stability of the reductive amination process for immobilizing proteins on agarose supports

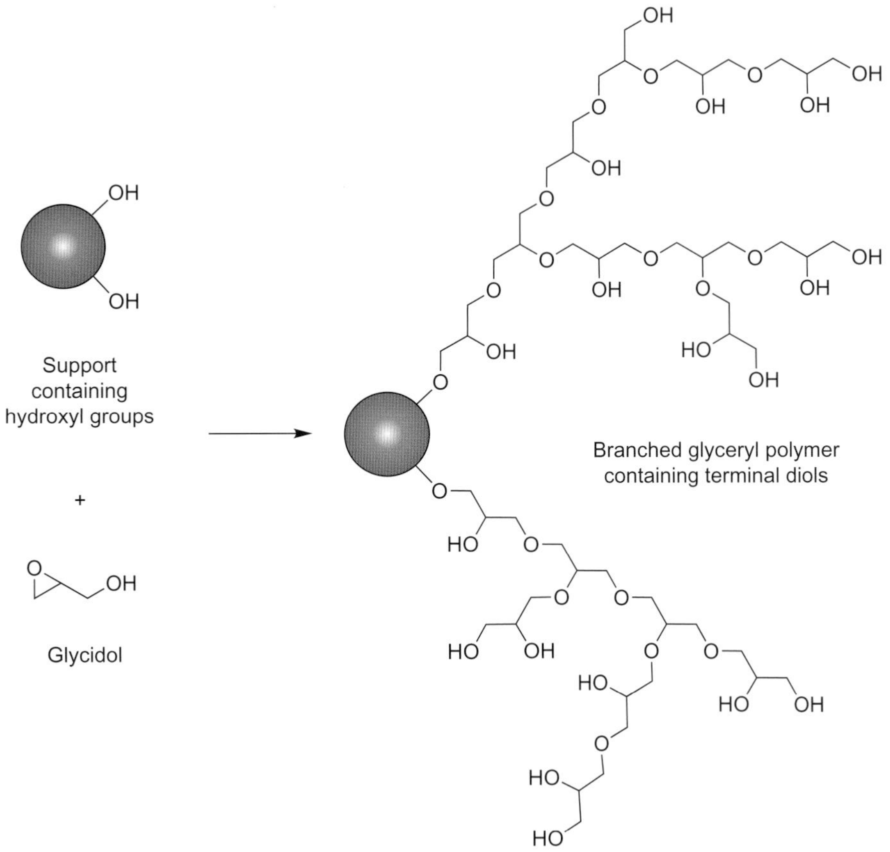

FIGURE 15.31 Glycidol reacts with hydroxyl-containing supports in a ring-opening process to create ether linkages, which in turn create additional hydroxyl groups. Glycidol can polymerize within the matrix to form branched polymers wherein each branch terminates in a diol, which may be periodate oxidized to form aldehyde groups.

are both excellent (Domen *et al.*, 1990). Proteins having abundant amino groups for immobilization typically have high coupling yields, which can exceed 95%. Yields can be enhanced by forming the initial Schiff base in a reaction medium having a more alkaline pH of 9 to 10 and then reducing the pH toward neutrality for subsequent reduction of the iminium double bond (Hornsey *et al.*, 1986). In addition, Ruzicka *et al.* (2006) showed that reductive amination reactions can be complete in just several minutes instead of the typical recommended reaction times of hours. They also contended that the reduction step with cyanoborohydride is unnecessary, as protein ligands immobilized *via* Schiff base formation at high pH appeared to have little ligand leakage even when treated with 0.1-*M* HCl. Despite these results on microliter quantities of support materials, care should be taken when coupling using much larger volumes of supports to determine the optimal reaction time for each step in the coupling process. Mixing and diffusion rates within bead slurries of quantities from hundreds of milliliters to multiple liters of gel can be quite different from reactions carried out in a micro flow cell with microliter amounts of gel. In addition, while Schiff base interactions can be more secure due to increased stability gained from multipoint attachment with proteins—given that the probability of all the associated Schiff base linkages hydrolyzing at once and releasing a protein molecule is low—it is still highly recommended that stabilization of these linkages be done through reduction to prevent the possibility of hydrolytic leakage of ligand over time. In particular, if an affinity ligand is immobilized through only a single amine to form an iminium Schiff base group, then failure to reduce that bond will result in significant ligand leakage over time by hydrolysis.

The following protocols involve various strategies to form the reactive aldehyde groups on support materials as well as the suggested coupling reactions for immobilizing amine-containing ligands. Note: The use of sodium cyanoborohydride is potentially toxic and dangerous. All operations should be performed in a fume hood while wearing appropriate personal protective equipment.

Periodate Activation Protocol

This protocol is designed to create aldehyde groups on diol-containing chromatography supports through

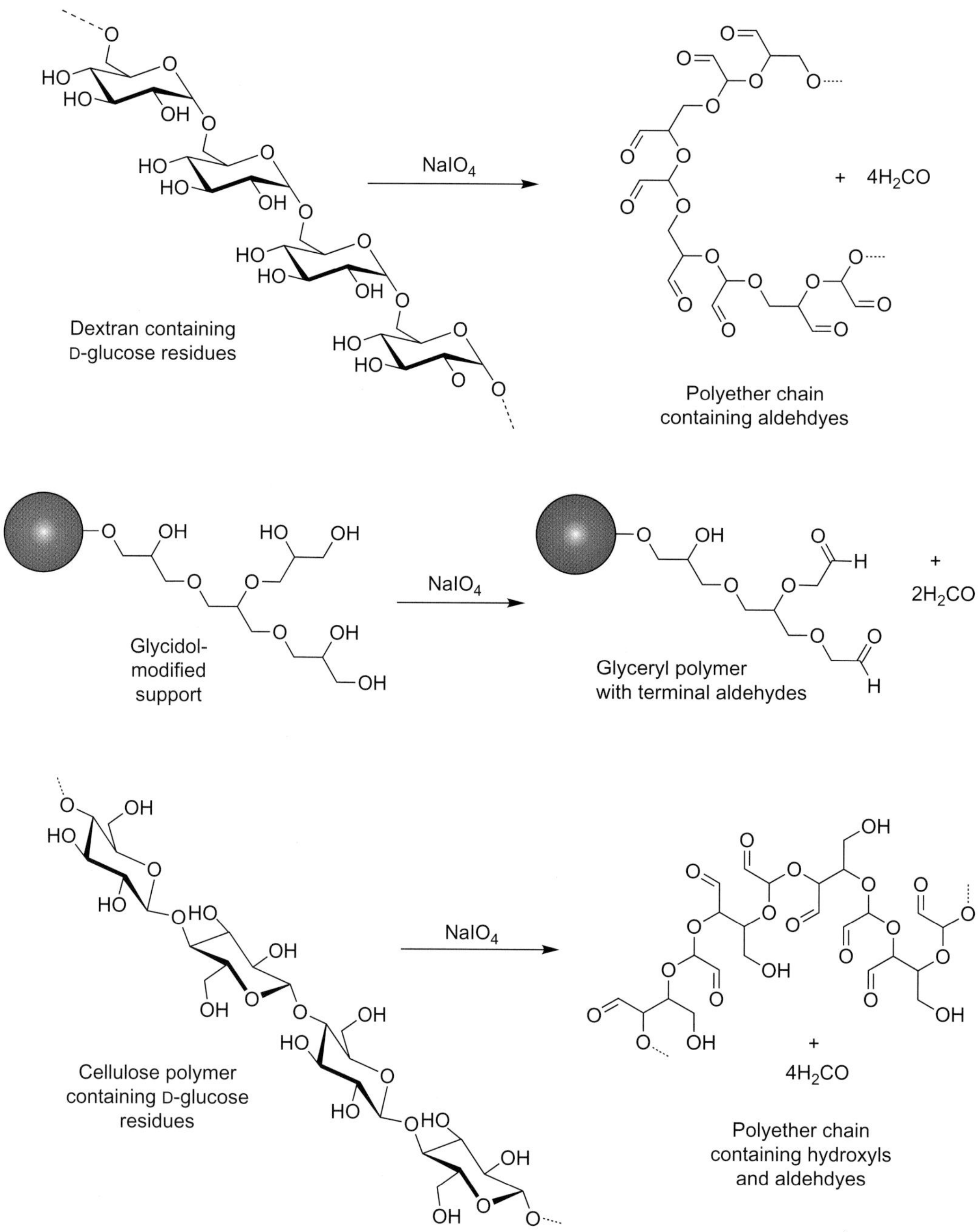

FIGURE 15.32 Diol groups in support materials may be oxidized with sodium periodate to form aldehydes, which can be used to immobilize amine-containing ligands by reductive amination. Saccharide residues containing diols, such as within dextran or cellulose polymers, will undergo ring cleavage and opening along with the formation of two aldehyde groups per sugar. Glycidol modified supports will undergo oxidation with the formation of a single aldehyde at the end of each branched-chain polymer.

oxidation with sodium periodate. This is useful for the activation of crosslinked agarose supports, cellulose- or dextran-containing beaded supports or membranes, and matrices that have been modified with glycidol or contain a hydrolyzed epoxide to create diol groups. Changes to the quantity of support material used from that described here may be done with corresponding proportional changes to the amounts of other reagents used.

1. Wash thoroughly 100 ml of a diol-containing chromatography support material (i.e., crosslinked agarose, dextran, cellulose, or a glycidol-modified support material) with at least several bed volumes of deionized water to remove any preservatives or other contaminants. The use of a sintered glass filter funnel works well for this purpose. Drain the support to a wet cake.

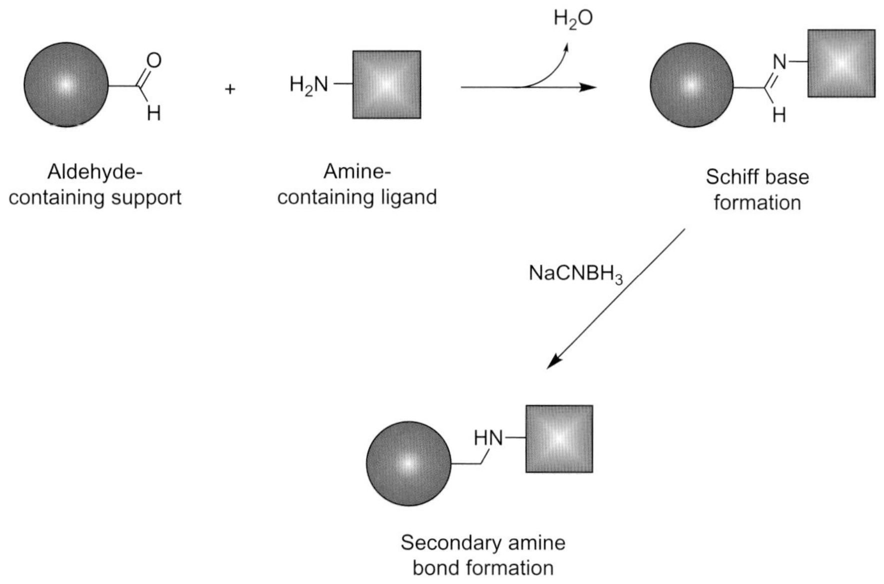

FIGURE 15.33 Aldehyde-containing supports can be reacted with amine-containing ligands in the presence of a reducing agent such as sodium cyanoborohydride to couple the ligand through a secondary amine bond.

2. Dissolve 4.28 g of sodium periodate ($NaIO_4$; MW = 213.89 g/mol) in 100 ml of water to prepare a 0.2-*M* solution. Mix the washed gel cake into the periodate solution to create a uniform suspension of the particles. Notes: At this level of periodate addition, the method will create a maximal level of aldehyde groups within the support as the internal diols are oxidized. If lower levels of aldehyde production are desired, then the amount of periodate addition can be controlled to provide this result. For instance, Guisán *et al.* (1997) determined that to create a support containing about 5 μmoles aldehydes/ml gel from a glycidol modified crosslinked agarose support (see the protocols which follow for glycidol modification), one could add a more dilute concentration of periodate to the modified support. To achieve similar moderated activation levels in this protocol, instead of using the recommended levels of periodate in step 2, dilute 5 ml of 0.1-*M* sodium periodate solution with 995 ml of water and then add this solution to the 100 ml of wet gel cake prepared in step 1. Conversely, to form about 75 μmoles aldehydes/ml gel on a glycidol modified agarose support, add 175 ml of 0.1-*M* sodium periodate to 825 ml of water and then add this solution to the wet gel cake. Stir these solutions for 60 min (for low-level activation) or 90 min (for higher activation) at room temperature to oxidize the supports to the desired level of activation, and then wash as in step 4. Another strategy that may control the level of aldehydes formed on a support is to regulate the initial glycidol modification reaction to control the amount of diols formed (Guisán, 1988).
3. Continue the reaction for 90 min at room temperature with constant mixing using an overhead paddle stirrer or a rotating mixer. Do not use a magnetic stir bar, as this will damage beaded supports. If oxidizing a membrane, the support material should be gently agitated throughout the reaction in a bath of the periodate solution. Periodate oxidation takes place rapidly to form aldehydes within the matrix, and such reactions on a small scale may be complete in as little as 10 to 15 min. For larger quantities of gel, longer reaction times may be appropriate to ensure complete mixing of the oxidant throughout the support. For some resins, such as non-crosslinked agarose or cellulose, extended oxidation may cause damage to the structure of the particles or may actually dissolve them. For such supports, a very brief 5- to 10-min exposure to periodate may aid in maintaining the matrix structure while still forming sufficient aldehyde groups for immobilization of ligands. For extremely sensitive supports, scale back both the concentration of periodate and the time of reaction.
4. Wash the oxidized support with at least 10 bed volumes of deionized water to stop the reaction and remove excess periodate and any formaldehyde that may have been formed during the reaction. The aldehyde groups created on the support by this process are stable in aqueous solution indefinitely if a preservative is added to the slurry to prevent microbial growth. Store the activated gel until use at 4°C as a 50% (v/v) slurry containing a preservative.

Use of Glycidol to Form Periodate-Oxidizable Diols

The following protocols can be used to modify primary or secondary hydroxyl groups on support materials to create diols susceptible to periodate oxidation. They can be used to modify supports that do not natively contain periodate-oxidizable diols, which are necessary to form aldehydes using oxidation. Polymeric support materials that can be used with success in this process include Trisacryl resins (containing primary hydroxyls from Tris-containing monomers) and Toyopearl resins (containing hydroxyls from pentaerythritol-containing monomers). In addition, other primary or secondary hydroxyl-containing supports such as those made from HEMA, PVA, and ethylene glycol methacrylate (EGMA) are candidates for modification with glycidol to create periodate oxidizable diols. Glycidol may also be used to modify amine-containing supports in a similar manner. The benefits of using hydrophilic modifications with glycidol to form polyglycerin spacers within a matrix for the coupling of affinity ligands have been documented (Rhemrev-Boom, 2009). Hyperbranched polyglycerols made from controlled polymerization of glycidol can form dendritic molecules as well as nanoparticles or microparticles containing pendent diols, which can be activated by periodate to form aldehydes for coupling ligands (Sunder *et al.*, 1999, 2000).

Glycidol has also been used to modify crosslinked agarose despite the fact that agarose already has periodate-oxidizable diols within its matrix structure. In this case, glycidol forms branched polyglycerin tethers that project off the agarose resin and are highly mobile within the porous structure of each particle. Periodate oxidation of the terminal diols on these polyglycerin groups forms aldehydes that can be used to couple enzymes at multiple lysine amino sites on their surface. Such multipoint attachment aids in enzyme stabilization to create immobilized reactors of high thermal stability and reusability for industrial processes (Ichikawa *et al.*, 2002; Ming *et al.*, 2006; Kuroiwa *et al.*, 2008). Site-directed mutagenesis of penicillin G acylase followed by immobilization on glycidyl-agarose resulted in increased stability and activity of the enzyme (Abian *et al.*, 2004). This technique also can be used on hydroxyl-containing monolithic supports, membranes, and on the surfaces of hydroxylic microfluidic channels to form reactive tethers for enzyme or affinity ligand immobilization (Albrecht *et al.*, 2010).

The following methods should be carried out in a fume hood using the appropriate personal protective equipment to avoid exposure to the epoxide-containing reactant, glycidol. The first two procedures describe aqueous reactions of glycidol with hydroxylic supports in the presence of base, while the second one involves a non-aqueous reaction using BF_3 etherate as catalyst for the epoxide reaction in dioxane.

(A) AQUEOUS GLYCIDOL REACTION FOR HIGH DENSITY MODIFICATIONS

The following protocol is designed to create a high concentration of glycidol polymer modifications throughout the hydroxylic support material. This will form numerous terminal diols for subsequent periodate oxidation, but the protocol also will create considerable polymer occlusion within the pore structure of some supports. This procedure has been determined to work well for certain polymeric supports, but the optimal level of glycidol modification for a specific application should be determined experimentally. See method (B) for a reaction that will yield a lower degree of glycidol modification.

1. Wash 100 ml of a support material, for instance Trisacryl GF-2000 or Toyopearl HW-65F, with 500 to 1000 ml of deionized water using a sintered glass filter funnel to remove the preservatives or storage solutions that are present in the commercial products. Finally, wash the gel with 100 ml of 1-*N* NaOH (caution: caustic solution) and drain to a wet cake, but do not allow it to dry.
2. Carefully add the washed gel to 100 ml of 1-*N* NaOH and resuspend it with stirring.
3. Add 100 ml of glycidol solution to the stirring gel suspension along with 1 g of sodium borohydride ($NaBH_4$). Note: With some matrices optimization of the amount of glycidol to add should be done to prevent unacceptable decreases in porosity due to internal glycidol polymer formation within the particles.
4. Continue the reaction overnight with constant stirring using an overhead paddle stirrer.
5. When the reaction is complete, carefully transfer the gel slurry to a sintered glass filter funnel, and wash extensively with water (1–2 l), 1-*M* NaCl (1 l), and again with water. The glycidol-modified gel can be stored until use at 4°C as a 50% aqueous slurry containing a preservative or oxidized immediately with periodate to form aldehyde groups according to the previous protocol.

(B) AQUEOUS GLYCIDOL REACTION FOR MEDIUM-DENSITY MODIFICATIONS

The following protocol will result in glycidol modifications within a hydroxylic support material at a much lower level than method (A) above. This process has been found to work well for the modification of agarose gels to create numerous periodate-oxidizable sites within the matrix and to form a more hydrophilic interior within the support (Guisan, 1988).

1. Wash 100 ml of a support material containing hydroxyl groups—for instance, agarose, Trisacryl

GF-2000, or Toyopearl HW-65F—with 500 to 1000 ml of deionized water using a sintered glass filter funnel to remove the preservatives or storage solutions, which usually are present in the commercial products. Finally, wash the gel with 100 ml of 0.375-*N* NaOH and drain to a wet cake, but do not allow it to dry.

2. Carefully add the washed gel to 100 ml of 0.375-*N* NaOH and resuspend it with stirring.
3. Very slowly add 34 ml of glycidol solution to the stirring gel suspension along with 1 g of sodium borohydride ($NaBH_4$). The glycidol should be added dropwise to prevent the solution from heating beyond about 25°C.
4. Continue the reaction overnight with constant stirring using an overhead paddle stirrer.
5. When the reaction is complete, carefully transfer the gel slurry to a sintered glass filter funnel, and wash extensively with water (1–2 l), 1-*M* NaCl (1 l), and again with water. The glycidol-modified gel can be stored until use at 4°C as a 50% aqueous slurry containing a preservative or oxidized immediately with periodate to form aldehyde groups according to the previous protocol.

(C) GLYCIDOL REACTION IN DIOXANE

An alternative glycidol modification method was described by Eriksson (1987), in which a high-percentage agarose gel that had been crosslinked with divinyl sulfone was further modified with glycidol in dioxane. This process results in a more hydrophilic agarose matrix when supports containing a high percentage of agarose are used (up to 20% agarose).

1. Wash 10 ml of a support material with 100 ml of deionized water to remove preservatives and other storage solution components. Next, sequentially wash the support into dioxane by using slowly increasing concentrations of the solvent in water until the agarose is in 100% solvent. This can be done by washing first with 50 ml of dioxane:water at a ratio of 1:3 (v/v), then washing with 50 ml of dioxane:water at 1:1, then washing with dioxane : water at 3 :1, and finally washing with 100 ml of 100% dioxane to fully remove the last traces of water.
2. Suspend the agarose support in an equal volume of dioxane and stir in a fume hood using a paddle stirrer or a rotating mixer. Add 3 ml of glycidol with mixing followed by 0.25 ml BF_3 etherate.
3. React for 1 h at room temperature with continuous mixing.
4. When the reaction is complete, carefully transfer the gel slurry to a sintered glass filter funnel and wash with 100 ml dioxane followed by a sequential transfer back into aqueous solution by reversing the dioxane:water ratios used in step 1. Finally, wash with 100 ml of water and store the gel at 4°C in an equal volume of water containing a preservative. The glycidol modified support may be periodate-oxidized to contain aldehyde groups according to the protocol described previously.

Immobilization of Ligands Using Reductive Amination

The following protocols are generalized procedures for the immobilization of amine-containing small ligands, spacer molecules, or proteins onto periodate-oxidized support materials, which contain aldehyde residues available for coupling by reductive amination. The reactions should be performed in a fume hood to avoid exposure to hazardous or toxic compounds, especially the cyanoborohydride used in the reduction step. Two methods are described for reductive amination: the first one using a reaction buffer at physiological pH and the second protocol using an initial incubation at pH 10 to form the Schiff bases followed by reduction at pH 7.2. The first method works well for the majority of ligands and should be used for proteins that are sensitive to the more alkaline pH reaction. However, if ligand coupling densities or coupling yields are lower than desired, especially using proteins, the pH 10 protocol typically provides increased coupling efficiency, which can exceed 95% yield.

The sodium cyanoborohydride used for the following reactions may be obtained as the pure solid or as a 5-*M* $NaCNBH_3$ solution in 1-*N* NaOH (Aldrich). The solution is more convenient to use, because the solid may be susceptible to static charge during weighing, which causes fine particles to fly away and increases the hazards of handling the compound. The only caveat in using the alkaline cyanoborohydride solution is that after addition the final pH of the reaction should be checked and adjusted back to pH 7.2 if necessary.

(A) COUPLING AT pH 7.2

1. Wash 100 ml of the periodate-oxidized support material containing aldehyde groups into coupling buffer (0.1-*M* sodium phosphate, pH 7.2) and drain to a wet cake. Other buffer components may be added to the coupling buffer, such as the addition of 0.15-*M* NaCl or alternative buffer salts; however, avoid amine-containing compounds like Tris, glycine, or imidazole, which will compete in the reaction. In general, avoid buffer compounds containing primary or secondary amines as well as any other nucleophile that could react with an aldehyde. In addition, certain detergents and other amphiphilic components

have been shown to decrease the stability of the intermediate Schiff base and these additives should be used only after validating that they have no affect on the immobilization reaction (Viguera *et al.*, 1990). Reductive amination coupling in aqueous solution has been shown to effectively occur between pH 4 and 10, with an optimal range for cyanoborohydride reduction of pH 6 to 8.

2. Suspend the washed support material containing aldehyde residues in an equal volume of 0.1-*M* sodium phosphate, pH 7.2, into which an amine-containing ligand has been dissolved. Notes: For protein ligands, a typical concentration range may be 3 to 5 mg/ml gel, but much higher concentrations may be used if a high density of coupled protein on the gel is desirable. In some cases, reactions containing protein at up to 20 mg/ml gel can be performed while still maintaining excellent coupling yields (>85%). For instance, high-capacity protein A supports often are reacted at levels exceeding 10 mg protein A per ml gel to obtain maximal binding capacity for immunoglobulins. However, optimization of ligand density on the affinity support should always be done to avoid having too much or too little ligand present, which could cause significant nonspecific binding at the high end or unacceptably low capacity at the low end. High densities of immobilized protein could also result in an affinity support in which it is difficult to effect elution of the target molecule after binding. For small amine-containing affinity ligands, a concentration of 2 to 3 mg ligand/ml gel for the immobilization reaction may be sufficient to obtain good binding capacity on the resultant affinity gel. Alternatively, a concentration representing 5 to 10 times the concentration of reactive aldehydes on the matrix may be used to ensure a high density of the final coupled ligand. The ultimate concentration of ligand used in the reaction should be optimized by performing a series of coupling reactions at different initial ligand concentrations and determining the performance of the affinity supports in capturing and eluting the desired target molecule. The coupling of diamine spacer molecules should be done at high concentration (i.e., 1-*M*) to avoid internal crosslinking or bead aggregation during the coupling reaction.
3. Stir the gel/ligand slurry in a fume hood using a paddle stirrer or using a rotator to maintain constant mixing.
4. Add to the gel slurry 0.63 g of solid sodium cyanoborohydride ($NaCNBH_3$; MW = 62.84) [toxic!] or 2 ml of 5-*M* $NaCNBH_3$ in 1-*N* NaOH with stirring. If the alkaline solution of cyanoborohydride is used, check the pH and readjust to 7.2, if required. Continue the reaction for 4 h to overnight at room temperature. The amount of cyanoborohydride added—whether as the pure solid compound or as the solution—results in a 50-m*M* solution in the final reaction slurry, which is in a total volume of 200 ml. A shorter coupling time may be sufficient for many ligands, but the optimal conditions to give acceptable yields for a particular ligand should be determined by doing a series of coupling reactions at different time points. Reactions at 4°C may also be done for thermally sensitive proteins or ligands, but reaction times likely will have to be extended to get the same level of coupling as a room temperature reaction.
5. Transfer the gel slurry to a sintered glass filter funnel (in the fume hood) and wash with several bed volumes of water to remove most of the unreacted ligand and reaction byproducts.
6. Unreacted aldehyde residues remaining on the support should be blocked by the addition of a small amine-containing compound, such as ethanolamine or Tris. Ethanolamine typically works best, and the compound is small enough to react with all remaining aldehydes to form hydrophilic terminal hydroxyl groups on the matrix upon coupling. Wash the support once with an equal volume of 1-*M* ethanolamine, pH 7.2, and then transfer the wet gel cake to a clean flask or vessel used for mixing the reaction. Add with stirring 100 ml of 1-*M* ethanolamine, pH 7.2, along with 0.63 g (or 2 ml of the 5-*M* $NaCNBH_3$ solution) of sodium cyanoborohydride. Readjust the pH if necessary and continue the blocking reaction for 30 min at room temperature.
7. Wash the support extensively with water, 1-*M* NaCl, and again with water to remove all unreacted components from the gel. Additional wash solutions may be utilized to completely remove ligands that may have some nonspecific binding potential and remain noncovalently bound to the immobilized ligand, such as the use of acidic and alkaline washes as well as washes containing denaturants (such as guanidine). After washing, the affinity gel may be stored as a 50% aqueous slurry containing a preservative at 4°C.

(B) COUPLING AT pH 10

This protocol may be used to increase the amount of ligand coupled to the support using reductive amination if the pH 7.2 protocol does not give sufficient coupling yields or does not result in high enough ligand density on the support. In this method, the formation of Schiff bases is first done at pH 10, where they occur more efficiently, and then the pH is adjusted to 7.2 for the reductive amination step. This two-step protocol typically results in much greater coupling yields than the previous procedure, especially when immobilizing proteins.

1. Wash 100 ml of a periodate-oxidized support into coupling buffer (0.1-*M* sodium carbonate, pH 10) and drain to a wet cake. Other buffer components may be added to the coupling buffer, such as the addition of 0.15-*M* NaCl or alternative buffer salts; however, avoid amine-containing compounds like Tris, glycine, or imidazole, which will compete in the reaction.
2. Suspend the washed support material containing aldehyde residues in an equal volume of 0.1-*M* sodium carbonate, pH 10, into which an amine-containing protein or ligand has been dissolved. See the notes in step 2 of the previous pH 7.2 coupling protocol that contains suggestions as to the amount of ligand to add to the coupling reaction.
3. Stir the gel/ligand slurry for 4 h to overnight at room temperature using a paddle stirrer or using a rotator to maintain constant mixing. For sensitive proteins, the reaction may be carried out at 4°C, but may require additional time to form the Schiff bases.
4. Transfer the reaction slurry to a sintered glass filter funnel, drain the gel of excess solution, and wash with 2 × 100 ml of 0.1-*M* sodium phosphate, pH 7.2, to reduce the pH for the reduction reaction. Drain to a wet cake.
5. Transfer the gel cake back to a clean vessel and suspend it in 100 ml of 0.1-*M* sodium phosphate, pH 7.2, and stir using a paddle stirrer. Add to the gel slurry 0.63 g of solid sodium cyanoborohydride ($NaCNBH_3$; MW = 62.84) [toxic!] or 2 ml of 5-*M* $NaCNBH_3$ in 1-*N* NaOH. If the alkaline solution of cyanoborohydride is used, subsequently check the pH of the solution and readjust to 7.2, if required.
6. React with mixing at room temperature for 4 h or at 4°C overnight.
7. Transfer the gel slurry to a sintered glass filter funnel (in the fume hood) and wash with several bed volumes of water to remove most of the unreacted ligand and reaction byproducts.
8. Aldehyde residues remaining on the support should be blocked by the addition of a small amine-containing compound, such as ethanolamine or Tris. Ethanolamine typically works best, and the compound is small enough to react with all remaining aldehydes to form hydrophilic terminal hydroxyl groups on the matrix upon coupling. Wash the support once with an equal volume of 1-*M* ethanolamine, pH 7.2, and then transfer the wet gel cake to a clean flask or vessel used for mixing the reaction. Add with stirring, 100 ml of 1-*M* ethanolamine, pH 7.2, along with 0.63 g (or 2 ml of the 5-*M* $NaCNBH_3$ solution) of sodium cyanoborohydride. Readjust the pH if necessary and continue the blocking reaction for 30 min at room temperature.
9. Wash the support extensively with water, 1-*M* NaCl, and again with water to remove all unreacted components from the gel. Additional wash solutions may be utilized to completely remove ligands that may have some nonspecific binding potential and remain noncovalently bound to the immobilized ligand, such as the use of acidic and alkaline washes as well as washes containing denaturants (such as guanidine). After washing, the affinity gel may be stored as a 50% aqueous slurry containing a preservative at 4°C.

Glutaraldehyde Activation

Glutaraldehyde is a 5-carbon, *bis*-aldehyde compound that can be used to activate amine- or hydrazide-containing supports to form reactive groups for the subsequent immobilization of amine-containing ligands. Given its structure, it may appear that the aldehydes in this crosslinker might undergo reactions with two amines much the same as that seen using similar aldehyde-containing supports for immobilization using reductive amination (described previously). It also might be expected that a bifunctional glutaraldehyde molecule would form two Schiff bases with two amine-containing molecules, which could then be reduced with cyanoborohydride to form secondary amine linkages. However, the nature of glutaraldehyde coupling is far more complex than a straightforward reductive amination process using its terminal aldehydes. Glutaraldehyde undergoes significant transformation reactions in aqueous solution (and even in some solvents), resulting in a range of potential derivative forms (Figure 15.34), which can create a variety of intermediary reactive species. These transitional forms can then have a number of potential reaction routes with amines beyond simple Schiff base formation. In this regard, glutaraldehyde activation of amine-containing chromatography supports and subsequent coupling to amine-containing ligands can potentially entail Schiff base formation with aldehydes, the Michael-type addition of amines to α,β-unsaturated sites within glutaraldehyde polymers, covalent reactions with a dimeric cyclic glutaraldehyde species, or ring formation through an aldol condensation product which subsequently can react with amines to form cyclic quaternary pyridinium complexes (see Chapter 2, Section 4.4, and Chapter 5, Section 6.2) (Monsan, 1978; Migneault *et al.*, 2004).

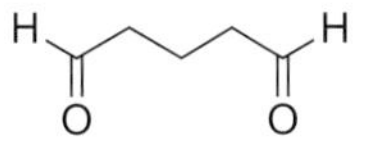

Glutaraldehyde

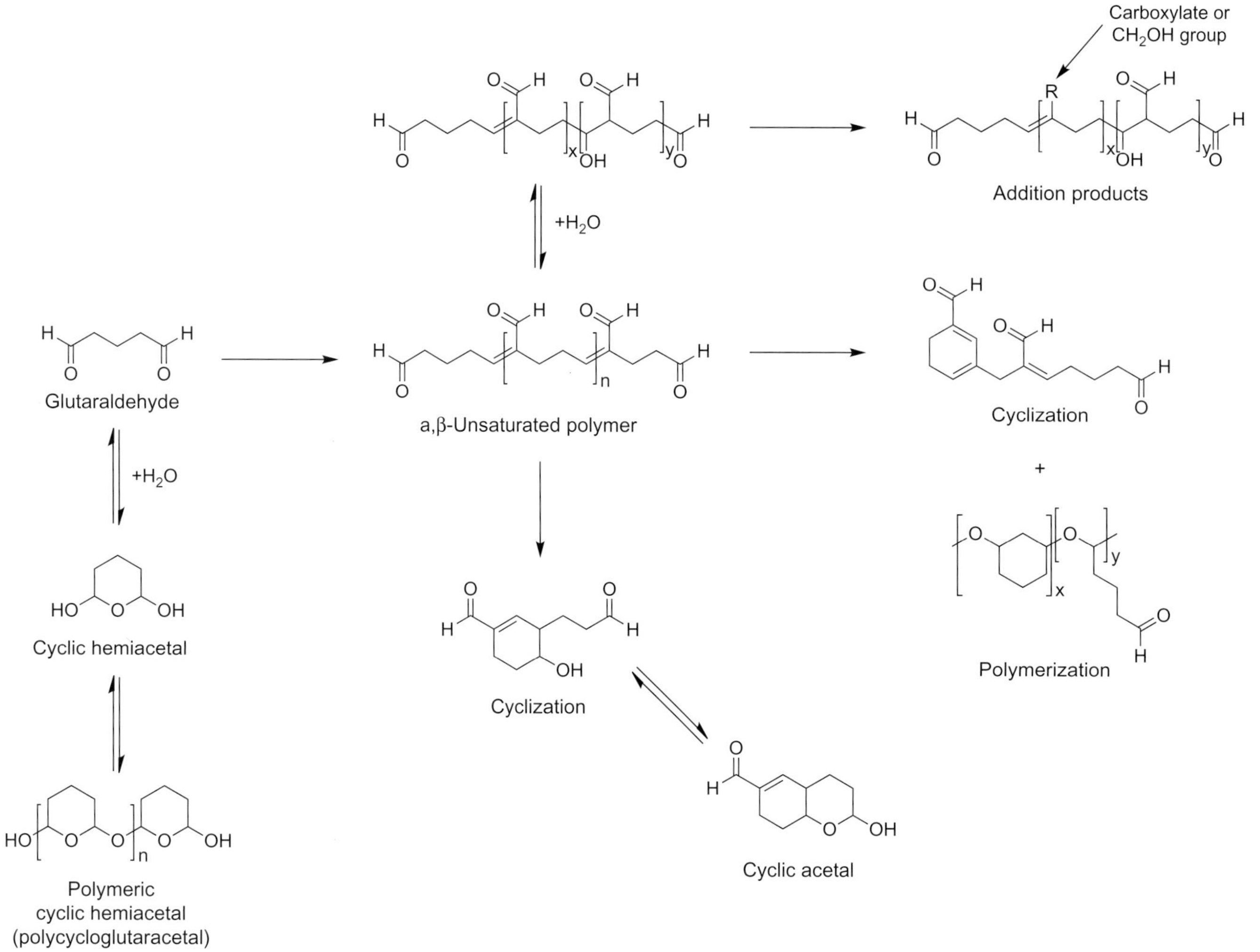

FIGURE 15.34 The potential transformations of glutaraldehyde in aqueous solution are complex and create many different reactive species. Once water reacts with one of the aldehydes of glutaraldehyde, the molecule can cyclize to form a hemiacetal derivative, which can also polymerize in solution to create polycycloglutaracetal. The acetal derivatives are reactive with amine-containing molecules through addition to the anomeric carbon atoms on each outer ring. Glutaraldehyde may also form a variety of unsaturated polymers at alkaline pH, which can undergo numerous transformations, including the creation of high-molecular-weight polymers that can precipitate out of solution. The cyclic acetal derivative can react with an amine-containing molecule by addition to the double bond on its left ring and through addition to the anomeric carbon atom on the right hemiacetal ring. Reactions with the unsaturated polymer form of glutaraldehyde typically occur through Michael-type addition to the double bonds along the polymer backbone.

The formation of aldol condensation unsaturated constructs and the potential for creating cyclic hemiacetal entities due to the intra- and intermolecular reactions of glutaraldehyde in aqueous alkaline solution both may result in nondescript oligomerization and layering of the active intermediates within and on a support material during activation. Formation of these polymeric glutaraldehyde species, once started, is not easy to control and could cause clogging of pore structures within chromatography supports as well as difficulty in reproducing the activation and immobilization reactions. In most cases, however, through careful adjustment of the initial glutaraldehyde activation conditions, side reactions such as polymerization can be limited and result in more defined reaction products that work well for immobilization of amine-containing affinity ligands.

In acidic or neutral pH conditions, glutaraldehyde can undergo an intramolecular cyclization reaction to form a hemiacetal ring (tetrahydro-2H-pyran-2,6-diol) that subsequently can polymerize through 6,6′-oxo linkages to form what some people call it by the nonstandard name "polycycloglutaracetal". The monomeric hemiacetal or the polymeric polycycloglutaracetal can react with amines by nucleophilic attack at the anomeric carbons bearing the hydroxyl groups (on both sides of a

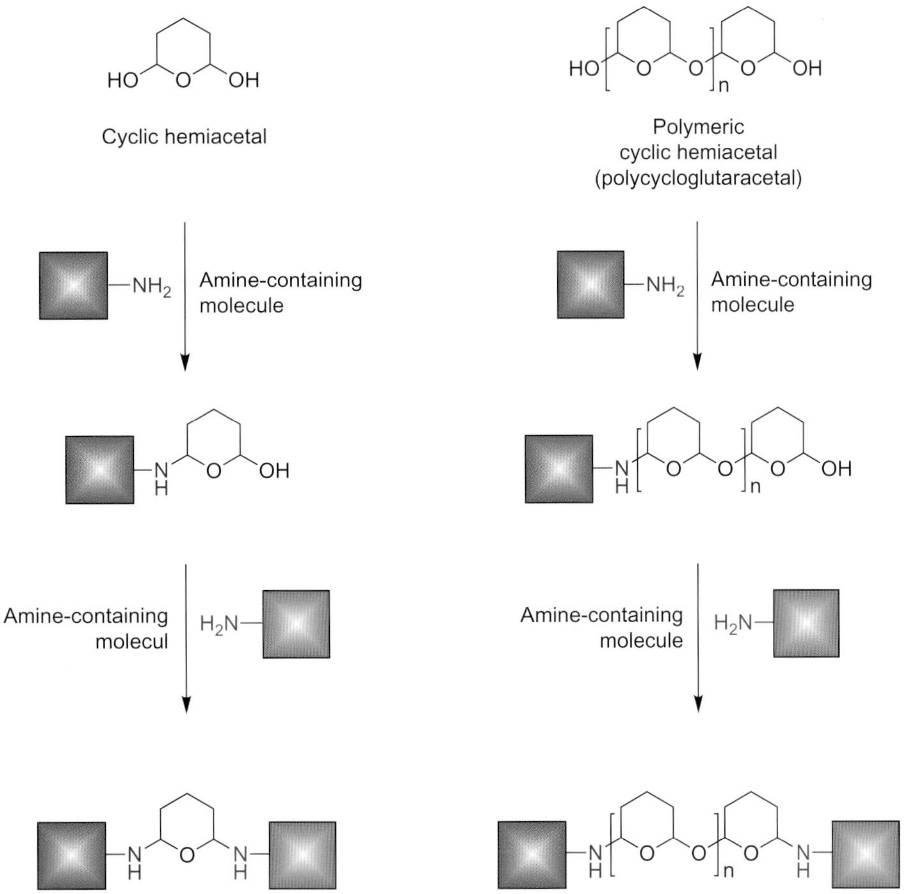

FIGURE 15.35 The glutaraldehyde cyclic hemiacetal can react with two amine-containing molecules to result in addition to the anomeric carbon atoms on both sides of its ring structure.

single cyclic acetal ring or at both ends of the polymer). Subsequent bond formation results in a substitution reaction that creates a secondary amine linkage along with the loss of one molecule of water (Figure 15.35). The reactions of monomeric glutaraldehyde hemiacetal may be one of the principal species responsible for stable ligand coupling without the need for reduction of Schiff bases, which normally would be necessary to stabilize aldehyde–amine complexes. This monomeric form may also be the primary type observed in immobilization reactions using an activation reaction with dilute glutaraldehyde done for a brief period of time (Betancor *et al.*, 2006) (Figure 15.36). In this case, the initial activation step of an amine-containing chromatography support, such as monoaminoethyl-*N*-aminoethyl (MANAE)–agarose, involves the reaction of one anomeric carbon of the cyclic acetal with the amine on the matrix to form a secondary amine bond. The other anomeric carbon atom of the acetal then can be used to couple amine-containing ligands after removal of excess glutaraldehyde species.

Under slightly more alkaline pH conditions (~pH 8.5), hemiacetal formation of glutaraldehyde can be accompanied by aldol dimerization to result in a bicyclic derivative, which can undergo reactions at two sites within the molecule (Tashima *et al.*, 1991) (Figure 15.37). Coupling of amine-containing molecules to this dimeric compound occurs by nucleophilic substitution at the anomeric hemiacetal carbon atom as well as through Michael-type addition to the double bond on the aldol addition ring product (Figure 15.38). This dimeric form of glutaraldehyde may be the predominant species observed in some activation and immobilization reaction protocols, wherein the activation step is done by the addition of a relatively high concentration of glutaraldehyde that is reacted for an extended period of time (Betancor *et al.*, 2006). This glutaraldehyde dimer is an extremely potent reactant with amines and it has been implicated as the form responsible for the best crosslinking performance in tissue fixation, protein conjugation, and in the immobilization of enzymes (Robertson and Schultz, 1970; Makino *et al.*, 1988).

Dimeric cyclic glutaraldehyde can be used to activate amine-containing chromatography supports, such as a MANAE–agarose matrix, to link initially to the α,β–unsaturated double bond to create an intermediate reactive derivative. The initial activation reaction is done at relatively high concentrations of glutaraldehyde at neutral pH and for an extended period to ensure that the reaction occurs with a glutaraldehyde dimer instead of a hemiacetal monomer (Betancor *et al.*, 2006). The activated support then can be coupled with amine-containing ligands through covalent attachment to the anomeric carbon of the hemiacetal ring with loss of one molecule of water to form a secondary amine linkage (Figure 15.39). Both of these reactions require no reducing step as is common with coupling aldehydes to amine-containing molecules, and the coupling occurs rapidly and in high yield.

In even more alkaline pH conditions, glutaraldehyde can undergo transformation by a series of aldol condensation reactions that creates, after dehydration and loss of the hydroxyl groups along with the adjacent hydrogens, numerous α,β-unsaturated sites within glutaraldehyde polymers of indeterminate length (Rembaum *et al.*, 1978). The resultant polymers reportedly also can contain a number of carboxylic, hydroxylic, and aldehyde groups along its length, although the exact composition is not well characterized (Margel and Rembaum, 1980). The presence of carboxylate functionalities presumably is due to further redox reactions that occur with the initial polymer in the presence of oxygen. Aldol polymerization of glutaraldehyde at pH 11 or above reportedly can be so severe that precipitation occurs in aqueous solution (Monsan *et al.*, 1975). Kawahara *et al.* (1992) indicated that extensive polymerization also can occur if the glutaraldehyde is dissolved in anhydrous solvent as well as high pH aqueous solutions. This form of polyglutaraldehyde long has been used to activate amine-containing microparticles for the subsequent coupling of amine-containing ligands (Rembaum *et al.*, 1978; Margel and Rembaum, 1980) (see also Chapter 14). However, it is likely that the other

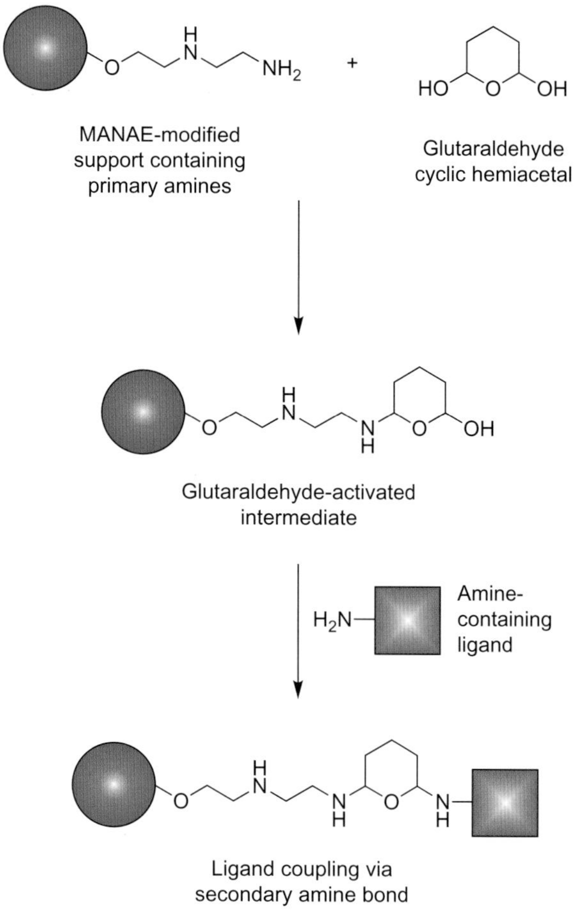

FIGURE 15.36 The activation of MANAE–agarose with glutaraldehyde monomer (under dilute glutaraldehyde conditions) and coupling of an amine-containing ligand.

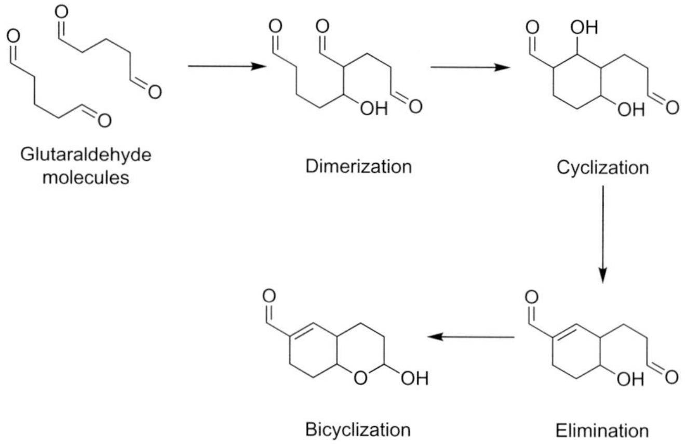

FIGURE 15.37 Formation of a dimeric, bicyclic glutaraldehyde.

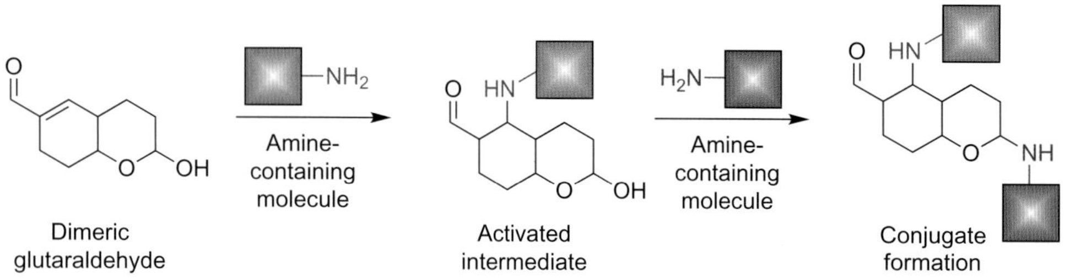

FIGURE 15.38 Reaction of glutaraldehyde dimer with two amine-containing compounds.

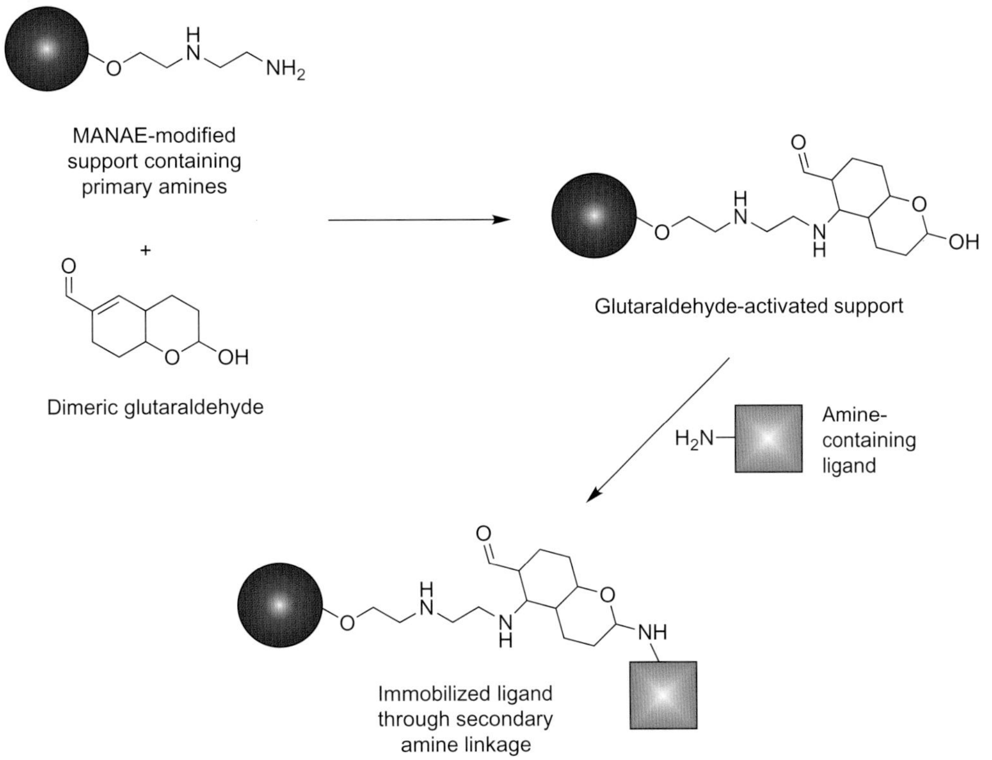

FIGURE 15.39 Activation of MANAE–agarose with dimeric glutaraldehyde and coupling of an amine-containing ligand.

monomer and dimer species described above may also contribute to the overall coupling reactions occurring with glutaraldehyde, depending on the exact activation procedure being used.

Aldol-mediated polyglutaraldehyde formation can be used to activate amine-containing supports by initially coupling through Michael-type addition reactions to various double bonds within the polymer strands. After activation with an excess of this polymeric form, the remaining uncoupled polyglutaraldehyde can be washed off the matrix and the activated intermediate used to immobilize an amine-containing ligand (Figure 15.40). Since polyglutaraldehyde contains both aldehydes and double bonds, activation of supports with this species likely involves two routes of amine reactivity. The Michael-type addition of amines to the double bond sites will result in secondary amine linkages, which have no requirement for further stabilization by reduction. However, amines may also link to the polymer through Schiff base formation with the aldehydes still present in each polymer strand. These Schiff bases may need a reductive amination reaction using sodium cyanoborohydride to stabilize them against hydrolysis and form secondary amine bonds. Particularly, if ligands are being immobilized that contain only a single amine group, then it is recommended that reduction be carried out to prevent ligand leakage through hydrolysis.

Glutaraldehyde and its derivatives that form in aqueous solution potentially can react with a number of nucleophilic groups on proteins or other ligands, such as amines, thiols, and imidazole nitrogens, as well as phenolic –OH groups (Habeeb and Hiramoto, 1968). The relative reactivity of glutaraldehyde toward some of the most predominant functional groups in proteins has been estimated to be ε-amine > α-amine > guanidinyl > secondary amine > hydroxyl groups (Migneault *et al.*, 2004). Thiol reactivity with glutaraldehyde-reactive species is also possible, but it has been reported to occur only in the presence of amine functionalities (Okuda *et al.*, 1991). The bifunctional nature of the glutaraldehyde monomeric or dimeric forms makes it suitable for activation of chromatography supports followed by immobilization of affinity ligands containing nucleophiles, such as amines. Due to the diverse reactions of glutaraldehyde in aqueous solution, the stability of coupling to amine-containing ligands typically is far greater than the stability of the initial Schiff bases created from aldehyde-containing supports as they react with amine-containing ligands (see previous section). Normally without reductive amination coupling of amino ligands to aldehyde-containing supports, the immobilized ligand will leach off the matrix as the Schiff bases slowly hydrolyze in aqueous solution. By contrast, glutaraldehyde immobilization is more resistant to ligand leakage even without a subsequent

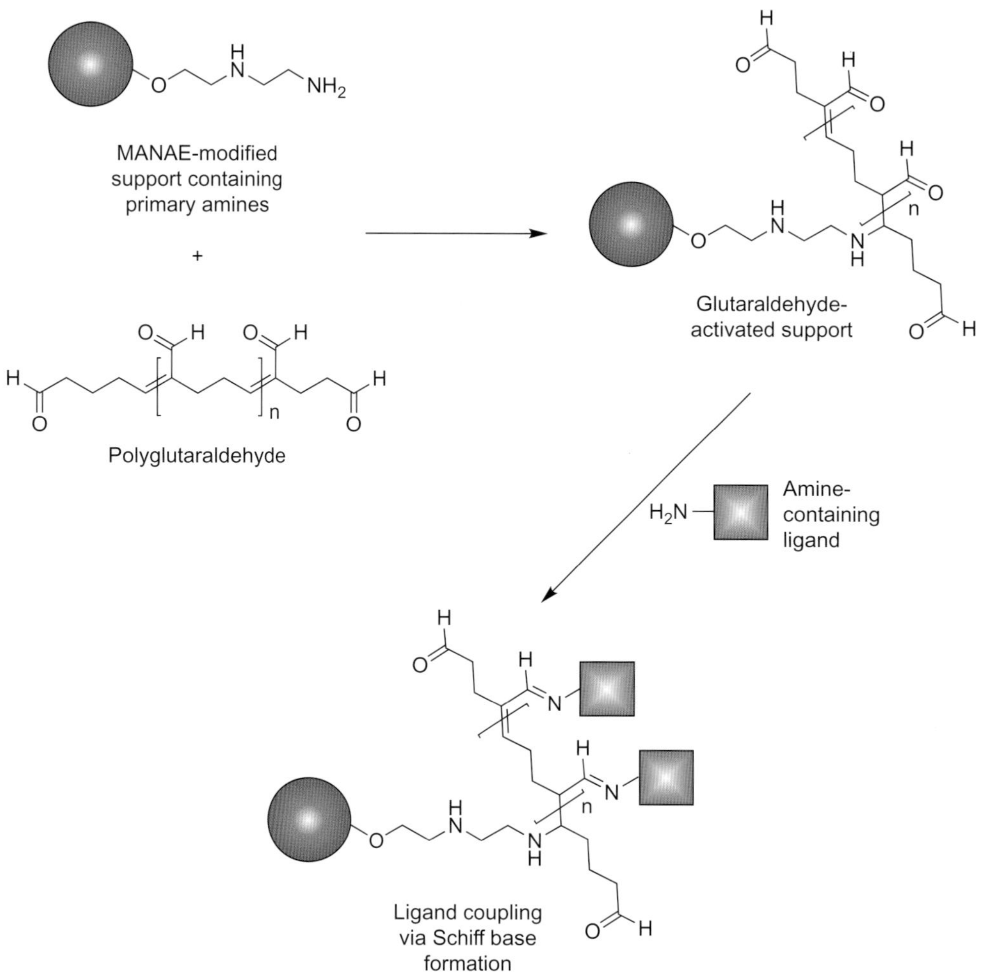

FIGURE 15.40 Activation of MANAE–agarose with polyglutaraldehyde (aldol-mediated) followed by coupling of an amine-containing ligand to form a secondary amine linkage.

reduction step, because most of the ligand coupling reactions involve addition reactions without the intermediate formation of a Schiff base.

Due to the diverse nature of glutaraldehyde reactions, the exact reactive species that is formed after activation may be less certain than when using other activation methods. Nevertheless, the immobilization reaction can be designed to take place with high efficiency and yield. In addition, recently it has been shown that by carefully controlling the level of glutaraldehyde activation of an amine-containing support at neutral pH, the result can be fairly well constrained to either a highly reactive dimeric species or a somewhat less reactive and predominantly monomeric glutaraldehyde derivative (Betancor *et al.*, 2006). The major reactive species created on an amine-containing support can be predetermined by the amount and concentration of glutaraldehyde addition as well as by the time of the reaction. By contrast, activation of an amine-containing support under highly alkaline conditions will result in a greatly polymerized glutaraldehyde intermediate that reacts with nucleophiles on ligands mainly through the addition to double bonds (although Schiff base formation may also occur).

The use of glutaraldehyde to activate and couple enzymes for the preparation of immobilized reactors remains one of the most popular methods for preparation of these catalytic supports (Burteau *et al.*, 1989; Van Aken *et al.*, 2000; Dos Reis-Costa *et al.*, 2003; Magnan *et al.*, 2004; Seyhan and Alptekin, 2004; Betancor *et al.*, 2006; Ma *et al.*, 2011; Matosevic *et al.*, 2011). Unlike reductive amination processes described in the previous section, immobilization using glutaraldehyde usually does not require the use of a reducing agent, such as the hazardous sodium cyanoborohydride. Especially when coupling multivalent protein molecules, immobilization onto a glutaraldehyde-activated support can create multiple attachment sites that firmly link the ligand with minimal potential for leakage. Thus, if care is taken to control the initial activation reaction, the resultant preparation of an affinity support can be performed reproducibly and give high activity in its intended application.

The following protocols can be used to activate amine-, amide-, or hydrazide-containing supports with glutaraldehyde for subsequent coupling of amine-containing ligands. The first two procedures are based on the method of Betancor *et al.* (2006). These activations can be performed using any chromatographic support material containing available primary amines, such as the use of an agarose support that has been modified with a diamine spacer to contain free terminal amino groups. One such support described in the literature is monoaminoethyl-*N*-aminoethyl (MANAE)–agarose, which contains low-pKa amino groups that are highly reactive using neutral pH buffers (Fernandez-Lafuente *et al.*, 1993). In one regard, the final level of glutaraldehyde activation can be controlled by regulating the amount of these amines produced initially on the support material.

Glutaraldehyde should be handled with care in a fume hood while using the appropriate personal protective equipment to prevent contact or exposure to the solutions or vapors. All waste should be treated as aqueous hazardous waste and disposed of according to environmental regulations and local policy.

ACTIVATION PROTOCOLS

(A) Glutaraldehyde Activation to Form a Monomeric-Reactive Species

This protocol forms a reactive intermediate with moderate rates of reactivity for coupling amine-containing ligands.

1. Wash 100 ml of an amine-containing support material using gentle vacuum filtration in a sintered glass filter funnel (such as MANAE–agarose, which can be prepared by methods described in this chapter for coupling diamine spacers to chromatography supports) with several bed volumes of deionized water to remove any storage solution. Finally, wash the support with 200 ml of activation buffer (0.2-*M* sodium phosphate, pH 7.0). The support material and all wash solutions as well as the reaction should be maintained at 25°C. Note: Using a buffer pH of >8 will result in glutaraldehyde polymerization and an uncontrolled reaction.
2. In a fume hood, add the moist gel cake to a clean vessel that can be used to stir or rotate the gel during the activation and coupling reactions. Add 200 ml of 0.5% (v/v) glutaraldehyde dissolved in coupling buffer with mixing to resuspend the support in the solution.
3. React with mixing for 1 h.
4. Remove the gel slurry from the reaction vessel and drain the excess glutaraldehyde solution into a suction filter flask using vacuum filtration in a sintered glass filter funnel (in a fume hood). Wash the activated support extensively with 25-m*M* sodium phosphate, pH 7.0, buffer and then with deionized water. The activated support can be stored at 4°C until ready for ligand coupling.

(B) Glutaraldehyde Activation to Form a Dimeric-Reactive Species

This protocol forms a reactive intermediate with extremely high reactivity for coupling amine-containing ligands.

1. Wash 100 ml of an amine-containing support material using gentle vacuum filtration in a sintered glass filter funnel (such as MANAE–agarose) with several bed volumes of deionized water to remove any storage solution. Finally, wash the support with 200 ml of activation buffer (0.2-*M* sodium phosphate, pH 7.0). The support material and all wash solutions as well as the reaction should be maintained at 25°C.
2. In a fume hood, add the moist gel cake to a clean vessel that can be used to stir or rotate the gel during the activation and coupling reactions. Add 200 ml of 15% (v/v) glutaraldehyde dissolved in coupling buffer with mixing to resuspend the support in the solution.
3. React with mixing for 15 h at 25°C.
4. Remove the gel slurry from the reaction vessel and drain the excess glutaraldehyde solution into a suction filter flask using vacuum filtration in a sintered glass filter funnel (in a fume hood). Wash the activated support extensively with 25-m*M* sodium phosphate, pH 7.0, buffer and then with deionized water. The activated support can be stored at 4°C until ready for ligand coupling.

IMMOBILIZATION OF LIGANDS ON GLUTARALDEHYDE-ACTIVATED SUPPORTS

The coupling of amine-containing ligands such as proteins or enzyme reactors to glutaraldehyde supports prepared according to the previous activation protocols will proceed rapidly under low ionic strength conditions (i.e., using 25-m*M* potassium phosphate, pH 7, with no additional salt added). Under high ionic strength conditions (25-m*M* potassium phosphate, pH 7, containing 0.5-*M* NaCl) the dimeric glutaraldehyde activation process (protocol (B), above) will react and couple ligand almost as fast as when using low ionic strength buffers; however, when using the monomeric glutaraldehyde activation process (protocol (A), above) it results in slower coupling rates, which could take many hours or even overnight to go to completion (Betancor *et al.*, 2006). The reason coupling takes place faster under low salt conditions is thought to be due to the formation of an initial ionic interaction between the protein molecules to be coupled and the underlying secondary amines to which the glutaraldehyde derivatives first are linked during activation. The initial ion-exchange character of the activated support facilitates recruitment of the protein to the immediate proximity of the reactive groups, thus accelerating immobilization. Adding high salt obviates these charge interactions and thus slows down the overall reaction process.

Protocol

1. Wash 100 ml of a glutaraldehyde-activated support containing dimeric glutaraldehyde (activation protocol (B), above) with 2 bed volumes of 25-m*M* potassium phosphate, pH 7.0 (coupling buffer), and drain to a wet cake. The addition of salt (NaCl) to the coupling buffer may be done for ligands that require it, but the use of additional ionic strength may also require increased reaction times to accommodate a somewhat slower reaction rate (see previous discussion). Transfer the support to a vessel that can be used to mix the gel suspension during the coupling reaction by end-over-end rotation or by using a paddle stirrer.
2. Dissolve an amine-containing ligand such as a protein, enzyme, or small ligand in 100 ml of coupling buffer. The ligand should be at a concentration of at least 2-fold greater than the quantity desired to be immobilized. For protein ligands, reacting at a concentration of 3 to 5 mg/ml resin is often sufficient. However, if a high density of immobilized protein or enzyme is desired, reactions can be done at concentrations of 10 to 20 mg/ml to obtain better coupling yields. For small ligands, the reaction should be done at a concentration that exceeds the level of glutaraldehyde activation or at least by using a 2-fold greater amount of ligand than the anticipated final immobilization level. To obtain an optimal affinity support for a given application, it may be necessary to perform a series of reactions using different ligand concentrations to determine the best coupling concentration to use.
3. Mix the ligand solution with the activated support and mix to resuspend the resin.
4. React with mixing for 2 h at room temperature or at 25°C. If a monomeric glutaraldehyde-activated support is used (activation protocol (A), above), then the reaction time should be extended by at least an hour. If high salt was added to the coupling buffer (e.g., 0.5-*M* NaCl) in step 1 and 2, then the reaction using dimeric glutaraldehyde activation should be extended to 3 h, and if monomeric glutaraldehyde activation was used with high salt, it should be extended to an overnight reaction.
5. Remove the reaction slurry and transfer to a filter flask. Drain the excess reaction solution and wash with water to remove the majority of uncoupled ligand. If the remaining active groups are not to be blocked, then the wash should be done extensively with water, 1-*M* NaCl, and again with water.
6. Reactive groups remaining on the support should be blocked by the addition of a small amine-containing compound, such as ethanolamine or Tris. Ethanolamine typically works best, and the compound is small enough to react with all remaining active sites to form hydrophilic terminal hydroxyl groups on the matrix upon coupling. In a fume hood, wash the support once with an equal volume of 1-*M* ethanolamine, pH 7.0, and then transfer the wet gel cake to a clean flask or vessel used for mixing the reaction. Add with stirring, 100 ml of 1-*M* ethanolamine, pH 7.0, along with 0.63 g of sodium cyanoborohydride (or 2 ml of a 5-*M* $NaCNBH_3$ solution in 1-*N* NaOH). Cyanoborohydride is extremely toxic and should be handled with care in a fume hood. The use of a reducing agent in the final blocking step is optional, but it is included here to ensure that if there are any aldehydes present in the activated support they too will be blocked by the ethanolamine addition (by reducing any Schiff bases formed by the reaction of the amine on ethanolamine with the aldehyde groups). Note that if the ligand is sensitive to the presence of cyanoborohydride, use the ethanolamine solution without adding the reducing agent. Readjust the pH if necessary and continue the blocking reaction for 30 min at room temperature.
7. Wash the support extensively with water, 1-*M* NaCl, and again with water to remove all unreacted components from the gel. Additional wash solutions may be utilized to completely remove ligands that may have some nonspecific binding potential and remain noncovalently bound to the immobilized ligand, such as the use of alternating acidic and alkaline washes as well as washes containing denaturants (such as guanidine). After washing, the affinity gel may be stored as a 50% aqueous slurry containing a preservative at 4°C.

NHS Ester and NHS Carbonate Activation

N-Hydroxysuccinimide (NHS) esters are activated carbonyl groups that may be formed from carboxylates through the use of a carbodiimide-mediated esterification reaction done under aqueous or nonaqueous conditions. They may also be formed through trans-esterification or transfer reactions using reagents such as Sakakibara's reagent, which is the NHS ester of trifluoroacetate, TSTU (*N*,*N*,*N*′,*N*′-tetramethyl(succinimido) uronium tetrafluoroborate), or disuccinimidyl carbonate

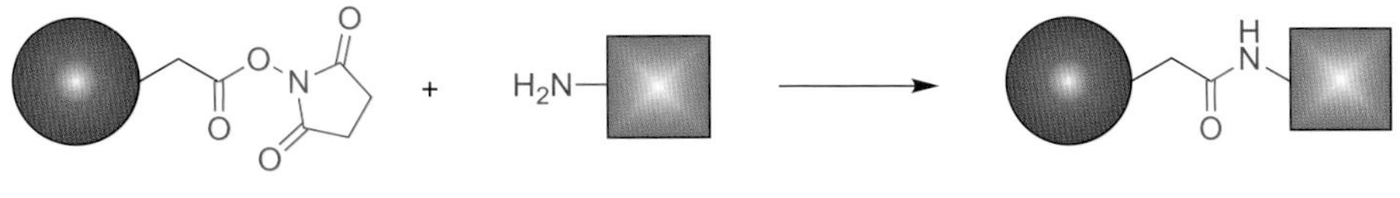

FIGURE 15.41 Reaction of an NHS ester-activated support with amine-containing ligands to form an amide bond.

(DSC) (Sakakibara and Inukai, 1965; Wilchek *et al.*, 1994). NHS esters are electrophilic reactive groups that can couple with nucleophiles such as amines to form stable amide bonds. They have been used extensively in the design of crosslinkers and modification reagents to form covalent conjugates with amine-containing proteins or other amine-containing molecules (Chapter 3, Section 1.4). NHS esters also can be used to immobilize amino-ligands onto chromatography supports in a rapid, single-step reaction that typically occurs in high yield (Prokudina *et al.*, 2011) (Figure 15.41).

Chromatography supports containing NHS esters are available commercially from several sources, and most contain a spacer arm that terminates in the activated carboxylate (e.g., Thermo Fisher, Bio-Rad, GE Healthcare). These supports thus are synthesized first by building a spacer arm off the chromatography support and then forming the NHS esters by activation of the carboxylates. A wide range of spacer molecules can be considered for this purpose, including hydrophobic aliphatic compounds such as 6-aminocaproic acid or hydrophilic spacers such as those containing PEG-based cross-bridges (i.e., amino–PEG_4–carboxylate; Thermo Fisher, Quanta BioDesign). Usually, the spacers are attached to the support through their amino end by coupling to an amine-reactive resin material, such as any of the amine-reactive immobilization chemistries described in this chapter. Once the carboxylate spacer derivative is formed on the support, it then can be activated to create the NHS ester groups using a number of different reaction routes.

When working with microparticles or nanoparticles, the formation of the NHS ester reactive group often is done in aqueous solution immediately prior to adding an amine-containing molecule to the particles for coupling (Chapter 14). This is accomplished by using a water soluble carbodiimide such as EDC (Chapter 4, Section 1) in the presence of NHS or sulfo-NHS (the sulfonated form promotes greater water solubility). The active NHS esters are formed despite the fact that the aqueous environment also causes continual hydrolysis of the esters back to unreactive carboxylates. This activation strategy works well with small nonporous particles if the ligand to be immobilized is added immediately after the activation step or even during the activation process. However, when working with chromatography supports that are made of larger porous particles, it is best to create the NHS ester groups in a nonaqueous environment so that competing hydrolysis is eliminated and maximal activation levels are obtained throughout the inner bead structures.

Although the use of a carbodiimide-mediated formation of NHS ester groups is common when working with carboxylate microparticles or nanoparticles, a major potential side reaction has been reported in the formation of NHS esters using carbodiimides on hydroxylic supports that have been modified to possess carboxylate spacers (Wilchek and Miron, 1987). If a reaction involving the carbodiimide DCC with added NHS is used to create the NHS ester groups on a carboxylate support under nonaqueous conditions, as is also typical for synthesizing most NHS ester-containing bioconjugation reagents, the formation of a *bis*-NHS derivative of β-alanine can be created at the same time that NHS esters are being formed with the carboxylate spacers on the matrix. This side reaction results from the initial interaction of DCC with the hydroxyl on NHS to form a carbodiimide-activated NHS group. This ability of carbodiimides to react with hydroxyl groups correlates to the reported activation of hydroxyls on chitosan using the carbodiimide EDC (Chiou and Wu, 2004). The DCC–NHS intermediate in this side reaction goes on to react with two additional equivalents of NHS, which then is followed by a Lossen rearrangement to create a reactive carbamate ester on the amino end of β-alanine and an NHS ester on the carboxylate end (Gross and Bilk, 1968) (Figure 15.42). This bifunctional compound then reacts with any remaining hydroxyl groups within the original matrix to create unstable ester linkages. The other end still can react with amine-containing ligands to form a covalent bond, but the immobilized product formed from this side reaction is unstable due to the potential for ester hydrolysis, and thus it can continually leach off the matrix in aqueous solution (Figure 15.43). This side reaction may be the main reason why many NHS ester-mediated immobilization reactions result in leaky supports.

To overcome these potential side reactions a two-step protocol might be used wherein the carboxylate

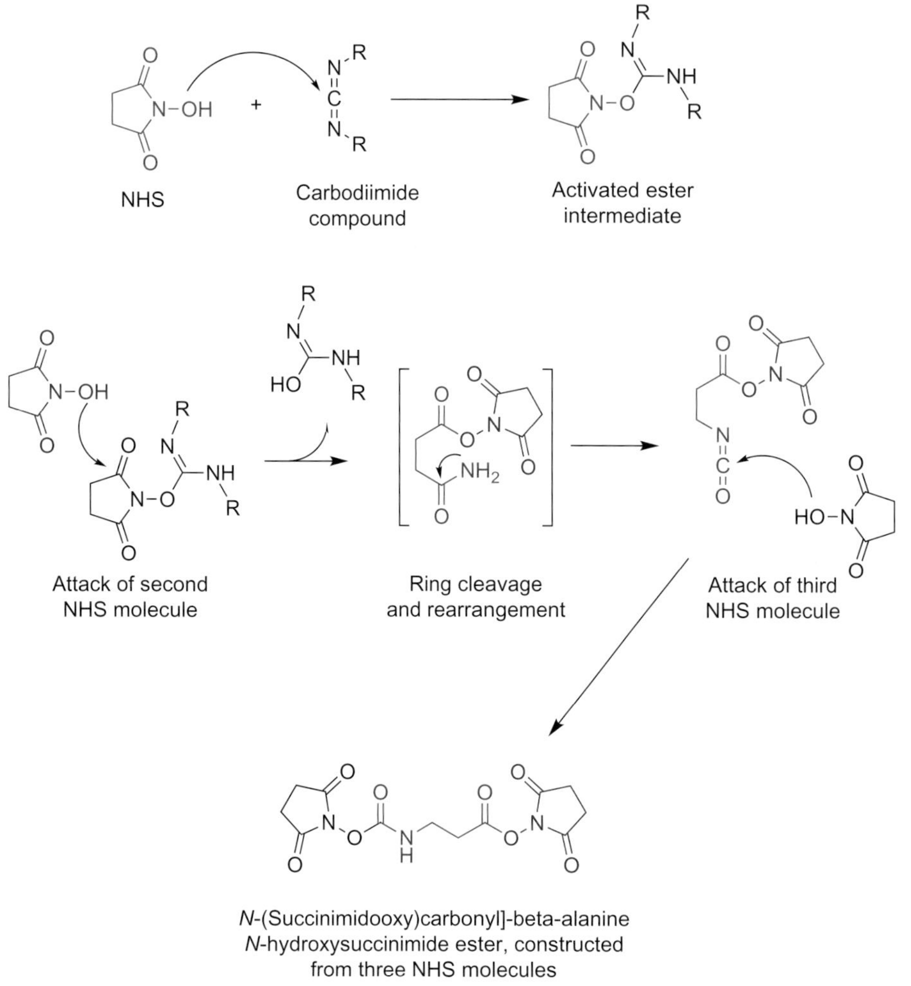

FIGURE 15.42 Reaction of NHS with DCC to form an activated β-alanine byproduct, which can form when using carbodiimides to create NHS esters in organic solvent.

support first is activated with a carbodiimide such as DCC in nonaqueous conditions, then the support is washed free of excess carbodiimide and reacted with NHS to form the final NHS esters. However, this procedure has been found to result in decreased activation yields and is cumbersome (Wilchek and Miron, 1987). Fortunately, there are now alternative strategies to creating NHS esters on hydroxylic support materials that have the ability to give highly stable reaction products without the potential side reactions that carbodiimides plus NHS are known to cause. One route was developed by Wilchek *et al.* (1994) specifically to avoid the problems of using carbodiimides with NHS and hydroxylic supports.

The coupling reagent TSTU originally was developed as an amide bond-forming agent for peptide synthesis (Knorr *et al.*, 1989; Bannwarth and Knorr, 1991). It is able to directly activate a carboxylate group to an NHS ester in a single step without the formation of detrimental side reaction products or other reactive components having crosslinking potential. Uronium coupling reagents react with carboxylate groups through an initial attack on the central positively charged carbon atom of the tetramethyl uronium component (Figure 15.44). This is followed by a rearrangement leading to the formation of an intermediary uronium ester with the carboxylate along with displacement of the leaving group attached to the uronium part of the reagent. In the case of TSTU, it is the NHS group that gets removed in this step. Next, the hydroxyl of the released NHS group in turn attacks the carbonyl carbon of this intermediate uronium ester, which causes displacement of the tetramethyl uronium group and formation of the final NHS ester on the carboxylate. Most TSTU activation reactions are done in nonaqueous solvent (e.g., dioxane or DMF), which preserves the

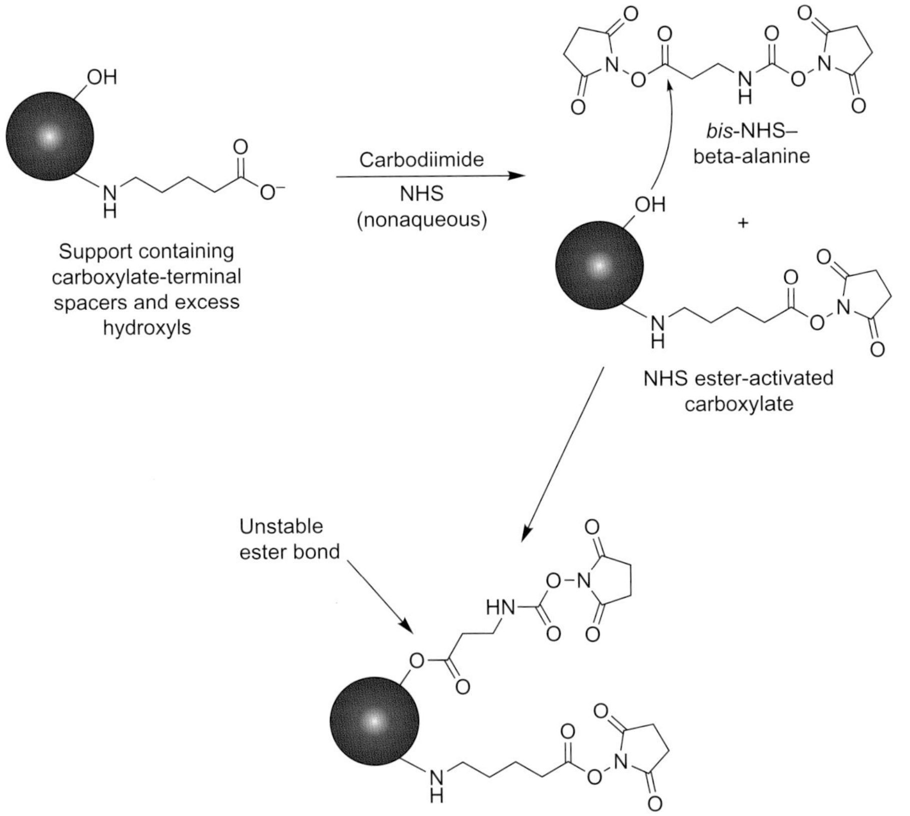

FIGURE 15.43 Reaction of activated β-alanine with the hydroxyl groups remaining on a support, which was modified to contain carboxylates.

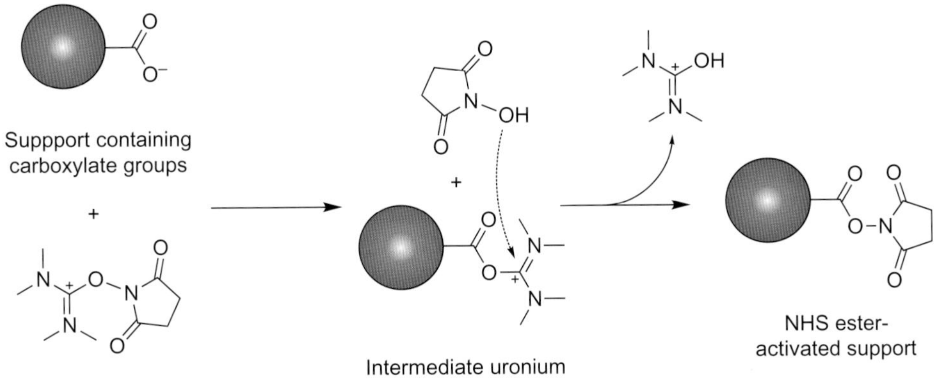

FIGURE 15.44 Mechanism of NHS ester formation using TSTU.

NHS ester from hydrolysis; however, the reagent may also be used in a mixture of organic solvent and water. In addition to Bannwarth and Knorr (1991) demonstrating an organic/water reaction environment consisting of a 2:2:1 mixture of DMF/dioxane/water, Andersson *et al.* (1993) used the coupling reagent to activate a carboxylate derivative of a saccharide in an aqueous/organic solvent mixture and used the NHS ester to conjugate the saccharide to a protein.

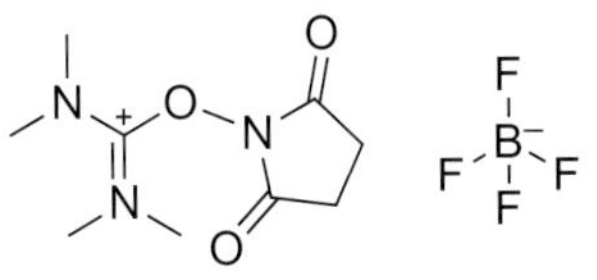

TSTU
2-succinimido-1,1,3,3-tetramethyl-uronium tetrafluoroborate
(tetrafluoroborate salt)

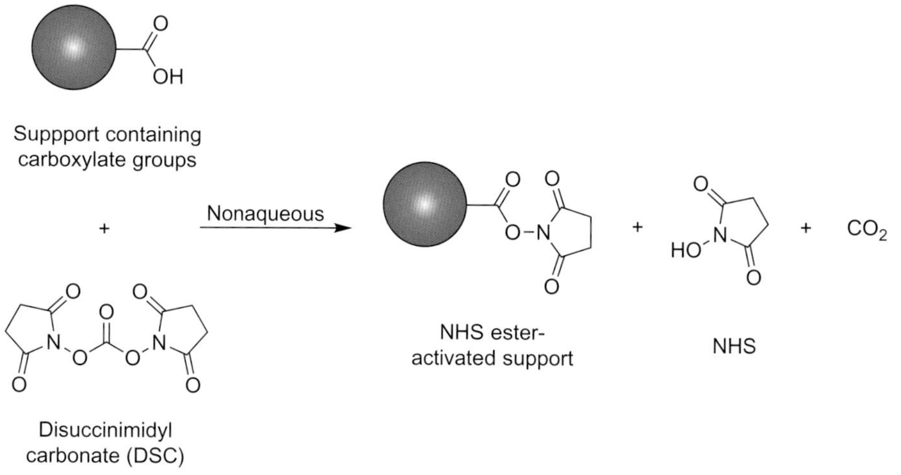

FIGURE 15.45 DSC activation of carboxylate supports to form NHS esters.

In addition to the use of TSTU to create NHS ester-activated derivatives, the bifunctional reagent *N,N'*-disuccinimidyl carbonate (DSC) also can be used to form these reactive groups on carboxylate-containing matrices (Ogura *et al.*, 1979; MacBeath and Schreiber, 2000). In this reaction, which is done in a nonaqueous environment, DSC reacts directly with the carboxylate to transfer one of its NHS groups to the acid in a transesterification process (Figure 15.45). The result is the single-step formation of an NHS ester on the carboxylate-containing support, which is then washed free of excess reactants and stored in nonaqueous solution to maintain stability of the ester until use. The byproducts of this process are CO_2 and NHS, both of which are completely innocuous. This reaction also avoids the potential side reactions of a carbodiimide/NHS reaction to form NHS esters. In similar methods, other *bis*-carbonates have been used to activate carboxylates, such as *tert*-butyl carbonates (Basel and Hassner, 2002). The formation of NHS esters on carboxylate supports with *bis*-carbonates can be performed in anhydrous DMF (or other appropriate water-miscible solvents) using 100-m*M* DSC along with an equal concentration of an organic base as a proton acceptor, such as *N,N*-diisopropylethylamine (DIEA).

As an alternative approach, reactive NHS esters also can be created directly on hydroxyl-containing chromatography supports without the need to add an intermediary carboxylate spacer. In this strategy, the hydroxylic resin is washed into nonaqueous solvent and activated using the reagent DSC. This compound reacts with a hydroxyl on the matrix with loss of one NHS group to form an *N*-succinimidyl carbonate reactive group, which is similar in reactivity toward amine nucleophiles as that of an NHS ester formed from a carboxylate (Wilchek and Miron, 1985). This same activation process has been used successfully to activate the hydroxyl groups on PEG compounds for subsequent coupling to amine-containing molecules (Chapter 18, Section 2.1). The immobilization of amine-containing ligands onto DSC-activated supports results in the formation of carbamate linkages (Figure 15.46), which are identical to the bonds formed from the reaction of amines with CDI-activated matrices (see this chapter, *CDI-Activated Supports*). Carbamates have been found to be very secure linkages that are similar to amide bonds in chemical stability. Another advantage of DSC activation of hydroxyl supports is that the hydroxyl is reformed by hydrolysis if an amine ligand is not coupled during the immobilization reaction. This leaves an uncharged, hydrophilic group on the matrix if hydrolysis occurs as opposed to a typical NHS ester, which upon hydrolysis, creates a carboxylate group bearing a negative charge. Thus, if a spacer arm on the matrix is not required to present the affinity ligand to its intended binding target, then direct activation of a hydroxylic support with DSC potentially is a better choice than an NHS ester formed from a carboxylate spacer. Even if a spacer arm is desirable to create an immobilized ligand, the use of a PEG spacer containing a terminal hydroxyl group could be created to subsequently activate the spacer with DSC. The use of PEG spacers would form a highly hydrophilic environment that will dramatically lower the nonspecific binding potential of the final affinity support. In addition, if a hydroxylic matrix first is modified to contain carboxylate spacers for conversion to NHS esters using DSC, it is likely that some remaining hydroxyls also are converted to the NHS carbonate in the activation process. Therefore, this result would produce both amide and carbamate bonds in the final immobilized ligand support, the relative proportion of which would depend on the ratio of activated carboxylates to activated hydroxyls.

A final option using DSC for the activation of chromatography supports is to use it to activate amine-containing

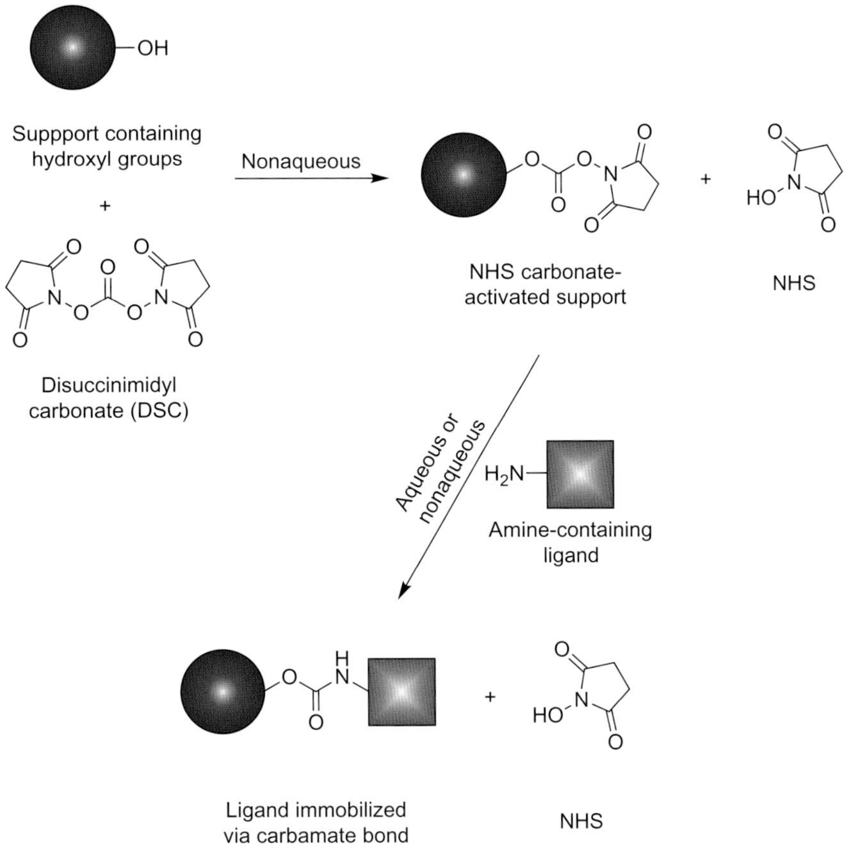

FIGURE 15.46 DSC activation of hydroxyl supports followed by coupling to amine-containing ligands to form carbamate bonds.

matrices. A beaded chromatography support modified to contain amine spacers can be created through the coupling of a diamine compound to an activated matrix, which then results in terminal primary amines on the support for further activation and coupling reactions (see section on coupling spacer molecules). DSC can be used to directly activate these amines to reactive NHS carbamates, which in turn can be coupled to amine-containing ligands to form an isourea linkage (a urethane bond) (Figure 15.47). This reaction has been used successfully to couple immobilized proteins with other amine-containing molecules (MacBeath and Schreiber, 2000) and to activate amine spacers on a slide surface for immobilization of amine-containing ligands (Niemeyer, 2004). This procedure offers a way of coupling amine-containing ligands to an amine-containing support without further modification of the amino spacers to convert them to a different functionality for activation. However, it should be kept in mind that the underlying amines on this support may create positively charged sites, especially with regard to any DSC-activated amines that do not couple to ligands in the immobilization reaction, because the hydrolysis product regenerates the original amine. Thus, using this method there may be some potential for nonspecificity in the final affinity support from ion-exchange effects due to leftover or uncoupled amines.

The following protocols represent methods that can be used to create amine-reactive NHS esters on chromatography supports. Carbodiimide-based methods using DCC in organic solvent can also be done, but are not included in the recommended protocols due to the likely side reactions involving the formation of the bifunctional β-alanine crosslinking agent described previously, which would result in an affinity support having a high potential for ligand leakage. Note that for protocols related to the activation and coupling of ligands to carboxylate-containing microparticles or carboxylated nanoparticles using the carbodiimide EDC with NHS to form intermediate NHS esters, the reader should refer to Chapter 14.

All solvents used in the following protocols should be as anhydrous as possible to avoid the potential for hydrolysis of the NHS ester groups during storage of the activated supports.

ACTIVATION PROTOCOLS

(A) NHS Ester Activation Procedure using TSTU

The following procedure is based on the method of Wilchek *et al.* (1994) using TSTU in a reaction with

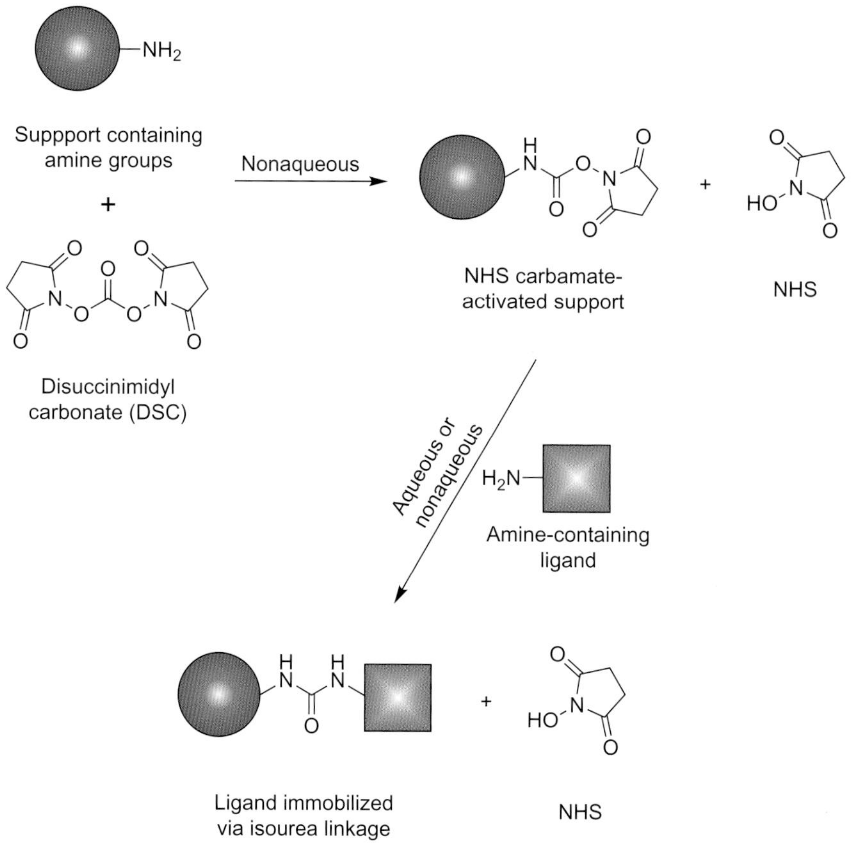

FIGURE 15.47 Activation of amine-containing supports with DSC and subsequent coupling of amine-containing ligands.

carboxylate groups on a chromatography support to form reactive NHS esters that can be used to couple with amine-containing ligands.

1. Wash 100 ml of a hydroxylic chromatography support material (such as crosslinked agarose) that has been modified to contain carboxylate groups with 0.3-*N* HCl and then with water (several bed volumes each) using a sintered glass filter funnel. In a fume hood, sequentially wash the support into increasing concentrations of dioxane-in-water (such as 25% (v/v), 50%, and 75% dioxane/water) and finally with 100% solvent for at least 10 to 20 bed volumes to remove the last traces of water. Drain to a wet gel cake and then pull a gentle vacuum on the filter funnel while breaking up the gel cake with a spatula to obtain moist, evenly divided pieces, which still contain dioxane within the internal pores of the beads. Remove the vacuum as soon as the gel is uniformly divided and appears as a white fluffy snow-like consistency. Be careful not to allow the support to dry, as this will cause particle collapse and irreversible damage to the pore structure.
2. In a fume hood, prepare a 100-ml solution of 0.1-*M* TSTU (*N,N,N′,N′*-tetramethyl(succinimido)uronium tetrafluoroborate) in DMF. Add the washed gel cake to the TSTU solution with stirring to resuspend the gel. Next add 0.244 g of 4-(dimethylamino)pyridine (DMAP) to the gel suspension to prepare a 0.2-*M* DMAP solution and mix to dissolve. A small amount of additional DMF may be added if necessary to aid in stirring the mixture.
3. Mix the reaction for 1 h by rotation or using an overhead paddle stirrer.
4. Wash the activated support with several bed volumes each of DMF, methanol, and isopropanol to remove excess reactants and reaction byproducts. The NHS ester-activated support can be stored as a 50% slurry in isopropanol at 4°C until use.

(B) NHS Ester Activation of Carboxylate Supports using DSC

1. Wash 100 ml of a carboxylate-containing chromatography support material (such as crosslinked agarose that has been modified to contain carboxylate groups) with water (at least several bed volumes) using a sintered glass filter funnel. In a fume hood, sequentially wash the support into increasing concentrations of

acetone-in-water (such as 25% (v/v), 50%, and 75% acetone/water) and finally with 100% dry acetone for at least 10 to 20 bed volumes to remove the last traces of water. Alternative solvents include dioxane, DMF, DMAC, or DMSO, which can be substituted for acetone depending on the compatibility of the matrix to a particular solvent. Drain to a wet gel cake and then pull a gentle vacuum on the filter funnel while breaking up the gel with a spatula to obtain moist, evenly divided pieces, which still contains acetone within the internal pores of the beads. Remove the vacuum as soon as the gel is uniformly divided and appears as a white fluffy snow-like consistency. Be careful not to allow the support to dry too much, as this will cause particle collapse and irreversible damage to the pore structure.

2. In the fume hood, prepare 100 ml of a solution consisting of 80 mg/ml DSC in dry acetone. Stir vigorously to dissolve. There may be some insoluble DSC left in the solution as a fine suspension, but this material will be driven into solution as the activation reaction proceeds.
3. Add the acetone wet gel cake to the DSC solution with stirring to uniformly resuspend the support. To this suspension, add 6.5 g of DMAP to provide a base to catalyze the reaction; alternatively, add 7.5 ml of anhydrous triethylamine (TEA). Mix the gel suspension using an overhead stirring paddle or by end-over-end rocking in the fume hood.
4. React for 1 h at room temperature with mixing.
5. In the fume hood, wash the activated support with acetone to thoroughly remove excess reactant and reaction byproducts. Wash with at least 10 bed volumes of solvent to ensure removal of the last traces of unreacted DSC.
6. Store the NHS ester-activated support as a 50% slurry in dry acetone at 4°C until use.

(C) NHS Carbonate Activation of Hydroxylic Supports using DSC

1. Wash 100 ml of a hydroxyl-containing chromatography support material (such as agarose, Toyopearl, or Trisacryl) with water (3–5 bed volumes) using a sintered glass filter funnel. In a fume hood, sequentially wash the support in increasing concentrations of acetone-in-water (such as 25% (v/v), 50%, and 75% acetone/water) and finally with 100% dry acetone for at least 10 to 20 bed volumes to remove the last traces of water. Alternative solvents include dioxane, DMF, DMAC, or DMSO, which can be substituted for acetone depending on the compatibility of the matrix to a solvent. Drain to a wet gel cake and then pull a gentle vacuum on the filter funnel while breaking up the gel cake with a spatula to obtain moist, evenly divided pieces, which still contain acetone within the internal pores of the beads. Internal pores of the beads, with a spatula to obtain moist, evenly divided pieces. Remove the vacuum. Be careful not to allow the support to dry too much, as this will cause particle collapse and irreversible damage to the pore structure.
2. In the fume hood, prepare 100 ml of a solution consisting of 80 mg/ml DSC in dry acetone. Stir vigorously to dissolve. There may be some insoluble DSC left in the solution as a fine suspension, but this material will be driven into solution as the activation reaction proceeds.
3. Add the acetone wet gel cake to the DSC solution with stirring to uniformly resuspend the support. To this suspension, add 6.5 g of DMAP (alternatively, add of 7.5 ml of anhydrous TEA) to provide a base to catalyze the reaction. Mix the gel suspension using an overhead stirring paddle or by end-over-end rocking in the fume hood.
4. React for 1 h at room temperature with mixing.
5. In the fume hood, wash the activated support with acetone to thoroughly remove excess reactant and reaction byproducts. Wash with 5 to 10 bed volumes of solvent to ensure removal of the last traces of unreacted DSC.
6. Store the NHS carbonate-activated support as a 50% slurry (the volume of the support is equal to half the total volume of the slurry) in dry acetone at 4°C until use. As an alternative storage solvent, the activated gel may be washed into dry isopropanol and stored in this solvent. Avoid the use of methanol, as some transesterification may occur with the NHS esters to form methyl esters upon storage.

The activation of an amine-containing support using DSC can be performed similar to methods described in previous protocols, except for substitution of a support modified to contain a terminal amino spacer for the hydroxylic or carboxylate supports described above.

LIGAND COUPLING TO NHS ESTER- OR NHS CARBONATE-ACTIVATED SUPPORTS

The following protocols describe general methods for coupling amine-containing ligands to NHS ester- or NHS carbonate-activated chromatography supports. This process can be used successfully to immobilize proteins or other amine-containing affinity ligands as well as amine-containing spacer molecules. During the coupling reaction, avoid introducing any other primary or secondary amine-containing components into the coupling buffer—such as Tris, glycine, imidazole, ammonium ions, or other small molecules containing a reactive amine—as these will interfere with the desired immobilization of ligand. Amine reactions with reactive ester supports will occur efficiently in aqueous solution at pH 7 to 9.0, so the coupling buffer pH may be modified somewhat to promote ligand stability or solubility if necessary; however, be aware that the higher the pH of the reaction, the greater the rate of hydrolysis of the NHS esters, so use care when reacting at the high end of this pH range. The immobilization reaction may also be done entirely in nonaqueous solution or in a combination of organic solvent/aqueous buffered solution to promote ligand solubility, when necessary. Nonaqueous conditions will eliminate the potential for hydrolysis and thus promote higher yields when immobilizing small amine-containing ligands. To perform immobilizations in 100% organic solvent, an organic base should be added as a proton acceptor to catalyze the coupling reaction, such as triethylamine (TEA), diisopropylethylamine (DIEA), or dimethylaminopyridine (DMAP).

(A) Aqueous Reaction Protocol

1. Prepare an amine-containing ligand solution consisting of a protein or small molecule dissolved in 100 ml of 0.1-*M* sodium phosphate, 0.15-*M* NaCl, pH 7.2 (coupling buffer), at a concentration of 1 to 20 mg protein/ml or 1 to 5 mg/ml of a small amine-containing molecule. Alternative coupling buffers for immobilization onto NHS-activated supports include 0.1-*M* MOPS (pH 7.0), 0.1- to 0.2-*M* phosphate (pH 7.2–7.5), 0.1 to 0.2-*M* $NaHCO_3$ (pH 8.0), or 0.1-*M* sodium borate (pH 8.5), which all may or may not contain NaCl. The pH of the ligand solution may be varied from approximately pH 7 to 9 to take into account differential reactivities of some ligands. The optimal concentration of ligand that is used for the immobilization reaction should be determined experimentally to obtain the best performance of the affinity matrix after coupling. This can be done using only 1 to 2 ml of activated support for each immobilization trial and proportionally reducing the quantities of reagents used for the reactions. Diamine spacer molecules should be coupled at much higher concentrations to avoid internal crosslinking while promoting coupling to only one end of the linker (i.e., 0.5–1.0 *M*). See the section on spacer arm coupling in this chapter for additional information.
2. In a fume hood, transfer 100 ml of an NHS ester- or NHS carbonate-activated support stored in isopropanol or acetone to a clean sintered glass filter funnel in the hood positioned over a suction filter flask. Drain the matrix of excess solvent by applying a vacuum and break up the resin with a plastic spatula into small pieces as the solvent is removed, then remove the vacuum to prevent gel drying. Wash the resin quickly with 2 to 3 bed volumes of deionized water to remove the majority of remaining solvent. When working with small quantities of activated support, this operation can be performed using spin columns for solvent removal and washings. Finally, wash the activated support with 100 ml of coupling buffer and drain to a wet cake. Perform all aqueous washing operations quickly to avoid extensive hydrolysis of the reactive groups prior to mixing with ligand solution.
3. Immediately add the washed support to the ligand solution with stirring to uniformly mix and suspend the resin within the medium. Continue mixing using an overhead paddle stirrer or by rotation using a sealed container. React for at least 1 h at room temperature. For sensitive ligands, all solutions can be cooled to 4°C and the reaction also performed in the cold; however, the reaction time should be at least doubled to accommodate a slower reaction rate. The majority of coupling will occur in the first 30 min at room temperature.
4. Upon completion of the reaction, transfer the support slurry to a clean sintered glass filter funnel suspended in a suction filter flask, drain excess ligand solution, and wash with several bed volumes of coupling buffer. The initial washes may be saved to determine the amount of ligand that coupled to the support. This is done by measuring the remaining ligand in washes and comparing this amount to the quantity initially reacted. The difference is the amount of ligand immobilized onto the support.
5. Block unreacted NHS ester groups by adding to the support an equal volume of 1-*M* ethanolamine in coupling buffer and mixing for 30 min. The amino end of ethanolamine will couple to the remaining NHS ester groups and leave hydrophilic hydroxyl groups on the support.
6. Wash the support extensively with water, 1-*M* NaCl, and water, using at least 5 bed volumes of wash for each solution. Store the immobilized ligand support at 4°C as a 50% aqueous slurry containing a preservative.

(B) Nonaqueous Coupling Protocol

1. In a fume hood, dissolve an amine-containing ligand in 100 ml of a dry organic solvent at a concentration of 1 to 5 mg/ml. Suitable solvents include acetone, dioxane, DMF, DMSO, DMAC, ethanol, and isopropanol. The use of methanol may also be done, but it could more easily than other alcohols undergo transesterification with the NHS ester-reactive groups and slow down or reduce the coupling yield. Other water-miscible but dry solvents may be used as long as the solvent does not harm or unduly collapse the particle structure of the activated chromatography support. Add to this solution a 2-times mole excess of an organic base over the

amount of ligand present, such as TEA, DIEA, or DMAP to catalyze the coupling reaction.

2. Drain 100 ml of an NHS ester-activated support of excess solvent by using a sintered filter funnel suspended in a suction filter flask in a fume hood. If the coupling reaction is to be done in a solvent other than the one in which the activated support was stored, then wash the matrix into the coupling solvent by initially pulling a vacuum on the support in the filter funnel while breaking up the resin into small pieces with a spatula. Do not let the support dry in the filter funnel during the process. Remove the vacuum and resuspend the matrix in an equivalent volume of the desired solvent. Then wash with several volumes of this solvent to fully exchange the matrix into it. Drain to remove excess solvent.
3. Add the activated support to the ligand solution with stirring to uniformly resuspend the resin. Mix in a fume hood by using a paddle stirrer or by rotation in a sealed container. Continue the reaction for 1 h at room temperature with constant mixing.
4. Transfer the gel reaction slurry to a clean sintered glass filter funnel and drain the excess ligand solution. Wash with several bed volumes of solvent to remove the majority of not-coupled ligand. Retain the washes to determine the amount of ligand that coupled to the matrix, if desired.
5. Block unreacted NHS ester groups by adding to the support an equal volume of 1-*M* ethanolamine in solvent and mixing for 30 min. The amino end of ethanolamine will couple to the remaining NHS ester groups and leave hydrophilic hydroxyl groups on the support.
6. Wash the support extensively with solvent to remove all traces of ethanolamine and any remaining uncoupled ligand (at least 5–10 bed volumes). Next, wash the support into aqueous solution by sequentially washing with increasing amounts of water in solvent until 100% water is used and all traces of solvent have been eliminated from the resin. The final affinity support may be stored at 4°C as a 50% slurry in water containing a preservative.

Carbonyl Diimidazole (CDI) Activation

N,N′-Carbonyl diimidazole (CDI; also called 1,1′-carbonyl diimidazole) is a highly reactive carbonylating compound originally developed for use in peptide synthesis, but which is also capable of activating carboxylates and hydroxyls on chromatography supports for the immobilization of amine-containing ligands (Paul and Anderson, 1960; Bartling *et al.*, 1973; Bethell *et al.*, 1979, 1987; Hearn *et al.*, 1981, 1983; Batista-Viera *et al.*, 2011). CDI is extremely sensitive to hydrolysis in water, where it decomposes to give two molecules of imidazole

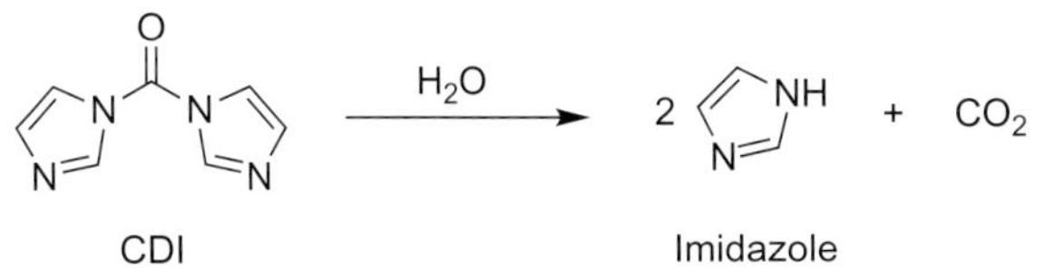

FIGURE 15.48 Hydrolysis of CDI in water to form CO_2 and imidazole.

with loss of CO_2. Placing CDI in an aqueous solution will result in vigorous fizzing as the CO_2 is released (Figure 15.48). All activation procedures, therefore, must be done in nonaqueous conditions using extremely dry solvents. Typically, water-miscible solvents are used so that the support can be washed in and out of aqueous solution by exchange with increasing or decreasing concentrations of the solvent in water. Solvents that work well in this process include acetone, dioxane, DMSO, DMF, and DMAC. Avoid alcohols such as methanol, ethanol, or isopropanol, as the hydroxyls on these compounds will react with CDI.

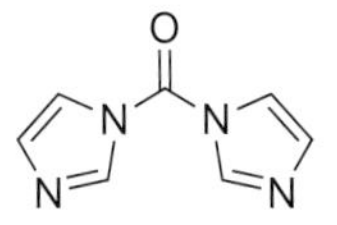

CDI
N,N′-Carbonyl diimidazole

The activation of a carboxylate group with CDI proceeds through transamidination to give a secondary amide with imidazole. This intermediate acyl imidazolide is reactive towards amines in proteins and other molecules, because the imidazole is a good leaving group. A coupling reaction involves attack of the nucleophilic amino group nitrogen on the carbonyl of the acyl imidazolide to displace the imidazole group and form an amide linkage with the ligand (Figure 15.49). Carboxylate-containing supports that can be activated with CDI include chromatography resins modified to contain a carboxylate-terminal spacer (see section on spacer arms, this chapter), membranes similarly modified with carboxylates, and small microparticles or nanoparticles that contain carboxylates (see Chapter 14, Section 4.2).

Hydroxyl-containing supports can also can be activated using CDI, including those that contain primary or secondary hydroxyls (Hearn *et al.*, 1983). In this case, CDI can be used to activate the hydroxyls to reactive imidazolyl carbamates, an intermediate that is similar in its reactions to NHS carbonates (described in the previous section). Hydroxylic resins activated in this manner can then be used to immobilize amino ligands through displacement of the imidazole leaving group on the carbonyl with subsequent formation of an alkyl carbamate linkage (Figure 15.50). The resulting urethane derivative

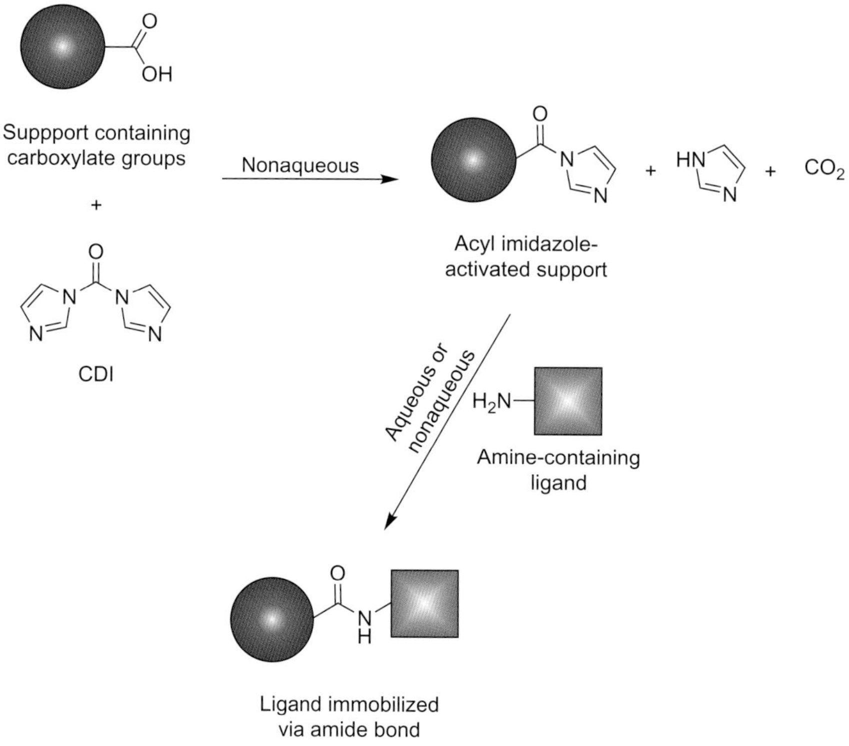

FIGURE 15.49 Activation of carboxylates with CDI and reaction with amine-containing ligands.

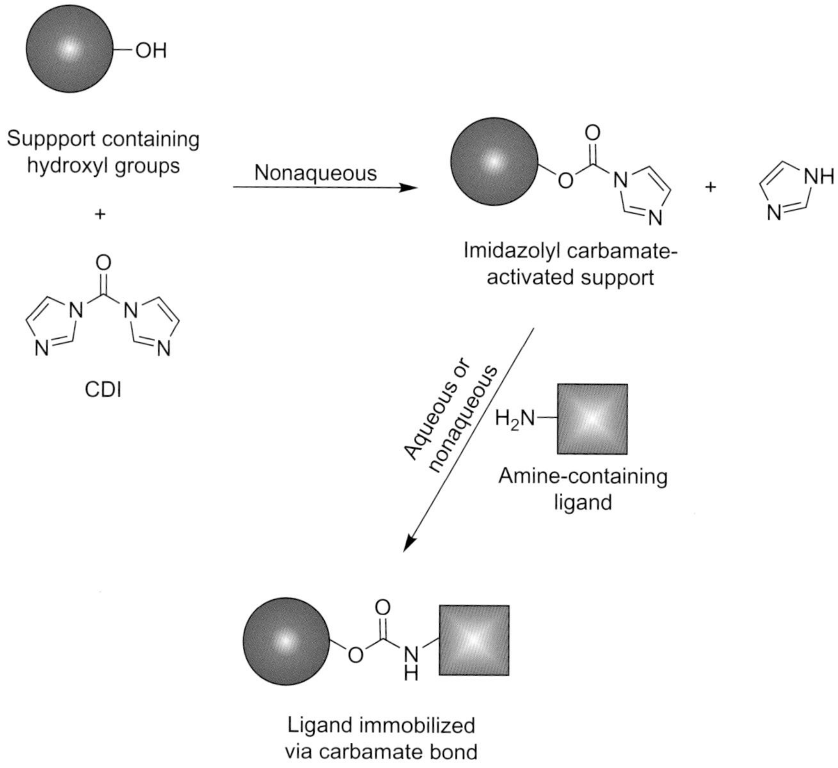

FIGURE 15.50 Activation of hydroxyls with CDI and reaction with amine-containing ligands.

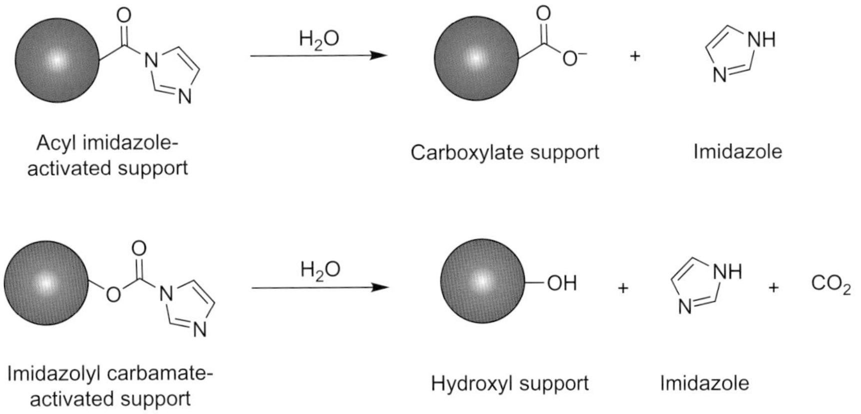

FIGURE 15.51 Hydrolysis of CDI-activated hydroxylic supports and CDI-activated carboxylate supports to give hydroxyls and carboxyls, respectively.

creates an uncharged bond on the support that has excellent chemical stability toward ligand leakage. While both the activation of carboxylates and the activation of hydroxyls with CDI produce an intermediate reactive carbonyl derivative, the respective bonds they form with amines are different, as are the hydrolysis products that are formed if a ligand does not couple in aqueous solution. If a CDI-activated carboxylate undergoes hydrolysis instead of ligand coupling, the result is a reformation of the original carboxylic acid group on the support. However, if a CDI-activated hydroxyl group is hydrolyzed, the result is loss of imidazole and CO_2 along with reformation of the original hydroxyl. Thus, if a carboxylate support that has been activated with CDI should hydrolyze during a coupling reaction, it can create negatively charged groups on the matrix, which may cause subsequent nonspecific binding during affinity separations; whereas, if a hydroxylic support activated with CDI undergoes hydrolysis, it reverts back to a hydrophilic and uncharged hydroxyl, which helps to reduce nonspecific interactions (Figure 15.51). These reactions and their hydrolysis products are similar to the activation of hydroxylic- or carboxylic-containing supports with DSC (described previously).

CDI-activated carboxylate supports or CDI-activated hydroxyl supports both will react nearly exclusively with primary amines on proteins and other ligands, even in the presence of secondary amines (Rannard and Davis, 2000). This surprising specificity can be used to advantage during the immobilization of spacer arms or small ligands, which may contain both primary and secondary amines within their structure. This may also be an important consideration in certain ligand immobilization needs in choosing CDI activation over the use of NHS ester- or NHS carbonate-activated supports, because they have the potential to react with secondary amines as well as primary amines.

Amine-containing supports may also be activated with CDI to give an intermediate imidazolyl isourea derivative. This reactive group also can be coupled with primary amines on ligands to result in an isourea (or urethane) bond (Figure 15.52). While the activation of amine-containing supports is less frequently performed than the activation of carboxylate- or hydroxyl-containing supports, it does open up an additional route of activation and ligand coupling that avoids having to convert the amine to some other functionality prior to activation. The only potential deficiency of this method is the fact that amine-containing supports typically have greater nonspecific binding with biomolecules due to their net positive charge at physiological pH. Hydrolysis rather than ligand coupling of the CDI-activated intermediate would cause a reversion to a primary amine on the support, which would be the major cause of a positively charged affinity support.

Hydroxylic supports (such as agarose, Trisacryl, or Toyopearl matrices) that are activated with CDI and stored in nonaqueous solvent are stable for years if kept in a sealed container at 4°C. In addition, the stability of the imidazolyl carbamate reactive groups is much greater toward hydrolysis in aqueous coupling buffer than NHS ester- or NHS carbonate-activated supports. At pH 8.5 to 9.0 in sodium borate buffer, a CDI-activated agarose support can take up to 30 h at room temperature to completely lose activity due to hydrolysis. By comparison NHS ester or NHS carbonate supports have half-lives due to hydrolysis measured in minutes in this pH range.

The immobilization of amine-containing ligands onto CDI-activated supports can be performed in nonaqueous or aqueous environments. It is particularly advantageous to couple small, organic soluble ligands in 100% organic solvent, as it eliminates the potential

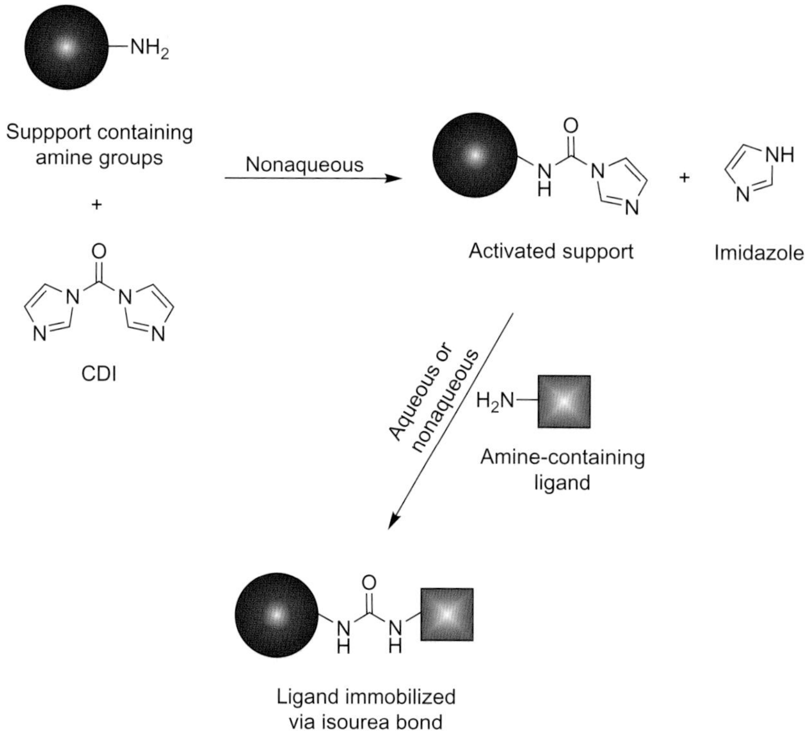

FIGURE 15.52 CDI activation of amine-containing supports and coupling to amine ligands.

for hydrolysis during the reaction, thus maximizing the coupling yield. For aqueous-phase reactions, as in the coupling of proteins or other amine-containing biomolecules, the optimal pH for immobilization often is dependent on the pI or pK_a of the ligand. For maximal efficiency of coupling, the reaction conditions should be controlled at a pH representing at least 1 pH unit above the ligand's pI or pK_a point. This ensures that a majority of amines on the ligand will be unprotonated and thus at their maximal potential nucleophilicity for reaction with the active groups on the matrix.

CDI ACTIVATION PROTOCOL

The following protocols describe the activation of supports using CDI followed by a number of coupling methods that can be used to immobilize affinity ligands. Regardless of the support being used, the activation reaction should be done in anhydrous organic solvent to avoid hydrolytic breakdown of CDI or the active intermediate formed on the matrix. Make certain to use extensive solvent washing protocols to completely remove any traces of water from a hydrated support prior to the addition of CDI. Finally, when working with solvents all operations should be carried out in a fume hood using personal protective equipment and precautions to avoid static electricity. Dispose of all solvents according to hazardous organic waste guidelines.

1. Wash 100 ml of a hydroxyl-, carboxyl-, or amine-containing chromatography support material (such as crosslinked agarose) with water (at least several bed volumes) to remove preservatives and storage solution components. In a fume hood, sequentially wash the support into increasing concentrations of acetone-in-water (such as 25% (v/v), 50%, and 75% acetone/water) and finally with 100% dry acetone for at least 10 to 20 bed volumes to remove the last traces of water. Drain the support to a wet gel cake and then pull a gentle vacuum on the filter funnel while breaking up the gel cake with a spatula to obtain moist, evenly divided pieces, which still contain acetone within the internal pores of the beads. Be careful not to allow the support to dry, as this will cause particle collapse and irreversible damage to the pore structure of some matrices. Remove the vacuum as soon as the support is fully divided into small pieces (it will look like moist, fluffy snow). Note: Alternative solvents can be used for the activation process including dioxane, DMF, DMAC, or DMSO, which can be substituted for acetone depending on the compatibility of the matrix to a particular solvent. For example, some supports such as dextran will maintain their swollen particle nature best in solvents such as DMSO or DMF, whereas acetone will cause them to

unacceptably shrink and restrict access to the inner pore structures.

2. In the fume hood, dissolve 10 g of CDI in 100 ml of dry acetone with stirring.
3. Add the acetone wet gel cake to the CDI solution with stirring to uniformly resuspend the support. Mix the gel suspension using an overhead paddle stirrer or by end-over-end rocking in a sealed container within the fume hood.
4. React for 1 h at room temperature with mixing.
5. In the fume hood, wash the activated support with acetone to thoroughly remove excess reactant and reaction byproducts. Wash with at least 10 bed volumes of solvent to ensure removal of the last traces of unreacted CDI.
6. Store the CDI-activated support as a 50% slurry in dry acetone at 4°C until use.

Using the above protocol, the CDI-activated support may contain as much as 100 µmoles of reactive groups per milliliter (using crosslinked agarose), which is more than sufficient for coupling primary amine-containing affinity ligands of all types at very high density, if required.

LIGAND COUPLING TO CDI-ACTIVATED SUPPORTS

The following ligand coupling protocols represent generalized methods that should be optimized for a particular ligand to obtain the best possible affinity support performance in its intended application. CDI-activated supports typically have slower reaction kinetics when coupling proteins as compared to NHS ester-activated supports and therefore extended reaction times often are necessary to realize the best coupling yields (i.e., 16–24 h). However, for coupling small amine-containing ligands or spacer molecules, maintaining a high concentration should give maximal yields within 1 to 2 h at room temperature. Take care, however, in creating an immobilized ligand at very high density on the support, because this may cause nonspecific binding or result in too high an affinity toward the target molecule, making it difficult to elute it from the support.

For ligands that are more soluble or stable in organic solvent, the entire coupling reaction can be performed under nonaqueous conditions using the same solvent used in the activation process, if appropriate. If a nonaqueous reaction is performed, the addition of an organic base (e.g., TEA, DIEA, DMAP) at a concentration that is 2 to 3 times that of the CDI activation level on the matrix will catalyze the process and increase the reaction rate. Reactions performed in organic solvent have the advantage of not undergoing hydrolysis, thus increasing potential yields of the ligand coupling process. Alternatively, a mixture of aqueous buffer with solvent can be used to maintain the solubility of some ligands. In this situation, the ligand may be first dissolved in organic solvent and then added to the reaction buffer just prior to adding the activated support. The final amount of solvent added to an aqueous/organic solvent reaction is often determined by the percentage of solvent that can be tolerated by the buffered solution before buffer salts begin to precipitate. The maximal level of organic solvent addition can be discovered by making a series of solutions with increasing amounts of organic solvent to an aqueous coupling buffer and then identifying which concentrations do not cause buffer precipitation.

(A) Aqueous Coupling Protocol

1. In a fume hood, drain 100 ml of CDI-activated support of solvent using a sintered glass filter funnel suspended in a suction filter flask. Gently pull a vacuum to remove most of the excess solvent while breaking up the support into small, finely divided pieces, but be careful not to allow the matrix to dry out. Stop using suction as soon as the support is divided into small pieces. When coupling ligands that might precipitate or be damaged by the presence of some residual solvent, the support should be quickly washed using several bed volumes of cold deionized water to remove the organic phase. Drain the support to a wet gel cake.
2. Dissolve the ligand to be coupled in 100 ml of a buffered solution at pH 8.5 to 11, or in a buffer having a pH at least 1 pH unit above the pK_a or pI of the ligand or protein. Proteins may be coupled at a concentration of 1 to 20 mg/ml or, for small molecules, at a concentration of 1 to 5 mg/ml of an amine-containing ligand. The optimal concentration of ligand that is used for the immobilization reaction should be determined experimentally to obtain the best performance of the affinity matrix after coupling. Suitable coupling buffers include 0.1-*M* sodium borate, pH 8.5 or 0.1- to 0.5-*M* sodium carbonate, pH 9 to 11. Avoid buffers or solution additives that contain amines, such as Tris, glycine, or imidazole, as these will compete with the ligand coupling reaction.
3. Add the activated wet gel cake to the ligand solution with stirring to fully resuspend the gel. Mix the reaction slurry for at least 24 h using an overhead paddle stirrer or by end-over-end rocking in a sealed container. When using conditions at a pH of less than 9, the reaction time should be extended to 30 h to obtain maximal coupling yields. The reaction may be performed at 4°C or at room temperature, depending on the stability of the ligand or protein being immobilized. Excess CDI-reactive groups

on the support may be blocked by the addition of ethanolamine to the reaction slurry at a final concentration of at least 0.1-*M*. Note: Titrate the ethanolamine solution to the proper pH by addition of HCl before adding it to the reaction solution. Continue to mix for 1 h at room temperature.

4. Filter and wash the affinity support to remove uncoupled ligand and reaction byproducts using coupling buffer, water, 1-*M* NaCl, and again with water. Depending on the ligand being coupled, other wash solutions may be used to completely remove unreacted ligand, such as detergents, denaturants, and high or low pH conditions. Finally, store the affinity support in water containing a preservative at 4°C until used.

(B) Nonaqueous Coupling Protocol

1. In a fume hood, drain 100 ml of CDI-activated support of excess acetone using a sintered glass filter funnel suspended in a suction filter flask.
2. Dissolve the amine-containing ligand to be coupled in 100 ml of acetone (or another water-miscible solvent) at a concentration of 1 to 5 mg/ml, which is normally sufficient for the immobilization of a small organic compound. The optimal concentration of ligand to be used in the coupling reaction should be determined experimentally on small quantities of activated support to identify the best affinity support performance in its intended application. Add an organic base to the ligand solution, such as DMAP, DIEA, or TEA, to make a final concentration of 2-m*M*.
3. Add the activated wet gel cake to the ligand solution with stirring to fully resuspend the gel. Mix the reaction slurry for 1 to 2 h using an overhead paddle stirrer or by end-over-end rocking in a sealed container. Longer reaction times may be used if appropriate. Excess CDI-reactive groups on the support may be blocked by the addition of ethanolamine to the reaction slurry at a final concentration of at least 0.1 *M*. Continue to mix for 1 h at room temperature.
4. Transfer the gel slurry to a sintered glass filter in the fume hood that is suspended in a suction filter flask and wash extensively (at least 10 bed volumes) with solvent to remove the remaining ligand and reaction byproducts. If the ligand is detectable, continue to wash the support until no further ligand is detected in the washings. Finally, drain the support of excess solvent by pulling a gentle vacuum on the filter flask while breaking up the support into small, finely divided pieces using a spatula, but be careful not to allow the matrix to dry out. Once the support is broken into small pieces, remove the vacuum and resuspend the gel in water with mixing. Continue to wash the support with water until all traces of solvent have been removed. Additional washes with 1-*M* NaCl as well as low and high pH conditions may be carried out as appropriate. Finally, wash with water and store the affinity support as a 50% slurry in water containing a preservative at 4°C until used.

Activation Using FMP (2-Fluoro-1-Methylpyridinium)

Ngo in 1986 first described the use of 2-fluoromethylpyridinium (FMP), as the toluene-4-sulfonate salt, in a simple process to activate primary and secondary hydroxyls on supports and surfaces for the coupling of amine- or thiol-containing ligands. Under nonaqueous conditions, FMP reacts with hydroxyl groups on chromatography supports to form an intermediate reactive methylpyridinium ether group, which can be coupled to ligands in aqueous or nonaqueous conditions to give stable secondary amine or thioether linkages. The immobilization reaction occurs in buffered solutions within the pH range of 5 to 10, but with greater reaction rates observed in the higher pH range. The coupling of ligands may also be performed in dry solvents with the addition of an organic base to catalyze the reaction (e.g., TEA, DIEA, DMAP) for ligands that are insoluble in aqueous solution.

The leaving group in the FMP coupling reaction (due to nucleophilic displacement with amino- or thiol-containing ligands) is 1-methyl-2-pyridone, which is a convenient chromophore that can be used to determine the activation level of an FMP-activated support. The molar extinction coefficient of this leaving group is $5900\,M^{-1}\,cm^{-1}$ at 297 nm, which also permits the coupling reaction to be followed provided the reaction is performed under nonaqueous conditions to eliminate simultaneous hydrolysis. Even in aqueous buffers, FMP-activated supports display relatively low hydrolysis rates. In acidic conditions, the activated support hydrolyzes extremely slowly and even after months of storage retains a significant number of reactive groups. Ngo (1988) found that in 10-m*M* phosphoric acid, an FMP-activated support stored at 4°C for months was still able to couple 15 mg/ml of BSA. Even under alkaline conditions the activated support hydrolyzes slowly with a half-life of about 130 h at pH 7 and about 35 h at pH 9. Long-term storage in aqueous solution, however, still is not recommended, because some reactive groups will degrade over time and the reproducibility of coupling ligands may be negatively affected. One benefit of FMP-mediated immobilization is that if the activated support does hydrolyze during ligand coupling, the result is to reform the original hydroxyls on the support, thus leaving behind no groups that potentially can cause nonspecific binding. In addition, successful ligand coupling

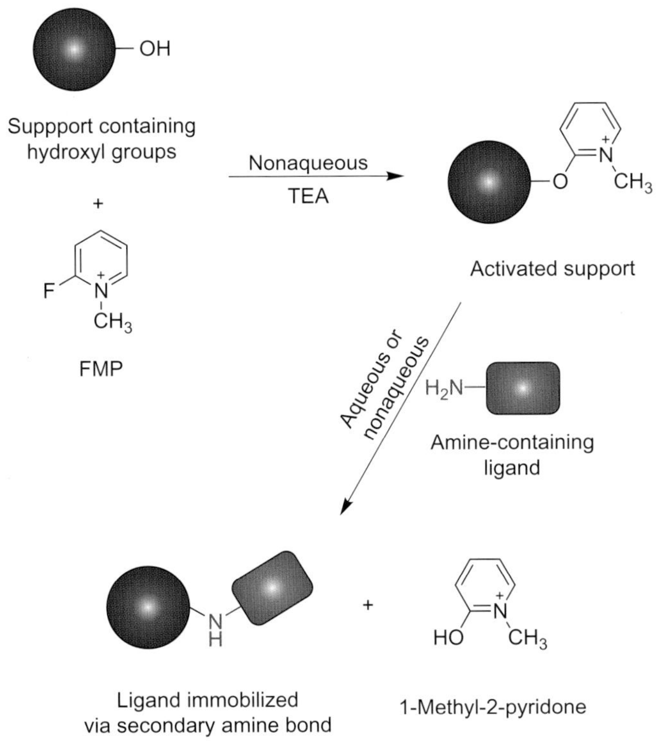

FIGURE 15.53 Activation of a hydroxylic support with FMP and coupling of amine- or thiol-containing ligands.

displaces the pyridone group, which bears the oxygen originating from the hydroxyl on the support material. Therefore, the net result of the immobilization reaction is to replace the hydroxyl group on the matrix with the nucleophile of the ligand, forming a new, chemically stable bond (Figure 15.53). This process is similar to that of tosyl and tresyl activation, which also replaces the activated hydroxyl with the amino group of the ligand (see tosyl and tresyl activation discussion, this chapter).

Amine- or thiol-containing ligands react with the FMP-activated support within the first several hours to result in at least 90% yields with regard to the amount of reactive groups that can react. Thiols appear to react at slightly greater rates due to their somewhat greater nucleophilicity. FMP-activated agarose or Toyopearl type supports should be able to immobilize proteins such as BSA at densities of 10 to 20 mg/ml gel.

After the coupling reaction is complete, the residual reactive groups should be blocked with a small molecule to ensure that no further reactivity remains that might couple molecules nonspecifically from subsequent samples applied to the affinity gel. This blocking step is especially important with FMP-activated matrices, because of the extended hydrolytic stability of the reactive groups. Blocking with ethanolamine, for example, will remove these active groups and result in numerous hydrophilic hydroxyls being created in their place.

FMP ACTIVATION PROTOCOL

The following protocols describe two methods of activating hydroxyl-containing supports with FMP in nonaqueous environments. The first one uses acetone and DMF as the solvents and the second method uses acetone and acetonitrile. Other solvents may be used for the activation step, but they should be water miscible so that water can be removed from the support by washing and anhydrous to prevent hydrolysis of the activation agent before the hydroxyls are activated.

(A) ACTIVATION OF TOYOPEARL SUPPORTS WITH FMP IN DMF

1. Wash 100 ml of a Toyopearl resin (such as HW-65 or HW-75) with water (at least several bed volumes) to remove preservatives and storage solution components. In a fume hood, sequentially wash the support into increasing concentrations of acetone-in-water (such as 25% (v/v), 50%, and 75% acetone/water) and finally wash with 100% anhydrous acetone for at least 10 to 20 bed volumes to remove the last traces of water. Drain the support to a wet gel cake and then pull a gentle vacuum on the filter funnel while quickly breaking up the gel cake with a spatula to obtain moist, evenly divided pieces, which still contains acetone within the internal pores of the beads. Be careful not to allow the support to dry. Remove the vacuum and add 200 ml of anhydrous DMF with stirring. Filter to remove excess DMF and then further wash the support with 500 ml of DMF to remove the acetone. Finally, filter off the excess solvent and keep as a moist gel cake.
2. In the fume hood, dissolve with stirring 7.1 g of FMP (5 mmoles; as the toluene-4-sulfonate salt) in 100 ml of dry DMF containing 6 mmoles TEA (3.03 g or 4.18 ml) as catalyst. Alternative organic base catalysts may be used, including DIEA or DMAP, which serve to accept protons during the course of the activation reaction.
3. Add the DMF wet gel cake to the FMP solution with stirring to uniformly resuspend the support. Mix the gel suspension using an overhead paddle stirrer or by end-over-end rocking in a sealed container within the fume hood.
4. React for 1 h at room temperature with mixing.
5. In the fume hood, wash the activated support with 500 ml of DMF to thoroughly remove excess reactant and reaction byproducts. Filter off excess DMF under a gentle vacuum and break up the gel cake into small pieces with a spatula. Resuspend the gel in acetone and then wash with at least 10 bed volumes of acetone.
6. Store the FMP-activated Toyopearl support as a 50% slurry in dry acetone at 4°C until use.

(B) ACTIVATION OF CROSSLINKED AGAROSE SUPPORTS WITH FMP IN ACETONITRILE

1. Wash 100 ml of a hydroxyl-containing, crosslinked agarose support (such as Sepharose CL-4B) with water (at least several bed volumes) to remove preservatives and storage solution components, then sequentially wash the gel in a fume hood with increasing concentrations of acetone-in-water (25% (v/v), 50%, and 75% acetone/water; 500 ml each). Finally, wash with 100% anhydrous acetone for at least 10 bed volumes to remove the last traces of water. Drain the support to a wet gel cake and then wash with at least 500 ml of dry acetonitrile. Finally, filter off the excess solvent and keep as a moist gel cake.
2. In a fume hood, dissolve 7.1 g of FMP (toluene-4-sulfonate salt) in 300 ml of acetonitrile containing 4.18 ml of TEA.
3. Add the FMP solution to the moist gel cake and stir to uniformly mix the suspension.
4. Continue to react with mixing for 1 h at room temperature using an overhead paddle stirrer or end-over-end rocker.
5. In the fume hood, wash the activated support with 500 ml of acetonitrile to thoroughly remove excess reactant and reaction byproducts. Filter off excess acetonitrile under a gentle vacuum and break up the gel cake into small pieces with a spatula. Resuspend the gel in acetone and then wash with at least 10 bed volumes of acetone to remove the acetonitrile.
6. Store the FMP-activated agarose support as a 50% slurry in dry acetone at 4°C until use.

LIGAND COUPLING TO FMP-ACTIVATED SUPPORTS

The following immobilization protocols are generalized methods for coupling amine- or thiol-containing affinity ligands to FMP-activated chromatography supports. The optimal protocol for a given affinity ligand should be discovered through experimentation using different coupling conditions and concentrations of ligand to produce the best-performing affinity support in the intended application. The first procedure is an aqueous coupling method that is appropriate for immobilizing proteins or small water soluble molecules. In the second procedure, a nonaqueous method is described that can be used to maximize the coupling yield (without competing hydrolysis) or for use with ligands that are not soluble in aqueous buffers.

(A) AQUEOUS COUPLING PROTOCOL

1. In a fume hood, drain 100 ml of the FMP-activated support of solvent using a sintered glass filter funnel suspended in a suction filter flask. Gently pull a vacuum to remove most of the excess solvent while breaking up the support into small, finely divided pieces using a spatula, while also being careful not to allow the matrix to dry out. Stop using suction as soon as the support is divided into small pieces. When coupling ligands that might precipitate or be damaged by the presence of some residual solvent, the support should be quickly washed using several bed volumes of deionized water to remove the majority of acetone still present within the gel. Drain the support to a wet gel cake.
2. Dissolve the ligand to be coupled in 100 ml of a buffered solution at pH 7 to 9. As a guideline, proteins may be coupled at a concentration of 1 to 20 mg/ml or for small molecules, at a concentration of 1 to 5 mg/ml for an amine-containing ligand (or thiol-containing ligand). The optimal concentration of ligand that is used for the immobilization reaction should be determined experimentally to obtain the best performance of the affinity matrix after coupling. Suitable coupling buffers include 0.1 *M* sodium borate, pH 8.5, or 0.1- to 0.5-*M* sodium carbonate, pH 8.5 to 9. More physiological coupling conditions may also be used, such as 0.1-*M* sodium phosphate, pH 7.5, especially for molecules sensitive to higher pH conditions. Avoid buffers or solution additives that contain amines or thiols, such as DTT, 2-mercaptoethanol, glutathione, Tris, glycine, or imidazole, as these will compete with the ligand coupling reaction.
3. Add the activated wet gel cake to the ligand solution with stirring to fully resuspend the gel. Mix the reaction slurry for at least 2 h using an overhead paddle stirrer or by end-over-end rocking in a sealed container. For ligands that have slower reaction kinetics with the FMP reactive group, the reaction time may be increased to overnight. In addition, when using conditions at a pH of less than 9, the reaction time can be extended to as much as 30 h to obtain maximal coupling yields. The reaction may be carried out at 4°C or at room temperature, depending on the stability of the ligand being immobilized.
4. Filter and wash the affinity support to remove uncoupled ligand and reaction byproducts using coupling buffer, water, 1-*M* NaCl, and again with water. Depending on the ligand being coupled, other wash solutions may be used to completely remove unreacted ligand, such as detergents, denaturants, and high or low pH conditions. Finally, store the affinity support in water containing a preservative at 4°C until used.

(B) NONAQUEOUS COUPLING PROTOCOL

1. In a fume hood, drain 100 ml of the FMP-activated support of excess acetone using a sintered glass filter funnel suspended in a suction filter flask.

2. Dissolve the amine-containing ligand to be coupled in 100 ml of acetone (or another water-miscible solvent) at a concentration of 1 to 5 mg/ml, which is normally sufficient for the immobilization of a small organic compound. If a bifunctional spacer molecule is to be coupled to the activated gel, such as a diamine compound, then much higher concentrations should be used (i.e., 0.5–1.5-*M*) to avoid internal crosslinking of the support during the reaction. The optimal concentration of ligand to be used in the coupling reaction should be determined experimentally on small quantities of activated support to identify the best affinity support performance in its intended application. Add an organic base to the ligand solution, such as DMAP, DIEA, or TEA, to make a final concentration of 2-m*M*.
3. In a suitable vessel, mix the activated wet gel cake with the ligand solution with stirring to fully resuspend the gel. Mix the reaction slurry for 1 to 2 h using an overhead paddle stirrer or by end-over-end rocking in a sealed container. Longer reaction times may be used if needed to obtain maximal yields, if appropriate.
4. Transfer the gel slurry to a clean sintered glass filter in the fume hood that is suspended in a suction filter flask and wash it extensively (at least 10 bed volumes) with solvent to remove the remaining ligand and reaction byproducts. Wash with considerably more solvent if the initial concentration of ligand or spacer molecule in the reaction medium was particularly high. If the ligand is easily detectable in the filtrates, continue to wash the support until no further ligand is detected in the washings. Finally, drain the support of excess solvent by pulling a gentle vacuum on the filter flask while breaking up the support into small, finely divided pieces using a spatula, but be careful not to allow the matrix to dry out. Once the support is broken into small pieces, remove the vacuum and resuspend the gel in water with mixing. Continue to wash the support with water until all traces of solvent have been removed. Additional washes with 1-*M* NaCl as well as low and high pH conditions may be done as appropriate to remove any noncovalently bound ligand molecules. Finally, wash with water and store the affinity support as a 50% slurry in water containing a preservative at 4°C until used.

Activation Using Organic Sulfonyl Chlorides

Organic sulfonyl chlorides are activating agents that can facilitate the conjugation of hydroxyl-containing compounds to other nucleophiles, particularly amine-containing ligands. They have also been used in organic synthesis for many years (Tipson, 1944; Brown *et al.*, 1967; Beard *et al.*, 1973; Kabalka *et al.*, 1986; Whitaker *et al.*, 2006) and were first applied to the immobilization of ligands onto hydroxylic chromatography supports by Nilsson and Mosbach (1980) (also see Nilsson and Mosbach, 1981, 1984; Nilsson *et al.*, 1981). Perhaps the most well-known organic sulfonyl chloride compounds used for synthesis are 4-toluenesulfonyl chloride, which also is referred to as tosyl chloride or TsCl (or TosCl), and methanesulfonyl chloride (also called mesyl chloride or MsCl). The sulfonyl chloride group of TsCl or MsCl can react with a hydroxyl or an amine to form a sulfonyl ester or a sulfonamide group, respectively (Carey and Sundberg, 2007) (Figure 15.54). The sulfonyl ester formed with hydroxyls is further reactive with other nucleophiles, such as amine-containing affinity ligands, to facilitate their covalent linkage to activated chromatography supports.

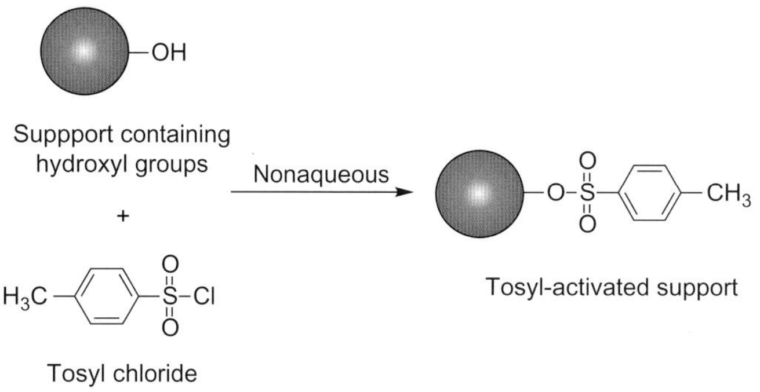

FIGURE 15.54 Activation of a hydroxylic matrix with TsCl to form reactive sulfonate esters.

Sulfonate esters that are formed from the activation of hydroxyl groups on chromatography supports with TsCl create particularly good leaving groups, which get displaced as an amine-containing ligand couples to the matrix (Figure 15.55). In this case, the reaction essentially converts the hydroxyl on the support to a secondary amine derivative that bonds the ligand to the matrix. Tosylates also can react with other nucleophiles, such as with hydroxyl groups under higher pH conditions (as the alkoxide, RO^-) to form ether linkages, with thiols (as the thiolate anion RS^-) to form thioether bonds, and also under alkaline conditions with OH^-, which causes hydrolysis back to the hydroxyl. Reaction with these groups in nonaqueous conditions requires the presence of an organic base to act as a proton acceptor to catalyze the coupling.

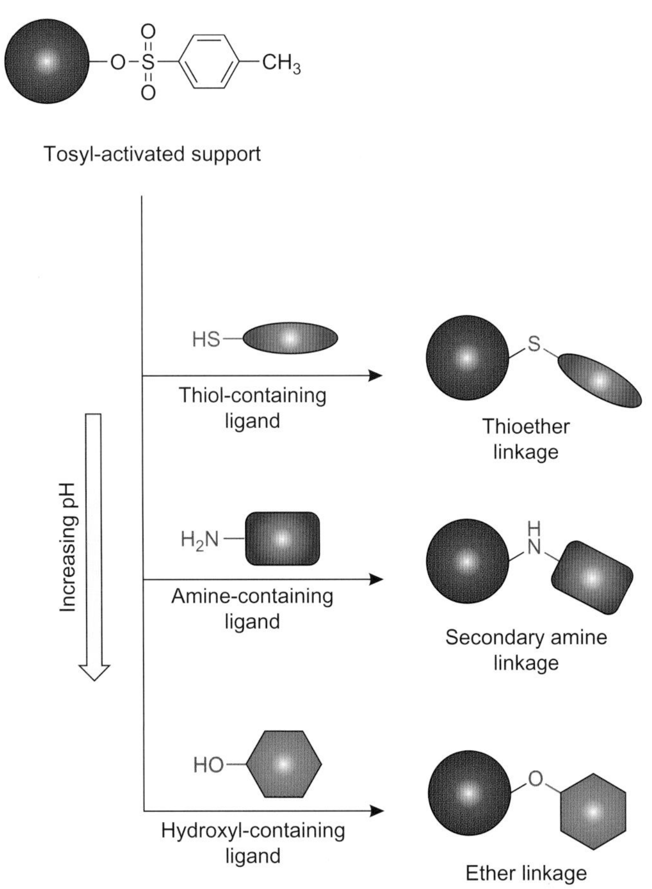

FIGURE 15.55 Reactions of active tosylates with amines, thiols, and hydroxyls.

The attack of a nucleophile on a sulfonate ester can occur through one of two reaction mechanisms, S_N1 or S_N2, the path of which may be governed by the solvent environment and the chemical groups immediately adjacent to the ester (Cremlyn, 1996; James and Cremlyn, 2002). The normal site of nucleophilic attack is on the carbon atom attached to the sulfonyl ester, which causes cleavage of the ester with concomitant bond formation of the nucleophile to the carbon atom. However, it has been demonstrated that nucleophilic attack can also occur onto the sulfur atom of the sulfonyl ester, which in this case may cause displacement of the constituent attached to the sulfonate opposite to the ester. The result of this reaction is the formation of a sulfonamide linkage with amine nucleophiles and retention of the sulfonate onto the originally activated hydroxyl group on the matrix (Figure 15.56).

Using TsCl or MsCl, the alternative reaction product consisting of a sulfonamide is an extremely minor side reaction that may not occur to any appreciable extent when immobilizing amine-containing ligands. Nevertheless, the reaction mechanism for coupling ligands using tresyl chloride (2,2,2-trifluoroethanesulfonyl chloride) has been found to be quite different. Tresyl chloride was first used for organic synthesis reactions by Crossland *et al.* (1971) and later employed for the activation of agarose chromatography supports by Mosbach and Nilsson (1981). It was Crossland *et al.* (1971) that demonstrated that the reaction of a tresyl ester with an amine compound resulted in the loss of 2,2,2-trifluoroethane sulfonate as the leaving group. This reaction process was believed to be true for tresyl-activated hydroxylic chromatography supports, as well (Nilsson and Mosbach, 1981). However, it subsequently was discovered that the actual reaction mechanism leads to the formation of a sulfonamide linkage on the matrix with an amine-containing ligand, not a secondary amine linkage as in the use of tosyl chloride coupling (Demiroglou and Jennissen, 1990; Demiroglou *et al.*, 1994). Careful elemental analysis indicated that the sulfur atom remains behind and the fluorine atoms are released in the immobilization process. In addition, the reaction of a thiol-containing ligand was found to yield a thiosulfate ester bond, which can have avid binding characteristics toward some proteins in affinity separations, particularly antibodies (Hutchens and Porath, 1986; Scoble and Scopes, 1997; Hansen *et al.*, 1998) (Figure 15.57).

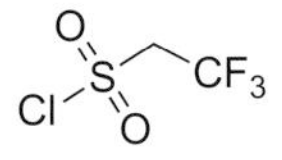

Tresyl chloride

The unique reaction mechanism of tresyl ester-activated supports may be due to the presence and strong electron-withdrawing properties of the trifluoro group. This may create a more powerful electrophilic center at the sulfur atom than that normally produced at the carbon atom when using tosyl ester-mediated activation and coupling. Thus, an attack by a nucleophile such as an amine would occur at the sulfur atom center of the sulfonate group instead of at the carbon atom immediately adjacent to the ester, which in turn would cause displacement of the trifluoroacetyl group. Under basic conditions with an abundance of OH^- ions, this group is further transformed after displacement within the reaction medium into acetic acid and three fluorine ions (F^-). In addition, the strong electron-withdrawing effects of the trifluoroacetyl group results in faster reaction kinetics for tresyl ester-activated supports than for tosyl ester-activated ones.

Sulfonyl chloride activation and coupling has been used for the immobilization of numerous affinity ligands onto hydroxylic supports of all types. In addition to its use in the activation of beaded chromatography supports, tosyl chloride has been used to activate partially hydrolyzed rayon/polyester cloth for the

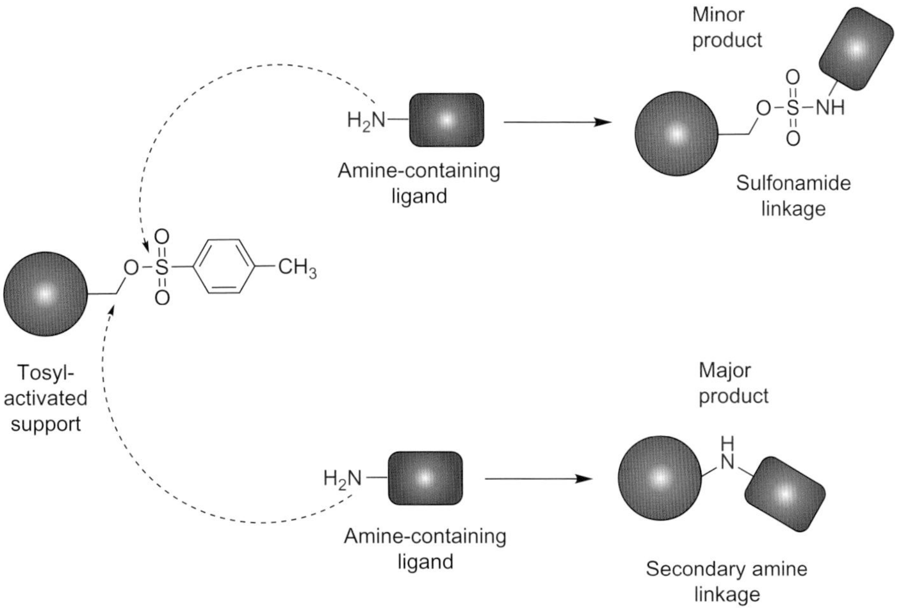

FIGURE 15.56 Alternative routes of nucleophilic attack on sulfonyl esters.

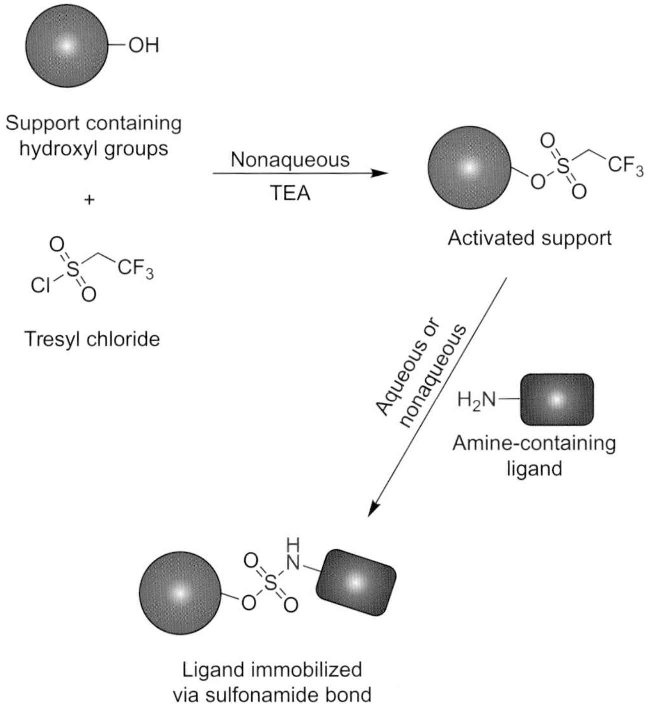

FIGURE 15.57 Tresyl chloride activation and coupling reactions.

coupling of antibodies and other proteins for purification of target molecules (Boyd and Yamazaki, 1993) as well as for the immobilization of Aspergillus oryzae β-galactosidase onto cotton cloth (Albayrak and Yang, 2002) and for the general coupling of amine nucleophiles onto cellulose (Heinze *et al.*, 2001). Even relatively inert membranes, such as polypropylene, have been modified using oxygen-plasma treatment to produce –OH functionalities and then activated with tosyl chloride for the coupling of peptides (Gérard *et al.*, 2011). In addition, tresyl chloride activation has been used to immobilize fibronectin onto titanium solid phases (Hayakawa *et al.*, 2003) and to couple antibodies onto PEG-modified particles for use in immunoassays (Chen *et al.*, 2009).

Still another alternative sulfonyl chloride compound to that of TsCl or tresyl chloride has been described for activation of chromatography supports. The reagent 4-fluorobenzenesulfonyl chloride (called fosyl chloride or fosCl) has been used to activate hydroxylic groups on polymeric supports for the coupling of proteins and other biomolecules (Chang *et al.*, 1992). It was found that the strong electron-withdrawing properties of the fluorobenzene group provides enhanced reaction rates for the activation and coupling steps over that of tosyl chloride.

SULFONYL CHLORIDE ACTIVATION PROTOCOL

The following activation protocol is a generalized procedure for the use of tosyl chloride to activate hydroxyl groups on beaded chromatography supports. Other sulfonyl chloride compounds may be used under similar conditions, such as tresyl chloride, mesyl chloride, or fosyl chloride. The activated support may be used to couple amine-containing affinity ligands and proteins as well as thiol-containing and hydroxyl-containing molecules. All procedures should be carried out in a well-ventilated fume hood to avoid exposure to potentially toxic or reactive compounds and solvent fumes. The use of a sintered glass filter funnel suspended in a suction filter flask can facilitate the washing steps.

1. Wash 100 ml of a hydroxyl-containing chromatography support (such as crosslinked agarose, Toyopearl, or Trisacryl resin) with water (at least several bed volumes) to remove preservatives and storage solution components. In a fume hood, sequentially wash the support into increasing concentrations of acetone-in-water (such as 25% (v/v), 50%, and 75% acetone/water). Finally, wash with 100% dry acetone for at least 10 to 20 bed volumes to remove the last traces of water. Drain the support to a wet gel cake and then pull a gentle vacuum on the filter funnel while breaking up the gel cake with a spatula to obtain moist, evenly divided pieces, which still contains acetone within the internal pores of the beads. Be careful not to allow the support to dry, as this will cause particle collapse and irreversible damage to the pore structure of some matrices. Remove the vacuum as soon as the support is fully divided into small pieces (it will look like moist, fluffy snow). Note: alternative solvents can be used for the activation process including dioxane, DMF, DMAC, or DMSO, which can be substituted for acetone depending on the compatibility of the matrix to a particular solvent. For example, some supports such as dextran will maintain their swollen particle nature best in solvents such as DMSO or DMF, whereas acetone will cause them to unacceptably shrink and restrict access to the inner pore structures for activation. Whichever solvent is used for the activation reaction, it should be anhydrous to prevent decomposition of the sulfonyl chloride by hydrolysis.
2. In the fume hood, dissolve 11.75 g of tosyl chloride (61.6 mmoles) in 100 ml of dry acetone with stirring. If an alternative sulfonyl chloride activating agent is to be used, add an equivalent mole quantity to the acetone solvent, and dissolve with mixing.
3. Add the acetone wet gel cake to the tosyl chloride solution with stirring to uniformly resuspend the support. Add 123 mmoles of an organic base as a proton acceptor to catalyze the activation by taking up generated HCl from the sulfonyl chloride as it reacts with the hydroxyls on the support. Suitable bases include pyridine, triethylamine (TEA), diisopropylethylamine (DIEA), or dimethylaminopyridine (DMAP). Mix the gel suspension using an overhead paddle stirrer or by end-over-end rocking in a sealed container within the fume hood.
4. React for 1 h at room temperature with mixing.
5. In the fume hood, wash the activated support with acetone to thoroughly remove excess reactant and reaction byproducts. Wash with at least 10 bed volumes of solvent to ensure removal of the last traces of unreacted tosyl chloride.
6. Store the tosyl ester-activated support as a 50% slurry in dry acetone at 4°C until use.

LIGAND COUPLING TO SULFONYL CHLORIDE-ACTIVATED SUPPORTS

The following methods are generalized procedures for the immobilization of amine-containing ligands onto sulfonyl chloride-activated supports. Nonaqueous reactions may be carried out for small organic molecules or spacer arms that are soluble in organic solvent. Aqueous reactions are more appropriate for the coupling or proteins or other biological molecules that are soluble and stable in a buffered medium. Thiol-containing or hydroxyl-containing ligands may also be coupled to sulfonyl chloride-activated supports using similar protocols. Thiol-containing molecules will typically react faster than amines and will require buffers at the lower end of the pH spectrum described below. Hydroxyl-containing ligands, however, will need conditions at the high end of the pH range, as they are less nucleophilic at lower pH values.

(A) AQUEOUS COUPLING PROTOCOL

1. In a fume hood, drain 100 ml of the sulfonyl ester-activated support of solvent using a sintered glass filter funnel suspended in a suction filter flask. Gently pull a vacuum to remove most of the excess solvent while breaking up the support into small, finely divided pieces, but be careful not to allow the matrix to dry out. Stop using suction as soon as the support is divided into small pieces. When coupling ligands that might precipitate or be damaged by the presence of some residual solvent, the support should be washed quickly using several bed volumes of cold deionized water to remove the organic phase. Drain the support to a wet gel cake.
2. Dissolve the ligand to be coupled in 100 ml of a buffered solution at pH 8.5 to 10. Proteins may be coupled at a concentration of 1 to 20 mg/ml or for small molecules at a concentration of 1 to 5 mg/ml of an amine-containing ligand. The optimal concentration of ligand that is used for the immobilization reaction should be determined experimentally to obtain the best performance of the affinity matrix after coupling. Coupling buffers that may be used in this reaction include 0.1-*M* sodium borate, pH 8.5, or 0.2- to 0.5-*M* sodium carbonate, pH 9 to 10. If the ligand to be coupled is stable at the higher alkaline pH conditions, then the use of 0.25-*M* sodium carbonate, pH 9.5 will provide excellent coupling yields. Avoid buffers or solution

additives that contain amines, such as Tris, glycine, or imidazole, as these will compete with the ligand coupling reaction. If a tresyl-activated support is being used to immobilize ligands, then the reaction can be done in a sodium phosphate buffer at pH 7.5 to 8.0 (such as 0.2-*M* sodium phosphate, pH 7.5), because the reaction rate is much greater even at lower pH values.

3. Add the activated wet gel cake to the ligand solution with stirring to fully resuspend the gel. Mix the reaction slurry for 24h using an overhead paddle stirrer or by end-over-end rocking in a sealed container. The reaction may be performed at at 4°C or at room temperature, depending on the stability of the ligand being immobilized. Note that the reaction rate using tresyl-activated supports should be much faster than tosyl reactions, thus maximal coupling levels may be reached within 2 to 4h. The exact time of the coupling reaction should be optimized for the best performance of the affinity support being produced.
4. Filter and wash the affinity support to remove uncoupled ligand and reaction byproducts using coupling buffer, water, 1-*M* NaCl, and again with water. Depending on the ligand being coupled, other wash solutions may be used to completely remove unreacted ligand, such as washing with detergent solutions, denaturants, and high or low pH conditions. Finally, store the affinity support in water containing a preservative at 4°C until used.

(B) Nonaqueous Coupling Protocol

1. In a fume hood, drain 100ml of a sulfonyl ester-activated support of excess acetone (or other solvent) using a sintered glass filter funnel suspended in a suction filter flask. Do not allow the support to dry.
2. Dissolve the amine-containing ligand to be coupled in 100ml of acetone (or another water-miscible solvent) at a concentration of 1 to 5mg/ml, which is normally sufficient for the immobilization of a small organic compound. The optimal concentration of ligand to be used in the coupling reaction should be determined experimentally on small quantities of activated resin to identify the best affinity support performance in its intended application. For organic solvent-based reactions, add an organic base to the ligand solution, such as DMAP, DIEA, or TEA, to make a final concentration of 2-m*M*.
3. Add the activated wet gel cake to the ligand solution with stirring to fully resuspend the gel. Mix the reaction slurry for 1 to 2h using an overhead paddle stirrer or by end-over-end rocking in a sealed container. Longer reaction times may be used if appropriate.
4. Transfer the gel slurry to a sintered glass filter in the fume hood that is suspended in a suction filter flask and wash extensively with at least 10 bed volumes of solvent to remove the remaining ligand and reaction byproducts. If the ligand is detectable, continue to wash the support until no further ligand is detected in the washings. Finally, drain the support of excess solvent by pulling a gentle vacuum on the filter flask while breaking up the support into small, finely divided pieces using a spatula, but be careful not to allow the matrix to dry out. Once the support is broken into small pieces, remove the vacuum and resuspend the gel in water with mixing. Continue to wash the support with water until all traces of solvent have been removed. Additional washes with 1-*M* NaCl as well as low and high pH conditions may be done as appropriate to remove any noncovalently bound ligand. Finally, wash with water and store the affinity support as a 50% slurry in water containing a preservative at 4°C until used.

Azlactone-Activated Supports

An oxazolone is a heterocyclic anhydride that can be prepared from an *N*-acyl amino acid derivative through a dehydration and cyclization reaction. A particular type of oxazolone, an oxazol-5(4H)-one, is also known as an azlactone and consists of a five-membered ring that contains nitrogen and oxygen within its heterocyclic structure (see structure). Oxazolones have been used for many years in organic synthesis and have also become important constituents within organic compounds as potential drug candidates for a variety of applications in medicinal chemistry (Rao and Filler, 1986; Bala *et al.*, 2011). The creation of azlactone groups can be accomplished from carboxylic acids through the reaction of α-methyl alanine in a two-step process using a condensing agent as the catalyst. This process is a variant of the classic Erlenmeyer–Plöchl azlactone synthesis, which was described in the late 1800s to prepare oxazolones (Plöchl, 1884; Erlenmeyer, 1893). Under anhydrous conditions, a carboxylate will react with a condensing agent to form an activated intermediate, which then goes on to react with the amine of α-methyl alanine to form an amide linkage (an *N*-acyl amino acid derivative). This intermediate can undergo a subsequent condensation reaction through cyclization of the amino acid carboxylate end *via* dehydration, which results in the azlactone functionality being formed (Figure 15.58). Suitable condensing agents to drive this reaction include anhydrides, alkyl chloroformates, and carbodiimides in a nonaqueous environment. A preferred cyclization agent is acetic anhydride, which can be used as the solvent to form the azlactone functional groups on a dry carboxylate-containing

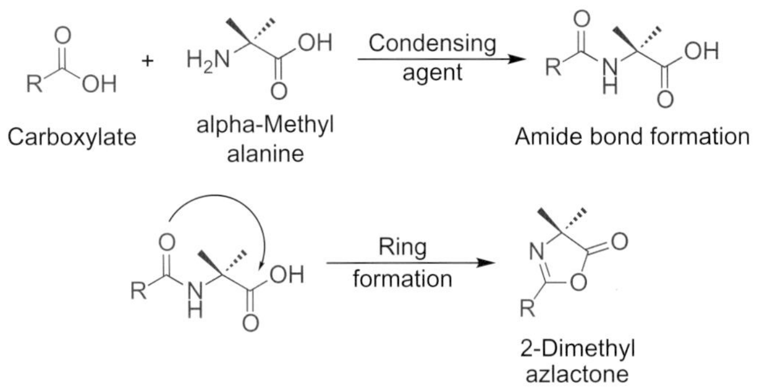

FIGURE 15.58 Formation of an azlactone from a carboxylate and α-methyl alanine in the presence of a cyclization catalyst.

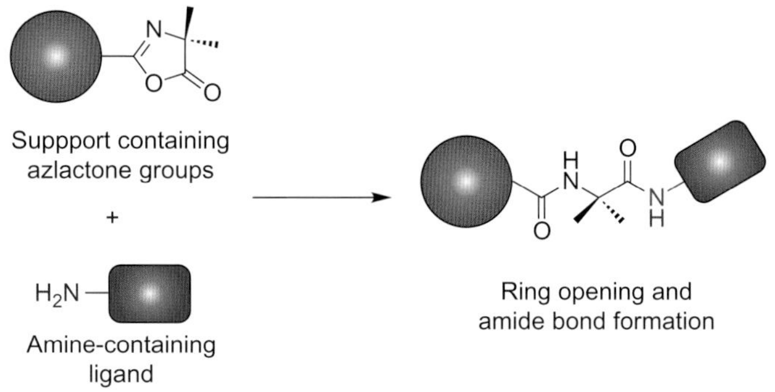

FIGURE 15.59 Reaction of an α-methyl alanine-containing azlactone with an amine.

support material. The reaction proceeds at 80 to 100°C by mixing for 2h.

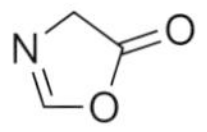

General structure of an azlactone (oxazol-5(4H)-one)

The formation of similar cyclic azlactones on carboxylate-containing chromatography supports can be used to create electrophilic reactive groups that can be exploited for the immobilization of amine-containing (nucleophilic) ligands. The creation of solid support materials containing azlactone reactive groups was first described by 3M Corporation (U.S. patents 4,871,824 and 4,737,560; Coleman *et al.*, 1990; Hermanson *et al.*, 1995). To eliminate the potential for nucleophilic attack on the azlactone C4 site, the amino acid used to form the azlactone should contain substitutions on the C4 carbon. A particularly convenient choice of amino acid in this regard is α-methyl alanine (or 2-dimethyl glycine). The two methyl groups prevent substitution at the C4 carbon after ring formation and therefore direct all nucleophile coupling reactions at the C5 position, which results in the desired ring opening process combined with amide bond formation with an amine-containing molecule (Figure 15.59).

Polymeric chromatography supports containing azlactone groups may be created more conveniently at the time of resin production through the use of an azlactone-containing vinyl monomer in a copolymerization reaction with other monomers. The reactive monomer vinyldimethyl azlactone (or 2-vinyl-4,4′-dimethylazlactone) has been used along with *N,N′*-methylene bisacrylamide as a crosslinking monomer to create polymeric beaded particles for affinity chromatography (Coleman *et al.*, 1990; Stanek *et al.*, 2005) (Figure 15.60). This route of polymer formation yields a reactive support immediately upon manufacture without the need to create a carboxylate support first and then activate the carboxylates to azlactones. The degree of reactivity in such polymers may be modulated by controlling the ratio of the azlactone-containing monomer to the other non-functionalized monomer(s) present in the final support. Using this

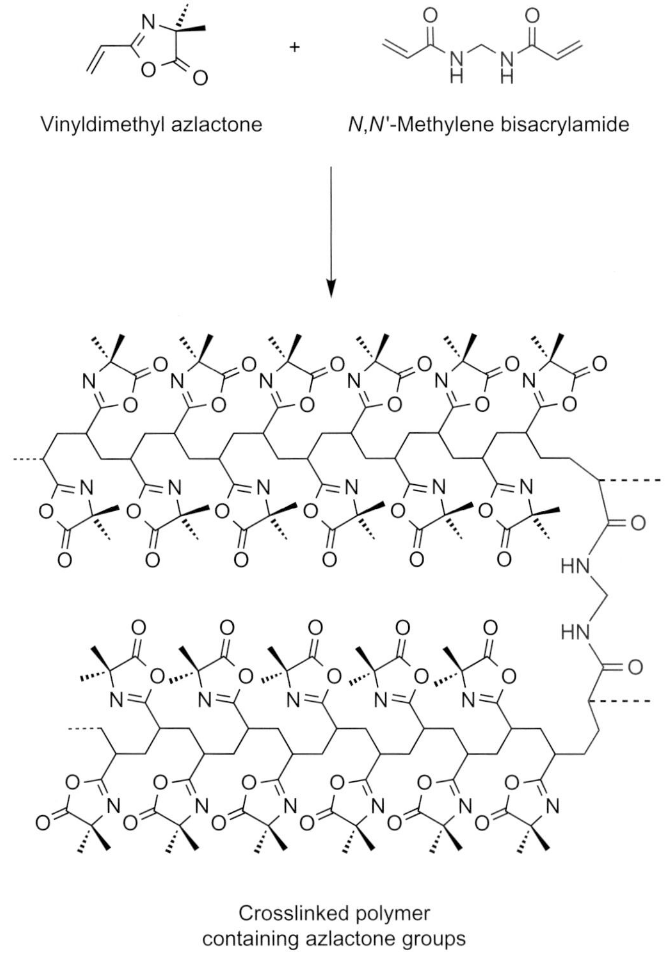

FIGURE 15.60 Synthesis of azlactone beads from vinyldimethyl azlactone and methylene bisacrylamide.

method, particles have been made that contain from approximately 100 µmoles/ml of azlactone groups up to a level of about 300 µmoles/ml, depending on the percentage of vinyldimethyl azlactone used in the copolymerization process. This type of support was previously commercialized under the Emphaze name through 3M and Pierce, and it now is available under the UltraLink name from Thermo Scientific (Pierce).

Supports prepared with azlactone groups should be stored in anhydrous organic solvent or kept dried as a stabilized particle powder to prevent hydrolysis of the reactive groups. Although the hydrolysis rate of the azlactone ring is slow during coupling reactions compared to the rate of ligand immobilization, the continuous exposure of the activated support to an aqueous environment will still degrade the reactive groups over time. Drying supports often requires the addition of excipients to stabilize the particles to prevent irreversible collapse or pore size damage. The UltraLink support is dried from a solution containing a small amount of detergent to decrease surface tension and allow rapid rehydration within the internal pore structures of the beads upon the addition of the dry support to a reaction medium. Therefore, ligand coupling reactions can be performed simply by means of the measured addition of the activated support to the appropriately buffered ligand solution and mixed to uniformly distribute the particles throughout the solution. Amine-containing ligands such as proteins can react with azlactone groups in a buffered environment over a pH range of 4 to 9, and even more extreme pH conditions may be used for coupling spacer arms or small organic molecules that are stable to highly alkaline conditions. For instance, coupling a diamine spacer to an azlactone-activated support can be done at a concentration of 0.5 to 1.5-*M* without pH adjustment (pH >12) to accelerate the reaction and increase immobilization densities. Ligand coupling to the azlactone groups results in the formation of amide bonds with formation of a short spacer arm created from the ring opening process (Figure 15.59). A comparison of the coupling of protein A, protein G, protein A/G, avidin, and streptavidin using two different buffers at pH 7.5 and pH 9.0 was made by Hermanson *et al.* (1995). In general, higher coupling yields were observed at pH 9.0 for three different ligand loading levels in each immobilization reaction.

An azlactone-activated support can have an initial degree of hydrophobic character due to the presence of the heterocyclic reactive groups. Many hydrophilic charged molecules such as proteins have difficulty approaching close enough to the activated support surfaces to facilitate efficient coupling using reasonable reaction times. This problem can be overcome through the use of buffer additives that are lyotropic in nature and which can have the effect of salting out and pushing the proteins closer to the surfaces for coupling. Suitable lyotropic salts for this purpose can be chosen from the Hofmeister series, such as the addition of sodium sulfate to the coupling buffer at a concentration of 0.8 to 1.5-*M*. The salt concentration should be adjusted to maintain protein solubility while maximizing the rate of immobilization to the support. Using these conditions, the immobilization reaction can be complete within 1 h and achieve maximal yields. Protein A has been coupled to azlactone-activated supports at densities of over 30 mg/ml and immunoglobulins have been immobilized to levels of over 20 mg/ml. After the coupling reaction is carried out, any excess reactive groups should be blocked using a small molecule such as ethanolamine. Once an affinity support is prepared using these methods, the support material reverts to a hydrophilic surface with relatively low nonspecific binding potential to the matrix itself.

Azlactone activation has been used for the coupling of affinity ligands to beaded chromatography supports (Coleman *et al.*, 1990; Hermanson *et al.*, 1995; Stanek *et al.*, 2005), for the immobilization of ligands onto various polymeric constructs (Laquièvre *et al.*, 2012), for modification and subsequent immobilization onto surfaces (Lokitz *et al.*, 2009), and to couple enzymes to monolithic supports for use in microfluidic devices (Logan, 2007). Surface polymer scaffolds have also been built to immobilize affinity ligands using multilayer copolymers incorporating vinyldimethyl azlactone groups (Barringer *et al.*, 2009). The use of the vinyl monomer is advantageous in creating copolymer constructs of many different structures as well as for photografting azlactone-reactive groups onto existing materials for immobilization of amine-containing biomolecules.

LIGAND COUPLING TO AZLACTONE-ACTIVATED SUPPORTS

The following procedure describes a generalized method for the coupling of protein affinity ligands to 10 ml of an azlactone-activated beaded chromatography support. Similar methods may be used to immobilize biomolecules onto activated membranes, surfaces, or monolithic supports containing azlactone groups. Two additional procedures are subsequently presented that describe methods for coupling small amine-containing molecules, such as organic ligands and spacer arms, using aqueous or nonaqueous conditions.

(B) Protein Immobilization onto Azlactone Supports

1. Dissolve the protein to be immobilized at a concentration of 1 to 20 mg/ml in 20 ml of coupling buffer (sufficient for coupling to 10 ml

of swollen azlactone beads) consisting of 0.1-*M* buffer concentration and at a pH range of 4 to 9. Most proteins react best at the high end of the recommended pH range, but some optimization might have to be carried out to obtain the best coupling yield and affinity ligand performance for a particular protein to be coupled. Also add to the coupling buffer a lyotropic salt to enhance the rate of protein coupling to the azlactone-activated support. A suggested coupling buffer is 0.1-*M* sodium carbonate, pH 9.0, with the addition of 0.6-*M* sodium citrate as the lyotropic agent. An alternative lyotrope that may be used is sodium sulfate at a concentration of 0.8-*M*. A more neutral pH coupling buffer that has been used successfully is 0.1-*M* sodium phosphate at pH 7.5, containing 0.6-*M* sodium citrate or 0.8-*M* sodium sulfate. The concentration of the lyotropic salt may be lowered for proteins that have solubility issues at the recommended concentrations; however, the recommended levels for sodium citrate or sodium sulfate have been found to work well for antibody (IgG) immobilizations, for avidin or streptavidin, as well as for the coupling of immunoglobulin binding proteins such as protein A or protein G to an azlactone support. In addition, other buffer components may be used to stabilize the protein in the reaction medium at different pH levels, but avoid additives that contain amines or other nucleophiles that would compete with the reaction—for example, Tris, imidazole, thiol reducing agents, glycine, and ammonium ions. If the activated particles are added as a dry powder, prepare enough ligand solution to equal at least twice the volume of the hydrated beads. This volume is required because half of the ligand solution volume will be used to hydrate and fill the pore structure of the particles, while the other half will create a 50% slurry for efficient mixing. If the support is in a hydrated state prior to addition of the ligand solution, then prepare an equal volume to the amount of swollen support used.

2. Measure out a quantity of dry azlactone-activated particles corresponding to exactly half the volume of the ligand solution prepared in step 1—in this case, weigh out the equivalent of 10 ml of beads. For instance, a typical batch of UltraLink azlactone-activated beads has a swelling ratio of approximately 120 mg/ml hydrated gel (it varies slightly for each individual lot of manufactured particles, so refer to the lot-specific information to be precise). To yield 10 ml of hydrated beads, therefore, add 1.2 g of azlactone beads to the ligand solution with stirring to fully hydrate the particles and initiate the coupling reaction.
3. React with mixing for 1 to 2 h at room temperature using an overhead paddle stirrer or end-over-end rotation in a sealed container. The majority of protein coupling reactions will reach maximal yield in 1 h.
4. Wash the support with several bed volumes of water to remove excess reactants by using a sintered glass filter funnel suspended in a suction filter flask. Note that if the UltraLink support is used, it is supplied dry from the manufacturer with a small amount of Triton X-100 added to promote rapid rehydration within the pores. The presence of the detergent in the initial washings after coupling will prevent the accurate measurement of protein concentration by absorbance at 280 nm. However, the protein concentration that did not couple to the support may be determined by use of the BCA assay, which is detergent tolerant (Smith *et al.*, 1985). The measurement of the non-coupled protein in the washings can facilitate the determination of how much protein has coupled to the support by the difference between the amount of protein not coupled and the initial amount of protein used in the coupling reaction. Drain the support to a wet cake.
5. To block the remaining azlactone reactive groups, add 1 bed volume (10 ml) of 1-*M* ethanolamine, pH 9.0. Mix for 30 min. Note: Adjust the pH of the ethanolamine solution in a fume hood using the slow addition of 50% HCl while maintaining the solution on ice. Adjust to room temperature before using.
6. Wash the support with coupling buffer without containing the lyotropic agent, then wash extensively with 1-*M* NaCl and finally with water to completely remove unreacted molecules. Additional washes may be carried out using acid or basic pH conditions as well as employing the use of detergents or denaturants to ensure complete removal of any proteins still bound by noncovalent interactions. These additional wash conditions should be done only if the protein immobilized is stable to such environments and only if necessary to minimize the leaching of ligand molecules in subsequent chromatographic operations. Finally, store the affinity support in water or buffer containing a preservative at 4°C.

(B) Aqueous Coupling of Small Ligands or Spacers to Azlactone Supports

Affinity molecules may also be small organic compounds that can be immobilized onto an azlactone support through an available amine group. Additionally, amine-containing spacer arms may be coupled to the support to facilitate further derivatization or to create a different chemical functionality at its terminal end for subsequent coupling to a ligand through something other than an amine (e.g., a thiol). The aqueous coupling of small molecules to an azlactone support is performed somewhat differently from the coupling methods typically used with proteins, because a lyotropic agent usually is not required to enhance the reaction rate and yield. Instead of using the salting out effect of a lyotrope to drive proteins toward the surface azlactone groups to enhance reaction rates, small-molecule coupling is improved through increased concentrations and higher pH conditions. Another caveat to consider is that if a molecule has more than one amine—as in diamine spacer arms, which makes multipoint attachment to the support more likely—then the concentration of ligand used during the reaction should be very high (i.e., 1.0–2.0 *M*) to prevent internal crosslinking or even bead-to-bead linking caused by both ends of the compound reacting with the support. In this case, the high concentration range encourages single point attachments by making it more likely that another molecule will react rather than a second site within the same molecule getting coupled. This is an especially important factor when coupling diamine spacers to any amine-reactive support, since the desired end product involves one amine getting immobilized and the other amine of the spacer remaining free for subsequent reactions. On the other hand, for spacer arms containing an amine on one end and another functional group, such as a carboxylate on the other end, the need for very high initial concentrations is not as important, since only one end of the molecule will be capable of coupling.

Coupling of Small Diamine Spacers

1. In a fume hood, dissolve a diamine spacer molecule, such as 3,3′-diaminodipropylamine (DADPA; also called *bis*(3-aminopropyl)amine), in water at a concentration of at least 1.5 *M*. When using diamines as the free base compounds (not as the hydrochloride salts), there is no need to adjust the pH once the solution is made. The pH will be extremely basic, but this will drive the coupling reaction to the azlactone groups on the support with higher efficiency (see additional information in the section on coupling spacer arms, this chapter). Prepare 20 ml of the diamine solution to react with 10 ml of hydrated azlactone beads.
2. Add 1.2 g of azlactone beads to the diamine solution with stirring to fully hydrate the particles and initiate the coupling reaction (this quantity is equivalent to 10 ml of particles once they are fully hydrated and swollen).
3. React for 1–2 h at room temperature with constant mixing by use of an overhead paddle stirrer or end-over-end mixing on a rotator (do not use a magnetic stirring bar).
4. Wash the amine-modified support using a sintered glass filter funnel in a fume hood. Wash extensively with water, 1-*M* NaCl, and again with water to remove excess diamine. Use 10 to 20 bed volumes of washes for each solution. The presence of amines on the support, which indicates successful coupling, can be monitored by reacting a small quantity of support with TNBS solution, which will give a bright orange color with primary aliphatic amines (see Chapter 2, Section 4.3). Store the amine support in water or buffer containing a preservative at 4°C.

Coupling Small, Water Soluble Amino Ligands

1. Dissolve an amine-containing ligand in 20 ml 0.1-*M* sodium carbonate, pH 9.0, at a concentration that will result in the desired density of affinity groups on the surface, which ultimately will be optimal for the chromatography application. For some ligands, such as peptides, the concentration may be in the range of 3 to 5 mg/ml of support. If a high density of coupling is desired, the concentration of ligand may be increased to at least double the amount of azlactone groups on the surface (i.e., 2 × 100 µmoles/ml).
2. Weigh out 1.2 g of azlactone beads (10 ml gel after hydration) and add them to the ligand solution with stirring.
3. React for 2 h to overnight at room temperature with constant stirring.
4. If the ligand was reacted at a level that would not couple to every azlactone group present on the support surface, then wash the support with several column volumes of water to remove the majority of remaining uncoupled ligand. Next, add to the support 10 ml of 1-*M* ethanolamine, pH 9.0, to block the remaining reactive groups. React for 1 h at room temperature with mixing.
5. Wash the support with coupling buffer and then with water, 1-*M* NaCl, and again with water to remove uncoupled ethanolamine and ligand. Wash with at least 10 to 20 bed volumes of each solution. Store the amine support in water or buffer containing a preservative at 4°C.

(C) Coupling Amine Containing Ligands in Organic Solvent

Some amine-containing ligands may be only sparingly soluble in aqueous buffer conditions. For this reason, it may be necessary to perform the immobilization reaction in organic solution to maximize the coupling yields and obtain an optimal affinity support. The following protocol describes an organic solvent reaction using azlactone particles (UltraLink). An important caveat is that the solvent chosen should maintain a swollen bead state so as not to restrict ligand access to the internal pore structures of the particles.

1. In a fume hood, dissolve an amine-containing ligand to be coupled in 20 ml of dry DMSO or DMF at a concentration of at least 2 to 5 mg/ml. The optimal

concentration of ligand in the immobilization reaction may have to be determined experimentally using small quantities of activated support and by varying the concentration of ligand to observe its affect on affinity chromatography performance. Other solvents may be used; however they should be water miscible and able to maintain a swollen bead state for the activated support.

2. Add 1.2 g of azlactone support (10 ml hydrated volume) to the ligand solution with stirring.
3. React for 1 to 2 h at room temperature with constant stirring.
4. Wash the support in a fume hood using a sintered glass filter suspended in a suction filter flask. Wash with at least 20 bed volumes of the organic solvent used for the coupling reaction. Continue to wash until no ligand is detected in the filtrate. Wash the support into water using progressively increasing concentrations of water in solvent (e.g., 30%, 70%) until 100% water is used. Wash with 100% water for at least 10 bed volumes to completely remove the last traces of solvent. Finally, store the affinity support as a 50% slurry in water or buffer at 4°C containing a preservative.

Cyanogen Bromide Activation

One of the first activation methods introduced for the coupling of affinity ligands to solid supports involves the use of cyanogen bromide (CNBr) to activate hydroxyl groups (Axen *et al.*, 1967). For many years, CNBr activation was the method of choice for coupling amine-containing ligands to agarose supports, especially for the immobilization of proteins. Pharmacia first commercialized CNBr-activated agarose in a dry form under the name CNBr Sepharose and the product is still sold by GE for affinity ligand immobilization. Meng *et al.* (2009) recently reported on methods for stabilizing CNBr-activated agarose in dry form, which preserves both the reactive group and the structure of the beaded support. Under alkaline conditions, CNBr reacts with the hydroxyl groups on a matrix to produce reactive cyanate esters and imidocarbonates (Figure 15.61). The relative yield of cyanate esters *versus* imidocarbonates may vary depending on the hydroxylic structure of the chromatography matrix. If closely spaced or adjacent hydroxyls mainly are present in the support, such as within dextran- or cellulose-containing matrices that contain many diols, then a cyclic imidocarbonate may become the predominate product. However, if primary hydroxyls or hydroxyls that are not immediately adjacent to one another are present within the support, such as in agarose matrices, then cyanate esters are the major product formed from the reaction (Kohn and Wilchek, 1982).

CNBr activation is highly versatile and has been used to activate hydroxyl groups on many different chromatography supports as well as on microparticles, nanoparticles, membranes, and even surfaces (Jurado *et al.*, 2002; Yavuz *et al.*, 2008; Arazawa *et al.*, 2012). The major disadvantages of using this coupling method are the high toxicity of the reagent (it emits hydrogen cyanide gas, which is deadly if inhaled), the potential for creating a positive charge on the support after ligand coupling due to the bond type that is formed (primarily an isourea linkage, which is positively charged at neutral pH), and the labile nature of the ligand linkage to the support (having a tendency to constantly leach immobilized ligands at levels higher than that observed using other coupling methods). The unfortunate disadvantages of using CNBr activation have led to a significant decrease in its use over the years and an increase in the use of alternative amine coupling chemistries that do not have these shortcomings.

CNBr-activated supports react with amine-containing ligands to potentially produce two linkages, depending on whether the starting activation form was

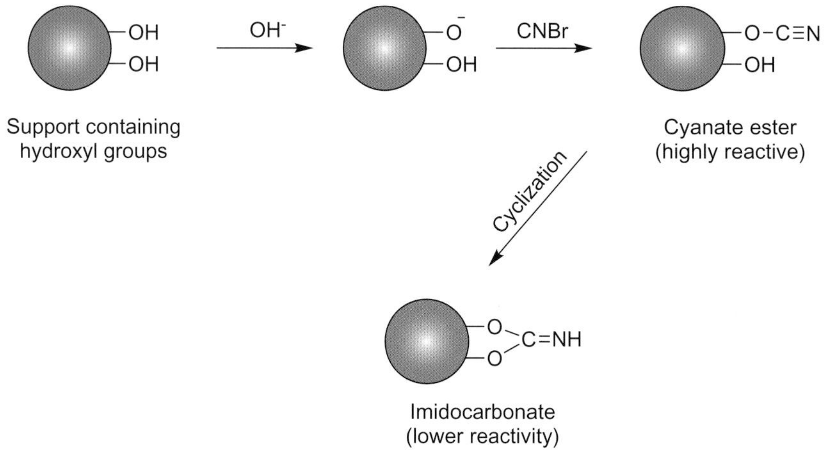

FIGURE 15.61 CNBr activation forming isocyanate and cyclic imidocarbonates.

a cyanate ester or a cyclic imidocarbonate. Cyanate esters rapidly react with amines under mildly alkaline pH conditions to form an isourea bond, whereas the cyclic imidocarbonate reactive group is far less reactive and couples with amines to create a substituted imidocarbonate linkage (Figure 15.62).

CNBr ACTIVATION PROTOCOL

The first procedure describes the original method reported by Cautrecasas (1970) of using titration with NaOH during the activation process to maintain the pH of the reaction. The second method describes the revised protocol of March *et al.* (1974), which involves reagent modifications that make the process easier to control. The reagent quantities may be proportionally scaled to activate different amounts of chromatography support materials. Read through the protocols thoroughly before setting up the activation reaction, because many of the steps are dependent on time, temperature, and pH, which all need to be controlled to result in successful and reproducible activation and coupling of an affinity ligand.

Caution: CNBr is highly toxic and may cause death if inhaled or ingested. Avoid contact with the solid compound and any solutions containing it. All operations should be carried out in a well-ventilated fume hood and using appropriate personal protective equipment. Dispose of all waste according to recommended safety protocols (see the product's MSDS data sheet for further details).

(A) Traditional Method: CNBr Activation using NaOH Titration (Cuatrecasas, 1970)

1. Wash the equivalent of 100 ml of settled chromatography support containing hydroxyl groups with 1 l of deionized water using a sintered glass filter funnel and a vacuum filter flask. Pull a gentle vacuum to facilitate the filtration process and collect all washes for proper disposal. After the wash, suction the support to a wet cake, stopping the filtration process by eliminating the vacuum at the point that the excess wash solution just enters the top of the gel. Do not allow the gel bed to get air within it or dry out at the top during the washing and filtering process. Note that if a monolithic support or a membrane is used in this procedure, then refer to the recommendations on handling these materials in Section 1 of this chapter.
2. Remove the washed and moist support from the filter funnel, transfer it to a beaker, and suspend it in 100 ml water by stirring. In a fume hood, set up an overhead paddle stirrer that will be used to mix the support during the activation procedure. Do not use a magnetic stir bar, as it will grind the support material and damage it. Insert a pH probe and a thermometer into the gel suspension to continuously monitor the pH and temperature during the activation reaction.
3. Prepare a 20% (w/w) solution of NaOH (i.e., 20 g/100 ml) by dissolving NaOH pellets in water or by dilution of a commercially available 50% solution. In addition, have available an ice bucket full of small, deionized ice chips to cool the reaction as it proceeds.
4. In the fume hood, weigh out 20 g of CNBr (**caution**: extremely toxic compound!) and add it to the stirring gel suspension. Maintain the pH of the reaction at pH 11 by dropwise addition of 20% NaOH solution. Also maintain the temperature of the reaction at about 25°C by periodic addition of crushed ice chips to the slurry.
5. Continue the activation reaction for 10 to 15 min at which point the CNBr should be completely

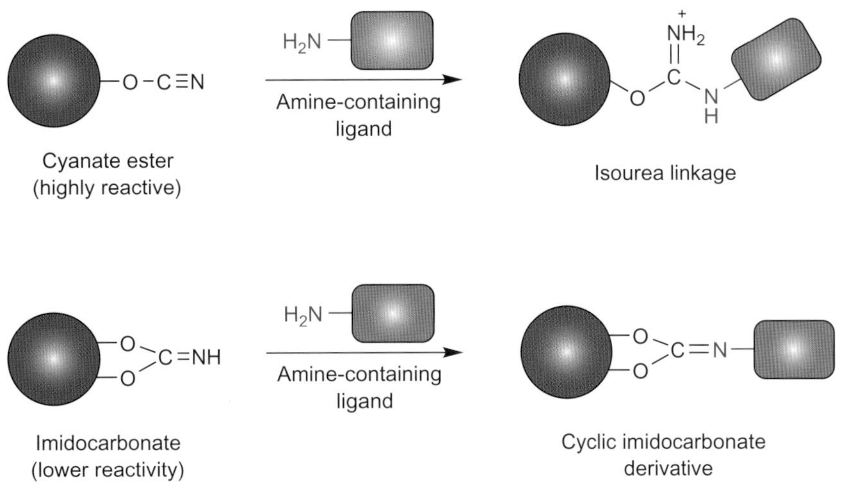

FIGURE 15.62 Ligand coupling to CNBr-activated supports.

dissolved and the rate of base consumption should be reduced. Transfer the activated support to a sintered glass filter funnel set up in the fume hood containing deionized ice chips to cool the reaction. Wash the support with 1 l of ice-cold deionized water followed by 500 ml of ice-cold 0.1-*M* sodium bicarbonate, pH 8.5 (coupling buffer). Drain the activated gel to a moist cake on the sintered glass filter pad.
6. Use the activated support to immediately couple an amine-containing ligand according to the immobilization protocol described below.

(B) Alternative Protocol: 2-min CNBr Activation using Carbonate Buffer at Room Temperature (March et al., 1974)

1. Wash the equivalent of 100 ml of settled chromatography support containing hydroxyl groups with 1 l of deionized water using a sintered glass filter funnel and a vacuum filter flask. Pull a gentle vacuum to facilitate the filtration process and collect all washes for proper disposal. Next wash the support with several bed volumes of 2-*M* sodium carbonate (no pH adjustment required). After the carbonate wash, suction the support to a wet cake, stopping the filtration process by eliminating the vacuum at the point that the excess wash solution just enters the top of the gel. Do not allow the gel bed to get air within it or dry out at the top during the washing and filtering process. Note that if a monolithic support or a membrane is used in this procedure, refer to the recommendations on handling these materials in Section 1 of this chapter.
2. Remove the washed and moist support from the filter funnel, transfer it to a beaker, and suspend it in 100 ml 2-*M* sodium carbonate buffer by stirring. In a fume hood, set up an overhead paddle stirrer that will be used to mix the support during the activation procedure. Do not use a magnetic stir bar, as it will grind the support material and damage it. Stir the gel slurry at a rate that keeps the particles well mixed and suspended in the activation buffer.
3. In a well-ventilated fume hood, weigh out 10 g of CNBr (**caution**: extremely toxic compound!) and dissolve it in 5 ml of acetonitrile.
4. Add the CNBr solution to the stirring gel slurry and react for exactly 2 min at room temperature.
5. Immediately transfer the gel to a sintered glass filter in the fume hood and wash with 1 l of ice-cold water followed by 500 ml of ice-cold coupling buffer (0.1-*M* sodium bicarbonate, pH 8.5). Drain the activated gel to a moist cake on the sintered glass filter pad.
6. Immediately use the activated support to couple an amine-containing ligand according to the following protocol.

LIGAND COUPLING TO CNBr-ACTIVATED SUPPORTS

The following protocol describes the general method for coupling amine-containing ligands to a CNBr-activated chromatography support material. This process can be used successfully to immobilize proteins or other amine-containing affinity ligands as well as amine-containing spacer molecules. Avoid introducing any other amine-containing components into the coupling buffer during the reaction, such as Tris, glycine, ammonium ions, or other small molecules containing a reactive amine, as these will interfere with the desired immobilization of ligand. Amine reactions with CNBr-activated supports will occur efficiently between pH 8 and pH 9.5, so the coupling buffer pH may be modified somewhat to promote ligand stability or solubility if necessary.

Protocol

1. Suspend the CNBr-activated support in an equal volume of 0.1-*M* sodium carbonate buffer, pH 8.5 (coupling buffer), into which an amine-containing affinity ligand has been dissolved. For many protein ligands, a suggested starting concentration is 3 to 6 mg/ml. For low-molecular-weight ligands use a concentration at least 3 times greater than the level of CNBr activation. For the activation of agarose with CNBr, the typical activation level is about 20 to 40 μmoles/ml gel.
2. Mix the reaction slurry using a paddle stirrer for 24 h at 4°C.
3. Using a sintered glass filter funnel, wash the gel extensively with coupling buffer, then with 1-*M* NaCl, and water to remove excess unreacted ligand, which did not get immobilized.
4. Excess reactive groups on the matrix can be blocked by the addition of 1-*M* ethanolamine, pH 9, or 1-*M* Tris, pH 9. Suspend the washed gel in an equal volume of the blocking solution and stir for 1 h at room temperature.
5. Remove excess blocking agent by washing the affinity support extensively with 1-*M* NaCl and water. The final washes can be tested for the presence of amines through reaction with TNBSA, which will turn orange in solution upon coupling to aliphatic amines (see Chapter 2, Section 4.3). Finally, store the affinity support as a 50% slurry in water or buffer at 4°C containing a preservative.

Trichloro-s-Triazine (TsT) Activation

Cyanuric chloride or trichloro-*s*-triazine (TsT) is a trifunctional, symmetrical, heterocyclic aromatic compound that has been used for many decades as a reactive group in the construction of dyes for staining fabrics and articles of clothing. Dyes such as the Procion series react through covalent bonding of the dye to the nucleophilic groups on the fabrics to create stable, nonfading, colored clothes (see structure of Procion Brilliant Blue). TsT can also be used to activate and form reactive groups on the surfaces of hydroxylic chromatography resins for the immobilization of affinity ligands. The three reactive chlorines of TsT potentially can covalently link to three separate nucleophilic groups on different molecules. The reaction of TsT in excess with the hydroxyl groups on chromatography support materials can result in two reactive sites remaining for coupling of amine-containing ligands. However, the reactivity of the triazine chlorine groups on TsT varies depending upon how many have already reacted. The first chlorine is the most highly reactive, and it can couple to the hydroxyls on supports under relatively mild conditions. The second reactive chlorine is able to couple with a nucleophile such as an amine, thiol, or even a hydroxyl group under slightly basic pH conditions with relatively rapid reaction kinetics. Finally, the last chlorine of TsT has the lowest reactivity, but it still can be used to couple with thiols or amines under mildly alkaline pH conditions and with hydroxyls under highly alkaline conditions. TsT activation has also been used to conjugate PEG molecules to proteins for the modulation of immunological properties (Abuchowski *et al.*, 1977) (Chapter 18, Section 2).

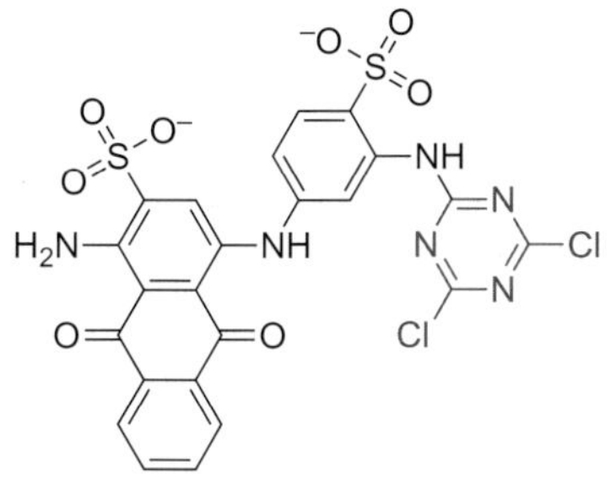

Procion Brilliant Blue

TsT activation of a hydroxyl-containing support such as agarose proceeds through loss of one chlorine (as HCl) with modification to the matrix through a hydroxyl by means of an ether linkage to the ring (Figure 15.63). If TsT is used in large excess, most of the active groups on the support will contain two remaining chlorines for coupling to affinity ligands. The activation process is best done under nonaqueous conditions to prevent hydrolysis of the TsT chlorines, and also while in the presence of two equivalents of organic base (e.g., DIEA) to accept the released protons produced during the process (Finlay *et al.*, 1978; Hodgins *et al.*, 1980). TsT-activated supports represent one of the few truly multi-purpose reactive groups—others being epoxides, iodoacetyl, and vinyl sulfones—that are able to couple with amine-, thiol-, and hydroxyl-containing ligands with success just by modulating the pH. In addition to the trichloro derivatives of triazine rings, the trifluoro derivatives have also been used with success for activation of support materials (Rerat *et al.*, 2010). The trifluorotriazine reactions proceed analogously to the use of TsT in the methods presented in this section.

After the initial activation of a support with TsT, the most reactive chlorine of the two remaining can be blocked with the relatively weak nucleophilic amine on aniline to create a monofunctional derivative. The last remaining chlorine of the cyanuric chloride ring is the most stable to hydrolysis, but it is still able to effectively couple to nucleophilic groups on affinity ligands under relatively mild conditions. The monofunctional derivative is best used for aqueous ligand coupling reactions with proteins or other biomolecules. However, ligands soluble in organic solvent may be coupled to the dichlorotriazinyl support (without blocking with aniline) and under nonaqueous conditions in the presence of an organic base to result in high-yield reactions devoid of any accompanying hydrolysis. Affinity ligands coupled

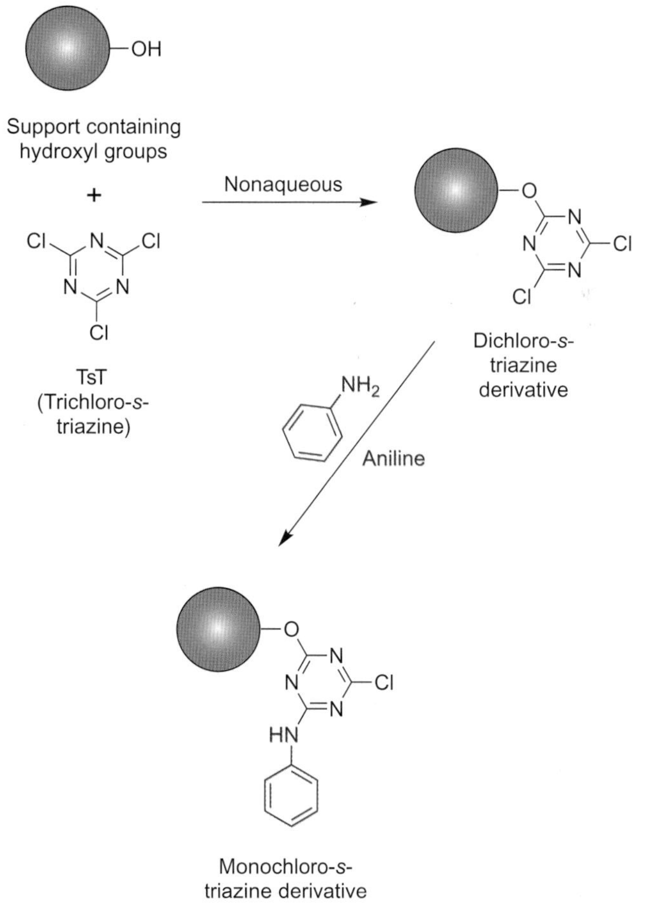

FIGURE 15.63 Activation of hydroxylic support with TsT and blocking with aniline to produce a mono-functional derivative.

using TsT activation result in more stable linkages relative to leakage than those coupled using CNBr activation (Hodgins *et al.*, 1980).

TsT mediated coupling has been used to immobilize a variety of biomolecules and synthetic organic ligands, including the coupling of peptides onto polypropylene membranes (Gérard *et al.*, 2011), a general activation method for the preparation of affinity membranes on polymers or carbohydrates (Avramescu *et al.*, 2008), as an activating agent of amino-substituted agarose to create rationally designed affinity ligands (Roque and Lowe, 2008), and as an activator of aminopropyl silica for the preparation of immobilized affinity ligands (Luong and Scouten, 2008).

TsT Activation Protocol

The following activation procedure should be carried out in a fume hood with the appropriate personal protective equipment to prevent contact or inhalation of solvents or reagents.

1. Wash 100 ml of a hydroxyl-containing chromatography support material (such as crosslinked agarose) with water (at least several bed volumes) to remove preservatives and storage solution components. In a fume hood, sequentially wash the support into increasing concentrations of acetone-in-water (such as 25% (v/v), 50%, and 75% acetone/water) and finally with 100% dry acetone for at least 10 to 20 bed volumes to remove the last traces of water. Note: Dioxane or acetonitrile can be substituted for acetone depending on the compatibility of the matrix to a particular solvent.
2. Drain the support to a wet gel cake, resuspend it in an equal volume of solvent by mixing, and transfer the slurry to a 500-ml, three-necked, round-bottom flask placed in a heating mantle within the fume hood. Mix the slurry using an overhead stirring motor with a paddle stirrer and add to the flask a water-jacketed condenser connected to cold flowing water and a thermometer.
3. In the fume hood, stir the gel slurry and heat to 50°C using the heating mantle to slowly increase the temperature. Use care not to overheat, as acetone boils at 56 to 57°C.
4. After the gel slurry has equilibrated at 50°C, add 20 ml of 2-*M* *N*,*N*-diisopropylethylamine (DIEA) in acetone. *N*,*N*-dimethylaniline may also be used as the proton acceptor, but avoid using organic bases such as TEA, pyridine, *N*-ethyl morpholine, or lutidine, because these will precipitate with TsT in the solvent (Hodgins *et al.*, 1980).
5. After 30 min of stirring, add 20 ml of 1-*M* TsT (highly purified) in acetone.
6. React for 1 h at 50°C with mixing.
7. In the fume hood, wash the activated support with acetone to thoroughly remove excess reactant and reaction byproducts. Wash with at least 10 to 20 bed volumes of solvent to ensure removal of the last traces of unreacted TsT and DIEA base.
8. Transfer the washed gel to a clean vessel and add 200 ml of 2-*M* aniline in acetone to block the most reactive chlorine on the triazine ring. Mix for 30 min at room temperature. Note: If the TsT-activated support is to be used to immobilize a solvent-miscible ligand, then it may be desirable not to block one of the two remaining acyl chlorides with aniline, because no hydrolysis will occur during the coupling reaction. If aniline blocking is not performed, then go to step 10.
9. Wash the TsT-activated support (now as the monochlorotriazine derivative) with at least 10 to 20 bed volumes of acetone to remove the last traces of unreacted aniline.
10. Store the activated support as a 50% slurry in dry acetone at 4°C until use.

TsT LIGAND-COUPLING PROTOCOL

TsT-activated supports that are stored in acetone or other anhydrous solvents may be used directly for coupling amine-containing ligands under nonaqueous conditions. In this case, a TsT-modified support that contains two remaining reactive chlorotriazine groups can be used (i.e., no blocking was done with aniline to eliminate the most reactive chlorine and leave only the least reactive one remaining). Alternatively, the activated gel may be filtered free of most solvent and reacted in aqueous buffer to couple proteins or other water soluble molecules. The following two protocols describe these methods in a general sense, but optimization should be performed to determine the best reaction conditions for a particular ligand and the optimal molar ratios for the best-performing affinity support. The reactions involved in coupling amine-containing ligands to TsT-activated supports are illustrated in Figure 15.64.

(A) Nonaqueous Coupling of Amine-Containing Ligands to TsT-Activated Supports

1. In a fume hood, drain 100 ml of the TsT-activated support of excess acetone (or other solvent) using a sintered glass filter funnel suspended in a suction filter flask. Do not allow the support to dry.
2. Dissolve the amine-containing ligand to be coupled in 100 ml of acetone (or another water-miscible solvent) at a concentration of 1 to 5 mg/ml, which is normally sufficient for the immobilization of a small organic compound. The optimal concentration of ligand to be used in the coupling reaction should be determined experimentally on small quantities of activated resin to identify the best affinity support

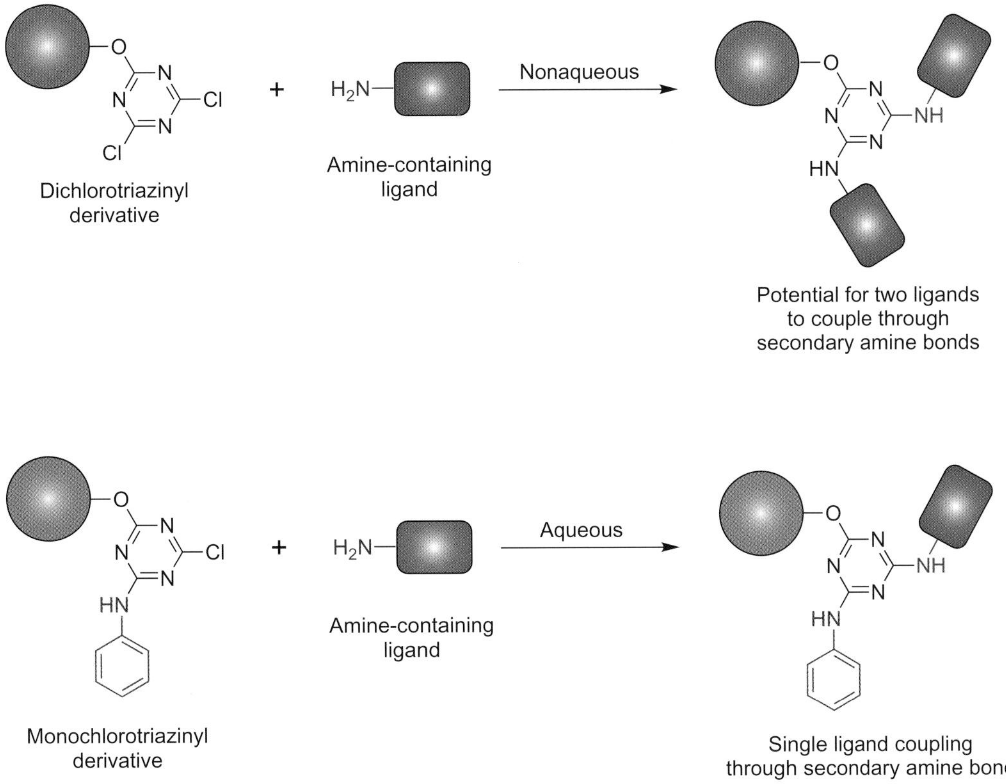

FIGURE 15.64 Coupling an amine-containing ligand to a TsT-activated support.

performance in its intended application. For organic solvent based reactions, add an organic base to the ligand solution, such as DIEA, to make a final concentration of 2-m*M*.

3. Add the activated wet gel cake to the ligand solution with stirring to fully resuspend the gel. Mix the reaction slurry for at least 2h using an overhead paddle stirrer or by end-over-end rocking in a sealed container. Longer reaction times may be done if appropriate to reach maximal yield of coupling.
4. Block unreacted active sites by the addition of ethanolamine (6.1g) to the solution to make a 1-*M* solution. Mix for an additional 30min.
5. Transfer the gel slurry to a sintered glass filter in the fume hood that is suspended in a suction filter flask and wash extensively (at least 10 bed volumes) with solvent to remove the remaining ligand and reaction byproducts. If the ligand is detectable, such as by using spectrophotometric methods, continue to wash the support until no further ligand is detected in the washings. Finally, drain the support of excess solvent by pulling a gentle vacuum on the filter flask while breaking up the support into small, finely divided pieces using a spatula, but be careful not to allow the matrix to dry out. Once the support is broken into small pieces, remove the vacuum and resuspend the gel in water with mixing. Continue to wash the support with water until all traces of solvent have been removed. Additional washes with 1-*M* NaCl as well as low and high pH conditions may be carried out as appropriate to remove any noncovalently adsorbed ligand. Finally, wash with water and store the affinity support as a 50% slurry in water containing a preservative at 4°C until used.

(B) Aqueous Coupling of Biomolecules to TsT-Activated Supports

1. Wash 100ml of a TsT-activated support into water and coupling buffer by suctioning off excess acetone storage solution and resuspending the gel into water. This can be done by pulling a gentle vacuum on the filter flask to filter off solvent while breaking up the support into small, finely divided pieces using a spatula, while being careful not to allow the matrix to dry out. Once the support is broken into small pieces, remove the vacuum and resuspend the gel in water with mixing. Continue to wash the support with water for at least 5 to 10 bed volumes. Finally, wash the support with 2 to 3 bed volumes of coupling buffer (0.1-*M* sodium borate, pH 8.5, containing 0.15-*M* NaCl).
2. Dissolve the protein or other amine-containing macromolecule to be immobilized in coupling buffer at a concentration of 1 to 20mg/ml, or for

small molecules, dissolve it at a concentration of 1 to 5 mg/ml. The optimal concentration of ligand that is used for the immobilization reaction should be determined experimentally to obtain the best performance of the affinity matrix after coupling. Avoid buffers or solution additives that contain amines, such as Tris, glycine, or imidazole, as these will compete with the ligand coupling reaction. Also avoid the presence of thiol-containing compounds such as DTT, because these will react with the chlorotriazine groups.

3. Add the activated wet gel cake to the ligand solution with stirring to fully resuspend the gel. Mix the reaction slurry for at least 24 h at room temperature using an overhead paddle stirrer or by end-over-end rocking in a sealed container. The reaction may be performed at 4°C or at room temperature, depending on the stability of the ligand being immobilized; however, the coupling yield at 4°C is about 50% less than that at room temperature. For small-molecule reactions where the compound is thermally stable, the temperature may be increased to higher levels (i.e., as high as 45°C) to further increase the efficiency of the immobilization process.
4. Filter and wash the support with several bed volumes of water to remove most of the unreacted ligand and reaction byproducts. Transfer the support to a clean vessel and add 100 ml of 1-*M* ethanolamine, pH 9, with mixing to block the remaining active groups. Stir the reaction for 1 h at room temperature.
5. Filter and wash the affinity support to remove uncoupled ligand, blocking agent, and reaction byproducts using coupling buffer, water, 1-*M* NaCl, and again with water. Depending on the ligand being coupled, other wash solutions may be used to completely remove unreacted ligand, such as detergents, denaturants, and high or low pH conditions. Finally, store the affinity support in water containing a preservative at 4°C until use.

2.2. Thiol-Reactive Immobilization Methods

Although the majority of methods used to immobilize affinity ligands onto chromatography supports involve reactive groups that target amines, the use of site-directed chemistry that can target other functional groups such as thiols can have significant benefit in certain instances (Domen *et al.*, 1990). Thiols are typically present in proteins at more limited locations than amines and they can, therefore, be used to covalently link at selective sites within a macromolecule. Cystine disulfides in proteins also can be mildly reduced to provide thiol groups for coupling even if a protein in its native state does not have an available free thiol, such as in the reduction of disulfides within antibody hinge regions or in $F(ab')_2$ fragments (see Chapter 20). This strategy can result in the immobilization of proteins in areas away from binding sites or active centers, thus avoiding the blocking of these regions by being forced down toward the matrix during coupling instead of facing outward and available to interact with molecules in the mobile phase. Peptides can also be synthesized with a cysteine residue at one end of their amino acid sequences, thereby providing a functional handle to orient the immobilized peptide with the important binding end facing out from the support. Thiols can even be purposely added to proteins and molecules through the use of special modification reagents to facilitate subsequent coupling to a thiol-reactive support (see Chapter 2, Section 4.1).

The many choices available in thiol-specific immobilization reactions described in this section can provide important options in designing an affinity support with the best possible performance for a particular application. Some of the methods form permanent or stable linkages with thiol-containing molecules by creating thioether bonds. Other methods are capable of forming reversible linkages using disulfide bonds, which subsequently can be reduced to elute off the ligand from the support along with any interacting molecules. The following sections describe the most common methods used for support activation and coupling to thiol-containing ligands for affinity chromatography. The reader is also directed toward the various other sections within this book that describe bioconjugate reagents for use with thiols to better understand the chemistry of thiol reactivity and coupling.

Iodoacetyl and Bromoacetyl Activation

Haloacetyl compounds have been used for decades in the crosslinking, modification, and immobilization of thiol-containing molecules, especially for covalently linking to cysteine-containing proteins and peptides (Narayan *et al.*, 2004; Wilhelmsen *et al.*, 2004; Handlogten, et al., 2005; Kim and Hage, 2006a; Mallik *et al.*, 2007). The reactivity of this group has its origin in the electron-withdrawing properties of the carbonyl oxygen of the carboxylate (note that the iodine atom of iodoacetyl compounds has an electronegativity approximately equivalent to that of the carbon to which it is attached). This effect causes an electrophilic center of partial positive charge on the carbon atom attached to the halogen. Nucleophiles such as thiols containing an unshared pair of electrons can attack this carbon resulting in a potential nucleophilic substitution reaction, which causes displacement of the halogen with simultaneous formation of a thioether bond with the thiol-containing compound.

Iodoacetyl groups in particular have been used to immobilize affinity ligands on all types of solid phases, including chromatography supports, nonporous particles, and planer surfaces for arrays (Camarero, 2006; Fu *et al.*, 2011). Once a support is activated to contain iodoacetyl groups it is relatively stable in aqueous solution, provided it is protected from light and reducing agents. Strong light exposure, especially sunlight, can cause the loss of the iodine atom (as HI) and rapidly eliminate coupling capacity. For this reason, activated supports should be stored in containers that prevent light transmittance to avoid this problem. Thiol-reducing agents can also react with the iodoacetyl groups and should be avoided, but even other reductants such as sodium cyanoborohydride, sodium borohydride, and phosphine reducing agents (e.g., TCEP) can liberate HI and destroy activity, as well.

Bromoacetyl groups can also can be used to create thiol-reactive support materials. All of the reactions described in this section relative to iodoacetyl groups and their synthesis apply to bromoacetyl chemistry, too. To use bromo- instead of iodo-derivatives, substitute bromoacetate for use of iodoacetate in the preparation of the activated support. The bromoacetyl group is more reactive than iodoacetyl, because the bromine atom is more electronegative than the iodine atom; however, in practice the two groups perform nearly identically in the immobilization of thiol-containing ligands onto chromatography supports. The iodoacetyl-activated supports are the most commonly cited in the literature for coupling thiol ligands.

Iodoacetyl groups can potentially react with any nucleophilic site in biomolecules or other ligands, including thiols, amines, and hydroxyl groups, depending on the conditions of the reaction. The product of the reaction is an alkylation of the ligand by the C2 carbon atom of the acetyl group, which forms thioether, secondary amine, or ether linkages with thiols, amines, or hydroxyls, respectively (Figure 15.65). Methionine thioether side chains are the most highly reactive toward alkylation and they can be modified even at acid pH conditions (pH 4.0) (Gundlach *et al.*, 1959). Reaction of an iodoacetyl group with methionine yields an unstable alkylation product, the carboxymethyl sulfonium salt of methionine, which subsequently can degrade by several routes into homocysteine, *S*-carboxymethyl homocysteine, or back to methionine (see Chapter 3, Section 2.1). Care should be taken using this method of coupling, therefore, if a methionine residue in a peptide or protein ligand represents a particularly important amino acid that must remain untouched by the immobilization process. Even if conditions are right to prevent amine or hydroxyl modification during the coupling of a thiol compound, the presence of methionine definitely may result in cross-reactions at this site.

Using iodo- or bromoacetyl-activated supports, thiols including cysteine will typically react under slightly alkaline conditions that involve the use of a buffer in the range of pH 8.0 to 8.5. In this range, amine reactivity will be very limited and hydroxyls will be virtually non-reactive. In fact, most protocols even make use of the amine-containing buffer Tris during the reaction, indicating that the specificity of the reaction toward cysteine thiols is very high. Increasing the alkalinity of the reaction can be used to purposely couple with amines (pH 10–12), while hydroxyls can be effectively targeted at pH > 12. Amine and hydroxyl group reactivity may also be accompanied by the covalent modification of one of the imidazole ring nitrogens of histidine as well as the phenolic hydroxyl group of tyrosine residues. Thus, iodoacetyl supports potentially are one of the few immobilization chemistries that can have multi-purpose immobilization selectivity—the others being epoxy, TsT, and vinyl sulfone, which are able to couple with a variety of nucleophilic functional groups depending on the conditions used during the reaction. However, the targeting of amines or hydroxyls in the presence of thiols is not possible, because thiols have greater reactivity and will always be modified under higher pH conditions (unless the thiols are protected beforehand). The greatest benefit of using iodoacetyl activation is the potential to chemoselectively target

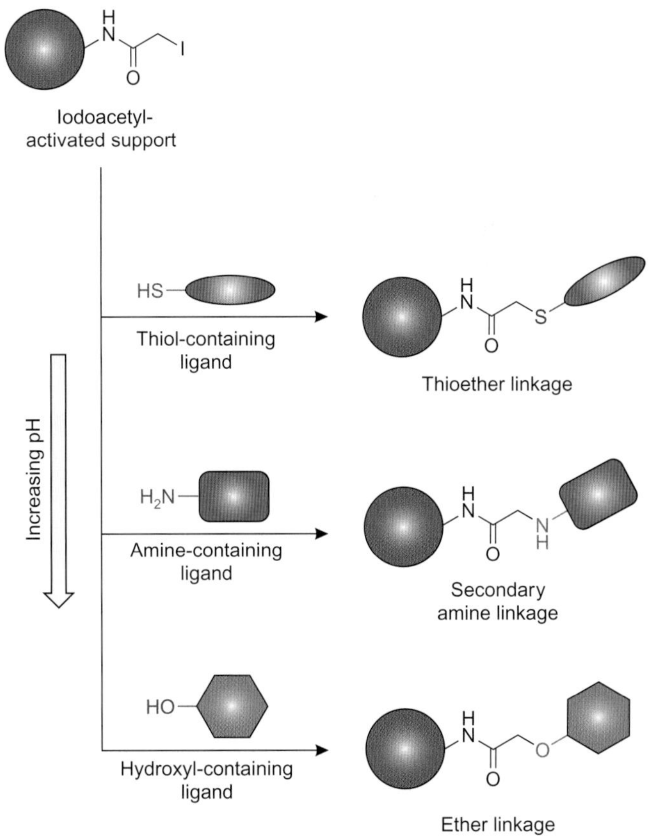

FIGURE 15.65 Reaction of an iodoacetyl-support with thiol-, amine-, and hydroxyl-containing supports.

only thiols (e.g., cysteines) at mild pH, thus leaving other functional groups alone.

Iodoacetyl supports are typically prepared from amine-containing matrices by coupling the carboxylate of iodoacetic acid to a terminal amine-containing spacer arm to form an amide bond. Thus, the initial step in the preparation of this support may be to modify an amine-reactive matrix with a diamine in large excess to form the terminal amino modifications suitable for reaction with iodoacetate. Many different diamine spacers can be used for this purpose, including diaminodipropylamine (DADPA), ethylene diamine (EDA; such as that used in the formation of MANAE supports), and the hydrophilic Jeffamine spacers containing short PEG groups (see the section on spacer arms, this chapter). The final coupling of the iodoacetate molecule onto the amino spacer can be performed using iodoacetic anhydride or NHS-iodoacetate, or through a carbodiimide (EDC) coupling procedure to attach the carboxylate to the amine through an amide bond. Mallik *et al.* (2007) prepared a thiol-reactive silica support using NHS–iodoacetate in the modification of an aminopropyl silane-coated silica particle (see Chapter 13, Section 2) to immobilize ligands for ultimate use in high-performance affinity chromatography (HPAC). The synthesis of a similar iodoacetate-agarose matrix made using the diamine spacer DADPA and attaching iodoacetate using EDC is described in the following protocol (Figure 15.66).

PREPARATION OF AN IODOACETYL-ACTIVATED SUPPORT

1. Prepare 100 ml of a DADPA-agarose support using periodate activation and diamine coupling through a reductive amination process as described elsewhere in this chapter in the section on the preparation of spacer arm derivatives. Wash the support with water and then with several bed volumes of 0.1-*M* MES buffer, pH 4.7 (coupling buffer). Drain to a wet cake, but do not allow the gel to dry.
2. Dissolve 12 g of iodoacetic acid into 100 ml of coupling buffer (makes a 0.645-*M* solution) and adjust the pH back to 4.7 with base. Protect the compound and the solution from light to prevent degradation of the iodoacetate. Note: Bromoacetic acid may be substituted for iodoacetic acid by adding an equal mole amount (8.96 g) to the reaction slurry.
3. Add the wet gel cake to the iodoacetate solution with stirring.
4. With stirring, slowly add 10 g of EDC to the slurry to dissolve. React for 2 h at room temperature with constant mixing using an overhead paddle stirrer or end-over-end rotation in a sealed container. Protect the slurry from light by wrapping the vessel in aluminum foil.
5. Filter off the excess reaction solution and wash the activated gel with water, 1-*M* NaCl, and water (at least 10 bed volumes each) to remove unreacted compound and reaction byproducts. Store the iodoacetyl–agarose support as a 50% slurry in water containing a preservative at 4°C until use. Protect the gel from light to avoid decomposition.

LIGAND COUPLING TO IODOACETYL-ACTIVATED SUPPORTS

The first method described below is a generalized protocol that can be used to immobilize thiol-containing ligands onto iodoacetyl-activated supports. Molecules containing free thiols are notoriously susceptible to

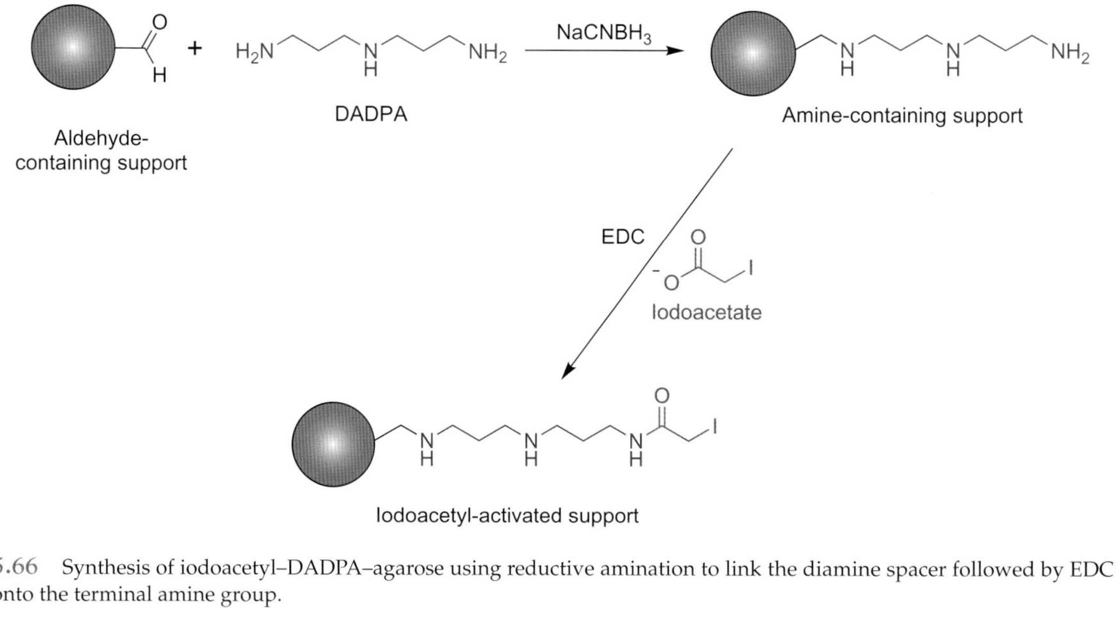

FIGURE 15.66 Synthesis of iodoacetyl–DADPA–agarose using reductive amination to link the diamine spacer followed by EDC coupling of iodoacetate onto the terminal amine group.

oxidation in solution. Cysteine-containing peptides should be dissolved in nitrogen-purged and vacuum-degassed buffer solutions containing EDTA (at least 5 to 10-m*M*) to avoid oxygen-catalyzed or metal-catalyzed oxidation of the thiols to disulfides. This oxidation process can take place very quickly in solution and result in virtually no coupling of ligand to the support material. In addition, after the reduction of protein disulfides to produce free thiols for coupling, the complete removal of the reducing agent is essential to prevent side reactions occurring with the iodoacetyl groups on the support.

The second procedure described below involves the conjugation of intact IgG or $F(ab')_2$ fragments to an iodoacetyl support after reduction of the disulfide linkages between the heavy chains (Figure 15.67). The immobilization of antibodies or antibody fragments through thiols can result in coupling to sites that are away from the antigen binding areas, thus potentially preserving activity better than when using amine-reactive strategies. Using intact antibodies, the reduction process will result in a number of free thiols being produced, the amount of which is dependent on the concentration of the reducing agent and the type of antibody. Often, antibodies are most susceptible to reduction at the disulfides in the hinge region between the heavy chains, but reduction may also occur between the heavy and light chains, which could result in disruption of the antigen binding site. High concentrations of reducing agents used with antibodies could result in the complete dissociation of the heavy and light chain fragments, and if this preparation is used during the immobilization reaction with an iodoacetyl support the result might be little to no specific binding activity after coupling. Therefore, controlling the amount of reductant added is important to the creation of thiols while not destroying the binding activity of the antibody toward antigen. After reduction, the amount of thiols present within the ligand to be immobilized can be determined using Ellman's reagent (see Chapter 2, Section 4.1).

(A) Coupling Thiol-Containing Ligands to Iodoacetyl-Activated Supports

1. Wash 10 ml of an iodoacetyl-activated support with water and then into coupling buffer using a sintered glass filter funnel suspended in a suction filter flask (coupling buffer: 50-m*M* Tris, 0.15-*M* NaCl, 10-m*M* EDTA, pH 8.5). The coupling buffer should be purged with nitrogen and degassed under vacuum to remove oxygen, which may oxidize the thiols and prevent coupling. Protect the activated resin and all solutions from light before and during the coupling reaction.

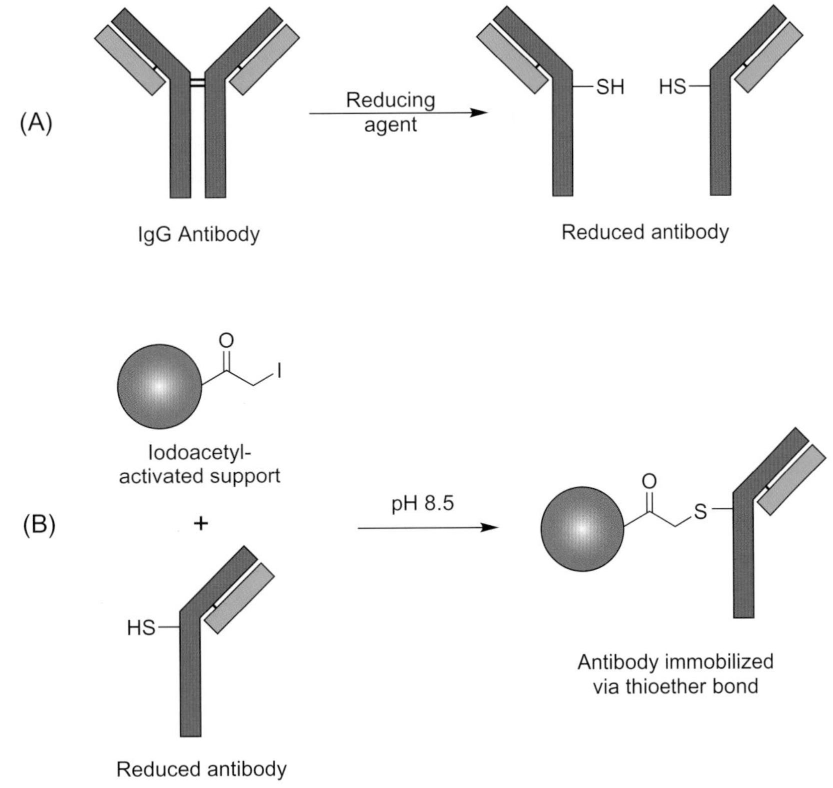

FIGURE 15.67 Reduction of antibody disulfides and coupling to an iodoacetyl-activated support.

2. Dissolve a thiol-containing ligand in 10 ml of coupling buffer at a concentration of at least 2 to 5 mg/ml. Determination of the optimal level of ligand concentration may be achieved experimentally by running a series of small reactions at different concentrations and determining which preparation performs best in the intended affinity chromatography application.
3. Add the ligand solution to the iodoacetyl-support and mix to thoroughly resuspend the resin in the solution. React for at least 1 h at room temperature (2–4 h at 4°C) with constant mixing.
4. Wash the gel with several bed volumes of coupling buffer to remove most of the excess uncoupled ligand. The amount of ligand not coupled may be determined by measuring the volume and concentration of the ligand in the pooled washes. This may be done using a protein assay or spectrophotometrically if the ligand has a characteristic spectral signature. The amount of ligand that has coupled to the matrix may then be determined by the difference. Drain the gel to a wet cake.
5. Block unreacted iodoacetyl active sites by adding to the gel 10 ml of a solution consisting of 50-m*M* cysteine (6.06 mg/ml) dissolved in coupling buffer. React for 30 min with stirring.
6. Wash the affinity resin thoroughly with water, 1-*M* NaCl, and again with water (at least 10 bed volumes each) to remove unreacted components. Store the support until use at 4°C as a 50% slurry in water containing a preservative.

(B) Coupling of Reduced IgG or F(ab′)$_2$ Fragments to Iodoacetyl-Activated Supports

1. Dissolve 1 to 10 mg of an IgG antibody or F(ab′)$_2$ fragment in 1 ml of 0.1-*M* sodium phosphate, 0.15-*M* NaCl, 5-m*M* EDTA, pH 6.0.
2. Add a reducing agent [such as DTT, 2-mercaptoethylamine (2-MEA), 2-mercaptoethanol (2-ME), or tris(carboxyethyl)phosphine (TCEP)] to the antibody solution to give a final concentration of at least 5-m*M*. Mix to dissolve and incubate for 1.5 h at 37°C. Note: Higher concentrations of the reducing agent are sometimes used (i.e., up to 50-m*M*), but the higher the concentration the more likely will be the reduction of disulfides between the heavy and light chains, which may disrupt the three-dimensional structure of the antibody and destroy antigen binding capability. Optimization of the reductant concentration may have to be done to ensure the best performance of the resultant immunoaffinity support.
3. Remove excess reducing agent from the antibody solution by desalting using size exclusion chromatography or by dialysis. During the chromatography or dialysis operation use coupling buffer consisting of 50-m*M* Tris, 0.15-*M* NaCl, 10-m*M* EDTA, pH 8.5, to equilibrate and elute protein through the gel filtration support or as the dialyzing solution. A gel filtration column should consist of at least 10 ml of a support having a molecular weight exclusion limit of no more than 5 to 10 kDa to ensure that the protein will come through in the void volume and be separated from the lower-molecular-weight reductant. A similarly sized dialysis membrane is appropriate. Complete removal of all reducing agent is essential to eliminate competition when coupling the reduced antibody to the iodoacetyl-activated support. Pool the fractions containing desalted protein peak from the gel filtration column or recover the dialyzed protein from the dialysis device. The solution may be concentrated if necessary using centrifugal concentrators to approximately 1 ml to maintain the desired concentration level of antibody in the coupling reaction.
4. Wash 1 ml of an iodoacetyl-activated support with water and then into coupling buffer by placing the resin into a drip column having a bottom porous frit and suspended in a test tube (coupling buffer: 50-m*M* Tris, 0.15-*M* NaCl, 10-m*M* EDTA, pH 8.5). If larger amounts of resin are used for the coupling reaction, then the use of a sintered glass filter funnel suspended in a suction filter flask is more appropriate for the washing steps. The coupling buffer should be purged with nitrogen and degassed under vacuum to remove oxygen, which may oxidize the thiols and prevent coupling. Protect the activated resin and all solutions from light before and during the coupling reaction.
5. Mix the washed iodoacetyl support with the reduced antibody solution and react in a sealed tube by gentle rotation for 1 h at room temperature. The tube should be wrapped in aluminum foil to prevent degradation of the iodoacetyl groups before the coupling reaction has occurred. If the antibody is sensitive to mixing, the tube can be rotated for the first 15 min and then allowed to incubate without continuous rotation. Every 5 min, gently resuspend the gel in the coupling solution to maintain a homogeneous slurry.
6. Wash the support with several bed volumes of coupling buffer to remove most of the not-coupled antibody. The washes may be analyzed *versus* the initial concentration of the antibody solution before coupling to determine the amount immobilized.
7. To block unreacted iodoacetyl groups, add to the washed resin 1 ml of 50-m*M* cysteine solution (6.06 mg/ml) prepared in coupling buffer. Mix and react for 30 min at room temperature.

8. Thoroughly wash the immunoaffinity support with coupling buffer, water, 1-*M* NaCl, and again with water to remove unreacted materials. Store the support until use at 4°C in water as a 50% slurry containing a preservative.

Maleimide Activation

Thiol-containing affinity ligands can be immobilized onto solid supports using maleimide-reactive groups. Maleimides long have been used in crosslinking and modification reagents to target thiols in cysteine-containing proteins for bioconjugation purposes (refer to the Index for many cross-references to this reactive group throughout this book). The specificity of the maleimide group toward thiols is excellent when reactions are performed at mildly basic pH. Side reactions might occur with amines if the pH is raised to a highly alkaline environment, but targeting thiols with maleimides is elegantly specific around physiological pH.

Maleimide groups can be formed on chromatography supports using an amine-containing spacer arm followed by reaction with a heterobifunctional crosslinker, which contains an amine-reactive NHS ester on one end and a maleimide group on the other end. Some examples of these reagents include sulfo-SMCC (Chapter 6, Section 1.3), sulfo-GMBS (Chapter 6, Section 7), or NHS–PEG_n–maleimide crosslinkers (Chapter 18, Section 1.2). An SMCC-based method was used to create maleimide groups on a silica support that had been modified with aminopropyl silane to form the requisite amino terminal spacers (Mallik *et al.*, 2007). EMCS-based linkers have also been described to immobilize affinity ligands containing thiols (Kim and Hage, 2006b). However, a superior alternative for modifying surfaces, particles, and chromatography supports with thiol-reactive maleimide groups is to use a PEG-based crosslinker, which can add hydrophilicity to the spacer group formed on the matrix and thus limit nonspecific binding by avoiding aliphatic linkers. Building hydrophilic amine-terminal spacers off of a glycidol-modified support (see section on periodate oxidation and reductive amination coupling, discussed previously) will also result in extremely biocompatible linkers, which can be used for modification with a maleimide heterobifunctional crosslinker. The reaction of NHS–PEG_4–maleimide with a MANAE–agarose amine-containing spacer arm prepared from a glycidol-modified precursor is shown in Figure 15.68. Other amine-terminal spacer groups may also be used for this purpose to create a thiol-reactive support. Similar chemical coupling strategies have been used successfully for immobilizing affinity ligands onto surfaces (Houseman *et al.*, 2003; Misra and Dwivedi, 2007).

The maleimide-reactive group undergoes rapid alkylation with a thiolate anion ($–S^-$) in the pH range of 5.5 to 8.5 and displays second-order reaction kinetics (Tournier *et al.*, 1998). Under these conditions, the major competing reaction is the potential of the maleimide ring to hydrolyze by opening to the maleamic acid derivative, which essentially inhibits effective coupling with thiols (see Chapter 19, Section 5). However, under standard coupling conditions the immobilization of a thiol ligand proceeds much faster than the rate of hydrolysis, yielding maximal density of ligand within 2 to 4h at room temperature (Figure 15.69). The reaction of the blocked amino acid *N*-acetylcysteine with the maleimide group of sulfo-SMCC was found to be 50% complete within 20min at pH 6.5 and totally complete within 10min at pH 8.5 (Tournier *et al.*, 1998). Allowing reactions to go for a longer time for immobilization reactions onto porous chromatography supports is good practice, as the diffusion of large protein ligands

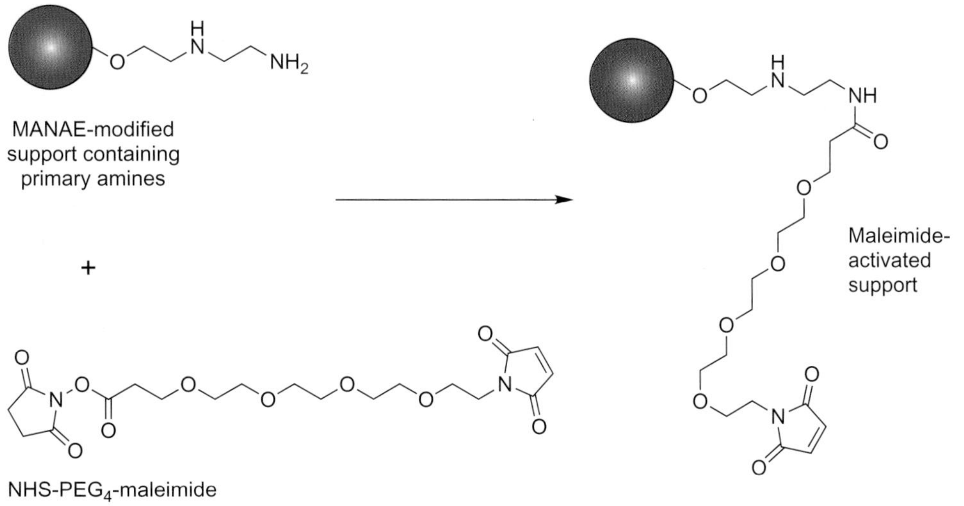

FIGURE 15.68 Preparation of maleimide-agarose by the reaction of NHS–PEG_4–maleimide with MANAE–agarose.

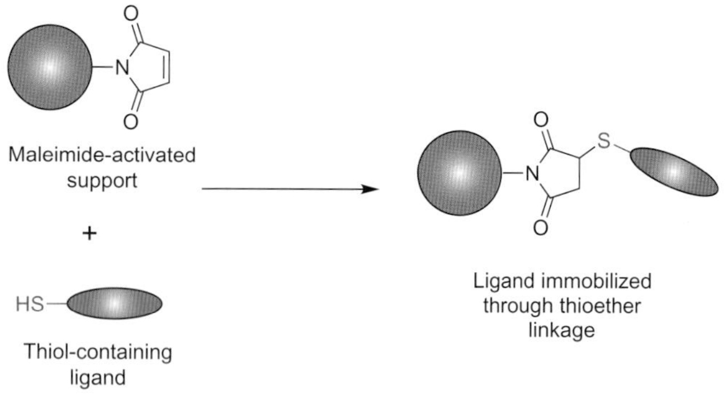

FIGURE 15.69 Coupling of thiol-containing ligands to a maleimide-activated support.

into the particles will increase the reaction times necessary to obtain optimal yields.

A survey of the literature on maleimide immobilization of thiol-containing affinity ligands indicates that this technique has mainly been used to activate and couple molecules to surfaces or nonporous particles (Hwang *et al.*, 2011; Cellet *et al.*, 2012). This is not a limitation of the maleimide group with respect to its utility, but a realization that the cost of materials to form a maleimide-reactive chromatography support may be prohibitive for large-scale use. The creation of other thiol-reactive supports, such as iodoacetyl, discussed previously, involves the use of reagents that are available in multi-gram quantities at relatively inexpensive price points. By contrast, maleimide crosslinkers used to form the support derivatives described in this section are much more expensive on a per-gram basis. Therefore, if small amounts of resin are being activated, this method is eminently appropriate; however, if quantities > 10 ml of support are to be prepared, then it might be better to consider another coupling method due to the price of the maleimide crosslinker.

PREPARATION OF A MALEIMIDE-ACTIVATED SUPPORT

The following protocol requires the use of an amine-containing chromatography support, which may be prepared by the methods discussed in the section on spacer arms elsewhere in this chapter.

Protocol

1. Wash 10 ml of an amine-containing support material (such as MANAE–agarose) with water to remove storage solutions and then with several bed volumes of 0.1-*M* sodium phosphate, pH 7.2 (activation buffer). Finally, suspend the support in an equal volume of activation buffer with mixing.
2. While mixing the suspended amine-containing gel, add 218 mg sulfo-SMCC (Thermo Scientific Pierce) (equivalent to about 50 μmoles/ml gel). Alternatively, an equivalent mole quantity of NHS–PEG_4–maleimide may be added instead of sulfo-SMCC to create a more hydrophilic support environment due to the presence of the PEG-based spacer arm.
3. React with constant mixing for 1 h at room temperature. The mixing may be done using an overhead stirring paddle or end-over-end rotation in a sealed container.
4. Quickly wash the activated support with several bed volumes of activation buffer, 1-*M* NaCl, and with water to remove unreacted crosslinker and reaction byproducts. Drain to a moist cake. To prevent hydrolysis of the maleimide groups, use the activated support immediately to immobilize a thiol-containing ligand according to the following protocol.

LIGAND COUPLING TO MALEIMIDE-ACTIVATED SUPPORTS

1. Prepare 10 ml of a ligand solution for immobilization by dissolving in coupling buffer (0.1-*M* sodium phosphate, 0.15-*M* NaCl, 10-m*M* EDTA, pH 7.2) a protein containing one or more free cysteine thiols or a small molecule ligand having an available thiol at a concentration of 1 to 20 mg/ml for the protein or 2 to 5 mg/ml for the small ligand. Alternatively, the small molecule may be reacted with the support at a concentration of 50 to 100 μmoles/ml gel. The thiols on proteins may be generated by the use of a reducing agent to cleave disulfides (see the protocol described previously for coupling reduced antibodies to iodoacetyl-activated supports) or through the use of a thiolation modification reagent (see Chapter 2, Section 4.1).
2. Add 10 ml of the washed, maleimide-activated support prepared above to the ligand solution with stirring to resuspend the matrix. React at room temperature for at least 2 h with constant mixing using an overhead paddle stirrer or end-over-end mixing in a sealed container.
3. Wash the support with several bed volumes of coupling buffer to remove most of the not coupled ligand. Drain to a moist cake. The washes may be analyzed *versus* the initial concentration of the ligand in the coupling solution before the reaction was initiated to determine the amount immobilized.
4. To block unreacted maleimide groups, add to the washed resin 10 ml of 50-m*M* cysteine solution (6.06 mg/ml) prepared in coupling buffer. Mix and react for an additional 30 min at room temperature.
5. Thoroughly wash the affinity support with coupling buffer, water, 1-*M* NaCl, and again with water to remove unreacted materials. Store the support until use at 4°C in water as a 50% slurry containing a preservative.

Divinyl Sulfone Activation

Divinyl sulfone (DVS) is a highly reactive homobifunctional compound that can be used to activate hydroxylic matrices for coupling to a wide variety of affinity ligands (Porath, 1974). It has also been used as a crosslinking agent to increase the physical strength of hydroxylic supports and as a gelling agent to form large-scale crosslinks in polymers for industrial applications (Kriegel, 2006). Activation of an agarose support with DVS in large excess occurs under alkaline pH conditions with modification of the hydroxyls present on the support. This process results in the reaction of one side of the DVS molecule with an available hydroxyl group on the support, which proceeds through a Michael-type addition to the double bond. At the same time, the formation of reactive vinyl sulfone groups occurs due to the presence of the unreacted end of the molecules remaining free (Figure 15.70). If the reaction is performed with a large excess of DVS, it will occur with minimal crosslinking within the matrix, although closely spaced hydroxyl groups may indeed link together and increase the overall structural rigidity within a resin at the same time as activation occurs. The resultant vinyl sulfone reactive groups are stable for storage in aqueous solution as long as no nucleophiles are present in the storage solution. This simple activation process combined with the excellent hydrolytic stability of DVS-activated supports (at 4°C) makes this method of ligand coupling highly attractive, especially for immobilizing ligands containing thiols or hydroxyls.

Supports activated to contain vinyl sulfone groups can be used to immobilize ligands containing nucleophilic thiol, amino, or hydroxyl groups with good linkage stability (Lihme *et al.*, 1986). Ubrich *et al.* (1992) also found that DVS-mediated coupling resulted in one of the most stable affinity supports for the immobilization of antibodies. Highly activated supports typically can couple proteins at densities of up to 30 to 50 mg/ml gel, depending on the quantity of protein initially reacted. Small-molecule ligands can also be coupled at high mole densities. Thiol-containing ligands are the most reactive and are able to couple with the vinyl sulfone groups at a pH of 6 to 8, while forming thioether linkages of excellent stability. Amine-containing ligands may also be immobilized onto vinyl sulfone supports within the somewhat higher pH range of 8 to 10, which results in a secondary amine bond upon addition to the double bond. Finally, hydroxyl groups on ligands such as those on polysaccharides, carbohydrates, and glycans may be linked to vinyl sulfone supports at pH values > 10 to create ether bonds (Figure 15.71). The pH modulation of vinyl sulfone to react with a single nucleophile type is only limited by the presence of other more nucleophilic groups. In other words, an amine cannot be targeted in the presence of a thiol without coupling also taking place at the thiol site. Similarly, hydroxyl functionalities cannot be targeted for coupling in the presence of other amines or thiols, because the more nucleophilic groups will react preferentially or in addition to the hydroxyl.

The yield of protein coupling to a vinyl sulfone-activated matrix may be enhanced by the addition of relatively high concentrations of PEG added to the reaction medium (5–7% PEG for antibodies, 7–10% PEG for most other proteins; Mini-Leak protocol, Kem-en-tec). In addition, similar enhanced coupling rates and

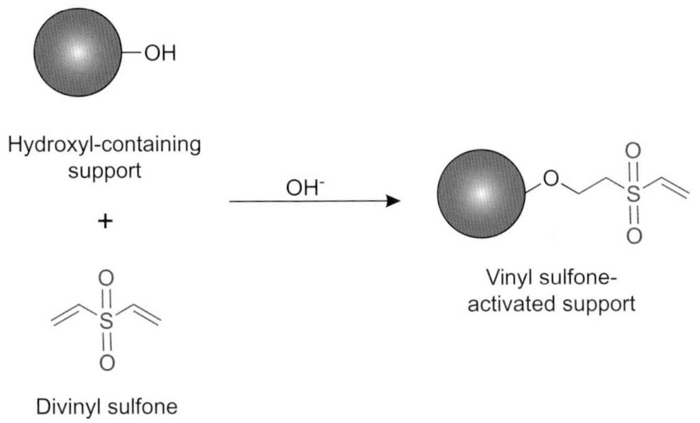

FIGURE 15.70 Activation of a hydroxylic support with divinyl sulfone.

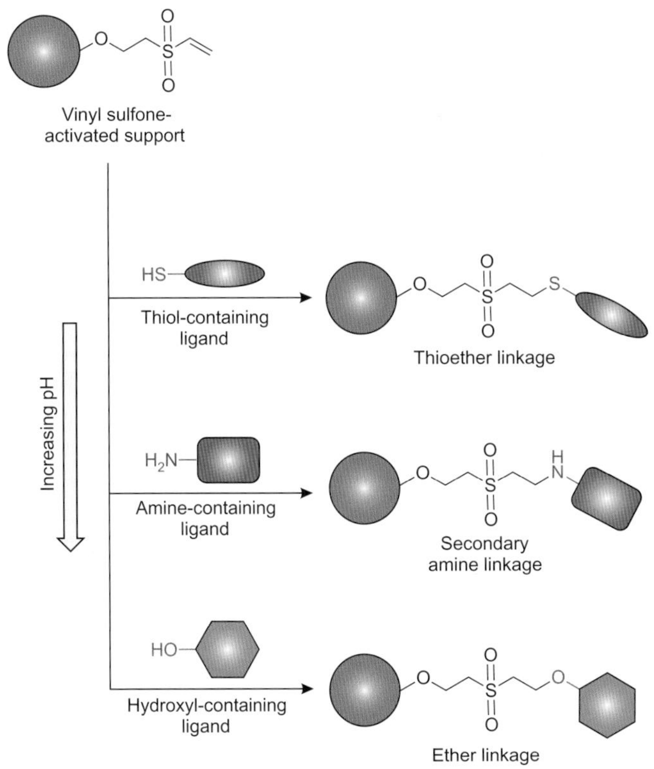

FIGURE 15.71 Reaction of vinyl sulfone-activated supports with thiol-, amine-, and hydroxyl-containing ligands.

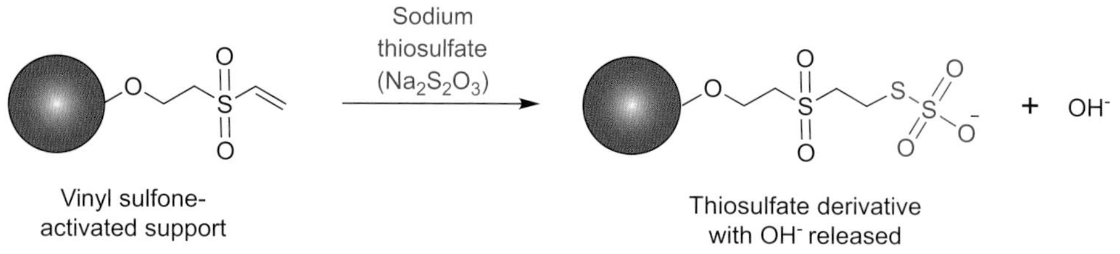

FIGURE 15.72 Reaction of sodium thiosulfate with vinyl sulfone to generate hydroxyl ions.

yields can be observed with the addition of a lyotropic salt (sodium sulfate, potassium sulfate, ammonium sulfate, or potassium phosphate, > 0.5-*M*). Both the PEG and high salt concentrations have the effect of pushing proteins out of solution and toward the surfaces of the activated support where they can more effectively react with the vinyl sulfone groups. This effect is also observed with azlactone-activated supports and with epoxide coupling reactions, described elsewhere in this chapter. Some optimization of the level of PEG or salt addition should be carried out to obtain the best rate of reaction for a given ligand to be immobilized.

The degree of vinyl sulfone-activated species within a chromatography support may be determined by reaction of a small sample of the gel with a large excess of sodium thiosulfate (Porath *et al.*, 1975; Lihme and Boenisch, 1996; Subramanian, 2000). The thiosulfate anion reacts with the vinyl groups and generates OH^- ions (Figure 15.72). Subsequent titration with HCl of the amount of hydroxide ions released provides a measurement of the quantity of vinyl sulfone groups originally present within the matrix.

Divinyl sulfone-activated supports are more reactive than epoxide-activated supports, but the abilities of both immobilization methods are similar in that they can be made to couple with thiols, amines, or hydroxyls simply by modulating the pH of the reaction. The resultant sulfone bridge created between the support and the ligand is stable under physiological conditions, but it can be purposely cleaved by exposure to highly alkaline conditions (see Chapter 9, Section 5). Retro-Michael cleavage of base labile sulfones can be accomplished by mixing the support as a 50% slurry in 0.1-*M* sodium carbonate (or sodium phosphate), pH 11.6, and heating the stirred gel suspension at 37°C for at least 2h (Zarling *et al.*, 1980). Longer incubation periods may be required for complete reversal of immobilized ligands. The reversible nature of the support indicates that DVS-coupled affinity ligands should not be exposed to conditions exceeding pH 8.5 for long periods of time to avoid potential ligand leakage through cleavage of the sulfone linker arm. However, under normal use and storage conditions the sulfone linker arm is very stable and provides a robust affinity support suitable for long term use.

DVS-activated supports have often been used to immobilize a wide variety of biological molecules to solid supports (Pepper, 1994; Morales-Sanfrutos *et al.*, 2010), including to covalently link glycans onto surfaces (Cheng *et al.*, 2011), to activate biodegradable polymers for coupling bioactive molecules (Wang *et al.*, 2011), in the activation of silica particles for the coupling of thiol-containing affinity ligands (Ortega-Munoz, 2010), and in the preparation of reactive polymers (Lihme and Boenisch, 1996).

DVS activation is also critical in the preparation of thiophilic adsorbents that are used in the isolation of immunoglobulins. The principle of thiophilic interaction chromatography involves an affinity for certain regions within the Fc fragment of antibodies by a ligand consisting of a sulfone in proximity to a thioether group, with or without the addition of an aromatic group nearby in the structure (Porath *et al.*, 1985; Hutchens and Porath, 1986, 1987a,b; Belew *et al.*, 1987; Porath, 1987; Porath and Belew, 1987; Nopper *et al.*, 1989; Lihme and Heegaard, 1991; Knudsen *et al.*, 1992; Hardouin *et al.*, 2007). The interaction was discovered entirely by accident, as the affinity does not involve a biospecific interaction, but a synthetic ligand which likely binds through hydrophobic contacts with pockets within the Fc domains of immunoglobulins. A thiophilic affinity resin can be prepared simply by coupling 2-mercaptoethanol to a DVS-activated support according to the following protocols (Figure 15.73). The resultant support can be used to capture antibodies in the presence of high concentrations of a lyotropic salt (at least 0.5-*M* potassium sulfate), which serves to enhance hydrophobic interactions. Subsequent elution of bound antibody is carried out merely by elimination of the salt from the buffer, thus making this method extremely mild for purifying IgG antibodies. The capacity of this affinity support for human IgG can be as high as 20mg/ml of thiophilic gel.

PREPARATION OF A DVS-ACTIVATED SUPPORT

The activation protocol described below should be done with caution, as divinyl sulfone is a highly reactive and toxic compound. All operations should be carried out in a fume hood and with the use of proper personal protective equipment.

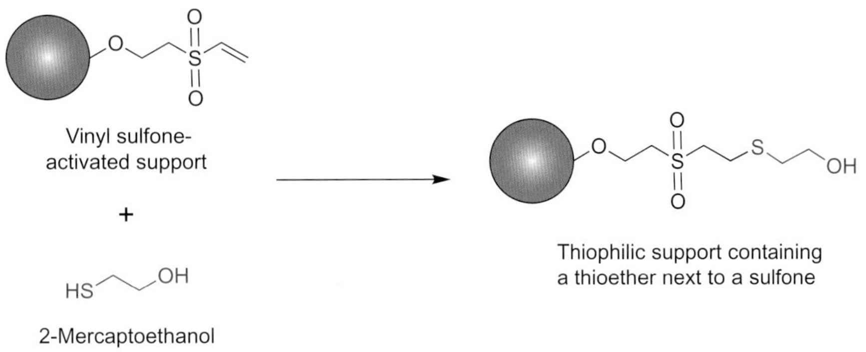

FIGURE 15.73 Preparation of a thiophilic adsorbent through coupling of 2-mercaptoethanol to DVS-activated supports.

Protocol

1. Wash 100 ml of a support material containing hydroxyl groups (such as agarose) with water (at least 5 bed volumes) to remove storage solutions and preservatives. Typically, a non-crosslinked agarose resin can be used for the activation, because crosslinking will occur during the activation step and create a more rigid matrix. However, other natural or synthetic polymeric, hydroxyl-containing supports or inorganic supports containing hydroxyl groups may be used as well, whether crosslinked or not. Use a sintered glass filter funnel suspended in a suction filter flask while pulling a gentle vacuum to facilitate the washing steps. After washing the gel, suction to a moist cake by removing excess water, but do not allow the support to dry.
2. Suspend the gel in 100 ml of 0.5-*M* sodium carbonate (no pH adjustment) and mix in a fume hood using an overhead paddle stirrer.
3. Add 10 ml of divinyl sulfone dropwise to the stirring gel slurry over a period of about 15 min.
4. Continue to react for 1 h at room temperature with constant mixing.
5. Wash the activated gel extensively with water to remove excess DVS (at least 20 bed volumes). Storage of the activated support may be achieved as a 50% slurry in water containing a preservative (e.g., 1,1,1-trichloro-2-methyl-2-propanol) at 4°C. Properly stored, the support should remain stable up to 2 years with minimal loss of vinyl sulfone coupling activity.

LIGAND COUPLING TO DVS-ACTIVATED SUPPORTS

The following ligand-coupling procedures describe the use of a DVS-activated support (10 ml) for the coupling of a thiol-, amine-, or hydroxyl-containing ligand. This may include the immobilization of proteins, carbohydrates, or other molecules that contain these functional groups. The immobilization through thiols may be performed in the presence of other amines or hydroxyl groups on the ligand; however, coupling through amines must be done while avoiding the presence of thiols, as they will preferentially react at a faster rate than amines. Similarly, the coupling of hydroxyls should be done while avoiding the presence of thiols or amines on the ligand.

(A) Coupling of Thiol-Containing Proteins or Ligands

1. Prepare 10 ml of a thiol-containing ligand solution by dissolving in coupling buffer (0.1-*M* sodium phosphate, 0.15-*M* NaCl, 10-m*M* EDTA, pH 7.5) a protein containing one or more free cysteine thiols or a small molecule ligand having an available thiol at a concentration of 1 to 20 mg/ml for the protein or 2 to 5 mg/ml for the small ligand. Notes: The EDTA chelator is present to inhibit metal-catalyzed oxidation of the thiols to disulfides. Alternatively, a small molecule may be reacted with the support at a concentration of 50 to 100 μmoles/ml gel. The thiols on proteins may be generated by the use of a reducing agent to cleave disulfides (see the protocol described previously for coupling reduced antibodies to iodoacetyl-activated supports) or through the use of a thiolation modification reagent (see Chapter 2, Section 4.1). Complete removal of any thiol-containing reducing agents must be performed before attempting to immobilize the reduced protein to avoid competition during the coupling reaction. The coupling buffer may also be formulated to contain a lyotropic salt, such as sodium sulfate, at a concentration of at least 0.5-*M* to increase the rate of reaction when immobilizing proteins. Alternatively, the addition of PEG (MW 20 kDa) to the coupling buffer may be used for this purpose at a concentration of 5 to 7% for antibodies or 7 to 10% for most other proteins.
2. Wash 10 ml of the DVS-activated support prepared above with water to remove any preservatives from the storage solution and then wash with coupling buffer (at least 2 bed volumes). Drain to a wet cake.
3. Add the gel to the ligand solution with stirring (total slurry volume: 20 ml). React overnight at room

temperature or for at least 18h with constant mixing using an overhead paddle stirrer or end-over-end mixing in a sealed container. The reaction may also be performed at 4°C for sensitive molecules or proteins, but the reaction time may have to be extended to obtain the same levels of coupling.

4. Wash the support with several bed volumes of coupling buffer (without the lyotropic salt or PEG present) to remove most of the not-coupled ligand. The washes may be analyzed *versus* the initial concentration of ligand in the coupling solution before the reaction was started to determine the amount that was immobilized. Continue to wash with several bed volumes of water and drain to a wet cake.
5. To block unreacted vinyl sulfone groups, add the washed resin to 10ml of 0.1-*M* cysteine solution (12.12mg/ml) prepared in 0.1-*M* sodium bicarbonate, pH 8.6. Alternatively, use a blocking solution consisting of 0.2-*M* ethanolamine prepared in the same buffer. Mix and react for an additional 2h at room temperature. Note: Do not block the support with 2-mercaptoethanol, because coupling this molecule may result in the creation of a number of thiophilic interaction sites within the matrix, which could cause nonspecific binding with antibodies and other molecules, depending on the conditions of the affinity chromatography being done.
6. Thoroughly wash the affinity support with 0.1-*M* sodium phosphate, pH 7.5, water, 1-*M* NaCl, and again with water to remove unreacted materials. Store the support until use at 4°C in water as a 50% slurry containing a preservative.

(B) Coupling of Amine-Containing Proteins or Ligands

1. Prepare 10ml of a amine-containing ligand solution by dissolving in coupling buffer (0.1-*M* sodium bicarbonate, pH 8.6) a protein containing one or more free amines or a small molecule ligand having an available amine at a concentration of 1 to 20mg/ml for the protein or 2 to 5mg/ml for the small ligand. Alternatively, the small molecule may be reacted with the support at a concentration of 50 to 100μmoles/ml gel. Note: The coupling buffer may also be made up to contain a lyotropic salt, such as sodium sulfate, at a concentration of at least 0.5-*M* to increase the rate of reaction when immobilizing proteins. Alternatively, the addition of PEG (MW 20kDa) to the coupling buffer may be used for this purpose at a concentration of 5 to 7% for antibodies or 7 to 10% for most other proteins.
2. Wash 10ml of the DVS-activated support prepared as described previously with water to remove any preservatives from the storage solution and then wash with coupling buffer (at least 2 bed volumes). Drain to a wet cake.
3. Add the gel to the ligand solution with stirring to resuspend the matrix (total slurry volume: 20ml). React overnight (for at least 18h) at room temperature or at 4°C with constant mixing using an overhead paddle stirrer or end-over-end mixing in a sealed container.
4. Wash the support with several bed volumes of coupling buffer (without the lyotropic salt or PEG present) to remove most of the not-coupled ligand. The washes may be analyzed *versus* the initial concentration of ligand in the coupling solution to determine the amount immobilized. Continue to wash with several bed volumes of water and drain to a wet cake.
5. To block unreacted vinyl sulfone groups, add the washed resin cake to 10ml of 0.2-*M* ethanolamine solution prepared in 0.1-*M* sodium bicarbonate, pH 8.6. Note: After the addition of ethanolamine to the buffer, adjust the pH back to 8.6 with 6-*N* HCl solution while maintaining the temperature near ambient using an ice bath if necessary. Mix and react for an additional 2h at room temperature.
6. Thoroughly wash the affinity support with 0.1-*M* sodium bicarbonate, water, 1-*M* NaCl, and again with water to remove unreacted materials. Store the support until use at 4°C in water as a 50% slurry containing a preservative.

(C) Coupling of Carbohydrates or Hydroxyl-Containing Ligands

1. Prepare 10ml of a hydroxyl-containing ligand solution (such as a carbohydrate, polysaccharide, or glycan) by dissolving it in coupling buffer (0.1-*M* sodium carbonate, pH 11.0) at a concentration of up to 100mg/ml for mono- or disaccharides, or at least 1 to 20mg/ml for larger polysaccharides or glycans. Alternatively, a hydroxyl-containing ligand may be reacted with the support at a concentration of 50 to 100μmoles/ml gel. Notes: The more concentrated the ligand solution, the better will be the coupling yield of the reaction and ultimate density of ligand on the support. However, for the immobilization of carbohydrates to be used in the affinity capture of lectins or other carbohydrate binding proteins, it may be optimal to immobilize the ligand at lower than maximal densities. This is due to the fact that many lectins contain more than one binding site for a carbohydrate, and multi-site interactions on the affinity support will have a tendency to create strong avidity, which may be difficult to reverse for the efficient elution of interacting proteins. Thus, some degree of experimental optimization of ligand density may have to be performed to obtain the best

performance of the affinity support in its intended application.

2. Wash 10 ml of the DVS-activated support prepared previously with water to remove any preservatives from the storage solution and then wash with coupling buffer (at least 2 bed volumes). Drain to a wet cake.
3. Add the gel to the ligand solution with stirring (total slurry volume: 20 ml). React overnight at room temperature (for at least 18 h) or at 4°C with constant mixing using an overhead paddle stirrer or end-over-end mixing in a sealed container. Reactions done at 4°C may take extended time periods to reach the same level of coupling yield as those done at room temperature.
4. Wash the support with several bed volumes of coupling buffer to remove most of the not coupled ligand. The washes may be pooled and analyzed *versus* the initial concentration of ligand in the coupling solution before the reaction was initialized to determine the amount of ligand immobilized. Drain the gel to a wet cake.
5. To block unreacted vinyl sulfone groups, add the washed resin cake to 10 ml of 0.2-*M* ethanolamine solution prepared in 0.1-*M* sodium carbonate, pH 11.0. (Note: After the addition of ethanolamine to the buffer, adjust the pH back to 11.0 with 6-*N* HCl, if necessary. Mix and react for an additional 2 h at room temperature.
6. Thoroughly wash the affinity support with 0.1-*M* sodium carbonate, water, 1-*M* NaCl, and again with water to remove unreacted materials. Store the support until use at 4°C in water as a 50% slurry containing a preservative.

(D) Coupling of 2-Mercaptoethanol to Make a Thiophilic Resin

The following protocol represents one of several methods available to form a thiophilic affinity support for the purification of immunoglobulins and other proteins able to interact with the ligand.

1. Wash 10 ml of the DVS-activated support prepared previously with water to remove any preservatives from the storage solution and then wash with at least 2 bed volumes of coupling buffer (0.1-*M* sodium carbonate, pH 9.0). Drain to a wet cake.
2. In a well-ventilated fume hood, suspend the washed activated support in 9 ml of coupling buffer and add 1 ml of 2-mercaptoethanol (stench!) with stirring.
3. React overnight (at least 18 h) at room temperature with constant mixing. For small volumes of gel, mixing may be accomplished by end-over-end rotation in a sealed container. For larger volumes of gel the use of an overhead paddle stirrer is appropriate; however, due to the unpleasant odor of 2-mercaptoethanol a sealed round-bottom flask should be used for the gel slurry during mixing operations.
4. Wash the resin thoroughly in the fume hood with coupling buffer, 1-*M* NaCl, and water. Continue the washing until no odor is detectable from 2-mercaptoethanol. Store the thiophilic support until use at 4°C in water as a 50% slurry containing a preservative.

Pyridyl Disulfide Activation

Pyridyl disulfide groups can react with thiol-containing molecules by disulfide exchange to form new mixed disulfide linkages. This group long has been used as a thiol-reactive end in crosslinking reagents and protein modification agents, such as the popular heterobifunctional compound SPDP (Chapter 6, Section 1.1). Pyridyl disulfide groups also can be formed on chromatography supports to immobilize thiol-containing ligands for affinity separations through a simple reaction process (Cuatrecasas, 1970; Egorov *et al.*, 1975; Carlsson *et al.*, 1976; Ngo, 1989). The similar reagents 4,4′-dipyridyl disulfide (Grassetti and Murray, 1967) or 2,2′-dipyridyl disulfide (Brocklehurst *et al.*, 1973) can be used as activation agents to form reactive pyridyl disulfide groups on resins. Reaction of these compounds in excess with a thiol-containing support material results in the formation of pyridyl disulfide groups and activation of the support for disulfide exchange reactions with a thiol-containing ligand (Figure 15.74). The requirement for activation using these reagents is to first form free thiol groups on a support using spacer arms that terminate in thiol groups. One method of creating such a thiol-containing matrix is to couple the *bis*-thiol compound DTT to an epoxy-activated support (see the section Epoxide Activation, this chapter). Another potential approach to forming terminal thiol groups is to modify an amine-containing support with a thiolation reagent

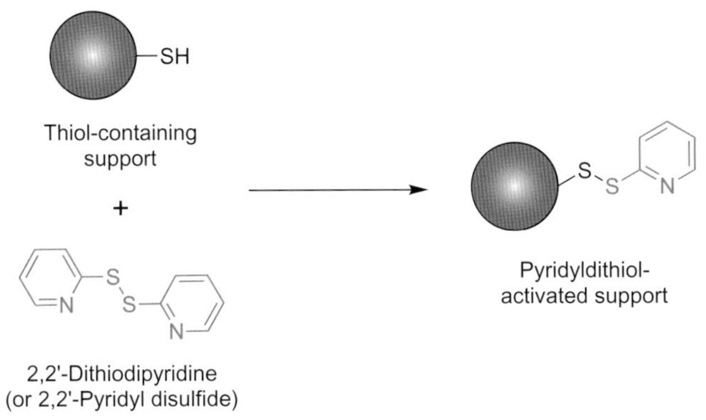

FIGURE 15.74 Activation of a thiol-containing support with 2,2′-pyridyl disulfide.

such as SATA, Traut's reagent, or homocysteine thiolactone (Chapter 2, Section 4.1), which form free or protected thiols upon reaction with the amines.

Amine-containing supports may also be modified with heterobifunctional crosslinking agents to form the pyridyl disulfide reactive groups directly from an amine-terminal spacer (Brogan and Schoenfisch, 2005). In this case, reaction of the terminal amines with the crosslinking reagent SPDP (Chapter 6, Section 1.1) through its NHS ester end creates amide bond linkages with the amines on the support, which then forms short spacer arms that terminate in pyridyl disulfide groups. An even better option is to use an NHS–PEG$_n$–pyridyl disulfide crosslinker, which has a hydrophilic spacer arm in its construction to create a matrix having low nonspecific binding characteristics (Chapter 18, Section 1.2) (Figure 15.75). This activation strategy may be the most straightforward, but it also is likely the most expensive due to the potential cost of the crosslinking reagent. For making small amounts of support material (i.e., <10 ml) the use of a pyridyl disulfide-containing crosslinker is cost effective and a simple route to production of the activated support. It also works well for activation of surfaces or small nonporous particles. For making larger quantities of resin materials, however, it may be best to use the methods that employ the dipyridyl disulfide activation reagents, because these compounds are much less expensive.

Perhaps the simplest method of introducing pyridyl disulfide groups into a chromatography support is to use the reagent PDEA, which is 2-(2-pyridinyldithio) ethaneamine. This small compound has an amine on one end and the thiol-reactive group on the other end. An amine-reactive chromatography support may be used to couple this reagent in high yield, thus forming the pyridyldithiol groups in a single step. Figure 15.76 illustrates the coupling of PDEA to a succinimidyl-carbonate-activated support, which reacts with amines within the pH range of 7 to 9 to give carbamate linkages (see the associated section, this chapter). Renberg *et al.* (2005) used PDEA to modify a sensor chip for the immobilization of thiol-containing Affibodies, which

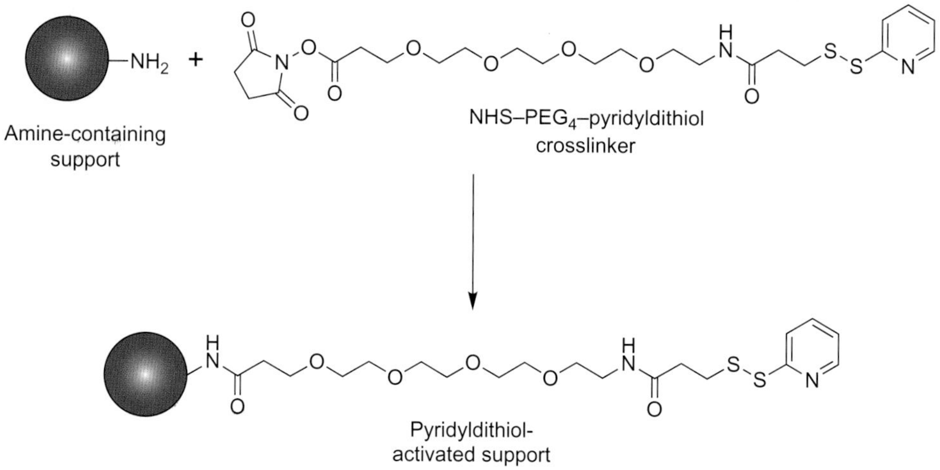

FIGURE 15.75 Use of NHS–PEG$_n$–pyridyl disulfide to create pyridyl disulfide groups from an amine-containing support.

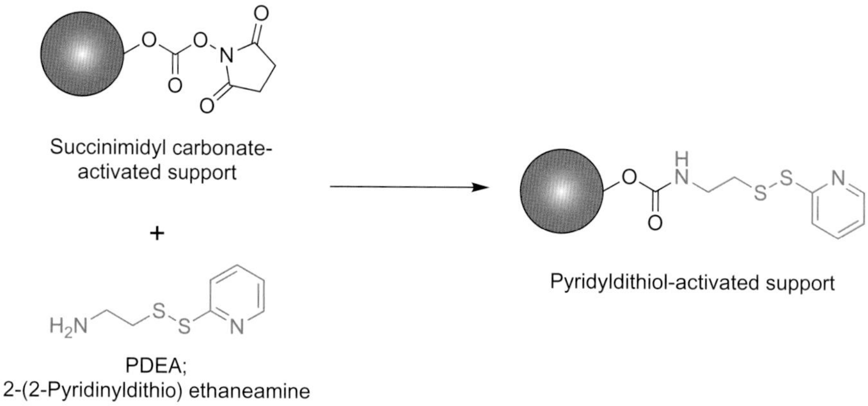

FIGURE 15.76 Use of PDEA to form pyridyl disulfide groups on an amine-reactive support.

are synthetic peptides having antibody-like binding properties toward antigens. The use of PDEA is also a recommended method of coupling thiol affinity ligands to a Biacore sensor chip, because the chip can be regenerated after use. PDEA-modified agarose supports have also been used to immobilize enzymes for use a bioreactors (Mansfeld and Ulbrich-Hofmann, 2000). The amino group of PDEA was coupled to CNBr-activated agarose or through an aminocaproic acid spacer using the EDC/NHS coupling reaction to yield the final thiol-reactive derivative (Chapter 4, Section 1).

Once pyridyl disulfide groups are formed on a support, they are stable for long-term storage in aqueous solution provided no reducing or oxidizing agents are present. The reaction of these groups with thiol-containing affinity ligands results in the formation of reversible disulfide bonds with release of the chromogenic leaving group pyridine-2-thione (Figure 15.77). Once the leaving group is displaced by a thiol ligand, it cannot go back and again react with the disulfide groups formed on the matrix, because the thione double bond makes the sulfur unreactive toward disulfide exchange. The release of this group also can be used to estimate the degree of coupling by measurement of its absorbance at 343 nm, provided the ligand does not absorb in this region. The reaction of a thiol-containing ligand with the pyridyl disulfide groups on the support will occur within the pH range of 4 to 8, with optimal reaction kinetics observed at pH 7 to 8 (Carlsson *et al.*, 1978).

Pyridyl disulfide immobilization is an alternative to the permanent coupling methods discussed previously for thiol ligands in that a coupled molecule can be subsequently removed from the support by treatment with a disulfide reducing agent. Disulfide immobilized ligands may be cleaved from the support by incubation with 25- to 50-m*M* DTT for 2 h at room temperature. This feature builds the potential for reversible covalent capture and release of ligand-target molecule interacting pairs under mild chromatography conditions. Therefore, this may be a valuable technique for co-immunoprecipitation (co-IP) assays in the study of protein interactions, because it allows for the isolation of these proteins under non-denaturing conditions. After a bound ligand is released, the support once again contains free thiol groups that then can be re-activated to form pyridyl disulfide groups for another immobilization reaction, thus allowing complete reusability of the support.

Pyridyl disulfide immobilization has also been used to couple thiol-containing antibodies, antibody fragments, proteins, and other molecules to polymers and solid phases of all types. For instance, Iwata *et al.* (2007) used pyridyl disulfide groups on polymer brushes to immobilize Fab′ fragments containing thiol groups

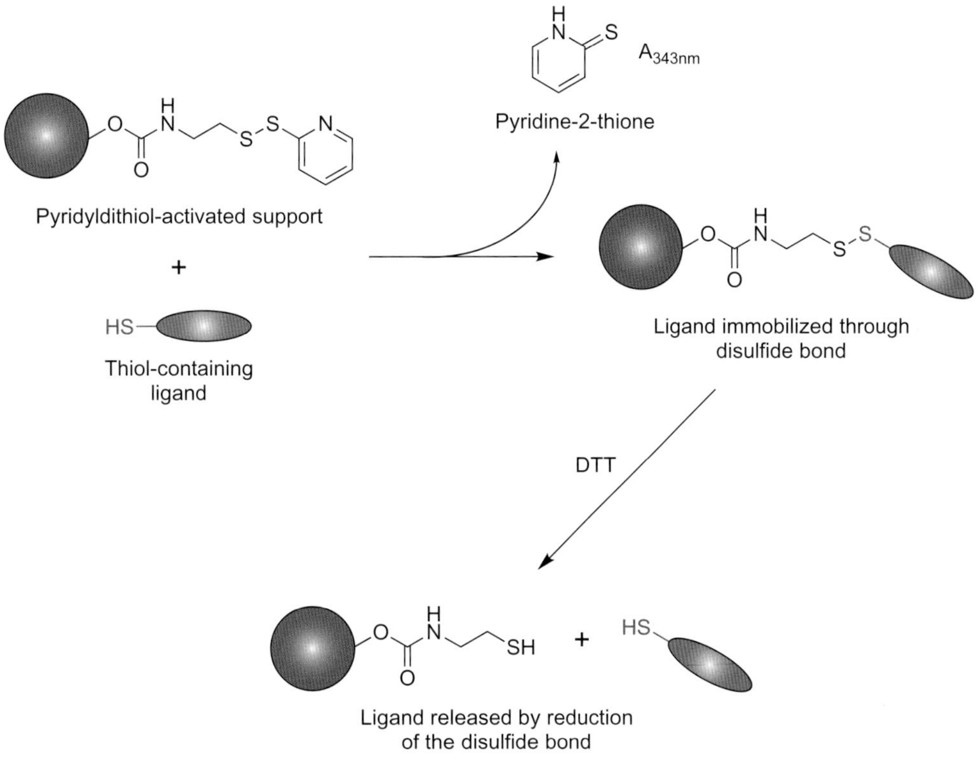

FIGURE 15.77 Coupling a thiol-containing ligand to a pyridyl disulfide-activated support with subsequent removal by treatment with a reducing agent.

within their hinge region. Vázquez-Dorbatt *et al.* (2009) formed these active groups within a glycopolymer by reaction of a pyridyl disulfide-containing initiator using atom transfer radical polymerization (ATRP) to create a copolymer matrix that could immobilize double-stranded siRNA, which had been previously modified to contain a terminal thiol group. Metal surfaces have also been activated with pyridyldithiol groups to allow affinity ligands to be coupled for molecular recognition force microscopy measurements (Brogan and Schoenfisch, 2005). In addition, a pyridyl disulfide-activated support can be used for the proteomics study of cysteine-containing peptides by reversible capture and elution with subsequent analysis by mass spec (Shen *et al.*, 2003).

PREPARATION OF A PYRIDYL DISULFIDE-ACTIVATED SUPPORT

The following methods for the preparation of a pyridyldithiol support represent different options for building this reactive group onto various base supports. The same general methods may be used to activate particles or surfaces with excellent results. The level of reactive pyridyl disulfide groups formed on a support may be controlled by adjusting the amount of functional groups initially created on the support, which are subsequently modified to create the reactive groups. The following methods use either amine- or thiol-terminal spacer arms on a support to form the final reactive pyridyl disulfides.

(A) Activation using 2,2′-Dipyridyl Disulfide

This protocol will work well using either 2,2′-dipyridyl disulfide or 4,4′-dipyridyl disulfide. These compounds may be obtained commercially (Acros, Aldrich) or synthesized from the corresponding mercaptopyridine precursor according to the method presented in Hermanson *et al.* (1992).

1. Prepare 10 ml of a support containing a thiol-terminal spacer using the methods described within the section on spacer arms in this chapter. The support should contain 10 to 50 μmoles –SH groups/ml of gel. Wash the support with water to remove storage solution components, if necessary, and drain to a wet cake.
2. In the fume hood, prepare 10 ml of an acetone:water (1:1, v/v) solution containing 1 g of 2,2′-dipyridyl disulfide dissolved into it.
3. Add the gel to the 2,2′-dipyridyl disulfide solution with stirring and react at room temperature for 30 min with constant mixing in the fume hood. For small amounts of gel, mixing may be accomplished by end-over-end rocking in a sealed container. If preparing larger quantities of activated gel, use an overhead paddle stirrer. During the course of the reaction, the release of pyridyl-2-thione molecules will turn the solution a yellow color.
4. Wash the activated gel with at least 10 bed volumes of acetone:water (1:1) to remove excess reactants and reaction byproducts. Continue to wash thoroughly with water to remove the last traces of acetone. Finally, wash the support with 1-m*M* EDTA (at neutral pH) to protect against metal-catalyzed oxidation. Store the pyridyl disulfide-activated support at 4°C as a 50% slurry in water containing 1-m*M* EDTA and a preservative until use.

(B) Activation using NHS–PEGn–Pyridyl Disulfide Crosslinkers

PEG-containing heterobifunctional crosslinkers containing an amine-reactive NHS ester on one end and a pyridyl disulfide group on the other end may be used to form thiol-reactive groups by modification of amine-containing supports according to the following protocol. Some optimization of the amount of crosslinker addition may have to be carried out to obtain the best density of pyridyldithiol groups on the support for a specific application. Refer to Chapter 18 for additional information on discrete PEG-based reagents. Other pyridyl disulfide-containing crosslinkers may also be used to activate an amine-support, such as SPDP or SMPT (Chapter 6, Sections 1.1 and 1.2).

1. Prepare 10 ml of a support containing an amine-terminal spacer using the methods described in the section on spacer arms in this chapter. The support should contain 10 to 50 μmoles $-NH_2$ groups per ml of gel. Wash the support with water to remove storage solution components, if necessary, and drain to a wet cake.
2. In a fume hood, dissolve 0.56 g SPDP–dPEG$_4$–NHS ester (also called PEG$_4$–SPDP) (Quanta Biodesign or Thermo Fisher) in 2 ml of dry DMSO, DMF, or DMAC solvent. This amount will provide 100 μmoles of crosslinker per milliliter of gel during the reaction.
3. Add the moist gel cake from step 1 to 8 ml of reaction buffer (0.1-*M* sodium phosphate, pH 7.5) and mix to resuspend the beads.
4. With stirring, add the 2 ml of dissolved crosslinker prepared in step 2 to the stirring support slurry prepared in step 3.
5. React with constant mixing at room temperature for 1 h. Mixing may be done using an overhead paddle stirrer or by using end-over-end rocking in a sealed container.
6. Wash the activated gel thoroughly with water, 1-*M* NaCl, and water to remove excess reactants and reaction byproducts. Finally, wash the support with 1-m*M* EDTA (at neutral pH) to protect against metal-catalyzed oxidation of the disulfide groups. Store the pyridyl disulfide-activated support at 4°C as a

50% slurry in water containing 1-m*M* EDTA and a preservative until use.

(C) Activation using PDEA

The following protocol can be used to create reactive pyridyl disulfide groups on an amine-reactive support. The procedure uses the compound PDEA, which is 2-(2-pyridinyldithio)ethaneamine, containing a free primary amine on one end and a pyridyldithiol group on the other end. PDEA may be obtained from GE Healthcare or synthesized by the reaction of 2 equivalents of 2,2′-dipyridyl disulfide (Aldrich) with 1 equivalent of cysteamine (or 2-mercaptoethylamine) in methanol according to the procedure of Li *et al.* (2008). The preparation of an amine-reactive support can be accomplished using any of the protocols previously described in this chapter; however, do not use an aldehyde-containing support to couple PDEA through reductive amination, because the presence of a reducing agent also will reduce the pyridyl disulfide groups and render the support inactive.

1. Prepare 10 ml of an amine-reactive support material according to the methods described in this chapter (e.g., CDI or an NHS ester-activated agarose). If the support is stabilized in an organic solvent (e.g., acetone), just before use filter off the excess solution using a sintered glass filter suspended in a vacuum filter flask within a fume hood. Draw a gentle vacuum until the excess solvent stops dripping from the filter while at the same time breaking up the support into finely divided pieces using a spatula, but do not allow it to dry. Release the vacuum.
2. In a fume hood, prepare a 20-m*M* PDEA solution in the appropriate solvent for coupling to the amine-reactive support of choice. For instance, if using a CDI- or NHS ester-activated support, to obtain maximal coupling yields dissolve 44 mg of PDEA in 10 ml of nonaqueous solvent, such as DMSO or DMF. Add to this solution, 2 mole equivalents (over the amount of PDEA added) of an organic base such as DIEA, TEA, or DMAP to catalyze the reaction by accepting the protons that are generated during the course of the reaction. Note: PDEA may also be coupled to amine-reactive supports in aqueous solution. The buffer composition and pH should be determined based on the suggested methods associated with the amine-reactive chemistries described earlier in this chapter. For aqueous coupling, mix the finely divided pieces of activated gel from step 1 with the PDEA buffer solution. The PDEA concentration may be increased for aqueous coupling to account for potential hydrolysis of reactive groups on the support. For instance, PDEA may be dissolved at 80-m*M* concentration in 50-m*M* sodium borate, pH 8.5, for amine-reactive immobilization methods compatible with this environment (i.e., for use with CDI- or NHS ester-activated supports).
3. With stirring and in the fume hood, mix the activated support material with the PDEA solution to resuspend it as a uniform slurry.
4. React with mixing for 1 h at room temperature using an overhead paddle stirrer or end-over-end rocking in a sealed container.
5. Wash the thiol-reactive support with solvent to remove excess reactants (at least 10–20 bed volumes). Wash the support into water by using sequentially increasing concentrations of water-in-solvent (e.g., 30% water, 70% water) until 100% water washes are used to completely remove the solvent. Note: If an aqueous reaction was done, then instead of washing with solvent wash with water, 1-*M* NaCl, and water to remove excess reactants. Finally, wash the support with 1-m*M* EDTA (at neutral pH) to protect against metal-catalyzed oxidation during storage. Store the pyridyl disulfide-activated support at 4°C as a 50% slurry in water containing 1-m*M* EDTA and a preservative until use.

LIGAND COUPLING TO PYRIDYL DISULFIDE-ACTIVATED SUPPORTS

The coupling of thiol-containing ligands to pyridyl-disulfide-activated supports can be performed over a broad pH range of approximately pH 4 to 8 with excellent results. The following immobilization protocols are generalized methods for coupling cysteine-containing proteins, peptides, or other thiol-containing molecules. The methods may also be used with proteins modified to contain –SH groups through the use of thiolation reagents (Chapter 2, Section 4.1). Some optimization may have to be carried out to determine a ligand density that is best for a particular affinity chromatography application. Preparing a series of small batches using different concentrations of ligand in each reaction medium will produce a range of ligand densities on the support for evaluation. In addition, after ligand coupling do not attempt to block excess reactive groups with a thiol-containing compound such as cysteine, as this may also cleave off any immobilized ligands.

(A) Coupling of Reduced IgG or F(ab′)$_2$ Fragments

1. Dissolve an IgG antibody or F(ab′)$_2$ fragment at a concentration of 1 to 10 mg/ml in 1 ml of 0.1-*M* sodium phosphate, 0.15-*M* NaCl, 10-m*M* EDTA, pH 7.5 (coupling buffer). The coupling buffer should be purged with nitrogen and degassed under vacuum to remove oxygen, which may oxidize the thiols and prevent coupling.

2. Add a reducing agent, such as DTT, 2-mercaptoethylamine (2-MEA), 2-mercaptoethanol (2-ME), or tris(carboxyethyl)phosphine (TCEP), to the antibody solution to give a final concentration of at least 5-m*M*. A highly concentrated stock solution of the reducing agent first may be prepared in coupling buffer and then a small aliquot of that solution added to the antibody to obtain the final desired concentration. Mix to dissolve and incubate for 1.5h at 37°C. Note: Higher concentrations of the reducing agent may be used (typically ≤50-m*M*), but the higher the concentration, the more likely the reduction of disulfides between the heavy and light chains, which may disrupt the three-dimensional structure of the antibody and destroy antigen binding capability. Optimization of the reductant concentration may have to be done to ensure the best performance of the resultant immunoaffinity support. An alternative to the use of a reducing agent to generate free thiols is to use a thiolation reagent to first modify the amines on an antibody to contain thiols. Various procedures on the use of thiolation reagents with proteins and antibodies can be found elsewhere in this book (Chapter 2, Section 4.1).
3. Remove excess reducing agent from the antibody solution by desalting using size exclusion chromatography or by dialysis. During the chromatography or dialysis operation, use coupling buffer to equilibrate and elute the protein through the gel filtration support or as the dialyzing solution. To desalt the 1-ml sample size, a gel filtration column should consist of at least 10ml of a support having a molecular weight exclusion limit of no more than 5 to 10kDa to ensure that the protein will come through first in the void volume and be separated from the salt peak containing the reducing agent. A similarly sized dialysis membrane having a size exclusion limit of 5 to 10kDa also is appropriate. Dialyze the antibody with numerous changes of the outside dialysis solution to ensure that the concentration of the reducing agent is decreased down to levels far below that of the reduced antibody. Complete removal of the reducing agent is essential to eliminate competition when coupling the reduced antibody to the pyridyl disulfide-activated support. Pool the fractions containing desalted protein from the gel filtration column or recover the dialyzed protein from the dialysis device. The solution may be concentrated if necessary using a centrifugal concentrator to approximately 1ml to maintain the desired concentration level of antibody in the coupling reaction.
4. Wash 1ml of a pyridyl disulfide-activated support with several bed volumes of water to remove storage solution components and then with two bed volumes of coupling buffer. Washing may be performed using spin columns or by using a small drip column. Drain to a moist cake.
5. Mix the washed pyridyl disulfide-activated gel with the reduced antibody solution and stir at room temperature (or 4°C) overnight or for at least 18h. Mixing may be achieved by end-over-end rocking in a sealed container. If the antibody is sensitive to extensive stirring or rocking, the mixture may be mixed for 2h and then left to sit with periodic resuspension and mixing every hour or so to maintain a somewhat homogeneous mixture of ligand throughout the resin.
6. Wash the support with several bed volumes of coupling buffer to remove most of the not-coupled antibody. The washes may be analyzed *versus* the initial concentration of the antibody solution before coupling to determine the amount immobilized.
7. Thoroughly wash the immunoaffinity support with coupling buffer, water, 1-*M* NaCl, and again with water to remove unreacted materials. Store the support until use at 4°C in water as a 50% slurry containing a preservative.

(B) Coupling Thiol-Containing Small Ligands

1. Wash 10ml of a pyridyl disulfide-activated support with water and then place into coupling buffer using a sintered glass filter funnel suspended in a suction filter flask (coupling buffer: 0.1-*M* sodium phosphate, 0.15-*M* NaCl, 10-m*M* EDTA, pH 7.5). The coupling buffer should be purged with nitrogen and degassed under vacuum to remove oxygen, which may oxidize the thiols and prevent coupling.
2. Dissolve a thiol-containing ligand (such as a cysteine-containing peptide) in 10ml of coupling buffer at a concentration of at least 2 to 5mg/ml. Determination of the optimal level of ligand concentration may be done experimentally by running a series of small reactions at different concentrations and determining which preparation performs best in the intended affinity chromatography application.
3. Add the ligand solution to the washed pyridyl disulfide-activated resin and mix to thoroughly resuspend the support in the solution. React overnight or for at least 18h at room temperature (or 4°C) with constant mixing.
4. Wash the gel with several bed volumes of coupling buffer to remove most of the excess uncoupled ligand. The amount of ligand not coupled may be determined by measuring the volume and concentration of the ligand in the pooled washes. This may be done using a protein assay or spectrophotometrically if the ligand has a characteristic spectral signature. The amount

of ligand that coupled to the matrix then may be determined by difference. Drain the gel to a wet cake.

5. Wash the affinity resin thoroughly with water, 1-*M* NaCl, and again with water (at least 10 bed volumes each) to remove unreacted components. Store the support until use at 4°C as a 50% slurry in water containing a preservative.

TNB–Thiol Activation

Ellman's reagent (5,5′-dithiobis(2-nitrobenzoic acid), or DTNB, long has been used as a chromogenic agent for the determination of free thiol groups, especially cysteine residues in proteins (Chapter 2, Section 4.1) (Ellman, 1958; Riddles *et al.*, 1979, 1983; Watabe *et al.*, 1982). The compound reacts with thiol groups in a disulfide exchange process to generate the chromogenic leaving group 5-thio-2-nitrobenzoic acid (TNB). The yellow color generated by the TNB anion at 412 nm can be used to quantify the amount of thiol groups present ($\varepsilon = 14{,}150$ at pH 8.0).

Ellman's reagent also can be used to create thiol-reactive groups on affinity supports for the coupling of sulfhydryl-containing ligands (Jayabaskaran *et al.*, 1987). Much like the reaction of 2,2′-dipyridyl disulfide discussed in the previous section, DTNB can be used to activate thiol-containing chromatography supports to contain reactive TNB–thiol groups (Figure 15.78). The TNB–thiol groups then can be used to immobilize thiol-containing proteins or other ligands through a disulfide exchange process leading to the formation of a disulfide linkage with the affinity ligand (Figure 15.79). The use of Ellman's reagent in this process provides advantages over the pyridyl disulfide chemistry in that the reactions of the TNB group can be followed visually and spectrophotometrically. The activation of a thiol-containing support proceeds with loss of one chromogenic TNB group per thiol, so the reaction can be quantified by measuring the amount of color released during the process. In addition, the immobilization of a thiol-containing protein or molecule can be quantified through the number of TNB groups generated during the coupling reaction.

The requirement for production of a TNB-activated support is to first prepare a spacer arm derivative that terminates in a thiol group. This can be done using the techniques discussed in this chapter on the preparation of spacer arm derivatives. Perhaps one of the easiest routes to accomplish this process is to use an amine-containing support (such as DADPA–agarose or MANEA–agarose) and modify it with Traut's reagent to spontaneously form a small spacer arm, which opens up to a free thiol on its terminus (Chapter 2, Section 4.1). This intermediate can then be immediately reacted with DTNB to create the activated TNB–thiol support (Figure 15.80).

PREPARATION OF A TNB–THIOL-ACTIVATED SUPPORT

The following protocol requires the prior preparation of an amine-containing support for the initial modification with a thiolation reagent to create thiols (see section on spacer arms in this chapter).

1. Wash 100 ml of an amine-containing support (such as DADPA–agarose or MANEA–agarose) with water to remove any storage components (at least 3 bed volumes) and then with 0.1-*M* sodium phosphate, 0.15-*M* NaCl, 10-m*M* EDTA, pH 8.0 (modification buffer). Use a sintered filter funnel suspended in a suction filter flask to facilitate the washing steps, while applying a gentle vacuum between each wash solution addition to accelerate flow through the resin. Drain to a wet cake.

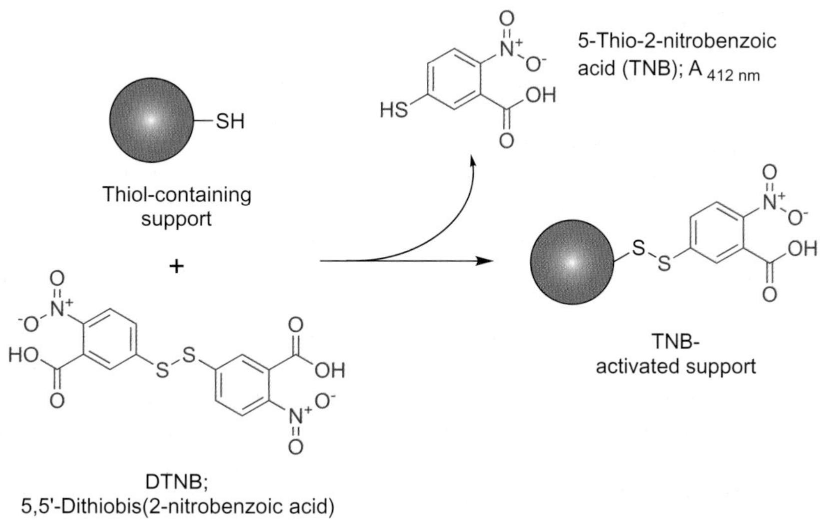

FIGURE 15.78 Activation of a thiol-containing matrix with DTNB.

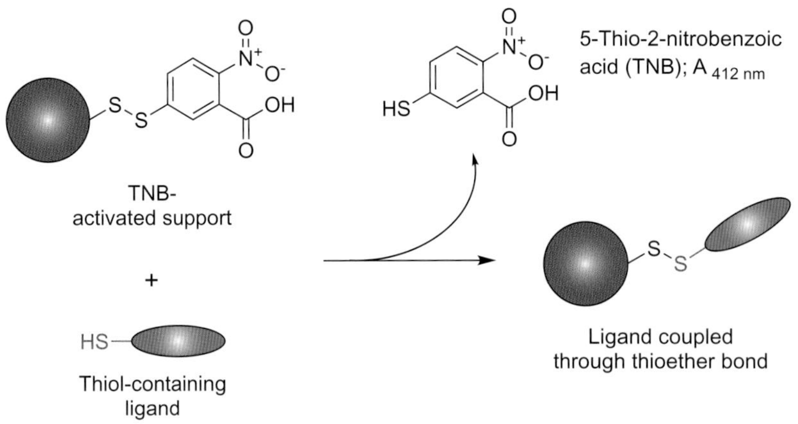

FIGURE 15.79 Coupling of a thiol-containing ligand to a TNB-thiol-activated support.

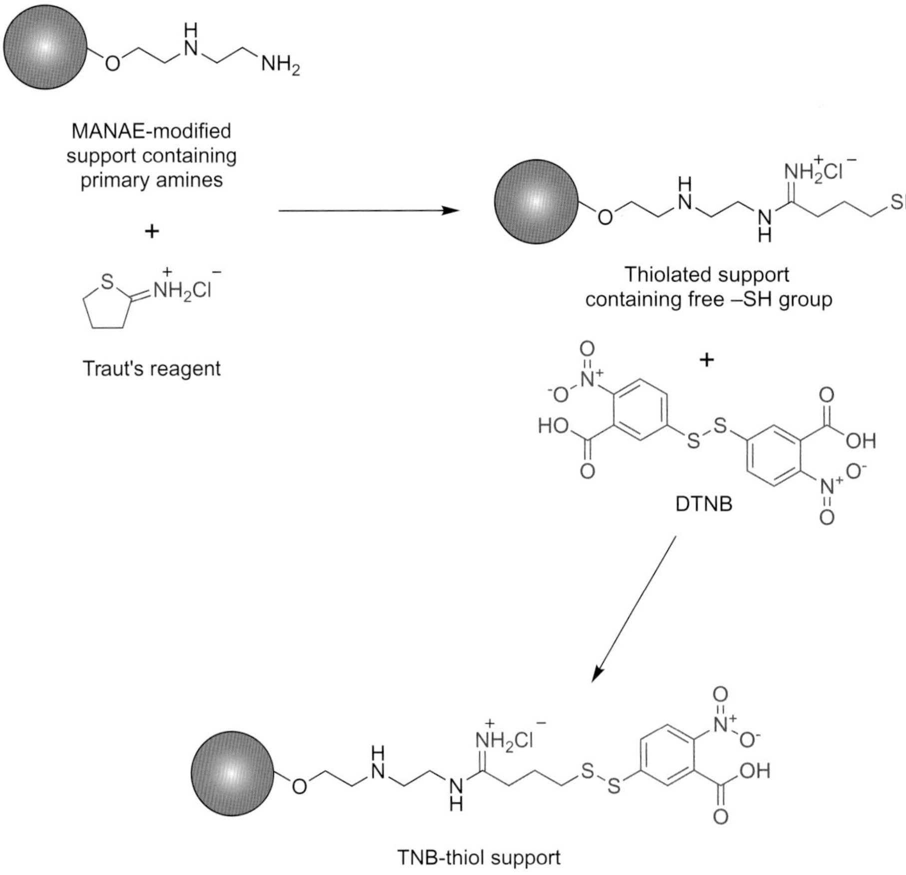

FIGURE 15.80 Reaction of Traut's reagent with MANEA–agarose and activation with DTNB.

2. Prepare 100 ml of a 20-m*M* 2-iminothiolane (Traut's reagent) solution in modification buffer and add the gel cake to it with stirring.
3. React for 1 h at room temperature with constant stirring using an overhead paddle stirrer or end-over-end rocking in a sealed container.
4. Quickly wash the support with several bed volumes of modification buffer to remove most of the reactant and then drain to a wet cake. Note: Do not store the thiolated support at this stage, because 2-iminothiolane modified species may undergo a recyclization reaction to an inactive state with the thiol group unavailable for further modification (refer to other sections on Traut's reagent within this book for additional information).

5. Prepare 100 ml of a 20-m*M* DTNB (Ellman's reagent) solution in modification buffer and add the gel cake to it with stirring.
6. React for 1 h at room temperature with constant mixing. During the reaction, the chromophoric TNB anion will be released and generates a yellow color within the gel slurry. This is indicative of a successful activation process.
7. Drain the activated gel of excess reaction solution and wash with an additional 2 bed volumes of modification buffer. These initial washes may be collected and pooled to determine the activation level of the support by measuring the absorbance of the released TNB anion at 412 nm ($\varepsilon = 14{,}150$ at pH 8.0). Continue to wash the activated support thoroughly with water, 1-*M* NaCl, and water to remove excess reactants and reaction byproducts. Finally, wash the support with 1-m*M* EDTA (at neutral pH) to protect against metal-catalyzed oxidation of the disulfide groups. Store the TNB–thiol-activated support at 4°C as a 50% slurry in water containing 1-m*M* EDTA and a preservative until use.

LIGAND COUPLING TO TNB–THIOL-ACTIVATED SUPPORTS

The following protocols describe the immobilization of a thiol-containing reduced antibody or a thiol-containing small ligand (such as a peptide containing a cysteine group) to a TNB–thiol-activated resin. The methods may also be used with proteins that have been modified to contain –SH groups through the use of thiolation reagents (Chapter 2, Section 4.1). Some optimization may have to be done to determine a ligand density that is best for a particular affinity chromatography application. Preparing a series of small batches using different concentrations of ligand in each reaction medium will produce a range of ligand densities on the support for evaluation. In addition, after ligand coupling do not attempt to block excess reactive groups with a thiol-containing compound such as cysteine, as this may also cleave off any immobilized ligands.

(A) Coupling of Reduced IgG or F(ab')$_2$ Fragments

1. Dissolve an IgG antibody or F(ab')$_2$ fragment at a concentration of 1 to 10 mg/ml in 1 ml of 0.1-*M* sodium phosphate, 0.15-*M* NaCl, 10-m*M* EDTA, pH 8.0 (coupling buffer). The coupling buffer should be purged with nitrogen and degassed under vacuum to remove oxygen, which may oxidize the thiols and prevent coupling.
2. Add a reducing agent, such as DTT, 2-mercaptoethylamine (2-MEA), 2-mercaptoethanol (2-ME), or tris(carboxyethyl)phosphine (TCEP), to the antibody solution to give a final concentration of at least 5-m*M*. A highly concentrated stock solution of the reducing agent may first be prepared in coupling buffer and then a small aliquot of that solution added to the antibody to obtain the final desired concentration. Mix to dissolve and incubate for 1.5 h at 37°C. Note: Higher concentrations of the reducing agent may be used (typically ≤50-m*M*), but the higher the concentration the more likely will be the reduction of disulfides between the heavy and light chains, which may disrupt the three-dimensional structure of the antibody and destroy antigen binding capability. Optimization of the reductant concentration may have to be done to ensure the best performance of the resultant immunoaffinity support. An alternative to the use of a reducing agent to generate free thiols is to use a thiolation reagent to modify the amines on an antibody to contain thiols. Various procedures on the use of thiolation reagents with proteins and antibodies can be found elsewhere in this book (Chapter 20, Section 1.1).
3. Remove excess reducing agent from the antibody solution by desalting using size exclusion chromatography or by dialysis. During the chromatography or dialysis operation, use coupling buffer to equilibrate and elute the protein through the gel filtration support or as the dialyzing solution. To desalt the 1-ml sample size, a gel filtration column should consist of at least 10 ml of a support having a molecular weight exclusion limit of no more than 5 to 10 kDa to ensure that the protein will come through first in the void volume and be separated from the salt peak containing the reducing agent. A similarly sized dialysis membrane having a size exclusion limit of 5 to 10 kDa is also appropriate. Dialyze the antibody with numerous changes of the outside dialysis solution to ensure a decrease in the concentration of the reducing agent down to levels far below that of the reduced antibody. Complete removal of the reducing agent is essential to eliminate competition when coupling the reduced antibody to the TNB–thiol-activated support. Pool the fractions containing desalted protein from the gel filtration column or recover the dialyzed protein from the dialysis device. The solution may be concentrated if necessary using a centrifugal concentrator to approximately 1 ml to maintain the desired concentration level of antibody in the coupling reaction.
4. Wash 1 ml of a TNB–thiol-activated support with several bed volumes of water to remove storage solution components and then wash with two bed volumes of coupling buffer. Washing may be carried out using spin columns or by using a small drip column. Drain to a moist cake.

5. Mix the washed TNB–thiol-activated gel with the reduced antibody solution and stir at room temperature (or 4°C) for 2 to 4 h. Mixing may be carried out by end-over-end rocking in a sealed container (which can consist of the sealed column used for washing). If the antibody is sensitive to extensive stirring or rocking, the mixture may be mixed for 2 h and then left to sit with only periodic resuspension and mixing every 30 min or so to maintain a somewhat homogeneous mixture of ligand throughout the resin.
6. Wash the support with several bed volumes of coupling buffer to remove most of the not-coupled antibody. The washes may be analyzed *versus* the initial concentration of the antibody solution before coupling to determine the amount immobilized by difference. During the course of the reaction, TNB groups will also be released as the thiol-containing antibody has coupled; therefore, measurement of the absorbance of this anion at 412 nm can provide an accurate determination of the amount of ligand coupled.
7. Thoroughly wash the immunoaffinity support with coupling buffer, water, 1-*M* NaCl, and again with water to remove unreacted materials. Store the support until use at 4°C in water as a 50% slurry containing a preservative.

(B) Coupling Thiol-Containing Small Ligands

1. Wash 10 ml of a TNB–thiol-activated support with water and then add into coupling buffer using a sintered glass filter funnel suspended in a suction filter flask (coupling buffer: 0.1-*M* sodium phosphate, 0.15-*M* NaCl, 10-m*M* EDTA, pH 8.0). The coupling buffer should be purged with nitrogen and degassed under vacuum to remove oxygen, which may oxidize the thiols and prevent coupling.
2. Dissolve a thiol-containing ligand (such as a cysteine-containing peptide) in 10 ml of coupling buffer at a concentration of at least 2 to 5 mg/ml. Determination of the optimal level of ligand concentration may be achieved experimentally by running a series of small reactions at different concentrations and determining which preparation performs best in the intended affinity chromatography application.
3. Add the ligand solution to the washed TNB–thiol-activated resin and mix to thoroughly suspend the support in the solution. React for 2 to 4 h at room temperature (or 4°C) with constant mixing.
4. Wash the gel with several bed volumes of coupling buffer to remove most of the excess uncoupled ligand. The amount of ligand not coupled may be determined by measuring the volume and concentration of the ligand in the pooled washes. This may be done using a protein assay or spectrophotometrically if the ligand has a characteristic spectral signature. The amount of ligand that coupled to the matrix may then be determined by the difference. Alternatively, monitoring the amount of the TNB anion released during the coupling reaction through its absorbance at 412 nm can provide an indirect, but accurate measurement of the degree of ligand coupling. Drain the gel to a wet cake.
5. Wash the affinity resin thoroughly with water, 1-*M* NaCl, and again with water (at least 10 bed volumes each) to remove unreacted components. Store the support until use at 4°C as a 50% slurry in water containing a preservative.

2.3. Hydroxyl-Reactive Immobilization Methods

This section describes the immobilization of hydroxyl-containing ligands to chromatography supports. Activated supports that are able to couple effectively with hydroxyl-containing ligands must be able to react with the hydroxyl groups while avoiding substantial inactivation by hydrolysis in aqueous solution. The following activation methods perform particularly well in this regard for coupling sugars, carbohydrates, polysaccharides, glycans, and other molecules containing one or more hydroxyl groups. One should keep in mind that in addition to these methods meant to first activate the matrix and then attach the ligand, hydroxyl-containing ligands could also be initially activated in solution with any of the activating reagents described in this chapter suitable for hydroxyls and then in a reverse manner reacted with a support material that contained the appropriate functional group for coupling to the activated ligand intermediate. In other words, a hydroxyl-containing molecule might be activated in organic solution with a limiting mole quantity of DSC, for example (see previous section on DSC activation, this chapter), and then the resultant NHS carbonate reactive group formed on the ligand could be coupled to an amine-containing support material to form a carbamate linkage. This strategy of reversing the normal sequence of support activation followed by ligand coupling to be instead ligand activation followed by coupling to a support can be performed with many organic-soluble small molecules containing a hydroxyl group. This alternative strategy therefore may open up many more options for designing a reaction sequence leading to the desired immobilized ligand.

Epoxide Activation

An epoxide group is a small three-member heterocyclic ring containing two carbon atoms and one oxygen. The combination of ring strain and the electron withdrawing effects of the oxygen atom cause the carbon atoms to carry a partial positive charge and therefore be good sites for potential nucleophilic attack. An epoxide reacts with a nucleophile through a ring opening process, which yields a bond to the nucleophilic group on one carbon and the formation of a hydroxyl group on the adjacent carbon as the epoxy ring opens. For epoxides that exist on terminal carbon atoms, the reaction with a nucleophile causes covalent linkage with the primary carbon atom and hydroxyl formation on the adjacent secondary carbon atom.

Depending on the pH and conditions of the reaction medium, epoxides can be made to react effectively with thiols, amines, or hydroxyl (even phenolic) groups (Figure 15.81). Similar to the relative reactivity of vinyl sulfone-activated supports described previously, epoxides efficiently react with thiols in the pH range of 7.5 to 8.5, amines react within the range of pH 9 to 11 (Ghazi *et al.*, 2005), and hydroxyl groups react at highly alkaline pH values (pH > 11). The reactions may be accelerated with heating (i.e., ≥37°C), which often is done during the activation of a hydroxyl-containing support with a *bis*-epoxide compound and during subsequent coupling reactions with carbohydrate hydroxyl groups. Epoxy groups also are known to react under certain conditions with carboxylic acids to form ester linkages. This reaction is important in many polymer resins that are designed to crosslink and form coatings, adhesives, moldings, and other synthetic solids for industrial and commercial applications (Blank *et al.*, 2001). Epoxy ester formation has also been investigated with amino acids as a mechanism in the preparation of crosslinked collagen-based materials, and it has found to occur best at slightly acidic pH (Zeeman, 1998). However, the reaction rate of carboxylate groups with epoxides is much slower than that of the reactions with thiols, amines, or hydroxyl groups under the recommended pH conditions described in this section. It is likely that if ester formation occurs with biological ligands that it is a minor product compared to the reactions commonly used with epoxy-activated supports.

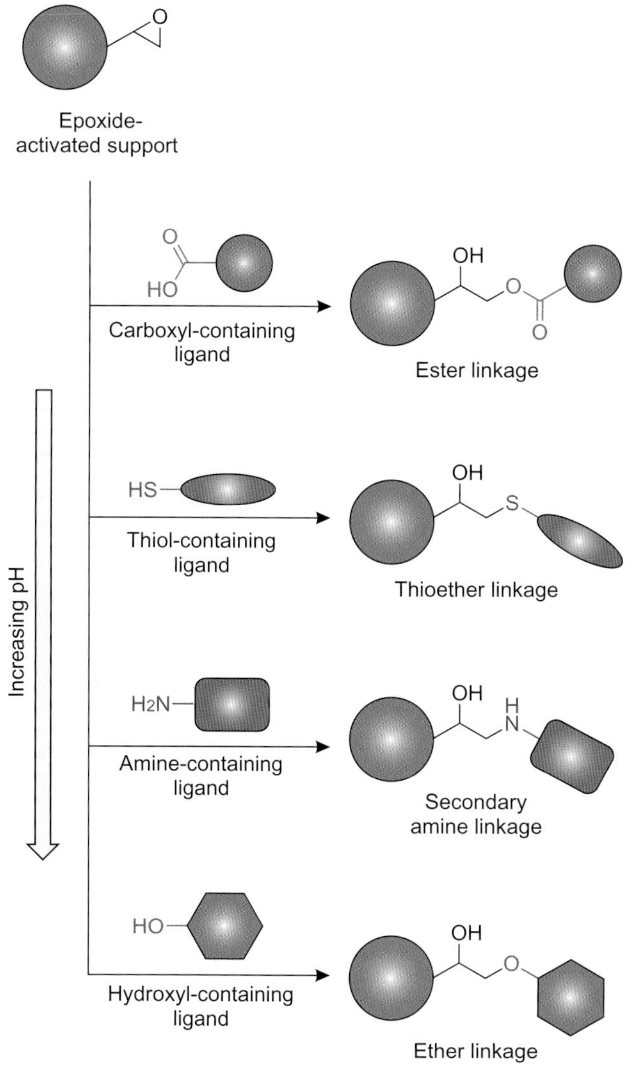

FIGURE 15.81 Reaction of an epoxide support with a thiol-, amine-, hydroxyl-, or carboxylate-containing compound.

Homobifunctional epoxide crosslinker compounds (bisoxiranes) are often used to activate hydroxyl-containing chromatography supports to result in the production of reactive epoxy groups for the immobilization of hydroxylic ligands. Perhaps the reagent most commonly used for this purpose is 1,4-butanediol diglycidyl ether (Sundberg and Porath, 1974), which contains a relatively hydrophilic 12-atom spacer group with epoxides at each end. At alkaline pH and in large excess, one end of this compound can be made to react with the hydroxyls on a matrix to create an ether linkage while forming a hydrophilic spacer which terminates in the second epoxide group for subsequent ligand immobilization (Figure 15.82).

The coupling of protein affinity ligands onto epoxide-activated supports can be significantly enhanced by the addition of certain salts to the reaction medium (Murthy and Moudgal, 1986; Wheatley and Schmidt, Jr., 1993, 1999). In particular, salts that increase the interactions of hydrophobic molecules in aqueous solution appear to promote protein adsorption onto the activated support prior to covalently reacting with the epoxy groups. Lyotropic salts such as sodium or potassium sulfate, sodium or potassium phosphate, and even ammonium sulfate can be used to drive soluble proteins toward the surface of the particles by a salting out effect, thus forcing them into proximity with the reactive groups. Surprisingly, even the presence of ammonium ions in ammonium sulfate salts does not interfere with the reaction with amines on proteins. Using ammonium sulfate concentrations of 0.4 to 2.5-*M*, most proteins will couple to the epoxide support in

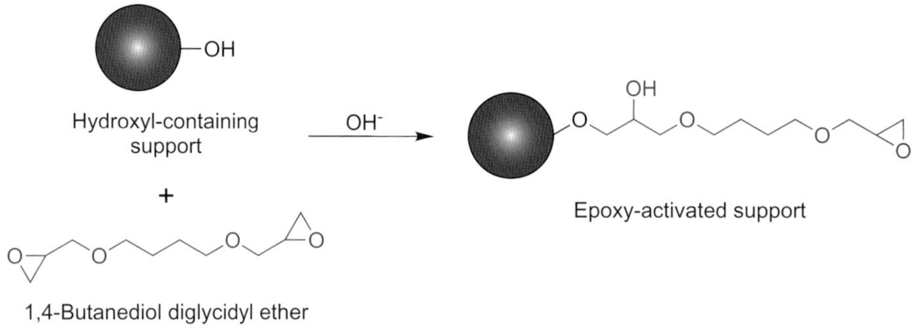

FIGURE 15.82 Activation of hydroxylic supports with 1,4-butanediol diglycidyl ether.

yields of 95 to 100% in 20h at room temperature. The other activated support that has this tendency to be enhanced for coupling proteins when in the presence of a lyotropic agent is the azlactone activation chemistry, described previously in this chapter. Presumably, the lyotropic-induced deposition of proteins onto the epoxide-activated surface occurs first, followed by the nucleophilic attack of protein amines onto the carbon of the epoxide ring, which causes covalent coupling.

Even the immobilization of some small affinity ligands onto epoxide-activated supports can be enhanced by the addition of a lyotropic salt to the coupling reaction. Bauer-Arnaz *et al.* (1998) found that various *S*-alkyl glutathione derivatives could be coupled in approximately twice the yield if a high concentration of potassium phosphate (2-*M*) was present in the reaction buffer. In this case, the coupling reaction was performed at pH 10.5, which indicates that even amine-containing small ligand coupling can be enhanced with salt if the compound structure contains hydrophobic regions that can interact with the activated support prior to covalent binding.

Mateo *et al.* (2002) developed salt-induced immobilization methods using the epoxide-activated support Sepabeads-EP. The coupling and stabilization of the enzyme penicillin G acylase (PGA) from *E. coli* was accomplished using an initially high concentration of sodium phosphate buffer (1-*M*) at pH 7. The effect of the salt to push the enzyme molecules toward the matrix surface near the epoxide active groups resulted in efficient immobilization even at neutral pH. Although the reaction took 24h to go to completion, the result was an increase of around 10,000-fold in stabilization of the enzyme compared to the soluble form.

Grazú *et al.* (2003) used an interesting combination of thiol/disulfide exchange coupling and epoxide reactions for enzyme immobilization to enhance the yield of epoxy coupling without the use of a lyotropic salt (see previous sections on pyridyl disulfide and TNB–thiol activation, this chapter). On a highly activated epoxide support, they reacted a limited amount of DTT with it to obtain a partially thiolated matrix, which contained both thiol and epoxide groups. Next, an enzyme was thiolated with SPDP followed by reduction and the thiols on the support were activated using 2,2′-dipyridyl disulfide to provide reactive pyridyldithiol groups. Mixture of the thiolated enzyme with the activated support resulted in the rapid formation of disulfide linkages with the thiol groups created on the protein followed by continued reaction with the epoxide groups to give a multi-point attachment with each enzyme.

Epoxy activation and coupling has been extensively used to couple metal chelating compounds to particles of all types for immobilized metal affinity chromatography (IMAC) applications (Novotna *et al.*, 2010). Often, an organic chelating molecule contains a secondary amine, such as in the example of iminodiacetic acid, which can be effectively coupled to a chromatography support by reaction with an epoxide (Figure 15.83). In addition, the coupling of glutathione is often done using epoxy-activated supports, because the most effective site of immobilization on the molecule is through its thiol group, which allows the ligand to interact with glutathione-*S*-transferase (GST) fusion proteins to capture and purify them after recombinant expression (Simons and Vander Jagt, 1977; Smith and Johnson, 1988; Stahl *et al.*, 2003).

Due to the versatility of epoxy-activated supports in coupling thiol-, amine-, or hydroxyl-containing ligands, the method has become a popular choice for immobilizing a wide range of affinity molecules. In addition, the activated support is extremely stable to hydrolysis even when stored in aqueous solution (neutral pH) for extended periods. The activation of chromatography supports, nonporous particles, and surfaces with the compound 1,4-butanediol diglycidyl ether creates a long spacer arm linking any potential ligand, which is also hydrophilic enough to minimize nonspecific binding.

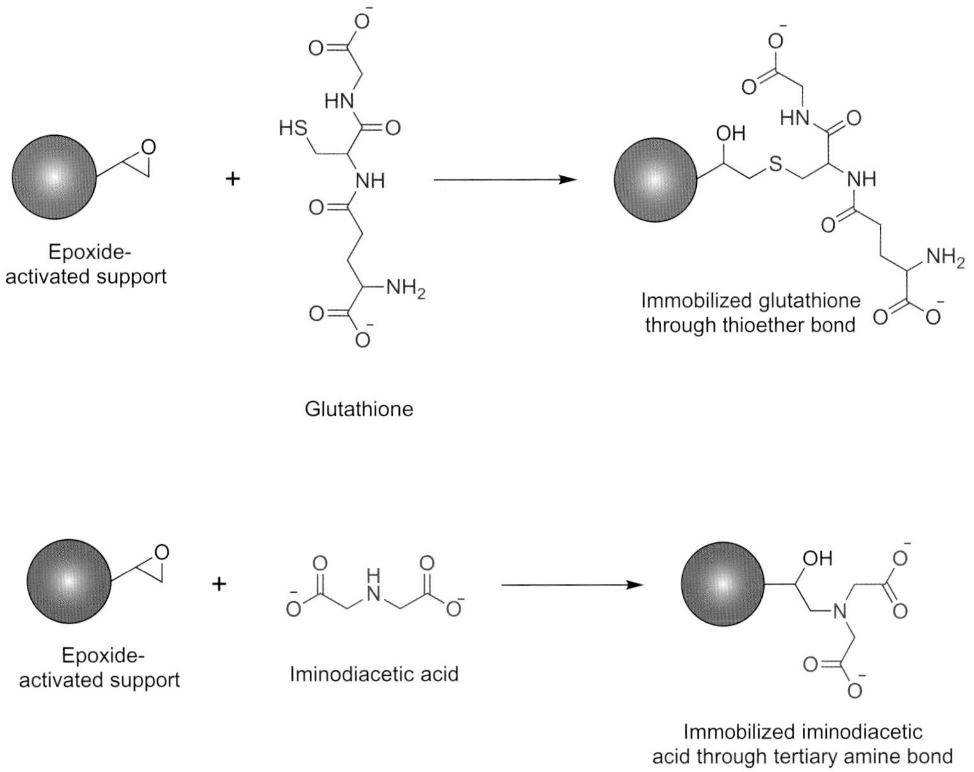

FIGURE 15.83 Immobilization of iminodiacetic acid and glutathione on epoxy-activated supports.

PREPARATION OF AN EPOXY-ACTIVATED SUPPORT

The following protocol describes the activation of a hydroxyl-containing support with the bis-epoxide compound 1,4-butanediol diglycidyl ether under highly alkaline conditions. All operations should be done in a fume hood while wearing the appropriate personal protective equipment. Avoid contact with all solutions and avoid breathing vapors of the epoxide compound or the solvents used during the operations.

1. Wash 100ml of a hydroxyl-containing support material (i.e., Sepharose (agarose), Toyopearl, or Trisacryl) with several bed volumes of water to remove storage solution and preservatives. The washing may be done using a sintered glass filter funnel suspended in a suction filter flask. Pull a gentle vacuum on the flask to facilitate the removal of the solutions from the gel after each addition of wash. Drain to a moist cake. Note: When activating agarose supports with a *bis*-epoxy compound, it is common to start with a non-crosslinked support (e.g., 4% agarose), because reactions with both ends of the epoxide will occur to some extent and form a crosslinked matrix.
2. Suspend the gel in 75ml of 0.6-*N* NaOH (caustic!) containing 150mg of sodium borohydride dissolved into it. Transfer the gel suspension into a three-necked, round-bottom flask and stir using an overhead paddle stirrer in the fume hood. Place the round-bottom flask in an open container into which water can be added to control the temperature of the reaction. Put a thermometer into one of the three necks of the flask to monitor the temperature throughout the activation reaction. Add water at room temperature to the surrounding bath up to the level of the inner gel slurry and have ice on hand if necessary to control any exothermic tendencies.
3. Slowly add to the stirring gel suspension 75ml of 1,4-butanediol diglycidyl ether over a period of about 20min. Monitor the temperature and add ice to the surrounding water bath if required to maintain a constant ambient temperature of about 20 to 25°C. There may be an exothermic temperature spike during the early stages of the reaction; however, do not cool the reaction below the recommended temperature range, as it will slow the activation process.
4. Stir the reaction for 8 to 10h or overnight at room temperature.
5. Transfer the gel slurry to the filter funnel and wash with 4 bed volumes of water to remove the NaOH solution and the borohydride. The excess epoxide

compound is more difficult to remove using only water washes. For this reason, sequentially wash the gel with increasing concentrations of acetone (or ethanol) (e.g., 30%, 70%) until 100% solvent is used. Continue to wash with 100% solvent for at least 10 bed volumes to remove the remaining traces of excess epoxide. Next, sequentially wash back into water by reversing the concentrations of solvent used previously. Finally, wash with 10 bed volumes of water to remove the last traces of solvent. Store the epoxy-activated support at 4°C as a 50% slurry in water and containing a preservative until use.

LIGAND COUPLING TO EPOXY-ACTIVATED SUPPORTS

The following protocols represent some common methods of coupling ligands to epoxy-activated supports. In the first example, a disaccharide (lactose) is immobilized through its hydroxyl groups onto epoxy-activated agarose at high pH. Other sugars or carbohydrates, including polysaccharides and glycans, may be coupled using similar procedures. In the second example, a protein is immobilized onto an epoxy-activated support at near physiological pH using a buffer containing a lyotropic agent to push the protein molecules into close proximity with the reactive groups, and thus facilitate efficient coupling while avoiding higher pH conditions to drive the reaction.

(A) Coupling of Carbohydrates

1. Wash 100 ml of an epoxy-activated support with several bed volumes of water to remove any storage solution and preservatives. Use a sintered glass filter funnel suspended in a vacuum filtration flask to facilitate the washing steps. Drain the gel to a moist cake.
2. Prepare a lactose ligand solution by dissolving 15 g of lactose in 100 ml of 0.1-*N* NaOH (makes a 0.438-*M* solution). Use care in handling the caustic NaOH solution and use the appropriate personal protective equipment during all operations.
3. Add the washed gel cake to the ligand solution with stirring and transfer the reaction slurry to a three-necked, round-bottom flask. Put the flask in a heating mantle and use an overhead paddle stirrer to mix the resin during the reaction. Add a thermometer to one of the openings in the flask and heat the reaction mixture with constant stirring to 40°C.
4. Stir the reaction for 24 h at 40°C.
5. Wash the support with several bed volumes of water to remove most of the reaction medium and then wash with 0.1-*M* sodium bicarbonate, pH 8.0, followed by an extensive water wash to remove the last traces of buffer and reactants. The immobilized lactose support can be stored until use at 4°C as a 50% slurry in water containing a preservative.

(b) Coupling of Proteins

Proteins containing cysteine thiol groups or amines may be immobilized onto epoxy-activated supports in the pH range of 7.0 to 11, with pH 7.5 to 8.5 being most effective for –SH groups and pH 9 to 11 optimal for lysine or N-terminal amines (Ghazi *et al.*, 2005). However, if a lyotropic salt is added to the reaction medium, then protein coupling to amines can be made to occur with excellent yield at pH 7.0 to 8.0, which is much more amenable to maintain protein stability. The following protocol describes the immobilization of a protein at mildly alkaline pH using the lyotropic salt sodium sulfate. Some optimization of the level of salt concentration in the reaction medium may have to be carried out for certain proteins to prevent precipitation while still promoting the best possible matrix-protein interactions leading to efficient reaction rates.

1. Wash 10 ml of an epoxy-activated support with several bed volumes of water to remove any storage solution and preservatives. Use a small sintered glass filter funnel suspended in a vacuum filtration flask to facilitate the washing steps. Drain the gel to a moist cake.
2. Prepare 10 ml of a protein ligand solution at a concentration of 1–20 mg/ml in coupling buffer. The coupling buffer may be formulated as 0.1-*M* sodium phosphate, pH 7.5, containing 0.5-*M* sodium sulfate as the lyotropic agent. Other lyotropic salts may also be used. For example, an alternative coupling buffer composition that was used with success by Mateo *et al.* (2002) to immobilize enzymes consisted of 1-*M* sodium phosphate, pH 7.0, in which the sodium phosphate functioned both as a buffer for pH stabilization and as the lyotropic agent. Despite the effect of a lyotropic agent to drive protein molecules toward the surface for reaction, some proteins may still require a higher-pH environment to efficiently couple. An alternative higher pH buffer is 0.1-*M* sodium carbonate, pH 9.0, containing 0.5-*M* sodium sulfate.
3. Add the washed epoxy-activated resin to the ligand solution with stirring.
4. React with constant mixing for at least 24 h at room temperature. Mixing may be done using an overhead paddle stirrer or by end-over-end rocking in a sealed container. Some proteins may require longer reaction times to reach maximal coupling yields for attachment to the support.
5. Wash the support with several bed volumes of coupling buffer and collect the washes. The amount of protein coupled may be determined by the difference between the initial amount of protein added to the reaction mixture and the total amount left in the not-coupled washings. Drain the support to a wet cake.
6. Block unreacted epoxide groups on the support using a small molecule containing a nucleophilic

group that will readily react with the leftover active sites. For instance, cysteine, ethanolamine, or glycine may be used for this purpose. Cysteine blocking will proceed faster than amine-containing blockers and at lower pH values, because of the increased reaction rate of epoxides with the highly nucleophilic thiol groups. However, avoid using thiol-containing compounds that are also strong disulfide reducing agents, such as DTT or 2-mercaptoethanol, if the protein that was immobilized is sensitive to reduction. Cysteine is much milder in this regard and can be used as a blocker without detrimentally affecting most proteins. Prepare at least a 0.1-*M* solution of the blocking agent in coupling buffer and mix with the washed resin.

7. React 2 to 4h with mixing (for a thiol-containing blocking agent, such as cysteine) or for 24h (if using an amine-containing blocking agent).
8. Wash the support with 10 bed volumes of coupling buffer without the lyotropic agent present (e.g., 0.1-*M* sodium phosphate, pH 7.5). Continue to wash with water, 1-*M* NaCl, and again with water to remove the last traces of all reactants and coupling buffer components. The immobilized protein can be stored until use at 4°C as a 50% slurry in water containing a preservative.

Divinyl Sulfone Activation

Divinyl sulfone (DVS) is a small bifunctional compound that can be used to activate hydroxyl-containing supports for the immobilization of hydroxyl-containing ligands, such as sugars, carbohydrates, polysaccharides, glycans, and small organic molecules containing an available –OH group. The chemistry of DVS activation and coupling is discussed in great detail in a previous section in this chapter, including a protocol for the coupling of hydroxylic ligands; see Thiol-Reactive Immobilization Methods for an overview and protocols for ligand immobilization.

Trichloro-s-Triazine Activation

The activation of hydroxyl-containing supports with trichloro-*s*-triazine (TsT) was presented in detail in the section on Amine-Reactive Immobilization Methods discussed earlier in this chapter. The cyanuric chloride reactive groups on a TsT-activated support can be used to immobilize thiol-, amine-, or hydroxyl-containing ligands with high efficiency. The coupling of hydroxyl-containing sugars, carbohydrates, polysaccharides, glycans, or other organic ligands containing hydroxyls can be done using a high pH environment (pH > 11), essentially using the same conditions that are recommended for the coupling of triazinyl dyes to hydroxyl matrices (Hermanson *et al.*, 1992; Chamani *et al.*, 2011).

LIGAND COUPLING TO TsT-ACTIVATED SUPPORTS

The following protocol assumes that 100ml of a TsT-activated support has been prepared as described previously in the section on Amine-Reactive Immobilization Methods.

1. Prepare 100ml of a hydroxyl-containing ligand solution in 0.5-*M* sodium carbonate (no pH adjustment necessary) by dissolving the ligand at a concentration range of 0.1 to 0.5-*M* (for a small sugar molecule or disaccharide) or about 5 to 10mg/ml (for larger polysaccharide or glycan ligands). Notes: The optimal concentration of a carbohydrate in the immobilization reaction may have to be determined by experimentation to obtain the best ligand loading for acceptable performance in the intended affinity chromatography application. For example, immobilized sugars used in the purification of lectins often are coupled at relatively low ligand levels (below the maximal coupling capacity of the gel), because a lower ligand density avoids the potential for multi-site binding with a lectin, which may result in high avidity interactions followed by difficulty in eluting the bound proteins from the support.
2. Wash 100ml of a TsT-activated support into water and coupling buffer by suctioning off excess acetone storage solution and resuspending the gel into water. This can be done by pulling a gentle vacuum on the filter flask to filter off solvent while breaking up the support into small, finely divided pieces using a spatula, at the same time as being careful not to allow the matrix to dry out. Once the support is broken into small pieces, remove the vacuum and resuspend the gel in water with mixing. Continue to wash the support with water for at least 5 to 10 bed volumes to remove the solvent. Finally, wash the support with 2 to 3 bed volumes of coupling buffer (0.5-*M* sodium carbonate, no pH adjustment).
3. Add the washed gel cake to the ligand solution with stirring and transfer the reaction slurry to a three-necked, round-bottom flask. Put the flask in a heating mantle and use an overhead paddle stirrer to mix the resin during the reaction. Add a thermometer to one of the openings in the flask and heat the reaction mixture with stirring to 40°C. After 2h of mixing, increase the temperature to 60°C with continued stirring.
4. Mix the reaction for 24h at 60°C.
5. Wash the support with several bed volumes of water to remove most of the reaction medium and then wash with 0.1-*M* sodium bicarbonate, pH 8.0, followed by an extensive water wash to remove the last traces of buffer and reactants. The immobilized carbohydrate support can be stored until use at 4°C as a 50% slurry in water containing a preservative.

2.4. Carbonyl-Reactive Immobilization Methods

Chromatography supports can be modified with reactive groups to specifically couple to carbonyls on affinity ligands or proteins and form stable covalent bonds. These carbonyl groups may consist of aldehydes, ketones, or sometimes even carboxylates on ligands that can be targeted under the right conditions to covalently link with certain reactive groups present on support materials. The following sections describe the principal methods of activating a support to couple with carbonyl-containing ligands. The reactive intermediates prepared in this section are somewhat unique, however, in that they typically do not spontaneously form covalent bonds with the functional groups encountered on most biomolecules. The hydrazide, aminooxy, and amine derivatives of supports described in this section will not immediately react with large complex macromolecules like proteins, nucleic acids, lipids, and most polysaccharides found within cells or living organisms. In this regard, they have more selective reactivity than most of the other electrophilic active groups described in this chapter for the immobilization of biomolecules. However, given the right circumstances where a ligand contains an aldehyde, ketone, or in some instances a carboxylate, the creation of these active intermediates can facilitate site-directed coupling to discrete locations on molecules that some of the other methods cannot accomplish.

Hydrazide or Hydrazine Supports for Coupling Aldehydes, Ketones, or Carboxylates

A hydrazine or hydrazide group is able to react with an aldehyde or ketone to form a dehydration product, called a hydrazone, which is a type of Schiff base having a double bond between the carbon atom of the original carbonyl group and the terminal hydrazino nitrogen. This product represents a more stable Schiff base than that formed between an amine and a carbonyl group (standard imine), because a hydrazone is less susceptible to hydrolysis back to the starting materials. In many cases, the hydrazone bond created with a carbonyl-containing ligand is strong enough not to need reduction with sodium cyanoborohydride to stabilize the linkage, as opposed to amino Schiff bases which almost always require reduction to prevent leakage. The hydrazone electrons can delocalize due to the electronegativity of the hydrazide carbonyl oxygen and therefore it stabilizes the bond much better than the linkage between an amine and an aldehyde group. Particularly, if there is a potential for more than one hydrazone attachment point with a ligand, such as in the immobilization of glycoconjugates, the likelihood of ligand leakage becomes negligible. In certain cases, however, hydrazone bonds should be reduced to prevent the possibility of slow ligand leaching over time. This is especially true if the point of attachment between the support and the ligand consists of only a single hydrazone bond.

The use of hydrazides or hydrazines created on a chromatography matrix to immobilize carbonyl-containing ligands was developed specifically to couple glycoproteins through their carbohydrates (O'Shannessy *et al.*, 1984; Bayer *et al.*, 1987; Hoffman and O'Shannessy, 1988; O'Shannessy and Wilchek, 1990). Limited oxidation of carbohydrates or glycans using sodium periodate can create aldehydes on either sialic acid groups alone or on any sugar of the glycan possessing adjacent hydroxyl groups (diols) (see Chapter 2, Section 2). Oxidized glycoproteins then may be reacted with a hydrazide-containing support to immobilize the proteins through hydrazone linkages, which only target the glycan portions of the molecules. This site-directed coupling method through the carbohydrate portion often aids in preserving active sites or binding regions within proteins. This can lead to better preservation of enzyme activity or better binding performance to a target molecule than using more random coupling methods like amine-reactive chemistries, which attach through lysine or N-terminal amines in many locations across the protein surface (Domen *et al.*, 1990).

Glycosylated antibodies, for instance, can be immobilized onto a hydrazide-containing support after mild periodate oxidation of the carbohydrate, which is usually located between the heavy chains within the Fc region. (Note: Some carbohydrate may be present on the Fab fragments near the antigen binding area in certain antibody types.) The use of hydrazone-mediated coupling might be used to avoid an attachment of the antibody to the support near the antigen binding regions. The kinetics of periodate oxidation with antibodies to generate aldehydes was investigated by Hage *et al.* (1997) and found to involve two distinct populations of oxidizable groups—one group that oxidized quickly within 5 to 10 min (likely sialic acid residues) and another group that took several hours to completely oxidize. The oxidation of rabbit polyclonal IgG with 10-m*M* periodate resulted in the production of approximately one aldehyde per immunoglobulin molecule after 10 min at 25°C, two aldehydes after 30 min, and about three aldehydes after 60 min. The rate of periodate oxidation of antibody carbohydrate also was found to be dependent on the pH of the reaction. From pH 3 to 7, the oxidation was found to decrease as the pH increased (Wolfe and Hage, 1995). After 30 min of oxidation at room temperature, two aldehydes were formed at pH 5 to 6, three were created at pH 4, and six resulted from the reaction at pH 3. Over-oxidation by extending the length of the periodate incubation procedure or by using concentrations of periodate exceeding

50-m*M* should be avoided to prevent damage to the avidity or immunoreactivity of antibodies (Abraham *et al.*, 1991). Normally, periodate oxidation reactions are not done for longer than about 15 to 30 min, depending on the sensitivity of the protein being treated. Treatment of antibodies with periodate at low concentrations for 30 min is thus sufficient to provide aldehydes on each antibody for efficient coupling while avoiding the damaging effects of over-oxidation.

Small-molecule aldehydes or ketones may also react with a hydrazide-activated support and end up being immobilized through hydrazone linkages. The reducing ends of sugars, carbohydrates, or even glycans that have been cleaved off of glycoproteins can be selectively immobilized through this process. The reducing ends of carbohydrates typically react with the hydrazide support at a slower rate due to the small amount of time most reducing sugars are in the open aldehyde form in solution as opposed to the cyclic hemiacetal form. Some biological small aldehyde- or ketone-containing molecules that may be present in some samples such as pyridoxal, glyceraldehyde, acetaldehyde, pyruvate, α-ketoglutarate, acetone, and retinal can also couple to the support quite effectively and block some of the hydrazides to further coupling to a glycoprotein. For this reason, exposure of a hydrazide-containing matrix to crude biological samples should be avoided. In addition, never wash a hydrazide support with acetone or any other organic solvent containing a ketone, because the support will become inactivated.

Hydrazide-containing supports can be prepared by the coupling of a *bis*-hydrazide compound to a matrix using at least two reaction strategies. A *bis*-hydrazide such as adipic dihydrazide or carbohydrazide can be added in large excess to the appropriately prepared matrix to form spacer arms, which then terminate in free hydrazide groups. This process can be done through direct coupling to an amine-reactive support or through the intermediary use of a carboxylate-terminal spacer arm to which the *bis*-hydrazide compound is then coupled using EDC. Early methods described for the preparation of a hydrazide-support involved the reaction of adipic acid dihydrazide with a periodate-oxidized agarose matrix (Hoffman and O'Shannessy, 1988). However, the reactivity of this aldehyde-containing support is so great toward the *bis*-hydrazide spacer that a large portion of the gel ends up being tremendously crosslinked, even to the extent of forming particle aggregates and damaged beads. This occurs even when adipic dihydrazide is added to the support in large mole excess, indicating that the potential for hydrazone formation is so high that both ends of the *bis*-hydrazide compound inevitably end up reacting with aldehydes on the support.

A more controlled approach to making a hydrazide-activated support is to build it from a terminal carboxylate spacer arm derivative, which is coupled to the matrix initially and then activated to attach the *bis*-hydrazide (see section on spacer arms, this chapter). Activated carboxylates will react with a hydrazide group on a *bis*-hydrazide compound to form a secondary amide (hydrazino) bond with a short spacer arm, which then terminates in a free hydrazide group (Figure 15.84). The reaction rate of coupling the hydrazide to the carboxylate is much less than it is with an aldehyde-containing support, so the matrix does not become highly crosslinked and damaged during the process. The reaction can be made to occur through activation of the carboxylate spacer using a carbodiimide such as EDC or through the use of an NHS ester-activated support (see previous discussion on amine reactive activation methods, this chapter). NHS may also be added to an EDC reaction to generate an intermediate NHS ester and enhance the reaction rate with the *bis*-hydrazide molecules. The result of these reactions is the preparation of a hydrazide-activated support that is ready to covalently couple to aldehyde-containing ligands.

Hydrazide-activated supports are relatively stable in aqueous solution if a preservative is added to prevent growth. The coupling of carbonyl-containing ligands to the matrix can be done over a broad pH range, but

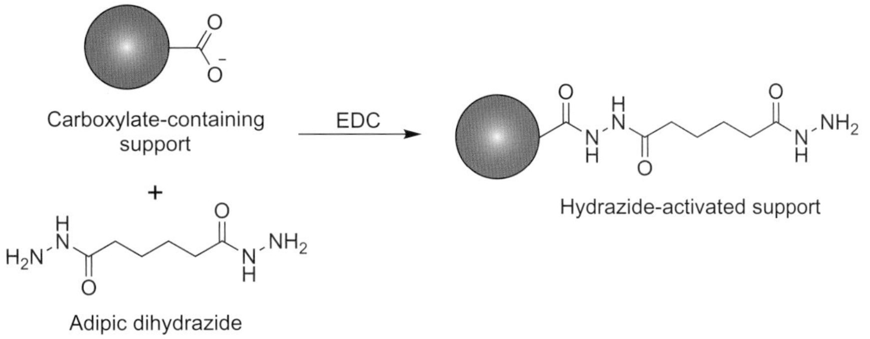

FIGURE 15.84 Preparation of a hydrazide support from a carboxylate spacer using EDC.

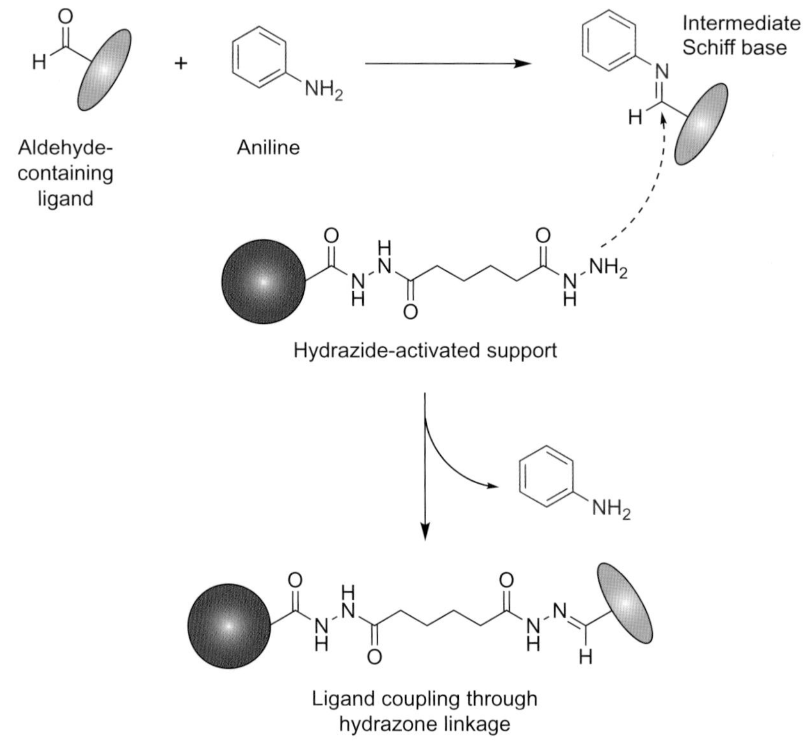

FIGURE 15.85 Immobilization of an aldehyde-containing ligand using an aniline catalyst.

is optimal under slightly acidic conditions (pH 5–6). It has also been found that the yield of hydrazone formation and the rate of the reaction can be increased dramatically by use of an aniline-catalyzed reaction process (Cordes and Jencks, 1962; Dirksen *et al.*, 2006a,b; Dirksen and Dawson, 2008; Byeon *et al.*, 2010). Aniline will react with an aldehyde group on the ligand to form an intermediate Schiff base between its aryl nitrogen and the carbon atom of the carbonyl group of the aldehyde. The aniline imine intermediate more rapidly gets protonated in solution than the initial oxygen of the aldehyde or ketone carbonyl, and it is this species that is the ideal form which can react with the hydrazide groups (Kohler, 2009). This aniline imine intermediate is then rapidly attacked by the hydrazides on the support, which results in displacement of aniline with associated hydrazone bond formation with the hydrazides (Figure 15.85). This two-step reaction process results in much faster hydrazone formation and ligand immobilization than reactions done without the aniline catalyst being present. Using aniline catalysis offers the potential of doing immobilization reactions under more neutral pH conditions while still maintaining high efficiency of hydrazone bond formation (Zeng *et al.*, 2009).

HYDRAZIDE ACTIVATION PROTOCOL

The following protocol describes the modification of a carboxylate-containing support with a *bis*-hydrazide compound to form a hydrazide-activated matrix. The prior creation of a carboxylate-containing support may be achieved by coupling a carboxylate-terminal spacer arm to a matrix using the methods described elsewhere in this chapter on spacer arm production. The *bis*-hydrazide compound used can be one of many; however, adipic dihydrazide probably is the most common. An alternative small bis-hydrazide reagent is carbohydrazide, which is just a single carbonyl group with two hydrazines. It is the most reactive dihydrazide available and can be used to minimize the amount of alkyl chains used in a spacer arm, especially to limit the potential for creating sites of hydrophobic binding and nonspecificity in the final matrix.

1. Wash 100 ml of a carboxylate-containing support material with water (at least 3 bed volumes) to remove storage buffers and preservatives. Wash the gel with an additional 2 bed volumes of 0.1-*M* MES, pH 4.75 (reaction buffer). The washing steps may be carried out using a sintered glass filter suspended in a suction filter flask and applying a gentle vacuum to facilitate drawing the wash solutions through the support. Drain the support to a moist cake.
2. Prepare 100 ml of a solution of 0.5-*M* adipic dihydrazide in the reaction buffer. Alternative *bis*-hydrazide compounds may be used at the same concentration with success. Readjust the pH if necessary.

3. Add the washed support to the dihydrazide solution with mixing. Continue to stir the support slurry by using an overhead paddle stirrer while adding 3 g of the carbodiimide EDC (fresh).
4. React for 3h at room temperature with constant stirring.
5. Wash the support extensively with water, 1-*M* NaCl, and again with water to remove unreacted dihydrazide and any reaction byproducts. Store the hydrazide-containing support until use at 4°C as a 50% slurry in water containing a preservative.

LIGAND COUPLING TO HYDRAZIDE-ACTIVATED SUPPORTS

The following protocols describe the periodate oxidation of an antibody followed by its aniline-catalyzed immobilization onto a hydrazide-containing support. A subsequent immobilization protocol describes the immobilization of sugars, polysaccharides, or glycans through their reducing ends. Similar protocols can be used to immobilize other glycoconjugates or carbohydrates after periodate oxidation. In addition, small ligands containing aldehydes may be coupled to a hydrazide-containing support using the identical buffers and aniline addition process. Larger quantities of immobilized ligands may be prepared by proportionally increasing the amount of reagents used in these procedures.

(A) Periodate Oxidation of Antibody or Glycoprotein

1. Dissolve 1 to 10mg of an IgG antibody that is glycosylated (or another glycoprotein) in 1ml of coupling buffer (0.1-*M* sodium acetate, 0.15-*M* NaCl, pH 5.5). Be careful that the antibody does not contain an amine-containing buffer in solution, such as glycine or Tris, which may alter the pH of the coupling buffer or interfere with the subsequent coupling reaction. Dialyze or desalt the antibody using coupling buffer if necessary to remove interfering components before proceeding.
2. Weigh out 2.1mg of sodium periodate into a small centrifuge tube and add the antibody solution to it with mixing using a vortex mixer. Continue to mix until the sodium periodate is completely dissolved and then wrap the tube in aluminum foil to protect it from light.
3. React for 30min at room temperature with periodic mixing. Do not allow the oxidation to continue longer than this time period or oxidative damage may occur to the protein structure.
4. Stop the reaction by desalting the antibody solution using a size exclusion chromatography support having a molecular weight exclusion limit of no more than 10kDa (i.e., a 10-ml column of Sephadex G-25 or the equivalent). Use coupling buffer to perform the chromatography and collect the protein peak, which will elute in the void volume before the salt peak. Spin columns may also be used for this operation, as they will result in less dilution of the protein solution during the separation and are quicker to use (e.g., Zeba Spin Desalting Columns from Thermo Fisher). Use the oxidized antibody or glycoprotein in the coupling reaction immediately to prevent the potential for protein crosslinking over time through Schiff base formation or Mannich reaction processes.

(B) Coupling an Oxidized Antibody to a Hydrazide Support

1. Wash 1ml of a hydrazide-containing support with water to remove storage solutions and preservatives. Then wash with several ml of coupling buffer (0.1-*M* sodium acetate, 0.15-*M* NaCl, pH 5.5). The washing of a small quantity of a chromatography support may be done in a drip column or a spin column without a top frit. Drain the gel to a wet cake and place the bottom cap on the column to stop the flow. The coupling reaction may be done in a sealed column or the washed gel transferred to a small plastic centrifuge tube able to hold at least 2.5ml of slurry.
2. Add the oxidized antibody or glycoprotein from part (A) to the washed gel cake and stir to resuspend the matrix.
3. In a fume hood, add 18µl of aniline catalyst to the gel slurry with stirring. This results in approximately a 0.1-*M* aniline solution in the coupling reaction mixture.
4. Continue the reaction for at least 2 to 4h with constant mixing. Proteins with higher glycosylation content will couple faster than proteins with lower amounts of carbohydrate. The mixing may be done by end-over-end rocking in a sealed container or column.
5. Blocking of excess hydrazide groups on the support may be done by the addition of glyceraldehyde to the reaction mixture at a final concentration of 0.1-*M* and continuing to mix for 30min. In most cases of glycoprotein immobilization, blocking of the unreacted hydrazides is not necessary; however, if this step is performed, do not add any reducing agents to the mixture or glyceraldehyde-protein adducts also will be formed (Acharya and Manning, 1980).
6. Wash the support with several bed volumes of coupling buffer and collect the washes. The amount of protein coupled may be determined by the difference in the amount of protein present in the reaction medium before coupling and after coupling, taking into account volume differences. Note that aniline may interfere with some methods of protein

concentration determination (e.g., absorbance at 280 nm) and may have to be removed. Continue to wash the support extensively with water, 1-*M* NaCl, and water to remove the last traces of all reactants.

7. The hydrazone linkages on the support may be reduced to stabilize them, if necessary. Often with coupling glycoproteins or glycosylated antibodies, the reduction step is not needed, because multi-point attachment takes place to the matrix and creates a high avidity bond. However, if reduction is deemed necessary due to a slow leakage of ligand from the support, then in a fume hood add to the washed gel 1 ml of coupling buffer containing 100-m*M* sodium cyanoborohydride (6.2 mg) (dangerous compound!). Reduce the hydrazone bonds for 1 h with constant mixing. Finally, wash the support as in step (6) and store the affinity support as a 50% slurry in water or buffer containing a preservative at 4°C.

(c) Coupling the Reducing end of Carbohydrates to Hydrazide Supports

Many carbohydrates, including monosaccharides, disaccharides, oligosaccharides, and glycans, contain a reducing sugar at their anomeric end that is in a cyclic hemiacetal form containing a masked aldehyde group. The aldehyde is only available for reaction when it is in the open form, which is a minority of time in aqueous solution, so immobilization reactions with hydrazide-containing supports usually take much longer to go to completion than reactions with freely available aldehydes. The coupling of reducing sugars onto hydrazide supports can be achieved with or without the presence of a reducing agent and the result will be two different linkage types (Figure 15.86). If a reductant such as sodium cyanoborohydride is used in the reaction medium, then coupling will take place at the anomeric carbon atom through ring opening and the formation of a reduced hydrazone linkage. However, if the reaction is performed without a reducing agent being present, the result will be the creation of a glycosylhydrazide bond with an intact ring structure at the reducing sugar (Rothenberg *et al.*, 1993; Toomre and Varki, 1994; Bigge *et al.*, 1995; Leteux *et al.*, 1998; Srikrishna *et al.*, 2001). It may be important for some affinity chromatography applications to maintain the conformation of a cyclic sugar structure at the reducing end in order to ensure interaction with certain lectins.

The addition of an aniline catalyst to the coupling reaction with a reducing sugar has been shown to dramatically accelerate the formation of the hydrazone bond (Thygesen *et al.*, 2010). The aniline initially traps the open ring form of the sugar's aldehyde through

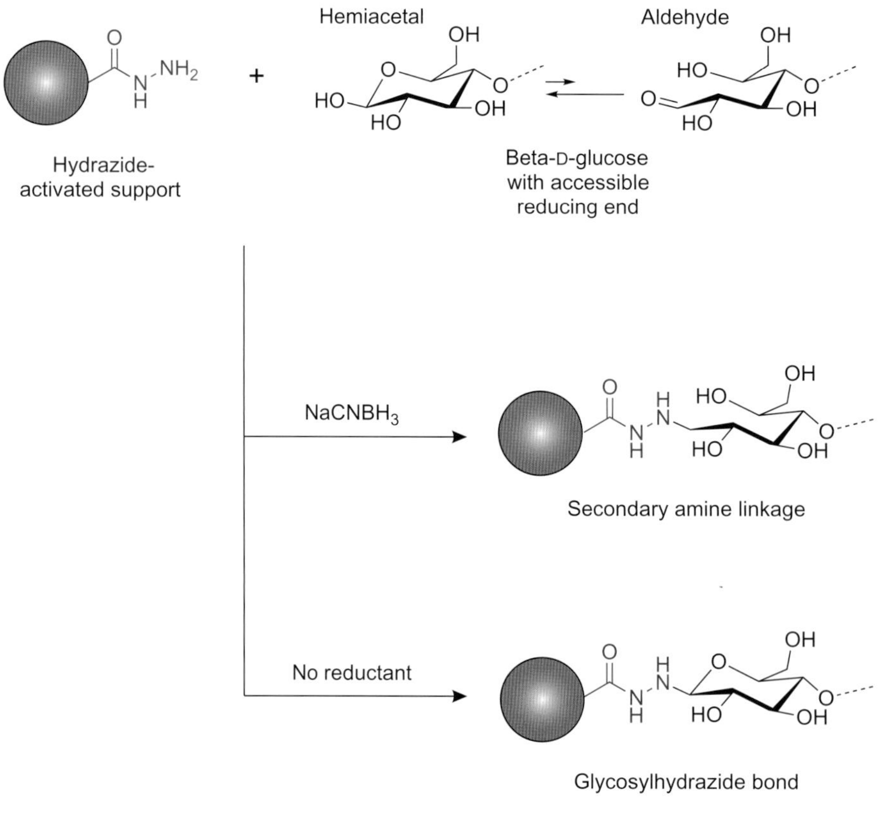

FIGURE 15.86 Coupling at the reducing end of a carbohydrate with and without a reducing agent.

the formation of an imine, which then gets attacked by the hydrazide to form the desired hydrazone with the matrix (Figure 15.87). The following protocol describes the catalyzed reaction with aniline being present in the coupling buffer.

1. Dissolve a carbohydrate, polysaccharide, or glycan (having a reducing end available) at a concentration of at least 5 to 100 μM in 1 ml of 0.1-*M* sodium acetate, 0.15 M NaCl, pH 5.5. Note: Higher concentrations may be used for more abundant carbohydrates as appropriate. The optimal concentration of the carbohydrate may have to be determined experimentally by making several small batches and varying the amount of ligand added to the reaction. Ultimately, the best performance of the immobilized carbohydrate in the intended affinity application should be used as the determining factor for the best concentration to use in the coupling reaction. An alternative coupling medium is a nonaqueous environment consisting of 30% glacial acetic acid in DMSO (v/v) or an acetic acid/pyridine mixture of 1:2 (v/v). Nonaqueous conditions may facilitate the dissolution of some carbohydrate molecules better than aqueous buffers.
2. Wash 1 ml of a hydrazide-containing support with water to remove storage solutions and preservatives. Then wash with several milliliters of coupling buffer (0.1-*M* sodium acetate, 0.15-*M* NaCl, pH 5.5). The washing of a small quantity of a chromatography support may be done in a drip column or a spin column without a top frit. Drain the gel to a wet cake and place the bottom cap on the column to stop the flow. Note: If an organic solvent medium is to be used for the immobilization reaction, sequentially wash the matrix into the acetic acid/solvent blend by first washing it into 100% solvent without the acid to completely remove water (e.g., DMSO) and then washing with the desired acetic acid/solvent mixture (e.g., 30% acetic acid/DMSO).
3. Add the carbohydrate solution to the washed gel cake and stir to resuspend the matrix. Transfer the slurry to a centrifuge tube that is large enough to hold a total volume of at least 4 ml.
4. In a fume hood, add 18 μl of aniline catalyst to the 2 ml gel slurry with stirring. This results in approximately a 0.1-*M* aniline solution in the coupling reaction mixture.
5. Seal the tube and continue the reaction for at least 4 h at room temperature and with constant mixing by end-over-end rocking.
6. The blocking of excess hydrazide groups on the support may be achieved by the addition of glyceraldehyde to the reaction mixture at a final concentration of 0.1-*M* and continuing to mix for 30 min.

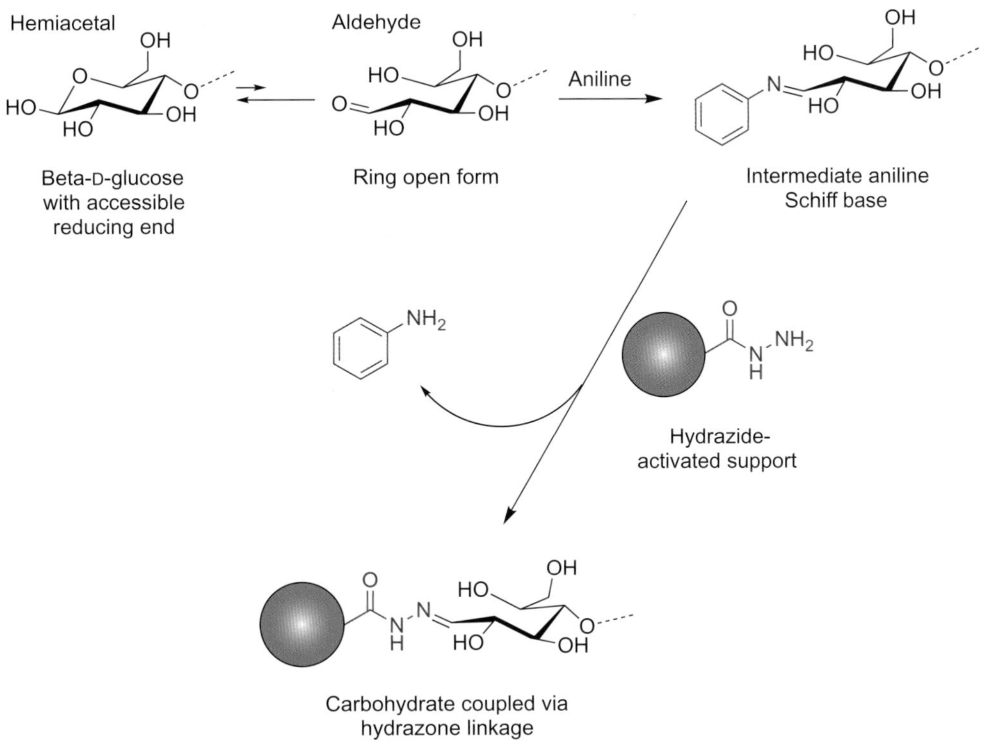

FIGURE 15.87 Immobilization through the reducing end of a carbohydrate using aniline.

7. The hydrazone linkages on the support may be reduced to stabilize them, if necessary. Often when coupling through a single hydrazone bond to a ligand, as in the reaction with the reducing end of a carbohydrate or glycan, the reduction step can be performed to prevent a slow leakage of ligand from the support. To the washed gel add 1 ml of coupling buffer containing 100-m*M* sodium cyanoborohydride (6.3 mg) (dangerous compound—use a fume hood!). Reduce the hydrazone bonds for 1 h with constant mixing.
8. In the fume hood, wash the support extensively with water, 1-*M* NaCl, and water to remove the last traces of all reactants and store the affinity support as a 50% slurry in water or buffer containing a preservative at 4°C.

Aminooxy Supports for Coupling Aldehydes or Ketones

The chemoselective ligation reaction of an aldehyde group with an aminooxy group (–ONH_2) to yield an oxime bond (aldoxime) has been described for use in many bioconjugation reactions, as well as in the coupling of ligands to insoluble supports including surfaces, for the organic synthesis of radiohalogenated compounds, and in the synthesis of drug candidates (Thumshirn *et al.*, 2003; Poethko *et al.*, 2004; Liu *et al.*, 2007; Colombo and Bianchi, 2010). This reaction with aminooxy groups also is quite efficient with ketones to form an oxime called a ketoxime. The simplest form of this reaction has been known for over a century and occurs with aldehydes or ketones as they react with the small compound hydroxylamine (Figure 15.88). The conversion of cyclohexanone to its oxime with hydroxylamine followed by subsequent treatment with sulfuric acid to yield caprolactam *via* a Beckmann rearrangement produces a raw material used in billions of kilogram quantities to create nylon-6 polymers by ring-opening polymerization.

Hydroxylamine derivatives that consist of an organic group linked to the oxygen also are very effective at

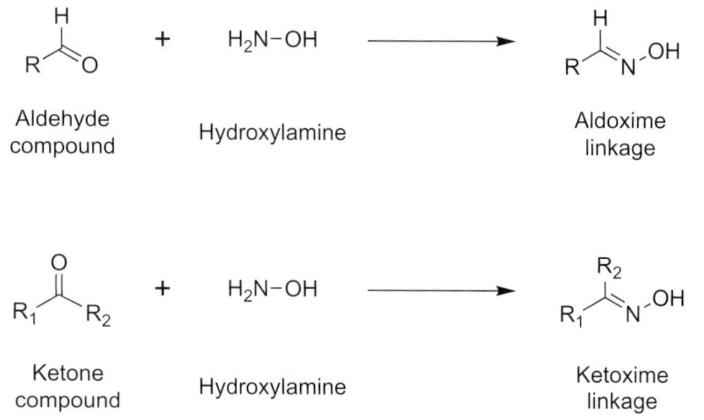

FIGURE 15.88 Basic reaction of an aldehyde or ketone with aminooxy group to give an oxime bond.

forming oxime bonds with carbonyl groups. In this case, the hydroxylamine derivative typically is called an aminooxy functional group, which is also known as an aminoxy or alkoxyamine derivative. The aminooxy compound of this reaction contains a substituent off the oxygen group of hydroxylamine, which can consist of virtually any organic compound or even a chromatography support or modified surface containing a spacer arm that terminates in the –ONH_2 group. The immobilization of aldehyde- or ketone-containing affinity ligands onto support materials that have been activated to contain aminooxy groups can be done similar to the methods described previously for hydrazide containing supports. However, oxime formation usually occurs more rapidly than the reaction between a hydrazide group and an aldehyde or ketone and it typically provides more stable oxime bonds than hydrazones.

The formation of an aminooxy group on a chromatography support can be achieved through the creation of the appropriate spacer arm. There are many methods that can be used to build a spacer on a support (see section on spacer arms, this chapter). In most cases, the creation of the spacer must be performed using a protected aminooxy group, because the terminal amino group will react in amine coupling reactions, which are typically performed to build the spacer molecules. For instance, the first step in making a support with aminooxy functionalities might be to couple a diamine spacer to an amine-reactive support (see section on amine reactive immobilization methods, described previously in this chapter). This would provide an amine-containing intermediate that could then be used to link a protected aminooxy-containing molecule, such as phthalimidooxy –$dPEG_{12}$–NHS ester (Quanta Biodesign). This crosslinker contains a phthalimide-protected aminooxy group on one end and an amine-reactive NHS ester on the other end which can be used to modify the amine-containing support. The result of reacting with the NHS ester ends will be the formation of amide bonds with the amine groups on the support, which will create extremely hydrophilic PEG_{12} chains that extend off the matrix and terminate in the protected aminooxy functionalities. Subsequent removal of the phthalimide-protecting groups with hydrazine in aqueous solution yields the free aminooxy-activated support (Figure 15.89).

Another approach to forming aminooxy functionalities on a resin is to make use of an activated support containing tosyl groups (see previous section on tosyl activation, this chapter). This reactive group is usually used to immobilize amine-containing ligands, but it can also be used to react with hydroxyl groups in nonaqueous conditions. The reaction of a tosyl-activated support with the compound *N*-hydroxyphthalimide results in coupling through the available hydroxyl group on the reagent. This intermediate will result in a protected aminooxy group immobilized on the support, which then can be deprotected by the addition of 0.5-*M* hydrazine in aqueous solution, thus producing the aminooxy-activated matrix (Figure 15.90).

Aminooxy groups have also been used on planar substrates to covalently array proteins to form nanopatterns on silicon wafers and other surface materials. In one example, the aminooxy functionalities were coated on the surface of wafers by Christman *et al.* (2008) using eight-arm PEG derivatives. The hydroxyl groups on the outer ends of the PEG arms were modified using a Mitsunobu-type reaction with *N*-hydroxyphthalimide and including triphenyl phosphine (PPh_3) with DIAD (diisopropyl azodicarboxylate) as the activators (Figure 15.91). Subsequent deprotection and removal of the phthalimide using hydrazine yielded the desired aminooxy groups.

The immobilization reaction using an aminooxy support with aldehyde- or ketone-containing ligands can be accelerated by the addition of an aniline catalyst. As in the formation of hydrazone linkages with aldehydes and hydrazide-activated supports, the formation of an oxime bond between an aldehyde or ketone and an aminoxy is dramatically increased in rate and yield by aniline (Cordes and Jencks, 1962; Dirksen *et al.*, 2006a,b; Dirksen and Dawson, 2008; Byeon *et al.*, 2010). The aromatic amine on aniline first reacts with the aldehyde or ketone group to form an intermediate imine, which then reacts with the aminooxy groups on the support to create the final oxime linkage (Figure 15.92). The intermediate aryl imine undergoes more rapid protonation in slightly acidic solution than does the oxygen of the aldehyde or ketone carbonyl, and it is this protonated species that can react with the aminooxy groups (Kohler, 2009).

The immobilization reaction on a surface that was activated with aminooxy groups to couple a peptide molecule was studied in detail to determine its kinetics (Lempens *et al.*, 2009). The peptide was prepared to contain an N-terminal aldehyde residue made *via* the pyridoxyl 5′-phosphate/sodium periodate method of Gilmore *et al.* (2006a,b) (also see Scheck and Francis, 2007; Scheck *et al.*, 2008). It was also found in this case that use of the aniline catalyst dramatically increased

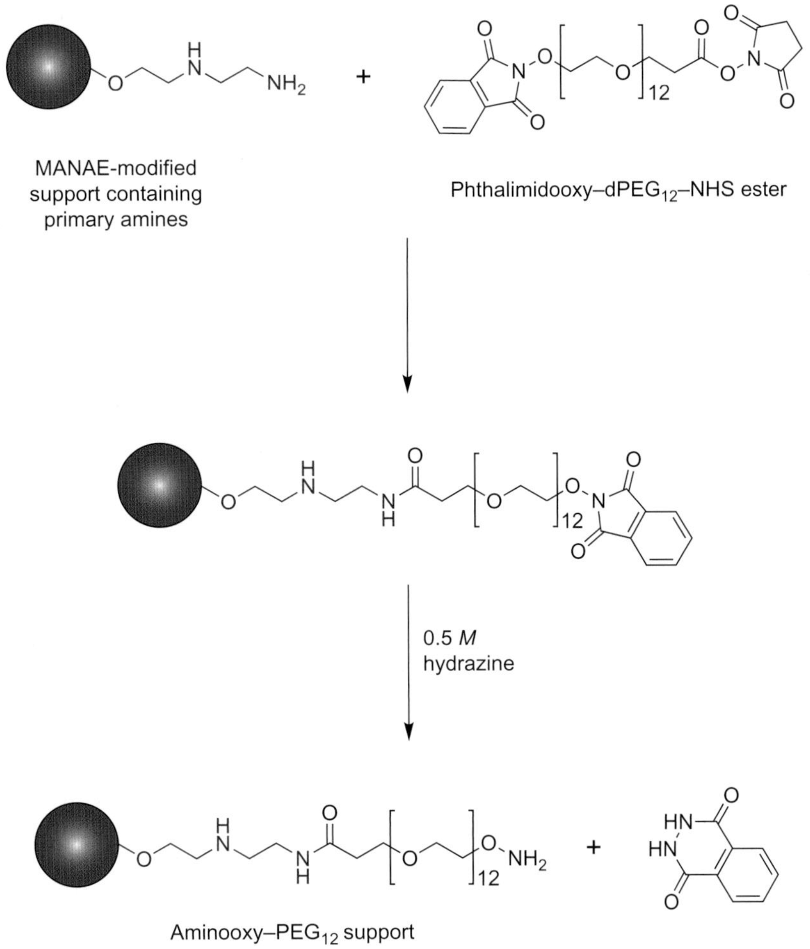

FIGURE 15.89 Preparation of an aminooxy support using PhthNO–dPEG$_{12}$–NHS ester.

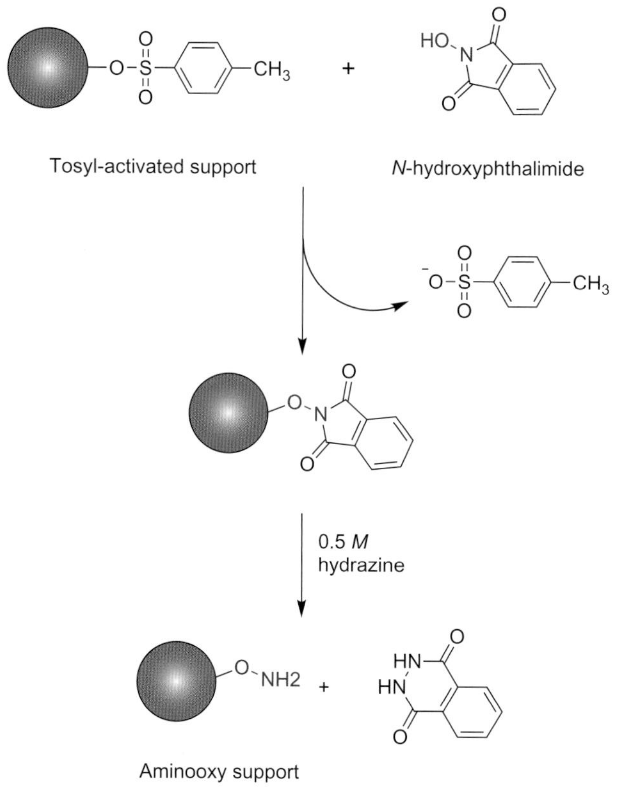

FIGURE 15.90 Formation of aminooxy groups on a support by reaction of *N*-hydroxyphthalimide with a tosyl-activated support.

the reaction rate and coupling efficiency at all pH values investigated, with the greatest rate and yield obtained at pH 4.5. The rate at pH 6.0 was less than half that observed at the lower pH value; however, the difference between a catalyzed and an uncatalyzed reaction was significant at all the acidic pH environments studied. As the reaction pH is increased from 4.5 to more of a neutral pH environment, the rate and yield of coupling significantly slow down; however, for molecules sensitive to more acidic pH environments the reaction may be performed with good results at higher pH. To compensate for slower kinetics at higher pH, the time of the reaction should be increased to realize acceptable immobilization yields.

AMINOOXY ACTIVATION PROTOCOL

The following protocol describes the modification of an amine-reactive support with a diamine spacer and the subsequent reaction of the amino-support with an NHS–PEG_{12}–oxyphthalimide crosslinker followed by deprotection of the phthalimide to form the aminooxy derivative. Other amine-containing supports may also be used in this procedure, such as can be prepared using the spacer arms described elsewhere in this chapter. Use a fume hood and personal protective equipment for all operations.

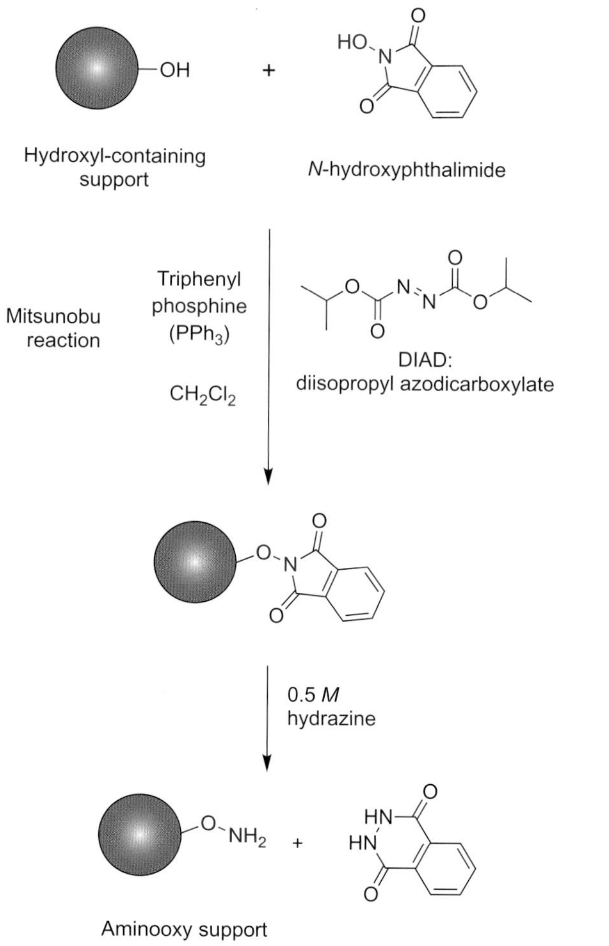

FIGURE 15.91 Aminooxy formation at the end of a PEG chain using the Mitsunobu reaction.

1. In a fume hood, drain 10 ml of a CDI-activated agarose support of acetone storage solution (see section on CDI activation). Other amine-reactive supports may also be used to couple the diamine; however, follow the respective protocols for coupling ligands if another reaction is used. All of the washing steps may be carried out using a sintered glass filter suspended in a suction filter flask and applying a slight vacuum to facilitate drawing the wash solutions through the support. Pull a gentle vacuum on the suction filter flask and break up the resin into small pieces as the acetone drains through, but do not allow the support to dry. Next, add to the resin approximately 2 bed volumes of the solvent DMAC (note that DMSO may be used as an alternative solvent) and stir the gel to resuspend it in the new solvent. Continue to wash the support with DMAC for at least 5 bed volumes to remove most of the remaining acetone.
2. Prepare 10 ml of a solution consisting of 1.0-*M* ethylenediamine in DMAC (or DMSO). Notes: Alternative diamine compounds may be used at the

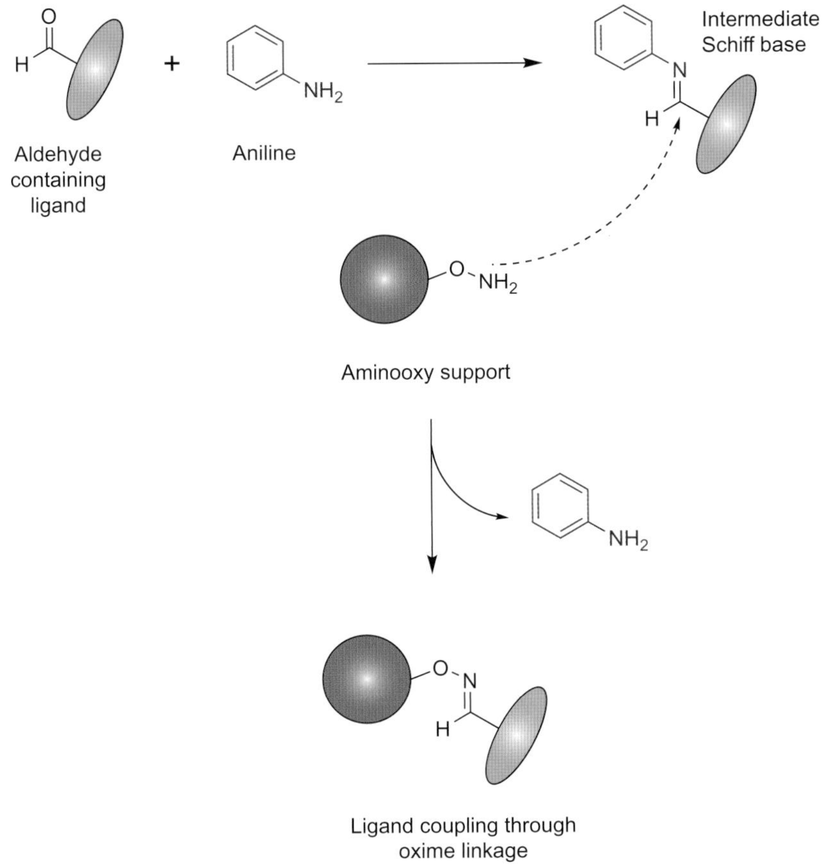

FIGURE 15.92 Immobilization of an aldehyde ligand on an aminooxy support using aniline catalysis.

same concentration with success; however, avoid the use of long aliphatic diamines as spacers, as this will create considerable hydrophobic character within the matrix and increase the potential for nonspecific interactions on the subsequent affinity support.

3. Add the washed support to the ethylenediamine solution with stirring. React for 1 h at room temperature with constant mixing. End-over-end rocking in a sealed container may be used to mix the support. Note: The addition of an organic base to the reaction to accept protons generated during the coupling of the spacer need not be done, because the free amine end of ethylenediamine will still be available.
4. Wash the amine-containing support extensively with DMAC to remove the last traces of diamine compound (use at least 20 bed volumes). Drain the washed support to a moist cake.
5. Prepare a solution of the heterobifunctional crosslinker PhthNO–dPEG$_{12}$–NHS ester (Quanta Biodesign) in 10 ml of DMAC dissolved at a concentration of at least 10 mg/ml (11.6 μmol/ml). Higher concentrations may be used to create higher densities of the final aminooxy groups on the matrix.
6. Add the washed support to the crosslinker solution and mix to resuspend the gel. To this suspension, add 0.65 g of dimethylaminopyridine (DMAP) as an organic base to catalyze the reaction and mix to dissolve (alternatively, the addition of 0.75 ml of anhydrous triethylamine (TEA) may be done or an equivalent mole amount of *N,N*-diisopropylethylamine (DIEA) may be added). React for 1 h at room temperature with constant mixing.
7. Extensively wash the modified support with DMAC to remove excess reactants and reaction byproducts (at least 20 bed volumes). Next, wash the support into water by sequentially washing with increasing concentrations of water in DMAC until 100% water is used. Continue washing with water until all traces of the solvent have been removed. Drain the gel to a wet cake.
8. In a fume hood, prepare 10 ml of an aqueous solution consisting of 0.5-*M* hydrazine (dangerous and toxic!) in 0.1-*M* MES, pH 6.0. If hydrazine dihydrochloride is used to prepare this solution, its dissolution in the buffer will not cause the pH to become highly alkaline. Avoid contact with hydrazine or its solutions by using the appropriate personal protective equipment, including being cautious to avoid contact with skin or eyes, and also avoiding the inhalation of dust. Prevent static charge buildup when dispensing and weighing the hydrazine compound by using the appropriate grounding precautions. Adjust the final pH of the solution back to 6.0, if necessary.

9. In the fume hood, add the washed gel containing the protected aminooxy groups with the hydrazine solution and mix by end-over-end rotation in a sealed container. React overnight to deprotect the aminooxy groups.
10. In the fume hood, drain the support of excess hydrazine deprotection reagent and wash the support extensively with water, 1-*M* NaCl, and water. Store the aminooxy-activated support until use as a 50% slurry in water or buffer containing a preservative at 4°C. (Note: Do not include any ketone- or aldehyde-containing compounds in the storage solution or the aminooxy groups will react and become inactive for coupling ligands.)

LIGAND COUPLING TO AMINOOXY-ACTIVATED SUPPORTS

The following protocols describe the periodate oxidation of sialic acid groups on glycoproteins followed by their aniline-catalyzed immobilization onto an aminooxy-containing support. A subsequent immobilization protocol describes the immobilization of sugars, polysaccharides, or glycans through their reducing ends. Similar protocols can be used to immobilize other glycoconjugates or carbohydrates after periodate oxidation or through their reducing ends, if available. In addition, small ligands containing aldehydes or ketones may also be coupled to an aminooxy-containing support using the identical buffers and an aniline catalysis process. Larger quantities of immobilized ligands may be prepared by proportionally increasing the reagent amounts used in these procedures.

(A) Periodate Oxidation of Sialic Acid Groups on Glycoproteins

1. Dissolve 1 to 10 mg of a glycoprotein containing sialic acid groups in 1 ml of coupling buffer (0.1-*M* sodium acetate, 0.15-*M* NaCl, pH 5.5). Be careful that the glycoprotein does not contain an amine-containing buffer in solution, such as glycine or Tris, which may alter the pH of the coupling buffer or interfere with the subsequent coupling reaction. For instance, dialyze or desalt a commercial antibody preparation containing Tris or glycine using coupling buffer to remove interfering components before proceeding. Chill the glycoprotein solution by placing it on ice.
2. Dissolve sodium periodate in water at a concentration of 10 mg/ml (46-m*M*). Continue to mix until the sodium periodate is completely dissolved and then wrap the tube in aluminum foil to protect it from light. Chill the periodate solution by placing it on ice.
3. Add 21.8 μl of the periodate solution to the 1 ml of glycoprotein solution and mix to dissolve (makes approximately 1-m*M* periodate final concentration in the glycoprotein solution). Maintain the solution on ice.
4. React for 30 min on ice with periodic mixing. Do not allow the oxidation to continue longer than this time or oxidation may occur at sites other than just sialic acid groups.
5. Stop the reaction by desalting the antibody solution using a size exclusion chromatography support having a molecular weight exclusion limit of no more than 10 kDa (i.e., at least 5 ml of Sephadex G-25 or the equivalent). Use cold coupling buffer to perform the chromatography and collect the protein peak, which will elute before the salt peak. Spin columns may also be used for this operation, as they will result in less dilution of the protein solution during the separation and are quicker to use (e.g., Zeba Spin Desalting Columns from Thermo Fisher). Use the oxidized glycoprotein in the coupling reaction immediately to prevent the potential for protein crosslinking over time through Schiff base formation or Mannich reaction processes.

(B) Coupling the Oxidized Glycoprotein to an Aminooxy Support

1. Wash 1 ml of an aminooxy-containing support with water to remove storage solutions and preservatives. Then wash with several milliliters of coupling buffer (0.1-*M* sodium acetate, 0.15-*M* NaCl, pH 5.5). The washing of a small quantity of a chromatography support may be done in a drip column or a spin column without a top frit. Drain the gel to a wet cake and place the bottom cap on the column to stop the flow. The coupling reaction may be done in a sealed column or the washed gel transferred to a small plastic centrifuge tube able to hold at least 2.5 ml of slurry. Note: The pH of the coupling buffer may be decreased to pH 4.5 to further enhance the reaction rate and yield of oxime bond formation.
2. Add the oxidized glycoprotein from part (A) to the washed gel cake and stir to resuspend.
3. In a fume hood, add 18 μl of aniline catalyst to the gel slurry with stirring. This results in approximately a 0.1-*M* aniline solution in the coupling reaction mixture.
4. Continue the reaction for at least 2 to 4 h with constant mixing. Proteins with higher glycosylation content will couple faster than proteins with lower amounts of carbohydrate. The mixing may be done by end-over-end rocking in a sealed container or column.
5. Blocking excess aminooxy groups on the support may be done by the addition of glyceraldehyde to the reaction mixture at a final concentration of 0.1-*M* and continuing to mix for 30 min. In most cases of glycoprotein immobilization, blocking the unreacted aminooxy groups is not necessary; however, if this step is performed, do not add any reducing agents to the mixture or glyceraldehyde–protein adducts will be formed (Acharya and Manning, 1980).

6. Wash the support with several bed volumes of coupling buffer and collect the washes. The amount of protein coupled to the resin may be determined by the difference in the amount of protein present in the reaction medium before coupling and that remaining after coupling, taking into account volume differences. Note that aniline may interfere with some methods of protein concentration determination (e.g., absorbance at 280 nm) and may have to be removed. Continue to wash the support extensively with water, 1-*M* NaCl, and water to remove the last traces of all reactants. Finally, wash the support as in step 6 and store the affinity support as a 50% slurry in water or buffer containing a preservative at 4°C.

(C) Coupling the Reducing end of Carbohydrates to Aminooxy Supports

Many carbohydrates contain a reducing sugar at their anomeric end that is in a cyclic hemiacetal form, which in reality is a masked aldehyde group. The aldehyde only is available for immobilization when it is in the open form, which is a minority of time in aqueous solution. For this reason, the immobilization reactions of reducing sugars and carbohydrates with aminooxy-containing supports usually take longer to go to completion than reactions with freely available aldehydes. The coupling of reducing sugars onto aminooxy supports is carried out very similarly to that described for hydrazide supports, described previously; however, a reducing step is optional to further stabilize the resultant bond. If the reaction is carried out without a reducing agent being present, the result will be the creation of an oxime bond yielding a mixture of an intact ring structure (glycosyl derivative) or acyclic oxime (open ring derivative) at the reducing sugar anomeric carbon. If a reductant is present during the reaction, only the acyclic derivative will result with the formation of a glycosyl-hydroxylamine linkage to the support (Peluso *et al.*, 2002) (Figure 15.93).

The addition of an aniline catalyst to the coupling reaction of an aminooxy group with a reducing sugar has been shown to dramatically accelerate the formation of the oxime bond (Thygesen *et al.*, 2010). The

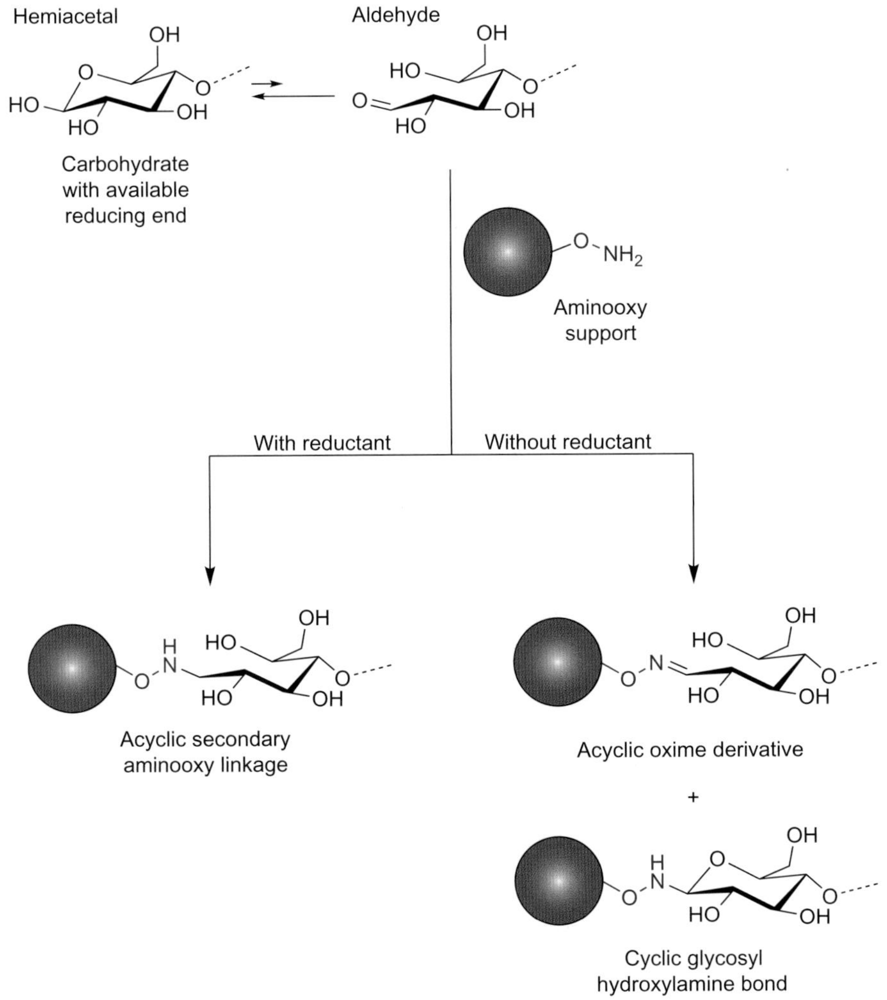

FIGURE 15.93 Immobilization of reducing carbohydrate onto an aminooxy support with and without the use of a reductant.

aniline initially traps the open ring aldehyde through the formation of an imine, which then gets attacked by the immobilized aminooxy group to form the desired oxime bond with the matrix (Figure 15.94). The following protocol describes the catalyzed reaction with aniline being present in the coupling buffer. Larger quantities of the immobilized ligand may be prepared by proportionally increasing the amount of reagents added at each step.

1. Dissolve a carbohydrate, polysaccharide, or glycan (having a reducing end available) at a concentration of at least 5 to 100 μ*M* in 1 ml of 0.1-*M* sodium acetate, 0.15-*M* NaCl, pH 4.5. Note: Higher concentrations may be used for more abundant carbohydrates as appropriate. The optimal concentration of the carbohydrate may have to be determined experimentally by making several small batches and varying the amount of ligand added to the reaction. Ultimately, the best performance of the immobilized carbohydrate in the intended affinity application should be the determining factor for the best coupling reaction conditions to use.

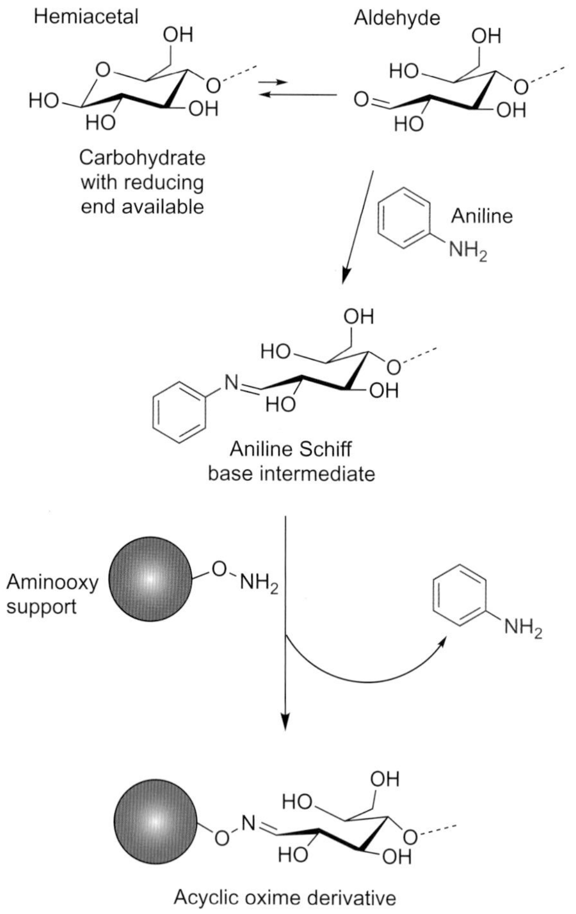

FIGURE 15.94 Immobilization through the reducing end of a carbohydrate using aniline.

2. Wash 1 ml of an aminooxy-containing support with water to remove storage solutions and preservatives. Then wash with several milliliters of coupling buffer (0.1-*M* sodium acetate, 0.15-*M* NaCl, pH 4.5). The washing of a small quantity of a chromatography support may be done in a drip column or a spin column without a top frit. Drain the gel to a wet cake and place the bottom cap on the column to stop the flow.
3. Add the carbohydrate solution to the washed gel cake and stir to resuspend the matrix. Transfer the slurry to a centrifuge tube that is large enough to hold a total volume of at least 4 ml.
4. In a fume hood, add 18 μl of aniline catalyst to the 2-ml gel slurry with stirring. This results in approximately a 0.1-*M* aniline solution in the coupling reaction mixture.
5. Seal the tube and continue the reaction for at least 4 h at room temperature and with constant mixing by end-over-end rocking.
6. The blocking of excess aminooxy groups on the support may be done, if necessary, by the addition of glyceraldehyde to the reaction mixture at a final concentration of 0.1-*M* and continuing to mix for 30 min.
7. In the fume hood, wash the support extensively (at least 20 bed volumes) with water, 1-*M* NaCl, and water to remove the last traces of all reactants and store the affinity support as a 50% slurry in water or buffer containing a preservative at 4°C.

Amine-Containing Supports for Coupling Aldehydes, Ketones, or Carboxylates

Supports that have available primary amines on them can be used to immobilize carbonyl-containing ligands with good efficiency. The amines may be formed from the construction of a spacer arm on a base support or created as a result of the polymerization of functional silane compounds or vinyl monomers containing amines. The amine groups can react with aldehydes or ketones on affinity ligands using a reductive amination process to yield secondary amine linkages to the support. Free amine groups can also be used to immobilize carboxylate-containing ligands by amide bond formation using a carbodiimide-mediated reaction. The reaction of an immobilized amine with an aldehyde-containing ligand is just the opposite of the immobilization of an amine-containing ligand on an aldehyde support, which is described under the section Amine Reactive Immobilization Methods earlier in this chapter; however, the reaction principles are identical.

Amine-containing spacer arms can be added to a support by the reaction of a diamine compound with an amine-reactive matrix. Using this method, a variety of spacer types may be used as desired to design particular properties into the resin, such as the use of short or long hydrophobic or hydrophilic cross-bridges, which may have an effect on subsequent affinity separations done on the support. Amines also can be added to particles using a functional silane reagent, such as aminopropyltrimethoxysilane, that can be used to coat the surface of particles and form the required primary amine groups. See the final section of this chapter for a discussion on the formation of spacer arms on supports and see also Chapter 13 for methods related to the use of functional silanes.

An amine-containing support can react with ligands containing aldehydes or ketones to form an initial Schiff base imine, which results from a dehydration reaction. This imine bond is highly reversible in aqueous environments and therefore must be stabilized by reduction to permanently link the ligand to the support (Figure 15.95). The reducing agent used in this process typically is sodium cyanoborohydride or other equivalent reagents that are capable of reducing the imine to a secondary amine, but will not reduce the initial aldehyde reactants. The use of an amine-containing support to immobilize aldehydes or ketones may not be the first or best choice, because hydrazide- or aminooxy-activated supports have better reaction characteristics and yield more stable linkages (see previous sections). However, if an amine-containing matrix is the desired starting point, the coupling of carbonyl ligands can be done with success using the cyanoborohydride reduction step to stabilize the linkage.

However, an amine-containing support is even more suitable for the immobilization of carboxylate-containing ligands. Carboxylates and amines can be made to react and form stable amide bonds. Although these two functional groups do not spontaneously react under normal conditions, the addition of an amide bond forming agent can be done to facilitate efficient coupling. The most common amide bond forming agents are carbodiimides, such as the water soluble EDC, as well as other such condensing agents that were originally developed to form bonds between amino acids in peptide synthesis applications. The water soluble reactants can be used in aqueous buffer to couple carboxylate-containing ligands that are soluble in water. EDC is a so-called "zero-length" crosslinker, since it mediates the formation of amide linkages without leaving behind a spacer molecule (Grabarek and Gergely, 1990). In addition, for carboxylate ligands that may be insoluble in aqueous conditions, amide bond forming agents that are soluble in organic solvents can be used. EDC reacts with the carboxylates on the ligand to form an intermediate reactive ester, which then goes on to react with the amines on the support surface to create amide bond linkages (Figure 15.96). See also Chapter 14 on the use of EDC for coupling of affinity ligands to microparticles and nanoparticles for additional information on the reactions involved with carbodiimide activation and immobilization.

In most cases, EDC-mediated amide bond formation is quite efficient and proceeds to completion within 2 to 4h. Staros *et al.* (1986) developed a modification of this reaction that incorporates the addition of NHS (or sulfo-NHS) into the reaction medium to form an intermediate NHS ester on the activated carboxylate. This two-stage reaction results in the creation of an amide bond through the reaction of an intermediate NHS ester with the amine groups on the support. An NHS ester undergoes fewer side reactions and is more efficient at forming amide bonds than the intermediate EDC ester, so the desired product is formed at an accelerated rate.

LIGAND COUPLING TO AMINE-CONTAINING SUPPORTS

The following protocols make use of an amine-containing resin made through the modification of an amine-reactive support with a diamine spacer molecule (see the last section in this chapter on spacer arm construction).

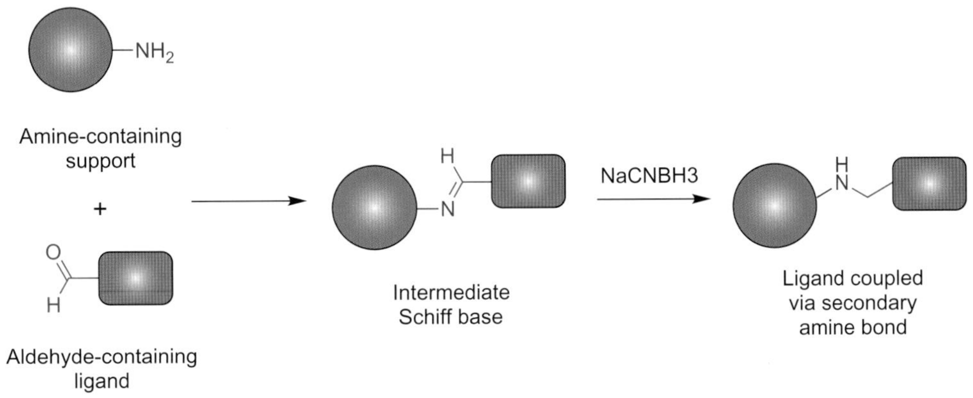

FIGURE 15.95 Immobilization of an aldehyde-containing ligand onto an amine-containing support.

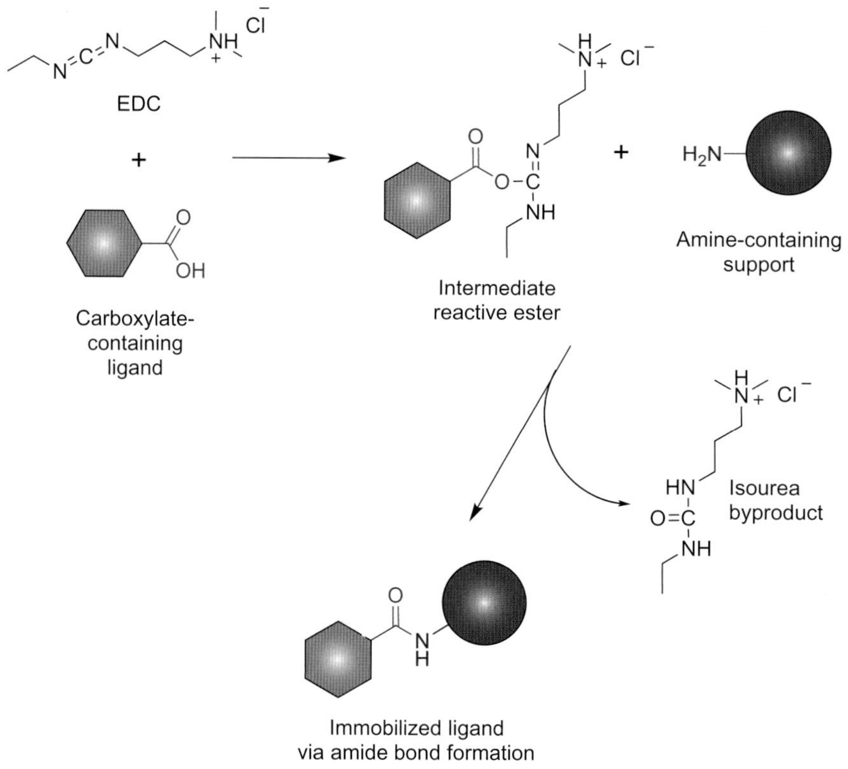

FIGURE 15.96 Immobilization of carboxylate ligands onto amine-containing supports using EDC.

Other amine-containing supports can also be used with success, such as polymeric supports prepared using amine-containing monomers during the polymerization process. Also see Chapter 14 for protocols related to the coupling of affinity molecules to micro- or nanoparticles.

Coupling Aldehydes or Ketones to Amine-Containing Supports

The following protocol makes use of sodium cyanoborohydride, which is a highly toxic compound that expels volatile cyanide into the atmosphere. For this reason, all operations using this compound as the solid or in solution should be done in a well-ventilated fume hood. Also use the appropriate personal protective equipment while doing this procedure, such as gloves, safety glasses, and a lab coat to prevent direct contact with the compound.

1. Wash 10 ml of the amine-containing support material into coupling buffer (0.1-*M* sodium phosphate, pH 7.2) and drain to a wet cake. Other buffer components may be added to the coupling buffer, such as the addition of 0.15-*M* NaCl or alternative buffer salts; however, avoid amine-containing compounds like Tris, glycine, or imidazole, which will compete in the reaction. In general, avoid buffer compounds containing primary or secondary amines as well as any other nucleophile that could react with an aldehyde on the ligand. In addition, certain detergents and other amphiphillic components have been shown to decrease the stability of the intermediate Schiff base and these additives should be used only after validating that they have no effect on the immobilization reaction (Viguera *et al.*, 1990). Reductive amination coupling in aqueous solution has been shown to effectively occur between pH 4 and 10, with an optimal range for cyanoborohydride reduction of pH 6 to 8.
2. Suspend the washed support material containing amine residues in an equal volume of 0.1-*M* sodium phosphate, pH 7.2, into which an aldehyde-containing ligand has been dissolved. Notes:
For glycoprotein ligands, a typical concentration range may be 3 to 5 mg/ml gel, but much higher concentrations may be used if a high density of coupled protein on the gel is desirable. In some cases, reactions containing protein at up to 20 mg/ml gel can be done, but the need for this density on the final affinity support is unusual. For small aldehyde-containing affinity ligands, a concentration of 2 to 3 mg ligand/ml gel in the immobilization reaction may be sufficient to obtain good binding capacity on the resultant affinity gel. Alternatively, a concentration representing 5 to 10 times the

concentration of reactive amines on the matrix may be used to ensure a high density of the final coupled ligand. The ultimate concentration of ligand used in the reaction should be optimized by performing a series of coupling reactions at different initial ligand concentrations and determining the performance of the affinity supports in capturing and eluting the desired target molecule.

3. Stir the gel/ligand slurry in a fume hood using a paddle stirrer or using a rotator to maintain constant mixing.
4. Add to the gel slurry 63 mg of solid sodium cyanoborohydride ($NaCNBH_3$; MW = 62.84) [toxic!] or 0.2 ml of 5-*M* $NaCNBH_3$ in 1-*N* NaOH (Sigma) with stirring. If the alkaline solution of cyanoborohydride is used, check the pH and readjust to 7.2, if necessary. Continue the reaction for 4 h to overnight at room temperature. The amount of cyanoborohydride added—whether as the pure solid compound or as the solution—results in a 50-m*M* solution in the final reaction slurry, which is in a total volume of 200 ml. A shorter coupling time may be sufficient for many ligands, but the optimal conditions to give acceptable yields for a particular ligand should be determined by doing a series of coupling reactions at different time points. Reactions at 4°C may also be done for thermally sensitive proteins or ligands, but reaction times may have to be extended to get the same level of coupling as a room temperature reaction.
5. Transfer the gel slurry to a sintered glass filter funnel (in the fume hood) and wash with several bed volumes of water to remove most of the unreacted ligand and reaction byproducts.
6. Unreacted amine residues remaining on the support can be blocked by the addition of a small aldehyde-containing compound, such as glyceraldehyde. Avoid the blocking of excess amine groups if a protein ligand or a ligand containing more than one amine has been immobilized, as the glyceraldehyde also will modify amines on the ligand. Wash the support once with an equal volume of 0.1-*M* glyceraldehyde dissolved in coupling buffer at pH 7.2, and then transfer the wet gel cake to a clean flask or vessel used for mixing the reaction. Add with stirring 10 ml of the 0.1-*M* glyceraldehyde solution along with 63 mg (or 0.2 ml of the 5-*M* $NaCNBH_3$ solution) of sodium cyanoborohydride. Readjust the pH if necessary and continue the blocking reaction for 30 min at room temperature.
7. Wash the support extensively with water, 1-*M* NaCl, and again with water to remove all unreacted components from the gel. Additional wash solutions may be utilized to completely remove ligands that may have some nonspecific binding potential to remain noncovalently bound to the immobilized ligand, such as the use of acidic and alkaline washes as well as washes containing denaturants (such as guanidine). After washing, the affinity gel may be stored as a 50% aqueous slurry containing a preservative at 4°C.

Coupling Carboxylate-Containing Ligands to Amine-Containing Supports

The following protocol may be used to immobilize a carboxylate-containing ligand onto a support material containing amines. The process uses the water soluble carbodiimide EDC, which activates the carboxylate to an intermediate ester that then reacts with the amines on the support to form amide bonds. This method is not recommended for coupling ligands that contain both carboxylates and amines, such as proteins, because the ligand will become oligomerized in solution upon the addition of EDC.

1. Wash 10 ml of an amine-containing support with several volumes of water to remove storage solutions and then wash with coupling buffer (0.1-*M* MES, pH 4.7). Drain to a wet cake. Note: The EDC reaction occurs efficiently at slightly acidic pH values with an optimal rate in the range of pH 4.5 to 6.0. However, buffers at physiological pH may also be used with success (e.g., 0.1-*M* sodium phosphate, pH 7.0–7.5). The reaction at the higher pH will be somewhat slower, and the reaction time should be extended to obtain the same yield of coupling. For neutral pH reactions, the addition of 2 equivalents of NHS (or sulfo-NHS) over the amount of EDC added can be done to further accelerate the reaction kinetics.
2. Prepare 10 ml of the carboxylate-containing ligand to be coupled in coupling buffer. For many small ligands, the use of 3 to 5 mg ligand/ml is a sufficient concentration; however, some optimization of ligand concentration may have to be done to obtain the best performance of the affinity resin in the intended application. Note: For ligands that are not very soluble in aqueous solution the carboxylate molecule may be first dissolved in a water-miscible solvent such as ethanol, DMSO, DMF, or DMAC and then added to the coupling buffer to obtain the final mixture. The final percentage of solvent in the coupling buffer should not exceed about 20% for DMSO, DMF, or DMAC or the buffer salts may begin to precipitate out of solution. For an ethanol-containing final solution, ethanol may be added up to 50% in the coupling buffer and still maintain buffer solubility. For ligands that are particularly insoluble in aqueous environments, there may be a micro-precipitate in the final solution, but the reaction still should proceed without difficulty.
3. Mix the washed gel with the ligand solution with stirring and add 0.3 g of EDC and mix to dissolve.

4. React with constant mixing for 2 to 4 h at room temperature. The gel slurry may be placed in a sealed container and mixed by end-over-end rocking.
5. Unreacted amines on the support may be blocked by the addition of a small hydrophilic carboxylate compound, such as glyceric acid (or 2,3-dihydroxypropanoic acid) or just by using sodium acetate. If a blocking step is desired, add the chosen blocking agent to the coupling reaction to make a final concentration of 0.1-*M* and readjust the pH, if necessary. Add an additional 0.3 g of EDC and react for 1 h with mixing.
6. Wash the support extensively with water, 1-*M* NaCl, and again with water to remove all unreacted components from the gel. Additional wash solutions may be utilized to completely remove ligands that may have some nonspecific binding potential to remain noncovalently bound to the immobilized ligand, such as the use of acidic and alkaline washes as well as washes containing denaturants (such as guanidine). After washing, the affinity gel may be stored as a 50% aqueous slurry containing a preservative at 4°C.

2.5. Streptavidin-Mediated Immobilization Methods

One of the most useful affinity interactions for biological techniques involves the tight binding of biotin (vitamin H) to the protein avidin (from egg whites) or streptavidin (from the bacterium *Streptomyces avidinii*). Both of these proteins are tetrameric in structure and each subunit contains a binding pocket for biotin. The deep binding pocket that biotin sits in contains amino acid residues able to form eight hydrogen bonding interactions plus additional van der Waals interactions with the bicyclic structure of biotin, thus creating one of the strongest noncovalent interactions known (dissociation constant, (K_d): ~10^{-14}–10^{-15} M). For this reason, the binding of biotin to avidin or streptavidin is similar to a chemoselective ligation reaction forming a covalent bond—highly specific and nearly irreversible.

Streptavidin conjugates are widely used as universal detection reagents in immunoassays, cellular imaging, flow cytometry, and other targeting and assay applications. Conjugates of streptavidin with a fluorescent molecule or an enzyme can provide sensitive detection of biotinylated antibodies and other affinity targeting molecules (see Chapter 11). Immobilized streptavidin can also be used to capture biotinylated molecules out of complex solutions, providing a mechanism for isolating interacting proteins or other biological complexes *via* immunoprecipitation (IP) (Figure 15.97).

For instance, an immobilized streptavidin support can be used to retrieve a biotinylated antibody that has specificity toward and has interacted with a desired protein target within a biological sample. As the affinity support captures the antibody–antigen complex

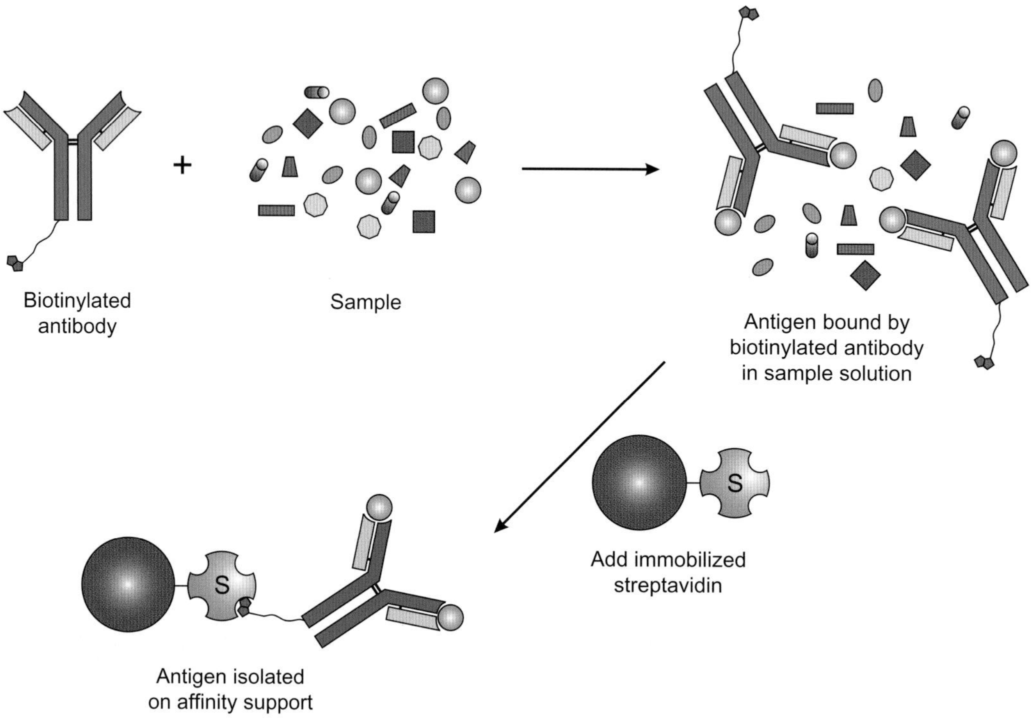

FIGURE 15.97 Use of immobilized streptavidin for IP and co-IP applications.

through the biotin modifications on the antibody, it may also pull out of solution any proteins or other biomolecules that have interacted with the antigen (a technique sometimes called a "pull-down" assay). Co-immunoprecipitation (co-IP) often is used to study interacting proteins if the protein complexes have sufficient affinity to remain intact during the affinity separation process (see Chapter 24). After washing off any non-interacting proteins in the sample, the co-IP isolated complexes then can be eluted from the affinity support using a variety of conditions, which can consist of an acid pH buffer (e.g., 0.1-*M* glycine, pH 2.8) to break the antigen–antibody interactions or much stronger elution agents such as 8-*M* guanidine hydrochloride, pH 1.5, or actually boiling the beads in SDS electrophoresis sample buffer, which will break the streptavidin–biotin interactions.

The preparation of an immobilized streptavidin affinity support can be achieved using amine-reactive resins, such as those described earlier in this chapter. For example, an aldehyde-containing matrix can be used to couple streptavidin by reductive amination and form a high-capacity affinity support for binding biotinylated proteins and other molecules. Coupling streptavidin to the support at a level of 5 to 10 mg/ml in the reaction medium will result in an excellent universal affinity gel for the noncovalent immobilization of biotinylated antibodies. Many such streptavidin affinity supports are available commercially (e.g., Thermo Fisher, Sigma).

Immobilization of Biotinylated Antibodies on Immobilized Streptavidin

The following protocol is a generalized method for the immobilization of biotinylated antibodies on a streptavidin-agarose affinity gel. Some optimization of the loading level on the support may have to be performed to obtain the best performance of the coupled antibody in its intended application. Biotinylation of the desired antibody, if not commercially available, can be done using methods described elsewhere in this book (Chapter 11 and Chapter 18).

1. Wash 1.0 ml of an immobilized streptavidin support with water to remove storage solutions and then with 0.1-*M* sodium phosphate, 0.15-*M* NaCl, pH 7.2 (binding buffer). Washing may be carried out in a small column containing a bottom disk to prevent the gel from escaping. Wash with at least several column volumes. Drain to a wet cake and place the bottom cap on the column to prevent further flow.
2. Prepare a biotinylated antibody solution by dissolving it at a concentration of about 2 to 5 mg/ml in binding buffer. If the amount of antibody being immobilized is far less than this amount, then it is best to use a proportionally lower amount of immobilized streptavidin.
3. Mix the biotinylated antibody solution with the washed streptavidin-agarose and stir to resuspend the gel in the solution. Place a top cap on the column and mix by rotation for 10 to 15 min at room temperature. For a reaction with microliter quantities of immobilized streptavidin, transfer the biotinylated antibody slurry to a small centrifuge tube, seal it, and mix by rotation for the recommended time.
4. Wash the immobilized antibody support extensively with binding buffer to remove any not-bound protein (at least 10 column volumes). After washing, the affinity gel may be stored as a 50% aqueous slurry containing a preservative at 4°C.

2.6. Protein A-, Protein G-, or Protein A/G-Mediated Antibody Immobilization Methods

The immunoglobulin binding proteins protein A, protein G, and protein A/G have been used extensively in immobilized form on chromatography supports for the purification of antibodies. Protein A is a 56-kDa cell wall constituent of the bacterium *Staphylococcus aureus* that contains five identical binding domains, each able to interact with the heavy chains of immunoglobulins in the region of the Fc fragment and in some cases within the region of the Fab fragments (Figure 15.98) (Graille *et al.*, 2000; Idusogie *et al.*, 2000). The recombinant form of protein A is typically truncated to about 45 kDa and is a robust, single polypeptide protein that can be immobilized onto chromatography supports through its lysine amine groups.

Protein G is another immunoglobulin binding protein that originates in group C and G *Streptococcal* bacteria. The native protein is expressed as a 56-kDa or 58-kDa polypeptide that contains multiple binding sites for immunoglobulins as well as a binding site for albumin. The recombinant form of the protein is a truncated version that has the albumin binding site removed, but retains the IgG binding capabilities of the native molecule. Protein G binds to antibodies through the heavy chains in the region of the Fc fragment, but at a different site than that of protein A. The differences in antibody binding between protein A and protein G translate into differential binding specificities and affinities and thus offer options in binding and purifying antibodies, depending on the type of antibody desired.

A chimeric fusion protein consisting of the combination of protein A and protein G, called protein A/G, merges the advantages of both protein specificities into one molecule. Protein A/G often is the immobilized immunoglobulin binding protein of choice for the purification of a wide range of antibody types from various species.

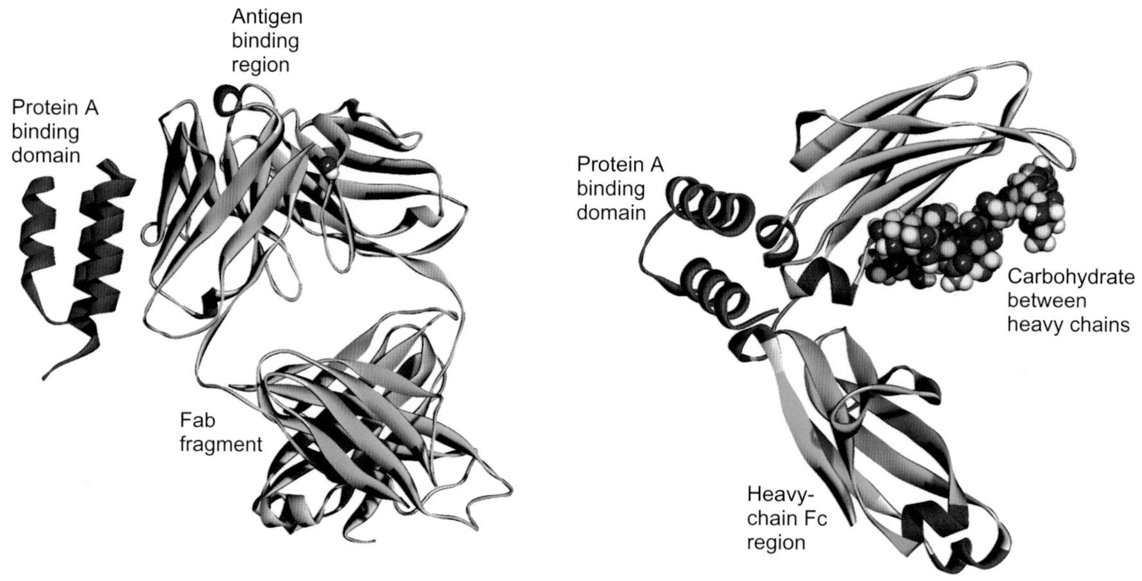

FIGURE 15.98 Three-dimensional molecular model of protein A binding to the Fc and Fab regions in antibodies (Graille *et al.*, 2000; Idusogie *et al.*, 2000; PDB IDs 1DEE and 1L6X.

However, the use of immobilized immunoglobulin binding proteins has been extended beyond just the affinity purification of antibodies. These supports can also be used to bind an antibody for subsequent targeting and immunoaffinity isolation of an antigen for which the antibody is designed to bind. Since most IgG type antibodies bind to protein A, G, or A/G in the region of their Fc fragments on the heavy chains, the antigen binding areas at the ends of the Fab fragments remain open to interact with antigens. This oriented binding of the antibody molecules places them in the ideal position to most effectively use both antigen binding sites and thus maximize the capacity of an immunoaffinity support. Other methods of antibody immobilization may result in almost random orientations of the antibody attached to the support, thus blocking some antigen binding sites by the antibody being positioned with the Fab ends facing the matrix instead of pointing out from it.

Schneider *et al.* (1983) described the use of immobilized protein A for the secondary immobilization of IgG type antibodies. Since the initial interaction of the antibody with protein A is potentially reversible just by changing the buffer conditions, this technique also uses a crosslinking agent to covalently trap the antibody on the support after binding to the protein A molecules. After applying the desired antibody to the protein A support and washing off excess not-bound material, the homobifunctional reagent DMP (dimethyl pimelimidate; see Chapter 5, Section 2.2) is incubated with the bound antibody–protein A complex. The imidoester ends of this crosslinker effectively locks the antibody onto the support by reacting with lysine amines on both the protein A and the antibody and forming covalent amidine bonds (Figure 15.99). This method has become quite popular but unfortunately it has one significant shortcoming: the amidine bonds formed from the reaction of the imidoesters with lysine amines are unstable and continually break down and leach antibody. This results in the presence of antibody in any isolated antigen preparations, which negatively affects the purity of most IP or co-IP experiments done.

Pierce (now Thermo Fisher) developed a modification of the Schneider method that solved the problem of continually leaching antibody. Instead of using DMP with its imidoester reactive groups to lock in place the antibody–protein A interaction, the homobifunctional crosslinker DSS (disuccinimidyl suberate; see Chapter 5, Section 1.2) is used. The NHS ester ends of DSS are much more able to effectively crosslink and stabilize the protein A–antibody interactions, and since the resultant linkages involve the formation of amide bonds the stability of the immobilized antibody is excellent. Essentially no leaching of antibody molecules is observed using this technique for IP or co-IP experiments. The crosslinker also stabilizes the antibody itself from breaking off the support and releasing light or heavy chains during the elution process from an experiment. In addition, this same crosslinking procedure can be used with immobilized protein A, protein G, or protein A/G supports to create immunoaffinity resins that are useful for the capture of any targeted protein or other antigen molecules from a complex biological solutions.

The downside of using an immunoglobulin-binding protein to immobilize an antibody with a crosslinker is that there may be nonspecific binding potentially

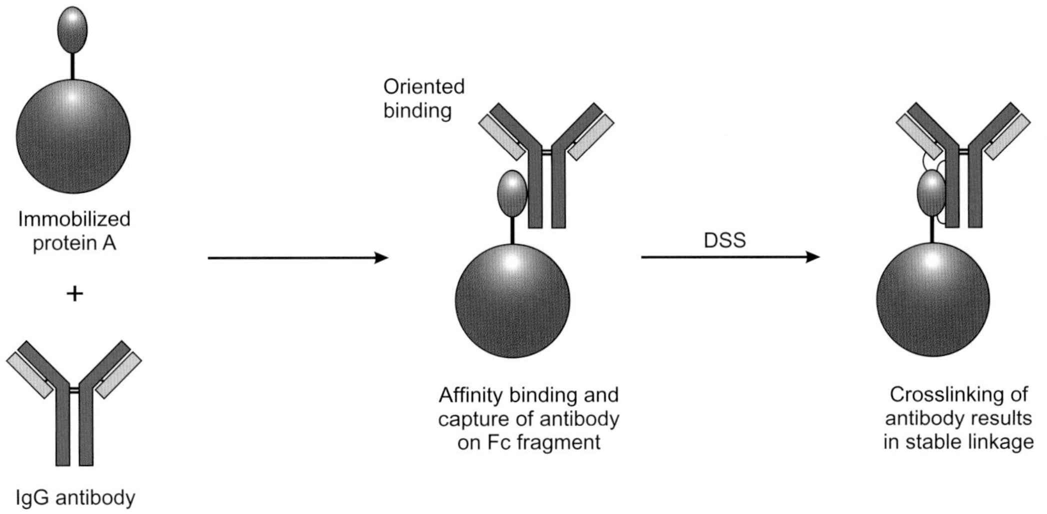

FIGURE 15.99 An immobilized protein A support can be first loaded with antibody through noncovalent affinity interactions with the Fc region of an IgG and then crosslinked with a homobifunctional reagent to stabilize the complex.

created by the presence of the other components besides the antibody. Protein A, for instance, may have a tendency to interact with unwanted proteins in a sample, or the presence of crosslinker modifications may cause some additional nonspecificity especially if the cross-bridge is somewhat hydrophobic. In particular, samples containing immunoglobulins should not be used with these supports, because the protein A on the matrix still has capacity to bind considerable IgG even with an antibody crosslinked onto its surface. Although the Schneider method and its various derivations are popular techniques to prepare an immunoaffinity support, if nonspecific binding is observed it may be advantageous to explore a method that allows direct coupling of the antibody to an amine-reactive, thiol-reactive, or carbonyl-reactive support as described previously in this chapter. Direct coupling would avoid any potential for nonspecific binding due to the protein A or the crosslinker used to stabilize the linkage.

Antibodies immobilized using a crosslinking process onto immunoglobulin binding supports have been used for many studies involving IP and co-IP procedures. Rubenstein *et al.* (2011) used an immobilized primary antibody on protein A/G–agarose to investigate the regulation of endogenous ENaC functional expression by CFTR (cystic fibrosis transmembrane conductance regulator) in airway epithelial cells. Qian *et al.* (2011) coupled a primary antibody to immobilized protein G on agarose to study the regulation of the alternative splicing of tau exon 10 by SC35 and Dyrk1A.

Immobilization of Antibodies on Protein A/G Supports

The following protocol is a generalized method for the binding and crosslinking of an IgG type antibody (polyclonal or monoclonal) onto an immobilized protein A/G support. Avoid the use of Fab fragments, as these will not interact as effectively with the protein A/G on the matrix surface, even though many Fab fragments do have a protein A binding site. Protein A/G is used instead of protein A due to its property of being a fusion protein and containing the combined properties of both protein A and protein G, thus providing excellent binding potential toward the widest variety of antibody species and subclasses. This procedure is adjusted to be appropriate for the immobilization of 100 to 200 μg of IgG antibody onto 100 μl of immobilized protein A/G; however, the amount of resin prepared may be scaled up or down by proportionally changing the quantity of each reagent used. The protein A/G support material may be obtained commercially (e.g., Thermo Fisher) or it can be prepared by reacting 5 to 10 mg protein A/G per ml of an amine-reactive support. See the section in this chapter on amine-reactive immobilization methods for additional information.

1. Wash 100 μl of immobilized protein A/G with several volumes of water and then equilibrate the gel with coupling buffer (0.1-*M* sodium phosphate, 0.15-*M* NaCl, pH 7.2). The washing steps may be done in an appropriately sized drip column or spin column. If a spin column is used, avoid centrifuging at higher than about 100 to 300× *g* to prevent gel damage or clumping. Drain to a wet gel cake and place the bottom cap onto the column.
2. Prepare 100 μg of the antibody to be coupled at a concentration of 1 μg/μl in 100 μl of coupling buffer. If the antibody is already in solution in another buffer, dialyze or buffer exchange it into coupling buffer using size exclusion chromatography.

3. Mix the antibody solution with the immobilized protein A/G support with stirring to resuspend the gel. Place the top cap on the column or alternatively, transfer the slurry to a small tube and seal the tube with a cap. Mix the reaction medium with end-over-end rocking for 30 to 60 min to ensure complete binding of the antibody to the immobilized protein A/G.
4. Wash the support with 3 to 5 bed volumes of coupling buffer to remove any not-bound antibody and drain the gel to a wet cake.
5. Weigh 2 mg of the crosslinker DSS into a microfuge tube and dissolve it in 217 µl of dry DMSO or DMF to prepare a 25-m*M* solution. Make a 1:9 dilution with solvent by taking 20 µl of the DSS solution and diluting it with 180 µl of additional solvent to result in a 2.5-m*M* solution.
6. Add 300 µl of coupling buffer to the washed gel cake and mix to resuspend the gel. With mixing, add 100 µl of the diluted DSS solution to the gel slurry. The total slurry volume will be 500 µl, including the hydrated matrix. This mixture makes a final DSS concentration of 500 µ*M* in the reaction, which will be sufficient to crosslink and stabilize the antibody onto the protein A/G attached to the resin.
7. React with constant mixing for 30 to 60 min at room temperature.
8. After the reaction is complete, wash the immunoaffinity resin with an elution buffer designed to remove antibodies from protein A/G affinity supports. A suggested buffer is 0.1-*M* glycine, pH 2.8. The acidic pH will break any remaining noncovalent protein A/G–antibody interactions and thus elute off non-crosslinked species. Wash with at least 5 column volumes of elution buffer to fully remove non-crosslinked antibody.
9. Wash the immunoaffinity support with water, 1-*M* NaCl, and water to completely remove the remaining buffers and reactants. The affinity gel may be stored as a 50% aqueous slurry containing a preservative at 4°C.

2.7. Reactive Hydrogen-Mediated Immobilization Methods

The methods commonly used to immobilize proteins or other biomolecules as well as couple organic compounds to chromatography supports usually relate to the covalent attachment of the ligand through a functional group, which most often includes amines, thiols, carboxylates, aldehydes, ketones, and occasionally hydrazides or aminooxy groups. However, sometimes a ligand does not contain any of these standard functional groups to facilitate easy immobilization using the activation methods thus far described in this chapter. In such cases, either the ligand must be derivatized to provide a functional group appropriate for coupling or another option must be used to facilitate immobilization.

One potential route to coupling ligands that do not have this common set of functionalities is to use a reactive hydrogen-mediated method, which targets certain replaceable hydrogen atoms within the compound's structure. These methods may work for compounds containing reactive aromatic hydrogens that may be present within the ligand structure, as is often the case with certain drugs, steroidal compounds, dyes, or other aromatic organic compounds.

The following sections describe two reaction strategies that might be used to immobilize ligands having none of the standard functional groups, as described in the previous sections of this chapter.

Diazonium Activation

Diazonium reactions have been used for many years in organic synthesis as well as for protein modification, crosslinking, and immobilization procedures (Higgins and Fraser, 1952; Phillips *et al.*, 1965; Inman and Dintzis, 1969; Cuatrecasas, 1970). Some of the earliest dyes contained diazo groups within their structure, which were formed from the reaction of a diazonium intermediate with a reactive hydrogen on another aromatic compound. The preparation of reactive chromatography supports containing diazonium groups can be done using an intermediary *p*-aminobenzylalkyl group, which in turn is created from the prior immobilization of a *p*-nitrophenyl derivative. A sequence of reactions is needed to form the required intermediate aminophenyl groups and then activate them to the diazonium derivatives. For immobilization reactions on chromatography supports, the diazonium groups need to be made immediately before use due to their high reactivity and instability.

Two routes to the formation of a reactive diazonium group on a support are shown in Figure 15.100. In one option, an amine-terminal spacer arm on a support is reacted with *p*-nitrobenzoyl chloride (Sigma-Aldrich) under nonaqueous conditions to create a nitrophenyl intermediate. In the second reaction option, the modification reagent *N*-succinimidyl-*p*-nitrophenylacetate (SNPA; Bachem) is coupled to the amine-terminal spacer arms on the support to form an amide bond linkage, which terminates in the nitrophenyl groups. In both options, the nitro groups are then reduced to aromatic amines through treatment with sodium dithionite in aqueous solution to form the necessary aminophenyl derivatives. This intermediate is stable for long-term storage until the support is needed to immobilize an affinity ligand. Just before the coupling reaction, the support is activated using an ice-cold solution of

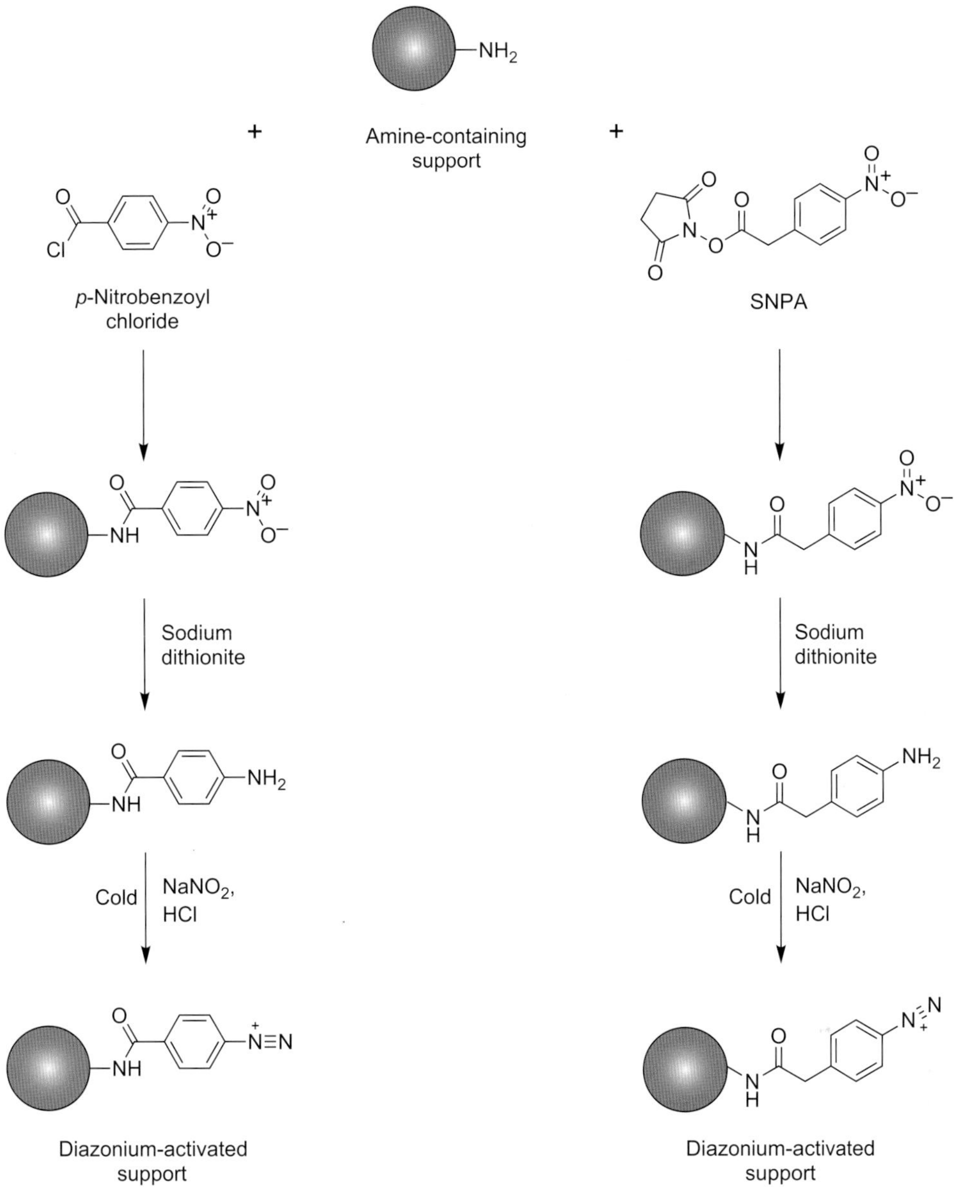

FIGURE 15.100 Modification of an amine-containing support with *p*-nitrobenzoyl chloride or SNPA with subsequent reduction with sodium dithionite to form aryl amines. These intermediates can then be treated with sodium nitrite in cold, acidic conditions to create reactive diazonium groups.

sodium nitrite in HCl to create the reactive diazoniums from the aminophenyl groups. The reaction is done in the cold to prevent the diazonium group from immediately reacting with water to form phenol and nitrogen gas.

Diazonium groups are relatively unstable and should be used immediately to couple a ligand. The rate of diazo bond formation is extremely rapid with ligands, but even if no ligand is added to the support, diazo bond formation can be observed to occur within the matrix due to intramolecular crosslinks. This is due to the reaction of a diazonium group with the aminophenyl precursor molecules and will result in extensive crosslinking and inactivation of the activated support within about an hour after activation. For this reason, ligand should be added quickly to limit the potential for the crosslinking side reactions.

Histidine and other imidazole-containing compounds can be coupled to a diazonium-activated support at pH 8, while tyrosine and phenolic compounds are best immobilized in the range of pH 8 to 10. Figure 15.101 shows the reactions associated with ligand immobilization to create the final diazo linkage with the support. Ligands coupled using this method often result in a highly colored resin due to the presence of the diazo bonds. During the reaction, the support

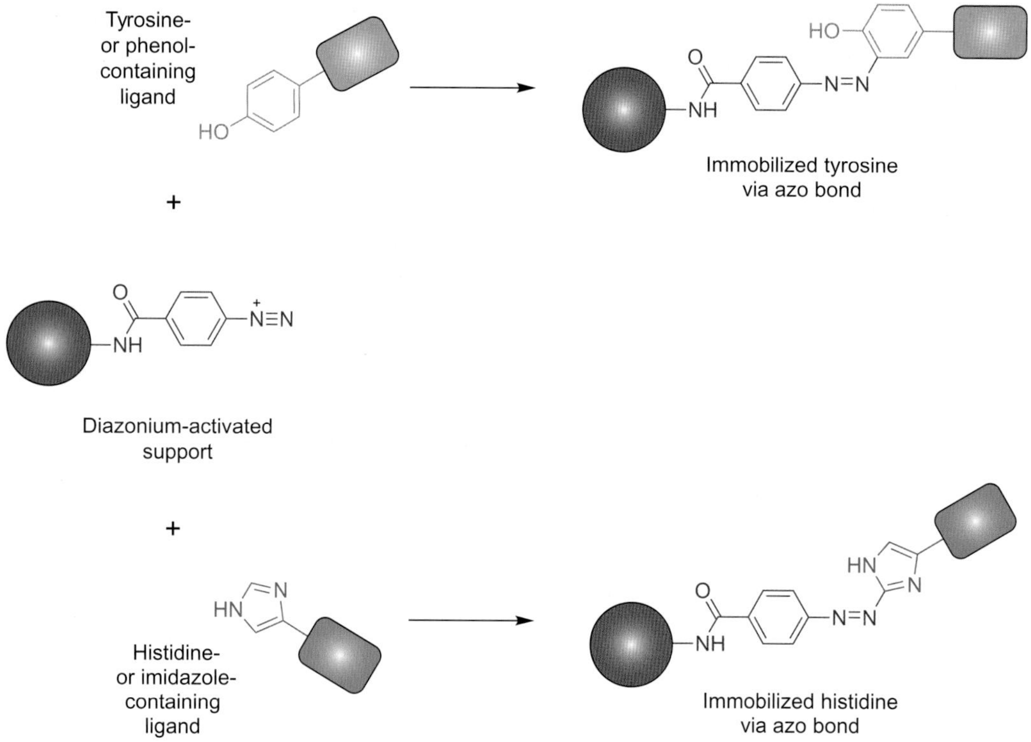

FIGURE 15.101 Immobilization of histidine (imidazole) or tyrosine (phenolic) ligands onto diazonium supports.

can turn dark-brown or even black as the ligand molecules are coupled. This color is normal and should not be viewed as unusual or a problem for subsequent use of the support for affinity separations. The diazo bonds also can be cleaved resulting in bond breakage and release of the immobilized ligand from the support. Treatment with 0.1-*M* sodium dithionite in 0.2-*M* sodium borate, pH 9.0, results in complete bond cleavage as evidenced by the disappearance of the diazo bond color within the support.

PREPARATION OF DIAZONIUM-ACTIVATED SUPPORTS

The following methods assume that an amine-containing support has previously been prepared according to the procedures outlined in the last section of this chapter. All operations should be done in a fume hood while wearing the appropriate personal protective equipment to prevent contact with chemical compounds and solvents.

(A) Preparation of Aminophenyl Groups on the Support

1. Wash 10 ml of an amine-containing resin with water to remove any storage solutions and then wash the support into DMF (dry) using sequentially increasing concentrations of DMF in water (e.g., 30%, 60%) until 100% DMF is used. Washing can be performed using a sintered glass filter funnel suspended in a vacuum filter flask in the fume hood. Continue to wash with at least 10 to 20 bed volumes of DMF until all remaining water is removed. Drain the support to a moist cake.
2. Prepare a solution of SNPA (Bachem) in DMF by dissolving 0.3 g of the modification reagent in 10 ml solvent. Mix thoroughly to dissolve.
3. With stirring, add the gel cake to the SNPA solution and mix the support to create a uniform suspension.
4. React with constant mixing for 1 h at room temperature. The mixing may be carried out in a sealed centrifuge tube by rotating it on an end-over-end mixer.
5. Wash the support extensively with DMF to remove excess reactants and reaction byproducts (at least 10 column volumes). Next, sequentially wash the support back into water by using decreasing concentration of DMF in water until pure water is used. Continue to wash with water (at least 10 column volumes) and then wash the support with several volumes of 0.1-*M* sodium borate, pH 9.0. Drain to a moist cake.
6. Prepare a solution of sodium dithionite in 0.1-*M* sodium borate, pH 9.0, by dissolving 1.2 g in 10 ml of buffer. Mix thoroughly to dissolve.
7. Add the moist gel cake to the dithionite solution with mixing. The dithionite will reduce the nitrophenyl

groups on the support to aminophenyl groups, the precursors used to form diazonium groups. React for 1 h at room temperature.

8. Wash the support thoroughly with 0.1-*M* sodium borate, pH 9.0, and then with water to remove all remaining reactants. The gel may be stored at this point before activation and coupling of ligand if desired. Storage should be done at 4°C as an aqueous slurry in the presence of a preservative.

(B) Activation to Form Diazonium Groups and Coupling of Ligands

Note that the following activation steps should be performed using ice-cold reagents and the reactions performed at 4°C. Do not allow the activated support to sit around for any length of time before coupling ligand, as the diazonium groups will quickly degrade and the support will lose coupling capacity. For this reason, the ligand solution (step 6) should be prepared prior to beginning the activation procedure.

1. Wash 10 ml of the aminophenyl-support prepared in (A) with ice-cold water and then with several bed volumes of cold 0.3-*N* HCl. Finally, suspend the gel in 10 ml of cold 0.3-*N* HCl using a 50-ml centrifuge tube or small vessel that can be sealed. Maintain the solution on ice.
2. Prepare 2.5 ml of a solution consisting of sodium nitrite ($NaNO_2$) in cold water by dissolving it at a concentration of 50 mg/ml.
3. Add the sodium nitrite solution to the gel slurry with stirring.
4. React with mixing for 15 min at 4°C on ice.
5. Quickly wash the activated gel with several bed volumes of cold 0.3-*N* HCl, cold water, and then with cold coupling buffer (0.1-*M* sodium phosphate, pH 8.0, for coupling histidine/imidazole-containing ligands or 0.1-*M* sodium borate, pH 9–10, for coupling tyrosine/phenolic compounds). Drain the gel to a moist cake.
6. Prepare a ligand solution in the chosen coupling buffer (cold) at a concentration of 3 to 10 mg/ml for proteins or 3 to 5 mg/ml for small peptides or organic molecules. Some optimization of the ligand concentration should be carried out to obtain the best performance in the intended application. For ligands which are not very soluble in aqueous solution, the borate coupling buffer may be made with up to 50% ethanol (v/v) to aid in solubility. The ligand may first be dissolved in ethanol and then diluted in coupling buffer to promote solubility in the reaction medium.
7. Immediately add the ligand solution to the washed, activated support with stirring to resuspend the gel. React overnight at 4°C in a sealed container with constant mixing (e.g., end-over-end rocking).
8. Wash the affinity support with coupling buffer (containing ethanol if that was used to aid in solubility during the reaction) and then with water (again containing ethanol, if used), and finally with water alone. The affinity support can be stored until use at 4°C as a 50% aqueous slurry in the presence of a preservative.

Mannich Condensation

The methods available to immobilize molecules are highly diverse and well characterized for ligands that contain a common functional group, which can easily be targeted and covalently linked to a support. Ligands that have at least one amine, carboxylate, aldehyde, ketone, thiol, or hydroxyl group can be coupled to a reactive solid support using one or more of the appropriate reactions discussed previously in this chapter. However, for molecules that possess no such functionalities the route toward successful immobilization may not be as clear. In some cases, ligands such as drugs, steroidal compounds, inhibitors, dyes, or other organic molecules simply do not have a convenient functional group for linking to a support. In other cases, functional groups that are present may have low reactivity or be sterically hindered to allow efficient coupling to an activated matrix.

Frequently, however, such difficult-to-immobilize molecules may contain reactive hydrogens (i.e., replaceable) within their structures to facilitate coupling in a Mannich condensation procedure. The classic Mannich reaction involves the condensation of formaldehyde (or another aldehyde compound) with ammonia (as its salt), and a third compound containing an active hydrogen. The product of this reaction, usually performed under acidic conditions, involves the replacement of the active hydrogen with the methylene group from formaldehyde, and then attached to this is an amine, which originates from the ammonia. Instead of using ammonia, the reaction can also be performed using an amino group (primary or secondary) to achieve the linkage of the amine-containing compound with the active hydrogen-containing molecule through the methylene (CH_2) bridge. An example of this reaction using acetophenone, formaldehyde, and an organic amine salt proceeds as follows (the active hydrogens shown in bold text):

$$C_6H_5COC\mathbf{H_3} + CH_2O + RNH_2 \cdot HCl \rightarrow C_6H_5COCH_2{-}CH_2{-}NH{-}R \cdot HCl$$

Hermanson *et al.* (1992) first described the use of the Mannich reaction to immobilize ligands that did not have any common functional groups, but had sufficiently active hydrogens to participate in this reaction. Chapter 19, Section 6, also describes the use of this reaction for the bioconjugation of active hydrogen-containing haptens to carrier proteins to produce immunogens suitable for immunization. The same reactions can be used to immobilize ligands onto amine-containing supports using formaldehyde as the third reactant.

The use of the Mannich reaction to couple active hydrogen-containing ligands assumes the prior preparation of a support that contains spacer arms that terminate in the primary amino groups required to participate in the process (see the final section in this chapter for spacer arm preparation). The coupling reaction is driven in aqueous buffer by mildly acidic conditions and moderate heat, which most ligands of this type can tolerate without degradation. In addition, since one component of the reaction is already coupled to the support, no potential is present for uncontrolled polymerization of the ligand in solution, as is often the case with molecules containing more than one reactive hydrogen undergoing Mannich condensation. Figure 15.102 illustrates the immobilization of a steroidal molecule onto an amine-containing support using this reaction process. The exact point of coupling to the ligand will be the most reactive or most replaceable hydrogen; however, there may be more than one orientation of ligand attachment if more than one active hydrogen site is present.

A modification of a Mannich coupling reaction for biomolecules was made by Joshi *et al.* (2004). This reaction uses aniline as the amine component similar to the methods used to enhance oxime or hydrazone formation (see previous sections, this chapter). However, instead of being a catalyst as it is in the reactions of aldehydes with hydrazides or aldehydes with aminooxy compounds, in the Mannich reaction aniline actually becomes incorporated into the final product. It first reacts with the formaldehyde (or another aldehyde) component to create an intermediate imine, which then goes on to specifically react with phenolic molecules in solution at positions *ortho* or *para* to the –OH group (Figure 15.103). The result is much more efficient end product formation even at room temperature than the standard Mannich reaction procedure, which is a benefit for coupling sensitive biomolecules.

This bioconjugation method of aniline-promoted Mannich condensation may also be extended to the immobilization of phenolic molecules by using immobilized aniline as the reactive group on the support. An

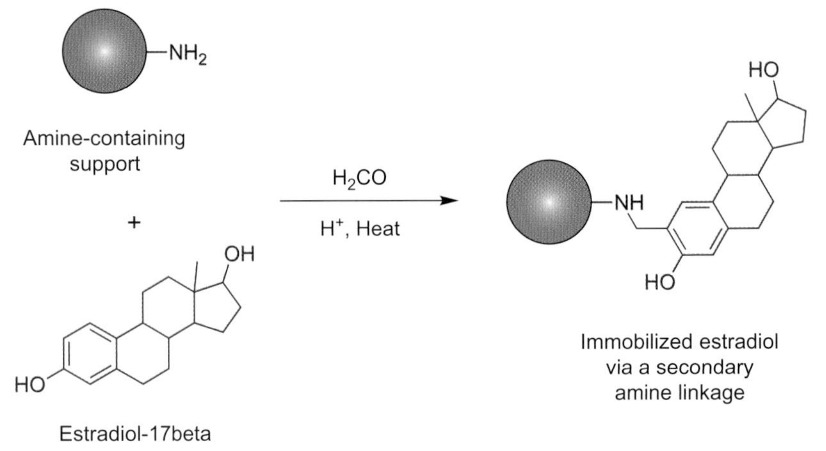

FIGURE 15.102 Immobilization of estradiol-17beta onto an amine-containing support using the Mannich reaction.

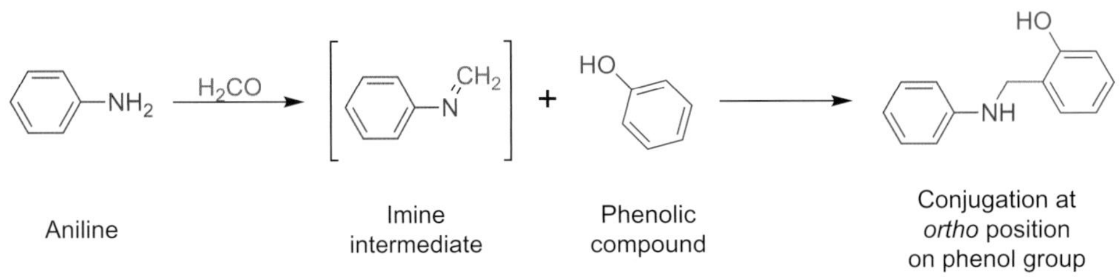

FIGURE 15.103 Mannich condensation reaction using aniline.

aniline-modified support was prepared in the previous section as a precursor to a diazonium-activated support. That same aniline intermediate may also be used in the Mannich reaction to couple with active hydrogens on phenolic compounds for the preparation of affinity chromatography supports in a mild reaction process. Indeed, even proteins or peptides containing a tyrosine residue may be coupled specifically through the phenolic side chain using this reaction without affecting any other functional group on the molecule. The only requirement for proteins to efficiently participate in this reaction is that the tyrosine groups must be surface accessible. The reaction proceeds at room temperature and is optimal at a pH range of 5.5 to 6.5, which is mild for most biomolecules. Figure 15.104 illustrates the immobilization of a tyrosine-containing peptide on a support containing aniline groups using the Mannich reaction.

Therefore, using the Mannich reaction two potential reaction paths are possible: (1) coupling of active-hydrogen containing molecules onto aliphatic amine-containing supports when a ligand contains no other available common functional groups to facilitate coupling; and (2) immobilization of tyrosine-containing proteins or peptides onto aniline-containing supports specifically through tyrosine's phenolic side chain while avoiding the reaction of other functional groups. These choices make the Mannich reaction a powerful alternative for the immobilization of certain affinity ligands.

Because of the versatility of the Mannich condensation method of immobilization it is being used more frequently for reactive hydrogen-containing ligands. Pyell and Stork (1992) used the method to prepare immobilized 8-hydroxyquinoline on an aminopropyl silane-modified silica support for use as a chelating agent. In addition, the phenol derivatives 4-(2-pyridylazo) resorcinol, 8-hydroxyquinoline, and 1-(2-pyridylazo)-2-naphthol were immobilized onto an amino-silica particle using the Mannich reaction (Tertykh *et al.*, 2000), while Pu *et al.* (1998) used a similar reaction process to couple 2-mercaptobenzothiazole to aminopropyltriethoxysilane-modified silica gel. In addition, Zhang *et al.* (2009) used the Mannich reaction to immobilize phenolphthalein onto an amine-modified polyacrylonitrile fiber for use as a halochromic fiber, which would change color with rapid response time depending on the pH of the environment. Members of this same group also used the Mannich reaction to produce a heavy-metal-detection fiber by coupling 4-(2-pyridylazo)-1,3-benzenediol onto ethylenediamine-modified polyacrylonitrile fibers (Li *et al.*, 2010).

COUPLING LIGANDS *VIA* MANNICH CONDENSATION

The following protocols describe the use of either an aliphatic amine-containing matrix to immobilize active-hydrogen-containing ligands or an aniline-containing support to couple tyrosine-containing (or phenolic) ligands. Both reactions involve the use of formaldehyde and should be carried out in a fume hood using the appropriate personal protective equipment to prevent contact with reactants or solvents.

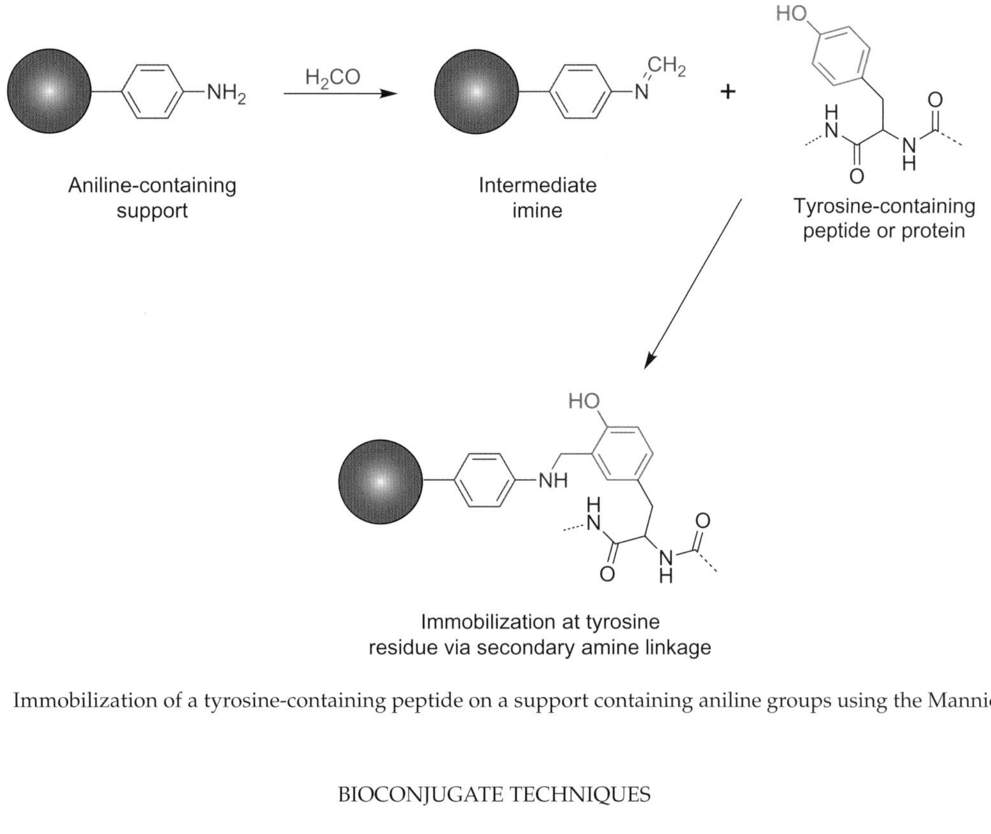

FIGURE 15.104 Immobilization of a tyrosine-containing peptide on a support containing aniline groups using the Mannich reaction.

(A) Mannich Reaction with Active Hydrogen-Containing Ligands

This protocol assumes the prior preparation of an aliphatic amine-containing support as described in the last section of this chapter on the modification of supports with spacer arms.

1. Wash 10 ml of a primary amine-containing support with water to remove any storage solution and then with 0.1-*M* MES, pH 4.7 (coupling buffer). Examples of primary amine-containing spacer arm gels that work well in this process include MANAE–agarose and DADPA–agarose. Wash with at least 3 to 5 bed volumes of the solutions. The washing steps may be carried out in a small sintered glass filter funnel suspended in a suction filter flask. After washing, drain the support to a moist cake.
2. Dissolve the ligand containing an active hydrogen site in 10 ml of coupling buffer. If the ligand is relatively insoluble in aqueous buffer, it may first be dissolved in ethanol and then an aliquot added to the final buffer solution, making the total volume up to 10 ml; however, the final concentration of ethanol in the MES buffer should not exceed 50% to avoid buffer salt precipitation.
3. Mix the ligand solution with the washed amine-support and stir to resuspend the gel. Place the slurry in a sealable container and in a fume hood add to it 1.0 ml of 37% formaldehyde solution with mixing. Seal the container.
4. React at 37 to 57°C with constant mixing for a minimum of 24 h. For compounds with a slower reaction rate in the Mannich reaction, a longer reaction time or a higher temperature may be required for effective coupling yields. Some degree of optimization may have to be done to determine the best reaction conditions for a given molecule.
5. Wash the support extensively with coupling buffer and then with water (both containing ethanol, if used) to remove excess unreacted ligand. If the ligand is especially soluble in ethanol, a wash with 100% ethanol may be done after the water/ethanol wash to completely remove the last traces of free ligand. Finally, wash sequentially back into water and then with pure water. The affinity support may be stored as a 50% aqueous slurry at 4°C containing a preservative.

(B) Mannich Reaction with Tyrosine-Containing Ligands

This protocol assumes the prior preparation of an aniline-containing support as described in the previous section on diazonium activation. Note that an alternative approach to creating an immobilized aniline intermediate is to immobilize 2-(4-aminophenyl)ethylamine directly onto an amine-reactive support through the aliphatic amino end. See the previous sections in this chapter on the preparation and use of amine-reactive supports for information on how to prepare this type of aniline-containing resin.

1. Wash 10 ml of an aniline-containing support with water to remove any storage solution and then with 0.1-*M* MES, pH 6.0 (coupling buffer). The washing steps may be done in a small sintered glass filter funnel suspended in a suction filter flask. After washing, drain the support to a moist cake.
2. Dissolve the ligand containing a tyrosine residue (i.e., protein, peptide, or a phenolic compound containing a free *ortho* or *para* position relative to its –OH group) in 10 ml of coupling buffer. If the ligand is relatively insoluble in aqueous buffer, it may first be dissolved in ethanol and then an aliquot added to the final buffer solution, making the total volume up to 10 ml; however, the final concentration of ethanol in the MES buffer should not exceed 50% to avoid buffer salt precipitation.
3. Mix the ligand solution with the washed aniline-support and stir to resuspend the gel. Place the slurry in a sealable container and in a fume hood add formaldehyde to make a final concentration of 25-m*M*. Mix well and seal the container.
4. React at room temperature (i.e., 20–25°C) with constant mixing for a minimum of 24 h.
5. Wash the support extensively with coupling buffer and then with water (both containing ethanol, if used during the coupling reaction) to remove excess unreacted ligand. If the ligand is especially soluble in ethanol, a wash with 100% ethanol may be done after the water/ethanol wash to completely remove the last traces of free ligand. Finally, wash sequentially back into water and then with pure water. The affinity support may be stored as a 50% aqueous slurry at 4°C containing a preservative.

2.8. Creating Spacer Arms on Chromatography Supports

The most common route of preparing an affinity chromatography support involves the direct activation of a solid-phase material followed by the coupling of the affinity ligand. Most of the methods described in the previous sections of this chapter use this path to ligand immobilization. Occasionally, however, it is useful to create an intermediate derivative on the support that includes the formation of a spacer arm, which might terminate in a functional group useful for subsequent reactions. Spacers are typically low-molecular-weight molecules that are most often linear and consist of alkyl or hetero-alkyl components that extend out from the matrix from about 2 to 20 atoms in length. Spacer arms can serve multiple purposes on chromatography supports. They can simply be a bridge between an initial functional group or reactive group on a support and the creation of another needed functionality at the other end of the spacer. They also can provide a long tether that extends the affinity ligand out from the matrix surface, thus providing greater steric accommodation for target molecules that may interact or dock with it. This is an especially important feature if the binding site for a ligand is a deep pocket within the target protein, such as is the case for avidin or streptavidin in binding to immobilized biotin. Even with much larger ligands, however, spacers can be used to extend the molecule away from the matrix surface for better accessibility to bind targets. Extremely long spacers such as PEG-based compounds can provide much greater freedom of motion for immobilized proteins, thus allowing greater potential for interactions to occur within the support. This may be especially important if an immobilized protein has more than one binding site for a target molecule, as is the case for many immunoglobulin-binding proteins (i.e., protein A, protein G, and protein A/G) or in the case of avidin or streptavidin used to capture biotinylated proteins.

Spacers can also create additional properties on a support, such as the introduction of increased hydrophilicity that can reduce nonspecific binding to the final affinity resin. An early report recognized the benefit of using hydrophilic linker arms in the design of affinity chromatography supports (O'Carra *et al.*, 1974). Spacers can also consist of branched constructs that create multi-point immobilization sites for an affinity ligand, which can stabilize some proteins or enzymes (e.g., glycidol modified supports, see previous corresponding section, this chapter). If desired, the cross-bridge of a spacer arm can also be chosen to be cleavable to permit chemical release of an affinity ligand after it has bound a target. Sometimes, in the design of an affinity support, it takes one or two spacer arm modifications to create the desired effects and functionalities on the matrix prior to the immobilization of the affinity ligand. In other cases, a spacer is formed during matrix activation or coupling of a ligand, such as in activation methods using a *bis*-epoxide compound or in the immobilization of an amine-containing ligand to an azlactone-activated support (see previous sections, this chapter).

The options available for adding spacer groups onto chromatography supports are numerous and mixing and matching these options can multiply the design choices dramatically. In the most fundamental spacer arm design, there are functional groups on both ends of a molecule that are separated by an aliphatic or hetero-aliphatic chain. Less frequently, there might be aromatic groups present in the bridging portion of the spacer, but this is less common due to the hydrophobicity such groups can add to the final support. Spacers can even be constructed from the use of homobifunctional or heterobifunctional crosslinking agents through the reaction of one end of the crosslinker to a functional group on the surface of the support, which leaves the other reactive group free to couple with a ligand or another spacer molecule. The epoxy activation of a support using the *bis*-epoxide compound 1,4-butanediol diglycidyl ether yields a hydrophilic 12-atom spacer, which terminates in a reactive epoxide group before a ligand is coupled.

It is also important to consider the overall chromatography effects in having a spacer arm in the design of an affinity resin. Some of the potential consequences of an inappropriate linker arm include the introduction of hydrophobic or ion-exchange interactions that can result in binding to nonrelevant molecules in a sample, resulting in lower purity for the desired isolated target molecule. For instance, long aliphatic chains can create considerable hydrophobicity on an otherwise fairly hydrophilic base support. Although spacer arms of long length may seem desirable, increasing the length of an aliphatic spacer may only serve to increase the hydrophobic interaction potential of the final affinity resin. A better alternative would be to choose a hydrophilic spacer instead of a hydrophobic one, which may provide the extended length needed while still maintaining or even improving the hydrophilicity, and therefore low nonspecific binding character, of a support. Similarly, some spacers may create charges on a support from ionized or protonated groups, thus generating the potential for nonspecific binding with oppositely charged molecules in a sample. Secondary amine-containing spacers are particularly notorious for creating positive charge on a matrix, which may lead to undesirable interactions. Spacers containing carboxylates or sulfonates may also generate nonspecific binding characteristics from creating negative charges on the support.

Figure 15.105 shows examples of common spacer molecules used in the preparation of immobilized affinity ligands, which is by no means exhaustive. Many of these molecules contain aliphatic or hydrophobic cross-bridges, which should be used with caution to avoid nonspecific binding issues in the final support. Even the spacers containing only a six-carbon aliphatic bridge can result in considerable hydrophobic interaction potential, even though they are used quite frequently with success in designing affinity supports. For hydrophilicity in the final support, it is important to choose a spacer composition containing polar groups within an alkyl chain, which can reduce or eliminate hydrophobic character. The best choices for water solubility and low nonspecific binding include

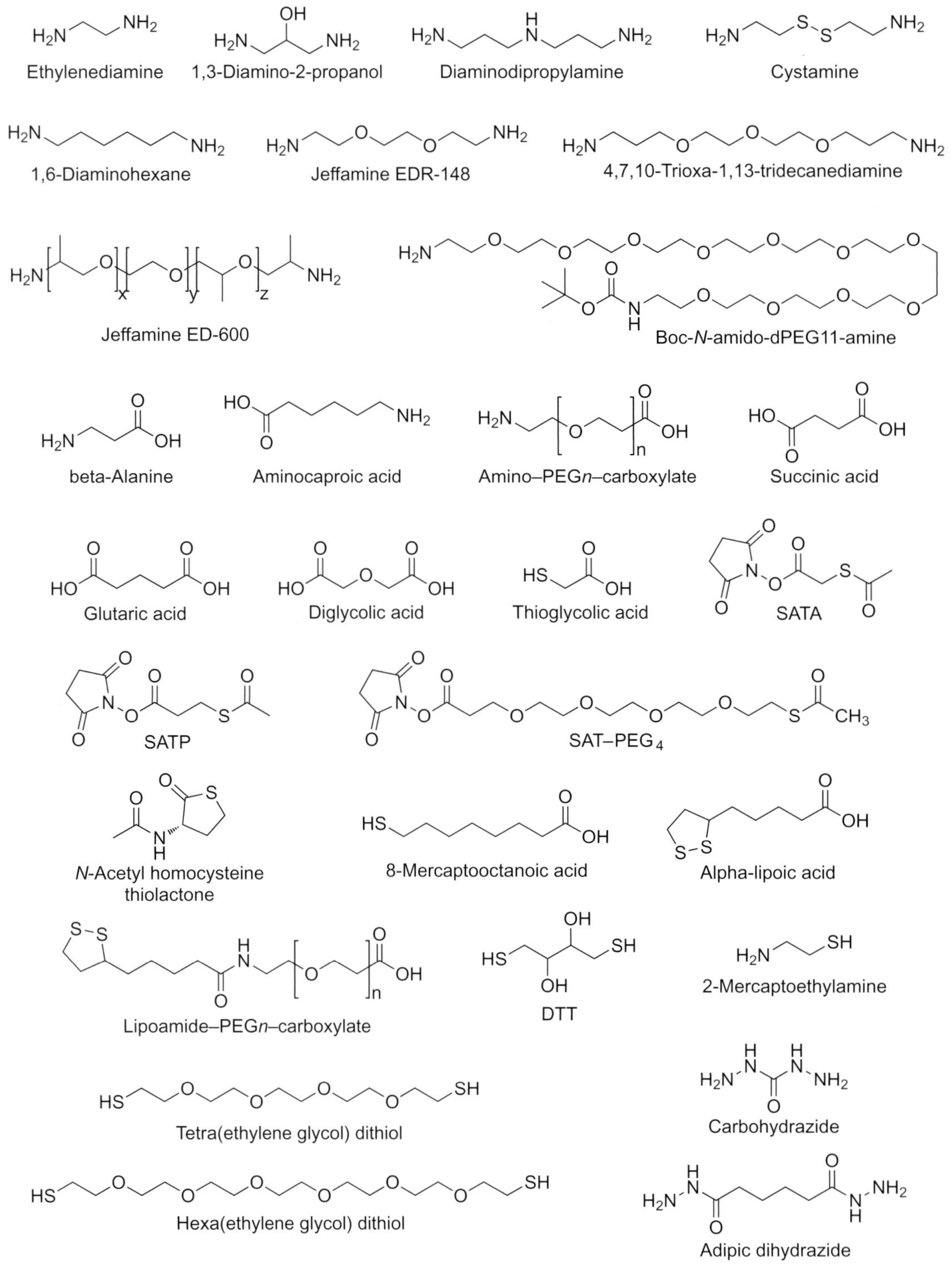

FIGURE 15.105 Small molecules used as spacers in affinity chromatography.

polar components such as secondary amines, carbonyl groups, amide bonds, ether groups, and hydroxyls. If possible, avoid charged groups, such as protonated secondary (or tertiary) amines or carboxylates, which may introduce ion-exchange effects. Perhaps a superior choice in this regard is to use a PEG-based spacer, which avoids both the issue of hydrophobicity and also does not generate within the cross-bridge any charged sites with ion-exchange potential. There are now many discrete PEG-based spacers and crosslinkers available for the design of virtually any immobilized ligand (see Chapter 18). A list of spacers and modification reagents that have been used to create linker molecules on solid phases and categorized by the type of spacer follows.

Diamine Spacers
- Ethylenediamine
- 1,3-Diamino-2-propanol
- DADPA
- Cystamine
- 1,6-Diaminohexane
- Jeffamine EDR-148
- Jeffamine ED-600
- 4,7,10-Trioxa-1,13-tridecanediamine
- Boc-*N*-amido-dPEG$_{11}$-amine
- Boc-*N*-amido-dPEG$_3$-amine

Amino–Carboxylate Spacers
- Beta-alanine
- Aminocaproic acid
- Amino–PEG$_n$–carboxylate compounds

***bis*-Carboxylate Spacers**
- Succinic acid (or succinic anhydride)
- Glutaric acid (or glutaric anhydride)
- Diglycolic acid (or diglycolic anhydride)

Thiol–Carboxylate Spacers
- Thioglycolic acid
- SATA
- SATP
- *N*-Acetyl homocysteine thiolactone
- 8-Mercaptooctanoic acid
- Alpha-lipoic acid
- Lipoamide–PEG$_n$–carboxylate compounds
- Thiol–PEG$_n$–carboxylate compounds
- NHS–PEG$_n$–acetylated thiol compounds (SATA-like)

***bis*-Thiol Spacers**
- DTT
- Tetra(ethylene glycol) dithiol
- Hexa(ethylene glycol) dithiol
- Poly(ethylene glycol) dithiol

Amino–Thiol Spacers
- 2-mercaptoethylamine

***bis*-Hydrazide Spacers**
- Adipic dihydrazide
- Carbohydrazide

Of course, some hydrophobic spacer molecules may be purposely chosen to result in a resin having hydrophobic character designed into the gel for certain separations. Various forms and lengths of hydrophobic linkers can be coupled to a support to create a resin useful for hydrophobic interaction chromatography. In this case, the longer the hydrophobic spacer arm, the greater will be the hydrophobic interaction potential. A 4-carbon chain, for instance, will have much weaker hydrophobic interaction potential with proteins than an 8-carbon chain. The extreme hydrophobicity of an 18-carbon aliphatic chain is often designed into HPLC chromatography resins, which is the length typically used for reverse phase separations.

The following sections describe some of the most popular spacer molecules used to make affinity resins and the reactions that can be used to prepare the derivatized supports. The choice of spacer for a particular immobilization reaction should be done based upon the functional groups required at both ends of the spacer and the length and physical properties of its cross-bridge. To design an optimal affinity resin containing a spacer molecule, it is often best to evaluate a number of spacer arm types to determine which one performs best for a given application. For instance, if there is a need for a diamine spacer in the construction of an immobilized ligand, then it is best if several different diamines are compared to see how changes in spacer structure and length affect the final affinity separation.

Diamine Spacers

Spacer molecules containing an amine at both ends often are used to create a primary amine on a support for further modification or for the coupling of carboxylate-containing ligands. They are also used as an intermediary in the creation of some activated supports, such as in the preparation of an iodoacetyl-support useful for coupling thiol-containing ligands (see earlier section, this chapter). To couple a diamine spacer to a support, an amine-reactive resin should first be created according to the methods described previously. The diamine is then reacted in large excess with the amine-reactive support to result in one end of the spacer coupling to the support and the other end remaining free. To reduce the potential for crosslinking that results in both ends of the spacer being linked to the support, the reaction needs to be performed at a minimum concentration of 0.5-*M* diamine. Alternatively, if the diamine spacer molecule is available with one end protected (one amine blocked), then a much lower concentration can be used for coupling; however, the final product then needs to be deblocked to reveal the terminal amine for further modification. Typical protecting groups for amines include carbobenzyloxy (Cbz), *tert*-butyloxycarbonyl (BOC), and 9-fluorenylmethyloxycarbonyl (FMOC), which allow one end of the diamine to be immobilized without the potential for crosslinking. Removal of the protecting groups is by catalytic hydrogenolysis (Cbz), strong acid such as trifluoroacetic acid (TFA) (BOC), or an organic base such as piperidine (FMOC). Use care to ensure that the deprotection step does not cause adverse effects on the base support. In this regard, deprotection of Cbz groups may not be compatible with porous chromatography supports, as the typical reagent used for this procedure is a particulate Pd/C suspension and hydrogen. The Pd/C particles may get entrapped within the matrix and not be easily removed. A better protecting group would be FMOC, because the addition of a 20% piperidine solution in DMF for the deprotection step would be tolerable for most support materials.

Diamine compounds that are in the free base form (not as salts) are extremely caustic in aqueous solution and upon dissolving will typically give a pH > 11. The dissolving process in water and the subsequent neutralization reaction with acid can be hazardous, as both are exothermic processes. If the dihydrochloride salt of a diamine is commercially available, then the neutralization step is not needed and the diamine typically is supplied as a solid and is much more convenient to work with. To neutralize diamines safely, always use a fume hood and proper personal protective equipment to avoid contact with solutions and inhalation of fumes. To limit the exothermic heat effect, a diamine compound may be first added to an excess of crushed, deionized ice (equal to a little less than the final desired solution volume) and then 6-*N* HCl added slowly with stirring. As the diamine is neutralized, the ice will melt and prevent the solution from becoming extremely hot. Once the diamine solution is neutralized, the appropriate buffer salt may be added and the pH adjusted to the proper value for the coupling reaction being done.

DIAMINODIPROPYLAMINE

One of the more common diamine spacer molecules used in affinity chromatography is diaminodipropylamine (DADPA), also known as 3,3′-iminobispropylamine (Figure 15.106). It is a nine-atom spacer with a central secondary amine and a primary amine at each of the ends. The central secondary amine aids in overall hydrophilicity, but may also be a potential site of positive charge due to protonation at physiological pH. In addition, the secondary amine might participate in some reactions that are targeted at the terminal primary amine, thus complicating the structural nature of the final immobilized ligand. DADPA is liquid at room temperature (d = 0.938 g/ml; mp = −14°C) and volatile with a strong amine odor; therefore, all solutions should be handled in a fume hood. The reagent is soluble in aqueous solution for buffered reactions and it is also soluble in many organic solvents suitable for work with chromatography supports (i.e., DMSO, DMAC, DMF, acetone, dioxane, and ethanol).

Amine-reactive supports may be modified with DADPA to produce spacer arms terminating in a primary amine. When using amine-reactive supports that are labile in aqueous environments (e.g., due to hydrolysis), then the DADPA coupling reaction can be done in organic solvent to eliminate the competing

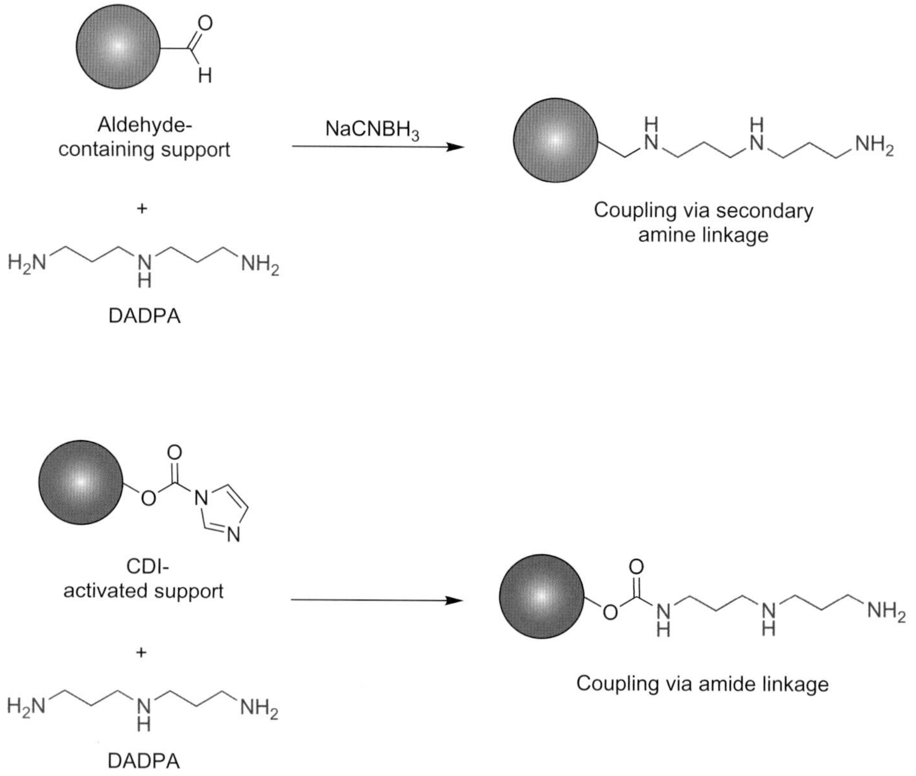

FIGURE 15.106 Coupling the diamine spacer DADPA to periodate-oxidized agarose and CDI-activated agarose.

hydrolysis reaction and maximize yields. DADPA reactions should be done with a large excess of diamine (at least 0.5-*M*) to limit the potential for crosslinking within the matrix and maximize the probability of only one end of each molecule reacting with the support.

Immobilized DADPA has been used to couple carboxylate-containing antigens using the carbodiimide EDC, which resulted in an affinity support for the purification of specific antibodies from immune serum (Tsai *et al.*, 1998). It has also been used as a precursor in the preparation of a maleimide-activated support by reacting the heterobifunctional crosslinker SMCC (Chapter 6, Section 1.3) with the amine on the support to yield a thiol-reactive derivative for the coupling of the rat GIP receptor epitope using a cysteine-terminal peptide. This affinity support was subsequently used to purify antibodies against the intact receptor protein (Lewis *et al.*, 2000). Shenouda *et al.* (2002) similarly used immobilized DADPA to couple a hapten peptide sequence (human urotensin-II) to the support using the EDC amide bond forming method, and the affinity support then was used to purify anti-urotensin-II antibodies from rabbit antiserum. In addition, Kost *et al.* (2011) used DADPA–agarose to couple the CNS-modulating compound clavulanic acid *via* its carboxylate using EDC and subsequently was able to isolate two proteins that specifically interacted with this ligand from neuronal cells.

PREPARATION OF A DADPA-MODIFIED SUPPORT The following protocols represent two methods of coupling DADPA to amine-reactive supports. Analogous methods may be used with other short diamine spacers, such as ethylene diamine or 1,3-diamino-2-propanol. In the first method, a periodate-oxidized agarose support is used in a reductive amination procedure in aqueous solution (see previous section on immobilization by reductive amination, this chapter). In the second protocol, DADPA is coupled to a CDI-activated support in a nonaqueous solution to eliminate the hydrolysis reaction, which would occur with the reactive support in aqueous solution. Both of these methods yield supports containing spacer arms that terminate in free primary amines for further reactions. The CDI coupling procedure will give a higher density of amines on the support, but both work quite well for subsequent immobilization reactions. All operations should be carried out in a fume hood.

(A) Coupling DADPA to Periodate-Oxidized Agarose

1. Wash 100 ml of periodate-oxidized agarose (or another aldehyde-containing support) with water to remove storage solutions and then into coupling buffer (0.1-*M* sodium phosphate, pH 7.2). Washing steps may be performed with a sintered glass filter funnel suspended in a vacuum filter flask. Wash with at least several bed volumes for each wash step. Drain to a moist cake.
2. In the fume hood, add 20 g (21.3 ml) of DADPA to about 100 ml of crushed, deionized ice (use personal protective equipment to prevent contact with the highly caustic diamine and the acid used for neutralization). Slowly add 8 to 10 ml of concentrated HCl to the DADPA/ice slurry with manual stirring. Add the acid slowly using a pipette. As the acid is added, the solution will warm and the ice will melt. After the acid has been added, stir the solution using a magnetic stir bar and continue to neutralize the solution to about pH 7 with acid. Add to this solution a quantity of sodium phosphate buffer salt to make the final concentration 0.1-*M* phosphate when the total volume is diluted to 100 ml. Stir and readjust the pH to 7.2 using acid or base. The final solution is 1.52-*M* DADPA, 0.1-*M* sodium phosphate, pH 7.2. Allow the solution to come to room temperature before continuing.
3. Add the washed periodate-oxidized agarose to the DADPA solution and mix to resuspend the gel.
4. In the fume hood weigh out and add 0.63 g of sodium cyanoborohydride (toxic!) to the gel slurry and mix for 2 to 4 h. Mixing may be done using an overhead paddle stirrer (not a stir bar) or in a sealed plastic container by end-over-end rocking. Avoid the use of sealed glass containers, because there is some gas evolution during the reductive amination process.
5. Extensively wash the DADPA–agarose support with water, 1-*M* NaCl, and water to completely remove unreacted diamine and reaction byproducts. The support may be stored as a 50% slurry containing a preservative at 4°C until use.

(B) Coupling DADPA to CDI-Activated Supports

1. In a fume hood, dissolve 20 g of DADPA (21.3 ml) in 100 ml of dry acetone with stirring (makes a 1.52-*M* DADPA solution).
2. Drain 100 ml of a CDI-activated support prepared in acetone of excess solvent using filtration on a sintered glass filter.
3. Add the drained support as an acetone-wet cake to the DADPA solution with mixing to resuspend the gel.
4. Stir the reaction slurry for 2 to 3 h using an overhead paddle stirrer or in a sealed container by end-over-end rocking.
5. Wash the modified support with 1 l of acetone to completely remove excess diamine. Next, sequentially wash the support into water by using increasing concentrations of water in acetone (e.g., 30%, 60%, and 100%). Continue to wash with water to completely remove the last traces of acetone, and

then with 1-*M* NaCl, and again with water. The support may be stored as a 50% slurry containing a preservative at 4°C until use.

1,6-DIAMINOHEXANE

The aliphatic diamine spacer 1,6-diaminohexane (DAH; also called hexamethylenediamine) has been used in a variety of applications in organic synthesis, including the formation of spacer arms on chromatography supports. DAH consists of a linear six-carbon linker arm with a primary amine on each end, and it is a solid at room temperature (mp 42–45°C) with hygroscopic properties. The crystals of DAH should be white or off-white in color, but occasionally may have a yellow tint to them. These are typically all right to use without further purification; however, avoid highly colored material as they are likely to be oxidized or contaminated. Nearly one billion kilograms of DAH are produced annually as a key ingredient for several polymers. The use of DAH as a spacer arm for the preparation of affinity supports dates to the earliest research on matrix preparation (O'Carra and Barry, 1972).

Since DAH does not contain any other polar constituents in its cross-bridge, it is considerably more hydrophobic than DADPA, discussed previously. For this reason, spacer arms built from DAH on chromatography supports will create more hydrophobic character on the matrix surface, even after an affinity ligand is attached. This can have detrimental effects on the specific binding potential of an affinity support, because hydrophobic molecules in samples may bind through hydrophobic interactions and result in lower selectivity for capturing the desired target molecule. In some cases, however, DAH has been found to be optimal in creating affinity supports with certain ligands immobilized, such as the dye Cibacron Blue 3GA (Suen and Tsai, 2000). In addition, for some applications of immobilized metal affinity chromatography (IMAC) it was found that medium-length hydrophobic spacers created a support with the best binding characteristics toward penicillin G acylase (Liu *et al.*, 2005). Recent novel applications of immobilized ligands in the field of supported photosensitizers (SPS) indicate that even longer aliphatic spacer arms showed better efficiency than those with shorter chains (Pineiro *et al.*, 2010).

DAH has also been used as a type of affinity ligand in the isolation of human IgG from serum and plasma, which involved a one-step purification process using HEPES buffer at pH 6.8 (de Souza *et al.*, 2010). Bayramoglu *et al.* (2006) used DAH as a spacer on polymer beads containing epoxy groups to subsequently couple L-histidine as an affinity ligand for the purification of immunoglobulins.

DAH dissolved in aqueous solution is extremely caustic and should be handled with care. Upon dissolution, the pH should be adjusted to the recommended point for coupling the spacer to an amine-reactive support and maintained with the appropriate buffer. Neutralization should be carried out using HCl and cooled on ice, similar to the procedure outlined in the previous section on DADPA. Alternatively, the dihydrochloride form of DAH can be used, which does not affect the pH of an aqueous solution, nor does it require titration with acid for neutralization.

PREPARATION OF A DAH-MODIFIED SUPPORT The following protocol describes the coupling of DAH to a tresyl- or tosyl-activated support, prepared according to the procedures described previously in this chapter (Figure 15.107). Other amine-reactive supports may also be used, such as the activated supports and general procedures outlined in the section on the spacer DADPA.

1. In a fume hood, dissolve 15.28 g of DAH dihydrochloride in 100 ml of coupling buffer (0.2-*M* sodium phosphate, pH 7.5) and mix to dissolve. Adjust the pH back to 7.5 if necessary. Note: Alternatively, dissolve 10 g of DAH (as the free base) in a minimum quantity (~70 ml) of coupling buffer. Cool the solution in an ice bath. Slowly titrate the pH back down to 7.5 using concentrated HCl (caution: highly so use personal protective equipment!), while maintaining the solution cool to control the heat of the exothermic process. After the proper pH is reached, adjust the total solution volume to 100 ml by the addition of coupling buffer.
2. Drain 100 ml of a tresyl- or tosyl-activated support of acetone using a sintered glass filter suspended in a suction filter flask in a fume hood. Pull a gentle vacuum on the support and break up the matrix into small pieces using a spatula. Do not allow the support to dry. Remove the vacuum and add an equivalent bed volume of water and mix to resuspend the support. Continue to quickly wash the support with 3 bed volumes of water and 2 volumes of coupling buffer. Drain to a moist cake.
3. Mix the washed support with the DAH solution and mix to resuspend the gel.
4. If a tresyl-activated support was used to couple DAH, react with constant mixing for 2 to 4 h at room temperature. If a tosyl-activated support was used, the reaction should be continued overnight at room temperature to realize maximal coupling yields. Mixing may be performed using an overhead paddle stirrer or in a sealed container by end-over-end rocking.
5. Wash the support with coupling buffer, 1-*M* NaCl, and again with water to remove excess uncoupled DAH. The support may be stored as a 50% slurry containing a preservative at 4°C until use.

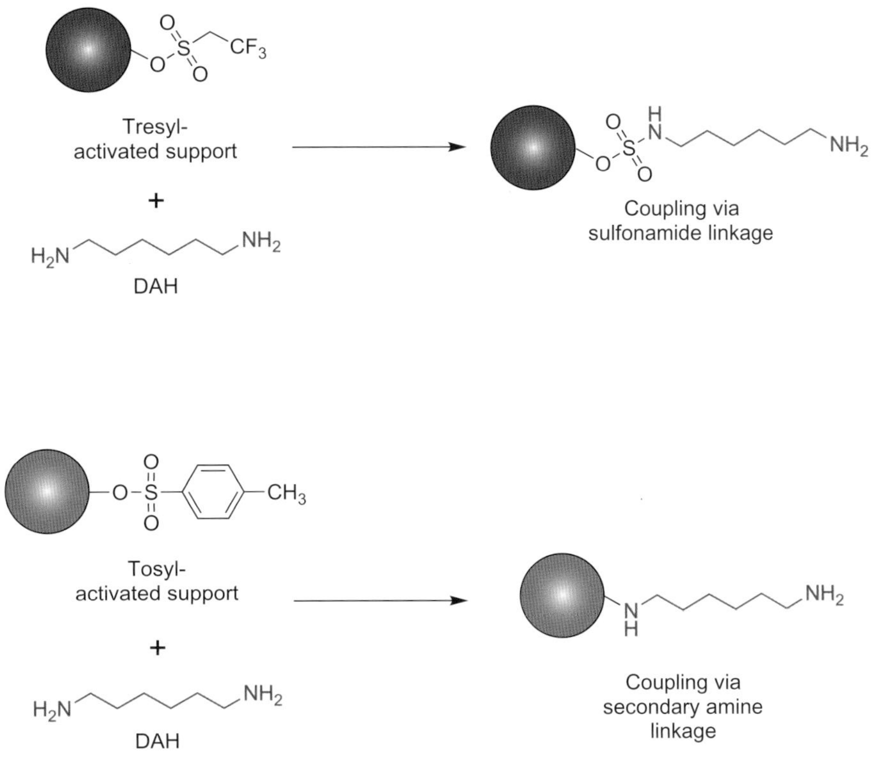

FIGURE 15.107 Coupling the diamine spacer DAH to tresyl- and tosyl-activated supports.

DIAMINE–PEG COMPOUNDS

Diamine spacers that contain a central PEG chain with amines on both ends are among the most hydrophilic bridges that can be created on a chromatography support. There are compounds available in this design that have a broad range of spacer arm lengths, from short PEG chains to those with high-molecular-weight long polymer chains. The discrete, short PEG compounds are useful for building amine functionality on a support while maintaining a hydrophilic and low nonspecific binding character on the base matrix. Longer PEG spacers might be chosen to further mask a more hydrophobic base support by developing a polyether coating that terminates in an amine. For the development of affinity supports for use with biomolecules, the choice of a diamine–PEG spacer may be an excellent starting point for high purity target molecule purification, as the spacer itself will not create hydrophobic or ionic interaction sites on the final support.

Diamine compounds containing a linear PEG bridge can be obtained from a number of sources. The shortest diamine–PEG spacers can be obtained from Aldrich [2,2′-(ethylenedioxy)*bis*(ethylamine)] or Huntsman (Jeffamine EDR-148) in the form of a PEG_2 length and having ethylamine groups on each end. This compound is available in bulk quantities from Huntsman (from 5-gallon pails to tank-car loads), and Aldrich offers it in 100-ml and 500-ml package sizes, which are more appropriate for small-scale affinity support production. This compound can easily be coupled to an amine-reactive support to yield a 10-atom spacer, which terminates in an amine. Another, slightly longer diamine–PEG compound is available from Aldrich that contains a PEG_3 bridge with propylamine groups on each end (4,7,10-Trioxa-1,13-tridecanediamine). This compound is still extremely hydrophilic and provides a 15-atom spacer once it is coupled with an amine-reactive support (Figure 15.108). Both the PEG_2 and PEG_3 diamine spacers need to be reacted in high molar excess to avoid the potential for both ends of the molecules reacting with the activated support and crosslinking within the matrix.

Longer PEG-based diamine spacers can be obtained from Huntsman with the additional incorporation of a propylene oxide-capped polyethylene glycol internal construct. For instance, the Jeffamine ED series (e.g., ED-600) (Figure 15.109) contains a core PEG_9 repeat with several propylene oxide groups at each end, which are terminally capped with a primary amine. The reagents are extremely hydrophilic and provide longer PEG-based spacers at relatively inexpensive price points. These are good choices if a longer spacer arm is needed to move an affinity ligand away from the base matrix to provide better accessibility for the docking of biomolecules.

Discrete diamine PEG–compounds are also available from Quanta Biodesign. Two of these spacers have

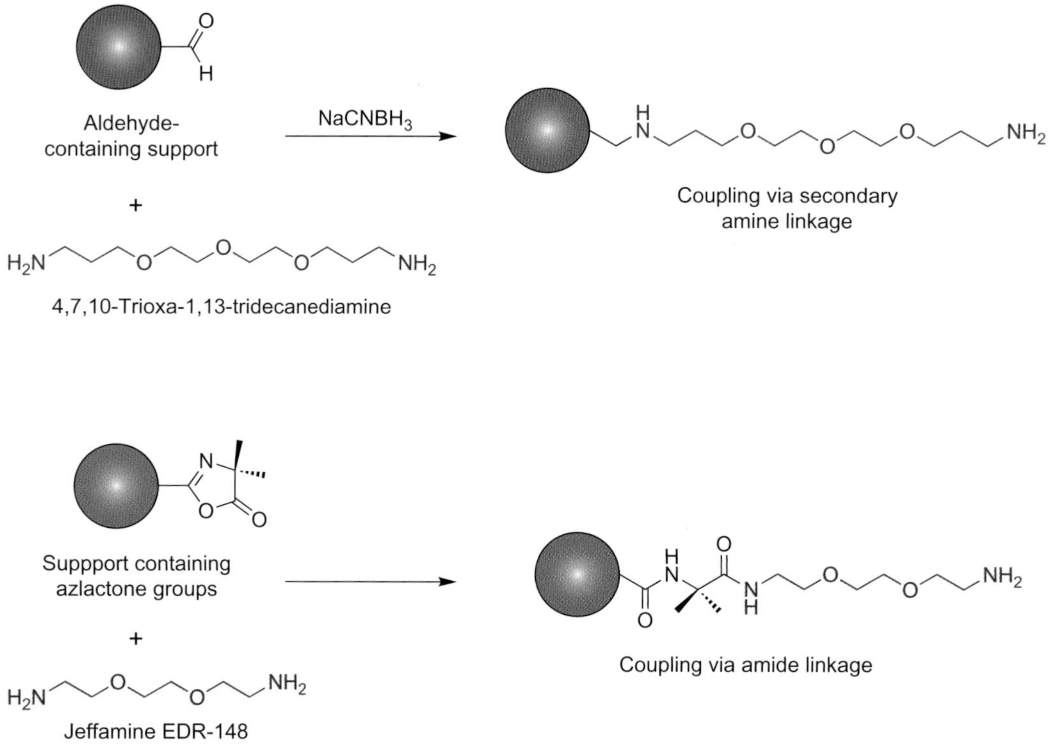

FIGURE 15.108 Coupling of Jeffamine EDR-148 or the diamine–PEG_3 compound to amine-reactive supports to yield terminal amine groups.

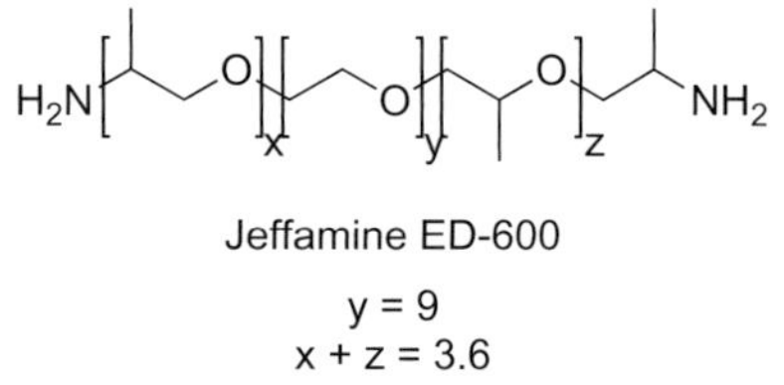

FIGURE 15.109 Jeffamine ED-600 structure.

one of their two amine groups protected to facilitate coupling to a support without the need to use a large molar excess to avoid crosslinking. Since only one amine will initially couple to the amine-reactive support the final density of amines on the matrix can be finely controlled to an optimal level for immobilizing whatever affinity ligand is desired. This avoids having a high density of amines on the support that could contribute to an ion-exchange character, especially if the excess amines are not blocked after coupling a ligand. The particularly long protected diamine–PEG reagent t-boc-*N*-amido–$dPEG_{11}$–amine contains a 37-atom crossbridge that is extremely hydrophilic with a free amine on one end and a BOC-protected amine on the other end. The use of a protected diamine–PEG spacer does require a deprotection step after coupling to a support to free the protected amine for further modifications. The type of protecting agent used with these reagents is a *tert*-butyloxycarbonyl (BOC) group, which typically requires strongly acidic conditions (TFA; 30–90%) in a water-immiscible solvent (dichloromethane) for deprotection. Some supports may not be able to tolerate these conditions without some degradation and washing in and out of a water immiscible solvent is problematic. Some modifications to this standard scheme may work for aqueous phase deprotection. Heating to 60 to 100°C in a neutral pH water solution was found to be efficient in one study and may be viable for use with crosslinked chromatography supports (Wang *et al.*, 2009). In addition, a modified acidic deprotection method was investigated by Han *et al.* (2001) that demonstrated fast deprotection using 4-*N* HCl in anhydrous dioxane. This method uses acidic cleavage of the protecting group, but it does so in a water-miscible solvent, which makes operations using chromatography supports more amenable to washing in and out of an aqueous environment. Two examples of discrete diamine–PEG compounds are shown in Figure 15.110, both of which use the BOC protecting group. The coupling and deprotection reaction for the shorter compound is illustrated in Figure 15.111.

COUPLING A DIAMINE–PEG-BASED SPACER The following protocols describe the use of diamine–PEG-based spacers to modify amine-reactive supports for additional reactions or for the coupling of

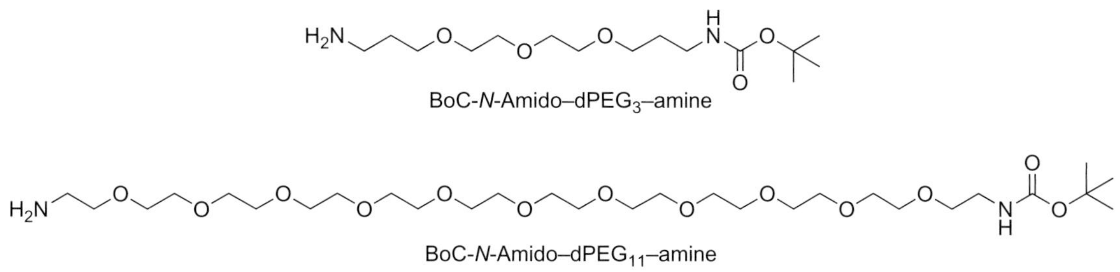

FIGURE 15.110 Diamine–PEG$_n$ compounds with one end blocked by a BOC group.

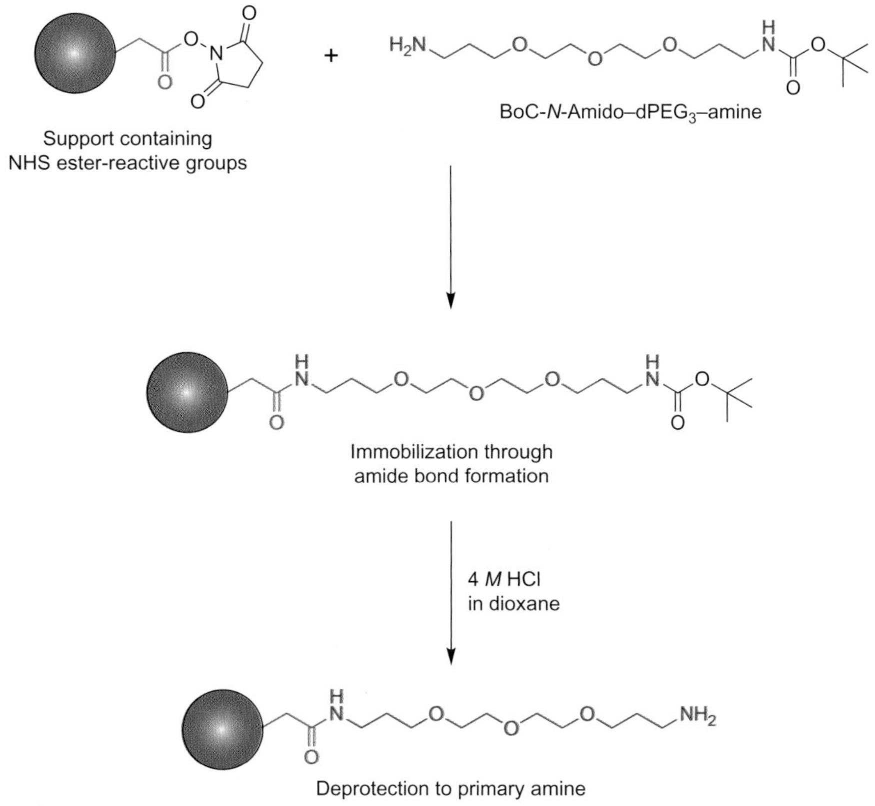

FIGURE 15.111 Coupling of a BOC-protected diamine–PEG compound to an amine-reactive support with subsequent deprotection with 4-*M* HCl in dioxane or heating to 60 to 100°C.

a carboxylate-containing ligand. The first procedure makes use of 4,7,10-Trioxa-1,13-tridecanediamine as the free amine (not protected and not as the dihydrochloride salt) in the modification of an aldehyde-containing support using reductive amination (see discussion on reductive amination, this chapter). The second protocol makes use of a discrete diamine–PEG compound that has one end protected with a BOC group.

(A) Coupling of 4,7,10-Trioxa-1,13-tridecanediamine to Periodate-Oxidized Agarose

1. Wash 100 ml of periodate-oxidized agarose (or another aldehyde-containing support) with water to remove storage solutions and then into coupling buffer (0.1-*M* sodium phosphate, pH 7.2). Washing steps may be performed with a sintered glass filter funnel suspended in a vacuum filter flask. Wash with at least several bed volumes for each wash step. Drain to a moist cake.
2. In the fume hood, add 33.5 g (33.4 ml) of 4,7,10-trioxa-1,13-tridecanediamine (Aldrich) to about 100 ml of crushed, deionized ice (use personal protective equipment to prevent contact with the highly caustic diamine and the acid used for neutralization). Slowly add 8 to 10 ml of concentrated HCl to the diamine/ice slurry with manual stirring. Add the acid slowly using a pipette. As the acid is added, the solution will warm and the ice will melt. After the acid has

been added, stir the solution using a magnetic stir bar and continue to neutralize the solution to about pH 7 with acid. Add to this solution a quantity of sodium phosphate buffer salt to make the final concentration 0.1-*M* phosphate when the total volume is diluted to 100 ml. Stir and readjust the pH to 7.2 using acid or base. The final solution is 1.52-*M* diamine, 0.1-*M* sodium phosphate, pH 7.2. Allow the solution to come to room temperature before continuing.
3. Add the washed periodate-oxidized agarose to the diamine solution and mix to resuspend the gel.
4. Add 0.63 g of sodium cyanoborohydride (toxic!) to the gel slurry and mix for 2 to 4 h. Mixing may be carried out using an overhead paddle stirrer (not a stir bar) or in a sealed plastic container by end-over-end rocking. Avoid the use of sealed glass containers, because there is some gas evolution during the reductive amination process.
5. Extensively wash the diamine-agarose support with water, 1-*M* NaCl, and water to completely remove unreacted diamine and reaction byproducts. The support may be stored as a 50% slurry containing a preservative at 4°C until use.

(B) Coupling of a BOC–Amido–PEG–Amine Spacer to CDI-Activated Supports

The following protocol may be used to couple a *bis*-amino–PEG_n compound that has one amino end blocked with a BOC protecting group. The support material used in this procedure must be able to withstand the deprotection conditions for removing the BOC group using formic acid in solvent. Small samples of a given support should be tested for stability in a formic acid/DMAC solution before proceeding to use this procedure. Many crosslinked or polymeric supports should be able to tolerate the deprotection step without difficulty.

1. In a fume hood, drain 10 ml of A CDI-activated support of excess acetone using a sintered glass filter funnel suspended in a suction filter flask (see Carbonyl Diimidazole (CDI) Activation in Section 2.1, this chapter). While pulling a gentle vacuum to remove the remaining excess acetone, break the support up into small pieces so that it resembles fluffy snow. Do not allow the support to dry. Remove the vacuum and resuspend the matrix in dry DMAC with mixing. Wash with DMAC to remove the last traces of acetone (at least 10 bed volumes). Drain to a moist cake.
2. Dissolve the blocked diamine compound BOC-*N*–amido–$dPEG_{11}$–amine (Quanta Biodesign) in 10 ml of DMAC at a concentration of 12.8 mg/ml, which will equal a level of 20 μmol/ml gel in the reaction medium. Control of the final density of amines on the support can be accomplished by adjusting the reaction concentration of the spacer. The optimal concentration of spacer to be used in the coupling reaction should be determined experimentally on small quantities of activated support to identify the best final affinity support performance in its intended application. Add an organic base to the ligand solution, such as DMAP, DIEA, or TEA, to make a final concentration of 2-m*M*.
3. Add the activated wet gel cake to the ligand solution with stirring to fully resuspend the gel. Mix the reaction slurry for 1 to 2 h using an overhead paddle stirrer or by end-over-end rocking in a sealed container. Longer reaction times may be used if appropriate.
4. Excess CDI reactive groups on the support may be blocked by the addition of ethanolamine to the reaction slurry at a final concentration of 0.1-*M*. Continue to mix for 1 h at room temperature.
5. Transfer the gel slurry to a sintered glass filter in the fume hood that is suspended in a suction filter flask and wash extensively (at least 10 bed volumes) with solvent to remove the remaining ligand and reaction byproducts. Finally, drain the support of excess solvent by pulling a gentle vacuum on the filter flask while breaking up the support into small, finely divided pieces using a spatula, but be careful not to allow the matrix to dry out. Once the support is broken into small pieces, remove the vacuum and resuspend the gel in neat formic acid with mixing (caution: highly corrosive; use a fume hood and personal protective equipment to avoid contact or inhalation of vapors). Stir for 1 h at room temperature.
6. Wash the support with 2 bed volumes of formic acid and then wash extensively with DMAC to completely remove all traces of remaining acid. Once the acid has been thoroughly removed, the support should be washed into water by sequential washes using increasing concentrations of water in DMAC until 100% water is attained. Continue to wash the support with water until all traces of solvent have been removed. Additional washes with 1-*M* NaCl as well as low and high pH conditions may be done as appropriate. Finally, wash with water and store the amine-containing support as a 50% slurry in water containing a preservative at 4°C until used.

AMINO–CARBOXYLATE SPACERS

Spacer arms containing an amine group on one end and a carboxylic acid group on the other end are popular choices for building many affinity supports. This type of spacer is typically used to modify a support to contain a terminal carboxylate for further coupling to amine-containing molecules. The spacer's cross-bridge

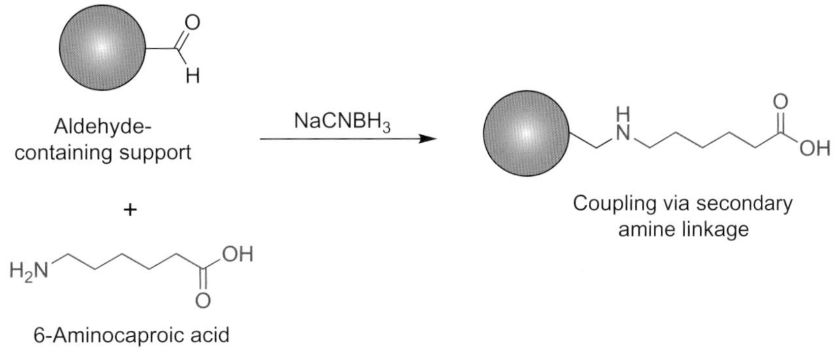

FIGURE 15.112 Coupling of 6-aminocaproic acid to an aldehyde-containing support via reductive amination.

can be designed to possess additional properties that may be beneficial to the final affinity support being created. For instance, short or long spacers can be chosen to modify the support, depending on the total length desired to tether a ligand away from the matrix surface. In addition, the spacer cross-bridge can be designed to be relatively hydrophobic (aliphatic) or hydrophilic (typically hetero-aliphatic construction).

The shortest amino–carboxylate spacer is β-alanine, which contains two central methylene groups with an amine on one end and a carboxylic acid group on the other end (3-aminopropanoic acid). This short spacer is used often in the design of modification and cross-linking agents, and it also can be used as a spacer on chromatography supports to provide a terminal carboxylate for subsequent coupling reactions. β-Alanine is extremely water soluble and will not contribute any hydrophobicity to a support, which may occur with a longer aliphatic cross-bridge; however, the molecule also is quite short, which means it cannot be used to extend a ligand out from the matrix to facilitate greater binding accessibility for a target molecule. It is mainly used just to form a carboxylate on the support for further modification reactions or for facilitating ligand immobilization.

One of the most common, medium length, aliphatic amino–carboxylate spacers is 6-aminocaproic acid, which contains a six-atom methylene cross-bridge between the two functional groups. This compound has been widely used to modify amine-reactive supports to contain carboxylates for further immobilization of an amine-containing ligand. Usually, the use of a spacer is done to move the ligand away from the matrix to allow efficient docking of target molecules to bind with the ligand. For instance, Bansal *et al.* (2006) coupled *p*-aminobenzamidine by its amine to a monolithic cryogel containing aminocaproic acid spacers and used the resultant affinity support to isolate urokinase from cell culture broth of human kidney cells. The hydrophobic nature of the cross-bridge in 6-aminocaproic acid can provide some enhancement of binding toward certain target molecules, depending on the circumstances of the interaction. The spacer has found use in other areas of bioconjugation as well, including its incorporation into the very popular biotinylation reagent, NHS–LC-biotin (see Chapter 11, Section 6.2). The 6-aminocaproic acid spacer has also been used to modify a matrix to create an NHS ester, amine-reactive support by forming the active ester on the terminal carboxylate after the spacer has been attached (Wilchek and Miron, 1987).

The spacer 6-aminocaproic acid is an inexpensive compound that has very good water solubility (50 mg/ml). It can be dissolved in aqueous buffers without significantly effecting the pH of the solution (unlike the diamine compounds described previously). The compound is an analog to the amino acid lysine and has antifibrinolytic properties *in vivo* by inhibiting plasminogen. The coupling of 6-aminocaproic acid to an aldehyde-containing support using reductive amination is illustrated in Figure 15.112. Other amine-reactive supports can be used in a similar process to create a carboxylate-terminal spacer.

COUPLING 6-AMINOCAPROIC ACID TO AN AMINE-REACTIVE SUPPORT The following protocol describes the modification of an aldehyde-containing support, such as periodate-oxidized agarose (see earlier section, this chapter) with 6-aminocaproic acid to provide a carboxylate derivative for subsequent immobilization of amine-containing ligands. The subsequent immobilization of an affinity ligand to form an amide bond may be performed using the methods previously described in this chapter, such as a carbodiimide-mediated coupling reaction (using EDC) or through NHS ester formation.

1. Wash 100 ml of periodate-oxidized agarose (or another aldehyde-containing support) with water to remove storage solutions and then into coupling buffer (0.1-*M* sodium phosphate, pH 7.2). Washing steps may be performed with a sintered glass filter

funnel suspended in a vacuum filter flask. Wash with at least several bed volumes for each wash step. Drain to a moist cake.

2. In the fume hood, add 10 g of 6-aminocaproic acid (Acros, Aldrich) to 100 ml of coupling buffer. Stir the solution to dissolve and readjust the pH to 7.2 using acid or base, if necessary. The final solution is 0.76-*M* aminocaproic acid, 0.1-*M* sodium phosphate, pH 7.2. Lower concentrations of 6-aminocaproic acid may be used for this reaction to create a lower density of carboxylates on the surface of the support. Some experimentation may have to be carried out to determine the optimal level of carboxylates for the intended affinity chromatography application.
3. Add the washed periodate-oxidized agarose to the aminocaproic acid solution and mix to resuspend the gel.
4. In a fume hood, add 0.63 g of sodium cyanoborohydride (toxic!) to the gel slurry and mix for 2–4 h. Mixing may be performed using an overhead paddle stirrer (not a stir bar) or in a sealed plastic container by end-over-end rocking. Avoid the use of sealed glass containers, because there is some gas evolution during the reductive amination process.
5. Extensively wash the aminocaproic acid-agarose support with water, 1-*M* NaCl, and water to completely remove unreacted reagent and reaction byproducts. The support may be stored as a 50% slurry containing a preservative at 4°C until use.

AMINO–PEG$_n$–CARBOXYLATE SPACERS

A version of an amine-carboxylate spacer that has a PEG-based cross-bridge is available that makes the linker extremely hydrophilic (Thermo Fisher, Quanta Biodesign). These spacers are particularly advantageous for creating support derivatives that have very low nonspecific binding character, unlike aliphatic spacers that may increase the nonspecificity of a support. PEG-based spacers can also increase the freedom of motion of tethered affinity ligands, thus making them more likely to be able to interact with a target molecule. A series of amino–PEG$_n$–carboxylate compounds are available depending on the spacer need, ranging from PEG$_4$ to PEG$_{36}$ in cross-bridge length (Quanta). The shortest is an amino–PEG$_4$–carboxylate that has a 16 atom spacer about 18 Å in length. A longer length amino–PEG$_8$–carboxylate compound has a 33.6-Å spacer and the amino–PEG$_{12}$–carboxylate reagent contains a 46.8-Å tether. The longest compound in this series has an internal PEG$_{36}$ chain with a linear cross-bridge of 132.7 Å, making it perhaps the longest discrete spacer available for modification of chromatography supports. The largest spacer arm is actually longer than the average diameter of a typical IgG antibody molecule (~110 Å), which ensures that any affinity ligand tethered at the end of it will be accessible to virtually any docking protein or biomolecule that may bind to it (Figure 15.113).

Amino–PEG$_n$–carboxylate spacers can be reacted with an amine-reactive support to give a covalent linkage forming an amide or secondary amine bond with the surface. The terminal carboxylate then can be used to build another reactive group for subsequent coupling of an affinity ligand, such as in the creation of an NHS ester or a hydrazide group (see previous sections, this chapter). The carboxylate may also be used to directly immobilize an amine-containing molecule through the use of a carbodiimide-mediated reaction sequence, which ultimately will form an amide bond with the ligand (see previous section, this chapter, on EDC coupling). A possible reaction sequence using these spacers is illustrated in Figure 15.114. The use of these long PEG spacer molecules will create an extremely hydrophilic layer on any solid-phase surface, which will lower nonspecific binding character and maximize the purity of any target molecule captured.

Any of the amine-reactive immobilization methods described previously in this chapter can be used to couple an amino–PEG$_n$–carboxylate spacer to a support. In the following protocol examples, first a DSC-activated hydroxylic support is used in nonaqueous solvent to couple the amino–PEG$_n$–carboxylate spacer to form an amide bond and in the second example a periodate-oxidized agarose support containing aldehyde groups is used to couple the spacer using reductive amination to form a secondary amine bond. Other reactive supports may be used by following the recommended coupling protocols found in previous sections.

COUPLING AMINO–PEG$_n$–CARBOXYLATE SPACERS

(A) Coupling Amino–PEG$_8$–Carboxylate to a DSC-Activated Hydroxyl-Support

1. In a fume hood, drain 10 ml of A DSC-activated support of excess acetone using a sintered glass filter funnel suspended in a suction filter flask. While pulling a gentle vacuum to remove the remaining excess acetone, break the support up into small pieces so that it resembles fluffy snow. Do not allow the support to dry. Remove the vacuum and resuspend the matrix in dry DMAC with mixing. Wash with DMAC to remove the last traces of acetone (at least 10 bed volumes). Drain to a moist cake.
2. Dissolve the amino-PEG$_8$-carboxylate spacer compound (Quanta BioDesign or Thermo Fisher) in 10 ml of DMAC at a concentration of 8.83 mg/ml, which translates into a level of 20 μmol/ml gel in the reaction medium. Control of the final density of amines on the support can be accomplished by adjusting the reaction concentration of the

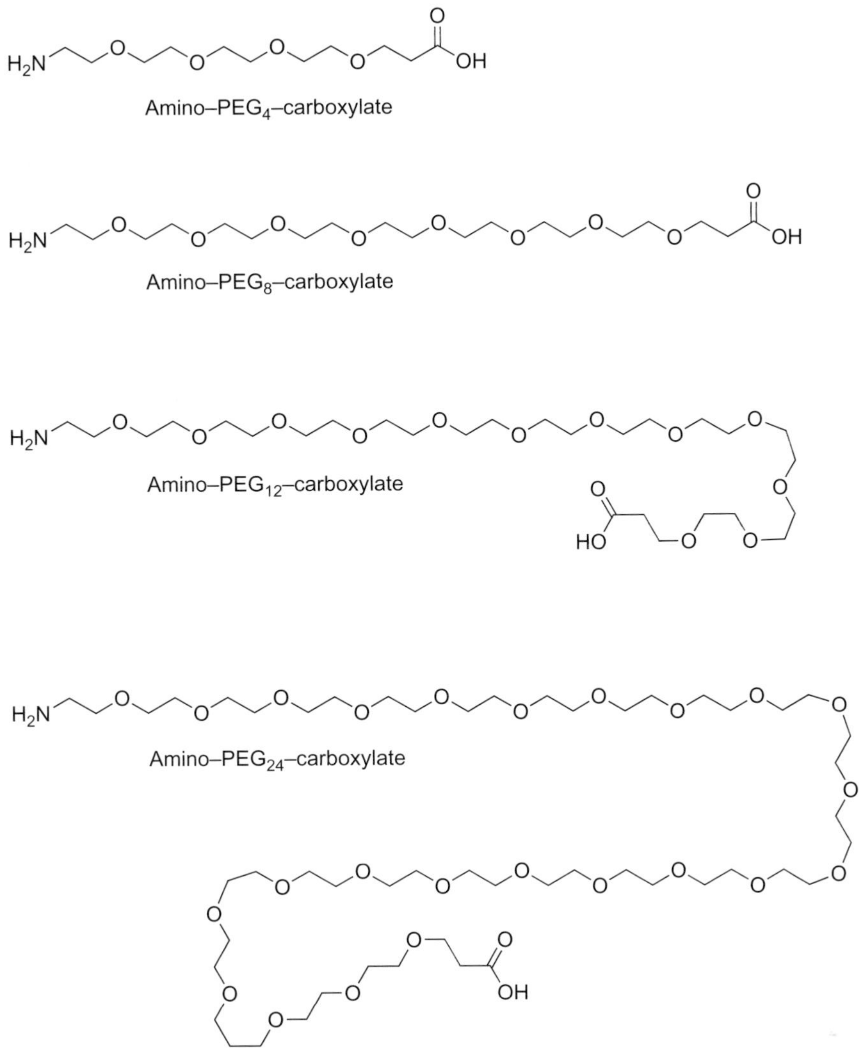

FIGURE 15.113 Examples of amino–PEG$_n$–carboxylate spacers.

spacer. The optimal concentration of spacer to be used in the coupling reaction should be determined experimentally on small quantities of activated support to identify the best final affinity support performance in its intended application. Add an organic base to the ligand solution, such as DMAP, DIEA, or TEA, to make a final concentration of 2-m*M*.

3. Add the activated wet gel cake to the ligand solution with stirring to fully resuspend the gel. Mix the reaction slurry for 1 h using an overhead paddle stirrer or by end-over-end rocking in a sealed container. Longer reaction times may be done if appropriate.
4. Excess NHS carbonate reactive groups on the support may be blocked by the addition of ethanolamine to the reaction slurry at a final concentration of 0.1-*M*. Continue to mix for 1 h at room temperature.
5. In the fume hood, transfer the gel slurry to a sintered glass filter that is suspended in a suction filter flask and wash extensively (at least 10 bed volumes) with solvent to remove the remaining ligand and reaction byproducts. Finally, drain the support of excess solvent by pulling a gentle vacuum on the filter flask while breaking up the support into small, finely divided pieces using a spatula, but be careful not to allow the matrix to dry out. Once the support is divided into small pieces, remove the vacuum and resuspend the gel in water with mixing. Continue to wash the support with water until all traces of solvent have been removed. Additional washes with 1-*M* NaCl as well as low and high pH conditions may be done as appropriate. Finally, wash with water and store the carboxylate-containing support as a 50% slurry in water containing a preservative at 4°C until used.

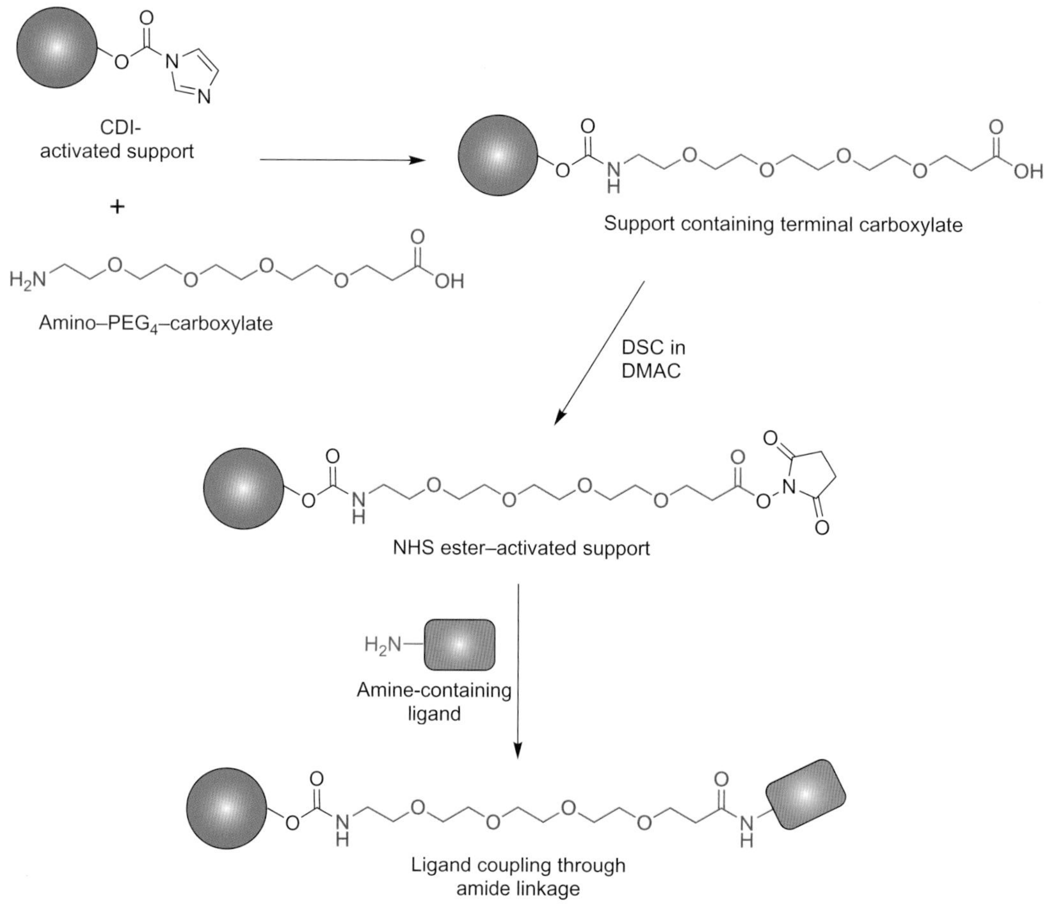

FIGURE 15.114 Coupling of an amino–PEG$_n$–carboxylate to an amine-reactive support followed by NHS ester formation and the immobilization of an amine-containing ligand.

(B) Coupling Amino–PEG$_8$–Carboxylate to a Periodate-Oxidized Agarose Support

1. Wash 10 ml of periodate-oxidized agarose (or another aldehyde-containing support, such as a glycidol/periodate-treated polymeric support) with water to remove storage solutions and then into coupling buffer (0.1-*M* sodium phosphate, pH 7.2). The washing steps may be done with a sintered glass filter funnel suspended in a vacuum filter flask. Wash with at least several bed volumes for each wash step. Drain to a moist cake.
2. In the fume hood, add 176 mg of amino–PEG$_8$–carboxylate spacer compound (Quanta BioDesign or Thermo Fisher) to 10 ml of coupling buffer. Stir the solution to dissolve and readjust the pH to 7.2 using acid or base, if necessary. The final solution is at a concentration of 40 μmol/ml of the spacer in 0.1-*M* sodium phosphate, pH 7.2. Higher or lower concentrations of amino–PEG$_8$–carboxylate may be used for this reaction to create custom densities of carboxylates on the surface of the support. Some experimentation may have to be done to determine the optimal level of carboxylates for the intended affinity chromatography application.
3. Add the washed periodate-oxidized agarose to the amino–PEG$_8$–carboxylate solution and mix to resuspend the gel.
4. Add 63 mg of sodium cyanoborohydride (toxic!) to the gel slurry and mix for 2 to 4 h. Mixing may be performed using an overhead paddle stirrer (not a stir bar) or in a sealed plastic container by end-over-end rocking. Avoid the use of sealed glass containers, because there is some gas evolution during the reductive amination process.
5. Extensively wash the carboxylate-agarose support with water, 1-*M* NaCl, and water to completely remove unreacted reagent and reaction byproducts. The support may be stored as a 50% slurry containing a preservative at 4°C until use.

BIS-CARBOXYLATE SPACERS

bis-Carboxylate spacer compounds are useful in extending an amine-containing support and converting the amine to a terminal carboxylate for further coupling

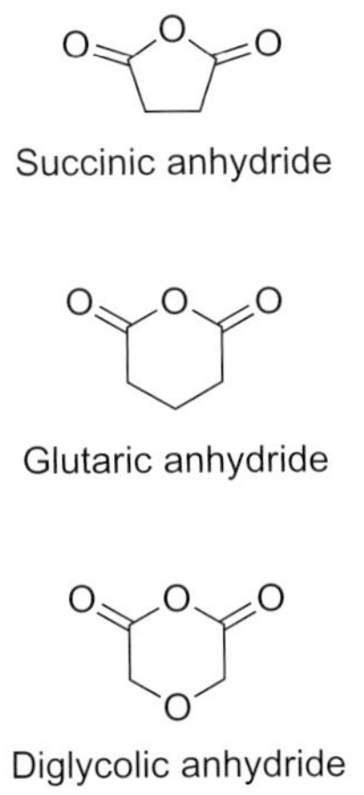

FIGURE 15.115 The structures of common anhydride compounds used to create terminal carboxylate groups on amine-containing supports.

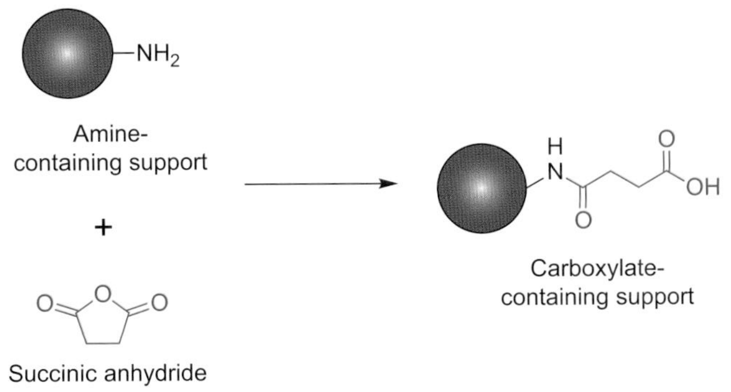

FIGURE 15.116 Reaction of succinic anhydride with an amine-containing support.

or immobilization reactions. Most often, a *bis*-carboxylate spacer is added to a support through reaction of its corresponding cyclic anhydride with the amino groups on a matrix to create amide bond linkages. Ring opening of the anhydride forms the short spacer group as the other carboxylate is freed at the uncoupled end. Several anhydride compounds are commonly used to form these spacers on supports (Figure 15.115). Succinic anhydride is the shortest of these spacers, containing a 4-carbon ring structure, which when reacted with an amine-containing support provides a 2-atom methylene bridge between the two terminal carboxylates. Glutaric anhydride is a 5-carbon compound that when opened results in a 3-atom methylene bridge between the two terminal carboxylates. Diglycolic anhydride is the most hydrophilic in this series, which after ring opening contains a 5-atom spacer between the two terminal carboxylate groups and also has a central ether group, which adds water solubility to the compound. All three anhydride compounds may be used in similar reaction schemes to modify an amine-containing support to contain terminal carboxylates. The reaction of succinic anhydride with an amine-containing support is illustrated in Figure 15.116.

The following protocol describes the modification of an amine-containing support with succinic anhydride to form a carboxylate-containing support for further ligand immobilization reactions. The other anhydride compounds may be used similarly.

Coupling Succinic Anhydride to Amine-Containing Supports

1. Wash 100 ml of an amine-containing support with water to remove storage solutions and then wash with several bed volumes of reaction buffer (1.0-*M* sodium bicarbonate, pH 8.0). Finally, suspend the support in an equal volume of reaction buffer. Other reaction buffers and pH conditions may be used over the range of about pH 6 to pH 9 with success. Some protocols just use water, but maintain the pH at the desired value with periodic titration with base (NaOH at 50%).
2. Stir the gel slurry using an overhead paddle stirrer and slowly add 10 g of succinic anhydride. Initially, the anhydride will not be fully soluble in the reaction mixture, but it will go into solution as it reacts and hydrolyzes in the buffered solution.
3. React with constant stirring for 1 h at room temperature.
4. Wash the succinylated support with reaction buffer, water, 1-*M* NaCl, and again with water to completely remove unreacted succinic anhydride or succinic acid. The carboxylic acid-containing support may be stored as a 50% slurry containing a preservative at 4°C until use.

Other anhydride compounds may be used in a similar reaction protocol with success; however, adjust the mole quantity of anhydride addition to be equivalent to the succinic anhydride protocol.

THIOL–CARBOXYLATE SPACERS

Spacer arms containing a thiol group on one end and a carboxylate group on the other end have been used extensively for some applications involving immobilized affinity ligands. Most of these applications have involved the use of a solid phase that contains a metallic surface, either planar or particulate in nature. In this case, the thiol end of the spacer is usually coupled to the metal by way of a dative bond, wherein the unshared pair of electrons on the sulfur atom are shared with the metal. The application of these spacers to modify metallic microparticles or nanoparticles is described in Chapter 14. Although chromatography supports typically do not use metallic solid phases, in certain instances it may be advantageous to use a thiol–carboxylate spacer to modify an amine-containing matrix with the carboxylate end, and thus create a thiol group on a support for subsequent immobilization of thiol-containing ligands. Terminal thiols created in this manner on a support can be activated to couple ligands through disulfide bonds, which then can be reversed to subsequently release interacting molecules (see previous section on thiol-reactive immobilization methods, this chapter). Of course, it also is possible to use a thiol–carboxylate spacer to modify a thiol-reactive support to create a terminal carboxylate, but this probably is less frequently done than coupling the spacer in the reverse way.

A number of thiol–carboxylate spacer arms are available to modify a support, including molecules that contain aliphatic cross-bridges as well as the more hydrophilic PEG-based linkers. The shortest thiol–carboxylate spacer is thioglycolic acid, which is 2-mercaptoacetic acid (note that this compound is not thioacetic acid, which is the acetylated thiol in a thioester form). An analog to thioglycolic acid is SATA (Chapter 2, Section 4.1), which has a protected thiol at one end and an NHS ester at the other end of the acetate bridge. SATA, or the similar compound SATP which is one carbon atom longer, can provide rapid modification of an amine-containing support to convert it to a protected thiol-containing support. The NHS ester end will react with the amines on the support to result in an amide bond, and the protected thiol end is stable in this form until needed for immobilization of ligands. The thiol can be deprotected with hydroxylamine under alkaline conditions to reveal the thiol. These short compounds can be useful if an amine-containing support already has a long spacer arm present and all that is desired is to convert a terminal amine into a thiol group.

There also are longer thiol–carboxylate spacers available to form tethers of greater length on supports for the coupling of affinity ligands. Both hydrophobic, aliphatic spacer molecules can be used as well as hydrophilic, PEG-based spacers. A PEG-containing spacer will provide greater hydrophilicity within the support and have a tendency to decrease the nonspecific binding potential of biomolecules. There is a wide selection of different spacer lengths available for these compounds with PEG groups ranging from PEG_4 to PEG_{20} in size (Quanta BioDesign, Thermo Fisher). However, the use of a thiol-carboxylate spacer without having the thiol group protected can be problematic. The reactions that have to occur to couple the carboxylate end to an amine-containing support are unfortunately interfered with by the thiol end. If an active ester is formed on the carboxylate to couple with the amines on the support without protecting the thiol, then the ester can potentially react with the thiol groups to form thioester bonds, causing polymerization of the reagent in solution. In addition, if the carbodiimide EDC is used to attach the carboxylate to the amino groups, then EDC can also react with the thiol end to form an irreversible complex. Fortunately, thiol-protected versions of these reagents are available that allow these reactions to occur without interference. These compounds are SATA-like in structure, having an NHS ester on one end and a protected (acetylated) thiol on the other end, but with the added feature of containing an internal PEG cross-bridge to provide increased hydrophilicity over that of the aliphatic reagent. Figure 15.117 illustrates the major thiol-carboxylate reagents available for use as spacers in chromatography supports and Figure 15.118 shows the reactions involved with coupling one of these reagents to an amine-containing support.

Coupling NHS–PEG$_4$–Thioacetyl to an Amine-Containing Support

The following protocol describes the use of a NHS–PEG_4–thioacetyl compound in the modification of an amine-containing support to provide an alteration that terminates in a protected thiol group. The reagents $SAT(PEG)_4$ from Thermo and $dPEG_4$-SATA from Quanta BioDesign are identical structurally and can be used in this method with success. This method describes the reaction of the NHS ester end with the amine groups on the support in a nonaqueous environment to eliminate the hydrolysis of the ester during coupling. All operations with solvent should be performed in a fume hood using the appropriate personal protective equipment. The reaction may also be carried out in an aqueous buffer (e.g., 0.1-*M* sodium phosphate, pH 7.2), but the amount of reagent added may have to be increased to account for some yield loss due to hydrolysis. The optimal density of protected thiols on the support should be investigated experimentally by assessing the performance of the ultimate affinity support in the intended application. Adjusting the concentration of the reagent in the reaction will control the density of the protected thiols after the reaction is complete, and so will adjusting the initial concentration or density of amines on the support.

1. Wash 10 ml of an amine-containing support with water to remove storage solution and preservatives. Then wash the support into DMAC by washing with sequentially increasing concentrations of solvent-in-water until 100% DMAC is used. Continue to wash with DMAC for at least 10 to 20 bed volumes to completely remove the last traces of water. Drain the support to a wet cake, but do not allow it to dry.
2. Dissolve 421 mg of the NHS–PEG_4–thioacetyl compound in 10 ml of DMAC. This will result in 100 µmoles of reagent per milliliter of amine-containing gel within the reaction mixture.
3. Add the washed gel to the reagent solution with mixing to resuspend the matrix. An organic base

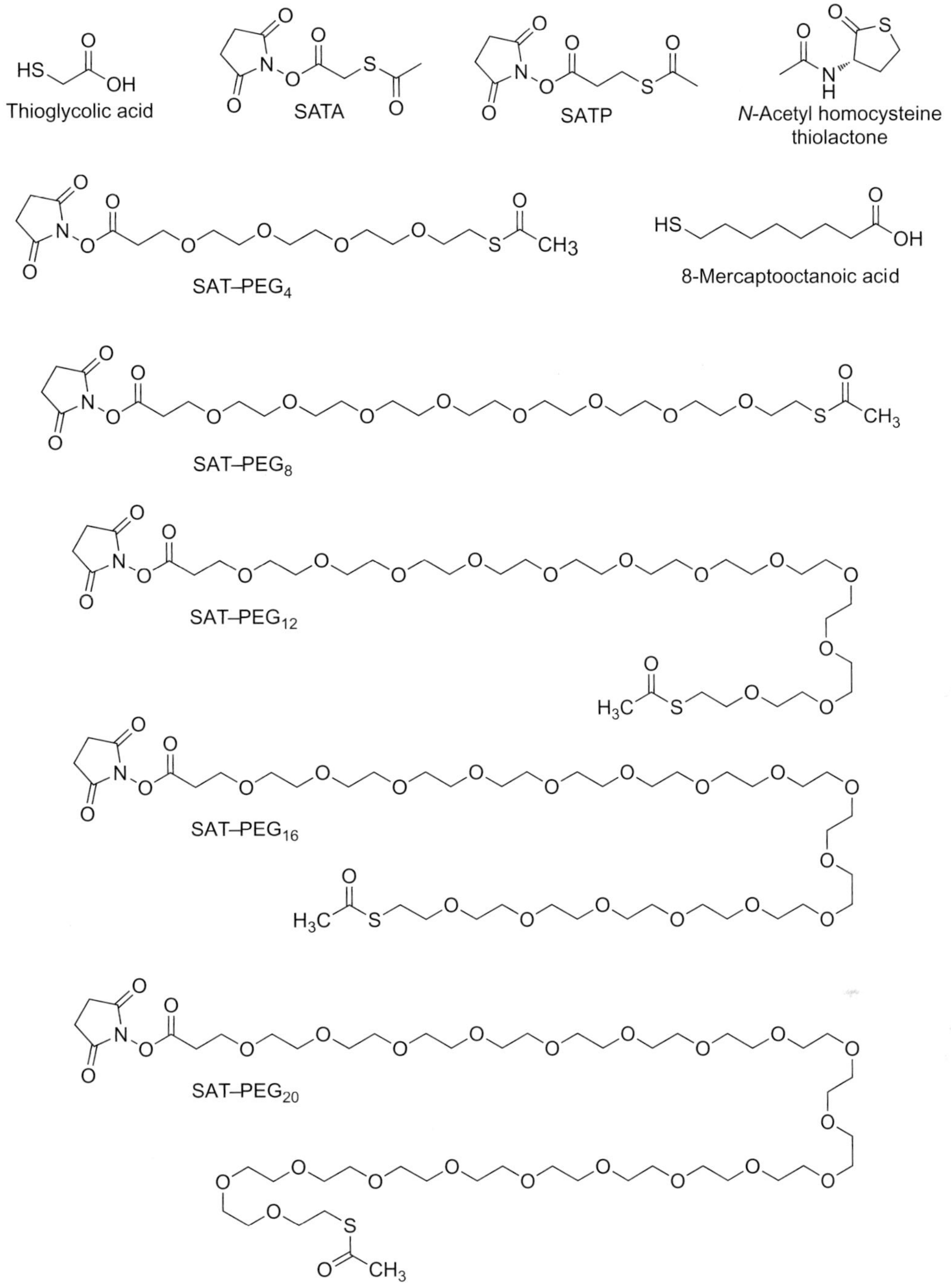

FIGURE 15.117 Thiol–carboxylate reagents for spacer arm applications.

may be added to the mixture to accelerate the reaction, such as DMAP, DIEA, or TEA, to make a final concentration of 2-m*M*. React for 1 h at room temperature with constant mixing. The mixing may be accomplished by use of a paddle and overhead stir motor or by end-over-end rocking in a sealed container.

4. Wash the support with DMAC to remove excess reagent (at least 10 column volumes) and then wash sequentially back into water by using increasing concentrations of water-in-solvent until 100% water is used. Continue to wash with water for 10 to 20 bed volumes to completely remove the solvent. The protected thiol-containing support may be stored at 4°C as a 50% slurry containing a preservative until use. As needed, the gel can be treated to deprotect the thiol groups prior to immobilizing a ligand or doing further reactions.

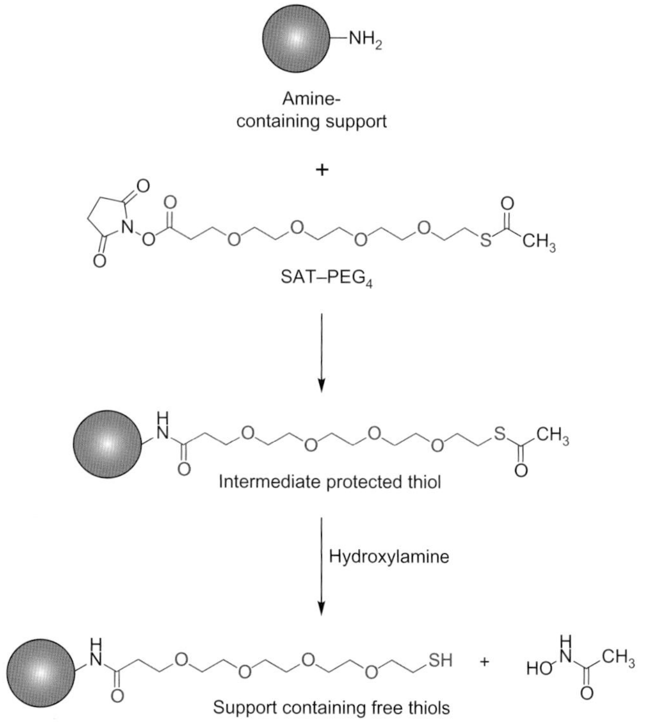

FIGURE 15.118 Coupling of a NHS–PEG$_n$–thioacetyl reagent to an amine-containing support with subsequent deprotection of the thiol using hydroxylamine.

Deprotection of the Thioacetyl-Containing Support

1. Prepare 20 ml (or 2 times the amount of protected thiol-containing resin to deprotect) of an aqueous solution containing 0.5-*M* hydroxylamine in 0.1-*M* sodium phosphate, pH 7.2, containing 25-m*M* EDTA.
2. Wash 10 ml of the protected thiol resin with several bed volumes of water to remove storage solution and then slowly wash with 10 ml of the hydroxylamine solution. Drain to a wet cake.
3. Add the washed support to the remaining 10 ml of hydroxylamine solution and mix to resuspend the gel. React for 2 h at room temperature with constant mixing to remove the acetyl protecting groups.
4. Wash the deprotected support extensively with water to completely remove the hydroxylamine solution and reaction byproducts. The thiol-containing matrix should be used immediately to immobilize a ligand or in another reaction to couple with the free –SH group. Avoid storage of the support in this form, because thiol oxidation likely will take place rapidly and degrade the amount of thiols available. EDTA (25-m*M*) may be added to the wash solutions or reaction buffers to prevent metal-catalyzed oxidation.

CHAPTER

16

Buckyballs, Fullerenes, and Carbon Nanotubes

1. BUCKYBALLS AND FULLERENES

1.1. Properties of Fullerenes

Carbon is an incredible element that is able to form structures having highly diverse properties depending on its bonding patterns and 3-dimensional organization. Natural allotropes of carbon include diamond, graphite, amorphous carbon, and several other known forms (Figure 16.1). Depending on the bond structure and atomic orientation that carbon takes on within the structure of an allotrope, the resultant characteristics can range from the hardest-known abrasive mineral, diamond, to the extremely soft graphite, which is used as a lubricant.

In 1985, the story of carbon allotropes took a dramatic turn with the discovery of C_{60}, which resulted in a new type of carbon structure, called fullerenes (Kroto *et al.*, 1985). This discovery earned the 1996 Nobel Prize in chemistry for Harold Kroto, Robert Curl, and Richard Smalley. Buckminsterfullerene (named after Buckminster Fuller for his geodesic dome architectural design) is a spherical cage of carbon having 60 atoms forming a truncated icosahedron of average diameter 0.72 nm, which contains 12 pentagons and 20 hexagons of bonded carbon (Figure 16.2). The shape is exactly the same as a modern soccer ball, with the pentagon and hexagon configurations clearly outlined on its surface.

There now are known to be a whole family of caged carbon structures having various numbers of carbon atoms, including C_{30}, C_{50}, C_{70}, C_{72}, C_{76}, C_{84}, and the huge C_{540}. The name "fullerene" has replaced the unwieldy, "Buckminsterfullerene" used to describe this general spheroid structure of carbon, although they still are referred to as "Buckyballs."

Of all the fullerene forms, the nearly spherical properties of C_{60} have attracted the greatest attention, especially in the field of bioconjugation (Montellano *et al.*, 2011). In addition to its physical properties, C_{60} fullerenes have unique photo-optical and electrochemical properties, which make them useful as carriers for biomedical research applications. For instance, upon exposure to light C_{60} will generate singlet oxygen, which can be used *in vivo* to cleave biological molecules, particularly DNA and RNA. Studies indicate that irradiation of C_{60} in solution can be used to destroy virus contamination (Kasermann and Kempf, 1997). Fullerene conjugates have been investigated for use as drug delivery platforms for targeting mammary carcinoma cells (Lucafò *et al.*, 2012), as folate conjugates to target upregulated folate receptors on tumor cells (Shi *et al.*, 2012), and as conjugates to 5-fluorouracil to enhance the drug's effectiveness as an antitumor agent (Dou *et al.*, 2012).

Fullerene C_{60} also functions efficiently as an antioxidant, actually being better than other lipid-soluble antioxidants at scavenging reactive oxygen species (ROS) (Wang *et al.*, 1999). Water soluble derivatives of C_{60}, such as a poly-hydroxyl form, are able to function in the same respect in aqueous environments.

Pure fullerenes are insoluble in aqueous environments and only sparingly soluble in many organic solvents. Solutions of Buckminsterfullerene are a deep-purple color, whereas other sizes of fullerenes display a variety of other colors. The greatest solubility is found in 1,2,4-trichlorobenzene (20 mg/ml), carbon disulfide (12 mg/ml), toluene (3.2 mg/ml), and benzene (1.8 mg/ml) (Wikipedia.org). Solubility calculations have been performed on C_{60} in 75 different organic solvents (Silvaraman *et al.*, 2001).

1.2. Modification of Fullerenes

C_{60} contains exactly 30 double bonds in its ring structures that are chemically and electronically equivalent. The carbons at the double bond junctions do not possess any hydrogens, so substitution reactions are not feasible at these sites; however, addition reactions and redox reactions can occur to yield derivatives of the fullerene molecule. Many chemical derivatization

Bioconjugate Techniques, Third Edition.
DOI: http://dx.doi.org/10.1016/B978-0-12-382239-0.00016-9

methods have been developed to afford fullerene solubility in particular environments and to provide functional handles for bioconjugation (Bosi *et al.*, 2003). The combination of adding polar groups and reactive functionalities to fullerenes, such as COOH, NH_2, and OH groups, provides water solubility and bioconjugation targets. Examples of these modifications include the method of Brettreich and Hirsch (1998) to add multiple carboxylates in a dendritic fashion and Wang *et al.* (1999), who added multiple pairs of carboxylates to the surface carbons. In addition, Cusan *et al.* (2002) developed a C_{60}–PEG dendrimer-based diamine derivative using a substituted fulleropyrrolidine modification linked to the surface. Polymer carriers have also been used to provide water solubility and sites of attachment. Cyclodextrins have been found to be excellent

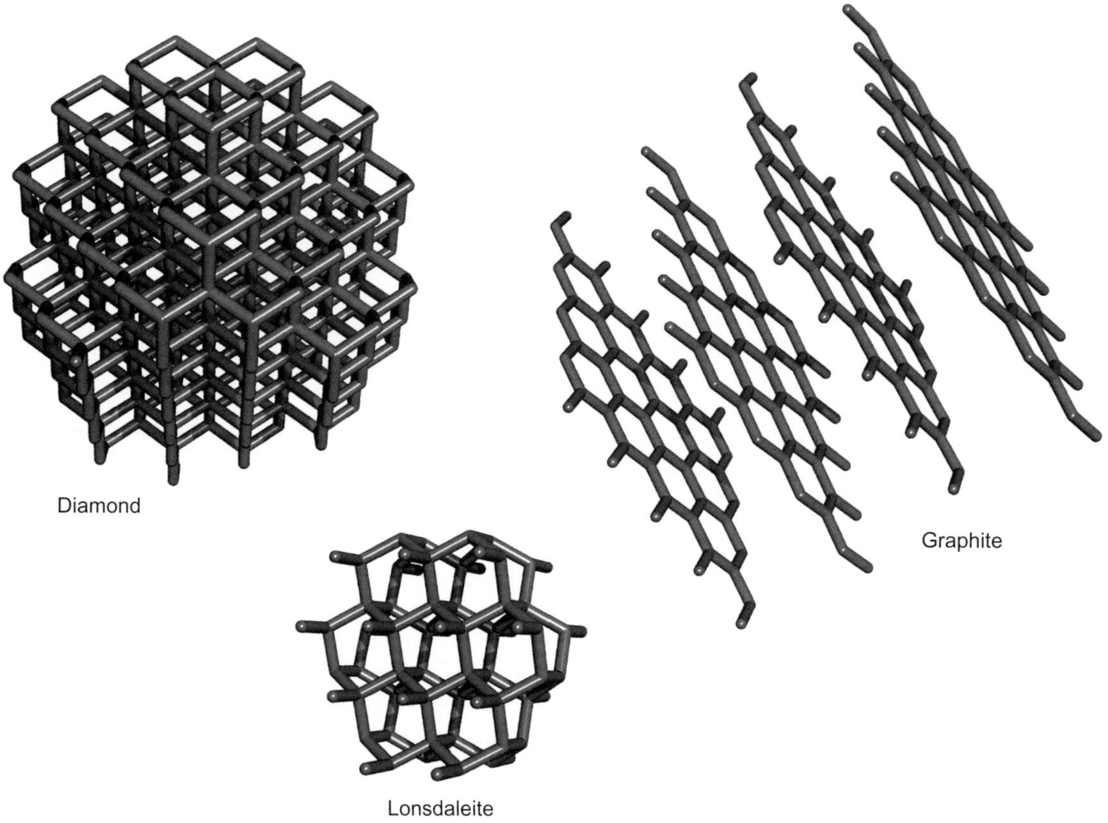

FIGURE 16.1 Three major allotropes of carbon.

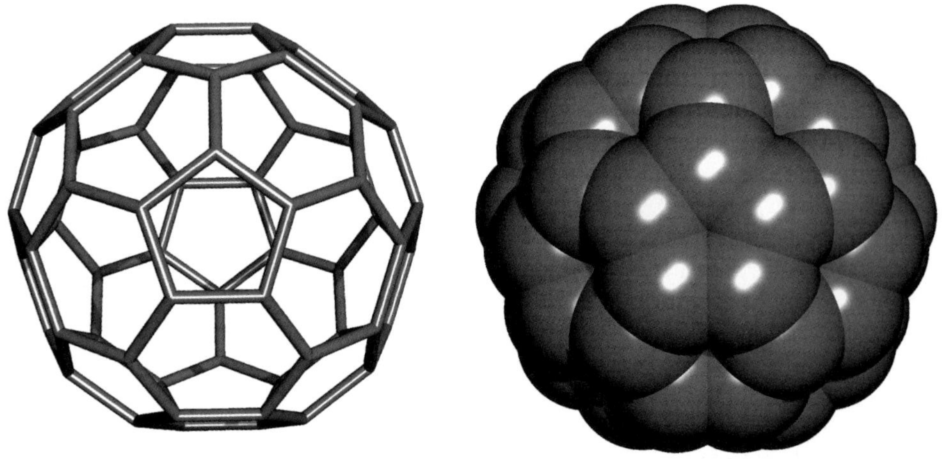

FIGURE 16.2 The structure of a C_{60} fullerene, also called a Buckyball.

carriers of C_{60} by holding the fullerene within its hydrophobic core (Andersson *et al.*, 1992; Braun, 1997; Filippone *et al.*, 2002; Samai and Geckeler, 2000) and covalent conjugates of C_{60} with alpha-cyclodextrin have been prepared (Xiao *et al.*, 2012) (Figure 16.3).

In many methods for derivatization of C_{60}, the initial modification is based on the reaction at a 6,6 ring junction on the fullerene with an azomethine ylide to form the 1,3-dipolar cycloaddition product (the "Prato reaction"), which creates a fulleropyrrolidine (Prato *et al.*, 1996; Bottari and Torres, 2011). The reaction is carried out overnight with heating to reflux in organic solvent. Typical reactants that combine with C_{60} in this reaction include an *N*-glycine derivative (with a constituent off the alpha-amino group) and an aldehyde derivative, which gives the fulleropyrrolidine compound according to Figure 16.4. By judicious choice of the right starting materials, the process provides a range of derivative possibilities to employ fullerenes in various bioconjugate applications. If the reactants are added in large excess over the concentration of the fullerene, then up to nine such pyrrolidine groups can be introduced per C_{60} molecule. The number of modifications actually ends up being a bell shaped curve from five to nine pyrrolidine derivatives, with a peak at seven modifications (Prato and Maggini, 1998). By controlling the length of the reaction, a mono-substituted product can be obtained in 40 to 50% yields.

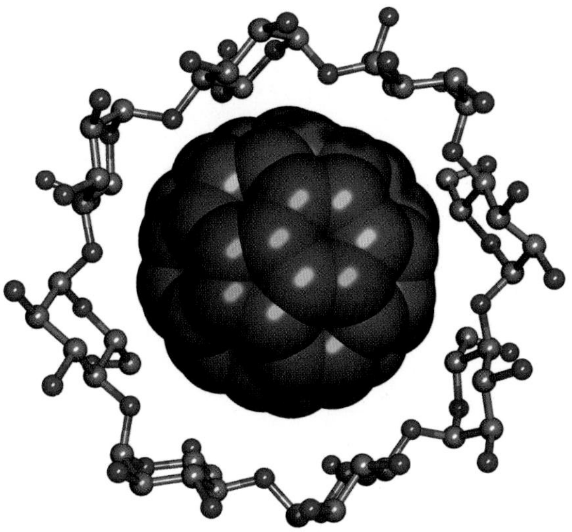

FIGURE 16.3 Gamma-cyclodextrin is able to accommodate a C_{60} fullerene within its core and solubilize it in aqueous solution.

In addition, the use of appropriate hydrophilic constituents on the aldehyde or glycine reactants can result in excellent water solubility of the C_{60} derivative. Two such modification arms can be added simultaneously to the pyrrolidine ring, thus providing a functional group for further conjugation and a hydrophilic arm for increased water solubility. PEG derivatives have been formed in this manner that create highly soluble fullerene derivatives.

The following procedure adapted from Prato *et al.* (1996) is an example of how glycine and formaldehyde derivatives may be used to create fullerene modifications for subsequent bioconjugation purposes.

Protocol

1. In a fume hood, prepare a solution of 100 mg of C_{60} dissolved in 100 ml of toluene.
2. Add to the fullerene solution with mixing, 25 mg of *N*-methyl glycine (sarcosine) and 20 mg of paraformaldehyde.
3. React by refluxing for 2 h with mixing.
4. Remove solvent under vacuum and purify the *N*-methyl-3,4-fulleropyrrolidine (Figure 16.5) by flash chromatography using toluene as eluent.

A similar preparative procedure can be used to form the *N*-(triphenylmethyl)-3,4-fulleropyrrolidine, which is a trityl protected amine derivative that may be deprotected using trifluoromethanesulfonic acid (TFMSA) to create a free secondary amine for conjugation. Kurz *et al.* (1998) used this C_{60} derivative to react with 3-maleimidopropionyl chloride to form a sulfhydryl-reactive fullerene. Although this derivative was not soluble in aqueous buffers, it was reactive enough as a micro-precipitate to be conjugated to thiol-containing proteins using the following procedure.

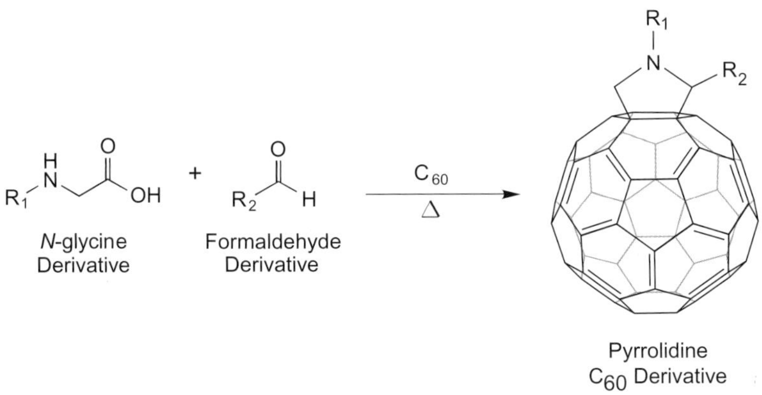

FIGURE 16.4 The reaction of a glycine derivative and a formaldehyde compound with a C_{60} molecule leads to the formation of a fulleropyrrolidine.

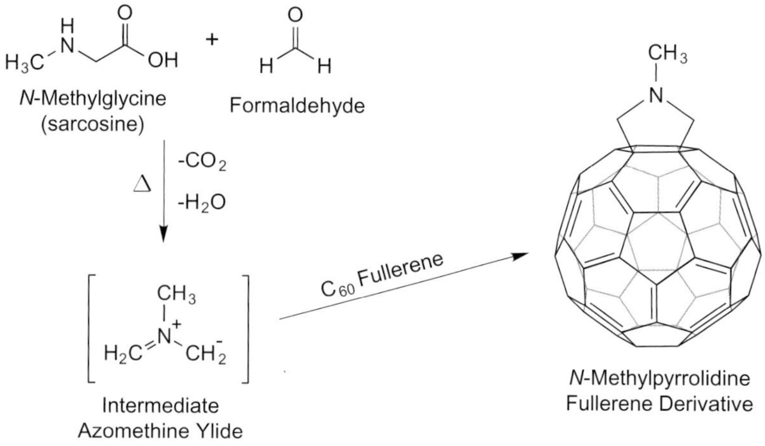

FIGURE 16.5 The synthesis of *N*-methylpyrrolidine–C_{60} proceeds through an intermediate azomethine ylide by the reaction of *N*-methylglycine plus formaldehyde upon heating.

Protocol

1. The thiol-containing protein is dissolved as a 60-μ*M* solution in 20-m*M* HEPES, pH 7.
2. Add the maleimide–C_{60} derivative to the protein solution at 100× mole excess with stirring. The addition of detergent to the solution may increase the solubility of the fullerene compound.
3. React with mixing at 4°C for 72 h.
4. Purify the labeled protein by size exclusion chromatography using a column with an exclusion limit of MW 5,000.

The nitrogen group in fulleropyrrolidines can be used for conjugation with crosslinking agents or hydrophilic biotinylation compounds for subsequent use in bioconjugation reactions. To create the free secondary amine group (NH) fulleropyrrolidine, an amine-protected starting material can be used in the reaction (Chai *et al.*, 2006). For instance, a trityl-oxazolidinone (using either triphenylmethyl- or, better, 4-methoxytriphenylmethyl-protecting groups) can be reacted with C_{60} to yield the trityl-protected pyrrolidine (Figure 16.6).

The following procedure for creating the NH fulleropyrrolidine is adapted from Maggini *et al.* (1994), Prato and Maggini (1998), and Prato *et al.* (1996). Extreme care should be taken when using TFMSA, as this acid is 30-times stronger than concentrated sulfuric acid.

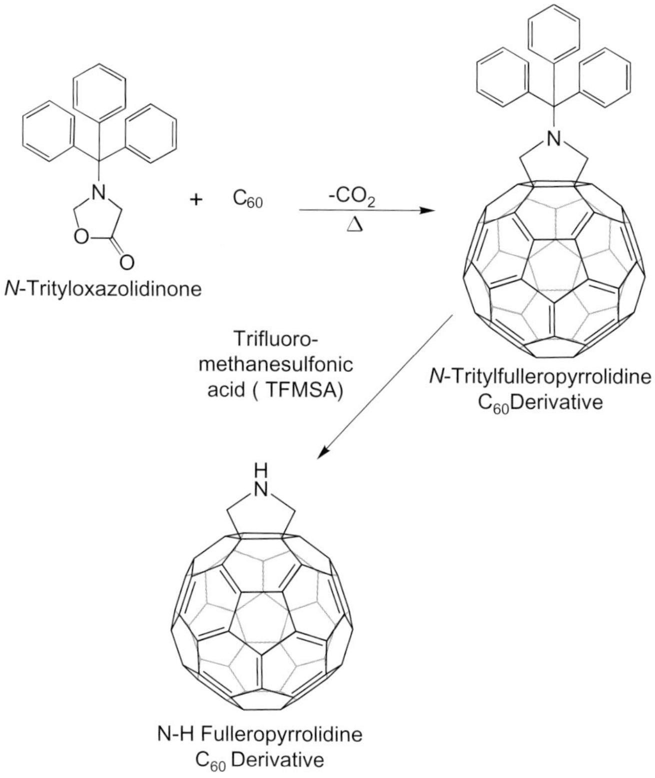

FIGURE 16.6 A trityl-protected pyrrolidine derivative of C_{60} can be prepared by the reaction of *N*-trityl-oxazolidinone with a fullerene. Deprotection of the trityl group using methanesulfonic acid gives the secondary amine, which can be used in further conjugation reactions.

Protocol

1. Dissolve 100 mg of C_{60} in 130 ml of chlorobenzene in a fume hood.
2. Add with mixing to the fullerene solution 54 mg of either *N*-(triphenylmethyl)-5 oxazolidinone or the same amount of the methoxytrityl oxazolidinone.
3. Reflux overnight in a fume hood with constant mixing.
4. Remove the solvent under vacuum and purify the residue using flash chromatography with an 8:2 mixture of petroleum ether/toluene as the eluent.
5. Remove the trityl-protecting group by dissolving the derivatized fulleropyrrolidine in 5 ml of dichloromethane in a fume hood and then adding 50 μl of TFMSA (caution!) with stirring.
6. React at room temperature for 1 h to remove the trityl protecting group, yielding the NH pyrrolidine derivative of C_{60}.

Capaccio *et al.* (2005) used the NH fulleropyrrolidine derivative to couple long-chain biotin to the C_{60} using the carbodiimide dicyclohexyl carbodiimide (DCC) in a pyridine/DMF/CH_2Cl_2 solvent mixture and reacting overnight at room temperature. This derivative was prepared using a hydrophobic, aliphatic long-chain biotin, but similar derivatives could be prepared using a biotin–PEG–carboxylate compound (e.g., NHS–PEG_4–biotin) to create a hydrophilic fullerene modification (Figure 16.7). The biotinylated fullerene could be used with streptavidin and another biotinylated enzyme to create a bioconjugate.

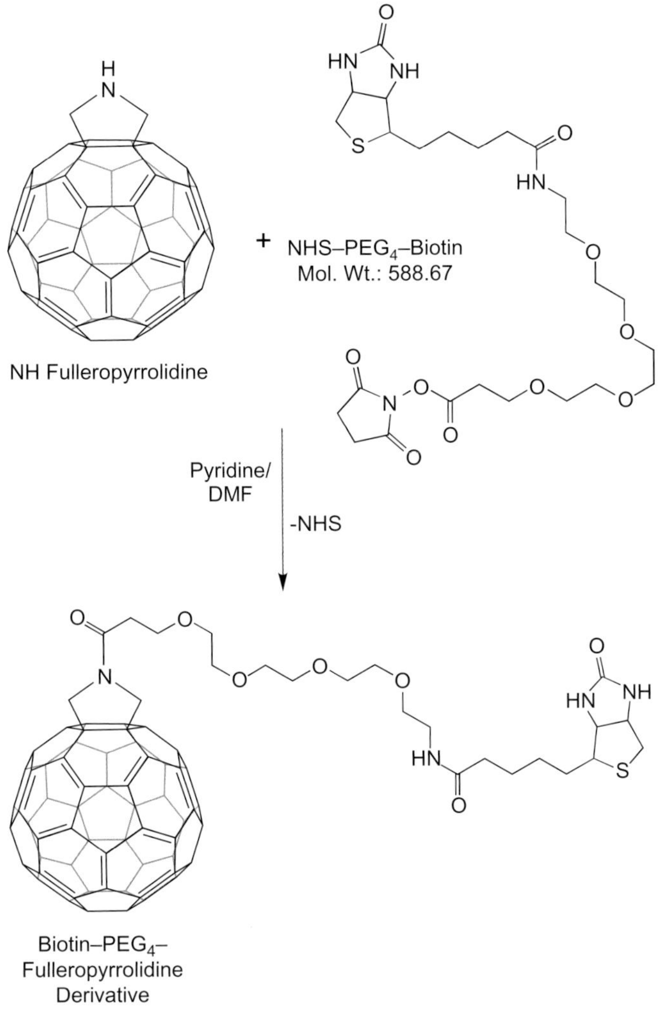

FIGURE 16.7 Reaction of an NH fulleropyrrolidine with NHS–PEG_4–biotin creates the biotinylated C_{60} derivative *via* an amide bond.

Similar to the fullerene modifications using either glycine/formaldehyde derivatives or oxazolidinone compounds, Maggini and Scorrano (1993) found that aziridines could yield similar pyrrolidine derivatives. Heating aziridine compounds in toluene was found to result in ring opening of the aziridine with subsequent covalent linking to the C_{60} ring system (Figure 16.8). Thus, synthesizing pyrrolidine derivatives of fullerenes can be done by several routes with success.

Another route to the preparation of C_{60} derivatives involves the reaction of a halogen derivative of diethylmalonate in the presence of a strong base with fullerenes, which results in a cyclopropanation product (Bingel, 1993; Isaacs *et al.*, 1993; Isaacs and Diederich, 1993). Using this reaction, various malonate derivatives of C_{60} can be made to facilitate production of bioconjugates (Zakharian, 2005). This reaction is often called the Bingel reaction (or Bingel–Hirsh addition) after the publication by the author in 1993 (see also Hirsch *et al.*, 1994). The reaction can yield carboxylate derivatives that then can be used to couple biomolecules through carbodiimide conjugation (Figure 16.9). Modifications of this reaction have been used frequently to produce important fullerene conjugates (Montellano *et al.*, 2011; Ruppert *et al.*, 2011; Tuktarov *et al.*, 2011; Xiao *et al.*, 2012).

The following protocol represents a C_{60} modification with an ethylmalonate derivative using the Bingel-type reaction, which creates an amine functional group on the fullerene surface, and is based on the method of Zakharian *et al.* (2005) (Figure 16.10). Ashcroft *et al.* (2006) used this method to create C_{60}

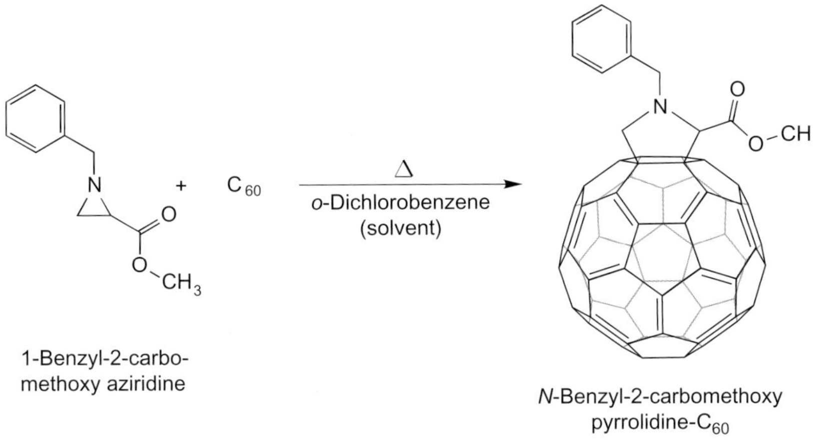

FIGURE 16.8 The reaction of aziridine derivatives with fullerenes can also give pyrrolidine derivatives useful for bioconjugation.

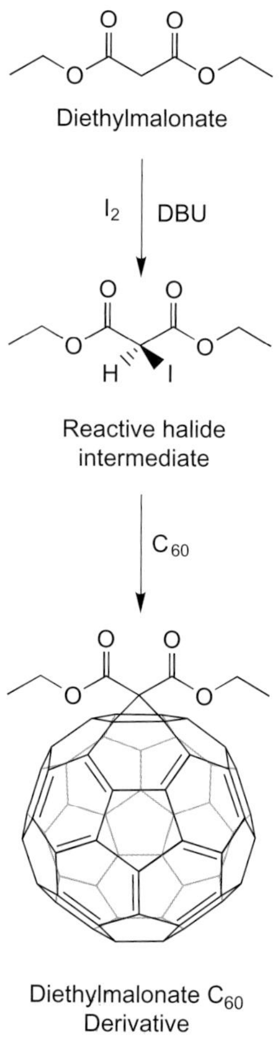

FIGURE 16.9 The Bingel reaction for the modification of fullerenes involves the *in situ* formation of a reactive halogen species in the presence of the strong base DBU. The cyclopropanation product can be used to create many bioconjugates.

immunoconjugates with the murine anti-gp240 melanoma antibody, which was made water soluble through the additional modification of the fullerene with hydrophilic malonodiserinolamide groups.

Protocol

1. In a fume hood, prepare a solution of 1.16 g *tert*-butyl-*N*-(3-hydroxypropyl)carbamate (Aldrich) in 100 ml of dry CH_2Cl_2 and add 1 ml of pyridine with stirring.
2. Cool the solution to 0°C and slowly add 1 g of ethylmalonyl chloride (Aldrich) under nitrogen with stirring.
3. React for 12 h with stirring at room temperature.
4. The resultant ethylmalonate–protected-amine compound is concentrated *in vacuo* and then purified using silica gel chromatography with a hexane/ethyl acetate (1:1) eluent.
5. In a fume hood, dissolve 400 mg of C_{60} in 700 ml of toluene with stirring.
6. Add to the C_{60} solution 100 mg of the purified ethylmalonate–protected-amine compound from step 4 along with 88 mg of I_2 and 105 mg of DBU (Aldrich).
7. React for 30 min at room temperature with mixing.
8. Remove the solvent *in vacuo* and purify the resultant mixture using silica gel chromatography. The initial eluent is toluene, which will remove the unreacted C_{60}, and then follow with a 10:1 mixture of toluene/ethyl acetate to elute the desired C_{60}–protected-amine derivative.
9. To remove the *tert*-butyl protecting group on the amine, in a fume hood add 190 mg of the C_{60}–protected-amine derivative from step 8 to 50 ml CH_2Cl_2 with stirring. Add to this solution with stirring 50 ml of TFA and mix for 30 minutes. Remove the solvent *in vacuo* to yield the final amine–C_{60} compound.

The amine group on the C_{60} fullerene may be used to couple carboxylate-containing spacers or affinity ligands using a carbodiimide conjugation reaction to form an amide linkage. The reaction may be performed in organic solvent or aqueous buffer, depending on the hydrophilicity of the fullerene derivative. A mixture of organic and aqueous solution also may be done if the ligand is more soluble in an aqueous environment.

A similar fullerene modification process developed by Hummelen *et al.* (1995) resulted in a number of phenyl C_{60} butyric acid methyl ester (PCBM) derivatives, which now are commercially available (Nano-C, Solenne). The reaction proceeds through the creation of a diazo derivative of methyl 4-benzoylbutyrate, which then is reacted with the aromatic ring system of C_{60} to give the cycloaddition product (Figure 16.11). The conjugate forms through the formation of both a 5,6 ring addition product (fulleroid) or a 6,6 ring addition product (methanofullerene). The fulleroid product results in breaking the 5,6 ring junction on the C_{60} molecule with dimeric bond formation to the aryl methyl group on PCBM.

The PCBM methyl ester can be used for coupling amine-containing ligands after removal of the methyl group and activation of the carboxylate using a number of different reaction strategies. Hummelen *et al.* (1995) successfully coupled cholestanol and histamine to the fullerene–PCBM derivative (after acid chloride formation) for use in fabrication of photodetectors and biological studies, respectively. For specific applications of PCBM-fullerenes, see Shaheen *et al.* (2001), Brabec *et al.* (2001), Yu *et al.* (1995), Mecher *et al.* (2002), Meijer *et al.* (2003), van Duren *et al.* (2004), and Anthopoulos *et al.* (2004).

Various commercial suppliers now offer fullerene derivatives with functionalities available for bioconjugation, including carboxylic and poly-hydroxylic derivatives, which are very hydrophilic and water soluble (BuckyUSA, NanoLab, NanoNB, Nano-C, and Aldrich).

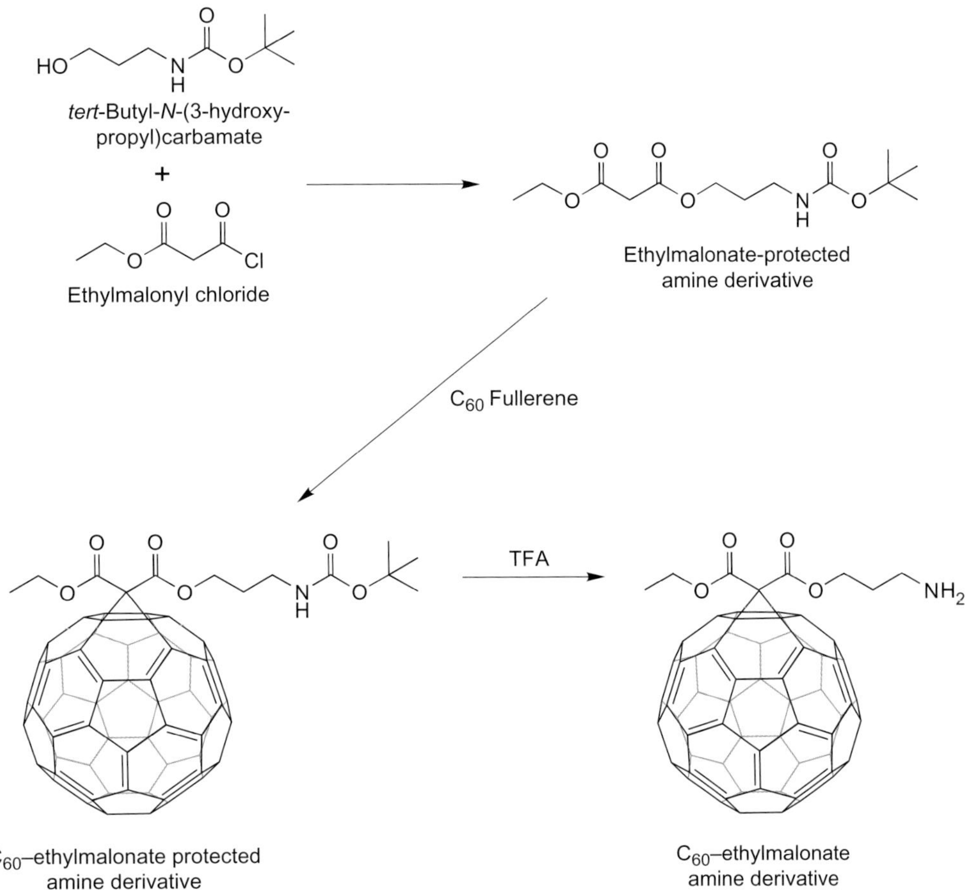

FIGURE 16.10 The reaction of the amine-blocked derivative of 3-hydroxypropylamine with ethylmalonyl chloride gives an ethylmalonate protected amine compound, which can be used in the Bingel reaction to create an amine group on a fullerene surface. Reaction with C_{60} in the presence of I_2 and DBU gives the cyclopropanation product that can be deprotected with TFA to yield the free amine.

2. CARBON NANOTUBES

2.1. Nanotube Properties

Cousins of the spheroidal fullerene molecules are carbon nanotubes. These are cylindrical graphene constructs with either open or closed ends and consist of carbon atoms arranged in a hexagonal pattern (Kit *et al.*, 2012). Similar to graphene, carbon nanotubes are being investigated and used as nanowires and sensors due to their electronic properties (Fam *et al.*, 2011; Zhang *et al.*, 2011 Tan *et al.*, 2012). There are two main families of carbon nanotubes that are distinguished by being either single-walled nanotubes (SWNTs) or multi-walled nanotubes (MWNTs). SWNTs are actually a graphene sheet that is seamlessly wound into a cylinder. MWNTs consist of multiple SWNTs that are concentrically wound around each other and nested together to create a tubes-within-tubes configuration (Figure 16.12). The type of nanotube is also determined by the manner in which the hexagonal pattern of carbon rings is arranged in the cylindrical structure.

In most publications, Iijima is given credit for the discovery in 1991 of the nanotube structure of carbon (Iijima, 1991; Bethune *et al.*, 1993; Iijima and Ichihashi, 1993). However, it has been said that Oberlin *et al.* (1976) also imaged carbon nanotubes, perhaps even SWNTs. Incredibly, nearly a century earlier, there was a study on the thermal decomposition of methane that resulted in the formation of long carbon strands, which were proposed at the time as a candidate for filaments in light bulbs (see Bacon and Bowman *et al.*, 1957).

SWNTs are typically only a few nanometers in diameter (0.4 to ~3 nm), but MWNTs can be from about 1.4 nm to over 100 nm in diameter, depending on the number of concentric nanotubes making up the bundle (Baughman *et al.*, 2007). However, carbon nanotubes can be from nanometers to millimeters or even microns in length, depending on how they are made. The length-to-diameter ratio typically exceeds 10,000 in most preparations. This unique molecular structure can result in fascinating properties, which include extremely high tensile strength, electrical conductivity

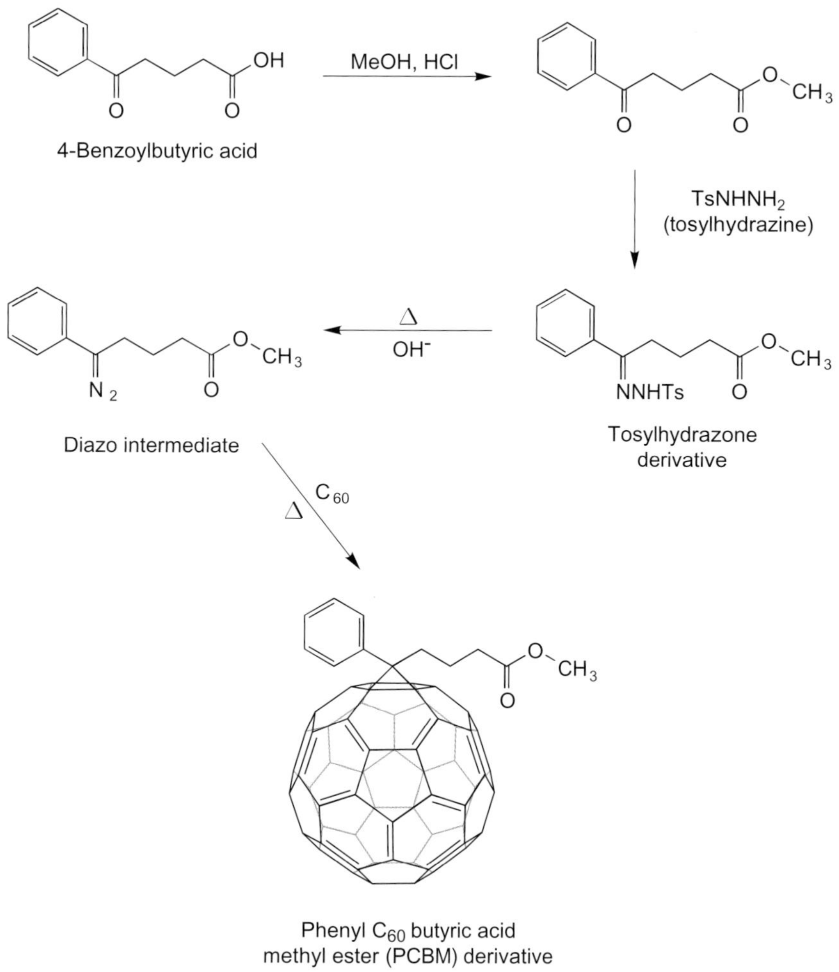

FIGURE 16.11 Fullerene–PCBM derivatives can be prepared using reactive diazo intermediates, which yield a cyclopropanation product similar to the Bingel reaction derivatives.

(or even semiconductor properties, depending on how the graphene sheet is wrapped), resistance to heat, and a great deal of chemical robustness. Nanotubes are being explored for use in applications ranging from electronics, optics, material science, biomedicine, biosensors, hydrogen storage, and nanoelectromechanical systems (NEMS) fabrication. The rate of growth in nanotube-related patents and publications has been increasing almost exponentially since the early 1990s, demonstrating the broad applications they can be used in (Baughman *et al.*, 2002; Park *et al.*, 2003).

The tensile strength of carbon nanotubes has been determined to be over 50 times that of high-carbon steel (Yu *et al.*, 2000). The strength of the bond structure in carbon nanotubes results from the fact that they are entirely sp^2 bonds, which are even stronger than the sp^3 carbon bonds in diamond. The addition of nanotubes to polymers, metals, and other structural materials has been used to add considerable strength to these materials. In fact, a carbon nanotube tether into low Earth orbit has been proposed as the only way of creating an elevator system into space.

The orbital bonding nature within carbon nanotubes creates unique electrical properties within a non-metallic molecule, which is a result of the delocalization of the pi-electron donated by each atom. Electrical conductivity can take place along the entire nanotube due to the freedom of pi-electron flow, making possible the design of circuits of extremely low-nanometer diameter.

Synthetic methods for the production of carbon nanotubes include arc discharge from graphite electrodes (Iijima, 1991; Collins and Avouris, 2000), pulsed laser ablation of a graphite substrate under high temperature (Guo *et al.*, 1995, 1995), chemical vapor deposition (José-Yacamán *et al.* 1993; Ren *et al.*, 1998), and high-pressure carbon monoxide (HiPco) method developed by Carbon Nanotechnologies, Inc. Methods for nanotube purification have been developed that result in 99.9% pure single-wall nanotubes (Chiang *et al.*, 2001).

Unlike the smaller, spheroidal fullerenes discussed previously, carbon nanotubes are not easily solubilized, even in organic solution. The reality is that all SWNTs and MWNTs are insoluble in all solvent systems. They also have a strong tendency to bind together and aggregate due to van der Waals attractive forces along the

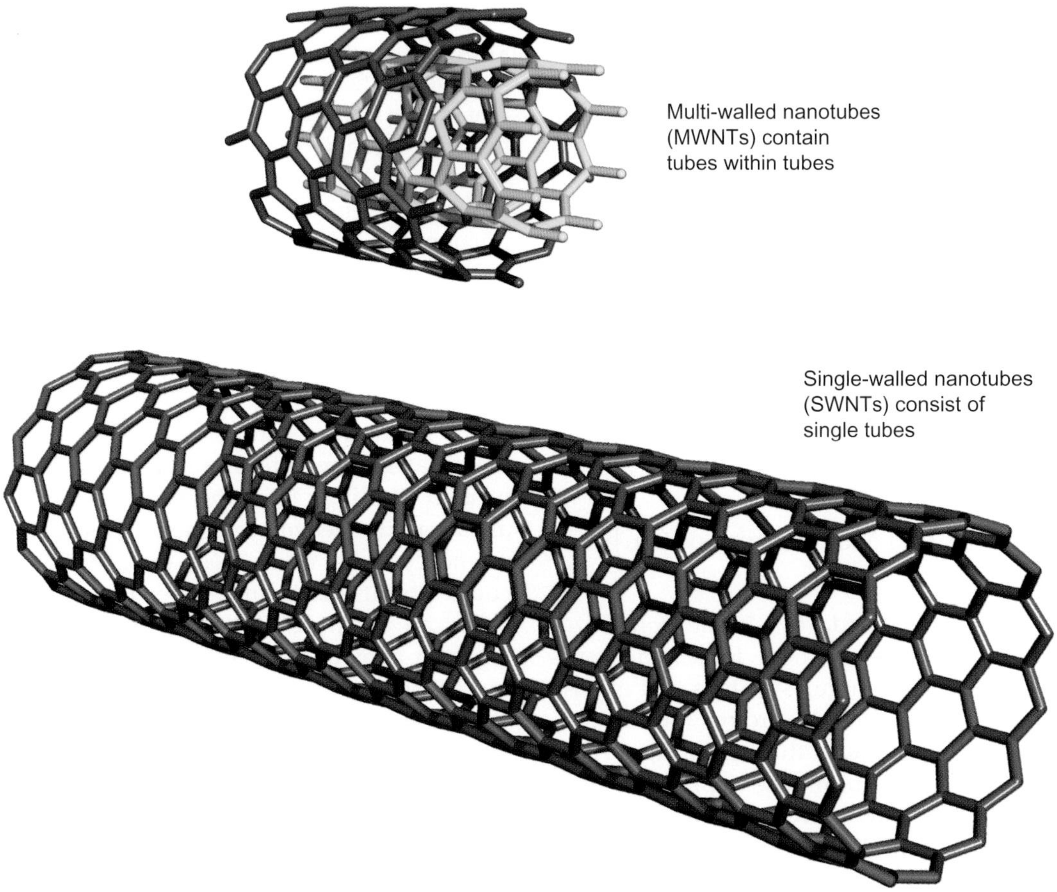
Multi-walled nanotubes
(MWNTs) contain
tubes within tubes
Single-walled nanotubes
(SWNTs) consist of
single tubes

FIGURE 16.12 An example of a single-walled carbon nanotube and a multi-walled carbon nanotube. Multi-walled varieties of CNTs can consist of numerous tubes within tubes.

length of the nanotubes. Since the length-to-diameter ratio is so high for nanotubes, bundles are often observed to be knotted and tangled masses, which are very difficult to unravel. The best that can be done is to disrupt the bundle and form a dispersion of the nanotubes in a solvent medium.

Some of the better solvents for pure SWNTs are the amide-containing ones, like DMF or *N*-methylpyrrolidone, but they still do not permit full dissolution, just dispersion (Boul *et al.*, 1999; Liu *et al.*, 1999). The addition of surfactants to carbon nanotube suspensions can aid in their solubilization and even permit their complete dispersion in aqueous solution. The hydrophobic tails of surfactant molecules quickly adsorb onto the surface of the carbon nanotubes, while the hydrophilic parts permit interaction with the surrounding polar solvent medium.

2.2. Nanotube Functionalization

An important route to solubilization of carbon nanotubes is to functionalize their surface to form groups that are more soluble in the desired solvent environment. It has been shown that acid treatment of nanotube bundles, particularly with HCl or HNO_3 at elevated temperatures, opens up the aggregate structure, reduces nanotube length, and facilitates dispersion (An *et al.*, 2004; Kordás *et al.*, 2006). Nitric acid treatment oxidizes the nanotubes at the defect sites of the outer graphene sheet, especially at the open ends (Hirsch, 2002; Álvaro *et al.*, 2004), and creates carbonyl, carboxyl, and hydroxyl groups, which aid in their solubility in polar solvents.

Such carbonyls may be further oxidized using potassium permanganate ($KMnO_4$) and perchloric acid ($HClO_4$) to convert all of these groups into carboxylic acids. Once functionalized in this manner, the nanotubes can be fully dispersed in aqueous systems. Kordás *et al.* (2006) used these derivatives to print nanotube patterns on paper or polymer surfaces to create conductive patterns for potential use in electronic circuitry. The carboxylates may also be used as conjugation sites to link other ligands or proteins to the nanotube surface using a carbodiimide reaction as previously discussed (Section 1, this chapter; Chapter 3, Section 1.12; Chapter 4, Section 1).

Untreated carbon nanotubes nonspecifically adsorb protein in an irreversible manner much like the noncovalent adsorption of protein onto hydrophobic surfaces (Chen *et al.*, 2003). It is essential, therefore, to modify the surfaces of nanotubes to prevent hydrophobic binding of biomolecules, especially if the resultant conjugates are to be used in biological assays or systems. Two main strategies have been used to make nanotubes biocompatible: (1) the noncovalent modification of the graphene surface with amphipathic molecules, which have functional groups that can be used for conjugation; or (2) the covalent modification of the outer nanotube surface to promote hydrophilicity and create functional groups for coupling other molecules.

2.3. Detergent or Lipid Modification of Carbon Nanotubes

Detergents have been used for simple solubilization of SWNTs in aqueous solution. Ionic detergents such as SDS will coat the nanotube surface and expose the negatively charged sulfonate groups to the surrounding aqueous environment, thus allowing SWNT dispersion in aqueous environments. Similarly, nonionic detergents such as Triton X-100 will coat the tubes and present their hydrophilic groups to the aqueous phase. Fischer *et al.* found that coating SWNTs with sodium dodecylbenzene sulfonate resulted in the best solubilization properties with long-term stability of the nanotubes in aqueous buffers (Zhou *et al.*, 2004). The detergent molecules do not merely lie in a random pattern on the nanotube surface; they coat the SWNT in a series of half-micelle structures, which create small knobs along its length (Figure 16.13).

Detergents have also been exploited in the noncovalent modification of carbon nanotubes by using modified detergents containing a coupled affinity ligand, which is linked to the hydrophilic part of the detergent molecule. Detergents that contain both a hydrophobic portion and a hydrophilic part with at least one terminal functional group for conjugation can be used in this process. The hydrophobic tail of many detergents will strongly adsorb to the nanotube's outer graphene cylinder, leaving the hydrophilic portions pointing outward and available for conjugation, if they contain an appropriate functional group. The result is a hydrophilic surface that is completely masked to prevent nonspecific protein binding. This approach to nanotube functionalization also leaves the chemical structure of the graphene cylinder unaffected, thus avoiding defects that could alter its electronic properties.

As an example of this strategy, Chen *et al.* (2003) used an activated Tween 20 detergent to coat SWNTs for subsequent conjugation to biotin, protein A, and U1A antigen. Tween 20 is a polyoxyethylene sorbitan monolaurate compound with 20 ethylene oxide units, 1 sorbitol unit, and 1 lauric acid group esterified as the hydrophobic tail. The fatty acid group avidly adsorbs to the carbon nanotube surface, leaving the three PEG arms, each containing terminal hydroxyl groups, sticking out from the coating to create an extremely biocompatible construct (Figure 16.14). The hydroxyl groups on the Tween molecules can be activated for coupling to biomolecules using many of the methods discussed for hydroxylic particles in Chapters 14 and 15.

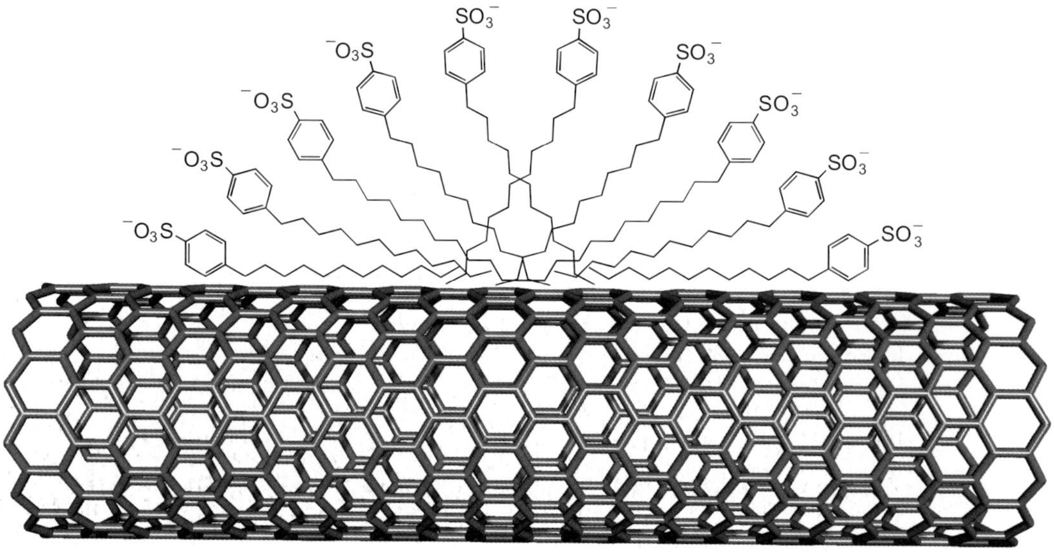

FIGURE 16.13 Detergent molecules can be used to solubilize carbon nanotubes by adsorption onto the surface through hydrophobic interactions and create half-micelle structures with the hydrophilic head groups facing outward into the aqueous environment.

For instance, activation of Tween 20 with carbonyl diimidazole (CDI) or disuccinimidyl carbonate (DSC) in a nonaqueous solution provides a reactive derivative suitable for coupling to amine-containing molecules, such as proteins.

The following protocol for the activation of Tween 20 with CDI and its subsequent use in modifying a carbon nanotube and coupling an affinity ligand is based on the method of Chen *et al.* (2003).

Protocol for Activation of Tween 20

1. Dissolve 5 mg of Tween 20 in 25 ml of dry DMSO with stirring.
2. Add 4 g of CDI with mixing and react for 1 h at room temperature.
3. Precipitate the CDI-activated Tween 20 by the addition of ethyl ether, and then isolate the precipitate using centrifugation or filtration. Redissolve the precipitated product in DMSO and

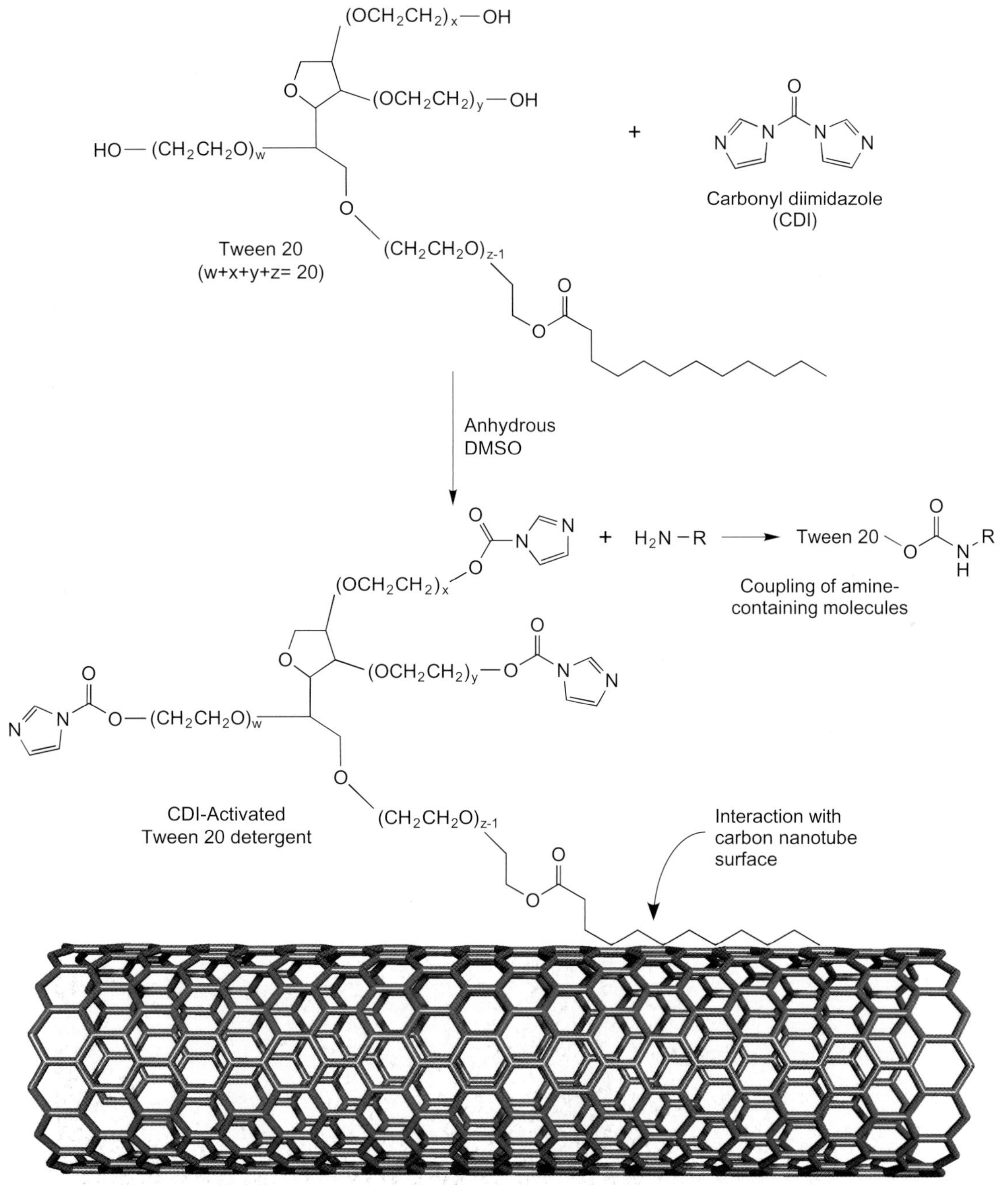

FIGURE 16.14 Tween-20 can be activated with CDI using its hydroxyl groups to create an amine-reactive imidazole carbamate intermediate that then can be used to coat a carbon nanotube. The result is an activated nanotube that can be used to couple proteins and other amine-containing molecules.

repeat the precipitation process two more times to ensure removal of excess reactants and reaction byproducts. After the final precipitation, dry the isolated precipitate overnight *in vacuo* to remove remaining solvent.

Protocol for Coating Nanotubes with CDI-Activated Tween 20

4. Suspend the SWNTs in 1% (w/w) CDI-activated Tween 20 solution in water using sonication and allow the detergent molecules to bind for 30 min at room temperature.
5. Quickly remove excess detergent by filtration on a 0.2-μm filter and washing the modified nanotubes with water.

Protocol for Coupling Activated Tween 20 to an Amine-Containing Ligand

6. Dissolve an amine-containing ligand in 0.1-*M* sodium carbonate, pH 9.5. If coupling a small ligand, such as a biotin–PEG–amine compound (Chapter 18, Section 1.3), then use a concentration of about 5 to 10-m*M* in the carbonate buffer. For proteins, concentrations of 10 n*M* to 1 μ*M* can be used with success.
7. Resuspend the activated nanotubes in the ligand-containing carbonate buffer and react with mixing overnight at room temperature or 4°C (e.g., for sensitive proteins).
8. Wash the coupled nanotubes with water or a suitable buffer using filtration and finally store them in buffer containing a preservative at 4°C.

In a similar approach to the noncovalent modification of carbon nanotubes with detergent molecules, Kam *et al.* (2005) used phospholipid derivatives to coat SWNTs for photo-therapeutic agents against tumor cells *in vivo*. A phospholipid containing a PEG–NH_2 group was used to couple folic acid as an affinity ligand (using EDC to form an amide bond), which preferentially can be taken up by cancer cells. SWNTs were modified by the lipid derivative in aqueous solution using sonication for 1 h and centrifuged to remove insoluble material. The aqueous fraction contained modified nanotubes, which contained the surface-adsorbed lipid derivative. After the modified SWNTs were incubated with cells, the solution was irradiated using an 808-nm laser. The nanotubes absorb light in this region of the spectrum and heat up to the point of causing cell death.

2.4. Pyrene Modification of Carbon Nanotubes

Another method for the noncovalent modification of carbon nanotubes involves the interaction of pyrene derivatives with the graphene sidewalls, presumably due to pi-stacking (Nakashima *et al.*, 2002) (Figure 16.15). This interaction forms tight complexes that completely coat the nanotube surface, and if the pyrene contains a hydrophilic portion, it can impart water solubility to the SWNT, as well. The additional presence of reactive groups or functional groups on a pyrene side chain permits conjugation to affinity ligands for biological applications.

Nakashima *et al.* (2002) found that a pyrene derivative containing a single side chain terminating in a positively

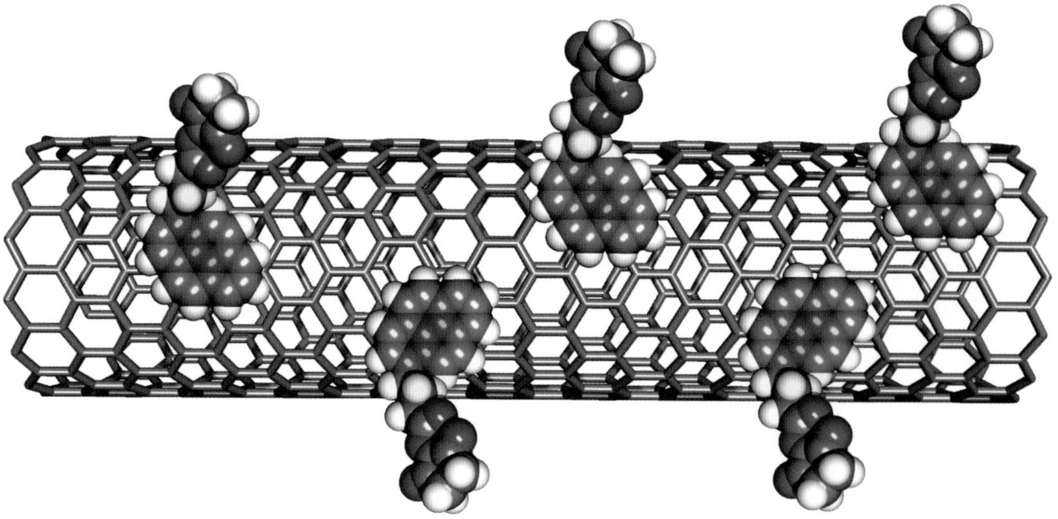

FIGURE 16.15 The NHS ester of a pyrene butyric acid derivative can be used to modify a carbon nanotube by adsorption of its rings onto the surface of the tube. The NHS ester groups then can be used to couple amine-containing molecules to form amide bonds.

charged quaternary amine imparted water solubility to SWNTs treated with the compound. The application of other pyrene derivatives can be performed to contribute both water solubility and an appropriate functionality for bioconjugation. Examples of pyrene derivatives that are suitable for the noncovalent modification of carbon nanotubes include those that have a single modification to the pyrene ring structure. Derivatives that contain multiple modifications off the pyrene group may affect its interaction potential for the SWNTs and should be avoided. For instance, water soluble poly-sulfonated derivatives of pyrene, such as the Cascade Blue fluorescent dyes described in Chapter 10, Section 5, should not be used, because the pyrene structure is too hydrophilic to associate with the nanotube surface due to the three negative charges contributed by the sulfonic acids.

Some commercially available pyrene compounds that may be used to functionalize a carbon nanotube by this method include 1-pyrenebutyric acid and 1-aminopyrene (from Acros or Aldrich) as well as *N*-(1-pyrenyl) maleimide, 2-(1-pyrenyl)ethyl chloroformate, 1-pyrenebutyric acid *N*-hydroxysuccinimide ester, 1-pyrenecarboxaldehyde, and 1-pyreneacetic acid (from Aldrich). Each of these compounds provides a single site of derivatization off the basic pyrene rings to contain either a functional group or reactive group for coupling ligands. Molecules modified with these pyrene derivatives may be used to treat a carbon nanotube to form a stable noncovalent complex.

The pyrene derivatives containing a carboxylate group, chloroformate, aldehyde, or an NHS ester can be used to couple to amine-containing ligands, including proteins. The maleimide–pyrene derivative may be used to couple with thiol-containing ligands, while the amine–pyrene compound may be used to conjugate with carboxylate-containing ligands. Also available are (1-pyrenyl)butyic acid hydrazide and pyrene-1-isothiocyanate from Molecular BioSciences, which react with aldehydes and amines, respectively. Once a ligand is conjugated to a pyrene derivative, the complex may be incubated with a carbon nanotube to produce the final noncovalent complex. Since the initial modification is done on a water-insoluble nanotube, it is best to perform the primary coating of the pyrene derivative in an organic solvent, such as DMF. It then is desirable to make the nanotube water soluble by linking a hydrophilic spacer arm to the pyrene–nanotube complex. If a hydrophilic spacer is built into the resultant pyrene conjugate, such as the use of a short-chain PEG compound, then the resultant SWNT complex will be completely water soluble (for example, see Figure 16.16). The PEG spacer chosen for this purpose should contain a terminal functional group for coupling to another molecule. At this point, the water soluble complex can be reacted with a protein or other affinity ligand in aqueous buffer to make the desired bioconjugate. This multi-step process will result in a biocompatible carbon nanotube that retains its electronic properties, is water soluble, and has added fluorescent properties due to the pyrene molecules coating its surface.

Maehashi *et al.* (2007) used pyrene adsorption to make carbon nanotubes labeled with DNA aptamers and incorporated them into a field effect transistor constructed to produce a label-free biosensor. The biosensor could measure the concentration of IgE in samples down to 250 p*M*, as the antibody molecules bound to the aptamers on the nanotubes. Felekis and Tagmatarchis (2005) used a positively charged pyrene compound to prepare water soluble SWNTs and then electrostatically adsorb porphyrin rings to study electron transfer interactions. Pyrene derivatives have also been used successfully to add a chromophore to carbon nanotubes using covalent coupling to an oxidized SWNT (Álvaro *et al.*, 2004). In this case, the pyrene ring structure was not used to adsorb directly to the nanotube surface, but a side-chain functional group was used to link it covalently to modified SWNTs.

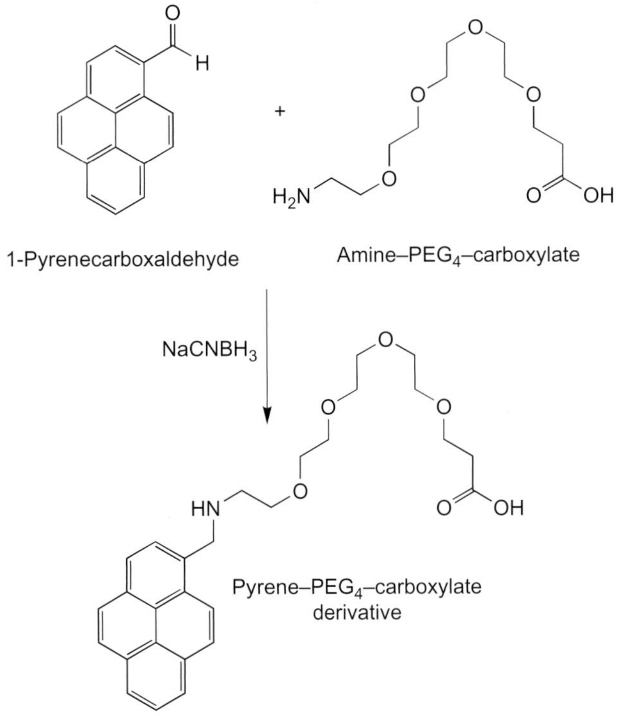

FIGURE 16.16 An aldehyde derivative of pyrene can be used to couple a hydrophilic amino–PEG–carboxylate spacer by reductive amination. The resultant derivative then can be used to coat a carbon nanotube through pyrene ring adsorption and result in a water soluble derivative containing terminal carboxylates for coupling amine-containing ligands.

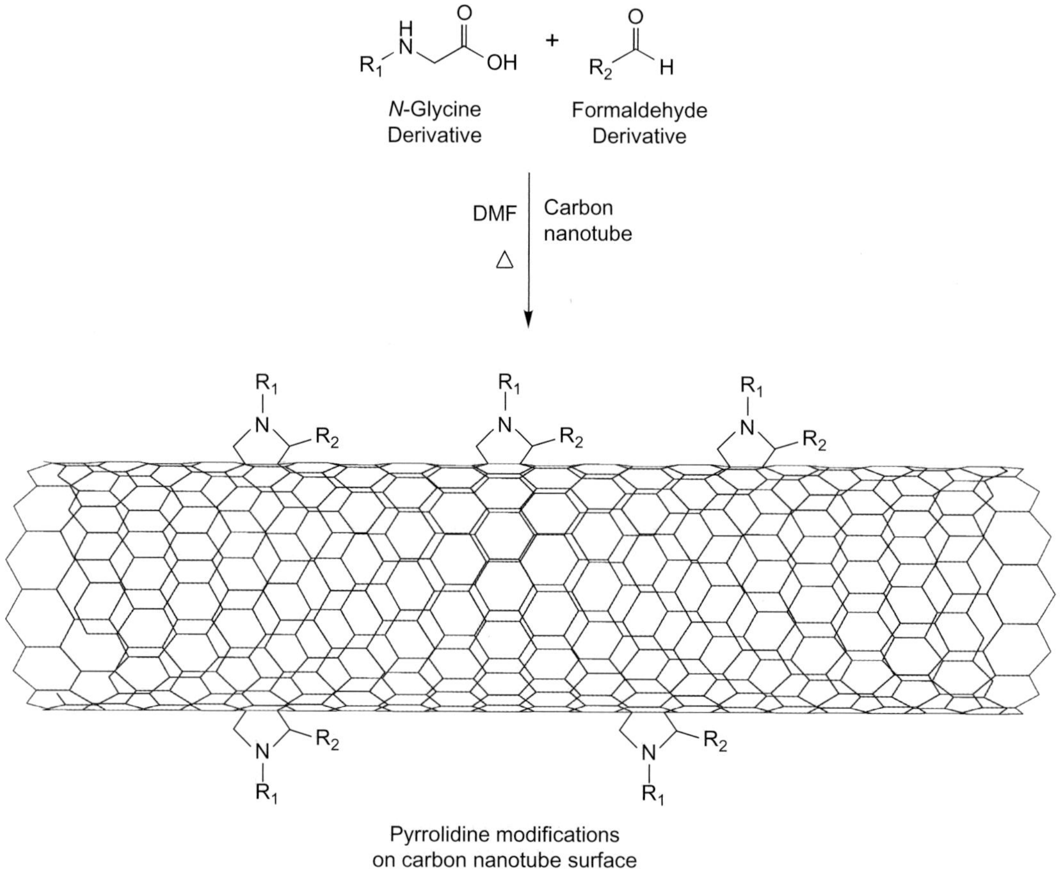

FIGURE 16.17 Some modification methods that are useful for fullerenes also can be used with carbon nanotubes. The reaction of an *N*-glycine compound with an aldehyde derivative can result in cycloaddition products, which create pyrrolidine modifications on the nanotube surface.

2.5. Modification of Carbon Nanotubes by Cycloaddition

The covalent methods previously discussed for fullerene modification using cycloaddition reactions also can be applied to carbon nanotubes (Kumar *et al.*, 2011). This strategy results in chemically linking molecules to the graphene rings on the outer surface of the cylinder, resulting in stable conjugates that can be designed to include hydrophilic groups for water solubilization. Georgakilas *et al.* (2002) described the use of a 1,3-dipolar cycloaddition process to carbon nanotubes with azomethine ylides, generated by condensation of an amino acid derivative and an aldehyde. The reaction occurs in organic solvent at high temperature over a time period of several days.

Typically, SWNTs are suspended in DMF using sonication and the aldehyde and glycine derivatives are added to the mixture with stirring. The alpha-amine derivative of glycine can include hydrophilic spacers to make the resultant nanotube water soluble as well as include protected functional groups to couple affinity ligands after deprotection (Kurz *et al.*, 1998). The aldehyde also can include R groups that add water solubility or functionality to the nanotube. A combination of a glycine derivative with an aldehyde derivative can result in both hydrophilicity and a functional group to conjugate ligands (Figure 16.17).

Felekis and Tagmatarchis (2005) used this cycloaddition process to prepare SWNT derivatives possessing photoactive components, such as the addition of ferrocene groups. They used a short PEG-type spacer on the glycine to impart water solubility at the same time.

Singh *et al.* (2006) also used cycloaddition to prepare carbon nanotubes containing indium-labeled diethylenetriamine pentaacetic acid (DTPA derivatives (Figure 16.18). In the initial modification, an SWNT was derivatized to contain a primary amine at the end of a short PEG spacer. The resultant water soluble nanotube then was reacted with DTPA to create a metal-chelating group at the end of the chain. Subsequent loading of the chelate with ^{111}In created a radionuclide–SWNT complex for *in vivo* biodistribution studies.

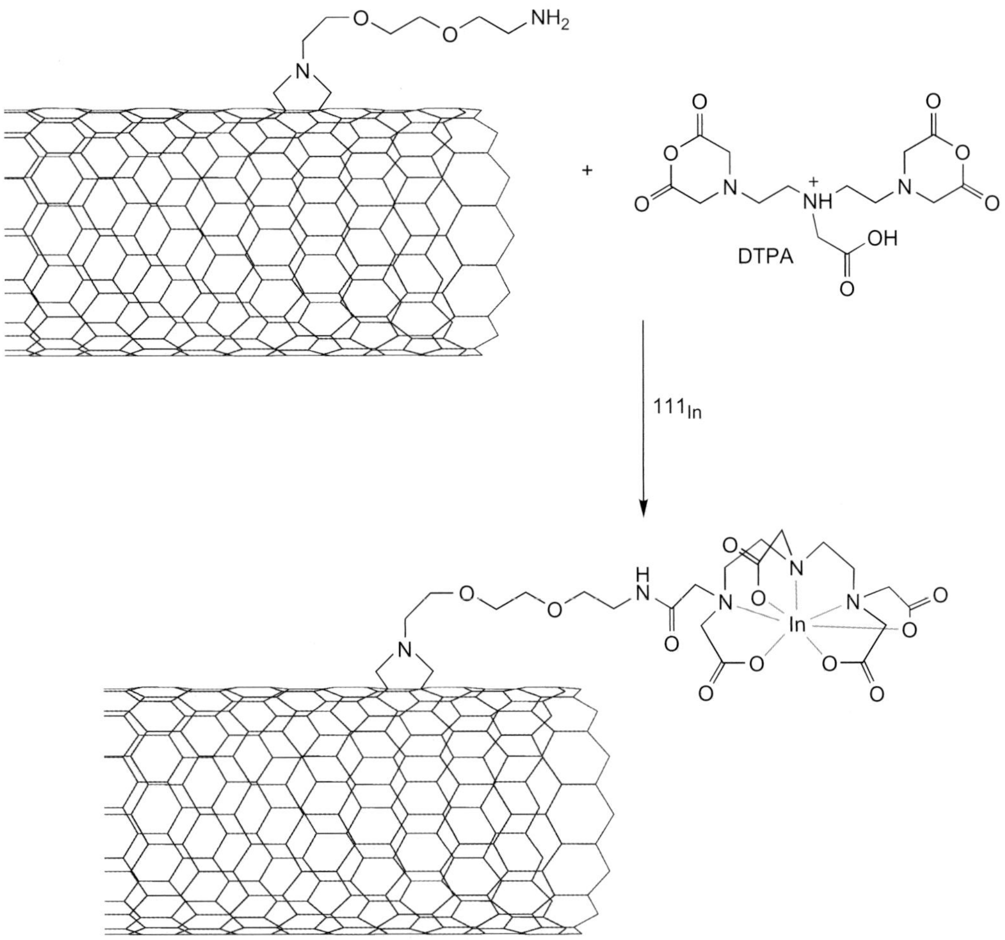

FIGURE 16.18 An amino–PEG–pyrrolidine derivative of carbon nanotubes can be used to couple metal-chelating groups, such as DTPA. Subsequent coordination of ^{111}In results in an indium chelate that can be used for imaging applications.

The demonstration that the 1,3-dipolar cycloaddition process with azomethine ylides works with nanotubes implies that similar reactions developed for use with fullerenes also will be successful with carbon nanotubes. In particular, the cyclopropanation reactions discussed previously for the modification of C_{60} also have been used for derivatization of SWNTs and MWNTs (Zakharian *et al.*, 2005; Umeyama *et al.*, 2011).

CHAPTER

17

Chemoselective Ligation; Bioorthogonal Reagents

The many dozens of reactions that are available for bioconjugation purposes generally are designed to work with biological molecules and the functionalities they contain. The main goal of most bioconjugate techniques is to use the functional groups on biomolecules to label with another type of biomolecule or link to synthetic probes. However, it is often desirable to couple one molecule to another without the potential for cross-reactivity with biomolecules. The term "chemoselective ligation" has been coined to describe the coupling of one reactive group specifically with another reactive group without side reactions in aqueous solution or in the presence of biological material (Lemieux and Bertozzi, 1998).

The incredible diversity of biomolecules in cells and organisms often presents problems for the goal of total chemoselectivity and bioorthogonality, as the number and variety of reactive sites on the molecules of life are extraordinary. Even the best crosslinkers or labeling reagents designed to be somewhat site specific in their reactions often display cross-reactivity with functional groups on biomolecules other than the ones intended for coupling. For instance, *N*-hydroxysuccinimide (NHS) esters usually are considered amine reactive, but they also can react with cysteine, histidine, serine, threonine, and tyrosine side chain groups. Maleimide groups too are touted as being thiol specific, but amines can also react with maleimides given the right conditions.

Bioorthogonal reagents ideally should contain a reactive group that will only react with another specific reactive group without any potential for cross-reactivity with biomolecule functionalities. In other words, a bioorthogonal reactive group could be added to a complex mixture of biological molecules in aqueous solution without reacting with any of them. Moreover, the ideal bioorthogonal system should be immune to instability in aqueous solutions, such as the tendency to hydrolyze or easily oxidize. True bioorthogonal reagents of this type will link to each other and only to each other in the presence of intracellular environments or in cell lysates or in defined biomolecule solutions.

The reality is that there are few options in this category of bioconjugate reagents and the systems or reactant pairs that have been developed for bioorthogonal applications have differing degrees of how well they perform in this task. The following sections describe the major chemoselective ligation reactions that can be considered to have a degree of bioorthogonal characteristics. In each system, the chemoselective pair of reactants can be separately linked or built into the design of crosslinkers or labeling reagents and used to modify biomolecules, surfaces, particles, or organic compounds. Subsequently, these labeled components can be brought together, even in complex solutions, to facilitate conjugation between the two bioorthogonal reacting species.

In addition, for several reactant strategies in chemoselective ligation, one of the reagent pairs can be designed into a biological monomer that can be utilized by cellular processes to become incorporated into biopolymers. This advantage provides a unique *in vivo* labeling capability through the feeding of monomer analogs to cells or organisms, such as modified amino acids or sugar derivatives, which then get selectively added into proteins, carbohydrates, or lipids. Thus, the ultimate application of bioorthogonal chemoselective ligation is to label specifically the molecules of life directly within living systems and without cross-reactions with other biological functional groups. For a review on the use of chemoselective reactions in living systems, see Prescher and Bertozzi (2005), as well as van Berkel *et al.* (2011).

Bioconjugate Techniques, Third Edition.
DOI: http://dx.doi.org/10.1016/B978-0-12-382239-0.00017-0

1. DIELS–ALDER REAGENT PAIRS

The Diels–Alder reaction has long been a staple for forming carbon–carbon bonds in organic synthesis (Smith and March, 2007). The typical reaction proceeds through the 2 + 4 cycloaddition of a double bond (alkene) and a diene to give a six-member ring product. The double-bond reactant is often called a dienophile, and electron-withdrawing constituents next to the alkene are used to accelerate the reaction (such as COOH, CHO, and COR groups, among others). Conversely, electron-donating groups on the diene are important for increasing reaction rates (Figure 17.1).

Some reports indicated that the Diels–Alder reaction could be done in aqueous environments with a potential for accelerated reaction rates under the right conditions (Rideout and Breslow, 1980; Blokzijl and Engberts, 1992; Pai and Smith, 1995; Otto *et al.*, 1996; Wijnen and Engberts, 1997), and the addition of $InCl_3$ was determined to act as a catalyst in aqueous environments (Loh *et al.*, 1996). For a review of organic reactions that can be done in aqueous media, see Li (2005).

In addition, it has been discovered that there are naturally occurring enzymes that facilitate Diels–Alder type reactions within certain metabolic pathways and that enzymes are also instrumental in forming polyketides, isoprenoids, phenylpropanoids, and alkaloids (de Araujo *et al.*, 2006). Agresti *et al.* (2002) identified ribozymes from RNA oligo libraries that catalyzed multiple-turnover Diels–Alder cycloaddition reactions.

In 2001, Hill *et al.* extended the use of aqueous-phase Diels–Alder reactions for the bioconjugation of diene-modified oligonucleotides. The dienophile that was used consisted of a simple maleimide derivative, which is present on a broad range of commercially available bioconjugation reagents. Modified oligonucleotides were prepared by solid-phase synthesis using a 3,5-hexadiene phosphoramidite derivative, which could be incorporated into the oligo at the 5′ end. Maleimide compounds investigated for oligo bioconjugation include *N*-ethylmaleimide, biotin–BMCC, fluorescein–maleimide, coumarin–maleimide, maleimide-PEG_2-biotin, and an mPEG–maleimide. Figure 17.2 shows the reaction of maleimide–PEG_2–biotin with the diene-modified oligo to yield the cycloaddition product.

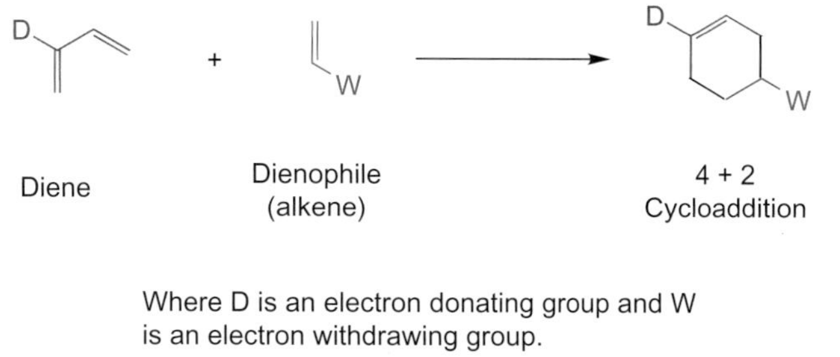

FIGURE 17.1 A general Diels Alder reaction consists of a 4 + 2 cycloaddition between a diene and an alkene, often called a dienophile. The reaction rate and yield increase if the diene contains an electron donating group and the alkene contains an electron withdrawing group.

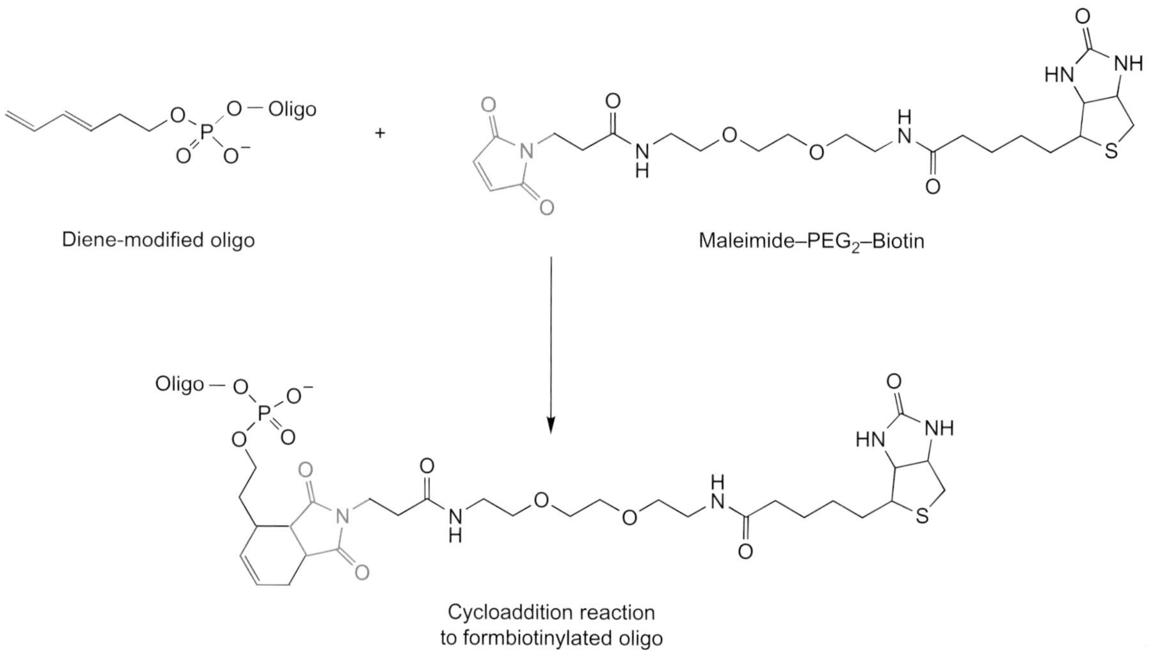

FIGURE 17.2 Maleimide groups provide good dienophiles for a Diels–Alder reaction. Biotin–PEG_2–maleimide can react with an oligo-diene molecule to form a covalent cycloaddition product, which adds the biotin tag to the oligo.

The reaction kinetics between a maleimide derivative and a 3,5-hexadiene derivative varies depending on the maleimide compound being reacted. Cycloaddition yields of greater than 80% and often as much as 90 to 95% can be expected within 1 to 18 h at room temperature or slightly elevated reaction conditions (e.g., 30°C).

The conjugation of oligonucleotides with peptides can also be carried out using Diels–Alder cycloadditions in water (Tona and Häner, 2005). Marchan *et al.* (2006) used the same 3,5-hexadiene phosphoramidite derivative as Hill *et al.* (2001), but in this case used a maleimide-modified peptide sequence. A mild, aqueous Diels–Alder reaction between them resulted in the formation of the cycloaddition product.

The Diels–Alder reaction for bioconjugation has been used for the chemoselective ligation of peptides and proteins in aqueous solution (de Araujo *et al.*, 2006). Peptides modified using a 2,4-hexadienyl ester were derivatized to contain a diene and were found to be reactive toward other peptides containing an N-terminal maleimide group to give the cycloaddition product in high yield. The hexadienyl group was also attached to biotin and allowed to interact with streptavidin, which could then be conjugated with peptides containing a maleimide group.

Diels–Alder cycloaddition reactions have also been used to link carbohydrates covalently to proteins (Pozsgay *et al.*, 2002), carbohydrates to surfaces (Beckmann *et al.*, 2012), for the immobilization of proteins onto surfaces (Chen *et al.*, 2011), as well as for the immobilization of oligonucleotides on glass surfaces to create arrays (Latham-Timmons *et al.*, 2003). In an application that used two chemoselective ligation reactions, Sun *et al.* (2006) employed sequential Diels–Alder and azide–alkyne (click chemistry) cycloaddition reactions to immobilize protein, biotin, or carbohydrate ligands on solid surfaces. In this case, glass slides containing maleimidocaproyl groups were used as the dienophiles and a PEG_4 spacer containing an alkyne on one end and a cyclopentadiene at the other end was the reactive linker. A Diels–Alder reaction coupled the maleimide groups to the cyclopentadiene groups on the spacer, while the alkyne groups at the other end were used in a click chemistry reaction to attach azide-containing ligands (Figure 17.3). The cycloaddition reaction between the maleimide groups on the slides and the cyclopentadiene group on the spacer was done in a 1:1 solution of water:*tert*-BuOH at room temperature for 12 h. After washing with the same water/solvent mixture, the slides contained hydrophilic spacers terminating in alkyne groups, which then could be coupled with the azide-containing ligands using click chemistry (see Section 6).

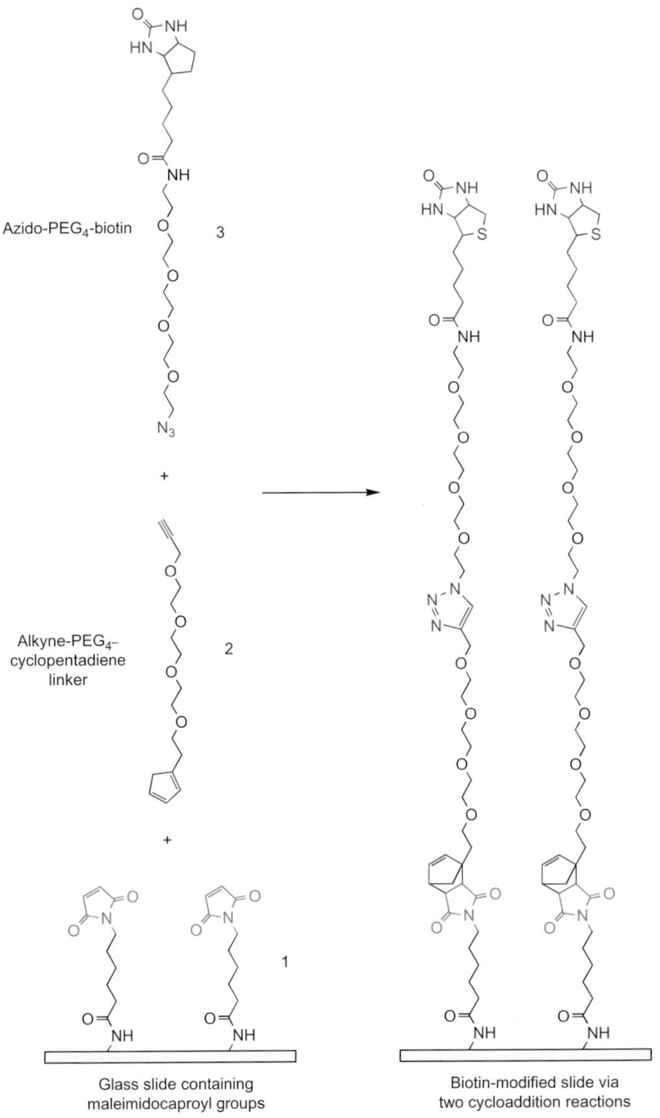

FIGURE 17.3 Maleimide-modified glass slides (1) can be derivatized using two chemoselective ligation reactions to create biotin modifications. In the first step, alkyne–PEG_4–cyclopentadiene linkers (2) are added to the maleimide groups using a Diels–Alder reaction. In the second reaction, an azido–PEG_4–biotin compound (3) is reacted with the terminal alkyne on the slide using click chemistry to result in another cycloaddition product, a triazole ring.

Chemoselective ligation reactions using the Diels–Alder cycloaddition process offer another bioconjugation route using a reactive component available commercially (maleimide-containing reagents). However, their use as true bioorthogonal reactants is limited due to the cross-reactivity of maleimides toward thiols. For instance, in the modification of a Rab protein, a cysteine residue had to be protected prior to cycloaddition using a maleimide compound (de Araujo *et al.*, 2006); otherwise, the maleimide would have coupled to the sulfhydryl, too.

2. HYDRAZINE–ALDEHYDE REAGENT PAIRS

The reaction between an aldehyde or ketone and a hydrazide or hydrazine derivative to form a hydrazone bond has been frequently used for bioconjugation purposes (Chapter 3, Section 5.1). The reaction is appealing from a bioorthogonal perspective, because natural biopolymers do not normally contain these reactive groups. Although aldehydes may form temporary Schiff base interactions with amines on proteins and other biomolecules, in aqueous solution they are fully reversible and will rapidly exchange with a hydrazide or hydrazine, if present. This also holds true of aminooxy (hydroxylamine) derivatives, which form stable oxime bonds with aldehydes, although the number of reagents available with an aminooxy functionality is limited, these reagents are growing in popularity.

The only potential problem of cross-reactivity for this chemoselective reaction pair with molecules of biological origin might occur from a hydrazine reacting with aldehyde- or ketone-containing metabolic intermediates, reducing sugars, or similar small organic molecules present within cells or cell lysates. However, if the hydrazine reagent is added in sufficient excess, the desired coupling reaction will still occur in such environments, as evidenced by the many successful bioconjugation reactions carried out with oxidized glycoproteins on cell surfaces or in lysates (Bayer and Wilchek, 1990; Bayer *et al.*, 1987a,b; 1990).

The hydrazine–aldehyde reaction has been used intracellularly to deliver nontoxic drug components, which when linked to form a hydrazone bond *in situ*, become cytotoxic (Rideout, 1986, 1994; Rideout *et al.*, 1990). This same approach has been used to generate enzyme inhibitors *in vivo*, wherein the hydrazine and aldehyde precursors are not active, but when coupled together within cells to form a hydrazone linkage, become active site binders (Rotenberg *et al.*, 1991).

Ketone-containing bio-monomer analogs have been used successfully to incorporate these functionalities into biopolymers, such as proteins and carbohydrates (glycans). For example, a phenylalanine analog containing a *para*-substituted aryl ketone group is able to be transformed into an aminoacyl t-RNA and then introduced with sequence specificity into a protein by cellular ribosomal machinery (Datta *et al.*, 2002). In addition, the post-translational glycosylation process within cells is tolerant of certain sugar analogs, and ketone derivatives of monosaccharides have been used to incorporate these functions into glycans. In particular, growing cells in the presence of *N*-levulinoyl-D-mannosamine (ManLev; Figure 17.4) was found to result in the incorporation of this ketone sugar into cell surface carbohydrates through enzymatic transformation into a keto sialic acid (Mahal *et al.*, 1997; Lemieux and Bertozzi, 1999). Since sialic acid residues are terminal sugars on most mammalian cell surface glycoproteins, the addition of pendent ketone groups allows for specific targeting of these glycans with hydrazine (hydrazide/aminooxy) based reagents.

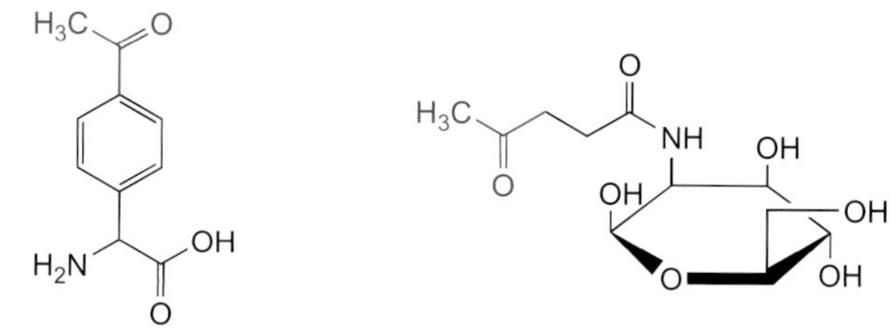

p-Acetylphenylalanine (an aryl ketone derivative) *N*-Levulinoyl-D-mannosamine (ManLev; a keto sugar derivative)

FIGURE 17.4 Ketone derivatives of phenylalanine and mannose can be fed to cells to incorporate the monomers into proteins and glycans. The resultant modifications can be probed using hydrazide-containing reagents.

There are many bioconjugate reagents and probes available with aldehyde or hydrazide functional groups that may be used for hydrazone bond formation. However, the best choices for making this reaction as chemoselective and bioorthogonal as possible are aromatic aldehyde and aromatic hydrazine derivatives due to their extremely low reactivity with biological functional groups, minimal nonspecific interactions with biomolecules (charge or Schiff base formation), and their highly efficient hydrazone bond formation in complex aqueous solutions (Solulink). Reagents of this type typically contain reactive groups consisting of either an aryl aldehyde (benzaldehyde) or a 6-hydrazinium nicotinate (Schwartz *et al.*, 1993, 1995).

These aldehyde and hydrazine functionalities can be associated with virtually any other reactive group or tag, such as heterobifunctional crosslinkers, fluorescent labels, or biotin compounds. Two reagents frequently used to chemically introduce the aldehyde and hydrazine groups include SANH (succinimidyl 4-hydrazinonicotinate acetone hydrazone) and SFB (succinimidyl 4-formylbenzoate) (Figure 17.5). Long-chain analogs of these two compounds also exist, which contain either a 6-aminocaproyl spacer arm in the cross-bridge or a short PEG spacer to promote increased water solubility. The SFB and SANH NHS ester compounds are amine-reactive and form amide bonds with the molecules modified by them (Figure 17.6). Other reactive groups or tags are also available having the benzaldehyde or hydrazinium nicotinate groups, including maleimide (thiol-reactive), biotin, and various fluorescent probes (Solulink).

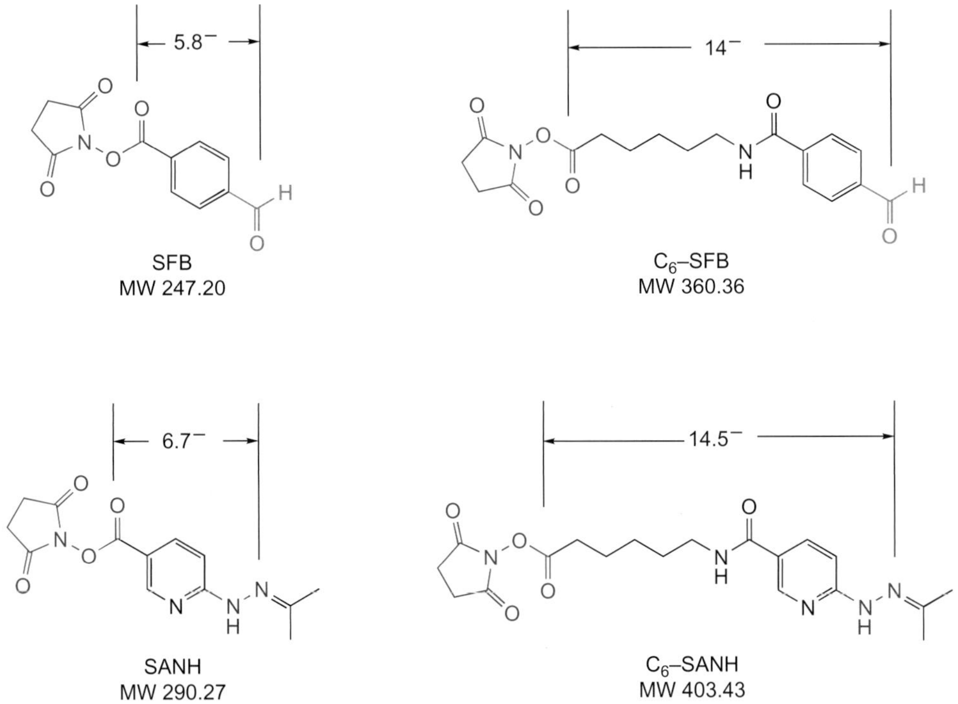

FIGURE 17.5 Structures of aldehyde-containing and hydrazine-containing heterobifunctional crosslinkers that can be used in chemoselective hydrazone conjugation procedures.

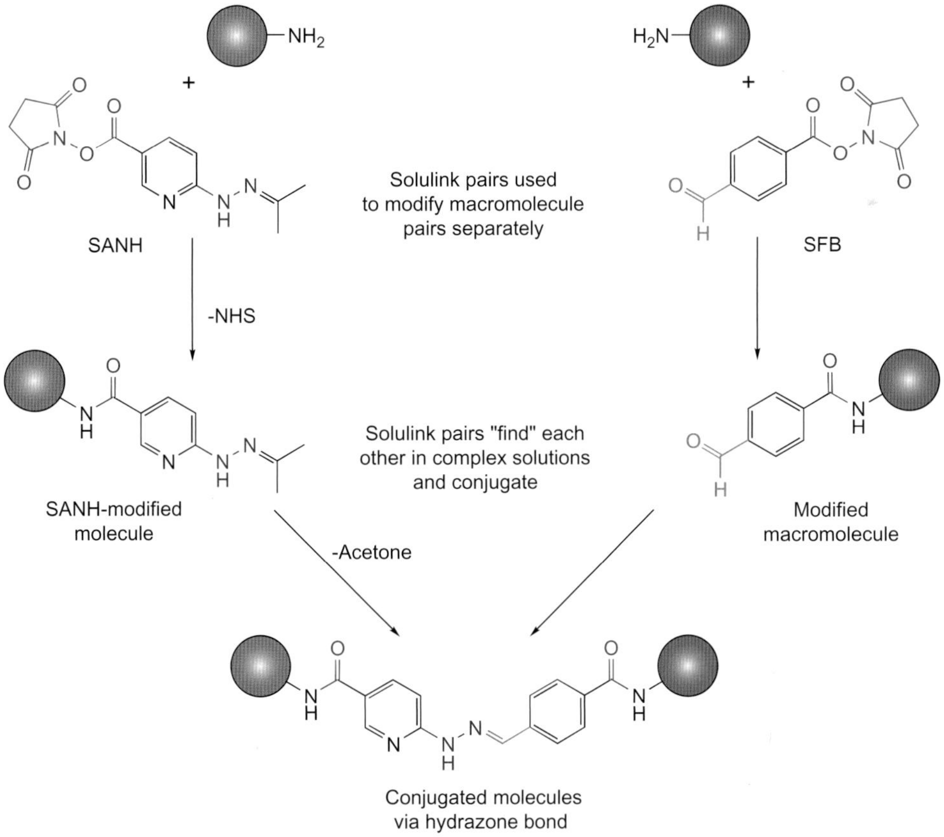

FIGURE 17.6 The reaction of SANH with amine-containing proteins or other molecules results in amide bond modifications containing terminal hydrazine groups. The reaction of SFB with amine-containing proteins or other molecules results in amide bond modifications containing terminal aldehyde groups. Subsequently, the two modified molecules can be reacted together to create a conjugate via hydrazone bond formation.

The hydrazinium nicotinate group on these reagents commonly is protected against reaction with the active ester by the addition of acetone to form the acetone hydrazone derivative. This hydrazone protective group is readily reversible at neutral or mildly acidic pH and will immediately exchange with a benzaldehyde on the corresponding chemoselective partner to form a stable hydrazone linkage.

Proteins, other molecules, or surfaces that are activated to contain either the benzaldehyde functionality or the hydrazinium nicotinate group can be quantified as to the number of these groups present using certain chromophoric reactants. For instance, 2-hydrazinopyridine can be reacted with a benzaldehyde-containing molecule to give a UV-absorbing derivative with $\lambda_{max} = 350$ nm and an extinction coefficient of 18,000 $M^{-1}cm^{-1}$. Conversely, the hydrazinium nicotinate-activated molecules can be reacted with *p*-nitrobenzaldehyde to create another UV detectable derivative having $\lambda_{max} = 390$ nm with an extinction coefficient of 24,000 $M^{-1}cm^{-1}$.

The aromatic nature of the SFB and SANH reactants create a hydrazone structure with unique absorptivity for monitoring the course of the ligation process. The aromatic hydrazone bond has a maximal absorbance at 354 nm and a molar extinction coefficient equal to 29,000/M. If the conjugation reaction is carried out with sufficient amounts of benzaldehyde and hydrazinium nicotinate reactants, then the yield of the reaction can be directly measured by monitoring the change in absorbance at this wavelength over time.

The *bis*-aryl hydrazone bond formed by this reaction is stable in aqueous solution over a broad pH range (pH 2–11) and under elevated temperature conditions (up to 94°C) (Solulink website).

Kozlov *et al.* (2004) used SANH and SFB to create DNA–antibody conjugates for highly sensitive detection in immunoassays using PCR amplification with fluorescently labeled primers. They also investigated hydrazone bond formation and its stability in aqueous solution using 3′- and 5′-labeled DNA pairs, wherein one oligonucleotide contained a benzaldehyde group and the other contained the hydrazinium nicotinate group. It was found that the most efficient pH for the creation of the hydrazone bond was pH 5.0 to 7.0, with an optimum at about pH 6 and yields dropping off at lower or higher pH. In addition, once the hydrazone bond was formed and the conjugate purified, it was stable in PBS buffer, pH 7.2, with or without 10 mg/ml BSA for 12 weeks at 4°C. The hydrazone linkage was also found to be stable at pH values between 2.3 and 11.3 without hydrolytic breakdown.

When reactions between the benzaldehyde and hydrazinium nicotinate groups are done in relatively dilute solution, for example using a modified antibody molecule at 1 mg/ml (~6 μ*M*) containing one of the two reactants, then to obtain maximal yield of the hydrazone conjugate, the second component should be added in sufficient excess to drive the reaction to completion. Kozlov *et al.* (2004) found that at least an 8-fold molar excess or above of the second reactant over the first reactant is necessary to obtain a yield of about 80% hydrazone bond formation.

The practical use of chemoselective ligation reactions for bioconjugation purposes involves the attachment of one of the reactants to a first biomolecule using a standard coupling chemistry (e.g., NHS ester) to an available functional group, while the second reactant is coupled to a second molecule or a surface, depending on the final conjugate desired. In this regard, the following protocols for modification of oligonucleotides or proteins can be carried out with either the benzaldehyde component or the hydrazine component and the opposite reactant is used to modify another molecule or surface that will be coupled in the final conjugation step.

Protocol for Modification of Amine–Oligo with SANH or SFB

1. Dissolve a 5′-amine modified oligonucleotide at a concentration of 0.5-m*M* in 0.1-*M* sodium phosphate, 0.15-*M* NaCl, pH 7.4.
2. Dissolve SANH (or SFB) in DMF at a concentration of 100-m*M*. Note: Both reagents are soluble in this solvent to about 50-mg/ml concentration.
3. Add a quantity of SANH or SFB solution to the oligo to provide a 40-fold molar excess of crosslinker over the amount of amine–oligo present. Mix well.
4. React at room temperature for 2 h to overnight with gentle mixing.
5. Purify the modified oligo from unreacted crosslinker and reaction byproducts using a micro-spin concentrator with a membrane having a molecular weight cutoff able to retain the size of the DNA being modified.

A similar method can be used to modify proteins, such as antibodies, to contain either benzaldehyde or hydrazinium nicotinate groups for subsequent conjugation with another molecule modified by the opposite functionality. The following protocol is based on the methods of Solulink, and Kozlov *et al.* (2004).

Protocol for Modification of Protein or Antibody with SANH or SFB

1. Dissolve the protein or antibody to be conjugated in 0.1-*M* sodium phosphate, 0.15-*M* NaCl, pH 7.4. The antibody solution should be as concentrated as possible, given the amount of antibody available for modification. This general protocol will work for protein or antibody concentrations ranging from

about 0.5 mg/ml to 10 mg/ml, but an increase in the molar excess of either SANH or SFB may have to be made at lower antibody concentrations to provide the same modification yields obtained at higher concentrations.

2. Dissolve SANH or SFB in DMF at a concentration of 2 mg/100 µl DMF.
3. Add a quantity of the crosslinker solution of choice (SANH or SFB) to the antibody solution to obtain the desired molar excess of reagent over the antibody. Typically, antibody modification procedures are done with 10- to 20-fold molar excess, but for dilute antibody concentrations, this may have to be doubled, depending on how many hydrazine or aldehyde groups are desired to be introduced on the modified antibody.
4. React for 2 to 3 h at room temperature with gentle mixing.
5. Purify the modified antibody by use of size exclusion chromatography or dialysis, using a molecular weight exclusion limit of 5000 to 10,000 Daltons. The modified antibody may be stored at 4°C or frozen until used to make a conjugate. The protected hydrazine on SANH is stable for several months and the SFB benzaldehyde group is stable indefinitely. For longer storage, the modified antibody can be lyophilized.

Once a molecule is modified with a hydrazine reagent and another molecule is modified with the benzaldehyde compound, they may be combined to form the final conjugate, which will result in a hydrazone linkage between the two molecules. In addition, chemoselective ligation using aldehyde/hydrazine reactions may be carried out to immobilize biomolecules. In this regard, one modified component may be a surface and the other one an antibody, protein, or oligonucleotide destined for immobilization onto the surface.

2.1. Conjugation using the Aldehyde/Hydrazine Reaction

1. Separately dissolve the SFB-modified molecule and the SANH-modified molecule in citrate buffer (100-m*M* sodium citrate, 150-m*M* NaCl, pH 6.0) at concentrations of at least 1 mg/ml. Note that pH 6 is an optimal pH for the formation of the hydrazone bond, but pH values slightly lower (to pH 4.7) or higher (to pH 7.4) may also work, but they will result in lower reaction rates and lower yields.
2. Mix a portion of the SFB-modified molecule with the SANH-modified molecule to obtain a molar ratio that will give the desired conjugate properties. The optimal ratio of reactants may have to be determined experimentally by performing a series of conjugations using different molar ratios and testing the performance of the final conjugate in the intended application.
3. React for at least 2 h at room temperature, longer if pH values other than pH 6 are used.
4. To purify the conjugate from reactants that did get incorporated into the conjugate, size exclusion chromatography may be used with resins having a molecular exclusion limit able to accommodate both the labeled molecules and the final conjugate.

3. AMINOOXY–ALDEHYDE REAGENT PAIRS

The chemoselective ligation of an aminooxy group (R–ONH_2) with an aldehyde is similar to that of the reaction of a hydrazide with an aldehyde (see previous section), except instead of giving a hydrazone bond it yields an oxime linkage. This aminooxy–aldehyde pair of reactants has been used in many bioconjugation reactions, including the coupling of ligands to surfaces, for the synthesis of radiohalogenated compounds, and in the synthesis of drug candidates (Thumshirn *et al.*, 2003; Poethko *et al.*, 2004; Liu *et al.*, 2007; Colombo and Bianchi, 2010; Mezö *et al.*, 2011). The reaction of aminooxy groups with ketones is also quite efficient and generally gives better yields than the reaction of hydrazide groups with ketones.

The simplest form of this reaction occurs with aldehydes or ketones as they react with the small molecule hydroxylamine (Figure 17.7). An organic hydroxylamine derivative is typically called an aminooxy group, which can also be termed an aminoxy or alkoxyamine group. The aminooxy compound of this reaction contains an organic substituent off the oxygen of hydroxylamine, which can consist of virtually any organic compound that terminates in the ONH_2 group. Oxime formation usually occurs more rapidly than in the reaction between a hydrazide group and an aldehyde or ketone and it typically provides a more stable oxime bond than a hydrazone bond. The bioorthogonal nature of this reaction is not absolute, however, as a number

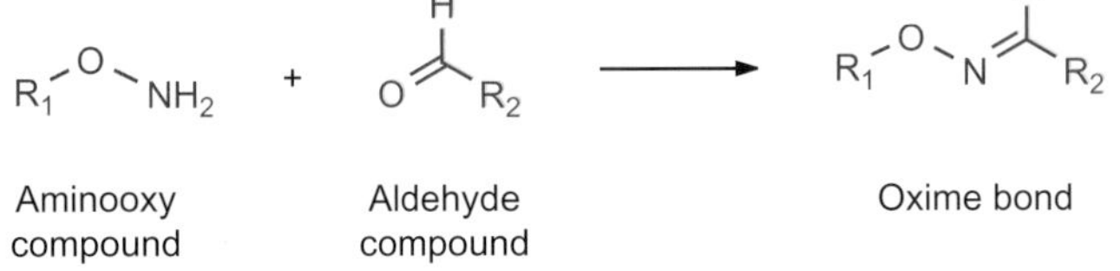

FIGURE 17.7 The reaction of an aminooxy derivative with an aldehyde (or ketone) compound forms a stable oxime linkage.

of small metabolites in biological systems contain aldehyde or ketone groups, which may participate in the reaction with an aminooxy group. The reaction can be used to probe unnatural aldehyde-containing sugar or amino acid derivatives in glycans, carbohydrates, or proteins within cells using aminooxy reagents.

The presence of an aminooxy group on a crosslinking reagent can be used to chemoselectively react with other compounds or biomolecules containing an aldehyde or ketone group. In particular, hydrophilic discrete PEG-based reagents are available that contain a protected aminooxy group, such as the phthalimidooxy–$dPEG_{12}$–NHS ester (Quanta Biodesign). This crosslinker has a phthalimide-protected aminooxy group on one end and an amine-reactive NHS ester on the other end that can be used to modify amine molecules to contain the protected aminooxy group. Without the protecting group being present, the NHS ester end of the compound would simply react with the aminooxy end. The phthalimide group prevents this from occurring until the NHS ester end is first used to label an amine-containing molecule. The result of this reaction will be the formation of an amide bond with the amine-containing molecule, which will create a derivative with hydrophilic PEG_{12} chains that terminate in the protected aminooxy function. The subsequent removal of the phthalimide protecting group can be done with 0.5-*M* hydrazine (toxic) in aqueous solution, which then yields the free aminooxy-modified molecule (Figure 17.8).

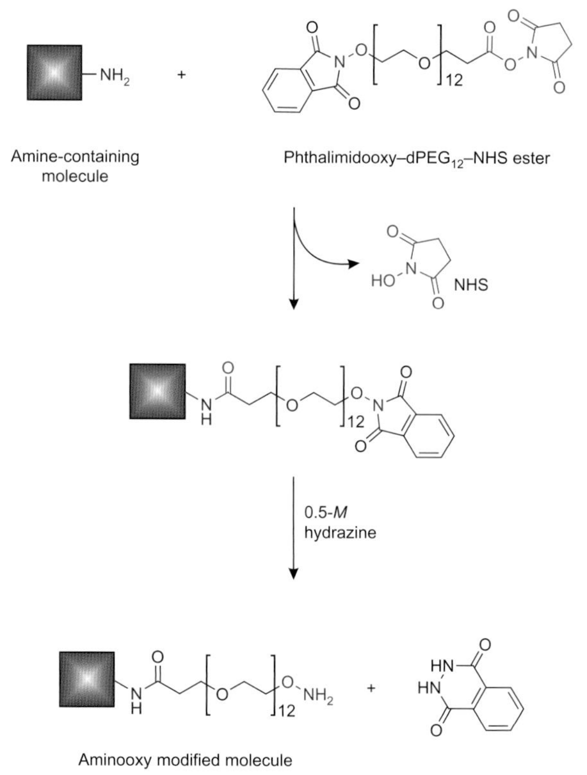

FIGURE 17.8 The modification of an amine-containing molecule with phthalimidooxy-$dPEG_{12}$-NHS ester proceeds through the reactive ester end to form an amide bond derivative. The protected aminooxy group can be removed using an aqueous solution of hydrazine to yield the aminooxy group available for further conjugation with aldehyde- or ketone-containing reagents.

The creation of aminooxy groups on surfaces or substrates has been used to covalently array proteins to form nano-patterns on silicon wafers and other surface materials. In one example, the aminooxy functionalities were coated on the surface of wafers using 8-arm PEG derivatives (Christman *et al.*, 2008). Coupling proteins to the surface aminooxy groups could be carried out after modification of the protein with an aryl aldehyde-containing crosslinker, such as 4-formyl-benzamido–$dPEG_{12}$–TFP ester (Quanta Biodesign). The reaction of this compound with the amine groups on proteins will create amide bonds through the reaction of the TFP ester side of the crosslinker and thus leave the aryl aldehyde groups available to react with an aminooxy-modified surface or another aminooxy-modified molecule (Figure 17.9). Both the aminooxy and aryl aldehyde groups have little to no reactivity with biological macromolecules, thus ensuring that a bioorthogonal reaction will occur only with each other to form the oxime linkage. In addition, aryl aldehyde groups only have a weak tendency to form Schiff base interactions with amines, so side reactions usually do not occur with this aldehyde.

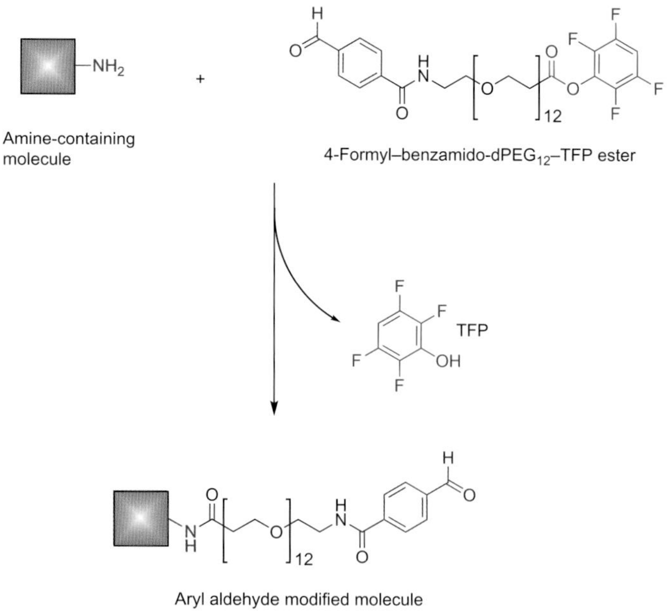

FIGURE 17.9 The modification of an amine-containing molecule with 4-formyl-benzamido–$dPEG_{12}$–TFP ester occurs by reaction at the ester end to form an amide bond derivative with loss of the TFP group. The aryl aldehyde can then be used to conjugate with aminooxy-containing reagents.

The reaction of an aminooxy group with an aldehyde- or ketone-containing molecule can be accelerated by the addition of an aniline catalyst. As in the formation of hydrazone linkages between aldehydes and hydrazides, the formation of an oxime bond between an aldehyde or ketone and an aminoxy is dramatically increased in rate and yield by the presence of aniline (Cordes and Jencks, 1962; Dirksen *et al.*, 2006a,b; Dirksen and Dawson, 2008; Byeon *et al.*, 2010). During this reaction, the aromatic amine on aniline first reacts with the aldehyde or ketone group to form an intermediate imine, which then reacts with the aminooxy group to create the final oxime linkage (Figure 17.10) (Kohler, 2009).

The aniline-catalyzed reaction was studied in detail to determine its kinetics under various conditions (Lempens *et al.*, 2009). A peptide was modified to contain an N-terminal aldehyde residue made via the pyridoxyl 5′-phosphate/sodium periodate method of Gilmore *et al.* (2006a,b) (see also Scheck and Francis, 2007; Scheck *et al.*, 2008). It was found that the aniline catalyst dramatically increased the reaction rate and coupling efficiency at all pH values studied, with the greatest rate and yield obtained at pH 4.5. The rate at pH 6.0 was less than half that observed at the lower pH condition; however, the positive difference between a catalyzed and an uncatalyzed reaction was significant at all the acidic pH environments studied.

The aminooxy–aldehyde chemoselective ligation reaction has been used to make peptide–drug conjugates (Mezö *et al.*, 2011), to label aldehyde-containing unnatural glycans with aminooxy reagents (Hudak *et al.*, 2011), for tethering bioactive molecules to platinum prodrugs for cancer therapy (Wong *et al.*, 2012), for immobilization of peptides and proteins on biosensor surfaces (Lempens *et al.*, 2011), and for the detection of sialic acid groups on glycans using an aminooxy derivative of nitrobenzoxadiazole fluorophore (Key *et al.*, 2012).

The conditions and methods for performing the reaction of an aminooxy-containing molecule with an aldehyde- or ketone-containing molecule can be found in Chapter 15, Section 2.4, including the use of aniline as a catalyst. The modification of proteins and other molecules

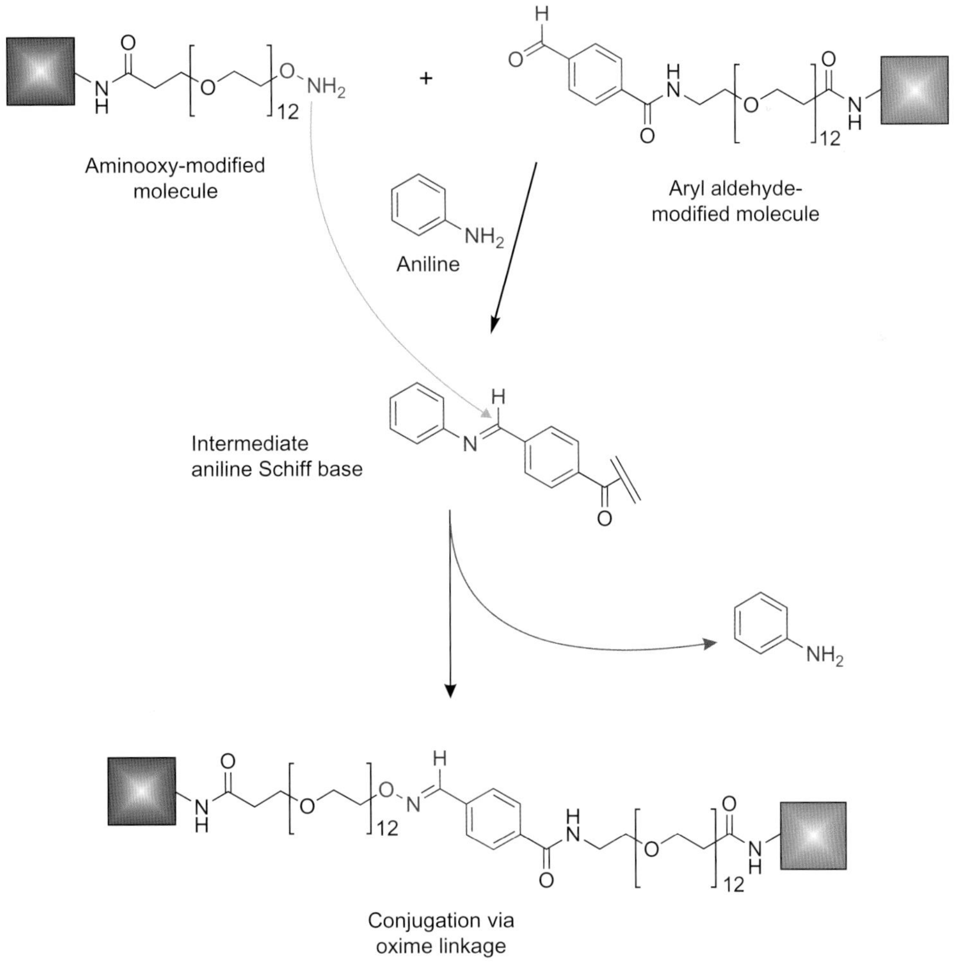

FIGURE 17.10 The chemoselective reaction between an aminooxy-modified molecule and an aryl aldehyde-modified molecule using aniline as the catalyst. With aniline in excess, the initial reaction is to form an imine intermediate (Schiff base) with the aniline amino group. This imine is subsequently displaced by the aminooxy group to form the oxime bonded conjugate in high yield.

using the crosslinking reagents 4-formyl-benzamido–dPEG$_{12}$–TFP ester and phthalimidooxy–dPEG$_{12}$–NHS ester can be found in Chapter 18, Section 1.2.

4. BORONIC ACID–SALICYLHYDROXAMATE REAGENT PAIRS

Phenylboronic acid (PBA) groups can interact with a variety of polar constituents on adjacent or nearby carbon atoms to result in a complex consisting of a 5- or 6-member heterocyclic ring. This process has been used for the affinity chromatographic purification of carbohydrates, glycoproteins, RNA, AMP (from cAMP), glycated proteins (such as glycated hemoglobin formed in diabetes; Klenk *et al.*, 1982), and a range of small molecules containing 1,2- or 1,3-diols, 1,2- or 1,3-hydroxy acids, 1,2- or 1,3-hydroxylamines, 1-2- or 1,3-hydroxyamide, 1,2- or 1,3-hydroxyoxime, as well as various sugars containing these species (Weith at al., 1970; Rosenberg and Gilham, 1971; Rosenberg *et al.*, 1972; Pace and Pace, 1980; Singhal *et al.*, 1980). For a review on the use of phenylboronic acid in affinity separations, see Scouten (1983). In addition, bioconjugate labeling reagents containing a PBA group have also been used as probes of these species in biological molecules, including fluorescent reagents for targeting glycans on cell surfaces (Burnett *et al.*, 1980; O'Shannessy and Quarles, 1987).

The interaction of PBA derivatives with molecular species having the above functional groups occurs optimally in the pH range of 8 to 9, but it is typically reversible at acid pH or in the presence of a high concentration of competing ligand. However, the heterocyclic boronic acid complex is relatively stable under optimal conditions of formation.

Stolowitz (1997) exploited this interaction potential in the design of a new chemoselective bioconjugation reagent pair consisting of a PBA group on one reagent and a salicylhydroxamic acid (SHA) group on a second reagent. Each reactant of the pair can be used to modify biomolecules, surfaces, or other compounds for subsequent conjugation or immobilization through specific PBA–SHA ring formation (Springer *et al.*, 2002). The major product of this reaction forms a 6-membered ring structure consisting of the PBA's boron atom along with one of its oxygens coordinated with the SHA's hydroxyl oxygen and hydroxamate nitrogen atoms (Figure 17.11). It also is possible that a 5-membered ring structure can form from interaction of the PBA boron with the hydroxamate hydroxyl and carbonyl oxygens on SHA, but this is a minor product as proven by NMR (Stolowitz *et al.*, 2001).

A significant advancement in the PBA reagent was to add another boronic acid group to the phenyl ring and thus allow two cycloaddition products to form from a single complexation. The phenyldiboronic acid (PDBA) group effectively increases the affinity constant of the interaction if reacting with SHA-modified molecules or surfaces that have more than one near-neighbor SHA group available. The resultant formation of two 6-membered rings per conjugation reaction assures that the bond will not hydrolyze even under high or low pH conditions. This is particularly useful for using the reaction to couple proteins or other molecules to surfaces for arrays, because multiple linkages maintain stability of the immobilized molecule without the possibility for leaching off. A single SHA–PBA bond reportedly has an affinity constant in the range of $10^6\,M^{-1}$, which is relatively weak for affinity binding properties. Two such bonds, however effectively raise the avidity to an observed affinity constant of $>10^{10}\,M^{-1}$ (Lonza product information). Note that multiple single PBA groups also can combine with multiple SHA groups when conjugating proteins together or proteins to surfaces and thus increase the effective avidity of the resultant bonding interaction. Springer *et al.* (2003) described the use of SHA membranes in this process for the immobilization

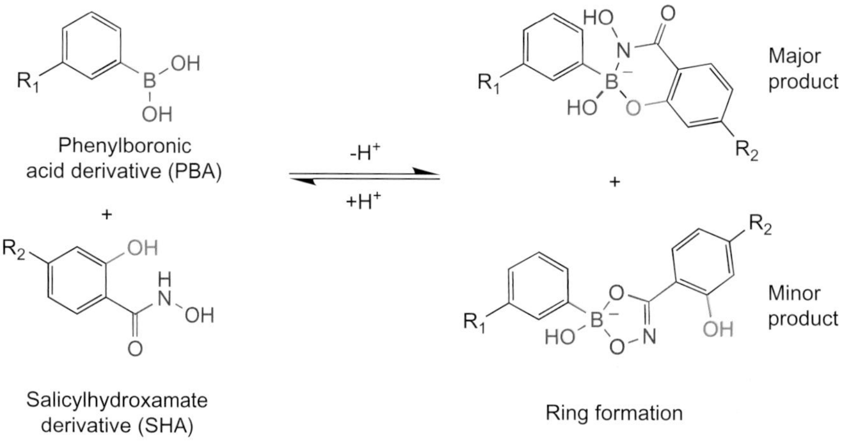

FIGURE 17.11 Phenylboronic acid derivatives react with salicylhydroxamate derivatives to form 5-membered or 6-membered ring structures, with the 6-membered ring the major product.

of PDBA-modified molecules, including nucleic acids and proteins.

Complex formation between PBA or PDBA groups and the SHA group occurs in a wide range of buffer types from pH 5 to 9, and it can tolerate high salt conditions (to 1.5-*M*) or the presence of moderate amounts of water-miscible solvents, detergents, chaotropes, and denaturing agents commonly used when working with protein solutions.

Ring formation between SHA groups coupled to solid-phase matrices and ligands containing either PBA or PDBA groups have been investigated for the immobilization of affinity molecules for protein purification (Wiley *et al.*, 2001). Using the dimeric PDBA group, immobilization of modified alkaline phosphatase onto SHA–agarose resulted in good retention of enzyme activity plus high stability of the coupled protein. The immobilized protein was also stable to elution conditions of acidic (pH 2.5) or basic (pH 11) buffers without cleavage of the boronic acid complexes or leakage of coupled protein. This indicates that multivalent attachment points using PDBA instead of PBA result in high tolerance for extreme environmental conditions without hydrolysis.

Bergseid *et al.* (2002) also reported on the use of SHA-activated chromatography supports for the coupling of boronate-containing affinity ligands. In this case, immobilized RNase A was used to purify anti-RNase antibodies from antiserum samples. RNase A was modified with an NHS–PDBA crosslinker at a molar ratio of 100:1 (crosslinker:protein), purified to remove excess crosslinker, and then coupled to an SHA-agarose support in 0.1-*M* sodium bicarbonate, pH 8.0.

Prolinx originally developed the technology utilizing the SHA–PBA interaction and later portions of it were commercialized by Invitrogen and then Lonza (Cambrex; Versalinx reagents). A number of PDBA crosslinking agents are now available to introduce the diboronic acid group into biomolecules. These include heterobifunctional compounds containing an NHS ester (amine reactive), a hydrazide group (aldehyde reactive), a maleimide group (thiol reactive), a pyridyl disulfide group for reversible linkage to thiols or for thiolation of target molecules, and an iodoacetyl group for creating thioether bonds. SHA crosslinkers include a hydrazide-containing compound, an NHS ester, and a *bis*-SHA compound containing a hydrazide group. Some examples of these reagents are shown in Figure 17.12.

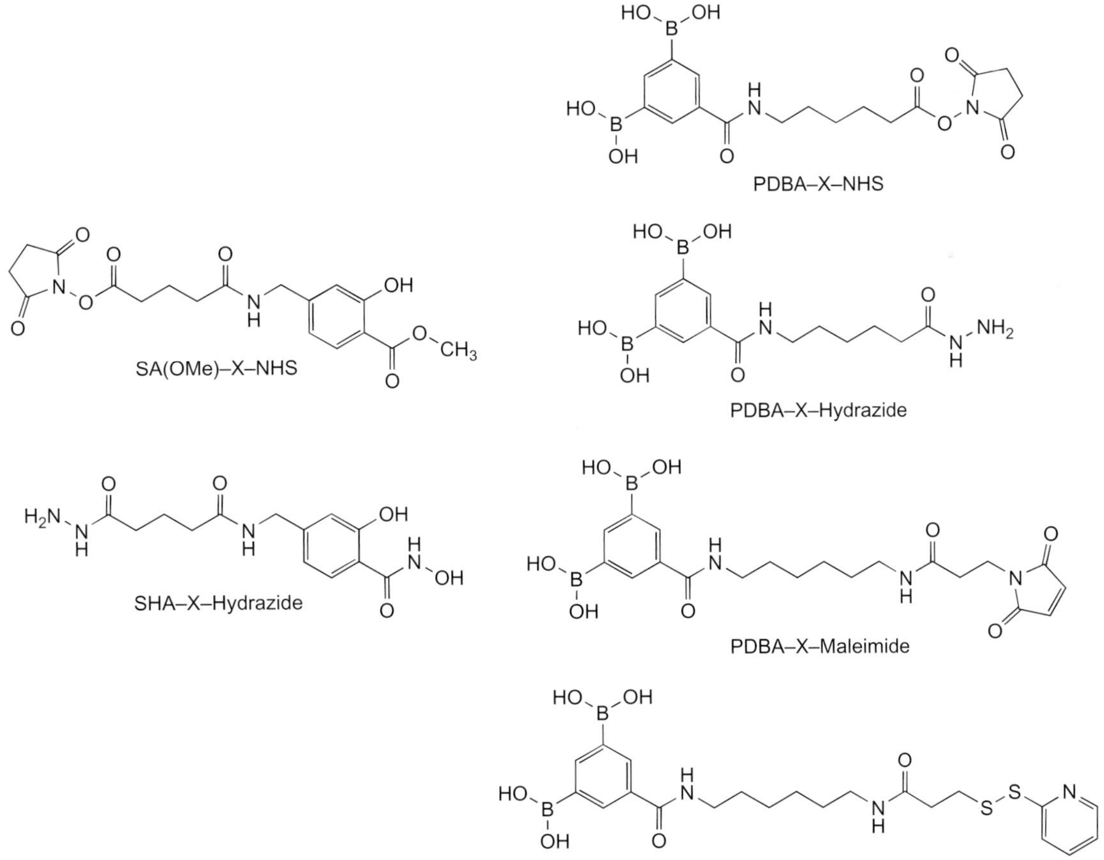

FIGURE 17.12 Heterobifunctional crosslinking agents containing the phenylboronic acid or salicylhydroxamate group for chemoselective conjugation purposes.

Conjugation reactions between phenylboronates and salicylhydroxamates entail the formation of a low-affinity interaction involving a reversible heterocyclic ring formation. Conjugates formed as a result of this reaction are relatively stable provided the pH is maintained within the optimal range and there are no competing species in solution, which may exchange for the SHA group. This may include sugars or carbohydrates containing diols or other organic constituents mentioned previously. The creation of dimeric or multivalent interactions with two or more SHA groups and a phenylboronate-modified molecule (using PBA or PDBA) will dramatically increase the stability of the linkage over that observed with a single PBA–SHA linkage. Wiley *et al.* (2001) found that modification of alkaline phosphatase with PBA or PDBA for subsequent immobilization on an SHA–agarose support resulted in stable immobilization if six PBA groups were present on each alkaline phosphatase molecule or the equivalent of three dimeric PDBA groups. This difference reflects the dimeric nature of the PDBA group in forming more than one boronate ring structure per modification.

The use of the PBA–SHA conjugation in a bioorthogonal mode will probably be problematic due to the potential side interactions that could occur involving the phenylboronate groups with biological molecules in cells or cell lysates. In addition, there are no reports of using SHA or PBA bio-monomer analogs (e.g., amino acids containing these groups) to incorporate into biopolymers *in vivo* as there are with other chemoselective reacting groups. However, the use of the PBA–SHA reaction as a chemoselective immobilization process for attaching proteins or other molecules to surfaces may have advantages over other bioconjugation reactive groups, because the reactive groups are very stable and will not hydrolyze in aqueous solution. In addition, the PBA–SHA reaction can be used with success for the conjugation of two pure biomolecules, such as in the creation of a protein–protein conjugate (Le Roch *et al.*, 2000).

The following protocol describes the immobilization of a protein on an amine-surface using the P(D)BA–SHA chemistry.

Protocol for Preparation of P(D)BA-Modified Protein

1. Dissolve a first protein to be modified at a concentration of 1 to 10 mg/ml in 0.1-*M* sodium bicarbonate buffer, pH 8.5. PBS buffer at physiological pH may also be used as the reaction medium.
2. Add a quantity of PDBA–NHS ester to the protein solution to provide the desired molar excess of crosslinker over the protein. A suggested starting point is to use a 10- to 15-fold molar excess of reagent, but the optimal amount to be added should be determined by experimentation to provide a final conjugate having the best possible properties for the intended application. The PDBA–NHS ester may first be dissolved in DMF as a concentrated stock solution and then an aliquot added to the reaction mixture. Mix well to dissolve.
3. React for at least 1 h at room temperature with gentle mixing.
4. Purify the modified protein by gel filtration or dialysis using a molecular weight cutoff that will be appropriate for the protein being modified.

Protocol for Preparation of SHA-Modified Surface

An amine-containing surface, such as an APTS-modified glass slide (see Chapter 13, Section 2), may be modified with a long-chain NHS–salicylic acid methyl ester derivative (Lonza), which then can be converted to the SHA group for coupling to a P(D)BA-modified protein. The following protocol describes this method.

1. Thoroughly wash the amine-containing slide with deionized water and then with 0.1-*M* sodium bicarbonate, pH 10 (coupling buffer).
2. Add a quantity of NHS–salicylic acid methyl ester reagent dissolved in coupling buffer to the slide surface to provide at least a 2 fold molar excess of crosslinker over the quantity of amines present on the surface. The surface of the slide may be coated with a minimum solution volume of the crosslinker by layering the solution over the surface. Slide masks or gaskets may be used to isolate only certain regions for modification. Alternatively, the slide may be immersed in the crosslinker solution. The NHS–salicylic acid methyl ester may first be dissolved in DMF as a concentrated stock solution and then an aliquot added to the reaction mixture to prepare the final concentration desired. Mix well to dissolve before immediately exposing the solution to the slide surface.
3. React for at least 1 h at room temperature with gentle mixing.
4. Wash the slide with 10 volumes of water.
5. Prepare a 1-*M* solution of hydroxylamine in coupling buffer sufficient to again treat the slide. The pH of the coupling buffer should be adjusted to pH 10 after dissolving the hydroxylamine into it. Expose the slide to the hydroxylamine solution in the same manner as the crosslinker treatment.

The hydroxylamine will react with the methyl ester groups on the salicylic acids and form hydroxamate functionalities suitable for conjugation with the P(D) BA-modified protein from above.

6. React the slide with the hydroxylamine solution for 16 to 24 h at room temperature.
7. Thoroughly wash the slide with coupling buffer and at least 10 volumes of water.

Protocol for Coupling PDBA-Modified Protein to SHA-Modified Slide

Proteins modified with boronic acid groups may be covalently linked to SHA-modified slides simply by spotting the PDBA-modified protein onto the slide surface in 0.1-*M* sodium bicarbonate buffer, pH 8.0. The arraying technique may use pin spotters, piezoelectric contactless printers, or even pipette dispensing into masked wells on the surface. The quantity of PDBA-modified protein placed on the surface should at least be in 2-fold molar excess to the theoretical density of SHA groups present. A large quantity of protein spotting may obviate a covalent attachment strategy by building up a "mountain" of dried protein, as is typically done using simple surface adsorption of protein to form array spots.

5. CLICK CHEMISTRY: CU(I)-PROMOTED AZIDE–ALKYNE [3 + 2] CYCLOADDITION

The 1,3-dipolar cycloaddition reactions to unsaturated carbon–carbon bonds have been known for quite some time and have become an important part of strategies for organic synthesis of many compounds (Smith and March, 2007). The 1,3-dipolar compounds that participate in this reaction include many of those that can be drawn having charged resonance hybrid structures, such as azides, diazoalkanes, nitriles, azomethine ylides, and aziridines, among others. The heterocyclic ring structures formed as the result of this reaction typically are triazoline, triazole, or pyrrolidine derivatives. In all cases, the product is a 5-membered heterocycle that contains components of both reactants and occurs with a reduction in the total bond unsaturation. In addition, this type of cycloaddition reaction can be carried out using carbon–carbon double bonds or triple bonds (alkynes).

The reaction between an azide and an alkyne has been referred to as the Huisgen cycloaddition reaction, after the name of its originator (Huisgen *et al.*, 1964). This type of reaction, shown in Figure 17.13 for both alkenes and alkynes, results in similar heterocycles only

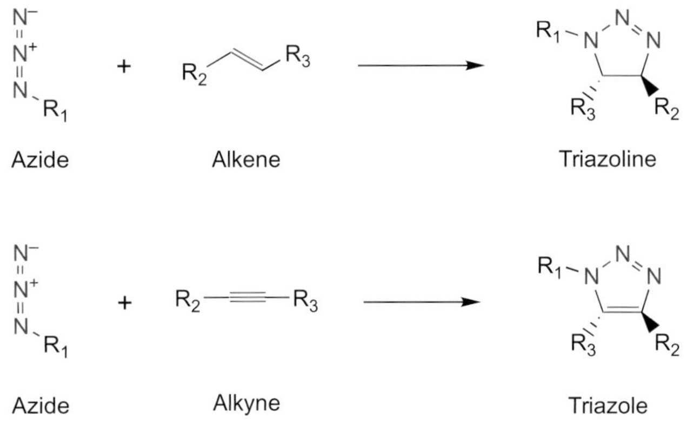

FIGURE 17.13 A general Huisgen reaction involves the cycloaddition of an azide with an alkene or an azide with an alkyne. The products of these reactions are a triazoline ring or a triazole ring, respectively.

differing in a single carbon–carbon double bond within the ring. The Sharpless group has coined the term "click chemistry" to describe these reactions, because of the seemingly "spring-loaded" nature of the electrophiles that participate in it (for review, see Kolb *et al.*, 2001). It was also noticed that such reactions appear to be accelerated in aqueous solution compared to the same reactions carried out in organic solvent.

Click chemistry reactions historically are done at elevated temperatures, and sometimes elevated pressures, to increase the rate of reaction and make the yield of heterocycle formation acceptable. However, it was discovered that in the presence of Cu(I), the reaction kinetics are dramatically accelerated to provide high yields even at room temperature and ambient pressures (Rostovtsev *et al.*, 2002; Tornøe *et al.*, 2002 Sharpless *et al.*, 2005). There was an early indication that Cu(I) could catalyze this process (L'Abbé, 1984), but it was not pursued further at that time, especially as a potentially useful bioconjugation method. In fact, in aqueous solution, the presence of only a small catalytic amount of Cu(I) in a click chemistry reaction can increase the rate of cycloaddition by about a million-fold, making the reaction biocompatible.

The copper-catalyzed azide–alkyne cycloaddition process has resulted in a proliferation of applications in organic synthesis and bioconjugation. There are hundreds of references to the use of this conjugation reaction for small molecule synthesis, protein conjugation, activity-based protein profiling, nucleic acid conjugation, surface modification, detection schemes, and *in vivo* targeting of molecules on cells. One of the primary reasons for the increasing popularity of the click chemistry reaction is the bioorthogonal nature of the two reacting groups.

Azides in particular are convenient electrophilic participants in the click chemistry reaction due to their ease of formation and stability. Alkyl azides undergo nearly no side reactions and are extremely stable in aqueous solution, even in the presence of complex biological material. Note that the aryl azides, which often are used as photoreactive crosslinking agents, are highly unstable to UV light or reducing agents and probably should be avoided for this purpose (although they may be perfectly good substrates for the click reaction).

Alkyne groups are also remarkably stable in biological solutions, provided they do not have an activating group adjacent to them, such as a carbonyl, which would make them predisposed to Michael-type addition reactions, especially with thiols. Adding an alkyne group to a modification reagent or a crosslinker can be as simple as coupling an activated carbonyl group with propargylamine, which forms the propargylamide linkage and creates a terminal acetylene group for conjugation. Link *et al.* (2004) synthesized a biotin–PEG–alkyne modification reagent using this strategy, which then could be used to modify proteins containing azide amino acids (Figure 17.14).

The functional groups used for click chemistry conjugations are completely unreactive toward biological molecules and virtually free of side reactions that otherwise would cause reagent instability in aqueous environments. This means that a molecule modified to

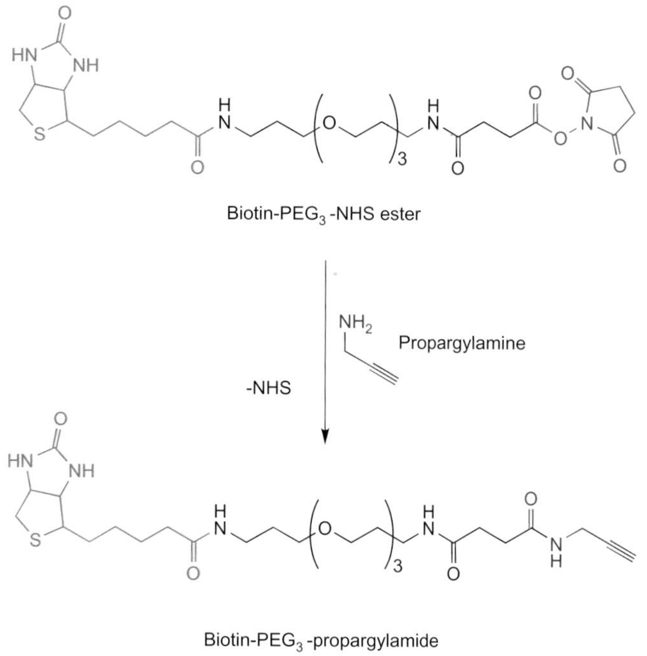

FIGURE 17.14 Propargylamine can be used to add an alkyne group to amine-reactive reagents, such as the NHS ester group on the biotin-PEG$_3$ compound.

contain an azide functionality would be able to react specifically with another molecule containing an alkyne group, even in the presence of biological fluids, cells, or cell lysates. In addition, without the presence of Cu(I), the azido-molecule and the alkyne-molecule would not react to an appreciable extent at room temperature even when placed together in solution. Only upon the addition of Cu(I) in sufficient concentration would the cycloaddition reaction take place and a triazole linkage be formed.

The Cu(I) source used to drive the click reaction can be generated in several ways. The use of CuBr can be carried out to add Cu(I) directly to the reaction; however, Cu(I) salts are relatively impure and in solution they are labile and may degrade by oxidation to a significant extent over the time course of the reaction. The use of Cu(I) salts has also been found to result in some degree of side reaction products as well as needing an organic base and organic co-solvent during the reaction to efficiently drive the cycloaddition process. For these reasons, the source for Cu(I) typically is generated *in situ* using Cu(II) in the presence of a reducing agent.

The Cu(II) salt, $CuSO_4$, is particularly convenient, as it is readily available and easily converted to Cu(I) with a reducing agent, such as sodium ascorbate or TCEP. In solution, Cu(II) is reduced to Cu(I) by ascorbate with concomitant oxidation of ascorbate to dehydroascorbate. For reactions between pure click chemistry components in solution, the amount of catalyst addition only has to be from about 0.25 mole percent to 2 mole percent relative to the amount of reactants present, with a 5-fold molar excess of ascorbate over the amount of Cu(II). Therefore the reaction is initiated by production of only a small amount of Cu(I), which catalytically gets oxidized and then regenerated by reduction during the cycloaddition process. The proposed mechanism for the click chemistry reaction has been illustrated as a catalytic cycle by both the Meldal group (Tornøe *et al.*, 2002) and the Sharpless group (Rostovtsev *et al.*, 2002), giving a cyclic intermediate azide–Cu(I)–alkyne complex, which then goes on to form the 5-membered triazole ring.

Another source for Cu(I) in the click reaction is to use elemental copper metal filings, which generate Cu(I) ion in solution slowly by oxidation. This last option, however, is considerably slower in generating the necessary Cu(I) than the other methods and will result in reactions needing to be done for at least 24 h.

For click reactions made in complex solutions, such as in the presence of biological molecules, the amount of Cu(II) and ascorbate addition typically is at a concentration of at least 0.1-m*M* $CuSO_4$ and 0.2-m*M* ascorbate. In this type of environment, the labeling reaction usually is done on azide or alkyne targets at very low concentration levels and for extended times. At this concentration of metal salt and ascorbate, cells may not remain viable for long periods and may die.

In some cases, click chemistry ligation reactions may not be appropriate for labeling within cells if continued cell viability is important. Live cell labeling requires that the conjugation chemistry not adversely affect cell viability or dramatically alter protein expression or pathway activation. Due to this limitation, the click chemistry reaction has been said to be undesirable for performing conjugations within a living cell, and only useful for labeling targets on live cell surfaces (Link and Tirrell, 2003; Prescher and Bertozzi, 2005).

However, some groups have worked around these issues and developed strategies for live cell labeling wherein the first step occurs *in vivo*, but then subsequent steps use *in vitro* cycloaddition for detection. Speers and Cravatt (2004a,b) used a click chemistry reactant to label enzymes *in vivo* at their active sites with an azide-substrate analog. Activity-based protein profiling (ABPP) typically involves using a binding probe along with a reactive group and a detectable tag, which is able to target specifically the binding site of an enzyme. The reactive group covalently links the affinity molecule to the active site, while the tag is used to image the enzyme *in vitro*. Using the click chemistry strategy, the active site binder in ABPP does not contain the detectable tag, but only possesses an azide group. The azide functionality is extremely stable *in vivo*, so the affinity reagent can be used in living cells or whole organisms. After incubation with the azide affinity component, the probe specifically interacts with the enzymes being targeted. Subsequently, the tissue or cells can be lysed (or fixed) and probed for bound enzyme using an alkyne-labeled reagent. This can be a fluorescent probe or an affinity handle, such as biotin, for purification.

In fact, most cell-based assays are done using fixed cells, not live cells, which makes click chemistry reactions eminently practicable. In this approach, a test population and a control population of cells is grown and after treating the test population of cells with a potential drug candidate or another modulator of cellular processes, they are compared relative to the expression of a biological component or the activity of a biomolecule. Most high-content screening assays are done on cells after a formaldehyde fixation step followed by a permeabilization process to allow passage of molecular probes into the cells (refer to Thermo Fisher Scientific, Cellomics). For these applications, the use of azide–alkyne reagents in a click chemistry strategy is entirely appropriate and may be the best choice of all conjugation reactions, because of its exquisite chemoselectivity, bioorthogonality, and excellent reaction kinetics.

The triazole ring generated by the reaction of an azide and an alkyne is a very stable linkage and not likely to undergo hydrolysis or any other breakdown reaction to cleave the linkage. Even under relatively extreme conditions used in some biological operations involving the addition of denaturants, detergents, chaotropic agents, organic solvents, or acidic or basic conditions, the triazole ring will survive and remain intact. Click chemistry reactions are thus highly chemoselective and result in strong conjugation bonds for use in any application.

Another important advantage to the use of click chemistry for cell-based targeting is the ability to create bio-monomer analogs containing either azido or alkyne functionalities, which then can be incorporated into biopolymers using a cell's native enzymatic machinery. Methionine and phenylalanine analogs containing side-chain alkyne or azide groups have been synthesized and proven able to be introduced into proteins at normal methionine sites in a sequence-specific manner (Link *et al.*, 2004; Prescher and Bertozzi, 2005). Kiick *et al.* (2002) also demonstrated that both azido and alkynyl amino acid derivatives could be used as methionine surrogates and get integrated into proteins with nearly the same efficiency as normal methionine. In addition, azide-sugar derivatives have been prepared that are capable of being incorporated into glycans and glycoconjugates using normal enzymatic biosynthetic pathways in cells (Saxon and Bertozzi, 2000). Thus, proteins and carbohydrates can be specifically tagged to contain non-canonical amino acids or sugars for subsequent bioconjugation using reagents containing the opposite click chemistry reactant (Figures 17.15 and 17.16). Cells grown in the presence of azido or alkyne monomer analogs will utilize these as building blocks for creating biopolymers. Proteins and glycans within the cells and on cell surfaces afterward display discrete bioconjugation targets for subsequent detection, crosslinking, or capture using click chemistry applications.

Lin *et al.* (2006) used click chemistry combined with site-specific labeling of recombinant protein using expressed protein ligation (EPL) (Muir, 2003) to couple proteins to array surfaces. A propargylamido-cysteine reagent was used to modify an expressed protein containing a thioester intein at its C-terminal. Reaction of the free thiol on the propargylamido–cysteine with the thioester linkage on the protein resulted in transthioesterification followed by an immediate S→N shift to give an amide bond between the alkyne compound and the protein (Figure 17.17).

Once the protein is modified to contain an alkynyl group at its C-terminal it can be used to covalently link to its click chemistry reactant partner, an azide on the surface of an array. Other azido molecules can also be

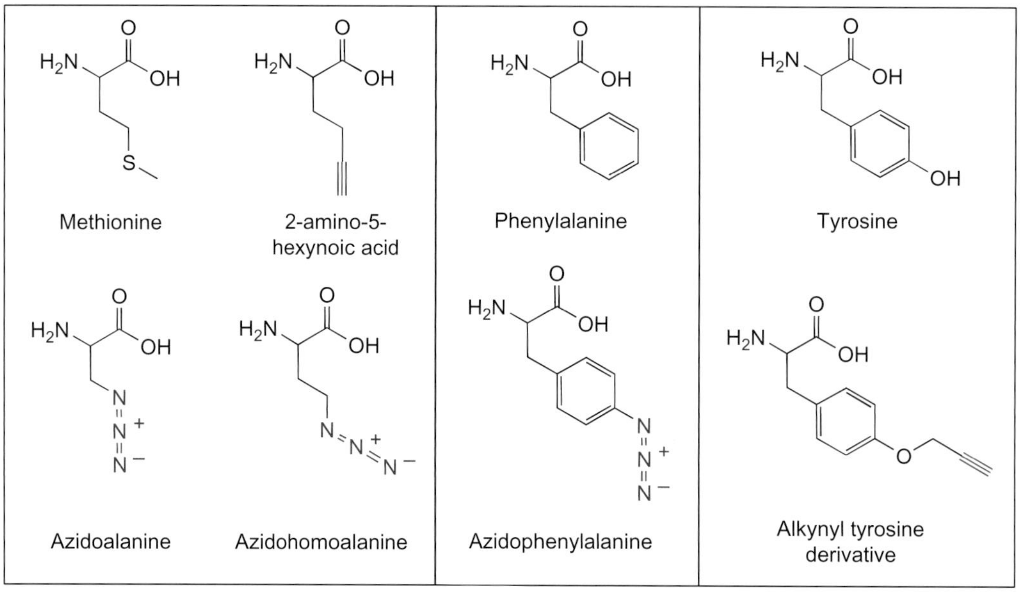

FIGURE 17.15 Amino acid analogs containing either azido or alkyne modifications can be fed to cells and these monomers incorporated into expressed proteins.

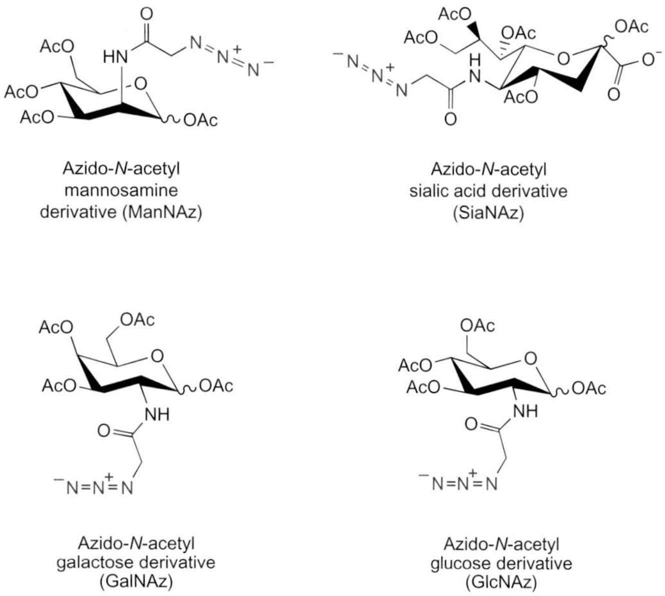

FIGURE 17.16 Azido derivatives of sugars can be used as monomers for glycan and carbohydrate synthesis by cells. Such modifications can be probed using click chemistry or Staudinger ligation reactions.

conjugated with an alkyne-labeled protein to facilitate the detection or capture of the protein using affinity techniques. For instance, an azido–fluorescein reagent can be used to detect fluorescently the expressed protein in complex samples or an azido–biotin derivative can be used to biotinylate the protein at its C-terminal for subsequent purification or detection (Figure 17.18).

In another application of coupling proteins to surfaces using click chemistry, Duckworth *et al.* (2006) carried out prenylation of a protein using a farnesyl azide derivative and the enzyme farnesyl transferase for subsequent chemoselective ligation to alkyne-functionalized agarose beads. The result is a highly discrete, site-specific attachment of the protein to the solid phase at a single location.

Bonnet *et al.* (2006) used the click chemistry reaction to synthesize receptor ligands containing fluorescent probes or biotin groups to specifically tag and detect or isolate receptor proteins. The efficiency of the click cycloaddition reaction catalyzed by Cu(I) presented many benefits over carrying out such synthesis by other routes. If receptor ligands can be modified to contain an alkyne group or an azide, then the opposite click chemistry reactant can be used on any number of probes to conjugate with the ligand for subsequent receptor probing applications.

Nanoparticles can also be modified with one of the click chemistry reactants to facilitate protein or ligand coupling to them in high yield. Brennan *et al.* (2006) modified gold nanoparticles with a thiol–azido spacer to produce azide functional groups on the particles for subsequent coupling of alkynyl–lipases. The thiol end formed a dative bond with the gold surface, which produced a long, hydrophilic, PEG-containing spacer terminating in an azide (Figure 17.19). Interaction of the alkynyl–lipase with the modified particles in the presence of Cu(I) resulted in triazole linkages, which efficiently immobilized the enzyme on the surface. The resultant enzyme–gold particles retained high enzymatic activity with an average of seven protein molecules conjugated per nanoparticle.

The following protocol adapted from Brennan *et al.* (2006) describes the coupling of alkyne-modified protein to 14-nm gold nanoparticles. The lipase enzyme

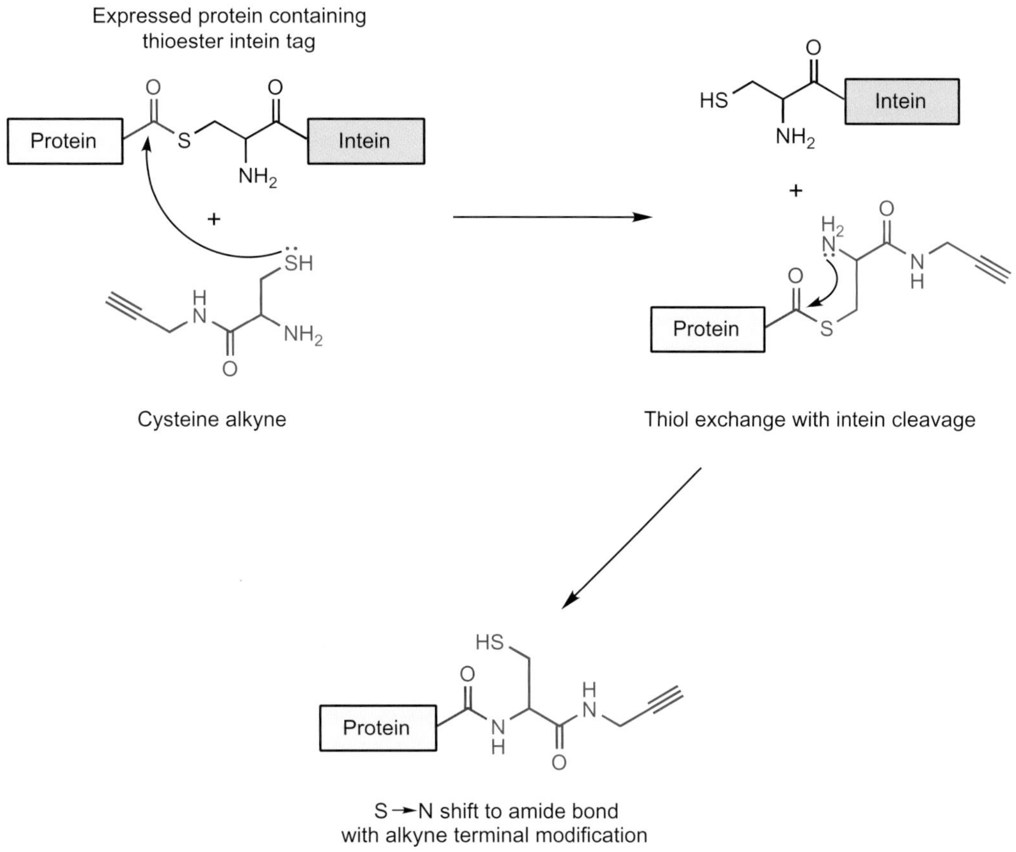

FIGURE 17.17 Expressed proteins containing a thioester intein tag can be specifically modified using a cysteine–alkyne derivative by trans-thioesterification followed by an internal S→N shift.

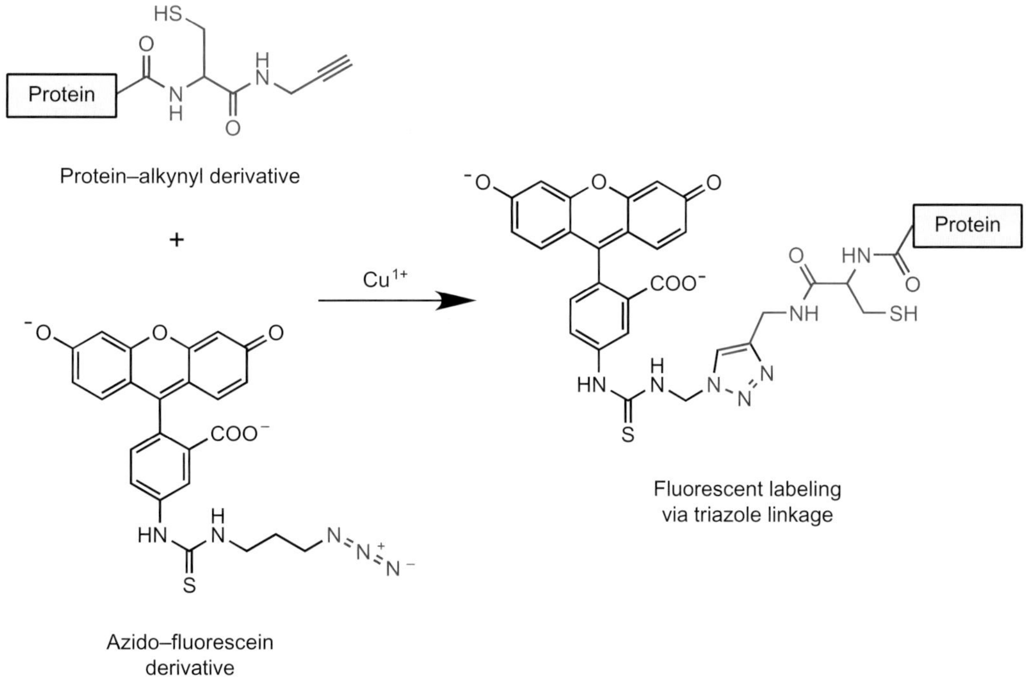

FIGURE 17.18 An expressed protein containing a thioester intein tag that was subsequently modified by native chemical ligation to contain an alkyne group can then be labeled using an azido–fluorescein probe by the click chemistry reaction in the presence of Cu^{1+}.

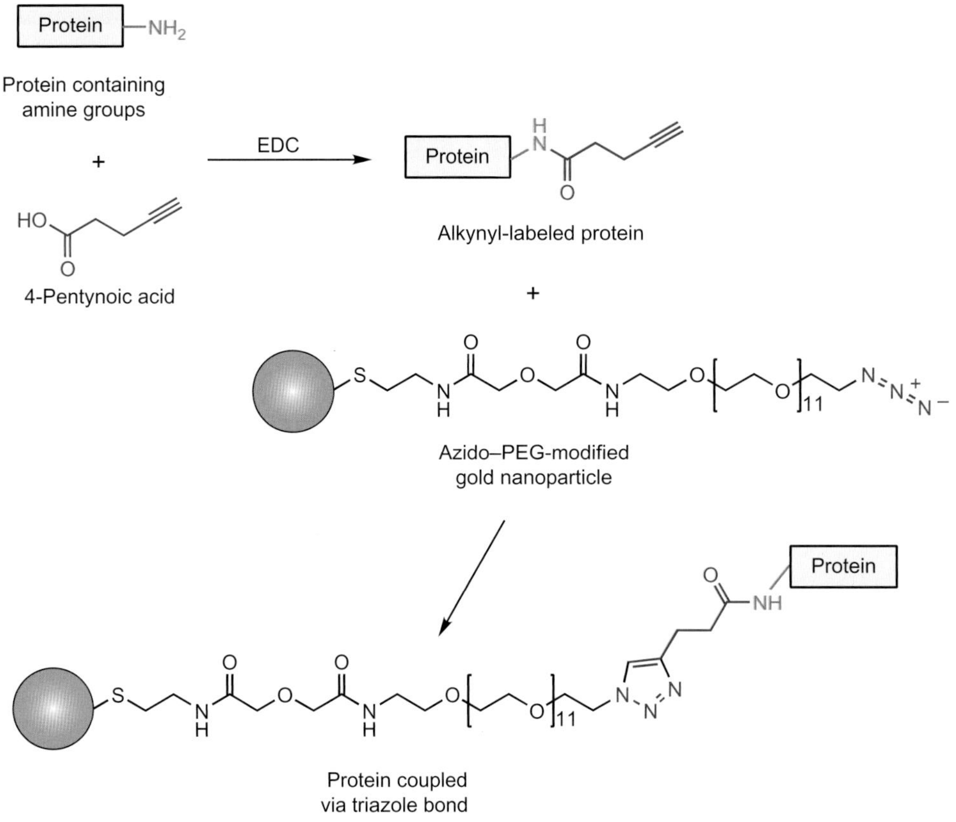

FIGURE 17.19 The small carboxylate–alkyne compound 4-pentynoic acid can be used to modify proteins at their amine groups with EDC to provide alkyne sites for click chemistry-mediated conjugation. The subsequent reaction of an azido–PEG-modified gold nanoparticle with the alkynyl-labeled protein in the presence of Cu^{1+} yields the triazole-coupled protein.

was engineered to contain a single lysine residue that was accessible to the aqueous environment for labeling with 4-pentynoic acid using carbodiimide coupling. This procedure is also applicable to other proteins containing lysine residues to add alkynyl groups for subsequent click chemistry conjugation procedures.

An alternative method of modifying proteins to contain alkyne groups is to use the propargyl–PEG_1–NHS ester compound described in Chapter 18, Section 1.2. This reagent will react spontaneously with available amine groups in proteins to form an amide bond without the need to use EDC, as in the following protocol.

Protocol for Modification of Protein with Alkynyl Groups

1. Dissolve a protein to be modified with an alkynyl group in 1.2 ml of 20-m*M* sodium phosphate, pH 7.0 (PBS), at a concentration of at least 20 μ*M*.
2. Prepare a stock solution of 4-pentynoic acid by dissolving it at a concentration of 100-m*M* in 50% THF/PBS. Add 6.7 μl of this solution to the protein solution with mixing.
3. Prepare a stock solution of EDC just prior to use by dissolving it in PBS at a concentration of 50-m*M*. Note: EDC is unstable in aqueous environments and should be used immediately after the solution is made. Quickly add 13 μl of the EDC solution to the protein solution with mixing.
4. Add to the reaction solution with mixing 150 μl of THF and 130 μl of PBS.
5. React with mixing in a fume hood overnight at room temperature. Note: This reaction may be complete within 2 to 4 h, as carbodiimide conjugations usually proceed to completion within this time frame.
6. Purify the alkynyl-protein by dialysis or gel filtration using PBS to remove excess reactants and solvent.

Protocol for Modification of Gold Nanoparticles with Thiol–PEG–Azide Linker

1. Gold nanoparticles (or other metallic or semiconductor particles) are functionalized with azide groups by suspending a 2.8-nm concentration of particles in 20 ml of water and adding a 20-μmole amount of a thiol–PEG–azide spacer ligand to the suspension with stirring.
2. Mix for 18 h at room temperature.
3. Purify the modified particles from excess linker by repeated centrifugation (at least three times at 15,000 *g*) and resuspending each time with water.

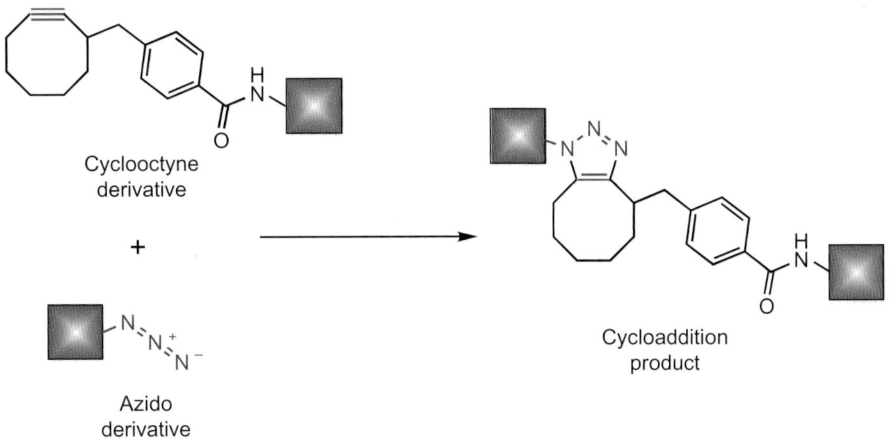

FIGURE 17.20 Cyclooctyne derivatives can be used as alternative click chemistry reactants, as they are capable of reacting with an azide group without the presence of Cu^{1+} to form a cycloaddition product. This reaction proceeds at a slower rate than the Cu^{1+}-catalyzed process, but it avoids the cytotoxic effects that copper addition can have on cells.

Protocol for Coupling Alkyne-Modified Protein to Azide-Nanoparticles

1. Suspend the azide-nanoparticles in water at a concentration of 13 nm.
2. With mixing, add a quantity of alkyne-modified protein to the particle suspension to provide at least a 10-fold molar excess over the quantity of azide groups present on the particles. The high molar excess is important to prevent particle aggregation if the modified protein has more than one alkyne group, which could crosslink more than one particle with a single protein, if the protein concentration is too low. However, if the modified protein only has a single alkyne modification, reacting it at a lower molar ratio is okay. For example, Brennan *et al.* (2006) added 36 µl of alkyne–lipase (69 µ*M*) in PBS buffer to 192 µl of azide-nanoparticles (13 nm).
3. Add 2.5 µl of 10-m*M* $CuSO_4.5H_2O$, 50-m*M* ascorbic acid dissolved in water per ml of the protein and particle mixture. Mix to dissolve.
4. React at room temperature for up to 3 days (or at 4°C, if the protein is not stable at ambient temperature). The optimal time of reaction is dependent on the protein being coupled and the number of alkyne reactive groups available. An alkynyl-protein added to the nanoparticles at a high molar ratio would probably reach maximal coupling yield in a matter of hours.
5. Purify the protein particles by repeated centrifugation (at least three times at 15,000 *g*) and resuspending each time with water.

In another application to couple ligands to surfaces, Sun *et al.* (2006) used two chemoselective ligation reactions, a Diels–Alder cycloaddition followed by an azide–alkyne (click chemistry) reaction, to immobilize protein, biotin, or carbohydrate ligands on glass slides. In this strategy, glass slides were prepared containing maleimidocaproyl groups, which were used in the Diels–Alder reaction to couple a PEG_4 spacer containing an alkyne on one end and a cyclopentadiene at the other end. The Diels–Alder reaction was used to link the cyclopentadiene to the maleimide groups, while the alkyne groups at the other end were used in the click chemistry reaction to attach the azide-containing ligands (see Figure 17.3).

Click chemistry reactant pairs used for surface immobilization have the advantage of being stable to aqueous conditions and long-term storage. Unlike many of the other coupling chemistries used with surfaces (e.g., NHS esters, EDC conjugation), which suffer from hydrolysis and degradation over time, the alkyne or azide components can be used to activate a surface and stored indefinitely until needed. A ligand modified with the opposite reactant can then be spotted on the array surface in the presence of Cu(I) to initiate covalent attachment through triazole ring formation.

Another version of the click chemistry azide/alkyne reaction has been developed to eliminate the requirement for Cu(I) to be added to catalyze the triazole ring formation. Agard *et al.* (2004) used a cyclooctyne ring to take the place of the typical linear alkynes used as the reactant partner for azides in the Cu(I) catalyzed reactions. This cyclic triple bond reactant is activated, because of ring strain, to produce better kinetics in the 3 + 2 cycloaddition reaction (Prescher and Bertozzi, 2005; Agard *et al.*, 2006). The reaction with an azide lessens the ring strain of the alkyne within the cyclooctyne structure, and thus drives the reaction without the addition of cytotoxic copper (Figure 17.20). This cycloaddition reaction is still much slower than a copper-catalyzed reaction, but its usefulness on living cells may give it an advantage in this application.

6. STAUDINGER LIGATION

Early in the last century, the Nobel Prize-winning chemist Hermann Staudinger discovered a reaction between phosphines and azides which became known as the Staudinger reaction (Staudinger and Meyer, 1919). Triphenylphosphine reacts with azides to form an intermediate iminophosphorane with the release of nitrogen gas. This intermediate quickly breaks down in aqueous environments to yield triphenylphosphine oxide and a primary amine (Figure 17.21). This reaction has since been used successfully to synthesize amines in countless numbers of organic compounds and still remains one of the most common organic reactions performed today. Often azides are thought of as hidden amines, because the azide is relatively inert to other reactants until it is revealed through the Staudinger reaction.

A significant modification to the Staudinger reaction was developed by Saxon and Bertozzi (2000) that effectively turns it into a covalent coupling reaction with bioconjugation potential (see also Saxon and Bertozzi, 2003, 2006). Termed "Staudinger ligation," this reaction is carried out using a triphenylphosphine derivative that contains an electrophilic group next to the phosphorus core. *Ortho* positioning of an electrophilic benzyl methyl ester group on one of the phenyl rings provides a reactive site for the nucleophilic nitrogen from an azide group that temporarily forms an aza–ylide interaction with the phosphorus atom core. Nucleophilic attack of the nitrogen on the carbonyl "electrophilic trap" releases methanol and forms a stable amide bond (Figure 17.22).

Thus, using the Staudvinger ligation process, a triphenylphosphine derivative containing a benzyl methyl ester group can be covalently conjugated to an azide derivative through the formation of an amide linkage. The two derivatives can have attached to them virtually any other molecules, such as proteins,

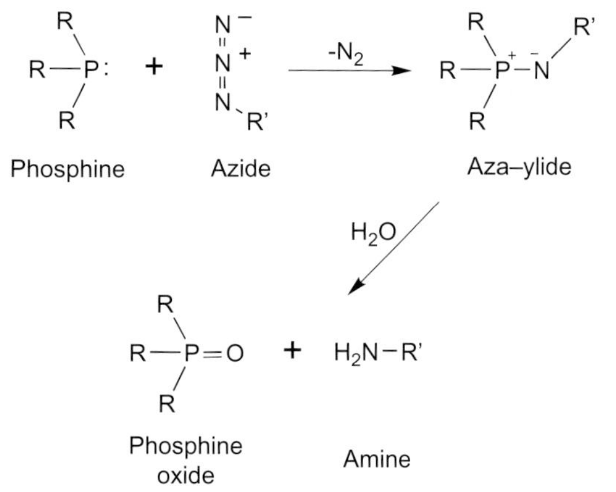

FIGURE 17.21 The Staudinger reaction involves the reduction of an azide to a primary amine with loss of N_2 and the concomitant oxidation of a phosphine derivative to a phosphine oxide.

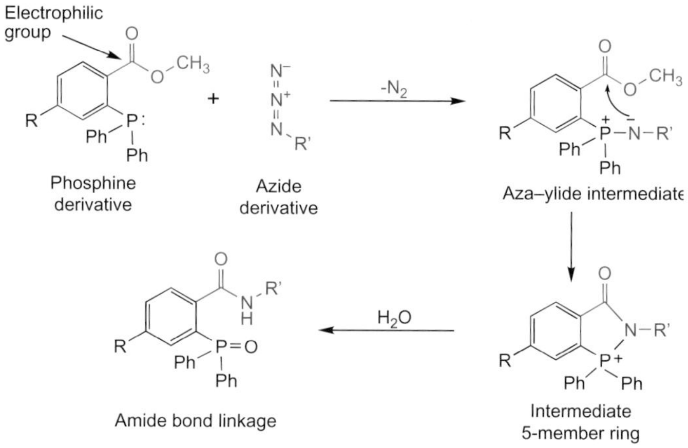

FIGURE 17.22 The Staudinger ligation reaction uses a modified phosphine derivative containing an electrophilic group that acts as a trap for the nucleophilic nitrogen in the intermediate aza–ylide. The resultant shift yields an amide bond derivative between the phosphine-containing molecule and the azide-containing molecule.

carbohydrates, other biological molecules, fluorescent tags, biotin groups, or other organic compounds, to create a conjugate between them.

The Staudinger ligation reactants are also extremely bioorthogonal. As discussed previously in the section on click chemistry, the (aliphatic) azide component is extremely stable in aqueous or biological solution and will not cross-react with other functional groups. The azide derivative is not appreciably reduced even given the reducing potential inside cells. In addition, the aryl derivative of triphenylphosphine used as a partner in this reaction has also been shown to be very stable in complex solutions, even to the point of not being an effective reducing agent for disulfides, as are many other phosphine compounds, including triphenylphosphine (Saxon and Bertozzi, 2000).

A number of alkyl–azide compounds that may be used in click chemistry reactions and the Staudinger ligation processes are now available. It is not recommended, however, to use aryl–azide compounds, as these are light sensitive and photoreactive as well as highly susceptible to reduction in the presence of thiols. More choices for obtaining commercial aryl–phosphine compounds to participate in this reaction are becoming available, as the applications for Staudinger ligation grow (Thermo Fisher).

The reaction between an alkyl–azide and the aryl–triphenylphosphine occurs without the requirement for other catalysts or activators, such as the need for Cu(I) in the click reaction. This provides an advantage for working with biological samples, since there is no possibility for toxicity or side reactions with added components being present. For this reason, the Staudinger ligation reaction is very amenable to being carried out with living cells without affecting cell viability.

Additionally, the azide group can be incorporated into amino acids, sugars, and lipids to label cellular molecules *in vivo* prior to reacting with a phosphine probe (see Figures 17.15 and 17.16). Kiick *et al.* (2002) demonstrated that both azido– and alkynyl–amino acid derivatives could be used as methionine surrogates and get integrated into proteins with nearly the same efficiency as normal methionine. Growing cells in the presence of one of these azido monomers will result in their incorporation into biopolymers at specific sites. In this manner, biomolecules can be purposely tagged using Staudinger ligation with a detectable phosphine probe, which is all done within living cells to track or locate them within their native environment. For instance, Prescher *et al.* (2004) used azido-sugar derivatives to modify cell surface glycans with azide groups at sialic acid residues, which are typically the outermost sugars on glycoproteins. It was found that cells took up these compounds most effectively if the hydroxyl groups were acetylated, presumably due to the easy transport of such hydrophobic derivatives through lipid membranes. The acetyl groups are removed within the cells by esterases, thereby providing the azido-sugar to the cell machinery for synthesizing glycoconjugates. The azido–glycan modifications produced on the cell surface using this process could then be labeled by the Staudinger ligation reaction with a phosphine–Flag tag derivative and subsequently detected using fluorescently labeled anti-Flag tag antibody and flow cytometry.

Due to the bioorthogonal nature of the reactants used for Staudinger ligation as well as the ability to incorporate azido analogs in biopolymers *in vivo* and the mild effect the reagents have on living cells, the process has been termed "a gift to chemical biology" (Köhn and Breinbauer, 2004). For the first time, efficient labeling of biomolecules within cells can be done with very small modifications to the biopolymers *in vivo*. Staudinger ligation permits discrete chemical tagging with detectable probes or affinity handles that facilitate purification.

The methods used for *in vivo* incorporation of azido monomers and performing a labeling reaction with live cells are relatively simple. The following protocol is based on the methods of Saxon and Bertozzi (2000), which use acetylated azidoacetylmannosamine as the azido-monomer source and a biotin–PEG–phosphine compound to biotinylate cell surface glycoproteins at the specific azide–sialic acid incorporation sites (Figure 17.23).

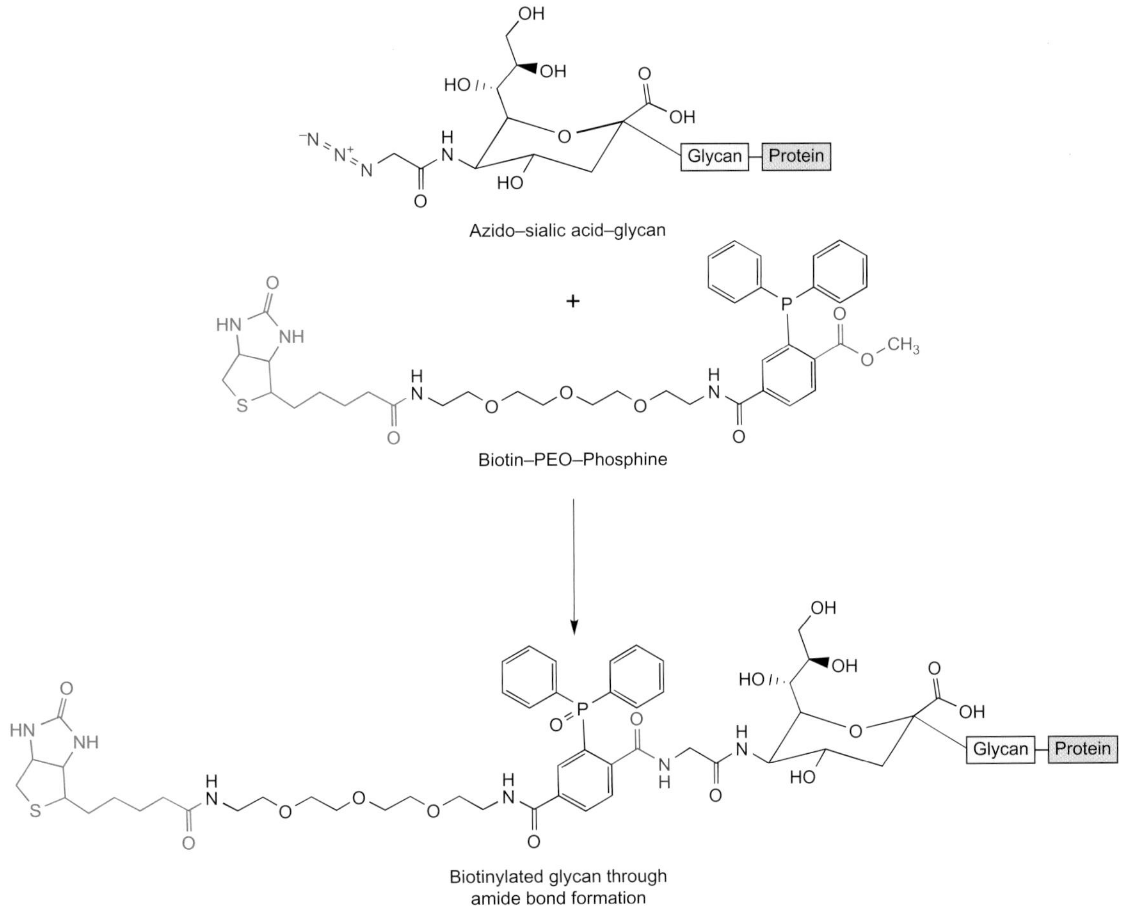

FIGURE 17.23 An azido–sialic acid derivative that gets incorporated into glycans in cells can be labeled specifically with a biotin–phosphine tag using the Staudinger ligation process. The result is an amide bond linkage with the glycan.

Protocol

1. Grow cells (~1×10^5 cells/ml) for 3 days in appropriate media containing a 20-μ*M* concentration of acetylated azidoacetylmannosamine.
2. Wash the cells at least twice with 0.1% fetal bovine serum in 10-m*M* sodium phosphate, 0.15-*M* NaCl, pH 7.4 (PBS), to remove excess azido-sugar.
3. Suspend the washed cells in 0.25 ml of PBS, pH 7.4.
4. Add to the washed cells 60 μl of a 5-m*M* concentration of the phosphine derivative to couple to the azido-sugar groups on the cell surface (e.g., biotin–PEG–phosphine).
5. Incubate for 1 h at room temperature with gentle mixing.
6. Wash the cells with PBS, pH 7.4, to remove excess biotinylation reagent.

The biotinylated glycans on the cell surfaces subsequently may be probed with (strept)avidin reagents to detect the azido–sialic acid modifications. Alternatively, the cells may be lysed and the glycoproteins isolated using an immobilized (strept)avidin or monomeric avidin affinity resin.

The same approach to *in vivo* labeling may be carried out using azide derivatives of sugars administered intraperitoneally in mice (once per day for 7 days) (Prescher *et al.*, 2004). Tissue samples can then be taken of particular organs and reacted *ex vivo* with a phosphine derivative to undergo Staudinger ligation. This process can facilitate probing of glycans on the cell surfaces of organs or cells within an animal. Various treatment procedures can be carried out, such as administering drug candidates to test animals, to determine the effect on glycosylation of proteins *in vivo*.

Staudinger ligation techniques can also be used to detect post-translational modification of proteins *in vivo*. Hang *et al.* (2007) developed a method to monitor fatty acid acylation of proteins using azido-fatty acids fed to cells. The two major types of fatty acid acylation, *N*-myristoylation and *S*-palmitoylation, could be detected with terminal (ω-)azido-labeled myristic acid or palmitic acid. Mammalian cells grown in the presence of these fatty acid azide derivatives resulted in certain proteins being modified to contain ω-azido-fatty acid modifications. Subsequent Staudinger ligation with a biotin–PEG–phosphine reagent resulted in covalent attachment to any post-translationally modified protein containing these groups (Figure 17.24). The biotin group could then be used for detection or purification of *N*-myristoylated or *S*-palmitoylated proteins using (strept)avidin reagents.

In a similar application, Kho *et al.* (2004) were able to detect post-translationally modified proteins using an azido–farnesyl analog. Cells grown in the presence of this derivative enzymatically incorporated the azido group into farnesylated proteins through the action of farnesyl transferase. These modifications could then be targeted through Staudinger ligation using the same biotin–PEG–phosphine reagent (Figure 17.25).

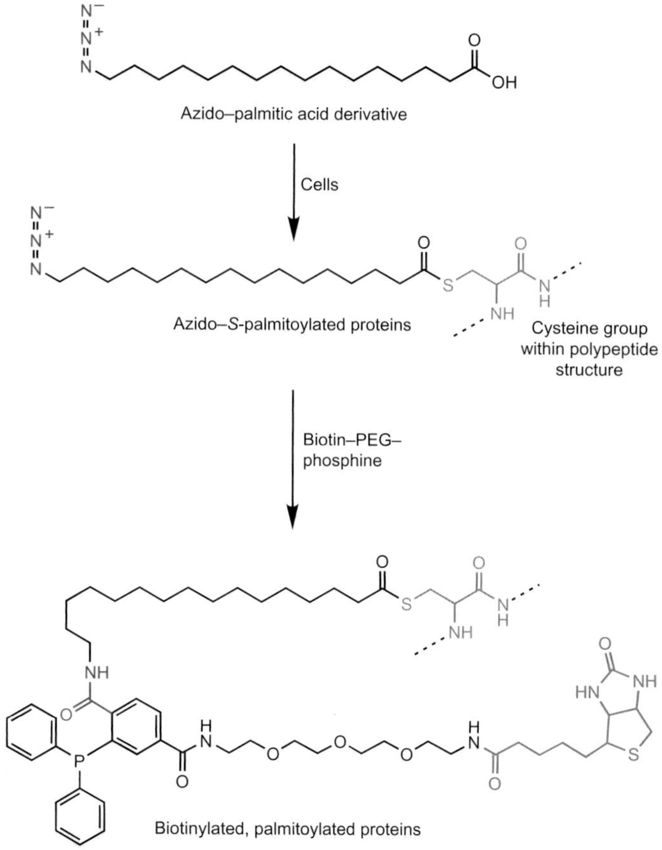

FIGURE 17.24 An azido–palmitic acid derivative can be added to cells to obtain palmitoylated proteins that contain an azide group able to participate in the Staudinger ligation reaction. Biotinylation of these post-translationally modified sites can then be carried out *in vivo* using a biotin–phosphine reagent.

Protein farnesyl transferase can also be used to add a geranylazide derivative to a synthetic peptide by incorporating the enzyme recognition sequence "CAAX" at the C-terminal of any peptide. This enzyme uses a farnesyl diphosphate derivative to transfer covalently the lipid to the cysteine residue via a thioether bond. Xu *et al.* (2006) found that the use of 6,7-dihydrogeranylazide diphosphate resulted in appending the terminal azido derivative onto such peptides, yielding an azido functionality for subsequent conjugation using Staudinger ligation. This process enabled targeted coupling through the C-terminal of any peptide or protein containing the CAAX box sequence.

Another important variation of the Staudinger ligation reaction described above involves the use of cleavable aryl groups on the triphenylphosphine component, which allows for a "traceless" ligation reaction to occur. This strategy results in a zero-length amide bond between the phosphine derivative and the azide derivative, thus removing the triphenylphosphine component from the final conjugate and linking the two molecules together directly (Saxon *et al.*, 2000; Nilsson *et al.*, 2000, 2001; Soellner *et al.*, 2002; Saxon and Bertozzi, 2003, 2006). The phosphanes useful in this process are built from acyl derivatives of compounds such as those shown in Figure 17.26. During

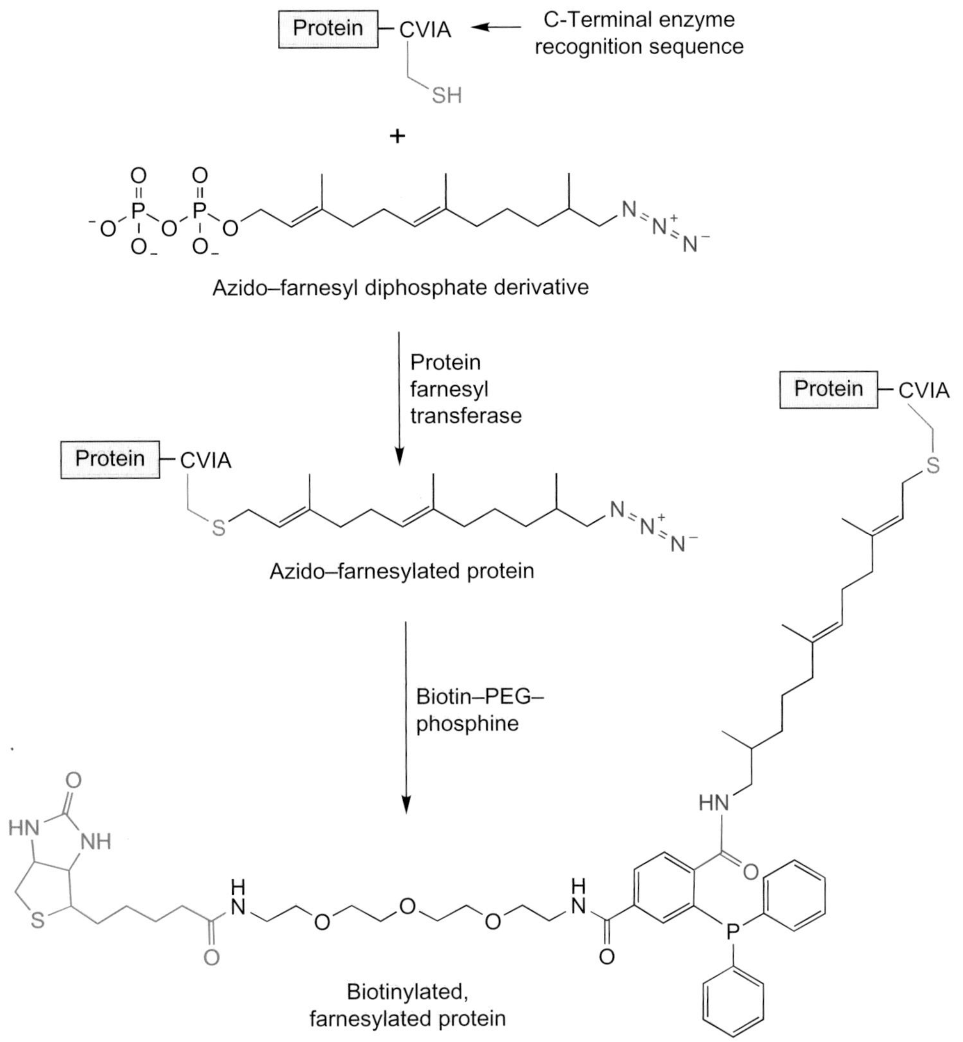

FIGURE 17.25 An azido–farnesyl diphosphate derivative can be added to cells to obtain farnesylated proteins containing terminal azide groups that can be targeted in a Staudinger ligation reaction. Biotinylation of these post-translationally modified proteins can be done *in vivo* using a biotin–phosphine derivative.

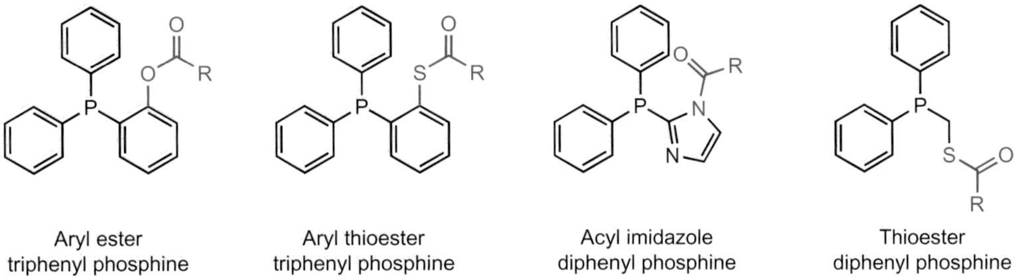

FIGURE 17.26 Certain unique phosphine derivatives can be used in the design of modification or conjugation reagents to create a traceless Staudinger ligation process, wherein the phosphine group is lost and an amide bond between an azide-containing molecule and the phosphine-containing molecule results.

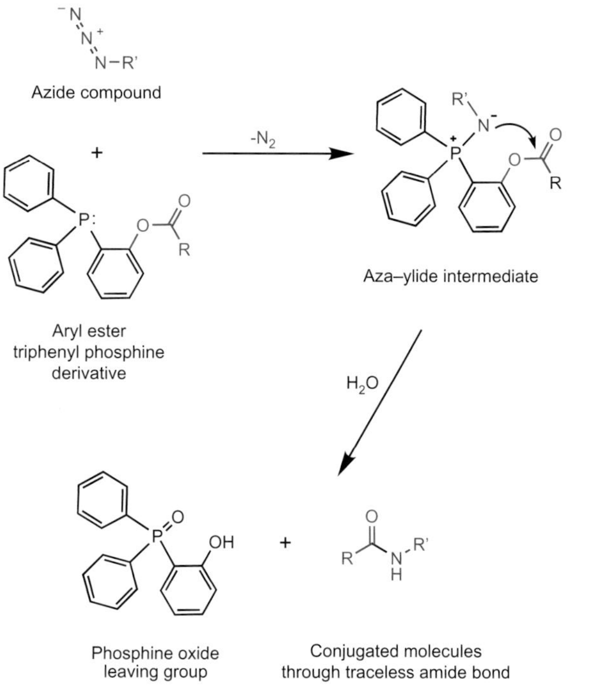

FIGURE 17.27 A traceless Staudinger ligation process involves the formation of an intermediate aza–ylide with subsequent attack of the nucleophilic nitrogen atom on the neighboring electrophilic group. The formation of an amide bond then occurs concomitant with the loss of the phosphine component, thus forming a zero-length crosslink between the two molecules.

the Staudinger ligation process, once the azide reactant forms the aza–ylide with the phosphine, electrophilic attraction induces the nitrogen to attack the electron deficient carbonyl, which in turn causes release of the phosphonium group and forms the amide bond (Figure 17.27).

The most useful phosphane derivatives in a traceless Staudinger ligation reaction include the acyl-modified 2-diphenylphosphanylphenol and the acyl-modified diphenylphosphanylmethanethiol. These two core structures provide the best rate of reaction and efficiently form the amide bond conjugate (Köhn and Breinbauer, 2004). The traceless Staudinger ligation method no doubt will become a popular choice to avoid retention of the bulky phosphane species in bioconjugates. The reaction can be used with success to produce long polypeptide chains by linking together azido-containing peptides with phosphine-containing peptides to form the appropriate biological peptide bond between them (Nilsson *et al.*, 2003). This makes it possible to synthetically create polypeptides that are too long to create using standard solid-phase peptide synthesis procedures (Figure 17.28). In addition, the traceless Staudinger ligation reaction can be used to link biomolecules containing azide groups to surfaces containing the traceless phosphane derivative (Soellner *et al.*, 2003).

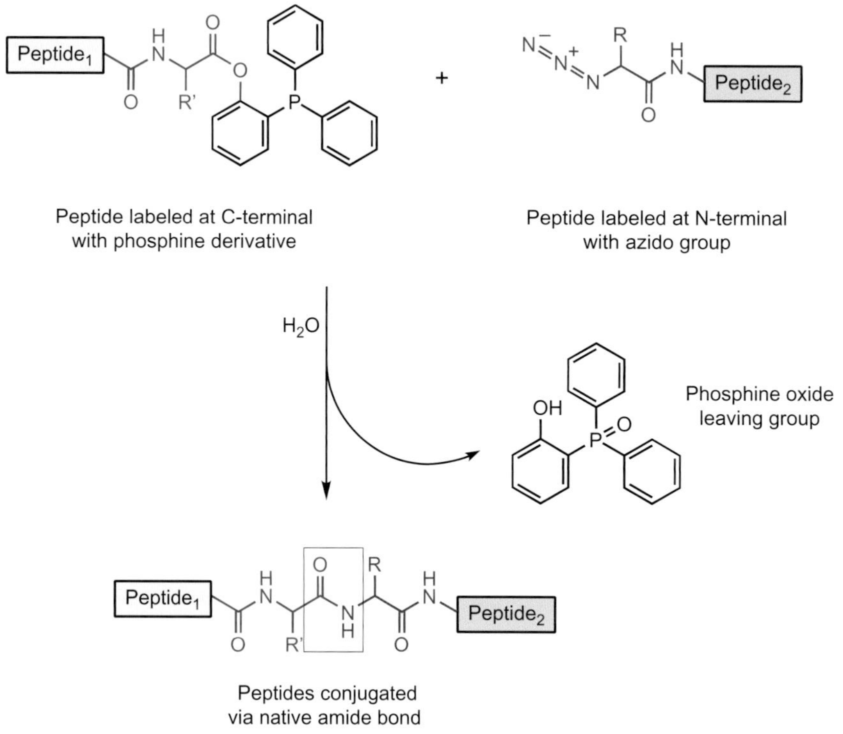

FIGURE 17.28 The traceless Staudinger reaction can be used to form larger peptides from smaller peptides, if one contains an azido group at the N-terminal and the other one contains a phosphine ester at its C-terminal. The reaction gives a native peptide (amide) bond with loss of the phosphine group.

7. NATIVE CHEMICAL LIGATION

Native chemical ligation is an important alternative chemoselective peptide conjugation technique to the previously discussed non-native methods (Dawson *et al.*, 1994). This system can provide discrete coupling of the N-terminal of one peptide to the C-terminal of another peptide using a unique reaction process, essentially extending a peptide chain while maintaining native sequence and bonding characteristics. The reactant partners are prepared through typical peptide synthetic procedures, wherein one peptide is made to contain an alpha-thioester on its C-terminal carboxylate and the other peptide is synthesized to contain a cysteine amino acid residue at its N-terminal. The reaction of these two derivatives proceeds through nucleophilic attack of the cysteine thiol of one peptide onto the carbonyl group of the alpha-thioester at the C-terminal of a second peptide. The result forms an intermediate thioester by transthioesterification, which then spontaneously rearranges by an S→N acyl transfer to form a native amide bond between the peptides with no foreign organic structure remaining (Figure 17.29). The reaction proceeds at physiological pH using fully unprotected (i.e., unblocked) amino acid functional groups, which is unusual in peptide synthetic strategies.

Peptides typically are prepared for this ligation process using alpha-alkyl thioesters, because they are simple to make at the time of peptide synthesis. However, due to the relatively slow reaction kinetics of alkyl thioesters, most native chemical ligation processes have been catalyzed through the use of thiol compound additives, such as benzyl mercaptan or thiophenol (Dawson *et al.*, 1997). These compounds react with the initial alpha-alkyl thioester to form another intermediate, an aryl thioester, which is more reactive toward the N-terminal cysteine on the other peptide to be coupled. A study of the rate of reaction for different thiol catalysts indicated that (4-carboxymethyl)thiophenol performed the best. The addition of this compound to a native chemical ligation reaction resulted in at least a 10-fold improvement in reaction rates (Johnson and Kent, 2006).

Another advantage to the use of a thiol additive is that the abundance of free thiol groups in the reaction environment will prevent the oxidation of the cysteine thiol at the N-terminal of the other peptide. Without added thiol transesterification catalysts, disulfide formation resulting in dimerization of the Cys peptide would be a dominant side reaction in aqueous, oxygenated buffer conditions.

Native chemical ligation has been used successfully to couple two unprotected peptides together during solid phase synthesis, wherein one of the peptides is attached to the resin using a thioester linkage and the other peptide is introduced containing a cysteine at its N-terminal (Camarero *et al.*, 1998). The ligation process could also form cyclic peptides, although the efficiency of the reaction was less than the conjugation of two separate peptides.

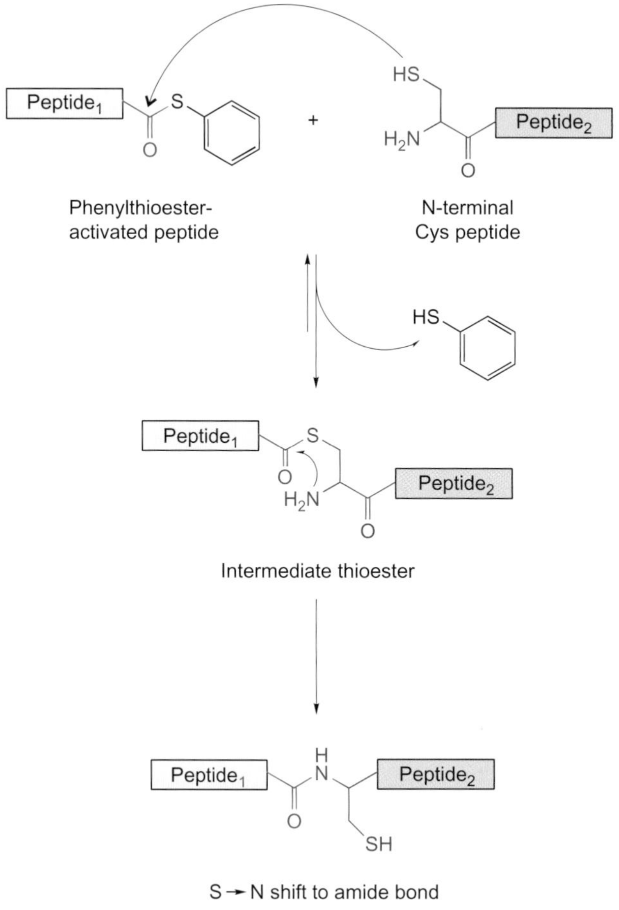

FIGURE 17.29 The native chemical ligation reaction can be used to form larger peptides from smaller peptides, if one contains a cysteine residue at its N-terminal and the other one contains a thioester on its C-terminal. Reaction of the peptide derivatives gives a native peptide (amide) bond.

So efficient is this method to link peptide sequences together in a native amide bonded state that complete proteins of various sizes have been synthetically made (Hackeng *et al.*, 1999), such as the preparation of a serine protease (Pal *et al.*, 2003), bovine pancreatic trypsin inhibitor (Lu *et al.*, 1998), cytochrome b562 (Low *et al.*, 2001), and the triple zinc finger protein Zif268 (Beligere and Dawson, 1999). Many other examples of full or partial protein synthesis can be found in the published literature (for reviews, see Dawson and Kent, 2000, and David *et al.*, 2004).

Native chemical ligation can also be extended to the conjugation of peptides or proteins to other molecules or surfaces. For instance, Reulen *et al.* (2007) prepared liposomes that contained cysteine–PEG–phospholipid derivatives and then coupled thioester-modified peptides or proteins to form a protein–liposome conjugate. Using this procedure, approximately 100 molecules of a collagen binding protein could be coupled to the cysteine-containing liposomes.

In addition, Dose and Seitz (2005) employed native chemical ligation to synthesize peptide nucleic acids (PNAs) by linking shorter segments of PNAs to make long contiguous strands, which could not be made through typical oligo synthesis procedures.

7.1. Expressed Protein Ligation and Inteins

Although there are a limited number of reports indicating that N-terminal cysteine-containing proteins can be expressed recombinantly, the *in vivo* production of a protein having a C-terminal thioester is considerably more difficult. That's why the initial development of native chemical ligation techniques was restricted to the use of peptide synthesis to generate the two peptide derivatives for ligation. However, the recombinant generation of C-terminal thioesters and N-terminal cysteine residues in peptides for native chemical ligation can now be achieved using intein technology. In the same frame that native chemical ligation reactions were being discovered and explored, intein technology was being investigated as a new *in vivo* native protein splicing and ligation mechanism.

Inteins are certain sequences of amino acids in precursor proteins that can catalyze a cleavage of their own internal peptide structure, resulting in the excision of the intein peptide segment and ligation together of the two peptide segments flanking the excised region (exteins). The exteins are thus ligated together with a native peptide (amide) bond. The immediate C-terminal side flanking the intein usually contains either a serine/threonine residue or a cysteine amino acid group at the splice junction. The N-terminal extein side next to the intein can also contain either a serine/threonine group or a cysteine residue, which are able to form an ester or thioester functionality, respectively. If a cysteine group is present at the N-terminal splice junction side of the intein, then the thioester intermediate that forms is very similar to the reaction intermediate of native chemical ligation.

The intein segment is excised through a number of steps that first involves an N→S shift at the C-terminal cysteine residue at the extein junction to create a thioester intermediate. Then nucleophilic attack of either the serine/threonine hydroxyl or the cysteine thiol of the extein on the N-terminal side of the intein proceeds to cleave the C-terminal extein and ligate it to the N-terminal extein thiol (or hydroxyl) through another (thio)ester bond. Therefore, the cleavage and ligation reaction proceeds through a second ester or thioester intermediate that involves either an O→N shift or an S→N shift, which cleaves off the intein segment entirely. The complete thioester reaction route is shown in Figure 17.30, which can be seen to be very similar to that of a native chemical ligation reaction, illustrated previously.

Muir *et al.* (1998) realized that the intein reaction could be used to facilitate a native chemical ligation with a synthetic N-terminal cysteine-containing peptide or cysteine-containing molecule. With the discovery of a mutant intein that could form an intermediate thioester but not go on to complete the splice and ligation reaction (Xu and Perler, 1996; Chong *et al.*, 1997), this expression technology now could be combined with the native chemical ligation reaction to facilitate a new expressed protein ligation (EPL) method.

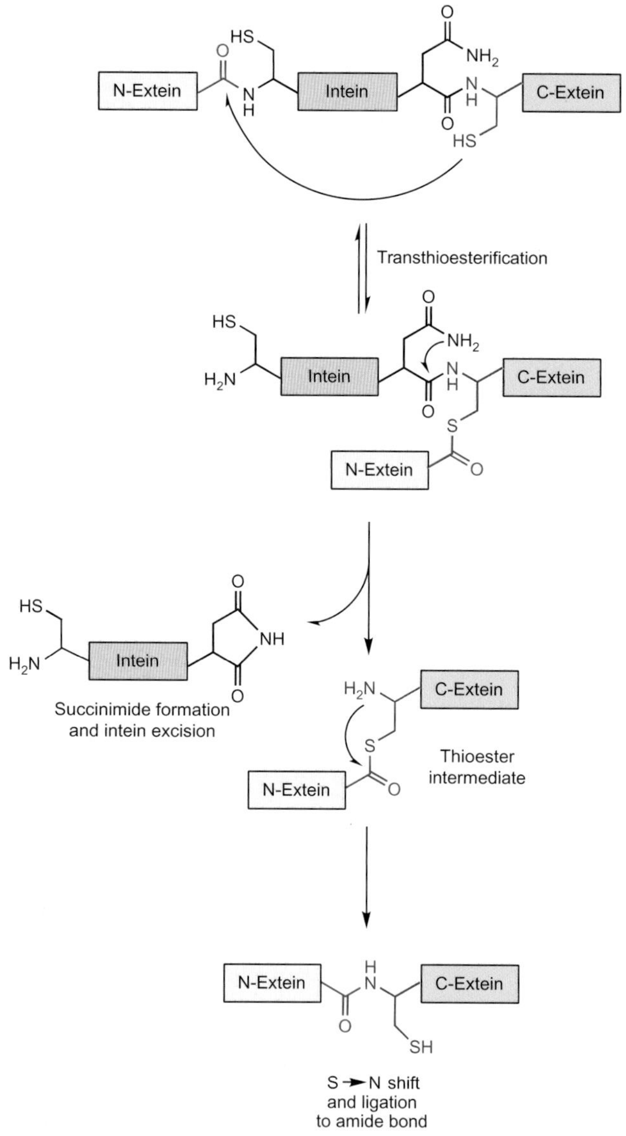

FIGURE 17.30 The native process leading to intein excision and ligation of extein fragments involves a sequence of reactions involving transthioesterification, cleavage of the intein fragment, and an S→N shift, which ligates the two extein peptides together via an amide bond.

Fusion vectors are available that combine a recombinant protein with a mutant mini-intein segment (not containing an endonuclease domain) and followed by a chitin binding domain (CBD) (Zhang *et al.*, 2001). These mutants typically also have an alanine substitution that replaces the cysteine or serine/threonine usually found on the C-extein splice junction. Alanine cannot facilitate attack on the thioester on the N-extein side and thus the intein is not cleaved. Using the CBD portion, the entire expressed fusion protein can be purified on a chitin column through specific affinity binding and then the N-extein segment induced to cleave using DTT, which results in release of the desired recombinant protein having a C-terminal thioester.

However, if the expressed protein is treated on the affinity support using thiophenol, this also will release the protein and result in a phenylthioester at its C-terminal, which is the reactive intermediate imminently suitable for native chemical ligation. Treatment of this activated thioester protein with a N-terminal cysteine peptide induces the native chemical ligation reaction and couples the peptide to the expressed protein through an amide bond (Severinov and Muir, 1998) (Figure 17.31).

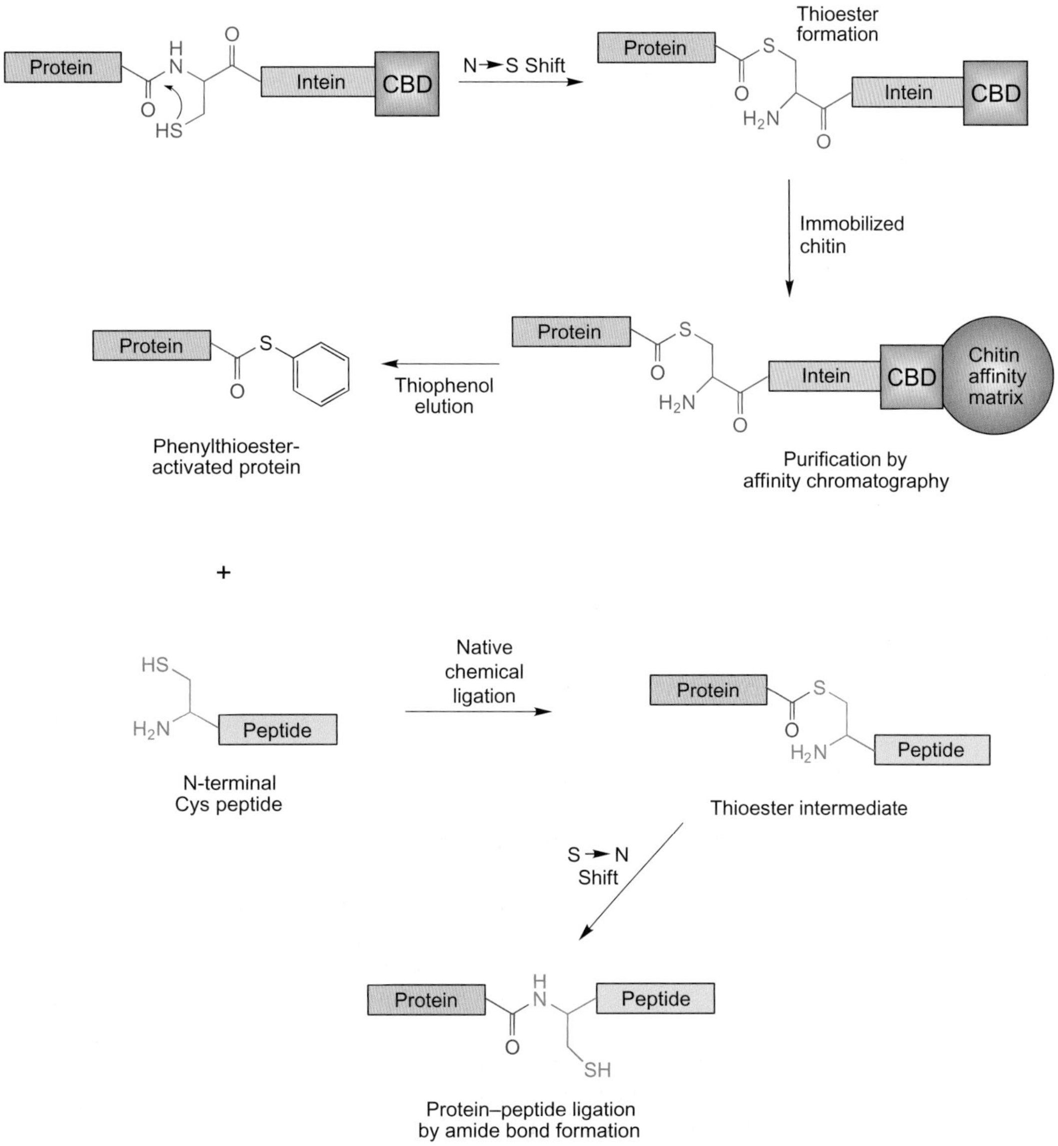

FIGURE 17.31 The expressed protein ligation process involves a fusion protein containing an intein tag plus a chitin binding domain. The fusion protein is captured on an immobilized chitin resin and after removal of contaminating proteins is eluted using thiophenol, which cleaves at the thioester bond between the intein and the desired expressed protein. This releases a phenylthioester-activated protein that can be used in the native chemical ligation reaction with another peptide containing an N-terminal cysteine residue. Conjugation results in a native amide (peptide) bond formed between them.

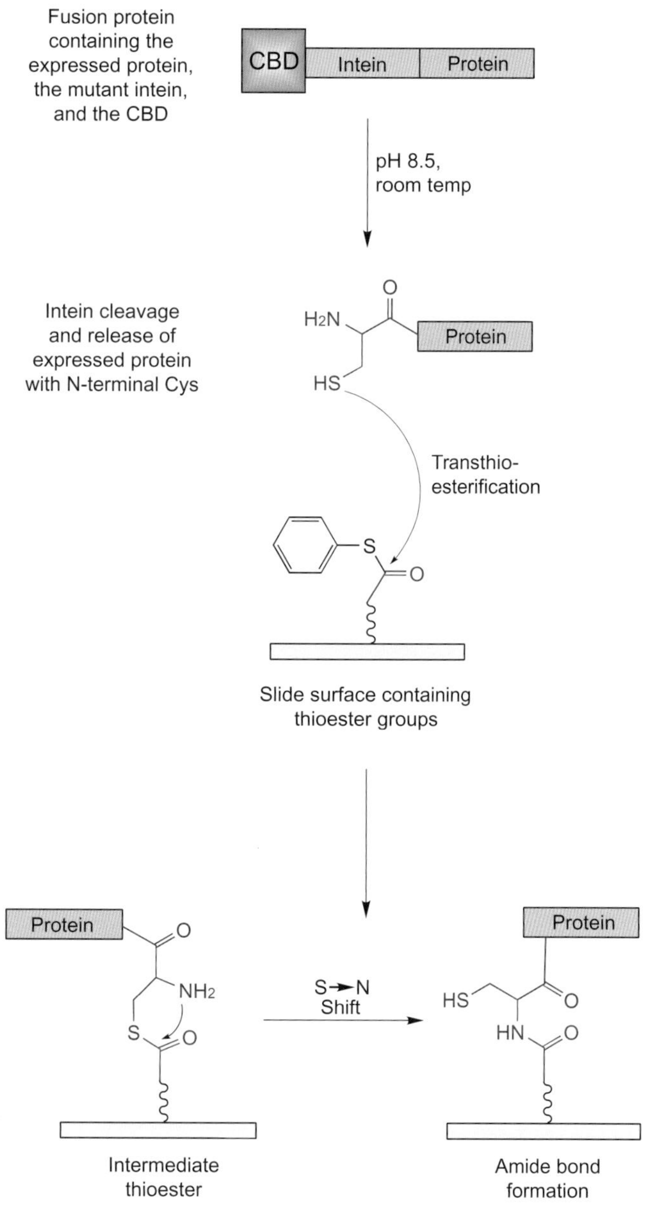

FIGURE 17.32 Expressed protein ligation reactions can be used to couple a fusion protein to a surface containing a thioester derivative. After cells are grown and the fusion protein expressed, a pH and temperature shift causes intein cleavage with release of the expressed protein with an N-terminal cysteine residue. Reaction with the thioester surface results in a native chemical ligation reaction that forms an amide bond linkage with the expressed protein.

Expressed protein ligation extends the applicability of native chemical ligation to recombinantly produced proteins using the mutant mini-intein vector system. Proteins being expressed using this method will contain a C-terminal thioester and therefore can be conjugated to any reagent or probe molecule containing a cysteine group with an available alpha-amine and thiol group. Expressed proteins can also be immobilized onto solid surfaces by coupling them solely at their C-terminal ends.

Girish *et al.* (2005) coupled proteins onto surfaces using this approach, but in this case expressing a protein containing an N-terminal Cys residue and then reacting it with a thioester group on a glass slide (Figure 17.32). A similar strategy has also been used with success to label specifically expressed proteins in live cells using thioester-containing probes (Nilsson *et al.*, 2003). Using the pTWIN vector (New England Biolabs), a recombinant system was developed to express proteins having an N-terminal Cys using the Ssp DnaB mini-intein segment (Yeo *et al.*, 2003). This intein contains a C-terminal asparagine residue that undergoes a self-cleaving reaction after a shift in pH and temperature. Growing cells in physiological pH conditions does not cleave the intein. However, after cell lysis, an increase to pH 8.5 under room temperature conditions will initiate the cleavage reaction. Thioester-containing probes, including thioester–fluorescent molecules or thioester–biotin, can be added directly to live cells to label these expressed proteins via native chemical ligation (Figure 17.33).

The following protocol for EPL, including purification using a CBD fusion tag followed by native chemical ligation, is based on the methods of Muir *et al.* (1998), Chong *et al.* (1997), Chong *et al.* (1998), Evans *et al.* (1998), Severinov and Muir (1998), and the NEB instruction manual for the IMPACT-TWIN system. The recombinant protein is recovered from the affinity column as the thioester derivative ready for reaction with a N-terminal Cys peptide or another tag containing a Cys residue.

Protocol

1. A gene encoding a recombinant protein of interest is cloned into a cleavable mutant intein–CBD plasmid vector and transfected into cells (e.g., using the IMPACT-TWIN system from New England Biolabs). The cells are grown in LB media with 200-μg/ml ampicillin at 37°C to an OD_{600} nm of about 0.5 to 0.8, induced to express the fusion protein complex by the addition of 1-m*M* IPTG, and incubated overnight.
2. Prepare a chitin affinity column by washing with at least 10 bed volumes of 25-m*M* HEPES, 250-m*M* NaCl, 1-m*M* EDTA, 0.1% Triton X-100, pH 7.0 (wash buffer).
3. Recover the cells by centrifugation, lyse them in wash buffer containing protease inhibitors (e.g., 20-μ*M* PMSF), and clarify the supernatant by centrifugation.
4. Apply the lysate onto the affinity column and wash with at least 10 column volumes of wash buffer to

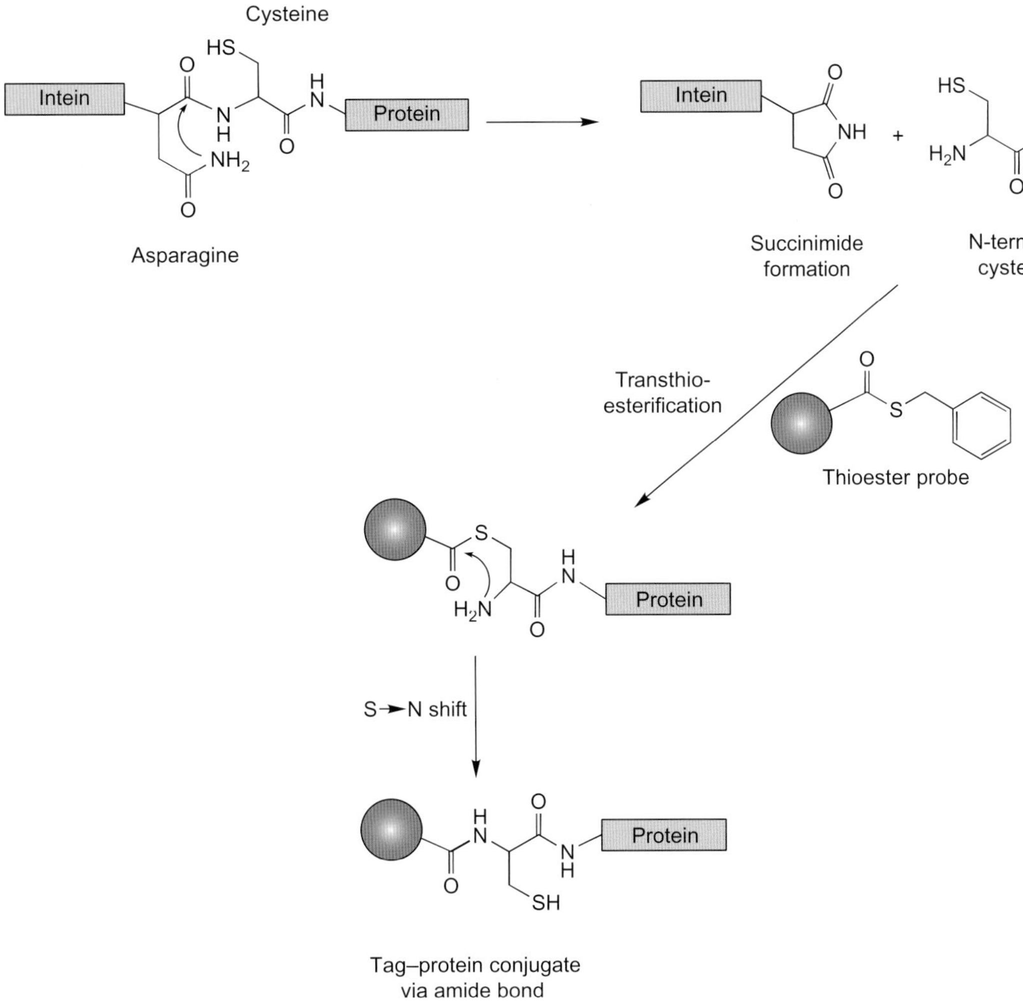

FIGURE 17.33 An expressed protein containing a mutant intein segment can undergo self cleavage to form an N-terminal cysteine residue, which can then be reacted with a thioester probe to specifically label the protein via an amide bond.

remove not-bound protein. Monitor the eluate by absorbance at 280 nm to ensure that baseline has been reached.

5. Wash the column with 2 to 3 column volumes of elution buffer (25-m*M* HEPES or Tris-HCl containing 500-m*M* NaCl, 1-m*M* EDTA, and 2% (v/v) thiophenol (or 50-m*M* 2-mercaptoethansulfonic acid), pH 8.5. Next, add 2 column volumes of elution buffer containing 1 to 2-m*M* of an N-terminal Cys peptide or Cys-containing tag. Stop the flow and incubate the column at room temperature for 24 h. The thiophenol will cleave the recombinant protein at the intein splice junction, while the cysteine-containing peptide will react with the intermediate ester through transthioesterification and an S→N shift to ligate the molecules via an amide bond.
6. Elute the conjugated molecules using wash buffer and purify using dialysis or gel filtration.

CHAPTER

18

PEGylation and Synthetic Polymer Modification

1. DISCRETE PEG REAGENTS

Poly(ethylene glycol) (PEG) has been used for many years as a modification and conjugation reagent for biological molecules (Roberts *et al.*, 2002; Veronese, 2009; Ikeda and Nagasaki, 2012). Part of the rationale for using PEG polymers in bioconjugation applications includes a dramatic increase in the water solubility of modified molecules, a decrease in immunogenicity due to the shielding of modified molecules from the immune system, protection of modified protein and peptides from digestion by proteolytic enzymes, and effectively increasing their hydrodynamic volume and decreasing clearance rates by renal filtration, all of which results in an increase in the serum half-life of modified molecules *in vivo*. Thus, hydrophobic drugs modified with PEG polymers become much more water soluble, foreign proteins and other immunogenic molecules are hidden from the circulating antibodies and cells of the immune system, labile peptides or proteins cannot be degraded by enzymes, and the reticuloendothelial system and kidneys cannot remove labeled drugs as quickly as their unlabeled counterparts. For these reasons, many drug candidates using PEG modification are currently in clinical trials and several are already on the market.

The majority of PEG applications involve the use of long-chain linear or branched PEG polymers having molecular masses of at least 2.5 kDa to over 50 kDa (see Section 2, this chapter, for selected applications of these long PEG polymers). PEG is typically made from ethylene oxide by an anionic ring-opening reaction, which results in long polymer molecules consisting of the general structure $HO(CH_2CH_2O)_nH$. The number of repeat units (n) in standard commercial PEG polymers created through this process can typically be anything from less than 50 to over 1100. However, in these conventional PEG polymers of any given size, there actually exists a distribution of chain lengths, as is typical for any polymer-based substance. The range of chain lengths for a given PEG size is approximately Gaussian in distribution, which means that most PEG reagents prepared by standard polymerization processes are fairly disperse. The level of polydispersity is usually indicated by Mw/Mn, which is called the polydispersity index (PDI) and is equal to the weight average molecular weight (Mw) divided by the number average molecular weight (Mn). If the PDI is equal to 1, the polymer is said to be monodisperse. For PEG polymers, the PDI is typically less than 1.2, which is quite good for polymeric materials, but this value probably reflects the concern over the use of polydisperse PEG for therapeutic applications. Even at this low level of PDI, the chain distribution often can be quite high. Kenworthy *et al.* (1995) reported that the mean number of repeat units and variance for PEG-2000 averages 53 units with a variance of 11, while a PEG-5000 polymer has an average 130 units with a variance of 20. Shorter chain PEG polymers have a tendency to have greater polydispersity than the larger polymers. For example, the commercially available PEG-1500 can have between 19 and 48 repeat units in a typical preparation, which correspond to a molecular weight distribution of 800 to 2100 Daltons (Davis and Crapps, 2006). Thus, most commercial sources of crude PEG polymers are highly disperse and they probably should be avoided entirely for critical bioconjugation work, unless they have been carefully purified to isolate a single chain length. This isolation process typically is done using size exclusion chromatography, which is a time-consuming and expensive procedure and still may not yield an entirely homogeneous PEG preparation.

However, true monodisperse PEG reagents have become available, which are made not by polymerizing small monomers, but by linking short, discrete PEG segments together to create pure polymers of known structure and purity. These discrete PEG molecules can be made in chain lengths from as little as 2 to over 24

Bioconjugate Techniques, Third Edition.
DOI: http://dx.doi.org/10.1016/B978-0-12-382239-0.00018-2

repeating ethylene oxide units, and theoretically, virtually any chain length can be produced by building up from smaller precursors (Davis and Crapps, 2006).

In the synthesis process, a short PEG segment containing a hydroxyl protecting group on one side and a free hydroxyl on the other end is reacted with another PEG segment containing a reactive group on one end and a protecting group on the other end. The reactive group must also be a good leaving group, so that once it reacts with the hydroxyl on the other PEG unit, a conjugation reaction occurs to form an ether bond. This results in the covalent linking of the two PEG molecules together to form a longer PEG compound equal to the combined length of the reactants. Deprotection of the ends can then be done to add additional functionality, such as a reactive group, functional group, or to form a hydroxyl or methoxy end (mPEG).

Prior to this synthetic method being developed, small PEG-containing compounds were limited to very short ethylene oxide segments, such as the commonly used reagents ethylene glycol and tetraethylene glycol. Now, using new discrete PEG reagents, the advantages that PEG compounds have provided for use in the modification and crosslinking of biomolecules can be incorporated at known polymer lengths into any bioconjugation reagent to enhance its properties.

Discrete PEG reagents have been reported that incorporate reactive groups, fluorescent probes, metal chelates, drug molecules, affinity ligands, biotin, and a host of other constituents. For instance, Wei *et al.* (2006) developed a PEG-functionalized texaphyrin derivative, which was shown to have enhanced solubility and anti-cancer activity *in vivo*. Four $mPEG_4$ chains decorating the central gadolinium(III) texaphyrin were found to convey dramatic anti-proliferative effects compared to the parent chelate without PEGs present. In another application using the chelator DOTA, Shi *et al.* (2011) created several ^{111}In-labeled cyclic RGD peptides with PEG_4 linkers to target integrin receptors on human gliomas cells. For the creation of real-time fluorescence imaging probes for MMP enzymes, Zhu *et al.* (2011) used amine derivatives of $mPEG_4$, $mPEG_{12}$, $mPEG_{24}$, and $mPEG_{67}$ attached to the C-terminus of the peptide sequence to improve water solubility and *in vivo* activity. The $mPEG_{12}$ probe provided the best specific MMP targeting ability toward SCC7 tumor cells in mice.

In another application of a PEG_4 spacer, Clevenger *et al.* (2004) prepared a biotinylated derivative of the antibiotic geldanamycin (GDA) to use as an inhibitor of the 90-kDa heat shock protein Hsp90. Use of the PEG linker in building such an organic drug complex has the advantage of adding a hydrophilic arm to an otherwise very hydrophobic probe. The biotin–PEG_4–GDA conjugate could be used to bind the active site of Hsp90 proteins and then affinity-purify them on a (strept)avidin-containing resin.

Similarly, Kruszynski *et al.* (2005) used the reagent NHS–PEG_4–biotin to make biotinylated analogs of human MCP-1. This compound, described later in this section, provides a long-chain biotin handle that has better solubility properties than the corresponding aliphatic reagent NHS–LC–biotin, which has been used in many applications (Chapter 11, Section 6.2). Kornilova *et al.* (2005) used the same PEG reagent to biotinylate various γ-secretase peptide inhibitors to create probes of this multi-protein complex.

Hydrophilic short biotin–PEG tags have also found their way into the design of multifunctional crosslinkers to study protein structures by mass spectrometry. Fujii *et al.* (2004) developed a homobifunctional NHS ester crosslinker that, in addition, has a PEG–biotin handle (Figure 18.1). The reagent is actually a trifunctional compound similar to the biotinylated PIR compound described in Chapter 24, Section 1.4. The NHS ester arms are of identical length to provide linkages of the same molecular distance in each direction, while the PEG group on the biotin arm avoids the hydrophobic collapse of alkyl chain spacers by providing an extremely hydrophilic linker with high freedom of motion in aqueous environments. Due to the increased water solubility provided by the PEG linker arm, the sensitivity of mass spectrometry analysis was enhanced over the use of more traditional aliphatic spacers. In particular, this discrete PEG compound was found to be optimal for performing mass spectrometry analysis in three dimensions (MS^3).

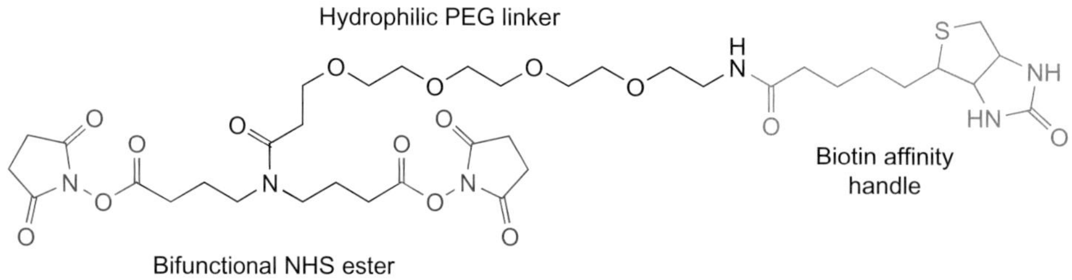

FIGURE 18.1 A trifunctional reagent for studying protein interactions by mass spectrometry. The *bis*-NHS ester arms crosslink interacting proteins, while the discrete PEG-containing biotin arm can be used to isolate or detect the conjugates using (strept)avidin reagents.

Coupling of affinity molecules to surfaces can also be enhanced by the use of discrete PEG linkers. Nishimura *et al.* (2005) modified an amino surface with a NHS–PEG_{12}–maleimide crosslinker to create a hydrophilic self-assembled monolayer (SAM) surface that was thiol reactive for the conjugation of sulfhydryl-modified RNAs. This array was then used to investigate the binding specificity of synthetic kanamycins with selected RNA sequences to prove the specific interaction of ribosomal RNA with this molecule. The PEG linkers on surfaces provide lower nonspecific binding character than alkyl linkers, when preparing SAM surfaces for affinity interactions.

The low nonspecificity of PEG layers was also used to eliminate biomolecule binding to certain areas of an array. Kidambi *et al.* (2004) patterned an mPEG–carboxylate molecule onto polyelectrolyte multilayers to mask portions of the surface. The extremely low binding character of PEG provides advantages for creating patterned surfaces that other modifiers using aliphatic alkyl linkers do not provide.

SAM surfaces on metals, such as gold particles or planar arrays, have been enhanced in their performance and properties through the use of PEG-containing modifications. Prime and Whitesides (1991) used an mPEG–thiol combined with a slightly longer thiol–PEG–carboxylate compound to create monolayers on gold through dative bonding with the thiol groups. The thiol–PEG–carboxylate compound is typically used at about 10% of the concentration of the thiol–mPEG compound to form a lawn of low-binding mPEG molecules which is interspersed with enough carboxylates to provide sites for covalent attachment of affinity ligands or antibodies. Spangler *et al.* (2004) improved upon this strategy by creating a dendritic structure branching off from a central phenyl ring core and containing two thiol arms and a single PEG arm containing either an mPEG group or a PEG–carboxylate (Figure 18.2). The presence of two thiol groups increases the strength of the linkage with the gold surface, thus providing resistance to oxidative cleavage of the SAM surface.

In another application involving a study of biotin compounds for potential use *in vivo*, it was found that the preferred structures included molecules that contained PEG spacers to provide increased water solubility and *in vivo* activity (Wilbur *et al.*, 2000). Fifteen biotin derivatives were conjugated to cyanocobalamin, which is a binder of cobalt that may be used in radiotherapy for cancer. Structures with enhanced activity were formed with PEG spacers of varying length between the cyanocobalamin and the biotin handle, which provided maximal hydrophilicity to the entire complex.

Biotinylation reagents have historically used aliphatic chains to provide a spacer between a modified molecule and the bicyclic ring of biotin. This allows for enough molecular distance for (strept)avidin to easily bind during detection or targeting applications. Even long spacers on biotin reagents have typically been constructed of hydrophobic alkyl chains for the biotinylation of proteins. This can be detrimental to protein solubility, because biotin itself is sparingly soluble in aqueous solution and adding an alkyl spacer to this group only decreases its water solubility further. In particular, antibodies modified with long hydrophobic biotin compounds often aggregate and lose activity, especially at high modification levels. Replacing the alkyl chain with a discrete PEG spacer, however, can dramatically increase water solubility and prevent antibody aggregation as well as significantly boost long-term stability.

Discrete PEG reagents can provide benefit for nearly any crosslinking compound or modification reagent designed for use in aqueous environments. Frequently, PEG chain lengths between 4 and 24 repeating ethylene oxide units create increasingly hydrophilic character for modified biomolecules or surfaces. The following sections describe a number of these PEG compounds, including crosslinkers, biotinylation compounds, and multi-armed PEG modification agents designed to block molecules and surfaces. In addition, PEG-based compounds are available that have functional groups on both ends for use as building blocks to create other PEG compounds or as spacers to build unique surface functionality. Most of these reagents are available commercially through Quanta BioDesign or Thermo Fisher Scientific.

Modification or crosslinking agents containing a discrete PEG spacer are often not powders or crystalline substances, but frequently they are thick, sticky, viscous liquids. Such materials are difficult to dispense by weighing out a small portion from a vial. For this reason, it may be best to dissolve an entire vial in a dry organic solvent prior to use or try to weigh out a much larger quantity than may be initially required. Stock solutions may be stored for weeks at <0°C, but long-term stability of active groups is dependent on the quality of the solvent used. For this reason, it is best to prepare solutions fresh.

Organic solvents that can be used with discrete PEG compounds include DMSO, DMF, DMAC (*N,N'*-dimethylacetamide), and methylene chloride. The compounds are also very soluble in many other commonly used organic solvents, which provide flexibility for doing reactions. DMAC is particularly convenient, because it is easily dried of contaminating water (using molecular sieves), it does not decompose like DMF (producing amines), and it does not have the odors of some of the other solvents. Methylene chloride can be used for water-insoluble molecules that are to be reacted with the PEG compounds, but do not require subsequent water miscibility.

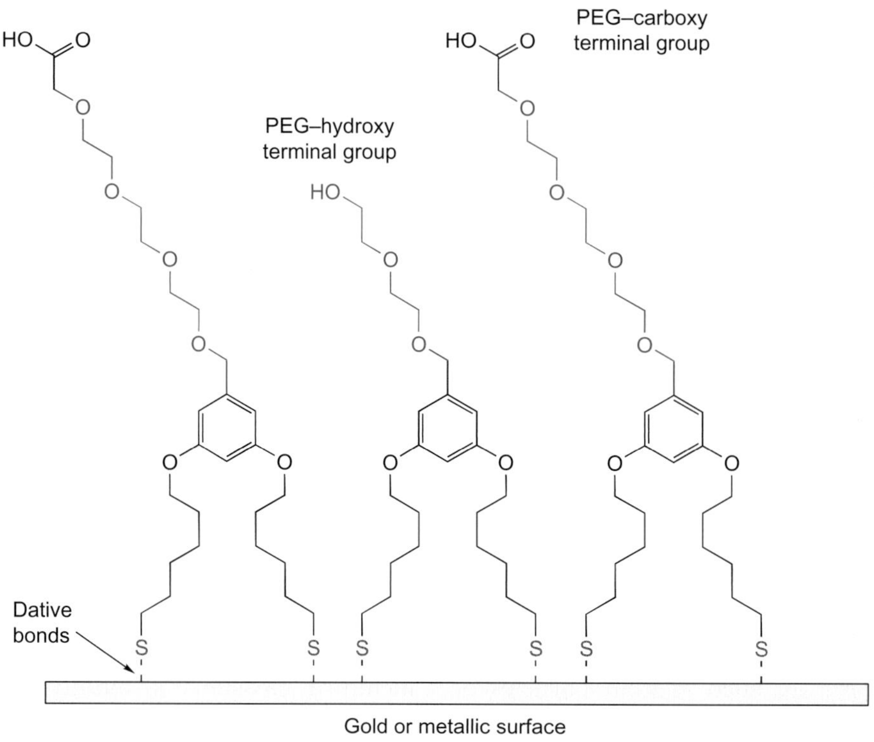

FIGURE 18.2 Dithiol linkers containing discrete PEG arms can be used to stably modify metallic surfaces for coupling biomolecules.

1.1. Homobifunctional PEG Crosslinkers

Compounds having the same functionality on both ends are homobifunctional in nature and can be conjugated with the same target functionality on biomolecules, surfaces, or other molecules. Chapter 5 describes traditional homobifunctional compounds in detail, but the discrete PEG-based reagents are described here, because of their unique hydrophilic properties.

bis-NHS Ester–PEG Compounds

Bifunctional NHS esters can be used to conjugate two amine-containing molecules together. The NHS ester groups react with amines to form amide linkages with loss of NHS. Two different PEG spacer lengths are available in this type of reagent, a PEG_5 compound and a PEG_9 derivative (Figure 18.3). Unlike the popular BS^3 reagent (Chapter 5, Section 1.2), which is initially hydrophilic and water soluble due to the presence of negatively charged sulfo-NHS esters on both ends but after reacting leaves behind a 6-carbon hydrophobic spacer, these PEG compounds attach two molecules together using a long and very hydrophilic bridge.

The *bis*-NHS–PEG_5 reagent provides a 21.7-Å spacer after attaching both ends to amines on two molecules, which is nearly twice the spacer distance provided by BS^3. The longer chain *bis*-NHS–PEG_9 reagent has a very long 35.8-Å spacer. All of these distances are measured between the carbonyl groups of the two NHS esters, which would form the amide bonds and remain behind after crosslinking two molecules together. The distances represent the energy minimized, maximal molecular distance of a linear structure and may not reflect how such structures exist in solution. All such molecular distance measurements are approximations and only serve to provide comparable information relative to the sizes of different reagents.

The reaction of the NHS esters occurs at physiological pH or under slightly basic conditions to couple rapidly with amines and form amide bonds (Figure 18.4). The NHS ester groups are also subject to hydrolysis in aqueous solution, and the rate of hydrolysis increases with increasing pH. The more hydrophilic the molecule, the greater the potential for hydrolysis. Nectar reported that for long-chain polymeric PEGs containing NHS esters that the half-life of hydrolysis effectively triples upon lowering the pH one unit. In addition, the molecular constituents immediately adjacent to the NHS ester affect the half-life of hydrolysis of activated PEG compounds. For instance, the addition of a 4-carbon aliphatic chain between the PEG polymer and the NHS ester results in a half-life of hydrolysis of about 44 min at pH 8. Reducing this to only a 2-carbon unit decreases the half-life of the NHS ester to about 3 min. Therefore, to aid in NHS ester stability maintaining a reaction pH in the range of 7.0 to 7.5 is optimal for most applications.

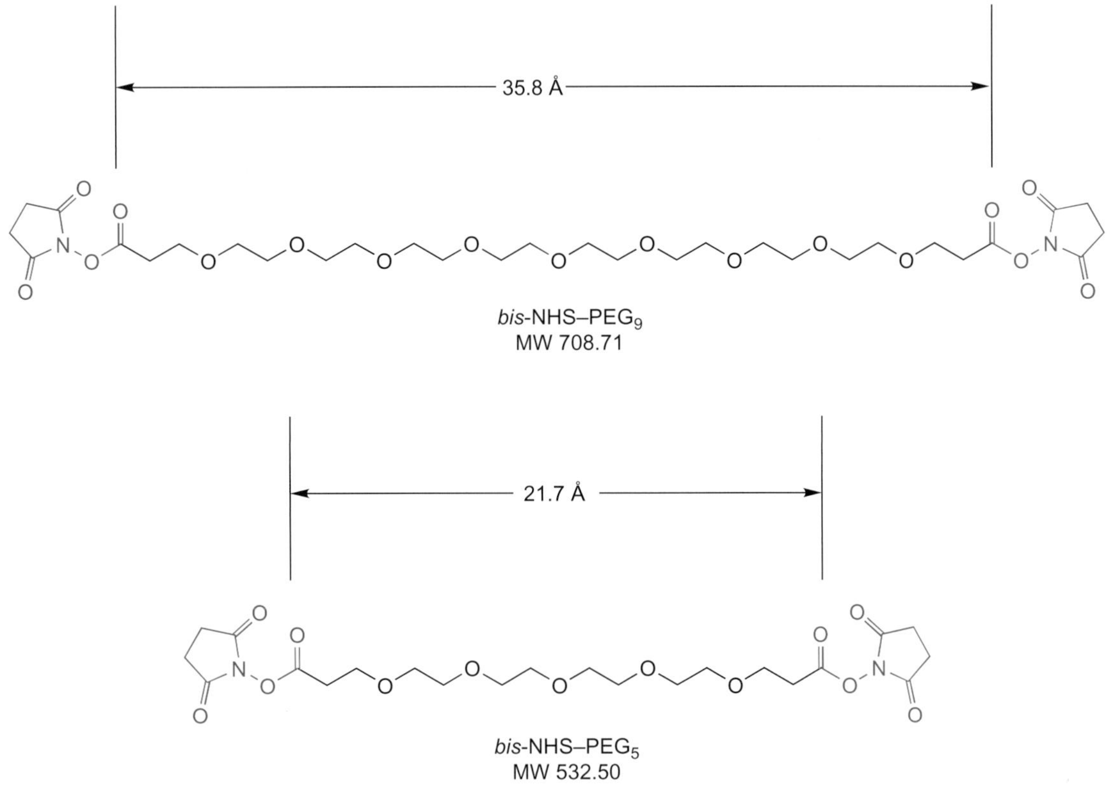

FIGURE 18.3 Homobifunctional NHS ester compounds containing PEG spacers for water solubility.

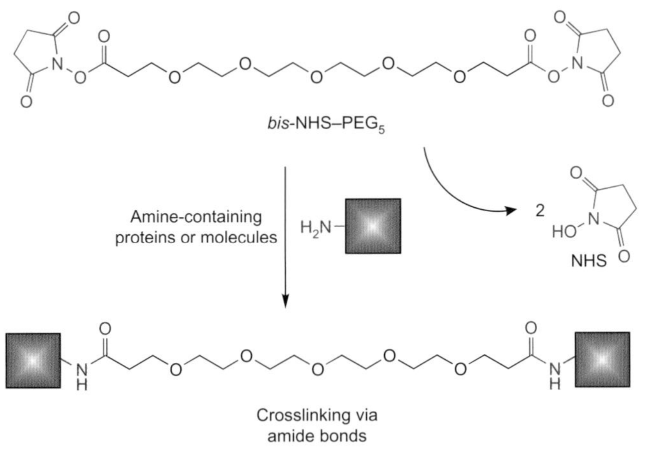

FIGURE 18.4 The reaction of *bis*-NHS–PEG$_5$ with amines on proteins yields amide bond linkages with amine-containing molecules.

The following protocol describes a general method for using *bis*-NHS ester–PEG compounds. Optimization of concentrations should be carried out for each application to ensure the best possible results. See also the protocol in Chapter 24, Section 1, which describes the use of homobifunctional NHS-ester compounds to study protein interactions. These *bis*-NHS–PEG compounds may provide a superior crosslinker for studying such interactions due to their water solubility and the fact that the PEG bridge will not get buried in hydrophobic pockets on proteins or within hydrophobic membrane structures.

Protocol

1. Dissolve the *bis*-NHS–PEG compound in a dry, water-miscible organic solvent to make a concentrated stock solution. To prepare a 10-m*M* solution of the *bis*-NHS–PEG$_9$ reagent, dissolve 7 mg per milliliter of DMAC; for the *bis*-NHS–PEG$_5$ reagent dissolve 5.3 mg/ml of DMAC.
2. Dissolve the molecules to be conjugated in 0.1-*M* sodium phosphate, pH 7.2 (for aqueous reactions), or in DMSO, DMAC, or methylene chloride (for organic reactions). If proteins are to be conjugated, a concentration of 1 to 10 mg/ml in buffer will work well in this protocol. For more dilute protein solutions, greater quantities of the *bis*-NHS–PEG compound may have to be added than recommended here to obtain similar levels of crosslinking.
3. Add a quantity of the crosslinker solution to the protein solution to provide a 1- to 10-fold molar excess of reagent over the concentration of protein. The use of lower molar ratios will limit the potential for oligomerization of proteins in solution. A series of reactions using different concentrations of crosslinker

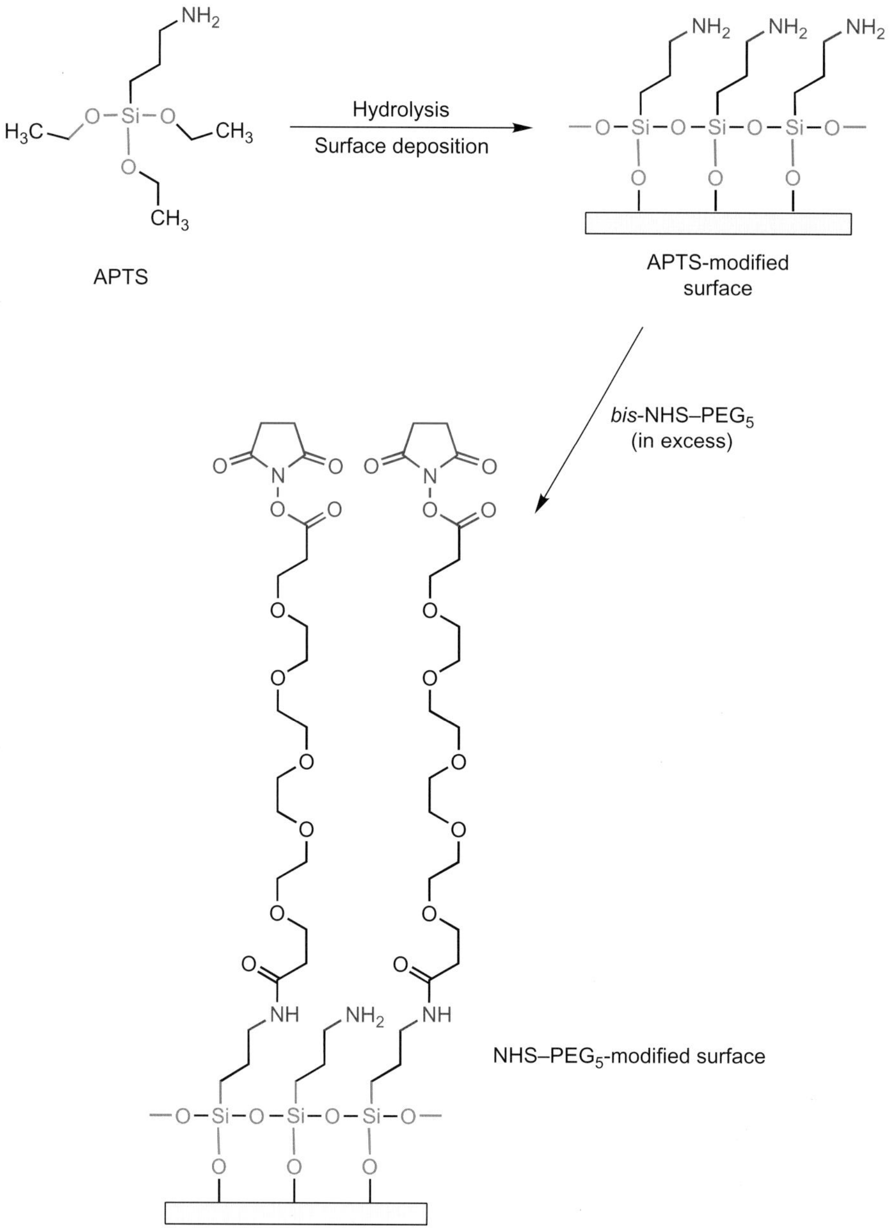

FIGURE 18.5 The modification of an APTS-modified surface containing amines with *bis*-NHS–PEG$_5$ yields hydrophilic spacers containing terminal NHS esters for coupling proteins.

may have to be performed to determine the optimal level to use for a particular application.

4. React for 30 to 60 min at room temperature.
5. Remove excess crosslinker by dialysis or gel filtration.

The *bis*-NHS–PEG compounds may also be used for modifying surfaces or particles that contain amine groups. If the *bis*-NHS–PEG reagent is reacted in large excess to the concentration of amines, then a single end of each crosslinker will couple to the amine groups on the surface and result in a PEG spacer terminating in a reactive NHS ester for further conjugation. Surfaces modified with an amino–silane group, such as 3-aminopropyltriethoxysilane (APTS), are particularly good for further modification with PEG-based reagents. For instance, APTS-modified glass slides or silica particles contain aminopropyl groups that can be modified with the *bis*-NHS–PEG compounds. This will create a hydrophilic monolayer having low nonspecific binding character and that can be used to covalently link to biomolecules *via* the terminal NHS ester group (Figure 18.5).

The following procedure can be used with planar or spherical surfaces containing amine groups. The reaction is performed in organic solvent to preserve the activity of the terminal NHS ester after modification of the surface. For particle modification, care should be taken in choosing a solvent that will not damage the particle core structure, such as the potential for certain organic solvents to dissolve polymeric particles. Silica particles, however, are very robust to solvent conditions and can be used without damage. See also Chapters 10, 14, and 15 for other particle conjugation methods.

Protocol

1. In a fume hood, wash the amine-containing surface with organic solvent to remove any contaminants or water (especially when working with particles). Suggested solvents to use for this reaction are highly pure and dry DMAC, DMSO, or DMF. Particles can be washed by repeated centrifugation and resuspension.
2. Dissolve *bis*-NHS–PEG_5 into the solvent of choice at a concentration of 1 mg/ml also containing an equal molar concentration of triethylamine as base. Add the crosslinker solution to the surface or to the particles to coat them fully. When working with particles, centrifuge them to remove solvent prior to resuspending in the crosslinker solution.
3. React for 30 to 60 min at room temperature with mixing.
4. Wash the surface or particles with solvent at least 3 times to remove excess crosslinker.

Planar surfaces activated with the NHS–PEG groups may be sealed in a pouch and stored dry in the presence of a desiccant. Activated particles may be stored as a suspension in dry solvent under a head of nitrogen at 4°C until used for further conjugation (e.g., in DMAC). The addition of an amine-containing protein or another amine-molecule will cause covalent coupling to the surface NHS ester groups to form amide bonds.

bis-Maleimide PEG Compounds

Homobifunctional crosslinkers containing thiol-reactive maleimides on each end of a PEG spacer are available in several sizes. These compounds are hydrophilic and react with sulfhydryls to produce thioether linkages, which are stable under most conditions. The following compounds can be obtained from Thermo Fisher or Quanta BioDesign.

BM(PEG)$_2$, BM(PEG)$_3$, and bis-MAL–dPEG$_3$

BM$(PEG)_2$ is 1,8-*bis*-maleimidodithyleneglycol, a hydrophilic reagent containing a 14.7-Å spacer arm. This compound is also called BM$(PEO)_2$ by Thermo Fisher, wherein PEO is the acronym for poly(ethylene oxide), but to maintain a consistent nomenclature throughout this book, it is referred to under the PEG name. Another pair of crosslinkers, BM$(PEG)_3$ and *bis*-MAL–$dPEG_3$ have one additional PEG unit and

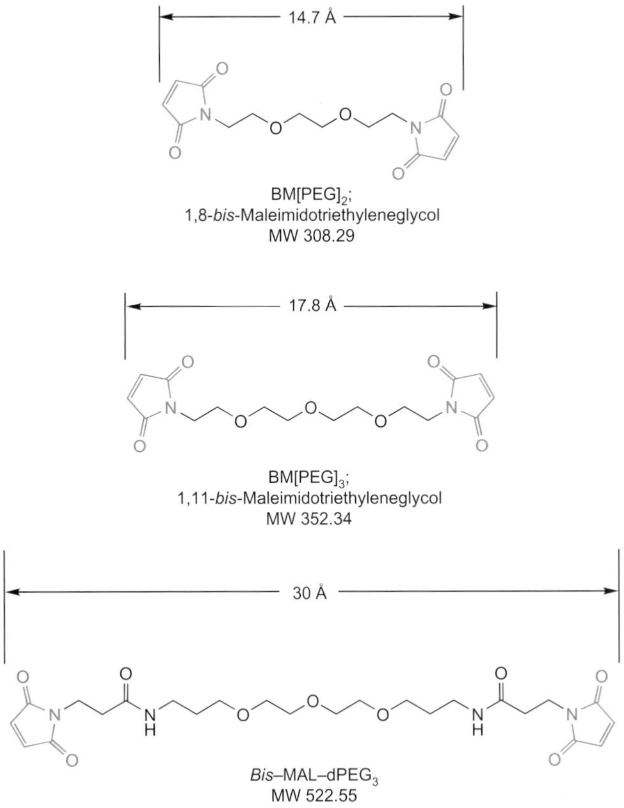

FIGURE 18.6 The structures of BM$(PEG)_2$, BM$(PEG)_3$, and *bis*-MAL–$dPEG_3$ contain thiol-reactive maleimide groups on the ends of a discrete PEG spacer arm.

provide similar bifunctional maleimides, but differ only in their structures leading from the PEG spacer to the maleimides (Figure 18.6). BM$(PEG)_5$, has a 17.8-Å spacer, while *bis*-MAL–$dPEG_3$ has a longer 30-Å length. These dimensions are measured between the outer hydrogens on the maleimide groups using linear three-dimensional structures after energy minimization. The actual structural configurations in aqueous solution almost certainly will not be linear and thus the molecular dimensions will differ from these values.

All of these bifunctional maleimides can be used to conjugate together two proteins containing available thiol groups (Figure 18.7). For instance, artificial bispecific antibodies have been created by coupling two disulfide-reduced antibodies together having different antigenic specificities using the thiol-reactive crosslinker SPDP (Chapter 6, Section 1.1). This was done by reducing the disulfides in the hinge region of antibodies to produce two heavy-/light-chain complexes containing one antigen binding site each and then conjugating the fragments from both antibodies together with SPDP (Foglesong *et al.*, 1989). Coupling one reduced antibody with another of a different specificity using a hydrophilic reagent such as BM$(PEG)_2$ will form a conjugate having two different antigen binding capabilities. The

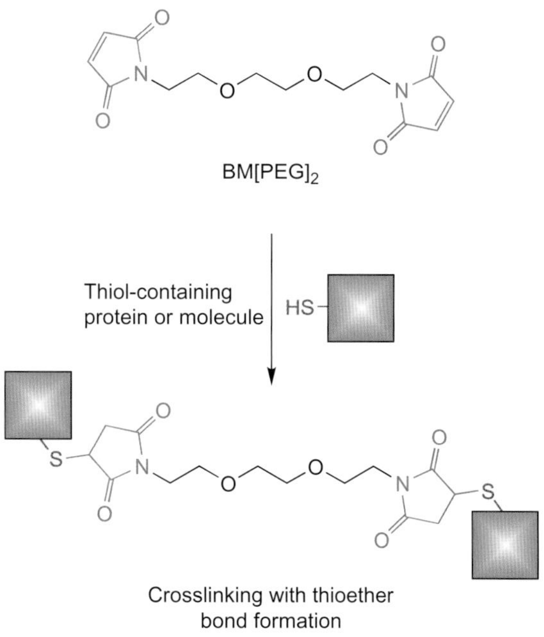

FIGURE 18.7 $BM(PEG)_2$ reaction with thiol-containing proteins forms crosslinks *via* thioether linkages.

advantage of using a PEG-based compound over SPDP is greater water-solubility of the resultant conjugate.

$BM(PEG)_2$ can also be used to determine protein–protein interactions or subunit interactions if there are free thiol groups available on each protein or subunit. The following protocol can be used to crosslink two thiol-containing proteins.

Protocol

1. Dissolve the thiol-containing proteins to be crosslinked in 50-m*M* sodium phosphate, pH 6.5–7.5, containing 10-m*M* EDTA to prevent metal-catalyzed sulfhydryl oxidation.
2. Dissolve $BM(PEG)_2$ in DMSO or DMF at a concentration of 10 to 20-m*M* (3.1 mg/ml to make a 10-m*M* solution).
3. Add a quantity of the crosslinker solution to the protein solution to provide a 1- to 10-fold molar excess of crosslinker to protein. Lower molar ratios will help prevent oligomerization of protein if the proteins contain more than one available thiol group. A series of different molar ratios may be studied to optimize the level of reagent addition.
4. React for 1 to 2 h at room temperature.
5. Purify the conjugate by dialysis or gel filtration.

1.2. Heterobifunctional PEG Reagents

Heterobifunctional crosslinkers contain different reactive groups or functionalities at each end of a spacer arm. Traditional reagents of this type are discussed in Chapter 6, but the PEG-based compounds are reviewed exclusively here, because of their unique water-soluble characteristics. The most popular aliphatic heterobifunctional compound is SMCC (or sulfo-SMCC), which contains an NHS ester on one end and a maleimide group on the other end. However, this compound suffers from a cross-bridge that is both water insoluble and immunogenic. Redesigning this crosslinker to have a PEG cross-bridge provides enhanced water solubility for modified proteins or other molecules as well as displaying very low immunogenicity. These benefits of discrete PEG spacers create a new generation of heterobifunctional compounds, which dramatically improve the performance of conjugates made from them. The following sections describe these reagents in more detail.

Maleimide–PEG$_n$–NHS Ester Compounds

One of the most useful types of crosslinkers ever invented is a heterobifunctional compound containing an NHS ester on one end and a maleimide group on the other end. The ability to link an amine-containing molecule to another molecule containing a thiol group provides control over the conjugation process, which avoids the potential for oligomerization that often occurs when using homobifunctional crosslinkers. Probably the most significant recent development in this type of reagent is the introduction of PEG spacers in their cross-bridge construction. Unlike the previous iteration of these compounds that all used hydrophobic spacers (Chapter 6, Section 1), the PEG-based reagents provide water-solubility both of the initial compound and of the cross-link itself after the reaction has taken place.

NHS–PEG$_n$–maleimide crosslinkers are available as a series of different chain lengths of PEG within the cross-bridge, including repeating ethylene oxide units of 2, 4, 6, 8, 12, and 24 (Quanta BioDesign and Thermo Fisher). This series provides a range of molecular lengths after conjugation from 17.6 Å to 95.2 Å, so that an optimal size can be determined for nearly any application (Figure 18.8). The longest crosslinker of the group, NHS–PEG$_{24}$–maleimide, has a total length that compares to almost the diameter of a typical immunoglobulin (IgG) antibody molecule. Longer-chain NHS–PEG–maleimide crosslinkers may be appropriate especially for the masking of surfaces to permit coupling of affinity ligands, while avoiding the potential for nonspecific interactions with the surface structure. In addition, these heterobifunctional PEG compounds are superior choices for linking haptens to carrier proteins to create immunogens. In this application, the PEG chain provides a high degree of freedom of motion and does not illicit an immune response itself, thus effectively generating antibodies *in vivo* with greater specificity toward the hapten, not the carrier or the crosslinker.

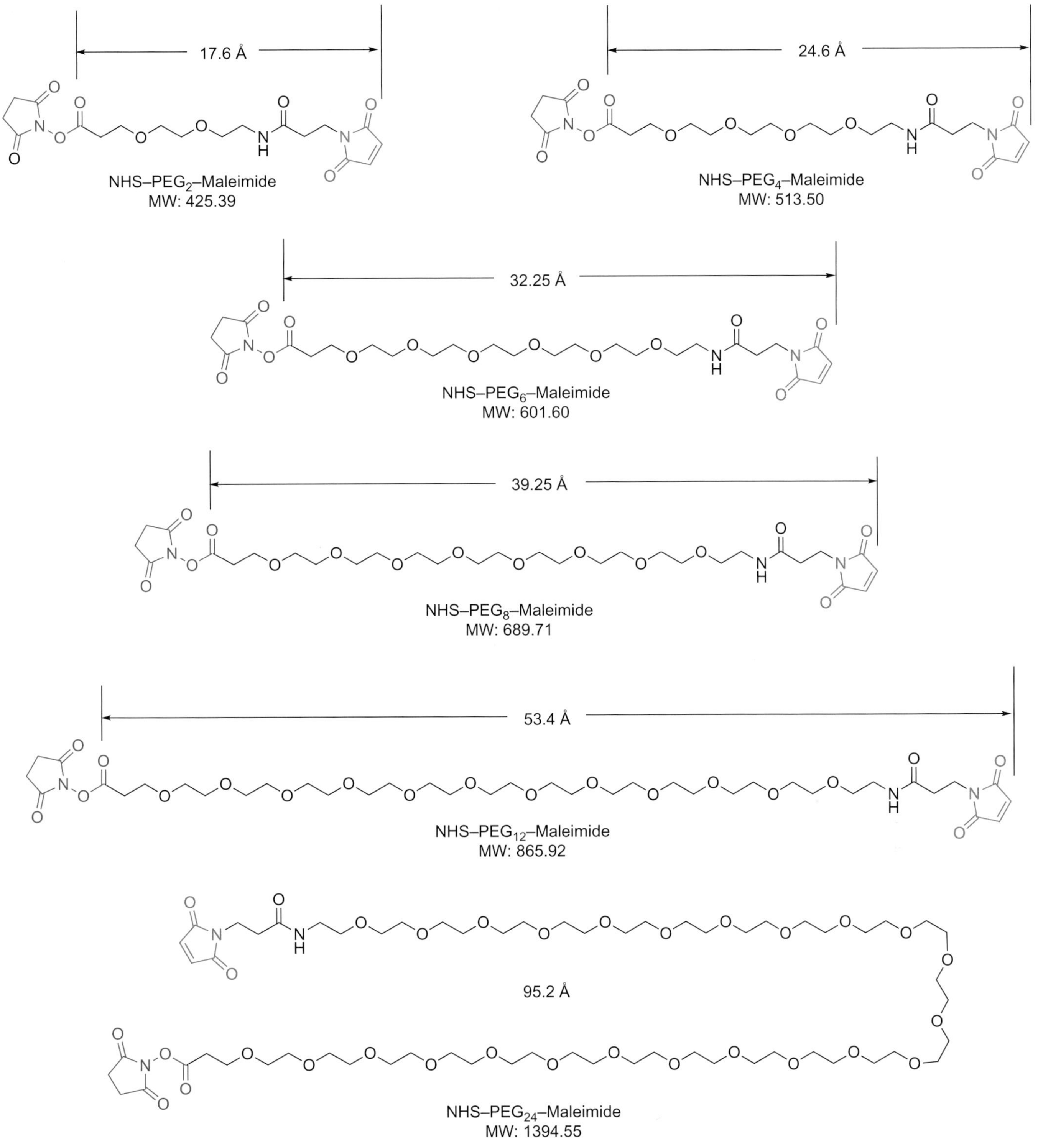

FIGURE 18.8 Series of NHS–PEG–maleimide crosslinkers.

In use, NHS–PEG–maleimides are first reacted in at least a 10-fold molar excess with an amine-containing molecule, such as a protein, to form an intermediate derivative with terminal maleimide groups. Depending on the number of maleimide modifications needed, the molar excess of crosslinker may be adjusted to provide the desired level of activation. The maleimide groups are more stable in aqueous solution than the NHS esters, so the modified protein can be purified from excess crosslinker before reacting it with

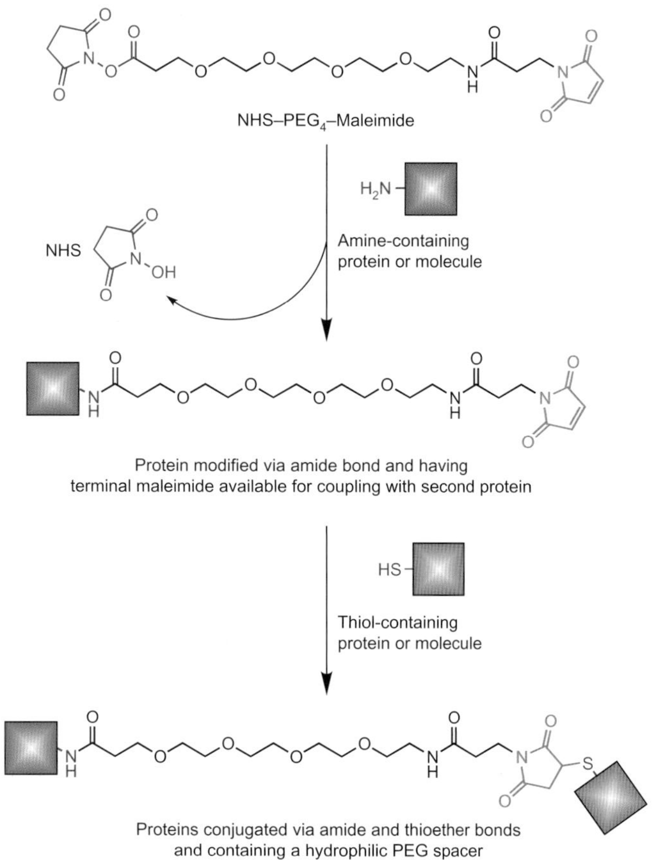

FIGURE 18.9 NHS–PEG$_4$–maleimide conjugation reactions are carried out in two steps involving modification of an amine-containing molecule with the NHS ester end with subsequent coupling of the maleimide end with a thiol-containing molecule.

a sulfhydryl-containing molecule or protein. After purification, the maleimide–PEG–protein derivative is reacted with a second protein containing sulfhydryls to form the final conjugate *via* a thioether bond (Figure 18.9). The higher the number of maleimide–PEG modifications on the first protein, the greater the number of potential sulfhydryl-containing proteins may be conjugated to it. The ratio of the reactants is often chosen to give at least several thiol-proteins conjugated to each of the maleimide–PEG–protein derivatives. Optimization of this ratio should be carried out to determine the best conjugate for a given application.

The intermediate maleimide–PEG–protein derivative maintains excellent solubility due to the presence of the PEG chains. Unlike a hydrophobic NHS–maleimide-type crosslinker made from an aliphatic cross-bridge, the PEG chains have a tendency to increase the solubility of modified proteins or other molecules. This effect is amplified when using PEG compounds of longer dimensions. The result is that modified proteins remain soluble, and even if not every maleimide group gets conjugated to a thiol-containing molecule, the presence of unreacted crosslinkers on the protein surface does not contribute to nonspecific interactions of the conjugates.

Most of these PEG crosslinkers come as thick, sticky, viscous liquids or low-melting solids (PEG$_2$). For this reason, weighing out a small sample of a compound can be difficult or impossible. It is usually best to dissolve an entire vial or a larger amount in organic solvent at a known concentration to permit accurate dispensing of a smaller amount into a reaction. Suitable solvents to prepare a stock solution include dry (molecular sieved) DMSO, DMF, DMAC, acetonitrile, or methylene chloride (for non-water-miscible reactions).

The following protocol is a general guide for using NHS–PEG$_n$–maleimide crosslinkers. This method may be used as a starting point for developing an optimized procedure for creating a unique conjugate.

Protocol

1. In a fume hood, dissolve the NHS–PEG$_n$–maleimide compound of choice in a dry, water-miscible organic solvent to make a concentrated stock solution. For instance, to prepare a 100-m*M* solution of the NHS–PEG$_6$–maleimide reagent (MW 601.6), dissolve an entire 100-mg vial of the crosslinker in 1.66 ml of DMAC (dry DMSO or DMF works well, too).
2. Dissolve the amine-containing protein to be activated in 0.1-*M* sodium phosphate, pH 7.2 (coupling buffer). A protein concentration of 1 to 10 mg/ml in buffer will work well in this protocol. For more dilute protein solutions, greater quantities of the NHS–PEG$_n$–maleimide compound may have to be added to obtain equivalent levels of modification.
3. Add a quantity of the crosslinker solution to the protein solution to provide at least a 10-fold molar excess of reagent over the amount of protein present. Higher levels of reagent to protein may be used, even up to a 50-fold molar excess, to obtain a large number of active groups for subsequent coupling to a thiol-containing protein. Conversely, the use of lower molar ratios will limit the number of maleimide–PEG modifications on each protein to just a few. A series of reactions using different concentrations of crosslinker may have to be done to determine the optimal level to use for a particular application.
4. React with gentle mixing for 30 min at room temperature or 60 min at 4°C.
5. Remove excess crosslinker by centrifugal dialysis or gel filtration using a molecular weight exclusion of 5000. This procedure should be carried out quickly due to the labile nature of the maleimide group in aqueous solution, which will hydrolyze to a ring-open maleamic acid that is ineffective at coupling to thiols.
6. Dissolve a sulfhydryl-containing protein in coupling buffer containing 10-m*M* EDTA at a concentration of at least 1 to 10 mg/ml. Thiols may be generated

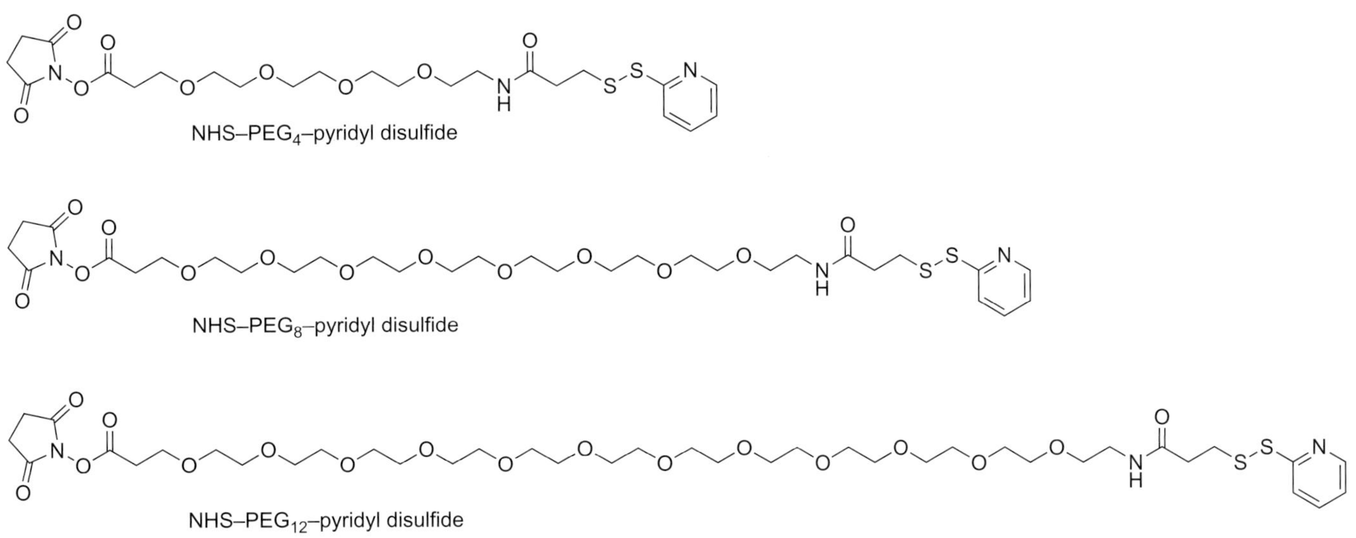

FIGURE 18.10 Examples of NHS–PEG$_n$–pyridyl disulfide compounds that can be used to crosslink amine-containing molecules with thiol-containing molecules.

in proteins by disulfide reduction or through use of a thiolation reagent (Chapter 2, Section 4.1). The amount of the thiol-containing protein needed is determined by the optimal molar ratio desired in the final conjugate of the thiol-containing protein to the maleimide-activated protein purified in step 5. Often, this means performing a reaction of the thiol-containing protein to the maleimide–PEG–protein in at least a 4-to-1 molar ratio, but higher ratios (e.g., 15 to 1) can also be used, depending on the application. Note that the number of thiol-containing protein molecules that can be conjugated with the maleimide–PEG–protein is limited by the number of reactive maleimide groups present and also by the molecular size of the two molecules being conjugated. Steric crowding will limit how many thiol-containing proteins can be attached to the maleimide-activated protein.

7. Add the thiol-containing protein to the maleimide–PEG–protein in the desired molar ratio to initiate the conjugation reaction.
8. React with gentle mixing for at least 2h at room temperature or overnight at 4°C.
9. The conjugate may be purified to remove unconjugated protein using gel filtration on a column of resin having a molecular weight cut-off able to accommodate the proteins being separated.

NHS–PEG$_n$–Pyridyl Disulfide Compounds

One of the earliest heterobifunctional reagents that were introduced for bioconjugation applications is SPDP (Chapter 6, Section 1.1). This compound has an NHS ester on one end for coupling with amine groups, forming amide bonds, and a pyridyl disulfide on the other end to couple with thiols, which forms disulfide bonds. The reagent has been used extensively to design cleavable conjugates with all types of molecules, especially to create antibody–toxin complexes, which can release a toxin payload after binding to tumor cells *in vivo* (Chapter 20, Section 3.2.1). SPDP has an aliphatic spacer that is hydrophobic and is of limited molecular length (6.8Å), which limits its usefulness for applications that might benefit from a longer cross-bridge. The longer analog to this reagent, NHS–LC–SPDP, contains an internal 6-aminocaproic acid spacer that provides greater length, but the increase in hydrophobicity due to the aliphatic portion can be problematic. A new alternative set of compounds built upon PEG cross-bridges now is available that solves the problems associated with hydrophobic heterobifunctionals (Thermo Fisher, Quanta BioDesign). These NHS–PEG$_n$–pyridyl disulfide compounds contain the same reactive groups on either end as SPDP, but the internal PEG spacer options provide water solubility and a dramatically decreased nonspecific binding characteristic in biological systems or samples. Quanta BioDesign calls these compounds by the name SPDP–dPEG®n–NHS ester and Thermo Fisher uses the general name PEGn–SPDP, where "n" in both names represents the number of repeating PEG units in the cross-bridge.

NHS–PEG$_n$–pyridyl disulfide reagents are available in options containing a PEG_4, PEG_8, PEG_{12}, PEG_{16}, PEG_{20}, PEG_{24}, or PEG_{36} internal cross-bridge, which provides molecular lengths of approximately 26Å, 37Å, 54Å, 64Å, 79Å, 97Å, or 138Å, respectively. Figure 18.10 shows the three shorter chain lengths in this series. The longest of these reagents provide chain lengths that equal or exceed the diameters of most biological macromolecules, such as proteins. These second-generation reagents are similar in reactivity to the original SPDP-type crosslinkers,

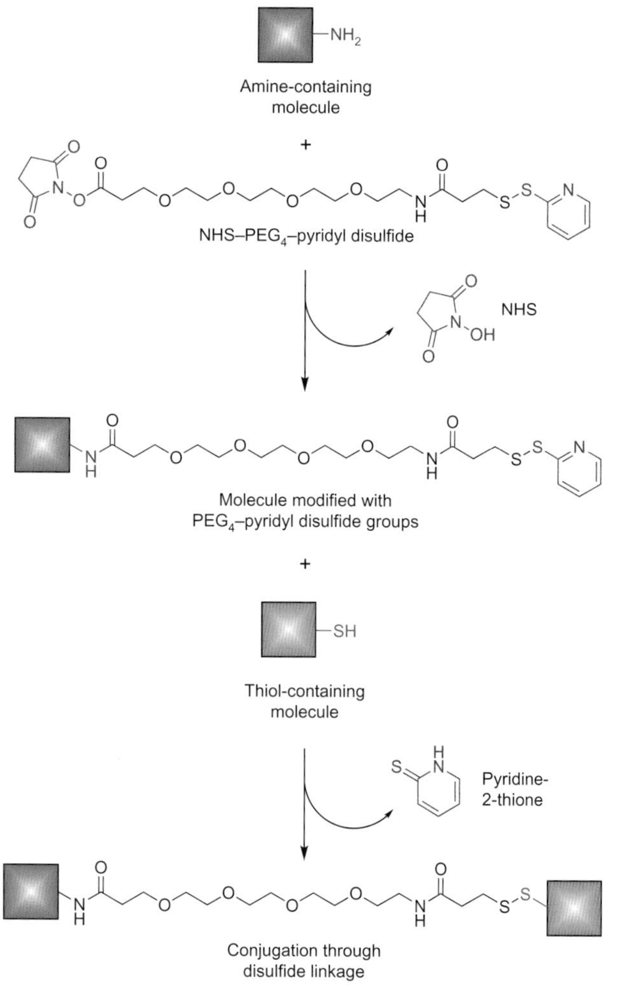

FIGURE 18.11 The reactions of NHS–PEG$_4$–pyridyl disulfide in conjugating an amine-containing molecule with a thiol-containing molecule. The disulfide bond formed in the final reaction is reversible using a disulfide reducing agent.

nevertheless they have superior properties related to hydrophilicity, spacer arm length, and decreased nonspecific binding.

Figure 18.11 illustrates the NHS–PEG$_4$–pyridyl disulfide compound and its reactions for conjugating an amine-containing molecule with a thiol-containing molecule. The NHS ester end is typically reacted first to form a stable amide bond with an amine-containing molecule. After purification of the modified molecule from excess reagent, the pyridyl disulfide end can next be coupled to a thiol-containing molecule to form a disulfide bond. The disulfide linkage is reversible by using reducing agents such as DTT, TCEP, or 2-mercaptoethylamine. The compounds can also be used as a thiolation agent to modify amines on molecules, particles, or surfaces to create terminal thiols for further modification or coupling of another molecule or affinity ligand. Thiolation reactions with SPDP-type reagents have been performed since their introduction and the similar PEG-based compounds may be used for this purpose using the same general protocols. For thiolation, the pyridyl disulfide end is simply reduced to create the free thiol on the modified component instead of immediately reacting it with another thiol-containing molecule. Thus, the same crosslinker can be used to create reactive pyridyl disulfide groups on one molecule and thiolate a second molecule to create SH groups. These two modified molecules then can be reacted together to form a bioconjugate containing the combination of two PEG internal cross-bridges linked by a central disulfide bond.

The following protocols for the conjugation of an antibody with a thiol-containing particle can be applied in a similar manner to the conjugation of a pyridyl disulfide-modified antibody with another protein that has been thiolated with this reagent. In this protocol, an antibody molecule is modified with NHS–PEG$_4$–pyridyl disulfide and a particle is thiolated to contain thiol groups using the same crosslinker. The two modified intermediates are then reacted to form the final conjugate, which is linked together by disulfide bonds. For modification of surfaces or particles, the reactions should be optimized to include enough of the reagent to thoroughly modify the surface amine groups and create the desired density of PEG–pyridyl disulfides for further conjugation. The use of long PEG cross-bridges in the coupling of proteins and other molecules to surfaces can have the added benefit of providing significant hydrophilicity to the substrate, thus increasing the freedom of motion in aqueous solution and decreasing the nonspecific binding potential of the final complex toward nonrelevant molecules in a sample.

Protocol for Activation of an Antibody with NHS–PEG$_4$–Pyridyl Disulfide

1. Dissolve 2 mg of an antibody (IgG) in 200 μl of 0.1-*M* sodium phosphate, 0.15-*M* NaCl, pH 7.5, containing 10-m*M* EDTA (coupling buffer). Note: This makes a 10-mg/ml solution of the antibody in coupling buffer. More dilute antibody solutions may require an increase in the amount of crosslinker added to the reaction to obtain the same level of modification as this protocol provides.
2. Dissolve 2 mg of NHS–PEG$_4$–pyridyl disulfide crosslinker in 200 μl of dry DMSO, DMF, or DMAC (makes a 17.9-m*M* solution).
3. Add 7.5 μl of the dissolved crosslinker to the antibody solution and immediately vortex mix to create a uniform solution. This amount of addition represents approximately a 10-fold mole excess of the crosslinker over the antibody in the solution. Note: Some optimization of the amount of NHS–PEG$_4$–pyridyl disulfide compound to add should be

performed to obtain the best possible final conjugate in the intended application. Experiments carried out using a range of different mole excesses of crosslinker over the antibody present can identify the best level of modification to use.

4. React for 30 min at room temperature with periodic mixing in a sealed container (e.g., microfuge tube).
5. Purify the modified antibody by dialysis or size exclusion chromatography using a 5-ml desalting column. Use coupling buffer as the dialysis or chromatography solution. Concentrate the antibody using centrifugal concentrators to approximately 200 μl, if necessary. The pyridyl disulfide-activated antibody may be used immediately to conjugate with a thiol-containing particle or protein.

Protocol for Thiolation of Amine-Containing Particles with NHS–PEG$_4$–Pyridyl Disulfide

1. Wash 10 mg of amine-containing particles with 50-m*M* sodium phosphate, pH 7.5, containing 10-m*M* EDTA, while using centrifugation to separate the particles or another appropriate method of exchanging them into coupling buffer. Finally, resuspend the particles in 1 ml of coupling buffer.
2. Dissolve 2 mg of NHS–PEG$_4$–pyridyl disulfide into 200 μl of dry DMSO, DMF, or DMAC (17.9-m*M* solution).
3. Add 55 μl of the NHS–PEG$_4$–pyridyl disulfide solution to the particle suspension with mixing.
4. React with mixing for 30 min at room temperature.
5. Wash the modified particles with coupling buffer at least 3 times using centrifugation and resuspend in 1 ml of the same buffer.
6. Reduce the pyridyl disulfide groups on the particles by the addition of DTT to a final concentration of 50-m*M* and incubate for 30 min with mixing at room temperature.
7. Thoroughly wash the thiolated particles with coupling buffer to remove excess reducing agent and the released pyridine-2-thione. After the final wash and centrifugation, resuspend the thiolated particles in 1 ml of coupling buffer and proceed to part (C).

Protocol for Conjugation of the Pyridyl Disulfide-Modified Antibody to the Thiolated Particles

1. Mix the thiolated particles prepared in part (B) with the activated antibody solution prepared in part (A) using vortex mixing.
2. React for 18 h at room temperature (or at 4°C) with constant mixing using an end-over-end rotator with the particle suspension contained in a sealed centrifuge tube.
3. Wash the particles using coupling buffer at least 3 to 5 times using centrifugation and resuspension.

The immobilized antibody may be stored until use at 4°C in water or buffer containing a preservative.

NHS–PEG$_n$–Azide/Alkyne Compounds for Chemoselective Ligation

Another family of heterobifunctional compounds containing an internal PEG spacer is represented by reagents with an NHS ester on one end and either an azide or alkyne group on the other end (Figure 18.12). These reagents react with amines on proteins or other molecules to create an amide bond derivative having a terminal functionality suitable for use in chemoselective ligation reactions, particularly click chemistry cycloaddition and Staudinger ligation (Chapter 17, Sections 5 and 6). Click chemistry is the reaction between an azide and an acetylene group in the presence of a catalytic amount of Cu(I) yielding a triazole ring linkage, which is a very efficient process for bioconjugation purposes. NHS ester heterobifunctional compounds containing these groups may be used to functionalize biomolecules, organic compounds, particles, or surfaces with azides or alkynes. Since both the azido and acetylene groups are highly stable in aqueous solution or in the presence of biomolecules, the modified substances or devices maintain nearly indefinitely the ability to conjugate molecules containing the opposite functionality.

The azide-containing PEG compounds can also be used in the Staudinger ligation process with a phosphine-containing group having an electrophilic trap. This reaction, like the click chemistry process, is chemoselective and highly bioorthogonal in that the functional groups will not react with typical functionalities present in biological solutions. The Staudinger ligation reaction with PEG–azide compounds proceeds to give an amide bond with either retention of the phosphine component or, in the traceless version of the reaction, with loss of the phosphine and formation of a zero-length crosslink between the two reacting molecules (Figure 18.13).

Click chemistry or Staudinger ligation-equipped heterobifunctional crosslinkers containing a hydrophilic PEG spacer provide a stable, yet highly efficient means to conjugate or immobilize biomolecules. The PEG spacer provides water solubility, low nonspecific binding character, and low immunogenicity.

Three PEG lengths are available in NHS–PEG$_n$–azide compounds, including n = 4, 8, or 12 ethylene oxide repeat units (Figure 18.12) (Quanta BioDesign and Thermo Fisher). The NHS–PEG$_4$–azide compound provides a 16-atom spacer, which is 17.7 Å long after reaction with an amine-containing molecule. The NHS–PEG$_8$–azide and the NHS–PEG$_{12}$–azide reagents provide longer spacer arms having molecular distances of 32.2 Å and 46.4 Å, respectively. All of these measurements are done using three-dimensional molecular models that

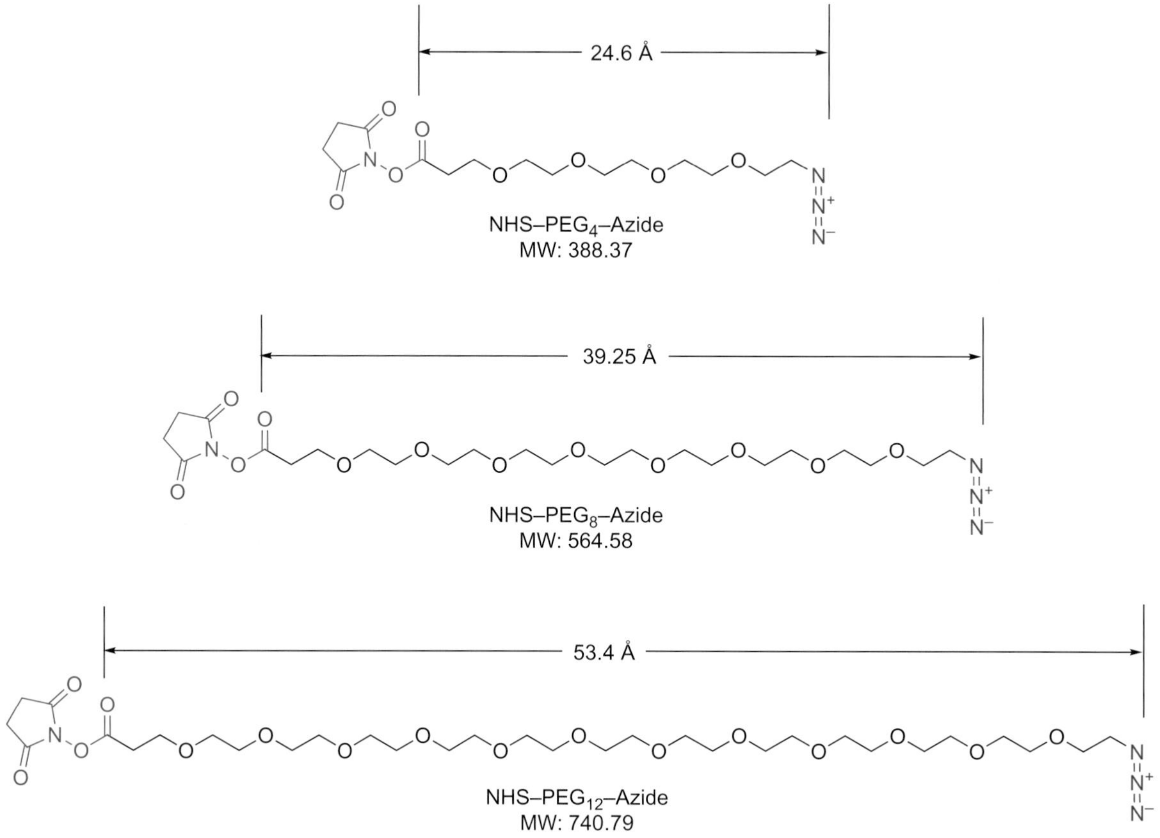

FIGURE 18.12 NHS–PEG–azide compounds can be used to modify amine-containing proteins or other molecules for subsequent conjugation using either the click chemistry reaction or Staudinger ligation.

are linearized structures with distances taken from the NHS ester carbonyl group to the first nitrogen atom of the azide. The actual molecular dimensions between crosslinked molecules will differ from these values due to the different conformations that the PEG chains can take in aqueous solution and due to the formation of the triazole ring after the click reaction has taken place.

Only one reactive compound is currently available for modifying amine-containing molecules with an acetylene group, which is the propargyl–PEG_1–NHS ester containing only a single ethylene oxide unit (Figure 18.14) (Quanta BioDesign, Thermo Fisher). Other simple alkyne-based raw materials are also commercially available with functional groups, such as propargylamine and various carboxylic acid derivatives (4-pentynoic acid, Aldrich), which can be used in a carbodiimide reaction to modify molecules or surfaces. Creative PEGWorks also offers longer $mPEG_n$–alkyne reagents and *bis*-alkyne–PEG compounds as well as tetra-alkynyl PEG dendrimers, but no heterobifunctional compounds of discrete PEG lengths.

The following protocol may be used to modify a protein for coupling to a surface using the NHS–PEG_n–azide and propargyl–PEG_1–NHS ester compounds. Similar procedures may be used to conjugate two molecules together, such as forming an antibody–enzyme conjugate.

Protocol for Modification of Protein 1 with NHS–PEG_n–Azide Groups

1. Dissolve a first protein to be modified with an azide group in 100-m*M* sodium phosphate, pH 7.2 (PBS), at a concentration of at least 1 to 10 mg/ml.
2. In a fume hood, prepare a stock solution of an NHS–PEG_n–azide by dissolving it at a concentration of 20-m*M* in DMAC, DMSO, or DMF (using highly pure and dry solvent).
3. Add a quantity of the crosslinker solution to the protein solution with mixing to provide at least a 10-fold molar excess of crosslinker over the amount of protein present.
4. React with mixing for 30 to 60 min at room temperature.
5. Purify the azido–PEG–protein by dialysis or gel filtration using PBS to remove excess reactants and solvent. The azide modifications are stable to aqueous conditions, so the protein derivative may be stored in this form until needed, provided the protein is stable.

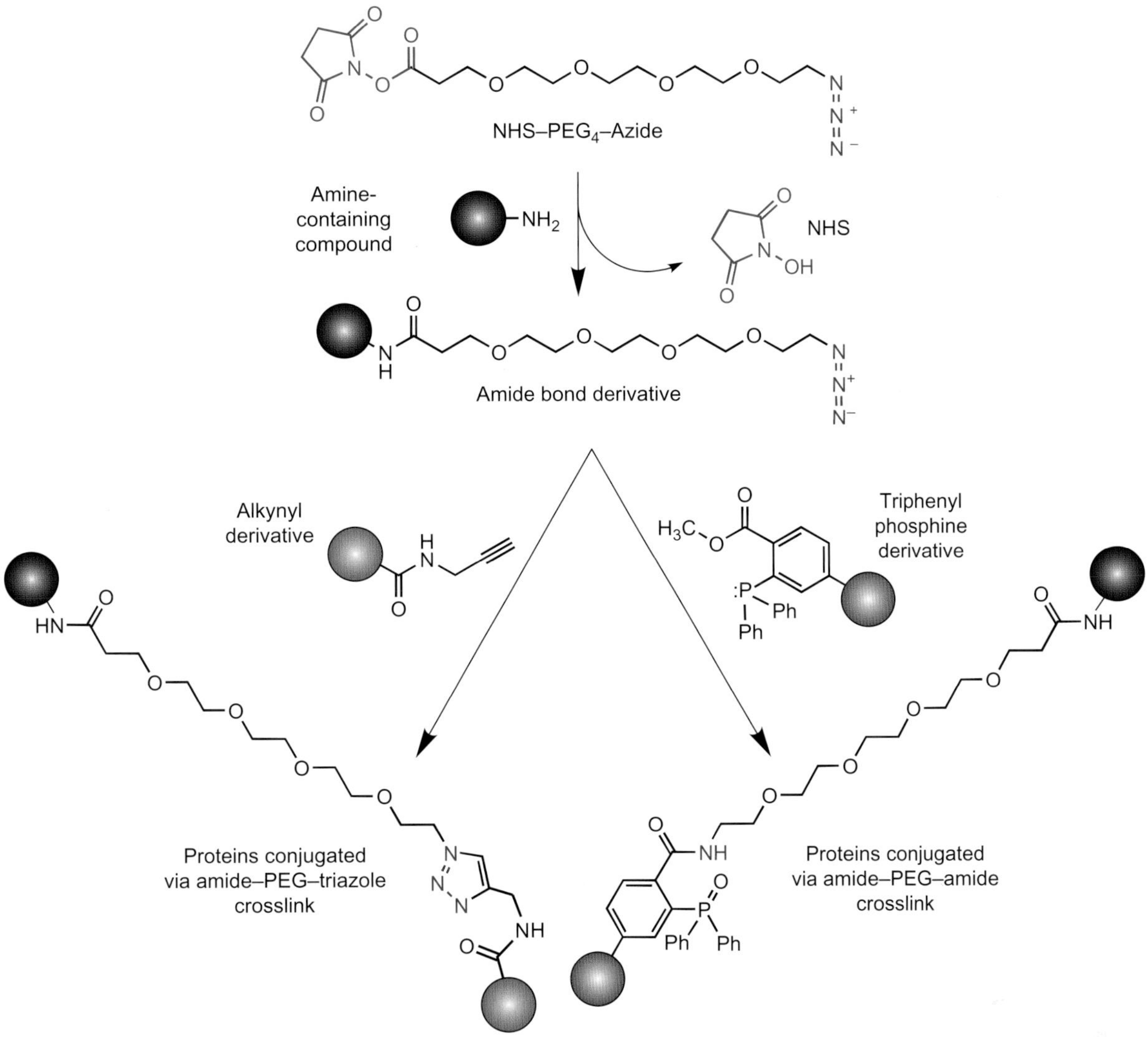

FIGURE 18.13 NHS–PEG$_4$–azide can be used to modify an amine-containing molecule to create an amide derivative terminating in azido groups. The azide modifications then can be used in a click chemistry reaction that forms a triazole linkage with an alkyne-containing molecule. Alternatively, the azide derivative can be used in a Staudinger ligation reaction with a phosphine derivative, which results in an amide bond linkage.

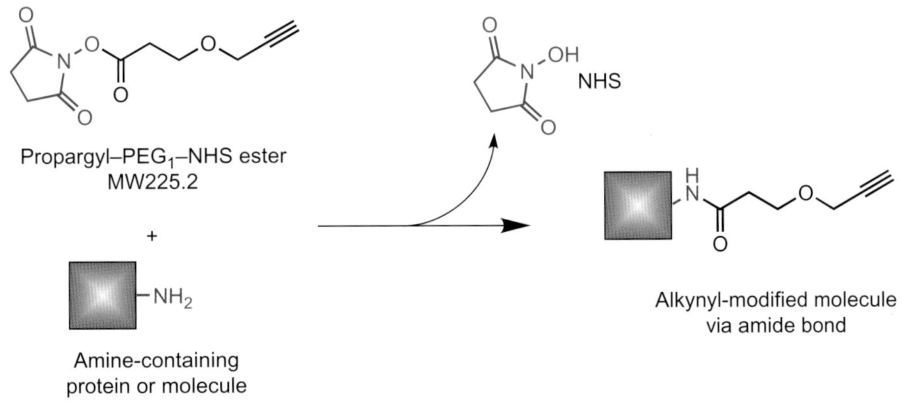

FIGURE 18.14 Progargyl–PEG$_1$–NHS ester can be used to react with amine-containing molecules to add short alkyne modification sites for subsequent click chemistry reactions with azide-containing molecules.

Protocol for Modification of Protein 2 with Propargyl–PEG$_1$–NHS Ester

1. Dissolve a second protein to be functionalized with alkyne groups at a concentration of 1 to 10 mg/ml in PBS.
2. In a fume hood, prepare a stock solution of propargyl–PEG$_1$–NHS ester in DMAC, DMSO, or DMF (highly pure and dry) at a concentration of 20-m*M* (4.5 mg/ml).
3. Add a quantity of the propargyl–PEG$_1$–NHS ester solution with stirring to the protein solution to obtain a 10-fold molar excess. Optimization of reactant ratios may be done to determine the best modification for a particular conjugation.
4. Gently mix for 30 to 60 min at room temperature.
5. Purify the alkyne-modified protein from excess linker by dialysis or gel filtration. The alkyne modification is stable in aqueous solution at this point.

Protocol for Coupling the Alkyne-Protein to the Azide-Modified Protein

1. With mixing, add a quantity of the azide-modified protein to the alkyne-modified protein to provide the desired molar excess over the quantity of protein and amount of alkyne groups present. Maintain the overall protein concentration at 1 to 10 mg/ml or greater to obtain the best reaction kinetics. Often, a one protein is reacted in 4-to 15-fold molar excess over a second protein to create a conjugate having several molecules of the first protein attached to the second protein. However, the appropriate molar excess may have to be determined by doing a series of reactions at different ratios and determining the best ratio for use in a particular application.
2. Add 2.5 μl of a solution containing 10-m*M* $CuSO_4.5H_2O$ and 50-m*M* ascorbic acid dissolved in water per ml of the protein mixture. Mix to dissolve.
3. React with mixing at room temperature for at least 4 h (or at 4°C overnight, if the protein is not stable at ambient temperature). The optimal time of reaction is dependent on the proteins being coupled together and the number of azide and alkyne reactive groups available on them. An alkynyl-modified protein added to the azide-modified protein at a high molar ratio probably would reach maximal coupling yield in a matter of hours.
4. Purify the protein conjugate by dialysis or gel filtration using a molecular weight cut-off appropriate for the sizes of the proteins being separated.

A similar protocol may be used to couple proteins or molecules modified with an azide or alkyne to a particle or surface modified with the other functionality.

Aminooxy- and Aryl Aldehyde PEG Reagents

Aminooxy groups, also called aminoxy or alkoxyamines, are compounds of the general formula R–ONH_2 that contain an oxygen-substituted hydroxylamine. This functional group contains a nucleophilic terminal amine that is able to undergo reactions with electrophilic reactive groups much the same as primary amines. However, the aminooxy group is relatively unreactive with the typical functional groups found on biological macromolecules. It even has a relatively low potential for interactions with negatively charged carboxylates or phosphate groups, because under physiological conditions it is not protonated and therefore does not carry a positive charge (pKa of aminooxy groups is 5–6, as opposed to amines, which is typically much higher (pKa~9)). Thus, aminooxy reagents largely are bioorthogonal with most macromolecules (Chapter 17) in that they have no inherent reactivity toward proteins, nucleic acids, or lipids typically found in biological samples. The exception to this statement is the potential reactivity to carbonyl sites on these molecules, if created by oxidative damage. This means that an aminooxy group is very reactive toward aldehydes and ketones, in which the formation of an oxime linkage occurs quite readily (see Chapter 15, Section 2.4, for additional details on this reaction).

In addition, aryl aldehyde groups are also particularly unreactive toward functional groups typically encountered in biological systems. Unlike aliphatic aldehydes, aryl aldehydes do not as readily form Schiff base interactions with amines in aqueous solution and thus do not spontaneously modify or crosslink proteins. The aryl aldehyde group may form weak Schiff base interactions, but these are quickly reversible under mild conditions, especially if another more reactive functional group is encountered. For instance, an aryl aldehyde rapidly reacts with an aminooxy group to form a stable oxime bond. This dehydration reaction occurs in mildly acidic aqueous buffers and is accelerated by the presence of an aniline catalyst, as described in Chapter 15, Section 2.4 (Cordes and Jencks, 1962; Dirksen *et al.*, 2006a,b; Dirksen and Dawson, 2008; Byeon *et al.*, 2010). Therefore, aminooxy reagents and aryl aldehyde reagents may be used in conjunction for bioconjugation strategies involving chemoselective and bioorthogonal reaction schemes, which can be used to crosslink, modify, or immobilize molecules in the presence of complex sample mixtures containing many other functional groups.

PEG-based aminooxy and aryl aldehyde crosslinking reagents now are available that combine the chemoselective nature of the two bioorthogonal reactive groups along with the hydrophilicity of discrete polyether spacer arms. These compounds contain one of these bioorthogonal groups at one end of a crosslinker

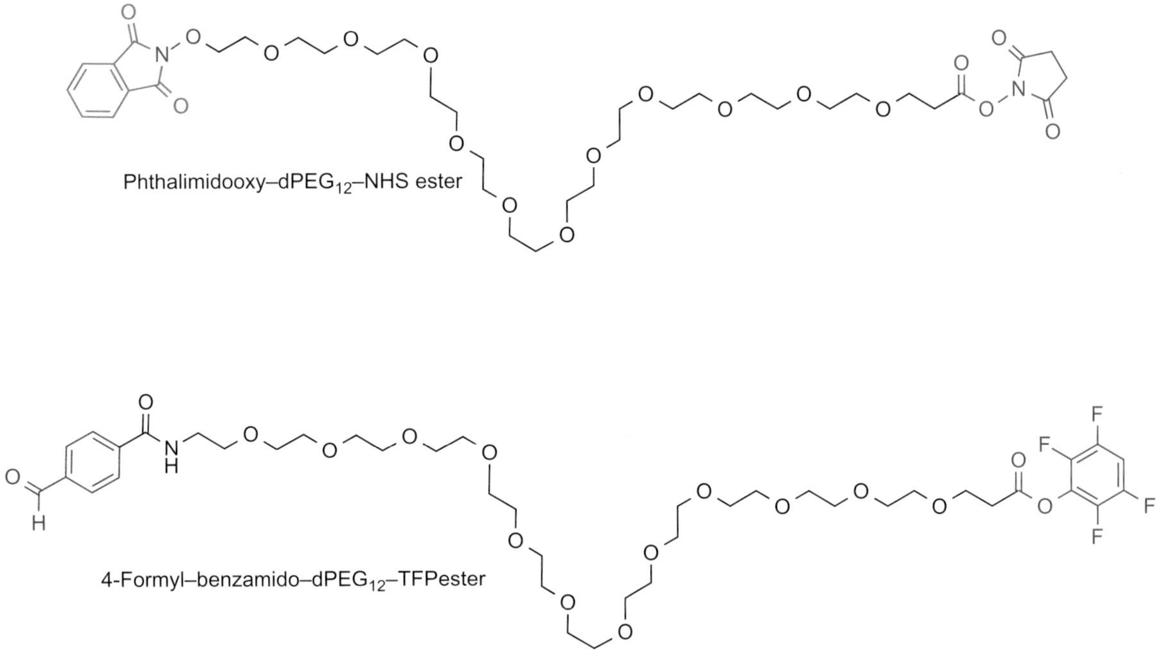

FIGURE 18.15 Aminooxy and aryl aldehyde crosslinking agents containing PEG_{12} spacer arms. The aminooxy compound has an amine-reactive NHS ester, while the aryl aldehyde compound has an amine-reactive TFP ester group.

and another type of reactive group or functional group at the other end—combined with a variety of different length PEG spacers as options for the central cross-bridge. Figure 18.15 illustrates two of these reagents, which may contain the aminooxy group as a protected phthalimide derivative to prevent its premature reaction with an electrophilic group at the opposite end. For instance, the phthalimidooxy–$dPEG_{12}$–NHS ester (Quanta BioDesign) has an amine-reactive group on one end for the initial modification of an amine-containing molecule or surface, while the aminooxy group is protected from reacting with the ester by use of the phthalimide protecting group.

Once an amine-containing molecule is modified with the PEG–phthalimidooxy group, the phthalimide can be removed using an aqueous solution of hydrazine to reveal the aminooxy group. The companion crosslinking reagents to the aminooxy compounds contain an aryl aldehyde group at one end, a PEG_{12} or PEG_{24} spacer group in the middle, and an amine-reactive tetrafluorophenyl (TFP) ester at the other end. Thus, two different molecules containing amines can be separately labeled with these reagents and later mixed to form a bioconjugate, which will get linked together by oxime bonds. Particles or surfaces may also be modified with one of these reagents to immobilize another molecule modified with the complementary bioorthogonal reagent. In this manner, a wide variety of bioconjugates may be prepared through the initial modification with these hydrophilic PEG-based reagents. The resultant oxime linkages are very stable and have no need to be reduced as is common with reactions between aldehydes and amines or aldehydes and hydrazides. Figure 18.16 shows the reactions that take place in the modification of two separate amine-containing molecules with the aminooxy and aryl aldehyde compounds and Figure 18.17 illustrates their subsequent conjugation to form a covalent complex.

The following protocols describe the reaction of the amine-reactive aminooxy and aryl aldehyde compounds with two separate amine-containing molecules (such as proteins) and their subsequent conjugation to form a complex linked together by oxime bonds. This general procedure can also be used with particles or surfaces containing amines to immobilize affinity ligands. Refer to Chapter 15, Section 2.4, for an alternative protocol using the phthalimidooxy–$dPEG_{12}$–NHS ester to immobilize aldehyde-containing affinity ligands on insoluble chromatography supports.

(A) Protocol for Modification of a Protein with PhthalimidooxyddPEG$_{12}$–NHS Ester

1. Dissolve an amine-containing protein, such as an antibody, at a concentration of 10 mg/ml in 0.1-*M* sodium phosphate, 0.15-*M* NaCl, pH 7.5 (coupling buffer). For instance, 2 mg of an antibody may be dissolved in 200 µl of the buffer.
2. Dissolve 2 mg of phthalimidooxy–$dPEG_{12}$–NHS ester in 200-µl DMSO, DMF, or DMAC.

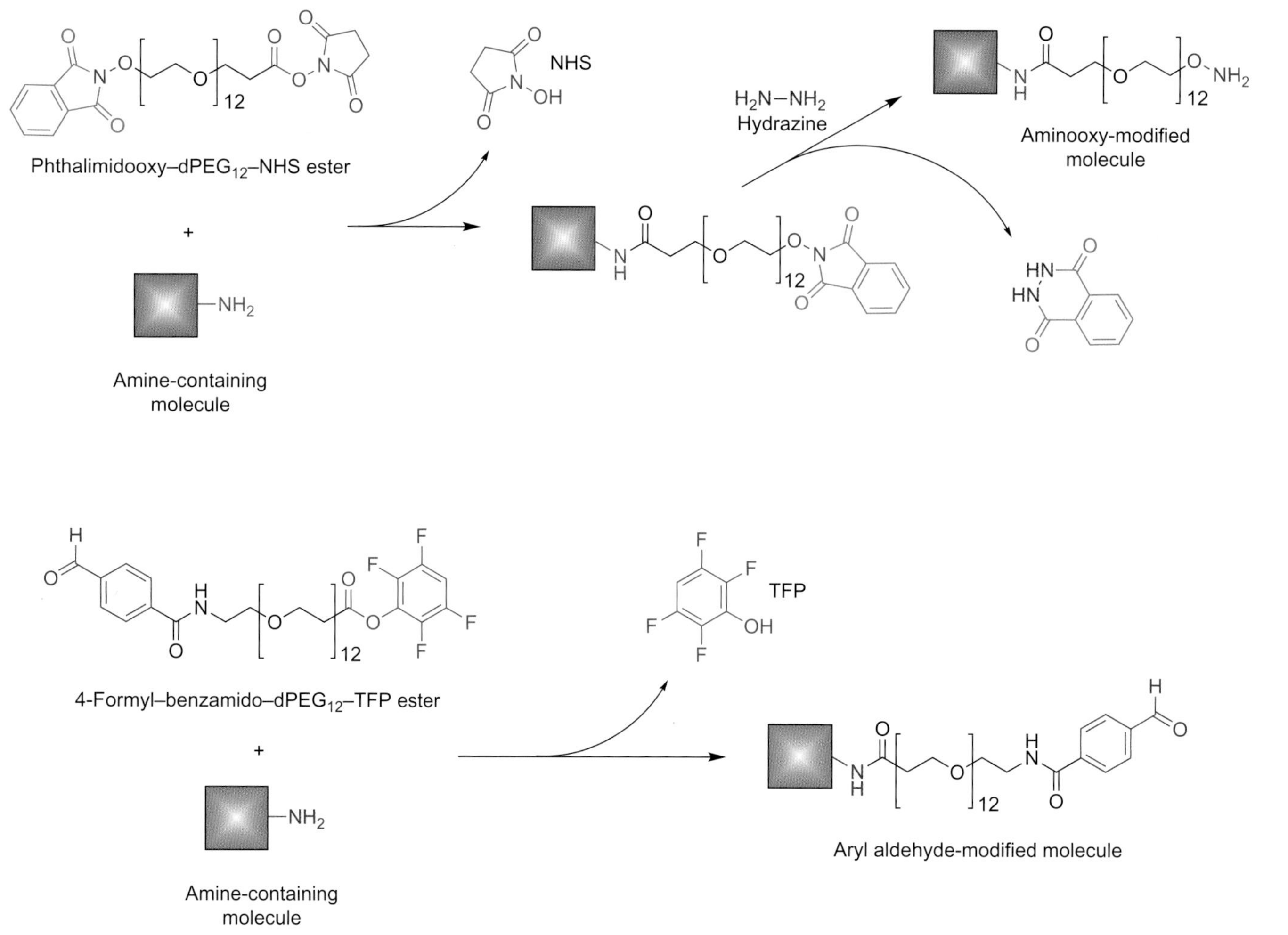

FIGURE 18.16 The reactions of the aminooxy and aryl aldehyde PEG-based crosslinkers in the modification of amine-containing molecules. The aminooxy group can be deprotected by the addition of hydrazine.

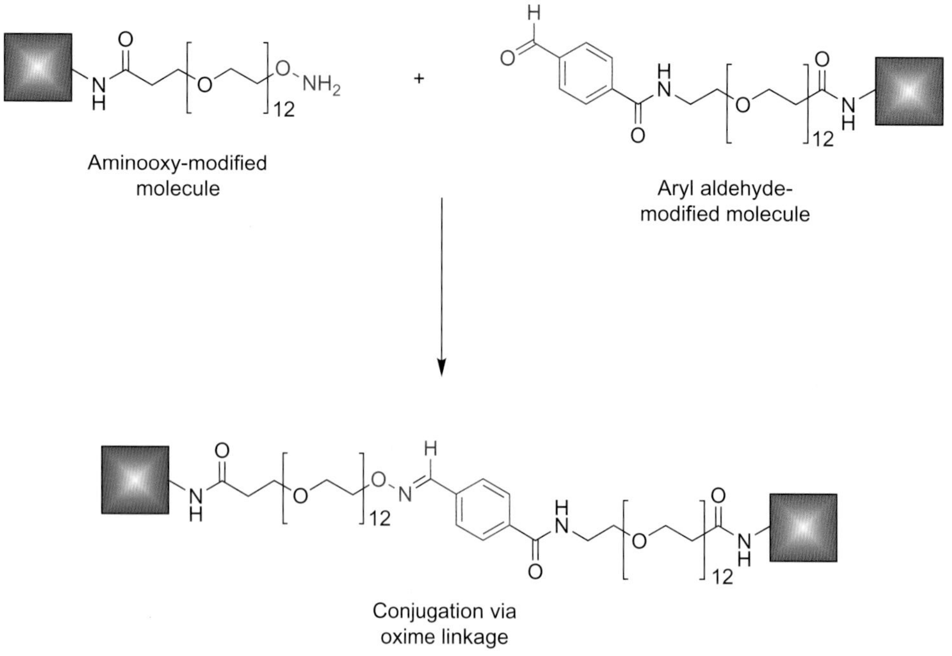

FIGURE 18.17 Molecules modified with the aminooxy and aryl aldehyde PEG-based crosslinkers can be conjugated through formation of a stable oxime linkage, which can be accelerated in aqueous solution by the addition of aniline as catalyst.

3. Add 17.5 µl of the crosslinker solution prepared in (2) to the antibody solution prepared in (1). This level of addition provides a 10-fold mole excess of the crosslinker over the quantity of antibody present. Note: Optimization of the amount to add of phthalimidooxy–dPEG$_{12}$–NHS ester compound should be carried out to obtain the best possible final conjugate in the intended application. Experiments performed using a range of different mole excesses of crosslinker over the antibody present can identify the best level of modification to use.
4. React for 30 min at room temperature with periodic mixing in a sealed container (e.g., microfuge tube).
5. Purify the modified antibody by dialysis or size exclusion chromatography using a 5-ml desalting column. Use 0.1-*M* sodium acetate, 0.15-*M* NaCl, pH 5.5 (conjugation buffer) as the dialysis or chromatography solution. Concentrate the antibody using centrifugal concentrators to approximately 200 µl, if necessary. The protected aminooxy-modified antibody may be stored until use at 4°C in buffer containing a preservative.

(B) Protocol for Deprotection of the Phthalimidooxy-Modified Antibody

1. In a fume hood and while wearing the appropriate personal protective equipment, add 3.1 µl of hydrazine (3.2 mg) to the phthalimidooxy-modified antibody solution to make a final solution that is 0.5-*M* hydrazine. Mix well to dissolve. Note: At much higher concentrations of hydrazine (> 20%) and at elevated temperatures (60°C or 95°C), protein structural reactions may occur, including deglycosylation, deamidation, cleavage of ester bonds to form hydrazides, and deguanidination (de la Burde *et al.*, 1963; Takasaki and Kobata, 1978; Patel *et al.*, 1992). At the recommended concentration of 0.5-*M* hydrazine needed to deprotect the phthalimidooxy groups, no such chemical modifications on proteins should occur. However, to be certain that a particular protein or antibody is not sensitive to hydrazine at this concentration, a separate test of the protein dissolved in 0.5-*M* hydrazine should be carried out to check stability.
2. Incubate at room temperature overnight with periodic mixing to remove the phthalimide group from the aminooxy end of the crosslinker modification sites.
3. Purify the aminooxy-modified antibody by dialysis or size exclusion chromatography using a 5-ml desalting column. Use 0.1-*M* sodium acetate, 0.15-*M* NaCl, pH 5.5 (conjugation buffer), as the dialysis or chromatography solution. Concentrate the antibody using centrifugal concentrators to approximately 200 µl, if necessary. Dispose of all hydrazine-containing solutions according to your facility's guidelines concerning hazardous waste disposal.

(C) Protocol for Modification of a Protein with 4-Formyl-Benzamido–dPEG$_{12}$–TFP Ester

1. Dissolve 2 mg of an amine-containing protein, at a concentration of 10 mg/ml in 200 µl of 0.1-*M* sodium phosphate, 0.15-*M* NaCl, pH 7.5 (coupling buffer).
2. Dissolve 2 mg of 4-formyl–benzamido-dPEG$_{12}$–TFP ester in 200 µl DMSO, DMF, or DMAC.
3. Add a quantity of the crosslinker solution prepared in step 2 to the protein solution prepared in step 1 to result in a 10-fold mole excess of crosslinker over the amount of protein present (the amount to add is dependent upon the MW of the protein being modified). Note: Optimization of the amount of the amount of crosslinker addition should be carried out to obtain the best possible final conjugate in the intended application. Experiments performed using a range of different mole excesses of crosslinker over the protein present can identify the best level of modification to use.
4. React for 30 min at room temperature with periodic mixing in a sealed container (e.g., microfuge tube).
5. Purify the modified protein by dialysis or size exclusion chromatography using a 5-ml desalting column. Use 0.1-*M* sodium acetate, 0.15-*M* NaCl, pH 5.5 (conjugation buffer) as the dialysis or chromatography solution. Concentrate the protein, if necessary, using centrifugal concentrators to approximately 200 µl. The aryl aldehyde-modified protein may be stored until use at 4°C in buffer containing a preservative.

Protocol for Conjugation of the Aminooxy-Modified Antibody with the Aryl Aldehyde-Modified Protein

1. Mix the deprotected aminooxy-modified antibody prepared in (B) with the aryl aldehyde-modified protein prepared in (C), with both proteins dissolved in conjugation buffer (0.1-*M* sodium acetate, 0.15-*M* NaCl, pH 5.5). Note: The conjugation reaction may be performed using an equal mole quantity of the modified proteins or it may be performed with one of the proteins in excess. Optimization of the reaction ratio of the two proteins may have to be carried out to determine the best resultant conjugate for the intended application. For instance, if the aryl aldehyde-modified protein happens to be an enzyme that can be used for detection purposes, it is typical to react the enzyme in excess over the amount of antibody present. Refer to Chapters 20 and 22 for suggested ratios of enzyme to antibody to use in particular conjugation strategies.

2. In a fume hood, add aniline to the conjugation reaction to obtain a final concentration of 0.1-*M*. Mix well to dissolve. Note: If the total solution volume is 400 μl, then add 7.2 μl of aniline to obtain the 0.1-*M* final concentration.
3. React for at least 2 to 4 h with periodic mixing at room temperature. Note: The blocking of excess aminooxy groups within the conjugate may be performed by the addition of 0.1-*M* glyceraldehyde and the blocking of excess aryl aldehyde groups may be performed using 0.1-*M* hydroxylamine, both blockers being dissolved in the conjugation buffer. However, these blocking steps should be carried out sequentially and with a purification step in between using dialysis or gel filtration to avoid the reaction of the blocking agents with each other.
4. Purify the conjugate with a buffer of choice using dialysis or size exclusion chromatography with a 10-ml desalting column. The final conjugate may be stored in buffer or water containing a preservative at 4°C.

1.3. Biotinylation Reagents Containing Discrete PEG Linkers

Biotin modification reagents are widely used to attach a biotin group to proteins or other molecules for subsequent use in avidin, streptavidin, or NeutrAvidin separations or assays. Traditional biotin compounds containing aliphatic or other hydrophobic linker arms are discussed in detail in Chapter 11, Section 6. In this chapter, the biotin–PEG compounds are exclusively discussed due to their unique hydrophilic properties, which include low nonspecific binding character and low immunogenicity.

Traditional biotinylation compounds include the very popular (sulfo)NHS–LC–biotin, which contains either an uncharged NHS ester or a negatively charged sulfo-NHS ester in addition to a C6 alkyl chain leading to the biotin group. Although the sulfo-NHS ester provides a degree of water solubility for the entire compound prior to using it to modify a protein, the resultant LC–biotin modification left on the protein is extremely hydrophobic. The result of adding such hydrophobic groups to a protein surface is often a balance between maintaining water solubility or tending toward protein aggregation and precipitation. Many antibodies that are biotinylated with NHS-LC–biotin undergo slow aggregation and loss of activity despite limiting the degree of biotinylation during production.

The use of discrete PEG spacers in the construction of biotinylation compounds not only increases the water solubility of the modification reagent itself, but significantly increases the hydrophilicity and stability of proteins modified with them. Even when high modification levels are used with PEG–biotin compounds, the resultant biotinylated protein is typically very water soluble and does not aggregate in solution.

The following compounds represent some of those that are commercially available for adding a PEG–biotin group to proteins and other molecules. Reactive group options include NHS esters for coupling to amines, maleimide groups for coupling to thiols, hydrazides for conjugation with carbonyl compounds, and a photoreactive benzophenone for nonselective insertion into molecular structures (Thermo Fisher, Quanta BioDesign, Solulink, and Molecular Biosciences). In addition, a chromogenic biotin compound containing a PEG spacer is available from Thermo Fisher and Solulink. This unique compound allows exact measurement of the amount of biotinylation by measuring the absorbance of the final purified conjugate.

NHS–PEG$_n$–Biotin Compounds

NHS ester biotinylation reagents are the most popular choice for adding a biotin group to another molecule that contains an available amine. The ester reacts with an amine at neutral or slightly alkaline pH values to form a stable amide bond with the amine-containing protein or molecule. Discrete PEG biotinylation compounds containing an NHS ester are available in chain lengths of 4 or 12 repeating poly(ethylene oxide) units and in one form that contains a cleavable disulfide in the cross-bridge.

The NHS ester compounds are sensitive to hydrolysis in aqueous solution, and they will likely hydrolyze faster than more hydrophobic biotinylation compounds due to their hydrophilicity. If a stock solution is made at a higher concentration to facilitate the addition of a small amount to a reaction solution, the initial solution should be made in a water-miscible organic solvent that is dried with a molecular sieve. Suitable solvents include DMAC, DMSO, or DMF. If using DMF, use only highly pure solvent, as it may contain amines that can react with the NHS ester groups (Figure 18.18).

Reactions performed with NHS–PEG$_n$–biotin compounds are typically done with the reagent in molar excess over the amount of protein being modified. The efficiency of the reaction is dependent upon the concentrations of reactants and the solvent exposed area of the amine groups on the protein. Reactions performed with a 10-fold molar excess of NHS–PEG$_n$–biotin usually will result in at least two to three biotin labels per protein, while doubling the molar excess should provide at least four to six biotin modifications; however, the yield of biotinylation varies depending on the protein being modified. The optimal number of biotin groups added to a particular protein should be determined experimentally to provide the best performance in the intended application. When using PEG-based biotinylation compounds, over-modifying an antibody

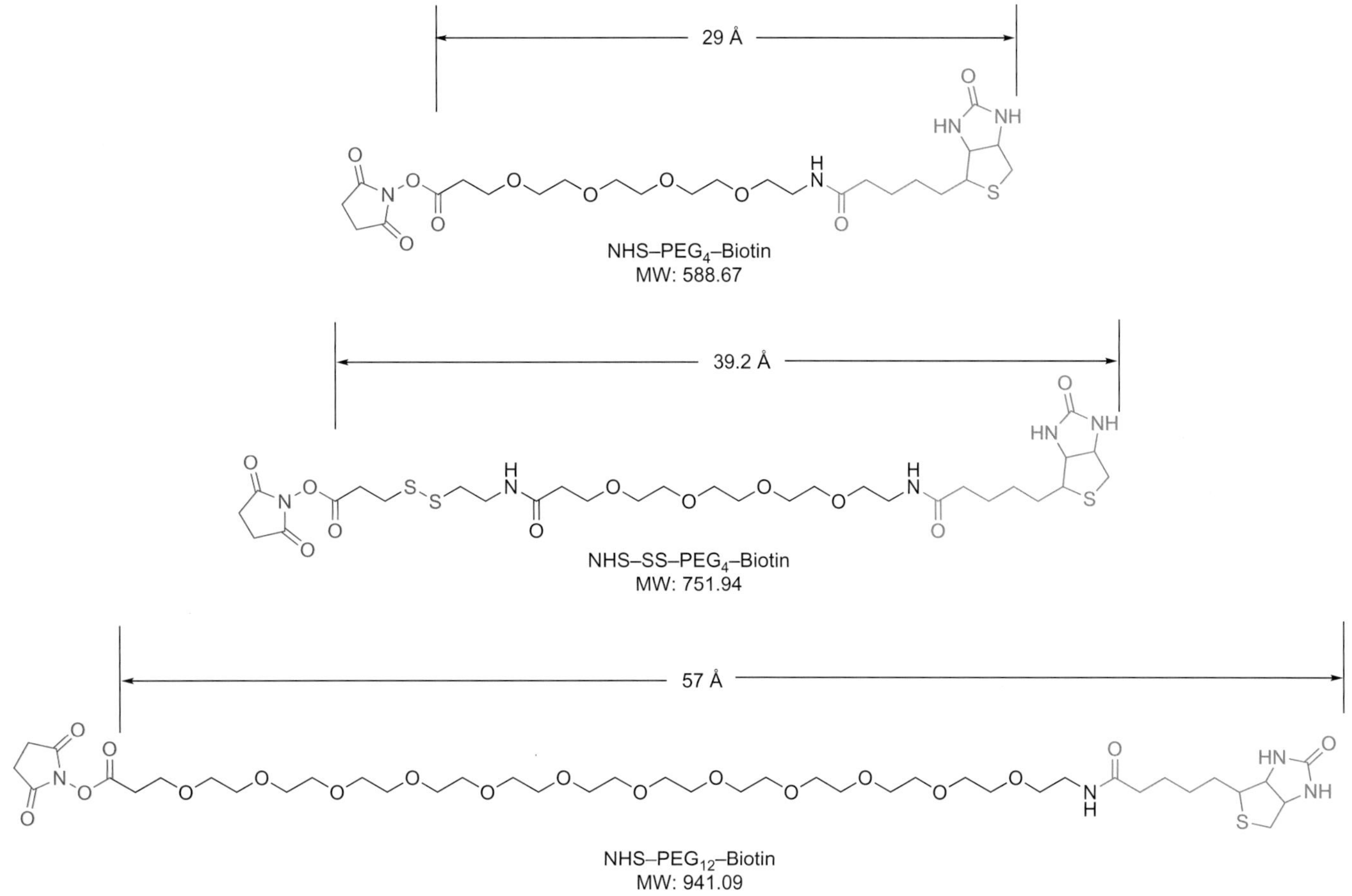

FIGURE 18.18 NHS–PEG$_n$–biotin compounds.

for example will not cause aggregation of the final conjugate as it would with aliphatic biotin reagents. Therefore, it is not as critical to maintain a low level of biotin modification to avoid this problem.

NHS–SS–PEG$_4$–biotin contains a cleavable disulfide bridge next to the PEG chain. This feature allows the biotin group to be released from the biotinylated molecule using a disulfide reducing agent, such as DTT or TCEP (Figure 18.19). This is useful for doing immunoprecipitation (IP) or co-immunoprecipitation (co-IP) assays, because a biotinylated antibody can be used to capture and affinity isolate a target protein on immobilized (strept)avidin and then elute the bound proteins by reduction (using 50-m*M* DTT for 2 h at room temperature or for 30 min at 50°C). This eliminates the severe denaturing conditions usually required to break the (strept)avidin–biotin bond, thus better preserving the native structure and activity of the captured proteins. Compounds of this type may also be used successfully in cell surface biotinylation, as described in Chapter 11, Section 6.2, under Sulfo-NHS–SS–Biotin.

The following protocol is based on the applications done at Thermo Fisher (formerly Pierce). The use of higher pH values for the NHS ester reaction than those recommended may result in lower biotinylation yields due to increased hydrolysis, especially when using an extremely hydrophilic PEG compound.

Protocol

1. Dissolve an antibody or protein to be modified at a concentration of 1 to 10 mg/ml in 0.1-*M* sodium phosphate, 0.15-*M* NaCl, pH 7.2–7.5. Lower concentrations of protein may result in decreased reaction yields and require increased quantities of reagent to obtain acceptable levels of biotinylation. Avoid amine-containing buffers or components, such as Tris or imidazole, which will react with the NHS ester and interfere with the biotinylation process.
2. Dissolve the NHS–PEG$_n$–biotin compound in DMAC, DMSO, or DMF (pure and dry) at a concentration of 10 to 20 m*M*. Prepare fresh in a fume hood.
3. With mixing, add an aliquot of the biotinylation stock solution to the protein solution to provide at least a 10-fold molar excess over the concentration of protein present. A series of reactions with different

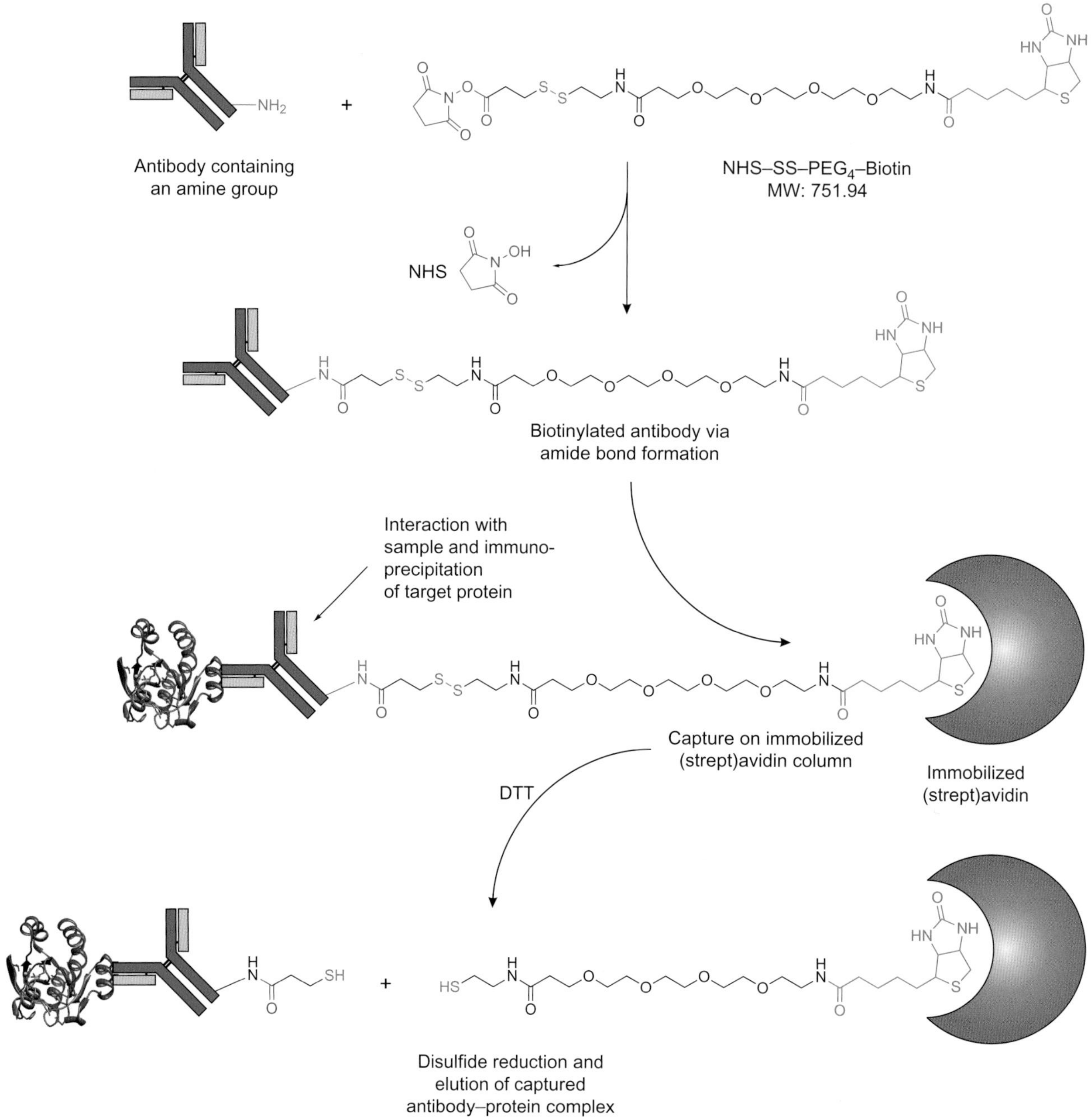

FIGURE 18.19 NHS–SS–PEG_4–biotin can be used to label a primary antibody molecule that has specificity for a protein of interest. Incubation of the biotinylated antibody with a sample, such as a cell lysate, allows the antibody to bind to its target. Capture of the antibody–antigen complex on an immobilized streptavidin reagent effectively isolates the targeted protein from the other proteins in the sample. The disulfide linkage in the spacer arm of the biotin tag permits elution of the immune complex from the streptavidin support using DTT and without using the strong denaturing condition typically required to break the streptavidin–biotin interaction.

molar amounts of the NHS–PEG_n–biotin compound may be performed to optimize the modification level.
4. React with gentle mixing for 30–60 min at room temperature or 2 h at 4°C.
5. Purify the biotinylated protein from excess reagent and reaction byproducts using dialysis or gel filtration (desalting resin).

NHS–Chromogenic–PEG_3–Biotin

A novel detectable biotinylation reagent containing a hydrophilic PEG spacer is NHS–chromogenic–PEG_3–biotin (also called chromogenic biotin; Solulink) (Figure 18.20). Next to the terminal NHS ester of this compound is a *bis*-aryl hydrazone group created from the reaction of a 6-hydrazinium nicotinate derivative and a benzaldehyde

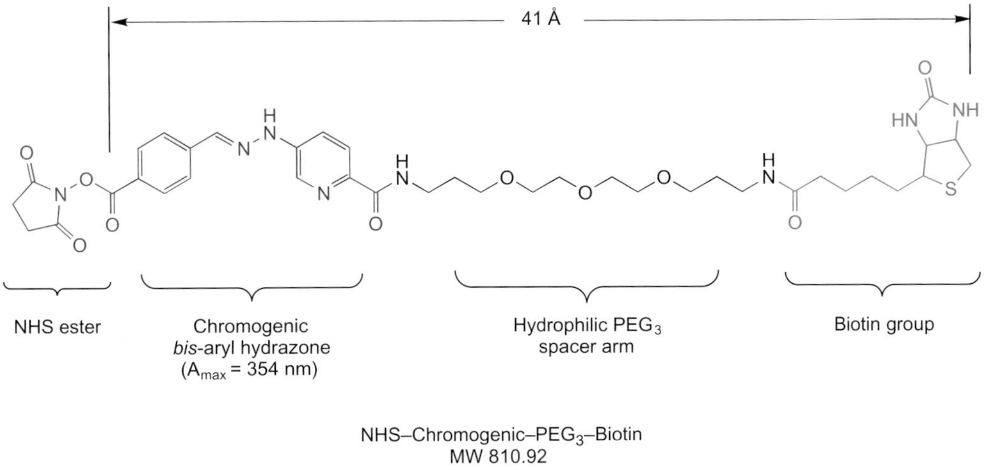

FIGURE 18.20 NHS–chromogenic–PEG_3–biotin contains an amine-reactive NHS ester that can be used to label biomolecules through an amide linkage. The chromogenic *bis*-aryl hydrazone group within the spacer arm of the reagent allows the degree of biotinylation to be quantified by measuring its absorbance at 354 nm. The compound also contains a hydrophilic PEG spacer, which provides greater water solubility.

group to form the chromogen having an absorbance at 354 nm ($\varepsilon = 29{,}000\,M^{-1}cm^{-1}$). The NHS ester end can be used to modify amine-containing molecules and form a stable amide linkage (Figure 18.21). The spacer arm contains a hydrophilic spacer made from three ethylene oxide units, which provide water solubility for the compound.

Proteins biotinylated with this reagent will have a characteristic absorbance band at 354 nm, which can be used to determine accurately the number of biotin groups per molecule. No other biotinylation compound has such built-in quantification capability. This feature eliminates the need to consume conjugate by doing a HABA assay to test for the level of biotin incorporation (Chapter 11, Section 6.1).

The following protocol is adapted from the manufacturers' recommendations.

Protocol

1. Dissolve a protein or other amine-containing molecule to be biotinylated in 0.1-*M* sodium phosphate, 0.15-*M* NaCl, pH 7.2 to 7.5, at a concentration of 1 to 10 mg/ml. Note that protein solutions that are more dilute than this may require higher levels of biotinylation reagent addition to achieve the same yield of modification.
2. In a fume hood, dissolve NHS–chromogenic–PEG_3–biotin in DMF at a concentration of 12.33-m*M* (2 mg/200 μl DMF). With mixing, add a quantity of the reagent to the protein solution to provide the desired molar excess (i.e., 10- to 20-fold excess).
3. React for 30 to 60 min at room temperature or 2 h at 4°C.
4. Purify the modified protein from unreacted biotinylation reagent and reaction byproducts using dialysis or gel filtration. Complete removal of the excess reagent is necessary to provide accurate measurement of the biotin incorporation level by absorptivity.

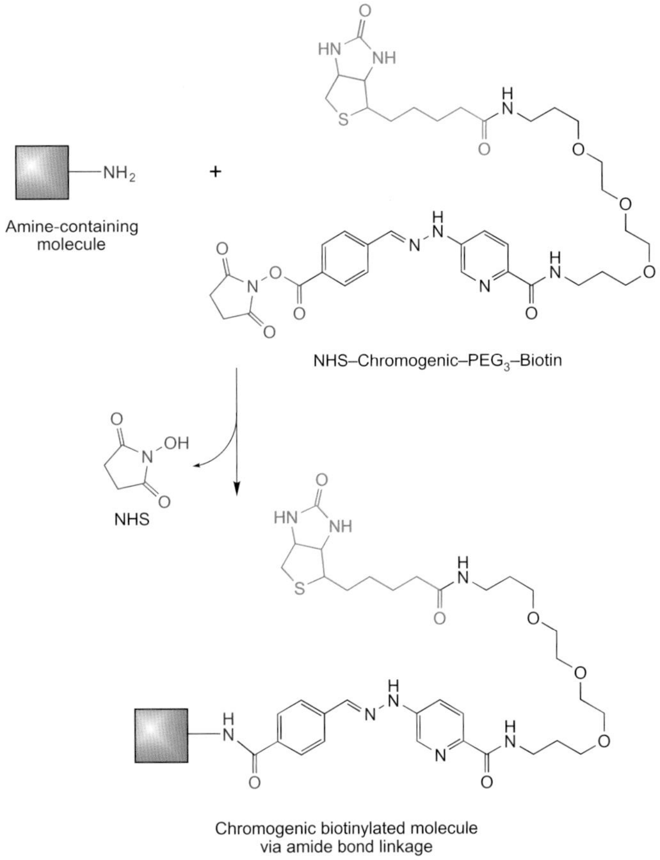

FIGURE 18.21 NHS–chromogenic–PEG_3–biotin reacts with amine groups in proteins or other molecules to form amide bond derivatives.

5. Measure the absorbance of the biotinylated protein solution at 354 nm. Use the molar extinction coefficient for the chromogenic group ($\varepsilon = 29{,}000\,M^{-1}cm^{-1}$) to determine the concentration of biotin present. To determine the molar ratio of

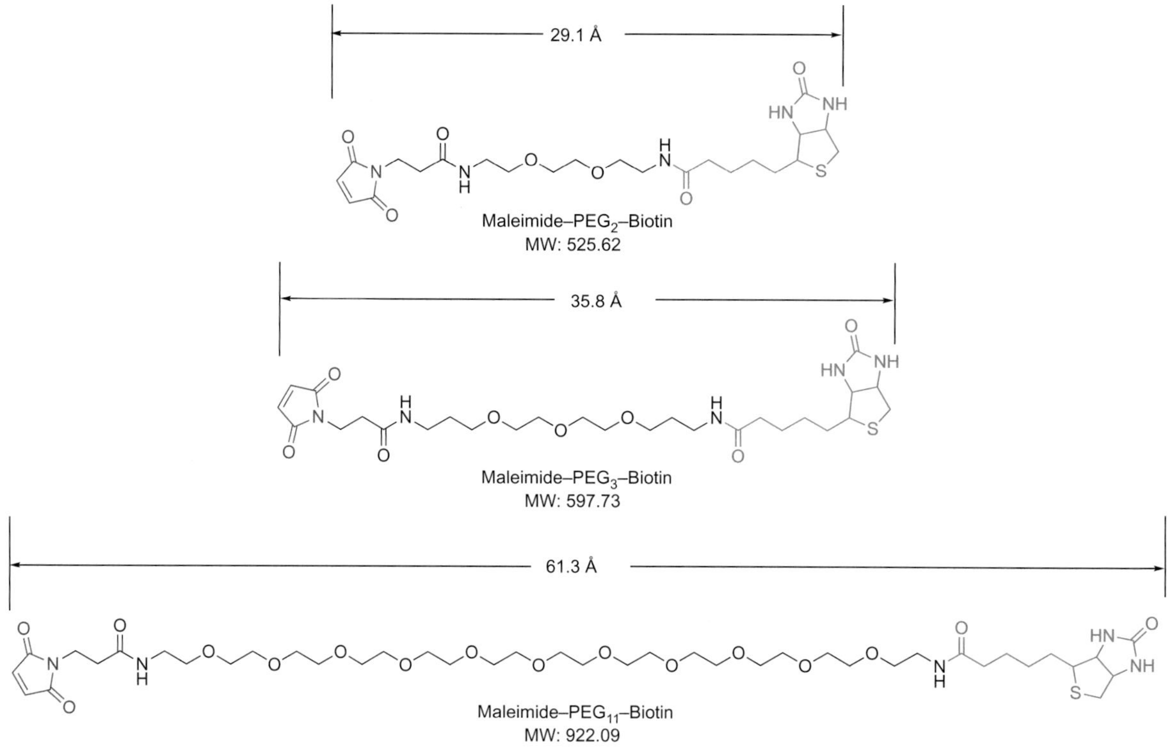

FIGURE 18.22 Maleimde–PEG$_n$–biotin compounds of three different discrete PEG sizes are available, including a PEG$_{11}$ chain that provides a molecular length of over 60 Å.

biotin to protein, divide the molar concentration of biotin by the molar concentration of protein present (which may be determined by using the Coomassie assay or the BCA assay methods).

Maleimide–PEG$_n$–Biotin Compounds

Discrete PEG–biotin compounds containing a terminal maleimide group may be used to label sulfhydryl-containing proteins and other molecules through thioether bond formation (Figure 18.22). The targeting of thiol groups in proteins is often used to direct the modification reaction away from binding sites or active centers in proteins, thus preserving activity. Maleimide reagents in general are the second most-popular reactive group used for bioconjugation purposes, second only to NHS esters. Unlike biotinylation compounds containing a hydrophobic hydrocarbon chain (Chapter 11), the discrete maleimide–PEG-based reagents provide increased hydrophilicity for modified molecules and maintain solution stability even at high substitution levels.

The maleimide group reacts with thiols in the pH range of 6.5 to 7.5 to form a stable thioether linkage with very little cross-reactivity with amines at this pH (Figure 18.23). However, the maleimide ring is subject to hydrolysis in aqueous solution, and since it is next to an extremely hydrophilic PEG chain in these reagents, this factor may increase the hydrolysis rate beyond that typically observed with hydrocarbon-based spacers (Chapter 3, Section 2.2). For this reason, stock solutions of a maleimide–PEG$_n$–biotin compound should be made in highly pure and dry organic solvent, which then can be added to an aqueous reaction medium to commence the biotinylation process.

The three maleimide–PEG$_n$–biotin compounds illustrated in this section provide short-, medium-, and very-long-chain spacer options, with the longer chains resulting in the greatest degree of hydrophilicity of modified molecules. The following protocol is adapted from general maleimide-based biotinylation methods, as discussed in Chapter 11, Section 6.3.

Protocol

1. Dissolve a sulfhydryl-containing protein or other thiol-molecule in a thiol-free buffer within a pH range of 6.5 to 7.5. The use of 20-m*M* sodium phosphate, 150-m*M* NaCl, pH 7.2, works well for this reaction. The concentration of protein should be in the range of 1 to 10 mg/ml. Lower concentrations of protein may result in the need to increase the molar excess of biotinylation reagent to obtain an acceptable level of modification. If a thiol is not present on the molecule to be biotinylated, one may be created by disulfide reduction or through the use of a thiolation reagent (Chapter 2, Section 4.1).
2. Prepare a stock solution of the maleimide–PEG$_n$–biotin compound in DMAC, DMSO, or DMF (pure and dry solvents only) at a concentration of 10 to 20 m*M*.

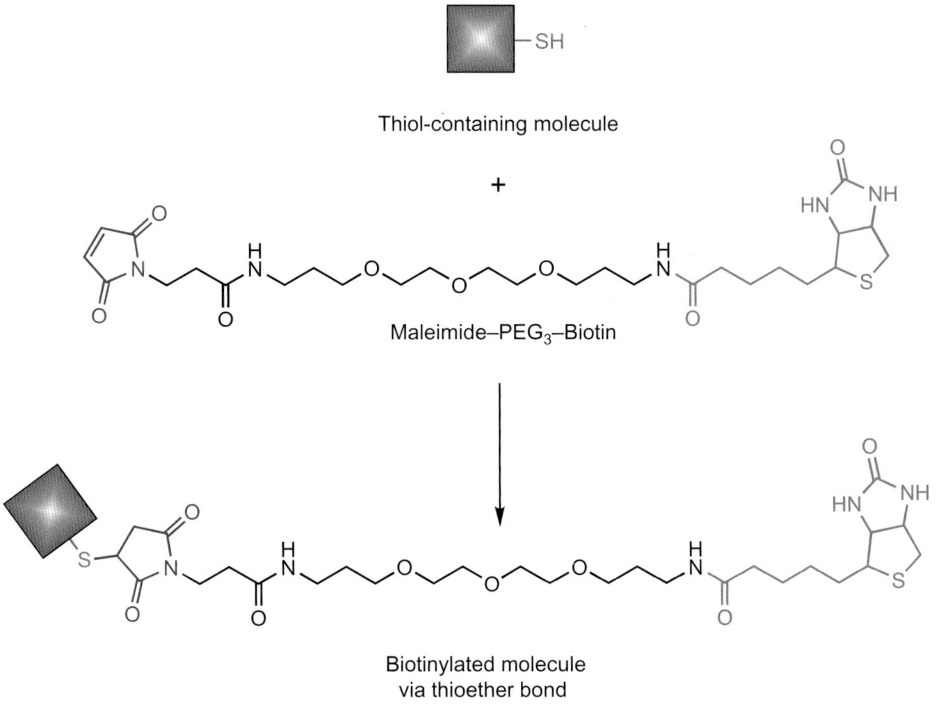

FIGURE 18.23 Maleimide–PEG_n–biotin compounds react with thiol-containing molecules to form thioether linkages.

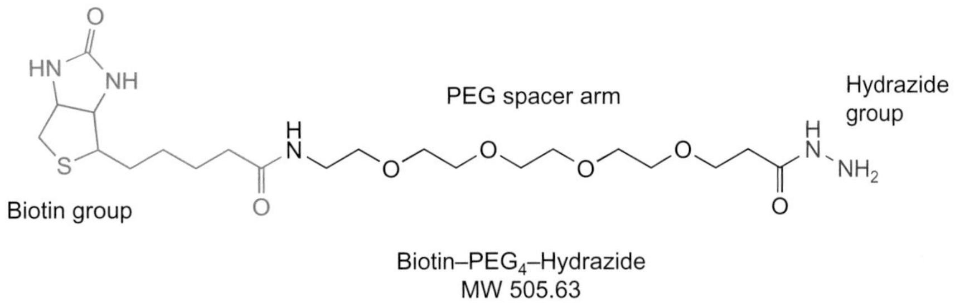

FIGURE 18.24 Biotin–PEG_4–hydrazide is a hydrophilic biotinylation reagent that can be used to modify glycans or carbohydrates at their reducing end or after periodate oxidation to create aldehydes.

3. With mixing, add an aliquot of the biotin solution to the protein solution to obtain at least a 10-fold molar excess over the quantity of protein present. As thiols are typically present in limiting amounts on proteins, the use of a high molar reagent ratio is not required to achieve acceptable yields of biotinylation.
4. React with gentle mixing for 2h at room temperature or 4h at 4°C.
5. Purify the biotinylated protein by dialysis or gel filtration using a desalting resin.

Hydrazide–PEG$_4$–Biotin

Hydrazide-containing PEG–biotinylation reagents provide reactivity with carbonyl groups (e.g., aldehydes) to label carbohydrates or glycoproteins via hydrazone bond formation (Figures 18.24 and 18.25). The hydrazide group may also be coupled with carboxylate-containing molecules using a carbodiimide reaction with EDC (Chapter 4, Section 1.1) or an active ester derivative (Chapter 3, Section 1.4). Like the other discrete PEG reagents, the hydrazide–PEG_4–biotin compound is very hydrophilic and will not promote aggregation or precipitation of labeled proteins. Aldehyde functionalities may be created on glycoproteins or other carbohydrates by oxidation using sodium periodate (Chapter 2, Sections 2.2, 4.4–4.6) or by modification with SFB (Chapter 17, Section 2). The reducing end of sugars or glycans may also be labeled with these hydrazide reagents to produce a biotin–carbohydrate that is modified at only a single site.

Hydrazide–PEG_4–biotin can be used to specifically label glycoproteins on cell surfaces after mild periodate oxidation of the glycan structures (Wilchek and Bayer, 1987). The hydrophilic nature of the PEG spacer will prevent the biotinylation reagent from easily penetrating

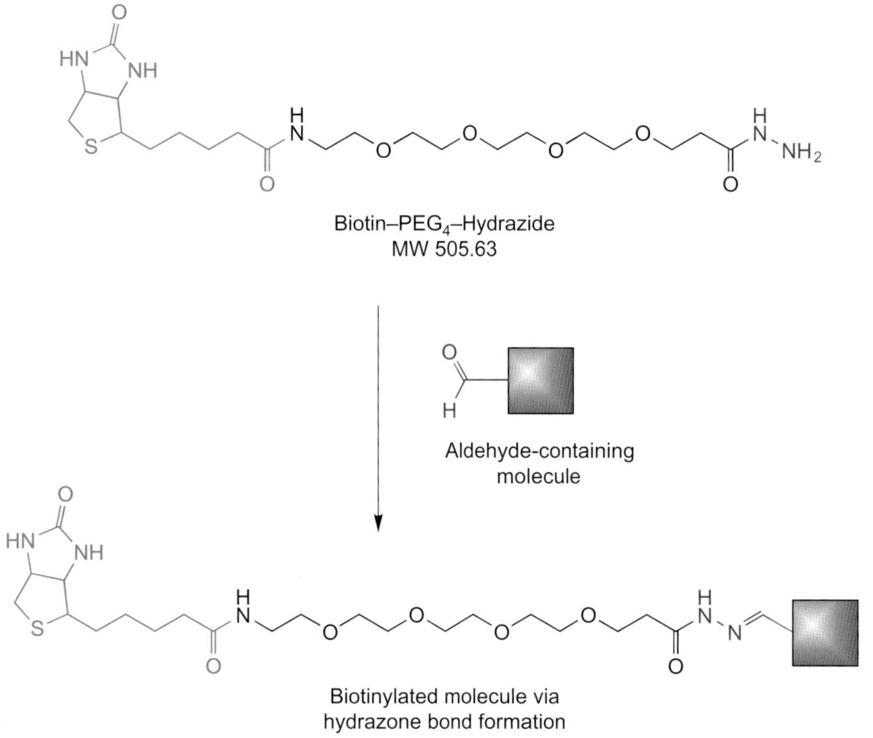

FIGURE 18.25 Biotin–PEG_4–hydrazide reacts with aldehyde-containing molecules to form a hydrazone linkage.

cell membranes, thus the labeling reaction is restricted to outer membrane glycoproteins. After biotinylation and cell lysis, the labeled proteins may be detected or isolated using (strept)avidin reagents (Jang and Hanash, 2003; Ding *et al.*, 2005; Handlogten *et al.*, 2005).

Hydrazide–PEG_4–biotin contains a 31.5-Å spacer consisting of 4 ethylene oxide units, which effectively imparts water solubility to the reagent. Other hydrazide biotinylation reagents that contain no spacer or a hydrocarbon spacer, such as biotin–hydrazide and biotin–LC–hydrazide, are water insoluble and actually will lower the solubility of modified molecules. Hydrazide–PEG_4–biotin can be used to modify molecules without the tendency for aggregation or precipitation. In addition, the compound is stable in aqueous environments, as it contains no groups that are easily hydrolysable. A stock solution may be prepared in a water-miscible organic solvent such as DMAC, DMSO, or DMF to facilitate transfer of a small amount to an aqueous reaction.

The following protocol describes a method for the periodate oxidation of a glycoprotein followed by biotinylation of the resultant aldehydes using hydrazide–PEG_4–biotin. Chapter 2, Section 4.6 describes an alternative protocol for the modification of glycans at their reducing ends with hydrazide compounds.

Protocol

1. Dissolve a glycoprotein to be oxidized in 0.1-*M* sodium acetate, pH 5.5 (oxidation buffer), at a concentration of 2 to 10 mg/ml. PBS at physiological pH may also be used for this reaction. The use of cold buffers for the oxidation step will limit the extent of carbohydrate oxidation and the potential for protein oxidation.
2. Dissolve sodium *meta*-periodate in oxidation buffer at a concentration of 20-m*M*. Protect from light.
3. Add an equal volume of the glycoprotein solution to the periodate solution with mixing.
4. React for 10 to 20 min with gentle mixing and protected from light.
5. Quench the oxidation reaction by the addition of at least a 4-fold molar excess of *N*-acetylmethionine or sodium sulfite over the concentration of periodate in the reaction mixture (e.g., 40-m*M*). Pre-dissolve the quencher in buffer at a higher concentration prior to adding an aliquot of it to the reaction solution. React for 10 min. Alternatively, the oxidation reaction may be stopped by the removal of excess periodate by gel filtration using a desalting column.
6. Prepare a 50-m*M* solution of hydrazide–PEG_4–biotin in DMAC, DMSO, or DMF. Add a quantity of this solution to the purified, oxidized protein to provide at least a 10-fold molar excess of biotinylation reagent over the concentration of protein present.
7. React with mixing for 2 h at room temperature.
8. The hydrazone bond can be reduced to stabilize the linkage by the addition of sodium cyanoborohydride to a final concentration of 50-m*M*. React for 30 min at room temperature with mixing. All operations with cyanoborohydride should be carried out in a fume

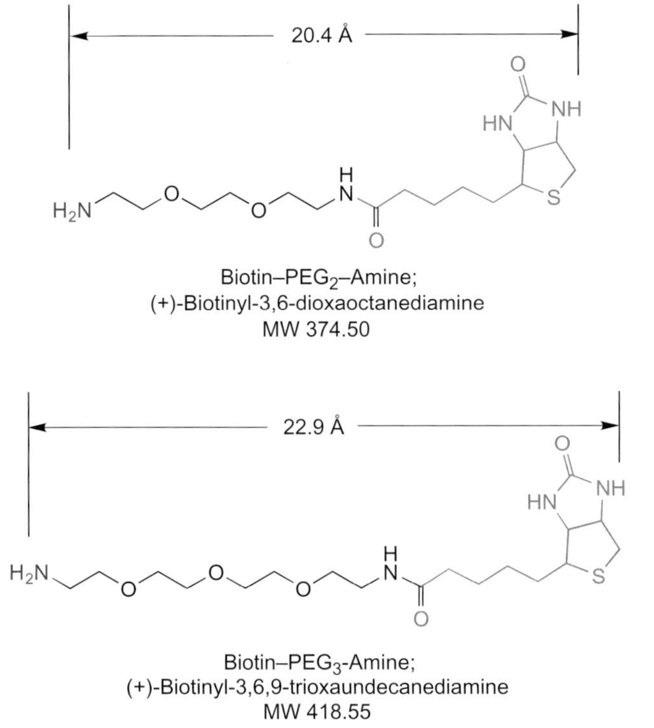

FIGURE 18.26 Biotin–PEG$_n$–amine compounds can be used to modify carboxylate- or aldehyde-containing compounds using a carbodiimide reaction.

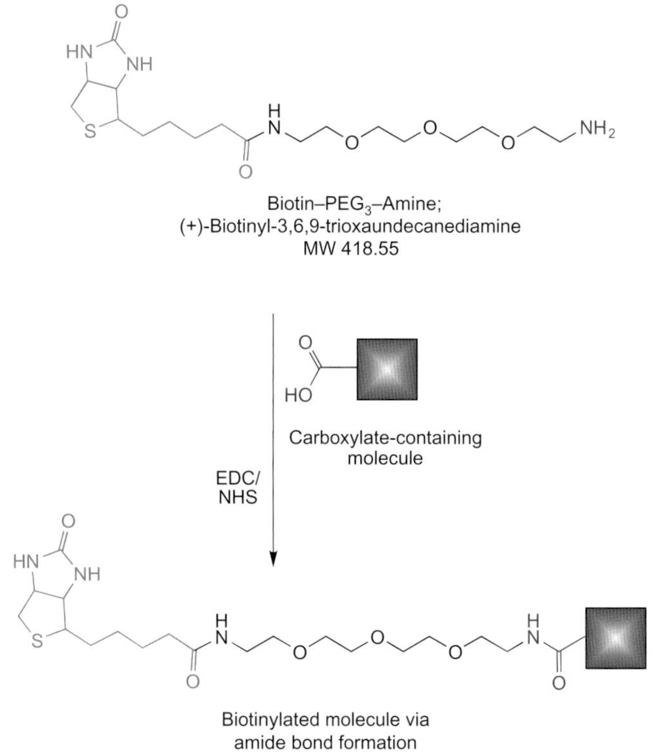

FIGURE 18.27 Biotin–PEG$_n$–amine can be used to add a biotin label to carboxylate-containing molecules using the EDC/(sulfo)NHS reaction, which forms a stable amide linkage.

hood. If the glycoprotein being modified is sensitive to disulfide reduction and potential denaturation, then this step should be avoided.

9. Purify the biotinylated glycoprotein by gel filtration or dialysis.

Biotin–PEG$_n$–Amine Compounds

Biotin compounds containing a PEG spacer that terminates in a primary amine can be used for the labeling of carboxylate molecules (Figure 18.26). Activated carboxylates, such as those containing an NHS ester, spontaneously react with the amines to give amide bond linkages. An active ester may also be formed *in situ* by the activation of carboxylates with EDC in the presence of NHS or sulfo-NHS (Chapter 4, Section 1.2) (Figure 18.27).

Biotin–PEG$_2$–amine contains a short, two-unit ethylene oxide cross-bridge that provides a 20.4-Å hydrophilic spacer, which has an amine on its end. Biotin–PEG$_3$–amine is identical except for one additional ethylene oxide unit. Both compounds are extremely water soluble and can be used to label organic molecules or biomolecules containing carboxylates. Bronfman *et al.* (2003) used the biotin–PEG$_n$–amine compounds to label nerve growth factor (NGF) peptide on its carboxylates using EDC-mediated amide bond formation. The biotinylated growth factor was then used to study receptor internalization in live cells by probing with fluorescent streptavidin conjugates.

The following protocol can be used to biotinylate carboxylate-containing molecules in aqueous solution using the EDC/sulfo-NHS reaction.

Protocol

1. Dissolve a carboxylate-containing peptide or other molecule in 0.1-*M* MES, pH 5.0 (reaction buffer). Ideally, this molecule should contain only one carboxylate with no amines to direct biotinylation to a single site and prevent polymerization of it during the conjugation process. However, if a peptide is to be biotinylated that also has amine groups, then the use of a very high molar excess of the biotin–PEG$_n$–amine reagent during the reaction will limit the potential for peptide–peptide linking. The concentration of the carboxylate molecule should be low if it also has amines present, but if it only has one or more carboxylates, then it can be prepared at higher concentration. For example, to use the biotin–PEG$_n$–amine compounds to biotinylate a protein, the concentration should be on the order of 1 to 2 mg/ml so that a large excess of biotinylation agent can be added. For molecules that are sparingly soluble in aqueous solution, they may be dissolved first in

ethanol and then added to the reaction buffer with mixing to make a final ethanol concentration of not more than 50%.

2. Dissolve the biotin–PEG_n–amine reagent in reaction buffer at a concentration of 25-m*M*.
3. Add a quantity of the biotin–PEG_n–amine solution to the solution containing the carboxylate molecule to achieve the desired molar excess. For molecules containing a single carboxylate to be modified, a 1.5- to 2-fold molar excess may be sufficient. However, for proteins or peptides that also contain competing amines, a much larger excess of biotin compound should be used (e.g., 100-fold excess). For instance, for protein biotinylation, add 120 µl of the biotin–PEG_n–amine solution per milliliter of the solution prepared in step 1.
4. Immediately before use, dissolve EDC in reaction buffer at a concentration of 25-m*M*. Add 12 µl of this solution per milliliter of the combined solution from step 2. Mix well.
5. React for 2 h at room temperature or 4 h at 4°C with gentle mixing.
6. Purify the biotinylated protein or molecule using dialysis or gel filtration. For small-molecule biotinylation where these separation methods may not be appropriate, other procedures may have to be developed, such as reverse-phase chromatography or organic precipitation techniques.

Aminooxy–PEG_n–Biotin Compounds

Aminooxy reagents are useful in forming covalent linkages with carbonyl compounds, such as aldehydes and ketones (see Section 1.2, this chapter, and Chapter 15, Section 2.4). Biotinylation reagents containing a PEG spacer with a terminal aminooxy group can be used to add a biotin affinity handle to molecules having these functionalities. For instance, glycoproteins and polysaccharides containing oxidizable diols can be reacted with sodium periodate to form aldehyde residues, which then can be biotinylated in a single step with an aminooxy–PEG_n–biotin reagent. In addition, carbohydrates and isolated glycans containing a reducing end may be specifically labeled with an aminooxy compound to form either cyclic or acyclic derivatives. The addition of a PEG spacer between the biotin end and the aminooxy end provides significant hydrophilicity and decreased nonspecific binding character to labeled molecules, thus increasing stability of biotinylated probes in solution and increasing the signal-to-noise of assays.

The reaction between an aminooxy compound and an aldehyde or ketone residue yields a stable oxime bond, which is a type of Schiff base but more stable than those formed between amines and carbonyls or hydrazides and carbonyls. This reaction is being used more often for bioconjugation purposes due to its chemoselectivity and its bioorthogonality (Chapter 17, Section 3) (Thumshirn *et al.*, 2003; Poethko *et al.*, 2004; Liu *et al.*, 2007; Colombo and Bianchi, 2010). The use of the aminooxy reactive group on biotinylation compounds is recent and includes various PEG-based cross-bridges, including a cleavable one containing an internal disulfide bond (Thermo Fisher and Quanta BioDesign).

Some of these aminooxy–PEG_n–biotin compounds are shown in Figure 18.28. They include different chain lengths of the internal PEG spacer, which provide options for spanning different molecular distances between the labeled molecule and the biotin end. With aliphatic biotinylation compounds, it has been shown to be beneficial to have an extended spacer arm to alleviate steric issues due to the docking of large streptavidin conjugates (see Chapter 11, Section 6.2). Aliphatic linker arms, however, create the potential for nonspecificity from hydrophobic interactions occurring with the cross-bridge. When using PEG-based spacers in biotin reagents, the increased length of a spacer can be more fully realized in aqueous solution due to the extreme hydrophilicity of the polyether arm. For this reason, glycoproteins and other molecules labeled with aminooxy-PEG_n–biotin compounds do not aggregate in solution like aliphatic biotin compounds typically do, especially at moderate to high biotinylation levels.

Aminooxy–PEG–biotin compounds can be used to biotinylate carbohydrates and glycoconjugates by an aniline-catalyzed reaction, which accelerates the formation of the oxime at slightly acid pH values (Cordes and Jencks, 1962; Dirksen *et al.*, 2006a,b; Dirksen and Dawson, 2008; Byeon *et al.*, 2010). The aniline first reacts with the aldehyde (or ketone) group to form an intermediate aryl imine, which then reacts more rapidly with the aminooxy group on the biotin reagent to form the oxime linkage (Kohler, 2009). Figure 18.29 illustrates this reaction in the biotinylation of an aldehyde-containing molecule with an aminooxy–PEG_4–biotin compound. The aldehyde-containing molecule may consist of a sugar, disaccharide, carbohydrate, or a glycan having an available reducing end or it may consist of a periodate-oxidized glycoprotein. For instance, polyclonal antibodies may be biotinylated at their associated carbohydrate after periodate oxidation (see Chapter 20, Section 1.3).

The following protocols describe the biotinylation of the reducing end of a carbohydrate using the aminooxy–PEG_4–biotin reagent or the biotinylation of a glycoprotein after periodate oxidation. Other aminooxy–PEG–biotin compounds may be substituted in these procedures with

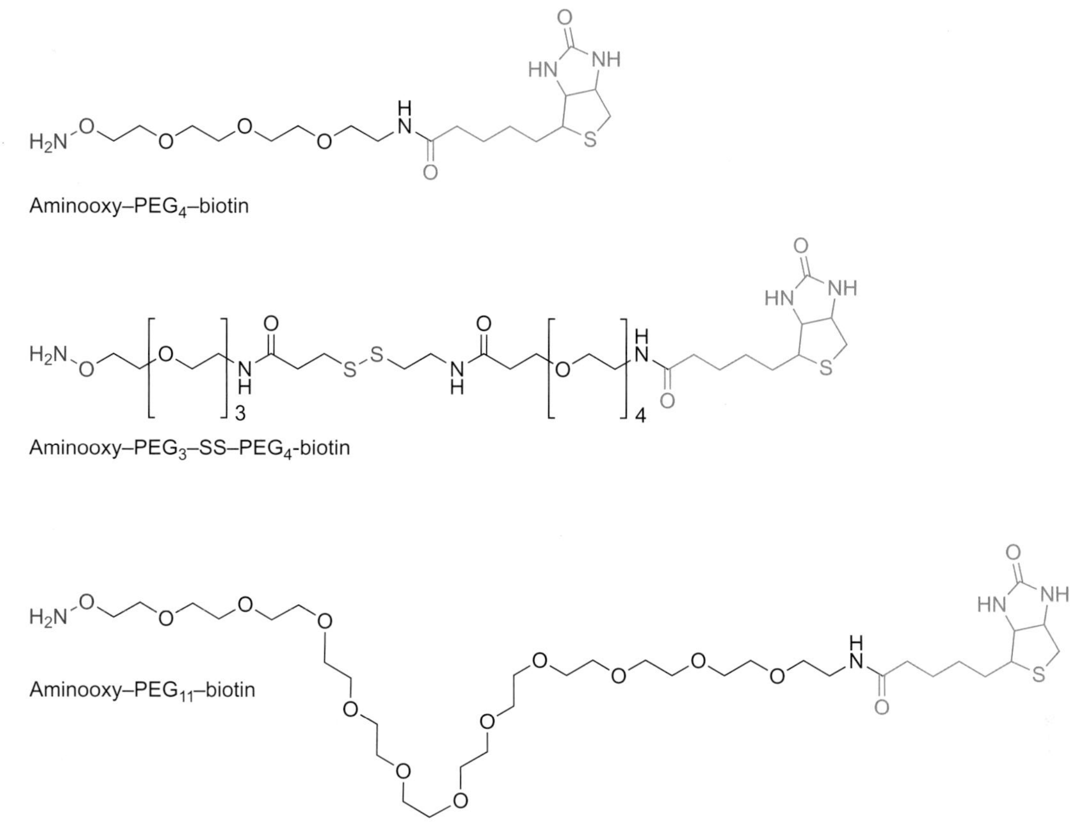

FIGURE 18.28 Aminooxy–PEG$_n$–biotin compounds.

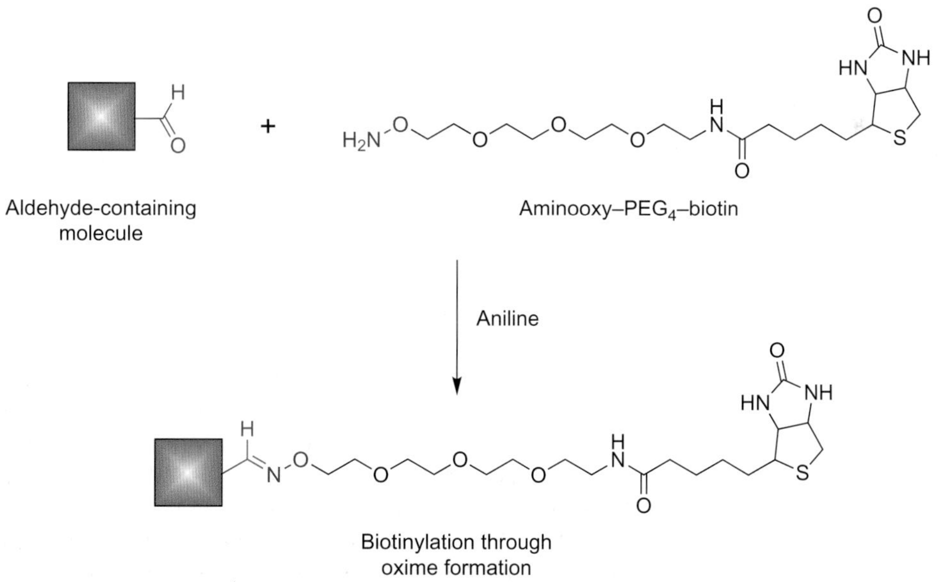

FIGURE 18.29 Biotinylation of an aldehyde-containing compound with aminooxy–PEG$_4$–biotin.

the appropriate changes in the amount of reagent added to a reaction due to differences in molecular weight. These protocols use the addition of aniline as catalyst to increase the reaction rates of oxime formation. For the biotinylation of a reducing carbohydrate, the protocol also results in a mixture of cyclic glycosyl bonds and acyclic oxime bonds to the reducing end of the carbohydrate (Peluso *et al.*, 2002).

(A) Protocol for Biotinylation of a Carbohydrate at its Reducing End

1. Dissolve a carbohydrate, polysaccharide, or glycan (having a reducing end available) at a concentration of at least 5–100 μ*M* in 1 ml of 0.1-*M* sodium acetate, 0.15-*M* NaCl, pH 4.5 (coupling buffer). Note: Higher concentrations of a reducing carbohydrate may be used, if soluble, as long as the total concentration does not exceed 0.1× the concentration of the biotinylation reagent used in the reaction (step 3). In addition, the buffer pH may be increased to the range of 5.5 to 6.5 and still obtain good yields of oxime bond formation.
2. In a fume hood, dissolve 50 mg of the aminooxy–PEG_4–biotin reagent (an entire vial; Thermo Fisher) in 460 μl of dry DMSO, DMF, or DMAC (makes a 250-m*M* solution). Note: PEG-based biotin reagents are often very hygroscopic and difficult to weigh out in small quantities. For this reason, it is typically easier to dissolve an entire vial of the reagent in a dry solvent rather than attempt to dispense it by weight.
3. Add 9.2 μl of the biotin solution from step 2 to the carbohydrate solution from step 1 with stirring to create approximately a 5-m*M* final concentration of aminooxy–PEG_4–biotin compound in the reaction medium.
4. In a fume hood, add 9 μl of aniline catalyst to the reaction mixture with stirring. This results in approximately a 0.1-*M* aniline solution in the reaction medium.
5. Seal the tube and continue the reaction for at least 4 h at room temperature and with constant mixing by end-over-end rocking.
6. Purify the biotinylated carbohydrate by dialysis or size exclusion chromatography using an exclusion limit for the membrane or for the chromatography support that excludes the molecular weight of the carbohydrate but allows the excess biotin compound and aniline catalyst to get into the pores and be removed. For the biotinylation of low-molecular-weight carbohydrates, this purification step may have to be altered to include separations such as HPLC chromatography on a reverse-phase column or differential precipitation steps to isolate the final biotinylated complex from excess reactants.

(B) Protocol for Biotinylation of a Glycoprotein after Periodate Oxidation

1. Dissolve a glycoprotein to be oxidized at a concentration of 2 mg/ml in 1 ml of 0.1-*M* sodium acetate, pH 5.5 (coupling buffer). Chill on ice.
2. Dissolve sodium periodate in 2 ml of coupling buffer at a concentration of 20-m*M*. Chill the solution on ice and protect from light.
3. To oxidize all the diols within the glycan portion of the glycoprotein to aldehydes, add 1 ml of the cold sodium periodate solution to the glycoprotein with mixing. To oxidize only the sialic acid groups within the carbohydrate add 50 μl of the sodium periodate solution to the glycoprotein solution with mixing.
4. Maintain the solution on ice and react for 30 min with periodic mixing and while protecting the solution from light.
5. Remove excess periodate from the oxidized glycoprotein by size exclusion chromatography on a column of Sephadex G-25 or the equivalent while using coupling buffer as the mobile phase. Concentrate the purified protein to approximately 1 ml using a centrifugal concentrator with a molecular weight exclusion limit of less than that of the glycoprotein (e.g., 10 kDa for most proteins).
6. In a fume hood, dissolve the aminooxy–PEG_4–biotin reagent at a concentration of 50-m*M* in DMSO, DMF, or DMAC. It may be simpler to dissolve an entire bottle of the biotin compound rather than weigh out a fraction of it, because of the hygroscopic or sticky nature of PEG-based substances.
7. Add 100 μl of the biotinylation reagent solution to the glycoprotein solution with mixing. This makes a final concentration of about 5-m*M* aminooxy–PEG_4–biotin in the reaction medium.
8. In a fume hood, add 9 μl of aniline to the reaction mixture and stir well. This makes approximately a 0.1-*M* solution with respect to the aniline catalyst.
9. React for 2 h at room temperature with periodic mixing.
10. Purify the biotinylated glycoprotein by dialysis or size exclusion chromatography to remove excess reactants and reaction by products.

Biotin–PEG_n–Azide and Biotin–PEG_3–Phosphine Compounds

Biotinylation reagents containing a PEG spacer arm and a chemoselective reacting group at the other end now are available to use in bioorthogonal reaction methods with alkyne-containing molecules or azide-containing molecules. The click chemistry reaction, which involves a copper-catalyzed cycloaddition of an azide group with an alkyne triple bond, has been used extensively to produce bioconjugates (Chapter 17, Section 5). In the presence of Cu–I, an alkyne will selectively couple with an azido group to give a triazole ring with very high yield and rapid reaction kinetics. This reaction is highly specific even in the presence of a wide variety of biomolecule functional groups within complex sample extracts. PEG-based biotinylation agents can label alkyne-modified substances within cells or homogenates to give biotin-modified targets that can be detected using streptavidin conjugates (Chapter 11).

In addition, PEG-based biotin reagents are also available that contain a reactive phosphine group on the other end, which can participate in the chemoselective

Staudinger ligation reaction (Chapter 17, Section 6). These compounds react in a bioorthogonal manner with azide-containing molecules to give an amide bonded conjugate. Much like the copper-catalyzed click reaction, biotinylation using biotin–PEG_3–phosphine is very specific for coupling to azido tags, even in the presence of other biomolecules; however, unlike the click reaction, the Staudinger ligation process occurs without the addition of a highly oxidizing substance (Cu–I). Thus, biotin–PEG_3–phosphine can be used to label live cells that have expressed azido-modified components, such as proteins, glycans, or lipids.

Figure 18.30 illustrates some of these biotinylation compounds that are useful in a broad range of chemoselective labeling processes. The central PEG spacer arms of these reagents aid in water solubility and decrease the tendency of the biotin tags to bind nonspecifically to hydrophobic pockets on biomolecules, which would definitely occur with the use of aliphatic cross-bridges. Even the properties of the phosphine biotin compounds are dramatically improved by the presence of the PEG spacer arms.

Figure 18.31 shows the reactions of the biotin–PEG_7–azide compound with an alkyne-labeled molecule Figure 18.32 shows the reaction of the biotin–PEG_3–phosphine compound with an azido-modified molecule. The reaction rate of the Staudinger ligation process is slower than the copper-catalyzed click reaction, but the advantage of being able to label and image in real time living cells is a major advantage of the phosphine/azide reaction.

The following protocol describes the use of biotin–PEG_3–phosphine to label azide-containing molecules by use of the Staudinger ligation reaction. The conjugation yields an amide bond to the carbonyl-containing phenyl ring on the triphenylphosphine group at the end of the biotin tag. This general reaction scheme can be used to label azido-proteins in solution or glycoproteins in cells that have been expressed with azido-sugar-containing glycans. Note that reducing agents should be avoided in the presence of azide-containing molecules, because the azido group might be degraded to form an amine with release of nitrogen gas, thus eliminating the activity of the azide to participate in the Staudinger ligation reaction.

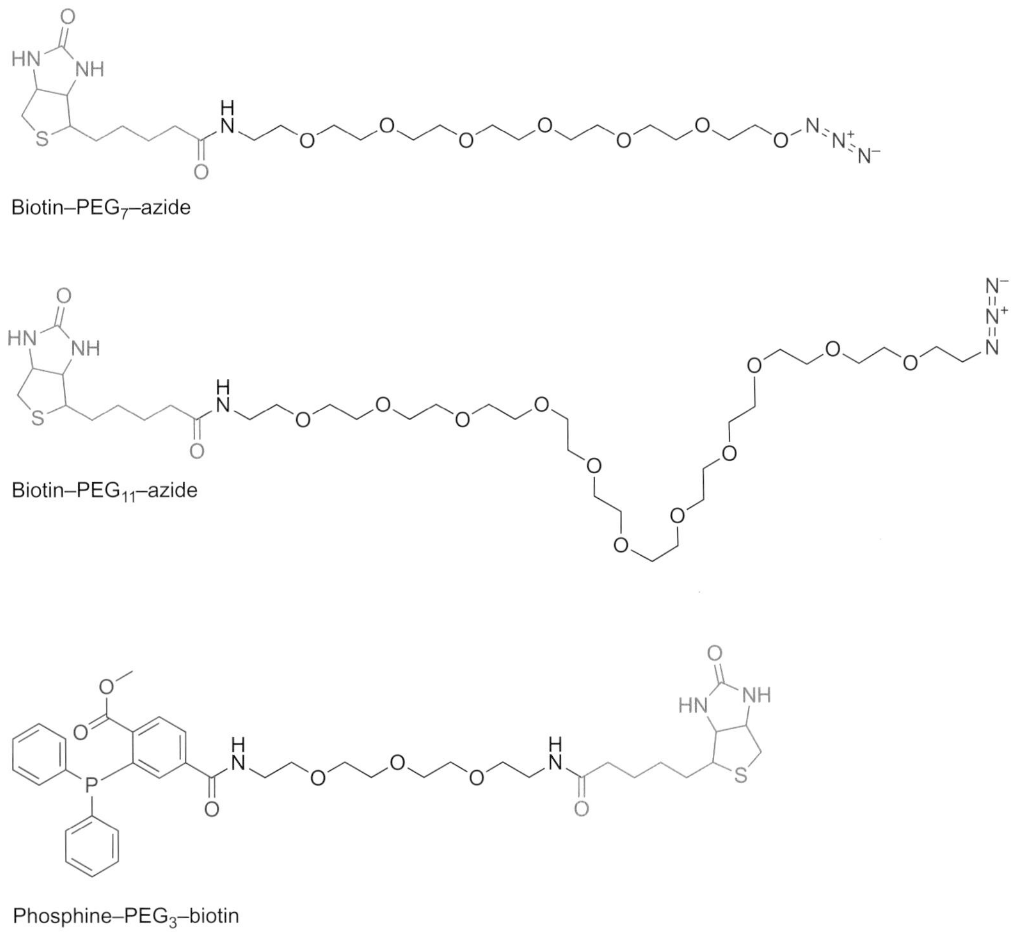

FIGURE 18.30 PEG-based biotin compounds that can be used in chemoselective ligation reactions. The first two azido compounds can be used in copper-catalyzed or copper-free click reactions with alkynes, while the third reagent contains a reactive phosphine group, which can be used in a Staudinger ligation reaction with an azido compound.

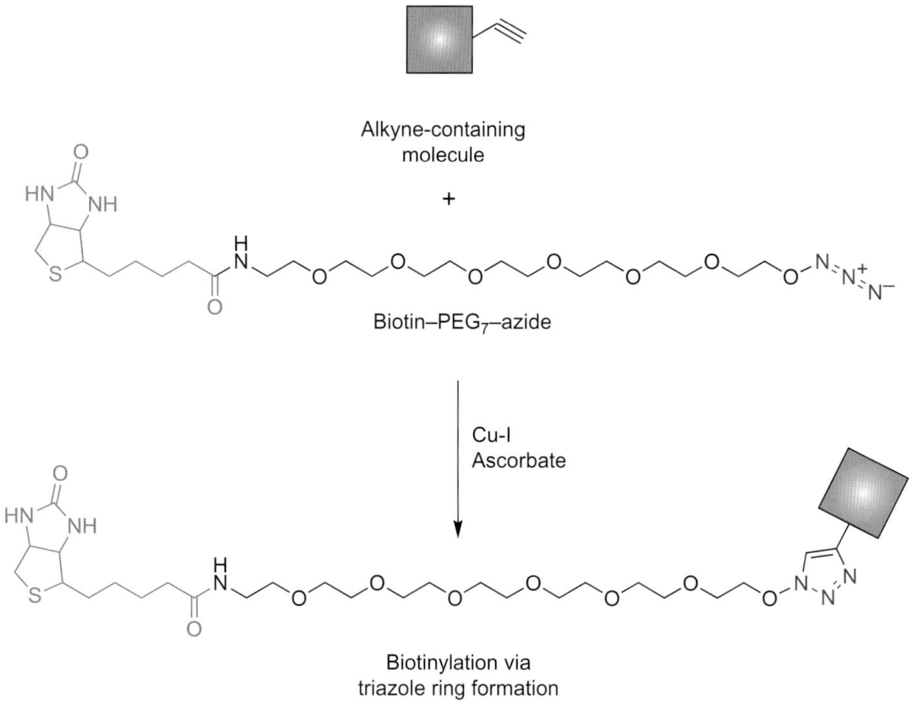

FIGURE 18.31 An alkyne-containing molecule can be biotinylated using an azide-containing PEG–biotin reagent to form a triazole bond.

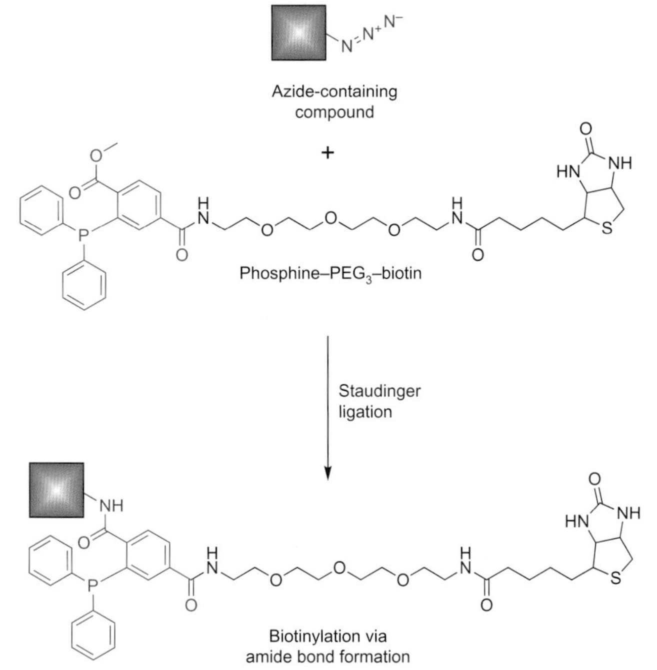

FIGURE 18.32 An azide-containing compound can be biotinylated with a phosphine-containing PEG–biotin reagent through amide bond formation.

For a suggested protocol on the use of the biotin–PEG$_n$–azide compounds to biotinylate alkyne-containing molecules, see the copper-catalyzed click chemistry reaction as described in the previous section entitled NHS–PEG$_n$–Azide/Alkyne Compounds for Chemoselective Ligation.

Protocol

1. Dissolve an azide-containing protein or other molecule in 50-m*M* sodium phosphate, pH 7.2, at a concentration of approximately 2–5 mg/ml. Notes: Higher protein concentrations may be used in this protocol provided the biotinylation reagent is adjusted in concentration as recommended in step 3. For probing living cells with the biotin–PEG$_3$–phosphine reagent, the use of physiological buffers compatible with cell viability may be used, provided that no reducing agents are present in the solution (e.g., 10-m*M* sodium phosphate, 0.15-*M* NaCl, pH 7.2).
2. Dissolve biotin–PEG$_3$–phosphine in anhydrous DMSO, DMF, or DMAC to make a final concentration of 10-m*M*. The stock solution may be stored frozen at −20°C for 6 months with no loss of activity.
3. Add an aliquot of the biotin–PEG$_3$–phosphine solution to the azido–protein solution to obtain a 2-fold molar excess of the biotinylation agent over the concentration of protein. For use of higher protein concentrations, the amount of biotin compound added to the reaction may be reduced to a 10-fold excess without affecting the yield of biotinylation.
4. Incubate the reaction at 37°C for 2–4 h or at room temperature for overnight (at least 16 h).
5. Remove excess reactant by size exclusion chromatography using a desalting column or by dialysis. If cells are being labeled, wash the cells with buffer at least several times to remove unreacted biotin compound.

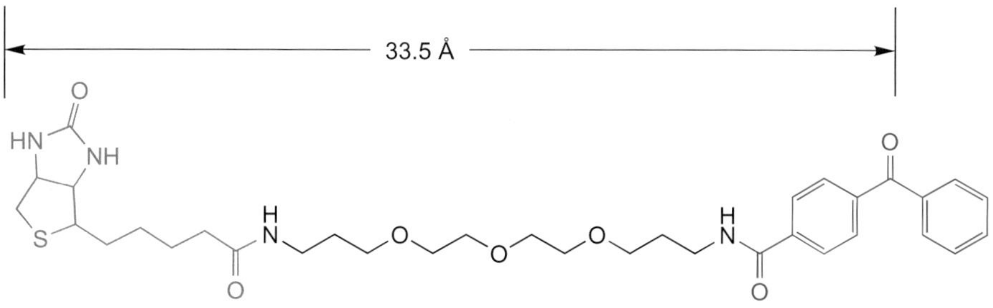

FIGURE 18.33 Biotin–PEG_3–benzophenone is a water soluble photoreactive biotinylation reagent that can be used to add a biotin group to surfaces or molecules containing no easily derivatized functional groups.

Biotin–PEG_3–Benzophenone

Biotin–PEG_3–benzophenone is a biotinylation reagent with a hydrophilic spacer containing three ethylene oxide units and a photoreactive group at its end (Quanta BioDesign) (Figure 18.33). The benzophenone is activated by UV light to an extremely reactive triplet-state ketone, which can insert into CH, NH, and other structures, resulting in a covalent bond (Chapter 3, Section 7.2). The reaction is one of the most efficient photoreactive conjugation mechanisms available (Campbell and Gioannini, 1979). Thus, this reagent provides a method of adding a biotin group to molecules that do not contain typical functionalities useful for bioconjugation. This may include polymeric surfaces or organic molecules lacking reactive targets.

The presence of the PEG_3 spacer in this compound provides water solubility to the biotin arm, whereas the benzophenone group should associate with more hydrophobic regions or surfaces, which may be ideal for the biotinylation photoreaction. The reagent can be used by dissolving it in an aqueous buffer suitable for use with whatever substance is to be biotinylated. After mixing this solution with the target molecule or surface, exposure to UV light will initiate the conjugation reaction. Unlike other photoreactive groups, a benzophenone does not undergo decomposition to an inactive form if it does not couple to target molecules. Instead, it degrades from the photo-excited state back to its initial state, so it can be once again photolyzed to an active state. This process increases the likelihood that the benzophenone will couple to a target molecule during the photoreaction. See Chapter 6, Section 4.3, for an illustration of the benzophenone-coupling reaction.

1.4. Discrete PEG Modification Reagents

Large-polymer PEG reagents having molecular weights > 2000 Daltons have been used for over 20 years as modification agents for biological molecules (Section 2.1). These compounds are often used with a reactive group on one end and a blocked hydroxyl group on the other end (e.g., as the methyl ether). In addition, large, branched PEG molecules have been created to add more bulk or exclusion volume at a modification site, thus increasing the protective effect of the PEG molecule toward the biological molecule.

Discrete PEG compounds have also been developed in various reactive forms with methyl ether blocking groups on the terminal end (Thermo Fisher, Quanta BioDesign). Unlike the original PEG polymer reagents that display polydispersity, these PEG compounds are pure and consist of only one chain length per reagent type. The chain lengths in discrete PEG modifiers include, for example, polyethylene oxide repeat units of 3, 4, 8, 12, and 24. NHS–$mPEG_n$ modification reagents can be used directly to couple with amine-containing molecules or proteins through amide bond formation. This reaction occurs in aqueous buffers at physiological pH or slightly alkaline pH conditions. Conversely, maleimide–$mPEG_n$ reagents are designed to conjugate with thiol-containing molecules, and these may be used to target reduced disulfide bonds in proteins or thiols created on molecules using a thiolation reagent.

In addition, branched-chain compounds have been developed consisting of a functional group or a reactive group followed by a PEG_4 chain, which then leads to three branches each having an $mPEG_{12}$ arm on them. Such compounds are expected to provide large exclusion volumes in aqueous solution to surround, protect, and solubilize modified molecules.

Another type of PEG modification reagent that has been developed contains a functional group on each end that can be used for conjugation purposes and which can be used to build structures on solid supports, surfaces, or on other molecules. Some of these reagents have been developed to contain an amine group on one end and a carboxylate on the other end. Such PEG-based amino acids can be used as spacer arms to mask surfaces or provide highly hydrophilic tethers for the attachment of affinity ligands. A thiol–PEG_n–carboxylate, for instance, can be used to modify metal particles or surfaces through dative binding of the thiol to the metal and then create PEG–carboxylic acids for further conjugation.

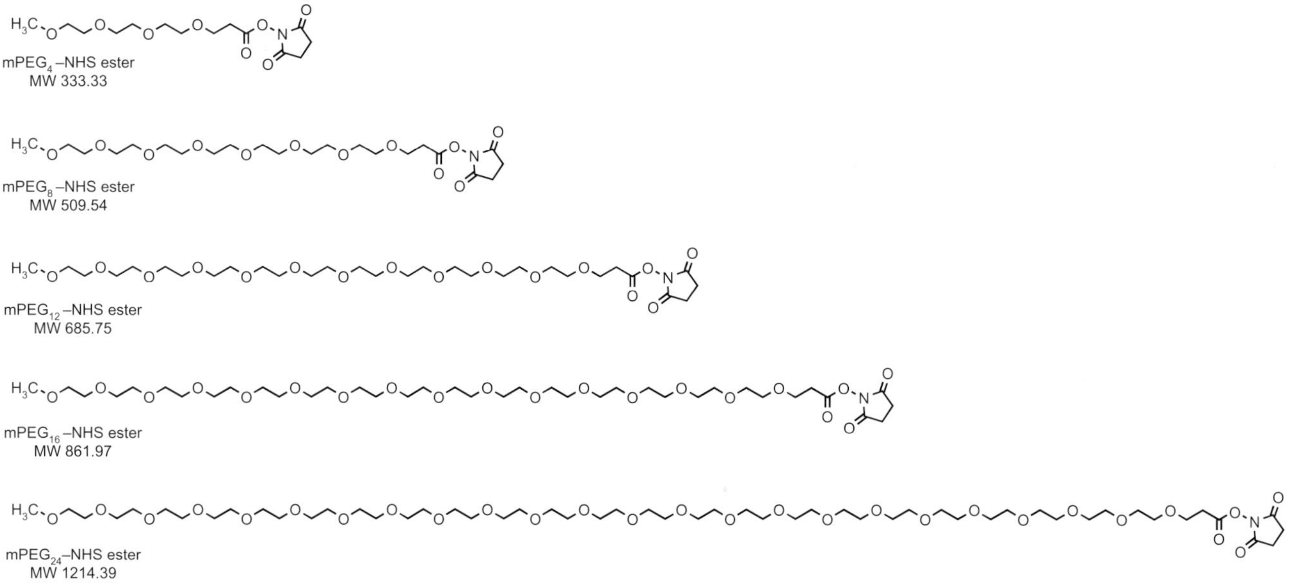

FIGURE 18.34 Discrete PEGylation reagents are available to provide a range of different chain lengths for adding mPEG modification arms to biomolecules. They can also be used to add water soluble mPEG groups to organic molecules that are normally not very soluble in aqueous solution. The NHS ester end of the mPEG compounds reacts with amine-containing molecules to form amide bonds, leaving the mPEG chain to interact with the aqueous environment.

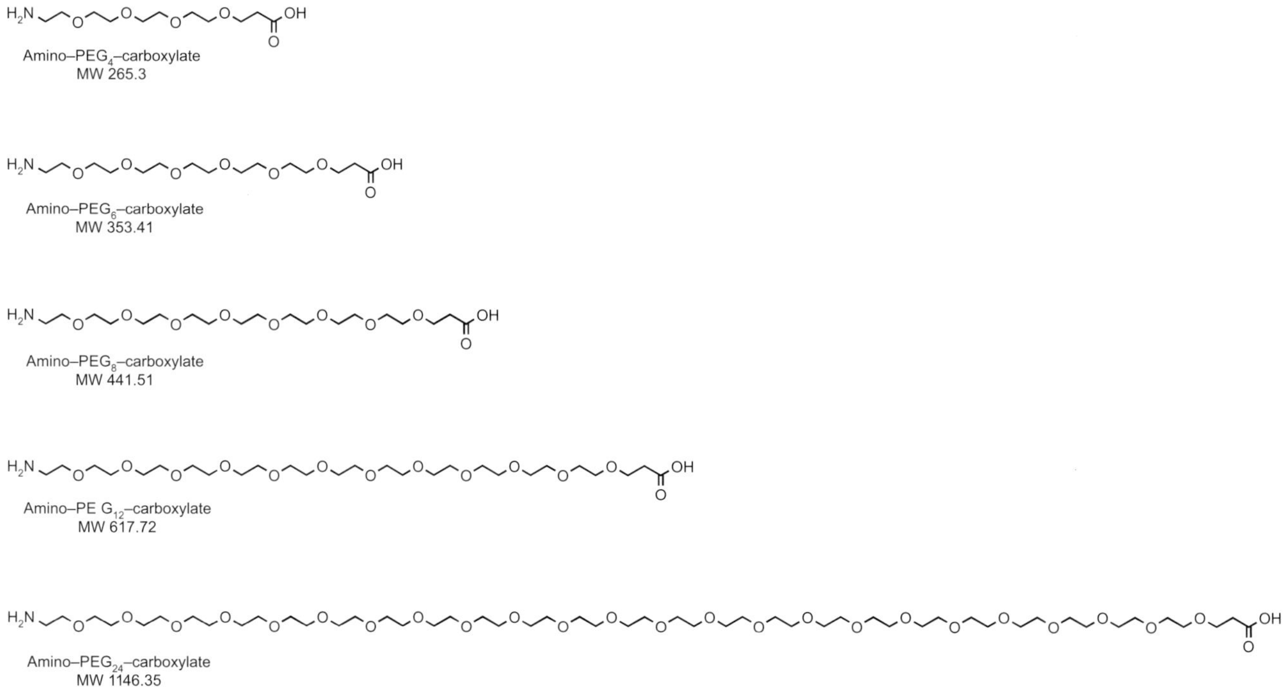

FIGURE 18.35 Amino–PEG$_n$–carboxylate compounds contain a primary amine on one end and a carboxylate group on the other end. They can be used to add water soluble spacer arms to molecules or surfaces. Using an amine-reactive group, the amino–PEG$_n$–carboxylate compound can be coupled *via* an amide bond, thus leaving the carboxylate end free for further conjugation reactions. Avoid the use of single-step EDC conjugation reactions, as this will polymerize the amino–PEG$_n$–carboxylate by reacting with both ends.

Figures 18.34, 18.35, and 18.36 illustrate these PEG-based modification reagents. The methods for their use follow the same general protocol guidelines as discussed in previous sections of this chapter for the corresponding reactive group or functional group. When modifying biomolecules with mPEG-based reagents, a series of modification levels should be investigated to determine the optimal performance in an intended application.

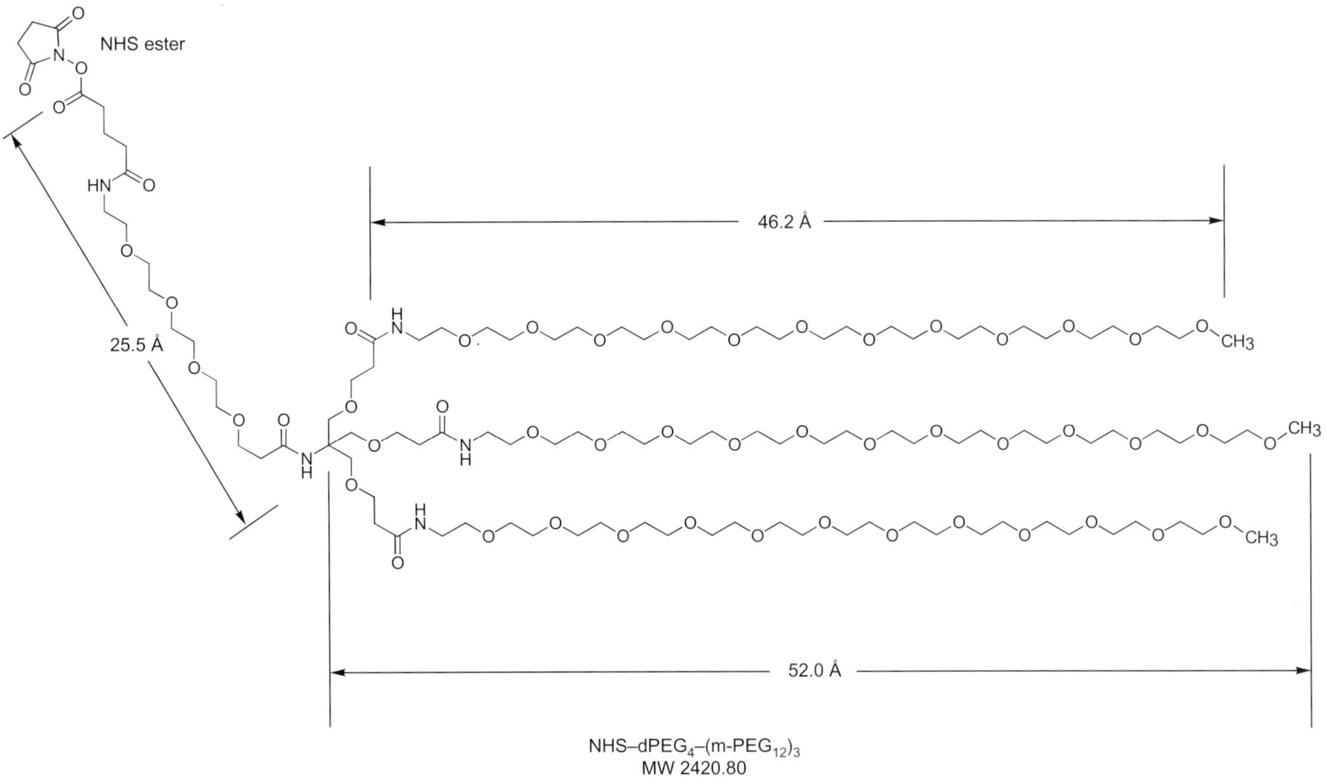

FIGURE 18.36 The branched PEGylation compound NHS–dPEG$_4$–(mPEG$_{12}$)$_3$ contains three mPEG arms, which provide an increased sphere of hydration around modified molecules compared to straight-chain PEGylation compounds.

For surface or particle modification, reference should be made to Chapter 14, especially the section on covalent coupling to particles.

2. POLYMER MODIFICATION REAGENTS

Modification or attachment of proteins or other molecules with synthetic polymers can provide many benefits for both *in vivo* and *in vitro* applications. Covalent coupling of polymers to large macromolecules can alter their surface and solubility properties, creating increased water solubility or even organic solvent solubility for molecules normally sparingly miscible in such environments. Polymer modification of foreign molecules can provide increased biocompatibility, reducing the immune response, increasing *in vivo* stability, and delaying clearance by the reticuloendothelial system. Modification of enzymes with polymers can dramatically enhance their stability in solution. Polymer attachment can provide cryoprotection for proteins sensitive to freezing. Polymers with multivalent reactive sites can be used to couple numerous small molecules for creating pharmacologically active agents that possess long half-lives in biological systems. Similar complexes can be formed to create highly potent immunogens consisting of hapten–polymer conjugates for induction of an antibody response toward the hapten. Polymer modification of surfaces can effectively mask the intrinsic character of the surface and thus prevent nonspecific protein adsorption. Finally, multifunctional polymers can serve as extended crosslinking agents for the conjugation of more than one molecule of one protein to multiple numbers of a second molecule, creating large complexes with increased sensitivity or activity in detecting or acting upon target analytes.

Many polymers have been studied for their usefulness in producing pharmacologically active complexes with proteins or drugs. Synthetic and natural polymers such as polysaccharides, poly(L-lysine) and other poly(amino acids), poly(vinyl alcohols), polyvinylpyrrolidinones, poly(acrylic acid) derivatives, various polyurethanes, and polyphosphazenes have been coupled with a diversity of substances to explore their properties (Duncan and Kopecek, 1984; Braatz *et al.*, 1993; Canal *et al.*, 2011; Gada *et al.*, 2012; Tong and Cheng, 2012). Copolymer preparations of two monomers have also been tried (Nathan *et al.*, 1993). See Chapter 1 for an in-depth discussion of polymeric scaffolds used in therapeutic bioconjugate applications.

The two polymers most often used in these applications are dextran and polyethylene glycol (PEG).

Both polymers consist of repeating units of a single monomer—glucose in the case of dextran and an ethylene oxide basic unit in the case of PEG. The polymers may be composed of linear strands or branched constructs. An additional similarity is that both types possess hydroxyl and ether linkages, lending significant hydrophilicity and water solubility to the molecules. Dextran and PEG can be activated through their hydroxyl groups by a number of chemical methods to allow efficient coupling of other molecules. Dextran can be activated at multiple sites throughout its chain, since each monomer contains hydroxyl residues. PEG, by contrast, only has hydroxyls at the termini of each polymer strand. Derivatives of both polymers are commercially available that incorporate other functional groups, including amines and carboxylates.

The following sections discuss the major properties and conjugation chemistries associated with the use of these polymers in modifying or conjugating proteins and other molecules.

2.1. Polymeric PEG Reagents

Since the first report by Abuchowski and co-workers in 1977 concerning the alteration of the immunological properties of bovine serum albumin (BSA) that had been modified with PEG, the interest in polymer modification of biological molecules has grown tremendously (Roberts *et al.*, 2012; Milla *et al.*, 2012). PEG coupled to other molecules can be used for altering solubility characteristics in aqueous or organic solvents (Inada *et al.*, 1986; Milla *et al.*, 2012 1986), for modulation of the immune response (Delgado *et al.*, 1992), to increase the stability of proteins and peptides in solution (Berger and Pizzo, 1988; Jain and Ashbaugh, 2011), to enhance the half-life of substances *in vivo* (Knauf *et al.*, 1988; Gunasekaran *et al.*, 2011), to aid in penetrating cell membranes, to alter pharmacological properties (Dunn and Ottenbrite, 1991; Liu, 2011), to increase biocompatibility, especially toward implanted foreign substances, and to prevent protein adsorption to surfaces (Chen *et al.*, 2009).

PEG consists of repeating units of ethylene oxide which terminate in hydroxyl groups on both ends of a linear chain. Some constructs have branches that have multiple linear strands emanating from the branch points. PEG is made from the anionic polymerization of ethylene oxide, resulting in the formation of polymer strands of various potential molecular weights, depending on the polymerization conditions. Thus, all polymeric PEGs are polydisperse and exist as distribution of multiple lengths and molecular weights. Most forms of PEG useful in bioconjugate applications have molecular weights less than about 20,000 and are soluble both in aqueous solution and in many organic solvents.

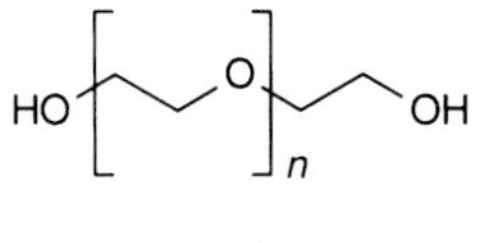

PEG
Poly(ethylene glycol)

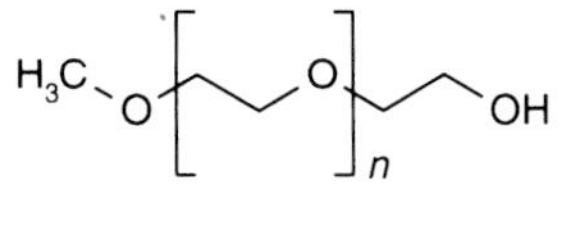

mPEG
Monomethoxy-
Poly(ethylene glycol)

Unlike the PEG molecules formed from anionic polymerization techniques, highly discrete forms of the polymer also exist, made by controlled addition of small PEG units to create chains of exact molecular size. These discrete PEGs have a single molecular weight and do not display the polydispersity of the traditional PEG polymers. See Section 1 for a complete discussion of discrete PEG-based reagents and their applications.

Since the polymer backbone of PEG is not of biological origin, it is not readily degraded by mammalian enzymes (although some bacterial enzymes will break it down). This property results in only slow degradation of the polymer when used *in vivo*, and this characteristic also causes an extension of the half-life of modified substances. PEG modification serves to mask any molecule that it is coupled to—the "PEGylated" molecule often being protected from immediate breakdown or from being complexed and inactivated by immunoglobulins in the bloodstream.

PEG's properties in solution are especially unusual, frequently displaying amphiphilic tendencies, having the ability to both solubilize in aqueous layers and in hydrophobic membranes or organic phases. The partitioning quality of PEG across membranes is important in aiding the formation of hybridomas in the production monoclonal antibodies (Goding, 1986b). The partitioning characteristics of PEG also create the ability to use it in aqueous two-phase systems for the purification of biological molecules (Johansson, 1992).

PEG in solution is a highly mobile molecule that creates a large exclusion volume for its molecular weight, much larger in fact than proteins of comparable size. Whether in solution or attached to other insoluble supports or surfaces, PEG has a tendency to exclude other polymers. This property forms a protein-rejecting region that is effective in preventing nonspecific protein binding (Bergstrom *et al.*, 1992). Conjugation with PEG can create the same exclusion effects surrounding a macromolecule, even preventing interaction between a ligand and its target (Klibanov *et al.*, 1991), an enzyme and its substrate (Berger and Pizzo, 1988), or the immune system and a foreign substance (Davis *et al.*, 1979). Thus, PEG-modified molecules display low immunogenicity, have good resistance to proteolytic digestion, and survive in the bloodstream for extended periods (Abuchowski *et al.*, 1977a; Dreborg and Akerblom, 1990; Roberts *et al.*, 2012).

PEG can be conjugated to other molecules through its two hydroxyl groups at the ends of each linear chain. This process is typically done by the creation of a reactive electrophilic intermediate that is capable of spontaneously coupling to nucleophilic residues on a second molecule. To prevent the potential for crosslinking when using a bifunctional polymer, monofunctional PEG polymers can be used which contain one end of each chain blocked with a methyl ether group. Monomethoxypolyethylene glycol (mPEG) contains only one hydroxyl group per linear chain, thus limiting activation and coupling to one site and preventing the crosslinking and polymerization of modified molecules. The mPEG derivative also stabilizes the blocked end to oxidation or degradation in solution.

Trichloro-s-Triazine Activation and Coupling

The most common activation methods for PEG create amine-reactive derivatives that can form amide or secondary amine linkages with proteins and other amine-containing molecules. The oldest method of PEG activation is through the use of trichloro-*s*-triazine (TsT; cyanuric chloride) (Abuchowski *et al.*, 1977). TsT is a symmetrical heterocyclic compound containing three reactive chlorines. This reagent and its derivatives are extensively used in industrial applications to form strong covalent bonds between dye molecules and fabrics. The compound has also been used to activate affinity chromatography supports for the coupling of amine-containing ligands (Finlay *et al.*, 1978) (see Chapter 15, Section 2.1). Reaction of TsT with PEG results in the formation of an activated derivative with an ether bond to the hydroxyl group of the polymer. If mPEG is used, TsT activation will be restricted to the one free hydroxyl, thus forming a monovalent intermediate that can be coupled to proteins without polymerization (Figure 18.37).

The three reactive chlorines on TsT have dramatically different reactivities toward nucleophiles in aqueous solution. The first chlorine is reactive toward hydroxyls as well as primary and secondary amine groups at 4°C and a pH of 9.0 (Abuchowski, 1977a; Mumtaz and Bachhawat, 1991). Once the first chlorine is coupled, the second one requires at least room temperature conditions at the same pH to react efficiently. If two chlorines are conjugated to nucleophilic groups, the third is even more difficult to couple, requiring at least 80°C at alkaline pH. After activation of mPEG with TsT, it is therefore, for all practical purposes, only possible to couple one additional component to the triazine ring.

TsT activation provides a simple route to an amine-reactive PEG derivative and it has been used extensively as an activation method for modifying proteins (Wieder *et al.*, 1979; Zalipsky and Lee, 1992; Gotoh *et al.*, 1993). The modification of primary amine-containing molecules such as proteins is pH dependent. At physiological pH values, the reaction will proceed slower than in a more alkaline pH environment. Optimal derivatization efficiency is reached at conditions equal to or above pH 9.0. However, TsT reactivity is not exclusive toward amines. TsT–mPEG modification of proteins can result in modifying other nucleophilic groups such as sulfhydryls and the phenolate ring of tyrosine. In addition, there is potential for toxicity associated with TsT and its derivatives—an especially important consideration for *in vivo* use.

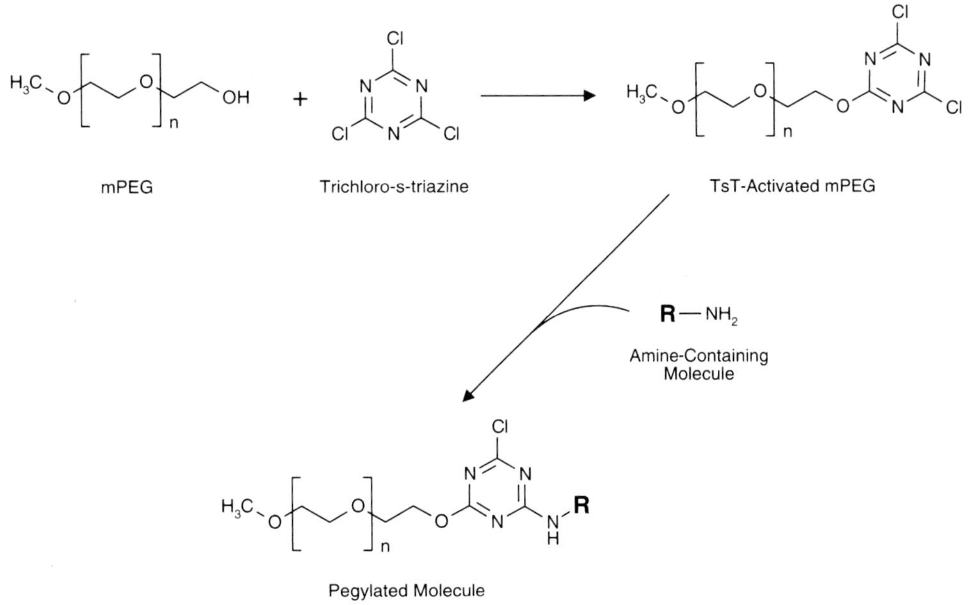

FIGURE 18.37 mPEG polymers may be activated by trichloro-*s*-triazine for the modification of amine-containing molecules.

The following protocol for mPEG activation using TsT and its coupling to proteins is based on the protocols of Abuchowski *et al.* (1977b) and Gotoh *et al.* (1993).

Protocol for the Activation of mPEG with TsT

Note: All operations should be carried out in a fume hood. Wear proper personal protective equipment and dispose of hazardous waste according to EPA guidelines.

1. Dissolve 5.5 g of TsT in 400 ml of anhydrous benzene which contains 10 g of anhydrous sodium carbonate.
2. Add to the TsT solution, 50 g of mPEG-5000 (monomethoxypolyethylene glycol having a molecular weight of 5000). Mix well to dissolve.
3. React overnight at room temperature with stirring.
4. Filter the solution through a glass-fiber filter pad and slowly add, with stirring, 600 ml of petroleum ether (bp 35–60°C).
5. Collect the precipitated product by filtration and redissolve it in 400 ml of benzene. Repeat steps 4 and 5 several times to ensure complete removal of unreacted TsT. The residual TsT may be detected by HPLC using a 250 × 3.2-mm LiChrosorb (5 μm particle size) column from E. Merck. The separation is performed using a mobile phase of hexane, and peaks are detected with a UV detector.
6. Remove excess solvents by rotary evaporation. The TsT–mPEG should be used immediately or stored in anhydrous conditions at 4°C.

Protocol for Coupling of TsT–mPEG to Proteins

1. Dissolve the protein to be modified with TsT–mPEG in ice-cold 0.1-*M* sodium borate, pH 9.4, at a concentration of 2 to 10 mg/ml. Other buffers at lower pH values (down to pH 7.2) can be used and still obtain modification, but the yield will be less. Avoid amine-containing buffers such as Tris or the presence of sulfhydryl-containing compounds, such as disulfide reductants.
2. Slowly add TsT–mPEG to the protein solution at a level of at least a 5-fold molar excess over the desired modification level. For example, Gotoh *et al.* (1993) added 100 mg of TsT–mPEG-5000 to 19 mg of protein dissolved in 6 ml of buffer. Add the polymer over a period of about 15 min with stirring at 4°C.
3. React for 1 h at 4°C.
4. Remove excess TsT–mPEG by dialysis or gel filtration using a column of Sephacryl S-300.

NHS Ester and NHS Carbonate Activation and Coupling

Carboxylate groups activated with *N*-hydroxysuccinimide (NHS) esters are highly reactive toward amine nucleophiles. In the mid-1970s, NHS esters were introduced as reactive ends of crosslinking reagents (Bragg and Hou, 1975; Lomant and Fairbanks, 1976). Their excellent reactivity at physiological pH quickly established NHS esters as the major amine-coupling chemistry in bioconjugate chemistry.

NHS ester-containing compounds react with nucleophiles to release the NHS leaving group and form an acylated product (Chapter 3, Section 1.4). The reaction of such esters with sulfhydryl or hydroxyl groups is possible, but does not yield stable conjugates, forming thioesters or ester linkages. Both of these bonds typically hydrolyze in aqueous environments or can undergo transesterification reactions. Histidine side chain nitrogens of the imidazolyl ring may also be acylated with an NHS ester reagent, but they too hydrolyze rapidly (Cuatrecasas and Parikh, 1972). Reaction with primary and secondary amines, however, creates stable amide and secondary amide linkages, respectively, that don't readily break down. In protein molecules, NHS ester groups primarily react with the α-amines at the N-terminals and the ε-amines of lysine side chains, due to their relative abundance.

PEG contains no carboxylate groups in its native state, but can be modified to possess them by reaction with anhydride compounds. Either PEG or mPEG may be acylated with anhydrides to yield ester derivatives terminating in free carboxylate groups. Modification of PEG with succinic anhydride or glutaric anhydride gives *bis*-modified products having carboxylates at both ends. Modification of mPEG yields the mono-substituted derivative containing a single carboxylate. Creation of the succinimidyl succinate and succinimidyl glutarate derivative of PEG was described by Abuchowski *et al.* (1984). A method for the succinylation of mPEG can be found later in Section 2.1. Subsequent formation of the NHS ester derivatives of these acylated PEG compounds produce highly reactive polymers that can be used to modify amine-containing molecules under mild conditions and with excellent yields (Figures 18.38 and 18.39). The main deficiency of the succinimidyl succinate or succinimidyl glutarate activation procedures is the potential for hydrolysis of the ester bond formed by acylation of the hydroxyl end groups of mPEG.

A modification of the anhydride-acylation route to obtaining reactive NHS ester–PEG compounds was introduced by Zalipsky *et al.* (1991, 1992). In this approach, the terminal hydroxyl group of mPEG is treated with phosgene to give a reactive intermediate, an mPEG–chloroformate compound. Next, the addition of *N*-hydroxysuccinimide (NHS) gives the succinimidyl carbonate derivative (Figure 18.40). Nucleophiles, such as the primary amino groups of proteins, can react with the succinimidyl carbonate to give stable carbamate (aliphatic urethane) bonds (Figure 18.41). The linkage is identical to that obtained through

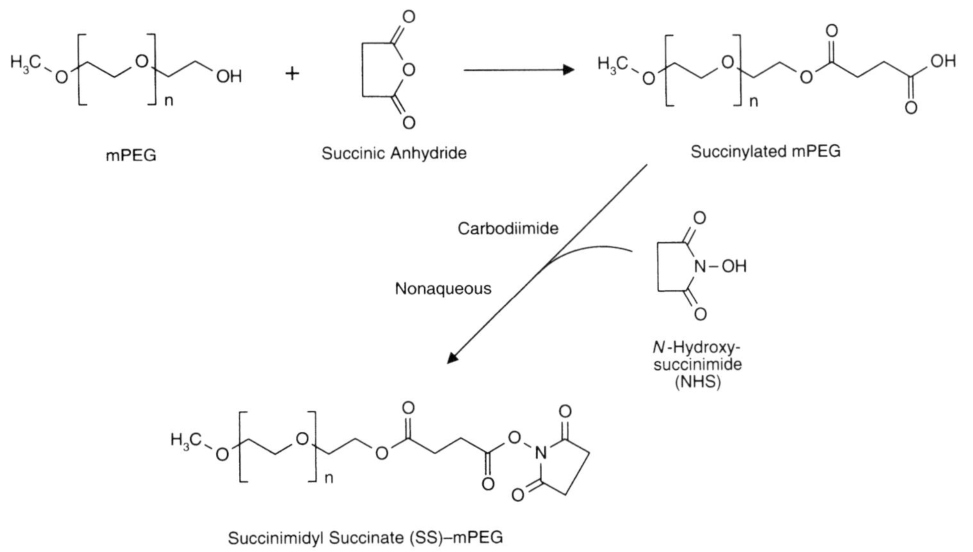

FIGURE 18.38 mPEG may be derivatized with succinic anhydride to produce a carboxylate end. A reactive NHS ester can be formed from this derivative by use of a carbodiimide-mediated reaction under nonaqueous conditions. The succinimidyl succinate–mPEG is highly reactive toward amine nucleophiles.

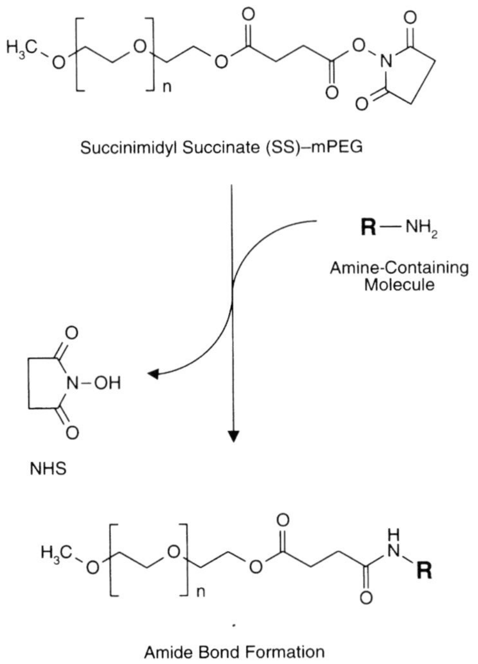

FIGURE 18.39 Succinimidyl succinate–mPEG may be used to modify amine-containing molecules to form amide bond derivatives. The ester bond of the succinylated mPEG, however, is subject to hydrolysis.

N,N′-carbonyldiimidazole (CDI) activation of hydroxyl groups with subsequent coupling of amines (Chapter 3, Section 3; Chapter 15, Section 2.1; and this chapter, Section 1.4). However, the reactivity of the succinimidyl carbonate is much greater than that of the imidazole carbamate formed as the active species in CDI activation.

Unlike the succinimidyl succinate or succinimidyl glutarate activation methods, succinimidyl carbonate chemistry does not suffer from the presence of a labile ester bond. The intermediate reactive NHS carbonate may hydrolyze in aqueous solution to release NHS and CO_2, essentially regenerating the underivatized PEG hydroxyl. After coupling to amine-containing molecules, however, the resultant carbamate linkage stabilizes the chemistry to the point that a modified molecule will not lose PEG by hydrolytic cleavage. For these reasons, the succinimidyl carbonate method of PEG activation and coupling has become the chemistry of choice for attaching the polymer to amine-containing proteins and other molecules.

A modification of Zalipsky's method by Miron and Wilchek (1993) simplifies the creation of the succinimidyl carbonate activated species. Instead of using highly toxic phosgene to form a chloroformate intermediate and then reacting with NHS, the new procedure utilizes either *N*-hydroxysuccinimidyl chloroformate or *N,N′*-disuccinimidyl carbonate (DSC; Chapter 5, Section 1.7) to produce the succinimidyl carbonate–PEG in one step (Figure 18.42). Since both activation reagents are commercially available, creating an amine-reactive PEG derivative has never been easier.

The following procedure is based on the Miron and Wilchek modification of Zalipsky's method.

Protocol for the Activation of PEG with N-Succinimidyl Chloroformate or N,N′-Disuccinimidyl Carbonate

Caution: The steps using flammable solvents, especially diethyl ether, should be done in a fume hood.

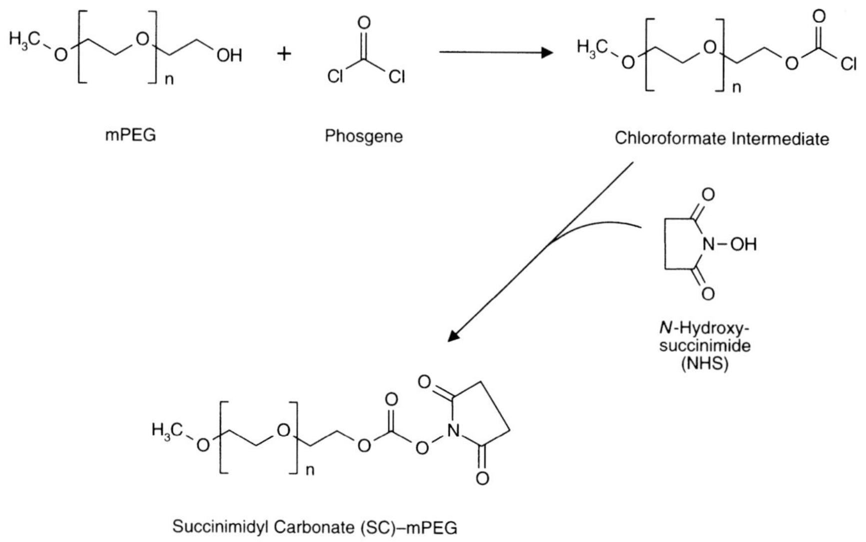

FIGURE 18.40 A succinimidyl carbonate derivative of mPEG was first prepared through the use of phosgene to form a chloroformate intermediate. Reaction with NHS gives the amine-reactive succinimidyl carbonate-mPEG.

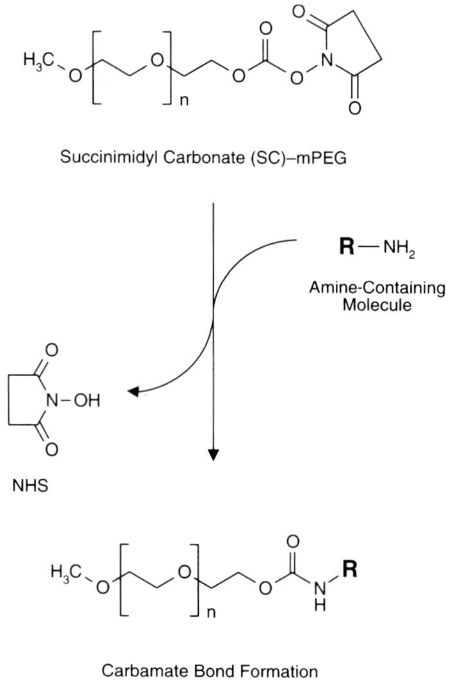

FIGURE 18.41 Succinimidyl carbonate-mPEG can be used to modify amine-containing molecules to form stable carbamate linkages.

1. Dissolve 5g PEG or mPEG (MW 5000; 1mmol) in 25ml of dry dioxane. Heating in a water bath may be necessary to solubilize fully the polymer. Cool to room temperature.
2. Dissolve 6mmol of either *N*-succinimidyl chloroformate or *N,N'*-disuccinimidyl carbonate (Aldrich) in 10ml of dry acetone.
3. Dissolve 6mmol of 4-(dimethylamino)pyridine (DMAP) in 10ml of dry acetone (base).
4. With stirring, add the solution prepared in step 2 to the PEG solution prepared in step 1. Next, slowly add the solution prepared in step 3.
5. React for 2h if activating with *N*-succinimidyl chloroformate or 6h if using *N,N'*-disuccinimidyl carbonate (DSC). Maintain stirring with a magnetic stirring bar.
6. If *N*-succinimidyl chloroformate was used, filter out the white precipitate of 4-(dimethylamino)pyridine hydrochloride using a glass-fiber filter pad. Collect the supernatant.
7. For either activation chemistry, precipitate the succinimidyl carbonate (SC)–PEG formed by addition of diethyl ether until no further precipitation is observed (typically 3–4 volumes of solvent).
8. Redissolve the precipitated product in acetone and precipitate again using diethyl ether. Repeat at least once more to remove completely excess reactants.
9. Dry the SC–PEG and store at 4°C.

Protocol for the Coupling of SC–mPEG to Proteins

1. Dissolve the protein to be PEGylated in cold 0.1-*M* sodium phosphate, pH 7.5, at a concentration of 1–10mg/ml.
2. With stirring, add a quantity of SC–mPEG to the protein solution at the molar ratio of polymer-to-protein desired. The ratio of activated polymer addition typically is expressed *versus* the molar quantity of primary amines present on the protein being modified. Ratios of SC–PEG to amines between

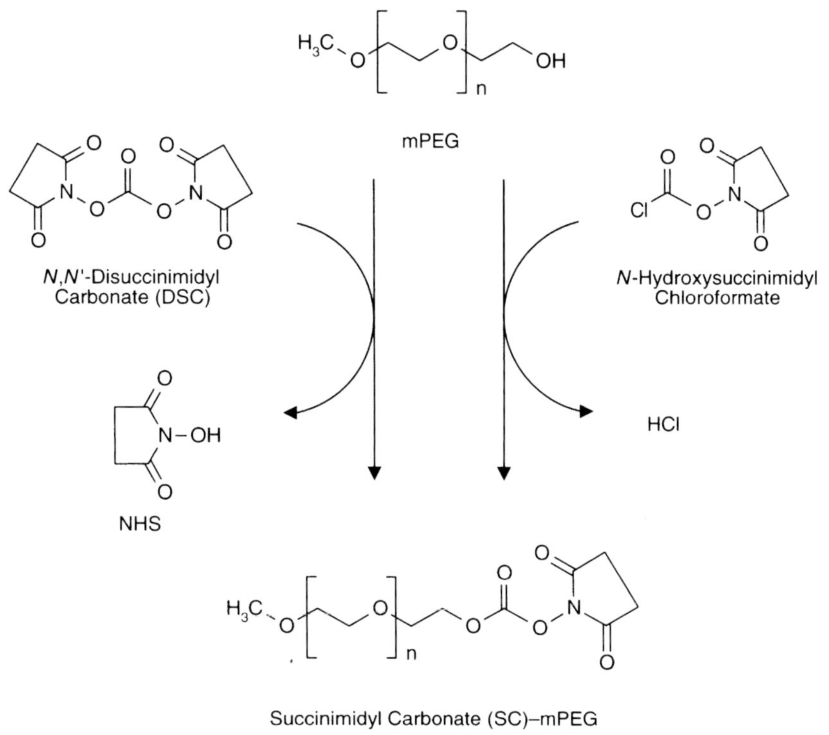

FIGURE 18.42 An alternative route to an succinimidyl carbonate derivative of mPEG can be accomplished by the reaction of the terminal hydroxyl group of the polymer with either *N,N'*-disuccinimidyl carbonate (DSC) or N-hydroxysuccinimidyl chloroformate.

0.3:1 and 8:1 were investigated by Miron and Wilchek (1993) for the derivatization of egg white lysozyme. The greater the ratio is of activated polymer to protein, the higher will be the molecular weight of the resultant complex. Experiments may have to be done using a number of different reaction ratios to determine the optimal PEGylation level for a particular protein.

3. React overnight at 4°C.
4. Remove excess SC–PEG by dialysis or gel filtration.

Carbodiimide Coupling of Carboxylate–PEG Derivatives

PEG contains only hydroxyl functional groups in its native state that need to be activated or modified in some manner to allow efficient conjugation to other molecules. These hydroxyls can be modified to possess carboxylates by reaction with anhydride compounds. Acylation of PEG with succinic anhydride or glutaric anhydride gives *bis*-modified products having carboxylates at both ends. Modification of mPEG yields the monosubstituted derivative containing a single carboxylate. Creation of these derivatives was first described by Abuchowski *et al.* (1984). Once the carboxylate–PEG modification is formed, it can be used to couple directly with amine-containing molecules by use of the carbodiimide reaction (Chapter 4, Section 1).

A carbodiimide may be used to activate the carboxylates to highly reactive *o*-acylisourea intermediates. When generated in the presence of an amine-containing protein or other molecule, these active esters will react with the nucleophiles to give amide bond derivatives (Figure 18.43). Atassi and Manshouri (1991) used this technique to PEGylate various peptides. In this instance, the reaction was carried out in dimethylformamide (DMF) due to the solubility of the peptides in this solvent. The organic-soluble carbodiimides dicyclohexyl carbodiimide (DCC) (Chapter 4, Section 1.4) or diisopropyl carbodiimide (DIC) (Chapter 4, Section 1.5) were used to perform the conjugation. However, aqueous-phase reactions can be performed using this approach just as easily as organic-based conjugations if the water soluble reagent EDC (1-ethyl-3-(3-dimethylaminopropyl)carbodiimide) is employed (Chapter 4, Section 1.1). The general protocols for using carbodiimides outlined in the referenced sections may be used to conjugate a carboxylate-containing PEG derivative to an amine-containing protein or other molecule. The formation of a carboxylate-containing PEG derivative can be done according to the following protocol (adapted from Atassi and Manshouri, 1991).

Protocol

1. Dissolve 1 g of mPEG (MW 5,000) in 5 ml of anhydrous pyridine by heating to 50°C.
2. To the stirring mPEG solution, add several 0.5-g aliquots of solid succinic anhydride over a period of several hours.

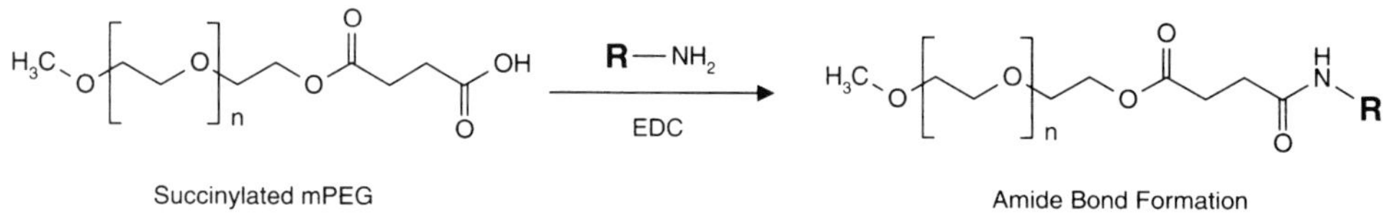

FIGURE 18.43 A succinylated mPEG derivative may be coupled to amine-containing molecules using a carbodiimide reaction to form an amide bond.

3. React for a further 2h at 50°C.
4. Evaporate the pyridine solvent by using a flash evaporator or a rotary evaporator under vacuum.
5. Redissolve the residue in water (the solution may have to be heated to fully dissolve) and again evaporate to dryness. Repeat until the odor of pyridine is nearly gone.
6. Remove remaining reactants by dialysis against water using a membrane having a molecular weight cut-off of 1000.

CDI Activation and Coupling

N,N'-Carbonyldiimidazole (CDI) is a highly reactive carbonylating compound which was first shown to be an excellent amide bond forming agent in peptide synthesis (Paul and Anderson, 1962). Later it was used to activate both carboxylic groups and hydroxyls in the immobilization of amine-containing ligands to prepare affinity chromatography supports (Bartling *et al.*, 1973; Hearn, 1987) (see Chapter 15, Section 2.1).

The activation of a carboxylate group with CDI proceeds to give an intermediate acyl derivative with imidazole as the active leaving group. In the presence of a primary amine-containing compound, the nucleophile attacks the electron-deficient carbonyl, displacing the imidazole and forming a stable amide bond.

For hydroxyl-containing compounds, CDI will react to form an intermediate imidazolyl carbamate which in turn can react with *N*-nucleophiles to give an *N*-alkyl carbamate linkage. Proteins normally couple through their N-terminals (α-amine) and lysine side chain (ε-amine) functional groups. The final bond is an uncharged, urethane derivative having excellent chemical stability (Figure 18.44). The result of CDI activation of PEG and subsequent coupling to a protein or other amine-containing molecule is a linkage identical to that obtained using succinimidyl carbonate chemistry, described previously (Section 1.2).

CDI-activated PEG is stable for years in a dried state or in organic solvents devoid of water or other nucleophiles. The activated polymer will also have an excellent half-life to hydrolysis even in the coupling environment. Unlike some activation chemistries which degrade rapidly and have half-lives on the order of minutes, imidazole carbamates have half-lives measured in hours. For instance, an agarose chromatography support activated with CDI will take up to 30h at pH 8.5 to 9.0 for complete loss of activity. The hydrolysis of CDI–PEG derivatives causes the release of CO_2 and imidazole. The hydrolyzed product thus reverts back to the original hydroxylic PEG compound, leaving no residual groups to cause potential sites for nonspecific interactions.

The optimal coupling condition for a CDI–PEG or CDI–mPEG reaction is in an alkaline pH environment, typically above pH 8.5. The coupling reaction proceeds at greatest efficiency when the target molecule is reacted at about 1 pH value above its pI or pK_a. The reaction can be performed directly in an organic solvent environment if the molecule to be modified demonstrates poor solubility in aqueous systems. The advantage of an organic coupling reaction is that there is no competing hydrolysis of the active groups, so very high modification yields of PEG can be realized.

There are a few precautions that should be noted when performing a CDI activation and coupling experiment. First, CDI itself is extremely unstable to aqueous environments, much more so than the active imidazolyl carbamate that is formed after PEG activation. Therefore, the activation step must be carried out in a solvent that is as free of water as possible. If unacceptable amounts of water are present, CDI will be immediately broken down to CO_2 and imidazole. The evolution of bubbles upon addition of CDI to a PEG solution is the tell-tale sign of high water content. Only freshly obtained solvents analyzed to be extremely low in moisture or those dried over a molecular sieve should be used. A water content of less than 0.1% in the solvent is usually acceptable for a CDI activation procedure.

A second precaution is to carry out the activation step in a fume hood away from sources of ignition. Most CDI activation protocols use flammable or toxic solvents and care should be taken in handling and disposing of them.

The coupling reaction using CDI–mPEG or CDI–PEG derivatives is slower than that obtained using NHS ester or succinimidyl carbonate coupling methods. Therefore, the reaction times used with CDI chemistry are typically on the order of 1 to 2 days at 4°C at a pH of about 8.5. Increasing the pH of the reaction to pH 9 or pH 10 will speed up the coupling. In addition, performing the reaction at room temperature also helps in

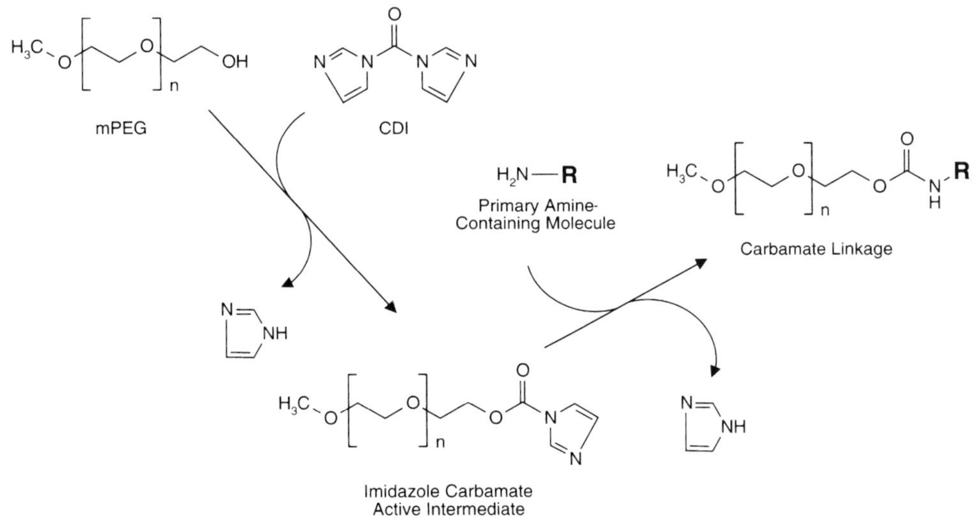

FIGURE 18.44 *N,N′*-Carbonyl diimidazole (CDI) may be used to activate the terminal hydroxyl of mPEG to an imidazole carbamate. Reaction of this intermediate reactive group with an amine-containing compound results in the formation of a stable carbamate linkage.

this regard. If the molecule to be modified is stable at alkaline pH values and room temperature, then these conditions may be used to decrease the time of the suggested protocol.

The following method is adapted from Beauchamp *et al.* (1983).

Protocol for the Activation of mPEG with CDI

1. Dissolve mPEG (MW 5000) in dry dioxane at a concentration of 50-m*M* (0.25 g/ml) by heating to 37°C.
2. Add solid CDI to a final concentration of 0.5-*M* (81 mg/ml).
3. React for 2 h at 37°C with stirring.
4. To remove excess CDI and reaction byproducts, Beauchamp *et al.* (1983) dialyzed against water at 4°C. However, the imidazole carbamate groups on mPEG formed during the activation process are subject to hydrolysis in aqueous environments. A better method may be to precipitate the activated mPEG with diethyl ether as in the protocol described previously for succinimidyl carbonate activation (Section 1.2).
5. Finally, dry the isolated product by lyophilization (if the water dialysis method is used) or by use of a rotary evaporator (if the ether precipitation method is used).

Protocol for the Coupling of CDI–mPEG to Proteins

1. Dissolve the protein to be PEGylated in 10-m*M* sodium borate, pH 8.5, at a concentration of 1–10 mg/ml. Higher pH values may be used to increase the reaction rate—for instance, 0.1-*M* sodium carbonate, pH 9 to 10.
2. Add CDI–mPEG to this solution with stirring to bring the final concentration of the activated polymer to 180-m*M*. Note: Other ratios of polymer-to-protein may be used, depending on the modification level desired. Some optimization of the derivatization level may have to be carried out to obtain conjugates having the best amount of polymer substitution with retention of protein activity.
3. React for 48 h at 4°C. If higher pH or room temperature conditions are used, the reaction time can be decreased to 24 h.
4. Remove unconjugated mPEG and reaction byproducts by dialysis or gel filtration.

Miscellaneous Coupling Reactions

PEG or mPEG may be conjugated to proteins or other molecules using other coupling chemistries in addition to the ones mentioned in the previous sections. Almost any activation method that can be built off of the terminal hydroxyl(s) of PEG may be employed to PEGylate target molecules. For instance, Bergstrüm *et al.* (1992) created an epoxy derivative of the polymer by reaction with epichlorohydrin under alkaline conditions. The reactive alkyl halogen end of epichlorohydrin first is coupled to the hydroxyls of PEG to give the terminal glycidyl ether derivative (Figure 18.45). The epoxy-functionalized polymer then could be used to covalently modify a poly(ethylene imine)-coated polystyrene surface to prevent nonspecific protein adsorption. This type of derivative could also be used to modify other amine-, hydroxyl-, or sulfhydryl-containing molecules (Chapter 3, Section 4.1).

Creation of a sulfhydryl-reactive PEG derivative was carried out by Goodson and Katre (1990) by reacting an

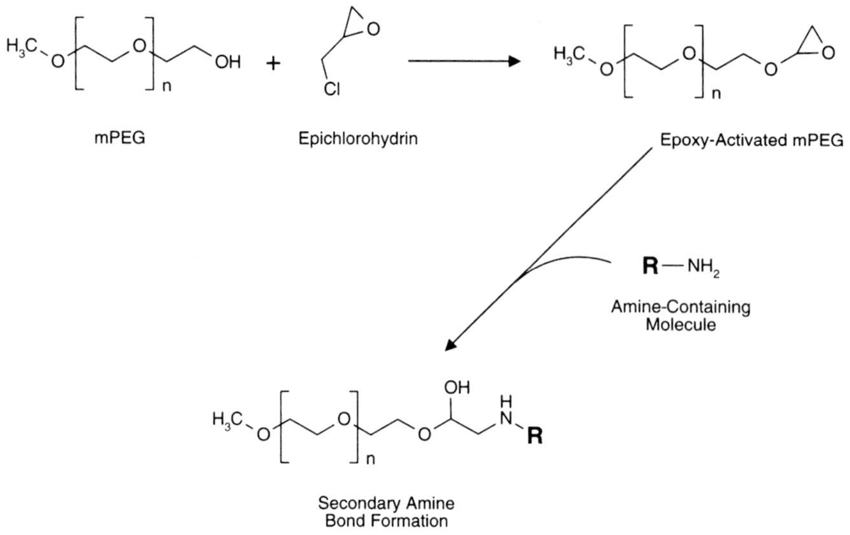

FIGURE 18.45 Epichlorohydrin can be used to activate the hydroxyl group of mPEG, creating an epoxy derivative. Reaction with amine-containing molecules yields secondary amine bonds.

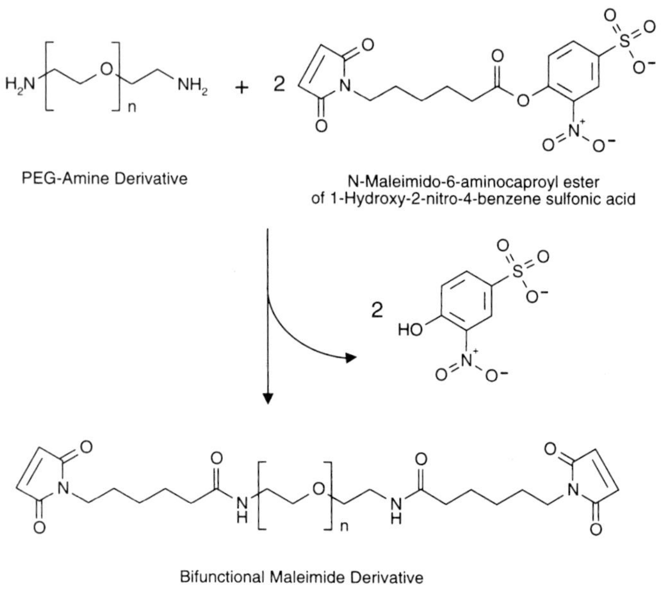

FIGURE 18.46 A PEG-diamine compound may be reacted with this heterobifunctional crosslinker to form amide bond derivatives terminating in maleimide groups. This results in a homobifunctional reagent capable of crosslinking thiol molecules. Subsequent reaction with sulfhydryl-containing molecules yields thioether linkages. Other amine- and thiol-reactive heterobifunctional crosslinkers may also be used in this manner to create bifunctional PEG derivatives.

active ester–maleimide heterobifunctional crosslinker with the amino groups of a PEG-amine polymer (Pillai and Mutter, 1980). Amine-terminal derivatives of discrete PEG-type polymers are available from Thermo Fisher or Quanta BioDesign (Chapter 15, Section 2.8). Reaction with the *N*-maleimido-6-aminocaproyl ester of 1-hydroxy-2-nitro-4-benzene sulfonic acid results in an amide bond derivative (Figure 18.46). This creates terminal maleimide groups on each PEG-amine molecule. Maleimide compounds can be used in a site-directed coupling procedure to specifically PEGylate at the sulfhydryl groups of proteins and other molecules (Chapter 3, Section 2.2).

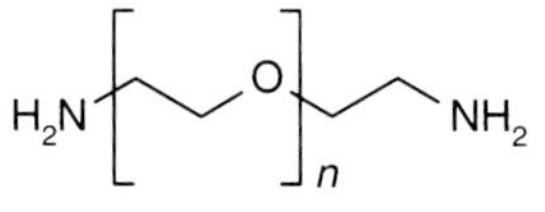

PEG-Amine Derivative
(Jeffamine Series from Texaco;
Various Polymer Lengths Available)

In another approach, Wirth *et al.* (1991) and Chamow *et al.* (1994) transformed the terminal hydroxyl group of mPEG into an aldehyde residue by the Moffatt oxidation procedure (Harris *et al.*, 1984). In this reaction, the hydroxyl is treated with acetic anhydride in dimethyl sulfoxide (DMSO) containing triethylamine, converting it to the aldehyde (Figure 18.47). After stirring at room temperature for 48h, the aldehyde derivative is isolated by precipitation with ether and ethyl acetate. An aldehyde can be conjugated to proteins or other amine-containing molecules by reductive amination using sodium cyanoborohydride (Chapter 4, Section 4). The advantage of an aldehyde-PEG derivative over the other amine-reactive methods described previously is that the active group will not hydrolyze or readily degrade before the coupling reaction is initiated. In addition, reductive amination is a reasonably mild conjugation technique well tolerated by most proteins.

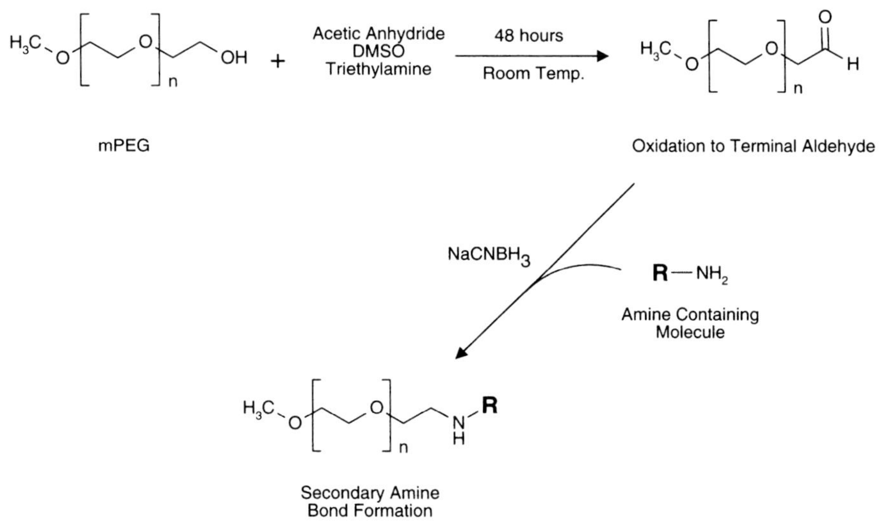

FIGURE 18.47 A terminal aldehyde function on mPEG may be formed through an oxidative process at elevated temperatures. This derivative may be used to modify amine-containing molecules by reductive amination.

2.2. Dextran and Carbohydrate Polymers

Dextran is a naturally occurring polymer that is synthesized in yeasts and bacteria for energy storage. It is mainly a linear polysaccharide consisting of repeating units of D-glucose linked together in glycosidic bonds (Chapter 2, Section 2.1), wherein the carbon-1 of one monomer is attached to the hydroxyl group at the carbon-6 of the next residue. This configuration is the same as that found in the α-1,6 linked disaccharide isomaltose. The same disaccharide is found at the branch points of glycogen and amylopectin. Occasional branch points may also be present in a dextran polymer, occurring as α-1,2, α-1,3, or α-1,4 glycosidic linkages. The branch type and degree of branching vary by species.

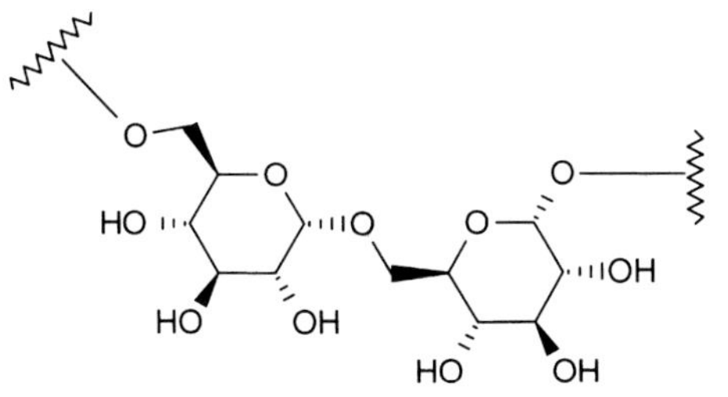

Isomaltose (a-1,6) Repeating
Unit of Dextran Polymer Chains

The hydroxylic content of the dextran sugar backbone makes the polymer very hydrophilic and easily modified for coupling to other molecules. Unlike PEG, discussed previously, which has modifiable groups only at the ends of each linear polymer, the hydroxyl functional groups of dextran are present on each monomer in the chain. The monomers contain at least three hydroxyls (four on the terminal units) that may undergo derivatization reactions. This multivalent nature of dextran allows molecules to be attached at numerous sites along the polymer chain.

Soluble dextran of molecular weight 10,000 to 500,000 has been used extensively as a modifying or crosslinking agent for proteins and other molecules (Goodarzi *et al.*, 2012). It has been used as a drug carrier to transport greater concentrations of antineoplastic pharmaceuticals to tumor sites *in vivo* (Bernstein *et al.*, 1978; Heindel *et al.*, 1990; Vaidya *et al.*, 2011; Jain *et al.*, 2012), as conjugated to biotin to make a sensitive anterograde tracer for neuroatomic studies (Brandt and Apkarian, 1992), as a hapten carrier to illicit an immune response against coupled molecules (Dintzis *et al.*, 1989; Shih *et al.*, 1991), as an inducer of B cell proliferation by coupling anti-Ig antibodies (Brunswick *et al.*, 1988), as a multifunctional linker to crosslink monoclonal antibody conjugates with chemotherapeutic agents (Heindel *et al.*, 1991), and as a stabilizer of enzymes and other proteins (Zlateva *et al.*, 1988; Nakamura *et al.*, 1990). As is true of PEG conjugates with proteins, dextran modification of macromolecules provides increased circulatory half-life *in vivo*, decreased immunogenicity, and a heat and protease protective effect when coupled at sufficient density (Mumtaz and Bachhawat, 1991).

The following sections describe the major activation and coupling methods used with dextran polymers. The reactive derivatives may be used to couple with proteins and other molecules containing the appropriate functional groups.

Polyaldehyde-Dextran; Activation and Coupling

The dextran polymer contains adjacent hydroxyl groups on each glucose monomer. These diols may be

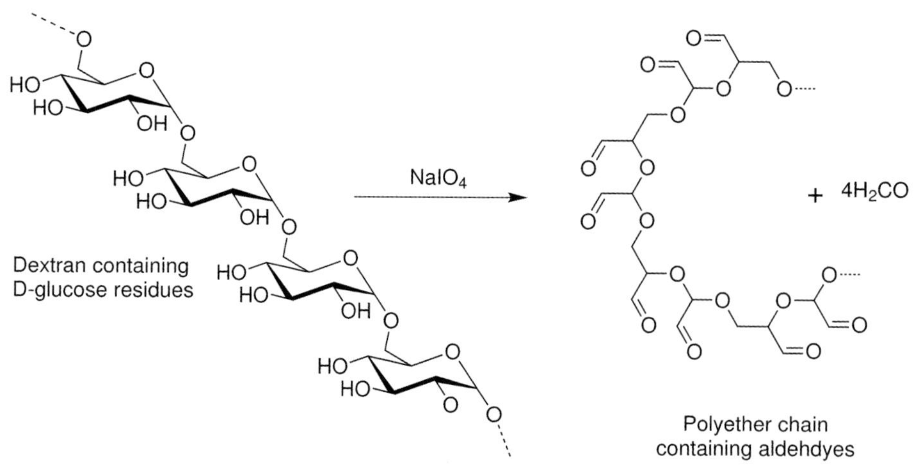

FIGURE 18.48 Dextran polymers can be oxidized with sodium periodate to create a polyaldehyde derivative. Note that additional oxidation may occur to cleave off another carbon atom (the C-3) and create an aldehyde on the adjacent C-4–OH group.

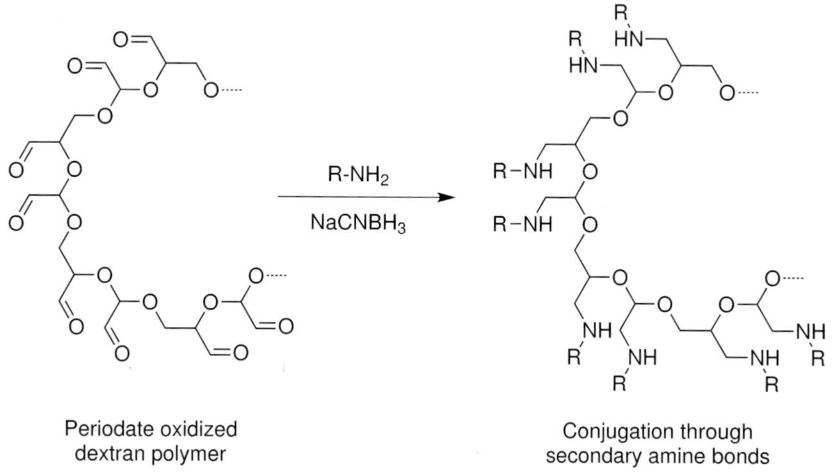

FIGURE 18.49 Polyaldehyde dextran may be used as a multifunctional crosslinking agent for the coupling of amine-containing molecules. Reductive amination creates secondary amine or tertiary amine linkages.

oxidized with sodium periodate to cleave the associated carbon–carbon bonds and produce aldehydes (Chapter 2, Section 4.4). This procedure results in two aldehyde groups formed per glucose monomer, thus producing a highly reactive, multifunctional polymer able to couple with numerous amine-containing molecules (Bernstein *et al.*, 1978) (Figure 18.48). Polyaldehyde dextran may be conjugated with amine groups by Schiff base formation followed by reductive amination to create stable secondary (or tertiary amine) linkages (Chapter 3, Section 5.4) (Figure 18.49).

Proteins may be modified with oxidized dextran polymers under mild conditions using sodium cyanoborohydride as the reducing agent. The reaction proceeds primarily through ε-amino groups of lysine located at the surface of the protein molecules. The optimal pH for the reductive amination reaction is an alkaline environment between pH 7 and 10. The rate of reaction is greatest at pH 8 to 9 (Kobayashi and Ichishima, 1991), reflecting the efficiency of Schiff base formation at this pH.

Polyaldehyde dextran can be used to couple many small molecules, such as drugs, to a targeting molecule like an antibody. The multivalent nature of the oxidized dextran backbone provides more sites for conjugation than possible using direct coupling of the drug with the antibody itself. Similarly, detection molecules such as fluorescent probes can be conjugated in greater amounts using a dextran carrier than is feasible with direct modification of a protein.

The following protocol for creating the polyaldehyde dextran derivative is based on the method of Berstein *et al.* (1978).

Protocol for Oxidizing Dextran with Sodium Periodate

1. Dissolve sodium periodate ($NaIO_4$) (Sigma) in 500 ml of deionized water at a concentration of 0.03-*M* (6.42 g). Protect from light.
2. Dissolve dextran (Polysciences) of molecular weight between 10,000 and 40,000 in the sodium periodate solution with stirring.

3. React overnight at room temperature in the dark.
4. Remove excess reactant by extensive dialysis against water. The purified polyaldehyde dextran may be lyophilized for long-term storage.

The degree of oxidation may be assessed by measurement of the aldehydes formed. Zhao and Heindel (1991) suggest derivatizing the polyaldehyde dextran with hydroxylamine hydrochloride and measuring the amount of HCl released by titration. However, this may be tedious and time consuming. A simpler method may be to take advantage of the fact that periodate-oxidized sugars are capable of reducing Cu^{2+} to Cu^{1+}, which can be detected using the bicinchoninic acid (BCA) reagent (Thermo Fisher) (Smith *et al.*, 1985). The formation of Cu^{1+} is in direct proportion to the amount of aldehydes present in the polymer. BCA will form a purple-colored complex with Cu^{1+} which can be measured at 562 nm.

Protocol for Coupling Polyaldehyde Dextran to Proteins

1. Dissolve or buffer-exchange the periodate-oxidized dextran in 0.1-*M* sodium phosphate, 0.15-*M* NaCl, pH 7.2, at a concentration of 10 to 25 mg/ml. Other buffers having a pH range of 7 to 10 may be used with success, as long as they do not contain competing amines (such as Tris). A reaction environment of pH 8–9 (0.1-*M* sodium bicarbonate) will give the greatest yield of reductive amination coupling.
2. Add 10 mg of the protein to be coupled to the dextran solution. Other ratios of dextran-to-protein may be used as appropriate. For instance, if more than one protein or a protein plus a smaller molecule are both to be conjugated to the dextran backbone, the amount of protein added initially may have to be scaled back to allow the second molecule to be coupled latter. Many times, a small molecule such as a drug will be coupled to the dextran polymer first, and then a targeting protein such as an antibody conjugated secondarily. The optimal ratio of components forming the dextran conjugate should be determined experimentally to obtain the best combination possible.
3. In a fume hood, add 0.2 ml of 1-*M* sodium cyanoborohydride (Aldrich) to each milliliter of the protein/dextran solution. Mix well. *Caution: cyanoborohydride is extremely toxic and should be handled only in well-ventilated fume hoods. Dispose of cyanide-containing solutions according to approved guidelines.*
4. React for at least 6 h at room temperature. Overnight reactions may also be done.
5. To block excess aldehydes, add 0.2 ml of 1-*M* Tris or 1-*M* ethanolamine, pH 8, to each milliliter of the reaction. Note: If a second molecule is to be coupled after the initial protein conjugation, don't block the remaining aldehydes until the second molecule is coupled.
6. React for an additional 2 h at room temperature.
7. Purify the protein–dextran conjugate from unconjugated protein and dextran by gel filtration using a column of Sephacryl S-200 or S-300. Small molecules may be removed from a dextran conjugate by dialysis.

Carboxyl, Amine, and Hydrazide Derivatives

Dextran derivatives containing carboxyl- or amine-terminal spacer arms may be prepared by a number of techniques. These derivatives are useful for coupling amine- or carboxylate-containing molecules through a carbodiimide-mediated reaction to form an amide bond (Chapter 4, Section 1). Amine-terminal spacers can also be used to create secondary reactive groups by modification with a heterobifunctional crosslinking agent (Chapter 6).

This type of modification process has been used to form sulfhydryl-reactive dextran polymers by coupling amine spacers with crosslinkers containing an amine-reactive end and a thiol-reactive end (Brunswick *et al.*, 1988; Noguchi *et al.*, 1992). The result was a multivalent sulfhydryl-reactive dextran derivative that could couple numerous sulfhydryl-containing molecules per polymer chain.

Several chemical approaches may be used to form the amine- or carboxyl-terminal dextran derivative. The simplest procedure may be to prepare polyaldehyde dextran according to the procedure of Section 2.1 and then make the spacer arm derivative by reductively aminating an amine-containing organic compound onto it. For instance, short diamine compounds such as ethylene diamine or diaminodipropylamine (3,3′-iminobispropylamine) can be reacted in large excess with polyaldehyde dextran to create numerous modifications along the polymer having terminal primary amines. Carboxyl-terminal derivatives may be prepared similarly by coupling molecules such as 6-aminocaproic acid or β-alanine to polyaldehyde dextran. Alternatively, an amine-terminal spacer may be reacted with succinic anhydride to form the carboxylate derivative (Chapter 2, Section 4.2).

Another approach uses reactive alkyl halogen compounds containing a terminal carboxylate group on the other end to form spacer arms off the dextran polymer from each available hydroxyl. In this manner, Brunswick *et al.* (1988) used chloroacetic acid to modify the hydroxyl groups to form the carboxymethyl derivative. The carboxylates then were aminated with ethylenediamine to create an amine-terminal derivative (Inman, 1985). Finally, the amines were modified with iodoacetate to form a sulfhydryl-reactive polymer (Figure 18.50).

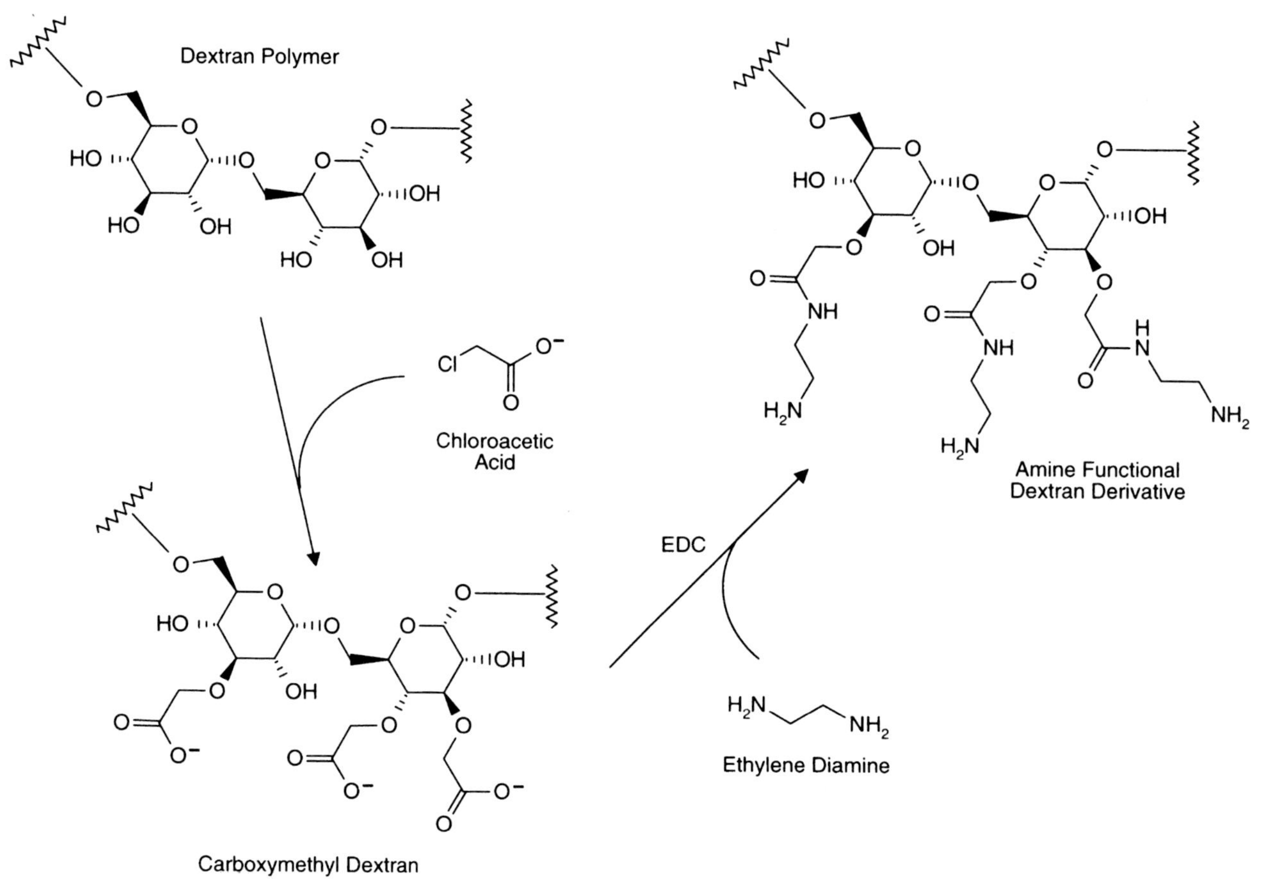

FIGURE 18.50 An amine derivative of dextran may be prepared through a two-step process involving the reaction of chloroacetic acid with the hydroxyl groups of the polymer to create carboxylates. Next, ethylene diamine is coupled in excess using a carbodiimide-mediated reaction to give the primary amine functional groups.

In a somewhat similar scheme, Noguchi *et al.* (1992) prepared a carboxylate spacer arm by reacting 6-bromohexanoic acid with a dextran polymer. The carboxylate then was aminated with ethylene diamine to form an amine-terminal spacer (Figure 18.51). This dextran derivative was finally reacted with *N*-succinimidyl 3-(2-pyridyldithio)propionate (SPDP) (Chapter 6, Section 1.1) to create the desired sulfhydryl-reactive polymer (Section 2.4, this chapter). The SPDP-activated polymer could then be used to prepare an immunoconjugate composed of an antibody against human colon cancer conjugated with the drug mitomycin-C.

Hydrazide derivatives may also be prepared from a periodate-oxidized dextran polymer or from a carboxyl-containing dextran derivative by reaction with *bis*-hydrazide compounds (Chapter 5, Section 8). A hydrazide terminal spacer provides reactivity toward aldehyde- or ketone-containing molecules. Thus, the hydrazide–dextran polymer can be used to conjugate specifically glycoproteins or other polysaccharide-containing molecules after they have been oxidized with periodate to form aldehydes (Chapter 2, Section 4.4).

The following protocols may be used to create carboxyl-, amine-, or hydrazide-containing derivatives of dextran.

Protocol for Preparation of Amine or Hydrazide Derivatives by Reductive Amination

1. Prepare polyaldehyde dextran according to the method of Section 2.1.
2. To make an amine derivative of dextran, dissolve ethylenediamine (or another suitable diamine) in 0.1-*M* sodium phosphate, 0.15-*M* NaCl, pH 7.2, at a concentration of 3M. Note: Use of the hydrochloride form of ethylenediamine is more convenient, since it avoids having to adjust the pH of the highly alkaline free-base form of the molecule. Alternatively, to prepare a hydrazide-dextran derivative, dissolve adipic acid dihydrazide (Chapter 5, Section 8.1) in the coupling buffer at a concentration of 30mg/ml (heating under a hot water tap may be necessary to completely dissolve the hydrazide compound). Adjust the pH to 7.2 with HCl and cool to room temperature.
3. Dissolve polyaldehyde dextran in the ethylenediamine (or adipic dihydrazide) solution at a concentration of 25mg/ml.

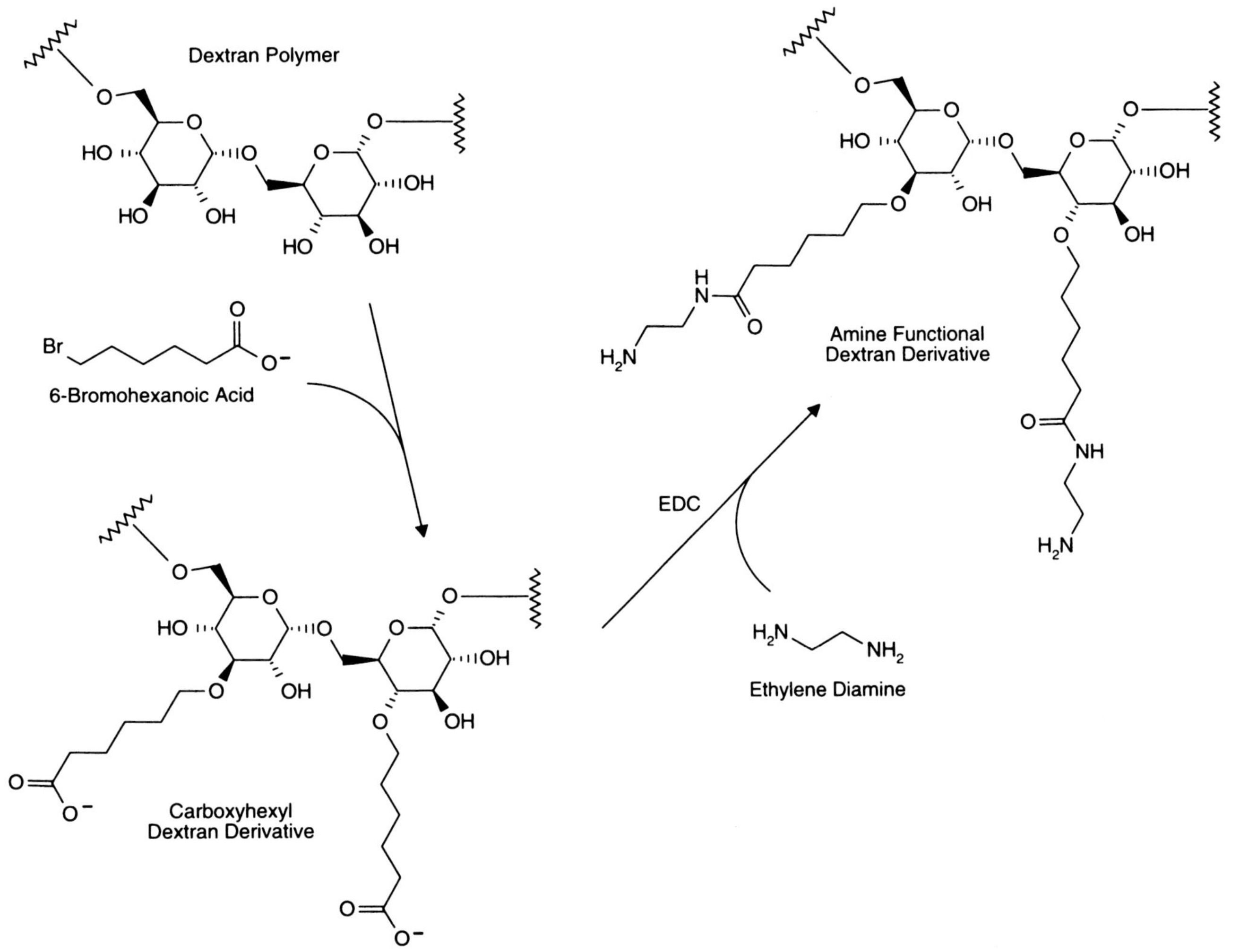

FIGURE 18.51 Amino-dextran derivatives may be prepared by the reaction of 6-bromohexanoic acid with the hydroxyl groups of the polymer followed by coupling of ethylene diamine using EDC.

4. In a fume hood, add 0.2 ml of 1-*M* sodium cyanoborohydride to each ml of the diamine/dextran solution. Mix well. *Caution: Cyanoborohydride is extremely toxic and should be handled only in well-ventilated fume hoods. Dispose of cyanide-containing solutions according to approved guidelines.*
5. React for at least 6 h at room temperature. Overnight reactions may also be done.
6. Remove excess diamine and reaction byproducts by dialysis.

The ethylenediamine–dextran derivative may be used for the coupling of carboxylate-containing molecules by the carbodiimide reaction, for the coupling of amine-reactive probes, or to modify further using heterobifunctional crosslinkers. The hydrazide–dextran derivative may be used to crosslink aldehyde-containing molecules, such as oxidized carbohydrates or glycoproteins.

Protocol for the Modification of Dextran with Chloroacetic Acid

1. In a fume hood, prepare a solution consisting of 1-*M* chloroacetic acid in 3-*N* NaOH.
2. Immediately add dextran polymer to a final concentration of 40 mg/ml. Mix well to dissolve.
3. React for 70 min at room temperature with stirring.
4. Stop the reaction by adding 4 mg/ml of solid NaH_2PO_4 and adjusting the pH to neutral with 6-*N* HCl.
5. Remove excess reactants by dialysis.

The carboxymethyl–dextran derivative may be used to couple amine-containing molecules by the carbodiimide reaction. Heindel *et al.* (1994) prepared the lactone derivative of carboxymethyl–dextran by refluxing for 5 h in toluene or other anhydrous solvents. The lactone derivative is highly reactive toward amine-containing molecules, thus creating a preactivated polymer for conjugation purposes.

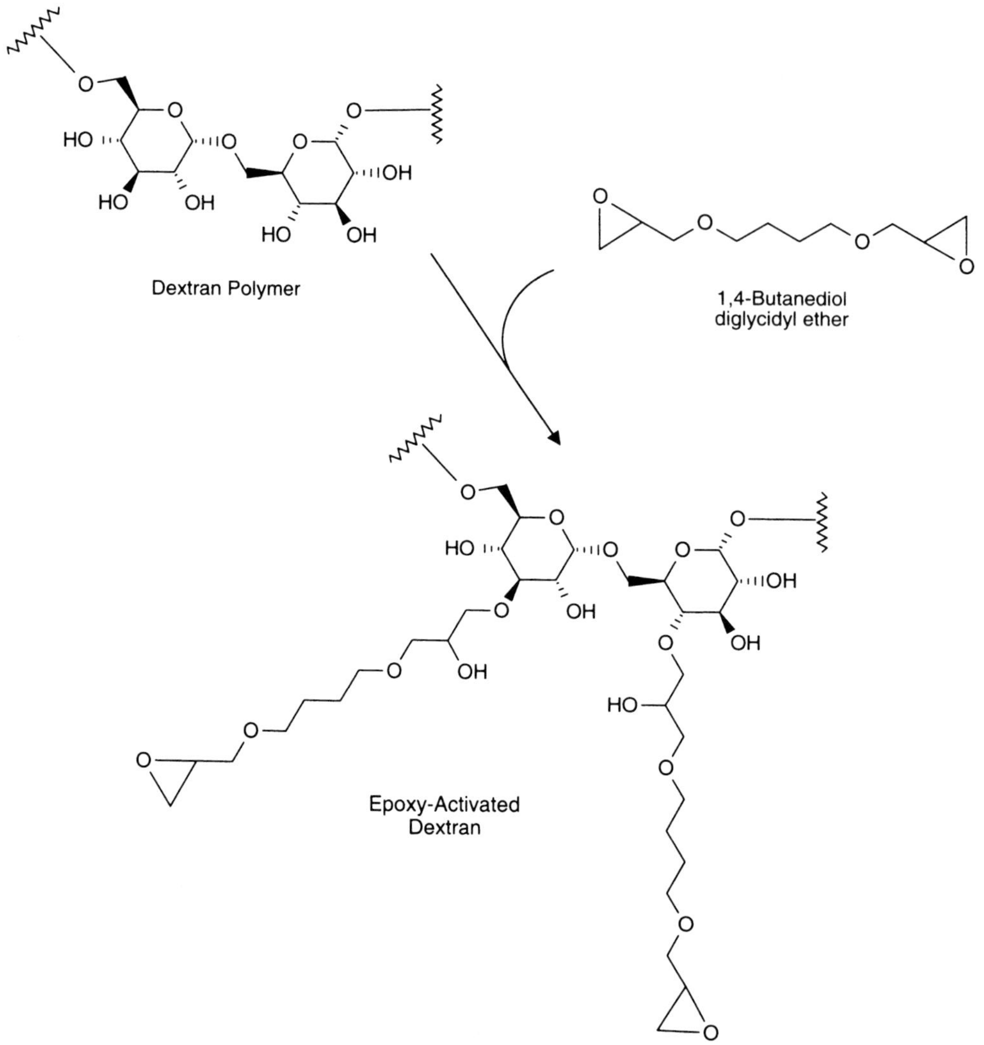

FIGURE 18.52 An epoxy-functional dextran derivative may be prepared by the reaction of 1,4-butanediol diglycidyl ether with the hydroxyl groups of the polymer.

Epoxy Activation and Coupling

Epoxy activation of hydroxylic polymers is commonly used as a means to immobilize molecules on solid-phase chromatographic supports that contain hydroxyl groups (Sundberg and Porath, 1974). *bis*-oxirane compounds can also be used to introduce epoxide groups into soluble dextran polymers in much the same manner (Bölkicke *et al.*, 1988; Böcher *et al.*, 1992). The epoxide group reacts with nucleophiles in a ring-opening process to form a stable covalent linkage. The reaction can take place with primary amines, sulfhydryls, or hydroxyl groups to create secondary amine, thioether, or ether bonds, respectively (Chapter 3, Section 1.8).

Modification of dextran polymers with 1,4-butanediol diglycidyl ether results in ether derivatives of the dextran hydroxyl groups, which then contain hydrophilic spacers with terminal epoxy functions (Figure 18.52).

Protocol

1. In a fume hood, mix 1 part 1,4-butanediol diglycidyl ether with 1 part 0.6-*N* NaOH containing 2 mg/ml sodium borohydride.
2. With stirring, add 5 mg of dextran to each milliliter of the *bis*-epoxide solution. Mix well to dissolve.
3. React for 12 h at 25°C or 3 to 4 h at 37°C.
4. Extensively dialyze the solution against water to remove excess reactants. The activated dextran may be lyophilized for long-term storage.

The epoxide-activated dextran may be used to conjugate amine-, sulfhydryl-, or hydroxyl-containing molecules. The reaction of the epoxide groups with hydroxyls requires high pH conditions, usually in the range of pH 11 to 12. Amine nucleophiles react at more moderate alkaline pH values, typically needing buffer environments of at least pH 9 to 10. Sulfhydryl groups

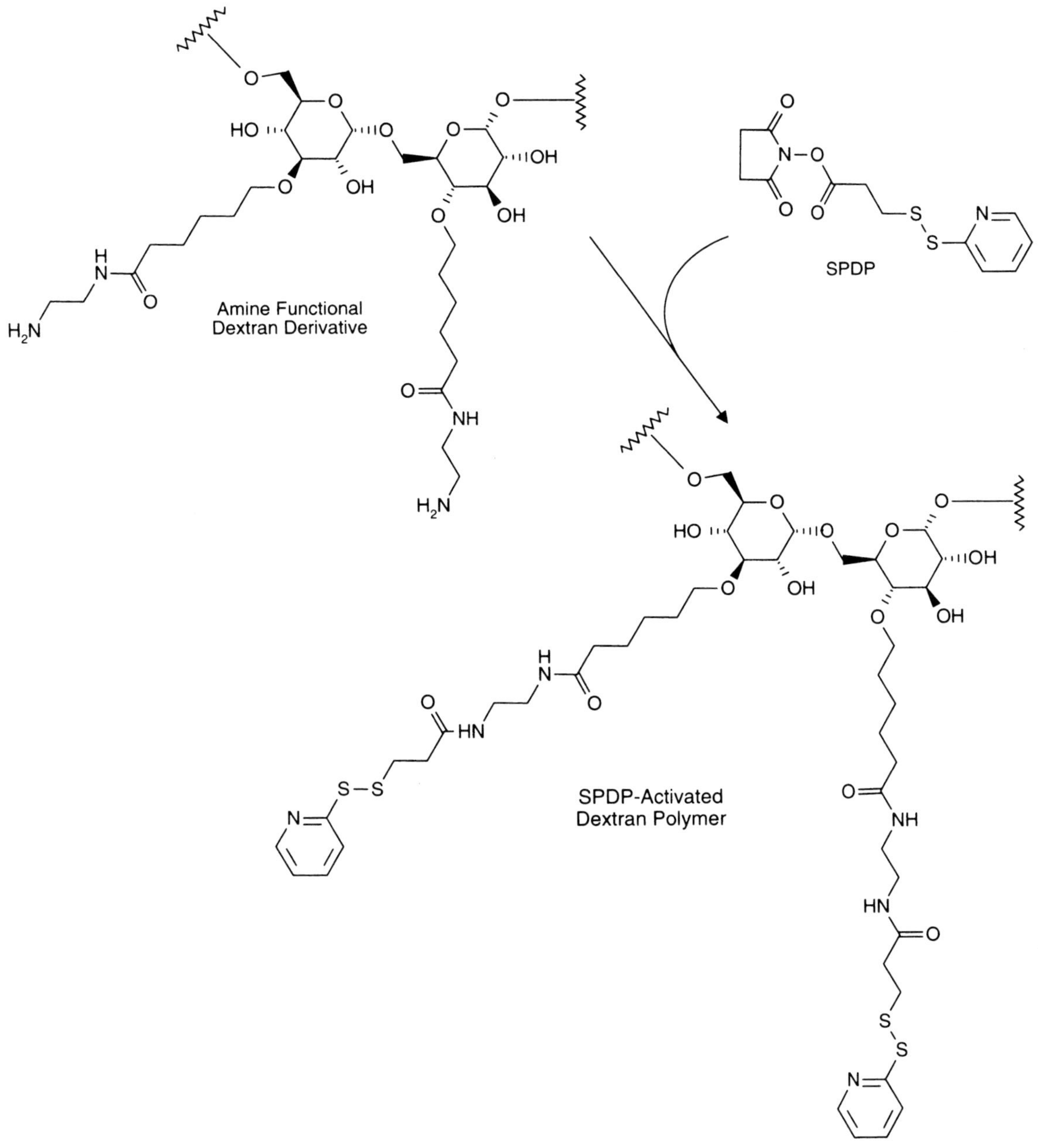

FIGURE 18.53 An amine-functionalized dextran derivative may be further reacted with SPDP to create a sulfhydryl-reactive product.

are the most highly reactive nucleophiles with epoxides, requiring a buffered system closer to the physiological range, pH 7.5 to 8.5, for efficient coupling.

Sulfhydryl-Reactive Derivatives

Sulfhydryl-reactive dextran derivatives may be prepared through the use of heterobifunctional crosslinking agents (Chapter 6). In particular, crosslinkers containing pyridyl disulfide, maleimide, or iodoacetyl groups on one end are quite effective in directing a conjugation reaction to thiols. Both maleimide and iodoacetyl activation procedures will yield nonreversible bonds with sulfhydryl-containing molecules. Pyridyl disulfide compounds, by contrast, react with thiols to form cleavable disulfide bonds that can be reversed by reduction.

Noguchi *et al.* (1992) used an amine-terminal spacer arm derivative of dextran to react with SPDP (Chapter 6, Section 1.1) in the creation of a pyridyl disulfide-activated polymer (Figure 18.53). Coupling of thiol-containing molecules to this activated dextran will result in the formation of disulfide linkages, which can be cleaved by the addition of 50-m*M* DTT. Brunswick *et al.* (1988) used a different amine-terminal spacer arm derivative of dextran and subsequently coupled iodoacetate

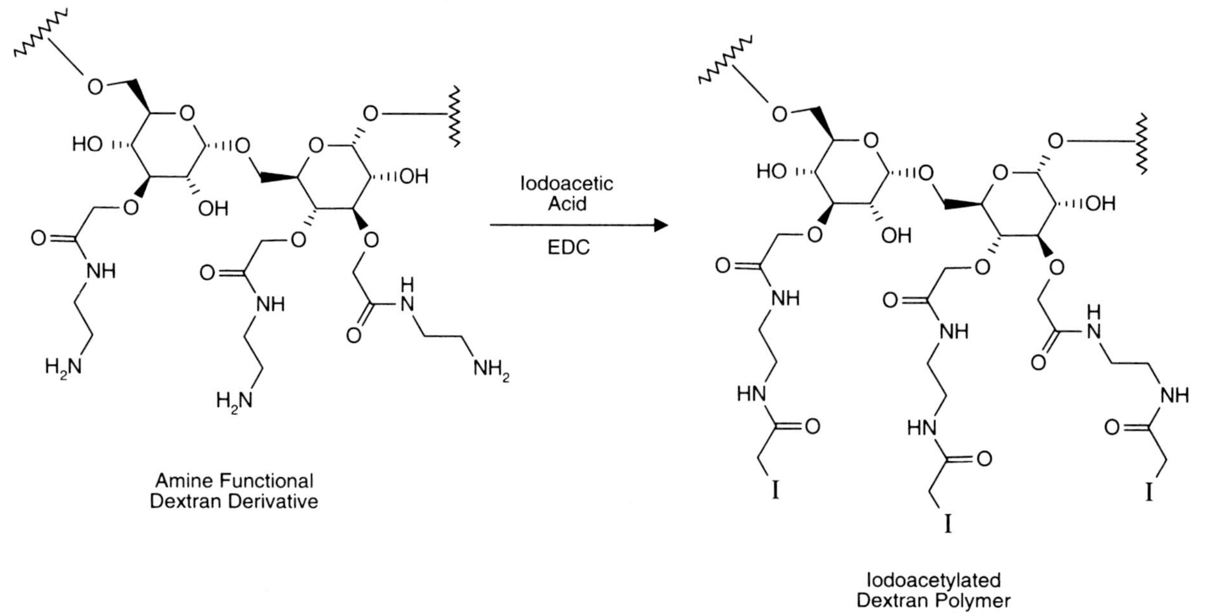

FIGURE 18.54 An amine-derivative of dextran may be coupled with iodoacetic acid using a carbodiimide reaction to produce a sulfhydryl-reactive iodoacetamide polymer.

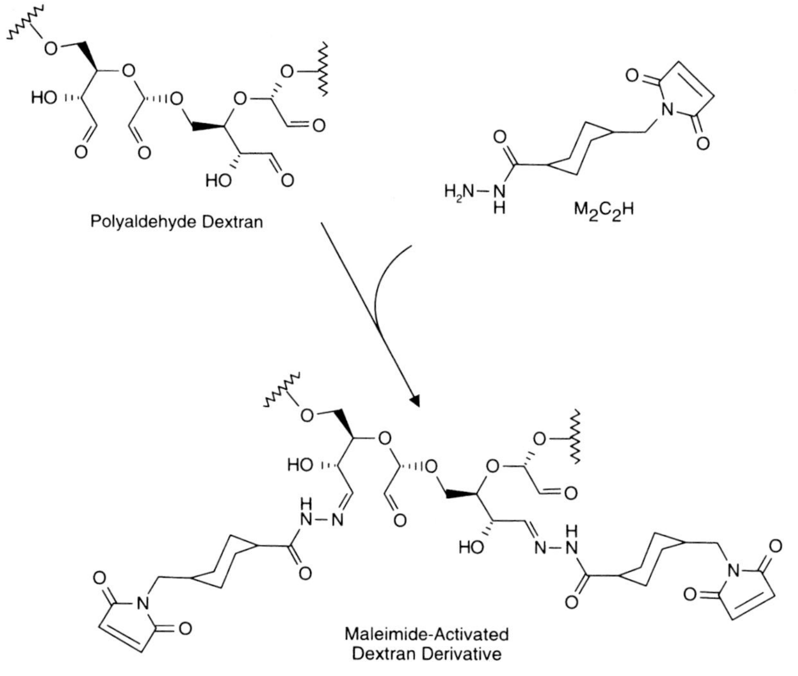

FIGURE 18.55 Polyaldehyde dextran may be modified with the hydrazide end of M_2C_2H to create a thiol-reactive polymer.

to form a sulfhydryl-reactive polymer (Figure 18.54). Conjugation to this activated polymer with thiol-containing molecules results in stable thioether linkages, which cannot be cleaved by reduction. Heindel *et al.* (1991) used a third approach to form a thiol-reactive polymer. They modified polyaldehyde dextran with a heterobifunctional crosslinker containing a hydrazide group on one end and a maleimide group on the other (Chapter 6, Section 2). The hydrazides reacted with the aldehyde groups to form hydrazone linkages, leaving the maleimide ends free to result in a thiol-reactive dextran derivative (Figure 18.55). Coupling thiol-containing molecules to this derivative will result in the formation of thioether bonds.

CHAPTER

19

Vaccines and Immunogen Conjugates

This chapter involves the design, preparation, and use of conjugate vaccines or hapten–carrier conjugates used to elicit an immune response toward a coupled hapten. The chemical reactions discussed for these conjugations are useful for coupling peptides, proteins, carbohydrates, oligonucleotides, and other small organic molecules to various carrier macromolecules. The resultant conjugates are important in antibody production, immune response research, and in the creation of vaccines.

1. THE BASIS OF IMMUNITY

The essence of adaptive immunity is the ability of an organism to react to the presence of foreign substances and produce components (antibodies and cells) capable of specifically interacting with and protecting the host from their invasion. An "antigen" or "immunogen" is the name given to a substance that is able to elicit this type of immune response and is also capable of interacting with the sensitized cells and antibodies that are manufactured against it (Abbas *et al.*, 2011; Murphy *et al.*, 2011).

The immune system has two basic components which respond to the challenge of a foreign substance: a cellular response mediated by T lymphocytes and a humoral response mediated by secreted proteins called antibodies produced by B lymphocytes, also called plasma cells. The B lymphocytes recognize antigens through cell-surface immunoglobulins that bind to discrete chemical and structural epitopes on the antigen molecule. Each B cell possesses surface immunoglobulin of a single type (i.e., is monoclonal) and has a binding capability that is directed against a discrete epitopic target.

Antigen binding by a complementary immunoglobulin molecule on the surface of B cells starts a process of cellular internalization of the foreign substance by pinocytosis. Once internalized by endosomes, systematic processing of the antigen takes place which breaks it down into smaller components.

At this point, the endosome may fuse with vesicles containing newly synthesized or recycling major histocompatibility complex (MHC) antigens. Some of the partially degraded antigenic fragments may form a complex with the MHC and be transported back to the cell surface. There they are "presented" to the circulating T helper (T_h) cells which contain receptors able to bind specifically to particular structural and chemical characteristics of the degraded antigen–MHC complex. If a T_h cell recognizes and binds to the presented antigen on the surface of the antigen-presenting cells (APCs), the T_h cell proliferates and begins to produce various lymphokines. Finally, the recognition and binding of the presented antigen by the T_h cells, coupled with the release of lymphokines, stimulates the associated B cells to proliferate and produce antibodies which recognizes the intact antigen (Germain, 1986; Pier *et al.*, 2004).

Antigens are usually macromolecules or macromolecular complexes that contain distinct antigenic sites or "epitopes," which can be recognized and interact with the various components of the immune system. They can exist as individual molecules composed of synthetic organic chemicals, proteins, lipoproteins, glycoproteins, RNA, DNA, and polysaccharides—or they may be parts of cellular structures (bacteria or fungi) or viruses (Male *et al.*, 1987; Harlow and Lane, 1988).

Small molecules like short peptides, although normally able to interact with the products of an immune response, often cannot cause a response on their own. These "haptens," as they are called, are actually incomplete antigens, and while not able by themselves to cause immunogenicity or to elicit antibody production, they can be made immunogenic by coupling them to a suitable carrier molecule (Figure 19.1). Carriers are typically antigens of higher molecular weight that are able to cause an immunological response when administered *in vivo*.

Antibodies are typically able to recognize peptide sequences as small as five to six amino acids in length. For instance, IgE auto-antibodies were found to have clinical significance in multiple sclerosis by binding

Bioconjugate Techniques, Third Edition.
DOI: http://dx.doi.org/10.1016/B978-0-12-382239-0.00019-4

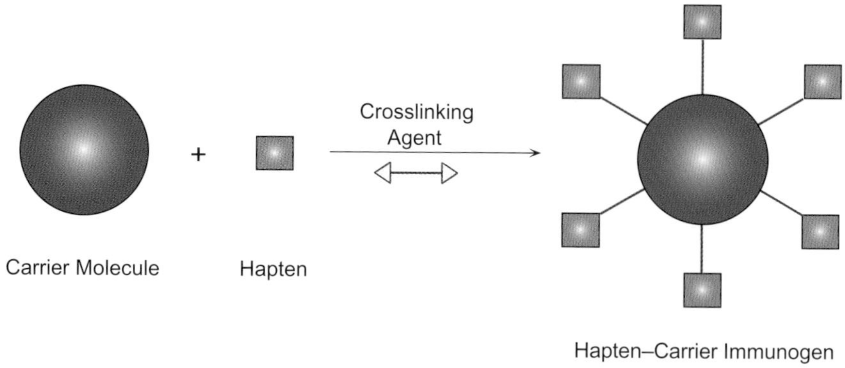

FIGURE 19.1 Immunogens are made by the conjugation of a hapten molecule to a carrier using a conjugation reagent.

specifically to short 5- and 6-amino acid epitopes on the surface of myelin proteins (Mikol *et al.*, 2006).

In an immune response, antibodies are produced and secreted by the B lymphocytes in conjunction with the T_h cells. In the majority of hapten–carrier systems, the B cells end up producing antibodies that are specific for both the hapten and the carrier. In these cases, the T lymphocytes will have specific binding domains on the carrier, but will not recognize the hapten alone. In a kind of synergism, the B and T cells cooperate to induce a hapten specific antibody response. After such an immune response has taken place, if the host is subsequently challenged with only the hapten, usually it will respond by producing hapten-specific antibodies from memory cells formed after the initial immunization. For a review of immunobiology, see Janeway (2004).

Synthetic haptens mimicking some critical epitopic structures on larger macromolecules are often conjugated to carriers to create an immune response to the larger "parent" molecule. For instance, short peptide segments can be synthesized from the known sequence of a viral coat protein and coupled to a carrier to induce immunogenicity toward the native virus. This type of synthetic approach to immunogen production has become the basis of much of the current research into the creation of vaccines.

The complete picture of the immune system is much more complex than this brief discussion can justly describe. In many instances, merely creating a B cell response by using synthetic peptide–carrier conjugates, however well designed, will not always guarantee complete protective immunity toward an intact antigen. The immune response generated by a short peptide epitope from, say, a larger viral particle or bacterial cell may only be sufficient to generate memory at the B cell level. In these cases, it is generally now accepted that a cytotoxic T cell response is a more important indicator of protective immunity. Designing peptide immunogens with the proper epitopic binding sites for both B cell and T cell recognition is one of the most challenging research areas in immunology today.

Hapten–carrier conjugates are also being used to produce highly specific monoclonal antibodies that can recognize discrete chemical epitopes on the coupled hapten. The resulting monoclonals are often used to investigate the epitopic structure and interactions between native proteins. In many cases, the haptens used to generate these monoclonals are again small peptide segments representing crucial antigenic sites on the surface of larger proteins. Monoclonals developed from known peptide sequences will interact in highly defined ways with the protein from which the sequence originated. These antibodies can then be used, for example, as competitors to the natural interactions between a receptor and its ligand. Thus, using antibodies generated from hapten–carrier conjugates, information can be obtained as to the precise sites of binding between macromolecules.

The preparation of hapten–carrier conjugates using peptide sequences can be controlled to produce immunogens that generate high-affinity antibodies when administered *in vivo*. Pedersen *et al.* (2006) determined that antibody titers increased in response to increasing the peptide-to-carrier ratio of conjugation. However, just the opposite effect was found for generating high affinity antibodies. The lower the peptide-to-carrier conjugation ratio, the higher the relative affinity of the antibodies produced. In addition, it was also found that coupling peptides to the carrier through a central amino acid residue caused higher antibody titers than using a terminal amino acid residue for conjugation. For this reason, for the preparation of particular immunogen conjugates, several ratios and methods of conjugation may have to be investigated to result in the optimal level and affinity of antibodies produced.

Also see Chapter 1, Section 3.6, for a review on the use of immunogen conjugates or conjugate vaccines for therapeutic applications, including the production of vaccines against infectious diseases and cancer.

2. TYPES OF IMMUNOGEN CARRIERS

The most commonly used carriers are typically immunogenic, large molecules that are capable of imparting immunogenicity to covalently coupled haptens. Some of the more popular ones are proteins, but other carriers may be composed of lipid bilayers (liposomes), synthetic or natural polymers (dextran, agarose, poly-L-lysine), synthetically designed organic molecules (i.e., dendrimers, see Chapter 8), and even inorganic particles (e.g., gold, see Chapter 14, Section 6). The liposome, polymer, and synthetic carriers may not be as immunogenic as the protein carriers, and this feature may be important for developing an immune response toward certain antigens which are difficult to get the immune system to recognize. The criteria for selecting a successful carrier molecule are the potential for imparting immunogenicity towards an associated hapten, the presence of suitable functional groups for conjugation, reasonable solubility properties even after derivatization (although this is not an absolute requirement, since precipitated molecules can be highly immunogenic), and lack of toxicity *in vivo*. The design of the appropriate vaccine delivery carrier is vitally important to developing a successful vaccine (Correia-Pinto *et al.*, 2012).

Some synthetic carriers are actually designed to have low immunogenicity on their own to minimize the potential for antibody production against them. When a hapten is coupled to these molecules, the immune response is directed principally toward the modification, not at the carrier. This design approach guides most of the immune response toward the desired target and minimizes the production of carrier-specific antibodies.

2.1. Protein Carriers

The first carrier molecules used for immunogen conjugation were proteins. A foreign protein administered *in vivo* by any one of a number of potential routes nearly ensured the elicitation of an immune response. In addition, protein carriers could be chosen to be highly soluble and possessed of abundant functional groups that could facilitate easy conjugation with a hapten molecule. When proteins are used as carriers in immunogen complex, the conjugates can be injected in any animal except the animal of origin for the carrier protein itself. In other words, the use of bovine serum albumin (BSA) would not be suitable for administration into cows, since self proteins would not be expected to elicit good immune responses, even when attached with hapten molecules.

The most common carrier proteins in use today are keyhole limpet hemocyanin (KLH; MW 4.5×10^5 to 1.3×10^7), bovine serum albumin (BSA; MW 67,000), aminoethylated (or cationized) BSA (cBSA), thyroglobulin (MW 660,000), ovalbumin (OVA; MW 43,000), and various toxoid proteins, including tetanus toxoid and diphtheria toxoid. Other proteins occasionally used include myoglobin, rabbit serum albumin, immunoglobulin molecules (particularly IgG) from bovine or mouse sera, tuberculin purified protein derivative, and synthetic polypeptides such as poly-L-lysine and poly-L-glutamic acid.

KLH

Perhaps the most popular carrier protein is KLH, which has been used as part of immunogen conjugates for antibody production, infectious disease vaccines, and anti-cancer vaccines (Chen *et al.*, 2011; Aarntzen *et al.*, 2012). The hemocyanin from keyhole limpets (the mollusk *Megathura crenulata*) is the oxygen-carrying protein of these primitive sea creatures. KLH is an extremely large, multi-subunit protein that contains chelated copper of non-heme origin. In concentrated solutions above pH 7.0, it displays a characteristic opalescent blue color that betrays its near insolubility and copper prosthetic groups. In acidic solutions, the blue color changes to green. At physiological pH, the protein exists in various subunit aggregate states of large molecular weight. For instance, in Tris buffer at pH 7.4 it is known to associate in five different aggregate forms (Senozan *et al.*, 1981). In highly alkaline or acidic environments, KLH disassociates into subunits (Hersckovits, 1988). The protein exhibits increased immunogenicity when it is disassociated into subunits, probably due to exposure of additional epitopic sites to the immune system (Bartel and Campbell, 1959). The intact protein usually creates considerable light-scattering or iridescent effects due to its size and almost colloidal nature in aqueous solutions. Subunits of KLH that are highly soluble in aqueous solution are available commercially (Thermo Fisher, Biosyn). KLH is a frequent choice for developing immunogen conjugates, especially for the treatment of cancer (Curigliano *et al.*, 2006; Sabbatini and Odunsi, 2007). The carrier has been used to conjugate sialyl-Tn, a glycan found on tumor cell-associated mucins, to treat metastatic breast cancer (the Theratope vaccine) (Miles *et al.*, 2011), and although the clinical trials of this vaccine did not prove efficacious, KLH continues to be used as an important carrier for anticancer agents. It is even used in the unconjugated form as an adjuvant to stimulate the immune system to respond to cancer (Aarntzen *et al.*, 2012).

Since keyhole limpets are marine creatures existing in a high-salt environment, native KLH maintains its best stability and solubility in buffers containing at least 0.9-*M* NaCl (not 0.9%). As the concentration of NaCl is decreased below about 0.6-*M*, the protein begins to precipitate and denature. Conjugation reactions using

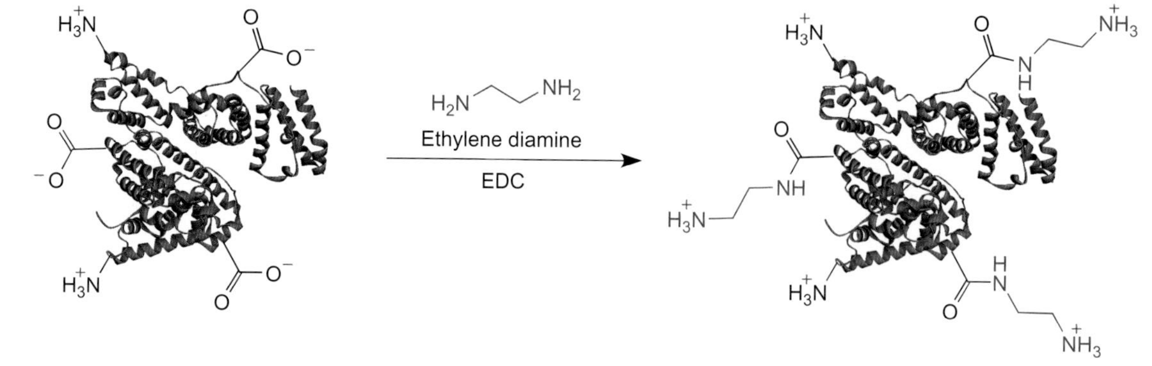

FIGURE 19.2 Cationized bovine serum albumin (cBSA) is formed by the reaction of ethylene diamine with native BSA using the water soluble carbodiimide EDC. Blocking of the carboxylate groups on the protein combined with the addition of terminal primary amines raises the pI of the molecule to highly basic values.

multi-subunit KLH, therefore, should be performed under high-salt conditions to preserve the solubility of the hapten–carrier complex. KLH used in the form of discrete subunits does not have this requirement of high salt to maintain solubility.

Native, multi-subunit KLH also should not be frozen. Freeze–thaw effects cause extensive denaturation and result in considerable amounts of insoluble material. Commercial preparations of native KLH are typically freeze-dried solids that no longer fully dissolve in aqueous buffers and do not display the protein's typical blue color due to loss of chelated copper. The partial denatured state of these products often makes conjugation reactions difficult.

KLH contains an abundance of functional groups available for conjugation with hapten molecules. On a per-mole basis (using an average multi-subunit MW of 5,000,000 Daltons), KLH has over 2000 amines from lysine residues, over 700 sulfhydryls from cysteine groups, and over 1900 tyrosine residues. Activation of the protein with succinimidyl-4-(*N*-maleimidomethyl) cyclohexane-1-carboxylate (SMCC) (Section 5, this chapter) typically results in 300 to 600 maleimide groups per molecule for coupling to sulfhydryl containing haptens.

The preparation of immunogen conjugates often requires the coupling of a sparingly soluble hapten to a carrier molecule. Pre-dissolving the hapten in an organic solvent and adding an aliquot of this solution to an aqueous reaction mixture is typically performed to maintain at least some solubility of the hapten in the conjugation solution. Dimethyl sulfoxide (DMSO) may be used for this purpose with KLH while maintaining very good solubility characteristics of the protein as well as the hapten. KLH is completely soluble in 50% (v/v) DMSO, becomes cloudy at a level of 60%, and definitely precipitates at 67%. Therefore, conjugation reactions may be performed by adding a volume of aqueous KLH to an equal volume of hapten dissolved in DMSO. Care should be taken, however, to avoid buffer salt precipitation upon addition of organic solvent.

BSA and cBSA

BSA (MW 67,000) and cationized BSA (cBSA) are highly soluble proteins containing numerous functional groups suitable for conjugation. Even after extensive modification with hapten molecules these carriers usually retain their solubility. The exception to this statement is when hydrophobic peptides or other sparingly soluble molecules are conjugated to the proteins. Modification of any carrier with numerous hydrophobic haptens may cause enough masking of the hydrophilic surface to result in precipitation. Depending on the degree of precipitation, such conjugates are often still useful in generating an immune response. To limit the production of insoluble complexes, however, the conjugation reaction can be scaled back to reduce the level of carrier modification and thus reduce or eliminate precipitation.

BSA possesses a total of 59 lysine ε-amine groups (with only 30–35 of these typically available for derivatization), one free cysteine sulfhydryl (with an additional 17 disulfides buried within its three-dimensional structure), 19 tyrosine phenolate residues, and 17 histidine imidazole groups. In addition, the presence of numerous carboxylate groups gives BSA its net negative charge (pI 5.1).

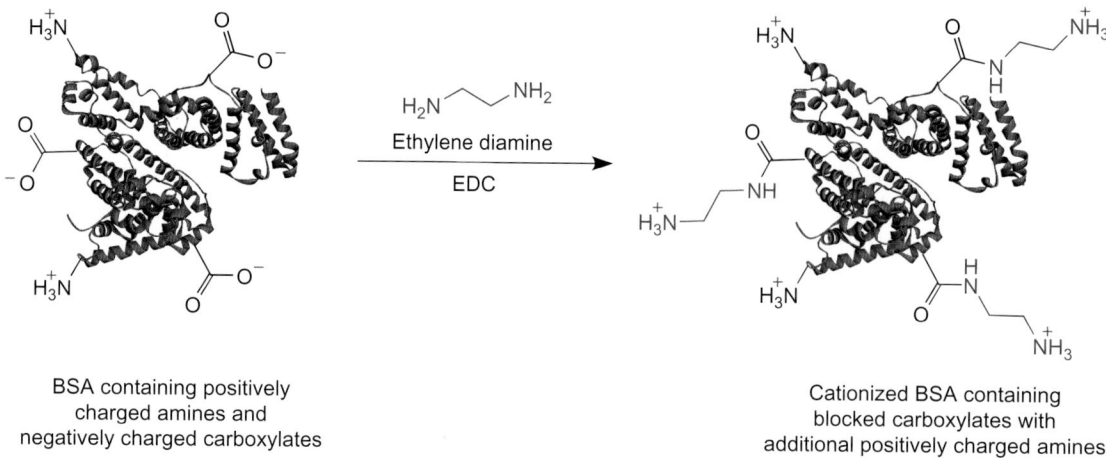

Cationized BSA is prepared by modification of its carboxylate groups with ethylenediamine (Chapter 2, Section 4.3) (Figure 19.2). Controlled aminoethylation using the water soluble carbodiimide EDC results in blocking many of BSA's aspartic and glutamic acid side chains (and possibly the C-terminal carboxylate), forming an amide bond with a 2-carbon spacer containing a terminal primary amine group. Since the negative charge contributions of the native carboxylates are masked and positively charged amines are created in their place, the result of this process is a significant rise in the protein's pI. Cationization performed according to published procedures alters the net charge of BSA from a pI of about 5.1 (Cohn *et al.*, 1947) to over pI 11.0 (Muckerheide *et al.*, 1987).

The highly positive charge of cBSA dramatically increases its immunogenicity. The positive character of the molecule aids in its binding to antigen-presenting cells (APCs) *in vivo*, the first step in antibody production. The protein thus gets incorporated into the APCs faster than molecules having lower pI values. It also gets processed at an accelerated rate, producing a quicker immune response, and one that occurs with greater concentrations of specific antibody (Domen *et al.*, 1987; Muckerheide *et al.*, 1987b; Apple *et al.*, 1988; Domen and Hermanson, 1992; Chen *et al.*, 2002).

Cationized BSA used as a carrier protein also induces a similar increase in the production of antibodies against any attached hapten molecules. Even when haptens are coupled through cBSA's amine residues, the overall charge of the molecule remains basic enough to augment the immune response beyond that usually obtained using other carriers. This augmentation occurs even when the attached molecule is not merely a hapten, but a larger antigen macromolecule. Conjugation of a complete antigen (a molecule able to generate an immune response on its own) to cBSA causes an increased immune response against the antigen beyond that normally obtainable for the native antigen administered in unconjugated form (Domen and Hermanson, 1992). Bioconjugates of cBSA have been developed for polysaccharide-based glycoconjugate vaccines (Burtnick *et al.*, 2012), for vaccines strategies in overcoming atherosclerosis (de Jager and Kuiper, 2011), and to make cationic cBSA-nanoparticles for treatment of Parkinson's disease (Rodríguez *et al.*, 2011).

The effectiveness of cBSA as a carrier for peptides was investigated using arginine vasopressin (AV) as the hapten. Figure 19.3 shows the antibody concentration resulting after injection of the AV–cBSA conjugate intraperitoneally (i.p.) into BDF_1 female mice. As a control, native BSA was similarly conjugated with AV and administered in a second set of mice under identical conditions. The antibody concentrations in the sera were monitored periodically by enzyme-linked-immunosorbent assay (ELISA). The antibody response resulting from a set of mice injected with unconjugated peptide was subtracted in all cases. All injections were performed using 100 μg of conjugate mixed with an equal volume of alum (22.5 mg/ml aluminum hydroxide) as adjuvant.

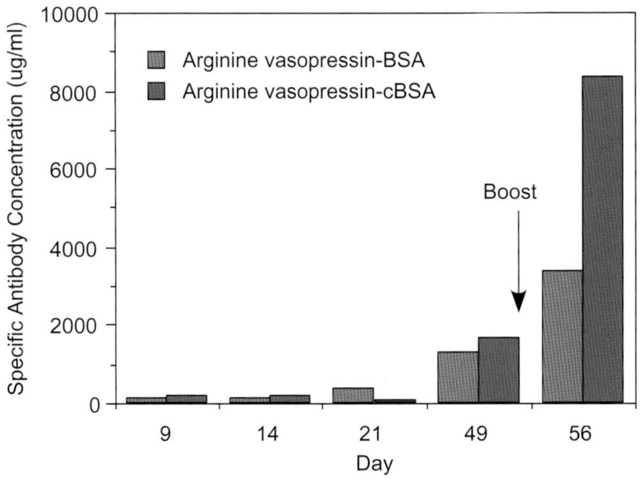

FIGURE 19.3 The effectiveness of a cationized carrier conjugate used as an immunogen can be seen by the comparison of specific antibody response in mice to arginine vasopressin (AV) coupled to both native BSA (nBSA) and cationized BSA (cBSA). The quantity injected was standardized according to the amount of arginine vasopressin (AV) present. The cationized carrier results in higher concentrations of antibody produced against the peptide than the immunogen made with native nBSA.

After the boost, the group of mice receiving the AV–cBSA conjugate generated over twice the antibody response as the group receiving the peptide conjugated to native BSA.

In a similar study, OVA conjugated to cBSA was compared to the same protein conjugated to native BSA (nBSA) and also OVA administered in an unconjugated form in mice. Figure 19.4 shows that before and after the boost, the OVA–cBSA conjugate resulted in much higher antibody concentrations than either the OVA–nBSA conjugate or OVA injected in an unconjugated form. Similar results were obtained for a conjugate of human IgG with cBSA (Figure 19.5).

A corollary to the use of cBSA as a carrier protein is that its increased immune response often abrogates the use of complete Freund's adjuvant, which is a source of concern because of its potential side-effects in animals. A relatively innocuous mixture with alum is usually all that is required as adjuvant to result in good antibody production.

As mentioned previously for KLH, DMSO may be used to solubilize hapten molecules that are rather insoluble in aqueous environments. Conjugation reactions may be performed in solvent/aqueous phase mixtures to maintain some solubility of the hapten once it is added to a buffered solution. BSA remains soluble in the presence of up to 35% DMSO, becomes slightly cloudy at 40%, and precipitates at 45% (v/v).

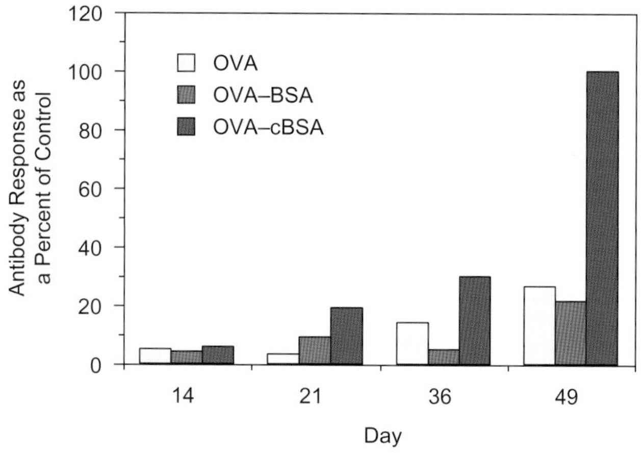

FIGURE 19.4 Cationized cBSA can even increase the specific antibody response to large proteins coupled to it. This graph shows a comparison of the relative antibody response in mice to injections of ovalbumin (OVA), either in an unconjugated form or conjugated to nBSA (native) or cationized (cBSA). The quantity injected was standardized according to the amount of ovalbumin OVA present. The highly basic cBSA molecule modulates the immune response to enhance the production of antibodies toward even full sized proteins conjugated with it.

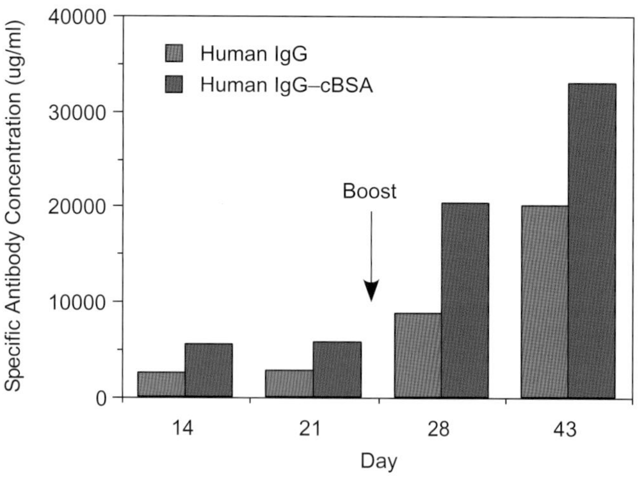

FIGURE 19.5 Human IgG was injected in mice in either an unconjugated form or crosslinked with cBSA. The quantity injected was standardized according to the amount of IgG present. A greater antibody response was obtained using the cBSA conjugate.

Thyroglobulin and OVA

Thyroglobulin and ovalbumin (OVA) are used less often as carriers, but they are particularly valuable as non-relevant carriers in ELISA tests designed to measure the antibody response after injection of an immunogen conjugate. Since an antibody response would be directed both against the carrier and the attached hapten, an ELISA performed to quantify specific antibody that only interacts with the hapten must not utilize the same carrier in the conjugate coated on the microplates. If the identical carrier conjugate is used for the ELISA as was used in the original immunization, the test results will be skewed by the contribution of carrier-specific antibodies. For this reason, a non-relevant carrier—one that is not recognized by the products of the immune response—must be coupled with hapten and used for the ELISA test.

Since OVA and BSA possess some immunologically similar epitopes, a population of the antibodies produced against one of them will often cross-react against the other. Therefore, OVA cannot function as a non-relevant carrier for BSA and *vice versa*. Either OVA or BSA, however, may be used as non-relevant carriers for KLH, thyroglobulin, or the various toxoid proteins used as immunogen conjugates.

OVA comprises about 75% of the total protein in hen egg whites. It is a phosphoprotein containing one *N*-glycosylation site and 386 amino acids. The protein contains 20 lysine residues, 14 aspartic acids, and 33 glutamic acid groups. This gives a total of 20 ε-amines, one N-terminal amine, 47 side-chain carboxylates, and one C-terminal carboxylate for conjugation reactions. The majority of acidic groups give the protein a pI of 4.63. Additional sites of modification include 4 sulfhydryl groups, 10 tyrosines, and 7 histidine residues. OVA is sensitive to temperature (above 56°C), electric fields, and vigorous mixing. Care should be taken in handling the protein to prevent denaturation and subsequent precipitation.

One advantage of OVA is its extreme solubility characteristics in the presence of DMSO. A sparingly soluble hapten molecule may be dissolved in this solvent and added to an aqueous OVA reaction mixture to maintain solubility of the molecule during conjugation. OVA is soluble at up to 70% DMSO, becomes cloudy at 75%, and precipitates at 80% (v/v).

Thyroglobulin is a prohormone protein which is synthesized and stored in the thyroid gland. Specific proteolytic action on the protein *in vivo* causes the release of triiodothyronine and thyroxine, low-molecular-weight amino acid derivatives that affect metabolic rate and oxygen consumption. Thyroglobulin is a large, multi-subunit protein composed of several polypeptide chains (MW 670,000). Its acidic pI (4.7) reflects the abundance of carboxylate groups. Thyroglobulin is also glycosylated, containing about 8 to 10% carbohydrate. Its use as an immunogen carrier protein is less frequent than that of KLH or BSA.

Tetanus and Diphtheria Toxoids

Toxoid proteins are biologically inactivated forms of native toxins. The most often used toxoid is tetanus toxoid, but diphtheria-derived toxoids and other proteins are also used occasionally (Anderson *et al.*, 1989). These carrier proteins have been used frequently for pertussis and influenza vaccines (Halperin *et al.*, 2011; McCormick, 2012). Tetanus toxoid (MW 150,000) has 106 amine groups, 10 sulfhydryls, 81 tyrosine residues, and 14 histidines that may participate in conjugation reactions with hapten molecules (Bizzini *et al.*, 1970). Diphtheria toxoid is derived from a protein secreted by certain strains of *Corynebacterium diphtheriae*. Its molecular weight is approximately 63,000 Daltons (Collier and Kandel, 1971). Both protein toxoids can be used to couple haptens through any of the chemical reactions described in this section. They generate strong immunological responses *in vivo*.

2.2. Liposome Carriers

Liposomes are artificial structures composed of phospholipid bilayers exhibiting amphiphilic properties (Chapter 21). In complex liposome morphologies, concentric spheres or sheets of lipid bilayers are usually separated by aqueous regions that are sequestered or compartmentalized from the surrounding solution. The phospholipid constituents of liposomes consist of hydrophobic lipid tails connected to a head constructed of various glycerylphosphate derivatives. The hydrophobic interaction between the fatty acid tails is the primary driving force for creating liposomal bilayers in aqueous solutions.

The morphology of a liposome may be classified according to the compartmentalization of aqueous regions between bilayer sheets. If the aqueous regions are sequestered by only one bilayer each, the liposomes are called unilamellar vesicles (ULV). If there is more than one bilayer surrounding each aqueous compartment, the liposomes are termed multilamellar vesicles (MLV). ULV forms are further classified as to their relative size, although rather crudely. Thus, there can be small unilamellar vesicles (SUV; usually less than 100nm in diameter) and large unilamellar vesicles (LUV; usually greater than 100nm in diameter). With regard to MLV, however, the bilayer structures cannot easily be classified due to the almost infinite number of ways each bilayer sheet can be associated and interconnected with the next one. MLVs typically form large complex honeycomb structures that are difficult to classify or reproduce.

The overall composition of a liposome—its morphology, composition (including a variety of potential phospholipids and the degree of its cholesterol content), charge, and any attached functional groups—can affect the antigenicity of the vesicle *in vivo* (Allison and Gregoriadis, 1974; Alving, 1987; Therien and Shahum, 1989). When liposomes are used as carriers for immunization purposes, the haptens or antigens are usually attached covalently to the head groups using various phospholipid derivatives and crosslinking strategies (Derksen and Scherphof, 1985). Most often, these derivatization reactions are performed off of phosphatidylethanolamine constituents within the liposomal mixture. The primary amine modification of the glycerylphosphate head group of phosphatidylethanolamine provides an ideal functional group for activation and subsequent coupling of hapten molecules (Shek and Heath, 1983). Stock preparations of activated liposomes may be prepared and lyophilized to be used as needed in coupling hapten molecules (Friede *et al.*, 1993). Conjugates of liposomes with peptides or other molecules have been used to target cells *in vivo* for disease therapy (Du *et al.*, 2007). Liposomes are becoming more popular as conjugate vaccines for the treatment of cancer (Butts *et al.*, 2011; Wu *et al.*, 2011). All of the amine-reactive conjugation methods discussed in this section may be used with phosphatidylethanolamine-containing liposomes; however, see Chapter 21 for a more complete discussion of the unique considerations associated with conjugation of molecules to liposomes.

2.3. Synthetic Carriers

Synthetic molecules may be used as immunogen carriers if they are designed with the appropriate functional groups to facilitate the coupling of hapten molecules. These carriers may consist of simple polymers such as poly-L-lysine, poly-L-glutamic acid, Ficoll, dextran, polyethylene glycol (PEG), dendrimers, or other synthetic organic constructs (Lee *et al.*, 1980; Fok *et al.*, 1982; Boyle *et al.*, 1983; Hopp, 1984; Wheat *et al.*, 1985). Coupling of hapten molecules to the principle functional groups of these polymers can produce immunogenic conjugates that may be injected in animals to generate a specific antibody response. The advantage of a synthetic vaccine carrier is the potential for having little to no immune response to the carrier itself. This may be important if the hapten is relatively low in immunogenicity, such as is often the case when using tumor-associated carbohydrate antigens. If the carrier is too immunogenic, it could overwhelm and prevent a response to the carbohydrate antigen (Huang *et al.*, 2012). Thus, synthetic carrier molecules are becoming more important for developing vaccines against cancer and infectious diseases when the antigen target is a glycan associated with tumor cells or a polysaccharide associated with pathogens.

Poly-L-lysine may be coupled to carboxylate-containing molecules using the carbodiimide conjugation procedure to yield amide linkages (Chapter 4, Section 1). Homobifunctional or heterobifunctional crosslinking agents may also be used with poly-L-lysine, such as in the use of sulfo-SMCC (Chapter 6, Section 1.3). The polymer can be used as well for coupling hapten molecules and subsequent coating of microplates for ELISA procedures (Gegg and Etzler, 1993). Conversely, poly-L-glutamic acid may be coupled to amine-containing haptens by the same carbodiimide protocol. Ficoll and dextran carriers may be activated by mild sodium periodate oxidation to generate reactive aldehyde groups (Chapter 2, Section 4.4, and Chapter 2, Section 2.1). Coupling to amine-containing haptens may then be performed by reductive amination (Chapter 3, Section 5). Polyethylene glycol chemistry involves alternate activation and coupling schemes that are addressed in Chapter 18.

A unique synthetic molecule that can be used as a carrier is the so-called multiple antigenic peptide (MAP) (Posnett *et al.*, 1988; Tam, 1988; Kowalczyk *et al.*, 2011). The MAP core structure is composed of a scaffolding of sequential levels of poly-L-lysine. The dendritic molecule is constructed from a divalent lysine compound to which two additional levels of lysine are attached. The final MAP compound consists of a symmetrical, octavalent primary amine surface to which hapten molecules may be attached. Coupling of up to eight peptide haptens to the MAP core yields a highly immunogenic complex having a molecular weight typically greater than 10,000. The nature of the MAP carrier makes it ideal for remarkably defined conjugates useful in vaccine development (Jayaprakash Babu *et al.*, 2011; Shreewastav *et al.*, 2012).

One particularly novel carrier was reported to consist of 50- to 70-nm colloidal gold particles of the type often used in cytochemical labeling techniques for microscopy (Pow and Crook, 1993) (Chapter 14, Section 6). Adsorption of peptide antigens onto gold and subsequent injection of the complex into rabbits in an adjuvant mixture resulted in rapid production of antibodies of extremely high titer. The resultant antibodies could be used in immunocytochemistry at dilutions from 1-in-250,000 down to 1-in-1,000,000, which is orders of magnitude beyond the dilutions typically used with lower-titer antibodies.

Gold nanoparticle vaccines are becoming more popular in developing effective anticancer and infectious disease treatments (Wang *et al.*, 2011; Barhate *et al.*, 2012; Gregory *et al.*, 2012; Lee *et al.*, 2012).

3. CARBODIIMIDE-MEDIATED HAPTEN–CARRIER CONJUGATION

The coupling chemistry used to prepare an immunogen from a hapten and carrier protein is an important consideration for the successful production and correct specificity of the resultant antibodies. The choice of crosslinking methodology is governed by the functional groups present on the carrier and the hapten as well as the orientation of the hapten desired for appropriate presentation to the immune system. An associated concern is the potential for antibody recognition and cross-reactivity toward the crosslinking reagent used to effect the conjugation. If antibodies are generated against the crosslinker bridge connecting the carrier with the hapten, then this may dilute the desired antibody response against the hapten. The use of a zero-length crosslinking procedure mediated by the water soluble carbodiimide EDC eliminates this problem, since no bridging molecule is introduced between the hapten and carrier.

The reactions involved in an EDC-mediated conjugation are discussed in Chapter 4, Section 1.1. (Note: EDC is 1-ethyl-3-(3-dimethylaminopropyl) carbodiimide hydrochloride, which has a molecular weight of 191.7 and is sometimes referred to as EDAC.) The carbodiimide first reacts with available carboxylic groups on either the carrier or hapten to form a highly reactive *O*-acylisourea intermediate. The activated carboxylic group can then react with a primary amine to form an amide bond, with release of the EDC mediator as a soluble isourea derivative. The reaction is quite efficient with no more than 2h required for it to go to completion and form a conjugated immunogen.

Since most peptide haptens contain either amines or carboxylic groups available for coupling, EDC-mediated immunogen formation may be the simplest method for the majority of hapten–carrier protein conjugations. Figure 19.6 shows the coupling of a carrier protein to a short peptide molecule through its amine terminus. It should be kept in mind, however, that this type of conjugation may occur at either the C- or N-terminal of the peptide or at any carboxyl- or amine-containing side chains. Therefore, this method probably should be avoided if a particularly interesting part of the peptide contains groups which may be blocked or undergo coupling using the carbodiimide reaction. Also, when using peptides rich in Lys, Glu, or Asp, an unacceptable level of hapten crosslinking may occur upon conjugation and thus change the antigenic structure of the resulting immunogen. However, some crosslinking or polymerization of the peptide on the surface of the carrier actually may be beneficial to the immunogenicity of the peptide, and thus create an even greater antibody response. Some investigators even advocate using no carrier protein when a peptide hapten is involved; merely polymerizing the peptide in the presence of EDC may result in a complex of high enough molecular weight to be immunogenic by itself. In general, EDC coupling is a very efficient, one-step method for forming a wide variety of peptide-carrier protein immunogens.

Figure 19.7 shows the results of an EDC conjugation study comparing a reaction performed at pH 4.7 (A) to one performed at pH 7.3 (B and C), with and without added sulfo-NHS (see Chapter 4, Section 1.2).

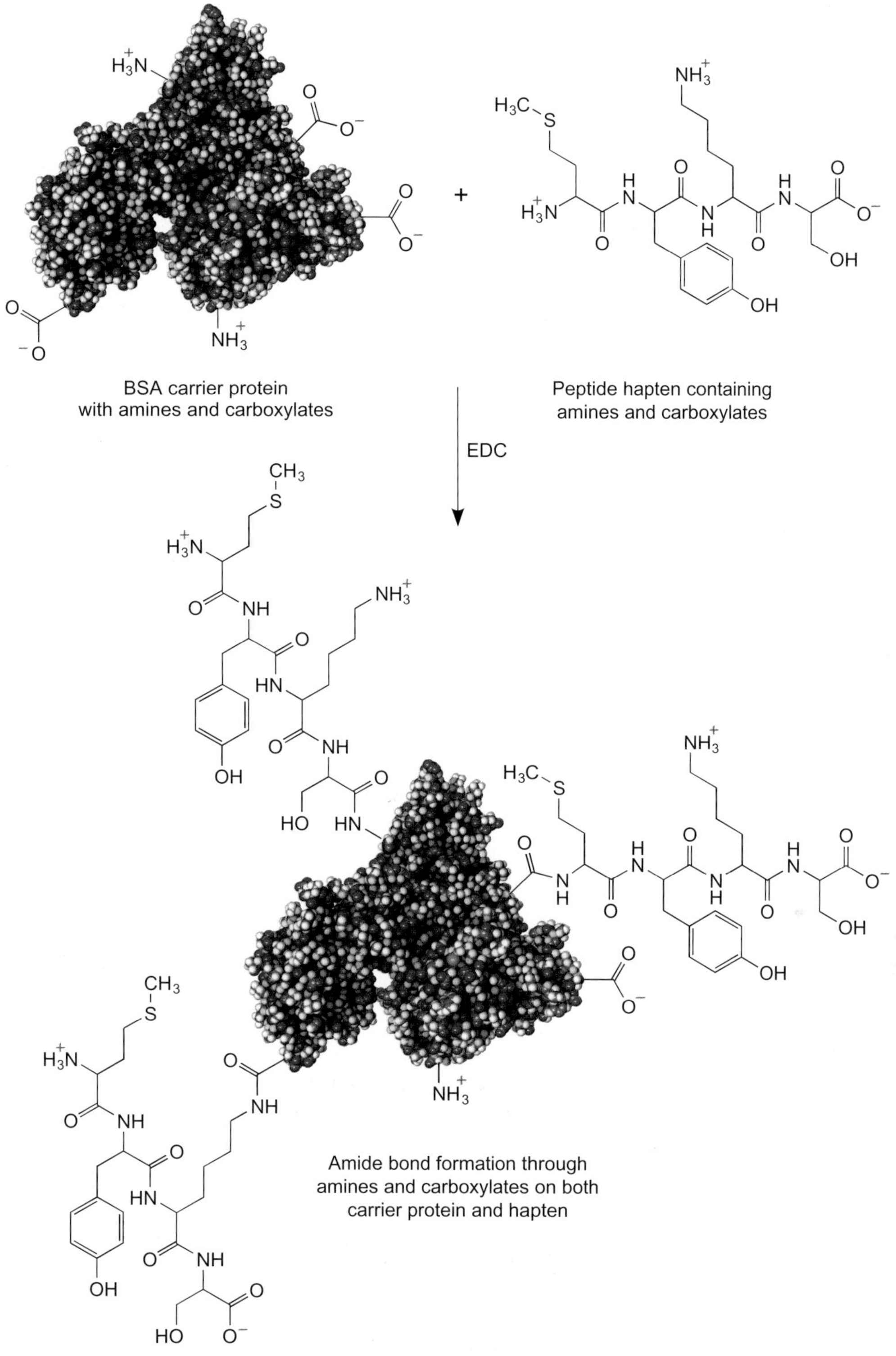

FIGURE 19.6 Peptide haptens are easily conjugated to carrier proteins using the water soluble carbodiimide EDC.

The graphs show the elution profiles of a gel filtration separation after conjugation. In each case, a blank run performed without the addition of EDC illustrates the separation of the protein carrier (the first peak) from the lower molecular weight peptide and reagent peak (the second peak). Decrease in the peptide peak is indicative of successful conjugation. Complete recovery of the total absorbance at 280 nm usually does not occur,

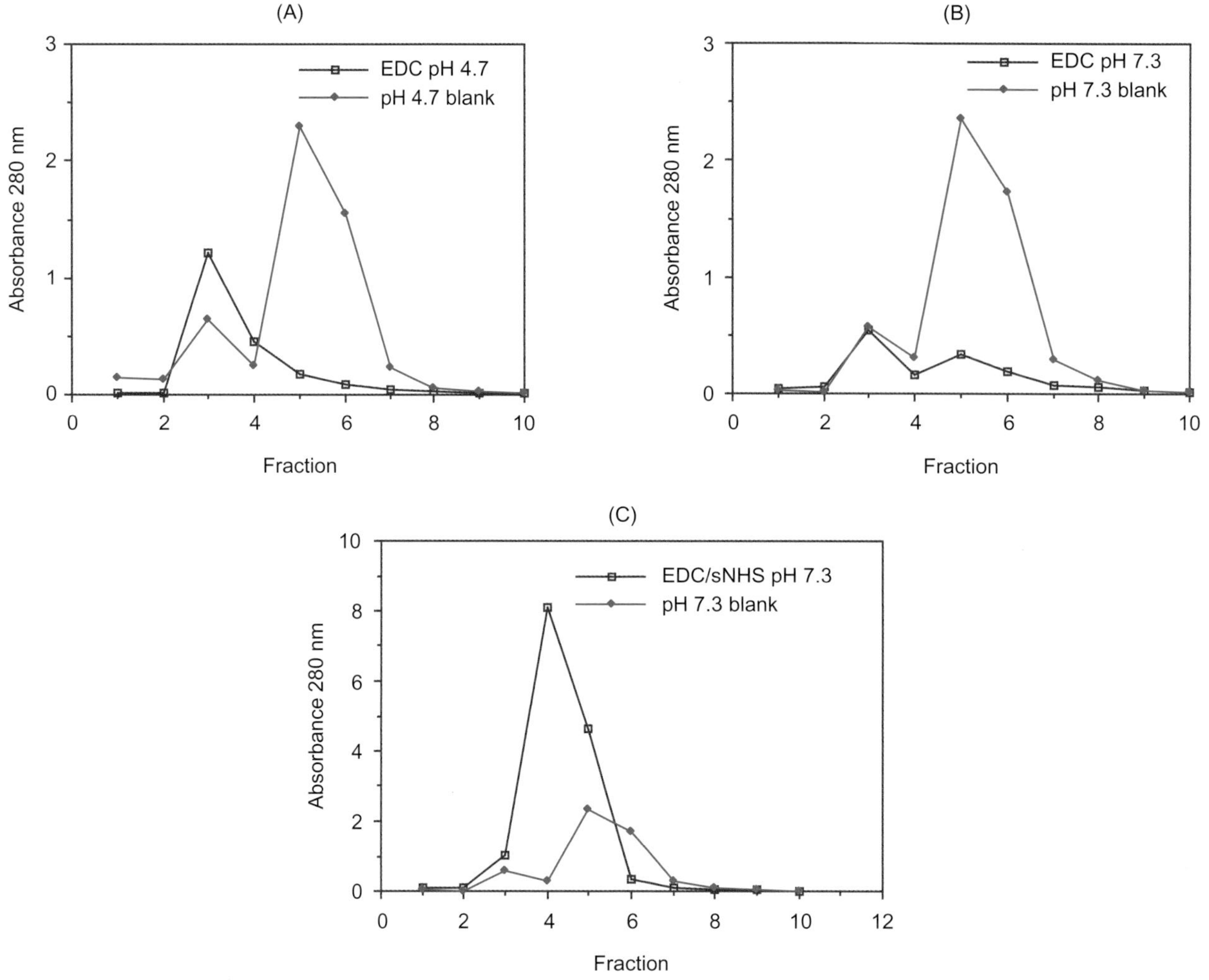

FIGURE 19.7 To assess the effectiveness of an EDC conjugation reaction of a peptide with a carrier protein, glycyl-tyrosine was coupled to BSA using various conditions. The graphs show a gel filtration profile on Sephadex G-25 after completion of the conjugation reaction. The first peak eluting off the column is the higher-molecular-weight, while the second peak is excess peptide. The elution profiles demonstrate that the carbodiimide reaction proceeds with nearly equal efficiency at pH 4.7 (A) or at pH 7.3 (B). In each graph, a comparison is shown between the separation of peptide and carrier without addition of EDC and the same mixture after reaction with EDC. Depletion of the peptide peak in the EDC-containing elution profiles indicates uptake of glycyl-tyrosine in the carrier conjugate. Some polymerization of peptide is also possible using this method, as evidenced by movement of the peptide peak toward higher-molecular-weight elution points. The addition of sulfo-NHS to the reaction caused precipitation problems as well as obscuring the analysis by size exclusion chromatography due to the absorbance of excess sulfo-NHS (C).

presumably due to a decrease in the peptide's absorptivity as it is conjugated or polymerized. Staros' method of adding sulfo-NHS to form an intermediate active ester that subsequently reacts with an amine to form the amide bond does not work as well due to excessive conjugation (causing precipitation in most cases) and interference from the eluting sulfo-NHS peak. The reaction proceeds with similar yields at either acid or neutral pH. Thus, the efficiency of an EDC conjugation reaction is approximately the same from pH 4.7 to physiological conditions.

Figure 19.8 shows the result of the conjugation of the dipeptide tyrosyl-lysine to BSA using various concentrations of EDC. Again, the elution profile shows the gel filtration pattern resulting after the reaction. The first peak is the protein carrier while the second is the peptide. Progressive decrease in the peptide peak with increasing amounts of EDC added to the reaction mixture correlates to increased conjugation yields. A side reaction to EDC conjugation of haptens that contain both an amine and a carboxylate group is hapten polymerization. This is revealed in the movement of the

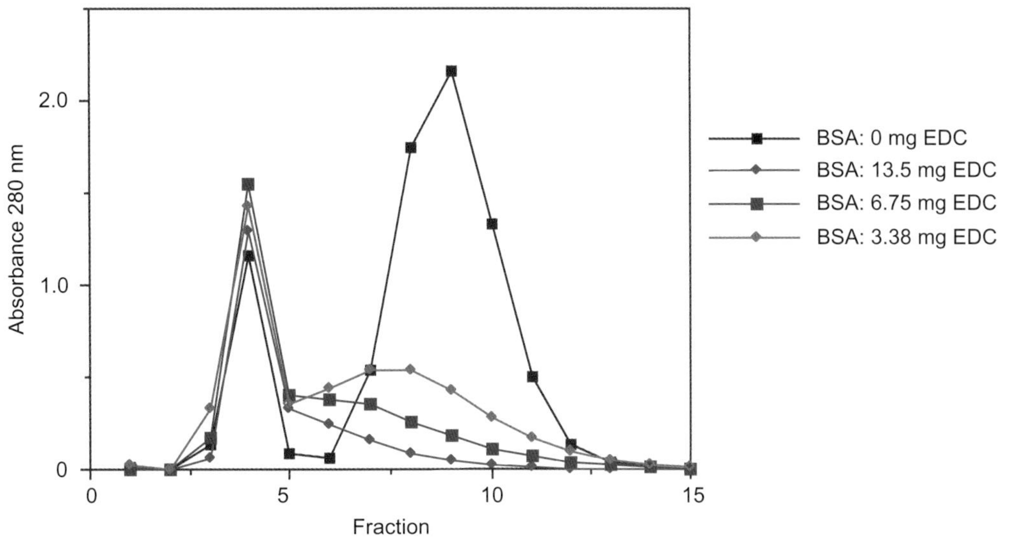

FIGURE 19.8 To study the conjugation of peptides to carriers using different levels of EDC, tyrosyl-lysine was conjugated to BSA and separated after the reaction by chromatography on a Sephadex G-25 column. As the EDC level was increased in the reaction, more peptide reacted and the peptide peak (the second peak) was depleted. The absorbance of the carrier peak (the first one) that elutes off the column was observed to increase as more peptide was conjugated.

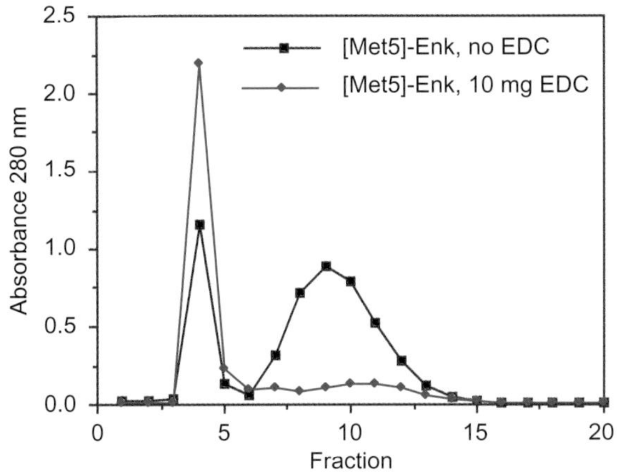

FIGURE 19.9 Conjugation of the biological peptide [Met5]-enkephalin to BSA using EDC. The graph shows the gel filtration profile (on Sephadex G-25) after completion of the conjugation reaction. A blank run with no added EDC was performed to illustrate the peak absorbance that would be obtained if no conjugation took place. With the addition of 10 mg of EDC to a reaction mixture consisting of 2 mg of BSA plus 2 mg of peptide, nearly complete conjugate formation was obtained.

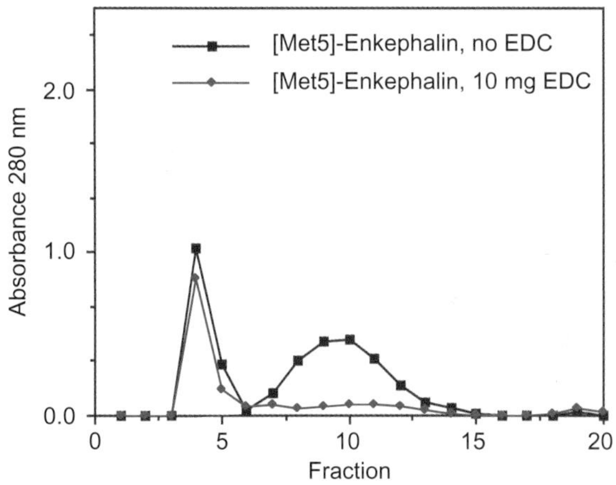

FIGURE 19.10 To illustrate the consistency of an EDC-mediated reaction, [Met5]-enkephalin was conjugated to ovalbumin (OVA) using conditions identical to those described for BSA in Figure 19.9. Note the similarity in the degree of conjugate formation.

peptide peak toward higher molecular weights (e.g., decreased time of elution) with increasing amounts of EDC added to the reaction.

Figure 19.9 illustrates the conjugation of [Met5]-enkephalin with BSA using EDC. The gel filtration profile after crosslinking reveals that the peptide peak effectively disappears upon complete conjugation with the carrier protein. With nicely soluble peptides such as this one, the immunogen remains freely soluble even at high modification levels. For less soluble peptides or haptens, reducing the amount of EDC addition may be necessary to maintain solubility in the conjugate.

To illustrate the similarity of an EDC conjugation reaction using a different carrier protein, but the same peptide, Figure 19.10 shows the gel filtration separation after conjugation of [Met5]-enkephalin to OVA. The uptake of peptide upon addition of EDC is almost

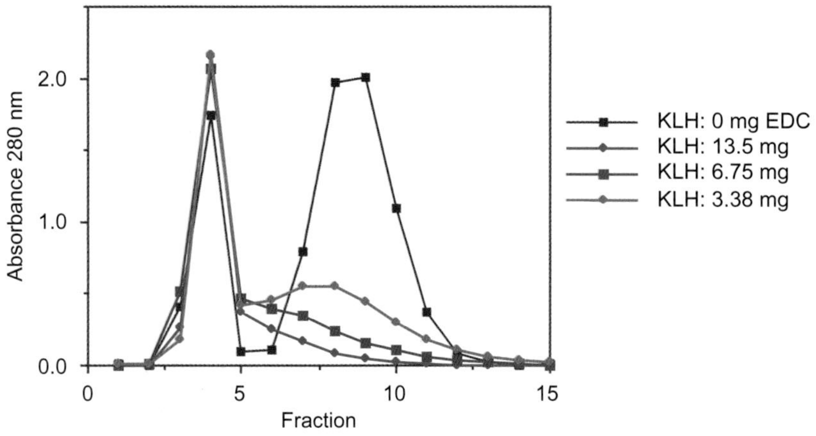

FIGURE 19.11 The EDC conjugation of tyrosyl-lysine to KLH is illustrated by the gel filtration pattern on Sephadex G-25 after the reaction. The first peak is the carrier protein and the second peak is the peptide. A blank containing no EDC is also shown to provide baseline peak heights that would be obtained if no crosslinking occurred. When more EDC was added, more peptide was conjugated, as evidenced by peptide peak depletion.

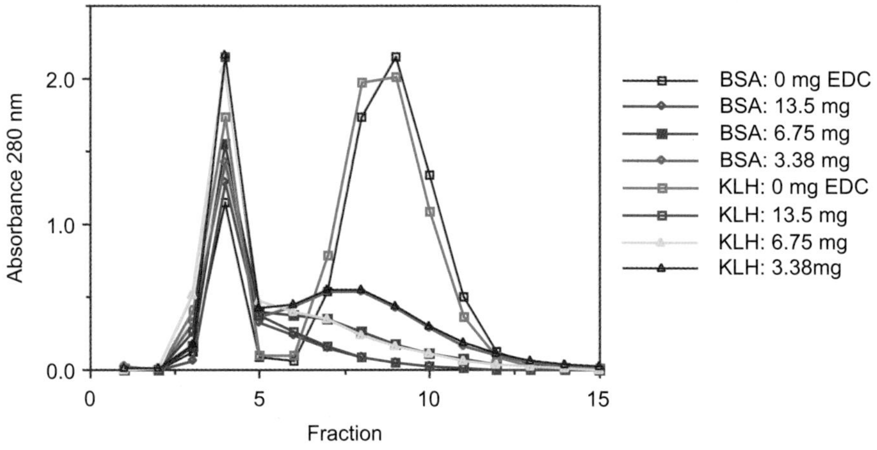

FIGURE 19.12 EDC conjugation reactions can be extraordinarily consistent using the same peptide crosslinked to two carrier proteins. This figure shows the gel filtration pattern on Sephadex G-25 after completion of the crosslinking reaction. Conjugation of tyrosyl-lysine to BSA and KLH are shown. The first peaks represent the carrier eluting from the column, while the second peaks are the excess peptide. Note the consistency of conjugation using the same levels of EDC addition.

identical to that observed when conjugating to BSA. This is logical, since on a per mass basis, there is very little difference between these proteins in the amount of amines or carboxylates available for conjugation.

Figure 19.11 shows the conjugation of tyrosyl-lysine to KLH using various concentrations of EDC. The elution profile shows the gel filtration pattern resulting after the reaction. Progressive decrease in the peptide peak (peak 2) with increasing amounts of EDC correlates to increased conjugation (or polymerization) yields. Despite the extremely high molecular weight of KLH compared to the other commonly used carriers, the conjugation reaction using EDC again proceeds with virtually identical results to the similar study shown in Figure 19.8 using BSA as the carrier. In fact, superimposing the two studies on the same graph demonstrates the reproducibility of an EDC-facilitated reaction (Figure 19.12).

Due to the high molecular weight of KLH and its solubility characteristics, the conjugation of this protein to some haptens can result in precipitation of the complex. This is especially true if the level of EDC addition is similar to the EDC concentrations used with lower molecular weight carriers such as BSA or OVA. Figure 19.13 shows the elution profile resulting from the gel filtration separation of KLH and the peptide [Met5]-enkephalin after an EDC reaction. To result in a soluble immunogen, only 0.1 to 0.2 times the amount of EDC was added as compared to similar BSA or OVA conjugation reactions. Even at levels this low, however, the

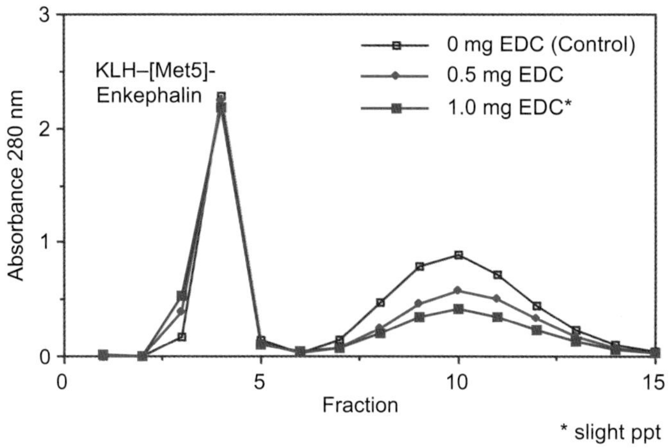

FIGURE 19.13 Conjugation to KLH can often cause precipitation due to the high-molecular weight of the carrier protein. The conjugation of [Met5]-enkephalin to KLH yields a soluble immunogen if the level of EDC addition is about 0.1-times that typically used with BSA as a carrier. This figure shows the gel filtration pattern on Sephadex G-25 after completion of the crosslinking reaction. The first peak is KLH and the second peak is excess peptide. Depletion of the peptide peak correlates to hapten–carrier conjugation.

coupling of peptide to the carrier is very efficient and results in an excellent immunogen.

These studies using EDC-facilitated conjugations were performed to develop an optimal protocol for the preparation of immunogens by carbodiimide cross-linking. For haptens (i.e., peptides) that display good solubility in aqueous solution, the level of reagent addition should result in a soluble immunogen conjugate; however, when using haptens that are sparingly soluble or insoluble in aqueous environments, the conjugation reaction may result in a precipitated complex. Precipitation can often be controlled by scaling back the level of EDC addition or limiting the time of the reaction. If a precipitated immunogen is not a problem (most precipitated, high-molecular-weight conjugates are very immunogenic), then the following protocol is applicable to the great majority of peptide-carrier protein conjugations. Thermo Fisher offers a kit containing all the reagents necessary for an EDC-mediated hapten-carrier conjugation.

Protocol

1. Dissolve the carrier protein in 0.1-*M* MES, 0.15-*M* NaCl, pH 4.7, at a concentration of 10 mg/ml. If using native, multi-subunit KLH, increase the NaCl concentration of all buffers to 0.9-*M* (yes, 0.9-*M*, not 0.9%) to maintain solubility of the protein. If using KLH subunits, the high salt concentration is not necessary. For neutral pH conjugations, substitute 0.1-*M* sodium phosphate, 0.15-*M* NaCl, pH 7.2, for the MES buffer.
2. Dissolve up to 4 mg of the peptide or hapten to be coupled in 1.0 ml of the reaction buffer chosen in step 1. If the peptide to be coupled is already in solution, it may be used directly if it is in a buffer containing no other amines or carboxylic acids and is at a pH between 4.7 and 7.2. Note: If an assessment of the degree of peptide coupling is desired, measure the absorbance at 280 nm of the 1.0-ml peptide solution before proceeding to step 3. In some cases, a dilution of the peptide solution may be necessary to keep the absorbance on scale for the spectrophotometer. If the peptide is sparingly soluble in aqueous solution, it may dissolve in DMSO and an aliquot can be added to the carrier solution to aid in dissolution. See the previous discussion on carrier proteins to determine the levels of DMSO compatible with carrier protein solubility. Peptide haptens should be at least five to seven amino acids in length to obtain suitable immunogenicity and correct specificity of the resulting antibodies produced against them. Many immunogen conjugates use peptides that are slightly longer than the desired epitopic sequence to provide enough size to allow interaction with the components of the immune system.
3. Add 500 μl of the peptide solution to 200 μl of carrier protein. For greater reaction volumes, keep the molar ratio of peptide-to-carrier addition the same and proportionally scale up the amount of EDC added in the next step. If the peptide is initially dissolved in DMSO, much less peptide volume compared to protein volume should be used to maintain solubility (see discussion in step 2).
4. For conjugations using relatively low molecular weight proteins, such as BSA or OVA, add the peptide/carrier solution to a vial containing 10 mg of EDC (Thermo Fisher) and gently mix to dissolve. For high-molecular-weight KLH immunogens, first dissolve one vial containing 10 mg of EDC in 1 ml of deionized water, and immediately transfer 50 μl of this solution to the carrier/peptide solution. Gently mix.
5. Allow the reaction to continue at room temperature for 2 h.

   Note: Although the conjugation protocols have been optimized by preparing a number of different peptide–carrier conjugates, some peptide sequences or other haptens may cause precipitation of the carrier upon coupling. This may occur as a result of changing the carrier's solubility characteristics through surface modification or due to polymerization. A small amount of precipitation is not a problem and can easily be removed by centrifugation before the gel filtration step. If severe precipitation occurs, however, the amount of EDC added to the reaction may have to be scaled back to

eliminate or reduce it. With BSA or OVA conjugates, this may mean using as little as 1 to 3 mg of EDC instead of the recommended 10 mg. With native, multi-subunit KLH as a carrier, reducing the EDC levels to 0.1 mg may be necessary.
6. Purify the hapten–carrier conjugate by gel filtration or dialysis.

4. NHS ESTER-MEDIATED HAPTEN–CARRIER CONJUGATION

Hapten–carrier conjugation may be accomplished by the use of homobifunctional reagents containing NHS ester groups on both ends. The active esters are highly reactive toward amines on proteins and other molecules to form stable amide bonds. Crosslinking agents of various lengths may be used for this conjugation strategy, including the sulfo-NHS ester analogs which are more water soluble than the NHS esters without a sulfonic acid group (Chapter 5, Section 1).

Using homobifunctional NHS esters, amine-containing haptens may be conjugated to amine-containing carriers in a single step (Figure 19.14). The carrier is dissolved in a buffer having a pH of 7–9 (0.1-*M* sodium phosphate, pH 7.2, works well). The hapten molecule is added to this solution at a suitable molar excess to ensure multipoint attachment of the hapten to the carrier. A molar excess of 20 to 30 times that of the carrier concentration is a good starting point. Next, the NHS ester crosslinker is added to the solution to provide at least a 3-fold molar excess over that of the hapten. For crosslinkers insoluble in aqueous solution, first solubilize them in dimethylformamide (DMF) or DMSO at higher concentration, and then add an aliquot of this stock solution to the hapten–carrier solution. The conjugation reaction is complete within 2 h at room temperature. Some adjustment of the level of hapten and crosslinker addition may be necessary to avoid extensive precipitation of the conjugate, especially when using rather hydrophobic hapten molecules.

Another method of NHS ester-mediated hapten–carrier conjugation is to create reactive sulfo-NHS esters directly on the carboxylates of the carrier protein using the EDC/sulfo-NHS reaction described in Chapter 4, Section 1.2. A carbodiimide reaction in the presence of sulfo-NHS activates the carboxylate groups on the carrier protein to form amine-reactive sulfo-NHS esters. The activation reaction is performed at pH 6.0, since the amines on the protein will be protonated and therefore be less reactive toward the sulfo-NHS esters that are formed. In addition, the hydrolysis rate of the esters is dramatically slower at acid pH. Thus, the active species may be isolated in a reasonable time frame without significant loss in conjugation potential. To quench unreacted EDC, 2-mercaptoethanol is added to form a stable complex with the remaining carbodiimide, according to Carraway and Triplett (1970). In the following protocol, a modification of Grabarek and Gergely's (1990) two-step method, sulfo-NHS is used instead of NHS so that active ester hydrolysis is slowed even more (Anjaneyulu and Staros, 1987; Thelen and Deuticke, 1988). Subsequent conjugation with amine-containing hapten molecules yields hapten–carrier conjugates created by amide bond formation (Figure 19.15).

Protocol

1. Dissolve the carrier protein to be activated in 0.05-*M* MES, 0.5-*M* NaCl, pH 6.0 (reaction buffer), at a concentration of 1 mg/ml.
2. Add to the solution in step 1 a quantity of EDC and sulfo-NHS (both from Thermo Fisher) to obtain a concentration of 2-m*M* EDC and 5-m*M* sulfo-NHS. To aid in aliquoting the correct amount of these reagents, they may be quickly dissolved in water at a higher concentration, and then immediately pipette a volume into the protein solution to obtain the proper molar quantities.
3. Mix and react for 15 min at room temperature to form the sulfo-NHS esters.
4. Add 2-mercaptoethanol to the reaction solution to obtain a final concentration of 20-m*M*. Mix and incubate for 10 min at room temperature. Note: If the protein being activated is sensitive to this level of 2-mercaptoethanol, instead of quenching the reaction chemically the activation may be terminated by rapid desalting (see step 5).
5. If the reaction was quenched by the addition of 2-mercaptoethanol, the activated protein may be added directly to an amine-containing hapten molecule for conjugation. Alternatively, or if no 2-mercaptoethanol was added, the activated protein may be purified from reaction byproducts by gel filtration using a desalting resin. The desalting operation should be performed rapidly to minimize hydrolysis and recover as much active ester functionality as possible. The use of centrifugal spin columns may afford the greatest speed in separation (Thermo Fisher). After purification, add the activated protein to the hapten for conjugation. The hapten molecule should be dissolved in 0.1-*M* sodium phosphate, pH 7.5.
6. React for at least 2 h at room temperature.
7. Remove excess reactants by gel filtration or dialysis.

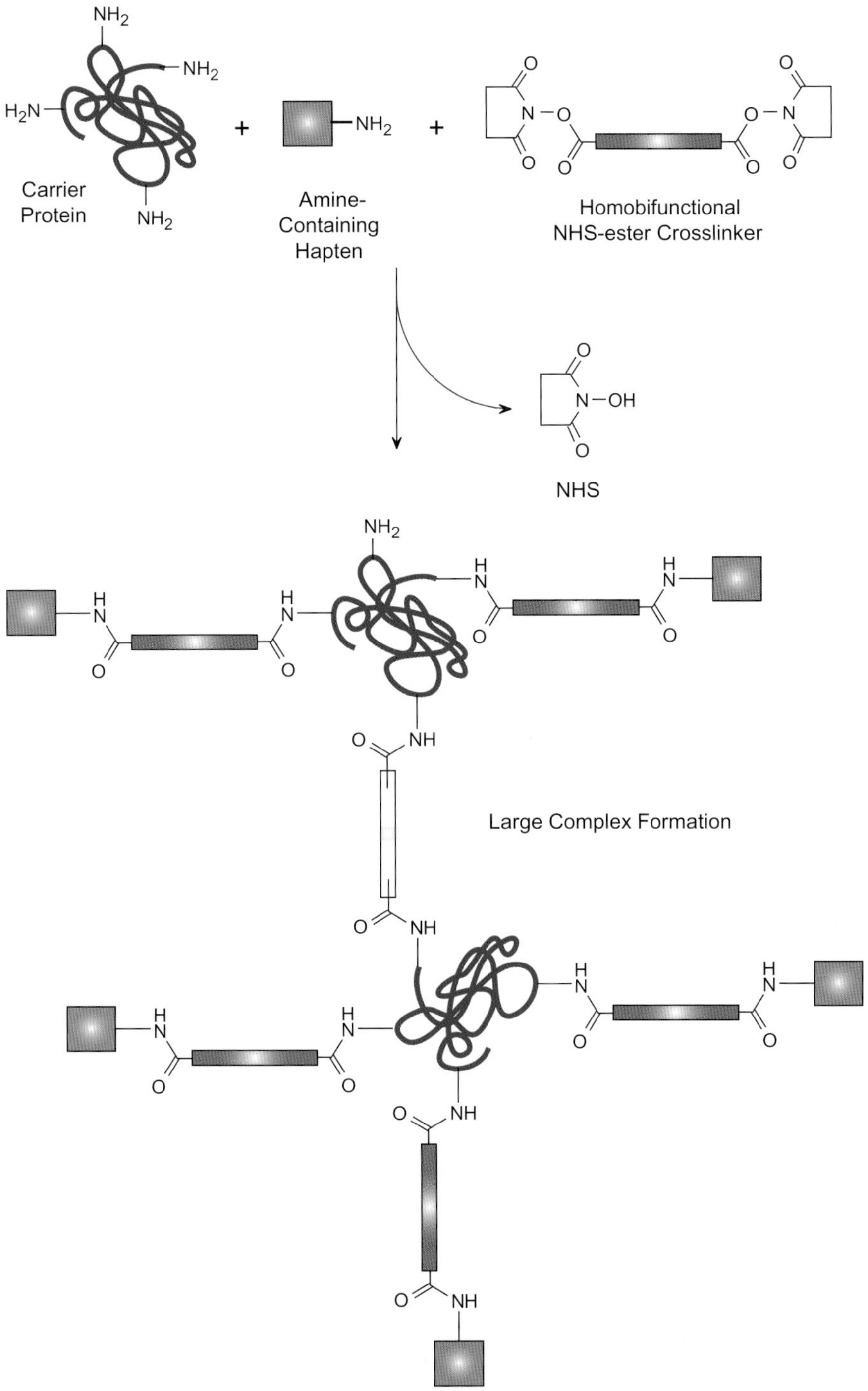

FIGURE 19.14 Hapten–carrier immunogen conjugates can be formed using homobifunctional NHS ester crosslinkers. The reaction may create large polymeric complexes, some of which could precipitate.

5. NHS ESTER–MALEIMIDE HETEROBIFUNCTIONAL CROSSLINKER-MEDIATED HAPTEN–CARRIER CONJUGATION

A common method for coupling haptens to carrier proteins involves the use of a heterobifunctional crosslinker containing an NHS ester and a maleimide group. This type of crosslinker allows better control over the coupling process than homobifunctional or zero-length conjugation methods by incorporating a two- or three-step reaction strategy directed against two different functional targets. In this approach, the carrier protein is first activated with the crosslinker through its

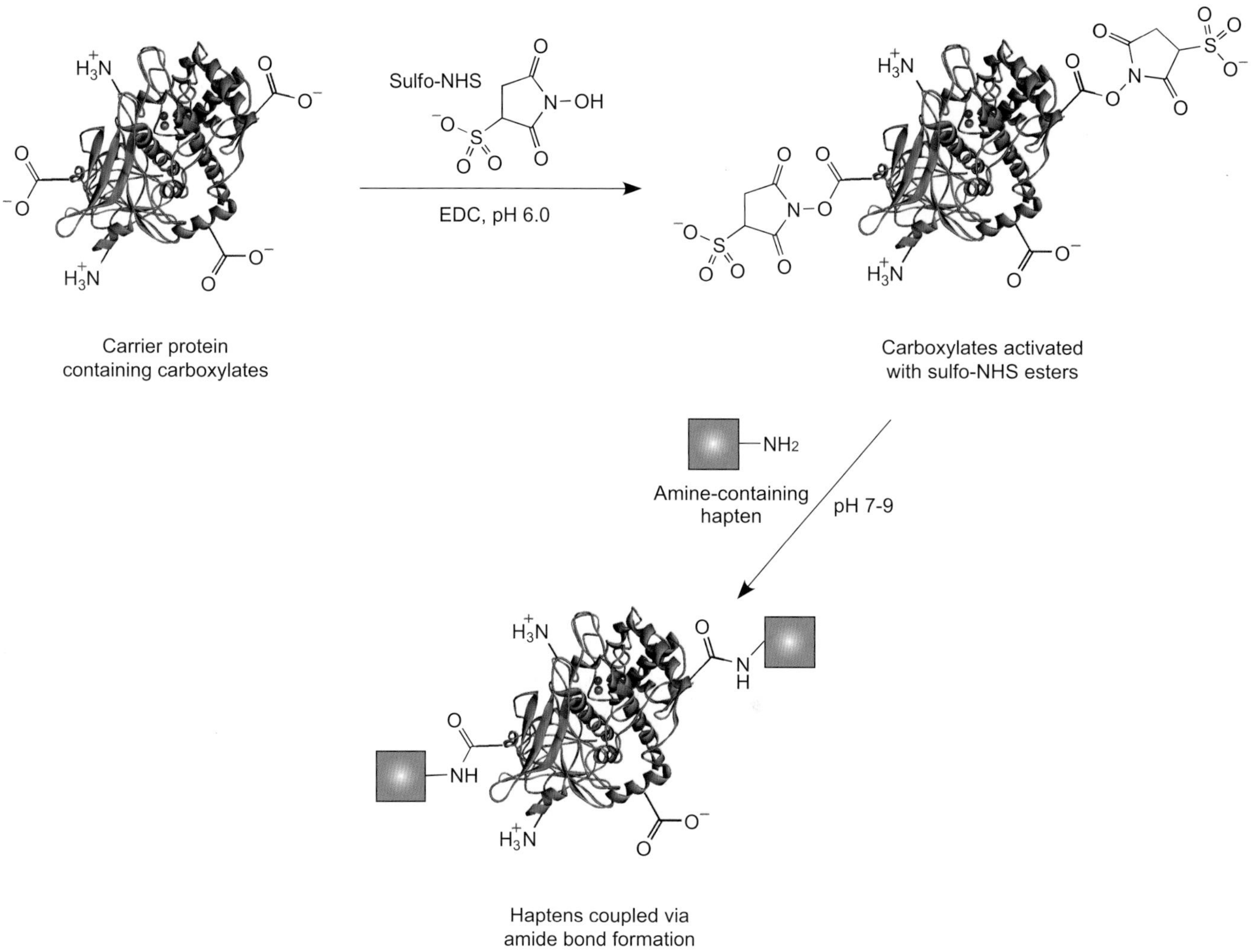

FIGURE 19.15 The carbodiimide EDC can be used in the presence of sulfo-NHS to create reactive sulfo-NHS ester groups on a carrier protein. Subsequent coupling with an amine-containing hapten can be performed to create amide bond linkages.

amine groups, purified to remove excess reactants, and then crosslinked to a hapten molecule containing a sulfhydryl group. One of the most useful reagents for this conjugation approach is sulfo-SMCC.

The reactions associated with a sulfo-SMCC conjugation are shown in Figure 19.16. (Note: sulfo-SMCC is sulfosuccinimidyl-4-(*N*-maleimidomethyl) cyclohexane-1-carboxylate; MW = 436.37; see Chapter 6, Section 1.3.) This crosslinking reagent mediates the conjugation of a carrier protein through its primary amine groups to a peptide or other hapten through sulfhydryl groups. The active *N*-hydroxysulfosuccinimide ester (sulfo-NHS) end of sulfo-SMCC is first reacted with available primary amine groups on the carrier protein. This reaction results in the formation of an amide bond between the protein and the crosslinker with the release of sulfo-NHS as a byproduct. The carrier protein is then isolated by gel filtration to remove excess reagents. At this stage, the purified carrier possesses modifications generated by the crosslinker resulting in a number of reactive maleimide groups projecting from its surface. The maleimide portion of sulfo-SMCC is a thiol-reactive group that can be used in a secondary step to conjugate with a free sulfhydryl (i.e., a cysteine residue) on a peptide or other hapten, resulting in a stable thioether bond.

The use of sulfo-SMCC over the other common maleimide-containing crosslinkers such as *m*-maleimidobenzoyl-*N*-hydroxysuccinimide ester (MBS) or succinimidyl-4-(*p*-maleimidophenyl)butyrate (SMPB) provides the advantages of initial water solubility during the activation step and increased stability of the maleimide group prior to conjugation with a peptide. The improved stability ensures that the majority of the maleimide groups substituted on the carrier will survive the subsequent purification process without degradation. The relatively good stability of the maleimide group of

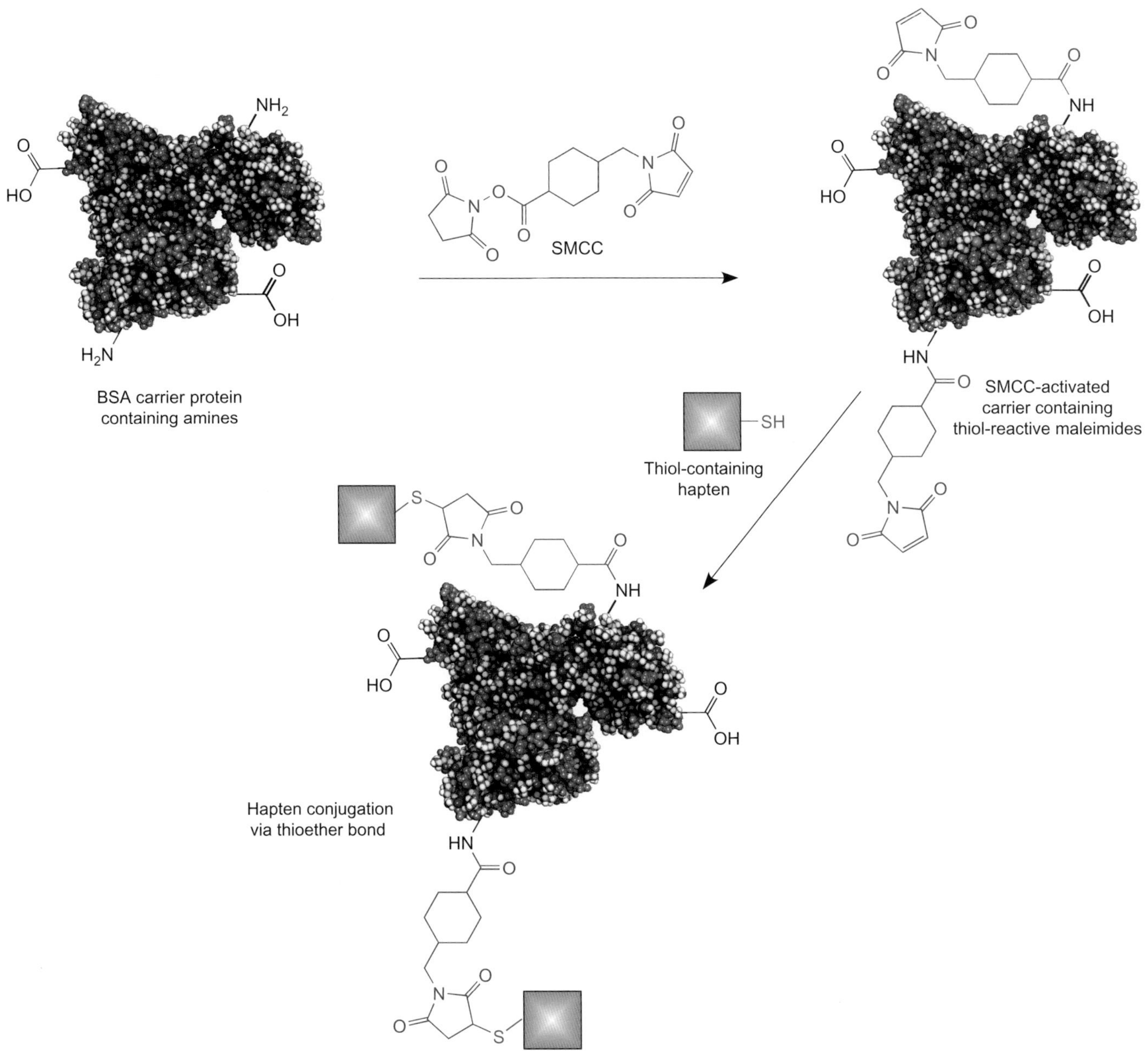

FIGURE 19.16 A common way of conjugating sulfhydryl-containing haptens to carrier proteins is to activate the carrier with sulfo-SMCC to create an intermediate maleimide derivative. The maleimide groups can then be coupled to thiol-containing haptens to form thioether bonds.

sulfo-SMCC is probably due to the neighboring steric effects of its cyclohexane ring. The faster hydrolysis rates of other maleimide type crosslinkers can be a significant problem, since they readily break down to the maleamic acid form, which is no longer reactive toward sulfhydryls (Figure 19.17).

A disadvantage to using SMCC or other NHS–maleimide type crosslinkers with hindered ring structures (such as MBS) is the relatively high immunogenicity of the cross-bridge. Studies have shown that a hapten-carrier

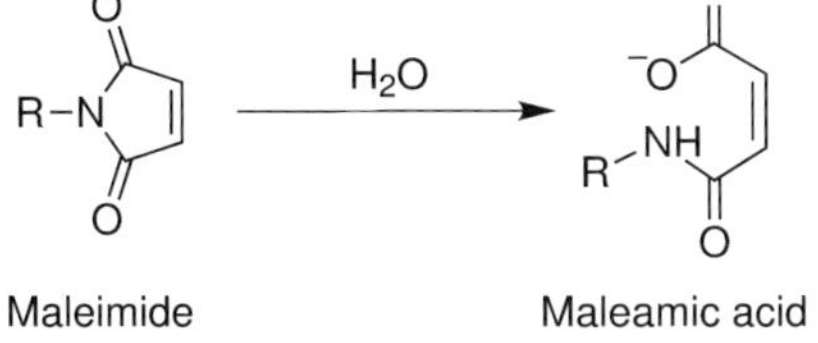

FIGURE 19.17 Maleimide groups may hydrolyze in aqueous solution to an open maleamic acid form that is unreactive with sulfhydryls.

complex formed from such crosslinkers generates significant antibody response against the spacer group itself, not just the hapten and carrier. To minimize the antibody population directed against the cross-bridge of the conjugate, the use of aliphatic straight-chain spacers will exhibit lower immunogenicity (Peeters *et al.*, 1989). However, perhaps a better choice for this type of conjugation is a polyethylene glycol (PEG)-based crosslinker containing an NHS ester on one end and a maleimide group on the other end (see Chapter 18, Section 1.2). A PEG group used as a cross-bridge in a heterobifunctional reagent to prepare immunogen conjugates will result in non-immunogenic modifications on the carrier protein and thus no antibody production against the polyether linker. Although SMCC (or sulfo-SMCC) is used in the following protocol, direct substitution of a PEG-based crosslinker will limit the immune response to the hapten and not generate unwanted antibodies to the cross-bridge.

Since many peptides do not naturally contain cysteine residues with free sulfhydryls, a terminal cysteine may be incorporated during peptide synthesis, or, where appropriate, disulfide groups may be reduced to generate them. Alternatively, thiolating reagents such as 2-iminothiolane (Traut's reagent) can be used to modify existing amino groups and introduce a sulfhydryl (see Chapter 2, Section 4.1). Caution must be taken when using this last technique, however, because multiple sites of modification may alter the immunogenic structure of the hapten.

If a terminal cysteine residue is added to a peptide during its synthesis, its sulfhydryl group provides a highly specific conjugation site for reacting with a sulfo-SMCC-activated carrier. All peptide molecules coupled using this approach will display the same basic conformation after conjugation. In other words, they will have a known and predictable orientation, leaving the majority of the molecule free to interact with the immune system. This method can therefore preserve the major epitopes on a peptide while still enhancing the immune response to the hapten by being covalently linked to a larger carrier protein. In addition, the well-known chemical reactivity of a sulfo-SMCC-mediated immunogen preparation permits covalent conjugation in a controllable fashion that can be highly defined for quality assurance purposes.

The process of carrier activation by sulfo-SMCC may be followed by performing a simple purification step after the reaction using a desalting resin. Figure 19.18 shows the gel filtration profiles for the separation of sulfo-SMCC-activated BSA and OVA. The first peak of both separations represents the elution point for the carrier protein, while the absorbance due to reaction by-products of the crosslinker is contained in the second peak (shown only as its leading edge). Activated proteins exhibit an increase in their absorbance at 280 nm over an identical sample with no added sulfo-SMCC due to their covalently attached maleimide groups. After isolation, the activated protein may be frozen and lyophilized to preserve maleimide coupling activity toward sulfhydryl-containing haptens. Thermo Fisher sells a number of maleimide-activated carrier proteins in lyophilized form for easy hapten conjugation.

After a carrier protein has been activated with sulfo-SMCC, it is often useful to measure the degree of maleimide incorporation prior to coupling an expensive hapten. Ellman's reagent may be used in an indirect method to assess the level of maleimide activity of sulfo-SMCC-activated proteins and other carriers. First, a sulfhydryl-containing compound such as 2-mercaptoethanol or cysteine is reacted in excess with the activated protein. The amount of unreacted sulfhydryls remaining in solution is then determined using the

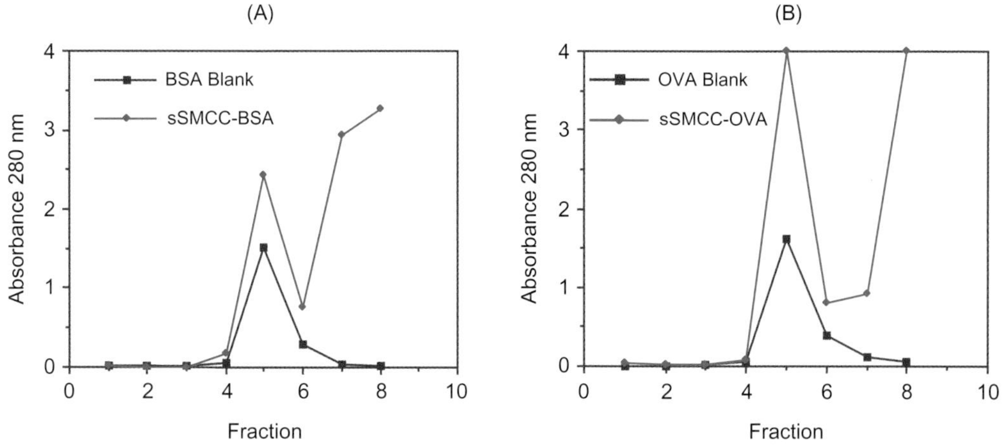

FIGURE 19.18 Carrier proteins may be activated with sulfo-SMCC to produce maleimide derivatives reactive with sulfhydryl-containing molecules. The graphs show the size exclusion separation on Sephadex G-25 of maleimide-activated BSA (A) and OVA (B) after reaction with sulfo-SMCC. The first peak is the protein and the second peak is excess crosslinker. The maleimide groups create increased absorbance at 280 nm in the activated proteins as compared with the blank separation performed without the addition of crosslinker.

Ellman's reaction (Chapter 2, Section 4.1). Comparison of the response of the sample to a blank reaction using the native, non-activated protein at the same concentration and a series of standards made from a serial dilution of the sulfhydryl compound employed in the assay gives the amount of sulfhydryl compound conjugated and thus an estimate of the original maleimide activity.

Figure 19.19 shows a plot of the results of such an assay performed to determine the maleimide content of activated BSA. This particular assay used 2-mercaptoethanol which is relatively unaffected by metal-catalyzed oxidation. For the use of cysteine or cysteine-containing peptides in the assay, however, the addition of EDTA is required to prevent disulfide formation. Without the presence of EDTA at 0.1-*M*, the metal contamination of some proteins (especially serum proteins such as BSA) is so great that disulfide formation proceeds preferential to maleimide coupling. Figure 19.20 shows a similar assay for maleimide-activated BSA using the more innocuous cysteine as the sulfhydryl-containing compound.

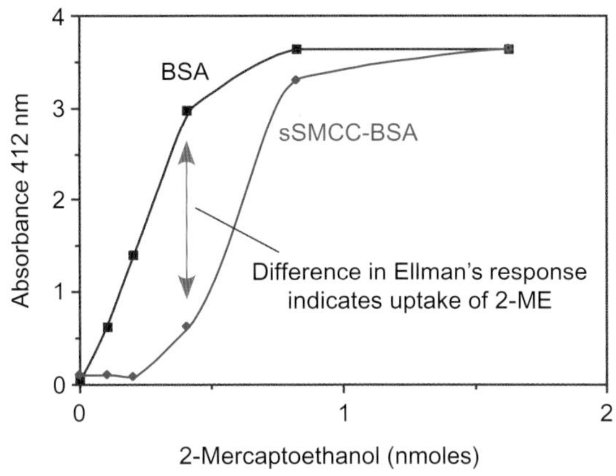

FIGURE 19.19 An Ellman's assay may be performed to determine the maleimide activation level of SMCC-derivatized proteins. Reaction of the activated carrier with different amounts of 2-mercaptoethanol will result in various levels of sulfhydryls remaining after the reaction. Detection of the remaining thiols using an Ellman's assay indirectly indicates the amount of sulfhydryl uptake into the activated carrier. Comparison of the Ellman's response to the same quantity of 2-mercaptoethanol plus an unactivated carrier indicates the absolute amount of sulfhydryl that reacted. Calculation of the maleimide activation level can then be performed.

Using this type of cysteine-uptake assay, it is possible to determine the percentage of maleimides that reacted over time. Thus, an indication of the reaction efficiency of a sulfhydryl-containing compound coupling with a maleimide-activated protein may be determined. Figure 19.21 shows the reaction rate for the coupling of cysteine to maleimide-activated BSA. Note that maximal coupling is obtained in less than 2h, and over 80% yield is achieved in less than 30min.

The following protocol describes the activation of a carrier molecule with sulfo-SMCC and its subsequent conjugation with a hapten. The preactivated carriers containing maleimide groups ready for coupling to a sulfhydryl-containing compound are commercially available in a stable freeze-dried form (Thermo Fisher). Substitution of GMBS (Chapter 6, Section 1.7) or a NHS–PEG_n–maleimide crosslinker (Chapter 18, Section 1.2) in the following protocol will provide straight-chain spacer arms with lower or no immunogenicity compared to the ring structure of SMCC's cross-bridge.

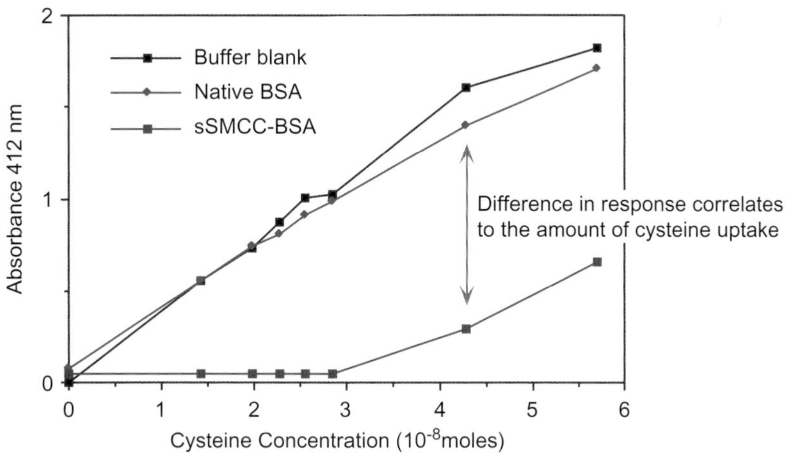

FIGURE 19.20 Cysteine may also be used in an Ellman's assay to determine the maleimide activation level of SMCC-derivatized proteins. Reaction of the activated carrier with different amounts of cysteine results in various levels of sulfhydryls remaining after the reaction. The coupling must be performed in the presence of EDTA to prevent metal-catalyzed oxidation of sulfhydryls. Detection of the remaining thiols using an Ellman's assay indirectly indicates the amount of sulfhydryl uptake into the activated carrier. Comparison of the Ellman's response to the same quantity of cysteine plus an unactivated carrier indicates the absolute amount of sulfhydryl that reacted. Calculation of the maleimide activation level then can be performed.

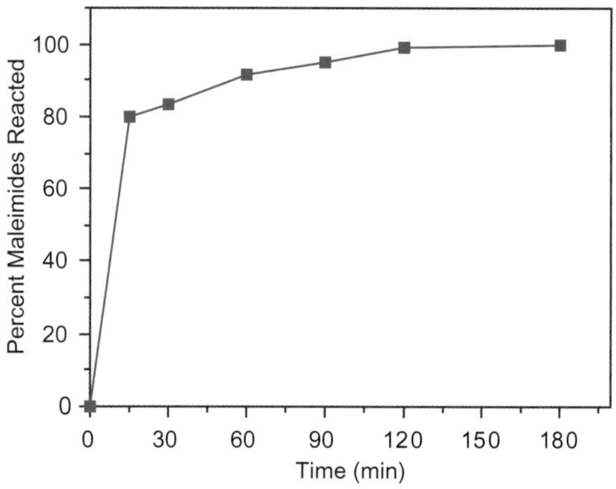

FIGURE 19.21 The rate of reaction of cysteine with maleimide-activated BSA was determined using an Ellman's assay to determine the remaining sulfhydryl groups after the reaction, which was performed according to Figure 19.20. Nearly all of the available maleimides are coupled with sulfhydryls within 2h.

Protocol

1. Dissolve the carrier of choice at a concentration of 10mg/ml in 0.1-*M* sodium phosphate, 0.15-*M* NaCl, pH 7.2 to 7.5 (activation buffer). Note: For use of native, multi-subunit KLH, increase the NaCl concentration to 0.9-*M*.
2. Dissolve sulfo-SMCC (Thermo Fisher) at a concentration of 10mg/ml in the activation buffer. Immediately transfer the appropriate amount of this crosslinker solution to the vial containing the dissolved carrier protein. Note: The amount of crosslinker solution to be transferred is dependent on the level of activation desired. Suitable activation levels can be obtained for the following proteins by adding the indicated quantities of the sulfo-SMCC solution. The degree of sulfo-SMCC modification often determines whether the carrier will maintain solubility after activation and coupling to a hapten. Multimeric KLH in particular, is sensitive to the amount of crosslinker addition. KLH usually retains solubility at about 0.1 to 0.2 times the mass of crosslinker added to BSA. This level of addition still results in excellent activation yields, since KLH is significantly larger than most of the other protein carriers. Add the following quantities of sulfo-SMCC solution to each milliliter of carrier protein solution:
   a. BSA: 500 µl
   b. cBSA: 200 µl
   c. OVA: 500 µl
   d. KLH: 100 µl
   Carriers having molecular weights similar to those of BSA or OVA may be activated at the same level with good success. Cationized BSA requires less crosslinker addition due to its greater quantity of amines present.
3. React for 1h at room temperature.
4. Immediately purify the activated carrier protein by gel filtration using a desalting resin with a bed volume equal to 15 times the volume of the activation reaction. To perform the chromatography use 0.1-*M* sodium phosphate, 0.15-*M* NaCl (0.9-*M* for KLH), 0.1-*M* EDTA, pH 7.2 to 7.5 (conjugation buffer). The EDTA is present to prevent metal-catalyzed sulfhydryl oxidation to disulfides, which will result in an inability to couple to the maleimide groups on the carrier. This is a particular problem when using BSA due to contaminating iron from hemolysis. Concentrations less than 0.1-*M* EDTA will not fully inhibit the oxidation reaction, especially if a cysteine-containing peptide is to be conjugated to the activated carrier. Apply the sSMCC/carrier reaction mixture to the column while collecting 0.5 to 1.0ml fractions. Pool the fractions containing the activated carrier (the first peak to elute from the column), and discard the fractions containing excess sulfo-SMCC (the second peak). The activated carrier should be used immediately or freeze-dried to maintain maleimide stability.
5. Dissolve a sulfhydryl-containing hapten or peptide to be conjugated at a concentration of 10mg/ml in the conjugation buffer. Other hapten concentrations may be used depending on its solubility. If an excess of the peptide solution is made at this time, an estimate of the degree of conjugation may be determined later (see section below). Add this solution to the pooled fractions containing the activated carrier at an equivalent mass ratio (1mg hapten per milligram of the carrier). Alternatively, the peptide may be added in solid form directly to the activated carrier solution if it is known to be freely soluble and can be weighed out in the appropriate quantity.
6. Allow the conjugation reaction to proceed for 2h at room temperature.
7. The hapten–carrier conjugate may now be mixed with adjuvant and used for injection purposes without further purification.

An estimate of the degree of conjugation may be made by assaying the amount of sulfhydryl present before and after the coupling reaction. A portion of the peptide solution before mixing with the activated carrier should be saved to compare with the reaction mixture after the conjugation is complete. The comparison is made using a solution of Ellman's reagent (5,5′-dithiobis-(2-nitrobenzoic acid)), which reacts with sulfhydryls to form a highly colored chromophore having an absorbance maximum at 412 nm (ε_{412} nm = $1.36 \times 10^4 M^{-1} cm^{-1}$) (Chapter 2, Section 4.1). A generalized procedure is presented here. Modifications to this guideline may have to be made for each individual peptide to obtain the appropriate response to the Ellman's reagent. For reactions performed with very small quantities of peptide, the Ellman's assay may not be sensitive enough to measure the degree of conjugation.

1. Using a microtiter plate (96 well) dispense 200 μl of 0.1-*M* sodium phosphate, 0.15-*M* NaCl, 0.1-*M* EDTA, pH 7.2 (conjugation buffer), into each well to be used.
2. Add 10 μl of the peptide solution before conjugation to the appropriate wells in duplicate.
3. Add 10 μl of the reaction mixture after the conjugation reaction is complete to another set of wells in duplicate.
4. Add 20 μl of Ellman's reagent (1 mg/ml dissolved in the gel filtration purification buffer) to each well including one containing only buffer (220 μl) to use as a blank.
5. Incubate for 15 min at room temperature.
6. Measure the absorbance of all wells using a microplate reader with a filter set at 410 nm.

A comparison of the blank corrected values before and after conjugation should give an indication of the percent of peptide coupled. To be more quantitative, a standard curve must be run to focus in on the linear response range of the peptide–Ellman's reaction. Using cysteine as a representative sulfhydryl compound (similar in Ellman's response to a peptide having one free sulfhydryl), it is possible to obtain very accurate determinations of the amount which coupled to the activated carrier. Figure 19.20, discussed previously in this section, shows the results of this type of assay.

6. ACTIVE-HYDROGEN-MEDIATED HAPTEN–CARRIER CONJUGATION

Conjugation chemistry for the coupling of haptens to carrier molecules is fairly well defined for compounds having common functional groups to facilitate such attachment. The types of functional groups generally useful for this operation include easily reactive components such as primary amines, carboxylic acids, aldehydes, or sulfhydryls.

However, for hapten molecules containing no easily reactive functional groups, conjugation can be difficult or impossible using current technologies. To solve this problem, demanding organic synthesis is frequently required to modify the hapten molecule to contain a suitable reactive portion. Particularly, certain drugs, steroidal compounds, dyes, or other organic molecules often have structures that contain no available "handles" for convenient crosslinking.

Frequently, these difficult-to-conjugate compounds do have certain sufficiently active hydrogens that can be reacted with a carrier molecule using specialized reactions designed for this purpose. This section describes two choices for this conjugation problem, the diazonium procedure and the Mannich reaction. Both of them are able to crosslink haptens through any available active hydrogen to carrier molecules, resulting in immunogens suitable for injection.

6.1. Diazonium Conjugation

Diazonium coupling procedures have been used for many years in organic synthesis and for crosslinking or immobilization of active-hydrogen-containing compounds (Inman and Dintzis, 1969; Cuatrecasas, 1970) (see Chapter 15, Section 2.7). Diazonium derivatives can couple with haptens containing available phenolic or, to a lesser extent, imidazole groups in an electrophilic substitution reaction (Riordan and Vallee, 1972). They may also undergo minor secondary reactions with sulfhydryl groups and primary amines (Glazer *et al.*, 1975).

The most important reaction of a diazonium group, however, is with available tyrosine and histidine residues within peptide haptens, rapidly creating diazo linkages. This method of conjugation is especially useful for site-directed crosslinking of tyrosine-containing peptides. Since tyrosine is usually present only in limited quantities in a given peptide, use of diazonium conjugation can crosslink and orient all peptide molecules in an identical fashion on a carrier. The result is excellent reproducibility in preparation of the immunogen, and a consistent presentation of the peptide on the surface of the carrier to the immune system for antibody production.

Derivatives of carbohydrate antigens have also been coupled to carrier proteins through the use of an intermediate diazonium group (McBroom *et al.*, 1976). In this case, an aminophenyl glycoside was prepared by reaction of the reducing end of the oligosaccharide with β-(*p*-aminophenyl)ethylamine and then forming the diazotized derivative with sodium nitrite (Zopf *et al.*, 1978). Upon mixing with carrier proteins containing tyrosine residues, the carbohydrate derivative is coupled via a diazo bond.

Creation of a diazonium group on phenolic compounds or tyrosine side chain groups is possible by forming an intermediate nitrophenol derivative. Reaction of tyrosine-containing proteins and peptides with tetranitromethane effectively nitrates the ring in the *ortho* position (Vincent *et al.*, 1970). Reduction of the nitro group to an amine is then performed using sodium dithionite (sodium hydrosulfite; $Na_2S_2O_4$)

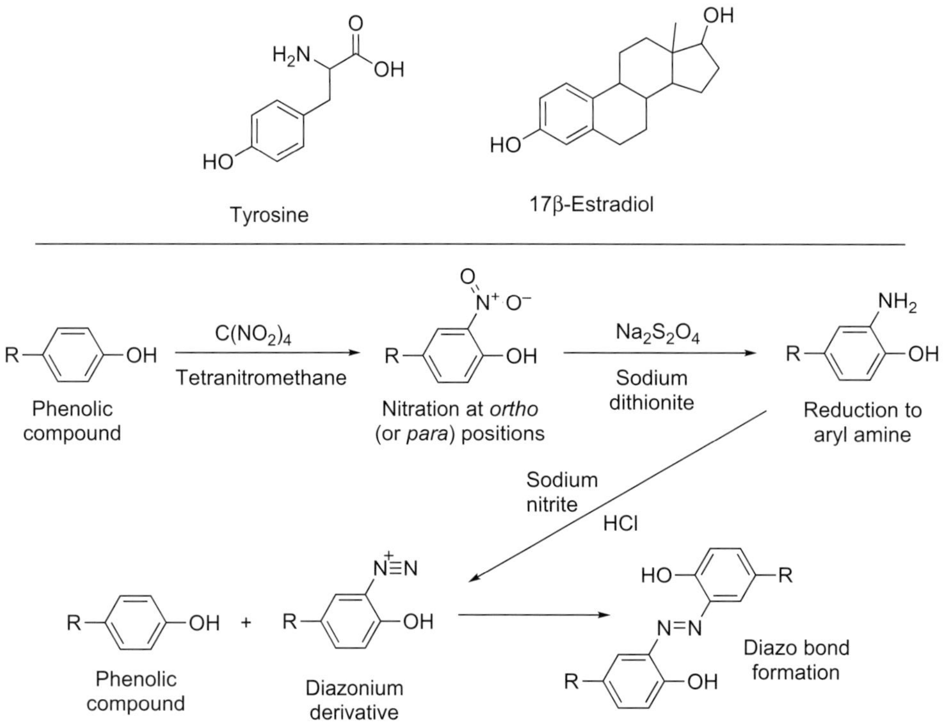

FIGURE 19.22 Phenolic compounds such as tyrosine and estradiol may be derivatized to contain reactive diazonium groups by nitration with tetranitromethane followed by reduction with sodium dithionite and finally diazotization with sodium nitrite in dilute HCl. The diazo groups rapidly react with active hydrogen containing molecules (such as other phenolic compounds) to create diazo linkages.

(Sokolovsky *et al.*, 1967) (Chapter 2, Section 4.3). The aminophenol derivative is finally reacted with sodium nitrite in acidic conditions to form the highly reactive diazonium group (Figure 19.22). Once created, the diazonium compound must be added immediately to the conjugation reaction, since the species is extremely unstable in aqueous environments.

The active diazonium is typically a colored compound, sometimes orange, dark brown, or even black in concentrated solutions. The conjugated immunogen is usually deeply colored due to the resultant diazo bond. The coupling reaction is performed at alkaline pH, optimally at pH 8 for histidine residues and pH 9 to 10 for tyrosine groups. In practice, however, it is not possible to target a histidine group in the presence of a tyrosine group without some cross-reactivity.

Diazo linkages are reversible bonds that may be cleaved by addition of 0.1-*M* sodium dithionite in 0.2-*M* sodium borate, pH 9. Release of the crosslinks can be followed by loss of the diazo bond color.

A simple, one-step conjugation reaction is possible with diazonium chemistry if a *bis*-aminophenyl compound is used as a homobifunctional crosslinking agent. Activation of the aminophenyl groups with sodium nitrite creates the requisite *bis*-diazonium derivative that can then couple with active-hydrogen-containing haptens and carriers. In this way, tyrosine-containing peptides can be conjugated with tyrosine-containing carrier proteins in a single step. Compounds useful for this procedure include *o*-tolidine and benzidine (Chapter 5, Section 9), both of which contain aromatic amines that can easily be diazotized (Figure 19.23).

From a practical perspective, however, any of the conjugation methods utilizing diazonium chemistry can be problematic. The rate of reaction of the diazonium species is so rapid that much of the total coupling potential can be lost through intramolecular crosslinking. As the diazonium groups are formed they may immediately crosslink to the active hydrogens present on the aminophenyl precursor molecules, even before addition of a second molecule to be conjugated. Even without addition of a second active-hydrogen-containing compound, the diazonium activated molecule will turn brown to black within an hour, indicating formation of diazo bonds and self-conjugation. For this reason, the reproducibility of conjugation reactions using this method can be poor.

The following protocol describes the use of diazotized *o*-tolidine for the crosslinking of active-hydrogen-containing haptens to active-hydrogen-containing carriers. Using a *bis*-diazonium compound is perhaps the simplest method of conjugation, but as in many

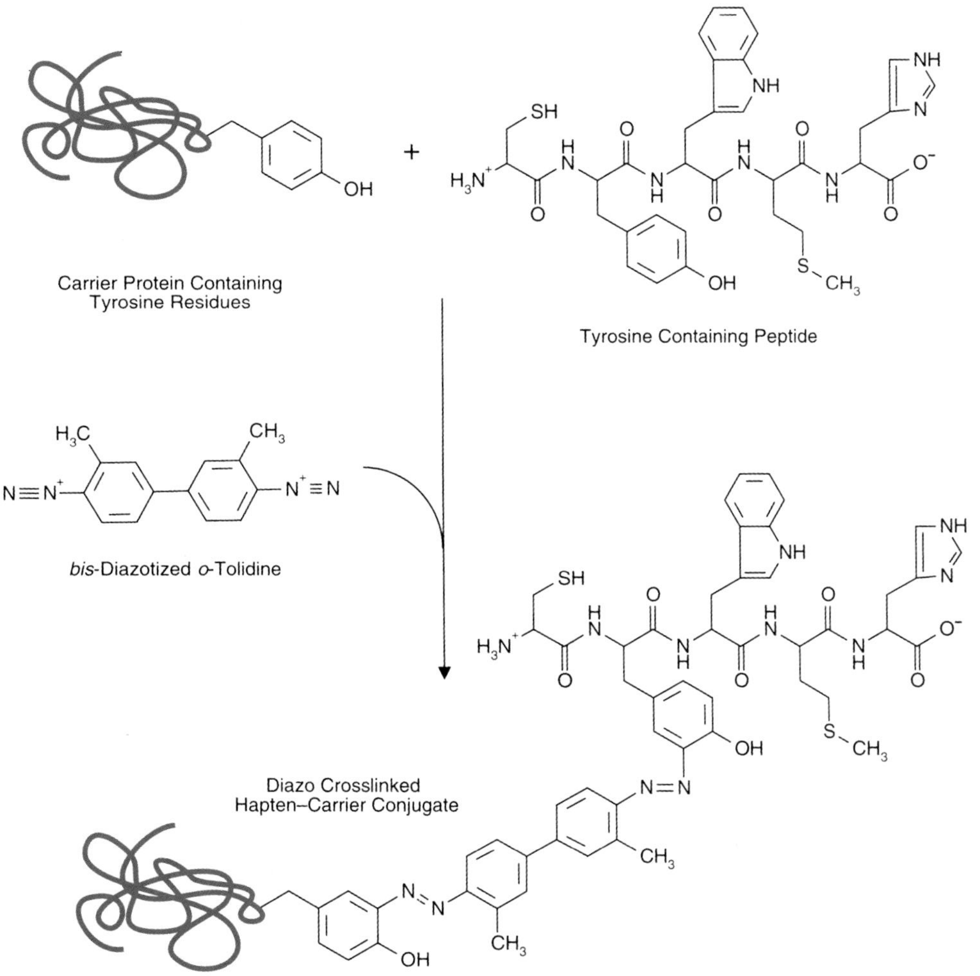

FIGURE 19.23 The conjugation of a tyrosine-containing carrier protein and a tyrosine-containing peptide may be performed using *bis*-diazotized tolidine to form diazo crosslinks.

one-step crosslinking procedures, it often results in some precipitation of the final product. Reaction conditions may have to be adjusted to prevent severe precipitation, however even an insoluble immunogen can be useful in generating an antibody response.

Caution: Both *o*-tolidine and benzidine are potential carcinogens. Protective clothing, gloves, and the use of a fume hood are recommended. Avoid all contact of the compounds with skin or clothing and do not inhale vapors or dust.

Protocol

1. Diazotization of *o*-tolidine: Weigh out 25 mg of *o*-tolidine and place in a small test tube or vial. Add 4.5 ml of 0.2-*N* HCl and mix to dissolve. Chill the solution on ice. Dissolve 17.5 mg of sodium nitrite into 0.5 ml of ice-cold deionized water, and add it to the vial containing the *o*-tolidine. The solution should begin to turn an orange color, progressively getting darker as the reaction continues. React for 1 h on ice, mixing periodically. At the completion of the diazotization reaction, aliquots of the solution may be stored at −20°C.
2. Dissolve 10 mg of carrier protein into 0.5 ml 0.15-*M* sodium borate, 0.15-*M* NaCl, pH 9.0.
3. Dissolve 5 to 10 mg of a peptide hapten containing at least one tyrosine residue per milliliter of 0.15-*M* sodium borate, 0.15-*M* NaCl, pH 9.0.
4. Mix 0.5 ml of the peptide solution with 0.5 ml of the carrier protein solution. Chill on ice. Add 0.4 ml of the *bis*-diazotized tolidine solution. There should be a color change from orange to red almost immediately. Continue the reaction for 2 h on ice in the dark.
5. Purify the conjugate by gel filtration or dialysis using PBS, pH 7.4. The preparation is now ready for immunization purposes.

6.2. Mannich Condensation

Another approach for crosslinking haptens to carriers when the hapten has no available common functional groups (amines, carboxylates, sulfhydryls, etc.), but does possess active hydrogens, is to use the Mannich reaction. Using this strategy an active hydrogen-containing compound can be condensed with formaldehyde and an amine in the Mannich reaction resulting in a stable alkylamine linkage. In particular, compounds containing replaceable hydrogens provided by the presence of certain activating chemical constituents can be aminoalkylated using this reaction (see Chapter 3, Section 5.5, and Chapter 5, Section 6.1 for additional information on active hydrogens).

In its simplest form, the Mannich reaction consists of the condensation of formaldehyde (or sometimes another aldehyde) with ammonia, in the form of its salt, and another compound containing an active hydrogen. Instead of using ammonia, however, this reaction can be performed with primary or secondary amines, or even with amides. An example is illustrated in the condensation of acetophenone, formaldehyde, and a secondary amine salt (the active hydrogens are shown underlined):

$$C_6H_5COC\underline{H}_3 + CH_2O + R_2NH \cdot HCl \rightarrow C_6H_5COCH_2CH_2NR_2 \cdot HCl + H_2O$$

The Mannich reaction provides a viable alternative to the diazonium conjugation method (discussed previously), because of the disadvantages inherent in the instability of both the diazonium group and the resultant diazo linkage. By contrast, conjugations performed through Mannich condensations will result in stable covalent bonds. The Mannich reaction has also been used successfully to immobilize affinity ligands onto chromatography supports and particles (see Chapter 15, Section 2.7) (Hermanson *et al.*, 1992).

The crosslinking scheme using this method can make use of the native ε- and N-terminal amines on carrier proteins as the source of primary amine for the condensation reaction. Formaldehyde and the desired hapten to be coupled containing an appropriately active hydrogen are then added to the conjugation reaction.

To increase the yield of conjugated hapten using this procedure, cationized BSA (cBSA) is used as the carrier protein in the method described below (see Section 2.1, this chapter for additional information on this carrier). The greater density of amine groups on cBSA available for participation in the Mannich reaction over that available on native proteins provides better results in coupling active hydrogen-containing haptens. The use of cBSA to facilitate Mannich-based conjugations has been used successfully in the study of the immune response to certain allergens (Liao *et al.*, 2011).

One note of caution should be regarded when using the Mannich reaction. The hapten to be coupled should not contain any amine groups or hapten polymerization may occur preferential to conjugation to the carrier. For instance, when performing site-directed coupling of tyrosine-containing peptides through their phenolic side chain, the diazonium reaction should be used instead of the Mannich procedure, otherwise peptide-to-peptide coupling may occur.

Protocol

1. In a vial or test tube are placed and mixed:
   a. 200 μl of a solution containing 10 mg/ml cationized BSA (cBSA, Thermo Fisher) in 0.1-*M* MES, 0.15-*M* NaCl, pH 4.7 (coupling buffer). The acidic conditions of this coupling buffer are optimal for the Mannich reaction.
   b. 200 μl of a solution consisting of 10 mg/ml of a hapten containing an active hydrogen. The solution can be made up in absolute ethanol in the case of water-insoluble haptens and is made up in coupling buffer in the case of water soluble haptens.
   c. 50 μl of additional absolute ethanol in the case of water-insoluble haptens.
   d. 50 μl of 37% formaldehyde (Sigma) solution. *Caution:* Use a fume hood and avoid contact or inhalation of vapors.
2. Incubate the reaction mixture in a water bath or oven at a temperature of 37 to 57°C for a period of 3 to 24 h.
3. To separate unconjugated hapten and formaldehyde from the synthesized conjugate, apply the entire volume of reactants to a desalting column containing a bed volume of at least 10 times the volume of the reaction mixture. PBS, pH 7.2, can be used for the desalting step. The purified conjugate is recovered in the void volume.

The yield of conjugation using the Mannich reaction is dependent on the reactivity of active hydrogens within the hapten molecule. It is often difficult to predict the relative reactivity of any given compound in this reaction. Thus, trial and error may be necessary to determine the suitability of the Mannich procedure.

Figure 19.24 shows the conjugation reaction of the dye phenol red to cBSA using the Mannich reaction. The active hydrogens that participate in the conjugation are *ortho* to the hydroxyl group on the phenol ring. After purification of the conjugate by gel filtration to remove any unconjugated dye and formaldehyde, a wavelength scan was performed to assess the degree of conjugate formation. Figure 19.25 shows the results of this scan. The protein solution appeared red after

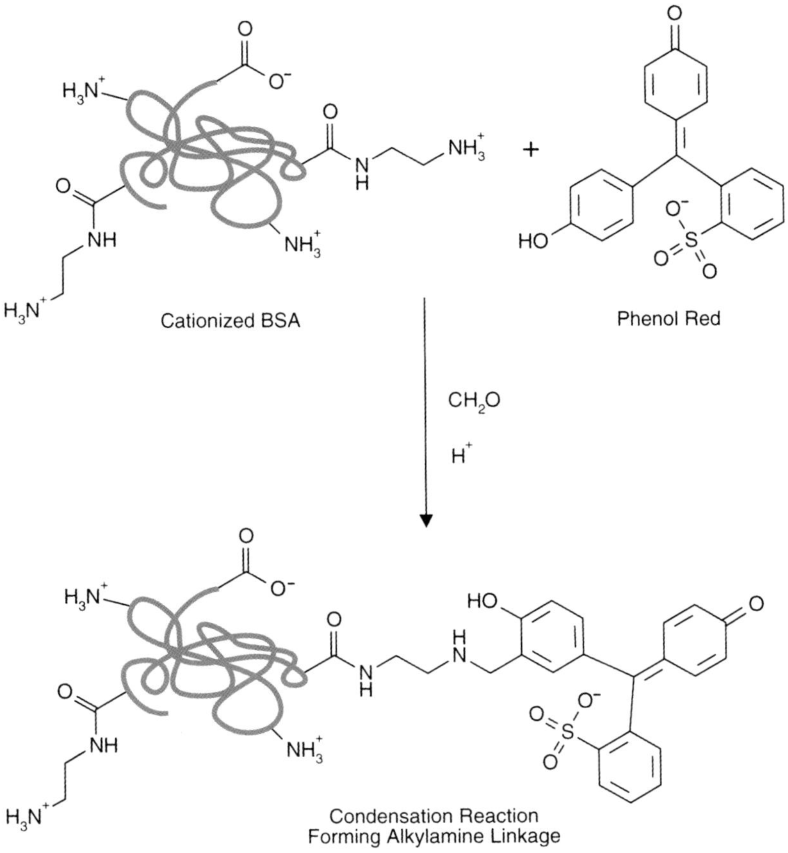

FIGURE 19.24 The conjugation of phenol red to cationized cBSA using the Mannich reaction.

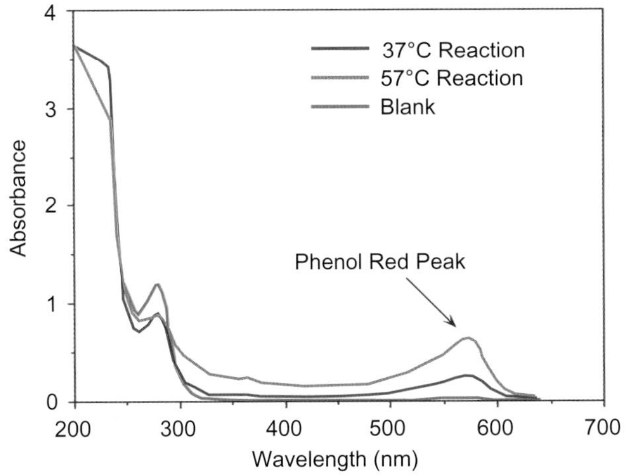

FIGURE 19.25 Absorbance scan comparing unconjugated cBSA with the same carrier that had been coupled with phenol red using the Mannich reaction. Two different reaction times are compared, indicating that extended reactions yield increased conjugate formation.

conjugation and desalting, indicating successful crosslinking had occurred.

The steroidal compound 17β-estradiol was also conjugated to cBSA using the Mannich reaction. Similar to phenol red, conjugation with estradiol occurs *ortho* to the hydroxyl group on its phenolic ring (Figure 19.26). After purification of the conjugate by gel filtration, it was injected in mice intraperitoneally using alum as adjuvant. Antibodies were successfully produced against the coupled estradiol. Controls consisting of unconjugated estradiol with and without mixed carrier molecules were also injected, but resulted in no antibody production.

7. GLUTARALDEHYDE-MEDIATED HAPTEN–CARRIER CONJUGATION

The homobifunctional crosslinking reagent glutaraldehyde can be used in a one- or two-step conjugation protocol to prepare hapten–carrier conjugates. Glutaraldehyde undergoes various cyclization and polymerization reactions in aqueous solution to form hemiacetal and aldol products (see Chapter 15, Section 2.1 for a detailed description of its reactions). These intermediates can react with primary amine groups on carrier proteins and haptens to covalently conjugate them together through direct linkages to its reactive hemiacetal forms or through double bond (Michael-type) addition products on aldol rings or polymers (Chapter 5,

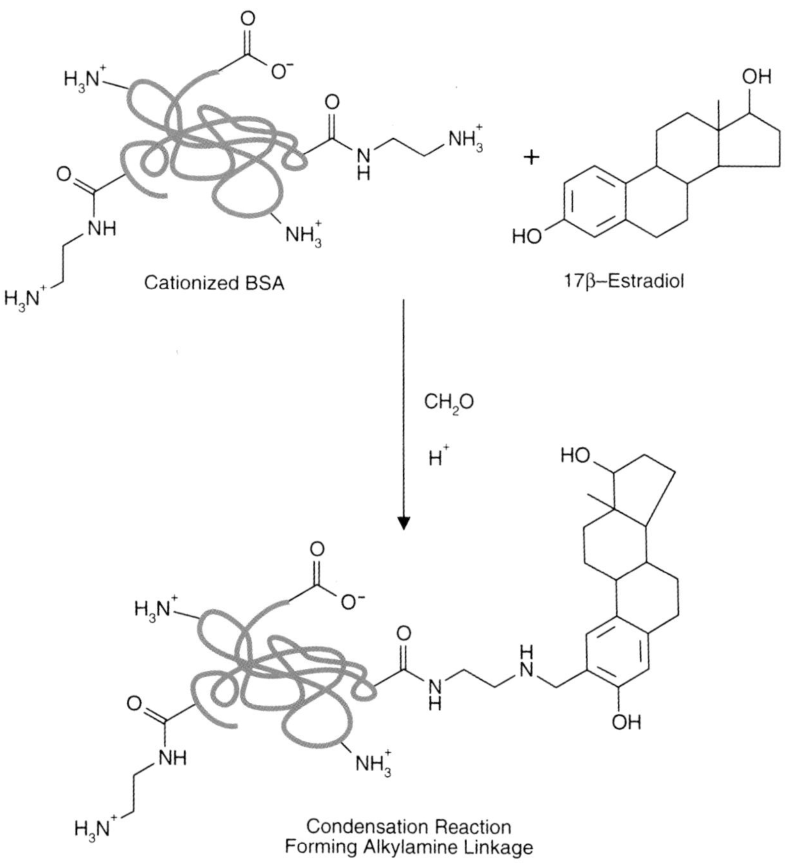

FIGURE 19.26 The conjugation of estradiol to cBSA using the Mannich reaction.

Section 6.2) (Korn *et al.*, 1972; Monsan *et al.*, 1975; Peters and Richards, 1977). One conjugation product, a quaternary pyridinium complex, can form as a crosslink between two lysine residues (Chapter 2, Section 4.4). Reduction of any Schiff bases formed during the reaction with sodium borohydride or sodium cyanoborohydride will result in stable secondary amine linkages.

The reaction of glutaraldehyde with protein carriers and peptide haptens involves mainly lysine ε-amine and N-terminal α-amino groups. The conjugates formed are usually of high molecular weight and may cause precipitation products. In addition, the orientation of the hapten on the carrier is indiscriminate with oligomers of the peptide predominating. However, despite the difficulties involved in controlling a glutaraldehyde-mediated crosslinking reaction, it still remains one of the most popular techniques for creating vaccine bioconjugates.

There are several different protocols commonly used in the literature to form glutaraldehyde conjugates. Some methods utilize a neutral pH environment in phosphate buffer (pH 6.8–7.5) while others use more alkaline pH conditions in carbonate buffer (pH 8–9) (Price *et al.*, 1993). In general, the higher pH conditions will result in aldol polymer formation of glutaraldehyde and create higher molecular weight conjugates. The concentration of glutaraldehyde in the reaction medium generally varies from 0.20% to 1% (Avrameas, 1969; Ford *et al.*, 1978; Jeanson *et al.*, 1988) with occasional use of very dilute solutions (0.05%). The lower concentrations of glutaraldehyde generate lower yields of conjugation and result in less stable conjugates (Briand *et al.*, 1985). In comparing the preparation of a peptide-carrier influenza A virus vaccine using glutaraldehyde conjugation *versus* an EDC-mediated conjugation process it was found that the glutaraldehyde vaccines were far more immunogenic (De Filette *et al.*, 2011).

The following procedure utilizes the one-step glutaraldehyde method. A two-step method may be used to somewhat limit polymerization of the conjugate (Chapter 20, Section 1.2). Varying the pH and the amount of glutaraldehyde added to the reaction can control the yield and molecular weight of the conjugates formed (see Chapter 15, Section 2.1).

Protocol

1. Dissolve the carrier protein (or another carrier that contains amine groups) in 0.1-*M* sodium carbonate, 0.15-*M* NaCl, pH 8.5, at a concentration of 2 mg/ml.
2. Add peptide hapten to the carrier solution to obtain a concentration of about 2 mg/ml. Alternatively, determine the molar ratio of peptide to carrier. Ratios

of 20:1 to 40:1 (peptide:carrier) usually result in good immunogens.

3. Add fresh glutaraldehyde to the peptide/carrier solution to obtain a 1% final concentration. Mix well. *Cautions:* Use of a fume hood is recommended when working with glutaraldehyde. Avoid contact with skin and clothing. Do not breathe vapors.
4. React for 2 to 4h at 4°C. Periodically mix the solution or use a gentle rocker.
5. The conjugate may be stabilized by addition of a reductant such as sodium borohydride or sodium cyanoborohydride. Usually sodium cyanoborohydride is recommended for specific reduction of any Schiff bases that may have formed with the aldehyde groups, but since the conjugate has already formed at this point, the use of sodium borohydride will both reduce the associated Schiff bases and eliminate any remaining aldehyde groups. Add sodium borohydride to a final concentration of 10mg/ml. Continue to react for 1h at 4°C.
6. Purify the conjugate to remove excess reagents by gel filtration using a desalting resin or by dialysis. The presence of high molecular weight conjugates may cause some precipitation in the final product. If turbidity is evident, instead of using gel filtration, dialyze against PBS, pH 7.4.

8. REDUCTIVE AMINATION-MEDIATED HAPTEN–CARRIER CONJUGATION

Hapten molecules containing aldehyde residues may be crosslinked to carrier molecules by use of reductive amination (Chapter 4, Section 4). At alkaline pH values, the aldehyde groups form intermediate Schiff bases with available amine groups on the carrier. Reduction of the resultant Schiff bases with sodium cyanoborohydride or sodium borohydride creates a stable conjugate held together by secondary amine bonds.

Oligosaccharide haptens are especially amenable for coupling to carriers by reductive amination. Carbohydrate molecules may contain reducing ends that can be utilized for this purpose (Chapter 2, Section 2.1) (Gray, 1978), or aldehyde residues may be specifically created from other functional groups (Chapter 2, Section 4.4). Often, mild oxidation using sodium periodate can be used to cleave adjacent diols on sugar residues, forming reactive aldehyde groups (Anderson *et al.*, 1989), but this process may alter the antigenic epitopes from that of the native carbohydrate. In addition, the reducing ends of glycans can be coupled to carrier molecules after their release from glycoproteins or other glycoconjugates (Chapter 2, Section 4.6). This technique can be used to create antibodies that are specific for binding the glycosylation sites on certain proteins.

If the reducing ends of oligosaccharide or glycan molecules are used for this technique, then the time necessary to obtain good yields of hapten–carrier conjugates may be from several days to several weeks, depending upon the reaction conditions used. The extended reaction period is due to the limited time reducing sugars are in their open, aldehyde form (usually far less than 1% of the available saccharide at any given time). By contrast, if periodate-oxidized carbohydrate is used, then the reaction time is reduced to only hours. It should be noted, however, that extensive periodate oxidation could modify antigenic determinants and no longer reflect the native structure and characteristics of the carbohydrate.

Protocol

1. Dissolve the carrier protein at a concentration of 10mg/ml in 0.1-*M* sodium phosphate, 0.15-*M* NaCl, pH 8.0.
2. Add the aldehyde-containing oligosaccharide to the carrier solution at a concentration sufficient to obtain at least a 20-fold molar excess of hapten to carrier. Adding a much greater molar excess of oligosaccharide to couple through reducing ends (i.e., up to 200-fold excess) will help to drive the conjugation reaction to completion.
3. Add sodium cyanoborohydride (Thermo Fisher) to the mixture to give a final concentration of 20mg/ml. *Caution*: Highly toxic! Use a fume hood and avoid inhalation of dust or vapors. Seal the reaction vessel with parafilm. Do not use a rigid sealing cap, since cyanoborohydride will liberate hydrogen gas bubbles over time and may rupture the vessel.
4. React at room temperature or at 37°C with periodic mixing. Reaction times can vary significantly depending on the reactivity of the aldehyde group. For coupling of the reducing ends of polysaccharide or glycan molecules, continue the reaction for 2 to 4 days. High density derivatization through the reducing ends may take up to 2 weeks. For coupling of periodate oxidized carbohydrate, where the aldehyde residues are more accessible, the reaction is complete within 4h. See Chapter 2, Section 4.6, for additional protocol options for working with glycans.
5. Purify the hapten–carrier conjugate to remove excess reductant by gel filtration or dialysis using a PBS, pH 7.4 buffer. Removal of unconjugated carbohydrate may be more difficult. If the oligosaccharide was of high molecular weight so that the unconjugated carbohydrate cannot be easily separated from the conjugate using typical desalting gels or small-porosity dialysis tubing, then a gel filtration matrix possessing greater exclusion limits may be used. However, often it is not necessary to remove unconjugated hapten from such preparations.

CHAPTER

20

Antibody Modification and Conjugation

The ability to conjugate an antibody to another protein or molecule is critically important for many applications in life science research, diagnostics, and therapeutics. Antibody conjugates have become one of the most important classes of biological agents associated with targeted therapy for cancer and other diseases. There literally are dozens of markers that have been identified on tumor cells to which monoclonal antibodies have been developed for targeted therapy (Carter *et al.*, 2004). The preparation of antibody conjugates to find and destroy cancer cells *in vivo* has become one of the leading strategies of research into investigational new drugs (McCarron *et al.*, 2005; Sievers and Senter, 2012). In most cases, the site-specific delivery of drugs involves the successful development of defined monoclonal antibody conjugates that can target diseased cells without affecting normal ones (see Chapter 1 for an extensive review of antibody conjugates used in the treatment of disease).

In addition, the use of antibody molecules in immunoassay or detection techniques encompasses a broad variety of applications affecting nearly every field of research. The availability of relatively inexpensive polyclonal and monoclonal antibodies of exacting specificity has made possible the design of reagent systems that can interact in high affinity with virtually any conceivable analyte. The directed specificity of purified immunoglobulins provides powerful tools for constructing immunological reagents. Using a number of conjugation and modification techniques, these specific antibodies can be modified to allow easy tracking in complex mixtures. For instance, an antibody molecule labeled with an enzyme, a fluorescent compound, or biotin provides a detectable complex able to be quantified or visualized through its tag. See Chapter 1 for an overview of how antibody conjugates are being used in a wide variety of applications in research, diagnostics, and therapeutics.

To maintain specificity in antibody conjugates derived from polyclonal antisera, only affinity-purified immunoglobulins should be used. Such purified preparations are isolated from antisera by affinity chromatography using the corresponding immobilized antigen (see Chapter 15). These preparations thus contain only that population of antibody molecules which has the desired antigenic specificity. Modification or conjugation of whole immunoglobulin fractions should be avoided, since other antibody populations will be present and cause considerable nonspecificity in the resultant activity of the reagent. Even secondary antibodies should be affinity purified and highly cross-adsorbed against immunoglobulins of other species' antibody types to prevent nonspecific interactions.

Monoclonal antibodies also should be purified by affinity chromatography prior to undergoing bioconjugation. This can be accomplished using an immobilized antigen or, if the antigen is not available in large enough quantities, an immobilized immunoglobulin binding protein (such as protein A) may be employed. Most monoclonals that can be successfully purified while maintaining activity also will be stable enough to withstand the rigors of chemical modification. Occasionally, however, a particular monoclonal will be partially or completely inactivated through the modification reaction. Sometimes this activity loss is caused by physically blocking the antigen binding sites during conjugation. In other cases, conformational changes in the complementarity-determining regions (CDRs) are the cause of the problem. If the antigen binding site is merely being blocked, then choosing an appropriate site-directed chemistry may solve the problem. On the other hand, some monoclonals are too labile to undergo modification reactions, regardless of the coupling method. Trial and error often is necessary when working with monoclonals to determine if modification will severely affect activity.

The unique structural characteristics of antibody molecules supply a number of choices for modification and conjugation schemes (Roitt, 1977; Goding, 1986; Harlow and Lane, 1988). The chemistry used to effect conjugate formation should be chosen to yield the best possible retention of antigen binding activity. A detailed

Bioconjugate Techniques, Third Edition.
DOI: http://dx.doi.org/10.1016/B978-0-12-382239-0.00020-0

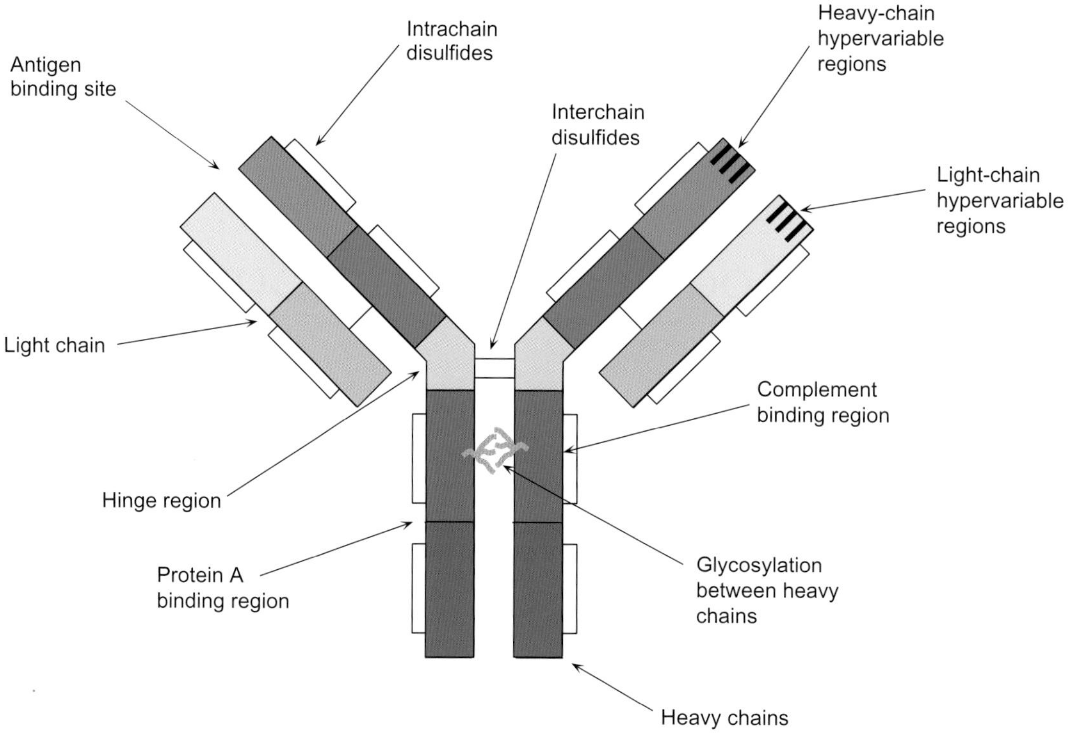

FIGURE 20.1 Detailed structure of an immunoglobulin G (IgG) antibody molecule.

illustration of antibody structure is shown in Figure 20.1. The most basic immunoglobulin G molecule is composed of two light and two heavy chains, held together by noncovalent interactions as well as a number of disulfide bonds. The light chains are disulfide-bonded to the heavy chains in the C_L and $C_H{}^1$ regions, respectively. The heavy chains are, in turn, disulfide-bonded to each other in the hinge region.

The two heavy chains in each immunoglobulin molecule are identical. Depending on the class of immunoglobulin, the molecular weight of these subunits ranges from about 50,000 to around 75,000. Similarly, the two light chains of an antibody are identical and have a molecular weight of about 25,000. For IgG molecules, the intact molecular weight representing all four subunits is in the range of 150,000 to 160,000.

There are two forms of light chains that may be found in antibodies. A single antibody will have light-chain subunits of either lambda (λ) or kappa (κ) variety, but not both types in the same molecule. The immunoglobulin class, however, is determined by an antibody's heavy-chain variety. A single antibody will also possess only one type of heavy chain (designated as γ, μ, α, ε, or δ). Thus, there are five major classes of antibody molecules, each determined from their heavy-chain type and designated as IgG, IgM, IgA, IgE, or IgD. Three of these antibody classes, IgG, IgE, and IgD, consist of the basic Ig monomeric structure containing two light and two heavy chains. By contrast, IgA molecules can exist as a singlet, doublet, or triplet of this basic Ig monomeric structure, while IgM molecules are large pentameric constructs (Figure 20.2). Both IgA and IgM contain an additional subunit, called the J chain—a very acidic polypeptide of molecular weight 15,000 that is very rich in carbohydrate. The heavy chains of immunoglobulin molecules also are glycosylated, typically in the $C_H{}^2$ domain within the Fc fragment region, but also may contain carbohydrate near the antigen binding sites.

There are two antigen binding sites on each of the basic Ig-type monomeric structures, formed by the heavy-/light-chain proximity in the N-terminal, hypervariable region at the tips of the "y" structure. The unique tertiary structure created by these subunit pairings produces the conformation necessary to interact with a complementary antigen molecule. The points of interaction on the immunoglobulin molecule with an antigen involve noncovalent forces that may encompass numerous non-sequential amino acids within the heavy and light chains. In other words, the binding site is formed not strictly from a linear sequence of amino acids on each chain, but from the unique orientation of these groups in three-dimensional space as the beta-sheet structure turns at the tips of the Fab regions. The binding site thus has affinity for a particular antigen molecule due to both structural complementarity as

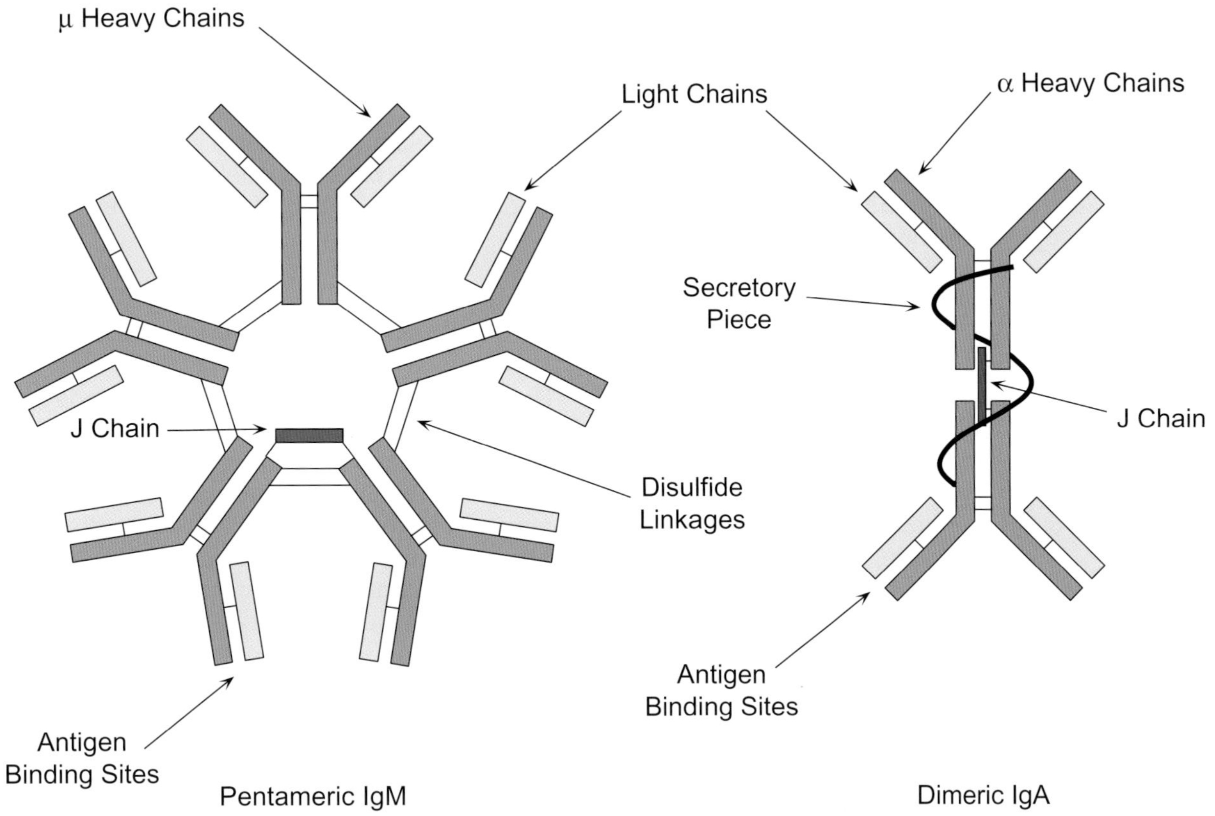

FIGURE 20.2 General structures of IgA and IgM antibodies.

well as the combination of van der Waals, ionic, hydrophobic, and hydrogen-bonding forces which may be created at each point of contact.

Useful enzymatic derivatives of antibody molecules may be prepared that still retain the antigen binding sites. Two principal digested forms of IgG antibodies are useful for creating immunological reagents. Enzymatic digestion with papain produces two small fragments of the immunoglobulin molecule, each containing an antigen binding site (called Fab fragments), and one larger fragment containing only the lower portions of the two heavy chains (called Fc, for "fragment crystallizable") (Section 1.4) (Coulter and Harris, 1983). Alternatively, pepsin cleavage produces one large fragment containing two antigen binding sites (called $F(ab')_2$) and many smaller fragments formed from extensive degradation of the Fc region (Rousseaux *et al.*, 1983). The $F(ab')_2$ fragment is held together by retention of the disulfide bonds in the hinge region. Specific reduction of these disulfides using 2-mercaptoethylamine (MEA), dithiothreitol (DTT), tris(2-carboxyethyl) phosphine (TCEP), or other reducing agents (Chapter 2, Section 4.1) produces two Fab′ fragments, each of which has one antigen binding site.

Antibody molecules possess a number of functional groups suitable for modification or conjugation purposes. Crosslinking reagents may be used to target lysine ε-amine and N-terminal α-amine groups. Carboxylate groups also may be coupled to another molecule using the C-terminal end as well as aspartic acid and glutamic acid residues. Although both amine and carboxylate groups are as plentiful in antibodies as they are in most proteins, the distribution of them within the three-dimensional structure of an immunoglobulin is nearly uniform throughout the surface topology. For this reason, conjugation procedures that utilize these functionalities will crosslink to the most accessible ones in a somewhat random fashion on nearly all parts of the antibody molecule. This, in turn, leads to a random orientation of the antibody within the conjugate structure, which may block the antigen binding sites against the surface of another coupled protein or molecule. Obscuring the binding sites in this manner results in decreased antigen binding activity in the conjugate compared to that observed for the unconjugated antibody. This effect also is dependent on how many points of modification are made on the antibody molecules. Controlling the level of modification and conjugation will typically lead to antibody conjugates having the highest possible activity in the anticipated application.

Conjugation chemistry done with antibody molecules may be more successful at preserving activity if the functional groups that are utilized are present in limiting quantities or only at discrete sites on the molecule. Such "site-directed conjugation" schemes make use of crosslinking reagents that can specifically react with residues that are only in certain positions on the immunoglobulin surface—usually chosen to be well removed from the antigen binding sites. By proper selection of the conjugation chemistry and knowledge of antibody structure, the immunoglobulin molecule can be oriented so that its bivalent binding potential for antigen remains maximally available.

Two site-directed chemical reactions are especially useful in this regard. The disulfides in the hinge region that hold the heavy chains together can be cleaved with a reducing agent (such as MEA, DTT, or TCEP) to reveal some thiols in the intact antibody or in some cases to create two half-antibody molecules that have thiols available, each containing an antigen binding site (Palmer and Nissonoff, 1963; Sun *et al.*, 2005) (Chapter 2, Section 4.1). Alternatively, smaller antigen binding fragments may be made from pepsin digestion (which makes $F(ab')_2$ fragments lacking most of the Fc region) and similarly reduced to form monovalent Fab′ molecules containing thiols. Both of these preparations contain free sulfhydryl groups that can be targeted for conjugation using thiol-reactive probes or crosslinkers. Conjugations done using hinge-region SH groups will orient the attached protein or other molecule away from the antigen binding sites, thus preventing blockage of these regions and better preserving activity.

The second method of site-directed conjugation of antibody molecules takes advantage of the glycan carbohydrate chains typically attached to the C_H2 domain between the heavy chains in the Fc region. Mild oxidation of the polysaccharide sugar residues with sodium periodate will generate aldehyde groups. A crosslinking or modification reagent containing a hydrazide or aminooxy functional group then can be used to specifically target these aldehydes for coupling to another molecule. Directed conjugation through antibody carbohydrate chains may avoid the antigen binding regions while allowing for use of intact antibody molecules. This method can result in the highest retention of antigen binding activity within the ensuing conjugate. However, care should be taken in using this method, because some antibody molecules can be glycosylated near the antigen binding area, thus potentially interfering with activity upon conjugate formation (Wright *et al.*, 1991).

Another limitation to the use of this strategy is the necessity for the antibody molecule to be glycosylated. Antibodies of polyclonal origin (from antisera) are usually glycosylated and work well in this procedure, but other antibody preparations may not possess polysaccharide. In particular, some monoclonals may not be post-translationally modified with carbohydrate after hybridoma synthesis. Recombinant antibodies grown in bacteria may also be devoid of carbohydrate. Before attempting to use a conjugation method that couples through polysaccharide regions, it is best to test the antibody to see if it contains carbohydrate—especially if the immunoglobulin is of hybridoma or recombinant origin.

1. PREPARATION OF ANTIBODY–ENZYME CONJUGATES

The most extensive application of antibody conjugation using crosslinking reagents is for the preparation of antibody–enzyme conjugates. Since the development of enzyme-linked immunosorbent assay (ELISA) systems, the ability to make conjugates of specific antibodies with enzymes has provided the means to quantify or detect hundreds, if not thousands, of important analytes. The use of enzymes as labels in immunoassay procedures surpassed radioactive tags as the means of detection, primarily due to the long-term stability potential of an enzyme system and the hazards and waste problems associated with radioisotopes. Designed properly, an antibody–enzyme conjugate assay system can be just as sensitive as a radiolabeled antibody system.

The development of viable methods for crosslinking antibody and enzyme molecules—methods that retain high antigen binding activity coupled with high enzymatic activity—has formed the basis for much of today's diagnostic industries, literally a multi-billion dollar enterprise with enormous impact on world health (see Chapter 1 for an extensive review). The conjugation chemistries that make this possible are designed around a knowledge of both antibody and enzyme structure. The best methods make use of definitive site-directed chemistries that target both molecules in regions removed from their respective active centers.

The major enzymes used in ELISA technology include horseradish peroxidase (HRP), alkaline phosphatase (AP), β-galactosidase (β-gal), and glucose oxidase (GO). See Chapter 22 for a detailed description of each enzyme's properties and activities. HRP is by far the most popular enzyme used in antibody–enzyme conjugates. One survey of enzyme use stated that HRP is incorporated in about 80% of all antibody conjugates, most of them utilized in diagnostic assay systems. AP is the second-most popular choice for antibody–enzyme conjugation, being used in almost 20% of all commercial enzyme-linked assays. Although β-gal and GO are used frequently in research labs and cited numerous times in the literature, their utilization for commercial ELISA applications represents less than 1% of the total assays available.

Conjugation methods for attaching these enzymes to antibody molecules vary according to the functional groups available. HRP is a glycoprotein and easily can be periodate oxidized for coupling *via* reductive amination to the amino groups on immunoglobulins. β-Gal contains abundant free sulfhydryl groups in its native state. The thiols can be utilized for coupling to the sulfhydryl-reactive end of heterobifunctional crosslinkers such as SMCC (Chapter 6, Section 1.3). Any of these enzymes can be conjugated through their amine groups using crosslinking agents such as glutaraldehyde or various heterobifunctional agents. The catalytic properties and activation methods often used with these enzymes are discussed in detail in Chapter 22.

The following sections describe the most common reagents and reactions used to create antibody–enzyme conjugates.

1.1. NHS Ester–Maleimide-Mediated Conjugation

Heterobifunctional reagents containing an amine-reactive NHS ester on one end and a sulfhydryl-reactive maleimide group on the other end generally have great utility for producing antibody–enzyme conjugates (see Chapter 6, Section 1). Crosslinking reagents possessing these reactive groups can be used in controlled, multi-step procedures that yield conjugates of defined composition and high activity. Among the most popular of these NHS ester–maleimide crosslinkers are SMCC (Chapter 6, Section 1.3), MBS (Chapter 6, Section 1.4), and GMBS (Chapter 6, Section 1.7). The use of any one of these crosslinkers in the following protocol can result in useful conjugates. However, SMCC and its water soluble analog, sulfo-SMCC, possess the most stable maleimide functionalities and are probably the most widely used crosslinkers of this type. This increased stability to hydrolysis of SMCC's hindered maleimide group allows activation of either enzyme or antibody *via* the amine-reactive NHS ester end, resulting in a maleimide-activated intermediate. The intermediate species can then be purified away from excess cross-linker and reaction byproducts before mixing with the second protein to be conjugated. The multi-step nature of this process limits polymerization of the conjugated proteins and provides control over the extent and sites of crosslinking. Antibody conjugation using SMCC or sulfo-SMCC continues to be a popular route to the formation of conjugates useful in detection and targeted therapeutic purposes (Nwe *et al.*, 2011; Lee *et al.*, 2012; Sukhanova *et al.*, 2012).

In addition, the PEG-based heterobifunctional cross-linkers described in Chapter 18, Section 1.2, provide enhanced water solubility for antibody conjugation applications. Conjugation of antibody molecules using a maleimide–PEG$_n$–NHS ester compound actually increases the solubility of the antibody and may help to maintain better stability for certain sensitive monoclonals than the traditional aliphatic crosslinkers. The methods described below for SMCC may be used with success for PEG-based reagents or other maleimide–NHS ester heterobifunctionals.

In protocols involving enzyme activation with SMCC and subsequent conjugation with an antibody molecule (the most common method of producing antibody–enzyme conjugates with this crosslinker), the antibody usually has to be prepared for coupling to the maleimide groups on the enzyme by introduction of sulfhydryl residues. Since antibodies typically do not contain free sulfhydryls accessible for conjugation, they must be fabricated by chemical means. Two main options are available for creating sulfhydryl functions on immunoglobulin molecules. The disulfide residues in the hinge region of the IgG structure may be reduced with DTT, TCEP, or MEA to cleave the immunoglobulin into two half-antibody molecules each possessing one antigen binding site and the requisite sulfhydryls. Alternatively, a thiolation reagent may be used to modify the intact antibody to contain sulfhydryls (Chapter 2, Section 4.1). Both options are described below. Although there are numerous thiolation reagents from which to choose, only SATA and Traut's reagent are discussed in this section, since they are the most popular.

Activation of Enzymes with NHS Ester–Maleimide Crosslinkers

The first step in conjugation of antibody molecules and enzymes using NHS ester–maleimide crosslinkers usually is modification of the enzyme with the NHS ester end of the reagent to produce a maleimide-activated derivative (Figure 20.3). The protocol described here uses sulfo-SMCC as the crosslinking agent due to the enhanced stability of its maleimide group and the water-solubility afforded by the negatively charged sulfonate on its sulfo-NHS ring. Other NHS ester–maleimide crosslinkers may be substituted without difficulty; however, water-insoluble varieties should be solubilized in DMSO or DMF prior to addition to the aqueous reaction mixture.

One note should be mentioned before proceeding: when conjugating antibody molecules with β-galactosidase, the antibody is usually activated with sulfo-SMCC first to take advantage of the indigenous sulfhydryl groups on the enzyme. Therefore, if β-gal is being used, substitute the antibody for the enzyme mentioned in this protocol, and then after the purification step add the enzyme in the desired molar excess to produce the final conjugation.

The following protocol describes the activation of horseradish peroxidase (HRP) with sulfo-SMCC.

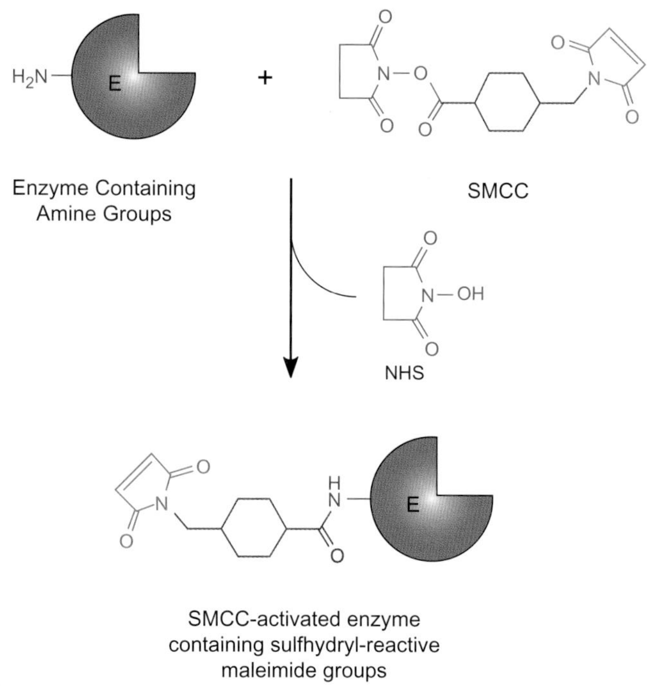

FIGURE 20.3 The reaction of SMCC with the amine groups on enzyme molecules yields a maleimide-activated derivative capable of coupling with sulfhydryl-containing antibody molecules.

Activation of other enzymes is performed similarly, with the appropriate adjustments in the mass of enzyme added to the reaction to account for differences in molecular weight.

The gel filtration column described in step 3 should be prepared and equilibrated prior to starting the modification reaction. Enzymes activated with sulfo-SMCC are available from Thermo Fisher.

Protocol

1. Dissolve 18 mg of HRP in 0.1-*M* sodium phosphate, 0.15-*M* NaCl, pH 7.2, at a concentration of 20 to 30 mg/ml. The more highly concentrated the enzyme solution, the more efficient the modification reaction will be. For conjugating smaller quantities of enzyme and antibody, proportionally decrease the amount of the reagents used, while attempting to maintain the same relative concentrations in solution.
2. Add 6 mg of sulfo-SMCC (Thermo Fisher) to the HRP solution. Mix to dissolve and react for 30 min at room temperature. Alternatively, two 3-mg additions of crosslinker may be performed—the second one after 15 min of incubation—to obtain even more efficient modification.
3. Immediately purify the maleimide-activated HRP away from excess crosslinker and reaction byproducts by gel filtration using a desalting resin. Use 0.1-*M* sodium phosphate, 0.15-*M* NaCl, pH 7.2, as the chromatography buffer. At this concentration, HRP can be observed visually as it flows through the column due to the color of its heme ring. Pool the fractions containing the HRP peak. After elution, adjust the HRP concentration to 10 mg/ml for the conjugation reaction. At this point, the maleimide-activated enzyme may be frozen and lyophilized to preserve its maleimide activity. The modified enzyme is stable for at least 1 year in a freeze-dried state. If kept in solution, the maleimide-activated HRP should be used immediately to conjugate with an antibody following one of the three options outlined below.

Conjugation with Reduced Antibodies

One method of introducing sulfhydryl residues into antibody molecules for conjugation with maleimide-activated enzymes is to reduce indigenous disulfide groups in the hinge region of the immunoglobulin structure. Reduction with low concentrations of DTT, TCEP, or 2-mercaptoethylamine (MEA) will principally cleave the disulfide bonds holding the heavy chains together, but the disulfides between the heavy and light chains may also be partially reduced to some extent. In a comparative study of disulfide reducing agents, it was determined that use of the relatively strong reductants DTT and TCEP required only 3.25 and 2.75 mole equivalents per mole equivalent of antibody molecule to achieve the reduction of two inter-chain disulfide bonds between the heavy chains of a monoclonal IgG (Sun *et al.*, 2005). This limited reduction strategy retains intact bispecific antibody molecules while providing discrete sites for conjugation to thiols. Using higher concentrations of DTT, TCEP, or MEA will result in complete cleavage of the disulfides between the heavy chains and formation of two half antibody molecules, each containing an antigen binding site. Under these conditions, some inter-chain cleavage also will occur and result in some smaller fragments being produced. Similar reduction can be done with $F(ab')_2$ fragments produced from pepsin digestion of IgG molecules. Either of these reduction steps creates half-antibody fragments, each containing one heavy and one light chain and one antigen binding site (Figure 20.4). The sulfhydryl groups produced by this reduction are able to couple with maleimide-activated enzymes without blocking the antigen binding area.

Antibody reduction is usually carried out in the presence of EDTA to prevent re-oxidation of the sulfhydryls by metal catalysis. In phosphate buffer at pH 6 to 7 and 4°C, one report stated that the number of available thiols decreased only by about 7% in the presence of EDTA over a 40-h time span. In the absence of EDTA, this sulfhydryl loss increased to 63 to 90% in the same period (Yoshitake *et al.*, 1979).

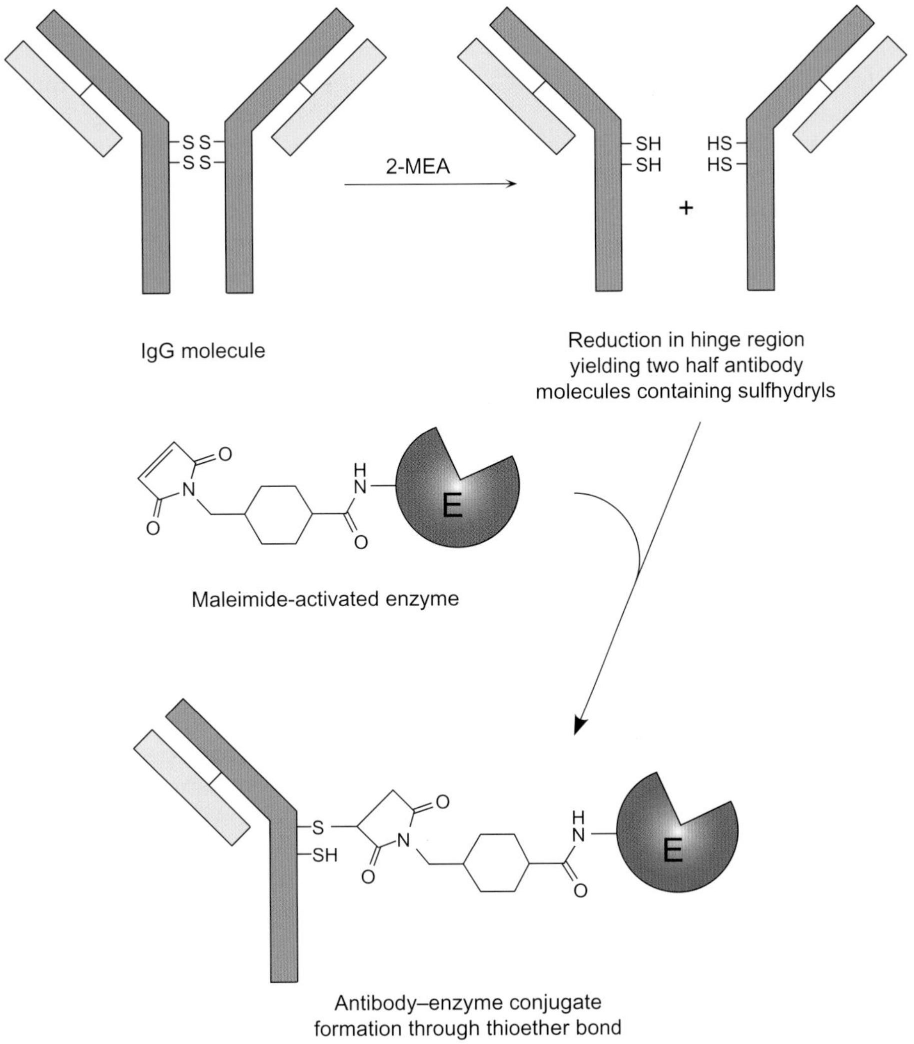

FIGURE 20.4 Reduction of the disulfide bonds within the hinge region of an IgG molecule can produce half-antibody molecules containing thiol groups. More limited reduction with low concentrations of reducing agents can be used to retain the intact antibody structure while revealing several thiols for conjugation. Reaction of these reduced antibodies with a maleimide-activated enzyme creates a conjugate through thioether bond formation.

In the following protocol, the most critical aspects are the concentration of reducing agent and EDTA in the reaction mixture. Good reduction of IgG will take place with 50- to 100-m*M* MEA and 1- to 100-m*M* EDTA. For DTT or TCEP, the concentration of reducing agent should be lowered to a 3-fold molar excess over the amount of antibody present. The pH of the reaction can vary from pH 6 to 9, with about pH 8 being optimal. The absolute concentration of antibody can vary and still yield acceptable results. With some monoclonals, however, reduction may not be completely efficient in cleaving the antibody between the heavy-chain pairs. Particularly, some subclasses of immunoglobulins contain structures with unusually high numbers of disulfides in the hinge region, and some of them may not be reduced except under much higher concentrations of reductant. Polyclonal populations typically work well in this procedure.

A final consideration is to provide adequate desalting of the reduced antibody molecule from excess reducing agent. If even a small amount of a thiol-containing reductant remains, subsequent conjugation with a maleimide-activated enzyme will be inhibited.

Protocol

1. Dissolve the IgG to be reduced at a concentration of 1 to 10 mg/ml in 0.1-*M* sodium phosphate, 0.15-*M* NaCl, pH 7.2, containing 10-m*M* EDTA.
2. Add 6 mg of MEA to each ml of antibody solution. Alternatively, add DTT or TCEP to a final concentration equal to 3 mole equivalents per mole equivalent of antibody present. Mix to dissolve.

3. Incubate for 90 min at 37°C.
4. Purify the reduced IgG by gel filtration using a desalting resin. Perform the chromatography using 0.1-*M* sodium phosphate, 0.15-*M* NaCl, pH 7.2, containing 10-m*M* EDTA as the buffer. To obtain efficient separation between the reduced antibody and excess reductant, the sample size applied to the column should be at a ratio of no more than 5% sample volume to column volume. Collect 0.5-ml fractions and monitor for protein at 280 nm. Since the reducing agents typically have no absorbance at 280 nm, the elution profile also may be monitored by use of the BCA Protein Assay method (Thermo Fisher). The BCA–copper reagent reacts with the reductants to produce a colored product. EDTA in the chromatography buffer will inhibit the BCA method somewhat, but a color response to the reducing agent peak will still be obtained. A micromethod for monitoring each fraction is as follows:
   a. Take 5 μl from each fraction collected and place in a separate well of a microtiter plate.
   b. Add 200 μl of BCA working reagent.
   c. Incubate at room temperature or 37°C for 15 to 30 min or until color develops. The color response may be measured visually or by absorbance at 562 nm. To ensure good separation between the antibody peak and excess MEA, at least one fraction of little or no color should separate the two peaks.
5. Pool the fractions containing antibody and immediately mix with an amount of maleimide-activated enzyme to obtain the desired molar ratio of antibody-to-enzyme in the conjugate. Use of a 4:1 (enzyme:antibody) molar ratio in the conjugation reaction usually results in high-activity conjugates suitable for use in many enzyme-linked immunoassay procedures. Higher molar ratios also have been used with success.
6. React for 30 to 60 min at 37°C or 2 h at room temperature. The conjugation reaction also may be carried out at 4°C overnight.
7. The conjugate may be further purified away from unconjugated enzyme by the procedures described in Section 1.5. For storage, the conjugate should be kept frozen, lyophilized, or sterile filtered and kept at 4°C. Stability studies may have to be carried out to determine the optimal method of long-term storage for a particular conjugate.

ETAC REAGENTS FOR DITHIOL CONJUGATION

The use of disulfide reduction within antibodies to create sites for bioconjugation has the potential deficiency of breaking down the subunit structure of an antibody. This may occur even when using limiting quantities of reducing agent, such as the method described in the previous protocol. Each time a cysteine disulfide is reduced to two thiols, the subsequent conjugation step usually involves the reaction of only one of the thiols with a reactive group, such as in the use of a maleimide, haloacetyl, vinyl sulfone, or epoxy group. Using these common reagents, one of the two sulfhydryls gets covalently linked to the crosslinker, but the other one may remain free without attachment. A broken disulfide linkage within the antibody may allow for subunit dissociation to occur even if the conjugation reaction is successful. However, this situation may be overcome if the conjugation reaction is able to link both thiols simultaneously using a single crosslinking agent.

Both thiols of a reduced disulfide within a protein can be targeted concurrently using a unique compound containing two reactive components consisting of an α,α-*bis*[(*p*-tolylsulfonyl)methyl]-*m*-aminoacetophenone group, termed an "equilibrium transfer alkylation crosslink" reagent (ETAC) (Liberatore *et al.*, 1990; Rosario *et al.*, 1990; Wilbur *et al.*, 1994). The two *p*-tolylsulfonyl groups on this compound react with neighboring thiols in a multi-step reaction sequence leading to the formation of two thioether bonds with a three-carbon bridge between the thiols (Figure 20.5). The molecular distance between the two thiols after disulfide reduction is nearly perfect for covalent conjugation with both reactive *p*-tolylsulfonyl groups, thus forming a linkage which maintains subunit association in proteins where disulfides occur between the subunits of multi-subunit proteins, such as within antibodies. The modification of hinge region disulfides with this reagent yields highly specific site-directed conjugation while maintaining the bifunctional nature of the intact antibody's antigen binding sites. Similarly, a $F(ab')_2$ fragment can be labeled with an ETAC reagent and maintain the bifunctional nature of the antibody, which normally does not occur when the reduced disulfides of the fragments are labeled using standard alkylation reagents.

ETAC derivatives have been used to biotinylate proteins and antibodies after disulfide reduction and to add PEG modifications to selected sites (Brocchini *et al.*, 2008; Choi *et al.*, 2009). At the time of writing, ETAC conjugation strategies are being applied to drug development and biotherapeutic conjugates, but no reagent systems are commercially available.

Conjugation with 2-Iminothiolane-Modified Antibodies

Traut's reagent, or 2-iminothiolane, is described in Chapter 2, Section 4.1. The reagent reacts with amine groups in proteins or other molecules in a ring-opening reaction to result in covalent modifications containing terminal sulfhydryl residues (Figure 20.6). Antibodies may be modified with Traut's reagent to create the

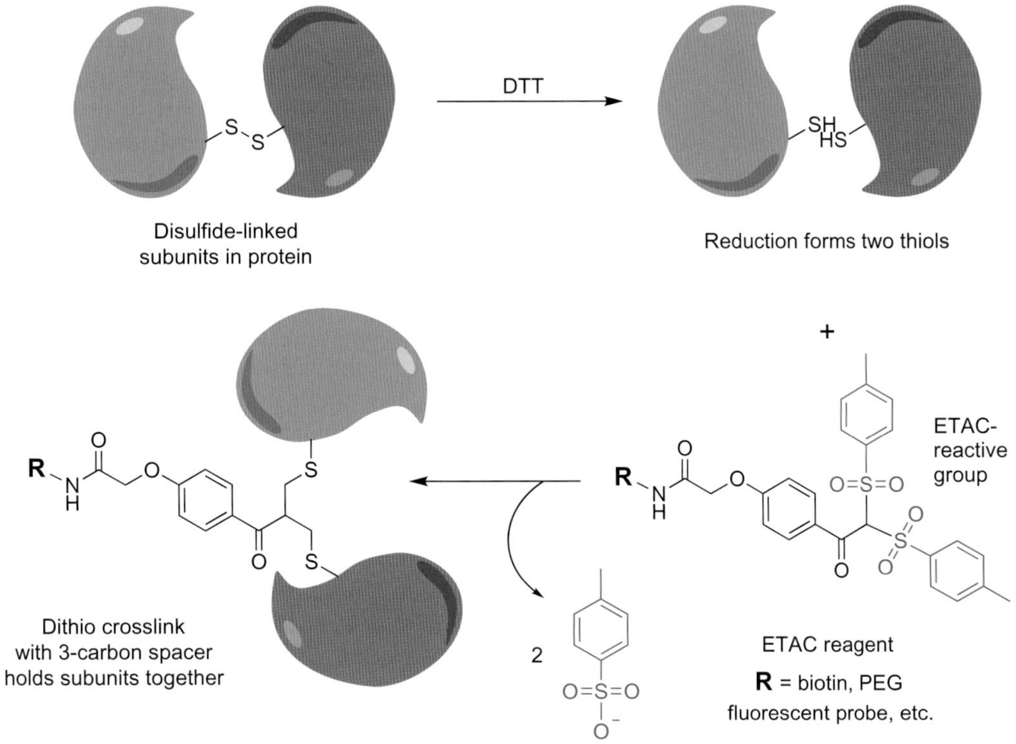

FIGURE 20.5 ETAC reagents can be used to form a covalent linkage between two thiols that were previously a part of a disulfide bond, thus retaining the subunit structure of reduced proteins or antibodies.

requisite sulfhydryls necessary for conjugation with a maleimide-activated enzyme. Unlike the disulfide reduction method described in the previous section, this protocol better retains the divalent nature of the antibody molecule. However, since amine modification of antibodies can take place at virtually any available lysine ε-amine location, the resultant sulfhydryls are distributed almost randomly over the immunoglobulin structure. Conjugation through these SH groups may result in a certain population of antibodies that have their antigen binding sites obscured or blocked by enzyme molecules. Typically, however, enough free antigen binding sites are available in the conjugate to result in high-activity complexes useful in ELISA procedures.

The number of sulfhydryls created on the immunoglobulin using thiolation procedures such as this one is more critical to the yield of conjugated enzyme molecules than the molar excess of maleimide-activated enzyme used in the conjugation reaction. Therefore, it is important to use a sufficient excess of Traut's reagent to obtain a sufficient number of available sulfhydryls. In addition, the thiolated antibody should be used immediately to prevent loss of thiols due to re-cyclization, which can tie up the available sulfhydryls in an undesired side reaction (see Chapter 2, Section 4.1).

Protocol

1. Dissolve the antibody to be modified at a concentration of 1 to 10 mg/ml in 0.1-*M* sodium phosphate, 0.15-*M* NaCl, pH 7.2, containing 10-m*M* EDTA. High levels of EDTA often are required to completely stop metal-catalyzed oxidation of sulfhydryl groups when working with serum proteins—especially polyclonal antibodies purified from antisera. Presumably, carry-over of iron from partially hemolyzed blood is the contaminating culprit.
2. Add 2-iminothiolane (Thermo Fisher) to this solution to give a molar excess of 20 to 40 times over the amount of antibody present (MW of Traut's reagent is 137.63). Solid 2-iminothiolane may be added despite the fact that the compound is relatively insoluble in aqueous solution. As the reagent reacts, it will be completely drawn into solution. Alternatively, a stock solution of Traut's may be made in DMF and an aliquot added to the antibody solution (not to exceed 10% DMF in the final solution).
3. React for 30 min at 37°C or 1 h at room temperature.
4. Purify the thiolated antibody by gel filtration using a desalting resin. Perform the chromatography using

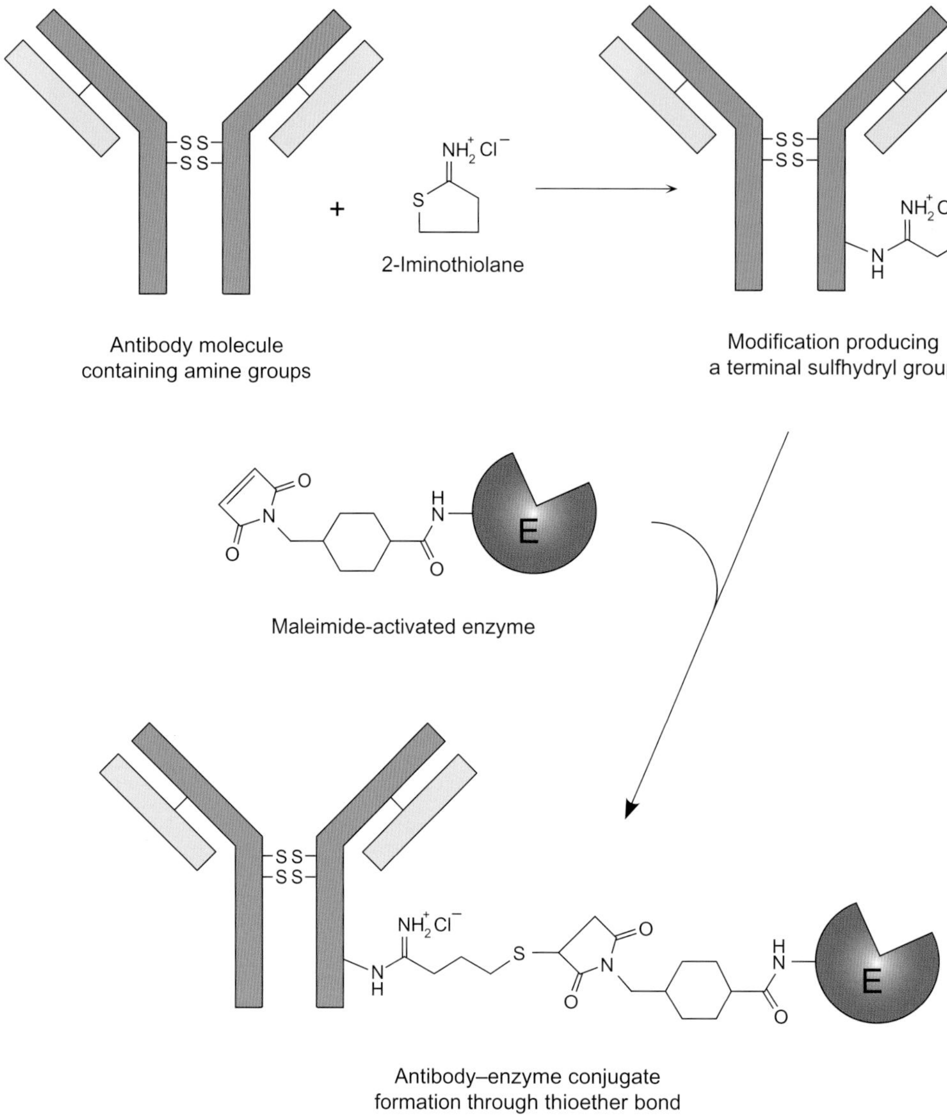

FIGURE 20.6 Antibodies may be modified with 2-iminothiolane at their amine groups to create sulfhydryls for conjugation with SMCC-activated enzymes. The maleimide groups on the derivatized enzyme react with the thiols on the antibody to form thioether bonds.

0.1-*M* sodium phosphate, 0.15-*M* NaCl, pH 7.2, containing 10-m*M* EDTA as the buffer. To obtain efficient separation between the reduced antibody and excess reductant, the sample size applied to the column should be at a ratio of no more than 5% sample volume to the total column volume. Collect 0.5-ml fractions and monitor for protein at 280 nm. To monitor the separation of the second peak (excess Traut's reagent), the BCA Protein Assay reagent (Thermo Fisher) may be used according to the procedure described in the previous section, protocol step 4.

5. Pool the fractions containing antibody and immediately mix with an amount of maleimide-activated enzyme to obtain the desired molar ratio of antibody-to-enzyme in the conjugate. Use of a 4:1 to 15:1 (enzyme:antibody) molar ratio in the conjugation reaction usually results in high-activity conjugates suitable for use in many enzyme-linked immunoassay procedures.
6. React for 30 to 60 min at 37°C or 2 h at room temperature. The conjugation reaction also may be performed at 4°C overnight.
7. The conjugate may be further purified away from unconjugated enzyme by the procedures described in Section 1.5. For storage, the conjugate should be kept frozen, lyophilized, or sterile filtered and kept at 4°C. Stability studies may have to be performed to determine the optimal method of long-term storage for a particular conjugate.

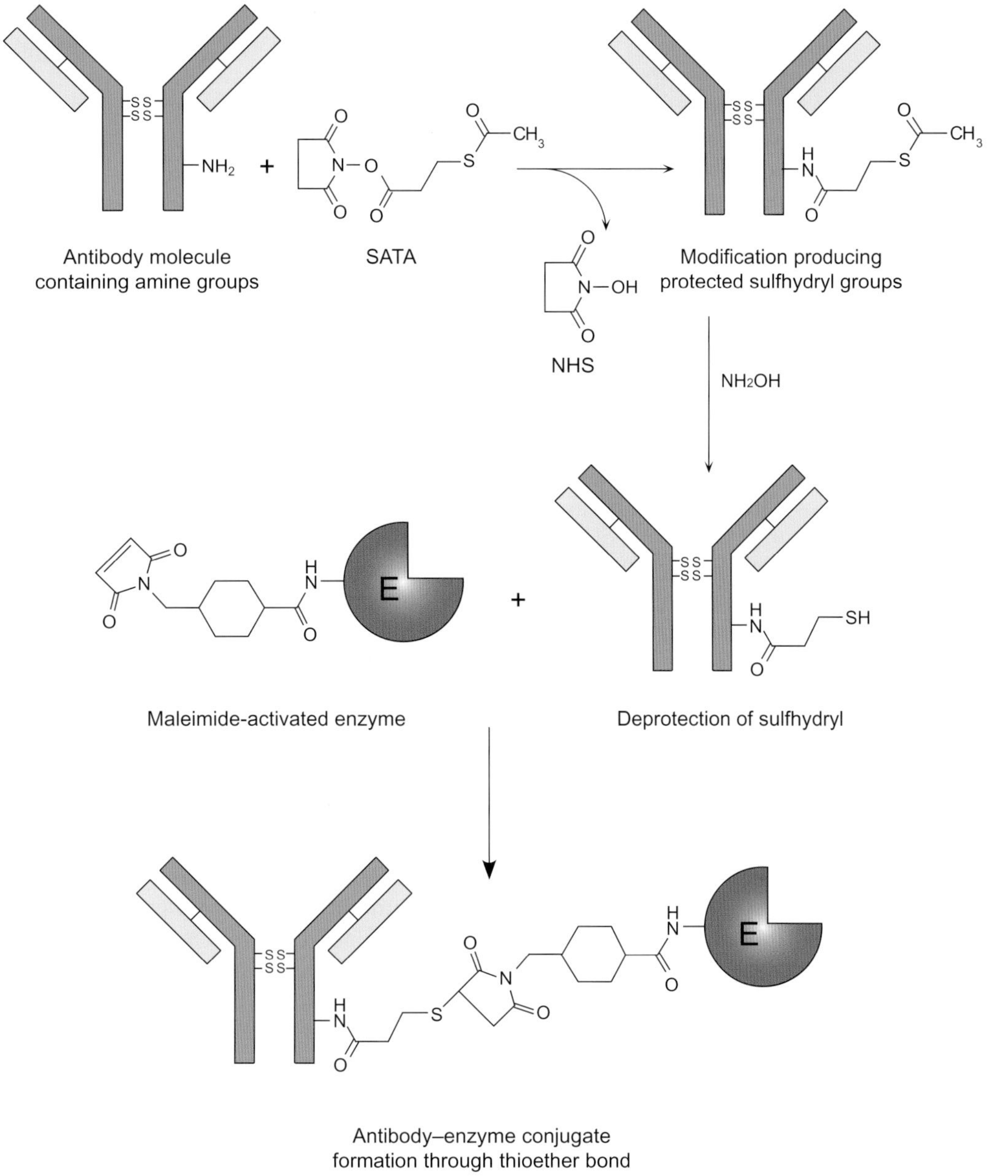

FIGURE 20.7 Available amine groups on an antibody molecule may be modified with the NHS ester end of SATA to produce amide bond derivatives containing terminal protected sulfhydryls. The acetylated thiols may be deprotected by treatment with hydroxylamine at alkaline pH. Reaction of the thiolated antibody with a maleimide-activated enzyme results in thioether crosslinks.

Conjugation with SATA-Modified Antibodies

N-Succinimidyl-*S*-acetylthioacetate (SATA) is a thiolation reagent described in detail in Chapter 2, Section 4.1. The compound reacts with primary amines *via* its NHS ester end to form stable amide linkages. The acetylated sulfhydryl group is stable until deacetylated with hydroxylamine. Thus, antibody molecules may be thiolated with SATA to create the sulfhydryl target groups necessary to couple with a maleimide-activated enzyme (Figure 20.7). Using this reagent, stock preparations of SATA-modified antibodies may be prepared and deacetylated as needed. Unlike thiolation procedures which immediately form a free sulfhydryl residue, the protected sulfhydryl group of SATA-modified proteins is stable to long-term storage without degradation.

Although amine-reactive protocols, such as SATA thiolation, result in nearly random attachment over the surface of the antibody structure, it has been shown that modification with up to six SATAs per antibody molecule typically results in no decrease in antigen binding activity (Duncan *et al.*, 1983). Even higher ratios of

SATA to antibody are possible with excellent retention of activity.

The following protocol should be compared to the method described for SATA thiolation in Chapter 2, Section 4.1. Although the procedures are slightly dissimilar, the differences indicate the flexibility inherent in the chemistry. For convenience, the buffer composition indicated here was chosen to be consistent throughout this section on enzyme–antibody conjugation using SMCC. Other buffers and alternate protocols can be found in the literature.

Protocol

1. Dissolve the antibody to be modified in 0.1-*M* sodium phosphate, 0.15-*M* NaCl, pH 7.2, at a concentration of 1 to 5 mg/ml. Note: Phosphate buffers at various pH values between 7.0 and 7.6 have been used successfully with this protocol. Other mildly alkaline buffers may be substituted for phosphate in this reaction, providing they do not contain extraneous amines (e.g., Tris) or promote hydrolysis of SATA's NHS ester (e.g., imidazole).
2. Prepare a stock solution of SATA (Thermo Fisher) by dissolving it in DMF or DMSO at a concentration of 8 mg/ml. Use a fume hood to handle the organic solvents.
3. Add 10 to 40 µl of the SATA stock solution per milliliter of 1 mg/ml antibody solution. This will result in a molar excess of approximately 12- to 50-fold of SATA over the antibody concentration (for an initial antibody concentration of 1 mg/ml). A 12-fold molar excess works well, but higher levels of SATA incorporation will potentially result in more maleimide-activated enzyme molecules able to couple to each thiolated antibody molecule. For higher concentrations of antibody in the reaction medium, proportionally increase the amount of SATA addition; however, do not exceed 10% DMF in the aqueous reaction medium.
4. React for 30 min at room temperature.
5. To purify the SATA-modified antibody, perform a gel filtration separation using desalting resin or by dialysis against 0.1-*M* sodium phosphate, 0.15-*M* NaCl, pH 7.2, containing 10-m*M* EDTA. Purification is not absolutely required, since the following deprotection step is performed using hydroxylamine at a significant molar excess over the initial amount of SATA added. Whether a purification step is carried out or not, at this point, the derivative is stable and may be stored under conditions which favor long-term antibody activity (i.e., sterile filtered at 4°C, frozen, or lyophilized).
6. Deprotect the acetylated sulfhydryl groups on the SATA-modified antibody according to the following protocol:
   a. Prepare a 0.5-*M* hydroxylamine (Thermo Fisher) solution in 0.1-*M* sodium phosphate, pH 7.2, containing 10-m*M* EDTA.
   b. Add 100 µl of the hydroxylamine stock solution to each ml of the SATA-modified antibody. Final concentration of hydroxylamine in the antibody solution is 50-m*M*.
   c. React for 2 h at room temperature.
   d. Purify the thiolated antibody by gel filtration on a desalting resin using 0.1-*M* sodium phosphate, 0.1-*M* NaCl, pH 7.2, containing 10-m*M* EDTA as the chromatography buffer. To obtain efficient separation between the thiolated antibody and excess hydroxylamine and reaction byproducts, the sample size applied to the column should be at a ratio of no more than 5% sample volume to the total column volume. Collect 0.5-ml fractions. Pool the fractions containing protein by measuring the absorbance of each fraction at 280 nm.
7. Immediately mix the thiolated antibody with an amount of maleimide-activated enzyme to obtain the desired molar ratio of antibody-to-enzyme in the conjugate. Use of a 4:1 to 15:1 (enzyme:antibody) molar ratio in the conjugation reaction usually results in high-activity conjugates suitable for use in many enzyme-linked immunoassay procedures.
8. React for 30 to 60 min at 37°C or 2 h at room temperature. The conjugation reaction also may be performed at 4°C overnight.
9. The conjugate may be further purified away from unconjugated enzyme by the procedures described in Section 1.5. For storage, the conjugate should be kept frozen, lyophilized, or sterile filtered and kept at 4°C. Stability studies may have to be carried out to determine the optimal method of long-term storage for a particular conjugate.

1.2. Glutaraldehyde-Mediated Conjugation

Glutaraldehyde was one of the first crosslinking agents used for creating antibody–enzyme conjugates, and it is still being used today in manufacturing processes despite its shortcomings. The crosslinking process using glutaraldehyde is believed to proceed by a number of mechanisms, including Schiff base formation with possible rearrangement to a stable product, addition to hemiacetal ring structures, or through a Michael-type addition reaction that takes place at points of double-bond unsaturation created by polymerization of the reagent in solution (see Chapter 2,

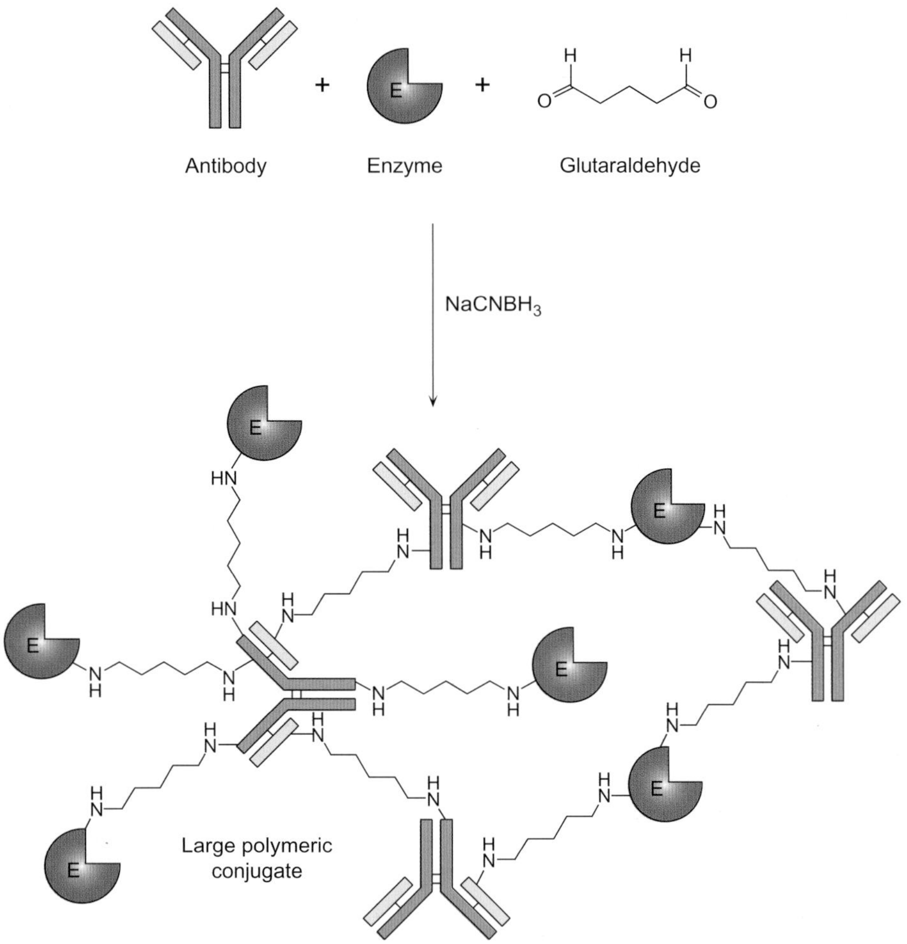

FIGURE 20.8 Glutaraldehyde antibody–enzyme crosslinking procedures usually produce a wide range of high-molecular-weight complexes, some of which may precipitate from solution. Note that there are many additional crosslinked forms that may occur upon crosslinking, including species that involve addition reactions to hemiacetal and aldol forms of glutaraldehyde which may involve high-molecular-weight polymers with multiple attachment points.

Section 4.4; Chapter 5, Section 6.2; Chapter 15, Section 2.1) (Avrameas, 1969). Reduction of Schiff base intermediates is also possible using sodium cyanoborohydride to form stable secondary amine linkages.

The problem of indeterminate reaction products is a deficiency that plagues all conjugations performed using glutaraldehyde. Part of this difficulty is due to the reagent's bifunctional nature, but a significant part of the problem is also due to the ambiguous nature of the commercial product. In aqueous solutions at alkaline pH, glutaraldehyde can undergo hemiacetal cyclization and aldol condensation reactions with itself to form a variety of structures, some of which contain α, β-unsaturated aldehydes (Hardy *et al.*, 1969, 1976). Another disadvantage of the reagent is the tendency to form high-molecular-weight conjugates due to uncontrollable polymerization during the crosslinking process. The resultant conjugates often have a significant amount of insoluble polymer which causes yield and activity losses in the preparation of antibody–enzyme conjugates. This is especially true when the conjugation is performed using the one-step method where glutaraldehyde is simply added to a solution containing the two proteins to be crosslinked (Figure 20.8). Enzymatic activity yields using this process can be as little as 10% in the final antibody–enzyme conjugate. To somewhat overcome the polymerization problem, a two-step procedure was developed which involves first activating one of the proteins with glutaraldehyde, purifying the intermediate from excess reagent, and then adding the second protein to effect the final conjugation. Unfortunately, even when using the two-step method, it results in significant formation of large-molecular-weight species that may precipitate out of solution. The only enzyme that the two-step method seems to work well with is HRP, since it only contains a limited number of available lysine amine groups.

Despite these deficiencies, antibody–enzyme conjugates are still being made using glutaraldehyde—particularly for many commercial diagnostic ELISA kits which were developed before the advent of more controllable, heterobifunctional crosslinking procedures. Today, choosing another method of producing antibody–enzyme conjugates will result in much better conjugates of higher activity and higher yield.

One-Step Glutaraldehyde Protocol

1. Prepare a solution containing 2 mg/ml antibody and 5 mg/ml enzyme in 0.02-*M* sodium phosphate, 0.15-*M* NaCl, pH 7.4, chilled to 4°C.
2. In a fume hood, add 10 µl of 25% glutaraldehyde (Sigma) per milliliter of antibody/enzyme solution. Mix well.
3. React for 2 h at 4°C.
4. To reduce the resultant Schiff bases and any excess aldehydes, add sodium borohydride (Aldrich) to a final concentration of 10 mg/ml. Note: Some protocols do not call for a reduction step. As an alternative to reduction, add 50 µl of 0.2-*M* lysine in 0.5-*M* sodium carbonate, pH 9.5, to each milliliter of the conjugation reaction to block excess reactive sites. Block for 2 h at room temperature. Other amine-containing small molecules may be substituted for lysine—such as glycine, Tris buffer, or ethanolamine.
5. Reduce for 1 h at 4°C.
6. To remove any insoluble polymers that may have formed, centrifuge the conjugate or filter it through a 0.45-µm filter. Purify the conjugate by gel filtration or dialysis using PBS, pH 7.4.

Two-Step Glutaraldehyde Protocol

1. Dissolve the enzyme at a concentration of 10 mg/ml in 0.1-*M* sodium phosphate, 0.15-*M* NaCl, pH 6.8.
2. Add glutaraldehyde to a final concentration of 1.25%.
3. React overnight at room temperature.
4. Purify the activated enzyme from excess glutaraldehyde by gel filtration using a desalting resin or by dialysis against PBS, pH 6.8.
5. Dissolve the antibody to be conjugated at a concentration of 10 mg/ml in 0.5-*M* sodium carbonate, pH 9.5. Mix the activated enzyme with the antibody at the desired molar ratio to effect the conjugation. Mixing the equivalent of 4 mg of enzyme per milligram of antibody usually results in acceptable conjugates.
6. React overnight at 4°C.
7. To reduce any Schiff bases that may have formed as well as reduce any excess aldehydes, add sodium borohydride to a final concentration of 10 mg/ml. Note: Some protocols avoid a reduction step. As an alternative to reduction, add 50 µl of 0.2-*M* lysine in 0.5-*M* sodium carbonate, pH 9.5 to each milliliter of the conjugation reaction to block excess reactive sites. Block for 2 h at room temperature. Other amine-containing small molecules may be substituted for lysine—such as glycine, Tris buffer, or ethanolamine.
8. Reduce for 1 h at 4°C.
9. To remove any insoluble polymers that may have formed, centrifuge the conjugate or filter it through a 0.45-µm filter. Purify the conjugate by gel filtration or dialysis using PBS, pH 7.4.

1.3. Reductive Amination-Mediated Conjugation

Oxidation of polysaccharide residues in glycoproteins with sodium periodate provides an efficient way of generating reactive aldehyde groups for subsequent conjugation with amine- or hydrazide-containing molecules *via* reductive amination (Chapter 2, Section 4.4, and Chapter 4, Section 4). Some selectivity of monosaccharide oxidation may be accomplished by regulating the concentration of periodate in the reaction medium. In the presence of 1-m*M* sodium periodate at approximately 0°C, sialic acid groups will be preferentially oxidized at their adjacent hydroxyl residues on the Nos. 7, 8, and 9 carbon atoms, cleaving off two molecules of formaldehyde and leaving one aldehyde group on the No. 7 carbon. At higher concentrations of sodium periodate (10-m*M* or greater) at room temperature, other sugar residues will be oxidized at points where adjacent carbon atoms contain hydroxyl groups.

Thus, glycoproteins such as horseradish peroxidase, glucose oxidase, or most antibody molecules can be activated for conjugation by brief treatment with periodate. Crosslinking with an amine-containing protein takes place under alkaline pH conditions through the formation of Schiff base intermediates. These relatively labile intermediates can be stabilized by reduction to a secondary amine linkage with sodium cyanoborohydride (Figure 20.9).

The use of periodate coupling chemistry for HRP first was introduced by Nakane and Kawaoi (1974; see also Nakane, 1975). In the first step of their protocol, the few amine groups on HRP were initially blocked with 2,4-dinitrofluorobenzene (DNFB) before periodate oxidation. The blocking step was designed to eliminate the possibility of self-conjugation of enzyme molecules during reductive amination with an immunoglobulin. However, Boorsma and Streefkerk (1976a,b) determined that HRP can still dimerize even after DNFB blocking, perhaps by a mechanism similar to Mannich condensation (Chapter 3, Section 6.2) or through aldol formation. In fact, amine-blocked, periodate-oxidized HRP will form insoluble complexes during storage after just

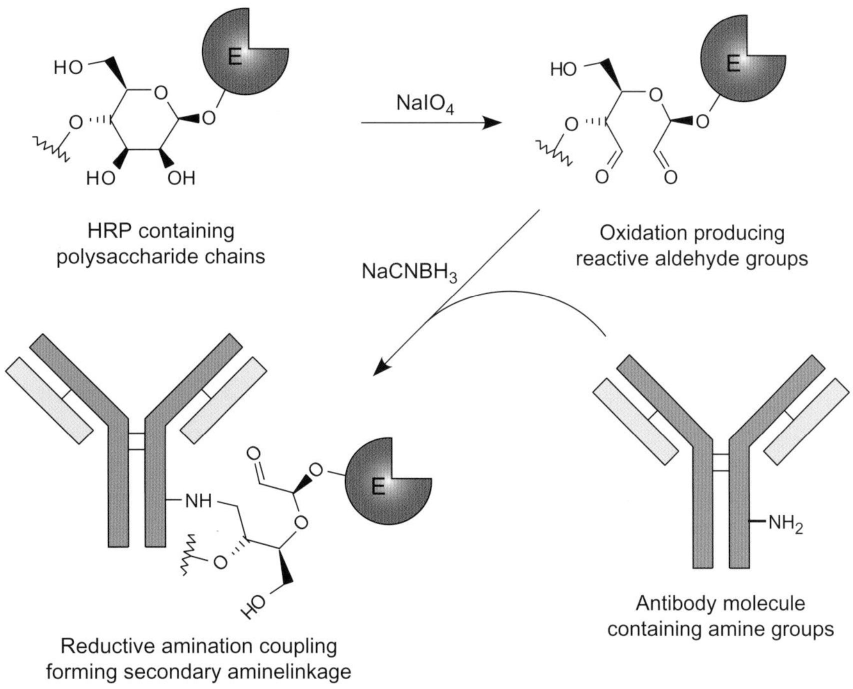

FIGURE 20.9 Enzymes that are glycoproteins like HRP may be oxidized with sodium periodate to produce reactive aldehyde residues. Conjugation with an antibody then may be done by reductive animation using sodium cyanoborohydride.

weeks in solution at room temperature or at 4°C, indicating that another route of conjugation is taking place.

Reductive amination crosslinking has been carried out using sodium borohydride or sodium cyanoborohydride; however, cyanoborohydride is the better choice since it is more specific for reducing Schiff bases and will not reduce aldehydes. Small blocking agents such as lysine, glycine, ethanolamine, or Tris can be added after conjugation to quench any unreacted aldehyde sites (Mannik and Downey, 1973; Mattiasson and Nilsson, 1977; Barbour, 1976). Ethanolamine and Tris are the best choices for blocking agents, since they contain hydrophilic hydroxyl groups with no charged functional groups.

The pH of the reductive amination reaction can be controlled to affect the efficiency of the crosslinking process and the size of the resultant antibody–enzyme complexes formed. At physiological pH, the initial Schiff base formation is less efficient and conjugates of lower molecular weight will result. At more alkaline pH values (i.e., pH 9–10), Schiff base formation occurs rapidly and with high efficiency, resulting in conjugates of higher molecular weight and greater incorporation of enzyme (when oxidized HRP is reacted in excess).

The ability to select the relative size of the antibody–enzyme complex is important depending on the assay application. Low-molecular-weight conjugates may be more optimal for immunohistochemical staining or blotting techniques where penetration of the complex through membrane barriers is an important consideration. Washing steps also more effectively remove excess reagent if the conjugate is of low molecular weight, thus maintaining a low background signal in an assay. By contrast, conjugates of high molecular weight are more appropriate for ELISA procedures in a microplate or array format, where high sensitivity is important, but washing off excess conjugate is not a problem.

The protocols appearing in the literature vary according to the amount of periodate used during polysaccharide oxidation, the type of reductant and blocking agent employed for reductive amination, and the pH at which the various reactions are done. This variability indicates considerable flexibility in the protocols, all of which yield usable antibody–enzyme conjugates. There are, however, several conclusions that can be drawn from these studies: Investigations performed using HRP indicate the optimal concentration of sodium periodate during oxidation to be approximately 4 to 8-m*M* (Tussen and Kurstak, 1984). This reaction should be performed in the dark to prevent periodate breakdown and for a limited period of time (no more than 15–30 min) to avoid loss of enzymatic activity. The conjugation reaction should be carried out at alkaline pH (7.2–9.5) in the presence of a reducing agent to stabilize the Schiff base intermediates. If sodium cyanoborohydride is used as the reductant, a blocking agent should be added at the completion of the conjugation reaction to cap excess aldehyde sites. The following protocol follows these general guidelines and works well especially in the preparation of HRP–antibody conjugates.

Activation of Enzymes with Sodium Periodate

Enzymes that are glycosylated (i.e., HRP and glucose oxidase) may be oxidized according to the following method to produce aldehyde groups for reductive amination coupling to an antibody molecule.

Protocol

1. Dissolve the enzyme to be oxidized in water or 0.01-*M* sodium phosphate, 0.15-*M* NaCl, pH 7.2, at a concentration of 10 to 20 mg/ml.
2. Dissolve sodium periodate in water at a concentration of 0.088 *M*. Protect from light.
3. Immediately add 100 μl of the sodium periodate solution to each milliliter of the enzyme solution. This ratio of addition results in an 8-m*M* periodate concentration in the reaction mixture. Mix to dissolve. Protect from light.
4. React in the dark for 15 to 20 min at room temperature. If HRP is the enzyme being oxidized, a color change will be apparent as the reaction proceeds—changing the brownish/gold color of concentrated HRP to green. Limiting the time of oxidation will help to preserve enzyme activity.
5. Immediately quench the reaction by the addition of 0.1 ml of glycerol per milliliter of reaction solution. Instead of glycerol, *N*-acetylmethionine may be added to quench the reaction, because the thioether of the methionine side chain will react with periodate to form sulfoxide or sulfone products (Geoghegan and Stroh, 1992). In addition, sodium sulfite (Na_2SO_3) was used by Stolowitz *et al.* (2001) to quench the periodate oxidation of HRP in solution. This may be the simplest route to stopping the reaction, as sulfite is inexpensive and the reduction does not form reactive byproducts. Add quenching reagent to provide at least a 2-times molar excess over the amount of periodate initially added to the reaction. Alternatively, the reaction may be stopped by immediate gel filtration on a desalting resin. If a dextran-based resin is used for the chromatography, the support itself will react with sodium periodate to quench excess reagent. Purify the oxidized enzyme by gel filtration using 0.01-*M* sodium phosphate, 0.15-*M* NaCl, pH 7.2. To obtain efficient separation between the oxidized enzyme and excess periodate (or quenching agent), the sample size applied to the column should be at a ratio of no more than 5% sample volume to the total column volume. Collect 0.5-ml fractions and monitor for protein at 280 nm. HRP also may be detected by its absorbance at 403 nm. When oxidizing large quantities of HRP, the fraction collection process may be performed visually—just pooling the main colored HRP peak as it comes off the column.
6. Pool the fractions containing protein. Adjust the enzyme concentration to 10 mg/ml for the conjugation step (see next section). The periodate-activated enzyme may be stored frozen or freeze-dried for extended periods without loss of activity. Do not store the preparation in solution at room temperature or 4°C, since precipitation will occur over time due to self-polymerization.

Activation of Antibodies with Sodium Periodate

Many immunoglobulin molecules are glycoproteins that can be periodate-oxidized to contain reactive aldehyde residues. Polyclonal IgG molecules often contain carbohydrate in the Fc portion of the molecule. This is sufficiently removed from the antigen binding sites to allow conjugation to take place through the polysaccharide chains without compromising activity. Occasionally, however, some antibodies may contain sites of glycosylation near the antigen binding regions, and in this situation conjugation through these sites may affect binding activity. Although antibody–enzyme conjugation by reductive amination is typically done by oxidation of the enzyme with subsequent crosslinking to an amine-containing antibody, oxidation of the antibody with subsequent conjugation to an amine- or hydrazide-containing molecule is also possible. It should be noted, however, that many monoclonal antibodies are not glycosylated and therefore can not be used in this protocol. Recombinant antibodies also do not contain carbohydrate. A given monoclonal should be checked to verify the presence of carbohydrate before attempting to use a periodate-mediated conjugation protocol.

Protocol

1. Dissolve the antibody to be periodate-oxidized at a concentration of 10 mg/ml in 0.01-*M* sodium phosphate, 0.15-*M* NaCl, pH 7.2.
2. Dissolve sodium periodate in water to a final concentration of 0.1-*M*. Protect from light.
3. Immediately add 100 μl of the sodium periodate solution to each milliliter of the antibody solution. Mix to dissolve. Protect from light.
4. React in the dark for 15 to 20 min at room temperature.
5. Immediately quench the reaction by the addition of sodium sulfite (Na_2SO_3) to provide a 2-times molar excess over the initial amount of periodate added. Purify the oxidized antibody by gel filtration using a desalting resin. The chromatography buffer is 0.1-*M* sodium phosphate, 0.15-*M* NaCl, pH 7.2. To obtain efficient separation between the oxidized

antibody and excess periodate, the sample size applied to the column should be at a ratio of no more than 5% sample volume to the total column volume. Collect 0.5-ml fractions and monitor for protein at 280 nm.

6. Pool the fractions containing protein. Adjust the antibody concentration to 10 mg/ml for the conjugation step. The oxidized antibody should be used immediately.

Conjugation of Periodate-Oxidized HRP to Antibodies by Reductive Amination

The following protocol assumes that HRP has already been periodate-oxidized by the method given in Section 1.3, above.

Protocol

1. Dissolve the IgG to be conjugated at a concentration of 10 mg/ml in 0.2-*M* sodium bicarbonate, pH 9.6, at room temperature. The high pH buffer will result in very efficient conjugation with the highest possible incorporation of enzyme molecules per antibody molecule. To produce lower molecular weight conjugates, dissolve the IgG at a concentration of 10 mg/ml in 0.1-*M* sodium phosphate, 0.15-*M* NaCl, pH 7.2.
2. The periodate-oxidized enzyme (HRP) prepared in Section 1.3 was finally purified using 0.01-*M* sodium phosphate, 0.15-*M* NaCl, pH 7.2. For conjugation using the lower-pH-buffered environment, this HRP preparation can be used directly at 10 mg/ml concentration. For conjugation using the higher pH carbonate buffer, dialyze the HRP solution against 0.2-*M* sodium carbonate, pH 9.6 for 2 h at room temperature prior to use. A volume of HRP solution equal to the volume of antibody solution will be required.
3. Mix the antibody solution with the enzyme solution at a ratio of 1:1 (v/v). Since an equal mass of antibody and enzyme is present in the final solution, this will result in a 3.75 molar excess of HRP over the amount of IgG. For conjugates consisting of greater enzyme-to-antibody ratios, proportionally increase the amount of enzyme solution as required. Typically, molar ratios of 4:1 to 15:1 (enzyme:antibody) give acceptable conjugates useful in a variety of ELISA techniques.
4. React for 2 h at room temperature.
5. In a fume hood, add 10 µl of 5-*M* sodium cyanoborohydride (Sigma) per milliliter of reaction solution. **Caution**: Cyanoborohydride is extremely toxic. All operations should be carried out with care in a fume hood. Also, avoid any contact with the reagent, as the 5-*M* solution is prepared in 1-*N* NaOH.
6. React for 30 min at room temperature with gentle mixing (in a fume hood).
7. Block unreacted aldehyde sites by addition of 50 µl of 1-*M* ethanolamine, pH 9.6, per milliliter of conjugation solution. Approximately a 1-*M* ethanolamine solution may be prepared by addition of 300 µl ethanolamine to 5 ml of deionized water. Adjust the pH of the ethanolamine solution by addition of concentrated HCl, keeping the solution cool on ice.
8. React for 30 min at room temperature.
9. Purify the conjugate from excess reactants by dialysis or gel filtration using a desalting resin. Use 0.01-*M* sodium phosphate, 0.15-*M* NaCl, pH 7.0, as the buffer for either operation. The conjugate may be further purified by removal of unconjugated enzyme using one of the methods described in Section 1.5.

Conjugation of Periodate-Oxidized Antibodies with Amine or Hydrazide Derivatives

The following protocol assumes that the antibody has already been periodate-oxidized by the method of Section 1.3 (above) to create reactive aldehyde groups suitable for coupling with amine-containing or hydrazide-containing molecules. This is an excellent method for directing the antibody modification reaction away from the antigen binding sites, if the antibody glycosylation points are solely in the Fc region of the molecule. For instance, biotinylation of intact antibodies can be performed after mild periodate treatment using biotin-hydrazide (Chapter 11, Section 6.4) (Figure 20.10). It should be noted, however, that periodate-oxidized antibodies can self-conjugate through their own amines if high-pH reductive amination is used. Conjugation with periodate-oxidized antibodies works best if the receiving molecule is modified to contain hydrazide groups and the reaction is carried out at more moderate pH values (e.g., slightly acidic to neutral pH).

1. For conjugation to hydrazide-containing proteins, dissolve the periodate-oxidized antibody at a concentration of 10 mg/ml in 0.1-*M* sodium phosphate, 0.15-*M* NaCl, pH 6.0 to 7.2. For conjugation to amine-containing molecules and proteins, dissolve the oxidized antibody at 10 mg/ml in 0.2-*M* sodium carbonate, pH 9.6.
2. Dissolve a hydrazide-containing enzyme or other protein at a concentration of 10 mg/ml in 0.1-*M* sodium phosphate, 0.15-*M* NaCl, pH 6.0 to 7.2. For the preparation of a hydrazide-activated enzyme, see Chapter 22, Section 2.4. For modification with a hydrazide-containing probe, such as biotin–hydrazide, use a concentration of 5-m*M* in the phosphate buffer. For conjugation through the amine groups of a secondary molecule, dissolve the

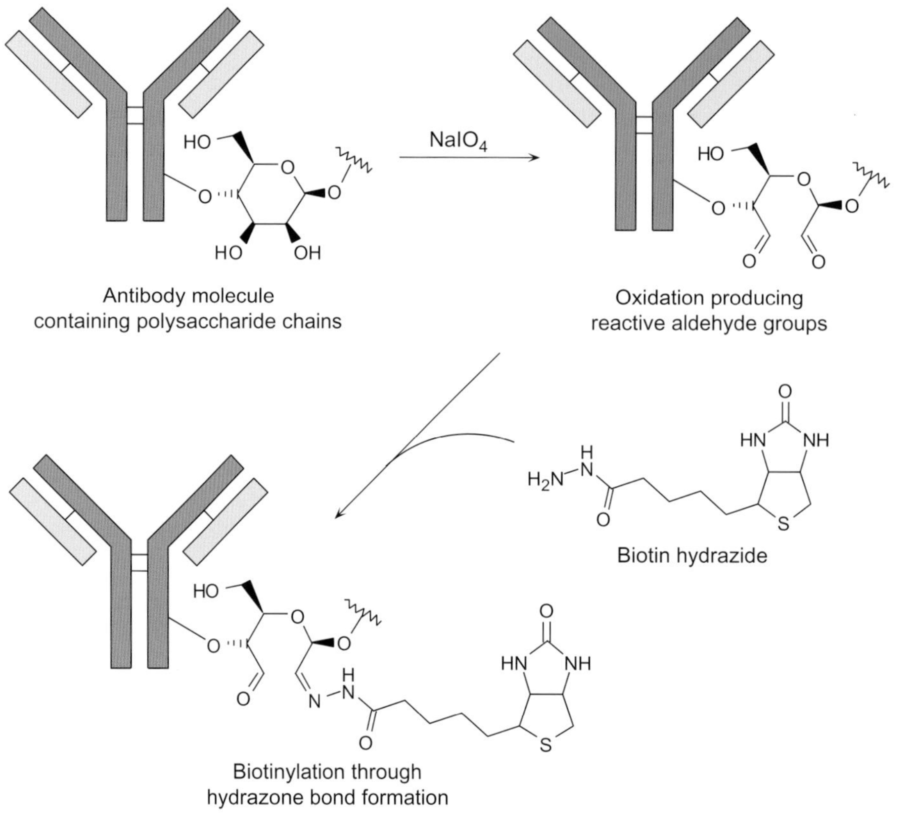

FIGURE 20.10 Polysaccharide groups on antibody molecules may be oxidized with periodate to create aldehydes. Modification with biotin–hydrazide results in hydrazone linkages. The sites of modification using this technique are often away from the antibody–antigen binding regions, thus preserving antibody activity.

amine-containing protein at 10 mg/ml in 0.2-*M* sodium carbonate, pH 9.6.

3. Mix the antibody solution from step 1 with the protein solution from step 2 in amounts necessary to obtain the desired molar ratio for conjugation. Often, the secondary molecule is reacted in approximately a 4- to 15-fold molar excess over the amount of antibody present.
4. React for 2 h at room temperature.
5. In a fume hood, add 10 μl of 5-*M* sodium cyanoborohydride (Sigma) per milliliter of reaction solution. **Caution**: Cyanoborohydride is extremely toxic. All operations should be carried out with care in a fume hood. Also, avoid any contact with the reagent, as the 5-*M* solution is prepared in 1-*N* NaOH. The addition of a reductant is necessary for stabilization of the Schiff bases formed between an amine-containing protein and the aldehydes on the antibody. For coupling to a hydrazide-activated protein, however, most protocols do not include a reduction step. Even so, hydrazone linkages may be further stabilized by cyanoborohydride reduction. The addition of a reductant during hydrazide/aldehyde reactions also increases the efficiency and yield of the reaction.
6. React for 30 min at room temperature (in a fume hood).
7. Block unreacted aldehyde sites by addition of 50 μl of 1-*M* ethanolamine, pH 9.6, per milliliter of conjugation solution. Approximately a 1-*M* ethanolamine solution may be prepared by addition of 300 μl ethanolamine to 5 ml of deionized water. Adjust the pH of the ethanolamine solution by addition of concentrated HCl, while keeping the solution cool on ice.
8. React for 30 min at room temperature.
9. Purify the conjugate from excess reactants by dialysis or gel filtration using desalting resin. Use 0.01-*M* sodium phosphate, 0.15-*M* NaCl, pH 7.0, as the buffer for either operation. The conjugate may be further purified by removal of unconjugated enzyme by one of the methods given in Section 1.5.

1.4. Conjugation Using Antibody Fragments

It is often advantageous to use antibody fragments in the preparation of antibody–enzyme conjugates. Selected fragmentation carried out by enzymatic digestion of intact immunoglobulins can yield lower-molecular-weight molecules still able to recognize and bind

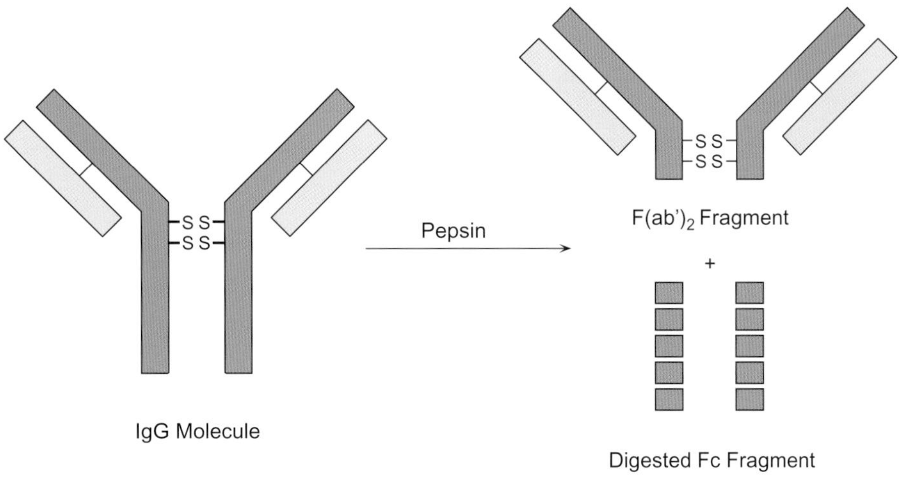

FIGURE 20.11 Digestion of IgG-class antibodies with pepsin results in heavy-chain cleavage below the disulfide groups in the hinge region. The bivalent fragments that are formed are called F(ab′)$_2$. The remaining Fc region normally is severely degraded into smaller peptide fragments.

antigen. Conjugation of these fragments with enzyme molecules can result in ELISA reagents that possess better characteristics than corresponding conjugates prepared with intact antibody. Such antibody fragment conjugates display less interference with various Fc binding proteins and also less immunogenicity (due to lack of the Fc region), more facile membrane penetration for immunohistochemical staining techniques (due to lower overall conjugate molecular weight) (Wilson and Nakane, 1978; Farr and Nakane, 1981), and lower nonspecific binding to surfaces or membranes (resulting in increased signal-to-noise ratios) (Hamaguchi *et al.*, 1979; Ishikawa *et al.*, 1981a,b).

Enzymatic digests of IgG can result in two particularly useful fragments called Fab and F(ab′)$_2$, prepared by the action of papain and pepsin, respectively. Most specific enzymatic cleavages of IgG occur in relatively unfolded regions between the major domains. Papain and pepsin, and similar enzymes including bromelain, ficin, and trypsin, cleave immunoglobulin molecules in the hinge region of the heavy-chain pairs. Depending on the location of cleavage, the disulfide groups holding the heavy chains together may or may not remain attached to the antigen-binding fragments that result. If the disulfide bonded region does remain with the antigen binding fragment, as in pepsin digestion, then a divalent molecule is produced (F(ab′)$_2$) which differs from the intact antibody by lack of an extended Fc portion. If the disulfide region is below the point of digestion, then the two heavy-/light-chain complexes that form the two antigen binding sites of an antibody are cleaved and released, forming individual dimeric fragments (Fab) containing one antigen binding site each (see Figure 20.4, discussed previously).

Methods for producing immobilized papain or pepsin for antibody fragmentation can be found in Hermanson *et al.* (1992). The following protocol describes the use of pepsin to cleave IgG molecules at the C-terminal side of the inter-heavy-chain disulfides in the hinge region, producing a bivalent antigen binding fragment, F(ab′)$_2$, with a molecular weight of about 105,000 (Figure 20.11). Using this enzyme, most of the Fc fragments undergo extensive degradation and cannot be recovered intact.

Protocol for Preparation of F(ab′)$_2$ Fragments Using Pepsin

1. Equilibrate by washing 0.25 ml of immobilized pepsin (Thermo Fisher) with 4 × 1 ml of 20-m*M* sodium acetate, pH 4.5 (digestion buffer). Finally, suspend the gel in 1 ml of digestion buffer.
2. Dissolve 1 to 10 mg of IgG in 1 ml digestion buffer and add it to the gel suspension.
3. Mix the reaction slurry in a shaker at 37°C for 2 to 48 h. The optimal time for complete digestion varies depending on the IgG subclass and species of origin. Mouse IgG1 antibodies are usually digested within 24 h, human antibodies are fragmented in 12 h, whereas some minor subclasses (e.g., mouse IgG2a) require a full 48-h digestion period.
4. After the digestion is complete, add 3 ml of 10-m*M* Tris–HCl, pH 8.0, to the gel suspension. Separate the gel from the antibody solution using filtration or by centrifugation.
5. Apply the fragmented IgG solution to an immobilized protein A column containing 2 ml of gel (Thermo Fisher) that was previously equilibrated with 10-m*M* Tris–HCl, pH 8.0.
6. After the sample has entered the gel, wash the column with 10-m*M* Tris–HCl, pH 8.0, while collecting 2-ml fractions. The fractions may be monitored for protein by measuring absorbance at

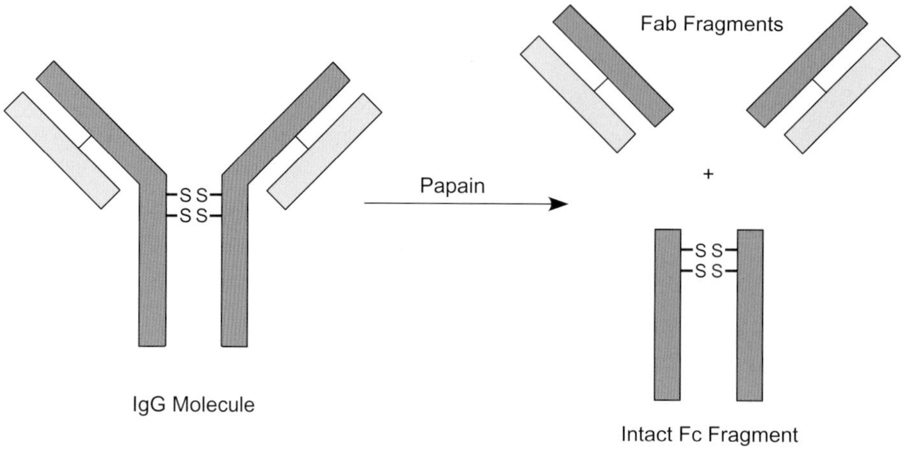

FIGURE 20.12 Papain digestion of IgG antibodies primarily results in cleavage in the hinge region above the interchain disulfides. This produces two heavy–light chain pairs, called Fab fragments, each containing one antigen binding site. The Fc region can normally be recovered intact.

280 nm. The protein peak eluting unretarded from the column is $F(ab')_2$.

7. Bound Fc or Fc fragments and any undigested IgG may be eluted from the column with 0.1-*M* glycine, pH 2.8.

Similarly, immobilized papain may be used to generate Fab fragments from immunoglobulin molecules. Papain is a sulfhydryl protease that is activated by the presence of a reducing agent. Cleavage of IgG occurs above the disulfides in the hinge region, creating two types of fragments, two identical Fab portions and one intact Fc fragment (Figure 20.12). For preparation of the immobilized papain gel used in the following protocol, see Hermanson *et al.* (1992). The gel is also commercially available from Thermo Fisher.

Protocol for Preparation of Fab Fragments using Papain

1. Wash 0.5 ml of immobilized papain (Thermo Fisher) with 4 × 2 ml of 20-m*M* sodium phosphate, 20-m*M* cysteine-HCl, 10-m*M* EDTA, pH 6.2 (digestion buffer), and finally suspend the gel in 1.0 ml of digestion buffer.
2. Dissolve 10 mg of human IgG solution in 1.0 ml of digestion buffer and add it to the immobilized papain gel suspension.
3. Mix the gel suspension in a shaker at 37°C for 4 to 48 h. Maintain the gel in suspension during mixing. The optimal time for complete digestion varies depending on the IgG subclass and species of origin. Mouse IgG_1 antibodies are usually digested within 27 h, whereas other mouse subclasses require only 4 h; human antibodies are fragmented in 4 h (IgG_1 and IgG_3), 24 h (IgG_4), or 48 h (IgG_2); and bovine, sheep, and horse antibodies are somewhat resistant to digestion and require a full 48 h.
4. After the required time of digestion, add 3.0 ml 10-m*M* Tris–HCl buffer, pH 8.0, to the gel suspension, mix, and then separate the digest solution from the gel by filtration or centrifugation at 2000 *g* for 5 min.
5. Apply the supernatant liquid to an immobilized protein A column (2 ml gel) (Thermo Fisher) which was previously equilibrated by washing with 20 ml of 10-m*M* Tris–HCl buffer, pH 8.0.
6. After the sample has entered the gel bed, wash the column with 15 ml of 10-m*M* Tris–HCl buffer, pH 8.0, while 2.0-ml fractions are collected. Monitor the fractions for protein by their absorbance at 280 nm. The protein eluted unretarded from the column is purified Fab.
7. Elute Fc and undigested IgG bound to the immobilized protein A column with 0.1-*M* glycine–HCl buffer, pH 2.8.

Conjugation of these fragments with enzymes is carried out using similar methods to those previously discussed for intact antibody molecules. $F(ab')_2$ fragments may be selectively reduced in the hinge region with DTT, TCEP, or MEA using the identical protocols outlined for whole antibody molecules (Chapter 2, Section 4.1, and Section 1.1, this chapter). Mild reduction results in cleaving the disulfides holding the heavy-chain pairs together at the central portion of the fragment, thus creating two F(ab′) fragments each containing one antigen binding site (Figure 20.13).

The amine groups on these fragments also may be modified with thiolating agents, such as SATA or 2-iminothiolane, to create sulfhydryl residues suitable for coupling to maleimide-activated enzymes (Section 1.1) (Figure 20.14). Amine groups may be further utilized in reductive amination coupling to periodate-oxidized glycoproteins, such as in the protocol outlined for HRP

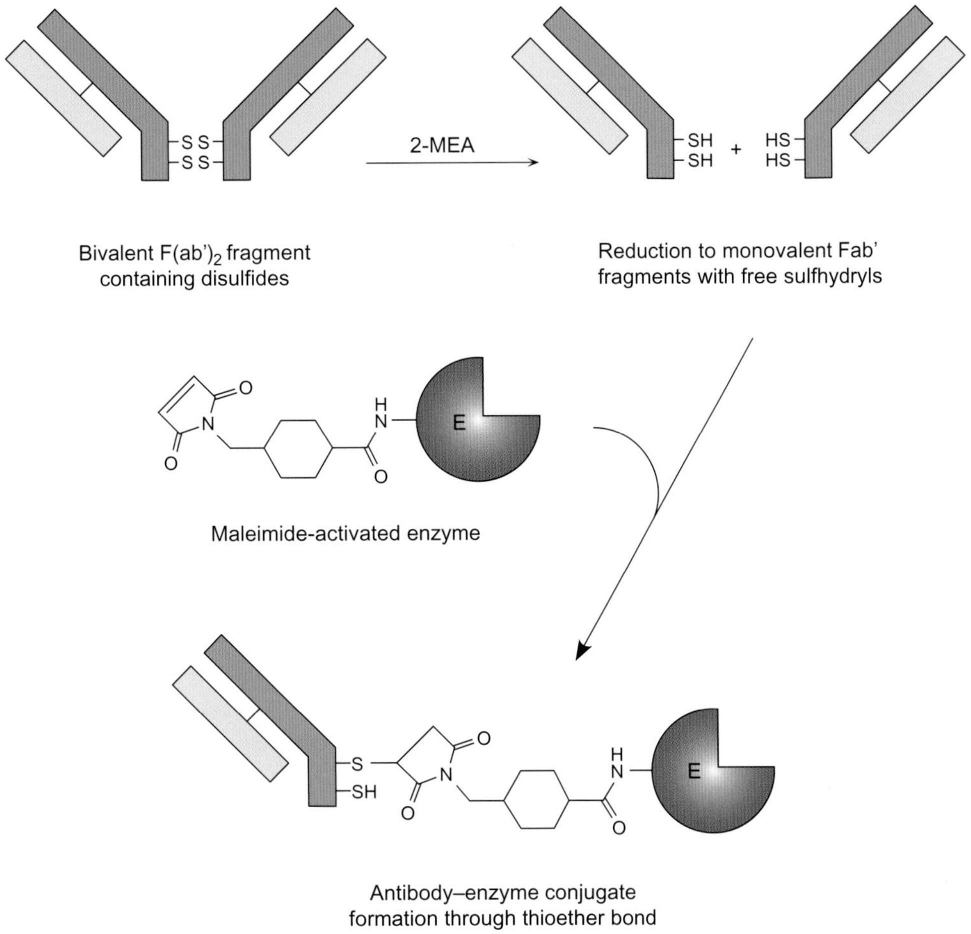

FIGURE 20.13 F(ab′)$_2$ fragments produced by pepsin digestion of IgG can be reduced at their heavy-chain disulfides using a reducing agent, such as 2-mercaptoethylamine (MEA), DTT, or TCEP. Conjugation then can be carried out with a maleimide-activated enzyme to produce relatively low-molecular-weight complexes linked by thioether bonds.

conjugation, previously (Section 1.3, this chapter) (Figure 20.15). Successful periodate oxidation of the fragments themselves, however, may not be possible unless they contain carbohydrate in the antigen binding region, which is true for some polyclonal antibodies. Finally, glutaraldehyde-mediated conjugation techniques will work with antibody fragments, but are not recommended due to the reasons discussed in Section 1.2.

The primary goal of any of these conjugation strategies using antibody fragments is to maintain the activity of the antigen binding site while limiting the size of the final complex with a second molecule. The use of heterobifunctional crosslinkers such as SMCC or reductive amination techniques allows sufficient control over the process to realize these goals.

1.5. Removal of Unconjugated Enzyme from Antibody–Enzyme Conjugates

Conjugates of antibodies and enzymes are essential components in immunoassay and detection systems. In the preparation of such conjugates, a molar excess of enzyme is typically crosslinked to a specific antibody to obtain a complex of high activity. As a result of this ratio, there is typically excess enzyme left unconjugated after completion of the reaction. The unconjugated enzyme confers nothing to the utility of the final product and can be detrimental if it contributes to increased backgrounds in assay procedures. The removal of this free enzyme component may be advantageous to improving the resultant signal-to-noise ratio in some immunoassays. Commercial preparations of antibody–enzyme conjugates usually are not purified to remove unconjugated enzyme. Frequently, the major proteinaceous part of these products is not active conjugate, but leftover enzyme that contributes nothing to the immunochemical activity of what was purchased.

Boorsma and Kalsbeek (1976) state that unconjugated HRP must be removed from antibody–enzyme conjugates to obtain optimal staining in immunoassay procedures. This is especially true in blotting techniques and cytochemical staining where free enzyme

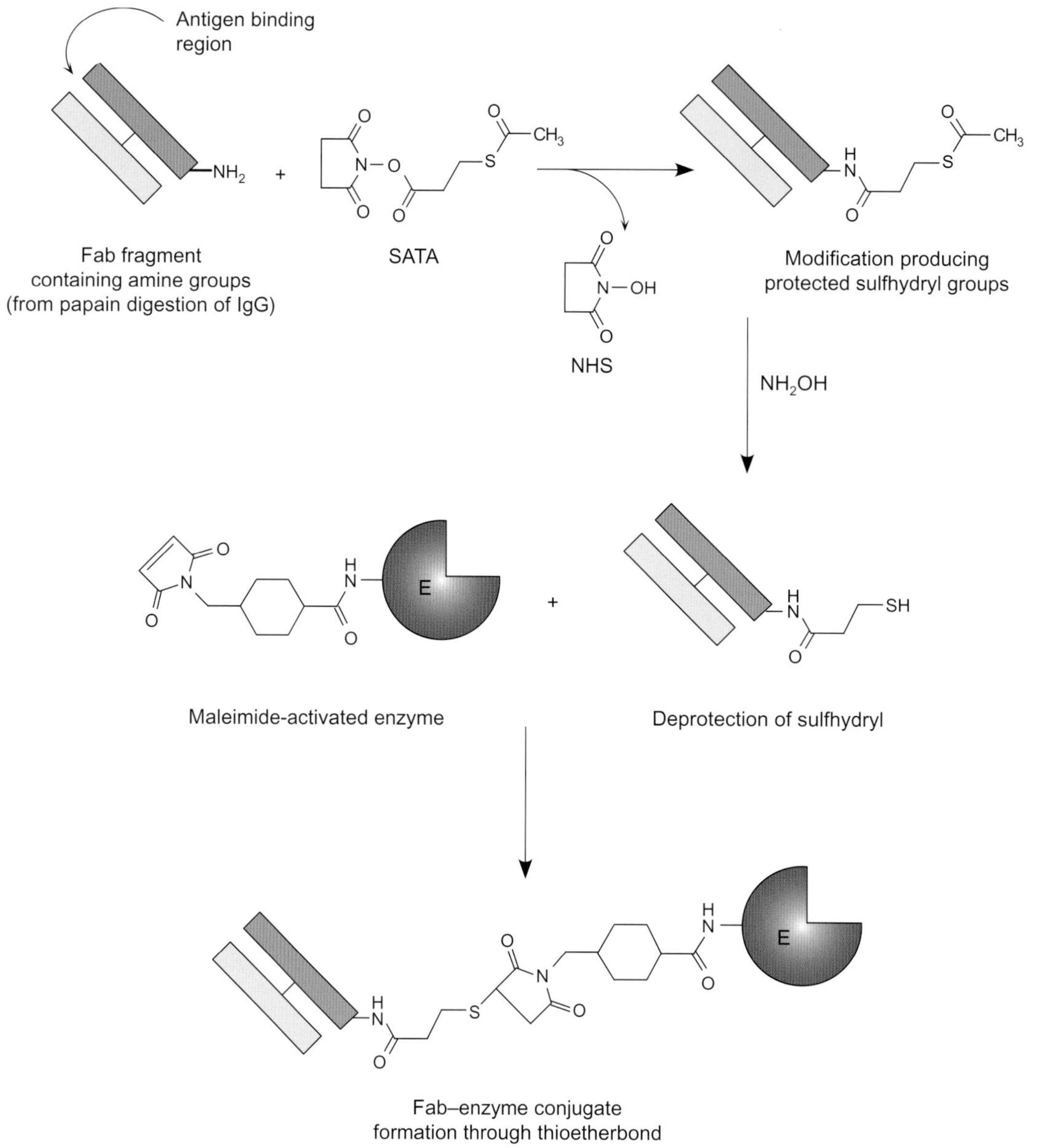

FIGURE 20.14 The thiolation reagent SATA can be used to create sulfhydryl groups on Fab fragments. After deprotection of the acetylated thiol of SATA with hydroxylamine, conjugation with a maleimide-activated enzyme can take place, producing thioether linkages.

may become entrapped nonspecifically within the membrane or cellular structures. The presence of this unconjugated enzyme leads to diffuse substrate noise that can obscure the immunospecific signal.

Several methods may be used to purify an antibody–enzyme conjugate and remove unconjugated enzyme. For instances where the enzyme molecular weight is significantly different than the conjugate molecular weight, separation may be achieved by gel filtration chromatography. Using the proper support with an exclusion limit and separation range able to accommodate all the proteins in the sample, the conjugate peak will elute before the enzyme peak, thus providing an efficient way of removing free enzyme. However, gel filtration procedures can be time consuming and of relatively low capacity for the amount of gel required. In addition, separation of higher-molecular-weight enzymes from antibody conjugates, such as in the case of AP (MW 140,000), is considerably less efficient or impossible. Gel filtration separation also becomes a problem if the conjugate itself consists of a broad range of molecular weights, as is often true when glutaraldehyde is used as the crosslinking agent.

The most effective methods of removing unconjugated enzyme all make use of affinity chromatography systems using specific ligands that can interact with the

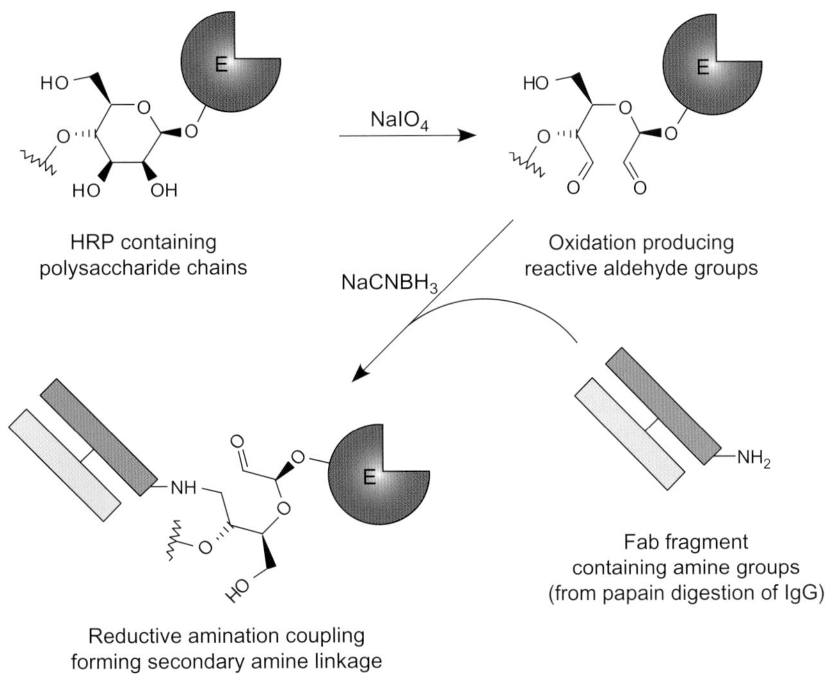

FIGURE 20.15 Periodate oxidation of HRP creates aldehyde groups on the carbohydrate chains of the enzyme. Reaction with a Fab fragment may then be done using reductive amination to produce a lower-molecular-weight complex than would be obtained using intact IgG antibodies.

antibody portion of the conjugate. Thus, the supports retain any unconjugated antibody (usually in very low percentage when the enzyme is reacted in excess) as well as the antibody–enzyme conjugate produced from the crosslinking reaction. The unconjugated enzyme, however, passes through such affinity columns unretarded. Two main methods are discussed below: (1) affinity chromatography, which makes use of immobilized immunoglobulin binding proteins or immobilized antigen molecules having specificity for the antibody used in the conjugate; and (2) nickel-chelate affinity chromatography, which binds the Fc region of antibody molecules.

The use of immunoaffinity techniques (whether antigen specific or immunoglobulin binding proteins such as protein A) allows strong binding of the antibody conjugate, but they have the significant disadvantage of requiring elution conditions that are often too severe for maintaining activity of the antibody or enzyme components. By contrast, nickel-chelate affinity techniques give excellent binding of the conjugate while allowing free enzyme to pass through the gel unretarded. It also has the significant advantage of having mild elution conditions which preserve the activity of the conjugate.

Immunoaffinity Chromatography

Immunoaffinity chromatography makes use of immobilized antigen molecules to bind and separate specific antibody from a complex mixture. After the preparation of an antibody–enzyme conjugate, the antibody binding capability of the crosslinked complex toward its complementary antigen ideally remains intact. This highly specific interaction can be used to purify the conjugate from excess enzyme if the antibody and enzyme can survive the conditions necessary for binding and elution from such a column. Binding conditions are typically mild physiological pH conditions which cause no difficulty. However, many elution conditions require acidic or basic conditions or the presence of a chaotropic agent to deform the antigen binding site. Sometimes these conditions can irreversibly damage the antigen binding recognition capability of the antibody or denature the active site of the enzyme, thus diminishing enzymatic activity. Activity losses for both the antibody and enzyme can be severe under such circumstances.

Another potential disadvantage of an immunoaffinity separation is the assumed abundance of the purified antigen in sufficient quantities to immobilize on a chromatography support (see Chapter 15 for common immobilization techniques). Protein antigens should be immobilized at densities of at least 2 to 3 mg/ml of affinity gel to produce supports of acceptable capacity for binding antibody. Often, the antigen is too expensive or scarce to obtain in the amounts needed.

However, if the antigen is abundant and inexpensive and the antibody–enzyme complex will survive the associated elution conditions, then immunoaffinity chromatography can provide a very efficient method of purifying a conjugate from excess enzyme. This method also ensures that the recovered antibody still retains its ability to bind specific target molecules (i.e., the antigen binding site was not blocked during conjugation). The preparation of immunoaffinity supports can be found in Hermanson *et al.* (1992) and in Chapter 15. A suggested method for performing immunoaffinity chromatography follows:

Protocol

1. Equilibrate the immunoaffinity column with 50-m*M* Tris, 0.15-*M* NaCl, pH 8.0 (binding buffer). Wash with at least 5 column volumes of buffer. The amount of gel used should be based on the total binding capacity of the support. A determination of binding capacity can be achieved by overloading a small-scale column, eluting, and measuring the amount of conjugate that bound. Such an experiment may be coupled with a determination of conjugate viability for using immunoaffinity as the purification method. The final column size should represent an amount of gel capable of binding at least 1.5-times more than the amount of conjugate that will be applied.
2. Apply the conjugate to the column in the binding buffer while taking 2-ml fractions.
3. Wash with binding buffer until the absorbance at 280 nm decreases back to baseline. The unbound protein flowing through the column will consist of mainly unconjugated enzyme. Some conjugate may also flow through if some of the conjugate is inactive or the column is overloaded.
4. Elute the bound conjugate with 0.1-*M* glycine, 0.15-*M* NaCl, pH 2.8, or another suitable elution buffer. A neutral pH alternative to this buffer is the Gentle Elution Buffer from Thermo Fisher. If acid pH conditions are used, immediately neutralize the fractions eluting from the column by the addition of 0.5 ml of 1-*M* Tris, pH 8.0, per fraction.

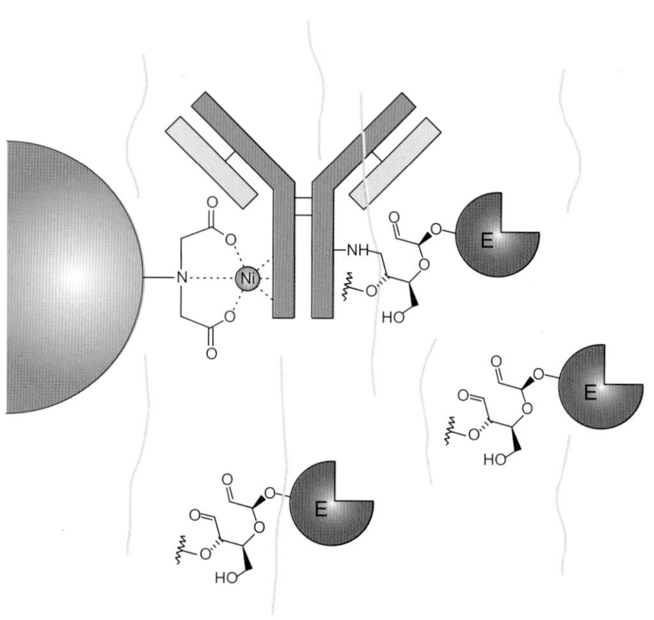

FIGURE 20.16 An affinity chromatography support containing iminodiacetic acid groups chelated with nickel may be used to remove excess enzyme after reactions to produce antibody–enzyme conjugates. The nickel chelate binds to the antibody in the Fc region, retaining the conjugate while allowing free enzyme to pass through the gel unretarded.

Nickel-Chelate Affinity Chromatography

Immobilized metal-chelate affinity chromatography (IMAC) is a powerful purification technique whereby proteins or other molecules can be separated based upon their ability to form coordination complexes with immobilized metal ions (Porath *et al.*, 1975; Lonnerdal and Keen, 1982; Porath and Belew, 1983; Porath and Olin, 1983; Sulkowski, 1985; Kagedal, 1989). The metal ions are stabilized on a matrix through the use of organic chelating compounds which usually have multivalent points of interaction with the metal atoms. To form useful affinity supports, these metal ion complexes must have some free or weakly associated and exchangeable coordination sites. These exchangeable sites can then form complexes with coordination sites on proteins or other molecules. Substances that are able to interact with the immobilized metals will bind and be retained on the column. Elution is typically accomplished by one or a combination of the following options: (1) lowering the pH; (2) raising the salt strength; and/or (3) inclusion of a competing chelating agent such as EDTA or imidazole in the buffer.

Sorensen (1993) reported that a nickel-chelate affinity column will specifically bind IgG class immunoglobulins while allowing certain enzymes to pass through the gel unretarded (Thermo Fisher). This phenomenon allows the separation of antibody–enzyme complexes containing, in particular, HRP or alkaline phosphatase conjugated to common polyclonal or monoclonal antibodies. The nickel chelate column binds the conjugate through the Fc region of the associated antibody, even if enzyme molecules are covalently attached. Any unconjugated enzyme will pass through the affinity column unretarded (Figure 20.16).

Elution of the bound antibody–enzyme conjugate occurs by only a slight shift in pH to acidic conditions or through the inclusion of a metal chelating agent like EDTA or imidazole in the binding buffer. Either method of elution is mild compared to most immunoaffinity separation techniques (discussed in the previous section). Thus, purification of the antibody–enzyme complex can be performed without damage to the activity of either component.

One limitation to this method should be noted. If the antibody–enzyme conjugate is prepared using antibody fragments such as Fab or $F(ab')_2$, then nickel chelate affinity chromatography will not work, since the requisite Fc portion of the antibody necessary for complexing with the metal is not present.

The preparation of a metal-chelate affinity support containing iminodiacetic acid functionalities may be found in Hermanson *et al.* (1992), or purchased from a commercial source (see also Chapter 15 for a discussion on the preparation of immobilized affinity ligands, including metal chelate resins). Any metal chelate support that is designed to bind His-tagged fusion proteins also will work well in this procedure. The following protocol is adapted from the instructions accompanying the nickel-chelate support. Thermo Fisher offers a kit based on this technology for the purpose of removing unconjugated enzyme from antibody–enzyme conjugates (called the FreeZyme™ Conjugate Purification Kit).

Protocol

1. Pack a column containing an immobilized iminodiacetic acid support (or another chelating agent designed to bind His-tagged proteins) (Thermo Fisher). The column size should be no less than 1.5 times that required to bind the anticipated amount of conjugate to be applied. The maximal capacity of such a column for binding antibody can be up to 50 mg/ml gel; however, best results are obtained if no more than 10 to 20 mg/ml of conjugate is applied.
2. Dissolve 50 mg of nickel ammonium sulfate per ml of deionized water. Apply 1 ml of nickel solution per ml of gel to the column. Note: The metal salt and all solutions containing it should be considered hazardous waste and disposed of according to relevant environmental regulations.
3. Wash the column with 10 volumes of water, then equilibrate the support with 2 volumes of 10-m*M* sodium phosphate, 0.15-*M* NaCl, pH 7.0 (binding buffer).
4. Dissolve or dialyze the conjugate into binding buffer. Apply the conjugate solution to the column while collecting 2-ml fractions.
5. Continue to wash the gel with 0.15-*M* NaCl (saline solution) until the absorbance at 280 nm is down to baseline. The protein eluting from the column at this point is unconjugated enzyme.
6. Elute the bound conjugate with 0.1-*M* sodium acetate, 0.5-*M* NaCl, pH 5.0. Pool the fractions containing protein, and dialyze the conjugate into 10-m*M* sodium phosphate, 0.15-*M* NaCl, pH 7.0, or other suitable storage buffers.

2. PREPARATION OF LABELED ANTIBODIES

In addition to labeling immunoglobulins with enzymes to provide detectability through their catalytic action on a substrate, antibody molecules can also be labeled or tagged with small compounds that can provide detectable properties, either directly or indirectly. The specificity of the antibody component of the conjugate can then be used to bind unique antigenic determinants, while the attached tag supplies the properties necessary for detection. Such small chemical labels are typically one of several types: intense fluorescent molecules, contrast agents, affinity tags, or unstable radioactive isotopes.

Radiolabeling antibodies with ^{125}I forms the basis for most of the original radioimmunoassays (RIA) that were first developed in the early days of immunoglobulin-mediated testing. The use of radioisotopes in tagging antibodies is used less often today for *in vitro* immunoassays due to the hazards associated with handling and disposal of radioactive compounds and the better sensitivity offered by using enzyme labels. However, radioisotopes are becoming very important as monoclonal antibody labels for *in vivo* diagnostic or therapeutic conjugates for cancer therapy or detection. Radioactive isotopes either directly attached to antibodies or attached through metal-chelating groups have become a significant tactic for the targeting and detection of tumor cells *in vivo* (Chapter 1 and Chapter 12) (van Dongen *et al.*, 2012). In addition, a radiolabel may have a distinct advantage over other chemical tags, because it is not influenced by conformational changes within the antibody molecule or by changes in its chemical environment, as enzymes or labels with unique spectral characteristics are often affected. Thus, in certain circumstances, radiolabels can still provide an important means of detection that approaches the most sensitive and reliable tags now available, particularly for *in vivo* imaging purposes.

Another form of label often used to tag antibody molecules is chemical modification with a biotin group. Biotinylation (see Chapter 11, Section 6) creates an affinity handle on an immunoglobulin with the ability to bind strongly either avidin or streptavidin in two of the most tightly held noncovalent interactions known. With a dissociation constant (K_d) on the order of 1.3×10^{-15}, the avidin–biotin interaction can be used to detect biotinylated molecules with extreme sensitivity. In this type of system, instead of the antibody being directly labeled with a detectable probe, the avidin (or streptavidin) molecules are modified to contain the detection complex—consisting of an enzyme, fluorescent probe, contrast agent, or radiolabel. Interaction of the

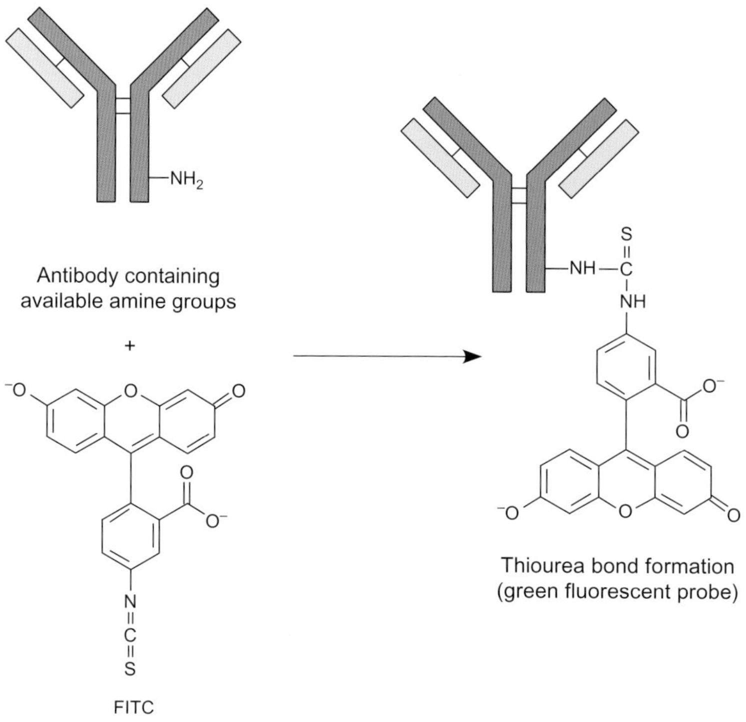

FIGURE 20.17 FITC may be used to label amine groups on antibody molecules, forming isothiourea bonds.

biotinylated antibody with its targeted antigen is then amplified and detected by the addition of such labeled avidin or streptavidin reagents.

The following three sections describe the preparation and properties of fluorescent, radiolabeled, and biotinylated antibodies. See also Chapter 1 for a review of labeled antibody reagents currently being used in a broad range of applications in research, diagnostics, and therapeutics as well as Chapters 10, 11, and 12 for discussions on the labels themselves.

2.1. Fluorescently Labeled Antibodies

Antibody molecules can be labeled with any one of more than a dozen different fluorescent probes currently available from commercial sources. Each probe option has its own characteristic spectral signals of excitation (or absorption) and emission (or fluorescence). Many derivatives of these fluorescent probes possess reactive functionalities convenient for covalently linking them to antibodies and other targeting molecules. Each of the main fluorophore families contains at least a few main choices in coupling chemistry to direct the modification reaction to selected functional groups on the molecule to be labeled. The most popular derivatives include amine-reactive, sulfhydryl-reactive, and carbonyl-reactive. Examples of some of the more useful varieties of fluorescent probes can be found in Chapter 10.

In addition to the wide range of commercial probes, many other fluorescent molecules have been synthesized and described in the literature. Only a handful, however, are generally used to label antibody molecules. Perhaps the most common fluorescent tags with application to immunoglobulin assays are reflected in the main derivatives produced by the prominent antibody manufacturing companies. These include derivatives of cyanine dyes, fluorescein, rhodamine, Texas red, aminomethylcoumarin (AMCA), and phycobiliproteins. Figure 20.17 shows the reaction of fluorescein isothiocyanate (FITC), one of the most common fluorescent probes, with an antibody molecule.

To a large degree, standardization has occurred in the use of these fluorescent probes due to the large body of published literature available on their successful application to antibody-based assays. As a result of this history, instrumentation has become widely available for measuring the fluorescence signals, including standard lasers and filter sets which match the common excitation and emission wavelengths. Such fluorescently labeled antibodies are used in immunohistochemical staining (IHC) (Osborn and Weber, 1982), in flow cytometry or cell sorting techniques (Ormerod, 1990; Watson, 1991), for tracking and localization of antigens, and in various double-staining methods (Kawamura, 1977). Extensive use of fluorescent antibodies in microscopy, pathology, and high content screening for drug discovery has made fluorescently labeled antibodies the most common antibody derivatives used in immunochemical detection techniques (Lichtman and Conchello, 2005; Luo *et al.*, 2011).

In choosing a fluorescent tag, the most important factors to consider are good absorption (high extinction coefficient), stable excitation without photobleaching, and efficient, high quantum yield of fluorescence. Some fluorophores, such as fluorescein, exhibit rapid photobleaching and undergo fluorescent quenching due to dye–dye interactions at even moderate dye substitution levels, which lowers the quantum yield and fluorescence signal. Up to 50% of the fluorescent intensity observed on a fluorescein-stained slide can be lost within one month in storage. AMCA and some cyanine dyes have much better stability, but all fluorophores lose some intensity upon exposure to light or upon storage. Exceptions to this rule are the use of fluorescent nanoparticles, such as dye-doped silica (Chapter 14, Section 5) and quantum dots (Chapter 10, Section 10).

In some cases, the preparation of a fluorescently labeled antibody is not even necessary. Particularly, if indirect methods are used to detect antibody binding to antigens, then preparing a fluorescently labeled primary antibody is not needed. Instead, the selection from a commercial source of a labeled secondary antibody having specificity for the species and class of primary antibody to be used is all that is required. However, if the primary antibody needs to be labeled and it is not manufactured commercially, then a custom labeling procedure will have to be performed.

Generalized protocols for the attachment of these fluorophores to protein molecules, including antibodies, can be found in Chapter 10 and Chapter 14, Section 5. The main consideration for the modification of immunoglobulins is to couple these probes at an optimal level to allow good detectability without high backgrounds. Too low a substitution level and the response of the fluorophore will yield low signal strength and poor sensitivity. Too high a substitution level and the fluorophore may self-quench through energy transfer, decrease the antibody's ability to bind target molecules by blocking the antigen binding sites, or cause nonspecific interactions resulting in high background or noise levels. In some cases, trial and error will be required to optimize this process.

For other examples of antibody labeling protocols see Goding (1976) and Harlow and Lane (1988).

2.2. Radiolabeled Antibodies

The attachment of a radioactive label onto an antibody molecule provides a powerful means of detection in immunoassay procedures, tracking of analytes, for *in vivo* diagnostic procedures, and, more recently, for the detection or therapy of numerous types of cancers. Originally, radiolabeling of antibodies merely meant modifying tyrosine residues with ^{125}I. Now, a number of different radioactive elements are being attached, both covalently and through specialized chelating compounds to provide imaging capabilities for the detection of primary tumors and metastases (Order, 1989).

Radioiodination can be carried out using any one of a number of techniques. Most of the procedures utilize ^{125}I as the unstable isotope of choice for *in vitro* use due to its easy availability, comparably long 60-day half-life, and relatively low-energy photon emissions. Radioactive ^{125}I usually is supplied as its sodium salt and must be oxidized to create an electrophilic species capable of modifying molecules. Commonly used oxidizing agents include chloramine-T, Iodogen, and Iodo-beads (Chapter 12, Section 2). When used in direct labeling techniques with antibodies and other proteins, these oxidants cause an iodination reaction to occur at available tyrosine or histidine residues within the polypeptide chain. If either of these amino acids is important to antibody activity and cannot be labeled, then certain crosslinking or modification reagents containing an activated aromatic ring may be used to radiolabel the molecule at other functional sites, particularly at lysine amines. An example of this technique is to use the Bolton–Hunter reagent (Chapter 12, Section 2.5) labeled with radioactive iodine to modify the primary amines within the antibody. This reagent also can be used to add an iodinatable site to molecules containing no tyrosine residues (Figure 20.18).

Reagent options and protocols for the radioiodination of antibodies and other molecules may be found in Chapter 12.

Another method of adding a radioactive tag to antibodies is to use a chelating compound capable of complexing metal isotopes. One of the most frequently used chelating reagents is diethylenetriamine pentaacetic acid (DTPA) (Chapter 12, Section 1.1). The reagent contains 2 anhydride groups that can be used to modify primary amines in proteins and other molecules. The reaction process involves ring opening and the formation of an amide bond. Ring opening also creates up to four free carboxylate groups which, combined with the three nitrogen atoms in the chelator, are able to form strong coordination complexes with metals such as indium-111 (Figure 20.19). Monoclonal antibodies labeled with bifunctional chelating agents containing radioisotopes can be used in targeting tumor cells *in vivo*. The detection sensitivity of radiolabeled antibodies has led to effective diagnostic procedures to monitor primary and secondary cancer growths. In addition, the intensity of radioactivity at the tumor site when labeled monoclonals are used therapeutically can be great enough to cause tumor cell death and remission.

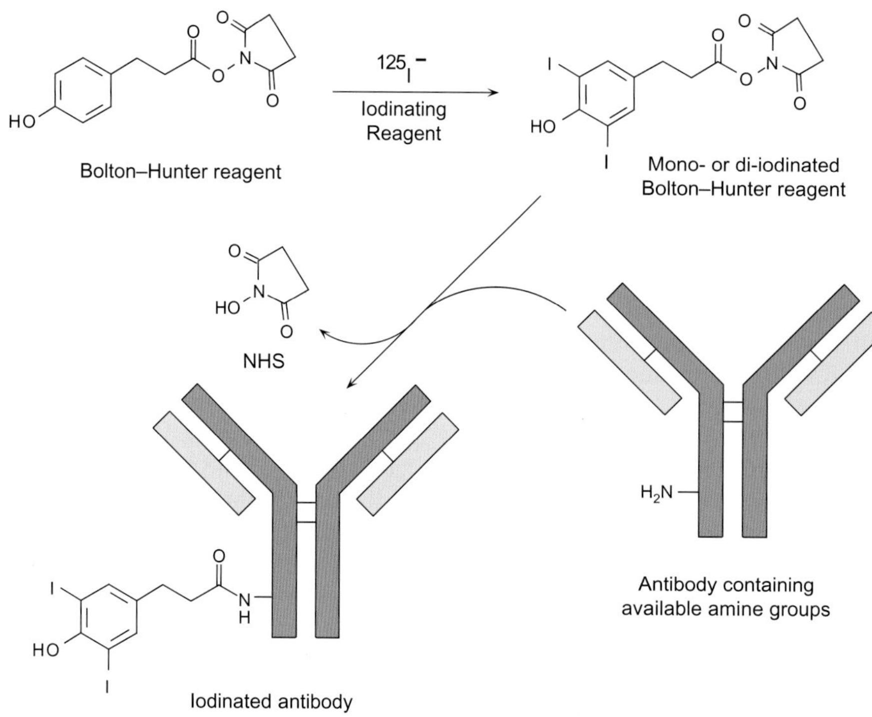

FIGURE 20.18 Bolton–Hunter reagent may be used to add radioactive iodine labels to antibody molecules by modification of amines.

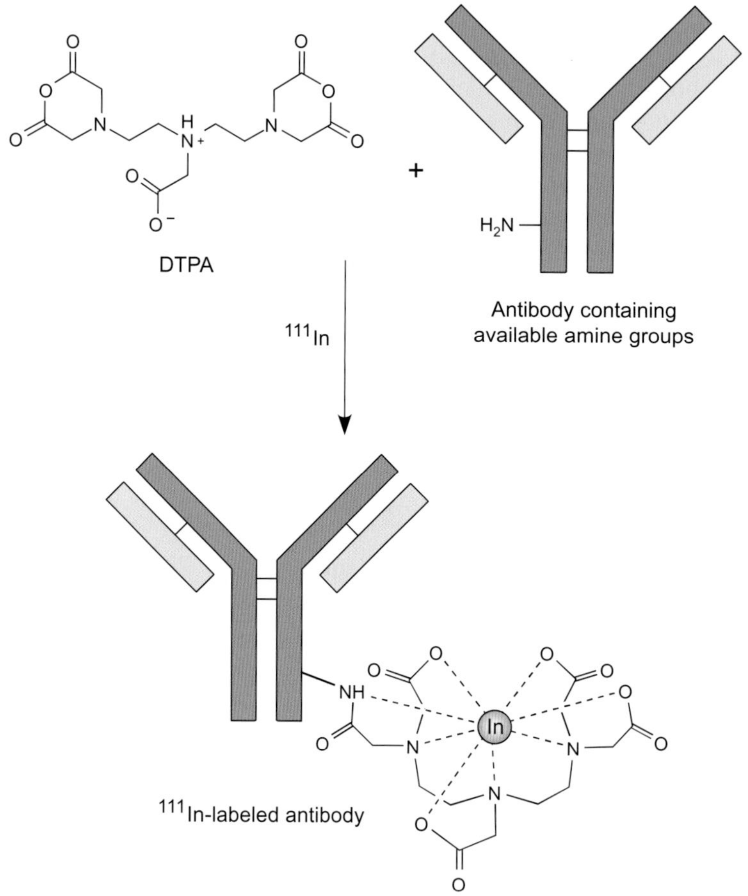

FIGURE 20.19 The bifunctional chelating reagent DTPA may be used to modify amine groups on antibody molecules, forming amide bond linkages. The isotope indium-111 may then be complexed to the chelator group to create a radiolabeled targeting reagent.

2.3. Biotinylated Antibodies

Another popular tag for use with immunoglobulins is biotin. Modification reagents that can covalently add a functional biotin group to proteins, nucleic acids, and other molecules now come in many shapes and reactivities (Chapter 11, Section 6, and Chapter 18, Section 1.3). Depending on the reactive group present on the biotinylation reagent, precise functional groups on antibodies may be modified to create an affinity tag capable of binding avidin or streptavidin derivatives. Amines, carboxylates, sulfhydryls, and carbohydrate groups can be specifically targeted for biotinylation through the appropriate choice of reactive biotin compound and modification procedure. Figure 20.20 shows the biotinylation of an antibody with NHS–LC-biotin, one of the most common biotinylation reagents.

Streptavidin conjugates are now used extensively to detect biotinylated antibodies after they have bound an antigen. A streptavidin probe labeled with an enzyme, fluorophore, high-contrast agent, or radiolabel can provide a very sensitive detection system to locate or assay antigens in many different applications. In addition, the potential for more than one labeled (strept)avidin conjugate to become attached to each biotinylated antibody through multiple biotin groups provides an increase in detectability over antibodies directly labeled with a detectable tag.

Several assay designs that use the enhanced sensitivity afforded through biotinylated antibodies have been developed. Most of these systems use conjugates of avidin or streptavidin with enzymes (such as HRP or AP), although other labels (such as fluorophores) can be used as well. In the simplest assay design, called the labeled avidin–biotin (LAB) system, a biotinylated antibody is allowed to incubate and bind with its target antigen. Next, an avidin–enzyme conjugate is introduced and allowed to interact with the available biotinylation sites on the bound antibody. Substrate development then provides the detectability necessary to quantify the antigen.

In a slightly more complex design, the bridged avidin–biotin system (BRAB) uses (strept)avidin's multiple biotin-binding sites to create an assay potentially of higher sensitivity than that of the LAB assay. Again, the biotinylated antibody is allowed to bind its target, but next unmodified (strept)avidin is introduced to bind with the biotin binding sites on the antibody. Finally, a biotinylated enzyme is added to provide a detection vehicle. Since the bound (strept)avidin still has additional biotin binding sites available, the potential exists for more than one biotinylated enzyme to interact with each bound (strept)avidin molecule. In some cases, sensitivity can be increased over that of the LAB technique by using the bridging ability of avidin or streptavidin (Chapter 11, Section 2).

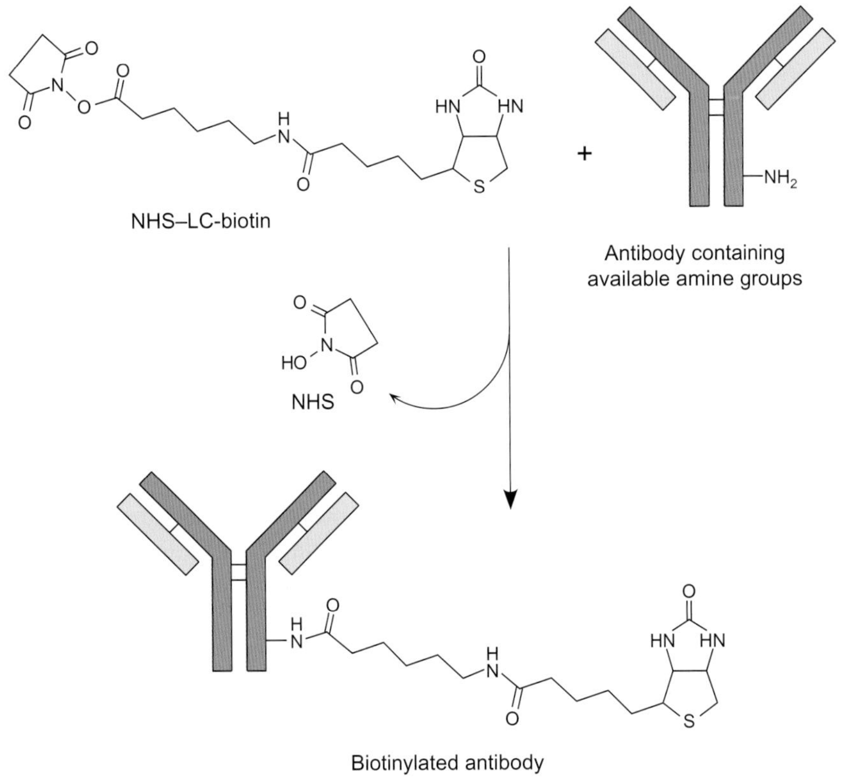

FIGURE 20.20 Biotinylated antibodies can be formed by reacting NHS–LC-biotin with available amine groups to create amide bonds.

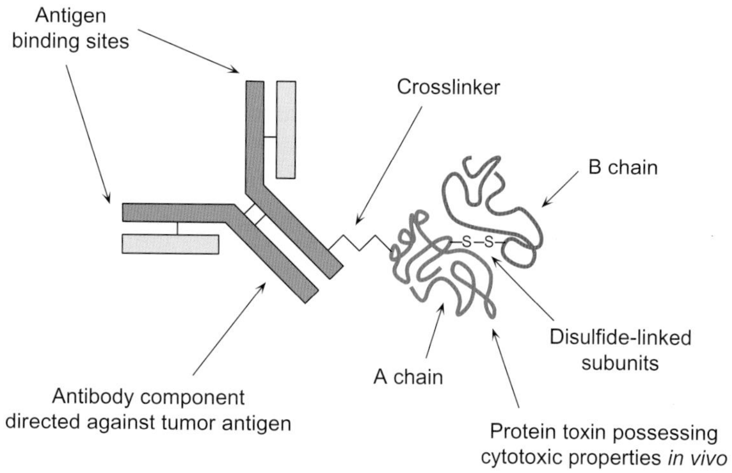

FIGURE 20.21 The basic design of an immunotoxin conjugate consists of an antibody-targeting component crosslinked to a toxin molecule. The complexation typically includes a disulfide bond between the antibody portion and the cytotoxic component of the conjugate to allow release of the toxin intracellularly. In this illustration, an intact A–B toxin protein provides the requisite disulfide, but the linkage may also be designed into the crosslinker itself.

A modification on this theme can be used to produce one of the most sensitive enzyme-linked assay systems known. The ABC system (for avidin–biotin complex) increases antigen detectability beyond that possible with either the LAB or BRAB designs by forming a polymer of biotinylated enzyme and (strept)avidin before addition to an antigen-bound, biotinylated antibody. When avidin and a biotinylated enzyme are mixed together in solution in the proper proportion, the multiple binding sites on (strept)avidin create a high-molecular-weight, multimeric complex. If the biotinylated enzyme is not in large enough excess to block the binding sites on all the (strept)avidin molecules, then additional sites will still be available on this complex to bind a biotinylated antibody bound to its complementary antigen. The large complex provides multiple enzyme molecules to dramatically enhance the sensitivity of detecting antigen. Thus, the ABC procedure is currently among the highest-sensitivity methods available for immunoassay work.

More detailed discussions of (strept)avidin–biotin systems as well as the process of adding a biotin affinity group to proteins, nucleic acids, and other biomolecules can be found in Chapter 11.

3. IMMUNOTOXIN CONJUGATES

Monoclonal antibodies directed against tumor antigens may be used as targeting agents for conducting certain cytotoxic substances to malignant cells for selective killing. Numerous cell surface markers are known to proliferate in human solid tumors (Boyer *et al.*, 1988; Carter *et al.*, 2004; Aplin *et al.*, 2011; Kelly *et al.*, 2011; Yang *et al.*, 2011). The ability to raise monospecific antibodies to these markers creates the capacity to discretely target the tumor, causing cell death while leaving healthy cells alone (Salmon, 1989; McCarron *et al.*, 2005).

The design of such cytotoxic antibodies is conceptually simple: attach a toxic substance or a mediator of toxicity to the appropriate monoclonal and you have a "magic bullet" that can find and eliminate the one-in-a-billion cells that have the requisite marker (Figure 20.21). The antibody provides the recognition and binding capacity, while the associated toxic component effects cellular alterations leading to cell death (Pastan *et al.*, 2006).

The approach to constructing antibody immunoconjugates for cancer has taken a number of forms (Vogel, 1987; Dosio *et al.*, 2011). One of the first designs used conjugates of monoclonal antibodies with toxins that were able to block protein synthesis at the ribosome level inside the cell (Lord *et al.*, 1988). Other conjugate forms used radioactive labels that killed cells by overexposure to radiation in proximity to where antibody docking occurred (Order, 1989). Drug conjugates were also constructed that combined the known benefits of chemotherapy with the targeting capability of monoclonals (Reisfeld *et al.*, 1989; Willner *et al.*, 1993; Bross *et al.*, 2001; Sievers and Senter, 2012). The expected result was the effective concentration of the chemotherapeutic agent at the location of the tumor—hopefully eliminating or minimizing the side-effects of traditional chemotherapy. In addition, conjugates of monoclonals with certain biological modulators such as lymphokines or growth factors were tried to affect malignant cell viability (Obrist *et al.*, 1988; Choudhary *et al.*, 2011).

Some immunoconjugates utilize intermediate carrier systems consisting of polymeric molecules such as polysaccharides, particularly dextran (Hurwitz *et al.*, 1978, 1980, 1983a,b, 1985; Manabe *et al.*, 1983; Sela and Hurwitz, 1987). The activated dextran is crosslinked to both the monoclonal and the cytotoxic agent, providing multivalent conjugation sites to create larger complexes (Section 2.3, this chapter, and Chapter 18, Section 2). Liposomes can be used in similar fashion by anchoring the antibody to its outer surface and charging the vesicle with cytotoxic compounds (Gregoriadis, 1984; Ho *et al.*, 1986; Matthay *et al.*, 1986; Singhal and Gupta, 1986; Kirpotin *et al.*, 2012). See Chapter 21 for a survey of liposome conjugation techniques. Also, dendrimer conjugates with antibody targeting molecules and cytotoxic components have been used to create multivalent immunotoxin conjugates (Chapter 8).

Other systems have been designed to use a two-stage approach where an antibody is conjugated with an intermediate agent, which, when combined with another factor, could elicit cytotoxicity. For instance, enzyme conjugates with monoclonals have been used that could transform an inactive pro-drug into a chemo-cytotoxic agent *in vivo* (Senter *et al.*, 1989). Antibody-directed enzyme-prodrug therapy (ADEPT) has become an important option in the design of antibody therapeutics (Tietze and Schmuck, 2011). See Chapter 1, Section 3.5, for current applications of ADEPT conjugates.

Monoclonals can also be tagged with a biotin label and a secondary, avidin–toxin conjugate used to target the antibody once it bound the tumor cell. Two-stage radiative treatment also has been tried through the use of boron neutron capture therapy (Holmberg and Meurling, 1993). In this case, a carrier molecule is modified to contain ^{10}B. After administration *in vivo* to target tumor cells, neutron bombardment yields an unstable intermediate ^{11}B which immediately undergoes a fission reaction to yield ^{7}Li and ^{4}He. The induced radiation from this decomposition then kills the surrounding cells.

Tumor-targeting conjugates which use biospecific agents other than monoclonal antibodies have been developed as well. The targeting component in these systems consists of any molecule that can function as a ligand having specific affinity for some receptor molecule on the surface of the tumor cells. Such an affinity molecule might be a hormone (typically called hormonotoxins) (Singh *et al.*, 1989; Singh *et al.*, 1993), a growth factor, an antigen specific for binding particular antibodies projecting from B cell surfaces, transferrin, α_2-macroglobulin, or anything else able to specifically interact with the targeted tumor cells. See Chapter 1 for a review of current strategies in the use of antibody conjugates and other conjugates in the treatment of cancer.

It became apparent very early in the development of such agents that their conception and design was much easier to imagine than to successfully implement. Monoclonal antibodies used *in vitro* could easily detect antigen molecules in complex mixtures with very little nonspecificity or cross-reactivity, even in cell-based assays. Very quickly following the invention of hybridoma technology, monoclonals were employed in numerous diagnostic assays and their use in detection and assay applications continues to grow. The use of monoclonals as therapeutic agents *in vivo*, however, was complicated by the body's natural immune response designed to prevent the invasion of foreign substances.

A major problem in the use of immunotoxins is that injection of conjugates prepared from mouse (or animal-based) monoclonals usually results in antibody production against the foreign protein. Antibodies of animal origin quickly get removed from circulation by the immune system, which prevents any therapeutic benefit from the conjugate drug. Sometimes allergic reactions can further complicate the side-effects, making continued therapy infeasible. Most often, however, induced host immunoglobulins will quickly bind the immunoconjugate and remove it from the circulatory system. Instead of finding the targeted tumor cells, the immunotoxin conjugate ends up sequestered and degraded in the liver or removed by the kidneys. Since the common culprit in this scenario is often the monoclonal, the acronym HAMA, for *human anti-mouse antibody*, is given to this response.

In an attempt to overcome the HAMA problem, "humanized" mouse monoclonals were designed where large portions of the murine antibody are substituted for their human counterparts. For instance, replacing a mouse Fc portion with the corresponding human ones can significantly decrease the immune response against such conjugates. Replacing everything but the hypervariable regions that code for antigen binding has been accomplished, too. Unfortunately, regardless of how much "humanization" is done, the remaining murine part still has the potential of causing immunological reactions. However, in the intervening years since the beginning of monoclonal antibody targeting in the late 1970s, the development of fully human antibodies and antibody–drug conjugates (ADCs) of many types has dramatically changed the effectiveness of such agents *in vivo* (Jakobovits, 1995; Kucherlapati *et al.*, 2012). The advent of recombinant antibody technology and transgenic mice producing antibodies of completely human origin has essentially overcome the problem of immune system response to the antibody component. As a result, there are now hundreds of antibody therapeutic agents in development and over a dozen antibody conjugates in late-stage clinical trials with excellent prospects for approval (see *Therapeutic Monoclonal Antibodies: World Market 2010–2025*, Visiongain, 2010). ADCs employing fully human antibodies attached to many different chemotherapeutic agents are now poised to revolutionize the use of antibody conjugates for the treatment of disease (Chari, 2008). See Chapter 1 for an extensive overview of antibody bioconjugates designed for therapeutic purposes.

Modification of immunotoxin conjugates with synthetic polymers also has been used to mask the complex from the host immune response. Particularly, polyethylene glycol (PEG; see Chapter 18) has been found to be quite successful in reducing or eliminating immune system reactions to modified proteins or antibodies (Roffler and Tseng, 1994). Modification of an antibody conjugate with two to four PEG molecules increases the serum half-life and improves tumor localization of the targeted reagent.

Other innovations in preparing targeted conjugates for cancer utilize recombinant DNA techniques to create antibody molecules that are entirely of human origin (Huse *et al.*, 1989; Orlandi *et al.*, 1989; Sastry *et al.*, 1989; Ward *et al.*, 1989). A completely human antibody molecule eliminates the immunological problems associated with mouse monoclonals. Intact antibodies, Fab fragments, small Fv fragments held together by synthetically designed amino acid segments, and short peptides representing the antigen binding site have all been developed by recombinant means. Although frequently the word "antibody" is used to describe these engineered proteins, many of the molecules are far removed from the traditional picture of an antibody molecule. The terms "single-chain antibody" or "single-chain Fv protein" are commonly used and more closely describe these new targeting molecules.

With the great diversity of targeted toxic agents being developed for cancer therapy, it would be difficult to characterize this section strictly as antibody conjugation. While many, if not most, studies utilize monoclonal antibodies of one form or another as the biospecific targeting component, the complete picture involves the crosslinking of a wide variety of molecules together to create the final conjugate. This section presents some of the most common methods of immunotoxin preparation. For the preparation of other unique targeted toxin conjugates, including.

ADCs, the methods found throughout this book for linking one particular functional group to another, can be followed with an excellent probability for success. When preparing ADC conjugates, it is often important to give great consideration to the cross-bridge and linkages formed between the antibody and the drug. Bonds that are cleavable *in vivo*, such as disulfides and hydrazone linkages, have proven to be essential in the successful design of effective ADCs (McCarron *et al.*, 2005; Beck, 2010).

3.1. Properties and Use of Immunotoxin Conjugates

Conjugates of monoclonal antibodies and protein toxins have been studied extensively for their usefulness in the treatment of cancer (Chandramohan *et al.*, 2012). Toxins of many different types can be used to create effective immunotoxin conjugates, including the proteins ricin from castor beans (*Ricinus communis*), abrin from *Abrus precatorius*, modeccin, gelonin from *Gelonium multiflorum* seeds, diphtheria toxin produced by *Corynebacterium diphtheriae*, pokeweed antiviral proteins (PAPs; three types: PAP, PAP II, and PAP-S) from *Phytolacca americana* seeds, cobra venom factor (CVF), *Pseudomonas* exotoxin, restrictocin from *Aspergillus restrictus*, momordin from *Momordica charantia* seeds, saporin from *Saponaria officinalis* seeds, as well as other ribosome-inactivating proteins (RIPs).

By far the most popular choices for the toxin component of protein-based immunotoxins are ricin, abrin, modeccin, and diphtheria toxin. The three plant toxins have lectin binding activity toward terminal β-galactosyl residues and they can be inhibited by the presence of simple sugars like galactose and lactose. The toxin proteins bind to cell-surface polysaccharide receptors with high affinity (K_a in the range of 10^7–10^8/M). Ricin, abrin, and modeccin consist of two subunits with remarkably similar structures and activities. The intact proteins have molecular weights of approximately 63–65,000 with each subunit of about equal size. The subunits are joined by disulfide linkages that are important reversible bonds in the mechanism of cytotoxicity. The A chain is called the effectomer and possesses ribosomal-inactivating properties. The B chain contains the carbohydrate binding site and it is termed the haptomer. While the intact toxin molecules have potent cytotoxic effects on cells, they exhibit no ribosomal inactivating activity on ribosomes in a cell-free system. By contrast, reduction of the toxins with a disulfide reducing agent creates the opposite effects. Reduced, dissociated toxin subunits inhibit ribosomal activity in cell-free systems, but they have no affect on intact cells.

The reason for these properties is due to the toxins' mode of action. Toxin molecules bind through saccharide recognition sites on the B chain to particular β-galactosyl-containing glycoprotein or glycolipid components on the surface of cell membranes. In animals sensitive to these toxins, the necessary polysaccharide ligands are present in large quantities on virtually all cell types (Cumber *et al.*, 1985). Upon binding of the protein dimer to the cell, the A chain enters the cell either by active transport into endocytic vesicles or through some mechanism of its own. Once inside the cell membrane, the A chain enters the cytoplasmic space, binding to and enzymatically inactivating the 60S subunit of ribosomes (Olsnes and Pihl, 1976, 1982). The result is cessation of protein synthesis and eventual cell death. Because the A chain's action is through enzymatic means, as little as one active toxin molecule is enough to seriously disrupt protein synthesis operations and probably sufficient to kill a target cell (Eiklid

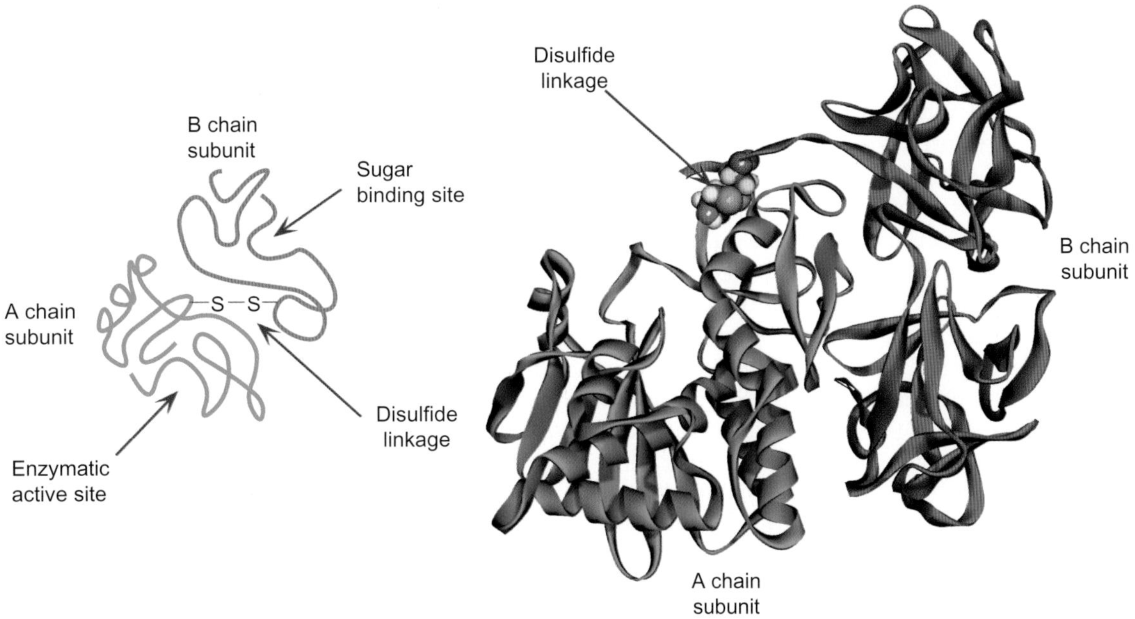

FIGURE 20.22 Conceptualized construction of an A–B subunit protein toxin (left). The B chain contains a binding region for docking onto cell surfaces, while the A chain contains a catalytic site that produces cytotoxic affects intracellularly. The two subunits are joined by a disulfide bond that is reductively cleaved at the cellular level to allow the A subunit to cause cell death. A molecular model of the protein toxin ricin is on the right.

et al., 1980). The turnover rate of one A chain molecule is about 1500 ribosomes inactivated per minute (Olsnes, 1978).

Diphtheria toxin also is a two-subunit protein, but it is initially synthesized by certain strains of *Corynebacterium diphtheriae* as a single polypeptide chain of molecular weight 63,000. Proteolytic processing results in the formation of a "nicked toxin" which is enzymatically inactive, but consists of two subunits bonded together by an interchain disulfide. Upon reduction of the disulfide, the A chain (MW 24,000) is released and manifests enzymatic activity toward ribosomal proteins (Collier and Cole, 1969; Sandvig and Olsnes, 1981). Its mode of action is different than that of the plant toxins. The A chain fragment of diphtheria toxin catalyzes the ADP-ribosylation of eukaryotic aminoacyl-transferase II (EF2) using NAD^+ (Honjo *et al.*, 1968; Gill *et al.*, 1969). The B chain, by contrast, possesses no enzymatic activity, but evidence points to the fact that a binding site on it recognizes certain cell surface receptors. As in the action of ricin, abrin, and modeccin, the B chain of diphtheria toxin is necessary for cytotoxicity (Colombatti *et al.*, 1986). There also is a role for the C-terminus cysteine residue of the B chain in cell penetration (Dell'Arciprete *et al.*, 1988).

Figure 20.22 illustrates the basic structure of these common two-subunit toxins, showing schematically their major characteristics. The molecular model of ricin is from Rutenber *et al.* (1991), RSCB structure No. 2aai.

Due to the extraordinary toxicity of intact ribosome-inactivating toxins like ricin, abrin, and modeccin, purification and handling of these proteins must be carried out with extreme care. Even dust from crude seed powders or lyophilized proteins should be considered dangerous. During the height of the Cold War days, a Soviet KGB agent killed a man from the West by injecting at most only milligram quantities of ricin into his leg using a modified umbrella tip. There even have been instances of worker deaths at companies that routinely purify these proteins. For this reason, all handling operations of intact toxin dimers and purified subunits should be performed in fume or laminar-flow hoods. Avoid, also, the use of laboratory tools that could lead to puncture wounds causing contaminating toxin injection.

Some potentially cytotoxic proteins contain only a single polypeptide chain, such as gelonin and PAPs. Such toxins manifest similar enzymatic ribosome-inactivating properties as the multi-subunit proteins like ricin, but do not possess the cell-recognition capacity that the B chain subunit contains. The result is the inability of these toxins to bind or affect intact cells. However, they do maintain the typical ribosome-inactivating properties in a cell-free system that the A chain of two-subunit toxins possess. If these proteins are conjugated with a cell-targeting agent, such as the B chain of ricin or a specific antibody that recognizes cell-surface epitopes, full cytotoxicity results.

Gelonin and PAPs are much more convenient to work with than ricin and the other two-subunit toxins. Most importantly, they are relatively nontoxic to cells unless conjugated with something that can facilitate cell binding and internalization (Stirpe *et al.*, 1980; Irvin, 1983). The pI of these toxins is in the basic range, and they each have a molecular weight of about 30,000 (Barbieri and Stirpe, 1982). These single-subunit proteins are very stable, especially to purification techniques, but also to most modification and crosslinking steps associated with preparing immunotoxins. Studies have shown (Lambert *et al.*, 1985) that modification of gelonin or PAPs can be performed with 2-iminothiolane (Traut's reagent; Chapter 2, Section 4.1) to create sulfhydryl groups without loss of activity. Previous studies, however, have determined that the use of SPDP to modify gelonin resulted in a 90% inactivation (Thorpe *et al.*, 1981). The difference in these results may be due to the retention of positive charge characteristics on the modified amine when using Traut's reagent, but neutralization of that charge when using SPDP. This is an example of how a slight difference in conjugation strategy can result in a dramatic difference in conjugate performance.

Immunotoxin conjugates consist of an antibody covalently crosslinked to a toxin molecule in a way that maintains the unique properties of both proteins. The antibody component consists of a monoclonal having specificity for an antigenic determinant on the surface of a particular cell type. Most often, the targeted cells are tumors that express a unique cell surface marker which can be recognized by the monoclonal. The role of the antibody, therefore, is to function as a passive taxi, carrying the toxin component to the targeted cells. Once at the tumor location, the toxin component effects its intended ribosome-inhibiting action, ultimately causing cell death and tumor destruction.

Since immunotoxin conjugates are destined to be used *in vivo*, their preparation involves more critical consideration of crosslinking methods than most of the other conjugation protocols described in this book. The following sections discuss the issues associated with toxin conjugates and the main crosslinking methods for preparing them.

3.2. Preparation of Immunotoxin Conjugates

It has become apparent that the method of crosslinking can dramatically affect the activity of an immunotoxin *in vivo*. This is true not only with regard to possible direct blocking by the crosslinker of the enzymatic active site which is responsible for inactivation of ribosomes, but the chemistry of conjugation is also an important factor in proper binding and entry of the conjugate into the cell. Preparation of the conjugate should maintain the antigen binding character of the attached antibody and at the same time not block the ribosome-inactivating activity of the toxin component.

Studies have been carried out to investigate the importance of using a cleavable linker between the antibody and the toxin. This configuration in the immunoconjugate would mimic the natural state of two-subunit toxins like ricin that are held together by disulfide bonds. There is evidence that disulfide reduction and cleavage of the A chain from the B chain is necessary for cytotoxicity in native toxins (Olsnes, 1978). There is similar evidence that the creation of cytotoxic immunotoxins using only A chain subunits requires that the conjugation be done with a monoclonal using a crosslinker that possesses a disulfide bond in its cross-bridge or creates a disulfide linkage upon coupling (Masuho *et al.*, 1982). Using disulfide-cleavable crosslinkers in the preparation of immunotoxins results in the antibody taking on the role of the B chain in recognizing and binding to antigenic determinants on the surface of cells. After binding, some mechanism internalizes the conjugate wherein the two components then are separated by disulfide reduction. The A chain subunit is then freed to enter the cytoplasmic space where enzymatic degradation of the ribosomal proteins occurs.

Other investigators, however, have demonstrated that conjugations of antibody with intact, two-subunit toxins can be carried out using non-cleavable crosslinkers such as NHS ester–maleimide heterobifunctionals (Chapter 6, Section 1) (Myers *et al.*, 1989). Presumably, the toxin is still able to release the A chain after the antibody has bound to the cell, since the conjugation process does not permanently attach the two toxin subunits together—only the toxin to the antibody.

Thus, two main strategies can be used in making immunotoxin conjugates (Figure 20.23). In the most often used method, the isolated A chain of two-subunit toxins (or the intact polypeptide of single-subunit toxins like gelonin) is conjugated to a monoclonal using a crosslinker that can introduce a disulfide bond. When using only purified A chain, it is common (but not absolutely required) to couple through the sulfhydryl that is freed during A–B chain cleavage by disulfide reduction. The single-chain toxins like gelonin, however, have no free sulfhydryls, so a thiolation agent such as 2-iminothiolane (Chapter 2, Section 4.1) may be used to create them (Lambert *et al.*, 1985).

When using ricin A chains, it has been found that chemical deglycosylation of the subunit prevents its nonspecific binding to receptors for mannose on certain cells of the reticuloendothelial system (Bitetta and Thorpe, 1985; Ghetie *et al.*, 1988, 1991; O'Hare *et al.*, 1988). Thus, immunotoxin conjugates consisting of deglycosylated ricin A chain (dgA) have been shown to survive longer *in vivo* and are more efficient

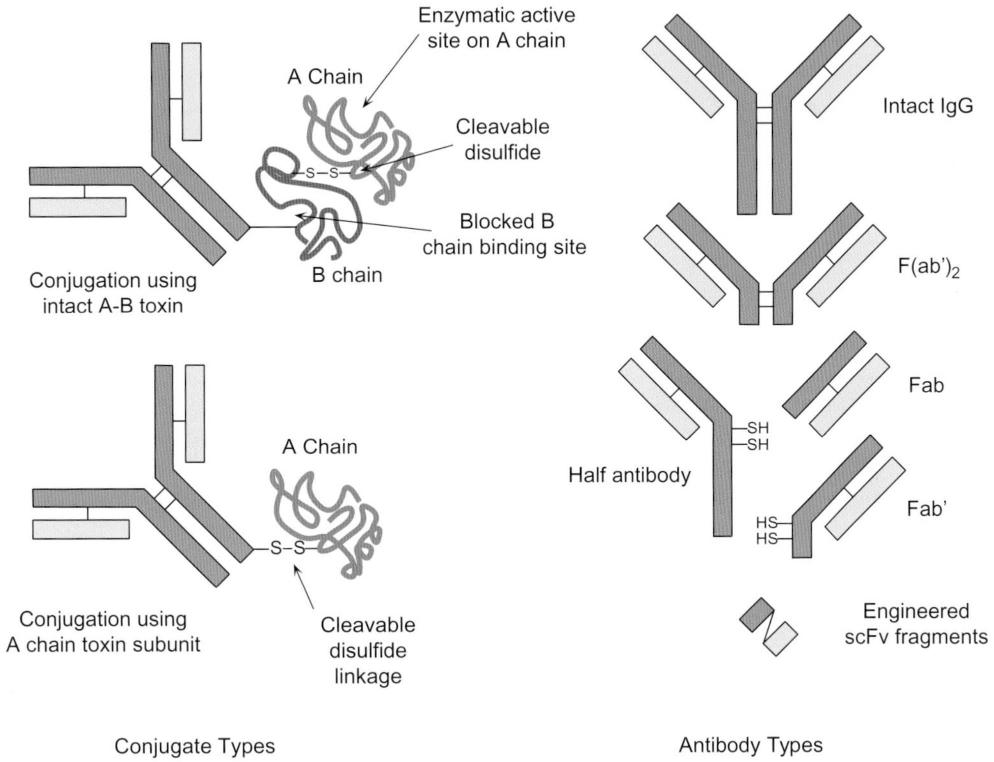

FIGURE 20.23 The strategies involved in creating an immunotoxin conjugate are numerous. Intact antibody molecules or enzymatic fragments may be used as the targeting component. Even small recombinant Fv fragments that are genetically engineered to limit host immune response can be employed. Conjugates can include two-subunit toxins or purified A chain components. If intact toxins are used, the B chain binding site must be blocked to prevent nonspecific cell death. If A chain subunits are used, to induce cytotoxic effects in the conjugate the crosslinking agent must generate a disulfide bond that can allow toxin release.

at reaching their intended target cells. In addition, if the antibody component does not contain Fc region, but consists of only $F(ab')_2$, Fab′, Fab, or smaller Fv fragments, then nonspecific binding of the immunotoxin *in vivo* will be reduced to a minimum. One study found that constructing immunotoxin conjugates with molar ratios of two dgA per antibody molecule resulted in a 7-fold increase in cytotoxicity over a 1:1 conjugate ratio (Ghetie *et al.*, 1993).

A chain immunotoxins, however, may not be quite as cytotoxic as conjugates formed from intact toxin molecules (Manske *et al.*, 1989). In an alternative approach to A chain use, the intact toxin of two-subunit proteins is directly conjugated to a monoclonal without isolation of the A chain. Conjugation of an antibody with intact A–B chain toxins can be performed without a cleavable linker, as long as the A chain can still separate from the B chain once it is internalized. Therefore, it is important to avoid intramolecular crosslinking during the conjugation process which can prevent release of the A–B complex. In addition, since the B chain possesses a recognition site for most cell surfaces, it still has the ability to nonselectively bind and kill non-tumor cells. To maintain antibody specificity in intact toxin conjugates toward only one cell type (and thus prevent nonspecific cell death), all cell binding capability within the toxin itself must be removed. Fortunately, a large proportion of the binding sites on the B chains are usually blocked during the conjugation process, and the galactose binding potential is significantly impaired. Further purification to remove conjugates that have remaining galactose binding potential can be carried out on an acid-treated agarose chromatography column (which contains galactose residues) or on a column of asialofetuin bound to agarose (Cumber *et al.*, 1985). Conjugate fractions which do not bind to both affinity gels contain no nonspecific binding potential toward non-targeted cells.

More elaborate methods of blocking or eliminating the B chain galactose binding site can also be done to prevent nonspecific cytotoxicity. For instance, the crosslinking agent may have a lactose molecule designed into it that can block B chain activity. The lactose portion possesses natural affinity for the B chain binding site, and thus it occupies that area while a nearby reactive group covalently attaches the crosslinker to neighboring functional groups on the protein.

Moroney *et al.* (1987) used this approach in creating a ricin conjugate. Lactose was modified at its reducing

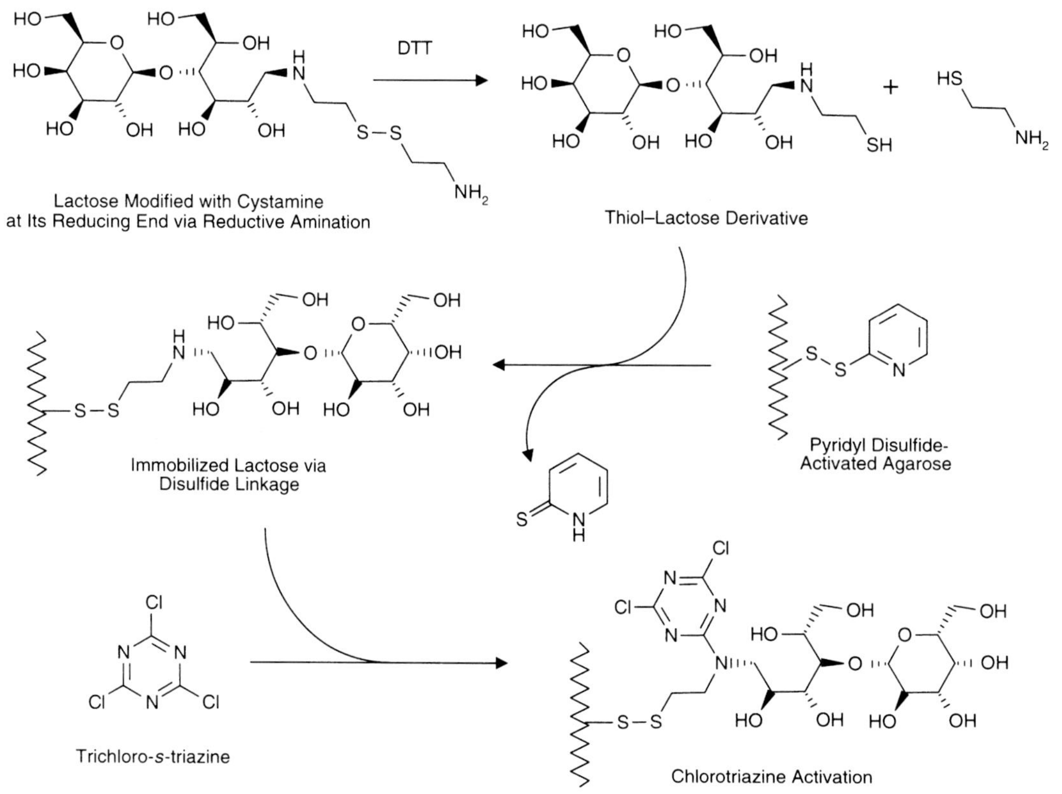

FIGURE 20.24 In an elaborate strategy to block the B chain binding site in the construction of immunotoxins using intact A–B subunit toxins, cystamine was first coupled to the reducing end of lactose by reductive amination. DTT was then used to reduce the cystamine disulfide group, revealing the free thiol. A pyridyl disulfide-activated agarose gel then was used to couple the lactose derivative through its sulfhydryl. Next, trichloro-*s*-triazine was reacted with the support to modify the secondary amine on the cystamine component, forming a reactive gel. Finally, addition of an intact toxin to the affinity support caused binding with the lactose group at the B chain binding site. Since the B chain was now in proximity to the chlorotriazine ring, covalent coupling occurred with available amines on the protein toxin, thus permanently blocking the binding pocket. After removal of the modified toxin from the gel using a disulfide reducing agent, the free thiol of the cystamine group was used to conjugate with an SMCC-activated immunoglobulin.

end with cystamine *via* reductive amination (Chapter 2, Section 5.3). The cystamine was reduced with dithiothreitol (DTT) and immobilized on an affinity gel which was activated with a pyridyl dithiol group (Chapter 3, Section 2.6). The coupled lactose was then modified with a chlorotriazine derivative through the secondary amine that was created by the reductive amination process. Next, the ricin molecule was immobilized by the additional reactive group on the chlorotriazine ring. Since this reactive group was immediately adjacent to the lactose residue, the ricin bound the sugar at its B chain binding site and it was covalently coupled through a nearby amine. This process effectively blocked and eliminated all nonspecific binding potential in the ricin dimer. After removal of the blocked ricin from the support by DTT, the free sulfhydryl group on the protein was conjugated to an SMCC-modified antibody, forming the final conjugate (Figure 20.24). Although this process worked in preparing an effective immunotoxin conjugate, most conjugation schemes are less elaborate.

Regardless of their method of preparation, the required and ideal characteristics of immunotoxin conjugates can be summarized in the following points:

1. The conjugation process must leave the antigen binding sites on the antibody component free to interact with its intended target. Crosslinker modification or blockage of these binding sites by the attached toxin must be kept to a minimum.
2. The toxin component of the conjugate must be able to elicit cytotoxicity by ribosomal damage as it could in its native state. This means that the cell penetration and enzymatic properties of the toxin remain unaltered, although an antibody molecule is conjugated to it.
3. The activity of toxin binding to cells through the B chain must be eliminated in the conjugate to prevent nonspecific binding and nonselective cell death. This may be accomplished by using only the A chain subunit or by blocking the B chain binding site in the intact toxin conjugate.

4. Avoid covalently linking the A and B chains together during the crosslinking process. This can be done by using heterobifunctional crosslinkers that are more controllable in their reactivity than homobifunctional reagents.
5. The crosslinking process must minimize polymerization of either the antibody or the toxin. Low-molecular-weight, 1:1 or 1:2 conjugates of antibody-to-toxin are best.
6. The crosslinker used to form the bond between the antibody and toxin must be able to survive *in vivo* and not be cleaved by enzymatic or reductive means before reaching the targeted cells.
7. The conjugate must reach its intended target without being intercepted, bound and destroyed by the host immune system.
8. Administration of the immunotoxin should result in cell death and complete elimination of all target cells.

The last two points are the most difficult to realize. The following methods describing the conjugation strategies used to prepare immunotoxins work well in creating complexes containing active antibody and toxin. The majority of research today is not so much concerned with further optimization of the crosslinking process, but primarily is directed at overcoming the host immune system and making the conjugates more effective in accomplishing complete targeted tumor destruction.

Preparation of Immunotoxin Conjugates via Disulfide Exchange Reactions

Since the cytotoxic potential of most common toxins relies on their subunit disulfide cleavability with subsequent release of associated A chains, most successful conjugation techniques for preparing immunotoxins involve the use of disulfide exchange reactions. Heterobifunctional crosslinking agents containing an amine-reactive group at one end and a disulfide bond with a good leaving group on the other end are common choices for making these conjugates (Chapter 6, Section 1, and Chapter 18, Section 1.2). The leaving group on the disulfide portion of the crosslinker permits efficient disulfide interchange with a free sulfhydryl on the antibody or toxin. The resultant covalent bond thus is a cleavable disulfide which mimics the native cleavability inherent in the toxin dimer.

PYRIDYL DISULFIDE REAGENTS

The most common reactive group for initiating disulfide interchange reactions is a pyridyl disulfide. Attack of a nucleophilic thiolate anion dissociates the pyridine-2-thione leaving group and forms a new disulfide bond with the incoming sulfhydryl compound. Several crosslinking reagents containing these groups are popular choices for producing antibody–toxin conjugates.

SPDP

SPDP, N-succinimidyl 3-(2-pyridyldithio)propionate, is by far the most popular heterobifunctional crosslinking agent used for immunotoxin conjugation (Chapter 6, Section 1.1). The activated NHS ester end of SPDP reacts with amine groups in one of the two proteins to form an amide linkage. The 2-pyridyldithiol group at the other end reacts with sulfhydryl groups in the other protein to form a disulfide linkage (Carlsson *et al.*, 1978). The result is a crosslinked antibody–toxin conjugate containing cleavable disulfide bonds that can emulate the activity of native two-subunit toxin molecules.

LC–SPDP (Chapter 6, Section 1.1) is an analog of SPDP containing a hexanoate spacer arm within its internal cross-bridge. The increased length of the extended crosslinker is important in some conjugations to avoid steric problems associated with closely linked macromolecules. However, for the preparation of immunotoxins, no advantages were observed for LC–SPDP over SPDP (Singh *et al.*, 1993).

SPDP is also useful in creating sulfhydryls in one of the two proteins being conjugated (Chapter 2, Section 4.1). Once modified with SPDP, the protein can be treated with DTT (or another disulfide reducing agent) to release the pyridine-2-thione leaving group and form the free sulfhydryl. The terminal SH group then can be used to conjugate with any crosslinking agents containing sulfhydryl-reactive groups, such as maleimide or iodoacetyl (for covalent conjugation) or 2-pyridyldithiol groups (for reversible conjugation).

In the preparation of immunotoxins, some procedures call for the modification of the antibody with SPDP to introduce reactive thiols (Cumber *et al.*, 1985). The NHS ester end of the crosslinker is reacted at slightly alkaline pH with the primary amines on the antibody. After removal of excess reagent by gel filtration, the pyridyl disulfide groups are reduced by DTT. The reductant causes the removal of pyridine-2-thione groups and the creation of sulfhydryl groups on the immunoglobulin. The reason the antibody is thiolated in this manner and not the toxin is to avoid exposing the intact toxin to reducing conditions that could disassociate the A and B subunits.

To activate the toxin, SPDP again can be used to modify the intact A–B component. After purification of the modified toxin from excess crosslinker, the SPDP–toxin is mixed with the thiolated antibody to effect the final conjugate (Figure 20.25).

This multi-step crosslinking method employing SPDP on both molecules has been used to prepare a number of immunotoxin conjugates (Edwards *et al.*, 1982; Thorpe *et al.*, 1982; Colombatti *et al.*, 1983; Wiels *et al.*, 1983; Vogel, 1987; Reiter and Fishelson, 1989).

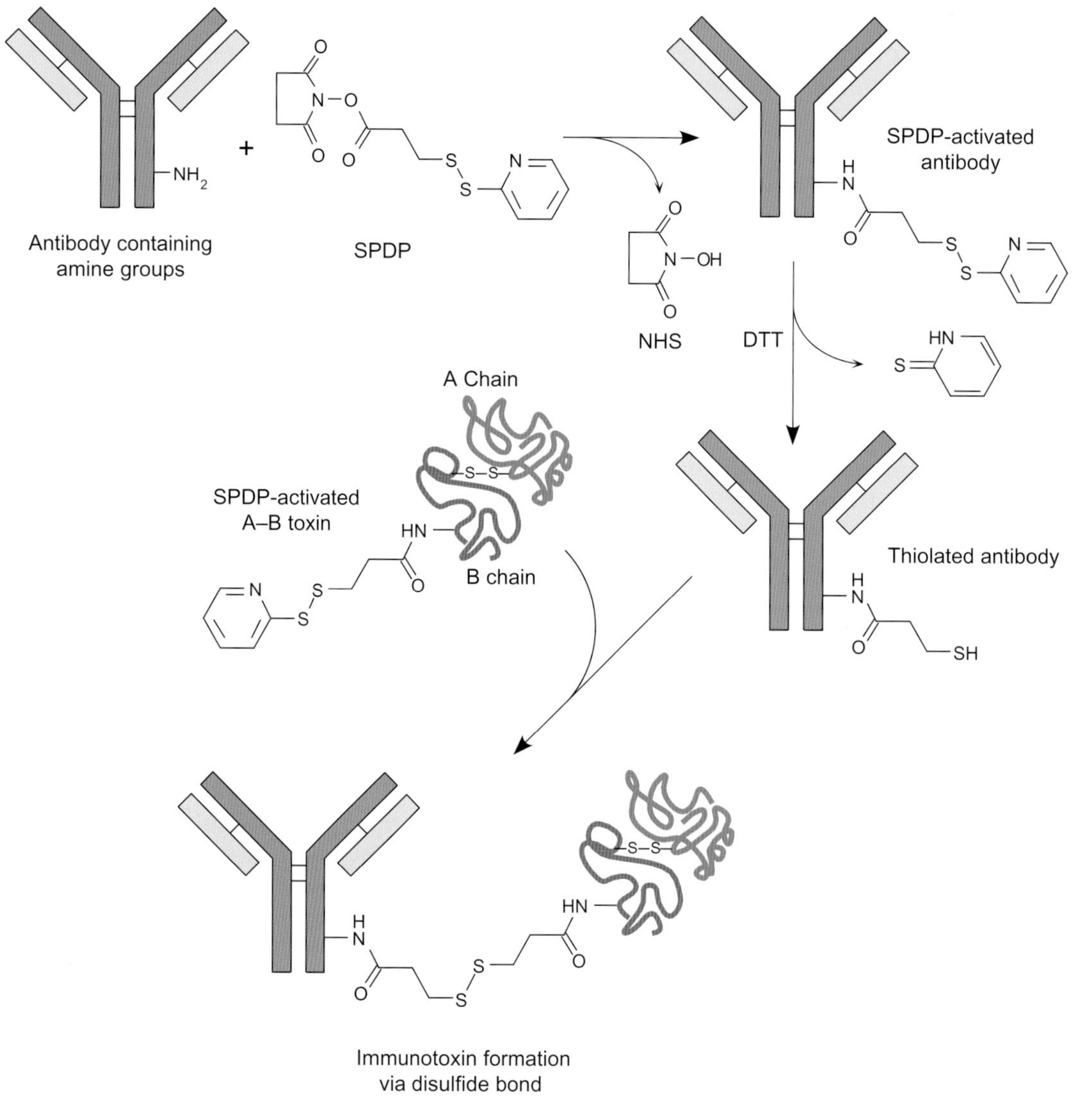

FIGURE 20.25 SPDP can be used to modify both an antibody and a toxin molecule for conjugation purposes. In this case, the antibody is thiolated to contain a sulfhydryl group by modification with SPDP followed by reduction with DTT. A toxin molecule is then activated with SPDP and reacted with the thiolated antibody to form the final conjugate through a disulfide bond.

While this method has worked well for many different toxins, its main potential disadvantage is exposure of the antibody to reducing conditions that potentially could cleave the disulfide bonds holding its heavy and light chains together. Alternative methods using SPDP in a non-reducing environment may result in better conjugates.

For instance, if toxin A chain–antibody conjugates are to be prepared, the antibody can be similarly activated with SPDP, but in this case not treated with reductant. After removal of excess crosslinker, the activated antibody can be directly mixed with isolated A chain to create the conjugate (Figure 20.26). This procedure makes use of the indigenous sulfhydryl residues produced during reductive separation of the A and B chains and therefore does not require crosslinker thiolation of one of the proteins.

Another way of utilizing SPDP is to again activate the antibody to create the pyridyl disulfide derivative, but this time thiolate the toxin component using 2-iminothiolane (Chapter 2, Section 4.1). 2-Iminothiolane reacts with primary amines in a ring-opening reaction that creates a terminal sulfhydryl group without reduction. Intact A–B toxins and toxins containing only one subunit, like gelonin, PAPs, and *Pseudomonas* toxin A, can be coupled to antibodies using this procedure (Lambert *et al.*, 1985; Bjorn *et al.*, 1986; Scott *et al.*, 1987; Lambert *et al.*, 1988; Ozawa *et al.*, 1989). Mixing the SPDP-activated antibody with the thiolated toxin effectively forms the final conjugation (Figure 20.27).

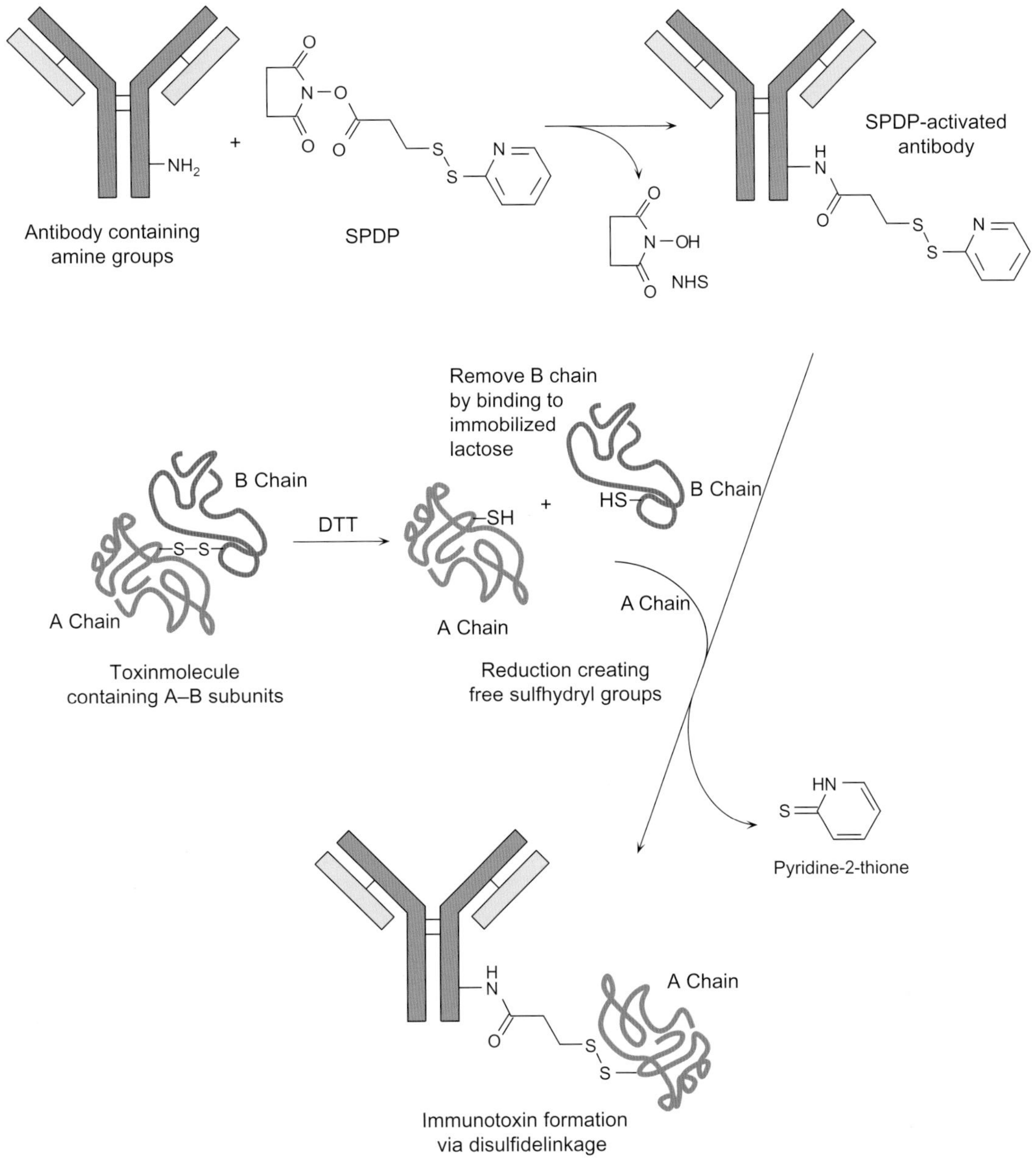

FIGURE 20.26 SPDP can be used to activate an antibody molecule through its available amine groups to form a sulfhydryl-reactive derivative. Toxin molecules containing disulfide-linked A–B chains may be reduced with DTT to isolate the A chain component containing a free thiol. The SPDP-activated antibody is then mixed with the reduced A chain to effect the final conjugate by disulfide bond formation.

SPDP can also be used to conjugate other targeting molecules to toxins, such as transferrin, epidermal growth factor, α_2-macroglobulin, and human chorionic gonadotropin (Fizgerald *et al.*, 1980; Helenius *et al.*, 1980; Keen *et al.*, 1982; Oeltmann, 1985). To create these conjugates, one of the two components must be activated with SPDP to generate the sulfhydryl-reactive pyridyl disulfide groups, while the other component must be modified to contain the SH functionality. Mixing these modified molecules together forms the toxin conjugate.

The following methods are generalized to provide an overview of how SPDP can be used in these conjugation techniques. The appropriate optimization for a particular toxin conjugate should be done.

Caution! Toxins are highly toxic even in very low amounts. Handle all toxin molecules and their isolated subunits with extreme care.

Protocol for Thiolation of Antibody with SPDP and Conjugation to an SPDP-Activated Toxin

Caution: Toxin molecules are dangerously toxic even in small amounts. Use extreme care in handling.

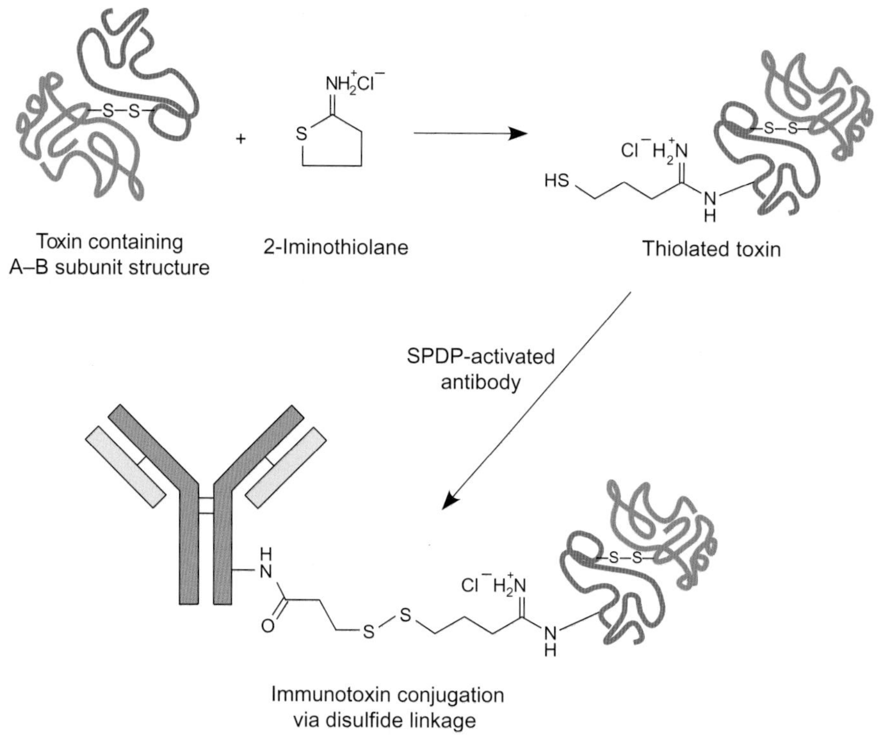

FIGURE 20.27 An intact A–B subunit toxin molecule may be activated with 2-iminothiolane with good retention of cytotoxic activity. The thiolated toxin may then be conjugated with SPDP-activated antibody to generate the immunotoxin conjugate through a disulfide bond.

Note: In this protocol, for every milligram of toxin employed, 2.5 mg of antibody are required to obtain the correct molar ratios in the final conjugate.

(A) ACTIVATION OF TOXIN WITH SPDP

1. Dissolve the toxin to be conjugated in 0.1-*M* sodium phosphate, 0.15-*M* NaCl, pH 7.5, at a concentration of 10 mg/ml. Some protocols use as an SPDP reaction buffer, 50-m*M* sodium borate, 0.3-*M* NaCl, 0.5% *n*-butanol, pH 9.0. Both buffer systems work well for the NHS ester modification reaction, although the pH 9 buffer is at the higher end of effective derivatization with active esters, since the hydrolysis rate is dramatically increased at this level of alkalinity.
2. Dissolve SPDP (Thermo Fisher) in DMF at a concentration of 3 mg/ml. Add 20 µl of this solution to each milliliter of the toxin solution. Gently mix to effect dissolution. Retain the SPDP stock solution for use in the antibody modification step, below.
3. React for 30 min at room temperature.
4. Purify the SPDP-activated toxin from excess reagents and reaction byproducts by gel filtration using a desalting resin. For the chromatography separation, use as a buffer 0.1-*M* sodium phosphate, 0.15-*M* NaCl, pH 7.5, containing 10-m*M* EDTA.
5. Concentrate the toxin to 10 mg/ml using a centrifugal concentrator with a molecular weight cutoff of 10,000. Retain this solution for the conjugation reaction.

(B) THIOLATION OF ANTIBODY WITH SPDP

1. Dissolve the antibody to be conjugated in 0.1-*M* sodium phosphate, 0.15-*M* NaCl, pH 7.5, at a concentration 10 mg/ml. Note: Some protocols use the borate buffer system described in step A. Use 2.5 mg of antibody per mg of toxin to be conjugated.
2. Dissolve SPDP in DMF at a concentration of 3 mg/ml. Add 24 µl of this solution to each milliliter of the antibody solution with gentle mixing to effect complete dissolution.
3. React for 30 min at room temperature.
4. Remove excess crosslinker by gel filtration using a desalting resin. Perform the chromatography using as the buffer 0.1-*M* sodium phosphate, 0.15-*M* NaCl, 10-m*M* EDTA, pH 7.5. The solution should be degassed under vacuum and nitrogen bubbled through it to remove oxygen. The presence of EDTA stabilizes the free sulfhydryls formed in the following steps against metal-catalyzed oxidation.
5. Concentrate the fractions containing protein from the gel filtration step to 10 mg/ml using a centrifugal concentrator (MW cutoff of 10,000).
6. To reduce the pyridyl dithiol groups and create reactive sulfhydryls, dissolve DTT in water at a

concentration of 17.2 mg/ml and immediately add 500 μl of this solution to each ml of concentrated antibody solution. Mix to dissolve and react for 30 min at room temperature.
7. Remove excess DTT by gel filtration using the same buffer as in step 4. Pool the fractions containing protein and concentrate to 10 mg/ml.

(C) CONJUGATION OF SPDP-ACTIVATED TOXIN WITH THIOLATED ANTIBODY

1. Immediately mix the concentrated, thiolated antibody solution from part B with the SPDP-activated toxin from part A.
2. React for 18 h at room temperature to form the final conjugate. Isolation of the ideal 1:1 or 1:2 antibody–toxin conjugate can be achieved through gel filtration separation using a column of Sephacryl S-300 or the equivalent.

(D) ACTIVATION OF ANTIBODY WITH SPDP AND CONJUGATION TO A TOXIN A CHAIN *Caution: Toxin molecules are dangerously toxic even in small amounts. Use extreme care in handling.*

Since the A chain of toxin molecules contains a free sulfhydryl group, there is no need in this conjugation strategy to thiolate one of the molecules. The following protocol calls for 1.73 mg of antibody per milligram of toxin A chain to produce a conjugate possessing the correct molar ratio of components. Best results for creating a highly cytotoxic immunotoxin will be obtained if deglycosylated ricin A chain is used.

1. Dissolve the antibody to be conjugated in 0.1-*M* sodium phosphate, 0.15-*M* NaCl, pH 7.5, at a concentration 10 mg/ml.
2. Dissolve SPDP (Thermo Fisher) in DMF at a concentration of 3.0 mg/ml. Add 30 μl of this solution to each milliliter of the antibody solution with gentle mixing to effect complete dissolution.
3. React for 30 min at room temperature.
4. Remove excess crosslinker by gel filtration using a desalting resin. Perform the chromatography using 0.1-*M* sodium phosphate, 0.15-*M* NaCl, 10-m*M* EDTA, pH 7.5.
5. Concentrate the fractions containing SPDP-activated antibody from the gel filtration step to 10 mg/ml using a centrifugal concentrator (MW cutoff of 10,000).
6. Mix the activated antibody solution with a solution of deglycosylated toxin A chain (dgA) dissolved in 0.1-*M* sodium phosphate, 0.15-*M* NaCl, pH 7.5, containing 10-m*M* EDTA. The ratio of mixing should equal 1.73 mg of antibody per milligram of A chain or 580 μl of A chain solution at 10 mg/ml per milliliter of activated antibody solution at 10 mg/ml. The A chain solution must not contain any reducing agents left over from the disassociation of the toxin subunits during the A subunit isolation. Reductants will compete for the SPDP activation sites on the antibody molecule.
7. React for 18 h at room temperature. Isolation of the 1:1 or 1:2 antibody–toxin conjugate can be achieved through a gel filtration separation using a column of Sephacryl S-200. Isolation of conjugates containing molar ratios of 1:2 antibody:dgA have resulted in greater cytotoxicity behavior *in vivo* (Ghetie *et al.*, 1993).

(E) CONJUGATION OF SPDP-ACTIVATED ANTIBODIES WITH 2-IMINOTHIOLANE-MODIFIED TOXINS *Caution: Toxin molecules are dangerously toxic even in small amounts. Use extreme care in handling.*

A third option for immunotoxin preparation is to again activate the antibody with SPDP, while this time thiolating a single-chain toxin molecule to conjugate with it. This method works especially well using 2-iminothiolane (Chapter 2, Section 4.1) to create sulfhydryls on gelonin or PAPs. Gelonin is a single-polypeptide toxin containing no free sulfhydryls. A number of options are available for thiolation; however, the use of SPDP to add sulfhydryl groups inactivates the toxin, while 2-iminothiolane preserves its activity, perhaps by maintaining the positive charge on the amines that are being modified (Lambert *et al.*, 1985).

(F) ACTIVATION OF ANTIBODY WITH SPDP

1. Dissolve the antibody to be conjugated in 0.1-*M* sodium phosphate, 0.15-*M* NaCl, pH 7.5, at a concentration of 10 mg/ml.
2. Dissolve SPDP (Thermo Fisher) in DMF at a concentration of 3.0 mg/ml. Add 30 μl of this solution to each milliliter of the antibody solution with gentle mixing to effect dissolution.
3. React for 30 min at room temperature.
4. Remove excess crosslinker by gel filtration using a desalting resin. Perform the chromatography using 0.1-*M* sodium phosphate, 0.15-*M* NaCl, 10-m*M* EDTA, pH 7.5.
5. Concentrate the fractions containing SPDP-activated antibody from the gel filtration step to 10 mg/ml using a centrifugal concentrator (MW cutoff of 10,000).

(G) THIOLATION OF GELONIN (OR OTHER SINGLE-POLYPEPTIDE TOXINS) WITH 2-IMINOTHIOLANE (TRAUT'S REAGENT)

1. Dissolve gelonin at a concentration of 10 mg/ml in 50-m*M* triethanolamine hydrochloride, pH 8.0, containing 10-m*M* EDTA. The buffer should be de-gassed under vacuum and bubbled with nitrogen to remove oxygen that may cause sulfhydryl oxidation after thiolation.

2. Dissolve 2-iminothiolane (Thermo Fisher) in degassed, nitrogen-bubbled deionized water at a concentration of 20 mg/ml (makes a 0.14-*M* stock solution). The solution should be used immediately. Add 70 μl of the 2-iminothiolane solution to each ml of the gelonin solution (final concentration is about 10-m*M*).
3. React for 1 h at 0°C (or on ice) under a nitrogen blanket.
4. Purify the thiolated protein from unreacted Traut's reagent by gel filtration on a desalting resin using 0.1-*M* sodium phosphate, 0.15-*M* NaCl, pH 7.5, containing 10-m*M* EDTA. The presence of EDTA in this buffer helps to prevent oxidation of the sulfhydryl groups and resultant disulfide formation. The degree of SH modification in the purified protein may be determined using the Ellman's assay (Chapter 2, Section 4.1).
5. Concentrate the thiolated toxin to 10 mg/ml using centrifugal concentrators. Immediately use the modified protein in the conjugation reaction to prevent inactivation of 2-iminothiolane-modified molecules by recyclization (Chapter 2, Section 4.1).

(H) CONJUGATION OF SPDP-ACTIVATED ANTIBODY WITH THIOLATED GELONIN

1. Mix the SPDP-activated antibody with the thiolated gelonin in equal mass quantities (or equal volumes if they are at the same concentration). This ratio results in about a 5-fold molar excess of toxin over the amount of antibody.
2. React for 20 h at 4°C under a nitrogen blanket.
3. To block unreacted sulfhydryl groups, add iodoacetamide to the solution to a final concentration of 2-m*M*.
4. React for an additional 1 h at room temperature.
5. Remove unconjugated gelonin by passage of the conjugate solution over a column of immobilized protein A (Thermo Fisher). Use 2 ml of the protein A column for each 10 mg of conjugate to be purified. Equilibrate the column with 50-m*M* sodium phosphate, 0.15-*M* NaCl, pH 7.5. Apply the conjugate sample and allow it to enter the gel. Continue to wash the column with equilibration buffer while taking 2-ml fractions until baseline is reached (monitored at an absorbance of 280 nm). Unconjugated gelonin will pass through the column unretarded. Elute bound conjugate with 0.1-*M* acetic acid, 0.15-*M* NaCl. Immediately add 0.1 ml of 1-*M* potassium phosphate, pH 7.5 to each bound fraction for neutralization. Alternatively, gel filtration may be used to isolate the conjugate from lower-molecular-weight antibody and gelonin. A column of Sephacryl S-200 works well for this purpose.

SMPT

Succinimidyloxycarbonyl-α-methyl-α-(2-pyridyldithio)toluene (SMPT) is a heterobifunctional crosslinking agent similar to SPDP that contains an amine-reactive NHS ester on one end and a sulfhydryl-reactive pyridyl disulfide group on the other (Chapter 6, Section 1.2). Reaction with a sulfhydryl-containing protein results in a cleavable disulfide linkage, important for immunotoxin activity. SMPT is an analog of SPDP that differs only in its cross-bridge, which contains an aromatic ring and a hindered disulfide group (Thorpe *et al.*, 1987; Ghetie *et al.*, 1990). The spacer arm of SMPT is slightly longer than SPDP, but the presence of the benzene ring and α-methyl group adjacent to the disulfide sterically hinders the structure sufficiently to provide increased half-life of immunotoxin conjugates *in vivo*.

SMPT is often used in place of SPDP for the preparation of immunotoxin conjugates. The hindered disulfide of SMPT has distinct advantages in this regard. Thorpe *et al.* (1987) showed that SMPT conjugates had approximately twice the half-life *in vivo* as SPDP conjugates. Antibody–toxin conjugates prepared with SMPT possess a half-life *in vivo* of up to 22 h, presumably due to the decreased susceptibility of the hindered disulfide toward reductive cleavage.

Ghetie *et al.* (1991) developed a large-scale preparation procedure for antibody-deglycosylated ricin A chain (dgA) conjugates utilizing this crosslinker. The following procedure describes a generalized method for using SMPT to prepare dgA–antibody conjugates. It is based on the Ghetie protocol, but using smaller quantities of reagents. Figure 20.28 illustrates the reactions involved in using SMPT.

Protocol

Caution: Toxin molecules are dangerously toxic even in small amounts. Use extreme care in handling.

The following method calls for mixing activated antibody with ricin A chain at a ratio of 2 mg antibody per milligram of A chain. Adjustments to the amount of antibody and A chain initially dissolved in the reaction buffers should be done to anticipate this ratio.

1. Dissolve the antibody to be conjugated in 0.1-*M* sodium phosphate, 0.15-*M* NaCl, pH 7.5, at a concentration of 10 mg/ml. If the antibody contains oligomers (as evidenced by nondenaturing electrophoresis or HPLC gel filtration analysis), then the monomeric IgG form should be isolated by gel filtration using a column of Sephacryl S-200HR. If no oligomers are present, then omit the chromatographic purification.
2. Dissolve SMPT (Thermo Fisher) in DMF at a concentration of 4.8 mg/ml. Add 27 μl of this solution

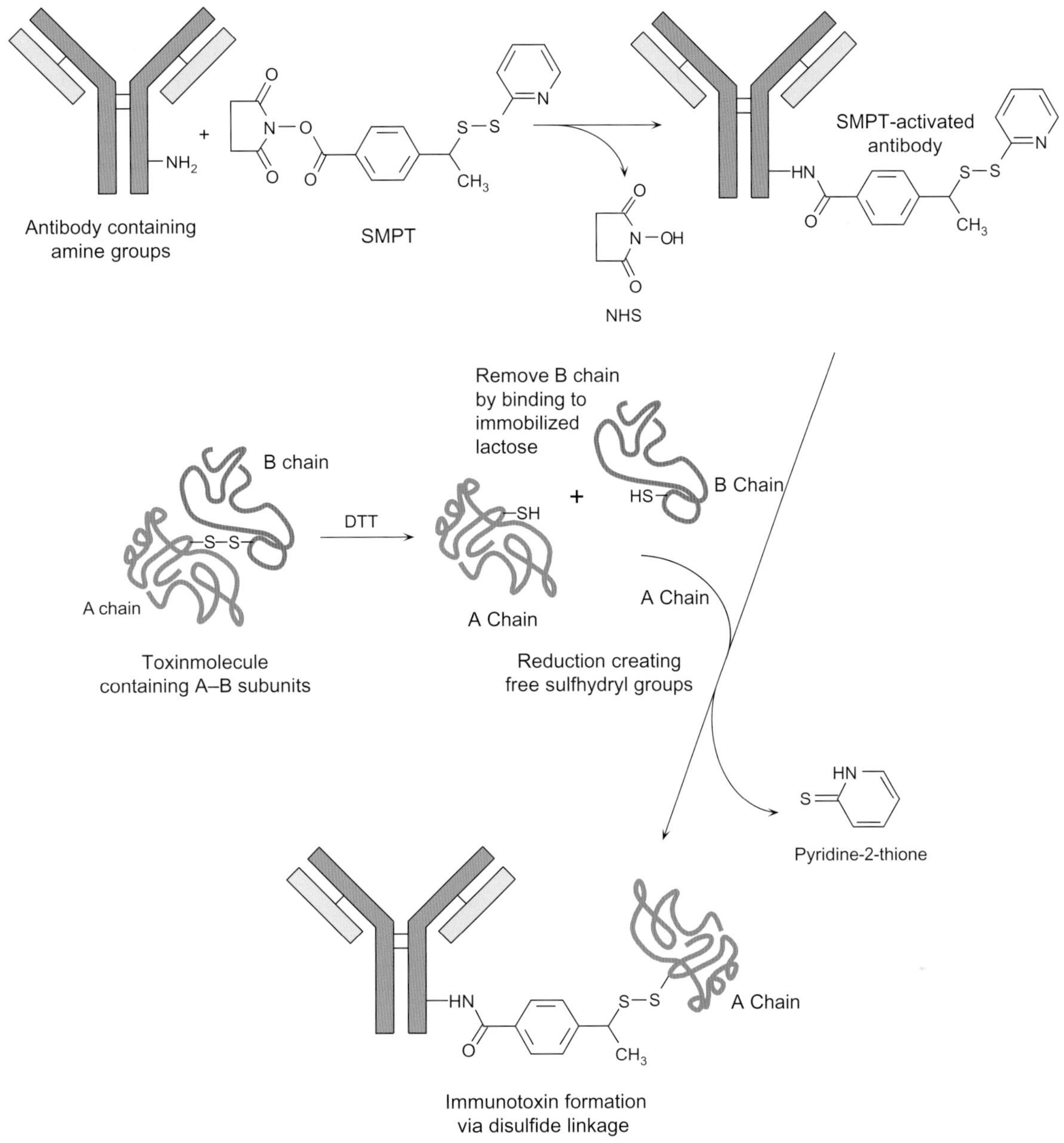

FIGURE 20.28 SMPT may be used to form immunotoxin conjugates by activation of the antibody component to form a thiol-reactive derivative. Reduction of an A–B toxin molecule with DTT can facilitate subsequent isolation of the A chain containing a free thiol. Mixing the A chain containing a sulfhydryl group with the SMPT-activated antibody causes immunotoxin formation through disulfide bond linkage. The hindered disulfide of an SMPT crosslink has been found to survive *in vivo* for longer periods than conjugates formed with SPDP.

to each ml of the antibody solution. Mix gently. The final concentration of SMPT in the reaction mixture is 0.13 mg/ml, which translates into about a 4.8-fold molar excess of crosslinker over the amount of antibody present.

3. React for 1 h at room temperature.
4. Remove unreacted SMPT and reaction byproducts by gel filtration on a desalting resin. Pool fractions containing SMPT-activated antibody (the first peak eluting from the column) and concentrate them to 10 mg/ml using centrifugal concentrators with a molecular weight cut-off of 10,000.
5. Dissolve deglycosylated ricin A chain (dgA) in 0.1-*M* sodium phosphate, 0.15-*M* NaCl, 10-m*M* EDTA, pH 7.5, at a concentration of 10 mg/ml. The buffer should be degassed under vacuum and nitrogen bubbled through it to remove oxygen. Prepare half the amount of A chain solution as the amount of antibody prepared in step 1. If the A chain preparation is done in bulk quantities or if the

protein has been stored for lengthy periods, it may be necessary to reduce the sulfhydryls with DTT prior to proceeding with the crosslinking reaction. If A chain sulfhydryl oxidation is suspected, add 2.5 mg of DTT per milliliter of A chain solution. React for 1 h at room temperature. Purify the reduced ricin A chain by gel filtration on a desalting resin using the PBS–EDTA buffer. Apply no greater volume of sample to the gel than is represented by 5% of the column volume to ensure good removal of excess DTT. Collect the protein and concentrate to 10 mg/ml using centrifugal concentrators.

6. Mix the reduced A chain solution with activated antibody solution at a ratio of 2 mg of antibody per milligram of A chain. Sterile filter the solution through a 0.22-μm membrane, and react at room temperature under nitrogen for 18 h.
7. To block excess pyridyl disulfide active sites on the antibody, add cysteine to a final concentration of 25 μg/ml. React for an additional 6 h at room temperature.
8. To isolate the conjugate, apply the immunotoxin solution to a column of Sephacryl S-200HR. Collect the peaks with molecular weights between 150,000 and 210,000. Further purification to remove excess unconjugated antibody can be done on a column of immobilized Cibacron Blue (available commercially from Thermo Fisher or for column preparation, see Hermanson *et al.*, 1992). Equilibration of the column with 50-m*M* sodium borate, 1-m*M* EDTA, pH 9.0, will cause binding of the conjugate, but not the free antibody. Elution of purified immunotoxin conjugate can be done with 50-m*M* sodium borate, 1-m*M* EDTA, 0.5-*M* NaCl, pH 9.0 (see Ghetie *et al.*, 1991).

3-(2-PYRIDYLDITHIO)PROPIONATE

A lesser-used reagent to introduce sulfhydryl-reactive pyridyl disulfide groups is 3-(2-pyridyldithio) propionate (PDTP), the acid precursor of SPDP containing no NHS ester group on the carboxylate. Sulfhydryl interchange reaction at the pyridyl dithiol end results in the formation of a disulfide linkage with SH containing molecules. The carboxylate end is not further derivatized to contain a reactive species, but may be coupled to amines by the carbodiimide reaction (Chapter 4, Section 1). Reaction of PDTP with an antibody molecule in the presence of 1-ethyl-3-(3-dimethylaminopropyl) carbodiimide (EDC) results in the formation of amide linkages with the active pyridyl disulfide groups still available for coupling to sulfhydryl-containing toxins (Figure 20.29). Mixing the PDTP–antibody with purified ricin A chain results in disulfide crosslinks identical to those obtained using SPDP as the crosslinker (Jansen *et al.*, 1980; Gros *et al.*, 1985). PDTP also has been used to activate transferrin to contain reactive pyridyl dithiol groups for conjugation to ricin A chain molecules (Raso and Basala, 1984, 1985).

Since activated molecules and crosslinks formed between two species are identical to those formed using SPDP, it is of little advantage to use PDTP. Furthermore, an EDC-mediated reaction of the carboxylate end of the crosslinker with amine groups on proteins can cause concomitant zero-length crosslinking and polymerization of protein molecules. For these reasons, SPDP is the better choice for preparing immunotoxin conjugates.

USE OF CYSTAMINE, ELLMAN'S REAGENT, OR S-SULFONATES

Other reagent systems can be used to form disulfide linkages between antibody and toxin molecules in immunotoxin conjugates. Cystamine can be incorporated into proteins by reaction of its terminal amines with the carboxylates on the proteins *via* the carbodiimide reaction (Chapter 4, Section 1). The resultant modifications contain disulfide linkages that can undergo disulfide interchange reactions with other sulfhydryl-containing molecules (Chapter 2, Section

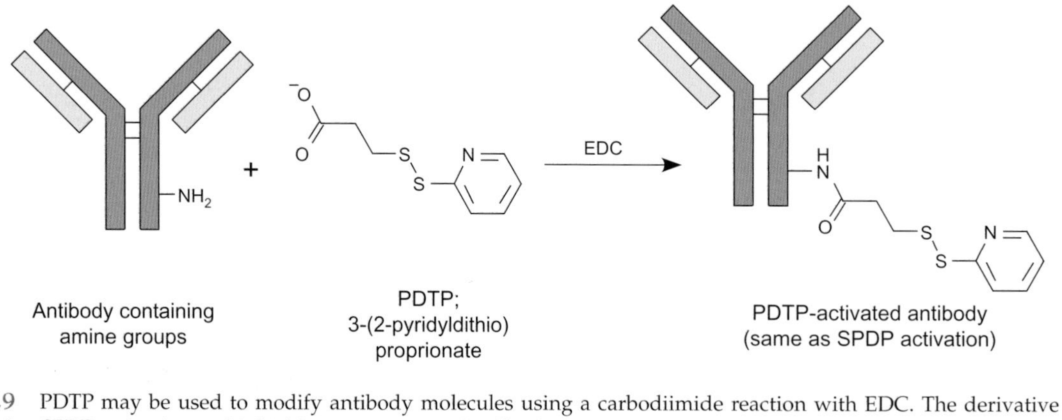

FIGURE 20.29 PDTP may be used to modify antibody molecules using a carbodiimide reaction with EDC. The derivative is the same as that obtained using SPDP activation and is highly reactive toward sulfhydryls.

4.1). For instance, a cystamine-modified targeting component, such as an antibody, can be mixed with the reduced A chain of a toxin molecule to cause conjugate formation through the creation of a disulfide bond (Figure 20.30) (Oeltmann and Forbes, 1981). Epidermal growth factor was modified with cystamine and coupled with reduced diphtheria toxin using this approach (Shimisu *et al.*, 1980).

Similarly, Ellman's reagent [5,5′-dithiobis(2-nitrobenzoic acid)] can be used to activate one thiol-containing molecule by disulfide exchange and subsequently used to couple to a second sulfhydryl-containing molecule by the same mechanism (Chapter 2, Section 5.2) (Figure 20.31). The disulfide of Elman's reagent readily undergoes disulfide exchange with a free sulfhydryl to form a mixed disulfide with simultaneous release of one molecule of the highly chromogenic 5-sulfido-2-nitrobenzoate, also called 5-thio-2-nitrobenzoic acid (TNB). The intense yellow color produced by the TNB anion can be measured by its absorbance at 412 nm. Thus, the efficiency of conjugation can be determined spectrophotometrically using this procedure (Pirker *et al.*, 1986; Fitzgerald *et al.*, 1988). A method for the large-scale conjugation of Fab′ fragments containing an available sulfhydryl group and deglycosylated ricin A chain (also containing an SH group) were developed using Ellman's reagent as the crosslinker (Ghetie *et al.*, 1988).

A final method of forming disulfide crosslinks between toxins and targeting molecules is the use of *S*-sulfonate formation using sodium sulfite (Na_2SO_3) in the presence of sodium tetrathionate ($Na_2S_4O_6$). Tetrathionate reacts with sulfhydryls to form

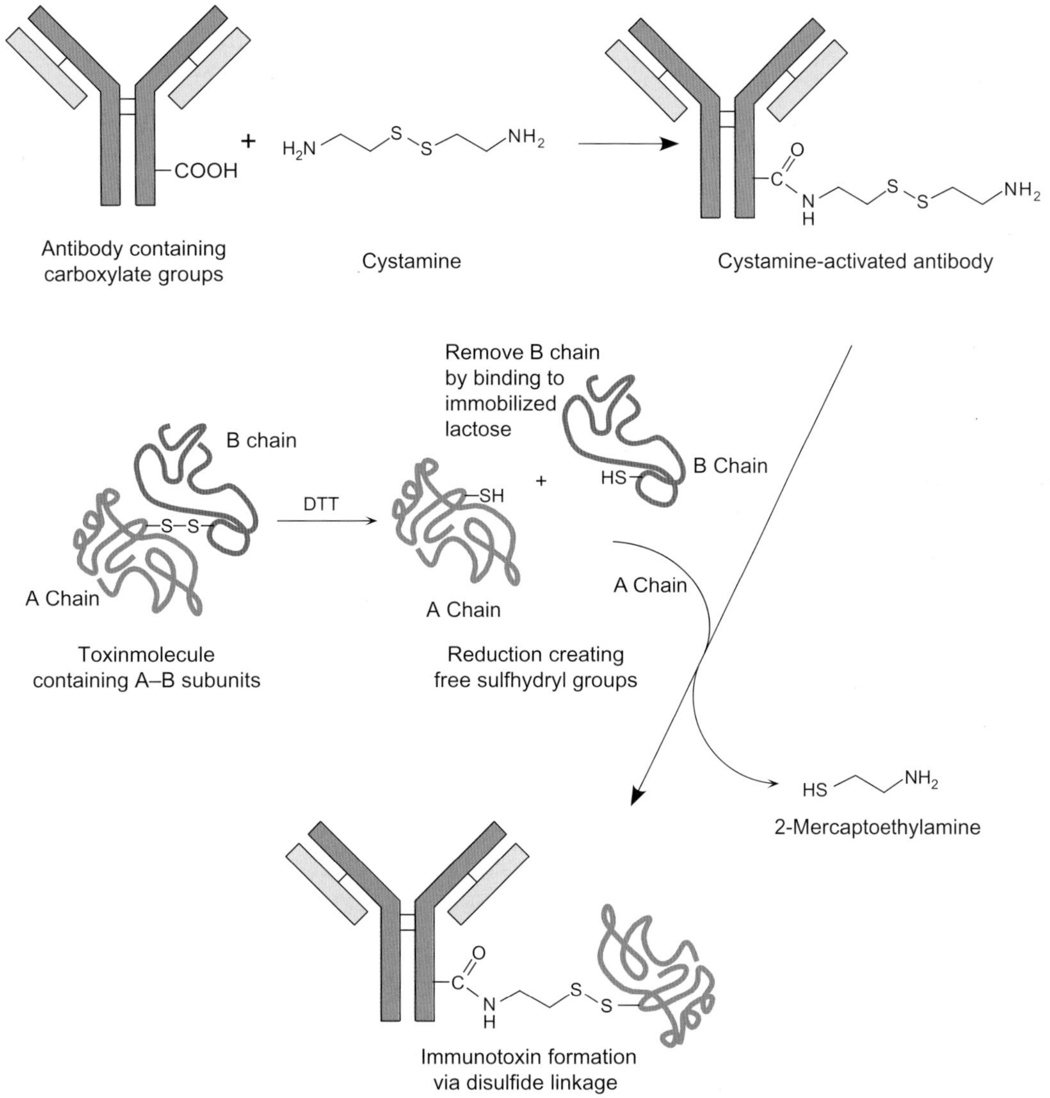

FIGURE 20.30 Cystamine may be used to make immunotoxin conjugates by a disulfide interchange reaction. Modification of antibody molecules using an EDC-mediated reaction creates a sulfhydryl-reactive derivative. A chain toxin subunits containing a free thiol can be coupled to the cystamine-modified antibody to form disulfide crosslinks.

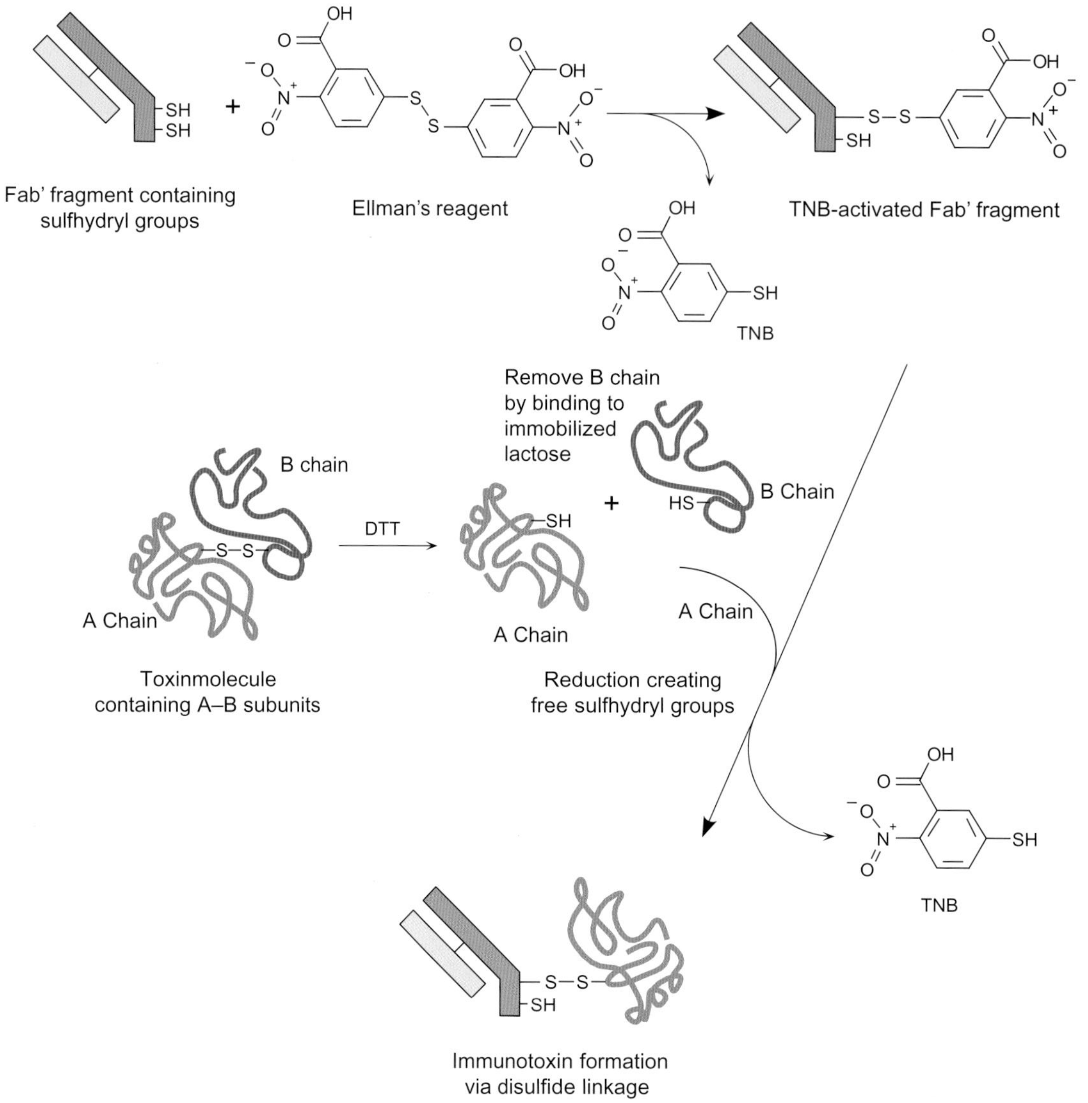

FIGURE 20.31 Fab′ antibody fragments containing free thiols can be activated with Ellman's reagent to form a sulfhydryl-reactive derivative. A chain toxin subunits that contain a free thiol group may be coupled to the activated Fab′ molecule to produce an immunotoxin complex.

sulfenylthiosulfate intermediates (Section 1.1). These derivatives are reactive toward other thiols to create disulfide linkages rapidly. Sulfite ions react with disulfides to form *S*-substituted thiosulfates, also known as *S*-sulfonates, and a thiol. The combination of these reagents results in the transformation of available thiols and disulfides into reactive *S*-sulfonates that can be used to crosslink with sulfhydryl-containing molecules. *S*-sulfonate conjugation can be used to conjugate the A chain of toxin molecules with sulfhydryl-containing Fab′ fragments with good efficiency (Masuho *et al.*, 1979).

Although the use of these alternative disulfide-generating agents has proven successful in some applications, pyridyl disulfide-containing crosslinkers, as discussed previously, are more common for producing immunotoxin conjugates.

Preparation of Immunotoxin Conjugates via Amine- and Sulfhydryl-Reactive Heterobifunctional Crosslinkers

Other forms of heterobifunctional crosslinkers that can be used for this purpose are the amine- and sulfhydryl-reactive agents that produce a thioether bond with SH containing molecules (Chapter 6, Section 1). The amine-reactive end of these crosslinkers is usually an NHS ester group that can form a stable amide bond with amine-containing proteins. One of two main reactive groups are usually used on the sulfhydryl-reactive end: an iodoacetyl group which couples to sulfhydryls with loss of HI or a maleimide group which undergoes a double-bond addition reaction with SH groups (see Chapter 3, Sections 2.1 and 2.2).

Since this type of crosslinker forms non-cleavable thioether bonds between toxin molecules and the targeting component of the conjugates, they are not appropriate for use with A chain or single-chain toxins. This is because the crosslinker will not allow the conjugated A chain to break free of the antibody by disulfide reduction after docking at the cellular target. Since release of the A chain is a prerequisite to ribosomal inactivation, such conjugates will prove to be ineffective cytotoxic agents. One report found a 1000-fold increase in cytotoxicity when an immunotoxin containing PAP or gelonin was prepared using a cleavable disulfide linker as opposed to a non-cleavable thioether linkage (Lambert *et al.*, 1985).

To make effective immunotoxin conjugates using the following crosslinkers, it is necessary to crosslink intact A–B toxins to antibodies, not single-chain or A chain toxins. Using intact two-subunit toxins allows the A chain to break free of the complex and perform its cytotoxic duties upon entering the target cell. Two main criteria are especially important when using A–B toxin conjugates: the B chain binding site must be blocked or inoperative in the final immunotoxin complex to prevent nonspecific cell death, and, second, the two subunits of the toxin must not be covalently crosslinked by the conjugation procedure, precluding them from being separated *in vivo*.

Fortunately, satisfying these criteria is not difficult. The heterobifunctional crosslinkers described in this section are sufficiently controllable as to prevent A–B chain crosslinking. In addition, during the conjugation process, the B chain binding site often becomes inactivated or physically blocked by the attached antibody molecule. Subsequent cleanup of the conjugate using affinity chromatography over a column containing an immobilized sugar can completely eliminate any potential nonspecificity contributed by the B chain in the final preparation.

It should be noted that the use of the following crosslinkers to create other forms of toxic conjugates for cancer therapy is not restricted by the disulfide bridge requirement (Trail *et al.*, 1993; Willner *et al.*, 1993). Drug–toxin conjugates, hormono-toxins, lymphokine– or growth factor–toxin conjugates all can be made using nonreversible thioether linkages without difficulty.

SIAB

N-Succinimidyl(4-iodoacetyl)aminobenzoate (SIAB) is a heterobifunctional crosslinker containing amine-reactive and sulfhydryl-reactive ends (Chapter 6, Section 1.5). The NHS ester on one end of the reagent can be used to couple with primary amine-containing molecules, forming stable amide linkages (Chapter 3, Section 1.4). The other end contains an iodoacetyl group that is specific for coupling to sulfhydryl residues, potentially creating stable thioether bonds (Chapter 3, Section 2.1).

Conjugations using SIAB to create immunotoxins can be performed by first reacting the NHS ester end of the crosslinker with available amine groups on the antibody and then coupling to a thiolated toxin dimer—or by first reacting it with the toxin and coupling to a thiolated antibody (Figure 20.32). Thiolation of the secondary component is usually done with SPDP or 2-iminothiolane. Other crosslinkers containing an iodoacetyl group can be used in a similar fashion.

Conjugations with iodoacetyl crosslinkers have been done using ricin and cobra venom factor (Thorpe *et al.*, 1984; Vogel, 1987; Myers *et al.*, 1989). The following generalized protocol for using SIAB is based on the method of Cumber *et al.* (1985).

Protocol

Caution: Toxin molecules are dangerously toxic even in small amounts. Use extreme care in handling.

To prepare an antibody–ricin conjugate using this protocol, 2.25 mg of antibody is needed for every milligram of toxin.

(A) ACTIVATION OF TOXIN WITH SIAB

1. Dissolve intact ricin in 0.1-*M* sodium phosphate, 0.15-*M* NaCl, pH 7.5, at a concentration of 10 mg/ml.
2. Dissolve SIAB (Thermo Fisher) in DMSO at a concentration of 1.4 mg/ml. Prepare fresh and protect from light to avoid breakdown of the active halogen group.
3. Add 160 μl (225 μg) of the SIAB solution to each milliliter of the ricin solution.
4. React for 30 min at room temperature in the dark.
5. Remove excess crosslinker from the activated ricin by gel filtration using a desalting resin.
6. Concentrate the purified, SIAB-activated toxin to 10 mg/ml using centrifugal concentrators with a molecular weight cut-off of 10,000. Protect the activated toxin from light to prevent degradation of the iodoacetyl reactive group.

(B) THIOLATION OF SPECIFIC ANTIBODY MOLECULE WITH SPDP

1. Dissolve the antibody to be conjugated in 0.1-*M* sodium phosphate, 0.15-*M* NaCl, pH 7.5, at a concentration 10 mg/ml. Use 2.25 mg of antibody per milligram of toxin to be conjugated.
2. Dissolve SPDP (Thermo Fisher) in DMF at a concentration of 3 mg/ml. Add 24 μl of this solution to each milliliter of the antibody solution with gentle mixing to effect complete dissolution.
3. React for 30 min at room temperature.
4. Remove excess crosslinker by gel filtration using a column of desalting resin. Perform the

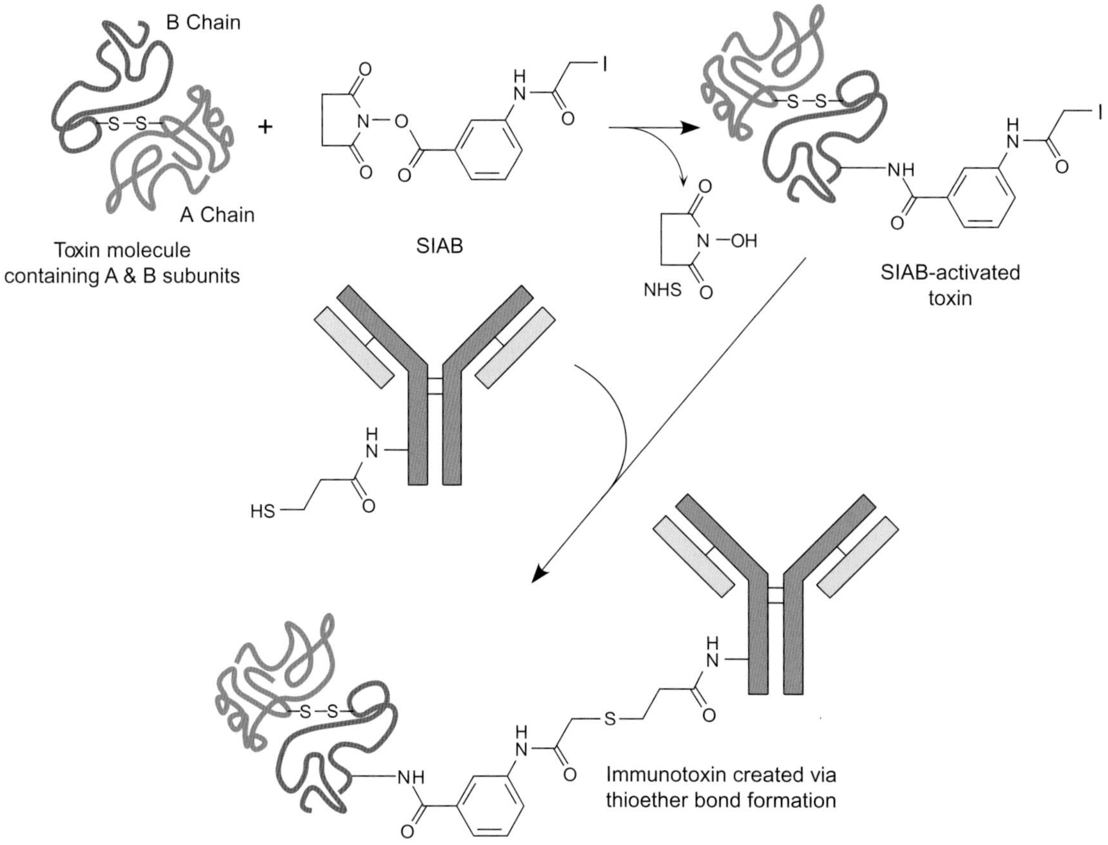

FIGURE 20.32 SIAB can be used to activate toxin molecules for coupling with sulfhydryl-containing antibodies. In this case, the antibody molecule is thiolated using SATA and deprotected to reveal the free sulfhydryl. Reaction with the SIAB-activated toxin forms the final conjugate by thioether bond formation.

chromatography using 0.1-*M* sodium phosphate, 0.15-*M* NaCl, 10-m*M* EDTA, pH 7.5. The buffer should be degassed under vacuum and nitrogen bubbled through it to remove oxygen. The presence of EDTA stabilizes the free sulfhydryls formed in the following steps against metal-catalyzed oxidation.

5. Concentrate the fractions containing protein from the gel filtration step to 10 mg/ml using a centrifugal concentrator (MW cut-off of 10,000).
6. To reduce the pyridyl dithiol groups and create reactive sulfhydryls, dissolve DTT (Thermo Fisher) in water at a concentration of 17.2 mg/ml and immediately add 500 μl of this solution to each milliliter of concentrated antibody solution. Mix to dissolve and react for 30 min at room temperature.
7. Remove excess DTT by gel filtration using the same buffer as in step 4. Pool the fractions containing protein and concentrate to 10 mg/ml.

(C) CONJUGATION OF SIAB-ACTIVATED TOXIN WITH THIOLATED ANTIBODY

1. Mix activated toxin from part A with thiolated antibody from part B at a ratio of 2.25 mg of antibody per milligram of toxin. Protect the solution from light.
2. React for 18 h at room temperature in the dark.
3. To block unreacted sulfhydryl groups, add iodoacetamide to the solution to a final concentration of 2 m*M*. React for an additional 1 h at room temperature.
4. Isolation of the ideal 1:1 or 1:2 antibody–toxin conjugate can be done by gel filtration separation using a column of Sephacryl S-300.

SMCC

Succinimidyl-4-(*N*-maleimidomethyl)cyclohexane-1-carboxylate (SMCC) is a crosslinker with significant utility in protein conjugation (Chapter 6, Section 1.3). It is a popular choice among heterobifunctional reagents, especially for the preparation of antibody–enzyme and hapten–carrier conjugates (Hashida and Ishikawa, 1985; Dewey *et al.*, 1987). The NHS ester end of the reagent can react with primary amine groups on proteins to form stable amide bonds. The maleimide end of SMCC is specific for coupling to sulfhydryls when the reaction pH is in the range of 6.5 to 7.5 (Smyth *et al.*, 1964). The

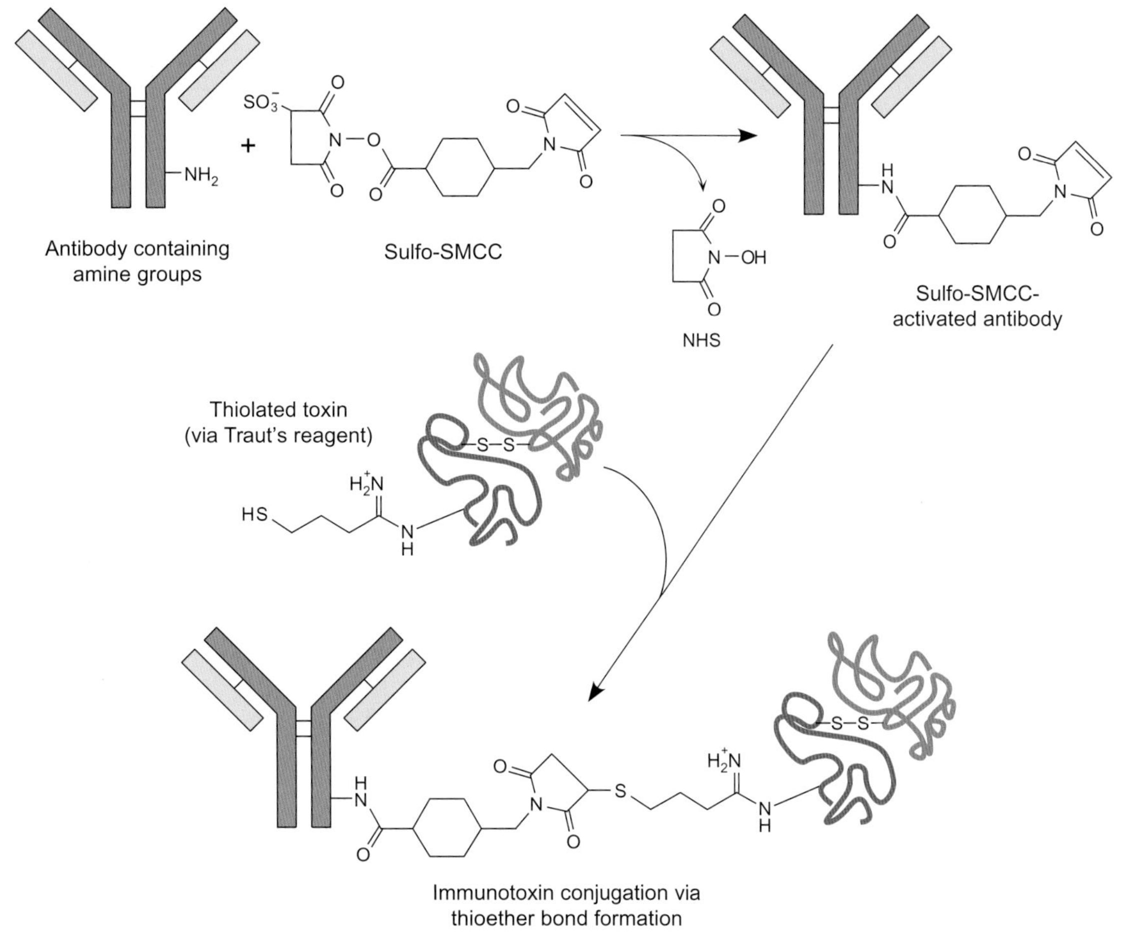

FIGURE 20.33 Sulfo-SMCC may be used to activate antibody molecules for coupling to thiolated toxin components. An intact A–B toxin molecule can be modified to contain sulfhydryls by treatment with 2-iminothiolane. Thiolation with this reagent retains the cytotoxic properties of the toxin while generating a sulfhydryl for conjugation. Reaction of the thiolated toxin with the maleimide-activated antibody creates the immunotoxin through thioether bond formation.

nature of the reactive groups of SMCC allow for highly controlled crosslinking procedures to be performed wherein the resulting products can be closely limited to a 1:1 ratio in the final complex. Thus, low-molecular-weight conjugates can be made which make ideal reagents for *in vivo* purposes.

However, since SMCC forms nonreversible thioether linkages with sulfhydryl groups, it can only be used in the preparation of immunotoxins if intact A–B toxins are employed in the conjugate. In such conjugates, the A chain still has the potential for reductive release from the B chain subunit after cellular docking and internalization. Immunotoxins prepared with A chain or single-subunit toxins will not display cytotoxicity if crosslinked with SMCC, since the crosslinker does not create cleavable disulfide bonds upon conjugation.

SMCC has been used to prepare immunotoxins with cobra venom factor (Vogel, 1987) and was compared to other crosslinkers in the preparation of gelonin and PAP conjugates (Lambert *et al.*, 1985).

The following protocol is a suggested method for the conjugation of SMCC-activated antibodies with 2-iminothiolane-modified, intact toxin molecules (Figure 20.33). It utilizes the water soluble analog of SMCC, sulfo-SMCC, which contains a negatively charged sulfonate group on its NHS ring.

Protocol

Caution: Toxin molecules are dangerously toxic even in small amounts. Use extreme care in handling.

Note: This protocol requires mixing activated antibody with thiolated toxin at a ratio of 2.25 mg of antibody per milligram of toxin. This ratio should be taken into account before starting the reactions.

(A) ACTIVATION OF ANTIBODY WITH SULFO-SMCC

1. Dissolve 10 mg of specific antibody in 1 ml of 0.1-*M* sodium phosphate, 0.15-*M* NaCl, pH 7.2.
2. Weigh out 2 mg of sulfo-SMCC (Thermo Fisher) and add it to the above solution. Mix gently to dissolve.

To aid in measuring the exact quantity of crosslinker, a concentrated stock solution may be made in water and an aliquot equal to 2 mg transferred to the reaction solution. If a stock solution is made, it should be dissolved rapidly and used immediately to prevent extensive hydrolysis of the active ester. Alternatively, a stock solution of sulfo-SMCC may be prepared in DMSO and an aliquot added to the aqueous reaction.
3. React for 1 h at room temperature.
4. Immediately purify the maleimide-activated protein by applying the reaction mixture to a desalting column. Do not dialyze the solution, since the maleimide activity will be lost over the time course required to complete the operation. To obtain good separation between the protein peak (eluting first) and the peak representing excess reagent and reaction byproducts (eluting second), the applied sample size should be no more than 5 to 8% of the column bed volume.
5. Collect the peak containing the activated antibody (eluting first) and concentrate to 10 mg/ml using centrifugal concentrators. Use immediately for conjugating to a thiolated toxin.

(B) THIOLATION OF INTACT A–B TOXIN

1. Dissolve the toxin (e.g., intact ricin) at a concentration of 10 mg/ml in 50-m*M* triethanolamine hydrochloride, pH 8.0, containing 10-m*M* EDTA. The buffer should be degassed under vacuum and bubbled with nitrogen to remove oxygen that may cause sulfhydryl oxidation after thiolation.
2. Dissolve 2-iminothiolane in degassed, nitrogen-bubbled deionized water at a concentration of 20 mg/ml (makes a 0.14-*M* stock solution). The solution should be used immediately. Add 70 μl of the 2-iminothiolane solution to each milliliter of the toxin solution (final concentration is about 10-m*M*).
3. React for 1 h at 0°C (on ice) under a nitrogen blanket.
4. Purify the thiolated toxin from unreacted Traut's reagent by gel filtration using 0.1-*M* sodium phosphate, 0.15-*M* NaCl, pH 7.5, containing 10-m*M* EDTA. The presence of EDTA in this buffer helps to prevent oxidation of the sulfhydryl groups with resultant disulfide formation. The degree of SH modification in the purified protein may be determined using the Ellman's assay (Chapter 2, Section 4.1).
5. Concentrate the thiolated toxin to 10 mg/ml using centrifugal concentrators.

(C) CONJUGATION OF SMCC-ACTIVATED ANTIBODY WITH THIOLATED TOXIN

1. Mix the thiolated toxin with SMCC-activated antibody at a ratio of 2.25 mg of antibody per milligram of toxin. Protect the solution from light.
2. React for 18 h at room temperature.
3. To block unreacted sulfhydryl groups, add iodoacetamide to the solution to a final concentration of 2 mM. React for an additional 1 h at room temperature.
4. Isolation of the ideal 1:1 antibody–toxin conjugate can be done by gel filtration separation using a column of Sephacryl S-300.
5. Removal of nonspecific binding potential in the B chain must be done before using an A–B intact toxin conjugate *in vivo*. A large proportion of the binding sites on the B chains are usually blocked during the above conjugation process, and the galactose binding potential is significantly impaired. Further purification to remove conjugates that have galactose binding potential can be performed on an acid-treated agarose chromatography column (which contains galactose residues) or on a column of asialofetuin bound to agarose (Cumber *et al.*, 1985). Conjugate fractions that do not bind to both affinity gels contain no nonspecific binding potential toward non-targeted cells.

MBS

m-Maleimidobenzoyl-*N*-hydroxysuccinimide ester (MBS) is a heterobifunctional crosslinking agent containing an NHS ester and a maleimide group. The NHS ester can react with primary amines in proteins and other molecules to form stable amide bonds, while the maleimide end reacts with sulfhydryl groups to create stable thioether linkages (Chapter 6, Section 1.4). The reagent can be used in many different conjugation protocols to crosslink amine-containing proteins with sulfhydryl-containing proteins. Since the thioether bond formed at the maleimide end is nonreversible, MBS can be used for immunotoxin preparation only if the conjugate involves crosslinking intact A–B toxins with antibody molecules. Using intact toxins (as opposed to single-chain or A chain isolates), the A chain still is able to release from the complex after cellular docking and inactivate ribosomal activity (Youle and Nevelle, 1980; Dell'Arciprete *et al.*, 1988; Myers *et al.*, 1989).

MBS contains a benzoic acid derivative as its cross-bridge. In many applications involving NHS–maleimide crosslinkers, non-aromatic cross-bridges are considered superior to aromatic ones. This is reflected in the stability of the maleimide group to hydrolysis prior to conjugating with a sulfhydryl group. For immunotoxin preparation, however, aromatic maleimides resulted in better conjugate yield and more potent cytotoxic effects when compared to aliphatic ones (Myers *et al.*, 1989). MBS, therefore, may be

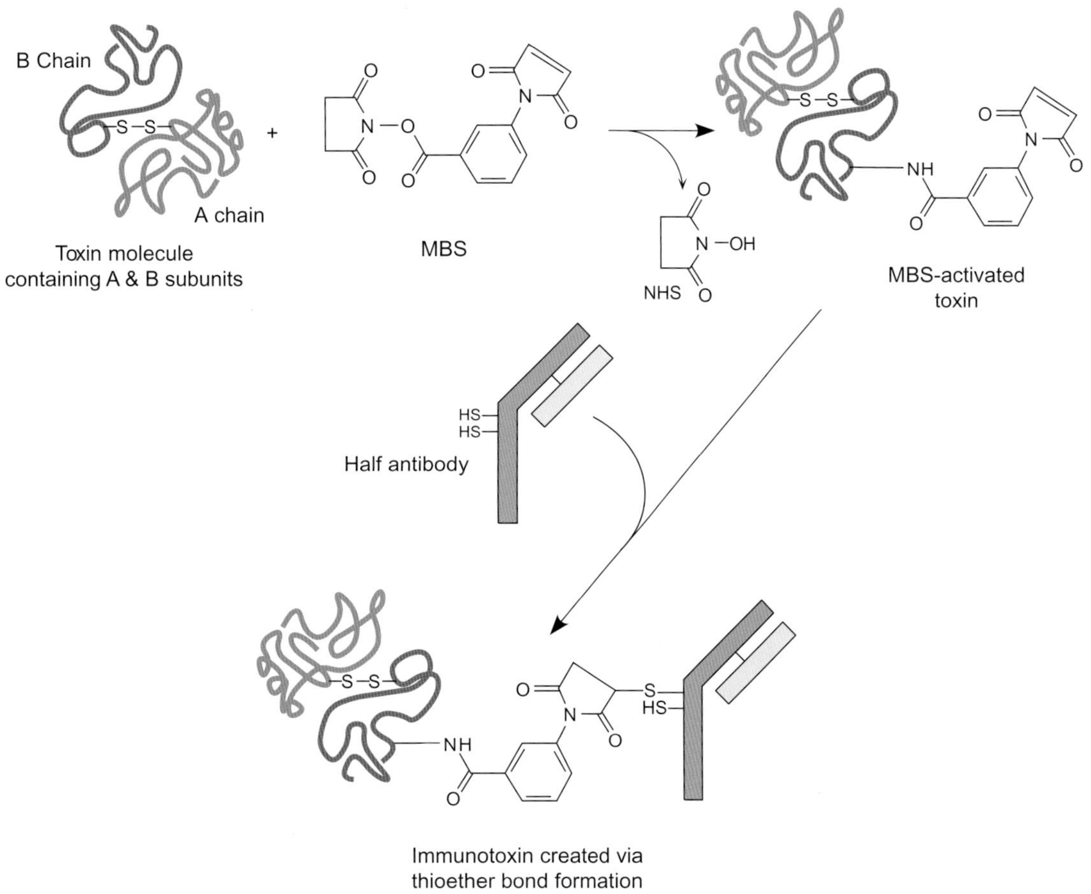

FIGURE 20.34 Activation of an intact A–B toxin molecule with MBS with subsequent conjugation with a reduced antibody fragment to produce an immunotoxin.

a crosslinker of choice when making conjugates with intact toxin molecules.

The following protocol is adapted from Myers *et al.* (1989). It involves activation of ricin with MBS and conjugation with a partially reduced antibody (Figure 20.34).

Protocol

Caution: Ricin molecules are dangerously toxic even in small amounts. Use extreme care in handling.

This method uses a molar ratio of 15:1 for ricin:antibody. This requires 6.24 mg of ricin per milligram of antibody. This ratio should be considered when determining how much starting materials to use for each step.

(A) ACTIVATION OF RICIN WITH MBS

1. Dissolve ricin at a concentration of 10 mg/ml in 0.1-*M* sodium phosphate, 0.15-*M* NaCl, pH 7.5.
2. Dissolve MBS (Thermo Fisher) in DMF at a concentration of 2 mg/ml.
3. Add 76 µl of the MBS solution to each ml of the ricin solution. This represents a 3:1 molar ratio of crosslinker to protein.
4. React for 30 min at room temperature.
5. Immediately purify the MBS-activated toxin by gel filtration using a column of desalting resin. Apply no more sample than represents 5 to 8% of the gel volume. Isolate the protein peak by its absorbance at 280 nm and concentrate to 10 mg/ml using centrifugal concentrators with a molecular weight cut-off of 10,000.

(B) PARTIAL REDUCTION OF ANTIBODY WITH DTT

1. Dissolve the antibody in 0.1-*M* sodium phosphate, 0.15-*M* NaCl, 10-m*M* EDTA, pH 7.5, at a concentration of 10 mg/ml.
2. Add DTT to a final concentration of 50 m*M*. Note: Lower concentrations of DTT may be used to avoid extensive reduction that may dissociate the antibody subunits. Concentrations in the range of 3- to 5-m*M* DTT have been shown to be appropriate for the reduction of just a few disulfides within the antibody structure (Sun *et al.*, 2005). Some optimization of the reductant concentration may have to be performed to obtain the best results for a particular antibody molecule.

3. Reduce for 30 min at room temperature.
4. Purify the reduced antibody using gel filtration on a column of Sephadex G-25. Concentrate the protein to 10 mg/ml using centrifugal concentrators.

(C) CONJUGATION OF MBS-ACTIVATED RICIN WITH PARTIALLY REDUCED ANTIBODY

1. Mix the MBS-activated ricin with the partially reduced antibody in a molar ratio of 15:1 (or 6.24 mg activated ricin per mg of reduced antibody). This represents a volume ratio (at 10 mg/ml for both proteins) of 1 ml ricin solution mixed with 160 μl antibody solution.
2. React for 18 h at room temperature.
3. Purification of the immunotoxin conjugate from unconjugated ricin can be performed using a column of TSK3000 SW (Toya Soda, Japan) according to the method of Myers *et al.* (1989).
4. Removal of nonspecific binding potential in the B chain must be done before using an A–B intact toxin conjugate *in vivo*. See step 5 of the MBS conjugation protocol discussed previous to this section.

SMPB

Succinimidyl-4-(*p*-maleimidophenyl)butyrate (SMPB) is a heterobifunctional analog of MBS containing an extended cross-bridge (Chapter 6, Section 1.6). The crosslinker has an amine-reactive NHS ester on one end and a sulfhydryl-reactive maleimide group on the other. Conjugates formed using SMPB thus are linked by stable amide and thioether bonds.

As in the case of MBS, discussed previously, SMPB was found to be more effective than aliphatic crosslinkers in producing immunotoxin conjugates with ricin that have high yields of cytotoxicity (Myers *et al.*, 1989). This was attributed to the reagent's aromatic ring structure. A comparison with SPDP produced immunotoxin conjugates concluded that SMPB formed more stable complexes that survive in serum for longer periods (Martin and Papahadjopoulos, 1982).

The method for the preparation of immunotoxins with SMPB is identical to that used for MBS (above). Since the thioether bonds formed with sulfhydryl-containing molecules are non-cleavable, A chain isolated fractions or single-chain toxin molecules cannot be conjugated with antibodies with retention of cytotoxicity. Only intact A–B toxin molecules may be used with this crosslinker, since the A chain is still capable of being reductively released from the complex.

3.3. Preparation of Immunotoxin Conjugates *via* Reductive Amination

Conjugations involving aldehyde groups and amine-containing molecules can be performed through Schiff base formation with subsequent reduction using sodium cyanoborohydride to form stable secondary amine linkages (Chapter 3, Section 5.3). Carbohydrates, glycoproteins, and other polysaccharide-containing molecules can be oxidized to contain aldehyde residues by sodium periodate or specific oxidases (Chapter 2, Section 4.4). Some antibodies and toxin molecules are glycoproteins and contain sufficient carbohydrate to be utilized for reductive amination crosslinking.

A second method of immunotoxin preparation by reductive amination involves the use of a polysaccharide spacer. Soluble dextran may be oxidized with periodate to form a multifunctional crosslinking polymer. Reaction with antibodies and cytotoxic molecules in the presence of a reducing agent forms multivalent immunotoxin conjugates. The following sections discuss these options.

Periodate Oxidation of Glycoproteins Followed by Reductive Conjugation

Antibody molecules usually contain carbohydrate in their Fc regions. Similarly, many toxins, such as ricin and abrin, are glycoproteins that contain abundant polysaccharide. These carbohydrate residues can be oxidized with 10-m*M* sodium periodate to form reactive aldehyde groups capable of being conjugated with primary amines (Chapter 2, Section 4.4). Mixing an aldehyde-containing glycoprotein with another amine-containing molecule in the presence of sodium borohydride or sodium cyanoborohydride reduces the intermediate Schiff bases that are formed to stable secondary amine bonds. Since functional groups on the antibody and the toxin components are the only ones necessary for this type of conjugation strategy, it is often referred to as a zero-length crosslinking procedure (Chapter 4). In other words, no additional crosslinking reagents are introduced into the site of the crosslink. This method of conjugation is used with great success in the formation of antibody–enzyme conjugates, especially using the glycosylated enzyme, horseradish peroxidase (HRP) (Chapter 22, Section 1.1).

The disadvantage of this type of conjugation approach for producing immunotoxins is that many of the monoclonal antibodies or antibody fragments used for immunotoxin conjugation are devoid of carbohydrate. Especially when using small Fv fragments or single-chain antibodies produced by recombinant techniques, there are typically no polysaccharide portions attached to them. In this case, creation of aldehydes on the targeting component is not possible. In addition, not all toxin molecules contain carbohydrate. Ricin, abrin, and cobra venom factor are glycoproteins and can be oxidized and coupled to antibodies without difficulty (Olsnes and Pihl, 1982a,b; Vogel and Muller-Eberhard, 1984). However, it is not well known whether

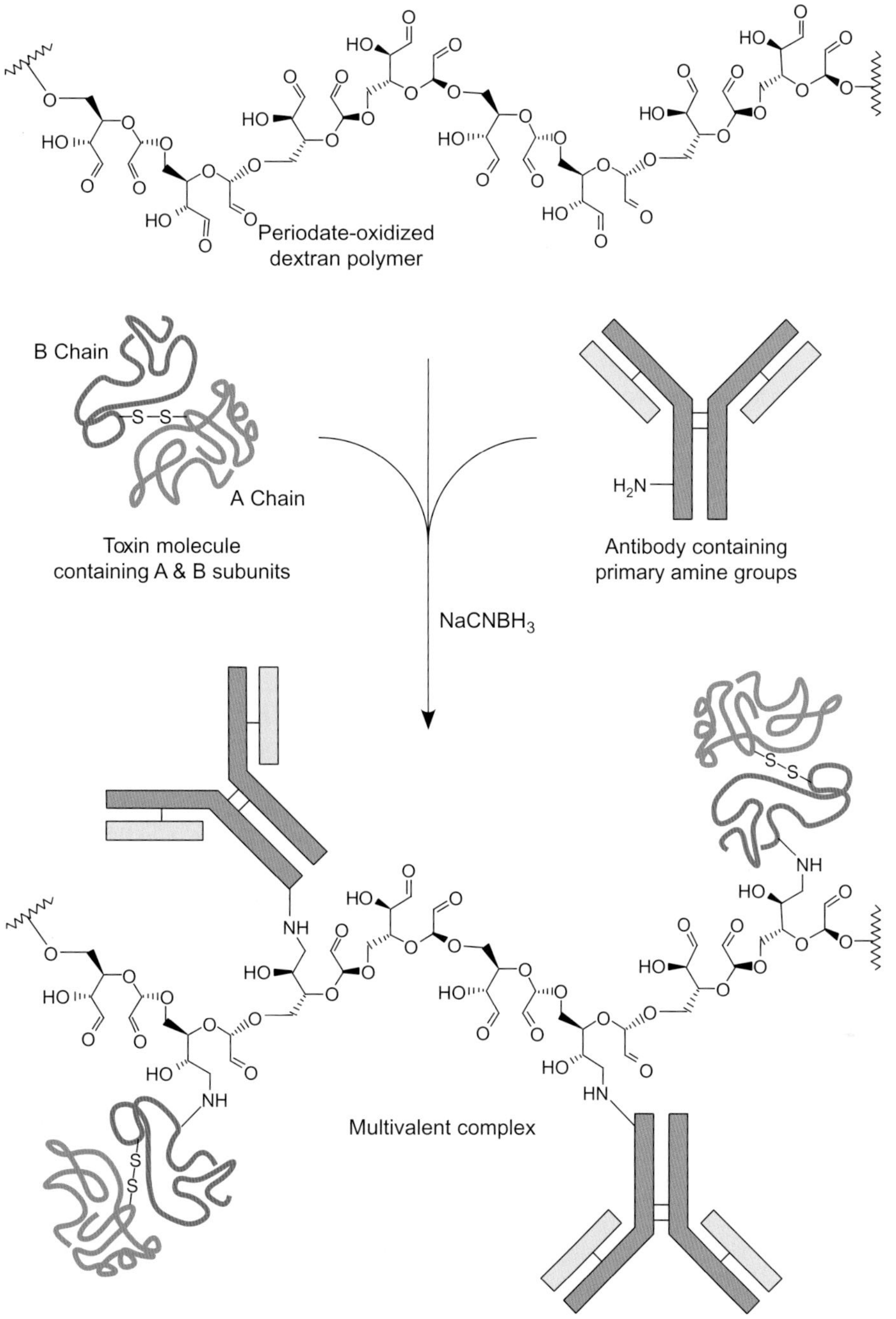

FIGURE 20.35 A periodate-oxidized dextran polymer may be reacted with both an antibody and an intact toxin component using reductive amination to form a multivalent immunotoxin complex.

immunotoxin conjugates formed by this procedure retain their ability to inhibit ribosomal activity.

Suggested procedures for using reductive amination techniques may be found in Chapter 2, Section 4.4 and Chapter 4, Section 4.

Periodate Oxidized Dextran as Crosslinking Agent

Dextran polymers consist of glucose residues bound together predominantly in α-1,6 linkages. The main repeating unit is an isomaltose group. Most preparations of dextran contain some branching, mainly incorporating 1,2, 1,3, and 1,4 linkages. The degree of branching is characteristic of its source—the strain and species of yeast or bacteria from which the dextran originated. The terminating monosaccharide in a dextran polymer is often a fructose group. Dextran polymers of molecular weight 10,000 to 40,000 provide long, hydrophilic arms that can accommodate multiple attachment points for macromolecules along their length. Soluble dextrans can be oxidized in aqueous solution to create numerous aldehyde residues suitable for use in reductive amination techniques (Hurwitz *et al.*, 1978, 1985; Manabe *et al.*, 1983; Sela and Hurwitz, 1987). Periodate oxidation results in the cleavage of the carbon–carbon bonds between the Nos. 2 and 3 carbons within each monosaccharide unit of the chain, transforming the associated hydroxyl groups into aldehydes (Chapter 2, Section 4.4).

Periodate oxidized dextran can be used as a protein modification or crosslinking agent (Chapter 18, Section 2.2). Conjugation of antibody molecules to toxins can be performed with dextran to produce immunotoxins suitable for *in vivo* administration. Mixing of the antibody and toxin together with the oxidized dextran under alkaline conditions will result in the formation of Schiff base interactions with the amines on both proteins. Reduction of these Schiff base linkages with sodium borohydride or sodium cyanoborohydride results in stable secondary amine bonds, covalently attaching multiple antibody and toxin molecules along the length of the polysaccharide chain (Figure 20.35).

Chemoimmunoconjugates or antibody–drug conjugates (ADCs) consisting of anticancer agents attached to antibody targeting molecules also can be formed using oxidized dextran carriers. Cancer therapeutic agents such as adriamycin, bleomycin, and daunomycin can be coupled to the oxidized dextran through their amine groups (also see Chapter 1, Section 3.5). After formation of Schiff base linkages between these drugs and the carrier, the antibody is added along with a reducing agent to create the final secondary amine linkages (Sela and Hurwitz, 1987). The dextran backbone provides the ability to couple many more drug molecules associated with each antibody than could be accomplished by direct conjugation to the antibody itself.

Although dextran can be a versatile crosslinking agent for the preparation of many forms of macromolecular conjugates, immunotoxin conjugation may be impeded by the non-reversibility of the multiple amide bond linkages formed during reductive amination. Certainly, only intact A–B toxins have a chance of succeeding with this method, since A chain or single-subunit toxins would not be capable of release from the complex after cellular docking. Even intact two-subunit toxins, however, may not be capable of releasing an A chain unit, due to the multivalent nature of the oxidized dextran linker. For this reason, activated dextran may be more useful for constructing antibody conjugates consisting of cytotoxic organic compounds, not protein toxins—for example, chemotherapeutic drugs, hormones, or radioactive complexes.

Methods for using oxidized dextran and reductive amination techniques can be found in Chapter 3, Section 4.4; Chapter 3; Section 5; and especially Chapter 18, Section 2.2. Reference should also be made to the use of dendrimers as carriers for making cytotoxic targeting complexes (Chapter 8) as well as Chapter 1, Section 3.5, which provides a broad overview of the types and applications of therapeutic bioconjugates used today.

CHAPTER

21

Liposome Conjugates and Derivatives

A fast growing field that heavily depends on bioconjugate technology involves the use of liposomes. At one time, liposomes were studied only for their interesting structural characteristics in solution. Their physicochemical properties were investigated extensively as models of membrane morphology. Now they are being put to use as macromolecular carriers for nearly every application of bioconjugate chemistry. They are used as delivery devices to encapsulate cosmetics, drugs, fluorescent detection reagents, and as vehicles to transport nucleic acids, peptides, and proteins to cellular sites *in vivo* (Sawant and Torchilin, 2011; Karchemiski *et al.*, 2012; Lu and Low, 2012). Targeting components such as antibodies can be attached to liposomal surfaces and used to create large antigen-specific complexes (Smith *et al.*, 2011; Kirpotin *et al.*, 2012). In this sense, liposomal derivatives are being used to target cancer cells *in vivo*, to enhance detectability in immunoassay systems, and as multivalent scaffolds in avidin–biotin-based assays. Covalent attachment of antigens to the surface of liposomes provides excellent immunogen complexes for the generation of specific antibodies or as vaccine carriers to elicit protective immunity (Amidi *et al.*, 2011).

The end-products of liposome technology are used in retail markets, for the diagnosis of disease, as therapeutic agents, as vaccines, and as important components in assays designed to either detect or quantify certain analytes.

The following sections discuss the properties and applications of liposome technology as well as the most common methods of preparing conjugates of them with proteins and other molecules.

1. PROPERTIES AND USE OF LIPOSOMES

1.1. Liposome Morphology

Liposomes are artificial structures primarily composed of phospholipid bilayers exhibiting amphiphilic properties. Other molecules, such as cholesterol or fatty acids may also be included in the bilayer construction. In complex liposome morphologies, concentric spheres or sheets of lipid bilayers are usually separated by aqueous regions that are sequestered or compartmentalized from the surrounding solution. The phospholipid constituents of liposomes consist of hydrophobic lipid "tails" connected to a "head" constructed of various glycerylphosphate derivatives. The hydrophobic interaction between the fatty acid tails to the exclusion of the aqueous phase is the primary driving force for creating liposomal bilayers in aqueous solution.

However, the organization of liposomes in aqueous solution may be highly complicated. The nature of the lipid constituents, the composition of the medium, and the temperature of the solution all affect the association and morphology of liposomal construction. Small "monomers" or groupings of lipid molecules may assemble to create larger structures having several main forms (Figure 21.1). Aggregation of these monomers may fuse them into spherical micelles, wherein the polar head groups are all facing outward toward the surrounding aqueous medium and the hydrophobic tails are all pointing inward, thus excluding water from the core. In addition, aggregation may result in bilayer construction. In this case, sheets of lipid molecules, all with their head groups facing one direction and their tails facing the other way, are fused with another lipid sheet having their tails and heads facing the opposite direction. Thus, the inside of the bilayer contains only hydrophobic tails from both sheets, while the outside contains the hydrophilic heads facing the outer aqueous environment.

The various configurations that the bilayers can assume can also be complex. A bilayer may be in a spherical form, having one layer of hydrophilic head groups pointing outward toward the surrounding solution and the second hydrophilic layer pointing inward toward a compartment of aqueous solution sequestered within the sphere. The morphology of a liposome may be classified according to the compartmentalization of aqueous regions between bilayer shells. If the

Bioconjugate Techniques, Third Edition.
DOI: http://dx.doi.org/10.1016/B978-0-12-382239-0.00021-2

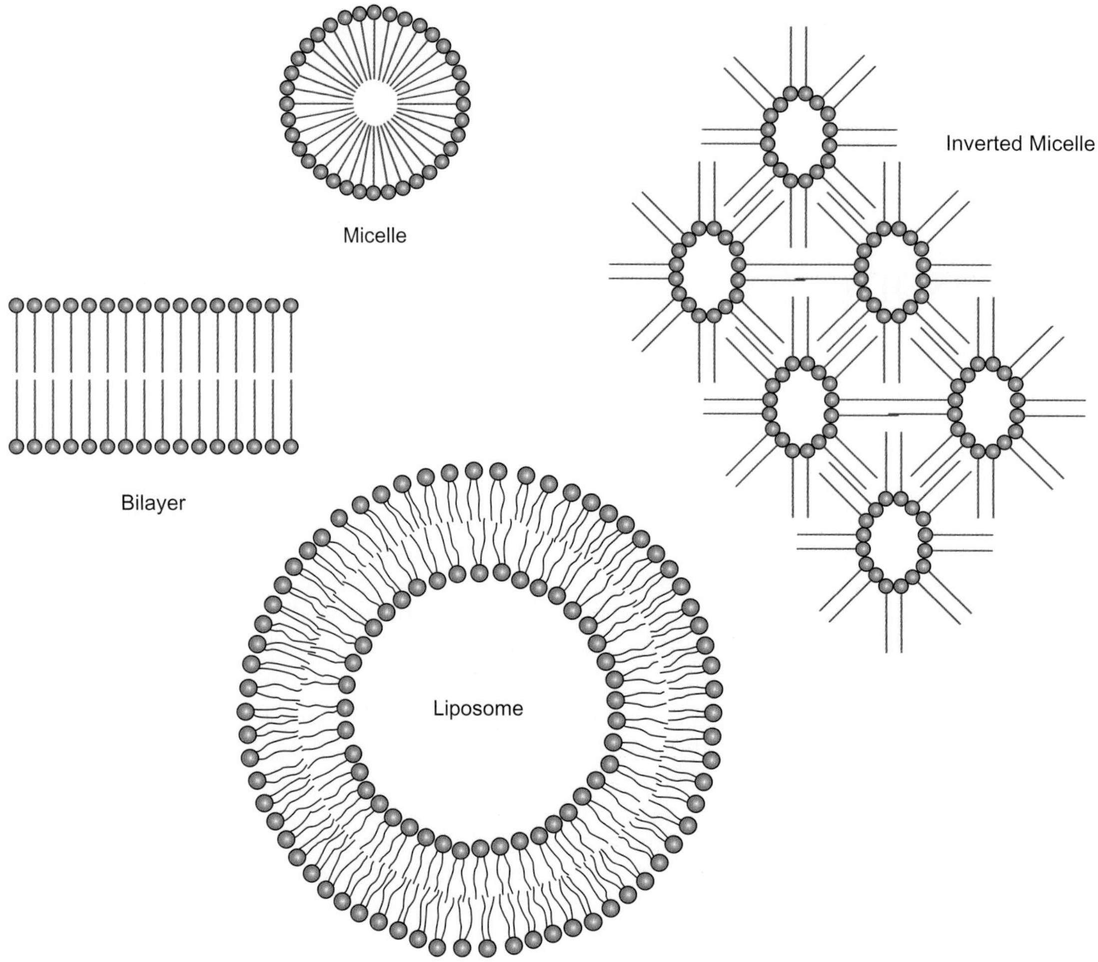

FIGURE 21.1 The amphiphilic nature of phospholipids in solution drives the formation of complex structures. Spherical micelles may form in aqueous solution, wherein the hydrophilic head groups all point out toward the surrounding water environment and the hydrophobic tails point inward to the exclusion of water. Larger lipid bilayers may form by similar forces, creating sheets, spheres, and other highly complex morphologies. Contiguous bilayer constructions, especially spheres, are called liposomes. In nonaqueous solution, inverted micelles may form, wherein the tails all point toward the outer hydrophobic region and the heads point inward forming hexagonal shapes. The yellow shaded areas represent the hydrophobic regions that are occupied by the aliphatic tails of the lipid molecules.

aqueous regions are segregated by only one bilayer each, the liposomes are called unilamellar vesicles (ULV) (Figure 21.2). If there is more than one bilayer surrounding each aqueous compartment, the liposomes are termed multilamellar vesicles (MLV). ULV forms are further classified as to their relative size, although rather crudely. Thus, there can be small unilamellar vesicles (SUV; usually thought of as less than 100 nm in diameter, with a minimum of about 25 nm) and large unilamellar vesicles (LUV; usually greater than 100 nm in diameter, with a maximal size of about 2500 nm). With regard to MLV, however, the bilayer structures cannot be as easily classified due to the almost infinite number of ways each bilayer sheet can be associated and interconnected with the next one. MLVs typically form large complex honeycomb structures that are difficult to categorize or exactly reproduce. However MLVs are the simplest to prepare, most stable, and easiest to scale up to large production levels.

Small lipid groupings or monomers may also fuse into bigger, inverted micelles, wherein their hydrophobic tails point outward toward other inverted micelle lipid tails. The individual structures are usually hexagonal in shape, but typically they exist as large groupings of inverted micelles, the outer edge of which contains partial inverted micelles, exposing their inner hydrophilic heads to the surrounding aqueous environment.

The most useful form of liposomes for bioconjugate applications consists of small, spherical ULVs that possess layers of hydrophilic head groups on their inner and outer surfaces. The inside of each vesicle can contain hydrophilic molecules that are protected from the outer environment by the lipid shell. The outside surface can be derivatized to contain covalently attached

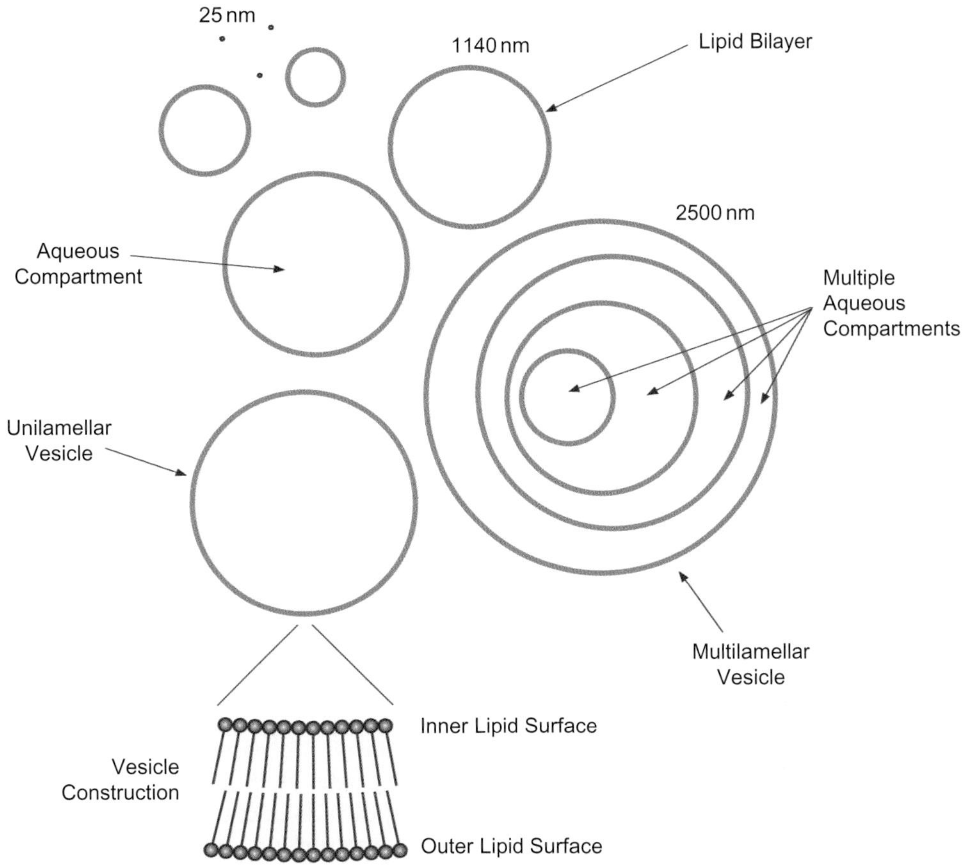

FIGURE 21.2 The highly varied morphologies of lipid bilayer construction.

molecules designed to target the liposome for specific interactions. Modification of the outer liposomal surface provides a multivalent scaffold that can accommodate many different molecules conjugated to it, including antibodies, detection molecules, high contrast agents, enzymes, drugs, or other molecules having some specific activity or function.

1.2. Preparation of Liposomes

Mixtures of phospholipids in aqueous solution will spontaneously associate to form liposomal structures. To prepare liposomes having morphologies useful for bioconjugation or delivery techniques, it is necessary to control this assemblage to create vesicles of the proper size and shape. Many methods are available to accomplish this goal, however all of them have several steps in common: (1) dissolving the lipid mixture in organic solvent, (2) dispersion of the lipid solution in an aqueous phase, and (3) fractionation to isolate the correct liposomal population.

In the first stage, the desired mix of lipid components is dissolved in organic solvent, usually chloroform:methanol (2:1 by volume), to create a homogeneous mixture. This mixture will include any phospholipid derivatized to contain reactive groups as well as other lipids used to form and stabilize the bulk of the liposomal structure. During all handling procedures using lipids or their derivatives, it is essential that the solutions be protected from oxidation and excessive exposure to light, especially the sun. Organic solvents should be maintained under a nitrogen or argon atmosphere to prevent introduction of oxygen. Water and buffers should be degassed using a vacuum and bubbled with inert gas before introducing lipid components.

The correct ratio of lipid constituents is important to form stable liposomes. For instance, a reliable liposomal composition for encapsulating aqueous substances may contain molar ratios of lecithin:cholesterol:negatively charged phospholipid (e.g., phosphatidyl glycerol (PG))) of 0.9:1:0.1. A composition that is typical when an activated phosphatidylethanolamine (PE) derivative is included may contain molar ratios of phosphatidylcholine (PC):cholesterol:PG:derivatized PE of 8:10:1:1. Another typical composition using a maleimide derivative of PE without PG is PC:maleimide-PE:cholesterol of 85:15:50 (Friede *et al.*, 1993). In general, to maintain membrane stability, the PE derivative should not exceed a concentration ratio of about 1 to 10 mol PE per 100 mol of total lipid.

An example of a lipid mixture preparation based on mass would be to dissolve 100 mg of PC, 40 mg of cholesterol, and 10 mg of PG in 5 ml of chloroform/methanol solution. When using activated PE components, inclusion of 10 mg of the PE derivative to this recipe will result in a stable liposome preparation.

Once the desired mixture of lipid components is dissolved and homogenized in organic solvent, one of several techniques may be used to disperse the liposomes in aqueous solution. These methods may be broadly classified as (1) mechanical dispersion, (2) detergent-assisted solubilization, and (3) solvent-mediated dispersion.

Probably the most popular option is mechanical dispersion, simply because the greatest number of methods that utilize it have been developed. When using mechanical means to form vesicles, the lipid solution is first dried to remove all traces of organic solvent prior to dispersion in an aqueous media. The dispersion process is the key to producing liposomal membranes of the correct morphology. This method uses mechanical energy to break up large lipid agglomerates into smaller vesicles having the optimal size and shape characteristics necessary for encapsulation or bioconjugation.

Mechanical dispersion methods involve adding an aqueous solution (which may contain substances to be encapsulated) to the dried, homogeneous lipid mixture and manipulating it to effect dispersion. Major methods of mechanical dispersion include simple shaking (Bangham *et al.*, 1965), non-shaken aqueous contact (Reeves and Dowben, 1969), high-pressure emulsification (Mayhew *et al.*, 1984), water-in-oil emulsions (Wang *et al.*, 2011), sonication (Huang, 1969), extrusion through small-pore membranes (Olson *et al.*, 1980; Morton *et al.*, 2012), various freeze–thaw techniques (Pick, 1981), and the use of supercritical solvents such as carbon dioxide (Aburai *et al.*, 2011). Some devices are available commercially which automate the mechanical dispersion process, usually by high-pressure emulsification or sonication (Branson Ultrasonics Corp.).

Most of these methods result in a population of vesicles ranging from SUVs of only 25-nm diameter to very large MLVs. Classification of the desired liposomal morphology may be performed by chromatographic means using columns of Sepharose 2B or Sepharose 4B, by density-gradient centrifugation using Ficoll or metrizamide gradients, or by dialysis.

Liposome formation by detergent-assisted solubilization utilizes the amphipathic nature of detergent molecules to bring the lipid components more effectively into the aqueous phase for dispersion. The detergent molecules presumably bind and mask the hydrophobic tails of lipids from the surrounding water molecules. Detergent treatment may take place from a dried lipid mixture or after formation of small vesicles. Usually, nonionic detergents such as the Triton X family, alkyl glycosides, or bile salts such as sodium deoxycholate are employed for this procedure. The immediate structures which form as the detergent molecules solubilize the lipids from a dried state are small micelles. Upon removal of the detergent from the solution, the lipid micelles aggregate to create larger liposome structures. Liposomes of up to 1000 Å containing a single bilayer may be formed using detergent-assisted methods (Enoch and Strittmatter, 1979; Richmond *et al.*, 2011). Unfortunately, some detergent-removal processes may also remove other molecules that were to be entrapped in the liposomes during formation.

Solvent-mediated dispersion techniques used to create liposomes first involve dissolving the lipid mixture in an organic solvent to create a homogeneous solution, and then introducing this solution into an aqueous phase. The solvent may or may not be soluble in the aqueous phase to effect this process. There may also be components dissolved in the aqueous phase to be encapsulated in the developing liposomes.

Perhaps the simplest solvent dispersion method is that developed by Batzri and Korn (1973). Phospholipids and other lipids to be a part of the liposomal membrane are first dissolved in ethanol. This ethanolic solution is then rapidly injected into an aqueous solution of 0.16-*M* KCl using a syringe, resulting in a maximum concentration of no more than 7.5% ethanol. Using this method, single bilayer liposomes of about 25 nm diameter can be created that are indistinguishable from those formed by mechanical sonication techniques. The main disadvantages of ethanolic injection are the limited solubility of some lipids in the solvent (about 40-m*M* for PC) and the dilute nature of the resultant liposome suspension. However, for the preparation of small quantities of SUVs, this method may be one of the best available.

Other solvent dispersion methods utilize solvents that are insoluble in the aqueous phase. The key to the production of liposomes by this procedure involves the formation of a "water-in-oil" emulsion. To create the proper reverse-phase emulsion, a small quantity of aqueous phase must be introduced into a large quantity of organic phase containing the dissolved liposomes. The result is a milky dispersion containing the "homogenized" liposomes. A number of techniques have been developed to perform this procedure (Kim and Martin, 1981; Kim *et al.*, 1983; Pidgeon *et al.*, 1986). The emulsification process in each of these solvent-dispersion techniques involves the use of mechanical means (shaking, stirring, or sonication) to effect the formation of small droplets of aqueous solution uniformly dispersed in the lipid–organic phase.

For the preparation of large quantities of liposomes, mechanical dispersion using a commercially available emulsifier is probably the best route. For limited quantities, the use of simple shaking or ethanolic dispersion techniques works well.

Regardless of their method of fabrication, most liposome preparations need to be further classified and purified before use. To remove excess aqueous components that were not encapsulated during the vesicle formation process, gel filtration using a column of Sephadex G-50 or dialysis can be employed. To fractionate the liposome population according to size, gel filtration using a column of Sepharose 2B or 4B should be performed.

Small liposome vesicles often aggregate upon standing to form larger, more complex structures. Therefore, long-term storage in aqueous solution is usually not possible without major transformations in liposome morphology. Freezing also fractures the liposomal membrane, releasing any entrapped substances. The inclusion of cryoprotectants such as sugars or poly-hydroxyl-containing compounds can overcome the structural degradation problems upon freezing (Harrigan *et al.*, 1990; Talsma *et al.*, 1991; Park and Huang, 1992). Presumably, the hydroxyl groups in cryoprotectants can take the place of water in hydrogen bonding activities, thus providing structural support even under conditions in which water is removed. A procedure by Friede *et al.* (1993) allows freezing and lyophilization of SUVs in the presence of 4% sorbitol with complete retention of liposome integrity upon reconstitution. Thus, freeze-drying may be the best method for the long-term storage of intact liposomes.

For a review of liposome manufacturing methods and their use in targeting biomarkers to deliver active agents *in vivo*, see Maherani *et al.* (2011).

1.3. Chemical Constituents of Liposomes

The overall composition of a liposome—its morphology, chemical constituents (including a large variety of phospholipids and other lipids or fatty acids), charge, and any attached functional groups—can affect the properties of the vesicle both *in vitro* and *in vivo* (Allison and Gregoriadis, 1974; Alving, 1987; Therien and Shahum, 1989). Although there are literally dozens of lipid components that potentially can be included in a liposomal recipe, only a handful are commonly used.

Phospholipids are the most important of these liposomal constituents. Being the major component of cell membranes, phospholipids are composed of a hydrophobic, fatty acid tail and a hydrophilic head group. The amphipathic nature of these molecules is the primary force that drives the spontaneous formation of bilayers in aqueous solution and holds the vesicles together.

Naturally occurring phospholipids can be isolated from a variety of sources. One of the most common phospholipid raw materials is egg yolk. However, since the composition of egg phospholipid is from a biological source and can vary considerably depending on age of the eggs, the diet of the chickens, and the method of processing, newer enzymatic and synthetic chemical methods are now being employed to manufacture the required phospholipid derivatives in higher purity and yield.

Two main forms of lipid derivatives exist biologically: molecules containing a glycerol backbone and those containing a sphingosine backbone. The most important type for liposomal construction is a phosphodiglyceride derivative, which consists of a glycerol backbone that links two fatty acid molecules with a polar head group (Figure 21.3). The fatty acids are acyl bonded in ester linkages to the Nos. 1 and 2 carbon

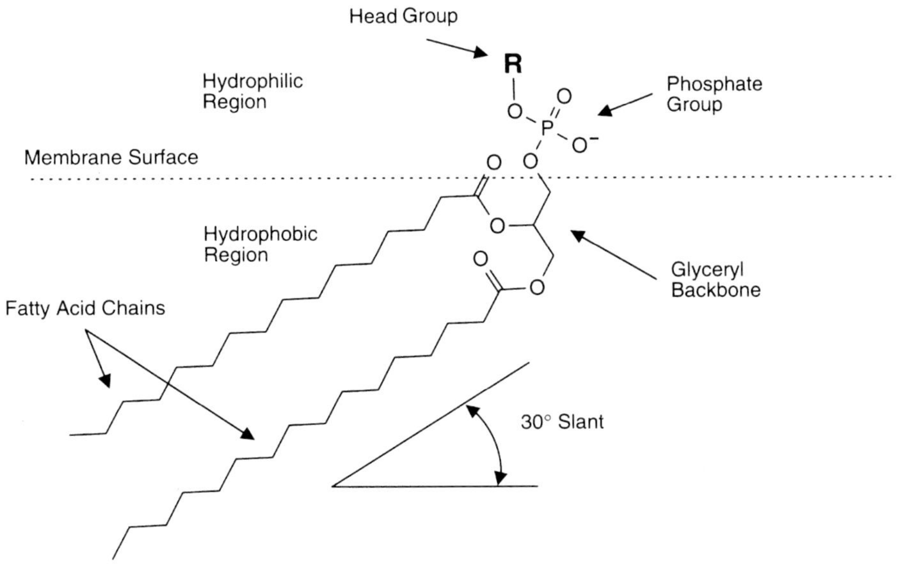

FIGURE 21.3 The basic construction of phosphodiglyceride molecules within lipid bilayers. The fatty acid chains are embedded in the hydrophobic inner region of the membrane, oriented at an angle to the plane of the membrane surface. The hydrophilic head group, including the phosphate portion, points out toward the hydrophilic aqueous environment. Complex lipid bilayers differ from this ideal presentation in that the fatty acid hydrophobic tails of each phospholipid molecule are entwined with other tails in a tightly crowded three-dimensional space within a highly dynamic membrane structure.

hydroxyls of the glycerol bridge. The No. 3 carbon hydroxyl of the glycerol group is phosphorylated and possesses a negative charge at physiological pH. This basic phosphodiglyceride construct of two fatty acids and one glycerylphosphate group is called phosphatidic acid. This is the simplest form of phospholipid available.

The fundamental phosphatidyl group can also be further derivatized at the phosphate to contain an additional polar constituent. Several common derivatives of phosphodiglycerides are naturally occurring, including PC (commonly called lecithin), phosphatidyl ethanolamine (PE), phosphatidyl serine (PS), phosphatidyl glycerol (PG), and phosphatidyl inositol (PI) (Figure 21.4). All of these phospholipids have polar groups that are linked to the phosphatidyl moiety in a phosphate ester bond. The most abundant of these derivatives in biological cell membranes is PC—the trimethyl derivative of PE, possessing a positive charge at physiological pH. Some or all of these phosphodiglyceride derivatives can be mixed to create a particular liposomal recipe.

The fatty acid constituents of phosphodiglycerides can vary considerably in nature among a number of different chain lengths and points of unsaturation. A given isolated phosphatidyl derivative from a biological source usually possesses a range of fatty acid components, varying in chain length from C16 to about C24. Some of the fatty acids may also contain points of unsaturation—one or more double bonds between certain carbon atoms within the chain (Matreya and Avanti Polar Lipids, suppliers). For instance, egg lecithin is not a single compound, but contains a mixture of PC containing about 31% saturated fatty acid having a chain length of 16 carbons, 16% saturated fatty acid with 18 carbons, about 48% also with 18 carbons but having at least one to two points of unsaturation, and the rest a variety of other fatty acid constituents. Naturally

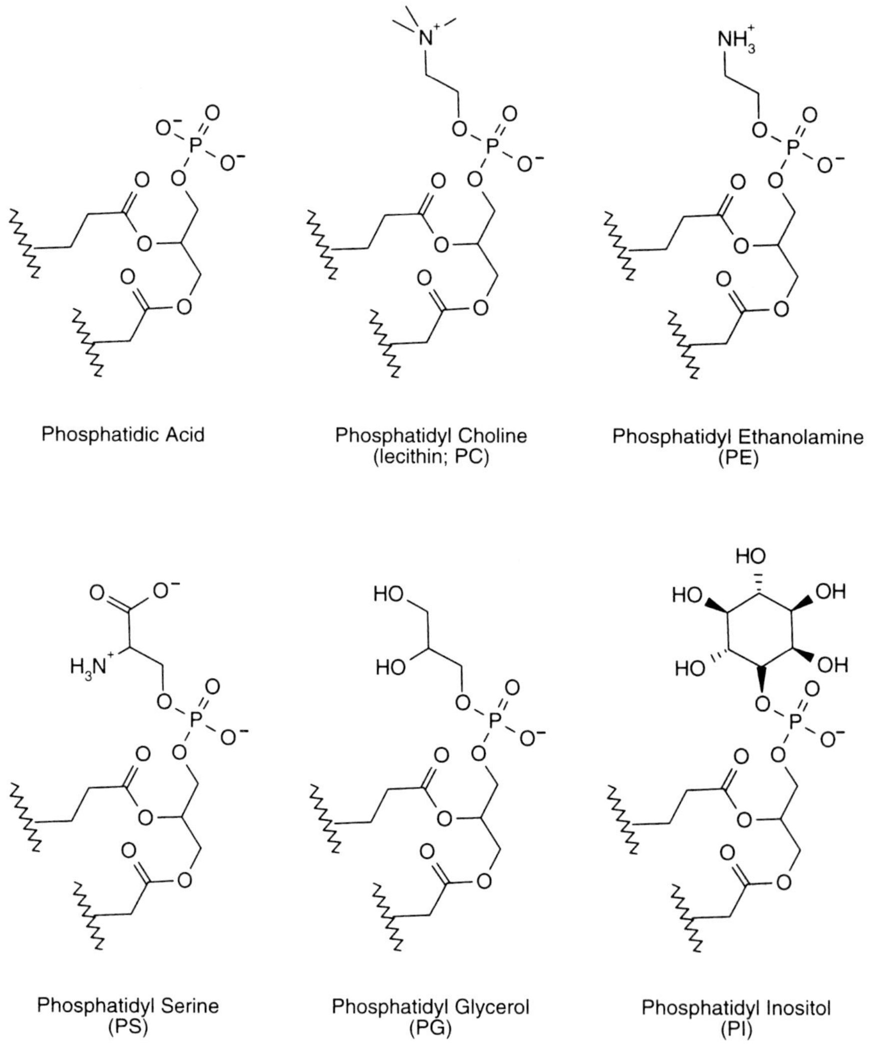

FIGURE 21.4 The head-group structures of the six commonly encountered phosphatidyl derivatives.

occurring, unsaturated fatty acids typically are of the *cis* conformation, not *trans*. The existence of unsaturation within a fatty acid is usually abbreviated as the chain length followed by a colon and the number of double bonds. For instance, *cis*-9-hexadecenoic acid (palmitoleic acid) contains one double bond at carbon 9 and it is abbreviated as C16:1.

By contrast, a given synthetic preparation of a major phospholipid possesses fatty acid constituents all of identical chain length and unsaturation. A synthetic PC derivative can be purchased that contains only, for instance, 1,2-dimyristoyl (C14) fatty acid substitutions on its glyceryl backbone (Genzyme). The use of synthetic rather than natural phospholipids for making liposomes thus produces reagents of known chemical purity, which is very important for regulatory requirements surrounding the introduction of products used topically or *in vivo*.

Despite the large variety of potential fatty acid components in natural-occurring phosphodiglycerides, only three major fatty acid derivatives of synthetic phospholipids are commonly used in liposome preparation: (1) myristic acid (n-tetradecanoic acid; containing 14 carbons), (2) palmitic acid (n-hexadecanoic acid; containing 16 carbons), and (3) stearic acid (n-octadecanoic acid; containing 18 carbons) (Figure 21.5).

The nomenclature for associating individual fatty acid groups with particular phosphodiglyceride derivatives is straightforward. For instance, a phosphatidic acid (PA) derivative which contains two myristic acid chains is commonly called dimyristoyl phosphatidic acid (DMPA). Likewise, a PC derivative containing two palmitate chains is called dipalmitoyl phosphatidyl choline (DPPC). Other phosphodiglyceride derivatives are similarly named.

The second form of lipid derivative that occurs naturally in membrane structures is derived from sphingosine. Unlike the phosphodiglyceride derivatives discussed above, sphingolipids contain no glycerol backbone. Instead, these lipids are constructed from a derivative of 4-sphingenine, containing an *N*-acyl-linked fatty acid group and possibly other constituents off the No. 1 carbon hydroxyl group (Figure 21.6). Sphingolipids are highly similar in their construction to glyceryl lipids, in that there are two hydrophobic tails present on a 3-carbon backbone (one of them contributed from 4-sphingenine itself and the other from the attached fatty acid), and there also exists a hydrophilic head group. This creates the typical amphipathic properties common to all lipid membrane components.

The simplest form of sphingolipid, ceramide, contains a fatty acid group, but no additional components on the No. 1 hydroxyl. Major derivatives of ceramide at the 1-hydroxyl position include a positively charged phosphocholine compound called sphingomyelin, a glucose derivative called glucosylcerebroside, and other complex carbohydrate derivatives termed gangliosides (Figure 21.7). Gangliosides are involved in various cellular recognition phenomena, including being part of the blood group determinants, A, B, and O, in humans.

The use of sphingolipids in liposome formation is possible due to the natural amphipathic properties of the molecules. Some sphingolipids can lend structural advantages to the integrity of liposomal membranes. Sphingomyelin, for example, is capable of hydrogen bonding with adjacent glyceryl lipids, thus increasing the order and stability of the vesicle construction. This stability may translate into a lower potential for passage of molecules through the membrane bilayer, forming vesicles that are better able to retain their contents than more fluid liposome constructions. The temperature of phase transition in sphingolipid-containing membranes is often greater than membranes

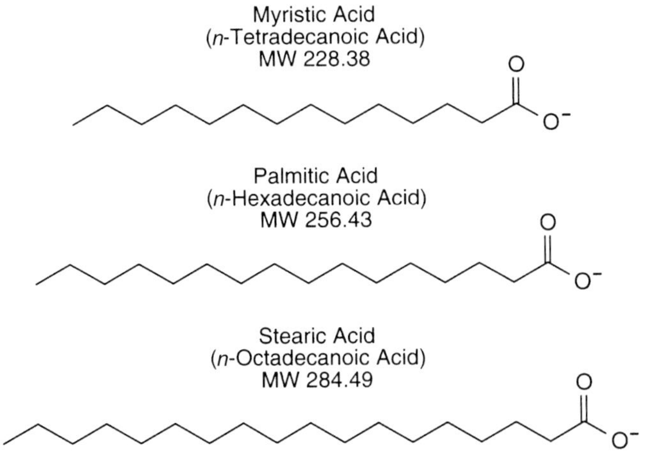

FIGURE 21.5 The three fatty acid components commonly used in liposome construction.

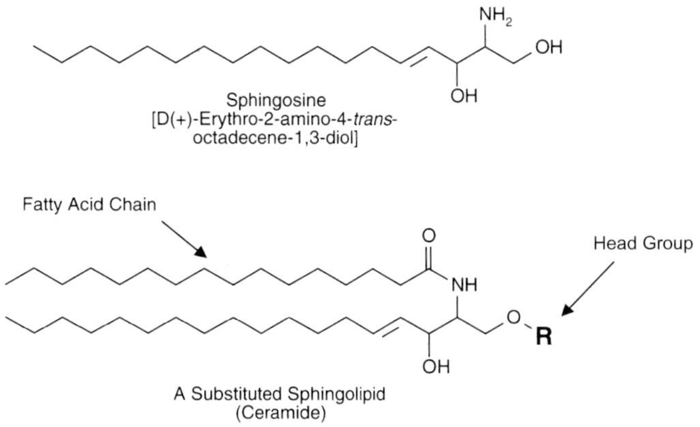

FIGURE 21.6 Sphingolipids are constructed of sphingosine derivatives containing an acylated fatty acid and a head group attached to the hydroxyl.

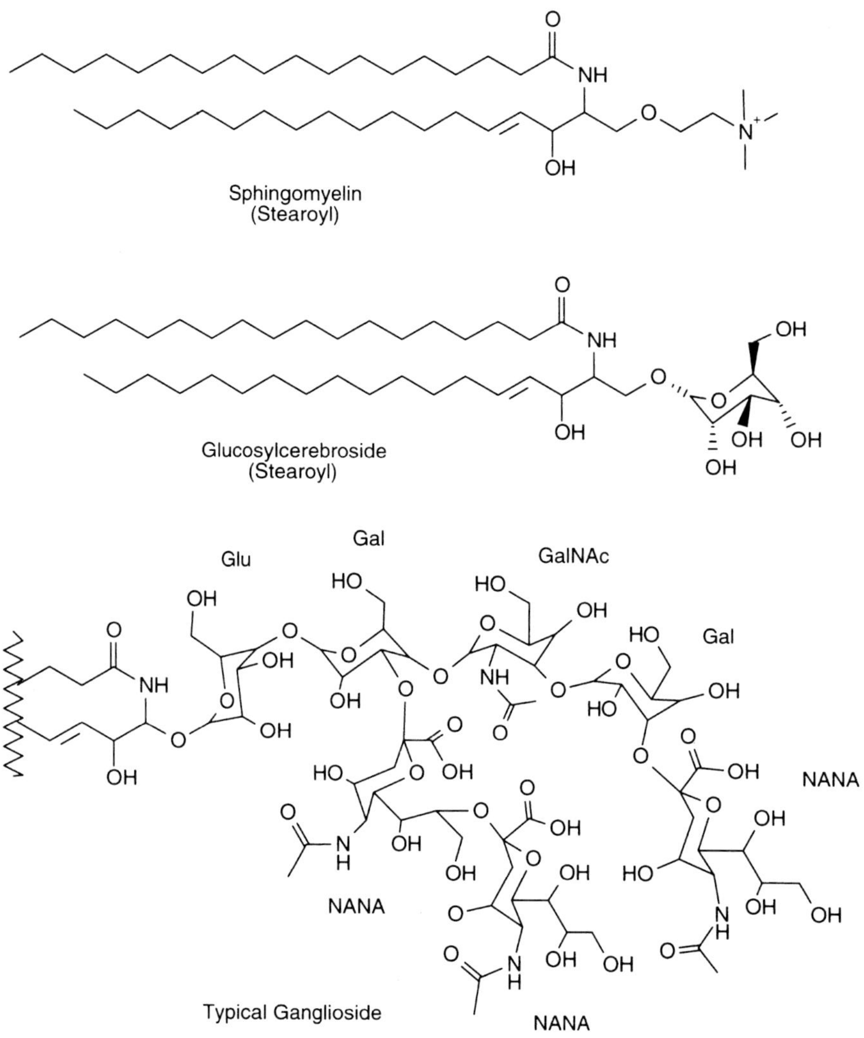

FIGURE 21.7 Common sphingolipid derivatives include small and highly complex head groups.

constructed of only phosphodiglyceride derivatives. Liposomes containing sphingomyelin or gangliosides also have prolonged lifetimes *in vivo* (Gregoriadis and Senior, 1980; Allen and Chonn, 1987) and may be advantageous for creating liposome immunogen complexes.

The main disadvantage of incorporating sphingolipids in liposomes is their high cost. Purified phosphodiglyceride derivatives may be obtained in bulk quantities and in highly defined synthetic preparations, whereas sphingolipid derivatives are not so readily available in similar purity.

Another significant component of many liposome preparations is cholesterol. In natural cell membranes, cholesterol makes up about 10 to 50% of the total lipid. For liposome preparation, it is typical to include a mole ratio of about 50% cholesterol in the total lipid recipe. The addition of cholesterol to phospholipid bilayers alters the properties of the resultant membrane in important ways. As it dissolves in the membrane, cholesterol orients itself with its polar hydroxyl group pointed toward the aqueous outer environment, approximately even, in a three-dimensional sense, with the glyceryl backbone of the bilayer's phosphodiglyceride components (Figure 21.8). Structurally, cholesterol is a rigid component in membrane construction, not having the same freedom of movement that the fatty acid tails of phosphodiglycerides possess. Adjacent phospholipid molecules are restricted in their freedom of movement throughout the length of their fatty acid chains that are abutting the cholesterol molecules. However, since the cholesterol components have the effect of creating spaces in the uniform hydrophobic morphology of the bilayer, the portion of the fatty acid chains below the abutted regions are increased in their freedom of movement.

Cholesterol's presence in liposome membranes has the effect of decreasing or even abolishing (at high

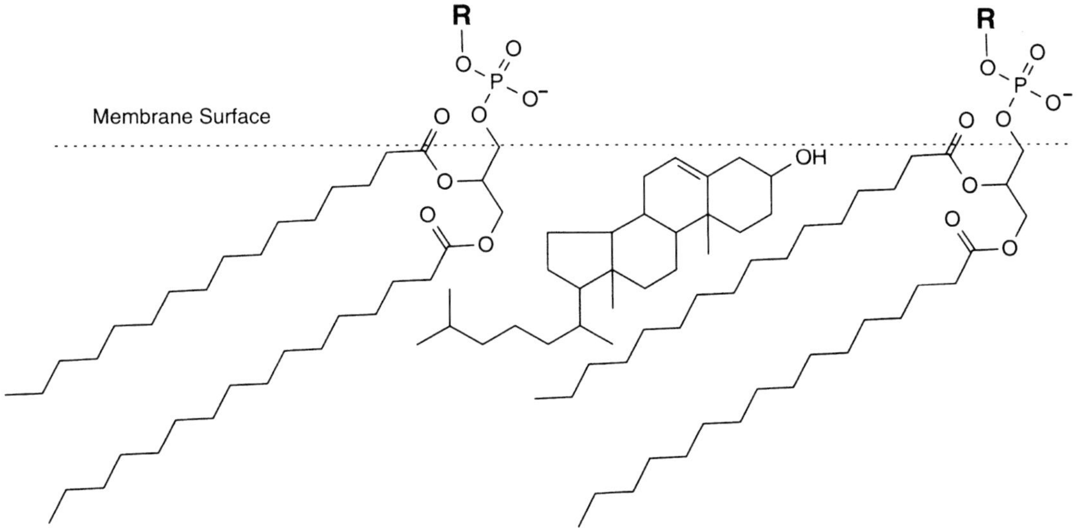

FIGURE 21.8 The orientation of cholesterol in phospholipid bilayers.

cholesterol concentrations) the phase transition from the gel state to the fluid or liquid crystal state that occurs with increasing temperature. It can also modulate the permeability and fluidity of the associated membrane—increasing both parameters at temperatures below the phase transition point and decreasing both above the phase transition temperature. Most liposomal recipes include cholesterol as an integral component in membrane construction.

1.4. Functional Groups of Phospholipids

For the production of liposomal conjugates, lipid derivatives must be incorporated into the bilayer construction that contain available functional groups that are able to be chemically crosslinked or modified. A number of phosphodiglyceride compounds can be employed for conjugation purposes. Each of these components contains a head group that can be directly derivatized or chemically modified to contain a reactive group. For instance, several lipid derivatives contain amine groups that can be utilized in nucleophilic reactions with crosslinkers or other modification reagents. These include phosphatidyl ethanolamine (PE), phosphatidyl serine (PS), and stearylamine. Carboxyl-containing molecules include all the individual fatty acids as well as PS. These can be coupled to amine-containing molecules by the use of the carbodiimide reaction (Chapter 4, Section 1). Hydroxyl-containing lipids include phosphatidyl glycerol (PG), fatty acid alcohols, phosphatidyl inositol (PI), and various gangliosides and cerebrosides of sphingolipid derivation. Lipids possessing hydroxyl groups on adjacent carbon atoms, such as those containing sugar constituents, may be oxidized with sodium periodate to produce reactive aldehyde residues (Chapter 2, Section 4.4). Coupling aldehydes to amine-containing molecules can be accomplished by reductive amination (Chapter 4, Section 4). Finally, the phosphate groups of phosphatidic acid (PA) residues may be used to conjugate with amine-containing molecules in a method similar to modification of the 5′-phosphates of DNA probes. This is performed through the use of the carbodiimide reaction with 1-ethyl-3-(3-dimethylaminopropyl) carbodiimide (EDC) (Chapter 23, Section 2.1). Figure 21.9 shows the structures and reactive sites of these lipid functional groups.

When liposomes are used as part of a conjugate system, the targeting molecules are usually attached covalently to these head group functionalities using standard crosslinking techniques (Derksen and Scherphof, 1985). Although all of the above-mentioned functional groups on lipid molecules can be used for conjugation procedures, most often the derivatization reactions are performed off the PE constituents within the liposomal mixture. The primary amine off the glycerylphosphate head of PE provides an ideal functional group for activation and subsequent coupling of targeting or detection molecules (Shek and Heath, 1983). Liposomes may be constructed with reactive groups already prepared on their PE constituents, all set to be conjugated with selected molecules having the correct functional group. Stock preparations of activated liposomes may even be prepared and lyophilized to be used as needed in coupling macromolecules (Friede *et al.*, 1993). All of the amine-reactive conjugation methods discussed in this section may be used with PE-containing liposomes.

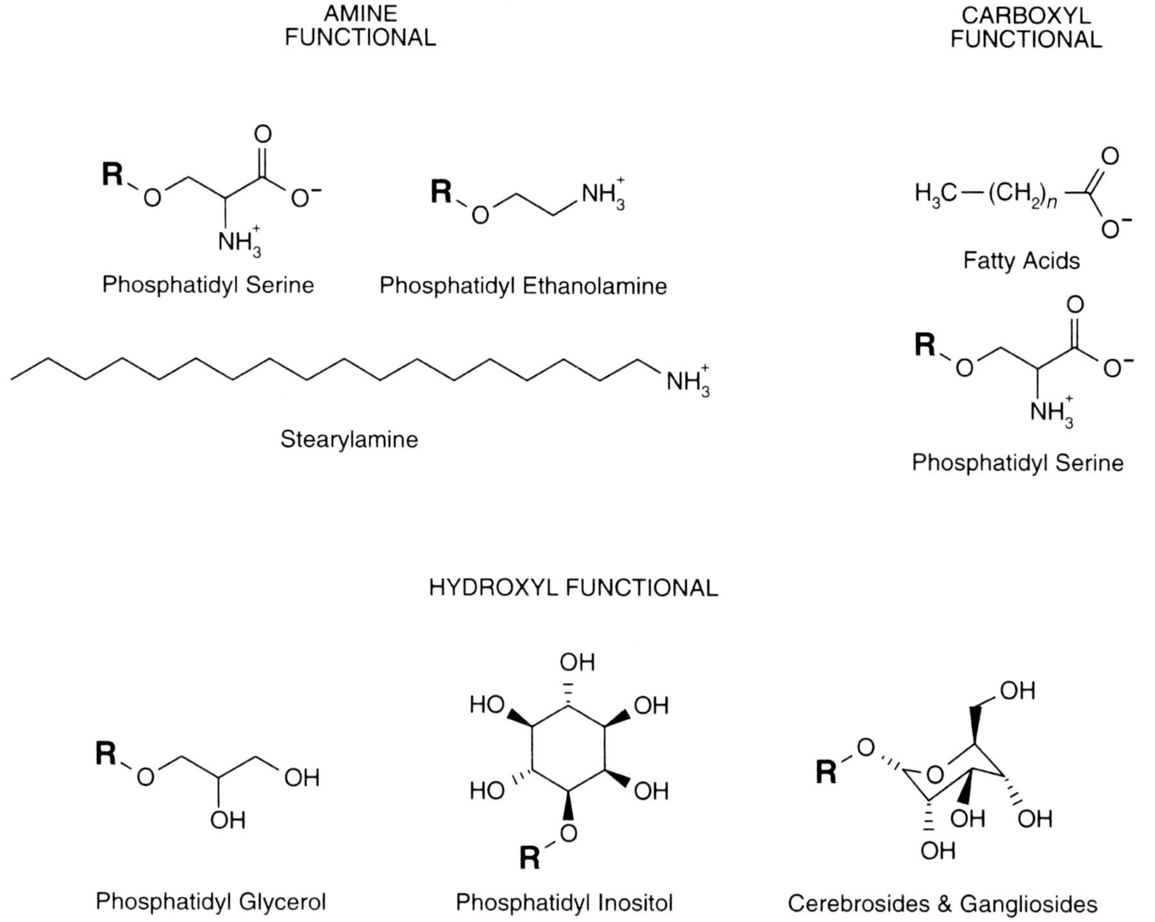

FIGURE 21.9 The major functional groups of lipids that may participate in bioconjugate techniques include amines, carboxylates, and hydroxyls.

2. DERIVATIZATION AND ACTIVATION OF LIPID COMPONENTS

Two approaches for the activation of lipid components may be used to create reactive groups in liposomes. A purified lipid may be activated prior to incorporation into the bilayer construction or the activation step may occur after formation of the intact liposome. Either way, the goal of the activation process is to provide a reactive species that can be used to couple with selected target groups on proteins or other molecules. While numerous crosslinking methods can be used with lipid functional groups, three main strategies are commonly used to conjugate proteins with liposomes: (1) reductive amination to couple aldehyde residues with amines; (2) carbodiimide-mediated coupling of an amine to a carboxylate or an amine to a phosphate group; and (3) multi-step, heterobifunctional crosslinker-mediated conjugation. Both reductive amination and heterobifunctional processes involve activation of particular lipid components.

2.1. Periodate Oxidation of Liposome Components

Reductive amination-mediated conjugation can be performed by periodate-oxidizing carbohydrate or glycerol groups on lipid components and using them to couple with amine-containing molecules. It may also be accomplished by using amines on liposomes (i.e., by the incorporation of PE or SA residues) and coupling them to aldehydes present on proteins or other molecules. Using the first approach, liposomes containing PG or glycosphingolipid residues are oxidized by sodium periodate, purified, and then used to conjugate with protein molecules in the presence of sodium cyanoborohydride (Figure 21.10) (Chapter 4, Section 4). A protocol for the formation of aldehyde groups on liposomes can be found in the method of Heath *et al.*, 1981.

Protocol

1. Prepare a 5-mg/ml liposome construction in 20-m*M* sodium phosphate, 0.15-*M* NaCl, pH 7.4,

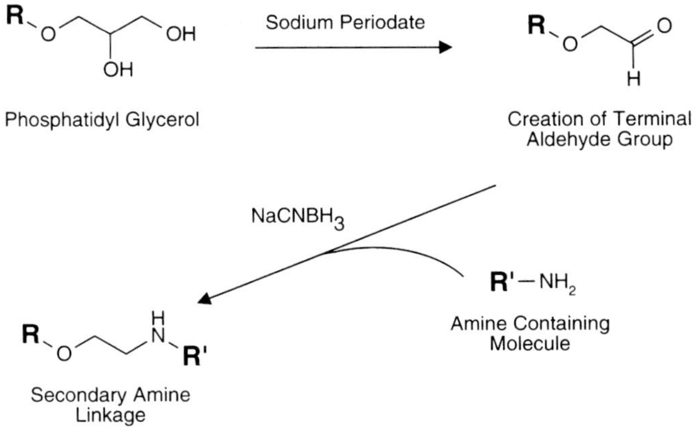

FIGURE 21.10 Hydroxyl-containing lipid components, such as phosphatidyl glycerol, can be oxidized with sodium periodate to produce aldehyde residues. Modification with amine-containing molecules can then take place using reductive amination.

containing, on a mole ratio basis, a mixture of PC:cholesterol:PG:other glycolipids of 8:10:1:2. The other glycolipids that can be incorporated include PI, lactosylceramide, galactose cerebroside, or various gangliosides. Other liposome compositions may be used, for example recipes without cholesterol, as long as a periodate-oxidizable component containing diols is present. Any method of liposome formation may be used.

2. Dissolve sodium periodate to a concentration of 0.6-*M* by adding 128 mg per ml of water. Add 200 μl of this stock periodate solution to each ml of the liposome suspension with stirring.
3. React for 30 min at room temperature in the dark.
4. Dialyze the oxidized liposomes against 20-m*M* sodium borate, 0.15-*M* NaCl, pH 8.4, to remove unreacted periodate. This buffer is ideal for the subsequent coupling reaction. Chromatographic purification using a column of Sephadex G-50 can also be performed.

The periodate-oxidized liposomes may be used immediately to couple with amine-containing molecules such as proteins (see Section 7.6, below), or they may be stored in a lyophilized state in the presence of sorbitol (Friede *et al.*, 1993) for later use.

2.2. Activation of PE Residues with Heterobifunctional Crosslinkers

The most common type of heterobifunctional reagent used for the activation of lipid components includes the amine- and sulfhydryl-reactive crosslinkers containing an *N*-hydroxysuccinimide (NHS) ester group on one end and either a maleimide, iodoacetyl, or pyridyl disulfide group on the other end (Chapter 6, Section 1). Principle reagents used to effect this activation process include SMCC (Chapter 6, Section 1.3), MBS (Chapter 6, Section 1.4), SMPB (Chapter 6, Section 1.6), SIAB (Chapter 6, Section 1.5), and SPDP (Chapter 6, Section 1.1). Other hydrophilic heterobifunctional crosslinkers containing a discrete polyethylene glycol (PEG) spacer may be used in a similar manner, which may provide superior characteristics in many liposomal conjugate applications (Chapter 18, Section 1).

Activation of PE residues with these crosslinkers can proceed by one of two routes: the purified PE phospholipid may be modified in organic solvent prior to incorporation into a liposome, or an intact liposome containing PE may be activated while suspended in aqueous solution. Most often, the PE derivative is prepared before the liposome is constructed. In this way, a stable, stock preparation of modified PE may be made and used in a number of different liposomal recipes to determine the best formulation for the intended application. However, it may be desirable to modify PE after formation of the liposomal structures to ensure that only the outer half of the lipid bilayer is altered. This may be particularly important if substances to be entrapped within the liposome are sensitive or may react with the PE derivatives.

Crosslinkers used to activate PE should contain at least moderately long spacer arms. The length of the spacer is important in providing enough distance from the liposome surface to accommodate the binding of another macromolecule. Short activating reagents often restrict protein accessibility to approach close enough to react with the functional groups on the bilayer surface. For instance, direct modification of PE with iodoacetate results in little or no sulfhydryl-modified IgG coupled to the associated liposomes. However, when an aminoethylthioacetyl spacer is used to move the iodoacetyl group farther away from the bilayer surface, good IgG coupling occurs (Hashimoto *et al.*, 1986). The use of longer discrete PEG-based crosslinkers may enhance the coupling of proteins to liposome surfaces, because the extreme water-solubility of the spacer provides greater aqueous phase access to approaching proteins, whereas hydrophobic linkers may tend to bury themselves within the bilayer structure. To use the PEG-based crosslinker analogs to the aliphatic crosslinker types described in the following activation reactions, refer to the corresponding compounds in Chapter 18, Section 1, having the same reactive groups on both ends.

For the activation of PE prior to liposome formation, it is best to employ a highly purified form of the molecule. While egg PE is abundantly available, it consists of a range of fatty acid derivatives—many of which are unsaturated—and is highly susceptible to oxidation. Synthetic PE, by contrast, can be obtained having a discrete fatty acid composition and is much more stable to oxidative degradation.

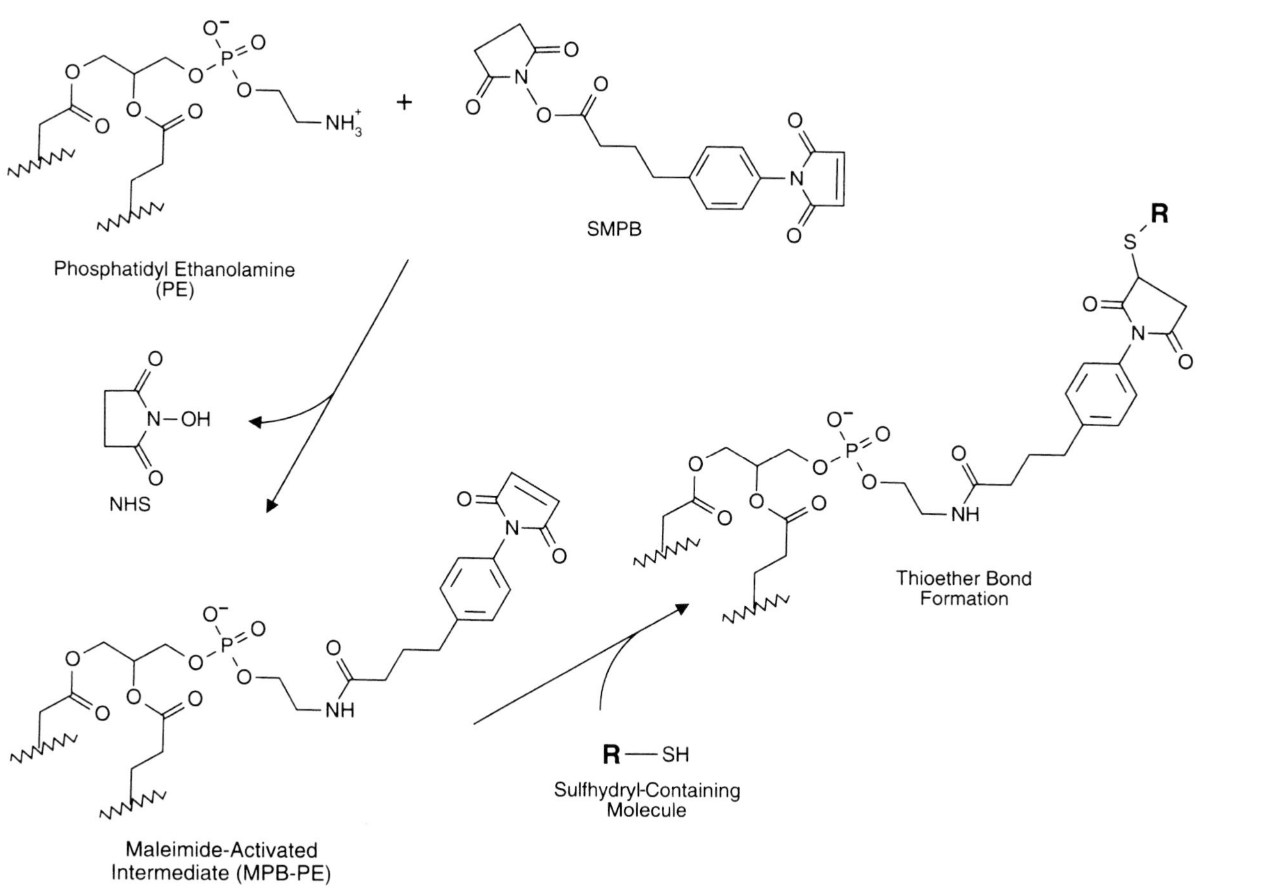

FIGURE 21.11 A sulfhydryl-reactive lipid derivative can be prepared through the reaction of SMPB with phosphatidyl ethanolamine (PE) to produce a maleimide-containing intermediate. Sulfhydryl-containing molecules may then be coupled to the modified phospholipid to form stable thioether linkages.

The following suggested protocols are modifications of those described by Martin and Papahadjopoulos (1982), Martin *et al.* (1990), and Hutchinson *et al.* (1989). Also see Sawant and Torchilin (2011) for a review of lipid-based bioconjugation methods. Although many methods were developed for use with SMPB, SPDP, and MBS, the same basic principles can be used to activate PE with any of the heterobifunctional crosslinkers mentioned above (Manjappa *et al.*, 2011). In addition, the use of hydrophilic NHS–PEG_n–maleimide compounds (Chapter 18, Section 1.2) may be a superior alternative to the use of crosslinkers with hydrophobic cross-bridges, as the PEG linkers will not dissolve as readily within the lipid bilayer structure. The reaction sequence for activation and coupling using SMPB is shown in Figure 21.11. The PE employed should be of a synthetic variety having fatty acid constituents of dimyristoyl-PE (DMPE), dipalmitoyl-PE (DPPE), or distearoyl-PE (DSPE) forms. For activation of pure PE, the heterobifunctional reagents should not be of the sulfo-NHS ester variety, since they are best used in aqueous reaction mediums and PE is activated under nonaqueous conditions. For activation of intact liposomes in aqueous suspension, then the sulfo-NHS variety of the crosslinkers may be the best choice, since they are very water soluble and relatively incapable of penetrating membranes, thus ensuring that only the outer surfaces of the vesicles will be modified. For PEG-based crosslinkers, typically only the NHS ester forms are available (not sulfo-NHS), but the reagents will perform well for activating PE derivatives within liposomes in aqueous solution.

Protocol for the Activation of DPPE with SMPB

1. Dissolve 100 μmoles of PE in 5 ml of argon-purged, anhydrous methanol containing 100 μmoles of triethylamine (TEA). Maintain the solution over an argon or nitrogen atmosphere. The reaction may also be performed in dry chloroform. Note: Methanol or chloroform and TEA should be handled in a fume hood.
2. Add 50 mg of SMPB (Thermo Fisher) to the PE solution. Mix well to dissolve.
3. React for 2 h at room temperature, while maintaining the solution under an argon or nitrogen atmosphere.

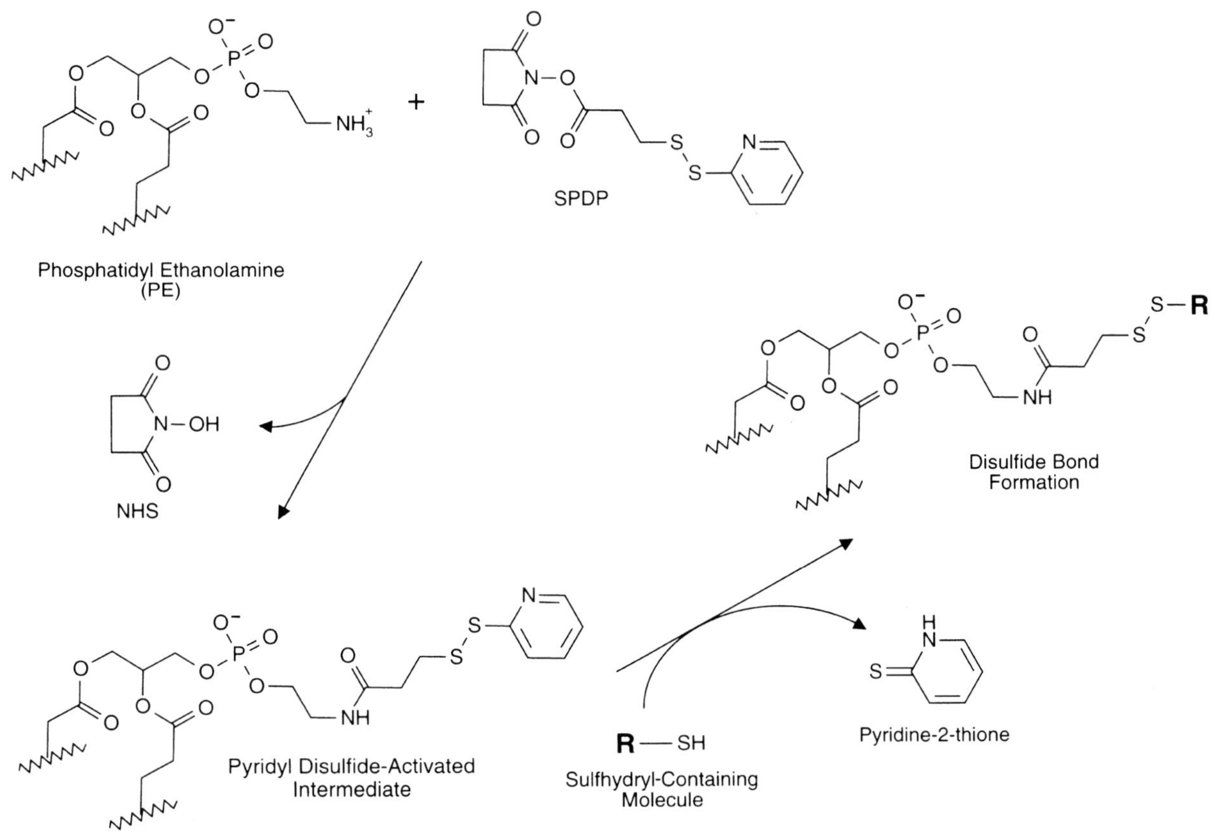

FIGURE 21.12 The reaction of SPDP with PE creates a maleimide derivative capable of coupling thiols. Reaction with a sulfhydryl-containing molecule forms a conjugate through a thioether linkage.

Reaction progress may be determined by thin-layer chromatography (TLC) using silica gel 60-F_{254} plates (Merck) and developed with a 65:25:4 (by volume) mixture of chloroform:methanol:water. The activated PE derivative will develop faster on TLC ($R_F = 0.52$ for MPB–PE) than the unmodified PE.

4. Remove the methanol from the reaction solution by rotary evaporation and redissolve the solids in chloroform (5 ml).
5. Extract the water soluble reaction byproducts from the chloroform with an equal volume of 1% NaCl. Extract twice.
6. Purify the MPB–PE derivative by chromatography on a column of silicic acid (Martin *et al.*, 1981). The following description is from Martin *et al.*, 1990. Add 2 g silicic acid to 10 ml of chloroform and pour the solution into a syringe barrel containing a plug of glass wool at the bottom. Apply the chloroform-dissolved lipids on the silicic acid column. Wash with 4 ml of chloroform, then elute with 4 ml each of the following series of chloroform:methanol mixtures: 4:0.25, 4:0.5, 4:0.75, and 4:1. During the chromatography, collect 2-ml fractions. The presence of purified MPB–PE may be monitored by TLC according to step 3.
7. Remove chloroform from the MBP–PE by rotary evaporation. Store the derivative at −20°C under a nitrogen atmosphere until use.

N-succinimidyl 3-(2-pyridyldithio)propionate (SPDP) may also be used to activate pure PE lipids in a similar manner to SMPB. The result will be a derivative containing pyridyl disulfide groups rather than maleimide groups (Figure 21.12). Pyridyl disulfides react with sulfhydryls to form disulfide linkages. Either the standard SPDP or the long-chain version, LC-SPDP, may be employed in the following protocol.

Protocol for Activation of PE with SPDP

1. Dissolve 20 μmol of PE (15 mg) in 2 ml of argon-purged, anhydrous methanol containing 20 μmol of triethylamine (TEA; 2 mg). Maintain the solution over an argon or nitrogen atmosphere. The reaction may also be performed in dry chloroform. Note: Methanol or chloroform and TEA should be handled in a fume hood.
2. Add 30 μmol (10 mg) of SPDP (Thermo Fisher) to the PE solution. Mix well to dissolve.
3. React for 2 h at room temperature, while maintaining the solution under an argon atmosphere. Reaction

progress may be determined by thin-layer chromatography (TLC) using silica gel plates developed with a 65:25:4 (by volume) mixture of chloroform:methanol:water. The activated PE derivative (PDP–PE) will develop faster on TLC than the unmodified PE.

4. Remove the methanol from the reaction solution by rotary evaporation and re-dissolve the solids in chloroform (5 ml).
5. Extract the water soluble reaction by-products from the chloroform with an equal volume of 1% NaCl. Extract twice.
6. Purify the PDP–PE derivative by chromatography on a column of silicic acid (Martin *et al.*, 1981). The following description is from Martin *et al.* (1990). Add 2 g silicic acid to 10 ml of chloroform and pour the solution into a syringe barrel containing a plug of glass wool at the bottom. Apply the chloroform-dissolved lipids on the silicic acid column. Wash with 4 ml of chloroform, then elute with 4 ml each of the following series of chloroform:methanol mixtures: 4:0.25, 4:0.5, 4:0.75, and 4:1. During the chromatography, collect 2 ml fractions. The presence of purified PDP–PE may be monitored by TLC according to step 3.
7. Remove chloroform from the PDP–PE by rotary evaporation. Store the derivative at −20°C under a nitrogen atmosphere until use.

Other heterobifunctional reagents containing an NHS ester end can be used to activate PE in a similar manner to those protocols described above. A somewhat abbreviated protocol (eliminating the silicic acid chromatography step) for the activation of DPPE with MBS (Chapter 6, Section 1.4), as adapted from Hutchinson *et al.* (1989), follows. The NHS ester of MBS reacts with PE's free amine group to create an amide bond. The maleimide end of the crosslinker then remains available for subsequent conjugation with a sulfhydryl-containing molecule after liposome formation (Figure 21.13).

Protocol for the Activation of DPPE with MBS

1. Dissolve 40 mg of DPPE in a mixture of 16 ml dry chloroform and 2 ml dry methanol containing 20 mg triethylamine. Maintain under nitrogen to prevent lipid oxidation.
2. Add 20 mg of MBS to the lipid solution and mix to dissolve.
3. React for 24 hours at room temperature under nitrogen.

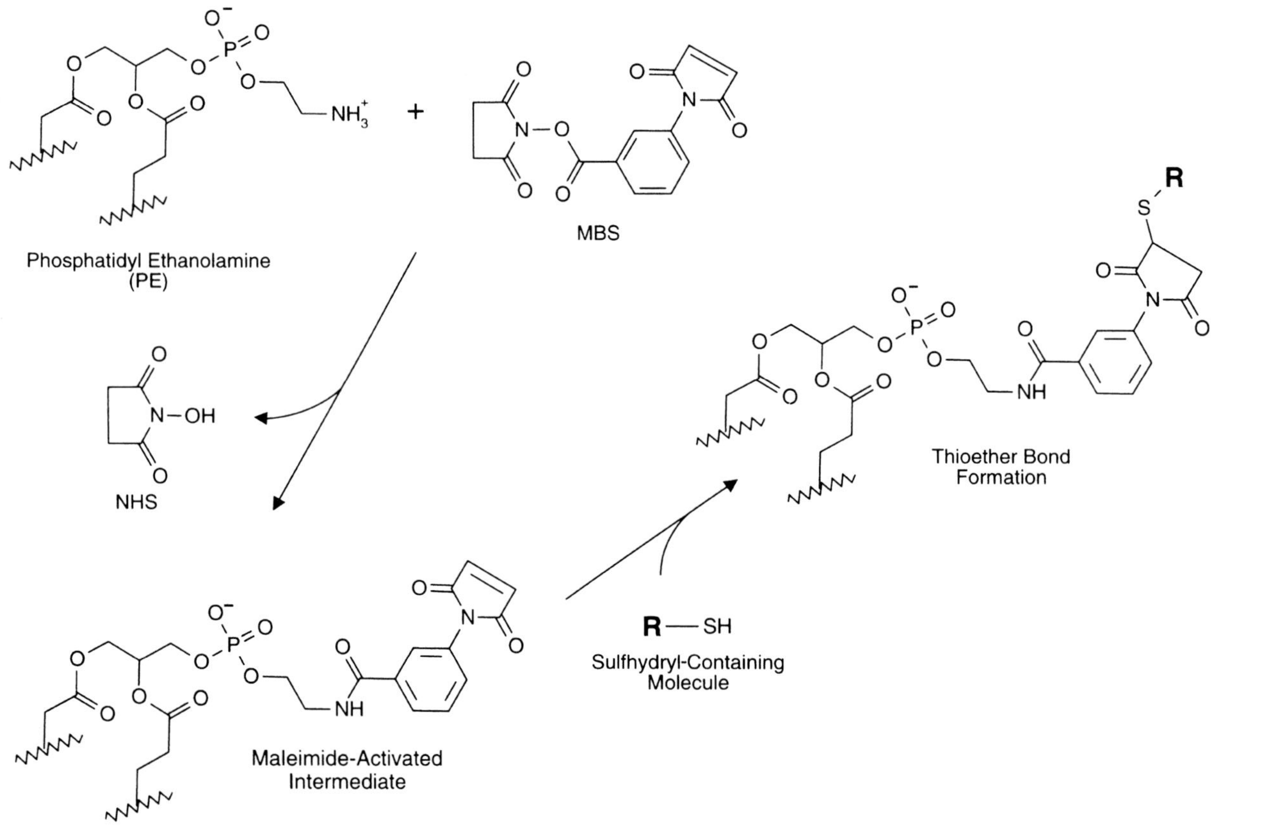

FIGURE 21.13 MBS reacted with PE produces a maleimide derivative that can couple to thiol compounds through a stable thioether bond.

4. Wash the organic phase three times with PBS, pH 7.3, to extract excess crosslinker and reaction by-products.
5. Remove the organic solvents by rotary evaporation under vacuum.
6. Analyze the MBS–DPPE derivative by TLC using a silica plate and developing with a solvent mix containing chloroform:methanol:glacial acetic acid in the volume ratio of 65:25:13. The R_f value of underivatized DPPE is 0.56, while that of the MBS–DPPE product is R_f 0.78.

The MBS–DPPE derivative can be stored dry under a nitrogen blanket at 4°C or dissolved in chloroform: methanol (9:1, v/v) under the same conditions.

If intact liposomes containing PE are to be activated with these crosslinkers, the methods employed are similar to those used to modify proteins and other macromolecules in aqueous solution. The following protocol is a generalized version for the activation of liposomes containing PE with SPDP (Figure 21.14).

Protocol for the Activation of Liposomes with SPDP

1. Prepare a 5 mg/ml liposome suspension in 0.1-*M* sodium phosphate, 0.15-*M* NaCl, pH 7.5, containing, for example, a mixture of PC:cholesterol:PG:PE at molar ratios of 8:10:1:1. Other lipid recipes may be used as long as they contain about this percentage of PE. In addition, if this level of cholesterol is maintained in the liposome, then the integrity of the bilayer will be stable up to a level of organic solvent addition of about 5%. This factor is important for adding an aliquot of the crosslinker to the liposome suspension as a concentrated stock solution in an organic solvent. Dispersion of the liposomes to the desired size and morphology may be performed by any common method (see Section 1.2, this chapter).
2. Dissolve SPDP (Thermo Fisher) at a concentration of 6.2 mg/ml in dimethylformamide (DMF) (makes a 20-m*M* stock solution). Alternatively, LC-SPDP may be used and dissolved at a concentration of 8.5 mg/ml in DMF (also makes a 20-m*M* solution). If the water soluble sulfo-LC-SPDP is used, a stock solution in water may be prepared just prior to adding an aliquot to the reaction. The sulfo-NHS form of the crosslinker contains a negatively charged sulfonate group which prevents the reagent from penetrating lipid bilayers. Thus, only the outer surfaces of the liposomes can be activated using sulfo-LC-SPDP. If this is desirable, prepare a 10-m*M* solution of sulfo-LC-SPDP by dissolving 5.2 mg/ml in water. Since an aqueous solution of the crosslinker will degrade by hydrolysis of the sulfo-NHS ester, it should be used quickly to prevent significant loss of activity. If a sufficiently large amount of liposomes will be modified, the solid sulfo-LC-SPDP may be added directly to the reaction mixture without preparing a stock solution in water.
3. Add 25 to 50 µl of the stock solution of either SPDP or LC-SPDP in DMF to each ml of the liposome suspension to be modified. If sulfo-LC-SPDP is used, add 50 to 100 µl of the stock solution in water to each ml of liposome suspension.
4. Mix and react for at least 30 min at room temperature. Longer reaction times, even overnight, will not adversely affect the modification.
5. Purify the modified liposomes from reaction byproducts by dialysis or gel filtration using a column of Sephadex G-50.

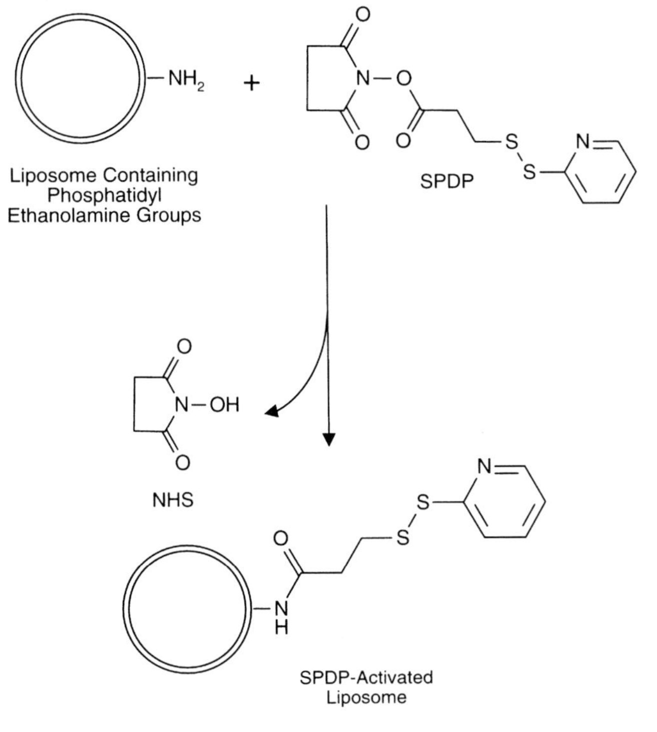

FIGURE 21.14 Intact liposomes containing PE components may be modified with SPDP to produce thiol reactive derivatives.

The SPDP-activated liposomes may be used immediately to couple with sulfhydryl-containing molecules such as proteins (see Section 7.7, this chapter), or they may be stored in a lyophilized state in the presence of sorbitol (Friede *et al.*, 1993) for later use.

3. USE OF GLYCOLIPIDS AND LECTINS TO EFFECT SPECIFIC CONJUGATIONS

Glycolipids are carbohydrate-containing molecules, usually of sphingosine derivation, possessing a hydrophobic, fatty acid tail that embeds them into membrane bilayers. The hydrophilic carbohydrate ends of these

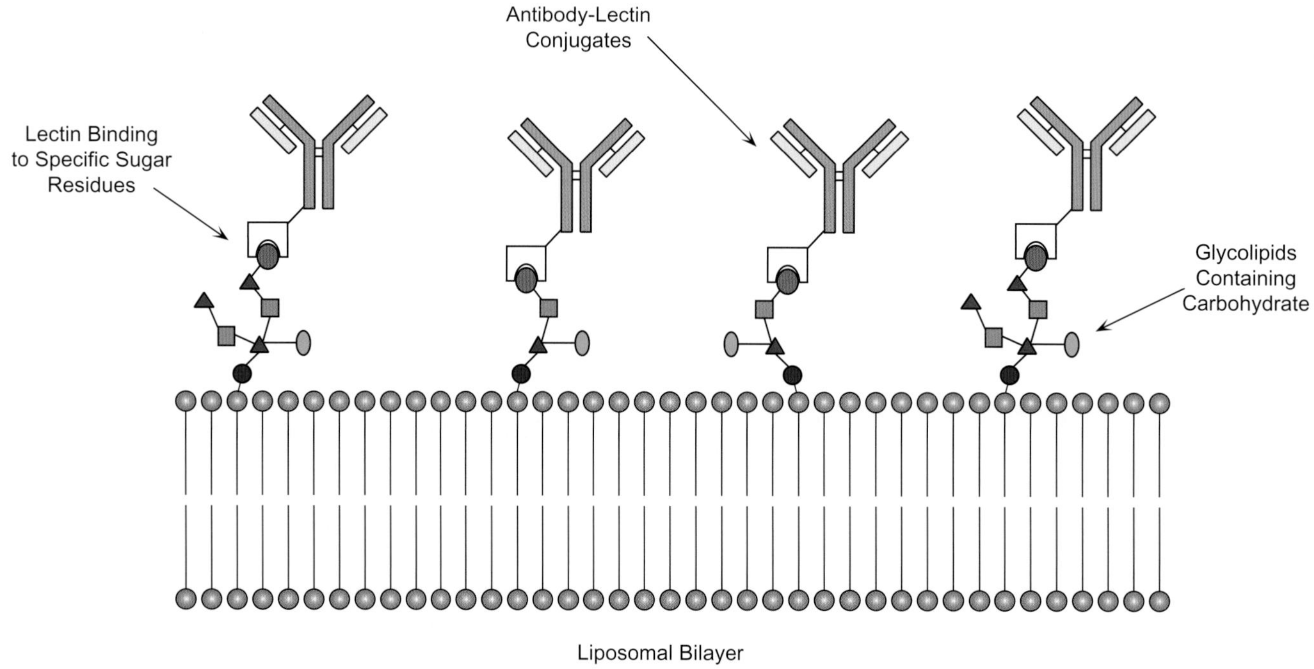

FIGURE 21.15 Glycolipids included in liposome construction may be used to couple antibody molecules by using conjugates of lectins with the proper specificity for binding the sugar groups.

amphipathic molecules orient toward the outer aqueous phase, protruding from the bilayer surface, and thus having the capability to interact with molecules dissolved in the surrounding environment. Sphingosine glycolipids may consist of the simple glucosylcerebroside molecules (containing a single glucose residue), lactosylceramide (containing up to four glucose and galactose residues), or complex gangliosides (containing elaborate oligosaccharides that may approach the complexity of those carbohydrate "trees" found on glycoproteins). In membranes of biological origin, glycoconjugates provide sites of cellular recognition for the binding or attachment of proteins and other molecules that possess binding sites able to interact with the particular saccharides present. Such proteins, called lectins, recognize unique sugars or polysaccharide sequences within the carbohydrates of glycoconjugates.

Liposomes partially constructed of glycolipids of known carbohydrate content may be targeted by lectin molecules possessing the requisite binding properties. The liposome may be labeled in this manner with a lectin conjugate, wherein the lectin possesses another molecule covalently attached to it having secondary detection or recognition properties. For instance, a liposome containing a glycolipid may be modified by a lectin–antibody complex, producing a conjugated antibody for specific antigen-targeting applications (Figure 21.15). In addition, since lectins are typically multivalent in character, having more than one binding site for a particular carbohydrate type, they can act as multifunctional crosslinking agents to agglutinate cells or liposomes containing the proper saccharide receptors.

The advantage of this approach to liposome conjugation is that the linkage between the lectin complex and the membrane bilayer is noncovalent and reversible. The addition of a saccharide containing the proper sequence or sugar type recognized by the lectin breaks the binding and releases the attached molecules. This property can be a significant disadvantage too, if stable linkages are desired.

Carbohydrate residues on the surface of liposomes may be used to bind selected receptor molecules on cell surfaces. This approach can provide a targeting ability for the vesicles *in vivo*, delivering drugs or toxic agents to intended cellular destinations (for review, see Leserman and Machy, 1987). Lectins may also be covalently conjugated to liposome surfaces to provide targeting capability toward cells or molecules containing the complementary carbohydrate needed for binding.

4. ANTIGEN OR HAPTEN CONJUGATION TO LIPOSOMES

Liposomes exert an adjuvant effect *in vivo*, and thus they may be used as carrier systems in the generation of a specific immune response directed against associated antigen or hapten molecules (Heath *et al.*, 1976). Since the main targets of liposomes are the reticulo-endothelial system, particularly macrophages, they naturally associate with the very cells important for mediating humoral immunity. The antigenicity of a liposomal vesicle is to a large degree determined by its overall composition. Its morphology, phospholipid composition, charge, and any attached functional groups all affect the resultant antibody response (Allison and Gregoriadis, 1974; Alving, 1987; Therien and Shahum, 1989).

For instance, incorporation of beef sphingomyelin instead of egg lecithin into liposomal antigen-carrier systems can increase the antibody response to the associated antigen (Yasuda *et al.*, 1977). Liposomal recipes using immune modulators such as lipid A (0.8 nmole of lipid A per μmol of phospholipid) or attaching muramyl dipeptide to the surface of the bilayer can also stimulate the immune response (Daemen *et al.*, 1989). In addition, the fatty acid composition of the phospholipid components can dramatically affect liposome immunogenicity. In general, the higher the transition temperature of the associated phospholipids, the greater will be the immune response to the liposome. Thus, the relative order of immunogenicity related to fatty acid composition is: distearoyl > dipalmitoyl > dimyristoyl. It is also possible that the presence of a positive charge on the bilayer surface may increase the resultant immune response (Domen *et al.*, 1987; Muckerheide *et al.*, 1987b; Apple *et al.*, 1988; Domen and Hermanson, 1992).

Two methods may be used to make immunogenic antigen–liposome or hapten–liposome complexes: (1) the molecule may be dissolved in solution and encapsulated within the vesicle construction; or (2) it may be covalently coupled to the phospholipid constituents using standard crosslinking reactions (Shek and Sabiston, 1982a,b). If the antigen molecules are not chemically coupled to the liposome, then they must be entrapped within them to effect an enhancement of the antibody response. If antigen is simply mixed with preformed liposomal vesicles, then there is no beneficial modulation of immunogenicity (Therien and Shahum, 1989). Encapsulation of soluble or particulate vaccines into giant liposomes provides a means of extending the half-life of the vaccine molecules *in vivo* and potentiating the immune response toward the vaccine (Antimisiaris *et al.*, 1993).

When haptens or antigens are covalently attached to liposomes, it is typically performed through the head groups using various phospholipid derivatives and crosslinking reagents as described previously (Derksen and Scherphof, 1985). Usually, these derivatization reactions are performed using PE constituents within the liposomal mixture. The primary amine on PE molecules provides an ideal functional group for activation and subsequent coupling of hapten molecules (Shek and Heath, 1983). All of the amine-reactive activation methods discussed in this section using heterobifunctional crosslinkers may be used with PE containing liposomes to prepare immunogen conjugates. In addition, the crosslinking methods in Chapter 19, dealing specifically with hapten–carrier conjugation, should be consulted for potential use with liposomes.

The following protocols provide suggested crosslinking strategies for producing an antigen or hapten complex with liposomes. The first procedure is a simple encapsulation of antigen molecules. The second method involves the coupling of a sulfhydryl-containing peptide hapten to a liposome that had been previously activated with SMPB (Figure 21.16) (see Section 2, this chapter). They are based on the methods of Van Regenmortel *et al.* (1988).

Protocol for the Encapsulation of Antigen into Liposomal Vesicles

1. Prepare a homogeneous lipid mixture by dissolving the desired components in chloroform. A suggested recipe may be to use a mixture of DPPC:DPPG:cholesterol at a molar ratio of 7.5:2.5:5. Evaporate the chloroform using a rotary evaporator under vacuum. Maintain lipids under nitrogen or argon to prevent air oxidation.
2. Dissolve the hapten or antigen to be encapsulated at a concentration of 21 μmol/ml in degassed, nitrogen-purged 10-m*M* HEPES, 0.15-*M* NaCl, pH 6.5.
3. Create liposomal vesicles using any established method (see Section 1, this chapter) by mixing the antigen solution with the lipid mixture to obtain a final concentration of 5 mg/ml lipid in the aqueous buffer. A suggested procedure may be to redissolve the lipids in a minimum quantity of diethyl ether, and then mix the buffer with the ether phase at a ratio necessary to give a 5 mg/ml concentration of lipid in the buffer. Use sonication to emulsify the liposomal preparation. Remove diethyl ether by vacuum evaporation. Periodically mix by vortexing to maintain a homogeneous suspension of liposomes.
4. Remove free antigen from encapsulated antigen by gel filtration using a column of Sephadex G-75 or by dialysis using 10-m*M* HEPES, 0.15-*M* NaCl, pH 6.5. Store the liposome preparation under an inert gas at 4°C in the dark until use.

Protocol for the Coupling of Peptide Haptens Containing Sulfhydryl Groups to Liposomal Vesicles

1. Prepare a homogeneous lipid mixture by dissolving in chloroform a mixture of DPPC:DPPG:cholesterol:MPB-DPPE at a mole ratio of 6.3:2.12:4.25:1.5. Preparation of MPB-activated DPPE can be found in Section 2, this chapter.

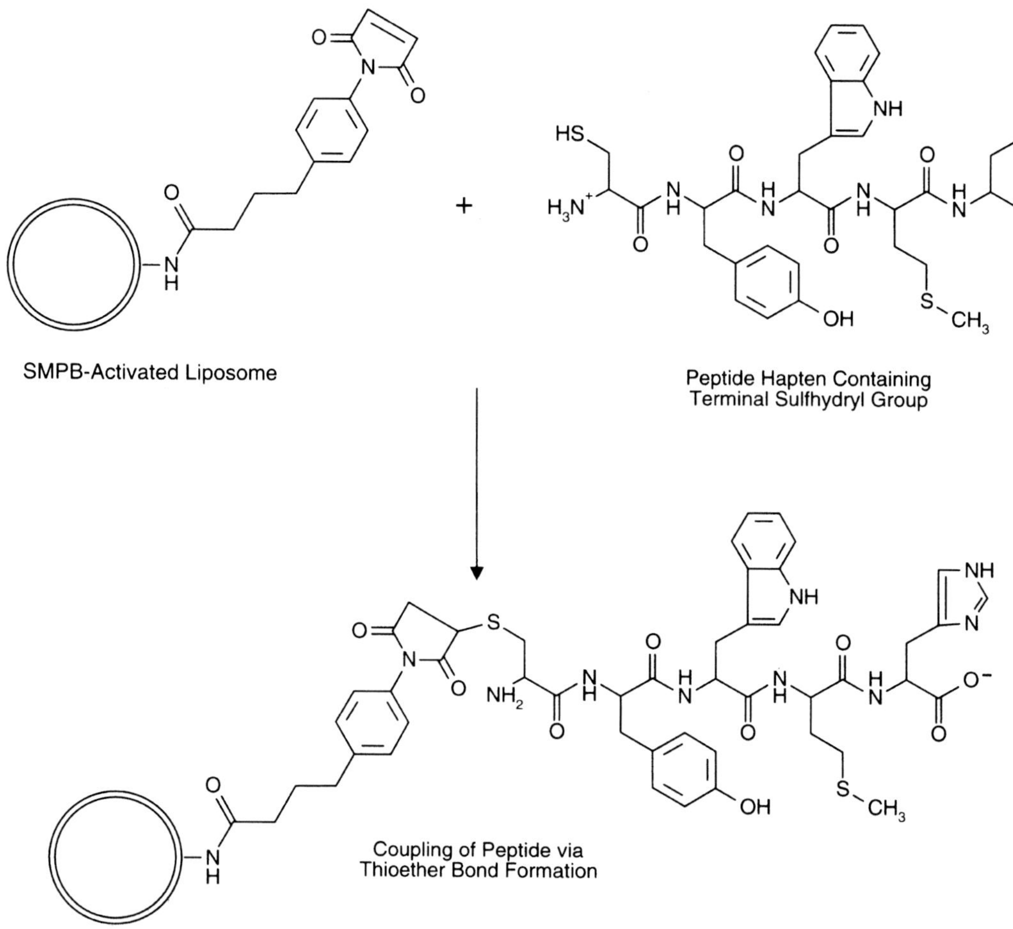

FIGURE 21.16 SMPB-activated liposomes may be modified with peptide hapten molecules containing cysteine thiol groups. The resultant immunogen may be used for immunization purposes to generate an antibody response against the coupled peptide.

Maintain all solutions under nitrogen or argon. The lipid derivative provides reactive maleimide groups for the coupling of sulfhydryl-containing molecules. Evaporate the chloroform using a rotary evaporator under vacuum.

2. Create liposomal vesicles using any established method (see Section 1, this chapter) by mixing the lipid mixture with degassed, nitrogen-purged 10-m*M* HEPES, 0.15-*M* NaCl, pH 7.0, to obtain a final concentration of 5 mg/ml lipid in the aqueous buffer. Use sonication to emulsify the liposomal preparation. Remove diethyl ether by vacuum evaporation. Periodically mix by vortexing to maintain a homogeneous suspension of liposomes.
3. Dissolve a sulfhydryl-containing peptide hapten at a concentration of 25 μmole/ml in degassed, nitrogen-purged 10-m*M* HEPES, 0.15-*M* NaCl, pH 7.0. Add the peptide solution to the liposome suspension at a molar ratio necessary to obtain at least a 5:1 excess of thiol groups to the amount of maleimide groups present (as MPB-DPPE).
4. React overnight at room temperature. Maintain an inert-gas blanket over the vessel to prevent lipid oxidation.
5. Purify the derivatized liposomes from excess peptide by gel filtration using a column of Sephadex G-75 or by dialysis. Store the immunogenic vesicles at 4°C under nitrogen or argon and protected from light until use.

5. PREPARATION OF ANTIBODY–LIPOSOME CONJUGATES

Covalent attachment of antibody molecules to liposomes can provide a targeting capacity to the vesicle that can modulate its binding to specific antigenic determinants on cells or to molecules in solution. Antibody-bearing liposomes may possess encapsulated components that can be used for detection or therapy (Figure 21.17). For instance, fluorescent molecules encapsulated within antibody-targeted vesicles

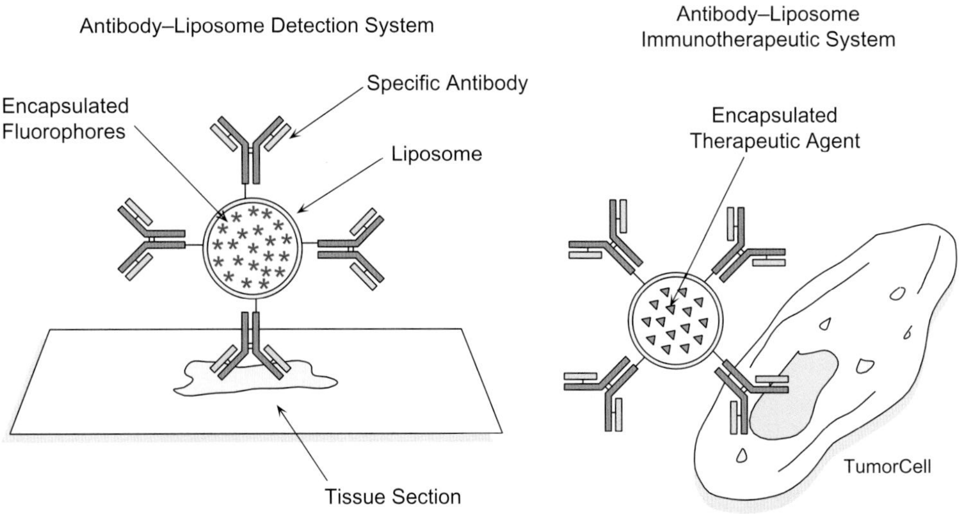

FIGURE 21.17 Antibody–liposome conjugates can be used as targeting reagents for detection or therapeutic applications. The liposome may be constructed to contain fluorescent molecules for detection purposes or bioactive agents for therapy. The coupled antibody creates the targeting component on the complex for binding to specific antigenic determinants.

can be used as imaging tools or in flow cytometry (Truneh *et al.*, 1987). Specific antibodies coupled to the vesicle surface can improve diagnostic assays involving agglutination of latex particles (Kung *et al.*, 1985). Liposomes possessing antibodies directed against tumor cell antigens can deliver encapsulated toxins or drugs to the associated cancer cells, effecting toxicity and cell death (Heath *et al.*, 1983, 1984; Matthay *et al.*, 1984; Straubinger *et al.*, 1988; Kirpotin *et al.*, 2012; Tong *et al.*, 2012).

Encapsulation of chemotherapeutic agents within lipid bilayers reduces systemic toxicity and local irritation often caused by anticancer drugs (Gabizon *et al.*, 1986; Crielaard *et al.*, 2012). The liposomal membrane acts as a slow-release agent so that cytotoxic components do not come into contact with non-tumor cells. Liposome binding to cells causes internalization and release of the encapsulated drugs. Antibody targeting can increase the likelihood of vesicle binding to the desired tumor cells.

However, there are problems associated with the use of antibody–liposome conjugates for drug delivery *in vivo*. Particularly, since lipid vesicles are huge compared to similar immunotoxin conjugates (Chapter 20, Section 3), their passage to particular tissue destinations may be difficult or impossible. Liposomes are almost entirely limited to the reticuloendothelial system. Their ability to pass through tissue barriers to target cells in other parts of the body is limited by their size. If liposomal conjugates can reach their intended destination, their contents are delivered to the cells by endocytosis. Endocytic vesicles arising from the surface of cells have diameters in the range of 1000 to 1500 Å. This limits the size of liposomes that can be used to small vesicles that can bind to the surface of a cell and be internalized efficiently. Large liposomes, by contrast, will not be internalized and therefore not be able to deliver their contents (Leserman and Machy, 1987).

The methods for coupling antibody molecules to liposomal surfaces are not unlike those described for general protein coupling (Section 7, this chapter), below. Antibodies may be coupled through sulfhydryl residues using liposomes containing PE groups that are derivatized with heterobifunctional crosslinkers such as SMCC (Chapter 6, Section 1.3), MBS (Chapter 6, Section 1.4), SMPB (Chapter 6, Section 1.6), SIAB (Chapter 6, Section 1.5), SPDP (Chapter 6, Section 1.1), and heterobifunctional PEG-based crosslinkers (Chapter 18, Section 1.2). They may also be coupled through their amine groups using reductive amination to periodate-oxidized glycolipids (Sections 2 and 7.6, this chapter).

6. PREPARATION OF BIOTINYLATED OR (STREPT)AVIDIN-CONJUGATED LIPOSOMES

Liposome conjugates may be used in various immunoassay procedures. The lipid vesicle can provide a multivalent surface to accommodate numerous antigen–antibody interactions and thus increase the sensitivity of an assay. At the same time, it can function as a vessel to carry encapsulated detection components needed for the assay system. This type of enzyme-linked immunosorbent assay (ELISA) is called a

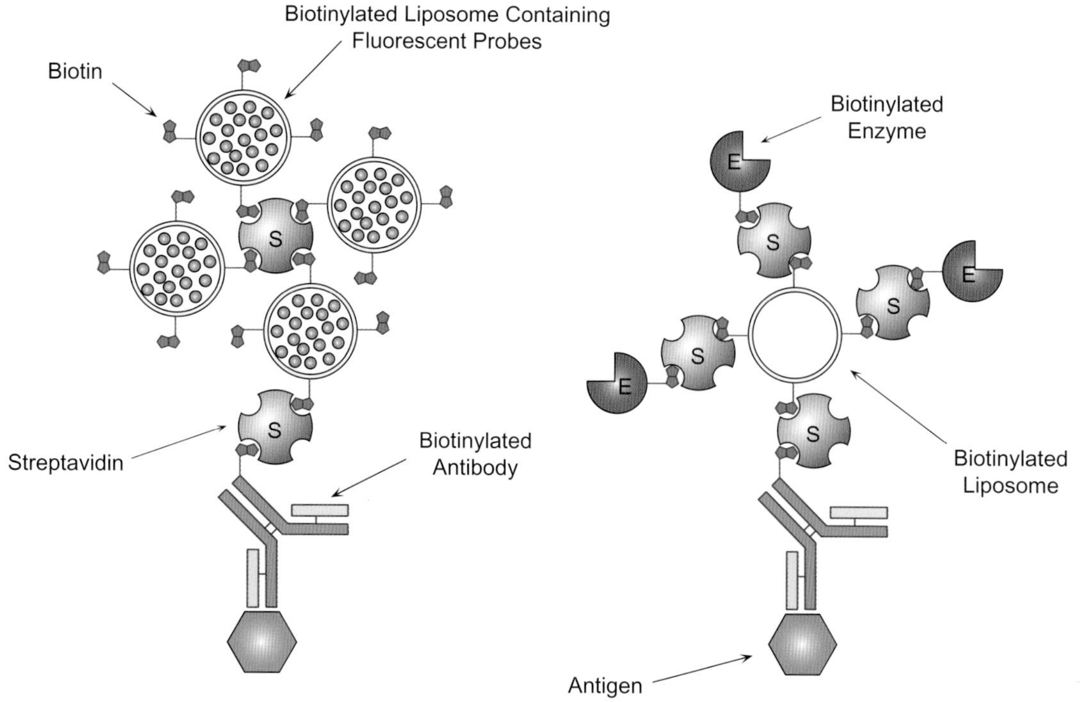

FIGURE 21.18 Biotinylated liposomes can be used in immunoassay systems to enhance the signal for detection or measurement of specific analytes. The liposome components may be constructed to include encapsulated fluorescent molecules to facilitate detection of antigens within tissue sections.

liposome immunosorbent assay or LISA. One method of using liposomes in an immunoassay is to modify the surface so that it can interact to form biotin–avidin or biotin–streptavidin complexes. The avidin–biotin interaction can be used to increase detectability or sensitivity in immunoassay tests (Chapter 11) (Savage *et al.*, 1992).

Liposomes containing biotinylated phospholipid components can be used in a bridging assay system with avidin and a biotinylated antibody molecule, creating large multivalent complexes able to bind antigen (Plant *et al.*, 1989) (Figure 21.18). The inside of the vesicles may contain fluorescent detection reagents that can be used to localize or quantify target analytes. One small liposome provides up to 10^5 molecules of fluorophore to allow excellent detectability of a binding event. LISA systems using biotinylated liposomes to detect antigen molecules can increase the sensitivity of an immunoassay up to 100-fold over that obtainable using traditional antibody–enzyme ELISAs.

Biotin–liposome conjugates are being used to target tumor cells and bring encapsulated drugs to the site of cancer cells for enhanced tumor accumulation of biotherapeutic agents. Soininen *et al.* (2012) studied the biodistribution of biotinylated doxorubicin bound to biotinylated liposomes using the avidin–biotin interaction. In addition, biotinylated liposomes can be used to target the lymphatic system for extended lymph node retention (Cai *et al.*, 2011).

Biotinylated liposomes are usually created by modification of PE components with an amine-reactive biotin derivative, for example the use of the popular reagent NHS–LC-Biotin (Chapter 11, Section 6.2). The NHS ester end of this compound reacts with the primary amine on PE residues, forming an amide bond linkage (Figure 21.19). However, since many of the traditional biotinylation reagents contain hydrophobic spacers, their use with amphipathic liposomal constructions may not be optimal. Hydrophobic spacers may have a tendency to dissolve into the lipid bilayer, thus decreasing their accessibility to the aqueous phase. A better choice of biotinylation agent may be to use one of the NHS–PEG_n–biotin compounds (Chapter 18, Section 1.3), because their hydrophilic PEG spacers provide better binding potential toward (strept)avidin conjugates than does a hydrophobic biotin spacer. A hydrophilic PEG-based biotin compound creates a water soluble biotin modification on the outer aqueous surface of the liposome bilayer. In addition, since the modification occurs at the hydrophilic end of the phospholipid molecule, after vesicle formation the biotin component can protrude out from the liposomal surface with high freedom of motion. In this configuration, the surface-immobilized

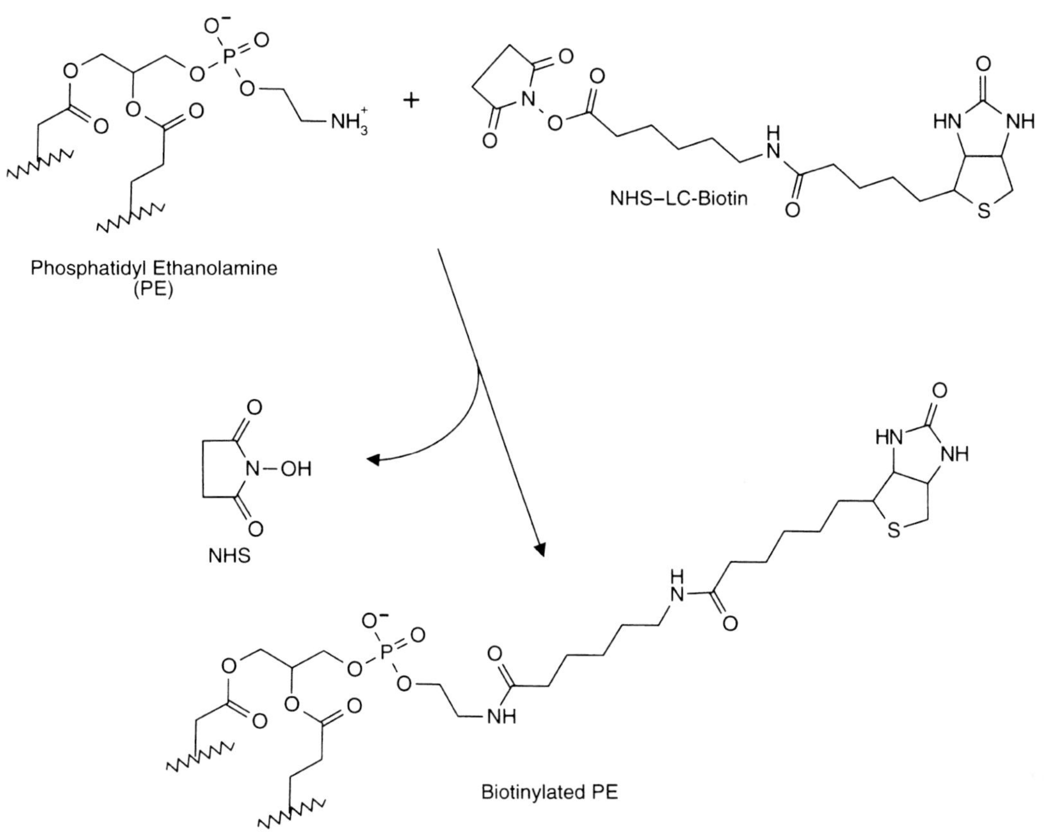

FIGURE 21.19 Biotinylated liposomes may be formed using biotinylated PE. Reaction of NHS–LC-biotin with the amine group of PE will result in an amide bond linkage with a long spacer arm terminating in a biotin group.

biotins are able to efficiently bind (strept)avidin-containing reagents added to the outer aqueous environment.

Biotinylation may be performed before or after liposome formation, but having a stock supply of biotin-modified PE is an advantage, since it can then be used to test a number of liposomal recipes. In addition, only a very small percentage of the total lipid should be biotinylated to prevent avidin-induced aggregation in the absence of antigen. It is difficult to precisely control the biotin content if direct biotinylation of intact liposomes is performed. Using pure biotinylated phospholipid allows incorporation of measured amounts of biotin binding sites into the final liposomal membrane. The preparation of biotinylated PE (B-PE) can be performed similarly to the methods described for activation of PE with SMPB (Section 2, this chapter) or it may be obtained commercially.

The following method for the formation of a biotinylated liposome is adapted from Plant *et al.* (1989). It assumes prior production of B-PE.

Protocol

1. Prepare a biotinylated liposome construct by first dissolving in chloroform, the lipids DMPC:chole sterol:dicetylphosphate (Sigma) at mole ratios of 5:4:1, and adding to this solution 0.1 mol% B-PE. Larger mole ratios of B-PE to total lipid may result in nonspecific aggregation of liposomes in the presence of (strept)avidin. Maintain all lipids under an inert atmosphere to prevent oxidation.
2. Evaporate 2 μmol of total homogenized lipid in chloroform using a stream of nitrogen or a rotary vacuum evaporator.
3. Re-dissolve the dried lipid in 50 μ of dry isopropanol.
4. Take the lipid solution up into a syringe and inject it into 1 ml of degassed, nitrogen-purged 20-m*M* Tris, 0.15-*M* NaCl, pH 7.4, which is being vigorously stirred using a vortex mixer. To encapsulate a fluorescent dye using this procedure, include 100-m*M* 5,6-carboxyfluorescein (or another suitable fluorophore) in the buffer solution before adding the lipids. The fluorescent probe used in the encapsulation procedure should be chemically nonreactive so that no lipid components are covalently modified during the process.

The biotinylated liposomes prepared by this procedure may be stored under an inert-gas atmosphere at 4°C for long periods without degradation.

7. CONJUGATION OF PROTEINS TO LIPOSOMES

Covalent attachment of proteins to the surface of liposomal bilayers is performed through reactive sites created on the head groups of phospholipids with the intermediary use of a crosslinker or other activating agent. The lipid functional groups described in Section 1 of this chapter are modified according to the methods discussed in Section 2 to be reactive toward specific target groups in proteins. Conjugation of liposomes with proteins may be performed with homobifunctional or heterobifunctional crosslinking reagents, carbodiimides, reductive amination, by NHS ester activation of carboxylates, or through the noncovalent use of the avidin–biotin interaction.

Characterizing the resultant complex for the amount of protein per liposome is somewhat more difficult than in other protein conjugation applications. The protein–liposome composition is highly dependent on the size of each liposomal particle, the amount of protein charged to the reaction, and the mole quantity of reactive lipid present in the bilayer construction. An approach to solving this problem is presented by Hutchinson *et al.* (1989). In analyzing at least 17 different protein–liposome preparations, the ratio of protein:lipid content (μg protein/μg lipid) in most of the complexes ranged from a low of about 4 to as much as 675. In some instances, however, up to 6000 molecules of a particular protein could be incorporated into each liposome.

Coupling of protein molecules to liposomes may occasionally induce vesicle aggregation. This may be due to the unique properties or concentration of the protein used, or it may be a result of liposome-to-liposome crosslinking during the conjugation process. Adjusting the amount of protein charged to the reaction as well as the relative amounts of crosslinking reagents employed may have to be performed to solve an aggregation problem.

The following sections will present suggested protocols for creating protein-bearing liposomes. Each method utilizes specific lipid modifications to form reactive groups capable of targeting amines, sulfhydryls, aldehydes, or carboxylates on the protein molecules.

7.1. Coupling via the NHS Ester of Palmitic Acid

Huang *et al.* (1980) coupled monoclonal antibodies to liposomes using an NHS ester modification of palmitic acid incorporated into the bilayer construction (Lapidot *et al.*, 1967). The NHS ester reacts with amine groups on the protein molecule, producing stable amide bond linkages (Figure 21.20). The specificity of the antibody-bearing liposomes for mouse L-929 cells was documented, illustrating the preservation of antibody binding activity.

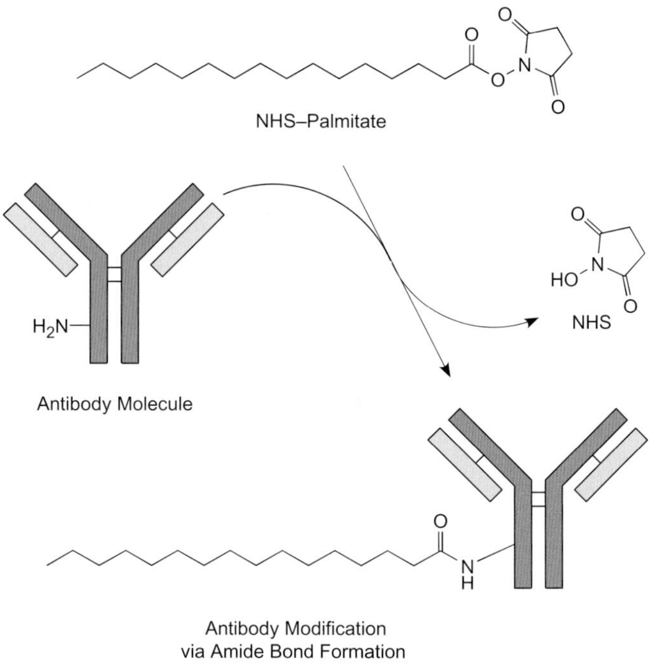

FIGURE 21.20 The NHS ester derivative of palmitic acid can be used to couple antibody molecules through amide bonds. These complexes then may be incorporated into liposomes.

Protocol

(A) PREPARATION OF NHS-PALMITATE

1. Dissolve 3.45 mg of *N*-hydroxysuccinimide (NHS) in 30 ml dry ethyl acetate.
2. Add 30 mMol of palmitic acid to the NHS solution. Maintain a nitrogen blanket over the solution.
3. Dissolve 6.18 gm of dicyclohexyl carbodiimide (DCC; Chapter 4, Section 1.4) in 10 ml of ethyl acetate and add it to the NHS/palmitic acid solution.
4. React overnight at room temperature under a nitrogen blanket.
5. Remove the insoluble dicyclohexyl urea (DCU) byproduct by filtration using a glass-fiber filter pad and vacuum.
6. Remove solvent from the filtered solution by using a rotary evaporator under vacuum.
7. The NHS–palmitate may be purified by recrystallization using ethanol. Dissolve the activated fatty acid in a minimum quantity of hot ethanol. Immediately upon dissolving, filter it through a filter funnel containing a fluted glass-fiber filter pad, both of which have been warmed to the same temperature as the ethanol solution. Allow the NHS–palmitate to recrystallize overnight at room temperature. Remove solvent from the recrystallized solid by filtration. Dry under vacuum in a desiccator.

8. Analyze the NHS–palmitate for purity using TLC on silica plates. Develop the chromatography using a solvent mixture of chloroform:petroleum diethyl ether (bp 40–60°C) of 8:2. Excess NHS and NHS–palmitate may be detected by staining with 10% hydroxylamine in 0.1-*N* NaOH, followed after 2 min by a 5% solution of $FeCl_3$ in 1.2-*N* HCl (creates red-colored spots).

(B) COUPLING OF PROTEIN TO NHS-PALMITATE

1. Add 2 mg of protein to 44 µg NHS–palmitate in 20-m*M* sodium phosphate, 0.15-*M* NaCl, pH 7.4, containing 2% deoxycholate.
2. Incubate at 37°C for 10 h.
3. Remove excess palmitic acid by chromatography on a column of Sephadex G-75. Use PBS, pH 7.4, containing 0.15% deoxycholate to perform the gel filtration. Collect the fractions containing derivatized protein, as monitored by absorbance at 280 nm.

(C) ADDITION OF PROTEIN–PALMITATE CONJUGATE TO LIPOSOMAL MEMBRANES

Since the protein–palmitate derivative cannot be dissolved in organic solvent during homogenization of lipid to form liposomal membranes, it must be inserted into intact liposomes by detergent dialysis.

1. Construct a liposome by dissolving the desired lipids in chloroform to fully homogenize the mixture, drying them to remove solvent, and using any established method of forming bilayer vesicles in aqueous solution (e.g., sonication; see Section 1.2, this chapter).
2. Add protein–palmitate conjugate to the formed liposomes in a ratio of 20:1 (w/w). Add concentrated deoxycholate to give a final concentration of 0.7%. Mix thoroughly using a vortex mixer.
3. Dialyze the liposome preparation against PBS, pH 7.4.
4. The liposome vesicles may be characterized for size by chromatography on a column of Sepharose 4B.

7.2. Coupling via Biotinylated PE Lipid Derivatives

Biotinylated PE (B-PE) incorporated into liposomal membranes can be used to interact non-covalently with (strept)avidin–protein conjugates or with other biotinylated proteins using (strept)avidin as a bridging molecule (Plant *et al.*, 1989). It is important that a long-chain spacer be used in constructing the B-PE derivative to allow enough spatial separation from the bilayer surface to accommodate avidin docking (Hashimoto *et al.*, 1986). Thus, any biotinylated protein can be coupled to the liposome surface through the strength of the (strept)avidin–biotin interaction. Section 6 (this chapter) describes the preparation of B-PE derivatives and their addition into vesicle construction. Incubation of (strept)avidin conjugates or (strept)avidin plus a biotinylated protein with the biotinylated liposome in PBS, pH 7.4, will form essentially nonreversible complexes, immobilizing the proteins to the outer surface of the bilayers. This method can be used to couple biotin-modified antibody molecules to liposomes (Figure 21.21). Removal of non-complexed protein may be performed using gel filtration chromatography on a column of Sephadex G-50 or G-75.

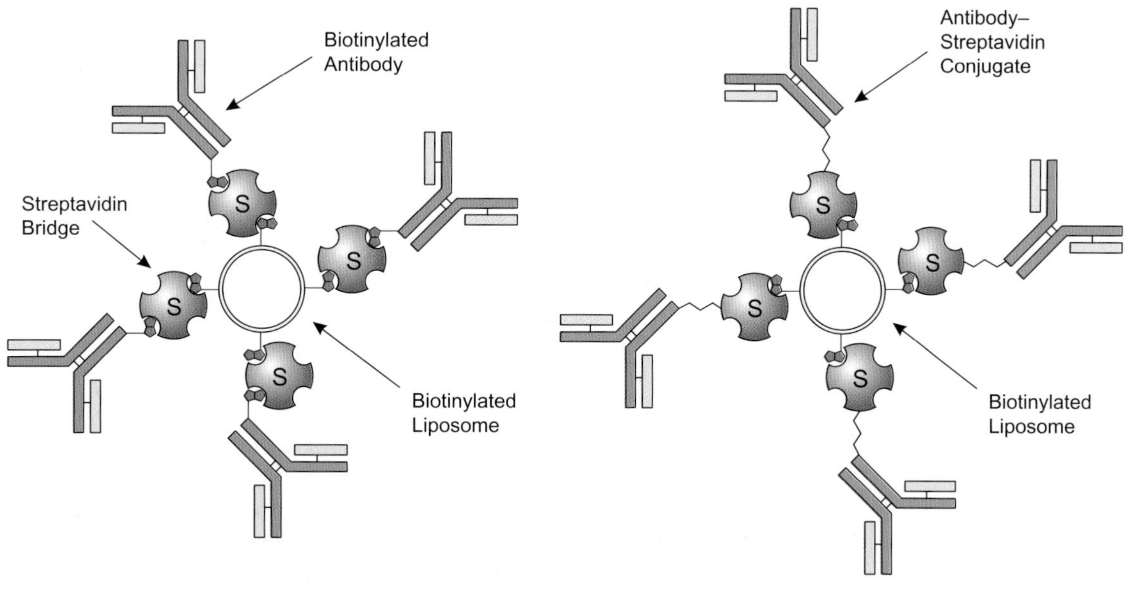

FIGURE 21.21 Antibodies can be conjugated to liposomes using an indirect approach by building a multilayered (strept)avidin–biotin complex on the surface. Biotinylated liposomes may be complexed with biotinylated antibodies using (strept)avidin as a bridging molecule or may be complexed with an antibody–(strept)avidin conjugate.

7.3. Conjugation via Carbodiimide Coupling to PE Lipid Derivatives

Underivatized PE in liposomal membranes contains an amine group that can participate in the carbodiimide reaction with carboxylate groups on proteins or other molecules (Dunnick *et al.*, 1975). The water soluble carbodiimide EDC (Chapter 4, Section 1.1) activates carboxylate groups to form active-ester intermediates that can react with PE to form an amide linkage. Unfortunately, EDC coupling of proteins to surfaces often results in considerable protein-to-protein crosslinking, since proteins contain an abundance of both amines and carboxylates. There is also potential for vesicle aggregation by proteins coupling to more than one liposome. Martin *et al.* (1990) suggested avoiding this polymerization problem by first blocking the amine groups of the protein with citraconic acid, which has been used successfully with antibodies (Jansons and Mallett, 1980).

However, even with the potential for protein–protein conjugation, carbodiimide coupling of peptides and proteins to liposomes can be performed with EDC without blocking polypeptide amines. The approach is similar to that described for the EDC conjugation of hapten molecules to carrier proteins to form immunogen complexes (Chapter 19, Section 3). Thus, this method may be used to prepare peptide hapten–liposome conjugates for immunization purposes (Figure 21.22). The procedure also works particularly well for coupling molecules containing only carboxylates to the amines on the liposomes.

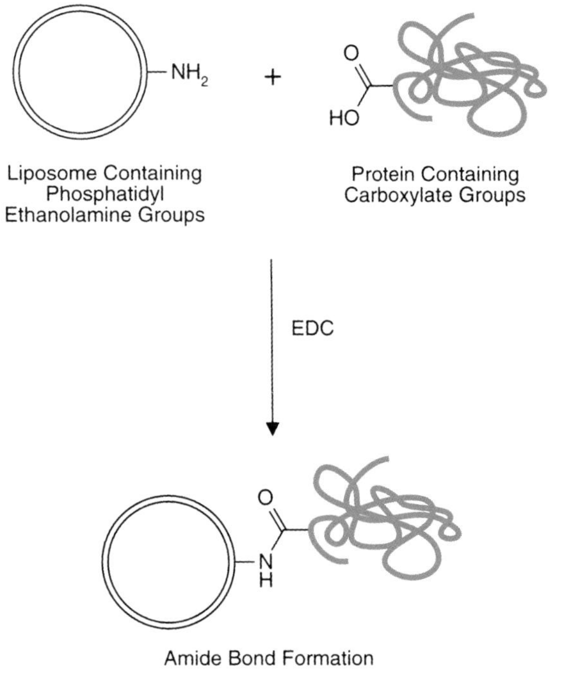

FIGURE 21.22 A protein may be conjugated with a liposome containing PE groups using a carbodiimide reaction with EDC.

Protocol

1. Prepare liposomes containing PE by any desired method. For instance, the common recipe mentioned in Section 1.2 (this chapter) that involves mixing PC:cholesterol:PG:PE in a molar ratio of 8:10:1:1 may be used. Thoroughly emulsify the liposome construction to obtain a good population of SUVs. The final liposome suspension should be in 20-m*M* sodium phosphate, 0.15-*M* NaCl, pH 7.2. Adjust the concentration to about 5 mg lipid/ml buffer.
2. Add the protein or peptide to be conjugated to the liposome suspension. The protein may be dissolved first in PBS, pH 7.2, and an aliquot added to the reaction lipid mixture. The amount of protein to be added can vary considerably, depending on the abundance of the protein and the desired final density required. Reacting from 1 mg protein per ml liposome suspension up to about 20 mg protein/ml can be performed.
3. Add 10 mg EDC per ml of lipid/protein mixture. Solubilize the carbodiimide using a vortex mixer.
4. React for 2 h at room temperature. If liposome aggregation or protein precipitation occurs during the crosslinking process, scale back the amount of EDC added to the reaction.
5. Purify the conjugate by gel filtration using a column of Sephadex G-75.

7.4. Conjugation via Glutaraldehyde Coupling to PE Lipid Derivatives

Glutaraldehyde is among the earliest homobifunctional crosslinkers employed for protein conjugation (Chapter 5, Section 6.2). It reacts with amine groups through several routes, including the formation of Schiff base linkages which can be reduced with borohydride or cyanoborohydride to create stable secondary amine bonds. Although very efficient in reacting with proteins, glutaraldehyde typically causes extensive polymerization accompanied by precipitation of high molecular weight oligomers during the reaction. Even with this significant disadvantage, the reagent is still used routinely in protein conjugation techniques. Some control over the glutaraldehyde conjugation processes may be accomplished using the protocols discussed in Chapter 15, Section 2.1.

Liposomes containing PE residues can be reacted with glutaraldehyde to form an activated surface possessing reactive aldehyde groups. A two-step glutaraldehyde reaction strategy is probably best when working with liposomes, since precipitated protein would be difficult to remove from a vesicle suspension.

The following protocol describes the two-step method wherein the liposome is glutaraldehyde-activated, purified away from excess crosslinker, and then coupled to a protein by reductive amination (Figure 21.23).

Protocol

1. Prepare a liposome suspension, containing PE, at a total-lipid concentration of 5 mg/ml in 0.1-*M* sodium phosphate, 0.15-*M* NaCl, pH 6.8. Maintain all lipid-containing solutions under an inert gas atmosphere. Degas all buffers and bubble them with nitrogen or argon prior to use.
2. Add glutaraldehyde to this suspension to obtain a final concentration of 1.25%.
3. React overnight at room temperature under a nitrogen blanket.
4. Purify the activated liposomes from excess glutaraldehyde by gel filtration (using Sephadex G-50) or by dialysis against PBS, pH 6.8.
5. Dissolve the protein or peptide to be conjugated at a concentration of 10 mg/ml in 0.5-*M* sodium carbonate, pH 9.5. Mix the activated liposome suspension with the polypeptide solution at the desired molar ratio to effect the conjugation. Mixing the equivalent of 4 mg of protein per mg of total lipid usually results in acceptable conjugates.
6. React overnight at 4°C under an atmosphere of nitrogen.
7. To reduce the resultant Schiff bases and any excess aldehydes, add sodium borohydride to a final concentration of 10 mg/ml.

7.5. Conjugation via DMS Crosslinking to PE Lipid Derivatives

Dimethyl suberimidate (DMS) is a homobifunctional crosslinking agent containing amine-reactive imidoester groups on both ends. The compound is reactive toward the ε-amine groups of lysine residues and N-terminal α-amines in the pH range of 7 to 10 (pH 8–9 is optimal). The resulting amidine linkages are positively charged at physiological pH, thus maintaining the positive charge contribution of the original amine.

bis-Imidoesters like DMS may be used to couple proteins to PE-containing liposomes by crosslinking with the amines on both molecules (Figure 21.24). However, single-step crosslinking procedures using homobifunctional reagents are particularly subject to uncontrollable polymerization of protein in solution. Polymerization is possible because the procedure is performed with the liposomes, protein, and crosslinker all in solution at the same time.

The reaction is carried out in 0.2-*M* triethanolamine, pH 8.2. DMS should be the limiting reagent in the reaction to avoid blocking all amines on both molecules with only one end of the crosslinker, thus eliminating any conjugation. The amounts of total lipid and protein

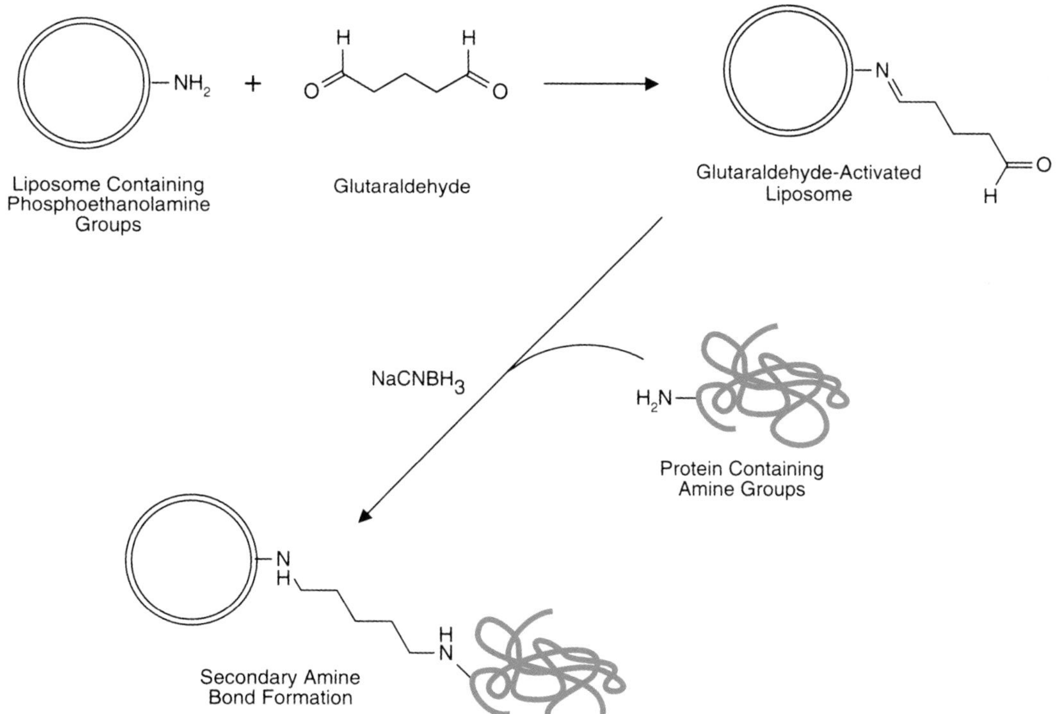

FIGURE 21.23 Glutaraldehyde activation of PE-containing liposomes may be used to couple protein molecules.

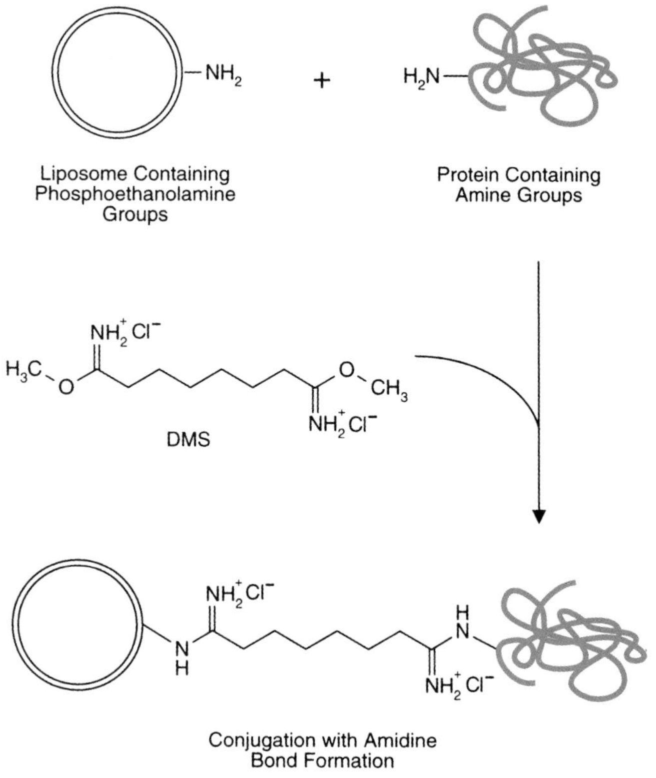

FIGURE 21.24 The homobifunctional crosslinker DMS may be used to conjugate PE-containing liposomes with proteins via amidine bond formation.

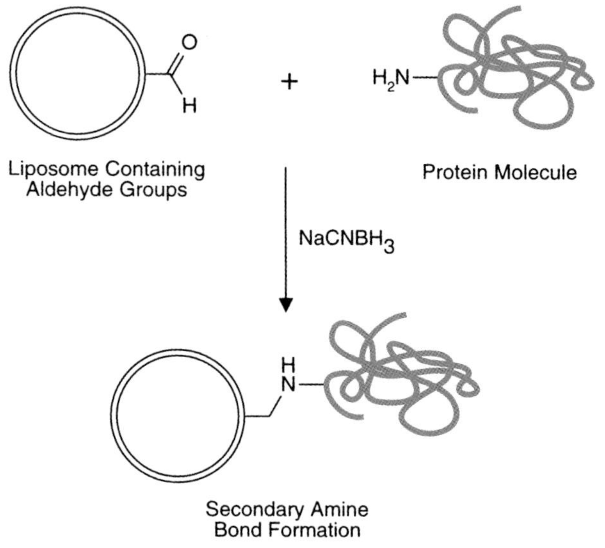

FIGURE 21.25 Glycolipids incorporated into liposomes may be oxidized with periodate to produce aldehydes suitable for coupling proteins via reductive amination.

in solution may have to be adjusted to optimize each conjugation reaction and avoid precipitation of protein or aggregation of liposomes.

7.6. Conjugation via Periodate Oxidation Followed by Reductive Amination

Periodate-oxidized liposomes which contain glycolipid moieties may be used to couple proteins and other amine-containing molecules by reductive amination. Section 2 (this chapter) describes the oxidative procedure that results in the formation of reactive aldehyde groups on the liposomal surface. Amine-containing polypeptides form Schiff base linkages with the aldehyde groups under alkaline conditions. The addition of a reducing agent such as borohydride or cyanoborohydride reduces the labile Schiff bases to form stable secondary amine bonds (Figure 21.25).

The following generalized method is based on the procedure described by Heath *et al.* (1981) for the coupling of immunoglobulins to liposomes containing glycosphingolipids.

Protocol

1. Prepare a periodate-oxidized liposome suspension containing glycolipid components according to Section 2 of this chapter. Adjust the concentration of total lipid to about 5 mg/ml.
2. Dissolve the protein to be coupled in 20-m*M* sodium borate, 0.15-*M* NaCl, pH 8.4, at a concentration of at least 10 mg/ml.
3. Add 0.5 ml of protein solution to each ml of liposome suspension with stirring.
4. Incubate for 2 h at room temperature to form Schiff base interactions between the aldehydes on the vesicles and the amines on the protein molecules.
5. In a fume hood, dissolve 125 mg of sodium cyanoborohydride in 1 ml water (makes a 2-*M* solution). *Caution: Highly toxic compound; handle with care.* This solution may be allowed to sit for 30 min to eliminate most of the hydrogen-bubble evolution that could affect the vesicle suspension.
6. Add 10 μl of the cyanoborohydride solution to each ml of the liposome reaction.
7. React overnight at 4°C.
8. Remove unconjugated protein and excess cyanoborohydride by gel filtration using a column of Sephadex G-50 or G-75.

7.7. Conjugation via SPDP-Modified PE Lipid Derivatives

N-Succinimidyl 3-(2-pyridyldithio)propionate (SPDP) is one of the most popular heterobifunctional crosslinking agents available, especially for protein conjugation (Chapter 6, Section 1.1). The activated NHS ester end of SPDP reacts with amine groups in proteins and other

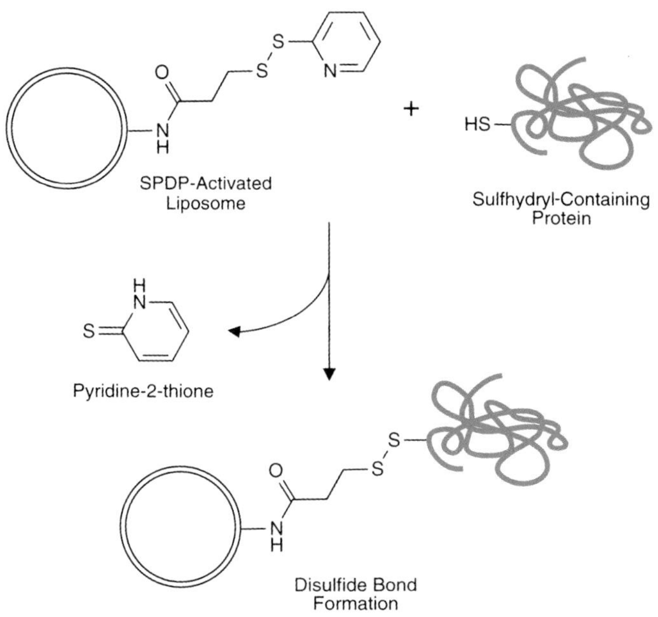

FIGURE 21.26 SPDP-activated liposomes can be used to couple sulfhydryl-containing proteins, forming disulfide linkages.

molecules to form an amide linkage (Figure 21.26). The 2-pyridyldithiol group at the other end reacts with sulfhydryl residues to form a disulfide linkage with sulfhydryl-containing molecules (Carlsson *et al.*, 1978).

SPDP is also a popular choice for coupling sulfhydryl-containing molecules to liposomes. PE residues in vesicles may be activated with this crosslinker to form pyridyl disulfide derivatives that can react with sulfhydryls to form disulfide linkages. Unlike the iodoacetyl- and maleimide-based crosslinkers discussed previously, the linkage formed with SPDP is reversible by simple disulfide reduction. Pure PE may be activated with SPDP prior to its incorporation into a liposome, or intact liposomes containing PE may be activated using the methods described in Section 2 (this chapter). Activation of PE with an SPDP crosslinker forms the intermediate reactive pyridyldithiopropionate-PE (PDP–PE) derivative. Stearylamine can also be activated with SPDP to be used in liposome conjugation (Goundalkar *et al.*, 1983). If the long-chain version, sulfo-LC-SPDP, is used with intact vesicles, the crosslinker will be water soluble and may be added directly to the buffered suspension without prior organic solvent dissolution. The negatively charged sulfonate group on its NHS ring prevents the reagent from penetrating the hydrophobic region of the lipid bilayer. Thus, only the outer surface of the liposomes will be modified. Using preactivated PE, both inner and outer surfaces end up containing reactive pyridyl disulfide groups. If liposome-sequestered components have the potential to react with this functional group or are sensitive, activation of intact liposomes with sulfo-LC-SPDP may be the better tactic.

Hydrophilic PEG-based heterobifunctional crosslinkers that are similar in design to SPDP are also available (Quanta BioDesign and Thermo Fisher). These compounds have the general design NHS–PEG_n–pyridyl disulfide and contain a water soluble spacer arm instead of the aliphatic spacers present in SPDP and sulfo-LC-SPDP (see Chapter 18, Section 1.2, and Chapter 15, Section 2.2). Use of the PEG-based reagents will result in better biocompatibility of the modified liposome surface and more accessibility of the reactive group for subsequent conjugation with proteins and other thiol-containing ligands. To use these reagents in the following coupling protocols, simply replace the SPDP compound in the activation procedure described previously (Section 2.2) with the corresponding PEG-based reagent by using an equivalent mole quantity for the reaction.

The following protocol is a suggested method for coupling sulfhydryl-containing proteins to SPDP-activated vesicles.

1. Prepare a 5 mg/ml liposome suspension containing a mixture of PC:cholesterol:PG:PDP–PE in molar ratios of 8:10:1:1. The PDP–PE activated lipid may be prepared according to the method described in Section 2.2, this chapter. The emulsification may be performed by any established method (Section 1, this chapter). Suspend the vesicles in 50-m*M* sodium phosphate, 0.15-*M* NaCl, 10-m*M* ethylenediamine triacetic acid (EDTA), pH 7.2.
2. Add at least 5 mg/ml of a sulfhydryl-containing protein or other molecule to the SPDP-modified vesicles to effect the conjugation reaction. Molecules lacking available sulfhydryl groups may be modified to contain them by a number of methods (Chapter 2, Section 4.1). The conjugation reaction should be performed in the presence of at least 10-m*M* EDTA to prevent metal-catalyzed sulfhydryl oxidation.
3. React overnight with stirring at room temperature. Maintain the suspension in a nitrogen or argon atmosphere to prevent lipid oxidation.
4. The modified liposomes may be separated from excess protein by gel filtration using Sephadex G-75 or by centrifugal floatation in a polymer gradient (Derksen and Scherphof, 1985).

7.8. Conjugation via SMPB-Modified PE Lipid Derivatives

Succinimidyl-4-(*p*-maleimidophenyl)butyrate (SMPB, Chapter 6, Section 1.6), is a heterobifunctional crosslinking agent that has an amine-reactive NHS ester on one end and a sulfhydryl-reactive maleimide group on the

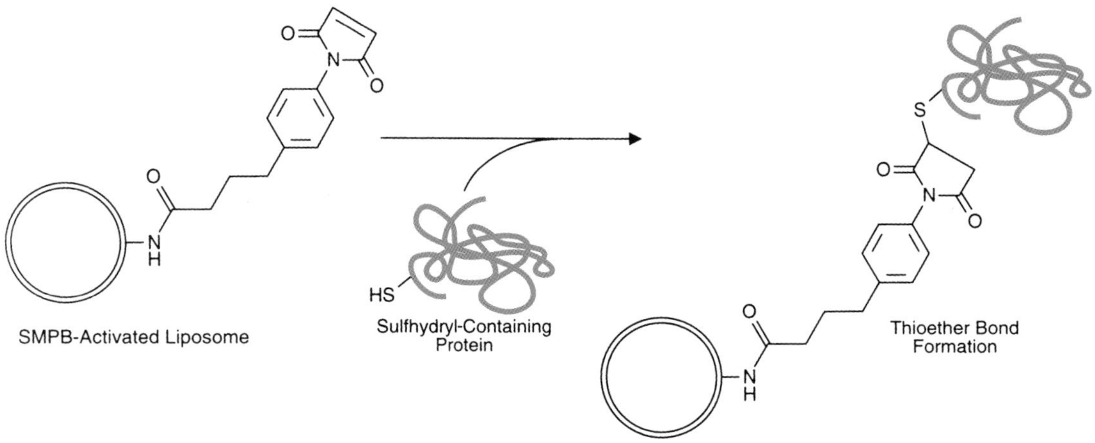

FIGURE 21.27 SMPB-activated liposomes may be used to couple thiol-containing protein molecules, forming stable thioether linkages.

other. Conjugates formed using SMPB are linked by stable amide and thioether bonds.

SMPB can be used to activate PE residues to contain sulfhydryl-reactive maleimide groups (Section 2, this chapter). Lipid vesicles formed with reactive maleimidophenylbutyrate–PE (MPB–PE) components can thus couple proteins through available SH groups, forming thioether linkages (Derksen and Scherphof, 1985) (Figure 21.27). A comparison with SPDP-produced conjugates concluded that SMPB formed more stable complexes that survived in serum for longer periods (Martin and Papahadjopoulos, 1982). The following protocol is a generalized method for the conjugation of proteins to SMPB-activated liposomes.

1. Prepare a liposome suspension, containing MPB–PE, at a total lipid concentration of 5 mg/ml in 0.05-*M* sodium phosphate, 0.15-*M* NaCl, pH 7.2. Activation of DPPE with SMPB is described in Section 3.4. A suggested lipid composition for vesicle formation is PC:cholesterol:PG:MPB–PE mixed at a molar ratio of 8:10:1:1. The presence of relatively high levels of cholesterol in the liposomal recipe dramatically enhances the conjugation efficiency of the component MPB–PE groups (Martin *et al.*, 1990). Any method of emulsification to create liposomes of the desired size and morphology may be used (Section 1.2, this chapter)
2. Dissolve a sulfhydryl-containing protein at a concentration of at least 5 mg/ml in 0.05-*M* sodium phosphate, 0.15-*M* NaCl, 10-m*M* EDTA, pH 7.2. The sulfhydryl groups on the protein molecule may be indigenous or created by any of the methods described in Chapter 2, Section 4.1.
3. Mix the protein solution with the liposome suspension in equal volume amounts.
4. React overnight at room temperature with stirring. Maintain an atmosphere of nitrogen over the reaction to prevent lipid oxidation.
5. Separate unreacted protein from modified liposomes by gel filtration using a column of Sephadex G-75 or by centrifugal floatation in a polymer gradient (Derksen and Scherphof, 1985).

7.9. Conjugation via SMCC-Modified PE Lipid Derivatives

Succinimidyl-4-(*N*-maleimidomethyl)cyclohexane-1-carboxylate (SMCC) is a heterobifunctional crosslinker with significant utility in crosslinking proteins, particularly in the preparation of antibody–enzyme (Chapter 20) and hapten–carrier (Chapter 19) conjugates (Hashida and Ishikawa, 1985; Dewey *et al.*, 1987). It is normally used in a two-step crosslinking procedure, wherein the NHS ester end of the reagent is first reacted with primary amine groups on proteins or other molecules to form stable amide bonds. This creates a reactive intermediate containing terminal maleimide groups on the modified molecule. The maleimide end is specific for coupling to sulfhydryls when the reaction pH is in the range of 6.5 to 7.5 (Smyth *et al.*, 1964). Addition of a sulfhydryl-containing protein forms a stable thioether linkage with the SMCC-activated molecule.

In a similar manner, PE may be activated through its head-group primary amine to possess reactive maleimide groups capable of coupling sulfhydryl-containing proteins to liposomes (Figure 21.28). The method of derivatizing DPPE with SMCC is essentially the same as that described for SMPB (Section 2, this chapter). SMCC, however, contains a more stable maleimide reactive group toward hydrolysis in aqueous reaction environments, due to the proximity of an aliphatic cyclohexane ring rather than the aromatic phenyl group of SMPB. In protein conjugation to liposomes, this stability may translate into higher activity and more efficient crosslinking. A general protocol for the coupling of sulfhydryl-containing proteins to liposomes containing

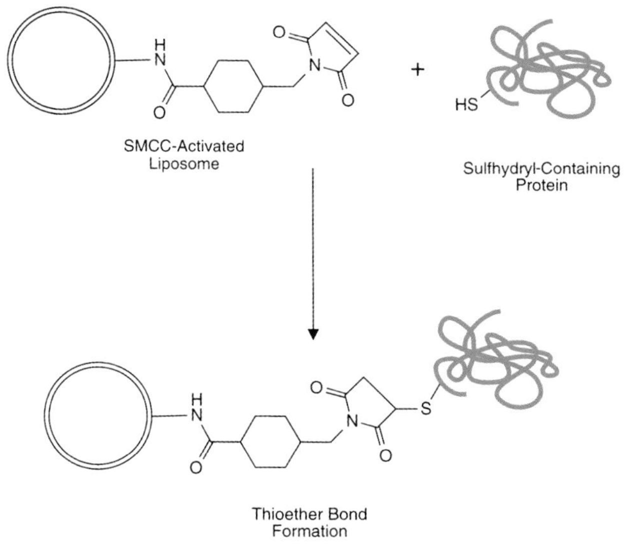

FIGURE 21.28 The reaction of an SMCC-activated liposome with a sulfhydryl-containing protein forms stable thioether bonds.

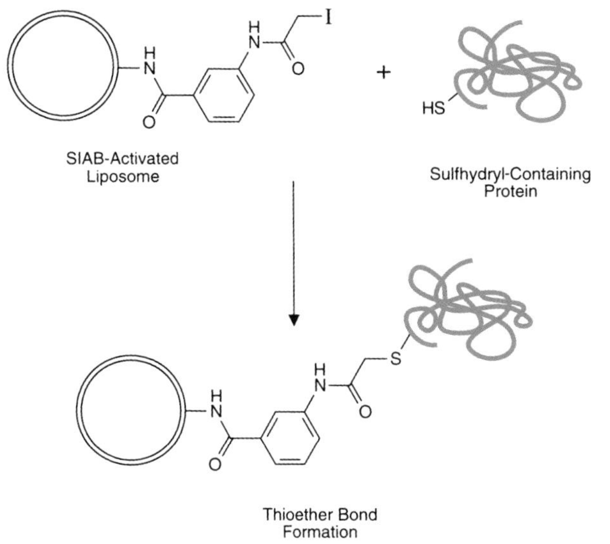

FIGURE 21.29 SIAB-activated liposomes can couple with sulfhydryl-containing proteins to produce thioether linkages.

SMCC–PE is essentially the same as that described previously for SMPB (Section 7.8, this chapter).

7.10. Conjugation via Iodoacetate-Modified PE Lipid Derivatives

Iodoacetate derivatives have been used for decades to block or crosslink sulfhydryl groups in proteins and other molecules (Chapter 2, Section 5.2). At mildly alkaline pH values (pH 8–8.5), iodoacetyl derivatives are almost entirely selective toward the cysteine SH groups in proteins. Disulfide reduction or thiolation reagents can be used to create the required sulfhydryl groups on proteins containing no free sulfhydryls.

Crosslinking reagents containing an amine-reactive NHS ester on one end and an iodoacetyl group on the other end are particularly useful for two-step protein conjugation. Heterobifunctional reagents like SIAB (Chapter 6, Section 1.5), SIAX (Chapter 6, Section 1.8), or SIAC (Chapter 6, Section 1.9) can be used to modify amine-containing molecules, resulting in iodoacetyl derivatives capable of coupling to sulfhydryl-containing molecules.

Liposomes containing PE lipid components may be activated with these crosslinkers to contain iodoacetyl derivatives on their surface (Figure 21.29). The reaction conditions described in Chapter 6, Section 1.5, may be used, substituting a liposome suspension for the initial protein being modified in that protocol. The derivatives are stable enough in aqueous solution to allow purification of the modified vesicles from excess reagent (by dialysis or gel filtration) without loss of activity. The only consideration is to protect the iodoacetyl derivative from light, which may generate iodine and reduce the activity of the intermediate. Finally, the modified liposome can be mixed with a sulfhydryl-containing molecule to effect the conjugation through a thioether bond.

Alternatively, pure PE may be derivatized to contain iodoacetyl groups prior to vesicle formation. This may be performed using heterobifunctional crosslinkers or through the use of iodoacetic anhydride according to Hashimoto *et al.* (1986). However, a single iodoacetyl group on PE was found not to be sufficiently extended from the vesicle surface to allow efficient protein coupling. Only after creating a longer spacer by reacting 2-mercaptoethylamine with the initial iodoacetamide derivative and then reacting a second iodoacetic anhydride to form an extended, reactive arm, did the derivative possess enough length to give it efficient conjugation ability (Figure 21.30). This example illustrates the importance of a longer spacer in avoiding steric problems during conjugation to liposomal vesicles surface.

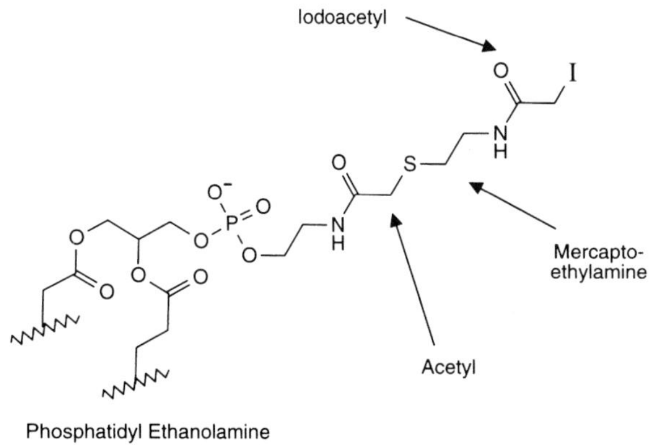

FIGURE 21.30 An iodoacetamide derivative of PE containing an extended spacer arm can be constructed through a carbodiimide coupling of iodoacetic acid to PE, followed by reaction with 2-mercaptoethylamine, and finally another reaction with iodoacetate.

CHAPTER

22

Enzyme Modification and Conjugation

Enzymes are widely used in bioconjugate chemistry as detection components in assay systems. The catalytic activity of an enzyme can be used to turn substrate molecules into chromogenic, fluorescent, or chemiluminescent products, which are easily detectable or quantifiable by imaging, microscopy, or spectroscopy. If an enzyme is conjugated to a targeting molecule specific for some analyte of interest, then an assay system can be constructed to localize or measure the analyte. The most common targeting molecule is an antibody having antigen-binding specificity for the substance to be measured. An enzyme conjugated to such an antibody can be used to visualize the presence of antigen. Due to the advantages of this simple concept, enzyme-linked immunoadsorbent assays (ELISAs) have become the most important type of immunoassay system available.

The rapid turnover rate of some enzymes allows ELISAs to be designed that far surpass the sensitivity of radiolabeling techniques. In addition, substrates can be chosen to produce soluble products that can be accurately quantified by their absorbance or fluorescence. Alternatively, substrates are available that form insoluble, highly colored precipitates, excellent for localizing antigens in blots, cells, or tissue sections. The flexibility of enzyme-based assay systems makes the chemistry of enzyme conjugation one of the most important application areas in bioconjugate techniques.

In addition, the immobilization of enzymes for use in catalytic transformations (called immobilized reactors) has also become an important field in the use of these proteins. Specialized immobilized reactors are being used to cleave or modify biological molecules, to synthesize complex organic compounds, to produce food products, and for the production of bioenergy molecules from biomass feedstock. See Chapter 1 for a review of immobilized reactor applications and Chapter 15 for the activation and coupling methods that can be used to covalently attach enzymes to various types of insoluble support materials.

The following sections briefly describe the principal enzymes used for conjugation with other protein molecules, particularly in the design of ELISA and other immunoassay systems.

1. PROPERTIES OF COMMON ENZYMES

1.1. Horseradish Peroxidase (HRP)

HRP (donor:hydrogen peroxide oxidoreductase; EC 1.11.1.7), derived from horseradish roots, is a enzyme of molecular weight 40,000 that can catalyze the reaction of hydrogen peroxide with certain organic, electron-donating substrates to yield highly colored products (Figure 22.1). The reaction of HRP with its fundamental substrate, H_2O_2, forms a stable intermediate that can dissociate in the presence of a suitable electron donor, oxidizing the donor and potentially creating a color change. The donor can consist of oxidizable molecules like ascorbate, cytochrome *c*, ferrocyanide, or the leuco forms of many dyes. A large variety of

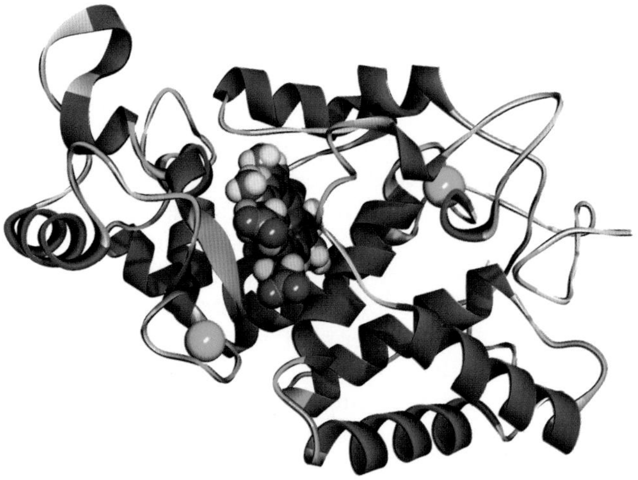

FIGURE 22.1 Horseradish peroxidase shown as the ribbon structure with the heme ring in its active center and two bound calcium ions. The molecular model is based on structure 1H58 in the RCSB Protein Databank by Berglund *et al.* (2002).

Bioconjugate Techniques, Third Edition.
DOI: http://dx.doi.org/10.1016/B978-0-12-382239-0.00022-4

electron-donating dye substrates are commercially available for use as HRP detection reagents. Some of them can be used to form soluble colored products for use in spectrophotometric detection systems, while other substrates form insoluble products that are especially appropriate for staining techniques. In addition, substrates are available that create fluorescent or chemiluminescent products upon oxidation with HRP. The chemiluminescent substrates are among the most sensitive of all detection reagents, facilitating the detection of as little as attogram quantities of many targeted analytes. The pH optimum for HRP is 7.0 although particular substrate detection reactions may be performed at pH values slightly different from neutrality.

The use of antibody–HRP or streptavidin–HRP conjugates in peroxidase-catalyzed enhanced chemiluminescent assays can result in one of the most sensitive detection methods for assaying targeted analytes in ELISA and western blotting applications. The reaction cascade that occurs during HRP catalysis can be dramatically improved by the addition of various enhancer molecules, which create oxidized intermediates leading to the oxidation and light emission of a chemiluminescent substrate such as luminol. Vdovenko *et al.* (2012) analyzed this reaction using a multi-factorial design of experiments (DOE) approach to identify the best combination and concentrations of H_2O_2, luminol, and two different enhancer compounds (3-(10′-phenothiazinyl) propane-1-sulfonate and 4-morpholinopyridine. This combination at an optimized concentration resulted in the best signal-to-noise ratio and the longest chemiluminescent emission.

HRP is a hemoprotein containing photohemin IX as its prosthetic group. The presence of the heme structure gives the enzyme its characteristic color and maximal absorptivity at 403 nm. The ratio of its absorbance in solution at 403 nm to its absorbance at 275 nm, called the RZ or Reinheitzahl ratio, can be used to approximate the purity of the enzyme. However, at least seven isoenzymes exist for HRP (Shannon *et al.*, 1966; Kay *et al.*, 1967; Strickland *et al.*, 1968), and their RZ values vary from 2.50 to 4.19. Thus, unless the RZ ratio is precisely known or determined for the particular isoenzyme of HRP utilized in the preparation of an antibody–enzyme conjugate, subsequent measurement after crosslinking would yield questionable results in the determination of the amount of HRP present in the conjugate.

HRP is a glycoprotein that contains significant amounts of carbohydrate. Its polysaccharide chains are often used in crosslinking reactions to couple the enzyme to targeting molecules. Mild oxidation of its associated glycan sugar residues with sodium periodate generates reactive aldehyde groups that can be used for conjugation to amine-containing molecules. Reductive amination of oxidized HRP to antibody molecules in the presence of sodium cyanoborohydride is perhaps the simplest method of preparing highly active conjugates with this enzyme (Chapter 4, Section 1.4, and Chapter 20, Section 1.3).

Other methods of HRP conjugation include the use of the homobifunctional reagent glutaraldehyde (Chapter 5, Section 6.2, and Chapter 15, Section 2.1) and the heterobifunctional crosslinker, SMCC (succinimidyl-4-(*N*-maleimidomethyl)cyclohexane-1-carboxylate) (Chapter 6, Section 1.3). Using glutaraldehyde, a two-step protocol is usually employed to try to limit the extent of oligomer formation. Even using the most highly controlled reactions, however, this method often causes unacceptable amounts of precipitated conjugate. Despite this disadvantage, glutaraldehyde conjugation is still routinely used, especially in the preparation of some antibody–enzyme reagents that go into established diagnostic assays. The use of the *N*-hydroxysuccinimide (NHS) ester–maleimide crosslinker, SMCC, provides far better control over the conjugation process. SMCC is usually reacted first with HRP to create a derivative containing sulfhydryl-reactive maleimide groups. HRP activation of the native enzyme should result in the modification of a maximum of about two amine groups on the protein, because HRP only contains two lysines. An increase in the activation level can be realized if the enzyme first is modified with ethylenediamine (EDA) using the carbodiimide EDC according to the methods described in Chapter 19 for the production of cationized bovine serum albumin (cBSA). The EDA-modified HRP is also more stable than the unmodified version, so cationization may have benefits in the retention of enzyme activity. The maleimide-activated enzyme can be purified and freeze-dried, providing a ready source of modified HRP to react with a sulfhydryl-containing antibody. Several preactivated forms of this enzyme are available from Thermo Fisher.

The size of HRP is an advantage in preparing antibody–enzyme conjugates, since the overall complex size also can be designed to be small. Relatively low-molecular-weight conjugates are able to penetrate cellular structures better than large, polymeric complexes. This is why HRP conjugates are often the best choice for immunohistochemical (IHC) and immunocytochemical staining techniques. Small conjugate size means greater accessibility to antigenic structures within tissue sections.

Another distinctive advantage of HRP is its robust nature and stability, especially under the conditions employed for crosslinking. HRP is stable for years in a freeze-dried state, and the purified enzyme can be stored in solution at 4°C for many months without significant loss of activity. The enzyme also retains excellent activity after being modified with a conjugation reagent or after being periodate-oxidized to form aldehyde groups on its polysaccharide chains. Depending on the methods used for crosslinking, HRP conjugates can be constructed to have a high ratio of enzyme to antibody or a low ratio—both retaining high specific activity.

The disadvantages associated with HRP are several. The enzyme only contains two available primary ε-amine groups—extraordinarily low for most proteins—thus limiting its ability to be activated with amine-reactive heterobifunctionals. HRP is sensitive to the presence of many antibacterial agents, especially azide. It is also reversibly inhibited by cyanide and sulfide (Theorell, 1951). Finally, while the enzymatic activity of HRP is extremely high, its useful life span or practical substrate development time is somewhat limited. After about an hour of substrate turnover, in some situations its activity can be decreased severely.

Nevertheless, HRP is by far the most popular enzyme used in antibody–enzyme conjugates. One survey of enzyme use stated that HRP is incorporated in about 80% of all antibody conjugates, most of them utilized in diagnostic assay systems.

1.2. Alkaline Phosphatase

Alkaline phosphatases [AP, orthophosphoric monoester phosphorylase (alkaline optimum); EC 3.1.3.1] represent a large family of almost ubiquitous isoenzymes found in organisms from bacteria to animals (Figure 22.2). In mammals, there are two forms of AP, one form present in a variety of tissues and another form found only in the intestines. They share common attributes in that the phosphatase activity is optimal at pH 8 to 10, is activated by the presence of divalent cations, and is inhibited by cysteine, cyanide compounds, arsenate various metal chelators, and phosphate ions.

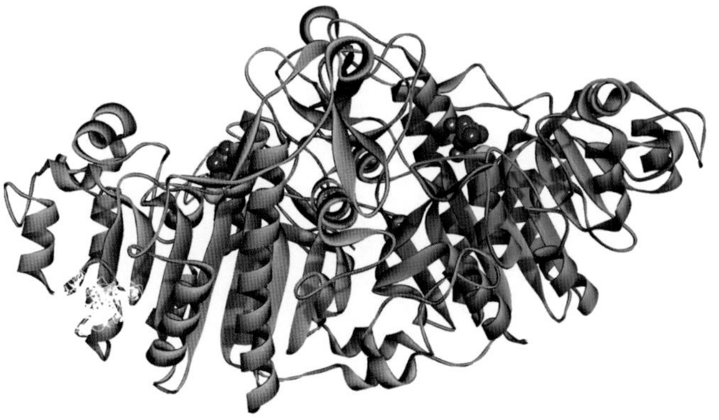

FIGURE 22.2 Alkaline phosphatase shown as the dimer ribbon structure with two molecules of phosphate bound in its active sites. The molecular model is based on structure 3TG0 in the RCSB Protein Databank by Bobyr *et al.* (2012).

Most conjugates created with AP utilize the form isolated from calf intestine.

AP isoenzymes can cleave associated phosphomonoester groups from a wide variety of substrates. The exact biological function of these enzymes is not fully understood, although they definitely function as the opposite of kinase enzymes—removing phosphate groups from phosphorylated proteins and thus affecting signal transduction processes within cells. They behave *in vivo* in their classic phosphohydrolase role at alkaline pH values, but at neutral pH AP isoenzymes can act as phosphotransferases. In this sense, suitable phosphate acceptor molecules can be utilized in solution to increase the reaction rates of AP on selected substrates. Typical phosphate acceptor additives include diethanolamine Tris, and 2-amino-2-methyl-1-propanol. The presence of these additives in substrate buffers can dramatically increase the sensitivity of AP-based ELISA determinations, even when the substrate reaction is performed in alkaline conditions.

Calf intestinal AP has a molecular weight of about 140,000. The active site of AP contains two zinc ions and a single magnesium ion, both of which are essential for activity (Kim and Wyckoff, 1991). Substrate development with AP should thus be carried out in buffered environments containing small concentrations of these divalent cations to maintain optimal active site conformation. Avoid the presence of metal chelators such as EDTA, since they may extract these ions from the enzyme and inhibit activity. The pH optimum for APs can vary from pH 8 to 10, depending on the type of isoenzyme. Calf intestinal AP peaks in activity at the higher pH values of this range, and substrate reactions are commonly performed in diethanolamine buffer at pH 9.8. The calf intestinal enzyme has the highest catalytic rate constant yet discovered for AP isoenzymes, and it is available commercially in high activity for bioconjugation applications.

Purified preparations of calf intestinal AP maintained in solution are usually stored in the presence of a stabilizer, which is typically 3-*M* NaCl. The enzyme may also be lyophilized, but it may experience activity loss with each freeze–thaw cycle. AP is not stable under acidic conditions. Lowering the pH of an AP solution to 4.5 reversibly inhibits the enzyme. It is recommended that all handling, storage, and use of AP be carried out under conditions > pH 7.0 to maintain the highest possible catalytic activity.

AP is often a difficult enzyme to work with when preparing enzyme conjugates. Activity losses may occur upon modification with a crosslinking agent or after coupling to an antibody molecule. Simply following established protocols for making antibody–AP conjugates does not always ensure retention of enzyme activity. Sometimes activity losses can be traced to particular batches or to certain suppliers of the enzyme. Using a highly purified, high-activity AP preparation helps to maintain good resultant activity in the conjugate.

Ironically, AP is the enzyme of choice for some applications due to its stability. Since it can withstand the moderately high temperatures associated with hybridization assays better than HRP, AP often is the enzyme of choice for labeling oligonucleotide probes (Kaatz *et al.*, 2012). AP is also capable of maintaining enzymatic activity for extended periods of substrate development. Increased sensitivity can be realized in ELISA procedures by extending the substrate incubation time to hours and sometimes even days, provided that the background interference is low. These properties make AP the second most popular choice for antibody–enzyme conjugation (behind HRP), being used in almost 20% of all commercial enzyme-linked assays.

Conjugation methods typically employed with AP include glutaraldehyde-mediated crosslinking (Chapter 5, Section 6.2) and the use of the heterobifunctional reagents SMCC (Chapter 6, Section 1.3) or SPDP (*N*-succinimidyl 3-(2-pyridyldithio)propionate) (Chapter 6, Section 1.1). Heterobifunctional crosslinkers provide the best control over the crosslinking process and typically result in antibody–enzyme conjugates of high activity. Many conjugation protocols incorporate a sodium phosphate buffer system to reversibly block the AP active site during chemical modification. This prevents derivatization from occurring in the catalytic site, thus better retaining activity in the resultant conjugate.

1.3. β-Galactosidase

β-Galactosidase (β-Gal; β-D-galactoside galactohydrolase; EC 3.2.1.23; also called lactase) catalyzes the hydrolysis of β-D-galactoside in the presence of water to galactose and alcohol. This type of enzyme is found widespread in many microorganisms, plants, and animals. β-Gal can be used to determine lactose in biological fluids and it is employed in food processing operations, particularly in immobilized form. The enzyme is often used as a reporter enzyme for monitoring gene activation and transcription (Beucher *et al.*, 2012; Ju *et al.*, 2012). β-Gal also has good characteristics when conjugated to antibody molecules or streptavidin for use in ELISA systems (Wallenfels and Weil, 1972; Byrne and Johnson, 1975; Kato *et al.*, 1975a,b).

β-Gal has a molecular weight of 540,000 and is composed of four identical subunits of MW 135,000, each with an independent active site (Melcher and Messer, 1973) (Figure 22.3). The enzyme has divalent metals as cofactors, with chelated Mg^{2+} ions required to maintain active site conformation. The presence of NaCl or dilute solutions (5%) of low-molecular-weight alcohols (methanol, ethanol, etc.) causes enhanced substrate turnover. β-Gal contains numerous sulfhydryl groups and is glycosylated.

Commercially available β-gal is usually isolated from *E. coli* and has a pH optimum at 7 to 7.5. By contrast, mammalian β-galactosidases usually have a pH optimum within the range of 5.5 to 6; thus, interference from endogenous β-gal during immunohistochemical staining can be avoided.

Due to the relatively high molecular weight of the enzyme, conjugates formed with antibodies and β-gal can be much bulkier than those associated with alkaline phosphatase or horseradish peroxidase. For this reason, antibody conjugates made with β-gal may have more difficulty penetrating tissue structures during immunohistochemical or immunocytochemical staining techniques than those made with the other enzymes.

Although numerous research articles have been written describing the preparation and use of antibody conjugates with β-gal, the enzyme remains a minor player in ELISA procedures. Less than 1% of all commercial ELISA products utilize this enzyme.

β-Gal may be conjugated to antibody molecules using the heterobifunctional reagent SMCC. This crosslinker is reacted first with an antibody through its amine-reactive NHS ester end to form a maleimide-activated derivative. This is in contrast with most antibody–enzyme conjugation schemes utilizing SMCC, wherein the enzyme is typically modified first and a sulfhydryl-containing antibody is coupled secondarily. However, since β-gal already contains abundant free sulfhydryl residues that can participate in coupling to a maleimide-activated protein, conjugations with this enzyme often are done with the antibody being the first

FIGURE 22.3 β-Galactosidase shown as the four-subunit biological assembly. The molecular model is based on structure 1BGL in the RCSB Protein Databank by Jacobson *et al.* (1994).

component modified. This route avoids having to create sulfhydryls on the antibody molecule, either by reduction or modification with a thiolation reagent. Thus, antibody–β-gal conjugates usually are simpler to make than using other enzymes.

1.4. Glucose Oxidase

Glucose oxidase (β-D-glucose: oxygen 1-oxidoreductase; EC 1.1.3.4; GO) is a flavoenzyme that catalyzes the oxidation of β-D-glucose to D-gluconolactone. The intermediate product of the catalysis is a reduced enzyme–$FADH_2$ complex that, in the presence of oxygen, gets oxidized back to enzyme–FAD with release of hydrogen peroxide. The enzyme consists of two identical subunits (MW 80,000 each) bound together by disulfide linkages (O'Malley and Weaver, 1972). GO contains two tightly bound flavin adenine dinucleotide (FAD) cofactors, one per subunit, which are critical to its oxidoreductase activity (Figure 22.4). Each subunit also contains one molecule of chelated iron. The intact protein consists of about 74% amino acids, 16% neutral carbohydrate, and 2% amino sugars (total molecular weight 160,000). GO operates under a relatively broad pH range of 4 to 7, but its pH optimum is 5.5. The commercially available preparation of GO is typically isolated from *Aspergillus niger*.

Glucose oxidase is widely used in diagnostic assays for the determination of glucose concentration in physiological fluids. Detectability of the oxidation products is done through an enzyme-coupled reaction wherein liberated H_2O_2 is reacted with peroxidase and a suitable chromogenic substrate. The development of substrate color thus is proportional to the amount of H_2O_2 released which is in turn related to the amount of glucose originally present. The production of hydrogen peroxide also can be quantified using a luminescence procedure with luminol to produce light in proportion to the glucose concentration (Williams *et al.*, 1976).

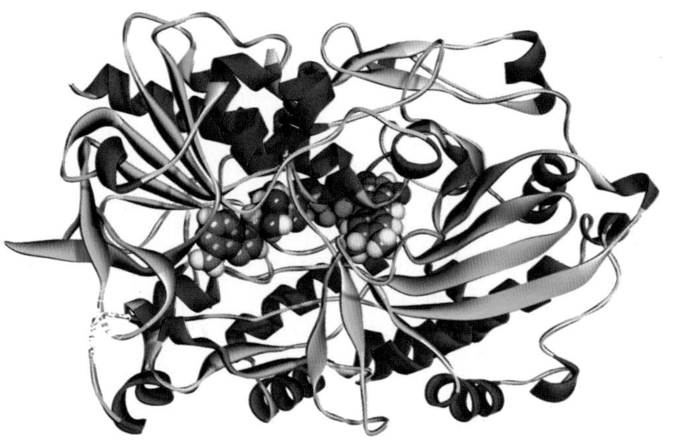

FIGURE 22.4 A single subunit of glucose oxidase shown as the ribbon structure with the $FADH_2$ cofactor. The biological assembly typically exists as the dimer. The molecular model is based on structure 3QVP in the RCSB Protein Databank by Kommoju *et al.* (2011).

GO is often used in solution phase reactions as well as being immobilized on "dip-sticks" and electrodes. Methods for the coupling of enzymes to solid supports can be found in Chapter 15. Although the overall clinical usage of GO is widespread, its use as conjugated to antibodies in enzyme-linked assay systems is minor compared to the popularity of other enzymes like HRP and AP. Of the total number of commercial diagnostic assays utilizing antibody–enzyme conjugates, GO is employed in less than 1% of clinical tests. The enzyme remains, however, an important tool in many assays developed for research use (Berron *et al.*, 2011; Holland *et al.*, 2011). One particular advantage to the enzyme is that there is no endogenous GO activity in mammalian tissues, making it an excellent choice for immunohistochemical staining procedures.

Antibody conjugates with GO can be made using the crosslinking agents glutaraldehyde (Chapter 5, Section 6.2) or SMCC (Chapter 6, Section 1.3). The heterobifunctional reagent SMCC provides the best control over the conjugation process and usually results in high-activity preparations. Also consider PEG-based heterobifunctional reagents as described in Chapter 18, Section 1.2, because the hydrophilicity of their crossbridge provides greater water solubility and less nonspecific binding in assay systems.

2. PREPARATION OF ACTIVATED ENZYMES FOR CONJUGATION

Enzymes may be modified to contain reactive groups useful for conjugation with other proteins. This operation may be carried out using homobifunctional (Chapter 5) or heterobifunctional (Chapters 6, 17, and 18) reagents that can covalently couple to some chemical target on the enzyme and result in a terminal reactive group that can crosslink with another molecule. Enzyme activation may also take advantage of the presence of polysaccharide constituents—oxidizing them with sodium periodate to form reactive aldehydes.

Whatever the method of conjugate creation, the most important considerations are retention of activity in the complex and prevention of extensive oligomer generation, which may cause precipitation. The following methods discuss some of the more common methods for producing enzyme conjugates. The list, however, is by no means exhaustive of every possible procedure used in the literature. Many other crosslinkers and reaction strategies as described in this book may be used with success, including the incorporation of polyvalent scaffolds to attach greater numbers of enzymes to a targeting molecule and thus enhance detectability, such as dendrimers (Chapter 8) and polymers (Chapter 18, Section 2).

2.1. Glutaraldehyde-Activated Enzymes

Glutaraldehyde is a homobifunctional crosslinker containing an aldehyde residue at both ends of a 5-carbon chain. Its primary reactivity is toward amine groups, but the reaction may occur by more than one mechanism. As discussed in Chapter 5, Section 6.2, and Chapter 15, Section 2.1, glutaraldehyde is able to crosslink proteins to create stable secondary amine linkages. Glutaraldehyde exists in a number of different forms in aqueous solution, such as hemiacetal and aldol ring structures as well as large α-, β-unsaturated polymers containing carbon–carbon double bonds. The double bonds of these polymers can undergo an addition reaction with amines that results in covalent bond formation even without a reductant being present. Fresh glutaraldehyde may contain little polymer formation, except if it is exposed to highly alkaline conditions. However, the older the preparation of glutaraldehyde, the more likely it is that it contains appreciable amounts of polymer. Thus, reactions with this crosslinking agent can result in indistinct conjugation products and may be difficult or impossible to reproduce or scale up as the conjugation need arises.

The high reactivity and indistinct forms of glutaraldehyde make it difficult to control the size and composition of the final conjugate. Proteins crosslinked with this reagent often form substantial amounts of precipitated products due to polymerization. The degree of oligomer formation can be moderated somewhat by using a two-step protocol, but the first protein activated with the molecule can still form large-molecular-weight complexes.

Despite the obvious disadvantages of glutaraldehyde-mediated conjugation, the crosslinker continues to be used to form enzyme–antibody complexes and to create bioconjugates used in other applications. Many diagnostic tests still utilize antibody–enzyme conjugates prepared through glutaraldehyde crosslinking procedures.

The one- and two-step procedures for enzyme activation and conjugation using glutaraldehyde can be found in Chapter 20, Section 1.2.

2.2. Periodate Oxidation Techniques

Molecules containing polysaccharide chains may be oxidized to possess reactive aldehyde residues by treatment with sodium periodate. Any adjacent carbon atoms containing hydroxyl groups will be affected, cleaving the carbon–carbon bond and transforming the hydroxyls into aldehydes. Glycoproteins may be oxidized in this manner to form reactive intermediates useful for crosslinking procedures involving reductive amination (Chapter 4, Section 4). This conjugation technique can direct the coupling process away from polypeptide active regions, thus helping to preserve catalytic activity or binding sites.

Enzymes that contain carbohydrate, such as HRP or GO, may be oxidized with periodate to create reactive derivatives that subsequently can be used to label antibodies or other targeting molecules at their amine groups. The aldehyde–HRP intermediate may be stored for extended periods in a frozen or lyophilized state without loss of activity (either enzymatic or coupling potential). Avoid, however, storage in a liquid state, since polymerization may occur—resulting in precipitation and loss of activity as Schiff base interactions between proteins builds up in solution.

The protocols for periodate oxidation of HRP and its conjugation with other proteins may be found in Chapter 20, Section 1.3.

2.3. SMCC-Activated Enzymes

The heterobifunctional crosslinker SMCC, or its water soluble analog sulfo-SMCC, can be used to activate enzymes through their amines, leaving terminal maleimide groups on the protein surface (Chapter 6, Section 1.3). The NHS ester end of the crosslinker reacts with ε-lysine or N-terminal amines to form amide bonds. The maleimide end of the reagent is stable enough in aqueous solution to allow purification of the activated enzyme prior to conjugation with a second protein. The maleimide group can react with sulfhydryl groups to create thioether linkages. A maleimide-activated enzyme may be stored in a lyophilized state for extended periods without loss of sulfhydryl-coupling capability.

The use of this type of heterobifunctional reagent allows controlled conjugations to take place, precisely regulating the exact ratio of each protein in the final complex and the size of the resultant conjugate. In addition, the second-stage conjugation through sulfhydryl groups provides directed coupling at discrete sites within a protein molecule, thus providing the potential to better avoid active centers or binding regions. For instance, antibody molecules can be coupled to enzymes in their hinge region after mild disulfide reduction to effect the crosslink in an area away from the antigen binding site.

Protocols for the activation of enzyme molecules with SMCC (or sulfo-SMCC) can be found in Chapter 20, Section 1.1. Conjugates formed using this method usually result in high-activity complexes giving excellent sensitivity for use in immunoassays or other applications. In addition, refer to Chapter 18, Section 1.2, for the use of similar NHS ester–maleimide crosslinkers that contain a discrete PEG cross-bridge to make the reagent much more hydrophilic. Antibody—enzyme conjugates formed from NHS–PEG$_n$–maleimide reagents perform better in assays than those made from crosslinkers with aliphatic cross-bridges, such as SMCC, because the hydrophobic nature of the aliphatic group often causes nonspecific binding of the conjugate to non-relevant components in a sample. Thus, antibody—enzyme conjugates formed using PEG-based reagents typically have better signal-to-noise ratios in assays, which can be very important when developing diagnostic or commercial assays.

2.4. Hydrazide-Activated Enzymes

Hydrazide groups can react with carbonyl groups to form stable hydrazone linkages. Derivatives of proteins formed from the reaction of their carboxylate side chains with adipic acid dihydrazide (Chapter 5, Section 8.1) and the water soluble carbodiimide EDC (Chapter 4, Section 1.1) create activated proteins that can covalently bind to aldehyde residues. Hydrazide-modified enzymes prepared in this manner can bind specifically to aldehyde groups formed by periodate oxidation of carbohydrates (Chapter 2, Section 4.4). These reagents can be used in assay systems to detect or measure glycoproteins in cells, tissue sections, or blots (Gershoni *et al.*, 1985).

Other molecules can be used in this type of assay approach, too. Hydrazide-modified (strept)avidin, lectins, biocytin, fluorescent probes, and other detectable molecules can be used to detect or image glycoconjugates in biological samples (Wilchek and Bayer, 1987).

The activation of enzymes using adipic acid dihydrazide and EDC is identical to the procedure outlined for the modification of (strept)avidin (Chapter 11, Section 5). Simply replace the (strept)avidin component for an equivalent mole amount of the enzyme of choice. Alternatively, hydrazide groups may be created on enzymes using the heterobifunctional chemoselective reagents described in Chapter 17, Section 2.

2.5. SPDP-Activated Enzymes

SPDP is a heterobifunctional crosslinker containing an NHS ester on one end and a pyridyl disulfide group on the other end (Chapter 6, Section 1.1). The NHS ester end can be used to modify amine groups on enzymes, forming amide bonds. The result of this procedure is to create sulfhydryl-reactive pyridyl disulfide groups on the surface of each enzyme molecule that are able to complex with thiol-containing proteins and other molecules. SPDP-activated enzymes may be purified and stored for extended periods without breakdown of the coupling capacity. The reaction with a sulfhydryl group forms a reversible disulfide linkage that can be cleaved with reducing agents.

The two-step nature of SPDP crosslinking provides control over the conjugation process. Complexes of defined composition can be constructed by adjusting the ratio of enzyme to secondary molecule in the reaction as well as the amount of SPDP used in the initial activation. The use of SPDP in conjugation applications is extensively cited in the literature, perhaps making it one of the more popular crosslinkers available. It is commonly used to form immunotoxins, antibody–enzyme conjugates, and enzyme-labeled DNA probes. A standard activation and coupling procedure using SPDP can be found in Chapter 6, Section 1.1.

3. PREPARATION OF BIOTINYLATED ENZYMES

Biotinylated enzymes can be used as detection reagents in (strept)avidin–biotin assay procedures. Particularly, in the bridged avidin–biotin (BRAB) approach or the ABC technique (Chapter 11, Section 2), a biotin-labeled enzyme is used as the signaling agent after the binding to an antigen of a biotinylated antibody and a (strept)avidin bridging molecule. The biotins on the surface of the enzyme can bind with extraordinary affinity to (strept)avidin–antibody complexes, providing near-covalent interaction potential with high specificity.

Adding a biotin label to an enzyme molecule is simple, given the wide variety of options available. A biotinylation reagent is chosen that has a reactive group that will couple to functional groups on the enzyme (Chapter 11, Section 6, and Chapter 18, Section 1.3). For instance, NHS–LC-biotin can be used to modify amine groups—a popular choice for many biotinylation procedures involving proteins (Chapter 11, Section 6.2). When free sulfhydryls are present, as in β-gal, a thiol-reactive biotin label may be more appropriate, such as biotin–BMCC (Chapter 11, Section 6.3). However, a better choice than these popular reagents may be to use hydrophilic biotinylation compounds containing a PEG spacer arm, which results in better solubility of the biotinylated enzyme (Chapter 18, Section 1.3). Biotin–PEG reactive compounds are available in different reactivities and spacer lengths to accommodate virtually any application. Enzymes and other proteins modified with these biotinylation reagents retain their water solubility better than when using aliphatic biotin compounds. The hydrophilic PEG spacer prevents aggregation of modified proteins and dramatically reduces nonspecific binding in assays.

CHAPTER

23

Nucleic Acid and Oligonucleotide Modification and Conjugation

Molecular biology techniques incorporating highly sensitive detection methods involving fluorescence, chemiluminescence, or chromogenic enzyme substrates are used widely for assaying oligonucleotide interactions. A major factor in the development of assay and detection systems for RNA and DNA measurement is the ability to modify a nucleic acid with a detectable component while not affecting base pairing. The attachment of a small label, such as a fluorescent molecule, or a large catalytic enzyme to an oligonucleotide probe forms the basis for constructing sensitive hybridization reagents that can be used to detect genomic sequences (Gottfried and Weinhold, 2011; Paredes *et al.*, 2011; Zohar and Muller, 2011). Unfortunately, the methods developed to crosslink or label proteins do not always apply to the modification of nucleic acids. The major reactive sites on proteins involve groups that are familiar and relatively easy to modify and derivatize, such as primary amines, sulfhydryls, glycans, or carboxylates. Oligonucleotides contain none of these groups and they may also be unreactive with many of the common electrophilic bioconjugation reagents discussed throughout this book.

To modify the unique chemical groups on nucleic acids, novel methods have been developed that allow derivatization through discrete sites on the available bases, sugars, or phosphate groups (see Chapter 2, Section 3 for a discussion of RNA and DNA structure). These chemical methods can be used to add a functional group or a label to an individual nucleotide or to one or more sites in oligonucleotide probes or full-sized genomic DNA or RNA polymers.

If an individual nucleotide is modified in the appropriate way, various enzymatic techniques can be used to polymerize the derivative into an existing oligonucleotide molecule. Alternatively, nucleotide polymers can be treated with chemical activators that can facilitate the attachment of a label at particular reactive sites. Thus, there are two main approaches to modifying DNA or RNA molecules: enzymatic or chemical. Both procedures can produce highly active conjugates for sensitive assays to quantify or localize the binding of an oligo probe to its complementary strand in a complex mixture.

The following sections describe the major enzymatic and chemical modification procedures used to label nucleic acids and oligonucleotides.

1. ENZYMATIC LABELING OF DNA

Enzymatic techniques can employ a variety of DNA- or RNA-specific enzymes, including polymerases, transferases, or ligases, to add controlled amounts of modified nucleotides to an existing oligonucleotide strand. The most common procedures utilize DNA polymerase I, terminal deoxynucleotide transferase (TdT), or T4 RNA ligase 1. The polymerase is most often used with a template to add modified nucleoside triphosphates to the end of a DNA molecule or to various sites within the middle of a sequence. The terminal transferase enzyme can add modified nucleotides to the 3′ end of a chain without a template. The T4 RNA ligase 1 enzyme similarly can add a single modified 5′ phosphoryl-terminated nucleotide to the 3′ hydroxyl-terminated end of an RNA strand. The following sections describe procedures for enzymatic labeling using random primed labeling, nick translation, PCR, 3′ tailing with terminal transferase, and 3′ labeling of RNA using T4 RNA ligase I.

1.1. Random-Primed Labeling

Three main procedures of enzyme labeling make use of a DNA polymerase to make a complementary modified oligo to a parent template: (1) random-primed labeling, (2) nick translation, and (3) polymerase

Bioconjugate Techniques, *Third Edition*.
DOI: http://dx.doi.org/10.1016/B978-0-12-382239-0.00023-6

chain reaction (PCR). In random-primed labeling, modified and unmodified deoxy nucleoside triphosphates (dNTPs) are added to a DNA template using a random mixture of hexa-deoxynucleotides to serve as 3′-OH primers. The form of polymerase typically used is the Klenow fragment which lacks the 5′–3′ exonuclease activity of intact *E. coli* DNA polymerase I (Feinberg and Vogelstein, 1983, 1984; Kessler *et al.*, 1990). The reaction creates a large selection of random complementary strands, which can be labeled in this process using virtually any affinity ligand or detection component attached to the dNTPs that is compatible with the polymerase enzyme. Random primed labeling is one of the most proficient methods of adding modified nucleotides in multiple positions along the length of an oligo strand. If it is used to incorporate ^{32}P radiolabeled nucleotides, which can be done at high density without affecting the resultant hybridization efficiency of the probe, then this method is likely the best route to making probes. It is also a simple way of tagging probes prepared from a restriction digest template with randomly incorporated, labeled nucleotides. When using biotin-labeled or fluorescently labeled nucleotides, however, the amount of labeling that is done on the complementary oligo needs to be controlled to not affect hybridization. The ratio of modified to unmodified dNTPs ultimately determines the degree of labeling within the growing complementary strands. The modified dNTPs that get incorporated into the labeled oligo can include detection molecules such as fluorescent dyes (Chapter 10) or affinity molecules such as a biotin (Chapter 11), which can be used with streptavidin conjugates to detect hybridization events. The preparation of non-self-quenching levels of fluorescent dye incorporation into DNA strands has been optimized to give highly sensitive probes (Yu *et al.*, 2012). Figure 23.1 illustrates the reactions that take place during a random primed labeling process.

Protocol for Random Primed Labeling of DNA

1. Denature 1 µg of probe DNA (single stranded) with 5 µg of random hexanucleotide primers by boiling for 5 min and then rapidly chill on ice. Incubate at least 10 min to allow the primers to hybridize to random sites within the probe DNA.
2. Add to a tube on ice, 5 µl of 10× random primed labeling buffer (0.5-*M* Tris, 0.1-*M* $MgCl_2$, 10-m*M* 2-mercaptoethanol, pH 6.6, containing 0.5 mg/ml BSA), 1 µl each of three types of unmodified deoxynucleoside triphosphates (dNTPs at 100-µ*M* concentration), 1 µl of a labeled dNTP (at 100-µ*M*), plus 48 µl of water. Then add 2 µl of DNA polymerase (5–10 units).
3. Combine the probe DNA/hexanucleotide preparation with the reaction solution and incubate for 2 h at 37°C.
4. Quench the reaction by the addition of 2 µl of 0.5-*M* EDTA, 2 µl of 10-mg/ml tRNA, and 150 µl of 10-m*M* Tris, pH 7.5.
5. Purify the labeled DNA from excess reactants by precipitation (or by use of a spin column—Qiagen). For the precipitation method, add 20 µl of 4-*M* LiCl and 500 µl of ethanol (chilled to −20°C). Mix well.
6. Store at −20°C for 30 min, and then separate the precipitated DNA by centrifugation at 12,000 *g*.
7. Remove the supernatant and wash the pellet with 70% and 100% ethanol, centrifuging after each wash.
8. Re-dissolve the labeled DNA pellet in water and store at −20°C until used.

1.2. Nick Translation Labeling

Nick translation labeling involves the use of a dual enzyme system acting on double-stranded DNA (Rigby *et al.*, 1977; Langer *et al.*, 1981; Höltke *et al.*, 1990). The enzymes pancreatic deoxyribonuclease I (DNase I) and *E. coli* DNA polymerase I act in tandem on a DNA helix to incorporate labeled nucleotides (dNTPs) into the sequence. DNase I is capable of breaking phosphodiester bonds in intact DNA double-stranded molecules. If it is used in the presence of magnesium ions, it limits the hydrolysis caused by the enzyme to a single strand at a time within the DNA helix. If DNase is further restricted in the amount added to the reaction, the number of breaks in the double helix can be controlled. The result of this reaction is the breaking of phosphodiester bonds along the DNA double helix, leaving an OH on the 3′ end and a phosphate on the 5′ end at the nick site. The addition of DNA polymerase I results in removal of the base at the 5′ end of the nick and replacement of it with a new nucleotide by attachment to the open 3′ OH group. By adding the appropriate labeled and unlabeled dNTPs to this reaction, the polymerase causes the breaks to be filled as quickly as they form from the added pool of nucleotides. Each time the polymerase enzyme adds a nucleotide, however, it attaches the new nucleotide at its 5′ phosphate end to the open 3′ OH group in a new phosphodiester bond, but it leaves the new 3′ OH end open. Thus, as a nucleotide is added into the gap, the nick in the strand moves toward the 3′ end of the DNA strand. Since a new nick opens in the chain each time a nucleotide is added, the reaction sequence repeats over and over again as the polymerase creates a new complementary strand and the nick moves (or is translated) down the DNA. The labeled and unlabeled nucleotides are incorporated into the growing sequence, resulting in the addition of a number of labels along the length of the new strand. Since a known quantity of labeled to unlabeled dNTPs is present during the

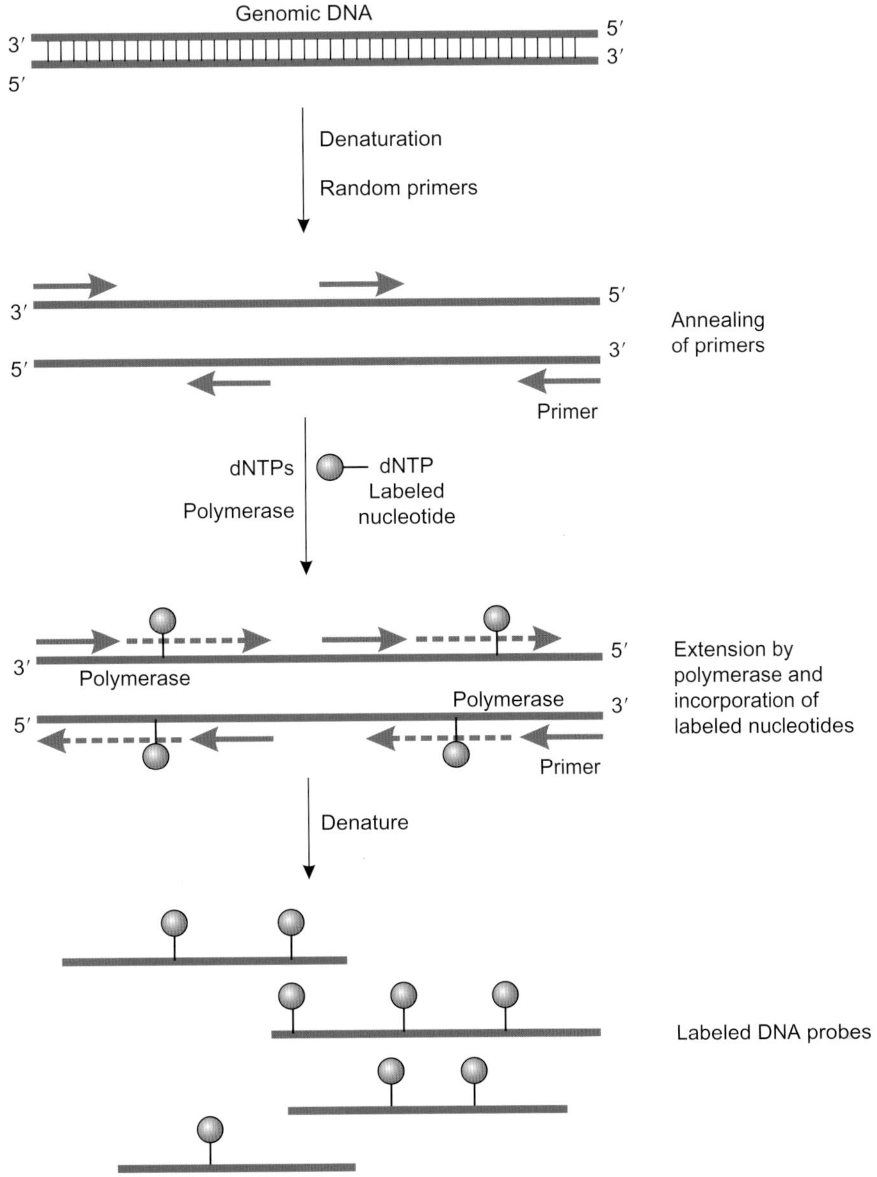

FIGURE 23.1 In a random primed labeling reaction a set of random primers is annealed to a denatured DNA sample and a complementary strand is synthesized using a polymerase enzyme in the presence of labeled and unlabeled dNTPs. The result is the generation of new oligonucleotides containing labeled nucleotides incorporated within their sequences.

reaction, the degree of labeling occurs at a predictable level as the parent DNA strands are modified (Figure 23.2).

Nick translation can be used to incorporate affinity tags such as biotin and fluorescent labels into an oligo probe for use in hybridization assays. Care should be taken to limit the number of modifications along the DNA strand so as not to affect the resultant hybridization efficiency with a target sequence. The labeling of a DNA strand with non-self-quenching fluorescent dyes has been optimized to result in the highest theoretical limit of incorporating modified nucleotides (Yu *et al.*, 2012).

Protocol for the Labeling of DNA by Nick Translation

1. In a tube kept cold on ice, add 10 μl of 10× nick-translation buffer (0.5-*M* Tris, 0.1-*M* $MgCl_2$, 0.08-*M* 2-mercaptoethanol, pH 7.5, containing 0.5 mg/ml bovine serum albumin (BSA), 0.5 μg of double-stranded probe DNA to be labeled, 1 μl of DNase I at a concentration of 2 ng/ml, 1 μl each of three types of unmodified deoxynucleoside triphosphates (dNTPs at 100-μ*M* concentration), 1 μl of a labeled dNTP (at 300-μ*M*), 32 μl water. Then add 1 μl of DNA polymerase containing 5 to 10 units of activity.

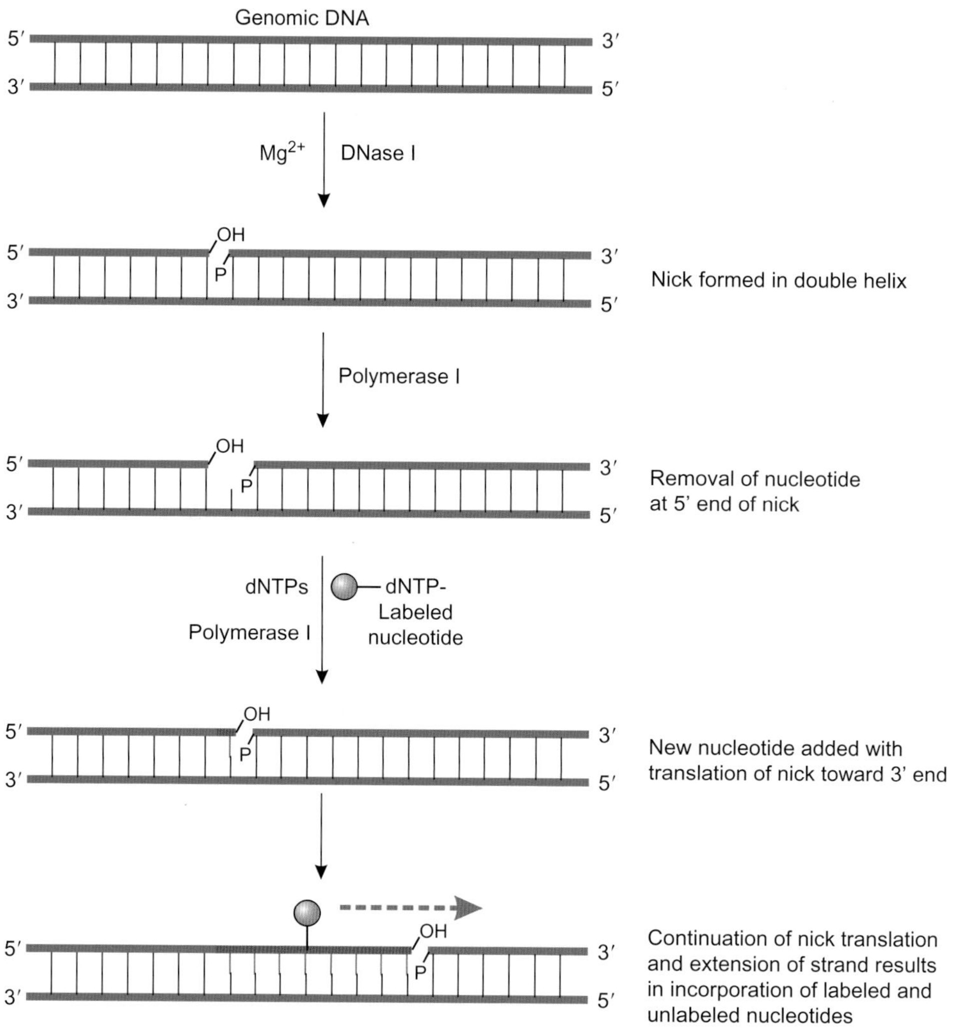

FIGURE 23.2 Nick translation labeling of DNA starts with the creation of defects within the sequence of existing DNA double-helix molecules by cleavage of phosphodiester bonds with DNase along the backbone of one strand. Polymerase then repairs these nicks beginning with the removal of the adjacent nucleotide and the immediate filling back in of those gaps with new nucleotides from the added dNTP pool. As each new nucleotide is added, the polymerase leaves the 3′ OH group open, thus translating the nick toward the 5′ end. As the reaction sequence is repeated, the polymerase enzyme continues to remove existing nucleotides and replace them with new ones at the site of the new nick. The result of these reactions is numerous labeled and unlabeled nucleotides being incorporated as a complementary sequence along the length of each DNA strand, starting at the site of the original nick.

2. React for 1 h at 15°C.
3. Quench the reaction by the addition of 4 µl of 0.25-*M* EDTA, 2 µl of 10-mg/ml tRNA, and 150 µl of 10-m*M* Tris, pH 7.5.
4. Purify the labeled DNA from excess reactants by precipitation (or by use of a spin column—Qiagen). For the precipitation method, add 20 µl of 4-*M* LiCl and 500 µl of ethanol (chilled to −20°C). Mix well.
5. Store at −20°C for 30 min, and then separate the precipitated DNA by centrifugation at 12,000 *g*.
6. Remove the supernatant and wash the pellet with 70% and 100% ethanol, centrifuging after each wash.
7. Re-dissolve the labeled DNA pellet in water and store at −20°C until used.

1.3. PCR Labeling

Enzymatic labeling of DNA by use of PCR techniques perhaps provides one of the most powerful ways of not only adding a label, but also of amplifying the labeled polymer to produce numerous copies of itself at the same time. First invented by Mullis (who went on to win the Nobel Prize; see Saiki *et al.*, 1985, 1988), PCR utilizes heat-stable forms of DNA polymerase, for example the commonly employed *Taq* polymerase isolated from thermophilic eubacterium (*Thermus aquaticus*). The stability of the enzyme allows repeated elevated-temperature denaturation of target DNA, followed by hybridization of primers onto the single strands. *Taq* DNA polymerase then

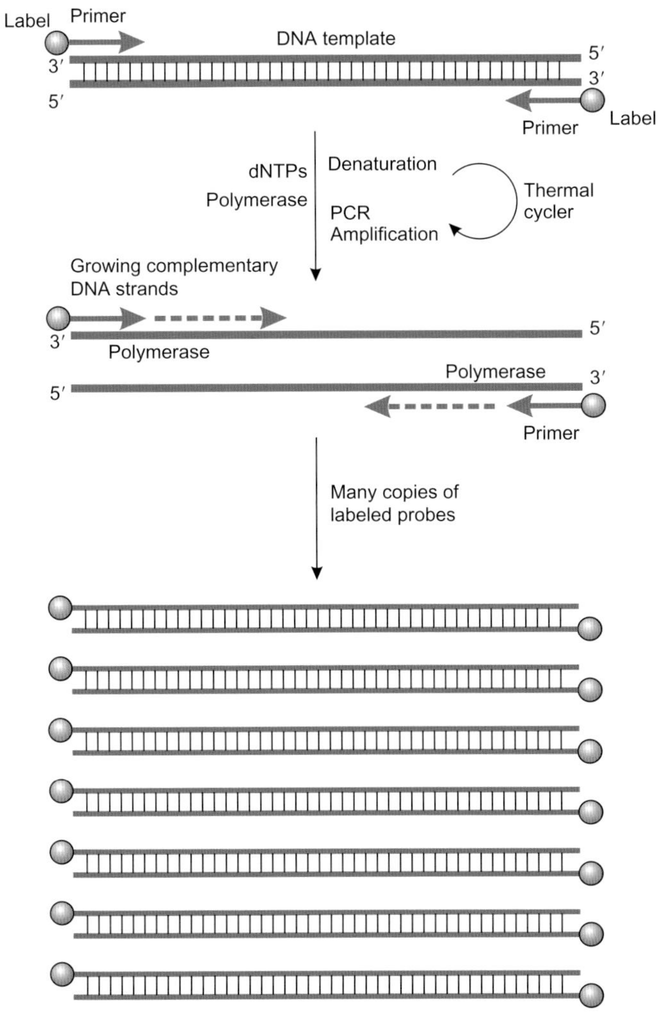

FIGURE 23.3 Labeling oligonucleotides by PCR can be carried out through the use of labeled primers or by the inclusion of labeled dNTPs within the reaction mixture. The polymerase enzyme simultaneously amplifies the original DNA template and incorporates labels within each of the copies.

creates a complementary sequence to the single strands by elongation of the primers. If repeated cycles of denaturation, hybridization, and elongation are done, the result is an exponential amplification of the original DNA strands (for a review of PCR, see van Pelt-Verkuil *et al.*, 2010; Innis *et al.*, 1990). Labeling of these amplified strands can be accomplished by one of two routes: using labeled primers that add the labels to the 3′ end regions of each new strand or by using labeled dNTPs that add labels to various sites within the complementary sequence. Either way, the *Taq* polymerase enzyme incorporates the labels into the growing DNA copies made during each PCR cycle (Figure 23.3). The fluorescent labeling and amplification of target DNA by PCR has resulted in extremely sensitive assays that can detect the target nearly down to the single molecule level (Hindson *et al.*, 2011).

Another popular technique for PCR labeling uses the enzyme Moloney murine leukemia virus (MMLV) reverse transcriptase to create a complementary labeled DNA (cDNA) probe from an RNA transcript. MMLV reverse transcriptase is an RNA-dependent DNA polymerase that can be used with labeled primers or labeled dNTPs to create the labeled cDNA oligo probe. RT-PCR-based labeling and detection has been used to create highly sensitive assays for viruses (Lung *et al.*, 2011).

There are many methods for performing PCR labeling that use different commercial enzyme and reagent formulations. The following method is a generalized one that can be customized to conform to any desired procedure. The preparation of the initial stock solutions is carried out in water.

Protocol for the Labeling of DNA Using PCR

1. Prepare a 10× PCR reaction buffer consisting of 100-m*M* Tris, 500-m*M* KCl, 20-m*M* $MgCl_2$, pH 8.4.
2. Prepare the following stock solutions in nuclease-free water:
   a. DNA template at a concentration of about 0.5 to 1.0 ng/μl.
   b. Primers at a concentration of about 25 to 50 μ*M*.
   c. dNTPs in a single stock solution at a concentration of 33-m*M*. Reduce the concentration of the particular unlabeled dNTP by the concentration of the labeled dNTP that is to be added to the reaction. For instance, if a biotin–dUTP derivative is to be incorporated into the PCR product, then reduce the concentration of dTTP in the nucleotide stock solution. Some optimization of the concentration of the labeled dNTP may have to be done to obtain the best labeled oligo for a particular application.
   d. Labeled dNTP solution at a concentration of 5-m*M*.
   e. BSA at 10 mg/ml.
   f. Taq polymerase at 5 U/μl.
   g. $MgCl_2$ at 50-m*M*.
3. Mix the following quantities of each reagent in a PCR reaction tube:
   a. 2.5 μl of 10× PCR reaction buffer
   b. 1 to 2 μl of the DNA template
   c. 1 μl of primers
   d. 0.5 μl of the unlabeled dNTP mixture
   e. 0.5 to 2.0 μl of the labeled dNTP
   f. 2 μl of the $MgCl_2$ solution
   g. 1 μl of the BSA solution
   h. 0.2 to 0.4 μl of the Taq polymerase solution
   i. Add water to make the total volume to 25 μl.
4. Perform PCR amplification/labeling using a thermal cycler to the desired number of cycles (e.g., 30 cycles).
5. Purify the labeled DNA by precipitation (as described in the previous protocol) or by use of a spin column (Qiagen).

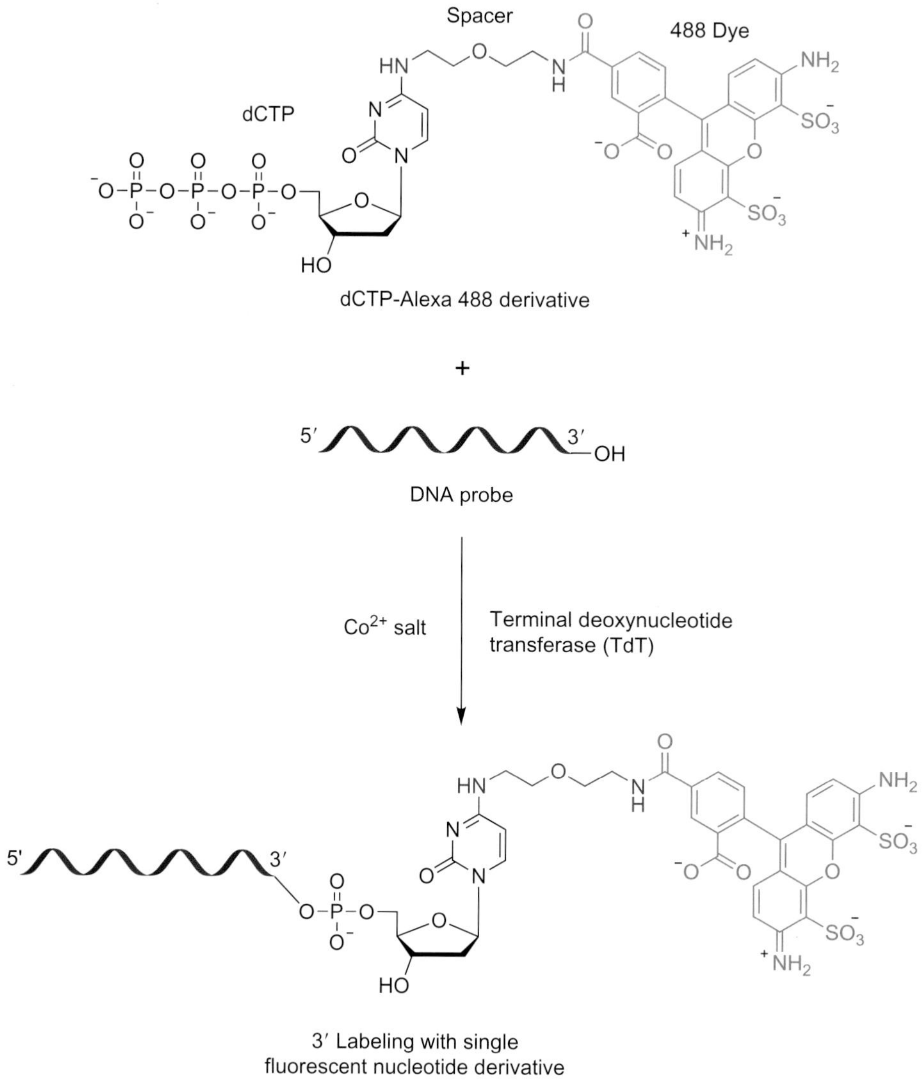

FIGURE 23.4 Oligonucleotide labeling by terminal transferase. Under controlled conditions, a single labeled nucleotide can be added to an existing DNA strand at its 3′ OH end without the need for a template. The Alexa 488 fluorescent dCTP derivative provides a bright green emission dye at the end of every oligo probe for detection purposes in hybridization assays. Other fluorescent dye derivatives may be used in similar labeling reactions.

1.4. Terminal Transferase Labeling

Enzymatic labeling using any of the polymerase-mediated methods (except when using labeled primers during PCR) results in derivatized dNTPs being incorporated at numerous locations within an oligonucleotide strand. These modifications can potentially interfere with the hybridization of a probe to a complementary sequence, especially if the level of labeling is high. Enzymatic labeling using terminal transferase is a way to avoid derivatization in the middle of a strand, and thus preserve sequence or targeting specificity. Terminal deoxynucleotidyl transferase (TdT) is a DNA polymerase enzyme commonly found in some lymphoid cell types. Under normal circumstances, TdT adds multiple dNTPs to the 3′ end of a DNA molecule. However, under the right conditions (in the presence of a Co^{2+} salt) the enzyme will add only a single base to the 3′ OH end (Horáková *et al.*, 2011) (Figure 23.4). It is also able to add the labeled nucleotide to the DNA without the need for a complementary template, thus making it capable of adding a labeled nucleotide to an existing single-stranded oligo probe. Since modification is limited to the end of the oligonucleotide and only one labeled nucleotide is added per strand, the main probe sequence is not disturbed by modifications that could possibly prevent hydrogen bonding interactions between base pairs. This process also is ideal for labeling a detection oligo at its 3′ end with biotin to detect hybridization events on oligo arrays. This discrete modification strategy ensures that each detection oligo can interact with a streptavidin conjugate at its 3′ end, which can be designed in the assay to be pointing up from the array surface.

Terminal transferase labeling was originally developed using radiolabeled (typically ^{32}P) nucleoside triphosphates (Roychoudhury *et al.*, 1979; Tu and Cohen, 1980). Later, the technique was extended to the use of nonradioactive nucleotide derivatives (Kumar *et al.*, 1988). The technique is especially convenient for adding a single biotin affinity tag to the 3′ end of an oligo probe to facilitate the interaction of streptavidin detection conjugates at a position that will not interfere with oligo probe binding to a target sequence (Leduc *et al.*, 2011).

Protocol for Labeling DNA at the 3′ End Using Terminal Transferase

1. Prepare 1 μg of purified DNA probe, either by restriction digestion or by synthetic means.
2. Add to the purified probe: (a) 20 μl of 0.5-*M* potassium cacodylate, 5-m*M* $CoCl_2$, 1-m*M* dithiothreitol (DTT), pH 7.0; (b) 100 μ*M* of a modified deoxynucleoside triphosphate, 4 μl of 5-m*M* dCTP, and 100 μl of water. Mix.
3. Add terminal transferase to a final concentration of 50 units in the reaction mixture.
4. React for 45 min at 37°C.
5. Isolate the labeled probe by alcohol precipitation as described previously for nick translation.

1.5. T4 RNA Ligase 1 Labeling

The labeling of the 3′ hydroxyl end of RNA using T4 RNA ligase 1 is another way of adding a single nucleotide derivative to an oligonucleotide probe, similar to the use of terminal transferase with a DNA probe (described previously). It was originally developed to incorporate a radiolabeled (^{32}P) *bis*-phosphoryl–cytidine derivative onto an RNA probe (England *et al.*, 1977; England and Uhlenbeck, 1978). However, other derivatives of cytidine bases formed by modification of the C-5 site can be used with the T4 RNA ligase enzyme with success (Opperman *et al.*, 2011; Paredes *et al.*, 2011). For example, biotinylated or fluorescently labeled *bis*-phosphoryl–cytosine derivatives have been synthesized and used to create highly sensitive oligo probes (Kore *et al.*, 2009). In addition, T4 RNA ligase can be used with N-6-labeled *bis*-phosphoryl–adenosine derivatives to add a single labeled adenosine nucleotide to the 3′ end of RNA (Richardson and Gumport, 1983). The enzymatic techniques for end labeling of DNA and RNA have an advantage over the PCR methods, because they add only a single label to each oligo probe at a position that will not affect hybridization efficiency with DNA or RNA targets.

Unlike the polymerase-based labeling methods described previously that use nucleotide triphosphate derivatives, the type of labeled nucleotide required for T4 RNA ligase labeling is a *bis*-phosphoryl derivative, wherein the 3′ and 5′ hydroxyl groups both have a single phosphate group on them. The *bis*-phosphoryl derivative is required for ligase recognition and incorporation onto the 3′ end of an existing RNA strand. This reaction is illustrated in Figure 23.5.

Protocol for Labeling DNA at the 3′ End Using T4 RNA Ligase

The following protocol is based on the method described by Thermo Fisher for their RNA 3′ end biotinylation kit.

1. Prepare an RNA probe to be labeled at a concentration of 125 nM in water.
2. Add to a small centrifuge tube 3 μl of nuclease-free water and 3 μl of 10× reaction buffer (0.5-*M* Tris, 0.1-*M* $MgCl_2$, 0.1-*M* DTT, 10-m*M* ATP, pH 7.8).
3. Add 1 μl of an RNase inhibitor to this solution to prevent RNA digestion (use the recommended final concentration of the particular inhibitor chosen).
4. Add 5 μl of the unlabeled RNA probe from step 1 to the reaction mixture to obtain a final concentration of 50 p*M*.
5. Prepare a labeled cytidine *bis*-phosphate derivative wherein the label is attached to the C-5 carbon atom (e.g., biotin labeled) at a concentration of 1 m*M* in water. Add 1 μl of this labeled nucleotide to the reaction solution with mixing.
6. Add to the reaction mixture 2 μl of a 20,000-U/ml solution of T4 RNA ligase to obtain a final activity of 40 U.
7. Add 300 μl of a 30% PEG solution to obtain a final concentration of 15% PEG in the reaction mixture.
8. Incubate the reaction for at least 2 h at 16°C. The reaction may be carried out overnight to increase the yield if necessary.
9. Add 70 μl of additional nuclease-free water to the reaction with mixing.
10. Extract and remove the RNA ligase enzyme from the reaction by the addition of 100 μl of chloroform : isoamyl alcohol and vortex mix thoroughly. Centrifuge the mixture to separate the phases and recover the top aqueous phase carefully from the tube.
11. Add to the aqueous phase 10 μl of 5-*M* NaCl followed by 1 μl of 20 mg/ml glycogen and 300 μl of ice-cold ethanol (100%).
12. Allow the labeled RNA to precipitate for 1 h at −20°C and centrifuge at 13,000 *g* for 15 min to pellet the precipitated oligo. Remove the supernatant and wash the pellet with an additional 300-μl quantity of ice-cold ethanol, then centrifuge a nd remove the ethanol. Allow the pellet to dry and redissolve the labeled RNA in nuclease-free water or a buffer for use.

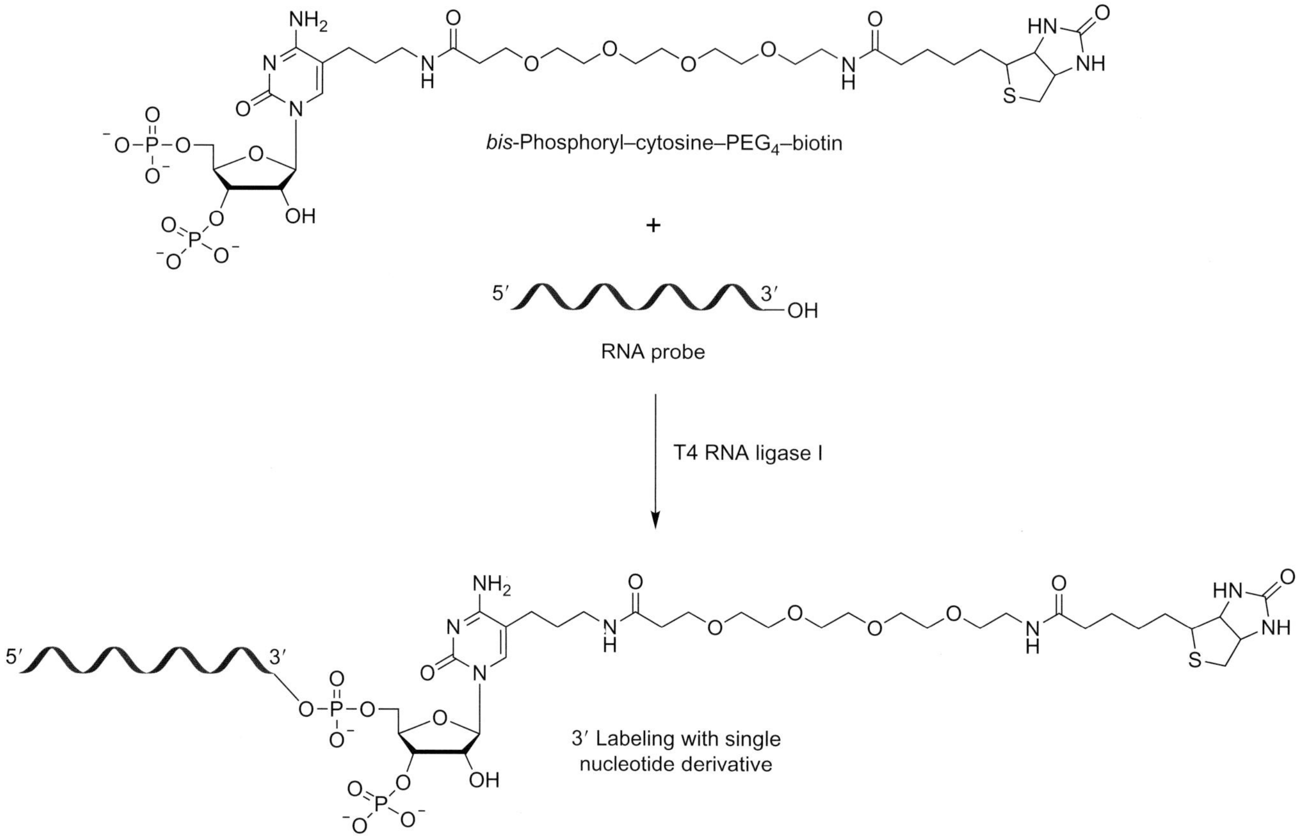

FIGURE 23.5 Labeling of an oligo probe can be done at the 3′ end of an existing RNA molecule by the use of T4 RNA ligase. The reaction attaches a single labeled *bis*-phosphoryl nucleotide to the 3′ OH without the need for a template. The *bis*-phosphoryl–cytosine–biotin derivative provides a biotinylated oligo probe that can be detected using streptavidin conjugates.

Regardless of the type of enzymatic labeling being done, it is important that the label be incorporated into the nucleotides or primers in a way that does not affect enzyme recognition and activity. Thus, every enzymatic labeling procedure for modifying RNA or DNA probes must start with chemical derivatization of individual nucleotides. Of the many chemical approaches that can be used to modify a nucleotide monomer, there are only a few that will result in a derivative still able to be enzymatically added to an existing oligonucleotide strand.

Of the purine nucleotides, dATP may be derivatized at its N-6 position using a long linker arm terminating in a detectable group without losing the ability to be enzymatically incorporated into DNA probes. By contrast, if modification is performed at the C-8 position of purine bases, DNA polymerase cannot be used to add the labeled monomer to an existing strand. C-8 derivatives of purines, however, can be added at the 3′ terminal using terminal transferase or T4 RNA ligase enzymes.

The pyrimidine nucleosides dUTP or dCTP can be modified at their C-5 position with a spacer arm containing a tag, such as a biotin group or a fluorescent molecule, and still remain good substrates for DNA polymerase. Enzymatic labeling with a biotin-modified pyrimidine-based dNTP is one of the most common methods of adding a detectable group to an existing DNA strand.

Figure 23.6 illustrates some of the common nucleoside triphosphate derivatives that can be used in enzymatic labeling processes. For a review of these methods in greater detail, see Gottfried and Weinhold (2011), Horakova *et al.* (2011), Kricka (1992), or Keller and Manak (1989). See Section 2 for a discussion of the chemical methods that can be used to label individual nucleic acids for incorporation into oligonucleotides by enzymatic means.

2. CHEMICAL MODIFICATION OF NUCLEIC ACIDS AND OLIGONUCLEOTIDES

The chemical modification of nucleic acids at specific sites within individual nucleotides or within

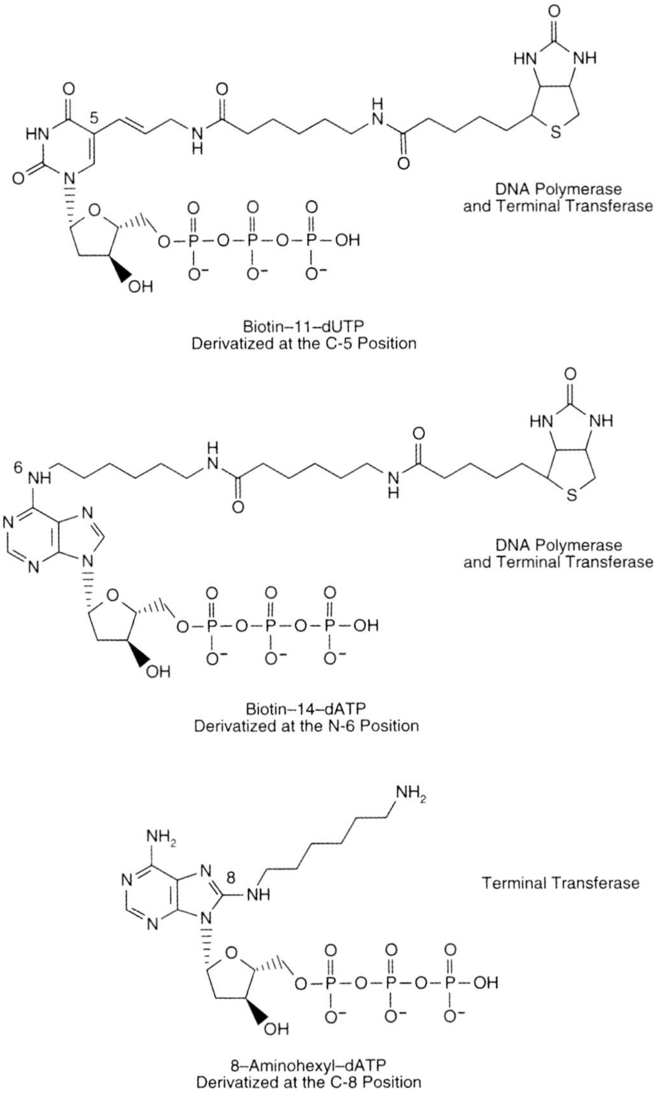

FIGURE 23.6 Three common nucleoside triphosphate derivatives that can be incorporated into oligonucleotides by enzymatic means. The first two are biotin derivatives of pyrimidine and purine bases, respectively, that can be added to an existing DNA strand using either polymerase or terminal transferase enzymes. Modification of DNA with these nucleosides results in a probe detectable with labeled avidin or streptavidin conjugates. The third nucleoside triphosphate derivative contains an amine group that can be added to DNA using terminal transferase. The modified oligonucleotide can then be labeled with amine-reactive bioconjugation reagents to create a detectable probe.

oligonucleotides allows various labels to be incorporated into DNA or RNA probes. This labeling process can produce conjugates having sensitive detection properties for the localization or quantification of oligo binding to a complementary strand using hybridization assays.

Some form of chemical labeling process must be used regardless of whether the final oligo conjugate is created by enzymatic or strictly chemical means. If enzymatic modification is to be performed, the initial label still must be incorporated into an individual nucleoside triphosphate, which is then polymerized into an existing oligonucleotide strand (Section 1). Fortunately, many useful modified nucleoside triphosphates are now available from commercial sources, often eliminating the need for custom derivatization of individual nucleotides.

Chemical modification may also be used to directly label an oligonucleotide, eliminating the enzymatic step altogether. The chemical modification of nucleic acids can encompass several strategies, ranging from creating intermediate derivatives containing a functional group needed for subsequent conjugation, producing a reactive group that can spontaneously couple to a desired biomolecule, or forming a final covalent linkage to a surface, such as a particle or planar array. The initial derivatization might only be done to add a spacer arm to a particular site on the nucleotide structure. The spacer typically contains a terminal functional group, such as an amine, that can be used to couple with another molecule. Another modification built upon this amine derivative might be to add a fluorescent probe to the end of the spacer, thus creating a detectable complex. The spacer may also be used to react with a cross-linking agent, such as a heterobifunctional compound (Chapter 6), which can facilitate the conjugation of a protein or another molecule to the modified nucleotide. It should be noted that if enzymatic methods are used to incorporate a small spacer into an oligonucleotide, subsequent chemical conjugation steps still will be needed to label the oligo at this site or link it with another molecule.

In some cases, if an oligonucleotide contains the appropriate functional group, a label may be directly incorporated into it using chemical methods. For instance, certain fluorescent molecules or biotin tags can be used to modify nucleotides without going through an initial derivatization step with a spacer arm. Such labels usually contain nucleophilic or photoreactive groups that can couple directly to the oligo using an intermediate activating agent or by exposure to UV light, respectively.

Many of the chemical derivatization methods employed in these strategies involve the use of an activation step that produces a reactive intermediary. The activated species can then be used to couple a molecule containing a nucleophile, such as a primary amine or a thiol group. The following sections describe the chemical modification methods suitable for derivatizing individual nucleic acids as well as oligonucleotide polymers.

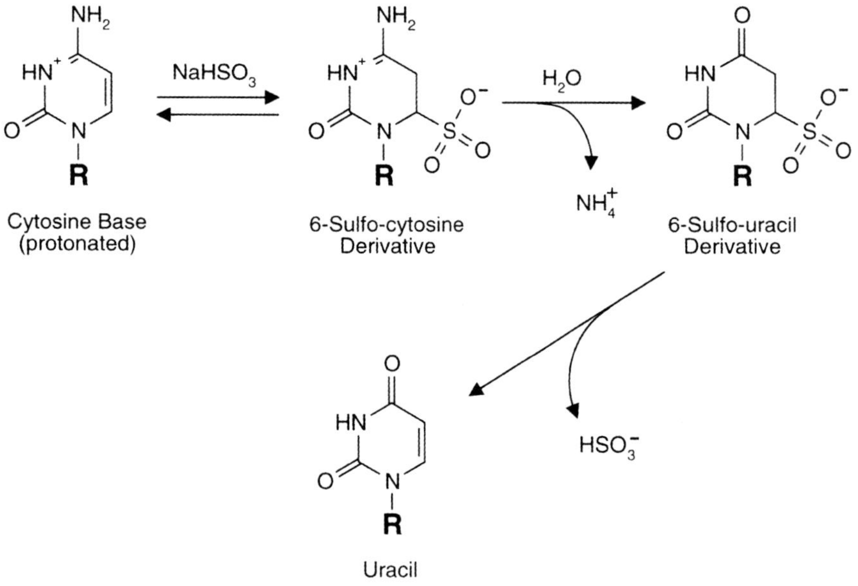

FIGURE 23.7 Treatment of cytosine bases with bisulfite results in a multistep deamination reaction, ultimately leading to uracil formation.

2.1. Diamine or *bis*-Hydrazide Modification of DNA

One of the more useful chemical modifications that can be performed on nucleic acids or oligonucleotides is to add an amine-terminal spacer arm using a diamine compound. The resultant amine derivative can be targeted by numerous amine-reactive crosslinkers or modification reagents to create a detectable conjugate. A similar approach can be used to modify a DNA probe with a *bis*-hydrazide compound (Chapter 5, Section 8) to produce terminal hydrazide group. The oligonucleotide derivative can then be coupled with aldehyde-containing molecules to form conjugates. The following methods utilize activation reagents which transform a particular site on nucleic acids into an amine-reactive or hydrazide-reactive intermediate. Coupling a diamine or *bis*-hydrazide compound to these activated species results in the formation of an alkyl spacer arm terminating in a primary amine group or a hydrazide functional group, respectively.

Conjugation **via** *Bisulfite Activation of Cytosine*

Single-stranded DNA molecules can react with sodium bisulfite, adding a sulfonate group across the 5,6-double bond of cytidine bases and creating 6-sulfo-cytosine derivatives. The reaction also catalyzes the deamination of cytosine to uracil by loss of the 4-amino group, which is the first step in assays that determine DNA methylation (Cheung *et al.*, 2012). Subsequent loss of HSO_3^- effectively forms the uracil base (Figure 23.7). This reaction sequence was recognized in the early 1970s as potential evidence for the mutagenicity of bisulfite (Shapiro *et al.*, 1973, 1974). Shapiro and Weisgras (1970) demonstrated that the bisulfite reaction also can cause transamination to occur at the N-4 position of cytosine. In the presence of an amine-containing molecule, such as a diamine, sodium bisulfite will cause the exchange of the N-4 amine for another amine-containing compound, effectively forming a new covalent linkage with release of ammonium ion (Figure 23.8). Draper and Gold (1980) used this reaction to produce primary amine groups on a poly(C) oligonucleotide by coupling 1,3-diaminopropane to a limited number of the cytosine residues. The amine derivative subsequently could be used to couple a fluorescent probe to the polymer, allowing sensitive studies of messenger RNA.

Bisulfite-catalyzed transamination can also be used to label oligonucleotide probes for application in nonradioisotopic hybridization assays. Viscidi *et al.* (1986) described a method for derivatizing cytosine groups in DNA probes using the short spacer, ethylenediamine. Other diamine molecules may also be used, such as 1,3-diaminopropane, 1,6-diaminohexane, or 3,3′-iminobispropylamine. The use of the long, hydrophilic Jeffamine molecules (Huntsman International; see Chapter 2, Section 4.3, and Chapter 15, Section 2.8) may be especially well suited for this type of modification due to the presence of a hydrophilic polyethylene glycol (PEG)-based spacer. Longer spacer arms may provide better steric accommodation for larger detection components without interfering substantially in the probe's ability to hybridize to a complementary DNA strand.

If an amine-containing fluorescent probe or hydrazide-containing compound is transaminated onto

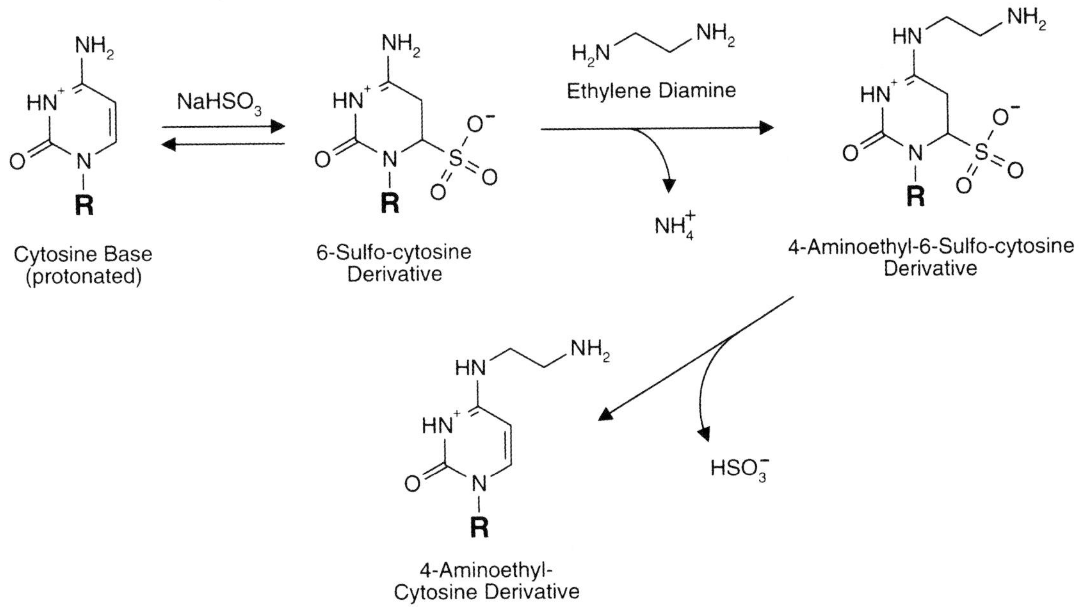

FIGURE 23.8 The reaction of cytosine with bisulfite in the presence of an excess of an amine nucleophile (such as a diamine compound) leads to transamination at the N-4 position. This process is a route to adding an amine functional group to cytosine residues in oligonucleotides.

an oligonucleotide using bisulfite, the labeling of nucleic acids can be done in a single step. An example of this approach is the coupling of biotin hydrazide (Chapter 11, Section 6.4) to cytosine residues, resulting in a biotinylated oligonucleotide suitable for (strept) avidin-based detection systems (Reisfeld *et al.*, 1987) (Chapter 11, Section 2).

Since the site of modification on cytosine bases is at a hydrogen bonding position in double-helix formation, the degree of bisulfite derivatization should be carefully controlled. Reaction conditions such as pH, diamine concentration, and incubation time and temperature affect the yield and type of products formed during the transamination process. At low concentrations of diamine, deamination and uracil formation dramatically exceed transamination. At high concentrations of diamine (3-*M*), transamination can approach 100% yield (Draper and Gold, 1980). Ideally, only about 30 to 40 bases should be modified per 1000 bases to ensure hybridization ability after derivatization.

Bisulfite modification of cytosine residues also can be used to permanently add a sulfone group to the C-6 position. In this scheme, the sulfone functions as a hapten recognizable by specific anti-sulfone antibodies. At high concentrations of bisulfite and in the presence of methylhydroxylamine, cytosines are transformed into N^4-methoxy-5,6-dihydrocytosine-6-sulphonate derivatives (Herzberg, 1984; Nur *et al.*, 1989). Labeled antibodies can then be used to detect the hybridization of such probes.

Protocol for Labeling Nucleic Acids by Bisulfite-Catalyzed Transamination

1. Prepare single-stranded DNA (denatured) at a concentration of 1 mg/ml.
2. Prepare bisulfite modification solution consisting of: 3-*M* concentration of a diamine (i.e., ethylenediamine), 1-*M* sodium bisulfite, pH 6. The use of the dihydrochloride form of the diamine avoids having to adjust the pH down from the severe alkaline pH of the free-base form. Note: The optimum pH is 4.5 for the transamination of biotin–hydrazide to cytosine residues using bisulfite (see Section 2.3).
3. Add 20 µl of the DNA to 180 µl of bisulfite modification solution. Mix well.
4. React for 3 h at 42°C.
5. Dialyze the solution against water overnight at 4°C to remove excess reactants.
6. The modified DNA may be recovered by alcohol precipitation according to the method in Section 1 described previously for nick-translation modification. Alternatively, dialysis or gel filtration may be carried out to remove excess reactants.

Conjugation via Bromine Activation of Thymine, Guanine, and Cytosine

The nucleotide bases of DNA and RNA can be activated with bromine to produce reactive intermediates capable of coupling to nucleophiles (Traincard *et al.*, 1983; Sakamoto *et al.*, 1987; Keller *et al.*, 1988;

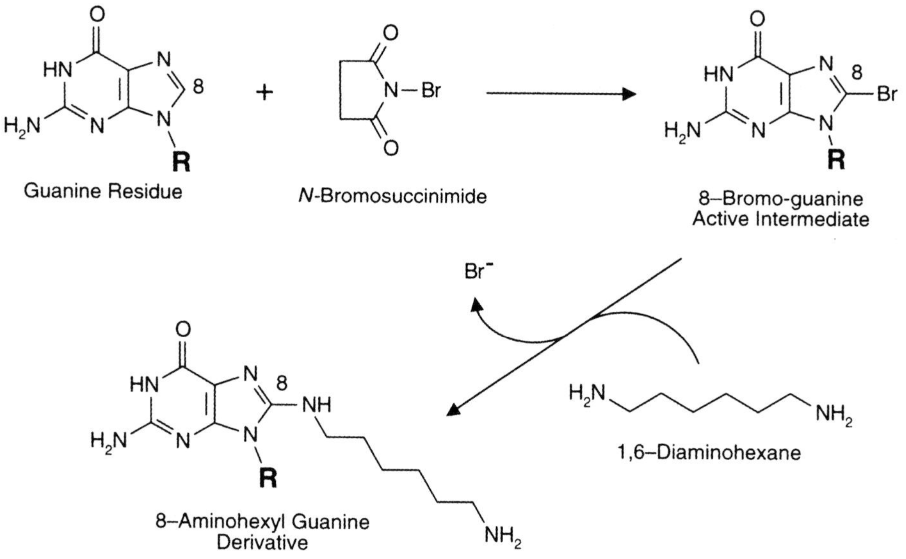

FIGURE 23.9 Reaction of guanine bases with N-bromosuccinimide causes bromination at the C-8 position of the ring. Amine nucleophiles can be coupled to this active derivative by nucleophilic displacement. Reaction of diamine compounds results in amine terminal spacers that can be further modified to contain detectable components.

Litosh *et al.*, 2011). Bromination occurs at the C-8 position of guanine residues and the C-5 of cytosine, yielding reactive derivatives which can be used to couple diamine spacer molecules by nucleophilic substitution (Figure 23.9). Other pyrimidine derivatives are also reactive to bromine compounds, but adenine residues are more resistant. However, even AMP can be immobilized through the introduction of an aminohexyl spacer at the C-8 position using bromination (Lowe, 1979). Either an aqueous solution of bromine or the compound *N*-bromosuccinimide can be used for this reaction. The alkaline modification proceeds rapidly but may be too severe for RNA molecules. Coupling of amine-containing molecules is done at elevated temperatures (50°C) to ensure good incorporation. Both amine-bearing spacers and probes may be coupled using this strategy. Moreover, the sites of derivatization using bromine activation are not involved in hydrogen bonding during base pairing, thus maintaining hybridization ability in the final conjugate.

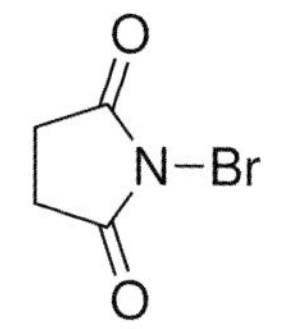

N-Bromosuccinimide
MW 177.98

Optimal bromination of a DNA probe is in the range of 30 to 35 bases per 1000 bases, a level which can be controlled by the amount of *N*-bromosuccinimide added. Over-labeling can prevent specific interactions with target DNA, even if the point of initial modification is not a hydrogen bonding site.

The major disadvantage with bromination is the extreme toxicity of bromine. Use a fume hood for all operations. Avoid the breathing of fumes or contact with skin or eyes. Protective clothing and gloves are recommended.

Protocol for Labeling Nucleic Acids by N-Bromosuccinimide Activation

(A) BROMINATION

1. Mix in a microfuge tube 20 μg of the DNA probe to be labeled, 20 μl of 1-*M* sodium bicarbonate, pH 9.6, and 196 μl of water. Chill on ice.
2. In a fume hood, dissolve *N*-bromosuccinimide (Thermo Fisher) in water at a concentration of 1.42 mg/ml.
3. Add 4 μl of the *N*-bromosuccinimide solution to the DNA solution (makes an 8-m*M* final concentration of brominating reagent). Mix well.
4. React on ice for 10 min. Use the bromine-activated DNA immediately.

(B) COUPLING A DIAMINE-CONTAINING SPACER OR PROBE

1. Dissolve a diamine spacer (i.e., ethylenediamine or 1,6-diaminohexane—Thermo Fisher) in water at a concentration of 80 to 100-m*M*. Caution: Amine-containing molecules such as diamines are highly corrosive if they are in the free-base form (not the

dihydrochloride). Work in a fume hood and wear gloves and other protective clothing. The pH of an aqueous solution of free-base diamine will be >pH 12 and may fume, especially during neutralization. The solution will also generate heat upon dissolution of the amine. Keeping it in an ice bath will help maintain a cool solution with less fuming. Using a dihydrochloride form of a diamine, if available, will avoid the problems associated with corrosiveness, heat, and fuming.

2. Add 25 µl of the diamine solution to the bromine-activated DNA solution prepared in protocol (A), above.
3. React for 1 h at 50°C.
4. The diamine-modified DNA may be isolated from excess reactants by ethanol precipitation according to steps 4 to 7 of the protocol described earlier for nick translation (Section 1). Alternatively, dialysis or gel filtration may be done to remove excess reactants.

Conjugation via Carbodiimide Reaction with the 5′ Phosphate of DNA (Phosphoramidate Formation)

The water soluble carbodiimide EDC (Chapter 4, Section 1.1) rapidly reacts with carboxylates or phosphates to form an active complex able to couple with amine-containing compounds. The carbodiimide activates an alkyl phosphate group to a highly reactive phosphodiester intermediate. Diamine spacer molecules or amine-containing probes then may react with this active species to form a stable phosphoramidate bond. Alternatively, *bis*-hydrazide compounds (Chapter 5, Section 8) may be coupled to DNA using this protocol to result in terminal hydrazide groups able to react with aldehyde-containing molecules (Ghosh *et al.*, 1989). Specific labeling of DNA probes only at the 5′ end is possible using these techniques.

Carbodiimide modification of the phosphomonoester end groups on DNA molecules was first used in Khorana's lab to determine nucleotide sequences (Ralph *et al.*, 1962). That early work used the water-insoluble reagent N,N′-dicyclohexylcarbodiimide (DCC) (Chapter 4, Section 1.4) in an organic/aqueous solvent system to effect the conjugations.

Chu *et al.* (1983, 1986) and Ghosh *et al.* (1990) described modified carbodiimide protocols using the water soluble reagent EDC instead of DCC. They also incorporate a second reactive intermediate, a phosphorimidazolide, created from the reaction of the phosphomonoester at the 5′-terminus of DNA with EDC in the presence of imidazole. A reactive phosphorimidazolide will rapidly couple to amine-containing molecules to form a phosphoramidate linkage (Figure 23.10). The chemistry had been used previously to cause the formation of phosphodiester linkages between short DNA strands (Shabarova *et al.*, 1983).

The formation of a phosphorimidazolide intermediate provides better reactivity toward amine nucleophiles than the EDC phosphodiester intermediate if EDC is used without added imidazole. The EDC phosphodiester intermediate also has a shorter half-life than the phosphorimidazolide in aqueous conditions due to hydrolysis. Although EDC alone will create nucleotide phosphoramidate conjugates with amine-containing molecules (Shabarova, 1988), the result of forming the secondary phosphorimidazolide-activated species is increased conjugation yield over carbodiimide-only reactions.

The downside of EDC conjugation with oligonucleotides is the potential for reaction of the carbodiimide at the guanosine N-1 site or with thymidine residues (von der Haar *et al.*, 1971). In practice, however, this cross-reactivity appears to be low enough to maintain complete biological activity and hybridization efficiency in the final conjugate, indicating most of the derivatization occurs at the 5′ phosphate group (Chu *et al.*, 1983).

Carbodiimide-mediated labeling of DNA or RNA probes at the 5′-phosphate end has been used as a general method of end-labeling RNA (Paredes *et al.*, 2011), to fabricate DNA-polymer brush arrays (Barbee *et al.*, 2011), for fluorescent tagging of small mRNAs (Vogel and Richert, 2012), and to fix siRNA molecules in cells by crosslinking to any nearby amines for subsequent probing by hybridization using the tyramide signal amplification (TSA) process (Shi *et al.*, 2011).

The following protocol describes the modification of DNA or RNA probes at their 5′-phosphate ends with a *bis*-hydrazide compound, such as adipic acid dihydrazide or carbohydrazide. A similar procedure for coupling the diamine compound cystamine can be found in Section 2.2.

Protocol

1. Weigh out 1.25 mg of the carbodiimide EDC (1-ethyl-3-(3-dimethylaminopropyl)carbodiimide hydrochloride; Thermo Fisher) into a microfuge tube.
2. Add to the tube 7.5 µl of RNA or DNA containing a 5′ phosphate group. The concentration of the oligonucleotide should be 7.5 to 15 nmol or total of about 57 to 115.5 µg. Also immediately add 5 µl of 0.25-*M bis*-hydrazide compound dissolved in 0.1-*M* imidazole, pH 6.0. Because EDC is labile in aqueous solutions, the addition of the oligo and *bis*-hydrazide/imidazole solutions should be performed quickly.
3. Mix by vortexing, then place the tube in a microcentrifuge and spin for 5 min at maximal rpm.

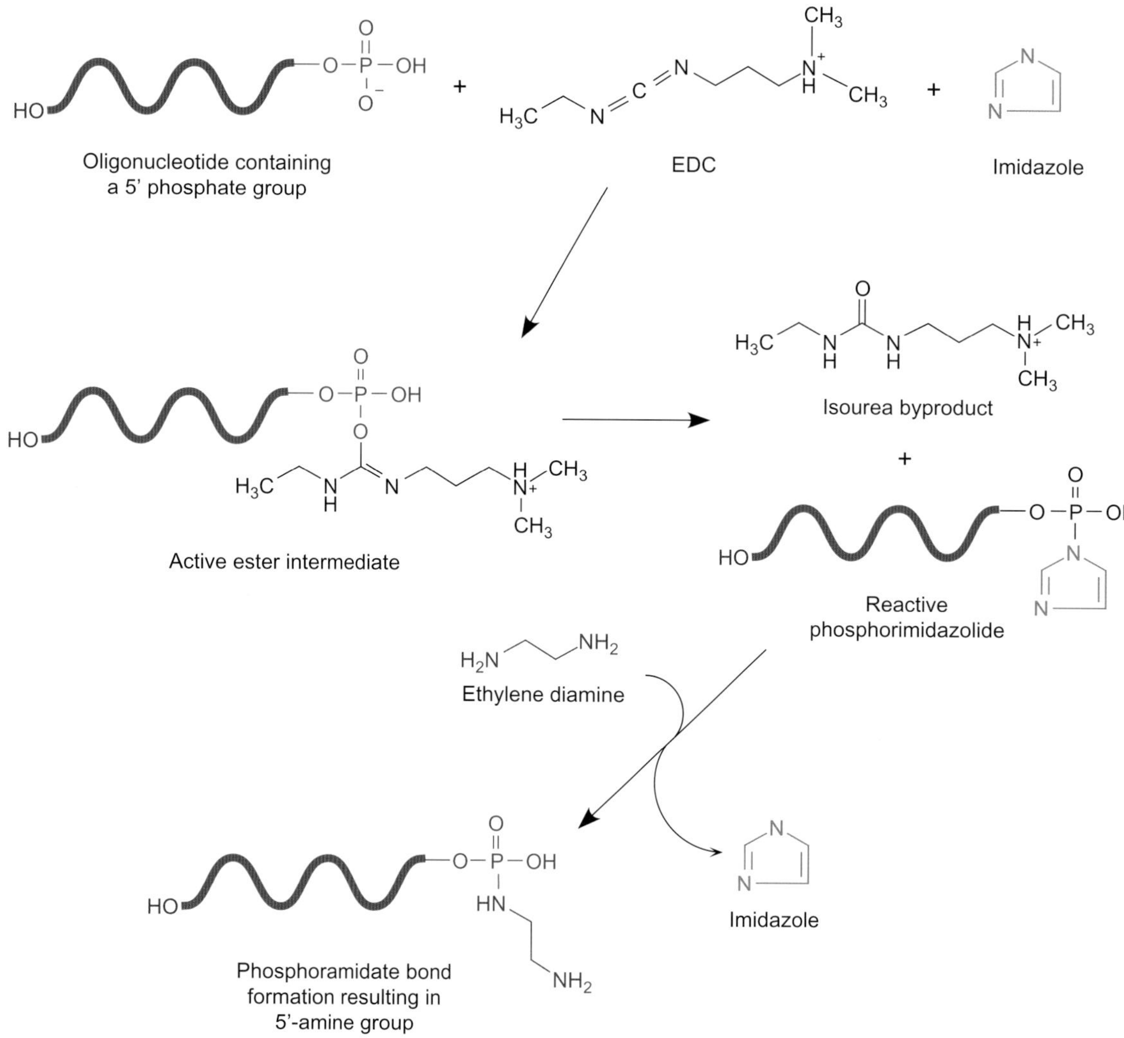

FIGURE 23.10 Oligonucleotides containing a 5′-phosphate group can be reacted with EDC in the presence of imidazole to form an active phosphorimidazolide intermediate. This derivative is highly reactive with amine nucleophiles, forming a phosphoramidate linkage. Diamines reacted with the phosphorimidazolide result in amine terminal spacers that can be modified with detectable components.

4. Add an additional 20 μl of 0.1-*M* imidazole, pH 6.0. Mix and react for 30 min at room temperature.
5. Purify the hydrazide-labeled oligo by gel filtration on desalting resin using 10-m*M* sodium phosphate, 0.15-*M* NaCl, 10-m*M* EDTA, pH 7.2. The probe may now be used to conjugate with an aldehyde-containing molecule.

2.2. Sulfhydryl Modification of DNA

Creating a sulfhydryl group on nucleic acid probes allows conjugation reactions to be performed with sulfhydryl-reactive heterobifunctional crosslinkers (Chapter 6), providing increased control over the derivatization process. Proteins can be activated with a crosslinking agent containing an amine-reactive and a sulfhydryl-reactive end, such as *N*-succinimidyl 3-(2-pyridyldithio)propionate (SPDP) (Chapter 6, Section 1.1), leaving the sulfhydryl-reactive portion free to couple with the modified DNA probe. Having a sulfhydryl group on the probe directs the coupling reaction to a discrete site on the nucleotide strand, thus better preserving hybridization ability in the final conjugate. In addition, heterobifunctional crosslinkers of this type allow two- or three-step conjugation procedures to be carried out, which result in better yield of the desired conjugate than when using homobifunctional reagents.

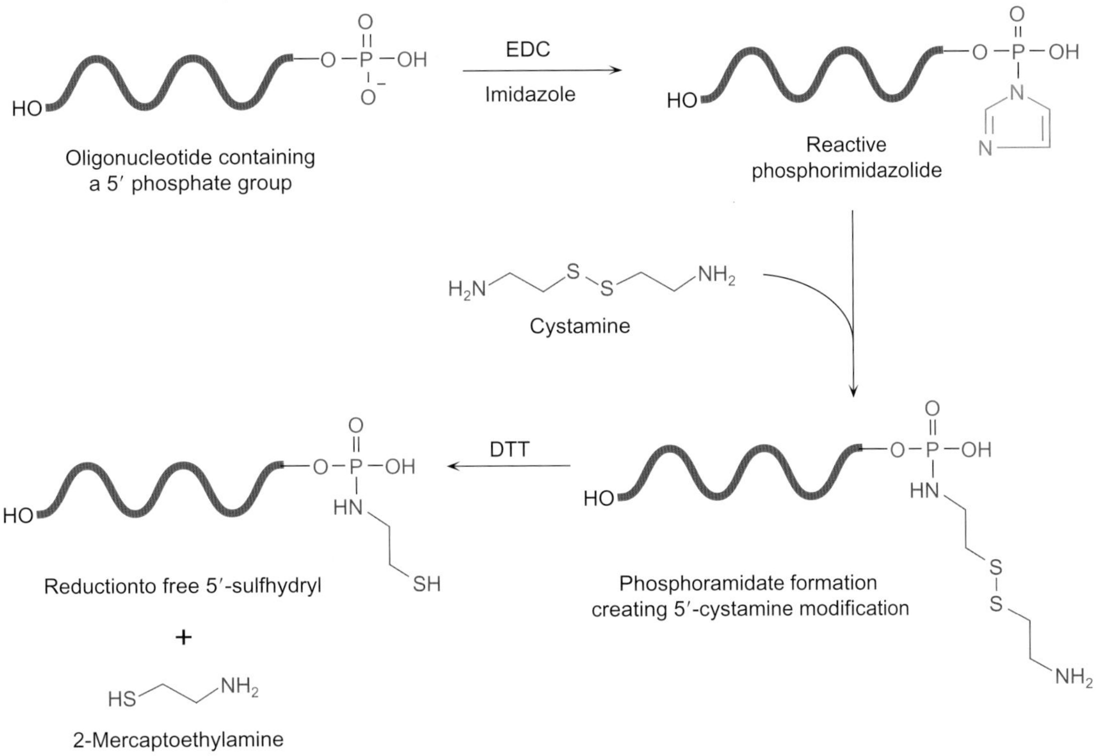

FIGURE 23.11 The 5′-phosphate group of oligonucleotides may be labeled with cystamine using the EDC/imidazole reaction. This results in the formation of an amine terminal spacer containing an internal disulfide group. Reduction of the disulfide provides a route for creating a free thiol for further derivatization.

Cystamine Modification of 5′ Phosphate Groups Using EDC

DNA or RNA may be modified with cystamine at the 5′ phosphate group using a carbodiimide reaction identical to that described previously (Section 2.1). In some procedures, the reaction is carried out in a two-step process by first forming a reactive phosphorimidazolide by EDC conjugation in an imidazole buffer. Next, cystamine is reacted with the activated oligonucleotide, causing the imidazole to be replaced by the amine and creating a phosphoramidate linkage (Chu *et al.*, 1986). An easier protocol was described by Ghosh *et al.* (1990) in which the oligo, cystamine, and EDC were all reacted together in an imidazole buffer. A modification of this method developed by Zanocco *et al.* (1993) is described below.

Reduction of the cystamine-labeled oligo using a disulfide reducing agent releases 2-mercaptoethylamine and creates a thiol group for conjugation (Figure 23.11). DNA probes labeled in this manner have been successfully coupled with SPDP-activated alkaline phosphatase (Chapter 22, Sections 1.2 and 2.5), maleimide-activated horseradish peroxidase (Chapter 22, Section 1.1), NHS–LC-biotin (Chapter 11, Section 6.2, and Section 2.3, this chapter), and the fluorescent tag AMCA–HPDP (Chapter 10, Section 3, and Section 2.5, this chapter).

Protocol

1. Weigh out 1.25 mg of the carbodiimide EDC (1-ethyl-3-(3-dimethylaminopropyl)carbodiimide hydrochloride; Thermo Fisher) into a microfuge tube.
2. Add to the tube 7.5 μl of RNA or DNA containing a 5′ phosphate group. The concentration of the oligonucleotide should be 7.5 to 15 nmol/μl or total of about 57 to 115.5 μg. Also immediately add 5 μl of 0.25-*M* cystamine in 0.1-*M* imidazole, pH 6.0. Because EDC is labile in aqueous solutions, the addition of the oligo and cystamine/imidazole solutions should be done quickly.
3. Mix by vortexing, then place the tube in a microcentrifuge and spin for 5 min at maximal rpm.
4. Add an additional 20 μl of 0.1-*M* imidazole, pH 6.0. Mix and react for 30 min at room temperature.
5. For reduction of the cystamine disulfides, add 20 μl of 1.0-*M* DTT and incubate at room temperature for 15 min. This will release 2-mercaptoethylamine from the cystamine modification site and create the free sulfhydryl on the 5′ terminus of the oligonucleotide.

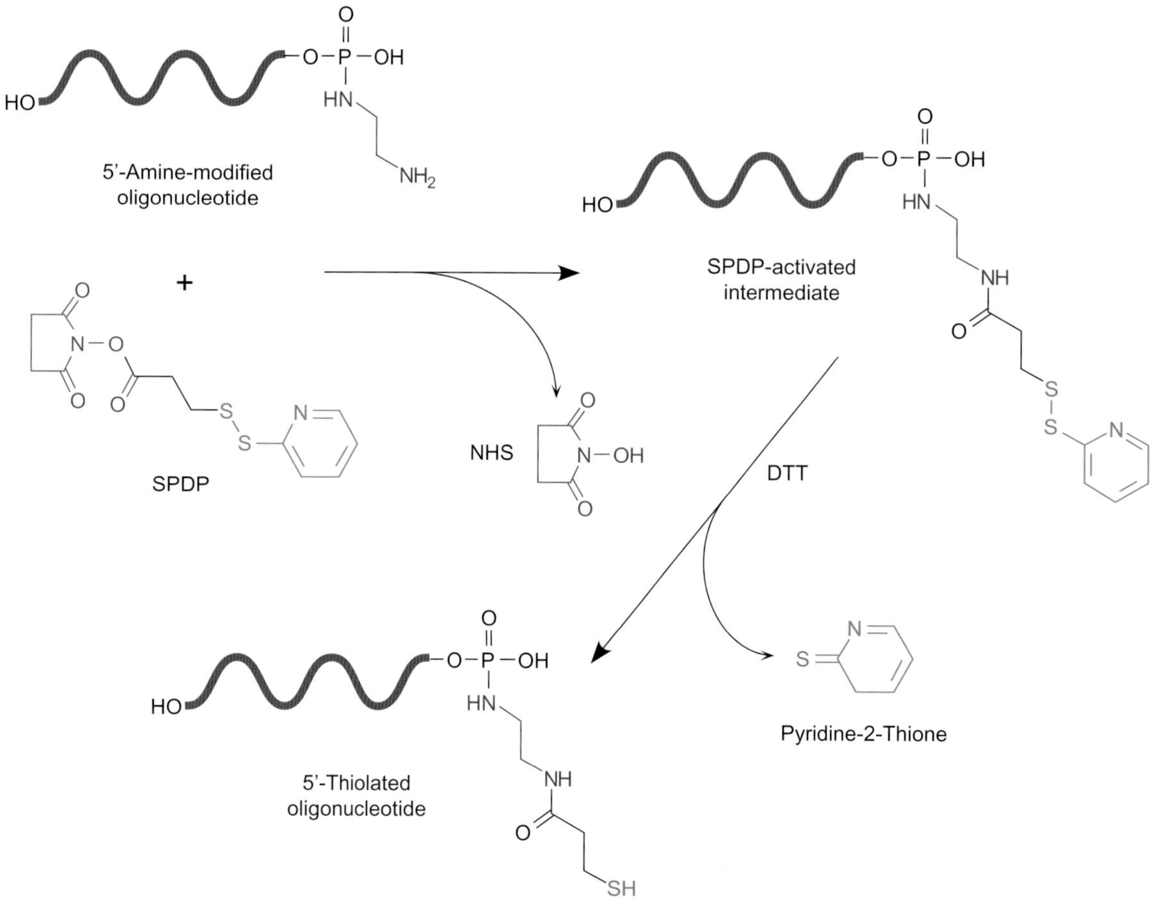

FIGURE 23.12 An oligonucleotide modified at its 5′-phosphate with a diamine compound may be reacted with SPDP and subsequently reduced to create a free sulfhydryl.

6. Purify the SH-labeled oligo by gel filtration on a desalting resin using 10-m*M* sodium phosphate, 0.15-*M* NaCl, 10-m*M* EDTA, pH 7.2. The probe now may be used to conjugate with an activated enzyme, biotin, fluorescent tag, or other molecules containing a sulfhydryl-reactive group.

SPDP Modification of Amines on Nucleotides

Oligonucleotide probes that have been modified with an amine-terminal spacer arm using any of the methods discussed in Sections 1 and 2 of this chapter may be thiolated to contain a sulfhydryl residue. Theoretically, any of the amine-reactive thiolation reagents described in Chapter 2, Section 4.1, may be used to convert an amino group on a DNA molecule into a thiol. One of the more common choices, both for crosslinking and for thiolation reactions, is the heterobifunctional reagent SPDP (Chapter 6, Section 1.1). The NHS ester end of SPDP reacts with primary amine groups to produce stable amide bonds. The other end of the crosslinker contains a thiol-reactive pyridyl disulfide group that also can be reduced with DTT to create a free sulfhydryl.

The reaction of a 5′-diamine-modified oligonucleotide probe with SPDP proceeds under mildly alkaline conditions (optimal pH 7–9) to give the pyridyl disulfide-activated intermediate (Figure 23.12). This derivative has dual functionality. It can be used to couple directly with sulfhydryl-containing detection reagents or enzymes, or it may be converted into a free sulfhydryl for coupling to thiol-reactive compounds (Gaur *et al.*, 1989; Gaur, 1991). In an alternative approach, Chu and Orgel (1988) used 2,2′-dipyridyldisulfide (Chapter 2, Section 5.2) to create reactive pyridyl disulfide groups on a reduced 5′-cystamine-labeled oligonucleotide probe. This derivative can then be used to couple with sulfhydryl-containing molecules, forming a disulfide bond.

Reduction of the pyridyl disulfide end after SPDP modification releases the pyridine-2-thione leaving group and generates a terminal–SH group. This procedure allows sulfhydryl-reactive derivatives such as maleimide-activated enzymes (Chapter 22, Section 2.3) to be conjugated with DNA probes for use in hybridization assays (Malcolm and Nicolas, 1984).

Protocol

1. Dissolve the amine-modified oligonucleotide to be thiolated in 250 µl of 50-m*M* sodium phosphate, pH 7.5.
2. Dissolve SPDP (Thermo Fisher) at a concentration of 6.2 mg/ml in DMSO (makes a 20-m*M* stock solution). Alternatively, LC-SPDP may be used and dissolved at a concentration of 8.5 mg/ml in DMSO (also makes a 20-m*M* solution). The 'LC' form of the crosslinker provides a longer spacer arm that often results in better probe activity after modification. If the water soluble sulfo-LC-SPDP is used, a stock solution in water may be prepared just prior to adding an aliquot to the thiolation reaction. In this case, prepare a 10-m*M* solution of sulfo-LC-SPDP by dissolving 5.2 mg/ml in water. Since an aqueous solution of the crosslinker will degrade by hydrolysis of the sulfo-NHS ester, it should be used quickly to prevent significant loss of activity.
3. Add 50 µl of the SPDP (or LC-SPDP) solution to the oligo solution. Add 100 µl of the sulfo-LC-SPDP solution, if the water soluble crosslinker is used. Mix.
4. React for 1 h at room temperature.
5. Remove excess reagents from the modified oligo by gel filtration on a desalting resin. The modified probe now may be used to conjugate with a sulfhydryl-containing molecule, or it may be reduced to create a thiol for conjugation with sulfhydryl-reactive molecules.
6. To release the pyridine-2-thione leaving group and form the free sulfhydryl, add 20 µl of 1.0-*M* DTT and incubate at room temperature for 15 min. If present in sufficient quantity, the release of pyridine-2-thione can be followed by its characteristic absorbance at 343 nm ($\varepsilon = 8.08 \times 10^3 M^{-1} cm^{-1}$). For many oligonucleotide modification applications, however, the leaving group will be present in too low a concentration to be detectable.
7. Purify the thiolated oligonucleotide from excess DTT by dialysis or gel filtration using 50-m*M* sodium phosphate, 1-m*M* EDTA, pH 7.2. The modified probe should be used immediately in a conjugation reaction to prevent sulfhydryl oxidation and formation of disulfide crosslinks.

SATA Modification of Amines on Nucleotides

Oligonucleotides containing amine groups introduced by enzymatic or chemical means may be modified with *N*-succinimidyl *S*-acetylthioacetate (SATA) (Chapter 2, Section 4.1), to produce protected sulfhydryl derivatives. The NHS ester end of SATA reacts with a primary amine to form a stable amide bond. After modification, the acetyl protecting group can be removed as needed by treatment with hydroxylamine under mildly alkaline conditions (Figure 23.13). The result is terminal sulfhydryl groups that can be used for subsequent labeling with thiol-reactive probes or activated-enzyme derivatives (Kumar and Malhotra, 1992).

The advantage of using SATA over disulfide-containing thiolation reagents such as SPDP (previous section) is that the introduction of sulfhydryl residues does not include the use of a disulfide reducing agent. Typically, the pyridyl dithiol group resulting from an SPDP thiolation must be reduced with a sulfhydryl-containing disulfide reducing compound like DTT to free the- SH group. With SATA, the sulfhydryl is freed by hydroxylamine cleavage, thus eliminating the need for removal of sulfhydryl reductants prior to a conjugation reaction.

Protocol

1. Dissolve the amine-modified oligonucleotide to be thiolated in 250 µl of 50-m*M* sodium phosphate, pH 8.0.
2. Dissolve SATA in DMF at a concentration of 8 mg/ml.
3. Add 250 µl of the SATA solution to the oligo solution. Mix.
4. React for 3 h at 37°C.
5. Remove excess reagents from the modified oligo by gel filtration.
6. To deprotect the acetylated thiol group, add 100 µl of 50-m*M* hydroxylamine hydrochloride, 2.5-m*M* EDTA, pH 7.5.
7. React for 2 h.
8. The sulfhydryl-containing oligonucleotide may be used immediately to conjugate with a sulfhydryl-reactive label, or it can be purified from excess hydroxylamine by gel filtration.

2.3. Biotin Labeling of DNA

Biotinylation of oligonucleotide probes provides a highly specific biological recognition site for detection of DNA using (strept)avidin conjugates. The preparation of biotin-labeled DNA can be performed by either enzymatic or chemical means. Enzyme-catalyzed reactions utilize biotinylated nucleoside triphosphates that can be incorporated into an oligonucleotide randomly or at the 3′ terminus (Section 1). Chemical derivatization methods make use of certain reactive biotin compounds that can couple to functionally modified probes or react with DNA with the use of an activating reagent. For a description of the wide range of biotinylation compounds available, see Chapter 11, Section 6. The preparation and use of avidin and streptavidin conjugates is discussed in Sections 1 to 5 of that chapter.

Biotin–LC-dUTP

Perhaps the most common method of DNA biotinylation is through enzymatic incorporation with the use of a biotin-labeled deoxynucleoside triphosphate. First reported by Langer *et al.* in 1981, the procedure

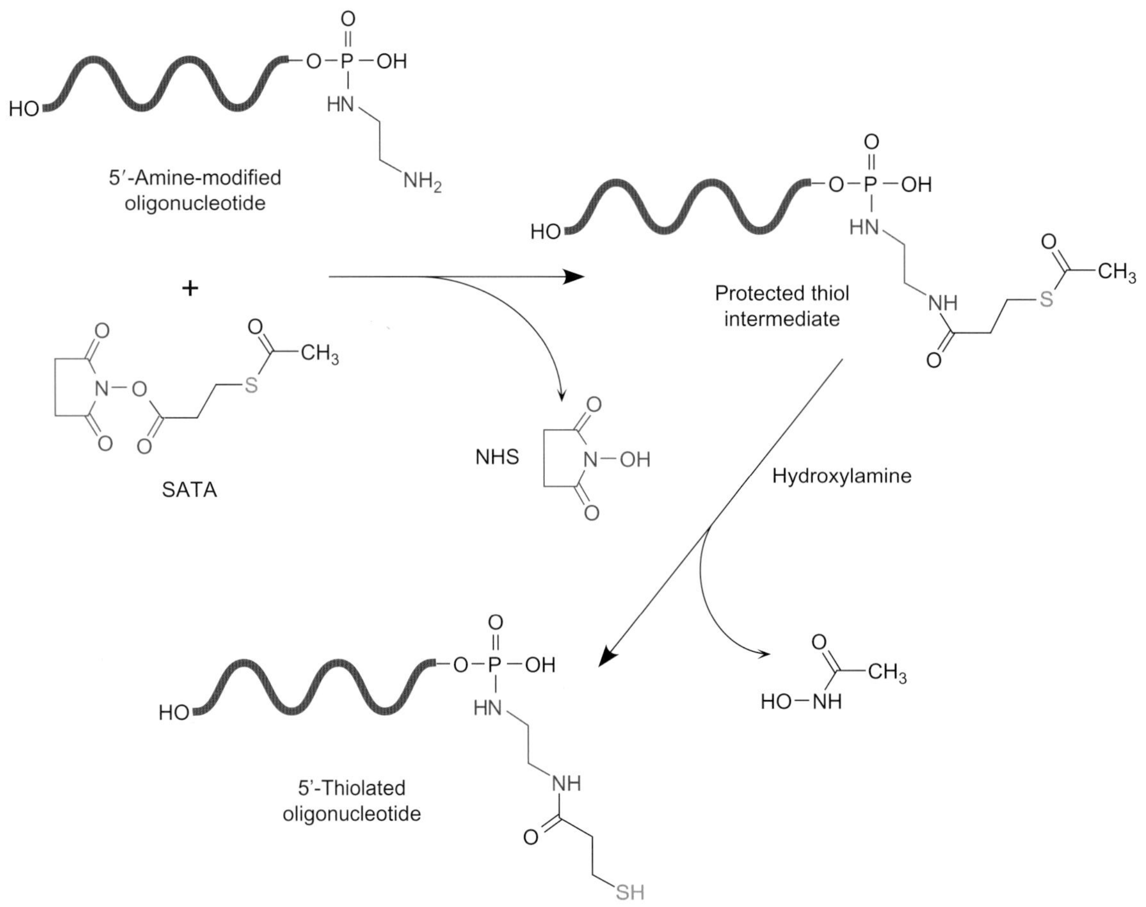

FIGURE 23.13 SATA can be used to modify a 5′-amine derivative of an oligonucleotide, forming a protected sulfhydryl. Deprotection with hydroxylamine results in generation of a free thiol.

is probably the most popular nonradioactive labeling technique reported for oligonucleotide probes. Although biotinylated derivatives of dCTP and dATP are reported in the literature, by far the most frequently employed derivative is biotin–dUTP prepared from the reaction of an amine-modified dUTP with an amine-reactive biotinylation reagent, such as NHS–LC-biotin (Li *et al.*, 2011; Kim *et al.*, 2012) (Chapter 11, Section 6.2).

Biotin–dUTP derivatives are formed by modification of the C-5 position of uridine. This location is not involved in hydrogen bonding activity with complementary DNA strands, thus hybridization efficiency is not immediately compromised. By contrast, biotin–dCTP or biotin–dATP derivatives involve modification of the bases at the N-4 position of cytosine and the N-6 position of adenine, locations directly involved in hydrogen bond formation with complementary bases. Thus, DNA biotinylation through the use of modified deoxynucleoside triphosphates to be incorporated into existing DNA strands may result in better activity of the probe if dUTP is used over dATP or dCTP.

The length of the spacer arm between the C-5 position of uridine and the biotin group is another important parameter for activity of the resulting conjugate. The spacer affects the incorporation efficiency into existing probes using DNA polymerases, and it also affects the ability of an (strept)avidin conjugate to effectively bind the biotinylated probe. Designation of spacer length is usually expressed as the number of atoms separating the nucleotide base from the biotin component. Thus, biotin–*n*–dUTP would describe a biotinylated deoxynucleoside triphosphate having a spacer arm *n* atoms long. The shorter the spacer arm, the better the derivative is able to be recognized and incorporated into DNA polymers using polymerase enzymes. Conversely, the longer the spacer arm, the better the biotinylated probe is able to hybridize to its target and still maintain the capacity to have a streptavidin conjugate be complexed with it. Thus, there is an optimal trade-off in spacer length between enzymatic incorporation efficiency and labeled-probe detectability. Studies have determined that this optimal range is

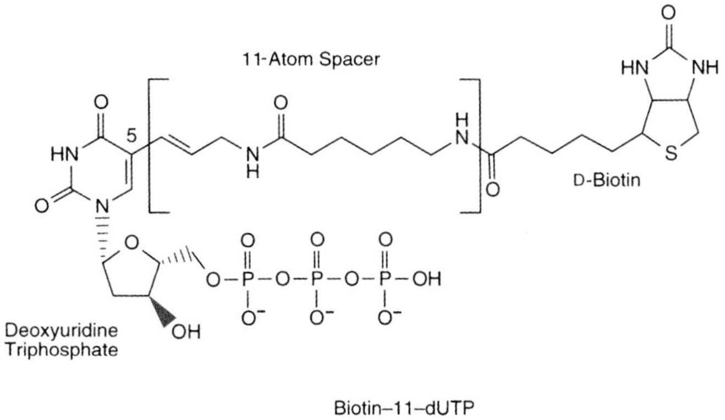

FIGURE 23.14 Biotin–11–dUTP is perhaps one of the most popular nucleotide derivatives used for enzymatic biotinylation of oligonucleotides. The "11" designation refers to the number of atoms in its spacer arm.

rather broad—between 7 atoms and 21 atoms in length. Perhaps the most common derivative is biotin–11–dUTP, wherein an 11-atom spacer is employed (Figure 23.14).

General protocols for the enzymatic incorporation of biotin–11–dUTP into DNA probes can be found in Section 1. A particularly interesting modification of the typical enzymatic incorporation protocol for biotin is described by Didenko (1993). The single-strand template is immobilized by adsorption onto membranes before synthesis of the biotinylated probe. After polymerase incorporation of biotin–11–dUTP, the labeled probe is removed by brief heating to 90°C in water. The result is highly pure probe with no contaminating complementary DNA strands.

Photobiotin Modification of DNA

The photoreactive biotin derivative, *N*-(4-azido-2-nitrophenyl)-aminopropyl-*N*′-(*N*-d-biotinyl-3-aminopropyl)-*N*′-methyl-1,3-propanediamine, simply called photoactivatable biotin or photobiotin (Forster *et al.*, 1985) contains a 9-atom diamine spacer group on the biotin valeric acid side chain on one end, while the other end of the spacer terminates in an phenyl azide group. The phenyl azide group can be photolyzed with UV light (350nm) resulting in the formation of a highly reactive nitrene intermediate. In most instances, this nitrene rapidly reacts via ring expansion to form a dehydroazepine that is reactive with nucleophiles, such as amine groups (Chapter 6, Section 3).

When photobiotin is irradiated in the presence of DNA the reaction process nonselectively couples a biotin label to every 100 to 200 base residues (Saito *et al.*, 2012). The result is an oligonucleotide probe detectable by the use of (strept)avidin conjugates. The use of photobiotin for DNA or RNA modification is summarized in Chapter 11, Section 6.5. Refer also to the same section of Chapter 11 for use of the compound psoralen–PEO_3–biotin for an alternative option in the photoreactive labeling of, in this case, double stranded DNA or RNA molecules.

The following protocol is based on the method of Forster *et al.* (1985). Some optimization may be necessary to obtain the best signal and activity for particular probes in hybridization assays.

Protocol for Labeling DNA Probes with Photobiotin

1. In subdued lighting conditions, dissolve photobiotin in water at a concentration of 1 μg/μl. Protect from light.
2. Dissolve the oligonucleotide probe in water or 0.1-m*M* EDTA, pH 7.0, at a concentration of 1 μg/μl.
3. Mix an equal volume of the photobiotin solution with the DNA probe solution.
4. Place the solution in an ice bath and irradiate from above (about 10cm away) for 15min using a sunlamp (such as Philips Ultrapnil MLU 300W, General Electric sunlamp RSM 275W, or National Self-Ballasted BHRF 240–250V 250W W-P lamp).
5. Add 50μl of 0.1-*M* Tris, pH 9, and increase the total volume of the solution to 100μl (if it is less than this amount).
6. To extract excess photobiotin, add 100μl of 2-butanol. Mix well and centrifuge. Discard the upper phase. Repeat this process two more times.
7. To recover the biotinylated DNA, add 75μl of 4-*M* NaCl and mix.
8. Add 100μl of ethanol and cool the sample in dry ice (CO_2) for 15min.
9. Centrifuge to collect the precipitated, biotinylated DNA.

Reaction of NHS–LC-Biotin with Diamine-Modified DNA Probes

NHS–LC-biotin is an extended spacer arm derivative of biotin containing an amine-reactive NHS ester (Chapter 11, Section 6.2). The compound is a popular choice for biotinylating a wide range of molecules containing primary amine groups, especially proteins. Oligonucleotides modified to contain amine-terminal spacer arms also can be modified with NHS–LC-biotin to create stable amide bond derivatives. Alternatively, a hydrophilic biotinylation compound containing a PEG spacer arm can be used to form a biotin–oligo derivative without the hydrophobic character of the alkyl chain within NHS–LC-biotin. Chapter 18, Section 1.3, describes the amine-reactive NHS–PEG_n–biotin compounds suitable for this purpose.

Whether an amine is incorporated into an oligo by enzymatic means or chemical derivatization, an NHS-ester-containing biotinylation reagent can be used to label the derivative in high yield. If an amine group is

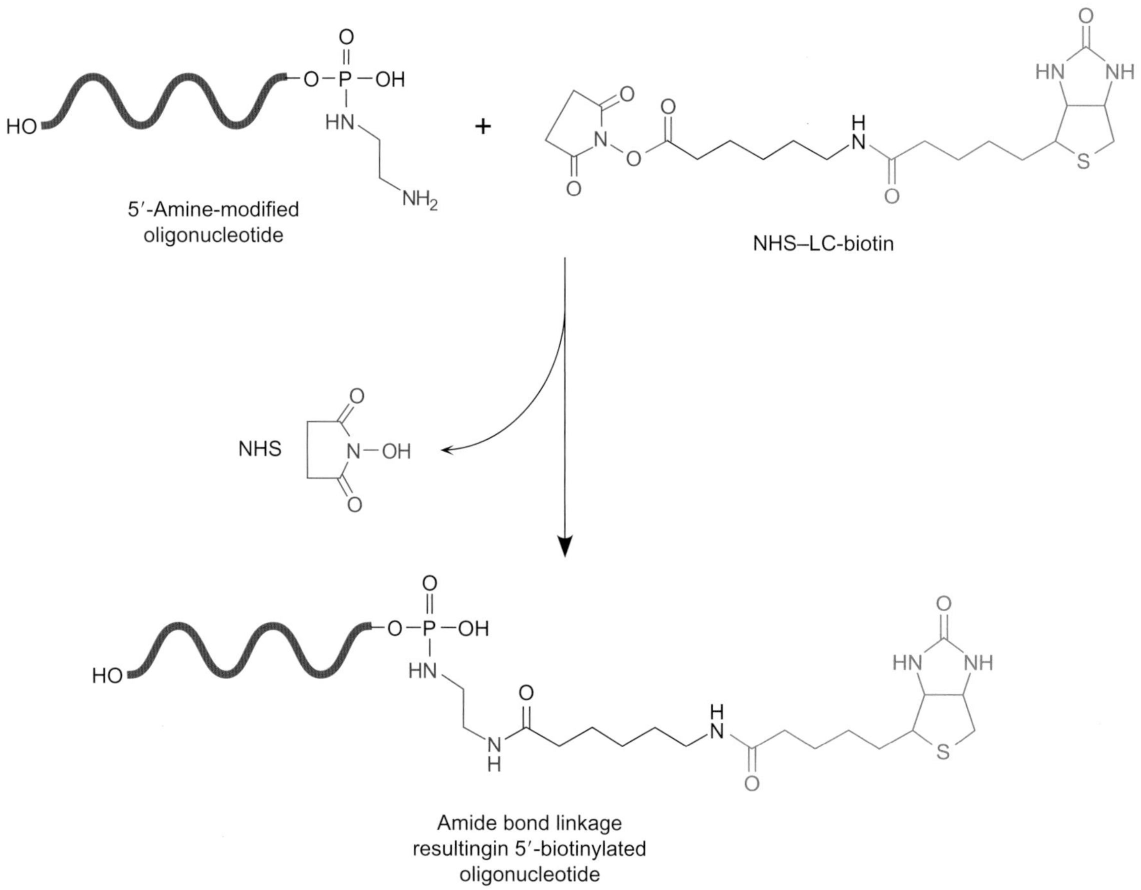

FIGURE 23.15 Biotinylation of oligonucleotides may be performed at the 5′-phosphate end using a diamine derivative and reacting with NHS–LC-biotin.

added to the 5′ end of a DNA probe by phosphoramidate formation (Section 2.1), then biotinylation of such molecules directs the label to a region totally removed from interfering in subsequent hybridization with a target DNA strand (Figure 23.15).

The following protocol assumes that the amine-containing oligo has already been synthesized by any of the methods discussed in Section 2.1.

Protocol

1. Prepare 10 to 20 µg of amine-containing oligonucleotide in 200 µl of water. Add to this solution, 20 µl of 1-*M* sodium bicarbonate, pH 9.0.
2. Dissolve NHS–LC-biotin (Thermo Fisher) in DMSO at a concentration of 10 mg/ml. Add 50 µl of the biotinylation solution to the oligo solution. Mix well.
3. React for 2 h at room temperature.
4. Isolate the biotinylated probe by ethanol/salt precipitation as described in Section 1 for nick-translation modification of DNA probes. Alternatively, dialysis, gel filtration, or *n*-butanol extraction may be used to remove excess reagents.

Biotin–Diazonium Modification of DNA

Diazonium groups are able to couple at the C-8 position of adenosine or guanosine residues, forming diazo bonds. *p*-Aminobenzoyl biocytin can be used in this reaction to add a biotin handle to purine bases within oligonucleotides (Chapter 11, Section 6.6). This biotinylation reagent contains a 4-aminobenzoic acid amide derivative off the α-amino group of biocytin's lysine residue (Thermo Fisher). The aromatic amine can be treated with sodium nitrite in dilute HCl to form a highly reactive diazonium derivative, which is able to couple with active hydrogen-containing compounds. A diazonium reacts rapidly with histidine or tyrosine residues within proteins, forming covalent diazo bonds (Wilchek *et al.*, 1986). It also can react with purine residues within DNA at position 8 of the bases (Lowe, 1979; Rothenberg and Wilchek, 1988) (Figure 23.16).

Protocol

1. Prepare diazotized *p*-aminobenzoyl biocytin by using the protocol outlined in Chapter 11, Section 6.6, but instead of starting with 2 mg the biotinylation reagent dissolved in 40 µl of 1-*N* HCl, use 9 mg in

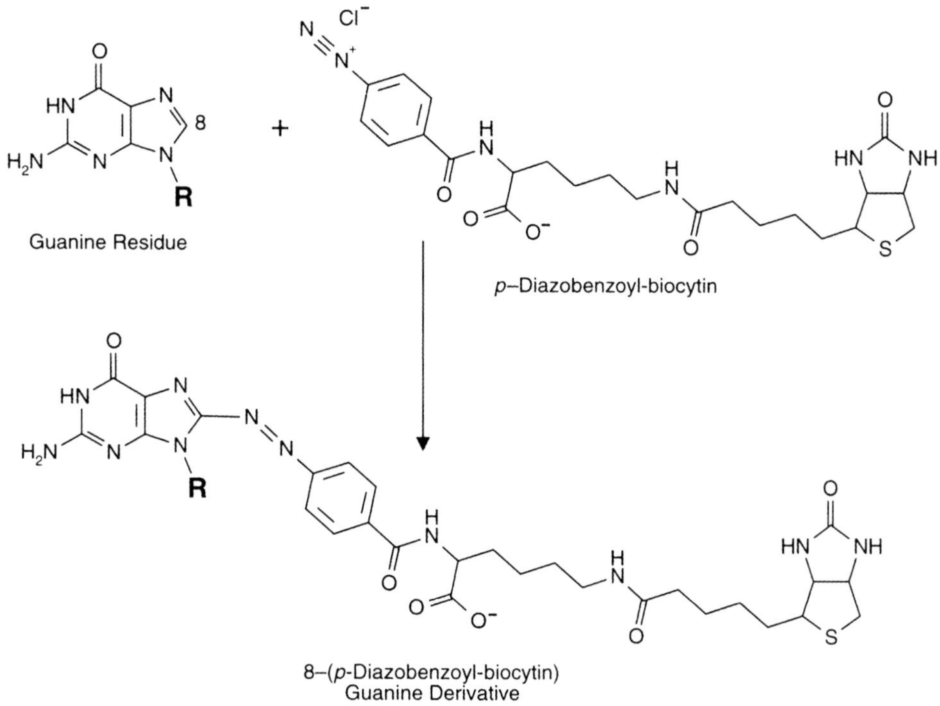

FIGURE 23.16 This diazo derivative of biocytin may be used to modify guanine bases at the C-8 position.

180 µl of 1-*N* HCl. Proportionally scale up the other reactant quantities used in the protocol. After the reaction is complete, immediately adjust the pH of the final solution to 9.

2. Add 1 µg of single-stranded DNA to the above solution.
3. React for 30 min at room temperature.
4. Purify the biotinylated DNA probe by ethanol precipitation, gel filtration, *n*-butanol extraction, or dialysis as discussed in previous sections.

Reaction of Biotin–BMCC with Sulfhydryl-Modified DNA

Biotin–BMCC is a sulfhydryl-reactive biotinylation reagent containing a maleimide group at the end of an extended spacer arm. The long spacer (32.6 Å) provides enough distance between modified oligonucleotides and the bicyclic biotin end to allow efficient binding of (strept) avidin probes, even when hybridized to target sequences. The reagent may be used to add a biotin label to DNA or RNA molecules after they have been modified to contain thiol groups. For instance, cystamine labeling at the 5′ phosphate group of DNA via carbodiimide-mediated phosphoramidate formation followed by disulfide reduction (Section 2.2) can create the required sulfhydryl groups. Subsequent reaction with biotin–BMCC results in a derivative labeled only at an end of the DNA probe (Figure 23.17), thus avoiding the potential for hydrogen bonding interference in hybridization assays.

Since maleimide groups are highly specific for coupling to thiols in the pH range of 6.5 to 7.5, side reaction products can be reasonably avoided. The reaction is complete within 2 h at room temperature.

Protocol

1. Prepare 10 to 20 µg of a sulfhydryl-containing oligonucleotide in 200 µl of 50-m*M* sodium phosphate, 10-m*M* EDTA, pH 7.2 (the methods outlined in Section 2.2 can be used to form the thiol group).
2. Dissolve biotin–BMCC in DMSO at a concentration of 5 mg/ml. Prepare fresh.
3. Add 50 µl of the biotinylation solution to the oligo. Mix well.
4. React for 2 h at room temperature.
5. Isolate the biotinylated probe by ethanol/salt precipitation as described in Section 1 for nick translation (this chapter).

Biotin–Hydrazide Modification of Bisulfite-Activated Cytosine Groups

Biotin–hydrazide is the hydrazine derivative of D-biotin prepared using its valeric acid carboxylate (Chapter 11, Section 6.4). The hydrazide group is typically used to react with aldehyde and ketone groups to give hydrazone linkages. However, the hydrazide compound can also undergo transamination reactions with cytosine residues *via* catalysis with bisulfite (Section 2.1)

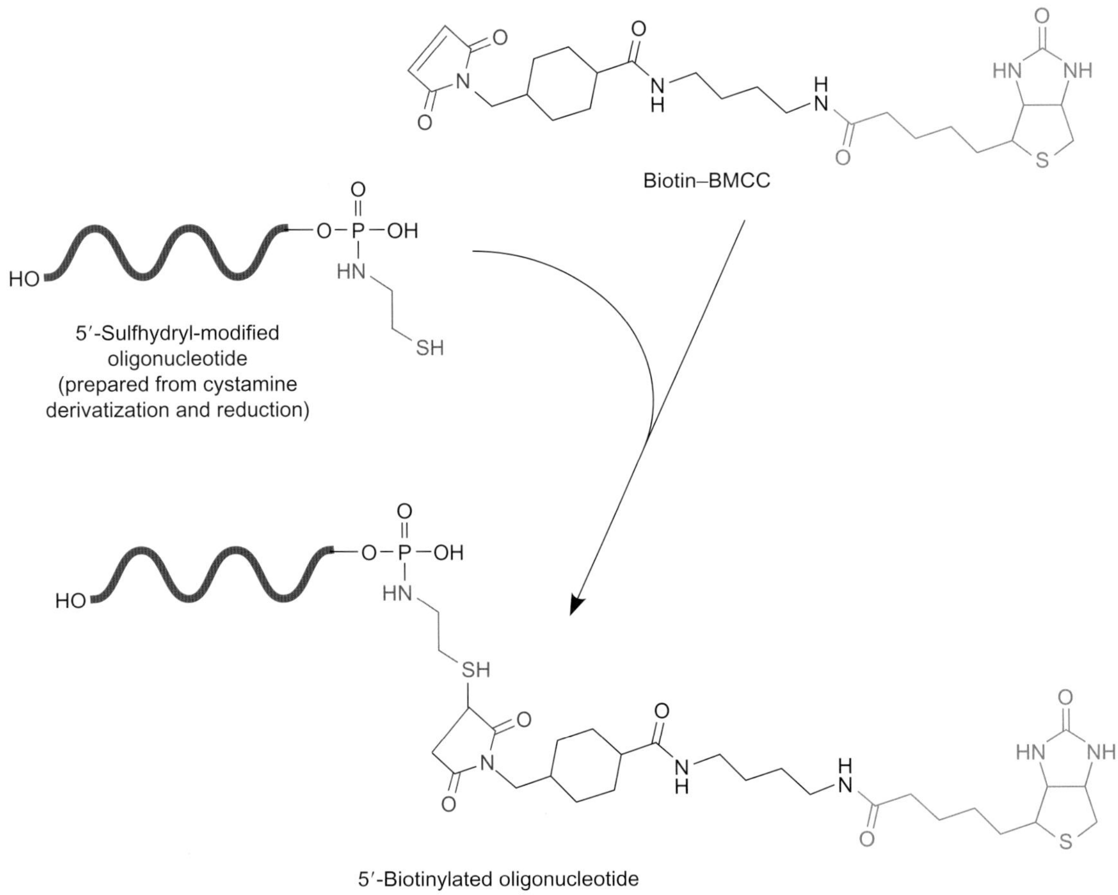

FIGURE 23.17 Biotin–BMCC can be used to modify a reduced, cystamine derivative of DNA, forming a thioether linkage.

(Figure 23.18). DNA or RNA probes containing cytosine groups may be modified to contain biotin labels using a simple, one-step procedure (Reisfeld *et al.*, 1987). The detection limit of DNA probes biotinylated using this technique can be less than 1 pg on blots, when analyzed using a streptavidin–HRP conjugate with chemiluminescent detection. A longer-chain analog of biotin–hydrazide, biotin–LC-hydrazide, may be used to create an extended spacer between the oligonucleotide and the bicyclic biotin group, thus increasing the binding efficiency of avidin or streptavidin conjugates. Alternatively, a hydrophilic biotin–PEG$_n$–hydrazide compound may be used in this reaction to provide greater water solubility for the label (Chapter 18, Section 1.3). Leary *et al.* (1983) reported that increasing the spacer arm length from 4 to 11 atoms when biotinylating DNA probes can increase the detectability of the target DNA approximately 4-fold.

The following method is adapted from Reisfeld *et al.* (1987).

Protocol

1. Prepare 50 μg of a single-stranded DNA probe in 300 μl of 50-m*M* sodium acetate, pH 4.5.
2. Dissolve biotin–hydrazide in water at a concentration of 10 mg/ml.
3. Add 300 μl of the biotin–hydrazide solution to the DNA solution.
4. Add sodium bisulfite to obtain a final concentration of 1-*M* in the reaction medium.
5. React for 24 h at 37°C.
6. Remove excess reactants by dialysis against water at 4°C.

2.4. Enzyme Conjugation to DNA

Enzymes useful for detection purposes in ELISA techniques (Chapter 22) can also be employed in the creation of highly sensitive DNA probes for hybridization assays. The attached enzyme molecule provides

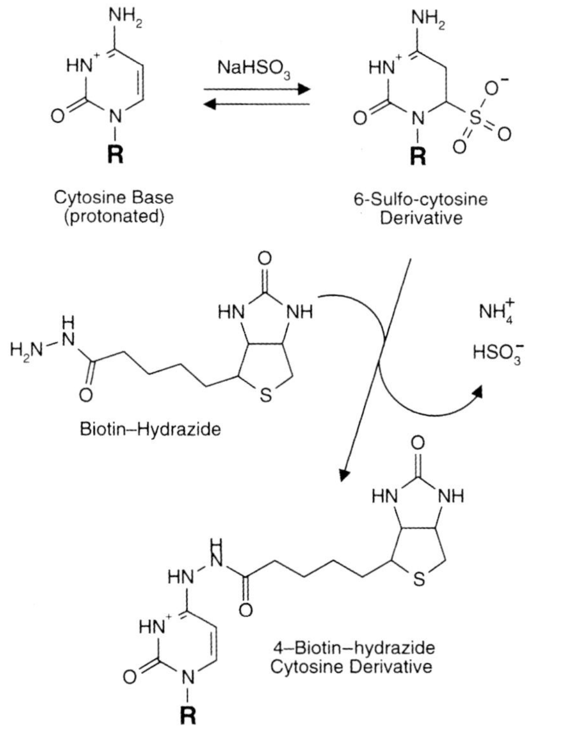

FIGURE 23.18 Biotin–hydrazide may be incorporated into cytosine bases using a bisulfite-catalyzed transamination reaction.

detectability for the oligonucleotide through turnover of substrates that can produce chromogenic or fluorescent products. Enzyme-based hybridization assays are perhaps the most common method of nonradioactive detection used in nucleic acid chemistry today. The sensitivity of enzyme-labeled probes can equal or exceed that of radiolabeled nucleic acids, thus eliminating the need for radioactivity in most assay systems.

The conjugation reactions involved in DNA–enzyme crosslinking are not unlike the methods used to form antibody–enzyme conjugates (Chapter 20, Section 1). Bifunctional crosslinkers can be used to couple a modified oligonucleotide to an enzyme molecule using the same basic principles effective in protein–protein conjugation. The only requirement is that the DNA molecule be modified to contain one or more suitable functional groups, such as nucleophiles like amines or sulfhydryls. The modification process used to create these functional groups can use enzymatic (Section 1, this chapter) or chemical (Section 2, this chapter) means and it can result in random incorporation of modification sites or be directed exclusively at one end of the DNA molecule, such as in 5′ phosphate coupling.

The following sections describe some of the more common procedures of preparing DNA–enzyme conjugates.

Alkaline Phosphatase Conjugation to Cystamine-Modified DNA Using Amine- and Sulfhydryl-Reactive Heterobifunctional Crosslinkers

A cystamine group added to the 5′ phosphate of DNA molecules using a carbodiimide reaction (Section 2.2) can be used in a heterobifunctional crosslinking scheme to conjugate with alkaline phosphatase. Crosslinking agents containing an amine-reactive portion and a sulfhydryl-reactive part work best in forming this type of conjugate. Perhaps one of the most common heterobifunctional reagents used for DNA–enzyme formation is SPDP (Chapter 6, Section 1.1). SPDP contains an NHS ester on one end able to create an amide bond linkage with amino groups on protein molecules. After modification of alkaline phosphatase with this crosslinker, the enzyme is activated to contain pyridyl disulfide groups for coupling to the sulfhydryls on a cystamine-modified DNA probe (Figure 23.19). The reaction forms disulfide bonds between the oligonucleotide and the alkaline phosphatase enzyme. Since the crosslink occurs only at the 5′ end of the DNA strand, the presence of an enzyme molecule does not adversely affect the ability of base pairing and hybridization to a target sequence.

The following protocol assumes that the labeling process used to create a sulfhydryl-modified DNA probe already has been done according to the method of Section 2.2 (this chapter). The modification procedure for activating alkaline phosphatase with SPDP may be done according to the protocol described in Chapter 6, Section 1.1 (see also Chapter 22, Section 1.2). To obtain efficient labeling of all the alkaline phosphatase added to the reaction medium, the modified oligo is reacted in a 10-fold molar excess. Reaction of the DNA probe in excess allows easy separation of not-coupled oligo from conjugated probe, thus eliminating any potential interference in hybridization assays due to unlabeled oligonucleotide.

Protocol

1. Dissolve a 5′-sulfhydryl-modified oligonucleotide in water or 10-m*M* EDTA at a concentration of 0.05 to 25 μg/μl. Calculate the total nanomoles of oligo present based upon its molecular weight.
2. Prepare SPDP-activated alkaline phosphatase in 50-m*M* sodium phosphate, 0.15-*M* NaCl, 10-m*M* EDTA, pH 7.2. Add to the oligo solution, an amount of the activated enzyme representing a 10-fold molar excess over the calculated amount of DNA present.
3. React at room temperature for 30 min with gentle mixing.
4. The alkaline phosphatase–DNA conjugate may be purified away from excess oligo by dialysis, gel filtration, or through the use of centrifugal concentrators. A simple way of removing unreacted

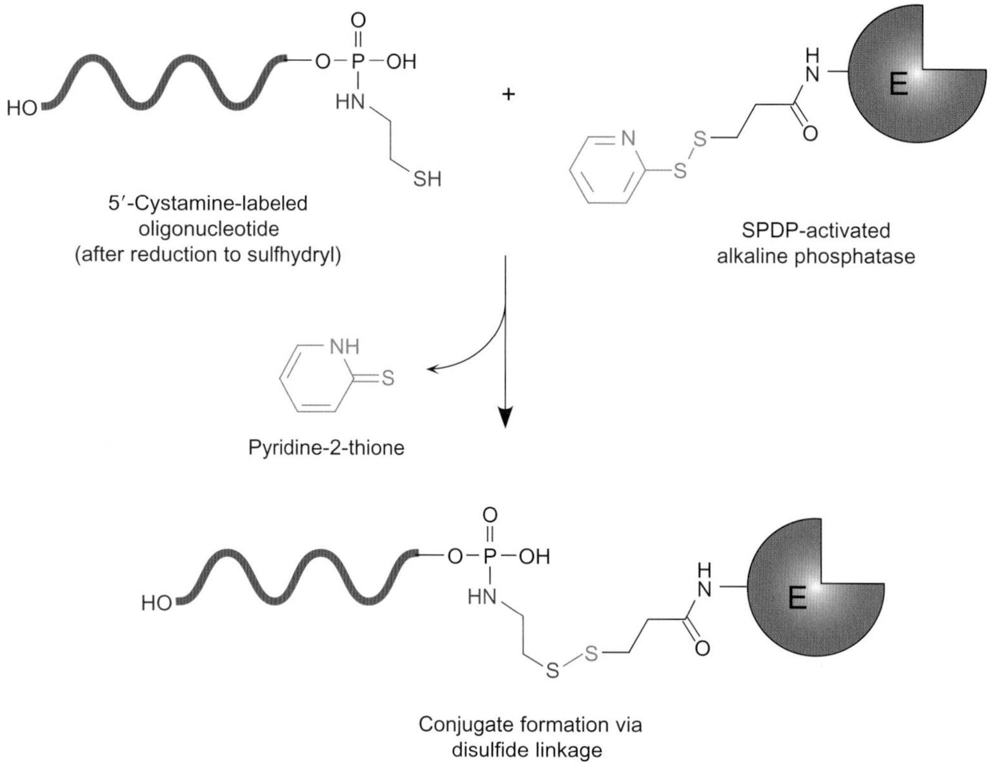

FIGURE 23.19 An oligonucleotide modified with cystamine and reduced to generate a free sulfhydryl may be conjugated with an SPDP-modified enzyme, forming a disulfide linkage.

oligo is to use Centricon-30 concentrators (Amicon) which have a molecular weight cutoff of 30,000. Since the enzyme molecular weight is approximately 140,000 and the conjugate is even higher, a relatively small DNA probe will pass through the membranes of these units while the conjugate will not. To purify the conjugate using Centricon-30s, add 2 ml of the phosphate buffer from step 2 to one concentrator unit, then add the reaction mixture to the buffer and mix. Centrifuge at 1000 *g* for 15 min or until the retentate volume is about 50 μl. Add another 2 ml of buffer and centrifuge again until the retentate is 50 μl. Invert the Centricon-30 unit and centrifuge to collect the retentate in the collection tube provided by the manufacturer.

Alkaline Phosphatase Conjugation to Diamine-Modified DNA Using DSS

Disuccinimidyl suberate (DSS) is a homobifunctional crosslinker containing an amine-reactive NHS ester at both ends (Chapter 5, Section 1.2). Reaction of the reagent in excess with diamine-modified DNA probes creates an activated intermediate able to conjugate with enzyme molecules through their available amine groups (Figure 23.20) (Jablonski *et al.*, 1986). The coupling reaction produces stable amide linkages under mildly alkaline conditions. Although the following protocol has been used to label oligonucleotides with success, it may be less efficient than the previous protocol at forming the desired conjugate due to the homobifunctional nature of the crosslinker. During the activation step, the modified DNA must be purified away from excess DSS. Since this is carried out under aqueous conditions, hydrolysis of the free NHS ester at the other end of the crosslinker takes place at the same time. Activity losses can be severe if the separation step is not done rapidly. In fact, the original protocol called for several hours of gel filtration chromatography and concentration before the conjugation reaction was performed. Farmer and Castaneda (1991) made a significant improvement to this procedure by including a faster separation step using alcohol extraction after activation of the oligo with DSS. The purification time decreased to minutes instead of hours. This does help to limit hydrolysis, but cannot completely avoid it.

The following protocol is based on the methods of Farmer and Castaneda (1991), Kiyama *et al.* (1992), and Ruth (1993).

Protocol

1. Prepare an amine-modified oligonucleotide according to any of the protocols discussed in Section 2.1. Dissolve or buffer-exchange the oligo into 0.1-*M* sodium borate, 2-m*M* EDTA, pH 8.25, at a concentration of 9 nmol (2.0 $A_{260\,nm}$ units) in 15 μl.

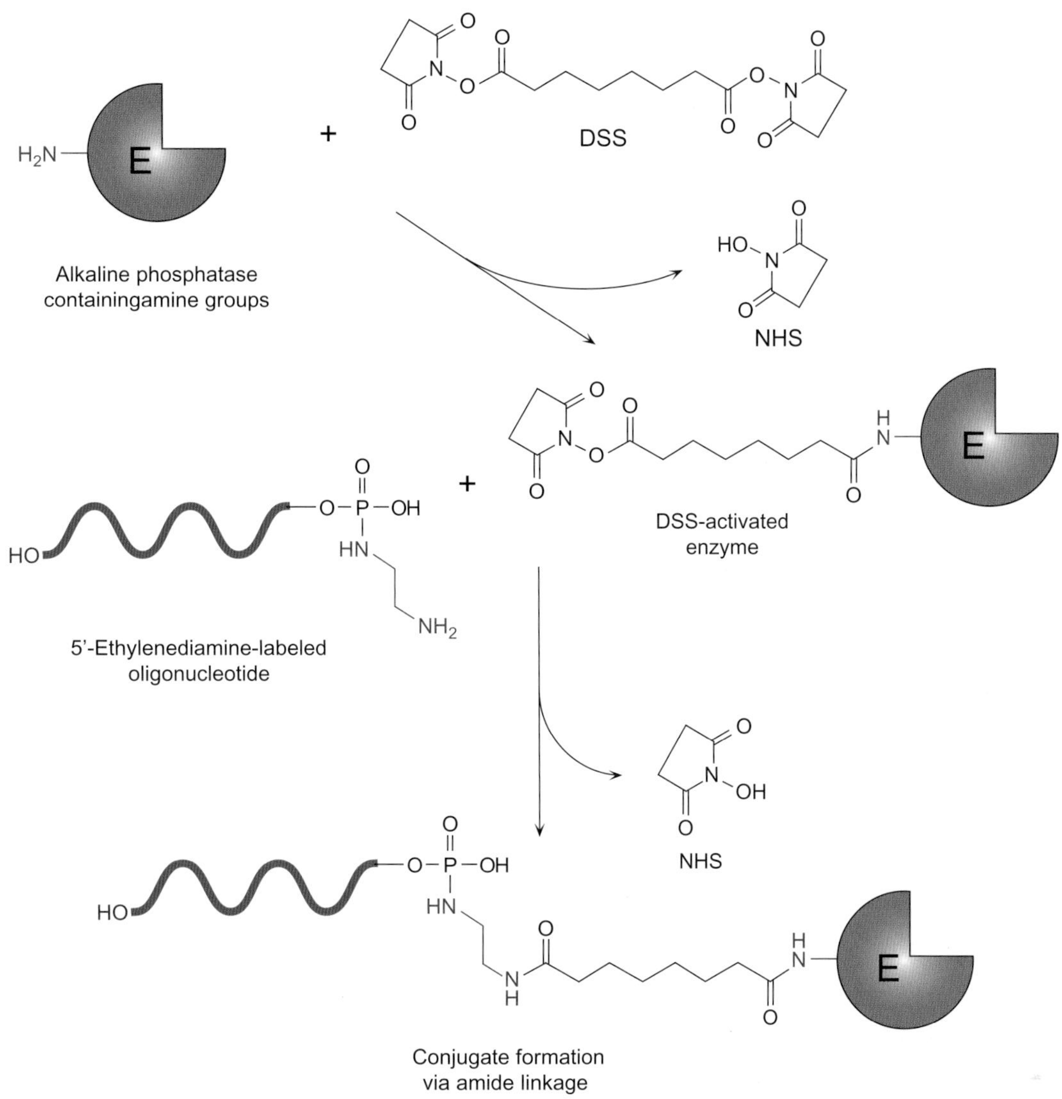

FIGURE 23.20 The homobifunctional crosslinker DSS may be used to conjugate an enzyme to a 5′-diamine-modified oligonucleotide. The NHS ester groups on DSS react with the amines to form amide bonds.

2. Dissolve DSS in dry DMSO at a concentration of 1 mg/100 μl. Prepare fresh.
3. Add 30 μl of the DSS solution to the oligo. Mix well.
4. React for 15 min at room temperature in the dark.
5. Immediately extract excess DSS and reaction by-products by the addition of 0.5 ml of *n*-butanol. Mix vigorously by vortexing and centrifuge (1 min, 15,000 rpm) to separate the two phases. Carefully remove the upper layer and discard. Extract two more times with *n*-butanol.
6. Chill the remaining sample on dry ice and lyophilize to remove the last traces of liquid. The drying period will only take 15 to 30 min. The dried, DSS-activated DNA is stable under anhydrous conditions.
7. Dissolve or dialyze alkaline phosphatase into 3-*M* NaCl, 30-m*M* triethanolamine, 1-m*M* $MgCl_2$, pH 7.6, at a concentration of 20 mg/ml.
8. Add 70 μl of the alkaline phosphatase to the dried, DSS-activated DNA. Mix gently to dissolve.
9. React overnight at 4°C in the dark.
10. Remove unconjugated oligo by using a Centricon-30 centrifugal concentrator according to step 4 of the protocol described in the previous section. Unconjugated enzyme may be removed by ion-exchange chromatography using a MonoQ-10 (0.5 × 5 cm) FPLC column (GE Healthcare) or the equivalent. Binding buffer is 20-m*M* Tris, pH 8. Elute using a linear gradient of 0–100% 20-m*M* Tris, 1-*M* NaCl, pH 8. Free enzyme will elute before the more negatively charged oligo–enzyme conjugate.

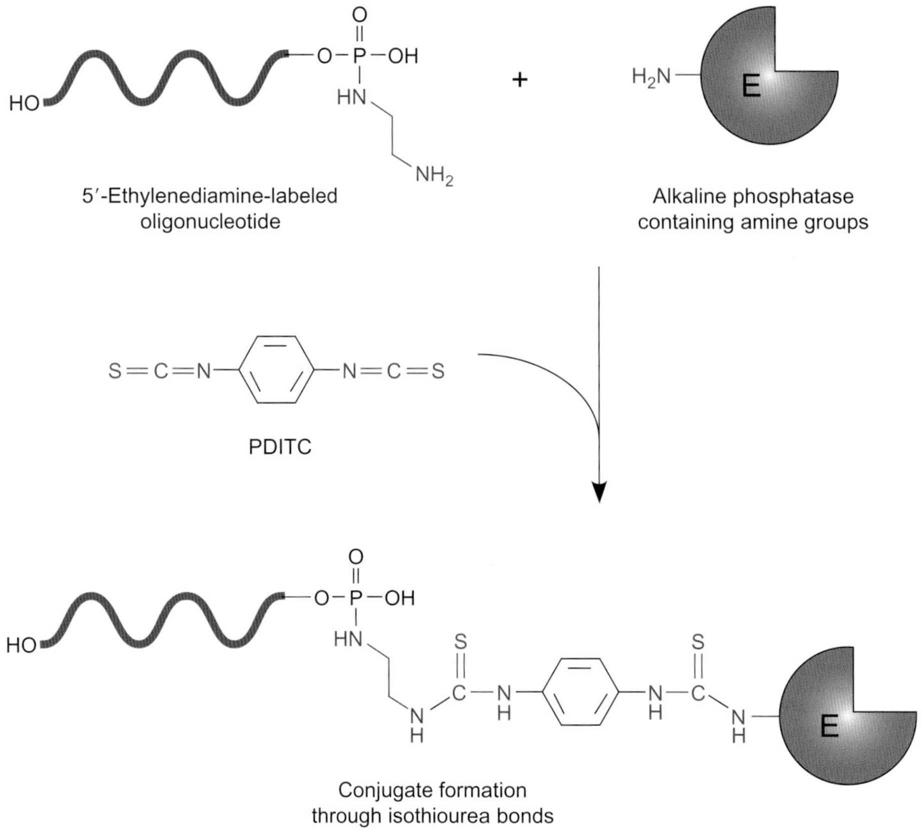

FIGURE 23.21 The homobifunctional crosslinker PDITC may be used to conjugate an enzyme to a 5′-diamine-modified oligonucleotide, creating isothiourea linkages.

Enzyme Conjugation to Diamine-Modified DNA Using PDITC

PDITC, 1,4-phenylene diisothiocyanate, is a homobifunctional crosslinker containing two amine-reactive isothiocyanate groups on a phenyl ring. Reaction in excess with an amine-modified oligonucleotide results in the formation of a thiourea linkage, leaving the second isothiocyanate group free to couple with amine-containing enzymes or other molecules (Urdea *et al.*, 1988) (Figure 23.21).

The following protocol is adapted from Keller and Manak (1989).

Protocol

1. Prepare an amine-modified oligonucleotide using any of the methods of Section 1 or 2.1. Dissolve or buffer exchange 70 µg of the oligo into 25 µl of 0.1-*M* sodium borate, pH 9.3.
2. Dissolve 10 mg of PDITC (Aldrich) in 500 µl of DMF. Add this solution to the oligo prepared in step 1.
3. React for 2 h at room temperature in the dark.
4. To extract excess reactant, add to the reaction medium, 3 ml of *n*-butanol and 3 ml of water. Mix well. Centrifuge the mixture to separate the two phases. Discard the upper yellow layer. Repeat the extraction process several times, and then dry the remaining solution containing activated oligo using lyophilization or a rotary evaporator. The PDITC-activated DNA is stable in a dried state.
5. Dissolve HRP (Chapter 22) in 200 µl 0.1-*M* sodium borate, pH 9.3, at a concentration of 10 mg/ml. If the HRP is supplied as an ammonium sulfate suspension, all ammonium ions must be removed by extensive dialysis prior to the conjugation reaction. Add the HRP solution to the activated oligo.
6. React overnight at room temperature in the dark.
7. Excess enzyme may be removed through isolation of the oligo–enzyme conjugate using electrophoresis separation under nondenaturing conditions. The reaction solution is applied to a 7% polyacrylamide gel using 90-m*M* Tris, 90-m*M* boric acid, 2.7-m*M* EDTA, pH 8.3, as the running buffer. The conjugate appears as a brown band in the middle of the gel.

Conjugation of SFB-Modified Alkaline Phosphatase to bis-Hydrazide-Modified Oligonucleotides

DNA probes modified to contain a 5′-terminal hydrazide group (Section 2.1) can be conjugated to aldehyde-containing molecules, resulting in the formation of a hydrazone bond. The crosslinking agent succinimidyl *p*-formylbenzoate (SFB; Chapter 17, Section 2) can be used to add aldehyde groups to proteins and other molecules that do not naturally contain them (Chapter 2, Section 4.4). Reaction of SFB-modified alkaline phosphatase with a hydrazide–DNA derivative can produce a conjugate having excellent sensitivity for use as a hybridization probe (Figure 23.22). Other enzymes and detection molecules modified to contain aldehydes may be coupled to hydrazide–DNA probes using similar methods. Using an SFB-modified HRP (or periodate-oxidized HRP) that contains aldehyde groups to prepare the DNA conjugate gives about a 40-fold less sensitivity in hybridization assays when compared to an alkaline phosphatase conjugate in this procedure (Ghosh *et al.*, 1989). However, this result may change if chemiluminescent detection using an HRP conjugate is used, as this method can result in femtogram detection limits.

The following protocol assumes the prior derivatization of an oligonucleotide at the 5′ end using a *bis*-hydrazide compound according to the protocol of Section 2.1 using a carbodiimide-mediated reaction.

Protocol

1. Dissolve alkaline phosphatase at a concentration of 10 mg/ml in 0.1-*M* sodium bicarbonate, 3-*M* NaCl, pH 8.5. Dialyze against this solution if the enzyme is already dissolved in another buffer. This protocol requires at least 0.4 ml of the enzyme solution.
2. Dissolve SFB in acetonitrile at a concentration of 50-m*M* (12.35 mg/ml). Make at least 100 μl.
3. Add 40 μl of the SFB solution to the 0.4-ml alkaline phosphatase solution with mixing.
4. React for 30 min at room temperature.
5. Remove excess reactants by dialysis against 50-m*M* MOPS, 0.1-*M* NaCl, pH 7.5.
6. Add the aldehyde-derivatized alkaline phosphatase to 8 nmol of a 5′-hydrazide oligonucleotide preparation made according to Section 2.1 using a carbodiimide coupling protocol.
7. React overnight at room temperature.
8. Remove unconjugated oligonucleotide using gel filtration on a column of Bio-Rad P-100 (1.5 × 65 cm). Use 50-m*M* Tris, pH 8.5, as the chromatography buffer. Pool the enzyme fractions and apply the sample to a 1 × 7 cm column of DEAE–cellulose equilibrated with the same buffer. After washing the column with 0.1-*M* Tris, pH 8.5, a salt gradient from 0- to 0.2-*M* NaCl in the same buffer is used to remove unconjugated enzyme. The enzyme–DNA conjugate is then eluted with 0.1-*M* Tris, 0.5-*M* NaCl, pH 8.5.

2.5. Fluorescent Labeling of DNA

Oligonucleotide probes may be labeled with small fluorescent molecules for detection of hybridization by luminescence. Fluorescent probes are widely used in assay systems involving biospecific interactions (Chapter 10). Receptors for ligands may be localized in tissues or cells by modification of the ligand with the appropriate fluorophore. Targeted molecules may be quantified through measurement or modulation of a fluorescent signal upon binding of a tagged ligand. The sensitivity of fluorescent assays can exceed that obtained using radiolabels, but enzyme assays typically are more sensitive than fluorescent detection without catalytic enhancement.

Fluorescently labeled DNA probes can be used for detection, localization, or quantification of target DNA sequences. *In situ* hybridization mapping of genomic DNA sequences can be carried out using fluorescent probes to target particular regions within chromosomes. Called FISH for fluorescent *in situ* hybridization, the technique is used extensively to identify marker chromosomes or chromosomal rearrangements. Since many genomic sequences are repeated, usually occurring in multiple copies within isolated regions of the chromosome, the fluorescent label on the DNA probe allows localization of targeted genes with high sensitivity. For a review of FISH, see Meyne (1993) and also refer to Chapter 1 for a discussion of current fluorescent DNA probe-based conjugates and methods.

Fluorescently labeled DNA probes also can be used in homogeneous assay systems to detect and quantify target complementary sequences. The majority of these systems use a process of energy transfer and fluorescent quenching to detect hybridization phenomena. The principle of these assays involves the labeling of two binding components that can specifically interact with a target DNA. One or both of the labels may be a luminescent compound. The luminescent quality of the first label may consist of a chemiluminescent probe that can be excited through specific chemical processes, producing light emission. Alternatively, the label may be a fluorescent probe that can absorb light of a particular wavelength and subsequently emit light at another wavelength.

The second label may also be a fluorescent compound, but does not necessarily have to be. As long as the second label can absorb the emission of the first label and modulate its signal, binding events can be

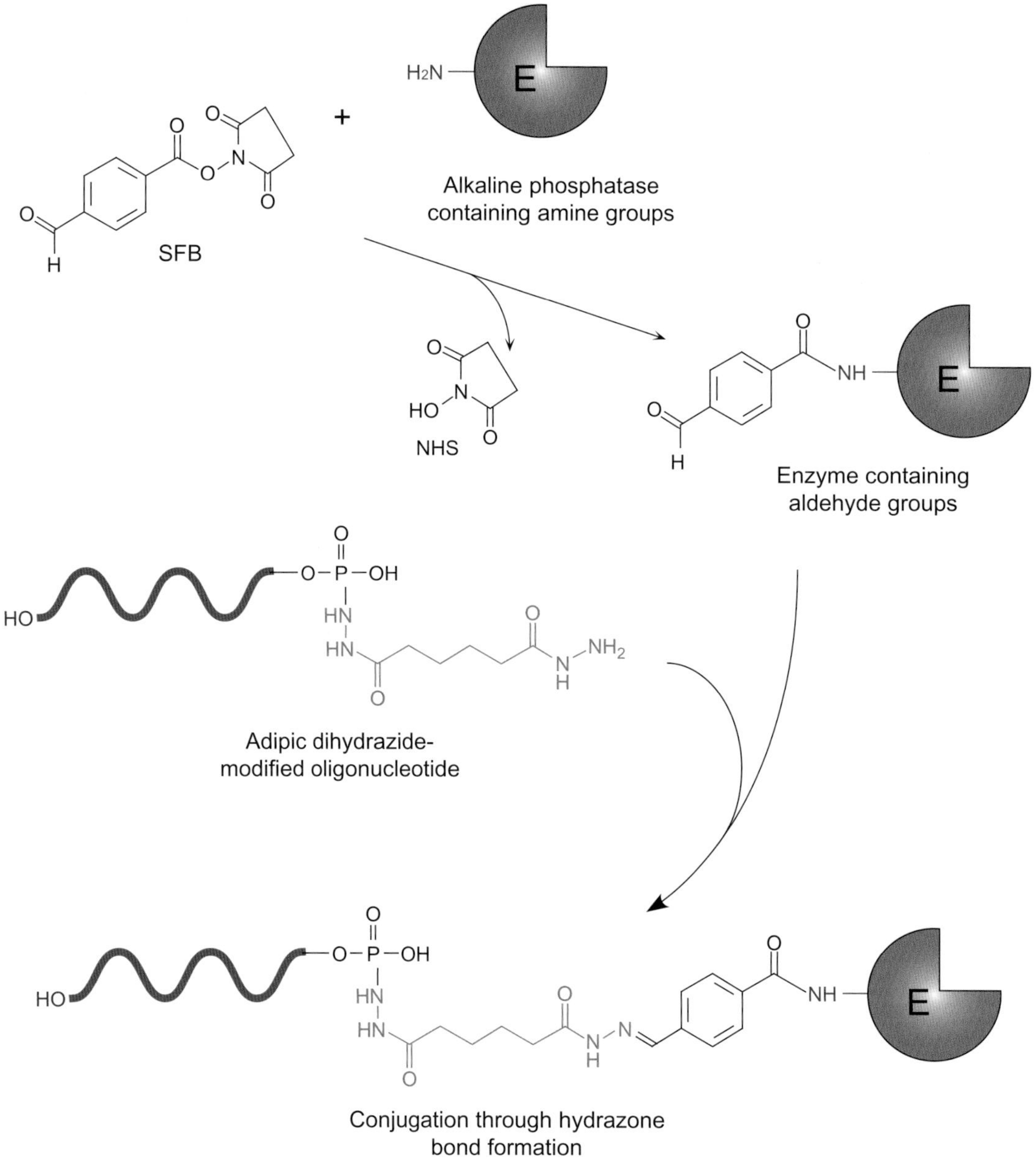

FIGURE 23.22 SFB may be used to create aldehyde groups on enzyme molecules for subsequent conjugation to a 5′-bishydrazide-modified oligonucleotide, forming hydrazone bonds.

observed. Thus, the two labeled DNA probes interact with each other to produce fluorescence modulation only after both have bound target DNA and are in enough proximity to initiate fluorescence resonance energy transfer (FRET). Common labels utilized in such assay techniques include the chemiluminescent probe *N*-(4-aminobutyl)-*N*-ethylisoluminol, and reactive fluorescent derivatives of fluorescein, rhodamine, and the cyanine dyes (Chapter 10). For a review of these techniques, see Morrison (1992) and Chapter 1.

To prepare labeled DNA molecules for use in fluorescent assays, the oligo must first be derivatized to contain a functional group. Any of the methods of Sections 2.1 and 2.2 (this chapter) may be used to add an amine or sulfhydryl residue to specific regions of the DNA polymer. Once modified in this manner, the oligo may be further reacted with a fluorescent probe to create the final derivative. However, many of the fluorescent quenching formats exclusively specify either 3′ or 5′ labeled DNA molecules. This is because discrete modification at just one end of the oligo ensures that the label on a hybridized probe will be near enough to a second hybridized-and-labeled DNA molecule to effect the luminescent modulation necessary to make the system viable. Multiple fluorescent labels on nucleotides within the DNA probe would not be affected to the same degree by a second label attached to another oligo hybridized some distance away on the target strand. Therefore, use of the terminal transferase method of adding a modified nucleoside triphosphate to the 3′ end (Section 1) or 5′-phosphate modification using a carbodiimide-mediated reaction (Section 2.1) work best for creating functionalized DNA derivatives for fluorescent modulation techniques.

Another method of fluorescent detection of DNA or RNA targets involves the modification of a targeting oligo at both ends. In this approach, one end is modified with a fluorescent molecule and the other end is modified with another label that can be a quencher of fluorescence or another fluorescent probe able to accept energy from the first label. These "molecular beacons," as they are called, contain terminal sequences that are able to hybridize the ends together in solution in the absence of target. Thus, if a fluorescent molecule is at one end of the molecular beacon and a quencher is at the other end, in solution without a target DNA present the fluorescent signal would be completely quenched. In the presence of the target DNA, the molecular beacon would hybridize to the target and open up the stem-and-loop structure of the probe, thus forcing the fluorophore and quencher too far away from each other to cause fluorescent quenching. The result is that specific target binding of the molecular beacon causes the generation of a fluorescent signal. The more fluorescence that occurs the more target is present in solution. For a review of molecular beacon technology, see Bratu (2006).

The following sections describe two methods of coupling fluorescent labels to functionalized DNA probes. Other fluorophores or quenching molecules may be attached using similar procedures with careful reference to the properties and reactivities of such labels as discussed in Chapter 10.

Conjugation of Amine-Reactive Fluorescent Probes to Diamine-Modified DNA

DNA modified with a diamine compound to contain terminal primary amines may be coupled with amine-reactive fluorescent labels. The most common fluorophores used for oligonucleotide labeling are the cyanine dyes and derivatives of fluorescein and rhodamine. However, any of the amine-reactive labels discussed throughout Chapter 10 are valid candidates for DNA applications.

Some fluorescent probes are relatively water insoluble and must be dissolved in an organic solvent prior to addition to an aqueous reaction medium containing the DNA to be labeled. Even water soluble fluorescent probes may be dissolved first in an organic solvent to permit easy addition to an aqueous reaction medium without hydrolysis of the reactive group. Suitable solvents are identified for each fluorophore, but mainly DMF, DMAC, or DMSO are used to prepare a stock solution. Some protocols utilize acetone when labeling DNA. However, avoid the use of DMSO for sulfonyl chloride compounds as this group reacts with the solvent. For oligonucleotide labeling, the amount of solvent added to the reaction mixture should not exceed more than 20% (although at least one protocol calls for a 50% acetone addition; see Nicolas *et al.*, 1992).

The following protocol is a generalized method for labeling amine-modified oligonucleotides with a fluorescent probe, such as FITC. It is based on the method of Morrison (1992).

Protocol

1. Prepare 10 nmol of a diamine-modified DNA probe using the chemical methods discussed in Section 2.1 or through enzymatic derivatization using an amine-containing nucleoside triphosphate (Section 1). Dissolve or buffer exchange the oligo into 1.0 ml of a suitable coupling buffer for the type of amine-reactive fluorophore utilized (see recommended reaction conditions for the particular fluorescent label in Chapter 10). For FITC, the appropriate buffer condition for the oligo is 0.1-*M* sodium carbonate, pH 9.0.
2. Dissolve the fluorophore in DMF or another suitable solvent at a concentration of 0.01-*M*. For FITC, this translates into a concentration of 3.89 mg/ml.
3. Add 50 μl of the FITC solution to the oligo solution and mix. For the use of NHS ester or sulfonyl chloride fluorescent probes, add up to 150 μl of the fluorophore solution to the DNA.
4. React overnight at room temperature.
5. Remove excess fluorophore from the labeled oligo using gel filtration on a desalting resin, dialysis, or a centrifugal concentrator.

Conjugation of Sulfhydryl-Reactive Fluorescent Probes to Sulfhydryl-Modified DNA

Fluorescent probes containing sulfhydryl-reactive groups can be coupled to DNA molecules containing thiol modification sites. The chemical derivatization methods outlined in Section 2.2 may be used to thiolate the oligo for subsequent modification with a fluorophore. Appropriate fluorescent compounds and their reaction conditions may be found in Chapter 10. The protocol discussed in the previous section can be used as a general guide for labeling DNA molecules.

CHAPTER

24

Bioconjugation in the Study of Protein Interactions

The study of protein interactions has become a vital research effort as a result of the sequencing of the human genome and the genomes of other organisms. The next great challenge beyond merely having knowledge of gene sequences is to understand the complex interplay of the resultant protein molecules within cells as they bind, interact, and affect cellular processes. From the many research papers that have appeared on this subject, it is becoming clear that each protein molecule interacts with other proteins and molecules not in isolated obscurity, but in a highly complex web of interactions, which can have far reaching effects on overall biological function (Braun and Gingras, 2012).

Protein interactions mediate virtually every cellular process. They are involved with transcription, translation, transport, cell cycle control, the determination of cell type and function, protein folding, post-translational modifications, signal transduction, metabolism and energy production, cell structure and motility, the formation of biological machines, cell and organism defense, and apoptosis (Katz *et al.*, 2011; Khan *et al.*, 2011). Genetic mutations or damaging modifications resulting in abnormal protein interactions are often the root cause of many diseases, especially tumorogenesis.

The general types of protein–protein interactions that occur in cells include receptor–ligand, enzyme–substrate, multimeric complex formation, structural scaffolds, and chaperones. However, proteins interact with more targets than just other proteins. Protein interactions can include protein–protein or protein–peptide, protein–DNA/RNA or protein–nucleic acid, protein–glycan or protein–carbohydrate, protein–lipid or protein–membrane, and protein–small molecule or protein–ligand. It is likely that every molecule within a cell has some kind of specific interaction with a protein (Roldos *et al.*, 2011; Smith, 2012).

The consequences of protein interactions in effecting cellular biochemistry include the synthesis, destruction, or recycling of biomolecules; the supply of energy for cellular processes; the generation of chemical signals; activation or inhibition of proteins and enzymes; changes in protein conformation and structure; the creation of new binding sites or active centers in proteins; the motility of cells, tissues, and organisms; transport of molecules within cells or into/out of cells; and the formation of cellular structures and compartments.

Protein interactions can be described in relative ways related to the strength of the binding that takes place between the two molecules and the time that the interaction lasts. Any interacting molecules can be characterized as having (1) either high affinity (strong) or low affinity (weak) binding, and (2) either stable (long lasting) or transient (short-lived) binding. Thus, a given protein interaction may be described in one of four possible ways: strong and stable, strong and transient, weak and stable, or weak and transient. Interactions of the weak and stable type may at first seem like an oxymoron, but often multiple weak interactions can take place simultaneously with a target and the combined "avidity" makes the resultant complex stable.

Of the large number of protein interactions that take place in cells, perhaps the vast majority may be described as transient. Most proteins that modify other molecules do so very rapidly and so interact only briefly with their substrates or binding partners (i.e., enzymes). In addition, since proteins within cells are highly compartmentalized, the affinity of most interactions does not have to be very great, because each potential binding partner is within short diffusion distances and the relative concentration of molecules within these small volumes is high.

Of course, the designations of strong, weak, stable, or transient are all subjective terms. They mainly result from the outcome of affinity capture experiments of binding partners on insoluble supports or the analytical determination of the kinetic parameters of specific

Bioconjugate Techniques, Third Edition.
DOI: http://dx.doi.org/10.1016/B978-0-12-382239-0.00024-8

binding interactions. In general, if the affinity constant of a protein interaction is strong enough to allow a binding partner to be captured and purified using an immobilized affinity ligand, then the interaction can be described as being reasonably strong. This usually correlates to an affinity constant of $>10^6$ or a dissociation constant of $<10^{-6}$ (with units dependent upon the number of molecules interacting and the resultant equilibrium equation). Conversely, interactions having affinity constants of $\leq 10^6$/M are often too weak to survive the washing steps needed to isolate the interacting protein on a solid phase affinity support. Quantitative measurement of the affinity constant between interacting proteins and the half-life of the interaction may be performed using surface plasmon resonance (SPR) techniques (Homola *et al.*, 1995, 1999).

The vast network of protein–protein interactions that have been deciphered in recent years has grown to include literally thousands of proteins in a sometimes chaotic dance of complexity. Using the two-hybrid method, for instance, which involves the use of split transcription factor fusion proteins that allows detection of interacting proteins by activation of reporter gene expression, many putative protein–protein interaction partners have been identified in yeast (Fields and Song, 1989; Chien *et al.*, 1991; Criekinge and Beyaert, 1999). A map of these interactions looks a lot like a picture of the myriad nodes of addresses on a vast network of computers, such as the interconnections that make up the Internet (Jeong *et al.*, 2001). Although most of the proteins in an organism have only a few partners that interact with them, major hubs in protein "interactomes" can have up to 10 to 20 links with other proteins, indicating that these key proteins are critical players in cell vitality. Many of these sentinel proteins are enzymes, such as kinases, that act on a variety of different protein substrates in turning on or turning off pathways within cells in response to metabolic or external stimuli. Figure 24.1 shows a small segment of the yeast interactome.

Through the growing knowledge of protein–protein interactions and their corresponding gene sequences major interaction domains on protein surfaces are being identified. These relatively conserved amino acid sequences create structural motifs that are designed to bind with certain targeted sequences in other proteins or molecules (Pawson and Nash, 2003; Ingham *et al.*, 2005; Ibrahim *et al.*, 2011). Using this information, many common binding regions on proteins can be identified just through their gene sequences. The role of protein interaction domains has been an important topic in identifying disease mechanisms (Ibrahim *et al.*, 2011).

However, it is much more difficult to characterize the interactions of proteins with no known interaction domains. The traditional "lock and key" approach to

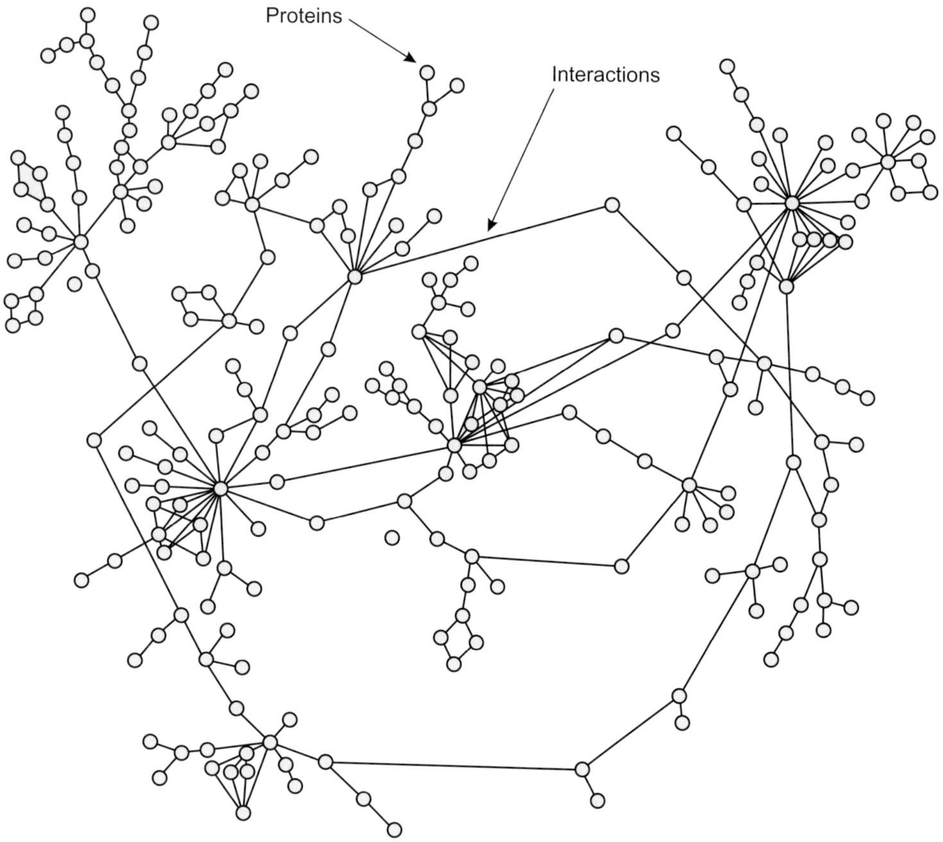

FIGURE 24.1 A small segment of the yeast interactome. The spheres represent proteins and the interconnecting lines are identified protein interactions. Many proteins are seen to interact with one or two other proteins, but some can have over a dozen other interacting partners.

conceptualizing binding pairs is far too simplistic to allow easy visual identification of interacting surfaces on the complex three-dimensional space making up the topology of protein molecules. Even in those instances where protein interacting partners have been crystallized together and their three-dimensional structures determined, it is obvious from the molecular models that it would be difficult or impossible to visually identify the site of interaction without having such structural data in place beforehand. For this reason, experimental schemes are needed that are more elaborate than just knowledge of genetic sequences or interaction domains to characterize the majority of specific interactions proteins undergo. Even when two-hybrid studies indicate the probability of a protein interaction occurring, it still does not provide information on the nature of the interaction, the binding sites, or its function.

The techniques developed to study protein interactions can be divided into a number of major categories (Table 24.1), including bioconjugation, protein interaction mapping, affinity capture, fusion protein labeling, protein probing, and instrumental analysis (i.e., NMR, crystallography, mass spectrometry, and surface plasmon resonance). Many of these methods are dependent on the use of an initial bioconjugation step to discern key information on protein interaction partners. For a review of the major methods used in the study of protein interactions, see Perrakis *et al.* (2011).

TABLE 24.1 **Common Reagents and Techniques Used to Investigate Protein Interactions**

Category	Technique
Bioconjugation	• Bifunctional crosslinking agents • Trifunctional label-transfer agents • Oxidative cross-linking • Photoreactive amino acids
Protein Interaction Mapping	• Mutagenesis • Site-directed modification • Chemical cutting agents • Enzymatic digestion
Affinity Capture	• Co-Immunoprecipitation • Pull-down assays • Streptavidin–biotin • Antibody & protein arrays • Tandem tag isolation • Affinity arrays
Fusion Protein Labeling	• Yeast two-hybrid techniques • Protein complementation assays • Fluorescent protein FRET assays • BRET assays
Protein Probing	• Far-western blot analysis • Labeled protein probes • Antibody probing of epitopes • Substrate or ligand inhibitors • Spin labels and photoaffinity probes • Fluorescent probes • Proximity ligation assays
NMR, Crystallography, Mass Spectrometry, SPR	• Mass tags and heavy atom labels • Molecular modeling of interacting pairs • Label-free detection

Many of the methods developed to study protein interactions use the bait/prey model to detect interacting partners (Phizicky and Fields, 1995; Archakov *et al.*, 2003; Piehler, 2005). The bait protein is a purified protein (often recombinant) that is used to lure and capture a putative interacting protein or biomolecule. The bait protein may be immobilized to a solid phase for affinity separations or used in solution. It may also be fusion tagged (i.e., GST or 6xHis) or labeled with a detectable molecule, such as a fluorescent probe. It is often the case that there exists an antibody specific for the bait protein to use for detection or in recovery of the interacting complexes.

The prey is a protein, protein complex, or other biomolecule that interacts with the bait protein. It can be captured by its specific affinity interaction with the bait. Since many protein interactions may involve low affinity binding events or are transient, the use of chemical crosslinking techniques can greatly facilitate analysis of an interacting prey protein. In fact, the use of bioconjugation to "fix" interacting proteins is a powerful route to capturing weak affinity or transiently interacting molecules, as the crosslinked complexes can be isolated and analyzed without loss of some components.

The following sections describe the use of bioconjugation reagents for the study of protein interactions. These reagents and techniques can be used to crosslink and capture interacting proteins, to investigate the binding sites involved with interactions, and to identify which peptide regions or amino acids participate in the binding event. In addition to the bioconjugation reagents described in this section, the reader is also referred to Chapter 12, Section 3, entitled Mass Tags and Isotope Tags, wherein some of those reagents can also be used to investigate interacting proteins.

1. HOMOBIFUNCTIONAL CROSSLINKING AGENTS

One of the first applications of bioconjugate techniques was the use of simple homobifunctional compounds (see Chapter 5) to capture interacting proteins in biological samples. As the name of this type of reagent implies, the compounds have reactive groups on each end of a spacer arm that are identical and react with the same type of functionality on different proteins. For instance, early development of crosslinking compounds resulted in the creation of homobifunctional amine-reactive reagents using either imidoesters or NHS esters as the reactive groups. Both of these reagent types could be reacted with

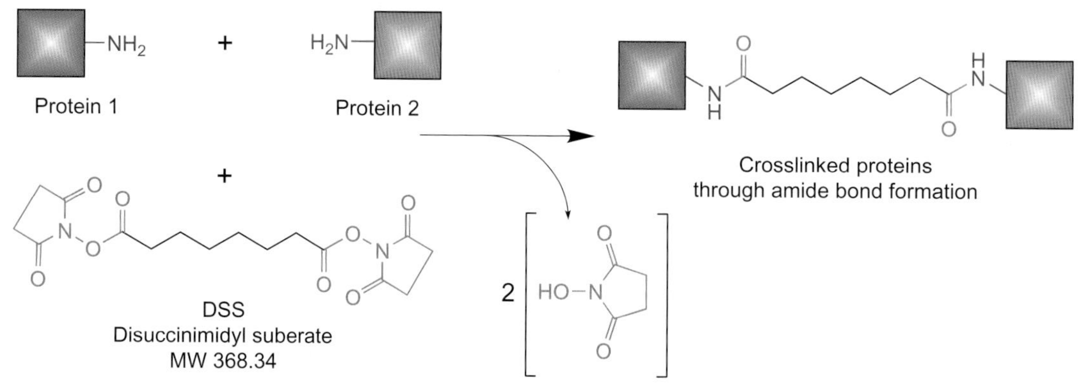

FIGURE 24.2 DSS can capture protein interacting partners through amide bond crosslinks.

a complex protein sample mixture to result in efficient conjugation of proteins through their available amine groups. The reactive groups on these compounds survive long enough in a physiological pH environment to effectively capture and covalently link proteins in proximity to one another. In this way, any proteins undergoing specific biological interactions can be "fixed" by chemical conjugation and thus link together permanently interacting complexes for analysis.

However, as one might expect, this strategy is more or less a crude "shotgun" approach, wherein every protein in the sample has the potential to be crosslinked with other proteins regardless of whether they are undergoing specific interactions or not. The use of homobifunctional reagents to study protein interactions may therefore result in high noise levels or many false positives, because of their non-selective crosslinking characteristics. The end result often makes it difficult to identify conjugated proteins that are specifically interacting from those proteins crosslinked just due to random collisions.

However, even acknowledging their disadvantages, homobifunctional crosslinkers have been used successfully to investigate many protein–protein interactions, within cells and within lysates or protein solutions. The key to capturing true interacting proteins while limiting the degree of nonspecific conjugation is to optimize the amount of crosslinker at the lowest possible concentration. Using too high a concentration will extensively conjugate all proteins, even those that are not specifically interacting at the moment. Using too low a concentration will not effectively conjugate even interacting proteins. Therefore, optimization is typically needed to determine the best concentration of homobifunctional crosslinker for a particular application.

1.1. DSS and BS^3

Two crosslinking agents that have been used extensively to study protein interactions are disuccinimidyl suberate (DSS) and *bis*-sulfosuccinimidyl suberate (BS^3) (see Chapter 5, Section 1.2). Both reagents contain an 8-carbon spacer arm built from suberate core and have reactive esters at each end that couple with amines to form amide bonds. They differ, however, in the fact that DSS contains NHS esters and BS^3 contains negatively charged sulfo-NHS esters. BS^3 therefore is water soluble and will not penetrate cell membranes, whereas DSS is hydrophobic, water insoluble, and membrane permeable. For this reason, BS^3 can be used to study cell surface–protein interactions (Friedrichson and Kurzchalia, 1998; Simons *et al.*, 1999), while DSS can be used to study intracellular protein interactions (Ishmael *et al.*, 2005). The reaction of DSS to capture and crosslink interacting proteins is shown in Figure 24.2.

The following protocol represents a generalized strategy for crosslinking interacting proteins using either DSS or BS^3. The buffer conditions and reagent amounts are gleamed from published procedures, but the exact quantities should be optimized for each protein interaction studied.

Protocol

1. Suspend cells at ~25×10^6cells/ml in PBS (pH 8.0).
2. Wash cells three times with ice-cold PBS (pH 8.0) to remove amine-containing culture media and extracellular proteins from the cells.
3. For cell–surface interaction studies, add ligands to the cells and incubate for 1 h at 4°C.
4. Dissolve DSS or BS^3 in dry DMSO at a concentration of 25-m*M*. Note: BS^3 may be added directly to PBS buffer or dissolved as a stock solution in DMSO.
5. Add an aliquot of the DSS or BS^3 solution to the reaction medium to obtain a final concentration of 0.5 to 5-m*M*. Note: Simons *et al.* (1999) successfully used a concentration of 0.5-m*M* BS^3 with Madin–Darby canine kidney (MDCK) cells permanently expressing a GPI-anchored form of growth hormone decay accelerating factor (GH-DAF) to crosslink the protein interaction complexes on the cell surfaces.

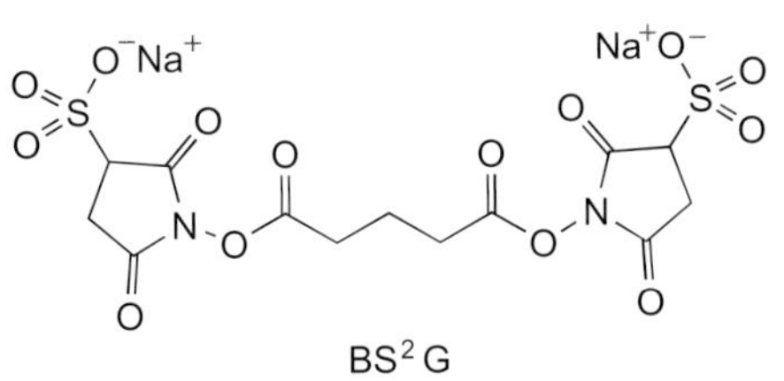

BS^2G
bis-Sulfosuccinimidyl glutarate
Mol. Wt.: 530.35

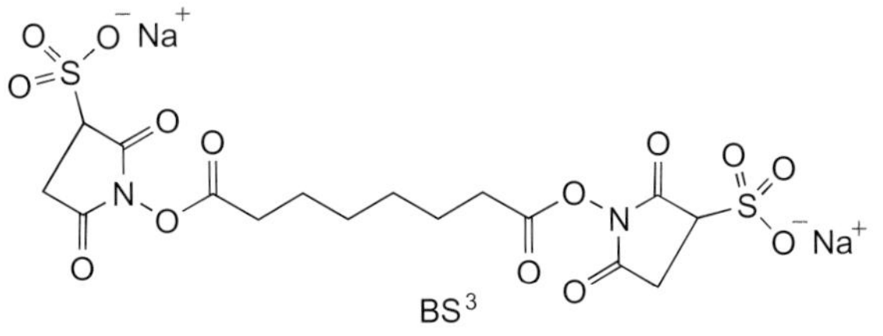

BS^3
bis-Sulfosuccinimidyl suberate
Mol. Wt.: 572.43

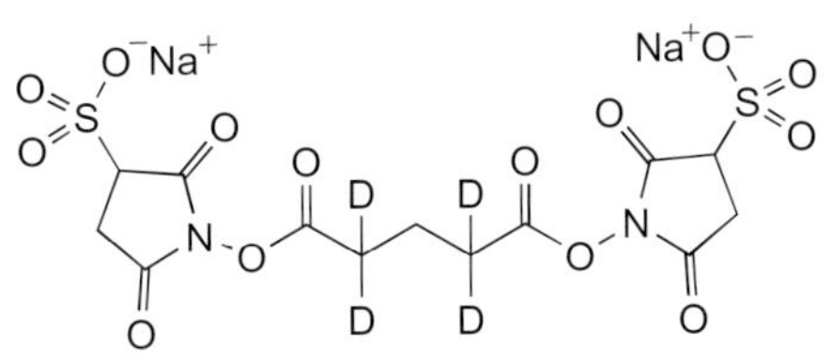

BS^2G-d_4
bis[Sulfosuccinimidyl] 2,2,4,4 glutarate-d_4
Mol. Wt.: 534.38

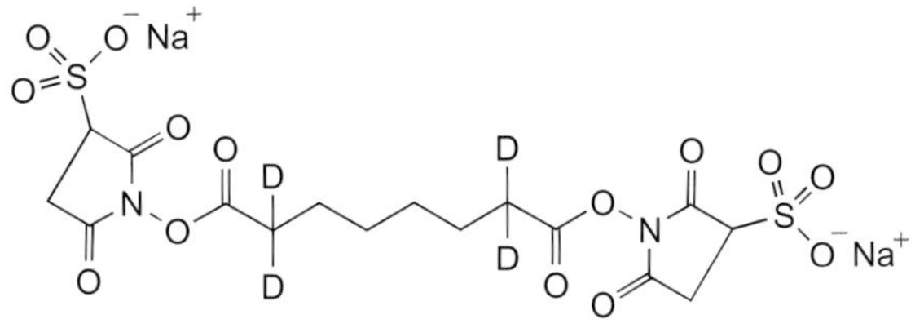

BS^3-d_4
bis[Sulfosuccinimidyl] 2,2,7,7 suberate-d_4
Mol. Wt.: 576.45

FIGURE 24.3 The homobifunctional crosslinkers BS^2G and BS^3 can be used to capture protein interactions through amide bond formation. The deuterium-labeled analogs of these reagents can be used to differentiate samples by mass spectrometry.

6. Incubate the reaction mixture for 30 min at room temperature. To reduce active internalization of BS^3 into cells, this incubation may be performed at 4°C.
7. Quench the reaction by adding an aliquot of 1-*M* Tris, pH 7.5, to give a final concentration of 10 to 20-m*M*.
8. Incubate the quenching reaction for 15 min at room temperature.
9. Lyse cells and analyze the protein interactions by electrophoresis, western blotting, and mass spectrometry.

1.2. Heavy Atom, Deuterated Crosslinking Agents

Another approach to the study of protein interactions using homobifunctional crosslinkers uses the incorporation of isotopes, such as deuterium, into the reagent structure to produce a heavy atom analog suitable for mass spectrometry analysis. In this technique, the normal or light H (hydrogen) version of a crosslinker is used in an equal molar ratio to a heavy D (deuterium) version to capture interacting proteins through covalent conjugation. The heavy and light analogs are reacted with a sample at the same time, so that each form of the reagent will have an equal chance of reacting with an interaction complex. Subsequent proteolytic digestion of the sample (i.e., with trypsin) creates peptide fragments, some of which will contain covalent crosslinks from the heavy or light crosslinkers. In addition, these crosslinked peptides will have an equal chance of having either a light atom linker or a heavy atom linker holding them together.

Mass spectrometer analysis of the peptide fragments formed by this process yields pairs of MS peaks differing only by the mass change caused by the substitution of deuterium atoms for hydrogen atoms in half of the crosslinks. Thus, searching for MS peaks in the data that differ by the number of deuterium substitutions will immediately identify peptides from the interacting proteins that have been captured by the crosslinking process.

Using this approach, homobifunctional crosslinking agents containing sulfo-NHS esters have been developed based on the core structures of glutaric and suberic acids. The suberate-based crosslinker is also known as BS^3 and has been described previously (Section 1.1, this chapter). These two standard amine-reactive reagents are then modified at two carbons of their respective cross-bridge structures to contain two pairs of deuterium atoms, increasing their molecular mass by exactly four from the normal hydrogen atom analogs.

These heavy atom reagent pairs are termed BS^2G-d_4 (*bis*[sulfosuccinimidyl] 2,2,4,4 glutarate-d_4) and BS^3-d_4 (*bis*[sulfosuccinimidyl] 2,2,7,7 suberate-d_4), and their light atom analogs are called BS^2G-d_0 (*bis*[sulfosuccinimidyl] glutarate-d_0) and BS^3-d_0 (*bis*[sulfosuccinimidyl] suberate-d_0) (Thermo Fisher) (Figure 24.3). The sulfo-NHS esters of these four compounds all react equally well with amine groups on proteins to form amide bonds. Therefore, there are virtually no differences in properties or reactivities between the heavy and light versions of these reagents.

The conjugation of interacting proteins with heavy/light crosslinkers potentially can result in a number of derivatives having unique atomic mass units (amu) observed by mass spectrometry. For instance, any of

TABLE 24.2 Potential Atomic Mass Unit Contributions Associated with the Heavy/Light Crosslinkers

Reagent	Both Ends Amide Bond Linked	One End Hydrolyzed to Carboxylate
BS^3-d_0	138.068 amu	157.079 amu
BS^3-d_4	142.090 amu	161.102 amu
BS^2G-d_0	98.102 amu	116.113 amu
BS^2G-d_4	102.124 amu	120.135 amu

these crosslinkers might react with amines on interacting proteins at both ends to form amide bonds. Alternatively, one end of these homobifunctional compounds may react with a protein to form an amide bond while the other end hydrolyzes, resulting in loss of the sulfo-NHS group to form a carboxylate. Thus, the heavy/light crosslinking pairs may have two potential amu results when peptides are analyzed by mass spectrometry. Table 24.2 shows the potential amu contributions for each of the possible products formed by the reactions of the heavy/light crosslinkers with proteins.

This technique has been described as a general method of studying protein–protein interactions as well as a method for investigating the three-dimensional structure of individual proteins (Muller *et al.*, 2001; Back *et al.*, 2003; Dihazi and Sinz, 2003; Sinz, 2003, 2006). It has also been used for the study of the interactions of cytochrome *c* and ribonuclease A (Pearson *et al.*, 2002), to investigate the interaction of calmodulin with a specific peptide binder (Kalkhof *et al.*, 2005a; Schmidt *et al.*, 2005) and for probing laminin self-interaction (Kalkhof *et al.*, 2005b).

In practice, sample concentrations are typically kept in the micromolar range to limit the widespread conjugation of molecules not specifically interacting or to limit intermolecular conjugation when studying the structure of a single protein. Therefore, most protein concentrations will be in the microgram/milliliter range prior to the addition of crosslinkers. Even in cases wherein a single purified protein is crosslinked to study its three-dimensional structure, limiting the amount of crosslinker in the reaction mixture will help to limit polymerization of the protein. However, successful results have been obtained by this method using up to a 200-fold excess of crosslinker over the concentration of protein in solution.

The following protocol is based on the published methods as well as the specific instructions provided with the heavy atom crosslinkers from Thermo Fisher Scientific. For new applications, the amount of crosslinkers added to the sample will have to be optimized to obtain useful information about the interactions.

Protocol

1. Dissolve the protein(s) to be studied in 20-m*M* HEPES buffer, pH 7.5, at a concentration of 5 to 10 μ*M*.
2. Dissolve together in one solution the desired heavy and light crosslinker analogs in dry DMSO at an equal concentration of up to 100-m*M*. Use either the pair BS^2G-d_4/BS^2G-d_0 or BS^3G-d_4/BS^3G-d_0, but do not mix the different sized crosslinkers together. The heavy and light analogs of the same type should always be dissolved at equivalent concentrations in DMSO to prepare the stock solution.
3. Add a quantity of the crosslinker solution to the protein solution to obtain at least a 10-fold molar excess of the crosslinkers over the concentration of the protein. Studies should be performed at several levels of crosslinker addition to determine the optimal conjugation conditions (e.g., 10-, 50-, 100-, and 200-fold excess).
4. Quench the reaction by the addition of NH_4HCO_3 to a final concentration of 20-m*M*. The optimal time course for the reaction should be determined by removing portions of the solution at different points, starting at about 5 to 10 min and extending out to 2 h in length.
5. Analyze the quenched reaction by SDS electrophoresis, western blotting, and mass spectrometer analysis.

1.3. Formaldehyde

By far the simplest bifunctional crosslinking agent is formaldehyde. Although structurally it appears to be a mono-functional aldehyde compound, formaldehyde reacts with proteins via a two-step reaction to yield a methylene bridge crosslink between two amines on proteins and other molecules, thus it behaves as though it is bifunctional in nature (Figure 24.4). The reaction of formaldehyde with proteins is rapid and efficient, and it has long been used to fix cells and tissue samples for preservation, staining, and probing. Formaldehyde quickly penetrates cells and at the right concentration, results in locking biomolecules in place, preventing diffusion, morphological changes, or loss of proteins through solubilization. It has long been used to inactivate viruses in the production of vaccines (Eckels and Putnak, 2003).

Formaldehyde can also be used at limiting concentrations to crosslink interacting proteins, while avoiding the extensive global crosslinking typically obtained when fixing cells. At relatively low concentrations, formaldehyde will link together only those proteins and other amine-containing molecules within proximity to one another and presumably, therefore, undergoing specific biological interactions (Prossnitz *et al.*, 1988;

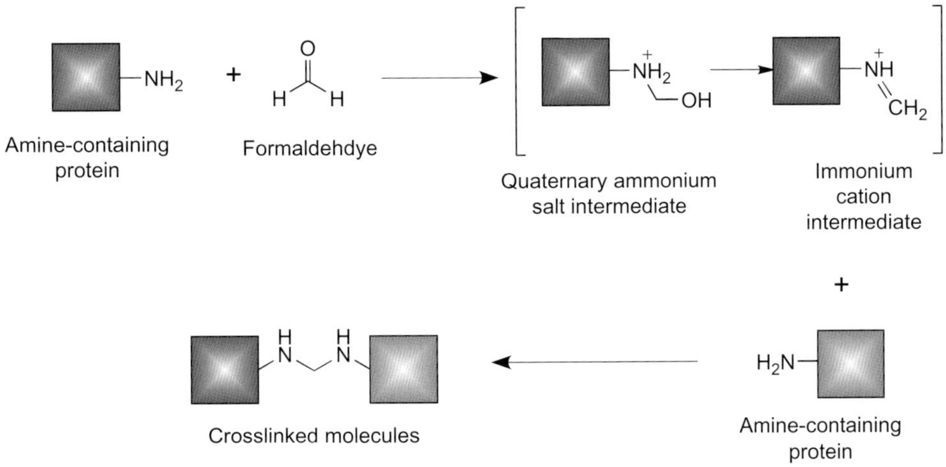

FIGURE 24.4 Formaldehyde can be used to capture protein interactions if it is used at low concentrations. The reaction proceeds through modification of a protein to create an intermediate immonium cation, which then goes on to react with a neighboring protein to form the crosslinked product via secondary amine bonds.

Skare *et al.*, 1993; Derouiche *et al.*, 1995; Orlando *et al.*, 1997; Orlando, 2000; Hall and Struhl, 2002; Vasilescu *et al.*, 2004).

Guerrero *et al.* (2006) used this technique along with the quantitative mass spectrometry strategy called SILAC (stable isotope labeling of amino acids in cell culture) (Ong *et al.*, 2002) to identify the yeast proteins that interact with the 26S proteasome. Formaldehyde has particular compatibility with mass spectrometry analysis, because the crosslinks have the option of being reversed by heating. A specific interaction complex from a sample can be isolated using co-immunoprecipitation (co-IP) by allowing a biotinylated antibody directed against one of the interacting proteins to interact with a lysed cell sample along with subsequent capture of the complex on an immobilized streptavidin column (Chapter 15, Section 2.5). The general process of immunoprecipitation (IP) is illustrated in Chapter 1, Figure 1.6. In a co-IP procedure, the protein pulled down by the antibody would have putative interacting partners crosslinked to it by glutaraldehyde (or by use of another crosslinker). Recombinant fusion tags have also been used with immobilized affinity ligands (Chapter 15 and Chapter 1, Table 1.2) for purification of protein–protein interaction complexes after crosslinking (Young *et al.*, 2012). The protein complexes thus isolated can be analyzed using mass spectrometry with or without reversing the crosslinks holding the complexes together. Without reversing the glutaraldehyde crosslinks, the peptide segments making up the interacting protein complex may be analyzed after proteolytic digestion to determine the exact sites of crosslinking.

The use of mass spectrometry has also aided in identifying the products of formaldehyde crosslinking with proteins, which potentially involves initial reactions with the N-terminal amines on peptide chains, the ε-amine of lysine side chains, the secondary amine of proline (if available), and tryptophan's ring nitrogen. These primary sites of glutaraldehyde reaction create a methylol imine at the modified nitrogens. The major reaction product occurs with lysine side-chain amines. In a second step, these initially modified amino acids can go on to react with one of several amino acids: a histidine side-chain nitrogen, an asparagine or glutamine side-chain amide nitrogen, addition to a tyrosine phenolic ring in the *ortho* position to the –OH group, a tryptophan side-chain nitrogen, or an arginine guanidinyl group (Sutherland *et al.*, 2008). If the formaldehyde crosslinks from this process are reversed prior to MS analysis, then the original peptides can be analyzed to determine the proteins from which they originated. A review of co-IP techniques was published by Geva and Sharan (2011).

Formaldehyde crosslinking can also be used to study the interactions of proteins with DNA or RNA molecules. This technique is typically called chromatin immunoprecipitation (ChIP) and the experiments can be performed using several methods (Bortz and Wamhoff, 2011; Truax and Greer, 2012). ChIP-based assays can be performed similarly to co-IP assays in that cell samples can be treated and fixed with dilute glutaraldehyde, lysed, and the targeted protein that may be interacting with the DNA or RNA segment isolated by affinity chromatography and then analyzed by use of a gel-shift assay (Yakhnin *et al.*, 2012) or by mass spectrometry. The ChIP approach can also be performed using immobilized oligonucleotides in an array (ChIP-on-chip) to probe for interacting proteins from lysed cell samples (Qin *et al.*, 2011). These methods are illustrated in Figures 1.20 and 1.21 in Chapter 1.

The following protocol is a generalized method that summarizes the publications on the use of formaldehyde for capturing interaction proteins. The concentration ranges indicated for the reactants and the time of the reaction need to be optimized for each protein interaction studied.

Protocol

1. Grow cells and wash into ice-cold 10-m*M* PBS buffer, pH 6.8.
2. Add formaldehyde to a final concentration of 0.125 to 1% (w/w) (optimize to find the best concentration level for the particular protein being studied).
3. React at room temperature to 37°C for 5 to 60 min (optimize to find the best reaction time for the protein being studied).
4. Wash cells and solubilize pellet in SDS-PAGE sample buffer.
5. Heat at 37°C for 10 min to fully solubilize and maintain crosslinked proteins, and then enrich specific complexes by immunoprecipitation using an immobilized antibody specific for the bait protein that was used. Alternatively, heat at 96°C for 20 min to solubilize and break all crosslinks (this may be used as a control).
6. Analyze interacting proteins by electrophoresis, western blotting, and mass spectrometry.

1.4. Protein Interaction Reporters

Standard homobifunctional crosslinkers, such as those described in the previous sections, can capture protein interactions effectively through covalent linkages, but they create severe challenges for analyzing exactly what proteins have been conjugated. This is not just the result of the large number of protein interacting partners that get crosslinked; it is also a result of the wide variety of products that can form from the process, including side reactions, nonspecific conjugations, and crosslinkers that only reacted with one protein. Attempting to deconvolute the identity of true protein interacting complexes created by large-scale crosslinking of complex samples is the major problem of most conjugation techniques used to study protein interactions.

A new type of crosslinking strategy may overcome this problem, as it takes advantage of definitive mass spectrometry identification after the chemical crosslinking of interacting proteins. The protein interaction reporter (PIR) reagent as described by Tang *et al.* (2005) is based on a bifunctional crosslinker concept, but having the additional feature of containing two mass spectrometry cleavable bonds within a specially designed spacer arm. The cleavable parts of the molecule consist of two RINK groups, which are amide-releasing, acid-cleavable components based on the solid-phase peptide synthesis resin first described by Rink (1987), and containing the 4-(2′,4′-dimethoxyphenylaminomethyl)phenoxy linker on each side of a central mass reporter tag (Figure 24.5). On both ends of the PIR compound, NHS esters provide amine reactivity to capture any interacting proteins in a complex sample, such as within a cell or lysate. In use, the PIR crosslinker is first reacted with a sample to form conjugates and then the proteins are subjected to proteolysis to create peptides. Some of the peptides that are formed will be conjugated together through the PIR crosslinks. This complex mixture of peptides is then subjected to mass spectrometry analysis using a mass spectometry capable of MS^2 or MS^3 separations.

Second-generation PIR reagents have been designed to include an affinity handle branching off from the central reporter group's free carboxylate. In one such design, a biotin group is built at the end of a

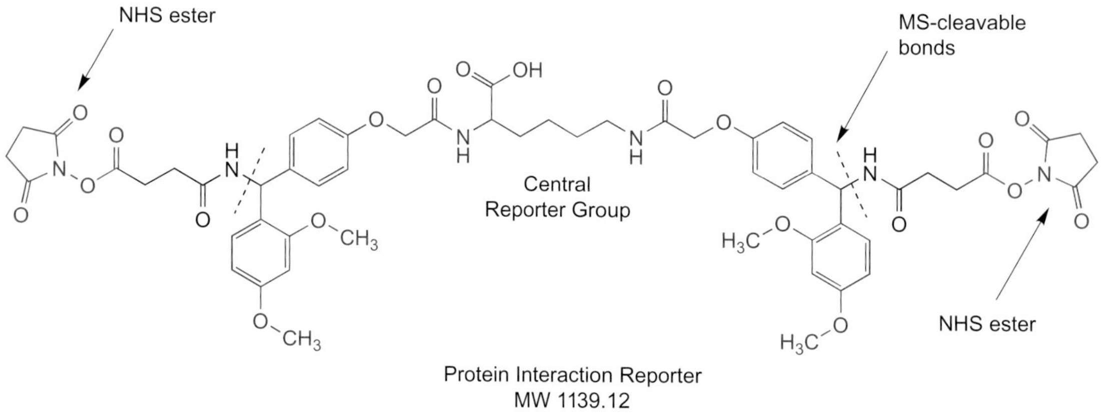

FIGURE 24.5 This PIR compound contains NHS esters at both ends to capture interacting proteins through amide bond formation. It also contains MS-cleavable bonds that release a central reporter group, which can be used to identify crosslinked peptides by mass spectrometry.

hydrophilic PEG spacer (Figure 24.6). The PEG arm provides increased water solubility to the overall reagent, which is otherwise rather hydrophobic. The biotin group can be used to detect or capture cross-linked interacting proteins out of complex solutions. For instance, immobilized streptavidin can be used to pull down any proteins modified with the biotin–PIR reagent. Alternatively, a complex solution may be separated by electrophoresis and the modified proteins identified after western blotting through the use of a streptavidin conjugate detection complex.

The use of a crosslinker containing MS labile bonds along with a mass reporter group can significantly reduce the complexity of finding and identifying coupled peptides in a huge amount of mass spectrometry data. In the first dimension of a MS separation, the overall mass of the PIR crosslinked peptides can be accurately determined as a single peak in the MS spectrum. The resultant mass correlates to the sum of the two crosslinked peptides plus the intervening PIR spacer, which links them together.

With bifunctional NHS ester reagents, one of three modification products can occur with proteins: (1) a dead-end linkage wherein one end of the crosslinker has attached to an amine group within a protein and the other end has hydrolyzed and not formed an attachment; (2) an intra-protein crosslink wherein the PIR reagent has been coupled at both ends to amines within the same protein; or (3) an inter-protein crosslink wherein both ends of the PIR reagent have been coupled with amines on two different protein molecules (Figure 24.7).

The second stage of an MS^2 analysis can be performed to bombard with more energy the mass products of the first stage separation in order to fragment the proteolytically digested complexes. At this point, the mass reporter group within the PIR crosslink is released due to breakdown of the two labile RINK bonds within the reagent structure (see Figure 24.5). Mass spectrometry cleavage of these groups and release of the reporter results in one of two potential mass signatures depending on if both ends of the PIR reagent have reacted (which gives a reporter mass = m/z 711) or if only one end has reacted (a dead-end; reporter mass = m/z 828). The resulting mass spectrum then shows either two or three peaks, depending on the type of modification formed from the initial PIR reaction. This process also releases the crosslinked peptides without disrupting the peptide backbone and thus leaves a small part of the PIR compound attached to each peptide fragment, minus the reporter group.

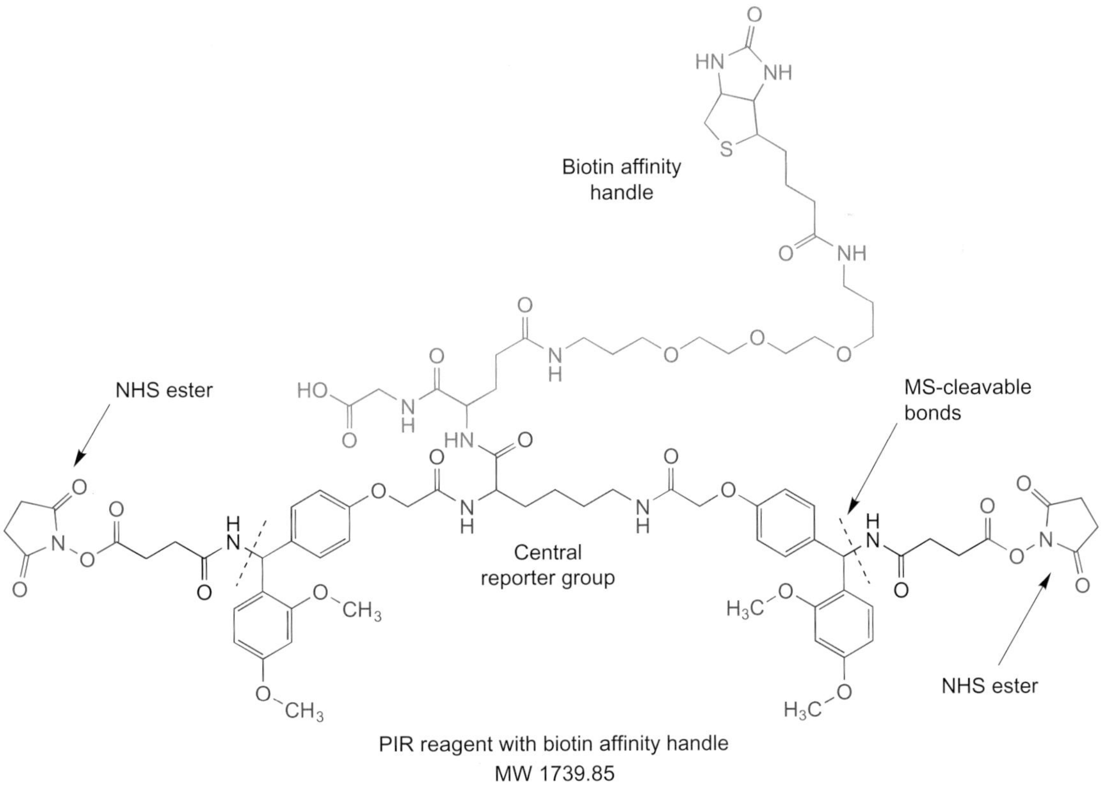

FIGURE 24.6 A trifunctional PIR compound that contains two NHS esters to capture interacting proteins through amide bond formation and a PEG–biotin arm to permit isolation of crosslinked proteins on (strept)avidin supports.

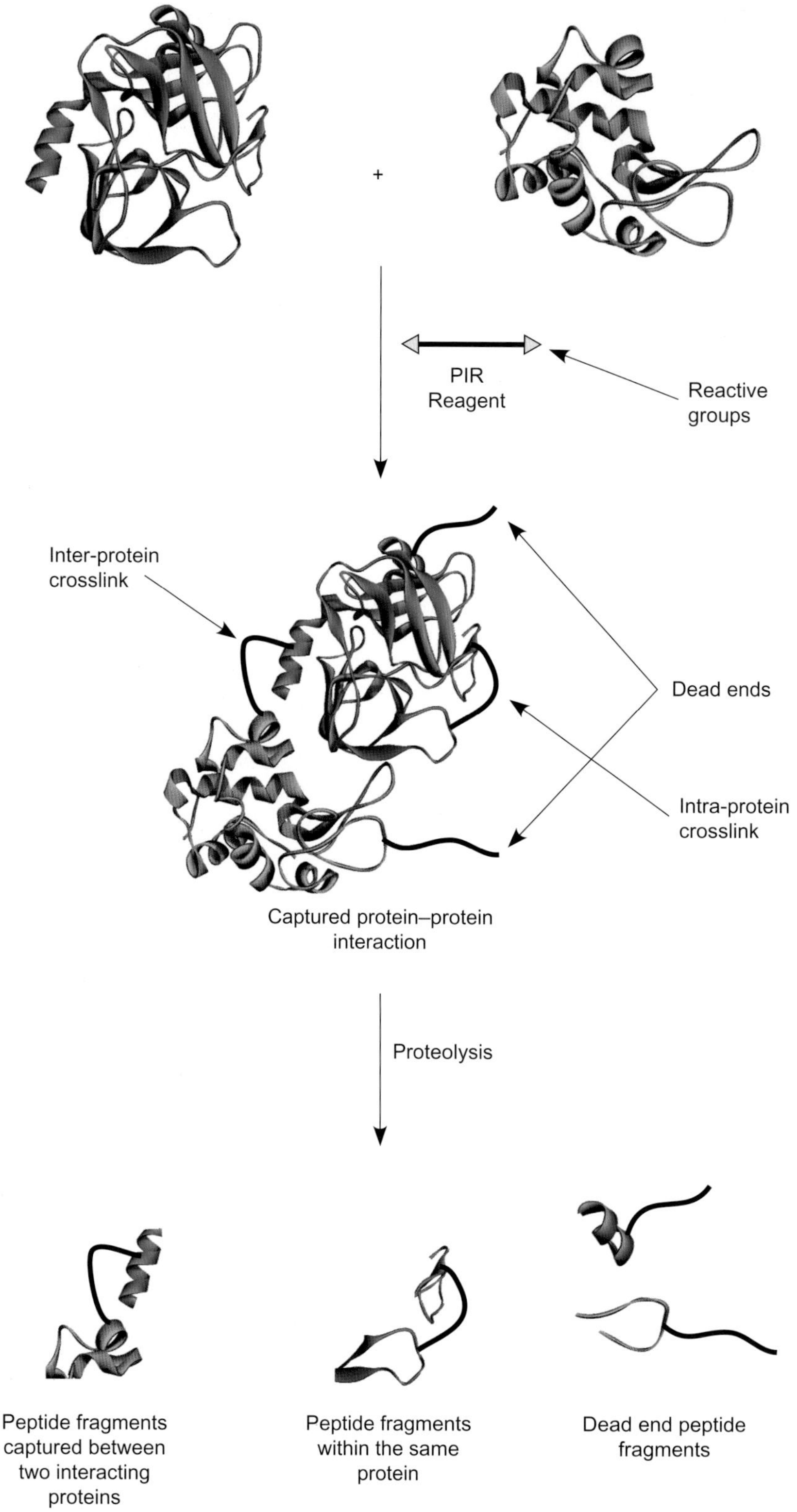

FIGURE 24.7 Reaction of a PIR compound with a protein sample can result in several products, all of which can be identified by mass spectrometer analysis. A true inter-protein crosslink can occur that links two interacting proteins together, which is the desired product. However, the crosslinking process may also result in intra-protein crosslinks or dead ends wherein only one end of the PIR reagent has coupled to a protein.

The specific fragmentation pattern observed at this stage distinguishes the kind of crosslink or modification that was initially formed. In this regard, a single modified peptide peak plus a reporter peak of mass *m/z* 828 indicates a dead-end modification with no value in determining protein interactions. Alternatively, a single peptide fragment peak plus a reporter group peak of mass *m/z* 711 indicates an intra-molecular crosslink made between regions of the same protein, which is also not of interest. However, a fragmentation pattern containing two labeled peptide peaks plus a reporter peak of mass *m/z* 711 indicates a successful conjugation between two protein molecules, which may be indicative of a true protein–protein interaction. Note that alternative designs of a PIR-type reagent containing other reporter structures, including those with a biotin handle and a PEG spacer, will result in different reporter fragmentation mass values than those stated here.

The use of PIR compounds to study protein interactions is a significant advance over the use of standard homobifunctional crosslinkers (Bruce, 2011; Yang *et al.*, 2012). The unique design of the PIR reagent facilitates deconvolution of putative protein interaction complexes through a simplified mass spectrometry analysis. The software can ignore all irrelevant peak data and just focus analysis on the two labeled peptide peaks, which accompany the reporter signal of appropriate mass. This greatly simplifies the bioinformatics of data analysis and provides definitive conformation of protein–protein crosslinks.

Finally, knowledge of the peptide masses that resulted from the PIR conjugation provides information to identify the parent proteins from which they originated. Peptide mass and sequence databases are now sufficiently developed to provide rapid confirmation of protein–protein interaction partners.

The following protocol is designed for treating cells with the PIR reagent to study protein interactions *in vivo*. It is based on the method of Tang *et al.* (2005). The use of the PIR compound to treat intact cells will result in the crosslinking of proteins both on the cell surface and within the cell, which indicates that the reagent is able to cross the cell membrane.

Protocol

1. Dissolve the PIR compound in dry DMSO to make a 100-m*M* stock solution.
2. Grow cells in media to a density of about $OD_{600\,nm} = 1.2$ and harvest in mid-log phase. Centrifuge cells at 3200 rpm to pellet them and wash three times with ice-cold PBS (150-m*M* sodium phosphate, 100-m*M* NaCl, pH 7.5).
3. Suspend the cells in 1 ml of PBS and add an aliquot of the dissolved PIR compound to bring the final concentration to 1-m*M*.
4. React at room temperature with gentle shaking for 5 min.
5. Quench the reaction by the addition of 50 µl of 1-*M* Tris, pH 7.5.
6. Wash the cells five times by centrifugation with cold PBS to remove excess PIR reagent and any secreted proteins.
7. Lyse the cells using a detergent lysis buffer suitable for the cell type being treated. Centrifuge the lysate at 15,000 rpm for 40 min at 4°C to remove insoluble material. Collect the supernatant and discard the pellet. At this point, the soluble protein fraction may be analyzed by electrophoresis, if desired.
8. Remove unreacted PIR reagent and reaction byproducts by gel filtration or dialysis.
9. Precipitate the protein with TCA to further remove any remaining salts and detergent, centrifuge to pellet the precipitated protein, wash the pellet with TCA, and centrifuge again. Re-dissolve the washed pellet in 100 µl of 100-m*M* NH_4HCO_3, pH 7.8, containing 8-*M* urea.
10. Reduce protein disulfides by adding dithiothreitol (DTT) to a final concentration of 5-m*M* and incubate for 30 min at 60°C. Add iodoacetamide to a final concentration of 25-m*M* to alkylate the thiols. React for 1 h in the dark.
11. Dilute the solution 4-fold with 100-m*M* 25-m*M* NH_4HCO_3, and then add 20 µg trypsin to digest the proteins. Incubate at 37°C for 4 h or overnight at 30°C with shaking.
12. Purify the tryptic peptides by chromatography on a C18 column to remove salts (follow the manufacturer's directions for peptide purification). Dry the eluent and re-dissolve the peptides in 20 µl 0.1% formic acid.
13. Analyze the purified peptides using 2D-LC/MS/MS.

2. USE OF PHOTOREACTIVE CROSSLINKERS TO STUDY PROTEIN INTERACTIONS

The use of crosslinking agents containing at least one photoreactive group provides reagents that can be activated at a desired point after bait proteins have been allowed to interact and bind to prey proteins. The availability of the first photoreactive homobifunctional or heterobifunctional compounds permitted protein interactions to be studied at a new level of specificity. Unlike the use of spontaneously reactive homobifunctional reagents, the incorporation of a photoreactive group helps to prevent nonselective crosslinking and uncontrolled polymerization of proteins that may or may not be specifically interacting.

In use, a bait protein is first modified with the spontaneously reactive (thermoreactive) end of a photoreactive heterobifunctional compound, while protecting the solution from light exposure. The modified bait protein is then allowed to interact with a sample containing other proteins and biomolecules, which may contain prey proteins able to interact with it. Finally, the sample mixture is exposed to UV light to activate the photoreactive end of the crosslinker, causing conjugation of this group with any nearby interacting molecules (Leitner, 2012). The rapid reaction rate of the activated photoreactive intermediates assures that they do not survive long enough to cause much conjugation to proteins just due to random collisions. However, interacting proteins that are in proximity to the modified bait protein are more likely to be captured through reaction with the activated photoreactive group.

The following sections discuss the application of several photoreactive heterobifunctional crosslinkers to the study of protein interactions.

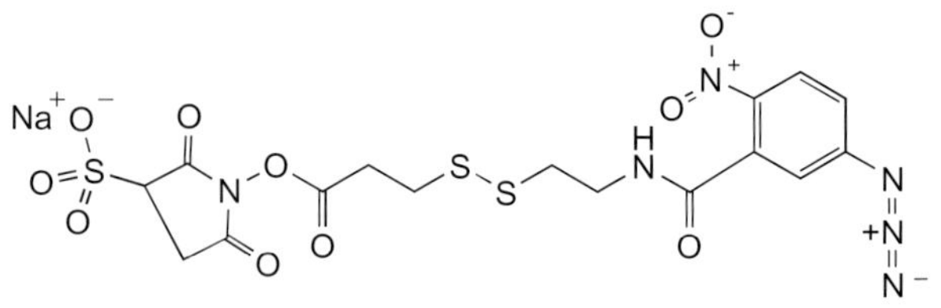

Sulfo-SAND
Sulfosuccinimidyl 2-(*m*-azido-*o*-nitrobenzamido) ethyl-1,3'-dithiopropionate
Mol. Wt.: 570.51

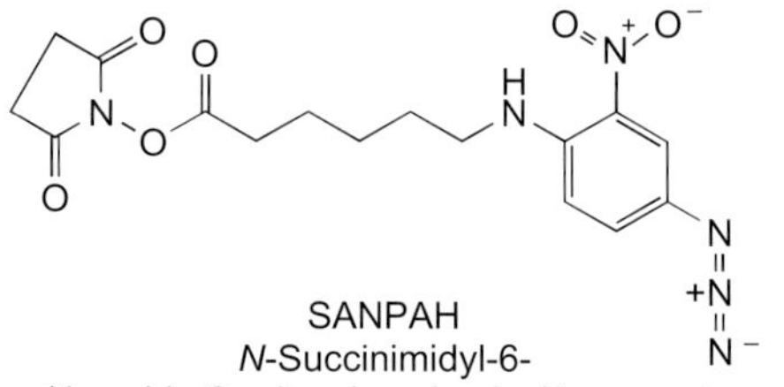

SANPAH
N-Succinimidyl-6-(4'-azido-2'-nitrophenylamino)hexanoate
Mol. Wt.:390.35

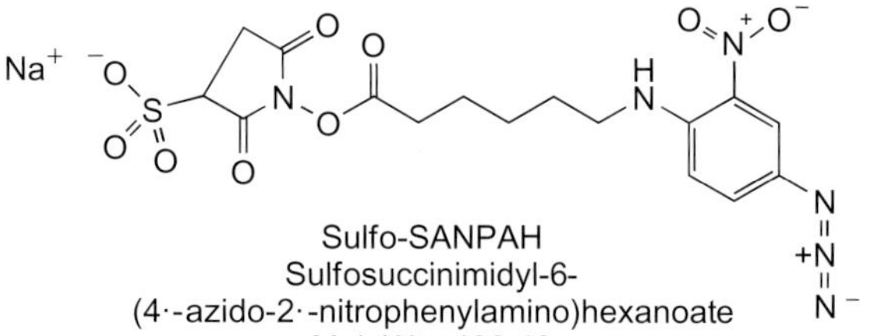

Sulfo-SANPAH
Sulfosuccinimidyl-6-(4'-azido-2'-nitrophenylamino)hexanoate
Mol. Wt.: 492.40

FIGURE 24.8 The heterobifunctional crosslinkers sulfo-SAND, SANPAH, and sulfo-SANPAH contain an amine-reactive (sulfo)NHS ester on one end and a photoreactive phenyl azide group on the other end. Sulfo-SAND allows release of conjugates by reduction of its internal disulfide bridge.

2.1. Sulfo-SAND, SANPAH, and Sulfo-SANPAH

These three photoreactive crosslinkers are described in terms of their properties and reactivities in Chapter 6, Section 3, Amine-Reactive and Photoreactive Crosslinkers. They represent early examples of the use of standard phenyl azide photoreactive compounds for the study of protein interactions. All of them contain either an amine-reactive NHS ester or the charged and water soluble analog, a sulfo-NHS ester (Figure 24.8). SANPAH is uncharged and hydrophobic and thus provides membrane permeability for studying intracellular protein interactions in whole cells (Mikhailov *et al.*, 2001; Kota and Ljungdahl, 2005). Sulfo-SAND and sulfo-SANPAH, however, possess a negatively charged sulfonate group, which prevents them from passing through cell membranes, and thus are membrane impermeable (Uckun *et al.*, 1995; Tiberi *et al.*, 1996; Gaudet *et al.*, 2003). The sulfonated compounds may be used with whole cells to study cell surface protein interactions, although, in use, their concentration should be limited to prevent internalization by active transport into cells.

Sulfo-SAND, SANPAH, and sulfo-SANPAH all contain a nitrated phenyl azide photoreactive group. The presence of the nitro group shifts the optimal wavelength for photoactivation to higher wavelengths, thus avoiding the potential for biomolecule damage due to UV irradiation (photoactivation occurs at 320-350 nm). Sulfo-SAND provides one further option for studying interacting proteins. It has a cleavable disulfide-containing cross-bridge that permits recovery of any prey proteins that have been captured by the photo-coupling process (McMahan and Burgess, 1994; Uchiyama *et al.*, 2002). Finally, each of these photoreactive crosslinkers has phenyl azide rings that undergo ring expansion to the 7-membered ring, dehydroazepine intermediate, which reacts primarily with amines on target molecules. Thus, the proteins that are captured by these crosslinkers are typically coupled through secondary amine linkages to the photoreactive end of the reagents. A general protocol for the use of these compounds to study protein interactions is given at the end of the next section.

2.2. Sulfo-SFAD

Sulfo-SFAD is sulfosuccinimidyl-[perfluoroazidobenzamido]-ethyl- 1,3′-dithiopropionate, an amine-reactive and photoreactive crosslinker with more advanced properties than the photoreactive reagents discussed above. This reagent contains a sulfo-NHS ester to provide increased water solubility prior to conjugation,

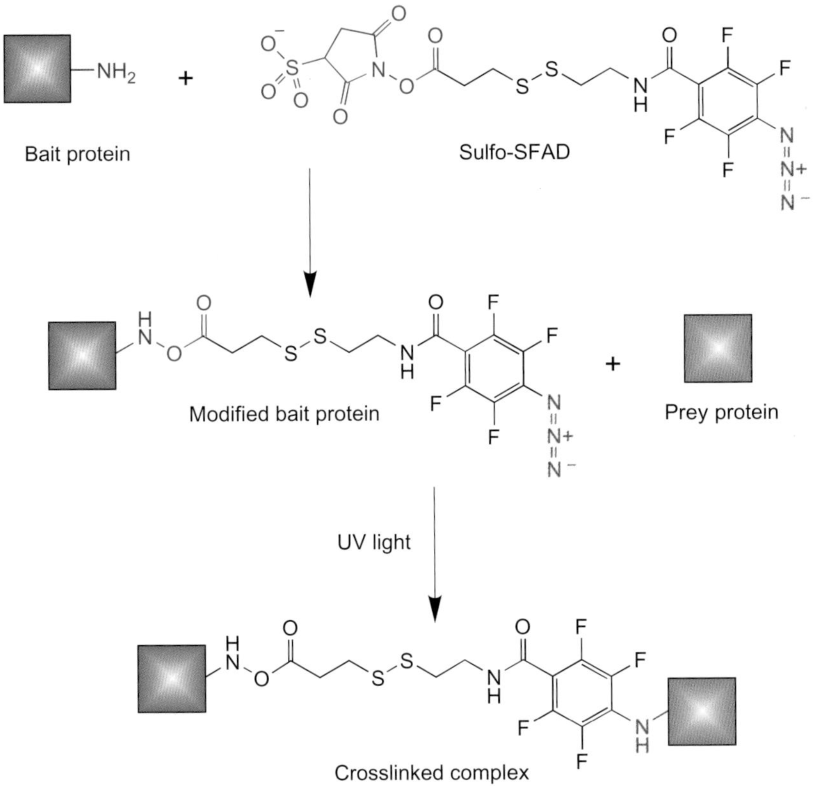

FIGURE 24.9 Sulfo-SFAD is an advanced heterobifunctional photoreactive crosslinker that contains an amine-reactive sulfo-NHS ester on one end and a tetrafluorophenyl azide group on the other end. The fluorine substitutions on the phenyl ring prevent the photoactivated nitrene from reacting with the ring, thereby providing greater yields in capturing interacting proteins. The disulfide-containing cross-bridge allows cleavage of conjugated molecules by reduction with DTT.

and this group will react with an amine on proteins and other biomolecules to form an amide linkage. Sulfo-SFAD also contains a perfluorophenyl azide group that has better photo-insertion capability than the original unsubstituted phenyl azide group. The reason for this enhanced yield is that, after photoactivation, the nitrene intermediate cannot react with the phenyl ring itself and undergo ring expansion to a dehydroazepine as happens with typical unsubstituted phenyl azides. The result is that the nitrene survives long enough to react with neighboring molecules in the immediate molecular vicinity, such as proteins that are interacting with the modified bait protein. Perfluorophenyl azides thus have higher yields of conjugation, and they can react by insertion into C–H bonds, N–H, unsaturated carbon–carbon bonds, and other structures in target molecules. By contrast, unsubstituted phenyl azides ring expand and react mainly with amines, and even then at low yield. The perfluorophenyl azide may be activated with UV light at 320nm.

Sulfo-SFAD also contains a disulfide bond in its cross-bridge, which provides subsequent cleavability of any crosslinks formed. Thus, prey proteins can be recovered from complexes for analysis by simple reduction of the conjugate with DTT. Once reduced, the perfluorophenyl ring group of sulfo-SFAD is transferred to the interacting prey protein. This group can be identified by ^{19}F NMR, and it also provides a traceable label for identification of peptide fragments by mass spectrometry. Figure 24.9 illustrates the reactions of sulfo-SFAD in capturing a bait–prey complex.

The following references provide further information on the use of sulfo-SFAD and perfluorophenyl azide photoreactive groups: Camperchioli *et al.* (2011); Pandurangi *et al.* (1995a,b, 1996, 1997a,b, 1998); Yan *et al.* (1994).

A general protocol that can serve as a guide for the use of heterobifunctional crosslinkers in the study of protein interactions is given below. Some optimization of concentrations may have to be performed depending on the particular type and properties of proteins being studied.

Protocol

1. Dissolve the photoreactive crosslinker in DMF or DMSO at a concentration of 10 to 25-m*M* (protect from light). Water soluble crosslinkers can be added to water or buffer, but should be used immediately.
2. Modify a purified bait protein by adding an aliquot of crosslinker to achieve a molar excess of 2 to 10 moles per mole of protein (protect from light). For NHS ester-containing reagents, react in 0.1-*M* sodium phosphate buffer at pH 7.2 to 7.4. Avoid using any other amine containing buffer components, such as Tris or imidazole, which will interfere with the reaction.

3. React for 30 to 60 min in the dark (room temperature or 4°C).
4. Remove excess crosslinker from the modified bait protein by gel filtration or dialysis in the dark.
5. Add the modified bait protein to a sample containing potentially interacting prey proteins and incubate for 1 h protected from light. The sample may be cells (for cell surface interaction studies), cell lysate, or various extracts from cells, tissues, or biological fluids.
6. Expose the sample to UV light to crosslink and capture interacting proteins via the photoreactive group.
7. Recover interaction complexes by affinity chromatography. This is typically performed using immobilized antibodies to the bait protein or using fusion tag binders, if the bait protein contains a fusion partner (e.g., immobilized glutathione for binding GST-labeled bait proteins).
8. Analyze the purified complexes by SDS-PAGE, western blotting, and mass spectrometry.

3. TRIFUNCTIONAL LABEL TRANSFER REAGENTS

The earliest examples of label transfer reagents were cleavable heterobifunctional compounds that incorporated a phenyl azide group, which also had a phenolic modification on the ring. The phenolic hydroxyl activates the ring for substitution reactions to occur *ortho* or *para* to its position. These compounds thus can be radioiodinated using typical oxidation reagents such as chloramine T or Iodobeads. Iodination of the crosslinker with ^{125}I prior to its use will result in a radioactive label transfer reagent that can tag an unknown interacting protein with a radiolabel after cleavage of the crosslinker's spacer arm.

Subsequent designs of label transfer reagents used non-radioactive labels to avoid the safety and regulatory issues posed by ^{125}I. Fluorescent constituents designed into cleavable photoreactive crosslinkers make possible transfer of a fluorescent label to an unknown interacting protein. An example of this type of reagent that incorporates a coumarin group is SAED, which has been derivatized with an azido group on the aromatic ring. The reagent is non-fluorescent prior to exposure to UV light, but upon photo-activation and coupling to interacting proteins, it becomes highly fluorescent. The reagent also has a disulfide bond that can be reduced, resulting in cleavage of the crosslinked proteins and transfer of the label to the unknown interacting species. In this case, the fluorescently labeled interacting proteins can be followed in cells to determine the site of interactions or the fate of the proteins after interacting.

The most advanced type of bioconjugation reagent that is designed to study interacting proteins is a trifunctional label transfer reagent, and it has become an important component in the reagent toolbox to capture protein complexes (Rutkowska and Schultz, 2012). These compounds are a special category of trifunctional crosslinkers (see Chapter 7), which possess two reactive groups and a third arm containing a label. Typically, one of the reactive groups is thermoreactive and can be used to label a bait protein, while the second reactive group is usually photoreactive for selective activation and coupling upon UV irradiation. The third arm of the reagent can be designed to have a fluorescent group for detection or terminate in a biotin group for both detection and purification (Figure 24.10).

An important feature of trifunctional label transfer reagents is the presence of a cleavable cross-bridge on the arm containing the thermoreactive group. After a bait protein has been conjugated with an interacting prey protein using this type of reagent, the complex can then be cleaved, which releases the interacting proteins and transfers the label over to the prey protein. The label can be used to detect prey proteins on a western blot or to purify them from the sample solution. Unlike the heterobifunctional compounds discussed previously for studying protein interactions, label transfer reagents allow specific detection and purification of unknown interacting prey proteins for subsequent analysis.

The use of trifunctional label transfer crosslinkers has involved the study of the glycolytic enzyme

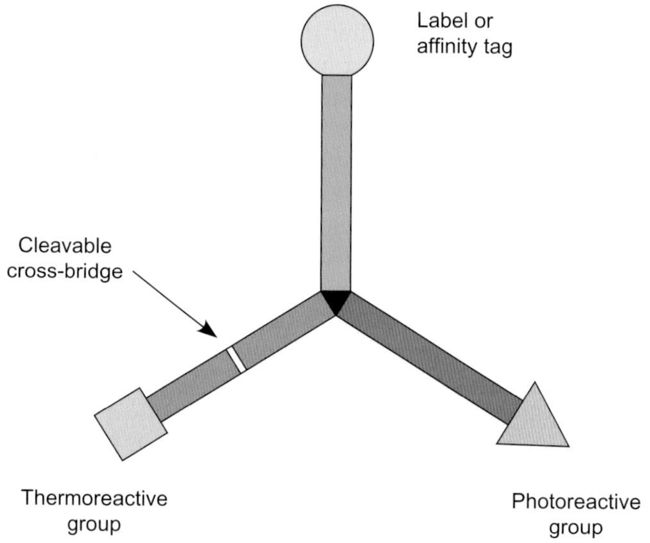

FIGURE 24.10 Trifunctional label transfer agents contain two arms with terminal reactive groups and a third arm with a label or affinity tag, such as a biotin group. One of the reactive groups typically is thermoreactive to couple with bait proteins, while the second reactive group usually is photoreactive. The thermoreactive arm has a cleavable cross-bridge to facilitate release of the captured protein and transfer of the label of affinity tag to it.

components (Puchulu-Campanella *et al.*, 2012), investigation into the interaction of human frataxin with ISU (Watson *et al.*, 2012), and to study the interaction complex between cell adhesion molecule L1 with matrix metalloproteinase 14 and adenine nucleotide translocator at the plasma membrane (Loers *et al.*, 2012).

The following sections describe three types of label transfer reagents, which are all built with a biotin handle. Note: A new version of a trifunctional label transfer agent was recently described for studying receptor–ligand interactions and termed the TRICEPTS reagent (Frei *et al.*, 2012). This crosslinker is described in Chapter 7, Section 5.

3.1. Sulfo-SBED

Sulfo-SBED is sulfosuccinimidyl-2-[6-(biotinamido)-2-(*p*-azidobenzamido) hexanoamido] ethyl-1,3′-dithiopropionate), a trifunctional reagent containing an amine reactive sulfo-NHS ester on one arm, a photoreactive phenyl azide group on the second arm, and a biotin handle on its third arm (Figure 24.11) (see Chapter 7, Section 2, for a general description and use). This compound has found great utility in capturing protein interactions and subsequently isolating them for analysis. A purified bait protein initially is labeled with sulfo-SBED by its sulfo-NHS ester to form amide linkages through reaction with lysine or N-terminal amines. The modified bait is then incubated with a sample and biomolecules are allowed to interact with it. Finally, the sample is exposed to UV light and the photoreactive group is able to couple with nearby amine groups on any interacting proteins. The resultant complex can be purified, identified, or detected through specific binding to the biotin groups (Figure 24.12 and 24.13).

Sulfo-SBED is soluble in organic solvents, such as DMSO (125-m*M*), DMF (170-m*M*), and methanol (12-m*M*), or to a lesser degree in pure water (~5-m*M*). The solubility of sulfo-SBED in buffered aqueous solutions may vary from about 0.1-m*M* to 3-m*M* (e.g., ~1-m*M* in 0.1-*M* PBS). However, to dissolve sulfo-SBED at higher concentrations in a reaction buffer, first dissolve it in a water-miscible organic solvent such as DMSO or DMF and transfer an aliquot to the aqueous buffer solution. Use no more than about 1 to 10% of solvent in the final buffered reaction mixture to prevent buffer precipitation and minimize the denaturation of proteins.

Bait proteins modified with sulfo-SBED may precipitate if the level of modification is too high, primarily due to the hydrophobic nature of the crosslinker and the biotin handle. To prevent precipitation or at least minimize it, adjust the molar excess of sulfo-SBED over the bait protein to a level where the protein remains in solution. Some precipitation may be removed by centrifugation or filtration prior to use.

A derivative of sulfo-SBED containing a thiol-reactive pyridyl disulfide group on its thermoreactive arm has

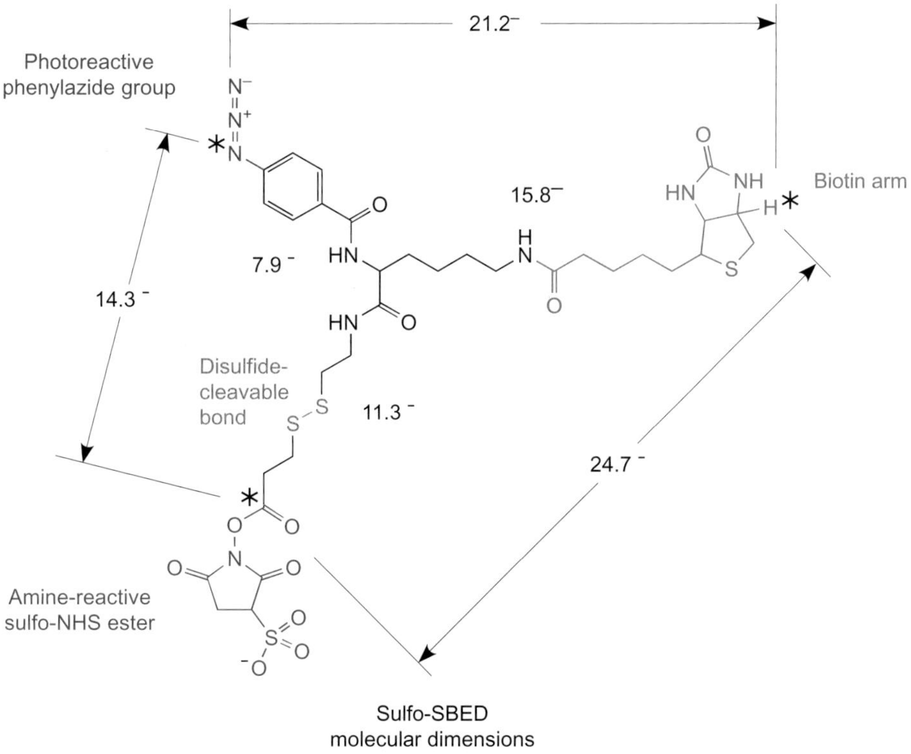

FIGURE 24.11 Sulfo-SBED is a label transfer agent that contains a water soluble sulfo-NHS ester to label bait proteins and a phenyl azide group for photoreactive capture of a prey protein. The biotin label can be used for detection or isolation of protein–protein conjugates using (strept) avidin reagents. The stars indicate the atoms that were used to measure the indicated molecular dimensions.

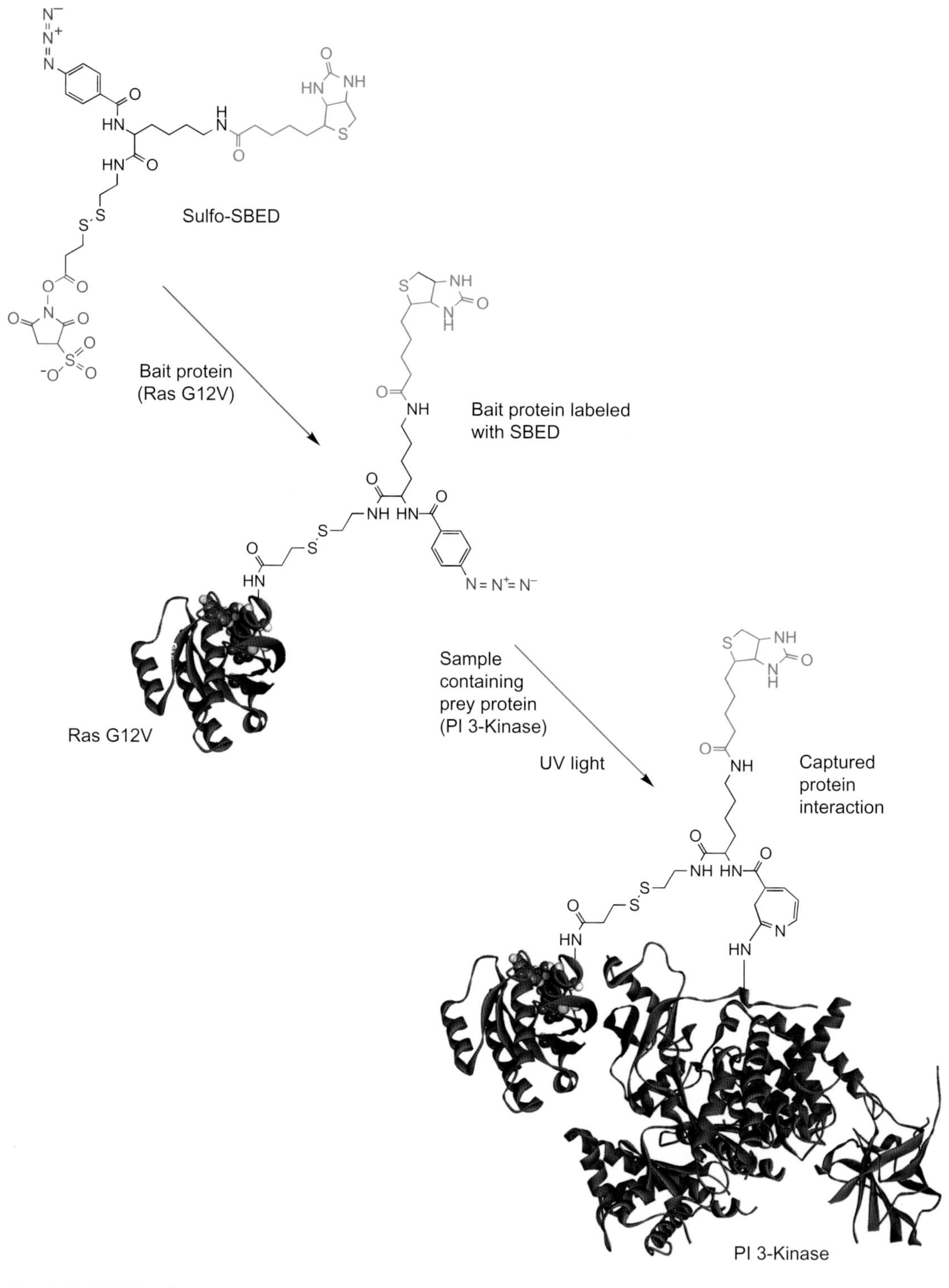

FIGURE 24.12 Sulfo-SBED is first used to label a bait protein through reaction of the sulfo-NHS ester with available amine groups on the protein, yielding an amide bond linkage. This labeled bait protein is then added to a sample containing proteins that potentially could interact with the bait. After an incubation period, the sample is exposed to UV light to photoactivate the phenyl azide group. This reaction causes any interacting prey proteins to be crosslinked with the bait protein, forming a complex containing a biotin affinity tag.

been reported for modification of bait proteins containing a cysteine residue. Reaction of this compound with a thiol results in the formation of a reversible disulfide linkage. After capturing interacting proteins with this reagent, reduction of the disulfide transfers the biotin label to the unknown prey protein(s). This compound was prepared from sulfo-SBED by reduction of the disulfide in its sulfo-NHS ester arm followed by reaction

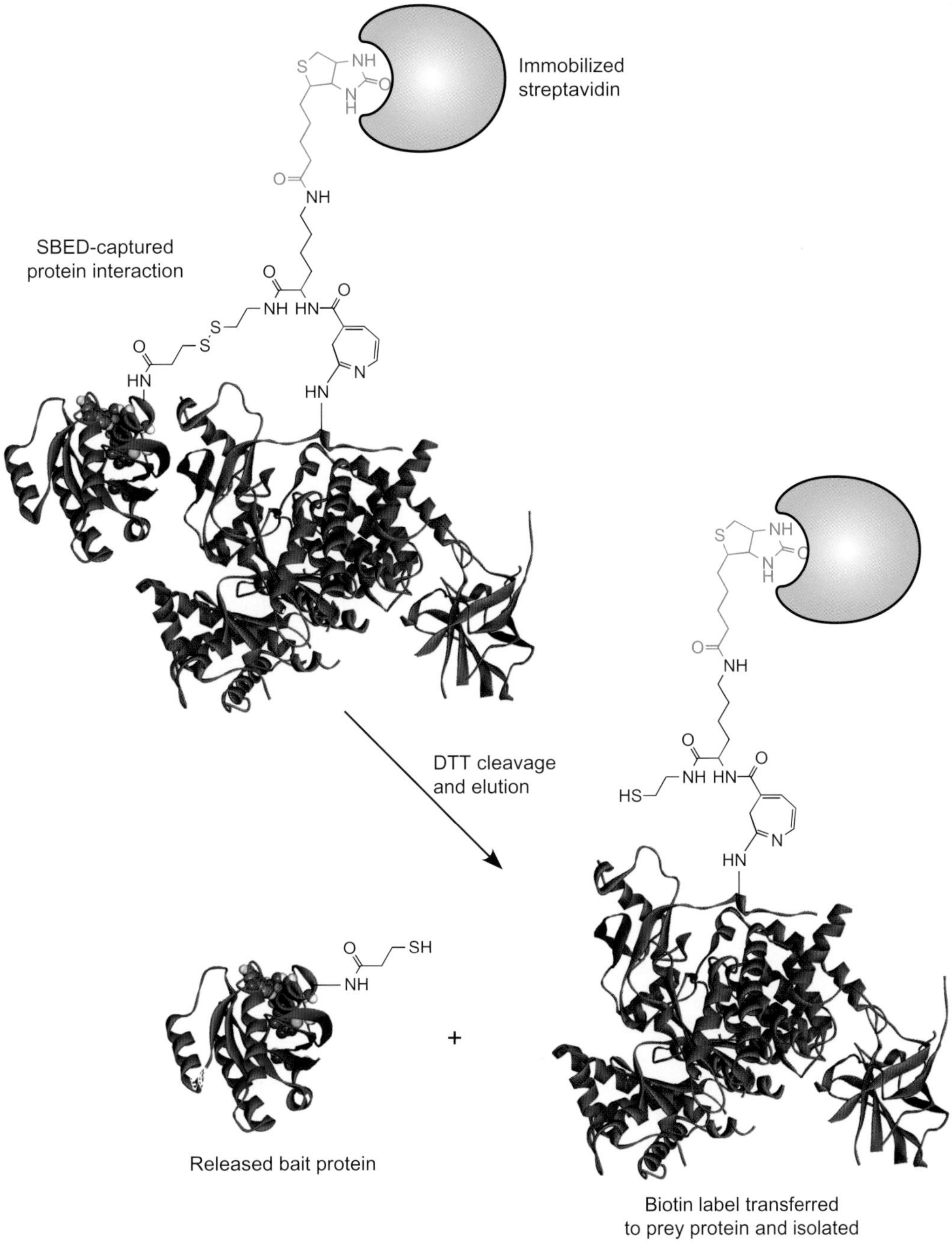

FIGURE 24.13 A sulfo-SBED-captured protein interaction can be released using DTT to cleave the disulfide within the cross-bridge leading to the bait protein. The result transfers the biotin label to the unknown interacting protein. The biotin tag thus allows the interacting protein to be detected or isolated using (strept)avidin reagents.

with 2,2′-dithiodipyridine to give the thiol-reactive compound, 1-[6-(biotinamido)-2-(*p*-azidobenzamido)-hexanoamido]-2-(2′-pyridyldithio)ethane (Figure 24.14) (Alley *et al.*, 2000; Ishmael *et al.*, 2001, 2002; Trakselis, 2005). This derivative has most of the features of sulfo-SBED in terms of being a trifunctional label transfer agent containing biotin, but extends its application to sulfhydryl-containing bait proteins.

Sulfo-SBED has been used to investigate protein–protein interactions in the following applications: to study the bacteriophage T4 replisome (Ishmael *et al.*, 2003), cell surface antigens of mycoplasma species bovine group 7 in binding and activation of plasminogen (Bower *et al.*, 2003), bacterium–host protein–carbohydrate interactions (Ilver *et al.*, 2003), transcription activator interactions with multiple SWI/SNF subunits (Neely *et al.*, 2002), binding

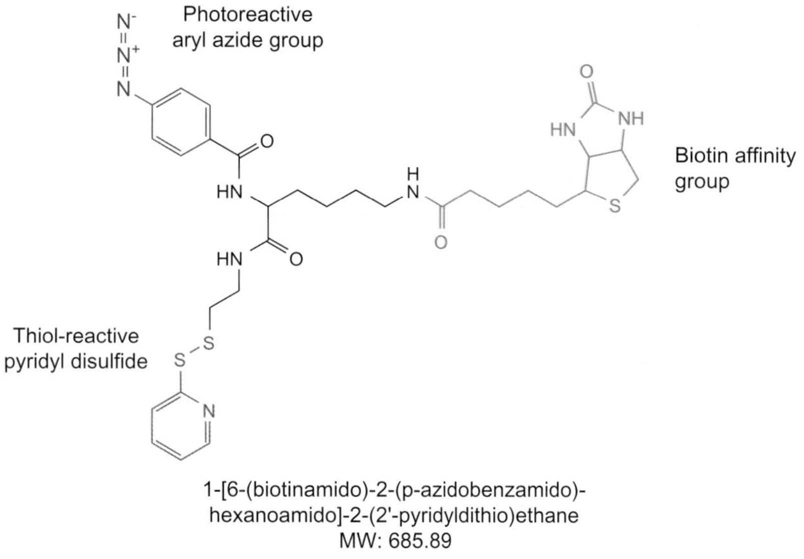

FIGURE 24.14 A trifunctional label-transfer reagent containing a thiol-reactive pyridyl disulfide group, a photoreactive phenyl azide, and a biotin affinity tag. This compound can be used to label bait proteins through available thiol groups and capture interacting prey proteins by photoreactive conjugation.

of protein D/E to the surface of rat epidermal sperm (Tubbs et al, 2002), and the gp41–gp59 complex in bacteriophage T4 helicase (Ishmael *et al.*, 2002); to identify a region in alcohol dehydrogenase that binds to α-crystallin during chaperone action (Santhoshkumar and Sharma, 2002), regions of the mouse CD14 molecule required for toll-like receptor 2- and 4-mediated activation of NF-κB (Muroi *et al.*, 2002), and active site residues of cyclophilin A that are important for signaling via CD147 (Yurchenko *et al.*, 2002); to map protein–protein interactions in the bacteriophage T4 DNA polymerase holoenzyme (Alley *et al.*, 2000); to study the import of adenovirus DNA involving the nuclear pore complex receptor CAN/ Nup214 and histone H1 interactions (Trotman *et al.*, 2001), insulin-like growth factor (IGF)-1 interaction with IgG-binding proteins (Horney *et al.*, 2001), 3F3/2 anti-phospho-epitope antibody binding to the mitotically phosphorylated anaphase-promoting complex/cyclosome (Daum *et al.*, 2000), SH3 binding sites of ZG29p in its interaction with amylase (Kleene *et al.*, 2000), and the proteasome activator PA28 in Hsp90-dependent protein refolding (Minami *et al.*, 2000); to investigate functional elements in α-crystallin (Sharma et al., 2000) and *Helicobacter pylori* adhesin binding to fucosylated histo-blood group antigens (Ilver *et al.*, 1998); to identify low abundance proteins by electrophoresis and MALDI-TOF MS (Bergstrom *et al.*, 1998) and molecular probes for muscarinic receptors (Jacobson *et al.*, 1995); to quantitate triple-helix formation (Geselowitz and Neumann, 1995) and to study prion–protein interactions with its receptor (Santuccione *et al.*, 2005); to study the activation of Hsp70 chaperones (Steel et al, 2004), the effect of oxidized βB3-crystallin peptide on lens βL-crystallin and its interaction with βB2- crystallin (Udupa and Sharma, 2005), and the effect of oxidized βB3-crystallin peptide on the thermal aggregation of bovine lens gamma-crystallins (Udupa and Sharma, 2005); for the mass spectrometric detection of affinity-purified crosslinked peptides (Hurst *et al.*, 2004); to map protein interfaces combined with MALDI-TOF and ESI-FTICR mass spectrometry (Sinz *et al.*, 2005); to study the activation of the antioxidant enzyme 1-CYS peroxiredoxin and its requirement for glutathionylation mediated by binding with GST (Manevich *et al.*, 2004); to investigate the recruitment of HAT complexes by direct activator interactions with the ATM-related Tra1 subunit (Brown *et al.*, 2001); to identify annexin A2 heterotetramer as a receptor for the plasmin-induced signaling in human peripheral monocytes (Laumonnier *et al.*, 2005); to investigate the role of proteasome activator PA28 in Hsp90-dependent protein refolding (Minami *et al.*, 2000); to confirm the interaction between human frataxin and the scaffold protein ISU with mass spectometry detection (Watson et al, 2012); to investigate the binding of receptor tyrosine kinase TrkB to the neural cell adhesion molecule (NCAM) (Cassens *et al.*, 2010); and to study an interaction complex at the plasma membrane (Loers *et al.*, 2012).

In addition to the use of sulfo-SBED to capture unknown prey proteins interacting with labeled bait proteins, the reagent may also be used to study the interaction interfaces between two known proteins that specifically interact. Sinz *et al.* (2005) used this approach to identify the interaction surfaces between calmodulin and M13, which is a short peptide from skeletal muscle light-chain kinase. In this application, calmodulin was labeled with sulfo-SBED and allowed to interact with purified M13. After an incubation period, the solution was exposed to UV light and the two proteins were crosslinked together. Next, the conjugated proteins were proteolytically digested with trypsin and the resultant biotinylated peptides were enriched on a column of immobilized monomeric avidin (Thermo Fisher). The crosslinked peptides come from one peptide segment of calmodulin and one peptide segment of M13. Finally, these biotinylated and crosslinked peptides were identified using nano-HPLC separation into a nano-ESI-FTICRMS with software-facilitated deconvolution of the peptide identities. Such analysis provides insight into the interaction surfaces involved with the binding event between the two proteins.

The following protocol represents a suggested method that will work well for many proteins. It is a blend of protocols used in the literature and recommended by Thermo Fisher in the sulfo-SBED instruction manual. Modifications to reaction conditions may be necessary in certain cases to maintain protein stability or solubility, depending on the properties of the particular bait protein being used.

Protocol

1. Dissolve a purified bait protein in 0.1-*M* sodium phosphate, 0.15-*M* NaCl, pH 7.2, or a similar buffer at neutral pH, which does not contain any competing amines (i.e., avoid Tris or imidazole). The bait protein may be at a concentration of anywhere from 0.1 mg/ml to 10 mg/ml. Prepare the solution in a dark tinted vial or wrap the vessel in foil to prevent light exposure when the crosslinker is added.
2. Prepare a solution of sulfo-SBED in dry DMF or DMSO at a concentration of 40 μg/ml. Protect from light.
3. Add a quantity of the sulfo-SBED solution to the bait protein solution so that a 1- to 5-fold molar excess of crosslinker over the bait protein results in the reaction mixture. Mix well. Using greater quantities of sulfo-SBED to the bait protein may result in precipitation due to the hydrophobic nature of crosslinker. In addition, over modification of the bait protein with the crosslinker may block sites of protein interaction, thus preventing complex formation. As a practical example, Horney *et al.* (2001) used a 1:1 molar ratio of sulfo-SBED to the bait protein IGF-1 with success.
4. React for 30 min at room temperature or 2 h at 4°C. Continue to protect the solution from light.
5. Remove excess crosslinker and reaction by-products by dialysis or size exclusion chromatography. For small quantities of bait proteins, dialysis may be the better choice, because gel filtration columns often bind enough protein nonspecifically to make recoveries unacceptably low.
6. Add the labeled bait protein to a sample containing the putative interacting prey proteins. The quantity of bait protein to be added to a given sample should be within the same concentration level as the amount of prey proteins present. The optimal level of addition may have to be determined by varying the amount of bait protein concentrations in a number of sample solutions to decide which concentration results in the best interaction complexes being formed.
7. Incubate the sample mixture for at least 1 h at room temperature or under conditions optimal for the interaction to be studied.
8. Expose the sample to UV light to photoactivate the aryl azide and cause the conjugation reaction to occur. For best results, use a UV lamp that irradiates in the range of 300 to 370 nm. Irradiate for 2 to 15 min, using a briefer exposure when using UV lamps of greater wattage.
9. Interaction complexes that were captured by the crosslinking reaction can be recovered or analyzed using the biotin groups. An immobilized streptavidin support, for instance, can be used to purify the conjugates away from other sample proteins. Alternatively, a streptavidin–HRP conjugate can be used to probe a western blot after electrophoresis separation of the sample. Reduction of the conjugate may be performed by cleavage of the disulfide bond with DTT. The typical addition of DTT to the electrophoresis sample buffer will facilitate disruption of the complexes and transfer of the biotin label to the interacting prey proteins. This reduction step can also be performed using 50-m*M* DTT or 50-m*M* TCEP in aqueous solution to permit recovery of the biotinylated prey proteins on immobilized streptavidin. Alternatively, the use of an affinity support having milder elution characteristics for biotinylated proteins, such as immobilized monomeric avidin, would facilitate isolation of the biotinylated complexes or prey proteins under non-denaturing conditions.

3.2. MTS–ATF–Biotin and MTS–ATF–LC-Biotin

MTS-ATF-biotin is 2-[N2-(4-Azido-2,3,5,6-tetrafluorobenzoyl)-N6-(6-biotinamidocaproyl)-L-lysinyl]ethyl methanethiosulfonate and MTS-ATF-LC-biotin is 2-{N2-[N6-(4-Azido-2,3,5,6-tetrafluorobenzoyl)-6-aminocaproyl]-N6-(6-biotinamidocaproyl)-L-lysinylamido}]ethyl methanethiosulfonate. Both reagents are trifunctional crosslinkers similar in design to sulfo-SBED discussed previously, but in addition to the biotin handle they contain a thiol-reactive group and an enhanced photoreactive, perfluorinated phenyl azide group. The two reagents differ in the length of the cross-bridge in the photoreactive arm, with the LC version containing an extended aminocaproyl spacer. Relative to the spacing possible between the reactive groups on these compounds, the LC version therefore provides nearly twice the maximal molecular distance over its shorter analog (21.8 Å *versus* 11.1 Å) (Figure 24.15). Thus, interacting proteins may be captured either through use of a long or short crosslink, depending on the optimal distances

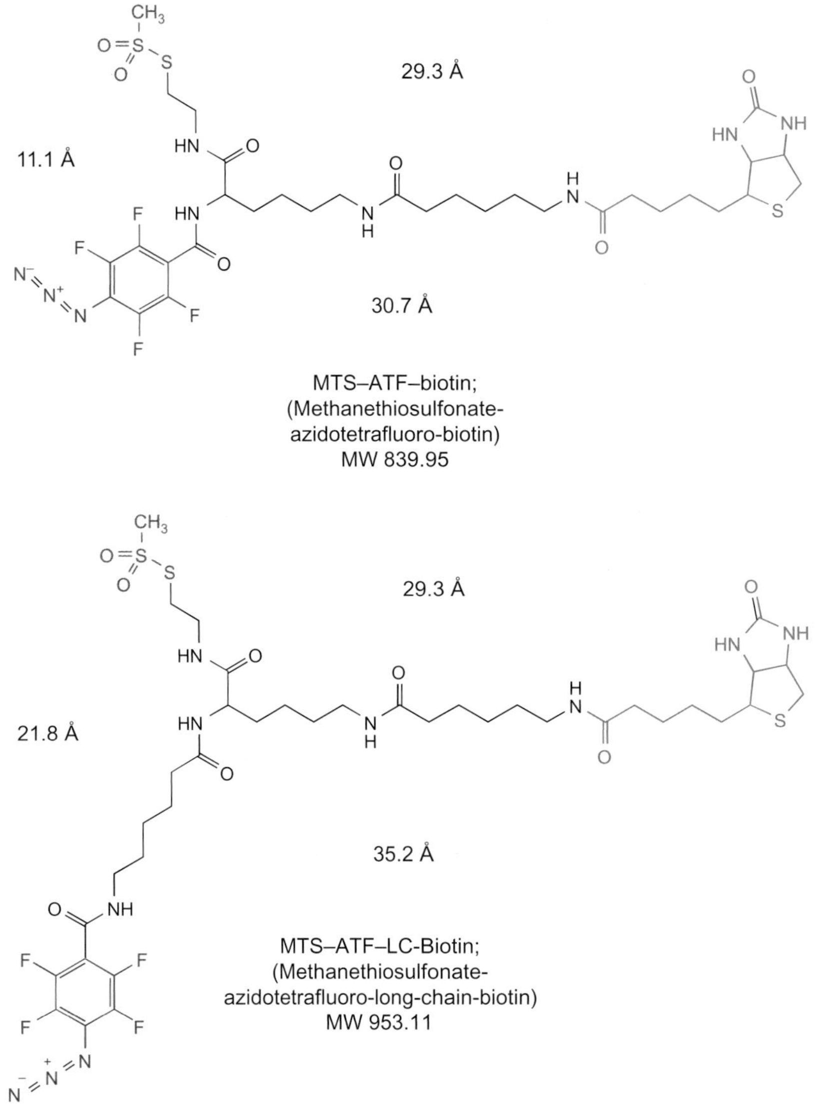

FIGURE 24.15 Two similar label transfer reagents containing a thiol-reactive methanethiolsulfonate group to label bait protein through available sulfhydryls, a tetrafluorophenyl azide group for high-efficiency photoreactive conjugation with interacting prey proteins, and a long biotin affinity tag. The lengths indicated in the figure represent the distances between the reactive linking site on one arm and the reactive group or biotin outer ring hydrogen atom on the adjacent arm. The point of measurement on the reactive groups is the last atom that is retained in any crosslink formed after the coupling reaction takes place.

between the proteins—or at least to the nearest thiol on the bait protein.

Both MTS–ATF–biotin and MTS–ATF–LC-biotin contain a methanethiolsulfonate group (MTS) on their thermoreactive arm which is able to couple with the sulfhydryl on a cysteine residue. This reaction proceeds with loss of the methyl sulfonate leaving group (sulfinic acid) to produce a disulfide linkage (Figure 24.16). Unlike a pyridyl disulfide group, however, which also reacts with thiols to form disulfide linkages, the MTS group is unstable to hydrolysis in aqueous solution, especially if other strong nucleophiles are present. Therefore, most MTS compounds dissolved or brought into PBS buffer at physiological pH will hydrolyze with a half-life on the order of 10 to 15 min. However, they also have very rapid reactivities with thiols (Stauffer and Karlin, 1994; Holmgren *et al.*, 1996; Liu *et al.*, 1996). If the parent MTS reagent is fully water soluble, the reaction with a thiol on a bait protein can take place with high yields in just a matter of minutes. However, both of these trifunctional label transfer compounds are hydrophobic, so their MTS reactivity in the aqueous phase may be somewhat slower than corresponding hydrophilic MTS reagents.

For use in studying protein interactions, these compounds are first reacted with a thiol on a purified bait protein to form a disulfide bond. Since the reagents are water insoluble, they must be dissolved in an organic solvent such as DMF or DMSO and then an aliquot

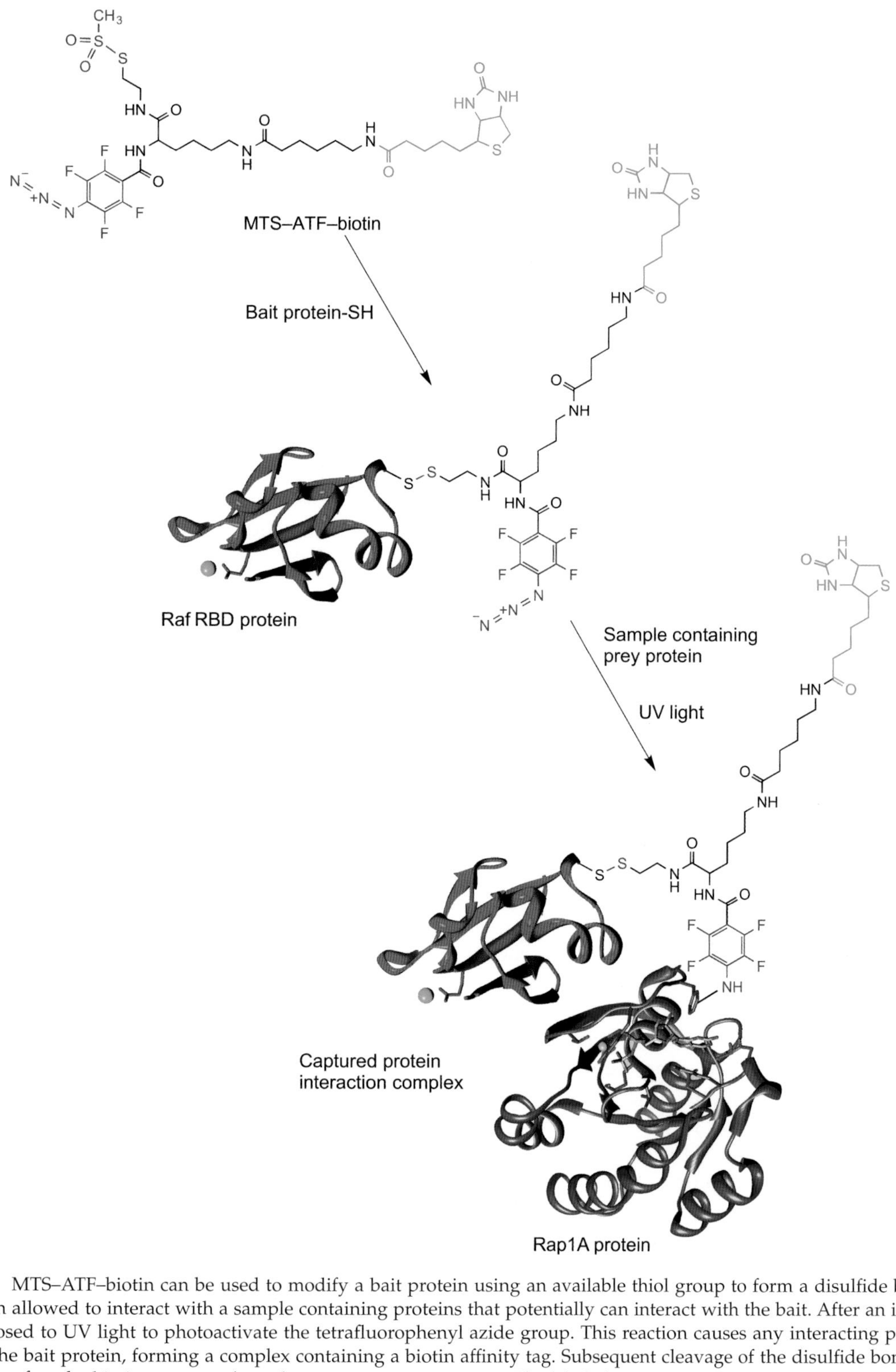

FIGURE 24.16 MTS–ATF–biotin can be used to modify a bait protein using an available thiol group to form a disulfide bond. The labeled bait protein is then allowed to interact with a sample containing proteins that potentially can interact with the bait. After an incubation period, the sample is exposed to UV light to photoactivate the tetrafluorophenyl azide group. This reaction causes any interacting prey proteins to be crosslinked with the bait protein, forming a complex containing a biotin affinity tag. Subsequent cleavage of the disulfide bond to the bait protein using DTT transfers the biotin group to the unknown interacting protein. The labeled prey protein can then be detected or isolated using (strept)avidin reagents.

added to the bait protein in an aqueous buffer to initiate the reaction. Once modified, the biotinylated bait protein is then incubated with a sample containing potential interactive prey proteins. After an incubation period, initiating the photoreaction by exposure to UV irradiation captures the interacting proteins. Any interaction complexes thus formed can be isolated or detected using the biotin handle. In addition, the disulfide bond formed with the bait protein during the crosslinking reaction can be reduced to cleave the conjugates and transfer the biotin label to the unknown interacting prey proteins. This is a powerful way of labeling unknown interacting proteins for subsequent analysis.

MTS–ATF–biotin and MTS–ATF–LC-biotin have been used to study the piccolo NuA4 acetylation of nucleosomal histones (Huang and Tan, 2012), to investigate protein interactions involving the SufBCD Fe-S scaffold complex binding with SufA for Fe-S cluster transfer (Chahal *et al.*, 2009), in the characterization of carriers and receptors of the Lewis^X glycan (Katagihallimath, 2008), and to study the piccolo NuA4-catalyzed acetylation of nucleosomal histones (Huang and Tan, 2012). The following protocol should be used as a suggested starting point to develop optimized methods for studying specific protein interactions. The bait protein to be modified with MTS–ATF–biotin or MTS–ATF–LC-biotin should be purified and contain at least one accessible thiol for reaction with the MTS group. Ellman's reagent (Chapter 2, Section 4.1) may be used to determine if a sulfhydryl is present and able to react with the crosslinkers. A free thiol may also be created by reduction of internal disulfides or through the use of a thiolation reagent. However, adding thiols by chemical derivatization may not be advantageous, because this will create numerous surface modifications, which may have detrimental effects on protein–protein interactions.

Protocol

1. Dissolve a thiol-containing bait protein in 10-m*M* HEPES, 0.15-*M* NaCl, pH 7.5 or in 0.1-*M* sodium phosphate, 0.15-*M* NaCl, pH 7.2. Other buffers around physiological pH will work well; however, avoid the presence of reducing agents containing thiols (i.e., DTT or 2-mercaptoethanol). Typical concentrations that will work with the suggested protocol range from 100 μg/ml to 10 mg/ml, but higher concentrations are preferred to give high reaction yields.
2. Dissolve MTS–ATF–biotin or MTS–ATF–LC–biotin in DMF or DMSO at a concentration of at least 10 mg/ml. Protect from light.
3. Add a quantity of the crosslinker solution to the bait protein solution to provide a 1 to 5 molar excess of the crosslinker over the quantity of protein to be modified. The addition of the organic solution containing the crosslinker should be performed so that the final reaction contains no more than 10% organic solvent. Even lower concentrations of solvent may be required for certain sensitive proteins.
4. React for at least 30 min at room temperature or 1 h at 4°C. Continue to protect the solution from light.
5. Remove excess crosslinker and reaction byproducts by dialysis or size exclusion chromatography.
6. Add the labeled bait protein to a sample containing the putative interacting prey proteins. The quantity of bait protein to be added to a given sample should be within the same concentration level as the amount of prey proteins present. The optimal level of addition may have to be determined by varying the amount of bait protein concentrations in a number of sample solutions to decide which concentration results in the best interaction complexes being formed.
7. Incubate the sample mixture for at least 1 h at room temperature or under conditions optimal for the interaction to be studied.
8. Expose the sample to UV light to photoactivate the aryl azide and cause the conjugation reaction to occur. For best results, use a UV lamp that irradiates in the range of 300 to 370 nm. Irradiate for 2 to 15 min, using briefer exposures when using UV lamps of higher wattage.
9. Interaction complexes that are captured by the crosslinking reaction can be recovered or analyzed using the biotin groups. For instance, immobilized streptavidin can be used to purify the conjugates or a streptavidin–HRP conjugate can be used to probe a western blot after electrophoresis separation of the sample. Reduction of conjugates may be performed by cleavage with DTT of the disulfide bond linking the bait protein. This can be performed using 50-m*M* DTT or 50-m*M* TCEP in aqueous solution to permit recovery of the biotinylated prey proteins on immobilized streptavidin. Alternatively, the use of an affinity support having milder elution characteristics for biotinylated proteins, such as immobilized monomeric avidin, would facilitate isolation of the biotinylated complexes or prey proteins under non-denaturing conditions.

4. METAL CHELATES IN THE STUDY OF PROTEIN INTERACTIONS

Certain bifunctional metal chelating agents have been used to investigate protein interactions by virtue of their ability to generate reactive oxygen species that affect protein structure in the immediate vicinity of their modification site. The following sections discuss two applications of such chelate labels, one of which cleaves peptide bonds while the other one causes covalent crosslinks to occur between interacting protein structures.

4.1. FeBABE for Protein Mapping Studies

Transition metals such as iron can catalyze oxidation reactions in aqueous solution, which are known to cause modification of amino acid side chains and damage to polypeptide backbones (see Chapter 2, Section 1.1) (Halliwell and Gutteridge, 1984; Kim *et al.*, 1985; Tabor and Richardson, 1987). These reactions can oxidize thiols, create aldehydes and other carbonyls on certain amino acids, and even cleave peptide bonds. The purposeful use of metal-catalyzed oxidation in the study of protein interactions has been applied to map interaction surfaces or identify which regions of biomolecules are in contact during specific affinity binding events.

Hydroxyl radical "footprinting," as the technique is called, can produce a picture of where ligand binding sites are located or map the interaction surfaces of protein complexes. An early example of this application was used to map the binding site of protein–DNA interactions (Tullius and Dombroski, 1986; Latham and Cech, 1989). Hydroxyl radicals are generated by an EDTA chelated iron(II) complex using a redox reaction involving hydrogen peroxide and ascorbic acid. In this case, the EDTA chelate is merely added to the solution of the interacting protein and DNA. Nonselective cleavage of the DNA strand (and protein) then occurs everywhere except at the region in which the interaction takes place, thus leaving a "footprint" of the interacting partners.

A second iteration of this reagent involves the attachment of affinity ligands to the EDTA chelator group to direct the hydroxyl radical generator to specific binding pockets on proteins and receptors. Hoyer *et al.* (1990) created a biotin–EDTA derivative that was able to bind to the biotin binding pocket on streptavidin and cause peptide cleavage in the region immediately surrounding the binding site. Similarly, Schepartz and Cuenoud (1990) formed a trifluoroperazine-based affinity reagent containing an EDTA group to map the binding site of the calcium binding protein calmodulin. Once affinity reagents of this sort bind to their receptor protein, hydroxyl radical footprinting can identify exactly where in the protein's three-dimensional structure the interaction took place.

Rana and Meares (1990, 1991) created a more versatile footprinting reagent by synthesizing a bifunctional metal chelating agent with an EDTA group on one end and a bromoacetyl group on the other end. This compound, Fe(III) (S)-1-(*p*-bromoacetamido-benzyl)ethylene diamine tetraacetic acid (or FeBABE, pronounced "iron-babe"), resulted in a universal modification reagent for protein interaction footprinting which could be used to study interacting protein complexes or other biomolecules (Figure 24.17). The bromoacetyl group of FeBABE can react with a thiol to form a thioether bond (Figure 24.18). Thus, proteins or other molecules containing a thiol can be covalently modified with the

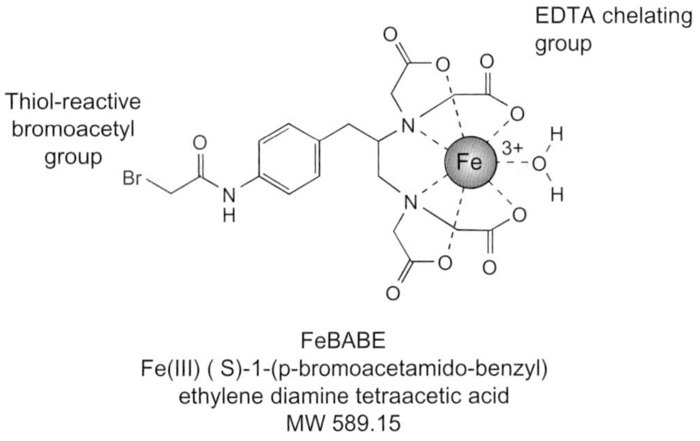

FIGURE 24.17 FeBABE is a bifunctional chelating agent containing an EDTA group on one side and a thiol-reactive bromoacetyl group on the other end. The EDTA group is coordinated with an iron ion.

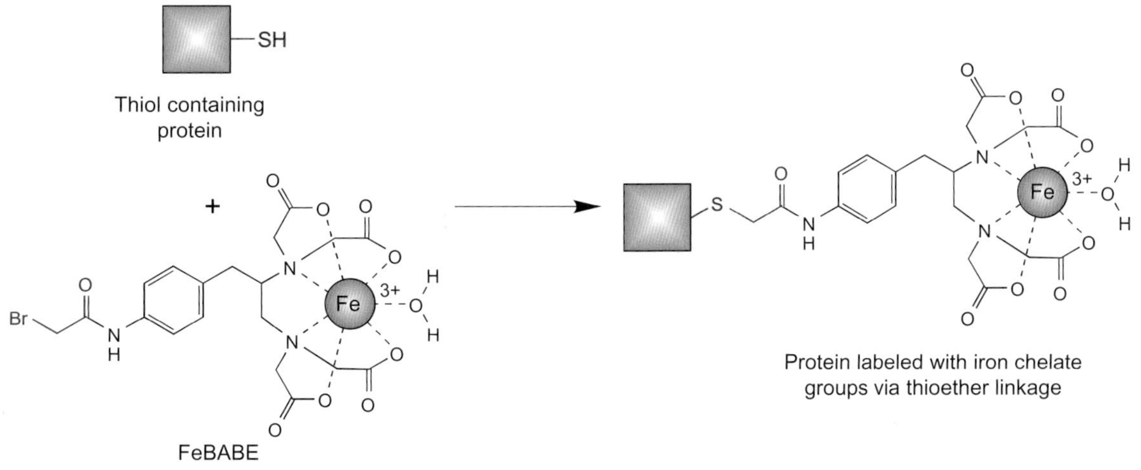

FIGURE 24.18 FeBABE can be coupled to an available thiol group on a bait protein using the bromoacetyl reactive group to form a thioether linkage.

FeBABE reagent forming a hydroxyl radical probe to footprint interacting domains. Such modified bait proteins are termed "artificial proteases" because of their ability to chemically cleave peptide bonds.

The reactions involved in the cleavage of biological molecules by FeBABE and similar iron–EDTA complexes have been theorized to involve two possible processes. Rana and Meares (1991) speculated that the reaction proceeds by the coordination of peroxo oxygen to the iron chelate group. They also found that O_2 might participate in this type of reaction, as evidenced by the successful incorporation of ^{18}O into the carboxyl terminus of cleaved peptide chains of human carbonic anhydrase. This peroxo–metal complex may facilitate subsequent nucleophilic attack of oxygen on the carbonyl carbon of a peptide bond, which then results in cleavage of the peptide bond through oxygen atom transfer to the carbonyl carbon, forming a C-terminal carboxylate. Figure 24.19 schematically shows part of this reaction sequence.

Another proposed mechanism involves the well known Fenton reaction, which is catalyzed effectively

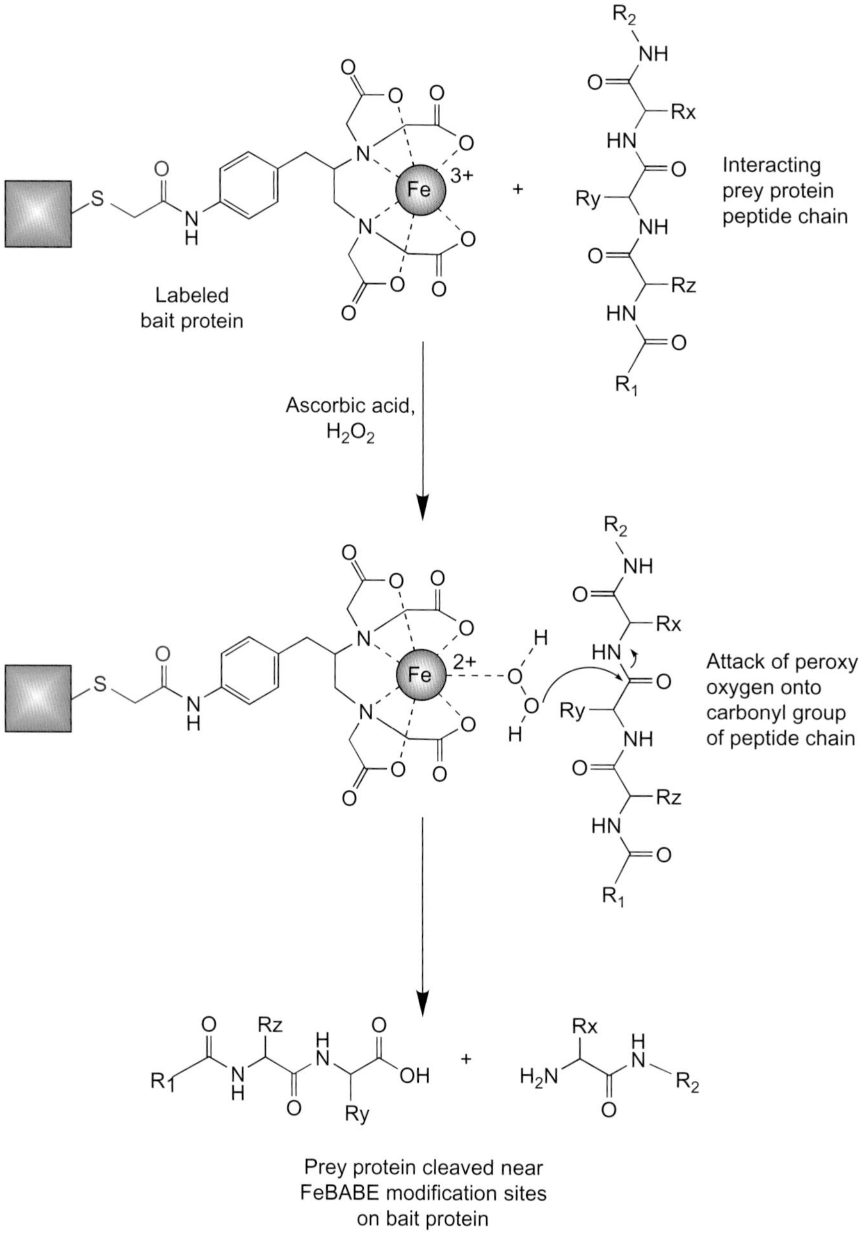

FIGURE 24.19 The cleavage reaction of FeBABE involves a catalytic process using peroxide and ascorbate to form reactive oxygen species. Any protein structure in the immediate vicinity of the FeBABE label on the bait protein will undergo peptide bond cleavage.

by Fe^{2+}–EDTA complexes. In this process, the ascorbate functions by reducing the FeBABE Fe^{3+} complex to Fe^{2+}, producing the ascorbate radical ($Asc^{\cdot-}$). The reduced iron can then form hydroxyl radicals through the Fenton reaction, which re-oxidizes the iron to the Fe^{3+} state. As additional Fe^{3+} is formed, ascorbate then regenerates the active Fe^{2+} chelate by reduction, resulting in the formation of the ascorbate radical. The entire process thus is catalytic and may occur numerous times at the site of one FeBABE modification, potentially resulting in a number of peptide bond fragmentations. Notice also that the FeBABE reagent can be used in an initial EDTA–Fe^{3+} form or in the reduced EDTA–Fe^{2+} form, as addition of the peroxide/ascorbate activators can catalyze the reaction process from either metal oxidation state.

$$\text{EDTA–Fe}^{2+} + H_2O_2 \rightarrow \text{EDTA–Fe}^{3+} + {}^{\cdot}OH + OH^- \quad \text{Fenton reaction}$$

$$\text{EDTA–Fe}^{3} + \text{ascorbate} \rightarrow \text{EDTA–Fe}^{2+} + \text{Asc}^{\cdot-} + H^+ \quad \text{Regeneration with ascorbate}$$

The FeBABE modifying group is approximately 12 Å in length. Peptide bond cleavage can occur anywhere on a protein surface that is within molecular distance of the iron chelate modifying group. Cleavage probably does not occur at distant sites, because the peroxo oxygen or hydroxyl radical intermediates are extremely reactive and will quickly react with the neighboring peptide structure or be quenched by ascorbate in solution. An interacting protein that is within the molecular distance of FeBABE's reach therefore is susceptible to peptide bond cleavage.

In use, a bait protein labeled with FeBABE is bound to an interacting prey protein and the peroxide/ascorbate cleavage reagents added to initiate the protein mapping reaction. After the cleavage process is complete, the peptide fragments on each protein are analyzed to gleam information about the interaction surfaces, which will be left relatively unaffected by the reaction. Obviously, to utilize this technique the two interacting proteins must be highly pure and their amino acid sequences known.

Typically, a bait protein is labeled with FeBABE at any available cysteine thiol group. If no free thiols are available, disulfides may be reduced to create sulfhydryls or a thiolation reagent may be used to add them to the protein surface (see Chapter 2, Section 4.1). Care should be taken when adding thiols through thiolation, however, because too high a substitution level could affect the ability of the bait protein to bind to the prey protein. One to three thiol substitutions on the surface of a bait protein should be sufficient to modify it with FeBABE and study protein interactions. A third option for adding a thiol is to change recombinantly one amino acid in the bait protein's sequence to a cysteine. Site-directed mutagenesis can create a thiol at a known region of the protein's surface, thus permitting the study of protein interactions through rational design of FeBABE modifications.

In order to facilitate analysis of FeBABE-produced fragments, the prey protein or biomolecule is labeled at one end with a tag that can be detected after electrophoresis, usually in a transfer blot. The tag can be a fusion tag, such as 6xHis, or any other group that can be targeted with an antibody and detected. Alternatively, radiolabels and fluorescent labels have been used with prey molecules, including the use of end-labeled DNA to study where DNA binding proteins dock onto the oligonucleotide sequence.

The fragments formed by FeBABE fragmentation are analyzed by comparing them to enzymatic or chemical cleavage patterns observed by treatment on the same prey protein. Since the cleaved prey protein is detected by its end-labeled tag, the only fragments detected are those that extend from the end labeled terminal to the point of cleavage. Comparison with known fragmentation patterns using protease digestion or chemical cleavage provides information as to the approximate size of the FeBABE fragment and its point of cleavage.

FeBABE has been used to study protein interactions involving the RNA polymerase RbpA (Hu *et al.*, 2012), in mapping protein interactions of the ATP-dependent chromatin remodelers (Hota *et al.*, 2012), to examine the binding of the Swi2/Snf2 remodeler protein Mot1 in complex with its substrate TBP (Wollmann *et al.*, 2011), and to identify the point of interaction between transcriptional factor SoxS and a subunit of RNA polymerase (Zafar *et al.*, 2011).

The following protocol assumes that the user has at least two pure proteins (or biomolecules) of known sequences which are able to interact specifically in solution. One of the two proteins (the prey) is end labeled with a fusion tag or another detectable component and the other protein contains at least one thiol group. All buffers and reagents used in this protocol should be of high purity and contain a very low metal content to prevent nonspecific cleavage reactions.

Protocol

This protocol is based on the procedure of Rana and Meares (1991) and the instruction booklet provided by Thermo Fisher for use with the FeBABE kit.

1. Dissolve the bait protein to be modified with the FeBABE reagent in 50-m*M* MOPS, 100-m*M* NaCl, 1-m*M* EDTA, 5% glycerol, pH 8.2 (coupling buffer).

The protein first may be treated by dialysis with EDTA to remove residual metals (30-m*M* MOPS, 4-m*M* EDTA, pH 8.2), and then dialyzed into the final coupling buffer for the reaction with FeBABE.

2. Dissolve FeBABE in DMSO at a concentration of 5 mg/ml (8.5 nmol/μl) and add an aliquot of this solution to the thiol-containing bait protein solution to obtain a 2- to 20-fold molar excess of the reagent over the protein concentration. Protect the FeBABE solution and the reaction mixture from light, as exposure to light will degrade the bromoacetyl reactive group. The optimal molar excess of FeBABE over the bait protein may have to be determined by experimentation. Since thiols are typically limited in number in most proteins, an excess of FeBABE reagent will not result in more labels than the total number of thiols present, provided the suggested reaction conditions are maintained.
3. Incubate the reaction mixture at room temperature or 37°C for 1 h.
4. Purify the modified bait protein from excess reactants by desalting or dialysis. During this process buffer exchange the labeled bait protein into 50-m*M* MOPS, 120-m*M* NaCl, 0.1-m*M* EDTA, 10-m*M* $MgCl_2$, 10% glycerol, pH 8.0. Components may be added or removed from this buffer to promote the protein interaction being studied as long as they do not interfere with the subsequent FeBABE cleavage reaction (i.e., they do not contain transition metals like iron or copper, oxidants, or reducing agents).
5. Dissolve or buffer exchange the prey protein into 50-m*M* MOPS, 120-m*M* NaCl, 0.1-m*M* EDTA, 10-m*M* $MgCl_2$, 10% glycerol, pH 8.0.
6. Mix the labeled bait protein with the prey protein at equal molar ratios and incubate for 30 min at room temperature to create protein interaction complexes.
7. Initiate the cleavage reaction by the addition of an 8-fold molar excess of ascorbate (over the concentration of protein) followed by an 8-fold molar excess of H_2O_2. Mix well.
8. React for 10 to 30s at room temperature.
9. Quench the cleavage reaction by the addition of an equal volume of SDS electrophoresis sample buffer containing up to 40% glycerol. The SDS will denature and disrupt the protein interaction and glycerol acts as a free radical scavenger, thus effectively quenching the reaction.
10. Analyze the cleaved proteins by SDS gel electrophoresis and western blotting, followed by specific detection of the prey protein label. Comparison of prey protein fragments formed by chemical and enzymatic digestion can yield information as to the site of the resultant FeBABE-mediated cleavage.

4.2. Ru(II)bpy$_3^{2+}$ for Light-Triggered, Zero-Length Crosslinking of Protein Interactions

In many instances within cells or on cell membrane surfaces, large protein complexes are formed containing many different proteins specifically interacting to effect biological processes. Protein machines of this type are difficult to capture and study using traditional cross-linking agents, because the resultant reactions may link some of the proteins, but not all of them in the complex (Friedrichson and Kurzchalia, 1998; Simons *et al.*, 1999). In addition, covalent modification of many nucleophilic groups in a protein by the electrophilic reactive groups of standard crosslinking reagents can decorate the surface of proteins with these compounds and affect non-specific protein interactions.

Fancy and Kodadek (1999) developed a novel protein crosslinking process using a metal chelating reagent that can be activated by visible light and effect zero-length crosslinks between proteins. Photo-induced crosslinking of unmodified proteins (PICUP) can capture protein interactions without the use of modifying agents such as organic crosslinkers or fusion tags, thus maintaining the interacting proteins in their native structure and conformation. The reaction is mediated by the chelate ruthenium(II) tris(bipyridine) (Ru(II)bpy$_3^{2+}$; pronounced "rūbippy"; see Figure 24.20) in a two-step reaction involving light absorption by the bipyridine groups followed by energy transfer and excitation of the Ru(II) metal to an excited state intermediate. Subsequent reaction of the excited-state Ru(II)bpy$_3^{2+}$ with persulfate anion results in the formation of a sulfate radical and Ru(III). The Ru(III) group is then postulated to undergo a one-electron transfer to tyrosine side-chain phenolic groups in proteins, resulting in the formation of a tyrosine radical. This highly reactive intermediate can rapidly react with neighboring nucleophilic groups in proteins or with another nearby tyrosine phenolic group, resulting in the formation of tyrosine dimers or a tyrosine heteroatom arene linkage with nearby

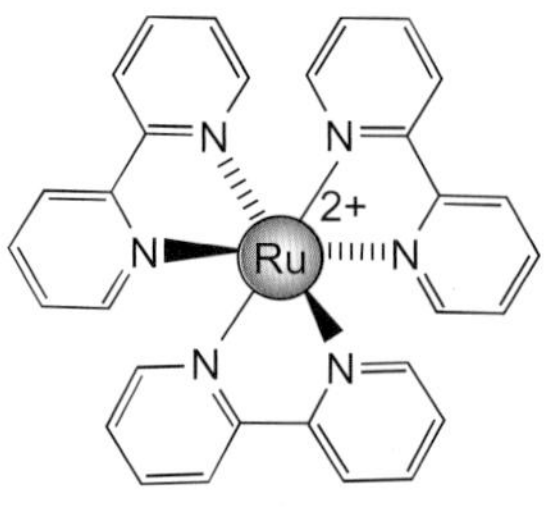

Ruthenium(II) tris(bipyridine) chelate
Ru(II)bpy$_3^{2+}$
MW 569.62

FIGURE 24.20 The structure of Ru(II)pby$_3^{2+}$ a chelated ruthenium ion coordinated with three bipyridine groups.

nucleophilic groups. Either route of reaction can result in the formation of covalent bonds between protein amino acids. If the linkages are formed between interacting proteins, then these complexes are captured for subsequent analysis. The proposed reactions of $Ru(II)bpy_3^{2+}$ in the light-mediated zero-length crosslinking of interacting proteins are shown in Figure 24.21.

The following protocol was successfully used to study the interactions of the 180-amino acid C-terminal domain of yeast TATA box binding protein (TBP), the Gal80 protein, and the Gal4 activation domain of the associated transcription factor complex (Fancy and Kodadek, 1999). It was also used for the study of the amyloid β-protein (αβ) assembly (Bitan *et al.*, 2003), to characterize the tRNA-specific adenosine deaminase (tadA) from *E. coli* (Wolf *et al.*, 2002) and to understand the nature of the interaction between MIP-1a and proteoglycans (Ottersbach and Graham, 2001). The reaction

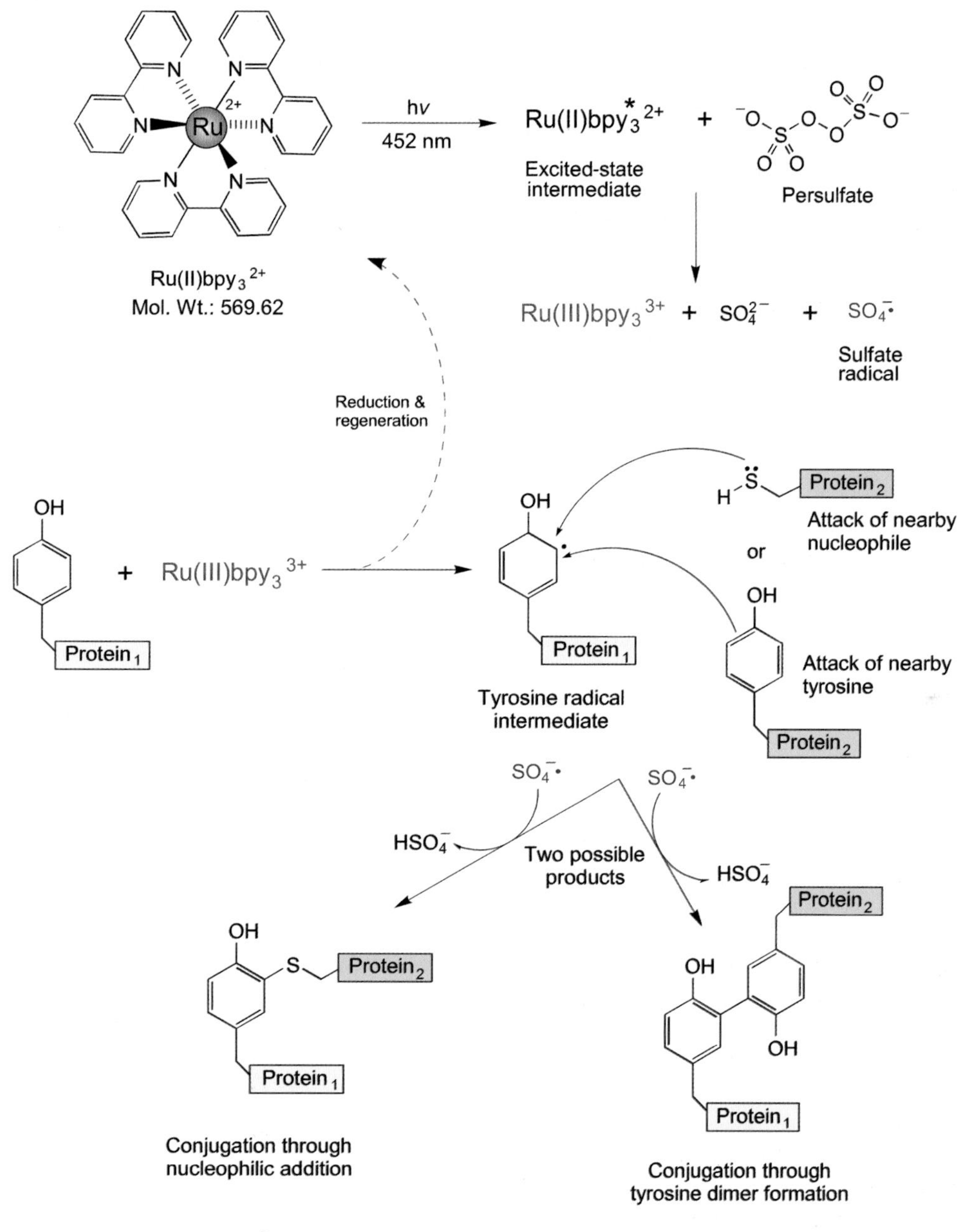

FIGURE 24.21 The reactions of $Ru(II)pby_3^{2+}$ are catalyzed by light at 452 nm that begins by forming an excited state intermediate. In the presence of persulfate, a sulfate radical is formed concomitant with the oxidative product $Ru(III)bpy_3^{3+}$. This form of the chelate is able to catalyze the formation of a radical on a tyrosine phenolic ring that can react along with the sulfate radical either with a nucleophile, such as a cysteine thiol, or with another tyrosine side chain to form a covalent linkage. The result of this reaction cascade is to cause protein crosslinks to form when a sample containing these components is irradiated with light.

is very efficient and typically can result in conjugation yields of 60% or better between interacting proteins. In fact, the conjugation efficiency may yield crosslinks between proteins that are not specifically interacting. To limit the degree of extraneous conjugation, the addition of a nucleophile in the reaction medium (e.g., histidine) can help to restrict crosslinks only to proteins that are in close contact, as in those that are forming interaction complexes at the time.

Protocol

1. Adjust the protein sample concentration to be between 0.01 μ*M* and 20 μ*M* by dilution in 20 μl of reaction buffer: 15-m*M* sodium phosphate, 150-m*M* NaCl, 0.125-m*M* $Ru(bpy)_3Cl_2$ (pH 7.5). The protein sample should not contain any components that are easily oxidized by the reaction, such as thiol reducing agents (e.g., DTT). The addition of histidine to the reaction buffer (7.5-m*M*) can be performed to modulate the crosslinking reaction and limit the degree of protein conjugation between non-interacting proteins. The presence of other nucleophiles will also inhibit the crosslinking reaction, such as Tris buffer or DTT.
2. Add an aliquot of ammonium persulfate (APS) to make a final concentration of 2.5-m*M*. Protect from light until ready to start the reaction.
3. Illuminate the sample in a small open tube or the well of a microplate by placing a white light source approximately 5 to 50 cm from the sample. If a weak light source is used, such as a flashlight, move the light source closer to the sample, but if a strong photoflood lamp is used, then use the greater distance measurement to avoid heating the sample. Illuminate the sample for 0.5 to 30s, depending again on the intensity of the light source.
4. Immediately after irradiation, stop the reaction by the addition of 7 μl of 4× SDS electrophoresis loading buffer or the equivalent (with a high concentration of reducing agent present): 0.2-*M* Tris, 8% SDS, 2.88-*M* β-mercaptoethanol, 40% glycerol, 0.4% xylene cyanol, 0.4% bromophenol blue. Heat the sample at 95°C for 5 min and analyze the complexes formed by electrophoresis.

References

Aarntzen, E.H., de Vries, I.J.M., Göertz, J.H., Beldhuis–Valkis, M., Brouwers, H.M., van de Rakt, M.W., et al., 2012. Humoral anti-KLH responses in cancer patients treated with dendritic cell-based immunotherapy are dictated by different vaccination parameters. Cancer Immunol. Immunother., 1–9.

Abate-Pella, D., Zeliadt, N.A., Ochocki, J.D., Warmka, J.K., Dore, T.M., Blank, D.A., et al., 2012. Photochemical modulation of ras-mediated signal transduction using caged farnesyltransferase inhibitors: activation by one-and two-photon excitation. ChemBioChem 13 (7), 1009–1016.

Abbas, A.K., Lichtman, A.H., Pillai, S., 2011. Cellular and Molecular Immunology, seventh ed., Elsevier Saunders, ISBN: 978-1-4377-1528-6.

Abdella, R.M., Smith, P.K., Royer, G.P., 1979. A new cleavable reagent for cross-linking and reversible immobilization of proteins. Biochem. Biophys. Res. Comm. 87, 734–742.

Abdelrahman, A.I., 2011. Lanthanide-encoded polystyrene microspheres for mass cytometry-based bioassays. Doctoral dissertation. University of Toronto.

Abdullah, N., Chase, H.A., 2005. Removal of poly-histidine fusion tags from recombinant proteins puriWed by expanded bed adsorption. Biotechnol. Bioeng. 92, 501–513.

Abian, O., Grazu, V., Hermoso, J., Gonzalez, R., Garcia, J.L., Fernandez-Lafuente, R., et al., 2004. Stabilization of penicillin G acylase from Escherichia coli: site-directed mutagenesis of the protein surface to increase multipoint covalent attachment. Appl. Environ. Microbiol. 70, 1249–1251.

Abonnenc, M., Mayr, M., 2012. Proteomics of atherosclerosis. Inflamm. Atheroscler., 249–266.

Abou-Samra, A.B., Freeman, M., Juppner, H., Uneno, S., Segre, G.V., 1990. Characterization of fully active biotinylated parathyroid hormone analogs. Applications to fluorescence activated cell sorting of parathyroid hormone receptor bearing cells. J. Biol. Chem. 265, 58–62.

Abraham, R., Moller, D., Gabel, D., Senter, P., Hellström, I., Hellström, K.E., 1991. The influence of periodate oxidation on monoclonal antibody avidity and immunoreactivity. J. Immunol. Methods 144 (1), 77–86.

Abuchowski, A., Kazo, G.M., Verhoest Jr., C.R., van Es, T., Kafkewitz, D., Nucci, M.L., et al., 1984. Cancer therapy with chemically modified enzymes. Anti-tumor properties of polyethylene glycol asparaginase conjugates. Cancer Biochem. Biophys. 7, 175–186.

Abuchowski, A., McCoy, J.R., Palczuk, N.C., van Es, T., Davis, F.F., 1977a. Effect of covalent attachment of polyethylene glycol on immunogenicity and circulating life of bovine liver catalase. J. Biol. Chem. 252, 3582–3586.

Abuchowski, A., van Es, T., Palczuk, N.C., Davis, F.F., 1977b. Alteration of immunological properties of bovine serum albumin by covalent attachment of polyethylene glycol. J. Biol. Chem. 252, 3578–3581.

Abuelyaman, A.S., Hudig, D., Woodard, S.L., Powers, J.C., 1994. Fluorescent derivatives of diphenyl [1-(N-Peptidylamino)alkyl] phosphonate esters: synthesis and use in the inhibition and cellular localization of serine proteases. Bioconjug. Chem. 5 (5), 400–405.

Aburai, K., Yagi, N., Yokoyama, Y., Okuno, H., Sakai, K., Sakai, H., et al., 2011. Preparation of liposomes modified with lipopeptides using a supercritical carbon dioxide reverse-phase evaporation method. J. Oleo Sci. 60 (5), 209–215.

Acharya, A.S., Manning, J.M., 1980. Reactivity of the amino groups of carbonmonoxyhemoglobin S with glyceraldehyde. J. Biol. Chem. 255, 1406–1412.

Adamczyk, M., Gebler, J.C., Mattingly, P.G., 1996. Characterization of protein–hapten conjugates. 2. Electrospray mass spectrometry of BSA–hapten conjugates. Bioconjug. Chem. 7, 475–481.

Adams, C.A., Kar, S.R., Hopper, J.E., Fried, M.G., 2004. Self-association of the amino-terminal domain of the yeast TATA-binding protein. J. Biol. Chem. 279, 1376–1382.

Adams, P.C., Powell, L.W., Halliday, J.W., 1988. Isolation of a human hepatic ferritin receptor. Hepatology 8 (4), 719–721.

Adams, R., Bachman, W.E., Fieser, L.F., Johnson, J.R., Snyder, H.R., 1942. Organic Reactions, vol. 1. Wiley, New York. p. 303.

Addy, M., Losey, B., Mohseni, R., Zlotnikov, E., Vasiliev, A., 2012. Adsorption of heavy metal ions on mesoporous silica-modified montmorillonite containing a grafted chelate ligand. Appl. Clay Sci. 59, 115–120.

Adessi, C., Matton, G., Ayala, G., Turcatti, G., Mermod, J.-J., Mayer, P., et al., 2000. Solid phase DNA amplification: characterisation of primer attachment and amplification mechanisms. Nucleic Acids Res. 28, e87.

Adolfson, R., Moudrianokis, E.N., 1976. Molecular polymorphism and mechanisms of activation and deactivation of the hydrolytic function of the coupling factor of oxidative phosphorylation. Biochemistry 15, 4164–4170.

Aebersold, R., 2003. Quantitative proteome analysis: methods and applications. J. Infect. Dis. 187 (Suppl. 2), S315–S320.

Agard, N.J., Prescher, J.A., Bertozzi, C.R., 2004. A strain-promoted [312] azide-alkyne cycloaddition for covalent modification of biomolecules in living systems. J. Am. Chem. Soc. 126, 15046–15047.

Agard, N.J., Baskin, J.M., Prescher, J.A., Lo, A., Bertozzi, C.R., 2006. A comparative study of bioorthogonal reactions with azides. ACS Chem. Biol. 1 (10) A–F. published online: November 10, 2006.

Agata, H., Yamazaki, M., Uehara, M., Hori, A., Sumita, Y., Tojo, A., et al., 2012. Characteristic differences among osteogenic cell populations of rat bone marrow stromal cells isolated from untreated, hemolyzed or Ficoll-treated marrow. Cytotherapy 14 (7), 791–801.

Agresti, J.J., Kelly, B.T., Jäschke, A., Griffiths, A.D., 2005. Selection of ribozymes that catalyse multiple-turnover Diels–Alder cycloadditions by using in vitro compartmentalization. Proc. Natl. Acad. Sci. USA 102, 16170–16175.

Aguirre, R., Gonsoulin, F., Cheung, H.C., 1986. Interaction of fluorescently labelled myosin subfragment 1 with nucleotide and actin. Biochemistry 25, 6827.

Ahamed, J., Versteeg, H.H., Kerver, M., Chen, V.M., Mueller, B.M., Hogg, P.J., et al., 2006. Disulfide isomerization switches tissue factor from coagulation to cell signaling. Proc. Natl. Acad. Sci. 103, 13932–13937.

Ahmad, H., Miah, M.A.J., Rahman, M.M., 2003. Preparation of micron-sized composite polymer particles containing hydrophilic 2-hydroxyethyl methacrylate and their biomedical applications. Colloid Polym. Sci. 281, 988–992.

Ahmad, S., Gromiha, M.M., Sarai, A., 2003. Real value prediction of solvent accessibility from amino acid sequence. Proteins 50 (4), 629–635.

Ahn, B., Rhee, S.G., Stadtman, E.R., 1987. Use of fluorescein hydrazide and fluorescein thiosemicarbazide reagents for the fluorometric determination of protein carbonyl groups and for the detection of oxidized proteins on polyacrylamide gels. Anal. Biochem. 161, 245–257.

Ahn, J., Shin, Y.B., Chang, W.S., Kim, M.G., 2011. Sequential patterning of two fluorescent streptavidins assisted by photoactivatable biotin on an aminodextran-coated surface. Colloids Surf. B Biointerfaces 87 (1), 67–72.

Aime, S., Anelli, P.L., Botta, M., Fedeli, F., Grandi, M., Paoli, P., et al., 1992. Synthesis, characterization, and l/T1 NMRD profiles of gadolinium (III) complexes of monoamide derivatives of DOTA-like ligands. X-ray structure of the 10-[2-[[2-hydroxy-1-(hydroxylmethyl)ethyl]amino]-1-[phenylmethoxy)methyl]-2-oxo-ethyl]-1,4,7,10-tetraaza-cyclododecane-1,4,7-triacetic acid-gadolinium (III) complex. Inorg. Chem. 31, 2422–2428.

Aithal, H.N., Knigge, K.M., Kartha, S., Czyewski, E.A., Toback, F.G., 1988. An alternate method utilizing small quantities of ligand for affinity purification of monospecific antibodies. J. Immunol. Methods 112, 63–70.

Aizawa, M., Ikariyama, M., Kobatake, E., Ogasawara, M., Tanaka, M., 2006. Luminescence by reacting an acridinium ester with superoxide. United States Patent RE39047.

Ajtai, K., 1992. Stereospecific reaction of muscle fiber proteins with the 59 or 69 iodoacetamido derivative of tetramethylrhodamine: only the 69 isomer is mobile on the surface of S1. Biophys. J. 61, A278, Abstract 1647.

Akgöl, S., Bayramoğlu, G., Kacar, Y., Denizli, A., Yakup Arıca, M., 2002. Poly(hydroxyethyl methacrylate-co-glycidyl methacrylate) reactive membrane utilized for cholesterol oxidase immobilization. Polym. Int. 51, 1316–1322.

Akin, M., Bongartz, R., Walter, J.G., Demirkol, D.O., Stahl, F., Timur, S., et al., 2012. PAMAM-functionalized water soluble quantum dots for cancer cell targeting. J. Mater. Chem. 22, 11529–11536.

Alagon, A.C., King, T.P., 1980. Activation of polysaccharides with 2-iminothiolane and its uses. Biochemistry 19, 4341–4345.

Alahakoon, T., Koh, J.W., Chong, X.W.C., Lim, W.T.L., 2012. Immobilization of cellulases on amine and aldehyde functionalized Fe_2O_3 magnetic nanoparticles. Prep. Biochem. Biotechnol. 42 (3), 234–248.

Alam, F., Soloway, A.H., Barth, R.F., Mafune, N., Adams, D.M., Knoth, W.H., 1989. Boron neutron capture therapy: linkage of a boronated macromolecule to monoclonal antibodies directed against tumor-associated antigens. J. Med. Chem. 32, 2326–2330.

Albayrak, N., Yang, S.-T., 2002. Immobilization of Aspergillus oryzae β-galactosidase on tosylated cotton cloth. Enzyme Microb. Technol. 31, 371–383.

Albertazzi, L., Brondi, M., Pavan, G.M., Sato, S.S., Signore, G., Storti, B., et al., 2011. Dendrimer-based fluorescent indicators: in vitro and in vivo applications. PLoS ONE 6 (12), e28450.10.1371/journal.pone.0028450.

Alberts, B., Johnson, A., Lewis, J., Raff, M., Roberts, K., Walters, P., 2002. The Shape and Structure of Proteins, Molecular Biology of the Cell, fourth ed., Garland Science, New York and London.

Albrecht, K., Moeller, M., Groll, J., 2010. Nano- and microgels through addition reactions of functional oligomers and polymersPich, A.Richtering, W. Chemical Design of Responsive Microgels, Advances in Polymer Science, vol. 234. Springer-Verlag, Berlin Heidelberg, ISBN: 2010936633.

Albrecht, R.M., Goodman, S.L., Simmons, S.R., 1989. Distribution and movement of membrane-associated platelet glycoproteins: use of colloidal gold with correlative video-enhanced light microscopy, low-voltage high-resolution scanning electron microscopy, and high-voltage transmission electron microscopy. Am. Anat. 185, 149–164.

Alderson, R.F., Toki, B.E., Roberge, M., Geng, W., Basler, J., Chin, R., et al., 2006. Characterization of a CC49-Based single-chain fragment-â-lactamase fusion protein for antibody-directed enzyme prodrug therapy (ADEPT). Bioconjug. Chem. 17, 410–418.

Alexander, J.P., Cravatt, B.F., 2006. The putative endocannabinoid transport blocker LY2183240 is a potent inhibitor of FAAH and several other brain serine hydrolases. J. Am. Chem. Soc. 128, 9699–9704.

Alexander, P., 1954. The reactions of carcinogens with macromolecules. Adv. Cancer Res. 2, 1.

Alexander, S., Gorboulev, V., Gorbunov, D., Keller, T., Volk, C., Schmitt, B.M., et al., 2007. Identification of cysteines in rat organic cation transporters rOCT1 (C322,C451) and rOCT2 (C451) critical for transport activity and substrate affinity. Am. J. Physiol. Renal Physiol. web preprint, June 13, 2007.

Aliosman, F., Caughlan, J., Gray, G.S., 1989. Diseased DNA intrastrand cross-linking and cytotoxicity induced in human brain tumor cells by 1,3-bis(2-chloroethyl)-1-nitrosourea after in vitro reaction with glutathione. Cancer Res. 49, 5954.

Alivisatos, A.P., 1996. Semiconductor clusters, nanocrystals, and quantum dots. Science 271, 933–937.

Allen, T.M., Chonn, A., 1987. Large unilamellar liposomes with low uptake into the reticuloendothelial system. FEBS Lett. 223, 42–46.

Allerson, C.R., Martinez, A., Yikilmaz, E., Rouault, T.A., 2003. A high-capacity RNA affinity column for the purification of human IRP1 and IRP2 overexpressed in Pichia pastoris. RNA 9, 364.

Alley, S.C., Ishmael, F.T., Jones, A.D., Benkovic, S.J., 2000. Mapping protein–protein interactions in the bacteriophage T4 DNA polymerase holoenzyme using a novel trifunctional photo-cross-linking and affinity reagent. J. Am. Chem. Soc. 122, 6126–6127.

Allison, A.C., Gregoriadis, G., 1974. Liposomes as immunological adjuvants. Nature (London) 252, 252.

Allmer, K., Hilborn, J., Larsson, P.H., Hult, A., Ranby, B., 1990. Surface modification of polymers. V. Biomaterial applications. J. Polym. Sci. Part A Polym. Chem. 28, 173–183.

Alston, K., Robinson, R.C., Park, S.S., Gelboin, H.V., Friedman, F.K., 1991. Interactions among cytochromes P-450 in the endoplasmic reticulum. Detection of chemically cross-linked complexes with monoclonal antibodies. J. Biol. Chem. 266, 735–739.

Álvaro, M., Atienzar, P., Bourdelande, J.L., García, H., 2004. An organically modified single wall carbon nanotube containing a pyrene chromophore: fluorescence and diffuse reflectance laser flash photolysis study. Chem. Phys. Lett. 384, 119–123.

Alvear, M., Jabalquinto, A.M., Cardemil, E., 1989. Inactivation of chicken liver mevalonate 5-diphosphate decarboxylase by sulfhydryl-directed reagents: evidence of a functional dithiol. Biochim. Biophys. Acta 994, 7–11.

Alving, C.R., 1987. Liposomes as carriers for vaccines. In: Ostro, M.J. (Ed.), Liposomes from Biophysics to Therapeutics. Dekker, New York, pp. 195–218.

Amato, R.J., 2008. Heat shock protein–peptide complex-96 (Vitespen) for the treatment of cancer. Oncol. Rev. 2 (1), 29–35.

Amessou, M., Fradagrada, A., Falguières, T., Lord, J.M., Smith, D.C., Roberts, L.M., et al., 2007. Syntaxin 16 and syntaxin 5 are required for efficient retrograde transport of several exogenous and endogenous cargo proteins. J. Cell Sci. 120, 1457–1468.

Amidi, M., de Raad, M., Crommelin, D.J., Hennink, W.E., Mastrobattista, E., 2011. Antigen-expressing immunostimulatory liposomes as a genetically programmable synthetic vaccine. Syst Synth Biol 5 (1), 21–31.

Amstad, P.A., Johnson, G.L., Lee, B.W., Dhawan, S., 2000. An in situ marker for the detection of activated caspases. Biotechnol. Lab. 18, 52–56.

An, K.H., Jeon, K.K., Moon, J.-M., Eum, S.J., Yang, C.W., Park, G.-S., et al., 2004. Transformation of singlewalled carbon nanotubes to multiwalled carbon nanotubes and onion-like structures by nitric acid treatment. Synth. Met. 140, 1–8.

Anderson, G.W., 1958. N,N′-Carbonyldiimidazole, a new reagent for peptide synthesis. J. Am. Chem. Soc. 80, 4323.

Anderson, P.W., Pichichero, M.E., Stein, E.C., Porcelli, S., Betts, R.F., Connuck, D.M., et al., 1989. Effect of oligosaccharide chain length, exposed terminal group, and hapten loading on the antibody response of human adults and infants to vaccines consisting of *Haemophilus influenzae* type b capsular antigen uniterminally coupled to the diphtheria protein CRMI97. J. Immunol. 142, 2464–2468.

Anderson, W.L., Wetlaufer, D.B., 1975. A new method for disulfide analysis of peptides. Anal. Biochem. 67, 493–502.

Andersson, M.A., Oscarson, S., Öhberg, L., 1993. Synthesis of oligosaccharides with oligoethylene glycol spacers and their conversion into glycoconjugates using N,N,N',N'-tetramethyl (succinimido)uronium tetrafluoroborate as a coupling reagent. Glycoconj. J. 10, 461–465.

Andersson, T., Nilsson, K., Sundahl, M., Westman, G., Wennerström, O., 1992. C60 embedded in c-cyclodextrin: a water-soluble fullerene. J. Chem. Soc. Comm., 604–606.

Ando, T., 1984. Fluorescence of fluorescein attached to myosin SH1 distinguishes the rigor state from the actin–myosin–nucleotide state. Biochemistry 23, 375.

André, S., Ortega, P.J.C., Perez, M.A., Roy, R., Gabius, H.J., 1999. Lactose-containing starburst dendrimers: influence of dendrimer generation and binding-site orientation of receptors (plant/animal lectins and immunoglobulins) on binding properties. Glycobiology 9 (11), 1253–1261.

Anikeeva, N., Lebedeva, T., Clapp, A.R., Goldman, E.R., Dustin, M.L., Mattoussi, H., et al., 2006. Quantum dot/peptide-MHC biosensors reveal strong CD8-dependent cooperation between self and viral antigens that augment the T cell response. Proc. Natl. Acad. Sci. USA 103 (45), 16846–16851.

Anin, M.F., Gaucheron, F., Leng, M., 1992. Lability of mono functional *cis* platinum adducts: role of DNA double helix. Nucleic Acids Res. 20, 4825–4830.

Anjaneyulu, P.S.R., Staros, J.V., 1987. Reactions of N-hydroxysulfosuccinimide active esters. Int. J. Pept. Protein Res. 30, 117–124.

Annunziato, M.E., Patel, U.S., Ranade, M., Palumbo, P.S., 1993. *p*-Maleimidophenyl isocyanate: a novel heterobifunctional linker for hydroxyl to thiol coupling. Bioconjug. Chem. 4, 212–218.

Anselme, M.J., Tedder, D.W., 1987. Characteristics of immobilized yeast reactors producing ethanol from glucose. Biotechnol. Bioeng. 30, 736–745.

Anthopoulos, T.D., Tanase, C., Setayesh, S., Meijer, E.J., Hummelen, J.C., Blom, P.W.M., et al., 2004. Ambipolar organic field-effect transistors based on a solution-processed methanofullerene. Adv. Mater. 16 (23–24), 2174–2179.

Antimisiaris, S.C., Jayasekera, P., Gregoriadis, C., 1993. Liposomes are vaccine carriers: incorpora-tion of soluble and particulate antigens in giant vesicles. J. Immunol. Methods 166, 271–280.

Aoi, K., Itoh, K., Okada, M., 1995. Globular carbohydrate macromolecule "sugar balls". 1. Synthesis of novel sugar-persubstituted poly(amido amine) dendrimers. Macromolecules 28, 5391–5393.

Aplin, J.D., Hey, N.A., Li, T.C., 2011. MUC1 as a cell surface and secretory component of endometrial epithelium: reduced levels in recurrent miscarriage. Am. J. Reprod. Immunol. 35 (3), 261–266.

Appel, E., Rabinkov, A., Neeman, M., Kohen, F., Mirelman, D., 2011. Conjugates of daidzein-alliinase as a targeted pro-drug enzyme system against ovarian carcinoma. J. Drug Target. 19 (5), 326–335.

Apple, R.J., Domen, P.L., Muckerheide, A., Michael, J.C., 1988. Cationization of protein antigens. IV. Increased antigen uptake by antigen-presenting cells. J. Immunol. 140, 3290–3295.

Arazawa, D.T., Oh, H., Ye, S.-H., Johnson Jr., C.A., Woolley, J.R., Wagner, W.R., et al., 2012. Immobilized carbonic anhydrase on hollow fiber membranes accelerates CO_2 removal from blood. J. Memb. Sci. 1, 25–31.

Archakov, A.I., Govorun, V.M., Dubanov, A.V., Ivanov, Y.D., Veselovsky, A.V., Lewi, P., et al., 2003. Protein–protein interactions as a target for drugs in proteomics. Proteomics 3 (4), 380–391.

Arima, H., Arizono, M., Higashi, T., Yoshimatsu, A., Ikeda, H., Motoyama, K., et al., 2012. Potential use of folate-polyethylene glycol (PEG)-appended dendrimer (G3) conjugate with α-cyclodextrin as DNA carriers to tumor cells. Cancer Gene Ther. 19, 358–366.

Armstrong, B., et al. 2000. Suspension arrays for high throughput, multiplexed single nucleotide polymorphism genotyping. Cytometry 40, 102–108.

Arnau, J., Lauritzen, C., Petersen, G.E., Pedersen, J., 2006. Current strategies for the use of affinity tags and tag removal for the purification of recombinant proteins. Protein Expr. Purif. 48, 1–13.

Arnaud, N., Georges, J., 2003. Comprehensive study of the luminescent properties and lifetimes of Eu(31) and Tb(31) chelated with various ligands in aqueous solutions: influence of the synergic agent, the surfactant and the energy level of the ligand triplet. Spectrochim. Acta A Mol. Biomol. Spectrosc. 59 (8), 1829–1840.

Arriagada, F.J., Osseo-Asare, K., 1995. Synthesis of nanosize silica in aerosol OT reverse microemulsions. J. Colloid Interface Sci. 170, 8–17.

Arroyo, I.J., Hu, R., Merino, G., Tang, B.Z., Peña-Cabrera, E., 2009. The smallest and one of the brightest. Efficient preparation and optical description of the parent borondipyrromethene system. J. Org. Chem. 74 (15), 5719–5722.

Arrua, R.D., Talebi, M., Causon, T.J., Hilder, E.F., 2012. Review of recent advances in the preparation of organic polymer monoliths for liquid chromatography of large molecules. Anal. Chim. Acta 738, 1–12.

Arshady, R., 1993. Micro-spheres for biomedical applications: preparation of reactive and labeled micro-spheres. Biomaterials 14, 5–15.

Arvidsson, P., Plieva, F.M., Savina, I.N., Lozinsky, V.I., Fexby, S., Bülow, L., et al., 2002. Chromatography of microbial cells using continuous supermacroporous affinity and ion-exchange columns. J. Chromatogr. A 977, 27–38.

Arvidsson, P., Plieva, F.M., Lozinsky, V.I., Galaev, I.Y., Mattiasson, B., 2003. Direct chromatographic capture of enzyme from crude homogenate using immobilized metal affinity chromatography on a continuous supermacroporous adsorbent. J. Chromatogr. A 986, 275–290.

Ashcroft, J.M., Tsyboulski, D.A., Hartman, K.B., Zakharian, T.Y., Marks, J.W., Weisman, R.B., et al., 2006. Fullerene (C60) immunoconjugates: interaction of water-soluble C60 derivatives with the murine anti-gp240 melanoma antibody. Chem. Comm., 3004–3006.

Atassi, M.Z., Manshouri, T., 1991. Synthesis of tolerogenic monomethoxypolyethylene glycol and polyvinyl alcohol conjugates of peptides. J. Protein Chem. 10, 623–627.

Atha, D.H., Brew, S.A., Ingham, K.C., 1964. Interactions and thermal stability of fluorescent labeled derivatives of thrombin and antithrombin III. Biochim. Biophys. Acta 785, 1.

Avigad, E., Amaral, D., Asensio, C., Horecker, B.L., 1962. The d-Galactose oxidase of Polyporus circinatus. J. Biol. Chem. 237, 2736–2743.

Avrameas, S., 1969. Coupling of enzyme to proteins with glutaraldehyde. Immunochemistry 6, 43–52.

Avrameas, S., Ternynck, T., 1969. The cross-linking of proteins with glutaraldehyde and its use for the preparation of immunosorbents. Immunochemistry 6, 53–66.

Avrameas, S., Ternynck, T., 1971. Peroxidase labelled antibody and Fab conjugates with enhanced intracellular penetration. Immunochemistry 8, 1175.

Avramescu, M.E., Borneman, Z., Wessling, M., 2008. In: Pabby, A.K., Rizvi, S.S.H., Sastre, A.M. (Eds.), Membrane Chromatography in Handbook of Membrane Separations; Chemical, Pharmaceutical, Food, and Biotechnological Applications. CRC Press, pp. 25–63. Print ISBN: 978-0-8493-9549-9.

Axen, R., Porath, J., Ernback, S., 1967. Chemical coupling of peptides and proteins to polysaccharides by means of cyanogen halides. Nature (London) 214, 1302–1304.

Ayhan, F., Yousefi Rad, A., Ayhan, H., 2002. Biocompatibility investigation and urea removal from blood by urease-immobilized

HEMA incorporated poly(ethyleneglycol dimethacrylate) microbeads. J. Biomed. Mater. Res. B Appl. Biomat. 64B (1), 13–18.

Azarkana, M., Huetb, J., Baeyens-Volanta, D., Loozeb, Y., Vandenbussche, G., 2007. Affinity chromatography: a useful tool in proteomics studies. J. Chromatogr. B 849 (Issues 1–2), 81–90.

Babu, P., Sinha, S., Surolia, A., 2007. Sugar-quantum dot conjugates for a selective and sensitive detection of lectins. Bioconjug. Chem. 18, 146–151.

Bacha, P., Murphy, J.R., Reichlin, S., 1983. Thyrotropin-releasing hormone-diphtheria toxin-related polypeptide conjugates. J. Biol. Chem. 258, 1565.

Back, J.W., et al. 2003. Chemical cross-linking and mass spectrometry for protein structural modeling. J. Mol. Biol. 331, 303–313.

Bäckman, C., Rose, G.M., Hoffer, B.J., Henry, M.A., Bartus, R.T., Friden, P., et al., 1996. Systemic administration of a nerve growth factor conjugate reverses age-related cognitive dysfunction and prevents cholinergic neuron atrophy. J. Neurosci. 16, 5437.

Bacon, R., Bowman, J.C., 1957. Production and properties of graphite whiskers. Bull. Am. Phys. Soc. 2, 131.

Baenziger, J.U., Fiete, D., 1982. Photoactivatable glycopeptide reagents for site-specific labeling of lectins. J. Biol. Chem. 257, 4421–4425.

Baeuerle, P.A., Reinhardt, C., 2009. Bispecific T-cell engaging antibodies for cancer therapy. Cancer Res. 69, 4941–4944.

Bagchi, P., Birnbaum, S.M., 1981. Effect of pH on the adsorption of immunoglobulin G on anionic poly(vinyltoluene) model latex particles. J. Colloid Interface Sci. 83, 460–478.

Bagshawe, K.D., Springer, C.J., Searle, F., Antoniw, P., Sharma, S.K., Melton, R.G., et al., 1988. A cytotoxic agent can be generated selectively at cancer sites. Br. J. Cancer 58, 700–703.

Baird, B.A., Hammes, G.G., 1976. Chemical cross-linking studies of chloroplast coupling factor 1. J. Biol. Chem. 251, 6953–6962.

Baird, B.A., Hammes, G.G., 1977. Chemical cross-linking studies of beef heart mitochondrial coupling factor 1. J. Biol. Chem. 252, 4743–4748.

Bakkus, M.H., Brakel-van Peer, K.M., Adriaansem, H.J., Wierenga-Wolf, A.F., van den Akker, T.W., Dicke-Evinger, M.J., et al., 1989. Detection of oncogene expression by fluorescent *in situ* hybridization in combination with immunofluorescent staining of cell surface markers. Oncogene 4, 1255–1262.

Bala, S., Saini, M., Kamboj, S., 2011. Methods for synthesis of oxazolones: a review. Int. J. Chemtech Res. 3, 1102–1118.

Bale, M.D., Danielson, S.J., Daiss, J.L., Goppert, K.E., Sutton, R.C., 1989. Influence of copolymer composition on protein adsorption and structural rearrangements at the polymer surface. J. Colloid Interface Sci. 132, 176–1874.

Ballmer-Hofer, K., Schlup, V., Burn, P., Burger, M.M., 1982. Isolation of in situ cross-linked ligand-receptor complexes using an anticrosslinker specific antibody. Anal. Biochem. 126, 246–250.

Balls, A.K., Wood, H.N., 1956. Acetyl chymotrypsin and its reaction with ethanol. J. Biol. Chem. 219, 245–256.

Balzani, V., Juris, A., Venturi, M., Campagna, S., Serroni, S., 1996. Luminescent and redox-active polynuclear transition metal complexes. Chem. Rev. 96, 759–833.

Balzani, V., Ceroni, P., Gestermann, S., Kauffmann, C., Gorka, M., Vögtle, F., 2000. Dendrimers as fluorescent sensors with signal amplification. Chem. Comm., 853–854.

Bangham, A.D., Standish, M.M., Watkins, J.C., 1965. Diffusion of univalent ions across the lamellae of swollen phospholipids. J. Mol. Biol. 13, 238.

Bangs, J.D., Andrews, N.W., Hart, G.W., Englund, P.T., 1986. Posttranslational modification and intracellular transport of a trypanosome variant surface glycoprotein. J. Cell Biol. 103, 255–263.

Bannwarth, W., Knorr, R., 1991. Formation of carboxamides with N,N,N‴-tetramethyl(succinimido)uronium tetrafluoroborate in aqueous/organic solvent systems. Tetrahedron Lett. 32, 1157–1160.

Bantscheff, M., Schirle, M., Sweetman, G., Rick, J., Kuster, B., 2007. Quantitative mass spectrometry in proteomics: a critical review. Anal. Bioanal. Chem. 389, 1017–1031.

Bantscheff, M., Lemeer, S., Savitski, M.M., Kuster, B., 2012. Quantitative mass spectrometry in proteomics: critical review update from 2007 to the present. Anal. Bioanal. Chem., 1–27.

Baranowska-Kortylewicz, J., Kassis, A.I., 1993a. Labeling of sulfhydryl groups in intact mammalian cells with coumarins. Bioconjug. Chem. 4, 305–307.

Baranowska-Kortylewicz, J., Kassis, A.I., 1993b. Labeling of immunoglobulins with bifunctional, sulfhydryl-selective, and photoreactive coumarins. Bioconjug. Chem. 4, 300–304.

Barany, C., Merrifield, R.B., 1980. In: Gross, E., Meienhofer, J. (Eds.), The Peptides. Academic Press, New York, pp. 1–284.

Baraquet, C., Théraulaz, L., Guiral, M., Lafitte, D., Méjean, V., Jourlin-Castelli, C., 2006. TorT, a member of a new periplasmic binding protein family, triggers induction of the Tor respiratory system upon trimethylamine N-oxide electron-acceptor binding in *Escherichia coli*. J. Biol. Chem. 281, 38189–38199.

Barbee, K.D., Chandrangsu, M., Huang, X., 2011. Fabrication of DNA polymer brush arrays by destructive micropatterning and rolling-circle amplification. Macromol. Biosci. 11 (5), 607–617.

Barbet, J., Machy, P., Leserman, L.D., 1981. Monoclonal antibody covalently coupled to liposomes: specific targeting to cells. J. Supramol. Struct. Cell Biochem. 16 (3), 243–258.

Barbieri, L., Stirpe, F., 1982. Dye affinity chromatography of ricin subunits. Cancer Surv. 1, 489–520.

Barbour, H.M., 1976. Development of an enzyme immunoassay for human placental lactogen using labelled antibodies. J. Immunol. Methods 11, 15.

Bargou, R., Leo, E., Zugmaier, G., et al. 2008. Tumor regression in cancer patients by very low doses of a T cell-engaging antibody. Science 321, 974–977.

Barhate, G., Gautam, M., Gairola, S., Jadhav, S., Pokharkar, V., 2012. Quillaja saponaria extract as mucosal adjuvant with chitosan functionalized gold nanoparticles for mucosal vaccine delivery: stability and immunoefficiency studies. Int. J. Pharm.10.1016/j.ijpharm.2012.10.033 in press.

Baron-Bodo, V., Doceur, P., Lefebvre, M.L., Labroquere, K., Defaye, C., Cambouris, C., et al., 2005. Anti-tumor properties of human-activated macrophages produced in large scale for clinical application. Immunobiology 210, 267–277.

Barrett, A.J., Kembhavi, A.A., Brown, M.A., Kirschke, H., Knight, C.G., Tamai, M., et al., 1982. L-trans-epoxysuccinyl-leucylamido (4-guanidino)butane (E-64) and its analogues as inhibitors of cysteine proteinases including cathepsins B, H and L. Biochem J. 201 (1), 189–198.

Barringer, J.E., Messman, J.M., Banaszek, A.L., Meyer III, H.M., Kilbey, S.M., 2009. Immobilization of biomolecules on poly(vinyldimethylazlactone)-containing surface scaffolds. Langmuir 25, 262–268.

Bartczak, D., Kanaras, A.G., 2011. Preparation of peptide-functionalized gold nanoparticles using one pot EDC/sulfo-NHS coupling. Langmuir 27, 10119–10123.

Bartel, A., Campbell, D., 1959. Some immunochemical differences between associated and dissociate hemocyanin. Arch. Biochem. Biophys. 82, 2332.

Barth, R.F., Soloway, A.H., 1994. Boron neutron capture therapy of primary and metastatic brain tumors. Mol. Chem. Neuropathol. 21, 139–154.

Barth, R.F., Adams, D.M., Soloway, A.H., Alam, F., Darby, M.V., 1994. Boronated starburst dendrimer–monoclonal antibody immunoconjugates: evaluation as a potential delivery system for neutron capture therapy. Bioconjug. Chem. 5, 58–66.

Barth, R.F., Coderre, J.A., Vicente, M.G.H., Blue, T.E., 2005. Boron neutron capture therapy of cancer: current status and future prospects. Clin. Cancer Res. 11, 3987–4002.

Bartling, G.J., Brown, H.D., Chattopadhyay, S.K., 1973. Synthesis of a matrix-supported enzyme in non-aqueous conditions. Nature (London) 243, 342–344.

Basel, Y., Hassner, A., 2002. Activation of carboxylic acids as their active esters by means of tert-butyl 3-(3,4-dihydrobenzotriazine-4-on)yl carbonate. Tetrahedron Lett. 43, 2529–2533.

Baskin, J.M., Prescher, J.A., Laughlin, S.T., Agard, N.J., Chang, P.V., Miller, I.A., et al., 2007. Copper-free click chemistry for dynamic in vivo imaging. Proc. Natl. Acad. Sci. 104, 16793–16797.

Baskin, L.S., Yang, C.S., 1980a. Cross-linking studies of cytochrome P-450 and reduced nicotinamide adenine dinucleotide phosphate-cytochrome P-450 reductase. Biochemistry 19, 2260–2264.

Baskin, L.S., Yang, C.S., 1980b. Microsomes, Drug Oxidations, and Chemical Carcinogenesis. Academic Press, New York. pp. 103–106.

Baskin, L.S., Yang, C.S., 1982. Cross-linking studies of the protein topography of rat liver microsomes. Biochim. Biophys. Acta 684, 263–271.

Batista-Viera, F., Janson, J.C., Carlsson, J., 2011. Affinity chromatography Protein Purification: Principles, High Resolution Methods, and Applications, in Methods of Biochemical Analysis, vol. 54. John Wiley & Sons, Inc., Chapter 9, pp. 221–258.

Battinelli, L., Cernia, E., Delbò, M., Ortaggi, G., Pala, A., Soro, S., 1996. New class of poly(vinyl alcohol) polymers as column-chromatography stationary phases for Candida rugosa lipase isoforms separation. J. Chromatogr. A 753 (1), 47–55.

Batzri, S., Korn, E.D., 1973. Single bilayer liposomes prepared without sonication. Biochim. Biophys. Acta 298, 1015–1019.

Bauer-Arnaz, K., Napolitano, E.W., Roberts, D.N., Montali, J.A., Hughes, B.R., Schmidt Jr., D.E., 1998. Salt-induced immobilization of small affinity ligands on an epoxide-activated affinity support. J. Chromatogr. A 803, 73–82.

Baues, R.J., Cray, G.R., 1977. Lectin purification on affinity columns containing reductively aminated disaccharides. J. Biol. Chem. 252, 57.

Baughman, R.H., Zakhidov, A.A., de Heer, W.A., 2002. Carbon nanotubes—the route toward applications. Science 297, 787–792.

Baxter, E.W., Reitz, A.B., 2002. Reductive aminations of carbonyl compounds with borohydride and borane reducing agents. Org. React., 1–714.

Bayer, E.A., Wilchek, M., 1980. The use of the avidin–biotin complex as a tool in molecular biology. Methods Biochem. Anal. 26, 1–45.

Bayer, E.A., Wilchek, M., 1990. Avidin- and streptavidin-containing probesWilchek, M.Bayer, E.A. Methods in Enzymology, vol. 184. Academic Press, San Diego CA, pp. 174–185.

Bayer, E.A., Wilchek, M., 1992. Labeling and detection of proteins and glycoproteins. In: Kessler, C. (Ed.), Nonradioactive Labeling and Detection of Biomolecules. Springer-Verlag, New York, pp. 98–99.

Bayer, E.A., Skutelsky, E., Wynne, D., Wilchek, M., 1976. Preparation of ferritin–avidin conjugates by reductive alkylation for use in electron microscopic cytochemistry. J. Histochem. Cytochem. 24, 933–939.

Bayer, E., Ben-Hur, H., Citlin, G., Wilchek, M., 1986. An improved method for the single-step purification of streptavidin. J. Biochem. Biophys. Methods 13, 103–112.

Bayer, E.A., Ben-Hur, H., Wilchek, M., 1987a. Enzyme-based detection of glycoproteins on blot transfers using avidin–biotin technology. Anal. Biochem. 161 (1), 123–131.

Bayer, E.A., Safars, M., Wilchek, M., 1987b. Selective labeling of sulfhydryls and disulfides on blot transfers using avidin–biotin technology: studies on purified proteins and erythrocyte membranes. Anal. Biochem. 161, 262–271.

Bayer, E.A., Ben-Hur, H., Wilchek, M., 1988. Biocytin hydrazide—a selective label for sialic acids, galactose, and other sugars in glycoconjugates using avidin–biotin technology. Anal. Biochem. 170, 271–281.

Bayer, E.A., Ben-Hur, H., Hiller, T., Wilchek, M., 1989. Postsecretory modifications of streptavidin. Biochem. J. 259, 369–376.

Bayer, E.A., Ben-Hur, H., Wilchek, M., 1990. Direct labeling of blotted glycoproteinsWilchek, M.Bayer, E.A. Methods in Enzymology, vol. 184. Academic Press, San Diego CA, pp. 427–429.

Bayne, S., Ottesen, M., 1976. Carlsberg Res. Commun. 41, 211–216.

Baynes, J.W., Thorpe, S.R., 2000. Glycoxidation and lipoxidation in atherogenesis. Free Radic. Biol. Med. 28, 1708–1716.

Bayramoglu, G., Senel, A.U.S., Arica, M.Y., 2006. Effect of spacer-arm and Cu(II) ions on performance of l-histidine immobilized on poly(GMA/MMA) beads as an affinity ligand for separation and purification of IgG. Sep. Purif. Technol. 50, 229–239.

Beard, C.D., Baum, K., Grakauskas, V., 1973. Synthesis of some novel trifluoromethanesulfonates and their reactions with alcohols. J. Org. Chem. 38, 3673–3677.

Beardsley, R.L., Reilly, J.P., 2002. Optimization of guanidination procedures for MALDI mass mapping. Anal. Chem. 74, 1884–1890.

Beatty, K.E., 2008. Imaging the proteome: metabolic tagging of newly synthesized proteins with reactive methionine analogues. Dissertation (Ph.D.), California Institute of Technology.

Beaucage, S.L., Iyer, R.P., 1993. The functionalization of oligonucleotides via phosphoramidite derivatives. Tetrahedron 49, 1925–1963.

Beauchamp, C.O., Gonias, S.L., Menapace, D.P., Pizzo, S.V., 1983. A new procedure for the synthesis of polyethylene glycol–protein adducts: effects on function, receptor recognition, and clearance of superoxide dismutase, lactoferrin, and a2macroglobulin. Anal. Biochem. 131, 25–33.

Beck, A., 2010. The next generation of antibody-drug conjugates comes of age. Discov. Med. 10 (53), 329–339.

Beckmann, H.S., Niederwieser, A., Wiessler, M., Wittmann, V., 2012. Preparation of carbohydrate arrays by using diels–alder reactions with inverse electron demand. Chem. A Eur. J. 18 (21), 6548–6554.

Bedner, E., Smolewski, P., Amstad, P.A., Darzynkiewicz, Z., 2000. Activation of caspases measured in situ by binding or fluorochrome-labeled inhibitors of caspases (FLICA): correlation with DNA fragmentation. Exp. Cell Res. 259, 308–313.

Behr, T.M., Behe, M., Stabin, M.G., Wehrmann, E., Apostolidis, C., Molinet, R., et al., 1999. High-linear energy transfer (LET) alpha versus low-LET beta emitters in radioimmunotherapy of solid tumors: therapeutic efficacy and dose-limiting toxicity of 213-Bi-versus 90-Y-labeled CO17-1A Fab′ fragments in human colonic cancer model. Cancer Res. 59, 2635–2643.

Beija, M., Charreyre, M.-T., Martinho, J.M.G., 2011. Dye-labelled polymer chains at specific sites: synthesis by living/controlled polymerization. Prog. Polym. Sci. 36, 568–602.

Belew, M., Juntti, N., Larsson, A., Porath, J., 1987. A one-step purification method for monoclonal antibodies based on salt-promoted adsorption chromatography on a 'thiophilic' adsorbent. J. Immunol. Methods 102, 173–182.

Beligere, G.S., Dawson, P.E., 1999. Synthesis of a three zinc finger protein, Zif268, by native chemical ligation. Biopolymers 51 (5), 363–369.

Bendayan, M., 1989. Ultrastructural localization of insulin and C-peptide antigenic sites in rat pancreatic B cell obtained by applying the quantitative high-resolution protein A-gold approach. Am. J. Anat. 185, 205–216.

Bendayan, M., Garzon, S., 1988. Protein G-gold complex: comparative evaluation with protein A-gold for high-resolution immunocytochemistry. J. Histochem. Cytochem. 36, 597–607.

Benesch, R., Benesch, R.E., 1956. Formation of peptide bonds by aminolysis of homocysteine thiolactones. J. Am. Chem. Soc. 78, 1597.

Benesch, R., Benesch, R.E., 1958. Thiolation of proteins. Proc. Natl. Acad. Sci. USA 44, 848.

Benesch, R.E., Kwong, S., 1988. Bis-pyridoxal polyphosphates: a new class of specific intramolecular cross-linking agents for hemoglobin. Biochem. Biophys. Res. Comm. 156, 9.

Beneteau, J., Renard, D., Marchoe, L., Douville, E., Lavenant, L., Rahboe, Y., et al., 2010. Binding properties of the N-acetylglucosamine and high-mannose N-glycan PP2-A1 phloem lectin in Arabidopsis. Plant Physiol. 153, 1345–1361.

Benhamou, N., Gilboa-Garber, N., Trudel, J., Asselin, A., 1988. A new lectin–gold complex for ultrastructural localization of galacturonic acids. J. Histochem. Cytochem. 36, 1403–1411.

Benoit, S.L., Mehta, N., Weinberg, M.V., Maier, C., Maier, R.J., 2007. Interaction between the *Helicobacter pylori* accessory proteins HypA and UreE is needed for urease maturation. Microbiology 153, 1474–1482.

Benters, R., Niemeyer, C.M., Drutschmann, D., Blohm, D., Wöhrle, D., 2002. DNA microarrays with PAMAM dendritic linker systems. Nucl. Acids Res. 30 (2), e10.

Berg, H.C., Diamond, J.M., Marfey, P.S., 1965. Erythrocyte membrane: chemical modification. Science 150, 64.

Berger, H., Pizzo, S.V., 1988. Preparation of polyethylene glycol–tissue plasminogen activator adducts that retain functional activity: characteristics and behavior in three different species. Blood 71, 1641–1647.

Bergin, J., 2009. Biologic imaging reagents: technologies and global markets. BCC Res. Market Res. Rep., 5.

Berglund, G.I., Carlsson, G.H., Smith, A.T., Szoke, H., Henridsen, A., Hajdu, J., 2002. The catalytic pathway of horseradish peroxidase at high resolution. Nature 417, 463–468.

Bergmann, K.E., Carlson, K.E., Katzenellenbogen, J.A., 1994. Hexestrol diazirine photo-affinity labeling reagent for the estrogen receptor. Bioconjug. Chem. 5, 141–150.

Bergmann-Leitner, E.S., Mease, R.M., Duncan, E.H., Khan, F., Waitumbi, J., Angov, E., 2008. Evaluation of immunoglobulin purification methods and their impact on quality and yield of antigen-specific antibodies. Malar. J. 7, 129.

Bergseid, M., Baytan, A.R., Wiley, J.P., Andener, W.M., Stolowitz, M.L., Hughes, K.A., et al., 2000. Small molecule-based chemical affinity system for the purification of proteins. BioTechniques 29, 1126–1133.

Bergstrom, J., et al., 1998. Identification of low abundance proteins by electrophoresis and MALDI-TOF MS. Poster available at <www.glycobiology.med.gu.se>.

Bergström, K., Holmberg, K., Safranj, A., Hoffman, A.S., Edgell, M.J., Kozlowski, A., et al., 1992. Reduction of fibrinogen adsorption on PEG-coated polystyrene surfaces. J. Biomed. Mater. Res. 26, 779–790.

Berning, D.E., et al. 1999. Chemical and biomedical motifs of the reactions of hydroxymethylphosphines with amines, amino acids and model peptide. J. Am. Chem. Soc. 121 (8), 1658–1864.

Bernstein, A., Hurwitz, E., Maron, R., Arnon, R., Sela, M., Wilchek, M., 1978. Higher antitumor efficacy of daunomycin when linked to dextran: in vivo and in vitro studies. J. Natl. Cancer Inst. 60, 379–384.

Berrio, J., Plou, F.J., Ballesteros, A., Martinez, N.T., Martinez, M.J., 2007. Immobilization of pycnoporus coccineus laccase on eupergit C: stabilization and treatment of olive oil mill wastewaters. Biocatal. Biotransformation 25 (2-4), 130–134.

Berron, B.J., Johnson, L.M., Ba, X., McCall, J.D., Alvey, N.J., Anseth, K.S., et al., 2011. Glucose oxidase-mediated polymerization as a platform for dual-mode signal amplification and biodetection. Biotechnol. Bioeng. 108 (7), 1521–1528.

Bertozzi, C.R., Kiessling, L.L., 2001. Chemical glycobiology, review in carbohydrates and glycobiology. Science 291, 2357–2364.

Bertrand, Y., Currie, J.C., Demeule, M., Régina, A., Ché, C., Abulrob, A., et al., 2010. Transport characteristics of a novel peptide platform for CNS therapeutics. J. Cell. Mol. Med. 14 (12), 2827–2839.

Betancor, L., L´opez-Gallego, F., Hidalgo, A., Alonso-Morales, N., Gisela Dellamora-Ortiz Cesar, M., Fern´andez-Lafuente, R., et al., 2006. Different mechanisms of protein immobilization on glutaraldehyde activated supports: effect of support activation and immobilization conditions. Enzyme Microb. Technol. 39, 877–882.

Beth, A.H., Conturo, T.E., Venkataramu, S.D., Staros, J.V., 1986. Dynamics and interactions of the anion channel in intact human erythrocytes: an electron paramagnetic resonance spectroscopic study employing a new membrane-impermeant bifunctional spin-label. Biochemistry 25, 3824–3832.

Bethell, G.S., Ayers, J.S., Hancock, W.S., Hearn, M.T.W., 1979. A novel method of activation of cross-linked agaroses with 1,1′-carbonyldiimidazole which gives a matrix for affinity chromatography devoid of additional charged groups. J. Biol. Chem. 254, 2572–2574.

Bethell, G.S., Ayers, J.S., Hearn, M.T.W., Hancock, W.S., 1987. Investigation of the activation of cross-linked agarose with carbonylating reagents and the preparation of matrices for affinity chromatography purifications. J. Chromatogr. A 219, 361–371.

Bethge, W.A., Lange, T., Meisner, C., von Harsdorf, S., Bornhaeuser, M., Federmann, B., et al., 2010. Radioimmunotherapy with yttrium-90-ibritumomab tiuxetan as part of a reduced- intensity conditioning regimen for allogeneic hematopoietic cell transplantation in patients with advanced non-Hodgkin lymphoma: results of a phase 2 study. Blood 116, 1795–1802.

Bethune, D.S., Kiang, C.H., De Vries, M.S., Gorman, G., Savoy, R., et al., 1993. Cobalt catalysed growth of carbon nanotubes with single-atomic-layer walls. Nature 363, 605–607.

Betzig, E., Patterson, G.H., Sougrat, R., Lindwasser, O.W., Olenych, S., Bonifacino, J.S., et al., 2006. Imaging intracellular fluorescent proteins at nanometer resolution. Science 313, 1642–1645.

Beutner, E.H., 1971. Defined immunofluorescent staining: past progress, present status, and future prospects for defined conjugates. Ann. NY Acad. Sci. 177, 506–526.

Bewley, T.A., Li, C.H., 1969. The reduction of protein disulfide bonds in the absence of denaturants. Int. J. Protein Res. 1 (2), 117–124.

Bewley, T.A., Dixon, J.S., Li, C.H., 1968. Human pituitary growth hormone. XVI. Reduction with dithiothreitol in the absence of urea. Biochim. Biophys. Acta 154, 420–422.

Biermann, C.J., McGinnis, G.D. (Eds.), 1989. Analysis of Carbohydrates by GLC and MS. CRC Press, Boca Raton, FL.

Bigelow, D.J., Inesi, G., 1991. Frequency-domain fluorescence spectroscopy resolves the location of maleimide-directed spectroscopic probes within the tertiary structure of the Ca-ATPase of sarcoplasmic reticulum. Biochemistry 30, 2113–2125.

Bigge, J.C., Patel, T.P., Bruce, J.A., Goulding, P.N., Charles, S.M., Parekh, R.B., 1995. Nonselective and efficient fluorescent labeling of glycans using 2-amino benzamide and anthranilic acid. Anal. Biochem. 230, 229–238.

Bingel, C., 1993. Cyclopropylation of fullerenes. Chem. Ber. 126 (8), 1957.

Binkley, R.W., 1988. Modern Carbohydrate Chemistry. Dekker, San Diego CA.

Birnbaumer, M.E., Schrader, W.T., O'Malley, B.W., 1979. Chemical cross-linking of chick oviduct progesterone-receptor subunits using a reversible bifunctional cross-linking agent. Biochem. J. 181, 201–213.

Bisagni, E., 1992. Synthesis of psoralens and analogs. J. Photochem. Photobiol. 14, 23–46.

Biswas, S., Dodwadkar, N.S., Sawant, R.R., Torchilin, V.P., 2011. Development of the novel PEG-PE-based polymer for the reversible attachment of specific ligands to liposomes: synthesis and in vitro characterization. Bioconjug. Chem. 22 (10), 2005–2013.

Biswas, S., Dodwadkar, N.S., Piroyan, A., Torchilin, V.P., 2012. Surface conjugation of triphenylphosphonium to target poly(amidoamine) dendrimers to mitochondria. Biomaterials 33, 4773–4782.

Bitan, G., Kirkitadze, M.D., Lomakin, A., Vollers, S.S., Benedek, G.B., Teplow, D.B., 2003. Amyloid B-protein (ab) assembly: ab40 and ab42 oligomerize through distinct pathways. Proc. Natl. Acad. Sci. USA 100, 330–335.

Biver, T., De Biasi, A., Secco, F., Venturini, M., Yarmoluk, S., 2005. Cyanine dyes as intercalating agents: kinetic and thermodynamic studies on the DNA/Cyan40 and DNA/CCyan2 systems. Biophys. J. 89, 374–383.

Bizzini, B., Blass, J., Turpin, A., Raynaud, M., 1970. Chemical characterization of tetanus toxin and toxoid amino acid composition, number of SH and S-S groups and N-terminal amino acid. Eur. J. Biochem. 17, 100–105.

Bjorn, M.J., Croetsema, G., Scalapino, L., 1986. Antibody–Pseudomonas exotoxin a conjugates cytotoxic to human breast cancer cells in vitro. Cancer Res. 46, 3262.

Blank, W.J., He, Z.A., Picci, M., 2001. Catalysis of the epoxy-carboxyl reaction. International Waterborne, High-Solids, and Powder Coatings Symposium, Paper 23, presented February 21-23, New Orleans, LA.

Blass, J., Bizzini, B., Raynaud, M., 1965. Mechanism of detoxication by formol. Compt. Rend. 261, 1448.

Blattler, W.A., Kuenzi, B.S., Lambert, J.M., Senter, P.D., 1985a. New heterobifunctional protein cross-linking reagent that forms an acid-labile link. Biochemistry 24, 1517–1524.

Blattler, W.A., Kuenzi, B.S., Lambert, J.M., Senter, P.D., 1985b. New heterobifunctional protein cross-linking reagents and their use in the preparation of antibody–toxin conjugates. Photochem. Photobiol. 42, 231.

Blixt, O., Head, S., Mondala, T., Scanlan, C., Huflejt, M.E., Alvarez, R., et al., 2004. Printed covalent glycan array for ligand profiling of diverse glycan binding proteins. Proc. Natl. Acad. Sci. USA 101 (49), 17033–17038.

Block, H., Maertens, B., Spriestersbach, A., Brinker, N., Kubicek, J., Fabis, R., et al., 2009. Immobilized-metal affinity chromatography (IMAC): a review. Methods Enzymol. 463, 439–473. Chapter 27.

Blokzijl, W., Engberts, J.B.F.N.J., 1992. Initial-state and transition-state effects on Diels-Alder reactions in water and mixed aqueous solvents. J. Am. Chem. Soc. 114, 5440–5442.

Bloxham, D.P., Cooper, C.K., 1982. Formation of a polymethylene bis(disulfide) inter-subunit cross-link between cys-281 residues in rabbit muscle glyceraldehyde-3-phosphate dehydrogenase using octamethylene bis-(methane[35]thiosulfonate. Biochemistry 21, 1807.

Bloxham, D.P., Sharma, R.P., 1979. The development of S,S9-polyme thylenebis(methanethiosulfonates) as reversible cross-linking reagent for thiol groups and their use to form stable catalytically active cross-linked dimers with glyceraldehyde-3-phosphate dehydrogenase. Biochem. J. 181, 355.

Blum, G., Mullins, S.R., Keren, K., Fonovicˇ, M., Jedeszko, C., Rice, M.J., et al., 2005. Dynamic imaging of protease activity with fluorescently quenched activity-based probes. Nat. Chem. Biol. 1 (4), 203–209.

Blum, G., Weimer, R.M., Edgington, L.E., Adams, W., Bogyo, M., 2009. Comparative assessment of substrates and activity based probes as tools for non-invasive optical imaging of cysteine protease activity. PLoS ONE 4 (7), e6374.10.1371/journal.pone.0006374.

Boas, M.A., 1927. The effect of desiccation upon the nutritive properties of egg-white. Biochem. J. 21, 712–724.

Boas, U., Christensen, J.B., Heegaard, P.M.H., 2006. Dendrimers in Medicine and Biotechnology: New Molecular Tools. Royal Society of Chemistry.

Bobbitt, J.M., 1956. Periodate oxidation of carbohydrates. Adv. Carbohydr. Chem. 11, 1–41.

Bobyr, E., Lassila, J.K., Wiersma-Koch, H.I., Fenn, T.D., Lee, J.J., Nikolic-Hughes, I., et al., 2012. High-resolution analysis of Zn(2+) coordination in the alkaline phosphatase superfamily by EXAFS and x-ray crystallography. J. Mol. Biol. 415, 102–117.

Böcher, M., Giersch, T., Schmid, R.D., 1992. Dextran, a hapten carrier in immunoassays for s-triazines. A comparison with ELISAs based on hapten–protein conjugates. J. Immunol. Methods 151, 1–8.

Bodanszky, A., Bodanszky, M., 1970. Sepharose–avidin column for the binding of biotin or biotin-containing peptides. Experientia 26, 327.

Boehl, M., Tietze, S., Sokoll, A., Madathil, S., Pfennig, F., Apostolakis, J., et al., 2007. Flavonoids affect actin functions in cytoplasm and nucleus. Biophys. J.10.1529/biophysj.107.107813.

Boens, N., Leen, V., Dehaen, W., 2012. Fluorescent indicators based on BODIPY. Chem. Soc. Rev. 41 (3), 1130–1172.

Boeva, V.I., Solovievb, A., Silva, C.J.R., Gomes, M.J.M., 2005. Incorporation of CdS nanoparticles from colloidal solution into optically clear ureasilicate matrix with preservation of quantum size effect. Solid State Sci. 8 (1), 50–58.

Bog-Hansen, T.C., Prahl, P., Lowenstein, H., 1978. A set of analytical electrophoresis experiments to predict the results of affinity chromatographic separations. Fractionation of allergens from cow's hair and dander. J. Immunol. Methods 22, 293.

Bogyo, M., Verhelst, S., Bellingard-Dubouchaud, V., Toba, S., Greenbaum, D., 2000. Selective targeting of lysosomal cysteine proteases with radiolabeled electrophilic substrate analogs. Chem. Biol. 7, 27–38.

Boi, C., Dimartino, S., Sarti, G.C., 2008. Performance of a new protein a affinity membrane for the primary recovery of antibodies. Biotechnol. Prog. 24, 640–647.

Böldicke, T., Kindt, S., Maywald, F., Fitzlaff, G., Böcher, M., Frank, R., et al., 1988. Production of specific monoclonal antibodies against the active sites of human pancreatic secretory trypsin inhibitor variants by in vitro immunization with synthetic peptides. Eur. J. Biochem. 175, 259–264.

Boller, T., Meier, C., Menzler, S., 2002. EUPERGIT Oxirane acrylic beads: how to make enzymes fit for biocatalysis. Org. Process Res. Dev. 6, 509–519.

Bolton, A.E., Hunter, W.M., 1973. The labeling of proteins to high specific radioactivities by conjugation to a 125I-containing acylating agent. Biochem. J. 133, 529–539.

Bolton, A.E., Hunter, W.M., 1986. Radioimmunoassay and related methods (fourth ed.)Weir, D.M. Handbook of Experimental Immunology. Immunochemistry, vol. 1. Blackwell, London, pp. 26.1–26.56.

Bonfield, T.L., John, N., Barna, B.P., Kavuru, M.S., Thomassen, M.J., Yen-Lieberman, B., 2005. Multiplexed particle-based anti-granulocyte macrophage colony stimulating factor assay used as pulmonary diagnostic test. Clin. Diagn. Lab. Immunol. 12, 821–824.

Bonnard, C., Papermaster, D.S., Kraehenbuhl, J.-P., 1984. The streptavidin–biotin bridge technique: application in light and electron microscope immunocytochemistry. In: Polak, J.M., Varndell, I.M. (Eds.), Immunolabelling for Electron Microscopy. Elsevier, New York, pp. 95–111.

Bonnet, D., Ilien, B., Galzi, J.-L., Riché, S., Antheaune, C., Hibert, M., 2006. A rapid and versatile method to label receptor ligands using "click" chemistry: validation with the muscarinic M1 antagonist pirenzepine. Bioconjug. Chem. 17, 1618–1623.

Boorsma, D.M., Kalsbeek, G.L., 1976. A comparative study of horseradish peroxidase conjugates prepared with a one-step and a two-step method. Histochem. Cytochem. 23, 200–207.

Boorsma, D.M., Streefkerk, J.G., 1976a. Peroxidase-conjugate chromatography. Isolation of conjugates prepared with glutaraldehyde or periodate using polyacrylamide-agarose gel. J. Histochem. Cytochem. 24, 481.

Boorsma, D.M., Streefkerk, J.G., 1976b. Some aspects of the preparation, analysis, and use of peroxidase-antibody conjugates in immunohistochemistry. Protides Biol. Fluids Proc. Colloq. 24, 795.

Bordo, D., Argos, P., 1991. Suggestions for "safe" residue substitutions in site-directed mutagenesis. J. Mol. Biol. 217, 721–729.

Borges, A.R., Schengrund, C.L., 2005. Dendrimers and antivirals: a review. Curr. Drug. Targets Infect. Disord. 5 (3), 247–254.

Borges, R.A., Puri, A., Krebs, F.C., Wigdahl, B., Blumenthal, R., Rawat, S.S., et al., 2005. Multivalent compounds functionalized with the carbohydrate headgroups of immune cell-surface GSLs as inhibitors of HIV-1 infection. Third IAS Conference on HIV Pathogenesis and Treatment, Rio de Janeiro, July 24–27.

Borneman, Z., 2006. Particle loaded membrane chromatography. Ph.D. thesis, University of Twente, The Netherlands. ISBN: 90-365-2433-4.

Borole, A., Dai, S., Chen, C.L., Rodriguez, M., Davison, B.H., 2004. Performance of chloroperoxidase stabilization in mesoporous sol-gel glass using in situ glucose oxidase peroxide generation. Appl. Biochem. Biotechnol. 113, 273–285.

Borque, L., Maside, C., Rus, C.J., 1994. Latex immunoassay of b2-microglobulin in serum and urine. J. Clin. Immunoassay 17, 160–165.

Bortz, P.D.S., Wamhoff, B.R., 2011. Chromatin Immunoprecipitation (ChIP): revisiting the efficacy of sample preparation, sonication, quantification of sheared DNA, and analysis via PCR. PloS one 6 (10), e26015.

Borucki, M.K., James Reynolds, D.R., Call, T.J., Ward, B.P., Kadushin, J., 2005. Suspension microarray with dendrimer signal amplification allows direct and high-throughput subtyping of listeria monocytogenes from Genomic DNA. J. Clin. Microbiol. 43, 3255–3259.

Borup, B., Weissenbach, K., 2010. In: Xanthos, M. (Ed.), Silane Coupling Agents, in Functional Fillers for Plastics (second ed.). Wiley-VCH Verlag GmbH & Co. KGaA, Weinheim, Germany, pp. 61–90. doi:10.1002/9783527629848.ch4.

Bos, F., 1981. Optimization of spectral coverage in an eight-cell oscillator-amplifier dye laser pumped at 308 nm. Appl. Opt. 20, 3553.

Bosi, S., Da Ros, T., Spalluto, G., Prato, M., 2003. Fullerene derivatives: an attractive tool for biological applications. Eur. J. Med. Chem. 38, 913–923.

Bosnjakovic, A., Mishra, M.K., Han, H.J., Romero, R., Kannan, R.M., 2012. A dendrimer-based immunosensor for improved capture and detection of tumor necrosis factor-α cytokine. Anal. Chim. Acta 720, 118–125.

Bottari, G., Torres, T., 2011. Towards collective physical properties in supramolecular organized phthalocyanine—C60-fullerene conjugates. Macroheterocycles 3, 16–18.

Bouizar, Z., Fouchereau-Person, M., Taboulet, J., Moukhtar, M.S., Milhaud, G., 1986. Purification and characterization of calcitonin receptors in rat kidney membranes by covalent cross-linking techniques. Eur. J. Biochem. 155, 141–147.

Boul, P.J., Liu, J., Mickelson, E.T., Huffman, C.B., Ericson, L.M., Chiang, I.W., et al., 1999. Chem. Phys. Lett. 310, 367–372.

Bourne, M.W., Margerun, L., Hylton, N., Campion, B., Lai, J.J., Derugin, N., et al., 1996. Evaluation of the effects of intravascular MR contrast media (gadolinium dendrimer) on 3D time of flight magnetic resonance angiography of the body. J. Magn. Reson. Imaging 6, 305–310.

Bower, K., Djordjevic, S.P., Andronicos, N.M., Ranson, M., 2003. Cell surface antigens of Mycoplasma species bovine group 7 bind to and activate plasminogen. Infect. Immun. 71, 4823–4827.

Boyd, S., Yamazaki, H., 1993. Tosyl chloride activation of a rayon/polyester cloth for protein immobilization. Biotechnol. Tech. 7, 277–282.

Boyer, C.M., Lidor, Y., Lottich, S.C., Bast Jr., R.C., 1988. Antigenic cell surface markers in human solid tumors. Antibody, Immunoconjugates, Radiopharm. 1, 105.

Boyer, T.D., 1986. Covalent labeling of the nonsubstrate ligand-binding site of glutathione S-transferase with bilirubin-woodward's reagent K. J. Biol. Chem. 261, 5363.

Boyle, R., 1966. The reaction of dimethyl sulfoxide and 5-dimethylaminonaphthalene-1-sulfonyl chloride. J. Org. Chem. 31, 3880–3882.

Boyle, W.J., Lipsick, J.S., Reddy, E.P., Baluda, M.A., 1983. Identification of the leukemogenic protein of avian myeloblastosis virus and of its normal cellular homologue. Proc. Natl. Acad. Sci. USA 80, 2834–2838.

Braatz, J.A., Yasuda, Y., Olden, K., Yamada, K.M., Heifetz, A.H., 1993. Functional peptide–polyurethane conjugates with extended circulatory half-lives. Bioconjug. Chem. 4, 262–267.

Brabec, C.J., Sariciftci, N.S., Hummelen, J.C., 2001. Plastic solar cells. Adv. Funct. Mater. 11, 15–26.

Bragg, P.D., Hou, C., 1975. Subunit composition, function, and spatial arrangement in the Ca21-and Mg21-activated adenosine triphosphatases of Escherichia coli and Salmonella typhimurium. Arch. Biochem. Biophys. 167, 311–321.

Bragg, P.D., Hou, C., 1980. A crosslinking study of the $Ca2^1$, $Mg2^1$-activated adenosine triphosphate of Escherichia coli. Eur. J. Biochem. 106, 495–503.

Brancia, F.L., Oliver, S.G., Gaskell, S.J., 2000. Improved matrix-assisted laser desorption/ionization mass spectrometric analysis of tryptic hydrolysates of proteins following guanidination of lysine-containing peptides. Rapid Comm. Mass Spectrom. 14, 2070–2073.

Brandon, D.L., 1980. Studies of sheep red blood cell membranes, using cleavable crosslinking reagents. Cell. Mol. Biol. 26, 569–573.

Brandt, H.M., Apkarian, A.V., 1992. Biotin–dextran: a sensitive anterograde tracer for neuroatomic studies in rat and monkey. J. Neurosci. Methods 45, 35–40.

Branham, W.S., Melvin, C.D., Han, T., Desai, V.G., Moland, C.L., Scully, A.T., et al., 2007. Elimination of laboratory ozone leads to a dramatic improvement in the reproducibility of microarray gene expression measurements. BMC Biotechnol. 7, 8.10.1186/1472-6750-7-8.

Brask, J., 2009. Immobilized Enzymes in Organic Synthesis. In: Tulla-Puche, J., Albericio, F. (Eds.), The Power of Functional Resins in Organic Synthesis. Wiley-VCH Verlag GmbH & Co. KGaA, Weinheim, Germany.

Bratkovic, T., Lunder, M., Popovic, T., Kreft, S., Turk, B., Strukelj, B., et al., 2005. Affinity selection to papain yields potent peptide inhibitors of cathepsins L, B, H, and K. Biochem. Biophys. Res. Commun. 332 (3), 897–903.

Bratu, D.P., 2006. Molecular beacons: fluorescent probes for detection of endogenous mRNAs in living cells. Methods Mol. Biol. 319, 1–14.

Braun, G.S., Kretzler, M., Heider, T., Floege, J., Holzman, L.B., Kriz, W., et al., 2007. Differentially spliced isoforms of FAT1 are asymmetrically distributed within migrating cells. J. Biol. Chem. 282, 22823–22833.

Braun, P., Gingras, A.-C., 2012. History of protein-protein interactions: from egg-white to complex networks. Proteomics 12, 1478–1498.

Braun, T., 1997. Water soluble fullerene-cyclodextrin suframolecular assemblies preparation, structure, properties (an annotated bibliography). Fullerene Sci. Technol. 5, 615–626.

Brechbiel, M.W., 2008. Bifunctional chelates for metal nuclides. The quarterly journal of nuclear medicine and molecular imaging: official publication of the Italian Association of Nuclear Medicine (AIMN)[and] the International Association of Radiopharmacology (IAR),[and] Section of the Society of., 52(2), 166-173.

Brechbiel, M.W., Gansow, O.A., 1991. Backbone-substituted DTPA ligands for 90Y radioimmunotherapy. Bioconjug. Chem. 2, 187–194.

Brechbiel, M.W., McMurry, T.J., Gansow, O.A., 1993. A direct synthesis of a bifunctional chelating agent for radiolabeling proteins. Tetrahedron Lett. 34, 3691–3694.

Bremer, C., Tung, C.H., Weissleder, R., 2001. In vivo molecular target assessment of matrix metalloproteinase inhibition. Nat Med 7, 743–748.

Brenna, O., Bianchi, E., 1994. Immobilised laccase for phenolic removal in must and wine. Biotechnol. Lett. 16, 35–40.

Brennan, J.F., Beutel, J., 1969. Quinone photochemistry. II. The mechanism of photoreduction of 9,10-phenanthrenequinone and 2-tert-butyl-9,10-anthraquinone in ethanol. J. Phys. Chem. 73, 3245–3249.

Brennan, J.L., Hatzakis, N.S., Tshikhudo, T.R., Dirvianskyte, N., Razumas, V., Patkar, S., et al., 2006. Bionanoconjugation via click chemistry: the creation of functional hybrids of lipases and gold nanoparticles. Bioconjug. Chem. 17, 1373–1375.

Brethauer, S., Wyman, C.E., 2010. Review: continuous hydrolysis and fermentation for cellulosic ethanol production. Bioresour. Technol. 101, 4862–4874.

Brettreich, M., Hirsch, A., 1998. Tetrahedron Lett. 39, 2731–2734.

Brew, K., Shaper, J.H., Olsen, K.W., Trayer, I.P., Hill, R.L., 1975. Cross-linking of the components of lactose synthetase with dimethylpimelimidate. J. Biol. Chem. 250, 1434–1444.

Brewer, C.F., Riehm, J.P., 1967. Evidence for possible nonspecific reactions between N-ethylmaleimide and proteins. Anal. Biochem. 18, 248.

Briand, J.P., Muller, S., Van Regenmortel, M.H.V., 1985. Synthetic peptides as antigens: pitfalls of conjugation methods. J. Immunol. Methods 78, 59–69.

Bright, F.V., 1988. Bioanalytical applications of fluorescence spectroscopy. Anal. Chem. 60, 1031A–1039A.

Brillhart, K.L., Ngo, T.T., 1991. Use of microwell plates carrying hydrazide groups to enhance antibody immobilization in enzyme immunoassays. J. Immunol. Methods 144, 19–25.

Brinkley, M., 1992. A brief survey of methods for preparing protein conjugates with dyes, haptens, and cross-linking reagents. Bioconjug. Chem. 3, 2.

Brocchini, S., Godwin, A., Balan, S., Choi, J.-W., Zloh, M., Shaunak, S., 2008. Disulfide bridge based PEGylation of proteins in Advanced Drug Delivery Reviews, Peptide and Protein PEGylation III: Advances in Chemistry and Clinical Applications. 60 (1), 3-12.

Brockhausen, I., Yang, J.-M., Burchell, J., Whitehouse, C., Taylor-Papadimitriou, J., 1995. Mechanisms underlying aberrant glycosylation of MUC1 mucin in breast cancer cells. Eur. J. Biochem. 233, 607–617.

Brocklehurst, K., Carlsson, J., Kierstan, M.P.J., Crook, E.M., 1973. Covalent chromatography. Preparation of fully active papain from dried papaya latex. Biochem. J. 133, 573–584.

Brocklehurst, K., Carlsson, J., Kierstan, M.P.J., Crook, E.M., 1974. Covalent chromatography by thiol-disulfide interchangeJakoby, W.B.Wilchek, M. Methods of Enzymology, vol. 34. Academic Press, New York, pp. 531–544.

Bromme, D., Smith, R.A., Coles, P.J., Kirschke, H., Storer, A.C., Krantz, A., 1994. Potent inactivation of cathepsins S and L by peptidyl (acyloxy) methyl ketones. Biol. Chem. Hoppe Seyler 375 (5), 343–347.

Bronckers, A.J.J., Gay, S., Finkelman, R.D., Butler, W.T., 1987. Immunolocalization of Gla proteins (osteocalcin) in rat tooth germs: comparison between indirect immunofluorescence, peroxidase-antiperoxidase, avidin–biotin–peroxidase complex, and avidin–biotin–gold complex with silver enhancement. J. Histochem. Cytochem. 35, 825–830.

Bronfman, F.C., Tcherpakov, M., Jovin, T.M., Fainzilber, M., 2003. Ligand-induced internalization of the p75 neurotrophin receptor: a slow route to the signaling endosome. J. Neurosci. 23 (8), 3209–3220.

Brooks, B.R., Klamerth, O.L., 1968. Interaction of DNA with bifunctional aldehydes. Eur. Biochem. 5, 178.

Bross, P.F., Beitz, J., Chen, G., Chen, X.H., Duffy, E., Kieffer, L., et al., 2001. Approval summary: gemtuzumab ozogamicin in relapsed acute myeloid leukemia. Clin. Cancer Res. 7 (6), 1490–1496.

Brouwers, A.H., Buijs, W.C.A.M., Oosterwijk, E., Boerman, O.C., Mala, C., De Mulder, P.H.M., et al., 2003. Targeting of metastatic renal cell carcinoma with the chimeric monoclonal antibody G250 labeled with 131-I or 111-In: an intrapatient comparison. Clin. Cancer Res. 9 (Suppl.), 3953s–3960s.

Brown, C.E., Howe, L., Sousa, K., Alley, S.C., Carrozza, M.J., Tan, S., et al., 2001. Recruitment of HAT complexes by direct activator interactions with the ATM-related tra1 subunit. Science 292, 2333.

Brown, D.M., 1974. Chemical reactions of polynucleotides and nucleic acidsTs'O, P.O.P. Basic Principles in Nucleic Acid Chemistry, vol. 2. Academic Press, New York, pp. 1–90.

Brown, H.C., Bernheimer, R., Kim, C.J., Scheppele, S.E., 1967. Structural effects in solvolytic reactions. II. Nature of the intermediates involved in the solvolysis of symmetrically substituted β-anisylethyl derivatives. J. Am. Chem. Soc. 89, 370378.

Browne, D.T., Kent, S.B.H., 1975. Formation of nonamidine products in the reaction of primary amines with imido esters. Biochem. Biophys. Res. Comm. 67, 126.

Browning, J., Ribolini, A., 1989. Studies on the differing effects of tumor necrosis factor and lymphotoxin on the growth of several human tumor lines. J. Immunol. 143, 1859–1867.

Bruce, I.J., Sen, T., 2005. Surface modification of magnetic nanoparticles with alkoxysilanes and their application in magnetic bioseparations. Langmuir 21, 7029–7035.

Bruce, J., 2011. Protein interactions and topologies in cells. Journal of Biomolecular Techniques: The fifty-nineth ASMA Conference on Mass Spectrometry, JBT, 22 (Suppl.), S13.

Bruchez, M., Moronne, M., Gin, P., Weiss, S., Alivisatos, A.P., 1998. Semiconductor nanocrystals as fluorescent biological labels. Science 281, 2013–2015.

Brunner, J., 1993. New photolabeling and cross-linking methods. Annu. Rev. Biochem. 62, 483–514.

Brunswick, M., Finkelman, F.D., Higher, P.F., Inman, J.K., Dintzis, H.M., Mond, J.J., 1988. Picogram quantities of anti-Ig antibodies coupled to dextran induce B cell proliferation. J. Immunol. 140, 3364–3372.

Bryant, L.H., Brechbiel, M.W., Wu, C., Bulte, J.W.M., Herynek, V., Frank, J.A., 1999. Synthesis and relaxometry of high-generation (G = 5, 7, 9, and 10) PAMAM dendrimer-DOTA-gadolinium chelates. J. Magn. Reson. Imaging 9, 348–352.

Buck, M.J., Lieb, J.D., 2004. ChIP-chip: considerations for the design, analysis, and application of genome-wide chromatin immunoprecipitation experiments. Genomics 83, 349–360.

Bugawan, T.L., Begovich, A.B., Erlich, H.A., 1990. Rapid HLA–DPB typing using enzymatically amplified DNA and nonradioactive sequence-specific oligonucleotide probes. Immunogenetics 32, 231–241.

Bull, H.B., 1956. Adsorption of bovine serum albumin on glass. Biochim. Biophys. Acta 19, 464–471.

Bulte, J.W.M., Modo, M.M.J., 2008. Nanoparticles in Biomedical Imaging: Emerging Technologies and Applications. Springer Science + Business Media, LLC, ISBN: 978-0-387-72026-5.

Bunnett, J.F., 1963. Nucleophilic reactivity. Annu. Rev. Phys. Chem. 14, 271.

Burnett, T.J., et al. 1980. Synthesis of a fluorescent boronic acid which reversibly binds to cell walls, etc. Biochem. Biophys. Res. Comm. 96, 157–162.

Burns, J.A., Butler, J.C., Moran, J., Whitesides, G.M., 1991. Selective reduction of disulfides by tris(2-carboxyethyl)phosphine. J. Org. Chem. 56, 2648–2650.

Burns, N.Z., Jacobsen, E.N., 2011. Mannich Reaction. ChemInform 42 (40), Wiley Online Library, DOI: 10.1002/chin.201140201.

Burteau, N., Burton, S., Crichton, R.R., 1989. Stabilisation and immobilisation of penicillin amidase. FEBS Lett. 258, 185–189.

Burtnick, L.D., 1984. Modification of actin with fluorescein isothiocyanate. Biochim. Biophys. Acta 791, 57.

Burtnick, M.N., Heiss, C., Roberts, R.A., Schweizer, H.P., Azadi, P., Brett, P.J., 2012. Development of capsular polysaccharide-based glycoconjugates for immunization against melioidosis and glanders. Front. Cell. Infect. Microbiol. 2, 108.10.3389/fcimb.2012.00108 published online.

Buss, H., Chan, T.P., Sluis, K.B., Domigan, N.M., Winterbourn, C.C., 1997. Protein carbonyl measurement by a sensitive ELISA method. Free Radic. Biol. Med. 23, 361–366.

Butler, J.E., 2000a. Solid supports in enzyme-linked immunosorbent assay and other solid-phase immunoassays. Methods 22, 4–23.

Butler, J.E., 2000b. Solid supports in enzyme-linked immunosorbent assay and other solid-phase immunoassays. Decker, J., Reischl, Methods in Molecular Medicine Molecular Diagnosis of Infectious Diseases, vol. 94. Humana Press, Inc., Totowa, NJ, pp. 333–372.

Butler, J.E., Ni, L., Nessler, R., Joshi, K.S., Suter, M., Rosenberg, B., et al., 1992. The physical and functional behavior of capture antibodies adsorbed on polystyrene. J. Immunol. Methods 150, 77–90.

Butler, J.E., Ni, L., Brown, W.R., Joshi, K.S., Chang, J., Rosenberg, B., et al., 1993. The immunochemistry of sandwich ELISAs—VI. Greater than 90% of monoclonal and 75% of polyclonal anti-fluorescyl capture antibodies (CAbs) are denatured by passive adsorption. Mol. Immunol. 30, 1165–1175.

Butler, J.E., Navarro, P., Sun, J., 1997. Adsorption-induced antigenic changes and their significance in ELISA and immunological disorders. Immunol. Invest. 26 (1–2), 39–54.

Butler, P.J.G., Harris, J.I., Hartley, B.S., Leberman, R., 1967. Use of maleic anhydride for the reversible blocking of amino groups in polypeptide chains. Biochem. J. 103, 78P.

Butts, C., Maksymiuk, A., Goss, G., Soulières, D., Marshall, E., Cormier, Y., et al., 2011. Updated survival analysis in patients with stage IIIB or IV non-small-cell lung cancer receiving BLP25 liposome vaccine (L-BLP25): phase IIB randomized, multicenter, open-label trial. J. Cancer Res. Clin. Oncol. 137 (9), 1337–1342.

Byeon, J.-Y., Limpoco, F.T., Bailey, R.C., 2010. Efficient bioconjugation of protein capture agents to biosensor surfaces using aniline-catalyzed hydrazone ligation. Langmuir 26 (19), 15430–15435.

Byrne, M.J., Johnson, D.B., 1975. Studies on the immobilization of b-galactosidase. Biochem. Soc. Trans. 2, 496.

Bystranowska, D., Siejda, B., Ożyhar, A., Kochman, M., 2012. The dityrosine cross-link as an intrinsic donor for assembling FRET pairs in the study of protein structure. Biophys. Chem. 170, 1–8.

Caamano, C.A., Fernandez, H.N., Paladani, A.C., 1983. Specificity of covalently stabilized complexes of 125I-labeled human somatotropin and components of the lactogenic binding sites of rat liver. Biochem. Biophys. Res. Comm. 115, 29–37.

Cabacungan, J.C., Ahmed, A.I., Feeney, R.E., 1982. Amine boranes as alternative reducing agents for reductive alkylation of proteins. Anal. Biochem. 124, 272–278.

Cai, H., Song, C., Endoh, I., Goyette, J., Jessup, W., Freedman, S.B., et al., 2007. Serum amyloid a induces monocyte tissue factor. J. Immunol. 178, 1852–1860.

Cai, L., Chen, Z.Z., Chen, M.Y., Tang, H.W., Pang, D.W., 2012. MUC-1 aptamer-conjugated dye-doped silica nanoparticles for MCF-7 cells detection. Biomaterials 34, 371–381.

Cai, S., Yang, Q., Bagby, T.R., Forrest, M.L., 2011. Lymphatic drug delivery using engineered liposomes and solid lipid nanoparticles. Adv. Drug Deliv. Rev. 63 (10), 901–908.

Cai, S.X., Glenn, D.J., Gee, K.R., Yan, M., Cotter, R.E., Reddy, N.L., et al., 1993. Chlorinated phenyl azides as photolabeling reagents. Synthesis of an ortho, ortho-dichlorinated arylazido PCP receptor ligand. Bioconjug. Chem. 4, 545–548.

Cai, T., Slebodnick, C., Xu, L., Harich, K., Glass, T.E., Chancellor, C., et al., 2006. A pirouette on a metallofullerene sphere: interconversion of isomers of N-tritylpyrrolidino Ih Sc3N@C80. J. Am. Chem. Soc. 128, 6486–6492.

Cai, W., Gambhir, S.S., Chen, X., 2005. Multimodality tumor imaging targeting integrin αvβ3. BioTechniques 39, S14–S25.

Cai, W., Zhang, X., Wu, Y., Chen, X., 2006. A thiol-reactive 18F-labeling agent, N-[2-(4-18F-fluorobenzamido)ethyl]maleimide, and synthesis of RGD peptide-based tracer for PET imaging of avb3 integrin expression. J. Nucl. Med. 47, 1172–1180.

Calleri, E., Temporini, C., Massolini, G., 2011. Frontal affinity chromatography in characterizing immobilized receptors. J. Pharm. Biomed. Anal. 54 (5), 911–925.

Camarero, J.A., 2006. New developments for the site-specific attachment of protein to surfaces. Biophys. Rev. Lett. 1, 1–28.

Camarero, J.A., Cotton, G.J., Adeva, A., Muir, T.W., 1998. Chemical ligation of unprotected peptides directly from a solid support. J. Pept. Res. 51 (4), 303–316.

Campana, V., Caputo, A., Sarnataro, D., Paladino, S., Tivodar, S., Zurzolo, C., 2007. Characterization of the properties and trafficking of an anchorless form of the prion protein. J. Biol. Chem. 282, 22747–22756.

Campbell, P., Gioannini, T.L., 1979. The use of benzophenone as a photoaffinity label. Labeling in*p*-benzoylphenylacetyl chymotrypsin at unit efficiency. Photochem. Photobiol. 29, 883.

Camperchioli, A., Mariani, M., Bartollino, S., Petrella, L., Persico, M., Orteca, N., et al., 2011. Investigation of the Bcl-2 multimerisation process: structural and functional implications. Biochimica et Biophysica Acta (BBA)-Mol. Cell Res. 1813 (5), 850–857.

Canal, F., Sanchis, J., Vicent, M.J., 2011. Polymer-drug conjugates as nano-sized medicines. Curr. Opin. Biotechnol. 22, 894–900.

Cantarero, L.A., Butler, J.E., Osborne, J.W., 1980. The adsorptive characteristics of proteins for polystyrene and their significance in solid-phase immunoassays. Anal. Biochem. 105, 375–382.

Cantin, G.T., Yi, W., Lu, B., Park, S.K., Xu, T., Lee, J.-D., et al., 2008. Combining protein-based IMAC, peptide-based IMAC, and MudPIT for efficient phosphoproteomic analysis. J. Proteome Res. 7, 1346–1351.

Cantor, C.R., 2012. Fluorescence studies of biopolymer structure. Trans. N. Y. Acad. Sci. 33, 576–585.

Cao, L., van Rantwijk, F., Sheldon, R.A., 2000. Cross-linked enzyme aggregates: a simple and effective method for the immobilization of penicillin acylase. Org. Lett. 2 (10), 1361–1364.

Capaccio, M., Gavalas, V.G., Meier, M.S., Anthony, J.E., Bachas, L.G., 2005. Coupling biomolecules to fullerenes through a molecular adapter. Bioconjug. Chem. 16, 241–244.

Cardoza, J.D., Kleinfeld, A.M., Stallcup, K.C., Mescher, M.F., 1984. Hairpin configuration of H-2Kk in liposomes formed by detergent dialysis. Biochemistry 23, 4401–4409.

Carey, F.A., Sundberg, R.J., 2007. Conversion of alcohols to alkylating agents; sulfonate esters Advanced Organic Chemistry, Part B: Reactions and Synthesis, Fifth ed., Springer Science + Business Media, New York, NY. p. 216.

Carlsson, J., Olsson, I., Axen, R., Drevin, H., 1976. A new method for the preparation of Jack-bean ureas involving covalent chromatography. Acta Chem. Scand. B30, 180–182.

Carlsson, J., Drevin, H., Axen, R., 1978. Protein thiolation and reversible protein–protein conjugation. N-Succinimidyl 3(2-pyridyldithio)propionate, a new heterobifunctional reagent. Biochem. J. 173, 723–737.

Carraway, K.L., Koshland Jr., D.E., 1968. Reaction of tyrosine residues in proteins with carbodiimide reagents. Biochim. Biophys. Acta 160, 272–274.

Carraway, K.L., Triplett, R.B., 1970. Reaction of carbodiimides with protein sulfhydryl groups. Biochim. Biophys. Acta 200, 564–566.

Carregal-Romero, S., Casula, M.F., Gutiérrez, L., Morales, M.P., Böhm, I.B., Heverhagen, J.T., et al., 2012. Biological applications of magnetic nanoparticles. Chem. Soc. Rev. 41, 4306–4334.

Carter, P., Smith, L., Ryan, M., 2004. Identification and validation of cell surface antigens for antibody targeting in oncology. Endocr. Relat. Cancer 11, 659–687.

Carter, P.J., 2006. Potent antibody therapeutics by design. Nat. Rev. Immunol. 6, 343–357.

Caruthers, S.D., Wickline, S.A., Lanza, G.M., 2007. Nanotechnological applications in medicine. Curr. Opin. Biotechnol. 18, 26–30.

Casanova, J., Horowitz, Z.D., Copp, R.P., Nclntyre, W.R., Pascual, A., Samuels, H.H., 1984. Photoaffinity labeling of thyroid hormone nuclear receptors. J. Biol. Chem. 259, 12084–12091.

Cassens, C., Kleene, R., Xiao, M.-F., Friedrich, C., Dityateva, G., Schafer-Nielsen, C., et al., 2010. Binding of the receptor tyrosine kinase TrkB to the neural cell adhesion molecule (NCAM) regulates phosphorylation of NCAM and NCAM-dependent neurite outgrowth. J. Biol. Chem. 285, 28959–28967.

Cassette, E., Helle, M., Bezdetnaya, L., Marchal, F., Dubertret, B., Pons, T., 2012. Design of new quantum dot materials for deep tissue infrared imaging. Adv. Drug Deliv. Rev.10.1016/j.addr.2012.08.016 Available online: 5 September 2012.

Cater, C.W., 1963. The evaluation of aldehydes and other difunctional compounds as cross-linking agents for collagen. J. Soc. Leather Trade Chem. 47, 259.

Caufield, M.P., Horiuchi, S., Tai, P.C., Davis, B.D., 1984. The 64-kilodalton membrane protein of Bacillus subtilis is also present as a multiprotein complex on membrane-free ribosomes. Biochemistry 81, 7772–7776.

Cellet, T.S.P., Guilherme, M.R., Silva, R., Pereira, G.M., Mauricio, M.R., Muniz, E.C., et al., 2012. Synthesis of a thermosensitive surface by construction of a thin layer of poly(N-isopropylacrylamide) on maleimide-immobilized polypropylene. J. Colloid and Interface Sci. 367, 494–501.

Ceroni, P., Bergamini, G., Marchioni, F., Balzani, V., 2005. Luminescence as a tool to investigate dendrimer properties. Prog. Polym. Sci. Vol., Dendrimers and Dendritic Polym. 30 (3–4), 453–473.

Chackerian, B., Schiller, J.T., 2012. Virus-like particles as antigen scaffolds. In: Morrow, W.J.W., Sheikh, N.A., Schmidt, C.S., Davies, D.H. (Eds.), Vaccinology: Principles and Practice. Wiley-Blackwell, Oxford, UK, doi: 10.1002/9781118345313.ch13.

Chahal, H.K., Dai, Y., Saini, A., Ayala-Castro, C., Outten, F.W., 2009. The SufBCD Fe-S scaffold complex interacts with SufA for Fe-S cluster transfer. Biochemistry 48, 10644–10653.

Chaiet, L., Wolf, F.J., 1964. The properties of streptavidin, a biotin-binding protein produced by Streptomycetes. Arch. Biochem. Biophys. 106, 1–5.

Chaiken, I.M., Wilchek, M., Parikh, I. (Eds.), 1983. Affinity Chromatography and Biological Recognition. Academic Press, San Diego p. 173.

Chakravorty, S., Aladegbami, B., Burday, M., Levi, M., Marras, S.A.E., Shah, D., et al., 2010. Rapid universal diagnosis of bacterial pathogens from clinical cultures using a novel sloppy molecular beacon melting temperature signature technique. J. Clin. Microbiol. 48, 258–267.

Chamani, M., Rasi, H.M., Panahi, H.A., Asghar, A., 2011. Pseudo-affinity chromatography of rumen microbial cellulase on Sepharose- Cibacron Blue F3GA. Afr. J. Biotechnol. 10 (24), 4926–4931.

Chamberlain, N.R., Deogny, L., Slaughter, C., Radolf, J.D., Norgard, M.V., 1989. Acylation of the 47-kilodalton major membrane immunogen of Treponema pallidum determines its hydrophobicity. Infect. Immun. 57, 2878–2885.

Chamow, S.M., Kogan, T.P., Peers, D.H., Hastings, R.C., Byrn, R.A., Ashkenazi, A., 1992. Conjugation of soluble CD4 without loss of biological activity via a novel carbohydrate-directed cross-linking reagent. Biol. Ghem. 267, 15916–15922.

Chamow, S.M., Kogan, T.P., Venuti, M., Gadek, T., Harris, R.J., Peers, D.H., et al., 1994. Modification of CD4 immunoadhesin with monomethoxypoly (ethylene glycol) aldehyde via reductive alkylation. Bioconjug. Chem. 5, 133–140.

Champagne, J., Delattre, C., Shanthi, C., Satheesh, B., Duverneuil, L., Vijayalakshmi, M.A., 2007. Pseudoaffinity chromatography using a convective interaction media® disk monolithic column. Chromatographia 65, 639–648.

Chan, J.K., Anderson, B.M., 1975. A novel diazonium-sulfhydryl reaction in the inactivation of yeast alcohol dehydrogenase by diazotized 3-aminopyridine adenine dinucleotide. J. Biol. Chem. 250, 67–72.

Chan, W.C.W., Nie, S., 1998. Quantum dot bioconjugates for ultrasensitive nonisotopic detection. Science 281, 2016–2018.

Chandra, S., Dietrich, S., Lang, H., Bahadur, D., 2012. Dendrimer–Doxorubicin conjugate for enhanced therapeutic effects for cancer. J. Mater. Chem. 21, 5729–5737.

Chandramohan, V., Sampson, J.H., Pastan, I., Bigner, D.D., 2012. Toxin-based targeted therapy for malignant brain tumors. Clin. Dev. Immunol. 2012, doi: 10.1155/2012/480429, Article ID 480429, 15 pages, 2012.

Chang, F.N., Flaks, J.C., 1972. Specific cross-linking of *Escherichia coli* 30S ribosomal subunit. J. Mol. Biol. 68, 177.

Chang, M., Colvin, M., Rembaum, A., 1986. Acrolein and 2-hydroxyethyl methacrylate copolymer microspheres. J. Polym. Sci. Part C: Polym. Lett. 24, 603–606.

Chang, M.-H., Chen, C.-J., Lai, M.-S., Hsu, H.-M., Wu, T.-C., Kong, M.-S., et al., 1997. Universal Hepatitis B Vaccination in Taiwan and the incidence of Hepatocellular Carcinoma in children. N. Engl. J. Med. 336 (26), 1855–1859.

Chang, N., Sutherland, C., Hesse, E., Winkfein, R., Wiehler, W.B., Pho, M., et al., 2007. Identification of a novel interaction between the $Ca2^{1}$-binding protein S100A11 and the Ca^{21}-and phospholipid-binding protein annexin A6. Am. J. Physiol. Cell Physiol. 292, C1417–C1430.

Chang, Y.-A., Gee, A., Smith, A., Lake, W., 1992. Activating hydroxyl groups of polymeric carriers using 4-fluorobenzenesulfonyl chloride. Bioconjug. Chem. 3, 200–202.

Chantler, P., Bower, S.M., 1988. Cross-linking between translationally equivalent sites on the heads of myosin: relationship to energy transfer results between the same pair of sites. J. Biol. Chem. 263, 938.

Chari, R.V., 2008. Targeted cancer therapy: conferring specificity to cytotoxic drugs. Acc. Chem. Res. 41 (1), 98–107.

Charleux, B., D'Agosto, F., Delaittre, G., 2011. Preparation of hybrid latex particles and core–shell particles through the use of controlled radical polymerization techniques in aqueous media. Hybrid Latex Particles, 125–183.

Chase, J.W., Merrill, B.M., Williams, K.P., 1983. F sex factor encodes a single-stranded DNA binding protein (SSB) with extensive sequence homology to *Escherichia coli* SSB. Proc. Natl. Acad. Sci. USA 80, 5480–5484.

Chattopadhyay, A., James, H.I., Fair, P.S., 1992. Molecular recognition sites on factor Xa which participate in the prothrombinase complex. J. Biol. Chem. 267, 12323–12329.

Chaudhuri, A.R., de Waal, E.M., Pierce, A., Van Remmen, H., Ward, W.F., Richardson, A., 2007. Detection of protein carbonyls in aging liver tissue: a fluorescence-based proteomic approach. Mech. Ageing Dev. 127 (11), 849–861.

Chaudhuri, D., Suriano, R., Mittelman, A., Tiwari, R.K., 2009. Targeting the immune system in cancer. Curr. Pharm. Biotechnol. 10 (2), 166–184.

Chazotte, B., Hackenbrock, C.R., 1991. Lateral diffusion of redox components in the mitochondrial inner membrane is unaffected by inner membrane folding and matrix density. J. Biol. Chem. 266, 5973.

Chehab, F.F., Kan, Y.W., 1989. Detection of specific DNA sequences by fluorescence amplification: a color complementation assay. Proc. Natl. Acad. Sci. USA 86, 9178.

Chelsky, D., Dahlquist, F.W., 1980. Chemotaxis in *Escherichia coli*: association of protein components. Biochemistry 19, 4633–4639.

Chen, A., Kozak, D., Battersby, B.J., Forrest, R.M., Scholler, N., Urban, N., et al., 2009. Antifouling surface layers for improved signal-to-noise of particle-based immunoassays. Langmuir 25, 13510–13515.

Chen, J., Selvin, P.R., 1999. Thiol-reactive luminescent chelates of terbium and europium. Bioconjug. Chem. 10, 311–315.

Chen, J., Arnold, M.A., Small, G.W., 2004. Comparison of combination and first overtone spectral regions for near-infrared calibration models for glucose and other biomolecules in aqueous solutions. Anal. Chem. 76, 5405–5413.

Chen, J.R., Ma, C., Wong, C.H., 2011. Vaccine design of hemagglutinin glycoprotein against influenza. Trends Biotechnol. 29 (9), 426–434.

Chen, K.K., Rose, C.L., Clowes, C.H.A., 1934. Am. J. Med. Sci. 188, 767.

Chen, R., Kang, V.H., Chen, J., Shope, J.C., Torabinejad, J., DeWald, D.B., et al., 2002. A monoclonal antibody to visualize PtdIns(3,4,5)P3 in cells. J. Histochem. Cytochem. 50, 697–708.

Chen, R.F., Knutson, J.R., 1988. Mechanism of fluorescent concentration quenching of carboxyfluorescein in liposomes: energy transfer to nonfluorescent dimers. Anal. Biochem. 172, 61.

Chen, R.J., Bangsaruntip, S., Drouvalakis, K.A., Shi Kam, N.W., Shim, M., Li, Y., et al., 2003. Noncovalent functionalization of carbon nanotubes for highly specific electronic biosensors. Proc. Natl. Acad. Sci. USA 100, 4984–4989.

Chen, W.-T., Thirumalai, D., Shih, T.T.-F., Chen, R.-C., Tu, S.-Y., Lin, C.-I., et al., 2010. Dynamic contrast-enhanced folate-receptor-targeted MR imaging using a Gd-loaded PEG-dendrimer–folate conjugate in a mouse xenograft tumor model. Mol. Imaging Biol. 12, 145–154.

Chen, X., Conti, P.S., Moats, R.A., 2004. In vivo near-infrared fluorescence imaging of integrin $\alpha v \beta 3$ in brain tumor xenografts. Cancer Res. 64, 8009–8014.

Chen, Y., Anderson, D.E., Rajagopalan, M., Erickson, H.P., 2007. Assembly dynamics of mycobacterium tuberculosis FtsZ. J. Biol. Chem. 10 doi: 1074/jbc.M703788200.

Chen, Y.X., Triola, G., Waldmann, H., 2011. Bioorthogonal chemistry for site-specific labeling and surface immobilization of proteins. Acc. Chem. Res. 44 (9), 762–773.

Cheng, F., Shang, J., Ratner, D.M., 2011. A versatile method for functionalizing surfaces with bioactive glycans. Bioconjug. Chem. 22, 50–57.

Cheng, F., Svensson, G., Fransson, L.Å., Mani, K., 2012. Non-conserved, S-nitrosylated cysteines in glypican-1 react with N-unsubstituted glucosamines in heparan sulfate and catalyze deaminative cleavage. Glycobiology 22, 1480–1486.

Cheng, Q., Brajter-Toth, A., 1992. Selectivity and sensitivity of self-assembled thioctic acid electrodes. Anal. Chem. 64, 1998–2000.

Cheng, Y., Dovichi, N.J., 1988. Subattomole amino acid analysis by capillary zone electrophoresis and laser-induced fluorescence. Science 242, 562.

Cheng, Y., Samia, A.C., Meyers, J.D., Panagopoulos, I., Fei, B., Burda, C., 2008. Highly efficient drug delivery with gold nanoparticle vectors for in vivo photodynamic therapy of cancer. J. Am. Chem. Soc. 130 (32), 10643–10647.

Cheng, Z., Wu, Y., Xiong, Z., Gambhir, S.S., Chen, X., 2005. Near-infrared fluorescent RGD peptides for optical imaging of integrin avb3 expression in living mice. Bioconjug. Chem. 16, 1433–1441.

Cheong, W.F., Prahl, S.A., Welch, A.J., 1990. A review of the optical properties of biological tissues. IEEE J. Quantum Electron. 26, 2166–2195.

Chetrit, P., Gaudin, V., de Courcel, A., Vedel, F., 1989. A cross-hybridization method for DNA mapping with photobiotin-labeled probes. Anal. Biochem. 178, 273–275.

Cheung, H.H., Lee, T.L., Rennert, O.M., Chan, W.Y., 2012. Methylation profiling using methylated DNA immunoprecipitation and tiling array hybridization. Methods Mol. Biol. (Clifton, NJ) 825, 115–126. Chapter 10.

Cheung, R.C.F., Wong, J.H., Ng, T.B., 2012. Immobilized metal ion affinity chromatography: a review on its applications. Appl. Microbiol. Biotechnol., 1–10.

Chi, K.R., 2009. Ever-increasing resolution; Overcoming the limitations of spatial and temporal resolution to image within a cell is no easy feat. Nature 462, 675–678.

Chiang, I.W., Brinson, B.E., Smalley, R.E., Margrave, J.L., Hauge, R.H., 2001. Purification and characterization of single-wall carbon nanotubes. J. Phys. Chem. B105, 1157–1161.

Chiang, Y.H., Wu, Y.J., Lu, Y.T., Chen, K.H., Lin, T.C., Chen, Y.K.H., et al., 2011. Simple and specific dual-wavelength excitable dye staining for glycoprotein detection in polyacrylamide gels and its application in glycoproteomics. J. Biomed. Biotechnol. doi: 10.1155/2011/780108, Article ID 780108, 8 pages.

Chien, C.-T., Bartel, P.L., Sternglanz, R., Fields, S., 1991. The two-hybrid system: a method to identify and clone genes for proteins that interact with a protein of interest. Proc. Natl. Acad. Sci. USA 88, 9578–9582.

Childs, C.V., Lloyd, J.M., Unabia, C., Gharib, S.D., Wierman, M.E., Chin, W.W., 1987. Detection of luteinizing hormone b messenger ribonucleic acid (RNA) in individual gonadotropes after castration: use of a new in situ hybridization method with a photobiotinylated complementary RNA probe. Mol. Endocrinol. 1, 926–932.

Chiou, S.-H., Wu, W.-T., 2004. Immobilization of Candida rugosa lipase on chitosan with activation of the hydroxyl groups. Biomaterials 25, 197–204.

Cho, K., Wang, X., Nie, S., Chen, Z.G., Shin, D.M., 2008. Therapeutic nanoparticles for drug delivery in cancer. Clin. Cancer Res. 14 (5), 1310–1316.

Choi, J.-W., Godwin, A., Balan, S., Bryant, P., Cong, Y., Pawlisz, E., et al., 2009. Rebridging disulfides: site-specific PEGylation of sequential bis-alkylation in PEGylated protein drugs: basic science and clinical applications. Milestones in Drug Ther., 47–73.

Choi, N., Kim, S.-M., Hong, K.S., Cho, G., Cho, J.-H., Lee, C., et al., 2012. The use of the fusion protein RGD-HSA-TIMP2 as a tumor targeting imaging probe for SPECT and PET. Biomaterials 32, 7151–7158.

Choi, S.K., Thomas, T.P., Li, M.-H., Desai, A., Kotlyar, A., Baker, J.R., 2012. Photochemical release of methotrexate from folate receptor-targeting PAMAM dendrimer nanoconjugate. Photochem. Photobiol. Sci. 11, 653–660.

Chong, S., Mersha, F.B., Comb, D.G., Scott, M.E., Landry, D., Vence, L.M., et al., 1997. Single-column purification of free recombinant proteins using a self-cleavable affinity tag derived from a protein splicing element. Gene 192, 271–281.

Chong, S., Montello, G.E., Zhang, A., Cantor, E.J., Liao, W., Xu, M.-Q., et al., 1998. Utilizing the C-terminal cleavage activity of a protein splicing element to purify recombinant proteins in a single chromatographic step. Nucl. Acids Res. 26, 5109–5115.

Chong, S.A.C., Lee, W., Arora, P.D., Laschinger, C., Young, E.W.K., Simmons, C.A., et al., 2007. Methylglyoxal inhibits the binding step of collagen phagocytosis. J. Biol. Chem. 282, 8510–8520.

Choudhary, S., Mathew, M., Verma, R.S., 2011. Therapeutic potential of anticancer immunotoxins. Drug Discov. Today 16 (11), 495–503.

Chowdhry, V., Vaughn, R., Westheimer, F.H., 1976. 2-Diazo-3,3,3-trifluoropropionyl chlorides: reagent for photoaffinity labeling. Proc. Natl. Acad. Sci. USA 73, 1406–1408.

Christman, K.L., Schopf, E., Broyer, R.M., Li, R.C., Chen, Y., Maynard, H.D., 2009. Positioning multiple proteins at the nanoscale with electron beam cross-linked functional polymers. J. Am. Chem. Soc. 131, 521–527.

Chu, B.C.F., Orgel, L.E., 1988. Ligation of oligonucleotides to nucleic acids or proteins via disulfide bonds. Nucleic Acids Res. 16, 3671.

Chu, B.C.F., Wahl, G.M., Orgel, L.E., 1983. Derivatization of unprotected polynucleotides. Nucleic Acids Res. 11, 6513–6529.

Chu, B.C.F., Kramer, F.R., Orgel, L.E., 1986. Synthesis of an amplifiable reporter RNA for bioassays. Nucleic Acids Res. 14, 5591–5603.

Chu, F.S., Ueno, I., 1977. Production of antibody against aflatoxin B1. Appl. Environ. Microbiol. 33, 1125–1128.

Chu, F.S., Fred, C.C., Hinsdill, R.D., 1976. Production of antibody against ochratoxin A. Appl. Environ. Microbiol. 31, 831–835.

Chu, F.S., Lau, H.P., Fan, T.S., Zhang, C.S., 1982. Ethylenediamine modified bovine serum albumin as protein carrier in the production of antibody against mycotoxins. J. Immunol. Methods 55, 73–78.

Cimino, C.P., Camper, H.B., Isaacs, S.T., Hearst, J.E., 1985. Psoralens as photoactive probes of nucleic acid structure and function: organic chemistry, photochemistry, and biochemistry. Annu. Rev. Biochem. 54, 1151–1193.

Clapp, A.R., Goldman, E.R., Mattoussi, H., 2006. Capping of CdSe-ZnS quantum dots with DHLA and subsequent conjugation with proteins. Nat. Protoc. 1 (3), 1258–1266.

Clarke, A.E., Wilson, I.A. (Eds.), 1988. Carbohydrate–Protein Interactions. Springer-Verlag, Heidelberg, Germany.

Clausen, J., 1988. Immunochemical techniques for the identification and estimation of macromolecules (third ed.)Burdon, R.H.Knippenberg, P.H. Laboratory Techniques in Biochemistry and Molecular Biology, vol. 1. Elsevier, New York. Part 3.

Cleland, W.W., 1964. Dithiothreitol, a new protective reagent for SH groups. Biochemistry 3, 480–482.

Clevenger, R.C., Raibel, J.M., Peck, A.M., Blagg, B.S.J., 2004. Biotinylated geldanamycin. J. Org. Chem. 69, 4375–4380.

Clonis, Y.D., Labrou, N.E., Kotsira, V.P., Mazitsos, C., Melissis, S., Gogolas, G., 2000. Biomimetic dyes as affinity chromatography tools in enzyme purification. Journal of Chromatography A 891, 33–44.

Cocco, L., Martelli, A., Billi, A., Matteucci, A., Vitale, M., Neri, L., et al., 1986. Changes in nucleosome structure and histone H3 accessibility. Iodoacetamidofluorescein labeling after treatment with phosphatidylserine vesicles. Exp. Cell Res. 166, 465–474.

Codelli, J.A., Baskin, J.M., Agard, N.J., Bertozzi, C.R., 2008. Second-Generation difluorinated cyclooctynes for copper-Free click chemistry. J. Am. Chem. Soc. 130, 11486–11493.

Cohn, E.J., et al. 1947. Preparation and properties of serum and plasma proteins. XIII. Crystallization of serum albumins from ethanol–water mixtures. J. Am. Chem. Soc. 69, 1753–1761.

Cohen, S., Cahan, R., Ben-Dov, E., Nisnevitch, M., Zaritsky, A., Michael, A.F., 2007. Specific targeting to murine myeloma cells of Cyt1Aa toxin from *Bacillus thuringiensis* subsp. Israelensis. J. Biol. Chem. doi: 10.1074/jbc.M703567200.

Cohen-Anisfeld, S.T., Lansbury Jr., P.T., 1993. A practical, convergent method for glycopeptide synthesis. J. Am. Chem. Soc. 115, 10531–10537.

Cole, R.D., 1967. S-AminoethylationHirs, C.H.W. Methods in Enzymology, vol. 11. Academic Press, New York, pp. 315–317.

Cole, R.D., Stein, W.H., Moore, S., 1958. On the cysteine content of human hemoglobin. J. Biol. Chem. 233, 1359–1363.

Coleman, P.L., Walker, M.M., Milhrath, D.S., Stauffer, D.M., Rasmussen, J.K., Krepski, L.R., et al., 1990. Immobilization of Protein A at high density on azlactone-functional polymeric beads and their use in affinity chromatography. J. Chromatogr. 512, 345–363.

Colland, F., Fujita, N., Ishihama, A., Kolb, A., 2002. The interaction between sigmaS, the stationary phase sigma factor, and the core enzyme of *Escherichia coli* RNA polymerase. Genes Cells 7, 233–247.

Collier, R.J., Cole, H.A., 1969. Diphtheria toxin subunit active in vitro. Science 164, 1179.

Collier, R.J., Kandel, J., 1971. Structure and activity of diphtheria toxin. J. Biol. Chem. 246, 1496–1503.

Collins, M.O., Yu, L., Coba, M.P., Husi, H., Campuzano, I., Blackstock, W.P., et al., 2005. Proteomic analysis of in vivo phosphorylated synaptic proteins. J. Biol. Chem. 280, 5972–5982.

Collins, P.G., Avouris, P., 2000. Nanotubes for electronics. Sci. Am., 67.

Collioud, A., Clemence, J.-F., Sänger, M., Sigrist, H., 1993. Oriented and covalent immobilization of target molecules to solid supports: synthesis and application of a light-activatable and thiol-reactive cross-linking reagent. Bioconjug. Chem. 4, 528–536.

Colman, R.F., 1969. The role of sulfhydryl groups in the catalytic function of isocitrate dehydrogenase. I. Reaction with 5,59-dithiobis (2-nitrobenzoic acid). Biochemistry 8, 888.

Colombatti, M., Nabholz, M., Gros, O., Brown, C., 1983. Selective killing of target cells by antibody-ricin A-chain or antibody-gelonin hybrid molecules: comparison of cytotoxic potency and use in immunoselection procedures. J. Immunol. 131, 3091.

Colombatti, M., Greenfield, L., Youle, R.J., 1986. Cloned fragment of diphtheria toxin linked to T cell-specific antibody identifies regions of B chain active in cell entry. J. Biol. Chem. 261, 3030.

Colvin, M., Smolka, A., Rembaum, A., Chang, M., 1988. The Covalent Binding of Enzymes and Immunoglobulins to Hydrophilic Microspheres in Microspheres: Medical and Biological Applications. CRC, Boca Raton, FL. pp. 1–13.

Comley, J., 2006. TR-FRET based assays–getting better with age. Drug Discov. World 7 (2), 22–38.

Conn, P.M., Rogers, D.C., Stewart, J.M., Niedel, J., Sheffield, T., 1982a. Conversion of a gonadotropin-releasing hormone antagonist to an agonist. Nature (London) 296, 633–655.

Conn, P.M., Rogers, P.C., McNeil, R., 1982b. Potency enhancement of a GnRH agonist: GnRH – receptor microaggregation stimulates gonadotropin release. Endocrinology (Baltimore) 111, 335–337.

Connock, M., Tubeuf, S., Malottki, K., Uthman, A., Round, J., Bayliss, S., et al., 2010. Certolizumab pegol (CIMZIA®) for the treatment of rheumatoid arthritis. Health Technol. Assess 14 (Suppl. 2), 1–10.

Cordes, E.H., Jencks, W.P., 1962. Nucleophilic catalysis of semicarbazone formation by anilines. J. Am. Chem. Soc. 84, 826–831.

Corneillie, T.M., Fisher, A.J., Meares, C.F., 2003. Crystal structures of two complexes of the rare-earth-DOTA-binding antibody 2d12.5: ligand generality from a chiral system. J. Am. Chem. Soc. 125, 15039–15048.

Corneillie, T.M., Lee, K.C., Whetstone, P.A., Wong, J.P., Meares, C.F., 2004. Irreversible engineering of the multielement-binding antibody 2D12.5 and its complementary ligands. Bioconjug. Chem. 15 (6), 1392–1402.

Correia-Pinto, J.F., Csaba, N., Alonso, M.J., 2012. Vaccine delivery carriers: insights and future perspectives. Int. J. Pharm. doi: 10.1016/j.ijpharm.2012.04.047, in press.

Coscoy, S., Lingueglia, E., Lazdunski, M., Barbry, P., 1998. The Phe-Met-Arg-Phe-amide-activated sodium channel is a tetramer. J. Biol. Chem. 273, 8317.

Coulter, A., Harris, R., 1983. Simplified preparation of rabbit Fab fragments. J. Immunol. Methods 59, 199–203.

Cover, J.A., Lambert, J.M., Norman, C.M., Traut, R.R., 1981. Identification of proteins at the subunit interface of the *Escherichia coli* ribosome by cross-linking with dimethyl 3,3′-dithiobis(propio nimidate). Biochemistry 20, 2843–2852.

Cox, J.P.L., Craig, A.S., Helps, L.M., Jandowski, K.J., Parker, P., Eaton, M.A.W., et al., 1990. Synthesis of C- and N-functionalized derivatives of 1,4,7-triazacyclononane-1,4,7-triyltriacetic acid (NOTA), 1,4,7,10-tetraazacyclododecane-1,4,7,10-tetrayltetraacetic acid (DOTA), and diethylenetriaminepentaacetic acid (DTPA): bifunctional complexing agents for the derivatization of antibodies. J. Chem. Soc. Perkin Trans. 1, 2567–2576.

Cramer, E.M., Beesley, J.E., Pulford, K.A.F., Breton-Gorius, J., Mason, D.Y., 1989. Colocalization of elastase and myeloperoxidase in human blood and bone marrow neutrophils using a monoclonal antibody and immunogold. Am. J. Pathol. 134, 1275–1284.

Crane, B.R., 2011. Structural biology by mass-spec: mapping protein interaction surfaces of membrane receptor complexes with ICAT. J. Mol. Biol. 409 (4), 481–482.

Cravatt, B.F., Wright, A.T, Kozarich, J.W., 2008. Activity-based protein profiling: from enzyme chemistry to proteomic chemistry. Annu. Rev. Biochem. 77, 383–414.

Cremlyn, R.J., 1996. An Introduction to Organosulfur Chemistry. Wiley, Chichester, UK. pp. 104–107.

Crestfield, A.M., Stein, W.H., Moore, S., 1963. Alkylation and identification of the histidine residues at the active site of ribonuclease. J. Biol. Chem. 238, 2413–2419.

Crestini, C., Crusianelli, M., Orlandi, M, Saladino, R., 2010. Oxidative strategies in lignin chemistry: a new environmental friendly approach for the functionalisation of lignin and lignocellulosic fibers. Catal. Today 156, 8–22.

Criekinge, W.V., Beyaert, R., 1999. Yeast two-hybrid: state of the art. Biol. Proced. Online 2 (1) <www.biologicalprocedures.com>.

Crielaard, B.J., van der Wal, S., Le, H.T., Bode, A.T., Lammers, T., Hennink, W.E., et al., 2012. Liposomes as carriers for colchicine-derived prodrugs: vascular disrupting nanomedicines with tailorable drug release kinetics. Eur. J. Pharm. Sci. 45 (4), 429–435.

Cristea, I.M., Gaskell, S.J., Whetton, A.D., 2004. Proteomics techniques and their application to hematology. Blood 103, 3624–3634.

Crowther, J.R., 2009. The ELISA guidebook, second ed., Methods in Molecular Biology, vol. 516, p. 566, ISBN: 978-1-60327-253-7.

Cruz, L.J., Iglesias, E., Aguilar, J.C., Quintana, D., Garay, H.E., Duarte, C., et al., 2001. Study of different coupling reagents in the conjugation of a V3-based synthetic MAP to carrier proteins. J. Pep. Sci. 7 (9), 511–518.

Cuatrecasas, P., 1970. Protein purification by affinity chromatography; derivatizations of agarose and polyacrylamide beads. J. Biol. Chem. 245, 3059–3065.

Cuatrecasas, P., 1972. Affinity chromatography of macromolecules. Adv. Enzymol. 36, 29–89.

Cuatrecasas, P., Parikh, I., 1972. Adsorbents for affinity chromatography. Use of N-hydroxysuccinimide esters of agarose. Biochemistry 11, 2291–2299.

Cuatrecasas, P., Wilchek, M., 1968. Single-step purification of avidin from egg white by affinity chromatography of biocytin–Sepharose columns. Biochem. Biophys. Res. Comm. 33, 235–246.

Cubie, H.A., Norval, M., 1989. Detection of human papilloma viruses in paraffin wax sections with biotinylated synthetic oligonucleotide probes and immunogold staining. J. Clin. Pathol. 42, 988–991.

Cui, Y., Chen, X., Li, Y., Liu, X., Lei, L., Xuan, S., 2012. Novel magnetic microspheres of P (GMA-b-HEMA): preparation, lipase immobilization and enzymatic activity in two phases. Appl. Microbiol. Biotechnol., 1–10.

Cumber, A.J., Forrester, J.A., Foxwell, B.M.J., Ross, W.C.J., Thorpe, P.E., 1985. Preparation of antibody–toxin conjugates, Widder, K.J.Green, R. Methods in Enzymology, vol. 112. Academic Press, New York, pp. 207–224.

Cummings, R.D., 2005. Apparatus and method for mixed-bed lectin chromatography. U.S. Patent Application 2005/0245737A1.

Cummings, R.D., Smith, D.F., 2005. The selectin family of carbohydrate-binding proteins: structure and importance of carbohydrate ligands for cell adhesion. BioEssays 14 (12), 849–856.

Curigliano, G., Spitaleri, G., Pietri, E., Rescigno, M., de Braud, F., Cardillo, A., et al., 2006. Breast cancer vaccines: a clinical reality or fairy tale? Ann. Onc. 17, 750–762.

Cursio, R., Colosetti, P., Auberger, P., Gugenheim, J., 2008. Liver apoptosis following normothermic ischemia-reperfusion: in vivo evaluation of caspase activity by FLIVO™ assay in rats. Transplant. Proc. 40, 2038–2041.

Cursio, R., Miele, C., Filippa, N., Colosetti, P., Auberger, P., Van Obberghen, E., et al., 2009. Tyrosine phosphorylation of insulin receptor substrates during ischemia/reperfusion-induced apoptosis in rat liver. Langenbecks Arch. Surg. 394, 123–131.

Curtis, J.M., Hahn, W.S., Stone, M.D., Inda, J.J., Droullard, D.J., Kuzmicic, J.P., et al., 2012. Protein carbonylation and adipocyte mitochondrial function. J. Biol. Chem. 287 (39), 32967–32980.

Cusan, C., Da Ros, T., Spalluto, G., Foley, S., Janot, J.-M., Seta, P., et al., 2002. A new multi-charged C60 derivative: synthesis and biological properties. Eur. J. Org. Chem. 17, 2928–2934.

Czerney, P., Wenzel, M., Schweder, B., Frank, W., 2005. Benzopyrylo-polymethine-based hydrophilic markers. U.S. Patent No. 6,924,372 B2.

Czworkowski, J., Odom, O.W., Hardesty, B., 1991. Study of the topology of messenger RNA bound to the 30S ribosomal subunit of *Escherichia coli*. Biochemistry 30, 4821.

Dabbousi, B.O., Rodriguez-Viejo, J., Mikulec, F.V., Heine, J.R., Mattouss, I.H., Ober, R., et al., 1997. (CdSe)ZnS core-shell quantum dots: synthesis and characterization of a size series of highly luminescent nanocrystallites. J. Phys. Chem. B101, 9463–9475.

Daemen, T., Veninga, A., Dijkstra, J., Scherphof, G., 1989. Differential effects of liposomeincorporation on liver macrophage activating potencies of rough lipopolysaccharide, lipid A, and muramyl dipeptide: differences in susceptibility to lysosomal enzymes. Immunol. 142, 2469–2474.

Dahan, M., Levi, S., Luccardini, C., Rostaing, P., Riveau, B., Triller, A., 2003. Diffusion dynamics of glycine receptors revealed by single-quantum dot tracking. Science 302, 442–445.

Dai, H., Smith, A., Meng, X.W., Schneider, P.A., Pang, Y.-P., Kaufmann, S.H., 2011. Transient binding of an activator BH3 domain to the Bak BH3-binding groove initiates Bak oligomerization. J. Cell Biol. 194, 39–48.

Dainiak, M.B., Plieva, F.M., Galaev, I.Y., Hatti-Kaul, R., Mattiasson, B., 2005. Cell chromatography. Separation of different microbial cells using IMAC supermacroporous monolithic columns. Biotechnol. Prog. 21, 644–649.

Dakin, H.D., 1906. The oxidation of amido-acids with the production of substances of biological importance. J. Biol. Chem. 1, 171–176.

Dalhoff, C., Hüben, M., Lenz, T., Poot, P., Nordhoff, E., Köster, H., et al., 2010. Synthesis of S-adenosyl-L-homocysteine capture compounds for selective photoinduced isolation of methyltransferases. ChemBioChem 11 (2), 256–265.

Damjanovich, S., Kleppe, K., 1966. Biochim. Biophys. Acta 122, 145.

Danscher, C., Rytter-Nörgaard, 1983. Light microscropic visualization of colloidal gold on resin-embedded tissue. J. Histochem. Cytochem. 31, 1394–1398.

Darbandi, M., Nann, T., 2005. Single quantum dots in silica spheres by microemulsion synthesis. Chem. Mater. 17, 5720–5725.

Darzynkiewicz, Z., Crissman, H.A., 1990. Flow cytometry. Methods Cell Biol.

Das, M., Fox, C.F., 1979. Chemical cross-linking in biology. Annu. Rev. Biophys. Bioeng. 8, 165–196.

Das, N., Valjavec-Gratian, M., Basuray, A.N., Fekete, R.A., Papp, P.P., Paulsson, J., et al., 2005. Multiple homeostatic mechanisms in the control of P1 plasmid replication. PNAS 102, 2856–2861.

Datta, D., Wang, P., Carrico, I.S., Mayo, S.L., Tirrell, D.A., 2002. A designed phenylalanyl-tRNA synthetase variant allows efficient in vivo incorporation of aryl ketone functionality into proteins. J. Am. Chem. Soc. 124, 5652–5653.

Daum, J.R., et al. 2000. The 3F3/2 anti-phosphoepitope antibody binds the mitotically phosphorylated anaphase-promoting complex/cyclosome. Curr. Biol. 10 (23) R850–R857. S1–S2.

David, R., Richter, M.P.O., Beck-Sickinger, A.G., 2004. Expressed protein ligation; method and applications. Eur. J. Biochem. 271, 663–677.

Davidson, R.S., Hilchenbach, M.M., 1990. The use of fluorescent probes in immunochemistry. Photochem. Photobiol. 52, 431.

Davies, C.E., Kaplan, J.G., 1972. Use of diimidoester cross-linking reagent to examine the subunit structure of rabbit muscle pyruvate kinase. Can. J. Biochem. 50, 416–422.

Davies, C.E., Palek, J., 1981. 125I-labeling of platelet proteins with Bolton–Hunter reagent. Anal. Biochem. 115, 383–387.

Davies, C.E., Stark, G.R., 1970. Use of dimethyl suberimidate, a cross-linking reagent, in studying the subunit structure of oligomeric proteins. Proc. Natl. Acad. Sci. USA 66, 651.

Davies, K.J.A., 1987. Protein damage and degradation by oxygen radicals. J. Biol. Chem. 262 (20), 9895–9901.

Davies, M.J., Fu, S., Wang, H., Dean, R.T., 1999. Stable markers of oxidant damage to proteins and their application in the study of human disease. Free Radic. Biol. Med. 27, 1151–1163.

Davis, B.G., 1999. Recent developments in glycoconjugates. J. Chem. Soc. Perkin Trans. 1, 3215–3237.

Davis, B.G., Jones, J.B., Bott, R.R., 2005. Synthesis and use of glycodendrimer reagents. US Patent Application 2005/0272670.

Davis, F.F., Van Es, T., Palczuk, N.C., 1979. Nonimmunogenic polypeptides. US Patent 4,179,337.

Davis, P.D., Crapps, E.C., 2006. Selective and specific preparation of discrete PEG compounds. US Patent Application Publication No. 2006/0020134.

Dawson, P.E., Kent, S.B., 2000. Synthesis of native proteins by chemical ligation. Annu. Rev. Biochem. 69, 923–960.

Dawson, P.E., Churchill, M., Ghadiri, M.R., Kent, S.B.H., 1997. Modulation of reactivity in native chemical ligation through the use of thiol additives. J. Am. Chem. Soc. 119, 4325–4329.

Dawson, P.E., Muir, T.W., Clark-Lewis, I., Kent, S.B.H., 1994. Synthesis of proteins by native chemical ligation. Science 266, 776–779.

Dayon, L., Sanchez, J.C., 2012. Relative protein quantification by MS/MS using the tandem mass tag technology. Methods Mol. Biol. (Clifton, NJ) 893, 115–127.

de Araujo, A.D., Palomo, J.M., Cramer, J., Seitz, O., Alexandrov, K., Waldmann, H., 2006. Diels–Alder ligation of peptides and proteins. Chemistry 12 (23), 6095–6109.

De Bruin, B., Kuhnast, B., Hinnen, F., Yaouancq, L., Amessou, M., Johannes, L., et al., 2005. 1-[3-(2-[18F]Fluoropyridin-3-yloxy)propyl] pyrrole-2,5-dione: design, synthesis and radiosynthesis of a new [18F]fluoropyridine-based maleimide reagent for the labeling of peptides and proteins. Bioconjug. Chem. 16, 406–420.

De Filette, M., Ysenbaert, T., Roose, K., Schotsaert, M., Roels, S., Goossens, E., et al., 2011. Antiserum against the conserved nine amino acid N-terminal peptide of influenza A virus matrix protein 2 is not immunoprotective. J. Gen. Virol. 92, 301–306.

de Graaf, A.J., Kooijman, M., Hennink, W.E., Mastrobattista, E., 2009. Nonnatural amino acids for site-specific protein conjugation. Bioconjug. Chem. 20 (7), 1281–1295.

De Grand, A.M., Frangioni, J.V., 2003. An operational near-infrared fluorescence imaging system prototype for large animal surgery. Technol. Cancer Res. Treat. 2, 553–562.

de Jager, S.C., Kuiper, J., 2011. Vaccination strategies in atherosclerosis. Thromb. Haemost. 106 (5), 796–803.

de la Burde, R., Peckham, L., Veis, A., 1963. The action of hydrazine on collagen. J. Biol. Chem. 238, 189–197.

De Mey, J., Moeremans, M., Ceuens, G., Nuydens, R., De Brabander, M., 1981. High resolution light and electron microscopic localization of tubulin with the IgS (immuno gold staining) method. Cell Biol. Int. Rep. 5, 889.

De Oliveira, V.L., Almeida, S.C., Soares, H.R., Crespo, A., Marshall-Clarke, S., Parkhouse, R.M., 2011. A novel TLR3 inhibitor encoded by African swine fever virus (ASFV). Arch. Virol. 156 (4), 597–609.

de Rosario, R.B., Wahl, R.L., Brocchini, S.J., Lawton, R.C., Smith, R.H., 1990. Sulfhydryl site-specific cross-linking and labeling of monoclonal antibodies by a fluorescent equilibrium transfer alkylation cross-link reagent. Bioconjug. Chem. 1, 51–59.

De Silva, B.S., Egodage, K.L., Wilson, G.S., 1999. Purified protein derivative (PPD) as an immunogen carrier elicits high antigen specificity to haptens. Bioconjug. Chem. 10 (3), 496–501.

de Souza, M.C.M., Bresolin, I.T.L., Bueno, S.M.A., 2010. Purification of human IgG by negative chromatography on ω-aminohexyl-agarose. J. Chromatogr. B 878, 557–566.

De Waele, M., Renmans, W., Segers, E., De Valck, V., Jochmans, K., van Camp, B., 1989. An immunogold-silver staining method for detection of cell surface antigens in cell smears. J. Histochem. Cytochem. 37, 1855–1862.

DeBlaquiere, J., Burgess, A.W., 1999. Affinity purification of the plasma membranes. J. Biomol. Tech. 10, 64–71.

Debs, J.E., Ebendorff-Heidepriem, H., Quinton, J.S., Monro, T.M., 2009. A fundamental study into the surface functionalization of soft glass microstructured optical fibers via silane coupling agents. J. Lightwave Technol. 27 (5), 576–582.

Debye, P., 1947. Molecular-weight determination by light scattering. J. Phys. Colloid Chem. 51, 18–32.

Degtyarenko, K., 2000. Bioinorganic motifs: towards functional classification of metalloproteins. Bioinformatics Rev. 16, 851–864.

del Campo, A., sen, T., Lellouche, J.P., Bruce, I.J., 2005. Multifunctional magnetite and silica-magnetite nanoparticles: synthesis, surface activation and application in life sciences. J. Magn. Magn. Mater. 293, 33–40.

Delaunay, N, Pichon, V, Hennion, MC, 2000. Immunoaffinity solid-phase extraction for the trace-analysis of low-molecular-mass analytes in complex sample matrices. J. Chromatogr. B Biomed. Sci. Appl. 745, 15–37.

Delcroix, M., Sajid, M., Caffrey, C.R., Lim, K.-C., Dvorák, J., Hsieh, I., et al., 2006. A multienzyme network functions in intestinal protein digestion by a platyhelminth parasite. J. Biol. Chem. 281, 39316–39329.

D'Eletto, M., Farrace, M.G., Rossin, F., Strappazzon, F., Di Giacomo, G., Cecconi, F., et al., 2012. Type 2 transglutaminase is involved in the autophagy-dependent clearance of ubiquitinated proteins. Cell Death & Differ. 19 (7), 1228–1238.

Delgado, C., Francis, G.F., Fisher, D., 1992. The uses and properties of PEG-linked proteins. Grit. Rev. Ther. Drug Carrier Syst. 9, 249–304.

Delgado-Martín, C., Riol-Blanco, L., Alonso-C, L.M., Rodríguez-Fernández, J.L., 2009. A protocol to detect apoptotic dendritic cells in murine lymph nodes using multiphoton microscopy. Nat. Protoc. doi: 10.1038/nprot.2009.133.

Dell'Arciprete, L., Colombatti, M., Rappuoli, R., Tridente, G., 1988. A C terminus cysteine of diphtheria toxin B chain involved in immunotoxin cell penetration and cytotoxicity. J. Immunol. 140, 2466–2471.

Della-Penna, D., Christofferson, R.E., Bennett, A.B., 1986. Biotinylated proteins as molecular weight standards on Western blots. Anal. Biochem. 152, 329–332.

DeMar Jr., J.C., Disher, R.M., Wensel, T.G., 1992. HPLC analysis of protein-linked fatty acids using fluorescence detection of 4-(diazomethyl)-7-diethylaminocoumarin derivatives Abstract 465. Biophys. J. 61 (A81).

Demiroglou, A., Jennissen, H.P., 1990. Synthesis and protein-binding properties of spacer-free thioalkyl agaroses. J. Chromatogr. 521, 1–17.

Demiroglou, A., Bandel-Schlesselmann, C., Jennissen, H.P., 1994. A novel reaction sequence for the coupling of nucleophiles to agarose with 2,2,-trifluoroethane-sulfonyl chloride. Angew. Chem. Int. Ed. Engl. 33, 120–123.

DeNardo, S.L., Kukis, D.L., Kroger, L.A., O'Donnell, R.T., Lamborn, K.R., Miers, L.A., et al., 1997. Synergy of Taxol and radioimmunotherapy with yttrium-90-labeled chimeric L6 antibody: efficacy and toxicity in breast cancer xenografts. Proc. Natl. Acad. Sci. USA 94, 4000–4004.

Deng, D., 2011. Methods and compositions for cell stabilization. U.S. Patent Application No. 2011/0027771 A1.

Denizli, A., Rad, A.Y., Piskin, E., 1995. Protein A immobilized polyhydroxyethylmethacrylate beads for affinity sorption of human immunoglobulin G. J. Chromatogr. B. Biomed. Appl. 668 (1), 13–19.

Denney, J.B., Blobel, G., 1984. 125I-Labeled cross-linking reagent that is hydrophilic, photoactivatable, and cleavable through an azo linkage. Proc. Natl. Acad. Sci. USA 81, 5286–5290.

Denti, G., Campagna, S., Serroni, S., Ciano, N., Balzani, V., 1992. Decanuclear homo- and heterometallic polypyridine complexes: synthesis, absorption spectra, luminescence, electrochemical oxidation, and intercomponent energy transfer. J. Am. Chem. Soc. 114, 2944–2950.

dePont, J.J., Schoot, B.M., Bonting, S.L., 1980. Use of mono- and bifunctional group-specific reagents in the study of the renal Na^1-K^1-ATPase. Int. J. Biochem. 12, 307–313.

DePont, J.J.H.H.M., 1979. Reversible inactivation of (Na^1)-ATPase by use of a cleavable bifunctional reagent. Biochim. Biophys. Acta 567, 247–256.

Derfus, A.M., Chan, W.C.W., Bhatia, S.N., 2004. Probing the cytotoxicity of semiconductor quantum dots. Nano. Lett. 4, 11–18.

Derjaguin, B.V., Landau, L., 1941. Theory of the stability of strongly charged lyophobic sols and of the adhesion of strongly charged particles in solutions of electrolytes. Acta Physiochim. U.R.S.S 14, 633–662.

Derksen, J.T.P., Scherphof, G.L., 1985. An improved method for the covalent coupling of proteins to liposomes. Biochim. Biophys. Acta 814, 151–155.

Dermer, O.C., Ham, G.E., 1969. Ethylenimine and Other Aziridines. Academic Press, New York. pp. 327–333.

Derouiche, R., Benedetti, H., Lazzaroni, J.C., Lazdunski, C., Lloubes, R., 1995. Protein complex within *Escherichia coli* inner membrane. TolA N-terminal domain interacts with TolQ and TolR proteins. J. Biol. Chem. 270, 11078–11084.

DeStefano, F., Pfeifer, D., Nohynek, H., 2008. Safety profile of pneumococcal conjugate vaccines: systematic review of pre- and postlicensure data. Bull. World Health Organ. 86, 373–380.

Detmers, F., Hermans, P., Jiao, J.-A., McCue, J.T., 2010. Novel affinity ligands provide for highly selective primary capture. BioProcess Int. 8, 50–54.

Devaraj, N.K., Keliher, E.J., Thurber, G.M., Nahrendorf, M., Weissleder, R., 2009. 18F Labeled Nanoparticles for in Vivo PET-CT Imaging. Bioconjug. Chem. 20 (2), 397–401.

Dewey, R.E., Timothy, D.H., Levings III, C.S., 1987. A mitochondrial protein associated with cytoplasmic male sterility in the T cytoplasm of maize. Proc. Natl. Acad. Sci. USA 84, 5374–5378.

Dewey, T.G. (Ed.), 1991. Biophysical and Biochemical Aspects of Fluorescence Spectroscopy. Plenum, New York.

Dhar, T.K., Dasgupta, S., Ray, D., Banerjee, M., 2012. A Filtration method for rapid preparation of conjugates for immunoassay. J. Immunol. Methods 385, 71–78.

Diamandis, E.P., 1993. Time-resolved fluorometry in nucleic acid hybridization and Western blotting techniques (Review). Electrophoresis 14, 866–875.

Diamandis, E.P., Christopoulos, T.K., 1990. Europium chelate labels in time-resolved fluorescence immunoassays and DNA hybridization assays (Review). Anal. Chem. 62, 1149–1157.

Didenko, V.V., 1993. Biotinylation of DNA on membrane supports: a procedure for preparation and easy control of labeling of nonradioactive single-stranded nucleic acid probes. Anal. Biochem. 213, 75–78.

Dihazi, G.H., Sinz, A., 2003. Mapping low-resolution three-dimensional protein structures using chemical cross-linking and Fourier transform ion-cyclotron resonance mass spectrometry. Rapid Comm. Mass Spectrom. 17, 2005–2014.

Ding, S., et al. 2005. Investigating the putative glycine hinge in shaker potassium channel. J. Gen. Physiol. 126, 213–226.

Dintzis, R.Z., Okajima, M., Middleton, M.H., Greene, G., Dintzis, H.M., 1989. The immunogenicity of soluble hapenated polymers is determined by molecular mass and hapten valence. J. Immunol. 143, 1239–1244.

Dirksen, A., Dawson, P.E., 2008. Rapid oxime and hydrazone ligations with aromatic aldehydes for biomolecular labeling. Bioconjug. Chem. 19, 2543–2548.

Dirksen, A., Dirksen, S., Hackeng, T.M., Dawson, P.E., 2006a. Nucleophilic catalysis of hydrazone formation and transamination: implications for dynamic covalent chemistry. J. Am. Chem. Soc. 128 (49), 15602–15603.

Dirksen, A., Hackeng, T.M., Dawson, P.E., 2006b. Nucleophilic catalysis of oxime ligation. Angew. Chem. Int. Ed. 45, 7581–7584.

Dixon, H.B.F., Perham, R.N., 1968. Reversible blocking of amino groups with citraconic anhydride. Biochem. J. 109, 312–314.

Domashevskiy, A.V., Miyoshi, H., Goss, D.J., 2012. Inhibition of pokeweed antiviral protein (PAP) by Turnip Mosaic virus genome-linked protein (VPg). J. Biol. Chem. 287, 29729–29738.

Dombrowski, T.R., Wilson, G.S., Thurman, E.M., 1998. Investigation of anion-exchange and immunoaffinity particle-loaded membranes for the isolation of charged organic analytes from water. Anal. Chem. 70, 1969–1978.

Domen, P.L., Hermanson, G.T., 1992. Cationized carriers for immunogen production. US Patent No. 5,142,027.

Domen, P.L., Muckerheide, A., Michael, J.G., 1987. Cationization of protein antigens III. Abrogation of oral tolerance. J. Immunol. 139, 3195–3198.

Domen, P.L., Nevens, J.R., Mallia, A.K., Hermanson, G.T., Klenk, D.C., 1990. Site-directed immobilization of proteins. J. Chromatogr. 510, 293–302.

Donnelly, E.T., Liu, Y., Fatunmbi, Y.O., Lee, I., Magda, D., Rockwell, S., 2004. Effects of texaphyrins on the oxygenation of EMT6 mouse mammary tumors. Int. J. Radiat. Oncol. Biol. Phys. 58 (5), 1570–1576.

Donovan, J.A., Jennings, M.L., 1986. *N*-Hydroxysulfosuccinimido active esters and the L-(1)-lactate transport protein in rabbit erythrocytes. Biochemistry 25, 1538–1545.

Dordick, J.S., July 1991. Enzymatic catalysis in organic media: fundamentals and selected applications. ASGSB Bull. 4 (2), 125–132.

Dorman, G., Prestwich, G.D., 2000. Using photolabile ligands in drug discovery and development. TIBTECH 18, 64–76.

Doronina, S.O., Mendelsohn, B.A., Bovee, T.D., Cerveny, C.G., Alley, S.C., Meyer, D.L., et al., 2006. Enhanced activity of monomethylauristatin F through monoclonal antibody delivery: effects of linker technology on efficacy and toxicity. Bioconjug. Chem. 17 (1), 114–124.

Doronina, S.O., Bovee, T.D., Meyer, D.W., Miyamoto, J.B., Anderson, M.E., Morris-Tilden, C.A., et al., 2008. Novel peptide linkers for highly potent antibody–Auristatin conjugate. Bioconjug. Chem. 19 (10), 1960–1963.

Dos Reis-Costa, L.S., Andreimar, M., França, S.C., Trevisan, H.C., Roberts, T.J.C., 2003. Immobilization of lipases and assay in continuous fixed bed reactor. Protein Peptide Lett. 10, 619–628.

Dose, C., Seitz, O., 2005. Convergent synthesis of peptide nucleic acids by native chemical ligation. Org. Lett. 7 (20), 4365–4368.

Dosio, F., Brusa, P., Cattel, L., 2011. Immunotoxins and anticancer drug conjugate assemblies: the role of the linkage between components. Toxins 3 (7), 848–883.

Dottavio-Martin, D., Ravel, J.M., 1978. Radiolabeling of proteins by reductive alkylation with [14C]-formaldehyde and sodium cyanoborohydride. Anal. Biochem. 87, 562.

Dou, Z., Xu, Y., Sun, H., Liu, Y., 2012. Synthesis of PEGylated fullerene/5-fluorouracil conjugate to enhance anti-tumor effect of 5-fluorouracil. Nanoscale 4, 4624–4630.

Dougan, M., Dranoff, G., 2012. Immunotherapy of cancer. Innate Immune Regulation and Cancer Immunotherapy, 391–414.

Douroumis, D., Onyesom, I., Maniruzzaman, M., Mitchell, J., 2012. Mesoporous silica nanoparticles in nanotechnology. Crit. Rev. Biotechnol., (00), 1–17, early online access, doi: 10.3109/07388551.2012.685860.

Dower, S.K., PeLisi, C., Titus, J.A., Segal, D.M., 1981. Mechanism of binding of multivalent immune complexes to Fc receptors. 1. Equilibrium binding. Biochemistry 20, 6326–6334.

Drafler, F.L., Marinetti, G.V., 1977. Synthesis of a photoaffinity probe for the b-adrenergic receptor. Biochem. Biophys. Res. Comm. 79, 1.

Dransfield, M.T., Nahm, M.H., Han, M.K., Harnden, S., Criner, G.J., Martinez, F.J., et al., 2009. Superior immune response to protein-conjugate versus free pneumococcal polysaccharide vaccine in chronic obstructive pulmonary disease. Am. J. Respir Crit. Care Med. 180, 499–505.

Draper, P.E., Gold, L., 1980. A method for linking fluorescent labels to polynucleotides: application to studies of ribosome–ribonucleic acid interactions. Biochemistry 19, 1774–1781.

Dreborg, S., Akerblom, E.B., 1990. Immunotherapy with monomethoxypolyethylene glycol modified allergens. Crit. Rev. Ther. Drug Carrier Syst. 6, 315–365.

Drisdel, RC, Green, WN, 2004. Labeling and quantifying sites of protein palmitoylation. Biotechniques 36 (2), 276–285.

du Vigneaud, V., Melville, D.B., Gyorgy, P., Rose, C.S., 1940. On the identity of vitamin H with biotin. Science 92, 62–63.

Du, S.-L., Pan, H., Lu, W.-Y., Wang, J., Wu, J., Wang, J.-Y., 2007. Cyclic Arg-Gly-Asp peptide-labeled liposomes for targeting drug therapy of hepatic fibrosis in rats. J. Pharmacol. Exp. Ther. 322, 560–568.

Duan, L., Wang, Y., Li, S.S.-C., Wan, Z., Zhai, J., 2005. Rapid and simultaneous detection of human hepatitis B virus and hepatitis C virus antibodies based on a protein chip assay using nano-gold immunological amplification and silver staining method. BMC Infect. Dis. 5, 53. <http://www.biomedcentral.com/1471-2334/5/53>.

Duband, J.-L., Nuckolls, G., Ishihara, A., Hasegawa, T., Yamada, K., Thiery, J.P., et al., 1988. Fibronectin receptor exhibits high lateral mobility in embryonic locomoting cells but is immobile in focal contacts and fibrillar streaks in stationary cells. J. Cell Biol. 107, 1385–1396.

Dubertret, B., Skourides, P., Norris, D.J., Noireaux, V., Brivanlou, A.H., Libchaber, A., 2002. In vivo imaging of quantum dots encapsulated in phospholipid micelles. Science 298, 1759–1762.

Duckworth, B.P., Xu, J., Taton, T.A., Guo, A., Distefano, M.D., 2006. Site-specific, covalent attachment of proteins to a solid surface. Bioconjug. Chem. 17, 967–974.

Ducry, L., Stump, B., 2010. Antibody-drug conjugates: linking cytotoxic payloads to monoclonal antibodies. Bioconjug. Chem. 21, 5–13.

Duijndam, W.A.L., Wiegant, J., Van Duijn, P., Haaijman, J.J., 1988. A simple method for labeling the carbohydrate moieties of antibodies with fluorochromes. J. Immunol. Methods 109, 289–290.

Dunbar, B.S., 1987. Two-Dimensional Electrophoresis and Immunological Techniques. Plenum, New York. pp. 229–335.

Duncan, R., 2003. The dawning era of polymer therapeutics. Nature Rev. Drug Discov. 2, 347–360.

Duncan, R., Kopecek, J., 1984. Soluble synthetic polymers as potential drug carriers. Adv. Polym. Sci. 57, 53–101.

Duncan, R., Gac-Breton, S., Keane, R., Musila, R., Sat, Y.N., Satchi, R., et al., 2001. Polymer–drug conjugates, PDEPT and PELT: basic principles for design and transfer from the laboratory to clinic. J. Control Release 74, 135–146.

Duncan, R.J.S., Weston, P.D., Wrigglesworth, R., 1983. A new reagent which may be used to introduce sulfhydryl groups into proteins, and its use in the preparation of conjugates for immunoassay. Anal. Biochem. 132, 68–73.

Dunn, B.M., Affinsen, C.B., 1974. Kinetics of Woodward's reagent K hydrolysis and reaction with staphylococcal nuclease. J. Biol. Chem. 249, 3717.

Dunn, R.L., Ottenbrite, R.M. (Eds.), 1991. Polymeric Drugs and Drug Delivery Systems. American Chemical Society, Washington, D. C.

Dunnick, J.K., McDougall, R., Aragon, S., Coris, M., Kriss, J., 1975. Vesicle interactions with polyamino acids and antibody: in vitro and in vivo studies. J. Nucl. Med. 16, 483–487.

Durán, N., Rosa, M.A., DAnnibale, A., Gianfreda, L., 2002. Applications of laccases and tyrosinases (phenoloxidases) immobilized on different supports: a review. Enzyme Microb. Technol. 31 (2002), 907–931.

Durand, R.E., Olive, P.L., 1983. Flow cytometry techniques for studying cellular thiols. Radiat. Res. 95, 456.

Duru, P.E., Bektas, S., Genç, ö., Patir, S., Denizli, A., 2001. Adsorption of heavy-metal ions on poly(ethylene imine)-immobilized poly(methyl methacrylate) microspheres. J. Appl. Polym. Sci. 81, 197–205.

Dutta, D., Pulsipher, A., Luo, W., Mak, H., Yousaf, M.N., 2011. Engineering cell surfaces via liposome fusion. Bioconjug. Chem. 22 (12), 2423–2433.

East, D.A., Mulvihill, D.P., Todd, M., Bruce, I.J., 2011. QD-antibody conjugates via carbodiimide-mediated coupling: a detailed study of the variables involved and a possible new mechanism for the coupling reaction under basic aqueous conditions. Langmuir 27, 13888–13896.

Eastman, A., 1987. The formation, isolation and characterization of DNA adducts produced by anticancer Platinium complexes. Pharmacol. Ther. 34, 155–166.

Ebrahim, H., Dakshinamurti, K., 1986. A fluorometric assay for biotinidase. Anal. Biochem. 154, 282–286.

Ebrahim, H., Dakshinamurti, K., 1987. Determination of biocytin. Anal. Biochem. 162, 319–324.

Eckels, K.H., Putnak, R., 2003. Formalin-inactivated whole virus and recombinant subunit flavivirus vaccines. Adv. Virus Res. 61, 395–418.

Edelhoch, H., Katchalsk, E., Maybury, R.H., Hughes Jr., W.L., Edsall, J.T., 1953. Pimenization of serum mercaptalbumin in the presence of mercunials. I. Kinetic and equilibrium studies with mercuric salts. J. Am. Chem. Soc. 75, 5058.

Edelman, G.M., Gall, W.E., Waxdal, M.J., Konigsberg, W.H., 1968. The covalent structure of a human gG-immunoglobulin. I. Isolation and characterization of the whole molecules, the polypeptide chains, and the tryptic fragments. Biochemistry 7, 1950–1958.

Edgington, L.E., Berger, A.B., Blum, G., Albrow, V.E., Paulick, M.G., Lineberry, N., et al., 2009. Noninvasive optical imaging of apoptosis by caspase-targeted activity-based probes. Nat. Med. 15, 967–974.

Edsall, J.T., Maybury, R.H., Simpson, R.B., Straessle, R., 1954. Dimerization of serum mercaptalbumin in the presence of mercurials. II. Studies with a bifunctional organic mercurial. J. Am. Chem. Soc. 76, 3131.

Edwards, D.C., Ross, W.C.J., Cumber, A.J., McIntosh, D., Smith, A., Thorpe, P.E., et al., 1982. A comparison of the in vitro and in vivo activities of conjugates of anti-mouse lymphocytes globulin and abrin. Biochim. Biophys. Acta 717, 272.

Edwards, J.O., Pearson, R.G., 1962. The factors determining nucleophilic reactivities. J. Chem. Soc. 84, 26.

Edwards, R.J., Singleton, A.M., Boobis, A.R., Davies, D.S., 1989. Cross-reaction of antibodies to coupling groups used in the production of anti-peptide antibodies. J. Immunol. Methods 117, 215–220.

Efros, A.L., Rosen, M., 1997. Random telegraph signal in the photoluminescence intensity of a single quantum dot. Phys. Rev. Lett. 78, 1110–1113.

Egorov, T.A., Svenson, A., Ryden, L., Carlsson, J., 1975. A rapid and specific method for isolation of thiol-containing peptides from large proteins by thiol-disulfide exchange on a solid support. Proc. Natl. Acad. Sci. USA 72, 3029–3033.

Ehses, S., Leonhardt, R.M., Hansen, G., Knittler, M.R., 2005. Functional role of C-terminal sequence elements in the transporter associated with antigen processing. J. Immunol. 174, 328–339.

Eichler, J., 2008. Peptides as protein binding site mimetics. Curr. Opin. Chem. Biol. 12, 707–713.

Eichman, J.D., Bielinska, A.U., Kukowska-Latallo, J.F., Donovan, B.W., Baker Jr., J.R., 2001. Bioapplications of PAMAM dendrimers in dendrimers and other dendritic polymers. In: Fréchet, M.J., Tomalia, D.A. (Eds.), Wiley Series in Polymer Science. John Wiley & Sons, Ltd, West Sussex, UK, pp. 441–461.

Eiklid, K., Olsnes, S., Pihl, A., 1980. Entry of lethal doses of abrin, ricin, and modeccin into the cytosol of Hela cells. Exp. Cell Res. 126, 321–326.

Eisen, H.N., Belman, S., Carsten, M.E., 1953. The reaction of 2,4-dinitrobenzenesulfonic acid with free amino groups of proteins. J. Am. Chem. Soc. 75, 4583.

Eisenhut, M., Mier, W., 2011. Radioiodination chemistry and radioiodinated compounds. In: Vértes, A., Nagy, S., Klencsár, Z., Lovas, R.G., Rösch, F. (Eds.), Handbook of Nuclear Chemistry (second ed.). Springer Science + Business Media B.V., pp. 2121–2141. Chapter 44, ISBN: 978-1-4419-0719-6.

Ekert, P.G., Silke, J., Vaux, D.L., 1999. Caspase inhibitors. Cell Death Differ. 6, 1081–1086.

ElBakri, A., Nelson, P.N., Abu Odeh, O., 2010. The state of antibody therapy. Hum. Immunol. 71, 1243–1250.

El-Bary, A.A., Amin, A.M., El-Wetery, A., Saad, S., Shoukry, M., 2012. Radioiodination of cephalexin with 125 I and its biological behavior in Mice. Pharmacol. Pharm. 3 (1), 97–102.

El-Boubbou, K., Zhu, D.C., Vasileiou, C., Borhan, B., Prosperi, D., Li, W., et al., 2010. Magnetic Glyco-nanoparticles: a tool to detect, differentiate, and unlock the glyco-codes of cancer via magnetic resonance imaging. J. Am. Chem. Soc. 132 (12), 4490–4499.

Eldjarn, L., Jellum, E., 1963. Organomercurial-polysaccharide, a chromatographic material for the separation and isolation of SH-proteins. Acta Chem. Scand. 17, 2610–2621.

Eldridge, G.M., Weiss, G.A., 2011. Hydrazide reactive peptide tags for site-specific protein labeling. Bioconjug. Chem. 22 (10), 2143–2153.

Ellerbroek, S.M., et al. 2001. Functional interplay between type I collagen and cell surface matrix metalloproteinase activity. J. Biol. Chem. 276 (27), 24833–24842.

Ellerman, D., Scheer, J.M., 2011. Generation of bispecific antibodies by chemical conjugation. In: Ellerman, D. (Ed.), Biospecific Antibodies. Springer-Verlag, Berlin Heidelberg, pp. 47–64. Chapter 3.

Ellis, I.O., Bell, J., Bancroft, J.D., 1988. An investigation of optimal gold particle size for immunohistological immunogold and immunogold-silver staining to be viewed by polarized incident light (EPI polarization) microscopy. J. Histochem. Cytochem. 36, 121–124.

Ellman, C.L., 1959. Tissue sulfhydryl groups. Arch. Biochem. Biophys. 82, 70–77.

Ellman, G.L., 1958. A colorimetric method for determining low concentrations of mercaptans. Arch. Biochem. Biophys. 74, 443–450.

Ellman, G.L., 1959. Tissue sulfhydryl groups. Arch. Biochem. Biophys. 82, 70–77.

Elsner, H.I., Mouritsen, S., 1994. Use of psoralens for covalent immobilization of biomolecules in solid phase assays. Bioconjug. Chem. 5 (5), 463–467.

Endo, T., Wright, A., Morrison, S.L., Kobata, A., 1995. Glycosylation of the variable region of immunoglobulin G-site specific maturation of the sugar chains. Mol. Immunol. 32 (13), 931–940.

Engin, S., Trouillet, V., Franz, C.M., Welle, A., Bruns, M., Wedlich, D., 2010. Benzylguanine Thiol self-assembled monolayers for the immobilization of SNAP-tag proteins on microcontact-printed surface structures. Langmuir 26 (9), 6097–6101.

England, T., Uhlenbeck, O., 1978. 3′-Terminal labelling of RNA with T4 RNA ligase. Nature 275, 560–562.

England, T., Gumport, R., Uhlenbeck, O., 1977. Dinucleoside pyrophosphates are substrates for T4-induced RNA ligase. Proc. Natl. Acad. Sci. USA 74, 4839–4842.

Englund, P.T., King, T.P., Craig, L.C., 1968. Studies on ficin. I. Its isolation and characterization. Biochemistry 7, 163–174.

Enoch, H.C., Strittmatter, P., 1979. Formation and properties of 100--diameter, singlebilayer phospholipid vesicles. Proc. Natl. Acad. Sci. USA 76, 145–149.

Entwistle, A., Noble, M., 2011. The use of lucifer yellow, BODIPY, FITC, TRITC, RITC and Texas Red for dual immunofluorescence visualized with a confocal scanning laser microscope. J. Microsc. 168 (3), 219–238.

Epps, P.E., et al. 1992. Spectral characterization of environment-sensitive adducts of interleukin 1b. J. Biol. Chem. 267, 3129.

Erathodiyil, N., Ying, J.Y., 2011. Functionalization of inorganic nanoparticles for bioimaging applications. Acc. Chem. Res. 44 (10), 925–935.

Eriksson, K.-O., 1987. Glycidol-modified gels for molecular-sieve chromatography. Surface hydrophilization and pore size reduction. J. Biochem. Biophys. Methods 15, 105–110.

Erlenmeyer, F., 1893. Ueber die Condensation der Hippursäure mit Phtalsäureanhydrid und mit Benzaldehyd. Justus Liebigs Annalen der Chemie 275, 1–8.

Ermacora, M.R., Delfino, J.M., Cuenoud, B., Schepartz, A., Fox, R.O., 1992. Conformation-dependent cleavage of staphylococcal nuclease with a disulfide-linked iron chelate. PNAS 89, 6383.

Erman, A., Zupancic, D., Jezernik, K., 2009. Apoptosis and desquamation of urothelial cells in tissue remodeling during rat postnatal development. J. Histochem. Cytochem. 57, 721–730.

Ernst, L.A., Gupta, R.K., Mujumdar, R.B., Waggoner, A.S., 1989. Cyanine dye labeling reagents for sulfhydryl groups. Cytometry 10 (1), 3–10.

Ernsting, N.P., Asimov, M., Shaefer, F.P., 1982. The electronic origin of the p-p* absorption of amino coumarins studied in a supersonically cooled free jet. Chem. Phys. Lett. 91, 231.

Eschrich, T.C., Morgan, T.J., 1985. Dye laser radiation in the 370–760 region pumped by a xenon monofluoride excimer laser. Appl. Opt. 24, 937.

Eton, O., Ross, M.I., East, M.J., Mansfield, P.F., Papadopoulos, N., Ellerhorst, J.A., et al., 2010. Autologous tumor-derived heat-shock protein peptide complex-96 (HSPPC-96) in patients with metastatic melanoma. J. Transl. Med. 8, 9. (published online by BioMed Central Ltd.).

Etzel, M.R., Bund, T., 2011. Monoliths for the purification of whey protein-dextran conjugates. J. Chromatogr. A 1218, 2445–2450.

Evans, T.C., Benner, J., Xu, M.-Q., 1998. Semisynthesis of cytotoxic proteins using a modified protein splicing element. Protein Sci. 7, 2256–2264.

Evident Technologies, 2005. Coupling of sulfhydryl-modified oligonucleotides to amine-EviTags using BMPA and EDC, web site quantum dot protocols.

Evstigneeva, R.P., Zaitsev, A.V., Luzgina, V.N., Ol'shevskaya, V.A., Shtil, A.A., 2003. Carboranylporphyrins for boron neutron capture therapy of cancer. Curr. Med. Chem. Anticancer Agents 3, 383–392.

Ewig, R.A.C., Kohn, K.W., 1977. DNA – protein cross-linking and DNA interstrand cross-linking by haloethylnitrosoureas in L1210 cells. Cancer Res. 38, 3197.

Fabris, L., 2012. Bottom-up optimization of SERS hot spots. Chem. Commun. 48, 9346–9348.

Fahien, L.A., Ruoho, A.E., Kmiotek, E., 1978. A study of glutamate dehydrogenase aminotransferase complexes with a bifunctional imidate. J. Biol. Chem. 253, 5745–5751.

Fairbank, M., Huang, K., El-Husseini, A., Nabi, I.R., 2012. RING finger palmitoylation of the endoplasmic reticulum Gp78 E3 ubiquitin ligase. FEBS Lett. 586, 2488–2493.

Falck, E., Groenhagen, A., Mühlisch, J., Hempel, G., Wünsch, B., 2011. Genome-wide DNA methylation level analysis by MEKC-LIF after treatment of cell lines with azacytidine and antifolates. Anal. Biochem. 421, 439–445.

Falke, J.J., Dernburg, A.F., Sternberg, D.A., Zalkin, N., Milligan, D.L., Koshland Jr., D.E., 1988. Structure of a bacterial sensory receptor. Biol. Chem. 263, 14850–14858.

Fam, D.W.H., Palaniappan, A., Tok, A.I.Y., Liedberg, B., Moochhala, S.M., 2011. A review on technological aspects influencing commercialization of carbon nanotube sensors. Sens. Actuators B: Chem. 157 (1), 1–7.

Fancy, D.A., Kodadek, T., 1997. Site-directed oxidative protein crosslinking. Tetrahedron 53, 11953–11960.

Fancy, D.A., Kodadek, T., 1998. A critical role for tyrosine residues in His6Ni-mediated protein crosslinking. Biochem. Biophys. Res. Commun. 247, 420–426.

Fancy, D.A., Kodadek, T., 1999. Chemistry for the analysis of protein–protein interactions: rapid and efficient cross-linking triggered by long wavelength light. Proc. Natl. Acad. Sci. USA 96, 6020–6024.

Fancy, D.A., Melcher, K., Johnston, S.A., Kodadek, T., 1996. New chemistry for the study of multiprotein complexes: the six-histidine tag as a receptor for a protein crosslinking reagent. Chem. Biol. 3, 551–559.

Farmer, J.C., Castaneda, M., 1991. An improved preparation and purification of oligonucleotide–alkaline phosphatase conjugates. Bio. Tech. 11, 588–589.

Farr, A.C., Nakane, P.K., 1981. Immunohistochemistry with enzyme labeled antibodies: a brief review. J. Immunol. Methods 47, 129–144.

Farries, T.C., Atkinson, J.P., 1989. Biosynthesis of properdin. J. Immunol. 142, 842–847.

Fasold, H., Groschel-Stewart, U., Turba, F., 1963. Azophenyl-dimaleimide als spaltbare peptidbrucken-bildende reagentien zwischen cysteinresten. Biochem. Z. 337, 425.

Fassina, G., Verdoliva, A., Odierna, M.R., Ruvo, M., Cassini, G., 1996. Protein A mimetic peptide ligand for affinity purification of antibodies. J. Mol. Recognit. 9, 564–569.

Fassina, G., Verdoliva, A., Palombo, G., Ruvo, M., Cassani, G., 1998. Immunoglobulin specificity of TG19318: a novel synthetic ligand for antibody affinity purification. J. Mol. Recognit. 11, 128–133.

Fassina, G., 2000. Protein A mimetic (PAM) affinity chromatography. Immunoglobulins purification. Methods Mol. Biol. 147, 57–68.

Faulk, W.P., Taylor, G.M., 1971. An immunocolloid method for the electron microscope. Immunochemistry 8, 1081–1083.

Fauq, A.H., Kache, R., Khan, M.A., Vega, I.E., 2006. Synthesis of acid-cleavable light isotope-coded affinity tags (ICAT-L) for potential use in proteomic expression profiling analysis. Bioconjug. Chem. 17, 248–254.

Faye, A., Esnous, C., Price, N.T., Onfray, M.A., Girard, J., Prip-Buus, C., 2007. Rat liver carnitine palmitoyltransferase 1 forms an oligomeric complex within the outer mitochondrial membrane. J. Biol. Chem. doi: 10.1074/jbc.M705418200.

Fearnley, C., Speakman, J.B., 1950. Cross-linkage formation in keratin. Nature (London) 166, 743.

Fein, M.L., Filachione, E.M., 1957. Tanning studies with aldehydes. J. Am. Leather Chem. Assoc. 52, 17.

Feinberg, A.P., Vogelstein, B., 1983. A technique for radiolabeling DNA restriction endonuclease fragments to high specific activity. Anal. Biochem. 132, 6–13.

Feinberg, A.P., Vogelstein, B., 1984. A technique for radiolabeling DNA restriction endonuclease fragments to high specific activity (Addendum). Anal. Biochem. 137, 266–267.

Felekis, T.A., Tagmatarchis, N., 2005. Single-walled carbon nanotube-based hybrid materials for managing charge transfer processes. Rev. Adv. Mater. Sci. 10, 272–276.

Feller, A.C., Parwaresch, M.R., Wacker, H.-H., Radzun, H.-J., Lennert, K., 1983. Combined immunohistochemical staining for surface IgD and T-lymphocyte subsets with monoclonal antibodies in human tonsils. Histochem. J. 15, 557–562.

Feng, T., Du, Y., Yang, J., Li, J., Shi, X., 2006. Immobilization of a non-specific chitosan hydrolytic enzyme for application in preparation of water-soluble low-molecular-weight chitosan. J. Appl. Polym. Sci. 101, 1334–1339.

Fenton, R.A., Brond, L., Nielsen, S., Praetorius, J., 2007. Cellular and subcellular distribution of the type II vasopressin receptor in kidney. Am. J. Physiol. Renal Physiol. doi: 10.1152/ajprenal.00316.2006.

Ferguson, B., Yang, D., 1986. Localization of noncovalently bound ethidium in free and methionyl-tRNA synthetase bound tRNA(fMet) by singlet – singlet energy transfer. Biochemistry 25, 5298.

Fernandes, F., Prieto, M., Loura, L.M., 2012. Advanced FRET methodologies: protein–lipid selectivity detection and quantification. Biochem. Roles Eukaryotic Cell Surf. Macromol., 171–185.

Fernandez-Lafuente, R., Rossell, C.M., Rodriguez, V., Santana, C., Soler, G., Bastida, A., et al., 1993. Preparation of activated supports containing low pK amino groups. A new tool for protein immobilization via the carboxyl coupling method. Enzym. Microb. Tech. 15, 546–550.

Ferreira, S.A., Correia, A., Madureira, P., Vilanova, M., Gama, F.M., 2012. Unraveling the uptake mechanisms of mannan nanogel in bone-marrow-derived macrophages. Macromol. Biosci. 12, 1172–1180.

Field, L., Harle, H., Owen, T.C., Ferretti, A., 1964. Preparation and oxidation of some asymmetrical dialkyl and alkyl pyridnium disulfides. J. Org. Chem. 29, 1632–1635.

Field, L., Owen, T.C., Crenshaw, R.R., Bryan, A.W., 1961. Thiosulfonates and disulfides containing 2-aminoethyl moieties. J. Am. Chem. Soc. 83, 4414–4417.

Fields, S., Song, O., 1989. A novel genetic system to detect protein–protein interactions. Nature (London) 340, 245–246.

Filippone, S., Heimann, F., Rassat, A., 2002. A highly water-soluble 2:1 beta-cyclodextrin-fullerene conjugate. Chem. Commun., 1508–1509.

Finlay, T.H., Troll, V., Levy, M., Johnson, A.J., Hodgins, L.T., 1978. New methods for the preparation of biospecific adsorbents and immobilized enzymes utilizing trichloro-s-triazine. Anal. Biochem. 87, 77–90.

Finlay, T.H., Troll, V., Levy, M., Johnson, A.J., Hodgins, L.T., 1978. New methods for the preparation of biospecific adsorbents and immobilized enzymes utilizing trichloro-s-triazine. Anal. Biochem. 87, 77–90.

Fioramonte, M., dos Santos, A.M., Mcllwain, S., Noble, W.S., Fanchini, K.G., Gozzo, F.C., 2012. Analysis of secondary structure in proteins by chemical cross-linking coupled to MS. Proteomics 12, 2746–2752.

Fischer, J.E., Johnson, A.T., Luzzi, D.E., Therien, M., Winey, K.I., Yodh, A.G., Carbon nanotube-derived materials: high-quality suspensions of single-wall carbon nanotubes. Poster, Materials Research Science and Engineering Center, University of Pennsylvania.

Fischer, J.J., Graebner Baessler, O.Y., Dalhoff, C., Michaelis, S., Schrey, A.K., Ungewiss, J., et al., 2010. Comprehensive identification of staurosporine-binding kinases in the hepatocyte cell line HepG2 using capture compound mass spectrometry (CCMS). J. Proteome Res. 9 (2), 806–817.

Fitzgerald, D.J., Willingham, M.C., Pastan, I., 1988. Pseudomonas exotoxin–immunotoxin. In: Frankel, A.E. (Ed.), Immunotoxins. Kluwer, Boston, pp. 161.

FitzGerald, S.P., Lamont, J.V., McConnell, R.I., Benchikh, E.O., 2005. Development of a high-throughput automated analyzer using biochip array technology. Clin. Chem. 51 (7), 1165–1176.

Fizgerald, D., Morris, R.E., Saelinger, C.B., 1980. Receptor-mediated internalization of pseudomonas toxin by mouse fibroblasts. Cell 21, 867.

Flenniken, M.L., Willits, D.A., Harmsen, A.L., et al., 2006. Melanoma and lymphocyte cell-specific targeting incorporated into a heat shock protein cage architecture. Chem. Biol. 13, 161–170.

Foglesong, P.D., Winkler, M.A., Price, J.O., Marshall, G.D., Reagh, S.H., Bush, D.A., et al., 1989. Preparation and analysis of bifunctional immunoconjugates containing monoclonal antibodies OKT3 and BABR1. Cancer Immunol. Immunother. 30 (3), 177–184.

Fok, K.-F., Ohga, K., Incefy, G.S., Erickson, B.W., 1982. Antigenic specificity of two antibodies directed against the thymic hormone serum thymic factor (FTS). Mol. Immunol. 19, 1667–1673.

Fonovic, M., Bogyo, M., 2007. Activity based probes for proteases: applications to biomarker discovery, molecular imaging and drug screening. Curr. Pharm. Des. 13, 253–261.

Fonovi⊠, M., Verhelst, S.H.L., Sorum, M.T., Bogyo, M., 2007. Proteomics evaluation of chemically cleavable activity-based probes. Mol. Cell. Proteomics 6, 1761–1770.

Ford, D.J., Radin, R., Pesce, A.J., 1978. Characterization of glutaraldehyde coupled alkaline phosphatase–antibody and lactoperoxidase–antibody conjugates. Immunochemistry 15, 237.

Forster, A.C., McInnes, J.L., Skingle, D.C., Symons, R.H., 1985. Non-radioactive hybridization probes prepared by the chemical labeling of DNA and RNA with a novel reagent, photobiotin. Nucleic Acid Res. 13, 745–761.

Forster, T., 1948. Zwischenmolekulare Energiewanderung Und Fluoreszenz. Ann. Physik. 2, 55–75.

Fraenkel-Conrat, H., 1959. Methods for investigating the essential groups for enzyme activityColowick, S.P.Kaplan, N.O. Methods in Enzymology, vol. 4. Academic Press, New York, pp. 247–269.

Fraker, P.J., Speck Jr., J.C., 1978. Protein and cell membrane iodinations with a sparingly soluble chloroamide, 1,3,4,6-tetrachloro-3a,6a-diphenylglycouril. Biochem. Biophys. Res. Commun. 80, 849–857.

Francis, R., Sharma, S.K., Springer, C., Green, A.J., Hope-Stone, L.D., Sena, L., et al., 2002. A Phase I trial of antibody directed enzyme prodrug therapy (ADEPT) in patients with advanced colorectal carcinoma or other CEA producing tumors. Br. J. Cancer 87, 600–607.

Frangioni, J.V., 2008. New technologies for human cancer imaging. J. Clin. Oncol. 26, 4012–4021.

Franz, K.J., Nitz, M., Imperiali, B., 2003. Lanthanide-binding tags as versatile protein coexpression probes. Chem. Bio. Chem. 4, 265–271.

Frawley Cass, S.M., Reid, G.E., Tepe, J.J., 2009. Synthesis of diazo functionalized solid supports and their application towards the enrichment of phosphorylated peptides. Org. Biomol. Chem. 7, 3291–3299.

Fréchet, J.M.J., 1994. Functional polymers and dendrimers: reactivity, molecular architecture, and interfacial energy. Science 263, 1710–1715.

Fréchet, J.M.J., Tomalia, D.A. (Eds.), 2002. Dendrimers and Other Dendritic Polymers (Wiley Series in Polymer Science). Wiley.

Fredriksson, S., Gullberg, M., Jarvius, J., Olsson, C., Pietras, K., Gustafsdottir, S.M., et al., 2002. Protein detection using proximity-dependent DNA ligation assays. Nat. Biotechnol. 20, 473–477.

Freedberg, W.B., Hardman, J.K., 1971. Structural and functional roles of the cysteine residues in the a-subunit of the *Escherichia coli* tryptophan synthetase. J. Biol. Chem. 246, 1439.

Freedman, M.H., Grossberg, A.L., Pressman, D., 1968. The effects of complete modification of amino groups on the antibody activity of antihapten antibodies. Reversible inactivation with maleic anhydride. Biochemistry 7, 1941–1950.

Frei, A.P., Jeon, O.-Y., Kilcher, S., Moest, H., Henning, L.M., Jost, C., et al., 2012. Direct identification of ligand-receptor interactions on living cells and tissues. Nat. Biotech. doi: 10.1038/nbt.2354, published online: 16 September 2012.

Freytag, J.W., Dickinson, J.C., Tseng, S.Y., 1984b. A highly sensitive affinity-column-mediated immunometric assay, as exemplified by digoxin. Clin. Chem. 30, 417–420.

Freytag, J.W., Lau, H.P., Wadsley, J.J., 1984a. Affinity-column-mediated immunoenzymometric assays: influence of affinity-column ligand and valency of antibody–enzyme conjugates. Gun. Chem. 30, 1494–1498.

Friden, P.M., Walus, L.R., Watson, P., Doctrow, S.R., Kozarich, J.W., Backman, C., et al., 1993. Blood–brain barrier penetration and in vivo activity of an NGF conjugate. Science 259, 373–378.

Fried, V.A., Ando, M.E., Bell, A.J., 1985. Protein quantitation at the picomole level: an *o*-phthaldialdehyde pre-TSK column-derivatization assay. Anal. Biochem. 146, 271–276.

Friede, M., Van Regenmortel, M.H.V., Schuber, F., 1993. Lyophilized liposomes as shelf items for the preparation of immunogenic liposome–peptide conjugates. Anal. Biochem. 211, 117–122.

Friedman, M.L., Ball Jr., W.J., 1989. Determination of monoclonal antibody-induced alterations in Na^1/K^1-ATPase conformations using fluorescein-labeled enzyme. Biochim. Biophys. Acta 995, 42.

Friedrich, K., Woolley, P., Steinhauser, K.G., 1988. Fluorimetric distance determination by resonance energy transfer. Eur. J. Biophys. 173, 233.

Friedrichson, T., Kurzchalia, T.V., 1998. Microdomains of GPI-anchored proteins in living cells revealed by crosslinking. Nature 394 (6695), 802–805.

Frohn, C., Fricke, L., Puchta, J.-C., Kirchner, H., 2001. The effect of HLA-C matching on acute renal transplant rejection. Nephrol. Dial. Transplant. 16 (2), 355–360.

Frohnert, B.I., Sinaiko, A.R., Serrot, F.J., Foncea, R.E., Moran, A., Ikramuddin, S., et al., 2011. Increased adipose protein carbonylation in human obesity. Obesity 19 (9), 1735–1741.

Frolov, A., Hoffmann, R., 2010. Identification and relative quantification of specific glycation sites in human serum albumin. Anal. Bioanal. Chem. 397, 2349–2356.

Fronczek, F.R., MGH., Vicente, 2005. Synthesis and cellular studies of an octa-anionic 5,10,15,20-tetra[3,5(nidocarboranylmethyl)phenyl] porphyrin (H2OCP) for application in BNCT. Bioorg. Med. Chem. 13, 1633–1640.

Frytak, S., Creagan, E.T., Brown, M.L., Salk, D., Nelp, W., 1993. A technetium labeled monoclonal antibody for imaging metastatic melanoma. Am. J. Clin. Oncol. 14, 156–161.

Fu, G., Yue, X., Dai, Z., 2011. Glucose biosensor based on covalent immobilization of enzyme in sol–gel composite film combined with Prussian blue/carbon nanotubes hybrid. Biosens. Bioelectron. 26 (9), 3973–3976.

Fu, J., Rienhold, J., Woodbury, N.W., 2011. Peptide-modified surfaces for enzyme immobilization. PLoS ONE 6 (4), e18692.10.1371/journal.pone.0018692.

Fu, L., Ferreira, R.A.S., Nobre, S.S., Carlos, L.D., Rocha, J., 2006. Organically modified silica-based xerogels derived from 3-aminopropyltrimethoxysilane and 3-isocyanatepropyltriethoxysilane through carboxylic acid solvolysis. Mater. Sci. Forum 514–516, 108–112.

Fu, Y., Létourneau, M., Chatenet, D., Dupuis, J., Fournier, A., 2011. Characterization of iodinated adrenomedullin derivatives suitable for lung nuclear medicine. Nucl. Med. Biol. 38 (6), 867–874.

Fuji, N., Akaji, K., Hayashi, Y., Yajima, H., 1985. Studies on peptides. CXXV. 3-(3-p-methoxybenzylthiopropionyl)-thiazolidine-2-thione and its analogs as reagents for the introduction of the mercapto group into peptides and proteins. Chem. Pharm. Bull. 33, 362–367.

Fujii, N., Jacobsen, R.B., Wood, N.L., Schoeniger, J.S., Guy, R.K., 2004. A novel protein crosslinking reagent for the determination of moderate resolution protein structures by mass spectrometry (MS3-D). Bioorg. Med. Chem. Lett. 14, 427–429.

Fujita, H., Hirao, T., Takahashi, M., 2007. A simple and non-invasive visualization for assessment of carbonylated protein in the stratum corneum. Skin Res. Technol. 13 (1), 84–90.

Fujiwara, K., Matsumoto, N., Yagisawa, S., Tanimori, H., Kitagawa, T., Hirota, M., et al., 1988. Sandwich enzyme immunoassay of tumor-associated antigen sialosylated Lewisx using b-d-galactosidase coupled to a monoclonal antibody of IgM isotype. J. Immunol. Methods 112, 77–83.

Fuller, D.N., Gemmen, G.J., Rickgauer, J.P., Dupont, A., Millin, R., Recouvreux, P., et al., 2006. A general method for manipulating DNA sequences from any organism with optical tweezers. Nucleic Acids Res. 34, e15.

Funovics, M., Weissleder, R., Tung, C.H., 2003. Protease sensors for bioimaging. Anal. Bioanal. Chem. 377, 956–963.

Gabizon, A., Meshorer, A., Barenholz, Y., 1986. Comparative long-term study of the toxicities of free and liposome-associated doxorubicin in mice after intravenous administration. J. Natl. Cancer Inst. 77, 459–469.

Gacal, B., Durmaz, H., Tasdelen, M.A., Hizal, G., Tunca, U., Yagci, Y., et al., 2006. Anthracene-maleimide-based diels-alder "Click Chemistry" as a novel route to graft copolymers. Macromolecules 39, 5330–5336.

Gada, K.S., Patil, V., Panwar, R., Hatefi, A., Khaw, B.-A., 2012. Bispecific antibody complex pre-targeted delivery of polymer-drug conjugates for cancer therapy. Drug Deliv. Transl. Res. 2, 65–76.

Gaertner, H.F., Offord, R.E., 1996. Site-specific attachment of functionalized poly(ethylene glycol) to the amino terminus of proteins. Bioconjug. Chem. 7, 38–44.

Gaertner, H.F., Offord, R.E., Cotton, R., Timms, D., Camble, R., Rose, K., 1994. Chemo-enzymic backbone engineering of proteins. Site-specific incorporation of synthetic peptides that mimic the 64–74 disulfide loop of granulocyte colony-stimulating factor. J. Biol. Chem. 269, 7224–7230.

Gaffney, B.J., Willingham, G.L., Schopp, R.S., 1983. Synthesis and membrane interactions of a spin-label bifunctional reagent. Biochemistry 22, 881.

Gahmberg, C.G., 1978. Tritium labeling of cell-surface glycoproteins and glycolipids using galactose oxidaseGinsburg, V. Methods in Enzymology, vol. 50. Academic Press, New York, pp. 204–206.

Gailit, J., 1993. Restoring free sulfhydryl groups in synthetic peptides. Anal. Biochem. 214, 334–335.

Galardy, R.E., et al. 1978. Biologically active derivatives of angiotensin for labeling cellular receptors. J. Med. Chem. 21, 1279.

Galardy, R.E., Craig, L.C., Jamieson, J.D., Printz, M.P., 1974. Photoaffinity labeling of peptide hormone binding sites. J. Biol. Chem. 249, 3510–3518.

Gallant, S.R., 2004. Immunoaffinity chromatography of proteins in HPLC of peptides and proteins. Methods Mol. Biol. 251 (I), 103–109.

Gambhir, S.S., Paulmurugan, R., 2005. Firefly luciferase enzyme fragment complementation for imaging in cells and living animals. Anal. Chem. 77, 1295–1302.

Ganta, S., Devalapally, H., Shahiwala, A., Amiji, M., 2008. A review of stimuli-responsive nanocarriers for drug and gene delivery. J. Control. Release 126, 187–204.

Gao, X., Cui, Y., Levenson, R.M., Chung, L.W., Nie, S., 2004. *In vivo* cancer targeting and imaging with semiconductor quantum dots. Nat. Biotechnol. 22, 969–976.

Gao, X.H., Cui, Y.Y., Levenson, R.M., Chung, L.W.K., Nie, S.M., 2004. In vivo cancer targeting and imaging with semiconductor quantum dots. Nat. Biotechnol. 22, 969–976.

Garrido-Medina, R., Puerta, A., Rivera-Monroy, Z., de Frutos, M., Guttman, A., Diez-Masa, J.C., 2012. Analysis of alpha-1-acid glycoprotein isoforms using CE-LIF with fluorescent thiol derivatization. Electrophoresis 33 (7), 1113–1119.

Gartmann, N., Schutze, C., Ritter, H., Bruhwiler, D., 2010. The effect of water on the functionalization of mesoporous silica with 3-aminopropyltriethoxysilane. J. Phys. Chem. Lett. 1, 379–382.

Gartner, C.A., Elias, J.E., Bakalarski, C.E., Gygi, S.P., 2007. Catch-and-release reagents for broadscale quantitative proteomics analyses. J. Proteome Res. 6 (4), 1482–1491.

Gaston, B.M., Carver, J., Doctor, A., Palmer, L.A., 2003. S-Nitrosylation Signaling in Cell Biology. Mol. Interv. Physiol. Roles for S-Nitrosylation 3 (5), 253–263.

Gaudet, C., et al. 2003. Influence of type I collagen surface density on fibroblast spreading, motility, and contractility. Biophys. J. 85, 3329–3335.

Gaur, R., Sharma, S., Gupta, K.C., 1989. A simple method for the introduction of thiol group at 59-termini of oligodeoxynucleotides. Nucleic Acids Res. 17, 4404.

Gaur, R.K., 1991. Introduction of 59-terminal amino and thiol groups into synthetic oligonucleotides. Nucleoside Nucleotides 10, 895–909.

Gautam, S., 2010. Biomimetic ligands for immunoglobulin-M purification. Thesis, Department of Chemical and Biomolecular Engineering, National University of Singapore.

Gautier, A., Juillerat, A., Heinis, C., Corrêa, I.R., Kindermann, M., Beaufils, F., et al., 2008. An engineered protein tag for multiprotein labeling in living cells. Chem. Biol. 15 (2), 128–136.

Ge, P., Selvin, P.R., 2004. Carbostyril derivatives as antenna molecules for luminescent lanthanide chelates. Bioconjug. Chem. 15, 1088–1094.

Gee, B., Warhol, M.J., Roth, J., 1991. Use of an anti-horseradish peroxidase antibody gold complex in the ABC technique. J. Histochem. Cytochem. 39, 863–870.

Gee, K.R., Archer, E.A., Kang, H.C., 1999. 4-Sulfotetrafluorophenyl (STP) esters: new water-soluble amine-reactive reagents for labeling biomolecules. Tetrahedron Lett. 40, 1471.

Gegg, C.V., Etzler, M.E., 1993. Directional coupling of synthetic peptides to poly-l-lysine and applications to the ELISA. Anal. Biochem. 210, 309–313.

Geiger, B., Singer, S.J., 1980. Association of microtubules and intermediate filaments in chicken gizzard cells as detected by double immunofluorescence. Proc. Natl. Acad. Sci. USA 77, 4769.

Gemeiner, L.E., Miller, I., Moroder, L., 1992. Immunomodulating activity of 1,2-difattyacyl-3-mercaptoglycerol adducts. Biol. Chem. Hooppe Seyler 373 (11), 1085–1094.

Geoghegan, K.F., Stroh, J.G., 1992. Site-directed conjugation of non-peptide groups to peptides and proteins via periodate oxidation of a 2-amino alcohol. Applications to modification at N-terminal serine. Bioconjug. Chem. 3, 138–146.

Geoghegan, K.F., Emery, M.J., Martin, W.H., McColl, A.S., Daumy, G.O., 1993. Site-directed double fluorescent tagging of human renin and collagenase (MMP-1) substrate peptides using the periodate oxidation of N-terminal serine. An apparently general strategy for provision of energy-transfer substrates for proteases. Bioconjug. Chem. 4, 537–544.

Geoghegan, W.D., 1988. The effect of three variables on adsorption of rabbit IgG to colloidal gold. J. Histochem. Cytochem. 36, 401–407.

Geoghegan, W.P., Ambegaonkar, N., Calvanico, N., 1980. Passive gold agglutination: an alternative to passive hemagglutination. J. Immunol. Methods 34, 11.

Georgakilas, V., Kordatos, K., Prato, M., Guldi, D.M., Holzinger, M., Hirsch, A., 2002. Organic functionalization of carbon nanotubes. J. Am. Chem. Soc. 124, 760–761.

George, S.P., Wang, Y., Mathew, S., Kamalakkannan, S., Seema, K., 2007. Dimerization and actin-bundling properties of villin and its role in the assembly of epithelial cell brush borders. J. Biol. Chem. doi: 10.1074/jbc.M703617200.

Gérard, E., Bessy, E., Salvagnini, C., Rerat, V., Momtaz, M., Hénard, G., et al., 2011. Surface modifications of polypropylene membranes used for blood filtration. Polymer 52, 1223–1233.

Gérard, E., Bessy, E., Salvagnini, C., Rerat, V., Momtaz, M., Henard, G., et al., 2011. Surface modifications of polypropylene membranes used for blood filtration. Polymer 52, 1223–1233.

Germain, R.N., 1986. The ins and outs of antigen processing and presentation. Nature (London) 322, 687–689.

Gershoni, J.M., Bayer, E.A., Wilchek, M., 1985. Blot analysis of glycoconjugates: enzyme-hydrazide—A novel reagent for the detection of aldehydes. Anal. Biochem. 146, 59–63.

Geselowitz, D.A., Neumann, R.D., 1995. Quantitation of triple-helix formation using a photo-cross-linkable aryl azide/biotin/oligonucleotide conjugate. Bioconjug. Chem. 6, 502–506.

Geva, G., Sharan, R., 2011. Identification of protein complexes from co-immunoprecipitation data. Bioinformatics 27 (1), 111–117.

Ghadban, A., Albertin, L., Moussavou Mounguengui, R.W., Peruchon, A., Heyraud, A., 2012. Synthesis of b-D-glucopyranuronosylamine in aqueous solution: kinetic study and synthetic potential. Carbohydr. Res. 346, 2384–2393.

Ghaim, J.B., Greiner, D.P., Meares, C.F., Gennis, R.B., 1995. Proximity mapping the surface of a membrane protein using an artificial protease: demonstration that the quinone-binding domain of subunit I is near the N-terminal region of subunit II of cytochrome bd. Biochemistry 34 (36), 11311–11315.

Ghazi, I., Gomez de Segura, A., Fernandez-Arrojo, L., Alcalde, M., Yates, M., Rojas-Cervantes, M.L., et al., 2005. Immobilisation of fructosyltransferase from Aspergillus aculeatus on epoxy-activated Sepabeads EC for the synthesis of fructo-oligosaccharides. J. Mol. Catal. B-Enzym. 35, 19–27.

Ghebrehiwet, B., Bossone, S., Erdei, A., Reid, K.B.M., 1988. Reversible biotinylation of Clq with a cleavable biotinyl derivative. Application in Clq receptor (ClqR) purification. J. Immunol. Methods 110, 251–260.

Ghesquiere, B., Johckheere, V., Colaert, N., Van Durme, J., Timmerman, E., Goethals, M., et al., 2011. Redox proteomics of protein-bound methionine oxidation. Mol. Cell. Proteomics, doi: 10.5, 10.1074/mcp.M110.006866–1.

Ghetie, V., Ghetie, M.-A., Uhr, J.W., Vitetta, E.S., 1988. Large scale preparation of immunotoxins constructed with the Fab' fragment of IgGl murine monoclonal antibodies and chemically deglycosylated ricin A chain. J. Immunol. Methods 112, 267–277.

Ghetie, V., Till, M.A., Ghetie, M.-A., Tucker, T., Porter, J., Patzer, E.J., et al., 1990. Preparation and characterization of conjugates of recombinant CD4 and deglycosylated ricin A chain using different crosslinkers. Bioconjug. Chem. 1, 24–31.

Ghetie, V., Thorpe, P., Ghetie, M.-A., Knowles, P., Uhr, J.W., Vitetta, E.S., 1991. The GLP large scale preparation of immunotoxins containing deglycosylated ricin A chain and a hindered disulfide bond. J. Immunol. Methods 142, 223–230.

Ghetie, V., Swindel, E., Uhr, J.W., Vitetta, E.S., 1993. Purification and properties of immunotoxins containing one vs. two deglycosylated ricin A chains. J. Immunol. Methods 166, 117–122.

Ghose, S., McNerney, T., Hubbard, B., 2007. Process Scale Bioseparations for the Biopharmaceutical Industry. CRC Press, Boca Raton, FL.

Ghosh, S.S., Kao, P.M., Kwoh, D.Y., 1989. Synthesis of 59-oligonucleotide hydrazide derivatives and their use in preparation of enzyme-nucleic acid hybridization probes. Anal. Biochem. 178, 43–51.

Ghosh, S.S., Kao, P.M., McCue, A.W., Chappelle, H.L., 1990. Use of maleimide-thiol coupling chemistry for efficient syntheses of oligonucleotide–enzyme conjugate hybridization probes. Bioconjug. Chem. 1, 71–76.

Giarelli, E., 2007. Cancer vaccines: a new frontier in prevention and treatment. Oncology (Williston Park) 21 (11 Suppl Nurse Ed), 11–17. discussion: 18.

Gilchrist, T.L., Rees, C.W., 1969. Carbenes, Nitrenes, and Arynes (Studies in Modern Chemistry). Nelson Publishers, London. p. 131.

Gill, D.M., Pappenheimer Jr, A.M., Brown, R., Kurnick, J.T., 1969. Studies on the mode of action of diphtheria toxin: VII. Toxin-stimulated hydrolysis of nicotinimide adenine dinucleotide in mammalian cell extracts. J. Exp. Med. 129, 1–21.

Gilles, M.A., Hudson, A.Q., Borders, C.L., 1990. Stability of water-soluble carbodiimides in aqueous solution. Anal. Biochem. 184, 244–248.

Gillitzer, R., Berger, R., Moll, H., 1990. A reliable method for simultaneous demonstration of two antigens using a novel combination of immunogold-silver staining and immunoenzymatic labeling. J. Histochem. Cytochem. 38, 307–313.

Gilmore, J.M., Scheck, R.A., Esser-Kahn, A.P., Joshi, N.S., Francis, M.B., 2006. N-terminal protein modification through a biomimetic transamination reaction. Angew. Chem. Int. Ed. 45 (32), 5307–5311.

Gingras, A.-C., Aebersold, R., Raught, B., 2005. Advances in protein complex analysis using mass spectrometry. J. Physiol. 563 (1), 11–21.

Gingras, A.-C., Gstaiger, M., Raught, B., Aebersold, R., 2007. Analysis of protein complexes using mass spectrometry. Nat. Rev. Mol. Cell Biol. 8, 645–654.

Gioux, S., De Grand, A.M., Lee, D.S., Yazdanfarc, S., Idoine, J.D., Lomnes, S.J., et al., 2005. Improved optical sub-systems for intraoperative near-infrared fluorescence imaging. In: Analoui M., Dunn D.A., (Ed.), Proceedings of SPIE (International Society for Optical Imaging). Optical Methods in Drug Discovery and Development Volume 6009, Oct 23–26; Boston, MA. Bellingham, WA; pp.39–48.

Girelli, Anna, Maria, Enrico, Mattei, Messina, Antonella, 2007. Immobilized tyrosinase reactor for on-line HPLC application: development and characterization. Sens. Actuators B: Chem. 121, 515–521.

Girish, A., Sun, H., Yeo, D.S.Y., Chen, G.Y.J., Chua, T.-K., Yao, S.Q., 2005. Site-specific immobilization of proteins in a microarray using intein-mediated protein splicing. Bioorg. Med. Chem. Lett. 15, 2447–2451.

Gitlin, G., Bayer, E.A., Wilchek, M., 1987. Studies on the biotin-binding site of avidin. Lysine residues involved in the active site. Biochem. J. 242, 923–926.

Gitlin, G., Bayer, E.A., Wilchek, M., 1988. Studies on the biotin-binding site of avidin. Tryptophan residues involved in the active site. Biochem. J. 250, 291–294.

Gitman, A.G., Kahane, L., Loyter, A., 1985a. Use of virus-attached antibndies or insulin molecules to mediate fusion between Sendai virus envelopes and neuraminidase-treated cells. Biochemistry 24, 2762–2768.

Gitman, A.G., Graessmann, A., Loyter, A., 1985b. Targeting of loaded Sendai virus envelopes by covalently attached insulin molecules to virus receptor-depleted cells: fusion-mediated microinjection of ricin A and simian 40 DNA. Proc. Natl. Acad. Sci. USA 82, 7209–7313.

Givens, R.S., Timberlake, G.T., Conrad II, P.G., Yousef, A.L., Weber, J.F., Amslinger, S., 2003. A photoactivated diazopyruvoyl cross-linking agent for bonding tissue containing type-I collagen. Photochem. Photobiol. 78 (1), 23–29.

Glacy, S., 1983. Subcellular distribution of rhodamine-actin microinjected into living fibroblastic cells. Cell Biol. 97, 1207.

Glazer, A.N., 1981. Photosynthetic accessory proteins with bilin prosthetic groups. Biochem. Plants 8, 51–96.

Glazer, A.N., 1985. Light harvesting by phycobilisomes. Annu. Rev. Biophys. Chem. 14, 47–77.

Glazer, A.N., Hixson, C.S., 1977. Subunit structure and chromophore composition of rhodophytan phycoerythrins. Porphyridium cruentum B-phycoerythrin and b-phycoerythrin. J. Biol. Chem. 252, 32–42.

Glazer, A.N., Stryer, L., 1983. Fluorescent tandem phycobiliprotein conjugates: emission wavelength shifting by energy transfer. Biophys. J. 43, 383–386.

Goda, N., Tenno, T., Inomata, K., Iwaya, N., Sasaki, Y., Shirakawa, M., et al., 2007. LBT/PTD dual tagged vector for purification, cellular protein delivery and visualization in living cells. Biochim. Biophys. Acta 1773, 141–146.

Goding, J.W., 1976. Conjugation of antibodies with fluorochromes: modifications to the standard methods. J. Immunol. Methods 13, 215–226.
Goding, J.W., 1986a. Monoclonal Antibodies: Principles and Practice. Academic Press, Orlando, FL. pp. 6–58.
Goding, J.W., 1986b. Monoclonal Antibodies: Principles and Practice. Academic Press, Orlando, FL. p. 35.
Godwin, J.T., Farr, L.E., Sweet, W.H., Robertson, J.S., 1955. Pathological study of eight patients with glioblastoma multiforme treated by neutron-capture therapy using boron10. Cancer 8, 601–6015.
Goebel, W.F., Babers, F.H., Avery, O.T., 1932. Chemo-immunological studies on conjugated carbohydrate-proteins: VI. The synthesis of p-aminophenol alpha-glucoside and its coupling with protein. J. Exp. Med. 55 (5), 761–767.
Göhr-Rosenthal, S., Schmitt-Willich, H., Ebert, W., Conrad, J., 1993. The demonstration of human tumors on nude mice using gadolinium-labelled monoclonal antibodies for magnetic resonance imaging. Invest Radiol. 28 (9), 789–795.
Gokmen, M.T., Du Prez, F.E., 2012. Porous polymer particles—A comprehensive guide to synthesis, characterization, functionalization and applications. Prog. Polym. Sci. 37 (3), 365–405.
Goldman, B., DeFrancesco, L., 2009. The cancer vaccine roller coaster. Nat. Biotechnol. 27, 129–139.
Goldman, E.R., et al. 2002a. Avidin: a natural bridge for quantum dot antibody conjugates. J. Am. Chem. Soc. 124, 6378–6382.
Goldman, E.R., et al., 2002b. Conjugation of luminescent quantum dots with antibodies using an engineered adaptor protein to provide new reagents for fluoroimmunoassays. Anal. Chem. 74, 841–847.
Goldman, E.R., Mattoussi, H., Anderson, G.P., Medintz, I.L., Mauro, J.M., 2005. Fluoroimmunoassays using antibody-conjugated quantum dots. Methods Mol. Biol. 303, 19–34.
Golds, E.E., Braun, P.E., 1978. Protein associations and basic protein conformation in the myelin membrane. J. Biol. Chem. 253, 8162–8170.
Goldshaid, L., Rubinstein, E., Brandis, A., Segal, D., Leshem, N., Brenner, O., et al., 2010. Novel design principles enable specific targeting of imaging and therapeutic agents to necrotic domains in breast tumors. Breast Cancer Res. 12, R29.
Gong, H., Kovar, J., Little, G., Chen, H., Olive, D.M., 2010. In Vivo imaging of Xenograft tumors using an epidermal growth factor receptor-specific affibody molecule labeled with a near-infrared fluorophore. Neoplasia 12, 139–149.
Goodarzi, N., Varshochian, R., Kamalinia, G., Atyabi, F., Dinarvand, R., 2012. A review of polysaccharide cytotoxic drug conjugates for cancer therapy. Carbohydrate Polym.10.1016/j.carbpol.2012.10.036 in press.
Goodfellow, V.S., Settineri, M., Lawton, R.G., 1989. p-Nitrophenyl 3-diazopyruvate and diazopyruvamides, a new family of photoactivatable cross-linking bioprobes. Biochemistry 28, 6346.
Goodlad, G.A.J., 1957. Cross-linking of collagen by sulfur- and nitrogen-mustards. Biochim. Biophys. Acta 25, 202.
Goodson, R.J., Katre, N.V., 1990. Site-directed pegylation of recombinant interleukin-2 at its glycosylation site. Bio. Technol. 8, 343–346.
Gorecki, M., Patchornik, A., 1973. Polymer-bound dihydrolipoic acid: a new insoluble reducing agent for disulfides. Biochim. Biophys. Acta 303, 36–43.
Gorecki, M., Patchornik, A., 1975. US Patent No. 3,914,205.
Gorecki, M., Wilchek, M., Patchornik, A., 1971. The conversion of 3-monoazotyrosine to 3-aminotyrosine in peptides and proteins. Biochim. Biophys. Acta 220, 590–595.
Gorin, G., Martin, P.A., Doughty, G., 1966. Kinetics of the reaction of N-ethylmaleimide with cysteine and some congeners. Arch. Biochem. Biophys. 115, 593.
Gorman, J.J., 1984. Fluorescent labeling of cysteinyl residues to facilitate electrophoretic isolation of proteins suitable for amino-terminal sequence analysis. Anal. Biochem. 160, 376.
Gorman, J.J., Folk, J.E., 1980. Transglutaminase amine substrates for photochemical labeling and cleavable cross-linking of proteins. J. Biol. Chem. 255, 1175.
Gorman, J.J., Corino, G.L., Mitchell, S.J., 1987. Fluorescent labeling of cysteinyl residues. Application to extensive primary structure analysis of protein on a microscale. Eur. J. Biochem. 168, 169–179.
Gotoh, I., Uekita, T., Seiki, M., 2007. Regulated nucleo-cytoplasmic shuttling of human aci-reductone dioxygenase (hADI1) and its potential role in mRNA processing. Genes Cells 12, 105–117.
Gotoh, Y., Tsukada, M., Minoura, N., 1993. Chemical modification of silk fibroin with cyanuric chloride-activated polyethylene glycol: analysis of reaction site by ^{1}H-NMR spectroscopy and conformation of the conjugates. Bioconjug. Chem. 4, 554–559.
Gottfried, A., Weinhold, E., 2011. Sequence-specific covalent labeling of DNA. Biochem. Soc. Trans. 39, 623–628.
Gounaris, A.D., Perlman, G.E., 1967. Succinylation of pepsinogen. J. Biol. Chem. 242, 2739.
Goundalkar, A., Chose, T., Mezei, M., 1983. Covalent binding of antibodies to liposomes using a novel lipid derivative. J. Pharm. Pharmacol. 36, 465–466.
Govardhan, P., 1999. Crosslinking of enzymes for improved stability and performance. Curr. Opin. Biotechnol. 10, 331–335.
Gowda, A.S.P., Madhunapantula, S.V., Achur, R.N., Valiyaveettil, M., Bhavanandan, V.P., Gowda, D.C., 2007. Structural basis for the adherence of plasmodium falciparum-infected erythrocytes to chondroitin 4-sulfate and design of novel photoactivable reagents for the identification of parasite adhesive proteins. J. Biol. Chem. 282, 916–928.
Grabarek, Z., Gergely, J., 1990. Zero-length cross-linking procedure with the use of active esters. Anal. Biochem. 185, 131–135.
Grabowski, J., Gantt, E., 1978. Photophysical properties of phycobiliproteins from phycobilisomes: fluorescence lifetimes, quantum yields, and polarization spectra. Photochem. Photobiol. 28, 39–45.
Gradishar, W.J., Tjulandin, S., Davidson, N., Shaw, H., Desai, N., Bhar, P., et al., 2005. Phase III trial of nanoparticle albumin-bound paclitaxel compared with polyethylated castor oil-based paclitaxel in women with breast cancer. J. Clin. Oncol. 23, 7794–7803.
Graille, M., Stura, E.A., Corper, A.L., Sutton, B.J., Taussig, M.J., Charbonnier, J.B., et al., 2000. Crystal structure of a Staphylococcus aureus protein A domain complexed with the Fab fragment of a human IgM antibody: structural basis for recognition of B-cell receptors and superantigen activity. Proc. Natl. Acad. Sci. USA 97 (10), 5399–5404.
Granata, A.R., Kitai, S.T., 1992. Intracellular analysis in vivo of different barosensitive bulbospinal neurons in the rat rostral ventrolateral medulla. J. Neurosci. 12, 1–20.
Grassetti, D.R., Murray, J.F., 1967. The effect of 2,29-dithiodipyridine on thiols and oxidizable substrates of Ehrlich ascites cells and of normal mouse tissues. Biochem. Pharmacol. 16 (12), 2387–2393.
Gray, G.R., 1974. The direct coupling of oligosaccharides to proteins and derivatized gels. Arch. Biochem. Biophys. 163, 426–428.
Gray, G.R., 1978. Antibodies to carbohydrates: preparation of antigens by coupling carbohydrates to proteins by reductive amination with cyanoborohydride, Ginsburg, V. Methods in Enzymology, vol. 50. Academic Press, New York, pp. 155–160.
Grayeski, M.L., DeVasto, J.K., 1987. Coumarin derivatizing agents for carboxylic acid detection using peroxyoxalate chemiluminescence with liquid chromatography. Anal. Chem. 59, 1203.
Grayson, S., Fréchet, J.M.J., 2001. Convergent dendrons and dendrimers: from synthesis to applications. Chem. Rev. 101, 3819–3868.
Grazú, V., Abian, O., Mateo, C., Batista-Viera, F., Fernández-Lafuente, R., Guisán, J.M., 2003. Novel bifunctional epoxy/thiol-reactive

support to immobilize thiol containing proteins by the epoxy chemistry. Biomacromolecules 4, 1495–1501.

Green, N.M., 1963. Stability at extremes of pH and dissociation into sub-units by guanidine hydrochloride. Biochem. J. 89, 609–620.

Green, N.M., 1965. A spectrophotometric assay for avidin and biotin based on binding of dyes by avidin. Biochem. J. 94, 23c–24c.

Green, N.M., 1975. Avidin. Adv. Protein Chem. 29, 85–133.

Green, N.M., Konieczny, L., Toms, E.J., Valentine, R.C., 1971. The use of bifunctional biotinyl compounds to determine the arrangement of subunits in avidin. Biochem. J. 125, 781–984.

Greenacre, S.A., Ischiropoulos, H., 2001. Tyrosine nitration: localisation, quantification, consequences for protein function and signal transduction. Free Radic. Res. 34, 541–581.

Greene, L.E., 1986. Cooperative binding of myosin subfragment one to regulated actin as measured by fluoresce changes of troponin 1 modified within different fluorophores. J. Biol. Chem. 261, 1279.

Greenwood, F.C., Hunter, W.M., Clover, J.S., 1963. The preparation of 131I-labelled human growth hormone of high specific radioactivity. Biochem. J. 89, 114.

Gregoriadis, G., 1984. Liposome technology Targeted Drug Delivery and Biological Interaction, vol. III. CRC Press, Boca Raton, Florida.

Gregoriadis, G., Senior, J., 1980. The phospholipid component of small unilamellar liposomes controls the rate of clearance of entrapped solutes from the circulation. FEBS Lett. 119, 43–46.

Gregory, A.E., Williamson, E.D., Prior, J.L., Butcher, W., Thompson, I.J., Shaw, A.M., et al., 2012. Conjugation of Y. pestis F1-antigen to gold nanoparticles improves immunogenicity. Vaccine 30, 6777–6782.

Gregory, J.D., 1955. The stability of N-ethylmaleimide and its reaction with sulfhydryl groups. J. Am. Chem. Soc. 77, 3922.

Greiner, D.P., Hughes, K.A., Gunasekera, A.H., Meares, C.F., 1996. Binding of the sigma-70 protein to the core subunits of *Escherichia coli* RNA polymerase, studied by iron-EDTA protein footprinting. Proc. Natl. Acad. Sci. USA 93, 71–75.

Greiner, D.P., Miyake, R., Moran, J.K., Jones, A.D., Negishi, T, Ishihama, A., et al., 1997. Synthesis of the protein cutting reagent iron (S)-1-(*p*-bromoacetamidobenzyl)ethylenediaminetetraacetate and conjugation to cysteine side chains. Bioconjug. Chem. 8 (1), 44–48.

Gretch, D.R., Suter, M., Stinski, M.F., 1987. The use of biotinylated monoclonal antibodies and streptavidin affinity chromatography to isolate herpesvirus hydrophobic proteins or glycoproteins. Anal. Biochem. 163, 270–277.

Griffin, R.J., Williams, B.W., Bischof, J.C., Olin, M., Johnson, G.L., Lee, B.W., 2007. Use of a fluorescently labeled poly-caspase inhibitor for in vivo detection of apoptosis related to vascular-targeting agent arsenic trioxide for cancer therapy. Technol. Cancer Res. Treat. 6, 651–654.

Grinberg, M., Schwarz, M., Zaltsman, Y., Eini, T., Niv, H., Pietrokovski, S., et al., 2005. Mitochondrial carrier homolog 2 is a target of tBID in cells signaled to die by tumor necrosis factor alpha. Mol. Cell. Biol. 25, 4579–4590.

Gros, O., Gros, P., Jansen, F.K., Vidal, H., 1985. Biochemical aspects of immunotoxin preparation. J. Immunol. Methods 81, 283.

Gross, H., Bilk, L., 1968. Zur reaction von N-hydroxysuccinimid mit dicyclohexylcarbodiimid. Tetrahedron 24, 6935–6939.

Grossman, S.H., Pyle, J., Steiner, R.J., 1981. Kinetic evidence for active monomers during the reassembly of denatured creatine kinase. Biochemistry 21, 6122.

Gruber, M., 2002. FRET compatible long-wavelength labels and their application in immunoassays and hybridization assays. Dissertation, Department of Chemistry and Pharmacy, University of Regensburg, Germany.

Grubor, N.M., Shinar, R., Jankowiak, R., Porter, M.D., Small, G.J., 2004. Novel biosensor chip for simultaneous detection of DNA-carcinogen adducts with low-temperature fluorescence. Biosens. Bioelectron. 19, 547–556.

Gudheti, M.V., Mlodzianoski, Ml., Hess, S.T., 2007. Imaging and shape analysis of giant unilamellar vesicles (GUVs) as model plasma membranes: effect of *trans*-DOPC (dielaidoyl phosphatidylcholine) on membrane properties. Biophys. J. doi: 10.1529/biophysj.106.103374.

Guerrero, C., Tagwerker, C., Kaiser, P., Huang, L., 2006. An integrated mass spectrometry-based proteomic approach: quantitative analysis of tandem affinity-purified in vivo cross-linked protein complexes (QTAX) to decipher the 26 S proteasome-interacting network. Mol. Cell. Proteomics 5, 366–378.

Guesdon, J.-L., Ternynck, T., Avrameas, S., 1979. The use of avidin–biotin interaction in Immunoenzymatic techniques. J. Histochem. Cytochem. 27, 1131–1139.

Guire, P., 1976. Stepwise thermophotochemical cross-linking agents for enzyme stabilization and immobilization. Fed. Proc. 35, 1632.

Guisán, J.M., 1988. Aldehyde-agarose gels as activated supports for immobilization-stabilization of enzymes. Enzyme Microb. Technol. 10, 375–382.

Guisán, J.M., Bastida, A., Blanco, R.M., Fernández-Lafuente, R., García-Junceda, E., 1997. Immobilization of enzymes on glyoxal-agarose: strategies for enzyme stabilization by multipoint attachment. In: Bickerstaff, G.R. (Ed.), Immobilization of Enzymes and Cells. Humana Press, Inc., Totowa, NJ, pp. 277–288. Methods in Biotechnology, Chapter 31.

Gullberg, M., Gustafsdottir, S.M., Schallmeiner, E., Jarvius, J., Bjarnegård, M., Betsholtz, C., et al., 2004. Cytokine detection by antibody-based proximity ligation. Proc. Natl. Acad. Sci. USA 101, 8420–8424.

Gulley, J.L., Madan, R.A., Arlen, P.M., 2007. Enhancing efficacy of therapeutic vaccinations by combination with other modalities. Vaccine 25 (Suppl. 2), B89–B96.

Gunasekaran, K., Nguyen, T.H., Maynard, H.D., Davis, T.P., Bulmus, V., 2011. Conjugation of siRNA with comb-type PEG enhances serum stability and gene silencing efficiency. Marcormol. Rapid Commun. 32, 654–659.

Gundlach, H.G., Moore, S., Stein, W.H., 1959. The reaction of iodoacetate with methionine. J. Biol. Chem. 234, 1761–1764.

Guo, J., Wang, S., Dai, N., Teo, Y.N., Kool, E.T., 2011. Multispectral labeling of antibodies with polyfluorophores on a DNA backbone and application in cellular imaging. Proc. Natl. Adac. Sci. (USA) 108, 3493–3498.

Guo, R., Shi, X., 2012. Dendrimers in cancer therapeutics and diagnosis. Curr. Drug Metab. 13, 1097–1109.

Guo, T., Nicolaev, P., Thess, A., Colbert, D.T., Smalley, R.E., 1995a. Catalytic growth of single walled nanotubes by laser vaporization. Chem. Phys. Lett. 243, 49–54.

Guo, T., Nikolaev, P., Rinzler, A.G., TomBnek, D., Colbert, D.T., Smalley, R.E., 1995b. Self-assembly of tubular fullerenes. J. Phys. Chem. 99, 10694–10697.

Gupta, H., Dai, L., Datta, G., Garber, D.W., Grenett, H., Li, Y., et al., 2005. Inhibition of lipopolysaccharide-induced inflammatory responses by an apolipoprotein AI mimetic peptide. Circ. Res. 97, 236–243.

Gurd, F.R.N., 1967. CarboxymethylationHirs, C.H.W. Methods in Enzymology, vol. 11. Academic Press, New York, pp. 532.

Gürsel, S.A., Gubler, L., Gupta, B., Scherer, G.G., 2008. Radiation grafted membranesAbe, A.Albertsson, A.C.Duncan, R.Dušek, K. Fuel Cells 1. Advances in Polymer Science, 215. Springer-Verlag, Berlin, pp. 157–217.

Gygi, S.P., Rist, B., Gerber, S.A., Turecek, F., Gelb, M.H., Abersold, R., 1999. Quantitative analysis of complex protein mixtures using isotope-coded affinity tags. Nat. Biotechnol. 10, 994–999.

Gygi, S.P., Corthals, G.L., Zhang, Y., Rochon, Y., Ruedi, A., 2000. Evaluation of two-dimensional gel electrophoresis-based proteome analysis technology. Curr. Opin. Biotechnol. 11 (4), 396–401.

Ha, Y.S., Lee, H.Y., An, G.I., Kim, J., Kwak, W., Lee, E.J., et al., 2012. Synthesis and evaluation of a radioiodinated bladder cancer specific peptide. Bioorg. Med. Chem. 20, 4330–4335.

Habeeb, A.F.S.A., 1966. Determination of free amino groups in protein by trinitrobenzene sulfonic acid. Anal. Biochem. 14, 328.

Habeeb, A.F.S.A., Atassi, M.Z., 1970. Enzymatic and immunochemical properties of lysozyme. Evaluation of several amino group reversible blocking reagents. Biochemistry 9, 4939–4944.

Habeeb, A.F.S.A., Hiramoto, R., 1968. Reaction of proteins with glutaraldehyde. Arch. Biochem. Biophys. 126, 16–26.

Habeeb, A.F.S.A., Cassidy, H.G., Singer, S.J., 1958. Molecular structural effects produced in proteins by reaction with succinic anhydride. Biochim. Biophys. Acta 29, 587.

Haberland, J., Becker, J., Gerke, V., 1997. The acidic C-terminal domain of rna1p is Required for the binding of Ran GTP and for RanGAP activity. J. Biol. Chem. 272, 24717–24726.

Habili, N., McInnes, J.K., Symons, R.H., 1987. Non-radioactive photobiotin-labelled DNA probes for the routine diagnosis of barley yellow dwarf virus. J. Virol. Methods 16, 225–237.

Hackeng, T.M., Mounier, C.M., Bon, C., Dawson, P.E., Griffin, J.H., Kent, S.B.H., 1997. Total chemical synthesis of enzymatically active human type II secretory phospholipase A2. Proc. Natl. Acad. Sci. USA 94, 7845–7850.

Hackeng, T.M., Griffin, J.H., Dawson, P.E., 1999. Protein synthesis by native chemical ligation: expanded scope by using straightforward methodology. Proc. Natl. Acad. Sci. USA 96, 10068–10073.

Haddad, L.C., Swenson, B.C., Bothof, C.A., Raghavachari, M., 2010. Materials, methods, and kits for reducing nonspecific binding of molecules to a surface. U.S. Patent 7,727,710.

Hadi, U.A.M., Malcoltne-Lawes, J., Oldham, G., 1979. Rapid radiohalogenations of small molecules-II. Radiobromination of tyrosine, uracil, and cytosine. Int. J. Appl. Radiat. Isot. 30, 709–712.

Hadisaputri, Y.E., Miyazaki, T., Suzuki, S., Yokobori, T., Kobayashi, T., Tanaka, N., et al., 2011. TNFAIP8 overexpression: clinical relevance to esophageal squamous cell carcinoma. Ann. Surg. Oncol., 1–8.

Hadjikakou, S.K., Tsipis, A.C., Kubicki, M., Bakas, T., Hadjiliadis, N., 2011. Inhibition of peroxidase-catalyzed iodination by thioamides: experimental and theoretical study of the antithyroid activity of thioamides. New J. Chem. 35 (1), 213–224.

Haensler, J., Szoka Jr., F.C., 1993. Polyamidoamine cascade polymers mediate efficient transfection of cells in culture. Bioconjug. Chem. 4, 372–379.

Haeuw, J.F., Caussanel, V., Beck, A., 2009. Immunoconjugates, drug-armed antibodies to fight against cancer. Med. Sci. (Paris) 25 (12), 1046–1052.

Hagan, A.K., Zuchner, T., 2011. Lanthanide-based time-resolved luminescence immunoassays. Anal. Bioanal. Chem. 400 (9), 2847–2864.

Hage, D.S., 1999. Affinity chromatography: a review of clinical applications. Clin. Chem. 45 (5), 593–615.

Hage, D.S., Wolfe, C.A.C., Oates, M.R., 1997. Development of a kinetic model to describe the effective rate of antibody oxidation by periodate. Bioconjug. Chem. 8, 914–920.

Hager, H.J., 1974. Latex polymer reagents for diagnostic tests. US Patent 3, 857, 931.

Hahn, C.D., Leitner, C., Weinbrenner, T., Schlapak, R., Tinazli, A., Tampé, R., et al., 2007. Self-assembled monolayers with latent aldehydes for protein immobilization. Bioconjug. Chem. 18, 247–253.

Hajdu, J., Pombradi, V., Bot, C., Friedrich, P., 1979. Structural changes in glycogen phosphorylase as revealed by cross-linking with bifunctional diimidates: phosphorylase b. Biochemistry 18, 4037–4041.

Halford, B., 2005. Dendrimers branch out. Chem. Eng. News 83 (24), 30–36.

Hall, D.B., Struhl, K., 2002. The VP16 activation domain interacts with multiple transcriptional components as determined by protein–protein cross-linking *in vivo*. J. Biol. Chem. 277, 46043–46050.

Hall, L.D., Yalpani, M., 1980. Synthesis of luminescent probe-sugar conjugates of either protected or unprotected sugars. Carbohydr. Res. 78, C4.

Halliwell, B., Gutteridge, J.M., 1984. Role of iron in oxygen radical reactions. Methods Enzymol. 105, 47–56.

Halliwell, B., Gutteridge, J.M.C., 1989. Free Radicals in Biology and Medicine, second ed., Clarendon Press, Oxford.

Halliwell, B., Gutteridge, J.M.C., 1990. Role of free radicals and catalytic metal ions in human diseases: an overview. Methods Enzymol. 186, 1.

Halperin, S.A., Scheifele, D., De Serres, G., Noya, F., Meekison, W., Zickler, P., et al., 2011. Immune responses in adults to revaccination with a tetanus toxoid, reduced diphtheria toxoid, and acellular pertussis vaccine 10 years after a previous dose. Vaccine, 30, 974–982.

Hamada, H., Tsuro, T., 1987. Determination of membrane antigens by a covalent cross-linking method with monoclonal antibodies. Anal. Biochem. 160, 483–488.

Hamaguchi, Y., Yoshitake, S., Ishikawa, E., Endo, Y., Ohtaki, S., 1979. Improved procedure for the conjugation of rabbit IgG and Fab′ antibodies with b-D-galactosidase from *Escherichia coli* using N,N9-o-phenylenedimaleimide. J. Biochem. (Tokyo) 85, 1289–1300.

Hamaker, K.H., Ladisch, M.R., 1996. Intraparticle flow and plate height effects in liquid chromatography stationary phases. Sep. Purif. Methods 25, 47–83.

Hamdan, M.H., Righetti, P.G., 2005. Proteomics Today: Protein Assessment and Biomarkers Using Mass Spectrometry, 2D Electrophoresis, and Microarray Technology (Wiley-Interscience Series on Mass Spectrometry). Wiley-Interscience, ISBN-13: 978-0471648178.

Han, B.N., Stevens, J.F., Maier, C.S., 2007. Design, synthesis, and application of a hydrazide-functionalized isotope-coded affinity tag for the quantification of oxylipid–protein conjugates. Anal. Chem. 79 (9), 3342–3354.

Han, G., Tamaki, M., Hruby, V.J., 2001. Fast, efficient and selective deprotection of the tert-butoxycarbonyl (Boc) group using HCl/dioxane (4 M). J. Pept. Res. 58, 338–341.

Han, M., Gao, X., Su, J.Z., Nie, S., 2001. Quantum-dot-tagged microbeads for multiplexed optical coding of biomolecules. Nat. Biotechnol. 19, 631–635.

Hanani, M., 2011. Lucifer Yellow—An angel rather than the devil. J. Cell. Mol. Med. 16, 22–31.

Handlogten, M.E., Hong, S.-P., Westhoff, C.M., Weiner, I.D., 2005. Apical ammonia transport by the mouse inner medullary collecting duct cell (mIMCD-3). Am. J. Physiol. Renal Physiol. 289, F347–F358.

Hang, H.C., Geutjes, E.-J., Grotenbreg, G., Pollington, A.M., Bijlmakers, M.J., Ploegh, H.L., 2007. Chemical probes for the rapid detection of fatty-acylated proteins in mammalian cells. J. Am. Chem. Soc. Published on Web 02/17/2007.

Hanke, S., Mann, M., 2009. The Phosphotyrosine Interactome of the Insulin Receptor Family and Its Substrates IRS-1 and IRS-2. Mol. Cell. Proteomics 8, 519–534.

Hansberry, D.R., Clark, P.M., 2012. Quantitative antibody immobilization using hetero-and homo-bifunctional crosslinkers for analytical biosensing. Bioengineering Conference (NEBEC), 38th Annual Northeast, Conference publication, pp. 89–90.

Hansen, K.C., Schmitt-Ulms, G., Chalkley, R.J., Hirsch, J., Baldwin, M.A., Burlingame, A.L., 2003. Mass spectrometric analysis of protein mixtures at low levels using cleavable 13C-isotope-coded affinity tag and multidimensional chromatography. Mol. Cell. Proteomics 2, 299–314.

Hansen, P., Scoble, J.A., Hanson, B., Hoogenraad, N.J., 1998. Isolation and purification of immunoglobulins from chicken eggs using thiophilic interaction chromatography. J. Immunol. Methods 215, 1–7.

Harding, C.V., Kihlberg, J., Elofsson, M., Magnusson, G., Unanue, E.R., 1993. Glycopeptides bind MHC molecules and elicit specific T cell responses. J. Immunol. 151, 2419–2425.

Harding, F.A., Liu, A.D., Stickler, M., Razo, O.J., Chin, R., Faravashi, N., et al., 2005. A β-lactamase with reduced immunogenicity for the targeted delivery of chemotherapeutics using antibody-directed enzyme prodrug therapy. Mol. Cancer Ther. 4, 1791–1800.

Hardouin, J., Duchateau, M., Canelle, L., Vlieghe, C., Joubert-Caron, R., Caron, M., 2007. Thiophilic adsorption revisited. J. Chromatogr. B 845, 226–231. 10th International Symposium on Biochromatography.

Hardy, P.M., Nicholls, A.C., Rydon, H.N., 1969. The nature of glutaraldehyde in aqueous solution. Chem. Commun. 65, 525.

Hardy, P.M., Nicholls, A.C., Rydon, H.N., 1976. The nature of the cross-linking of proteins by glutaraldehyde. Interaction of glutaraldehyde with the amino-groups of 6-ami-nohexanoic acid and of b-*N*-acetyl-lysine. J. Chem. Soc., Perk Trans. 1, 958.

Harlow, E., Lane, D., 1988. Antibodies: A Laboratory Manual. Cold Spring Harbor Laboratory Press, Cold Spring Harbor, NY. ISBN: 0-87969-314-2.

Harlow, E., Lane, D., 1988a. Antibodies: A Laboratory Manual. Cold Springs Harbor Laboratory, Cold Spring Harbor, NY. pp. 23–135.

Harlow, E., Lane, D., 1988b. Antibodies: A Laboratory Manual. Cold Spring Harbor Laboratory, Cold Spring Harbor, NY. pp. 7–22.

Harlow, E., Lane, D., 1988c. Antibodies: A Laboratory Manual. Cold Spring Harbor Laboratory, Cold Spring Harbor, NY. pp. 319–358.

Harlow, E., Lane, D., 1999. Using Antibodies: A Laboratory Manual. Cold Spring Harbor Laboratory Press, Cold Spring Harbor, NY. ISBN: 0-87969-543-9.

Härmä, H., Tarkkinen, P., Soukka, T., Lövgren, T., 2000. Miniature single-particle immunoassay for prostate-specific antigen in serum using recombinant Fab' fragments. Clin. Chem. 46, 1755–1761.

Harper, G.R., Davis, S.S., Davies, M.C., Norman, M.E., Tadros, T.F., Taylor, D.C., et al., 1995. Influence of surface coverage with poly(ethylene oxide) on attachment of sterically stabilized microspheres to rat Kupffer cells in vitro. Biomaterials 16, 427–439.

Harrigan, P.R., Madden, T.D., Cullis, P.R., 1990. Protection of liposomes during dehydration or freezing. Chem. Phys. Lipids 52, 139–149.

Harris, J.M., Stuck, E.C., Case, M.G., Paley, M.S., Yalpani, M., van Alstine, J.M., et al., 1984. Synthesis and characterization of PEG derivatives. J. Polym. Sci. Polym. Chem. 22, 341–352.

Harrison, J.K., Lawton, R.G., Cnegy, M.E., 1989. Development of a novel photoreactive calmodulin derivative: cross-linking of purified adenylate cyclase from bovine brain. Biochemistry 28, 6023.

Hartman, F.C., Wold, F., 1966. Bifunctional reagents. Cross-linking of pancreatic ribonuclease with a diimido ester. J. Am. Chem. Soc. 88, 3890–3891.

Hartman, F.C., Wold, F., 1967. Cross-linking of bovine pancreatic ribonuclease A with dimethyl adipimidate. Biochemistry 6, 2439–2448.

Hartmann, D.M., Heller, M., Esener, S.C., Schwartz, D., Tu, G.S., 2002. DNA attachment of micro- and nanoscale particles to substrates. J. Mater. Res. 17 (2), 473–478.

Harvey, D.J., 2011. Derivatization of carbohydrates for analysis by chromatography; electrophoresis and mass spectrometry. J. Chromatogr. B 879 (17), 1196–1225.

Hashida, S., Ishikawa, E., 1985. Use of normal IgG and its fragments to lower the nonspecific binding of Fab'-enzyme conjugates in sandwich enzyme immunoassay. Anal. Lett. 18 (B9), 1143–1155.

Hashimoto, K., Loader, J.E., Kinsky, S.C., 1986. Iodoacetylated and biotinylated liposomes: effect of spacer length on sulfbydryl ligand binding and avidin precipitability. Biochim. Biophys. Acta 856, 556–565.

Hashimoto, N., Takatsu, K., Masuho, Y., Kishida, K., Hara, T., Hamaoka, T., 1984. Selective elimination of a B cell subset having acceptor site(s) for T cell-replacing factor (TRF) with biotinylated antibody to the acceptor site(s) and avidin-ricin A chain conjugate. J. Immunol. 132, 129–135.

Hassell, J., Hand, A., 1974. Tissue fixation with diimidoesters as an alternative to aldehydes. I. Comparison of cross-linking and ultrastructure obtained with dimethylsuberimidate and glutaraldehyde. J. Histochem. Cytochem. 22, 223–239.

Hata, T., Nakayam, M., 2007. Rapid single-tube method for small-scale affinity purification of polyclonal antibodies using HaloTag® technology. J. Biochem. Biophys. Methods 70, 679–682.

Hatakeyama, T., Kohzake, H., Yamasaki, N., 1992. A microassay for proteases using succinylcasein as a substrate. Anal. Biochem. 204, 181–184.

Haubner, R., Kuhnast, B., Mang, C., Weber, W.A., Kessler, H., Wester, H.-J., et al., 2004. [18F]Galacto-RGD: synthesis, radiolabeling, metabolic stability, and radiation dose estimates. Bioconjug. Chem. 15, 61–69.

Haugaard, N., Cutler, J., Ruggieri, M.R., 1981. Use of *N*-ethylmaleimide to prevent interference by sulfhydryl reagents with the glucose oxidase assay for glucose. Anal. Biochem. 116, 341–343.

Haugland, R.P., 1991. Fluorescent labels. In: Wise, D.L., Wingard, L.B. (Eds.), Biosensors with Fiberoptics. Humana Press, Totowa, NJ, pp. 85–109.

Haugland, R.P., Bhalgat, M.K., 2008. Preparation of avidin conjugatesMcMahon, Methods in Molecular Biology, vol. 418. Humana Press, pp. 1–12. Chapter 1.

Hauser, N.C., Martinez, R., Jacob, A., Rupp, S., Hoheisel, J.D., Matysiak, S., 2006. Utilising the left-helical conformation of L-DNA for analysing different marker types on a single universal microarray platform. Nucleic Acids Res. 34, 5101–5111.

Hawker, C.J., Fréchet, J.M.J., 1990. Preparation of polymers with controlled molecular architecture. A new convergent approach to dendritic macromolecules. J. Am. Chem. Soc. 112, 7638–7647.

Hay, R.W., Porter, D.S., 1999. The reaction of sulphur and nitrogen nucleophiles with [Pt(dien)Cl]1. Transit. Met. Chem 24, 186–188.

Hayakawa, T., Yoshinari, M., Nemoto, K., 2003. Direct attachment of fibronectin to tresyl chloride-activated titanium. J. Biomed. Mat. Res. Part A 67A, 684–688.

Hayashi, H., Naoi, S., Nakagawa, T., Nishikawa, T., Fukuda, H., Imajoh-Ohmi, S., et al., 2012. Sorting Nexin 27 Interacts with Multidrug Resistance-associated Protein 4 (MRP4) and Mediates Internalization of MRP4. J. Biol. Chem. 287 (18), 15054–15065.

He, X., McMahon, S., Rasooly, R., 2012. Evaluation and comparison of three enzyme-linked immunosorbent assay formats for the detection of ricin in milk and serum. Biocatal. Agr. Biotech. 1, 105–109.

Hearn, M.T.W., 1987. 1,19-Carbonyldiimidazole-mediated immobilization of enzymes and affinity ligandsMosbach, K. Methods in Enzymology, vol. 135. Academic Press, Orlando, FL, pp. 102–117.

Hearn, M.T.W., Bethell, G.S., Ayers, J.S., Hancock, W.S., 1979. Application of 1,19-carbonyldiimidazole-activated agarose for the purification of proteins. J. Chromatogr. 185, 463–470.

Hearn, M.T.W., Harris, E., Bethell, G.S., Hancock, W.S., Ayers, J.A., 1981. Application of 1,1′-carbonyldiimidazole-activated matrices for the purification of proteins : III. The use of 1,1′-carbonyldiimidazole-activated agaroses in the biospecific affinity chromatographic isolation of serum antibodies. J. Chromatogr. A 218, 509–518.

Hearn, M.T.W., Smith, P.K., Mallia, A.K., Hermanson, G.T., 1983. Preparative and analytical applications of CDI-mediated affinity chromatography Affinity Chromatography and Biological Recognition. Academic Press, New York. pp. 191–196.

Hearn, S.A., 1987. Electron microscopic localization of chromogranin A in osmium-fixed neuroendocrine cells with a protein A-gold technique. J. Histochem. Cytochem. 35, 795–801.

Heath, T.D., Edwards, D.C., Ryman, B.E., 1976. The adjuvant properties of liposomes. Biochem. Soc. Trans. 4, 129.

Heath, T.D., Macher, B.A., Papahadjopoulos, D., 1981. Covalent attachment of immunoglobulins to liposomes via glycosphingolipids. Biochim. Biophys. Acta 640, 66–81.

Heath, T.D., Montgomery, J.A., Piper, J.R., Papahadjopoulos, D., 1983. Antibody-targeted liposomes: increase in specific toxicity of methotrexate-g-aspartate. Proc. Natl. Acad. Sci. USA 80, 1377–1381.

Heath, T.D., Bragman, K.S., Matthay, K.K., Lopez-Straubinger, N.G., Papahadjopoulos, D., 1984. Antibody-directed liposomes: the development of a cell-specific cytotoxic agent. Biochem. Soc. Trans. 12, 340.

Hebert, G.A., Pittman, B., Cherry, W.B., 1967. Factors affecting the degree of nonspecific staining given by fluorescent isothiocyanate labeled globulins. J. Immunol. 98, 1204–1212.

Heetebrij, R.J., Talman, E.G., van Velzen, M.A., van Gijlswijk, R.P., Snoeijers, S.S., Schalk, M., et al., 2003. Platinum (II)-based coordination compounds as nucleic acid labeling reagents: synthesis, reactivity, and applications in hybridization assays. Chem. Bio. Chem. 4, 573–583.

Heilemann, M., Margeat, E., Kasper, R., Sauer, M., Tinnefeld, P., 2005. Carbocyanine Dyes as efficient reversible single-molecule optical switch. J. Am. Chem. Soc. 127, 3801–3806.

Heilmann, H.D., Holzner, M., 1981. The spatial organization of the active sites of the bifunctional oligomeric enzyme tryptophan synthetase: cross-linking by a novel method. Biochem. Biophys. Res. Commun. 99, 1146.

Heindel, N.D., Zhao, H., Leiby, J., VanDongen, J.M., Lacey, C.J., Lima, D.A., et al., 1990. Hydrazide pharmaceuticals as conjugates to polyaldehyde dextran: syntheses, characterization, and stability. Bioconjug. Chem. 1, 77–82.

Heindel, N.D., Zhao, H., Egolf, R.A., Chang, C.-H., Schray, K.J., Emrich, J.G., et al., 1991. A novel heterobifunctional linker for formyl to thiol coupling. Bioconjug. Chem. 2, 427–430.

Heindel, N.D., Kauffman, M.A., Akyea, E.K., Engel, S.A., Frey, M.F., Lacey, C.J., et al., 1994. Carboxymethyldextran lactone: a preactivated polymer for amine conjugations. Bioconjug. Chem. 5, 98–100.

Heinmark, R.L., Hershey, J.W.B., Traut, R.R., 1976. Cross-linking of initiation factor IF2 to proteins L7/L12 in 70S ribosomes of *Escherichia coli*. J. Biol. Chem. 251, 7779–7784.

Heinze, T., Koschella, A., Magdaleno-Maiza, L., Ulrich, A.S., 2001. Nucleophilic displacement reactions on tosyl cellulose by chiral amines. Polym. Bull. 46, 7–13.

Heitz, J.R., Anderson, C.D., Anderson, B.M., 1968. Inactivation of yeast alcohol dehydrogenase by N-alkylmaleimides. Arch. Biochem. Biophys. 127, 627.

Helenius, A., Kartenbech, J., Simons, K., Fries, E., 1980. On the entry of Semliki forest virus into BHK-21 cells. J. Cell Biol. 84, 404.

Helmeste, D.M., Hammonds Jr., R.G., Li, C.H., 1986. Preparation of [125I-Tyr27, Leu5]bh-endorphin and its use for cross-linking of opioid binding sites in human striatum and NG108-15 neuroblastoma-glioma cells. Proc. Natl. Acad. Sci. USA 83, 4622–4625.

Helsens, K., Martens, L., Vandekerckhove, J., Gevaert, K., 2011. Mass spectrometry-driven proteomics: an introduction. Methods Mol. Biol. 753, 1–27.

Hemmila, I., 1985. Fluoroimmunoassays and immunofluorometric assays [Review]. Clin. Chem. 31, 359–370.

Hemmila, I., 1998. Lanthanides as probes for time-resolved fluorometric immunoassays [Review]. Scand. J. Clin. Lab. Invest. 48, 389–399.

Hemmilä, I., Laitala, V., 2011. Sensitized bioassays. Lanthanide Lumin., 361–380.

Henderson, W., Olsen, G.M., Bonnington, L.S., 1994. Immobilized phosphines incorporating the chiral biopolymers chitosan and chitin. J. Chem. Soc. Comm., 1863–1864.

Henle, E.S., Han, Z., Tang, N., Rai, P., Luo, Y., Linn, S., 1999. Sequence-specific DNA cleavage by Fe21-mediated Fenton reactions has possible biological implications. J. Biol. Chem. 274, 962.

Henriksen, U., Buchardt, O., Nielsen, P., 1991. Azidobenzoyl-, azidoacridinyl-, diazocyclopentadienyl-carbonyl-, and 8-propyloxypsoralen photobiotinylation reagents. Syntheses and photoreactions with DNA and protein. Photochem. Photobiol. A: Chem. 57, 331–342.

Herbener, G.H., 1989. Use of the protein A-gold immunocytochemical and enzyme-gold cytochemical techniques in studies of vitellogenesis. Am. J. Anat. 185, 244–254.

Hermanson, G.T., Mallia, A.K., Smith, P.K., 1992. Immobilized Affinity Ligand Techniques. Academic Press, San Diego.

Hermanson, G.T., Mattson, G.R., Krohn, R.I., 1995. Preparation and use of immunoglobulin-binding affinity supports on Emphaze beads. J. Chrom. A 691, 113–122.

Hermansyah, H., Wijanarko, A., Gozan, M., Arbianti, R., Utami, T.S., Kubo, M., et al., 2007. Consecutive reaction model for triglyceride hydrolysis using lipase. J. Teknologi 2 (XXI), 151–157.

Herriott, R.M., 1947. Reactions of native proteins with chemical reagents. Adv. Protein Chem. 3, 169.

Hersckovits, T., 1988. Recent aspects of the subunit organization and dissociation of hemocyanins. Comp. Biochem. Physiol. 91B, 597–611.

Herzberg, M., 1984. Molecular genetic probe, assay technique, and a kit using this molecular genetic probe. Eur. Patent Appl., 0128018.

Hess, S.T., Girirajan, T.P.K., Mason, M.D., 2006. Ultra-High resolution imaging by fluorescence photoactivation localization microscopy. Biophys. J. 91, 4258–4272.

Heuck, A.P., Savva, C.G., Holzenburg, A., Johnson, A.E., 2007. Conformational changes that effect oligomerization and initiate pore formation are triggered throughout perfringolysin O upon binding to cholesterol. J. Biol. Chem. 282, 22629–22637.

Higano, C.S., Schellhammer, P.F., Small, E.J., Burch, P.A., Nemunaitis, J., Yuh, L., et al., 2009. Integrated data from 2 randomized, double-blind, placebo-controlled, phase 3 trials of active cellular immunotherapy with sipuleucel-T in advanced prostate cancer. Cancer 115, 3670–3679.

Higgins, H.G., Fraser, D., 1952. The reaction of amino acids and proteins with diazonium compounds. Aust. J. Sci. Res. Ser. A: Phys. Sci. 5, 736–752.

Higuchi, T., 1989. Mechanisms of lignin degradation by lignin peroxidase and laccase of white-rot fungi. In: Lewis, N.G., Paice, M.G. (Ed.), Plant Cell Wall Polymers: Biogenesis and Biodegradation, vol. 399, ACS Symposium Series, pp. 482–502.

Higuchi, Y., Wu, C., Chang, K.-L., Irie, K., Kawakami, S., Yamashita, F., et al., 2011. Polyamidoamine dendrimer-conjugated quantum-dots for efficient labeling of primary cultured mesenchymal stem cells. Biomaterials 32, 6676–6682.

Hilderbrand, A., Weissleder, Ralph, 2010. Near-infrared fluorescence: application to in vivo molecular imaging. Curr. Opin. Chem. Biol. 14, 71–79.

Hileman, F.D., Sievers, R.E., Hess, G.G., Ross, W.D., 1973. In situ preparation and evaluation of open pore polyurethane foams in chromatography. Anal. Chem. 45, 1126–1130.

Hill, K.W., Taunton-Rigby, J., Carter, J.D., Kropp, E., Vagle, K., Pieken, W., et al., 2001. Diels–Alder bioconjugation of diene-modified oligonucleotides. J. Org. Chem. 66, 5352–5358.

Hillel, Z., Wu, C.W., 1977. Subunit topography of RNA polymerase from *Escherichia coli*. A cross-linking study with bifunctional reagents. Biochemistry 16, 3334–3342.

Hiltensperger, G., Jones, N.G., Niedermeier, S., Stich, A., Kaiser, M., Jung, J., et al., 2012. Synthesis and structure–activity relationships of new quinolone-type molecules against trypanosoma brucei. J. Med. Chem. 55 (6), 2538–2548.

Hindson, B.J., Ness, K.D., Masquelier, D.A., Belgrader, P., Heredia, N.J., Makarewicz, A.J., et al., 2011. High-throughput droplet digital PCR system for absolute quantitation of DNA copy number. Anal. Chem. 83 (22), 8604–8610.

Hines, K., 1992. Pierce Chem. unpublished observations.

Hines, M.A., Guyot-Sionnest, P., 1996. Synthesis of strongly luminescing ZnS-capped CdSe nanocrystals. J. Phys. Chem. B100, 468–471.

Hirabayashi, J., 2008. Concept, strategy and realization of lectin-based glycan profiling. J. Biochem. 144 (2), 139–147.

Hiratsuka, T., 1987. Nucleotide-induced change in the interaction between the 20- and 26-kilodalton heavy-chain segments of myosin adenosine triphosphatase revealed by chemical cross-linking via the reactive thiol SH2. Biochemistry 26, 3168.

Hiratsuka, T., 1988. Cross-linking of three heavy-chain domains of myosin adenosine triphosphatase with a trifunctional alkylating agent. Biochemistry 27, 4110.

Hirsch, A., 2002. Functionalization of single-walled carbon nanotubes. Angew. Chem. Int. Ed. 41, 1853–1859.

Hirsch, A., Lamparth, I., Grdsser, T., 1994. Regiochemistry of multiple additions to the fullerene core: synthesis of a T′-symmetric hexakisadduct of Ca with bis(ethoxycarbonyl)methylene. J. Am. Chem. Soc. 116, 9385–9386.

Hirsch, R.E., Zukin, R.S., Nagel, R.L., 1986. Steady-state fluorescence emission from the fluorescent probe 5-iodoacetamido-fluorescein, bound to hemoglobin. Biochem. Biophys. Res. Commun. 138, 4889.

Hnatowich, D.J., 1990. Antibody radiolabeling, problems and promises. Nucl. Med. Biol. 17, 49–55.

Hnatowich, D.J., Layne, W.W., Childs, R.L., 1982. The preparation and labeling of PTPA-coupled albumin. Int. J. Appl. Radiat. Isot. 33, 327–332.

Hnatowich, D.J., Virzi, F., Rusckowski, M., 1987. Investigations of avidin and biotin for imaging applications. J. Nucl. Med. 28, 1294–1302.

Ho, R.J.Y., Rouse, B.T., Huang, L., 1986. Target-sensitive immunoliposomes: preparation and characterization. Biochemistry 25, 5500.

Hoare, D.G., Koshland, D.E., 1966. A procedure for the selective modification of carboxyl groups in proteins. J. Am. Chem. Soc. 88, 2057.

Hoare, D.G., Koshland Jr., D.E., 1967. A method for the quantitative modification and estimation of carboxylic acid groups in proteins. J. Biol. Chem. 242, 2447–2453.

Hochman, J.H., Shimizu, Y., PeMars, R., Edidin, M., 1988. Specific associations of fluorescent B-2 microglobulin with cell surfaces. J. Immunol. 140, 2322–2329.

Hodge, P., 1993. Polymer science branches out. Nature 362, 18–19.

Hodgins, L.T., Finlay, T.H., Johnson, A.J., 1980. Preparation of Trichloro-s-triazine activated supports for coupling ligands. U.S. Patent 4,229,537.

Hoefsmit, E.C.M., Korn, C., Blijleven, N., Ploem, J.S., 2011. Light microscopical detection of single 5 and 20 nm gold particles used for immunolabelling of plasma membrane antigens with silver enhancement and reflection contrast. J. Microsc. 143 (2), 161–169.

Hoffman, W.L., O'Shannessy, D.J., 1988. Site-specific immobilization of antibodies by their oligosacchanide moieties to new hydrazide derivatized solid supports. J. Immunol. Methods 112, 113–120.

Hofmann, K., Finn, F.M., Friesen, H.-J., Diaconescu, C., Zahn, H., 1977. Biotinylinsulins as potential tools for receptor studies. Proc. Natl. Acad. Sci. USA 74, 2697–2700.

Hofmann, K., Wood, S.W., Brinton, C.C., Montibeller, J.A., Finn, F.M., 1980. Iminobiotin affinity columns and their application to retrieval of streptavidin. Proc. Natl. Acad. Sci. USA 77, 4666–4668.

Hohng, S., Ha, T., 2004. Near-complete suppression of quantum dot blinking in ambient conditions. J. Am. Chem. Soc. 126, 1324–1325.

Holgate, C., Jackson, P., Cowen, P., Bird, C., 1983. Immunogold-silver staining: new method of immunostaining with enhanced sensitivity. J. Histochem. Cytochem. 31, 938.

Holland, J.T., Lau, C., Brozik, S., Atanassov, P., Banta, S., 2011. Engineering of glucose oxidase for direct electron transfer via site-specific gold nanoparticle conjugation. J. Am. Chem. Soc. 133 (48), 19262–19265.

Holmberg, A., Meurling, L., 1993. Preparation of sulfhydrylhorane-dextran conjugates for boron neutron capture therapy. Bioconjug. Chem. 4, 570–573.

Holmgren, M., Jurman, M.E., Yellen, G., 1996. On the use of thiol-modifying agents to determine channel topology. Neuropharmacology 35, 797–804.

Höltke, H.-J., Seibl, R., Burg, J., Mühlegger, K., Kessler, C., 1990. Non-radioactive labeling and detection of nucleic acids: II. Optimization of the digoxigenin system. Mol. Gen. Hoppe Seyler 371, 929–938.

Homola, J., Yee, S.S., Gauglitz, G., 1995. Biosensing with surface plasmon resonance – how it all started. Biosens. Bioelectron. 10, i–ix.

Homola, J., Yee, S.S., Gauglitz, G., 1999. Surface plasmon resonance sensors: review. Sens. Actuator B54, 3–15.

Hong, M.-Y., Yoon, H.C., Kim, H.-S., 2003a. Protein–ligand interactions at poly(amidoamine) dendrimer monolayers on gold. Langmuir 19, 416–421.

Hong, M.-Y., Lee, D., Yoon, H.C., Kim, H.-S., 2003b. Patterning biological molecules onto poly(amidoamine) dendrimer on gold and glass. Bull. Korean Chem. Soc. 24 (8), 1197–1202.

Hong, M.-Y., Kim, Y.-J., Lee, J.W., Kim, K., Lee, J.-H., Yoo, J.-S., et al., 2004. Synthesis and characterization of tri(ethylene oxide)-attached poly(amidoamine) dendrimer layers on gold. J. Colloid Interface Sci. 274, 41–48.

Hong, S., Leroueil, P.R., Majoros, I.J., Orr, B.G., Baker Jr., J.R., Holl, M.M.B., 2007. The binding avidity of a nanoparticle-based multivalent targeted drug delivery platform. Chem. Biol. 14 (1), 107–115.

Honisberger, M., 1979. Evaluation of colloidal gold as a cytochemical marker for transmission and scanning electron microscope. Biol. Cell 36, 253–258.

Honisbenger, M., Rosset, J., 1977. Colloidal gold, a useful marker for transmission and scanning electron microscopy. J. Histochem. Cytochem. 25, 295.

Honisberger, M., Tacchini-Vonlanthen, M., 1983. Ultrastructural localization of Kunitz inhibitor on thin sections of Glycine max (soybean) cv. Maple Arrow by the gold method. Histochemistry, 77, 37–50.

Honisberger, M., Rosset, J., Bauer, H., 1975. Colloidal gold granules as markers for cell surface receptors in the scanning electron microscope. Experientia 31, 1147–1149.

Honjo, J., Nishizuka, Y., Hayaishi, O., Kato, I., 1968. Diphtheria toxin-dependent adenosine diphosphate ribosylation of aminoacyl transferase II and inhibition of protein synthesis. J. Biol. Chem. 243, 3553–3555.

Hopman, A.H.N., Wiegant, J., Tesser, G.I., Van Duijn, P., 1986. A non-radioactive in situ hybridization method based on mercurated nucleic acid probes and sulfhydryl-hapten hgands. Nucleic Acids Res. 14, 6471–6488.

Hopp, T.P., 1984. Immunogenicity of a synthetic HBsAg peptide: enhancement by conjugation to a fatty acid carrier. Mol. Immunol. 21, 13–16.

Hoppmann, S., Miao, Z., Liu, S., Liu, H., Ren, G., Bao, A., et al., 2011. Radiolabeled affibody-albumin bioconjugates for HER2 positive cancer targeting. Bioconjug. Chem. 16, 413–421.

Hopwood, D., 1969. Comparison of the cross-linking abilities of glutaraldehyde, formaldehyde, and a-hydroxyadipaldehyde with bovine serum albumin and casein. Histochemie 17, 151.

Horak, D., Karpisek, M., Turkova, J., Benes, M., 1999. Hydrazide-functionalized poly(2-hydroxyethyl methacrylate) microspheres for immobilization of horseradish Peroxidase. Biotechnol. Prog. 15, 208–215.

Horák, D., Španová, A., Tvrdíková, J., Rittich, B., 2011. Streptavidin-modified magnetic poly (2-hydroxyethyl methacrylate-co-glycidyl methacrylate) microspheres for selective isolation of bacterial DNA. Eur. Polym. J. 47 (5), 1090–1096.

Horáková, P., Macíčková-Cahová, H., Pivoňková, H., Špaček, J., Havran, L., Hocek, M., et al., 2011. Tail-labelling of DNA probes using modified deoxynucleotide triphosphates and terminal deoxynucleotidyl tranferase. Application in electrochemical DNA hybridization and protein-DNA binding assays. Org. Biomol. Chem. 9 (5), 1366–1371.

Hordern, J.S., Leonard, J.D., Scraba, D.G., 1979. Structure of the mengo virion. Virology 97, 131–140.

Horikawa, K., Armstrong, W.E., 1988. A versatile means of intracellular labeling: injection of biocytin and its detection with avidin conjugates. J. Neurosci. Methods 25, 1–11.

Horisberger, M., Clerc, M.F., 1985. Labeling of colloidal gold with protein A. Histochemistry 82, 219.

Horlacher, Tim, Seeberger, H., 2006. The utility of carbohydrate microarrays in glycomics. OMICS: A J. Integr. Biol. 10 (4), 490–498.

Horney, M.J., et al. 2001. Synthesis and characterization of insulin-like growth factor (IGF)-1 photoprobes selective for the IgG-binding proteins (IGFBPs); photoaffinity labeling of the IGF-binding domain on IGFBP-2. J. Biol. Chem. 276 (4), 2880–2889.

Hornsey, V.S., Pnowse, C.V., Pepper, D.S., 1986. Reductive amination for solid-phase coupling of protein. A practical alternative to cyanogen bromide. J. Immunol. Methods 93, 83–88.

Hota, S.K., Dechassa, M.L., Prasad, P., Bartholomew, B., 2012. Mapping protein-DNA and protein-protein interactions of ATP-dependent chromatin remodelers. Methods Mol. Biol. (Clifton, NJ) 809, 381–409.

Houseman, B.T., Gawalt, E.S., Mrksich, M., 2003. Maleimide-functionalized self-assembled monolayers for the preparation of peptide and carbohydrate biochips. Langmuir 19, 1522–1531.

Howard, A., de La Baume, S., Gioannini, T.L., Hiller, J.M., Simon, E.J., 1985. Covalent labeling of opioid receptors with human b-endorphin. J. Biol. Chem. 260, 10833–10839.

Hoyer, D., Cho, H., Schultz, P.G., 1990. New strategy for selective protein cleavage. J. Am. Chem. Soc. 112, 3249–3250.

Hoyer, L.W., Shainoff, J.R., 1980. Factor VIII-related protein circulates in normal human plasma as high molecular weight multimers. Blood 55, 1056–1059.

Hrmova, M., Farkas, V., Lahnstein, J., Fincher, G.B., 2007. A barley xyloglucan xyloglucosyl transferase covalently links xyloglucan, cellulosic substrates, and (1,3;1,4)-D-glucans. J. Biol. Chem. 282, 12951–12962.

Hu, Y., Morichaud, Z., Chen, S., Leonetti, J.P., Brodolin, K., 2012. Mycobacterium tuberculosis RbpA protein is a new type of transcriptional activator that stabilizes the σA-containing RNA polymerase holoenzyme. Nucleic Acids Res. 40 (14), 6547–6557.

Huang, A., Huang, L., Kennel, S.J., 1980. Monoclonal antibody covalently coupled with fatty acid. J. Biol. Chem. 255, 8015–8018.

Huang, B., Tomalia, D.A., 2005. Dendronization of gold and CdSe/cdS (core–shell) quantum dots with tomalia type, thiol core, functionalized poly(amidoamine) (PAMAM) dendrons. J. Lumin. 111, 215–223.

Huang, B., Bates, M., Zhuang, X., 2009. Super-resolution fluorescence microscopy. Annu. Rev. Biochem. 78, 993–1016.

Huang, B., Kukowska-Latallo, J.F., Tang, S., Zong, H., Johnson, K.B., Desai, A., et al., 2012. The facile synthesis of multifunctional PAMAM dendrimer conjugates through copper-free click chemistry. Bioorg. Med. Chem. Lett. 22, 3152–3156.

Huang, C., 1969. Studies of phospholipid vesicles. Formation and physical characteristics. Biochemistry 8, 344.

Huang, J., Tan, S., 2012. Piccolo NuA4 catalyzed acetylation of nucleosomal histones: critical roles of an Esa1 Tudor/chromo barrel loop and an Epl1 Enhancer of Polycomb A (EPcA) basic region. Mol. Cell. Biol.10.1128/MCB.01131-12 Published ahead of print 29 October 2012.

Huang, K.-H., Fainclough, R.H., Cantor, C.R., 1975. Singlet energy transfer studies of the arrangement of proteins in the 30S Escherichia coli ribosome. J. Mol. Biol. 97, 443.

Huang, Z.H., Shi, L., Ma, J.W., Sun, Z.Y., Cai, H., Chen, Y.X., et al., 2012. A totally synthetic, self-assembling, adjuvant-free MUC1 glycopeptide vaccine for cancer therapy. J. Am. Chem. Soc. 134 (21), 8730–8733.

Hudak, J.E., Yu, H.H., Bertozzi, C.R., 2011. Protein glycoengineering enabled by the versatile synthesis of aminooxy glycans and the genetically encoded aldehyde tag. J. Am. Chem. Soc. 133 (40), 16127–16135.

Hudson, E.N., Weber, G., 1973. Synthesis and characterization of two fluorescent sulfhydryl reagents. Biochemistry 12, 4154.

Hudson, P.J., Souriau, C., 2007. Engineered antibodies. Nat. Med. 9, 129–134.

Hudson, J., Souriau, Christelle, 2003. Engineered antibodies. Nat. Med. 9, 129–134.

Hudson, S.D., Jung, H.Y., Percec, V., Cho, W.D., Johansson, G., Ungar, G., et al., 1997. Direct visualization of individual cylindrical and spherical supramolecular dendrimers. Science 278, 449–452.

Hughes, C.E., Pollitt, A.Y., Mori, J., Eble, J.A., Tomlinson, M.G., Hartwig, J.H., et al., 2010. CLEC-2 activates Syk through dimerization. Blood 115, 2947–2955.

Hughes, W.L., Straessle, R., 1950. Preparation and properties of serum and plasma proteins. XXIV. Iodination of human serum albumin. J. Am. Chem. Soc. 72, 452–457.

Huhtinen, P., Kivela, M., Kuronen, O., Hagren, V., Takalo, H., Tenhu, H., et al., 2005. Synthesis, characterization, and application of Eu(III), Tb(III), Sm(III), and Dy(III) lanthanide chelate nanoparticle labels. Anal. Chem. 77 (8), 2643–2648.

Huisgen, R., Grashey, R., Sauer, J., 1964. Chemistry of Alkenes. Interscience, New York. pp. 806–877.

Huisman, Han, Wynveen, Paul, Setter, W., 2009. Studies on the immune response and preparation of antibodies against a large panel of conjugated neurotransmitters and biogenic amines: specific polyclonal antibody response and tolerance. J. Neurochem. 112, 829–841.

Hummelen, J.C., Knight, B.W., LePeq, F., Wudl, F., Yao, J., Wilkins, C.L., 1995. Preparation and characterization of fulleroid and methanofullerene derivatives. J. Org. Chem. 60, 532–538.

Hunter, M.J., Ludwig, M.L., 1962. The reaction of imidoesters with protein and related small molecules. J. Am. Chem. Soc. 84, 3491.

Huq, I., Tamilarasu, N., Rana, T.M., 1999. Visualizing tertiary folding of RNA and RNA-protein interactions by a tethered iron chelate: analysis of HIV-1 Tat-TAR complex. Nucleic Acids Res. 27, 1084–1093.

Hurst, G.B., Lankford, T.K., Kennel, S.J., 2004. Mass spectrometric detection of affinity purified cross-linked peptides. J. Am. Soc. Mass Spectrom. 15 (6), 832–839.

Hurwitz, E., Maron, R., Arnon, R., Wilchek, M., Sela, M., 1978. Daunomycin immunoglobulin conjugates, uptake and activity in vitro. Eur. J. Cancer 14, 1213.

Hurwitz, E., Wilchek, M., Phita, J., 1980. Soluble macromolecules as carriers for daunomycin. J. Appl. Biochem. 2, 25.

Hurwitz, E., Arnon, R., Sahar, E., Danon, Y., 1983a. A conjugate of adriamycin and monoclonal antibodies to Thy-1 antigen inhibits human neuroblastoma cells in vitro. Ann. N.Y. Acad. Sci. 417, 125.

Hurwitz, E., Kashi, R., Burowsky, D., Arnon, R., Haimovich, J., 1983b. Site-directed chemotherapy with a drug bound to antiidiotypic antibody to a lymphoma cell-surface IgM. Int. J. Cancer 31, 745.

Hurwitz, E., Kashi, R., Arnon, R., Wilchek, M., Sela, M., 1985. The covalent linking of two nucleotide analogues to antibodies. J. Med. Chem. 28, 137.

Husain, Q., Husain, M., Kulshrestha, Y., 2009. Remediation and treatment of organopollutants mediated by peroxidases: a review. Crit. Rev. Biotech. 29, 94–119.

Husain, S.S., Lowe, G., 1968. Evidence for histidine in the active sites for ficin and stembromelain. Biochem. J. 110, 53.

Huse, K., Böhme, H.-J., Scholz, G.H., 2002. Purification of antibodies by affinity chromatography. J. Biochem. Biophys. Methods 51, 217–231.

Huse, W.D., Sastry, L., Iverson, S.A., Kang, A.S., Alting-Mees, M., Burton, D.R., et al., 1989. Generation of a large combinatorial library of the immunoglobulin repertoire in phage lambda. Science 246, 1275–1281.

Hussien, H., Goud, A.A., Amin, A.M., El-Sheikh, R., Seddik, U., 2011. Comparative study between chloramine-T and iodogen to prepare radioiodinated etodolac for inflammation imaging. J. Radioanal. Nucl. Chem. 288 (1), 9–15.

Hutchens, T.W., Porath, J., 1986. Thiophilic adsorption of immunoglobulins—Analysis of conditions optimal for selective immobilization and purification. Anal. Biochem. 159, 217–226.

Hutchens, T.W., Porath, J., 1987a. Thiophilic adsorption: a comparison of model protein behavior. Biochemistry 26, 7199–7204.

Hutchens, T.W., Porath, J., 1987b. Protein recognition of immobilized ligands: promotion of selective adsorption. Clin. Chem. 3319, 1502–1508.

Hutchinson, F.J., Francis, S.E., Lyle, I.G., Jones, M.N., 1989. The characterization of liposomes with covalently attached proteins. Biochim. Biophys. Acta 978, 17–24.

Hwang, K.S., Lee, M.H., Lee, J., Yeo, W.-S., Lee, J.H., Kim, K.-M., et al., 2011. Peptide receptor-based selective dinitrotoluene detection using a microcantilever sensor. Biosens. Bioelectron. 30, 249–254.

Hynes, R.O., 1987. Integrins: a family of cell surface receptors. Cell (Cambridge, Mass.) 48, 549–554.

Ibrahim, S.S., Eldeeb, M.A.R., Rady, M.A.H., Hady, K.M.A., Lotfy, M.S., Farag, N.S., et al., 2011. The role of protein interaction domains in the human cancer network. Netw. Biol. 1, 59–71.

Ichikawa, S., Takano, K., Kuroiwa, T., Hiruta, O., Sato, S., Mukataka, S., 2002. Immobilization and stabilization of chitosanase by multipoint attachment to agar gel support. J. Biosci. Bioeng. 93, 201–206.

Idusogie, E.E., Presta, L.G., Gazzano-Santoro, H., Totpal, K., Wong, P.Y., Ultsch, M., et al., 2000. Mapping of the C1q binding site on rituxan, a chimeric antibody with a human IgG1 Fc. J. Immunol. 164 (8), 4178–4184.

Iijima, M., Sato, N., Wuled Lenggoro, I., Kamiya, H., 2009. Surface modification of $BaTiO_3$ particles by silane coupling agents in different solvents and their effect on dielectric properties of $BaTiO_3$/epoxy composites. Colloids Surf. A Physicochem. Eng. Asp. 352 (1), 88–93.

Iijima, S., 1991. Helical microtubules of graphite carbon. Nature 354, 56–58.

Ijima, S., Ichihashi, T., 1993. Single-shell carbon nanotubes of 1-nm diameter. Nature 363, 603–605.

Ikai, A., Yanagita, Y., 1980. A cross-linking study of apo-low density lipoprotein. J. Biochem. (Tokyo) 88, 1359–1364.

Ikeda, Y., Nagasaki, Y., 2012. PEGylation technology in nanomedicine in Polymers in nanomedicineKunugi, S.Yamaoka, T. Advances in Polymer Science, vol. 247. Springer-Verlag, Berlin Heidelberg, pp. 115–140. ISBN: 0065-3195.

Illum, L., Jones, P.D.E., 1985. Attachment of monoclonal antibodies to microspheres. Meth. Enzymol. 112, 67–84. Academic Press, Inc.

Ilver, D., et al. 1998. Helicobactor pylori adhesin binding fucosylated histo-blood group antigens revealed by re-tagging. Science 279 (5349), 373–377.

Ilver, D., et al. 2003. Bacterium-host protein-carbohydrate interactions. Meth. Enzymol. 363, 134–157.

Imagawa, M., Yoshitake, S., Hamguchi, Y., Ishikawa, E., Niitsu, Y., Urushizaki, I., et al., 1982. Characteristics and evaluation of antibody-horseradish peroxidase conjugates prepared by using a maleimide compound, glutaraldehyde, and periodate. J. Appl. Biochem. 4, 41–57.

Inada, Y., Takahashi, K., Yoshimoto, T., Ajima, A., Matsushima, A., Saito, Y., 1986. Applications of polyethylene glycol-modified enzymes in biotechnological processes: organic solvent-soluble enzymes. Trends Biotechnol. 4, 190–194.

Ingalls, H.M., Goodloe-Holland, C.M., Luna, E.J., 1986. Junctional plasma membrane domains isolated from aggregating Dictyostelium discoideum amebae. Proc. Natl. Acad. Sci. USA 83, 4779–4783.

Ingham, R.J., Colwill, K., Howard, C., Dettwiler, S., Lim, C.S.H., Yu, J., et al., 2005. WW domains provide a platform for the assembly of multiprotein networks. Mol. Cell. Biol. 25, 7092–7106.

Inman, J.K., 1985. Functionalization of agarose beads via carboxymethylation and aminoethylamide formation. In: Dean, P.D.G., Johnson, W.S., Middle, F.A. (Eds.), Affinity Chromatography–A Practical Approach. IRL Press, Washington, DC, pp. 53–59.

Inman, J.K., Dintzis, N.D., 1969. Derivatization of cross-linked polyacrylamide beads. Controlled introduction of functional groups for the preparation of special-purpose, biochemical adsorbents. Biochemistry 8, 4074–4082.

Inman, J.K., Highet, P.F., Kolodny, N., Robey, F.A., 1991. Synthesis of N alpha-(tert-butoxycarbonyl)-N epsilon-[N-(bromoacetyl)-beta-alanyl]-L-lysine: its use in peptide synthesis for placing a bromoacetyl cross-linking function at any desired sequence position. Bioconjug. Chem. 2, 458–463.

Innis, M.A., Gelfand, D.H., Sninsky, J.J., White, T.J., 1990. PCR Protocols. A Guide to Methods and Applications. Academic Press, New York.

Inouye, S., Sato, J.I., Sasaki, S., Sahara, Y., 2011. Streptavidin-Aequorin fusion protein for bioluminescent immunoassay. Biosci. Biotechnol. Biochem. 75 (3), 568–571.

Irvin, J.D., 1983. Pokeweed antiviral protein. Pharmacol. Ther. 21, 371–387.

Isaacs, B.S., Husten, E.J., Esmon, C.T., Johnson, A.E., 1986. A domain of membrane-bound blood coagulation factor Va is located far from the phospholipid surface. A fluorescence energy transfer measurement. Biochemistry 25, 4958–4969.

Isaacs, L., Diederich, F., 1993. Structures and chemistry of methanofullerenes: a versatile route into N-[(methanofullerene) carbonyl]-substituted amino acids. Helv. Chim. Acta 76, 2454–2464.

Isaacs, L., Wehrsig, A., Diederich, F., 1993. Improved purification of C60 and formation of &- and n-homoaromatic methano-bridged fullerenes by reaction with alkyl diazoacetates. Helv. Chim. Acta 76, 1231–1250.

Ishi, Y., Lehrer, S.S., 1986. Effects of the state of the succinimido-ring on the fluorescence and structural properties of pyrene maleimide-labeled an-tropomyosin. Biophys. J. 50, 75–80.

Ishida-Yamamoto, A., Kishibe, M., Murakami, M., Honma, M., Takahashi, H., Iizuka, H., 2012. Lamellar granule secretion starts before the establishment of tight junction barrier for paracellular tracers in mammalian epidermis. PloS ONE 7 (2), e31641.

Ishikawa, E., Imagawa, M., Hashida, S., 1983b. Ultra sensitive enzyme immunoassay using fluorogenic, luminogenic, radioactive and related substances and factors to limit the sensitivity. Proceedings of the 2nd International Symposium on Immunoenzymatic Technology.

Ishikawa, E., Imagawa, M., Hashida, S., Yoshitake, S., Hamaguchi, Y., Ueno, T., 1983a. Enzyme-labeling of antibodies. J. Immunoassay 4, 209–327.

Ishikawa, E., Yamada, Y., Yoshitake, S., 1981a. Enzyme labeling with N,N9-*o*-phenylenedimaleimide. In: Ishikawa, E., Kawai, T., Miyazi, K. (Eds.), Enzyme Immunoassay, Tokyo, pp. 67–80.

Ishikawa, E., Yamada, Y., Yoshitake, S., Hamaguchi, Y., 1981b. A more stable maleimide, N-(4-carboxycyclohexylmethyl)maleimide, for enzyme labeling. In: Ishikawa, E., Kawai, T., Miyazi, K. (Eds.), Enzyme Immunoassay, Tokyo, pp. 90–105.

Ishmael, F.T., Alley, S.C., Benkovic, S.J., 2001. Identification and mapping of protein-protein interactions between gp32 and gp59 by cross-linking. J. Biol. Chem. 276, 25236–25242.

Ishmael, F.T., Alley, S.C., Benkovic, S.J., 2002. Assembly of the bacteriophage T4 helicase-architecture and stoichiometry of the gp41-gp59 complex. J. Biol. Chem. 277, 20555–20562.

Ishmael, F.T., Shier, V.K., Ishmael, S.S., Bond, J.S., 2005. Intersubunit and domain interactions of the meprin B metalloproteinase: disulfide bonds and protein-protein interactions in the MAM and TRAF domains. J. Biol. Chem. 280, 13895–13901.

Ishmael, F.T., Trakselis, M.A., Benkovic, S.J., 2003. Protein-protein interactions in the bacteriophage T4 replisome. The leading strand holoenzyme is physically linked to the lagging strand holoenzyme and the primosome. J. Biol. Chem. 278, 3145–3152.

Islam, M.T., Majoros, I.J., Baker Jr., J.R., 2005. HPLC analysis of PAMAM dendrimer based multifunctional devices. J. Chrom. B822, 21–26.

Ismaili, N., Sha, M., Gustafson, E.H., Konarska, M.M., 2001. The 100-kda U5 snRNP protein (hPrp28p) contacts the 59 splice site through its ATPase site. RNA 7, 182.

Iso, M., Chen, B., Eguchi, M., Kudo, T., Shrestha, S., 2001. Production of biodiesel fuel from triglycerides and alcohol using immobilized lipase. J. Mol. Catal. B: Enzymatic 16, 53–58.

Itagaki, T., Kusano, H., Miyata, E., Tashiro, T., 1989. Porous crosslinked polyvinyl alcohol particles, process for producing the same, and separating agent composed of the same. U.S. Patent 4,863,972.

Ito, K., Maruyama, J., 1983. Studies on stable diazoalkanes as potential fluorogenic reagents. I. 7-substituted 4-diazomethylcoumarins. Chem. Pharm. Bull. 31, 3014.

Ito, K., Sawanobori, J., 1982. 4-Piazomethyl-7-methoxycoumarin as a new type of stable aryldiazomethane reagent. Synth. Comm. 12, 665.

Iversen, L., Cherouati, N., Berthing, T., Stamou, D., Martinez, K.L., 2008. Templated protein assembly on micro-contact-printed surface patterns. Use of the SNAP-tag protein functionality. Langumuir 17, 6375–6381.

Iwai, K., Fukuoka, S.-I., Fushiki, T., Kido, K., Sengoku, Y., Semba, T., 1988. Preparation of a verifiable peptide-protein immunogen: direction-controlled conjugation of a synthetic fragment of the monitor peptide with myoglobin and application for sequence analysis. Anal. Biochem. 171, 277–282.

Iwata, R., Iwasaki, Y., Akiyoshi, K., 2007. Site-directed immobilization of antibodies on well-defined polymer brushes. Eur. Cell. Mater. 14 (Suppl. 3), 66.

Izzo, P.N., 1991. A note on the use of biocytin in anterograde tracing studies in the central nervous system: application at both light and electron microscopic level. J. Neurosci. Methods 36, 155–166.

Jablonski, E., Moomaw, E.W., Tullis, R.H., Ruth, J.L., 1986. Preparation of oligo-deoxynucleotide-alkaline phosphatase conjugates and their use as hybridization probes. Nucleic Acids Res. 14, 6115–6129.

Jackson, P., Dockey, D.A., Lewis, F.A., Wells, M., 1990. Application of 1-nm gold probes on paraffin wax sections for *in situ* hybridization histochemistry. J. Clin. Pathol. 43, 810–812.

Jacobsen, R.B., Sale, K.L., Ayson, M.J., Novak, P., Hong, J., Lane, P., et al., 2006. Structure and dynamics of dark-state bovine rhodopsin revealed by chemical cross-linking and high-resolution mass spectrometry. Protein Sci. 15, 1303–1317.

Jacobson, K.A., et al. 1995. Molecular probes for muscarinic receptors: functionalized congeners of selective muscarinic antagonists. Life Sci. 56 (11/12), 823–830.

Jacobson, R.H., Zhang, X.J., DuBose, R.F., Matthews, B.W., 1994. Three-dimensional structure of beta-galactosidase from E. coli. Nature 369, 761–766.

Jaffe, C.L., Lis, H., Sharon, N., 1980. New cleavable photoreactive heterobifunctional cross-linking reagents for studying membrane organization. Biochemistry 19, 4423.

Jaffrey, S.R., Snyder, S.H., 2001. The biotin switch method for the detection of S-Nitrosylated proteins. Sci. STKE, l1.

Jagannath, C., Sehgal, S., 1989. Enhancement of the antigen-binding capacity of incomplete IgG antibodies to Brucella melitensis through Fc region interactions with Staphylococcal protein A. J. Immunol. Meth. 124, 251–257.

Jahn, O., Eckart, K., Brauns, O., Tezval, H., Spiess, J., 2002. The binding protein of corticotropin-releasing factor: ligand-binding site and subunit structure. PNAS 99, 12055–12060.

Jain, A., Ashbaugh, H.S., 2011. Helix stabilization of poly(ethylene glycol)-peptide conjugates. Biomacromolecules 12, 2729–2734.

Jain, K., Kesharwani, P., Gupta, U., Jain, N.K., 2012. A review of glycosylated carriers for drug delivery. Biomaterials 33, 4166–4186.

Jaiswal, J.K., Goldman, E.R., Mattoussi, H., Simon, S.M., 2004. Use of quantum dots for live cell imaging. Nat. Methods 1 (1), 73–78.

Jaiswal, J.K., Mattoussi, H., Mauro, J.M., Simon, S.M., 2003. Long-term multiple color imaging of live cells using quantum dot bioconjugates. Nat. Biotechnol. 21 (1), 47–51.

Jakobovits, A., 1995. Production of fully human antibodies by transgenic mice. Curr. Opin. Biotechnol. 6, 561–566.

Jal, P.K., Patel, S., Mishra, B.K., 2004. Chemical modification of silica surface by immobilization of functional groups for extractive concentration of metal ions. Talanta 62, 1005–1028.

James, R., Cremlyn, W., 2002. Reactions of organic sulfonyl chlorides, in Chlorosulfonic acid: a versatile reagent. The Royal Society of Chemistry, Cambridge, UK. pp. 22–34.

Janeway, C., 2004. Immunobiology, sixth ed., Garland Science, Pub., ISBN No. 0815341016.

Jang, J.H., Hanash, S., 2003. Profiling of the cell surface proteome. Proteomics 3, 1947–1954.

Jansen, F.L., Blythman, H.E., Carriere, D., Casellas, P., Diaz, J., Gros, P., et al., 1980. High specific cytotoxicity of antibody-toxin hybrid molecules (immunotoxins) for target cells. Immunol. Lett. 2, 97.

Jansen, J.F.G.A., Meijer, E.W., 1995. The dendritic box: shape-selective liberation of encapsulated guests. J. Am. Chem. Soc. 117, 4417–4418.

Jansen, J.F.G.A., de Brabander van den Berg, E.M.M., Meijer, E., 1994. Encapsulation of guest molecules into a dendritic box. Science 266, 1226–1229.

Jansons, V.K., Mallet, P.L., 1980. Targeted liposomes: a method for preparation and analysis. Anal. Biochem. 111, 54–59.

Jarvius, J., Melin, J., Göransson, J., Stenberg, J., Fredriksson, S., Gonzalez-Rey, C., et al., 2006. Digital quantification using amplified single-molecule detection. Nat. Methods 3, 725–727.

Jarvius, M., Paulsson, J., Weibrecht, I., Leuchowius, K.J., Andersson, A.C., Wählby, C., et al., 2007. In situ detection of phosphorylated platelet-derived growth factor receptor β using a generalized proximity ligation method. Mol. Cell. Proteomics 6, 1500–1509.

Jayabaskaran, C., Davison, P.F., Paulus, H., 1987. Facile preparation and some applications of an affinity matrix with a cleavable connector arm containing a disulfide bond. Prep. Biochem. 17, 121–141.

Jayaprakash Babu, U.S., Bhat, A.A., Sah, A., Rao, D.N., 2011. Enhanced humoral and mucosal immune responses after intranasal immunization with chimeric multiple antigen peptide of LcrV antigen epitopes of Yersinia pestis coupled to palmitate in mice. Vaccine 29, 9352–9360.

Jeanloz, R.W., 1963. Mucopolysaccharides (acidic glycosaminoglycans) Florkin, M.Stotz, E. Comprehensive Biochemistry, vol. 3. Elsevier, New York, pp. 266–267.

Jeanson, A., Cloes, J.M., Bouchet, M., Rentier, B., 1988. Preparation of reproducible alkaline phosphatase-antibody conjugates for enzyme immunoassay using a heterobifunctional linking agent. Anal. Biochem. 172, 392.

Jedrychowski, L., Penninks, A.H., Kaczmarski, M., Cudowska, B., Korotkiewicz-Kaczmarska, E., Harrer, A., et al., 2009. Methods for detection of food allergens. In: Jedrychowski, L., Wichers, H.J. (Eds.), Chemical and Biological Properties of Food Allergens. CRC Press, Boca Raton, FL, pp. 83–168.

Jeffrey, A.M., Zopf, D.A., Ginsburg, V., 1975. Affinity chromatography of carbohydrate-specific immunoglobulins: coupling of oligosaccharides to sepharose. Biochem. Biophys. Res. Comm. 62, 608–613.

Jegannathan, K.R., Abang, S., Poncelet, D., Chan, E.S., Ravindra, P., 2008. Production of biodiesel using immobilized lipase—a critical review. Crit. Rev. Biotechnol. 28, 253–264.

Jellum, E., 1964. The prevention of thiol autoxidation in biological systems by means of thiolated Sephadex. Acta Chem. Scand. 18, 1887–1895.

Jemmerson, R., Agre, M., 1987. Monoclonal antibodies to different epitopes on a cellsurface enzyme, human placental alkaline phosphatase, effect different patterns of labeling with protein A-colloidal gold. J. Histochem. Cytochem. 35, 1277–1284.

Jenkins, D.W., Hudson, S.M., 2001. Review of vinyl graft copolymerization featuring recent advances toward controlled radical-based reactions and illustrated with Chitin/Chitosan trunk polymers. Chem. Rev. 101 (11), 3245–3274.

Jennings, M.L., Nicknish, J.S., 1985. Localization of a site of intermolecular cross-linking in human red blood cell band 3 protein. J. Biol. Chem. 260, 5472–5479.

Jennissen, H.P., Laub, M., 2007. Development of an universal affinity fusion tag (Poly-DOPA) for immobilizing recombinant proteins on biomaterials. Materwiss. Werksttech. 38, 1035–1039.

Jenny, R.J., Mann, K.G., Lundblad, R.L., 2003. A critical review of the methods for cleavage of fusion proteins with thrombin and factor Xa. Protein Expr. Purif. 31, 1–11.

Jentoft, N., 1990. Why are proteins O-glycosylated? Trends Biochem. Sci. 15, 291–294.

Jentoft, N., Dearborn, D.G., 1979. Labeling of proteins by reductive methylation using sodium cyanoborohydride. J. Biol. Chem. 254, 4359–4365.

Jeon, W.M., Lee, K.N., Birckbichler, P.J., Conway, E., Patterson Jr., M.K., 1989. Colorimetric assay for cellular transglutaminase. Anal. Biochem. 182, 170–175.

Jeong, H., Mason, S.P., Barabasi, A.-L., Oltvai, Z.N., 2001. Lethality and centrality in protein networks. Nature 411, 41–42.

Jeong, K.J., Wang, L., Stefanescu, C.F., Lawlor, M.W., Polat, J., Dohlman, C.H., et al., 2011. Polydopamine coatings enhance biointegration of a model polymeric implant. Soft Matter 7, 8305–8312.

Jerng, H.H., Kunjilwar, K., Pfaffinger, P.J., 2005. Multiprotein assembly of Kv4.2, KChIP3 and DPP10 produces ternary channel complexes with ISA-like properties. J. Physiol. 568, 767–788.

Jessani, N., Niessen, S., Wei, B.Q., Nicolau, M., Humphrey, M., Ji, Y., et al., 2005. A streamlined platform for high-content functional proteomics of primary human specimens. Nat. Methods 2, 691–697.

Jevevar, S., Kunstelj, M., Porekar, V.G., 2010. PEGylation of therapeutic proteins. Biotechnol. J. 5 (1), 113–128.

Ji, I., Ji, T.H., 1981. Both a and b subunits of human choniogonadotropin photoaffinity label the hormone receptor. Proc. Natl. Acad. Sci. USA 78, 5465–5469.

Ji, I., Shin, J., Ji, T.H., 1985. Radioiodination of a photoactivatable heterobifunctional reagent. Anal. Biochem. 151, 348–349.

Ji, S., Xu, J., Zhang, B., Yao, W., Xu, W., Wu, W., et al., 2012. RGD-conjugated albumin nanoparticles as a novel delivery vehicle in pancreatic cancer therapy. Cancer Biol. Ther. 13, 206–215.

Ji, T.H., 1979. The application of chemical cross-linking for studies of cell membrane and the identification of surface reporters. Biochim. Biophys. Acta 559, 39.

Ji, T.H., 1983. Bifunctional reagents. Meth. Enzymol. 91, 580.

Ji, T.H., Ji, I., 1982. Macromolecular photoaffinity labeling with radioactive photoactivatable heterobifunctional reagents. Anal. Biochem. 121, 286–289.

Ji, X., von Rosenvinge, E.C., Johnson, W.W., Tomarev, S.I., Piatigorsky, J., Armstrong, R.N., et al., 1995. Three-dimensional structure, catalytic properties, and evolution of a sigma class glutathione transferase from squid, a progenitor of the lens S-crystallins of cephalopods. Biochemistry 34 (16), 5317–5328.

Jobbagy, A., Jobbagy, G.M., 1973. Examination of FITC preparations. I. Measurements of the dye content of fluorescein isothiocyanate preparations. J. Immunol. Meth. 2, 159–168.

Jobbagy, A., Kiraly, K., 1966. Chemical characterization of fluorescein isothiocyanate–protein conjugates. Biochim. Biophys. Acta 124, 166.

Johansson, G., 1992. Affinity partitioning in PEG-containing two-phase systems. In: Harris, J.M. (Ed.), Poly(Ethylene Glycol) Chemistry: Biotechnical and Biomedical Applications. Plenum, New York, pp. 73–84.

Johnson, A.R., Dekker, E.E., 1996. Woodward's reagent K inactivation of Escherichia coli L-threonine dehydrogenase: increased absorbance at 340–350 nm is due to modification of cysteine and histidine residues, not aspartate or glutamate carboxyl groups. Protein Sci. 5, 382–390.

Johnson, E.C., Kent, S.B., 2006. Insights into the mechanism and catalysis of the native chemical ligation reaction. J. Am. Chem. Soc. 128 (20), 6640–6646.

Johnson, J.D., Collins, J.H., Potter, J.D., 1978. Dansylaziridine labeled troponin C. A fluorescent probe of calcium ion binding to the calcium ion-specific regulatory sites. J. Biol. Chem. 253, 6451.

Joiris, E., Basin, B., Thornback, J.A., 1991. A new method of labeling of monoclonal antibodies and their fragments with 99mTc. Nucl. Med. Biol. 18, 353–356.

Jokerst, J.V., Lobovkina, T., Zare, R.N., Gambhir, S.S., 2011. Nanoparticle PEGylation for imaging and therapy. Nanomedicine 6 (4), 715–728.

Jones, B.N., Gilligan, J.P., 1983. o-Phthaldialdehyde precolumn derivatization and reversed phase high-performance liquid chromatography of polypeptide hydrolysates and physiological fluids. J. Chromatogr. 266, 471–482.

Jones, D., 2004. Pharmaceutical applications of polymers for drug delivery in Rapra Review Reports, vol. 15, No. 6, Published by Rapra Technology, Shropshire, UK.

Jones, G., Bergmark, W.R., Jackson, W.R., 1984. Products of photodegradation for coumarin laser dyes. Opt. Commun. 50, 320.

Jones, G., Jackson, W.R., Choi, C.Y., Bergmark, W.R., 1985. Solvent effects on emission yields and lifetime for coumarin laser dyes. Requirements for the rotatory decay mechanism. J. Phys. Chem. 89, 294.

Jones, O.T., Kunze, D.L., Angelides, K.J., 1989. Localization and mobility of w-conotoxinsensitive Ca21 channels in hippocampal CAl neurons. Science 244, 1189.

Jongsma, M.A., Litjens, R.H., 2006. Self-assembling protein arrays on DNA chips by auto-labeling fusion proteins with a single DNA address. Proteomics 6 (9), 2650–2655.

Joralemon, M.J., O'Reilly, R.K., Matson, J.B., Nugent, A.K., Hawker, C.J., Wooley, K.L., 2005. Dendrimers clicked together divergently. Macromolecules 38 (13), 5436–5443.

Joseph, S., Weiser, B., Noller, H.F., 1997. Mapping the inside of the ribosome with an RNA helical ruler. Science 278, 1093–1098.

Jose-Yacaman, M., Miki-Yoshida, M., Rendon, L., Santiesteban, J.G., 1993. Catalytic growth of carbon microtubules with fullerene structure. Appl. Phys. Lett. 62, 657–659.

Joshi, N.S., Whitaker, L.R., Francis, M.B., 2004. A three-component Mannich-type reaction for selective tyrosine bioconjugation. J. Am. Chem. Soc. 126, 15942–15943.

Joshi, S., Burrows, R., 1990. ATP synthase complex from bovine heart mitochondria. J. Biol. Chem. 265, 14518–14525.

Jothikumar, P., Hill, V., Narayanan, J., 2009. Design of FRET-TaqMan probes for multiplex real-time PCR using an internal positive control. Biotechniques 46 (7), 519–524.

Ju, Q., Tu, D., Liu, Y., Li, R., Zhu, H., Chen, J., et al., 2011. Amine-Functionalized Lanthanide-Doped KGdF4 nanocrystals as potential optical/magnetic multimodal bioprobes. J. Am. Chem. Soc. 134 (2), 1323–1330.

Ju, S.M., Goh, A.R., Kwon, D.J., Youn, G.S., Kwon, H.J., Bae, Y.S., et al., 2012. Extracellular HIV-1 Tat induces human beta-defensin-2 production via NF-kappaB/AP-1 dependent pathways in human B cells. Mol. Cells, 1–7.

Jue, R., Lambert, J.M., Pierce, L.R., Traut, R.R., 1978. Addition of sulfhydryl groups to *Escherichia coli* ribosomes by protein modification with 2-iminothiolane (methyl 4-mercap-tobutyrimidate). Biochemistry 17, 5399–5405.

Jun, H.Y., Yin, H.-H., Kim, S.-H., Park, S.H., Kim, H.S., Yoon, K.-H., 2010. Visualization of tumor angiogenesis using MR imaging contrast agent Gd-DTPA-anti-VEGF receptor 2 antibody conjugate in a mouse tumor model. Korean J. Radiol. 11 (4), 449–456.

Jung, S.K., Wilson, G.S., 1996. Polymeric mercaptosilane-modified platinum electrodes for elimination of interferants in glucose biosensors. Anal. Chem. 68 (4), 591–596.

Jung, S.M., Moroi, M., 1983. Cross-linking of platelet glycoprotein Ib by N-succinimidyl-(4-azidophenyldithio)propionate and 3,39-dit hiobis(sulfosuccinimidyl propionate). Biochim. Biophys. Acta 761, 152–162.

Jungbauer, A., 1996. Insights into the chromatography of proteins provided by mathematical modeling. Curr. Opin. Biotechnol. 7 (2), 210–218.

Jungbauer, A., Hahn, R., 2008. Polymethacrylate monoliths for preparative and industrial separation of biomolecular assemblies. J. Chromatogr. A 1184, 62–79.

Jurado, L.A., Mosley, J., Jarrett, H.W., 2002. Cyanogen bromide activation and coupling of ligands to diol-containing silica for high-performance affinity chromatography optimization of conditions. J. Chromatogr. A 97, 95–104.

Jüse, U., van de Wal, Y., Koning, F., Sollid, L.M., Fleckenstein, B., 2010. Design of new high-affinity peptide ligands for human leukocyte antigen-DQ2 using a positional scanning peptide library. Hum. Immunol. 71, 475–481.

Juskowiak, B., 2011. Nucleic acid-based fluorescent probes and their analytical potential. Anal. Bioanal. Chem. 399, 3157–3176.

Kaatz, M., Schulze, H., Ciani, I., Lisdat, F., Mount, A.R., Bachmann, T.T., 2012. Alkaline phosphatase enzymatic signal amplification for fast, sensitive impedimetric DNA detection. Analyst 137 (1), 59–63.

Kabalka, G.W., Varma, M., Varma, P.C., Srivastava, P.C., Knapp Jr., F.F., 1986. The tosylation of alcohols. J. Org. Chem. 51, 2386–2388.

Kagedal, L., 1989. Protein purification: principles. In: Janson, J.-C., Ryden, L. (Eds.), High Resolution Methods and Applications. VCH Publishers, New York, pp. 227.

Kainz, Q.M., Schatz, A., Zopfl, A., Stark, W.J., Reiser, O., 2011. Combined covalent and noncovalent functionalization of nanomagnetic carbon surfaces with dendrimers and BODIPY fluorescent dye. Chem. Mater. 23, 3606–3613.

Kakade, M.L., Liener, I.E., 1969. Anal. Biochem. 27 (2), 273–280.

Kakinoki, A., Kaneo, Y., Tanaka, T., Hosokawa, Y., 2008. Synthesis and evaluation of water-soluble poly(vinyl alcohol)-paclitaxel conjugate as a macromolecular prodrug. Biol. Pharm. Bull. 31 (5), 963–969.

Kale, A.A., Torchilin, V.P., 2007. Design, synthesis and characterization of pH-sensitive PEG-PE conjugates for stimuli-sensitive pharmaceutical nano-carriers: the effect of substitutes at the hydrazone linkage on the pH-stability of PEG-PE conjugates. Bioconjug. Chem. 18, 363–370.

Kalkhof, S., Sinz, A., 2008. Chances and pitfalls of chemical cross-linking with amine-reactive N -hydroxysuccinimide esters. Anal. Bioanal. Chem. 392, 305–312.

Kalkhof, S., et al. 2005a. Chemical cross-linking and high-performance fourier transform ion cyclotron resonance mass spectrometry for protein interaction analysis: application to a calmodulin/target peptide complex. Anal. Chem. 77, 495–503.

Kalkhof, S., Haehn, S., Ihling, C., Smyth, N., Sinz, A., 2005b. Probing laminin self-interaction using isotope-labeled cross-linkers and ESI-FTICR mass spectrometry. Poster, Pierce Biotechnology web site.

Kallin, E., Lonn, H., Norberg, T., Elofsson, M., 1989. Derivatization procedures for reducing oligosaccharides, Part 3: preparation of oligosaccharide glycosylamines, and their conversion into oligosaccharide-acrylamide copolymers. J. Carbohydr. Chem. 8, 597–611.

Kam, C.M., Abuelyaman, A.S., Li, Z., Hudig, D., Powers, J.C., 1993. Biotinylated isocoumarins, new inhibitors and reagents for detection, localization, and isolation of serine proteases. Bioconjug. Chem. 4 (6), 560–567.

Kam, N.W.S., O'Connell, M., Wisdom, J.A., Dai, H., 2005. Carbon nanotubes as multifunctional biological transporters and near-infrared agents for selective cancer cell destruction. Proc. Natl. Acad. Sci. 102, 11600–11605.

Kamkaew, A., Lim, S.H., Lee, H.B., Kiew, L.V., Chung, L.Y., Burgess, K., 2012. BODIPY dyes in photodynamic therapy. Chem. Soc. Rev. Advance Article, DOI: 10.1039/C2CS35216H.

Kanan, S.M., Tze, W.T.Y., Tripp, C.P., 2002. Method to double the surface concentration and control the orientation of adsorbed (3-aminopropyl)dimethylethoxysilane on silica powders and glass slides. Langmuir 18 (17), 6623–6627.

Kang, J., Tarscafalvi, A., Fujimoto, E., Shahrokh, Z., Shohet, S., Ikemoto, N., 1991. Specific labeling of the foot protein moiety of the triad with a novel fluorescent probe: application to the studies of conformational changes of the foot protein (Abstract). Biophys. J. 59, 249a. Tu-Pos 81.

Kantoff, P.W., Higano, C.S., Shore, N.D., Berger, E.R., Small, E.J., Penson, D.F., et al., 2010a. Sipuleucel-T immunotherapy for castration-resistant prostate cancer. N. Engl. J. Med. 363 (5), 411–422.

Kantoff, P.W., Schuetz, T.J., Blumenstein, B.A., Glode, L.M., Bilhartz, D.L., Wyand, M., et al., 2010b. Overall survival analysis of a phase II randomized controlled trial of a poxviral-based PSA-targeted immunotherapy in metastatic castration-resistant prostate cancer. J. Clin. Oncol. 28 (7), 1099–10105.

Kapanidisa, A.N., Weiss, S., 2002. Fluorescent probes and bioconjugation chemistries for single-molecule fluorescence analysis of biomolecules. J. Chem. Phys. 117, 10953–10964.

Kaplan, M.R., Calef, E., Bercovici, T., Gitler, C., 1983. The selective detection of cell surface determinants by means of antibodies and acetylated avidin. Biochim. Biophys. Acta 728, 112–120.

Karaman, M.W., Herrgard, S., Treiber, D.K., Gallant, P., Atteridge, C.E., Campbell, B.T., et al., 2008. A quantitative analysis of kinase inhibitor selectivity. Nat. Biotechnol. 26, 127–132.

Karchemski, F., Zucker, D., Barenholz, Y., Regev, O., 2012. Carbon nanotubes-liposomes conjugate as a platform for drug delivery into cells. J. Control. Release 160, 339–345.

Kareva, V.V., Dobrovol'sky, A.B., Baratova, L.A., Friedrich, P., Gusev, N.B., 1986. Ca21-induced structural change in the Ca21/Mg21 domain of troponin C detected by cross-linking. Biochim. Biophys. Acta 869, 322.

Karlsson, H.J., Eriksson, M., Perzon, E., Äkerman, B., Lincoln, P., Westman, G., 2003. Groove-binding unsymmetrical cyanine dyes for staining of DNA: syntheses and characterization of the DNA-binding. Nucleic Acids Res. 31 (21), 6227–6234.

Karsten, U., Serttas, N., Paulsen, H., Danielczyk, A., Goletz, S., 2004. Binding patterns of DTR-specific antibodies reveal a glycosylation-conditioned tumor-specific epitope of the epithelial mucin (MUC1). Glycobiology 14 (8), 681–692.

Karty, J.A., Reilly, J.P., 2005. Deamidation as a consequence of beta-elimination of phosphopeptides. Anal. Chem. 77 (14), 4673–4676.

Kasermann, F., Kempf, C., 1997. Photodynamic inactivation of enveloped viruses by buckminsterfullerene. Antivir. Res. 34, 65–70.

Kasina, S., Rao, T.N., Srinivasan, A., Sanderson, J.A., Fitzner, J.N., Reno, J.M., et al., 1991. Development and biologic evaluation of a kit for preformed chelate 99mTc. J. Nucl. Med. 32, 1445–1451.

Katagihallimath, N., 2008. Characterization of carriers and receptors of the Lewis^X glycan in the nervous system of mice. Dissertation, Center for Molecular Neurobiology, Hamburg, Germany.
Kato, D., Boatright, K.M., Berger, A.B., Nazif, T., Blum, G., Ryan, C., et al., 2005. Activity-based probes that target diverse cysteine protease families. Nat. Chem. Biol. 1 (1), 33–38.
Kato, K., Hamaguchi, Y., Fukui, H., Ishikawa, E., 1975a. Enzyme-linked immunoassay. I. Novel method for synthesis of the insulin-b-d-galactosidase conjugate and its applicability for insulin assay. J. Biochem. (Tokyo) 78, 235.
Kato, K., Hamaguchi, Y., Fukui, H., Ishikawa, E., 1975b. Enzyme-linked immunoassay. II. A simple method for synthesis of the rabbit antibody-b-d-galactosidase complex and its general applicability. J. Biochem. (Tokyo) 78, 423.
Katsuhiko, M., Owens, J.T., Belyaeva, T.A., Meares, C.F., Busby, S.J.W., Ishihama, A., 1997. Positioning of two alpha subunit carboxy-terminal domains of RNA polymerase at promoters by two transcription factors. PNAS 94, 11274–11278.
Katti, K.V., 1996. Recent advances in the chemistry of water-soluble phosphines–catalytic and biomedical aspects. Curr. Sci. 70 (3), 219–225.
Katti, K.V., Karra, S.R., Berning, D.E., Smith, C.J., Volkert, W.A., Ketring, A.R., 1999. Hydroxymethyl phosphine compounds for use as diagnostic and therapeutic pharmaceuticals and method of making same. US patent No. 5,855,867.
Katz, C., Levy-Beladev, L., Rotem-Bamberger, S., Rito, T., Rudiger, S.G.D., Friedler, A., 2011. Studying protein-protein interactions using peptide arrays. Chem. Soc. Rev. 40, 2131–2145.
Katz, M.J., Lasek, R.J., Osdoby, P., Whittaker, J.R., Caplan, A.I., 1982. Bolton-Hunter reagent as a vital stain for developing systems. Dev. Biol. 90, 419–429.
Kawa, M., Fréchet, J.M.J., 1998. Self-assembled lanthanide-cored dendrimer complexes: enhancement of the luminescence properties of lanthanide ions through site-isolation and antenna effects. Chem. Mater. 10, 286–296.
Kawamura Jr., A. (Ed.), 1977. Fluorescent Antibody Techniques and Their Application. University of Tokyo Press, Baltimore, Maryland.
Kay, C.M., Edsall, J.T., 1956. Dimerization of mercaptalbumin in the presence of mercurials. III. Bovine mercaptalbumin in water and in concentrated urea solutions. Arch. Biochem. Biophys. 65, 354.
Kay, E., Shannon, L.M., Lew, J.Y., 1967. Peroxidase isozymes from horseradish roots. II. Catalytic properties. J. Biol. Chem. 242, 2470.
Keana, J.F.W., Cai, S.X., 1990. New reagents for photoaffinity labeling: synthesis and photolysis of functionalized perfluorophenyl azides. J. Org. Chem. 55, 3640.
Keck, C., Kobierski, S., Mauludin, R., Muller, R.H., 2008. Second generation of drug nanocrystals for delivery of poorly soluble drugs: smart crystals technology. DSOIS 24, 124–128.
Keen, J.H., Maxfield, F.R., Hardegree, M.C., Habig, W.H., 1982. Receptor-mediated endocytosis of diphtheria toxin by cell in culture. Proc. Natl. Acad. Sci. USA 79, 2912.
Keller, G.H., Manak, M.M., 1989. DNA Probes. Stockton, New York.
Keller, G.H., Cumming, C.U., Huang, D.P., Manak, M.M., Ting, R., 1988. A chemical method for introducing haptens onto DNA probes. Anal. Biochem. 170, 441–450.
Keller, G.H., Huang, D.-P., Manak, M.M., 1989. Labeling of DNA probes with a photoactivatable hapten. Anal. Biochem. 177, 392–395.
Kellogg, D.R., Michison, T.J., Alberts, B.M., 1988. Behavior of microtubules and actin filaments in living Drosophila embryos. Development (Cambridge, UK) 103, 675.
Kelly, O.G., Chan, M.Y., Martinson, L.A., Kadoya, K., Ostertag, T.M., Ross, K.G., et al., 2011. Cell-surface markers for the isolation of pancreatic cell types derived from human embryonic stem cells. Nat. Biotechnol. 29 (8), 750–756.
Kenny, J.W., Fanning, T.G., Lambert, J.M., Traut, R.R., 1979. The subunit interface of the Escherichia coli ribosome. Cross-linking of 30S protein S9 to proteins of the 50S subunit. J. Mol. Biol. 135, 151–170.
Kensinger, R.D., Catalone, B.J., Krebs, F.C., Wigdahl, B., Schengrund, C.-L., 2004b. Novel polysulfated galactose-derivatized dendrimers as binding antagonists of human immunodeficiency virus type 1 infection. Antimicrob. Agents Chemother. 48 (5), 1614–1623.
Kensinger, R.D., Yowler, B.C., Benesi, A.J., Schengrund, C.-L., 2004a. Synthesis of novel, multivalent glycodendrimers as ligands for HIV-1 gp120. Bioconjug. Chem. 15, 349–358.
Kent, O.A., Ritchie, D.B., MacMillan, A.M., 2005. Characterization of a U2AF-independent commitment complex (E′) in the mammalian spliceosome assembly pathway. Mol. Cell. Biol. 25, 233–240.
Kenworthy, A.K., Hristova, K., Needham, D., McIntosh, T.J., 1995. Range and magnitude of the steric pressure between bilayers containing phospholipids with covalently attached poly(ethylene glycol). Biophys. J. 68, 1921–1936.
Kessler, C., Höltke, H.-J., Seibl, R., Burg, J., Muhlegger, K., 1990. Nonradioactive labeling and detection of nucleic acids: I. A novel DNA labeling and detection system based on digoxigenin: anti-digoxigenin ELISA principle (digoxigenin system). Mol. Gen. Hoppe Seyler 371, 917–927.
Key, J.A., Li, C., Cairo, C.W., 2012. Detection of cellular sialic acid content using nitrobenzoxadiazole carbonyl-reactive chromophores. Bioconjug. Chem. 23 (3), 363–371.
Khalfan, H., Abuknesha, R., Rand-Weaver, M., Price, R.G., Robinson, D., 1986. Aminomethyl coumarin acetic acid: a new fluorescent labeling reagent for proteins. Histochem. J. 18, 497–499.
Khan, A.M., Wright, P.J., 1987. Detection of flavivirus RNA in infected cells using photobiotin-labelled hybridization probes. J. Virol. Methods 15, 121–130.
Khan, S.H., Ahmad, F., Ahmad, N., Flynn, D.C., Kumar, R., 2011. Protein-protein interactions: principles, techniques, and their potential role in new drug development. J. Biomol. Struct. Dyn. 28, 929–938.
Khandare, J., Minko, T., 2006. Polymer–drug conjugates: progress in polymeric prodrugs. Prog. Polym. Sci. 31, 359–397.
Khanna, P.L., Ullman, E.F., 1980. 49,59-dimethoxy-6-carboxyfluorescein: a novel dipole-dipole coupled fluorescence energy transfer acceptor useful for fluorescence immunoassays. Anal. Biochem. 108, 156.
Khatiwala, C.B., Peyton, S.R., Putnam, A.J., 2006. Intrinsic mechanical properties of the extracellular matrix affect the behavior of pre-osteoblastic MC3T3-E1 cells. Am. J. Physiol. Cell Physiol. 290, C1640–C1650.
Kho, Y., Kim, S.C., Jiang, C., Barma, D., Kwon, S.W., Cheng, J., et al., 2004. A tagging-via-substrate technology for detection and proteomics of farnesylated proteins. Proc. Natl. Acad. Sci. USA 101, 12479–12484.
Khosroshahi, M.E., Ghazanfari, L., 2012. Synthesis and functionalization of SiO_2 coated Fe_3O_4 nanoparticles with amine groups based on self-assembly. Mater. Sci. Eng. C: Biomim. Supramol. Syst. 32 (5), 1043–1049.
Khullar, O., Frangioni, J.V., Grinstaff, M., Colson, Y.L., 2009. Image-guided sentinel lymph node mapping and nanotechnology-based nodal treatment in lung cancer using invisible near-infrared fluorescent light. Semin. Thorac. Cardiovasc. Surg. 21, 309–315.
Kidambi, S., Chan, C., Lee, I., 2004. Selective depositions on polyelectrolyte multilayers: self-assembled monolayers of m-dPEG acid as molecular template. J. Am. Chem. Soc. 126, 4697–4703.
Kidd, D., Liu, Y., Cravatt, B.F., 2001. Profiling serine hydrolase activities in complex proteomes. Biochemistry 40, 4005–4015.
Kiehm, D., Ji, T.H., 1977. Photochemical cross-linking of cell membranes. J. Biol. Chem. 252, 8524–8531.
Kihlberg, J., Magnusson, G., 1996. Use of carbohydrates and peptides in studies of adhesion of pathogenic bacteria and in efforts to generate carbohydrate-specific T cells. Pure Appl. Chem. 68 (11), 2119–2128.

Kiick, K.L., Saxon, E., Tirrell, D.A., Bertozzi, C.R., 2002. Incorporation of azides into recombinant proteins for chemoselective modification by the Staudinger ligation. Proc. Natl. Acad. Sci. USA 99, 19–24.

Kim, C.G., Sheffrey, M., 1990. Physical characterization of the affinity purified CCAAT transcription, a-CP1. J. Biol. Chem. 265, 13362–13369.

Kim, E.E., Wyckoff, H.W., 1991. Reaction mechanism of alkaline phosphatase based on crystal structures. Two metal ion catalysis. J. Mol. Biol. 218, 449–464.

Kim, H.S., Hage, D.S., 2006a. In: Hage, D.S. (Ed.), Handbook of Affinity Chromatography. CRC Press, Boca Raton, pp. 35–78.

Kim, H.S., Hage, D.S., 2006b. In: Hage, D.S. (Ed.), Handbook of Affinity Chromatography. CRC Press, Boca Raton, pp. 60–62.

Kim, K., Rhee, S.G., Stadtman, E.R., 1985. Nonenzymatic cleavage of proteins by reactive oxygen species generated by dithiothreitol and iron. J. Biol. Chem. 260, 15394–15397.

Kim, M.I., Ham, H.O., Oh, S.-D., Park, H.G., Chang, H.N., Choi, S.-H., 2006. Immobilization of Mucor javanicus lipase on effectively functionalized silica nanoparticles. J. Mol. Catal. B: Enzymatic 39, 62–68.

Kim, S., Martin, G.M., 1981. Preparation of cell-size unilamellar liposomes with high captured volume and defined size distribution. Biochim. Biophys. Acta 646, 1–9.

Kim, S., Tuker, M.S., Chi, E.Y., Sela, S., Martin, G.M., 1983. Preparation of multivesicular liposomes. Biochim. Biophys. Acta 728, 339–348.

Kim, Y.T., Chen, Y., Choi, J.Y., Kim, W.J., Dae, H.M., Jung, J., et al., 2012. Integrated microdevice of reverse transcription-polymerase chain reaction with colorimetric immunochromatographic detection for rapid gene expression analysis of influenza A H1N1 virus. Biosens. Bioelectron. 33, 88–94.

Kim, Y., Klutz, A.M., Jacobson, K.A., 2008a. Systematic investigation of polyamidoamine dendrimers surface-modified with poly(ethylene glycol) for drug delivery applications: synthesis, characterization, and evaluation of cytotoxicity. Bioconjug. Chem. 19 (8), 1660–1672.

Kim, Y.-G., Shin, D.-S., Yang, Y.-H., Gil, G.-C., Park, C.-G., Mimura, Y., et al., 2008b. High-throughput screening of glycan-binding proteins using miniature pig kidney N-glycan-immobilized beads. Chem. Biol. 15, 215–223.

Kimura, A., Orn, A., Holmquist, G., Wizzell, H., Ersson, B., 1979. Unique lectin-binding characteristics of cytotoxic T-lymphocytes allowing their distribution from natural killer cells and "K" cells. Eur. J. Immunol. 9, 575.

Kimura, Y., Yamatsugua, K., Kanaia, M., Echigob, N., Kuzuharab, T., Shibasaki, M., 2009. Design and synthesis of immobilized Tamiflu analog on resin for affinity chromatography. Tetrahedron Lett. 50, 3205–3208.

King, M.A., Louis, P.M., Hunter, B.E., Walker, D.W., 1989. Biocytin: a versatile anterograde neuroanatomical tract-tracing alternative. Brain Res. 497, 361–367.

King, P., Li, Y., Kochoumian, L., 1978. Preparation of protein conjugates via intermolecular disulfide bond formation. Biochemistry 17, 1499.

Kingsley, J.D., Dou, H., Morehead, J., Rabinow, B., Gendelman, H.E., Destache, C.J., 2006. Nanotechnology: a focus on nanoparticles as a drug delivery system. J. Neuroimmune Pharmacol. 1, 340–350.

Kipp, R.T., McNeel, D.G., 2007. Immunotherapy for prostate cancer—recent progress in clinical trials. Clin. Adv. Hematol. Oncol. 5, 465–479.

Kirley, T.L., 1989. Reduction and fluorescent labeling of cyst(e)ine-containing proteins for subsequent structural analysis. Anal. Biochem. 180, 231–236.

Kirpotin, D.B., Noble, C.O., Hayes, M.E., Huang, Z., Kornaga, T., Zhou, Y., et al., 2012. Building and characterizing antibody-targeted lipidic nanotherapeutics. Methods Enzymol. 502, 139–166.

Kit, O.O., Tallinen, T., Mahadevan, L., Timonen, J., Koskinen, P., 2012. Twisting graphene nanoribbons into carbon nanotubes. Phys. Rev. B 85 9 pages, DOI: 10.1103/PhysRevB.85.085428.

Kitagawa, T., Aikawa, T., 1976. Enzyme coupled immunoassay of insulin using a novel coupling reagent. J. Biochem. (Tokyo) 79, 233–236.

Kitagawa, T., Fujitake, T., Taniyama, H., Aikawa, T., 1978. Enzyme immunoassay of viomycin. J. Biochem. (Tokyo) 83, 1493–1501.

Kitagawa, T., Kawasaki, T., Munechika, H., 1982. Enzyme immunoassay of blasticidin S with high sensitivity: a new and convenient method for preparation of immunogenic (hapten-protein) conjugates. J. Biochem. (Tokyo) 92, 585–590.

Kiyama, H., Emson, P.C., Tokyama, M., 1992. In situ hybridization histochemistry using alkaline phosphatase-labeled oligodeoxynucleotide probeLongstaff, A.Revest, P. Methods in Molecular Biology, Protocols in Molecular Neurobiology, vol. 13. Humana Press, Totowa, New Jersey, pp. 167–179.

Klapper, M.H., Klotz, I.M., 1972. Acylation with dicarboxylic acid anhydridesHirs, C.H.W.Timasheff, S.N. Methods in Enzymology, vol. 25. Academic Press, New York, pp. 531–552.

Klar, T.A., Jakobs, S., Dyba, M., Egner, A., Hell, S.W., 2000. Fluorescence microscopy with diffraction resolution barrier broken by stimulated emission. Proc. Natl. Acad. Sci. USA 97, 8206–8210.

Kleene, R., et al. 2000. SH3 binding sites of ZG29p mediate an interaction with amylase and are involved in condensation sorting in the exocrine rat pancreas. Biochemistry 39, 9893–9900.

Klemm, P.J., Floyd III, W.C., Smiles, D.E., Frechet, J.M.J., Raymond, K.N., 2012. Improving T1 and T2 magnetic resonance imaging contrast agents through the conjugation of an esteramide dendrimer to high-water-coordination Gd(III) hydroxypyridinone complexes. Contrast Media Mol. Imaging 7, 95–99.

Klenk, D.C., Hermanson, G.T., Krohn, R.I., Fujimoto, E.K., Mallia, A.K., Smith, P.K., et al., 1982. Determination of glycosylated hemoglobin by affinity chromatography: comparison with colorimetric and ion-exchange methods, and effects of common interferences. Clin. Chem. 28 (10), 2088–2094.

Klibanov, A.L., Maruyama, K., Beckerleg, A.M., Torchilin, V.P., Huang, L., 1991. Activity of amphipathic poly(ethylene glycol) 5000 to prolong the circulation time of liposomes depends on the liposome size and is unfavorable for immunoliposome binding to target. Biochim. Biophys. Acta 1062, 142–148.

Klotz, I.M., 1967. SuccinylationHits, C.H.W. Methods in Enzymology, vol. 11. Academic Press, New York, pp. 576.

Klotz, I.M., Heiney, R.E., 1962. Introduction of sulfhydryl groups into proteins using acetylmercaptosuccinic anhydride. Arch. Biochem. Biophys. 96, 605.

Klotz, I.M., Keresztes-Nagy, S., 1962. Dissociation of proteins into subunits by succinylation: haemerythrin. Nature (London) 195, 900.

Knauf, M.J., Bell, D.P., Hirtzer, P., Luo, Z.-P., Young, J.D., Katre, N.V., 1988. Relationship of effective molecular size to systemic clearance in rats of recombinant interleukin-2 chemically modified with water-soluble polymers. J. Biol. Chem. 263, 15064–15070.

Knezevic, I., Predescu, D., Bardita, C., Wang, M., Sharma, T., Keith, B., et al., 2011. Regulation of dynamin-2 assembly–disassembly and function through the SH3A domain of intersectin-1s. J. Cell. Mol. Med. 15 (11), 2364–2376.

Knoller, S., Shpungin, S., Pick, E., 1991. The membrane-associated component of the amphiphile-activated, cytosol-dependent superoxide-forming NADPH oxidase of macrophages is identical to cytochrome b559. J. Biol. Chem. 266, 2795–2804.

Knorr, R., Trzeciak, A., Bannwarth, W., Gillessen, D., 1989. New coupling reagents in peptide chemistry. Tetrahedron Lett. 30, 1927–1930.

Knudsen, K.L., Hansen, M.B., Henriksen, L.R., Andersen, B.K., Lihme, A., 1992. Sulfone-aromatic ligands for thiophilic adsorption chromatography: purification of human and mouse immunoglobulins. Anal. Biochem. 201, 170–177.

Ko, B.J., Brodbelt, J.S., 2011. Ultraviolet photodissociation of chromophore-labeled oligosaccharides via reductive amination and hydrazide conjugation. J. Mass Spectrom. 46 (4), 359–366.

Kobayashi, H., Kawamoto, S., Saga, T., Sato, N., Hiraga, A., Konishi, J., et al., 2001. Micro-MR angiography of normal and intratumoral vessels in mice using dedicated intravascular MR contrast agents with high generation of polyamidoamine dendrimer core: reference to pharmacokinetic properties of dendrimer-based MR contrast agents. J. Magn. Reson. Imag. 14, 705–713.

Kobayashi, H., Kawamoto, S., Star, R.A., Waldmann, T.A., Tagaya, Y., Brechbiel, M.W., 2003. Micro-magnetic resonance lymphangiography in mice using a novel dendrimer-based magnetic resonance imaging contrast agent. Can. Res. 63, 271–276.

Kobayashi, H., Sato, N., Saga, T., Nakamoto, Y., Ishimori, T., Toyama, S., et al., 2000. Monoclonal antibody-dendrimer conjugates enable radiolabeling of antibody with markedly high specific activity with minimal loss of immunoreactivity. Eur. J. Nucl. Med. 27, 1334–1339.

Kobayashi, H., Wu, C., Kim, M.K., Paik, C.H., Carrasquillo, J.A., Brechbiel, M.W., 1999. Evaluation of the in vivo biodistribution of indium-111 and yttrium-88 labeled dendrimer-1B4M-DTPA and its conjugation with anti-Tac monoclonal antibody. Bioconjug. Chem. 10, 103–111.

Kobayashi, H., Kawamoto, S., Star, R.A., Waldmann, T.A., Tagaya, Y., Brechbiel, M.W., 2003. Micro-magnetic resonance lymphangiography in mice using a novel dendrimer-based magnetic resonance imaging contrast agent. Cancer Res. 63, 271–276.

Kobayashi, M., Ichishima, E., 1991. Application of periodate oxidized glucans to biochemical reactions. J. Carbohydr. Chem. 10, 635–644.

Koch, T., Jacobsen, N., Fensholdt, J., Boas, U., Fenger, M., Jakobsen, M.H., 2000. Photochemical immobilization of anthraquinone conjugated oligonucleotides and PCR amplicons on solid surfaces. Bioconjug. Chem. 11, 474–483.

Koehler, L., Graf, F., Bergmann, R., Steinbach, J., Pietzsch, J., Wuest, F., 2010. Radiosynthesis and radiopharmacological evaluation of cyclin-dependent kinase 4 (Cdk4) inhibitors. Eur. J. Med. Chem. 45 (2), 727–737.

Koganty, R.R., Reddish, M.A., Longenecker, B.M., 1996. Glycopeptide- and carbohydrate-based synthetic vaccines for the immunotherapy of cancer. Drug Discov. Today 1, 190–198.

Koguma, I., Sugita, K., Saito, K., Sugo, T., 2000. Multilayer binding of proteins to polymer chains grafted onto porous hollow-fiber membranes containing different anion-exchange groups. Biotechnol. Prog. 16, 456–461.

Kohler, J.J., 2009. Aniline: a catalyst for sialic acid detection. ChemBioChem 10, 2147–2150.

Kohn, J., Wilchek, M., 1982. A new approach (cyano-transfer) for cyanogen bromide activation of Sepharose at neutral pH, which yields activated resins, free of interfering nitrogen derivatives. Biochem. Biophys. Res. Commun. 107, 878–884.

Kohn, J., Wilchek, M., 1982. Mechanism of activation of Sepharose and Sephadex by cyanogen bromide. Enzyme Microb. Technol. 4, 161–163.

Kohn, K.W., Spears, C.L., Dory, P., 1966. Interstrand cross-linking of DNA by nitrogen mustard. J. Mol. Biol. 19, 87.

Köhn, M., Breinbauer, R., 2004. The staudinger ligation–a gift to chemical biology. Angew. Chem. Int. 43, 3106–3116.

Kojima, C., Kono, K., Maruyama, K., Takagishi, T., 2000. Synthesis of polyamidoamine dendrimers having poly(ethylene glycol) grafts and their ability to encapsulate anticancer drugs. Bioconjug. Chem. 11, 910–917.

Kolb, H.C., Finn, M.G., Sharpless, K.B., 2001. Click chemistry: diverse chemical function from a few good reactions. Angew. Chem. Int. 40, 2004–2021.

Kolemen, S., Bozdemir, O.A., Cakmak, Y., Barin, G., Erten-Ela, S., Marszalek, M., et al., 2011. Optimization of distyryl-Bodipy chromophores for efficient panchromatic sensitization in dye sensitized solar cells. Chem. Sci. 2 (5), 949–954.

Koller, D., et al. 2004. Selective inactivation of adrenomedullin over calcitonin gene-related peptide receptor function by the deletion of amino acids 14–20 of the mouse calcitonin-like receptor. J. Biol. Chem. 279, 20387–20391.

Kolonin, M.G., Bover, L., Sun, J., Zurita, A.J., Do, K.-A., Lahdenranta, J., et al., 2006. Ligand-directed surface profiling of human cancer cells with combinatorial peptide libraries. Cancer Res. 66, 34–40.

Kolupaeva, V.G., Lomakin, I.B., Pestova, T.V., Hellen, C.U.T., 2003. Eukaryotic initiation factors 4G and 4A mediate conformational changes downstream of the initiation codon of the encephalomyocarditis virus internal ribosomal entry site. Mol. Cell. Biol. 23, 687–698.

Kommoju, P.R., Chen, Z.W., Bruckner, R.C., Mathews, F.S., Jorns, M.S., 2011. Probing oxygen activation sites in two flavoprotein oxidases using chloride as an oxygen surrogate. Biochemistry 50, 5521–5534.

Konigsberg, W., 1972. Reduction of disulfide bonds in proteins with dithiothreitolHirs, C.H.W.Timaseff, S.N. Methods in Enzymology, vol. 25. Academic Press, New York, pp. 185.

Koning, G.A., Morselt, H.W., Velinova, M.J., Donga, J., Gorter, A., Allen, T.M., et al., 1999. Selective transfer of a lipophilic prodrug of 5-fluorodeoxyuridine from immunoliposomes to colon cancer cells. Biochim. Biophys. Acta 1420 (1–2), 153–167.

Konishi, K., Fujioka, M., 1987. Chemical modification of a functional arginine residue of rat liver glycine methyltransferase. Biochemistry 26, 8496–8502.

Konno, K., Morales, M.F., 1985. Exposure of actin thiols by the removal of tightly held calcium ions. Proc. Natl. Acad. Sci. USA 82, 7904–7908.

Konno, K., Liu, M., Fréchet, J.M.J., 1999. Design of dendritic macromolecules containing folate or methotrexate residues. Bioconjug. Chem. 10, 1115–1121.

Koo, O.M., Rubinstein, I., Onyuksel, H., 2005. Role of nanotechnology in targeted drug delivery and imaging: a concise review. Nanomedicine 1, 193–212.

Kopaciewicz, W., Sheer, D.G., Arnold, T.E., Goel, V., 2000. Cast membrane structures for sample preparation. U.S. Patent 6,048,457.

Kordás, K., Mustonen, T., Tóth, G., Jantunen, H., Lajunen, M., Soldano, C., et al., 2006. Inkjet printing of electrically conductive patterns of carbon nanotubes. Small 2 (8–9), 1021–1025.

Kore, A.R., Charles, I., Yang, L., Kuersten, S., 2009. Synthesis and activity of modified cytidine 5′-monophosphate probes for T4 RNA ligase 1. Nucleosides Nucleotides Nucleic Acids 28, 292–302.

Korn, A.H., Feairheller, S.H., Filachione, E.M., 1972. Glutaraldehyde: nature of the reagent. J. Mol. Biol. 65, 525–529.

Kornblatt, J.A., Lake, D.F., 1980. Cross-linking of cytochrome oxidase subunits with difluorodinitrobenzene. Can. J. Biochem. 58, 219–224.

Kornfield, R., Kornfield, S., 1985. Assembly of asparagine-linked oligosaccharides. Annu. Rev. Biochem. 54, 631.

Kornilova, A.Y., Bihel, F., Das, C., Wolfe, M.S., 2005. The initial substrate-binding site of c-secretase is located on presenilin near the active site. Proc. Natl. Acad. Sci. USA 102, 3230–3235.

Kossaczka, Z., Szu, S.C., Robbins, J.B., Schneerson, R., Shiloach, J., 2012. Method of immunizing humans against Salmonella typhi using a VI-REPA conjugate vaccine. U.S. Patent No. 8,202,520B2.

Kost, G.C., Selvaraj, S., Lee, Y.B., Kim, D.J., Anh, C.-H., Singh, B.B., 2011. Clavulanic acid increases dopamine release in neuronal cells through a mechanism involving enhanced vesicle trafficking. Neurosci. Lett. 504, 170–175.

Kostal, J., Prabhukumar, G., Loi Lao, U., Chen, A., Matsumoto, M., Mulchandani, A., et al., 2005. Customizable biopolymers for heavy metal remediation. J. Nanoparticle Res. 7, 517–523.

Köster, H., Little, D.P., Luan, P., Muller, R., Siddiqi, S.M., Marappan, S., et al., 2007. Capture compound mass spectrometry: a technology for the investigation of small molecule protein interactions. Assay Drug Dev. Technol. 5 (3), 381–390.

Kota, J., Ljungdahl, P.O., 2005. Specialized membrane-localized chaperones prevent aggregation of polytopic proteins in the ER. J. Cell Biol. 168, 79–88.

Kota, U., Chien, K.-Y., Goshe, M.B., 2009. Isotope-labeling and affinity enrichment of phosphopeptides for proteomic analysis using liquid chromatography–tandem mass spectrometry Proteomics, Methods in Molecular Biology, vol. 564. Humana Press, Chapter 17, 303–321.

Kotite, N.J., Staros, J.V., Cunningham, L.W., 1984. Interaction of specific platelet membrane proteins with collagen: evidence from chemical cross-linking. Biochemistry 23, 3099–3104.

Kovacic, P., Hem, R.W., 1959. Cross-linking of polymers with dimaleimide. J. Am. Chem. Soc. 81, 1187.

Kowalczyk, W., Monsó, M., de la Torre, B.G., Andreu, D., 2011. Synthesis of multiple antigenic peptides (MAPs)—strategies and limitations. J. Pept. Sci. 17 (4), 247–251.

Kozlov, I.A., Melnyk, P.C., Stromsborg, K.E., Chee, M.S., Barker, D.L., Zhao, C., 2004. Efficient strategies for the conjugation of oligonucleoitdes to antibodies enabling highly sensitive protein detection. Biopolymers 73 (5), 621–630.

Kozulic, B., Barbaric, S., Ries, B., Mildner, P., 1984. Study of the carbohydrate part of yeast acid phosphatase. Biochem. Biophys. Rev. Comm. 122, 1083.

Kraehenbuhl, J.P., Calardy, R.E., Jamieson, J.D., 1974. Preparation and characterization of an immunoelectron microscope tracer consisting of a heme-octapeptide coupled to Fab. J. Exp. Med. 139, 208.

Krager, K.J., Sarkar, M., Twait, E.C., Lill, N.L., Koland, J.G., 2012. A novel biotinylated lipid raft reporter for electron microscopic imaging of plasma membrane microdomains. J. Lipid Res. 53 (10), 2214–2225.

Krasnoperov, L.N., Marras, S.A.E., Kozlov, M., Wirpsza, L., Mustaev, A., 2010. Luminescent probes for ultrasensitive detection of nucleic acids. Bioconjug. Chem. 21, 319–327.

Krausz, Y., Freedman, N., Rubinstein, R., Lavie, E., Orevi, M., Tshori, S., et al., 2011. 68 Ga-DOTA-NOC PET/CT imaging of neuroendocrine tumors: comparison with 111 In-DTPA-Octreotide (OctreoScan®). Mol. Imaging Biol. 13 (3), 583–593.

Krenkova, J., Svec, F., 2009. Less common applications of monoliths: IV. Recent developments in immobilized enzyme reactors for proteomics and biotechnology. J. Sep. Sci. 32 (5–6), 706–718.

Kricka, L.J., 1992. Nonisotopic DNA Probe Techniques. Academic Press, New York.

Kricka, L.J., 1999. Nucleic acid detection technologies—labels, strategies, and formats. Clin. Chem. 45 (4), 453–458.

Krieg, U.C., Walter, P., Johnson, A.E., 1986. Photocross-linking of the signal sequence of nascent preprolactin to the 54-kilodalton polypeptide of the signal recognition particle. Proc. Natl. Acad. Sci. USA 83, 8604–8608.

Kriegel, R.M., 2006. Divinyl sulfone crosslinking agents and methods of use in subterranean applications. U.S. Patent 7,131,492.

Krishna, O.D., Jha, A.K., Jia, X., Kiick, K.L., 2011. Integrin-mediated adhesion and proliferation of human MSCs elicited by a hydroxyproline-lacking, collagen-like peptide. Biomaterials 32 (27), 6412–6424.

Krock, L., Esposito, D., Castagner, B., Wang, C.-C., Bindschadler, P., Seeberger, P.H., 2012. Streamlined access to conjugation-ready glycans by automated synthesis. Chem. Sci. 3, 1617–1622.

Kroll, F., Glinski, M., Dreger, M., Dülsner, E., Köster, H., 2009. CCMS: a New Era in toxicoproteomics. Innov. Pharm. Technol. 29, 38–41.

Kronick, M.N., 1986. The use of phycobiliproteins as fluorescent labels in immunoassay. J. Immunol. Meth. 92, 1–13.

Kroto, H.W., Heath, J.R., O'Brien, S.C., Curl, R.F., Smalley, R.E., 1985. C60: Buckminsterfullerene. Nature 318.

Kruszynski, M., Tsui, P., Stowell, N., Luo, J., Nemeth, J.F., Das, A.M., et al., 2005. Synthetic, site-specific biotinylated analogs of human MCP-1. J. Pept. Sci. 12 (5), 354–360.

Kubin, M., Spacek, P., Chromecek, R., 1967. Gel permeation chromatography on porous poly(Ethylene Glycol Methacrylate). Collect. Czech. Chem. Commun. 32, 3881–3887.

Kubitzki, T., Noll, T., Lütz, S., 2008. Immobilization of bovine enterokinase and application of the immobilized enzyme in fusion protein cleavage. Bioprocess Biosyst. Eng. 31, 173–182.

Kucherlapati, R., Jakobovits, A., Brenner, D.G., Capon, D.J., Klapholz, S., 2012. Human antibodies derived from immunized xenomice. U.S. Patent Application 2012/0117669.

Kuhn, D., Coates, C., Daniel, K., Chen, D., Bhuiyan, M., Kazi, A., et al., 2004. Beta-lactams and their potential use as novel anticancer chemotherapeutics drugs. Front. Biosci. 9, 2605–2617.

Kuhnast, B., de Bruin, B., Hinnen, F., Tavitian, B., Dolle, F., 2004. Design and synthesis of a new [18F]fluoropyridinebased haloacetamide reagent for the labeling of oligonucleotides: 2-bromo-N-[3-(2-[18F]fluoropyridin-3-yloxy)propyl]acetamide. Bioconjug. Chem. 15, 617–627.

Kuhnast, B., Hinnen, F., Boisgard, R., Tavitian, B., Dolle, F., 2003. Fluorine-18 labelling of oligonucleotides: prosthetic labelling at the 5′-end using the N-(4-[18F]fluorobenzyl)-2-bromoacetamide reagent. J. Labelled Compd. Radiopharm. 46, 1093–1103.

Kukowska-Latallo, J.F., Candido, K.A., Cao, Z., Nigavekar, S.S., Majoros, I.J., Thomas, T.P., et al., 2005. Nanoparticle targeting of anticancer drug improves therapeutic response in animal model of human epithelial cancer. Cancer Res. 65 (12), 5317–5324.

Kulin, S., Kishore, R., Hubbard, J.B., Helmerson, K., 2002. Real-time measurement of spontaneous antigen-antibody dissociation. Biophys. J. 83, 1965–1973.

Kull Jr., F.C., Jacobs, S., Cuatrecasas, P., 1985. Cellular receptor for 125I-labeled tumor necrosis factor: specific binding, affinity labeling, and relationship to sensitivity. Proc. Natl. Acad. Sci. USA 82, 5756–5760.

Kumacheva, E., Garstecki, P., 2011. Methods for the Generation of Polymer Particles. Microfluidic Reactors for Polymer Particles, 7–15.

Kumakura, M., Kaetsu, I., 1984. Polymeric microspheres by radiation copolymerization of acrolein and various monomers at low temperatures. Colloid Polym. Sci. 262, 450–454.

Kumar, A., Malhotra, S., 1992. A simple method for introducing SH group at 59 OH terminus of oligonucleotides. Nucleosides Nucleotides 11, 1003–1007.

Kumar, A., Plieva, F.M., Galaev, I., Mattiasson, B., 2003. Affinity fractionation of lymphocytes using supermacroporous monolithic cryogel. J. Immunol. Methods 283, 185–194.

Kumar, A., Tchen, P., Roullet, F., Cohen, J., 1988. Nonradioactive labeling of synthetic oligonucleotide probes with terminal deoxynucleotidyl transferase. Anal. Biochem. 169, 376–382.

Kumar, I., Rana, S., Cho, J.W., 2011. Cycloaddition reactions: a controlled approach for carbon nanotube functionalization. Chem.-A Eur. J. 17 (40), 11092–11101.

Kumar, P., Agarwal, S.K., Gupta, K.C., 2004. N-(3-Trifluoroethanesulfonyloxypropyl)-anthraquinone-2-carboxamide: a new heterobifunctional reagent for immobilization of biomolecules on a variety of polymer surfaces. Bioconjug. Chem. 15, 7–11.

KumaraSwamy, G., Kumar, J.M.R., Sheshagirirao, J.V.L.N., Kumar, V., 2011. Affinity chromatography: a review. J. Pharm. Res. 4, 1567–1574.

Kung, V.T., Maxim, P.E., Veltri, R.W., Martin, F.J., 1985. Antibody-bearing liposomes improve agglutination of latex particles used in clinical diagnostic assays. Biochim. Biophys. Acta 839, 105–109.

Kuntner, C., Wanek, T., Hoffer, M., Dangl, D., Hornof, M., Kvaternik, H., et al., 2011. Radiosynthesis and assessment of ocular pharmacokinetics of 124 I-Labeled chitosan in rabbits using small-animal PET. Mol. Imaging Biol. 13 (2), 222–226.

Küper, M.A., Schüle, R., Mayer, P., Königsrainer, A., Beckert, S., 2012. Factor XIII Val34Leu polymorphism is associated with increased factor XIII activation and decreased transcutaneous oxygen readings in patients with diabetic foot ulcers. Diabetic Medicine, accepted manuscript, DOI: 10.1111/j.1464-5491.2012.03707.x.

Kuramitz, H., Miyagaki, S., Ueno, E., Hata, N., Taguchi, S., Sugawara, K., 2011. Binding assay for cholera toxin based on sequestration electrochemistry using lactose labeled with an electroactive compound. Analyst 136, 2373–2378.

Kuroiwa, T., Noguchi, Y., Nakajima, M., Sato, S., Mukataka, S., Ichikawa, S., 2008. Production of chitosan oligosaccharides using chitosanase immobilized on amylose-coated magnetic nanoparticles. Process Biochem. 43, 62–69.

Kurz, A., Halliwell, C.M., Davis, J.J., Allen, H., Hill, O., Canters, G.W., 1998. A fullerene-modified protein. Chem. Comm. 3, 433–434.

Kurzchalia, T.V., Wiedmann, M., Breter, H., Zimmermann, W., Bauschke, E., Rapoport, T.A., 1988. tRNA-mediated labeling of proteins with biotin. A nonradioactive method for the detection of cell-free translation products. Eur. J. Biochem. 172, 663–668.

Kuwata, K., Uebori, M., Yamada, K., Yamazaki, Y., 1982. Liquid chromatographic determination of alkylthiols via derivatization with 5,59-dithiobis(2-nitrobenzoic acid). Anal. Chem. 54, 1082–1087.

Kwon, G.S. (Ed.), 2005. Polymeric Drug Delivery Systems. Marcel Dekker, Inc., New York (Drugs and the Pharmaceutical Sciences, Vol. 148).

L'abbé, G., 1984. Are azidocumulenes accessible? Bull. Soc. Chim. Belg. 93 (7), 579–592.

Labbe, J.P., Mornet, D., Roseau, G., Kassab, R., 1982. Cross-linking of F-actin to skeletal muscle myosin subfragment 1 with bis(imido esters): further evidence for the interaction of myosin-head heavy chain with an actin dimer. Biochemistry 21, 6897–6902.

Laburthe, M., Breant, B., Rouyer-Fessard, C., 1984. Molecular identification of receptors for vasoactive intestinal peptide in rat intestinal epithelium by covalent cross-linking. Eur. J. Biochem. 139, 181–187.

Lacey, B., Grant, W.N., 1987. Photobiotin as a sensitive probe for protein labeling. Anal. Biochem. 163, 151–158.

Laemmli, U.K., 1970. Cleavage of structural proteins during the assembly of the head of the bacteriophage T4. Nature (London) 277, 680–685.

Lakowicz, J.R., 1999. Principles of Fluorescence Spectroscopy, second ed., Springer.

Lakowicz, J.R., 2006. Principles of Fluorescence Spectroscopy, third ed., Springer Science & Business Media, LLC, New York, NY. ISBN-10: 0-387-31278-1.

Lakowicz, J.R. Topics in Fluorescence Spectroscopy, vols. 1–3. Plenum, New York.

Lam, C.F., Giddens, A.C., Chand, N., Webb, V.L., Copp, B.R., 2012. Semi-synthesis of bioactive fluorescent analogues of the cytotoxic marine alkaloid discorhabdin C. Tetrahedron 68 (15), 3187–3194.

Lambert, J.M., Boileau, G., Cover, J.A., Traut, R.R., 1983. Cross-links between ribosomal proteins of 30S subunits in 70S tight couples and in 30S subunits. Biochemistry 22, 3913–3920.

Lambert, J.M., Senter, P.D., Yau-Young, A., Blattler, W.A., Goldmacher, V.S., 1985. Purified immunotoxins that are reactive with human lymphoid cells: monoclonal antibodies conjugated to the ribosome-inactivating proteins gelonin and the pokeweed antiviral proteins. J. Biol. Chem. 260, 12035–12041.

Lambert, J.M., Blattler, W.A., McIntyre, G.D., Golmacher, V.S., Scott Jr., C.F., 1988. Immunotoxins containing single chain ribosome-inactivating proteins. In: Frankel, A.E. (Ed.), Immunotoxins. Kluwer, Boston, pp. 175.

Lamture, J.B., Wensel, T.G., 1995. Intensely luminescent immunoreactive conjugates of proteins and dipicolinate-based polymeric Tb (III) chelates. Bioconjug. Chem. 6 (1), 88–92.

Lane, C.F., 1972. Sodium cyanoborohydride — a highly selective reducing agent for organic functional groups. Synthesis, 135–146.

Lange, V., Picotti, P., Domon, B., Aebersold, R., 2008. Selected reaction monitoring for quantitative proteomics: a tutorial. Mol. Syst. Biol. 4, 222.

Langer, P.R., Waldrop, A.A., Ward, D.C., 1981. Enzymatic synthesis of biotin-labeled polynucleotides: novel nucleic acid affinity probes. Proc. Natl. Acad. Sci. USA 78, 6633–6637.

Langone, J.J., 1980. Radioiodination by use of the Bolton–Hunter and related reagentsVan Vunakis, H.Langone, J.J. Methods in Enzymology, vol. 70. Academic Press, New York, pp. 221–243.

Langone, J.J., 1981. Radioiodination by use of the Bolton-Hunter and related reagentsLangone, J.J.Van Vunakis, H. Methods in Enzymology, vol. 73. Academic Press, New York, pp. 113–127.

Lanier, L.L., Recktenwald, D.J., 1991. Multicolor immunofluorescence and flow cytometry. Methods (San Diego) 2, 192.

Lansdell, T.A., Tepe, J.J., 2004. Isolation of phosphopeptides using solid phase enrichment. Tetrahedron Lett. 45, 91–93.

Lante, A., Crapisi, A., Krastanov, A., Spettoli, P., 2000. Biodegradation of phensols by laccase immobilised in a membrane reactor. Process Biochem. 36, 51–58.

Lanteigne, D., Hnatowich, D.J., 1984. The labeling of DTPA-coupled proteins with 99mTc. Int. J. Appl. Radiat. Isot. 35, 617–621.

Lapidot, Y., Rappoport, S., Wolman, Y., 1967. J. Lipid Res. 8, 142.

Laquièvre, A., Allaway, N.S., Lyskawa, J., Woisel, P., Lefebvre, J.-M., Fournier, D., 2012. Highly efficient ring-opening reaction of azlactone-based copolymer platforms for the design of functionalized materials. Macromol. Rapid Comm. 33, 848–855.

LaRochelle, W.J., Froehner, S.C., 1986a. Determination of the tissue distributions and relative concentrations of the postsynaptic 43-kDa protein and the acetylcholine receptor in Torpedo. J. Biol. Chem. 261, 5270–5274.

LaRochelle, W.J., Froehner, S.C., 1986b. Immunochemical detection of proteins biotinylated on nitrocellulose replicas. J. Immunol. Methods 92, 65–71.

Larsson, P.-O., Mosbach, K., 1971. Biotechnol. Bioeng. 13, 393.

Lateef, S.S., Gupta, S., Jayathilaka, L.P., Krishnanchettiar, S., Huang, J.-S., Lee, B.-S., 2007. An improved protocol for coupling synthetic peptides to carrier proteins for antibody production using DMF to solubilize peptides. J. Biomol. Tech. 18, 173–176.

Latham, J.A., Cech, T.R., 1989. Defining the inside and outside of a catalytic RNA molecule. Science 245, 276–282.

Latham-Timmons, H.A., Wolter, A., Roach, J.S., Giare, R., Leuck, M., 2003. Novel method for the covalent immobilization of oligonucleotides via diels–alder bioconjugation. Nucleos. Nucleot. Nucleic Acids 22, 1495–1497.

Lattuada, L., Barge, A., Cravotto, G., Giovenzana, G.B., Tei, L., 2011. The synthesis and application of polyamino polycarboxylic bifunctional chelating agents. Chem. Soc. Rev. 40 (5), 3019–3049.

Laughlin, S.T., Baskin, J.M., Amacher, S.L., Bertozzi, C.R., 2008. In vivo imaging of membrane-associated glycans in developing zebrafish. Science 320 (5876), 664–667.

Laumonnier, Y., Syrovets, T., Burysek, L., Simmet, T.,2005. Identification of the annexin A2 heterotetramer as a receptor for the plasmin-induced signaling in human peripheral monocytes. Blood 10.1182/blood-2005-07-2840.

Launay, N., Caminade, A.M., Lahana, R., Majoral, J.P., 1994. A general synthetic strategy for neutral phosphorus-containing dendrimers. Int. Ed. Engl. 33, 1589–1592.

Launay, N., Caminade, A.M., Majoral, J.P., 1997. Synthesis of bowl-shaped dendrimers from generation 1 to generation 8. J. Organomet. Chem. 529, 51–53.

Law, B.K., et al. 2002. Rapamycin potentiates transforming growth factor b-induced growth arrest in nontransformed, oncogene-transformed, and human cancer cells. Mol. Cell. Biol. 22, 8184–8198.

Le Berre, V., Trévisiol, E., Dagkessamanskaia, A., Sokol, S., Caminade, A.-M., Majoral, J.P., et al., 2003. Dendrimeric coating of glass slides for sensitive DNA microarrays analysis. Nucleic Acids Res. 31 (16), e88.

Le Roch, K., Sestier, C., Dorin, D., Waters, N., Kappesi, B., Chakrabarti, D., et al., 2000. Activation of a plasmodium falciparum cdc2-related kinase by heterologous p25 and cyclin H. J. Biol. Chem. 275 (12), 8952–8958.

Le Trong, I., Wang, Z., Hyre, D.E., Lybrand, T.P., Stayton, P.S., Stenkamp, R.E.,2011. Streptavidin and its biotin complex at atomic resolution. Acta Crystallogr., Sect.D 67, 813-821, Protein Database entry 3RY1.

Le, U.M., Tran, H., Pathak, Y., 2012. Methods for polymeric nanoparticle conjugation to monoclonal antibodies. In: Pathak, Y., Benita, S. (Eds.), Antibody-Mediated Drug Delivery Systems, Concepts, Technology, and Applications. John Wiley & Sons, Inc., Hoboken, New Jersey, pp. 351–361. Chapter 16.

Leary, J.J., Brigati, D.J., Ward, D.C., 1981. Enzymatic synthesis of biotin-labelled nucleotides: novel nucleic acid affinity probes. Proc. Natl. Acad. Sci. USA 80, 4045–4049.

Leary, J.J., Waldrop, A.A., Ward, D.C., 1983. Rapid and sensitive colorimetric method for visualizing biotin-labeled DNA probes hybridized to DNA or RNA immobilized on nitrocellulose: bioblots. Proc. Natl. Acad. Sci. USA 80, 4045–4049.

Leduc, F., Faucher, D., Nkoma, G.B., Grégoire, M.C., Arguin, M., Wellinger, R.J., et al., 2011. Genome-wide mapping of DNA strand breaks. PloS one 6 (2), e17353.

Lee, A.C.J., Powell, J.E., Tregear, G.W., Niall, H.D., Stevens, V.C., 1980. Mol. Immunol. 17, 749.

Lee, C.Y., Farha, O.K., Hong, B.J., Sarjeant, A.A., Nguyen, S.T., Hupp, J.T., 2011. Light-harvesting Metal–Organic Frameworks (MOFs): efficient strut-to-strut energy transfer in bodipy and porphyrin-based MOFs. J. Am. Chem. Soc. 133 (40), 15858–15861.

Lee, D.S.C., Criffiths, B.W., 1984. Comparative studies of iodo-bead and chloramine-T methods for the radioiodination of human alpha-fetoprotein. J. Immunol. Methods 74, 181–189.

Lee, H., Scherer, N.F., Messersmith, P.B., 2006. Single-molecule mechanics of mussel adhesion. Proc. Natl. Acad. Sci. U.S.A 103, 12999–13003.

Lee, I.H., Kwon, H.K., An, S., Kim, D., Kim, S., Yu, M.K., et al., 2012. Imageable antigen-presenting gold nanoparticle vaccines for effective cancer immunotherapy in vivo. Angew. Chem. 124 (35), 8930–8935.

Lee, J., Owens, J.T., Hwang, I., Meares, C., Kustu, S., 2000. Phosphorylation-induced signal propagation in the response regulator NtrC. J. Bacteriol. 182, 5188–5195.

Lee, J., Kwon, Y.J., Choi, Y., Kim, H.C., Kim, K., Kim, J., et al., 2012. Quantum dot-based screening system for discovery of g protein-coupled receptor agonists. ChemBioChem 13 (10), 1503–1508.

Lee, J.A., Fortes, P.A.G., 1985. Labeling of the glycoprotein subunit of (Na,K)ATPase with fluorescent probes. Biochemistry 24, 322–330.

Lee, J.H., Baker, T.J., Mahal, L.K., Zabner, J., Bertozzi, C.R., Wiemer, D.F., et al., 1999. Engineering novel cell surface receptors for virus-mediated gene transfer. J. Biol. Chem. 274, 21878.

Lee, J.W., Kim, B.-Ku., 2005. A facile route to triazole dendrimers via click chemistry linking tripodal acetylene and dendrons. Bull. Kor. Chem. Soc. 26 (4), 658–660.

Lee, J.W., Kim, J.H., Kim, H.J., Han, S.C., Kim, J.H., Shin, W.S., et al., 2007. Synthesis of symmetrical and unsymmetrical PAMAM dendrimers by fusion between azide- and alkyne-functionalized PAMAM dendrons. Bioconjug. Chem. 18, 579–584.

Lee, K.N., Maxwell, M.D., Patterson Jr., M.K., Birckbichler, P.J., Conway, E., 1992. Identification of transglutaminase substrates in HT29 colon cancer cells: use of 5-(bio-tinamido)pentylamine as a transglutaminase-specific probe. Biochim. Biophys. Acta 1136, 12–16.

Lee, K.Y., Birckbichler, P.J., Patterson Jr., M.K., 1988. Colorimetric assay of blood coagulation factor XIII in plasma. Clin. Chem. 34, 906–910.

Lee, L., Kelly, R.E., Pastra-Landis, S.C., Evans, D.R., 1985. Oligomeric structure of the multifunctional protein CAP that initiates pyrimidine biosynthesis in mammalian cells. Proc. Natl. Acad. Sci. USA 82, 6802–6806.

Lee, P.Y., Bae, K.H., Jeong, D.G., Chi, S.W., Moon, J.H., Kang, S., et al., 2011. The S-nitrosylation of glyceraldehyde-3-phosphate dehydrogenase 2 is reduced by interaction with glutathione peroxidase 3 in Saccharomyces cerevisiae. Mol. Cells 31 (3), 255–259.

Lee, R.T., Lin, P., Lee, Y.C., 1984. New synthetic cluster ligands for galactose/N-acetylgalactosamine-specific lectin of mammalian liver. Biochemistry 23, 4255–4261.

Lee, S., Young, N.L., Whetstone, P.A., Cheal, S.M., Benner, W.H., Lebrilla, C.B., et al., 2006. A method to site-specifically identify and quantitate carbonyl end products of protein oxidation using oxidation-dependent element coded affinity tags (O-ECAT) and nanoLiquid chromatography Fourier transform mass spectrometry. J. Proteome Res. 5 (3), 539–547.

Lee, S.H., Mok, H., Lee, Y., Park, T.G., 2011. Self-assembled siRNA–PLGA conjugate micelles for gene silencing. J. Control. Release 152, 152–158.

Lee, S.-K., Siefert, A., Beloor, J., Fahmy, T.M., Kumar, P., 2012. Cell-specific siRNA delivery by peptides and antibodies, Chapter 5, in protein engineering for therapeutics: part AWittrup, K.D.Verdine, G.L. Methods in Enzymology, vol. 502. Elsevier Inc., San Diego, CA, pp. 91–118.

Lee, W.T., Conrad, D.H., 1984. The murine lymphocyte receptor for IgE. II. Characterization of the multivalent nature of the B lymphocyte receptor for IgE. J. Exp. Med. 159, 1790–1795.

Lee, W.T., Conrad, D.H., 1985. The murine lymphocyte receptor for IgE. III. Use of chemical cross-linking reagents to further characterize the B lymphocyte Fce receptor. J. Immunol. 134, 518–525.

LeFebvre, A.K., Korneeva, N.L., Trutschl, M., Cvek, U., Duzan, R.D., Bradley, C.A., et al., 2006. Translation initiation factor eIF4G-1 binds to eIF3 through the eIF3e subunit. J. Biol. Chem. 281, 22917–22932.

Leffak, I.M., 1983. Decreased protein staining after chemical cross-linking. Anal. Biochem. 135, 95–101.

Lei, Z., Maeda, T., Tamura, A., Nakamura, T., Yamazaki, Y., Shiratori, H., et al., 2012. EpCAM contributes to formation of functional tight junction in the intestinal epithelium by recruiting claudin proteins. Dev. Biol. 371, 136–145.

Leitner, A., Lindner, W., 2004. Current chemical tagging strategies for proteome analysis by mass spectrometry. J. Chrom. B813, 1–26.

Leitner, A., Lindner, W., 2009. Chemical Tagging Strategies for Mass Spectrometry-Based Phospho-proteomics in Methods in Molecular Biology, vol. 527. Humana Press, Pub, (229–243).

Leitner, A., Reischl, R., Walzthoeni, T., Herzog, F., Bohn, S., Förster, F., et al., 2012. Expanding the chemical cross-linking toolbox by the use of multiple proteases and enrichment by size exclusion chromatography. Mol. Cell. Proteomics 11 (3)10.1074/mcp.M111.014126.

Lemieux, G.A., Bertozzi, C.R., 1998. Chemoselective ligation reactions with proteins, oligosaccharides and cells. Trends Biotechnol. 16 (12), 506–513.

Lemieux, G.A., Bertozzi, C.R., 1999. Exploiting differences in sialoside expression for selective targeting of MRI contrast reagents. J. Am. Chem. Soc. 121, 4278–4279.

Lemmon, C.A., Ohashi, T., Erickson, H.P., 2011. Probing the folded state of fibronectin type III domains in stretched fibrils by measuring buried cysteine accessibility. J. Biol. Chem. 286 (30), 26375–26382.

Lempens, E.H., Helms, B.A., Merkx, M., 2011. Chemoselective protein and peptide immobilization on biosensor surfaces. Methods Mol. Biol. 751, 401–420.

Lempens, E.H.M., Helms, B.A., Merkx, M., Meijer, E.W., 2009. Efficient and chemoselective surface immobilization of proteins by using aniline-catalyzed oxime chemistry. ChemBioChem 10, 658–662.

Lennarz, W.J. (Ed.), 1980. The Biochemistry of Glycoproteins and Proteoglycans. Plenum, New York.

Leon, J.W., Kawa, M., Fréchet, J.M.J., 1996. Isophthalate ester-terminated dendrimers: versatile nanoscopic building blocks with readily modifiable surface functionalities. J. Am. Chem. Soc. 118, 8847–8859.

Leopoldo, M., Lacivita, E., Berardi, F., Perrone, R., 2009. Developments in fluorescent probes for receptor research. Drug Discov. Today 14, 816–822.

Lequin, R.M., 2005. Enzyme Immunoassay (EIA)/Enzyme-Linked Immunosorbent Assay (ELISA). Clin. Chem. 51 (12), 2415–2418.

Lerner, R.A., Green, N., Alexander, H., Liu, F.-T., Sutcliffe, J.G., Shinnick, T.M., 1981. Chemically synthesized peptides predicted from the nucleotide sequence of the hepatitis B virus genome elicit antibodies reactive with the native envelope protein of pane particles. Proc. Natl. Acad. Sci. USA 78, 3403–3407.

Leserman, L., Machy, P., 1987. Ligand targeting of liposomes. In: Ostro, M.J. (Ed.), Liposomes: From Biophysics to Therapeutics. Dekker, New York, pp. 157–194.

Lesniak, W., Bielinska, A.U., Sun, K., Janczak, K.W., Shi, X., Baker Jr., J.R., et al., 2005. Silver/dendrimer nanocomposites as biomarkers: fabrication, characterization, in vitro toxicity, and intracellular detection. Nano Lett. 5, 2123–2130.

Leteux, C., Childs, R.A., Chai, W., Stoll, M.S., Kogelberg, H., Feizi, T., 1998. Biotinyl-L-3-(2-naphthyl)-alanine hydrazide derivatives of *N*-glycans: versatile solid-phase probes for carbohydrate-recognition studies. Glycobiology 8, 227–236.

Lethias, C., Hartmann, D.J., Masmejean, M., Ravazzola, M., Sabbagh, I., Ville, G., et al., 1987. Ultrastructural immunolocalization of elastic fibers in rat blood vessels using the protein A-gold technique. J. Histochem. Cytochem. 35, 15–21.

Leung, W.-Y., Cheung, C.-Y., Yue, S.,2005. Modified carbocyanine dyes and their conjugates. US Patent No. 6,977, 305.

Levison, M.E., Josephson, A.S., Kirschenbaum, D.M., 1969. Reduction of biological substances by water-soluble phosphines: gamma-globulin (IgG). Experientia 25, 126–127.

Lewis, J.T., Dayanandan, B., Habener, J.F., Kieffer, T.J., 2000. Glucose-dependent insulinotropic polypeptide confers early phase insulin release to oral glucose in rats: demonstration by a receptor antagonist. Endocrinology 141, 3710–3716.

Lewis, R.V., Roberts, M.F., Dennis, E.A., Allison, W.S., 1977. Photoactivated heterobifunctional cross-linking reagents which demonstrate the aggregation state of phospholipase A2. Biochemistry 16, 5650–5654.

Li, C., Zolotarevsky, E., Thompson, I., Anderson, M.A., Simeone, D.M., Casper, J.M., et al., 2011. A multiplexed bead assay for profiling glycosylation patterns on serum protein biomarkers of pancreatic cancer. Electrophoresis 32, 2028–2035.

Li, C.H., 1945. Iodination of tyrosine groups in serum albumin and pepsin. J. Am. Chem. Soc. 67, 1065–1069.

Li, C.J., 2005. Organic reactions in aqueous media with a focus on carbon–carbon bond formations. A decade update. Chem. Rev. 105, 3095–3165.

Li, G., Zhang, L., Li, Z., Zhang, W., 2010. PAR immobilized colorimetric fiber for heavy metal ion detection and adsorption. J. Hazard. Mat. 177, 983–989.

Li, H., Mao, T., Zhang, Z., Yuan, M., 2007. The AtMAP65-1 cross-bridge between microtubules is formed by one dimer. Plant Cell. Physiol. 48, 866–874.

Li, J., Ng, C.K., 2012, In: Pathak, Benita, Methods for Nanoparticle Conjugation to Monoclonal Antibodies, in Antibody-Mediated Drug Delivery Systems: Concepts, Technology, and Applications. John Wiley & Sons, Inc., Hoboken, NJ, pp. 191–208. Chapter 10.

Li, J., Steen, H., Gygi, S.P., 2003. Protein profiling with cleavable isotope-coded affinity tag (cICAT) reagents. The yeast salinity stress response. Mol. Cell. Proteomics 2, 1198–1204.

Li, M., Meares, C.F., 1993. Synthesis, metal chelate stability studies, and enzyme digestion of a peptide-liked DOTA derivatives and its corresponding radiolabeled immunoconjugates. Bioconjug. Chem. 4, 275–283.

Li, M., Selvin, P.R., 1997. Amine-reactive forms of a luminescent DTPA chelate of terbium and europium: attachment to DNA and energy transfer measurements. Bioconjug. Chem. 8 (2), 127–132.

Li, R., Dowd, V., Stewart, D.J., Burton, S.J., Lowe, C.R., 1998. Design, synthesis, and application of a protein a mimetic. Nat. Biotechnol. 16, 190–195.

Li, W., Backlund, P.S., Boykins, R.A., Wang, G., Chen, H.-C., 2003. Susceptibility of the hydroxyl groups in serine and threonine to β-elimination/michael addition under commonly used moderately high-temperature conditions. Anal. Biochem. 323, 94–102.

Li, W., Liu, M.L., Cai, J.H., Tang, Y.X., Zhai, L.Y., Zhang, J., 2012. Effect of the combination of a cyclooxygenase-1 selective inhibitor and taxol on proliferation, apoptosis and angiogenesis of ovarian cancer in vivo. Oncol. Lett. 4 (1), 168–174.

Li, X., Abell, C., Cooper, M.A., 2008. Single step biocompatible coating for sulfhydryl coupling of receptors using 2-(pyridinyldithio) ethylcarbamoyl dextran. Colloids Surf. B Biointerfaces 61, 113–117.

Li, Y., Zhou, J., Liu, C., Li, H., 2012. Composite quantum dots detect Cd (II) in living cells in a fluorescence "turning on" mode. J. Mater. Chem. 22 (6), 2507–2511.

Li, Z., He, L., He, N., Deng, Y., Shi, Z., Wang, H., et al., 2011. Polymerase chain reaction coupling with magnetic nanoparticles-based biotinavidin system for amplification of chemiluminescent detection signals of nucleic acid. J. Nanosci. Nanotechnol. 11 (2), 1074–1078.

Liang, A.M., Claret, E., Ouled-Diaf, J., Jean, A., Vogel, D., Light, D.R., et al., 2007. Development of a homogeneous time-resolved fluorescence leukotriene B4 assay for determining the activity of leukotriene A4 hydrolase. J. Biomol. Screen. 12, 536–545.

Liao, G., Li, W., He, S., Zhang, Z., 2011. Modeling a murine model of immunoglobulin-E (IgE)-mediated qingkailing injection anaphylaxis. Afr. J. Pharm Pharmacol 5 (8), 1106–1114.

Liao, H., Chen, D., Yuan, L., Zheng, M., Zhu, Y., Liu, X.,2010. Immobilized cellulase by polyvinyl alcohol/Fe2O3 magnetic nanoparticle to degrade microcrystalline cellulose. Carbohydrate Polymers, Article in Press.

Liao, J.-L., Hjerten, S.,2002. large-pore chromatographic beads prepared by suspension polymerization. U.S. Patent 6,423,666.

Liberatore, F.A., Comeau, R.D., McKearin, J.M., Pearson, D.A., Belonga III, B.Q., Brocchini, S.J., et al., 1990. Site-directed chemical modification and cross-linking of a monoclonal antibody using equilibrium transfer alkylating cross-link reagents. Bioconjug. Chem. 7, 36–50.

Lichtman, J.W., Conchello, J.-A., 2005. Fluorescence microscopy. Nat. Methods 2, 910–919.

Liehr, T. (Ed.), 2009. Fluorescence in situ hybridization (FISH) – Application guide, Springer protocols. Springer Berlin Heidelberg, Pub. ISBN: 3642089526.

Liener, I.E., Friedenson, B., 1970. FicinPerlmann, G.E.Lorand, L. Methods in Enzymology, vol. 19. Academic Press, New York, pp. 261–273.

Lihme, A. Boenisch, T.,1996. Water-soluble, polymer-based reagents and conjugates comprising moieties derived from divinyl sulfone. U.S. Patent 5,543,332.

Lihme, A., Heegaard, P.M.H., 1991. Thiophilic adsorption chromatography: the separation of serum proteins. Anal. Biochem. 192, 64–69.

Lihme, A., Schafer-Nielsen, C., Larsen, K.P., Muller, K.G., Boq-Hansen, T.C., 1986. Divinylsulphone-activated agarose. Formation of stable and non-leaking affinity matrices by immobilization of immunoglobulins and other proteins. J. Chromatogr. 376, 299–305.

Lihme, A.O.F. Boenisch, T.,1996. Water-soluble, polymer-based reagents and conjugates comprising moieties derived from divinyl sulfone. U.S. Patent 5,543,332.

Likhosherstov, L.M., Novikova, O.S., Derevitskaja, V.A., Kochetkov, N.K., 1986. A new simple synthesis of amino sugar b-D-glycosylamines. Carbohydr. Res. 146, C1–C5.

Lillich, M., Chen, X., Weil, T., Barth, H., Fahrer, J., 2012. Streptavidin-conjugated c3 protein mediates the delivery of mono-biotinylated rnase a into macrophages. Bioconjug. Chem. 23 (7), 1426–1436.

Lim, S., Franklin, S.J., 2004. Lanthanide-binding peptides and the enzymes that might have been. Cell. Mol. Life Sci. 61 (17), 2184–2188.

Lin, J., Chen, H., Ji, Y., Zhang, Y., 2012. Functionally modified monodisperse core-shell silica nanoparticles: silane coupling agent as capping and size tuning agent. Colloids Surf. A: Physicochem. Eng. Asp. 411, 111–121.

Lin, O.H., Akil, H.M., Mohd Ishak, Z.A., 2011. Surface-activated nanosilica treated with silane coupling agents/polypropylene composites: mechanical, morphological, and thermal studies. Polym. Comp. 32 (10), 1568–1583.

Lin, P.-C., Ueng, S.-H., Tseng, M.-C., Ko, J.-L., Huang, K.-T., Yu, S.-C., et al., 2006. Site-specific protein modification through CuI-catalyzed 1,2,3-triazole formation and its implementation in protein microarray fabrication. Angew. Chem. Int. Ed. 45, 4286–4290.

Lindersson, E., Lundvig, D., Petersen, C., Madsen, P., Nyengaard, J.R., Højrup, P., et al., 2005. p25 Stimulates synuclein aggregation and is co-localized with aggregated synuclein in synucleinopathies. J. Biol. Chem. 280, 5703–5715.

Lindley, H., 1956. A new synthetic substrate for trypsin and its application to the determination of the amino acid sequence of proteins. Nature (London) 178, 647.

Link, A.J., Tirrell, D.A., 2003. Cell surface labeling of *Escherichia coli* via copper(I)-catalyzed [3+2] cycloaddition. J. Am. Chem. Soc. 125, 11164–11165.

Link, A.J., Vink, M.K.S., Tirrell, D.A., 2004. Presentation and detection of azide functionality in bacterial cell surface proteins. J. Am. Chem. Soc. 126, 10598–10602.

Litosh, V.A., Wu, W., Stupi, B.P., Wang, J., Morris, S.E., Hersh, M.N., et al., 2011. Improved nucleotide selectivity and termination of 3′-OH unblocked reversible terminators by molecular tuning of 2-nitrobenzyl alkylated HOMedU triphosphates. Nucleic Acids Res. 39 (6), e39.

Liu, B., 2011. Peptide PEGylation: the next generation. Pharm. Technol. 35, s26–s28.

Liu, F.-T., Zinnecker, M., Hamaoka, T., Katz, D.H., 1979. New procedures for preparation and isolation of conjugates of proteins and a synthetic copolymer of D-amino acids and immunochemical characterization of such conjugates. Biochemistry 18, 690–697.

Liu, G., He, Y., 2012. Facile synthesis of nanocrystal encoded fluorescent silica microspheres. J. Colloid Interface Sci. 388, 86–91.

Liu, H., Ren, G., Miao, Z., Zhang, X., Tang, X., Han, P., et al., 2010. Molecular optical imaging with radioactive probes. PLoS ONE 5 (3), e9470.10.1371/journal.pone.0009470.

Liu, J., Casavant, M.J., Cox, M., Walters, D.A., Boul, P., Lu, W., et al., 1999. Controlled deposition of individual single-walled carbon nanotubes on chemically functionalized templates. Chem. Phys. Lett. 303, 125–129.

Liu, J., Liu, J., Chu, L., Wang, Y., Duan, Y., Feng, L., et al., 2011. Novel peptide–dendrimer conjugates as drug carriers for targeting nonsmall cell lung cancer. Int. J. Nanomedicine 6, 59–69.

Liu, J., Zhao, Y., Guo, Q., Wang, Z., Wang, H., Yang, Y., et al., 2012. TAT-modified nanosilver for combating multidrug-resistant cancer. Biomaterials 33, 6155–6161.

Liu, M., Kono, K., Fréchet, J.M.J., 2000. Water-soluble dendritic unimolecular micelles: their potential as drug delivery agents. J. Contr. Release 65, 121–131.

Liu, S.L., Tian, Z.Q., Zhang, Z.L., Wu, Q.M., Zhao, H.S., Ren, B., et al., 2012. High-efficiency dual labeling of influenza virus for single-virus imaging. Biomaterials 33, 7828–7833.

Liu, S., Edwards, D.S., Barrett, J.A., 1997. 99mTc labeling of highly potent small peptides. Bioconjug. Chem. 8, 621–636.

Liu, S.C., Fairbanks, G., Palek, J., 1977. Spontaneous reversible protein cross-linking in the human erythrocyte membrane. Temperature and pH dependence. Biochemistry 16, 4066.

Liu, S.-H., Cheng, H.-H., Huang, S.-Y., Yiu, P.-C., Chang, Y.-C., 2006. Studying the protein organization of the postsynaptic density by a novel solid phase- and chemical cross-linking-based technology. Mol. Cell. Proteomics 5, 1019–1032.

Liu, S., 2004. The role of coordination chemistry in the development of target-specific radiopharmaceuticals. Chem. Soc. Rev. 33, 445–461.

Liu, T., Wu, L.Y., Choi, J.K., Berkman, C.E., 2009a. In vitro targeted photodynamic therapy with a pyropheophorbide-a conjugated inhibitor of prostate specific membrane antigen. Prostate 69 (6), 585–594.

Liu, Y., Wu, C., 1991. Radiolabeling monoclonal antibodies with metal chelates. Pure Appl. Chem. 63, 427–463.

Liu, Y., Feize, T., Campanero-Rhodes, M.A., Childs, R.A., Zhang, Y., Mulloy, B., et al., 2007. Neoglycolipid probes prepared via oxime ligation for microarray analysis of oligosaccharide-protein interactions. Chem. Biol. 14, 847–859.

Liu, Y., Jurman, M.E., Yellen, G., 1996. Dynamic rearrangement of the outer mouth of a K1 channel during gating. Neuron 16, 859–867.

Liu, Y., Patricelli, M.P., Cravatt, B.F., 1999. Activity-based protein profiling: the serine hydrolases. Proc. Natl. Acad. Sci. USA 96, 14694–14699.

Liu, Y., Lipowsky, R., Dimova, R., 2012. Concentration dependence of the interfacial tension for aqueous two-phase polymer solutions of dextran and polyethylene glycol. Langmuir 28, 3831–3839.

Liu, Y.-C., Suen, S.-Y., Huang, C.-W., ChangChien, C.-C., 2005. Effects of spacer arm on penicillin G acylase purification using immobilized metal affinity membranes. J. Memb. Sci. 251, 201–207.

Liz-Marzan, L.M., Giersig, M., Mulvaney, P., 1996. Synthesis of nanosized gold-silica core-shell particles. Langmuir 12, 4329–4335.

Lloyd, R.V., Jin, L., Fields, K., 1990. Detection of chromogranins A and B in endocrine tissues with radioactive and biotinylated oligonucleotide probes. Am. I. Surg. Pathol. 14, 35–43.

Lo Conte, M., Robb, M.J., Hed, Y., Marra, A., Malkoch, M., Hawker, C.J., et al., 2011. Exhaustive glycosylation, PEGylation, and glutathionylation of a [G4]-ene48 dendrimer via photoinduced thiol-ene coupling. J. Polym. Sci., Part A: Polym. Chem. 49, 4468–4475.

Loers, G., Makhina, T., Bork, U., Dörner, A., Schachner, M., Kleene, R., 2012. The interaction between cell adhesion molecule L1, matrix metalloproteinase 14, and adenine nucleotide translocator at the plasma membrane regulates L1-mediated neurite outgrowth of murine cerebellar neurons. J. Neurosci. 32 (11), 3917–3930.

Logan, T.C.,2007. Immobilized enzymes in microfluidic systems. Dissertation, University of California, Berkeley.

Loh, T.P., Pei, J., Lin, M., 1996. Indium trichloride (InCl 3) catalysed diels–alder reaction in water. Chem. Comm., 2315–2316.

Loken, M.R., Keij, J.F., Kelley, K.A., 1987. Comparison of helium-neon and dye lasers for the excitation of allophycocyanin. Cytometry 8, 96.

Lokitz, B.S., Messman, J.M., Hinestrosa, J.P., Alonzo, J., Verduzco, R., Brown, R.H., et al., 2009. Dilute solution properties and surface attachment of RAFT polymerized 2-vinyl-4,4-dimethyl azlactone (VDMA). Macromolecules 42, 9018–9026.

Lomant, A.J., Fairbanks, G., 1976. Chemical probes of extended biological structures: synthesis and properties of the cleavable cross-linking reagent [35S] dithiobis(succinimidyl propionate). J. Mol. Biol. 104, 243–261.

Longshaw, V.M., et al., 2004. Nuclear translocation of the Hsp70/Hsp90 organizing protein mSTI1 is regulated by cell cycle kinases. J. Cell Sci. 117, 701–710.

Lonnerdal, B., Keen, C.L., 1982. Metal chelate affinity chromatography of proteins. J. Appl. Biochem. 4, 203–208.

Lopatin, S.A., Varlamov, V.P., 1995. New trends in immobilized metal affinity chromatography of proteins. Appl. Biochem. Microbiol. 31, 221–227.

López-Sánchez, L.M., López-Pedrera, C., Rodríguez-Ariza, A., 2012. Proteomics insights into deregulated protein S-nitrosylation and disease. Expert Rev. Proteomics 9 (1), 59–69.

Lord, J.M., Spooner, R.A., Hussain, K., Roberts, L.M., 1988. Immunotoxins: properties, applications, and current limitations. Adv. Drug Deliv. Rev. 2, 297.

Lord, S.J., Conley, N.R., Lee, H.-L.D., Samuel, R., Liu, N., Twieg, R.J., et al., 2008. A photoactivatable push–pull fluorophore for single-molecule imaging in live cells. J. Am. Chem. Soc. 130 (29), 9204–9205.

Lord, S.J., Conley, N.R., Lee, H.-L.D., Nishimura, S.Y., Pomerantz, A.K., Willets, K.A., et al., 2009a. DCDHF fluorophores for single-molecule imaging in cells. Chemphyschem. 10 (1), 55–65.

Lord, S.J., Lee, H.-L.D., Samuel, R., Weber, R., Liu, N., Conley, N.R., et al., 2009b. Azido push-pull fluorogens photoactivate to produce bright fluorescent labels. J. Phys. Chem. B (Articles ASAP (Web): October 27, 2009).

Lotan, R., Pebray, H., Cacan, M., Cacan, R., Sharon, N., 1975. Labeling of soybean agglutinin by oxidation with sodium periodate followed by reduction with [3H] borohydride. J. Biol. Chem. 250, 1955–1957.

Loudet, A., Burgess, K., 2007. BODIPY dyes and their derivatives: syntheses and spectroscopic properties. Chem. Rev. 107 (11), 4891–4932.

Louis, C.F., Saunders, M.J., Holroyd, J.A., 1977. The cross-linking of rabbit skeletal muscle sarcoplasmic reticulum protein. Biochim. Biophys. Acta 493, 78–92.

Low, D.W., Hill, M.G., Carrasco, M.R., Kent, S.B.H., Botti, P., 2001. Total synthesis of cytochrome b562 by native chemical ligation using a removable auxiliary. Proc. Natl. Acad. Sci. USA 98, 6554–6559.

Lowder, B.J., Duyvesteyn, M.D., Blair, D.F., 2005. FliG subunit arrangement in the flagellar rotor probed by targeted cross-linking. J. Bacteriol. 187, 5640–5647.

Lowe, C.R., 1979. Immobilized nucleotides and coenzymes for affinity chromatography. Pure Appl. Chem. 51, 1429–1441.

Lowe, C.R., Dean, P.D.G., 1971. Affinity chromatography of enzymes on insolubilized cofactors. FEBS Lett. 14, 313–316.

Lowe, C.R., Dean, P.D.G., 1974. Affinity Chromatography. Wiley, New York. (pp. 228–229).

Lowe, C.R., Harvey, M.J., Craven, D.B., Dean, P.D.G., 1973. Biochem. J. 133, 499.

Lowe, C.R., Burton, S.J., Pearson, J.C., Clonis, Y.D., Stead, V., 1986. Design and application of bio-mimetic dyes in biotechnology. J Chromatogr. 376, 121–130.

Lowe, C.R., Lowe, A.R., Gupta, G., 2001. New developments in affinity chromatography with potential application in the production of biopharmaceuticals. J. Biochem. Biophys. Methods 49, 561–574.

Lu, F., Hu, Z., Sinard, J., Garen, A., Adelman, R.A., 2009. Factor VII–verteporfin for targeted photodynamic therapy in a rat model of choroidal neovascularization. Invest. Ophthalmol. Vis. Sci. 50 (8), 3890–3896.

Lu, R.C., Wong, A., 1989. Glutamic acid-88 is close to SH-1 in the tertiary structure of myosin subfragment-1. Biochemistry 28, 4826.

Lu, W., Starovasnik, M.A., Kent, S.B., 1998. Total chemical synthesis of bovine pancreatic trypsin inhibitor by native chemical ligation. FEBS Lett. 429 (1), 31–35.

Lu, Y., Low, P.S., 2012. Folate-mediated delivery of macromolecular anticancer therapeutic agents. Adv. Drug Deliv. Rev.10.1016/j.addr.2012.09.020 in press.

Lu, Y., Bottari, P., Turecek, F., Aebersold, R., Gelb, M.H., 2004. Absolute quantification of specific proteins in complex mixtures using visible isotope-coded affinity tags. Anal. Chem. 76, 4104–4111.

Lucafò, M., Pacor, S., Fabbro, C., Da Ros, T., Zorzet, S., Prato, M., et al., 2012. Study of a potential drug delivery system based on carbon nanoparticles: effects of fullerene derivatives in MCF7 mammary carcinoma cells. J. Nanopart. Res. 14 (4), 1–13.

Lucker, B.F., Behal, R.H., Qin, H., Siron, L.C., Taggart, W.D., Rosenbaum, J.L., et al., 2005. Characterization of the intraflagellar transport complex B core: direct interaction of the IFT81 and IFT74/72 subunits. J. Biol. Chem. 280, 27688–27696.

Luduena, R.F., Roach, M.C., Trcka, P.P., Weintraub, S., 1982. Bioiodoacetyldithioethylamine: a reversible cross-linking reagent for protein sulfhydryl group. Anal. Biochem. 117, 76.

Ludwig, F.R., Jay, F.A., 1985. Reversible chemical cross-linking of the light-harvesting polypeptides of Rhodopseudomonas viridis. Eur. J. Biochem. 151, 83–87.

Lundblad, R., 1991. Chemical Reagents for Protein Modification. CRC Press, Boca Raton, Florida.

Lunder, M., Bratkovic, T., Doljak, B., Kreft, S., Urleb, U., Strukelj, B., et al., 2005. Comparison of bacterial and phage display peptide libraries in search of target-binding motif. Appl. Biochem. Biotechnol. 127 (2), 125–131.

Lung, O., Fisher, M., Beeston, A., Hughes, K.B., Clavijo, A., Goolia, M., et al., 2011. Multiplex RT-PCR detection and microarray typing of vesicular disease viruses. J. Virol. Methods 175 (2), 236–245.

Luo, S., Zhang, E., Su, Y., Cheng, T., Shi, C., 2011. A review of NIR dyes in cancer targeting and imaging. Biomaterials 32, 7127–7138.

Luo, Y., Fischer, J.J., Baessler, O.Y., Schrey, A.K., Ungewiss, J., Glinski, M., et al., 2009. GDP-capture compound–a novel tool for the profiling of GTPases in pro- and eukaryotes by capture compound mass spectrometry (CCMS). J. Proteomics 73 (4), 815–819.

Luo, Y., Blex, C., Baessler, O., Glinski, M., Dreger, M., Sefkow, M., et al., 2009b. The cAMP capture compound mass spectrometry as a novel tool for targeting cAMP-binding proteins: from protein kinase a to potassium/sodium hyperpolarization-activated cyclic nucleotide-gated channels. Mol. Cell Proteomics 8 (12), 2843–2856.

Luong, J.H.T., Scouten, W.H.,2008. Affinity purification of natural ligands. Current Protocols in Protein Science, Unit Number: Unit 9.3, Wiley Online Library, doi: 10.1002/0471140864.ps0903s52.

Leonard, M., Fournier, C., Dellacherie, E., 1995. Polyvinyl alcohol-coated macroporous polystyrene particles as stationary phases for the chromatography of proteins. J. Chromatogr. B: Biomed. Sci. Appl. 664, 39–46.

Ma, J., Zhang, L., Liang, Z., Zhang, W., Zhang, Y., 2009. Recent advances in immobilized enzymatic reactors and their applications in proteome analysis. Anal. Chim. Acta 632, 1–8.

Ma, J., Zhang, L., Liang, Z., Shan, Y., Zhang, Y., 2011. Immobilized enzyme reactors in proteomics. TrAC, Trends Anal. Chem. 30 (5), 691–702.

Ma, K., Temiakov, D., Jiang, M., Anikin, M., McAllister, W.T., 2002. Major conformational changes occur during the transition from an initiation complex to an elongation complex by T7 RNA polymerase. J. Biol. Chem. 277, 43206–43215.

MacBeath, G., 2007. Protein Arrays: Labeling the Compounds and Probing the Array for Protein-Small Molecule Interactions. CSH Protocols, 2007: pdb.prot4631.

MacBeath, G., Schreiber, S.L., 2000. Printing proteins as microarrays for high-throughput function determination. Science 289, 1760–1763.

MacLean, L.M., O'Toole, P.J., Stark, M., Marrison, J., Seelenmeyer, C., Nickel, W., et al., 2012. Trafficking and release of Leishmania metacyclic HASPB on macrophage invasion. Cell. Microbiol. 14, 740–761.

Mädler, S., Bich, C., Touboul, D., Zenobi, R., 2009. Chemical cross-linking with NHS esters: a systematic study on amino acid reactivities. J. Mass Spectrom. 44, 694–706.

Maeda, H., Fujita, N., Ishihama, A., 2000. Competition among seven *Escherichia coli* subunits: relative binding affinities to the core RNA polymerase. Nucleic Acids Res. 28, 3497–3503.

Maehashi, K., Katsura, T., Kerman, K., Takamura, Y., Matsumoto, K., Tamiya, E., 2007. Label-free protein biosensor based on aptamer-modified carbon nanotube field-effect transitors. Anal. Chem. 79, 782–787.

Maetzel, D., Denzel, S., Mack, B., et al., 2009. Nuclear signalling by tumour-associated antigen EpCAM. Nat. Cell Biol. (Published Online,Jan 11).

Magalhães, P.O., Lopes, A.M., Mazzola, P.G., Rangel-Yagui, C., Penna, T.C.V., Adalberto Jr, P., 2007. Methods of endotoxin removal from biological preparations: a review. J. Pharm. Pharmaceut. Sci. 10, 388–404.

Maggini, M., Scorrano, G., 1993. Addition of azomethine ylides to CM: synthesis, characterization, and functionalization of fullerene pyrrolidines. J. Am. Chem. Soc. 115, 9798–9799.

Maggini, M., Karlsson, A., Pasimeni, L., Scorrano, G., Prato, M., Vallid, L., 1994. Synthesis of N-acylated fulleropyrrolidines: new materials for the preparation of langmuir–blodgett films containing fullerenes. Tetrahedron Lett. 35, 2985–2988.

Magnan, E., Catarino, I., Paolucci-Jeanjean, D., Preziosi-Belloy, L., Belleville, M.P., 2004. Immobilization of lipase on a ceramic membrane: activity and stability. J. Membr. Sci., 161–166.

Mahal, L.K., Yarema, K.J., Bertozzi, C.R., 1997. Engineering chemical reactivity on cell surfaces through oligosaccharide biosynthesis. Science 276, 1125–1128.

Mahan, D.E., Morrison, L., Watson, L., Haugneland, L.S., 1987. Phase change enzyme lmmunoassay. Anal. Biochem. 162, 163–170.

Maherani, B., Arab-Tehrany, E., R Mozafari, M., Gaiani, C., Linder, M., 2011. Liposomes: a review of manufacturing techniques and targeting strategies. Curr. Nanosci. 7 (3), 436–452.

Mahrus, S., Craik, C.S., 2005. Selective chemical functional probes of granzymes A and B reveal granzyme B is a major effector of natural killer cell-mediated lysis of target cells. Chem. Biol. 12, 567–577.

Maia, J., Carvalho, R.A., Coelho, J.F.J., Simoes, P.N., Gil, M.H., 2011. Insight on the periodate oxidation of dextran and its structural vicissitudes. Polymer 52, 258–265.

Maiden, M.C., Spratt, B.G., 1999. Meningococcal conjugate vaccines: new opportunities and new challenges. Lancet 354, 615–616.

Maiden M.C.J., Stuart J.M. (for the UK Meningococcal Carriage Group),2002. Carriage of serogroup C meningococci one year after meningococcal C conjugate polysaccharide vaccination. Lancet 359: 1829–1830.

Major, M., Feuillerat, S., Schofield, M., Rosenblatt, M., Rogers, M., Opperman, K., et al., 2009. Exploration of the phospho-tyrosine proteome using SH2 domains. In: Proceedings of the 100th Annual Meeting of the American Association for Cancer Research; Apr 18-22; Denver, CO. AACR; 2009. Abstract No. 2625.

Majoros, I., Thomas, T.P., Mehta, C.B., Baker Jr., J.R., 2005. Poly(amidoamine) dendrimer-based multifunctional engineered nanodevice for cancer therapy. J. Med. Chem. 48, 5892–5899.

Makarova, O.V., Ostafin, A.E., Miyoshi, H., Norris Jr., J.R., Meisel, D., 1999. Adsorption and encapsulation of fluorescent probes in nanoparticles. J. Phys. Chem. B103, 9080–9084.

Maki, C.G.,2011. p53 Localization, Chapter 8, in Molecular Biology Intelligence Unit, Volume 1, 117-126, doi: 10.1007/978-1-4419-8231-5_8.

Makino, K., Maruo, S., Morita, Y., Takeuchi, T., 1988. Biotechnol. Bioeng. 31, 617.

Malcolm, A.D.B., Nicolas, J.L.,1984. Detecting a polynucleotide sequence and labelled polynucleotides useful in this method. WO Patent Appl. 8403520.

Male, D., Champion, B., Cooke, A., 1987. Advanced Immunology (Section 8.1–8.8). Lippincott Gower Medical Publ, London.

Mallia, A., Krishna, G.T., Hermanson, R.I., Krohn, E.K., Fujimoto, P.K., Smith, 1981. Preparation and use of a boronic acid affinity support for separation and quantitation of glycosylated hemoglobins. Anal. Lett. 14, 649–661.

Mallia, A.K., 1992. Pierce Chemical, personal communications.

Mallik, R., Wa, C., Hage, D.S., 2007. Development of sulfhydryl-reactive silica for protein immobilization in high-performance affinity chromatography. Anal. Chem. 15, 1411–1424.

Malmström, J., Lee, H., Aebersold, R., 2007. Advances in proteomic workflows for systems biology. Curr. Opin. Biotechnol. 18, 378–384.

Mamede, M., Saga, T., Kobayashi, H., Ishimori, T., Higashi, T., Sato, N., et al., 2003. Radiolabeling of avidin with very high specific activity for internal radiation therapy of intraperitoneally disseminated tumors. Clin. Cancer Res. 9, 3756–3762.

Manabe, Y., Tsubota, T., Haruta, Y., Okazaki, M., Haisa, S., Nakamura, K., et al., 1983. Production of monoclonal antibody–bleomycin conjugate utilizing dextran T40 and the antigen-targeting cytotoxicity of the conjugate. Biochem. Biophys. Res. Comm. 115, 1009.

Manchester, M., Singh, P., 2006. Virus-based nanoparticles (Vnanoparticles): platform technologies for diagnostic imaging. Adv. Drug Deliv. Rev. 58, 1505–1522.

Manduchi, E., Scearce, M., Brestelli, J.E., Grant, G.R., Kaestner, K.H., Stoeckert, C.J., 2002. Comparison of different labeling methods for two-channel high-density microarray experiments. Physiol. Genom. 10, 169–179.

Mandy, W.J., Rivers, M.M., Nisonoff, A., 1961. Recombination of univalent subunits derived from rabbit antibody. J. Biol. Chem. 236, 3221.

Manevich, Y., Feinstein, S.I., Fisher, A.B., 2004. Activation of the antioxidant enzyme 1-CYS peroxiredoxin requires glutathionylation mediated by heterodimerization with GST. Proc. Natl. Acad. Sci. USA 101, 3780–3785.

Manger, I.D., Rademacher, T.W., Dwek, R.A., 1992. 1-N-Glycyl *P*-oligosaccharide derivatives as stable intermediates for the formation of glycoconjugate probes. Biochemistry 31, 10724–10732.

Manjappa, A.S., Chaudhari, K.R., Venkataraju, M.P., Dantuluri, P., Nanda, B., Sidda, C., et al., 2011. Antibody derivatization and conjugation strategies: application in preparation of stealth immunoliposome to target chemotherapeutics to tumor. J. Control. Release 150, 2–22.

Mann, W.A., Meyer, N., Weber, W., Meyer, S., Greten, H., Beisiegel, U., 1995. Apolipoprotein E isoforms and rare mutations: parallel reduction in binding to cells and to heparin reflects severity of associated type III hyperlipoproteinemia. J. Lipid Res. 36, 517.

Mannik, M., Downey, W., 1973. Studies on the conjugation of horseradish peroxidase to fab fragments. J. Immunol. Methods 3, 233.

Mannweiler, K., Hohenberg, H., Bohn, W., Rutter, G., 2011. Protein-A gold particles as markers in replica immunocytochemistry: high resolution electron microscope investigations of plasma membrane surfaces. J. Microsc. 126 (2), 145–149.

Mansfeld, J., Ulbrich-Hofmann, R., 2000. Site-specific and random immobilization of thermolysin-like proteases reflected in the thermal inactivation kinetics. Biotechnol. Appl. Biochem. 32, 189–195.

Manske, J.M., Buchsbaum, D.J., Vallera, D.A., 1989. The role of ricin B chain in the intracellular trafficking of anti-CD5 immunotoxins. J. Immunol. 142, 1755–1766.

Marcelino-Cruz, A.M., Bhattacharya, M., Anselmo, A.C., Tessier, P.M., 2011. Site-specific structural analysis of a yeast prion strain with species-specific seeding activity. Prion 5 (3), 208–214.

March, S.C., Parikh, I., Cuatrecasas, P., 1974. A simplified method for cyanogen bromide activation of agarose for affinity chromatography. Anal. Biochem. 60, 149–152.

Marchán, V., Ortega, S., Pulido, D., Pedroso, E., Grandas, A., 2006. Diels–alder cycloadditions in water for the straightforward preparation of peptide-oligonucleotide conjugates. Nucleic Acids Res. 34 (3), e24.

Marcholonis, J.J., 1969. Biochem. J. 113, 299.

Marcus, S.L., Balbinder, E., 1972. Use of affinity matrices in determining steric requirements for substrate binding: binding of anthranilate 5-phosphoribosyl-pyrophosphate phosphoribosyltransferase from salmonella typhimurium to sepharose-anthranilate derivatives. Anal. Biochem. 48, 448–459.

Marfey, S.P., Tsai, K.H., 1975. Cross-linking of phospholipids in human erythrocyte membrane. Biochem. Biophys. Res. Comm. 65, 31–38.

Margaret, A.L., Nardin, A., Foley, P.L., Solga, M.D., Bankovich, A.J., Martin, E.N., et al., 2001. Targeting of pseudomonas aruginosa in the bloodstream with bispecific monoclonal antibodies. J. Immunol. 167, 2240–2249.

Margel, S., Rembaum, A., 1980. Synthesis and characterization of poly(glutaraldehyde). A potential reagent for protein immobilization and cell separation. Macromolecules 13, 19–24.

Margel, S., Zisblatt, S., Rembaum, A., 1979. Polyglutaraldehyde: a new reagent for coupling proteins to microspheres and for labeling cell-surface receptions. II. Simplified labeling method by means of non-magnetic and magnetic polyglutaraldehyde microspheres. J. Immunol. Methods 28, 341–353.

Marincean, S., Smith, M.R., Beltz, L., Borhan, B., 2012. Selectivity of labeled bromoethylamine for protein alkylation. J. Mol. Model. 18, 4547–4556.

Markell, C.G., Hagen, D.F., Hansen, P.E., Baumann, N.R.,1994. Particle-loaded nonwoven fibrous article for separations and purifications. U.S. Patent 5,328,758.

Markwell, M.A.K., 1982. A new solid-state reagent to iodinate proteins: conditions for the efficient labeling of antiserum. Anal. Biochem. 125, 427–432.

Markwell, M.A.K., Fox, C.F., 1978. Surface-specific iodination of membrane proteins of viruses and eukaryotic cells using 1,3,4,6-tetrachloro-3a,6a-diphenylglycouril. Biochemistry 17, 4807–4817.

Markwell, M.A.K., Fox, C.F., 1980. Protein-protein interactions within paramyxoviruses identified by native disulfide bonding or reversible chemical cross-linking. J. Virol. 33, 152–166.

Marquez, B.V., Beck, H.E., Aweda, T.A., Phinney, B., Holsclaw, C., Jewell, W., et al., 2012. Enhancing peptide ligand binding to vascular endothelial growth factor by covalent bond formation. Bioconjug. Chem. 23, 1080–1089.

Marr, M.T., Datwyler, S.A., Meares, C.F., Roberts, J.W., 2001. Restructuring of an RNA polymerase holoenzyme elongation complex by lambdoid phage Q proteins. PNAS 98, 8972–8978.

Marr, M.T., Roberts, J.W., Brown, S.E., Klee, M., Gussin, G.N., 2004. Interactions among CII protein, RNA polymerase and the PRE promoter: contacts between RNA polymerase and the—35 region of PRE are identical in the presence and absence of CII protein. Nucleic Acids Res. 32, 1083–1090.

Marrari, A., Iero, M., Pilla, L., Villa, S., Salvioni, R., VAldagni, R., et al., 2007. Vaccination therapy in prostate cancer. Cancer Immunol. Immunother. 56, 429–445.

Marsh, I.R., Smith, H.K., Leblanc, C., Bradley, M., 1996. Mol. Divers. 2, 165–170.

Martin, F.J., Papahadjopoulos, D., 1982. Irreversible coupling of immunoglobulin fragments to preformed vesicles. J. Biol. Chem. 257, 286–288.

Martin, F.J., Heath, T.D., New, R.R.C., 1990. Covalent attachment of proteins to liposomes. In Liposomes, a Practical Approach. IRL Press, New York. (163–182).

Martin, F.J., Hubbell, W., Papahyadjopoulos, D., 1981. Immunospecific targeting of liposomes to cells: a novel and efficient method for covalent attachment of Fab′ fragments via disulfide bonds. Biochemistry 20, 4229–4238.

Martin, L.J., Sculimbrene, B.R., Nitz, M., Imperiali, B., 2005. Rapid combinatorial screening of peptide libraries for the selection of lanthanide-binding tags (LBTs). QSAR Combin. Sci. 24 (10), 1149–1157.

Martina, M., Luk, C., Py, C., Martinez, D., Comas, T., Monette, R., et al., 2011. Recordings of cultured neurons and synaptic activity using patch-clamp chips. J. Neural. Eng. 8 (3), 034002.

Martinez-Ramon, A., Knecht, E., Rubio, V., Grisolia, S., 1990. Levels of carbamoyl phosphate synthetase I in livers of young and old rats assessed by activity and immunoassays and by electron microscopic immunogold procedures. J. Histochem. Cytochem. 38, 371–376.

Martini, C., Lucacchini, A., Ronca, G., Hrelia, S., Rossi, C.A., 1982. Isolation of putative benzodiazepine receptors from rat brain membranes by affinity chromatography. J. Neurochem. 38 (1), 15–19.

Maruk, A.Y., Bruskin, A.B., Kodina, G.E., 2011. Novel 99m Tc radiopharmaceuticals with bifunctional chelating agents. Radiochemistry 53 (4), 341–353.

Marzi, S., Knight, W., Brandi, L., Caserta, E., Soboleva, N., Hill, W.E., et al., 2003. Ribosomal localization of translation initiation factor IF2. RNA 9, 958–969.

Masamune, S., Palmer, M.A.J., Gamboni, R., Thompson, S., Davis, J.T., Williams, S.F., et al., 1989. Bio-claisen condensation catalyzed by thiolase from zoogloea ramigera. Active site cysteine residues. Chemtracts Org. Chem. 2, 247–251.

Mason, R.W., Wilcox, D., Wikstrom, P., Shaw, E.N., 1989. The identification of active forms of cysteine proteinases in kirsten-virus-transformed mouse fibroblasts by use of a specific radiolabelled inhibitor. Biochem. J. 257 (1), 125–129.

Masri, M.S., Friedman, M., 1988. Protein reactions with methyl and ethyl vinyl sulfones. J. Protein Chem. 7, 49–54.

Massague, J., Guillette, B.J., Czech, M.P., Morgan, C.J., Bradshaw, R.A., 1981. Identification of a nerve growth factor receptor protein in sympathetic ganglia membranes by affinity labeling. J. Biol. Chem. 256, 9419–9424.

Masuho, Y., Hara, T., Noguchi, T., 1979. Preparation of hybrid of fragment fab′ of antibody and fragment A of diphtheria toxin and its cytotoxicity. Biochem. Biophys. Res. Comm. 90, 320.

Masuho, Y., Kishida, K., Saito, M., Umemoto, N., Hara, T., 1982. Importance of the antigen-binding valency and the nature of the cross-linking bond in ricin A-chain conjugates with antibody. J. Biochem. (Tokyo) 91, 1583.

Mateo, C., Abian, O., Fernandez-Lorente, G., Pedroche, J., Fernandez-Lorente, R., Guisan, J.M., 2002. Epoxy sepabeads: a novel epoxy support for stabilization of industrial enzymes via very intense multipoint covalent attachment. Biotechnol. Prog. 18, 629–634.

Mathis, G., Bazin, H., 2011. Stable luminescent chelates and macrocyclic compounds. Lanthanide Luminescence, 47–88.

Matosevic, S., Szita, N., Baganz, F., 2011. Fundamentals and applications of immobilized microfluidic enzymatic reactors. J. Chem. Technol. Biotechnol. 86 (3), 325–334.

Matteucci, M.D., Caruthers, M.H., 1980. The synthesis of oligodeoxypyrimidines on a polymer support. Tetrahedron Lett. 21, 719–722.

Matthay, K.K., Heath, T.D., Papahadjopoulos, D., 1984. Specific enhancement of drug delivery to AKR lymphoma by antibody-targeted small unilamellar vesicles. Cancer Res. 44, 1880–1886.

Matthay, K.K., Heath, T.D., Badger, C.C., Bernstein, I.D., Papahadjopoulos, D., 1986. Antibody-directed liposomes: comparison of various ligands for association, endocytosis and drug delivery. Cancer Res. 46, 4904.

Matthews, J.A., Kricka, I.J., 1988. Analytical strategies for the use of DNA probes. Anal. Biochem. 169, 1–25.

Mattiasson, B., Nilsson, H., 1977. An enzyme immunoelectrode. FEBS Lett. 78, 251.

Mattoussi, H., Mauro, J.M., Goldman, E.R., Anderson, G.P., Sundar, V.C., Mikulec, F.V., et al., 2000. Self-assembly of CdSe-ZnS quantum dot bioconjugates using an engineered recombinant protein. J. Am. Chem. Soc. 122, 12142–12150.

Mattu, T.S., Pleass, R.J., Willis, A.C., Kilian, M., Wormald, M.R., Lellouch, A.C., et al., 1998. The glycosylation and structure of human serum IgA1, fab, and Fc regions and the role of

N-glycosylation on Fc alpha receptor interactions. J. Biol. Chem. 273 (4), 2260–2272.

Mawas, F., Niggemann, J., Jones, C., Corbel, M.J., Kamerling, J.P., Vliegenthart, J.F.G., 2002. Immunogenicity in a mouse model of a conjugate vaccine made with a synthetic single repeating unit of type 14 pneumococcal polysaccharide coupled to CRM197. Infect. Immun. 70 (9), 5107–5114.

Mayer, A., Sharma, S.K., Tolner, B., Minton, N.P., Purdy, D., Amlot, P., et al., 2004. Modifying an immunogenic epitope on a therapeutic protein: a step towards an improved system for antibody-directed enzyme prodrug therapy (ADEPT). Br. J. Cancer 90, 2402–2410.

Mayhew, E., Lazo, R., Vail, W.J., King, J., Green, A.M., 1984. Characterization of liposomes prepared using a microemulsifier. Biochim. Biophys. Acta 775, 169–174.

Mazaitis, J.K., Francis, B.E., Eckelman, W.C., Gibson, R.E., Reba, R.C., Barnes, J.W., et al., 1981. No-carrier-added bromination of estrogens with chloramine-T and Na77. Br. J. Labelled Compd. Radiopharm. 18, 1033–1038.

McAllister, L.A., Hixon, M.S., Schwartz, R., Kubitz, D.S., Janda, K.D., 2007. Synthesis and application of a novel ligand for affinity chromatography based removal of endotoxin from antibodies. Bioconjug. Chem. 18, 559–566.

McBroom, C.R., Samanen, C.H., Goldstein, I.J., 1976. Carbohydrate antigens: coupling of carbohydrates to proteins by diazonium and phenylisothiocyanate reactionsJakoby, W.B. Methods in Enzymology, vol. 2. Academic Press, New York, pp. 212.

McCafferty, J., Griffiths, A.D., Winter, G., Chiswell, D.J., 1990. Phage antibodies: filamentous phage displaying antibody variable domains. Nature 348 (6301), 552–554.

McCarron, P.A., Olwill, S.A., Marouf, W.M.Y., Buick, R.J., Walker, B., Scott, C.J., 2005. Antibody conjugates and therapeutic strategies. Mol. Interv. 5 (6), 368–380.

McCarron, P.A., Olwill, S.A., Marouf, W.M.Y., Buick, R.J., Walker, B., Scott, C.J., 2005. Antibody conjugates and therapeutic strategies. Mol. Interv. 5, 368–380.

McCarthy, J.R., Sazonova, I.Y., Erdem, S.S., Hara, T., Thompson, B.D., Patel, P., et al., 2012. Multifunctional nanoagent for thrombus-targeted fibrinolytic therapy. Nanomedicine 7, 1017–1028.

McCleary, B.V., Matheson, N.K., 1986. Enzymic analysis of polysaccharide structure. Adv. Carbohydr. Chem. Biochem. 44, 147–276.

McCormick, C.M., 2012. Tetanus Toxoid, Reduced Diphtheria Toxoid and Acellular Pertussis Vaccination and Influenza Vaccination of Pregnant and Postpartum Women. <http://corescholar.libraries.wright.edu/mph/82>.

McDevitt, M.R., Chattopadhyay, D., Kappel, B.J., Jaggi, J.S., Schiffman, S.R., Antczak, C., et al., 2007. Tumor targeting with antibody-functionalized, radiolabeled carbon nanotubes. J. Nucl. Med. 48, 1180–1189.

McGown, L.B., Warner, I.M., 1990. Molecular fluorescence, phosphorescence, and chemiluminescence spectroscopy. Anal. Chem. 190, 255R.

McInnes, J.L., Dalton, S., Vize, P.D., Robins, A.J., 1987. Non-radioactive photobiotinlabeled probes detect single copy genes and low abundance mRNA. Biotechnology 5, 269–272.

McIntyre, J.O., Fingleton, B., Wells, K.S., Piston, D.W., Lynch, C.C., Gautam, S., et al., 2004. Development of a novel fluorogenic proteolytic beacon for in vivo detection and imaging of tumor-associated matrix metalloproteinase-7 activity. Biochem. J. 377, 617–628.

McKinney, R.M., Spillane, J.T., Pearce, G.W., 1964. Factors affecting the rate of reaction of fluorescein isothiocyanate with serum proteins. J. Immunol. 93, 232–242.

McMahan, S.A., Burgess, R.R., 1994. Use of aryl azide cross-linkers to investigate protein–protein interactions: an optimization of important conditions as applied to *Escherichia coli* RNA polymerase and localization of a p70-a cross-link to the C-terminal region of a. Biochemistry 33, 12092–12099.

Means, G.E., Feeney, R.E., 1971. Chemical Modification of Proteins. Holden-Day, San Francisco. (20).

Meares, C.F., 1986. Chelating agents for the binding of metal ions to antibodies. Nucl. Med. Biol. 13, 311–318.

Meares, C.F., Whetstone, P.A., Corneillie, T.M., Butlin, N.G.,2007. Element-coded affinity tags. US Patent No. 7, 214, 545.

Mecher, E., Gallego-Gómez, F., Tillmann, H., Hörhold, H.H., Hummelen, J.C., Meerholz, K., 2002. Near-infrared sensitivity enhancement of photorefractive polymer composites by pre-illumination. Nature 418, 959–964.

Medintz, I.L., Clapp, A.R., Mattoussi, H., Goldman, E.R., Fisher, B., Mauro, J.M., 2003. Self-assembled nanoscale biosensors based on quantum dot FRET donors. Nat. Mater. 2, 630–638.

Medintz, I.L., Konnert, J.H., Clapp, A.R., Stanish, I., Twigg, M.E., Mattoussi, H., et al., 2004. A fluorescence resonance energy transfer-derived structure of a quantum dot-protein bioconjugate nanoassembly. Proc. Natl. Acad. Sci. USA 101 (26), 9612–9617.

Mei, Y., Kumar, A., Gross, R., 2003. Kinetics and mechanism of candida antarctica lipase B catalyzed solution polymerization of e-caprolactone. Macromolecules 36, 5530–5536.

Meige, J.B., Wang, Y.-L., 1986. Reorganization of alpha-actin and vinculin induced by a phorboll ester in living cells. J. Cell Biol. 102, 1430.

Meighen, F.A., Nicolim, M.Z., Hustings, J.W., 1971. Hybridization of bacterial luciferase with a variant produced by chemical modification. Biochemistry 10, 4062.

Meijer, E.J., de Leeuw, D.M., Setayesh, S., Van Veenendaal, E., Huisman, B.-H., Blom, P.W.M., et al., 2003. Solution-processed ambipolar organic field-effect transistors and inverters. Nat. Mater. 2, 678–682.

Meijer, F.W., Nijhuis, S., Vroonhoven, F.C.B.M., 1988. Poly-1,2-azepines by the photo-polymerization of phenyl azides. Precursors for conducting polymer films. J. Am. Chem. Soc. 110, 7209–7210.

Melchers, F., Messer, W., 1973. The activity of individual molecules of hybrid b-galactos-idase reconstituted from the wild-type and an inactive-mutant enzyme. Eur. J. Biochem. 34, 228.

Melnikov, V.Y., Faubel, S., Siegmund, B., Lucia, M.S., Ljubanovic, D., Edelstein, C.L., 2002. Neutrophil-independent mechanisms of caspase-1– and IL-18–mediated ischemic acute tubular necrosis in mice. J. Clin. Invest. 110, 1083–1091.

Meng, Q.-Q., Wang, J.-X., Ma, G.-H., Su, Z.-G., 2009. Lyophilization of CNBr-activated agarose beads with lactose and PEG. Process Biochemistry 44, 562–571.

Menjoge, A.R., Kannan, R.M., Tomalia, 2010. Dendrimer-based drug and imaging conjugates: design considerations for nanomedical applications. Drug Deliv. Today 15, 171–185.

Menon, V., Prakash, G., Prabhune, A., Rao, M., 2010. Biocatalytic approach for the utilization of hemicellulose for ethanol production from agricultural residue using thermostable xylanase and thermotolerant yeast. Bioresour. Technol. 101, 5366–5373.

Mentinova, M., Barefoot, N.Z., McLuckey, S.A., 2012. Solution versus gas-phase modification of peptide cations with NHS-ester reagents. J. Am. Soc. Mass Spectrom., 1–8.

Mentzer Jr., W.C., Lubin, B.H., 1979. The effect of cross-linking agents on red-cell shape. Semin. Hematol. 16, 115–127.

Mentzer Jr., W.C., Lewis, S., Pennathur-Das, R., Halpin, R., Cerrone, K.L., Lubin, B., et al., 1982. Formation of 5-carbomethoxyvaleramidine during hydrolysis of the protein cross-linking agent dimethyl adipimidate. J. Protein Chem. 1, 141–155.

Mercier, F., J. Paris, G. Kaisin, D. Thonon, J. Flagothier, N. Teller, et al., 2011. General Method for Labeling siRNA by Click Chemistry with Fluorine-18 for the Purpose of PET Imaging.

Meredith, G.D., Wu, H.Y., Allbritton, N.L., 2004. Targeted protein functionalization using His-tags. Bioconjug. Chem. 15, 969–982.

Messerli, S.M., Prabhakar, S., Tangy, Yi, Shah, K., Cortes, M.L., Murthy, V., et al., 2004. A novel method for imaging apoptosis

using a caspase-1 near-infrared fluorescent probe. Neoplasia 6, 95–105.
Metz, D.H., Brown, G.L., 1969. The investigation of nucleic acid secondary structure by means of chemical modification with a carbodiimide reagent. I. The reaction between N-cyclohexyl-N9-b-(4-methylmorpholinium)ethyl carbodiimide and model nucleotides. Biochemistry 8, 2312–2328.
Meyne, J., 1993. Chromosome mapping by fluorescent in situ hybridization. In: Howard, G.C. (Ed.), Methods in Nonradioactive Detection. Appleton & Lange, Norwalk, Connecticut, pp. 263–268.
Mezö, G., Szabó, I., Kertész, I., Hegedüs, R., Orbán, E., Leurs, U., et al., 2011. Efficient synthesis of an (aminooxy) acetylated-somatostatin derivative using (aminooxy) acetic acid as a 'carbonyl capture'reagent. J. Pept. Sci. 17 (1), 39–46.
Miao, Z., Ren, G., Liu, H., Jiang, L., Cheng, Z., 2010. Cy5.5-labeled affibody molecule for near-infrared fluorescent optical imaging of epidermal growth factor receptor positive tumors. J. Biomed. Opt. 15, 036007.10.1117/1.3432738.
Michel, R.B., Ochakovskaya, R., Mattes, M.J., 2002. Antibody localization to B-cell lymphoma xenografts in immunodeficient mice: importance of using residualizing radiolabels. Clin. Cancer Res. 8, 2632–2639.
Michnick, S.W., 2003. Protein fragment complementation strategies for biochemical network mapping. Curr. Opin. Biotechnol. 14, 610–617.
Michnick, S.W., Ear, P.H., Manderson, E.N., Remy, I., Stefan, E., 2007. Universal strategies in research and drug discovery based on protein-fragment complementation assays. Nat. Rev. 6, 569–582.
Mielke, S., Rezvani, K., Savani, B.N., Nunes, R., Yong, A.S.M., Schindler, J., et al., 2007. Reconstitution of foxp3+ regulatory T cells (Tregs) after CD25-depleted allotransplantion in elderly patients and association with acute graft-versus-host disease (GvHD). Blood, 10.1182/blood-2007-03-079160.
Migneault, I., Dartiguenave, C., Bertrand, M.J., Waldron, K.C., 2004. Glutaraldehyde: behavior in aqueous solution, reaction with proteins, and application to enzyme crosslinking. Biotechniques 37, 798–802.
Mikhailov, V., et al., 2001. Bcl-2 prevents Bax oligomerization in the mitochondrial outer membrane. J. Biol. Chem. 276, 18361–18374.
Mikkelsen, R.B., Wallach, D.F.H., 1976. Photoactivated cross-linking of protein within the erythrocyte membrane core. J. Biol. Chem. 251, 7413.
Miklis, P., Cagin, T., Goddard III, W.A., 1997. Dynamics of bengal rose encapsulated in the Meijer dendrimer box. J. Am. Chem. Soc. 119, 7458–7462.
Mikol, D.D., Ditlow, C., Usatin, D., Biswas, P., Kalbfleisch, J., Milner, A., et al., 2006. Serum IgE reactive against small myelin protein-derived peptides is increased in multiple sclerosis patients. J. Neuroimmunol. 180, 40–49.
Mikolajczyk, S.D., Meyer, D.L., Starling, J.J., Law, K.L., Rose, K., Dufour, B., et al., 1994. High yield, site-specific coupling of N-terminally modified beta-lactamase to a proteolytically derived single-sulfhydryl murine fab. Bioconjug. Chem. 5, 636–646.
Miles, D., Roché, H., Martin, M., Perren, T.J., Cameron, D.A., Glaspy, J., et al., 2011. Phase III multicenter clinical trial of the sialyl-TN (STn)-keyhole limpet hemocyanin (KLH) vaccine for metastatic breast cancer. Oncologist 16 (8), 1092–1100.
Milla, P., Dosio, F., Cattel, L., 2012. PEGylation of proteins and liposomes: a powerful and flexible strategy to improve the drug delivery. Curr. Drug Metab. 13, 105–119.
Millar, J.B., Rozengur, E., 1990. Chronic desensitization to bombesin by progressive down-regulation of bombesin receptors in swiss 3T3 cells. J. Biol. Chem. 265, 12052–12058.
Miller, E., Salisbury, D., Ramsay, M., 2001. Planning, registration, and implementation of an immunisation campaign against meningococcal serogroup C disease in the UK: a success story. Vaccine 20 (Suppl. 1), S58–S67.
Miller, M.D., Hata, S., De Waal Malefyt, R., Krangel, M.S., 1989. A novel polypeptide secreted by activated human T lymphocytes. J. Immunol. 143, 2907–2916.
Minami, Y., et al. 2000a. A critical role for the proteasome activator PA28 in the Hsp90-dependent protein refolding. J. Biol. Chem. 275 (12), 9055–9061.
Minami, Y., Kawasaki, H., Minami, M., Tanahashi, N., Tanaka, K., Yahara, I., 2000b. A critical role for the proteasome activator PA28 in the Hsp90-dependent protein refolding. J. Biol. Chem. 275, 9055.
Minard-Basquin, C., Weil, T., Hohner, A., Radler, J.O., Mullen, K., 2003. A polyphenylene dendrimer-detergent complex as a highly fluorescent probe for bioassays. J. Am. Chem. Soc. 125 (19), 5832–5838.
Ming, M., Kuroiwa, T., Ichikawa, S., Sato, S., Mukataka, Sukekuni, 2006. Production of chitosan oligosaccharides by chitosanase directly immobilized on an agar gel-coated multidisk impeller. Biochem. Eng. J. 28, 289–294.
Minussi, R.C., Rossi, M., Bologna, L., Rotilio, D., Pastore, G.M., Durán, N., 2007. Phenols removal in musts: Strategy for wine stabilization by laccase. J. Mol. Catal B: Enzymatic 45 (Issues 3-4), 102–107.
Miron, T., Wilchek, M., 1985. Activation of trisacryl gels with chloroformates and their use for affinity chromatography and protein immobilization. Appl. Biochem. Biotechnol. 11, 445–456.
Miron, T., Wilchek, M., 1993. A simplified method for the preparation of succinimidyl carbonate polyethylene glycol for coupling to proteins. Bioconjug. Chem. 4, 568–569.
Mirzadeh, S., Brechbiel, M.W., Atcher, R.W., Gansow, O.A., 1990. Radiometal labeling of immunoproteins: covalent linkage of 2-(4-isothiocyanatobenzyl) diethylenetriaminepentaacetic acid ligands to immunoglobulin. Bioconjug. Chem. 1, 59–65.
Misawa, J., Moriwaki, S.I., Kohno, E., Hirano, T., Tokura, Y., Takigawa, M., 2005. The role of low-density lipoprotein receptors in sensitivity to killing by Photofrin-mediated photodynamic therapy in cultured human tumor cell lines. J. Dermatol. Sci. July 20.
Mishra, B., Patel, B.B., Tiwari, S., 2010. Colloidal nanocarriers: a review on formulation technology, types, and applications toward targeted drug delivery. Nanomedicine 6, 9–24.
Miskimins, W.K., Shimizu, N., 1979. Synthesis of cytotoxic insulin cross-linked to diphtheria toxin fragment a capable of recognizing insulin receptors. Biochem. Biophys. Res. Comm. 91, 143.
Misra, A., Dwivedi, P., 2007. Immobilization of oligonucleotides on glass surface using an efficient heterobifunctional reagent through maleimide–thiol combination chemistry. Anal. Biochem. 369, 248–255.
Mittal, B., Sanger, J.M., Sanger, J.W., 1987. Visualization of myosin in living cells. J. Cell Biol. 105, 1753–1760.
Mittal, K.L. Silanes and other Coupling Agents, vol. 5. Brill, ISBN-10: 9004165916.
Miyagi, M., Rao, K.C.S., 2007. Proteolytic 18O-labeling strategies for quantitative proteomics. Mass Spectrom. Rev. 26 (1), 121–136.
Miyakawa, T., Takemoto, L.J., Fox, C.F., 1978. J. Supramol. Struct. 8, 303–310.
Miyake, R., Murakami, K., Owens, J.T., Greiner, D.P., Ozoline, O.N., Ishihama, A., et al., 1998. Dimeric association of Escherichia coli RNA polymerase alpha subunits, studied by cleavage of single-cysteine alpha subunits conjugated to iron-(S)-1-[*p*-(bromoacetamido)benzyl]ethylenediaminetetraacetate. Biochemistry 37 (5), 1344–1349.
Miyazaki, M., Maeda, H., 2006. Microchannel enzyme reactors and their applications for processing. Trends Biotechnol. 24, 463–470.
Mochizuki, M., Yano, M., Oda, T., Tateishi, H., Kobayashi, S., Yamamoto, T., et al., 2007. Scavenging free radicals by low-dose carvedilol prevents redox-dependent Ca^{21} leak via stabilization of ryanodine receptor in heart failure. J. Am. Coll. Cardiol. 49, 1722–1732.

Modjtahedi, H., Ali, S., Essapen, S., 2012. Therapeutic application of monoclonal antibodies in cancer: advances and challenges. Br. Med. Bull. doi: 10.1093/bmb/lds032 (published online).

Moebius, F.F., Hanner, M., Knaus, H.G., Weber, F., Striessnig, J., Glossmann, H., 1994. Purification and amino-terminal sequencing of the high affinity phenylalkylamine Ca21 antagonist binding protein from guinea pig liver endoplasmic reticulum. J. Biol. Chem. 269, 29314–29320.

Moeremans, M., Daneels, G., Van Dijck, A., Langanger, G., De Mey, J., 1984. Sensitive visualization of antigen-antibody reactions in dot and blot immuno overlay assays with the immunogold and immunogold/silver staining. J. Immunol. Methods 74, 353–360.

Moi, M.K., Meares, C.F., 1988. The peptide way to macrocyclic bifunctional chelating agents: synthesis of 2-(p-nitrobenzyl)-1,4,7,10-tetraazacyclododecane-N,N′,N″,N‴-tetraacetic acid and study of its yttrium(III) complex. J. Am. Chem. Soc. 110, 6266–6267.

Moi, M.K., Meares, C.F., McCall, M.J., Cole, W.C., DeNardo, S.J., 1985. Copper chelates as probes of biological systems: stable copper complexes with a macrocyclic bifunctional chelating agent. Anal. Biochem. 148, 249–253.

Mojazi Amiri, B., Adams, T.E., Doroshov, S.I., Paktinat, M., 2012. Use of a mammalian gonadotropin-releasing hormone (GnRH) agonist to characterize pituitary GnRH receptors in white sturgeon (Acipenser transmontanus Richardson). J. Appl. Ichthyology 28, 687–691.

Mojović, L., Rakin, M., Vukašinović, M., Nikolić, S., Pejin, J., Pejin, D., 2010. Production of bioethanol by simultaneous saccharification and fermentation of corn meal by immobilized yeast. Chem. Eng. Trans. 21, 1333–1338.

Mokotoff, M., Mocarski, Y.M., Gentsch, B.L., Miller, M.R., Zhou, J.H., Chen, J., et al., 2001. Caution in the use of 2-iminothiolane (Traut's reagent) as a cross-linking agent for peptides. The formation of N-peptidyl-2-iminothiolanes with bombesin (BN) antagonist (D-Trp(6),Leu(13)-psi[CH(2)NH]-Phe(14))BN(6-14) and D-Trp-Gln-Trp-NH(2). J. Pept. Res. 57 (5), 383–389.

Molina-Bolívar, J.A., Galisteo-González, F., lvarez, R.H., 1998. Anomalous colloidal stability of latex–protein systems. J. Colloid Interface Sci. 206, 518–526.

Møller, I.M., Rogowska-Wrzesinska, A., Rao, R.S.P., 2011. Protein carbonylation and metal-catalyzed protein oxidation in a cellular perspective. J. Proteomics 74 (11), 2228–2242.

Monaghan, S., Griffith-Johnson, D., Matthews, I. Bradley, M.,2001. Solid-phase synthesis of peptide–dendrimer conjugates for an investigation of integrin binding. Arkivoc, 2001, (Part (x)), 46-53. <http://eprints.soton.ac.uk/19576/>.

Monfardini, C., Veronese, F.M., 1998. Stabilization of substances in circulation. Bioconjug. Chem. 9, 418–450.

Monsan, P., 1978. Optimization of glutaraldehyde activation of a support for enzyme immobilization. J. Mol. Catal. 3, 371–384.

Monsan, P., Puzo, G., Mazarguil, H., 1975. Mechanism of glutaraldehyde-protein bond formation. Biochimie 57, 1281–1292.

Montellano, A., Da Ros, T., Bianco, A., Prato, M., 2011. Fullerene C60 as a multifunctional system for drug and gene delivery. Nanoscale 3 (10), 4035–4041.

Montesano, L., Cawley, D., Herschman, H.R., 1982. Disuccinimidyl suberate cross-linked ricin does not inhibit cell-free protein synthesis. Biochem. Biophys. Res. Comm. 109, 7–13.

Moon, J.J., Suh, H., Bershteyn, A., Stephan, M.T., Liu, H., Huang, B., et al., 2012. Interbilayer-crosslinked multilamellar vesicles as synthetic vaccines for potent humoral and cellular immune responses. Nat. Mater. 10, 243–251.

Moore, J.F., Ward, W.H., 1956. Cross-linking of bovine plasma albumin with wool keratin. J. Am. Chem. Soc. 78, 2414.

Morag, E., Bayer, E.A., Wilchek, M., 1996. Immobilized nitro-avidin and nitro-streptavidin as reusable affinity matrices for application in avidin-biotin technology. Anal. Biochem. 243 (2), 257–263.

Morales-Sanfrutos, J., Lopez-Jaramillo, J., Ortega-Muñoz, M., Megia-Fernandez, A., Perez-Balderas, F., Hernandez-Mateo, F., et al., 2010. Vinyl sulfone: a versatile function for simple bioconjugation and immobilization. Org. Biomol. Chem. 8, 667–675.

Moreland, R.B., Smith, P.K., Fujimoto, E.K., Pockter, M.E., 1982. Synthesis and characterization of *N*-(4-azidophenylthio)-phthalimide. Anal. Biochem. 121, 321.

Morgan, C.J., Stanley, E.R., 1984. Chemical crosslinking of the mononuclear phagocyte specific growth factor CSF-1 to its receptor at the cell surface. Biochem. Biophys. Res. Comm. 119, 35–41.

Moroder, L., Nyfeler, R., Gemeiner, M., Kalbacher, H., Wfinsch, E., 1983. Immunoassays of peptide hormones and their chemical aspects. BioPolymers 22, 481–486.

Moroder, L., Bovermann, G., Mourier, G., Göhring, W., Gemeiner, M., Wünsch, E., 1987. Studies on immunoassays of peptide factors. III. Gastrin/iso-1-cytochrome c as immunogen for raising anti-gastrin antisera. Biol. Chem. Hoppe Seyler 368, 839–848.

Moroney, J.V., Warncke, K., McCarthy, R.F., 1982. The distance between thiol groups in the gamma subunit of coupling factor 1 influences the protein permeability of thylakoid membranes. J. Bioenerg. Biomembr. 14, 347.

Moroney, S.E., D'Alarcao, L.J., Goldmacher, V.S., Lambert, H.M., Blattler, W.A., 1987. Modification of the binding site(s) of lectins by an affinity column carrying an activated galactose-terminated ligand. Biochemistry 26, 8390.

Morpurgo, M., Veronese, F.M., Kachensky, D., Harris, J.M., 1996. Preparation and characterization of poly(ethylene glycol)vinyl sulfone. Bioconjug. Chem. 7, 363–368.

Morris, M.C., Deshayes, S., Heitz, F., Divita, G., 2008. Cell-penetrating peptides: from molecular mechanisms to therapeutics. Biol. Cell 100, 201–217.

Morris, R.E., Saelinger, C.B., 1984. Visualization of intracellular trafficking: use of biotinylated ligands in conjunction with avidin–gold colloids. J. Histochem. Cytochem. 32, 124–128.

Morrison, L.E., 1992. Detection of energy transfer and fluorescence quenching. In: Kricka, L.J. (Ed.), Nonisotopic DNA Probe Techniques. Academic Press, New York, pp. 311–352.

Morrison, M., Bayse, G.S., 1970. Catalysis of iodination by lactoperoxidase. Biochemistry 9, 2995–3000.

Morrison, S.L., 2007. Two heads are better than one. Nat. Biotechnol. 25, 1233–1234.

Morton, L.A., Saludes, J.P., Yin, H., 2012. Constant pressure-controlled extrusion method for the preparation of nano-sized lipid vesicles. J. Vis. Exp. (64)10.3791/4151.

Mossberg, K., Ericsson, M., 1990. Detection of doubly stained fluorescent specimens using confocal microscopy. J. Microsc. 158, 215.

Motta-Hennessy, C., Eccles, S.A., Dean, C., Coghlan, G., 1985. Preparation of 67Ga-labeled human IgG and its fab fragments using deferoxamine as chelating agent. Eur. J. Nucl. Med. 11, 240–245.

Moulton, H.M., 2012. Cell-penetrating peptides enhance systemic delivery of antisense morpholino oligomers, Chapter 26, in exon skipping. Methods Mol. Biol. 867 (Part 3), 407–414.

Mourey, T.H., Turner, S.R., Rubenstein, M., Fréchet, J.M.J., Hawker, C.J., Wooley, K.L., 1992. Unique behavior of dendritic macromolecules: intrinsic viscosity of polyether dendrimers. Macromolecules 25, 2401–2406.

Muckerheide, A., Apple, R.J., Pesce, A.J., Michael, J.G., 1987a. Cationization of protein antigens. I. Alteration of immunogenic properties. J. Immunol. 138, 833–837.

Muckerheide, A., Domen, P.L., Michael, J.G., 1987b. Cationization of protein antigens. II. Alteration of regulatory properties. J. Immunol. 138, 2800–2804.

Mudd, J.A., Swanson, R.F., 1978. In situ crosslinking of vesicular stomatitis virus proteins with reversible agents. Virology 88, 263–280.

Mueller, J.E., Borrow, R., Gessner, B.D., 2006. Meningococcal serogroup W135 in the African meningitis belt: epidemiology, immunity and vaccines. Expert Rev. Vaccines 5, 319–336.

Muir, T.W., 2003. Semisynthesis of proteins by expressed protein ligation. Ann. Rev. Biochem. 72, 249–289.

Muir, T.W., Dawson, P.E., Kent, S.B.H., 1997. Protein synthesis by chemical ligation of unprotected peptides in aqueous-solution. Methods Enzymol. 289, 266–298.

Muir, T.W., Sondhi, D., Cole, P.A., 1998. Expressed protein ligation: A general method for protein engineering. Proc. Natl. Acad. Sci. USA 95, 6705–6710.

Mujumdar, R.B., Ernst, L.A., Mujumdar, S.R., Lewis, C.J., Waggoner, A.S., 1993. Cyanine dye labeling reagents: sulfoindocyanine succinimidyl esters. Bioconjug. Chem. 4 (2), 105–111.

Mukherjee, S., Brieba, L.G., Sousa, R., 2002. Structural transitions mediating transcription initiation by T7 RNA polymerase. Cell 110 (1), 81–91.

Mukkala, V.-M., Mikola, H., Hemmila, I., 1989. The synthesis and use of activated *N*-benzyl derivatives of diethylenetriaminetetraacetic acids: alternative reagents for labeling of antibodies with metal ions. Anal. Biochem. 176, 319–325.

Muller, D.R., et al. 2001. Isotope-tagged cross-linking reagents. A new tool in mass spectrometric protein interaction analysis. Anal. Chem. 73, 1927–1934.

Muller, W., 1990. New ion exchangers for the chromatography of biopolymers. J. Chromatogr. 510, 133–140.

Mumtaz, S., Bachhawat, B.K., 1991. Conjugation of proteins and enzymes with hydrophilic polymers and their applications. Indian J. Biochem. Biophys. 28, 346–351.

Murachi, T., 1976. Bromelain enzymesLorand, L. Methods in Enzymology, vol. 45. Academic Press, New York, pp. 475–485.

Muramoto, K., Kamiya, H., Kawauchi, H., 1984. The application of fluorescein isothiocyanate and high performance liquid chromatography for the microsequencing of proteins and peptides. Anal. Biochem. 141, 446.

Murayama, Y., Satoh, S., Oka, T., Imanishi, J., Noishiki, Y., 1988. Reduction of the antigenicity and immunogenicity of xenografts by a new cross-linking reagent. ASAIO Trans. 34, 546.

Muroi, M., et al. 2002. Regions of the mouse CD14 molecule required for toll-like receptor 2-and 4-mediated activation of NF-kB. J. Biol. Chem. 277, 42372–42379.

Murphy, F.R., Jorgensen, F.D., Cantor, C.R., 1982. Kinetics of histone endocytosis in Chinese hamster cells. A flow cytofluorometric analysis. J. Biol. Chem. 257, 1895.

Murphy, K., Travers, P., Walport, M., Janeway, C., 2011. Janeway's Immunobiology. Taylor & Francis.

Murray, C.B., Norris, D.J., Bawendi, M.G., 1993. Synthesis and characterization of nearly monodisperse CdE (E=S, Se, Te) semiconductor nanocrystallites. J. Am. Chem. Soc. 115, 8706–8715.

Murthy, G.S., Moudgal, N.R., 1986. Use of epoxy sepharose for protein immobilization. J. Biosci. 10, 351–358.

Myers, D.E., Uckun, F.M., Swaim, S.E., Vallera, D.A., 1989. The effects of aromatic and aliphatic maleimde cross-linkers on anti-CP5 ricin immunotoxins. J. Immunol. Methods 121, 129–142.

Nadeau, O.W. Carlson, G.M.,2007. Protein Interactions Captured by Chemical Cross-linking: Two-Step Cross-linking with ANB-NOS. CSH Protocols, April 2007; 2007: pdb.prot4635.

Nagai, Y., Yoshida, T., Hamaguchi, M., Linuma, M., Maeno, K., Matsumoto, T., 1978. Cross-linking of Newcastle disease virus (NDV) proteins. Arch. Virol. 58, 15–28.

Nagels, L.J., Maes, P.C., 1995. Post-column enzyme reactors for the HPLC determination of carbohydrates, in carbohydrate analysis; high performance liquid chromatography and capillary electrophoresisRassi, Z.E. Journal of Chromatography Library, vol. 58. Elsevier, Amsterdam, The Netherlands.

Nakai, T., Ono, K., Kuroda, S., Tanizawa, K., Okajima, T., 2012. An unusual subtilisin-like serine protease is essential for biogenesis of quinohemoprotein amine dehydrogenase. J. Biol. Chem. 287, 6530–6538.

Nakajima, M., Ito, N., Nishi, K., Okamura, Y., Hirota, T., 1988. Cytochemical localization of blood group substances in human salivary glands using lectin–gold complexes. J. Histochem. Cytochem. 36, 337–348.

Nakajima, N., Ikada, Y., 1995. Mechanism of amid formation by carbodiimides for bioconjugation in aqueous media. Bioconjug. Chem. 6, 123–130.

Nakamura, S., Kato, A., Kobayashi, K., 1990. Novel bifunctional lysozyme—dextran conjugate that acts on both Gram-negative and Gram-positive bacteria. Agric. Biol. Chem. 54, 3057–3059.

Nakane, P.K., 1975. Recent progress in the peroxidase-labeled antibody method. Ann. N.Y. Acad. Sci. 254, 203.

Nakane, P.K., Kawaoi, A., 1974. Peroxidase-labeled antibody. A new method of conjugation. J. Histochem. Cytochem. 22, 1084–1091.

Nakano, M., Kakehi, K., Taniguchi, N., Kondo, A., 2011. Capillary electrophoresis and capillary electrophoresis–mass spectrometry for structural analysis of N-glycans derived from glycoproteins. Capillary Electroph. Carb., 205–235.

Nakashima, N., Tomonari, Y., Murakami, H., 2002. Water-soluble single-walled carbon nanotubes via noncovalent sidewall-functionalization with a pyrene-carrying ammonium ion. Chem. Lett. 31, 638–639.

Nakayama, H., Ishihara, K., Akiba, S., Uenishi, J., 2011. Synthesis of N-[2-(2,4-difluorophenoxy)trifluoromethyl-3-pyridyl]-sulfonamides and their inhibitory activities against secretory phospholipase A2. Chem. Pharm. Bull. 59, 1069–1072.

Nalvarte, I., T. Schwend, J.-A. Gustafsson,2010. Proteomic analysis of the estrogen receptor alpha receptosome. MCP Papers in Press. Published on March 27, 2010 as Manuscript M900457-MCP200.

Namimatsu, S., Ghazizadeh, M., Sugisaki, Y., 2005. Reversing the effects of formalin fixation with citraconic anhydride and heat: a universal antigen retrieval method. J. Histochem. Cytochem. 53 (1), 3–11.

Napier, M.P., Sharma, S.K., Springer, C.J., Bagshawe, K.D., Green, A.J., Martin, J., et al., 2000. Antibodydirected enzyme prodrug therapy: efficacy and mechanism of action in colorectal carcinoma. Clin. Cancer Res. 6, 765–772.

Narayan, S.B., Pastor, J.V., Mitchison, H.M., Bennett, M.J., 2004. CLN3L, a novel protein related to the batten disease protein, is overexpressed in Cln3 mice and in batten disease. Brain 127, 1748–1754.

Nath, J., Johnson, K.L., 2000. A review of fluorescence in situ hybridization (FISH): current status and future prospects. Biotech. Histochem. 75, 54–78.

Nathan, A., Zalipsky, S., Ertel, S.I., Agathos, S.N., Yarmush, M.L., Kohn, J., 1993. Copolymers of lysine and polyethylene glycol: a new family of functionalized drug carriers. Bioconjug. Chem. 4, 54–62.

Nathan, C.F., Cohn, Z.A., 1981. Antitumor effects of hydrogen peroxide in vivo. J. Exp. Med. 154, 1539–1553.

Nayak, T.K., Garmestani, K., Baidoo, K.E., Milenic, D.E., Brechbiel, M.W., 2011. PET imaging of tumor angiogenesis in mice with VEGF-A-targeted 86Y-CHX-A"-DTPA-bevacizumab. Int. J. Cancer 128 (4), 920–926.

Neely, K.E., et al. 2002. Transcription activator interactions with multiple SWI/SNF subunits. Mol. Cell. Biol. 22 (6), 1615–1625.

Neff, S., Jungbauer, A., 2011. Monolith peptide affinity chromatography for quantification of immunoglobulin M. J. Chromatogr. A 1218 (17), 2374–2380.

Nelissen, N., Van Laere, K., Thurfjell, L., Owenius, R., Vandenbulcke, M., Koole, M., et al., 2009. Phase 1 study of the pittsburgh compound B derivative 18F-flutemetamol in healthy volunteers and patients with probable alzheimer disease. J Nucl Med. 50, 1251–1259.

Neri, D., Carnemolla, B., Nissim, A., et al., 1997. Targeting by affinity-matured recombinant antibody fragments of an angiogenesis associated fibronectin isoform. Nat. Biotechnol. 15, 1271–1275.

Ness, J.M., Akhtar, R.S., Latham, C.B., Roth, K.A., 2003. Combined tyramide signal amplification and quantum dots for sensitive and photostable immunofluorescence detection. J. Histochem. Cytochem. 51, 981–987.

Neville, D.C., Rozanas, C.R., Price, E.M., Gruis, D.B., Verkman, A.S., Townsend, R.R., 1997. Evidence for phosphorylation of serine 753 in CFTR using a novel metal-ion affinity resin and matrix-assisted laser desorption mass spectrometry. Protein Sci. 6, 2436–2445.

Newhall, J., Sawyer, W.D., Haak, R.A., 1980. Cross-linking analysis of the outer membrane proteins of neisseria gonorrhoeae. Infect. Immun. 28, 785–791.

Newkome, G.R., Moorefield, C.N., Keith, J.M., Baker, G.R., Escamilla, G.H., 1994. Chemistry of micelles. 37. Internal chemical transformations in a precursor of a unimolecular micelle: boron supercluster via site-specific addition of B10H14 to cascade molecules. Angew. Chem., Int. Ed. Engl. 33, 666–668.

Newkome, G.R., Yao, Z., Baker, G.R., Gupta, V.K., 1985. Cascade molecules: a new approach to micelles. A [27]-arborol. J. Org. Chem. 50, 2003–2004.

Ngo, T.T., 1988. Efficient coupling of amino and thiol ligands to solid-phase hydroxyl groups activated by 2-fluoro-1-methylpyrindinium salts (FMP) in Chemically modified surfaces in science and industry: Proceedings of the chemically modified surfaces symposium, Fort Collins, CO, June 17-19, pp. 49-66.

Ngo, T.T., 1989. Preparation of polymeric thiol gels for covalent bonding of biologically active ligands. U.S. Patent 4,886,755.

Ngo, T.T., Yam, C.F., Lenhoff, H.M., Ivy, J., 1981. *p*-Azidophenylglyoxal: a heterobifunctional photoactivatable cross-linking reagent selective for arginyl residues. J. Biol. Chem. 256, 11313–11318.

Nguyen, T.M. (1999) Characterization of the vitamin B-12 receptor in Salmonella typhimuium. Dissertation, Texas Tech University.

Nickel, S., Kaschani, F., Colby, T., van der Hoorn, R.A.L., Kaiser, M., 2012. A para-nitrophenol phosphonate probe labels distinct serine hydrolases of arabidopsis. Bioorg. Med. Chem. 20, 601–606.

Nicolas, J.-C., Balaguer, P., Terouanne, B., Villebrun, M.A., Boussioux, A.-M., 1992. Detection of glucose 6-phosphate dehydrogenase by bioluminescence. In: Kricka, L.J. (Ed.), Nonisotopic DNA Probe Techniques. Academic Press, New York, pp. 207.

Nicoli, R., Bartolini, M., Rudaz, S., Andrisano, V., Veuthey, J.-L., 2008. Development of immobilized enzyme reactors based on human recombinant cytochrome P450 enzymes for phase I drug metabolism studies. J. Chromatogr. A 1206, 2–10.

Nicolson, G.L., 1978. Ultrastructural localization of lectin receptors. In: Koehler, M. (Ed.), Advanced Techniques in Biological Electron Microscopy. Springer-Verlag, New York, pp. 1.

Nie, S. Bailey, R.E.,2007. Alloyed semiconductor quantum dots and concentration-gradient alloyed quantum dots, series comprising the same and methods related thereto. US Patent Application 2007/0111324 A1.

Nielsen, M.H., Bastholm, L., Chatterjee, S., Koga, J., Norrild, B., 1989. Simultaneous triple-immunogold staining of virus and host cell antigens with monoclonal antibodies of virus and host cell antigens in ultrathin cryosections. Histochemistry 92, 89–93.

Niemeyer, C.M. Methods in Molecular Biology, vol. 283. Humana Press, Totowa, New Jersey, pp. 286–293.

Niemi, P., Koskinen, S., Reisto, T., 1991. Tissue relaxation enhancement after intravenous administration of (ITCB-DTPA)-gadolinum conjugated albumin, an intravascular magnetic resonance imaging contrast agent. Invest. Radiol. 26 (7), 674–680.

Nillson, K., Mosbach, K., 1984. Immobilization of ligands with organic sulfonyl chlorides. Methods Enzymol. 104, 56–69.

Nilsen, T.W., Grayzel, J., Prensky, W., 1997. Dendritic nucleic acid structures. J. Theor. Biol. 187, 273–284.

Nilsson, B.L., Kiessling, L.L., Raines, R.T., 2000. Staudinger ligation: a peptide from a thioester and azide. Org. Lett. 2, 1939–1941.

Nilsson, B.L., Kiessling, L.L., Raines, R.T., 2001. High-yielding staudinger ligation of a phosphinothioester and azide to form a peptide. Org. Lett. 3, 9–12.

Nilsson, B.L., Hondal, R.J., Soellner, M.B., Raines, R.T., 2003a. Protein assembly by orthogonal chemical ligation methods. J. Am. Chem. Soc. 125, 5268–5269.

Nilsson, B.L., Soellner, M.B., Raines, R.T., 2003b. In: Chorev, M., Sawyer, T.K. (Eds.), Protein Assembly Using the Staudinger Ligation. Peptide Revolution: Genomics, Proteomics & Therapeutics. American Peptide Society.

Nilsson, K., Mosbach, K., 1980. p-Toluenesulfonyl chloride as an activating agent of agarose for the preparation of immobilized affinity ligands and proteins. Eur. J. Biochem. 112, 397–402.

Nilsson, K., Mosbach, K., 1981. Immobilization of enzymes and affinity ligands to various hydroxyl group carrying supports using highly reactive sulfonyl chlorides. Biochem. Biophys. Res. Commun. 102, 449–457.

Nilsson, K. Mosbach, K.,1984. Immobilization of ligands with organic sulfonyl chlorides. In: W.B. Jakoby, (Ed.),Methods in Enzymology Enzyme Purification and Related Techniques, vol. 104, Part C, Academic Press, Inc., Orlando, Florida, pp. 56-69.

Nilsson, K., Norrlöw, O., Mosbach, K., 1981. p-Toluenesulfonyl chloride as an activating agent of agarose for the preparation of immobilized affinity ligands and proteins. Optimization of conditions for activation and coupling. Acta Chem. Scand. B 35, 19–27.

Niman, H.L., Thompson, A.M.H., Yu, A., Markman, M., Willems, J.J., Herwig, K.R., et al., 1985. Anti-peptide antibodies detect oncogene-related proteins in urine. Proc. Natl. Acad. Sci. USA 82, 7924–7928.

Nirmal, M., Dabbousi, B.O., Bawendi, M.G., Macklin, J.J., Trautman, J.K., Harris, T.D., et al., 1996. Fluorescence intermittency in single cadmium selenide nanocrystals. Nature 383, 802–804.

Nishida, S., Nada, T., Terazima, M., 2004. Kinetics of intermolecular interaction during protein folding of reduced cytochrome C. Biophys. J. 87, 2663–2675.

Nishimura, Y., Adachi, H., Kyo, M., Murakami, S., Hattori, S., Ajito, K., 2005. A proof of the specificity of kanamycin–ribosomal RNA interaction with designed synthetic analogs and the antibacterial activity. Bioorg. Med. Chem. Lett. 15, 2159–2162.

Nita-Lazar, A., 2011. Quantitative analysis of phosphorylation-based protein signaling networks in the immune system by mass spectrometry. Syst. Biol. Med. 3, 368–376.

Nithipatikom, K., McGown, L.B., 1987. Homogeneous immunochemical technique for determination of human lactoferrin using excitation tranfer and phase-resolved fluorometry. Anal. Chem. 59, 423.

Nitz, M., Franz, K.J., Maglathlin, R.L., Imperiali, B., 2003. A powerful combinatorial screen to identify high-affinity terbium(III)-binding peptides. Chem. Bio. Chem. 4, 272–276.

Noel, R., Sanderson, A., Spark, L., 1993. In: Kennedy, J.F., Phillips, G.O., Williams, P.A. (Eds.), Cellulosice: Materials for Selective Separations and other Technologies. Horwood, New York, pp. 17–24.

Noguchi, A., Takahashi, T., Yamaguchi, T., Kitamura, K., Takakura, Y., Hashida, M., et al., 1992. Preparation and properties of the immunoconjugate composed of anti-human colon cancer monoclonal antibody and mitomycin C—dextran conjugate. Bioconjug. Chem. 3, 132–137.

Noh, S.M., Brayton, K.A., Brown, W.C., Norimine, J., Munske, G.R., Davitt, C.M., et al., 2008. Composition of the surface proteome of anaplasma marginale and its role in protective immunity induced by outer membrane immunization. Infect. Immun. 76, 2219–2226.

Nomura, T., Katunuma, N., 2005. Involvement of cathepsins in the invasion, metastasis and proliferation of cancer cells. J. Med. Invest. 52, 1–9.

Nopper, B., Kohen, F., Wilchek, M., 1989. A thiophilic adsorbent for the one-step high-performance liquid chromatography purification of monoclonal antibodies. Anal. Biochem. 180, 66–71.

Norde, W., 1986. Adsorption of proteins from solution at the solid–liquid interface. Adv. Colloid Interface Sci. 25, 267.

Noureddini, Hossein, Gao, X., Philkana, R.S., 2005. Immobilized pseudornonas cepacia lipase for biodiesel fuel production from soybean oil. Bioresour. Technol. 96, 769–777.

Novak-Hofer, I., Siegenthaler, P., 1978. Chemical cross-linking of neighboring thylakoid membrane polypeptides. Plant Physiol. 62, 368–372.

Novick, S., Quastel, M.R., Marcus, S., Chipman, D., Shani, G., Barth, R.F., et al., 2002. Linkage of boronated polylysine to glycoside moieties of polyclonal antibody; boronated antibodies as potential delivery agents for neutron capture therapy. Nucl. Med. Biol. 29, 93–101.

Novick, D., Orchansky, P., Revel, M., Rubenstein, M., 1987. The human interferon-g receptor. J. Biol. Chem. 262, 8483–8487.

Novitsky, T.J. Sloyer Jr., J.L.,2012. Fluorescence polarization assay for bacterial endotoxin. U.S. Patent Application 2012/0252137.

Novotna, L., Emmerova, T., Horak, D., Kucerova, Z., Ticha, M., 2010. Iminodiacetic acid-modified magnetic poly(2-hydroxyethyl methacrylate)-based microspheres for phosphopeptide enrichment. J. Chromatogr. A 1217, 8032–8040.

Nut, I., Reinhartz, A., Hyman, H.C., Razin, S., Herzberg, M., 1989. Chemiprobe, a nonradioactive system for labeling nucleic acid. Ann. Biol. Clin. 47, 601–606.

Nwagu, T.N., Okolo, B.N., Aoyagi, H., 2012. Stabilization of a raw starch digesting amylase from aspergillus carbonarius via immobilization on activated and non-activated agarose gel. World J. Microbiol. Biotechnol. 28 (1), 335–345.

Nwe, K., Milenic, D.E., Ray, G.L., Kim, Y.S., Brechbiel, M.W., 2011. Preparation of cystamine core dendrimer and antibody–dendrimer conjugates for MRI angiography. Mol. Pharm. 9 (3), 374–381.

Nyffenegger, R., Quellet, C., Ricka, J., 1993. Synthesis of fluorescent, monodisperse, colloidal silica particles. J. Colloid Interface Sci. 159, 150–157.

Nygrena, P.-Å., Skerra, A., 2004. Binding proteins from alternative scaffolds. J Immunol Methods 290, 3–28.

O'Keefe, E.T., Mordick, T., Bell, J.E., 1980. Bovine galactosyltransferase: Interaction with a-lactalbumin and the role of a-lactalbumin in lactose synthase. Biochemistry 19, 4962–4966.

O'Malley, J.J., Weaver, J.L., 1972. Subunit structure of glucose oxidase from Aspergillus niger. Biochemistry 11, 3527.

O'Shannessy, D., Dobersen, M.J., Quarles, R.H., 1984. A novel procedure for labeling immunoglobulins by conjugation to oligosaccharide moieties. Immunol. Lett. 8, 273–277.

O'Shannessy, D.J., Quarles, R.H., 1985. Specific conjugation reactions of the oligosaccharide moieties of immunoglobulins. J. Appl. Biochem. 7, 347–355.

O'Shannessy, D.J., Quarles, R.H., 1987. Labeling of the oligosaccharide moieties of immunoglobulins. J. Immunol. Meth. 99, 153–161.

O'Shannessy, D.J., Wilchek, M., 1990. Immobilization of glycoconjugates by their oligosaccharides: use of hydrazido-derivatized matrices. Anal. Biochem. 191, 1–8.

O'Shannessy, D.J., Doberson, M.J., Quarles, R.H., 1984. A novel procedure for labeling immunoglobulins by conjugation to oligosaccharide moieties. Immunol. Lett. 8, 273–277.

O'Shannessy, D.J., Voorstad, P.J., Quarles, R.H., 1987. Quantitation of glycoproteins on electroblots using the biotin—Streptavidin complex. Anal. Biochem. 163, 204–209.

O'Sullivan, M., Gnemmi, E., Morris, D., Chieregatti, G., Simmonds, A., Simmons, M., et al., 1979. Comparison of two methods of preparing enzyme—antibody conjugates: application of these conjugates for enzyme immunoassay. Anal. Biochem. 100, 100–108.

Oberlin, A., Endo, M., Koyama, T., 1976. Filamentous growth of carbon through benzene decomposition. J. Cryst. Growth. 32, 335–349.

Obrist, R., Schmidli, J., Obrecht, J.P., 1988. Chemotactic monoclonal antibody conjugates: a comparison of four different f-Met—Peptide conjugates. Biochem. Biophys. Res. Comm. 155, 1139–1144.

O'Carra, P., Barry, S., 1972. Affinity chromatography of lactate dehydrogenase Model studies demonstrating the potential of the technique in the mechanistic investigation as well as in the purification of multi-substrate enzymes. FEBS Lett. 21, 281–285.

O'Carra, P., Barry, S., Griffin, T., 1974. Spacer arms in affinity chromatography: use of hydrophilic arms to control or eliminate nonspecific adsorption effects. FEBS Lett. 43, 169–175.

Oda, Y., Nagasu, T., Chait, B.T., 2001. Enrichment analysis of phosphorylated proteins as a tool for probing the phosphoproteome. Nature Biotech. 19, 379–382.

Odom Jr., O.W., Robins, D.J., Lynch, J., Dottavio-Martin, D., Kramer, G., Hardesty, B., 1980. Distances between 39 ends of ribosomal ribonucleic acids reassembled into *Escherichia coli* ribosomes. Biochemistry 19, 5947–5954.

Odom, O.W., Dabbs, E.R., Dionne, C., Muller, M., Hardesty, B., 1984. The distance between Si, S21, and the 39 end of 16S RNA in 30S ribosomal subunits. The effect of poly(uridylic acid) and 50S subunits on these distances. Eur. J. Biochem. 142, 261.

Odom, O.W., Picking, W.D., Hardesty, B., 1990. Movement of tRNA but not the nascent peptide during peptide bond formation on ribosomes. Biochemistry 29, 10734–10744.

Odorico, M., Teulon, J.-M., Bessou, T., Vidaud, C., Bellanger, L., Chen, S.-W., et al., 2007. Energy landscape of chelated uranyl: antibody interactions by dynamic force spectroscopy. Biophys. J. 93, 645–654.

Oeltmann, T.N., 1985. Synthesis and in vitro activity of a hormone-diphtheria toxin fragment a hybrid. Biochem. Biophys. Res. Comm. 133, 430.

Oeltmann, T.N., Forbes, J.T., 1981. Inhibition of mouse spleen cell function by diphtheria toxin fragment A coupled to anti-mouse Thy-1.2 and by ricin A chain coupled to anti-mouse IgM. Arch. Biochem. Biophys. 209, 362.

Ogawa-Goto, K., Tanaka, K., Ueno, T., Tanaka, K., Kurata, T., Sata, T., et al., 2007. p180 is involved in the interaction between the endoplasmic reticulum and microtubules through a novel microtubule-binding and bundling domain. Mol. Biol. Cell, doi: 10.1091/mbc. E06-12-1125.

Ogura, H., Kobayashi, T., Shimizu, K., Kawabe, K., Takeda, K., 1979. A novel active ester synthesis reagent (N,N′-disuccinimidyl carbonate. Tetrahedron Lett. 20, 4745–4746.

Oi, V.T., Glazer, A.N., Stryer, L., 1982. Fluorescent phycobiliprotein conjugates for analyses of cells and molecules. J. Cell Biol. 93, 981–986.

Okerberg, ES, Wu, J, Zhang, B, Samii, B, Blackford, K, Winn, D.T., et al., 2005. High-resolution functional proteomics by active-site peptide profiling. Proc. Natl. Acad. Sci. USA 102, 4996–5001.

Okuda, K., Urabe, I., Yamada, Y., Okada, H., 1991. Reaction of glutaraldehyde with amino and thiol compounds. J. Ferment. Bioeng. 71, 100–105.

Olajos, M., Szekroenyes, A., Hajos, P., Gjerde, D.T., Guttman, A., 2010. Boronic acid lectin affinity chromatography (BLAC). 3. Temperature dependence of glycoprotein isolation and enrichment. Anal. Bioanal. Chem. 397, 2401–2407.

Olekhnovich, I.N., Kadner, R.J., 2002. Mutational scanning and affinity cleavage analysis of UhpA-binding sites in the *Escherichia coli* uhpT promoter. J. Bacteriol. 184, 2682–2691.

Olenjnik, J., Sonar, S., Krzymanska-Olejnik, E., Rothschild, K.J., 1995. Photocleavable biotin derivatives: a versatile approach for the isolation of biomolecules. Proc. Natl. Acad. Sci. USA 92, 7590–7594.

Olsnes, S., 1978. Binding, entry, and action of abrin, ricin, and modeccin. In: Silverstein, S.C. (Ed.), Transport of Macromolecules in Cellular Systems. Dahlem Konferenzen, Berlin, pp. 103–116.

Olsnes, S., Pihl, A., 1976. Abrin, ricin, and their associated agglutinins. In: Cuatrecasas, P., (Ed.), The Specificity of Animal, Bacterial and Plant Toxins. Receptors and Recognition, Series B, vol. 1. Chapman & Hall, London, pp. 129–173.

Olsnes, S., Pihl, A., 1982a. Cytotoxic proteins with intracellular site of action: mechanism of action and anti-cancer properties. Cancer Surv. 3, 467–487.

Olsnes, S., Pihl, A., 1982b. Chimeric toxins. Pharmacol. Ther. 15, 355.

Olsnes, S., Pihl, A., 1982c. Toxic lectins and related proteins. In: Cohen, P., van Heynigen, S. (Eds.), Molecular Action of Toxins and Viruses. Elsevier, New York, pp. 51.

Olson, E., Nievera, C.J., Liu, E., Yueh-Luen Lee, A., Chen, L., Wu, X., 2007. The Mre11 complex mediates the S-phase checkpoint through an interaction with RPA. Mol. Cell. Biol.10.1128/MCB.00532-07.

Olsson, L., Carl Fredrik Mandenius, J.V., 1990. Determination of monosaccharides in cellulosic hydrolyzates using immobilized pyranose oxidase in a continuous amperometric analyzer. Anal. Chem. 62 (24), 2688–2691.

Omer-Mizrahi, M., Margel, S., 2009. Synthesis and characterization of magnetic and non-magnetic core-shell polyepoxide micrometer-sized particles of narrow size distribution. J. Colloid Interface Sci. 329 (2), 228–234.

Ondetti, M.A., Thomas, P.L., 1965. Synthesis of a peptide lactone related to vernamycin Ba. J. Am. Chem. Soc. 87, 4373–4380.

Ong, S.-E., Blagoev, B., Kratchmarova, I., Kristensen, D.B., Steen, H., Pandey, A., et al., 2002. Stable isotope labeling by amino acids in cell culture, SILAC, as a simple and accurate approach to expression proteomics. Mol. Cell. Proteomics 1, 376–386.

Oparka, K.J., Hawes, C., 2011. Vacuolar sequestration of fluorescent probes in plant cells: a review. J. Microsc. 166 (1), 15–27.

Oparka, K.J., Murant, E.A., Wright, K.M., Prior, D.A.M., Harris, N., 1991. The drug probenecid inhibits the vacuolar accumulation of fluorescent anions in onion epidermal cells. J. Cell Sci. 99, 557–563.

Opperman, K., Kaboord, B.J., Schultz, J.-S., Etienne, C.L., Hermanson, G.T., 2011. Modified nucleotides. U.S. Patent Application No. 2011/0262917.

Order, S.E., 1982. Monoclonal antibodies potential in radiation therapy and oncology. Int. J. Radiat. Oncol. Biol. Phys. 8, 1193–1201.

Order, S.E., 1989. Therapeutic use of radioimmunoconjugates. Antibody, Immunoconjugates, Radiopharm. 2, 235.

Orlandi, R., Gussow, D.H., Jones, P.T., Winter, G., 1989. Cloning immunoglobulin variable domains for expression by the polymerase chain reaction. Proc. Natl. Acad. Sci. USA 86, 3833–3837.

Orlando, V., 2000. Mapping chromosomal proteins in vivo by formaldehyde-crosslinked-chromatin immunoprecipitation. Trends Biochem. Sci. 3, 99–104.

Orlando, V., Strutt, H., Paro, R., 1997. Analysis of chromatin structure by in vivo formaldehyde cross-linking. Methods 11, 205–214.

Ormerod, M.G. (Ed.), 1990. A Practical Approach. IRL Press, New York.

Orr, G.A., 1981. The use of the 2-iminobiotin—Avidin interaction for the selective retrieval of labeled plasma membrane components. J. Biol. Chem. 256, 761–766.

Ortega-Munoz, M., Morales-Sanfrutos, J., Megia-Fernandez, A., Lopez-Jaramillo, F.J., Hernandez-Mateo, F., Santoyo-Gonzalez, F., 2010. Vinyl sulfone functionalized silica: a "ready to use" pre-activated material for immobilization of biomolecules. J. Mater. Chem. 20, 7189–7196.

Osborn, M., Weber, K., 1982. Immunofluorescence and immunocytochemical procedures with affinity purified antibodies: tubulin-containing structures. Meth. Cell Biol. 24, 97–132.

Oser, A., Roth, W.K., Valet, G., 1988. Sensitive non-radioactive dot-blot hybridization using DNA probes labeled with chelate group substituted psoralen and quantitative detection by europium ion fluorescence. Nucleic Acids Res. 16, 1181–1196.

Ota, S., Miyazaki, S., Matsuoka, H., Morisato, K., Shintani, Y., Nakanishi, K., 2007. High-throughput protein digestion by trypsin-immobilized monolithic silica with pipette-tip formula. J. Biochem. Biophys. Methods 70, 57–62.

Otsuka, F.L., Welch, M.J., 1987. Methods to label monoclonal antibodies for use in tumor imaging. Nucl. Med. Biol. 14, 243–249.

Otsuka, H., Nagasaki, Y., Kataoka, K., 2012. PEGylated nanoparticles for biological and pharmaceutical applications. Adv. Drug Deliv. Rev.10.1016/j.addr.2012.09.022 (in press).

Ottersbach, K., Graham, G.J., 2001. Aggregation-independent modulation of proteoglycan binding by neutralization of C-terminal acidic residues in the chemokine macrophage inflammatory protein 1alpha. Biochem. J. 354, 447–453.

Otto, S., Blandamer, M.J., Engberts, J.B.F.N., 1996. J. Am. Chem. Soc. 118, 7702–7707.

Ow, H., Larson, D.R., Srivastava, M., Baird, B.A., Webb, W.W., Wiesner, U., 2005. Bright and stable core-shell fluorescent silica nanoparticles. Nano Lett. 5, 113–117.

Owens, J.T., Miyake, R., Murakami, K., Chmura, A.J., Fujita, N., Ishihama, A., et al., 1998. Mapping the sigma(70) subunit contact sites on *Escherichia coli* RNA polymerase with a sigma(70)-conjugated chemical protease. Proc. Nat. Acad. Sci. USA 95, 6021–6026.

Ozawa, H., 1967. Bridging reagent for protein. II. The reaction of N,N′-polymethylenebis(iodoacetamide) with cysteine and rabbit muscle aldolase. J. Biochem. (Tokyo) 62, 531.

Ozawa, S., Ueda, M., Ando, N., Abe, O., Minoshima, S., Shimizu, N., 1989. Selective killing of squamous carcinoma cells by an immunotoxin that recognizes the EGF receptor. Int. J. Cancer 43, 152.

Pace, B., Pace, N.R., 1980. The chromatography of RNA and oligoribonucleotides on boronate-substituted agarose and polyacrylamide. Anal. Biochem. 107, 128–135.

Packman, L.C., Perham, R.N., 1982. Quaternary structure of the pyruvate dehydrogenase multienzyme complex of *Bacillus stearothermophilus* studied by a new reversible cross-linking procedure with bis(imidoesters). Biochemistry 21, 5171–5175.

Paganelli, G., Riva, P., Deleide, G., Clivio, A., Chiolerio, F., Scassellati, G.A., et al., 1988. In vivo labeling of biotinylated monoclonal antibodies by radioactive avidin: a strategy to increase tumor radiolocalization. Int. J. Cancer 2, 121–125.

Pagano, J.M., Farley, B.M., McCoig, L.M., Ryder, S.P., 2007. Molecular basis of RNA recognition by the embryonic polarity determinant MEX-5. J. Biol. Chem. 282, 8883–8894.

Pai, C.K., Smith, M.B., 1995. Rate enhancement in dilute salt solutions of aqueous ethanol: the Diels-Alder reaction. J. Org. Chem. 60, 3731–3735.

Pal, G., Santamaria, F., Kossiakoff, A.A., Lu, W., 2003. The first semi-synthetic serine protease made by native chemical ligation. Protein Expr. Purif. 29 (2), 185–192.

Palmer, J.L., Nissonoff, A., 1963. Reduction and reoxidation of a critical disulfide bond in the rabbit antibody molecule. J. Biol. Chem. 238, 2393–2398.

Pan, Y., Zhang, Y., Jia, T., Zhang, K., Li, J., Wang, L., 2012. Development of a microRNA delivery system based on bacteriophage MS2 virus-like particles. FEBS J. 279, 1198–1208.

Pan, Z., Jeffery, D.A., Chehade, K., Beltman, J., Clark, J.M., Grothaus, P., et al., 2006. Development of activity-based probes for trypsin-family serine proteases. Bioorg. Med. Chem. Lett. 16, 2882–2885.

Pandurangi, R.S., et al. 1995a. Photolabeling of human serum albumin by 4-azido-2-(14C-Methylamino) trifluorobenzonitrile.

A high-efficiency long wavelength photolabel. App. Rad. Isot. 46 (4), 233–239.

Pandurangi, R.S., et al. 1995b. High efficiency photolabeling of human serum albumin and human g-globulin with [^{14}C]Methyl 4-azido-2,3,5,6-tetrafluorobenzoate. Bioconjug. Chem. 6, 630–634.

Pandurangi, R.S., et al. 1996. Preservation of immunoreactivity in the photolabeling of B72.3 human antibody. Photochem. Photobiol. 64 (1), 100–105.

Pandurangi, R.S., et al. 1997a. Recent trends in the evaluation of photochemical insertion characteristics of heterobifunctional perfluoroaryl azide chelating agents: biochemical implications in nuclear medicine. Photochem. Photobiol. 65 (2), 208–221.

Pandurangi, R.S., et al. 1997b. Chemistry of bifunctional photoprobes: Part 1. Perfluoro azido functionalized phosphorus hydrazides as novel photoreactive heterobifunctional chelating agents: high efficiency nitrene insertion on model solvents and proteins. J. Org. Chem. 62 (9), 2798–2807.

Pandurangi, R.S., et al. 1998. Chemistry of bifunctional photoprobes: Part 4. Synthesis of the chromogenic, cleavable, water soluble and heterobifunctional (N-Methyl amino perfluoroaryl azide benzamido)-ethyl-1,3-dithiopropionyl sulfosuccinimide: an efficient protein cross-linking agent. Bioorg. Chem. 26 (4), 201–212.

Pansare, V.J., Hejazi, S., Faenza, W.J., Prud'homme, R.K., 2012. Review of long-wavelength optical and NIR imaging materials: contrast agents, fluorophores, and multifunctional nano carriers. Chem. Mater. 24, 812–827.

Paredes, E., Evans, M., Das, S.R., 2011. RNA labeling, conjugation and ligation. Methods 54 (2), 251–259.

Park, B., Lee, S., Kim, E., Ahn, K., 2003. A single polymorphic residue within the peptide-binding cleft of MHC class I molecules determines spectrum of tapasin dependence. J. Immunol. 170, 961.

Park, J.W., Melisko, M.E., Esserman, L.J., Jones, L.A., Wollan, J.B., Sims, R., 2007. Treatment with autologous antigen-presenting cells activated with the HER-2–based antigen lapuleucel-T: results of a phase I study in immunologic and clinical activity in HER-2–overexpressing breast cancer. J. Clin. Oncol. 25, 3680–3687.

Park, K.D., Liu, R., Kohn, H., 2009. Useful tools for biomolecule isolation, detection, and identification: acylhydrazone-based cleavable linkers. Chem. Biol. 16, 763–772.

Park, K.H., Chhowalla, M., Iqbal, Z., Sesti, F., 2003. Single-walled carbon nanotubes are a new class of ion channel blockers. J. Biol. Chem. 278, 50212–50216.

Park, L.S., Friend, D., Gillis, S., Urdal, D.L., 1986. Characterization of the cell surface receptor for a multi-lineage colony-stimulating factor (CSF-2a). J. Biol. Chem. 261, 205–210.

Park, Y.S., Huang, L., 1992. Cryoprotective activity of synthetic glycophospholipids and their interactions with trehalose. Biochim. Biophys. Acta 1124, 241–248.

Parker, D.J., Allison, W.S., 1969. The mechanism of inactivation of glyceraldehyde 3-phosphate dehydrogenase by tetrathionate, o-iodosobenzoate, and iodine monochloride. J. Biol. Chem. 244, 180–189.

Parsons, B.J., 1980. Psoralen photochemistry. Photochem. Photobiol. 32, 813–821.

Partis, M.D., Griffiths, D.G., Roberts, G.C., Beechey, R.B., 1983. Cross-linking of protein by w-maleimido alkanoyl *N*-hydroxysuccinimido esters. J. Protein Chem. 2, 263–277.

Pascual, A., Casanova, J., Samuels, H.H., 1982. Photoaffinity labeling of thyroid hormone nuclear receptors in intact cells. J. Biol. Chem. 257, 9640–9647.

Pascual, J.M., Karlin, A., 1998. State-dependent accessibility and electrostatic potential in the channel of the acetylcholine receptor. Inferences from rates of reaction of thiosulfonates with substituted cysteines in the M2 segment of the alpha subunit. J. Gen. Physiol. 111 (6), 717–739.

Pashov, A, Monzavi-Karbassi, B, Raghava, GP, Kieber-Emmons, T., 2010. Bridging innate and adaptive antitumor immunity targeting glycans. J. Biomed. Biotechnol. 2010 Article ID: 354068, 19 pages.

Pastan, I., Hassan, R., Fitzgerald, D.J., Kreitman, R.J., 2006. Immunotoxin therapy of cancer. Nat. Rev. Cancer 6 (7), 559–565.

Patel, T., Bruce, J., Merry, A.H., Bigge, J.C., Wormald, M.R., Parekh, R.B., 1992. The use of hydrazine to release in intact and unreduced form both N- and O-linked oligosaccharides from glycoproteins. Biochemistry 32, 679–693.

Pathak, M.A., 1984. Mechanisms of psoralen photosensitization reactions. J. Natl. Cancer Inst. Monogr. 66, 41–46.

Pathy, L., Smith, E.L., 1975. Reversible modification of arginine residues: application to sequence studies by restriction of tryptic hydrolysis to lysine residues. J. Biol. Chem. 250, 557.

Patri, A.K., Majoros, I.J., Baker Jr., J.R., 2002. Curr. Opin. Chem. Biol. 6, 466.

Patri, A.K., Myc, A., Beals, J., Thomas, T.P., Bander, N.H., Baker Jr., J.R., 2004. Synthesis and in vitro testing of J591 antibody-dendrimer conjugates for targeted prostate cancer therapy. Bioconjug. Chem. 15, 1174–1181.

Patricelli, M.P., Giang, D.K., Stamp, L.M., Burbaum, J.J., 2001. Direct visualization of serine hydrolase activities in complex proteomes using fluorescent active site-directed probes. Proteomics 1, 1067–1071.

Patricelli, M.P., Szardenings, A.K., Liyanage, M., Nomanbhoy, T.K., Wu, M., Weissig, H., et al., 2007. Functional Interrogation of the kinome using nucleotide acyl phosphates. Biochemistry 46, 350–358.

Patsenker, L.D., Tatarets, A.L., Povrozin, Y.A., Terpetschnig, E.A., 2011. Long-wavelength fluorescence lifetime labels. Bioanal. Rev., 1–23.

Pattnaik, A., 2011. Synthesis and characterization of amine-functionalized magnetic silica nanoparticle. Doctoral Dissertation, National Institute of Technology, Rourkela, Orissa.

Paul, R., Anderson, G.W., 1960. N,N′-Carbonyldiimidazole, a new peptide forming reagent. J. Am. Chem. Soc. 82, 4596–4600.

Paul, R., Anderson, G.W., 1962. N,N′-Carbonyldiimidazole in peptide synthesis. III. A synthesis of isoleucine-5-angiotensin II amide-1. J. Org. Chem. 27, 2094–2099.

Pawley, J., Albrecht, R., 1988. Imaging colloidal gold labels in LVSEM. Scan. Microsc. 10, 184–189.

Pawson, T., Nash, P., 2003. Assembly of cell regulatory systems through protein interaction domains. Science 300, 445–452.

Pearson, K.M., et al. 2002. Intramolecular cross-linking experiments on cytochrome C and ribonuclease A using an isotope multiplet method. Rapid Comm. Mass Spectrom. 16, 149–159.

Pearson, R.G., Sobel, H., Songstad, J., 1968. Nucleophilic reactivity constants toward methyl iodide and trans-[Pt(py)2Cl2]. J. Am. Chem. Soc. 90, 319–326.

Pedersen, J.,C., Lauritzen, M.T., Madsen, S.W., 1999. Dahl, Removal of N-terminal polyhistidine tags from recombinant proteins using engineered aminopeptidases. Protein Expr. Purif. 15, 389–400.

Pedersen, M.K., Sorensen, N.S., Heegaard, P.M., Beyer, N.H., Bruun, L., 2006. Effect of different hapten-carrier conjugation ratios and molecular orientations on antibody affinity against a peptide antigen. J. Immunol. Meth. 311 (1–2), 198–206.

Peeters, J.M., Hazendonk, T.G., Beuvery, E.C., Tesser, G.I., 1989. Comparison of four bifunctional reagents for coupling peptides to proteins and the effect of the three moieties on the immunogenicity of the conjugates. J. Immunol. Meth. 120, 133–143.

Peethambaram, P.P., Melisko, M.E., Rinn, K.J., Alberts, S.R., Provost, N.M., Jones, L.A., et al., 2009. A Phase I Trial of Immunotherapy with Lapuleucel-T (APC8024) in Patients with Refractory Metastatic Tumors that Express HER-2/neu. Clin. Cancer Res. 15, 5937–5944.

Peltola, H, 2000. Worldwide Haemophilus influenzae type b disease at the beginning of the 21st century: global analysis of the disease burden 25 years after the use of the polysaccharide vaccine and a decade after the advent of conjugates. Clin. Microbiol. Rev. 13, 302–317.

Peluso, S., Ufret, M. de L., O'Reilly, M.K., Imperiali, B., 2002. Neoglycopeptides as inhibitors of oligosaccharyl transferase: insight into negotiating product inhibition. Chem. Biol. 9, 1323–1328.

Peña, M.M.O., Puig, S., Thiele, D.J., 2000. Characterization of the Saccharomyces cerevisiae high affinity copper transporter Ctr3. J. Biol. Chem. 275, 33244–33251.

Penchovsky, R., Birch-Hirschfeld, E., McCaskill, J.S., 2000. End-specific covalent photo-dependent immobilisation of synthetic DNA to paramagnetic beads. Nucleic Acids Res. 28, e98.

Peng, H.-H., Chen, J.-W., Yang, T.-P., Kuo, C.-F., Wang, Y.-J., Lee, M.-W., 2011. Polygalacturonic acid hydrogel with short-chain hyaluronate cross-linker to prevent postoperative adhesion. J. Bioact. Compat. Polym. 26, 552–564.

Peng, L., Calton, G.J., Burnett, J.W., 1987. Effect of borohydride reduction on antibodies. Appl. Biochem. Biotechnol. 14, 91–99.

Pennathur-Das, R., Heath, R., Mentzer, W.C., Lubin, B., 1982. Modification of hemoglobin s with dimethyl adipimidate. Contribution of individual reacted subunits to changes in properties. Biochim. Biophys. Acta 704, 389–397.

Pepinsky, R.B., Cappiello, D., Wilkowski, C., Vogt, V.M., 1980. Chemical crosslinking of proteins in avian sarcoma and leukemia viruses. Virology 102, 205–210.

Pepper, D.S., 1994. Some alternative coupling chemistries for affinity chromatography. Mol. Biotechnol. 2, 157–178.

Percec, V., Cho, W.D., Mosier, P.E., Ungar, G., Yeardley, D.J.P., 1998. Structural analysis of cylindrical and spherical supramolecular dendrimers quantifies the concept of monodendron shape control by generation number. J. Am. Chem. Soc. 120, 11061–11070.

Perez-Almodovar, E.X., Carta, G., 2009. IgG adsorption on a new protein A adsorbent based on macroporous hydrophilic polymers. I. Adsorption equilibrium and kinetics. J. Chromatogr. A 1216, 8339–8347.

Perham, R.N., Jones, G.M.T., 1967. The determination of the order of lysine-containing tryptic peptides of proteins by diagonal paper electrophoresis. Eur. J. Biochem. 2, 84–89.

Perham, R.N., Thomas, J.O., 1971. Reaction of tobacco mosaic virus with a thiol-containing imidoester and a possible application to X-ray diffraction analysis. J. Mol. Biol. 62, 415–418.

Perler, F.B., 2000. InBase, the intein database. Nucleic Acids Res. 28, 344–345.

Perrakis, A., Musacchio, A., Cusack, S., Petosa, C., 2011. Investigating a macromolecular complex: the toolkit of methods. J. Struct. Biol. 175 (2), 106–112.

Petach, H.H., Henderson, W., Olsen, G.M., 1994. P(CH2OH)3—A new coupling reagent for the covalent immobilisation of enzymes. J. Chem. Soc., Chem. Comm., 2181–2182.

Peters, K., Richards, F.M., 1977. Chemical cross-linking: reagents and problems in studies of membrane structure. Annu. Rev. Biochem. 46, 523–551.

Petrotchenko, E.V., Olkhovik, V.K., Borchers, C.H., 2005. Isotopically coded cleavable cross-linker for studying protein–protein interaction and protein complexes. Mol. Cell. Proteom. 4 (8), 1167–1179.

Petrou, P.S., Georgiou, S., Christofidis, I., Kakabakos, S.E., 2002. Increased sensitivity of heterogeneous fluoroimmunoassays employing fluorescein-labeled antibodies by simple treatment of the wells with glycerin solution. J. Immunol. Meth. 266, 175–179.

Petruzzelli, L., Herrer, R., Garcia-Arenas, R., Rosen, R.M., 1985. Acquisition of insulin-dependent protein tyrosine kinase activity during Drosophila embryogenesis. J. Biol. Chem. 226, 16072–16075.

Pfeuffer, E., Dreher, R.-M., Pfeuffer, T., 1985. Catalytic unit of adenylate cyclase purification and identification by affinity cross-linking. Proc. Natl. Acad. Sci. USA 82, 3086–3090.

Phillips, J.H., Robrish, S.T., Bates, C., 1965. High efficiency coupling of diazonium ions to proteins and amino acids. J. Biol. Chem. 240, 699–704.

Philomin, V., Vessieres, A., Jaouen, G., 1994. New applications of carbonylmetallo-immunoassay (CMIA): a nonradioisotopic approach to cortisol assay. J. Immunol. Meth. 171, 201–210.

Phizicky, E.M., Fields, S., 1995. Protein–protein interactions: methods for detection and analysis. Microbiol. Rev. 59, 94–123.

Pichot, C., Charleux, B., Charreyre, M.T., Revilla, J., 2011. Recent developments in the design of functionalised polymeric microspheres. In Macromolecular Symposia, vol. 88, No. 1. Hüthig & Wepf Verlag, pp. 71–87.

Pick, U., 1981. Liposomes with a large trapping capacity prepared by freezing and thawing of sonicated phospholipid mixtures. Arch. Biochem. Biophys. 212, 186.

Pidgeon, C., Hunt, A.H., Dittrich, K., 1986. Formation of multilayered vesicles from water/organic-solvent (W/O) emulsions: theory and practice. Pharm. Res. 3, 23–34.

Piehler, J., 2005. New methodologies for measuring protein interactions in vivo and in vitro. Curr. Opin. Struct. Biol. 15 (1), 4–14.

Piehler, J., Brecht, A., Valiokas, R., Liedberg, B., Gauglitz, G., 2000. A high-density poly(ethylene glycol) polymer brush for immobilization on glass-type surfaces. Biosens. Bioelectron. 15, 473–481.

Pier, G.B., Lyczak, J.B., Wetzler, L.M., 2004. Immunology, Infection, and Immunity. ASM Press, ISBN 1-55581-246-5.

Pietraszkiewicz, M., Karpiuk, J., Rout, A.K., 1993. Lanthanide complexes of macrocyclic and macrobicyclic N-oxides; light-converting supramolecular devices. Pure Appl. Chem. 65 (3), 563–566.

Pihl, A., Lange, R., 1962. The interaction of oxidized glutathione, cystamine mono-sulfoxide, and tetrathionate with the –SH groups of rabbit muscle D-glyceraldehyde 3-phosphate. J. Biol. Chem. 237, 1356–1362.

Pikuleva, I.A., Turko, I.V., 1989. A new method of preparing hemin conjugate with rabbit IgC. Bioorg. Khim. 15, 1480.

Pillai, V.N.R., Mutter, M., 1980. New, easily removable polyethylene glycol supports for liquid phase method of peptide synthesis. J. Org. Chem. 45, 5364–5367.

Pimm, M.V., Raiput, R.S., Frier, M., Cribben, S.J., 1991. Anomalies in reduction-mediated technetium-99m labeling of monoclonal antibodies. Eur. J. Nucl. Med. 18, 973–976.

Pinaud, F., Clarke, S., Assa Sittner, A., Dahan, M., 2010. Probing cellular events, one quantum dot at a time. Nature Methods 7, 275–285.

Pineiro, M., Ribeiro, S.M., Serra, A.C., 2010. The influence of the support on the singlet oxygen quantum yields of porphyrin supported photosensitizers. ARKIVOC 2010, ARKAT USA, Inc., p. 51–63. ISSN: 1551-7012.

Pinkel, D., Landegent, J., Collins, C., Fuscoe, J., Segraves, R., Lucas, J., et al., 1988. Fluorescence in situ hybridization with human chromosome-specific libraries: detection of trisomy 21 and translocations of chromosome 4. Proc. Natl. Acad. Sci. USA 85, 9138–9142.

Pirker, R., Fitzgerald, D.J.P., Hamilton, T., Ozols, R.F., Laird, W., Frankel, A.E., et al., 1986. Characterization of immunotoxins active against ovarian cancer cell lines. J. Clin. Invest. 76, 1261.

Pitt, M., Gallegos, M.-T., Buck, M., 2000. Single amino acid substitution mutants of Klebsiella pneumoniae 54 defective in transcription. Nucleic Acids Res. 28, 4419–4427.

Plank, L., Ware, B.R., 1987. Acanthamoeba profiln binding to fluorescein-labelled actin. Biophys. J. 51, 985.

Plant, A.L., Brizgys, M.V., Lacasio-Brown, L., Durst, R.A., 1989. Generic liposome reagent for immunoassays. Anal. Biochem. 176, 420–426.

Plapp, B.V., Raftery, M.A., Cole, R.D., 1967. The tryptic digestion of S-aminoethylated ribonuclease. J. Biol. Chem. 242, 265–270.

Plieva, FM, Mattiasson, B, 2008. Macroporous gel particles as novel sorption medium: rational design. Ind. Eng. Chem. Res. 47, 4131–4141.

Plöchl, J., 1884. Über einige Derivate der Benzoylimdozimtsäure. Ber. 17, 1616–1624.

Ploem, J.S., Tanke, H.J., 1987. Introduction to Fluorescence Microscopy. Oxford University Press, London.

Plotz, P.H., Rifai, A., 1982. Stable, soluble, model immune complexes made with a versatile multivalent affinity-labeling antigen. Biochemistry 21, 301.

Plückthun, A., 2009. Alternative scaffolds: expanding the options of antibodies. In: Little, M. (Ed.), Recombinant Antibodies for Immunotherapy. Cambridge Press, pp. 243–272. ISBN: 9780521887328, (Chapter 18).

Plueddemann, E.P., 1991. Silane Coupling Agents, second ed., Plenum Press, NY.

Podhradsky, D., Drobnica, L., Kristian, P., 1979. Reactions of cysteine, its derivatives, glutathione, coenzyme A, and dihydrolipoic acid with isothiocyanates. Experientia 35, 154.

Poethko, T., Schottelius, M., Thumshirn, G., Hersel, U., Herz, M., Henriksen, G., et al., 2004. Two-step methodology for high-yield routine radiohalogenation of peptides: 18F-Labeled RGD and octreotide analogs. J. Nucl. Med. 45, 892–902.

Pojer, P.M., 1979. Reduction of imines and cleavage of oximes by sodium dithionite. Aust. J. Chem. 32, 201–204.

Politz, S.M., Noller, H.F., McWhirter, P.D., 1981. Ribonucleic acid-protein cross-linking in Escherichia coli ribosomes: (4-azidophenyl) glyoxal, a novel heterobifunctional reagent. Biochemistry 20, 372–378.

Pollard, A.J., Perrett, K.P., Beverley, P.C., 2009. Maintaining protection against invasive bacteria with protein–polysaccharide conjugate vaccines. Nature Rev., Immunol. 9, 213–220.

Polson, A.G., Yu, S.-F., Elkins, K., Zheng, B., Clark, S., Ingle, G.S., et al., 2007. Antibody-drug conjugates targeted to CD79 for the treatment of non-Hodgkin lymphoma. Blood 110, 616–623.

Porath, J., 1974. General methods and coupling procedures. Meth. Enzymol. 34, 13–30.

Porath, J., 1976. General methods and coupling proceduresJakoby, W.B.Wilchek, M. Methods in Enzymology, vol. 34. Academic Press, New York, pp. 13.

Porath, J., 1987. Metal ion – hydrophobic, thiophilic and II-electron governed interactions and their application to salt-promoted protein adsorption chromatography. Biotechnol. Prog. 3, 14–21.

Porath, J., 1992. Immobilized metal ion affinity chromatography. Protein Expr. Purif. 3, 263–281.

Porath, J., Axén, R., 1976. Immobilization of enzymes to agar, agarose, and sephadex supportMosbach, K. Methods in Enzymology, vol. 44. Academic Press, New York, NY, pp. 19–45.

Porath, J., Belew, M., 1987. 'Thiophilic' interaction and the selective adsorption of proteins. Trends Biotechnol. 5, 225–229.

Porath, J., Olin, B., 1983. Immobilized metal ion affinity adsorption and immobilized metal ion affinity chromatography of biomaterials. Serum protein affinities for gel-immobilized iron and nickel ions. Biochemistry 22, 1621–1630.

Porath, J., Carlsson, J., Olsson, I., Belfrage, G., 1975a. Metal chelate affinity chromatography, a new approach to protein fractionation. Nature (London) 258, 598–599.

Porath, J., Laas, T., Janson, J.C., 1975b. Agar derivatives for chromatography, electrophoresis, and gel-bound enzymes. 111. Rigid agarose gels cross-linked with divinyl sulfone (DVS). J. Chromatogr. 103, 49–62.

Porath, J., Maisano, F., Belew, M., 1985. Thiophilic adsorption – a new method for protein fractionation. FEBS Lett. 185, 306–310.

Porstmann, B., Porstmann, T., Nugel, E., Evers, U., 1985. Which of the commonly used marker enzymes gives the best results in colorimetric and fluorimetric enzyme immunoassays: horseradish peroxidase, alkaline phosphatase or b-galactosidase. J. Immunol. Meth. 79, 27–37.

Portnoy, E, Lecht, S, Lazarovici, P, Danino, D, Magdassi, S., 2011. Cetuximab-labeled liposomes containing near-infrared probe for in vivo imaging. Nanomedicine January 25 [Epub ahead of print].

Pöselt, E., Schmidtke, C., Fischer, S., Peldschus, K., Salamon, J., Kloust, H., et al., 2012. Tailor-made quantum dot and iron oxide based contrast agents for in vitro and in vivo tumor imaging. ACS Nano 6 (4), 3346–3355.

Posewitz, M.C., Tempst, P., 1999. Immobilized gallium(III) affinity chromatography of phosphopeptides. Anal. Chem. 71 (14), 2883–2892.

Posnett, D.N., McGrath, H., Tam, J.P., 1988. A novel method for producing anti-peptide antibodies: production of site-specific antibodies to the T-cell antigen receptor b-chain. J. Biol. Chem. 263, 1719–1725.

Pow, D.V., Crook, D.K., 1993. Extremely high titre polyclonal antisera against small neurotransmitter molecules: rapid production, characterization and use in light- and electron-microscopic immunocytochemistry. J. Neurosci. Methods 48, 51–63.

Pow, D.V., Morris, J.F., 1991. Membrane routing during exocytosis and endocytosis in neuroendocrine neurons and endocrine cells: use of colloidal gold particles and immunocytochemical discrimination of membrane compartments. Cell Tissue Res. 264, 299–316.

Powsner, E.R., 1994. Basic principles of radioactivity and its measurement. In: Burtis, C.A., Ashwood, E.R. (Eds.), Tietz Textbook of Clinical Chemistry. Saunders, Pennsylvania, Philadelphia, pp. 256–282.

Pozsgay, V., 1998. Synthesis of glycoconjugate vaccines against Shigella dysenteriae type 1. J. Org. Chem. 63, 5983–5999.

Pozsgay, V., Vieira, N.E., Yergey, A., 2002. A method for bioconjugation of carbohydrates using Diels-Alder cycloaddition. Org. Lett. 4, 3191–3194.

Prabaharan, M., Grailer, J.J., Pilla, S., Steeber, D.A., Gong, S., 2009. Folate-conjugated amphiphilic hyperbranched block copolymers based on Boltorn® H40, poly(l-lactide) and poly(ethylene glycol) for tumor-targeted drug delivery. Biomaterials 30, 3009–3019.

Prasuhn, D.E., Susumu, K., Medintz, I.L., 2011. Multivalent conjugation of peptides, proteins, and DNA to semiconductor quantum dots. Methods Mol. Biol. (Clifton, NJ) 726, 95–110.

Prato, M., Maggini, M., 1998. Fulleropyrrolidines: a family of full-fledged fullerene derivatives. Accounts Chem. Res. 31, 519–526.

Prato, M., Li, Q., Wudl, F., Lucchini, V.J., 1993. Addition of azides to fullerene C60: synthesis of azafulleroids. J. Am. Chem. Soc. 115, 1148–1150.

Prato, M., Maggini, M., Giacometti, C., Scorrano, G., Sandonh, G., Farnia, G., 1996. Synthesis and electrochemical properties of substituted fulleropyrrolidines. Tetrahedron 52 (14), 5221–5234.

Predescu, S.A., Predescu, D.N., Palade, G.E., 2001. Endothelial transcytotic machinery involves supramolecular protein–lipid complexes. Mol. Biol. Cell 12, 1019–1033.

Preechakasedkit, P., Pinwattana, K., Dungchai, W., Siangproh, W., Chaicumpa, W., Tongtawe, P., et al., 2011. Development of a one-step immunochromatographic strip test using gold nanoparticles for the rapid detection of Salmonella typhi in human serum. Biosens. Bioelectron. 31, 562–566.

Preis, J., Stumpf, P.K., Conn, E.E. (Eds.), 1980. Carbohydrates: Structure and Function. Volume 3 of The Biochemistry of Plants: A Comprehensive Treatise. Academic Press, New York.

Prescher, J.A., Bertozzi, C.R., 2005. Chemistry in living systems. Nat. Chem. Biol. 1, 13–21.

Pressman, D., Keighley, G., 1948. The zone of activity of antibodies as determined by the use of radioactive tracers. The zone of activity of nephritoxic antikidney serum. J. Immunol. 59, 141–146.

Prestayko, A.W., Cooke, S.T., Carter, S.K. (Eds.), 1980. Cisplatin Current Status and New Developments. Academic Press, New York pp. 285–304.

Prestayko, A.W., Baker, L.H., Crooke, S.T., Carter, S.K., Schein, P.S., 1981. Nitrosoureas. Current Status and New Developments. Academic Press, New York. (Chapter 4).

Price, M.R., Sekowski, M., Hooi, D.S.W., Durrant, L.G., Hudecz, F., Tendler, S.J.B., 1993. Measurement of antibody binding to antigenic peptides conjugated in situ to albumin-coated microtitre plates. J. Immunol. Meth. 159, 277–281.

Prime, K.L., Whitesides, G.M., 1991. Self-assembled organic monolayers: model systems for studying adsorption of proteins at surfaces. Science 252, 1164.

Prokudina, E.A., Lanková, P., Koblovská, R., Al-Maharik, N., Lapčík, O., 2011. Development of sorbents for immunoaffinity extraction of isoflavonoids. Phytochem. Lett. 4 (2), 113–117.

Prossnitz, E., Nikaido, K., Ulbrich, S.J., Ames, G.F., 1988. Formaldehyde and photoactivatable cross-linking of the periplasmic binding protein to a membrane component of the histidine transport system of Salmonella typhimurium. J. Biol. Chem. 263, 17917–17920.

Pu, Q., Su, Z., Hu, Z., Chang, X., Yang, M., 1998. 2-Mercaptobenzothiazole-bonded silica gel as selective adsorbent for preconcentration of gold, platinum and palladium prior to their simultaneous inductively coupled plasma optical emission spectrometric determination. Anal. Atomic Spectrom. 13, 249–253.

Puchulu-Campanella, E., Chu, H., Anstee, D.J., Galan, J.A., Tao, W.A., Low, P.S., 2012. Identification of the components of a glycolytic enzyme metabolon on the human red blood cell membrane. J. Biol. Chem. published online: November 13, 2012, doi: 10.1074/jbc. M112.428573.

Puig, O., Caspary, F., Rigaut, G., Rutz, B., Bouveret, E., Bragado-Nilsson, E., et al., 2001. The Tandem Affinity Purification (TAP) method: a general procedure of protein complex purification. Methods 24, 218–229.

Pulliam, M.W., Boyd, L.F., Baylan, N.C., Bradshaw, R.A., 1975. Specific binding of covalently cross-linked mouse nerve growth factor to responsive peripheral neurons. Biochem. Biophys. Res. Comm. 67, 1281–1289.

Pushpanathan, M., Rajendhran, J., Jayashree, S., Sundarakrishnan, B., Jayachandran, S., Gunasekaran, P., 2012. Direct cell penetration of the antifungal peptide, MMGP1, in Candida albicans. J. Pept. Sci. 18, 657–660.

Pyell, U., Stork, G., 1992. Preparation and properties of an 8-hydroxyquinoline silica gel, synthesized via Mannich reaction. Fresenius J. Anal. Chem. 342, 281–286.

Qian, W., Liang, H., Shi, J., Jin, N., Grundke-Iqbal, I., Iqbal, K., et al., 2011. Regulation of the alternative splicing of tau exon 10 by SC35 and Dyrk1A. Nucleic Acids Res. Advance Access published Apr. 5, 2011. Doi: 10.1093/nar/gkr195.

Qiu, G., Li, Y., 2000. Immobilization of cellulase on magnetic agarose composite microspheres. Jingxi Huagong 17, 115–117.

Qualmann, B., Kessels, M.M., Musiol, H.J., Sierralta, W.D., Jungblut, P.W., Moroder, L., 1996. Synthesis of boron-rich lysine dendrimers as protein labels in electron microscopy. Angew. Chem. Int. Ed. Engl. 35, 909–911.

Quash, G., Roch, A.M., Niveleau, A., Grange, J., Keolouangkhot, T., Huppert, J., 1978. The preparation of latex particles with covalently bound polyamines, IgG and measles agglutinins and their use in visual agglutination tests. J. Immunol. Meth. 22, 165–174.

Qui, G., Li, Y., 2001. Studies on the preparation and characterization of magnetic gelatin microspheres and immobilization of cellulase. Yaowu Shengwu Jushi 8, 197–199.

Quintana, A., Raczka, E., Piehler, L., Lee, I., Myc, A., Majoros, I., et al., 2002. Design and function of a dendrimer-based therapeutic nanodevice targeted to tumor cells through the folate receptor. Pharm. Res. 19, 1310–1316.

Qureshi, N., Perera, P.-Y., Shen, J., Zhang, G., Lenschat, A., ASplitter, G., et al., 2003. The proteasome as a lipopolysaccharide-binding protein in macrophages: differential effects of proteasome inhibition on lipopolysaccharide-induced signaling events. J. Immunol. 171, 1515.

Rabinovici, G.D., Furst, A.J., O'Neil, J.P., Racine, C.A., Mormino, E.C., Baker, S.L., et al., 2007. 11C-PIB PET imaging in Alzheimer disease and frontotemporal lobar degeneration. Neurology 68, 1205–1212.

Radioisotopes in Medicine, published by the World Nuclear Association, October 15, 2010; <http://www.world-nuclear.org/info/inf55.html/>.

Raftery, M.A., Cole, R.D., 1963. Tryptic cleavage at cysteinyl peptide bonds. Biochem. Biophys. Res. Comm. 10, 467–472.

Raftery, M.A., Cole, R.D., 1966. On the aminoethylation of proteins. J. Biol. Chem. 241, 3457–3461.

Ragupathi, G., Park, T.K., Zhang, S., Kim, I.J., Graber, L., Adluri, S., et al., 1997. Immunization of mice with conjugates of fully synthetic hexasaccharide globo-H results in antibodies against human cancer cells. Angew. Chem. Int. Ed. Engl. 36, 125–128.

Rahman, M.S., Kabashima, T., Yasmin, H., Shibata, T., Kai, M., 2012. A novel fluorescence reaction for N-terminal Ser-containing peptides and its application to assay caspase activity. Anal. Biochem. Available online: 22 October 2012, doi: 10.1016/j.ab.2012.10.018.

Rahman, N.A., Hasan, M., Hussain, M.A., Jahim, J., 2008. Determination of glucose and fructose from glucose isomerization process by high-performance liquid chromatography with UV detection. Mod. Appl. Sci. 2 (4), 151–154.

Rai, P., Mallidi, S., Zheng, X., Rahmanzadeh, R., Mir, Y., Elrington, S., et al., 2010. Development and applications of photo-triggered theranostic agents. Adv. Drug Deliv. Rev. 62, 1094–1124.

Raju, K., Doulias, P.T., Tenopoulou, M., Greene, J.L., Ischiropoulos, H., 2012. Strategies and tools to explore protein S-nitrosylation. Biochim. Biophys. Acta-Gen. Subj. 1820 (6), 684–688.

Ralph, R.K., Young, R.J., Khorana, H.G., 1962. The labeling of phosphomonoester end groups in amino acid acceptor ribonucleic acids and its use in the determination of nucleotide sequences. J. Am. Chem. Soc. 84, 1490–1491.

Ramakrishnan, M., Wengenack, T.M., Kandimalla, K.K., Curran, G.L., Gilles, E.J., Ramirez-Alvarado, M., et al., 2008. Selective contrast enhancement of individual alzheimer's disease amyloid plaques using a polyamine and Gd-DOTA conjugated antibody fragment against fibrillar Aβ42 for magnetic resonance molecular imaging. Pharm. Res. 25, 1861–1872.

Ramanathan, H.N., Chung, D.-H., Plane, S.J., Sztul, E., Chu, Y.-K., Guttieri, M.C., et al., 2007. Dynein-dependent transport of the Hantaan virus nucleocapsid protein to the endoplasmic reticulum-golgi intermediate compartment. J. Virol. 81, 8634–8647.

Ramjiawan, B, Maiti, P, Aftanas, A, Kaplan, H, Fast, D, Mantsch, HH, et al., 2000. Noninvasive localization of tumors by immunofluorescence imaging using a single chain Fv fragment of a human monoclonal antibody with broad cancer specificity. Cancer 89, 1134–1144.

Rana, T.M., Meares, C.F., 1990a. N-Terminal modification of immunoglobulin polypeptide chains tagged with isothiocyanato chelates. Bioconjug. Chem. 1, 357–362.

Rana, T.M., Meares, C.F., 1990b. Specific cleavage of a protein by an attached iron chelate. J. Am. Chem. Soc. 112, 2457–2458.

Rana, T.M., Meares, C.F., 1991a. Iron chelate-mediated proteolysis: protein structure dependence. J. Am. Chem. Soc. 113, 1859–1861.

Rana, T.M., Meares, C.F., 1991b. Transfer of oxygen from an artificial protease to peptide carbon during proteolysis. PNAS 88, 10578.

Ranadive, G.N., Rosenzweig, H.S., Epperly, M.W., Seskey, T., Bloomer, W.D., 1993. A new method of technetium-99m labeling of monoclonal antibodies through sugar residues, a study with TAG-72 specific CC-49 antibody. Nucl. Med. Biol. 20, 719–726.

Rannard, S.P., Davis, N.J., 2000. The selective reaction of primary amines with carbonyl imidazole containing compounds: selective amide and carbamate synthesis. Org. Lett. 2, 2117–2120.

Ranzinger, R., Herget, S., von der Lieth, C.-W., Frank, M., 2010. GlycomeDB—a unified database for carbohydrate structures. Nucl. Acids Res. 39 (Suppl. 1), D373–D376.

Rao, A., Martin, P., Reithmeier, R.A.F., Cantley, L.C., 1979. Location of the stilbenedisulfonate binding site of the human erythrocyte anion-exchange system by resonance energy transfer. Biochemistry 18, 4505–4516.

Rao, J., Dragulescu-Andrasi, A., Yao, H., 2007. Fluorescence imaging in vivo: recent advances. Curr. Opin. Biotechnol. 18, 17–25.

Rao, R., Møller, I.M., 2011. Pattern of occurrence and occupancy of carbonylation sites in proteins. Proteomics 11 (21), 4166–4173.

Rao, Y.S., Filler, R., 2008. OxazolonesTurchi, I.J. Chemistry of Heterocyclic Compounds: Oxazoles, vol. 45. John Wiley & Sons, Inc., Hoboken, NJ, USA, doi: 10.1002/9780470187289.ch3.

Raphael, M.P., Rappole, C.A., Kurihara, L.K., Christodoulides, J.A., Qadri, S.N., Byers, J.M., 2011. Iminobiotin binding induces large fluorescent enhancements in avidin and streptavidin fluorescent conjugates and exhibits diverging pH-dependent binding affinities. J. Fluoresc. 21 (2), 647–652.

Rashidbaigi, A., Langer, J.A., Jung, V., Jones, C., Morse, R.G., Tischfield, J.A., et al., 1986. The gene for the human immune interferon receptor is located on chromosome 6. Proc. Natl. Acad. Sci. USA 83, 384–388.

Raso, V., Basala, M., 1984. A highly cytotoxic human transferrin – Ricin A chain conjugate used to select receptor-modified cells. J. Biol. Chem. 259, 1143.

Raso, V., Basala, M., 1985. Study of the transferrin receptor using a cytotoxic human transferrin-ricin A chain conjugate, Gregoriadis, G. Receptor-Mediated Targeting of Drugs, vol. 2. Plenum, New York, pp. 73.

Ratelade, J., Bennett, J.L., Verkman, A.S., 2011. Evidence against cellular internalization in vivo of NMO-IgG, aquaporin-4, and excitatory amino acid transporter 2 in neuromyelitis optica. J. Biol. Chem. 286 (52), 45156–45164.

Rattanakiat, S., Nishikawa, M., Takakura, Y., 2012. Self-assembling CpG DNA nanoparticles for efficient antigen delivery and immunostimulation. Eur. J. Pharm. Sci. 47, 352–358.

Rebek, J., Feitler, D., 1974. Mechanism of the carbodiimide reaction. II. Peptide synthesis on the solid phase. J. Am. Chem. Soc. 96, 1606–1607.

Rector, E.S., Schwenk, R.J., Tse, K.S., Sehon, A.H., 1978. A method for the preparation of protein-protein conjugates of predetermined composition. J. Immunol. Methods 24, 321–336.

Reedijk, J., Fichtinger-Schepman, A.M.J., van Oosterom, A.T., van de Putte, P., 1987. Platinum amine coordination compounds as anti-tumor drugs. Molecular aspects of the mechanism of action. Struct. Bonding (Berlin) 68, 53–72.

Reese, C.B., 1973. In: McOmie, Protective Groups in Organic Chemistry. Plenum, New York, pp. 95.

Reeves, J.P., Dowben, R.M., 1969. Formation and properties of thin-walled phospholipid vesicles. J. Cell Physiol. 73, 49.

Reeves, R., Nissen, M.S., 1993. Interaction of high mobility group-I (Y) nonhistone proteins with nucleosome core particles. J. Biol. Chem. 268, 21137–21146.

Rege, K., Patel, S.J., Megeed, Z., Yarmush, M.L., 2007. Amphipathic peptide-based fusion peptides and immunoconjugates for the targeted ablation of prostate cancer cells. Cancer Res. 67, 6368–6375.

Regnier, F.E., Noel, R., 1976. Glycerolpropylsilane bonded phases in the steric exclusion chromato-graphy of biological macromolecules. J. Chromatogr. Sci. 14, 316–320.

Regnier, F.E., Riggs, L., Zhang, R., Xiong, L., Liu, P., Chakraborty, A., et al., 2002. Comparative proteomics based on stable isotope labeling and affinity selection. Int. J. Mass Spectrom. 37, 133–145.

Regoeczi, E., 1984. Methods of protein iodination Iodine-labeled Plasma Proteins, vol. 1. CRC Press, New York. pp. 35–102.

Reisfeld, A., Rothenberg, J.M., Bayer, E.A., Wilchek, M., 1987. Nonradioactive hybridization probes prepared by the reaction of biotin hydrazide with DNA. Biochem. Biophys. Res. Comm. 142, 519–526.

Reisfeld, R.A., Yang, H.M., Muller, B., Wargalla, U.C., Schrappe, M., Wrasidlo, W., 1989. Promises, problems, and prospects of monoclonal antibody-drug conjugates for cancer therapy. Antibody, Immunoconjugates, Radiopharm. 2, 217–224.

Reiter, Y., Fishelson, Z., 1989. Targeting of complement to tumor cells by heteroconjugates composed of antibodies and of the complement component C3b. J. Immunol. 142, 2771.

Rembaum, A., Yen, S.P.S., Cheong, E., Wallace, S., Molday, R.S., Gordon, I.L., et al., 1976. Functional polymeric microspheres based on 2-hydroxyethyl methacrylate for immunochemical studies. Macromolecules 9, 328–336.

Rembaum, A., Margel, S., Levy, J., 1978. Polyglutaraldehyde: a new reagent for coupling proteins to microspheres and for labeling cell-surface receptors. J. Immunol. Meth. 24, 239–250.

Ren, Z.F., Huang, Z.P., Xu, J.W., Wang, J.H., Bush, P., Siegal, M.P., et al., 1998. Synthesis of large arrays of well-aligned carbon nanotubes on glass. Science 282, 1105–1107.

Renberg, B., Shiroyama, I., Engfeldt, T., Nygren, P.-A., Karlstrom, A.E., 2005. Affibody protein capture microarrays: synthesis and evaluation of random and directed immobilization of affibody molecules. Anal. Biochem. 341, 334–343.

Renn, O., Meares, C.F., 1992. Large scale synthesis of the bifunctional chelating agent 2-*p*-nitrobenzyl-1,4,7,10-tetraazacyclododecane-N,N9,N0N09-tetraacetic acid and the determination of its enantiomeric purity by chiral chromatography. Bioconjug. Chem. 3, 563–569.

Requena, J.S., Chao, C.-C., Levine, R.L., Stadtman, E.R., 2001. Glutamic and aminoadipic semialdehydes are the main carbonyl products of metal-catalyzed oxidation of proteins. Proc. Natl. Acad. Sci 98, 69–74.

Rerat, V., Pourcelle, V., Devouge, S., Nysten, B., Marchand-Brynaert, J., 2010. Surface grafting on poly(ethylene terephthalate) track-etched microporous membrane by activation with trifluorotriazine: application to the biofunctionalization with GRGDS peptide. J. Polym. Sci., Part A: Polym. Chem. 48, 195–208.

Rescigno, M, Avogadri, F, Curigliano, G., 2007. Challenges and prospects of immunotherapy as cancer treatment. Biochim. Biophys. Acta 1776, 108–123.

Reukov, V., Maximov, V., Vertegel, A., 2011. Proteins conjugated to poly(butyl cyanoacrylate) nanoparticles as potential neuroprotective agents. Biotechnol. Bioeng. 108, 243–252.

Reulen, S.W.A., Brusselaars, W.W.T., Langereis, S., Mulder, W.J.M., Breurken, M., Merkx, M., 2007. Protein–liposome conjugates using cysteine–lipids and native chemical ligation. Bioconjug. Chem. 18 (2), 590–596.

Reynolds, A.M., Xia, W., Holmes, M.D., Hodge, S.J., Danilov, S., Curiel, D.T., et al., 2007. Bone morphogenetic protein type 2 receptor gene therapy attenuates hypoxic pulmonary hypertension. Am. J. Physiol. Lung Cell Mol. Physiol. 292, L1182–L1192.

Reynolds, V.L., McGovern, J.P., Hurley, L.H., 1986. The chemistry, mechanism of action, and biological properties of CC-1065, a potent antitumor antibiotic. J. Antibiot. 33, 319–329.

Rhemrev-Boom, M.M., 2009. Improved chromatography resin, and methods and devices related thereto. European Patent Application EP 2 090 361 A1.

Rhode, B.M., Hartmuth, K., Urlaub, H., Lührmann, R., 2003. Analysis of site-specific protein–RNA cross-links in isolated RNP complexes, combining affinity selection and mass spectrometry. RNA 9, 1542.

Rhodes, B.A., 1991. Direct labeling of proteins with 99mTc. Nucl. Med. Biol. 18, 667–676.

Rich, D.H., Gesellchen, P.D., Tong, A., Cheung, A., Buckner, C.K., 1975. Alkylating derivatives of amino acids and peptides. Synthesis of N-maleoyl amino acids, 1-[N-maleoylglycyl-cysteinyl]-oxytocin and 1-[N-maleoyl-11-aminoundecanoyl-cysteinyl]-oxytocin. Effects on vasopressin stimulated water loss from isolated toad bladder. J. Med. Chem. 18, 1004–1010.

Richa, R.R., Kumari, S., Singh, K.L., Kannaujiya, V.K., Singh, G., Kesheri, M., et al., 2011. Biotechnological potential of mycosporine-like amino acids and phycobiliproteins of cyanobacterial origin. Biotechnol. Bioinf. Bioeng. 1, 159–171.

Richard, F.M., Knowles, J.R., 1968. Glutaraldehyde as a protein cross-linking reagent. J. Mol. Biol. 37, 231.

Richardson, R.W., Gumport, R.I., 1983. Biotin and fluorescent labeling of RNA using T4 RNA ligase. Nucleic Acids Res. 11, 6167–6184.

Richmond, D.L., Schmid, E.M., Martens, S., Stachowiak, J.C., Liska, N., Fletcher, D.A., 2011. Forming giant vesicles with controlled membrane composition, asymmetry, and contents. Proc. Natl. Acad. Sci. 108 (23), 9431–9436.

Riddles, P.W., Blakeley, R.L., Zerner, B., 1979. Ellman's reagent: 5,59-dithiobis(2-nitrobenzoic acid)— A reexamination. Anal. Biochem. 94, 75–81.

Riddles, P.W., Blakeley, R.L., Zerner, B., 1983. Reassessment of Ellman's reagent. Meth. Enzymol. 91, 49–60.

Rideout, D., 1986. Self-assembling cytotoxins. Science 233, 561–563.

Rideout, D., 1994. Self-assembling drugs: a new approach to biochemical modulation in cancer chemotherapy. Cancer Invest. 12, 189–202.

Rideout, D., Calogeropoulou, T., Jaworski, J., McCarthy, M., 1990. Synergism through direct covalent bonding between agents: a strategy for rational design of chemotherapeutic combinations. Biopolymers 29, 247–262.

Rideout, D.C., Breslow, R.J., 1980. Hydrophobic acceleration of Diels–Alder reactions. Am. Chem. Soc. 102, 7816–7817.

Riehm, J.P., Scheraga, H.A., 1965. Structural studies of ribonuclease. XVII. A reactive carboxyl group in ribonuclease. Biochemistry 4, 772.

Rifai, A., Wong, S.S., 1986. Preparation of phosphorylcholine-conjugated antigens. J. Immunol. Meth. 94, 25.

Rigaut, G., Shevchenko, A., Rutz, B., Wilm, M., Mann, M., Séraphin, B., 1999. A generic protein purification method for protein complex characterization and proteome exploration. Nat. Biotechnol. 17, 1030–1032.

Rigby, P.W.J., Dieckmann, M., Rhodes, C., Berg, P., 1977. Labeling deoxyribonucleic acid to high specific activity in vitro by nick translation with DNA polymerase I. J. Mol. Biol. 113, 237–251.

Rimon, A., Tzubery, T., Padan, E., 2007. Monomers of nhaa NA^1/H^1 antiporter of *Escherichia coli* are fully functional yet dimers are beneficial under extreme stress conditions at alkaline ph in the presence of NA^1 or LI^1. J. Biol. Chem. doi: 10.1074/jbc.M704469200.

Rinaudo, M., 2010. Periodate oxidation of methylcellulose: characterization and properties of oxidized derivatives. Polymers 2, 505–521.

Rink, H., 1987. Solid-phase synthesis of protected peptide fragments using a trialkoxy-diphenyl-methylester resin. Tetrahedron Lett. 28, 3787–3790.

Riordan, J.F., Vallee, B.L., 1963. Acetylcarboxypeptidase. Biochemistry 2, 1460.

Riordan, J.F., Vallee, B.L., 1964. Succinylcarboxy peptidase. Biochemistry 3, 1768.

Riordan, J.F., Vallee, B.L., 1972. Diazonium salts as specific reagents and probes of protein conformation, Hirs, C.H.W.Timasheff, S.N. Methods in Enzymology, vol. 25. Academic Press, New York, pp. 521.

Ritzefeld, M., Sewald, N., 2012. Real-time analysis of specific protein-DNA interactions with surface plasmon resonance. J. Amino Acids Article ID 816032, 19 pages, doi:10.1155/2012/816032.

Roberts, J.C., Adams, Y.E., Tomalia, D., Mercer-Smith, J.A., Lavallee, D.K., 1990. Using starburst dendrimers as linker molecules to radiolabel antibodies. Bioconjug. Chem. 1 (5), 305–308.

Roberts, M.J., Bentley, M.D., Harris, J.M., 2002. Chemistry for peptide and protein PEGylation. Adv. Drug Deliv. Rev. 54, 459–476.

Roberts, M.J., Bentley, M.D., Harris, J.M., 2012. Chemistry for peptide and protein PEGylation. Adv. Drug Deliv. Rev. Available online: 13 September 2012, doi: 10.1016/j.addr.2012.09.025.

Robertson, E.A., Schultz, R.L.J., 1970. J. Ultrastruct. Res. 30, 275.

Rocklage, S.M., Watson, A.D., 1993. Chelates of gadolinium and dysprosium as contrast agents for MR imaging. J. Magn. Reson. Imaging 3, 167–178.

Rodrigues, M.L., Carter, P., Wirth, C., Mullins, S., Lee, A., Blackburn, B.K., 1995. Synthesis and beta-lactamase-mediated activation of a cephalosporin-taxol prodrug. Chem. Biol. 2, 223–227.

Rodrigues, R.C., Berenguer-Murcia, A., Fernandez-Lafuente, R., 2011. Coupling chemical modification and immobilization to improve the catalytic performance of enzymes. Adv. Synth. Catal. 353, 2216–2238.

Rodríguez, M., Brito-Armas, J.M., Castro, R., 2011. Gene Therapy for Parkinson's Disease: Towards Non Invasive Approaches, published online, DOI: 10.5772/20631.

Roepstorff, P., 2012. Mass spectrometry based proteomics, background, status and future needs. Protein Cell 3 (9), 641–647.

Roffler, S.R., and Tseng, T.-L., 1994. Enhanced serum half-life and tumor localization of PEG-modified antibody–enzyme conjugates for targeted prodrug activation. Antibody Engineering Conference. San Diego, California.

Roffman, E., Spiegel, Y., Wilchek, M., 1980. Ferritin hydrazide, a novel conalent electron dense reagent for the ultrastructural localization of glycoconjugates. Biochem. Biophys. Res. Comm. 97, 1192–1198.

Rogach, A.L., Katsikas, L., Kornowski, A., Su, D., Eychmuller, A., Weller, H., 1996. Synthesis and characterization of thiol-stabilized CdTe nanocrystals. Ber. Bunsenges. Phys. Chem. 100, 1772–1778.

Roitt, I., 1977. Essential Immunology. Blackwell, London. p. 21.

Roldos, V., Canada, F.J., Jimenez-Barbero, J., 2011. Carbohydrate-protein interactions: a 3D view by NMR. ChemBioChem 12, 990–1005.

Romaniouk, A.V., Silva, A., Feng, J., Vijay, I.K., 2004. Synthesis of a novel photoaffinity derivative of 1-deoxynojirimycin for active site-directed labeling of I. Glycobiology 14, 301–310.

Rong, G., Reinhard, B.M., 2012. Monitoring the size and lateral dynamics of ErbB1 enriched membrane domains through live cell plasmon coupling microscopy. PLoS ONE 7 (3), e34175. doi: 10.1371/journal.pone.0034175.

Rooseboom, M., Commandeur, J.N.M., Vermeulen, N.P.E., 2004. Enzyme-catalysed activation of anticancer drugs. Pharm. Rev. 56, 53–102.

Roque, A.C.A., Lowe, C.R., 2008. Rationally designed ligands for use in affinity chromatography. In: Zachariou, M., (Ed.), Affinity Chromatography, Methods and Protocols, second ed., Methods in Molecular Biology 421, Humana Press, Totowa, NJ, pp. 93–102. ISBN: 978-1-58829-659-7.

Roque, A.C.A., Silva, C.S.O., Ângela Taipa, M., 2007. Affinity-based methodologies and ligands for antibody purification: advances and perspectives. J. Chromatogr. A 1160, 44–55.

Rosa, J., Sabelli, M., Soriano, E.R., 2010. Prefilled certolizumab pegol (Cimzia®) syringes for self-use in the treatment of rheumatoid arthritis. Med. Devices: Evid. Res. 3, 25–31.

del Rosario, R.B., Wahl, R.L., Brocchini, S.J., Lawton, R.G., Smith, R.H., 1990. Sulfhydryl site-specific cross-linking and labeling of monoclonal antibodies by a fluorescent equilibrium transfer alkylation cross-link reagent. Bioconjug. Chem. 1990 (1), 51–59.

Rosenberg, M., Gilham, P.T., 1971. The isolation of 39-terminal polynucleotides from RNA molecules. Biochim. Biophys. Acta 246, 337–340.

Rosenberg, M., Wiebers, J.L., Gilham, P.T., 1972. Interactions of nucleotides, polynucleotides, and nucleic acids with dihydroxyboryl-substituted celluloses. Biochemistry 11, 3623–3628.

Rosenberg, M.B., Hawrot, E., Breakefield, X.O., 1986. Receptor binding activities of biotinylated derivatives of b-nerve growth factor. J. Neurochem. 46, 641–648.

Rosenthal, E., Poizot-Martin, I., Saint-Marc, T., Spano, J.P., Cacoub, P., GroupDNXS, 2002. Phase IV study of liposomal daunorubicin (DaunoXome) in AIDS-related Kaposi sarcoma. Am. J. Clin. Oncol. 25, 57–59.

Ross, J.F., Chaudhuri, P.K., Ratnam, M., 1994. Differential regulation of folate receptor isoforms in normal and malignant tissues in vivo and established cell lines. Physiologic and clinical implications. Cancer 73, 2432–2443.

Ross, P.L., Huang, Y.N., Marchese, J.N., Williamson, B., Parker, K., Hattan, S., et al., 2004. Multiplexed protein quantitation in Saccharomyces cerevisiae using amine-reactive isobaric tagging reagents. Mol. Cell. Proteomics 3, 1154–1169.

Ross, S.E., Carson, S.D., Fink, L.M., 1986. Effects of detergents on avidin–biotin interaction. Biotechniques 4, 350–354.

Ross, W.C.J., 1953. The chemistry of cytotoxic alkylating agents. Adv. Cancer Res. 1, 397.

Ross, W.D., Jefferson, R.T., 1970. In situ formed open pore polyurethane as chromatography support. J. Chromatogr. Sci. 8, 386–389.

Rostovtsev, V.V., Green, L.G., Fokin, V.V., Sharpless, K.B., 2002. A stepwise Huisgen cycloaddition process: copper(I)-catalyzed regioselective "Ligation" of azides and terminal alkynes. Angew. Chem. 114 (14), 2708–2711.

Rotenberg, S.A., Calogeropoulou, T., Jaworski, J., Weinstein, I.B., Rideout, D.A., 1991. Self-assembling protein kinase C inhibitor. Proc. Natl. Acad. Sci. USA 88, 2490–2494.

Roth, J., 1983. Application of lectin—Gold complexes for electron microscopic localization of glycoconjugates on thin sections. J. Histochem. Cytochem. 31, 987.

Roth, J., 2011. Post-embedding cytochemistry with gold-labelled reagents: a review. J. Microsc. 143 (2), 125–137.

Roth, J., Binder, M., 1978. Colloidal gold, ferritin, and peroxidase as markers for electron microscopic double labeling lectin techniques. J. Histochem. Cytochem. 26, 163.

Roth, J., Taatjes, D.J., Warhol, M.J., 1989. Prevention of non-specific interactions of gold-labeled reagents on tissue sections. Histochemistry 92, 47–56.

Rothenberg, B.E., Hayes, B.K., Toomre, D., Manzi, A.E., Varki, A., 1993. Biotinylated diaminopyridine: an approach to tagging oligosaccharides and exploring their biology. PNAS 90, 11939–11943.

Rothenberg, J.M., Wilchek, M., 1988. p-Diazobenzoyl-biocytin: a new biotinylating reagent for DNA. Nucleic Acids Res. 16, 7197–7198.

Rother, D., Sen, T., East, D., Bruce, I.J., 2011. Silicon, silica and its surface patterning with alkoxysilanes for nanomedical applications. Nanomedicine 6, 281–300.

Rothfus, J.A., Smith, E.L., 1963. Glycopeptides. IV. The periodate oxidation of glycopeptides from human gamma-globulin. J. Biol. Chem. 238, 1402–1410.

Rothschild, K.J., Sonar, S.M., Olejnik, J., 1999. Photocleavable agents and conjugates for the detection and isolation of biomolecules. U.S. Patent No. 5,986,076.

Rouhi, M., 2004. High-yield path to dendrimers. Copper-catalyzed reaction offers easy, efficient route to globular molecules. Chem. Eng. News 82 (28), 5.

Rousseaux, J., Rousseaux-Prevost, R., Bazin, H., 1983. Optimal conditions for the preparation of Fab and F(ab')2 fragments from monoclonal IgG of different rat JgC subclasses. J. Immunol. Meth. 64, 141–146.

Roychoudhury, R., Tu, C.-P.D., Wu, R., 1979. Influence of nucleotide sequence adjacent to duplex DNA termini on 39-terminal labeling by terminal transferase. Nucleic Acids Res. 6, 1323–1333.

Rozema, D.B., Lewis, D.L., Wakefield, D.H., Wong, S.C., Klein, J.J., Roesch, P.L., et al., 2007. Dynamic polyconjugates for targeted in vivo delivery of siRNA to hepatocytes. PNAS, doi: 10.1073/pnas.0703778104.

Rubenstein, R.C., Lockwood, S.R., Lide, E., Bauer, R., Suaud, L., Grumbach, Y., 2011. Regulation of endogenous ENaC functional expression by CFTR and F508-CFTR in airway epithelial cells. Am. J. Physiol. Lung Cell Mol. Physiol. 300, L88–L101.

Rudd, P.M., Joao, H.C., Coghill, E., Fiten, P., Saunders, M.R., Opdenakker, G., et al., 1994. Glycoforms modify the dynamic stability and functional activity of an enzyme. Biochemistry 33 (1), 17–22.

Ruegg, U.T., Rudingder, J., 1977. Reductive cleavage of cystine disulfides with tributylphosphineHirs, C.H.W.Timasheff, S.N. Methods in Enzymology, vol. 47. Academic Press, New York, pp. 111.

Ruiz-Carrillo, A., Allfrey, V.G., 1973. A method for the purification of histone fraction F3 by affinity chromatography. Arch. Biochem. Biophys. 154, 185–191.

Ruppert, M., Bauer, W., Hirsch, A., 2011. Microenvironment Engineering in ortho-and para-Dendronized Metalloporphyrin–Fullerene Conjugates Involving a trans-2-Bisaddition Pattern. Chemistry-A Eur. J. 17 (31), 8714–8725.

Rust, M.J., Bates, M., Zhuang, X., 2006. Sub-diffraction-limit imaging by stochastic optical reconstruction microscopy (STORM). Nat. Methods 3, 793–795.

Rutenber, E., Katzin, B.J., Ernst, S., Collins, E.J., Mlsna, D., Ready, M.P., et al., 1991. Crystallographic refinement of ricin to 2.5 A. Proteins 10, 240–250.

Ruth, J.L., 1993. Direct attachment of enzymes to DNA probes. In: Howard, G.C. (Ed.), Methods in Nonradioactive Detection. Appleton and Lange, Norwalk, Connecticut, pp. 153–177.

Rutkowska, A., Schultz, C., 2012. Protein tango: the toolbox to capture interacting partners. Angew. Chem. Int. Ed. 51 (33), 8166–8176.

Ryu, D., Loh, K.J., Ireland, R., Karimzada, M., Yaghmaie, F., Gusman, A.M., 2011. In situ reduction of gold nanoparticles in PDMS matrices and applications for large strain sensing. Smart Struct. Syst. 8 (5), 471–486.

Sabbatini, P., Odunsi, K., 2007. Immunologic approaches to ovarian cancer treatment. J. Clin. Oncol. 25, 2884–2893.

Sabidó, E., Selevsek, N., Aebersold, R., 2011. Mass spectrometry-based proteomics for systems biology. Curr. Opin. Biotechnol. 23, 591–597.

Saghatelian, A., Jessani, N., Joseph, A., Humphrey, M., Cravatt, B.F., 2004. Activity-based probes for the proteomic profiling of metalloproteases. Proc. Natl. Acad. Sci. USA 101 (27), 10000–10005.

Saha, A.K., Kross, K., Kloszewski, E.D., Upson, D.A., Toner, J.L., Snow, R.A., et al., 1993. Time-resolved fluorescence of a new europium-chelate complex: demonstration of highly sensitive detection of protein and DNA samples. J. Am. Chem. Soc. 115, 11032–11033.

Saiki, R.K., Scharf, S., Faloona, F., Mullis, K.B., Horn, G.T., Erlich, H.A., et al., 1985. Enzymatic amplification of beta-globin genomic sequences and restriction site analysis for diagnosis of sickle cell anemia. Science 230, 1350–1354.

Saiki, R.K., Gelfand, D.H., Stoffel, S., Scharf, S.J., Higuchi, R., Horn, G.T., et al., 1988. Primer-directed enzymatic amplification of DNA with a thermostable DNA polymerase. Science 239, 487–491.

Saito, M., Shinozaki-Kuwahara, N., Takada, K., 2012. Gibbsiella dentisursi sp. nov., isolated from the bear oral cavity. Microbiol. Immunol. 56 (8), 506–512.

Sakaguchia, M., Shingo, T., Shimazaki, T., Okano, H.J., Shiwa, M., Ishibashi, S., et al., 2006. A carbohydrate-binding protein, Galectin-1, promotes proliferation of adult neural stem cells. Proc. Natl. Acad. Sci. (USA) 103, 7112–7117.

Sakakibara, S., Inukai, N., 1965. The trifluoroacetate method of peptide synthesis. I. The synthesis and use of trifluoroacetate reagents. Bull. Chem. Soc. Jpn. 38, 1979–1984.

Sakamoto, H., Traincard, F., Vo-Quang, T., Ternynck, T., Guesdon, J.L., Avrameas, S., 1987. 5-Bromodeoxyuridin in vivo labeling of M13 DNA, and its use as a nonradioactive probe for hybridization experiments. Mol. Cell. Probes 1, 109–120.

Sakato, M., Sakakibara, H., King, S.M., 2007. Chlamydomonas outer arm dynein alters conformation in response to Ca^{21}. Mol. Biol. Cell, doi: 10.1091/mbc.E06-10-0917.

Sakharov, D.V., Jie, A.F.H., Bekkers, M.E.A., Emeis, J.J., Rijken, D.C., 2001. Polylysine as a vehicle for extracellular matrix-targeted local drug delivery, providing high accumulation and long-term retention within the vascular wall. Aterioscler. Thromb. Vasc. Biol. 21, 943–948.

Saleh, M.A., Sandoval, R.M., Rhodes, G.J., Campos-Bilderback, S.B., Molitoris, B.A., Pollock, D.M., 2012. Chronic endothelin-1 infusion elevates glomerular sieving coefficient and proximal tubular albumin reuptake in the rat. Life Sci. 91, 634–637.

Saleh, S.M., Ali, R., Wolfbeis, O.S., 2011. New silica and polystyrene nanoparticles labeled with longwave absorbing and fluorescent chameleon dyes. Microchim. Acta 174 (3), 429–434.

Salmain, M., Vessiéres, A., Butler, I., Jaouen, G., 1991. *N*-Succinimidyl (4-pentynoate)hexacarbonyldicobalt: a transition-metal carbonyl complex having similar uses to the Bolton–Hunter reagent. Bioconjug. Chem. 2, 13–15.

Salmain, M., Vessiéres, A., Brossier, P., Butler, I.S., Jaouen, G., 1992. Carbonylmetallo-immunoassay (CMIA) a new type of nonradioisotopic immunoassay. Principles and application to phenobarbital assay. J. Immunol. Meth. 148, 65–75.

Salmain, M., Fischer-Durand, N., Cavalier, L., Rudolf, B., Zakrzewski, J., Jaouen, G., 2002. Transition metal-carbonyl labeling of biotin and avidin for use in solid-phase carbonyl metallo immunoassay (CMIA). Bioconjug. Chem. 13, 693–698.

Salmon, S.E., 1989. Monoclonal antibody immunoconjugates for cancer. Antibody, Immunoconjugates, Radiopharm. 2, 63–70.

Samaha, R.R., Joseph, S., O'Brien, B., O'Brien, T.W., Noller, H.F., 1999. Site-directed hydroxyl radical probing of 30S ribosomal subunits by using Fe(II) tethered to an interruption in the 16S rRNA chain. PNAS 96, 366.

Samal, S., Geckeler, K.E., 2000. Cyclodextrin–fullerenes: a new class of water-soluble fullerenes. J. Chem. Soc. Chem. Comm., 1101–1102.

Samarakoon, K., Jeon, Y.J., 2012. Bio-functionalities of proteins derived from marine algae–A review. Food Res. Int. 48, 948–960.

Sammes, P.G., Yahioglu, G., 1996. Modern bioassays using metal chelates as luminescent probes. Nat. Prod. Rep. 13, 1–28.

Sampathkumar, P., Mak, M.W., Fisher-Witholt, S.J., Guigard, E., Kay, C.M., Lemieux, M.J., 2012. Oligomeric state study of prokaryotic rhomboid proteases. Biochim. Biophys. Acta (BBA) – Biomembranes 1818, 3090–3097.

Sánchez-Pomales, G., Morris, T.A., Falabella, J.B., Tarlov, M.J., Zangmeister, R.A., 2012. A lectin-based gold nanoparticle assay for probing glycosylation of glycoproteins. Biotechnol. Bioeng. 109, 2240–2249.

Sanderson, C.J., Wilson, D.V., 1971. A simple method for coupling proteins to insoluble polysaccharides. Immunology 20, 1061–1065.

Sands, M.J., Levitin, A., 2004. Basics of magnetic resonance imaging. Semin. Vasc. Surg. 17, 66–82.

Sandvig, K., Olsnes, S., 1981. Rapid entry of nicked diphtheria toxin into cells at low pH. Characterization of the entry process and effects of low pH on the toxin molecule. J. Biol. Chem. 256, 9068.

Sangwung, P., Greco, T.M., Wang, Y., Ischiropoulos, H., Sessa, W.C., Iwakiri, Y., 2012. Proteomic identification of S-nitrosylated Golgi proteins: new insights into endothelial cell regulation by eNOS-derived NO. PLoS ONE 7 (2), e31564. doi: 10.1371/journal.pone.0031564.

Santhoshkumar, P., Sharma, K.K., 2002. Identification of a region in alcohol dehydrogenase that binds to a-crystallin during chaperone action. Biochim. Biophys. Acta 1589, 115–121.

Santos, S.G., Campbell, E.C., Lynch, S., Wong, V., Antoniou, A.N., Powis, S.J., 2007. Major histocompatibility complex class I–ERp57–tapasin interactions within the peptide-loading complex. J. Biol. Chem. 282 (24), 17587–17593.

Santra, S., Wang, K., Tapec, R., Tan, W., 2001. Development of novel dye-doped silica nanoparticles for biomarker application. J. Biomed. Optics 6, 160–166.

Santuccione, A., Sytnyk, V., Leshchyns'ka, I., Schachner, M., 2005. Prion protein recruits its neuronal receptor NCAM to lipid rafts to activate p59fyn and to enhance neurite outgrowth. J. Cell Biol. 169, 341–354.

Sapsford, K.E., Pons, T., Medintz, I.L., Mattoussi, H., 2006. Biosensing with luminescent semiconductor quantum dots. Sensors 6, 925–953.

Sashidhar, R.B., Capoor, A.K., Ramana, P., 1994. Quantitation of e-amino group using amino acids as reference standards by trinitrobenzene sulfonic acid. J. Immunol. Meth. 167, 121–127.

Sastry, L., Alting-Mees, M., Huse, W.D., Short, J.M., Sorge, J.A., Hay, B.N., et al., 1989. Cloning of the immunological repertoire in *Escherichia coli* for generation of monoclonal catalytic antibodies: construction of a heavy chain variable region-specific cDNA library. Proc. Natl. Acad. Sci. USA 86, 5728–5732.

Sasuga, Y., Tani, T., Hayashi, M., Yamakawa, H., Ohara, O., Harada, Y., 2006. Development of a microscopic platform for real-time monitoring of biomolecular interactions. Genome Res. 16, 132–139.

Sato, N., Kobayashi, H., Saga, T., Nakamoto, Y., Ishimori, T., Togashi, K., et al., 2001. Tumor targeting and imaging of intraperitoneal tumors by use of antisense oligo-DNA complexed with dendrimers and/or avidin in mice. Clin. Cancer Res. 7, 3606–3612.

Sato, S., Nakao, M., 1981. Cross-linking of intact erythrocyte membrane with a newly synthesized cleavable bifunctional reagent. J. Biochem. (Tokyo) 90, 1177.

Savage, D., Mattson, G., Desai, S., Nielander, G., Morgensen, S., Conklin, E., 1992. Avidin-Biotin Chemistry: A Handbook. Pierce Chemical Company, Rockford, IL.

Savina, I.N., Mattiasson, B., Galaev, IYu., 2006. Graft polymerization of vinyl monomers inside macroporous polyacrylamide gel, cryogel, in aqueous and aqueous-organic media initiated by diperiodatocuprate(III) complexes. J. Polymer Sci.: Part A: Polymer Chem. 44, 1952–1963.

Savina, I.N., Chudde, V., D'Hollander, S., van Hoorebeke, L., Mattiasson, B., Galaev, IYu., et al., 2007. Cryogels from poly (2-hydroxyethyl methacrylate): macroporoous, interconnected materials with potential as cell scaffolds. Soft Matter 3, 1176–1184.

Sawant, R.R., Torchilin, V.P., 2011. Design and synthesis of novel functional lipid-based bioconjugates for drug delivery and other applications. Methods Mol. Biol. (Clifton, NJ) 751, 357–378.

Sawant, R.R., Sriraman, S.K., Navarro, G., Biswas, S., Valvi, R.A., Torchilin, V.P., 2012. Polyethyleneimine-lipid conjugate-based pH-sensitive micellar carrier for gene delivery. Biomaterials 33, 3942–3951.

Sawin, K.E., Mitchison, T.J., 1991. Mitotic spindle assembly by two different pathways in vitro. J. Cell Biol. 112, 925.

Sawyer, S.T., Krantz, S.B., Luna, J., 1987. Identification of the receptor for erythropoietin by cross-linking to Friend virus-infected erythroid cells. Proc. Natl. Acad. Sci. USA 84, 3690–3694.

Saxena, C., Higgs, R.E., Zhen, E., Hale, J.E., 2009. Small-molecule affinity chromatography coupled mass spectrometry for drug target deconvolution. Expert Opin. Drug Discov. 4 (7), 701–714.

Saxon, E., Bertozzi, C.R., 2000. Cell surface engineering by a modified Staudinger reaction. Science 287, 2007–2010.

Saxon, E., Carolyn, R.B., 2003. Chemoselective ligation. US Patent No. 6,570,040.

Saxon, E., Carolyn, R.B., 2006. Chemoselective ligation. US Patent No. 7,122,703.

Saxon, E., Armstrong, J.I., Bertozzi, C.R., 2000. A "Traceless" Staudinger ligation for the chemoselective synthesis of amide bonds. Org. Lett. 2, 2141–2143.

Sayre, L.M., Smith, M.A., Perry, G., 2001. Chemistry and biochemistry of oxidative stress in neurodegenerative disease. Curr. Med. Chem. 8, 721–738.

Scheck, R.A., Francis, M.B., 2007. Regioselective labeling of antibodies through N-terminal transamination. ACS. Chem. Biol. 2 (4), 247–251.

Scheck, R.A., Dedeo, M.T., Iavarone, A.T., Francis, M.B., 2008. Optimization of a biomimetic transamination reaction. J. Am. Chem. Soc. 130 (35), 11762–11770.

Schepartz, A., Cuenoud, B., 1990. Site-specific cleavage of the protein calmodulin using a trifluoperazine-based affinity reagent. J. Am. Chem. Soc. 112, 3247–3249.

Scherson, T., Kreis, T.E., Schlessinger, J., Littauer, U., Borisy, G.G., Geiger, B., 1984. Dynamic interactions of fluorescently labeled microtubule-associated proteins in living cells. J. Cell Biol. 99, 425–434.

Schewale, J.G., Brew, K., 1982. Effects of Fe^{31} binding on the microenvironments of individual amino groups in human serum transferrin as determined by different kinetic labeling. J. Biol. Chem. 257, 9406.

Schiel, J.E., Mallik, R., Soman, S., Joseph, K.S., Hage, D.S., 2006. Applications of silica supports in affinity chromatography. J. Sep. Sci. 29, 719–737.

Schimitschek, E.J., Trias, J.A., Hammond, P.R., Atkins, R.L., 1974. Laser performance and stability of fluorinated coumarin dyes. Opt. Comm. 11, 352.

Schlom, J., 1986. Basic principles and applications of monoclonal antibodies in the management of carcinomas. Cancer Res. 46, 3225.

Schmer, G., 1972. Hoppe-Seyler's Z. Physiol. Chem. 353, 810.

Schmidt, T.G.M., Skerra, A., 2007. The Strep-tag system for one-step purification and high-affinity detection or capturing of proteins. Nat. Protoc. 2, 1528–1535.

Schmidt, A., Kalkhof, S., Ihling, C., Schulz, D.M., Beck-Sickinger, A.G., Cooper, D.M.F., et al., 2005. Studying calmodulin/adenylyl cyclase 8 interaction using isotope-labeled cross-linkers and FTICR mass spectrometry. Poster, Pierce Biotechnology web site.

Schmitt, A., Hinkeldey, B., Wild, M., Jung, G., 2009. Synthesis of the core compound of the BODIPY dye class: 4,4′-Difluoro-4-bora-(3a,4a)-diaza-s-indacene. J. Fluoresc. 19 (4), 755–759.

Schmitt, M., Painter, R.G., Jesaitis, A.J., Preissner, K., Sklar, L.A., Cochrane, C.G., 1983. Photoaffinity labeling of the *N*-formyl peptide receptor binding site of intact human polymorphonuclear leukocytes. J. Biol. Chem. 258, 649–654.

Schmitz, G.G., Walter, T., Seibl, R., Kessler, C., 1991. Nonradioactive labeling of oligonucleotides in vitro with the hapten digoxigenin by tailing with terminal transferase. Anal. Biochem. 192, 222–231.

Schnapp, K.A., Platz, M.S., 1993. A laser flash photolysis study of di-, tri- and tetrafluorinated phenylnitrenes; implications for photoaffinity labeling. Bioconjug. Chem. 4, 178–183.

Schnapp, K.A., Poe, R., Leyva, E., Soundararajan, N., Platz, M.S., 1993. Exploratory photochemistry of fluorinated aryl azides. Implications for the design of photoaffinity labeling reagents. Bioconjug. Chem. 4, 172–177.

Schneede, J., Ueland, P.M., 1992. Formation in an aqueous matrix and properties and chromatographic behavior of 1-pyrenyldiazomethane derivatives of methylmalonic acid and other short-chain dicarboxylic acids. Anal. Chem. 64, 315.

Schneider, C., Newman, R.A., Sutherland, D.R., Asser, U., Greaves, M.F., 1982. A one-step purification of membrane proteins using a high efficiency immunomatrix. J. Biol. Chem. 257 (18), 10766–10769.

Schneider, L.V., Hall, M.P., 2005. Stable isotope methods for high-precision proteomics. Drug Discov. Today 10 (5), 353–363.

Schröder, J., Benink, H., Dyba, M., Los, G.V., 2009. In vivo labeling method using a genetic construct for nanoscale resolution microscopy. Biophys. J. 96, L1–3.

Schroeder, W.A., Shelton, J.R., Robberson, B., 1967. Modification of methionyl residues during aminoethylation. Biochim. Biophys. Acta 147, 590–592.

Schuberth, H.-J., et al. 1996. Biotinylation of cell surface MHC molecules: a complementary tool for the study of MHC class II polymorphism in cattle. J. Immunol. Meth. 189, 89–98.

Schulz, C., Köhn, M., 2008. Simultaneous protein tagging in two colors. Chem. Biol. 15, 91–92.

Schulz, A., Woolley, R., Tabarin, T., McDonagh, C., 2011. Dextran-coated silica nanoparticles for calcium-sensing. Analyst 136 (8), 1722–1727.

Schwartz, B.A., Gray, G.R., 1977. Proteins containing reductively aminated disaccharides. Synthesis and chemical characterization. Arch. Biochem. Biophys. 181, 542–549.

Schwartz, D.A., Abrams, M.J., Giadomenico, C.M., Zubieta, J.A., 1993. Certain pyridyl hydrazines and hydrazides useful for protein labeling. US Patent No. 5, 206, 370.

Schwartz, D.A., Abrams, M.J., Giadomenico, C.M., Zubieta, J.A., 1995. Protein labelling utilizing certain pyridyl hydrazines, hydrazides and derivatives. US Patent No. 5, 420, 285.

Schwartz, W.E., Smith, P.K., Royer, G.P., 1980. *N*-(b-iodoethyl)trifluoroacetamide: a new reagent for the aminoethylation of thiol groups in proteins. Anal. Biochem. 106, 43–48.

Schwinghamer, E.A., 1980. A method for improved lysis of gram-negative bacteria. FEMS Microbiol. Lett. 7, 157–162.

Scoble, J.A., Scopes, R.K., 1997. Ligand structure of the divinylsulfone-based T-gel. J. Chromatogr. 787, 47–54.

Scopes, R., 1982. Protein Purification. Springer-Verlag, New York. p. 30.

Scorilas, A., Bjartell, A., Lilja, H., Moller, C., Diamandis, E.P., 2000. Streptavidin–polyvinylamine conjugates labeled with a europium chelate: applications in immunoassay, immunohistochemistry, and microarrays. Clin. Chem. 46, 1450–1455.

Scott Jr., C.J., Goldmacher, V.S., Lambert, J.M., Chari, R.V., Bolender, S., Gauthier, M.N., et al., 1987. The antileukemic efficacy of an immunotoxin composed of a monoclonal anti-Thy-1 antibody disulfide linked to the ribosome-inactivating protein gelonin. Cancer Immunol. Immunother. 25, 31.

Scott, D.C., Newton, S.M.C., Klebba, P.E., 2002. Surface loop motion in FepA. J. Bacteriol. 184, 4906–4911.

Scouten, W.H., 1983. In: Scouten, W.H. (Ed.), Solid Phase Biochemistry. John Wiley and Sons, New York, pp. 149–187.

Scouten, W.H., Van den Tweel, W., 1984. Chromophonic sulfonyl chloride agarose for immobilizing bioligands. Annal. N.Y. Acad. Sci. 434, 249.

Sculimbrene, B.R., Imperiali, B., 2006. Lanthanide-binding tags as luminescent probes for studying protein interactions. J. Am. Chem. Soc. 128 (22), 7346–7352.

Seela, F., Waldeck, S., 1975. Agarose linked adenosine and guanosine-59-monophosphate; a new general method for the coupling of ribonucleotides to polymers through their *cis*-diols. Nucleic Acids Res. 2, 2343–2349.

Segal, D.M., Hurwitz, E., 1976. Dimers and trimers of immunoglobulin G covalently cross-linked with a bivalent affinity label. Biochemistry 15, 5253.

Segal, E., Pan, H., Benayoun, L., Kopeckova, P., Shaded, Y., Kopecek, J., et al., 2011. Enhanced anti-tumor activity and safety profile of targeted nano-scaled HPMA copolymer-alendronate-TNP-470 conjugate in the treatment of bone malignances. Biomaterials 32, 4450–4463.

Sela, M., Hurwitz, E., 1987. Conjugates of antibodies with cytotoxic drugs. In: Vogel, C.-W. (Ed.), Immunoconjugates: Antibody Conjugates in Radioimaging and Therapy of Cancer. Oxford University Press, New York, pp. 189.

Seligsberger, L., Sadlier, C., 1957. New developments in tanning with aldehydes. J. Am. Leather Chem. Assoc. 52, 2.

Selvin, P.R., 2000. The renaissance of fluorescence resonance energy transfer. Nat. Struct. Biol. 7 (9), 730–734.
Selvin, P.R., 2002. Principles and biophysical applications of luminescent lanthanide probes. Annu. Rev. Biophys. Biomol. Struct. 31, 275–302.
Selvin, P.R., 2003. Lanthanide-labeled DNA, Lakowicz, J. In: Topics in Fluorescence Spectroscopy, Vol. 7. Kluwer Academic/Plenum, New York, pp. 177–212.
Sen, T., Bruce, I.J. 2012. Surface engineering of nanoparticles in suspension for particle based bio-sensing. Scientific Reports, 2, Article No. 564.
Senozan, N., et al. 1981. Hemocyanin of the giant keyhold limpet, Megathura crenulata. In: Lamy, J., Lamy, J. (Eds.), Invertebrate Oxygen Binding Proteins: Structure, Active Sites, and Function. Dekker, New York, pp. 703–717.
Senter, P.D., 1990. Activation of prodrugs by antibody-enzyme conjugates: a new approach to cancer therapy. FASEB J. 4, 188–193.
Senter, P.D., Saulnier, M.G., Schreiber, G.J., Hirschberg, D.L., Brown, J.P., Hellstrom, I., et al., 1988. Anti-tumor effects of antibody-alkaline phosphatase conjugates in combination with etoposide phosphate. Proc. Natl. Acad. Sci. USA 85, 4842.
Senter, P.D., Su, P.C., Katsuragi, T., Sakai, T., Cosand, W.L., Hellstrom, I., et al., 1991. Generation of 5-fluorouracil from 5-fluorocytosine by monoclonal antibody-cytosine deaminase conjugates. Bioconjug. Chem. 2, 447–451.
Séraphin, B., Puig, O., Bouveret, E., Rutz, B., Caspary, F., 2002. Tandem Affinity Purification to Enhance Interacting Protein Identification, in Protein–Protein Interactions: A Molecular Cloning Manual, by Cold Spring Harbor Laboratory Press, Chapter 17, pp. 313–328.
Sessler, J.L., Hemmi, G.W., Mody, T.D., 1995. Water soluble texaphyrin metal complexes for singlet oxygen production. United States Patent 5, 439, 570.
Sessler, J.L., Miller, R.A., 2000. Texaphyrins: new drugs with diverse clinical applications in radiation and photodynamic therapy. Biochem. Pharmacol. 59, 733–739.
Sessler, J.L., Tomat, E., 2007. Transition metal complexes of expanded porphyrins. Acc. Chem. Res. 40 (5), 371–379.
Seth, D., Stamler, J.S., 2011. The SNO-proteome: causation and classifications. Curr. Opin. Chem. Biol. 15 (1), 129–136.
Severinov, K., Muir, T.W., 1998. Expressed protein ligation, a novel method for studying protein–protein interactions in transcription. J. Biol. Chem. 273 (26), 16205–16209.
Seydack, M., 2008. Immunoassays: basic concepts, physical chemistry and validation. Standardization and Quality Assurance in Fluorescence Measurements II, 401–428.
Seyhan, T.S., Alptekin, O., 2004. Immobilization and kinetics of catalase onto magnesium silicate. Process Biochem. 39, 2149–2155.
Seymour, L.W., Ulbrich, K., Strahalm, J., Kopecek, J., Duncan, R., 1990. Pharmacokinetics of polymer-bound adriamycin. Biochem. Pharmacol. 39, 1125–1131.
Sgro, J., Jacrot, B., Chroboczek, J., 1986. Identification of regions of brome mosaic virus coat protein chemically cross-linked in situ to viral RNA. Eur. J. Biochem. 154, 69–76.
Shabarova, Z.A., 1988. Chemical development in the design of oligonucleotide probes for binding to DNA and RNA. Biochimie 70, 1323–1334.
Shabarova, Z.A., Ivanovskaya, M.G., Isaguliants, M.G., 1983. DNA-like duplexes with repetitions: efficient template-guided polycondensation of decadeoxyribonucleotide imidazolide. FEBS Lett. 154, 288–292.
Shacter, E., 2000. Quantification and significance of protein oxidation in biological samples. Drug Metab. Rev. 32, 307–326.
Shaheen, S., Brabec, C.J., Sariciftci, N.S., Padinger, F., Fromherz, T., Hummelen, J.C., 2001. 2.5% efficient organic plastic solar cells. Appl. Phys. Lett. 78, 841–846.
Shainoff, J.R., 1980. Zonal immobilization of proteins. Biochem. Biophys. Res. Comm. 95, 690–695.
Shaked, H., Shiff, I., Kott-Gutkowski, M., Zahava Siegfried, Z., Haupt, Y., Simon, I., 2008. Chromatin immunoprecipitation–on-chip reveals stress-dependent p53 occupancy in primary normal cells but not in established cell lines. Cancer Res. 68, 9671–9677.
Shaltiel, S., 1967. Thiolysis of some dinitrophenyl derivatives of amino acids. Biochem. Biophys. Res. Comm. 29, 178.
Shanahan, M.F., Wadzinski, B.E., Lowndes, J.M., Ruoho, A.E., 1985. Photoaffinity labeling of the human erythrocyte monosaccharide transporter with an aryl azide derivative of D-glucose. J. Biol. Chem. 260, 10897–10900.
Shannon, L.M., Kay, E., Lew, J.Y., 1966. Peroxidase isozymes from horseradish roots. I. Isolation and physical properties. J. Biol. Chem. 241, 2166.
Shao, J., Sun, H., Guo, H., Ji, S., Zhao, J., Yuan, X., et al., 2011. A highly selective red-emitting FRET fluorescent molecular probe derived from BODIPY for the detection of cysteine and homocysteine: an experimental and theoretical study. Chem. Sci. 3, 1049–1061.
Shapiro, R., Weisgras, J.M., 1970. Bisulfite-catalyzed transamination of cytosine and cytidine. Biochem. Biophys. Res. Comm. 40, 839–843.
Shapiro, R., Braverman, B., Louis, J.B., Servis, R.E., 1973. Nucleic acid reactivity and conformation. II. Reaction of cytosine and uracil with sodium bisulfite. J. Biol. Chem. 248, 4060–4064.
Shapiro, R., DiFate, V., Welcher, M., 1974. Deamination of cytosine derivatives by bisulfite. Mechanism of the reaction. J. Am. Chem. Soc. 96, 906–912.
Sharkey, R.M., Rossi, E.A., Chang, C.-H., Goldenberg, D.M., 2010. Improved cancer therapy and molecular imaging with multivalent, multispecific antibodies. Biother. Radiopharm. 25 (1), 1–12.
Sharma, K.K., Kumar, R.S., Kumar, G.S., Quinn, P.T., 2000. Synthesis and characterization of a peptide identified as a functional element in α-crystallin. J. Biol. Chem. 275 (6), 3767–3771.
Sharma, R.K., Gulati, S., Mehta, S., 2012. Preparation of gold nanoparticles using tea: a green chemistry experiment. J. Chem. Educ. 89, 1316–1318.
Sharon, N., 2007. Lectins: carbohydrate-specific reagents and biological recognition molecules. J. Biol. Chem. 282, 2753–2764.
Sharon, N., Lis, H., 1989. Lectins as cell recognition molecules. Science 246, 227–234.
Sharpless, B.K., Fokin, V., Rostovtsev, V.V., Green, L., Himo, F., 2005. Copper-catalysed ligation of azides and acetylenes. US Patent Application: 2005/0222427 A1, published: October 6, 2005, filed May 30, 2003; provisional application filed May 30, 2002.
Shawler, D.L., Bartholomew, R.M., Smith, L.M., Dillman, R.O., 1985. Human immune response to multiple injections of murine monoclonal IgG. J. Immunol. 135, 1530–1535.
Shechter, Y., Mironchik, M., Rubinraut, S., Saul, A., Tsubery, H., Fridkin, M., 2005. Albumin–insulin conjugate releasing insulin slowly under physiological conditions: a new concept for long-acting insulin. Bioconjug. Chem. 16, 913–920.
Sheehan, D.C., Hrapchak, B.B., 1980. Theory and Practice of Histotechnology, second ed., Mosby, St. Louis, MO.
Sheehan, J.C., Hess, G.P., 1955. A new method of forming peptide bonds. J. Am. Chem. Soc. 77, 1067–1068.
Sheehan, J.C., Hlavka, J.J., 1956. The use of water-soluble and basic carbodiimides in peptide synthesis. J. Org. Chem. 21, 439–441.
Sheehan, J.C., Cruickshank, P.A., Boshart, G.L., 1961. A convenient synthesis of water-soluble carbodiimides. J. Org. Chem. 26, 2525–2528.
Sheehan, J.C., Preston, J., Cruickshank, P.A., 1965. A rapid synthesis of oligonucleotide derivatives without isolation of intermediates. J. Am. Chem. Soc. 87, 2492–2493.
Shek, P.N., Heath, T.D., 1983. Immune response mediated by liposome-associated protein antigens. III. Immunogenicity of bovine serum albumin covalently coupled to vesicle surface. Immunology 50, 101.

Shek, P.N., Sabiston, B.H., 1982a. Immune response mediated by liposome-associated protein antigens. I. Potentiation of the plaque-forming cell response. Immunology 45, 349.

Shek, P.N., Sabiston, B.H., 1982b. Immune response mediated by liposome-associated protein antigens. II. Comparison of the effectiveness of vesicle-entrapped and surface-associated antigens in immunopotentiation. Immunology 47, 627.

Shen, M., Guo, L., Wallace, A., Fetzner, J., Eisenman, J., Jacobson, E., et al., 2003. Isolation and isotope labeling of cysteine- and methionine-containing tryptic peptides, application to the study of cell surface proteolysis. Mol. Cell. Proteomics 2, 315–324.

Shephard, E.G., DeBeer, F.C., von Holt, C., Hapgood, J.P., 1988. The use of sulfosuccinimidyl-2-(p-azidosalicylamido)-1,39-dithiopropionate as a cross-linking reagent to identify cell surface receptors. Anal. Biochem. 168, 306–313.

Sherry, A.D., Caravan, P., Lenkinski, R.E., 2009. Primer on gadolinium chemistry. J. Magn. Reson. Imaging 30, 1240–1248.

Sherry, A.D., Brown III, R.D., Geraldes, C.F.C., Koeng, S.H., Kuan, K.-T., 1989. Synthesis and characterization of the gadolinium (31) complex of DOTA-propylamide: a model DOTA–protein conjugate. Inorg. Chem. 28, 620–622.

Sheshagirirao, J.V.L.N., Kumar, V., 2011. Affinity chromatography: a review. J. Pharm. Res. 4 (5), 1567–1574.

Shetty, J.K., Kinsella, J.E., 1980. Ready separation of proteins from nucleoprotein complexes by reversible modification of lysine residues. Biochem. J. 191, 269–272.

Shetty, K.J., Rao, M.S.N., 1978. Effect of succinylation on the oligomeric structure of arachin. Int. J. Pept. Protein Res. 11, 305.

Shi, B., Keough, E., Matter, A., Leander, K., Young, S., Carlini, E., et al., 2011. Biodistribution of small interfering RNA at the organ and cellular levels after lipid nanoparticle-mediated delivery. J. Histochem. Cytochem. 59 (8), 727–740.

Shi, J., Zhang, H., Wang, L., Li, L., Wang, H., Wang, Z., et al., 2012. PEI-derivatized fullerene drug delivery using folate as a homing device targeting to tumor. Biomaterials 34, 251–261.

Shi, J., Zhou, Y., Chakraborty, S., Kim, Y.-S., Jia, B., Wang, F., et al., 2011. Evaluation of ^{111}In-labeled cyclic RGD peptides: effects of peptide and linker multiplicity on their tumor uptake, excretion kinetics and metabolic stability. Theranostics 1, 322–340.

Shi, X., Bányai, I., Islam, M.T., Lesniak, W., Davis, D.Z., Baker Jr., J.R., et al., 2005. Generational, skeletal and substitutional diversities in generation one poly(amidoamine) dendrimers. Polymer 46 (9), 3022–3034.

Shi, X., Wang, S., Sun, H., Baker Jr., J.R., 2007. Improved biocompatibility of surface functionalized dendrimer-entrapped gold nanoparticles. Soft Matter 3, 71–74.

Shiao, D.D.F., Lumry, R., Rejender, S., 1972. Modification of protein properties by change in charge. Succinylated chymotrypsinogen. Eur. J. Biochem. 29, 377.

Shibata, S., Raubitschek, A., Leong, L., Koczywas, M., Williams, L., Zhan, J., et al., 2009. A phase I study of a combination of Yttrium-90–labeled anti–carcinoembryonic antigen (CEA) antibody and gemcitabine in patients with CEA-producing advanced malignancies. Clin. Cancer Res. 15, 2935–2941.

Shih, L.B., Goldenberg, D.M., Xuan, H., Lu, H., Sharkey, R.M., Hall, T.C., 1991. Anthracycline immunoconjugates prepared by a site-specific linkage via an aminodextran intermediate carrier. Cancer Res. 51, 4192–4198.

Shimada, Y., Watanabe, Y., Samukawa, T., Sugihara, A., Noda, H., Fukuda, H., et al., 1999. Conversion of vegetable oil to biodiesel using immobilized Candida antarctica lipase. JAOCS Vol. 76, 789–793.

Shimisu, N., Nickimins, W.K., Shimizu, Y., 1980. A cytotoxic epidermal growth factor cross-linked to diphtheria toxin A-fragment. FEBS Lett. 118, 274.

Shimkus, M., Levy, J., Herman, T., 1985. A chemically cleavable biotinylated nucleotide: usefulness in the recovery of protein–DNA complexes from avidin affinity columns. Proc. Natl. Acad. Sci. USA 82, 2593–2597.

Shimomura, S., Fukui, T., 1978. Characterization of the pyridoxal phosphate site in glycogen phosphorylase b from rabbit muscle. Biochemistry 17, 5359.

Shin, M.J., Tan, L., Jeong, M.H., Kim, J.H., Choe, W.S., 2011. Monolith-based immobilized metal affinity chromatography increases production efficiency for plasmid DNA purification. J. Chromatogr. A 1218 (31), 5273–5278.

Shinohara, H., Matsubayashi, Y., 2007. Functional immobilization of plant receptor-like kinase onto microbeads towards receptor array construction and receptor-based ligand fishing. Plant J. 52 (1), 175–184.

Shivdasani, R.A., Thomas, D.W., 1988. Molecular associations of IA antigens after T–B cell interactions. J. Immunol. 141, 1252–1260.

Shoseyov, O., Shani, Z., Levy, I., 2006. Carbohydrate binding modules: biochemical properties and novel applications. Microbiol. Mol. Biol. Rev. 70 (2), 283–295.

Shreewastav, R.K., Ali, R., Babu, J., Rao, D.N., 2012. Cell–mediated immune response to epitopic MAP (Multiple antigen peptide) construct of LcrV antigen of yersinia pestis in murine model. Cell. Immunol. 278, 55–62.

Shukla, S., Wu, G., Chatterjee, M., Yang, W., Sekido, M., Diop, L.A., et al., 2003. Synthesis and biological evaluation of folate receptor-targeted boronated PAMAM dendrimers as potential agents for neutron capture therapy. Bioconjug. Chem. 14, 158–167.

Shuvaev, V.V., Tliba, S., Pick, J., Arguiri, E., Christofidou-Solomidou, M., Albelda, S.M., et al., 2011. Modulation of endothelial targeting by size of antibody–antioxidant enzyme conjugates. J. Control. Release 149 (3), 236–241.

Sia, C.L., Horecker, B.L., 1968. Dissociation of protein subunits by maleylation. Biochem. Biophys. Res. Comm. 31, 731–737.

Sievers, E.L., Senter, P.D., 2012. Antibody-Drug Conjugates in Cancer Therapy. Annual review of medicine, published online ahead of print.

Siezen, R.J., Bindels, J.C., Hoenders, H.J., 1980. The quaternary structure of bovine a-crystallin. Chemical cross-linking with bifunctional imido esters. Eur. J. Biochem. 107, 243–249.

Silman, H.I., Albu-Weissenberg, M., Katchalski, E., 1966. Some water-insoluble papain derivatives. Biopolymers 4, 441–448.

Silva, S.S., Ferreira, R.A.S., Fu, L., Carlos, L.D., Mano, J.F., Reis, R.L., et al., 2005. Functional nanostructured chitosan–siloxane hybrids. J. Mater. Chem. 15, 3952–3961.

Silvius, J.R., Leventis, R., Brown, P.M., Zuchermann, M., 1987. Novel fluorescent phospholipids for assays of lipid mixing between membranes. Biochemistry 26, 4279–4287.

Simister, N.E., Rees, A.R., 1985. Isolation and characterization of an Fc receptor from neonatal rat small intestine. Eur. J. Immunol. 15 (7), 733–738.

Simon, J.R., Taylor, D.L., 1988. Preparation of a fluorescent analog: acetamidofluoresceinyl labeled dictyostelium discoideum a-actin, Vallee, R.B. Methods in Enzymology, Vol. 134. Academic Press, San Diego, pp. 47.

Simon, S.R., Konigsberg, W.H., 1966. Chemical modification of hemoglobins: a study of conformation restraint by internal bridging. Proc. Natl. Acad. Sci. USA 56, 749.

Simons, M., Friedrichson, T., Schulz, J.B., Pitto, M., Masserini, M., Kurzchalia, T.V., 1999. Exogenous administration of gangliosides displaces GPI-anchored proteins from lipid microdomains in living cells. Mol. Biol. Cell 10, 3187–3196.

Simons, P.C., Vander Jagt, D.L., 1977. Purification of glutathione S-transferases from human liver by glutathione-affinity chromatography. Anal. Biochem. 82 (2), 334–341.

Singaravelu, R., Blais, D.R., McKay, C.S., Pezacki, J.P., 2010. Activity-based protein profiling of the hepatitis C virus replication in Huh-7 hepatoma cells using a non-directed active site probe. Proteome Sci. 8, 5.

Singer, J.M., Plotz, C.M., 1956. The latex fixation test. I. Application to the serologic diagnosis of rheumatoid arthritis. Am. J. Med. 21, 888.

Singer, S.J., Fothergill, J.E., Shainoff, J.R., 1960. A general method for the isolation of antibodies. J. Am. Chem. Soc. 82, 565.

Singh, P., 1998. Terminal groups in starburst dendrimers: activation and reactions with proteins. Bioconjug. Chem. 9, 54–63.

Singh, P., Gupta, U., Asthana, A., Jain, N.K., 2008. Folate and folate-PEG-PAMAM dendrimers: synthesis, characterization, and targeted anticancer drug delivery potential in tumor bearing mice. Bioconjug. Chem. 2008 (19), 2239–2252.

Singh, R., Kats, L., Blattler, W.A., Lambert, J.M., 1996. Formation of *N*-substituted 2-iminothiolanes when amino groups in proteins and peptides are modified by 2-iminothiolane. Anal. Biochem. 236 (1), 114–125.

Singh, R., Barden, A., Mori, T., Beilin, L., 2001. Advanced glycation end-products: a review. Diabetologia 44 (2), 129–146.

Singh, R., Pantarotto, D., Lacerda, L., Pastorin, G., Klumpp, C., Prato, M., et al., 2006. Tissue biodistribution and blood clearance rates of intravenously administered carbon nanotube radiotracers. Proc. Natl. Acad. Sci. USA 103, 3357–3362.

Singh, V., Mavila, A.K., Kar, S.K., 1993. Comparison of the cytotoxic effect of hormonotoxins prepared with the use of heterobifunctional cross-linking agents *N*-succinimidyl 3-(2-pyridyldithio)propionate and *N*-succinimidyl 6-[3-(2-pyridyldithio) propionamido]-hexanoate. Bioconjug. Chem. 4, 473–482.

Singh, V., Sairam, M.R., Bhargavi, G.N., Akhras, R.G., 1989. Hormonotoxins: preparation and characterization of ovine luteinizing hormone–gelonin conjugate. J. Biol. Chem. 264, 3089–3095.

Singhal, A., Gupta, C.M., 1986. Antibody-mediated targeting of liposomes to red cells in vivo. FEBS Lett. 201, 321.

Singhal, R.P., Bajaj, R.K., Buess, G.M., Smoll, D.B., Vakharia, V.N., 1980. Reversed-phase boronate chromatography for the separation of O-methylribose nucleosides and aminoacyl-tRNAs. Anal. Biochem. 109, 1–11.

Sinz, A., 2003. Chemical cross-linking and mass spectrometry for mapping three-dimensional structures of proteins and protein complexes. J. Mass Spectrom. 38, 1225–1237.

Sinz, A., 2006. Chemical cross-linking and mass spectrometry to map three-dimensional protein structures and protein–protein interactions. Mass Spectrom. Rev. 25, 663–682.

Sinz, A., Kalkhof, S., Ihling, C., 2005. Mapping protein interfaces by a trifunctional cross-linker combined with MALDI-TOF and ESI-FTICR mass spectrometry. J. Am. Soc. Mass Spectrom. 16 (12), 1921–1931.

Sipe, D.M., Jesurum, A., Murphy, R.F., 1991. Absence of Na^1, K^1-ATPase regulation of endosomal acidification in K562 erythroleukemia cells. J. Biol. Chem. 266, 3469.

Sippel, T.O., 1981. New fluorochromes for thiols: maleimide and iodoacetamide derivatives of 3-phenylcoumarin fluorophore. J. Histochem. Cytochem. 29, 314.

Sissona, T.H., Castor, C.W., 1990. An improved method for immobilizing IgG antibodies on protein A-agarose. J. Immunol. Methods Vol. 127, 215–220.

Siuzdak, G., 2006. The Expanding Role of Mass Spectrometry in Biotechnology, second ed., MCC Press, ISBN-13: 978-0974245126.

Sivakoff, S.I., Janes, C.J., 1988. Automated high performance gel-filtration chromatography (HPGFC) processing of avidin coupled b-galactosidase. Biochromatography 3, 62–68.

Sivaraman, N., Srinivasan, T., Vasudeva Rao, P., Natarajan, R., 2001. QSPR modeling for solubility of fullerene (C60) in organic solvents. J. Chem. Inf. Comput. Sci. 41, 1067–1074.

Skare, J.T., Ahmer, B.M., Seachord, C.L., Darveau, R.P., Postle, K., 1993. Energy transduction between membranes. TonB, a cytoplasmic membrane protein, can be chemically cross-linked in vivo to the outer membrane receptor FepA. J. Biol. Chem. 268, 16302–16308.

Skold, S.-E., 1983. Chemical crosslinking of elongation factor G to the 23S RNA in 70S ribosomes from *Escherichia coli*. Nucleic Acids Res 11, 4923.

Slatkin, D.N., 1991. A history of boron neutron capture therapy of brain tumors; postulation of a brain radiation dose tolerance limit. Brain 114 (4), 1609–1629.

Slaughter, T.F., Achyuthan, K.E., Lai, T.-S., Greenberg, C.S., 1992. A microtiter plate transglutaminase assay utilizing 5-(biotinamido) pentylamine as substrate. Anal. Biochem. 205, 1–6.

Sletten, E.M., Bertozzi, C.R., 2009. Bioorthogonal chemistry: fishing for selectivity in a sea of functionality. Angew. Chem. Int. Ed. 48, 6974–6998.

Slinkin, M.A., Klibanov, A.L., Torchilin, V.P., 1991. Terminal-modified polylysine-based chelating polymers: highly efficient coupling to antibody with minimal loss in immunoreactivity. Bioconjug. Chem. 2, 342–348.

Slinkin, M.A., Klibanov, A.L., Khaw, B.A., Torchilin, V.P., 1990. Succinylated polylysine as a possible link between an antibody molecule and deferoxamine. Bioconjug. Chem. 1, 291–295.

Slomkowski, S., Miksa, B., Chehimi, M.M., Delamar, M., Cabet-Deliry, M., Majoral, J.P., et al., 1999. Inorganic–organic systems with tailored properties controlled on molecular, macromolecular and microscopic level. React. Funct. Polym. 41, 45–57.

Slot, J.W., Geuze, H.J., 1984. Cold markers for single and double immunolabeling of ultrathin cryosections. In: Polak, J.M., Varndess, I.M. (Eds.), Immunolabeling for Electron Microscopy. Elsevier, New York, pp. 139.

Smirnov, E.A., Meledina, M.A., Garshev, A.V., Chelpanov, V.I., Frost, S., Wienecke, J.U., & et al., (2012). Grafting of titanium dioxide microspheres with a temperature-responsive polymer via surface-initiated atom transfer radical polymerization without the use of silane coupling agents. Polym. Int. doi: 10.1002/pi.4377.

Smith, B.J., 2002. Chemical cleavage of proteins at asparaginyl-glycyl peptide bonds. In: Walker, J.M. (Ed.), The Protein Protocols Handbook (second ed.). Humana Press Inc., Totowa, New Jersey, USA, pp. 507.

Smith, G.P., 1985. Filamentous fusion phage: novel expression vectors that display cloned antigens on the virion surface. Science 228, 1315–1317.

Smith, B., Lyakhov, I., Loomis, K., Needle, D., Baxa, U., Yavlovich, A., et al., 2011. Hyperthermia-triggered intracellular delivery of anticancer agent to HER2 cells by HER2-specific affibody (ZHER2-GS-Cys)-conjugated thermosensitive liposomes (HER2 affisomes). J. Control. Release 153 (2), 187–194.

Smith, C.H., Barker, J.N.W.N., 2006. Psoriasis and its management. BMJ 333, 380–384.

Smith, D.B., Johnson, K.S., 1988. Single-site purification of polypeptides expressed in Escherichia coli as fusions with glutathione S-transferase. Gene 67, 31–40.

Smith, D.F., Cummings, R.D., 2009. Glycan-binding proteins and glycan microarrays. In: Cummings, R.D., Pierce, J.M. (Eds.), Handbook of Glycomics. Academic Press, London, UK, pp. 139–160.

Smith, D.F., Zopf, D.A., Ginsburg, V., 1978. Carbohydrate antigens: coupling of oligosacchanide phenethylamine-isothiocyanate derivatives to bovine serum albumin, Ginsburg, V. Methods in Enzymology, Vol. 50. Academic Press, New York, pp. 169–171.

Smith, L.M., Fung, S., Hunkapiller, M.W., Hunkapiller, T.J., Hood, L.E., 1985. The synthesis of oligonucleotides containing an aliphatic amino group at the 59 terminus: synthesis of fluorescent

DNA primers for use in DNA sequence analysis. Nucleic Acids Res. 13, 2399–2412.

Smith, M.B., March, J., 2007. March's Advanced Organic Chemistry: Reactions, Mechanisms, and Structure, sixth ed., John Wiley and Sons, Inc., Hobokan, New Jersey.

Smith, P.K., Krohn, R.I., Hermanson, G.T., Malhia, A.K., Gartner, F.H., Provenzano, M.D., et al., 1985. Measurement of protein using bicinchoninic acid. Anal. Biochem. 150, 76–85.

Smith, R.A.G., Knowles, J.R., 1973. Aryldiazinines. Potential reagents for photolabeling of biological receptor sites. J. Am. Chem. Soc. 95, 5072–5073.

Smith, R.J., Capaldi, R.A., Muchmone, D., Dahlquist, F., 1978. Cross-linking of ubiquinone cytochnome c reductase (complex III) with periodate-cleavable bifunctional reagents. Biochemistry 17, 3719–3723.

Smith, S.M., Bomgarden, R.D., Deshpande, A., Farooqui, R., Kaboord, B.J., 2008. Comparison of small GTPase activation over time after growth factor stimulation in 3T3 cells. Mol. Biol. Cell 19 (Suppl.) abstract 1243/B452.

Smith, S.W., 2012. Lipid-protein interactions in biological membranes: a dynamic perspective. Biochim. Biophys. Acta – Biomembr. 1818, 172–177.

Smolewski, P., Bedner, E., Du, L., Hsieh, T.-C., Wu, J., Phelps, J.D., et al., 2001. Detection of caspase activation by fluorochrome-labeled inhibitors: multiparameter analysis by laser scanning cytometry. Cytometry 44, 73–82.

Smyth, D.G., 1967. Acetylation of amino and tyrosine hydroxyl groups. J. Biol. Chem. 242, 1592–1598.

Smyth, D.G., Nagamatsu, A., Fruton, J.S., 1960. Reactions of N-ethylmaleimide. J. Am. Chem. Soc. 82, 4600.

Smyth, D.G., Blumenfeld, O.O., Konigsberg, W., 1964. Reaction of N-ethylmaleimide with peptides and amino acids. Biochem. J. 91, 589.

Snyder, B., Hammes, G.G., 1984. Structural mapping of chloroplast coupling factor. Biochemistry 23, 5787–5795.

Snyder, B., Hammes, G.G., 1985. Structural organization of chloroplast coupling factor. Biochemistry 24, 2324–2331.

Snyder, H.R., Reedy, A.J., Lennarz, W.J., 1958. Synthesis of aromatic boronic acids, aldehydo boronic acids and a boronic acid analog of tyrosine. J. Am. Chem. Soc. 80, 835–838.

Söderberg, O., Gullberg, M., Jarvius, M., Ridderstråle, K., Leuchowius, K.J., Jarvius, J., et al., 2006. Direct observation of individual endogenous protein complexes in situ by proximity ligation. Nat. Methods 3 (12), 995–1000.

Soellner, M.B., Nilsson, B.L., Raines, R.T., 2002. Staudinger ligation of a-azido acids retains stereo-chemistry. J. Org. Chem. 67, 4993–4996.

Soellner, M.B., Dickson, K.A., Nilsson, B.L., Raines, R.T., 2003. Site-specific protein immobilization by Staudinger ligation. J. Am. Chem. Soc. 125, 11790–11791.

Soini, E., Kojola, H., 1983. Time-resolved fluorometer for lanthanide chelates—A new generation of monisotopic immunoassays. Clin. Chem. 29, 65–68.

Soini, E., Lövgren, T., 1987. Time-resolved fluorescence of lanthanide probes and applications in biotechnology. CRC Crit. Rev. Anal. Chem. 18, 104–154.

Soini, E., Hemmila, I., Dahlen, P., 1990. Time-resolved fluorescence in biospecific assays (Review). Ann. Biol. Clin. 48, 567–571.

Soininen, S.K., Lehtolainen-Dalkilic, P., Karppinen, T., Puustinen, T., Dragneva, G., Kaikkonen, M.U., et al., 2012. Targeted delivery via avidin fusion protein: intracellular fate of biotinylated doxorubicin derivative and cellular uptake kinetics and biodistribution of biotinylated liposomes. Eur. J. Pharm. Sci. 47, 848–856.

Sokolovsky, M., Riordan, J.F., Vallee, B.L., 1967. Conversion of 3-nitrotyrosine to 3-aminotyrosine in peptides and proteins. Biochem. Biophys. Res. Comm. 27, 20.

Solomon, S.R., Mielke, S., Savani, B.N., Montero, A., Wisch, L., Childs, R., et al., 2005. Selective depletion of alloreactive donor lymphocytes: a novel method to reduce the severity of graft-versus-host disease in older patients undergoing matched sibling donor stem cell transplantation. Blood 106, 1123–1129.

Song, X., Heimburg-Molinaro, J., Smith, D.F., Cummings, R.D., 2011. Derivatization of free natural glycans for incorporation onto glycan arrays: derivatizing glycans on the microscale for microarray and other applications. Curr. Protoc. Chem. Biol. 3, 53–63.

Sorensen, K., 1993. Method for isolation and purification of enzyme–antibody conjugates. US Patent No. 5,266,686.

Sorenson, P., Farber, N.M., Krystal, G., 1986. Identification of the interleukin-3 receptor using an iodinatable, cleavable, photoreactive cross-linking agent. J. Biol. Chem. 261, 9094–9097.

Soundararajan, N., Liu, S.H., Soundararajan, S., Platz, M.S., 1993. Synthesis and binding of new polyfluoninated aryl azides to a-chymotrypsin. New reagents for photoaffinity labeling. Bioconjug. Chem. 4, 256–261.

Southerland, B., Kulkarni-Datar, K., Keoni, C., Bricker, R., Grunwald, W.C., Ketcha, D.M., et al., 2010. Q-Ve-Oph, a negative control for O-phenoxy-conjugated caspase Inhibitors. J. Cell Death 3, 33–40.

Southwick, P.L., Ernst, L.A., Tauriello, E.W., Parker, S.R., Mujumdar, R.B., Mujumdar, S.R., et al., 2005. Cyanine dye labeling reagents—carboxymethylindocyanine succinimidyl esters. Cytometry 11 (3), 418–430.

Souza, E.D., Ginsberg, M.H., Lam, S., Plow, E.F., 1988. Chemical cross-linking of arginylglycyl-aspartic acid peptides to an adhesion receptor on platelets. J. Biol. Chem. 263, 3943–3951.

Spangler, C.W., Spangler, B.D., Tarter, E.S., Suo, Z., 2004. Design and synthesis of dendritic tethers for the immobilization of antibodies for the detection of class A bioterror pathogens. Poly. Preprints 45 (1), 524.

Spasic, D., Raemaekers, T., Dillen, K., Declerck, I., Baert, V., Serneels, L., et al., 2007. Rer1p competes with APH-1 for binding to nicastrin and regulates–secretase complex assembly in the early secretory pathway. J. Cell Biol. 176, 629–640.

Speers, A.E., Cravatt, B.F., 2004a. Chemical strategies for activity-based proteomics. Chem. Bio. Chem. 5, 41–47.

Speers, A.E., Cravatt, B.F., 2004b. Profiling enzyme activities in vivo using click chemistry methods. Chem. Biol. 11, 535–546.

Speers, A.E., Adam, G.C., Cravatt, B.F., 2003. Activity-based protein profiling in Vivo using a Copper(I)-Catalyzed Azide-Alkyne [3+2] Cycloaddition. J. Am. Chem. Soc. 125, 4686–4687.

Spektor, T.M., Congdon, L.M., Verappan, C.S., Rice, J.C., 2011. The UBC9 E2 SUMO conjugating enzyme binds the PR-Set7 histone methyltransferase to facilitate target gene repression. PLoS ONE 6 (7), e22785. doi: 10.1371/journal.pone.0022785.

Spiegel, S., Skutelsky, E., Bayer, E.A., Wilchek, M., 1982. A novel approach for the topographical localization of glycolipids on the cell surface. Biochim. Biophys. Acta 687, 27–34.

Spiegel, S., Wilchek, M., Fishman, P.H., 1983. Fluorescent labeling of cell surface glycoconjugates with Lucifer Yellow CH. Biochem. Biophys. Res. Comm. 112, 872–877.

Spiegel, S., Yamada, K.M., Hom, B.E., Moss, J., Fishman, P.H., 1985. Fluorescent gangliosides as probes for the retention and organization of fibronectin by ganglioside-deficient mouse cells. J. Cell Biol. 100, 721–726.

Springer, A.L., Gall, A.S., Hughes, K.A., Kaiser, R.J., Li, G., Lucas, D.D., et al., 2002. Affinity-based immobilization tools for functional genomics. Presented at Transcriptome 2002: From Functional to Systems Biology, 10–13 March 2002, Seattle, WA.

Springer, A.L., Gall, A.S., Hughes, K.A., Kaiser, R.J., Li, G., Lund, K.P., 2003. Salicylhydroxamic acid functionalized affinity membranes for specific immobilization of proteins and oligonucleotides. J. Biomol. Tech. 14 (3), 183–190.

Srikant, P., Singh, A.K., McElhanon, J.R., Dentinger, P.M., 2004. Dendrimer-activated surfaces for high density and high activity protein chip applications. Langmuir 20 (15), 6075–6079.

Srikrishna, G., Toomre, D.K., Manzi, A., Panneerselvam, K., Freeze, H.H., Varki, A., et al., 2001. A novel anionic modification of *N*-glycans on mammalian endothelial cells is recognized by activated neutrophils and modulates acute inflammatory responses. J. Immunol. 166, 624–632.

Srinivasachar, K., Neville Jr., D.M., 1989. New protein cross-linking reagents that are cleaved by mild acid. Biochemistry 28, 2501.

St. Clair, N.L., Navia, M.A., 1992. Cross-linked enzyme crystals as robust biocatalysts. J. Am. Chem. Soc. 114, 7314–7316.

Stadtman, E.R., 1992. Protein oxidation and aging. Science 257, 1220.

Stadtman, E.R., Levine, R.L., 2000. Protein oxidation. Ann. NY. Acad. Sci. 899, 191–208.

Stahl, S., Hober, S., Nilsson, J., Uhlen, M., Nygren, P.-A., 2003. Genetic approaches to facilitate protein purification. In: Hatti-Kaul, R., Mattiasson, B. (Eds.), Isolation and Purification of Proteins. Marcel Dekker AG, Basel, Switzerland.

Stahlberg, T., Markela, E., Mikola, H., Mottram, P., Hemmila, I., 1993. Europium and samarium in time-resolved fluoroimmunoassays. Am. Lab., 15–20.

Stamler, J.S., Lamas, S., Fang, F.C., 2001. Nitrosylation: the prototypic redox-based signaling mechanism. Cell 106, 675–683.

Stanek, L.G., Heilmann, S.M., Gleason, W.B., 2005. Synthesis and characterization of copolymers containing N,N-dimethylacrylamide and 2-vinyl-4,4′-dimethylazlactone. Polym. Bull. 55, 393–402.

Staneloudi, C., Smith, K.A., Hudson, R., Malatesti, N., Savoie, H., Boyle, R.W., et al., 2007. Development and characterization of novel photosensitizer: scFv conjugates for use in photodynamic therapy of cancer. Immunology 120 (4), 512–517.

Staros, J.V., 1982. N-hydroxysulfosuccinimide active esters: Bis(N-hydroxysulfosuccinimide) esters of two dicarboxylic acids are hydrophilic, membrane impermeant, protein cross-linkers. Biochemistry 21, 3950–3955.

Staros, J.V., 1988. Membrane-impermeant cross-linking reagents: probes of the structure and dynamics of membrane proteins. Acc. Chem. Res. 21, 435–441.

Staros, J.V., Kakkad, B.P., 1983. Cross-linking and chymotryptic digestion of the extracytoplasmic domain of the anion exchange channel in intact human erythrocytes. J. Membr. Biol. 74, 247–254.

Staros, J.V., Bayley, H., Standring, D.N., Knowles, J.R., 1978. Reduction of aryl azides by thiols: implication for the use of photoaffinity reagents. Biochem. Biophys. Res. Comm. 80, 568.

Staros, J.V., Wright, R.W., Swinghe, D.M., 1986. Enhancement by N-hydroxysulfosuccinimide of water-soluble carbodiimide-mediated coupling reactions. Anal. Biochem. 156, 220–222.

Staros, J.V., Lee, W.T., Conrad, D.H., 1987. Membrane-impermeant cross-linking reagents: application to the study of the cell surface receptor for IgEDi Sabato, G. Methods in Enzymology, Vol. 150. Academic Press, Orlando, FL, pp. 503–512.

Stasiuk, G.J., Faulkner, S., Long, N.J., 2012. Novel imaging chelates for drug discovery. Curr. Opin. Pharmacol. 12, 576–582.

Staudinger, H., Meyer, J., 1919. Über neue organische phosphorverbindungen III. Phosphinmethyl-enderivate und phosphinimine. Helv. Chim. Acta 2, 635–646.

Stauffer, D.A., Karlin, A., 1994. Electrostatic potential of the acetylcholine binding sites in the nicotinic receptor probed by reaction of binding-site cysteines with charged methanethiosulfonates. Biochemistry 33, 6840–6849.

Stears, R.L., Getts, R.C., Gullans, S.R., 2000. A novel, sensitive detection system for high-density microarrays using dendrimer technology. Physiol. Genomics 3, 93–99.

Steel, G.J., Fullerton, D.M., Tyson, J.R., Stirling, C.J., 2004. Coordinated activation of Hsp70 chaperones. Science 303, 98–101.

Steer, C.J., Ashwell, G., 1986. Hepatic membrane receptors for glycoproteins. Prog. Liver Dis. 8, 99–123.

Stefflova, K., Chen, J., Zheng, G., 2007. Using molecular beacons for cancer imaging and treatment. Front. Biosci. 12, 4709–4721.

Sternberger, L.A., 1986. Immunocytochemistry. Wiley, New York.

Stewart, J.M., Young, J.D., 1984. Solid Phase Peptide Synthesis, second ed., Pierce Chemical Company, Rockford, IL. p. 31.

Stewart, W.W., 1978. Functional connections between cells as revealed by dye-coupling with a highly fluorescent naphthahimide tracer. Cell 14, 741.

Stewart, W.W., 1981a. Lucifer dyes—Highly fluorescent dyes for biological tracing. Nature (London) 292, 17.

Stewart, W.W., 1981b. Synthesis of 3,6-disulfonate 4-aminonaphthalimides. J. Am. Chem. Soc. 103, 7615.

Stickel, S.K., Wang, Y.-L., 1988. Synthetic peptide GRGDS induces dissociation of alpha-actin and vinculin from the sites of focal contacts. J. Cell Biol. 107, 1231.

Stirpe, F., Olsnes, S., Pihl, A., 1980. Gelonin, a new inhibitor of protein synthesis, nontoxic to intact cells. Isolation, characterization, and preparation of cytotoxic complexes with concanavahin A. J. Biol. Chem. 255, 6947–6953.

Stöber, W., Fink, A., Bohn, E.J., 1968. Controlled growth of monodisperse silica spheres in the micron size range. Colloid Interface Sci. 26, 62–69.

Stocking, E.M., Williams, R.M., 2002. Chemistry and biology of biosynthetic Diels–Alder reactions. Angew. Chem. Int. Ed. Engl. 42 (27), 3078–3115.

Stolowitz, M.L., 1997. Phenylboronic acid complexes for bioconjugate preparation. US Patent No. 5, 594, 111.

Stolowitz, M.L., Ahlem, C., Hughes, K.A., Kaiser, R.J., Kesicki, E.A., Li, G., et al., 2001. Phenylboronic acid–salicylhydroxamic acid bioconjugates. 1. A novel boronic acid complex for protein immobilization. Bioconjug. Chem. 12, 229–239.

Strassberger, V., Trüssel, S., Fugmann, T., Neri, D., Roesli, C., 2011. A novel reactive ester derivative of biotin with reduced membrane permeability for in vivo biotinylation experiments. Proteomics-Clin. Appl. 5 (3–4), 194.

Straubinger, R.M., Lopez, N.G., Debs, R.J., Hong, K., Papahajopoulos, D., 1988. Liposome-based therapy of human ovarian cancer: parameters determining potency of negatively charged and antibody-targeted liposomes. Cancer Res. 48, 5237–5245.

Strickland, E., Hardin, E.K., Shannon, L.M., Horwitz, J., 1968. Peroxidase isoenzymes from horseradish roots. III. Circular dichroism of isoenzymes and apoisoenzymes. J. Biol. Chem. 243, 3560.

Striebel, H.M., Birch-Hirschfeld, E., Egerer, R., Foldes-Papp, Z., Tilz, G.P., Stelzner, A., 2004. Enhancing sensitivity of human herpes virus diagnosis with DNA microarrays using dendrimers. Exp. Mol. Pathol. 77, 89–97.

Strijkers, G.J., Mulder, W.J., van Tilborg, G.A., Nicolay, K., 2007. MRI contrast agents: current status and future perspectives. Anticancer Agents Med. Chem. 7, 291–305.

Strottmann, J.M., Robinson Jr., J.B., Stehlwagen, E., 1983. Advantages of preelectrophoretic conjugation of polypeptides with fluorescent dyes. Anal. Biochem. 132, 334–337.

Stuchbury, T., Shipton, M., Norris, R., Malthouse, J.P.G., Brocklehurst, K., 1975. Reporter groups delivery system with both absolute and selective specificity for thiol groups and an improved fluorescent probe containing the 7-nitrobenzo-2-oxa-1,3-diazole moiety. Biochem. J. 151, 417–432.

Stults, N.L., Asta, L.M., Lee, Y.C., 1989. Immobilization of proteins on oxidized crosslinked Sepharose preparations by reductive amination. Anal. Biochem. 180, 114–119.

Stump, R.F., Pfeiffer, J.R., Schneebeck, M.C., Seagrave, J.C., Oliver, J.M., 1989. Mapping gold-labeled receptors on cell surfaces by backscattered electron imaging and digital image analysis: studies of the IgE receptor on mast cells. Am. J. Anat. 185, 128–141.

Stutz, A., Bertheloot, D., Latz, E., 2011. Innate immune receptors for nucleic acidsRast, J.P.Booth, J.W.D. Methods in Molecular Biology, 748. Humana Press, Clifton, NJ, pp. 69–82.

Subbiahdoss, G., Sharifi, S., Grijpma, D.W., Laurent, S., van der Mei, H.C., Mahmoudi, M., et al., 2012. Magnetic targeting of surface-modified superparamagnetic iron oxide nanoparticles yields antibacterial efficacy against biofilms of gentamicin-resistant staphylococci. Acta Biomater. 8, 2047–2055.

Subramanian, A., 2000. Purification of immunoglobulins from serum using thiophilic cellulose beads. Chemical and Biomolecular Engineering Research and Publications. DigitalCommons@ University of Nebraska – Lincoln, Paper 24.

Subramanian, C., Woo, J., Cai, X., Xu, X., Servick, S., Johnson, C.H., et al., 2006. A suite of tools and application notes for in vivo protein interaction assays using bioluminescence resonance energy transfer (BRET). Plant J. 48, 138–152.

Subramanian, R., Meares, C.F., 1991. Bifunctional chelating agents for radiometal-labeled monoclonal antibodies. In: Coldenberg, D.M. (Ed.), Cancer Imaging with Radiolabeled Antibodies. Kluwer, Boston, MA, pp. 183–199.

Suchanek, M., Radzikowska, A., Thiele, C., 2005. Photo-leucine and photo-methionine allow identification of protein–protein interactions in living cells. Nat. Methods 2 (4), 261–267.

Suen, S.-Y., Tsai, Y.-D., 2000. Comparison of ligand density and protein adsorption on dye-affinity membranes using different spacer arms. Sep. Sci. Technol. 35, 69–87.

Suh, C.W., Choi, G.S., Lee, E.K., 2003. Enzymic cleavage of fusion protein using immobilized urokinase covalently conjugated to glyoxyl-agarose. Biotechnol. Appl. Biochem. 37, 149–155.

Sukhanova, A., Even-Desrumeaux, K., Kisserli, A., Tabary, T., Reveil, B., Millot, J.M., et al., 2012. Oriented conjugates of single-domain antibodies and quantum dots: toward a new generation of ultrasmall diagnostic nanoprobes. Nanomedicine 8 (4), 516–525.

Sulkowski, E., 1985. Purification of proteins by IMAC. Trends Biotechnol. 3, 1–7.

Sun, J.-S., Francois, J.-C., Lavery, R., Saison-Behmoaras, T., Montenay-Garestier, T., Thuong, N.T., et al., 1988. Sequence-targeted cleavage of nucleic acids by oligo-a-thymidylate–phenanthroline conjugates: parallel and antiparallel double helices are formed with DNA and RNA, respectively. Biochemistry 27, 6039–6045.

Sun, M.M.C., Beam, K.S., Cerveny, C.G., Hamblett, K.J., Blackmore, R.S., Torgov, M.Y., et al., 2005. Reduction–alkylation strategies for the modification of specific monoclonal antibody disulfides. Bioconjug. Chem. 16, 1282–1290.

Sun, T.T., Bollen, A., Kahan, L., Traut, R.R., 1974. Topography of ribosomal proteins of the Escherichia coli 30S subunit as studied with the reversible cross-linking reagent methyl 4-mercaptobutyrimidate. Biochemistry 13, 2334–2340.

Sun, X.-L., Stabler, C.L., Cazalis, C.S., Chaikof, E.L., 2006. Carbohydrate and protein immobilization onto solid surfaces by sequential diels–alder and azide–alkyne cycloadditions. Bioconjug. Chem. 17, 52–57.

Sundberg, L., Porath, J., 1974. Preparation of adsorbents for biospecific affinity chromatography. I. Attachment of group containing ligands to insoluble polymers by means of bufunctional oxiranes. J. Chromatogr. 90, 87–98.

Sunder, A., Hanselmann, R., Frey, H., Mülhaupt, R., 1999. Controlled synthesis of hyperbranched polyglycerols by ring-opening multibranching polymerization. Macromolecules 32, 4240–4246.

Sunder, A., Mülhaupt, R., Haag, R., Frey, H., 2000. Chiral hyperbranched dendron-analogs. Macromolecules 33, 253–254.

Susaki, H., Suzuki, K., Ikeda, M., Yamada, H., Watanabe, H.K., 1998. Renal drug targeting using a vector "alkylglycoside". Chem. Pharm. Bull. 46, 1530–1537.

Suter, M., Butler, J.E., 1986. The immunochemistry of sandwich ELISAs. II. A novel system prevents the denaturation of capture antibodies. Immunol. Lett. 13, 313–316.

Sutherland, B.W., Toews, J., Kast, J., 2008. Utility of formaldehyde cross-linking and mass spectrometry in the study of protein–protein interactions. J. Mass Spectrom. 43 (6), 699–715.

Sutoh, K., Hiratsuka, T., 1988. Spatial proximity of the glycine-rich loop and the SH2 thiol in myosin subfragment 1. Biochemistry 27 (8), 2964–2969.

Sutoh, K., Yamamoto, K., Wakabayashi, T., 1984. Electron microscopic visualization of the SH1 thiol of myosin by the use of an avidin–biotin system. J. Mol. Biol. 178, 323–339.

Suttiponparnit, K., Jiang, J., Sahu, M., Suvachittanont, S., Charinpanitkul, T., Biswas, P., 2011. Role of surface area, primary particle size, and crystal phase on titanium dioxide nanoparticle dispersion properties. Nanoscale Res. Lett. 6 (1), 27.

Svec, F., Frechet, J.M.J., 1999. Molded rigid monolithic porous polymers: an inexpensive, efficient, and versatile alternative to beads for the design of materials for numerous applications. Ind. Eng. Chem. Res. 38, 34–48.

Svec, F., 2002. Organic polymer support materials. In: Gooding, K.M., Regnier, F.E. (Eds.), HPLC of Biological Macromolecules. Marcel Dekker, Inc., New York, NY, pp. 17–48.

Svec, F., Tennikova, T.B., Deyl, Z. (Eds.), 2003. Monolithic Materials: Preparation, Properties and Applications (Journal of Chromatography Library). Elsevier Science B.V., Amsterdam, The Netherlands.

Swali, V., Wells, N.J., Langley, G.J., Bradley, M., 1997. Solid-phase dendrimer synthesis and the generation of super-high-loading resin beads for combinatorial chemistry. J. Org. Chem. 62, 4902–4903.

Swanson, J., Burke, E., Silverstein, S.C., 1987. Tubular lysosomes accompany stimulated pinocytosis in macrophages. J. Cell Biol. 104, 1217.

Swanson, S.D., Kukowska-Latallo, J.F., Patri, A.K., Chen, C., Ge, S., Cao, Z., et al., 2008. Targeted gadolinium-loaded dendrimer nanoparticles for tumor-specific magnetic resonance contrast enhancement. Int. J. Nanomed. 3 (2), 201–210.

Swanson, S.J., Lin, B.-F., Mullenix, M.C., Mortensen, R.F., 1991. A synthetic peptide corresponding to the phosphorylcholine (PC)-binding region of human C-reactive protein possesses the TEPC-15 myeloma PC-idiotype. J. Immunol. 146, 1596–1601.

Swanton, E., Holland, A., High, S., Woodman, P., 2005. Disease-associated mutations cause premature oligomerization of myelin proteolipid protein in the endoplasmic reticulum. PNAS 102, 4342–4347.

Sweeley, C.C., Nunez, H.A., 1985. Structural analysis of glycoconjugates by mass spectrometry and nuclear magnetic resonance spectroscopy. Annu. Rev. Biochem. 54, 765–801.

Sykaluk, L., 1994. unpublished data, Pierce Chemical.

Szabo, J., Kruger, S.R., Beall, G.N., 1989. Detection of cells producing anti-idiotypic antibody to thyroid stimulating hormone-reactive antibodies. Immunol. Invest. 18, 879–884.

Szoka, F., Olson, F., Heath, T., Vail, W., Mayhew, E., Paphadjopoulos, D., 1980. Preparation of unilamellar liposomes of intermediate size (0.1–0.2μmol) by a combination of reverse phase evaporation and extrusion through polycarbonate membranes. Biochim. Biophys. Acta 601, 559–571.

Szwergold, B.S., Howell, S., Beisswenger, P.J., 2001. Human fructosamine-3-kinase; purification, sequencing, substrate specificity, and evidence of activity *in vivo*. Diabetes 50, 2139–2147.

Tabor, S., Richardson, C.C., 1987. Selective oxidation of the exonuclease domain of bacteriophage T7 DNA polymerase. J. Biol. Chem. 262, 15330–15333.

Tada, H., Higuchi, H., Wanatabe, T.M., Ohuchi, N., 2007. *In vivo* real-time tracking of single quantum dots conjugated with monoclonal anti-HER2 antibody in tumors of mice. Cancer Res. 67, 1138–1144.

Tadayoni, B.M., Friden, P.M., Walus, L.R., Musso, G.F., 1993. Synthesis *in vitro* kinetics, and *in vivo* studies on protein conjugates of

AZT: evaluation as a transport system to increase brain delivery. Bioconjug. Chem. 4, 139–145.
Tager, H.S., 1976. Coupling of peptides to albumin with difluorodinitrobenzene. Anal. Biochem. 71, 367–375.
Taherzadeh, M.J., Karimi, K., 2007. Enzyme- based hydrolysis processes for ethanol from lignocellulosic materials: a review. BioResources 2, 707–738.
Takadate, A., Irikura, M., Suehiro, T., Fujino, H., Goya, S., 1985. New labeling reagents for alcohols in fluorescence high-performance liquid chromatography. Chem. Pharm. Bull. 33, 1164–1169.
Takahashi, K., 1968. The reaction of phenylglyoxal with arginine residues in proteins. J. Biol. Chem. 243, 6171–6179.
Takasaki, S., Kobata, A., 1978. Microdetermination of sugar composition of radioisotope labeling. Meth. Enzymol. 50, 50–54.
Taldone, T., Gomes-DaGama, E.M., Zong, H., Sen, S., Alpaugh, M.L., Zatorska, D., et al., 2011. Synthesis of purine-scaffold fluorescent probes for heat shock protein 90 with use in flow cytometry and fluorescence microscopy. Bioorg. Med. Chem. Lett. 21 (18), 5347–5352.
Talsma, H., van Steenberg, M.J., Crommelin, D.J.A., 1991. The cryopreservation of liposomes: 3. Almost complete retention of a water-soluble marker in small liposomes in a cryoprotectant containing dispersion after a freezing/thawing cycle. Int. J. Pharm. 77, 119–126.
Tam, J.P., 1988. Synthetic peptide vaccine design: synthesis and properties of a high-density multiple antigenic peptide system. Proc. Natl. Acad. Sci. USA 85, 5409–5413.
Tam, P., Lu, Y.-A., Liu, C.-F., Shao, J., 1995. Peptide synthesis using unprotected peptides through orthogonal coupling methods. Proc. Natl. Acad. Sci. USA 92, 12485–12489.
Tan, C.W., Tan, K.H., Ong, Y.T., Mohamed, A.R., Zein, S.H.S., Tan, S.H., 2012. Energy and environmental applications of carbon nanotubes. Environ. Chem. Lett., 1–9.
Tanaka, E.M., Kirschner, M.W., 1991. Microtubule behavior in the growth cones of living neurons during axon elongation. J. Cell Biol. 115, 345.
Tanford, C., Hauenstein, J.D., 1956. Hydrogen ion equilibria of ribonuclease. J. Am. Chem. Soc. 78, 5287–5291.
Tang, X., Munske, G.R., Siems, W.F., Bruce, J.E., 2005. Mass spectrometry identifiable cross-linking strategy for studying protein–protein interactions. Anal. Chem. 77, 311–318.
Tani, T., Murofushi, M., 1994. Silver microclusters on silver halide grains as latent image and reduction sensitization centers. J. Imag. Sci. Technol. 98, 1.
Tao, T., Lamkin, M., Schemer, C., 1984. Studies on the proximity relationships between thin filament proteins using benzophenone-4-maleimide as a site-specific photoreactive crosslinker. Biophys. J. 45, 261.
Tao, W.A., Wollscheid, B., O'Brien, R., Eng, J.K., Li, X.-J., Bodenmiller, B., et al., 2005. Quantitative phosphoproteome analysis using a dendrimer conjugation chemistry and tandem mass spectrometry. Nat. Methods 2 (8), 591–598.
Tardioli, P.W., Vieira, M.F., Vieira, A.M.S., Zanin, G.M., Betancor, L., Mateo, C., et al., 2011. Immobilization–stabilization of glucoamylase: chemical modification of the enzyme surface followed by covalent attachment on highly activated glyoxyl-agarose supports. Process Biochem. 46 (1), 409–412.
Tarentino, A.L., Phelan, A.W., Plummer Jr., T.H., 1993. 2-Iminothiolane: a reagent for the introduction of sulfhydryl groups into oligosaccharides derived from asparagine-linked glycans. Glycobiology 3, 279–285.
Tarvers, R.C., Noyes, C.M., Roberts, H.R., Lundblad, R.L., 1982. Influence of metal ions on prothrombin self-association. J. Biol. Chem. 257, 10708–10714.
Tashima, T., Imai, M., Kuroda, Y., Yagi, S., Nakagawa, T., 1991. Structure of a new oligomer of glutaraldehyde produced by aldol condensation reaction. J. Org. Chem. 56, 694–697.
Tateno, H., Uchiyama, N., Kuno, A., Togayachi, A., Sato, T., Narimatsu, H., et al., 2007. A novel strategy for mammalian cell surface glycome profiling using lectin microarray. Glycobiology 17 (10), 1138–1146.
Tauer, K., Imroz Ali, A.M., Sedlak, M., 2005. On the preparation of stable poly(2-hydroxyethyl methacrylate) nanoparticles. Colloid Polym. Sci. 283, 351–358.
Tavares, A.J., Chong, L., Petryayeva, E., Algar, W.R., Krull, U.J., 2011. Quantum dots as contrast agents for in vivo tumor imaging: progress and issues. Anal. Bioanal. Chem. 399 (7), 2331–2342.
Tawney, P.O., Snyder, R.H., Conger, R.P., Leibbrand, K.A., Stiteler, C.H., Williams, A.R., 1961. Maleimide and derivatives. II. Maleimide and N-methylmaleimide. J. Org. Chem. 26, 15.
Taylor, H.J., 1935. The disintegration of boron by neutrons. Proc. Phys. Soc. 47, 873.
Taylor, K.E., Wu, Y.C., 1980. A thiolation reagent for cell surface carbohydrate. Biochem. Int. 1, 353.
Taylor, M.E., Drickamer, K., 2009. Structural insights into what glycan arrays tell us about how glycan-binding proteins interact with their ligands. Glycobiology 19, 1155–1162.
Taylor, R.M., Riesselman, M.H., Lord, C.I., Gripentrog, J.M., Jesaitis, A.J., 2012. Anionic lipid-induced conformational changes in human phagocyte flavocytochrome b precede assembly and activation of the NADPH oxidase complex. Arch. Biochem. Biophys. 521 (1), 24–31.
Teale, J.M., Kearney, J.R., 1986. Clonotypic analysis of the fetal B cell repertoire: evidence for an early and predominant expression of idiotypes associated with the VH 36–60 family. J. Mol. Cell. Immunol. 2, 283–292.
Teke, A.B., Baysal, Ş.H., 2007. Immobilization of urease using glycidyl methacrylate grafted nylon-6-membranes. Process Biochem. 42, 439–443.
Tellechea, E., Wilson, K.J., Bravo, E., Hamad-Schifferli, K., 2012. Engineering the interface between glucose oxidase and nanoparticles. Langmuir 28 (11), 5190–5200.
Teng, S.F., Sproule, K., Husain, A., Lowe, C.R., 2000. Affinity chromatography on immobilized "'biomimetic"' ligands. Synthesis, immobilization and chromatographic assessment of an immunoglobulin G-binding ligand. J. Chromatogr. B: Biomed. Sci. Appl. 740, 1–15.
Tepe, J.J., Pinnavaia, T.J., 2008. Method for the enrichment and characterization of phosphorylated peptides or proteins. United States Patent No. 7, 423, 132.
Tertykh, V.A., Yanishpolskii, V.V., Panova, O.Y., 2000. Covalent attachment of some Phenol derivatives to the silica surface by use of single-stage aminomethylation. J. Therm. Anal. Calorim. 62, 545–549.
Thakur, M.L., DeFulvio, J.D., 1991. Technetium-99m labeled monoclonal antibodies for immunoscintigraphy. J. Immunol. Methods 137, 217–224.
Thanh, N.T.K., Green, L.A.W., 2010. Functionalization of nanoparticles for biomedical applications. Nano Today 5, 213–230.
Thanou, M., Duncan, R., 2003. Polymer-protein and polymer-drug conjugates in cancer therapy. Curr. Opin. Investig. Drugs 4, 701–709.
Theis, F.V., Freeland, M.R., 1941. Thrombo-angiitis obliterans: clinical observations and arterial blood oxygen studies during treatment of the disease with sodium tetrathionate and sodium thiosulfates. Ann. Surg. 113 (3), 411–423.
Theisen, S., Hänsch, R., Kothe, L., Leist, U., Galensa, R., 2010. A fast and sensitive HPLC method for sulfite analysis in food based on a plant sulfite oxidase biosensor. Biosens. Bioelectron. 26, 175–181.
Thelen, P., Deuticke, B., 1988. Chemo-mechanical leak formation in human erythrocytes upon exposure to a water-soluble carbodiimide followed by very mild shear stress. II. Chemical modifications involved. Biochim. Biophys. Acta 944, 297–307.
Theorell, H., 1951. The iron-containing enzymes. B. Catalases and peroxidases. HydroperoxidasesSumner, J.B.Myrback, K. The Enzymes, vol. 2. Academic Press, New York, pp. 397. Part 1.

Therien, H.-M., Shahum, E., 1989. Importance of physical association between antigen and liposomes in liposome adjuvanticity. Immunol. Lett. 22, 253–258.

Thevenin, B., Shahrokh, Z., Williard, R., Fujimoto, E., Ikemoto, N., Shohet, S., 1991. A novel reagent for functionally-directed site-specific fluorescent labeling of proteins, Abstract. Biophys. J. 59, 358a. Tu-Pos 476.

Thiruppathiraja, C., Kumar, S., Murugan, V., Adaikkappan, P., Sankaran, K., Alagar, M., 2011. An enhanced immuno-dot blot assay for the detection of white spot syndrome virus in shrimp using antibody conjugated gold nanoparticles probe. Aquaculture 318 (3), 262–267.

Thomas, D.D., Carlsen, W.F., Stryer, L., 1978. Fluorescence energy transfer in the rapid-diffusion limit. Proc. Natl. Acad. Sci. USA 75 (12), 15746–15750.

Thomas, T.P., Patri, A.K., Myc, A., Myaing, M.T., Ye, J.Y., Norris, T.B., et al., 2004. In vitro targeting of synthesized antibody-conjugated dendrimer nanoparticles. Biomacromolecules 5, 2269–2274.

Thompson, A., Prescott, M., Chelebi, N., Smith, J., Brown, T., Schmidt, G., 2007. Electrospray ionisation-cleavable tandem nucleic acid mass tag–peptide nucleic acid conjugates: synthesis and applications to quantitative genomic analysis using electrospray ionisation-MS/MS. Nucleic Acids Res. 35 (4), e28.

Thompson, A., Schäfer, J., Kuhn, K., Kienle, S., Schwarz, J., Schmidt, G., et al., 2003. Tandem mass tags: a novel quantification strategy for comparative analysis of complex protein mixtures by MS/MS. Anal. Chem. 75 (8), 1895–1904.

Thompson, R.E., Larson, D.R., Webb, W.W., 2002. Precise nanometer localization analysis for individual fluorescent probes. Biophys. J. 82, 2775–2783.

Thorpe, P.E., Brown, A.N.F., Ross, W.C.J., Cumber, A.J., Detre, S.I., Edwards, D.C., et al., 1981. Cytotoxicity acquired by conjugation of an anti-Thy 1.1 monoclonal antibody and the ribosome-inactivating protein, gelonin. Eur. J. Biochem. 116, 447–454.

Thorpe, P.E., Mason, D.W., Brown, A.N.F., Simmonds, S.J., Ross, W.C.J., Cumber, A.J., et al., 1982. Selective killing of malignant cells in a leukaemic rat bone marrow using an antibody–ricin conjugate. Nature (London) 297, 594.

Thorpe, P.E., Ross, W.C.J., Brown, A.N.F., Myers, C.D., Cumber, A.J., Foxwell, B.M.J., et al., 1984. Blockade of the galactose-binding sites of ricin by its linkage to antibody. Specific cytotoxic effects of the conjugates. Eur. J. Biochem. 140, 63–71.

Thorpe, P.E., Wallace, P.M., Knowles, P.P., Relf, M.G., Brown, A.N., Watson, G.L., et al., 1987. New coupling agents for the synthesis of immunotoxins containing a hindered disulfide bond with improved stability *in vivo*. Cancer Res. 47, 5924–5931.

Thumshirn, G., Hersel, U., Goodman, S.L., Kessler, H., 2003. Multimeric cyclic RGD peptides as potential tools for tumor targeting: solid phase peptide synthesis and chemoselective oxime ligation. Chem. Eur. J. 9, 2717–2725.

Thygesen, M.B., Munch, H., Sauer, J., Clo, E., Jorgensen, M.R., Hindsgaul, O., et al., 2010. Nucleophilic catalysis of carbohydrate oxime formation by anilines. J. Org. Chem. 75, 1752–1755.

Tian, B., Yang, J., Brasier, A.R., 2012. Two-step crosslinking for analysis of protein-chromatin interactions in transcriptional regulation. Methods Mol. Biol. 809 (Part 1), 105–120.

Tian, J., Zhou, L., Zhao, Y., Wang, Y., Peng, Y., Zhao, S., 2012. Multiplexed detection of tumor markers with multicolor quantum dots based on fluorescence polarization immunoassay. Talanta 92, 72–77.

Tian, Y., 2012. Förster Resonance Energy Transfer (FRET) Between Phycobiliproteins and Tandem Conjugates (Doctoral dissertation, Texas Tech University).

Tian, Y., Pappas, D., 2011. Energy transfer and light tolerance studies in a fluorescent tandem phycobiliprotein conjugate. Appl. Spectrosc. 65 (9), 991–995.

Tietze, L.F., Schmuck, K., 2011. Prodrugs for targeted tumor therapies: recent developments in ADEPT, GDEPT and PMT. Curr. Pharm. Des. 17 (32), 3527–3547.

Timberlake, G.T., Yousef, A.L., Chiles, S.R., Moses, R.A., Givens, R.S., 2005. Bonding corneal tissue: applications of photoactivated diazopyruvoyl cross-linking agent. Photochem. Photobiol. 81 (5), 1180–1185.

Tinette, S., René Feyereisen, A.R., 2006. Approach to systematic analysis of serine/threonine phosphoproteome using Beta elimination and subsequent side effects: intramolecular linkage and/or racemisation. J. Cell. Biochem. 100, 875–882.

Tinglu, G., Ghosh, A., Ghosh, B.K., 1984. Subcellular localization of alkaline phosphatase in Bacillus licheniformis 749/C by immunoelectron microscopy with colloidal gold. J. Bacteriol. 159, 668.

Tipson, R.S., 1944. On esters of p-toluenesulfonic acid. J. Org. Chem. 9, 235–241.

Titus, J.A., Haugland, R.P., Sharrow, D.M., Segal, J., 1982. Texas Red, a hydrophilic, red-emitting fluorophore for use with fluorescein in dual parameter flow microfluorometric and fluorescence microscopic studies. J. Immunol. Methods 50, 193–204.

Tjerneld, F., 1992. Aqueous two-phase partitioning on an industrial scale. In: Harris, J.M. (Ed.), Poly(Ethylene Glycol) Chemistry: Biotechnical and Biomedical Applications. Plenum, New York, pp. 85–102.

Tokuyasu, K.T., 1983. Present state of immunocryoultramicrotomy. J. Histochem. Cytochem. 31, 164.

Tolmachev, V., Orlova, A., Pehrson, R., Galli, J., Baastrup, B., Andersson, K., et al., 2007. Radionuclide therapy of HER2-positive microxenografts using a 177Lu-labeled HER2-specific Affibody molecule. Cancer Res. 67 (6), 2773–2782.

Tomalia, D.A., Dewald, J.R., Dense star polymers having core, core branches, terminal groups. US Patent 4, 507, 466, filed January 7, 1983, published March: 26, 1985.

Tomalia, D.A., Baker, H., Dewald, J., Hall, J.M., Kallos, G., Martin, R., et al., 1985. Polym. J. 17, 117–132.

Tomalia, D.A., Baker, J.R., Cheng, R.C., Bielinska, A.U., Fazio, M.J., Hedstrand, D.M., et al., 1998. Bioactive and/or targeted dendrimer conjugates. US Patent No. 5, 714, 166.

Tona, R., Häner, R., 2005. Synthesis and bioconjugation of diene-modified oligonucleotides. Bioconjug. Chem. 16, 837–842.

Tong, R., Cheng, J., 2012. Drug-initiated, controlled ring-opening polymerization for the synthesis of polymer-drug conjugates. Macromolecules 45, 2225–2232.

Tong, R., Tang, L., Cheng, J., 2012. Development and application of anticancer nanomedicine. Multifunctional Nanoparticles for Drug Delivery Applications, 31–46.

Toomre, D.K., Varki, A., 1994. Advances in the use of biotinylated diaminopyridine (BAP) as a versatile fluorescent tag for oligosaccharides. Glycobiology 4, 653–663.

Topp, M., Goekbuget, N., Kufer, P., et al., 2008. Treatment with anti-CD19 BiTE antibody blinatumomab (MT103/MEDI-538) is able to eliminate minimal residual disease (MRD) in patients with B-precursor acute lmphoblastic leukemia (ALL): first results of ongoing phase 2 study [ASH Annual Meeting Abstract]. Blood 112, 1926.

Torchilin, V.P., Klibanov, A.L., Slinkin, M.A., Danilov, S.M., Levitsky, D.O., Khow, B.A., 1989. Antibody-linked chelating polymers for immunoimaging *in vivo*. J. Control. Release 11, 297–303.

Torchilin, V.P., Trubetskoy, V.S., Narula, J., Khaw, B.A., Klibanov, A.L., Slinkin, M.A., 1993. Chelating polymer modified monoclonal antibodies for radioimmunodiagnostics and radioimmunotherapy. J. Control. Release 24, 111–118.

Tornøe, C.W., Christensen, C., Meldal, M., 2002. Peptidotriazoles on solid phase: [1,2,3]-triazoles by regiospecific copper(I)-catalyzed 1,3-dipolar cycloadditions of terminal alkynes to azides. J. Org. Chem. 67, 3057–3064.

Tosti, G., di Pietro, A., Ferrucci, P.F., Testori, A., 2009. HSPPC-96 vaccine in metastatic melanoma patients: from the state of the art to a possible future. Expert Rev. Vaccines 8 (11), 1513–1526.

Tournier, E.J.M., Wallach, J., Blond, P., 1998. Sulfosuccinimidyl 4-(N-maleimidomethyl)-1-cyclohexane carboxylate as a bifunctional immobilization agent. Optimization of the coupling conditions. Anal. Chim. Acta 361, 33–44.

Toyokuni, T., Singhal, A.K., 1995. Synthetic carbohydrate vaccines based on tumour-associated antigens. Chem. Soc. Rev. 24, 231–242.

Tozawa, T., Itoh, K., Yaoi, T., Tando, S., Umekage, M., Dai, H., et al., 2012. The shortest isoform of dystrophin (Dp40) interacts with a group of presynaptic proteins to form a presumptive novel complex in the mouse brain. Mol. Neurobiol. 45, 287–297.

Trail, P.A., Willner, D., Lasch, S.J., Henderson, A.J., Hofstead, S., Casazza, A.M., et al., 1993. Cure of xenografted human carcinomas by BR96-doxorubicin immunoconjugates. Science 261, 212–215.

Traincard, F., Ternynck, T., Danchin, A., Avrameas, S., 1983. An immunoenzymic procedure for the demonstration of nucleic acid molecular hybridization. Ann. Immunol. 134, 339–405.

Trakselis, M.A., Alley, S.C., Ishmael, F.T., 2005. Identification and mapping of protein–protein interactions by a combination of cross-linking, cleavage, and proteomics. Bioconjug. Chem. 16 (4), 741–750.

Tram, K., Yan, H., Jenkins, H.A., Vassiliev, S., Bruce, D., 2009. The synthesis and crystal structure of unsubstituted 4,4-difluoro-4-bora-3a, 4a-diaza-s-indacene (BODIPY). Dyes and Pigments 82 (3), 392–395.

Tran, L., Baars, J., Damen, C., Beijnen, J., Huitema, A., 2011. Three spectroscopic techniques evaluated as a tool to study the effects of iodination of monoclonal antibodies, exemplified by rituximab. J. Pharm. Biomed. Anal. 56 (3), 609–614.

Traut, R.R., Bollen, A., Sun, R.R., Hershey, J.W.B., Sundberg, J., Pierce, L.R., 1973. Methyl 4-mercaptobutyrimidate as a cleavable cross-linking reagent and its application to the Escherichia coli 30s ribosome. Biochemistry 12, 3266–3273.

Traut, R.R., Casiano, C., Zecherle, N., 1989. Cross-linking of protein subunits and ligands by the introduction of disulfide bonds. In: Creighton, T.E. (Ed.), Protein Function—A Practical Approach. IRL Press at Oxford University, Oxford, pp. 101–133.

Traviglia, S.L., Datwyler, S.A., Meares, C.F., 1999. Mapping protein-protein interactions with a library of tethered cutting reagents: the binding site of sigma (70) on Escherichia coli RNA polymerase. Biochemistry 38, 4259–4265.

Trindade, T., da Silva, A.L.D. (Eds.), 2011. Nanocomposite Particles for Bio-Applications: Materials and Bio-Interfaces. Pan Stanford Publishing.

Tripathi, S.K., Goyal, R., Gupta, K.C., 2011. Surface modification of crosslinked dextran nanoparticles influences transfection efficiency of dextran–polyethylenimine nanocomposites. Soft Matter 7, 11360–11371.

Trivedi, N., Jung, P., Brown, A., 2007. Neurofilaments switch between distinct mobile and stationary states during their transport along axons. J. Neurosci. 27, 507–516.

Trotman, L.C., et al. 2001. Import of adenovirus DNA involves the nuclear pole complex receptor CAN/Nup214 and histone H1. Nat. Cell Biol. 3, 1092–1100.

Trotter, C.L., Ramsay, M.E., 2007. Vaccination against meningococcal disease in Europe: review and recommendations for the use of conjugate vaccines. FEMS Microbiol. Rev. 31, 101–107.

Troyan, S.L., Kianzad, V., Gibbs-Strauss, S.L., Gioux, S., Matsui, A., Oketokoun, R., et al., 2009. The FLARE™ intraoperative near-infrared fluorescence imaging system: a first-in-human clinical trial in breast cancer sentinel lymph node mapping. Ann. Surg. Oncol. 16 (10), 2943–2952.

Troyanovsky, R.B., Sokolov, E.P., Troyanovsky, S.M., 2006. Endocytosis of cadherin from intracellular junctions is the driving force for cadherin adhesive dimer disassembly. Mol. Biol. Cell 17, 3484–3493.

Truax, A.D., Greer, S.F., 2012. ChIP and Re-ChIP assays: investigating interactions between regulatory proteins, histone modifications, and the DNA sequences to which they bind. Methods Mol. Biol. (Clifton, NJ) 809, 175–188.

Trubetskoy, V.S., Narula, J., Khaw, B.A., Torchilin, V.P., 1993. Chemically optimized antimyosin Fab conjugates with chelating polymers: importance of the nature of the protein–polymer single site covalent bond for biodistribution and infarction localization. Bioconjug. Chem. 4, 251–255.

Truneh, A., Machy, P., Horan, P.K., 1987. Antibody-bearing liposomes as multicolor immunofluorescent markers for flow cytometry and imaging. J. Immunol. Methods 100, 59–71.

Tsai, H.H., Huang, C.-H., Tessmer, I., Erie, D.A., Chen, C.W., 2011. Linear Streptomyces plasmids form superhelical circles through interactions between their terminal proteins. Nucleic Acids Res. 39, 2165–2174.

Tsai, L., Szweda, P.A., Vinogradova, O., Szweda, L.I., 1998. Structural characterization and immunochemical detection of a fluorophore derived from 4-hydroxy-2-nonenal and lysine. Proc. Natl. Acad. Sci. USA 95, 7975–7980.

Tsai, Y.-C., Du, D., Dominguez-Malfavon, L., Dimastrogiovanni, D., Cross, J., Callaghan, A.J., et al., 2012. Recognition of the 70S ribosome and polysome by the RNA degradosome in Escherichia coli. Nucleic Acids Res. 40, 10417–10431.

Tsao, I.-Fu, Shipman Jr., C., Wang, H.Y., 1988. The removal of adventitious viruses and virus-infected cells using a cellular adsorbent: a feasibility study. Nat. Biotechnol. 6, 1330–1333.

Tsien, R.Y., Waggoner, A., 1990. Fluorophores for confocal microscopy: photophysics and photochemistry. In: Pawley, J.B. (Ed.), Handbook of Biological Confocal Microscopy. Plenum, New York, pp. 169.

Tsomides, T.J., Walker, B.D., Eisen, H.N., 1991. An optimal viral peptide recognized by CD8 + T cells binds very tightly to the restricting class I major histocompatibility complex protein on intact cells but not to the purified class I protein. Proc. Natl. Acad. Sci. USA 88, 11276–11280.

Tsudo, M., Kozak, R.W., Goldman, C.K., Waldmann, T.A., 1987. Demonstration of a non-Tac peptide that binds interleukin 2: a potential participant in a multichain interleukin 2 receptor complex. Proc. Natl. Acad. Sci. USA 83, 9694–9698.

Tsukamoto, Y., Wakil, S.J., 1988. Isolation and mapping of the b-hydroxyacyl dehydratase activity of chicken liver fatty acid synthase. J. Biol. Chem. 263, 16225–16229.

Tu, C.-P.C., Cohen, S., 1980. 39-End labeling of DNA with [a-32P] cordycepin-59-triphosphate. Gene 10, 177–183.

Tubbs, C.E., et al. 2002. Binding of protein D/E to the surface of rat epidermal sperm before ejaculation and after deposition in the female reproductive tract. J. Androl. 23 (4), 512–521.

Tuktarov, A.R., Khuzina, L.L., Dzhemilev, U.M., 2011. Covalent binding of fullerene C 60 to pharmacologically important compounds. Russ. Chem. Bull. 60 (4), 662–666.

Tullius, T.D., Dombroski, B.A., 1986. Hydroxyl radical "footprinting": high-resolution information about DNA-protein contacts and application to repressor and Cro protein. PNAS 83, 5469–5473.

Tung, C.H., 2004. Fluorescent peptide probes for in vivo diagnostic imaging. Biopolymers 76, 391–403.

Turecek, F., 2002. Mass spectrometry in coupling with affinity capture-release and isotope-coded affinity tags for quantitative protein analysis. J. Mass Spectrom. 37, 1–14.

Turro, N.J., Ramamurthy, V., Scaiano, J.C., 2009. Principles of Molecular Photochemistry: An Introduction. Chapter 5, Section 5.29, pp. 312–313, University Science Books, ISBN: 978-1-891389-57-3.

Tussen, P., Kurstak, E., 1984. Highly efficient and simple methods for the preparation of peroxidase and active peroxidase—antibody conjugates for enzyme immunoassays. Anal. Biochem. 136, 451–457.

Ubrich, N., Hubert, P., Regnault, V., Dellacherie, E., Rivat, C., 1992. Compared stability of sepharose-based immunoadsorbents prepared by various activation methods. J. Chromatogr. B Biomed. Sci. Appl. 584, 17–22.

Uchegbu, I.F., Schatzlein, A.G. (Eds.), 2006. Polymers in Drug Delivery. CRC Press, Taylor & Francis Group, Boca Raton, FL.

Uchino, O., Mizunami, T., Maida, M., Miyazoe, Y., 1979. Efficient dye lasers pumped by an XeC1 excimer laser. Appl. Phys. 19, 35.

Uchiutni, T., Terao, K., Ogata, K., 1980. Identification of neighboring protein pairs in rat liver 60S ribosomal subunits cross-linked with dimethyl suberimidate or dimethyl 3,39-dithiobispropionimidate. J. Biochem. (Tokyo) 88, 1033–1044.

Uchiyama, K., et al. 2002. VCIP135, a novel essential factor for p97/p47-mediated membrane fusion, is required for Golgi and ER assembly in vivo. J. Cell Biol. 159, 855–866.

Uckun, F.M., et al. 1995. Biotherapy of B-cell precursor leukemia by targeting genistein to CD19-associated tyrosine kinases. Science 267, 886–891.

Udupa, E.G.P., Sharma, K.K., 2005. Effect of oxidized ßB3-crystallin peptide on lens ßL-crystallin: interaction with ßB2-crystallin. Invest. Ophthalmol. Vis. Sci. 46, 2514–2521.

Udupa, P.E., Sharma, K.K., 2005. Effect of oxidized betaB3-crystallin peptide (152–166) on thermal aggregation of bovine lens gamma-crystallins: identification of peptide interacting sites. Exp. Eye. Res. 80 (2), 185–196.

Ueda, K., Fukase, Y., Katagiri, T., Ishikawa, N., Irie, S., Sato, T.-A., et al., 2009. Targeted serum glycoproteomics for the discovery of lung cancer-associated glycosylation disorders using lectin-coupled proteinchip arrays. Proteomics 9 (8), 2182–2192.

Uhlemann, A.-C., Wittlin, S., Hugues Matile, H., Bustamante, L.Y., Krishna, S., 2007. Mechanism of antimalarial action of the synthetic trioxolane RBX11160 (OZ277). Antimicrob. Agents Chemother. 51 (2), 667–672.

Ulbricht, M., 2006. Advanced functional polymer membranes. Polymer 47 (2006), 2217–2262.

Umeyama, T., Fueno, H., Kawabata, E., Kobayashi, Y., Tanaka, K., Tezuka, N., et al., 2011. Density functional theory studies on chemical functionalization of single-walled carbon nanotubes by bingel reaction. Bull. Chem. Soc. Jpn. 84 (7), 748–753.

Uraki, Z., Terminiello, L., Bier, M., Nord, F.F., 1957. On the mechanism of enzyme action. LXIII. Specificity of acetylation of proteins with C14 anhydride. Arch. Biochem. Biophys. 69, 644–652.

Urdea, M.S., Warner, B.D., Running, J.A., Stempien, M., Clyne, J., Horn, T., 1988. A comparison of non-radioactive hybridization assay methods using fluorescent, chemiluminescent and enzyme-labeled synthetic oligodeoxyribonucleotide probes. Nucleic Acids Res. 16, 4937–4956.

Urge, L., Kollat, E., Hollosi, M., Laczko, I., Wroblewski, K., Thurin, J., et al., 1991. Solid-phase synthesis of glycopeptides: synthesis of Na-fluorenylmethoxycarbonyl L-asparagine Nb-glycosides. Tetrahedron Lett. 32, 3445–3448.

Urge, L., Otvos Jr., L., Lang, E., Wroblewski, K., Laczko, I., Hollosi, M., 1992. Fmoc-protected, glycosylated asparagines potentially useful as reagents in the solid-phase synthesis of *N*-glycopeptides. Carbohydr. Res. 235, 83–93.

Ushakov, D.S., Caorsi, V., Ibanez-Garcia, D., Manning, H.B., Konitsiotis, A.D., West, T.G., et al., 2011. Response of rigor cross-bridges to stretch detected by fluorescence lifetime imaging microscopy of myosin essential light chain in skeletal muscle fibers. J. Biol. Chem. 286 (1), 842–850.

Usuda, J., Kato, H., Okunaka, T., Furukawa, K., Tsutsui, H., Yamada, K., et al., 2006. Photodynamic therapy (PDT) for lung cancers. J. Thorac. Oncol. 1, 489–493.

Uto, I., Ishimatsu, T., Hirayama, H., Ueda, S., Tsuruta, J., Kambara, T., 1991. Determination of urinary Tamm–Horsfall protein by ELISA using a maleimide method for enzyme–antibody conjugation. J. Immunol. Methods 138, 87–94.

Uyeda, H.T., Medintz, I.L., Mattoussi, H., 2003. Design of water-soluble quantum dots with novel surface ligands for biological applications. Mater. Res. Soc. Proc., 789. Symposium N.

Uyeda, K., 1969. Reaction of phosphofructokinase with maleic anhydride, succinic anhydride, and pyridoxal 59-phosphate. Biochemistry 8, 2366–2373.

Vaidya, A., Agarwal, A., Jain, A., Agrawal, K., Ram, K., Jain, S., 2011. Bioconjugation of polymers: a novel platform for targeted drug delivery. Curr. Pharm. Des. 17, 1108–1125.

Vainauskas, S., Cortes, L.K., Taron, C.H., 2012. In vivo incorporation of an azide-labeled sugar analog to detect mammalian glycosylphosphatidylinositol molecules isolated from the cell surface. Carbohydr. Res. 362, 62–69.

Van Aken, B.P., Henry, L., Spiros, N., Agathos, N., 2000. Co-immobilization of manganese peroxidase from Phlebia radiata on porous silica beads. Biotechnol. Lett. 8, 641–646.

van Belkum, A., Linkels, E., Jelsma, T., van den Berg, F.M., Quint, W., 1994. Non-isotopic labeling of DNA by newly developed hapten-containing platinum compounds. Biotechniques 16 (1), 148–153.

van Berkel, S.S., van Eldijk, M.B., van Hest, J., 2011. Staudinger ligation as a method for bioconjugation. Angew. Chem. Int. Ed. 50 (38), 8806–8827.

van Blaaderen, A., Vrij, A., 1992. Synthesis and characterization of colloidal dispersions of fluorescent, monodisperse silica spheres. Langmuir 8, 2921–2931.

van Dalen, J.P.R., Haaijman, J.J., 1974. Determination of the molar absorbance coefficient of bound tetramethyl rhodamine isothiocyanate relative to fluorescein isothiocyanate. J. Immunol. Methods 5, 103–106.

van den Brink, W., van der Loos, C., Volkers, H., Lauwen, R., van den Berg, F., Houthoff, H.-J., et al., 1990. Combined b-galactosidase and immunogold/silver staining for immunohistochemistry and DNA in situ hybridization. J. Histochem. Cytochem. 38, 325–329.

van der Horst, G.T.J., Mancini, G.M.S., Brossmer, R., Rose, U., Verheijen, F.W., 1990. Photoaffinity labeling of a bacterial sialidase with an aryl azide derivative of sialic acid. J. Biol. Chem. 265, 10801–10804.

van Dongen, G.A.M.S., Poot, A.J., Vugts, D.J., 2012. PET imaging with radiolabeled antibodies and tryorsine kinase inhibitors: immuno-PET and TKI-PET. Tumor. Biology. 33, 607–615.

van Dongen, J.J.M., Hooijkaas, H., Comans-Bitter, W.M., Benne, K., van Os, T.M., de Josselin de Jong, J., 1985. Triple immunological staining with colloidal gold, fluorescein, and rhodamine as labels. J. Immunol. Methods 80, 1.

van Duren, J.K.J., Yang, X.N., Loos, J., Bulle-Lieuwma, C.W.T., Sieval, A.B., Hummelen, J.C., et al., 2004. Relating the morphology of poly(p-phenylene vinylene)/methanofullerene blends to solar-cell performance. Adv. Funct. Mater. 14, 425–434.

Van Lenten, L., Ashwell, G., 1971. Studies on the chemical and enzymatic modification of glycoproteins. A general method for the tritiation of sialic acid-containing glycoproteins. J. Biol. Chem. 46, 1889–1894.

van Oijen, A.M., Kohler, J., Schmidt, J., Muller, M., Brakenhoff, G., 1998. 3-Dimensional super-resolution by spectrally selective imaging. J. Chem. Phys. Lett. 292, 183–187.

Van Pelt-Verkuil, E., van Belkum, A., Hays, J.P., 2010. Principles and Technical Aspects of PCR Amplification. Springer Science & Business Media B.V., ISBN-10: 9048175798.

Van Regenmortal, M.H.V., Briand, J.P., Muller, S., Plaue, S., 1988. Synthetic polypeptides as antigens. Lab. Tech. Biochem. Mol. Biol. 19, 121–125.

Van Reis, R., Zydney, A.L., 2007. Bioprocess membrane technology. J. Memb. Sci. 297, 16–50.

Vandelen, R.L., Arcuri, K.E., Napier, M.A., 1985. Identification of a receptor for atrial natriuretic factor in rabbit aorta membranes by affinity cross-linking. J. Biol. Chem. 260, 10889–10892.

Vandenberghe, R., Koen Van Laere, A., Ivanoiu, E., Salmon, C., Bastin, E., Triau, S., et al., 2010. 18F-flutemetamol amyloid imaging in Alzheimer disease and mild cognitive impairment: a phase 2 trial. Ann. Neurol. 68, 319–329.

VanDerVoort, P., et al. 1996. Silylation of the silica surface a review. J. Liq. Chrom. Relat. Tech. 19, 2723–2752.

Vanin, E.F., Ji, T.H., 1981. Synthesis and application of cleavable photoactivatable heterobifunctional reagents. Biochemistry 20, 6754–6760.

Varenne, A., Salmain, M., Brisson, C., Jaouen, G., 1992. Transition metal carbonyl labeling of proteins. a novel approach to a solid-phase two-site immunoassay using Fourier transform infrared spectroscopy. Bioconjug. Chem. 3, 471–476.

Varenne, A., Vessières, A., Salmain, M., Brossier, P., Jaouen, G., 1995. Production of specific antibodies and development of a nonisotopic immunoassay for carbamazepine by the carbonylmetalloimmunoassay (CMIA) method. J. Immunol. Methods 186, 195–204.

Varki, A., Cummings, R.D., Esko, J.D., Freeze, H.H., Hart, G.W., Etzler, M.E. (Eds.), 2008. Essentials of Glycobiology (second ed.). Cold Spring Harbor Laboratory Press.

Varnavski, O., Ispasoiu, R.G., Balogh, L., Tomalia, D., Goodson, T., 2001. Ultrafast time-resolved photoluminescence from novel metal–dendrimer nanocomposites. J. Chem. Phys. 114, 1962–1965.

Vashist, S.K., Tewari, R., Bajpai, R.P., Bharadwaj, L.M., Raiteri, R., 2006. Review of quantum dot technologies for cancer detection and treatment. J. Nanotechnol. Online 2, 1–14. <http://www.azonano.com/nanotechnology.asp>.

Vasicek, L., O'Brien, J.P., Browning, K.S., Tao, Z., Liu, H.-W., Brodbelt, J.S., 2012. Mapping protein surface accessibility via an electron transfer dissociation selectively cleavable hydrazone probe. Available online, Molecular & Cellular Proteomics 11, O111.015826.

Vasilescu, J., Guo, X., Kast, J., 2004. Identification of protein–protein interactions using in vivo cross-linking and mass spectrometry. Proteomics 4, 3845–3854.

Vázquez-Dorbatt, V., Tolstyka, Z.P., Chang, C.-W., Maynard, H.D., 2009. Synthesis of a pyridyl disulfide end-functionalized glycopolymer for conjugation to biomolecules and patterning on gold surfaces. Biomacromolecules 10, 2207–2212.

Vdovenko, M.M., Demiyanova, A.S., Chemleva, T.A., Sakharov, I.Y., 2012. Optimization of horseradish peroxidase-catalyzed enhanced chemiluminescence reaction by full factorial design. Talanta 94, 223–226.

Verhaegh, N.A.M., van Blaaderen, A., 1994. Dispersions of rhodamine-labeled silica spheres: synthesis, characterization, and fluorescence confocal scanning laser microscopy. Langmuir 10, 1427–1438.

Veronese, F.M. (Ed.), 2009. PEGylated Protein Drugs: Basic Science and Clinical Applications. Birkhauser Verlag, Basel, Switzerland, ISBN: 978-3-7643-8678-8.

Veronese, F.M., Morpurgo, M., 1999. Bioconjugation in pharmaceutical chemistry (review). Il Farmaco 54, 497–516.

Verwey, E.J.W., Overbeek, J.T.G., 1948. Theory of the Stability of Lyophobic Colloids. Elsevier, New York.

Vessières, A., Kowalski, K., Zakrzewski, J., Stepien, A., Grabowski, M., Jaouen, G., 1999. Synthesis of CpFe(CO)-(L) complexes of hydantoin anions (Cp) eta5-C5H5 (L) CO (PPh3), and the use of the 5,5-diphenylhydantoin anion complexes as tracers in the nonisotopic immunoassay CMIA of this antiepileptic drug. Bioconjug. Chem. 10, 379–385.

Vetter, D., Tate, E.M., Gallop, M.A., 1995. Strategies for the synthesis and screening of glycoconjugates. 2. Covalent immobilization for flow cytometry. Bioconjug. Chem. 6, 319–322.

Vicente, M., Graça, H., Wickramasinghe, A., Nurco, D.J., Wang, H.J.H., Nawrocky, M.M., et al., 2003. Syntheses, toxicity and biodistribution of two 5,15-di[3,5-(nido-carboranyl-methyl)phenyl] porphyrin in EMT-6 tumor bearing mice. Bioorg. Med. Chem. 11, 3101–3108.

Vigers, G.P.A., Cone, J.R., Mcintosh, J., 1988. Fluorescent microtubules break up under illumination. J. Cell Biol. 107, 1011.

Viguera, A.-R., Villa, M.-J., Goni, F.M., 1990. A water-soluble polylysine-retinaldehyde Schiff base; stability in aqueous and nonaqueous environments. J. Biol. Chem. 265, 2527–2532.

Vilja, P., 1991. One- and two-step non-competitive avidin-biotin immunoassays for monomeric and heterodimeric antigen. J. Immunol. Methods 136, 77.

Vincent, J.P., Lazdunski, M., Delaage, M., 1970. Use of tetranitromethane as a nitration reagent. Reaction of phenol sidechains in bovine and porcine trypsinogens and trypsins. Eur. J. Biochem. 12, 250.

Viscidi, R.P., Connelly, C.J., Yolken, R.H., 1986. Novel chemical method for the preparation of nucleic acids for nonisotopic hybridization. J. Clin. Microbiol. 23, 311–317.

Vitetta, E.S., Thorpe, P.E., 1985. Immunotoxins containing ricin A or B chains with modified carbohydrate residues act synergistically in killing neoplastic B cells *in vitro*. Cancer Drug Deliv. 2, 191.

Vithayathil, P.J., Richards, F.M., 1960. Modification of the methionine residue in the peptide component of ribonuclease-S. J. Biol. Chem. 235, 2343–2351.

Vitols, K.S., Haag-Zeino, B., Baer, T., Montejano, Y.D., Huennekens, F.M., 1995. Methotrexate-alpha-phenylalanine: optimization of methotrexate prodrug for activation by carboxypeptidase A–monoclonal antibody conjugate. Cancer Res. 55, 478–481.

Vliegenthart, J.F.G., Dorland, L., van Halbeek, H., 1983. High-resolution, 1H-nuclear magnetic resonance spectroscopy as a tool in the structural analysis of carbohydrates related to glycoproteins. Adv. Carbohydr. Chem. Biochem. 41, 209–374.

Vljayalakshmi, M.A., 1998. Antibody purification methods. Appl. Biochem. Biotechnol. 75, 93–102.

Vogel, C.-W., 1987. Antibody conjugate without inherent toxicity: the targeting of cobra venom factor and other biological response modifiers. In: Vogel, C.-W. (Ed.), Immunoconjugates: Antibody Conjugates in Radioimaging and Therapy of Cancer. Oxford University Press, New York, pp. 170.

Vogel, C.-W., Muller-Eberhard, H.J., 1984. Cobra venom factor: improved method for purification and biochemical characterization. J. Immunol. Methods 73, 203.

Vogel, H., Richert, C., 2012. Labeling small RNAs through chemical ligation at the 5′ terminus: enzyme-free or combined with enzymatic 3′-Labeling. Chem. Bio. Chem. 13, 1474–1482.

Vogelbacker, H.H., Getts, R.C., Tian, N., Labaczewski, R., Nilsen, T.W., 1997. DNA dendrimers: assembly and signal amplification. Polymeric Materials Science and Engineering. ACS Spring Meeting, San Francisco, CA, pp. 458–460.

Vojinovi⊠, V., Esteves, F.M.F., Cabral, J.M.S., Fonseca, L.P., 2006. Bienzymatic analytical microreactors for glucose, lactate, ethanol, galactose and l-amino acid monitoring in cell culture media. Anal. Chim. Acta 565, 240–249.

Vollmers, H.P., Br¨andlein, S., 2009. Natural antibodies and cancer. N. Biotechnol. 25 (5), 294–298.

von der Haar, F., Schlimme, E., Gauss, D.H., 1971.Cantoni, G.L.Davies, D.R. Proceedings of Nucleic Acids Research, vol. 2. Harper & Row, New York, pp. 643–664.

Vrudhula, V.M., Svensson, H.P., Senter, P.D., 1995. Cephalosporin derivatives of doxorubicin as prodrugs for activation by monoclonal antibodybeta-lactamase conjugates. J. Med. Chem. 38, 1380–1385.

Vrudhula, V.M., Svensson, H.P., Senter, P.D., 1997. Immunologically specific activation of a cephalosporin derivative of mitomycin C by monoclonal antibody beta-lactamase conjugates. J. Med. Chem. 40, 2788–2792.

Vuignier, K., Guillarme, D., Veuthey, J.L., Carrupt, P.A., Schappler, J., 2012. High performance affinity chromatography (HPAC) as a high-throughput screening tool in drug discovery to study drug-plasma protein interactions. J. Pharm. Biomed. Anal.10.1016/j.jpba.2012.10.030 (in press).

Wade, D.P., Knight, B.L., Soutar, A.K., 1985. Detection of the low-density lipoprotein receptor with biotin-low-density lipoprotein. A rapid new method for ligand blotting. Biochem. J. 229, 785–790.

Waentig, L., Jakubowski, N., Hayen, H., Roos, P.H., 2011. Iodination of proteins, proteomes and antibodies with potassium triodide for LA-ICP-MS based proteomic analyses. J. Anal. At. Spectrom. 26 (8), 1610–1618.

Waggoner, A.S., 1990. Fluorescent probes for cytometry. In: Melamed, M.R., Lindmo, T., Mendelsohn, M.L. (Eds.), Flow Cytometry and Sorting (second ed.). Wiley-Liss, New York, pp. 209–225.

Waggoner, A.S., Ernst, L.A., Mujumdar, R.B., 1993. Method for labeling and detecting materials employing arylsulfonate cyanine dyes. US Patent No. 5, 268, 486.

Wai, N. K. Thet, W. N. Nway Oo, M. K. Thu and Mya Mya Oo, Isolation, Characterization and Screening of Thermotolerant, Ethanol Tolerant Indigenous Yeasts and Study on the Effectiveness of Immobilized Cell for Ethanol Production. GMSARN International Conference on Sustainable Development: issues and Prospects for the GMS, Nov. 12-14, 2008.

Waldmann, T.A., 1991. Monoclonal antibodies in diagnosis and therapy. Science 252, 1657.

Wallace, C.S., Strike, S.A., Truskey, G.A., 2007. Smooth muscle cell rigidity and extraellular matrix organization influence endothelial cell spreading and adhesion formation in co-culture. Am. J. Physiol. Heart. Circ. Physiol. doi: 10.1152/ajpheart. 00618.2007.

Wallace, P.M., Senter, P.D., 1991. *In vitro* and *in vivo* activities of monoclonal antibody-alkaline phosphatase conjugates in combination with phenol mustard phosphate. Bioconjug. Chem. 2, 349–352.

Wallenfels, K., Weil, R., 1972. (third ed.)Boyer, P.D. The Enzymes, vol. 7. Academic Press, New York, pp. 617.

Walling, C., Gibian, M.J., 1965. Hydrogen abstraction reactions by the triplet states of ketones. J. Am. Chem. Soc. 87, 3361.

Walter, M.V., Malkroch, M., 2012. Simplifying the synthesis of dendrimers: accelerated approaches. Chem. Soc. Rev. 41, 4593–4609.

Walus, L.R., Pardridge, W.M., Starzyk, R.M., Friden, P.M., 1996. Enhanced uptake of rsCD4 across the rodent and primate blood–brain barrier after conjugation to anti-transferrin receptor antibodies. J. Pharmacol. Exp. Ther. 277, 1067.

Wang, A.Z., Langer, R., Farokhzad, O.C., 2012. Nanoparticle delivery of cancer drugs. Annu. Rev. Med. 63, 185–198.

Wang, C., Yan, Q., Liu, H.B., Zhou, X.H., Xiao, S.J., 2011. Different EDC/NHS activation mechanisms between PAA and PMAA brushes and the following amidation reactions. Langmuir 27 (19), 12058–12068.

Wang, D., Wilson, G., Moore, S., 1976. Preparation of cross-linked dimers of pancreatic ribonuclease. Biochemistry 15, 660–665.

Wang, G., Strang, C., Pfaffinger, P.J., Covarrubias, M., 2007. Zn21-dependent redox switch in the intracellular T1-T1 interface of a Kv channel. J. Biol. Chem. 282, 13637–13647.

Wang, H., Shao, N., Qiao, S., Cheng, Y., 2012. Host–Guest chemistry of dendrimer–cyclodextrin conjugates: selective encapsulations of guests within dendrimer or cyclodextrin cavities revealed by NOE NMR techniques. J. Phys. Chem. B 116, 11217–11224.

Wang, I.C., Tai, L., Lee, D., Kanakamma, P., Shen, C.-F., Luh, T.-Y., et al., 1999. C60 and water-soluble fullerene derivatives as antioxidants against radical-initiated lipid peroxidation. J. Med. Chem. 42, 4614–4620.

Wang, J., Liang, Y.-L., Qu, J., 2009. Boiling water-catalyzed neutral and selective N-boc deprotection. Chem. Commun., 5144–5146.

Wang, K., Richards, F., 1974. An approach to nearest neighbor analysis of membrane proteins. Application to the human erythrocyte membrane of a method employing cleavable cross-linkages. J. Biol. Chem. 249, 8005–8018.

Wang, K., Richards, F., 1975. Reaction of dimethyl-3,39-dithiobispropionimidate with intact human erythrocytes. Cross-linking of membrane proteins and of hemoglobin. J. Biol. Chem. 250, 6622–6626.

Wang, L., Wang, K., Santra, S., Zhao, X., Hilliard, L.R., Smith, J.E., et al., 2006. Glow in the biological world. Anal. Chem. February, 646–654.

Wang, L., Zhao, W., O'Donoghue, M.B., Tan, W., 2007. Fluorescent nanoparticles for multiplexed bacteria monitoring. Bioconjug. Chem. 18, 297–301.

Wang, P., Dai, S., Waezsada, S.D., Tsao, A.Y., Davison, B.H., 2001. Enzyme stabilization by covalent binding in nanoporous sol-gel glass for nonaqueous biocatalysis. Biotechnol. Bioeng. 74, 249–255.

Wang, Q., Chan, T.R., Hilgraf, R., Fokin, V.V., Sharpless, K.B., 2003. Bioconjugation by copper(I)-catalyzed azide-alkyne [3+2] cycloaddition. J. Am. Chem. Soc. 125, 3192–3193.

Wang, R., Chen, W., Meng, F., Cheng, R., Deng, C., Feijen, J., et al., 2011. Unprecedented access to functional biodegradable polymers and coatings. Macromolecules 44, 6009–6016.

Wang, S.S., Carpenter, F.H., 1968. Kinetic studies at high pH of the trypsin-catalyzed hydrolysis of Na-benzoyl derivatives of L-arginamide, L-lysinamide, and S-2-aminoethyl-L-cysteinamide and related compounds. J. Biol. Chem. 243, 3702–3710.

Wang, T., Wang, N., Wang, T., Sun, W., Li, T., 2011. Preparation of submicron liposomes exhibiting efficient entrapment of drugs by freeze-drying water-in-oil emulsions. Chem. Phys. Lipids 164 (2), 151–157.

Wang, Y.T., Lu, X.M., Zhu, F., Huang, P., Yu, Y., Zeng, L., et al., 2011. The use of a gold nanoparticle-based adjuvant to improve the therapeutic efficacy of hNgR-Fc protein immunization in spinal cord-injured rats. Biomaterials 32 (31), 7988–7998.

Wang, Y., Gibney, P.A., West, J.D., Morano, K.A., 2012. The yeast Hsp70 Ssa1 is a sensor for activation of the heat shock response by thiol-reactive compounds. Mol. Biol. Cell 23 (17), 3290–3298.

Wang, Y.-L., 1985. Exchange of actin subunits at the leading edge of living fibroblasts: possible role of treadmilling. J. Cell Biol. 101, 597.

Wang, Y., Yuan, H., Wright, S.C., Wang, H., Larrick, J.W., 2001. Synthesis and preliminary cytotoxicity study of a cephalosporin-CC-1065 analogue prodrug. BMC Chem. Biol. 1, 4. <http://www.biomedcentral.com/1472-6769/1/4>.

Wang, Z.V., Schraw, T.D., Kim, J.-Y., Khan, T., Rajala, M.W., Follenzi, A., et al., 2007. Secretion of the adipocyte-specific secretory protein adiponectin critically depends on thiol-mediated protein retention. Mol. Cell. Biol. 27, 3716–3731.

Ward, W.S., Schmidt, W.N., Schmidt, C.A., Hnilica, L.S., 1989. Crosslinking of Novikoff ascites hepatoma cytokeratin filaments. Biochemistry 24 (16), 4429–4434.

Warwood, S., Mohammed, S., Cristea, I.M., Evans, C., Whetton, A.D., Gaskell, S.J., 2006. Guanidination chemistry for qualitative and quantitative proteomics. Rapid Commun. Mass Spectrom. 20, 3245–3256.

Watabe, S., Ochiai, Y., Hashimoto, K., 1982. Identification of 5,5'-dithio-bis-2nitrobenzoic acid (DTNB) and alkali light chains of piscine myosin. B. Jpn. Soc. Sci. Fish. 48, 827–832.

Waterman, M.R., Yanaoka, K., Chuang, A.H., Cottam, G.L., 1975. Antisickling nature of dimethyl adipimidate. Biochem. Biophys. Res. Comm. 63, 580–587.

Watson, H.M., Gentry, L.E., Asuru, A.P., Wang, Y., Marcus, S., Busenlehner, L.S., 2012. Heterotrifunctional chemical crosslinking mass spectrometry confirms physical interaction between human frataxin and ISU. Biochemistry 51, 6889–6891.

Watson, J.V., 2004. Introduction to Flow Cytometry. Cambridge University Press.

Waugh, S.M., DiBella, E.E., Pilch, P.F., 1989. Isolation of a proteolitically derived domain of the insulin receptor containing the major site of cross-linking/binding. Biochemistry 28, 3448–3455.

Webster, R., Didier, E., Harris, P., Siegel, N., Stadler, J., Tilbury, L., et al., 2007. PEGylated proteins: evaluation of their safety in the absence of definitive metabolism studies. Drug Metab. Dispos. 35, 9–16.

Wedekind, F., Baer-Pontzen, K., Bala-Mohan, S., Choli, D., Zahn, H., Brandenburg, D., 1989. Hormone binding site of the insulin receptor: analysis using photoaffinity-mediated avidin complexing. Biol. Chem. Hoppe-Seyler 370, 251–258.

Wei, W.-H., Fountain, M., Magda, D., Wang, Z., Lecane, P., Mesfin, M., et al., 2005. Gadolinium texaphyrin–methotrexate conjugates. Towards improved cancer chemotherapeutic agents. Org. Biomol. Chem. 3, 3290–3296.

Wei, W.-H., Wang, Z., Mizuno, T., Cortez, C., Fu, L., Sirisawad, M., et al., 2006. New polyethyleneglycol-functionalized texaphyrins: synthesis and in vitro biological studies. Dalton Trans., 1934–1942.

Wei, Y., Chen, S.S., 2005. Mass spectrometry-based quantitative proteomic profiling. Brief. Funct. Genomic. Proteomic. 4, 1–12.

Wei, Z., Huang, W., Li, J., Hou, G., Fang, J., Yuan, Z., 2007. Studies on endotoxin removal mechanism of adsorbents with amino acid ligands. J. Chromatogr. B 852, 288–292.

Weiner, R.E., Thakur, M.L., 2002. Radiolabeled peptides in the diagnosis and therapy of oncological diseases. Appl. Radiat. Isot. 57, 749–763.

Weissleder, R., Tung, C.-H., Mahmood, U., Bogdanov Jr., A., 1999. In vivo imaging of tumors with proteaseactivated near-infrared fluorescent probes. Nat. Biotechnol. 17, 375–378.

Weith, H.L., Wiebers, J.L., Gilham, P.T., 1970. Synthesis of cellulose derivatives containing the dihydroxyboryl group and a study of their capacity to form specific complexes with sugars and nucleic acid components. Biochemistry 9, 4396–4401.

Wellman, A., Meares, C.F., 1991. Footprint of the sigma protein: a re-examination. Biochem. Biophys. Res. Comm. 177, 140–144.

Wells, J.A., Knoeber, C., Sheldon, M.C., Werber, M.M., Yount, R.G., 1980. Cross-linking of myosin subfragment 1. Nucleotide-enhanced modification by a variety of bifunctional reagents. J. Biol. Chem. 255, 11135.

Wells, N.J., Basso, A., Bradley, M., 1998. Solid-phase dendrimer synthesis. Biopolymers 47, 381–396.

Weltman, J.K., Hohnson, S.-A., Langevin, J., Riester, E.F., 1983. *N*-Succinimidyl(4-iodoacetyl)aminobenzoate: a new heterobifunctional cross-linker. BioTech. 1, 148–152.

Weng, Y.J., Jing, F.J., Chen, J.Y., Huang, N., 2011. Construction of heparinylated multilayer films on Ti-O via streptavidin/biotin interaction. Appl. Surf. Sci. 258, 5947–5954.

Wessels, B.W., Rogus, R.D., 1984. Radionuclide selection and model absorbed dose calculations for radiolabeled tumor associated antibodies. Med. Phys. 11, 638–645.

Wessendorf, M.W., 1990. Characterization and use of multi-color fluorescence microscopic techniques, Björklund, A.Hökfelt, T. Handbook of Chemical Neuroanatomy, vol. 8. Elsevier, Amsterdam (Chapter 1).

Weston, P.D., Devries, J.A., Wrigglesworth, R., 1980. Conjugation of enzymes to immunoglobulins using dimaleimides. Biochim. Biophys. Acta 612, 40–49.

Wheat, T., Shelton, J.A., Conzales-Prevatt, V., Goldberg, E., 1985. The antigenicity of synthetic peptide fragments of lactate dehydrogenase C4. Mol. Immunol. 22, 1195–1199.

Wheatley, J.B., Schmidt Jr., D.E., 1993. Salt-induced immobilization of proteins on a high-performance liquid chromatographic epoxide affinity support. J. Chromatogr. A 644, 11–16.

Wheatley, J.B., Schmidt Jr., D.E., 1999. Salt-induced immobilization of affinity ligands onto epoxide-activated supports. J. Chromatogr. A 849, 1–12.

Whetstone, P.A., Butlin, N.G., Corneillie, T.M., Meares, C.F., 2003. Element-coded affinity tags for peptides and proteins. Bioconjug. Chem. 15, 3–6.

Whitaker, J.E., Haugland, R.P., Moore, P.L., Hewitt, P.C., Reese, M., Haugland, R.P., 1991. Cascade blue derivatives: water soluble, reactive, blue emission dyes evaluated as fluorescent labels and tracers. Anal. Biochem. 198, 119–130.

Whitaker, D.T., Whitaker, K.S., Johnson, C.R., Haas, J., 2006. p-Toluenesulfonyl Chloride in E-EROS Encyclopedia of Reagents for Organic Synthesis. John Wiley & Sons, Ltd., New York.

Wieder, K.J., Palczuk, N.C., van Es, T., Davis, F.F., 1979. Some properties of polyethylene glycol: phenylalanine ammonia-lyase adducts. J. Biol. Chem. 254, 12579–12587.

Wiels, J., Junqua, S., Dujardin, P., Le Pecq, J.B., Tursz, T., 1984. Properties of immunotoxins against a glycolipid antigen associated with Burkitt's lymphoma. Cancer Res. 44, 129.

Wiesner, U., Ow, H., Larson, D.E., Webb, W.W., United States patent application publication, Publication No.: US 2006/0183246 A1, August 17, 2006.

Wigneshweraraj, S.R., Fujita, N., Ishihama, A., Buck, M., 2000. Conservation of sigma-core RNA polymerase proximity relationships between the enhancer-independent and enhancer-dependent sigma classes. EMBO J. 19 (12), 3038–3048.

Wijnen, J.W., Engberts, J.B.F.N., 1997. Retro-diels-alder reaction in aqueous solution: toward a better understanding of organic reactivity in water. J. Org. Chem. 62, 2039–2044.

Wilbur, D., Scott, J.E., Stray, D.K., Hamlin, D.K., Curtis, Vessella, R.L., 1994. Monoclonal antibody fab′ fragment cross-linking using equilibrium transfer alkylation reagents. A strategy for site-specific conjugation of diagnostic and therapeutic agents with F(ab′)2 fragments. Bioconjug. Chem. 1994 (5), 220–235.

Wilbur, D., Scott, C., Ming-Kuan, P., Pradip, M., Hamlin, D.K., 2000. Biotin reagents for antibody pretargeting. 4. Selection of biotin conjugates for *in vivo* application based on their dissociation rate. Bioconjug. Chem. 11, 569–583.

Wilbur, D.S., 1992. Radiohalogenation of proteins: an overview of radionuclides, labeling methods, and reagents for conjugate labeling. Bioconjug. Chem. 3, 433–470.

Wilbur, S.D., Pathare, P.M., Hamlin, D.K., Buhler, K.R., Vessella, R.L., 1998. Biotin reagents for antibody pretargeting. 3. Synthesis, radioiodination, and evaluation of biotinylated starburst dendrimers. Bioconjug. Chem. 9, 813–825.

Wilchek, M., Bayer, E.A., 1987. Labeling glycoconjugates with hydrazide reagentsGinsburg, V. Methods in Enzymology, vol. 138. Academic Press, Orlando Florida, pp. 429–442.

Wilchek, M., Bayer, E.A., 1988. The avidin–biotin complex in bioanalytical applications. Anal. Biochem. 171, 1–32.

Wilchek, M., Givol, D., 1977. Affinity cross-linking of heavy and light chainsJakoby, W.B.Wilchek, M. Methods in Enzymology, vol. 46. Academic Press, New York, pp. 501.

Wilchek, M., Miron, T., 1982. Immobilization of enzymes and affinity ligands onto agarose via stable and uncharged carbamate linkages. Biochem. Int. 4, 629–635.

Wilchek, M., Miron, T., 1985. Activation of sepharose with N,N′-disuccinimidyl carbonate. Appl. Biochem. Biotechnol. 11, 191–193.

Wilchek, M., Miron, T., 1987. Limitations of N-hydroxysuccinimide esters in affinity chromatography and protein immobilization. Biochemistry 26, 2155–2161.

Wilchek, M., Spiegel, S., Spiegel, Y., 1980. Fluorescent reagents for the labeling of glycoconjugates in solution and on cell surfaces. Biochem. Biophys. Res. Comm. 92, 1215.

Wilchek, M., Ben-Hur, H., Bayer, E.A., 1986. p-Diazobenzoyl biocytin—a new biotinylating reagent for the labeling of tyrosines and histidines in proteins. Biochem. Biophys. Res. Comm. 138, 872–879.

Wilchek, M., Knudsen, K.L., Miron, T., 1994. Improved method for preparing N-hydroxysuccinimide ester-containing polymers for affinity chromatography. Bioconjug. Chem. 5, 491–492.

Wildling, L., Unterauer, B., Zhu, R., Rupprecht, A., Haselqrubler, T., Rankl, C., et al., 2011. Linking sensor molecules to amino groups to amino-functionalized AFM tips. Bioconjug. Chem. 22, 1239–1248.

Wiley, D.C., Skehel, J.J., Waterfield, M., 1977. Evidence from studies with a cross-linking reagent that the haemagglutinin of influenza virus is a trimer. Virology 79, 446–448.

Wiley, J.P., Hughes, K.A., Kaiser, R.J., Kesicki, E.A., Lund, K.P., Stolowitz, M.L., 2001. Phenylboronic acid-salicylhydroxamic acid bioconjugates. 2. Polyvalent immobilization of protein ligands for affinity chromatography. Bioconjug. Chem. 12, 240–250.

Wilhelmsen, K., Copp, J., Glenn, G., Hoffman, R.C., Tucker, P., van der Geer, P., 2004. Purification and identification of protein-tyrosine kinase-binding proteins using synthetic phosphopeptides as affinity reagents. Proteomics 3, 887–895.

Willey, T.M., Vance, A.L., Bostedt, C., van Buuren, T., Meulenberg, R.W., Terminello, L.J., et al., 2004. Surface structure and chemical switching of thioctic acid adsorbed on au(111) as observed using absorption fine structure. Langmuir 20 (12), 4939–4944.

Williams, A., Ibrahim, I.A., 1981. A mechanism involving cyclic tautomers for the reaction with nucleophiles of the water-soluble peptide coupling reagent 1-ethyl-3-(3-dimethyla-minopropyl) carbodiimide (EDC). J. Am. Chem. Soc. 103, 7090–7095.

Williams, D.C., Huff, G.F., Seitz, W.R., 1976. Glucose oxidase chemiluminescence measurement of glucose in urine compared with the hexokinase method. Clin. Chem. 22, 372.

Williams, E.H., Davydov, A.V., Motayed, A., Sundaresan, S.G., Bocchini, P., Richter, L., et al., 2012. Immobilization of streptavidin on 4H-SiC for biosensor development. Appl. Surf. Sci. 258, 6056–6063.

Willingham, G.L., Gaffney, B.J., 1983. Reactions of spin-label cross-linking reagents with red blood cell proteins. Biochemistry 22, 892.

Willner, D., Trail, P.A., Hofstead, S.J., King, H.D., Lasch, S.J., Braslawsky, G.R., et al., 1993. (6-Maleimidocaproyl) hydrazone of doxorubicin—a new derivative for the preparation of immunoconjugates of doxorubicin. Bioconjug. Chem. 4, 521–527.

Wilson, M.B., Nakane, P.K., 1978. Recent developments in the periodate method of conjugating horseradish peroxidase (HRPO) to antibodies. In: Knapp, W., Holuber, K., Wick, G. (Eds.), Immunofluorescence and Related Staining Techniques. Elsevier/North-Holland Biomedical Press, Amsterdam, pp. 215–224.

Wilson, R., Akhavan-Tafti, H., DeSilva, R., Schaap, A.P., 2001. Comparison between acridan ester, luminol, and ruthenium chelate electrochemiluminescence. Electroanalysis 13, 1083–1092.

Winterbourn, C.C., Buss, I.H., 1999. Protein carbonyl measurement by enzyme-linked immunosorbent assay. Methods Enzymol. 300, 106–111.

Winterbourn, C.C., Kettle, A.J., 2000. Biomarkers of myeloperoxidase-derived hypochlorous acid. Free Radic. Biol. Med. 29, 403–409.

Winzerling, J.J., Berna, P., Porath, J., 1992. How to use immobilized metal ion affinity chromatography. Methods 4, 4–13.

Wirth, P., Souppe, J., Tritsch, D., Biellmann, J.F., 1991. Chemical modification of horseradish peroxidase with ethanal-MePEG: solubility in organic solvents, activity and properties. Bioorg. Chem. 19, 133–142.

Wissing, J., Jänsch, L., Nimtz, M., Dieterich, G., Hornberger, R., Kéri, G., et al., 2007. Proteomics analysis of protein kinases by target class-selective prefractionation and tandem mass spectrometry. Mol. Cell. Proteomics 6, 537–547.

Wittig, I., Karas, M., Schägger, H., 2007. High resolution clear native electrophoresis for in-gel functional assays and fluorescence studies of membrane protein complexes. Mol. Cell. Proteomics 6, 1215–1225.

Witzig, T.E., Gordon, L.I., Cabanillas, F., Czuczman, M.S., Emmanouilides, C., Joyce, R., et al., 2002. Randomized controlled trial of yttrium-90-labeled ibritumomab tiuxetan radioimmunotherapy versus rituximab immunotherapy for patients with relapsed or refractory low-grade, follicular, or transformed B-cell non-Hodgkin's lymphoma. J. Clin Oncol. 20, 2453–2463.

Wöhnert, J., Franz, K.J., Nitz, M., Imperiali, B., Schwalbe, H., 2003. Protein alignment by a coexpressed lanthanide-binding tag for the measurement of residual dipolar couplings. J. Am. Chem. Soc. 125, 13338–13339.

Wojchowski, D.M., Sytkowski, A.J., 1986. Hybridoma production by simplified avidin-mediated electrofusion. J. Immunol. Methods 90, 173–177.

Wold, F., 1961. Reaction of bovine serum albumin with the bifunctional reagent p,p9-difluoro-m,m9-dinitrodiphenylsulfone. J. Biol. Chem. 236 (106).

Wold, F., 1972. Bifunctional reagentsHirs, C.H.W.Timasheff, S.N. Methods in Enzymology, vol. 25. Academic Press, New York, pp. 623.

Wolf, J., Gerber, A.P., Keller, W., 2002. tadA, an essential tRNA-specific adenosine deaminase from escherichia coli. EMBO J. 21 (14), 3841–3851.

Wolfe, C.A.C., Hage, D.S., 1995. Studies on the rate and control of antibody oxidation by periodate. Anal. Biochem. 231, 123–130.

Wolfe, L.A., Mullin, R.J., Laethem, R., Blumenkopf, T.A., Cory, M., Miller, J.F., et al., 1999. Antibody-directed enzyme prodrug therapy with the T268G mutant of human carboxypeptidases A1: in vitro and in vivo studies with prodrugs of methotrexate and the thymidylate synthase inhibitors GW1031 and GW1843. Bioconjug. Chem. 10, 38–48.

Wollenweber, H.-W., Morrison, D.C., 1985. Synthesis and biochemical characterization of a photoactivatable, iodinatable, cleavable bacterial lipopolysaccharide derivative. J. Biol. Chem. 260, 15068–15074.

Woller, E., Cloninger, M.J., 2001. Mannose functionalization of a sixth generation dendrimer. Biomacromolecules 2, 1052–1054.

Woller, E.K., Cloninger, M.J., 2002. The lectin-binding properties of six generations of mannose-functionalized dendrimers. Org. Lett. 4, 7–10.

Woller, E.K., Walter, E.D., Morgan, J.R., Singel, D.J., Cloninger, M.J., 2003. Altering the strength of lectin binding interactions and controlling the amount of lectin clustering using mannose/hydroxyl-functionalized dendrimers. J. Am. Chem. Soc. 125, 8820–8826.

Wollmann, P., Cui, S., Viswanathan, R., Berninghausen, O., Wells, M.N., Moldt, M., et al., 2011. Structure and mechanism of the swi2/snf2 remodeller mot1 in complex with its substrate TBP. Nature 475 (7356), 403–407.

Wong, S.S., 1991. Chemistry of Protein Conjugation and Cross-Linking. CRC Press.

Wong, D.Y.Q., Lau, J.Y., Ang, W.H., 2012. Harnessing chemoselective imine ligation for tethering bioactive molecules to platinum (iv) prodrugs. Dalton. Trans. 41 (20), 6104–6111.

Wood, C.L., O'Dorisio, M.S., 1985. Covalent cross-linking of vasoactive intestinal polypeptide to its receptors on intact human lymphoblasts. J. Biol. Chem. 260, 1243–1247.

Woodward, M.P., Young, W.W., Bloodgood, R.A., 1985. Detection of monoclonal antibodies specific for carbohydrate epitopes using periodate oxidation. J. Immunol. Methods 78, 143–153.

Woodward, R.B., Olofson, R.A., 1961. The reaction of isoxazolium salts with bases. J. Am. Chem. Soc. 83, 1010.

Woodward, R.B., Olofson, R.A., Mayer, H., 1961. A new synthesis of peptides. J. Am. Chem. Soc. 83, 1007–1009.

Wright, A., Tao, M.-h., Kabat, E.A., Morrison, S.L., 1991. Antibody variable region glycosylation: position effects on antigen binding and carbohydrate structure. EMBO J. 10, 2717–2723.

Wright, A., Taol, M.-h., Kabat, E.A., Morrison, S.L., 1991. Antibody variable region glycosylation: position effects on antigen binding and carbohydrate structure. EMBO J. 10 (10), 2717–2723.

Wright, B.S., Tyler, G.A., O'Brien, R., Coporale, L.H., Rosenblatt, M., 1987. Immunoprecipitation of the parathyroid hormone receptor. Proc. Natl. Acad. Sci. USA 84, 26–30.

Wu, C.-W., Yarbrough, L.R., Wu, F.Y.-H., 1976. N-(1-pyrene) maleimide: a fluorescent cross-linking reagent. Biochemistry 15, 2863–2867.

Wu, F., Yu, J., 2007. Novel biomimetic affinity ligands for human tissue plasminogen activator. Biochem. Biophys. Res. Comm. 355, 673–678.

Wu, G., Barth, R.F., Yang, W., Chatterjee, M., Tjarks, W., Ciesielski, M.J., et al., 2004. Site-specific conjugation of boron containing dendrimers to anti-EGF receptor monoclonal antibody cetuximab (IMCC225) and its evaluation as a potential delivery agent for neutron capture therapy. Bioconjug. Chem. 15, 185–194.

Wu, P., Feldman, A.K., Nugent, A.K., Hawker, C.J., Scheel, A., Voit, B., et al., 2004. Efficiency and fidelity in a click-chemistry route to triazole dendrimers by the copper(I)-catalyzed ligation of azides and alkynes. Angew. Chem. Int. Ed. 43 (30), 3928–3932.

Wu, P., Malkoch, M., Hunt, J.N., Vestberg, R., Kaltgrad, E., Finn, M.G., et al., 2005. Multivalent, bifunctional dendrimers prepared by click chemistry. Chem. Comm., 5775–5777.

Wu, M., Zhang, X., Bian, Q., Taylor, A., Liang, J.J., Ding, L., et al., 2012. Oligomerization with wt αA-and αB-crystallins reduces proteasome-mediated degradation of C-terminally truncated αA-crystallin. Invest. Ophthalmol. Vis. Sci. 53 (6), 2541–2550.

Wu, S.Y., Yuen, S.M., Ma, C., Huang, Y.L., Teng, C.C., 2011. Molecular motion, morphology and properties of 3-isocyanato-propyltriethoxysilane-modified multi-walled carbon nanotube/epoxy composites. Micro Nano Lett., IET 6 (6), 463–467.

Wu, X., Liu, H., Liu, J., Haley, K.N., Treadway, J.A., Larson, J.P., et al., 2003. Immunofluorescent labeling of cancer marker Her2 and other cellular targets with semiconductor quantum dots. Nat. Biotechnol. 21, 41–46.

Wu, Y.L., Park, K., Soo, R.A., Sun, Y., Tyroller, K., Wages, D., et al., 2011. Inspire: a phase III study of the BLP25 liposome vaccine (L-BLP25) in Asian patients with unresectable stage III non-small cell lung cancer. BMC cancer 11 (1), 430. doi: 10.1186/1471-2407-11-430.

Wu, Y.-Z., Manevich, Y., Baldwin, J.L., Dodia, C., Yu, K., Feinstein, S.I., et al., 2005. Interaction of surfactant protein A with peroxiredoxin 6 regulates phospholipase A2 activity. J. Biol. Chem. doi: 10.1074/jbc.M504525200.

Wudl, F., 1992. The chemical properties of buckminsterfullerene (C60) and the birth and infancy of fulleroids. Acc. Chem. Res. 25, 157–161.

Wygrecka, M., Morty, R.E., Markart, P., Kanse, S.M., Andreasen, P.A., Wind, T., et al., 2007. Plasminogen activator inhibitor-1 (PAI-1) is an inhibitor of factor VII-activating protease in patients with acute respiratory distress syndrome. J. Biol. Chem.10.1074/jbcM610748200 (published online ahead of print).

Xavier, F., Malcata, H.R., Reyes, H.S., Garcia, C.G.H., Amundson, C.H., 1990. Immobilized lipase reactors for modification of fats and oils—a review. J. Am. Oil Chem. Soc. 67, 890–910.

Xia, H., Jin, X., Wu, P., Zheng, Z., 2011. Porous ceramic/agarose composite adsorbents for fast protein liquid chromatography. J. Chromatogr. A 1223, 126–130.

Xiang, C.C., et al. 2004. Using DSP, a reversible crosslinker, to fix tissue sections for immunostaining, microdissection and expression profiling. Nucleic Acids Res. 32, e185.

Xiao, Y., Barker, P.E., 2004. Semiconductor nanocrystal probes for human metaphase chromosomes. Nucleic Acids Res. 32 e28/1–e28/5.

Xiao, S.L., Wang, Q., Yu, F., Peng, Y.Y., Yang, M., Sollogoub, M., et al., 2012. Conjugation of cyclodextrin with fullerene as a new class of HCV entry inhibitors. Bioorg. Med. Chem. 20, 5616–5622.

Xie, J., Gao, J., Michalski, M., Chen, X., 2011, In: Chen, Z. (Ed.), Nanoparticle Surface Modification and Bioconjugation, in Nanoplatform-Based Molecular Imaging. John Wiley & Sons, Inc., pp. 47–74.

Xu, M.-Q., Perler, F.B., 1996. The mechanism of protein splicing and its modulation by mutation. EMBO J. 15, 5146–5153.

Xu, J., DeGraw, A.J., Duckworth, B.P., Lenevich, S., Tann, C.-M., Henson, E.C., et al., 2006. Synthesis and reactivity of 6,7-dihydrogeranylazides: reagents for primary azide incorporation into peptides and subsequent staudinger ligation. Chem. Biol. Drug Des. 68, 85–96.

Xu, J., Huo, S., Yuan, Z., Zhang, Y., Xu, H., Guo, Y., et al., 2011. Characterization of direct cellulase immobilization with superparamagnetic nanoparticles. Biocatal. Biotransformation 29 (2-3), 71–76.

Yakhnin, A.V., Yakhnin, H., Babitzke, P., 2012. Gel mobility shift assays to detect protein–RNA interactions. Methods Mol. Biol. 905, 201.

Yamada, H., Imoto, T., Fujita, K., Okazaki, K., Motomura, M., 1981. Selective modification of aspartic acid-101 in lysozyme by carbodiimide reaction. Biochemistry 20, 4836–4842.

Yamaguchi, H., Miyazaki, M., Kawazumi, H., Maeda, H., 2010. Multidigestion in continuous flow tandem protease-immobilized microreactors for proteomic analysis. Anal. Biochem. (Article in Press).

Yamamoto, K., Sekine, T., Sutoh, K., 1984. Spatial relationship between SH1 and the actin binding site on myosin subfragment-1 surface. FEBS Lett. 176, 75–78.

Yamasaki, R.B., Shimer, D.A., Feeney, R.E., 1981. Colorimetric determination of arginine residues in proteins by p-nitrophenylglyoxal. Anal. Biochem. 111, 220.

Yan, M., Cai, S.X., Wybourne, M.N., Keana, J.F.W., 1994. N-Hydroxysuccinimide ester functionalized perfluorophenyl azides as novel photoactivatable heterobifunctional cross-linking reagents. The covalent immobilization of biomolecules to polymer surfaces. Bioconjug. Chem. 5, 151–157.

Yan, W., Chen, S.S., 2005. Mass spectrometry-based quantitative proteomic profiling. Brief. Funct. Genomics proteomics 4, 1–12.

Yanes, R.E., Tamanoi, F., 2012. Development of mesoporous silica nanomaterials as a vehicle for anticancer drug delivery. Therapeutic Deliv. 3 (3), 389–404.

Yang, L., Zheng, C., Weisbrod, C.R., Tang, X., Munske, G.R., Hoopmann, M.R., et al., 2012. In vivo application of photocleavable protein interaction reporter technology. J. Proteome Res. 11 (2), 1027–1041.

Yang, M.Y., Chaudhary, A., Seaman, S., Dunty, J., Stevens, J., Elzarrad, M.K., et al., 2011. The cell surface structure of tumor endothelial marker 8 (TEM8) is regulated by the actin cytoskeleton. Biochim. Biophys. Acta (BBA)-Mol. Cell. Res. 1813 (1), 39–49.

Yang, Q., Adrus, N., Tomicki, F., Ulbricht, M., 2011. Composites of functional polymeric hydrogels and porous membranes. J. Mater. Chem. 21, 2783–2811.

Yang, S., Papagiakoumou, E., Guillon, M., de Sars, V., Tang, C.M., Emiliani, V., 2011. Three-dimensional holographic photostimulation of the dendritic arbor. J. Neural. Eng. 8 (4), 046002.

Yang, S.J., Zhang, H., 2012. Glycan analysis by reversible reaction to hydrazide beads and mass spectrometry. Anal. Chem. 84, 2232–2238.

Yang, W., Cheng, Y., Xu, T., Wang, X., Wen, L.-P., 2009. Targeting cancer cells with biotin–dendrimerconjugates. Eur. J. Med. Chem. 44, 862–868.

Yang, W., Barth, R.F., Wu, G., Tjarks, W., Binns, P., Riley, K., 2009. Boron neutron capture therapy of EGFR or EGFRvIII positive gliomas using either boronated monoclonal antibodies or epidermal growth factor as molecular targeting agents. Appl. Radiat. Isot. 67, S328–S331.

Yano, M., Okuda, S., Oda, T., Tokuhisa, T., Tateishi, H., Mochizuki, M., et al., 2005. Correction of defective interdomain interaction within ryanodine receptor by antioxidant is a new therapeutic strategy against heart failure. Circulation 112, 3633–3643.

Yao, J., Larson, D.R., Vishwasrao, H.D., Zipfel, W.R., Webb, W.W., 2005. Blinking and nonradiant dark fraction of water-soluble quantum dots in aqueous solution. PNAS 102 (40), 14284–14289.

Yao, G., Wang, L., Wu, Y., Smith, J., Xu, J., Zhao, W., et al., 2006. FloDots: luminescent nanoparticles. Anal. Bioanal. Chem. 385, 518–524.

Yaoi, T., Chamnongpol, S., Jiang, X., Li, X., 2006. SH2 domain-based high-throughput assays for profiling downstream molecules in receptor tyrosine kinase pathways. Mol. Cell. Proteomics 5 (5), 959–968.

Yasuda, T., Dancey, G.F., Kinsky, S.C., 1977. Immunogenicity of liposomal model membranes in mice: dependence on phospholipid composition. Proc. Natl. Acad. Sci. USA 74, 1234–1236.

Yavuz, H., Ozden, K., Kin, E.P., Denizli, A., 2008. Concanavalin A binding on PHEMA beads and their interactions with myeloma cells. J. Macromol. Sci. Part A: Pure Appl. Chem. 46, 163–169.

Ye, Y., Bloch, S., Xu, B., Achilefu, S., 2006. Design, synthesis, and evaluation of near infrared fluorescent multimeric RGD peptides for targeting tumors. J. Med. Chem. 49, 2268–2275.

Ye, J., Zhang, X., Young, C., Zhao, X., Hao, Q., Cheng, L., et al., 2010. Optimized IMAC-IMAC protocol for phosphopeptide recovery from complex biological samples. J. Proteome Res. 9, 3561–3573.

Yegneswaran, k, Kojima, Y., Nguyen, P.M., Gale, A.J., Heeb, M.J., Griffin, J.H., 2007. Factor Va residues 311–325 represent an activated protein C binding region. J. Biol. Chem.10.1074/jbc. M704316200.

Yem, A.W., et al., 1992. Site-specific chemical modification of interleukin lb by acrylodan at cysteine 8 and lysine 103. J. Biol. Chem. 267, 3122.

Yen, S.P.S., Rembaum, A., Molday, R.W., Dreyer, W.J., 1979. Emulsion polymerization ACS Symp. Ser., vol. 24. American Chemical Society, Washington, D.C. 236.

Yeo, D.S.Y., Srinivasan, R., Uttamchandani, M., Chen, G.Y.J., Zhu, Q., Yao, S.Q., 2003. Cell-permeable small molecule probes for site-specific labeling of proteins. Chem. Commun. 23, 2870–2871.

Yeung, C.W.T., Moule, M.L., Yip, C.C., 1980. Photoaffinity labeling of insulin receptor with an insulin analogue selectively modified at the amino terminal of the B chain. Biochemistry 19, 2196–2203.

Yi, Y., Song, Y., Loscalzo, J., 2007. Regulation of the protein disulfide proteome by mitochondria in mammalian cells. PNAS 104, 10813–10817.

Yildiz, A., Forkey, J.N., McKinney, S.A., Ha, T., Goldman, Y.E., Selvin, P.R., 2003. Myosin V walks hand-over-hand: single fluorophore imaging with 1.5-nm localization. Science 300, 2061–2065.

Yim, E.K.F., Darling, E.M., Kulanqara, K., Guilak, F., Leong, K.W., 2010. Nanotopography-induced changes in focal adhesions, cytoskeletal organization, and mechanical properties of human mesenchymal stem cells. Biomaterials 31, 1299–1306.

Yokota, S., 1988. Effect of particle size on labeling density for catalase in protein A–gold immunocytochemistry. J. Histochem. Cytochem. 36, 107–109.

Yong-Sam, Kim, Yoo, H.S., Ko, J.H.e.o.n, 2009. Implication of aberrant glycosylation in cancer and use of lectin for cancer biomarker discovery. Protein Pept. Lett. 16, 499–507.

Yoo, H., Juliano, R.L., 2000. Enhanced delivery of antisense oligonucleotides with fluorophore-conjugated PAMAM dendrimers. Nucleic Acids Res. 28, 4225–4231.

Yoo, H., Sazani, P., Juliano, R.L., 1999. PAMAM dendrimers as delivery agents for antisense oligonucleotides. Pharm. Res. (N.Y.) 16, 1799–1804.

Yoon, H.C., Lee, D., Kim, H.-S., 2002. Reversible affinity interactions of antibody molecules at functionalized dendrimer monolayer: affinity-sensing surface with reusability. Anal. Chim. Acta 456, 209–218.

Yoshitake, S., Yamada, Y., Ishikawa, E., Masseyeff, R., 1979. Conjugation of glucose oxidase from aspergillus niger and rabbit antibodies using *N*-hydroxysuccinimide ester of *N*-(4-carboxycyclohexylmethyl)maleimide. Eur. J. Biochem. 101, 395–399.

Yoshitake, S., Imagawa, M., Ishikawa, E., Niitsu, Y., Urushizaki, I., Nishiura, M., et al., 1982a. Mild and efficient conjugation of rabbit fab′ and horseradish peroxidase using a maleimide compound and its use for enzyme immunoassay. J. Biochem. (Tokyo) 92, 1413–1424.

Yoshitake, S., Imagawa, M., Ishikawa, E., 1982b. Efficient preparation of rabbit fab′-horseradish peroxidase conjugates using maleimide compounds and its use for enzyme lmmunoassay. Anal. Lett. 15 (B2), 147–160.

Yoshimura, S.H., Khan, S., Maruyama, H., Nakayama, Y., Takeyasu, K., 2011. Fluorescence labeling of carbon nanotubes and visualization of a nanotube– protein hybrid under fluorescence microscope. Biomacromolecules 12 (4), 1200–1204.

Youle, R.J., Nevelle Jr., D.M., 1980. Anti-Thy 1.2 monoclonal antibody linked to ricin is a potent cell-type-specific toxin. Proc. Natl. Acad. Sci. USA 77, 5483–5486.

Young, J.L., 1979. The effect of dimethyl 3,39-dithiobispropionimidate on the adenylate cyclase activity of bovine corpus luteum. FEBS Lett. 104, 294–296.

Young, C.L., Britton, Z.T., Robinson, A.S., 2012. Recombinant protein expression and purification: a comprehensive review of affinity tags and microbial applications. Biotechnol. J. 7, 620–634.

Young, S.W., Qing, F., Harriman, A., Sessler, J.L., Dow, W.C., Mody, T.D., et al., 1996. Gadolinium(III) texaphyrin: a tumor selective radiation sensitizer that is detectable by MRI. Proc. Natl. Acad. Sci. USA 93, 6610–6615.

Young, J.-J., Cheng, K.-M., Tsou, T.-L., Liu, H.-W., Wang, H.-J., 2004. Preparation of cross-linked hyaluronic acid film using 2-chloro-1-methylpyridinium iodide or water-soluble 1-ethyl-(3,3-dimethylaminopropyl)carbodiimide. J. Biomater. Sci. Polymer Edn. 15, 767–780.

Yu, G., Gao, J., Hummelen, J.C., Wudl, F., Heeger, A.J., 1995. Polymer photovoltaic cells: enhanced efficiencies via a network of internal donor-acceptor heterojunctions. Science 270, 1789–1791.

Yu, C.C., Kuo, Y.Y., Liang, C.F., Chien, W.T., Wu, H.T., Chang, T.C., et al., 2012. Site-Specific immobilization of enzymes on magnetic nanoparticles and their use in organic synthesis. Bioconjug. Chem. 23 (4), 714–724.

Yu, M.-F., et al., 2000. Strength and breaking mechanism of multiwalled carbon nanotubes under tensile load. Science 287, 637–640.

Yu, R., Schweinberger, F., 1979. In vitro crosslinking of phytochrome to its putative receptor with Bi-imidoesters. Z. Pflanzenphysiol 94, 135–142.

Yu, S., Fu, L., Zhou, Y., Su, H., 2011. Novel bifunctional magnetic–near-infrared luminescent nanocomposites: near-infrared emission from Nd and Yb. Photochem. Photobiol. Sci. 10 (4), 548–553.

Yu, S.-H., Wands, A.M., Kohler, J.J., 2012. Photoaffinity probes for studying carbohydrate biology. J. Carbohydr. Chem. 31, 325–352.

Yu, Y., Liu, J., Zhao, Z., Ng, K.M., Luo, K.Q., Tang, B.Z., 2012. Facile preparation of Non-self-quenching fluorescent DNA strands with degree of labeling up to theoretic limit. Chem. Commun. 48, 6360–6362.

Yuan, K.H., Li, Q., Yu, W.L., Huang, Z., 2009. Photodynamic therapy in treatment of port wine stain birthmarks–recent progress. Photodiagn. Photodyn. Ther. 6 (3-4), 189–194.

Yurchenko, V., et al., 2002. Active site residues of cyclophilin A are crucial for its signaling activity via CD147. J. Biol. Chem. 277, 22959–22965.

Yurt, A., Daaboul, G.G., Connor, J.H., Goldberg, B.B., Ünlü, M.S., 2012. Single nanoparticle detectors for biological applications. Nanoscale 4 (3), 715–726.

Zafar, M.A., Sanchez-Alberola, N., Wolf, R.E., 2011. Genetic evidence for a novel interaction between transcriptional activator SoxS and region 4 of the σ^{70} subunit of RNA polymerase at class II SoxS-dependent promoters in *Escherichia coli*. J. Mol. Biol. 407 (3), 333–353.

Zahn, H., Lumper, L., 1968. Specificity of bifunctional sulfhydryl reagents and synthesis of a defined dimer of bovine serum albumin. Hoppe-Seyler's Z. Physiol. Chem. 349, 485.

Zahn, H., Meinhoffer, J., 1958. Reactions of 1,5-difluoro-2,4-dinitrobenzene with insulin. Makromol. Chem. 26, 153.

Zakharian, T.Y., Seryshev, A., Sitharaman, B., Gilbert, B.E., Knight, V., Wilson, L.J., 2005. A fullerene-paclitaxel chemotherapeutic: synthesis, characterization, and study of biological activity in tissue culture. J. Am. Chem. Soc. 127, 12508–12509.

Zalipsky, S., Lee, C., 1992. Use of functionalized poly(ethylene glycol)s for modification of polypeptides. In: Harris, J.M. (Ed.), Poly(Ethylene Glycol) Chemistry: Biotechnical and Biomedical Applications. Plenum, New York, pp. 347–370.

Zalipsky, S., Seltzer, R., Nho, K., 1991. Succinimidyl carbonates of polyethylene glycol: useful reactive polymers for preparation of protein conjugates. In: Dunn, R.L., Ottenbrite, R.M. (Eds.), Polymeric Drugs and Drug Delivery Systems. American Chemical Society, Washington, D.C., pp. 91–100.

Zalipsky, S., Seltzer, R., Menon-Rudolph, S., 1992. Evaluation of a new reagent for covalent attachment of polyethylene glycol to proteins. Biotechnol. Appl. Biochem. 15, 100–114.

Zaman, N.T., Tan, F.E., Joshi, S.M., Ying, J.Y., 2006. Targeted stimuli-responsive dextran conjugates for doxorubicin delivery to hepatocytes. Mol. Eng. Biol. Chem. Syst. <http://hdl.handle.net/1721.1/30394>.

Zanocco, J., Krohn, R., Sykaluk, L., Olson, B., 1993. Unpublished observations. Pierce Chemical.

Zappacosta, F., Annan, R.S., 2004. N-terminal isotope tagging strategy for quantitative proteomics: results-driven analysis of protein abundance changes. Anal. Chem. 76, 6618–6627.

Zappacosta, F., Collingwood, T.S., Huddleston, M.J., Annan, R.S., 2006. A quantitative results-driven approach to analyzing multisite protein phosphorylation. Mol. Cell. Proteomics 5 (11), 2019–2030.

Zara, J.J., et al. 1991. A carbohydrate-directed heterobifunctional cross-linking reagent for the synthesis of immunoconjugates. Anal. Biochem. 194, 156–162.

Zarling, D.A., Watson, A., Bach, F.H., 1980. Mapping of lymphocyte surface polypeptide antigens by chemical cross-linking with BSOCOES. J. Immunol. 124, 913–920.

Zarling, D.A., Miskimen, J.A., Fan, D.P., Fujimoto, E.K., Smith, P.K., 1982. Association of sendai virion envelope and a mouse surface membrane polypeptide on newly infected cells: lack of association with H-2K/D or alteration of viral immunogenicity. J. Immunol. 128, 251–257.

Zearfoss, N.R., Ryder, S.P., 2012. End-labeling oligonucleotides with chemical tags after synthesis. Methods Mol. Biol. 941, 181–193.

Zecherle, G.N., 1990. The ribosomal location and conformation of Escherichia coli protein L7/L12 studied by cysteine site directed mutagenesis and crosslinking. Doctoral Dissertation. University of California at Davis.

Zeeman, R., 1998. Cross-linking of collagen-based materials. Thesis University of Twente, Enschede, The Netherlands. ISBN: 90 365 1207 7.

Zeheb, R., Chang, V., Orr, G.A., 1983. An analytical method for the selective retrieval of iminobiotin-derivatized plasma membrane proteins. Anal. Biochem. 129, 156–161.

Zeng, Y., Ramya, T.N.C., Dirksen, A., Dawson, P.E., Paulson, P.C., 2009. High-efficiency labeling of sialylated glycoproteins on living cells. Nat. Methods 6 (3), 207–209.

Zeng, S., Yong, K.T., Roy, I., Dinh, X.Q., Yu, X., Luan, F., 2011. A review on functionalized gold nanoparticles for biosensing applications. Plasmonics 6 (3), 491–506.

Zeng, Z., Hincapie, M., Pitteri, S.J., Hanash, S., Schalkwijk, J., Hogan, J.M., et al., 2011. A proteomics platform combining depletion, Multi-lectin Affinity Chromatography (M-LAC) and isoelectric focusing to study the breast cancer proteome. Anal. Chem. 83 (12), 4845–4854.

Zhang, A., Gonzalez, S.M., Cantor, E.J., Chong, S., 2001. Construction of a mini-intein fusion system to allow both direct monitoring of soluble protein expression and rapid purification of target proteins. Gene 275, 241–252.

Zhang, H., Yan, W., Aebersold, R., 2004. Chemical probes and tandem mass spectrometry: a strategy for the quantitative analysis of proteomes and subproteomes. Curr. Opin. Chem. Biol. 8, 66–75.

Zhang, H., Wu, B., Hu, W., Liu, Y., 2011. Separation and/or selective enrichment of single-walled carbon nanotubes based on their electronic properties. Chem. Soc. Rev. 40 (3), 1324–1336.

Zhang, J., Walker, G.C., 1996. Identification of elements of the peptide binding site of dnaK by peptide cross-linking. J. Biol. Chem. 271, 19668.

Zhang, L., Gao, H., Chen, L., Wu, B., Zheng, Y., Liao, R., et al., 2008. Tumor targeting of vincristine by mBAFF-modified PEG liposomes in B lymphoma cells. Cancer Lett. 269, 26–36.

Zhang, L., Li, Z., Chang, R., Chen, Y., Zhang, W., 2009. Synthesis and characterization of novel phenolphthalein immobilized halochromic fiber. React. Funct. Polym. 69, 234–239.

Zhang, L., Su, Y., Zheng, Y., Jiang, Z., Shi, J., Zhu, Y., et al., 2010. Sandwich-structured enzyme membrane reactor for efficient conversion of maltose into isomaltooligosaccharides. Bioresour. Technol. (in press).

Zhang, M., Yao, Z., Garmestani, K., Yu, S., Goldman, C.K., Paik, C.H., et al., 2009. Preclinical evaluation of an Anti-CD25 Monoclonal antibody, 7G7/B6, armed with the β-Emitter, yttrium-90, as a radioimmunotherapeutic agent for treating lymphoma. Cancer Biother. Radiopharm. 24 (3), 303–309.

Zhang, Q., Tang, N., Brock, J.W.C., Mottaz, H.M., Ames, J.M., Baynes, J.W., et al., 2007. Enrichment and analysis of Non-enzymatically glycated peptides: boronate affinity chromatography coupled with electron transfer dissociation mass spectrometry. J. Proteome Res. 6 (6), 2323–2330.

Zhang, Y., Pardridge, W.M., 2005. Delivery of b-Galactosidase to mouse brain via the blood–brain barrier transferrin receptor. J. Pharmacol. Exp. Ther. 313, 1075–1081.

Zhang, Y., Milam, V.T., Graves, D.J., Hammer, D.A., 2006. Differential adhesion of microspheres mediated by DNA hybridization I: experiment. Biophys. J. 90, 4128–4136.

Zhang, Y., Hong, H., Engle, J.W., Bean, J., Yang, Y., Leigh, B.R., et al., 2011. Positron emission tomography imaging of CD105 expression with a ^{64}Cu-labeled monoclonal antibody: NOTA is superior to DOTA. PloS one 6 (12), e28005.

Zhang, Y., Wang, Z., Zhang, X., Zhou, W., Huang, L., 2011. One-pot fluorescent labeling of saccharides with fluorescein-5-thiosemicarbazide for imaging polysaccharides transported in living cells. Carbohydr. Res. 346 (14), 2156–2164.

Zhang, Z., Fan, J., Cheney, P.P., Berezin, M.Y., Edwards, W.B., Akers, W.J., et al., 2009. Activatable molecular systems using homologous near-infrared fluorescent probes for monitoring enzyme activities in vitro, in cellulo, and *in vivo*. Mol. Pharm. 6 (2), 416–427.

Zhao, H., Heindel, N.D., 1991. Determination of degree of substitution of formyl groups in polyaldehyde dextran. Pharm. Res. 8, 400–402.

Zhao, J., Milanova, M., Warmoeskerken, M.M., Dutschk, V., 2011. Surface modification of TiO_2 nanoparticles with silane coupling agents. Colloids Surf., A 413, 273–279.

Zhao, K., Zhou, H., Zhao, X., Wolff, D.W., Tu, Y., Liu, H., et al., 2012. Phosphatidic acid mediates the targeting of tBid to induce lysosomal membrane permeabilization and apoptosis. J. Lipid Res. 53, 2102–2114.

Zhao, L., Peralta-Videa, J.R., Ren, M., Varela-Ramirez, A., Li, C., Hernandez-Viezcas, J.A., et al., 2012. Transport of Zn in a sandy loam soil treated with ZnO NPs and uptake by corn plants: electron microprobe and confocal microscopy studies. Chem. Eng. J. 184, 1–8.

Zhao, X., Hilliard, L.R., Mechery, S.J., Wang, Y., Bagwe, R.P., Jin, S., et al., 2004. A rapid bioassay for single bacterial cell quantitation using bioconjugated nanoparticles. Proc. Natl. Acad. Sci. USA 101, 15027–15032.

Zhao, Y., Pérez-Segarra, W., Shi, Q., Wei, A., 2005. Dithiocarbamate assembly on gold. J. Am. Chem. Soc. 127 (20), 7328–7329.

Zheng, J., Dickson, R.M., 2002. Individual water-soluble dendrimer-encapsulated silver nanodot fluorescence. J. Am. Chem. Soc. 124, 13982–13983.

Zheng, Y., Yu, B., Weecharangsan, W., Piao, L., Darby, M., Mao, Y., et al., 2010. Transferrin-conjugated lipid-coated PLGA nanoparticles for targeted delivery of aromatase inhibitor 7α-APTADD to breast cancer cells. Int. J. Pharm. 390, 234–241.

Zhong, X., Reynolds, R., Kidd, J.R., Kidd, K.K., Jenison, R., Marlar, R.A., et al., 2003. Single-nucleotide polymorphism genotyping on optical thin-film biosensor chips. PNAS 100, 11559–11564.

Zhou, H., Ranish, J.A., Watts, J.D., Aebersold, R., 2002. Quantitative proteome analysis by solid-phase isotope tagging and mass spectrometry. Nat. Biotech. 19, 512–515.

Zhou, W., Merrick, B.A., Khaledi, M.G., Tomer, K.B., 2000. Detection and sequencing of phosphopeptides affinity bound to immobilized metal ion beads by matrix-assisted laser desorption/ionization mass spectrometry. J. Am. Soc. Mass. Spectrom. 11 (4), 273–282.

Zhou, W., Islam, M.F., Wang, H., Ho, D.L., Yodh, A.G., Winey, K.I., et al., 2004. Small angle neutron scattering from single-wall carbon nanotube suspensions: evidence for isolated rigid rods and rod networks. Chem. Phys. Lett. 384, 185–189.

Zhou, Q.H., Lu, J.Z., Hui, E.K.W., Boado, R.J., Pardridge, W.M., 2011. Delivery of a peptide radiopharmaceutical to brain with an IgG–avidin fusion protein. Bioconjug. chem. 22 (8), 1611–1618.

Zhu, J., He, J., Liu, Y., Simeone, D.M., Lubman, D.M., 2012. Identification of glycoprotein markers for pancreatic cancer CD24+ CD44+ stem-like cells using Nano-LC–MS/MS and tissue microarray. J. Proteome Res. 11 (4), 2272–2281.

Zhu, J.-X., Goldoni, S., Bix, G., Owens, R.T., McQuillan, D.J., Reed, C.C., et al., 2005. Decorin evokes protracted internalization and degradation of the epidermal growth factor receptor via caveolar endocytosis. J. Biol. Chem. 280, 32468–32479.

Zhu, L., Xie, J., Swierczewska, M., Zhang, F., Quan, Q., Ma, Y., et al., 2011. Real-Time video imaging of protease expression *in vivo*. Theranostics 1, 18–27.

Zhu, S., Qian, L., Hong, M., Zhang, L., Pei, Y., Jiang, Y., 2011. RGD-modified PEG-PAMAM-DOX conjugate: in vitro and in vivo targeting to both tumor neovascular endothelial cells and tumor cells. Adv. Mat. 23, H84–H89.

Zielinski, R., Lyakhov, I., Jacobs, A., Chertov, O., Kramer-Marek, G., Francella, N., et al., 2009. Affitoxin--a novel recombinant, HER2-specific, anticancer agent for targeted therapy of HER2-positive tumors. J. Immunother 32 (8), 817–825.

Zlateva, T.P., Krysteva, M., Balajthy, Z., Elodi, P., 1988. Properties of chymotrypsin bound covalently to dextran. Acta Biochim. Biophys. Hung. 23, 225–230.

Zmak, P.M., Podgornik, H., Jancar, J., Podgornik, A., Strancar, A., 2003. Transferof gradient chromatographic methods for protein separation to Convective Interaction Media monolithic columns. J. Chromatogr. A 1006, 195–205.

Zohar, H., Muller, S.J., 2011. Labeling DNA for single-molecule experiments: methods of labeling internal specific sequences on double-stranded DNA. Nanoscale 3, 3027–3039.

Zola, H., Neoh, S.H., Bantzioris, B.X., Webster, J., Loughman, M.S., 1990. Detection by immunofluorescence of surface molecules present in low copy numbers. J. Immunol. Methods 135, 247–255.

Zopf, D.A., Smith, D.F., Drzeniek, Z., Tsai, C.-M., Ginsburg, V., 1978. Affinity purification of antibodies using oligosaccharide-phenethylamine derivatives coupled to SepharoseGinsburg, V. Methods in Enzymology, vol. 50. Academic Press, New York, pp. 171–175.

Zopf, D.A., Tsai, C.-M., Ginsburg, V., 1978b. Carbohydrate antigens: coupling of oligosaccharide-phenethylamine derivatives to edestin by diazotization and characterization of antibody specificity by radioimmunoassay Ginsburg, V. Methods in Enzymology, vol. 50. Academic Press, New York, pp. 163–169.

Zsigmondy, R., 1905. Zur Erkenntnis der Kolloide. Jena, Germany.

Zuberbühler, K., Casi, G., Bernardes, G.J., Neri, D., 2012. Fucose-specific conjugation of hydrazide derivatives to a vascular-targeting monoclonal antibody in IgG format. Chem. Commun. 48, 7100–7102.

Index

Note: Page numbers followed by "*f*" and "*t*" refer to figures and tables, respectively.

C

D

I

J

K

L

N

Q

R

T

U

V